Developmental Mathematics: College Mathematics and Introductory Algebra

7th EDITION

Marvin L. Bittinger
Indiana University Purdue University Indianapolis

Judith A. Beecher
Indiana University Purdue University Indianapolis

Boston San Francisco New York
London Toronto Sydney Tokyo Singapore Madrid
Mexico City Munich Paris Cape Town Hong Kong Montreal

Publisher	Greg Tobin
Editor in Chief	Maureen O'Connor
Acquisitions Editor	Randy Welch
Executive Project Manager	Kari Heen
Assistant Editor	Joanna Doxey
Editorial Assistant	Antonio Arvelo
Production Manager	Ron Hampton
Cover Designer	Dennis Schaefer
Digital Assets Manager	Marianne Groth
Media Producer	Ceci Fleming
Software Development	Mary Durnwald, TestGen; Rebecca Williams, MathXL
Marketing Manager	Jay Jenkins
Marketing Coordinator	Alexandra Waibel
Senior Author Support/ Technology Specialist	Joseph K. Vetere
Senior Prepress Supervisor	Caroline Fell
Senior Media Buyer	Ginny Michaud
Rights and Permissions	Dana Weightman
Manufacturing Manager	Evelyn M. Beaton
Art and Design Services	Geri Davis/The Davis Group, Inc.
Production Coordination	Martha K. Morong/Quadrata, Inc.
Composition	BeaconPMG
Illustrations	William Melvin and Network Graphics
Cover Photo	© Chris Madeley/Photo Researchers, Inc.

About the cover: The aurora borealis, or northern lights, as photographed in the skies over British Columbia, Canada.

Photo credits appear on p. xxxii.

Library of Congress Cataloging-in-Publication Data

Bittinger, Marvin L.
Developmental mathematics.—7th ed./Marvin L. Bittinger, Judith A. Beecher.
p. cm.
Includes indexes.
ISBN-13: 978-0-321-33191-5
ISBN-10: 0-321-33191-5 (student version: alk. paper)
1. Arithmetic. 2. Algebra I. Beecher, Judith A. II. Title.
QA107.2.B58 2007
513'.14—dc22 2006050704

6 7 8 9 10–CKV–10 09

Contents

2 FRACTION NOTATION

3 DECIMAL NOTATION

4 PERCENT NOTATION

5 DATA, GRAPHS, AND STATISTICS

GEOMETRY

INTRODUCTION TO REAL NUMBERS AND ALGEBRAIC EXPRESSIONS

8 SOLVING EQUATIONS AND INEQUALITIES

9 GRAPHS OF LINEAR EQUATIONS

10 POLYNOMIALS: OPERATIONS

11 POLYNOMIALS: FACTORING

12 RATIONAL EXPRESSIONS AND EQUATIONS

SYSTEMS OF EQUATIONS

RADICAL EXPRESSIONS AND EQUATIONS

15

QUADRATIC EQUATIONS

APPENDIXES

Index of Applications

Agriculture

Astronomy

Automotive

Biology

Business

Chemistry

Construction

Consumer

Domestic

Economics

Education

Engineering

Environment

Finance

Food

Geometry

Government

Health/Medicine

Labor

Miscellaneous

Physics

Social Sciences

Sports/Entertainment

Statistics/Demographics

Technology

Transportation

Index of Study Tips

Preface

It is with great pride and excitement that we present to you the seventh edition of *Developmental Mathematics: College Mathematics and Introductory Algebra*. The text has evolved dramatically over the past 24 years through your comments, responses, and opinions. This feedback, combined with our overall objective of presenting the material in a clear and accurate manner, drives each revision. It is our hope that *Developmental Mathematics*, Seventh Edition, and the supporting supplements will help provide an improved teaching and learning experience by matching the needs of instructors and successfully preparing students for their future.

This text is intended for use by students needing a review in arithmetic skills before covering introductory algebra topics, and is part of a complete series that includes the following:

Bittinger: *Fundamental Mathematics,* Fourth Edition

Bittinger/Penna: *Basic Mathematics with Early Integers*

Bittinger: *Basic Mathematics,* Tenth Edition

Bittinger/Ellenbogen/Johnson: *Prealgebra,* Fifth Edition

Bittinger: *Introductory Algebra,* Tenth Edition

Bittinger: *Intermediate Algebra,* Tenth Edition

Bittinger/Ellenbogen: *Prealgebra and Introductory Algebra,* Second Edition

Bittinger/Beecher: *Introductory and Intermediate Algebra,* Third Edition

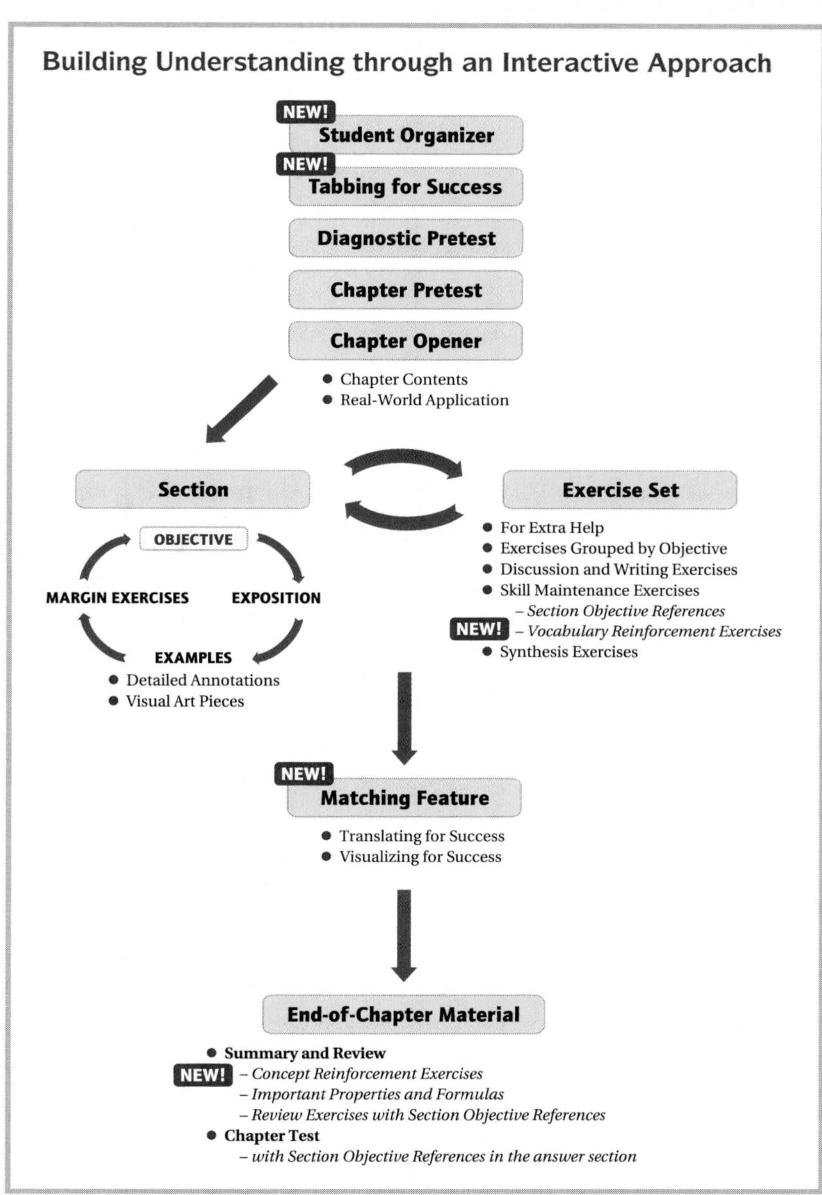

Building Understanding through an Interactive Approach

The pedagogy of this text is designed to provide an interactive learning experience between the student and the exposition, annotated examples, art, margin exercises, and exercise sets. This unique approach, which has been developed and refined over ten editions of the series and is illustrated at right, provides students with a clear set of learning objectives, involves them with the development of the material, and provides immediate and continual reinforcement and assessment through the margin exercises.

Let's Visit the Seventh Edition

The style, format, and approach of the seventh edition have been strengthened in a number of ways. However, the accuracy that the Bittinger books are known for has not changed. This edition, as with all editions, has gone through an exhaustive checking process to ensure accuracy in the problem sets, mathematical art, and accompanying supplements. We know what a critical role the accuracy of a book plays in student learning, and we value the reputation for accuracy that we have earned.

Guidelines from many states and educational institutions were considered while planning the revision, including those on the Texas Higher Education Assessment Test (THEA) (as outlined in Appendix Q) and the majority of topics on other state-level mathematics tests.

NEW! IN THE SEVENTH EDITION

Each revision gives us the opportunity to incorporate new elements and refine existing elements to provide a better experience for students and teachers alike. Below are six **new** features designed to help students succeed.

- *Student Organizer*
- *Tabbing for Success*
- *Translating for Success* matching exercises
- *Visualizing for Success* matching exercises
- *Vocabulary Reinforcement* exercises
- *Concept Reinforcement* exercises

These features, along with the hallmark features of this book, are discussed in the pages that follow.

In addition, the seventh edition has been designed to be open and flexible, helping students focus their attention on details that are critical at this level through prominent headings, boxed definitions and rules, and clearly labeled objectives.

NEW! CHANGES IN THE TABLE OF CONTENTS

In response to reviewers' suggestions, the two graphing chapters of the sixth edition—Chapter 9, "Graphs of Linear Equations," and Chapter 13, "Graphs, Slope, and Applications"—have been combined into one chapter—Chapter 9, "Graphs of Linear Equations"—in the seventh edition. Sections 9.1 and 9.2 of the sixth edition are now Section 9.1 in the seventh edition. Sections 9.3, 9.4, and 13.1–13.4 of the sixth edition are now Sections 9.2–9.7 of the seventh edition. Section 13.5 of the sixth edition is now Section 12.9 in the seventh edition. These changes were requested primarily so that instructors can cover the topics of graphing and slope in a more efficient manner. In addition, Appendix G ("Finding Equations of Lines: Point–Slope Equation") has been added to extend the coverage of equations of lines in Section 9.4.

Chapter Pretests, now located along with the Diagnostic Pretest in the *Printed Test Bank* and in MyMathLab, diagnose at the section and objective level and can be used to place students in a specific section, or objective, of the chapter, allowing them to concentrate on topics with which they have particular difficulty. Answers to these pretests are available in the *Printed Test Bank* and in MyMathLab.

NEW! STUDENT ORGANIZER AND TABBING FOR SUCCESS

Along with study tips found throughout the text, a pull-out schedule card helps students stay organized. Students can schedule time on the card for study, commuting, work, family, and relaxation, and can also record important dates, useful contact information, and technology references.

In addition to the new Student Organizer, the new Tabbing for Success page provides students with forty color-coded reusable tabs to quickly locate key examples, review important summaries, flag text topics, and highlight areas where they need help. Together, these features will help students better use their time and their textbooks to succeed in the course.

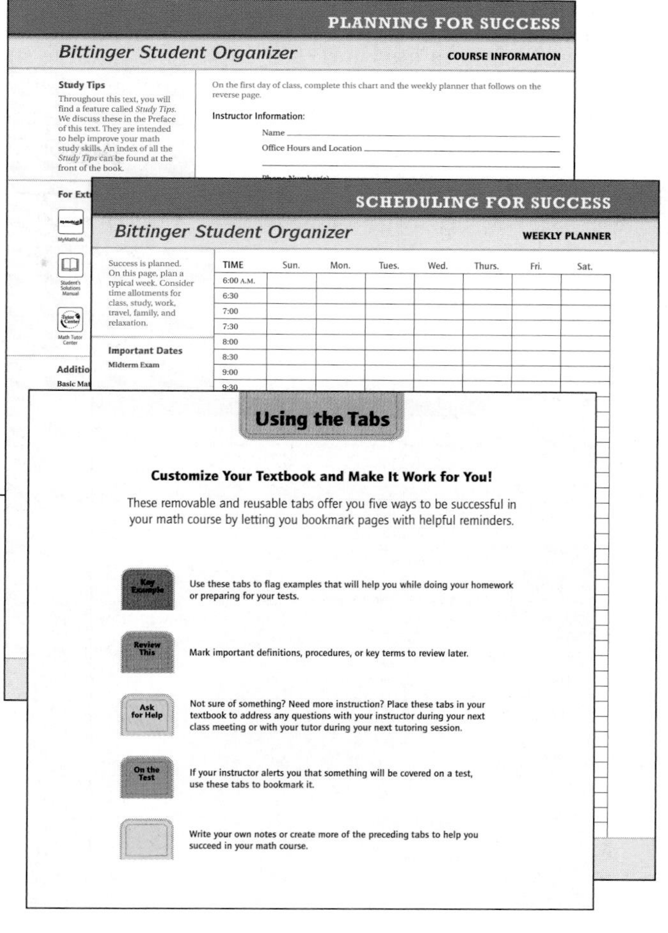
PLANNING FOR SUCCESS

Bittinger Student Organizer — COURSE INFORMATION

Study Tips
Throughout this text, you will find a feature called *Study Tips*. We discuss these in the Preface of this text. They are intended to help improve your math study skills. An index of all the *Study Tips* can be found at the front of the book.

On the first day of class, complete this chart and the weekly planner that follows on the reverse page.

Instructor Information:
Name
Office Hours and Location

SCHEDULING FOR SUCCESS

Bittinger Student Organizer — WEEKLY PLANNER

Success is planned. On this page, plan a typical week. Consider time allotments for class, study, work, travel, family, and relaxation.

Important Dates
Midterm Exam

TIME	Sun.	Mon.	Tues.	Wed.	Thurs.	Fri.	Sat.
6:00 A.M.							
6:30							
7:00							
7:30							
8:00							
8:30							
9:00							
9:30							

Using the Tabs

Customize Your Textbook and Make It Work for You!

These removable and reusable tabs offer you five ways to be successful in your math course by letting you bookmark pages with helpful reminders.

Key Example — Use these tabs to flag examples that will help you while doing your homework or preparing for your tests.

Review This — Mark important definitions, procedures, or key terms to review later.

Ask for Help — Not sure of something? Need more instruction? Place these tabs in your textbook to address any questions with your instructor during your next class meeting or with your tutor during your next tutoring session.

On the Test — If your instructor alerts you that something will be covered on a test, use these tabs to bookmark it.

Write your own notes or create more of the preceding tabs to help you succeed in your math course.

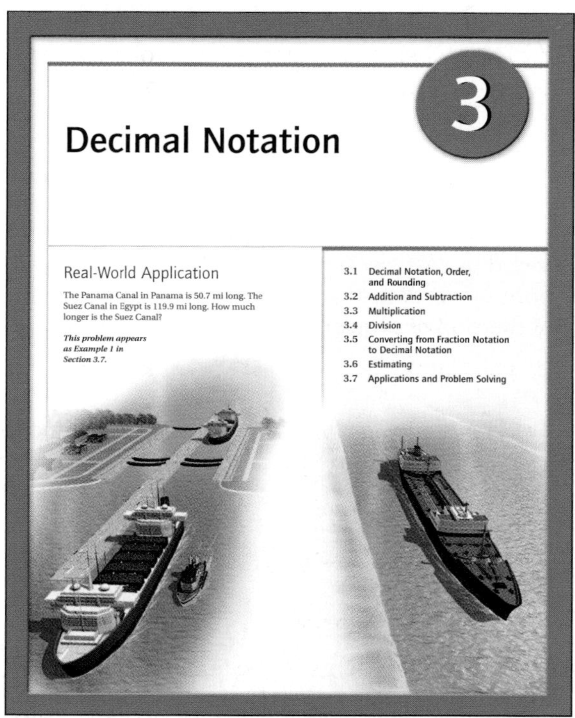
3

Decimal Notation

Real-World Application

The Panama Canal in Panama is 50.7 mi long. The Suez Canal in Egypt is 119.9 mi long. How much longer is the Suez Canal?

This problem appears as Example 1 in Section 3.7.

3.1 Decimal Notation, Order, and Rounding
3.2 Addition and Subtraction
3.3 Multiplication
3.4 Division
3.5 Converting from Fraction Notation to Decimal Notation
3.6 Estimating
3.7 Applications and Problem Solving

CHAPTER OPENERS

To engage students and prepare them for the upcoming chapter material, gateway chapter openers are designed with exceptional artwork that is tied to a motivating real-world application. (See pages 1, 101, and 247.)

OBJECTIVE BOXES

At the beginning of each section, a boxed list of objectives is keyed by letter not only to section subheadings, but also to the exercises in the Pretest (located in the *Printed Test Bank* and MyMathLab), the section exercise sets, and the Summary and Review exercises, as well as to the answers to the questions in the Chapter Tests. This correlation enables students to easily find appropriate review material if they need help with a particular exercise or skill at the objective level. (See pages 2, 174, and 248.)

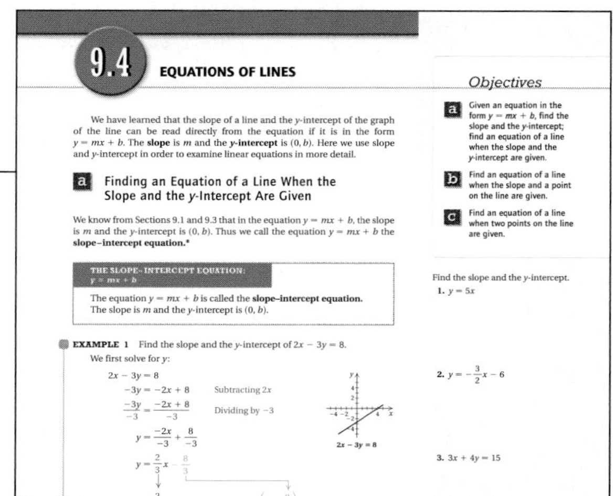
9.4 EQUATIONS OF LINES

We have learned that the slope of a line and the y-intercept of the graph of the line can be read directly from the equation if it is in the form $y = mx + b$. The **slope** is m and the **y-intercept** is $(0, b)$. Here we use slope and y-intercept in order to examine linear equations in more detail.

a Finding an Equation of a Line When the Slope and the y-Intercept Are Given

We know from Sections 9.1 and 9.3 that in the equation $y = mx + b$, the slope is m and the y-intercept is $(0, b)$. Thus we call the equation $y = mx + b$ the **slope–intercept equation.***

THE SLOPE–INTERCEPT EQUATION: $y = mx + b$

The equation $y = mx + b$ is called the **slope–intercept equation.** The slope is m and the y-intercept is $(0, b)$.

EXAMPLE 1 Find the slope and the y-intercept of $2x - 3y = 8$.

We first solve for y:

$$2x - 3y = 8$$
$$-3y = -2x + 8 \quad \text{Subtracting } 2x$$
$$\frac{-3y}{-3} = \frac{-2x + 8}{-3} \quad \text{Dividing by } -3$$
$$y = \frac{-2x}{-3} + \frac{8}{-3}$$
$$y = \frac{2}{3}x - \frac{8}{3}$$

Objectives

a Given an equation in the form $y = mx + b$, find the slope and the y-intercept; find an equation of a line when the slope and the y-intercept are given.

b Find an equation of a line when the slope and a point on the line are given.

c Find an equation of a line when two points on the line are given.

Find the slope and the y-intercept.

1. $y = 5x$

2. $y = -\frac{3}{2}x - 6$

3. $3x + 4y = 15$

ANNOTATED EXAMPLES

Detailed annotations and color highlights lead the student through the structured steps of the examples. The level of detail in these annotations is a significant reason for students' success with this book. (See pages 148, 290, and 558.)

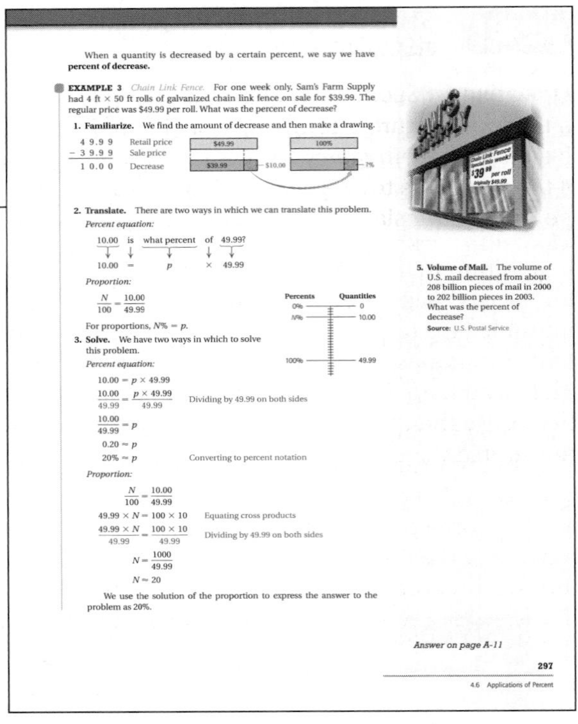

When a quantity is decreased by a certain percent, we say we have **percent of decrease.**

EXAMPLE 3 *Chain Link Fence.* For one week only, Sam's Farm Supply had 4 ft × 50 ft rolls of galvanized chain link fence on sale for \$39.99. The regular price was \$49.99 per roll. What was the percent of decrease?

1. **Familiarize.** We find the amount of decrease and then make a drawing.

4 9.9 9	Retail price
− 3 9.9 9	Sale price
1 0.0 0	Decrease

2. **Translate.** There are two ways in which we can translate this problem.

Percent equation:

10.00 is what percent of 49.99?

$10.00 = p \times 49.99$

Proportion:

$\frac{N}{100} = \frac{10.00}{49.99}$

For proportions, $N\% = p$.

3. **Solve.** We have two ways in which to solve this problem.

Percent equation:

$10.00 = p \times 49.99$

$\frac{10.00}{49.99} = \frac{p \times 49.99}{49.99}$ Dividing by 49.99 on both sides

$\frac{10.00}{49.99} = p$

$0.20 \approx p$

$20\% \approx p$ Converting to percent notation

Proportion:

$\frac{N}{100} = \frac{10.00}{49.99}$

$49.99 \times N = 100 \times 10$ Equating cross products

$\frac{49.99 \times N}{49.99} = \frac{100 \times 10}{49.99}$ Dividing by 49.99 on both sides

$N = \frac{1000}{49.99}$

$N \approx 20$

We use the solution of the proportion to express the answer to the problem as 20%.

5. Volume of Mail. The volume of U.S. mail decreased from about 208 billion pieces of mail in 2000 to 202 billion pieces in 2003. What was the percent of decrease?

Source: U.S. Postal Service

Answer on page A-11

297

4.6 Applications of Percent

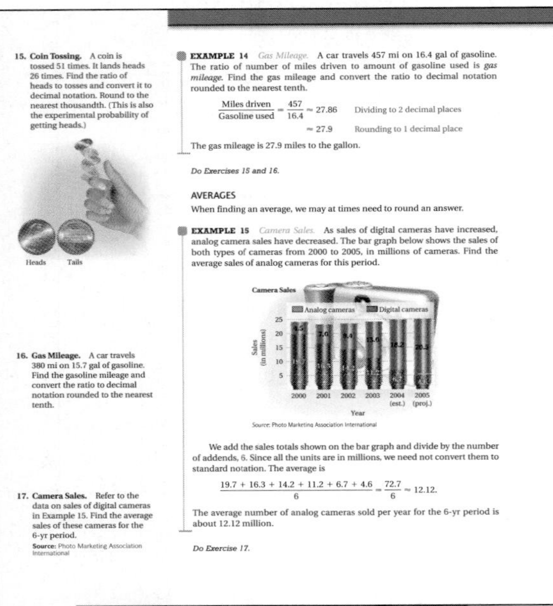

15. Coin Tossing. A coin is tossed 51 times. It lands heads 26 times. Find the ratio of heads to tosses and convert it to decimal notation. Round to the nearest thousandth. (This is also the experimental probability of getting heads.)

Heads Tails

EXAMPLE 14 *Gas Mileage.* A car travels 457 mi on 16.4 gal of gasoline. The ratio of number of miles driven to amount of gasoline used is *gas mileage*. Find the gas mileage and convert the ratio to decimal notation rounded to the nearest tenth.

$\frac{\text{Miles driven}}{\text{Gasoline used}} = \frac{457}{16.4} \approx 27.86$ Dividing to 2 decimal places

≈ 27.9 Rounding to 1 decimal place

The gas mileage is 27.9 miles to the gallon.

Do Exercises 15 and 16.

AVERAGES

When finding an average, we may at times need to round an answer.

EXAMPLE 15 *Camera Sales.* As sales of digital cameras have increased, analog camera sales have decreased. The bar graph below shows the sales of both types of cameras from 2000 to 2005, in millions of cameras. Find the average sales of analog cameras for this period.

16. Gas Mileage. A car travels 380 mi on 15.7 gal of gasoline. Find the gasoline mileage and convert the ratio to decimal notation rounded to the nearest tenth.

We add the sales totals shown on the bar graph and divide by the number of addends, 6. Since all the units are in millions, we need not convert them to standard notation. The average is

$\frac{19.7 + 16.3 + 14.2 + 11.2 + 6.7 + 4.6}{6} = \frac{72.7}{6} \approx 12.12.$

The average number of analog cameras sold per year for the 6-yr period is about 12.12 million.

Do Exercise 17.

17. Camera Sales. Refer to the data on sales of digital cameras in Example 15. Find the average sales of these cameras for the 6-yr period.

Source: Photo Marketing Association International

214

CHAPTER 3: Decim

MARGIN EXERCISES

Throughout the text, students are directed to numerous margin exercises that provide immediate practice and reinforcement of the concepts covered in each section. Answers are provided at the back of the book so students can immediately self-assess their understanding of the skill or concept at hand. (See pages 146, 250, and 282.)

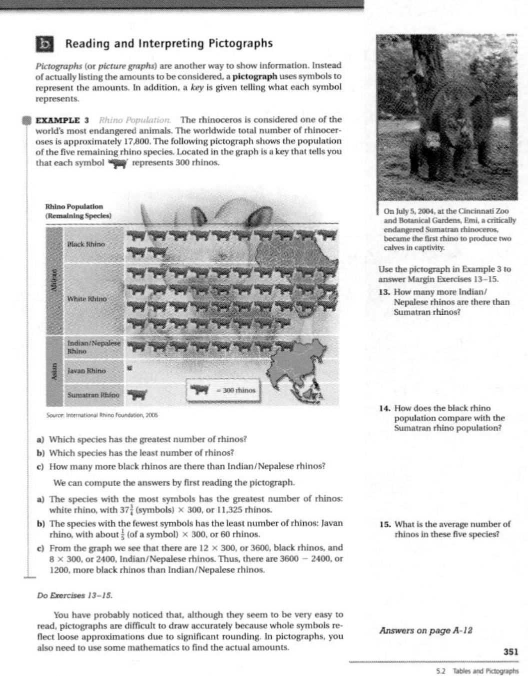

b Reading and Interpreting Pictographs

Pictographs (or *picture graphs*) are another way to show information. Instead of actually listing the amounts to be considered, a **pictograph** uses symbols to represent the amounts. In addition, a *key* is given telling what each symbol represents.

EXAMPLE 3 *Rhino Population.* The rhinoceros is considered one of the world's most endangered animals. The worldwide total number of rhinoceroses is approximately 17,800. The following pictograph shows the population of the five remaining rhino species. Located in the graph is a key that tells you that each symbol represents 300 rhinos.

a) Which species has the greatest number of rhinos?
b) Which species has the least number of rhinos?
c) How many more black rhinos are there than Indian/Nepalese rhinos?

We can compute the answers by first reading the pictograph.

a) The species with the most symbols has the greatest number of rhinos: white rhino, with $37\frac{3}{4}$ (symbols) × 300, or 11,325 rhinos.
b) The species with the fewest symbols has the least number of rhinos: Javan rhino, with about $\frac{1}{5}$ (of a symbol) × 300, or 60 rhinos.
c) From the graph we see that there are 12 × 300, or 3600, black rhinos, and 8 × 300, or 2400, Indian/Nepalese rhinos. Thus, there are 3600 − 2400, or 1200, more black rhinos than Indian/Nepalese rhinos.

Do Exercises 13–15.

You have probably noticed that, although they seem to be very easy to read, pictographs are difficult to draw accurately because whole symbols reflect loose approximations due to significant rounding. In pictographs, you also need to use some mathematics to find the actual amounts.

On July 5, 2004, at the Cincinnati Zoo and Botanical Gardens, Emi, a critically endangered Sumatran rhinoceros, became the first rhino to produce two calves in captivity.

Use the pictograph in Example 3 to answer Margin Exercises 13–15.

13. How many more Indian/Nepalese rhinos are there than Sumatran rhinos?

14. How does the black rhino population compare with the Sumatran rhino population?

15. What is the average number of rhinos in these five species?

Answers on page A-12

351

5.2 Tables and Pictographs

REAL-DATA APPLICATIONS

This text encourages students to see and interpret the mathematics that appears every day in the world around them. Throughout the writing process, an extensive and energetic search for real-data applications was conducted, and the result is a variety of examples and exercises that connect the mathematical content with the real world. A large number of the applications are new to this edition, and many are drawn from the fields of business and economics, life and physical sciences, social sciences, and areas of general interest such as sports and daily life. To further encourage students to understand the relevance of mathematics, many applications are enhanced by graphs and drawings similar to those found in today's newspapers and magazines, and feature source lines as well. (See pages 152, 287, and 321.)

NEW! TRANSLATING FOR SUCCESS VISUALIZING FOR SUCCESS

Translating for Success The goal of the matching exercises in this new feature is to practice step two, *Translate,* of the five-step problem-solving process. Students translate each of ten problems to an equation or an inequality and select the correct translation from fifteen given equations and inequalities. This feature appears once in each chapter that contains problems that can be solved using the five-step problem-solving process and reviews skills and concepts with problems from all preceding chapters. (See pages 300, 371, and 541.)

Visualizing for Success This new feature provides students with an opportunity to match an equation with its graph by focusing on characteristics of the equation or inequality and the corresponding attributes of the graph. This feature appears at least once in each chapter that contains graphing instruction and reviews graphing skills and concepts with exercises from all preceding chapters. (See pages 702, 774, and 1115.)

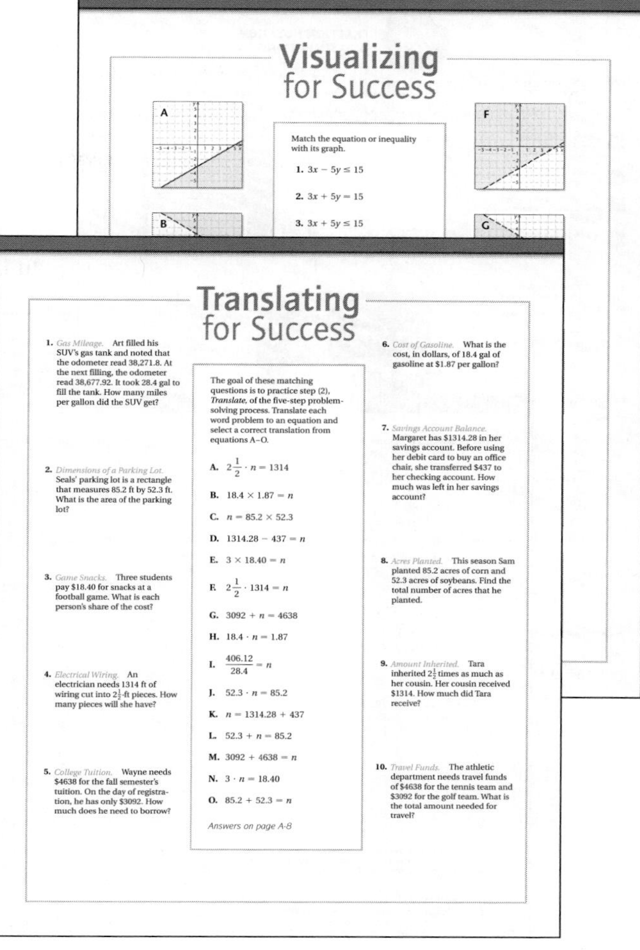
Visualizing for Success

Match the equation or inequality with its graph.

1. $3x - 5y \le 15$
2. $3x + 5y = 15$
3. $3x + 5y \le 15$

Translating for Success

1. *Gas Mileage.* Art filled his SUV's gas tank and noted that the odometer read 38,271.8. At the next filling, the odometer read 38,677.92. It took 28.4 gal to fill the tank. How many miles per gallon did the SUV get?

2. *Dimensions of a Parking Lot.* Seals' parking lot is a rectangle that measures 85.2 ft by 52.3 ft. What is the area of the parking lot?

3. *Game Snacks.* Three students pay $18.40 for snacks at a football game. What is each person's share of the cost?

4. *Electrical Wiring.* An electrician needs 1314 ft of wiring cut into $2\frac{1}{2}$-ft pieces. How many pieces will she have?

5. *College Tuition.* Wayne needs $4638 for the fall semester's tuition. On the day of registration, he has only $3092. How much does he need to borrow?

The goal of these matching questions is to practice step (2), *Translate,* of the five-step problem-solving process. Translate each word problem to an equation and select a correct translation from equations A–O.

A. $2\frac{1}{2} \cdot n = 1314$
B. $18.4 \times 1.87 = n$
C. $n = 85.2 \times 52.3$
D. $1314.28 - 437 = n$
E. $3 \times 18.40 = n$
F. $2\frac{1}{2} \cdot 1314 = n$
G. $3092 + n = 4638$
H. $18.4 \cdot n = 1.87$
I. $\frac{406.12}{28.4} = n$
J. $52.3 \cdot n = 85.2$
K. $n = 1314.28 + 437$
L. $52.3 + n = 85.2$
M. $3092 + 4638 = n$
N. $3 \cdot n = 18.40$
O. $85.2 + 52.3 = n$

Answers on page A-8

6. *Cost of Gasoline.* What is the cost, in dollars, of 18.4 gal of gasoline at $1.87 per gallon?

7. *Savings Account Balance.* Margaret has $1314.28 in her savings account. Before using her debit card to buy an office chair, she transferred $437 to her checking account. How much was left in her savings account?

8. *Acres Planted.* This season Sam planted 85.2 acres of corn and 52.3 acres of soybeans. Find the total number of acres that he planted.

9. *Amount Inherited.* Tara inherited $2\frac{1}{2}$ times as much as her cousin. Her cousin received $1314. How much did Tara receive?

10. *Travel Funds.* The athletic department needs travel funds of $4638 for the tennis team and $3092 for the golf team. What is the total amount needed for travel?

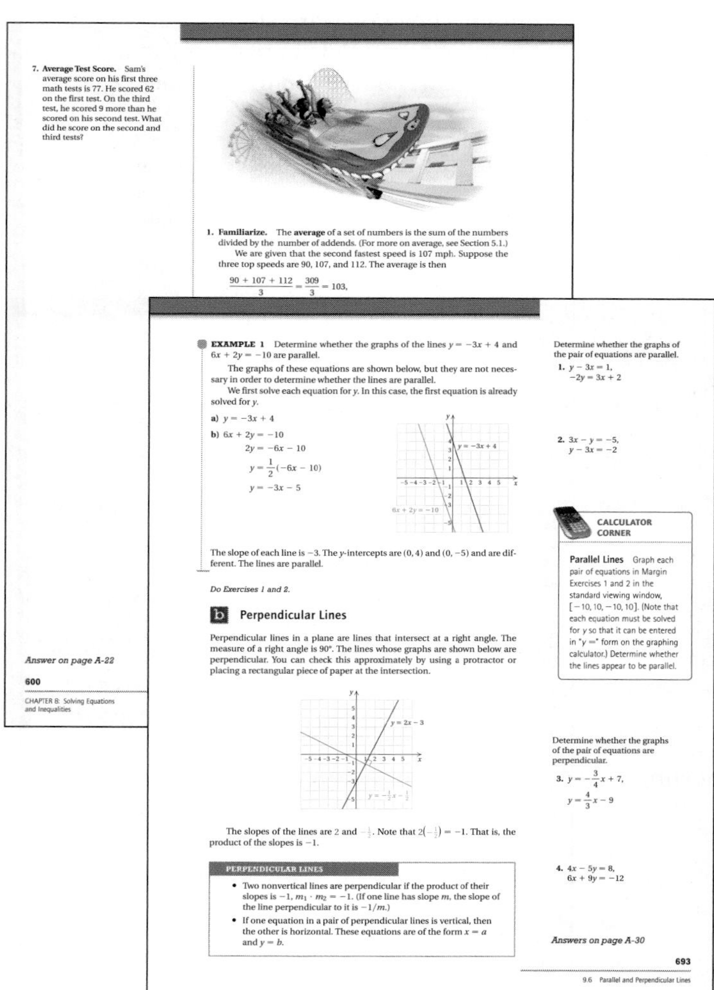
7. **Average Test Score.** Sam's average score on his first three math tests is 77. He scored 62 on the first test. On the third test, he scored 9 more than he scored on his second test. What did he score on the second and third tests?

1. **Familiarize.** The **average** of a set of numbers is the sum of the numbers divided by the number of addends. (For more on average, see Section 5.1.) We are given that the second fastest speed is 107 mph. Suppose the three top speeds are 90, 107, and 112. The average is then

$\frac{90 + 107 + 112}{3} = \frac{309}{3} = 103,$

Answer on page A-22

600 CHAPTER 8: Solving Equations and Inequalities

EXAMPLE 1 Determine whether the graphs of the lines $y = -3x + 4$ and $6x + 2y = -10$ are parallel.

The graphs of these equations are shown below, but they are not necessary in order to determine whether the lines are parallel.

We first solve each equation for y. In this case, the first equation is already solved for y.

a) $y = -3x + 4$

b) $6x + 2y = -10$
$2y = -6x - 10$
$y = \frac{1}{2}(-6x - 10)$
$y = -3x - 5$

The slope of each line is -3. The y-intercepts are $(0, 4)$ and $(0, -5)$ and are different. The lines are parallel.

Do Exercises 1 and 2.

b Perpendicular Lines

Perpendicular lines in a plane are lines that intersect at a right angle. The measure of a right angle is 90°. The lines whose graphs are shown below are perpendicular. You can check this approximately by using a protractor or placing a rectangular piece of paper at the intersection.

The slopes of the lines are 2 and $-\frac{1}{2}$. Note that $2(-\frac{1}{2}) = -1$. That is, the product of the slopes is -1.

PERPENDICULAR LINES
- Two nonvertical lines are perpendicular if the product of their slopes is -1, $m_1 \cdot m_2 = -1$. (If one line has slope m, the slope of the line perpendicular to it is $-1/m$.)
- If one equation in a pair of perpendicular lines is vertical, then the other is horizontal. These equations are of the form $x = a$ and $y = b$.

Determine whether the graphs of the pair of equations are parallel.

1. $y - 3x = 1,$
$-2y = 3x + 2$

2. $3x - y = -5,$
$y - 3x = -2$

CALCULATOR CORNER

Parallel Lines Graph each pair of equations in Margin Exercises 1 and 2 in the standard viewing window, [−10, 10, −10, 10]. (Note that each equation must be solved for y so that it can be entered in "$y =$" form on the graphing calculator.) Determine whether the lines appear to be parallel.

Determine whether the graphs of the pair of equations are perpendicular.

3. $y = -\frac{3}{4}x + 7,$
$y = \frac{4}{3}x - 9$

4. $4x - 5y = 8,$
$6x + 9y = -12$

Answers on page A-30

693 9.6 Parallel and Perpendicular Lines

ART PROGRAM

Today's students are often visually oriented and their approach to a printed page is no exception. The art program is designed to improve the visualization of the mathematical concepts and to enhance the real-data applications. (See pages 226, 234, and 346.)

The use of color is carried out in a methodical and precise manner so that it conveys a consistent meaning, which enhances the readability of the text. For example, the use of both red and blue in mathematical art increases understanding of the concepts. When two lines are graphed using the same set of axes, one is usually red and the other blue. Note that equation labels are the same color as the corresponding line to aid in understanding.

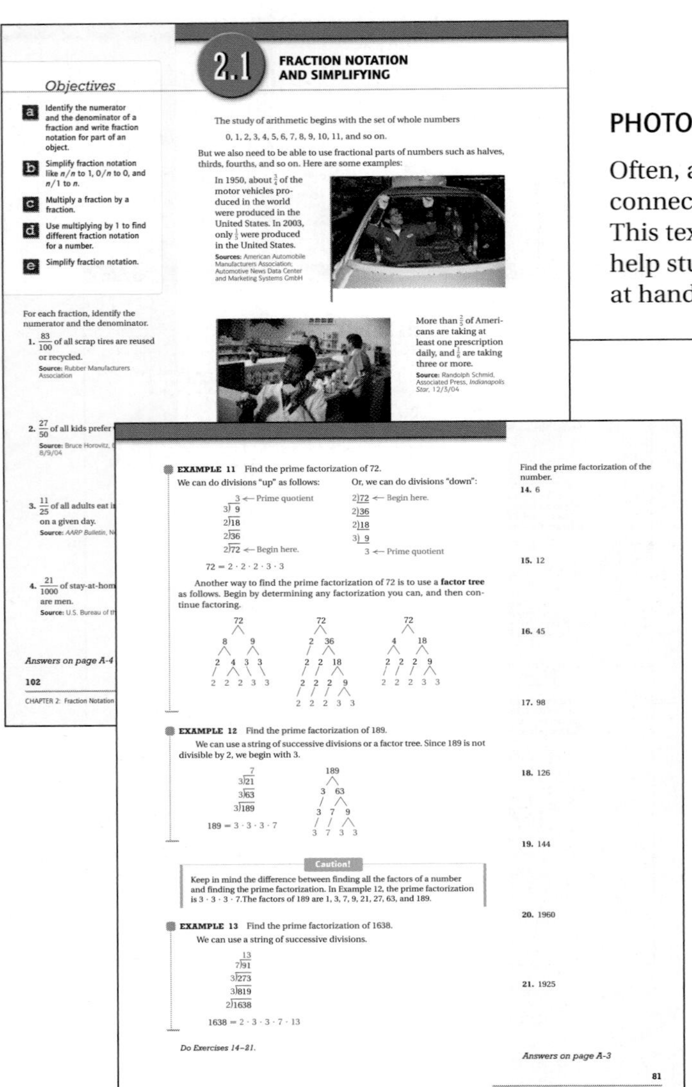

PHOTOGRAPHS

Often, an application becomes relevant to students when the connection to the real world is illustrated with a photograph. This text has numerous photographs throughout in order to help students see the relevance and visualize the application at hand. (See pages 102, 156, and 269.)

CAUTION BOXES

Found at relevant points throughout the text, boxes with the "Caution!" heading warn students of common misconceptions or errors made in performing a particular mathematics operation or skill. (See pages 81, 117, and 516.)

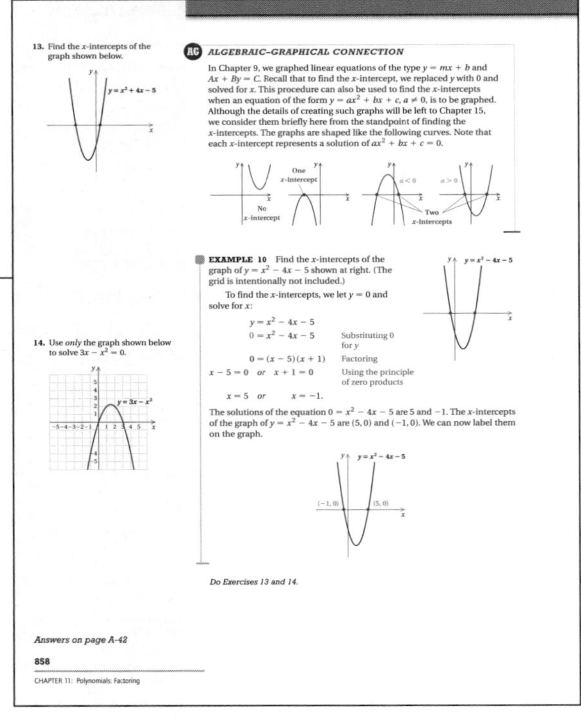

ALGEBRAIC–GRAPHICAL CONNECTIONS

To provide a visual understanding of algebra, algebraic–graphical connections are included in each chapter beginning with Chapter 9. This feature gives the algebra more meaning by connecting it to a graphical interpretation. (See pages 645–646, 858, and 920.)

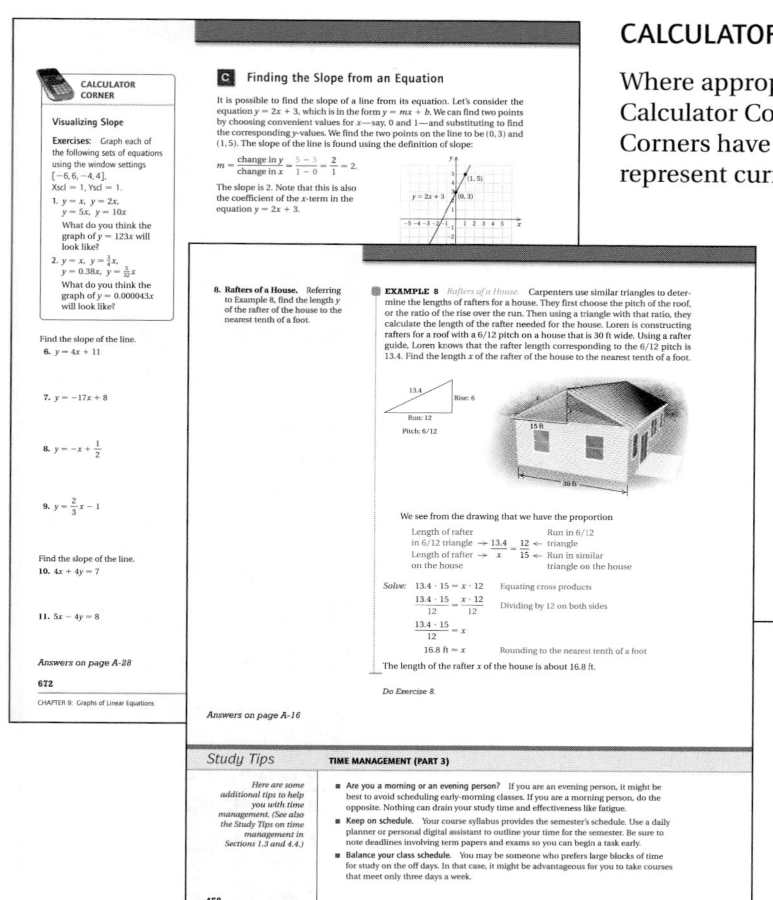

CALCULATOR CORNERS

Where appropriate throughout the text, students will find optional Calculator Corners. Popular in the Sixth Edition, these Calculator Corners have been revised to be more accessible to students and to represent current calculators. (See pages 540, 647, 701, and 922.)

STUDY TIPS

A variety of Study Tips throughout the text give students pointers on how to develop good study habits as they progress through the course. At times short snippets and at other times more lengthy discussions, these Study Tips encourage students to get involved in the learning process. (See pages 72, 143, and 229.)

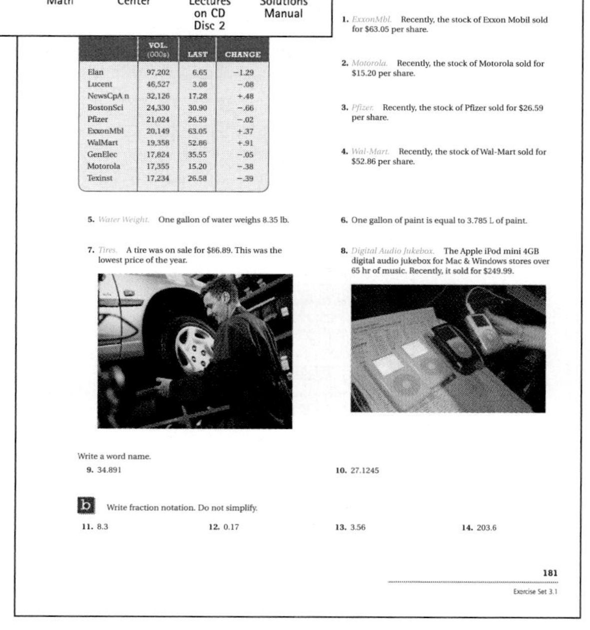

EXERCISE SETS

The exercise sets are a critical part of any math book. To give students ample opportunity to practice what they have learned, each section is followed by an extensive exercise set designed to reinforce the section concepts. In addition, students also have the opportunity to synthesize the objectives from the current section with those from preceding sections.

For Extra Help Many valuable study aids accompany this text. Located before each exercise set, "For Extra Help" references list appropriate video lectures on CD, tutorial, and Web resources so that students can easily find related support materials.

Exercises Grouped by Objective Exercises in the section exercise sets are keyed by letter to the section objectives for easy review and remediation. This reinforces the objective-based structure of the book. (See pages 7, 189, and 399.)

Discussion and Writing Exercises Designed to help students develop a deeper comprehension of critical concepts, Discussion and Writing exercises (indicated by D_W) are suitable for individual or group work. These exercises encourage students to both think and write about key mathematical ideas in the chapter. (See pages 10, 191, and 499.)

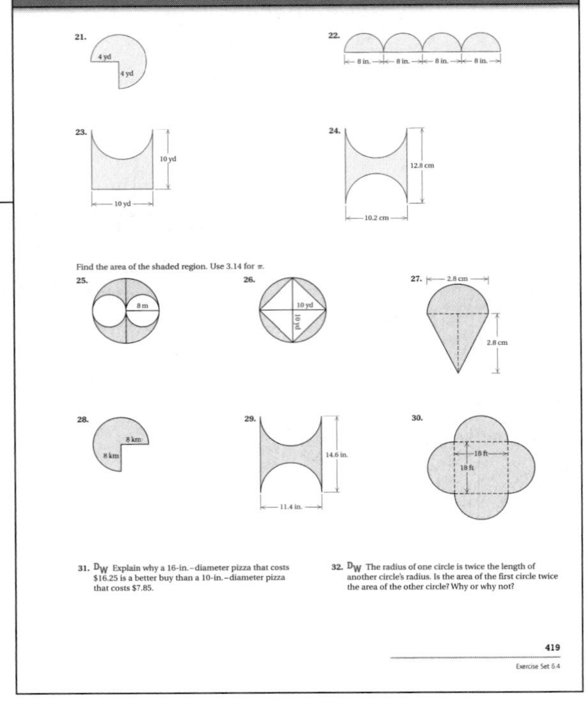

21. 4 yd, 4 yd

22. 8 in., 8 in., 8 in., 8 in.

23. 10 yd, 10 yd

24. 12.8 cm, 10.2 cm

Find the area of the shaded region. Use 3.14 for π.

25. 8 m

26. 10 yd, 10 yd

27. 2.8 cm, 2.8 cm

28. 8 km, 8 km

29. 14.6 in., 11.4 in.

30. 18 ft, 18 ft

31. D_W Explain why a 16-in.–diameter pizza that costs \$16.25 is a better buy than a 10-in.–diameter pizza that costs \$7.85.

32. D_W The radius of one circle is twice the length of another circle's radius. Is the area of the first circle twice the area of the other circle? Why or why not?

419

Exercise Set 6.4

SKILL MAINTENANCE

Evaluate. [1.6b]

33. 2^4 34. 17^2 35. 5^3 36. 8^2

Convert to percent notation. [4.3a]

37. $\frac{3}{8}$ 38. $\frac{5}{8}$ 39. $\frac{2}{3}$ 40. $\frac{1}{5}$

41. Find the rate of discount. [4.7c]

Cookware Set \$58[99]
Lowest Price of the Year
Original price \$100

42. Jack's commission is increased according to how much he sells. He receives a commission of 6% for the first \$3000 and 10% on the amount over \$3000. What is the total commission on sales of \$8500? [4.7b]

43. If 2 cans of tomato paste cost \$1.49, how many cans of tomato paste can you buy for \$7.45? [4.1e]

44. The weight of a human brain is 2.5% of total body weight. A person weighs 200 lb. What does the brain weigh? [4.6a]

SYNTHESIS

45. $\pi \approx \frac{3927}{1250}$ is another approximation for π. Find decimal notation using a calculator.

46. The distance from Kansas City to Indianapolis is 500 mi. A car was driven this distance using tires with a radius of 14 in. How many revolutions of each tire occurred on the trip? Use $\frac{22}{7}$ for π.

47. *Tennis Balls.* Tennis ... vertically three in a ca... the diameter of a tenn... stack of balls. Find the... is greater? Explain.

420

CHAPTER 6: Geometry

Skill Maintenance Exercises Found in each exercise set, these exercises review concepts from other sections in the text to prepare students for their final examination. Section and objective codes appear next to each Skill Maintenance exercise for easy reference. (See pages 95, 591, and 691.)

81. $\dfrac{|2^3 - 3^2| + |12 \cdot 5|}{-32 \div (-16) \div (-4)}$

82. $\dfrac{|3 - 5|^2 - |7 - 13|}{|12 - 9| + |11 - 14|}$

83. D_W Jake keys in $18/2 \cdot 3$ on his calculator and expects the result to be 3. What mistake is he making?

84. D_W Determine whether $|-x|$ and $|x|$ are equivalent. Explain.

SKILL MAINTENANCE

VOCABULARY REINFORCEMENT

In each of Exercises 85–92, fill in the blank with the correct term from the given list. Some of the choices may not be used and some may be used more than once.

85. The set of ______ is $\{\dots, -5, -4, -3, -2, -1, 0, 1, 2, 3, \dots\}$. [7.2a]

86. Two numbers whose sum is 0 are called ______ of each other. [7.3b]

87. The ______ of addition says that $a + b = b + a$ for any real numbers a and b. [7.7b]

88. The ______ states that for any real number a, $a \cdot 1 = 1 \cdot a = a$. [7.7a]

89. The ______ of addition says that $a + (b + c) = (a + b) + c$ for any real numbers a, b, and c. [7.7b]

90. The ______ of multiplication says that $a(bc) = (ab)c$ for any real numbers a, b, and c. [7.7b]

91. Two numbers whose product is 1 are called ______ of each other. [7.6b]

92. The equation $y + 0 = y$ illustrates the ______. [7.7a]

natural numbers
multiplicative inverses
distributive law
associative law
whole numbers
additive inverses
identity property of 0
property of −1
integers
commutative law
identity property of 1
real numbers

SYNTHESIS

Find an equivalent expression by enclosing the last three terms in parentheses preceded by a minus sign.

93. $6y + 2x - 3a + c$ 94. $x - y - a - b$ 95. $6m + 3n - 5m + 4b$

Simplify.

96. $z - \{2z - [3z - (4z - 5z) - 6z] - 7z\} - 8z$

97. $\{x - [f - (f - x)] + [x - f]\} - 3x$

98. $x - \{x - 1 - [x - 2 - (x - 3 - \{x - 4 - [x - 5 - (x - 6)]\})]\}$

99. Use your calculator to do the following.
a) Evaluate $x^2 + 3$ when $x = 7$, when $x = -7$, and when $x = -5.013$.
b) Evaluate $1 - x^2$ when $x = 5$, when $x = -5$, and when $x = -10.455$.

100. Express $3^3 + 3^3 + 3^3$ as a power of 3.

Find the average.

101. −15, 20, 50, −82, −7, −2

102. −1, 1, 2, −2, 3, −8, −10

545

Exercise Set 7.8

NEW! **Vocabulary Reinforcement Exercises** This new feature checks and reviews students' understanding of the vocabulary introduced throughout the text. It appears once in every chapter, in the Skill Maintenance portion of an exercise set, and is intended to provide a continuing review of the terms that students must know in order to be able to communicate effectively in the language of mathematics. (See pages 43, 158, and 225.)

Synthesis Exercises In most exercise sets, Synthesis exercises help build critical-thinking skills by requiring students to synthesize or combine learning objectives from the current section as well as from preceding text sections. (See pages 166, 431, and 696.)

END-OF-CHAPTER MATERIAL

At the end of each chapter, students can practice all they have learned as well as tie the current chapter content to material covered in earlier chapters.

SUMMARY AND REVIEW

A three-part *Summary and Review* appears at the end of each chapter. The first part includes the Concept Reinforcement Exercises described below. The second part is a list of important properties and formulas, when applicable, and the third part provides an extensive set of review exercises.

NEW! **Concept Reinforcement Exercises** Found in the Summary and Review of every chapter, these true/false exercises are designed to increase understanding of the concepts rather than merely assess students' skill at memorizing procedures. (See pages 96, 167, and 241.)

Important Properties and Formulas A list of the important properties and formulas discussed in the chapter is provided for students in an organized manner to help them prioritize topics learned and prepare for chapter tests. This list is only provided in those chapters in which new properties or formulas are presented. (See pages 546, 630, and 795.)

Review Exercises At the end of each chapter, students are provided with an extensive set of review exercises. Reference codes beside each exercise or direction line allow students to easily refer back to specific, objective-level content for remediation. (See pages 96, 167, and 241.)

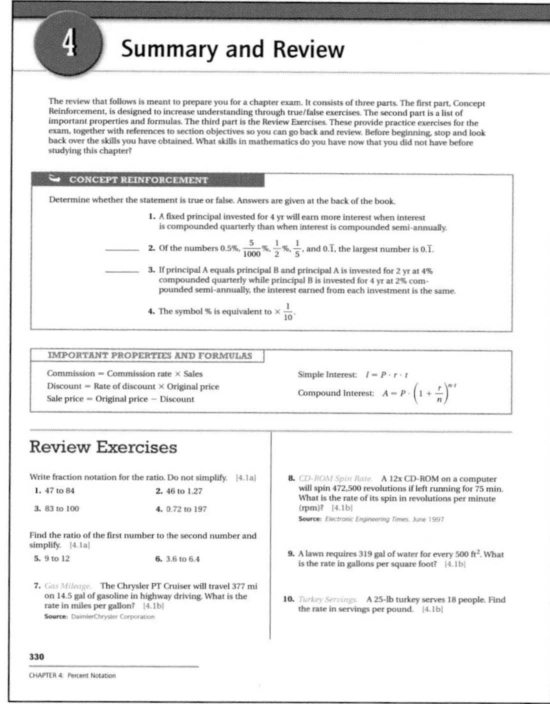

4 Summary and Review

The review that follows is meant to prepare you for a chapter exam. It consists of three parts. The first part, Concept Reinforcement, is designed to increase understanding through true/false exercises. The second part is a list of important properties and formulas. The third part is the Review Exercises. These provide practice exercises for the exam, together with references to section objectives so you can go back and review. Before beginning, stop and look back over the skills you have obtained. What skills in mathematics do you have now that you did not have before studying this chapter?

CONCEPT REINFORCEMENT

Determine whether the statement is true or false. Answers are given at the back of the book.

1. A fixed principal invested for 4 yr will earn more interest when interest is compounded quarterly than when interest is compounded semi-annually.
2. Of the numbers 0.5%, $\frac{5}{1000}$%, $\frac{1}{2}$%, $\frac{1}{5}$, and $0.\overline{1}$, the largest number is $0.\overline{1}$.
3. If principal A equals principal B and principal A is invested for 2 yr at 4% compounded quarterly while principal B is invested for 4 yr at 2% compounded semi-annually, the interest earned from each investment is the same.
4. The symbol % is equivalent to $\times \frac{1}{10}$.

IMPORTANT PROPERTIES AND FORMULAS

Commission = Commission rate × Sales
Discount = Rate of discount × Original price
Sale price = Original price − Discount
Simple Interest: $I = P \cdot r \cdot t$
Compound Interest: $A = P \cdot \left(1 + \frac{r}{n}\right)^{n \cdot t}$

Review Exercises

Write fraction notation for the ratio. Do not simplify. [4.1a]

1. 47 to 84
2. 46 to 1.27
3. 83 to 100
4. 0.72 to 197

Find the ratio of the first number to the second number and simplify. [4.1a]

5. 9 to 12
6. 3.6 to 6.4

7. *Gas Mileage.* The Chrysler PT Cruiser will travel 377 mi on 14.5 gal of gasoline in highway driving. What is the rate in miles per gallon? [4.1b]
Source: DaimlerChrysler Corporation

8. *CD-ROM Spin Rate.* A 12x CD-ROM on a computer will spin 472,500 revolutions if left running for 75 min. What is the rate of its spin in revolutions per minute (rpm)? [4.1b]
Source: *Electronic Engineering Times*, June 1997

9. A lawn requires 319 gal of water for every 500 ft^2. What is the rate in gallons per square foot? [4.1b]

10. *Turkey Servings.* A 25-lb turkey serves 18 people. Find the rate in servings per pound. [4.1b]

330

CHAPTER 4: Percent Notation

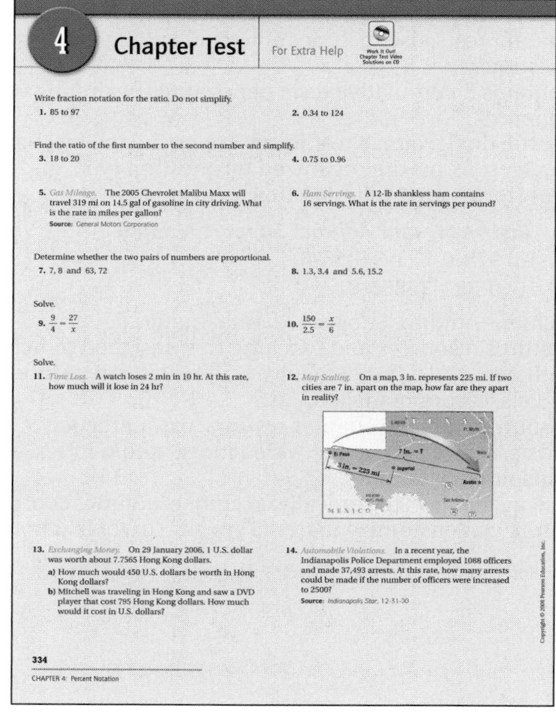

4 Chapter Test

For Extra Help

Work It Out! Chapter Test Video Solutions on CD

Write fraction notation for the ratio. Do not simplify.

1. 85 to 97
2. 0.34 to 124

Find the ratio of the first number to the second number and simplify.

3. 18 to 20
4. 0.75 to 0.96

5. *Gas Mileage.* The 2005 Chevrolet Malibu Maxx will travel 319 mi on 14.5 gal of gasoline in city driving. What is the rate in miles per gallon?
Source: General Motors Corporation

6. *Ham Servings.* A 12-lb shankless ham contains 16 servings. What is the rate in servings per pound?

Determine whether the two pairs of numbers are proportional.

7. 7, 8 and 63, 72
8. 1.3, 3.4 and 5.6, 15.2

Solve.

9. $\frac{9}{4} = \frac{27}{x}$
10. $\frac{150}{2.5} = \frac{x}{6}$

Solve.

11. *Time Loss.* A watch loses 2 min in 10 hr. At this rate, how much will it lose in 24 hr?

12. *Map Scaling.* On a map, 3 in. represents 225 mi. If two cities are 7 in. apart on the map, how far are they apart in reality?

13. *Exchanging Money.* On 29 January 2006, 1 U.S. dollar was worth about 7.7565 Hong Kong dollars.
a) How much would 450 U.S. dollars be worth in Hong Kong dollars?
b) Mitchell was traveling in Hong Kong and saw a DVD player that cost 795 Hong Kong dollars. How much would it cost in U.S. dollars?

14. *Automobile Violations.* In a recent year, the Indianapolis Police Department employed 1088 officers and made 37,493 arrests. At this rate, how many arrests could be made if the number of officers were increased to 2500?
Source: *Indianapolis Star*, 12-31-00

334

CHAPTER 4: Percent Notation

Copyright © 2008 Pearson Education, Inc.

CHAPTER TEST

Following the Review Exercises, a sample Chapter Test allows students to review and test comprehension of chapter skills prior to taking an instructor's exam. Answers to all questions in the Chapter Test are given at the back of the book. Section and objective references for each question are included with the answers. (See pages 99, 171, and 244.)

Ancillaries

The following ancillaries are available to help both instructors and students use this text more effectively.

Student Supplements

Student's Solutions Manual

(ISBN-13: 978-0-321-34858-6)
(ISBN-10: 0-321-34858-3)

- By Judith A. Penna, *Indiana University Purdue University Indianapolis*
- Contains completely worked-out solutions with step-by-step annotations for all the odd-numbered exercises in the text, with the exception of the discussion and writing exercises, as well as completely worked-out solutions to all the exercises in the Chapter Reviews and Chapter Tests.

Collaborative Learning Activities Manual

(ISBN-13: 978-0-321-34854-8)
(ISBN-10: 0-321-34854-0)

- Features group activities tied to text sections and includes the focus, time estimate, suggested group size and materials, and background notes for each activity.
- Available as a stand-alone supplement sold in the bookstore, as a textbook bundle component for students, or as a classroom activity resource for instructors.

Video Lectures on CD

(ISBN-13: 978-0-321-34853-1)
(ISBN-10: 0-321-34853-2)

- Complete set of digitized videos on CD-ROMs for student use at home or on campus.
- To tie student learning to the pedagogy of the text, lectures are organized by objectives, which are indicated on the screen at the start of each new objective.
- Present a series of lectures correlated directly to the content of each section of the text.
- Feature an engaging team of instructors who present material in a format that stresses student interaction, often using examples and exercises from the text.
- Ideal for distance learning or supplemental instruction.
- Includes an expandable window that shows text captioning. Captions can be turned on or off.

NEW! ***Work It Out! Chapter Test Video Solutions on CD***

(ISBN-13: 978-0-321-49674-4)
(ISBN-10: 0-321-49674-4)

- Presented by Judith A. Penna and Barbara Johnson
- Provides step-by-step solutions to every exercise in each Chapter Test from the text.
- Helps students prepare for chapter tests and synthesize content.

Instructor Supplements

Annotated Instructor's Edition

(ISBN-13: 978-0-321-34830-2)
(ISBN-10: 0-321-34830-3)

- Includes answers to all exercises printed in blue on the same page as those exercises.

Instructor's Solutions Manual

(ISBN-13: 978-0-321-34857-9)
(ISBN-10: 0-321-34857-5)

- By Judith A. Penna, *Indiana University Purdue University Indianapolis*
- Contains brief solutions to the even-numbered exercises in the exercise sets, answers to all discussion and writing exercises, and completely worked-out solutions to all the exercises in the Chapter Reviews and Chapter Tests.

Printed Test Bank

(ISBN-13: 978-0-321-34829-6)
(ISBN-10: 0-321-34829-X)

- By Mark Stevenson, *Oakland Community College*
- Contains one diagnostic test.
- Contains one pretest for each chapter.
- Provides 13 new test forms for every chapter and 8 new test forms for the final exam.
- For the chapter tests, 5 test forms are modeled after the chapter tests in the text, 3 test forms are organized by topic order following the text objectives, 3 test forms are designed for 50-minute class periods and organized so that each objective in the chapter is covered on one of the tests, and 2 test forms are multiple-choice. Chapter tests also include more challenging synthesis questions.
- Contains 2 cumulative tests per chapter beginning with Chapter 2.
- For the final exam, 3 test forms are organized by chapter, 3 test forms are organized by question type, and 2 test forms are multiple-choice.

NEW! ***Instructor and Adjunct Support Manual***

(ISBN-13: 978-0-321-34855-5)
(ISBN-10: 0-321-34855-9)

- Includes Adjunct Support Manual material.
- Features resources and teaching tips designed to help both new and adjunct faculty with course preparation and classroom management.
- Resources include chapter reviews, extra practice sheets, conversion guide, video index, audio index, and transparency masters.
- Also available electronically so course/adjunct coordinators can customize material specific to their schools.

Student Supplements

Math Study Skills for Students Video on CD

(ISBN-13: 978-0-321-29745-7)
(ISBN-10: 0-321-29745-8)

- Presented by author Marvin Bittinger
- Designed to help students make better use of their math study time and improve their retention of concepts and procedures taught in classes from basic mathematics through intermediate algebra.
- Through carefully crafted graphics and comprehensive on-camera explanation, focuses on study skills that are commonly overlooked.

Addison-Wesley Math Tutor Center

www.aw-bc.com/tutorcenter

- The Addison-Wesley Math Tutor Center is staffed by qualified mathematics instructors who provide students with tutoring on examples and odd-numbered exercises from the textbook. Tutoring is available via toll-free telephone, toll-free fax, e-mail, or the Internet. White Board technology allows tutors and students to actually see problems worked while they "talk" in real time over the Internet during tutoring sessions.

MathXL® Tutorials on CD

(ISBN-13: 978-0-321-34849-4)
(ISBN-10: 0-321-34849-4)

- Provides algorithmically generated practice exercises that correlate at the objective level to the content of the text.
- Includes an example and a guided solution to accompany every exercise and video clips for selected exercises.
- Recognizes student errors and provides feedback; generates printed summaries of students' progress.

Instructor Supplements

TestGen with Quizmaster

(ISBN-13: 978-0-321-34856-2)
(ISBN-10: 0-321-34856-7)

- Enables instructors to build, edit, print, and administer tests.
- Features a computerized bank of questions developed to cover all text objectives.
- Algorithmically based content allows instructors to create multiple but equivalent versions of the same question or test with a click of a button.
- Instructors can also modify test-bank questions or add new questions by using the built-in question editor, which allows users to create graphs, input graphics, and insert math notation, variable numbers, or text.
- Tests can be printed or administered online via the Internet or another network. Quizmaster allows students to take tests on a local area network.
- Available on a dual-platform Windows/Macintosh CD-ROM.

PowerPoint Lecture Presentation

- Classroom presentation software geared specifically to this textbook.
- Available within MyMathLab or on the Addison-Wesley catalog www.aw-bc.com/math.

MathXL® www.mathxl.com

MathXL is a powerful online homework, tutorial, and assessment system that accompanies Addison-Wesley textbooks in mathematics or statistics. With MathXL, instructors can create, edit, and assign online homework and tests using algorithmically generated exercises correlated at the objective level to the textbook. They can also create and assign their own online exercises and import TestGen tests for added flexibility. All student work is tracked in MathXL's online gradebook. Students can take chapter tests in MathXL and receive personalized study plans based on their test results. The study plan diagnoses weaknesses and links students directly to tutorial exercises for the objectives they need to study and retest. Students can also access supplemental animations and video clips directly from selected exercises. MathXL is available to qualified adopters. For more information, visit our Web site at www.mathxl.com or contact your Addison-Wesley sales representative.

MyMathLab™ www.mymathlab.com

MyMathLab is a series of text-specific, easily customizable online courses for Addison-Wesley textbooks in mathematics and statistics. Powered by CourseCompass™ (Pearson Education's online teaching and learning environment) and MathXL® (our online homework, tutorial, and assessment system), MyMathLab gives instructors the tools they need to deliver all or a portion of their course online, whether students are in a lab setting or working from home. MyMathLab provides a rich and flexible set of course materials, featuring free-response exercises that are algorithmically generated for unlimited practice and mastery. Students can also use online tools, such as video lectures, animations, and a multimedia textbook, to independently improve their understanding and performance. Instructors can use MyMathLab's homework and test managers to select and assign online exercises correlated directly to the textbook, and they can also create and assign their own online exercises and import TestGen tests for added flexibility. MyMathLab's online gradebook—designed specifically for mathematics and statistics—automatically tracks students' homework and test results and gives the instructor control over how to calculate final grades. Instructors can also add offline (paper-and-pencil) grades to the gradebook. MyMathLab is available to qualified adopters. For more information, visit our Web site at www.mymathlab.com or contact your Addison-Wesley sales representative.

InterAct Math® Tutorial Web site www.interactmath.com

Get practice and tutorial help online! This interactive tutorial Web site provides algorithmically generated practice exercises that correlate directly to the exercises in the textbook. Students can retry an exercise as many times as they like with new values each time for unlimited practice and mastery. Every exercise is accompanied by an interactive guided solution that provides helpful feedback for incorrect answers, and students can also view a worked-out sample problem that steps them through an exercise similar to the one they're working on.

ADDISON-WESLEY MATH ADJUNCT SUPPORT CENTER

The Addison-Wesley Math Adjunct Support Center is staffed by qualified mathematics instructors with over 50 years of combined experience at both the community college and university level. Assistance is provided for faculty in the following areas:

- Suggested syllabus consultation
- Tips on using materials packaged with your book
- Book-specific content assistance
- Teaching suggestions including advice on classroom strategies

For more information, visit www.aw-bc.com/tutorcenter/math-adjunct.html

Acknowledgments and Reviewers

Many of you helped to shape *Developmental Mathematics,* Seventh Edition, by reviewing and spending time with us on your campuses. Our deepest appreciation to all of you and in particular to the following:

Louis Audet, *Augusta Technical College*
Gail Burkett, *Palm Beach Community College*
Jimmy Bullock, *Wilson Technical Community College*
Sally Clark, *Ozarks Technical Community College*
Lucy Dechéne, *Fitchburg State College*
Penny Deggelman, *Lane Community College*
Joe Jordan, *John Tyler Community College*

Susan Meshulam, *Indiana University Purdue University Indianapolis*
Molly Misko, *Gadsden State Community College*
Michael Montano, *Riverside Community College City Campus*
Robert Payne, *Stephen F. Austin State University*
Cassonda Thompson, *York Technical College*
Mary Woestman, *Broome Community College*
Alice Wong, *Miami-Dade College*

We wish to express our heartfelt appreciation to a number of people who have contributed in special ways to the development of this textbook. Our editors, Jennifer Crum and Randy Welch, and marketing manager, Jay Jenkins, encouraged our vision and provided marketing insight. Kari Heen, the project manager, deserves special recognition for overseeing every phase of the project and keeping it moving. The unwavering support of the Developmental Math group, including Antonio Arvelo, editorial assistant, Joanna Doxey, assistant editor, Ron Hampton, production manager, Dennis Schaefer, cover designer, and Sharon Smith and Ceci Fleming, media producers, and the endless hours of hard work by Martha Morong and Geri Davis have led to products of which we are immensely proud.

Other strong support has come from Patty LaGree and Jeremy Pletcher for their accuracy checking of the manuscript. We also wish to recognize those who wrote scripts, presented lessons on camera, and checked the accuracy of the videos.

To the Student

As your authors, we would like to welcome you to this study of *Developmental Mathematics*. Whatever your past experiences, we encourage you to look at this mathematics course as a fresh start. Approach this course with a positive attitude about mathematics. Mathematics is a base for life, for many majors, for personal finances, for most careers, or just for pleasure.

You are the most important factor in the success of your learning. In earlier experiences, you may have allowed yourself to sit back and let the instructor "pour in" the learning, with little or no follow-up on your part. But now you must take a more assertive and proactive stance. This may be the first adjustment you make in college. As soon as possible after class, you should thoroughly read the textbook and the supplements and do all you can to learn on your own. In other words, rid yourself of former habits and take responsibility for your own learning. Then, with all the help you have around you, your hard work will lead to success.

One of the most important suggestions we can make is to allow yourself enough *time* to learn. You can have the best book, the best instructor, and the best supplements, but if you do not give yourself time to learn, how can they be of benefit? Many other helpful suggestions are presented in the Study Tips that you will find throughout the book. You may want to read through all the Study Tips before you begin the text. An Index of Study Tips can be found at the front of the book.

We wish you success.

M.L.B.
J.A.B.

Photo Credits

7, © Ed Quinn, CORBIS **8,** Tony Stone/Getty **9 (left),** © Jeff Greenberg/The Image Works **9 (right),** © CLARO CORTES IV/Reuters/Corbis **26,** © Kin Cheung/Reuters/Corbis **42,** © Michael S. Yamshita/CORBIS **61,** © Pete Saloutos/CORBIS **64,** CORBIS **102 (top),** © Keith Dannemiller/CORBIS **102 (bottom),** CORBIS **142,** © Bill Varie/CORBIS **146,** © Joseph Sohm; ChromoSohn, Inc./CORBIS **147,** John Deere **154 (top),** photodisc **154 (bottom),** © Royalty-Free/Corbis **156 (left),** © Ted Spiegel/CORBIS **157 (right),** © Royalty-Free/Corbis **160,** © Annie Griffiths Belt/CORBIS **169,** Greg Probst/Stone/Getty **170,** CORBIS **175 (left),** Plush Studios/Getty **175 (right),** © Rick Rickman/NewSport **181 (left),** DigitalVision/Getty **181 (right),** © Sion Touhig/Corbis **195,** Bettman/CORBIS **213,** Mark Gibson, DigitalVision/Getty **217,** Chuck Savage/CORBIS **218, 221, 224,** © Reuters/CORBIS **232,** Think-stock/Getty **241,** © Tom Nebbia/Corbis Stock Market **242,** © Royalty-Free/Corbis **248,** © Matt A. Brown/NewSport **250,** apwire photos **251,** Eyewire Collection **256 (left),** Brian Spurlock **256 (right),** Buccina Studios, photodisc **258,** Nicholas Devore III/Photographers Aspen **259 (left),** © Royalty-Free/Corbis **259 (right),** Albert J. Copley, photodisc **260 (top),** © JOE GIZA/Reuters/Corbis **260 (bottom left),** Jeremy Woodhouse, photodisc **260 (bottom right),** © Joe McDonald/CORBIS **262 (left),** Keith Brofsky, photodisc **262 (right),** © Kevin Fleming, Corbis **263,** © Royalty-Free/Corbis **264,** photodisc **265,** © Turbo/zefa/CORBIS **269 (top left), 269 (top right), 269 (bottom), 270,** photodisc **272,** © Guzelian Asadour/CORBIS **281, 282, 283, 287, 294,** © Royalty-Free/CORBIS **298,** DigitalVision **301,** photodisc **302,** © David H. Wells/CORBIS **317,** Christopher Bissell, Stone/Getty Images **321,** photodisc **323, 324,** © Royalty-Free/CORBIS **331 (left),** Patricia Clark, photodisc **331 (right),** Phil Schmermeister, Corbis **335,** © Royalty-Free/CORBIS **340,** Hulton Archive/Getty Images **341,** Reportage/Getty Images **347,** © Royalty-Free/CORBIS **351,** Cincinnati Zoo & Botanical Garden, Photo, S. David Jenike **352,** Stone/Getty Images **369,** © Larry Williams/CORBIS **399,** © Michael Prince/CORBIS **418,** © Nancy Dudley/Stock Boston **429,** © Richard Cummins/CORBIS **430,** Scott T. Baxer/photodisc **431,** © 1988 Al Satterwhite **474,** © Rob Howard/CORBIS **481,** AP Wire Services **489,** © Buddy Mays/CORBIS **498,** © Douglas Peebles/CORBIS **547,** Comstock Images **587,** © Jim Craigmyle/CORBIS **590 (top),** PhotoDisc **590 (bottom),** Brian Spurlock **592,** Appalachian Trail Conference **595,** Corbis **598,** © Patrik Giardino/CORBIS **601,** Doug Menuez, PhotoDisc **604,** EyeWire **605,** © Ed Bock/CORBIS **607,** EyeWire **622,** © Jeremy Birch **625,** © Charles O'Rear/CORBIS **628,** Andy Sacks, Stone/Getty **632,** © Lawrence Migdale, Tony Stone **670,** Karen Bittinger **676 (left),** Mary Clay/Tom Stack and Associates **676 (right),** Terje Rakke/The Image Bank **708,** Federal Highway Administration and Washington Infrastructure Services, Inc. **710,** Copyright © by Klev Schoening **729 (left),** © Dave G. Houser/CORBIS **729 (middle),** PhotoDisc **729 (right),** EyeWire **733 (left),** © Digital Art/CORBIS **733 (right),** PhotoDisc **736 (right),** Hoby Finn, PhotoDisc **737,** Bob Daemmrich/The Image Works **748,** Ron Chapple, FPG International **779,** Brian Spurlock **783,** © Nathan Benn/CORBIS **835,** PhotoDisc **865,** © LE SEGRETAIN PASCAL/CORBIS SYGMA **868,** PhotoDisc Collection **871 (top),** Duncan Smith/PhotoDisc **871 (bottom),** EyeWire **928,** © Kelly-Mooney Photography/CORBIS **933 (left),** © 1990 Glenn Randall **933 (right),** PhotoDisc **934,** © Anne Ryan/NewSport/Corbis **939,** PhotoLine/PhotoDisc **940 (left),** Geostock/PhotoDisc **940 (right),** Tom Turpin, Purdue University **941 (left),** Reuters/Anthony P. Bolante/Reuters NewMedia **941 (right),** © Reuters/CORBIS **942,** Joseph Sohm, ChromoSohn Inc./CORBIS **955,** PhotoDisc **956,** Betsie Van de Meer/Stone/Getty **957 (left),** © W. Cody/CORBIS **957 (right),** © Norbert Schaefer/Corbis Stock Market **960,** © ML Sinibaldi/CORBIS **962,** Hisham E. Ibrahim, PhotoDisc **963,** Steve Mason, PhotoDisc **992,** PhotoDisc **993,** © Richard Cummins/CORBIS **997,** Keith Brofsky/PhotoDisc **1001 (top),** © Dave Bartruff/CORBIS **1001 (bottom),** © Jeff Smith/Getty **1003,** CORBIS **1011,** Bettman/Corbis **1023,** Andre Lichenberg/FPG International **1024,** AP/Photo **1033,** Doug Menuez/PhotoDisc **1057,** © Galen Rowell/CORBIS **1107,** © Royalty-Free/Corbis **1108,** PhotoLink **1125,** Doug Menuez/PhotoDisc **1131,** © Joseph Sohm, ChromoSohn, Inc./CORBIS **1145,** © Owen Franken/CORBIS **1155,** © Tom and Dee Ann McCarthy/Corbis Stock Market **1156,** Digital Vision/photodisc **1158,** David Young-Wolff/PhotoEdit **1168,** © Galen Rowell/Corbis **1208,** Warren Morgan/Corbis

Bittinger Student Organizer

COURSE INFORMATION

Study Tips

Throughout this text, you will find a feature called *Study Tips*. We discuss these in the Preface of this text. They are intended to help improve your math study skills. An index of all the *Study Tips* can be found at the front of the book.

For Extra Help

MyMathLab

MathXL

Student's Solutions Manual

Video Lectures on CD

Math Tutor Center

InterAct Math

Additional Resources

Basic Math Review Card
(ISBN-10: 0-321-39476-3)
(ISBN-13: 978-0-321-39476-7)

Algebra Review Card
(ISBN-10: 0-321-39473-9)
(ISBN-13: 978-0-321-39473-6)

Math for Allied Health Reference Card
(ISBN-10: 0-321-39474-7)
(ISBN-13: 978-0-321-39474-3)

Graphing Calculator Reference Card
(ISBN-10: 0-321-39475-5)
(ISBN-13: 978-0-321-39475-0)

Spanish Basic Math Study Card
(ISBN-10: 0-321-43858-2)
(ISBN-13: 978-0-321-43854-4)

Math Study Skills for Students Video on CD
(ISBN-10: 0-321-29745-8)
(ISBN-13: 978-0-321-29745-7)

Go to www.aw-bc.com/math for more information.

On the first day of class, complete this chart and the weekly planner that follows on the reverse page.

Instructor Information:

Name ____________________

Office Hours and Location ____________________

____

Phone Number(s) ____________________

Fax Number ____________________

E-mail Address ____________________

Find the names of two students whom you could contact for class information or study questions:

1. Name ____________________

Phone Number(s) ____________________

E-mail Address ____________________

IM Name/Host ____________________

2. Name ____________________

Phone Number(s) ____________________

E-mail Address ____________________

IM Name/Host ____________________

Math Lab on Campus:

Location ____________________

Hours ____________________

Phone Number(s) ____________________

Tutoring:

Campus Location ____________________

Hours ____________________

To order the Addison-Wesley Math Tutor Center, call 1-888-777-0463.

(*See the Preface for important information concerning this tutoring.*)

Important Supplements: (*See the Preface for a complete list of available supplements.*)

Supplements recommended by the instructor ____________________

____

Online Log-in Information (*include access code, password, Web address, etc.*)

____

____

For more information about this text or any of its supplements, visit www.aw-bc.com/math

Bittinger Student Organizer

WEEKLY PLANNER

Success is planned. On this page, plan a typical week. Consider time allotments for class, study, work, travel, family, and relaxation.

Important Dates

Midterm Exam

____

Final Exam

____

Holidays

____
____
____
____
____

Other

____
____
____
____
____

TIME	Sun.	Mon.	Tues.	Wed.	Thurs.	Fri.	Sat.
6:00 A.M.							
6:30							
7:00							
7:30							
8:00							
8:30							
9:00							
9:30							
10:00							
10:30							
11:00							
11:30							
12:00 P.M.							
12:30							
1:00							
1:30							
2:00							
2:30							
3:00							
3:30							
4:00							
4:30							
5:00							
5:30							
6:00							
6:30							
7:00							
7:30							
8:00							
8:30							
9:00							
9:30							
10:00							
10:30							
11:00							
11:30							
12:00 A.M.							

Whole Numbers

1

Real-World Application

Races in which runners climb the steps inside a building are called "run-up" races. There are 2058 steps in the International Towerthon, Kuala Lumpur, Malaysia. Write expanded notation for the number of steps.

This problem appears as Exercise 13 in Section 1.1.

1.1 STANDARD NOTATION; ORDER

Objectives

a Give the meaning of digits in standard notation.

b Convert from standard notation to expanded notation.

c Convert between standard notation and word names.

d Use < or > for □ to write a true sentence in a situation like 6 □ 10.

We study mathematics in order to be able to solve problems. In this section, we study how numbers are named. We begin with the concept of place value.

a Place Value

Consider the numbers in the following table.

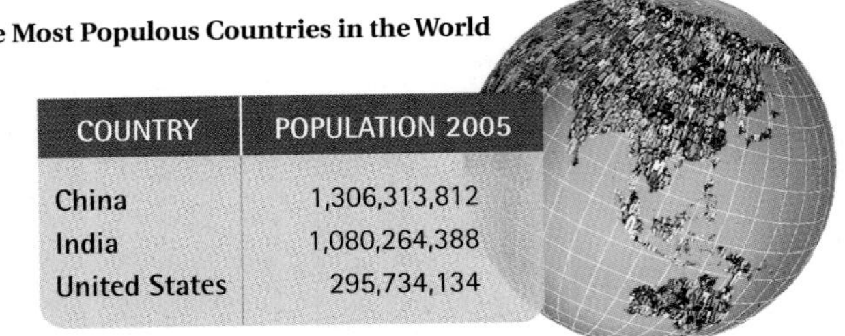

Three Most Populous Countries in the World

COUNTRY	POPULATION 2005
China	1,306,313,812
India	1,080,264,388
United States	295,734,134

Source: The World Factbook, July 2005 estimates

A **digit** is a number 0, 1, 2, 3, 4, 5, 6, 7, 8, or 9 that names a place-value location. For large numbers, digits are separated by commas into groups of three, called **periods.** Each period has a name: *ones, thousands, millions, billions, trillions,* and so on. To understand the population of China in the table above, we can use a **place-value chart,** as shown below.

PLACE-VALUE CHART

Periods →	Trillions			Billions			Millions			Thousands			Ones		
						1	3	0	6	3	1	3	8	1	2
	Hundreds	Tens	Ones	Hundreds	Tens	Ones	Hundreds	Tens	Ones	Hundreds	Tens	Ones	Hundreds	Tens	Ones
				1 billion			306 millions			313 thousands			812 ones		

EXAMPLES What does the digit 8 mean in each number?

1. 278,342 — 8 thousands
2. 872,342 — 8 hundred thousands
3. 28,343,399,223 — 8 billions
4. 1,023,850 — 8 hundreds
5. 98,413,099 — 8 millions
6. 6328 — 8 ones

Do Margin Exercises 1–6.

What does the digit 2 mean in each number?

1. 526,555

2. 265,789

3. 42,789,654

4. 24,789,654

5. 8924

6. 5,643,201

To the student:

In the preface, at the front of the text, you will find a Student Organizer card. This pullout card will help you keep track of important dates and useful contact information. You can also use it to plan time for class, study, work, and relaxation. By managing your time wisely, you will provide yourself the best possible opportunity to be successful in this course.

Answers on page A-1

EXAMPLE 7 *American Red Cross.* In 2003, private donations to the American Red Cross totaled about $587,492,000. What does each digit name?

Sources: *The Chronicle of Philanthropy*

Do Exercise 7.

b Converting from Standard Notation to Expanded Notation

To answer questions such as "How many?", "How much?", and "How tall?", we use whole numbers. The set, or collection, of **whole numbers** is

0, 1, 2, 3, 4, 5, 6, 7, 8, 9, 10, 11, 12, ...

The set goes on indefinitely. There is no largest whole number, and the smallest whole number is 0. Each whole number can be named using various notations. The set 1, 2, 3, 4, 5, . . . , without 0, is called the set of **natural numbers.**

Let's look at the data from the bar graph shown here.

Fewer Computer Majors

Source: Computing Research Association Taulbee Survey

The number of computer majors in 2003 was 17,706. **Standard notation** for the number of computer majors is 17,706. We write **expanded notation** for 17,706 as follows:

17,706 = 1 ten thousand + 7 thousands
+ 7 hundreds + 0 tens + 6 ones.

7. Presidential Library. In the first year of operation, 280,219 people visited the Ronald Reagan Presidential Library in Simi Valley, California. What does each digit name?

Sources: *USA Today* research by Bruce Rosenstein; National Archives & Records Administration; Associated Press

Write expanded notation.

8. 1895

9. 23,416, the number of computer majors in 2000

10. 3031 mi (miles), the diameter of Mercury

Answers on page A-1

11. 4180 mi, the length of the Nile River, the longest river in the world

12. 154,616, the number of Labrador retrievers registered in 2002
Source: The American Kennel Club

Write a word name. (Refer to the figure at right.)

13. 49, the total number of medals won by Australia

14. 16, the number of silver medals won by Germany

15. 38, the number of bronze medals won by Russia

Answers on page A-1

EXAMPLE 8 Write expanded notation for 4218 mi, the diameter of Mars.

$$4218 = 4 \text{ thousands} + 2 \text{ hundreds} + 1 \text{ ten} + 8 \text{ ones}$$

EXAMPLE 9 Write expanded notation for 3400.

$$3400 = 3 \text{ thousands} + 4 \text{ hundreds} + 0 \text{ tens} + 0 \text{ ones}, \quad \text{or}$$
$$3 \text{ thousands} + 4 \text{ hundreds}$$

EXAMPLE 10 Write expanded notation for 563,384, the population of Washington, D.C.

$$563{,}384 = 5 \text{ hundred thousands} + 6 \text{ ten thousands} + 3 \text{ thousands} + 3 \text{ hundreds} + 8 \text{ tens} + 4 \text{ ones}$$

Do Exercises 8–12. (Exercises 8–10 are on the preceding page.)

C Converting Between Standard Notation and Word Names

We often use **word names** for numbers. When we pronounce a number, we are speaking its word name. Russia won 92 medals in the 2004 Summer Olympics in Athens, Greece. A word name for 92 is "ninety-two." Word names for some two-digit numbers like 27, 39, and 92 use hyphens. Others like 17 use only one word, "seventeen."

TOP COUNTRIES IN SUMMER OLYMPICS 2004	MEDAL COUNT			
	GOLD	SILVER	BRONZE	TOTAL
United States of America	35	39	29	103
Russia	27	27	38	92
People's Republic of China	32	17	14	63
Australia	17	16	16	49
Germany	14	16	18	48

Source: 2004 Olympics, Athens, Greece

EXAMPLES Write a word name.

11. 35, the total number of gold medals won by the United States

Thirty-five

12. 17, the number of silver medals won by the People's Republic of China

Seventeen

Do Exercises 13–15 on the preceding page.

For word names for larger numbers, we begin at the left with the largest period. The number named in the period is followed by the name of the period; then a comma is written and the next period is named.

EXAMPLE 13 Write a word name for 46,605,314,732.

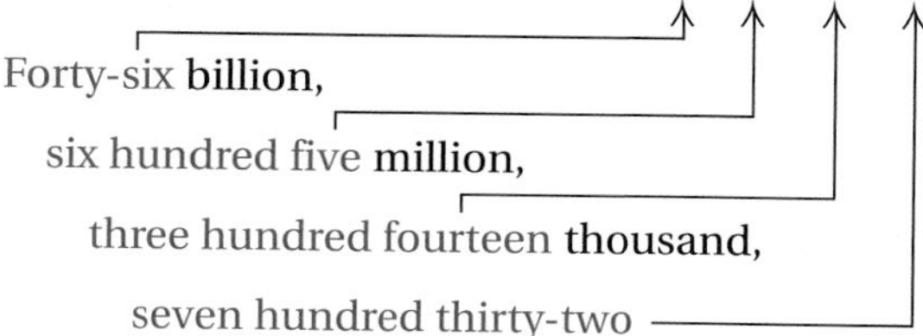

The word "and" *should not* appear in word names for whole numbers. Although we commonly hear such expressions as "two hundred *and* one," the use of "and" is not, strictly speaking, correct in word names for whole numbers. For decimal notation, it is appropriate to use "and" for the decimal point. For example, 317.4 is read as "three hundred seventeen *and* four tenths."

Do Exercises 16–19.

EXAMPLE 14 Write standard notation.

Five hundred six million,
three hundred forty-five thousand,
two hundred twelve

Standard notation is 506,345,212.

Do Exercise 20.

Write a word name.

16. 204

17. $44,155, the average salary in 2001 for those who have a bachelor's degree or more

Source: U.S. Bureau of the Census

18. 1,879,204

19. 6,449,000,000, the world population in 2005

Source: U.S. Bureau of the Census

20. Write standard notation.

Two hundred thirteen million, one hundred five thousand, three hundred twenty-nine

Answers on page A-1

Use < or > for $\square$ to write a true sentence. Draw a number line if necessary.

21. 8 $\square$ 12

22. 12 $\square$ 8

23. 76 $\square$ 64

24. 64 $\square$ 76

25. 217 $\square$ 345

26. 345 $\square$ 217

Answers on page A-1

d Order

We know that 2 is not the same as 5. We express this by the sentence $2 \neq 5$. We also know that 2 is less than 5. We symbolize this by the expression $2 < 5$. We can see this order on the number line: 2 is to the left of 5. The number 0 is the smallest whole number.

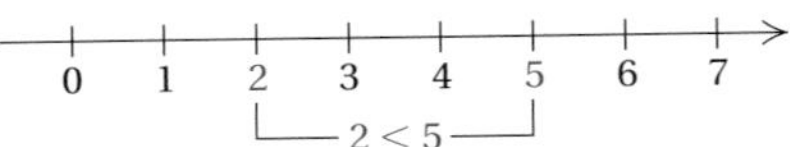

ORDER OF WHOLE NUMBERS

For any whole numbers a and b:

1. $a < b$ (read "a is less than b") is true when a is to the left of b on the number line.
2. $a > b$ (read "a is greater than b") is true when a is to the right of b on the number line.

We call < and > **inequality symbols.**

EXAMPLE 15 Use < or > for $\square$ to write a true sentence: 7 $\square$ 11.

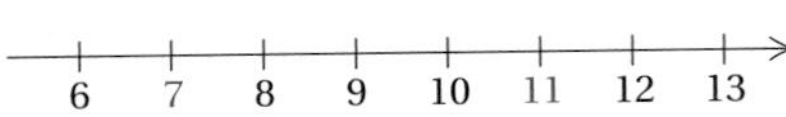

Since 7 is to the left of 11 on the number line, $7 < 11$.

EXAMPLE 16 Use < or > for $\square$ to write a true sentence: 92 $\square$ 87.

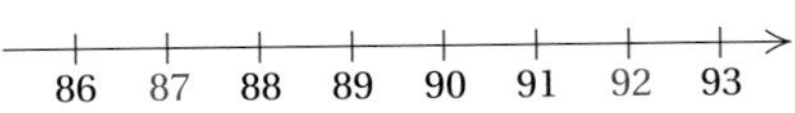

Since 92 is to the right of 87 on the number line, $92 > 87$.

A sentence like $8 + 5 = 13$ is called an **equation.** It is a *true* equation. The equation $4 + 8 = 11$ is a *false* equation. A sentence like $7 < 11$ is called an **inequality.** The sentence $7 < 11$ is a *true* inequality. The sentence $23 > 69$ is a *false* inequality.

Do Exercises 21–26.

EXERCISE SET

For Extra Help

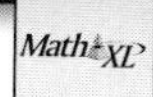 MathXL  MyMathLab  InterAct Math  Math Tutor Center Video Lectures on CD Disc 1  Student's Solutions Manual

 What does the digit 5 mean in each case?

1. 235,888

2. 253,777

3. 1,488,526

4. 500,736

Used Cars. 1,582,370 certified used cars were sold in 2004 in the United States. **Source:** *Motor Trend,* April 2005, p. 26

In the number 1,582,370, what digit names the number of:

5. thousands?

6. ones?

7. millions?

8. hundred thousands?

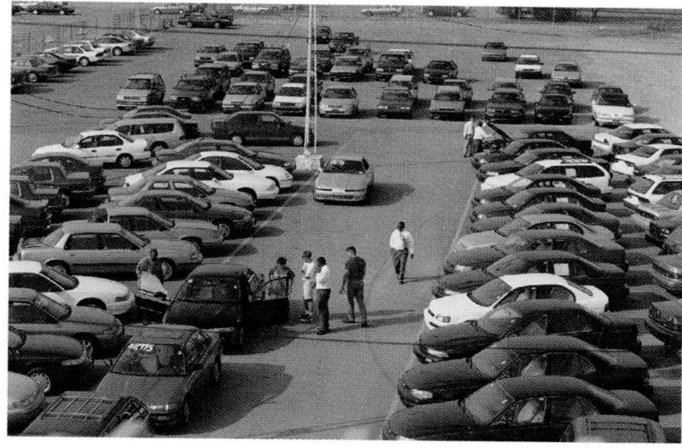

b Write expanded notation.

9. 5702

10. 3097

11. 93,986

12. 38,453

Step-Climbing Races. Races in which runners climb the steps inside a building are called "run-up" races. The graph below shows the number of steps in four buildings. In Exercises 13–16, write expanded notation for the number of steps in each race.

Step-Climbing Races

Source: New York Road Runners Club

13. 2058 steps in the International Towerthon, Kuala Lumpur, Malaysia

14. 1776 steps in the CN Tower Run-Up, Toronto, Ontario, Canada

15. 1268 steps in the World Financial Center, New York City, New York

16. 1081 steps in the Skytower Run-Up, Auckland, New Zealand

Overseas Travelers. The chart below shows the residence and number of overseas travelers to the United States in 2003. In Exercises 17–22, write expanded notation for the number of travelers from each country.

OVERSEAS TRAVELERS TO U.S., 2003

Country	Travelers
Australia	405,698
Brazil	348,945
Germany	1,180,212
India	272,161
Japan	3,169,682
Spain	284,031

Source: U.S. Department of Commerce, ITA, Office of Travel and Tourism Industries, 2004

17. 405,698 from Australia

18. 3,169,682 from Japan

19. 272,161 from India

20. 284,031 from Spain

21. 1,180,212 from Germany

22. 348,945 from Brazil

C Write a word name.

23. 85

24. 48

25. 88,000

26. 45,987

27. 123,765

28. 111,013

29. 7,754,211,577

30. 43,550,651,808

Write standard notation.

31. Two million, two hundred thirty-three thousand, eight hundred twelve

32. Three hundred fifty-four thousand, seven hundred two

33. Eight billion

34. Seven hundred million

Write a word name for the number in the sentence.

35. *Great Pyramid.* The area of the base of the Great Pyramid in Egypt is 566,280 square feet.

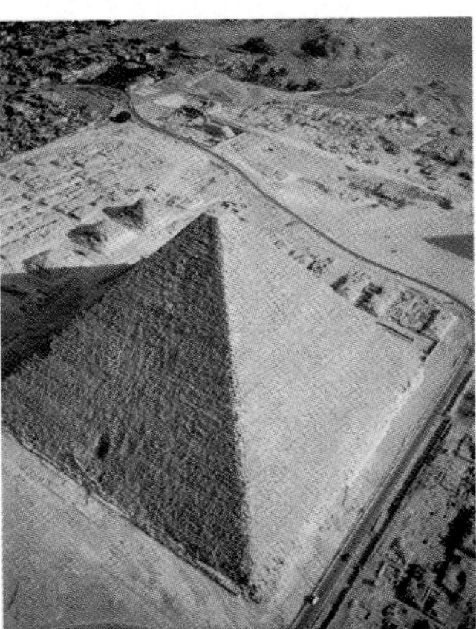

36. *Population of the United States.* The population of the United States in July 2005 was estimated to be 295,734,134.

Source: *The World Factbook*

37. *Busiest Airport.* In 2003, the world's busiest airport, Hartsfield in Atlanta, had 76,086,792 passengers.

Source: Airports Council International World Headquarters, Geneva, Switzerland

38. *Prisoners.* There was a total of 2,078,570 prisoners, federal, state, and local, in the United States in 2003.

Source: *Prison and Jail Inmates at Midyear 2003*, U.S. Bureau of Justice Statistics

Write standard notation for the number in the sentence.

39. Light travels nine trillion, four hundred sixty billion kilometers in one year.

40. The distance from the sun to Pluto is three billion, six hundred sixty-four million miles.

41. *Pacific Ocean.* The area of the Pacific Ocean is sixty-four million, one hundred eighty-six thousand square miles.

42. *Internet Users.* In a recent year, there were fifty-four million, five hundred thousand Internet users in China.

Source: Computer Industry Almanac, Inc.

d Use $<$ or $>$ for $\square$ to write a true sentence. Draw a number line if necessary.

43. 0 $\square$ 17

44. 32 $\square$ 0

45. 34 $\square$ 12

46. 28 $\square$ 18

47. 1000 $\square$ 1001

48. 77 $\square$ 117

49. 133 $\square$ 132

50. 999 $\square$ 997

51. 460 $\square$ 17

52. 345 $\square$ 456

53. 37 $\square$ 11

54. 12 $\square$ 32

Daily Newspapers. The top three daily newspapers in the United States are *USA Today,* the *Wall Street Journal,* and the *New York Times.* The daily circulation of each as of September 30, 2002, is listed in the table below. Use this table when answering Exercises 55 and 56.

NEWSPAPER	DAILY CIRCULATION
USA Today	2,136,068
Wall Street Journal	1,800,607
New York Times	1,113,000

Source: Editor and Publisher International Year Book 2003

55. Use an inequality to compare the daily circulation of the *Wall Street Journal* and *USA Today.*

56. Use an inequality to compare the daily circulation of *USA Today* and the *New York Times.*

57. *Pedestrian Fatalities.* The annual number of pedestrians killed when hit by a motor vehicle has declined from 6482 in 1990 to 4827 in 2003. Use an inequality to compare these annual totals of pedestrian fatalities.

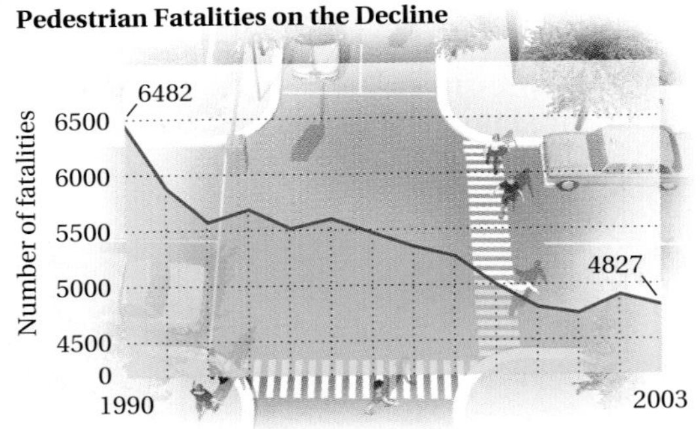

Source: National Center for Statistics and Analysis

58. *Life Expectancy.* The life expectancy of a female in 2010 is predicted to be about 82 yr and of a male about 76 yr. Use an inequality to compare these life expectancies.

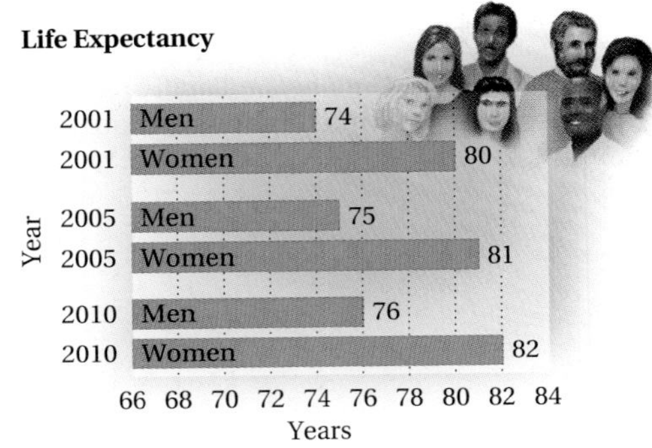

Source: U.S. Bureau of the Census

To the student and the instructor: The Discussion and Writing exercises are meant to be answered with one or more sentences. They can be discussed and answered collaboratively by the entire class or by small groups. Because of their open-ended nature, the answers to these exercises do not appear at the back of the book. They are denoted by the symbol **DW**.

59. **DW** Explain why we use commas when writing large numbers.

60. **DW** Write an English sentence in which the number 370,000,000 is used.

SYNTHESIS

To the student and the instructor: The Synthesis exercises found at the end of every exercise set challenge students to combine concepts or skills studied in that section or in preceding parts of the text. Exercises marked with a 🖩 symbol are meant to be solved using a calculator.

61. How many whole numbers between 100 and 400 contain the digit 2 in their standard notation?

62. 🖩 What is the largest number that you can name on your calculator? How many digits does that number have? How many periods?

1.2 ADDITION AND SUBTRACTION

Objectives

a Add whole numbers.

b Use addition in finding perimeter.

c Convert between addition sentences and subtraction sentences.

d Subtract whole numbers.

a Addition of Whole Numbers

Addition of whole numbers corresponds to combining or putting things together.

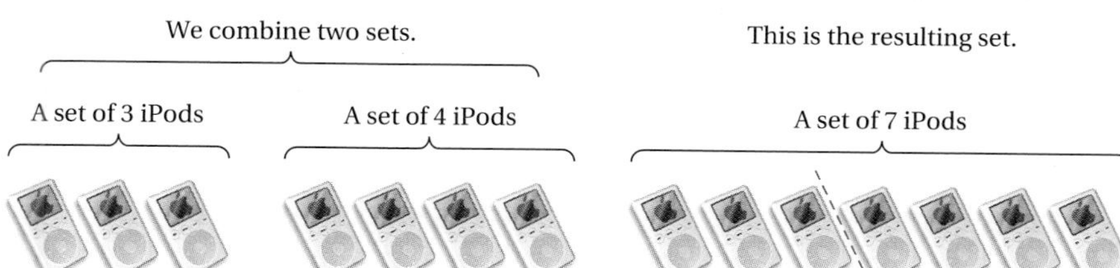
We combine two sets.
This is the resulting set.
A set of 3 iPods
A set of 4 iPods
A set of 7 iPods

The addition that corresponds to the figure above is $3 + 4 = 7$. We say that the **sum** of 3 and 4 is 7. The numbers added are called **addends.**

Addition also corresponds to moving distances on a number line. The number line at right is marked with tick marks at equal distances of 1 *unit.* The sum $3 + 4$ is shown. We first move 3 units from 0, and then 4 more units, and end up at 7. The addition that corresponds to the situation is $3 + 4 = 7$.

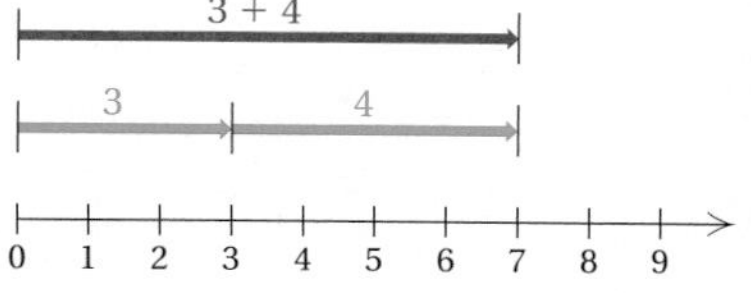
3 + 4
3
4
0 1 2 3 4 5 6 7 8 9

To add whole numbers, we add the ones digits first, then the tens, then the hundreds, then the thousands, and so on. Adding 0 to a number does not change the number: $a + 0 = 0 + a = a$. That is, $6 + 0 = 0 + 6 = 6$, or $198 + 0 = 0 + 198 = 198$. We say that 0 is the **additive identity.**

EXAMPLE 1 Add: $6878 + 4995$.

Place values are lined up in columns.

$$\begin{array}{r} {}^{1} \\ 6\,8\,7\,8 \\ +\;4\,9\,9\,5 \\ \hline 3 \end{array}$$

Add ones. We get 13 ones, or 1 ten + 3 ones. Write 3 in the ones column and 1 above the tens. This is called *carrying,* or *regrouping.*

$$\begin{array}{r} {}^{1}\;{}^{1} \\ 6\,8\,7\,8 \\ +\;4\,9\,9\,5 \\ \hline 7\,3 \end{array}$$

Add tens. We get 17 tens, or 1 hundred + 7 tens. Write 7 in the tens column and 1 above the hundreds.

$$\begin{array}{r} {}^{1}\;{}^{1}\;{}^{1} \\ 6\,8\,7\,8 \\ +\;4\,9\,9\,5 \\ \hline 8\,7\,3 \end{array}$$

Add hundreds. We get 18 hundreds, or 1 thousand + 8 hundreds. Write 8 in the hundreds column and 1 above the thousands.

Moving to the thousands, we have

$$\begin{array}{r} {}^{1}\;{}^{1}\;{}^{1} \\ 6\,8\,7\,8 \\ +\;4\,9\,9\,5 \\ \hline 1\,1\,8\,7\,3 \end{array}$$

Add thousands. We get 11 thousands.

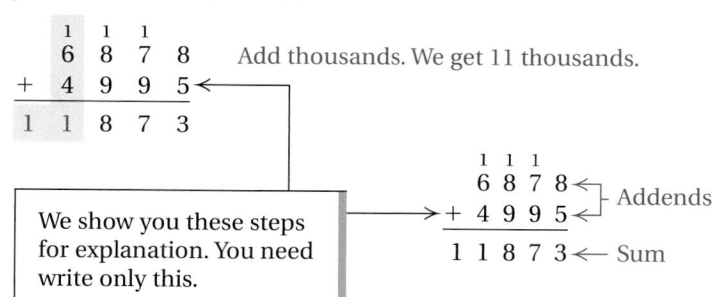

We show you these steps for explanation. You need write only this.

$$\begin{array}{r} {}^{1}\;{}^{1}\;{}^{1} \\ 6\,8\,7\,8 \\ +\;4\,9\,9\,5 \\ \hline 1\,1\,8\,7\,3 \end{array}$$

(6878 and 4995: Addends; 11873: Sum)

Add.

1. $\begin{array}{r} 7\,9\,6\,8 \\ +\;5\,4\,9\,7 \\ \hline \end{array}$

2. $6203 + 3542$

Answers on page A-1

3. $\begin{array}{r} 9804 \\ +\ 6378 \\ \hline \end{array}$

4. $\begin{array}{r} 1932 \\ 6723 \\ 9878 \\ +\ 8941 \\ \hline \end{array}$

Answers on page A-1

EXAMPLE 2 Add: 391 + 276 + 789 + 498.

$$\begin{array}{r} \overset{2}{3}\,9\,1 \\ 2\,7\,6 \\ 7\,8\,9 \\ +\ 4\,9\,8 \\ \hline 4 \end{array}$$

Add ones. We get 24, so we have 2 tens + 4 ones. Write 4 in the ones column and 2 above the tens.

$$\begin{array}{r} \overset{3}{3}\,\overset{2}{9}\,1 \\ 2\,7\,6 \\ 7\,8\,9 \\ +\ 4\,9\,8 \\ \hline 5\,4 \end{array}$$

Add tens. We get 35 tens, so we have 30 tens + 5 tens. This is also 3 hundreds + 5 tens. Write 5 in the tens column and 3 above the hundreds.

$$\begin{array}{r} \overset{3}{3}\,\overset{2}{9}\,1 \\ 2\,7\,6 \\ 7\,8\,9 \\ +\ 4\,9\,8 \\ \hline 1\,9\,5\,4 \end{array}$$

Add hundreds. We get 19 hundreds.

Do Exercises 1–4. (Exercises 1 and 2 are on the preceding page.)

CALCULATOR CORNER

Adding Whole Numbers *To the student and the instructor:* This is the first of a series of *optional* discussions on using a calculator. A calculator is *not* a requirement for this textbook. There are many kinds of calculators and different instructions for their usage. We have included instructions here for a minimum-cost calculator. Be sure to consult your user's manual as well. Also, check with your instructor about whether you are allowed to use a calculator in the course.

To add whole numbers on a calculator, we use the [+] and [=] keys. For example, to add 57 and 34, we press [5] [7] [+] [3] [4] [=]. The calculator displays [91], so 57 + 34 = 91. To find 314 + 259 + 478, we press [3] [1] [4] [+] [2] [5] [9] [+] [4] [7] [8] [=]. The display reads [1051], so 314 + 259 + 478 = 1051.

Exercises: Use a calculator to find each sum.

1. 19 + 36
2. 73 + 48
3. 925 + 677
4. 276 + 458
5. $\begin{array}{r} 826 \\ 415 \\ +\ 691 \\ \hline \end{array}$
6. $\begin{array}{r} 253 \\ 490 \\ +\ 121 \\ \hline \end{array}$

b Finding Perimeter

Addition can be used when finding perimeter.

PERIMETER

The distance around an object is its **perimeter.**

EXAMPLE 3 Find the perimeter of the figure.

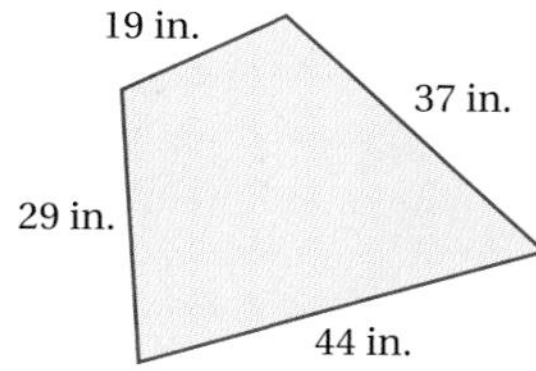

We add the lengths of the sides:

Perimeter = 29 in. + 19 in. + 37 in. + 44 in.

We carry out the addition as follows.

$$\begin{array}{r} {}^{2} \\ 29 \\ 19 \\ 37 \\ +\ 44 \\ \hline 129 \end{array}$$

The perimeter of the figure is 129 in.

Do Exercises 5 and 6.

EXAMPLE 4 Find the perimeter of the octagonal (eight-sided) resort swimming pool.

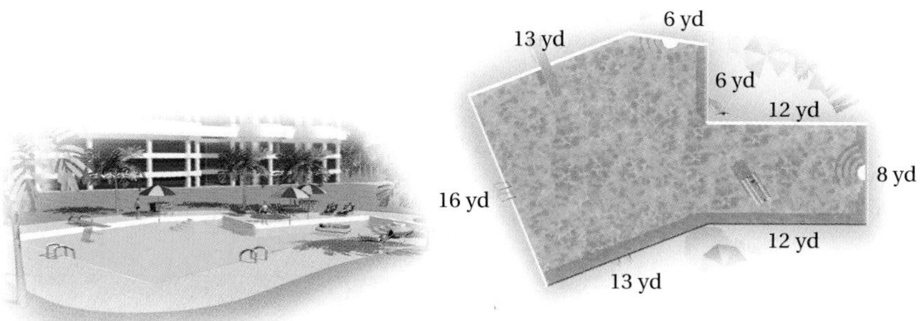

Perimeter = 13 yd + 6 yd + 6 yd + 12 yd + 8 yd + 12 yd + 13 yd + 16 yd

The perimeter of the pool is 86 yd.

Do Exercise 7.

Find the perimeter of the figure.

5.

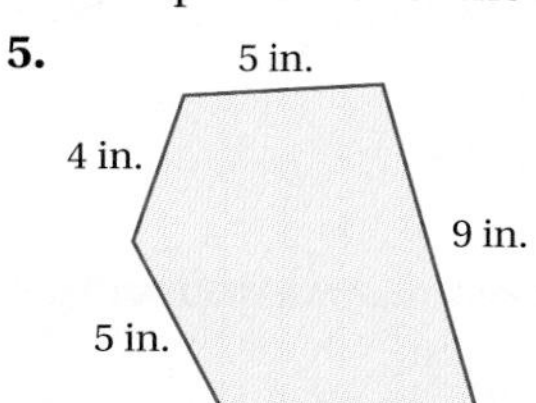

6.

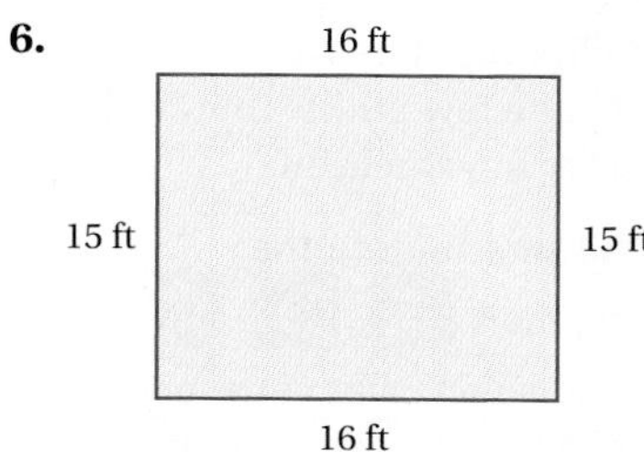

Solve.

7. Index Cards. Two standard sizes for index cards are 3 in. (inches) by 5 in. and 5 in. by 8 in. Find the perimeter of each card.

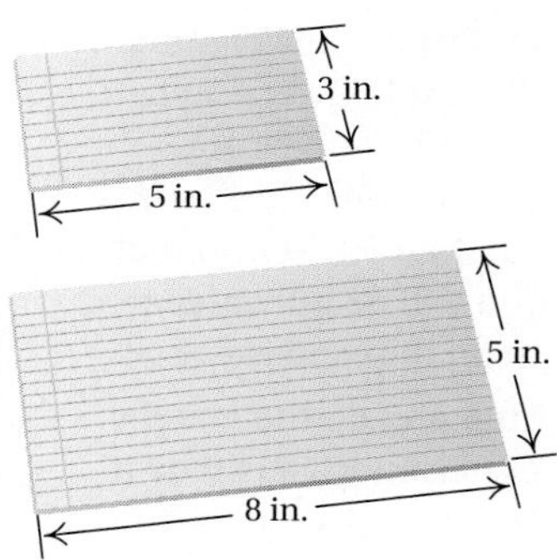

Answers on page A-1

c Subtraction and Related Sentences

TAKE AWAY

Subtraction of whole numbers applies to two kinds of situations. The first is called "take away." Consider the following example.

A bowler starts with 10 pins and knocks down 8 of them.

From 10 pins, the bowler "takes away" 8 pins. There are 2 pins left. The subtraction is $10 - 8 = 2$.

10

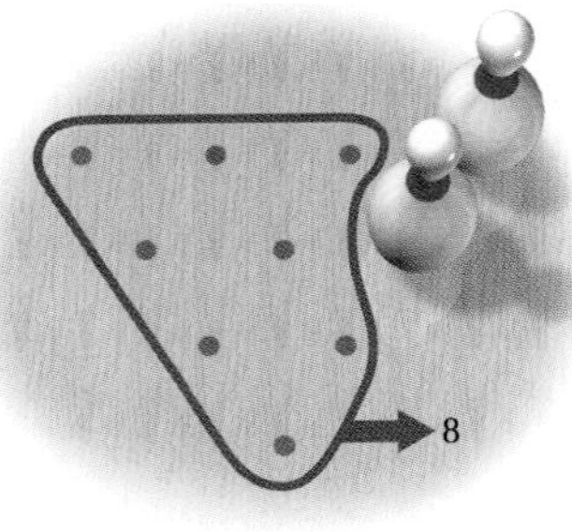

$10 - 8 = 2$

We use the following terminology with subtraction:

10	−	8	=	2 .
↓		↓		↓
Minuend		Subtrahend		Difference

The **minuend** is the number from which another number is being subtracted. The **subtrahend** is the number being subtracted. The **difference** is the result of subtracting the subtrahend from the minuend.

Subtraction also corresponds to moving distances on a number line. The number line below is marked with tick marks at equal distances of 1 unit. The difference $10 - 8$ is shown. We first move from 0 right 10 units, and then left 8 units, and end up at 2. The subtraction that corresponds to the situation is $10 - 8 = 2$.

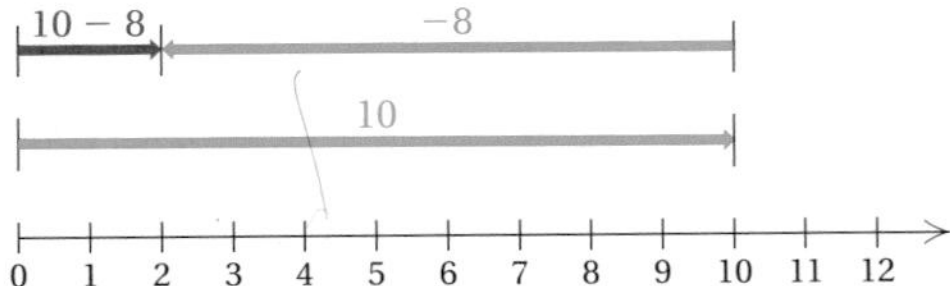

This leads us to the following definition of subtraction.

> **SUBTRACTION**
>
> The difference $a - b$ is that unique whole number c for which $a = c + b$.

For example, $13 - 4$ is the number 9 since $13 = 9 + 4$.

Study Tips

USING THIS TEXTBOOK

Throughout this textbook, you will find a feature called "Study Tips." One of the most important ways in which to improve your math study skills is to learn the proper use of the textbook. Here we highlight a few points that we consider most helpful.

- **Be sure to note the symbols a, b, c, and so on, that correspond to the objectives you are to master in each section.** The first time you see them is in the margin at the beginning of the section; the second time is in the subheadings of each section; and the third time is in the exercise set for the section.
- **Read and study each step of each example.** The examples include important side comments that explain each step. These examples and annotations have been carefully chosen so that you will be fully prepared to do the exercises.
- **Stop and do the margin exercises as you study a section.** This gives you immediate reinforcement of each concept as it is introduced and is one of the most effective ways to master the mathematical skills in this text. Don't deprive yourself of this benefit!

RELATED SENTENCES

Subtraction is defined in terms of addition. For example, $5 - 2$ is that number which, when added to 2, gives 5. Thus for the subtraction sentence

$5 - 2 = 3$, Taking away 2 from 5 gives 3.

there is a *related addition sentence*

$5 = 3 + 2$. Putting back the 2 gives 5 again.

In fact, we know that answers we find to subtractions are correct only because of the related addition, which provides a handy way to *check* a subtraction.

EXAMPLE 5 Write a related addition sentence: $8 - 5 = 3$.

$8 - 5 = 3$

This number gets added.

$8 = 3 + 5$

By the commutative law of addition, there is also another addition sentence:

$8 = 5 + 3$.

The related addition sentence is $8 = 3 + 5$.

Do Exercises 8 and 9.

EXAMPLE 6 Write two related subtraction sentences: $4 + 3 = 7$.

$4 + 3 = 7$

This addend gets subtracted from the sum.

$4 = 7 - 3$

(7 take away 3 is 4.)

$4 + 3 = 7$

This addend gets subtracted from the sum.

$3 = 7 - 4$

(7 take away 4 is 3.)

The related subtraction sentences are $4 = 7 - 3$ and $3 = 7 - 4$.

Do Exercises 10 and 11.

HOW MANY MORE

The second kind of situation to which subtraction can apply is called "how many more." You have 2 notebooks, but you need 7. You can think of this as "how many do I need to add to 2 to get 7?" Finding the answer can be thought of as finding a missing addend, and can be found by subtracting 2 from 7.

Need 7 notebooks

Have 2 notebooks

5 notebooks

What must be added to 2 to get 7? The answer is 5.

Missing addend: $2 + \square = 7$

Difference: $7 - 2 = \square$

Write a related addition sentence.

8. $7 - 5 = 2$

9. $17 - 8 = 9$

Write two related subtraction sentences.

10. $5 + 8 = 13$

11. $11 + 3 = 14$

Answers on page A-1

Let's look at the following example in which a missing addend occurs: Jillian wants to buy the roll-on luggage shown in this ad. She has $30. She needs $79. How much more does she need in order to buy the luggage?

Thinking of this situation in terms of a missing addend, we have:

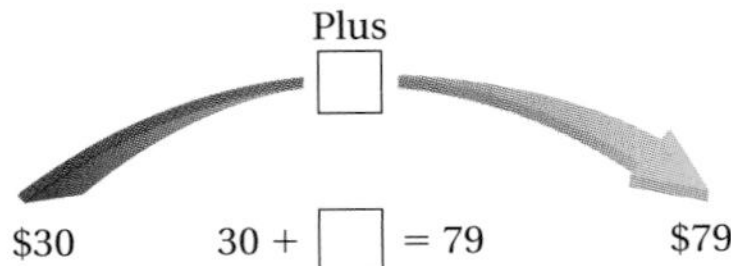

To find the answer, we think of the related subtraction sentence:

$$30 + \square = 79$$

$$\square = 79 - 30.$$

12. Subtract.

$$\begin{array}{r} 7\,8\,9\,3 \\ -\;4\,0\,9\,2 \\ \hline \end{array}$$

Answer on page A-1

d Subtraction of Whole Numbers

To subtract numbers, we subtract the ones digits first, then the tens digits, then the hundreds, then the thousands, and so on.

EXAMPLE 7 Subtract: 9768 − 4320.

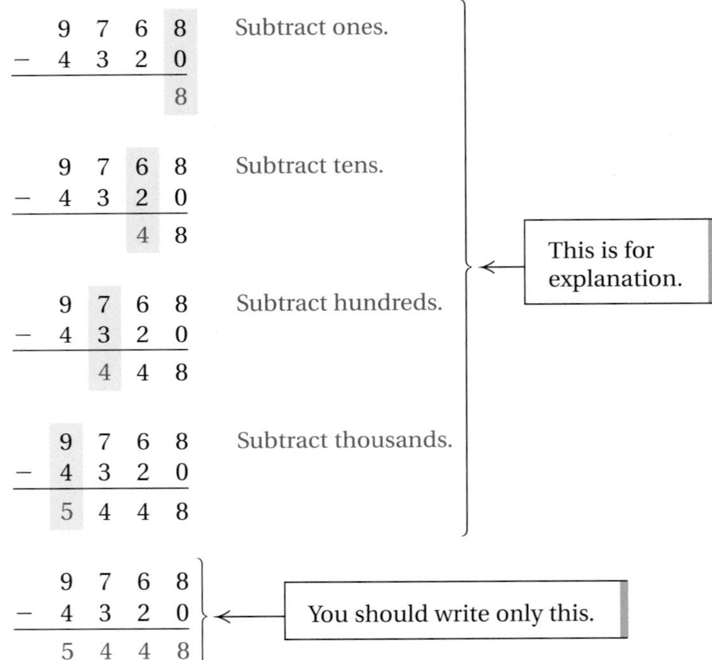

9 7 6 8 − 4 3 2 0 = 8 Subtract ones.
9 7 6 8 − 4 3 2 0 = 4 8 Subtract tens.
9 7 6 8 − 4 3 2 0 = 4 4 8 Subtract hundreds.
9 7 6 8 − 4 3 2 0 = 5 4 4 8 Subtract thousands.
This is for explanation.
9 7 6 8 − 4 3 2 0 = 5 4 4 8
You should write only this.

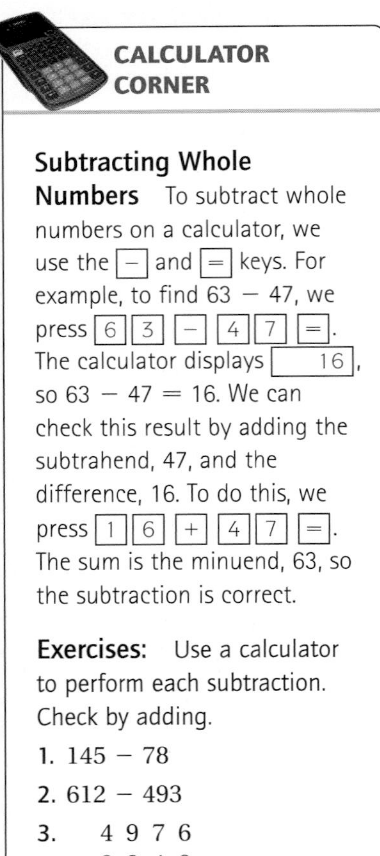

CALCULATOR CORNER

Subtracting Whole Numbers To subtract whole numbers on a calculator, we use the [−] and [=] keys. For example, to find 63 − 47, we press [6] [3] [−] [4] [7] [=]. The calculator displays [16], so 63 − 47 = 16. We can check this result by adding the subtrahend, 47, and the difference, 16. To do this, we press [1] [6] [+] [4] [7] [=]. The sum is the minuend, 63, so the subtraction is correct.

Exercises: Use a calculator to perform each subtraction. Check by adding.

1. 145 − 78
2. 612 − 493
3. $\begin{array}{r} 4\,9\,7\,6 \\ -\;2\,8\,4\,8 \\ \hline \end{array}$
4. $\begin{array}{r} 1\,2{,}4\,0\,6 \\ -\;9\,8\,1\,3 \\ \hline \end{array}$

We have considered the subtraction 9768 − 4320 = □. That is, we have found the missing addend in the sentence 9768 = 4320 + □. If 5448 is indeed the missing addend, then if we add it to 4320, the answer should be 9768. The related addition sentence is the basis for adding as a *check*.

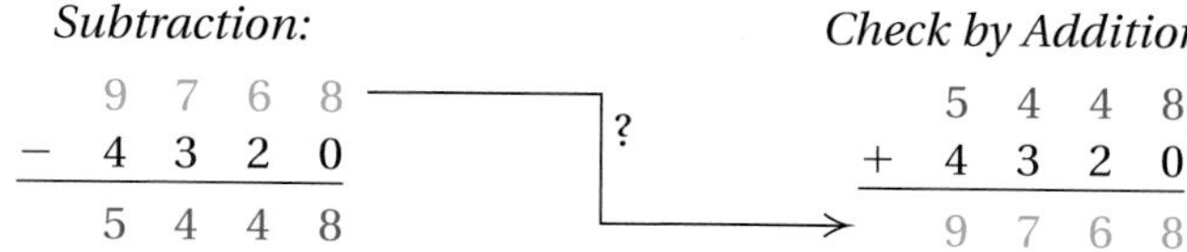

Do Exercise 12 on the preceding page.

EXAMPLE 8 Subtract: 348 − 165.

We have

3 hundreds + 4 tens + 8 ones	=	2 hundreds + 14 tens + 8 ones
− 1 hundred − 6 tens − 5 ones	=	− 1 hundred − 6 tens − 5 ones
	=	1 hundred + 8 tens + 3 ones
	=	183.

Note that in this case, although we can subtract the ones (8 − 5 = 3), we cannot do so with the tens, because 4 − 6 is *not* a whole number. To see why, consider

4 − 6 = □ and the related addition sentence 4 = □ + 6.

There is no whole number that when added to 6 gives 4. To complete the subtraction, we must *borrow* 1 hundred from 3 hundreds and regroup it with 4 tens. Then we can do the subtraction 14 tens − 6 tens = 8 tens. Below we consider a shortened form.

```
  3 4 8
− 1 6 5
      3
```
Subtract ones.

```
  2 14
  3̸ 4̸ 8
− 1  6 5
       3
```
Borrow one hundred. That is, 1 hundred = 10 tens, and 10 tens + 4 tens = 14 tens. Write 2 above the hundreds column and 14 above the tens.

```
  2 14
  3̸ 4̸ 8
− 1  6 5
  1  8 3
```
Subtract tens; subtract hundreds.

This is what you should write. →

```
  2 14                 1
  3̸ 4̸ 8   Check:      1 8 3
− 1  6 5  ——————→  + 1 6 5
  1  8 3             3 4 8 ←
```

The answer checks because this is the top number in the subtraction.

Study Tips

LEARNING RESOURCES

Are you aware of all the learning resources that exist for this textbook? Many details are given in the Preface.

- The *Student's Solutions Manual* contains worked-out solutions for the odd-numbered exercises in the exercise sets and for *all* the end-of-chapter Review Exercises and Chapter Test exercises.
- An extensive set of *videos* supplements this text. These are available on CD-ROM by calling 1-800-282-0693.
- *Tutorial software* called Math XL Tutorials on CD is available with this text. If it is not available in the campus learning center, you can order it by calling 1-800-282-0693.
- The Addison-Wesley *Math Tutor Center* has experienced instructors available to help with the odd-numbered exercises. You can order this service by calling 1-800-824-7799.
- Extensive help is available online via MyMathLab and/or MathXL. Ask your instructor for information about these or visit MyMathLab.com and MathXL.com.

Subtract. Check by adding.

13. $\begin{array}{r} 8686 \\ -\ 2358 \\ \hline \end{array}$

14. $\begin{array}{r} 7145 \\ -\ 2398 \\ \hline \end{array}$

EXAMPLE 9 Subtract: 6246 − 1879.

$$\begin{array}{rcccc} & & & 3 & 16 \\ & 6 & 2 & \cancel{4} & \cancel{6} \\ - & 1 & 8 & 7 & 9 \\ \hline & & & & 7 \end{array}$$

We cannot subtract 9 ones from 6 ones, but we can subtract 9 ones from 16 ones. We borrow 1 ten to get 16 ones.

$$\begin{array}{rcccc} & & & 13 & \\ & & 1 & \cancel{3} & 16 \\ & 6 & \cancel{2} & \cancel{4} & \cancel{6} \\ - & 1 & 8 & 7 & 9 \\ \hline & & & 6 & 7 \end{array}$$

We cannot subtract 7 tens from 3 tens, but we can subtract 7 tens from 13 tens. We borrow 1 hundred to get 13 tens.

$$\begin{array}{rcccc} & & 11 & 13 & \\ & 5 & \cancel{1} & \cancel{3} & 16 \\ & \cancel{6} & \cancel{2} & \cancel{4} & \cancel{6} \\ - & 1 & 8 & 7 & 9 \\ \hline & 4 & 3 & 6 & 7 \end{array}$$

We cannot subtract 8 hundreds from 1 hundred, but we can subtract 8 hundreds from 11 hundreds. We borrow 1 thousand to get 11 hundreds.

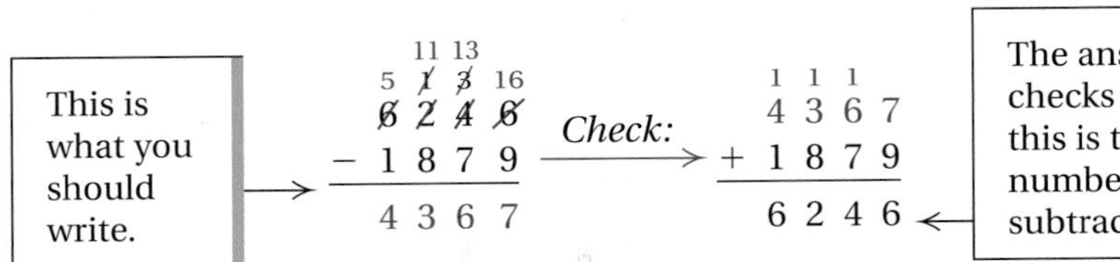

$$\begin{array}{rcccc} & & 11 & 13 & \\ & 5 & \cancel{1} & \cancel{3} & 16 \\ & \cancel{6} & \cancel{2} & \cancel{4} & \cancel{6} \\ - & 1 & 8 & 7 & 9 \\ \hline & 4 & 3 & 6 & 7 \end{array} \quad \text{Check:} \quad \begin{array}{rcccc} & & 1 & 1 & 1 \\ & 4 & 3 & 6 & 7 \\ + & 1 & 8 & 7 & 9 \\ \hline & 6 & 2 & 4 & 6 \end{array}$$

Do Exercises 13 and 14.

Subtract.

15. $\begin{array}{r} 70 \\ -\ 14 \\ \hline \end{array}$ **16.** $\begin{array}{r} 503 \\ -\ 298 \\ \hline \end{array}$

EXAMPLE 10 Subtract: 902 − 477.

$$\begin{array}{rccc} & 8 & 9 & 12 \\ & \cancel{9} & \cancel{0} & \cancel{2} \\ - & 4 & 7 & 7 \\ \hline & 4 & 2 & 5 \end{array}$$

We cannot subtract 7 ones from 2 ones. We have 9 hundreds, or 90 tens. We borrow 1 ten to get 12 ones. We then have 89 tens.

Do Exercises 15 and 16.

Subtract.

17. $\begin{array}{r} 7007 \\ -\ 6349 \\ \hline \end{array}$ **18.** $\begin{array}{r} 6000 \\ -\ 3149 \\ \hline \end{array}$

19. $\begin{array}{r} 9035 \\ -\ 7489 \\ \hline \end{array}$

EXAMPLE 11 Subtract: 8003 − 3667.

$$\begin{array}{rcccc} & 7 & 9 & 9 & 13 \\ & \cancel{8} & \cancel{0} & \cancel{0} & \cancel{3} \\ - & 3 & 6 & 6 & 7 \\ \hline & 4 & 3 & 3 & 6 \end{array}$$

We have 8 thousands, or 800 tens. We borrow 1 ten to get 13 ones. We then have 799 tens.

EXAMPLES

12. Subtract: 6000 − 3762.

$$\begin{array}{rcccc} & 5 & 9 & 9 & 10 \\ & \cancel{6} & \cancel{0} & \cancel{0} & \cancel{0} \\ - & 3 & 7 & 6 & 2 \\ \hline & 2 & 2 & 3 & 8 \end{array}$$

13. Subtract: 6024 − 2968.

$$\begin{array}{rcccc} & & & 11 & \\ & 5 & 9 & \cancel{1} & 14 \\ & \cancel{6} & \cancel{0} & \cancel{2} & \cancel{4} \\ - & 2 & 9 & 6 & 8 \\ \hline & 3 & 0 & 5 & 6 \end{array}$$

Do Exercises 17–19.

Answers on page A-1

1.2 EXERCISE SET

For Extra Help

MathXL

MyMathLab

InterAct Math

Math Tutor Center

Video Lectures on CD Disc 1

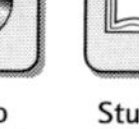
Student's Solutions Manual

a Add.

1. $\begin{array}{r} 364 \\ +\ \ 23 \\ \hline \end{array}$

2. $\begin{array}{r} 1521 \\ +\ \ 348 \\ \hline \end{array}$

3. $\begin{array}{r} 1716 \\ +\ 3482 \\ \hline \end{array}$

4. $\begin{array}{r} 7503 \\ +\ 2683 \\ \hline \end{array}$

5. 8113 + 390

6. 271 + 3338

7. 356 + 4910

8. 280 + 34,702

9. $\begin{array}{r} 99 \\ +\ \ 1 \\ \hline \end{array}$

10. $\begin{array}{r} 999 \\ +\ \ 11 \\ \hline \end{array}$

11. $\begin{array}{r} 5093 \\ +\ 3217 \\ \hline \end{array}$

12. $\begin{array}{r} 3654 \\ +\ 2700 \\ \hline \end{array}$

13. $\begin{array}{r} 4825 \\ +\ 1783 \\ \hline \end{array}$

14. $\begin{array}{r} 6775 \\ +\ 1432 \\ \hline \end{array}$

15. $\begin{array}{r} 23,443 \\ +\ 10,989 \\ \hline \end{array}$

16. $\begin{array}{r} 67,654 \\ +\ 98,786 \\ \hline \end{array}$

17. $\begin{array}{r} 12,070 \\ 2,954 \\ +\ \ 3,400 \\ \hline \end{array}$

18. $\begin{array}{r} 42,487 \\ 83,141 \\ +\ 36,712 \\ \hline \end{array}$

19. $\begin{array}{r} 327 \\ 428 \\ 569 \\ 787 \\ +\ 209 \\ \hline \end{array}$

20. $\begin{array}{r} 989 \\ 566 \\ 834 \\ 920 \\ +\ 703 \\ \hline \end{array}$

b Find the perimeter of (the distance around) the figure.

21.

14 mi
13 mi
8 mi
22 mi
10 mi
47 mi

22.

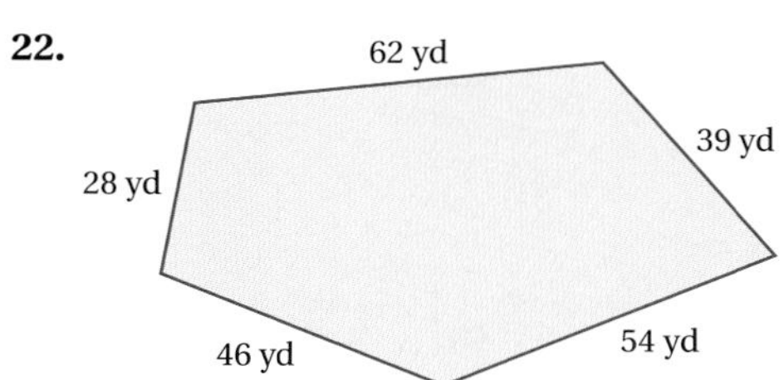

23. Find the perimeter of a standard hockey rink.

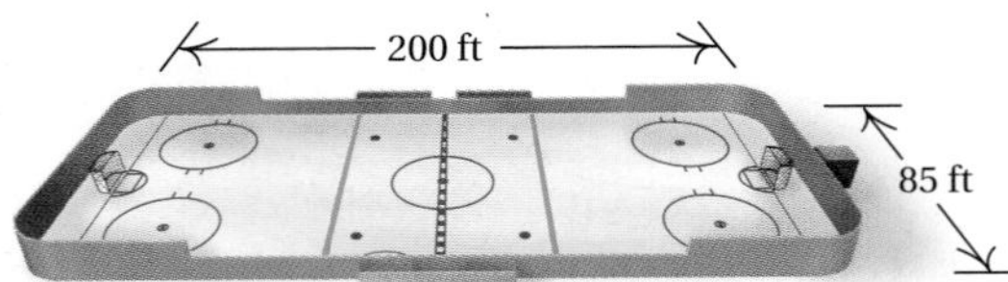

24. In Major League baseball, how far does a batter travel in circling the bases when a home run has been hit?

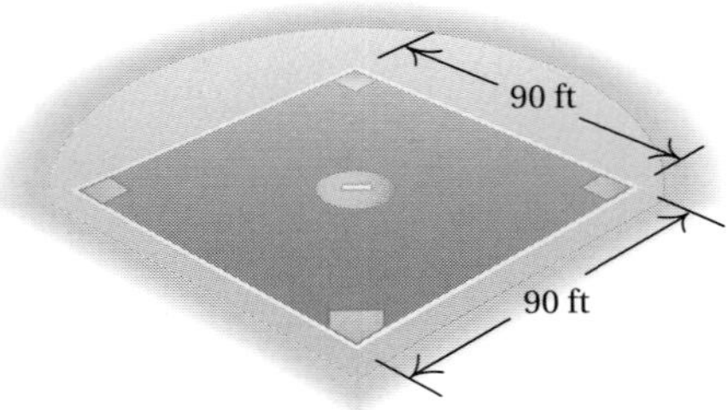

c Write a related addition sentence.

25. $7 - 4 = 3$

26. $12 - 5 = 7$

27. $43 - 16 = 27$

28. $51 - 18 = 33$

Write two related subtraction sentences.

29. $6 + 9 = 15$

30. $7 + 9 = 16$

31. $23 + 9 = 32$

32. $42 + 10 = 52$

d Subtract.

33. $\begin{array}{r} 65 \\ -\ 21 \\ \hline \end{array}$

34. $\begin{array}{r} 87 \\ -\ 34 \\ \hline \end{array}$

35. $\begin{array}{r} 4547 \\ -\ 3421 \\ \hline \end{array}$

36. $\begin{array}{r} 6875 \\ -\ 2111 \\ \hline \end{array}$

37. $\begin{array}{r} 7769 \\ -\ 2387 \\ \hline \end{array}$

38. $\begin{array}{r} 6431 \\ -\ 2896 \\ \hline \end{array}$

39. $\begin{array}{r} 7640 \\ -\ 3809 \\ \hline \end{array}$

40. $\begin{array}{r} 8003 \\ -\ 599 \\ \hline \end{array}$

41. $10{,}002 - 7834$

42. $23{,}048 - 17{,}592$

43. $90{,}237 - 47{,}209$

44. $84{,}703 - 298$

45. $\begin{array}{r} 140 \\ -\ 56 \\ \hline \end{array}$

46. $\begin{array}{r} 470 \\ -\ 188 \\ \hline \end{array}$

47. $\begin{array}{r} 690 \\ -\ 236 \\ \hline \end{array}$

48. $\begin{array}{r} 803 \\ -\ 418 \\ \hline \end{array}$

49. $\begin{array}{r} 903 \\ -\ 132 \\ \hline \end{array}$

50. $\begin{array}{r} 6408 \\ -\ 258 \\ \hline \end{array}$

51. $\begin{array}{r} 8092 \\ -\ 1073 \\ \hline \end{array}$

52. $\begin{array}{r} 6007 \\ -\ 1589 \\ \hline \end{array}$

53. 5843 − 98

54. 15,017 − 7809

55. 21,043 − 8909

56. 83,907 − 89

57. $\begin{array}{r} 7000 \\ -\ 2794 \\ \hline \end{array}$

58. $\begin{array}{r} 8001 \\ -\ 6543 \\ \hline \end{array}$

59. $\begin{array}{r} 48{,}000 \\ -\ 37{,}695 \\ \hline \end{array}$

60. $\begin{array}{r} 17{,}043 \\ -\ 11{,}598 \\ \hline \end{array}$

61. **D**W Describe two situations that correspond to the subtraction \$20 − \$17, one "take away" and one "missing addend."

62. **D**W Describe a situation that corresponds to this mathematical expression:

80 mi + 245 mi + 336 mi.

SKILL MAINTENANCE

The exercises that follow begin an important feature called *Skill Maintenance exercises.* These exercises provide an ongoing review of any preceding objective in the book. You will see them in virtually every exercise set. It has been found that this kind of extensive review can significantly improve your performance on a final examination.

63. What does the digit 8 mean in 486,205? [1.1a]

64. Write a word name for the number in the following sentence: [1.1c]

In fiscal year 2004, Starbucks Corporation had total net revenues of \$5,294,247,000.

Source: Starbucks Corporation

SYNTHESIS

65. A fast way to add all the numbers from 1 to 10 inclusive is to pair 1 with 9, 2 with 8, and so on. Use a similar approach to add all numbers from 1 to 100 inclusive.

66. Fill in the missing digits to make the subtraction true:

9,□48,621 − 2,097,□81 = 7,251,140.

1.3 MULTIPLICATION AND DIVISION; ROUNDING AND ESTIMATING

Objectives

a. Multiply whole numbers.

b. Use multiplication in finding area.

c. Convert between division sentences and multiplication sentences.

d. Divide whole numbers.

e. Round to the nearest ten, hundred, or thousand.

f. Estimate sums, differences, and products by rounding.

a Multiplication of Whole Numbers

REPEATED ADDITION

The multiplication 3×5 corresponds to this repeated addition:

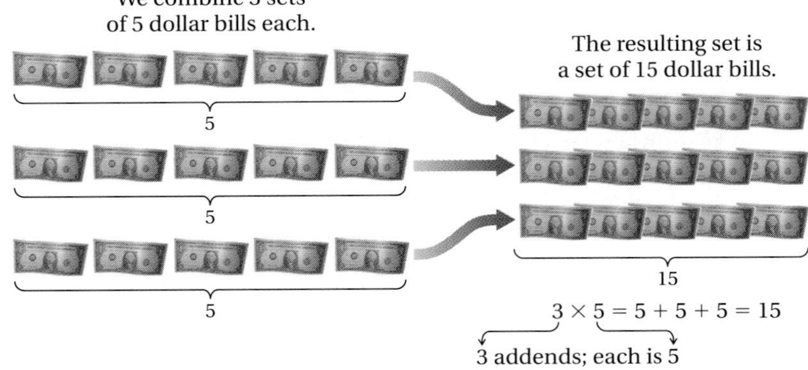

We combine 3 sets of 5 dollar bills each.
5
5
5
The resulting set is a set of 15 dollar bills.
15
3 × 5 = 5 + 5 + 5 = 15
3 addends; each is 5

The numbers that we multiply are called **factors.** The result of the multiplication is called a **product.**

$$\underset{\text{Factor}}{3} \times \underset{\text{Factor}}{5} = \underset{\text{Product}}{15}$$

RECTANGULAR ARRAYS

Multiplications can also be thought of as rectangular arrays. Each of the following corresponds to the multiplication 3×5.

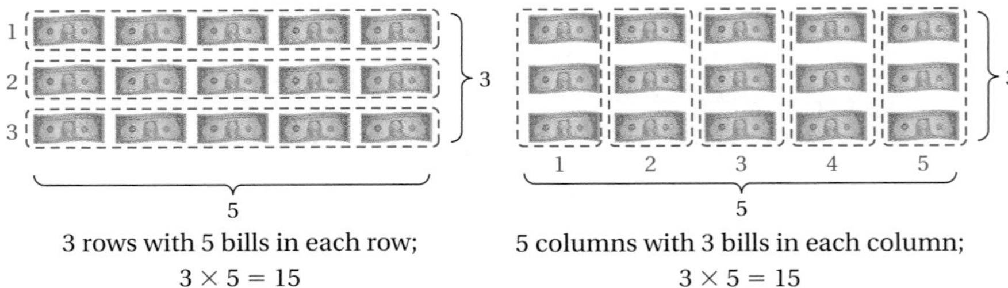

1
2
3
3
5
1 2 3 4 5
3
5

3 rows with 5 bills in each row; $3 \times 5 = 15$

5 columns with 3 bills in each column; $3 \times 5 = 15$

When you write a multiplication sentence corresponding to a real-world situation, you should think of either a rectangular array or repeated addition. In some cases, it may help to think both ways.

We have used an "×" to denote multiplication. A dot "·" is also commonly used. (Use of the dot is attributed to the German mathematician Gottfried Wilhelm von Leibniz in 1698.) Parentheses are also used to denote multiplication. For example,

$$3 \times 5 = 3 \cdot 5 = (3)(5) = 3(5) = 15.$$

The product of 0 and any whole number is 0: $0 \cdot a = a \cdot 0 = 0$. Multiplying a number by 1 does not change the number: $1 \cdot a = a \cdot 1 = a$. We say that 1 is the **multiplicative identity.** For example, $0 \cdot 3 = 3 \cdot 0 = 0$ and $1 \cdot 3 = 3 \cdot 1 = 3$.

EXAMPLE 1 Multiply: 5×734.

We have

$$\begin{array}{r} 734 \\ \times \quad 5 \\ \hline 20 \\ 150 \\ 3500 \\ \hline 3670 \end{array}$$

- 2 0 ← Multiply the 4 ones by 5: $5 \times 4 = 20$.
- 1 5 0 ← Multiply the 3 tens by 5: $5 \times 30 = 150$.
- 3 5 0 0 ← Multiply the 7 hundreds by 5: $5 \times 700 = 3500$.
- 3 6 7 0 ← Add.

Instead of writing each product on a separate line, we can use a shorter form.

$$\begin{array}{r} {}^{2} \\ 734 \\ \times \quad 5 \\ \hline 0 \end{array}$$

Multiply the ones by 5: $5 \cdot (4 \text{ ones}) = 20 \text{ ones} = 2 \text{ tens} + 0 \text{ ones}$. Write 0 in the ones column and 2 above the tens.

$$\begin{array}{r} {}^{1\,2} \\ 734 \\ \times \quad 5 \\ \hline 70 \end{array}$$

Multiply the 3 tens by 5 and add 2 tens: $5 \cdot (3 \text{ tens}) = 15 \text{ tens}$, $15 \text{ tens} + 2 \text{ tens} = 17 \text{ tens} = 1 \text{ hundred} + 7 \text{ tens}$. Write 7 in the tens column and 1 above the hundreds.

$$\begin{array}{r} {}^{1\,2} \\ 734 \\ \times \quad 5 \\ \hline 3670 \end{array}$$

Multiply the 7 hundreds by 5 and add 1 hundred: $5 \cdot (7 \text{ hundreds}) = 35 \text{ hundreds}$, $35 \text{ hundreds} + 1 \text{ hundred} = 36 \text{ hundreds}$.

$$\begin{array}{r} {}^{1\,2} \\ 734 \\ \times \quad 5 \\ \hline 3670 \end{array}$$

You should write only this.

Do Exercises 1–4.

Let's find the product

$$\begin{array}{r} 54 \\ \times 32 \\ \hline \end{array}$$

To do this, we multiply 54 by 2, then 54 by 30, and then add.

$$\begin{array}{r} 54 \\ \times \quad 2 \\ \hline 108 \end{array} \qquad \begin{array}{r} {}^{1} \\ 54 \\ \times \quad 30 \\ \hline 1620 \end{array}$$

Since we are going to add the results, let's write the work this way.

$$\begin{array}{r} 54 \\ \times 32 \\ \hline 108 \\ 1620 \\ \hline 1728 \end{array}$$

- 1 0 8 Multiplying by 2
- 1 6 2 0 Multiplying by 30
- 1 7 2 8 Adding to obtain the product

Multiply.

1. $\begin{array}{r} 58 \\ \times \quad 2 \\ \hline \end{array}$

2. $\begin{array}{r} 37 \\ \times \quad 4 \\ \hline \end{array}$

3. $\begin{array}{r} 823 \\ \times \quad 6 \\ \hline \end{array}$

4. $\begin{array}{r} 1348 \\ \times \quad 5 \\ \hline \end{array}$

Answers on page A-2

The fact that we can do this is based on a property called the **distributive law.** It says that to multiply a number by a sum, $a \cdot (b + c)$, we can multiply each addend by a and then add like this: $(a \cdot b) + (a \cdot c)$. Thus, $a \cdot (b + c) = (a \cdot b) + (a \cdot c)$. Applied to the example above, the distributive law gives us

$$54 \cdot 32 = 54 \cdot (30 + 2) = (54 \cdot 30) + (54 \cdot 2).$$

EXAMPLE 2 Multiply: 43×57.

$$\begin{array}{rrrr} & & {\scriptstyle 2} & \\ & & 5 & 7 \\ & \times & 4 & 3 \\ \hline & 1 & 7 & 1 \end{array}$$

Multiplying by 3

$$\begin{array}{rrrr} & & {\scriptstyle 2} & \\ & & {\scriptstyle 2} & \\ & & 5 & 7 \\ & \times & 4 & 3 \\ \hline & 1 & 7 & 1 \\ 2 & 2 & 8 & 0 \end{array}$$

Multiplying by 40. (We write a 0 and then multiply 57 by 4.)

> You may have learned that such a 0 does not have to be written. You may omit it if you wish. If you do omit it, remember, when multiplying by tens, to start writing the answer in the tens place.

$$\begin{array}{rrrr} & & {\scriptstyle 2} & \\ & & {\scriptstyle 2} & \\ & & 5 & 7 \\ & \times & 4 & 3 \\ \hline & 1 & 7 & 1 \\ 2 & 2 & 8 & 0 \\ \hline 2 & 4 & 5 & 1 \end{array}$$

Adding to obtain the product

Do Exercises 5 and 6.

EXAMPLE 3 Multiply: 457×683.

$$\begin{array}{rrrr} & {\scriptstyle 5} & {\scriptstyle 2} & \\ & 6 & 8 & 3 \\ \times & 4 & 5 & 7 \\ \hline 4 & 7 & 8 & 1 \end{array}$$

Multiplying 683 by 7

$$\begin{array}{rrrrr} & & {\scriptstyle 4} & {\scriptstyle 1} & \\ & & {\scriptstyle 5} & {\scriptstyle 2} & \\ & & 6 & 8 & 3 \\ & \times & 4 & 5 & 7 \\ \hline & 4 & 7 & 8 & 1 \\ 3 & 4 & 1 & 5 & 0 \end{array}$$

Multiplying 683 by 50

$$\begin{array}{rrrrrr} & & & {\scriptstyle 3} & {\scriptstyle 1} & \\ & & & {\scriptstyle 4} & {\scriptstyle 1} & \\ & & & {\scriptstyle 5} & {\scriptstyle 2} & \\ & & & 6 & 8 & 3 \\ & & \times & 4 & 5 & 7 \\ \hline & & 4 & 7 & 8 & 1 \\ & 3 & 4 & 1 & 5 & 0 \\ 2 & 7 & 3 & 2 & 0 & 0 \\ \hline 3 & 1 & 2, & 1 & 3 & 1 \end{array}$$

Multiplying 683 by 400

Adding

Do Exercises 7 and 8.

Multiply.

5. $\begin{array}{r} 45 \\ \times\ 23 \\ \hline \end{array}$

6. 48×63

Multiply.

7. $\begin{array}{r} 746 \\ \times\ \ 62 \\ \hline \end{array}$

8. 245×837

Multiply.

9. $\begin{array}{r} 472 \\ \times\ 306 \\ \hline \end{array}$

10. 408×704

11. $\begin{array}{r} 2344 \\ \times\ 6005 \\ \hline \end{array}$

Answers on page A-2

EXAMPLE 4 Multiply: 306×274.

Note that $306 = 3$ hundreds $+ 6$ ones.

```
    2 7 4
  × 3 0 6
  -------
  1 6 4 4     Multiplying by 6
  8 2 2 0 0   Multiplying by 3 hundreds. (We write 00
  ---------   and then multiply 274 by 3.)
  8 3,8 4 4   Adding
```

Do Exercises 9–11 on the preceding page.

EXAMPLE 5 Multiply: 360×274.

Note that $360 = 3$ hundreds $+ 6$ tens.

```
      2 7 4
  ×   3 6 0
  ---------
  1 6 4 4 0   Multiplying by 6 tens. (We write 0 and
              then multiply 274 by 6.)
  8 2 2 0 0   Multiplying by 3 hundreds. (We write 00
  ---------   and then multiply 274 by 3.)
  9 8,6 4 0   Adding
```

Do Exercises 12–15.

Multiply.

12.
```
    4 7 2
  × 8 3 0
```

13.
```
    2 3 4 4
  × 7 4 0 0
```

14. 100×562

15. 1000×562

Answers on page A-2

CALCULATOR CORNER

Multiplying Whole Numbers To multiply whole numbers on a calculator, we use the [×] and [=] keys. For example, to find 13×47, we press [1] [3] [×] [4] [7] [=]. The calculator displays 611, so $13 \times 47 = 611$.

Exercises: Use a calculator to find the product.

1. 56×8
2. 845×26
3. $5 \cdot 1276$
4. $126(314)$
5.
```
    3 7 6 0
  ×     4 8
```
6.
```
    5 2 1 8
  ×   4 5 3
```

Study Tips — TIME MANAGEMENT (PART 1)

Time is the most critical factor in your success in learning mathematics. Have reasonable expectations about the time you need to study math.

- **Juggling time.** Working 40 hours per week and taking 12 credit hours is equivalent to working two full-time jobs. Can you handle such a load? Your ratio of number of work hours to number of credit hours should be about 40/3, 30/6, 20/9, 10/12, or 5/14.
- **A rule of thumb on study time.** Budget about 2–3 hours for homework and study for every hour you spend in class each week.
- **Scheduling your time.** Use the Student Organizer card to make an hour-by-hour schedule of your typical week. Include work, school, home, sleep, study, and leisure times. Try to schedule time for study when you are most alert. Choose a setting that will enable you to focus and concentrate. Plan for success and it will happen!

"You cannot increase the quality or quantity of your achievement or performance except to the degree in which you increase your ability to use time effectively."

Brian Tracy, motivational/inspirational speaker

16. **Table Tennis.** Find the area of a standard table tennis table that has dimensions of 9 ft by 5 ft.

b Finding Area

The area of a rectangular region can be considered to be the number of square units needed to fill it. Here is a rectangle 4 cm (centimeters) long and 3 cm wide. It takes 12 square centimeters (sq cm) to fill it.

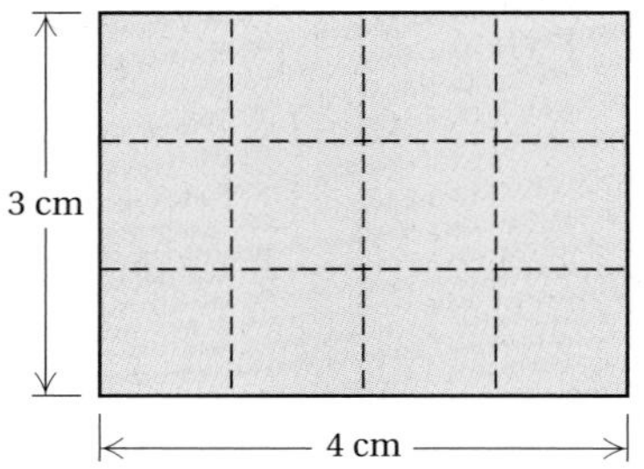

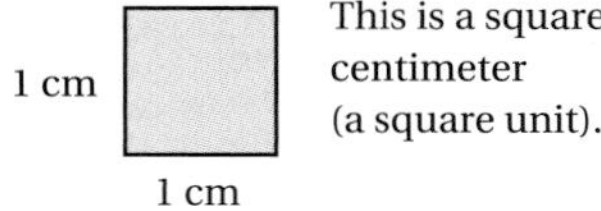

This is a square centimeter (a square unit).

In this case, we have a rectangular array of 3 rows, each of which contains 4 squares. The number of square units is given by $3 \cdot 4$, or 12. That is, $A = l \cdot w = 3 \text{ cm} \cdot 4 \text{ cm} = 12$ sq cm.

EXAMPLE 6 *Professional Pool Table.* The playing area of a standard pool table has dimensions of 50 in. by 100 in. (There are rails 6 in. wide on the outside that are not included in the playing area.) Find the playing area.

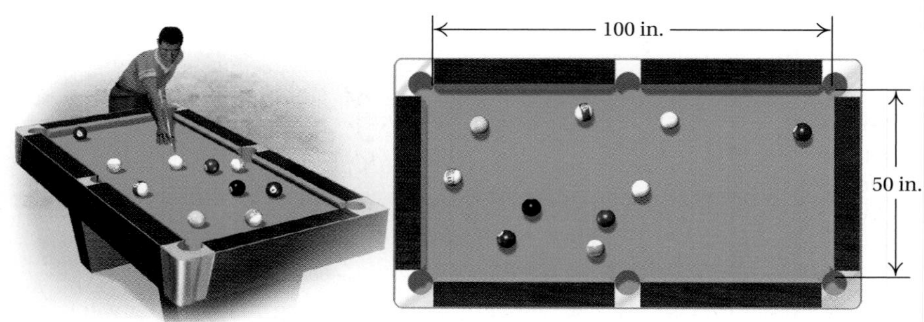

If we think of filling the rectangle with square inches, we have a rectangular array. The length $l = 100$ in. and the width $w = 50$ in. Thus the area A is given by the formula

$$A = l \cdot w = 100 \cdot 50 = 5000 \text{ sq in.}$$

Do Exercise 16.

Answer on page A-2

C Division and Related Sentences

REPEATED SUBTRACTION

Division of whole numbers applies to two kinds of situations. The first is repeated subtraction. Suppose we have 20 notebooks in a pile, and we want to find out how many sets of 5 there are. One way to do this is to repeatedly subtract sets of 5 as follows.

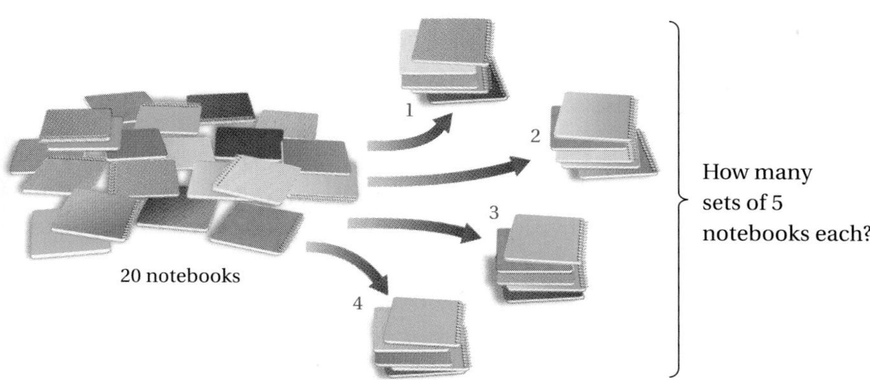

Since there are 4 sets of 5 notebooks each, we have

$$\underset{\text{Dividend}}{20} \div \underset{\text{Divisor}}{5} = \underset{\text{Quotient}}{4}.$$

The division $20 \div 5$, read "20 divided by 5," corresponds to the figure above. We say that the **dividend** is 20, the **divisor** is 5, and the **quotient** is 4. We divide the *dividend* by the *divisor* to get the *quotient.*

We can also express the division $20 \div 5 = 4$ as

$$\frac{20}{5} = 4 \quad \text{or} \quad 5\overline{)20}\,\text{ with quotient } 4$$

Do Exercise 17.

17. Consider $54 \div 6 = 9$. Express this division in two other ways.

RECTANGULAR ARRAYS

We can also think of division in terms of rectangular arrays. Consider again the pile of 20 notebooks and division by 5. We can arrange the notebooks in a rectangular array with 5 rows and ask, "How many are in each row?"

We can also consider a rectangular array with 5 notebooks in each column and ask, "How many columns are there?" The answer is still 4.

In each case, we are asking, "What do we multiply 5 by in order to get 20?"

$$5 \cdot \underbrace{\square}_{\text{Missing factor}} = 20 \qquad 20 \div 5 = \underbrace{\square}_{\text{Quotient}}$$

Answer on page A-2

Write a related multiplication sentence.

18. $15 \div 3 = 5$

19. $72 \div 8 = 9$

Write two related division sentences.

20. $6 \cdot 2 = 12$

21. $7 \cdot 6 = 42$

Answers on page A-2

This leads us to the following definition of division.

> **DIVISION**
>
> The quotient $a \div b$, where $b \neq 0$, is that unique number c for which $a = b \cdot c$.

RELATED SENTENCES

By looking at rectangular arrays, we can see how multiplication and division are related. The array of notebooks on the preceding page shows that $5 \cdot 4 = 20$. The array also shows the following:

$$20 \div 5 = 4 \quad \text{and} \quad 20 \div 4 = 5.$$

The division $20 \div 5$ is defined to be the number that when multiplied by 5 gives 20. Thus, for every division sentence, there is a related multiplication sentence.

$20 \div 5 = 4$ Division sentence

$20 = 4 \cdot 5$ Related multiplication sentence

> To get the related multiplication sentence, we use
> Dividend = Quotient · Divisor.

EXAMPLE 7 Write a related multiplication sentence: $12 \div 6 = 2$.

We have

$12 \div 6 = 2$ Division sentence

$12 = 2 \cdot 6.$ Related multiplication sentence

The related multiplication sentence is $12 = 2 \cdot 6$.

> By the commutative law of multiplication, there is also another multiplication sentence: $12 = 6 \cdot 2$.

Do Exercises 18 and 19.

For every multiplication sentence, we can write related division sentences.

EXAMPLE 8 Write two related division sentences: $7 \cdot 8 = 56$.

We have

$7 \cdot 8 = 56$ — This factor becomes a divisor. — $7 = 56 \div 8.$

$7 \cdot 8 = 56$ — This factor becomes a divisor. — $8 = 56 \div 7.$

The related division sentences are $7 = 56 \div 8$ and $8 = 56 \div 7$.

Do Exercises 20 and 21.

d Division of Whole Numbers

Before we consider division with remainders, let's recall four basic facts about division.

DIVIDING BY 1

Any number divided by 1 is that same number:

$$a \div 1 = \frac{a}{1} = a.$$

DIVIDING A NUMBER BY ITSELF

Any nonzero number divided by itself is 1:

$$\frac{a}{a} = 1, \quad a \neq 0.$$

DIVIDENDS OF 0

Zero divided by any nonzero number is 0:

$$\frac{0}{a} = 0, \quad a \neq 0.$$

EXCLUDING DIVISION BY 0

Division by 0 is not defined. (We agree not to divide by 0.)

$\frac{a}{0}$ is **not defined.**

Why can't we divide by 0? Suppose the number 4 could be divided by 0. Then if $\square$ were the answer,

$$4 \div 0 = \square$$

and since 0 times any number is 0, we would have

$$4 = \square \cdot 0 = 0. \qquad \text{False!}$$

Thus, $a \div 0$ would be some number $\square$ such that $a = \square \cdot 0 = 0$. So the only possible number that could be divided by 0 would be 0 itself.

But such a division would give us any number we wish, for

$$\left.\begin{aligned} 0 \div 0 = 8 \quad &\text{because} \quad 0 = 8 \cdot 0; \\ 0 \div 0 = 3 \quad &\text{because} \quad 0 = 3 \cdot 0; \\ 0 \div 0 = 7 \quad &\text{because} \quad 0 = 7 \cdot 0. \end{aligned}\right\} \quad \text{All true!}$$

We avoid the preceding difficulties by agreeing to exclude division by 0.

Study Tips

HIGHLIGHTING

- **Highlight important points.** You are probably accustomed to highlighting key points as you study. If that works for you, continue to do so. But you will notice many design features throughout this book that already highlight important points. Thus you may not need to highlight as much as you generally do.
- **Highlight points that you do not understand.** Use a unique mark to indicate trouble spots that can lead to questions to be asked during class, in a tutoring session, or when calling or contacting the AW Math Tutor Center.

Divide by repeated subtraction. Then check.

22. 54 ÷ 9

23. 61 ÷ 9

24. 53 ÷ 12

25. 157 ÷ 24

Answers on page A-2

Suppose we have 18 cans of soda and want to pack them in cartons of 6 cans each. How many cartons will we fill? We can determine this by repeated subtraction. We keep track of the number of times we subtract. We stop when the number of objects remaining, the **remainder,** is smaller than the divisor.

EXAMPLE 9 Divide by repeated subtraction: 18 ÷ 6.

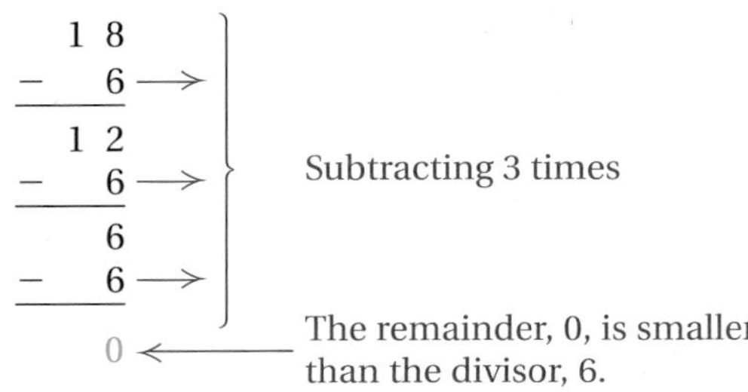

18
− 6
12
− 6
6
− 6
0
Subtracting 3 times
The remainder, 0, is smaller than the divisor, 6.

Thus, 18 ÷ 6 = 3.

Suppose we have 22 cans of soda and want to pack them in cartons of 6 cans each. We end up with 3 cartons with 4 cans left over.

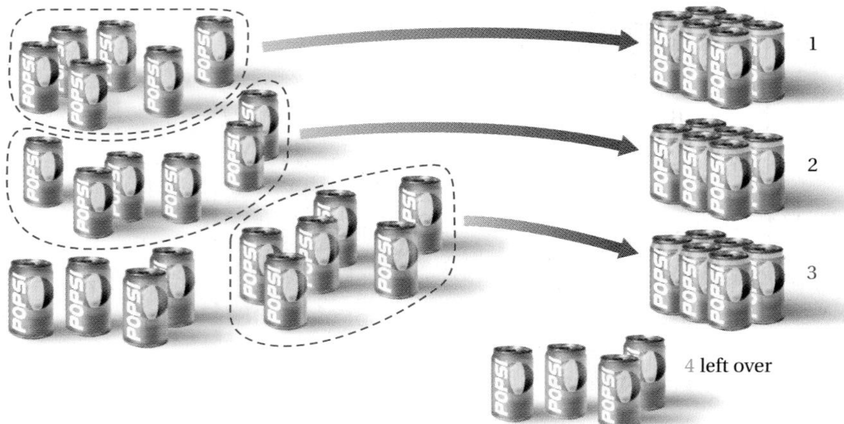

EXAMPLE 10 Divide by repeated subtraction: 22 ÷ 6.

$$\begin{array}{r} 22 \\ -\ 6 \\ \hline 16 \\ -\ 6 \\ \hline 10 \\ -\ 6 \\ \hline 4 \end{array}$$

Subtracting 3 times (the three subtractions of 6); 4 ← Remainder

Check: $3 \cdot 6 = 18$, $18 + 4 = 22$.

We can write this division as follows.

$$\begin{array}{r} 3 \\ 6\overline{)22} \\ \underline{18} \\ 4 \end{array}$$

Note that

Quotient · Divisor + Remainder = Dividend.

We write answers to a division sentence as follows.

$22 \div 6 = 3 \text{ R } 4$

Dividend (22), Divisor (6), Quotient (3), Remainder (4)

Do Exercises 22–25 on the preceding page.

EXAMPLE 11 Divide and check: $3642 \div 5$.

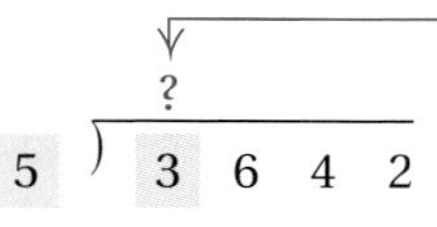

1. Find the number of thousands in the quotient. Consider 3 thousands ÷ 5 and think 3 ÷ 5. Since 3 ÷ 5 is not a whole number, move to hundreds.

```
        7
   5 ) 3 6 4 2
       3 5 0 0
         1 4 2  <— The remainder is larger than the divisor.
```

2. Find the number of hundreds in the quotient. Consider 36 hundreds ÷ 5 and think 36 ÷ 5. The estimate is about 7 hundreds. Multiply 700 by 5 and subtract.

```
        7 2
   5 ) 3 6 4 2
       3 5 0 0
         1 4 2
         1 0 0
           4 2  <— The remainder is larger than the divisor.
```

3. Find the number of tens in the quotient using 142, the first remainder. Consider 14 tens ÷ 5 and think 14 ÷ 5. The estimate is about 2 tens. Multiply 20 by 5 and subtract. (If our estimate had been 3 tens, we could not have subtracted 150 from 142.)

```
        7 2 8
   5 ) 3 6 4 2
       3 5 0 0
         1 4 2
         1 0 0
           4 2
           4 0
             2
```

4. Find the number of ones in the quotient using 42, the second remainder. Consider 42 ones ÷ 5 and think 42 ÷ 5. The estimate is about 8 ones. Multiply 8 by 5 and subtract. The remainder, 2, is less than the divisor, 5, so we are finished.

The remainder is less than the divisor.

> You may have learned to divide like this, not writing the extra zeros. You may omit them if desired.

```
      7 2 8
 5 ) 3 6 4 2
     3 5
       1 4
       1 0
         4 2
         4 0
           2
```

Check: $728 \cdot 5 = 3640$, $3640 + 2 = 3642$. The answer is 728 R 2.

Do Exercises 26–28.

To do division of whole numbers:

a) Estimate.
b) Multiply.
c) Subtract.

Divide and check.

26. $4 \overline{)239}$

27. $6 \overline{)8855}$

28. $5 \overline{)5075}$

Answers on page A-2

Divide.

29. $6\overline{)4846}$

30. $7\overline{)7616}$

ZEROS IN QUOTIENTS

EXAMPLE 12 Divide: 6341 ÷ 7.

$$\begin{array}{r} 9 \\ 7\overline{)6341} \\ \underline{6300} \\ 41 \end{array}$$

← *Think*: 63 hundreds ÷ 7. Estimate 9 hundreds. Multiply 900 · 7 and subtract.

$$\begin{array}{r} 90 \\ 7\overline{)6341} \\ \underline{6300} \\ 41 \end{array}$$

← *Think*: 4 tens ÷ 7. The tens digit of the quotient is 0.

$$\begin{array}{r} 905 \\ 7\overline{)6341} \\ \underline{6300} \\ 41 \\ \underline{35} \\ 6 \end{array}$$

41 ← *Think*: 41 ones ÷ 7. Estimate 5 ones. Multiply 5 · 7 and subtract.

6 ← The remainder, 6, is less than the divisor, 7.

The answer is 905 R 6.

Do Exercises 29 and 30.

Divide.

31. $27\overline{)9724}$

32. $56\overline{)44,847}$

EXAMPLE 13 Divide: 8889 ÷ 37.

We round 37 to 40.

$$\begin{array}{r} 2 \\ 37\overline{)8889} \\ \underline{7400} \\ 1489 \end{array}$$

← *Think:* 37 ≈ 40; 88 hundreds ÷ 40. Estimate 2 hundreds. Multiply 200 ? 37 and subtract.

$$\begin{array}{r} 24 \\ 37\overline{)8889} \\ \underline{7400} \\ 1489 \\ \underline{1480} \\ 9 \end{array}$$

1489 ← *Think:* 148 tens ÷ 40. Estimate 4 tens. Multiply 40 ? 37 and subtract.

$$\begin{array}{r} 240 \\ 37\overline{)8889} \\ \underline{7400} \\ 1489 \\ \underline{1480} \\ 9 \end{array}$$

9 ← The remainder, 9, is less than the divisor, 37.

The answer is 240 R 9.

Do Exercises 31 and 32.

Answers on page A-2

CALCULATOR CORNER

Dividing Whole Numbers: Finding Remainders To divide whole numbers on a calculator, we use the [÷] and [=] keys. For example, to divide 711 by 9, we press [7] [1] [1] [÷] [9] [=]. The display reads [79], so 711 ÷ 9 = 79.

When we enter 453 ÷ 15, the display reads [30.2]. Note that the result is not a whole number. This tells us that there is a remainder. The number 30.2 is expressed in decimal notation. The symbol "." is called a decimal point. (Decimal notation will be studied in Chapter 3.) The number to the left of the decimal point, 30, is the quotient. We can use the remaining part of the result to find the remainder. To do this, first subtract 30 from 30.2. Then multiply the difference by the divisor, 15. We get 3. This is the remainder. Thus, 453 ÷ 15 = 30 R 3. The steps that we performed to find this result can be summarized as follows:

$$453 \div 15 = 30.2,$$
$$30.2 - 30 = .2,$$
$$0.2 \times 15 = 3.$$

To follow these steps on a calculator, we press [4] [5] [3] [÷] [1] [5] [=] and write the number that appears to the left of the decimal point. This is the quotient. Then we continue by pressing [−] [3] [0] [=] [×] [1] [5] [=]. The last number that appears is the remainder. In some cases, it will be necessary to round the remainder to the nearest one.

To check this result, we multiply the quotient by the divisor and then add the remainder.

$$30 \times 15 = 450,$$
$$450 + 3 = 453$$

Exercises: Use a calculator to perform the division. Check the results with a calculator also.

1. 92 ÷ 27
2. $19\overline{)532}$
3. $6\overline{)746}$
4. 3817 ÷ 29
5. $126\overline{)35{,}715}$
6. $308\overline{)259{,}831}$

e Rounding

We round numbers in various situations when we do not need an exact answer. For example, we might round to see if we are being charged the correct amount in a store. We might also round to check if an answer to a problem is reasonable or to check a calculation done by hand or on a calculator.

To understand how to round, we first look at some examples using number lines, even though this is not the way we generally do rounding.

EXAMPLE 14 Round 47 to the nearest ten.

Here is a part of a number line; 47 is between 40 and 50. Since 47 is closer to 50, we round up to 50.

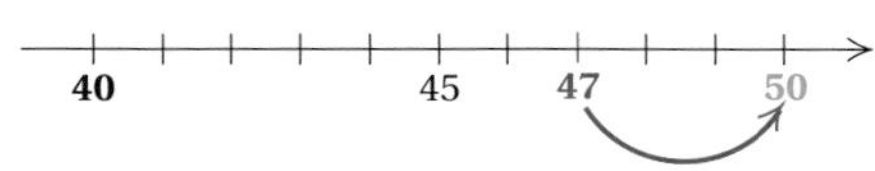

EXAMPLE 15 Round 42 to the nearest ten.

42 is between 40 and 50. Since 42 is closer to 40, we round down to 40.

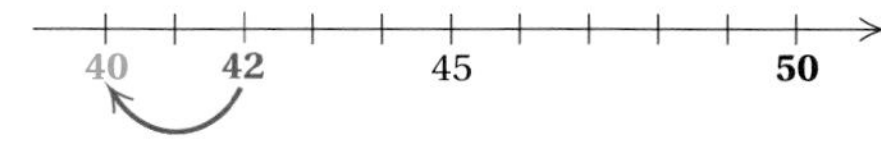

Do Exercises 33–36.

Round to the nearest ten.

33. 37

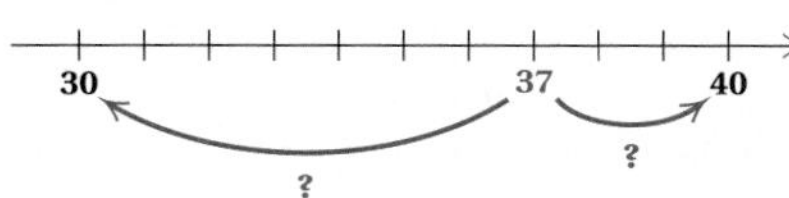

34. 52

35. 73

36. 98

Answers on page A-2

EXAMPLE 16 Round 45 to the nearest ten.

45 is halfway between 40 and 50. We could round 45 down to 40 or up to 50. We agree to round up to 50.

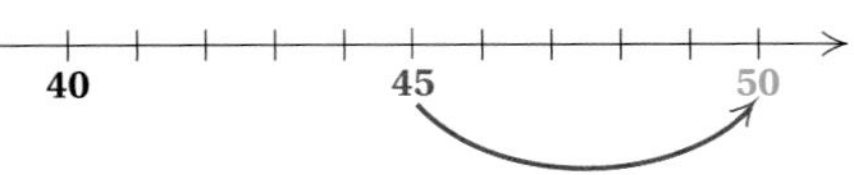

When a number is halfway between rounding numbers, round up.

Do Exercises 37–39.

Here is a rule for rounding.

ROUNDING WHOLE NUMBERS

To round to a certain place:

a) Locate the digit in that place.
b) Consider the next digit to the right.
c) If the digit to the right is 5 or higher, round up. If the digit to the right is 4 or lower, round down.
d) Change all digits to the right of the rounding location to zeros.

EXAMPLE 17 Round 6485 to the nearest **(a)** ten; **(b)** hundred; **(c)** thousand.

a) Locate the digit in the tens place. It is 8.

6 4 8 5
    ↑

The next digit to the right is 5, so we round up. The answer is 6490.

b) Locate the digit in the hundreds place. It is 4.

6 4 8 5
  ↑

The next digit to the right is 8, so we round up. The answer is 6500.

c) Locate the digit in the thousands place. It is 6.

6 4 8 5
↑

The next digit to the right is 4, so we round down. The answer is 6000.

Do Exercises 40–51.

Caution!

7000 is not a correct answer to Example 17. It is incorrect to round from the ones digit over, as follows:

6485, → 6490, → 6500, → 7000.

Note that 6485 is closer to 6000 than it is to 7000.

Round to the nearest ten.

37. 35

38. 75

39. 85

Round to the nearest ten.

40. 137

41. 473

42. 235

43. 285

Round to the nearest hundred.

44. 641

45. 759

46. 750

47. 9325

Round to the nearest thousand.

48. 7896

49. 8459

50. 19,343

51. 68,500

Answers on page A-2

f Estimating

Estimating can be done in many ways. In general, an estimate made by rounding to the nearest ten is more accurate than one rounded to the nearest hundred, and an estimate rounded to the nearest hundred is more accurate than one rounded to the nearest thousand, and so on.

In the following example, we see how estimation can be used in making a purchase.

EXAMPLE 18 *Estimating the Cost of an Automobile Purchase.* Ethan and Olivia Benson are shopping for a new car. They are considering a Saturn ION. There are three basic models of this car, and each has options beyond the basic price, as shown in the chart below. Ethan and Olivia have allowed themselves a budget of $16,500. They look at the list of options and want to make a quick estimate of the cost of model ION·2 with all the options.

Estimate by rounding to the nearest hundred the cost of the ION·2 with all the options and decide whether it will fit into their budget.

MODEL ION·1 SEDAN (4 DOOR) 2.2-LITER ENGINE, 4-SPEED AUTOMATIC TRANSMISSION	MODEL ION·2 SEDAN (4 DOOR) 2.2-LITER ENGINE, 5-SPEED MANUAL TRANSMISSION	MODEL ION·3 SEDAN (4 DOOR) 2.2-LITER ENGINE, 5-SPEED MANUAL TRANSMISSION
Base Price: $12,975	Base Price: $14,945	Base Price: $16,470
Each of these vehicles comes with several options. Note that some of the options are standard on certain models. Others are not available for all models.		
Antilock Braking System with Traction Control:	$400	
Head Curtain Side Air Bags:	$395	
Power Sunroof (Not available for ION·1):	$725	
Rear Spoiler (Not available for ION·1):	$250	
Air Conditioning with Dust and Pollen Filtration (Standard on ION·2 and ION·3):	$960	
CD/MP3 Player with AM/FM Stereo and 4 Coaxial Speakers (Standard on ION·3):	ION·1—$510 ION·2—$220	
Power Package: Power Windows, Power Exterior Mirrors, Remote Keyless Entry, and Cruise Control (Not available for ION·1 and Standard for ION·3):	$825	

Source: Saturn

Refer to the chart on this page to answer Margin Exercises 52 and 53.

52. By eliminating options, find a way that Ethan and Olivia can buy the ION·2 and stay within their $16,500 budget.

53. Tara and Alex are shopping for a new car. They are considering a Saturn ION·3 and have allowed a budget of $19,000.

a) Estimate by first rounding to the nearest hundred the cost of an ION·3 with all the options.

b) Can they afford this car with a budget of $19,000?

Answers on page A-2

54. Estimate the sum by first rounding to the nearest ten. Show your work.

$$\begin{array}{r} 74 \\ 23 \\ 35 \\ +\ 66 \\ \hline \end{array}$$

55. Estimate the difference by first rounding to the nearest hundred. Show your work.

$$\begin{array}{r} 9285 \\ -\ 6739 \\ \hline \end{array}$$

56. Estimate the difference by first rounding to the nearest thousand. Show your work.

$$\begin{array}{r} 23{,}278 \\ -\ 11{,}698 \\ \hline \end{array}$$

57. Estimate the product by first rounding to the nearest ten and to the nearest hundred. Show your work.

$$\begin{array}{r} 837 \\ \times 245 \\ \hline \end{array}$$

Answers on page A-2

First, we list the base price of the ION·2 and then the cost of each of the options. We then round each number to the nearest hundred and add.

$$\begin{array}{r} 14{,}945 \\ 400 \\ 395 \\ 725 \\ 250 \\ 220 \\ +\quad 825 \\ \hline \end{array} \qquad \begin{array}{r} 14{,}900 \\ 400 \\ 400 \\ 700 \\ 300 \\ 200 \\ +\quad 800 \\ \hline 17{,}700 \end{array} \leftarrow \text{Estimated answer}$$

Air conditioning is standard on the ION·2, so we do not include that cost. The estimated cost is \$17,700. Since Ethan and Olivia have allowed themselves a budget of \$16,500 for the car, they will need to forgo some options.

Do Exercises 52 and 53 on the preceding page.

EXAMPLE 19 Estimate the difference by first rounding to the nearest thousand: $9324 - 2849$.

We have

$$\begin{array}{r} 9324 \\ -\ 2849 \\ \hline \end{array} \qquad \begin{array}{r} 9000 \\ -\ 3000 \\ \hline 6000 \end{array}$$

EXAMPLE 20 Estimate the following product by first rounding to the nearest ten and to the nearest hundred: 683×457.

Nearest ten	*Nearest hundred*	*Exact*
$\begin{array}{r} 680 \\ \times\quad 460 \\ \hline 40800 \\ 272000 \\ \hline 312{,}800 \end{array}$	$\begin{array}{r} 700 \\ \times\quad 500 \\ \hline 350{,}000 \end{array}$	$\begin{array}{r} 683 \\ \times\quad 457 \\ \hline 4781 \\ 34150 \\ 273200 \\ \hline 312{,}131 \end{array}$

Do Exercises 54–57.

The sentence $7 - 5 = 2$ says that $7 - 5$ is the same as 2. When we round, the result is rarely the same as the number we started with. Thus we use the symbol $\approx$ when rounding. This symbol means "**is approximately equal to.**" For example, when 687 is rounded to the nearest ten, we can write $687 \approx 690$.

EXERCISE SET

For Extra Help

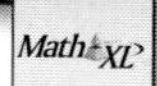
MathXL

MyMathLab

InterAct Math

Math Tutor Center

Video Lectures on CD Disc 1

Student's Solutions Manual

a Multiply.

1. $\begin{array}{r} 94 \\ \times \quad 6 \\ \hline \end{array}$

2. $\begin{array}{r} 76 \\ \times \quad 9 \\ \hline \end{array}$

3. $\begin{array}{r} 2340 \\ \times 1000 \\ \hline \end{array}$

4. $\begin{array}{r} 800 \\ \times \quad 70 \\ \hline \end{array}$

5. $3 \cdot 509$

6. $7 \cdot 806$

7. $7(9229)$

8. $4(7867)$

9. $90(53)$

10. $60(78)$

11. $(47)(85)$

12. $(34)(87)$

13. $\begin{array}{r} 640 \\ \times \quad 72 \\ \hline \end{array}$

14. $\begin{array}{r} 777 \\ \times \quad 77 \\ \hline \end{array}$

15. $\begin{array}{r} 444 \\ \times \quad 33 \\ \hline \end{array}$

16. $\begin{array}{r} 509 \\ \times \quad 88 \\ \hline \end{array}$

17. $\begin{array}{r} 509 \\ \times 408 \\ \hline \end{array}$

18. $\begin{array}{r} 432 \\ \times 375 \\ \hline \end{array}$

19. $\begin{array}{r} 853 \\ \times 936 \\ \hline \end{array}$

20. $\begin{array}{r} 346 \\ \times 650 \\ \hline \end{array}$

21. $\begin{array}{r} 6428 \\ \times 3224 \\ \hline \end{array}$

22. $\begin{array}{r} 8928 \\ \times 3172 \\ \hline \end{array}$

23. $\begin{array}{r} 3482 \\ \times \quad 104 \\ \hline \end{array}$

24. $\begin{array}{r} 6408 \\ \times 6064 \\ \hline \end{array}$

25. $\begin{array}{r} 5006 \\ \times 4008 \\ \hline \end{array}$

26. $\begin{array}{r} 6789 \\ \times 2330 \\ \hline \end{array}$

27. $\begin{array}{r} 5608 \\ \times 4500 \\ \hline \end{array}$

28. $\begin{array}{r} 4560 \\ \times 7890 \\ \hline \end{array}$

b What is the area of the region?

29.

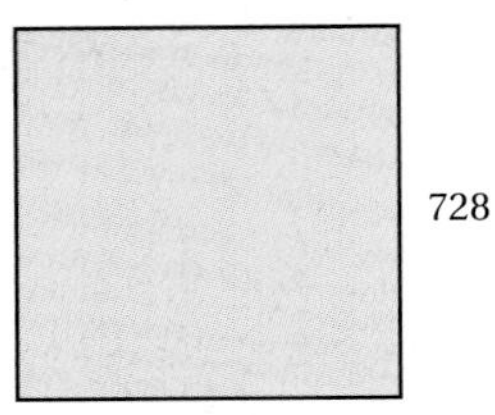

30.

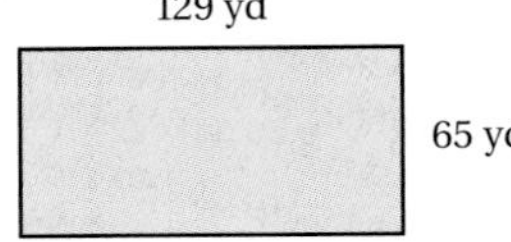

31. Find the area of the region formed by the base lines on a Major League baseball diamond.

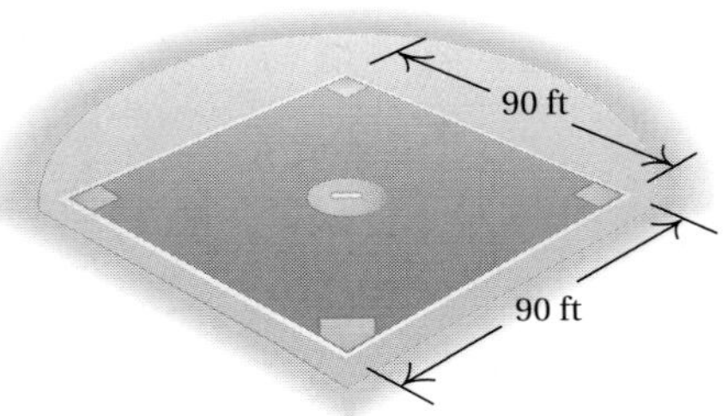

32. Find the area of a standard-sized hockey rink.

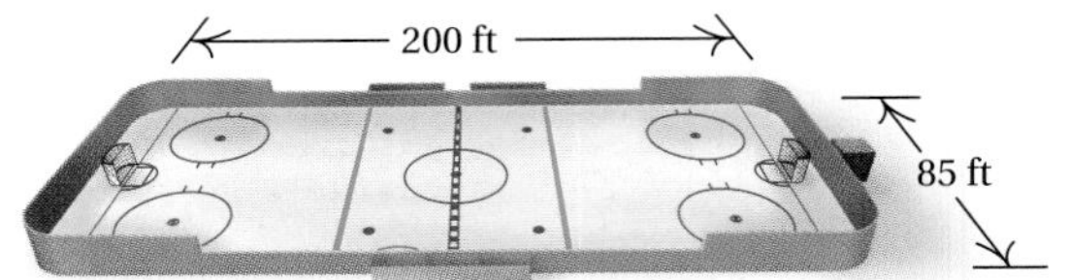

c Write a related multiplication sentence.

33. $18 \div 3 = 6$

34. $72 \div 9 = 8$

35. $22 \div 22 = 1$

36. $32 \div 1 = 32$

Write two related division sentences.

37. $9 \times 5 = 45$

38. $2 \cdot 7 = 14$

39. $37 \cdot 1 = 37$

40. $4 \cdot 12 = 48$

d Divide, if possible. If not possible, write "not defined."

41. $72 \div 6$

42. $54 \div 9$

43. $\frac{23}{23}$

44. $\frac{37}{37}$

45. $22 \div 1$

46. $\frac{56}{1}$

47. $\frac{16}{0}$

48. $74 \div 0$

Divide.

49. $277 \div 5$

50. $699 \div 3$

51. $864 \div 8$

52. $869 \div 8$

53. $4\overline{)1228}$

54. $3\overline{)2124}$

55. $738 \div 8$

56. $881 \div 6$

57. $5\overline{)8515}$

58. $3\overline{)6027}$

59. $127{,}000 \div 1000$

60. $4260 \div 10$

61. $30\overline{)875}$

62. $40\overline{)987}$

63. $852 \div 21$

64. $942 \div 23$

65. $85\overline{)7672}$

66. $54\overline{)2729}$

67. $111\overline{)3219}$

68. $102\overline{)5612}$

69. $5\overline{)8047}$

70. $9\overline{)7273}$

71. $5\overline{)5036}$

72. $7\overline{)7074}$

73. $1058 \div 46$

74. $7242 \div 24$

75. $3425 \div 32$

76. $48\overline{)4899}$

77. $24\overline{)8880}$

78. $36\overline{)7563}$

79. $28\overline{)17{,}067}$

80. $36\overline{)28{,}929}$

81. $80\overline{)24{,}320}$

82. $90\overline{)88{,}560}$

83. $285\overline{)999{,}999}$

84. $306\overline{)888{,}888}$

85. $456\overline{)3{,}679{,}920}$

86. $803\overline{)5{,}622{,}606}$

e Round to the nearest ten.

87. 48

88. 532

89. 467

90. 8945

91. 731

92. 17

93. 895

94. 798

Round to the nearest hundred.

95. 146

96. 874

97. 957

98. 650

99. 9079

100. 4645

101. 32,850

102. 198,402

Round to the nearest thousand.

103. 5876

104. 4500

105. 7500

106. 2001

107. 45,340

108. 735,562

109. 373,405

110. 6,713,855

Estimate the sum or difference by first rounding to the nearest ten. Show your work.

111. $\begin{array}{r} 78 \\ +\ 97 \\ \hline \end{array}$

112. $\begin{array}{r} 62 \\ 97 \\ 46 \\ +\ 88 \\ \hline \end{array}$

113. $\begin{array}{r} 8074 \\ -\ 2347 \\ \hline \end{array}$

114. $\begin{array}{r} 673 \\ -\ \ 28 \\ \hline \end{array}$

Estimate the sum or difference by first rounding to the nearest hundred. Show your work.

115. $\begin{array}{r} 7348 \\ +\ 9247 \\ \hline \end{array}$

116. $\begin{array}{r} 568 \\ 472 \\ 938 \\ +\ 402 \\ \hline \end{array}$

117. $\begin{array}{r} 6852 \\ -\ 1748 \\ \hline \end{array}$

118. $\begin{array}{r} 9438 \\ -\ 2787 \\ \hline \end{array}$

Estimate the sum or difference by first rounding to the nearest thousand. Show your work.

119.
```
   9643
   4821
   8943
 + 7004
```

120.
```
   7648
   9348
   7842
 + 2222
```

121.
```
   92,149
 − 22,555
```

122.
```
   84,890
 − 11,110
```

Planning a Kitchen. Perfect Kitchens offers custom kitchen packages with three choices for each of four items: cabinets, countertops, appliances, and flooring. The chart below lists the price for each choice.

Perfect Kitchens lets you customize your kitchen with one choice from each of the following four features:

CABINETS	TYPE	PRICE
(a)	Oak	$7450
(b)	Cherry	8820
(c)	Painted	9630
COUNTERTOPS	**TYPE**	**PRICE**
(d)	Laminate	$1595
(e)	Solid surface	2870
(f)	Granite	3528
APPLIANCES	**PRICE RANGE**	**PRICE**
(g)	Low	$1540
(h)	Medium	3575
(i)	High	6245
FLOORING	**TYPE**	**PRICE**
(j)	Vinyl	$ 625
(k)	Ceramic tile	985
(l)	Hardwood	1160

123. Estimate the cost of remodeling a kitchen with choices (a), (d), (g), and (j) by rounding to the nearest hundred dollars.

124. Estimate the cost of a kitchen with choices (c), (f), (i), and (l) by rounding to the nearest hundred dollars.

125. Sara and Ben are planning to remodel their kitchen and have a budget of $17,700. Estimate by rounding to the nearest hundred dollars the cost of their kitchen remodeling project if they choose options (b), (e), (i), and (k). Can they afford their choices?

126. The Davidsons must make a final decision on the kitchen choices for their new home. The allotted kitchen budget is $16,000. Estimate by rounding to the nearest hundred dollars the kitchen cost if they choose options (a), (f), (h), and (l). Does their budget allotment cover the cost?

127. Suppose you are planning a new kitchen and must stay within a budget of $14,500. Decide on the options you would like and estimate the cost by rounding to the nearest hundred dollars. Does your budget support your choices?

128. Suppose you are planning a new kitchen and must stay within a budget of $18,500. Decide on the options you would like and estimate the cost by rounding to the nearest hundred dollars. Does your budget support your choices?

Estimate the product by first rounding to the nearest ten. Show your work.

129. $\begin{array}{r} 45 \\ \times\ 67 \\ \hline \end{array}$

130. $\begin{array}{r} 51 \\ \times\ 78 \\ \hline \end{array}$

131. $\begin{array}{r} 34 \\ \times\ 29 \\ \hline \end{array}$

132. $\begin{array}{r} 63 \\ \times\ 54 \\ \hline \end{array}$

Estimate the product by first rounding to the nearest hundred. Show your work.

133. $\begin{array}{r} 876 \\ \times\ 345 \\ \hline \end{array}$

134. $\begin{array}{r} 355 \\ \times\ 299 \\ \hline \end{array}$

135. $\begin{array}{r} 432 \\ \times\ 199 \\ \hline \end{array}$

136. $\begin{array}{r} 789 \\ \times\ 434 \\ \hline \end{array}$

137. *Toyota Sienna.* A pharmaceutical company buys a Toyota Sienna for each of its 112 sales representatives. Each car costs $27,896 plus an additional $540 per car in destination charges.

a) Estimate the total cost of the purchase by rounding the cost of each car, the destination charge, and the number of sales representatives to the nearest hundred.

b) Estimate the total cost of the purchase by rounding the cost of each car to the nearest thousand and the destination charge and the number of reps to the nearest hundred.

Source: Toyota

138. A travel club of 176 people decides to fly from Los Angeles to Tokyo. The cost of a round-trip ticket is $643.

a) Estimate the total cost of the trip by rounding the cost of the airfare and the number of travelers to the nearest ten.

b) Estimate the total cost of the trip by rounding the cost of the airfare and the number of travelers to the nearest hundred.

139. D_W Is division associative? Why or why not? Give an example.

140. D_W Suppose a student asserts that "$0 \div 0 = 0$ because nothing divided by nothing is nothing." Devise an explanation to persuade the student that the assertion is false.

SKILL MAINTENANCE

VOCABULARY REINFORCEMENT

In each of Exercises 141–148, fill in the blank with the correct term from the given list. Some of the choices may not be used.

141. The distance around an object is its _______________. [1.2b]

142. A sentence like $10 - 3 = 7$ is called a(n) _______________; a sentence like $31 < 33$ is called a(n) _______________. [1.1d]

143. For large numbers, _______________ are separated by commas into groups of three, called _______________. [1.1a]

144. The number 0 is called the _______________ identity. [1.2a]

145. In the sentence $28 \div 7 = 4$, the _______________ is 28. [1.3c]

146. In the sentence $10 \times 1000 = 10{,}000$, 10 and 1000 are called _______________ and 10,000 is called the _______________. [1.3a]

147. The _______________ is the number from which another number is being subtracted. [1.2c]

148. The number 1 is called the _______________ identity. [1.3a]

addends | subtrahend
factors | digits
perimeter | periods
dividend | additive
quotient | multiplicative
product | equation
minuend | inequality

SYNTHESIS

149. Complete the following table.

a	b	$a \cdot b$	$a + b$
	68	3672	
84			117
		32	12

150. Find a pair of factors whose product is 36 and:

a) whose sum is 13.
b) whose difference is 0.
c) whose sum is 20.
d) whose difference is 9.

151. A group of 1231 college students is going to take buses for a field trip. Each bus can hold only 42 students. How many buses are needed?

152. ▦ Fill in the missing digits to make the equation true:

$34{,}584{,}132 \div 76\square = 4\square{,}386.$

153. ▦ An 18-story office building is box-shaped. Each floor measures 172 ft by 84 ft with a 20-ft by 35-ft rectangular area lost to an elevator and a stairwell. How much area is available as office space?

1.4 SOLVING EQUATIONS

Objectives

a Solve simple equations by trial.

b Solve equations like $x + 28 = 54$, $28 \cdot x = 168$, and $98 \cdot 2 = y$.

Find a number that makes the sentence true.

1. $8 = 1 + \square$

2. $\square + 2 = 7$

3. Determine whether 7 is a solution of $\square + 5 = 9$.

4. Determine whether 4 is a solution of $\square + 5 = 9$.

Answers on page A-2

a Solutions by Trial

Let's find a number that we can put in the blank to make this sentence true:

$$9 = 3 + \square.$$

We are asking "9 is 3 plus what number?" The answer is 6.

$$9 = 3 + 6$$

Do Exercises 1 and 2.

A sentence with = is called an **equation.** A **solution** of an equation is a number that makes the sentence true. Thus, 6 is a solution of

$9 = 3 + \square$ because $9 = 3 + 6$ is true.

However, 7 is not a solution of

$9 = 3 + \square$ because $9 = 3 + 7$ is false.

Do Exercises 3 and 4.

We can use a letter instead of a blank. For example,

$$9 = 3 + n.$$

We call n a **variable** because it can represent any number. If a replacement for a variable makes an equation true, it is a **solution** of the equation.

SOLUTIONS OF AN EQUATION

A **solution** is a replacement for the variable that makes the equation true. When we find all the solutions, we say that we have **solved** the equation.

EXAMPLE 1 Solve $y + 12 = 27$ by trial.

We replace y with several numbers.

If we replace y with 13, we get a false equation: $13 + 12 = 27$.

If we replace y with 14, we get a false equation: $14 + 12 = 27$.

If we replace y with 15, we get a true equation: $15 + 12 = 27$.

No other replacement makes the equation true, so the solution is 15.

EXAMPLES Solve.

2. $7 + n = 22$
(7 plus what number is 22?)
The solution is 15.

3. $63 = 3 \cdot x$
(63 is 3 times what number?)
The solution is 21.

Do Exercises 5–8 on the following page.

b Solving Equations

We now begin to develop more efficient ways to solve certain equations. When an equation has a variable alone on one side, it is easy to see the solution or to compute it. When a calculation is on one side and the variable is alone on the other, we can find the solution by carrying out the calculation.

EXAMPLE 4 Solve: $x = 245 \times 34$.

To solve the equation, we carry out the calculation.

$$\begin{array}{r} 2\ 4\ 5 \\ \times \quad 3\ 4 \\ \hline 9\ 8\ 0 \\ 7\ 3\ 5\ 0 \\ \hline 8\ 3\ 3\ 0 \end{array} \qquad \begin{aligned} x &= 245 \times 34 \\ x &= 8330 \end{aligned}$$

The solution is 8330.

Do Exercises 9–12.

Look at the equation

$x + 12 = 27.$

We can get x alone by writing a related subtraction sentence:

$x = 27 - 12$ — 12 gets subtracted to find the related subtraction sentence.

$x = 15.$ — Doing the subtraction

It is useful in our later study of algebra to think of this as "subtracting 12 *on both sides.*" Thus

$x + 12 - 12 = 27 - 12$ — Subtracting 12 on both sides

$x + 0 = 15$ — Carrying out the subtraction

$x = 15.$

SOLVING $x + a = b$

To solve $x + a = b$, subtract a on both sides.

If we can get an equation in a form with the variable alone on one side, we can "see" the solution.

EXAMPLE 5 Solve: $t + 28 = 54$.

We have

$t + 28 = 54$

$t + 28 - 28 = 54 - 28$ — Subtracting 28 on both sides

$t + 0 = 26$

$t = 26.$

Solve by trial.

5. $n + 3 = 8$

6. $x - 2 = 8$

7. $45 \div 9 = y$

8. $10 + t = 32$

Solve.

9. $346 \times 65 = y$

10. $x = 2347 + 6675$

11. $4560 \div 8 = t$

12. $x = 6007 - 2346$

Answers on page A-2

Solve. Be sure to check.

13. $x + 9 = 17$

14. $77 = m + 32$

15. Solve: $155 = t + 78$. Be sure to check.

Solve. Be sure to check.

16. $4566 + x = 7877$

17. $8172 = h + 2058$

Answers on page A-2

To check the answer, we substitute 26 for t in the original equation.

Check:

$$\begin{array}{c|c} t + 28 & 54 \\ \hline 26 + 28 \;?\; & 54 \\ 54 & \end{array} \quad \text{TRUE}$$

The solution is 26.

Do Exercises 13 and 14.

EXAMPLE 6 Solve: $182 = 65 + n$.

We have

$$\begin{aligned} 182 &= 65 + n \\ 182 - 65 &= 65 + n - 65 && \text{Subtracting 65 on both sides} \\ 117 &= 0 + n && \text{65 plus } n \text{ minus 65 is } 0 + n. \\ 117 &= n. \end{aligned}$$

Check:

$$\begin{array}{c|c} 182 & 65 + n \\ \hline 182 \;?\; & 65 + 117 \\ & 182 \end{array} \quad \text{TRUE}$$

The solution is 117.

Do Exercise 15.

EXAMPLE 7 Solve: $7381 + x = 8067$.

We have

$$\begin{aligned} 7381 + x &= 8067 \\ 7381 + x - 7381 &= 8067 - 7381 && \text{Subtracting 7381 on both sides} \\ x &= 686. \end{aligned}$$

The check is left to the student. The solution is 686.

Do Exercises 16 and 17.

We now learn to solve equations like $8 \cdot n = 96$. Look at

$$8 \cdot n = 96.$$

We can get n alone by writing a related division sentence:

$$\begin{aligned} n &= 96 \div 8 = \frac{96}{8} && \text{96 is divided by 8.} \\ n &= 12. && \text{Doing the division} \end{aligned}$$

It is useful in our later study of algebra to think of the preceding as "dividing by 8 *on both sides*." Thus,

$$\begin{aligned} \frac{8 \cdot n}{8} &= \frac{96}{8} && \text{Dividing by 8 on both sides} \\ n &= 12. && \text{8 times } n \text{ divided by 8 is } n. \end{aligned}$$

SOLVING $a \cdot x = b$

To solve $a \cdot x = b$, divide by a on both sides.

EXAMPLE 8 Solve: $10 \cdot x = 240$.

We have

$$10 \cdot x = 240$$

$$\frac{10 \cdot x}{10} = \frac{240}{10} \quad \text{Dividing by 10 on both sides}$$

$$x = 24.$$

Check:

$$\begin{array}{c|c} 10 \cdot x & 240 \\ \hline 10 \cdot 24 \ ? & 240 \\ 240 & \end{array} \quad \text{TRUE}$$

The solution is 24.

EXAMPLE 9 Solve: $5202 = 9 \cdot t$.

We have

$$5202 = 9 \cdot t$$

$$\frac{5202}{9} = \frac{9 \cdot t}{9} \quad \text{Dividing by 9 on both sides}$$

$$578 = t.$$

The check is left to the student. The solution is 578.

Do Exercises 18–20.

EXAMPLE 10 Solve: $14 \cdot y = 1092$.

We have

$$14 \cdot y = 1092$$

$$\frac{14 \cdot y}{14} = \frac{1092}{14} \quad \text{Dividing by 14 on both sides}$$

$$y = 78.$$

The check is left to the student. The solution is 78.

EXAMPLE 11 Solve: $n \cdot 56 = 4648$.

We have

$$n \cdot 56 = 4648$$

$$\frac{n \cdot 56}{56} = \frac{4648}{56} \quad \text{Dividing by 56 on both sides}$$

$$n = 83.$$

The check is left to the student. The solution is 83.

Do Exercises 21 and 22.

Solve. Be sure to check.

18. $8 \cdot x = 64$

19. $144 = 9 \cdot n$

20. $5152 = 8 \cdot t$

Solve. Be sure to check.

21. $18 \cdot y = 1728$

22. $n \cdot 48 = 4512$

Answers on page A-2

1.4 EXERCISE SET

For Extra Help

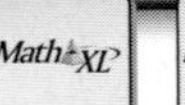

MathXL | MyMathLab | InterAct Math | Math Tutor Center | Video Lectures on CD Disc 1 | Student's Solutions Manual

a Solve by trial.

1. $x + 0 = 14$

2. $x - 7 = 18$

3. $y \cdot 17 = 0$

4. $56 \div m = 7$

b Solve. Be sure to check.

5. $13 + x = 42$

6. $15 + t = 22$

7. $12 = 12 + m$

8. $16 = t + 16$

9. $3 \cdot x = 24$

10. $6 \cdot x = 42$

11. $112 = n \cdot 8$

12. $162 = 9 \cdot m$

13. $45 \cdot 23 = x$

14. $23 \cdot 78 = y$

15. $t = 125 \div 5$

16. $w = 256 \div 16$

17. $p = 908 - 458$

18. $9007 - 5667 = m$

19. $x = 12{,}345 + 78{,}555$

20. $5678 + 9034 = t$

21. $3 \cdot m = 96$

22. $4 \cdot y = 96$

23. $715 = 5 \cdot z$

24. $741 = 3 \cdot t$

25. $10 + x = 89$

26. $20 + x = 57$

27. $61 = 16 + y$

28. $53 = 17 + w$

29. $6 \cdot p = 1944$

30. $4 \cdot w = 3404$

31. $5 \cdot x = 3715$

32. $9 \cdot x = 1269$

33. $47 + n = 84$

34. $56 + p = 92$

35. $x + 78 = 144$

36. $z + 67 = 133$

37. $165 = 11 \cdot n$

38. $660 = 12 \cdot n$

39. $624 = t \cdot 13$

40. $784 = y \cdot 16$

41. $x + 214 = 389$

42. $x + 221 = 333$

43. $567 + x = 902$

44. $438 + x = 807$

45. $18 \cdot x = 1872$

46. $19 \cdot x = 6080$

47. $40 \cdot x = 1800$

48. $20 \cdot x = 1500$

49. $2344 + y = 6400$

50. $9281 = 8322 + t$

51. $8322 + 9281 = x$

52. $9281 - 8322 = y$

53. $234 \cdot 78 = y$

54. $10{,}534 \div 458 = q$

55. $58 \cdot m = 11{,}890$

56. $233 \cdot x = 22{,}135$

57. ^D_W Describe a procedure that can be used to convert any equation of the form $a \cdot b = c$ to a related division equation.

58. ^D_W Describe a procedure that can be used to convert any equation of the form $a + b = c$ to a related subtraction equation.

SKILL MAINTENANCE

59. Write two related subtraction sentences: $7 + 8 = 15$. [1.2c]

60. Write two related division sentences: $6 \cdot 8 = 48$. [1.3c]

Use > or < for □ to write a true sentence. [1.1d]

61. 123 □ 789

62. 342 □ 339

63. 688 □ 0

64. 0 □ 11

Divide. [1.3d]

65. $1283 \div 9$

66. $1278 \div 9$

67. $17\overline{)5678}$

68. $17\overline{)5689}$

SYNTHESIS

Solve.

69. [calculator] $23{,}465 \cdot x = 8{,}142{,}355$

70. [calculator] $48{,}916 \cdot x = 14{,}332{,}388$

1.5 APPLICATIONS AND PROBLEM SOLVING

Objective

a Solve applied problems involving addition, subtraction, multiplication, or division of whole numbers.

a A Problem-Solving Strategy

Applications and problem solving are the most important uses of mathematics. To solve a problem using the operations on the whole numbers, we first look at the situation. We try to translate the problem to an equation. Then we solve the equation. We check to see if the solution of the equation is a solution of the original problem. We are using the following five-step strategy.

FIVE STEPS FOR PROBLEM SOLVING

1. *Familiarize* yourself with the situation.
   a) Carefully read and reread until you understand *what* you are being asked to find.
   b) Draw a diagram or see if there is a formula that applies to the situation.
   c) Assign a letter, or *variable,* to the unknown.
2. *Translate* the problem to an equation using the letter or variable.
3. *Solve* the equation.
4. *Check* the answer in the original wording of the problem.
5. *State* the answer to the problem clearly with appropriate units.

EXAMPLE 1 *International Adoptions.* The top ten countries of origin for United States international adoptions in 2003 and 2002 are listed in the table below. Find the total number of adoptions from China, South Korea, India, and Vietnam in 2003.

INTERNATIONAL ADOPTIONS

	2003		2002	
RANK	COUNTRY OF ORIGIN	NUMBER	COUNTRY OF ORIGIN	NUMBER
1	China (mainland)	6859	China (mainland)	5053
2	Russia	5209	Russia	4939
3	Guatemala	2328	Guatemala	2219
4	South Korea	1790	South Korea	1779
5	Kazakhstan	825	Ukraine	1106
6	Ukraine	702	Kazakhstan	819
7	India	472	Vietnam	766
8	Vietnam	382	India	466
9	Colombia	272	Colombia	334
10	Haiti	250	Bulgaria	260

Source: U.S. Department of State

1. **Familiarize.** We can make a drawing or at least visualize the situation.

6859	+	1790	+	472	+	382
from China		from South Korea		from India		from Vietnam

Since we are combining numbers of adoptions, addition can be used. First, we define the unknown. We let $n =$ the total number of adoptions from China, South Korea, India, and Vietnam.

2. **Translate.** We translate to an equation:

$$6859 + 1790 + 472 + 382 = n.$$

3. **Solve.** We solve the equation by carrying out the addition.

$$\begin{array}{r} \scriptstyle 2\ 3\ 1 \\ 6\ 8\ 5\ 9 \\ 1\ 7\ 9\ 0 \\ 4\ 7\ 2 \\ +\quad 3\ 8\ 2 \\ \hline 9\ 5\ 0\ 3 \end{array} \qquad \begin{aligned} 6859 + 1790 + 472 + 382 &= n \\ 9503 &= n \end{aligned}$$

4. **Check.** We check 9503 in the original problem. There are many ways in which this can be done. For example, we can repeat the calculation. (We leave this to the student.) Another way is to check whether the answer is reasonable. In this case, we would expect the total to be greater than the number of adoptions from any of the individual countries, which it is. We can also estimate by rounding. Here we round to the nearest hundred.

$$\begin{aligned} 6859 + 1790 + 472 + 382 &\approx 6900 + 1800 + 500 + 400 \\ &= 9600 \end{aligned}$$

Since $9600 \approx 9503$, we have a partial check. If we had an estimate like 4800 or 7500, we might be suspicious that our calculated answer is incorrect. Since our estimated answer is close to our calculation, we are further convinced that our answer checks.

5. **State.** The total number of adoptions from China, South Korea, India, and Vietnam in 2003 is 9503.

Do Exercises 1–3.

Refer to the table on the preceding page to answer Margin Exercises 1–3.

1. Find the total number of adoptions from China, Russia, Kazakhstan, and India in 2002.

2. Find the total number of adoptions from Russia, Guatemala, Vietnam, Colombia, and Haiti in 2003.

3. Find the total number of adoptions from the top ten countries in 2003.

Answers on page A-2

4. Checking Account Balance. The balance in Heidi's checking account is $2003. She uses her debit card to buy the same RotoZip Spiral Saw Combo, featured in Example 2, that Tyler did. Find the new balance in her checking account.

Answer on page A-2

EXAMPLE 2 *Checking Account Balance.* The balance in Tyler's checking account is $528. He uses his debit card to buy the RotoZip Spiral Saw Combo shown in this ad. Find the new balance in his checking account.

Source: Reproduced with permission from the copyright owner (©2004 by Robert Bosch Tool Corporation). Further reproductions strictly prohibited.

1. **Familiarize.** We first make a drawing or at least visualize the situation. We let $M =$ the new balance in his account. This gives us the following:

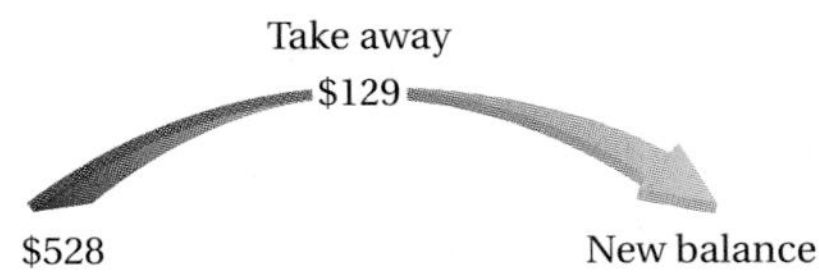

2. **Translate.** We can think of this as a "take-away" situation. We translate to an equation.

Money in the account	minus	Money spent	is	New balance
↓	↓	↓	↓	↓
528	$-$	129	$=$	M

3. **Solve.** This sentence tells us what to do. We subtract.

$$\begin{array}{r} \scriptstyle 11 \\ \scriptstyle 4\ \not{1}\ 18 \\ \not{5}\ \not{2}\ \not{8} \\ -\ 1\ 2\ 9 \\ \hline 3\ 9\ 9 \end{array} \qquad \begin{aligned} 528 - 129 &= M \\ 399 &= M \end{aligned}$$

4. **Check.** To check our answer of $399, we can repeat the calculation. We note that the answer should be less than the original amount, $528, which it is. We can add the difference, 399, to the subtrahend, 129: $129 + 399 = 528$. We can also estimate:

$$528 - 129 \approx 530 - 130 = 400 \approx 399.$$

5. **State.** Tyler has a new balance of $399 in his checking account.

Do Exercise 4.

In the real world, problems may not be stated in written words. You must still become familiar with the situation before you can solve the problem.

EXAMPLE 3 *Travel Distance.* Abigail is driving from Indianapolis to Salt Lake City to interview for a news anchor position. The distance from Indianapolis to Salt Lake City is 1634 mi. She travels 1154 mi to Denver. How much farther must she travel?

1. **Familiarize.** We first make a drawing or at least visualize the situation. We let $x =$ the remaining distance to Salt Lake City.

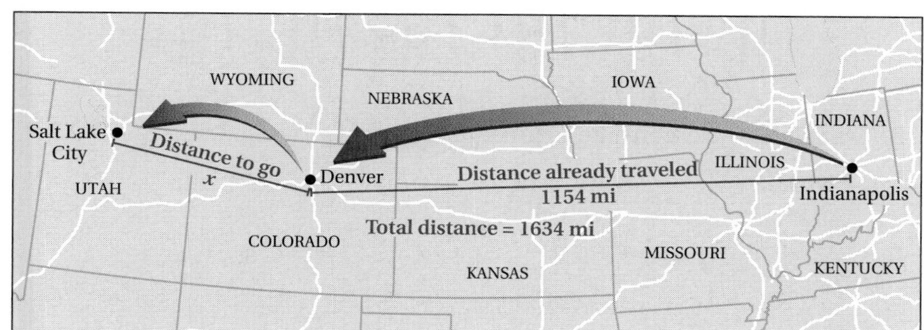

2. **Translate.** We see that this is a "how much more" situation. We translate to an equation.

Distance already traveled	plus	Distance to go	is	Total distance of trip
↓	↓	↓	↓	↓
1154	$+$	x	$=$	1634

3. **Solve.** To solve the equation, we subtract 1154 on both sides:

$$1154 + x = 1634$$
$$1154 + x - 1154 = 1634 - 1154$$
$$x = 480.$$

$$\begin{array}{r} \overset{5\,13}{1\,\cancel{6}\,\cancel{3}\,4} \\ -\ 1\,1\,5\,4 \\ \hline 4\,8\,0 \end{array}$$

4. **Check.** We check our answer of 480 mi in the original problem. This number should be less than the total distance, 1634 mi, which it is. We can add the difference, 480, to the subtrahend, 1154: $1154 + 480 = 1634$. We can also estimate:

$$1634 - 1154 \approx 1600 - 1200$$
$$= 400 \approx 480.$$

The answer, 480 mi, checks.

5. **State.** Abigail must travel 480 mi farther to Salt Lake City.

Do Exercise 5.

5. **Home Vacuum.** William has \$228. He wants to purchase the home vacuum shown in the ad below. How much more does he need?

Answer on page A-2

6. Total Cost of Gas Grills. What is the total cost of 14 gas grills, each with 520 sq in. of total cooking surface and porcelain cast-iron cooking grates, if each one costs $398?

EXAMPLE 4 *Total Cost of Chairs.* What is the total cost of 6 Logan side chairs from Restoration Hardware if each one costs $210?

1. **Familiarize.** We first make a drawing or at least visualize the situation. We let $T =$ the cost of 6 chairs. Repeated addition works well in this case.

2. **Translate.** We translate to an equation.

Number of chairs	times	Cost of each chair	is	Total cost
↓	↓	↓	↓	↓
6	$\times$	$210	$=$	T

3. **Solve.** This sentence tells us what to do. We multiply.

$$\begin{array}{r} 2\ 1\ 0 \\ \times \quad\ \ 6 \\ \hline 1\ 2\ 6\ 0 \end{array} \qquad \begin{aligned} 6 \times 210 &= T \\ 1260 &= T \end{aligned}$$

4. **Check.** We have an answer, 1260, that is much greater than the cost of any individual chair, which is reasonable. We can repeat our calculation. We can also check by estimating:

$$6 \times 210 \approx 6 \times 200 = 1200 \approx 1260.$$

The answer checks.

5. **State.** The total cost of 6 chairs is $1260.

Do Exercise 6.

EXAMPLE 5 *Truck Bed Cover.* The dimensions of a fiberglass truck bed cover for a pickup truck are 79 in. by 68 in. What is the area of the cover?

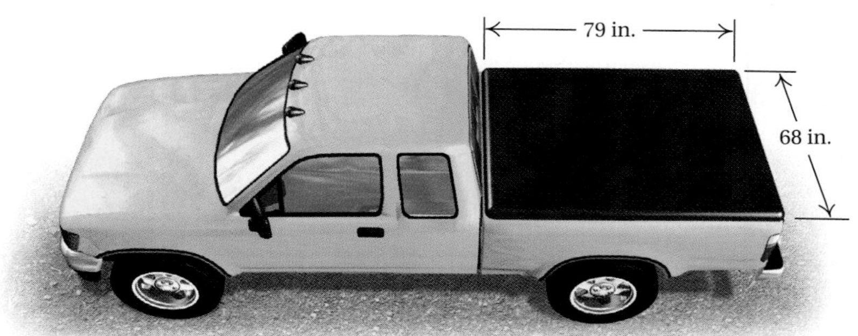

Answer on page A-2

1. **Familiarize.** The truck bed cover is a rectangle that measures 79 in. by 68 in. We let $A =$ the area and use the area formula $A = l \cdot w$.
2. **Translate.** Using this formula, we have

$$A = \text{length} \cdot \text{width} = l \cdot w = 79 \cdot 68.$$

3. **Solve.** We carry out the multiplication.

$$\begin{array}{r} 7\ 9 \\ \times\ 6\ 8 \\ \hline 6\ 3\ 2 \\ 4\ 7\ 4\ 0 \\ \hline 5\ 3\ 7\ 2 \end{array} \qquad \begin{aligned} A &= 79 \cdot 68 \\ A &= 5372 \end{aligned}$$

4. **Check.** We repeat our calculation. We also note that the answer is greater than either the length or the width, which it should be. (This might not be the case if we were using fractions or decimals.) The answer checks.
5. **State.** The area of the truck bed cover is 5372 sq in.

Do Exercise 7.

EXAMPLE 6 *Cartons of Soda.* A bottling company produces 3304 cans of soda. How many 12-can cartons can be filled? How many cans will be left over?

1. **Familiarize.** We first make a drawing. We let $n =$ the number of 12-can cartons that can be filled. The problem can be considered as repeated subtraction, taking successive sets of 12 cans and putting them into n cartons.

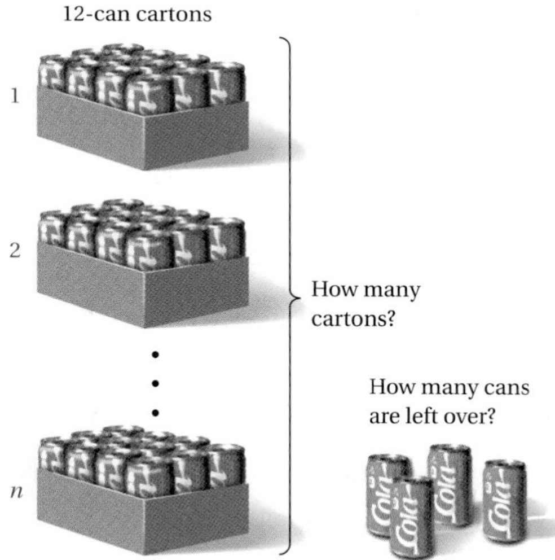

12-can cartons
1
2
n
How many cartons?
How many cans are left over?

2. **Translate.** We translate to an equation.

Number of cans	divided by	Number in each carton	is	Number of cartons
↓	↓	↓	↓	↓
3304	÷	12	=	n

7. **Bed Sheets.** The dimensions of a flat sheet for a queen-size bed are 90 in. by 102 in. What is the area of the sheet?

Answer on page A-2

8. Cartons of Soda. The bottling company in Example 6 also uses 6-can cartons. How many 6-can cartons can be filled with 2269 cans of cola? How many will be left over?

Answer on page A-2

3. Solve. We solve the equation by carrying out the division.

$$\begin{array}{r} 275 \\ 12\overline{)3304} \\ \underline{2400} \\ 904 \\ \underline{840} \\ 64 \\ \underline{60} \\ 4 \end{array} \qquad \begin{aligned} 3304 \div 12 &= n \\ 275 \text{ R } 4 &= n \end{aligned}$$

4. Check. We can check by multiplying the number of cartons by 12 and adding the remainder, 4:

$12 \cdot 275 = 3300,$

$3300 + 4 = 3304.$

5. State. Thus, 275 twelve-can cartons can be filled. There will be 4 cans left over.

Do Exercise 8.

EXAMPLE 7 *Automobile Mileage.* The Pontiac G6 GT gets 21 miles to the gallon (mpg) in city driving. How many gallons will it use in 3843 mi of city driving?

Source: General Motors

1. Familiarize. We first make a drawing. It is often helpful to be descriptive about how we define a variable. In this case, we let g = the number of gallons ("g" comes from "gallons").

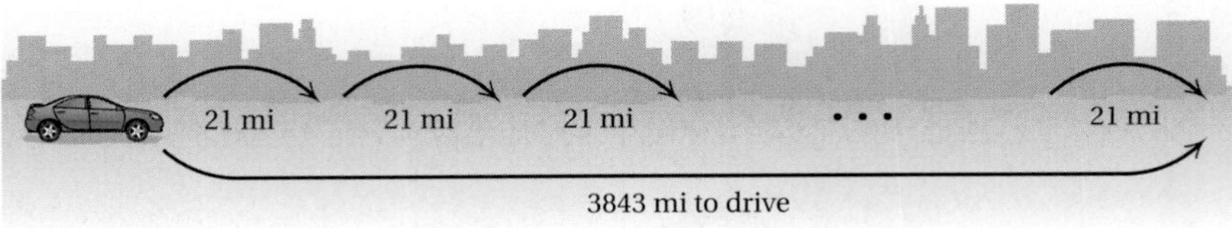

2. Translate. Repeated addition applies here. Thus the following multiplication applies to the situation.

Number of miles per gallon	times	Number of gallons needed	is	Number of miles to drive
↓	↓	↓	↓	↓
21	$\cdot$	g	$=$	3843

3. Solve. To solve the equation, we divide by 21 on both sides.

$$\begin{aligned} 21 \cdot g &= 3843 \\ \frac{21 \cdot g}{21} &= \frac{3843}{21} \\ g &= 183 \end{aligned} \qquad \begin{array}{r} 183 \\ 21\overline{)3843} \\ \underline{2100} \\ 1743 \\ \underline{1680} \\ 63 \\ \underline{63} \\ 0 \end{array}$$

4. **Check.** To check, we multiply 183 by 21: $21 \times 183 = 3843$.
5. **State.** The Pontiac G6 GT will use 183 gal.

Do Exercise 9.

Multistep Problems

Sometimes we must use more than one operation to solve a problem, as in the following example.

EXAMPLE 8 *Aircraft Seating.* Boeing Corporation builds commercial aircraft. A Boeing 767 has a seating configuration with 4 rows of 6 seats across in first class and 35 rows of 7 seats across in economy class. Find the total seating capacity of the plane.
Sources: The Boeing Company; Delta Airlines

1. **Familiarize.** We first make a drawing.

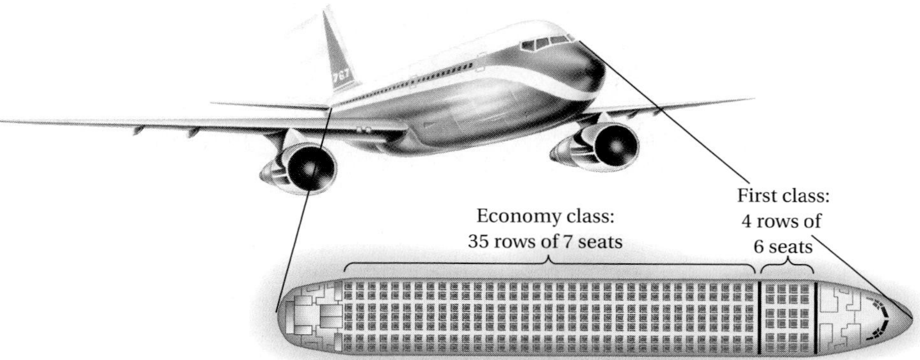

2. **Translate.** There are three parts to the problem. We first find the number of seats in each class. Then we add.

First-class: Repeated addition applies here. Thus the following multiplication corresponds to the situation. We let $F =$ the number of seats in first class.

Number of rows	times	Seats in each row	is	Total number
↓	↓	↓	↓	↓
4	$\cdot$	6	=	F

Economy class: Repeated addition applies here. Thus the following multiplication corresponds to the situation. We let $E =$ the number of seats in economy class.

Number of rows	times	Seats in each row	is	Total number
↓	↓	↓	↓	↓
35	$\cdot$	7	=	E

We let $T =$ the total number of seats in both classes.

Number of seats in first class	plus	Number of seats in economy class	is	Total number of seats on plane
↓	↓	↓	↓	↓
F	+	E	=	T

9. **Automobile Mileage.** The Pontiac G6 GT gets 29 miles to the gallon (mpg) in highway driving. How many gallons will it use in 2291 mi of highway driving?
Source: General Motors

Answer on page A-2

10. Aircraft Seating. A Boeing 767 used for foreign travel has three classes of seats. First class has 3 rows of 5 seats across; business class has 6 rows with 6 seats across and 1 row with 2 seats on each of the outside aisles. Economy class has 18 rows with 7 seats across. Find the total seating capacity of the plane.

Sources: The Boeing Company; Delta Airlines

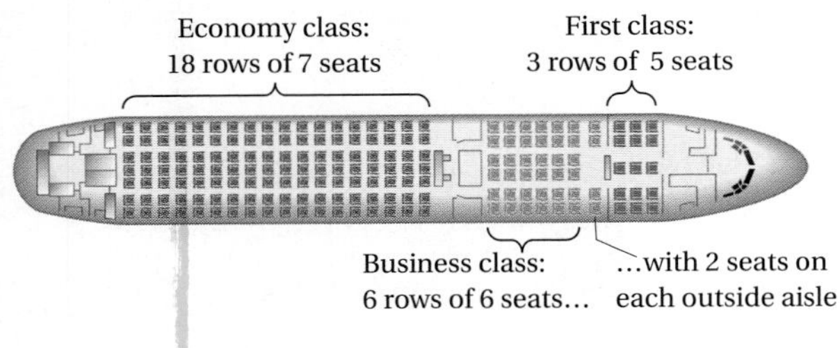

Answer on page A-2

Study Tips

EXERCISES

- **Odd-numbered exercises.** Usually an instructor assigns some odd-numbered exercises as homework. When you complete these, you can check your answers at the back of the book. If you miss any, check your work in the *Student's Solutions Manual* or ask your instructor for guidance.
- **Even-numbered exercises.** Whether or not your instructor assigns the even-numbered exercises, always do some on your own. Remember, there are no answers given for the chapter tests, so you need to practice doing exercises without answers. Check your answers later with a friend or your instructor.

3. Solve. We solve each equation and add the solutions.

$$4 \cdot 6 = F \qquad 35 \cdot 7 = E \qquad F + E = T$$
$$24 = F \qquad 245 = E \qquad 24 + 245 = T$$
$$269 = T$$

4. Check. To check, we repeat our calculations. (We leave this to the student.) We could also check by rounding, multiplying, and adding.

5. State. There are 269 seats in a Boeing 767.

Do Exercise 10.

As you consider the following exercises, here are some words and phrases that may be helpful to look for when you are translating problems to equations.

KEY WORDS, PHRASES, AND CONCEPTS

ADDITION (+)	SUBTRACTION (−)
add added to sum total plus more than increased by	subtract subtracted from difference minus less than decreased by take away how much more missing addend
MULTIPLICATION (·)	**DIVISION (÷)**
multiply multiplied by product times of repeated addition rectangular arrays	divide divided by quotient repeated subtraction missing factor finding equal quantities

Translating for Success

1. *Brick-mason Expense.* A commercial contractor is building 30 two-unit condominiums in a retirement community. The brick-mason expense for each building is $10,860. What is the total cost of bricking the buildings?

2. *Heights.* Dean's sons are on the high school basketball team. Their heights are 73 in., 69 in., and 76 in. How much taller is the tallest son than the shortest son?

3. *Account Balance.* You have $423 in your checking account. Then you deposit $73 and use your debit card for purchases of $76 and $69. How much is left in your account?

4. *Purchasing Camcorder.* A camcorder is on sale for $423. Jenny has only $69. How much more does she need to buy the camcorder?

5. *Purchasing Coffee Makers.* Sara purchases 8 coffee makers for the newly remodeled bed-and-breakfast hotel that she manages. If she pays $52 for each coffee maker, what is the total cost of her purchase?

The goal of these matching questions is to practice step (2), *Translate,* of the five-step problem-solving process. Translate each word problem to an equation and select a correct translation from equations A–O.

A. $8 \cdot 52 = n$

B. $69 \cdot n = 76$

C. $73 - 76 - 69 = n$

D. $423 + 73 - 76 - 69 = n$

E. $30 \cdot 10{,}860 = n$

F. $15 \cdot n = 195$

G. $69 + n = 423$

H. $n = 10{,}860 - 300$

I. $n = 423 \div 69$

J. $30 \cdot n = 10{,}860$

K. $15 \cdot 195 = n$

L. $n = 52 - 8$

M. $69 + n = 76$

N. $15 \div 195 = n$

O. $52 + n = 60$

Answers on page A-2

6. *Hourly Rate.* Miller Auto Repair charges $52 per hour for labor. Jackson Auto Care charges $60 per hour. How much more per hour does Jackson charge than Miller?

7. *College Band.* A college band with 195 members marches in a 15-row formation in the homecoming halftime performance. How many members are in each row?

8. *Shoe Purchase.* A professional football team purchases 15 pairs of shoes at $195 a pair. What is the total cost of this purchase?

9. *Loan Payment.* Kendra borrows $10,860 for a new boat. The loan is to be paid off in 30 payments. How much is each payment?

10. *College Enrollment.* At the beginning of the fall term, the total enrollment in Lakeview Community College was 10,860. By the end of the first two weeks, 300 students withdrew. How many students were then enrolled?

For Extra Help

MathXL · MyMathLab · InterAct Math · Math Tutor Center · Video Lectures on CD Disc 1 · Student's Solutions Manual

a Solve.

Longest Broadway Run. The bar graph below lists the five broadway shows with the greatest number of performances as of May 3, 2004.

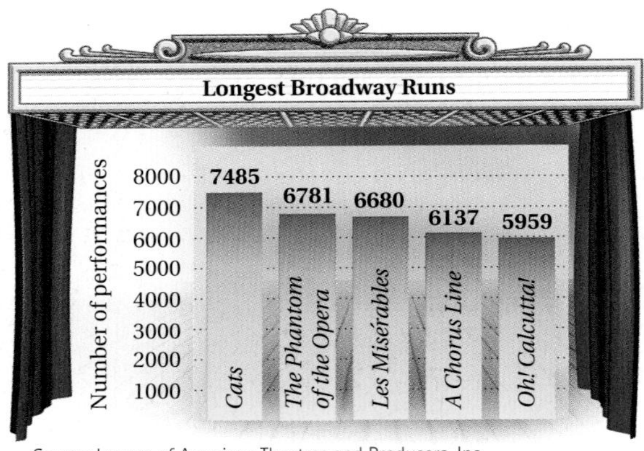

Source: League of American Theatres and Producers, Inc.

1. What was the total number of performances of all five shows?

2. What was the total number of performances of the top three shows?

3. How many more performances of *Cats* were there than performances of *A Chorus Line*?

4. How many more performances of *Les Misérables* were there than performances of *Oh! Calcutta!*?

5. *Boundaries Between Countries.* The boundary between mainland United States and Canada including the Great Lakes is 3987 mi long. The length of the boundary between the United States and Mexico is 1933 mi. How much longer is the Canadian border?

Source: U.S. Geological Survey

6. *Caffeine.* Hershey's 6-oz milk chocolate almond bar contains 25 milligrams (mg) of caffeine. A 20-oz bottle of Coca-Cola has 32 more milligrams of caffeine than the Hershey bar. How many milligrams of caffeine does the 20-oz bottle of Coca-Cola have?

Source: *National Geographic*, "Caffeine," by T. R. Reid, January 2005

7. A carpenter drills 216 holes in a rectangular array in a pegboard. There are 12 holes in each row. How many rows are there?

8. Lou works as a CPA. He arranges 504 entries on a spreadsheet in a rectangular array that has 36 rows. How many entries are in each row?

Bachelor's Degree. The line graph below illustrates data about bachelor's degrees awarded to men and women from 1970 to 2002. Use this graph when answering Exercises 9–12.

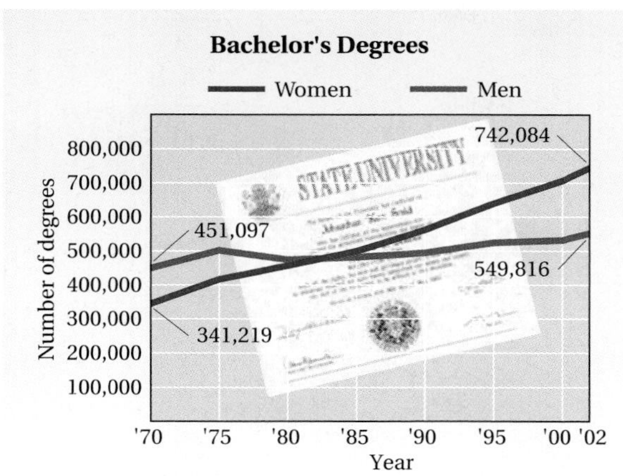

Source: U.S. Department of Education

9. Find the total number of bachelor's degrees awarded in 1970 and the total number awarded in 2002.

10. Determine how many more bachelor's degrees were awarded in 2002 than in 1970.

11. How many more bachelor's degrees were awarded to women than to men in 2002?

12. How many more bachelor's degrees were awarded to men than to women in 1970?

13. *Median Mortgage Debt.* The median mortgage debt in 2001 was $29,475 more than the median mortgage debt in 1989. The debt in 1989 was $39,802. What was the median mortgage debt in 2001?

Sources: Federal Reserve Board Survey of Consumer Finances; Consumer Bank Association 2003 Home Equity Study; Freddie Mac

14. *Olympics in Athens.* In the first modern Olympics in Athens, Greece, in 1896, there were 43 events. There were 258 more events in the 2004 Summer Olympics in Athens. How many events were there in 2004?

Source: *USA Today* research; The Olympic Games: Athens 1896–Athens 2004

15. *Longest Rivers.* The longest river in the world is the Nile in Egypt at 4100 mi. The longest river in the United States is the Missouri–Mississippi at 3860 mi. How much longer is the Nile?

16. *Speeds on Interstates.* Recently, speed limits on interstate highways in many Western states were raised from 65 mph to 75 mph. By how many miles per hour were they raised?

17. There are 24 hours (hr) in a day and 7 days in a week. How many hours are there in a week?

18. There are 60 min in an hour and 24 hr in a day. How many minutes are there in a day?

19. *Crossword.* The *USA Today* crossword puzzle is a rectangle containing 15 rows with 15 squares in each row. How many squares does the puzzle have altogether?

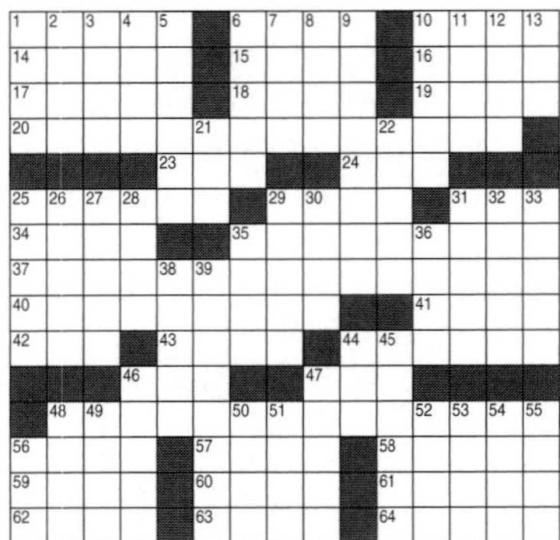

20. *Pixels.* A computer screen consists of small rectangular dots called *pixels.* How many pixels are there on a screen that has 600 rows with 800 pixels in each row?

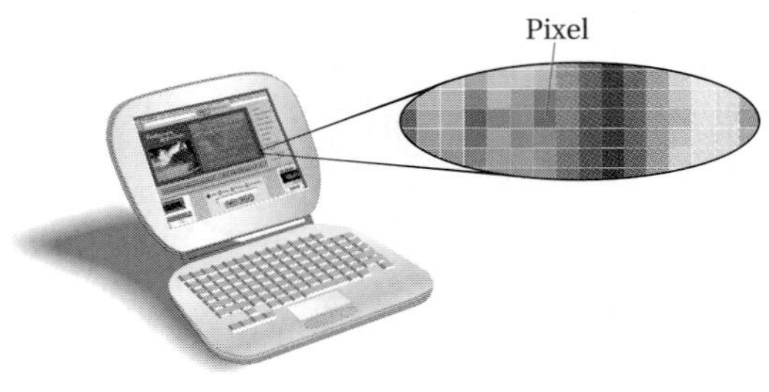

21. *Refrigerator Purchase.* Gourmet Deli has a chain of 24 restaurants. It buys a commercial refrigerator for each store at a cost of $1019 each. Find the total cost of the purchase.

22. *Microwave Purchase.* Bridgeway College is constructing new dorms in which each room has a small kitchen. It buys 96 microwave ovens at $88 each. Find the total cost of the purchase.

23. *"Seinfeld" Episodes.* "Seinfeld" was a long-running television comedy with 177 episodes. A local station picks up the syndicated reruns. If the station runs 5 episodes per week, how many full weeks will pass before it must start over with past episodes? How many episodes will be left for the last week?

24. A lab technician separates a vial containing 70 cubic centimeters (cc) of blood into test tubes, each of which contains 3 cc of blood. How many test tubes can be filled? How much blood is left over?

25. *Automobile Mileage.* The 2005 Hyundai Tucson GLS gets 26 miles to the gallon (mpg) in highway driving. How many gallons will it use in 6136 mi of highway driving?

Source: Hyundai

26. *Automobile Mileage.* The 2005 Volkswagen Jetta (5 cylinder) gets 24 miles to the gallon (mpg) in city driving. How many gallons will it use in 3960 mi of city driving?

Source: Volkswagen of America, Inc.

Boeing Jets. Use the chart below to compare the Boeing 747 jet with the Boeing 777 jet in Exercises 27–32.

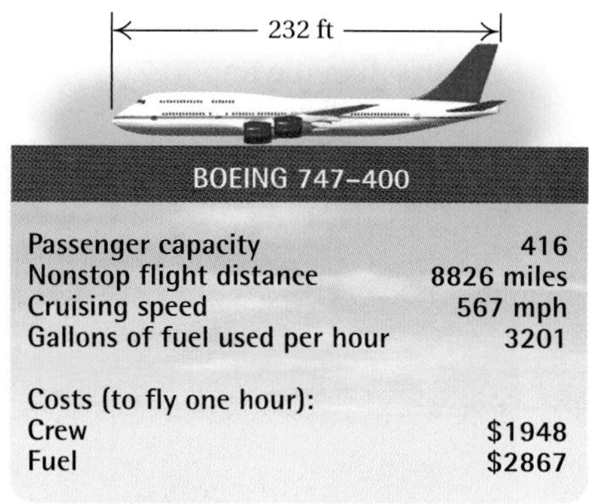

BOEING 747-400	
Passenger capacity	416
Nonstop flight distance	8826 miles
Cruising speed	567 mph
Gallons of fuel used per hour	3201
Costs (to fly one hour):	
Crew	$1948
Fuel	$2867

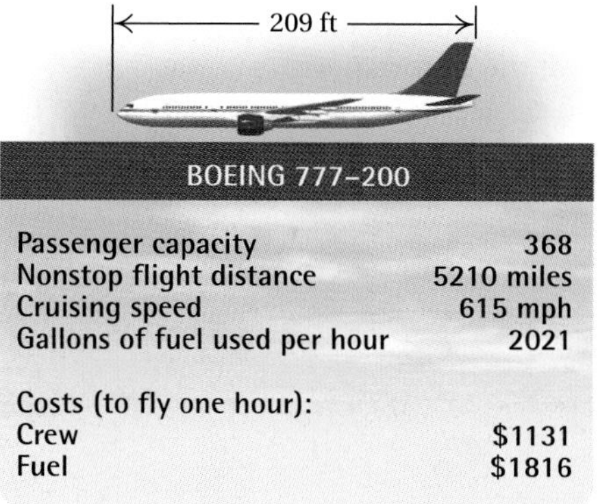

BOEING 777-200	
Passenger capacity	368
Nonstop flight distance	5210 miles
Cruising speed	615 mph
Gallons of fuel used per hour	2021
Costs (to fly one hour):	
Crew	$1131
Fuel	$1816

Source: Éclat Consulting; Boeing

27. How much greater is the nonstop flight distance of the Boeing 747 jet than the nonstop flight distance of the Boeing 777 jet?

28. How much larger is the passenger capacity of the Boeing 747 jet than the passenger capacity of the Boeing 777 jet?

29. How many gallons of fuel are needed for a 4-hr flight of the Boeing 747?

30. How much longer is the Boeing 747 than the Boeing 777?

31. What is the total cost for the crew and fuel for a 3-hr flight of the Boeing 747 jet?

32. What is the total cost for the crew and fuel for a 2-hr flight of the Boeing 777 jet?

33. Dana borrows $5928 for a used car. The loan is to be paid off in 24 equal monthly payments. How much is each payment (excluding interest)?

34. A family borrows $7824 to build a sunroom on the back of their home. The loan is to be paid off in equal monthly payments of $163 (excluding interest). How many months will it take to pay off the loan?

35. *High School Court.* The standard basketball court used by high school players has dimensions of 50 ft by 84 ft.

a) What is its area?
b) What is its perimeter?

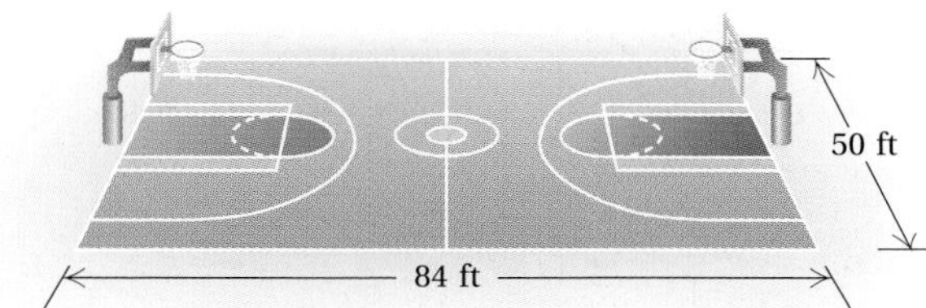

36. *NBA Court.* The standard basketball court used by college and NBA players has dimensions of 50 ft by 94 ft.

a) What is its area?
b) What is its perimeter?
c) How much greater is the area of an NBA court than a high school court? (See Exercise 35.)

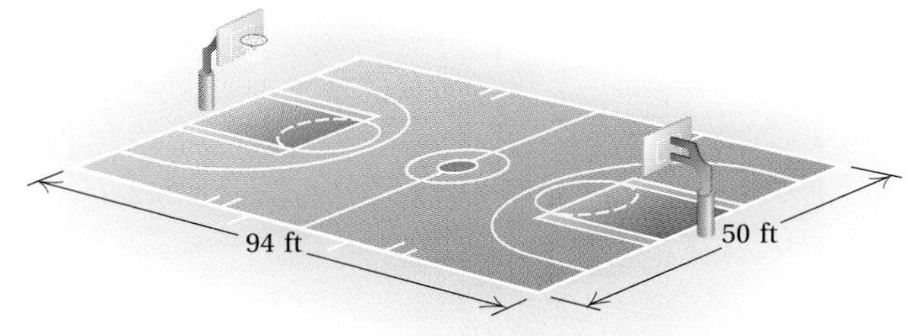

37. *Clothing Imports and Exports.* In the United States, the exports of clothing in 2003 totaled $2,596,000,000 while the imports totaled $31,701,000,000. How much more were the imports than the exports?

Source: U.S. Bureau of the Census, Foreign Trade Division

38. *Corn Imports and Exports.* In the United States, the exports of corn in 2003 totaled $2,264,000,000 while the imports totaled $130,000,000. How much more were the exports than the imports?

Source: U.S. Bureau of the Census, Foreign Trade Division

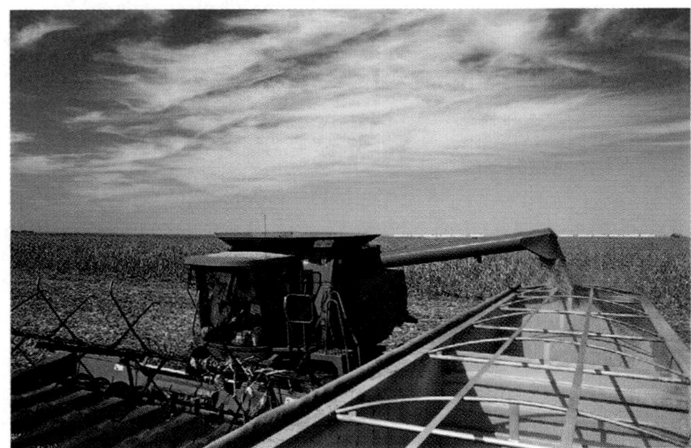

39. *Colonial Population.* Before the establishment of the U.S. Census in 1790, it was estimated that the Colonial population in 1780 was 2,780,400. This was an increase of 2,628,900 from the population in 1680. What was the Colonial population in 1680?

Source: *Time Almanac,* 2005

40. *Deaths by Firearms.* Deaths by firearms totaled 29,573 in 2001. This was a decrease of 10,022 from the number of deaths by firearms in 1993. How many deaths by firearms were there in 1993?

Source: Centers for Disease Control and Prevention, *National Vital Statistics Reports,* vol 52, no. 3, September 18, 2003

41. *Hershey Bars.* Hershey Chocolate USA makes small, fun-size chocolate bars. How many 20-bar packages can be filled with 11,267 bars? How many bars will be left over?

42. *Reese's Peanut Butter Cups.* H. B. Reese Candy Co. makes small, fun-size peanut butter cups. The company manufactures 23,579 cups and fills 1025 packages. How many cups are in a package? How many cups will be left over?

43. *Map Drawing.* A map has a scale of 64 mi to the inch. How far apart *in reality* are two cities that are 6 in. apart on the map? How far apart *on the map* are two cities that, in reality, are 1728 mi apart?

44. *Map Drawing.* A map has a scale of 150 mi to the inch. How far apart *on the map* are two cities that, in reality, are 2400 mi apart? How far apart *in reality* are two cities that are 13 in. apart on the map?

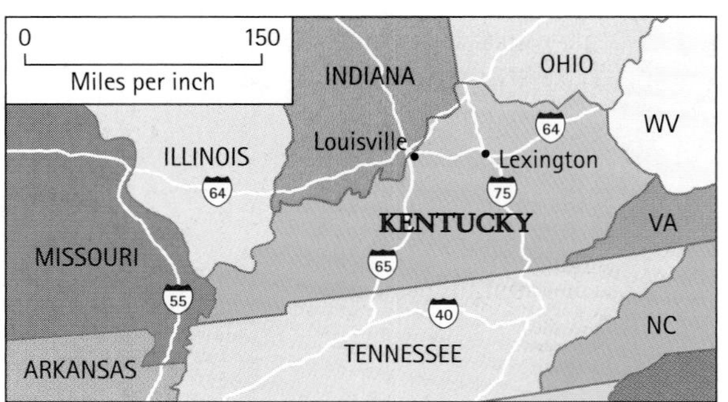

45. *Crossword.* The *Los Angeles Times* crossword puzzle is a rectangle containing 441 squares arranged in 21 rows. How many columns does the puzzle have?

46. *Sheet of Stamps.* A sheet of 100 stamps typically has 10 rows of stamps. How many stamps are in each row?

47. Copies of this book are generally shipped from the warehouse in cartons containing 24 books each. How many cartons are needed to ship 1355 books?

48. According to the H. J. Heinz Company, 16-oz bottles of ketchup are generally shipped in cartons containing 12 bottles each. How many cartons are needed to ship 528 bottles of ketchup?

49. Elena buys 5 video games at $64 each and pays for them with $10 bills. How many $10 bills does it take?

50. Pedro buys 5 video games at $64 each and pays for them with $20 bills. How many $20 bills does it take?

51. You have $568 in your bank account. You use your debit card for purchases totaling $46, $87, and $129. Then you deposit $94 back in the account after the return of some books. How much is left in your account?

52. The balance in your bank account is $749. You use your debit card for purchases totaling $34 and $65. Then you make a deposit of $123 from your paycheck. What is your new balance?

Weight Loss. Many Americans exercise for weight control. It is known that one must burn off about 3500 calories in order to lose one pound. The chart shown here details how much of certain types of exercise is required to burn 100 calories. Use this chart for Exercises 53–56.

To burn off 100 calories, you must:

- Run for 8 minutes at a brisk pace, or
- Swim for 2 minutes at a brisk pace, or
- Bicycle for 15 minutes at 9 mph, or
- Do aerobic exercises for 15 minutes.

53. How long must you run at a brisk pace in order to lose one pound?

54. How long must you swim in order to lose one pound?

55. How long must you do aerobic exercises in order to lose one pound?

56. How long must you bicycle at 9 mph in order to lose one pound?

57. *Bones in the Hands and Feet.* There are 27 bones in each human hand and 26 bones in each human foot. How many bones are there in all in the hands and feet?

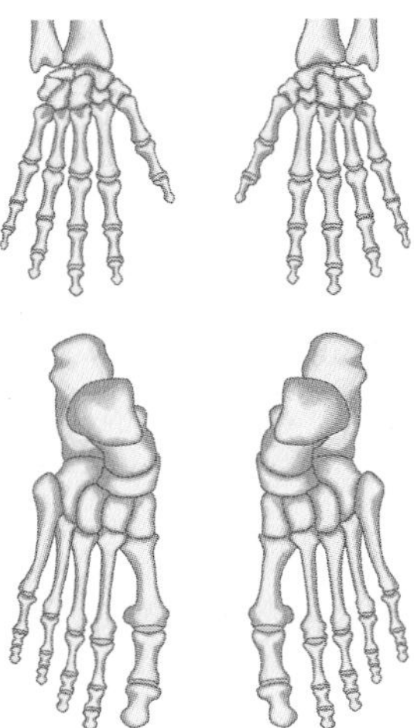

58. *Subway Travel.* The distance to Mars is about 303,000,000 mi. The number of miles that people in the United States traveled on subways in 2003 approximately equaled 22 round trips to Mars. What was the total distance traveled on subways?

Source: American Public Transportation Association, NASA

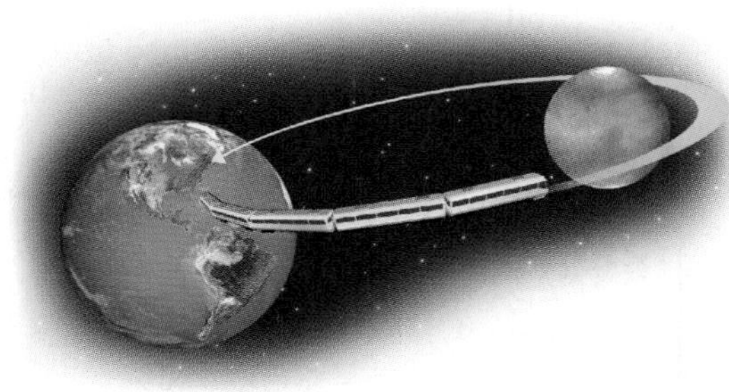

59. *Index Cards.* Index cards of dimension 3 in. by 5 in. are normally shipped in packages containing 100 cards each. How much writing area is available if one uses the front and back sides of all the cards in one package?

60. An office for adjunct instructors at a community college has 6 bookshelves, each of which is 3 ft wide. The office is moved to a new location that has dimensions of 16 ft by 21 ft. Is it possible for the bookshelves to be put side by side on the 16-ft wall?

61. D_W In the newspaper article "When Girls Play, Knees Fail," the author discusses the fact that female athletes have six times the number of knee injuries that male athletes have. What information would be needed if you were to write a math problem based on the article? What might the problem be?

Source: *The Arizona Republic,* 2/9/00, p. C1

62. D_W Write a problem for a classmate to solve. Design the problem so that the solution is "The driver still has 329 mi to travel."

SKILL MAINTENANCE

Round 234,562 to the nearest: [1.3e]

63. Hundred.

64. Ten.

65. Thousand.

Estimate the computation by rounding to the nearest thousand. [1.3f]

66. $2783 + 4602 + 5797 + 8111$

67. $28{,}430 - 11{,}977$

68. $2100 + 5800$

69. $5800 - 2100$

Estimate the product by rounding to the nearest hundred. [1.3f]

70. $787 \cdot 363$

71. $887 \cdot 799$

72. $10{,}362 \cdot 4531$

SYNTHESIS

73. 🖩 *Speed of Light.* Light travels about 186,000 miles per second (mi/sec) in a vacuum, as in outer space. In ice it travels about 142,000 mi/sec, and in glass it travels about 109,000 mi/sec. In 18 sec, how many more miles will light travel in a vacuum than in ice? than in glass?

74. Carney Community College has 1200 students. Each professor teaches 4 classes and each student takes 5 classes. There are 30 students and 1 teacher in each classroom. How many professors are there at Carney Community College?

1.6 EXPONENTIAL NOTATION AND ORDER OF OPERATIONS

Objectives

a Write exponential notation for products such as $4 \cdot 4 \cdot 4$.

b Evaluate exponential notation.

c Simplify expressions using the rules for order of operations.

d Remove parentheses within parentheses.

a Writing Exponential Notation

Consider the product $3 \cdot 3 \cdot 3 \cdot 3$. Such products occur often enough that mathematicians have found it convenient to create a shorter notation, called **exponential notation,** for them. For example,

$\underbrace{3 \cdot 3 \cdot 3 \cdot 3}_{\text{4 factors}}$ is shortened to 3^4 ← exponent (3 is the base)

We read exponential notation as follows.

NOTATION	WORD DESCRIPTION
3^4	"three to the fourth power," or "the fourth power of three"
5^3	"five-cubed," or "the cube of five," or "five to the third power," or "the third power of five"
7^2	"seven squared," or "the square of seven," or "seven to the second power," or "the second power of seven"

The wording "seven squared" for 7^2 comes from the fact that a square with side s has area A given by $A = s^2$.

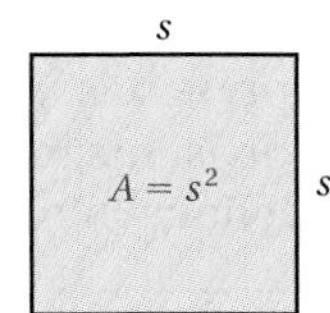

An expression like $3 \cdot 5^2$ is read "three times five squared," or "three times the square of five."

EXAMPLE 1 Write exponential notation for $10 \cdot 10 \cdot 10 \cdot 10 \cdot 10$.

Exponential notation is 10^5. 5 is the *exponent.* 10 is the *base.*

EXAMPLE 2 Write exponential notation for $2 \cdot 2 \cdot 2$.

Exponential notation is 2^3.

Do Exercises 1–4.

Write exponential notation.

1. $5 \cdot 5 \cdot 5 \cdot 5$

2. $5 \cdot 5 \cdot 5 \cdot 5 \cdot 5$

3. $10 \cdot 10$

4. $10 \cdot 10 \cdot 10 \cdot 10$

Evaluate.

5. 10^4

6. 10^2

7. 8^3

8. 2^5

Answers on page A-2

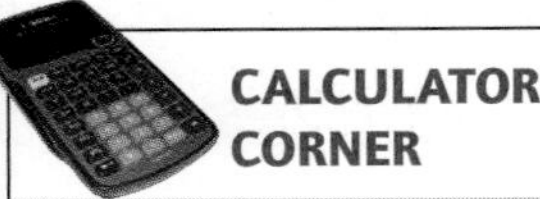

CALCULATOR CORNER

Exponential Notation Many calculators have a y^x or x^y key for raising a base to a power. To find 16^3, for example, we press 1 6 y^x 3 = or 1 6 x^y 3 =. The result is 4096.

Exercises: Use a calculator to find each of the following.

1. 3^5
2. 5^6
3. 12^4
4. 2^{11}

Simplify.

9. $93 - 14 \cdot 3$

10. $104 \div 4 + 4$

11. $25 \cdot 26 - (56 + 10)$

12. $75 \div 5 + (83 - 14)$

Answers on page A-2

b Evaluating Exponential Notation

We evaluate exponential notation by rewriting it as a product and computing the product.

EXAMPLE 3 Evaluate: 10^3.

$$10^3 = 10 \cdot 10 \cdot 10 = 1000$$

Caution!

10^3 does not mean $10 \cdot 3$.

EXAMPLE 4 Evaluate: 5^4.

$$5^4 = 5 \cdot 5 \cdot 5 \cdot 5 = 625$$

Do Exercises 5–8 on the preceding page.

c Simplifying Expressions

Suppose we have a calculation like the following:

$$3 + 4 \cdot 8.$$

How do we find the answer? Do we add 3 to 4 and then multiply by 8, or do we multiply 4 by 8 and then add 3? In the first case, the answer is 56. In the second, the answer is 35. We agree to compute as in the second case.

Consider the calculation

$$7 \cdot 14 - (12 + 18).$$

What do the parentheses mean? To deal with these questions, we must make some agreement regarding the order in which we perform operations. The rules are as follows.

RULES FOR ORDER OF OPERATIONS

1. Do all calculations within parentheses ( ), brackets [ ], or braces { } before operations outside.
2. Evaluate all exponential expressions.
3. Do all multiplications and divisions in order from left to right.
4. Do all additions and subtractions in order from left to right.

It is worth noting that these are the rules that computers and most scientific calculators use to do computations.

EXAMPLE 5 Simplify: $16 \div 8 \cdot 2$.

There are no parentheses or exponents, so we start with the third step.

$$\begin{aligned} 16 \div 8 \cdot 2 &= 2 \cdot 2 \\ &= 4 \end{aligned}$$ Doing all multiplications and divisions in order from left to right

EXAMPLE 6 Simplify: $7 \cdot 14 - (12 + 18)$.

$7 \cdot 14 - (12 + 18) = 7 \cdot 14 - 30$ Carrying out operations inside parentheses

$= 98 - 30$ Doing all multiplications and divisions

$= 68$ Doing all additions and subtractions

Do Exercises 9–12 on the preceding page.

EXAMPLE 7 Simplify and compare: $23 - (10 - 9)$ and $(23 - 10) - 9$.

We have

$23 - (10 - 9) = 23 - 1 = 22;$

$(23 - 10) - 9 = 13 - 9 = 4.$

We can see that $23 - (10 - 9)$ and $(23 - 10) - 9$ represent different numbers. Thus subtraction is not associative.

Do Exercises 13 and 14.

EXAMPLE 8 Simplify: $7 \cdot 2 - (12 + 0) \div 3 - (5 - 2)$.

$7 \cdot 2 - (12 + 0) \div 3 - (5 - 2) = 7 \cdot 2 - 12 \div 3 - 3$ Carrying out operations inside parentheses

$= 14 - 4 - 3$ Doing all multiplications and divisions in order from left to right

$= 7$ Doing all additions and subtractions in order from left to right

Do Exercise 15.

EXAMPLE 9 Simplify: $15 \div 3 \cdot 2 \div (10 - 8)$.

$15 \div 3 \cdot 2 \div (10 - 8) = 15 \div 3 \cdot 2 \div 2$ Carrying out operations inside parentheses

$= 5 \cdot 2 \div 2$
$= 10 \div 2$
$= 5$
} Doing all multiplications and divisions in order from left to right

Do Exercises 16–18.

Simplify and compare.

13. $64 \div (32 \div 2)$ and $(64 \div 32) \div 2$

14. $(28 + 13) + 11$ and $28 + (13 + 11)$

15. Simplify:

$9 \times 4 - (20 + 4) \div 8 - (6 - 2)$.

Simplify.

16. $5 \cdot 5 \cdot 5 + 26 \cdot 71 - (16 + 25 \cdot 3)$

17. $30 \div 5 \cdot 2 + 10 \cdot 20 + 8 \cdot 8 - 23$

18. $95 - 2 \cdot 2 \cdot 2 \cdot 5 \div (24 - 4)$

Answers on page A-3

Simplify.

19. $5^3 + 26 \cdot 71 - (16 + 25 \cdot 3)$

20. $(1 + 3)^3 + 10 \cdot 20 + 8^2 - 23$

21. $81 - 3^2 \cdot 2 \div (12 - 9)$

22. Simplify: $2^3 \cdot 2^8 \div 2^9$.

EXAMPLE 10 Simplify: $4^2 \div (10 - 9 + 1)^3 \cdot 3 - 5$.

$4^2 \div (10 - 9 + 1)^3 \cdot 3 - 5$

$= 4^2 \div (1 + 1)^3 \cdot 3 - 5$ — Subtracting inside parentheses

$= 4^2 \div 2^3 \cdot 3 - 5$ — Adding inside parentheses

$= 16 \div 8 \cdot 3 - 5$ — Evaluating exponential expressions

$= 2 \cdot 3 - 5$
$= 6 - 5$ — Doing all multiplications and divisions in order from left to right

$= 1$ — Subtracting

Do Exercises 19–21.

EXAMPLE 11 Simplify: $2^9 \div 2^6 \cdot 2^3$.

$2^9 \div 2^6 \cdot 2^3 = 512 \div 64 \cdot 8$ — There are no parentheses. Evaluating exponential expressions

$= 8 \cdot 8$
$= 64$ — Doing all multiplications and divisions in order from left to right

Do Exercise 22.

CALCULATOR CORNER

Order of Operations To determine whether a calculator is programmed to follow the rules for order of operations, we can enter a simple calculation that requires using those rules. For example, we enter [3] [+] [4] [×] [2] [=]. If the result is 11, we know that the rules for order of operations have been followed. That is, the multiplication $4 \times 2 = 8$ was performed first and then 3 was added to produce a result of 11. If the result is 14, we know that the calculator performs operations as they are entered rather than following the rules for order of operations. That means, in this case, that 3 and 4 were added first to get 7 and then that sum was multiplied by 2 to produce the result of 14. For such calculators, we would have to enter the operations in the order in which we want them performed. In this case, we would press [4] [×] [2] [+] [3] [=].

Many calculators have parenthesis keys that can be used to enter an expression containing parentheses. To enter $5(4 + 3)$, for example, we press [5] [(] [4] [+] [3] [)] [=]. The result is 35.

Exercises: Simplify.

1. $84 - 5 \cdot 7$
2. $80 + 50 \div 10$
3. $3^2 + 9^2 \div 3$
4. $4^4 \div 64 - 4$
5. $15 \cdot 7 - (23 + 9)$
6. $(4 + 3)^2$

Answers on page A-3

AVERAGES

In order to find the average of a set of numbers, we use addition and then division. For example, the average of 2, 3, 6, and 9 is found as follows.

$$\text{Average} = \frac{2 + 3 + 6 + 9}{4} = \frac{20}{4} = 5$$

The number of addends is 4.

Divide by 4.

The fraction bar acts as a pair of grouping symbols so

$$\frac{2 + 3 + 6 + 9}{4} \text{ is equivalent to } (2 + 3 + 6 + 9) \div 4.$$

Thus we are using order of operations when we compute an average.

AVERAGE

The **average** of a set of numbers is the sum of the numbers divided by the number of addends.

EXAMPLE 12 *Average Number of Career Hits.* The number of career hits of five Hall of Fame baseball players are given in the bar graph below. Find the average number of career hits of all five players.

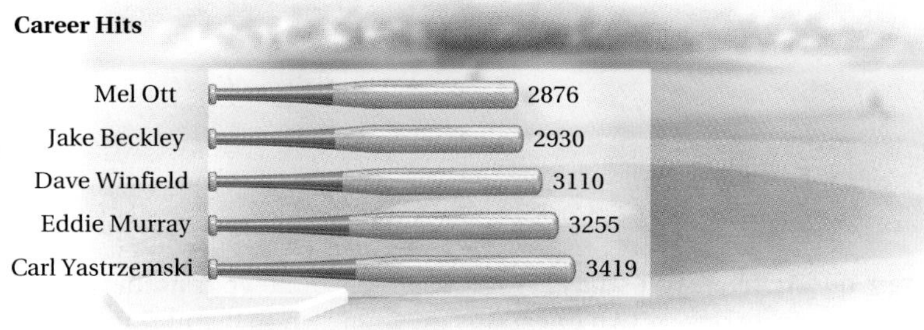

Sources: Associated Press; Major League Baseball

The average is given by

$$\frac{3419 + 3255 + 3110 + 2930 + 2876}{5} = \frac{15{,}590}{5} = 3118.$$

Thus the average number of career hits of these five Hall of Fame players is 3118.

Do Exercise 23.

d Removing Parentheses within Parentheses

When parentheses occur within parentheses, we can make them different shapes, such as [ ] (also called "brackets") and { } (also called "braces"). All of these have the same meaning. When parentheses occur within parentheses, computations in the innermost ones are to be done first.

23. **World's Tallest Buildings.** The heights, in feet, of the four tallest buildings in the world are given in the bar graph below. Find the average height of these buildings.

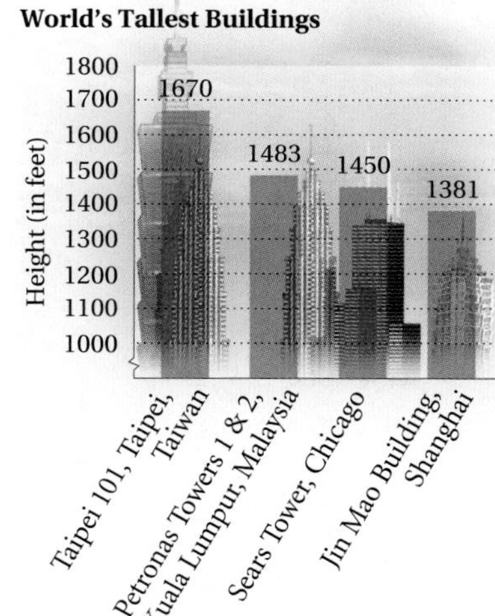

Source: Council on Tall Buildings and Urban Habitat, Lehigh University, 2004

Answer on page A-3

Simplify.

24. $9 \times 5 + \{6 \div [14 - (5 + 3)]\}$

25. $[18 - (2 + 7) \div 3] - (31 - 10 \times 2)$

Answers on page A-3

EXAMPLE 13 Simplify: $[25 - (4 + 3) \cdot 3] \div (11 - 7)$.

$[25 - (4 + 3) \cdot 3] \div (11 - 7)$

$= [25 - 7 \cdot 3] \div (11 - 7)$ Doing the calculations in the innermost parentheses first

$= [25 - 21] \div (11 - 7)$ Doing the multiplication in the brackets

$= 4 \div 4$ Subtracting

$= 1$ Dividing

EXAMPLE 14 Simplify: $16 \div 2 + \{40 - [13 - (4 + 2)]\}$.

$16 \div 2 + \{40 - [13 - (4 + 2)]\}$

$= 16 \div 2 + \{40 - [13 - 6]\}$ Doing the calculations in the innermost parentheses first

$= 16 \div 2 + \{40 - 7\}$ Again, doing the calculations in the innermost parentheses

$= 16 \div 2 + 33$ Subtracting inside the braces

$= 8 + 33$ Doing all multiplications and divisions in order from left to right

$= 41$ Adding

Do Exercises 24 and 25.

Study Tips

TEST PREPARATION

You are probably ready to begin preparing for your first test. Here are some test-taking study tips.

- **Make up your own test questions as you study.** After you have done your homework over a particular objective, write one or two questions on your own that you think might be on a test. You will be amazed at the insight this will provide.
- **Do an overall review of the chapter, focusing on the objectives and the examples.** This should be accompanied by a study of any class notes you may have taken.
- **Do the review exercises at the end of the chapter.** Check your answers at the back of the book. If you have trouble with an exercise, use the objective symbol as a guide to go back and do further study of that objective.
- **Call the AW Math Tutor Center at 1-888-777-0463 if you need extra help.**
- **Take the chapter test at the end of the chapter.** Check the answers and use the objective symbols at the back of the book as a reference for where to review.
- **Ask former students for old exams.** Working such exams can be very helpful and allows you to see what various professors think is important.
- **When taking a test, read each question carefully and try to do all the questions the first time through, but pace yourself.** Answer all the questions, and mark those to recheck if you have time at the end. Very often, your first hunch will be correct.
- **Try to write your test in a neat and orderly manner.** Very often, your instructor tries to give you partial credit when grading an exam. If your test paper is sloppy and disorderly, it is difficult to verify the partial credit. Doing your work neatly can ease such a task for the instructor.

1.6 EXERCISE SET

For Extra Help

MathXL | MyMathLab | InterAct Math | Math Tutor Center | Video Lectures on CD Disc 1 | Student's Solutions Manual

a Write exponential notation.

1. $3 \cdot 3 \cdot 3 \cdot 3$

2. $2 \cdot 2 \cdot 2 \cdot 2 \cdot 2$

3. $5 \cdot 5$

4. $13 \cdot 13 \cdot 13$

5. $7 \cdot 7 \cdot 7 \cdot 7 \cdot 7$

6. $10 \cdot 10$

7. $10 \cdot 10 \cdot 10$

8. $1 \cdot 1 \cdot 1 \cdot 1$

b Evaluate.

9. 7^2

10. 5^3

11. 9^3

12. 10^2

13. 12^4

14. 10^5

15. 11^2

16. 6^3

c Simplify.

17. $12 + (6 + 4)$

18. $(12 + 6) + 18$

19. $52 - (40 - 8)$

20. $(52 - 40) - 8$

21. $1000 \div (100 \div 10)$

22. $(1000 \div 100) \div 10$

23. $(256 \div 64) \div 4$

24. $256 \div (64 \div 4)$

25. $(2 + 5)^2$

26. $2^2 + 5^2$

27. $(11 - 8)^2 - (18 - 16)^2$

28. $(32 - 27)^3 + (19 + 1)^3$

29. $16 \cdot 24 + 50$

30. $23 + 18 \cdot 20$

31. $83 - 7 \cdot 6$

32. $10 \cdot 7 - 4$

33. $10 \cdot 10 - 3 \cdot 4$

34. $90 - 5 \cdot 5 \cdot 2$

35. $4^3 \div 8 - 4$

36. $8^2 - 8 \cdot 2$

37. $17 \cdot 20 - (17 + 20)$

38. $1000 \div 25 - (15 + 5)$

39. $6 \cdot 10 - 4 \cdot 10$

40. $3 \cdot 8 + 5 \cdot 8$

41. $300 \div 5 + 10$

42. $144 \div 4 - 2$

43. $3 \cdot (2 + 8)^2 - 5 \cdot (4 - 3)^2$

44. $7 \cdot (10 - 3)^2 - 2 \cdot (3 + 1)^2$

45. $4^2 + 8^2 \div 2^2$

46. $6^2 - 3^4 \div 3^3$

47. $10^3 - 10 \cdot 6 - (4 + 5 \cdot 6)$

48. $7^2 + 20 \cdot 4 - (28 + 9 \cdot 2)$

49. $6 \cdot 11 - (7 + 3) \div 5 - (6 - 4)$

50. $8 \times 9 - (12 - 8) \div 4 - (10 - 7)$

51. $120 - 3^3 \cdot 4 \div (5 \cdot 6 - 6 \cdot 4)$

52. $80 - 2^4 \cdot 15 \div (7 \cdot 5 - 45 \div 3)$

53. $2^3 \cdot 2^8 \div 2^6$

54. $2^7 \div 2^5 \cdot 2^4 \div 2^2$

55. Find the average of \$64, \$97, and \$121.

56. Find the average of four test grades of 86, 92, 80, and 78.

57. Find the average of 320, 128, 276, and 880.

58. Find the average of \$1025, \$775, \$2062, \$942, and \$3721.

d Simplify.

59. $8 \times 13 + \{42 \div [18 - (6 + 5)]\}$

60. $72 \div 6 - \{2 \times [9 - (4 \times 2)]\}$

61. $[14 - (3 + 5) \div 2] - [18 \div (8 - 2)]$

62. $[92 \times (6 - 4) \div 8] + [7 \times (8 - 3)]$

63. $(82 - 14) \times [(10 + 45 \div 5) - (6 \cdot 6 - 5 \cdot 5)]$

64. $(18 \div 2) \cdot \{[(9 \cdot 9 - 1) \div 2] - [5 \cdot 20 - (7 \cdot 9 - 2)]\}$

65. $4 \times \{(200 - 50 \div 5) - [(35 \div 7) \cdot (35 \div 7) - 4 \times 3]\}$

66. $15(23 - 4 \cdot 2)^3 \div (3 \cdot 25)$

67. $\{[18 - 2 \cdot 6] - [40 \div (17 - 9)]\} + \{48 - 13 \times 3 + [(50 - 7 \cdot 5) + 2]\}$

68. $(19 - 2^4)^5 - (141 \div 47)^2$

69. D_W Consider the problem in Example 8 of Section 1.5. How can you translate the problem to a single equation involving what you have learned about order of operations? How does the single equation relate to how we solved the problem?

70. D_W Consider the expressions $9 - (4 \cdot 2)$ and $(3 \cdot 4)^2$. Are the parentheses necessary in each case? Explain.

SKILL MAINTENANCE

Solve. [1.4b]

71. $x + 341 = 793$

72. $4197 + x = 5032$

73. $7 \cdot x = 91$

74. $1554 = 42 \cdot y$

75. $3240 = y + 898$

76. $6000 = 1102 + t$

77. $25 \cdot t = 625$

78. $10{,}000 = 100 \cdot t$

Solve. [1.5a]

79. *Colorado.* The state of Colorado is roughly the shape of a rectangle that is 273 mi by 382 mi. What is its area?

80. On a long four-day trip, a family bought the following amounts of gasoline for their motor home:

23 gallons, 24 gallons,
26 gallons, 25 gallons.

How much gasoline did they buy in all?

SYNTHESIS

Each of the answers in Exercises 81–83 is incorrect. First find the correct answer. Then place as many parentheses as needed in the expression in order to make the incorrect answer correct.

81. $1 + 5 \cdot 4 + 3 = 36$

82. $12 \div 4 + 2 \cdot 3 - 2 = 2$

83. $12 \div 4 + 2 \cdot 3 - 2 = 4$

84. Use one occurrence each of 1, 2, 3, 4, 5, 6, 7, 8, and 9 and any of the symbols $+$, $-$, $\times$, $\div$, and ( ) to represent 100.

1.7 FACTORIZATIONS

Objectives

a Determine whether one number is a factor of another, and find the factors of a number.

b Find some multiples of a number, and determine whether a number is divisible by another.

c Given a number from 1 to 100, tell whether it is prime, composite, or neither.

d Find the prime factorization of a composite number.

In Chapter 2, we will begin our work with fractions and fraction notation. Certain skills make such work easier. For example, in order to simplify $\frac{12}{32}$, it is important that we be able to *factor* 12 and 32, as follows:

$$\frac{12}{32} = \frac{4 \cdot 3}{4 \cdot 8}.$$

Then we "remove" a factor of 1:

$$\frac{4 \cdot 3}{4 \cdot 8} = \frac{4}{4} \cdot \frac{3}{8} = 1 \cdot \frac{3}{8} = \frac{3}{8}.$$

Thus factoring is an important skill in working with fractions.

a Factors and Factorization

In Sections 1.7 and 1.8, we consider only the **natural numbers** 1, 2, 3, and so on.

Let's look at the product $3 \cdot 4 = 12$. We say that 3 and 4 are **factors** of 12. When we divide 12 by 3, we get a remainder of 0. We say that the divisor 3 is a **factor** of the dividend 12.

FACTOR

- In the product $a \cdot b$, a and b are called **factors.**
- If we divide Q by d and get a remainder of 0, then the divisor d is a **factor** of the dividend Q.

EXAMPLE 1 Determine by long division whether 6 is a factor of 72.

$$\begin{array}{r} 12 \\ 6\overline{)72} \\ \underline{60} \\ 12 \\ \underline{12} \\ 0 \end{array}$$

The remainder is 0, so 6 is a factor of 72. We sometimes say that 6 divides 72 "evenly" because there is a remainder of 0.

EXAMPLE 2 Determine by long division whether 15 is a factor of 7894.

$$\begin{array}{r} 526 \\ 15\overline{)7894} \\ \underline{7500} \\ 394 \\ \underline{300} \\ 94 \\ \underline{90} \\ 4 \end{array} \leftarrow \text{Not } 0$$

The remainder is *not* 0, so 15 is not a factor of 7894.

Do Exercises 1 and 2 on the following page.

Consider $12 = 3 \cdot 4$. We say that $3 \cdot 4$ is a **factorization** of 12. Similarly, $6 \cdot 2$, $12 \cdot 1$, $2 \cdot 2 \cdot 3$, and $1 \cdot 3 \cdot 4$ are also factorizations of 12. Since $a = a \cdot 1$, every number has a factorization, and every number has factors. For some numbers, the factors consist of only the number itself and 1. For example, the only factorization of 17 is $17 \cdot 1$, so the only factors of 17 are 17 and 1.

EXAMPLE 3 Find all the factors of 70.

We find as many "two-factor" factorizations as we can. We check sequentially the numbers 1, 2, 3, and so on, to see if we can form any factorizations:

70

$1 \cdot 70$
$2 \cdot 35$
$5 \cdot 14$
$7 \cdot 10.$

Note that all but one of the factors of a natural number are *less* than the number.

Note that 3, 4, and 6 are not factors. If there are additional factors, they must be between 7 and 10. Since 8 and 9 are not factors, we are finished. The factors of 70 are 1, 2, 5, 7, 10, 14, 35, and 70.

Do Exercises 3–6.

b Multiples and Divisibility

A **multiple** of a natural number is a product of that number and some natural number. For example, some multiples of 2 are:

2 (because $2 = 1 \cdot 2$);
4 (because $4 = 2 \cdot 2$);
6 (because $6 = 3 \cdot 2$);
8 (because $8 = 4 \cdot 2$);
10 (because $10 = 5 \cdot 2$).

Note that all but one of the multiples of a number are *larger* than the number.

We find multiples of 2 by counting by twos: 2, 4, 6, 8, and so on. We can find multiples of 3 by counting by threes: 3, 6, 9, 12, and so on.

EXAMPLE 4 Show that each of the numbers 8, 12, 20, and 36 is a multiple of 4.

$8 = 2 \cdot 4 \qquad 12 = 3 \cdot 4 \qquad 20 = 5 \cdot 4 \qquad 36 = 9 \cdot 4$

Do Exercises 7 and 8.

EXAMPLE 5 Multiply by 1, 2, 3, and so on, to find ten multiples of 7.

$1 \cdot 7 = 7$	$6 \cdot 7 = 42$
$2 \cdot 7 = 14$	$7 \cdot 7 = 49$
$3 \cdot 7 = 21$	$8 \cdot 7 = 56$
$4 \cdot 7 = 28$	$9 \cdot 7 = 63$
$5 \cdot 7 = 35$	$10 \cdot 7 = 70$

Do Exercise 9.

Determine whether the second number is a factor of the first.

1. 72; 8

2. 2384; 28

Find all the factors of the number.

3. 10

4. 45

5. 62

6. 24

7. Show that each of the numbers 5, 45, and 100 is a multiple of 5.

8. Show that each of the numbers 10, 60, and 110 is a multiple of 10.

9. Multiply by 1, 2, 3, and so on, to find ten multiples of 5.

Answers on page A-3

10. Determine whether 16 is divisible by 2.

11. Determine whether 125 is divisible by 5.

12. Determine whether 125 is divisible by 6.

Answers on page A-3

DIVISIBILITY

The number a is **divisible** by another number b if there exists a number c such that $a = b \cdot c$. The statements "a is **divisible** by b," "a is a **multiple** of b," and "b is a **factor** of a" all have the same meaning.

Thus 27 is *divisible* by 3 because $27 = 3 \cdot 9$. We can also say that 27 is a *multiple* of 3, and 3 is a *factor* of 27.

EXAMPLE 6 Determine whether 45 is divisible by 9.

We divide 45 by 9 to determine whether 9 is a factor of 45. If so, then 45 is divisible by 9.

$$\begin{array}{r} 5 \\ 9\overline{)45} \\ \underline{45} \\ 0 \end{array}$$

Because the remainder is 0, 45 is divisible by 9.

EXAMPLE 7 Determine whether 98 is divisible by 4.

We divide 98 by 4:

$$\begin{array}{r} 24 \\ 4\overline{)98} \\ \underline{80} \\ 18 \\ \underline{16} \\ 2 \end{array} \leftarrow \text{Not } 0$$

Since the remainder is not 0, 98 is *not* divisible by 4.

Do Exercises 10–12.

CALCULATOR CORNER

Divisibility and Factors We can use a calculator to determine whether one number is divisible by another number or whether one number is a factor of another number. For example, to determine whether 387 is divisible by 18, we first press [3] [8] [7] [÷] [1] [8] [=]. The display reads [21.5]. Note that the result is not a natural number. (For a brief discussion of decimal notation, see the Calculator Corner on page 33. Decimal notation will be studied in detail in Chapter 3.) Thus we know that 387 is not a multiple of 18; that is, 387 is not divisible by 18 and 18 is not a factor of 387.

When we divide 387 by 9, the result is [43]. Since 43 is a natural number, we know that 387 is a multiple of 9; that is, $387 = 43 \cdot 9$. Thus, 387 is divisible by 9, and 9 is a factor of 387.

Exercises: For each pair of numbers, determine whether the first number is divisible by the second number.

1. 722; 19
2. 845; 7
3. 1047; 14
4. 5283; 9

For each pair of numbers, determine whether the second number is a factor of the first number.

5. 502; 8
6. 651; 21
7. 3875; 25
8. 8464; 12
9. 32,768; 256
10. 32,768; 864

c Prime and Composite Numbers

PRIME AND COMPOSITE NUMBERS

- A natural number that has exactly two *different* factors, only itself and 1, is called a **prime number.**
- The number 1 is *not* prime.
- A natural number, other than 1, that is not prime is **composite.**

EXAMPLE 8 Determine whether the numbers 1, 2, 3, 4, 5, 6, 7, 9, 10, 11, and 63 are prime, composite, or neither.

The number 1 is not prime. It does not have *two* different factors.

The number 2 is prime. It has only the factors 2 and 1.

The numbers 3, 5, 7, and 11 are prime. Each has only two factors, itself and 1.

The number 4 is not prime. It has the factors 1, 2, and 4 and is composite.

The numbers 6, 9, 10, and 63 are composite. Each has more than two factors.

Thus we have:

Prime: 2, 3, 5, 7, 11;
Composite: 4, 6, 9, 10, 63;
Neither: 1.

The number 2 is the *only* even prime number. It is also the smallest prime number. The number 0 is also neither prime nor composite, but 0 is *not* a natural number and thus is not considered here. We are considering only natural numbers.

Do Exercise 13.

The following is a table of the prime numbers from 2 to 157. There are more extensive tables, but these prime numbers will be the most helpful to you in this text.

A TABLE OF PRIMES FROM 2 TO 157

2, 3, 5, 7, 11, 13, 17, 19, 23, 29, 31, 37, 41, 43, 47, 53, 59, 61, 67, 71, 73, 79, 83, 89, 97, 101, 103, 107, 109, 113, 127, 131, 137, 139, 149, 151, 157

d Prime Factorizations

When we factor a composite number into a product of primes, we find the **prime factorization** of the number. To do this, we consider the primes

2, 3, 5, 7, 11, 13, 17, 19, 23, and so on,

and determine whether a given number is divisible by the primes.

13. Tell whether each number is prime, composite, or neither.

1, 2, 6, 12, 13, 19, 41, 65, 73, 99

Answer on page A-3

EXAMPLE 9 Find the prime factorization of 39.

a) We divide by the first prime, 2.

$$\begin{array}{r} 19 \quad R = 1 \\ 2\overline{)39} \\ \underline{38} \\ 1 \end{array}$$

Because the remainder is not 0, 2 is not a factor of 39, and 39 is not divisible by 2.

b) We divide by the next prime, 3.

$$\begin{array}{r} 13 \quad R = 0 \\ 3\overline{)39} \end{array}$$

The remainder is 0, so we know that $39 = 3 \cdot 13$. Because 13 is a prime, we are finished. The prime factorization is

$$39 = 3 \cdot 13.$$

EXAMPLE 10 Find the prime factorization of 220.

a) We divide by the first prime, 2.

$$\begin{array}{r} 110 \quad R = 0 \\ 2\overline{)220} \end{array} \qquad 220 = 2 \cdot 110$$

b) Because 110 is composite, we start with 2 again.

$$\begin{array}{r} 55 \quad R = 0 \\ 2\overline{)110} \end{array} \qquad 220 = 2 \cdot 2 \cdot 55$$

c) Since 55 is composite and is not divisible by 2 or 3, we divide by the next prime, 5.

$$\begin{array}{r} 11 \quad R = 0 \\ 5\overline{)55} \end{array} \qquad 220 = 2 \cdot 2 \cdot 5 \cdot 11$$

Because 11 is prime, we are finished. The prime factorization is

$$220 = 2 \cdot 2 \cdot 5 \cdot 11.$$

We abbreviate our procedure as follows.

$$\begin{array}{r} 11 \\ 5\overline{)55} \\ 2\overline{)110} \\ 2\overline{)220} \end{array}$$

$$220 = 2 \cdot 2 \cdot 5 \cdot 11$$

Multiplication is commutative so a factorization such as $2 \cdot 2 \cdot 5 \cdot 11$ could also be expressed as $5 \cdot 2 \cdot 2 \cdot 11$ or $2 \cdot 5 \cdot 11 \cdot 2$ (or, in exponential notation, as $2^2 \cdot 5 \cdot 11$ or $11 \cdot 2^2 \cdot 5$), but the prime factors are the same. For this reason, we agree that any of these is "the" prime factorization of 220.

Every number has just one (unique) prime factorization.

EXAMPLE 11 Find the prime factorization of 72.

We can do divisions "up" as follows:

$$\begin{array}{r} 3 \leftarrow \text{Prime quotient} \\ 3\overline{)\ 9} \\ 2\overline{)18} \\ 2\overline{)36} \\ 2\overline{)72} \leftarrow \text{Begin here.} \end{array}$$

Or, we can do divisions "down":

$$\begin{array}{l} 2\underline{)72} \leftarrow \text{Begin here.} \\ 2\underline{)36} \\ 2\underline{)18} \\ 3\underline{)\ 9} \\ \quad 3 \leftarrow \text{Prime quotient} \end{array}$$

$72 = 2 \cdot 2 \cdot 2 \cdot 3 \cdot 3$

Another way to find the prime factorization of 72 is to use a **factor tree** as follows. Begin by determining any factorization you can, and then continue factoring.

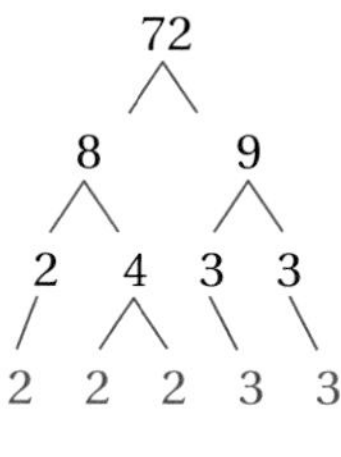

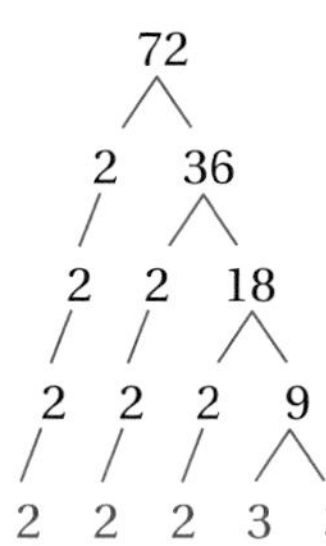

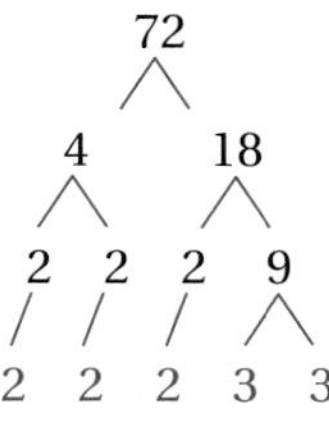

EXAMPLE 12 Find the prime factorization of 189.

We can use a string of successive divisions or a factor tree. Since 189 is not divisible by 2, we begin with 3.

$$\begin{array}{r} 7 \\ 3\overline{)21} \\ 3\overline{)63} \\ 3\overline{)189} \end{array}$$

$189 = 3 \cdot 3 \cdot 3 \cdot 7$

189
3 63
3 7 9
3 7 3 3

Caution!

Keep in mind the difference between finding all the factors of a number and finding the prime factorization. In Example 12, the prime factorization is $3 \cdot 3 \cdot 3 \cdot 7$. The factors of 189 are 1, 3, 7, 9, 21, 27, 63, and 189.

EXAMPLE 13 Find the prime factorization of 1638.

We can use a string of successive divisions.

$$\begin{array}{r} 13 \\ 7\overline{)91} \\ 3\overline{)273} \\ 3\overline{)819} \\ 2\overline{)1638} \end{array}$$

$1638 = 2 \cdot 3 \cdot 3 \cdot 7 \cdot 13$

Do Exercises 14–21.

Find the prime factorization of the number.

14. 6

15. 12

16. 45

17. 98

18. 126

19. 144

20. 1960

21. 1925

Answers on page A-3

1.7 EXERCISE SET

For Extra Help

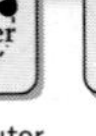

MathXL | MyMathLab | InterAct Math | Math Tutor Center | Video Lectures on CD Disc 1 | Student's Solutions Manual

a Determine whether the second number is a factor of the first.

1. 52; 14
2. 52; 13
3. 625; 25
4. 680; 16

Find all the factors of the number.

5. 18
6. 16
7. 54
8. 48

9. 4
10. 9
11. 1
12. 13

13. 98
14. 100
15. 255
16. 120

b Multiply by 1, 2, 3, and so on, to find ten multiples of the number.

17. 4
18. 11
19. 20
20. 50

21. 3
22. 5
23. 12
24. 13

25. 10
26. 6
27. 9
28. 14

29. Determine whether 26 is divisible by 6.

30. Determine whether 48 is divisible by 8.

31. Determine whether 1880 is divisible by 8.

32. Determine whether 4227 is divisible by 3.

33. Determine whether 256 is divisible by 16.

34. Determine whether 102 is divisible by 4.

35. Determine whether 4227 is divisible by 9.

36. Determine whether 200 is divisible by 25.

37. Determine whether 8650 is divisible by 16.

38. Determine whether 4143 is divisible by 7.

c Determine whether the number is prime, composite, or neither.

39. 1
40. 2
41. 9
42. 19

43. 11
44. 27
45. 29
46. 49

d Find the prime factorization of the number.

47. 8 **48.** 16 **49.** 14 **50.** 15 **51.** 42

52. 32 **53.** 25 **54.** 40 **55.** 50 **56.** 62

57. 169 **58.** 140 **59.** 100 **60.** 110 **61.** 35

62. 70 **63.** 72 **64.** 86 **65.** 77 **66.** 99

67. 2884 **68.** 484 **69.** 51 **70.** 91 **71.** 1200

72. 1800 **73.** 273 **74.** 675 **75.** 1122 **76.** 6435

77. D_W Is every natural number a multiple of 1? Explain.

78. D_W Explain a method for finding a composite number that contains exactly two factors other than itself and 1.

SKILL MAINTENANCE

Multiply. [1.3a]

79. $2 \cdot 13$ **80.** $8 \cdot 32$ **81.** $17 \cdot 25$ **82.** $25 \cdot 168$

Divide. [1.3d]

83. $0 \div 22$ **84.** $22 \div 1$ **85.** $22 \div 22$ **86.** $66 \div 22$

Solve. [1.5a]

87. Find the total cost of 7 shirts at $48 each and 4 pairs of pants at $69 each.

88. Sandy can type 62 words per minute. How long will it take her to type 12,462 words?

SYNTHESIS

89. *Factors and Sums.* The top number in each column of the table below can be factored as a product of two numbers whose sum is the bottom number in the column. For example, in the first column, 56 has been factored as $7 \cdot 8$, and $7 + 8 = 15$. Fill in the blank spaces in the table.

Product	56	63	36	72	140	96		168	110			
Factor	7									9	24	3
Factor	8						8	8		10	18	
Sum	15	16	20	38	24	20	14		21			24

1.8 DIVISIBILITY

Objective

 Determine whether a number is divisible by 2, 3, 4, 5, 6, 8, 9, or 10.

Suppose you are asked to find the simplest fraction notation for

$$\frac{117}{225}.$$

Since the numbers are quite large, you might feel that the task is difficult. However, both the numerator and the denominator are divisible by 9. If you knew this, you could factor and simplify quickly as follows:

$$\frac{117}{225} = \frac{9 \cdot 13}{9 \cdot 25} = \frac{9}{9} \cdot \frac{13}{25} = 1 \cdot \frac{13}{25} = \frac{13}{25}.$$

How did we know that both numbers have 9 as a factor? There is a simple test for determining this.

In this section, we learn quick ways to determine whether a number is divisible by 2, 3, 4, 5, 6, 8, 9, or 10. This will make simplifying fraction notation much easier.

a Rules for Divisibility

DIVISIBILITY BY 2

You may already know the test for divisibility by 2.

> **BY 2**
>
> A number is **divisible by 2** (is *even*) if it has a ones digit of 0, 2, 4, 6, or 8 (that is, it has an even ones digit).

Let's see why. Consider 354, which is

3 hundreds + 5 tens + 4.

Hundreds and tens are both multiples of 2. If the last digit is a multiple of 2, then the entire number is a multiple of 2.

EXAMPLES Determine whether the number is divisible by 2.

1. 355 is *not* a multiple of 2; 5 is *not* even.
2. 4786 is a multiple of 2; 6 is even.
3. 8990 is a multiple of 2; 0 is even.
4. 4261 is *not* a multiple of 2; 1 is *not* even.

Do Exercises 1–4.

Determine whether the number is divisible by 2.

1. 84

2. 59

3. 998

4. 2225

Answers on page A-3

DIVISIBILITY BY 3

BY 3

A number is **divisible by 3** if the sum of its digits is divisible by 3.

Let's illustrate why the test for divisibility by 3 works. Consider 852; since $852 = 3 \cdot 284$, 852 is divisible by 3.

$$852 = 8 \cdot 100 + 5 \cdot 10 + 2 \cdot 1$$
$$= 8(99 + 1) + 5(9 + 1) + 2(1)$$
$$= 8 \cdot 99 + 8 \cdot 1 + 5 \cdot 9 + 5 \cdot 1 + 2 \cdot 1$$

Using the distributive law, $a(b + c) = a \cdot b + a \cdot c$

Since 99 and 9 are each a multiple of 3, we see that $8 \cdot 99$ and $5 \cdot 9$ are multiples of 3. This leaves $8 \cdot 1 + 5 \cdot 1 + 2 \cdot 1$, or $8 + 5 + 2$. If $8 + 5 + 2$, the sum of the digits, is divisible by 3, then 852 is divisible by 3.

EXAMPLES Determine whether the number is divisible by 3.

5. 18 $1 + 8 = 9$
6. 93 $9 + 3 = 12$
7. 201 $2 + 0 + 1 = 3$

Each is divisible by 3 because the sum of its digits is divisible by 3.

8. 256 $2 + 5 + 6 = 13$

The sum of the digits, 13, is *not* divisible by 3, so 256 is *not* divisible by 3.

Do Exercises 5–8.

Determine whether the number is divisible by 3.

5. 111

6. 1111

7. 309

8. 17,216

DIVISIBILITY BY 6

A number divisible by 6 is a multiple of 6. But $6 = 2 \cdot 3$, so the number is also a multiple of 2 and 3. Thus we have the following.

BY 6

A number is **divisible by 6** if its ones digit is 0, 2, 4, 6, or 8 (is even) and the sum of its digits is divisible by 3.

EXAMPLES Determine whether the number is divisible by 6.

9. 720

Because 720 is even, it is divisible by 2. Also, $7 + 2 + 0 = 9$, so 720 is divisible by 3. Thus, 720 is divisible by 6.

720 (Even) $7 + 2 + 0 = 9$ (Divisible by 3)

10. 73

73 is *not* divisible by 6 because it is *not* even.

11. 256

Although 256 is even, it is *not* divisible by 6 because the sum of its digits, $2 + 5 + 6$, or 13, is *not* divisible by 3.

Do Exercises 9–12.

Determine whether the number is divisible by 6.

9. 420

10. 106

11. 321

12. 444

Answers on page A-3

DIVISIBILITY BY 9

The test for divisibility by 9 is similar to the test for divisibility by 3.

BY 9

A number is **divisible by 9** if the sum of its digits is divisible by 9.

EXAMPLES Determine whether the number is divisible by 9.

12. 6984

Because $6 + 9 + 8 + 4 = 27$ and 27 is divisible by 9, 6984 is divisible by 9.

13. 322

Because $3 + 2 + 2 = 7$ and 7 is *not* divisible by 9, 322 is *not* divisible by 9.

Do Exercises 13–16.

DIVISIBILITY BY 10

BY 10

A number is **divisible by 10** if its ones digit is 0.

We know that this test works because the product of 10 and *any* number has a ones digit of 0.

EXAMPLES Determine whether the number is divisible by 10.

14. 3440 is divisible by 10 because the ones digit is 0.

15. 3447 is *not* divisible by 10 because the ones digit is not 0.

Do Exercises 17–20.

DIVISIBILITY BY 5

BY 5

A number is **divisible by 5** if its ones digit is 0 or 5.

EXAMPLES Determine whether the number is divisible by 5.

16. 220 is divisible by 5 because the ones digit is 0.

17. 475 is divisible by 5 because the ones digit is 5.

18. 6514 is *not* divisible by 5 because the ones digit is neither 0 nor 5.

Do Exercises 21–24.

Let's see why the test for 5 works. Consider 7830:

$$7830 = 10 \cdot 783 = 5 \cdot 2 \cdot 783.$$

Since 7830 is divisible by 10 and 5 is a factor of 10, 7830 is divisible by 5.

Determine whether the number is divisible by 9.

13. 16

14. 117

15. 930

16. 29,223

Determine whether the number is divisible by 10.

17. 305

18. 300

19. 847

20. 8760

Determine whether the number is divisible by 5.

21. 5780

22. 3427

23. 34,678

24. 7775

Answers on page A-3

Consider 6734:

6734 = 673 tens + 4.

Tens are multiples of 5, so the only number that must be checked is the ones digit. If the last digit is a multiple of 5, the entire number is. In this case, 4 is *not* a multiple of 5, so 6734 is *not* divisible by 5.

DIVISIBILITY BY 4

The test for divisibility by 4 is similar to the test for divisibility by 2.

BY 4

A number is **divisible by 4** if the number named by its last *two* digits is divisible by 4.

EXAMPLES Determine whether the number is divisible by 4.

19. 8212 is divisible by 4 because 12 is divisible by 4.

20. 5216 is divisible by 4 because 16 is divisible by 4.

21. 8211 is *not* divisible by 4 because 11 is *not* divisible by 4.

22. 7538 is *not* divisible by 4 because 38 is *not* divisible by 4.

Do Exercises 25–28.

To see why the test for divisibility by 4 works, consider 516:

516 = 5 hundreds + 16.

Hundreds are multiples of 4. If the number named by the last two digits is a multiple of 4, then the entire number is a multiple of 4.

DIVISIBILITY BY 8

The test for divisibility by 8 is an extension of the tests for divisibility by 2 and 4.

BY 8

A number is **divisible by 8** if the number named by its last *three* digits is divisible by 8.

EXAMPLES Determine whether the number is divisible by 8.

23. 5648 is divisible by 8 because 648 is divisible by 8.

24. 96,088 is divisible by 8 because 88 is divisible by 8.

25. 7324 is *not* divisible by 8 because 324 is *not* divisible by 8.

26. 13,420 is *not* divisible by 8 because 420 is *not* divisible by 8.

Do Exercises 29–32.

A NOTE ABOUT DIVISIBILITY BY 7

There are several tests for divisibility by 7, but all of them are more complicated than simply dividing by 7. So if you want to test for divisibility by 7, simply divide by 7, either by hand or using a calculator.

Determine whether the number is divisible by 4.

25. 216

26. 217

27. 5862

28. 23,524

Determine whether the number is divisible by 8.

29. 7564

30. 7864

31. 17,560

32. 25,716

Answers on page A-3

1.8 EXERCISE SET

For Extra Help

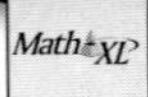
MathXL

MyMathLab

InterAct Math

Math Tutor Center

Video Lectures on CD Disc 1

Student's Solutions Manual

a To answer Exercises 1–8, consider the following numbers.

46	300	85	256
224	36	711	8064
19	45,270	13,251	1867
555	4444	254,765	21,568

1. Which of the above are divisible by 2?

2. Which of the above are divisible by 3?

3. Which of the above are divisible by 4?

4. Which of the above are divisible by 5?

5. Which of the above are divisible by 6?

6. Which of the above are divisible by 8?

7. Which of the above are divisible by 9?

8. Which of the above are divisible by 10?

To answer Exercises 9–16, consider the following numbers.

56	200	75	35
324	42	812	402
784	501	2345	111,111
55,555	3009	2001	1005

9. Which of the above are divisible by 3?

10. Which of the above are divisible by 2?

11. Which of the above are divisible by 5?

12. Which of the above are divisible by 4?

13. Which of the above are divisible by 9?

14. Which of the above are divisible by 6?

15. Which of the above are divisible by 10?

16. Which of the above are divisible by 8?

To answer Exercises 17–24, consider the following numbers.

305	313,332	876	64,000
1101	7624	1110	9990
13,025	111,126	5128	126,111

17. Which of the above are divisible by 2?

18. Which of the above are divisible by 3?

19. Which of the above are divisible by 6?

20. Which of the above are divisible by 5?

21. Which of the above are divisible by 9?

22. Which of the above are divisible by 8?

23. Which of the above are divisible by 10?

24. Which of the above are divisible by 4?

25. $\mathbf{D_W}$ How can the divisibility tests be used to find prime factorizations?

26. $\mathbf{D_W}$ Which of the years from 2000 to 2020, if any, also happen to be prime numbers? Explain at least two ways in which you might go about solving this problem.

SKILL MAINTENANCE

Solve. [1.4b]

27. $56 + x = 194$

28. $y + 124 = 263$

29. $3008 = x + 2134$

30. $18 \cdot t = 1008$

31. $24 \cdot m = 624$

32. $338 = a \cdot 26$

Divide. [1.3d]

33. $2106 \div 9$

34. $45\overline{)180{,}135}$

Solve. [1.5a]

35. An automobile with a 5-speed transmission gets 33 mpg in city driving. How many gallons of gas will it use to travel 1485 mi?

36. There are 60 min in 1 hr. How many minutes are there in 72 hr?

SYNTHESIS

Find the prime factorization of the number. Use divisibility tests where applicable.

37. 7800

38. 2520

39. 2772

40. 1998

41. ▦ Fill in the missing digits of the number

95, □ □ 8

so that it is divisible by 99.

42. A passenger in a taxicab asks for the driver's company number. The driver says abruptly, "Sure—you can have my number. Work it out: If you divide it by 2, 3, 4, 5, or 6, you will get a remainder of 1. If you divide it by 11, the remainder will be 0 and no driver has a company number that meets these requirements and is smaller than this one." Determine the number.

1.9 LEAST COMMON MULTIPLES

Objective

a Find the least common multiple, LCM, of two or more numbers.

1. By examining lists of multiples, find the LCM of 9 and 15.

In Chapter 2, we will study addition and subtraction using fraction notation. Suppose we want to add $\frac{2}{3}$ and $\frac{1}{2}$. To do so, we rewrite the fractions with a common denominator. The number we choose for the common denominator is the least common multiple of the denominators, 6; $\frac{2}{3} + \frac{1}{2} = \frac{4}{6} + \frac{3}{6}$. Then we add the numerators and keep the common denominator. In order to do this, we must be able to find the **least common denominator (LCD),** or **least common multiple (LCM)** of the denominators. (A review of Section 1.7b might be helpful.)

a Finding Least Common Multiples

LEAST COMMON MULTIPLE, LCM

The **least common multiple,** or LCM, of two natural numbers is the smallest number that is a multiple of both numbers.

EXAMPLE 1 Find the LCM of 20 and 30.

First we list some multiples of 20 by multiplying 20 by 1, 2, 3, and so on:

20, 40, 60, 80, 100, 120, 140, 160, 180, 200, 220, 240, ...

Then we list some multiples of 30 by multiplying 30 by 1, 2, 3, and so on:

30, 60, 90, 120, 150, 180, 210, 240, ...

Now we determine the smallest number common to both lists. The LCM of 20 and 30 is 60.

Do Exercise 1.

Next we develop three methods that are more efficient for finding LCMs. You may choose to learn only one method (consult with your instructor), but if you are going to study algebra, you should definitely learn method 2.

METHOD 1: FINDING LCMS USING ONE LIST OF MULTIPLES

One method for finding LCMs uses *one* list of multiples. Let's consider finding the LCM of 9 and 12. The largest number, 12, is not a multiple of 9, so we check multiples of 12 until we find a number that is also a multiple of 9:

$1 \cdot 12 = 12$, not a multiple of 9;
$2 \cdot 12 = 24$, not a multiple of 9;
$3 \cdot 12 = 36$, a multiple of 9: $4 \cdot 9 = 36$.

The LCM of 9 and 12 is 36.

Answer on page A-3

Method 1. To find the LCM of a set of numbers using a list of multiples:

a) Determine whether the largest number is a multiple of the others. If it is, it is the LCM. That is, if the largest number has the others as factors, the LCM is that number.
b) If it is not, check multiples of the largest number until you get one that is a multiple of each of the others.

EXAMPLE 2 Find the LCM of 12 and 15.

a) 15 is not a multiple of 12.
b) Check multiples of 15:

$1 \cdot 15 = 15,$ Not a multiple of 12. When we divide 15 by 12, we get a nonzero remainder.

$2 \cdot 15 = 30,$ Not a multiple of 12

$3 \cdot 15 = 45,$ Not a multiple of 12

$4 \cdot 15 = 60.$ A multiple of 12: $5 \cdot 12 = 60$

The LCM = 60.

Do Exercise 2.

EXAMPLE 3 Find the LCM of 4 and 14.

a) 14 is not a multiple of 4.
b) Check multiples of 14:

$1 \cdot 14 = 14,$

$2 \cdot 14 = 28.$ A multiple of 4: $7 \cdot 4 = 28$

The LCM = 28.

EXAMPLE 4 Find the LCM of 8 and 32.

a) 32 is a multiple of 8 ($4 \cdot 8 = 32$), so the LCM = 32.

EXAMPLE 5 Find the LCM of 10, 100, and 1000.

a) 1000 is a multiple of 10 ($100 \cdot 10 = 1000$) and of 100 ($10 \cdot 100 = 1000$), so the LCM = 1000.

Do Exercises 3–6.

METHOD 2: FINDING LCMS USING PRIME FACTORIZATIONS

A second method for finding LCMs uses prime factorizations. Consider again 20 and 30. Their prime factorizations are $20 = 2 \cdot 2 \cdot 5$ and $30 = 2 \cdot 3 \cdot 5$. Let's look at these prime factorizations in order to find the LCM. Any multiple of 20 will need to have *two* 2's as factors and *one* 5 as a factor. Any multiple of 30 will need to have *one* 2, *one* 3, and *one* 5 as factors. The smallest number satisfying these conditions is

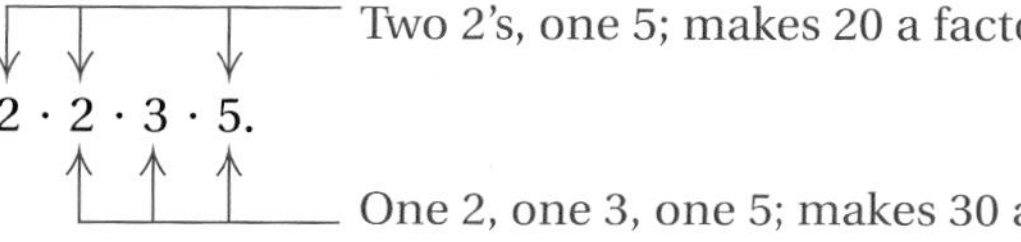
Two 2's, one 5; makes 20 a factor
2 · 2 · 3 · 5.
One 2, one 3, one 5; makes 30 a factor

2. By examining lists of multiples, find the LCM of 8 and 10.

Find the LCM.

3. 10, 15

4. 6, 8

5. 5, 10

6. 20, 40, 80

Answers on page A-3

Use prime factorizations to find the LCM.

7. 8, 10

8. 18, 40

9. 32, 54

Answers on page A-3

Thus the LCM of 20 and 30 is $2 \cdot 2 \cdot 3 \cdot 5$, or 60. It has all the factors of 20 and all the factors of 30, but the factors are not repeated when they are common to both numbers.

Observe that the greatest number of times 2 occurs as a factor of either 20 or 30 is two, and the LCM has 2 as a factor twice. The greatest number of times 3 occurs as a factor of either 20 or 30 is one, and the LCM has 3 as a factor once. The greatest number of times 5 occurs as a factor of either 20 or 30 is one, and the LCM has 5 as a factor once.

Method 2. To find the LCM of a set of numbers using prime factorizations:

a) Find the prime factorization of each number.
b) Create a product of factors, using each factor the greatest number of times that it occurs in any one factorization.

EXAMPLE 6 Find the LCM of 6 and 8.

a) Find the prime factorization of each number.

$6 = 2 \cdot 3, \qquad 8 = 2 \cdot 2 \cdot 2$

b) Create a product by writing factors that appear in the factorizations of 6 and 8, using each the greatest number of times that it occurs in any one factorization.

Consider the factor 2. The greatest number of times that 2 occurs in any one factorization is three. We write 2 as a factor three times.

$2 \cdot 2 \cdot 2 \cdot ?$

Consider the factor 3. The greatest number of times that 3 occurs in any one factorization is one. We write 3 as a factor one time.

$2 \cdot 2 \cdot 2 \cdot 3 \cdot ?$

Since there are no other prime factors in either factorization, the

LCM is $2 \cdot 2 \cdot 2 \cdot 3$, or 24.

EXAMPLE 7 Find the LCM of 24 and 36.

a) Find the prime factorization of each number.

$24 = 2 \cdot 2 \cdot 2 \cdot 3, \qquad 36 = 2 \cdot 2 \cdot 3 \cdot 3$

b) Create a product by writing factors, using each the greatest number of times that it occurs in any one factorization.

Consider the factor 2. The greatest number of times that 2 occurs in any one factorization is three. We write 2 as a factor three times:

$2 \cdot 2 \cdot 2 \cdot ?$

Consider the factor 3. The greatest number of times that 3 occurs in any one factorization is two. We write 3 as a factor two times:

$2 \cdot 2 \cdot 2 \cdot 3 \cdot 3 \cdot ?$

Since there are no other prime factors in either factorization, the

LCM is $2 \cdot 2 \cdot 2 \cdot 3 \cdot 3$, or 72.

Do Exercises 7–9.

EXAMPLE 8 Find the LCM of 27, 90, and 84.

a) Find the prime factorization of each number.

$27 = 3 \cdot 3 \cdot 3, \qquad 90 = 2 \cdot 3 \cdot 3 \cdot 5, \qquad 84 = 2 \cdot 2 \cdot 3 \cdot 7$

b) Create a product by writing factors, using each the greatest number of times that it occurs in any one factorization.

Consider the factor 2. The greatest number of times that 2 occurs in any one factorization is two. We write 2 as a factor two times:

$2 \cdot 2 \cdot ?$

Consider the factor 3. The greatest number of times that 3 occurs in any one factorization is three. We write 3 as a factor three times:

$2 \cdot 2 \cdot 3 \cdot 3 \cdot 3 \cdot ?$

Consider the factor 5. The greatest number of times that 5 occurs in any one factorization is one. We write 5 as a factor one time:

$2 \cdot 2 \cdot 3 \cdot 3 \cdot 3 \cdot 5 \cdot ?$

Consider the factor 7. The greatest number of times that 7 occurs in any one factorization is one. We write 7 as a factor one time:

$2 \cdot 2 \cdot 3 \cdot 3 \cdot 3 \cdot 5 \cdot 7 \cdot ?$

Since there are no other prime factors in any of the factorizations, the

LCM is $2 \cdot 2 \cdot 3 \cdot 3 \cdot 3 \cdot 5 \cdot 7$, or 3780.

Do Exercise 10.

EXAMPLE 9 Find the LCM of 7 and 21.

We find the prime factorization of each number. Because 7 is prime, it has no prime factorization.

$7 = 7, \qquad 21 = 3 \cdot 7$

Note that 7 is a factor of 21. We stated earlier that if one number is a factor of another, the LCM is the larger of the numbers. Thus the LCM is $7 \cdot 3$, or 21.

Do Exercises 11 and 12.

EXAMPLE 10 Find the LCM of 8 and 9.

We find the prime factorization of each number.

$8 = 2 \cdot 2 \cdot 2, \qquad 9 = 3 \cdot 3$

Note that the two numbers, 8 and 9, have no common prime factor. When this is the case, the LCM is just the product of the two numbers. Thus the LCM is $2 \cdot 2 \cdot 2 \cdot 3 \cdot 3$, or $8 \cdot 9$, or 72.

Do Exercises 13 and 14.

10. Find the LCM of 24, 35, and 45.

Find the LCM.

11. 3, 18

12. 12, 24

Find the LCM.

13. 4, 9

14. 5, 6, 7

Answers on page A-3

1.9 EXERCISE SET

For Extra Help

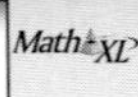
MathXL

MyMathLab

InterAct Math

Math Tutor Center

Video Lectures on CD Disc 1

Student's Solutions Manual

a Find the LCM of the set of numbers.

1. 2, 4

2. 3, 15

3. 10, 25

4. 10, 15

5. 20, 40

6. 8, 12

7. 18, 27

8. 9, 11

9. 30, 50

10. 24, 36

11. 30, 40

12. 21, 27

13. 18, 24

14. 12, 18

15. 60, 70

16. 35, 45

17. 16, 36

18. 18, 20

19. 32, 36

20. 36, 48

21. 2, 3, 5

22. 5, 18, 3

23. 3, 5, 7

24. 6, 12, 18

25. 24, 36, 12

26. 8, 16, 22

27. 5, 12, 15

28. 12, 18, 40

29. 9, 12, 6

30. 8, 16, 12

31. 180, 100, 450, 60

32. 18, 30, 50, 48

33. 8, 48

34. 16, 32

35. 5, 50

36. 12, 72

37. 11, 13

38. 13, 14

39. 12, 35

40. 23, 25

41. 54, 63

42. 56, 72

43. 81, 90

44. 75, 100

45. 36, 54, 80

46. 22, 42, 51

47. 39, 91, 108, 26

48. 625, 75, 500, 25

Applications of LCMs: Planet Orbits. The earth, Jupiter, Saturn, and Uranus all revolve around the sun. The earth takes 1 yr, Jupiter 12 yr, Saturn 30 yr, and Uranus 84 yr to make a complete revolution. On a certain night, you look at those three distant planets and wonder how many years it will take before they have the same position again. (*Hint*: To find out, determine the LCM of 12, 30, and 84. It will be that number of years.)

Source: *The Handy Science Answer Book*

49. How often will Jupiter and Saturn appear in the same direction in the night sky as seen from the earth?

50. How often will Jupiter and Uranus appear in the same direction in the night sky as seen from the earth?

51. How often will Saturn and Uranus appear in the same direction in the night sky as seen from the earth?

52. How often will Jupiter, Saturn, and Uranus appear in the same direction in the night sky as seen from the earth?

53. D_W Use both methods 1 and 2 to find the LCM of each of the following sets of numbers.

a) 6, 8 **b)** 6, 7 **c)** 6, 21 **d)** 24, 36

Which method do you consider more efficient? Explain why.

54. D_W Is the LCM of two numbers always larger than either number? Why or why not?

SKILL MAINTENANCE

Solve.

55. Use $<$ or $>$ for $\square$ to write a true sentence: [1.1d]

$9001 \square 10{,}001.$

56. *Broadway Runs.* As of May 3, 2004, *Cats* had the longest Broadway run with 7485 performances from October 1982 to September 2000. From June 1988 to the present, *The Phantom of the Opera* has had 6781 performances. How many more performances of *The Phantom of the Opera* are needed to equal the record set by the *Cats* show? [1.5a]

Source: League of American Theatres and Producers, Inc.

57. Add: $23{,}456 + 5677 + 4002$. [1.2a]

58. Subtract: $10{,}007 - 3068$. [1.2d]

59. Write expanded notation for 24,605. [1.1b]

60. Write a word name for 102,960. [1.1c]

SYNTHESIS

61. Find the LCM of 27, 90, 84, 210, 108, and 50.

62. Find the LCM of 18, 21, 24, 36, 63, 56, and 20.

63. A pencil company uses two sizes of boxes, 5 in. by 6 in. and 5 in. by 8 in. These boxes are packed in bigger cartons for shipping. Find the width and the length of the smallest carton that will accommodate boxes of either size without any room left over. (Each carton can contain only one type of box and all boxes must point in the same direction.)

64. Consider 8 and 12. Determine whether each of the following is the LCM of 8 and 12. Tell why or why not.

a) $2 \cdot 2 \cdot 3 \cdot 3$

b) $2 \cdot 2 \cdot 3$

c) $2 \cdot 3 \cdot 3$

d) $2 \cdot 2 \cdot 2 \cdot 3$

1 Summary and Review

The review that follows is meant to prepare you for a chapter exam. It consists of two parts. The first part, Concept Reinforcement, is designed to increase understanding of the concepts through true/false exercises. The second part is the Review Exercises. These provide practice exercises for the exam, together with references to section objectives so you can go back and review. Before beginning, stop and look back over the skills you have obtained. What skills in mathematics do you have now that you did not have before studying this chapter?

CONCEPT REINFORCEMENT

Determine whether the statement is true or false. Answers are given at the back of the book.

________ **1.** The product of two natural numbers is always greater than either of the factors.

________ **2.** The number 1 is not prime.

________ **3.** Each member of the set of natural numbers is a member of the set of whole numbers.

________ **4.** The sum of two natural numbers is always greater than either of the addends.

________ **5.** If a number is not divisible by 6, then it is not divisible by 3.

________ **6.** The number 0 is the smallest natural number.

Review Exercises

The review exercises that follow are for practice. Answers are given at the back of the book. If you miss an exercise, restudy the objective indicated in red next to the exercise or direction line that precedes it.

1. What does the digit 8 mean in 4,678,952? [1.1a]

2. In 13,768,940, what digit tells the number of millions? [1.1a]

Write expanded notation. [1.1b]

3. 2793

4. 56,078

5. 4,007,101

Write a word name. [1.1c]

6. 67,819

7. 2,781,427

8. 1,065,070,607, the population of India in 2004

Source: U.S. Bureau of the Census, International Database

Write standard notation. [1.1c]

9. Four hundred seventy-six thousand, five hundred eighty-eight

10. *e-books.* The publishing industry predicts that sales of digital books will reach two billion, four hundred thousand by 2005.

Source: Andersen Consulting

Add. [1.2a]

11. 7304 + 6968

12. 27,609 + 38,415

13. 2703 + 4125 + 6004 + 8956

14.

$$\begin{array}{r} 91{,}426 \\ +\ \ 7{,}495 \\ \hline \end{array}$$

15. Write a related addition sentence: [1.2c]

$10 - 6 = 4.$

16. Write two related subtraction sentences: [1.2c]

$8 + 3 = 11.$

Subtract. [1.2d]

17. 8045 − 2897

18. 9001 − 7312

19. 6003 − 3729

20.
$$\begin{array}{r} 37,405 \\ -\ 19,648 \\ \hline \end{array}$$

Round 345,759 to the nearest: [1.3e]

21. Hundred.

22. Ten.

23. Thousand.

24. Hundred thousand.

Estimate the sum, difference, or product by first rounding to the nearest hundred. Show your work. [1.3f]

25. 41,348 + 19,749

26. 38,652 − 24,549

27. 396 · 748

Use < or > for □ to write a true sentence. [1.1d]

28. 67 □ 56

29. 1 □ 23

Multiply. [1.3a]

30. 17,000 · 300

31. 7846 · 800

32. 726 · 698

33. 587 · 47

34.
$$\begin{array}{r} 8305 \\ \times\ \ 642 \\ \hline \end{array}$$

35. Write a related multiplication sentence: [1.3c]

56 ÷ 8 = 7.

36. Write two related division sentences: [1.3c]

13 · 4 = 52.

Divide. [1.3d]

37. 63 ÷ 5

38. 80 ÷ 16

39. $7\overline{)6394}$

40. 3073 ÷ 8

41. $60\overline{)286}$

42. 4266 ÷ 79

43. $38\overline{)17,176}$

44. 52,668 ÷ 12

Solve. [1.4b]

45. $46 \cdot n = 368$

46. $47 + x = 92$

47. Write exponential notation: 4 · 4 · 4. [1.6a]

Evaluate. [1.6b]

48. 10^4

49. 6^2

Simplify. [1.6c, d]

50. $8 \cdot 6 + 17$

51. $10 \cdot 24 - (18 + 2) \div 4 - (9 - 7)$

52. $7 + (4 + 3)^2$

53. $7 + 4^2 + 3^2$

54. $(80 \div 16) \times [(20 - 56 \div 8) + (8 \cdot 8 - 5 \cdot 5)]$

55. Find the average of 157, 170, and 168.

Solve. [1.5a]

56. *Computer Workstation.* Natasha has $196 and wants to buy a computer workstation for $698. How much more does she need?

Computer Workstation
Raised monitor platform, sliding keyboard shelf, and mobile CPU shelf. Locking 3-drawer file cabinet. Maple and honey finish with durable melamine work surface.
Workstation just...
$698

57. Toni has $406 in her checking account. She is paid $78 for a part-time job and deposits that in her checking account. How much is then in her account?

58. *Lincoln-Head Pennies.* In 1909, the first Lincoln-head pennies were minted. Seventy-three years later, these pennies were first minted with a decreased copper content. In what year was the copper content reduced?

59. A beverage company packed 228 cans of soda into 12-can cartons. How many cartons did they fill?

60. An apple farmer keeps bees in her orchard to help pollinate the apple blossoms. The bees from an average beehive can pollinate 30 surrounding trees during one growing season. A farmer has 420 trees. How many beehives does she need to pollinate them all?

Source: Jordan Orchards, Westminster, PA

61. An apartment builder bought 13 gas stoves at $425 each and 13 refrigerators at $620 each. What was the total cost?

62. A family budgets $7825 for food and clothing and $2860 for entertainment. The yearly income of the family was $38,283. How much of this income remained after these two allotments?

63. A chemist has 2753 mL of alcohol. How many 20-mL beakers can be filled? How much will be left over?

64. *Olympic Trampoline.* Shown below is an Olympic trampoline. Find the area and the perimeter of the trampoline. [1.2b], [1.3b]

Source: International Trampoline Industry Association, Inc.

Determine whether the number is prime, composite, or neither. [1.7c]

65. 37 **66.** 1 **67.** 91

Find the prime factorization of the number. [1.7d]

68. 70 **69.** 30

70. 45 **71.** 150

72. 648 **73.** 5250

Determine whether: [1.8a]

74. 2532 is divisible by 6. **75.** 182 is divisible by 4.

76. 4344 is divisible by 9. **77.** 4344 is divisible by 8.

Find the LCM. [1.9a]

78. 12 and 18 **79.** 18 and 45

80. 3, 6, and 30 **81.** 26, 36, and 54

82. DW Use the number 9432 to explain why the test for divisibility by 9 works. [1.8a]

SYNTHESIS

83. Determine the missing digit d. [1.3a]

$$\begin{array}{r} 9\,d \\ \times\ \ d\,2 \\ \hline 8\,0\,3\,6 \end{array}$$

84. Determine the missing digits a and b. [1.3d]

$$2\,b\,1\,\overline{)\,2\,3\,6{,}4\,2\,1}\quad \text{quotient: } 9\,a\,1$$

85. A mining company estimates that a crew must tunnel 2000 ft into a mountain to reach a deposit of copper ore. Each day the crew tunnels about 500 ft. Each night about 200 ft of loose rocks roll back into the tunnel. How many days will it take the mining company to reach the copper deposit? [1.5a]

86. A prime number that becomes a prime number when its digits are reserved is called a **palindrome prime.** For example, 17 is a palindrome prime because both 17 and 71 are primes. Which of the following numbers are palindrome primes? [1.7c]

13, 91,16, 11, 15, 24, 29, 101, 201, 37

1 Chapter Test

For Extra Help

Work It Out!
Chapter Test Video
Solutions on CD

1. In the number 546,789, which digit tells the number of hundred thousands?

2. Write expanded notation: 8843.

3. Write a word name: 38,403,277.

Add.

4. $\begin{array}{r} 6811 \\ +\ 3178 \\ \hline \end{array}$

5. $\begin{array}{r} 45{,}889 \\ +\ 17{,}902 \\ \hline \end{array}$

6. $\begin{array}{r} 12 \\ 8 \\ 3 \\ 7 \\ +\ 4 \\ \hline \end{array}$

7. $\begin{array}{r} 6203 \\ +\ 4312 \\ \hline \end{array}$

Subtract.

8. $\begin{array}{r} 7983 \\ -\ 4353 \\ \hline \end{array}$

9. $\begin{array}{r} 2974 \\ -\ 1935 \\ \hline \end{array}$

10. $\begin{array}{r} 8907 \\ -\ 2059 \\ \hline \end{array}$

11. $\begin{array}{r} 23{,}067 \\ -\ 17{,}892 \\ \hline \end{array}$

Multiply.

12. $\begin{array}{r} 4568 \\ \times\ 9 \\ \hline \end{array}$

13. $\begin{array}{r} 8876 \\ \times\ 600 \\ \hline \end{array}$

14. $\begin{array}{r} 65 \\ \times\ 37 \\ \hline \end{array}$

15. $\begin{array}{r} 678 \\ \times\ 788 \\ \hline \end{array}$

Divide.

16. $15 \div 4$

17. $420 \div 6$

18. $89 \overline{)\,8633}$

19. $44 \overline{)\,35{,}428}$

Solve.

20. *Hostess Ding Dongs®.* Hostess packages its Ding Dong® snack products in 12-packs. How many 12-packs can it fill with 22,231 cakes? How many will be left over?

21. *Largest States.* The following table lists the five largest states in terms of their land area. Find the total land area of these states.

STATE	AREA (in square miles)
Alaska	571,951
Texas	261,797
California	155,959
Montana	145,552
New Mexico	121,356

Source: Department of Commerce, Bureau of the Census

22. *Pool Tables.* The Hartford™ pool table made by Brunswick Billiards comes in three sizes of playing area, 50 in. by 100 in., 44 in. by 88 in., and 38 in. by 76 in.

a) Find the perimeter and the area of the playing area of each table.

b) By how much does the largest area exceed the smallest area?

Source: Brunswick Billiards

23. *Voting Early.* In the 2004 Presidential Election, 345,689 Nevada voters voted early. This was 139,359 more than in the 2000 election. How many voted early in 2000?

Source: National Association of Secretaries of State

24. A bag of oranges weighs 27 lb. A bag of apples weighs 32 lb. Find the total weight of 16 bags of oranges and 43 bags of apples.

25. A box contains 5000 staples. How many staplers can be filled from the box if each stapler holds 250 staples?

Solve.

26. $28 + x = 74$

27. $169 \div 13 = n$

28. $38 \cdot y = 532$

29. $381 = 0 + a$

Round 34,572 to the nearest:

30. Thousand.

31. Ten.

32. Hundred.

Estimate the sum, difference, or product by first rounding to the nearest hundred. Show your work.

33. $\begin{array}{r} 23{,}649 \\ +\ 54{,}746 \\ \hline \end{array}$

34. $\begin{array}{r} 54{,}751 \\ -\ 23{,}649 \\ \hline \end{array}$

35. $\begin{array}{r} 824 \\ \times\ 489 \\ \hline \end{array}$

36. Write exponential notation: $12 \cdot 12 \cdot 12 \cdot 12$.

Evaluate.

37. 7^3

38. 10^5

Use $<$ or $>$ for $\square$ to write a true sentence.

39. $34 \ \square\ 17$

40. $117 \ \square\ 157$

Simplify.

41. $35 - 1 \cdot 28 \div 4 + 3$

42. $10^2 - 2^2 \div 2$

43. $(25 - 15) \div 5$

44. $2^4 + 24 \div 12$

45. $8 \times \{(20 - 11) \cdot [(12 + 48) \div 6 - (9 - 2)]\}$

46. Find the average of 97, 98, 87, and 86.

Find the LCM.

47. 16 and 12

48. 15, 40, and 50

Determine whether the number is prime, composite, or neither.

49. 41

50. 14

Find the prime factorization of the number.

51. 18

52. 60

53. Determine whether 1784 is divisible by 8.

54. Determine whether 784 is divisible by 9.

55. Determine whether 5552 is divisible by 5.

56. Determine whether 2322 is divisible by 6.

SYNTHESIS

57. An open cardboard container is 8 in. wide, 12 in. long, and 6 in. high. How many square inches of cardboard are used?

58. Use trials to find the single-digit number a for which

$$359 - 46 + a \div 3 \times 25 - 7^2 = 339.$$

Fraction Notation

2

Real-World Application

Arie Luyendyk won the Indianapolis 500 in 1990 with an average speed of about 186 mph. This record high speed is about $2\frac{12}{25}$ times the average speed of the first winner, Ray Harroun, in 1911. What was the average speed in the first Indianapolis 500?

Source: Indianapolis Motor Speedway

This problem appears as Example 11 in Section 2.5.

2.1 FRACTION NOTATION AND SIMPLIFYING

Objectives

a Identify the numerator and the denominator of a fraction and write fraction notation for part of an object.

b Simplify fraction notation like n/n to 1, $0/n$ to 0, and $n/1$ to n.

c Multiply a fraction by a fraction.

d Use multiplying by 1 to find different fraction notation for a number.

e Simplify fraction notation.

For each fraction, identify the numerator and the denominator.

1. $\frac{83}{100}$ of all scrap tires are reused or recycled.
Source: Rubber Manufacturers Association

2. $\frac{27}{50}$ of all kids prefer white bread.
Source: Bruce Horovitz, *USA Today*, 8/9/04

3. $\frac{11}{25}$ of all adults eat in restaurants on a given day.
Source: *AARP Bulletin*, November 2004

4. $\frac{21}{1000}$ of stay-at-home parents are men.
Source: U.S. Bureau of the Census

Answers on page A-4

The study of arithmetic begins with the set of whole numbers

0, 1, 2, 3, 4, 5, 6, 7, 8, 9, 10, 11, and so on.

But we also need to be able to use fractional parts of numbers such as halves, thirds, fourths, and so on. Here are some examples:

In 1950, about $\frac{3}{4}$ of the motor vehicles produced in the world were produced in the United States. In 2003, only $\frac{1}{5}$ were produced in the United States.

Sources: American Automobile Manufacturers Association; Automotive News Data Center and Marketing Systems GmbH

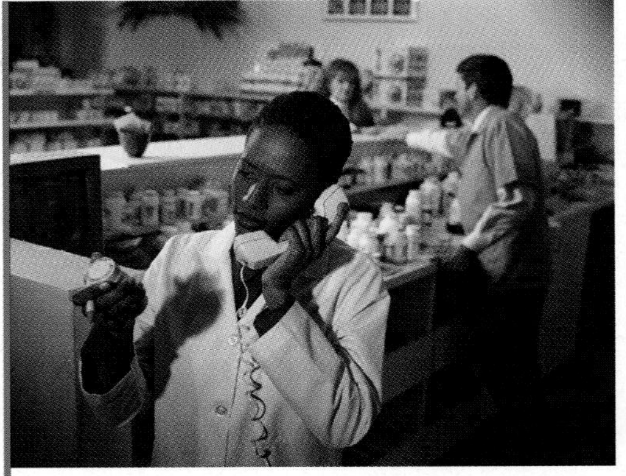

More than $\frac{2}{5}$ of Americans are taking at least one prescription daily, and $\frac{1}{6}$ are taking three or more.

Source: Randolph Schmid, Associated Press, *Indianapolis Star*, 12/3/04

a Fractions and the Real World

Numbers like those above and the ones below are written in **fraction notation.** The top number is called the **numerator** and the bottom number is called the **denominator.**

$$\frac{1}{2},\quad \frac{3}{4},\quad \frac{8}{5},\quad \frac{11}{23}$$

EXAMPLE 1 Identify the numerator and the denominator.

$$\frac{7}{8}\begin{matrix}\longleftarrow \text{Numerator}\\ \longleftarrow \text{Denominator}\end{matrix}$$

Do Exercises 1–4.

Let's look at various situations that involve fractions.

FRACTIONS AS A PARTITION OF AN OBJECT DIVIDED INTO EQUAL PARTS

Consider a candy bar divided into 5 equal sections. If you eat 2 sections, you have eaten $\frac{2}{5}$ of the candy bar.

The denominator 5 tells us the unit, $\frac{1}{5}$. The numerator 2 tells us the number of equal parts we are considering, 2.

EXAMPLE 2 What part is shaded?

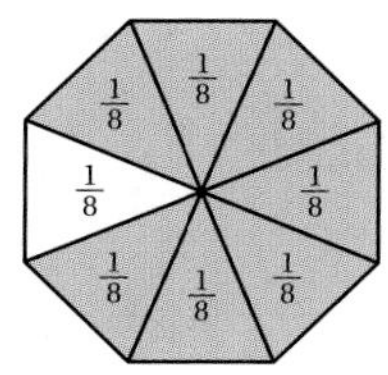

The equal parts are eighths. This tells us the unit, $\frac{1}{8}$. The *denominator* is 8. We have 7 of the units shaded. This tells us the *numerator*, 7. Thus,

$$\frac{7}{8} \begin{array}{l} \leftarrow \text{7 units are shaded.} \\ \leftarrow \text{The unit is } \frac{1}{8}. \end{array}$$

is shaded.

Do Exercises 5–9.

The markings on a ruler use fractions.

EXAMPLE 3 What part of an inch is highlighted?

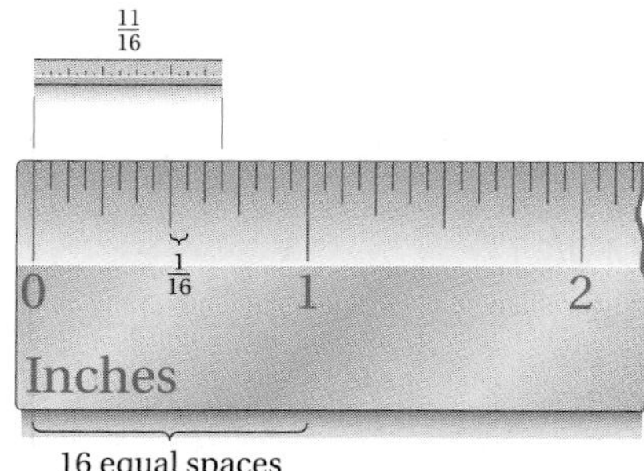

Each inch on the ruler shown above is divided into 16 equal parts. The highlighting extends to the 11th mark. Thus, $\frac{11}{16}$ of an inch is highlighted.

Do Exercise 10.

What part is shaded?

5.

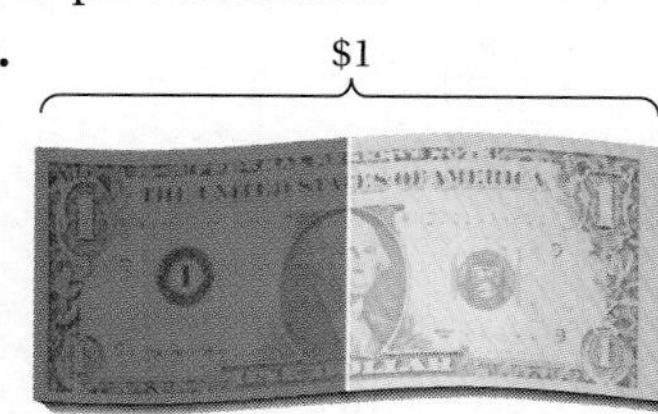

6.

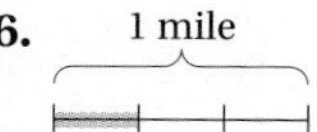

7.

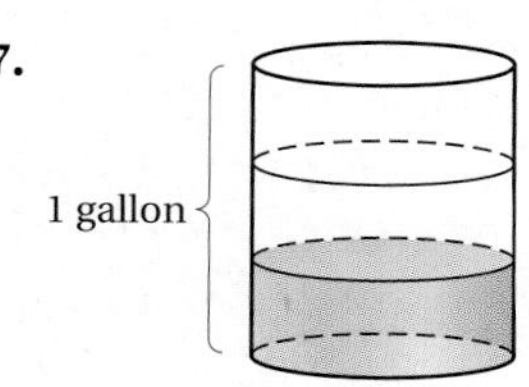

8.

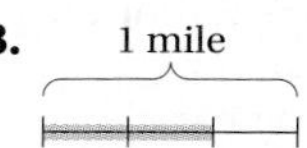

9.

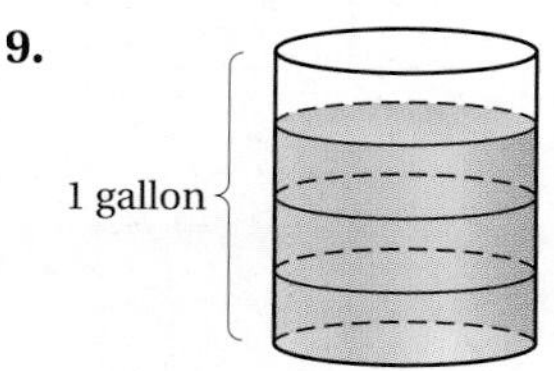

10. What part of an inch is highlighted?

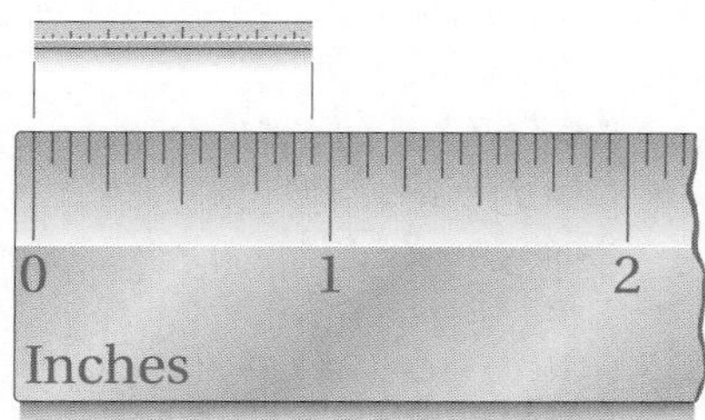

Answers on page A-4

What part is shaded?

11.

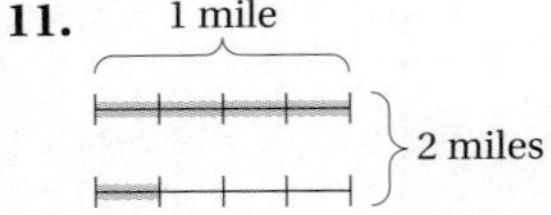

12.

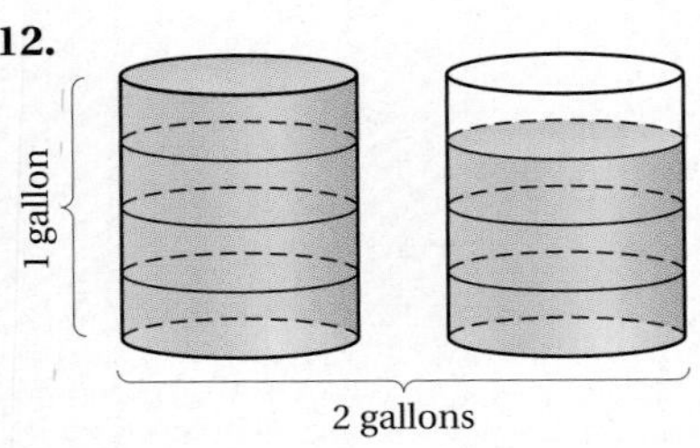

Answers on page A-4

Fractions larger than or equal to 1, such as $\frac{13}{6}$ or $\frac{9}{9}$, are sometimes referred to as "improper" fractions. We will not use this terminology because notation such as $\frac{27}{8}$, $\frac{11}{3}$, and $\frac{4}{4}$ is quite "proper" and very common in algebra.

Study Tips

STUDY TIPS REVIEW

The following Study Tips were presented in Chapter 1. If you have not studied them, go back and do so now.

- Using This Textbook (Section 1.2)
- Learning Resources (Section 1.2)
- Time Management (Part 1) (Section 1.3)
- Highlighting (Section 1.3)
- Exercises (Section 1.5)
- Test Preparation (Section 1.6)

Fractions greater than 1, such as $\frac{10}{3}$ and $\frac{5}{4}$, correspond to situations like the following.

EXAMPLE 4 What part is shaded?

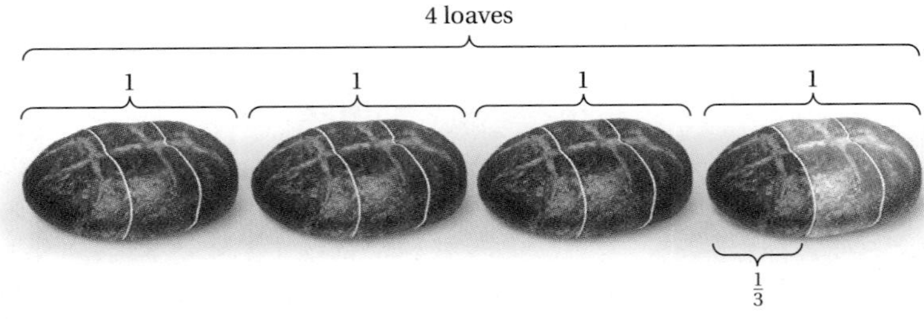

Each loaf of bread is divided into 3 equal parts. The unit is $\frac{1}{3}$. The *denominator* is 3. We have 10 of the units shaded. This tells us the *numerator* is 10. Thus, $\frac{10}{3}$ are shaded.

EXAMPLE 5 What part is shaded?

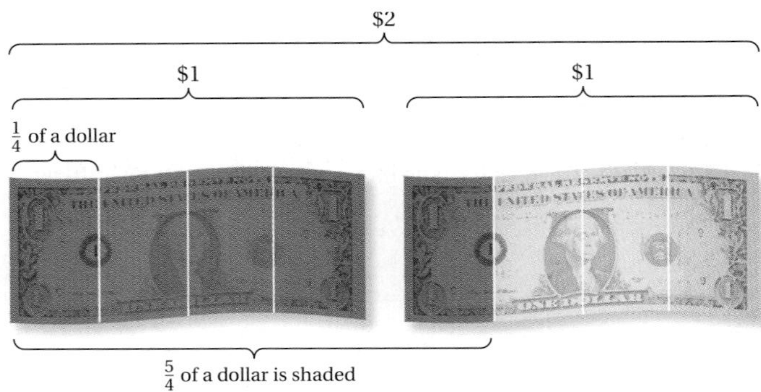

We can regard this as two objects of 4 parts each and take 5 of those parts. The unit is $\frac{1}{4}$. The denominator is 4, and the numerator is 5. Thus $\frac{5}{4}$ are shaded.

Do Exercises 11 and 12.

b Some Fraction Notation for Whole Numbers

FRACTION NOTATION FOR 1

The number 1 can correspond to situations like those shown here.

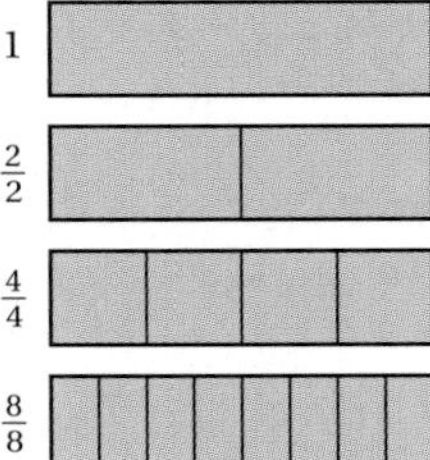

If we divide an object into n parts and take n of them, we get all of the object (1 whole object).

THE NUMBER 1 IN FRACTION NOTATION

$\frac{n}{n} = 1$, for any whole number n that is not 0.

EXAMPLES Simplify.

6. $\frac{5}{5} = 1$

7. $\frac{9}{9} = 1$

8. $\frac{23}{23} = 1$

Do Exercises 13–18.

FRACTION NOTATION FOR 0

Consider the fraction $\frac{0}{4}$. This corresponds to dividing an object into 4 parts and taking none of them. We get 0.

THE NUMBER 0 IN FRACTION NOTATION

$\frac{0}{n} = 0$, for any whole number n that is not 0.

EXAMPLES Simplify.

9. $\frac{0}{1} = 0$

10. $\frac{0}{9} = 0$

11. $\frac{0}{23} = 0$

Fraction notation with a denominator of 0, such as $n/0$, is meaningless because we cannot speak of an object being divided into *zero* parts. See also the discussion of excluding division by 0 in Section 1.3.

A DENOMINATOR OF 0

$\frac{n}{0}$ is not defined for any whole number n.

Do Exercises 19–24.

OTHER WHOLE NUMBERS

Consider the fraction $\frac{4}{1}$. This corresponds to taking 4 objects and dividing each into 1 part. (We do not divide them.) We have 4 objects.

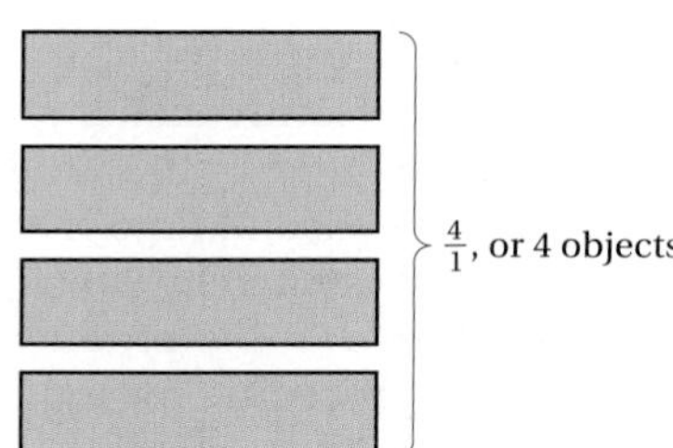

Simplify.

13. $\frac{1}{1}$

14. $\frac{4}{4}$

15. $\frac{34}{34}$

16. $\frac{100}{100}$

17. $\frac{2347}{2347}$

18. $\frac{103}{103}$

Simplify, if possible.

19. $\frac{0}{1}$

20. $\frac{0}{8}$

21. $\frac{0}{107}$

22. $\frac{4-4}{567}$

23. $\frac{15}{0}$

24. $\frac{0}{3-3}$

Answers on page A-4

Simplify.

25. $\frac{8}{1}$

26. $\frac{10}{1}$

27. $\frac{346}{1}$

28. $\frac{24 - 1}{23 - 22}$

29. Draw a diagram like the one at right to show the multiplication $\frac{2}{3} \cdot \frac{4}{5}$.

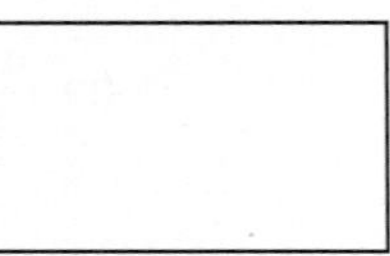

Answers on page A-4

ANY WHOLE NUMBER IN FRACTION NOTATION

Any whole number divided by 1 is the whole number. That is,

$$\frac{n}{1} = n, \quad \text{for any whole number } n.$$

EXAMPLES Simplify.

12. $\frac{2}{1} = 2$

13. $\frac{9}{1} = 9$

14. $\frac{34}{1} = 34$

Do Exercises 25–28.

C Multiplication Using Fraction Notation

Let's visualize the product of two fractions. We consider the multiplication

$$\frac{3}{5} \cdot \frac{3}{4}.$$

This is equivalent to finding $\frac{3}{5}$ of $\frac{3}{4}$. We first consider an object and take $\frac{3}{4}$ of it. We divide it into 4 equal parts using vertical lines and take 3 of them. That is shown by the shading in the figure at right.

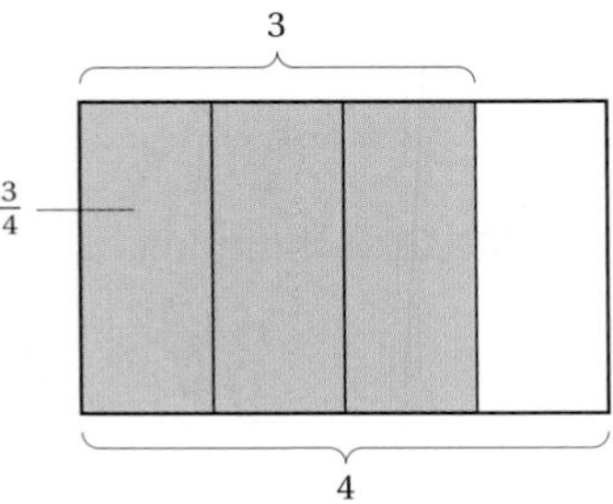

Next, we take $\frac{3}{5}$ of the shaded area. We divide it into 5 equal parts using horizontal lines and take 3 of them. That is shown by the darker shading in the figure below.

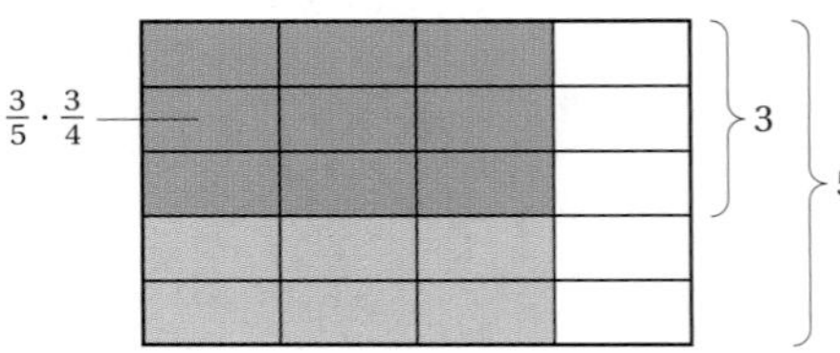

The entire object has been divided into 20 parts, and we have shaded 9 of them for a second time. Thus we see that $\frac{3}{5}$ of $\frac{3}{4}$ is $\frac{9}{20}$, or

$$\frac{3}{5} \cdot \frac{3}{4} = \frac{9}{20}.$$

The figure above shows a rectangular array inside a rectangular array. The number of pieces in the entire array is $5 \cdot 4$ (the product of the denominators). The number of pieces shaded a second time is $3 \cdot 3$ (the product of the numerators). And the product is represented by 9 pieces out of a set of 20, or $\frac{9}{20}$, or the product of the numerators over the product of the denominators. This leads us to a statement of the procedure for multiplying a fraction by a fraction.

Do Exercise 29.

We find a product such as $\frac{9}{7} \cdot \frac{3}{4}$ as follows.

To multiply a fraction by a fraction,

a) multiply the numerators to get the new numerator, and

$$\frac{9}{7} \cdot \frac{3}{4} = \frac{9 \cdot 3}{7 \cdot 4} = \frac{27}{28}$$

b) multiply the denominators to get the new denominator.

EXAMPLES Multiply.

15. $\frac{5}{6} \times \frac{7}{4} = \frac{5 \times 7}{6 \times 4} = \frac{35}{24}$

Skip writing this step whenever you can.

16. $\frac{3}{5} \cdot \frac{7}{8} = \frac{3 \cdot 7}{5 \cdot 8} = \frac{21}{40}$

17. $\frac{3}{5} \cdot \frac{3}{4} = \frac{9}{20}$

18. $\frac{1}{4} \cdot \frac{1}{3} = \frac{1}{12}$

19. $6 \cdot \frac{4}{5} = \frac{6}{1} \cdot \frac{4}{5} = \frac{24}{5}$

Do Exercises 30–33.

Multiply.

30. $\frac{3}{8} \cdot \frac{5}{7}$

31. $\frac{4}{3} \times \frac{8}{5}$

32. $\frac{3}{10} \cdot \frac{1}{10}$

33. $7 \cdot \frac{2}{3}$

Answers on page A-4

d Multiplying by 1

Recall the following:

$$1 = \frac{1}{1} = \frac{2}{2} = \frac{3}{3} = \frac{4}{4} = \frac{10}{10} = \frac{45}{45} = \frac{100}{100} = \frac{n}{n}.$$

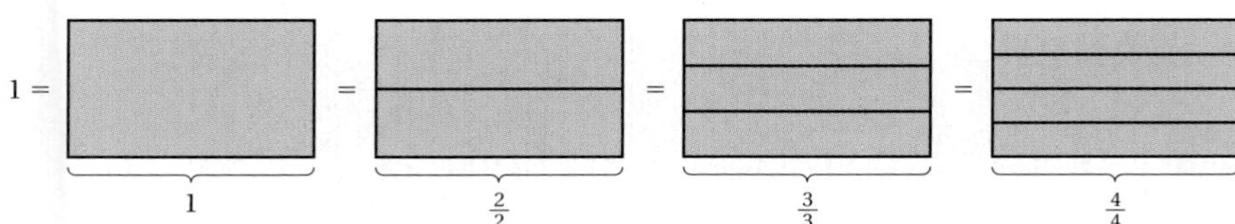

Any nonzero number divided by itself is 1. (See Section 1.3.)

Now recall the multiplicative identity from Section 1.3. For any whole number a, $1 \cdot a = a \cdot 1 = a$. This holds for fractions as well.

MULTIPLICATIVE IDENTITY FOR FRACTIONS

When we multiply a number by 1, we get the same number:

$$\frac{3}{5} = \frac{3}{5} \cdot 1 = \frac{3}{5} \cdot \frac{4}{4} = \frac{12}{20}.$$

Since $\frac{3}{5} = \frac{12}{20}$, we know that $\frac{3}{5}$ and $\frac{12}{20}$ are two names for the same number. We also say that $\frac{3}{5}$ and $\frac{12}{20}$ are **equivalent.**

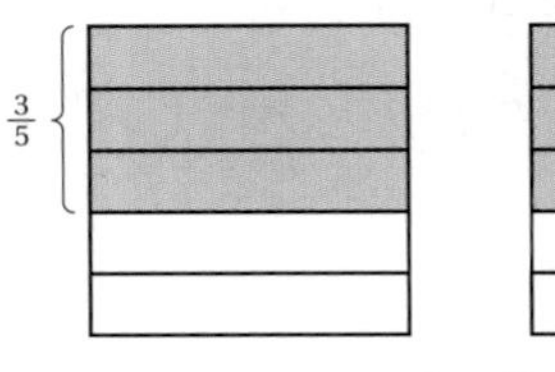

$\frac{3}{5} = \frac{12}{20}$

Do Exercises 34–37.

Suppose we want to find a name for $\frac{2}{3}$, but one that has a denominator of 9. We can multiply by 1 to find equivalent fractions:

$$\frac{2}{3} = \frac{2}{3} \cdot \frac{3}{3} = \frac{2 \cdot 3}{3 \cdot 3} = \frac{6}{9}.$$

Since $9 = 3 \cdot 3$, we chose $\frac{3}{3}$ for 1 in order to get a denominator of 9.

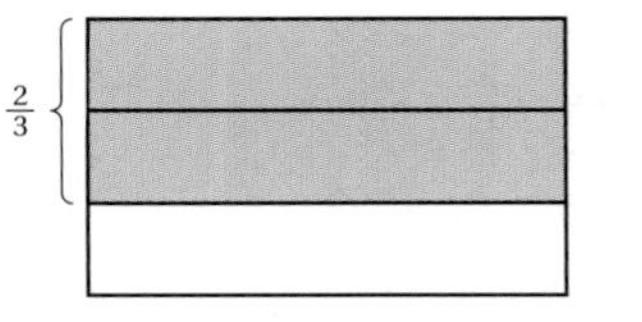

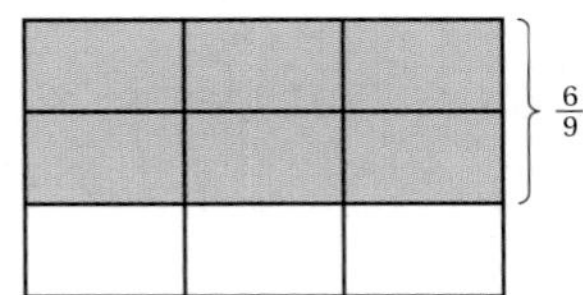

$\frac{2}{3} = \frac{6}{9}$

EXAMPLE 20 Find a name for $\frac{1}{4}$ with a denominator of 24.

Since $4 \cdot 6 = 24$, we multiply by $\frac{6}{6}$:

$$\frac{1}{4} = \frac{1}{4} \cdot \frac{6}{6} = \frac{1 \cdot 6}{4 \cdot 6} = \frac{6}{24}.$$

EXAMPLE 21 Find a name for $\frac{2}{5}$ with a denominator of 35.

Since $5 \cdot 7 = 35$, we multiply by $\frac{7}{7}$:

$$\frac{2}{5} = \frac{2}{5} \cdot \frac{7}{7} = \frac{2 \cdot 7}{5 \cdot 7} = \frac{14}{35}.$$

Do Exercises 38–42.

e Simplifying Fraction Notation

All of the following are names for three-fourths:

$$\frac{3}{4}, \frac{6}{8}, \frac{9}{12}, \frac{12}{16}, \frac{15}{20}.$$

We say that $\frac{3}{4}$ is **simplest** because it has the smallest numerator and the smallest denominator. That is, the numerator and the denominator have no common factor other than 1.

Multiply.

34. $\frac{1}{2} \cdot \frac{8}{8}$

35. $\frac{3}{5} \cdot \frac{10}{10}$

36. $\frac{13}{25} \cdot \frac{4}{4}$

37. $\frac{8}{3} \cdot \frac{25}{25}$

Find another name for the number, but with the denominator indicated. Use multiplying by 1.

38. $\frac{4}{3} = \frac{?}{9}$

39. $\frac{3}{4} = \frac{?}{24}$

40. $\frac{9}{10} = \frac{?}{100}$

41. $\frac{3}{15} = \frac{?}{45}$

42. $\frac{8}{7} = \frac{?}{49}$

Answers on page A-4

To simplify, we reverse the process of multiplying by 1:

$$\frac{12}{18} = \frac{2 \cdot 6}{3 \cdot 6} \quad \begin{array}{l}\leftarrow \text{Factoring the numerator} \\ \leftarrow \text{Factoring the denominator}\end{array}$$

$$= \frac{2}{3} \cdot \frac{6}{6} \qquad \text{Factoring the fraction}$$

$$= \frac{2}{3} \cdot 1 \qquad \frac{6}{6} = 1$$

$$= \frac{2}{3}. \qquad \text{Removing a factor of 1: } \frac{2}{3} \cdot 1 = \frac{2}{3}$$

EXAMPLES Simplify.

22. $\frac{8}{20} = \frac{2 \cdot 4}{5 \cdot 4} = \frac{2}{5} \cdot \frac{4}{4} = \frac{2}{5}$

23. $\frac{2}{6} = \frac{1 \cdot 2}{3 \cdot 2} = \frac{1}{3} \cdot \frac{2}{2} = \frac{1}{3}$

The number 1 allows for pairing of factors in the numerator and the denominator.

24. $\frac{30}{6} = \frac{5 \cdot 6}{1 \cdot 6} = \frac{5}{1} \cdot \frac{6}{6} = \frac{5}{1} = 5$

We could also simplify $\frac{30}{6}$ by doing the division $30 \div 6$. That is, $\frac{30}{6} = 30 \div 6 = 5$.

Do Exercises 43–46.

The use of prime factorizations can be helpful for simplifying when numerators and/or denominators are larger numbers.

EXAMPLE 25 Simplify: $\frac{90}{84}$.

$$\frac{90}{84} = \frac{2 \cdot 3 \cdot 3 \cdot 5}{2 \cdot 2 \cdot 3 \cdot 7} \qquad \text{Factoring the numerator and the denominator into primes}$$

$$= \frac{2 \cdot 3 \cdot 3 \cdot 5}{2 \cdot 3 \cdot 2 \cdot 7} \qquad \text{Changing the order so that like primes are above and below each other}$$

$$= \frac{2}{2} \cdot \frac{3}{3} \cdot \frac{3 \cdot 5}{2 \cdot 7} \qquad \text{Factoring the fraction}$$

$$= 1 \cdot 1 \cdot \frac{3 \cdot 5}{2 \cdot 7}$$

$$= \frac{3 \cdot 5}{2 \cdot 7} \qquad \text{Removing factors of 1}$$

$$= \frac{15}{14}$$

The tests for divisibility (Section 1.8) are very helpful in simplifying fraction notation. We could have shortened the preceding example had we noted that 6 is a factor of both the numerator and the denominator. Then we have

$$\frac{90}{84} = \frac{6 \cdot 15}{6 \cdot 14} = \frac{6}{6} \cdot \frac{15}{14} = \frac{15}{14}.$$

Simplify.

43. $\frac{2}{8}$

44. $\frac{10}{12}$

45. $\frac{40}{8}$

46. $\frac{24}{18}$

Simplify.

47. $\frac{35}{40}$

48. $\frac{801}{702}$

49. $\frac{24}{21}$

50. $\frac{75}{300}$

51. $\frac{280}{960}$

52. $\frac{1332}{2880}$

Answers on page A-4

53. Simplify each fraction in this circle graph.

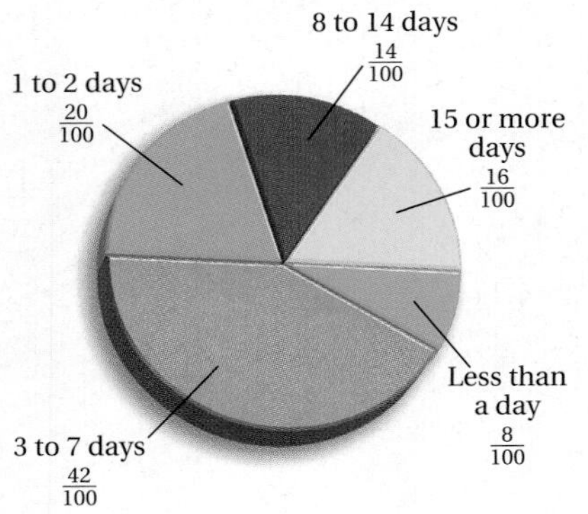

Answers on page A-4

CALCULATOR CORNER

Simplifying Fraction Notation Fraction calculators are equipped with a key, often labeled [$a^b/_c$], that allows for simplification with fraction notation. To simplify

$$\frac{208}{256}$$

with such a fraction calculator, the following keystrokes can be used.

[2] [0] [8] [$a^b/_c$]
[2] [5] [6] [=].

The display that appears

[13 ⌋ 16.]

represents simplified fraction notation $\frac{13}{16}$.

Exercises: Use a fraction calculator to simplify each of the following.

1. $\frac{84}{90}$
2. $\frac{35}{40}$
3. $\frac{690}{835}$
4. $\frac{42}{150}$

EXAMPLE 26 Simplify: $\frac{603}{207}$.

At first glance this looks difficult. But note, using the test for divisibility by 9 (sum of digits divisible by 9), that both the numerator and the denominator are divisible by 9. Thus we can factor 9 from both numbers:

$$\frac{603}{207} = \frac{9 \cdot 67}{9 \cdot 23} = \frac{9}{9} \cdot \frac{67}{23} = \frac{67}{23}.$$

EXAMPLE 27 Simplify: $\frac{660}{1140}$.

Using the tests for divisibility, we have

$$\frac{660}{1140} = \frac{10 \cdot 66}{10 \cdot 114} = \frac{10}{10} \cdot \frac{66}{114} = \frac{66}{114}$$ Both 660 and 1140 are divisible by 10.

$$= \frac{6 \cdot 11}{6 \cdot 19} = \frac{6}{6} \cdot \frac{11}{19} = \frac{11}{19}.$$ Both 66 and 114 are divisible by 6.

Do Exercises 47–53. (Exercises 47–52 are on the preceding page.)

CANCELING

Canceling is a shortcut that you may have used for removing a factor of 1 when working with fraction notation. With *great* concern, we mention it as a possibility for speeding up your work. Canceling may be done only when removing common factors in numerators and denominators. Each such pair allows us to remove a factor of 1 in a product.

Our concern is that canceling be done with care and understanding. In effect, slashes are used to indicate factors of 1 that have been removed. For instance, Example 25 might have been done faster as follows:

$$\frac{90}{84} = \frac{2 \cdot 3 \cdot 3 \cdot 5}{2 \cdot 2 \cdot 3 \cdot 7}$$ Factoring the numerator and the denominator

$$= \frac{\not{2} \cdot \not{3} \cdot 3 \cdot 5}{2 \cdot \not{2} \cdot \not{3} \cdot 7}$$ When a factor of 1 is noted, it is "canceled" as shown: $\frac{2 \cdot 3}{2 \cdot 3} = 1$.

$$= \frac{3 \cdot 5}{2 \cdot 7} = \frac{15}{14}.$$

Caution!

The difficulty with canceling is that it is often applied incorrectly in situations like the following:

$\frac{\not{2} + 3}{\not{2}} = 3$; Wrong! $\frac{\not{4} + 1}{\not{4} + 2} = \frac{1}{2}$; Wrong! $\frac{1\not{5}}{\not{5}4} = \frac{1}{4}$. Wrong!

The correct answers are

$\frac{2 + 3}{2} = \frac{5}{2}$; $\frac{4 + 1}{4 + 2} = \frac{5}{6}$; $\frac{15}{54} = \frac{3 \cdot 5}{3 \cdot 18} = \frac{3}{3} \cdot \frac{5}{18} = \frac{5}{18}$.

In each situation, the number canceled was not a factor of 1. Factors are parts of products. For example, in $2 \cdot 3$, 2 and 3 are factors, but in $2 + 3$, 2 and 3 are *not* factors. Canceling may not be done when sums or differences are in numerators or denominators, as shown here. **If you cannot factor, you cannot cancel! If in doubt, do not cancel!**

2.1 EXERCISE SET

For Extra Help

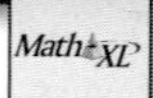
MathXL

MyMathLab

InterAct Math

Math Tutor Center

Video Lectures on CD Disc 1

Student's Solutions Manual

a Identify the numerator and the denominator.

1. $\frac{3}{4}$

2. $\frac{9}{10}$

3. $\frac{11}{2}$

4. $\frac{18}{5}$

5. $\frac{0}{7}$

6. $\frac{1}{13}$

What part of the object or set of objects is shaded?

7.

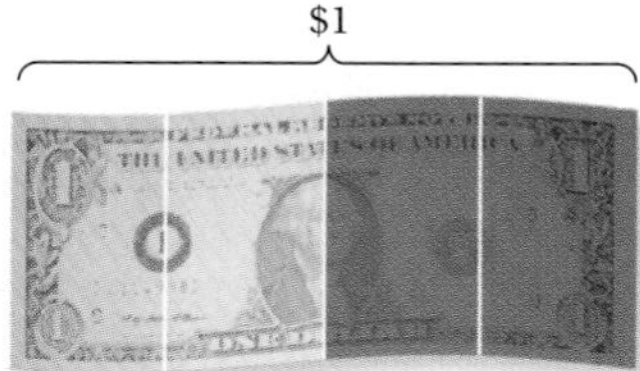

8.

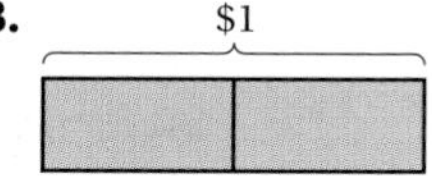

9.

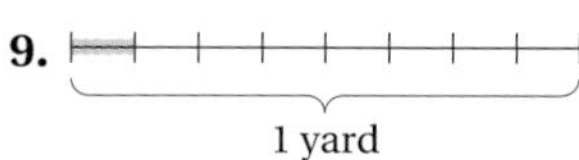

10.

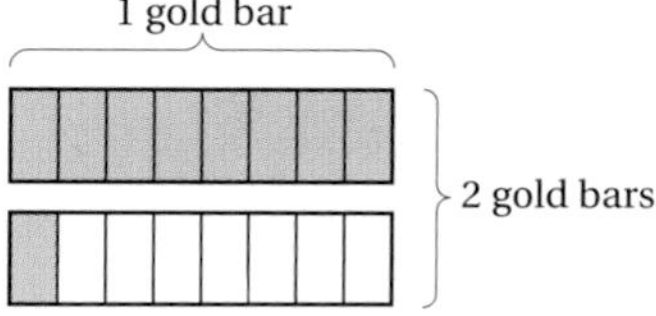

11.

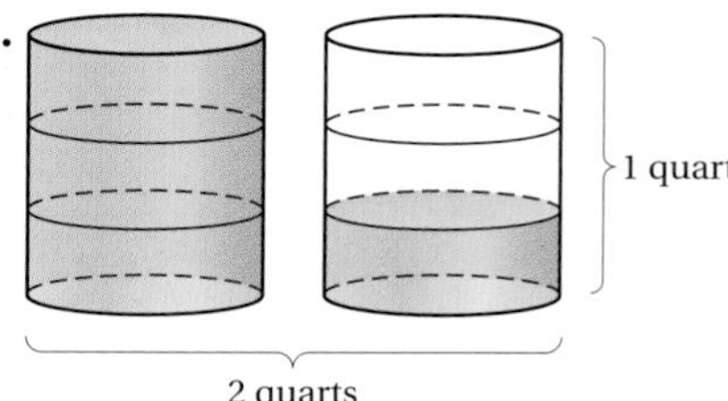

12.

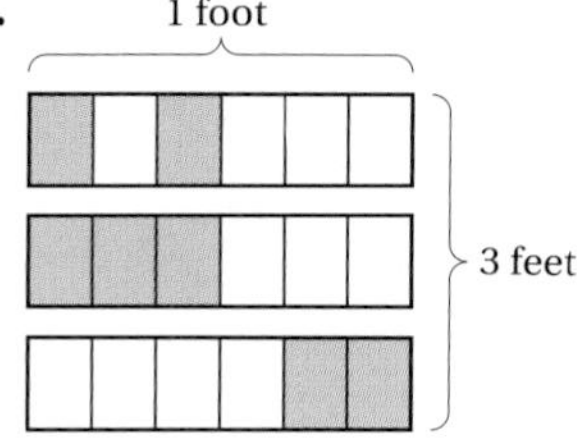

What part of an inch is highlighted?

13.

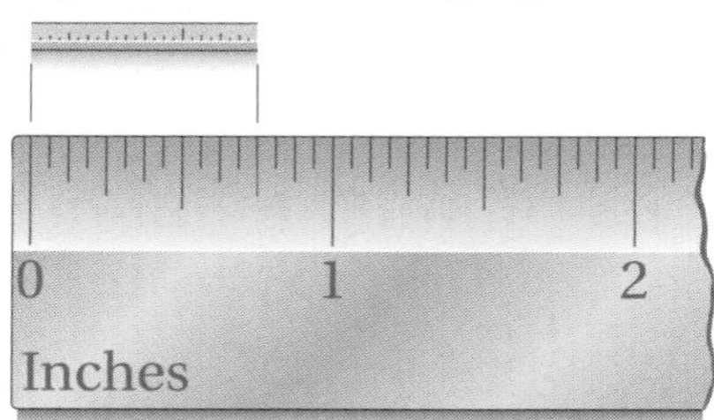

14.

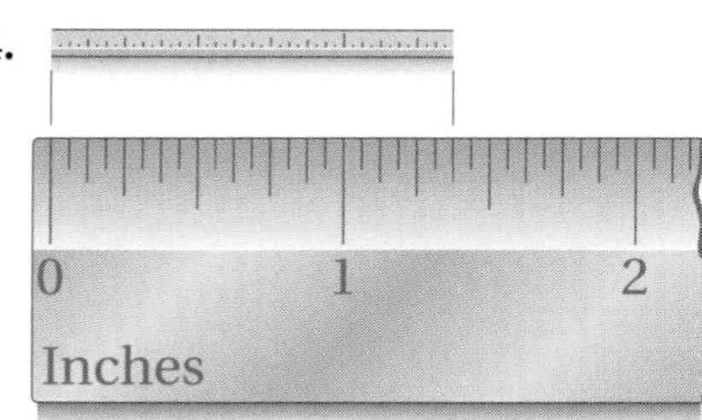

15.

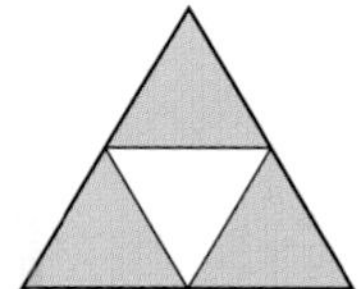

16.

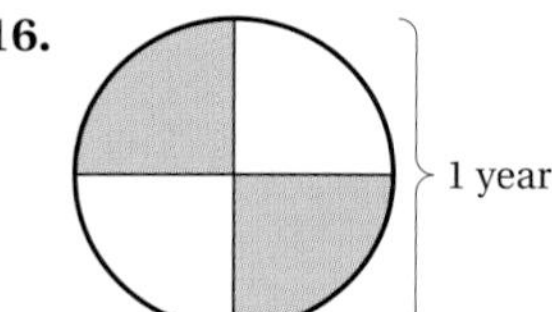

17.

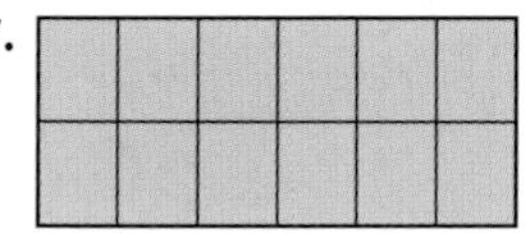

18.

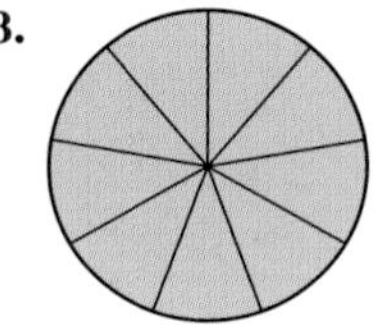

19. 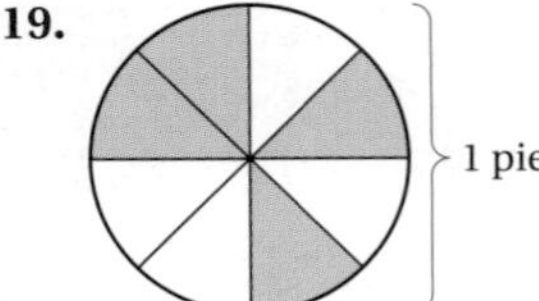

20. 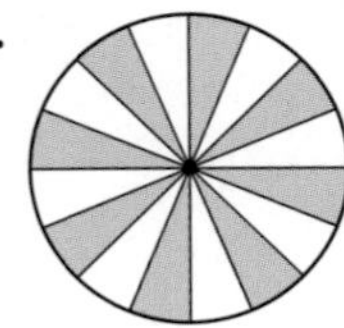

21. 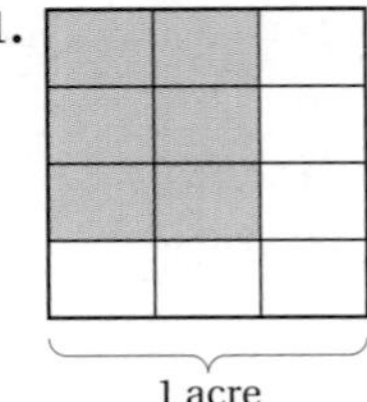

22. 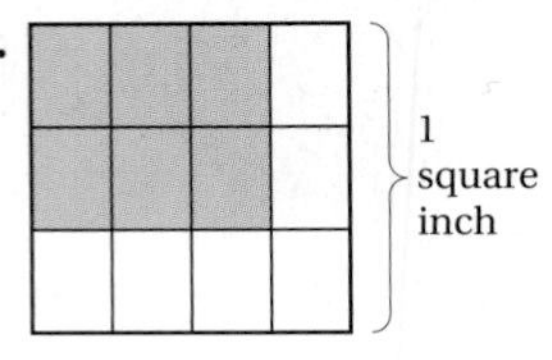

For each of Exercises 23–26, give fraction notation for the amount of gas **(a)** in the tank and **(b)** used from a full tank.

23. 

24. 

25. 

26. 

b Simplify.

27. $\frac{8}{0}$

28. $\frac{25}{1}$

29. $\frac{19}{19}$

30. $\frac{2-2}{8}$

31. $\frac{0}{87}$

32. $\frac{313}{313}$

33. $\frac{403}{1}$

34. $\frac{38}{0}$

35. $\frac{5}{6-6}$

36. $\frac{0}{9-8}$

37. $\frac{56}{56}$

38. $\frac{7}{7}$

39. $\frac{8-8}{1247}$

40. $\frac{20-5}{11-10}$

c Multiply.

41. $\frac{1}{2} \cdot \frac{1}{3}$

42. $\frac{1}{6} \cdot \frac{1}{4}$

43. $5 \times \frac{1}{8}$

44. $4 \times \frac{1}{5}$

45. $\frac{2}{3} \times \frac{1}{5}$

46. $\frac{3}{5} \times \frac{1}{5}$

47. $\frac{2}{5} \cdot \frac{2}{3}$

48. $\frac{3}{4} \cdot \frac{3}{5}$

49. $\frac{3}{4} \cdot \frac{3}{4}$

50. $\frac{3}{7} \cdot \frac{4}{5}$

51. $\frac{2}{3} \cdot \frac{7}{13}$

52. $\frac{3}{11} \cdot \frac{4}{5}$

53. $7 \cdot \frac{3}{4}$

54. $7 \cdot \frac{2}{5}$

55. $\frac{7}{8} \cdot \frac{7}{8}$

56. $\frac{3}{10} \cdot \frac{7}{100}$

d Find another name for the given number, but with the denominator indicated. Use multiplying by 1.

57. $\frac{1}{2} = \frac{?}{10}$

58. $\frac{1}{6} = \frac{?}{18}$

59. $\frac{5}{8} = \frac{?}{32}$

60. $\frac{2}{9} = \frac{?}{18}$

61. $\frac{5}{3} = \frac{?}{45}$

62. $\frac{11}{5} = \frac{?}{30}$

63. $\frac{7}{22} = \frac{?}{132}$

64. $\frac{10}{21} = \frac{?}{126}$

e Simplify.

65. $\frac{6}{8}$

66. $\frac{8}{12}$

67. $\frac{3}{15}$

68. $\frac{8}{10}$

69. $\frac{24}{8}$

70. $\frac{36}{9}$

71. $\frac{18}{24}$

72. $\frac{42}{48}$

73. $\frac{14}{16}$

74. $\frac{15}{25}$

75. $\frac{150}{25}$

76. $\frac{19}{76}$

77. $\frac{17}{51}$

78. $\frac{425}{525}$

79. **DW** Explain in your own words when it *is* possible to "cancel" and when it *is not* possible to "cancel."

80. **DW** On p. 106, we explained, using words and pictures, why $\frac{3}{5} \cdot \frac{3}{4}$ equals $\frac{9}{20}$. Present a similar explanation of why $\frac{2}{3} \cdot \frac{4}{7}$ equals $\frac{8}{21}$.

SYNTHESIS

What part of the object is shaded?

81.

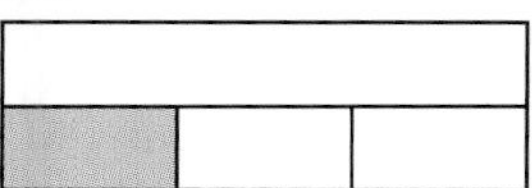

82.

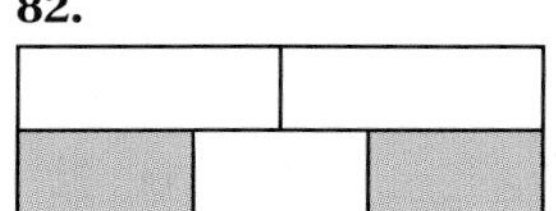

83.

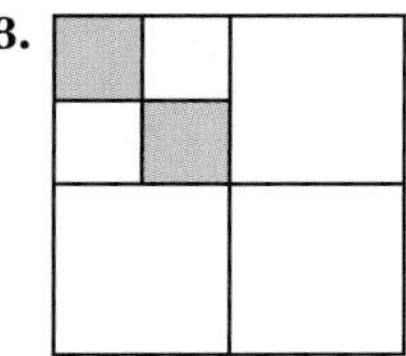

84.

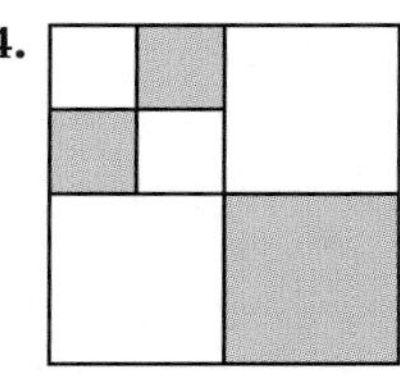

2.2 MULTIPLICATION AND DIVISION

Objectives

a Multiply and simplify using fraction notation.

b Find the reciprocal of a number.

c Divide and simplify using fraction notation.

d Solve equations of the type $a \cdot x = b$ and $x \cdot a = b$, where a and b may be fractions.

To the student:

In the preface, at the front of the text, you will find a Student Organizer card. This pullout card will help you keep track of important dates and useful contact information. You can also use it to plan time for class, study, work, and relaxation. By managing your time wisely, you will provide yourself the best possible opportunity to be successful in this course.

a Multiplying and Simplifying Using Fraction Notation

It is often possible to simplify after we multiply. To make such simplifying easier, it is generally best not to carry out the products in the numerator and the denominator, but to factor and simplify first. Consider the product

$$\frac{3}{8} \cdot \frac{4}{9}.$$

We proceed as follows:

$$\frac{3}{8} \cdot \frac{4}{9} = \frac{3 \cdot 4}{8 \cdot 9}$$ We write the products in the numerator and the denominator, but we do not carry them out.

$$= \frac{3 \cdot 2 \cdot 2}{2 \cdot 2 \cdot 2 \cdot 3 \cdot 3}$$ Factoring the numerator and the denominator

$$= \frac{3 \cdot 2 \cdot 2 \cdot 1}{2 \cdot 2 \cdot 2 \cdot 3 \cdot 3}$$ Using the identity property of 1 to insert the number 1 as a factor

$$= \frac{3 \cdot 2 \cdot 2}{3 \cdot 2 \cdot 2} \cdot \frac{1}{2 \cdot 3}$$ Factoring the fraction

$$= 1 \cdot \frac{1}{2 \cdot 3}$$

$$= \frac{1}{2 \cdot 3}$$ Removing a factor of 1

$$= \frac{1}{6}.$$

The procedure could have been shortened had we noticed that 4 is a factor of the 8 in the denominator:

$$\frac{3}{8} \cdot \frac{4}{9} = \frac{3 \cdot 4}{8 \cdot 9} = \frac{3 \cdot 4}{4 \cdot 2 \cdot 3 \cdot 3} = \frac{3 \cdot 4}{3 \cdot 4} \cdot \frac{1}{2 \cdot 3} = 1 \cdot \frac{1}{2 \cdot 3} = \frac{1}{2 \cdot 3} = \frac{1}{6}.$$

To multiply and simplify:

a) Write the products in the numerator and the denominator, but do not carry out the products.
b) Factor the numerator and the denominator.
c) Factor the fraction to remove a factor of 1.
d) Carry out the remaining products.

EXAMPLES Multiply and simplify.

1. $\frac{2}{3} \cdot \frac{9}{4} = \frac{2 \cdot 9}{3 \cdot 4} = \frac{2 \cdot 3 \cdot 3}{3 \cdot 2 \cdot 2} = \frac{2 \cdot 3}{2 \cdot 3} \cdot \frac{3}{2} = 1 \cdot \frac{3}{2} = \frac{3}{2}$

2. $\frac{6}{7} \cdot \frac{5}{3} = \frac{6 \cdot 5}{7 \cdot 3} = \frac{3 \cdot 2 \cdot 5}{7 \cdot 3} = \frac{3}{3} \cdot \frac{2 \cdot 5}{7} = 1 \cdot \frac{2 \cdot 5}{7} = \frac{2 \cdot 5}{7} = \frac{10}{7}$

3. $40 \cdot \frac{7}{8} = \frac{40 \cdot 7}{8} = \frac{8 \cdot 5 \cdot 7}{8 \cdot 1} = \frac{8}{8} \cdot \frac{5 \cdot 7}{1} = 1 \cdot \frac{5 \cdot 7}{1} = \frac{5 \cdot 7}{1} = 35$

Caution!

Canceling can be used as follows for Examples 1–3.

1. $\frac{2}{3} \cdot \frac{9}{4} = \frac{2 \cdot 9}{3 \cdot 4} = \frac{\not{2} \cdot \not{3} \cdot 3}{\not{3} \cdot \not{2} \cdot 2} = \frac{3}{2}$ Removing a factor of 1: $\frac{2 \cdot 3}{2 \cdot 3} = 1$

2. $\frac{6}{7} \cdot \frac{5}{3} = \frac{6 \cdot 5}{7 \cdot 3} = \frac{\not{3} \cdot 2 \cdot 5}{7 \cdot \not{3}} = \frac{2 \cdot 5}{7} = \frac{10}{7}$ Removing a factor of 1: $\frac{3}{3} = 1$

3. $40 \cdot \frac{7}{8} = \frac{40 \cdot 7}{8} = \frac{\not{8} \cdot 5 \cdot 7}{\not{8} \cdot 1} = \frac{5 \cdot 7}{1} = 35$ Removing a factor of 1: $\frac{8}{8} = 1$

Remember, if you can't factor, you can't cancel!

Do Exercises 1–4.

b Reciprocals

Look at these products:

$$8 \cdot \frac{1}{8} = \frac{8 \cdot 1}{8} = \frac{8}{8} = 1; \qquad \frac{2}{3} \cdot \frac{3}{2} = \frac{2 \cdot 3}{3 \cdot 2} = \frac{6}{6} = 1.$$

RECIPROCALS

If the product of two numbers is 1, we say that they are **reciprocals** of each other. To find a reciprocal of a fraction, interchange the numerator and the denominator.

$$\text{Number} — \frac{3}{4} \rightarrow \frac{4}{3} — \text{Reciprocal}$$

EXAMPLES Find the reciprocal.

4. The reciprocal of $\frac{4}{5}$ is $\frac{5}{4}$. $\frac{4}{5} \cdot \frac{5}{4} = \frac{20}{20} = 1$

5. The reciprocal of $\frac{8}{7}$ is $\frac{7}{8}$. $\frac{8}{7} \cdot \frac{7}{8} = \frac{56}{56} = 1$

6. The reciprocal of 24 is $\frac{1}{24}$. Think of 24 as $\frac{24}{1}$: $\frac{24}{1} \cdot \frac{1}{24} = \frac{24}{24} = 1$.

7. The reciprocal of $\frac{1}{3}$ is 3. $\frac{1}{3} \cdot 3 = \frac{3}{3} = 1$

Do Exercises 5–8.

Multiply and simplify.

1. $\frac{2}{3} \cdot \frac{7}{8}$

2. $\frac{4}{5} \cdot \frac{5}{12}$

3. $16 \cdot \frac{3}{8}$

4. $\frac{5}{8} \cdot 4$

Find the reciprocal.

5. $\frac{2}{5}$

6. $\frac{10}{7}$

7. 9

8. $\frac{1}{5}$

Answers on page A-4

Does 0 have a reciprocal? If it did, it would have to be a number x such that

$$0 \cdot x = 1.$$

But 0 times any number is 0. Thus we have the following.

0 HAS NO RECIPROCAL

The number 0, or $\frac{0}{n}$, has no reciprocal. $\left(\text{Recall that } \frac{n}{0} \text{ is not defined.}\right)$

C Dividing Using Fraction Notation

Consider the division $\frac{3}{4} \div \frac{1}{8}$. We are asking how many $\frac{1}{8}$'s are in $\frac{3}{4}$. We can answer this by looking at the figure below.

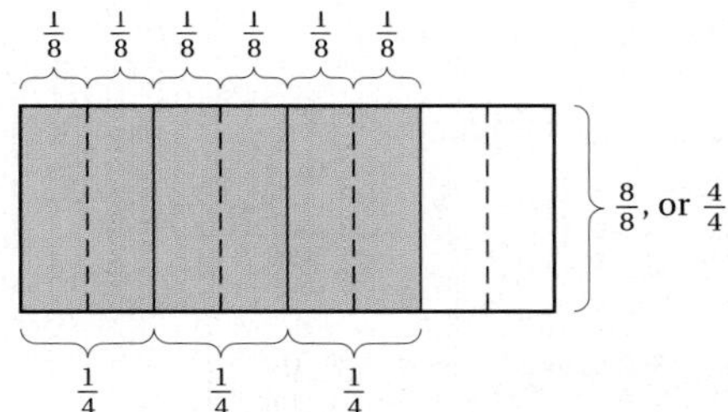

We see that there are six $\frac{1}{8}$'s in $\frac{3}{4}$. Thus,

$$\frac{3}{4} \div \frac{1}{8} = 6.$$

We can check this by multiplying:

$$6 \cdot \frac{1}{8} = \frac{6}{8} = \frac{3}{4}.$$

Here is a faster way to do this division:

$$\frac{3}{4} \div \frac{1}{8} = \frac{3}{4} \cdot \frac{8}{1} = \frac{24}{4} = 6.$$ Multiplying by the reciprocal of the divisor

To divide fractions, multiply the dividend by the reciprocal of the divisor:

$$\frac{2}{5} \div \frac{3}{4} = \frac{2}{5} \cdot \frac{4}{3} = \frac{2 \cdot 4}{5 \cdot 3} = \frac{8}{15}.$$

Multiply by the reciprocal of the divisor.

EXAMPLES Divide and simplify.

8. $\frac{5}{6} \div \frac{2}{3} = \frac{5}{6} \cdot \frac{3}{2} = \frac{5 \cdot 3}{6 \cdot 2} = \frac{5 \cdot 3}{3 \cdot 2 \cdot 2} = \frac{3}{3} \cdot \frac{5}{2 \cdot 2} = \frac{5}{2 \cdot 2} = \frac{5}{4}$

9. $\frac{7}{8} \div \frac{1}{16} = \frac{7}{8} \cdot 16 = \frac{7 \cdot 16}{8} = \frac{7 \cdot 2 \cdot 8}{8 \cdot 1} = \frac{8}{8} \cdot \frac{7 \cdot 2}{1} = \frac{7 \cdot 2}{1} = 14$

10. $\frac{2}{5} \div 6 = \frac{2}{5} \cdot \frac{1}{6} = \frac{2 \cdot 1}{5 \cdot 6} = \frac{2 \cdot 1}{5 \cdot 2 \cdot 3} = \frac{2}{2} \cdot \frac{1}{5 \cdot 3} = \frac{1}{5 \cdot 3} = \frac{1}{15}$

11. $\frac{3}{5} \div \frac{1}{2} = \frac{3}{5} \cdot 2 = \frac{3 \cdot 2}{5} = \frac{6}{5}$

Caution!

Canceling can be used as follows for Examples 8–10.

8. $\frac{5}{6} \div \frac{2}{3} = \frac{5}{6} \cdot \frac{3}{2} = \frac{5 \cdot 3}{6 \cdot 2} = \frac{5 \cdot \cancel{3}}{\cancel{3} \cdot 2 \cdot 2} = \frac{5}{2 \cdot 2} = \frac{5}{4}$ Removing a factor of 1: $\frac{3}{3} = 1$

9. $\frac{7}{8} \div \frac{1}{16} = \frac{7}{8} \cdot 16 = \frac{7 \cdot 16}{8} = \frac{7 \cdot \cancel{8} \cdot 2}{\cancel{8} \cdot 1} = \frac{7 \cdot 2}{1} = 14$ Removing a factor of 1: $\frac{8}{8} = 1$

10. $\frac{2}{5} \div 6 = \frac{2}{5} \cdot \frac{1}{6} = \frac{2 \cdot 1}{5 \cdot 6} = \frac{\cancel{2} \cdot 1}{5 \cdot \cancel{2} \cdot 3} = \frac{1}{5 \cdot 3} = \frac{1}{15}$ Removing a factor of 1: $\frac{2}{2} = 1$

Remember, if you can't factor, you can't cancel!

Do Exercises 9–13.

What is the explanation for multiplying by a reciprocal when dividing? Let's consider $\frac{2}{3} \div \frac{7}{5}$. We multiply by 1. The name for 1 that we will use is (5/7)/(5/7); it comes from the reciprocal of $\frac{7}{5}$.

$\frac{2}{3} \div \frac{7}{5} = \frac{\frac{2}{3}}{\frac{7}{5}}$ Writing fraction notation for the division

$= \frac{\frac{2}{3}}{\frac{7}{5}} \cdot 1$ Multiplying by 1

$= \frac{\frac{2}{3}}{\frac{7}{5}} \cdot \frac{\frac{5}{7}}{\frac{5}{7}}$ Multiplying by 1; $\frac{5}{7}$ is the reciprocal of $\frac{7}{5}$ and $\frac{\frac{5}{7}}{\frac{5}{7}} = 1$

$= \frac{\frac{2}{3} \cdot \frac{5}{7}}{\frac{7}{5} \cdot \frac{5}{7}}$ Multiplying the numerators and the denominators

$= \frac{\frac{2}{3} \cdot \frac{5}{7}}{1}$ After we multiplied, we got 1 for the denominator. The numerator shows the multiplication by the reciprocal.

$= \frac{2}{3} \cdot \frac{5}{7} = \frac{10}{21}$

Divide and simplify.

9. $\frac{6}{7} \div \frac{3}{4}$

10. $\frac{2}{3} \div \frac{1}{4}$

11. $\frac{4}{5} \div 8$

12. $60 \div \frac{3}{5}$

13. $\frac{3}{5} \div \frac{3}{5}$

Answers on page A-4

14. Divide by multiplying by 1:

$$\frac{\frac{4}{5}}{\frac{6}{7}}.$$

Solve.

15. $\frac{5}{6} \cdot y = \frac{2}{3}$

16. $\frac{3}{4} \cdot n = 24$

Answers on page A-4

Thus,

$$\frac{2}{3} \div \frac{7}{5} = \frac{2}{3} \cdot \frac{5}{7} = \frac{10}{21}.$$

Do Exercise 14.

d Solving Equations

Now let's solve equations $a \cdot x = b$ and $x \cdot a = b$, where a and b may be fractions. We proceed as we did with equations involving whole numbers. We divide by a on both sides.

EXAMPLE 12 Solve: $\frac{4}{3} \cdot x = \frac{6}{7}$.

We have

$$\frac{4}{3} \cdot x = \frac{6}{7}$$

$$\frac{\frac{4}{3} \cdot x}{\frac{4}{3}} = \frac{\frac{6}{7}}{\frac{4}{3}} \qquad \text{Dividing by } \tfrac{4}{3} \text{ on both sides}$$

$$x = \frac{6}{7} \cdot \frac{3}{4} \qquad \text{Multiplying by the reciprocal}$$

$$= \frac{6 \cdot 3}{7 \cdot 4} = \frac{2 \cdot 3 \cdot 3}{7 \cdot 2 \cdot 2} = \frac{2}{2} \cdot \frac{3 \cdot 3}{7 \cdot 2} = \frac{3 \cdot 3}{7 \cdot 2} = \frac{9}{14}.$$

The solution is $\frac{9}{14}$.

EXAMPLE 13 Solve: $t \cdot \frac{4}{5} = 80$.

Dividing by $\frac{4}{5}$ on both sides, we get

$$t = 80 \div \frac{4}{5} = 80 \cdot \frac{5}{4}$$

$$= \frac{80 \cdot 5}{4} = \frac{4 \cdot 20 \cdot 5}{4 \cdot 1} = \frac{4}{4} \cdot \frac{20 \cdot 5}{1} = \frac{20 \cdot 5}{1} = 100.$$

The solution is 100.

Do Exercises 15 and 16.

2.2 EXERCISE SET

For Extra Help

MathXL | MyMathLab | InterAct Math | Math Tutor Center | Video Lectures on CD Disc 1 | Student's Solutions Manual

a Multiply and simplify.

Don't forget to simplify!

1. $\frac{2}{3} \cdot \frac{1}{2}$

2. $\frac{3}{8} \cdot \frac{1}{3}$

3. $\frac{1}{4} \cdot \frac{2}{3}$

4. $\frac{4}{6} \cdot \frac{1}{6}$

5. $\frac{12}{5} \cdot \frac{9}{8}$

6. $\frac{16}{15} \cdot \frac{5}{4}$

7. $\frac{10}{9} \cdot \frac{7}{5}$

8. $\frac{25}{12} \cdot \frac{4}{3}$

9. $9 \cdot \frac{1}{9}$

10. $4 \cdot \frac{1}{4}$

11. $\frac{7}{5} \cdot \frac{5}{7}$

12. $\frac{2}{11} \cdot \frac{11}{2}$

13. $24 \cdot \frac{1}{6}$

14. $16 \cdot \frac{1}{2}$

15. $12 \cdot \frac{3}{4}$

16. $18 \cdot \frac{5}{6}$

17. $\frac{7}{10} \cdot 28$

18. $\frac{5}{8} \cdot 34$

19. $240 \cdot \frac{1}{8}$

20. $150 \cdot \frac{1}{5}$

21. $\frac{4}{10} \cdot \frac{5}{10}$

22. $\frac{7}{10} \cdot \frac{34}{150}$

23. $\frac{8}{10} \cdot \frac{45}{100}$

24. $\frac{3}{10} \cdot \frac{8}{10}$

25. $\frac{11}{24} \cdot \frac{3}{5}$

26. $\frac{15}{22} \cdot \frac{4}{7}$

27. $\frac{10}{21} \cdot \frac{3}{4}$

28. $\frac{17}{18} \cdot \frac{3}{5}$

b Find the reciprocal.

29. $\frac{5}{6}$

30. $\frac{7}{8}$

31. 6

32. 4

33. $\frac{1}{6}$

34. $\frac{1}{4}$

35. $\frac{10}{3}$

36. $\frac{17}{4}$

c Divide and simplify.

Don't forget to simplify!

37. $\frac{3}{5} \div \frac{3}{4}$

38. $\frac{2}{3} \div \frac{3}{4}$

39. $\frac{3}{5} \div \frac{9}{4}$

40. $\frac{6}{7} \div \frac{3}{5}$

41. $\frac{4}{3} \div \frac{1}{3}$

42. $\frac{10}{9} \div \frac{1}{3}$

43. $\frac{1}{3} \div \frac{1}{6}$

44. $\frac{1}{4} \div \frac{1}{5}$

45. $\frac{3}{8} \div 3$

46. $\frac{5}{6} \div 5$

47. $\frac{12}{7} \div 4$

48. $\frac{18}{5} \div 2$

49. $12 \div \frac{3}{2}$

50. $24 \div \frac{3}{8}$

51. $28 \div \frac{4}{5}$

52. $40 \div \frac{2}{3}$

53. $\frac{5}{8} \div \frac{5}{8}$

54. $\frac{2}{5} \div \frac{2}{5}$

55. $\frac{8}{15} \div \frac{4}{5}$

56. $\frac{6}{13} \div \frac{3}{26}$

57. $\frac{9}{5} \div \frac{4}{5}$

58. $\frac{5}{12} \div \frac{25}{36}$

59. $120 \div \frac{5}{6}$

60. $360 \div \frac{8}{7}$

d Solve.

61. $\frac{4}{5} \cdot x = 60$

62. $\frac{3}{2} \cdot t = 90$

63. $\frac{5}{3} \cdot y = \frac{10}{3}$

64. $\frac{4}{9} \cdot m = \frac{8}{3}$

65. $x \cdot \frac{25}{36} = \frac{5}{12}$

66. $p \cdot \frac{4}{5} = \frac{8}{15}$

67. $n \cdot \frac{8}{7} = 360$

68. $y \cdot \frac{5}{6} = 120$

69. $\mathbf{D_W}$ Without performing the division, explain why $5 \div \frac{1}{7}$ is a greater number than $5 \div \frac{2}{3}$.

70. $\mathbf{D_W}$ A student incorrectly insists that $\frac{2}{5} \div \frac{3}{4}$ is $\frac{15}{8}$. What mistake is he probably making?

SKILL MAINTENANCE

Divide. [1.3d]

71. $268 \div 4$

72. $268 \div 8$

73. $6842 \div 24$

74. $8765 \div 85$

Solve. [1.4b]

75. $4 \cdot x = 268$

76. $4 + x = 268$

77. $y + 502 = 9001$

78. $56 \cdot 78 = T$

SYNTHESIS

79. If $\frac{1}{3}$ of a number is $\frac{1}{4}$, what is $\frac{1}{2}$ of the number?

80. $\left(\frac{9}{10} \div \frac{2}{5} \div \frac{3}{8}\right)^2$

2.3 ADDITION AND SUBTRACTION; ORDER

Objectives

a Add using fraction notation.

b Subtract using fraction notation.

c Use $<$ or $>$ with fraction notation to write a true sentence.

d Solve equations of the type $x + a = b$ and $a + x = b$, where a and b may be fractions.

a Addition Using Fraction Notation

LIKE DENOMINATORS

Addition using fraction notation corresponds to combining or putting like things together, just as addition with whole numbers does. For example,

We combine two sets, each of which consists of fractional parts of one object that are the same size.

This is the resulting set.

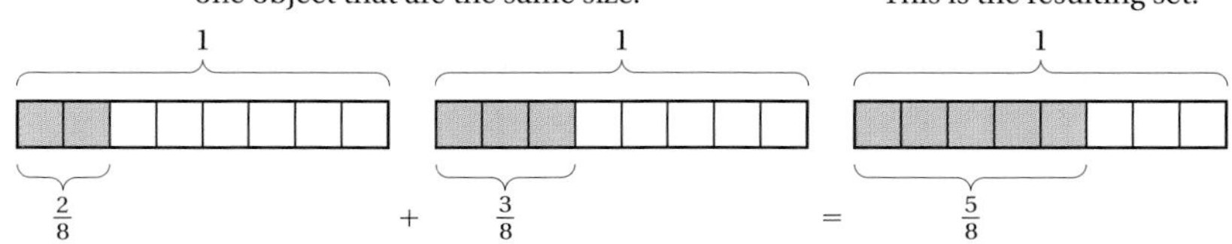

$$2 \text{ eighths} + 3 \text{ eighths} = 5 \text{ eighths},$$

or $\quad 2 \cdot \frac{1}{8} + 3 \cdot \frac{1}{8} = 5 \cdot \frac{1}{8}, \quad$ or $\quad \frac{2}{8} + \frac{3}{8} = \frac{5}{8}.$

We see that to add when denominators are the same, we add the numerators and keep the denominator.

Do Exercise 1.

1. Find $\frac{1}{5} + \frac{3}{5}$.

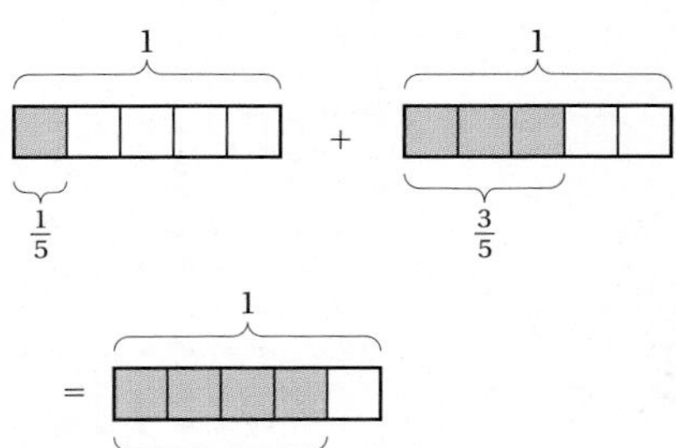

To add when denominators are the same,

a) add the numerators,

b) keep the denominator, and

c) simplify, if possible.

$$\frac{2}{6} + \frac{5}{6} = \frac{2+5}{6} = \frac{7}{6}$$

EXAMPLES Add and simplify.

1. $\frac{2}{4} + \frac{1}{4} = \frac{2+1}{4} = \frac{3}{4}$ No simplifying is possible.

2. $\frac{11}{6} + \frac{3}{6} = \frac{11+3}{6} = \frac{14}{6} = \frac{2 \cdot 7}{2 \cdot 3} = \frac{2}{2} \cdot \frac{7}{3} = 1 \cdot \frac{7}{3} = \frac{7}{3}$ Here we simplified.

3. $\frac{3}{12} + \frac{5}{12} = \frac{3+5}{12} = \frac{8}{12} = \frac{4 \cdot 2}{4 \cdot 3} = \frac{4}{4} \cdot \frac{2}{3} = 1 \cdot \frac{2}{3} = \frac{2}{3}$

Do Exercises 2–4.

Add and simplify.

2. $\frac{1}{3} + \frac{2}{3}$

3. $\frac{5}{12} + \frac{1}{12}$

4. $\frac{9}{16} + \frac{3}{16}$

DIFFERENT DENOMINATORS

What do we do when denominators are different? We can find a common denominator by multiplying by 1. Consider adding $\frac{1}{6}$ and $\frac{3}{4}$. There are many common denominators that can be obtained. Let's look at two possibilities.

Answers on page A-5

5. Add. (Find the least common denominator.)

$$\frac{2}{3} + \frac{1}{6}$$

6. Add: $\frac{3}{8} + \frac{5}{6}$.

A. $\frac{1}{6} + \frac{3}{4} = \frac{1}{6} \cdot 1 + \frac{3}{4} \cdot 1$

$= \frac{1}{6} \cdot \frac{4}{4} + \frac{3}{4} \cdot \frac{6}{6}$

$= \frac{4}{24} + \frac{18}{24}$

$= \frac{22}{24}$

$= \frac{11}{12}$

B. $\frac{1}{6} + \frac{3}{4} = \frac{1}{6} \cdot 1 + \frac{3}{4} \cdot 1$

$= \frac{1}{6} \cdot \frac{2}{2} + \frac{3}{4} \cdot \frac{3}{3}$

$= \frac{2}{12} + \frac{9}{12}$

$= \frac{11}{12}$

We had to simplify in (A). We didn't have to simplify in (B). In (B), we used the least common multiple of the denominators, 12. That number is called the **least common denominator, or LCD.** As we will see in Example 6, we may still need to simplify when using the LCD, but it is usually easier.

To add when denominators are different:

a) Find the least common multiple of the denominators. That number is the least common denominator, LCD.
b) Multiply by 1, using an appropriate notation, n/n, to express each number in terms of the LCD.
c) Add the numerators, keeping the same denominator.
d) Simplify, if possible.

EXAMPLE 4 Add: $\frac{3}{4} + \frac{1}{8}$.

The LCD is 8. 4 is a factor of 8, so the LCM of 4 and 8 is 8.

$\frac{3}{4} + \frac{1}{8} = \frac{3}{4} \cdot 1 + \frac{1}{8}$ ← This fraction already has the LCD as its denominator.

$= \frac{3}{4} \cdot \frac{2}{2} + \frac{1}{8}$ *Think*: $4 \times \square = 8$. The answer is 2, so we multiply by 1, using $\frac{2}{2}$.

$= \frac{6}{8} + \frac{1}{8} = \frac{7}{8}$

Do Exercise 5.

EXAMPLE 5 Add: $\frac{1}{9} + \frac{5}{6}$.

The LCD is 18. $9 = 3 \cdot 3$ and $6 = 2 \cdot 3$, so the LCM of 9 and 6 is $2 \cdot 3 \cdot 3$, or 18.

$\frac{1}{9} + \frac{5}{6} = \frac{1}{9} \cdot 1 + \frac{5}{6} \cdot 1 = \frac{1}{9} \cdot \frac{2}{2} + \frac{5}{6} \cdot \frac{3}{3}$

Think: $6 \times \square = 18$. The answer is 3, so we multiply by 1 using $\frac{3}{3}$.

Think: $9 \times \square = 18$. The answer is 2, so we multiply by 1 using $\frac{2}{2}$.

$= \frac{2}{18} + \frac{15}{18} = \frac{17}{18}$

Do Exercise 6.

Answers on page A-5

EXAMPLE 6 Add: $\frac{5}{9} + \frac{11}{18}$.

The LCD is 18.

$$\frac{5}{9} + \frac{11}{18} = \frac{5}{9} \cdot \frac{2}{2} + \frac{11}{18} = \frac{10}{18} + \frac{11}{18}$$

$$= \frac{21}{18} = \frac{3 \cdot 7}{3 \cdot 6} = \frac{3}{3} \cdot \frac{7}{6} = \frac{7}{6} \quad \text{Simplifying}$$

Do Exercise 7.

7. Add: $\frac{1}{6} + \frac{7}{18}$.

EXAMPLE 7 Add: $\frac{1}{10} + \frac{3}{100} + \frac{7}{1000}$.

Since 10 and 100 are factors of 1000, the LCD is 1000. Then

$$\frac{1}{10} + \frac{3}{100} + \frac{7}{1000} = \frac{1}{10} \cdot \frac{100}{100} + \frac{3}{100} \cdot \frac{10}{10} + \frac{7}{1000}$$

$$= \frac{100}{1000} + \frac{30}{1000} + \frac{7}{1000} = \frac{137}{1000}.$$

Do Exercise 8.

8. Add: $\frac{4}{10} + \frac{1}{100} + \frac{3}{1000}$.

When denominators are large, we most often use the prime factorization of each denominator to find the LCD. This is shown in Example 8. Using the prime factorization in this manner is similar to what is done in algebra.

EXAMPLE 8 Add: $\frac{13}{70} + \frac{11}{21} + \frac{6}{15}$.

We have

$$\frac{13}{70} + \frac{11}{21} + \frac{6}{15} = \frac{13}{2 \cdot 5 \cdot 7} + \frac{11}{3 \cdot 7} + \frac{6}{3 \cdot 5}. \quad \text{Factoring denominators}$$

The LCD is $2 \cdot 3 \cdot 5 \cdot 7$, or 210. Then

$$\frac{13}{70} + \frac{11}{21} + \frac{6}{15} = \frac{13}{2 \cdot 5 \cdot 7} \cdot \frac{3}{3} + \frac{11}{3 \cdot 7} \cdot \frac{2 \cdot 5}{2 \cdot 5} + \frac{6}{3 \cdot 5} \cdot \frac{7 \cdot 2}{7 \cdot 2}$$

The LCM of 70, 21, and 15 is $2 \cdot 3 \cdot 5 \cdot 7$. In each case, think of which factors are needed to get the LCD. Then multiply by 1 to obtain the LCD in each denominator.

$$= \frac{13 \cdot 3}{2 \cdot 5 \cdot 7 \cdot 3} + \frac{11 \cdot 2 \cdot 5}{3 \cdot 7 \cdot 2 \cdot 5} + \frac{6 \cdot 7 \cdot 2}{3 \cdot 5 \cdot 7 \cdot 2}$$

$$= \frac{39}{3 \cdot 5 \cdot 7 \cdot 2} + \frac{110}{3 \cdot 5 \cdot 7 \cdot 2} + \frac{84}{3 \cdot 5 \cdot 7 \cdot 2}$$

$$= \frac{233}{3 \cdot 5 \cdot 7 \cdot 2}$$

$$= \frac{233}{210}. \quad \text{We left 210 factored until we knew we could not simplify.}$$

Do Exercises 9 and 10.

Add.

9. $\frac{7}{10} + \frac{2}{21} + \frac{1}{7}$

10. $\frac{7}{18} + \frac{5}{24} + \frac{11}{36}$

Answers on page A-5

Subtract and simplify.

11. $\frac{7}{8} - \frac{3}{8}$

12. $\frac{10}{16} - \frac{4}{16}$

13. $\frac{8}{10} - \frac{3}{10}$

Answers on page A-5

b Subtraction Using Fraction Notation

LIKE DENOMINATORS

We can consider the difference $\frac{4}{8} - \frac{3}{8}$ as we did before, as either "take away" or "how many more." Let's consider "take away."

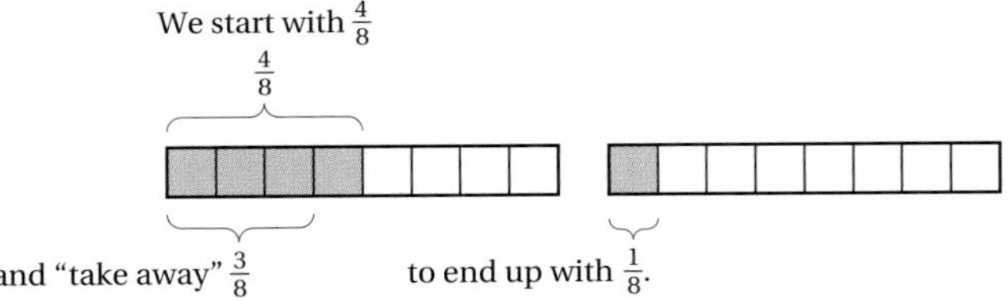

We start with $\frac{4}{8}$
$\frac{4}{8}$
and "take away" $\frac{3}{8}$
to end up with $\frac{1}{8}$.

We start with 4 eighths and take away 3 eighths:

4 eighths − 3 eighths = 1 eighth,

or $\quad 4 \cdot \frac{1}{8} - 3 \cdot \frac{1}{8} = \frac{1}{8}, \quad$ or $\quad \frac{4}{8} - \frac{3}{8} = \frac{1}{8}.$

To subtract when denominators are the same,

a) subtract the numerators,

b) keep the denominator, and

c) simplify, if possible.

$$\frac{7}{10} - \frac{4}{10} = \frac{7-4}{10} = \frac{3}{10}$$

EXAMPLES Subtract and simplify.

9. $\frac{7}{10} - \frac{3}{10} = \frac{7-3}{10} = \frac{4}{10} = \frac{2 \cdot 2}{5 \cdot 2} = \frac{2}{5} \cdot \frac{2}{2} = \frac{2}{5} \cdot 1 = \frac{2}{5}$

10. $\frac{8}{9} - \frac{2}{9} = \frac{8-2}{9} = \frac{6}{9} = \frac{2 \cdot 3}{3 \cdot 3} = \frac{2}{3} \cdot \frac{3}{3} = \frac{2}{3} \cdot 1 = \frac{2}{3}$

11. $\frac{32}{12} - \frac{25}{12} = \frac{32 - 25}{12} = \frac{7}{12}$

Do Exercises 11–13.

DIFFERENT DENOMINATORS

The procedure for subtraction with different denominators is similar to the procedure for addition.

To subtract when denominators are different:

a) Find the least common multiple of the denominators. That number is the least common denominator, LCD.
b) Multiply by 1, using an appropriate notation, n/n, to express each number in terms of the LCD.
c) Subtract the numerators, keeping the same denominator.
d) Simplify, if possible.

EXAMPLE 12 Subtract: $\frac{5}{6} - \frac{7}{12}$.

Since 12 is a multiple of 6, the LCM of 6 and 12 is 12. The LCD is 12.

$$\frac{5}{6} - \frac{7}{12} = \frac{5}{6} \cdot \frac{2}{2} - \frac{7}{12}$$
$$= \frac{10}{12} - \frac{7}{12} = \frac{10 - 7}{12} = \frac{3}{12}$$
$$= \frac{3 \cdot 1}{3 \cdot 4} = \frac{3}{3} \cdot \frac{1}{4} = \frac{1}{4}$$

Do Exercises 14 and 15.

EXAMPLE 13 Subtract: $\frac{17}{24} - \frac{4}{15}$.

We have

$$\frac{17}{24} - \frac{4}{15} = \frac{17}{3 \cdot 2 \cdot 2 \cdot 2} - \frac{4}{5 \cdot 3}.$$

The LCD is $3 \cdot 2 \cdot 2 \cdot 2 \cdot 5$, or 120. Then

$$\frac{17}{24} - \frac{4}{15} = \frac{17}{3 \cdot 2 \cdot 2 \cdot 2} \cdot \frac{5}{5} - \frac{4}{5 \cdot 3} \cdot \frac{2 \cdot 2 \cdot 2}{2 \cdot 2 \cdot 2}$$

The LCM of 24 and 15 is $2 \cdot 2 \cdot 2 \cdot 3 \cdot 5$. In each case, we multiply by 1 to obtain the LCD.

$$= \frac{17 \cdot 5}{3 \cdot 2 \cdot 2 \cdot 2 \cdot 5} - \frac{4 \cdot 2 \cdot 2 \cdot 2}{5 \cdot 3 \cdot 2 \cdot 2 \cdot 2}$$
$$= \frac{85}{120} - \frac{32}{120} = \frac{53}{120}.$$

Do Exercise 16.

C Order

We see from this figure that $\frac{4}{5}$ is greater than $\frac{3}{5}$, and $\frac{3}{5}$ is less than $\frac{4}{5}$. That is, $\frac{4}{5} > \frac{3}{5}$, and $\frac{3}{5} < \frac{4}{5}$.

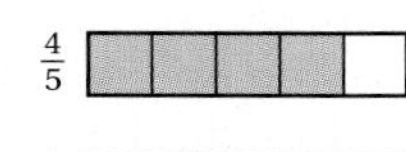

To determine which of two numbers is greater when there is a common denominator, compare the numerators:

$$\frac{4}{5}, \frac{3}{5}; \quad 4 > 3; \quad \frac{4}{5} > \frac{3}{5}.$$

Do Exercises 17 and 18.

Subtract.

14. $\frac{5}{6} - \frac{1}{9}$

15. $\frac{4}{5} - \frac{3}{10}$

16. Subtract: $\frac{11}{28} - \frac{5}{16}$.

17. Use $<$ or $>$ for $\square$ to write a true sentence:

$$\frac{3}{8} \square \frac{5}{8}.$$

18. Use $<$ or $>$ for $\square$ to write a true sentence:

$$\frac{7}{10} \square \frac{6}{10}.$$

Answers on page A-5

Use < or > for $\square$ to write a true sentence.

19. $\frac{2}{3} \square \frac{5}{8}$

20. $\frac{3}{4} \square \frac{8}{12}$

21. $\frac{5}{6} \square \frac{7}{8}$

Solve.

22. $x + \frac{2}{3} = \frac{5}{6}$

23. $\frac{3}{5} + t = \frac{7}{8}$

Answers on page A-5

When denominators are different, we cannot compare numerators. We multiply by 1 to make the denominators the same.

EXAMPLE 14 Use < or > for $\square$ to write a true sentence: $\frac{2}{5} \square \frac{3}{4}$.

The LCD is 20. We have

$$\frac{2}{5} \cdot \frac{4}{4} = \frac{8}{20}; \qquad \text{We multiply by 1 using } \frac{4}{4} \text{ to get the LCD.}$$

$$\frac{3}{4} \cdot \frac{5}{5} = \frac{15}{20}. \qquad \text{We multiply by 1 using } \frac{5}{5} \text{ to get the LCD.}$$

Now that the denominators are the same, 20, we can compare the numerators. Since $8 < 15$, it follows that $\frac{8}{20} < \frac{15}{20}$, so

$$\frac{2}{5} < \frac{3}{4}.$$

EXAMPLE 15 Use < or > for $\square$ to write a true sentence: $\frac{9}{10} \square \frac{89}{100}$.

The LCD is 100. We write $\frac{9}{10}$ with a denominator of 100 to make the denominators the same.

$$\frac{9}{10} \cdot \frac{10}{10} = \frac{90}{100} \qquad \text{We multiply by } \frac{10}{10} \text{ to get the LCD.}$$

Since $90 > 89$, it follows that $\frac{90}{100} > \frac{89}{100}$, so

$$\frac{9}{10} > \frac{89}{100}.$$

Do Exercises 19–21.

d Solving Equations

Now let's solve equations of the form $x + a = b$ or $a + x = b$, where a and b may be fractions. Proceeding as we have before, we subtract a on both sides of the equation.

EXAMPLE 16 Solve: $x + \frac{1}{4} = \frac{3}{5}$.

$$x + \frac{1}{4} - \frac{1}{4} = \frac{3}{5} - \frac{1}{4} \qquad \text{Subtracting } \frac{1}{4} \text{ on both sides}$$

$$x + 0 = \frac{3}{5} \cdot \frac{4}{4} - \frac{1}{4} \cdot \frac{5}{5} \qquad \text{The LCD is 20. We multiply by 1 to get the LCD.}$$

$$x = \frac{12}{20} - \frac{5}{20}$$

$$x = \frac{7}{20}$$

Do Exercises 22 and 23.

2.3 EXERCISE SET

For Extra Help

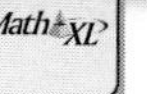

MathXL | MyMathLab | InterAct Math | Math Tutor Center | Video Lectures on CD Disc 1 | Student's Solutions Manual

a Add and simplify.

1. $\frac{7}{8} + \frac{1}{8}$

2. $\frac{2}{5} + \frac{3}{5}$

3. $\frac{1}{8} + \frac{5}{8}$

4. $\frac{3}{10} + \frac{3}{10}$

5. $\frac{2}{3} + \frac{5}{6}$

6. $\frac{5}{6} + \frac{1}{9}$

7. $\frac{1}{8} + \frac{1}{6}$

8. $\frac{1}{6} + \frac{3}{4}$

9. $\frac{4}{5} + \frac{7}{10}$

10. $\frac{3}{4} + \frac{1}{12}$

11. $\frac{5}{12} + \frac{3}{8}$

12. $\frac{7}{8} + \frac{1}{16}$

13. $\frac{3}{20} + \frac{3}{4}$

14. $\frac{2}{15} + \frac{2}{5}$

15. $\frac{5}{6} + \frac{7}{9}$

16. $\frac{5}{8} + \frac{5}{6}$

17. $\frac{3}{10} + \frac{1}{100}$

18. $\frac{9}{10} + \frac{3}{100}$

19. $\frac{5}{12} + \frac{4}{15}$

20. $\frac{3}{16} + \frac{1}{12}$

21. $\frac{9}{10} + \frac{99}{100}$

22. $\frac{3}{10} + \frac{27}{100}$

23. $\frac{7}{8} + \frac{0}{1}$

24. $\frac{0}{1} + \frac{5}{6}$

25. $\frac{3}{8} + \frac{1}{6}$

26. $\frac{5}{8} + \frac{1}{6}$

27. $\frac{5}{12} + \frac{7}{24}$

28. $\frac{1}{18} + \frac{7}{12}$

29. $\frac{3}{16} + \frac{5}{16} + \frac{4}{16}$

30. $\frac{3}{8} + \frac{1}{8} + \frac{2}{8}$

31. $\frac{8}{10} + \frac{7}{100} + \frac{4}{1000}$

32. $\frac{1}{10} + \frac{2}{100} + \frac{3}{1000}$

33. $\frac{3}{8} + \frac{5}{12} + \frac{8}{15}$

34. $\frac{1}{2} + \frac{3}{8} + \frac{1}{4}$

35. $\frac{15}{24} + \frac{7}{36} + \frac{91}{48}$

36. $\frac{5}{7} + \frac{25}{52} + \frac{7}{4}$

b Subtract and simplify.

37. $\frac{5}{6} - \frac{1}{6}$

38. $\frac{5}{8} - \frac{3}{8}$

39. $\frac{11}{12} - \frac{2}{12}$

40. $\frac{17}{18} - \frac{11}{18}$

41. $\frac{3}{4} - \frac{1}{8}$

42. $\frac{2}{3} - \frac{1}{9}$

43. $\frac{1}{8} - \frac{1}{12}$

44. $\frac{1}{6} - \frac{1}{8}$

45. $\frac{4}{3} - \frac{5}{6}$

46. $\frac{7}{8} - \frac{1}{16}$

47. $\frac{3}{4} - \frac{3}{28}$

48. $\frac{2}{5} - \frac{2}{15}$

49. $\frac{3}{4} - \frac{3}{20}$

50. $\frac{5}{6} - \frac{1}{2}$

51. $\frac{3}{4} - \frac{1}{20}$

52. $\frac{3}{4} - \frac{4}{16}$

53. $\frac{5}{12} - \frac{2}{15}$

54. $\frac{9}{10} - \frac{11}{16}$

55. $\frac{6}{10} - \frac{7}{100}$

56. $\frac{9}{10} - \frac{3}{100}$

57. $\frac{7}{15} - \frac{3}{25}$

58. $\frac{18}{25} - \frac{4}{35}$

59. $\frac{99}{100} - \frac{9}{10}$

60. $\frac{78}{100} - \frac{11}{20}$

61. $\frac{2}{3} - \frac{1}{8}$

62. $\frac{3}{4} - \frac{1}{2}$

63. $\frac{3}{5} - \frac{1}{2}$

64. $\frac{5}{6} - \frac{2}{3}$

65. $\frac{5}{12} - \frac{3}{8}$

66. $\frac{7}{12} - \frac{2}{9}$

67. $\frac{7}{8} - \frac{1}{16}$

68. $\frac{5}{12} - \frac{5}{16}$

69. $\frac{17}{25} - \frac{4}{15}$

70. $\frac{11}{18} - \frac{7}{24}$

71. $\frac{23}{25} - \frac{112}{150}$

72. $\frac{89}{90} - \frac{53}{120}$

c Use < or > for □ to write a true sentence.

73. $\frac{5}{8}$ □ $\frac{6}{8}$

74. $\frac{7}{9}$ □ $\frac{5}{9}$

75. $\frac{1}{3}$ □ $\frac{1}{4}$

76. $\frac{1}{8}$ □ $\frac{1}{6}$

77. $\frac{2}{3}$ □ $\frac{5}{7}$

78. $\frac{3}{5}$ □ $\frac{4}{7}$

79. $\frac{4}{5}$ □ $\frac{5}{6}$

80. $\frac{3}{2}$ □ $\frac{7}{5}$

81. $\frac{19}{20}$ □ $\frac{4}{5}$

82. $\frac{5}{6}$ □ $\frac{13}{16}$

83. $\frac{19}{20}$ □ $\frac{9}{10}$

84. $\frac{3}{4}$ □ $\frac{11}{15}$

85. $\frac{31}{21}$ □ $\frac{41}{13}$

86. $\frac{12}{7}$ □ $\frac{132}{49}$

d Solve.

87. $x + \frac{1}{30} = \frac{1}{10}$

88. $y + \frac{9}{12} = \frac{11}{12}$

89. $\frac{2}{3} + t = \frac{4}{5}$

90. $\frac{2}{3} + p = \frac{7}{8}$

91. $x + \frac{1}{3} = \frac{5}{6}$

92. $m + \frac{5}{6} = \frac{9}{10}$

93. D_W A fellow student made the following error:

$$\frac{8}{5} - \frac{8}{2} = \frac{8}{3}.$$

Find at least two ways to convince him of the mistake.

94. D_W To add numbers with different denominators, a student consistently uses the product of the denominators as a common denominator. Is this correct? Why or why not?

SKILL MAINTENANCE

Presidential Elections. The results of the five closest presidential elections are listed in the table at right. Use these data for Exercises 95–100. [1.2d]

DATE	PRESIDENT	ELECTORAL VOTES	POPULAR VOTES
1916	Woodrow Wilson (D)	277	9,129,606
	Charles E. Hughes (R)	254	8,538,221
1960	John F. Kennedy (D)	303	34,226,731
	Richard M. Nixon (R)	219	34,108,157
1968	Richard M. Nixon (R)	301	31,785,480
	Hubert H. Humphrey (D)	191	31,275,166
	George C. Wallace (American Independent)	46	9,906,473
1976	Jimmy Carter (D)	297	40,830,763
	Gerald R. Ford (R)	240	39,147,973
2000	George W. Bush (R)	271	50,455,156
	Albert A. Gore (D)	266	50,992,335

Source: Time Almanac 2005, p. 55

95. How many more popular votes did Albert A. Gore have than George W. Bush in the 2000 presidential election?

96. How many more electoral votes did George W. Bush have than Albert A. Gore in the 2000 presidential election?

97. In the 1960 presidential election, how many more electoral votes did John F. Kennedy have than Richard M. Nixon?

98. In the 1968 presidential election, how many more popular votes did Richard M. Nixon have than Hubert H. Humphrey?

99. How much greater was the total of the popular vote counts in the 2000 presidential election than in the 1976 election?

100. How much greater was the total of the popular vote counts in the 2000 presidential election than in the 1960 election?

Divide, if possible. If not possible, write "not defined." [2.1b]

101. $\frac{38}{38}$

102. $\frac{38}{0}$

103. $\frac{124}{0}$

104. $\frac{124}{31}$

SYNTHESIS

Simplify. Use the rules for order of operations given in Section 1.6.

105. $\frac{7}{8} - \frac{1}{10} \times \frac{5}{6}$

106. $\frac{2}{5} + \frac{1}{6} \div 3$

107. $\left(\frac{2}{3}\right)^2 + \left(\frac{3}{4}\right)^2$

108. $5 \times \frac{3}{7} - \frac{1}{7} \times \frac{4}{5}$

2.4 MIXED NUMERALS

Objectives

a Convert between mixed numerals and fraction notation.

b Add using mixed numerals.

c Subtract using mixed numerals.

d Multiply using mixed numerals.

e Divide using mixed numerals.

Convert to a mixed numeral.

1. $1 + \frac{2}{3} = \square\frac{\square}{\square}$

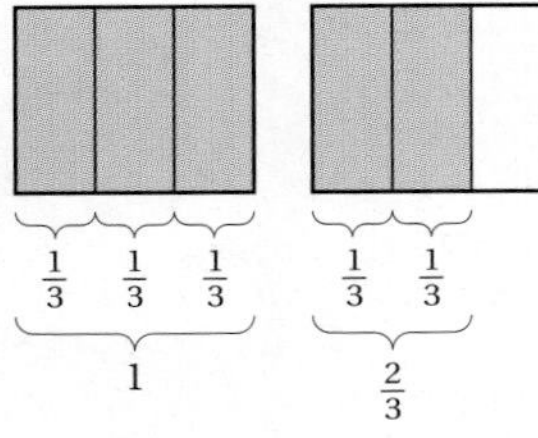

2. $2 + \frac{3}{4} = \square\frac{\square}{\square}$

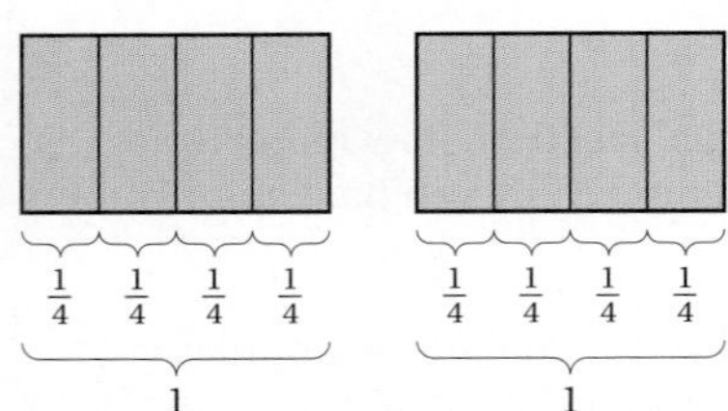

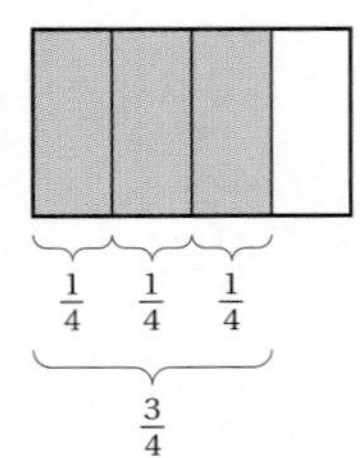

Answers on page A-5

a Mixed Numerals

The following figure illustrates the use of a **mixed numeral.** The bolt shown is $2\frac{3}{8}$ in. long. The length is given as a whole-number part, 2, and a fractional part less than 1, $\frac{3}{8}$. We can represent the measurement of the bolt with fraction notation as $\frac{19}{8}$, but the notation $2\frac{3}{8}$ makes the length easier to visualize and is thus more descriptive.

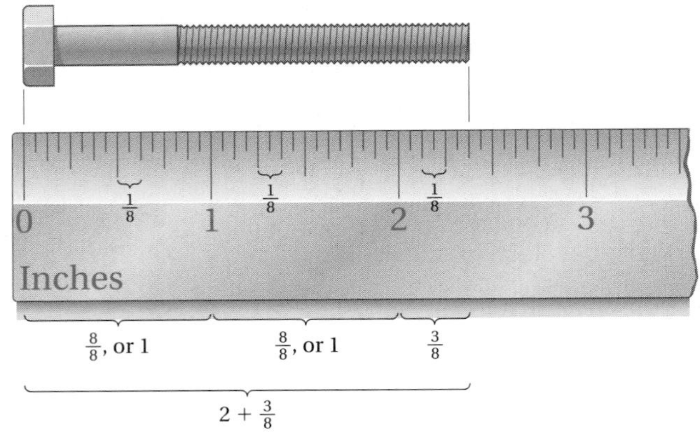

A mixed numeral $2\frac{3}{8}$ represents a sum:

$$2\frac{3}{8} \quad \text{means} \quad 2 + \frac{3}{8}$$

This is a whole number. This is a fraction less than 1.

EXAMPLES Convert to a mixed numeral.

1. $7 + \frac{2}{5} = 7\frac{2}{5}$

2. $4 + \frac{3}{10} = 4\frac{3}{10}$

Do Exercises 1–4. (Exercises 3 and 4 are on the following page.)

The notation $2\frac{3}{4}$ means $2 + \frac{3}{4}$. To aid in understanding, we sometimes write the missing plus sign.

EXAMPLES Convert to fraction notation.

3.
$$\begin{aligned} 2\frac{3}{4} &= 2 + \frac{3}{4} && \text{Inserting the missing plus sign} \\ &= \frac{2}{1} + \frac{3}{4} && 2 = \frac{2}{1} \\ &= \frac{2}{1}\cdot\frac{4}{4} + \frac{3}{4} && \text{Finding a common denominator} \\ &= \frac{8}{4} + \frac{3}{4} = \frac{11}{4} \end{aligned}$$

4. $4\frac{3}{10} = 4 + \frac{3}{10} = \frac{4}{1} + \frac{3}{10} = \frac{4}{1}\cdot\frac{10}{10} + \frac{3}{10} = \frac{40}{10} + \frac{3}{10} = \frac{43}{10}$

Do Exercises 5 and 6.

We now use Example 4 to illustrate a faster method for converting a mixed numeral to fraction notation.

To convert from a mixed numeral to fraction notation:

ⓐ Multiply the whole number by the denominator: $4 \cdot 10 = 40$.

ⓑ Add the result to the numerator: $40 + 3 = 43$.

ⓒ Keep the denominator.

$$4\frac{3}{10} = \frac{43}{10}$$

EXAMPLES Convert to fraction notation.

5. $6\frac{2}{3} = \frac{20}{3}$ $\quad 6 \cdot 3 = 18, 18 + 2 = 20$

6. $8\frac{2}{9} = \frac{74}{9}$ $\quad 8 \cdot 9 = 72, 72 + 2 = 74$

7. $10\frac{7}{8} = \frac{87}{8}$ $\quad 10 \cdot 8 = 80, 80 + 7 = 87$

Do Exercises 7–9.

WRITING MIXED NUMERALS

We can find a mixed numeral for $\frac{5}{3}$ as follows:

$$\frac{5}{3} = \frac{3}{3} + \frac{2}{3} = 1 + \frac{2}{3} = 1\frac{2}{3}.$$

In terms of objects, we can think of $\frac{5}{3}$ as $\frac{3}{3}$, or 1, plus $\frac{2}{3}$, as shown below.

$\frac{5}{3} =$ $\frac{3}{3}$, or 1 $+$ $\frac{2}{3}$

Fraction symbols like $\frac{5}{3}$ also indicate division; $\frac{5}{3}$ means $5 \div 3$. Let's divide the numerator by the denominator.

$$\begin{array}{r} 1 \\ 3\overline{)5} \\ \underline{3} \\ 2 \end{array} \leftarrow 2 \div 3 = \frac{2}{3}$$

Thus, $\frac{5}{3} = 1\frac{2}{3}$.

To convert from fraction notation to a mixed numeral, divide.

$\frac{13}{5}$

$$\begin{array}{r} 2 \\ 5\overline{)13} \\ \underline{10} \\ 3 \end{array} \qquad 2\frac{3}{5}$$

The divisor

The quotient

The remainder

3. $8 + \frac{3}{4}$

4. $12 + \frac{2}{3}$

Convert to fraction notation.

5. $4\frac{2}{5}$

6. $6\frac{1}{10}$

Convert to fraction notation. Use the faster method.

7. $4\frac{5}{6}$

8. $9\frac{1}{4}$

9. $20\frac{2}{3}$

Answers on page A-5

Convert to a mixed numeral.

10. $\frac{7}{3}$

11. $\frac{11}{10}$

12. $\frac{110}{6}$

13. Add.

$$\begin{array}{r} 2\frac{3}{10} \\ +\ 5\frac{1}{10} \\ \hline \end{array}$$

14. Add.

$$\begin{array}{r} 8\frac{2}{5} \\ +\ 3\frac{7}{10} \\ \hline \end{array}$$

Answers on page A-5

EXAMPLES Convert to a mixed numeral.

8. $\frac{69}{10}$

$$\begin{array}{r} 6 \\ 10\overline{)69} \\ \underline{60} \\ 9 \end{array} \qquad \frac{69}{10} = 6\frac{9}{10}$$

9. $\frac{122}{8}$

$$\begin{array}{r} 15 \\ 8\overline{)122} \\ \underline{80} \\ 42 \\ \underline{40} \\ 2 \end{array} \qquad \frac{122}{8} = 15\frac{2}{8} = 15\frac{1}{4}$$

Do Exercises 10–12.

b Addition Using Mixed Numerals

To find the sum $1\frac{5}{8} + 3\frac{1}{8}$, we first add the fractions. Then we add the whole numbers.

$$\begin{array}{rcl} 1\frac{5}{8} & = & 1\frac{5}{8} \\ +\ 3\frac{1}{8} & = & +\ 3\frac{1}{8} \\ \hline \frac{6}{8} & & 4\frac{6}{8} = 4\frac{3}{4} \end{array}$$

Add the fractions. Add the whole numbers. Simplifying

Do Exercise 13.

Sometimes we must write the fractional parts with a common denominator before we can add.

EXAMPLE 10 Add: $5\frac{2}{3} + 3\frac{5}{6}$. Write a mixed numeral for the answer.

The LCD is 6.

$$\begin{array}{rcl} 5\ \frac{2}{3}\cdot\frac{2}{2} & = & 5\frac{4}{6} \\ +\ 3\ \frac{5}{6} & = & +\ 3\frac{5}{6} \\ \hline & & 8\frac{9}{6} = 8 + \frac{9}{6} \\ & & = 8 + 1\frac{1}{2} \\ & & = 9\frac{1}{2} \end{array}$$

To find a mixed numeral for $\frac{9}{6}$, we divide:

$$\begin{array}{r} 1 \\ 6\overline{)9} \\ \underline{6} \\ 3 \end{array} \qquad \frac{9}{6} = 1\frac{3}{6} = 1\frac{1}{2}$$

Do Exercise 14.

EXAMPLE 11 Add: $10\frac{5}{6} + 7\frac{3}{8}$.

The LCD is 24.

$$\begin{array}{rl} 10\frac{5}{6}\cdot\frac{4}{4} & = 10\frac{20}{24} \\ +\ 7\frac{3}{8}\cdot\frac{3}{3} & = +\ 7\frac{9}{24} \\ \hline & 17\frac{29}{24} = 17 + \frac{29}{24} \end{array}$$

$$= 17 + 1\frac{5}{24}$$ Writing $\frac{29}{24}$ as a mixed numeral, $1\frac{5}{24}$

$$= 18\frac{5}{24}$$

Do Exercise 15.

C Subtraction Using Mixed Numerals

Subtraction is a lot like addition; we subtract the fractions and then the whole numbers.

EXAMPLE 12 Subtract: $7\frac{3}{4} - 2\frac{1}{4}$.

$$\begin{array}{rcrl} 7\frac{3}{4} & = & 7\frac{3}{4} & \\ -\ 2\frac{1}{4} & = & -\ 2\frac{1}{4} & \\ \hline \frac{2}{4} & & 5\frac{2}{4} & = 5\frac{1}{2} \end{array}$$

Subtract the fractions. Subtract the whole numbers. Simplifying

EXAMPLE 13 Subtract: $9\frac{4}{5} - 3\frac{1}{2}$.

The LCD is 10.

$$\begin{array}{rl} 9\frac{4}{5}\cdot\frac{2}{2} & = 9\frac{8}{10} \\ -\ 3\frac{1}{2}\cdot\frac{5}{5} & = -\ 3\frac{5}{10} \\ \hline & 6\frac{3}{10} \end{array}$$

Do Exercises 16 and 17.

EXAMPLE 14 Subtract: $7\frac{1}{6} - 2\frac{1}{4}$.

The LCD is 12.

$$\begin{array}{rl} 7\frac{1}{6}\cdot\frac{2}{2} & = 7\frac{2}{12} \\ -\ 2\frac{1}{4}\cdot\frac{3}{3} & = -\ 2\frac{3}{12} \\ \hline \end{array}$$

We borrow 1, or $\frac{12}{12}$, from 7:
$7\frac{2}{12} = 6 + 1 + \frac{2}{12} = 6 + \frac{12}{12} + \frac{2}{12} = 6\frac{14}{12}$.

We cannot subtract $\frac{3}{12}$ from $\frac{2}{12}$.

We can write this as

$$\begin{array}{rcr} 7\frac{2}{12} & = & 6\frac{14}{12} \\ -\ 2\frac{3}{12} & = & -\ 2\frac{3}{12} \\ \hline & & 4\frac{11}{12} \end{array}$$

Do Exercise 18.

15. Add.

$$\begin{array}{r} 9\frac{3}{4} \\ +\ 3\frac{5}{6} \\ \hline \end{array}$$

Subtract.

16. $$\begin{array}{r} 10\frac{7}{8} \\ -\ 9\frac{3}{8} \\ \hline \end{array}$$

17. $$\begin{array}{r} 8\frac{2}{3} \\ -\ 5\frac{1}{2} \\ \hline \end{array}$$

18. Subtract.

$$\begin{array}{r} 8\frac{1}{9} \\ -\ 4\frac{5}{6} \\ \hline \end{array}$$

Answers on page A-5

19. Subtract.

$$\begin{array}{r} 5 \\ -\ 1\frac{1}{3} \\ \hline \end{array}$$

20. Multiply: $6 \cdot 3\frac{1}{3}$.

21. Multiply: $2\frac{1}{2} \cdot \frac{3}{4}$.

Answers on page A-5

Study Tips

FORMING A STUDY GROUP

Consider forming a study group with some of your fellow students. Exchange e-mail addresses, telephone numbers, and schedules so that you can coordinate study time for homework and tests.

EXAMPLE 15 Subtract: $12 - 9\frac{3}{8}$.

$$\begin{array}{rcr} 12 & = & 11\frac{8}{8} \\ -\ 9\frac{3}{8} & = & -\ 9\frac{3}{8} \\ \hline & & 2\frac{5}{8} \end{array}$$

$$12 = 11 + 1 = 11 + \frac{8}{8} = 11\frac{8}{8}$$

Do Exercise 19.

d Multiplication Using Mixed Numerals

Carrying out addition and subtraction with mixed numerals is usually easier if the numbers are left as mixed numerals. With multiplication and division, however, it is easier to convert the numbers to fraction notation first.

MULTIPLICATION USING MIXED NUMERALS

To multiply using mixed numerals, first convert to fraction notation and multiply. Then convert the answer to a mixed numeral, if appropriate.

EXAMPLE 16 Multiply: $6 \cdot 2\frac{1}{2}$.

$$6 \cdot 2\frac{1}{2} = \frac{6}{1} \cdot \frac{5}{2} = \frac{6 \cdot 5}{1 \cdot 2} = \frac{2 \cdot 3 \cdot 5}{2 \cdot 1} = \frac{2}{2} \cdot \frac{3 \cdot 5}{1} = 15$$

Note that fraction notation is needed to carry out the multiplication.

Do Exercise 20.

EXAMPLE 17 Multiply: $3\frac{1}{2} \cdot \frac{3}{4}$.

$$3\frac{1}{2} \cdot \frac{3}{4} = \frac{7}{2} \cdot \frac{3}{4} = \frac{21}{8} = 2\frac{5}{8}$$

Here we write fraction notation.

Do Exercise 21.

EXAMPLE 18 Multiply: $8 \cdot 4\frac{2}{3}$.

$$8 \cdot 4\frac{2}{3} = \frac{8}{1} \cdot \frac{14}{3} = \frac{112}{3} = 37\frac{1}{3}$$

Do Exercise 22.

EXAMPLE 19 Multiply: $2\frac{1}{4} \cdot 3\frac{2}{5}$.

$$2\frac{1}{4} \cdot 3\frac{2}{5} = \frac{9}{4} \cdot \frac{17}{5} = \frac{153}{20} = 7\frac{13}{20}$$

Do Exercise 23.

Caution!

$2\frac{1}{4} \cdot 3\frac{2}{5} \neq 6\frac{2}{20}$. A common error is to multiply the whole numbers and then the fractions. This does not give the correct answer, $7\frac{13}{20}$, which is found by converting to fraction notation first.

22. Multiply: $2 \cdot 6\frac{2}{5}$.

23. Multiply: $3\frac{1}{3} \cdot 2\frac{1}{2}$.

e Division Using Mixed Numerals

The division $1\frac{1}{2} \div \frac{1}{6}$ is shown here. *Think*: "How many $\frac{1}{6}$'s are in $1\frac{1}{2}$?" We see that the answer is 9.

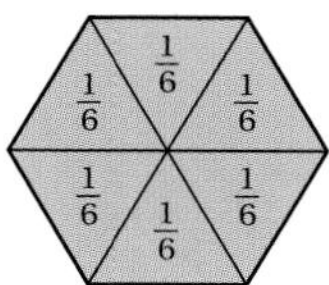

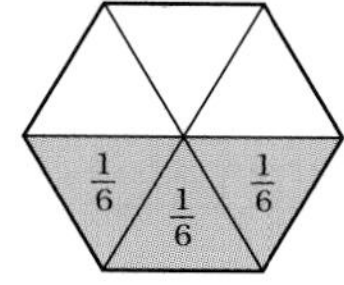

When we divide using mixed numerals, we convert to fraction notation first.

$$1\frac{1}{2} \div \frac{1}{6} = \frac{3}{2} \div \frac{1}{6} = \frac{3}{2} \cdot 6$$

$$= \frac{3 \cdot 6}{2} = \frac{3 \cdot 3 \cdot 2}{2 \cdot 1} = \frac{3 \cdot 3}{1} \cdot \frac{2}{2} = \frac{3 \cdot 3}{1} \cdot 1 = 9$$

DIVISION USING MIXED NUMERALS

To divide using mixed numerals, first write fraction notation and divide. Then convert the answer to a mixed numeral, if appropriate.

24. Divide: $84 \div 5\frac{1}{4}$.

EXAMPLE 20 Divide: $32 \div 3\frac{1}{5}$.

$$32 \div 3\frac{1}{5} = \frac{32}{1} \div \frac{16}{5}$$ Writing the mixed numeral in fraction notation

$$= \frac{32}{1} \cdot \frac{5}{16} = \frac{32 \cdot 5}{1 \cdot 16} = \frac{2 \cdot 16 \cdot 5}{1 \cdot 16} = \frac{16}{16} \cdot \frac{2 \cdot 5}{1} = 1 \cdot \frac{2 \cdot 5}{1} = 10$$

↑ Remember to multiply by the reciprocal.

Do Exercise 24.

Answers on page A-5

25. Divide: $26 \div 3\frac{1}{2}$.

EXAMPLE 21 Divide: $35 \div 4\frac{1}{3}$.

$$35 \div 4\frac{1}{3} = \frac{35}{1} \div \frac{13}{3} = \frac{35}{1} \cdot \frac{3}{13} = \frac{105}{13} = 8\frac{1}{13}$$

Do Exercise 25.

Divide.

26. $2\frac{1}{4} \div 1\frac{1}{5}$

27. $1\frac{3}{4} \div 2\frac{1}{2}$

EXAMPLE 22 Divide: $2\frac{1}{3} \div 1\frac{3}{4}$.

$$2\frac{1}{3} \div 1\frac{3}{4} = \frac{7}{3} \div \frac{7}{4} = \frac{7}{3} \cdot \frac{4}{7} = \frac{7 \cdot 4}{7 \cdot 3} = \frac{7}{7} \cdot \frac{4}{3} = 1 \cdot \frac{4}{3} = \frac{4}{3} = 1\frac{1}{3}$$

Caution!

The reciprocal of $1\frac{3}{4}$ is not $1\frac{4}{3}$!

EXAMPLE 23 Divide: $1\frac{3}{5} \div 3\frac{1}{3}$.

$$1\frac{3}{5} \div 3\frac{1}{3} = \frac{8}{5} \div \frac{10}{3} = \frac{8}{5} \cdot \frac{3}{10} = \frac{8 \cdot 3}{5 \cdot 10} = \frac{2 \cdot 4 \cdot 3}{5 \cdot 2 \cdot 5} = \frac{2}{2} \cdot \frac{4 \cdot 3}{5 \cdot 5} = \frac{12}{25}$$

Do Exercises 26 and 27.

Answers on page A-5

CALCULATOR CORNER

Operations on Fractions and Mixed Numerals Fraction calculators can add, subtract, multiply, and divide fractions and mixed numerals. The [$a^{b}/_{c}$] key is used to enter fractions and mixed numerals. To find $\frac{3}{4} + \frac{1}{2}$, for example, we press [3] [$a^{b}/_{c}$] [4] [+] [1] [$a^{b}/_{c}$] [2] [=]. Note that 3/4 and 1/2 appear on the display as [3 ⌟ 4] and [1 ⌟ 2], respectively. The result is given as the mixed numeral $1\frac{1}{4}$ and is displayed as [1 ⌟ 1 ⌟ 4]. Fraction results that are greater than 1 are always displayed as mixed numerals. To express this result as a fraction, we press [SHIFT] [d/c]. We get [5 ⌟ 4], or 5/4.

To find $3\frac{2}{3} \cdot 4\frac{1}{5}$, we press [3] [$a^{b}/_{c}$] [2] [$a^{b}/_{c}$] [3] [×] [4] [$a^{b}/_{c}$] [1] [$a^{b}/_{c}$] [5] [=]. The calculator displays [15 ⌟ 2 ⌟ 5], so the product is $15\frac{2}{5}$.

Some calculators are capable of displaying mixed numerals in the way in which we write them, as shown below.

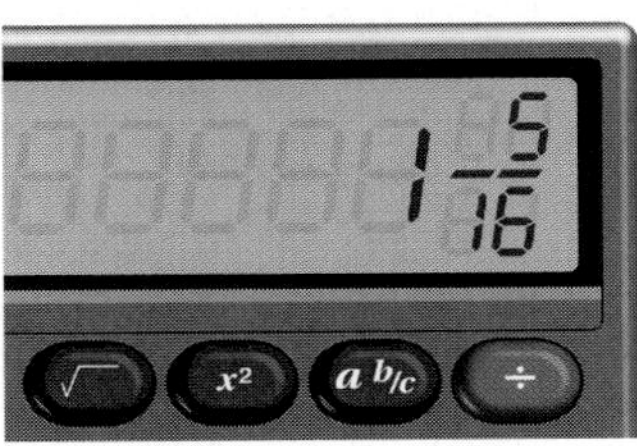

Exercises: Perform each calculation. Give the answer in fraction notation.

1. $\frac{1}{3} + \frac{1}{4}$
2. $\frac{7}{5} - \frac{3}{10}$
3. $\frac{15}{4} \cdot \frac{7}{12}$
4. $\frac{4}{5} \div \frac{8}{3}$

Perform each calculation. Give the answer as a mixed numeral.

5. $4\frac{1}{3} + 5\frac{4}{5}$
6. $9\frac{2}{7} - 8\frac{1}{4}$
7. $2\frac{1}{3} \cdot 4\frac{3}{5}$
8. $10\frac{7}{10} \div 3\frac{5}{6}$

2.4 EXERCISE SET

For Extra Help: MathXL, MyMathLab, InterAct Math, Math Tutor Center, Video Lectures on CD Disc 1, Student's Solutions Manual

a

1. *Alarm Clock Dimensions.* A world time clock/radio with NOAA (National Oceanic and Atmospheric Administration) Weather Band measures $14\frac{1}{4}$ in. × $6\frac{3}{4}$ in. × $2\frac{1}{4}$ in. Convert $14\frac{1}{4}$, $6\frac{3}{4}$, and $2\frac{1}{4}$ to fraction notation.

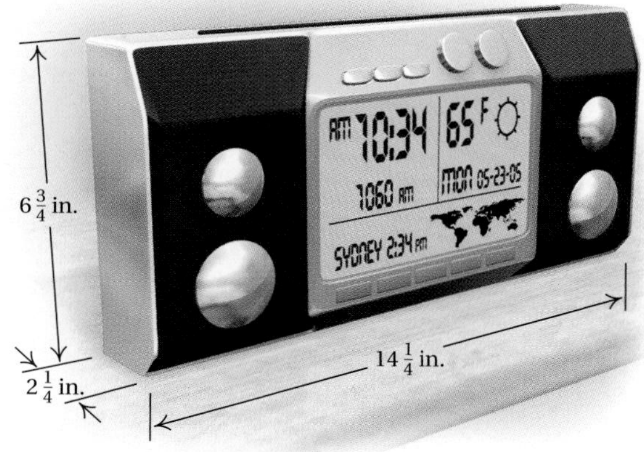

2. *Carpentry.* Dick Bonewitz, master carpenter, is making a display case according to the design below. Convert each mixed numeral to fraction notation.

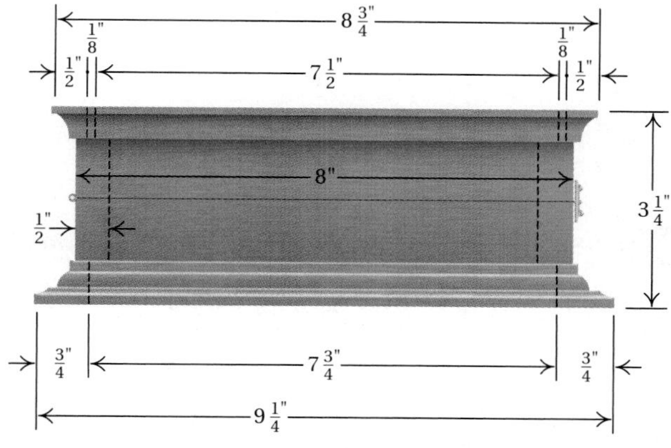

Convert to fraction notation.

3. $5\frac{2}{3}$

4. $20\frac{1}{5}$

5. $9\frac{5}{6}$

6. $1\frac{3}{5}$

7. $12\frac{3}{4}$

8. $33\frac{1}{3}$

Convert to a mixed numeral.

9. $\frac{18}{5}$

10. $\frac{17}{4}$

11. $\frac{57}{10}$

12. $\frac{50}{8}$

13. $\frac{345}{8}$

14. $\frac{467}{100}$

b

Add. Write a mixed numeral for the answer.

15. $\begin{array}{r} 2\frac{7}{8} \\ +\ 3\frac{5}{8} \\ \hline \end{array}$

16. $\begin{array}{r} 4\frac{5}{6} \\ +\ 3\frac{5}{6} \\ \hline \end{array}$

17. $1\frac{1}{4} + 1\frac{2}{3}$

18. $4\frac{1}{3} + 5\frac{2}{9}$

19. $8\frac{3}{4} + 5\frac{5}{6}$

20. $4\frac{3}{8} + 6\frac{5}{12}$

21. $12\frac{4}{5} + 8\frac{7}{10}$

22. $15\frac{5}{8} + 11\frac{3}{4}$

23. $14\frac{5}{8} + 13\frac{1}{4}$

24. $16\frac{1}{4} + 15\frac{7}{8}$

25. $7\frac{1}{8} + 9\frac{2}{3} + 10\frac{3}{4}$

26. $45\frac{2}{3} + 31\frac{3}{5} + 12\frac{1}{4}$

c Subtract. Write a mixed numeral for the answer.

27. $4\frac{1}{5} - 2\frac{3}{5}$

28. $5\frac{1}{8} - 2\frac{3}{8}$

29. $6\frac{3}{5} - 2\frac{1}{2}$

30. $7\frac{2}{3} - 6\frac{1}{2}$

31. $34 - 18\frac{5}{8}$

32. $23 - 19\frac{3}{4}$

33. $21\frac{1}{6} - 13\frac{3}{4}$

34. $42\frac{1}{10} - 23\frac{7}{12}$

35. $14\frac{1}{8} - \frac{3}{4}$

36. $28\frac{1}{6} - 5$

37. $25\frac{1}{9} - 13\frac{5}{6}$

38. $23\frac{5}{16} - 14\frac{7}{12}$

d Multiply. Write a mixed numeral for the answer.

39. $8 \cdot 2\frac{5}{6}$

40. $5 \cdot 3\frac{3}{4}$

41. $3\frac{5}{8} \cdot \frac{2}{3}$

42. $6\frac{2}{3} \cdot \frac{1}{4}$

43. $3\frac{1}{2} \cdot 2\frac{1}{3}$

44. $4\frac{1}{5} \cdot 5\frac{1}{4}$

45. $3\frac{2}{5} \cdot 2\frac{7}{8}$

46. $2\frac{3}{10} \cdot 4\frac{2}{5}$

47. $4\frac{7}{10} \cdot 5\frac{3}{10}$

48. $6\frac{3}{10} \cdot 5\frac{7}{10}$

49. $20\frac{1}{2} \cdot 10\frac{1}{5} \cdot 4\frac{2}{3}$

50. $21\frac{1}{3} \cdot 11\frac{1}{3} \cdot 3\frac{5}{8}$

e Divide. Write a mixed numeral for the answer.

51. $20 \div 3\frac{1}{5}$

52. $18 \div 2\frac{1}{4}$

53. $8\frac{2}{5} \div 7$

54. $3\frac{3}{8} \div 3$

55. $4\frac{3}{4} \div 1\frac{1}{3}$

56. $5\frac{4}{5} \div 2\frac{1}{2}$

57. $1\frac{7}{8} \div 1\frac{2}{3}$

58. $4\frac{3}{8} \div 2\frac{5}{6}$

59. $5\frac{1}{10} \div 4\frac{3}{10}$

60. $4\frac{1}{10} \div 2\frac{1}{10}$

61. $20\frac{1}{4} \div 90$

62. $12\frac{1}{2} \div 50$

63. $\mathbf{D_W}$ Write a problem for a classmate to solve. Design the problem so that its solution is found by performing the multiplication $4\frac{1}{2} \cdot 33\frac{1}{3}$.

64. $\mathbf{D_W}$ Under what circumstances is a pair of mixed numerals more easily added than multiplied?

SKILL MAINTENANCE

65. Round to the nearest hundred: 45,765. [1.3e]

66. Round to the nearest ten: 45,765. [1.3e]

Determine whether the first number is divisible by the second. [1.8a]

67. 9993 by 3

68. 9993 by 9

69. 2345 by 9

70. 2345 by 5

71. 2335 by 10

72. 7764 by 6

73. 18,888 by 8

74. 18,888 by 4

Subtract. [1.2d]

75. $\begin{array}{r} 3\,0\,0\,4 \\ -\ 2\,9\,5\,7 \\ \hline \end{array}$

76. $\begin{array}{r} 1\,1\,1\,1 \\ -\ \ \ 2\,2\,2 \\ \hline \end{array}$

77. $\begin{array}{r} 1\,0{,}0\,1\,3 \\ -\ \ \ 9\,9\,8\,8 \\ \hline \end{array}$

78. $\begin{array}{r} 5\,0\,7{,}9\,8\,7 \\ -\ \ \ 3\,0{,}0\,0\,9 \\ \hline \end{array}$

SYNTHESIS

Multiply. Write the answer as a mixed numeral whenever possible.

79. 🖩 $15\frac{2}{11} \cdot 23\frac{31}{43}$

80. 🖩 $17\frac{23}{31} \cdot 19\frac{13}{15}$

Simplify.

81. $8 \div \frac{1}{2} + \frac{3}{4} + \left(5 - \frac{5}{8}\right)^2$

82. $\frac{7}{8} - 1\frac{1}{8} \times \frac{2}{3} + \frac{9}{10} \div \frac{3}{5}$

2.5 APPLICATIONS AND PROBLEM SOLVING

Objective

a Solve applied problems involving addition, subtraction, multiplication, and division using fraction notation and mixed numerals.

a We solve applied problems using fraction notation and mixed numerals in the same way that we do when using whole numbers. The five steps for problem solving on p. 50 should be reviewed.

Many problems that can be solved by multiplying fractions can be thought of in terms of rectangular arrays.

EXAMPLE 1 A real estate developer owns a plot of land that measures 1 square mile. He plans to use $\frac{4}{5}$ of the plot for a small strip mall and parking lot. Of this, $\frac{2}{3}$ will be needed for the parking lot. What part of the plot will be used for parking?

1. **Familiarize.** We first make a drawing to help familiarize ourselves with the problem. The land may not be rectangular. It could be in a shape like A or B below. But to think out the problem, we can think of it as a rectangle, as shown in shape C.

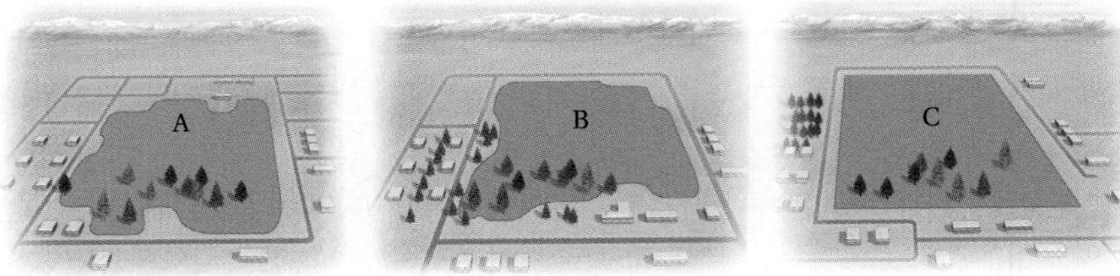

The strip mall including the parking lot uses $\frac{4}{5}$ of the plot. We shade $\frac{4}{5}$.

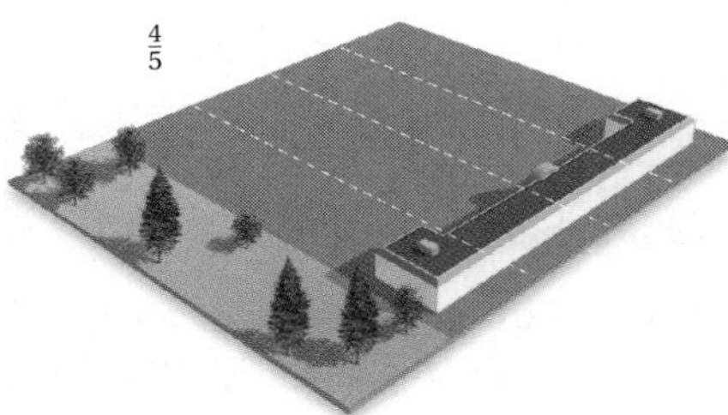

The parking lot alone takes $\frac{2}{3}$ of the preceding part. We shade that.

2. **Translate.** We let n = the part of the plot that is used for parking. We are taking "two-thirds of four-fifths." Recall from Section 1.5 that the word "of" corresponds to multiplication. Thus the following multiplication sentence corresponds to the situation:

$$n = \frac{2}{3} \cdot \frac{4}{5}.$$

3. **Solve.** The number sentence tells us what to do. We multiply:

$$n = \frac{2}{3} \cdot \frac{4}{5} = \frac{2 \cdot 4}{3 \cdot 5} = \frac{8}{15}.$$

4. **Check.** We can check partially by noting that the answer is smaller than the original area, 1, which we expect since the developer is using only part of the original plot of land. Thus, $\frac{8}{15}$ is a reasonable answer. We can also check this in the figure above, where we see that 8 of 15 parts have been shaded a second time.

5. **State.** The parking lot takes $\frac{8}{15}$ of the square mile of land.

Do Exercise 1.

1. A resort hotel uses $\frac{3}{4}$ of its extra land for recreational purposes. Of that, $\frac{1}{2}$ is used for swimming pools. What part of the land is used for swimming pools?

Example 1 and the preceding discussion indicate that the area of a rectangular region can be found by multiplying length by width. Remember, the area of a rectangular region is given by the formula

$$A = l \cdot w.$$

EXAMPLE 2 *Area of an Ice-Skating Rink.* The length of a rectangular ice-skating rink in the atrium of a shopping mall is $\frac{7}{100}$ mi. The width is $\frac{3}{100}$ mi. What is the area of the rink?

1. **Familiarize.** Recall that area is length times width. We make a drawing and let $A =$ the area of the ice-skating rink.

2. **Translate.** Then we translate.

Area	is	Length	times	Width
↓	↓	↓	↓	↓
A	$=$	$\frac{7}{100}$	$\times$	$\frac{3}{100}$

3. **Solve.** The sentence tells us what to do. We multiply:

$$A = \frac{7}{100} \cdot \frac{3}{100} = \frac{7 \cdot 3}{100 \cdot 100} = \frac{21}{10{,}000}.$$

4. **Check.** We check by repeating the calculation. This is left to the student.

5. **State.** The area is $\frac{21}{10{,}000}$ mi^2.

Do Exercise 2.

2. **Area of a Ceramic Tile.** The length of a rectangular ceramic tile on an inlaid ceramic counter is $\frac{4}{9}$ ft. The width is $\frac{2}{9}$ ft. What is the area of one tile?

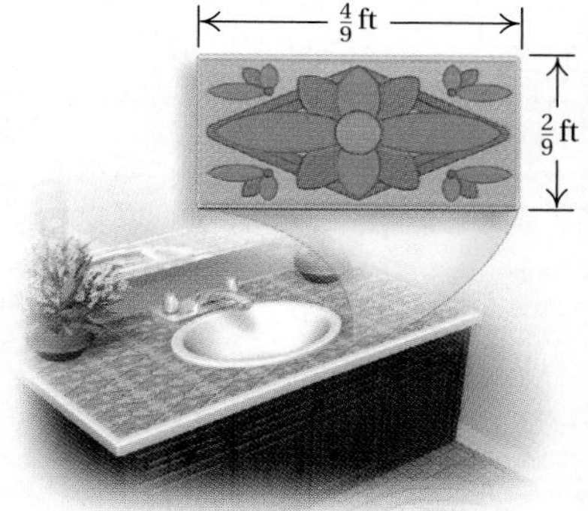

Answers on page A-5

3. A landscaper uses $\frac{2}{3}$ lb of peat moss for a rosebush. How much will be needed for 21 rosebushes?

EXAMPLE 3 Elite Elegance is preparing souvenir favors for a charity fund-raising dinner. How many pounds of caramels will be needed to fill 235 boxes if each box contains $\frac{2}{5}$ lb?

1. **Familiarize.** We first make a drawing or at least visualize the situation. Repeated addition will work here.

We let $n =$ the number of pounds of caramels.

2. **Translate.** The problem translates to the following equation:

$$n = 235 \cdot \frac{2}{5}.$$

3. **Solve.** To solve the equation, we carry out the multiplication:

$$\begin{aligned} n = 235 \cdot \frac{2}{5} &= \frac{235 \cdot 2}{5} && \text{Multiplying} \\ &= \frac{5 \cdot 47 \cdot 2}{5 \cdot 1} \\ &= \frac{5}{5} \cdot \frac{47 \cdot 2}{1} \\ &= 94. && \text{Simplifying} \end{aligned}$$

4. **Check.** We check by repeating the calculation. (The check is left to the student.) We can also think about the reasonableness of the answer. We are putting less than a pound of caramels in each box, so the answer should be less than 235. Since 94 is less than 235, we have a partial check of the reasonableness of the answer. The number 94 checks.

5. **State.** Thus, 94 lb of caramels will be needed.

Do Exercise 3.

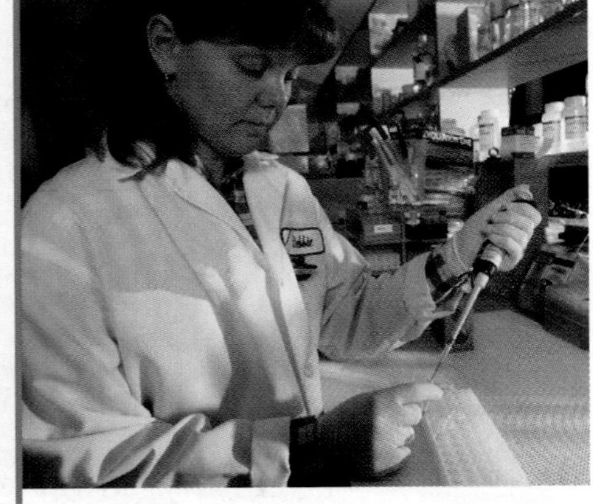

EXAMPLE 4 *Syringes.* How many syringes, each containing $\frac{3}{5}$ mL, can a clinical pharmacist fill from a 30-mL vial?

1. **Familiarize.** We are asking the question, "How many $\frac{3}{5}$'s are in 30?" Repeated addition will apply here. We make a drawing. We let $n =$ the number of syringes in all.

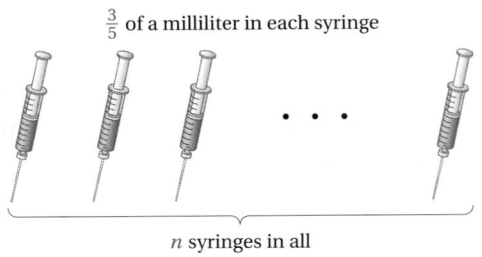

$\frac{3}{5}$ of a milliliter in each syringe

. . .

n syringes in all

Answer on page A-5

2. Translate. The equation that corresponds to the situation is

$$n = 30 \div \frac{3}{5}.$$

3. Solve. We solve the equation by carrying out the division:

$$n = 30 \div \frac{3}{5} = 30 \cdot \frac{5}{3} = \frac{30 \cdot 5}{3} = \frac{3 \cdot 10 \cdot 5}{3 \cdot 1}$$
$$= \frac{3}{3} \cdot \frac{10 \cdot 5}{1} = 50.$$

4. Check. We check by repeating the calculation.

5. State. The pharmacist can fill 50 syringes.

Do Exercise 4.

EXAMPLE 5 *Hand-Knit Scarves.* Nayah knits winter scarves for the Quick-Knit Boutique. After she knits 63 in. of a scarf, $\frac{7}{8}$ of the scarf is complete. What is the total length of this scarf?

1. Familiarize. We ask: "63 in. is $\frac{7}{8}$ of what length?" We make a drawing or at least visualize the problem. We let s = the length of the scarf.

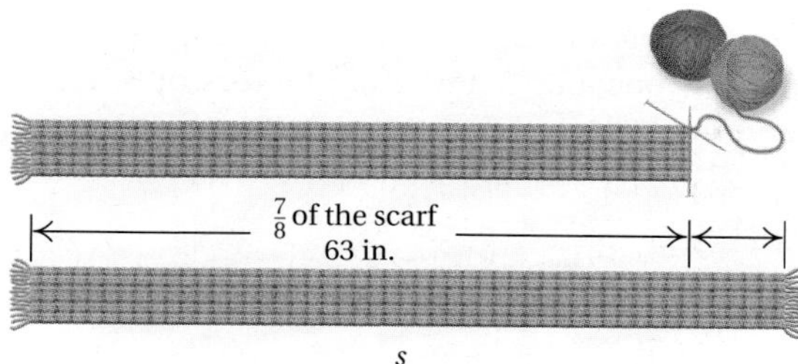

2. Translate. We translate to an equation.

Fraction completed	of	Total length of scarf	is	Amount already knitted
$\frac{7}{8}$	$\cdot$	s	$=$	63

3. Solve. The equation that corresponds to the situation is $\frac{7}{8} \cdot s = 63$. We divide by $\frac{7}{8}$ on both sides and carry out the division:

$$s = 63 \div \frac{7}{8} = 63 \cdot \frac{8}{7} = \frac{63 \cdot 8}{7} = \frac{7 \cdot 9 \cdot 8}{7} = \frac{7}{7} \cdot \frac{9 \cdot 8}{1} = 72.$$

4. Check. Since $\frac{7}{8} \cdot 72 = \frac{7 \cdot 8 \cdot 9}{8} = \frac{8}{8} \cdot \frac{7 \cdot 9}{1} = 63$, the answer, 72 checks.

5. State. The completed scarf is 72 in. long.

Do Exercise 5.

4. Each loop in a spring uses $\frac{3}{8}$ in. of wire. How many loops can be made from 120 in. of wire?

5. Sales Trip. Miles Lanosga sells soybean seeds to seed companies. After he had driven 210 mi, $\frac{5}{6}$ of his sales trip was completed. How long was the total trip?

Answers on page A-5

Study Tips

STUDY RESOURCES

The new mathematical skills and concepts presented in class will be most easily learned if you begin the homework assignment as soon as possible after the lecture while they are fresh in your mind. Then if you have difficulty with any of the exercises, you have time to access supplementary resources such as:

- *Student's Solutions Manual*
- Video Lectures
- MathXL Tutorials on CD
- AW Math Tutor Center
- MathXL
- MyMathLab

6. Sally jogs for $\frac{4}{5}$ mi, rests, and then jogs for another $\frac{1}{10}$ mi. How far does she jog in all?

EXAMPLE 6 *Subflooring.* Unique Builders and Developers requires their subcontractors to use two layers of subflooring under a ceramic tile floor. First the contractors install a $\frac{3}{4}$-in. ($\frac{3}{4}$") layer of oriented strand board (OSB). Then a $\frac{1}{2}$-in. ($\frac{1}{2}$") sheet of cement board is mortared to the OSB. The mortar is $\frac{1}{8}$-in. thick. What is the total thickness of the two installed subfloors?

1. **Familiarize.** We first make a drawing. We let $T =$ the total thickness of the subfloors.

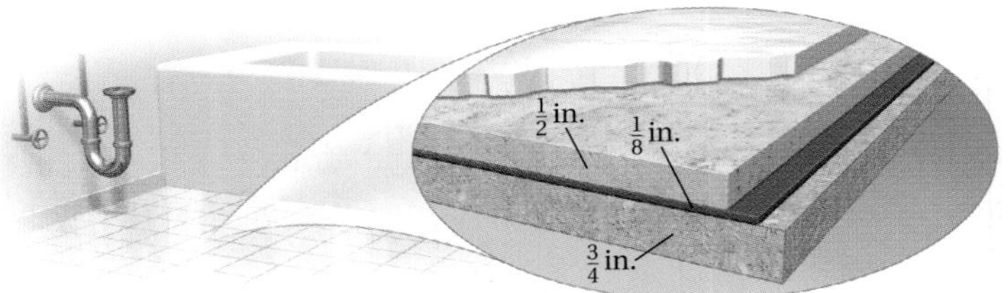

2. **Translate.** The problem can be translated to an equation as follows.

OSB	plus	Mortar	plus	Cement board	is	Total thickness
↓	↓	↓	↓	↓	↓	↓
$\frac{3}{4}$	$+$	$\frac{1}{8}$	$+$	$\frac{1}{2}$	$=$	T

3. **Solve.** To solve the equation, we carry out the addition. The LCM of the denominators is 8 because 2 and 4 are factors of 8. We multiply by 1 in order to obtain the LCD:

$$\frac{3}{4} + \frac{1}{8} + \frac{1}{2} = T$$
$$\frac{3}{4} \cdot \frac{2}{2} + \frac{1}{8} + \frac{1}{2} \cdot \frac{4}{4} = T$$
$$\frac{6}{8} + \frac{1}{8} + \frac{4}{8} = T$$
$$\frac{11}{8} = T.$$

4. **Check.** We check by repeating the calculation. We also note that the sum should be larger than any of the individual measurements, which it is. This tells us that the answer is reasonable.

5. **State.** The total thickness of the installed subfloors is $\frac{11}{8}$ in.

Do Exercise 6.

EXAMPLE 7 *Pendant Necklace.* At one time, Coldwater Creek offered the pendant necklace illustrated on the following page. The sterling silver capping at the top measures $\frac{11}{32}$ in. and the total length of the pendant is $\frac{7}{8}$ in. Find the length, or diameter, w of the pearl ball on the pendant.

1. **Familiarize.** We let $w =$ the length of the pearl on the pendant.

Answer on page A-5

2. **Translate.** We see that this is a "how many more" situation. We can translate to an equation.

$$\underbrace{\text{Length of silver capping}}_{\frac{11}{32}} \quad \underbrace{\text{plus}}_{+} \quad \underbrace{\text{Length of pearl ball}}_{w} \quad \underbrace{\text{is}}_{=} \quad \underbrace{\text{Total length of pendant}}_{\frac{7}{8}}$$

3. **Solve.** To solve the equation, we subtract $\frac{11}{32}$ on both sides:

$$\frac{11}{32} + w = \frac{7}{8}$$

$$\frac{11}{32} + w - \frac{11}{32} = \frac{7}{8} - \frac{11}{32}$$ Subtracting $\frac{11}{32}$ on both sides

$$w + 0 = \frac{7}{8} \cdot \frac{4}{4} - \frac{11}{32}$$ The LCD is 32. We multiply by 1 to obtain the LCD.

$$w = \frac{28}{32} - \frac{11}{32}$$

$$= \frac{17}{32}.$$

4. **Check.** To check, we return to the original problem and add:

$$\frac{11}{32} + \frac{17}{32} = \frac{28}{32} = \frac{7}{8} \cdot \frac{4}{4} = \frac{7}{8}.$$

5. **State.** The length of the pearl ball on the pendant is $\frac{17}{32}$ in.

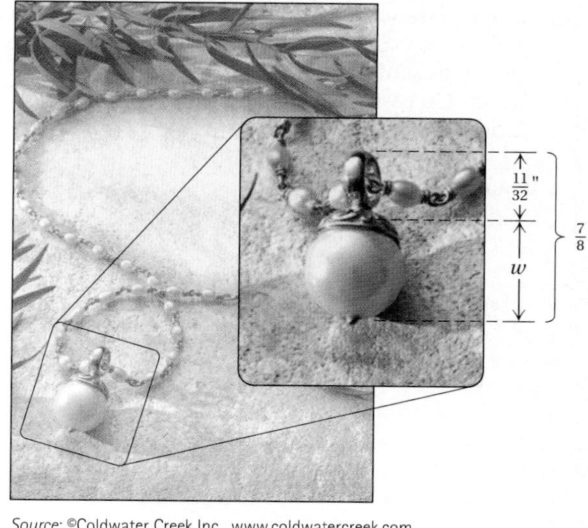

Source: ©Coldwater Creek Inc. www.coldwatercreek.com

Do Exercise 7.

7. Natasha has run for $\frac{2}{3}$ mi and will stop when she has run for $\frac{7}{8}$ mi. How much farther does she have to run?

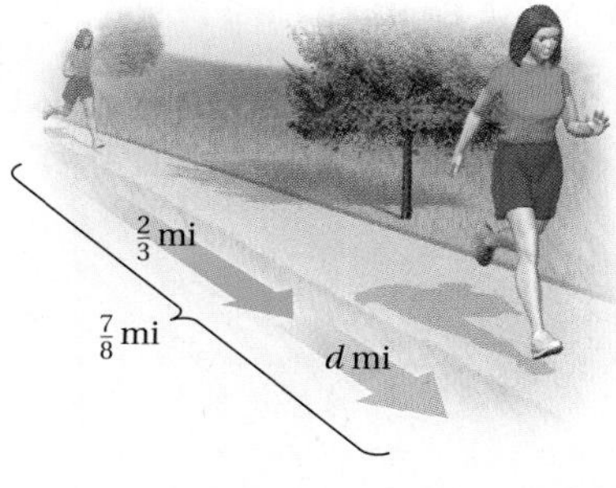

Answer on page A-5

EXAMPLE 8 *Widening a Driveway.* Sherry and Woody are widening their existing $17\frac{1}{4}$-ft driveway by adding $5\frac{9}{10}$ ft on one side. What is the width of the new driveway?

1. **Familiarize.** We let w = the width of the new driveway.
2. **Translate.** We translate as follows.

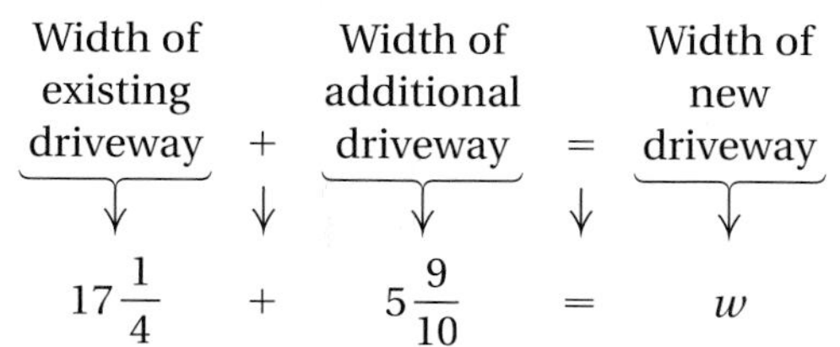

$$\underbrace{\text{Width of existing driveway}}_{17\frac{1}{4}} \quad \underbrace{+}_{+} \quad \underbrace{\text{Width of additional driveway}}_{5\frac{9}{10}} \quad \underbrace{=}_{=} \quad \underbrace{\text{Width of new driveway}}_{w}$$

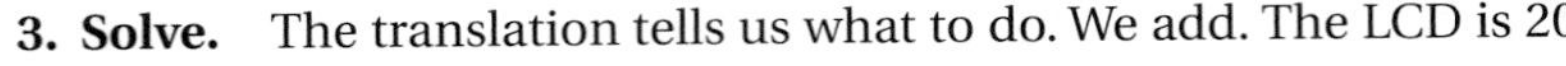

8. **Travel Distance.** Chrissy Jenkins is a college textbook sales representative for Addison-Wesley. On two business days, Chrissy drove $144\frac{9}{10}$ mi and $87\frac{1}{4}$ mi. What was the total distance Chrissy drove?

3. **Solve.** The translation tells us what to do. We add. The LCD is 20.

$$\begin{array}{rcrcr} 17\frac{1}{4} & = & 17\frac{1}{4}\cdot\frac{5}{5} & = & 17\frac{5}{20} \\ +\ 5\frac{9}{10} & = & +\ 5\frac{9}{10}\cdot\frac{2}{2} & = & +\ 5\frac{18}{20} \\ \hline & & & & 22\frac{23}{20} = 23\frac{3}{20} \end{array}$$

Thus, $w = 23\frac{3}{20}$.

4. **Check.** We check by repeating the calculation. We also note that the answer is larger than either of the widths, which means that the answer is reasonable.
5. **State.** The width of the new driveway is $23\frac{3}{20}$ ft.

Do Exercise 8.

EXAMPLE 9 *Communication Cable.* Celebrity Cable has two crews who install Internet communication cable. Crew A can install $38\frac{1}{8}$ ft per hour. Crew B can install $31\frac{2}{3}$ ft per hour. How many fewer feet can crew B install per hour than crew A?

1. **Familiarize.** The phrase "how many fewer" indicates subtraction. We let c = the difference in the numbers of feet per hour.
2. **Translate.** We translate as follows.

$$\underbrace{\text{Crew A: ft/hr}}_{38\frac{1}{8}} - \underbrace{\text{Crew B: ft/hr}}_{31\frac{2}{3}} = \underbrace{\text{Difference in ft/hr}}_{c}$$

$$38\frac{1}{8} - 31\frac{2}{3} = c$$

3. **Solve.** To solve the equation, we carry out the subtraction. The LCD = 24.

$$\begin{array}{rcrcrcr} 38\frac{1}{8} & = & 38\frac{1}{8}\cdot\frac{3}{3} & = & 38\frac{3}{24} & = & 37\frac{27}{24} \\ -31\frac{2}{3} & = & -31\frac{2}{3}\cdot\frac{8}{8} & = & -31\frac{16}{24} & = & -31\frac{16}{24} \\ \hline & & & & & & 6\frac{11}{24} \end{array}$$

Thus, $c = 6\frac{11}{24}$.

4. **Check.** To check, we add the difference, $6\frac{11}{24}$, to crew B's rate:

$$31\frac{2}{3} + 6\frac{11}{24} = 31\frac{16}{24} + 6\frac{11}{24} = 37\frac{27}{24} = 38\frac{3}{24} = 38\frac{1}{8}.$$

This checks.

5. **State.** Crew B installs $6\frac{11}{24}$ fewer feet per hour than crew A.

Do Exercise 9.

9. **NCAA Football Goal Posts.** In college football, the distance between goal posts was reduced from $23\frac{1}{3}$ ft to $18\frac{1}{2}$ ft in 1991. By how much was it reduced?

Source: NCAA

Answers on page A-5

EXAMPLE 10 *Carpentry.* The following diagram shows the layout for the construction of a desk drawer. Find the missing length a.

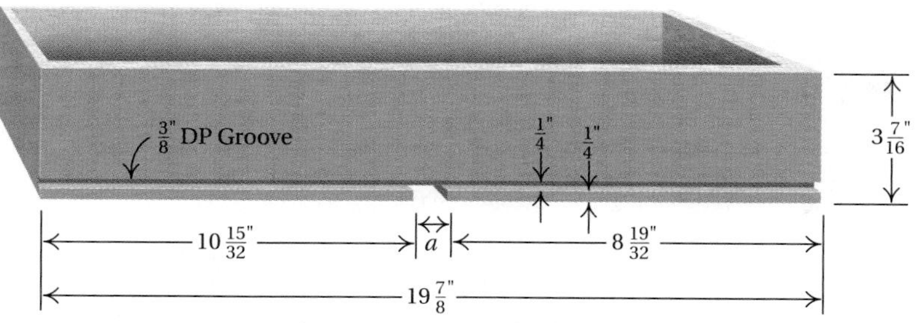

Middle Drawer / Back Layout

1. **Familiarize.** The length a is shown in the drawing.
2. **Translate.** From the drawing, we see that the length a is the full length of the drawer minus the sum of the lengths $10\frac{15}{32}$" and $8\frac{19}{32}$". Thus we have

$$a = 19\frac{7}{8} - \left(10\frac{15}{32} + 8\frac{19}{32}\right).$$

3. **Solve.** This is a two-step problem.

a) We first add the two lengths, $10\frac{15}{32}$ and $8\frac{19}{32}$.

$$\begin{array}{r} 10\frac{15}{32} \\ +\ 8\frac{19}{32} \\ \hline 18\frac{34}{32} = 18\frac{17}{16} = 19\frac{1}{16} \end{array}$$

b) Next, we subtract $19\frac{1}{16}$ from $19\frac{7}{8}$.

$$\begin{array}{rcr} 19\frac{7}{8} & = & 19\frac{14}{16} \\ -19\frac{1}{16} & = & -19\frac{1}{16} \\ \hline & & \frac{13}{16} \end{array}$$

Thus, $a = \frac{13}{16}$.

4. **Check.** We check by repeating the calculation, or adding the three measurements:

$$10\frac{15}{32} + 8\frac{19}{32} + \frac{13}{16} = 10\frac{15}{32} + 8\frac{19}{32} + \frac{26}{32}$$
$$= 18\frac{60}{32} = 19\frac{28}{32} = 19\frac{7}{8}.$$

5. **State.** The length a in the diagram is $\frac{13}{16}$".

Do Exercise 10.

10. Liquid Fertilizer. There are $283\frac{5}{8}$ gal of liquid fertilizer in a fertilizer application tank. After applying $178\frac{2}{3}$ gal to a soybean field, the farmer requests that Braden's Farm Supply deliver an additional 250 gal. How many gallons of fertilizer are in the tank after the delivery?

Answer on page A-5

11. Kyle's pickup truck travels on an interstate highway at 65 mph for $3\frac{1}{2}$ hr. How far does it travel?

12. Holly's minivan travels 302 mi on $15\frac{1}{10}$ gal of gas. How many miles per gallon does it get?

Answers on page A-5

EXAMPLE 11 *Average Speed in Indianapolis 500.* Arie Luyendyk won the Indianapolis 500 in 1990 with an average speed of about 186 mph. This record high speed is about $2\frac{12}{25}$ times the average speed of the first winner, Ray Harroun, in 1911. What was the average speed in the first Indianapolis 500?

Source: Indianapolis Motor Speedway

1. **Familiarize.** We ask the question, "186 is $2\frac{12}{25}$ times what number?" We let $s =$ the average speed in 1911. Then the average speed in 1990 was $2\frac{12}{25} \cdot s$.

2. **Translate.** The problem can be translated to an equation as follows.

$$\underbrace{\text{Average speed in 1990}}_{186} \quad \underbrace{\text{is}}_{=} \quad \underbrace{2\frac{12}{25}}_{2\frac{12}{25}} \quad \underbrace{\text{times}}_{\cdot} \quad \underbrace{\text{Average speed in 1911}}_{s}$$

3. **Solve.** To solve the equation, we divide on both sides.

$186 = \frac{62}{25} \cdot s$ Converting $2\frac{12}{25}$ to fraction notation

$\frac{186}{\frac{62}{25}} = \frac{\frac{62}{25} \cdot s}{\frac{62}{25}}$ Dividing by $\frac{62}{25}$ on both sides

$\frac{186}{\frac{62}{25}} = s$ Factoring and removing a factor of 1: $((62/25)/(62/25)) = 1$

$186 \cdot \frac{25}{62} = s$ Multiplying by the reciprocal

$75 = s$ Simplifying: $186 \cdot \frac{25}{62} = 3 \cdot 25 = 75$

4. **Check.** If the average speed in 1911 was about 75 mph, we find the average speed in 1990 by multiplying 75 by $2\frac{12}{25}$:

$$2\frac{12}{25} \cdot 75 = \frac{62}{25} \cdot 75 = \frac{62 \cdot 75}{25} = \frac{62 \cdot 25 \cdot 3}{25} = 62 \cdot 3 = 186.$$

The answer checks.

5. **State.** The average speed in the first Indianapolis 500 was about 75 mph.

Do Exercises 11 and 12.

EXAMPLE 12 *Mirror Area.* The mirror-backed candle shelf, shown below with a carpenter's diagram, was designed and built by Harry Cooper. Such shelves were popular in Colonial times because the mirror provided extra lighting from the candle. A rectangular walnut board is used to make the back of the shelf. Find the area of the original board and the amount left over after the opening for the mirror has been cut out.

Source: *Popular Science Woodworking Projects*

1. **Familiarize.** Refer to the figure at right. We let $h =$ the height of the back of the shelf and $B =$ the area of the original board. We know the width of the original board, $8\frac{1}{2}$". (Remember, $8\frac{1}{2}$" means $8\frac{1}{2}$ in.) We let $A =$ the area left over after the mirror has been cut out.

2. **Translate.** This is a multistep problem. To find B, which equals $8\frac{1}{2} \cdot h$, we first need to calculate h. We read the dimensions $5\frac{3}{8}$", $11\frac{1}{2}$", and $6\frac{3}{8}$" from the diagram and add them to find h:

$$h = 5\frac{3}{8} + 11\frac{1}{2} + 6\frac{3}{8}.$$

The dimensions of the mirror are $11\frac{1}{2}$" and $5\frac{1}{2}$". Then A is the area of the original board minus the area of the mirror. That is,

$$A = B - 11\frac{1}{2} \cdot 5\frac{1}{2}.$$

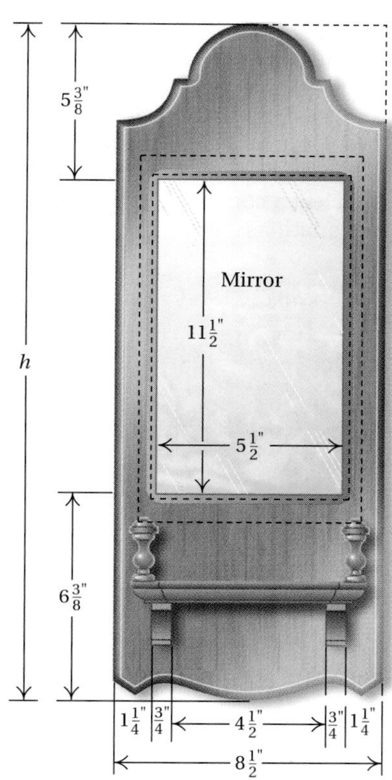

Front View

3. **Solve.** We carry out each calculation as follows:

$$\begin{aligned} h &= 5\frac{3}{8} + 11\frac{1}{2} + 6\frac{3}{8} \\ &= 5\frac{3}{8} + 11\frac{4}{8} + 6\frac{3}{8} \\ &= 22\frac{10}{8} = 22\frac{5}{4} = 23\frac{1}{4}; \end{aligned}$$

$$\begin{aligned} B &= 8\frac{1}{2} \cdot h \\ &= 8\frac{1}{2} \cdot 23\frac{1}{4} = \frac{17}{2} \cdot \frac{93}{4} \\ &= \frac{1581}{8} = 197\frac{5}{8}; \end{aligned}$$

$$\begin{aligned} A &= B - 11\frac{1}{2} \cdot 5\frac{1}{2} = 197\frac{5}{8} - 11\frac{1}{2} \cdot 5\frac{1}{2} \\ &= 197\frac{5}{8} - \frac{23}{2} \cdot \frac{11}{2} = 197\frac{5}{8} - \frac{253}{4} \\ &= 197\frac{5}{8} - 63\frac{1}{4} = 197\frac{5}{8} - 63\frac{2}{8} = 134\frac{3}{8}. \end{aligned}$$

4. **Check.** We perform a check by repeating the calculations.

5. **State.** The area of the original board is $197\frac{5}{8}$ in^2. The area left over is $134\frac{3}{8}$ in^2.

Do Exercise 13.

13. A room measures $22\frac{1}{2}$ ft by $15\frac{1}{2}$ ft. A 9-ft by 12-ft Oriental rug is placed in the center of the room. How much area is not covered by the rug?

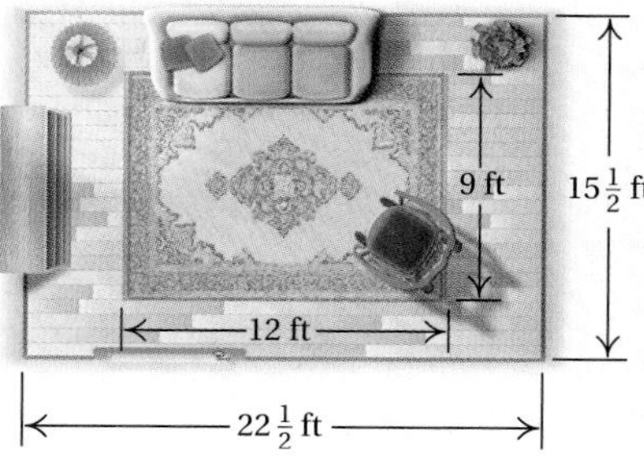

Answer on page A-5

Translating for Success

The goal of these matching questions is to practice step (2), *Translate,* of the five-step problem-solving process. Translate each word problem to an equation and select a correct translation from equations A–O.

1. *Raffle Tickets.* At the Happy Hollow Camp Fall Festival, Rico and Becca, together, spent \$270 on raffle tickets that sell for $\$\frac{9}{20}$ each. How many tickets did they buy?

2. *Irrigation Pipe.* Jed uses two pipes, one of which measures $5\frac{1}{3}$ ft, to repair the irrigation system in the Buxtons' lawn. The total length of the two pipes is $8\frac{7}{12}$ ft. How long is the other pipe?

3. *Vacation Days.* Together, Helmut and Claire have 36 vacation days a year. Helmut has 22 vacation days per year. How many does Claire have?

4. *Enrollment in Japanese Classes.* Last year at the Lakeside Community College, 225 students enrolled in basic mathematics. This number is $4\frac{1}{2}$ times as many as the number who enrolled in Japanese. How many enrolled in Japanese?

5. *Bicycling.* Cole rode his bicycle $5\frac{1}{3}$ mi on Saturday and $8\frac{7}{12}$ mi on Sunday. How far did he ride on the weekend?

A. $13\frac{11}{12} = x + 5\frac{1}{3}$

B. $\frac{3}{4} \cdot x = 1\frac{2}{3}$

C. $\frac{20}{9} \cdot x = 270$

D. $225 = 4\frac{1}{2} \cdot x$

E. $98 \div 2\frac{1}{3} = x$

F. $22 + x = 36$

G. $x = 4\frac{1}{2} \cdot 225$

H. $x = 5\frac{1}{3} + 8\frac{7}{12}$

I. $22 \cdot x = 36$

J. $x = \frac{3}{4} \cdot 1\frac{2}{3}$

K. $5\frac{1}{3} + x = 8\frac{7}{12}$

L. $\frac{9}{20} \cdot 270 = x$

M. $1\frac{2}{3} + \frac{3}{4} = x$

N. $98 - 2\frac{1}{3} = x$

O. $\frac{9}{20} \cdot x = 270$

Answers on page A-5

6. *Deli Order.* For a promotional open house for contractors last year, the Bayside Builders Association ordered 225 turkey sandwiches. Due to increased registrations this year, $4\frac{1}{2}$ times as many sandwiches will be needed. How many sandwiches should be ordered?

7. *Dog Ownership.* In Sam's community, $\frac{9}{20}$ of the households own at least one dog. There are 270 households. How many own dogs?

8. *Magic Tricks.* A magic trick requires a piece of rope $2\frac{1}{3}$ ft long. Gerry, a magician, has 98 ft of rope and needs to divide it into $2\frac{1}{3}$-ft pieces. How many pieces can be cut from the rope?

9. *Painting.* Laura needs $1\frac{2}{3}$ gal of paint to paint the ceiling of the exercise room and $\frac{3}{4}$ gal of the same paint for the bathroom. How much paint does Laura need?

10. *Chocolate Fudge Bars.* A recipe for chocolate fudge bars that serves 16 includes $1\frac{2}{3}$ cups of sugar. How much sugar is needed for $\frac{3}{4}$ of this recipe?

2.5 EXERCISE SET

For Extra Help

MathXL | MyMathLab | InterAct Math | Math Tutor Center | Video Lectures on CD Disc 1 | Student's Solutions Manual

a Solve.

1. If each piece of pie is $\frac{1}{6}$ of a pie, how much of the pie is $\frac{1}{2}$ of a piece?

2. It takes $\frac{2}{3}$ yd of ribbon to make a bow. How much ribbon is needed to make 5 bows?

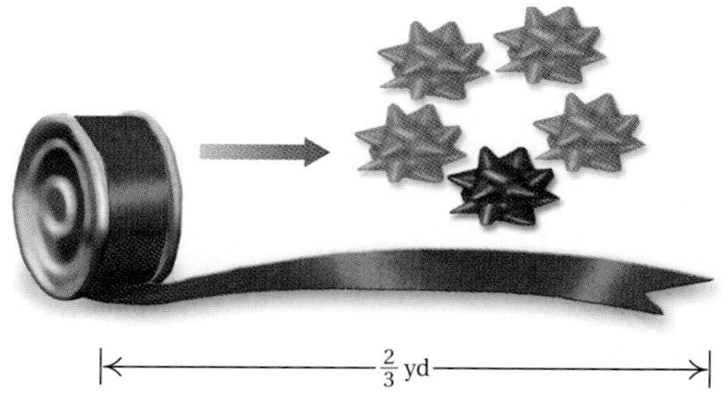

3. A gasoline can holds $\frac{7}{8}$ liter (L). How much will the can hold when it is $\frac{1}{2}$ full?

4. *Floor Tiling.* The floor of a room is being covered with tile. An area $\frac{3}{5}$ of the length and $\frac{3}{4}$ of the width is covered. What fraction of the floor has been tiled?

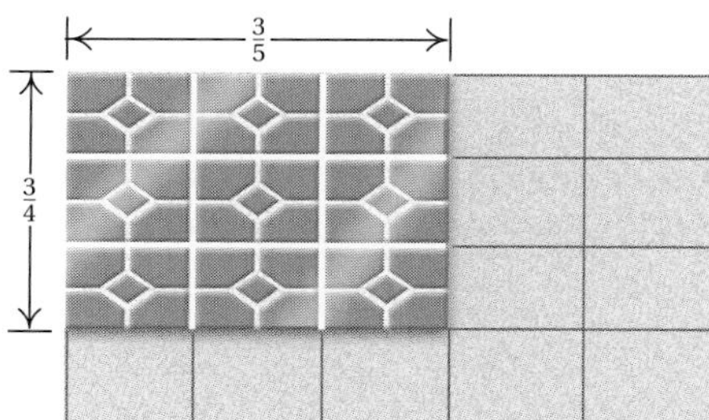

5. How many $\frac{2}{3}$-cup sugar bowls can be filled from 16 cups of sugar?

6. How many $\frac{2}{3}$-cup cereal bowls can be filled from 10 cups of cornflakes?

The *pitch* of a screw is the distance between its threads. With each complete rotation, the screw goes in or out a distance equal to its pitch. Use this information to answer Exercises 7 and 8.

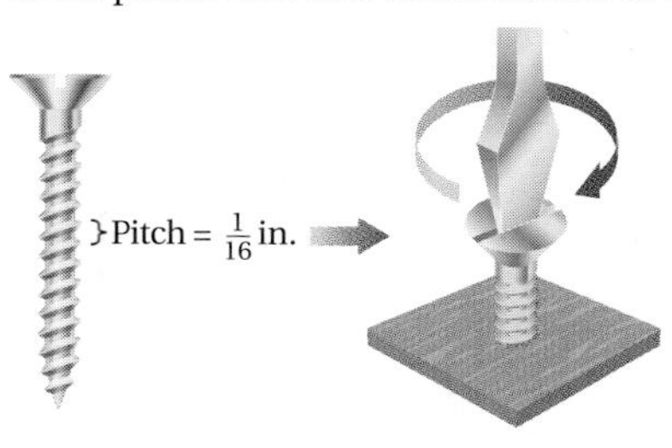

7. The pitch of a screw is $\frac{1}{16}$ in. How far will it go into a piece of oak when it is turned 10 complete rotations clockwise?

8. The pitch of a screw is $\frac{3}{32}$ in. How far will it go out of a piece of plywood when it is turned 10 complete rotations counterclockwise?

9. *Basement Carpet.* A basement floor is being covered with carpet. An area $\frac{7}{8}$ of the length and $\frac{3}{4}$ of the width is covered by lunch time. What fraction of the floor has been completed?

10. *Basketball: High School to Pro.* One of 35 high school basketball players plays college basketball. One of 75 college players plays professional basketball. What fractional part of high school players play professional basketball?

Source: National Basketball Association

11. *Running Speeds.* The maximum running speed of an elk over approximately quarter-mile distances is about 45 mph. The maximum speed of a grizzly bear is about $\frac{2}{3}$ that of an elk. Find the running speed of a grizzly bear.

Source: *Natural History* magazine, © American Museum of Natural History

12. *Map Scaling.* On a map, 1 in. represents 120 mi. How much does $\frac{3}{4}$ in. represent?

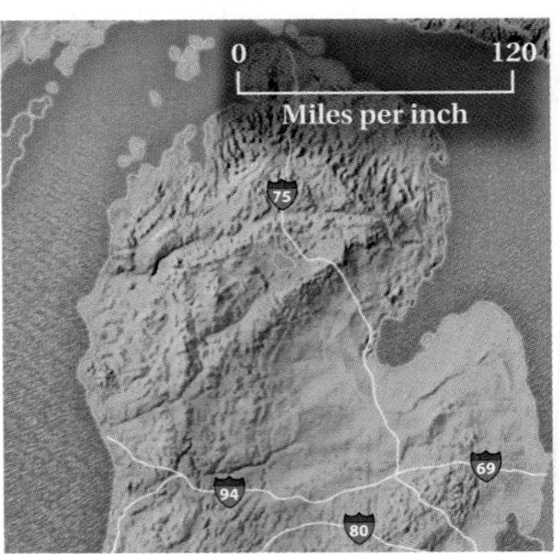

13. A house worth $154,000 is assessed for $\frac{3}{4}$ of its value. What is the assessed value of the house?

14. Roxanne's tuition was $4600. A loan was obtained for $\frac{3}{4}$ of the tuition. How much was the loan?

15. *Household Budgets.* A family has an annual income of $36,000. Of this, $\frac{1}{4}$ is spent for food, $\frac{1}{5}$ for housing, $\frac{1}{10}$ for clothing, $\frac{1}{9}$ for savings, $\frac{1}{4}$ for taxes, and the rest for other expenses. How much is spent for each?

16. *Household Budgets.* A family has an annual income of $29,700. Of this, $\frac{1}{4}$ is spent for food, $\frac{1}{5}$ for housing, $\frac{1}{10}$ for clothing, $\frac{1}{9}$ for savings, $\frac{1}{4}$ for taxes, and the rest for other expenses. How much is spent for each?

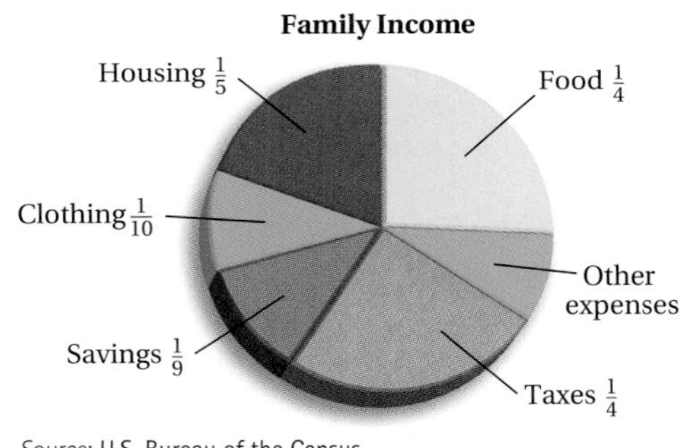

Source: U.S. Bureau of the Census

17. A baker used $\frac{1}{2}$ lb of flour for rolls, $\frac{1}{4}$ lb for donuts, and $\frac{1}{3}$ lb for cookies. How much flour was used?

18. A tile $\frac{5}{8}$ in. thick is glued to a board $\frac{7}{8}$ in. thick. The glue is $\frac{3}{32}$ in. thick. How thick is the result?

19. Benny uses $\frac{2}{5}$ gram (g) of toothpaste each time he brushes his teeth. If Benny buys a 30-g tube, how many times will he be able to brush his teeth?

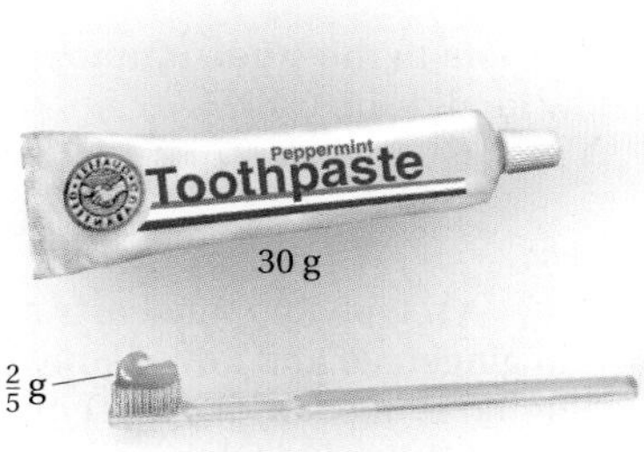

20. *Gasoline Tanker.* A tanker that delivers gasoline to gas stations had 1400 gal of gasoline when it was $\frac{7}{9}$ full. How much could the tanker hold when it is full?

21. A pair of basketball shorts requires $\frac{3}{4}$ yd of nylon. How many pairs of shorts can be made from 24 yd of nylon?

22. A child's baseball shirt requires $\frac{5}{6}$ yd of fabric. How many shirts can be made from 25 yd of the fabric?

23. A bucket had 12 L of water in it when it was $\frac{3}{4}$ full. How much could it hold altogether?

24. A tank had 20 L of gasoline in it when it was $\frac{4}{5}$ full. How much could it hold altogether?

25. Yoshi Teramoto sells hardware tools. After driving 180 kilometers (km), he completes $\frac{5}{8}$ of a sales trip. How long is the total trip? How many kilometers are left to drive?

26. A piece of coaxial cable $\frac{4}{5}$ meter (m) long is to be cut into 8 pieces of the same length. What is the length of each piece?

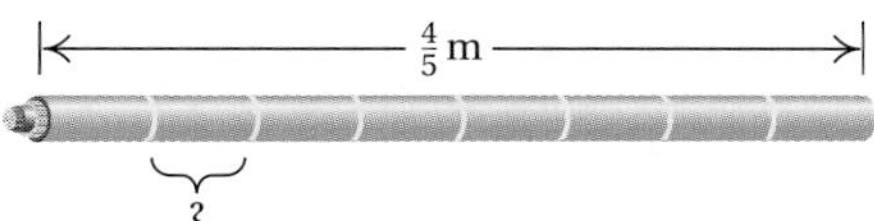

27. *Carpentry.* To cut expenses, a carpenter sometimes glues two kinds of plywood together. He glues a $\frac{1}{4}$" piece of walnut plywood to a $\frac{3}{8}$" piece of less expensive plywood. What is the total thickness of these pieces?

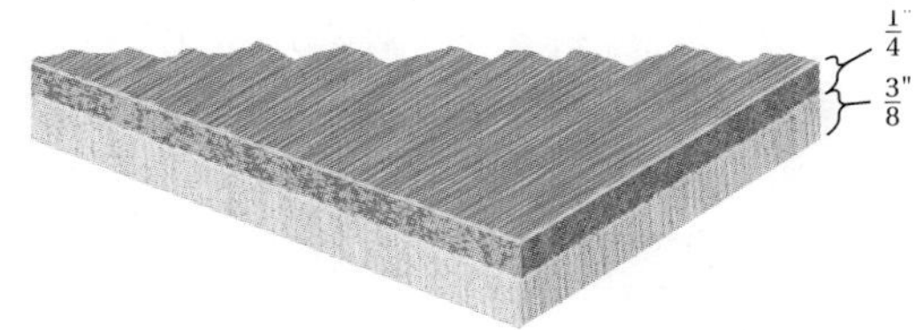

28. *Iced Cookies.* A chef prepared cookies for the freshmen orientation reception. He frosted the $\frac{5}{16}$" cookies with a $\frac{5}{32}$" layer of icing. What is the thickness of the iced cookie?

29. A Subway franchise is owned by Alicia, Erica, and Hannah. Alicia owns $\frac{7}{12}$ of the business and Erica owns $\frac{1}{6}$. What part of the business does Hannah own?

30. Jovan has an $\frac{11}{10}$-lb mixture of cashews and peanuts that includes $\frac{3}{5}$ lb of cashews. How many pounds of peanuts are in the mixture?

31. Rene bought $\frac{1}{3}$ lb of orange pekoe tea and $\frac{1}{2}$ lb of English cinnamon tea. How many pounds of tea did he buy?

32. Stan bought $\frac{1}{4}$ lb of gumdrops and $\frac{1}{2}$ lb of caramels. How many pounds of candy did he buy?

33. Russ walked $\frac{7}{6}$ mi to a friend's dormitory, and then $\frac{3}{4}$ mi to class. How far did he walk?

34. Elaine walked $\frac{7}{8}$ mi to the student union, and then $\frac{2}{5}$ mi to class. How far did she walk?

35. *Concrete Mix.* A cubic meter of concrete mix contains 420 kilograms (kg) of cement, 150 kg of stone, and 120 kg of sand. What is the total weight of a cubic meter of the mix? What part is cement? stone? sand? Add these fractional amounts. What is the result?

36. *Punch Recipe.* A recipe for strawberry punch calls for $\frac{1}{5}$ quart (qt) of ginger ale and $\frac{3}{5}$ qt of strawberry soda. How much liquid is needed? If the recipe is doubled, how much liquid is needed? If the recipe is halved, how much liquid is needed?

37. Jaci spent $\frac{3}{4}$ hr listening to DVDs of Maroon 5 and U2. She spent $\frac{1}{3}$ hr listening to Maroon 5. How many hours were spent listening to U2?

38. As part of a fitness program, Deb swims $\frac{1}{2}$ mi every day. One day she had already swum $\frac{1}{5}$ mi. How much farther should Deb swim?

39. A server has a bowl containing $\frac{11}{12}$ cup of grated Parmesan cheese and puts $\frac{1}{8}$ cup on a customer's spaghetti and meatballs. How much remains in the bowl?

40. *Tire Tread.* A long-life tire has a tread depth of $\frac{3}{8}$ in. instead of a more typical depth of $\frac{11}{32}$ in. How much deeper is the long-life tread depth?

Source: *Popular Science*

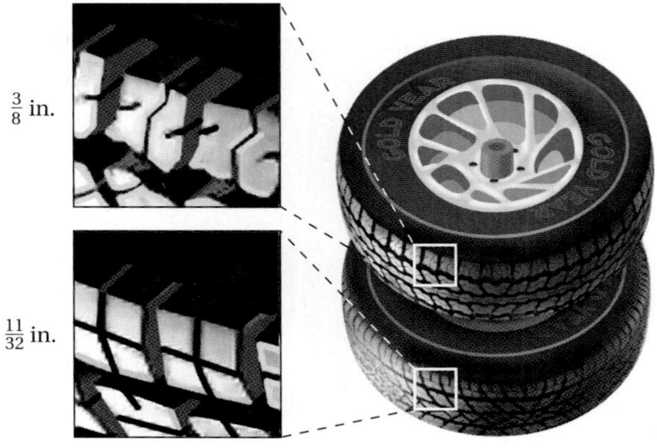

41. *Carpentry.* When cutting wood with a saw, a carpenter must take into account the thickness of the saw blade. Suppose that from a piece of wood 36 in. long, a carpenter cuts a $15\frac{3}{4}$-in. length with a saw blade that is $\frac{1}{8}$ in. thick. How long is the piece that remains?

42. *Painting.* A painter used $1\frac{3}{4}$ gal of paint for the Garcias' living room and $1\frac{1}{3}$ gal for their family room. How much paint was used in all?

43. *Sewing from a Pattern.* Suppose you want to make an outfit in size 8. Using 45-in. fabric, you need $1\frac{3}{8}$ yd for the dress, $\frac{5}{8}$ yd of contrasting fabric for the band at the bottom, and $3\frac{3}{8}$ yd for the jacket. How many yards of 45-in. fabric are needed to make the outfit?

44. Marsha's Butcher Shop sold packages of sliced turkey breast weighing $1\frac{1}{3}$ lb and $4\frac{3}{5}$ lb. What was the total weight of the meat?

45. Tara is 66 in. tall and her son, Tom, is $59\frac{7}{12}$ in. tall. How much taller is Tara?

46. A Boeing 767 flew 640 mi on a nonstop flight. On the return flight, it landed after having flown $320\frac{3}{10}$ mi. How far was the plane from its original point of departure?

47. Find the perimeter of (distance around) the figure.

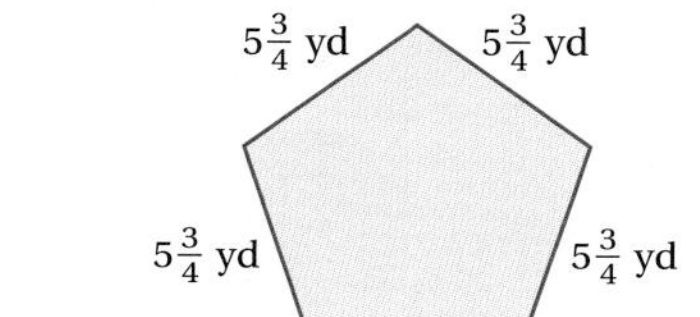

48. Find the length d in the figure.

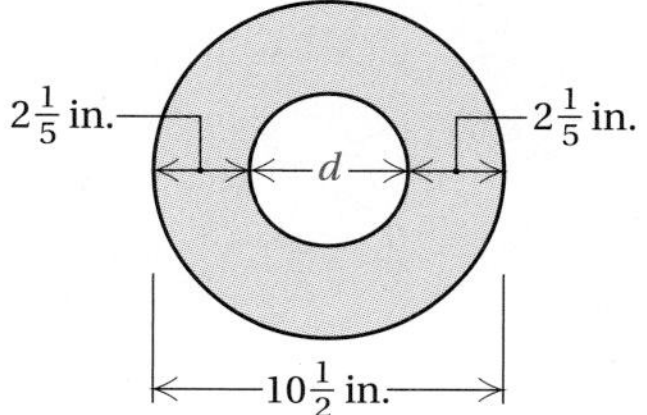

49. *Cutco Knives.* The Kitchen Classics knife set sold by Cutco contains three basic knives: $7\frac{5}{8}$" Petite Chef, $4\frac{3}{4}$" Trimmer, and $2\frac{3}{4}$" Paring Knife. How much larger is the blade in the Petite Chef than in the Trimmer? than in the Paring Knife?

Source: Cutco Cutlery Corporation

50. *Upholstery Fabric.* Executive Car Care sells 45-in. upholstery fabric for car restoration. Art buys $9\frac{1}{4}$ yd and $10\frac{5}{6}$ yd for two car projects. How many yards did Art buy in all?

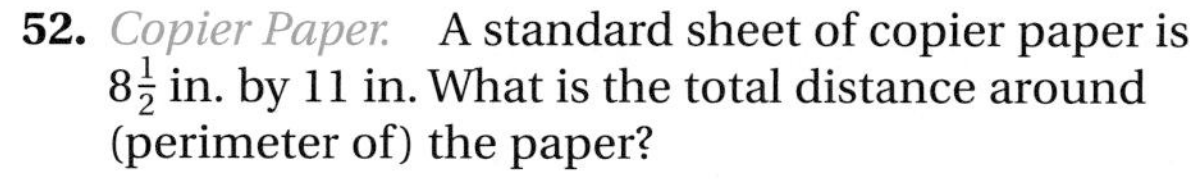

51. Find the smallest length of a bolt that will pass through a piece of tubing with an outside diameter of $\frac{1}{2}$ in., a washer $\frac{1}{16}$ in. thick, a piece of tubing with a $\frac{3}{4}$-in. outside diameter, another washer, and a nut $\frac{3}{16}$ in. thick.

52. *Copier Paper.* A standard sheet of copier paper is $8\frac{1}{2}$ in. by 11 in. What is the total distance around (perimeter of) the paper?

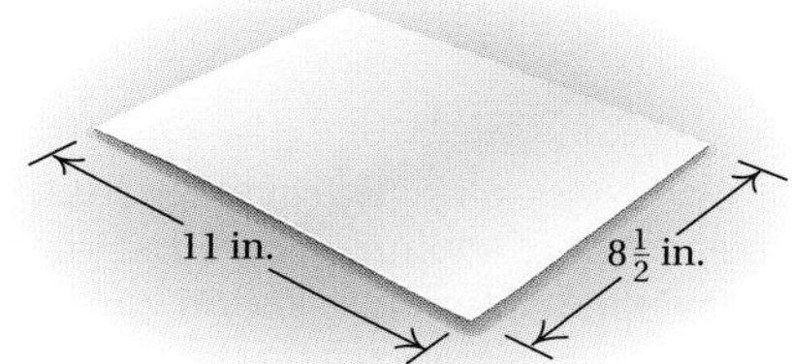

53. A car traveled 213 mi on $14\frac{2}{10}$ gal of gas. How many miles per gallon did it get?

54. A car traveled 385 mi on $15\frac{4}{10}$ gal of gas. How many miles per gallon did it get?

55. *Temperatures.* Fahrenheit temperature can be obtained from Celsius (centigrade) temperature by multiplying by $1\frac{4}{5}$ and adding 32°. What Fahrenheit temperature corresponds to a Celsius temperature of 20°?

56. *Weight of Water.* The weight of water is $62\frac{1}{2}$ lb per cubic foot. What is the weight of $2\frac{1}{4}$ cubic feet of water?

57. *Beagles.* There are about 155,000 Labrador retrievers registered with The American Kennel Club. This is $3\frac{4}{9}$ times the number of beagles registered. How many beagles are registered?

Source: The American Kennel Club

58. *Exercise.* At one point during an aerobics class, Kea's bicycle wheel was completing $76\frac{2}{3}$ revolutions per minute. How many revolutions did the wheel complete in 6 min?

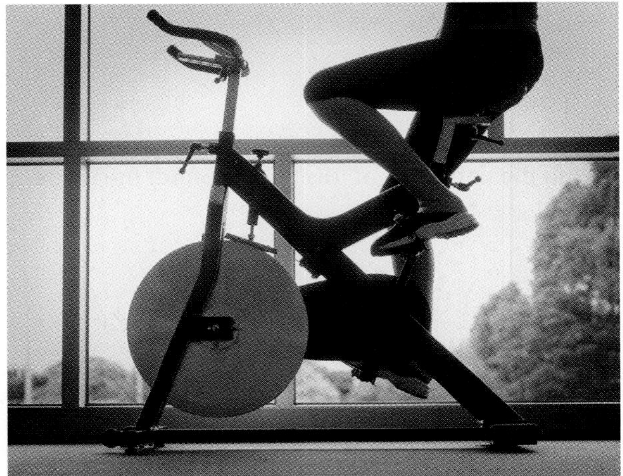

59. *Population.* The population of Louisiana is $2\frac{1}{2}$ times the population of West Virginia. The population of West Virginia is approximately 1,800,000. What is the population of Louisiana?

Source: U.S. Bureau of the Census

60. *Population.* The population of New York is $3\frac{1}{3}$ times that of Missouri. The population of New York is approximately 19,000,000. What is the population of Missouri?

Source: U.S. Bureau of the Census

61. *Mural.* Cecilia hired an artist to paint a mural on the wall in her twin sons' bedroom. The dimensions of the mural must be $6\frac{2}{3}$ ft by $9\frac{3}{8}$ ft. What is the area of the mural?

62. *Sidewalk.* A sidewalk alongside a garden at the conservatory is to be $14\frac{2}{5}$ yd long. Rectangular stone tiles that are each $1\frac{1}{8}$ yd long are used to form the sidewalk. How many tiles are used?

63. *Sodium Consumption.* The average American woman consumes $1\frac{1}{3}$ tsp of sodium each day. How much sodium does the average American woman consume in 10 days?

Source: *Nutrition Action Health Letter,* March 1994, p. 6. 1875 Connecticut Ave., N.W., Washington, DC 20009-5728

64. *Aeronautics.* Most space shuttles orbit the earth once every $1\frac{1}{2}$ hr. How many orbits are made every 24 hr?

65. *Weight of Water.* The weight of water is $62\frac{1}{2}$ lb per cubic foot. How many cubic feet would be occupied by 250 lb of water?

66. *Weight of Water.* The weight of water is $62\frac{1}{2}$ lb per cubic foot. How many cubic feet would be occupied by 375 lb of water?

Find the area of the shaded region.

67.

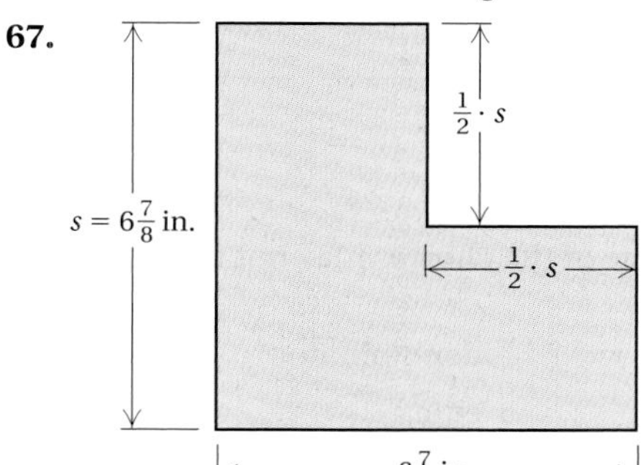

68.

$10\frac{1}{2}$ ft

$8\frac{1}{2}$ ft

4 ft

$10\frac{1}{2}$ ft

69. *Construction.* A rectangular lot has dimensions of $302\frac{1}{2}$ ft by $205\frac{1}{4}$ ft. A building with dimensions of 100 ft by $25\frac{1}{2}$ ft is built on the lot. How much area is left over?

70. *Word Processing.* Kelly wants to create a table using Microsoft® Word software for word processing. She needs to have two columns, each $1\frac{1}{2}$ in. wide, and five columns, each $\frac{3}{4}$ in. wide. Will this table fit on a piece of standard paper that is $8\frac{1}{2}$ in. wide? If so, how wide will each margin be if her margins on each side are to be of equal width?

71. *Creamy Peach Coffee Cake.* Listed below are the ingredients for creamy peach coffee cake. What are the ingredients for $\frac{1}{2}$ cake? for 4 cakes?

Source: Reprinted with permission from *Taste of Home* magazine, www.tasteofhome.com

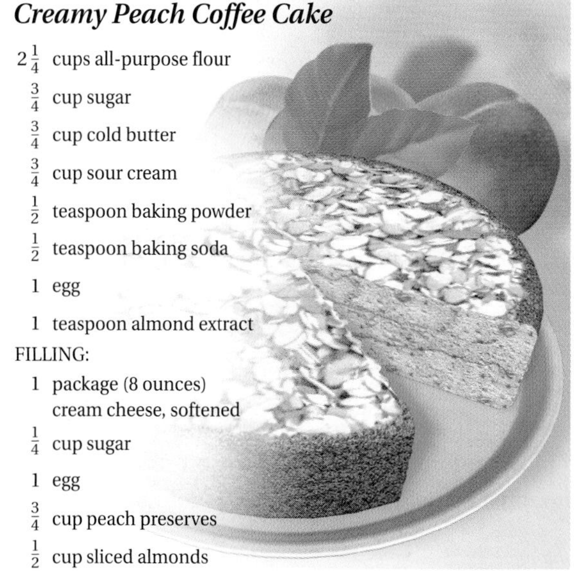

Creamy Peach Coffee Cake
$2\frac{1}{4}$ cups all-purpose flour
$\frac{3}{4}$ cup sugar
$\frac{3}{4}$ cup cold butter
$\frac{3}{4}$ cup sour cream
$\frac{1}{2}$ teaspoon baking powder
$\frac{1}{2}$ teaspoon baking soda
1 egg
1 teaspoon almond extract
FILLING:
1 package (8 ounces) cream cheese, softened
$\frac{1}{4}$ cup sugar
1 egg
$\frac{3}{4}$ cup peach preserves
$\frac{1}{2}$ cup sliced almonds

72. *Butterscotch Hard Candy.* Listed below are the ingredients for butterscotch hard candy. What are the ingredients for $\frac{1}{2}$ batch? for 3 batches?

Source: Reprinted with permission from *Taste of Home* magazine, www.tasteofhome.com

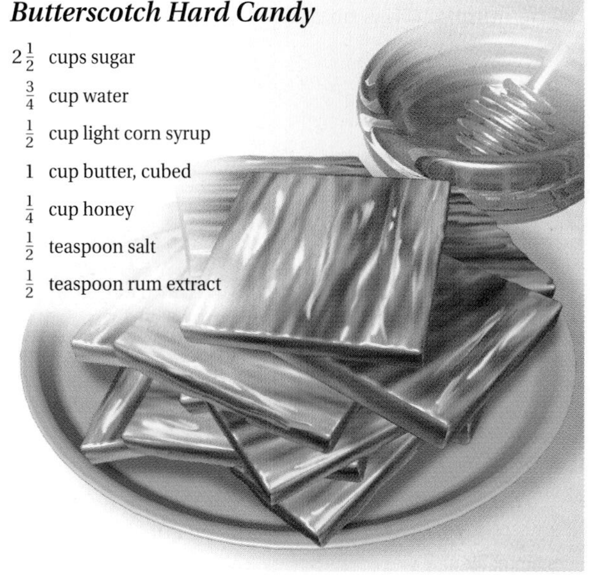

73. DW Write a problem for a classmate to solve. Design the problem so the solution is "The larger package holds $4\frac{1}{2}$ oz more than the smaller package."

74. DW Write a problem for a classmate to solve. Design the problem so that its solution is found by performing the multiplication $4\frac{1}{2} \cdot 33\frac{1}{3}$.

SKILL MAINTENANCE

VOCABULARY REINFORCEMENT

In each of Exercises 75–82, fill in the blank with the correct term from the given list. Some of the choices may not be used and some may be used more than once.

75. In the equation $420 \div 60 = 7$, 60 is the _______________, 7 is the _______________, and 420 is the _______________. [1.3d]

76. When denominators are the same, we say that fractions have a _______________ denominator. [2.3a]

77. The numbers 91, 95, and 111 are examples of _______________ numbers. [1.7c]

78. The number 22,223,133 is _______________ by 9 because the sum of its digits is _______________ by 9. [1.8a]

79. When simplifying $24 \div 4 + 4 \times 12 - 6 \div 2$, do all _______________ and _______________ in order from left to right before doing all _______________ and _______________ in order from left to right. [1.6c]

80. In the equation $2 + 3 = 5$, 2 and 3 are the _______________. [1.2a]

81. In the expression $\frac{c}{d}$, c is the _______________. [2.1a]

82. The number 0 has no _______________. [2.2c]

identity	denominator
reciprocal	equal
divisor	common
dividend	prime
quotient	composite
additions	product
subtractions	divisible
multiplications	digits
divisions	factors
numerator	addends

SYNTHESIS

83. A guitarist's band is booked for Friday and Saturday nights at a local club. The guitarist is part of a trio on Friday and part of a quintet on Saturday. Thus the guitarist is paid one-third of one-half the weekend's pay for Friday and one-fifth of one-half the weekend's pay for Saturday. What fractional part of the band's pay did the guitarist receive for the weekend's work? If the band was paid $1200, how much did the guitarist receive?

84. *Microsoft Interview.* The following is a question taken from an employment interview with Microsoft. Try to answer it. "Given a gold bar that can be cut exactly twice and a contractor who must be paid one-seventh of a gold bar every day for seven days, how should the bar be cut?"
Source: *Fortune* magazine, January 22, 2001

2.6 ORDER OF OPERATIONS; ESTIMATION

Objectives

a Simplify expressions using the rules for order of operations.

b Estimate with fraction notation and mixed numerals.

a Order of Operations; Fraction Notation and Mixed Numerals

The rules for order of operations that we use with whole numbers (see Section 1.6) apply when we are simplifying expressions involving fraction notation and mixed numerals. For review, these rules are listed below.

RULES FOR ORDER OF OPERATIONS

1. Do all calculations within parentheses before operations outside.
2. Evaluate all exponential expressions.
3. Do all multiplications and divisions in order from left to right.
4. Do all additions and subtractions in order from left to right.

EXAMPLE 1 Simplify: $\frac{1}{6} + \frac{2}{3} \div \frac{1}{2} \cdot \frac{5}{8}$.

$\frac{1}{6} + \frac{2}{3} \div \frac{1}{2} \cdot \frac{5}{8} = \frac{1}{6} + \frac{2}{3} \cdot \frac{2}{1} \cdot \frac{5}{8}$ — Doing the division first by multiplying by the reciprocal of $\frac{1}{2}$

$= \frac{1}{6} + \frac{2 \cdot 2 \cdot 5}{3 \cdot 1 \cdot 8}$ — Doing the multiplications in order from left to right

$= \frac{1}{6} + \frac{\not{2} \cdot \not{2} \cdot 5}{3 \cdot 1 \cdot \not{2} \cdot \not{2} \cdot 2}$ — Factoring in order to simplify

$= \frac{1}{6} + \frac{5}{6}$ — Removing a factor of 1: $\frac{2 \cdot 2}{2 \cdot 2} = 1$; simplifying

$= \frac{6}{6}$, or 1 — Doing the addition

Do Exercises 1 and 2.

Simplify.

1. $\frac{2}{5} \cdot \frac{5}{8} + \frac{1}{4}$

2. $\frac{1}{3} \cdot \frac{3}{4} \div \frac{5}{8} - \frac{1}{10}$

EXAMPLE 2 Simplify: $\frac{2}{3} \cdot 24 - 11\frac{1}{2}$.

$\frac{2}{3} \cdot 24 - 11\frac{1}{2} = \frac{2 \cdot 24}{3} - 11\frac{1}{2}$ — Doing the multiplication first

$= \frac{2 \cdot \not{3} \cdot 8}{\not{3}} - 11\frac{1}{2}$ — Factoring the numerator

$= 2 \cdot 8 - 11\frac{1}{2}$ — Removing a factor of 1: $\frac{3}{3} = 1$

$= 16 - 11\frac{1}{2}$ — Completing the multiplication

$= 4\frac{1}{2}$, or $\frac{9}{2}$ — Doing the subtraction

Do Exercise 3.

3. Simplify: $\frac{3}{4} \cdot 16 + 8\frac{2}{3}$.

Answers on page A-5

4. After two weeks, Kurt's tomato seedlings measure $9\frac{1}{2}$ in., $10\frac{3}{4}$ in., $10\frac{1}{4}$ in., and 9 in. tall. Find their average height.

5. Find the average of

$$\frac{1}{2}, \frac{1}{3}, \text{ and } \frac{5}{6}.$$

6. Find the average of $\frac{3}{4}$ and $\frac{4}{5}$.

7. Simplify:

$$\left(\frac{2}{3}+\frac{3}{4}\right) \div 2\frac{1}{3} - \left(\frac{1}{2}\right)^3.$$

Answers on page A-5

EXAMPLE 3 Melody has triplets. Their birth weights were $3\frac{1}{2}$ lb, $2\frac{3}{4}$ lb, and $3\frac{1}{8}$ lb. What was the average weight of her babies?

Recall that to compute an **average,** we add the numbers and then divide the sum by the number of addends (see Section 1.6). We have

$$\frac{3\frac{1}{2}+2\frac{3}{4}+3\frac{1}{8}}{3}.$$

We first add in the numerator:

$$\frac{3\frac{1}{2}+2\frac{3}{4}+3\frac{1}{8}}{3} = \frac{3\frac{4}{8}+2\frac{6}{8}+3\frac{1}{8}}{3}$$

$$= \frac{8\frac{11}{8}}{3} = \frac{\frac{75}{8}}{3}$$ Doing the additions; converting $8\frac{11}{8}$ to fraction notation, $\frac{75}{8}$

$$= \frac{75}{8}\cdot\frac{1}{3} = \frac{75}{24} = \frac{25}{8}$$ Multiplying by the reciprocal and simplifying

$$= 3\frac{1}{8}.$$ Converting to a mixed numeral

The average weight of the babies is $3\frac{1}{8}$ lb.

Do Exercises 4–6.

EXAMPLE 4 Simplify: $\left(\frac{7}{8}-\frac{1}{3}\right)\times 48 + \left(13+\frac{4}{5}\right)^2$.

$$\left(\frac{7}{8}-\frac{1}{3}\right)\times 48 + \left(13+\frac{4}{5}\right)^2$$

$$= \left(\frac{7}{8}\cdot\frac{3}{3}-\frac{1}{3}\cdot\frac{8}{8}\right)\times 48 + \left(13\cdot\frac{5}{5}+\frac{4}{5}\right)^2$$ Carrying out operations inside parentheses first. To do so, we first multiply by 1 to obtain the LCD.

$$= \left(\frac{21}{24}-\frac{8}{24}\right)\times 48 + \left(\frac{65}{5}+\frac{4}{5}\right)^2$$

$$= \frac{13}{24}\times 48 + \left(\frac{69}{5}\right)^2$$ Completing the operations within parentheses

$$= \frac{13}{24}\times 48 + \frac{4761}{25}$$ Evaluating the exponential expression next

$$= 26 + \frac{4761}{25}$$ Doing the multiplication

$$= 26 + 190\frac{11}{25}$$ Converting to a mixed numeral

$$= 216\frac{11}{25}, \text{ or } \frac{5411}{25}$$ Adding

Answers can be given using either fraction notation or mixed numerals.

Do Exercise 7.

b Estimation with Fraction Notation and Mixed Numerals

We now estimate with fraction notation and mixed numerals.

EXAMPLES Estimate each of the following as 0, $\frac{1}{2}$, or 1.

5. $\dfrac{2}{17}$

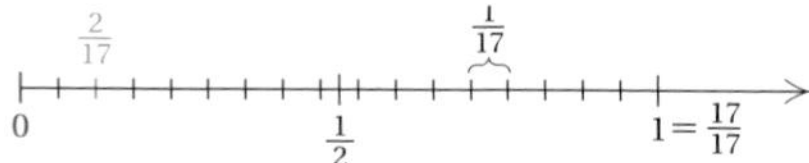

A fraction is close to 0 when the numerator is small in comparison to the denominator. Thus, 0 is an estimate for $\frac{2}{17}$ because 2 is small in comparison to 17. Thus, $\frac{2}{17} \approx 0$.

6. $\dfrac{11}{23}$

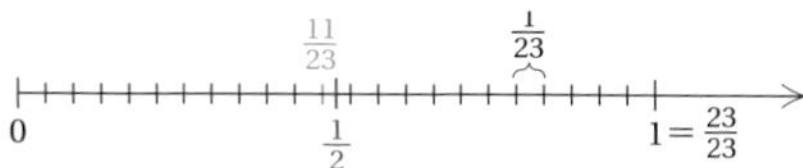

A fraction is close to $\frac{1}{2}$ when the denominator is about twice the numerator. Thus, $\frac{1}{2}$ is an estimate for $\frac{11}{23}$ because $2 \cdot 11 = 22$ and 22 is close to 23. Thus, $\frac{11}{23} \approx \frac{1}{2}$.

7. $\dfrac{37}{38}$

A fraction is close to 1 when the numerator is nearly equal to the denominator. Thus, 1 is an estimate for $\frac{37}{38}$ because 37 is nearly equal to 38. Thus, $\frac{37}{38} \approx 1$.

8. $\dfrac{43}{41}$

As in the preceding example, the numerator 43 is nearly equal to the denominator 41. Thus, $\frac{43}{41} \approx 1$.

Do Exercises 8–11.

EXAMPLE 9 Find a number for the blank so that $\dfrac{9}{\square}$ is close to but less than 1. Answers may vary.

If the number in the blank were 9, we would have 1, so we increase 9 to 10. An answer is 10; $\frac{9}{10}$ is close to 1. The number 11 would also be a correct answer; $\frac{9}{11}$ is close to 1.

Do Exercises 12 and 13.

Estimate each of the following as 0, $\frac{1}{2}$, or 1.

8. $\dfrac{3}{59}$

9. $\dfrac{61}{59}$

10. $\dfrac{29}{59}$

11. $\dfrac{57}{59}$

Find a number for the blank so that the fraction is close to but less than 1.

12. $\dfrac{11}{\square}$

13. $\dfrac{\square}{33}$

Answers on page A-5

Study Tips

SKILL MAINTENANCE EXERCISES

It is never too soon to begin reviewing for the final examination. The Skill Maintenance exercises found in each exercise set review and reinforce skills taught in earlier sections. Include all of these exercises in your weekly preparation. Answers to both odd-numbered and even-numbered exercises along with section references appear at the back of the book.

EXAMPLE 10 Find a number for the blank so that $\frac{9}{\square}$ is close to but less than $\frac{1}{2}$. Answers may vary.

If we double 9 to get 18 and use it for the blank, we have $\frac{1}{2}$. If we increase that denominator by 1, to get 19, and use it for the blank, we get a number less than $\frac{1}{2}$ but close to $\frac{1}{2}$. Thus, $\frac{9}{19} \approx \frac{1}{2}$.

Do Exercises 14 and 15.

EXAMPLE 11 Find a number for the blank so that $\frac{\square}{50}$ is close to but greater than 0.

Since 50 is rather large, any small number such as 1, 2, or 3 will make the fraction close to 0. For example, $\frac{1}{50} \approx 0$.

Do Exercises 16 and 17.

EXAMPLE 12 Estimate $16\frac{8}{9} + 11\frac{2}{13} - 4\frac{22}{43}$ as a whole number or as a mixed numeral where the fractional part is $\frac{1}{2}$.

We estimate each fraction as 0, $\frac{1}{2}$, or 1. Then we calculate:

$$16\frac{8}{9} + 11\frac{2}{13} - 4\frac{22}{43} \approx 17 + 11 - 4\frac{1}{2}$$
$$= 28 - 4\frac{1}{2}$$
$$= 23\frac{1}{2}.$$

Do Exercises 18–20.

Find a number for the blank so that the fraction is close to but less than $\frac{1}{2}$.

14. $\frac{13}{\square}$

15. $\frac{\square}{31}$

Find a number for the blank so that the fraction is close to but greater than 0.

16. $\frac{\square}{37}$

17. $\frac{13}{\square}$

Estimate each part of the following as a whole number or as a mixed numeral where the fractional part is $\frac{1}{2}$.

18. $5\frac{9}{10} + 26\frac{1}{2} - 10\frac{3}{29}$

19. $10\frac{7}{8} \cdot \left(25\frac{11}{13} - 14\frac{1}{9}\right)$

20. $\left(10\frac{4}{5} + 7\frac{5}{9}\right) \div \frac{17}{30}$

Answers on page A-5

2.6 EXERCISE SET

For Extra Help

MathXL | MyMathLab | InterAct Math | Math Tutor Center | Video Lectures on CD Disc 1 | Student's Solutions Manual

a Simplify.

1. $\frac{1}{2} \cdot \frac{1}{3} \cdot \frac{1}{4}$

2. $\frac{1}{3} \cdot \frac{1}{4} \cdot \frac{1}{5}$

3. $6 \div 3 \div 5$

4. $12 \div 4 \div 8$

5. $\frac{2}{3} \div \frac{4}{3} \div \frac{7}{8}$

6. $\frac{5}{6} \div \frac{3}{4} \div \frac{2}{5}$

7. $\frac{5}{8} \div \frac{1}{4} - \frac{2}{3} \cdot \frac{4}{5}$

8. $\frac{4}{7} \cdot \frac{7}{15} + \frac{2}{3} \div 8$

9. $\frac{3}{4} - \frac{2}{3} \cdot \left(\frac{1}{2} + \frac{2}{5}\right)$

10. $\frac{3}{4} \div \frac{1}{2} \cdot \left(\frac{8}{9} - \frac{2}{3}\right)$

11. $28\frac{1}{8} - 5\frac{1}{4} + 3\frac{1}{2}$

12. $10\frac{3}{5} - 4\frac{1}{10} - 1\frac{1}{2}$

13. $\frac{7}{8} \div \frac{1}{2} \cdot \frac{1}{4}$

14. $\frac{7}{10} \cdot \frac{4}{5} \div \frac{2}{3}$

15. $\left(\frac{2}{3}\right)^2 - \frac{1}{3} \cdot 1\frac{1}{4}$

16. $\left(\frac{3}{4}\right)^2 + 3\frac{1}{2} \div 1\frac{1}{4}$

17. $\frac{1}{2} - \left(\frac{1}{2}\right)^2 + \left(\frac{1}{2}\right)^3$

18. $1 + \frac{1}{4} + \left(\frac{1}{4}\right)^2 - \left(\frac{1}{4}\right)^3$

19. Find the average of $\frac{2}{3}$ and $\frac{7}{8}$.

20. Find the average of $\frac{1}{4}$ and $\frac{1}{5}$.

21. Find the average of $\frac{1}{6}$, $\frac{1}{8}$, and $\frac{3}{4}$.

22. Find the average of $\frac{4}{5}$, $\frac{1}{2}$, and $\frac{1}{10}$.

23. Find the average of $3\frac{1}{2}$ and $9\frac{3}{8}$.

24. Find the average of $10\frac{2}{3}$ and $24\frac{5}{6}$.

25. *Birth Weights.* The Piper quadruplets of Great Britain weighed $2\frac{9}{16}$ lb, $2\frac{9}{32}$ lb, $2\frac{1}{8}$ lb, and $2\frac{5}{16}$ lb at birth. Find their average birth weight.

Source: *The Guinness Book of Records,* 1998

26. *Vertical Leaps.* Eight-year-old Zachary registered vertical leaps of $12\frac{3}{4}$ in., $13\frac{3}{4}$ in., $13\frac{1}{2}$ in., and 14 in. Find his average vertical leap.

27. *Acceleration.* The results of a *Motor Trend* road acceleration test for five cars are given in the graph below. The test measures the time, in seconds, required to go from 0 mph to 60 mph. What was the average time?

Source: *Motor Trend,* March 2005, pp. 134–142

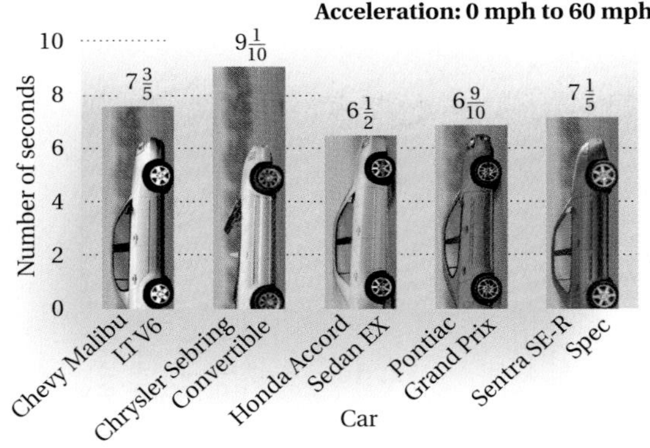

28. *Manufacturing.* A test of five light bulbs showed that they burned for the lengths of time given on the graph below. For how many days, on average, did the bulbs burn?

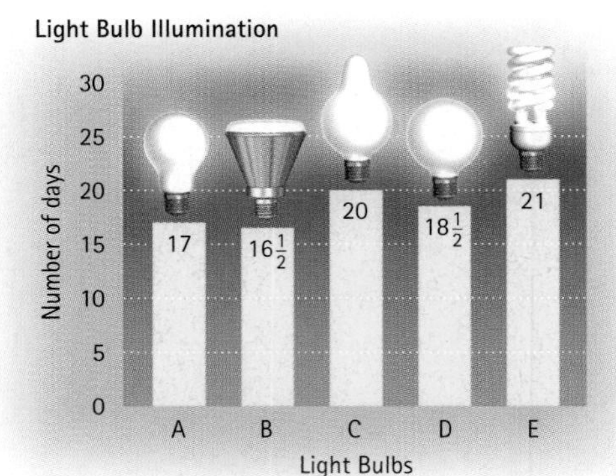

Simplify.

29. $\left(\frac{2}{3} + \frac{3}{4}\right) \div \left(\frac{5}{6} - \frac{1}{3}\right)$

30. $\left(\frac{3}{5} - \frac{1}{2}\right) \div \left(\frac{3}{4} - \frac{3}{10}\right)$

31. $\left(\frac{1}{2} + \frac{1}{3}\right)^2 \cdot 144 - \frac{5}{8} \div 10\frac{1}{2}$

32. $\left(3\frac{1}{2} - 2\frac{1}{3}\right)^2 + 6 \cdot 2\frac{1}{2} \div 32$

b Estimate each of the following as 0, $\frac{1}{2}$, or 1.

33. $\frac{2}{47}$

34. $\frac{4}{5}$

35. $\frac{1}{13}$

36. $\frac{7}{8}$

37. $\frac{6}{11}$

38. $\frac{10}{13}$

39. $\frac{7}{15}$

40. $\frac{1}{16}$

41. $\frac{7}{100}$

42. $\frac{5}{9}$

43. $\frac{19}{20}$

44. $\frac{5}{12}$

Find a number for the blank so that the fraction is close to but greater than $\frac{1}{2}$. Answers may vary.

45. $\frac{\square}{11}$

46. $\frac{\square}{8}$

47. $\frac{\square}{23}$

48. $\frac{\square}{35}$

49. $\frac{10}{\square}$

50. $\frac{51}{\square}$

Find a number for the blank so that the fraction is close to but greater than 1. Answers may vary.

51. $\frac{7}{\square}$

52. $\frac{11}{\square}$

53. $\frac{13}{\square}$

54. $\frac{27}{\square}$

55. $\frac{\square}{15}$

56. $\frac{\square}{100}$

Estimate each part of the following as a whole number, as $\frac{1}{2}$, or as a mixed numeral where the fractional part is $\frac{1}{2}$.

57. $2\frac{7}{8}$

58. $1\frac{1}{3}$

59. $12\frac{5}{6}$

60. $26\frac{6}{13}$

61. $\frac{4}{5} + \frac{7}{8}$

62. $\frac{1}{12} \cdot \frac{7}{15}$

63. $\frac{2}{3} + \frac{7}{13} + \frac{5}{9}$

64. $\frac{8}{9} + \frac{4}{5} + \frac{11}{12}$

65. $\frac{43}{100} + \frac{1}{10} - \frac{11}{1000}$

66. $\frac{23}{24} + \frac{37}{39} + \frac{51}{50}$

67. $7\frac{29}{60} + 10\frac{12}{13} \cdot 24\frac{2}{17}$

68. $5\frac{13}{14} - 1\frac{5}{8} + 1\frac{23}{28} \cdot 6\frac{35}{74}$

69. $24 \div 7\frac{8}{9}$

70. $43\frac{16}{17} \div 11\frac{2}{13}$

71. $76\frac{3}{14} + 23\frac{19}{20}$

72. $76\frac{13}{14} \cdot 23\frac{17}{20}$

73. $16\frac{1}{5} \div 2\frac{1}{11} + 25\frac{9}{10} - 4\frac{11}{23}$

74. $96\frac{2}{13} \div 5\frac{19}{20} + 3\frac{1}{7} \cdot 5\frac{18}{21}$

75. **DW** A student insists that $3\frac{2}{5} \cdot 1\frac{3}{7} = 3\frac{6}{35}$. What mistake is he making and how should he have proceeded?

76. **DW** A student insists that $5 \cdot 3\frac{2}{7} = (5 \cdot 3) \cdot \left(5 \cdot \frac{2}{7}\right)$. What mistake is she making and how should she have proceeded?

SKILL MAINTENANCE

77. Multiply: $27 \cdot 126$. [1.3a]

78. Multiply: $132 \cdot 7865$. [1.3a]

79. Divide: $7865 \div 132$. [1.3d]

Multiply and simplify. [2.1c], [2.2a]

80. $\frac{2}{3} \cdot 522$

81. $\frac{3}{2} \cdot 522$

Divide and simplify. [2.2c]

82. $\frac{4}{5} \div \frac{3}{10}$

83. $\frac{3}{10} \div \frac{4}{5}$

84. Classify the given numbers as prime, composite, or neither. [1.7c]

1, 5, 7, 9, 14, 23, 43

Solve.

85. *Luncheon Servings.* Ian purchased 6 lb of cold cuts for a luncheon. If Ian has allowed $\frac{3}{8}$ lb per person, how many people did he invite to the luncheon? [2.5a]

86. *Cholesterol.* A 3-oz serving of crabmeat contains 85 milligrams (mg) of cholesterol. A 3-oz serving of shrimp contains 128 mg of cholesterol. How much more cholesterol is in the shrimp? [1.5a]

SYNTHESIS

87. **a)** Find an expression for the sum of the areas of the two rectangles shown here.
b) Simplify the expression.
c) How is the computation in part (b) related to the rules for order of operations?

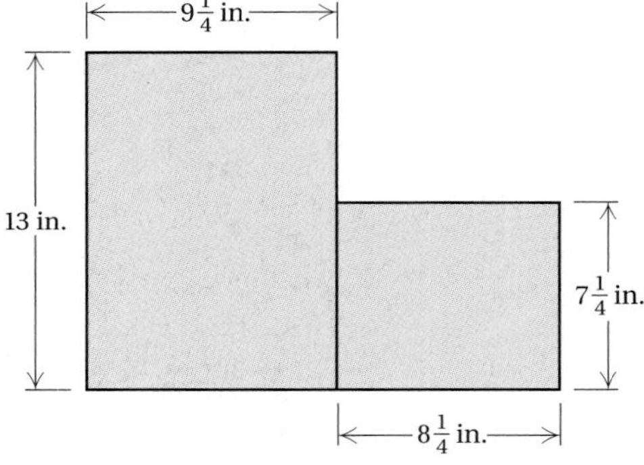

88. Find r if

$$\frac{1}{r} = \frac{1}{100} + \frac{1}{150} + \frac{1}{200}.$$

89. In the sum below, a and b are digits. Find a and b.

$$\frac{a}{17} + \frac{1b}{23} = \frac{35a}{391}$$

90. Consider only the numbers 3, 4, 5, and 6. Assume each can be placed in a blank in the following.

$$\square + \frac{\square}{\square} \cdot \square = ?$$

What placement of the numbers in the blanks yields the largest number?

91. Consider only the numbers 2, 3, 4, and 5. Assume each is placed in a blank in the following.

$$\frac{\square}{\square} + \frac{\square}{\square} = ?$$

What placement of the numbers in the blanks yields the largest sum?

92. Use a standard calculator. Arrange the following in order from smallest to largest.

$$\frac{3}{4}, \frac{17}{21}, \frac{13}{15}, \frac{7}{9}, \frac{15}{17}, \frac{13}{12}, \frac{19}{22}$$

2 Summary and Review

The review that follows is meant to prepare you for a chapter exam. It consists of two parts. The first part, Concept Reinforcement, is designed to increase understanding of the concepts through true/false exercises. The second part is the Review Exercises. These provide practice exercises for the exam, together with references to section objectives so you can go back and review. Before beginning, stop and look back over the skills you have obtained. What skills in mathematics do you have now that you did not have before studying this chapter?

CONCEPT REINFORCEMENT

Determine whether the statement is true or false. Answers are given at the back of the book.

________ **1.** A number a is divisible by another number b if b is a factor of a.

________ **2.** The least common multiple of two natural numbers is the smallest number that is a factor of both.

________ **3.** If $\frac{a}{b} > \frac{c}{b}$, $b \neq 0$, then $a > c$.

________ **4.** All mixed numerals represent numbers larger than 1.

________ **5.** The fraction $\frac{13}{7}$ is larger than the fraction $\frac{13}{6}$.

Review Exercises

1. Identify the numerator and the denominator of $\frac{2}{7}$. [2.1a]

What part of the object or set of objects is shaded? [2.1a]

2.

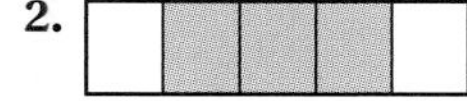

3.

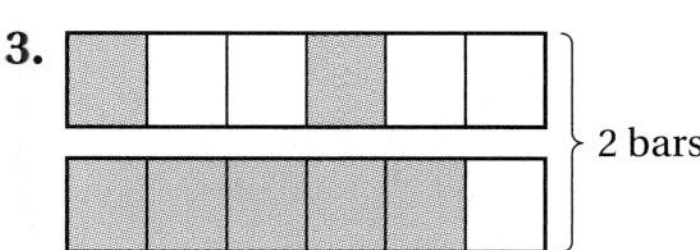

Simplify. [2.1b, e]

4. $\frac{21}{21}$ **5.** $\frac{0}{25}$ **6.** $\frac{12}{30}$

7. $\frac{18}{1}$ **8.** $\frac{32}{8}$ **9.** $\frac{9}{27}$

10. $\frac{18}{0}$ **11.** $\frac{5}{8-8}$ **12.** $\frac{88}{184}$

13. $\frac{140}{490}$ **14.** $\frac{1170}{1200}$ **15.** $\frac{288}{2025}$

16. Simplify, if possible, the fractions on this circle graph. [2.1e]

How the Business Travel Dollar Is Spent

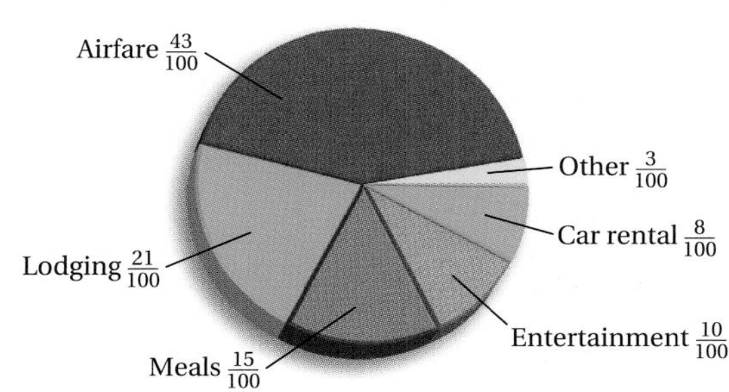

Multiply and simplify. [2.2a]

17. $4 \cdot \frac{3}{8}$

18. $\frac{7}{3} \cdot 24$

19. $9 \cdot \frac{5}{18}$

20. $\frac{6}{5} \cdot 20$

21. $\frac{3}{4} \cdot \frac{8}{9}$

22. $\frac{5}{7} \cdot \frac{1}{10}$

23. $\frac{3}{7} \cdot \frac{14}{9}$

24. $\frac{1}{4} \cdot \frac{2}{11}$

25. $\frac{4}{25} \cdot \frac{15}{16}$

26. $\frac{11}{3} \cdot \frac{30}{77}$

Find the reciprocal. [2.2b]

27. $\frac{4}{5}$

28. 3

29. $\frac{1}{9}$

30. $\frac{47}{36}$

Divide and simplify. [2.2c]

31. $6 \div \frac{4}{3}$

32. $\frac{5}{9} \div \frac{5}{18}$

33. $\frac{1}{6} \div \frac{1}{11}$

34. $\frac{3}{14} \div \frac{6}{7}$

35. $\frac{1}{4} \div \frac{1}{9}$

36. $180 \div \frac{3}{5}$

37. $\frac{23}{25} \div \frac{23}{25}$

38. $\frac{2}{3} \div \frac{3}{2}$

Add and simplify. [2.3a]

39. $\frac{6}{5} + \frac{3}{8}$

40. $\frac{5}{16} + \frac{1}{12}$

41. $\frac{6}{5} + \frac{11}{15} + \frac{3}{20}$

42. $\frac{1}{1000} + \frac{19}{100} + \frac{7}{10}$

Subtract and simplify. [2.3b]

43. $\frac{5}{9} - \frac{2}{9}$

44. $\frac{7}{8} - \frac{3}{4}$

45. $\frac{11}{27} - \frac{2}{9}$

46. $\frac{5}{6} - \frac{2}{9}$

Use $<$ or $>$ for $\square$ to write a true sentence. [2.3c]

47. $\frac{4}{7} \square \frac{5}{9}$

48. $\frac{8}{9} \square \frac{11}{13}$

Convert to fraction notation. [2.4a]

49. $7\frac{1}{2}$

50. $8\frac{3}{8}$

51. $4\frac{1}{3}$

52. $10\frac{5}{7}$

Convert to a mixed numeral. [2.4a]

53. $\frac{7}{3}$

54. $\frac{27}{4}$

55. $\frac{63}{5}$

56. $\frac{7}{2}$

Solve. [2.2d], [2.3d]

57. $\frac{5}{4} \cdot t = \frac{3}{8}$

58. $x \cdot \frac{2}{3} = 160$

59. $x + \frac{2}{5} = \frac{7}{8}$

60. $\frac{1}{2} + y = \frac{9}{10}$

Add. Write a mixed numeral for the answer. [2.4b]

61. $5\frac{3}{5} + 4\frac{4}{5}$

62. $8\frac{1}{3} + 3\frac{2}{5}$

63. $5\frac{5}{6} + 4\frac{5}{6}$

64. $2\frac{3}{4} + 5\frac{1}{2}$

Subtract. Write a mixed numeral for the answer where appropriate. [2.4c]

65. $12 - 4\frac{2}{9}$

66. $9\frac{3}{5} - 4\frac{13}{15}$

67. $10\frac{1}{4} - 6\frac{1}{10}$

68. $24 - 10\frac{5}{8}$

Multiply. Write a mixed numeral for the answer where appropriate. [2.4d]

69. $6 \cdot 2\frac{2}{3}$

70. $5\frac{1}{4} \cdot \frac{2}{3}$

71. $2\frac{1}{5} \cdot 1\frac{1}{10}$

72. $2\frac{2}{5} \cdot 2\frac{1}{2}$

Divide. Write a mixed numeral for the answer where appropriate. [2.4e]

73. $27 \div 2\frac{1}{4}$

74. $2\frac{2}{5} \div 1\frac{7}{10}$

75. $3\frac{1}{4} \div 26$

76. $4\frac{1}{5} \div 4\frac{2}{3}$

Solve. [2.5a]

77. A road crew repaves $\frac{1}{12}$ mi of road each day. How long will it take the crew to repave a $\frac{3}{4}$-mi stretch of road?

78. After driving 600 km, the Buxton family has completed $\frac{3}{5}$ of their vacation. How long is the total trip?

79. Molly is making a pepper steak recipe that calls for $\frac{2}{3}$ cup of green bell peppers. How much would be needed to make $\frac{1}{2}$ recipe? 3 recipes?

80. *Sewing.* Gloria wants to make a dress and jacket. She needs $1\frac{5}{8}$ yd of 60-in. fabric for the dress and $2\frac{5}{8}$ yd for the jacket. How many yards in all does Gloria need to make the outfit?

81. *Corn Production.* In 2003, the United States produced approximately $\frac{2}{5}$ of the world production of corn. The total world corn production was about 640,000,000 metric tons. How much corn did the United States produce?

Source: UN Food and Agriculture Organization

82. What is the sum of the areas in the figure below?

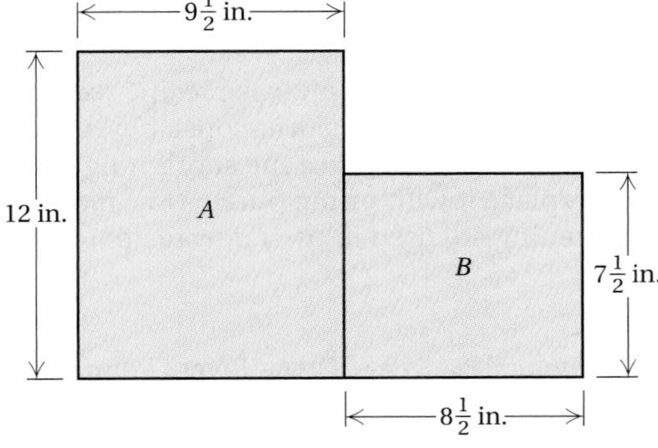

83. In the figure above, how much larger is the area of rectangle *A* than the area of rectangle *B*?

84. *Carpentry.* A board $\frac{9}{10}$ in. thick is glued to a board $\frac{8}{10}$ in. thick. The glue is $\frac{3}{100}$ in. thick. How thick is the result?

85. *Turkey Servings.* Turkey contains $1\frac{1}{3}$ servings per pound. How many pounds are needed for 32 servings?

86. *Weightlifting.* In 1998, Sun Tianni of China snatched 111 kg. This amount was about $1\frac{3}{5}$ times her body weight. How much did Tianni weigh?

Source: *The Guinness Book of Records*, 2000

87. *Cake Recipe.* A wedding-cake recipe requires 12 cups of shortening. Being calorie-conscious, the wedding couple decides to reduce the shortening by $3\frac{5}{8}$ cups and replace it with prune purée. How many cups of shortening are used in their new recipe?

88. *Firefighters' Pie Sale.* Green River's Volunteer Fire Department recently hosted its annual ice cream social. Each of the 83 pies donated was cut into 6 pieces. At the end of the evening, they had sold 382 pieces of pie. How many pies did they sell? How many were left over? Express your answers in mixed numerals.

Simplify the expression using the rules for order of operations. [2.6a]

89. $\frac{1}{8} \div \frac{1}{4} + \frac{1}{2}$

90. $\frac{4}{5} - \frac{1}{2} \cdot \left(1 + \frac{1}{4}\right)$

91. $20\frac{3}{4} - 1\frac{1}{2} \times 12 + \left(\frac{1}{2}\right)^2$

92. Find the average of $\frac{1}{2}$, $\frac{1}{4}$, $\frac{1}{3}$, and $\frac{1}{5}$. [2.6a]

Estimate each of the following as 0, $\frac{1}{2}$, or 1. [2.6b]

93. $\frac{29}{59}$

94. $\frac{2}{59}$

95. $\frac{61}{59}$

Estimate each of the following as a whole number or as a mixed numeral where the fractional part is $\frac{1}{2}$. [2.6b]

96. $6\frac{7}{8}$

97. $10\frac{2}{17}$

98. $\frac{3}{10} + \frac{5}{6} + \frac{31}{29}$

99. $32\frac{14}{15} + 27\frac{3}{4} - 4\frac{25}{28} \cdot 6\frac{37}{76}$

100. **DW** A student claims that "taking $\frac{1}{2}$ of a number is the same as dividing by $\frac{1}{2}$." Explain the error in this reasoning. [2.2c]

101. **DW** Discuss the role of least common multiples in adding and subtracting with fraction notation. [2.3a, b]

SYNTHESIS

102. In the division below, find a and b. The quotient has not been simplified. [2.2c]

$$\frac{19}{24} \div \frac{a}{b} = \frac{187,853}{268,224}$$

103. Place the numbers 3, 4, 5, and 6 in the boxes in order to make a true equation: [2.3a]

$$\frac{\square}{\square} + \frac{\square}{\square} = 3\frac{1}{4}.$$

2 Chapter Test

For Extra Help

Work It Out!
Chapter Test Video
Solutions on CD

1. What part is shaded?

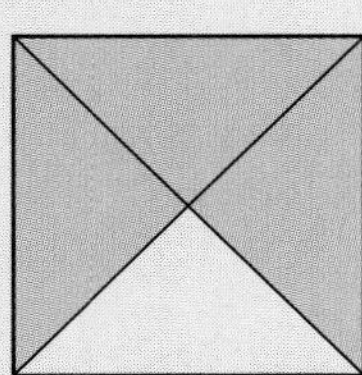

2. Use < or > for □ to write a true sentence:

$\frac{6}{7} \square \frac{21}{25}$.

Simplify.

3. $\frac{12}{12}$

4. $\frac{0}{16}$

5. $\frac{2}{28}$

6. $\frac{9}{0}$

Multiply and simplify.

7. $\frac{4}{3} \cdot 24$

8. $\frac{2}{3} \cdot \frac{15}{4}$

9. $\frac{3}{5} \cdot \frac{1}{6}$

10. $\frac{22}{15} \cdot \frac{5}{33}$

Find the reciprocal.

11. $\frac{5}{8}$

12. $\frac{1}{4}$

13. 18

Divide and simplify.

14. $\frac{3}{8} \div \frac{5}{4}$

15. $\frac{1}{5} \div \frac{1}{8}$

16. $12 \div \frac{2}{3}$

Add and simplify.

17. $\frac{1}{2} + \frac{5}{2}$

18. $\frac{7}{8} + \frac{2}{3}$

19. $\frac{7}{10} + \frac{19}{100} + \frac{31}{1000}$

Subtract and simplify.

20. $\frac{5}{6} - \frac{3}{6}$

21. $\frac{5}{6} - \frac{3}{4}$

22. $\frac{17}{24} - \frac{1}{15}$

Solve.

23. $\frac{7}{8} \cdot x = 56$

24. $x + \frac{2}{3} = \frac{11}{12}$

25. Convert to fraction notation: $3\frac{1}{2}$.

26. Convert to a mixed numeral: $\frac{74}{9}$.

Add, subtract, multiply, or divide. Write a mixed numeral for the answer in Exercises 27–30.

27. $\begin{array}{r} 6\frac{2}{5} \\ +\,7\frac{4}{5} \\ \hline \end{array}$

28. $\begin{array}{r} 10\frac{1}{6} \\ -\,5\frac{7}{8} \\ \hline \end{array}$

29. $6\frac{3}{4} \cdot \frac{2}{3}$

30. $2\frac{1}{3} \div 1\frac{1}{6}$

31. There are 7000 students at La Poloma College, and $\frac{5}{8}$ of them live in dorms. How many live in dorms?

32. A strip of taffy $\frac{9}{10}$ m long is cut into 12 equal pieces. What is the length of each piece?

33. *Weightlifting.* In 2002, Hossein Rezazadeh of Iran did a clean and jerk of 263 kg. This amount was about $2\frac{1}{2}$ times his body weight. How much did Rezazadeh weigh?

Source: *The Guinness Book of Records,* 2005

34. *Book Order.* An order of books for a math course weighs 220 lb. Each book weighs $2\frac{3}{4}$ lb. How many books are in the order?

35. *Carpentry.* The following diagram shows a middle drawer support guide for a cabinet drawer. Find each of the following.

a) The short length a across the top

b) The length b across the bottom

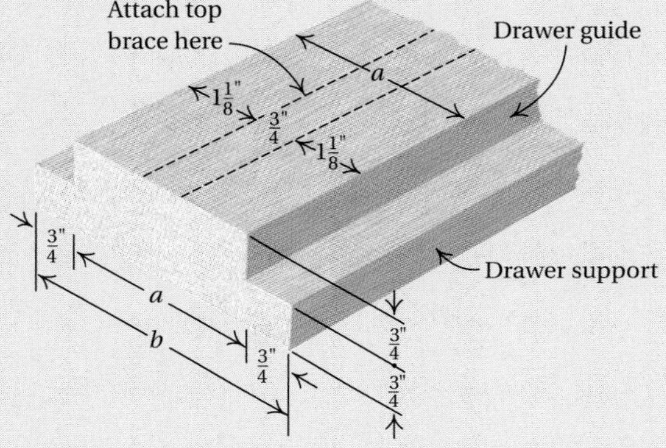

36. *Women's Dunks.* The first three women in the history of college basketball able to dunk a basketball are listed below. Their names, heights, and universities are:

Michelle Snow, $6\frac{5}{12}$ ft, Tennessee;
Charlotte Smith, $5\frac{11}{12}$ ft, North Carolina;
Georgeann Wells, $6\frac{7}{12}$ ft, West Virginia.

Find the average height of these women.

Source: *USA Today,* 11/30/00, p. 3C

37. Simplify: $1\frac{1}{2} - \frac{1}{2}\left(\frac{1}{2} \div \frac{1}{4}\right) + \left(\frac{1}{2}\right)^2$.

Estimate each of the following as 0, $\frac{1}{2}$, or 1.

38. $\frac{3}{82}$

39. $\frac{93}{91}$

Estimate each of the following as a whole number or as a mixed numeral where the fractional part is $\frac{1}{2}$.

40. $18\frac{9}{17}$

41. $256 \div 15\frac{19}{21}$

SYNTHESIS

42. Rebecca walks 17 laps at her health club. Trent walks 17 laps at his health club. If the track at Rebecca's health club is $\frac{1}{7}$ mi long, and the track at Trent's is $\frac{1}{8}$ mi long, who walks farther? How much farther?

43. Grandma Hammons left $\frac{2}{3}$ of her $\frac{7}{8}$-acre apple farm to Karl. Karl gave $\frac{1}{4}$ of his share to his oldest daughter, Eileen. How much land did Eileen receive?

3

Decimal Notation

Real-World Application

The Panama Canal in Panama is 50.7 mi long. The Suez Canal in Egypt is 119.9 mi long. How much longer is the Suez Canal?

This problem appears as Example 1 in Section 3.7.

3.1 DECIMAL NOTATION, ORDER, AND ROUNDING

Objectives

a Given decimal notation, write a word name.

b Convert between fraction notation and decimal notation.

c Given a pair of numbers in decimal notation, tell which is larger.

d Round decimal notation to the nearest thousandth, hundredth, tenth, one, ten, hundred, or thousand.

The set of **arithmetic numbers,** or **nonnegative rational numbers,** consists of the whole numbers 0, 1, 2, 3, 4, 5, 6, 7, 8, 9, 10, and so on, and fractions like $\frac{1}{2}, \frac{2}{3}, \frac{7}{8}, \frac{17}{10}$, and so on. Note that we can write the whole numbers using fraction notation. For example, 3 can be written as $\frac{3}{1}$. We studied the use of fraction notation for arithmetic numbers in Chapter 2.

In Chapter 3, we will study the use of *decimal notation*. The word *decimal* comes from the Latin word *decima,* meaning a tenth part. Although we are using different notation, we are still considering the nonnegative rational numbers. Using decimal notation, we can write 0.875 for $\frac{7}{8}$, for example, or 48.97 for $48\frac{97}{100}$.

a Decimal Notation and Word Names

A 5-quart stand mixer sells for \$249.98. The dot in \$249.98 is called a **decimal point.** Since \$0.98, or 98¢, is $\frac{98}{100}$ of a dollar, it follows that

$$\$249.98 = 249 + \frac{98}{100} \text{ dollars.}$$

Also, since \$0.98, or 98¢, has the same value as

9 dimes + 8 cents

and 1 dime is $\frac{1}{10}$ of a dollar and 1 cent is $\frac{1}{100}$ of a dollar, we can write

$$249.98 = 2 \cdot 100 + 4 \cdot 10 + 9 \cdot 1 + 9 \cdot \frac{1}{10} + 8 \cdot \frac{1}{100}.$$

This is an extension of the expanded notation for whole numbers that we used in Chapter 1. The place values are 100, 10, 1, $\frac{1}{10}, \frac{1}{100}$, and so on. We can see this on a **place-value chart.** The value of each place is $\frac{1}{10}$ as large as the one to its left.

Let's consider decimal notation using a place-value chart to represent 26.3385 min, the world record in the men's 10,000-m run, held by Kenenisa Bekele from Ethiopia.

PLACE-VALUE CHART							
Hundreds	Tens	Ones	Ten*ths*	Hundred*ths*	Thousand*ths*	Ten-Thousand*ths*	Hundred-Thousand*ths*
100	10	1	$\frac{1}{10}$	$\frac{1}{100}$	$\frac{1}{1000}$	$\frac{1}{10,000}$	$\frac{1}{100,000}$
	2	6 .	3	3	8	5	

Study Tips

QUIZ–TEST FOLLOW-UP

You may have just completed a chapter quiz or test. Immediately after each chapter quiz or test, write out a step-by-step solution of each question you missed. Visit your instructor or tutor for help with problems that are still giving you trouble. When the week of the final examination arrives, you will be glad to have the excellent study guide these corrected tests provide.

The decimal notation 26.3385 means

$$20 + 6 + \frac{3}{10} + \frac{3}{100} + \frac{8}{1000} + \frac{5}{10{,}000}, \text{ or } 26\frac{3385}{10{,}000}.$$

We read both 26.3385 and $26\frac{3385}{10{,}000}$ as

"Twenty-six and three thousand three hundred eighty-five ten-thousandths."

We can also read 26.3385 as

"Two six *point* three three eight five, or twenty-six *point* three three eight five."

To write a word name from decimal notation,

a) write a word name for the number named to the left of the decimal point, — 397.685 → Three hundred ninety-seven

b) write the word "and" for the decimal point, and — 397.685 → Three hundred ninety-seven and

c) write a word name for the number named to the right of the decimal point, followed by the place value of the last digit. — 397.685 → Three hundred ninety-seven and six hundred eighty-five *thousandths*

EXAMPLE 1 *Ice Cream.* Write a word name for the number in this sentence: The average person eats 26.3 servings of ice cream a year.

Source: NPD Group; J. M. Hirsch, Associated Press

Twenty-six and three tenths

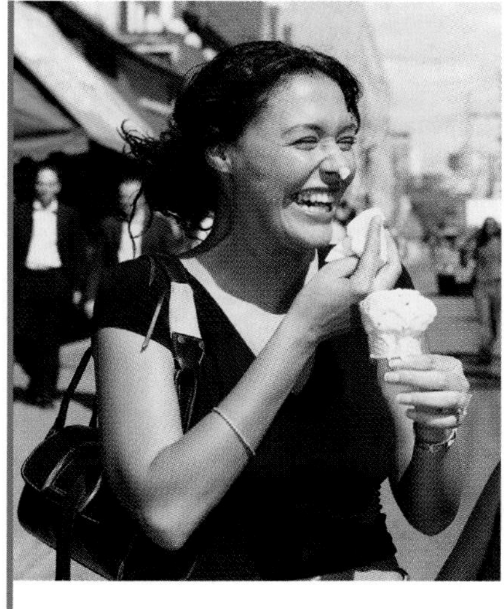

EXAMPLE 2 Write a word name for 410.87.

Four hundred ten and eighty-seven hundredths

Write a word name for the numbers.

1. Life Expectancy. In 2004, the life expectancy at birth in Switzerland was 80.31. In the United States, it was 77.43.

Source: *CIA World Factbook*, 2004

2. Kentucky Derby. In 2004, racehorse Smarty Jones won the Kentucky Derby in a time of 2.06767 min.

Source: *Time Almanac*, 2005

Answers on page A-6

3. 245.89

4. 34.0064

5. 31,079.764

Answers on page A-6

EXAMPLE 3 *5000-m Run.* Write a word name for the number in this sentence: The world outdoor record in the women's 5000-m run is 14.4113 min, held by Elvan Abeylegesse of Turkey.

Fourteen and four thousand one hundred thirteen ten-thousandths

EXAMPLE 4 Write a word name for 1788.405.

One thousand, seven hundred eighty-eight and four hundred five thousandths

Do Exercises 1–5. (Exercises 1 and 2 are on the preceding page.)

b Converting Between Decimal Notation and Fraction Notation

Given decimal notation, we can convert to fraction notation as follows:

$$\begin{aligned} 9.875 &= 9 + \frac{8}{10} + \frac{7}{100} + \frac{5}{1000} \\ &= 9 \cdot \frac{1000}{1000} + \frac{8}{10} \cdot \frac{100}{100} + \frac{7}{100} \cdot \frac{10}{10} + \frac{5}{1000} \\ &= \frac{9000}{1000} + \frac{800}{1000} + \frac{70}{1000} + \frac{5}{1000} = \frac{9875}{1000}. \end{aligned}$$

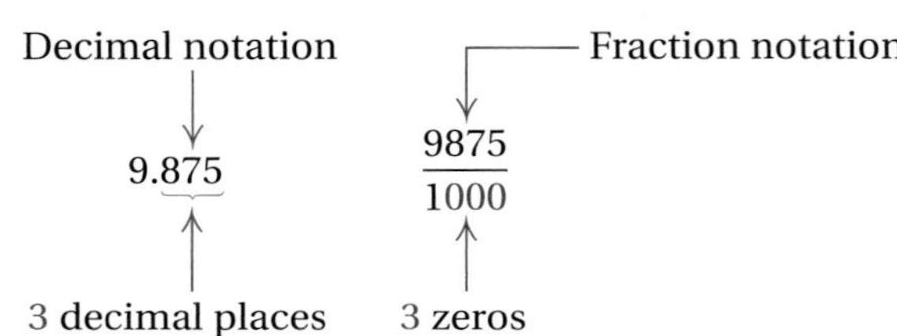

To convert from decimal notation to fraction notation,

a) count the number of decimal places, — 4.98 — 2 places

b) move the decimal point that many places to the right, and — 4.98. — Move 2 places.

c) write the result over a denominator with a 1 followed by that number of zeros. — $\frac{498}{100}$ — 2 zeros

EXAMPLE 5 Write fraction notation for 0.876. Do not simplify.

$$0.876 \qquad 0.876. \qquad 0.876 = \frac{876}{1000}$$

3 places — 3 zeros

For a number like 0.876, we generally write a 0 before the decimal point to draw attention to the presence of the decimal point.

EXAMPLE 6 Write fraction notation for 56.23. Do not simplify.

$$56.23 \qquad 56.23. \qquad 56.23 = \frac{5623}{100}$$

2 places — 2 zeros

EXAMPLE 7 Write fraction notation for 1.5018. Do not simplify.

$$1.5018 \qquad 1.5018. \qquad 1.5018 = \frac{15{,}018}{10{,}000}$$

4 places — 4 zeros

Do Exercises 6–9.

If fraction notation has a denominator that is a power of ten, such as 10, 100, 1000, and so on, we reverse the procedure we used before.

> To convert from fraction notation to decimal notation when the denominator is 10, 100, 1000, and so on,
>
> **a)** count the number of zeros, and $\frac{8679}{1000}$ ← 3 zeros
>
> **b)** move the decimal point that number of places to the left. Leave off the denominator. 8.679. Move 3 places.
>
> $$\frac{8679}{1000} = 8.679$$

EXAMPLE 8 Write decimal notation for $\frac{47}{10}$.

$$\frac{47}{10} \qquad 4.7. \qquad \frac{47}{10} = 4.7$$

1 zero — 1 place

Write fraction notation.

6. 0.896

7. 23.78

8. 5.6789

9. 1.9

Answers on page A-6

Write decimal notation.

10. $\frac{743}{100}$

11. $\frac{406}{1000}$

12. $\frac{67,089}{10,000}$

13. $\frac{9}{10}$

14. $\frac{57}{1000}$

15. $\frac{830}{10,000}$

Write decimal notation.

16. $4\frac{3}{10}$

17. $283\frac{71}{100}$

18. $456\frac{13}{1000}$

Answers on page A-6

EXAMPLE 9 Write decimal notation for $\frac{123,067}{10,000}$.

$\frac{123,067}{10,000}$ ← 4 zeros 12.3067. 4 places $\frac{123,067}{10,000} = 12.3067$

EXAMPLE 10 Write decimal notation for $\frac{13}{1000}$.

$\frac{13}{1000}$ ← 3 zeros 0.013. 3 places $\frac{13}{1000} = 0.013$

EXAMPLE 11 Write decimal notation for $\frac{570}{100,000}$.

$\frac{570}{100,000}$ ← 5 zeros 0.00570. 5 places $\frac{570}{100,000} = 0.0057$

Do Exercises 10–15.

When denominators are numbers other than 10, 100, and so on, we will use another method for conversion. It will be considered in Section 3.5.

If a mixed numeral has a fractional part with a denominator that is a power of ten, such as 10, 100, or 1000, and so on, we first write the mixed numeral as a sum of a whole number and a fraction. Then we convert to decimal notation.

EXAMPLE 12 Write decimal notation for $23\frac{59}{100}$.

$$23\frac{59}{100} = 23 + \frac{59}{100} = 23 \text{ and } \frac{59}{100} = 23.59$$

EXAMPLE 13 Write decimal notation for $772\frac{129}{10,000}$.

$$772\frac{129}{10,000} = 772 + \frac{129}{10,000} = 772 \text{ and } \frac{129}{10,000} = 772.0129$$

Do Exercises 16–18.

c Order

To understand how to compare numbers in decimal notation, consider 0.85 and 0.9. First note that $0.9 = 0.90$ because $\frac{9}{10} = \frac{90}{100}$. Then $0.85 = \frac{85}{100}$ and $0.90 = \frac{90}{100}$. Since $\frac{85}{100} < \frac{90}{100}$, it follows that $0.85 < 0.90$. This leads us to a quick way to compare two numbers in decimal notation.

COMPARING NUMBERS IN DECIMAL NOTATION

To compare two numbers in decimal notation, start at the left and compare corresponding digits moving from left to right. If two digits differ, the number with the larger digit is the larger of the two numbers. To ease the comparison, extra zeros can be written to the right of the last decimal place.

EXAMPLE 14 Which of 2.109 and 2.1 is larger?

Think.
2.109 ⟶ 2.109
2.1 ⟶ 2.100
Same ↑ Different; $9 > 0$

Thus, 2.109 is larger than 2.1. That is, $2.109 > 2.1$.

EXAMPLE 15 Which of 0.09 and 0.108 is larger?

Think.
0.09 ⟶ 0.090
0.108 ⟶ 0.108
Same ↑ Different; $1 > 0$

Thus, 0.108 is larger than 0.09. That is, $0.108 > 0.09$.

Do Exercises 19–24.

d Rounding

Rounding is done as for whole numbers. To understand, we first consider an example using the number line. It might help to review Section 1.3.

EXAMPLE 16 Round 0.37 to the nearest tenth.

Here is part of the number line.

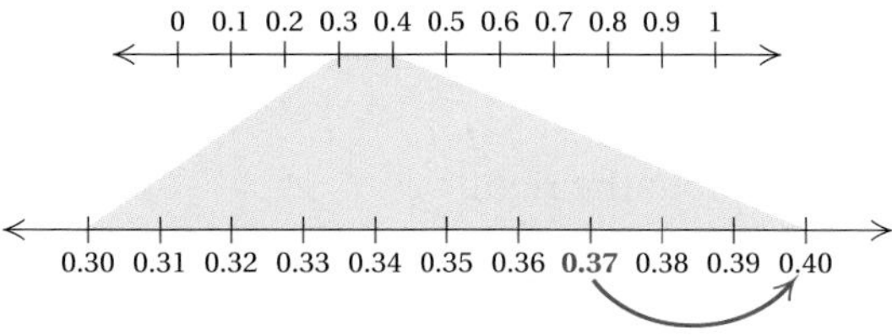

We see that 0.37 is closer to 0.40 than to 0.30. Thus, 0.37 rounded to the nearest tenth is 0.4.

Which number is larger?

19. 2.04, 2.039

20. 0.06, 0.008

21. 0.5, 0.58

22. 1, 0.9999

23. 0.8989, 0.09898

24. 21.006, 21.05

Answers on page A-6

Round to the nearest tenth.

25. 2.76 **26.** 13.85

27. 234.448 **28.** 7.009

Round to the nearest hundredth.

29. 0.636 **30.** 7.834

31. 34.675 **32.** 0.025

Round to the nearest thousandth.

33. 0.9434 **34.** 8.0038

35. 43.1119 **36.** 37.4005

Round 7459.3548 to the nearest:

37. Thousandth.

38. Hundredth.

39. Tenth.

40. One.

41. Ten. (*Caution:* "Tens" are not "tenths.")

42. Hundred.

43. Thousand.

Answers on page A-6

ROUNDING DECIMAL NOTATION

To round to a certain place:

a) Locate the digit in that place.
b) Consider the next digit to the right.
c) If the digit to the right is 5 or higher, round up; if the digit to the right is 4 or lower, round down.

EXAMPLE 17 Round 3872.2459 to the nearest tenth.

a) Locate the digit in the tenths place, 2.

3 8 7 2.2 4 5 9
↑

b) Consider the next digit to the right, 4.

3 8 7 2.2 4 5 9
↑

c) Since that digit, 4, is less than 5, round down.

3 8 7 2.2 ← This is the answer.

Caution!

3872.3 is not a correct answer to Example 17. It is *incorrect* to round from the ten-thousandths digit over to the tenths digit, as follows:

3872.246→3872.25→3872.3.

EXAMPLE 18 Round 3872.2459 to the nearest thousandth, hundredth, tenth, one, ten, hundred, and thousand.

Thousandth:	3872.246	Ten:	3870
Hundredth:	3872.25	Hundred:	3900
Tenth:	3872.2	Thousand:	4000
One:	3872		

EXAMPLE 19 Round 14.8973 to the nearest hundredth.

a) Locate the digit in the hundredths place, 9. 1 4.8 9 7 3 (↑ under 9)

b) Consider the next digit to the right, 7. 1 4.8 9 7 3 (↑ under 9)

c) Since that digit, 7, is 5 or higher, round up. When we make the hundredths digit a 10, we carry 1 to the tenths place.

The answer is 14.90. Note that the 0 in 14.90 indicates that the answer is correct to the nearest hundredth.

EXAMPLE 20 Round 0.008 to the nearest tenth.

a) Locate the digit in the tenths place, 0. 0.0 0 8 (↑ under the first 0 after the point)

b) Consider the next digit to the right, 0. 0.0 0 8 (↑ under the first 0 after the point)

c) Since that digit, 0, is less than 5, round down.

The answer is 0.0.

Do Exercises 25–43.

3.1 EXERCISE SET

For Extra Help

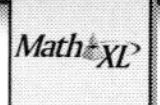

MathXL | MyMathLab | InterAct Math | Math Tutor Center | Video Lectures on CD Disc 2 | Student's Solutions Manual

a Write a word name for the number in the sentence.

NEW YORK STOCK EXCHANGE INDEX
MOST ACTIVE: SHARE VOLUME

	VOL. (000s)	LAST	CHANGE
Elan	97,202	6.65	−1.29
Lucent	46,527	3.08	−.08
NewsCpA n	32,126	17.28	+.48
BostonSci	24,330	30.90	−.66
Pfizer	21,024	26.59	−.02
ExxonMbl	20,149	63.05	+.37
WalMart	19,358	52.86	+.91
GenElec	17,824	35.55	−.05
Motorola	17,355	15.20	−.38
Texinst	17,234	26.58	−.39

1. *ExxonMbl.* Recently, the stock of Exxon Mobil sold for $63.05 per share.

2. *Motorola.* Recently, the stock of Motorola sold for $15.20 per share.

3. *Pfizer.* Recently, the stock of Pfizer sold for $26.59 per share.

4. *Wal-Mart.* Recently, the stock of Wal-Mart sold for $52.86 per share.

5. *Water Weight.* One gallon of water weighs 8.35 lb.

6. One gallon of paint is equal to 3.785 L of paint.

7. *Tires.* A tire was on sale for $86.89. This was the lowest price of the year.

8. *Digital Audio Jukebox.* The Apple iPod mini 4GB digital audio jukebox for Mac & Windows stores over 65 hr of music. Recently, it sold for $249.99.

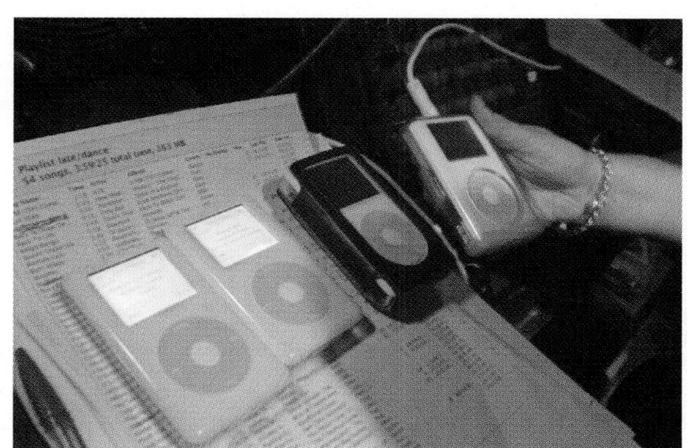

Write a word name.

9. 34.891

10. 27.1245

b Write fraction notation. Do not simplify.

11. 8.3

12. 0.17

13. 3.56

14. 203.6

15. 46.03

16. 1.509

17. 0.00013

18. 0.0109

19. 1.0008

20. 2.0114

21. 20.003

22. 4567.2

Write decimal notation.

23. $\frac{8}{10}$

24. $\frac{51}{10}$

25. $\frac{889}{100}$

26. $\frac{92}{100}$

27. $\frac{3798}{1000}$

28. $\frac{780}{1000}$

29. $\frac{78}{10,000}$

30. $\frac{56,788}{100,000}$

31. $\frac{19}{100,000}$

32. $\frac{2173}{100}$

33. $\frac{376,193}{1,000,000}$

34. $\frac{8,953,074}{1,000,000}$

35. $99\frac{44}{100}$

36. $4\frac{909}{1000}$

37. $3\frac{798}{1000}$

38. $67\frac{83}{100}$

39. $2\frac{1739}{10,000}$

40. $9243\frac{1}{10}$

41. $8\frac{953,073}{1,000,000}$

42. $2256\frac{3059}{10,000}$

c Which number is larger?

43. 0.06, 0.58

44. 0.008, 0.8

45. 0.905, 0.91

46. 42.06, 42.1

47. 0.0009, 0.001

48. 7.067, 7.054

49. 234.07, 235.07

50. 0.99999, 1

51. 0.004, $\frac{4}{100}$

52. $\frac{73}{10}$, 0.73

53. 0.432, 0.4325

54. 0.8437, 0.84384

d Round to the nearest tenth.

55. 0.11

56. 0.85

57. 0.49

58. 0.5794

59. 2.7449

60. 4.78

61. 123.65

62. 36.049

Round to the nearest hundredth.

63. 0.893

64. 0.675

65. 0.6666

66. 6.529

67. 0.995

68. 207.9976

69. 0.094

70. 11.4246

Round to the nearest thousandth.

71. 0.3246 **72.** 0.6666 **73.** 17.0015 **74.** 123.4562

75. 10.1011 **76.** 0.1161 **77.** 9.9989 **78.** 67.100602

Round 809.4732 to the nearest:

79. Hundred. **80.** Tenth. **81.** Thousandth.

82. Hundredth. **83.** One. **84.** Ten.

Round 34.54389 to the nearest:

85. Ten-thousandth. **86.** Thousandth. **87.** Hundredth.

88. Tenth. **89.** One. **90.** Ten.

91. D$_W$ Describe in your own words a procedure for converting from decimal notation to fraction notation.

92. D$_W$ A fellow student rounds 236.448 to the nearest one and gets 237. Explain the possible error.

SKILL MAINTENANCE

Round 6172 to the nearest: [1.3e]

93. Ten. **94.** Hundred. **95.** Thousand.

96. Find the LCM of 18, 27, and 54. [1.9a]

97. Subtract and simplify: $24 - 17\frac{2}{5}$. [2.4c]

Find the prime factorization. [1.7d]

98. 2000 **99.** 1530 **100.** 2002 **101.** 4312

SYNTHESIS

102. Arrange the following numbers in order from smallest to largest.

0.99, 0.099, 1, 0.9999, 0.89999, 1.00009, 0.909, 0.9889

103. Arrange the following numbers in order from smallest to largest.

2.1, 2.109, 2.108, 2.018, 2.0119, 2.0302, 2.000001

Truncating. There are other methods of rounding decimal notation. To round using **truncating,** we drop off all decimal places past the rounding place, which is the same as changing all digits to the right to zeros. For example, rounding 6.78093456285102 to the ninth decimal place, using truncating, gives us 6.780934562. Use truncating to round each of the following to the fifth decimal place, that is, the hundred-thousandth.

104. 6.78346623 **105.** 6.783465902 **106.** 99.999999999 **107.** 0.030303030303

3.2 ADDITION AND SUBTRACTION

Objectives

a Add using decimal notation.

b Subtract using decimal notation.

c Solve equations of the type $x + a = b$ and $a + x = b$, where a and b may be in decimal notation.

d Balance a checkbook.

a Addition

Adding with decimal notation is similar to adding whole numbers. First we line up the decimal points so that we can add corresponding place-value digits. Then we add digits from the right. For example, we add the thousandths, then the hundredths, and so on, carrying if necessary. If desired, we can write extra zeros to the right of the decimal point so that the number of places is the same in all of the addends.

EXAMPLE 1 Add: 56.314 + 17.78.

$$\begin{array}{r} 56.314 \\ +\ 17.780 \\ \hline \end{array}$$

Lining up the decimal points in order to add
Writing an extra zero to the right of the decimal point

$$\begin{array}{r} 56.314 \\ +\ 17.780 \\ \hline 4 \end{array}$$

Adding thousandths

$$\begin{array}{r} 56.314 \\ +\ 17.780 \\ \hline 94 \end{array}$$

Adding hundredths

$$\begin{array}{r} \overset{1}{5}6.314 \\ +\ 17.780 \\ \hline .094 \end{array}$$

Adding tenths
We get 10 tenths = 1 one + 0 tenths, so we carry the 1 to the ones column.
Write a decimal point in the answer.

$$\begin{array}{r} \overset{1}{5}\overset{1}{6}.314 \\ +\ 17.780 \\ \hline 4.094 \end{array}$$

Adding ones
We get 14 ones = 1 ten + 4 ones, so we carry the 1 to the tens column.

$$\begin{array}{r} \overset{1}{5}\overset{1}{6}.314 \\ +\ 17.780 \\ \hline 74.094 \end{array}$$

Adding tens

Do Exercises 1 and 2.

EXAMPLE 2 Add: 3.42 + 0.237 + 14.1.

$$\begin{array}{r} 3.420 \\ 0.237 \\ +\ 14.100 \\ \hline 17.757 \end{array}$$

Lining up the decimal points and writing extra zeros

Adding

Do Exercises 3–5.

Add.

1.
$$\begin{array}{r} 0.847 \\ +\ 10.07 \\ \hline \end{array}$$

2.
$$\begin{array}{r} 2.1 \\ 0.739 \\ +\ 31.3689 \\ \hline \end{array}$$

Add.

3. 0.02 + 4.3 + 0.649

4. 0.12 + 3.006 + 0.4357

5. 0.4591 + 0.2374 + 8.70894

Answers on page A-7

Consider the addition 3456 + 19.347. Keep in mind that any whole number has an "unwritten" decimal point at the right that can be followed by zeros. For example, 3456 can also be written 3456.000. When adding, we can always write in the decimal point and extra zeros if desired.

Add.

6. 789 + 123.67

EXAMPLE 3 Add: 3456 + 19.347.

```
      1
  3 4 5 6.0 0 0    Writing in the decimal point and extra zeros
+       1 9.3 4 7  Lining up the decimal points
-----------------
  3 4 7 5.3 4 7    Adding
```

Do Exercises 6 and 7.

7. 45.78 + 2467 + 1.993

b Subtraction

Subtracting with decimal notation is similar to subtracting whole numbers. First we line up the decimal points so that we can subtract corresponding place-value digits. Then we subtract digits from the right. For example, we subtract the thousandths, then the hundredths, the tenths, and so on, borrowing if necessary.

EXAMPLE 4 Subtract: 56.314 − 17.78.

```
  5 6.3 1 4    Lining up the decimal points in order to subtract
− 1 7.7 8 0    Writing an extra 0
-----------

  5 6.3 1 4    Subtracting thousandths
− 1 7.7 8 0
-----------
          4

      2 11
  5 6.3 1 4    Borrowing tenths to subtract hundredths
− 1 7.7 8 0
-----------
        3 4

      12
    5 2 11
  5 6.3 1 4    Borrowing ones to subtract tenths
− 1 7.7 8 0
-----------
     .5 3 4    Writing a decimal point

    15 12
  4 5 2 11
  5 6.3 1 4    Borrowing tens to subtract ones
− 1 7.7 8 0
-----------
    8.5 3 4

    15 12
  4 5 2 11
  5 6.3 1 4    Subtracting tens
− 1 7.7 8 0
-----------
  3 8.5 3 4
```

```
          1 1 1
Check:    3 8.5 3 4
        + 1 7.7 8 0
        -----------
          5 6.3 1 4
```

Do Exercises 8 and 9.

Subtract.

8. 37.428 − 26.674

9.
```
  0.3 4 7
− 0.0 0 8
---------
```

Answers on page A-7

Subtract.

10. 1.2345 − 0.7

11. 0.9564 − 0.4392

12. 7.37 − 0.00008

EXAMPLE 5 Subtract: 13.07 − 9.205.

$$\begin{array}{r} 1\,3.0\,7\,0 \\ -\;\;9.2\,0\,5 \\ \hline 3.8\,6\,5 \end{array}$$

Writing an extra zero (borrow marks: 12, 10, 6, 10)

Subtracting

EXAMPLE 6 Subtract: 23.08 − 5.0053.

$$\begin{array}{r} 2\,3.0\,8\,0\,0 \\ -\;\;5.0\,0\,5\,3 \\ \hline 1\,8.0\,7\,4\,7 \end{array}$$

Writing two extra zeros (borrow marks: 1, 13, 7, 9, 10)

Subtracting

Check by addition:

$$\begin{array}{r} 1\,8.0\,7\,4\,7 \\ +\;\;5.0\,0\,5\,3 \\ \hline 2\,3.0\,8\,0\,0 \end{array}$$

The answer checks because this is the top number in the subtraction.

Do Exercises 10–12.

When subtraction involves a whole number, again keep in mind that there is an "unwritten" decimal point that can be written in if desired. Extra zeros can also be written in to the right of the decimal point.

EXAMPLE 7 Subtract: 456 − 2.467.

$$\begin{array}{r} 4\,5\,6.0\,0\,0 \\ -\;\;2.4\,6\,7 \\ \hline 4\,5\,3.5\,3\,3 \end{array}$$

Writing in the decimal point and extra zeros (borrow marks: 5, 9, 9, 10)

Subtracting

Do Exercises 13 and 14.

Subtract.

13. 1277 − 82.78

14. 5 − 0.0089

CALCULATOR CORNER

Addition and Subtraction with Decimal Notation To use a calculator to add and subtract with decimal notation, we use the [·], [+], [−], and [=] keys. To find 47.046 − 28.193, for example, we press [4] [7] [·] [0] [4] [6] [−] [2] [8] [·] [1] [9] [3] [=]. The display reads [18.853], so 47.046 − 28.193 = 18.853.

Exercises:

Use a calculator to add.

1. 274.159 + 43.486
2. 19.805 + 486.748
3. 1.7 + 14.56 + 0.89
4. 3.4 + 45 + 0.68

Use a calculator to subtract.

5. 9.2 − 4.8
6. 52.34 − 18.51
7. 489 − 34.26
8. 6.09 − 5.1

Answers on page A-7

c Solving Equations

Now let's solve equations $x + a = b$ and $a + x = b$, where a and b may be in decimal notation. Proceeding as we have before, we subtract a on both sides.

EXAMPLE 8 Solve: $x + 28.89 = 74.567$.

We have

$$x + 28.89 - 28.89 = 74.567 - 28.89 \qquad \text{Subtracting 28.89 on both sides}$$
$$x = 45.677.$$

$$\begin{array}{r} \overset{6}{\not7}\,\overset{13}{\not4}.\overset{14}{\not5}\,\overset{16}{\not6}\,7 \\ -\;2\;8.8\;9\;0 \\ \hline 4\;5.6\;7\;7 \end{array}$$

The solution is 45.677.

EXAMPLE 9 Solve: $0.8879 + y = 9.0026$.

We have

$$0.8879 + y - 0.8879 = 9.0026 - 0.8879 \qquad \text{Subtracting 0.8879 on both sides}$$
$$y = 8.1147.$$

$$\begin{array}{r} \overset{8}{\not9}.\overset{9}{\not0}\,\overset{9}{\not0}\,\overset{11}{\not2}\,\overset{16}{\not6} \\ -\;0.8\;8\;7\;9 \\ \hline 8.1\;1\;4\;7 \end{array}$$

The solution is 8.1147.

Do Exercises 15 and 16.

EXAMPLE 10 Solve: $120 + x = 4380.6$.

We have

$$120 + x - 120 = 4380.6 - 120 \qquad \text{Subtracting 120 on both sides}$$
$$x = 4260.6$$

$$\begin{array}{r} 4\;3\;8\;0.6 \\ -\quad 1\;2\;0.0 \\ \hline 4\;2\;6\;0.6 \end{array}$$

The solution is 4260.6.

Do Exercise 17.

d Balancing a Checkbook

Let's use addition and subtraction with decimals to balance a checkbook.

EXAMPLE 11 Find the errors, if any, in the balances in this checkbook.

20____ RECORD ALL CHARGES OR CREDITS THAT AFFECT YOUR ACCOUNT

DATE	CHECK NUMBER	TRANSACTION DESCRIPTION	√ T	(−) PAYMENT/ DEBIT	(+ OR −) OTHER	(+) DEPOSIT/ CREDIT	BALANCE FORWARD
							8767 73
8/16	432	Burch Laundry		23 56			8744 16
8/19	433	Rogers TV		20 49			8764 65
8/20		Deposit				85 00	8848 65
8/21	434	Galaxy Records		48 60			8801 05
8/22	435	Electric Works		267 95			8533 09

Solve.

15. $x + 17.78 = 56.314$

16. $8.906 + t = 23.07$

17. Solve: $241 + y = 2374.5$.

Answers on page A-7

18. Find the errors, if any, in this checkbook.

20___ RECORD ALL CHARGES OR CREDITS THAT AFFECT YOUR ACCOUNT

DATE	CHECK NUMBER	TRANSACTION DESCRIPTION	√ T	(−) PAYMENT/ DEBIT	(+ OR −) OTHER	(+) DEPOSIT/ CREDIT	BALANCE FORWARD
							3078 92
12/1	888	H.H. Gregg Appliances		340 69			2738 23
12/3	889	Marie Callendar's Pies		78 56			2659 66
12/5		Deposit <Paycheck>				230 80	2890 46
12/6	890	Chili's Restaurant		13 14			2877 32
12/8	891	Stonycreek Golf Course		48 00			2829 32
12/8		Deposit <Molly>				39 58	2868 90
12/10	892	Galyan's Trading Post		102 87			2766 83
12/14	893	Goody's Music		68 59			2697 45
12/15	894	Salvation Army		100 00			2497 45

Answer on page A-7

There are two ways to determine whether there are errors. We assume that the amount $8767.73 in the "Balance forward" column is correct. If we can determine that the ending balance is correct, we have some assurance that the checkbook is correct. But two errors could offset each other to give us that balance.

METHOD 1

a) We add the debits: $23.56 + 20.49 + 48.60 + 267.95 = 360.60$.

b) We add the deposits/credits. In this case, there is only one deposit, 85.00.

c) We add the total of the deposits to the balance brought forward:

$$8767.73 + 85.00 = 8852.73.$$

d) We subtract the total of the debits: $8852.73 - 360.60 = 8492.13$.

The result should be the ending balance, 8533.09. We see that $8492.13 \neq 8533.09$. Since the numbers are not equal, we proceed to method 2.

METHOD 2

We successively add or subtract deposit/credits and debits, and check the result in the "Balance forward" column.

$$8767.73 - 23.56 = 8744.17.$$

We have found our first error. The subtraction was incorrect. We correct it and continue, using 8744.17 as the corrected balance forward:

$$8744.17 - 20.49 = 8723.68.$$

It looks as though 20.49 was added instead of subtracted. Actually, we would have to correct this line even if it had been subtracted, because the error of 1¢ in the first step has been carried through successive calculations. We correct that balance line and continue, using 8723.68 as the balance and adding the deposit 85.00:

$$8723.68 + 85.00 = 8808.68.$$

We make the correction and continue subtracting the last two debits:

$$8808.68 - 48.60 = 8760.08.$$

Then

$$8760.08 - 267.95 = 8492.13.$$

The corrected checkbook is shown below.

20___ RECORD ALL CHARGES OR CREDITS THAT AFFECT YOUR ACCOUNT

DATE	CHECK NUMBER	TRANSACTION DESCRIPTION	√ T	(−) PAYMENT/ DEBIT	(+ OR −) OTHER	(+) DEPOSIT/ CREDIT	BALANCE FORWARD	
							8767 73	
8/16	432	Burch Laundry		23 56			~~8744 16~~	→ 8744.17
8/19	433	Rogers TV		20 49			~~8764 65~~	→ 8723.68
8/20		Deposit				85 00	~~8848 65~~	→ 8808.68
8/21	434	Galaxy Records		48 60			~~8801 05~~	→ 8760.08
8/22	435	Electric Works		267 95			~~8533 09~~	→ 8492.13

Do Exercise 18.

There are other ways in which errors can be made in checkbooks, such as forgetting to record a transaction or writing the amounts incorrectly, but we will not consider those here.

3.2 EXERCISE SET

For Extra Help

MathXL

MyMathLab

InterAct Math

Math Tutor Center

Video Lectures on CD Disc 2

Student's Solutions Manual

a Add.

1. $\begin{array}{r} 316.25 \\ +\ \ 18.12 \\ \hline \end{array}$

2. $\begin{array}{r} 641.803 \\ +\ \ 14.935 \\ \hline \end{array}$

3. $\begin{array}{r} 659.403 \\ +\ 916.812 \\ \hline \end{array}$

4. $\begin{array}{r} 4203.28 \\ +\ \ \ \ 3.39 \\ \hline \end{array}$

5. $\begin{array}{r} 9.104 \\ +\ 123.456 \\ \hline \end{array}$

6. $\begin{array}{r} 81.008 \\ +\ \ 3.409 \\ \hline \end{array}$

7. 20.0124 + 30.0124

8. 0.687 + 0.9

9. 39 + 1.007

10. 2.3 + 0.729 + 23

11. $\begin{array}{r} 47.8 \\ 219.852 \\ 43.59 \\ +\ 666.713 \\ \hline \end{array}$

12. $\begin{array}{r} 13.72 \\ 9.112 \\ 6542.7908 \\ +\ \ \ 23.901 \\ \hline \end{array}$

13. 0.34 + 3.5 + 0.127 + 768

14. 17 + 3.24 + 0.256 + 0.3689

15. 99.6001 + 7285.18 + 500.042 + 870

16. 65.987 + 9.4703 + 6744.02 + 1.0003 + 200.895

Subtract.

17. $\begin{array}{r} 51.31 \\ -\ \ 2.29 \\ \hline \end{array}$

18. $\begin{array}{r} 44.345 \\ -\ \ 3.105 \\ \hline \end{array}$

19. $\begin{array}{r} 92.341 \\ -\ \ 6.42\ \ \\ \hline \end{array}$

20. $\begin{array}{r} 97.01 \\ -\ \ 3.15 \\ \hline \end{array}$

21. $\begin{array}{r} 2.5\ \ \ \ \\ -\ 0.0025 \\ \hline \end{array}$

22. $\begin{array}{r} 39.0\ \ \\ -\ \ 0.28 \\ \hline \end{array}$

23. $\begin{array}{r} 3.4\ \ \ \\ -\ 0.003 \\ \hline \end{array}$

24. $\begin{array}{r} 2.8\ \ \\ -\ 2.08 \\ \hline \end{array}$

25. 28.2 − 19.35

26. 100.16 − 0.118

27. 34.07 − 30.7

28. 36.2 − 16.28

29. $8.45 - 7.405$

30. $3.801 - 2.81$

31. $6.003 - 2.3$

32. $9.087 - 8.807$

33. $1 - 0.0098$

34. $2 - 1.0908$

35. $100 - 0.34$

36. $624 - 18.79$

37. $7.48 - 2.6$

38. $18.4 - 5.92$

39. $3 - 2.006$

40. $263.7 - 102.08$

41. $19 - 1.198$

42. $2548.98 - 2.007$

43. $65 - 13.87$

44. $45 - 0.999$

45. $\begin{array}{r} 32.7978 \\ -\ \ 0.0592 \\ \hline \end{array}$

46. $\begin{array}{r} 0.49634 \\ -\ 0.12678 \\ \hline \end{array}$

47. $\begin{array}{l} \ \ 6.07 \\ -\ 2.0078 \\ \hline \end{array}$

48. $\begin{array}{l} \ \ 1.0 \\ -\ 0.9999 \\ \hline \end{array}$

C Solve.

49. $x + 17.5 = 29.15$

50. $t + 50.7 = 54.07$

51. $17.95 + p = 402.63$

52. $w + 1.3004 = 47.8$

53. $13{,}083.3 = x + 12{,}500.33$

54. $100.23 = 67.8 + z$

55. $x + 2349 = 17{,}684.3$

56. $1830.4 + t = 23{,}067$

d Find the errors, if any, in the checkbook.

57.

20____ RECORD ALL CHARGES OR CREDITS THAT AFFECT YOUR ACCOUNT

DATE	CHECK NUMBER	TRANSACTION DESCRIPTION	√ T	(−) PAYMENT/ DEBIT	(+ OR −) OTHER	(+) DEPOSIT/ CREDIT	BALANCE FORWARD
							9704 56
8/8	342	Bill Rydman		27 44			9677 12
8/9		Deposit <Beauty Contest>				1000 00	10,677 12
8/12	343	Chuck Taylor		123 95			10,553 17
8/14	344	Jennifer Crum		124 02			10,677 19
8/22	345	Neon Johnny's Pizza		12 43			10,664 76
8/24		Deposit <Bowling Tournament>				2500 00	13,164 76
8/29	346	Border's Bookstore		137 78			13,302 54
9/2		Deposit <Bodybuilder Contest>				18 88	13,283 66
9/3	347	Fireman's Fund		2800 00			10,483 66

58.

20____ RECORD ALL CHARGES OR CREDITS THAT AFFECT YOUR ACCOUNT

DATE	CHECK NUMBER	TRANSACTION DESCRIPTION	√ T	(−) PAYMENT/ DEBIT	(+ OR −) OTHER	(+) DEPOSIT/ CREDIT	BALANCE FORWARD
							1876 43
4/1	500	Bart Kaufman		500 12			1376 31
4/3	501	Jim Lawler		28 56			1347 75
4/3		Deposit <State Lottery>				10,000 00	11,347 75
4/3	502	Victoria Montoya		464 00			10,883 75
4/3		Deposit <Jewelry Sale>.				2500 00	8383 75
4/4	503	Baskin & Robbins		1600 00			6783 75
4/8	504	Golf Galaxy		1349 98			5433 77
4/12	505	Don Mitchell Pro Shops		658 97			4774 80
4/13		Deposit <Publisher's Clearing House>				100000 00	104,774 80
4/15	506	American Airlines		6885 58			98,889 22

59. $\mathbf{D_W}$ Explain the error in the following:
Add.

$$13.07 + 9.205 = 10.512$$

60. $\mathbf{D_W}$ Explain the error in the following:
Subtract.

$$73.089 - 5.0061 = 2.3028$$

SKILL MAINTENANCE

61. Find the LCM of 32 and 85. [1.9a]

62. Find the prime factorization of 228. [1.7d]

Subtract.

63. $\frac{13}{24} - \frac{3}{8}$ [2.3b]

64. $\frac{8}{9} - \frac{2}{15}$ [2.3b]

65. $8805 - 2639$ [1.2d]

66. $8005 - 2639$ [1.2d]

Solve.

67. A serving of filleted fish is generally considered to be about $\frac{1}{3}$ lb. How many servings can be prepared from $5\frac{1}{2}$ lb of flounder fillet? [2.5a]

68. A photocopier technician drove $125\frac{7}{10}$ mi away from Scottsdale for a repair call. The next day he drove $65\frac{1}{2}$ mi back toward Scottsdale for another service call. How far was the technician from Scottsdale? [2.5a]

SYNTHESIS

69. A student presses the wrong button when using a calculator and adds 235.7 instead of subtracting it. The incorrect answer is 817.2. What is the correct answer?

3.3 MULTIPLICATION

Objectives

a Multiply using decimal notation.

b Convert from notation like 45.7 million to standard notation, and between dollars and cents.

a Multiplication

Let's find the product

$$2.3 \times 1.12.$$

To understand how we find such a product, we first convert each factor to fraction notation. Next, we multiply the whole numbers 23 and 112, and then divide by 1000.

$$2.3 \times 1.12 = \frac{23}{10} \times \frac{112}{100} = \frac{23 \times 112}{10 \times 100} = \frac{2576}{1000} = 2.576$$

Note the number of decimal places.

$$\begin{array}{rl} 1.12 & \text{(2 decimal places)} \\ \times\ \ 2.3 & \text{(1 decimal place)} \\ \hline 2.576 & \text{(3 decimal places)} \end{array}$$

Now consider

$$0.011 \times 15.0002 = \frac{11}{1000} \times \frac{150{,}002}{10{,}000} = \frac{1{,}650{,}022}{10{,}000{,}000} = 0.1650022.$$

Note the number of decimal places.

$$\begin{array}{rl} 15.0002 & \text{(4 decimal places)} \\ \times\ \ 0.011 & \text{(3 decimal places)} \\ \hline 0.1650022 & \text{(7 decimal places)} \end{array}$$

To multiply using decimals: 0.8×0.43

a) Ignore the decimal points and multiply as though both factors were whole numbers.

$$\begin{array}{rl} \overset{2}{0.43} & \\ \times\ \ 0.8 & \text{Ignore the decimal} \\ \hline 344 & \text{points for now.} \end{array}$$

b) Then place the decimal point in the result. The number of decimal places in the product is the sum of the numbers of places in the factors (count places from the right).

$$\begin{array}{rl} 0.43 & \text{(2 decimal places)} \\ \times\ \ 0.8 & \text{(1 decimal place)} \\ \hline 0.344 & \text{(3 decimal places)} \end{array}$$

EXAMPLE 1 Multiply: 8.3×74.6.

a) Ignore the decimal points and multiply as though factors were whole numbers.

$$\begin{array}{r} \scriptstyle 3\ 4 \\ \scriptstyle 1\ 1 \\ 74.6 \\ \times\ \ 8.3 \\ \hline 2238 \\ 59680 \\ \hline 61918 \end{array}$$

b) Place the decimal point in the result. The number of decimal places in the product is the sum of the number of places in the factors, 1 + 1, or 2.

```
      7 4.6    (1 decimal place)
×       8.3    (1 decimal place)
    2 2 3 8
  5 9 6 8 0
  6 1 9.1 8    (2 decimal places)
```

Do Exercise 1.

EXAMPLE 2 Multiply: 0.0032×2148.

```
       2 1 4 8    (0 decimal places)
× 0.0 0 3 2       (4 decimal places)
       4 2 9 6
     6 4 4 4 0
    6.8 7 3 6     (4 decimal places)
```

EXAMPLE 3 Multiply: 0.14×0.867.

```
      0.8 6 7    (3 decimal places)
×       0.1 4    (2 decimal places)
      3 4 6 8
      8 6 7 0
  0.1 2 1 3 8    (5 decimal places)
```

Do Exercises 2 and 3.

MULTIPLYING BY 0.1, 0.01, 0.001, AND SO ON

Now let's consider some special kinds of products. The first involves multiplying by a tenth, hundredth, thousandth, or ten-thousandth. Let's look at those products.

$$0.1 \times 38 = \frac{1}{10} \times 38 = \frac{38}{10} = 3.8$$

$$0.01 \times 38 = \frac{1}{100} \times 38 = \frac{38}{100} = 0.38$$

$$0.001 \times 38 = \frac{1}{1000} \times 38 = \frac{38}{1000} = 0.038$$

$$0.0001 \times 38 = \frac{1}{10{,}000} \times 38 = \frac{38}{10{,}000} = 0.0038$$

Note in each case that the product is *smaller* than 38. That is, the decimal point in each product is farther to the left than the unwritten decimal point in 38.

1. Multiply.

```
  8 5.4
×   6.2
```

Multiply.

2.

```
     1 2 3 4
× 0.0 0 4 1
```

3.

```
   4 2.6 5
× 0.8 0 4
```

Answers on page A-7

Multiply.

4. 0.1×3.48

5. 0.01×3.48

6. 0.001×3.48

7. 0.0001×3.48

To multiply any number by 0.1, 0.01, 0.001, and so on,

a) count the number of decimal places in the tenth, hundredth, or thousandth, and so on, and
b) move the decimal point that many places to the left.

0.001×34.45678 → 3 places

$0.001 \times 34.45678 = 0.034.45678$

Move 3 places to the left.

$0.001 \times 34.45678 = 0.03445678$

EXAMPLES Multiply.

4. $0.1 \times 14.605 = 1.4605$ 1.4.605

5. $0.01 \times 14.605 = 0.14605$

6. $0.001 \times 14.605 = 0.014605$
We write an extra zero.

7. $0.0001 \times 14.605 = 0.0014605$
We write two extra zeros.

Do Exercises 4–7.

MULTIPLYING BY 10, 100, 1000, AND SO ON

Next, let's consider multiplying by 10, 100, 1000, and so on. Let's look at those products.

$$10 \times 97.34 = 973.4$$
$$100 \times 97.34 = 9734$$
$$1000 \times 97.34 = 97{,}340$$
$$10{,}000 \times 97.34 = 973{,}400$$

Note in each case that the product is *larger* than 97.34. That is, the decimal point in each product is farther to the right than the decimal point in 97.34.

Multiply.

8. 10×3.48

9. 100×3.48

10. 1000×3.48

11. $10{,}000 \times 3.48$

To multiply any number by 10, 100, 1000, and so on,

a) count the number of zeros, and
b) move the decimal point that many places to the right.

1000×34.45678 → 3 zeros

$1000 \times 34.45678 = 34.456.78$

Move 3 places to the right.

$1000 \times 34.45678 = 34{,}456.78$

EXAMPLES Multiply.

8. $10 \times 14.605 = 146.05$ 14.6.05

9. $100 \times 14.605 = 1460.5$

10. $1000 \times 14.605 = 14{,}605$

11. $10{,}000 \times 14.605 = 146{,}050$ 14.6050.   We write an extra zero.

Answers on page A-7

Do Exercises 8–11 on the preceding page.

b Applications Using Multiplication with Decimal Notation

NAMING LARGE NUMBERS

We often see notation like the following in newspapers and magazines and on television.

- The largest building in the world is the Pentagon, which has 3.7 million square feet of floor space.
- In 2003, consumers spent $51.7 billion on online retail products. Total spending in specific categories is listed in the following table.

ONLINE CONSUMER SPENDING
(in billions of dollars)

CATEGORY	2003	CATEGORY	2003
1. PCs	$8.8	7. Event tickets	$2.7
2. Peripherals	$2.5	8. Apparel	$6.1
3. Software	$2.9	9. Jewelry	$1.3
4. Consumer electronics	$2.5	10. Toys	$1.3
5. Books	$3.1	11. Housewares/ small appliances	$1.9
6. Music	$1.3	12. Office products	$1.7

Source: Jupiter Media Metrix, Inc.

To understand such notation, consider the information in the following table.

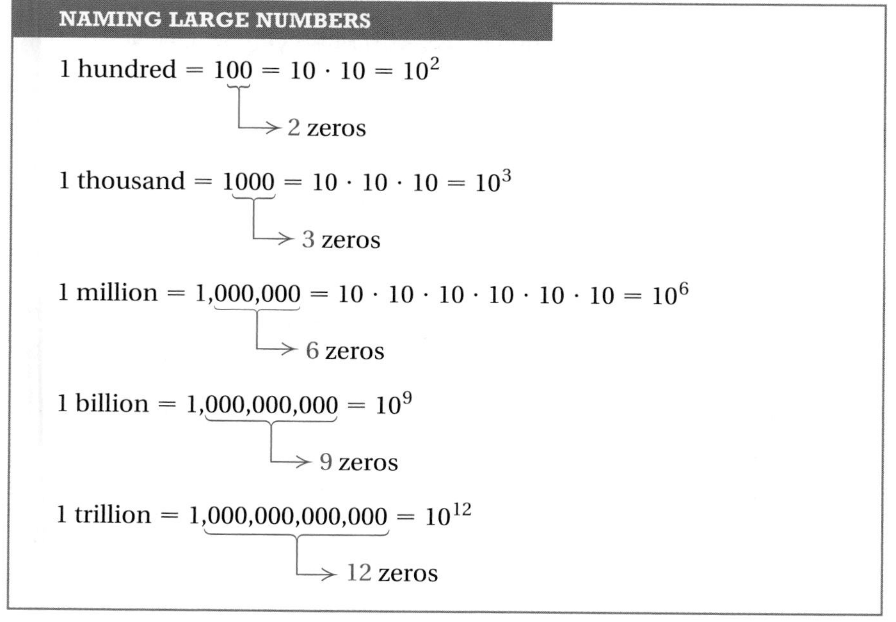

NAMING LARGE NUMBERS

1 hundred = $100 = 10 \cdot 10 = 10^2$
→ 2 zeros

1 thousand = $1000 = 10 \cdot 10 \cdot 10 = 10^3$
→ 3 zeros

1 million = $1{,}000{,}000 = 10 \cdot 10 \cdot 10 \cdot 10 \cdot 10 \cdot 10 = 10^6$
→ 6 zeros

1 billion = $1{,}000{,}000{,}000 = 10^9$
→ 9 zeros

1 trillion = $1{,}000{,}000{,}000{,}000 = 10^{12}$
→ 12 zeros

CALCULATOR CORNER

Multiplication with Decimal Notation To use a calculator to multiply with decimal notation, we use the [·], [×], and [=] keys. To find 4.78 × 0.34, for example, we press [4] [·] [7] [8] [×] [·] [3] [4] [=]. The display reads [1.6252], so 4.78 × 0.34 = 1.6252.

Exercises: Use a calculator to multiply.

1. 5.4×9
2. 415×16.7
3. 17.63×8.1
4. 0.04×12.69
5. 586.4×13.5
6. 4.003×5.1

Convert the number in the sentence to standard notation.

12. The largest building in the world is the Pentagon, which has 3.7 million square feet of floor space.

13. Online Apparel Sales. In 2003, consumers spent \$6.1 billion for apparel.
Source: Jupiter Media Metrix, Inc.

14. Online Ticket Sales. In 2003, consumers spent \$2.7 billion for event tickets.
Source: Jupiter Media Metrix, Inc.

Convert from dollars to cents.

15. \$15.69

16. \$0.17

Convert from cents to dollars.

17. 35¢

18. 577¢

Answers on page A-7

To convert a large number to standard notation, we proceed as follows.

EXAMPLE 12 Convert the number in this sentence to standard notation: In 1999, the U.S. Mint produced 11.6 billion pennies.
Source: U.S. Mint

$$\begin{aligned} 11.6 \text{ billion} &= 11.6 \times 1 \text{ billion} \\ &= 11.6 \times \underbrace{1{,}000{,}000{,}000}_{9 \text{ zeros}} \\ &= 11{,}600{,}000{,}000 \qquad \text{Moving the decimal point 9 places to the right} \end{aligned}$$

Do Exercises 12–14.

MONEY CONVERSION

Converting from dollars to cents is like multiplying by 100. To see why, consider \$19.43.

\$19.43 = 19.43 × \$1	We think of \$19.43 as 19.43 × 1 dollar, or 19.43 × \$1.
= 19.43 × 100¢	Substituting 100¢ for \$1: \$1 = 100¢
= 1943¢	Multiplying

DOLLARS TO CENTS

To convert from dollars to cents, move the decimal point two places to the right and change the \$ sign in front to a ¢ sign at the end.

EXAMPLES Convert from dollars to cents.

13. \$189.64 = 18,964¢

14. \$0.75 = 75¢

Do Exercises 15 and 16.

Converting from cents to dollars is like multiplying by 0.01. To see why, consider 65¢.

65¢ = 65 × 1¢	We think of 65¢ as 65 × 1 cent, or 65 × 1¢.
= 65 × \$0.01	Substituting \$0.01 for 1¢: 1¢ = \$0.01
= \$0.65	Multiplying

CENTS TO DOLLARS

To convert from cents to dollars, move the decimal point two places to the left and change the ¢ sign at the end to a \$ sign in front.

EXAMPLES Convert from cents to dollars.

15. 395¢ = \$3.95

16. 8503¢ = \$85.03

Do Exercises 17 and 18.

3.3 EXERCISE SET

For Extra Help

MathXL | MyMathLab | InterAct Math | Math Tutor Center | Video Lectures on CD Disc 2 | Student's Solutions Manual

a Multiply.

1. 8.6×7

2. 5.7×0.8

3. 0.84×8

4. 9.4×0.6

5. 6.3×0.04

6. 9.8×0.08

7. 87×0.006

8. 18.4×0.07

9. 10×23.76

10. 100×3.8798

11. 1000×583.686852

12. 0.34×1000

13. 7.8×100

14. 0.00238×10

15. 0.1×89.23

16. 0.01×789.235

17. 0.001×97.68

18. 8976.23×0.001

19. 78.2×0.01

20. 0.0235×0.1

21. 32.6×16

22. 9.28×8.6

23. 0.984×3.3

24. 8.489×7.4

25. 374×2.4

26. 865×1.08

27. 749×0.43

28. 978×20.5

29. 0.87×64

30. 7.25×60

31. 46.50×75

32. 8.24×703

33. $\begin{array}{r} 81.7 \\ \times\ 0.612 \\ \hline \end{array}$

34. $\begin{array}{r} 31.82 \\ \times\ 7.15 \\ \hline \end{array}$

35. $\begin{array}{r} 10.105 \\ \times\ 11.324 \\ \hline \end{array}$

36. $\begin{array}{r} 151.2 \\ \times\ 4.555 \\ \hline \end{array}$

37. $\begin{array}{r} 12.3 \\ \times\ 1.08 \\ \hline \end{array}$

38. $\begin{array}{r} 7.82 \\ \times\ 0.024 \\ \hline \end{array}$

39. $\begin{array}{r} 32.4 \\ \times\ 2.8 \\ \hline \end{array}$

40. $\begin{array}{r} 8.09 \\ \times\ 0.0075 \\ \hline \end{array}$

41. $\begin{array}{r} 0.00342 \\ \times\ 0.84 \\ \hline \end{array}$

42. $\begin{array}{r} 2.0056 \\ \times\ 3.8 \\ \hline \end{array}$

43. $\begin{array}{r} 0.347 \\ \times\ 2.09 \\ \hline \end{array}$

44. $\begin{array}{r} 2.532 \\ \times\ 1.067 \\ \hline \end{array}$

45. $\begin{array}{r} 3.005 \\ \times\ 0.623 \\ \hline \end{array}$

46. $\begin{array}{r} 16.34 \\ \times\ 0.000512 \\ \hline \end{array}$

47. 1000×45.678

48. 0.001×45.678

b Convert from dollars to cents.

49. $28.88

50. $67.43

51. $0.66

52. $1.78

Convert from cents to dollars.

53. 34¢

54. 95¢

55. 3445¢

56. 933¢

Convert the number in the sentence to standard notation.

57. *Wal-Mart.* In 2003, Wal-Mart stores had revenues of about $258.7 billion.

Source: Fortune 500

58. *Magazine Circulation.* The total paid circulation in 2003 of the magazine *Better Homes and Gardens* was about 7.6 million copies.

Source: Audit Bureau of Circulations, tabulated by Magazine Publishers of America

59. *Broadway.* Broadway productions grossed $748.9 million in 2004.

Source: League of American Theatres and Producers

60. *Rail Transportation.* People in the United States traveled 13.6 billion miles on city rail transportation in 2003.

Source: American Public Transportation Association, NASA

61. D$_W$ If two rectangles have the same perimeter, will they also have the same area? Experiment with different dimensions. Be sure to use decimals. Explain your answer.

62. D$_W$ A student insists that $346.708 \times 0.1 = 3467.08$. How could you convince him that a mistake had been made without checking on a calculator?

SKILL MAINTENANCE

Calculate.

63. $2\frac{1}{3} \cdot 4\frac{4}{5}$ [2.4d]

64. $2\frac{1}{3} \div 4\frac{4}{5}$ [2.4e]

65. $4\frac{4}{5} - 2\frac{1}{3}$ [2.4c]

66. $4\frac{4}{5} + 2\frac{1}{3}$ [2.4b]

Divide. [1.3d]

67. $24\overline{)8208}$

68. $4\overline{)348}$

69. $7\overline{)31{,}962}$

70. $18\overline{)22{,}626}$

71. $40\overline{)3480}$

72. $17\overline{)20{,}006}$

SYNTHESIS

Consider the following names for large numbers in addition to those already discussed in this section:

$$1 \text{ quadrillion} = 1{,}000{,}000{,}000{,}000{,}000 = 10^{15};$$
$$1 \text{ quintillion} = 1{,}000{,}000{,}000{,}000{,}000{,}000 = 10^{18};$$
$$1 \text{ sextillion} = 1{,}000{,}000{,}000{,}000{,}000{,}000{,}000 = 10^{21};$$
$$1 \text{ septillion} = 1{,}000{,}000{,}000{,}000{,}000{,}000{,}000{,}000 = 10^{24}.$$

Find each of the following. Express the answer with a name that is a power of 10.

73. (1 trillion) · (1 billion)

74. (1 million) · (1 billion)

75. (1 trillion) · (1 trillion)

76. Is a billion millions the same as a million billions? Explain.

3.4 DIVISION

Objectives

a Divide using decimal notation.

b Solve equations of the type $a \cdot x = b$, where a and b may be in decimal notation.

c Simplify expressions using the rules for order of operations.

Divide.

1. $9\overline{)5.4}$

2. $15\overline{)22.5}$

3. $82\overline{)38.54}$

Answers on page A-7

a Division

WHOLE-NUMBER DIVISORS

We use the following method when we divide a decimal quantity by a whole number.

To divide by a whole number,

a) place the decimal point directly above the decimal point in the dividend, and

b) divide as though dividing whole numbers.

```
                 0.8 4 <- Quotient
Divisor -> 7 ) 5.8 8 <- Dividend
               5 6 0
               -----
                 2 8
                 2 8
                 ---
                   0 <- Remainder
```

EXAMPLE 1 Divide: 379.2 ÷ 8.

```
         4 7.4      <- Place the decimal point.
  8 ) 3 7 9.2
      3 2 0 0
      -------
        5 9 2
        5 6 0       Divide as though dividing whole numbers.
        -----
          3 2
          3 2
          ---
            0
```

EXAMPLE 2 Divide: 82.08 ÷ 24.

```
          3.4 2     <- Place the decimal point.
  2 4 ) 8 2.0 8
        7 2 0 0
        -------
        1 0 0 8
          9 6 0     Divide as though dividing whole numbers.
        -------
            4 8
            4 8
            ---
              0
```

Do Exercises 1–3.

Suppose the dividend is a whole number. We can think of it as having a decimal point at the end with as many 0's as we wish after the decimal point. For example,

$$12 = 12. = 12.0 = 12.00 = 12.000, \text{ and so on.}$$

EXAMPLE 3 Divide: $30 \div 8$.

$$\begin{array}{r} 3. \\ 8\,\overline{)\,30.} \\ \underline{24} \\ 6 \end{array}$$

Place the decimal point and divide to find how many ones.

$$\begin{array}{r} 3. \\ 8\,\overline{)\,30.0} \\ \underline{24} \\ 60 \end{array}$$

Write an extra zero.

$$\begin{array}{r} 3.7 \\ 8\,\overline{)\,30.0} \\ \underline{24} \\ 60 \\ \underline{56} \\ 4 \end{array}$$

Divide to find how many tenths.

$$\begin{array}{r} 3.7 \\ 8\,\overline{)\,30.00} \\ \underline{24} \\ 60 \\ \underline{56} \\ 40 \end{array}$$

Write another zero.

$$\begin{array}{r} 3.75 \\ 8\,\overline{)\,30.00} \\ \underline{24} \\ 60 \\ \underline{56} \\ 40 \\ \underline{40} \\ 0 \end{array}$$

Divide to find how many hundredths.

Check:
$$\begin{array}{r} \scriptstyle 6\ 4 \\ 3.75 \\ \underline{\times\quad 8} \\ 30.00 \end{array}$$

EXAMPLE 4 Divide: $4 \div 25$.

$$\begin{array}{r} 0.16 \\ 25\,\overline{)\,4.00} \\ \underline{25} \\ 150 \\ \underline{150} \\ 0 \end{array}$$

Check:
$$\begin{array}{r} \scriptstyle 1 \\ \scriptstyle 3 \\ 0.16 \\ \underline{\times\quad 25} \\ 80 \\ \underline{320} \\ 4.00 \end{array}$$

Do Exercises 4–6.

Divide.

4. $25\,\overline{)\,8}$

5. $4\,\overline{)\,15}$

6. $86\,\overline{)\,21.5}$

Answers on page A-7

7. a) Complete.

$$\frac{3.75}{0.25} = \frac{3.75}{0.25} \times \frac{100}{100}$$
$$= \frac{(\ \ )}{25}$$

b) Divide.

$0.25\overline{)3.75}$

Divide.

8. $0.83\overline{)4.067}$

9. $3.5\overline{)44.8}$

Answers on page A-7

DIVISORS THAT ARE NOT WHOLE NUMBERS

Consider the division

$0.24\overline{)8.208}$

We write the division as $\frac{8.208}{0.24}$. Then we multiply by 1 to change to a whole-number divisor:

$$\frac{8.208}{0.24} = \frac{8.208}{0.24} \times \frac{100}{100} = \frac{820.8}{24}.$$

The division $0.24\overline{)8.208}$ is the same as $24\overline{)820.8}$.

The divisor is now a whole number.

To divide when the divisor is not a whole number,

a) move the decimal point (multiply by 10, 100, and so on) to make the divisor a whole number,

b) move the decimal point (multiply the same way) in the dividend the same number of places, and

c) place the decimal point directly above the new decimal point in the dividend and divide as though dividing whole numbers.

$0.24\overline{)8.208}$ Move 2 places to the right.

$0.24\overline{)8.208}$ Move 2 places to the right.

```
          3 4.2
0.2 4 ) 8.2 0^8
        7 2 0 0
        1 0 0 8
          9 6 0
            4 8
            4 8
              0
```

(The new decimal point in the dividend is indicated by a caret.)

EXAMPLE 5 Divide: $5.848 \div 8.6$.

$8.6\overline{)5.848}$ Multiply the divisor by 10 (move the decimal point 1 place). Multiply the same way in the dividend (move 1 place).

```
         0.6 8
8.6 ) 5.8^4 8
        5 1 6 0
          6 8 8
          6 8 8
              0
```

Place a decimal point above the new decimal point and then divide.

Note: $\frac{5.848}{8.6} = \frac{5.848}{8.6} \cdot \frac{10}{10} = \frac{58.48}{86}$.

Do Exercises 7–9.

Sometimes it helps to write extra zeros to the right of the decimal point in the dividend.

EXAMPLE 6 Divide: $12 \div 0.64$.

$0.64\overline{)12.}$ — Place a decimal point at the end of the whole number.

$0.64\overline{)12.00}$ — Multiply the divisor by 100 (move the decimal point 2 places). Multiply the same way in the dividend (move 2 places).

$$\begin{array}{r} 18.75 \\ 0.64\overline{)12.00{}_\wedge00} \\ \underline{640} \\ 560 \\ \underline{512} \\ 480 \\ \underline{448} \\ 320 \\ \underline{320} \\ 0 \end{array}$$

Place a decimal point above and then divide.

Do Exercise 10.

10. Divide.

$1.6\overline{)25}$

DIVIDING BY 10, 100, 1000, AND SO ON

It is often helpful to be able to divide quickly by a ten, hundred, or thousand, or by a tenth, hundredth, or thousandth. Each procedure we use is based on multiplying by 1. Consider the following example:

$$\frac{23.789}{1000} = \frac{23.789}{1000} \cdot \frac{1000}{1000} = \frac{23{,}789}{1{,}000{,}000} = 0.023789.$$

We are dividing by a number greater than 1: The result is *smaller* than 23.789.

To divide by 10, 100, 1000, and so on,

a) count the number of zeros in the divisor, and

$\frac{713.49}{100}$ → 2 zeros

b) write the quotient by moving the decimal point in the dividend that number of places to the left.

$\frac{713.49}{100}$, 7.13.49 (2 places to the left), $\frac{713.49}{100} = 7.1349$

EXAMPLE 7 Divide: $\frac{0.0104}{10}$.

$\frac{0.0104}{10}$ (1 zero), 0.0.0104, (1 place to the left), $\frac{0.0104}{10} = 0.00104$

CALCULATOR CORNER

Division with Decimal Notation To use a calculator to divide with decimal notation, we use the [.], [÷], and [=] keys. To find $237.12 \div 5.2$, for example, we press [2] [3] [7] [.] [1] [2] [÷] [5] [.] [2] [=]. The display reads [45.6], so $237.12 \div 5.2 = 45.6$.

Exercises: Use a calculator to divide.

1. $12.4\overline{)177.32}$
2. $49\overline{)125.44}$
3. $3.2\overline{)640}$
4. $16\overline{)12}$
5. $14 \div 0.7$
6. $1.6 \div 25$
7. $474.14 \div 30.2$
8. $518.472 \div 6.84$

Answer on page A-7

Study Tips

HOMEWORK TIPS

Prepare for your homework assignment by reading the explanations of concepts and following the step-by-step solutions of examples in the text. The time you spend preparing will save valuable time when you do your assignment.

Divide.

11. $\frac{0.1278}{0.01}$

12. $\frac{0.1278}{100}$

13. $\frac{98.47}{1000}$

14. $\frac{6.7832}{0.1}$

Answers on page A-7

DIVIDING BY 0.1, 0.01, 0.001, AND SO ON

Now consider the following example:

$$\frac{23.789}{0.01} = \frac{23.789}{0.01} \cdot \frac{100}{100} = \frac{2378.9}{1} = 2378.9.$$

We are dividing by a number less than 1: The result is *larger* than 23.789.
We use the following procedure.

> To divide by 0.1, 0.01, 0.001, and so on,
>
> **a)** count the number of decimal places in the divisor, and
>
> $\frac{713.49}{0.001}$ → 3 places
>
> **b)** write the quotient by moving the decimal point in the dividend that number of places to the right.
>
> $\frac{713.49}{0.001}$, 713.490. (3 places to the right) $\frac{713.49}{0.001} = 713{,}490$

EXAMPLE 8 Divide: $\frac{23.738}{0.001}$.

$\frac{23.738}{0.001}$, (3 places) 23.738. (3 places to the right to change 0.001 to 1) $\frac{23.738}{0.001} = 23{,}738$

Do Exercises 11–14.

b Solving Equations

Now let's solve equations of the type $a \cdot x = b$, where a and b may be in decimal notation. Proceeding as before, we divide by a on both sides.

EXAMPLE 9 Solve: $8 \cdot x = 27.2$.

We have

$$\frac{8 \cdot x}{8} = \frac{27.2}{8} \qquad \text{Dividing by 8 on both sides}$$

$$x = 3.4.$$

```
     3.4
8 ) 2 7.2
    2 4 0
    -----
      3 2
      3 2
      ---
        0
```

The solution is 3.4.

EXAMPLE 10 Solve: $2.9 \cdot t = 0.14616$.

We have

$$\frac{2.9 \cdot t}{2.9} = \frac{0.14616}{2.9} \qquad \text{Dividing by 2.9 on both sides}$$

$$t = 0.0504.$$

```
         0.0 5 0 4
2.9 ) 0.1^4 6 1 6
         1 4 5 0 0
             1 1 6
             1 1 6
                 0
```

The solution is 0.0504.

Do Exercises 15 and 16.

Solve.

15. $100 \cdot x = 78.314$

16. $0.25 \cdot y = 276.4$

C Order of Operations: Decimal Notation

The same rules for order of operations used with whole numbers and fraction notation apply when simplifying expressions with decimal notation.

RULES FOR ORDER OF OPERATIONS

1. Do all calculations within grouping symbols before operations outside.
2. Evaluate all exponential expressions.
3. Do all multiplications and divisions in order from left to right.
4. Do all additions and subtractions in order from left to right.

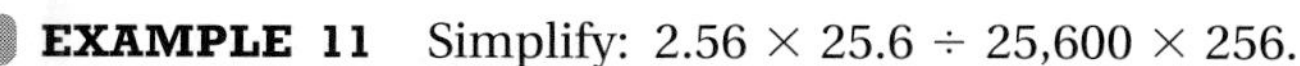

EXAMPLE 11 Simplify: $2.56 \times 25.6 \div 25{,}600 \times 256$.

There are no exponents or parentheses, so we multiply and divide from left to right:

$$\begin{aligned} 2.56 \times 25.6 \div 25{,}600 \times 256 &= 65.536 \div 25{,}600 \times 256 \\ &= 0.00256 \times 256 \\ &= 0.65536. \end{aligned}$$

Doing all multiplications and divisions in order from left to right

EXAMPLE 12 Simplify: $(5 - 0.06) \div 2 + 3.42 \times 0.1$.

$(5 - 0.06) \div 2 + 3.42 \times 0.1 = 4.94 \div 2 + 3.42 \times 0.1$ Carrying out the operation inside parentheses

$= 2.47 + 0.342$ Doing all multiplications and divisions in order from left to right

$= 2.812$

Answers on page A-7

Simplify.

17. $625 \div 62.5 \times 25 \div 6250$

18. $0.25 \cdot (1 + 0.08) - 0.0274$

19. $20^2 - 3.4^2 + \{2.5[20(9.2 - 5.6)] + 5(10 - 5)\}$

20. **Mountains in Peru.** Refer to the figure in Example 14. Find the average height, in meters, of the mountains.

Answers on page A-7

EXAMPLE 13 Simplify: $10^2 \times \{[(3 - 0.24) \div 2.4] - (0.21 - 0.092)\}$.

$10^2 \times \{[(3 - 0.24) \div 2.4] - (0.21 - 0.092)\}$

$= 10^2 \times \{[2.76 \div 2.4] - 0.118\}$ Doing the calculations in the innermost parentheses first

$= 10^2 \times \{1.15 - 0.118\}$ Again, doing the calculations in the innermost parentheses

$= 10^2 \times 1.032$ Subtracting inside the parentheses

$= 100 \times 1.032$ Evaluating the exponential expression

$= 103.2$

Do Exercises 17–19.

EXAMPLE 14 *Mountains in Peru.* The following figure shows a range of very high mountains in Peru, together with their altitudes, given in both feet and meters. Find the average height, in feet, of these mountains.

Source: *National Geographic,* July 1968, p. 130

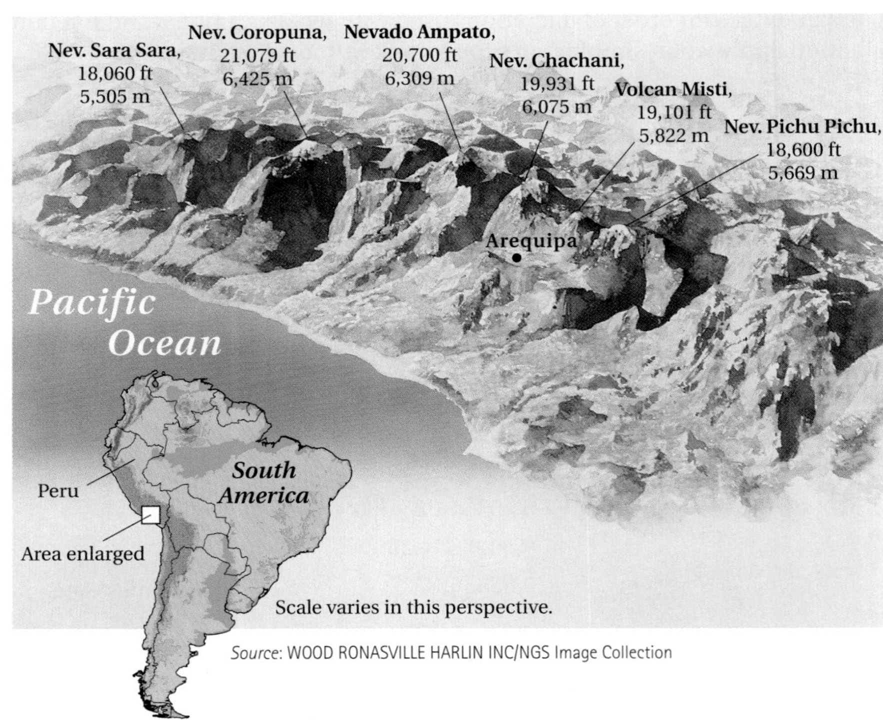

Source: WOOD RONASVILLE HARLIN INC/NGS Image Collection

The **average** of a set of numbers is the sum of the numbers divided by the number of addends. (See Section 1.6.) We find the sum of the heights divided by the number of addends, 6:

$$\frac{18{,}060 + 21{,}079 + 20{,}700 + 19{,}931 + 19{,}101 + 18{,}600}{6} = \frac{117{,}471}{6} = 19{,}578.5.$$

Thus the average height of these mountains is 19,578.5 ft.

Do Exercise 20.

3.4 EXERCISE SET

For Extra Help

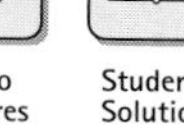

MathXL | MyMathLab | InterAct Math | Math Tutor Center | Video Lectures on CD Disc 2 | Student's Solutions Manual

a Divide.

1. $2\overline{)5.98}$

2. $5\overline{)18}$

3. $4\overline{)95.12}$

4. $8\overline{)25.92}$

5. $12\overline{)89.76}$

6. $23\overline{)25.07}$

7. $33\overline{)237.6}$

8. $12.4 \div 4$

9. $9.144 \div 8$

10. $4.5 \div 9$

11. $12.123 \div 3$

12. $7\overline{)5.6}$

13. $5\overline{)0.35}$

14. $0.04\overline{)1.68}$

15. $0.12\overline{)8.4}$

16. $0.36\overline{)2.88}$

17. $3.4\overline{)68}$

18. $0.25\overline{)5}$

19. $15\overline{)6}$

20. $12\overline{)1.8}$

21. $36\overline{)14.76}$

22. $52\overline{)119.6}$

23. $3.2\overline{)27.2}$

24. $8.5\overline{)27.2}$

25. $4.2\overline{)39.06}$

26. $4.8\overline{)0.1104}$

27. $8\overline{)5}$

28. $8\overline{)3}$

29. $0.47\overline{)0.1222}$

30. $1.08\overline{)0.54}$

31. $4.8\overline{)75}$

32. $0.28\overline{)63}$

33. $0.032\overline{)0.07488}$

34. $0.017\overline{)1.581}$

35. $82\overline{)38.54}$

36. $34\overline{)0.1462}$

37. $\dfrac{213.4567}{1000}$

38. $\dfrac{213.4567}{100}$

39. $\dfrac{213.4567}{10}$

40. $\dfrac{100.7604}{0.1}$

41. $\dfrac{1.0237}{0.001}$

42. $\dfrac{1.0237}{0.01}$

b Solve.

43. $4.2 \cdot x = 39.06$

44. $36 \cdot y = 14.76$

45. $1000 \cdot y = 9.0678$

46. $789.23 = 0.25 \cdot q$

47. $1048.8 = 23 \cdot t$

48. $28.2 \cdot x = 423$

c Simplify.

49. $14 \times (82.6 + 67.9)$

50. $(26.2 - 14.8) \times 12$

51. $0.003 + 3.03 \div 0.01$

52. $9.94 + 4.26 \div (6.02 - 4.6) - 0.9$

53. $42 \times (10.6 + 0.024)$

54. $(18.6 - 4.9) \times 13$

55. $4.2 \times 5.7 + 0.7 \div 3.5$

56. $123.3 - 4.24 \times 1.01$

57. $9.0072 + 0.04 \div 0.1^2$

58. $12 \div 0.03 - 12 \times 0.03^2$

59. $(8 - 0.04)^2 \div 4 + 8.7 \times 0.4$

60. $(5 - 2.5)^2 \div 100 + 0.1 \times 6.5$

61. $86.7 + 4.22 \times (9.6 - 0.03)^2$

62. $2.48 \div (1 - 0.504) + 24.3 - 11 \times 2$

63. $4 \div 0.4 + 0.1 \times 5 - 0.1^2$

64. $6 \times 0.9 + 0.1 \div 4 - 0.2^3$

65. $5.5^2 \times [(6 - 4.2) \div 0.06 + 0.12]$

66. $12^2 \div (12 + 2.4) - [(2 - 1.6) \div 0.8]$

67. $200 \times \{[(4 - 0.25) \div 2.5] - (4.5 - 4.025)\}$

68. $0.03 \times \{1 \times 50.2 - [(8 - 7.5) \div 0.05]\}$

69. Find the average of \$1276.59, \$1350.49, \$1123.78, and \$1402.58.

70. Find the average weight of two wrestlers who weigh 308 lb and 296.4 lb.

71. *Income Taxes.* The amount paid in United States individual income taxes in a 5-yr period is shown in the bar graph below. Find the average amount paid per year in individual income taxes over the 5-yr period.

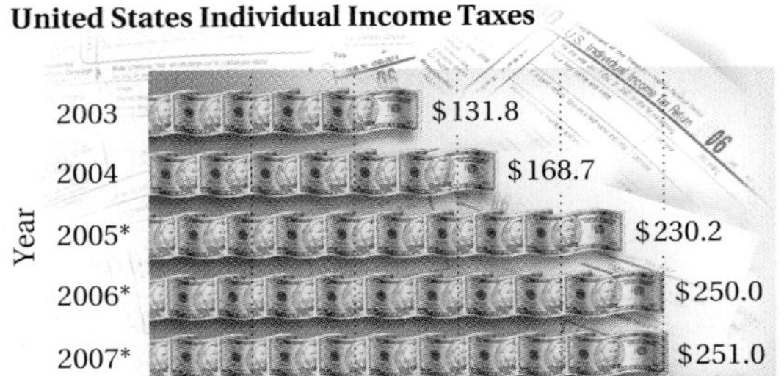

72. *Herbal Supplement.* Biologically based therapies in complementary and alternative medicine use substances found in nature such as herbs, food, and vitamins. The graph below shows spending on the herbal supplement echinacea, also known as purple coneflower, which is thought to strengthen the immune system. Find the average amount spent per year from 1999 to 2003.

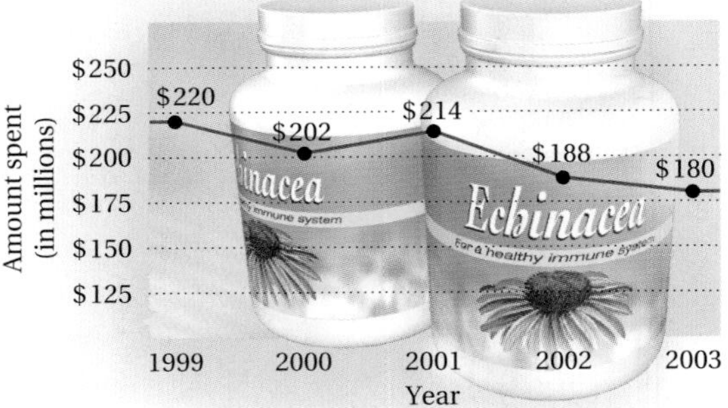

73. D_W How is division with decimal notation similar to division of whole numbers? How is it different?

74. D_W Kayla made these two computational mistakes:

$0.247 \div 0.1 = 0.0247$; $\quad 0.247 \div 10 = 2.47$.

In each case, how could you convince her that a mistake has been made?

SKILL MAINTENANCE

Simplify. [2.1e]

75. $\frac{36}{42}$

76. $\frac{56}{64}$

77. $\frac{38}{146}$

78. $\frac{114}{438}$

Find the prime factorization. [1.7d]

79. 684

80. 162

81. 2007

82. 2005

83. Add: $10\frac{1}{2} + 4\frac{5}{8}$. [2.4b]

84. Subtract: $10\frac{1}{2} - 4\frac{5}{8}$. [2.4c]

SYNTHESIS

Simplify.

85. 🖩 $9.0534 - 2.041^2 \times 0.731 \div 1.043^2$

86. 🖩 $23.042(7 - 4.037 \times 1.46 - 0.932^2)$

In Exercises 87–90, find the missing value.

87. $439.57 \times 0.01 \div 1000 \times \square = 4.3957$

88. $5.2738 \div 0.01 \times 1000 \div \square = 52.738$

89. $0.0329 \div 0.001 \times 10^4 \div \square = 3290$

90. $0.0047 \times 0.01 \div 10^4 \times \square = 4.7$

3.5 CONVERTING FROM FRACTION NOTATION TO DECIMAL NOTATION

a Fraction Notation to Decimal Notation

When a denominator has no prime factors other than 2's and 5's, we can find decimal notation by multiplying by 1. We multiply to get a denominator that is a power of ten, like 10, 100, or 1000.

EXAMPLE 1 Find decimal notation for $\frac{3}{5}$.

$$\frac{3}{5} = \frac{3}{5} \cdot \frac{2}{2} = \frac{6}{10} = 0.6$$ We use $\frac{2}{2}$ for 1 to get a denominator of 10.

EXAMPLE 2 Find decimal notation for $\frac{7}{20}$.

$$\frac{7}{20} = \frac{7}{20} \cdot \frac{5}{5} = \frac{35}{100} = 0.35$$ We use $\frac{5}{5}$ for 1 to get a denominator of 100.

EXAMPLE 3 Find decimal notation for $\frac{87}{25}$.

$$\frac{87}{25} = \frac{87}{25} \cdot \frac{4}{4} = \frac{348}{100} = 3.48$$ We use $\frac{4}{4}$ for 1 to get a denominator of 100.

EXAMPLE 4 Find decimal notation for $\frac{9}{40}$.

$$\frac{9}{40} = \frac{9}{40} \cdot \frac{25}{25} = \frac{225}{1000} = 0.225$$ We use $\frac{25}{25}$ for 1 to get a denominator of 1000.

Do Exercises 1–4.

We can always divide to find decimal notation.

EXAMPLE 5 Find decimal notation for $\frac{3}{5}$.

$$\frac{3}{5} = 3 \div 5 \qquad \begin{array}{r} 0.6 \\ 5\overline{)3.0} \\ \underline{3\,0} \\ 0 \end{array} \qquad \frac{3}{5} = 0.6$$

EXAMPLE 6 Find decimal notation for $\frac{7}{8}$.

$$\frac{7}{8} = 7 \div 8 \qquad \begin{array}{r} 0.8\,7\,5 \\ 8\overline{)7.0\,0\,0} \\ \underline{6\,4} \\ 6\,0 \\ \underline{5\,6} \\ 4\,0 \\ \underline{4\,0} \\ 0 \end{array} \qquad \frac{7}{8} = 0.875$$

Do Exercises 5 and 6.

Objectives

a Convert from fraction notation to decimal notation.

b Round numbers named by repeating decimals in problem solving.

c Calculate using fraction notation and decimal notation together.

Find decimal notation. Use multiplying by 1.

1. $\frac{4}{5}$

2. $\frac{9}{20}$

3. $\frac{11}{40}$

4. $\frac{33}{25}$

Find decimal notation.

5. $\frac{2}{5}$

6. $\frac{3}{8}$

Answers on page A-7

Find decimal notation.

7. $\frac{1}{6}$

8. $\frac{2}{3}$

Find decimal notation.

9. $\frac{5}{11}$

10. $\frac{12}{11}$

Answers on page A-7

In Examples 5 and 6, the division *terminated,* meaning that eventually we got a remainder of 0. A **terminating decimal** occurs when the denominator has only 2's or 5's, or both, as factors, as in $\frac{17}{25}$, $\frac{5}{8}$, or $\frac{83}{100}$. This assumes that the fraction notation has been simplified.

If division does *not* lead to a remainder of 0, but instead leads to a repeating pattern of nonzero remainders, we have a **repeating decimal.**

EXAMPLE 7 Find decimal notation for $\frac{5}{6}$.

$$\frac{5}{6} = 5 \div 6 \qquad \begin{array}{r} 0.833 \\ 6\overline{)5.000} \\ \underline{48} \\ 20 \\ \underline{18} \\ 20 \\ \underline{18} \\ 2 \end{array}$$

Since 2 keeps reappearing as a remainder, the digits repeat and will continue to do so; therefore,

$$\frac{5}{6} = 0.83333\ldots.$$

The red dots indicate an endless sequence of digits in the quotient. When there is a repeating pattern, the dots are often replaced by a bar to indicate the repeating part—in this case, only the 3:

$$\frac{5}{6} = 0.8\overline{3}.$$

Do Exercises 7 and 8.

EXAMPLE 8 Find decimal notation for $\frac{4}{11}$.

$$\frac{4}{11} = 4 \div 11 \qquad \begin{array}{r} 0.3636 \\ 11\overline{)4.0000} \\ \underline{33} \\ 70 \\ \underline{66} \\ 40 \\ \underline{33} \\ 70 \\ \underline{66} \\ 4 \end{array}$$

Since 7 and 4 keep repeating as remainders, the sequence of digits "36" repeats in the quotient, and

$$\frac{4}{11} = 0.363636\ldots, \quad \text{or} \quad 0.\overline{36}.$$

Do Exercises 9 and 10.

EXAMPLE 9 Find decimal notation for $\frac{5}{7}$.

$$\begin{array}{r} 0.714285 \\ 7\,\overline{)\,5.000000} \\ \underline{4\,9} \\ 1\,0 \\ \underline{7} \\ 3\,0 \\ \underline{2\,8} \\ 2\,0 \\ \underline{1\,4} \\ 6\,0 \\ \underline{5\,6} \\ 4\,0 \\ \underline{3\,5} \\ 5 \end{array}$$

Since 5 appears as a remainder, the sequence of digits "714285" repeats in the quotient, and

$$\frac{5}{7} = 0.714285714285\ldots, \quad \text{or} \quad 0.\overline{714285}.$$

11. Find decimal notation for $\frac{3}{7}$.

Round each to the nearest tenth, hundredth, and thousandth.

12. $0.\overline{6}$

13. $0.\overline{80}$

14. $6.\overline{245}$

Answers on page A-7

The length of a repeating part can be very long—too long to find on a calculator. An example is $\frac{5}{97}$, which has a repeating part of 96 digits.

Do Exercise 11.

b Rounding in Problem Solving

In applied problems, repeating decimals are rounded to get approximate answers. To round a repeating decimal, we can extend the decimal notation at least one place past the rounding digit, and then round as before.

EXAMPLES Round each of the following to the nearest tenth, hundredth, and thousandth.

	Nearest tenth	*Nearest hundredth*	*Nearest thousandth*
10. $0.8\overline{3} = 0.83333\ldots$	0.8	0.83	0.833
11. $0.\overline{09} = 0.090909\ldots$	0.1	0.09	0.091
12. $0.\overline{714285} = 0.714285714285\ldots$	0.7	0.71	0.714

Do Exercises 12–14.

CONVERTING RATIOS TO DECIMAL NOTATION

When solving applied problems, we often convert ratios to decimal notation.

EXAMPLE 13 *Forest Fires.* In October 2003, 15 devastating forest fires burned in Southern California for two weeks, burning 800,000 acres. Find the ratio of number of acres burned to number of fires and convert it to decimal notation. Round to the nearest thousandth.

Source: National Forest Service

We have

$$\frac{\text{Acres burned}}{\text{Number of fires}} = \frac{800{,}000 \text{ acres}}{15 \text{ fires}} \approx 53{,}333.\overline{3}.$$

There were about 53,333.333 acres burned per fire.

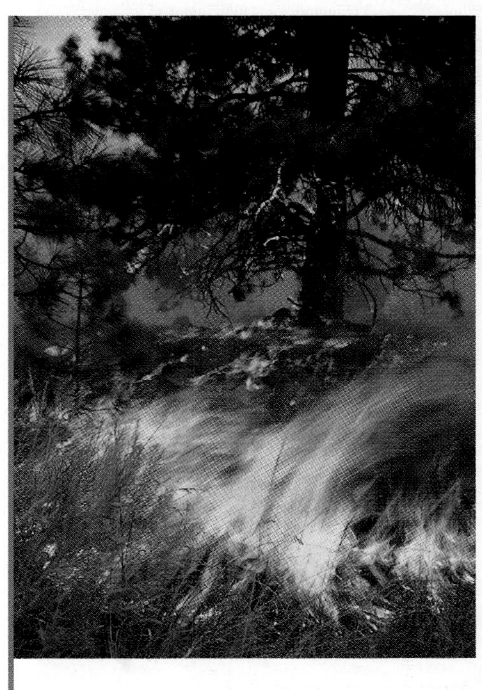

15. **Coin Tossing.** A coin is tossed 51 times. It lands heads 26 times. Find the ratio of heads to tosses and convert it to decimal notation. Round to the nearest thousandth. (This is also the experimental probability of getting heads.)

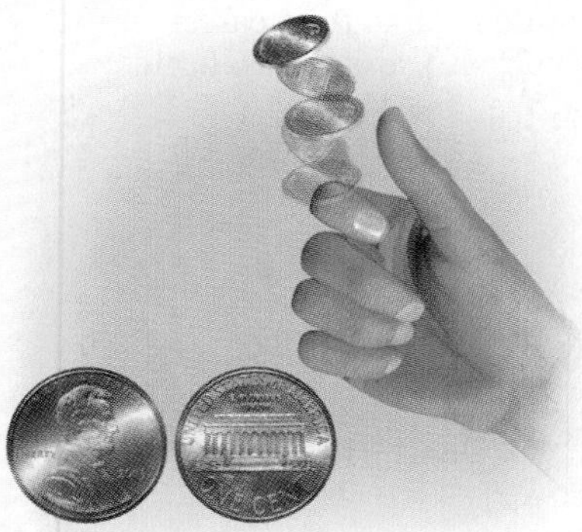

Heads Tails

16. **Gas Mileage.** A car travels 380 mi on 15.7 gal of gasoline. Find the gasoline mileage and convert the ratio to decimal notation rounded to the nearest tenth.

17. **Camera Sales.** Refer to the data on sales of digital cameras in Example 15. Find the average sales of these cameras for the 6-yr period.

Source: Photo Marketing Association International

Answers on page A-7

EXAMPLE 14 *Gas Mileage.* A car travels 457 mi on 16.4 gal of gasoline. The ratio of number of miles driven to amount of gasoline used is *gas mileage.* Find the gas mileage and convert the ratio to decimal notation rounded to the nearest tenth.

$$\frac{\text{Miles driven}}{\text{Gasoline used}} = \frac{457}{16.4} \approx 27.86 \qquad \text{Dividing to 2 decimal places}$$

$$\approx 27.9 \qquad \text{Rounding to 1 decimal place}$$

The gas mileage is 27.9 miles to the gallon.

Do Exercises 15 and 16.

AVERAGES

When finding an average, we may at times need to round an answer.

EXAMPLE 15 *Camera Sales.* As sales of digital cameras have increased, analog camera sales have decreased. The bar graph below shows the sales of both types of cameras from 2000 to 2005, in millions of cameras. Find the average sales of analog cameras for this period.

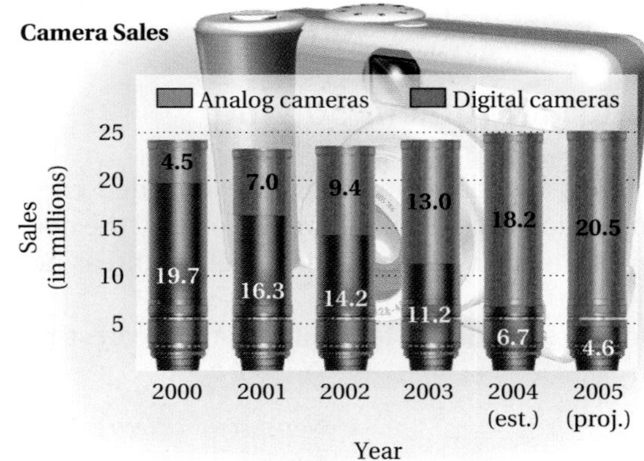

Source: Photo Marketing Association International

We add the sales totals shown on the bar graph and divide by the number of addends, 6. Since all the units are in millions, we need not convert them to standard notation. The average is

$$\frac{19.7 + 16.3 + 14.2 + 11.2 + 6.7 + 4.6}{6} = \frac{72.7}{6} \approx 12.12.$$

The average number of analog cameras sold per year for the 6-yr period is about 12.12 million.

Do Exercise 17.

C Calculations with Fraction Notation and Decimal Notation Together

In certain kinds of calculations, fraction notation and decimal notation might occur together. In such cases, there are at least three ways in which we might proceed.

EXAMPLE 16 Calculate: $\frac{2}{3} \times 0.576$.

METHOD 1 One way to do this calculation is to convert the fraction notation to decimal notation so that both numbers are in decimal notation. Since $\frac{2}{3}$ converts to repeating decimal notation, it is first rounded to some chosen decimal place. We choose three decimal places. Then, using decimal notation, we multiply.

$$\frac{2}{3} \times 0.576 = 0.\overline{6} \times 0.576 \approx 0.667 \times 0.576 = 0.384192$$

METHOD 2 A second way to do this calculation is to convert the decimal notation to fraction notation so that both numbers are in fraction notation. The answer can be left in fraction notation and simplified, or we can convert to decimal notation and round, if appropriate.

$$\begin{aligned}\frac{2}{3} \times 0.576 &= \frac{2}{3} \cdot \frac{576}{1000} = \frac{2 \cdot 576}{3 \cdot 1000}\\ &= \frac{2 \cdot 2 \cdot 2 \cdot 2 \cdot 2 \cdot 2 \cdot 2 \cdot 3 \cdot 3}{2 \cdot 2 \cdot 2 \cdot 3 \cdot 5 \cdot 5 \cdot 5}\\ &= \frac{2 \cdot 2 \cdot 2 \cdot 3}{2 \cdot 2 \cdot 2 \cdot 3} \cdot \frac{2 \cdot 2 \cdot 2 \cdot 2 \cdot 3}{5 \cdot 5 \cdot 5}\\ &= 1 \cdot \frac{2 \cdot 2 \cdot 2 \cdot 2 \cdot 3}{5 \cdot 5 \cdot 5}\\ &= \frac{2 \cdot 2 \cdot 2 \cdot 2 \cdot 3}{5 \cdot 5 \cdot 5} = \frac{48}{125}, \text{ or } 0.384\end{aligned}$$

METHOD 3 A third way to do this calculation is to treat 0.576 as $\frac{0.576}{1}$. Then we multiply 0.576 by 2, and divide the result by 3.

$$\frac{2}{3} \times 0.576 = \frac{2}{3} \times \frac{0.576}{1} = \frac{2 \times 0.576}{3} = \frac{1.152}{3} = 0.384$$

Note that we get an exact answer with methods 2 and 3, but method 1 gives an approximation since we rounded decimal notation for $\frac{2}{3}$.

Do Exercise 18.

EXAMPLE 17 Calculate: $\frac{2}{3} \times 0.576 + 3.287 \div \frac{4}{5}$.

We use the rules for order of operations, doing first the multiplication and then the division. Then we add.

$$\begin{aligned}\frac{2}{3} \times 0.576 + 3.287 \div \frac{4}{5} &= 0.384 + 3.287 \cdot \frac{5}{4}\\ &= 0.384 + 4.10875\\ &= 4.49275\end{aligned}$$

Method 3:
$\frac{2}{3} \times \frac{0.576}{1} = 0.384$;
$\frac{3.287}{1} \times \frac{5}{4} = 4.10875$

Do Exercises 19 and 20.

18. Calculate: $\frac{5}{6} \times 0.864$.

Calculate.

19. $\frac{1}{3} \times 0.384 + \frac{5}{8} \times 0.6784$

20. $\frac{5}{6} \times 0.864 + 14.3 \div \frac{8}{5}$

Answers on page A-7

3.5 EXERCISE SET

For Extra Help

MathXL | MyMathLab | InterAct Math | Math Tutor Center | Video Lectures on CD Disc 2 | Student's Solutions Manual

a Find decimal notation.

1. $\frac{23}{100}$ **2.** $\frac{9}{100}$ **3.** $\frac{3}{5}$ **4.** $\frac{19}{20}$ **5.** $\frac{13}{40}$ **6.** $\frac{3}{16}$

7. $\frac{1}{5}$ **8.** $\frac{4}{5}$ **9.** $\frac{17}{20}$ **10.** $\frac{11}{20}$ **11.** $\frac{3}{8}$ **12.** $\frac{7}{8}$

13. $\frac{39}{40}$ **14.** $\frac{31}{40}$ **15.** $\frac{13}{25}$ **16.** $\frac{61}{125}$ **17.** $\frac{2502}{125}$ **18.** $\frac{181}{200}$

19. $\frac{1}{4}$ **20.** $\frac{1}{2}$ **21.** $\frac{29}{25}$ **22.** $\frac{37}{25}$ **23.** $\frac{19}{16}$ **24.** $\frac{5}{8}$

25. $\frac{4}{15}$ **26.** $\frac{7}{9}$ **27.** $\frac{1}{3}$ **28.** $\frac{1}{9}$ **29.** $\frac{4}{3}$ **30.** $\frac{8}{9}$

31. $\frac{7}{6}$ **32.** $\frac{7}{11}$ **33.** $\frac{4}{7}$ **34.** $\frac{14}{11}$ **35.** $\frac{11}{12}$ **36.** $\frac{5}{12}$

b

37.–47. Odds. Round each answer of the odd-numbered Exercises 25–35 to the nearest tenth, hundredth, and thousandth.

38.–48. Evens. Round each answer of the even-numbered Exercises 26–36 to the nearest tenth, hundredth, and thousandth.

Round each to the nearest tenth, hundredth, and thousandth.

49. $0.\overline{18}$ **50.** $0.\overline{83}$ **51.** $0.2\overline{7}$ **52.** $3.5\overline{4}$

53. For this set of people, what is the ratio, in decimal notation rounded to the nearest thousandth, where appropriate, of:

a) women to the total number of people?
b) women to men?
c) men to the total number of people?
d) men to women?

54. For this set of nuts and bolts, what is the ratio, in decimal notation rounded to the nearest thousandth, where appropriate, of:

a) nuts to bolts?
b) bolts to nuts?
c) nuts to the total number?
d) total number to nuts?

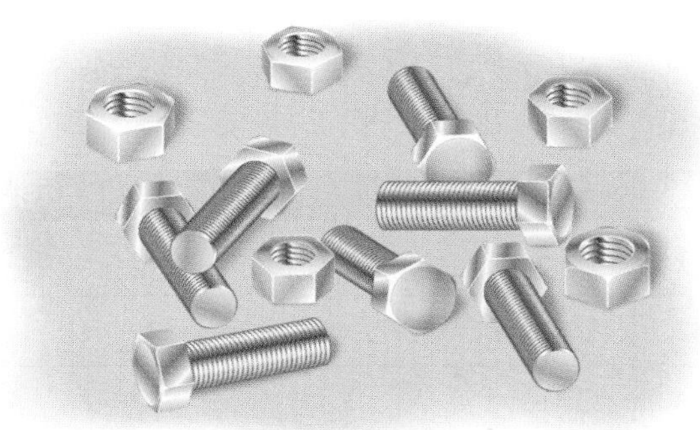

Gas Mileage. In each of Exercises 55–58, find the gas mileage rounded to the nearest tenth.

55. 285 mi; 18 gal

56. 396 mi; 17 gal

57. 324.8 mi; 18.2 gal

58. 264.8 mi; 12.7 gal

59. *Windy Cities.* Although nicknamed the Windy City, Chicago is not the windiest city in the United States. Listed in the table below are the six windiest cities and their average wind speeds. Find the average of these wind speeds and round your answer to the nearest tenth.

Source: *The Handy Geography Answer Book*

CITY	AVERAGE WIND SPEED (in miles per hour)
Mt. Washington, NH	35.3
Boston, MA	12.5
Honolulu, HI	11.3
Dallas TX	10.7
Kansas City, MO	10.7
Chicago, IL	10.4

60. *Areas of the New England States.* The table below lists the areas of the New England states. Find the average area and round your answer to the nearest tenth.

Source: *The New York Times Almanac*

STATE	TOTAL AREA (in square miles)
Maine	33,265
New Hampshire	9,279
Vermont	9,614
Massachusetts	8,284
Connecticut	5,018
Rhode Island	1,211

Stock Prices. At one time stock prices were given using mixed numerals. The Securities and Exchange Commission has mandated the use of decimal notation. Thus a price of $\$23\frac{13}{16}$ is now converted to decimal notation rounded to the nearest hundredth, that is, \$23.81. Complete the following table.

Sources: *The Indianapolis Star,* 1/30/01; www.yahoo.com

	STOCK	PRICE PER SHARE	DECIMAL NOTATION	ROUNDED TO NEAREST HUNDREDTH
61.	Pfizer	$\$24\frac{9}{16}$		
62.	Lucent	$\$3\frac{31}{64}$		
63.	Qwest	$\$3\frac{47}{64}$		
64.	Merck	$\$32\frac{5}{8}$		
65.	Exxon Mobil	$\$59\frac{7}{8}$		
66.	Hewlett Packard	$\$21\frac{7}{16}$		

Source: New York Stock Exchange

C Calculate.

67. $\frac{7}{8} \times 12.64$

68. $\frac{4}{5} \times 384.8$

69. $2\frac{3}{4} + 5.65$

70. $4\frac{4}{5} + 3.25$

71. $\frac{47}{9} \times 79.95$

72. $\frac{7}{11} \times 2.7873$

73. $\frac{1}{2} - 0.5$

74. $3\frac{1}{8} - 2.75$

75. $4.875 - 2\frac{1}{16}$

76. $55\frac{3}{5} - 12.22$

77. $\frac{5}{6} \times 0.0765 + \frac{5}{4} \times 0.1124$

78. $\frac{3}{5} \times 6384.1 - \frac{3}{8} \times 156.56$

79. $\frac{4}{5} \times 384.8 + 24.8 \div \frac{8}{3}$

80. $102.4 \div \frac{2}{5} - 12 \times \frac{5}{6}$

81. $\frac{7}{8} \times 0.86 - 0.76 \times \frac{3}{4}$

82. $17.95 \div \frac{5}{8} + \frac{3}{4} \times 16.2$

83. $3.375 \times 5\frac{1}{3}$

84. $2.5 \times 3\frac{5}{8}$

85. $6.84 \div 2\frac{1}{2}$

86. $8\frac{1}{2} \div 2.125$

87. $\mathbf{D_W}$ When is long division *not* the fastest way to convert from fraction notation to decimal notation?

88. $\mathbf{D_W}$ Examine Example 16 of this section. How could the problem be changed so that method 1 would give a result that is completely accurate?

SKILL MAINTENANCE

Multiply. [2.4d]

89. $9 \cdot 2\frac{1}{3}$

90. $10\frac{1}{2} \cdot 22\frac{3}{4}$

Divide. [2.4e]

91. $84 \div 8\frac{2}{5}$

92. $8\frac{3}{5} \div 10\frac{2}{5}$

Add. [2.4b]

93. $17\frac{5}{6} + 32\frac{3}{8}$

94. $14\frac{3}{5} + 16\frac{1}{10}$

Subtract. [2.4c]

95. $16\frac{1}{10} - 14\frac{3}{5}$

96. $32\frac{3}{8} - 17\frac{5}{6}$

Solve. [2.5a]

97. A recipe for bread calls for $\frac{2}{3}$ cup of water, $\frac{1}{4}$ cup of milk, and $\frac{1}{8}$ cup of oil. How many cups of liquid ingredients does the recipe call for?

98. A board $\frac{7}{10}$ in. thick is glued to a board $\frac{3}{5}$ in. thick. The glue is $\frac{3}{100}$ in. thick. How thick is the result?

Find the LCM. [1.9a]

99. 15, 27, and 30

100. 8, 11, and 36

SYNTHESIS

Find decimal notation.

101. $\frac{1}{7}$

102. $\frac{2}{7}$

103. $\frac{3}{7}$

104. $\frac{4}{7}$

105. $\frac{5}{7}$

106. From the pattern of Exercises 101–105, guess the decimal notation for $\frac{6}{7}$. Check on your calculator.

Find decimal notation.

107. $\frac{1}{9}$

108. $\frac{1}{99}$

109. $\frac{1}{999}$

110. From the pattern of Exercises 107–109, guess the decimal notation for $\frac{1}{9999}$. Check on your calculator.

3.6 ESTIMATING

Objective

a Estimate sums, differences, products, and quotients.

a Estimating Sums, Differences, Products, and Quotients

Estimating has many uses. It can be done before a problem is even attempted in order to get an idea of the answer. It can be done afterward as a check, even when we are using a calculator. In many situations, an estimate is all we need. We usually estimate by rounding the numbers so that there are one or two nonzero digits, depending on how accurate we want our estimate. Consider the following advertisements for Examples 1–3.

EXAMPLE 1 Estimate by rounding to the nearest ten the total cost of one all-in-one printer/copier and one TV/DVD combo.

We are estimating the sum

$199.98 + $141.99 = Total cost.

The estimate found by rounding the addends to the nearest ten is

$200 + $140 = $340. (Estimated total cost)

Do Exercise 1.

1. Estimate by rounding to the nearest ten the total cost of one refrigerator and one printer/copier. Which of the following is an appropriate estimate?

a) $500 **b)** $510
c) $5100 **d)** $51

EXAMPLE 2 About how much more does the refrigerator cost than the TV/DVD combo? Estimate by rounding to the nearest ten.

We are estimating the difference

$309.95 − $141.99 = Price difference.

The estimate to the nearest ten is

$310 − $140 = $170. (Estimated price difference)

Do Exercise 2.

2. About how much more does the printer/copier cost than the TV/DVD combo? Estimate by rounding to the nearest ten. Which of the following is an appropriate estimate?

a) $60 **b)** $600
c) $340 **d)** $6000

Answers on page A-8

Study Tips

ASKING QUESTIONS

Don't be afraid to ask questions in class. Most instructors welcome this and encourage students to ask them. Other students probably have the same questions you do.

"Better to ask twice than lose your way once."

Danish Proverb

EXAMPLE 3 Estimate the total cost of 4 TV/DVD combos.

We are estimating the product

$4 \times \$141.99 =$ Total cost.

The estimate is found by rounding 141.99 to the nearest ten:

$4 \times \$140 = \$560.$

Do Exercise 3.

EXAMPLE 4 About how many Xbox™ Online Ready Video Game Systems at $149.99 each can be purchased for $1480?

We estimate the quotient

$\$1480 \div \$149.99.$

Since we want a whole-number estimate, we choose our rounding appropriately. Rounding $149.99 to the nearest one, we get $150. Since $1480 is close to $1500, which is a multiple of 150, we estimate

$\$1500 \div \$150,$

so the answer is 10.

Do Exercise 4.

EXAMPLE 5 Estimate: 4.8×52. Do not find the actual product. Which of the following is an appropriate estimate?

a) 25 **b)** 250 **c)** 2500 **d)** 360

We round 4.8 to the nearest one and 52 to the nearest ten:

$5 \times 50 = 250.$ (Estimated product)

Thus an approximate estimate is (b).

Other estimates we might have used in Example 5 are

$5 \times 52 = 260$ or $4.8 \times 50 = 240.$

The estimate in Example 5, $5 \times 50 = 250$, is the easiest to do because it has the fewest nonzero digits. You could probably do it mentally. In general, we try to round so that a computation has as few nonzero digits as possible while still keeping the estimated value close to the original value.

Do Exercises 5–10.

3. Estimate the total cost of 6 refrigerators. Which of the following is an appropriate estimate?

a) $1200 **b)** $186
c) $18,600 **d)** $1860

4. About how many Xbox™ Game Systems can be purchased for $4530? Which of the following is an appropriate estimate?

a) 90 **b)** 30
c) 300 **d)** 45

Estimate the product. Do not find the actual product. Which of the following is an appropriate estimate?

5. 2.4×8

a) 16 **b)** 34
c) 125 **d)** 5

6. 24×0.6

a) 200 **b)** 5
c) 110 **d)** 20

7. 0.86×0.432

a) 0.04 **b)** 0.4
c) 1.1 **d)** 4

8. 0.82×0.1

a) 800 **b)** 8
c) 0.08 **d)** 80

9. 0.12×18.248

a) 180 **b)** 1.8
c) 0.018 **d)** 18

10. 24.234×5.2

a) 200 **b)** 120
c) 12.5 **d)** 234

Answers on page A-8

Estimate the quotient. Which of the following is an appropriate estimate?

11. 59.78 ÷ 29.1

a) 200 b) 20
c) 2 d) 0.2

12. 82.08 ÷ 2.4

a) 40 b) 4.0
c) 400 d) 0.4

13. 0.1768 ÷ 0.08

a) 8 b) 10
c) 2 d) 20

14. Estimate: 0.0069 ÷ 0.15. Which of the following is an appropriate estimate?

a) 0.5 b) 50
c) 0.05 d) 0.004

Answers on page A-8

EXAMPLE 6 Estimate: 82.08 ÷ 24. Which of the following is an appropriate estimate?

a) 400 **b)** 16 **c)** 40 **d)** 4

This is about 80 ÷ 20, so the answer is about 4. Thus an appropriate estimate is (d).

EXAMPLE 7 Estimate: 94.18 ÷ 3.2. Which of the following is an appropriate estimate?

a) 30 **b)** 300 **c)** 3 **d)** 60

This is about 90 ÷ 3, so the answer is about 30. Thus an appropriate estimate is (a).

EXAMPLE 8 Estimate: 0.0156 ÷ 1.3. Which of the following is an appropriate estimate?

a) 0.2 **b)** 0.002 **c)** 0.02 **d)** 20

This is about 0.02 ÷ 1, so the answer is about 0.02. Thus an appropriate estimate is (c).

Do Exercises 11–13.

In some cases, it is easier to estimate a quotient directly rather than by rounding the divisor and the dividend.

EXAMPLE 9 Estimate: 0.0074 ÷ 0.23. Which of the following is an appropriate estimate?

a) 0.3 **b)** 0.03 **c)** 300 **d)** 3

We estimate 3 for a quotient. We check by multiplying.

$0.23 \times 3 = 0.69$

We make the estimate smaller. We estimate 0.3 and check by multiplying.

$0.23 \times 0.3 = 0.069$

We make the estimate smaller. We estimate 0.03 and check by multiplying.

$0.23 \times 0.03 = 0.0069$

This is about 0.0074, so the quotient is about 0.03. Thus an appropriate estimate is (b).

Do Exercise 14.

3.6 EXERCISE SET

For Extra Help

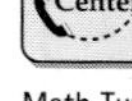

MathXL | MyMathLab | InterAct Math | Math Tutor Center | Video Lectures on CD Disc 2 | Student's Solutions Manual

a Consider the following advertisements for Exercises 1–8. In each exercise, estimate the sum, difference, product, or quotient involved and indicate which of the given choices is an appropriate estimate.

Weatherband Radio X4402
Rechargeable Two-Way Radios with weather band and up to a five-mile range.
$79 95
For Two
Satellite Radio
Receiver and car dock bundle. Over 120 digital channels of music, sports, entertainment and news including 65 channels of 100% commercial-free music.
$149 99
Upright Vacuum Cleaner
Features a lifetime HEPA filter and 17 ft quick draw hose.
$279

1. Estimate the total cost of one vacuum cleaner and one satellite radio.

a) \$43 **b)** \$4300 **c)** \$360 **d)** \$430

2. Estimate the total cost of one satellite radio and one set of two-way radios.

a) \$230 **b)** \$23 **c)** \$2300 **d)** \$400

3. About how much more does the vacuum cleaner cost than the satellite radio?

a) \$1300 **b)** \$200 **c)** \$130 **d)** \$13

4. About how much more does the satellite radio cost than a set of two-way radios?

a) \$7000 **b)** \$70 **c)** \$130 **d)** \$700

5. Estimate the total cost of 6 sets of two-way radios.

a) \$480 **b)** \$48 **c)** \$240 **d)** \$4800

6. Estimate the total cost of 4 vacuum cleaners.

a) \$1200 **b)** \$1120 **c)** \$11,200 **d)** \$600

7. About how many sets of two-way radios can be purchased for \$830?

a) 120 **b)** 100 **c)** 10 **d)** 1000

8. About how many vacuum cleaners can be purchased for \$5627?

a) 200 **b)** 20 **c)** 1800 **d)** 2000

Estimate by rounding as directed.

9. $0.02 + 1.31 + 0.34$; nearest tenth

10. $0.88 + 2.07 + 1.54$; nearest one

11. $6.03 + 0.007 + 0.214$; nearest one

12. $1.11 + 8.888 + 99.94$; nearest one

13. $52.367 + 1.307 + 7.324$; nearest one

14. $12.9882 + 1.0115$; nearest tenth

15. 2.678 − 0.445; nearest tenth

16. 12.9882 − 1.0115; nearest one

17. 198.67432 − 24.5007; nearest ten

Estimate. Choose a rounding digit that gives one or two nonzero digits. Indicate which of the choices is an appropriate estimate.

18. 234.12321 − 200.3223

a) 600 **b)** 60
c) 300 **d)** 30

19. 49 × 7.89

a) 400 **b)** 40
c) 4 **d)** 0.4

20. 7.4 × 8.9

a) 95 **b)** 63
c) 124 **d)** 6

21. 98.4 × 0.083

a) 80 **b)** 12
c) 8 **d)** 0.8

22. 78 × 5.3

a) 400 **b)** 800
c) 40 **d)** 8

23. 3.6 ÷ 4

a) 10 **b)** 1
c) 0.1 **d)** 0.01

24. 0.0713 ÷ 1.94

a) 3.5 **b)** 0.35
c) 0.035 **d)** 35

25. 74.68 ÷ 24.7

a) 9 **b)** 3
c) 12 **d)** 120

26. 914 ÷ 0.921

a) 10 **b)** 100
c) 1000 **d)** 1

27. *Fence Posts.* A zoo plans to construct a fence around its proposed African wildlife exhibit. The perimeter of the area to be fenced is 1760 ft. Estimate the number of wooden fence posts needed if the posts are placed 8.625 ft apart.

28. *Ticketmaster.* Recently, Ticketmaster stock sold for $8.63 per share. Estimate how many shares could be purchased for $27,000.

29. D_W Describe a situation in which an estimation is made by rounding to the nearest 10,000 and then multiplying.

30. D_W A roll of fiberglass insulation costs $21.95. Describe two situations involving estimating and the cost of fiberglass insulation. Devise one situation so that $21.95 is rounded to $22. Devise the other situation so that $21.95 is rounded to $20.

SKILL MAINTENANCE

VOCABULARY REINFORCEMENT

In each of Exercises 31–38, fill in the blank with the correct term from the given list. Some of the choices may not be used and some may be used more than once.

31. The decimal $0.57\overline{3}$ is an example of a ______________ decimal. [3.5a]

32. The least common ______________ of two natural numbers is the smallest number that is a multiple of both. [1.9a]

33. In the product $10 \cdot \frac{3}{4}$, 10 and $\frac{3}{4}$ are ______________. [1.7a]

34. A ______________ of an equation is a replacement for the variable that makes the equation true. [1.4a]

35. The number 1 is the ______________ identity. [1.3a]

36. The product of 6 and $\frac{1}{6}$ is 1; we say that 6 and $\frac{1}{6}$ are ______________ of each other. [2.2b]

37. The least common ______________ of two or more fractions is the least common ______________ of their denominators. [2.3a]

38. The number 3728 is ______________ by 4 if the number named by the last two digits is ______________ by 4. [1.8a]

additive
multiplicative
numerator
denominator
reciprocals
factors
solution
divisible
terminating
repeating
multiple
factor

SYNTHESIS

The following were done on a calculator. Estimate to determine whether the decimal point was placed correctly.

39. $178.9462 \times 61.78 = 11{,}055.29624$

40. $14{,}973.35 \div 298.75 = 501.2$

41. $19.7236 - 1.4738 \times 4.1097 = 1.366672414$

42. $28.46901 \div 4.9187 - 2.5081 = 3.279813473$

43. Use one of $+$, $-$, $\times$, and $\div$ in each blank to make a true sentence.

a) $(0.37 \square\ 18.78) \square\ 2^{13} = 156{,}876.8$

b) $2.56 \square\ 6.4 \square\ 51.2 \square\ 17.4 = 312.84$

3.7 APPLICATIONS AND PROBLEM SOLVING

Objective

a Solve applied problems involving decimals.

a Solving Applied Problems

Solving applied problems with decimals is like solving applied problems with whole numbers. We translate first to an equation that corresponds to the situation. Then we solve the equation.

EXAMPLE 1 *Canals.* The Panama Canal in Panama is 50.7 mi long. The Suez Canal in Egypt is 119.9 mi long. How much longer is the Suez Canal?

1. **Familiarize.** This is a "how much more" situation. We let $l =$ the distance in miles that the length of the longer canal differs from the length of the shorter canal.
2. **Translate.** We translate as follows, using the given information.

Length of Panama Canal, the shorter canal	plus	Additional length	is	Length of Suez Canal, the longer canal
↓	↓	↓	↓	↓
50.7 mi	$+$	l	$=$	119.9 mi

3. **Solve.** We solve the equation by subtracting 50.7 mi on both sides:

$$50.7 + l = 119.9$$
$$50.7 + l - 50.7 = 119.9 - 50.7$$
$$l = 69.2.$$

4. **Check.** We can check by adding 69.2 to 50.7 to get 119.9.
5. **State.** The Suez Canal is 69.2 mi longer than the Panama Canal.

Do Exercise 1.

1. **Photofinishing Market.** Because an increasing number of people are printing their own digital photographs at home, the commercial photofinishing industry's revenue is declining. Between 1994 and 2005, the total revenue from film processing ranged in value from a low of \$3.7 billion to a high of \$6.2 billion. By how much did the high value differ from the low value?

Photofinishing Market

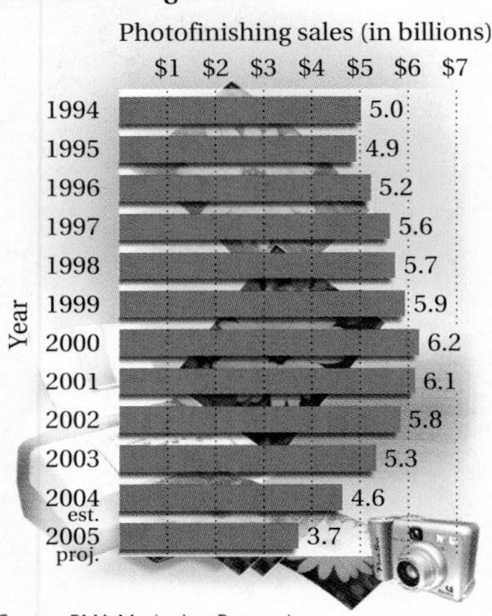

Source: PMA Marketing Research

Answer on page A-8

EXAMPLE 2 *Injections of Medication.* A patient was given injections of 2.8 mL, 1.35 mL, 2.0 mL, and 1.88 mL over a 24-hr period. What was the total amount of the injections?

1. **Familiarize.** We make a drawing or at least visualize the situation. We let $t =$ the amount of the injections.

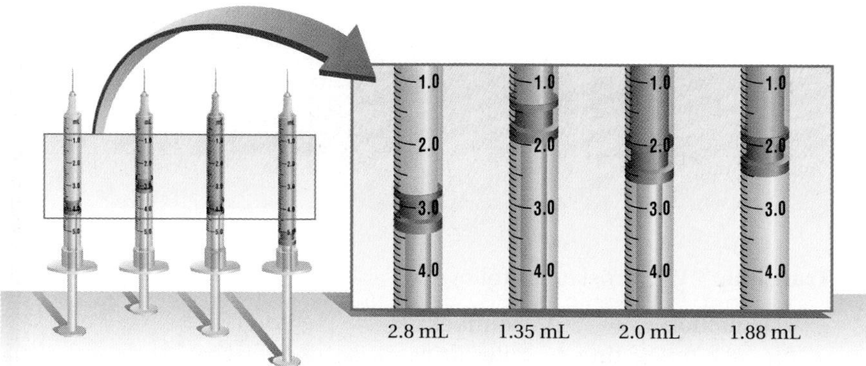

2. **Translate.** Amounts are being combined. We translate to an equation.

$$\underbrace{\text{First}}_{2.8} \text{ plus } \underbrace{\text{Second}}_{1.35} \text{ plus } \underbrace{\text{Third}}_{2.0} \text{ plus } \underbrace{\text{Fourth}}_{1.88} \text{ is } \underbrace{\text{Total}}_{t}$$

$$2.8 + 1.35 + 2.0 + 1.88 = t$$

3. **Solve.** To solve, we carry out the addition.

$$\begin{array}{r} {\scriptstyle 2\ 1} \\ 2.8\,0 \\ 1.3\,5 \\ 2.0\,0 \\ +\ 1.8\,8 \\ \hline 8.0\,3 \end{array}$$

Thus, $t = 8.03$.

4. **Check.** We can check by repeating our addition. We can also see whether our answer is reasonable by first noting that it is indeed larger than any of the numbers being added. We can also partially check by rounding:

$$2.8 + 1.35 + 2.0 + 1.88 \approx 3 + 1 + 2 + 2$$
$$= 8 \approx 8.03.$$

If we had gotten an answer like 80.3 or 0.803, then our estimate, 8, would have told us that we did something wrong, like not lining up the decimal points.

5. **State.** The total amount of the injections was 8.03 mL.

Do Exercise 2.

2. **Meat and Seafood Consumption.** In 2002, the average American consumed about 64.5 lb of beef, 48.2 lb of pork, 56.8 lb of chicken, 14.0 lb of turkey, and 15.6 lb of seafood. What is the total amount of meat and seafood that the average American consumed?

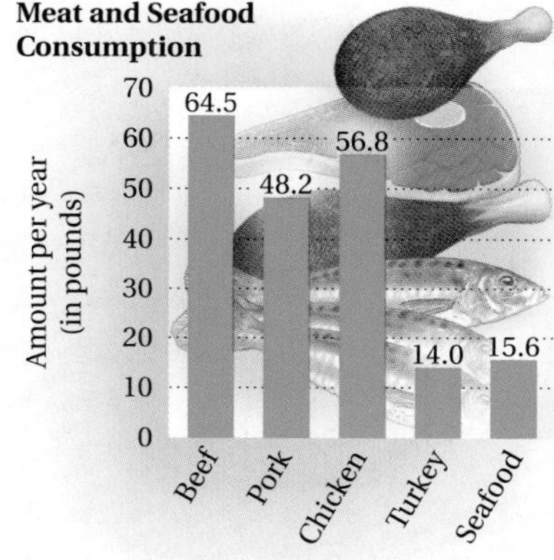

Source: Statistical Abstract of the United States, 2004

Answer on page A-8

3. **Printing Costs.** At a printing company, the cost of copying is 11 cents per page. How much, in dollars, would it cost to make 466 copies?

EXAMPLE 3 *IRS Driving Allowance.* In 2004, the Internal Revenue Service allowed a tax deduction of 37.5¢ per mile for mileage driven for business purposes. What deduction, in dollars, would be allowed for driving 9143 mi?

Source: Internal Revenue Service

1. **Familiarize.** We first make a drawing or at least visualize the situation. Repeated addition fits this situation. We let d = the deduction, in dollars, allowed for driving 9143 mi.

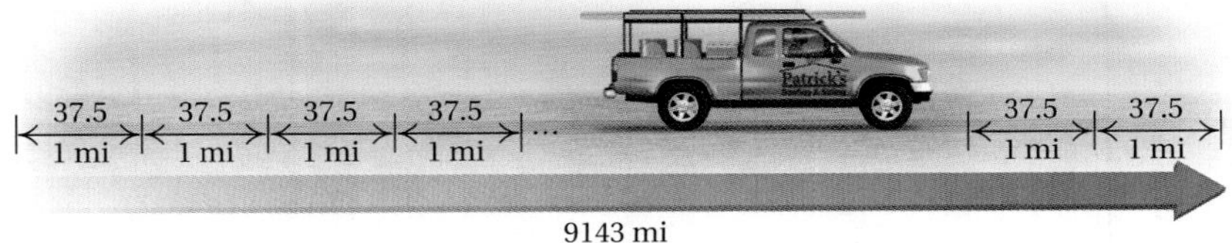

2. **Translate.** We translate as follows.

Deduction for each mile	times	Number of miles driven	is	Total deduction
$0.375	×	9143	=	d

Converting 37.5 cents to dollars gives us $0.375.

3. **Solve.** To solve the equation, we carry out the multiplication.

$$\begin{array}{r} 9\,1\,4\,3 \\ \times \quad 0.3\,7\,5 \\ \hline 4\,5\,7\,1\,5 \\ 6\,4\,0\,0\,1\,0 \\ 2\,7\,4\,2\,9\,0\,0 \\ \hline 3\,4\,2\,8.6\,2\,5 \end{array}$$

Thus, $d = 3428.625 \approx \$3428.63$.

4. **Check.** We can obtain a partial check by rounding and estimating:

$$9143 \times 0.375 \approx 9000 \times 0.4 = 3600 \approx 3428.63.$$

5. **State.** The total allowable deduction would be $3428.63.

Do Exercise 3.

EXAMPLE 4 *Loan Payments.* A car loan of $7382.52 is to be paid off in 36 monthly payments. How much is each payment?

1. **Familiarize.** We first make a drawing. We let n = the amount of each payment.

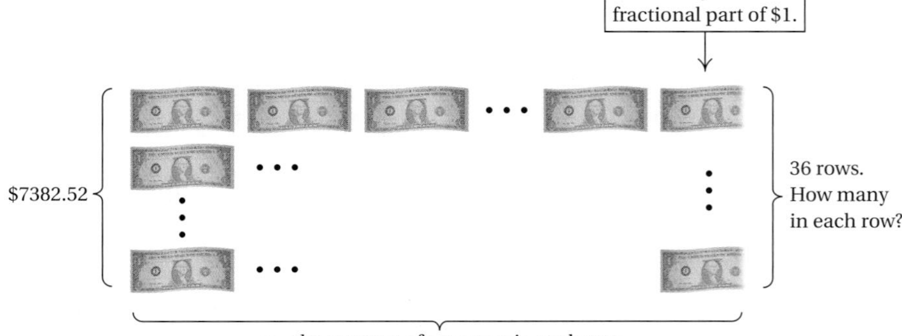

Answer on page A-8

2. **Translate.** The problem can be translated to the following equation, thinking that

$$(\text{Total loan}) \div (\text{Number of payments}) = \text{Amount of each payment}$$
$$\$7382.52 \div 36 = n.$$

3. **Solve.** To solve the equation, we carry out the division.

$$\begin{array}{r@{}l} & 205.07 \\ 36 \,\big)\!\!\! & \overline{7382.52} \\ & \underline{720000} \\ & 18252 \\ & \underline{18000} \\ & 252 \\ & \underline{252} \\ & 0 \end{array}$$

Thus, $n = 205.07$.

4. **Check.** A partial check can be obtained by estimating the quotient: $\$7382.56 \div 36 \approx 8000 \div 40 = 200 \approx 205.07$. The estimate checks.

5. **State.** Each payment is $205.07.

4. Loan Payments. A loan of $4425 is to be paid off in 12 monthly payments. How much is each payment?

Answer on page A-8

Do Exercise 4.

EXAMPLE 5 *Jackie Robinson Poster.* A limited-edition poster by sports artist Leroy Neiman commemorates Jackie Robinson's entrance into Major League Baseball in 1947. (Robinson was the first African-American player in the major leagues.) The dimensions of the poster are 19.3 in. by 27.4 in. Find the area.

Source: Barton L. Kaufman, private collection

1. **Familiarize.** We first make a drawing. We let $A =$ the area.

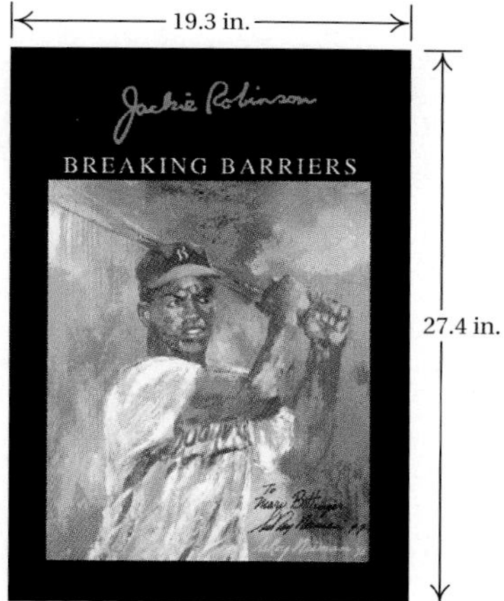

Study Tips

FIVE STEPS FOR PROBLEM SOLVING

Are you using the five steps for problem solving that were developed in Section 1.5?

1. **Familiarize** yourself with the situation.
   a) Carefully read and reread until you understand what you are being asked to find.
   b) Draw a diagram or see if there is a formula that applies.
   c) Assign a letter, or *variable,* to the unknown.
2. **Translate** the problem to an equation using the letter or variable.
3. **Solve** the equation.
4. **Check** the answer in the original wording of the problem.
5. **State** the answer to the problem clearly with appropriate units.

5. Index Cards. A standard-size index card measures 12.7 cm by 7.6 cm. Find its area.

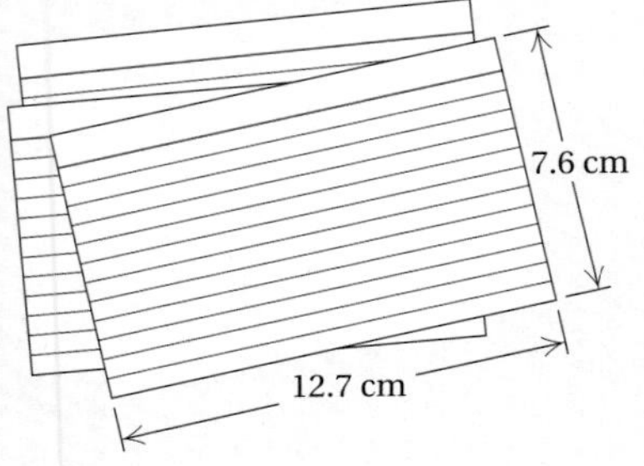

2. Translate. We use the formula $A = l \cdot w$ and substitute.

$$A = l \cdot w$$
$$A = 27.4 \times 19.3.$$

3. Solve. We solve by carrying out the multiplication.

$$\begin{array}{r} 27.4 \\ \times\ 19.3 \\ \hline 822 \\ 24660 \\ 27400 \\ \hline 528.82 \end{array}$$

4. Check. We obtain a partial check by estimating the product:

$$A = 27.4 \times 19.3 \approx 25 \times 20 = 500.$$

This estimate is close to 528.82 so it's a good check.

5. State. The area of the Jackie Robinson poster is 528.82 in^2.

Do Exercise 5.

EXAMPLE 6 *Digital Camera Purchase.* Dawson Real Estate spent \$10,399.74 on a set of 26 Nikon COOLPIX 5200 cameras, so that its realtors can place photos of properties on the firm's Web site. How much did each camera cost?

1. Familiarize. We let c = the cost of each camera.

2. Translate. We translate as follows.

Cost of each camera	is	Total cost of purchase	divided by	Number of cameras purchased
↓	↓	↓	↓	↓
c	$=$	\$10,399.74	$\div$	26

3. Solve. To solve, we carry out the division.

$$\begin{array}{r} 399.99 \\ 26\,\overline{)\,10{,}399.74} \\ 78 \\ \hline 259 \\ 234 \\ \hline 259 \\ 234 \\ \hline 257 \\ 234 \\ \hline 234 \\ 234 \\ \hline 0 \end{array}$$

4. Check. We check by estimating

$$10{,}399.74 \div 26 \approx 10{,}400 \div 25 = 416.$$

Since 416 is close to 399.99, the answer is probably correct.

5. State. The cost of each camera was \$399.99.

Do Exercise 6.

6. One pound of lean boneless ham contains 4.5 servings. It costs \$5.99 per pound. What is the cost per serving? Round to the nearest cent.

Answers on page A-8

MULTISTEP PROBLEMS

EXAMPLE 7 *Gas Mileage.* A driver filled the gasoline tank and noted that the odometer read 67,507.8. After the next filling, the odometer read 68,006.1. It took 16.5 gal to fill the tank. How many miles per gallon did the driver get?

1. Familiarize. We first make a drawing.

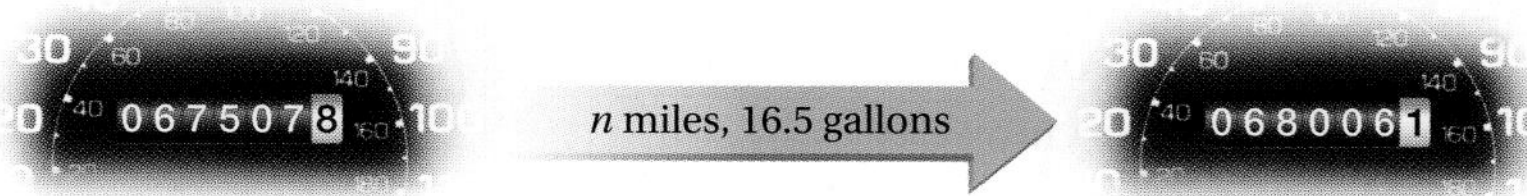

This is a two-step problem. First, we find the number of miles that have been driven between fillups. We let n = the number of miles driven.

2., 3. Translate and **Solve.** This is a "how many more" situation. We translate and solve as follows.

First odometer reading	plus	Number of miles driven	is	Second odometer reading
↓	↓	↓	↓	↓
67,507.8	+	n	=	68,006.1

To solve the equation, we subtract 67,507.8 on both sides:

$$n = 68{,}006.1 - 67{,}507.8$$
$$= 498.3.$$

$$\begin{array}{r} 68{,}006.1 \\ -\ 67{,}507.8 \\ \hline 498.3 \end{array}$$

Second, we divide the total number of miles driven by the number of gallons. This gives us m = the number of miles per gallon—that is, the mileage. The division that corresponds to the situation is

$$498.3 \div 16.5 = m.$$

To find the number m, we divide.

$$\begin{array}{r} 30.2 \\ 16.5\,\overline{)\,498.3_{\wedge}0} \\ \underline{495\,0} \\ 33\,0 \\ \underline{33\,0} \\ 0 \end{array}$$

Thus, $m = 30.2$.

4. Check. To check, we first multiply the number of miles per gallon times the number of gallons:

$$16.5 \times 30.2 = 498.3.$$

Then we add 498.3 to 67,507.8:

$$67{,}507.8 + 498.3 = 68{,}006.1.$$

The mileage 30.2 checks.

5. State. The driver got 30.2 miles per gallon.

Do Exercise 7.

7. Gas Mileage. A driver filled the gasoline tank and noted that the odometer read 38,320.8. After the next filling, the odometer read 38,735.5. It took 14.5 gal to fill the tank. How many miles per gallon did the driver get?

Answer on page A-8

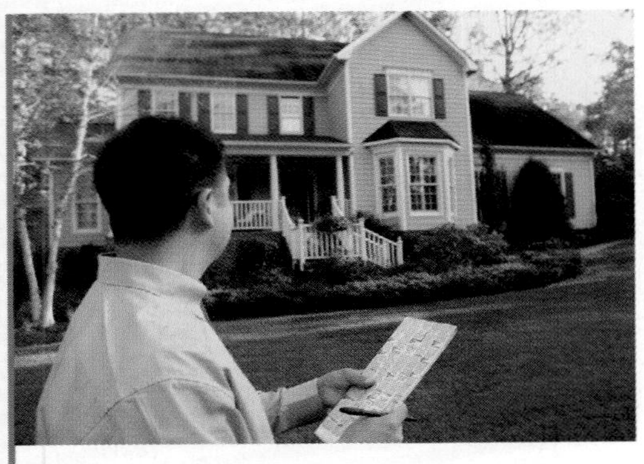

EXAMPLE 8 *Home-Cost Comparison.* Suppose you own a home like the one shown here and it is valued at $225,000 in Tulsa, Oklahoma. What would it cost to buy a similar (replacement) home in Boston, Massachusetts? To find out, we can use an index table prepared by Coldwell Banker Real Estate Corporation. (For a complete index table, contact your local representative.) We use the following formula:

$$\left(\begin{array}{c}\text{Cost of your}\\ \text{home in new city}\end{array}\right) = \frac{\left(\begin{array}{c}\text{Value of}\\ \text{your home}\end{array}\right) \times \left(\begin{array}{c}\text{Index of}\\ \text{new city}\end{array}\right)}{\left(\begin{array}{c}\text{Index of}\\ \text{your city}\end{array}\right)}.$$

Find the cost of your Tulsa home in Boston. Round to the nearest one.

STATE	CITY	INDEX	STATE	CITY	INDEX
Massachusetts	Boston	297	Texas	Dallas	67
Illinois	Chicago	215	Oklahoma	Tulsa	39
California	Palo Alto Sacramento San Diego	342 98 171	Florida	Miami/Coral Gables Orlando Key West	143 69 214
Alaska	Juneau	119	Colorado	Boulder	125
Kentucky	Louisville	63	North Carolina	Charlotte	54

Source: Coldwell Banker Real Estate Corporation

Refer to the table in Example 8 to answer Margin Exercises 8 and 9.

8. **Home-Cost Comparison.** Find the cost of a $180,000 home in Charlotte if you were to try to replace it when moving to Chicago. Round to the nearest one.

9. Find the cost of a $315,000 home in Boulder if you were to try to replace it when moving to Louisville. Round to the nearest one.

1. **Familiarize.** We let $C =$ the cost of the home in Boston. We use the table and look up the indexes of the city in which you now live and the city to which you are moving.
2. **Translate.** Using the formula, we translate to the following equation:

$$C = \frac{\$225{,}000 \times 297}{39}.$$

3. **Solve.** To solve, we carry out the computations using the rules for order of operations (see Section 3.4).

$$C = \frac{\$225{,}000 \times 297}{39}$$

$$= \frac{66{,}825{,}000}{39} \qquad \text{Carrying out the multiplication first}$$

$$\approx \$1{,}713{,}462. \qquad \text{Carrying out the division and rounding to the nearest one}$$

On a calculator, the computation could be done in one step.

4. **Check.** We can repeat our computations.
5. **State.** A home that sells for $225,000 in Tulsa would cost about $1,713,462 in Boston.

Do Exercises 8 and 9.

Answers on page A-8

Translating for Success

1. *Gas Mileage.* Art filled his SUV's gas tank and noted that the odometer read 38,271.8. At the next filling, the odometer read 38,677.92. It took 28.4 gal to fill the tank. How many miles per gallon did the SUV get?

2. *Dimensions of a Parking Lot.* Seals' parking lot is a rectangle that measures 85.2 ft by 52.3 ft. What is the area of the parking lot?

3. *Game Snacks.* Three students pay $18.40 for snacks at a football game. What is each person's share of the cost?

4. *Electrical Wiring.* An electrician needs 1314 ft of wiring cut into $2\frac{1}{2}$-ft pieces. How many pieces will she have?

5. *College Tuition.* Wayne needs $4638 for the fall semester's tuition. On the day of registration, he has only $3092. How much does he need to borrow?

The goal of these matching questions is to practice step (2), *Translate,* of the five-step problem-solving process. Translate each word problem to an equation and select a correct translation from equations A–O.

A. $2\frac{1}{2} \cdot n = 1314$

B. $18.4 \times 1.87 = n$

C. $n = 85.2 \times 52.3$

D. $1314.28 - 437 = n$

E. $3 \times 18.40 = n$

F. $2\frac{1}{2} \cdot 1314 = n$

G. $3092 + n = 4638$

H. $18.4 \cdot n = 1.87$

I. $\frac{406.12}{28.4} = n$

J. $52.3 \cdot n = 85.2$

K. $n = 1314.28 + 437$

L. $52.3 + n = 85.2$

M. $3092 + 4638 = n$

N. $3 \cdot n = 18.40$

O. $85.2 + 52.3 = n$

Answers on page A-8

6. *Cost of Gasoline.* What is the cost, in dollars, of 18.4 gal of gasoline at $1.87 per gallon?

7. *Savings Account Balance.* Margaret has $1314.28 in her savings account. Before using her debit card to buy an office chair, she transferred $437 to her checking account. How much was left in her savings account?

8. *Acres Planted.* This season Sam planted 85.2 acres of corn and 52.3 acres of soybeans. Find the total number of acres that he planted.

9. *Amount Inherited.* Tara inherited $2\frac{1}{2}$ times as much as her cousin. Her cousin received $1314. How much did Tara receive?

10. *Travel Funds.* The athletic department needs travel funds of $4638 for the tennis team and $3092 for the golf team. What is the total amount needed for travel?

3.7 EXERCISE SET

For Extra Help

MathXL | MyMathLab | InterAct Math | Math Tutor Center | Video Lectures on CD Disc 2 | Student's Solutions Manual

a Solve.

1. *Hurricanes.* Before hurricane Katrina struck the Gulf Coast in 2005, the five most costly hurricanes in the United States were those listed in the following table. The costs, adjusted to 2003 dollars by the Insurance Information Institute, ranged from a low of \$2.5 billion to a high of \$20.3 billion. By how much did the high value differ from the low value?

COSTLY HURRICANES

RANK	HURRICANE	DATE	IN 2003 DOLLARS (in billions)
1	Andrew	Aug. 23–26, 1992	\$20.3
2	Charley	Aug. 13–15, 2004	\$6.8
3	Hugo	Sept. 17–22, 1989	\$6.2
4	Georges	Sept. 21–28, 1998	\$3.3
5	Opal	Oct. 4, 1995	\$2.5

Sources: Insurance Services Office; Insurance Information Institute

2. *Compact Discs.* Because of increased sales of DVDs, the sales of recorded music in compact disc (CD) format are declining. Between 1996 and 2003, the total sales of recorded music on CDs ranged from a high of 942.5 million to a low of 745.9 million. By how much did the high value differ from the low value?

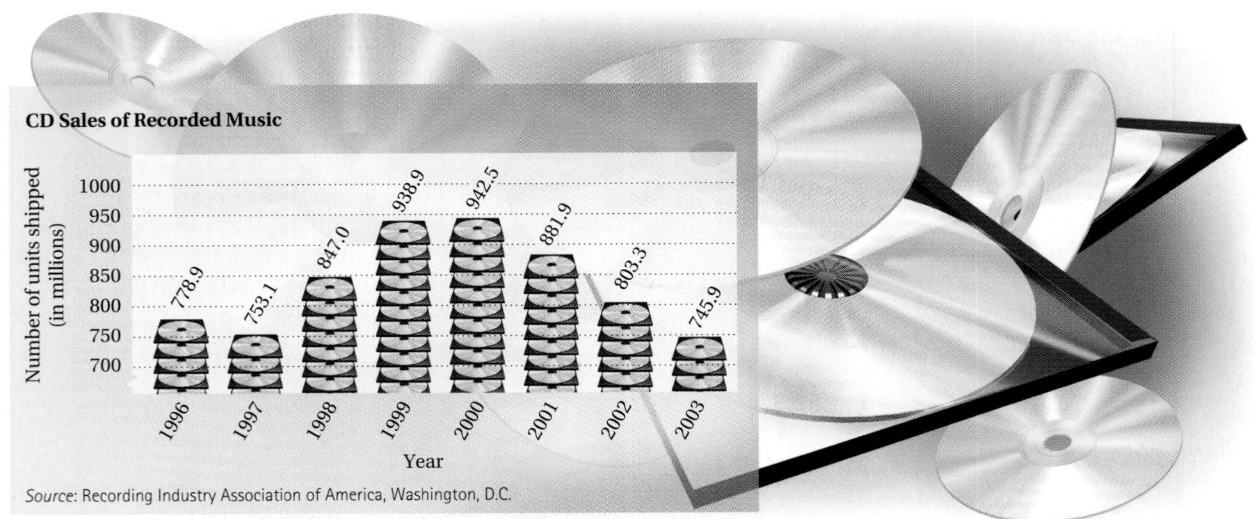

3. Andrew bought a DVD of the complete ninth season of the TV show "Friends" for \$29.24 plus \$1.61 sales tax. He paid for it with a \$50 bill. How much change did he receive?

4. Jaden bought the CD "Feels Like Home" by Norah Jones for \$13.49 plus \$0.81 sales tax. She paid for it with a \$20 bill. How much change did she receive?

5. *Body Temperature.* Normal body temperature is 98.6°F. During an illness, a patient's temperature rose 4.2°. What was the new temperature?

6. *Gasoline Cost.* What is the cost, in dollars, of 20.4 gal of gasoline at $2.13 per gallon? Round the answer to the nearest cent.

7. *Lottery Winnings.* The largest lotto jackpot ever won in California totaled $193,000,000 and was shared equally by 3 winners. How much was each winner's share? Round to the nearest cent.

Source: California State Lottery

8. *Lunch Costs.* A group of 4 students pays $47.84 for lunch. What is each person's share of the cost?

9. *Stamp.* Find the area and the perimeter of the stamp shown here.

10. *Pole Vault Pit.* Find the area and the perimeter of the landing area of the pole vault pit shown here.

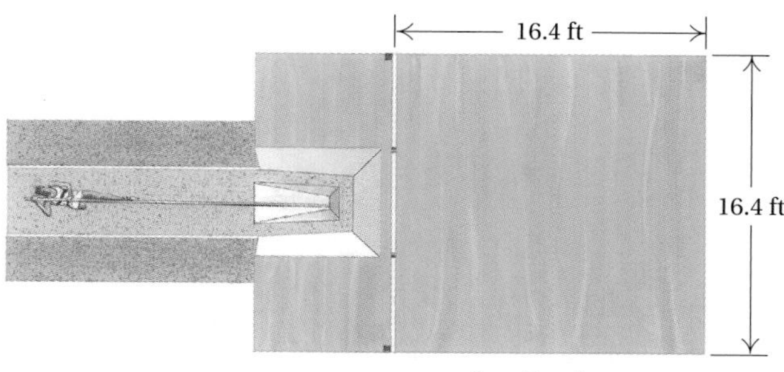

11. *Odometer Reading.* A family checked the odometer before starting a trip. It read 22,456.8 and they know that they will be driving 234.7 mi. What will the odometer read at the end of the trip?

12. *Miles Driven.* Petra bought gasoline when the odometer read 14,296.3. At the next gasoline purchase, the odometer read 14,515.8. How many miles had been driven?

13. *Gas Mileage.* Peggy filled her van's gas tank and noted that the odometer read 26,342.8. After the next filling, the odometer read 26,736.7. It took 19.5 gal to fill the tank. How many miles per gallon did the van get?

14. *Gas Mileage.* Henry filled his Honda's gas tank and noted that the odometer read 18,943.2. After the next filling, the odometer read 19,306.2. It took 13.2 gal to fill the tank. How many miles per gallon did the car get?

15. *Cost of Video Game.* A certain video game costs 75 cents and runs for 1.5 min. Assuming a player does not win any free games and plays continuously, how much money, in dollars, does it cost to play the video game for 1 hr?

16. *Property Taxes.* The Colavitos own a house with an assessed value of $184,500. For every $1000 of assessed value, they pay $7.68 in taxes. How much do they pay in taxes?

17. *Chemistry.* The water in a filled tank weighs 748.45 lb. One cubic foot of water weighs 62.5 lb. How many cubic feet of water does the tank hold?

18. *Highway Routes.* You can drive from home to work using either of two routes:

Route A: Via interstate highway, 7.6 mi, with a speed limit of 65 mph.

Route B: Via a country road, 5.6 mi, with a speed limit of 50 mph.

Assuming you drive at the posted speed limit, which route takes less time? (Use the formula *Distance* = *Speed* × *Time*.)

Find the perimeter of the figure.

19.

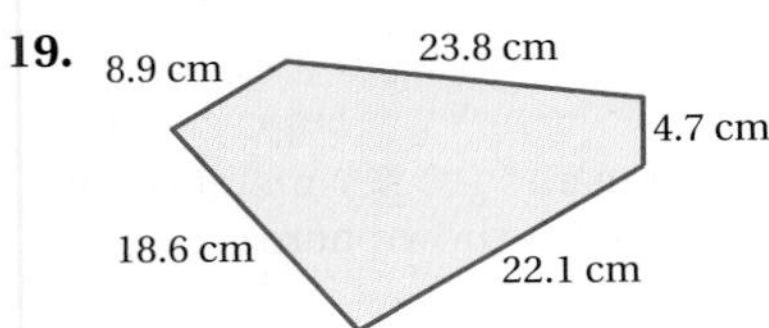

20.

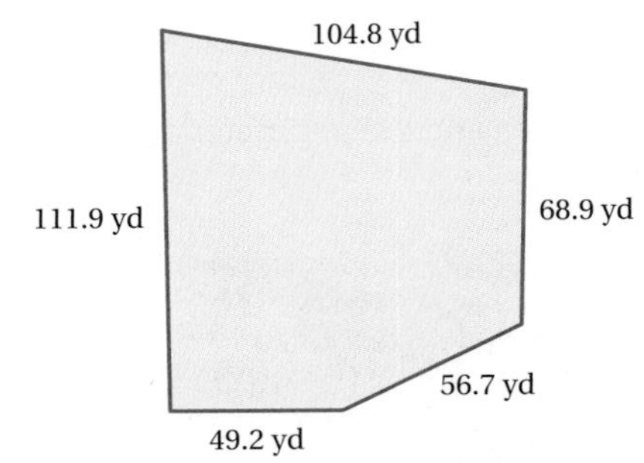

21.

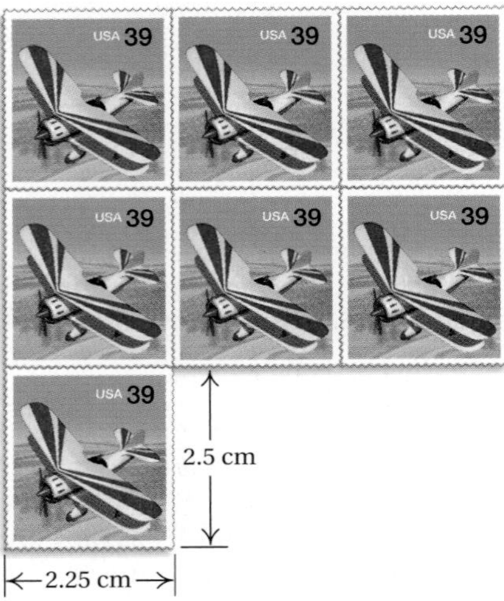

22.

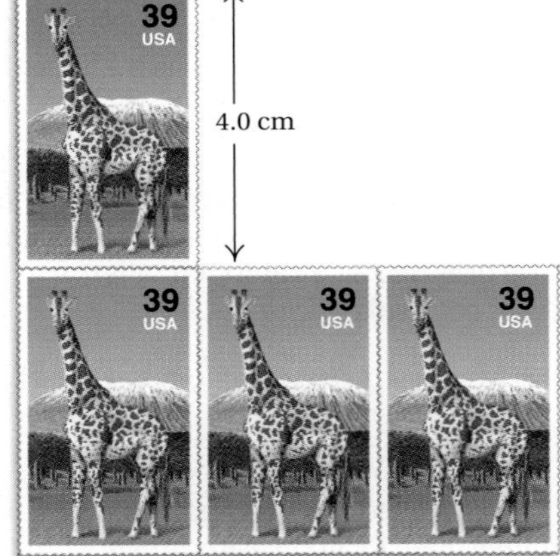

Find the length d in the figure.

23.

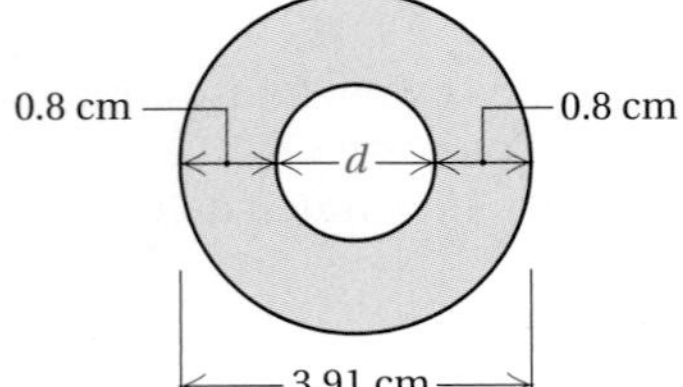

24.

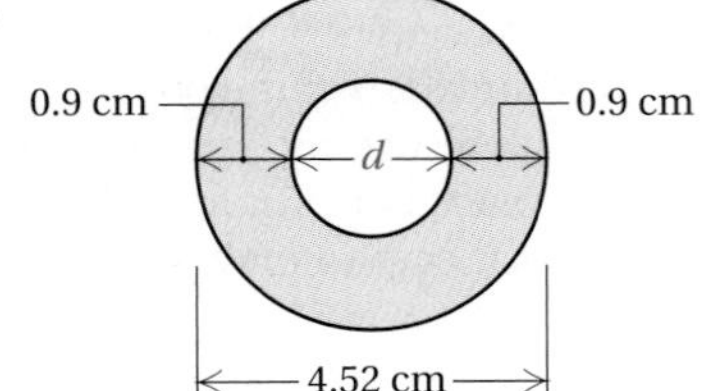

25. *Calories Burned Mowing.* A person who weighs 150 lb burns 7.3 calories per minute while mowing a lawn with a power lawnmower. How many calories would be burned in 2 hr of mowing?

Source: *The Handy Science Answer Book*

26. Lot A measures 250.1 ft by 302.7 ft. Lot B measures 389.4 ft by 566.2 ft. What is the total area of the two lots?

27. Holly had $1123.56 in her checking account. She used her debit card to pay bills of $23.82, $507.88, and $98.32. She then deposited a bonus check of $678.20. How much is in her account after these changes?

28. Natalie had $185.00 to spend for fall clothes: $44.95 was spent on shoes, $71.95 for a jacket, and $55.35 for pants. How much was left?

29. A rectangular yard is 20 ft by 15 ft. The yard is covered with grass except for an 8.5-ft square flower garden. How much grass is in the yard?

30. Rita earns a gross paycheck (before deductions) of $495.72. Her deductions are $59.60 for federal income tax, $29.00 for FICA, and $29.00 for medical insurance. What is her take-home paycheck?

31. *Batting Average.* In the 2005 baseball season, Derrek Lee of the Chicago Cubs led the National League in batting with 199 hits in 594 times at bat. What part of his at-bats were hits? Use decimal notation rounded to the nearest thousandth. (This is a player's *batting average.*)

Source: Major League Baseball

32. *Batting Average.* In the 2005 baseball season, Michael Young of the Texas Rangers led the American League in batting with 221 hits in 668 times at bat. What part of his at-bats were hits? Use decimal notation rounded to the nearest thousandth.

Source: Major League Baseball

33. *Verizon Wireless.* In 2005, Verizon Wireless offered its America's Choice cell-phone plan with 450 anytime minutes per month for a monthly access fee of $39.99. Minutes in excess of 450 were charged at the rate of $0.45 per minute. In June, Leila used her cell phone for 479 min. What was she charged?

Source: www.vzwshop.com

34. *Verizon Wireless.* In 2005, Verizon Wireless offered its America's Choice cell-phone plan with 900 anytime minutes per month for a monthly access fee of $59.99. Minutes in excess of 900 were charged at the rate of $0.40 per minute. One month Jeff used his cell phone for 946 min. What was the charge?

Source: www.vzwshop.com

35. *Field Dimensions.* The dimensions of a World Cup soccer field are 114.9 yd by 74.4 yd. The dimensions of a standard football field are 120 yd by 53.3 yd. How much greater is the area of a World Cup soccer field?

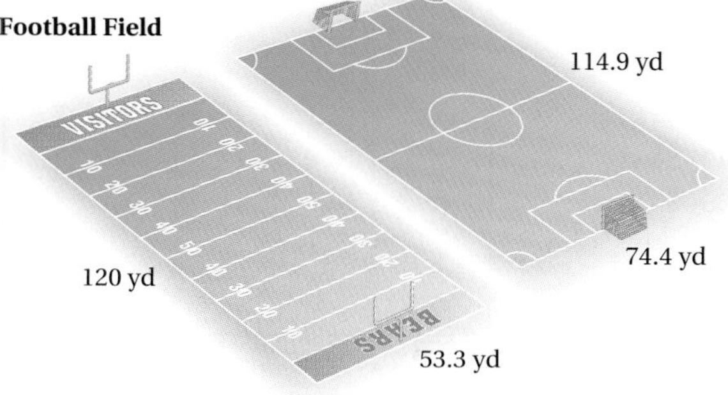

36. *Loan Payment.* In order to make money on loans, financial institutions are paid back more money than they loan. You borrow $120,000 to buy a house and agree to make monthly payments of $880.52 for 30 yr. How much do you pay back altogether? How much more do you pay back than the amount of the loan?

37. *Egg Costs.* A restaurant owner bought 20 dozen eggs for $25.80. Find the cost of each egg to the nearest tenth of a cent (thousandth of a dollar).

38. *Weight Loss.* A person who weighs 170 lb burns 8.6 calories per minute while mowing a lawn. One must burn about 3500 calories in order to lose 1 lb. How many pounds would be lost by mowing for 2 hr? Round to the nearest tenth.

39. *Construction Pay.* A construction worker is paid $18.50 per hour for the first 40 hr of work, and time and a half, or $27.75 per hour, for any overtime exceeding 40 hr per week. One week she works 46 hr. How much is her pay for the week?

40. *Summer Work.* Zachary worked 53 hr during a week one summer. He earned $6.50 per hour for the first 40 hr and $9.75 per hour for overtime (hours exceeding 40). How much did Zachary earn during the week?

41. *Projected World Population.* Using the information in the following bar graph, determine the average population of the world for the years 1950 through 2040.

World Population (historical and projected)

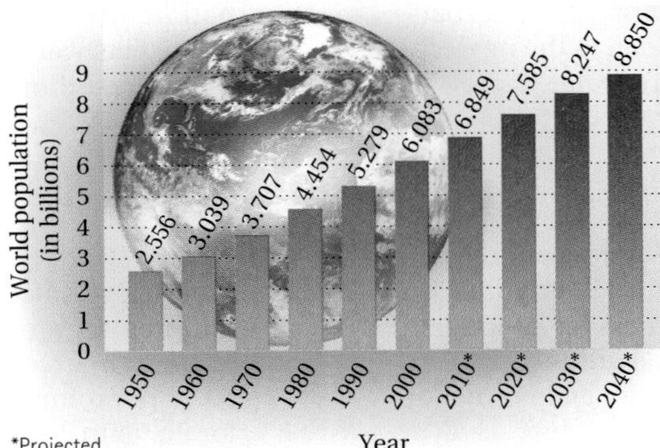

*Projected

Source: U.S. Bureau of the Census

42. *Volunteer Hours.* The table at right lists the average number of hours per month that volunteers of specific age groups spend doing volunteer work. What is the average number of volunteer hours of all age groups?

Source: *Statistical Abstract of the United States*

HOURS SPENT VOLUNTEERING

AGE	AVERAGE HOURS PER MONTH
21–24 yr	12.1
25–34 yr	15.9
35–44 yr	16.1
45–54 yr	14.7
55–64 yr	12.1
65–74 yr	14.1
75 yr and older	19.5

43. *Body Temperature.* Normal body temperature is 98.6°F. A baby's bath water should be 100°F. How many degrees above normal body temperature is this?

44. *Body Temperature.* Normal body temperature is 98.6°F. The lowest temperature at which a person has survived is 69°F. How many degrees below normal is this?

Home-Cost Comparison. Use the table and formula from Example 8. In each of the following cases, find the value of the house in the new location.

Source: Coldwell Banker Real Estate Corporation

	VALUE	PRESENT LOCATION	NEW LOCATION	NEW VALUE
45.	$125,000	Dallas	Miami/Coral Gables	
46.	$180,000	San Diego	Key West	
47.	$96,000	Orlando	Juneau	
48.	$300,000	Palo Alto	Dallas	
49.	$240,000	Louisville	Boston	
50.	$160,000	Charlotte	Sacramento	

Powells.com logo used with permission.

Comparison Shopping. The Internet now provides many sites to shop for a product rather than shopping in a local store. The following lists various Web sites for purchasing the recent best-selling novel *The Da Vinci Code* by Dan Brown.

51. D_W Complete the total costs in the table below for each merchant. (Assume you live in a state in which sales taxes are not charged on Internet purchases.) Then decide which site you would use to make a purchase. Discuss possible ways in which answers might vary.

MERCHANT	PRICE	SHIPPING AND HANDLING	SALES TAX	TOTAL COST
Amazon.com www.amazon.com	$14.97	$3.00 per order, plus $0.99 per item	$0	?
Barnes & Noble www.bn.com	$14.97	$3.00 per order, plus $0.99 per item	$0	?
Powell's Books www.powells.com	$24.95	$2.50 per order, plus $1.00 per item	$0	?
Costco Wholesale www.costco.com	$21.99	$5.00	$0	?
Overstock.com www.overstock.com	$13.47	$1.40	$0	?
Borders Bookstore, local store	$17.46	$0	$1.05	?

52. $\mathbf{D_W}$ *Internet Project.* Consider using the Internet to buy a copy of the book *Freakonomics* by Steven D. Levitt and Stephen J. Dubner. Then compare your costs with buying it at a local bookstore. Decide which way you would make a purchase. Discuss possible ways in which answers might vary.

SKILL MAINTENANCE

Add.

53. $4569 + 1766$ [1.2a]

54. $\frac{2}{3} + \frac{5}{8}$ [2.3a]

55. $4\frac{1}{3} + 2\frac{1}{2}$ [2.4b]

56. $\frac{5}{6} + \frac{7}{10}$ [2.3a]

Subtract.

57. $\frac{2}{3} - \frac{5}{8}$ [2.3b]

58. $4569 - 1766$ [1.2d]

59. $\frac{5}{6} - \frac{7}{10}$ [2.3b]

60. $4\frac{1}{3} - 2\frac{1}{2}$ [2.4c]

Simplify. [2.1e]

61. $\frac{3225}{6275}$

62. $\frac{125}{400}$

63. $\frac{325}{625}$

64. $\frac{625}{475}$

Solve.

65. If a water wheel made 469 revolutions at a rate of $16\frac{3}{4}$ revolutions per minute, how long did it rotate? [2.5a]

66. If a bicycle wheel made 480 revolutions at a rate of $66\frac{2}{3}$ revolutions per minute, how long did it rotate? [2.5a]

67. *Calories in Pie.* A piece of pecan pie ($\frac{1}{8}$ of a 9-in. pie) has 502 calories. A piece of pumpkin pie has 316 calories. How many more calories does a piece of pecan pie have than a piece of pumpkin pie? [1.5a]

68. *Calories in Turkey.* Dark meat turkey contains 187 calories per 3.5-oz serving. White meat turkey contains 157 calories per 3.5-oz serving. How many more calories per 3.5-oz serving does the dark turkey contain than the white turkey? [1.5a]

SYNTHESIS

69. Reggie bought a half-dozen packs of basketball cards with a dozen cards in each pack. The cost was twelve dozen cents for each half-dozen cards. How much did Reggie pay for the cards?

3 Summary and Review

The review that follows is meant to prepare you for a chapter exam. It consists of two parts. The first part, Concept Reinforcement, is designed to increase understanding of the concepts through true/false exercises. The second part is the Review Exercises. These provide practice exercises for the exam, together with references to section objectives so you can go back and review. Before beginning, stop and look back over the skills you have obtained. What skills in mathematics do you have now that you did not have before studying this chapter?

CONCEPT REINFORCEMENT

Determine whether the statement is true or false. Answers are given at the back of the book.

______ **1.** In the number 308.00567, the digit 6 names the hundreds place.

______ **2.** To multiply any number by a multiple of 10, count the number of zeros and move the decimal point that many places to the right.

______ **3.** One thousand billions is one trillion.

______ **4.** The number of decimal places in the product of two numbers is the product of the number of places in the factors.

______ **5.** When writing a word name for decimal notation, we write the word "and" for the decimal point.

Review Exercises

Convert the number in the sentence to standard notation.

1. Russia has the largest total area of any country in the world, at 6.59 million square miles. [3.3b]

2. The total weight of the turkeys consumed by Americans during the Thanksgiving holidays is about 6.9 million pounds. [3.3b]

Write a word name. [3.1a]

3. 3.47

4. 0.031

5. 27.00011

6. 0.000007

Write fraction notation. [3.1b]

7. 0.09

8. 4.561

9. 0.089

10. 3.0227

Write decimal notation. [3.1b]

11. $\frac{34}{1000}$

12. $\frac{42,603}{10,000}$

13. $27\frac{91}{100}$

14. $867\frac{6}{1000}$

Which number is larger? [3.1c]

15. 0.034, 0.0185

16. 0.91, 0.19

17. 0.741, 0.6943

18. 1.038, 1.041

Round 17.4287 to the nearest: [3.1d]

19. Tenth.

20. Hundredth.

21. Thousandth.

22. One.

Add. [3.2a]

23.
$$\begin{array}{r} 2.048 \\ 65.371 \\ +\ 507.1 \\ \hline \end{array}$$

24.
$$\begin{array}{l} 0.6 \\ 0.004 \\ 0.07 \\ +0.0098 \\ \hline \end{array}$$

25. $219.3 + 2.8 + 7$

26. $0.41 + 4.1 + 41 + 0.041$

Subtract. [3.2b]

27.
$$\begin{array}{r} 30.0 \\ -\ 0.7908 \\ \hline \end{array}$$

28.
$$\begin{array}{r} 845.08 \\ -\ 54.79 \\ \hline \end{array}$$

29. $37.645 - 8.497$

30. $70.8 - 0.0109$

Multiply. [3.3a]

31.
$$\begin{array}{r} 48 \\ \times\ 0.27 \\ \hline \end{array}$$

32.
$$\begin{array}{r} 0.174 \\ \times\ 0.83 \\ \hline \end{array}$$

33. 100×0.043

34. 0.001×24.68

Divide. [3.4a]

35. $8\overline{)60}$

36. $52\overline{)23.4}$

37. $2.6\overline{)117.52}$

38. $2.14\overline{)2.18708}$

39. $\dfrac{276.3}{1000}$

40. $\dfrac{13.892}{0.01}$

Solve. [3.2c], [3.4b]

41. $x + 51.748 = 548.0275$

42. $3 \cdot x = 20.85$

43. $10 \cdot y = 425.4$

44. $0.0089 + y = 5$

Solve. [3.7a]

45. Working as a coronary intensive care nurse, Stacia earned $620.80 during a recent 40-hr week. What is her hourly wage?

46. *Nutrition.* The average person eats 683.6 lb of fruits and vegetables in a year. What is the average consumption in one day? (Use 1 year = 365 days.) Round to the nearest tenth of a pound.

47. Derek had $1283.67 in his checking account. He used $370.99 to buy a Sony home theater system with his debit card. How much was left in his account?

48. *Verizon Wireless.* In 2005, Verizon Wireless offered its America's Choice cell-phone plan with 1350 anytime minutes per month for a monthly access fee of $79.99. Minutes in excess of 1350 were charged $0.35 per minute. One month Maria used her cell phone for 2000 min. What was the charge?

Source: www.vzwshop.com

49. *Gas Mileage.* A driver wants to estimate the gas mileage per gallon for her vehicle. At 36,057.1 mi, the tank is filled with 10.7 gal. At 36,217.6 mi, the tank is filled with 11.1 gal. Find the number of miles driven per gallon. Round to the nearest tenth.

50. *Seafood Consumption.* The following graph shows the annual consumption, in pounds, of seafood per person in the United States in recent years. [3.4c]

a) Find the total consumption per person for the seven years.

b) Find the average annual consumption per person for the seven years.

Seafood Consumption

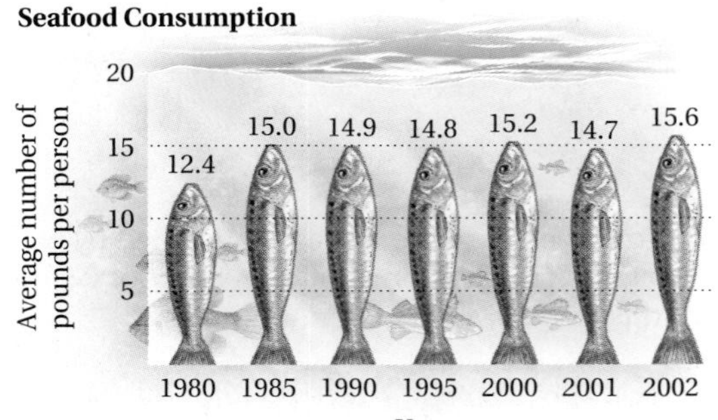

Estimate each of the following. [3.6a]

51. The product 7.82×34.487 by rounding to the nearest one

52. The difference $219.875 - 4.478$ by rounding to the nearest one

53. The quotient $82.304 \div 17.287$ by rounding to the nearest ten

54. The sum $45.78 + $78.99 by rounding to the nearest one

Find decimal notation. Use multiplying by 1. [3.5a]

55. $\frac{13}{5}$ **56.** $\frac{32}{25}$ **57.** $\frac{11}{4}$

Find decimal notation. Use division. [3.5a]

58. $\frac{13}{4}$ **59.** $\frac{7}{6}$ **60.** $\frac{17}{11}$

Round the answer to Exercise 60 to the nearest: [3.5b]

61. Tenth. **62.** Hundredth. **63.** Thousandth.

Convert from cents to dollars. [3.3b]

64. 8273 cents **65.** 487 cents

Convert from dollars to cents. [3.3b]

66. $24.93 **67.** $9.86

Calculate. [3.4c], [3.5c]

68. $(8 - 1.23) \div 4 + 5.6 \times 0.02$

69. $(1 + 0.07)^2 + 10^3 \div 10^2 + [4(10.1 - 5.6) + 8(11.3 - 7.8)]$

70. $\frac{3}{4} \times 20.85$

71. $\frac{1}{3} \times 123.7 + \frac{4}{9} \times 0.684$

72. D_W Consider finding decimal notation for $\frac{44}{125}$. Discuss as many ways as you can for finding such notation and give the answer. [3.5a]

73. D_W Explain how we can use fraction notation to understand why we count decimal places when multiplying with decimal notation. [3.3a]

SYNTHESIS

74. In each of the following, use one of +, −, ×, and ÷ in each blank to make a true sentence. [3.4c]

a) 2.56 ☐ 6.4 ☐ 51.2 ☐ 17.4 ☐ 89.7 = 72.62
b) (11.12 ☐ 0.29) ☐ 3^4 = 877.23

75. Find repeating decimal notation for 1 and explain. Use the following hints. [3.5a]

$$\frac{1}{3} = 0.33333333\ldots,$$

$$\frac{2}{3} = 0.66666666\ldots$$

76. Find repeating decimal notation for 2. [3.5a]

77. In the subtraction below, a and b are digits. Find a and b. [3.2b]

$$\begin{array}{r} b876.a4321 \\ -1234.a678b \\ \hline 8641.b7a32 \end{array}$$

3 Chapter Test

For Extra Help

Work It Out! Chapter Test Video Solutions on CD

Convert the number in the sentence to standard notation.

1. The annual sales of antibiotics in the United States is \$8.9 billion.

Source: IMS Health

2. There are 3.756 million people enrolled in bowling organizations in the United States.

Source: *Bowler's Journal International,* December 2000

Write a word name.

3. 2.34

4. 105.0005

Write fraction notation.

5. 0.91

6. 2.769

Write decimal notation.

7. $\frac{74}{1000}$

8. $\frac{37,047}{10,000}$

9. $756\frac{9}{100}$

10. $91\frac{703}{1000}$

Which number is larger?

11. 0.07, 0.162

12. 0.078, 0.06

13. 0.09, 0.9

Round 5.6783 to the nearest:

14. One.

15. Hundredth.

16. Thousandth.

17. Tenth.

Calculate.

18.
$$\begin{array}{r} 0.7 \\ 0.08 \\ 0.009 \\ +\ 0.0012 \\ \hline \end{array}$$

19. $102.4 + 6.1 + 78$

20. $0.93 + 9.3 + 93 + 930$

21.
$$\begin{array}{r} 52.678 \\ -\ \ 4.321 \\ \hline \end{array}$$

22.
$$\begin{array}{r} 20.0 \\ -\ \ 0.9099 \\ \hline \end{array}$$

23. $234.6788 - 81.7854$

24.
$$\begin{array}{r} 0.125 \\ \times\ \ 0.24 \\ \hline \end{array}$$

25. 0.001×213.45

26. 1000×73.962

27. $4\overline{)19}$

28. $3.3\overline{)\,100.32}$

29. $82\overline{)\,15.58}$

30. $\dfrac{346.89}{1000}$

31. $\dfrac{346.89}{0.01}$

Solve.

32. $4.8 \cdot y = 404.448$

33. $x + 0.018 = 9$

34. *Verizon Wireless.* In 2005, Verizon Wireless offered its America's Choice cell-phone plan with 2000 anytime minutes per month for a monthly access fee of \$99.99. Minutes in excess of 2000 were charged \$0.25 per minute. One month Trey used his cell phone for 2860 min. What was the charge?

Source: www.vzwshop.com

35. *Gas Mileage.* Tina wants to estimate the gas mileage (miles per gallon) for her economy car. At 76,843 mi, the tank is filled with 14.3 gal of gasoline. At 77,310 mi, the tank is filled with 16.5 gal of gasoline. Find the number of miles driven per gallon. Round to the nearest tenth.

36. *Checking Account Balance.* Nicholas has a balance of \$10,200 in his checking account before making purchases of \$123.89, \$56.68, and \$3446.98 with his debit card. What was the balance after making the purchases?

37. The Drake, Smith, and Nicholas law firm buys 7 cases of copy paper at \$25.99 per case. What is the total cost?

38. *Airport Passengers.* The following graph shows the number of passengers in 2003 who traveled through the country's busiest airports. Find the average number of passengers through these airports.

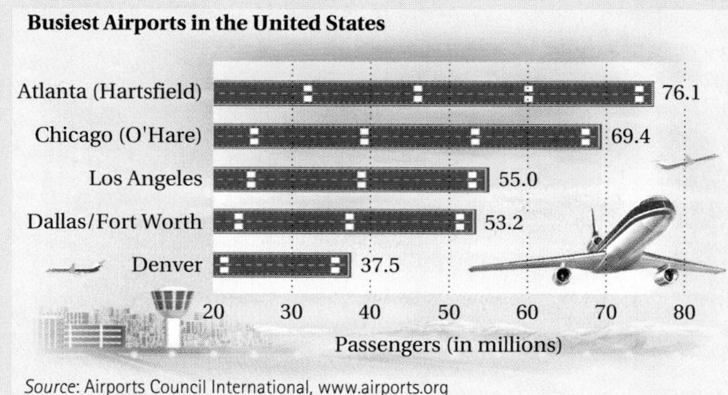

Source: Airports Council International, www.airports.org

Estimate each of the following.

39. The product 8.91×22.457 by rounding to the nearest one

40. The quotient $78.2209 \div 16.09$ by rounding to the nearest ten

Find decimal notation. Use multiplying by 1.

41. $\frac{8}{5}$

42. $\frac{22}{25}$

43. $\frac{21}{4}$

Find decimal notation. Use division.

44. $\frac{3}{4}$

45. $\frac{11}{9}$

46. $\frac{15}{7}$

Round the answer to Exercise 46 to the nearest:

47. Tenth.

48. Hundredth.

49. Thousandth.

50. Convert from cents to dollars: 949 cents.

Calculate.

51. $256 \div 3.2 \div 2 - 1.56 + 78.325 \times 0.02$

52. $(1 - 0.08)^2 + 6[5(12.1 - 8.7) + 10(14.3 - 9.6)]$

53. $\frac{7}{8} \times 345.6$

54. $\frac{2}{3} \times 79.95 - \frac{7}{9} \times 1.235$

SYNTHESIS

55. The Silver's Health Club generally charges a $79 membership fee and $42.50 a month. Allise has a coupon that will allow her to join the club for $299 for six months. How much will Allise save if she uses the coupon?

56. Arrange from smallest to largest.

$$\frac{2}{3}, \frac{15}{19}, \frac{11}{13}, \frac{5}{7}, \frac{13}{15}, \frac{17}{20}$$

4

Percent Notation

Real-World Application

There are 1,168,195 people in the United States in active military service. The numbers in the four armed services are as follows: Air Force, 314,477; Army, 391,126; Marine Corps, 135,324; Navy, 327,268. What percent of the total does each branch represent? Round the answers to the nearest tenth of a percent.

Source: U.S. Department of Defense

This problem appears as Exercise 2 in Section 4.6.

4.1 RATIO AND PROPORTION

Objectives

a Find fraction notation for ratios.

b Give the ratio of two different measures as a rate.

c Determine whether two pairs of numbers are proportional.

d Solve proportions.

e Solve applied problems involving proportions.

1. Find the ratio of 5 to 11.

2. Find the ratio of 57.3 to 86.1.

3. Find the ratio of $6\frac{3}{4}$ to $7\frac{2}{5}$.

4. **Rainfall.** The greatest amount of rainfall ever recorded for a 12-month period was 739 in. in Kukui, Maui, Hawaii, from December 1981 to December 1982. Find the ratio of rainfall to time in months.
Source: *The Handy Science Answer Book*

Answers on page A-9

a Ratios

RATIO

A **ratio** is the quotient of two quantities.

In the 2004–2005 season, the Detroit Pistons basketball team averaged 93.3 points per game and allowed their opponents an average of 89.5 points per game. The *ratio* of points earned to points allowed is given by the fraction notation

$$\frac{\text{Points earned} \rightarrow 93.3}{\text{Points allowed} \rightarrow 89.5} \quad \text{or by the colon notation} \quad 93.3:89.5.$$

We read both forms of notation as "the ratio of 93.3 to 89.5," listing the numerator first and the denominator second.

RATIO NOTATION

The **ratio** of a to b is given by the fraction notation $\frac{a}{b}$, where a is the numerator and b is the denominator, or by the colon notation $a:b$.

EXAMPLE 1 Find the ratio of 31.4 to 100.

The ratio is $\frac{31.4}{100}$, or $31.4:100$.

Do Exercises 1–3.

In most of our work, we will use fraction notation for ratios.

EXAMPLE 2 *Batting.* In the 2005 Major League Baseball season, Vladimir Guerrero of the Los Angeles Angels of Anaheim got 165 hits in 520 at-bats. What was the ratio of hits to at-bats? of at-bats to hits?
Source: Major League Baseball

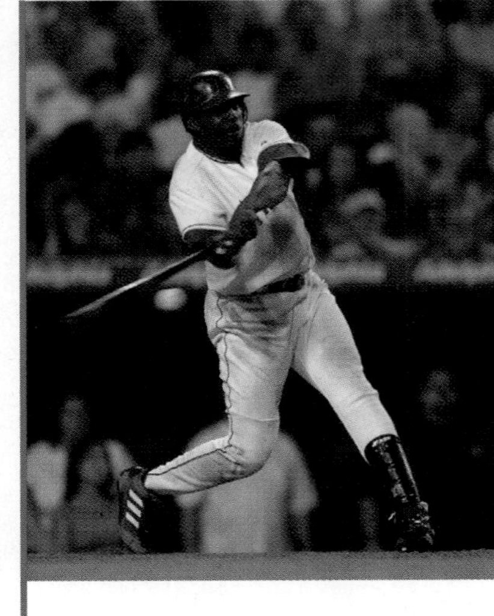

The ratio of hits to at-bats is

$$\frac{165}{520}.$$

The ratio of at-bats to hits is

$$\frac{520}{165}.$$

Do Exercises 4–6. (Exercises 5 and 6 are on the following page.)

EXAMPLE 3 Refer to the triangle below.

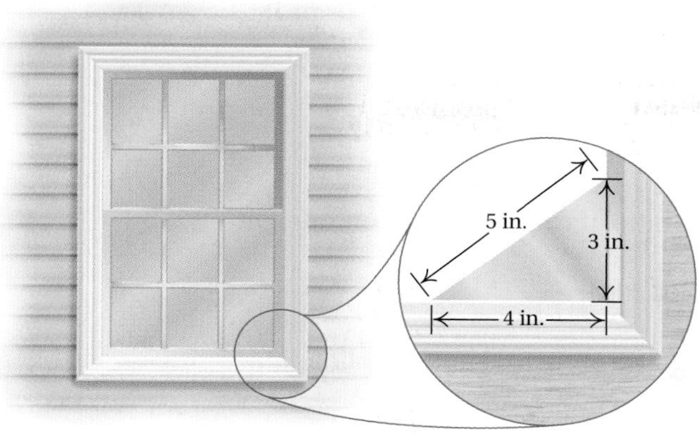

a) What is the ratio of the length of the longest side to the length of the shortest side?

$$\frac{5}{3}$$

b) What is the ratio of the length of the shortest side to the length of the longest side?

$$\frac{3}{5}$$

Do Exercise 7.

EXAMPLE 4 *Veterinary Medicine.* The number of women attending Purdue University's veterinary school of medicine has grown to surpass the number of men in the past three decades.

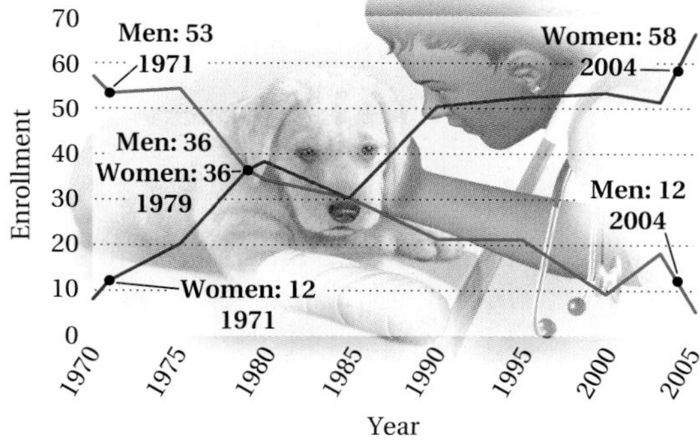

Sources: Purdue University School of Veterinary Medicine; *Indianapolis Star*

a) What was the ratio of women to men in 1971? in 1979? in 2004?

b) What was the ratio of men to women in 1971? in 1979? in 2004?

c) What was the ratio of women to total enrollment in 2004?

d) What was the ratio of men to total enrollment in 2004?

5. **Fat Content.** In one serving ($\frac{1}{2}$-cup) of fried scallops, there are 12 g of fat. In one serving ($\frac{1}{2}$-cup) of fried oysters, there are 14 g of fat. What is the ratio of grams of fat in one serving of scallops to grams of fat in one serving of oysters?

Source: *Better Homes and Gardens: A New Cook Book*

6. **Earned Runs.** In the 2005 baseball season, Roger Clemens of the Houston Astros gave up 44 earned runs in 211.1 innings pitched. What was the ratio of earned runs to innings pitched? of innings pitched to earned runs?

Source: Major League Baseball

7. In the triangle below, what is the ratio of the length of the shortest side to the length of the longest side?

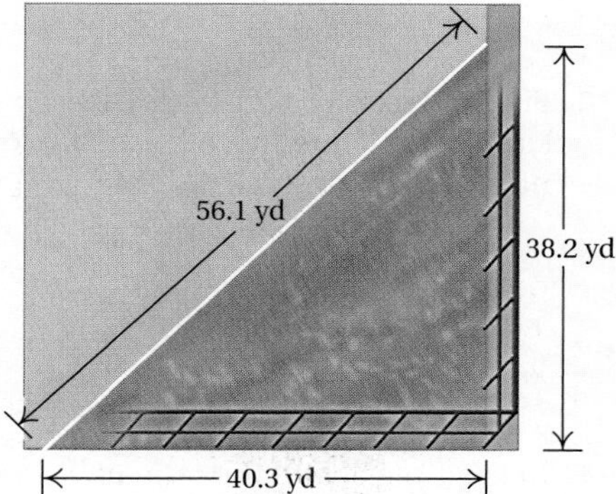

Answers on page A-9

a) The ratio of women to men

in 1971: $\dfrac{12}{53}$; in 1979: $\dfrac{36}{36}$; in 2004: $\dfrac{58}{12}$.

b) The ratio of men to women

in 1971: $\dfrac{53}{12}$; in 1979: $\dfrac{36}{36}$; in 2004: $\dfrac{12}{58}$.

c) The ratio of women to total enrollment

in 2004: $\dfrac{58}{70}$.

d) The ratio of men to total enrollment

in 2004: $\dfrac{12}{70}$.

Do Exercise 8.

EXAMPLE 5 Find the ratio of 2.4 to 10. Then simplify and find two other numbers in the same ratio.

We first write the ratio in fraction notation. Next, we multiply by 1 to clear the decimal from the numerator. Then we simplify.

$$\frac{2.4}{10} = \frac{2.4}{10} \cdot \frac{10}{10} = \frac{24}{100} = \frac{4 \cdot 6}{4 \cdot 25} = \frac{4}{4} \cdot \frac{6}{25} = \frac{6}{25}$$

Thus, 2.4 is to 10 as 6 is to 25.

Do Exercises 9 and 10.

EXAMPLE 6 A standard HDTV screen with a width of 40 in. has a height of $22\frac{1}{2}$ in. Find the ratio of width to height and simplify.

$$\text{The ratio is } \frac{40}{22\frac{1}{2}} = \frac{40}{22.5} = \frac{400}{225}$$

$$= \frac{25 \cdot 16}{25 \cdot 9} = \frac{25}{25} \cdot \frac{16}{9}$$

$$= \frac{16}{9}.$$

Thus we can say that the ratio of width to height is 16 to 9, which can also be expressed as 16:9.

Do Exercise 11.

8. Soap Box Derby. Participation in the All-American Soap Box Derby World Championship has increased by more than 300 competitors since 1985. In 2004, there were 483 participants, 278 boys and 205 girls. What was the ratio of girls to boys? of boys to girls? of boys to total number of participants?

Source: All-American Soap Box Derby

9. Find the ratio of 3.6 to 12. Then simplify and find two other numbers in the same ratio.

10. Find the ratio of 1.2 to 1.5. Then simplify and find two other numbers in the same ratio.

11. In Example 6, find the ratio of the height of the HDTV screen to the width and simplify.

Answers on page A-9

b Rates

A 2005 Kia Sportage EX 4WD can travel 414 mi on 18 gal of gasoline. Let's consider the ratio of miles to gallons:

Source: Kia Motors America, Inc.

$$\frac{414 \text{ mi}}{18 \text{ gal}} = \frac{414}{18}\,\frac{\text{miles}}{\text{gallon}} = \frac{23}{1}\,\frac{\text{miles}}{\text{gallon}}$$
$$= 23 \text{ miles per gallon} = 23 \text{ mpg}.$$

"per" means "division," or "for each."

The ratio $\frac{414 \text{ mi}}{18 \text{ gal}}$, or $\frac{414}{18}\,\frac{\text{mi}}{\text{gal}}$, or 23 mpg is called a **rate.**

RATE

When a ratio is used to compare two different kinds of measure, we call it a **rate.**

Suppose David says his car travels 392.4 mi on 16.8 gal of gasoline. Is the mpg (mileage) of his car better than that of the Kia Sportage above? To determine this, it helps to convert the ratio to decimal notation and perhaps round. Then we have

$$\frac{392.4 \text{ mi}}{16.8 \text{ gal}} = \frac{392.4}{16.8} \text{ mpg} \approx 23.357 \text{ mpg}.$$

Since $23.357 > 23$, David's car gets better mileage than the Kia Sportage does.

EXAMPLE 7 It takes 60 oz of grass seed to seed 3000 sq ft of lawn. What is the rate in ounces per square foot?

$$\frac{60 \text{ oz}}{3000 \text{ sq ft}} = \frac{1}{50}\,\frac{\text{oz}}{\text{sq ft}}, \quad \text{or} \quad 0.02\,\frac{\text{oz}}{\text{sq ft}}$$

EXAMPLE 8 A cook buys 10 lb of potatoes for \$3.69. What is the rate in cents per pound?

$$\frac{\$3.69}{10 \text{ lb}} = \frac{369 \text{ cents}}{10 \text{ lb}}, \quad \text{or} \quad 36.9\,\frac{\text{cents}}{\text{lb}}$$

EXAMPLE 9 A pharmacy student working as a pharmacist's assistant earned \$3690 for working 3 months one summer. What was the rate of pay per month?

The rate of pay is the ratio of money earned per length of time worked, or

$$\frac{\$3690}{3 \text{ mo}} = 1230\,\frac{\text{dollars}}{\text{month}}, \quad \text{or}$$
\$1230 per month.

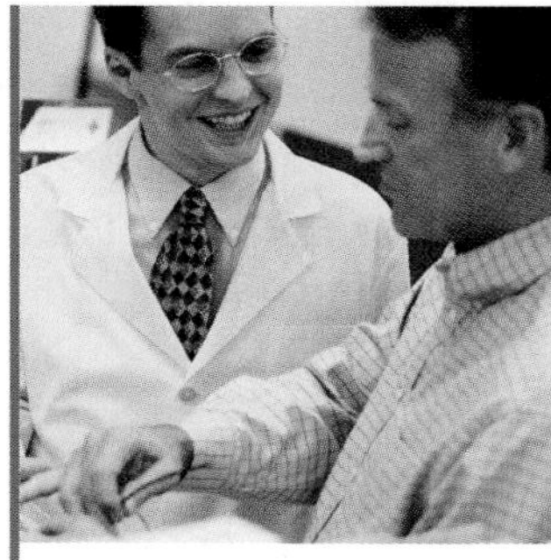

Do Exercises 12–19. (Exercises 18 and 19 are on the following page.)

A ratio of distance traveled to time is called *speed.* What is the rate, or speed, in miles per hour?

12. 45 mi, 9 hr

13. 120 mi, 10 hr

14. 89 km, 13 hr (Round to the nearest hundredth.)

What is the rate, or speed, in feet per second?

15. 2200 ft, 2 sec

16. 52 ft, 13 sec

17. 242 ft, 16 sec

Answers on page A-9

18. A well-hit golf ball can travel 500 ft in 2 sec. What is the rate, or speed, of the golf ball in feet per second?

19. **Babe Ruth.** In his entire career, Babe Ruth had 1330 strikeouts and 714 home runs. What was his home-run to strikeout rate?

Source: Major League Baseball

Answers on page A-9

C Proportions

Suppose we want to compare $\frac{3}{6}$ and $\frac{2}{4}$. We find a common denominator and compare numerators. To do this, we multiply each fraction by 1, using the denominator of the other fraction to form the symbol for 1.

The "unit" is $\frac{1}{6}$.

$$\frac{3}{6} = \frac{3}{6} \cdot \frac{4}{4} = \frac{3 \cdot 4}{6 \cdot 4} = \frac{12}{24}$$

$$\frac{2}{4} = \frac{2}{4} \cdot \frac{6}{6} = \frac{2 \cdot 6}{4 \cdot 6} = \frac{12}{24}$$

The "unit" is $\frac{1}{4}$.

Both "units" are $\frac{1}{24}$.

We multiply $\frac{3}{6}$ by $\frac{4}{4}$ and $\frac{2}{4}$ by $\frac{6}{6}$. We see that $\frac{3}{6} = \frac{2}{4}$.

Note in the preceding that if

$$\frac{3}{6} = \frac{2}{4}, \quad \text{then} \quad 3 \cdot 4 = 6 \cdot 2.$$

We need check only the products $3 \cdot 4$ and $6 \cdot 2$ to compare the fractions.

A TEST FOR EQUALITY

We multiply these two numbers: $3 \cdot 4$.

We multiply these two numbers: $6 \cdot 2$.

$$\frac{3}{6} \quad \frac{2}{4}$$

We call $3 \cdot 4$ and $6 \cdot 2$ **cross products.** Since the cross products are the same, that is, $3 \cdot 4 = 6 \cdot 2$, we know that

$$\frac{3}{6} = \frac{2}{4}.$$

When two pairs of numbers (such as 3, 6 and 2, 4) have the same ratio, we say that they are **proportional.** The equation

$$\frac{3}{6} = \frac{2}{4}$$

states that the pairs 3, 6 and 2, 4 are proportional. Such an equation is called a **proportion.** We sometimes read $\frac{3}{6} = \frac{2}{4}$ as "3 is to 6 as 2 is to 4."

EXAMPLE 10 Determine whether 1, 2 and 3, 6 are proportional.

We can use cross products:

$$1 \cdot 6 = 6 \qquad \frac{1}{2} \stackrel{?}{=} \frac{3}{6} \qquad 2 \cdot 3 = 6.$$

Since the cross products are the same, $6 = 6$, we know that $\frac{1}{2} = \frac{3}{6}$, so the numbers are proportional.

EXAMPLE 11 Determine whether 2, 5 and 4, 7 are proportional.

We can use cross products:

$$2 \cdot 7 = 14 \qquad \frac{2}{5} \stackrel{?}{=} \frac{4}{7} \qquad 5 \cdot 4 = 20.$$

Since the cross products are not the same, $14 \neq 20$, we know that $\frac{2}{5} \neq \frac{4}{7}$, so the numbers are not proportional.

Do Exercises 20–22.

Determine whether the two pairs of numbers are proportional.

20. 3, 4 and 6, 8

21. 1, 4 and 10, 39

22. 1, 2 and 20, 39

d Solving Proportions

Let's now look at solving proportions. Consider the proportion

$$\frac{x}{3} = \frac{4}{6}.$$

One way to solve a proportion is to use cross products. Then we can divide on both sides to get the variable alone:

$x \cdot 6 = 3 \cdot 4$ Equating cross products (finding cross products and setting them equal)

$\frac{x \cdot 6}{6} = \frac{3 \cdot 4}{6}$ Dividing by 6 on both sides

$x = \frac{3 \cdot 4}{6} = \frac{12}{6} = 2.$

We can check that 2 is the solution by replacing x with 2 and using cross products:

$$2 \cdot 6 = 12 \qquad \frac{2}{3} \stackrel{?}{=} \frac{4}{6} \qquad 3 \cdot 4 = 12.$$

Since the cross products are the same, it follows that $\frac{2}{3} = \frac{4}{6}$; so the numbers 2, 3 and 4, 6 are proportional, and 2 is the solution of the equation.

SOLVING PROPORTIONS

To solve $\frac{x}{a} = \frac{c}{d}$, equate *cross products* and divide on both sides to get x alone.

Do Exercise 23.

23. Solve: $\frac{x}{63} = \frac{2}{9}$.

Answers on page A-9

Study Tips

WRITING ALL THE STEPS

Take the time to include all the steps when working your homework problems. Doing so will help you organize your thinking and avoid computational errors. If you find a wrong answer, having all the steps allows easier checking of your work. It will also give you complete, step-by-step solutions of the exercises that can be used to study for an exam.

Writing down all the steps and keeping your work organized may also give you a better chance of getting partial credit.

"Success comes before work only in the dictionary."

Anonymous

EXAMPLE 12 Solve: $\frac{x}{7} = \frac{5}{3}$. Write a mixed numeral for the answer.

We have

$$\frac{x}{7} = \frac{5}{3}$$

$$x \cdot 3 = 7 \cdot 5 \quad \text{Equating cross products}$$

$$\frac{x \cdot 3}{3} = \frac{7 \cdot 5}{3} \quad \text{Dividing by 3}$$

$$x = \frac{7 \cdot 5}{3}$$

$$x = \frac{35}{3}, \text{ or } 11\frac{2}{3}.$$

The solution is $11\frac{2}{3}$.

Do Exercise 24 on the following page.

EXAMPLE 13 Solve: $\frac{7.7}{15.4} = \frac{y}{2.2}$.

We have

$$\frac{7.7}{15.4} = \frac{y}{2.2}$$

$$7.7 \times 2.2 = 15.4 \times y \quad \text{Equating cross products}$$

$$\frac{7.7 \times 2.2}{15.4} = \frac{15.4 \times y}{15.4} \quad \text{Dividing by 15.4}$$

$$\frac{7.7 \times 2.2}{15.4} = y$$

$$\frac{16.94}{15.4} = y \quad \text{Multiplying}$$

$$1.1 = y. \quad \text{Dividing: } 15.4\overline{)16.94} = 1.1;\ 1540,\ 154,\ 154,\ 0$$

The solution is 1.1.

Do Exercise 25 on the following page.

EXAMPLE 14 Solve: $\frac{8}{x} = \frac{5}{3}$. Write decimal notation for the answer.

We have

$$\frac{8}{x} = \frac{5}{3}$$

$$8 \cdot 3 = x \cdot 5 \quad \text{Equating cross products}$$

$$\frac{8 \cdot 3}{5} = \frac{x \cdot 5}{5}. \quad \text{Dividing by 5}$$

CALCULATOR CORNER

Solving Proportions Note in Examples 12–15 that when we solve a proportion, we equate cross products and then we divide on both sides to isolate the variable on one side of the equation. We can use a calculator to do the calculations in this situation. In Example 15, for instance, after equating cross products and dividing by 3.4 on both sides, we have

$$n = \frac{4.93 \times 10}{3.4}.$$

To find n on a calculator, we can press [4] [.] [9] [3] [×] [1] [0] [÷] [3] [.] [4] [=]. The result is 14.5, so $n = 14.5$.

Exercises: Solve each proportion.

1. $\frac{15.75}{20} = \frac{a}{35}$
2. $\frac{32}{x} = \frac{25}{20}$
3. $\frac{t}{57} = \frac{17}{64}$
4. $\frac{71.2}{a} = \frac{42.5}{23.9}$

Then

$$\frac{8 \cdot 3}{5} = x$$

$$\frac{24}{5} = x \qquad \text{Multiplying}$$

$$4.8 = x. \qquad \text{Simplifying}$$

The solution is 4.8.

Do Exercise 26.

EXAMPLE 15 Solve: $\dfrac{3.4}{4.93} = \dfrac{10}{n}$.

We have

$$\frac{3.4}{4.93} = \frac{10}{n}$$

$$3.4 \times n = 4.93 \times 10 \qquad \text{Equating cross products}$$

$$\frac{3.4 \times n}{3.4} = \frac{4.93 \times 10}{3.4} \qquad \text{Dividing by 3.4}$$

$$n = \frac{4.93 \times 10}{3.4}$$

$$n = \frac{49.3}{3.4} \qquad \text{Multiplying}$$

$$n = 14.5. \qquad \text{Dividing}$$

The solution is 14.5.

Do Exercise 27.

e Applications and Problem Solving

Proportions have applications in such diverse fields as business, chemistry, health sciences, and home economics, as well as to many areas of daily life. Proportions are useful in making predictions.

EXAMPLE 16 *Predicting Total Distance.* Donna drives her delivery van 800 mi in 3 days. At this rate, how far will she drive in 15 days?

1. **Familiarize.** We let d = the distance traveled in 15 days.
2. **Translate.** We translate to a proportion. We make each side the ratio of distance to time, with distance in the numerator and time in the denominator.

$$\text{Distance in 15 days} \rightarrow \frac{d}{15} = \frac{800}{3} \leftarrow \text{Distance in 3 days}$$
$$\text{Time} \rightarrow 15 \qquad 3 \leftarrow \text{Time}$$

It may help to verbalize the proportion above as "the unknown distance d is to 15 days as the known distance 800 mi is to 3 days."

24. Solve: $\dfrac{x}{9} = \dfrac{5}{4}$.

25. Solve: $\dfrac{21}{5} = \dfrac{n}{2.5}$.

26. Solve: $\dfrac{6}{x} = \dfrac{25}{11}$.

27. Solve: $\dfrac{0.4}{0.9} = \dfrac{4.8}{t}$.

Answers on page A-9

28. Calories Burned. Marv Bittinger, one of the authors of this book, generally exercises for 2 hr each day. The readout on an exercise machine tells him that if he exercises for 24 min, he will burn 356 calories. How many calories will he burn if he exercises for 30 min?

Source: Star Trac Treadmill

Answer on page A-9

3. **Solve.** Next, we solve the proportion:

$3 \cdot d = 15 \cdot 800$ Equating cross products

$\frac{3 \cdot d}{3} = \frac{15 \cdot 800}{3}$ Dividing by 3 on both sides

$d = \frac{15 \cdot 800}{3}$

$d = 4000.$ Multiplying and dividing

4. **Check.** We substitute into the proportion and check cross products:

$\frac{4000}{15} = \frac{800}{3};$

$4000 \cdot 3 = 12{,}000;$ $15 \cdot 800 = 12{,}000.$

The cross products are the same.

5. **State.** Donna will drive 4000 mi in 15 days.

Do Exercise 28.

EXAMPLE 17 *Recommended Dosage.* To control a fever, a doctor suggests that a child who weighs 28 kg be given 320 mg of Tylenol. If the dosage is proportional to the child's weight, how much Tylenol is recommended for a child who weighs 35 kg?

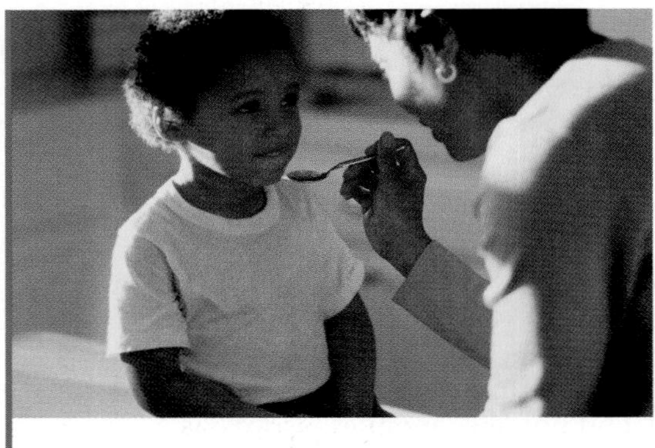

1. **Familiarize.** We let $t =$ the number of milligrams of Tylenol.
2. **Translate.** We translate to a proportion, keeping the amount of Tylenol in the numerators.

Tylenol suggested → 320 / Child's weight → 28 = t ← Tylenol suggested / 35 ← Child's weight

$$\frac{320}{28} = \frac{t}{35}$$

3. **Solve.** Next, we solve the proportion:

$320 \cdot 35 = 28 \cdot t$ Equating cross products

$\frac{320 \cdot 35}{28} = \frac{28 \cdot t}{28}$ Dividing by 28 on both sides

$\frac{320 \cdot 35}{28} = t$

$400 = t.$ Multiplying and dividing

4. Check. We substitute into the proportion and check cross products:

$$\frac{320}{28} = \frac{400}{35};$$

$320 \cdot 35 = 11{,}200; \qquad 28 \cdot 400 = 11{,}200.$

The cross products are the same.

5. State. The dosage for a child who weighs 35 kg is 400 mg.

Do Exercise 29.

29. Determining Paint Needs. Lowell and Chris run a summer painting company to support their college expenses. They can paint 1600 ft^2 of clapboard with 4 gal of paint. How much paint would be needed for a building with 6000 ft^2 of clapboard?

Problems involving proportion can be translated in more than one way. In Example 16, any one of the following is a correct translation:

$$\frac{800}{3} = \frac{d}{15}, \quad \frac{15}{d} = \frac{3}{800}, \quad \frac{15}{3} = \frac{d}{800}, \quad \frac{800}{d} = \frac{3}{15}.$$

Equating the cross products in each equation gives us the equation $3 \cdot d = 15 \cdot 800$.

EXAMPLE 18 *Construction Plans.* Architects make blueprints of projects being constructed. These are scale drawings in which lengths are in proportion to actual sizes. The Hennesseys are constructing a rectangular deck just outside their house. The architectural blueprints are rendered such that $\frac{3}{4}$ in. on the drawing is actually 2.25 ft on the deck. The width of the deck on the drawing is 4.3 in. How wide is the deck in reality?

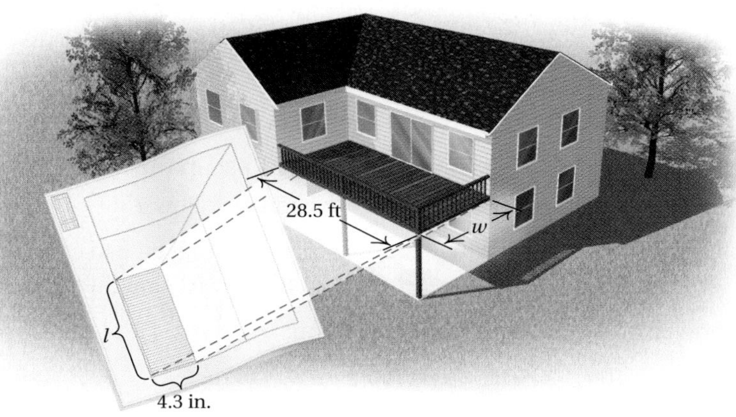

1. Familiarize. We let w = the width of the deck.

2. Translate. Then we translate to a proportion, using 0.75 for $\frac{3}{4}$ in.

$$\begin{matrix}\text{Measure on drawing} \rightarrow \\ \text{Measure on deck} \rightarrow\end{matrix} \frac{0.75}{2.25} = \frac{4.3}{w} \begin{matrix}\leftarrow \text{Width of drawing} \\ \leftarrow \text{Width of deck}\end{matrix}$$

3. Solve. Next, we solve the proportion:

$0.75 \times w = 2.25 \times 4.3$ Equating cross products

$\dfrac{0.75 \times w}{0.75} = \dfrac{2.25 \times 4.3}{0.75}$ Dividing by 0.75 on both sides

$w = \dfrac{2.25 \times 4.3}{0.75}$

$w = 12.9.$

Answer on page A-9

30. Construction Plans. In Example 18, the length of the actual deck is 28.5 ft. What is the length of the deck on the blueprints?

31. Estimating a Deer Population. To determine the number of deer in a forest, a conservationist catches 612 deer, tags them, and releases them. Later, 244 deer are caught, and it is found that 72 of them are tagged. Estimate how many deer are in the forest.

Answer on page A-9

4. Check. We substitute into the proportion and check cross products:

$$\frac{0.75}{2.25} = \frac{4.3}{12.9};$$

$$0.75 \times 12.9 = 9.675; \qquad 2.25 \times 4.3 = 9.675.$$

The cross products are the same.

5. State. The width of the deck is 12.9 ft.

Do Exercise 30.

EXAMPLE 19 *Estimating a Wildlife Population.* To determine the number of fish in a lake, a conservationist catches 225 fish, tags them, and throws them back into the lake. Later, 108 fish are caught, and it is found that 15 of them are tagged. Estimate how many fish are in the lake.

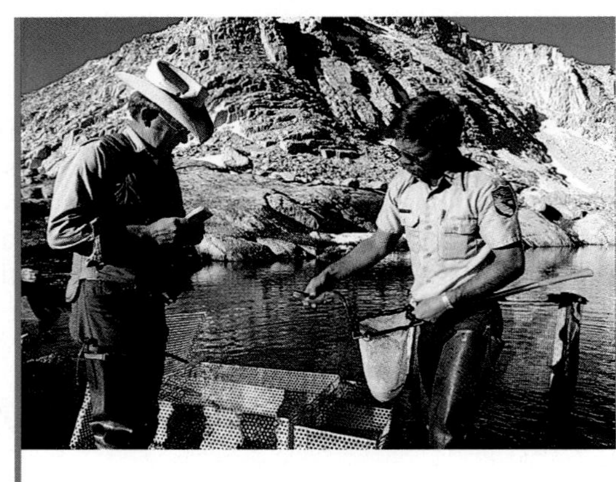

1. Familiarize. We let $F =$ the number of fish in the lake.

2. Translate. We translate to a proportion as follows:

$$\begin{array}{r} \text{Fish tagged originally} \rightarrow \\ \text{Fish in lake} \rightarrow \end{array} \frac{225}{F} = \frac{15}{108}. \begin{array}{l} \leftarrow \text{Tagged fish caught later} \\ \leftarrow \text{Fish caught later} \end{array}$$

3. Solve. Next, we solve the proportion:

$$225 \cdot 108 = F \cdot 15 \qquad \text{Equating cross products}$$

$$\frac{225 \cdot 108}{15} = \frac{F \cdot 15}{15} \qquad \text{Dividing by 15 on both sides}$$

$$\frac{225 \cdot 108}{15} = F$$

$$1620 = F. \qquad \text{Multiplying and dividing}$$

4. Check. We substitute into the proportion and check cross products:

$$\frac{225}{1620} = \frac{15}{108};$$

$$225 \cdot 108 = 24{,}300; \qquad 1620 \cdot 15 = 24{,}300.$$

The cross products are the same.

5. State. We estimate that there are 1620 fish in the lake.

Do Exercise 31.

4.1 EXERCISE SET

For Extra Help

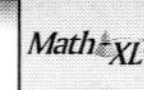

MathXL | MyMathLab | InterAct Math | Math Tutor Center | Video Lectures on CD Disc 2 | Student's Solutions Manual

a Find fraction notation for the ratio. You need not simplify.

1. 4 to 5

2. 329 to 967

3. 56.78 to 98.35

4. $10\frac{1}{2}$ to $43\frac{1}{4}$

5. *Physicians.* In 2001, there were 356 physicians in Connecticut per 100,000 residents. In Wyoming, there were 173 physicians per 100,000 residents. Find the ratio of the number of physicians to the number of residents in Connecticut and in Wyoming.

Sources: U.S. Bureau of the Census; American Medical Association

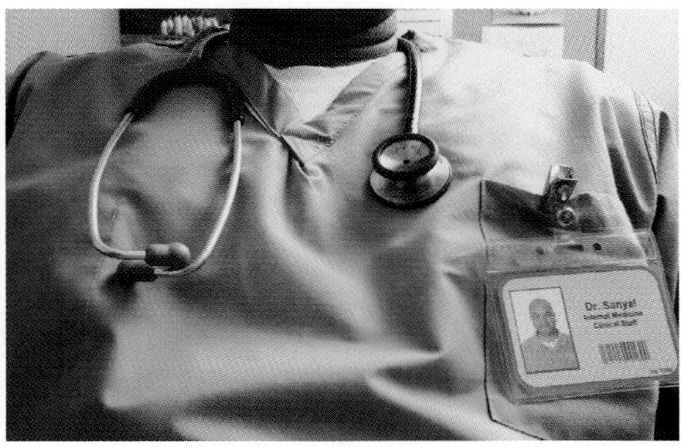

6. *Silicon in the Earth's Crust.* Of every 100 tons of the earth's crust, there is about 28 tons of silicon in its content. What is the ratio of silicon to the weight of crust? of the weight of crust to the weight of silicon?

Source: *The Handy Science Answer Book*

Find the ratio of the first number to the second and simplify.

7. 18 to 24

8. 5.6 to 10

9. 2.8 to 3.6

10. 0.32 to 0.96

11. In this rectangle, find the ratios of length to width and of width to length.

12. In this right triangle, find the ratios of shortest length to longest length and of longest length to shortest length.

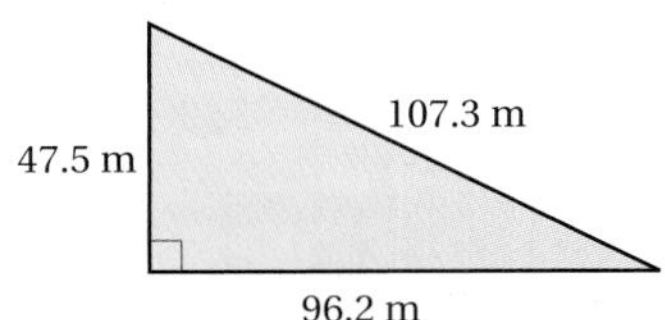

b In Exercises 13–16, find the rate, or speed, as a ratio of distance to time. Round to the nearest hundredth where appropriate.

13. 120 km, 3 hr

14. 18 mi, 9 hr

15. 217 mi, 29 sec

16. 443 m, 48 sec

17. *Audi A6 4.2 Quattro—Highway Driving.* A 2005 Audi A6 4.2 Quattro will travel 448.5 mi on 19.5 gal of gasoline in highway driving. What is the rate in miles per gallon?

Source: *Car and Driver,* May 2005, p. 52

18. *Audi A6 4.2 Quattro—City Driving.* A 2005 Audi A6 4.2 Quattro will travel 246.5 mi on 14.5 gal of gasoline in city driving. What is the rate in miles per gallon?

Source: *Car and Driver,* May 2005, p. 52

19. *Points per Game.* In the 2004–2005 season, Yao Ming of the Houston Rockets scored 1465 points in 80 games. What was the rate in points per game?

Source: National Basketball Association

20. *Population Density of Australia.* The continent of Australia, with the island state of Tasmania, has an area of 2,967,893 sq mi and a population of 19,913,144 people. What is the rate of number of people per square mile? The rate per square mile is called the *population density.* Australia has one of the lowest population densities in the world.

Source: *Time Almanac, 2005*

21. *Lawn Watering.* To water a lawn adequately requires 623 gal of water for every 1000 ft^2. What is the rate in gallons per square foot?

22. A student eats 3 hamburgers in 15 min. What is the rate in hamburgers per minute? in minutes per hamburger?

23. *Elephant Heartbeat.* The heart of an elephant, at rest, will beat an average of 1500 beats in 60 min. What is the rate in beats per minute?

Source: *The Handy Science Answer Book*

24. A black racer snake can travel 4.6 km in 2 hr. What is its rate, or speed, in kilometers per hour?

c Determine whether the two pairs of numbers are proportional.

25. 5, 6 and 7, 9

26. 7, 5 and 6, 4

27. 1, 2 and 10, 20

28. 7, 3 and 21, 9

29. 2.4, 3.6 and 1.8, 2.7

30. 4.5, 3.8 and 6.7, 5.2

31. $5\frac{1}{3}, 8\frac{1}{4}$ and $2\frac{1}{5}, 9\frac{1}{2}$

32. $2\frac{1}{3}, 3\frac{1}{2}$ and 14, 21

d Solve.

33. $\frac{18}{4} = \frac{x}{10}$

34. $\frac{x}{45} = \frac{20}{25}$

35. $\frac{t}{12} = \frac{5}{6}$

36. $\frac{12}{4} = \frac{x}{3}$

37. $\frac{2}{5} = \frac{8}{n}$

38. $\frac{10}{6} = \frac{5}{x}$

39. $\frac{16}{12} = \frac{24}{x}$

40. $\frac{7}{11} = \frac{2}{x}$

41. $\frac{t}{0.16} = \frac{0.15}{0.40}$

42. $\frac{x}{11} = \frac{7.1}{2}$

43. $\frac{100}{25} = \frac{20}{n}$

44. $\frac{35}{125} = \frac{7}{m}$

45. $\frac{\frac{1}{4}}{\frac{1}{2}} = \frac{\frac{1}{2}}{x}$

46. $\frac{5\frac{1}{5}}{6\frac{1}{6}} = \frac{y}{3\frac{1}{2}}$

47. $\frac{1.28}{3.76} = \frac{4.28}{y}$

48. $\frac{10.4}{12.4} = \frac{6.76}{t}$

 Solve.

49. *Cap'n Crunch's Peanut Butter Crunch® Cereal.* The nutritional chart on the side of a box of Quaker Cap'n Crunch's Peanut Butter Crunch® Cereal states that there are 110 calories in a $\frac{3}{4}$-cup serving. How many calories are there in 6 cups of the cereal?

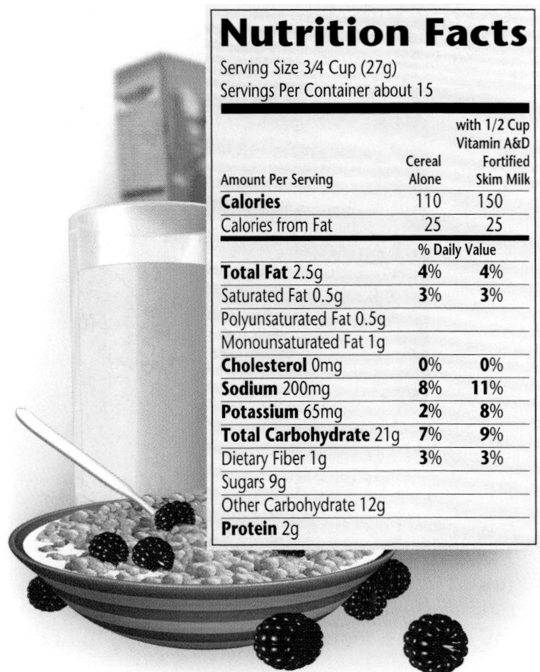

Nutrition Facts
Serving Size 3/4 Cup (27g)
Servings Per Container about 15

Amount Per Serving	Cereal Alone	with 1/2 Cup Vitamin A&D Fortified Skim Milk
Calories	110	150
Calories from Fat	25	25
	% Daily Value	
Total Fat 2.5g	4%	4%
Saturated Fat 0.5g	3%	3%
Polyunsaturated Fat 0.5g		
Monounsaturated Fat 1g		
Cholesterol 0mg	0%	0%
Sodium 200mg	8%	11%
Potassium 65mg	2%	8%
Total Carbohydrate 21g	7%	9%
Dietary Fiber 1g	3%	3%
Sugars 9g		
Other Carbohydrate 12g		
Protein 2g		

50. *Rice Krispies® Cereal.* The nutritional chart on the side of a box of Kellogg's Rice Krispies® Cereal states that there are 120 calories in a $1\frac{1}{4}$-cup serving. How many calories are there in 5 cups of the cereal?

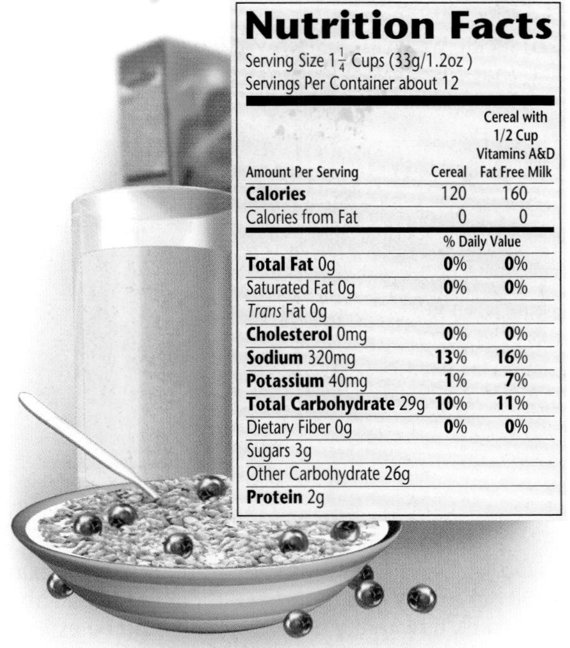

Nutrition Facts
Serving Size $1\frac{1}{4}$ Cups (33g/1.2oz)
Servings Per Container about 12

Amount Per Serving	Cereal	Cereal with 1/2 Cup Vitamins A&D Fat Free Milk
Calories	120	160
Calories from Fat	0	0
	% Daily Value	
Total Fat 0g	0%	0%
Saturated Fat 0g	0%	0%
Trans Fat 0g		
Cholesterol 0mg	0%	0%
Sodium 320mg	13%	16%
Potassium 40mg	1%	7%
Total Carbohydrate 29g	10%	11%
Dietary Fiber 0g	0%	0%
Sugars 3g		
Other Carbohydrate 26g		
Protein 2g		

51. *Overweight Americans.* A study recently confirmed that of every 100 Americans, 60 are considered overweight. There were 295 million Americans in 2005. How many would be considered overweight?

Source: U.S. Centers for Disease Control

52. *Study Time and Test Grades.* A mathematics instructor asserted that students' test grades are directly proportional to the amount of time spent studying. Brent studies for 15 hr for a particular test and gets a score of 85. At this rate, what score would he have received if he had studied for 16 hr?

53. *Mileage.* Jean bought a new car. In the first 8 months, it was driven 9000 mi. At this rate, how many miles will the car be driven in 1 yr?

54. *Bicycling.* Roy bicycled 234 mi in 14 days. At this rate, how far would Roy travel in 42 days?

55. *Metallurgy.* In a metal alloy, the ratio of zinc to copper is 3 to 13. If there are 520 lb of copper, how many pounds of zinc are there?

56. *Class Size.* A college advertises that its student-to-faculty ratio is 14 to 1. If 56 students register for Introductory Spanish, how many sections of the course would you expect to see offered?

57. *Gas Mileage.* A 2005 Ford Mustang GT Convertible will travel 372 mi on 15.5 gal of gasoline in highway driving.

a) How many gallons of gasoline will it take to drive 2690 mi from Boston to Phoenix?

b) How far can the car be driven on 140 gal of gasoline?

Source: Ford Motor Company

58. *Gas Mileage.* A 2005 Volkswagen Passat will travel 462 mi on 16.5 gal of gasoline in highway driving.

a) How many gallons of gasoline will it take to drive 1650 mi from Pittsburgh to Albuquerque?

b) How far can the car be driven on 130 gal of gasoline?

Source: Volkswagen of America, Inc.

59. *Exchanging Money.* On 29 January 2006, 1 U.S. dollar was worth about 10.4555 Mexican pesos.

a) How much would 120 U.S. dollars be worth in Mexican pesos?

b) Jackie was traveling in Mexico and bought a watch that cost 3600 Mexican pesos. How much would it cost in U.S. dollars?

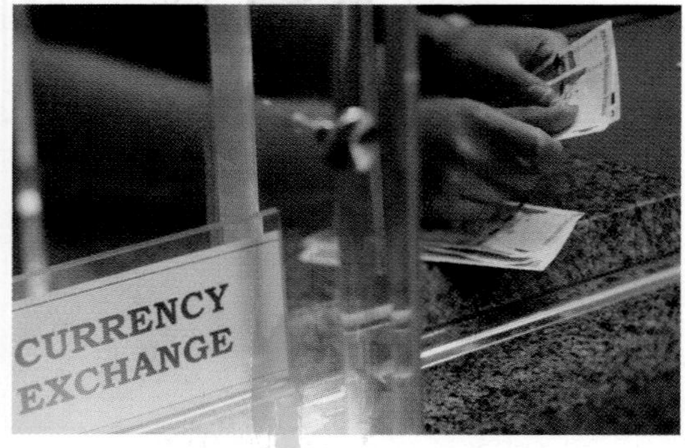

60. *Sugaring.* When 38 gal of maple sap are boiled down, the result is 2 gal of maple syrup. How much sap is needed to produce 9 gal of syrup?

61. *Quality Control.* A quality-control inspector examined 100 lightbulbs and found 7 of them to be defective. At this rate, how many defective bulbs can be expected in a lot of 2500?

62. *Coffee Production.* Coffee beans from 14 trees are required to produce the 17 lb of coffee that the average person in the United States drinks each year. How many trees are required to produce 375 lb of coffee?

63. *Painting.* Fred uses 3 gal of paint to cover 1275 ft^2 of siding. How much siding can Fred paint with 7 gal of paint?

64. *Waterproofing.* Bonnie can waterproof 450 ft^2 of decking with 2 gal of sealant. How many gallons should Bonnie buy for a 1200-ft^2 deck?

65. *Grass-Seed Coverage.* It takes 60 oz of grass seed to seed 3000 ft^2 of lawn. At this rate, how much would be needed for 5000 ft^2 of lawn?

66. *Snow to Water.* Under typical conditions, $1\frac{1}{2}$ ft of snow will melt to 2 in. of water. To how many inches of water will $5\frac{1}{2}$ ft of snow melt?

67. *Estimating a Deer Population.* To determine the number of deer in a game preserve, a forest ranger catches 318 deer, tags them, and releases them. Later, 168 deer are caught, and it is found that 56 of them are tagged. Estimate how many deer are in the game preserve.

68. *Map Scaling.* On a map, $\frac{1}{4}$ in. represents 50 mi. If two cities are $3\frac{1}{4}$ in. apart on the map, how far apart are they in reality?

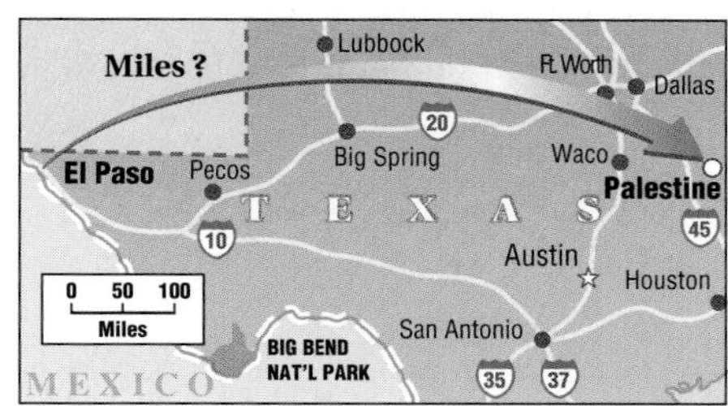

69. DW An instructor predicts that a student's test grade will be proportional to the amount of time the student spends studying. What is meant by this? Write an example of a proportion that involves the grades of two students and their study times.

70. DW *Earned Run Average.* In baseball, the average number of runs given up by a pitcher in nine innings is his *earned run average,* or *ERA.* Set up a formula for determining a player's ERA. Then verify it using the fact that in the 2000 season, Daryl Kile of the St. Louis Cardinals gave up 101 earned runs in $232\frac{1}{3}$ innings to compile an ERA of 3.91. Then use your formula to determine the ERA of Randy Johnson of the Arizona Diamondbacks, who gave up 73 earned runs in $248\frac{2}{3}$ innings. Is a low ERA considered good or bad?

Source: Major League Baseball

SKILL MAINTENANCE

Divide. Write decimal notation for the answer. [3.4a]

71. $200 \div 4$

72. $95 \div 10$

73. $232 \div 16$

74. $342 \div 2.25$

Solve. [2.5a]

75. Rocky is $187\frac{1}{10}$ cm tall and his daughter is $180\frac{3}{4}$ cm tall. How much taller is Rocky?

76. Aunt Louise is $168\frac{1}{4}$ cm tall and her son is $150\frac{7}{10}$ cm tall. How much taller is Aunt Louise?

SYNTHESIS

77. *Baseball Statistics.* Cy Young, one of the greatest baseball pitchers of all time, gave up an average of 2.63 earned runs every 9 innings. Young pitched 7356 innings, more than anyone in the history of baseball. How many earned runs did he give up?

78. *Real-Estate Values.* According to Coldwell Banker Real Estate Corporation, a home selling for $189,000 in Austin, Texas, would sell for $665,795 in San Francisco. How much would a $450,000 home in San Francisco sell for in Austin? Round to the nearest $1000.

Source: Coldwell Banker Real Estate Corporation

4.2 PERCENT NOTATION

Objectives

a Write three kinds of notation for a percent.

b Convert between percent notation and decimal notation.

a Understanding Percent Notation

Of all the surface area of the earth, 70% is covered by water. What does this mean? It means that of every 100 square miles of the earth's surface area, 70 square miles are covered by water. Thus, 70% is a ratio of 70 to 100, or $\frac{70}{100}$.

Source: *The Handy Geography Answer Book*

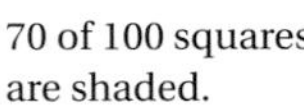

70 of 100 squares are shaded.

70% or $\frac{70}{100}$ or 0.70 of the large square is shaded.

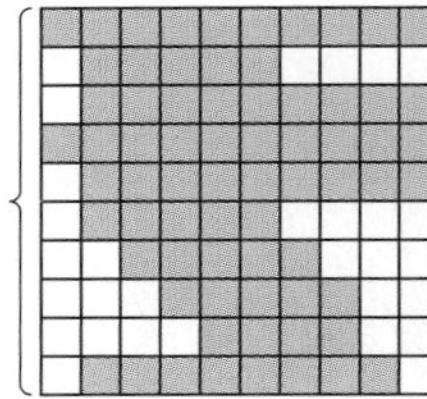

Percent notation is used extensively in our everyday lives. Here are some examples:

44% of all Americans use at least one prescribed medicine.

35.5% of Massachusetts residents age 25 and over have a bachelor's degree.

18% of household personal vehicles are pickup trucks.

50.3% of all paper used in the United States is recycled.

54% of all kids prefer white bread.

0.08% blood alcohol level is a standard used by some states as the legal limit for drunk driving.

Percent notation is often represented in circle graphs, or pie charts, to show how the parts of a quantity are related. For example, the circle graph below illustrates the percentage of vehicles manufactured in the most popular car colors during 2003 in North America.

Most Popular Car Colors, 2003 (SUV/Truck/Van)

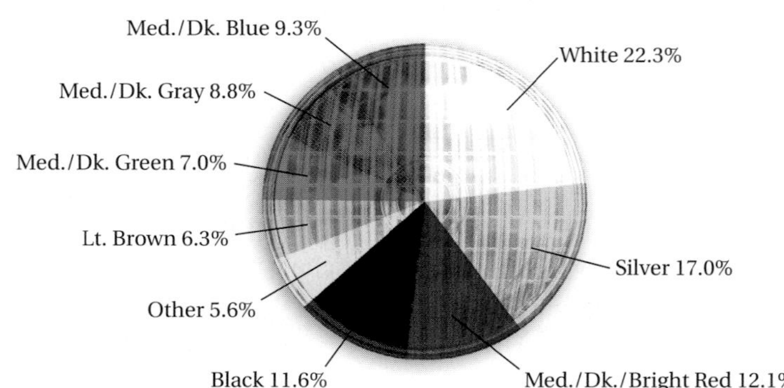

Source: DuPont Herberts Automotive Systems, Troy, Michigan, 2003, DuPont Automotive Color Survey Results

Write three kinds of notation as in Examples 1 and 2.

1. 70%

2. 23.4%

3. 100%

It is thought that the Roman emperor Augustus created percent notation by taxing goods sold at a rate of $\frac{1}{100}$. In time, the symbol "%" evolved by interchanging the parts of the symbol "100" to "0/0" and then to "%."

Answers on page A-9

PERCENT NOTATION

The notation ***n*%** means "n per hundred."

This definition leads us to the following equivalent ways of defining percent notation.

NOTATION FOR n%

Percent notation, *n*%, can be expressed using:

ratio → $n\% =$ the ratio of n to 100 $= \frac{n}{100}$,

fraction notation → $n\% = n \times \frac{1}{100}$, or

decimal notation → $n\% = n \times 0.01$.

From 1998 to 2008, the number of jobs for professional chefs will increase by 13.4%.

Source: *Handbook of U.S. Labor Statistics*

EXAMPLE 1 Write three kinds of notation for 35%.

Using ratio: $35\% = \frac{35}{100}$ A ratio of 35 to 100

Using fraction notation: $35\% = 35 \times \frac{1}{100}$ Replacing % with $\times \frac{1}{100}$

Using decimal notation: $35\% = 35 \times 0.01$ Replacing % with $\times$ 0.01

EXAMPLE 2 Write three kinds of notation for 67.8%.

Using ratio: $67.8\% = \frac{67.8}{100}$ A ratio of 67.8 to 100

Using fraction notation: $67.8\% = 67.8 \times \frac{1}{100}$ Replacing % with $\times \frac{1}{100}$

Using decimal notation: $67.8\% = 67.8 \times 0.01$ Replacing % with $\times$ 0.01

Do Exercises 1–3 on the preceding page.

b Converting Between Percent Notation and Decimal Notation

Consider 78%. To convert to decimal notation, we can think of percent notation as a ratio and write

$$78\% = \frac{78}{100} \quad \text{Using the definition of percent as a ratio}$$
$$= 0.78. \quad \text{Dividing}$$

Similarly,

$$4.9\% = \frac{4.9}{100} \quad \text{Using the definition of percent as a ratio}$$
$$= 0.049. \quad \text{Dividing}$$

We could also convert 78% to decimal notation by replacing "%" with "$\times$ 0.01" and write

$$78\% = 78 \times 0.01 \quad \text{Replacing \% with} \times 0.01$$
$$= 0.78. \quad \text{Multiplying}$$

CALCULATOR CORNER

Converting from Percent Notation to Decimal Notation Many calculators have a [%] key that can be used to convert from percent notation to decimal notation. This is often the second operation associated with a particular key and is accessed by first pressing a [2nd] or [SHIFT] key. To convert 57.6% to decimal notation, for example, you might press [5] [7] [.] [6] [2nd] [%] or [5] [7] [.] [6] [SHIFT] [%]. The display would read [0.576], so 57.6% = 0.576. Read the user's manual to determine whether your calculator can do this conversion.

Exercises: Use a calculator to find decimal notation.

1. 14%
2. 0.069%
3. 43.8%
4. 125%

Find decimal notation.

4. 34%

5. 78.9%

6. $6\frac{5}{8}\%$

Find decimal notation for the percent notation in the sentence.

7. Pickup Trucks. Of all household personal vehicles in the United States, 18% are pickup trucks.

Source: Department of Transportation, 2001 National Household Travel Survey

8. Blood Alcohol Level. A blood alcohol level of 0.08% is a standard used by some states as the legal limit for drunk driving.

Answers on page A-9

Similarly,

$$4.9\% = 4.9 \times 0.01 \quad \text{Replacing \% with } \times 0.01$$
$$= 0.049. \quad \text{Multiplying}$$

Dividing by 100 amounts to moving the decimal point two places to the left, which is the same as multiplying by 0.01. This leads us to a quick way to convert from percent notation to decimal notation—we drop the percent symbol and move the decimal point two places to the left.

To convert from percent notation to decimal notation,	36.5%	
a) replace the percent symbol % with × 0.01, and	36.5 × 0.01	
b) multiply by 0.01, which means move the decimal point two places to the left.	0.36.5 36.5% = 0.365	Move 2 places to the left.

EXAMPLE 3 Find decimal notation for 99.44%.

a) Replace the percent symbol with × 0.01. 99.44 × 0.01

b) Move the decimal point two places to the left. 0.99.44

Thus, 99.44% = 0.9944.

EXAMPLE 4 The interest rate on a $2\frac{1}{2}$-year certificate of deposit is $6\frac{3}{8}\%$. Find decimal notation for $6\frac{3}{8}\%$.

a) Convert $6\frac{3}{8}$ to decimal notation and replace the percent symbol with × 0.01. $6\frac{3}{8}\%$ 6.375 × 0.01

b) Move the decimal point two places to the left. 0.06.375

Thus, $6\frac{3}{8}\% = 0.06375$.

Do Exercises 4–8.

To convert 0.38 to percent notation, we can first write fraction notation, as follows:

$$0.38 = \frac{38}{100} \quad \text{Converting to fraction notation}$$
$$= 38\%. \quad \text{Using the definition of percent as a ratio}$$

Note that 100% = 100 × 0.01 = 1. Thus to convert 0.38 to percent notation, we can multiply by 1, using 100% as a symbol for 1.

$$0.38 = 0.38 \times 1$$
$$= 0.38 \times 100\%$$
$$= 0.38 \times 100 \times 0.01 \quad \text{Replacing 100\% with } 100 \times 0.01$$
$$= (0.38 \times 100) \times 0.01 \quad \text{Using the associative law of multiplication}$$

Then

$38 \times 0.01 = 38\%$. Replacing "$\times$ 0.01" with the % symbol

Even more quickly, since $0.38 = 0.38 \times 100\%$, we can simply multiply 0.38 by 100 and write the % symbol.

To convert from decimal notation to percent notation, we multiply by 100%—that is, we move the decimal point two places to the right and write a percent symbol.

To convert from decimal notation to percent notation, multiply by 100%. That is,	$0.675 = 0.675 \times 100\%$
a) move the decimal point two places to the right, and	0.67.5 Move 2 places to the right.
b) write a % symbol.	67.5%
	$0.675 = 67.5\%$

EXAMPLE 5 Find percent notation for 1.27.

a) Move the decimal point two places to the right. 1.27.

b) Write a % symbol. 127%

Thus, $1.27 = 127\%$.

EXAMPLE 6 Of the time that people declare as sick leave, 0.21 is actually used for family issues. Find percent notation for 0.21.

Source: CCH Inc.

a) Move the decimal point two places to the right. 0.21.

b) Write a % symbol. 21%

Thus, $0.21 = 21\%$.

EXAMPLE 7 Find percent notation for 5.6.

a) Move the decimal point two places to the right, adding an extra zero. 5.60.

b) Write a % symbol. 560%

Thus, $5.6 = 560\%$.

EXAMPLE 8 Of those who play golf, 0.149 play 8–24 rounds per year. Find percent notation for 0.149.

Source: U.S. Golf Association

a) Move the decimal point two places to the right. 0.14.9

b) Write a % symbol. 14.9%

Thus, $0.149 = 14.9\%$.

Do Exercises 9–13.

Find percent notation.

9. 0.24

10. 3.47

11. 1

Find percent notation for the decimal notation in the sentence.

12. High School Graduate. The highest level of education for 0.321 of persons 25 and older in the United States is high school graduation.

Source: U.S. Department of Commerce, Bureau of the Census, *Current Population Survey*

13. Golf. Of those who play golf, 0.253 play 25–49 rounds per year.

Source: U.S. Golf Association

Answers on page A-9

EXERCISE SET

For Extra Help

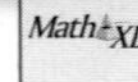
MathXL

MyMathLab

InterAct Math

Math Tutor Center

Video Lectures on CD Disc 2

Student's Solutions Manual

a Write three kinds of notation as in Examples 1 and 2 on p. 265.

1. 90%

2. 58.7%

3. 12.5%

4. 130%

b Find decimal notation.

5. 67%

6. 17%

7. 45.6%

8. 76.3%

9. 59.01%

10. 30.02%

11. 10%

12. 80%

13. 1%

14. 100%

15. 200%

16. 300%

17. 0.1%

18. 0.4%

19. 0.09%

20. 0.12%

21. 0.18%

22. 5.5%

23. 23.19%

24. 87.99%

25. $14\frac{7}{8}\%$

26. $93\frac{1}{8}\%$

27. $56\frac{1}{2}\%$

28. $61\frac{3}{4}\%$

Find decimal notation for the percent notation(s) in the sentence.

29. *Pediatricians.* By 2020, the population of children in the United States is expected to increase by about 9% while the number of pediatricians will leap 58%.

Source: Study by Dr. Scott Shipman, an Oregon Health and Science University pediatrician

30. *Cancer Survival.* In 2005, 78.9% of children with cancer survive at least 5 years compared to about 60% in 1975.

Source: National Cancer Institute

31. *Eating Out.* On a given day, 44% of all adults eat in a restaurant.

Source: *AARP Bulletin*, November 2004

32. Of those who play golf, 18.6% play 100 or more rounds per year.

Source: U.S. Golf Association

33. *Knitting and Crocheting.* Of all women ages 25 to 34, 36% know how to knit or crochet.

Source: Research Inc. for Craft Yarn Council of America

34. According to a recent survey, 95.1% of those asked to name what sports they participate in chose swimming.

Source: Sporting Goods Manufacturers

Find percent notation.

35. 0.47

36. 0.87

37. 0.03

38. 0.01

39. 8.7

40. 4

41. 0.334

42. 0.889

43. 0.75

44. 0.99

45. 0.4

46. 0.5

47. 0.006

48. 0.008

49. 0.017

50. 0.024

51. 0.2718

52. 0.8911

53. 0.0239

54. 0.00073

Find percent notation for the decimal notation(s) in the sentence.

55. *Hours of Sleep.* In 2005, only 0.26 of people got 8 or more hours of sleep a night on weekdays. This rate has declined from 0.38 in 2001.

Source: National Sleep Foundation's 2005 Sleep in America Poll

56. According to a recent survey, 0.526 of those asked to name what sports they participate in chose bowling.

Source: Sporting Goods Manufacturers

57. *Bachelor's Degrees.* For 0.177 of the United States population 25 and older, the bachelor's degree is the highest level of education attainment.

Source: U.S. Department of Commerce, Bureau of the Census, *Current Population Survey;* National Center for Educational Statistics, *Digest of Education Statistics,* 2003

58. About 0.69 of all newspapers are recycled.

Sources: American Forest and Paper Association; Newspaper Association of America

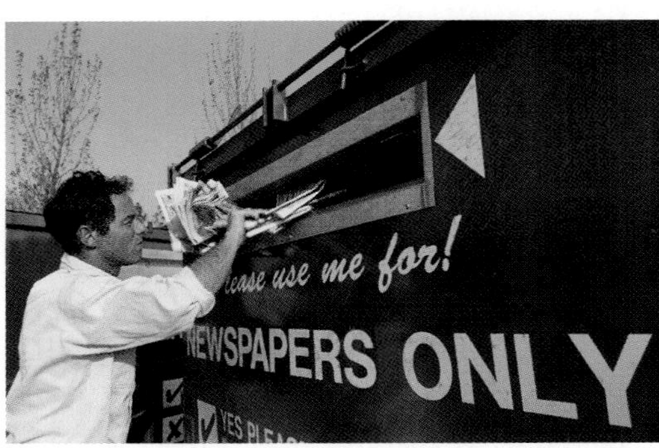

59. *65 and Older.* In Clearwater, Florida, 0.215 of the residents are 65 and older.

Source: U.S. Bureau of the Census

60. Of those living in North Carolina, 0.1134 will die of heart disease.

Source: American Association of Retired Persons

61. D_W *Winning Percentage.* During the 2005 regular baseball season, the Chicago White Sox won 99 of 162 games and went on to win the World Series. Find the ratio of number of wins to total number of games played in the regular season and convert it to decimal notation. Such a rate is often called a "winning percentage." Explain why.

62. D_W Athletes sometimes speak of "giving 110%" effort. Does this make sense? Explain.

SKILL MAINTENANCE

Convert to a mixed numeral. [2.4a]

63. $\frac{100}{3}$

64. $\frac{75}{2}$

65. $\frac{75}{8}$

66. $\frac{297}{16}$

67. $\frac{567}{98}$

68. $\frac{2345}{21}$

Convert to decimal notation. [3.5a]

69. $\frac{2}{3}$

70. $\frac{1}{3}$

71. $\frac{5}{6}$

72. $\frac{17}{12}$

73. $\frac{8}{3}$

74. $\frac{15}{16}$

4.3 PERCENT AND FRACTION NOTATION

Objectives

a Convert from fraction notation to percent notation.

b Convert from percent notation to fraction notation.

a Converting from Fraction Notation to Percent Notation

Consider the fraction notation $\frac{7}{8}$. To convert to percent notation, we use two skills we already have. We first find decimal notation by dividing:

$$\frac{7}{8} = 0.875 \qquad \begin{array}{r} 0.875 \\ 8\overline{)7.000} \\ \underline{64} \\ 60 \\ \underline{56} \\ 40 \\ \underline{40} \\ 0 \end{array}$$

Then we convert the decimal notation to percent notation. We move the decimal point two places to the right

$$0.87.5$$

and write a % symbol:

$$\frac{7}{8} = 87.5\%, \text{ or } 87\frac{1}{2}\%.$$

To convert from fraction notation to percent notation, $\frac{3}{5}$ Fraction notation

a) find decimal notation by division, and

$$\begin{array}{r} 0.6 \\ 5\overline{)3.0} \\ \underline{3\,0} \\ 0 \end{array}$$

b) convert the decimal notation to percent notation.

$0.6 = 0.60 = 60\%$ Percent notation

$\frac{3}{5} = 60\%$

EXAMPLE 1 Find percent notation for $\frac{9}{16}$.

a) We first find decimal notation by division.

$$\begin{array}{r} 0.5625 \\ 16\overline{)9.0000} \\ \underline{80} \\ 100 \\ \underline{96} \\ 40 \\ \underline{32} \\ 80 \\ \underline{80} \\ 0 \end{array} \qquad \frac{9}{16} = 0.5625$$

Study Tips

BEING A TUTOR

Try being a tutor for a fellow student. Understanding and retention of concepts can be maximized for yourself if you explain the material to someone else.

Find percent notation.

1. $\frac{1}{4}$

2. $\frac{5}{8}$

b) Next, we convert the decimal notation to percent notation. We move the decimal point two places to the right and write a % symbol.

0.56.25

$\frac{9}{16} = 56.25\%$, or $56\frac{1}{4}\%$

Don't forget the % symbol.

Do Exercises 1 and 2.

CALCULATOR CORNER

Converting from Fraction Notation to Percent Notation A calculator can be used to convert from fraction notation to percent notation. We simply perform the division on the calculator and then use the percent key. To convert $\frac{17}{40}$ to percent notation, for example, we press [1] [7] [÷] [4] [0] [2nd] [%], or [1] [7] [÷] [4] [0] [SHIFT] [%]. The display reads [42.5], so $\frac{17}{40} = 42.5\%$. Read the user's manual to determine whether your calculator can do this conversion.

Exercises: Use a calculator to find percent notation. Round to the nearest hundredth of a percent.

1. $\frac{13}{25}$
2. $\frac{5}{13}$
3. $\frac{43}{39}$
4. $\frac{12}{7}$
5. $\frac{217}{364}$
6. $\frac{2378}{8401}$

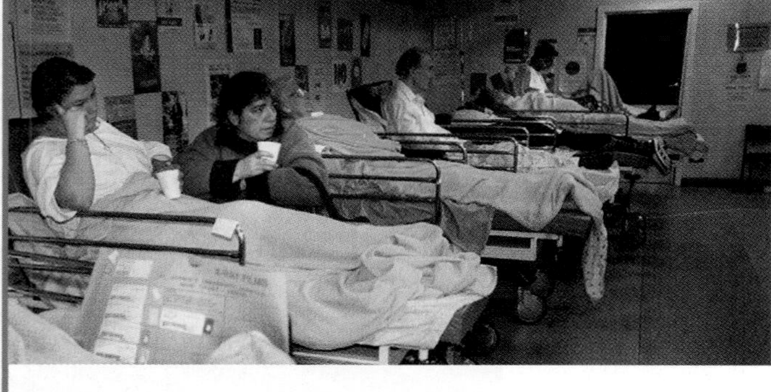

EXAMPLE 2 *Without Health Insurance.* Approximately $\frac{1}{6}$ of all people in the United States are without health insurance. Find percent notation for $\frac{1}{6}$.

Source: U.S. Bureau of the Census, *Current Population Survey*, March 2003

a) Find decimal notation by division.

$$\begin{array}{r} 0.1\,6\,6 \\ 6\,\overline{)\,1.0\,0\,0} \\ \underline{6} \\ 4\,0 \\ \underline{3\,6} \\ 4\,0 \\ \underline{3\,6} \\ 4 \end{array}$$

We get a repeating decimal: $0.16\overline{6}$.

b) Convert the answer to percent notation.

$0.16.\overline{6}$

$\frac{1}{6} = 16.\overline{6}\%$, or $16\frac{2}{3}\%$

Answers on page A-9

Do Exercises 3 and 4.

In some cases, division is not the fastest way to convert. The following are some optional ways in which conversion might be done.

EXAMPLE 3 Find percent notation for $\frac{69}{100}$.

We use the definition of percent as a ratio.

$$\frac{69}{100} = 69\%$$

EXAMPLE 4 Find percent notation for $\frac{17}{20}$.

We multiply by 1 to get 100 in the denominator. We think of what we have to multiply 20 by in order to get 100. That number is 5, so we multiply by 1 using $\frac{5}{5}$.

$$\frac{17}{20} \cdot \frac{5}{5} = \frac{85}{100} = 85\%$$

Note that this shortcut works only when the denominator is a factor of 100.

Do Exercises 5 and 6.

b Converting from Percent Notation to Fraction Notation

To convert from percent notation to fraction notation,	30%	Percent notation
a) use the definition of percent as a ratio, and	$\frac{30}{100}$	
b) simplify, if possible.	$\frac{3}{10}$	Fraction notation

EXAMPLE 5 Find fraction notation for 75%.

$$75\% = \frac{75}{100} \quad \text{Using the definition of percent}$$

$$= \frac{3 \cdot 25}{4 \cdot 25} = \frac{3}{4} \cdot \frac{25}{25} \quad \text{Simplifying}$$

$$= \frac{3}{4}$$

3. Water is the single most abundant chemical in the body. The human body is about $\frac{2}{3}$ water. Find percent notation for $\frac{2}{3}$.

4. Find percent notation: $\frac{5}{6}$.

Find percent notation.

5. $\frac{57}{100}$

6. $\frac{19}{25}$

Answers on page A-9

Find fraction notation.

7. 60%

8. 3.25%

9. $66\frac{2}{3}\%$

10. Complete this table.

FRACTION NOTATION	$\frac{1}{5}$		
DECIMAL NOTATION		$0.83\overline{3}$	
PERCENT NOTATION			$37\frac{1}{2}\%$

Answers on page A-9

Study Tips

MEMORIZING

Memorizing is a very helpful tool in the study of mathematics. Don't underestimate its power as you memorize the table of decimal, fraction, and percent notation on the inside back cover. We will discuss memorizing more later.

EXAMPLE 6 Find fraction notation for 62.5%.

$$62.5\% = \frac{62.5}{100} \quad \text{Using the definition of percent}$$

$$= \frac{62.5}{100} \times \frac{10}{10} \quad \text{Multiplying by 1 to eliminate the decimal point in the numerator}$$

$$= \frac{625}{1000}$$

$$\left.\begin{aligned} &= \frac{5 \cdot 125}{8 \cdot 125} = \frac{5}{8} \cdot \frac{125}{125} \\ &= \frac{5}{8} \end{aligned}\right\} \quad \text{Simplifying}$$

EXAMPLE 7 Find fraction notation for $16\frac{2}{3}\%$.

$$16\frac{2}{3}\% = \frac{50}{3}\% \quad \text{Converting from the mixed numeral to fraction notation}$$

$$= \frac{50}{3} \times \frac{1}{100} \quad \text{Using the definition of percent}$$

$$\left.\begin{aligned} &= \frac{50 \cdot 1}{3 \cdot 50 \cdot 2} = \frac{1}{6} \cdot \frac{50}{50} \\ &= \frac{1}{6} \end{aligned}\right\} \quad \text{Simplifying}$$

The table on the inside back cover lists decimal, fraction, and percent equivalents used so often that it would speed up your work if you memorized them. For example, $\frac{1}{3} = 0.\overline{3}$, so we say that the **decimal equivalent** of $\frac{1}{3}$ is $0.\overline{3}$, or that $0.\overline{3}$ has the **fraction equivalent** $\frac{1}{3}$.

EXAMPLE 8 Find fraction notation for $16.\overline{6}\%$.

We can use the table on the inside back cover or recall that $16.\overline{6}\% = 16\frac{2}{3}\% = \frac{1}{6}$. We can also recall from our work with repeating decimals in Chapter 3 that $0.\overline{6} = \frac{2}{3}$. Then we have $16.\overline{6}\% = 16\frac{2}{3}\%$ and can proceed as in Example 7.

Do Exercises 7–10.

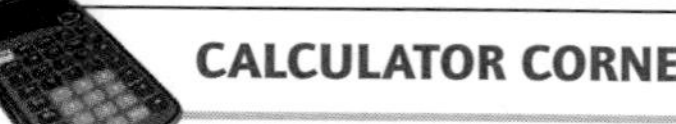

CALCULATOR CORNER

Applications of Ratio and Percent: The Price–Earnings Ratio and Stock Yields

The Price–Earnings Ratio If the total earnings of a company one year were \$5,000,000 and 100,000 shares of stock were issued, the earnings per share was \$50. At one time, the price per share of Coca-Cola was \$58.125 and the earnings per share was \$0.76. The **price-earnings ratio,** *P/E,* is the price of the stock divided by the earnings per share. For the Coca-Cola stock, the price-earnings ratio, *P/E,* is given by

$$\frac{P}{E} = \frac{58.125}{0.76} \approx 76.48.$$ Dividing, using a calculator, and rounding to the nearest hundredth

Stock Yields At one time, the price per share of Coca-Cola stock was \$58.125 and the company was paying a yearly dividend of \$0.68 per share. It is helpful to those interested in stocks to know what percent the dividend is of the price of the stock. The percent is called the **yield.** For the Coca-Cola stock, the yield is given by

$$\text{Yield} = \frac{\text{Dividend}}{\text{Price per share}} = \frac{0.68}{58.125} \approx 0.0117$$ Dividing and rounding to the nearest ten-thousandth

$$= 1.17\%$$ Converting to percent notation

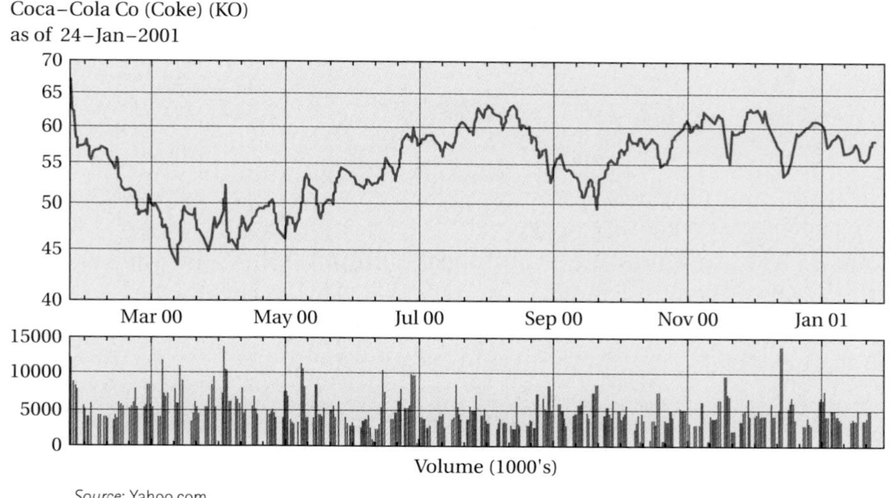

Source: Yahoo.com

Exercises: Compute the price–earnings ratio and the yield for each stock listed below.

	STOCK	PRICE PER SHARE	EARNINGS	DIVIDEND	P/E	YIELD
1.	Pepsi (PEP)	\$42.75	\$1.40	\$0.56		
2.	Pearson (PSO)	\$25.00	\$0.78	\$0.30		
3.	Quaker Oats (OAT)	\$92.375	\$2.68	\$1.10		
4.	Texas Insts (TEX)	\$42.875	\$1.62	\$0.43		
5.	Ford Motor Co (F)	\$27.5625	\$2.30	\$1.19		
6.	Wendy's Intl (WEN)	\$25.75	\$1.47	\$0.23		

4.3 EXERCISE SET

For Extra Help

MathXL | MyMathLab | InterAct Math | Math Tutor Center | Video Lectures on CD Disc 2 | Student's Solutions Manual

a Find percent notation.

1. $\frac{41}{100}$
2. $\frac{36}{100}$
3. $\frac{5}{100}$
4. $\frac{1}{100}$
5. $\frac{2}{10}$
6. $\frac{7}{10}$
7. $\frac{3}{10}$
8. $\frac{9}{10}$
9. $\frac{1}{2}$
10. $\frac{3}{4}$
11. $\frac{7}{8}$
12. $\frac{1}{8}$
13. $\frac{4}{5}$
14. $\frac{2}{5}$
15. $\frac{2}{3}$
16. $\frac{1}{3}$
17. $\frac{1}{6}$
18. $\frac{5}{6}$
19. $\frac{3}{16}$
20. $\frac{11}{16}$
21. $\frac{13}{16}$
22. $\frac{7}{16}$
23. $\frac{4}{25}$
24. $\frac{17}{25}$
25. $\frac{1}{20}$
26. $\frac{31}{50}$
27. $\frac{17}{50}$
28. $\frac{3}{20}$

Find percent notation for the fraction notation in the sentence.

29. *Driving While Drowsy.* Almost half of U.S. drivers say they have driven while drowsy. Of this group, $\frac{2}{5}$ fight off sleep by opening a window, and $\frac{9}{50}$ do so by drinking a caffeinated beverage.

Source: Harris Interactive for Tylenol PM

30. *Paved Roads.* About $\frac{13}{20}$ of the roads and streets in the United States are paved.

Source: U.S. Department of Transportation, *Highway Statistics*

In Exercises 31–36, write percent notation for the fractions in this pie chart.

Time Workers Spend Sorting Unsolicited e-mail and Spam

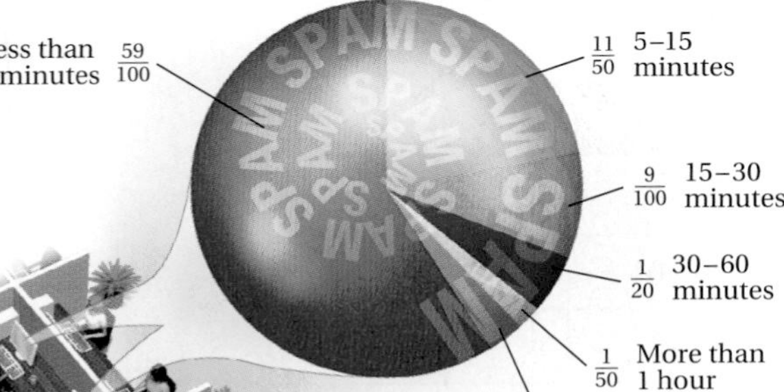

Source: Data from InsightExpress

31. $\frac{11}{50}$
32. $\frac{3}{100}$
33. $\frac{1}{20}$
34. $\frac{59}{100}$
35. $\frac{9}{100}$
36. $\frac{1}{50}$

b Find fraction notation. Simplify.

37. 85%

38. 55%

39. 62.5%

40. 12.5%

41. $33\frac{1}{3}\%$

42. $83\frac{1}{3}\%$

43. $16.\overline{6}\%$

44. $66.\overline{6}\%$

45. 7.25%

46. 4.85%

47. 0.8%

48. 0.2%

49. $25\frac{3}{8}\%$

50. $48\frac{7}{8}\%$

51. $78\frac{2}{9}\%$

52. $16\frac{5}{9}\%$

53. $64\frac{7}{11}\%$

54. $73\frac{3}{11}\%$

55. 150%

56. 110%

57. 0.0325%

58. 0.419%

59. $33.\overline{3}\%$

60. $83.\overline{3}\%$

Find fraction notation for the percent notation in the following bar graph.

Organ Transplants in the United States, 2003

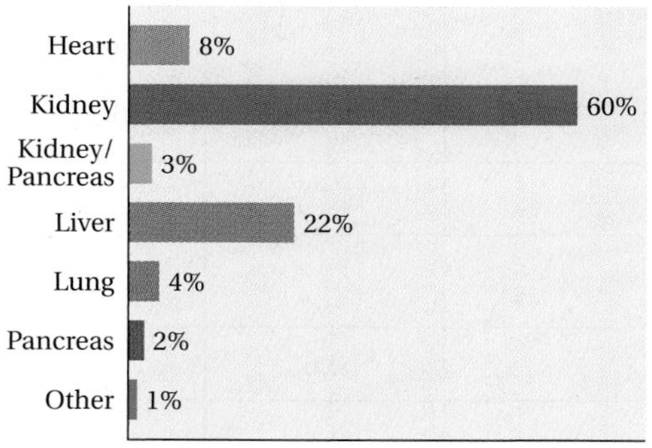

Heart	8%
Kidney	60%
Kidney/Pancreas	3%
Liver	22%
Lung	4%
Pancreas	2%
Other	1%

Source: National Organ Procurement and Transplantation Network

61. 8%

62. 22%

63. 60%

64. 4%

65. 2%

66. 3%

Find fraction notation for the percent notation in the sentence.

67. A 1.9-oz serving of Raisin Bran Crunch® cereal with $\frac{1}{2}$ cup of skim milk satisfies 35% of the minimum daily requirements for Vitamin B_{12}.

Source: Kellogg's USA, Inc.

68. A 1-cup serving of Wheaties® cereal with $\frac{1}{2}$ cup of skim milk satisfies 15% of the minimum daily requirements for calcium.

Source: General Mills Sales, Inc.

69. Of all those who are 85 and older, 47% have Alzheimer's disease.

Source: Alzheimer's Association

70. In 2003, 24.4% of Americans 18 and older smoked cigarettes.

Source: U.S. Centers for Disease Control and Prevention

Complete the table.

71.

FRACTION NOTATION	DECIMAL NOTATION	PERCENT NOTATION
$\frac{1}{8}$		12.5%, or $12\frac{1}{2}\%$
$\frac{1}{6}$		
		20%
	0.25	
		$33.\overline{3}\%$, or $33\frac{1}{3}\%$
		37.5%, or $37\frac{1}{2}\%$
		40%
$\frac{1}{2}$		

72.

FRACTION NOTATION	DECIMAL NOTATION	PERCENT NOTATION
$\frac{3}{5}$		
	0.625	
$\frac{2}{3}$		
	0.75	75%
$\frac{4}{5}$		
$\frac{5}{6}$		$83.\overline{3}\%$, or $83\frac{1}{3}\%$
$\frac{7}{8}$		87.5%, or $87\frac{1}{2}\%$
		100%

73.

FRACTION NOTATION	DECIMAL NOTATION	PERCENT NOTATION
	0.5	
$\frac{1}{3}$		
		25%
		$16.\overline{6}\%$, or $16\frac{2}{3}\%$
	0.125	
$\frac{3}{4}$		
	$0.8\overline{3}$	
$\frac{3}{8}$		

74.

FRACTION NOTATION	DECIMAL NOTATION	PERCENT NOTATION
		40%
		62.5%, or $62\frac{1}{2}\%$
	0.875	
$\frac{1}{1}$		
	0.6	
	$0.\overline{6}$	
$\frac{1}{5}$		

75. $\mathbf{D_W}$ What do the following have in common? Explain.

$\frac{23}{16}$, $1\frac{875}{2000}$, 1.4375, $\frac{207}{144}$, $1\frac{7}{16}$, 143.75%, $1\frac{4375}{10,000}$

76. $\mathbf{D_W}$ Is it always best to convert from fraction notation to percent notation by first finding decimal notation? Why or why not?

SKILL MAINTENANCE

Solve.

77. $13 \cdot x = 910$ [1.4b]

78. $15 \cdot y = 75$ [1.4b]

79. $0.05 \times b = 20$ [3.4b]

80. $3 = 0.16 \times b$ [3.4b]

81. $\frac{24}{37} = \frac{15}{x}$ [4.1d]

82. $\frac{17}{18} = \frac{x}{27}$ [4.1d]

83. $\frac{9}{10} = \frac{x}{5}$ [4.1d]

84. $\frac{7}{x} = \frac{4}{5}$ [4.1d]

Convert to a mixed numeral. [2.4a]

85. $\frac{100}{3}$

86. $\frac{75}{2}$

87. $\frac{250}{3}$

88. $\frac{123}{6}$

89. $\frac{345}{8}$

90. $\frac{373}{6}$

91. $\frac{75}{4}$

92. $\frac{67}{9}$

Convert from a mixed numeral to fraction notation. [2.4a]

93. $1\frac{1}{17}$

94. $20\frac{9}{10}$

95. $101\frac{1}{2}$

96. $32\frac{3}{8}$

SYNTHESIS

Write percent notation.

97. $\frac{41}{369}$

98. $\frac{54}{999}$

99. $2.5\overline{74631}$

100. $3.2\overline{93847}$

Write decimal notation.

101. $\frac{14}{9}\%$

102. $\frac{19}{12}\%$

103. $\frac{729}{7}\%$

104. $\frac{637}{6}\%$

4.4 SOLVING PERCENT PROBLEMS USING PERCENT EQUATIONS

Objectives

a Translate percent problems to percent equations.

b Solve basic percent problems.

a Translating to Equations

To solve a problem involving percents, it is helpful to translate first to an equation. To distinguish the method in Section 4.4 from that of Section 4.5, we will call these *percent equations*.

EXAMPLE 1 Translate:

23% of 5 is what?

$23\% \cdot 5 = a$ This is a *percent equation.*

KEY WORDS IN PERCENT TRANSLATIONS

"**Of**" translates to "·", or "×".
"**What**" translates to any letter.
"**Is**" translates to "=".
"**%**" translates to "$\times \frac{1}{100}$" or "× 0.01".

EXAMPLE 2 Translate:

What is 11% of 49?

$a = 11\% \cdot 49$ Any letter can be used.

Do Exercises 1 and 2.

EXAMPLE 3 Translate:

3 is 10% of what?

$3 = 10\% \cdot b$

EXAMPLE 4 Translate:

45% of what is 23?

$45\% \times b = 23$

Do Exercises 3 and 4.

EXAMPLE 5 Translate:

10 is what percent of 20?

$10 = p \times 20$

Translate to an equation. Do not solve.

1. 12% of 50 is what?

2. What is 40% of 60?

Translate to an equation. Do not solve.

3. 45 is 20% of what?

4. 120% of what is 60?

Answers on page A-10

EXAMPLE 6 Translate:

What percent	of	50	is	7?
↓	↓	↓	↓	↓
p	$\cdot$	50	=	7

Do Exercises 5 and 6.

Translate to an equation. Do not solve.

5. 16 is what percent of 40?

6. What percent of 84 is 10.5?

b Solving Percent Problems

In solving percent problems, we use the *Translate* and *Solve* steps in the problem-solving strategy used throughout this text.

Percent problems are actually of three different types. Although the method we present does *not* require that you be able to identify which type you are solving, it is helpful to know them.

We know that

15 is 25% of 60, or $15 = 25\% \times 60$.

We can think of this as:

Amount = Percent number × Base.

Each of the three types of percent problems depends on which of the three pieces of information is missing.

1. **Finding the *amount* (the result of taking the percent)**

Example:	**What**	is	25%	of	60?
	↓	↓	↓	↓	↓
Translation:	a	=	25%	$\cdot$	60

2. **Finding the *base* (the number you are taking the percent of)**

Example:	15	is	25%	of	**what number?**
	↓	↓	↓	↓	↓
Translation:	15	=	25%	$\cdot$	b

3. **Finding the *percent number* (the percent itself)**

Example:	15	is	**what percent**	of	60?
	↓	↓	↓	↓	↓
Translation:	15	=	p	$\cdot$	60

7. Solve:

What is 12% of 50?

Answers on page A-10

FINDING THE AMOUNT

EXAMPLE 7 What is 4.6% of 105,000,000?

Translate: $a = 4.6\% \times 105{,}000{,}000$.

Solve: The letter is by itself. To solve the equation, we just convert 4.6% to decimal notation and multiply:

$$a = 4.6\% \times 105{,}000{,}000$$
$$= 0.046 \times 105{,}000{,}000 = 4{,}830{,}000.$$

Thus, 4,830,000 is 4.6% of 105,000,000. The answer is 4,830,000.

Do Exercise 7.

There are 105,000,000 households in the United States and approximately 4.6% own a pet bird. How many households own at least one pet bird?

8. Solve:

64% of $55 is what?

In a survey of a group of people, it was found that 5%, or 20 people, chose strawberry as their favorite ice cream. How many people were surveyed?

Source: International Ice Cream Association

Solve.

9. 20% of what is 45?

10. $60 is 120% of what?

Answers on page A-10

EXAMPLE 8 120% of $42 is what?

Translate: $120\% \times 42 = a$.

Solve: The letter is by itself. To solve the equation, we carry out the calculation:

$$
\begin{aligned}
a &= 120\% \times 42 \\
&= 1.2 \times 42 \qquad 120\% = 1.2 \\
&= 50.4.
\end{aligned}
$$

Thus, 120% of $42 is $50.40. The answer is $50.40.

Do Exercise 8.

FINDING THE BASE

EXAMPLE 9 5% of what is 20?

Translate: $5\% \times b = 20$.

Solve: This time the letter is *not* by itself. To solve the equation, we divide by 5% on both sides:

$$
\begin{aligned}
\frac{5\% \times b}{5\%} &= \frac{20}{5\%} \qquad \text{Dividing by 5\% on both sides} \\
b &= \frac{20}{0.05} \qquad 5\% = 0.05 \\
b &= 400.
\end{aligned}
$$

Thus, 5% of 400 is 20. The answer is 400.

EXAMPLE 10 $3 is 16% of what?

Translate: $3 is 16% of what?

$$
\begin{array}{ccccc}
\$3 & \text{is} & 16\% & \text{of} & \text{what?} \\
\downarrow & \downarrow & \downarrow & \downarrow & \downarrow \\
3 & = & 16\% & \times & b.
\end{array}
$$

Solve: Again, the letter is *not* by itself. To solve the equation, we divide by 16% on both sides:

$$
\begin{aligned}
\frac{3}{16\%} &= \frac{16\% \times b}{16\%} \qquad \text{Dividing by 16\% on both sides} \\
\frac{3}{0.16} &= b \qquad 16\% = 0.16 \\
18.75 &= b.
\end{aligned}
$$

Thus, $3 is 16% of $18.75. The answer is $18.75.

Do Exercises 9 and 10.

FINDING THE PERCENT NUMBER

In solving these problems, you *must* remember to convert to percent notation after you have solved the equation.

EXAMPLE 11 13 is what percent of 260?

Translate: 13 is what percent of 260?

$$13 = p \times 260.$$

Solve: To solve the equation, we divide by 260 on both sides and convert the result to percent notation:

$p \cdot 260 = 13$

$\frac{p \cdot 260}{260} = \frac{13}{260}$ Dividing by 260 on both sides

$p = 0.05$ Converting to decimal notation

$p = 5\%.$ Converting to percent notation

Thus, 13 is 5% of 260. The answer is 5%.

Do Exercise 11.

EXAMPLE 12 What percent of \$50 is \$16?

Translate: What percent of \$50 is \$16?

$$p \times 50 = 16.$$

Solve: To solve the equation, we divide by 50 on both sides and convert the answer to percent notation:

$\frac{p \times 50}{50} = \frac{16}{50}$ Dividing by 50 on both sides

$p = \frac{16}{50}$

$p = 0.32$

$p = 32\%.$ Converting to percent notation

Thus, 32% of \$50 is \$16. The answer is 32%.

Do Exercise 12.

Caution!

When a question asks "what percent?", be sure to give the answer in percent notation.

The average worker in the United States gets 13 vacation days per year. What percent of the 260 work days per year are vacation days?

11. Solve:

16 is what percent of 40?

12. Solve:

What percent of \$84 is \$10.50?

Answers on page A-10

CALCULATOR CORNER

Using Percents in Computations Many calculators have a [%] key that can be used in computations. (See the Calculator Corner on page 265.) For example, to find 11% of 49, we press [1] [1] [2nd] [%] [×] [4] [9] [=], or [4] [9] [×] [1] [1] [SHIFT] [%]. The display reads [5.39], so 11% of 49 is 5.39.

In Example 9, we perform the computation 20/5%. To use the [%] key in this computation, we press [2] [0] [÷] [5] [2nd] [%] [=], or [2] [0] [÷] [5] [SHIFT] [%]. The result is 400.

We can also use the [%] key to find the percent number in a problem. In Example 11, for instance, we answer the question "13 is what percent of 260?" On a calculator, we press [1] [3] [÷] [2] [6] [0] [2nd] [%] [=], or [1] [3] [÷] [2] [6] [0] [SHIFT] [%]. The result is 5, so 13 is 5% of 260.

Exercises: Use a calculator to find each of the following.

1. What is 5% of 24?
2. What is 12.6% of $40?
3. What is 19% of 256?
4. 140% of $16 is what?
5. 0.04% of 28 is what?
6. 33% of $90 is what?
7. 8% of what is 36?
8. $45 is 4.5% of what?
9. 23 is what percent of 920?
10. What percent of $442 is $53.04?

Study Tips

TIME MANAGEMENT (PART 2)

Here are some additional tips to help you with time management. (See also the Study Tips on time management in Sections 1.3 and 6.8.)

- **Avoid "time killers."** We live in a media age, and the Internet, e-mail, television, and movies all are time killers. Allow yourself a break to enjoy some college and outside activities. But keep track of the time you spend on such activities and compare it to the time you spend studying.
- **Prioritize your tasks.** Be careful about taking on too many college activities that fall outside of academics. Examples of such activities are decorating a homecoming float, joining a fraternity or sorority, and participating on a student council committee. Any of these is important but keep your involvement to a minimum to be sure that you have enough time for your studies.
- **Be aggressive about your study tasks.** Instead of worrying over your math homework or test preparation, do something to get yourself started. Work a problem here and a problem there, and before long you will accomplish the task at hand. If the task is large, break it down into smaller parts, and do one at a time. You will be surprised at how quickly the large task can then be completed.

4.4 EXERCISE SET

For Extra Help

MathXL | MyMathLab | InterAct Math | Math Tutor Center | Video Lectures on CD Disc 2 | Student's Solutions Manual

a Translate to an equation. Do not solve.

1. What is 32% of 78?

2. 98% of 57 is what?

3. 89 is what percent of 99?

4. What percent of 25 is 8?

5. 13 is 25% of what?

6. 21.4% of what is 20?

b Translate to an equation and solve.

7. What is 85% of 276?

8. What is 74% of 53?

9. 150% of 30 is what?

10. 100% of 13 is what?

11. What is 6% of $300?

12. What is 4% of $45?

13. 3.8% of 50 is what?

14. $33\frac{1}{3}$% of 480 is what? (*Hint*: $33\frac{1}{3}\% = \frac{1}{3}$.)

15. $39 is what percent of $50?

16. $16 is what percent of $90?

17. 20 is what percent of 10?

18. 60 is what percent of 20?

19. What percent of $300 is $150?

20. What percent of $50 is $40?

21. What percent of 80 is 100?

22. What percent of 60 is 15?

23. 20 is 50% of what?

24. 57 is 20% of what?

25. 40% of what is \$16?

26. 100% of what is \$74?

27. 56.32 is 64% of what?

28. 71.04 is 96% of what?

29. 70% of what is 14?

30. 70% of what is 35?

31. What is $62\frac{1}{2}\%$ of 10?

32. What is $35\frac{1}{4}\%$ of 1200?

33. What is 8.3% of \$10,200?

34. What is 9.2% of \$5600?

35. **D**W Write a question that could be translated to the equation

$$25 = 4\% \times b.$$

36. **D**W Suppose we know that 40% of 92 is 36.8. What is a quick way to find 4% of 92? 400% of 92? Explain.

SKILL MAINTENANCE

Write fraction notation. [3.1b]

37. 0.09

38. 1.79

39. 0.875

40. 0.125

41. 0.9375

42. 0.6875

Write decimal notation. [3.1b]

43. $\frac{89}{100}$

44. $\frac{7}{100}$

45. $\frac{3}{10}$

46. $\frac{17}{1000}$

SYNTHESIS

Solve.

47. What is 7.75% of \$10,880?
Estimate ______________
Calculate ______________

48. 50,951.775 is what percent of 78,995?
Estimate ______________
Calculate ______________

49. \$2496 is 24% of what amount?
Estimate ______________
Calculate ______________

50. What is 38.2% of \$52,345.79?
Estimate ______________
Calculate ______________

51. 40% of $18\frac{3}{4}\%$ of \$25,000 is what?

4.5 SOLVING PERCENT PROBLEMS USING PROPORTIONS*

Objectives

a Translate percent problems to proportions.

b Solve basic percent problems.

a Translating to Proportions

A percent is a ratio of some number to 100. For example, 47% is the ratio $\frac{47}{100}$. The numbers 68,859,700 and 146,510,000 have the same ratio as 47 and 100.

$$\frac{47}{100} = \frac{68,859,700}{146,510,000}$$

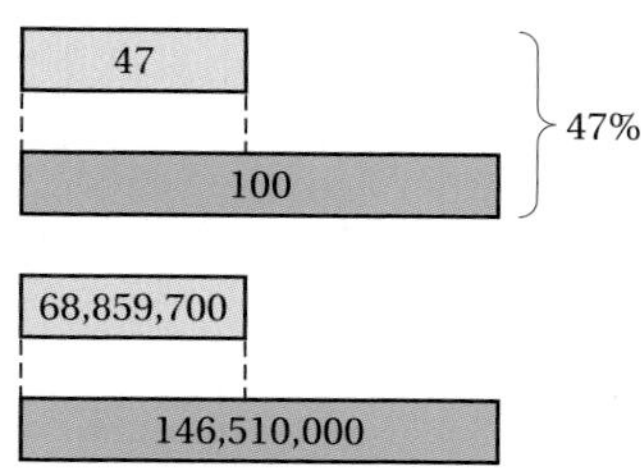

In the United States, 47% of the labor force are women. In 2003, there were 146,510,000 people in the labor force. This means that 68,859,700 were women.

Sources: U.S. Department of Labor, Bureau of Labor Statistics

To solve a percent problem using a proportion, we translate as follows:

$$\begin{array}{l} \text{Number} \rightarrow \\ 100 \longrightarrow \end{array} \frac{N}{100} = \frac{a}{b} \begin{array}{l} \leftarrow \text{Amount} \\ \leftarrow \text{Base} \end{array}$$

You might find it helpful to read this as "part is to whole as part is to whole."

For example, 60% of 25 is 15 translates to

$$\frac{60}{100} = \frac{15}{25}. \begin{array}{l} \leftarrow \text{Amount} \\ \leftarrow \text{Base} \end{array}$$

A clue in translating is that the base, b, corresponds to 100 and usually follows the wording "percent of." Also, $N\%$ always translates to $N/100$. Another aid in translating is to make a comparison drawing. To do this, we start with the percent side and list 0% at the top and 100% near the bottom. Then we estimate where the specified percent—in this case, 60%—is located. The corresponding quantities are then filled in. The base—in this case, 25—always corresponds to 100% and the amount—in this case, 15—corresponds to the specified percent.

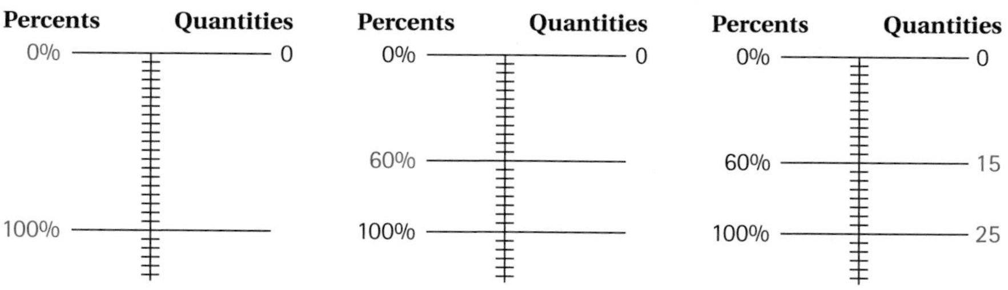

The proportion can then be read easily from the drawing: $\frac{60}{100} = \frac{15}{25}$.

*Note: This section presents an alternative method for solving basic percent problems. You can use either equations or proportions to solve percent problems, but you might prefer one method over the other, or your instructor may direct you to use one method over the other.

Translate to a proportion. Do not solve.

1. 12% of 50 is what?

2. What is 40% of 60?

3. 130% of 72 is what?

Translate to a proportion. Do not solve.

4. 45 is 20% of what?

5. 120% of what is 60?

Answers on page A-10

EXAMPLE 1 Translate to a proportion.

23% of 5 is what?

$$\frac{23}{100} = \frac{a}{5}$$

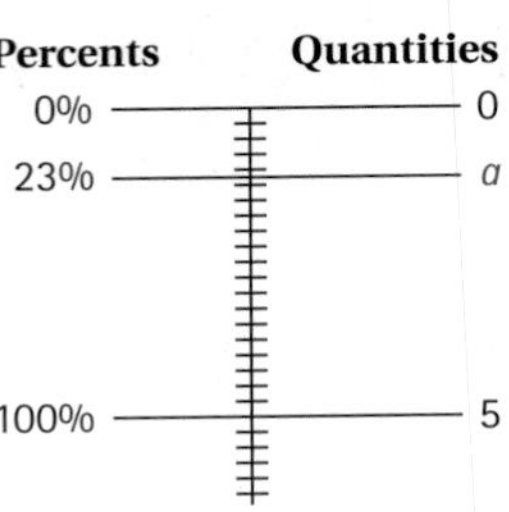

EXAMPLE 2 Translate to a proportion.

What is 124% of 49?

$$\frac{124}{100} = \frac{a}{49}$$

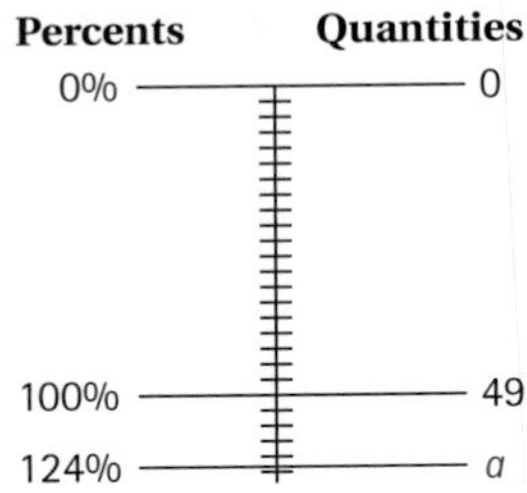

Do Exercises 1–3.

EXAMPLE 3 Translate to a proportion.

3 is 10% of what?

$$\frac{10}{100} = \frac{3}{b}$$

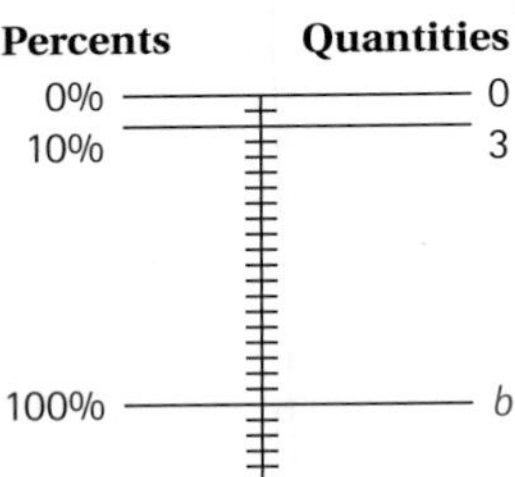

EXAMPLE 4 Translate to a proportion.

45% of what is 23?

$$\frac{45}{100} = \frac{23}{b}$$

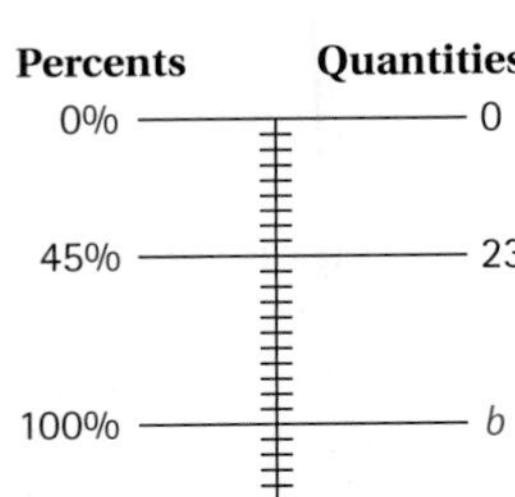

Do Exercises 4 and 5.

EXAMPLE 5 Translate to a proportion.

10 is what percent of 20?

$$\frac{N}{100} = \frac{10}{20}$$

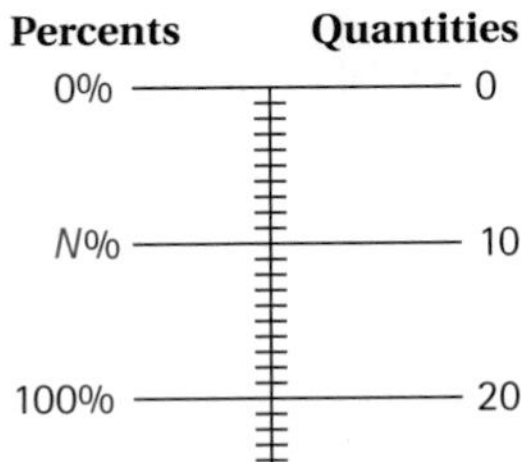

EXAMPLE 6 Translate to a proportion.

What percent of 50 is 7?

$$\frac{N}{100} = \frac{7}{50}$$

Percents Quantities
0% 0
N% 7
100% 50

Do Exercises 6 and 7.

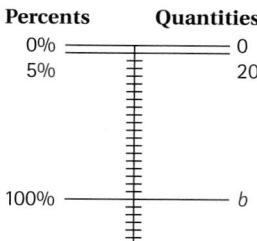

Translate to a proportion. Do not solve.

6. 16 is what percent of 40?

7. What percent of 84 is 10.5?

b Solving Percent Problems

After a percent problem has been translated to a proportion, we solve as in Section 4.1.

EXAMPLE 7 5% of what is $20?

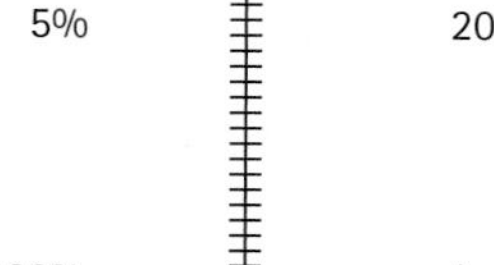

Translate: $\frac{5}{100} = \frac{20}{b}$

Solve: $5 \cdot b = 100 \cdot 20$ Equating cross products

$\frac{5 \cdot b}{5} = \frac{100 \cdot 20}{5}$ Dividing by 5

$b = \frac{2000}{5}$

$b = 400$ Simplifying

Thus, 5% of $400 is $20. The answer is $400.

Do Exercise 8.

8. Solve:

20% of what is $45?

EXAMPLE 8 120% of 42 is what?

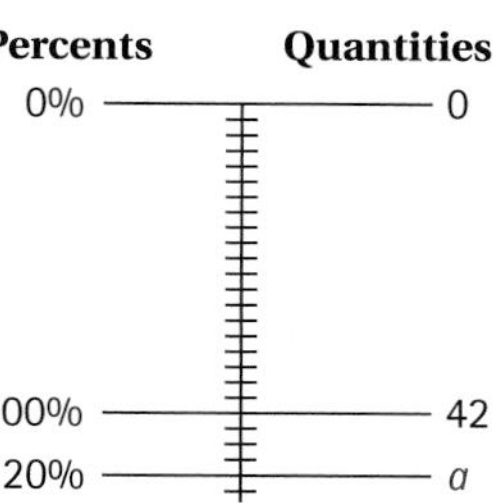

Translate: $\frac{120}{100} = \frac{a}{42}$

Solve: $120 \cdot 42 = 100 \cdot a$ Equating cross products

$\frac{120 \cdot 42}{100} = \frac{100 \cdot a}{100}$ Dividing by 100

$\frac{5040}{100} = a$

$50.4 = a$ Simplifying

Thus, 120% of 42 is 50.4. The answer is 50.4.

Do Exercises 9 and 10.

Solve.

9. 64% of 55 is what?

10. What is 12% of 50?

Answers on page A-10

11. Solve:

60 is 120% of what?

12. Solve:

$12 is what percent of $40?

13. Solve:

What percent of 84 is 10.5?

Answers on page A-10

Study Tips

RECORDING YOUR LECTURES

Consider recording your lectures and playing them back when convenient, say, while commuting to campus. (Be sure to get permission from your instructor before doing so, however.) Important points can be emphasized verbally.

EXAMPLE 9 3 is 16% of what?

Translate: $\frac{3}{b} = \frac{16}{100}$

Solve: $3 \cdot 100 = b \cdot 16$ Equating cross products

$\frac{3 \cdot 100}{16} = \frac{b \cdot 16}{16}$ Dividing by 16

$\frac{300}{16} = b$ Multiplying and simplifying

$18.75 = b$ Dividing

Thus, 3 is 16% of 18.75. The answer is 18.75.

Percents	Quantities
0%	0
16%	3
100%	b

Do Exercise 11.

EXAMPLE 10 $10 is what percent of $20?

Translate: $\frac{10}{20} = \frac{N}{100}$

Solve: $10 \cdot 100 = 20 \cdot N$ Equating cross products

$\frac{10 \cdot 100}{20} = \frac{20 \cdot N}{20}$ Dividing by 20

$\frac{1000}{20} = N$ Multiplying and simplifying

$50 = N$ Dividing

Thus, $10 is 50% of $20. The answer is 50%.

Percents	Quantities
0%	0
N%	$10
100%	$20

Do Exercise 12.

EXAMPLE 11 What percent of 50 is 16?

Translate: $\frac{N}{100} = \frac{16}{50}$

Solve: $50 \cdot N = 100 \cdot 16$ Equating cross products

$\frac{50 \cdot N}{50} = \frac{100 \cdot 16}{50}$ Dividing by 50

$N = \frac{1600}{50}$ Multiplying and simplifying

$N = 32$ Dividing

Thus, 32% of 50 is 16. The answer is 32%.

Percents	Quantities
0%	0
N%	16
100%	50

Do Exercise 13.

4.5 EXERCISE SET

For Extra Help

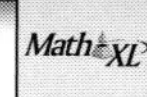
MathXL

MyMathLab

InterAct Math

Math Tutor Center
Video Lectures on CD Disc 2

Student's Solutions Manual

a Translate to a proportion. Do not solve.

1. What is 37% of 74?

2. 66% of 74 is what?

3. 4.3 is what percent of 5.9?

4. What percent of 6.8 is 5.3?

5. 14 is 25% of what?

6. 133% of what is 40?

b Translate to a proportion and solve.

7. What is 76% of 90?

8. What is 32% of 70?

9. 70% of 660 is what?

10. 80% of 920 is what?

11. What is 4% of 1000?

12. What is 6% of 2000?

13. 4.8% of 60 is what?

14. 63.1% of 80 is what?

15. $24 is what percent of $96?

16. $14 is what percent of $70?

17. 102 is what percent of 100?

18. 103 is what percent of 100?

19. What percent of $480 is $120?

20. What percent of $80 is $60?

21. What percent of 160 is 150?

22. What percent of 33 is 11?

23. $18 is 25% of what?

24. $75 is 20% of what?

25. 60% of what is 54?

26. 80% of what is 96?

27. 65.12 is 74% of what?

28. 63.7 is 65% of what?

29. 80% of what is 16?

30. 80% of what is 10?

31. What is $62\frac{1}{2}\%$ of 40?

32. What is $43\frac{1}{4}\%$ of 2600?

33. What is 9.4% of $8300?

34. What is 8.7% of $76,000?

35. $\mathbf{D_W}$ In your own words, list steps that a classmate could use to solve any percent problem in this section.

36. $\mathbf{D_W}$ In solving Example 10, a student simplifies $\frac{10}{20}$ before solving. Is this a good idea? Why or why not?

SKILL MAINTENANCE

Solve. [4.1d]

37. $\frac{x}{188} = \frac{2}{47}$

38. $\frac{15}{x} = \frac{3}{800}$

39. $\frac{4}{7} = \frac{x}{14}$

40. $\frac{612}{t} = \frac{72}{244}$

41. $\frac{5000}{t} = \frac{3000}{60}$

42. $\frac{75}{100} = \frac{n}{20}$

43. $\frac{x}{1.2} = \frac{36.2}{5.4}$

44. $\frac{y}{1\frac{1}{2}} = \frac{2\frac{3}{4}}{22}$

Solve. [2.5a]

45. A recipe for muffins calls for $\frac{1}{2}$ qt of buttermilk, $\frac{1}{3}$ qt of skim milk, and $\frac{1}{16}$ qt of oil. How many quarts of liquid ingredients does the recipe call for?

46. The Ferristown School District purchased $\frac{3}{4}$ ton (T) of clay. If the clay is to be shared equally among the district's 6 art departments, how much will each art department receive?

SYNTHESIS

Solve.

47. ▦ What is 8.85% of $12,640?
Estimate ________________
Calculate ________________

48. ▦ 78.8% of what is 9809.024?
Estimate ________________
Calculate ________________

4.6 APPLICATIONS OF PERCENT

Objectives

a Solve applied problems involving percent.

b Solve applied problems involving percent of increase or decrease.

a Applied Problems Involving Percent

Applied problems involving percent are not always stated in a manner easily translated to an equation. In such cases, it is helpful to rephrase the problem before translating. Sometimes it also helps to make a drawing.

EXAMPLE 1 *Presidential Deaths in Office.* George W. Bush was inaugurated as the 43rd President of the United States in 2001. Since Grover Cleveland was both the 22nd and the 24th presidents, there have been only 42 different presidents. Of the 42 presidents, 8 have died in office: William Henry Harrison, Zachary Taylor, Abraham Lincoln, James A. Garfield, William McKinley, Warren G. Harding, Franklin D. Roosevelt, and John F. Kennedy. What percent have died in office?

1. **Familiarize.** The question asks for a percent of the presidents who have died in office. We note that 42 is approximately 40 and 8 is $\frac{1}{5}$, or 20%, of 40, so our answer is close to 20%. We let $p =$ the percent who have died in office.

2. **Translate.** There are two ways in which we can translate this problem.

   Percent equation (see Section 4.4):

   $$\underbrace{8}_{\downarrow} \ \underbrace{\text{is}}_{\downarrow} \ \underbrace{\text{what percent}}_{\downarrow} \ \underbrace{\text{of}}_{\downarrow} \ \underbrace{42?}_{\downarrow}$$
   $$8 \ = \ p \ \cdot \ 42$$

   Proportion (see Section 4.5):

   $$\frac{N}{100} = \frac{8}{42}$$

   For proportions, $N\% = p$.

Percents	Quantities
0%	0
N%	8
100%	42

Study Tips

MAKING APPLICATIONS REAL

Newspapers and magazines are full of mathematical applications. Some of the easiest ones to find in the area of Basic Mathematics are about percent. Find such an application and share it with your class. As you obtain more skills in mathematics, you will find yourself observing the world from a different perspective, seeing mathematics everywhere. Math courses become more interesting when we connect the concepts to the real world.

1. **Presidential Assassinations in Office.** Of the 42 U.S. presidents, 4 have been assassinated in office. These were Garfield, McKinley, Lincoln, and Kennedy. What percent have been assassinated in office?

Answer on page A-11

3. **Solve.** We now have two ways in which to solve this problem.

Percent equation (see Section 4.4):

$$8 = p \cdot 42$$

$$\frac{8}{42} = \frac{p \cdot 42}{42} \quad \text{Dividing by 42 on both sides}$$

$$\frac{8}{42} = p$$

$$0.190 \approx p \quad \text{Finding decimal notation and rounding to the nearest thousandth}$$

$$19.0\% \approx p \quad \text{Remember to find percent notation.}$$

Note here that the solution, p, includes the % symbol.

Proportion (see Section 4.5):

$$\frac{N}{100} = \frac{8}{42}$$

$$N \cdot 42 = 100 \cdot 8 \quad \text{Equating cross products}$$

$$\frac{N \cdot 42}{42} = \frac{800}{42} \quad \text{Dividing by 42 on both sides}$$

$$N = \frac{800}{42}$$

$$N \approx 19.0 \quad \text{Dividing and rounding to the nearest tenth}$$

We use the solution of the proportion to express the answer to the problem as 19.0%. Note that in the proportion method, $N\% = p$.

4. **Check.** To check, we note that the answer 19.0% is close to 20%, as estimated in the *Familiarize* step.
5. **State.** About 19.0% of the U.S. presidents have died in office.

Do Exercise 1.

EXAMPLE 2 *Transportation to Work.* In the 15 largest cities in the United States, there are about 130,000,000 workers 16 years and older. Approximately 76.3% drive to work alone. How many workers in these 15 cities drive to work alone?

Transportation to Work (in the 15 Largest Cities in the United States)

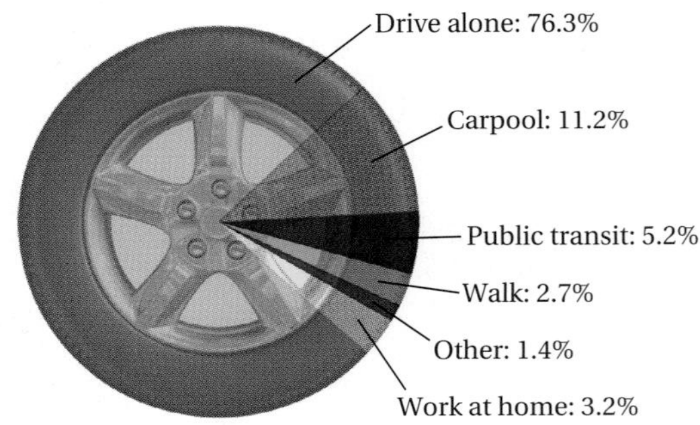

Source: U.S. Bureau of the Census

1. **Familiarize.** We can make a drawing of a pie chart to help familiarize ourselves with the problem. We let $b =$ the total number of workers who drive to work alone.

Transportation to Work

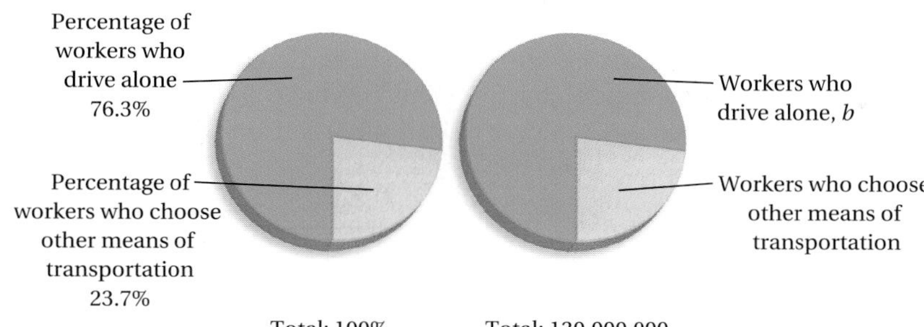

2. **Translate.** There are two ways in which we can translate this problem.

Percent equation:

What number is 76.3% of 130,000,000?

$$b = 76.3\% \cdot 130{,}000{,}000$$

Proportion:

$$\frac{76.3}{100} = \frac{b}{130{,}000{,}000}$$

Percents	Quantities
0%	0
76.3%	b
100%	130,000,000

3. **Solve.** We now have two ways in which to solve this problem.

Percent equation:

$$b = 76.3\% \cdot 130{,}000{,}000$$

We convert 76.3% to decimal notation and multiply:

$$b = 0.763 \times 130{,}000{,}000 = 99{,}190{,}000.$$

Proportion:

$$\frac{76.3}{100} = \frac{b}{130{,}000{,}000}$$

$$76.3 \times 130{,}000{,}000 = 100 \cdot b \qquad \text{Equating cross products}$$

$$\frac{76.3 \cdot 130{,}000{,}000}{100} = \frac{100 \cdot b}{100} \qquad \text{Dividing by 100}$$

$$\frac{9{,}919{,}000{,}000}{100} = b$$

$$99{,}190{,}000 = b \qquad \text{Simplifying}$$

4. **Check.** To check, we can repeat the calculations. We can also do a partial check by estimating. Since 76.3% is about 80%, or $\frac{4}{5}$, and $\frac{4}{5}$ of 130,000,000 is 104,000,000 and 104,000,000 is close to 99,190,000, our answer is reasonable.

5. **State.** The number of workers in the 15 largest cities who drive to work alone is 99,190,000.

Do Exercise 2.

2. **Transportation to Work.** There are about 130,000,000 workers 16 years and older in the 15 largest cities in the United States. Approximately 11.2% carpool to work. How many workers in these 15 cities carpool to work?

Source: U.S. Bureau of the Census

Answer on page A-11

3. **Percent of Increase.** The value of a car is \$36,875. The price is increased by 4%.

a) How much is the increase?

b) What is the new price?

4. **Percent of Decrease.** The value of a car is \$36,875. The car depreciates in value by 25% after one year.

a) How much is the decrease?

b) What is the depreciated value of the car?

Answers on page A-11

b Percent of Increase or Decrease

Percent is often used to state increase or decrease. Let's consider an example of each, using the price of a car as the original number.

PERCENT OF INCREASE

One year a car sold for \$20,455. The manufacturer decides to raise the price of the following year's model by 6%. The increase is $0.06 \times \$20,455$, or \$1227.30. The new price is $\$20,455 + \$1227.30$, or \$21,682.30. The *percent of increase* is 6%.

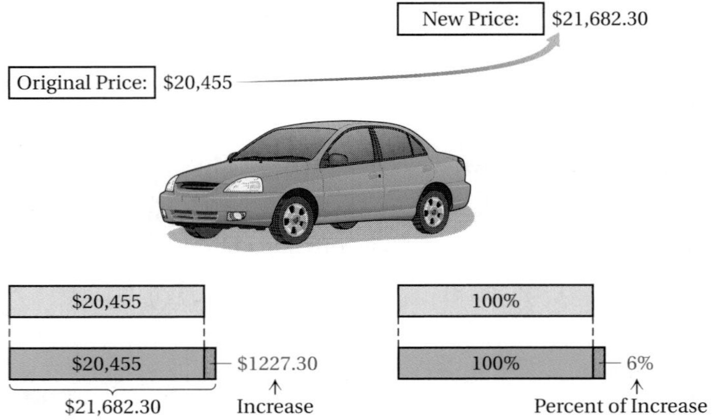

PERCENT OF DECREASE

Lisa buys the car listed above for \$20,455. After one year, the car depreciates in value by 25%. This is $0.25 \times \$20,455$, or \$5113.75. This lowers the value of the car to

$$\$20,455 - \$5113.75, \quad \text{or} \quad \$15,341.25.$$

Note that the new price is thus 75% of the original price. If Lisa decides to sell the car after a year, \$15,341.25 might be the most she could expect to get for it. The *percent of decrease* is 25%, and the decrease is \$5113.75.

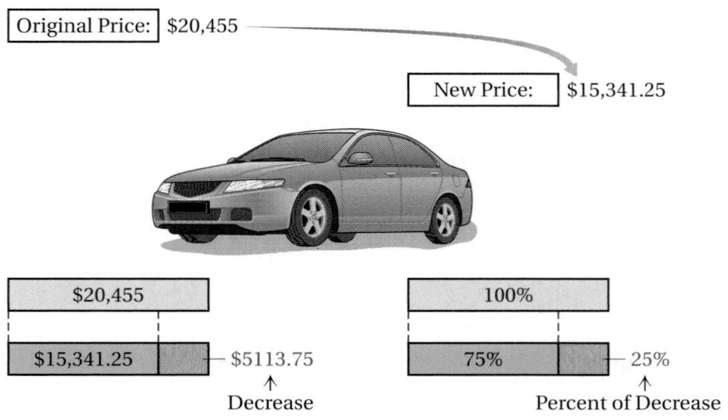

Do Exercises 3 and 4.

When a quantity is decreased by a certain percent, we say we have **percent of decrease.**

EXAMPLE 3 *Chain Link Fence.* For one week only, Sam's Farm Supply had 4 ft × 50 ft rolls of galvanized chain link fence on sale for \$39.99. The regular price was \$49.99 per roll. What was the percent of decrease?

1. **Familiarize.** We find the amount of decrease and then make a drawing.

4 9.9 9	Retail price
− 3 9.9 9	Sale price
1 0.0 0	Decrease

\$49.99 | 100%
\$39.99 — \$10.00 | ?%

2. **Translate.** There are two ways in which we can translate this problem.

Percent equation:

10.00 is what percent of 49.99?

$$10.00 = p \times 49.99$$

Proportion:

$$\frac{N}{100} = \frac{10.00}{49.99}$$

Percents	Quantities
0%	0
N%	10.00
100%	49.99

For proportions, $N\% = p$.

3. **Solve.** We have two ways in which to solve this problem.

Percent equation:

$$10.00 = p \times 49.99$$

$$\frac{10.00}{49.99} = \frac{p \times 49.99}{49.99} \quad \text{Dividing by 49.99 on both sides}$$

$$\frac{10.00}{49.99} = p$$

$$0.20 \approx p$$

$$20\% \approx p \quad \text{Converting to percent notation}$$

Proportion:

$$\frac{N}{100} = \frac{10.00}{49.99}$$

$$49.99 \times N = 100 \times 10 \quad \text{Equating cross products}$$

$$\frac{49.99 \times N}{49.99} = \frac{100 \times 10}{49.99} \quad \text{Dividing by 49.99 on both sides}$$

$$N = \frac{1000}{49.99}$$

$$N \approx 20$$

We use the solution of the proportion to express the answer to the problem as 20%.

5. **Volume of Mail.** The volume of U.S. mail decreased from about 208 billion pieces of mail in 2000 to 202 billion pieces in 2003. What was the percent of decrease?

Source: U.S. Postal Service

Answer on page A-11

4. **Check.** To check, we note that, with a 20% decrease, the reduced (or sale) price should be 80% of the retail (or original) price. Since

$$80\% \times 49.99 = 0.80 \times 49.99 = 39.992 \approx 39.99,$$

our answer checks.

5. **State.** The percent of decrease in the price of the roll of fence was 20%.

Do Exercise 5 on the preceding page.

When a quantity is increased by a certain percent, we say we have **percent of increase.**

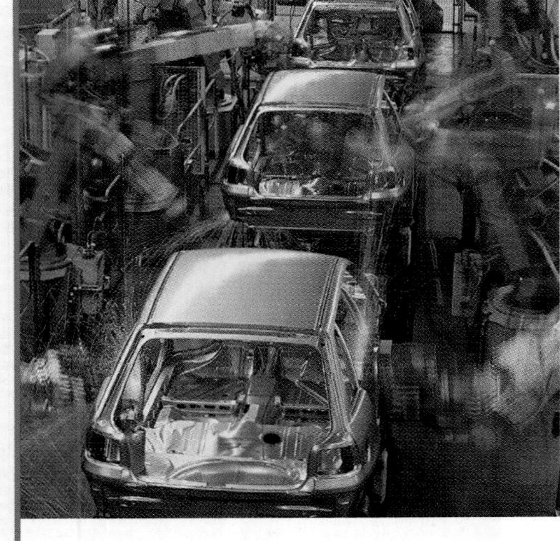

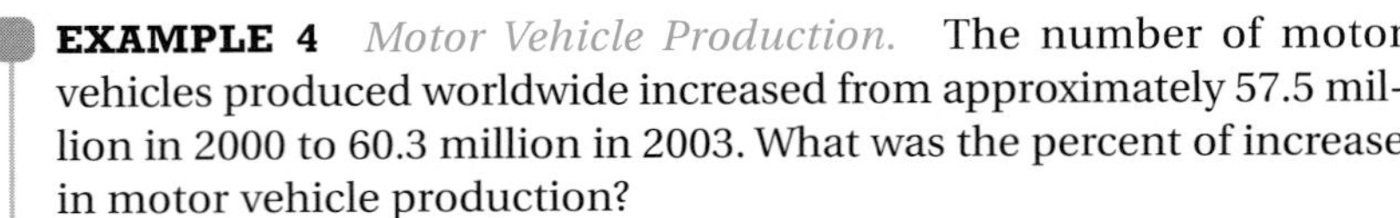

EXAMPLE 4 *Motor Vehicle Production.* The number of motor vehicles produced worldwide increased from approximately 57.5 million in 2000 to 60.3 million in 2003. What was the percent of increase in motor vehicle production?

Source: Ward's Communications, *Ward's Motor Vehicle Facts & Figures,* 2004

1. **Familiarize.** We first note that the increase in the number of vehicles produced was 60.3 − 57.5 million, or 2.8 million. A drawing can help us visualize the situation.

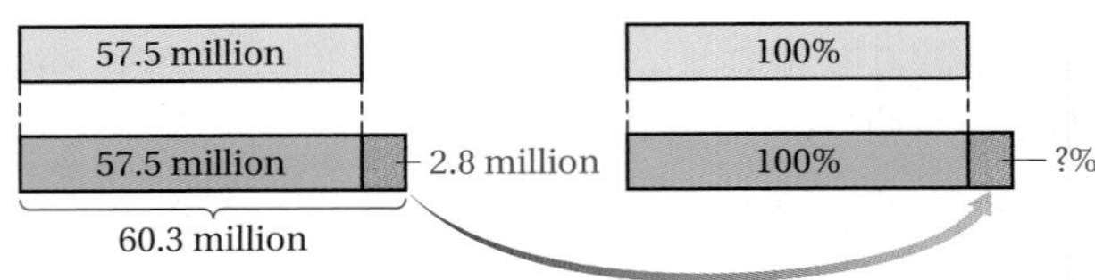

We are asking this question: The increase is what percent of the *original* amount? We let $p =$ the percent of increase.

2. **Translate.** There are two ways in which we can translate this problem.

Percent equation:

2.8 million	is	what percent	of	57.5 million?
↓	↓	↓	↓	↓
2.8	=	p	·	57.5

Proportion:

$$\frac{N}{100} = \frac{2.8}{57.5}$$

For proportions, $N\% = p$.

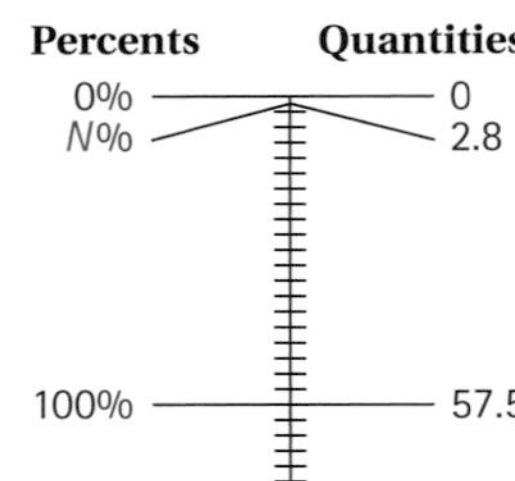

3. **Solve.** We have two ways in which to solve this problem.

Percent equation:

$$2.8 = p \cdot 57.5$$

$$\frac{2.8}{57.5} = \frac{p \times 57.5}{57.5} \qquad \text{Dividing by 57.5 on both sides}$$

$$\frac{2.8}{57.5} = p$$

$$0.049 \approx p$$

$$4.9\% \approx p \qquad \text{Converting to percent notation}$$

Proportion:

$$\frac{N}{100} = \frac{2.8}{57.5}$$

$$57.5 \times N = 100 \times 2.8 \qquad \text{Equating cross products}$$

$$\frac{57.5 \times N}{57.5} = \frac{100 \times 2.8}{57.5} \qquad \text{Dividing by 57.5 on both sides}$$

$$N = \frac{280}{57.5}$$

$$N \approx 4.9$$

We use the solution of the proportion to express the answer to the problem as 4.9%.

4. **Check.** To check, we take 4.9% of 57.5:

$$4.9\% \times 57.5 = 0.049 \times 57.5 = 2.8175.$$

Since we rounded the percent, this approximation is close enough to 2.8 to be a good check.

5. **State.** The percent of increase in the number of motor vehicles produced is 4.9%.

Do Exercise 6.

6. **Patents Issued.** The number of patents issued per year by the U.S. government increased from 107,332 in 1993 to 189,597 in 2003. What was the percent of increase over the 10-yr period?

Source: U.S. Patent and Trademark Office

Answer on page A-11

Translating for Success

1. *Distance Walked.* After knee replacement, Alex walked $\frac{1}{8}$ mi each morning and $\frac{1}{5}$ mi each afternoon. How much farther did he walk in the afternoon?

2. *Stock Prices.* A stock sold for \$5 per share on Monday and only \$$2\frac{1}{8}$ per share on Friday. What was the percent of decrease from Monday to Friday?

3. *SAT Score.* After attending a class called *Improving Your SAT Scores,* Jacob raised his total score from 884 to 1040. What was the percent of increase?

4. *Change in Population.* The population of a small farming community decreased from 1040 to 884. What was the percent of decrease?

5. *Lawn Mowing.* During the summer, brothers Steve and Rob earned money for college by mowing lawns. The largest lawn they mowed was $2\frac{1}{8}$ acres. Steve can mow $\frac{1}{5}$ acre per hour, and Rob can mow only $\frac{1}{8}$ acre per hour. Working together, how many acres did they mow per hour?

The goal of these matching questions is to practice step (2), *Translate,* of the five-step problem-solving process. Translate each word problem to an equation and select a correct translation from equations A–O.

A. $x + \frac{1}{5} = \frac{1}{8}$

B. $250 = x \cdot 1040$

C. $884 = x \cdot 1040$

D. $\frac{250}{16.25} = \frac{1000}{x}$

E. $156 = x \cdot 1040$

F. $16.25 = 250 \cdot x$

G. $\frac{1}{5} + \frac{1}{8} = x$

H. $2\frac{1}{8} = x \cdot 5$

I. $5 = 2\frac{7}{8} \cdot x$

J. $\frac{1}{8} + x = \frac{1}{5}$

K. $1040 = x \cdot 884$

L. $\frac{250}{16.25} = \frac{x}{1000}$

M. $2\frac{7}{8} = x \cdot 5$

N. $x \cdot 884 = 156$

O. $x = 16.25 \cdot 250$

Answers on page A-11

6. *Land Sale.* Cole sold $2\frac{1}{8}$ acres from the 5 acres he inherited from his uncle. What percent did he sell?

7. *Travel Expenses.* A magazine photographer is reimbursed 16.25¢ per mile for business travel up to 1000 mi per week. In a recent week, he traveled only 250 mi. What was the total reimbursement for travel?

8. *Trip Expenses.* The total expenses for Claire's recent business trip were \$1040. She put \$884 on her charge card and paid the balance in cash. What percent did she place on her charge card?

9. *Cost of Copies.* During the first summer session at a community college, the campus copy center advertised 250 copies for \$16.25. At this rate, what is the cost of 1000 copies?

10. *Cost of Insurance.* Following a raise in the cost of health insurance, 250 of a company's 1040 employees dropped their health coverage. What percent of the employees canceled their insurance?

4.6 EXERCISE SET

For Extra Help

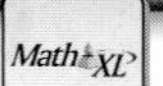

MathXL | MyMathLab | InterAct Math | Math Tutor Center | Video Lectures on CD Disc 2 | Student's Solutions Manual

a Solve.

1. *Wild Horses.* There are 27,369 wild horses on land managed by the Federal Bureau of Land Management. It is estimated that 48.4% of this total is in Nevada. How many wild horses are in Nevada?
Source: Bureau of Land Management, 2005

2. *U.S. Armed Forces.* There are 1,168,195 people in the United States in active military service. The numbers in the four armed services are listed in the table below. What percent of the total does each branch represent? Round the answers to the nearest tenth of a percent.

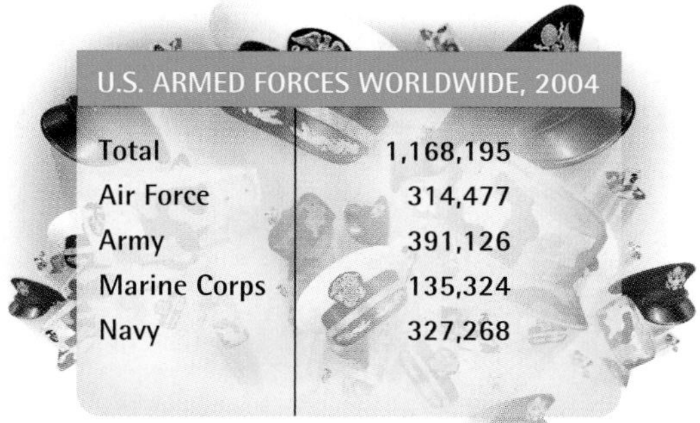

U.S. ARMED FORCES WORLDWIDE, 2004

Total	1,168,195
Air Force	314,477
Army	391,126
Marine Corps	135,324
Navy	327,268

Source: U.S. Department of Defense

3. *Car Value.* The base price of a Nissan 350Z is $34,000. This vehicle is expected to retain 62% of its value at the end of 3 yr and 52% at the end of 5 yr. What is the value of a Nissan 350Z after 3 yr? after 5 yr?
Source: November/December *Kelley Blue Book Residual Values Guide*

4. *Panda Survival.* Breeding the much-loved panda bear in captivity has been quite difficult for zookeepers.
   a) From 1964 to 1997, of 133 panda cubs born in captivity, only 90 lived to be one month old. What percent lived to be one month old?
   b) In 1999, Mark Edwards of the San Diego Zoo developed a nutritional formula on which 18 of 20 newborns lived to be one month old. What percent lived to be one month old?

5. *Overweight and Obese.* Of the 294 million people in the United States, 60% are considered overweight and 25% are considered obese. How many are overweight? How many are obese?
Source: U.S. Centers for Disease Control

6. *Smoking.* Of the 294 million people in the United States, 26% are smokers. How many are smokers?
Source: SAMHSA, Office of Applied Studies, National Survey on Drug Use and Health

7. A lab technician has 680 mL of a solution of water and acid; 3% is acid. How many milliliters are acid? water?

8. A lab technician has 540 mL of a solution of alcohol and water; 8% is alcohol. How many milliliters are alcohol? water?

9. *Mississippi River.* The Mississippi River, which extends from Minneapolis, Minnesota, to the Gulf of Mexico, is 2348 mi long. Approximately 77% of the river is navigable. How many miles of the river are navigable?

Source: National Oceanic and Atmospheric Administration

10. *Immigrants.* In 2003, 705,827 immigrants entered the United States. Of this total, 16.4% were from Mexico and 7.1% were from India. How many immigrants came from Mexico? from India?

Source: U.S. Department of Justice, *2003 Yearbook of Immigration Statistics*

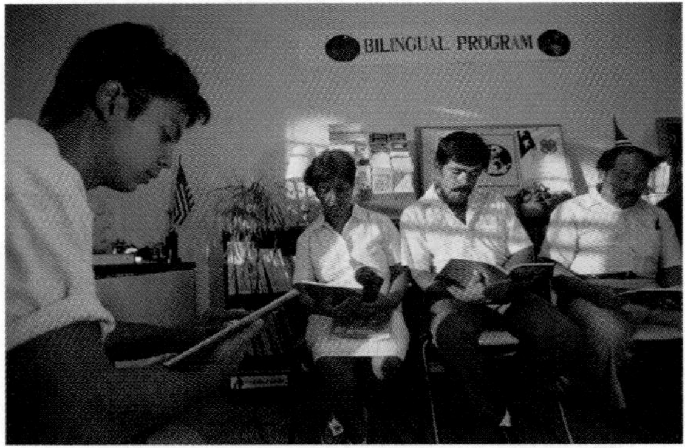

11. *Hispanic Population.* The Hispanic population is growing rapidly in the United States. In 2003, the population of the United States was about 291,000,000 and 13.7% of this total was Hispanic. How many Hispanic people lived in the United States in 2003?

Source: U.S. Bureau of the Census

12. *Age 65 and Older.* By 2010, it is predicted that 13.2% of the U.S. population will be 65 and older. If the population of the United States in 2010 is 307,000,000, how many people will be 65 and older?

Source: U.S. Bureau of the Census

13. *Test Results.* On a test of 40 items, Christina got 91% correct. (There was partial credit on some items.) How many items did she get correct? incorrect?

14. *Test Results.* On a test of 80 items, Pedro got 93% correct. (There was partial credit on some items.) How many items did he get correct? incorrect?

15. *Test Results.* On a test, Maj Ling got 86%, or 81.7, of the items correct. (There was partial credit on some items.) How many items were on the test?

16. *Test Results.* On a test, Juan got 85%, or 119, of the items correct. How many items were on the test?

17. *TV Usage.* Of the 8760 hr in a year, most television sets are on for 2190 hr. What percent is this?

18. *Colds from Kissing.* In a medical study, it was determined that if 800 people kiss someone who has a cold, only 56 will actually catch a cold. What percent is this?

Source: U.S. Centers for Disease Control

19. *Maximum Heart Rate.* Treadmill tests are often administered to diagnose heart ailments. A guideline in such a test is to try to get you to reach your *maximum heart rate,* in number of beats per minute. The maximum heart rate is found by subtracting your age from 220 and then multiplying by 85%. What is the maximum heart rate of someone whose age is 25? 36? 48? 55? 76? Round to the nearest one.

20. It costs an oil company $40,000 a day to operate two refineries. Refinery A accounts for 37.5% of the cost, and refinery B for the rest of the cost.

a) What percent of the cost does it take to run refinery B?

b) What is the cost of operating refinery A? refinery B?

Solve.

21. *Savings Increase.* The amount in a savings account increased from $200 to $216. What was the percent of increase?

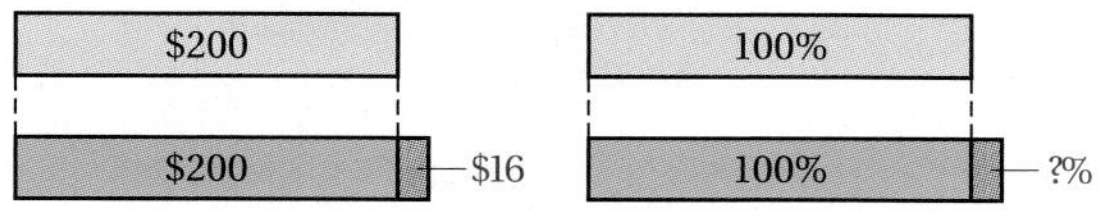

22. *Population Increase.* The population of a small mountain town increased from 840 to 882. What was the percent of increase?

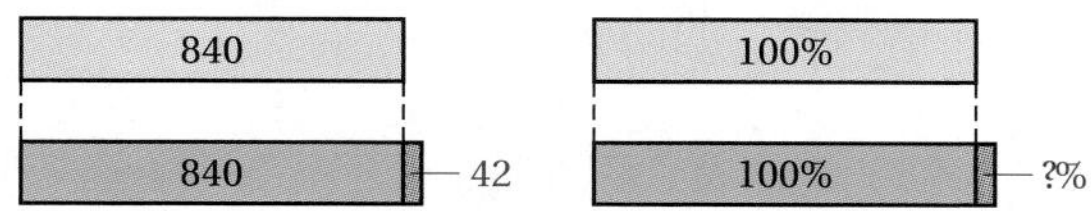

23. During a sale, a dress decreased in price from $90 to $72. What was the percent of decrease?

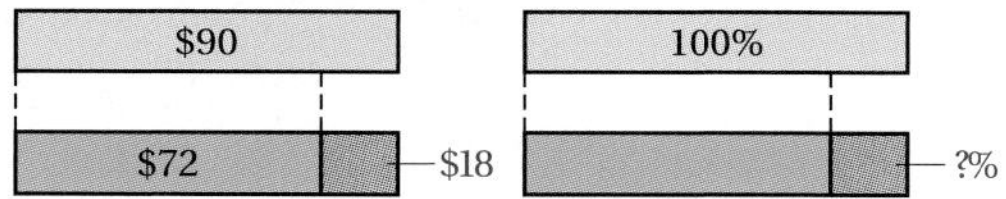

24. A person on a diet went from a weight of 125 lb to a weight of 110 lb. What was the percent of decrease?

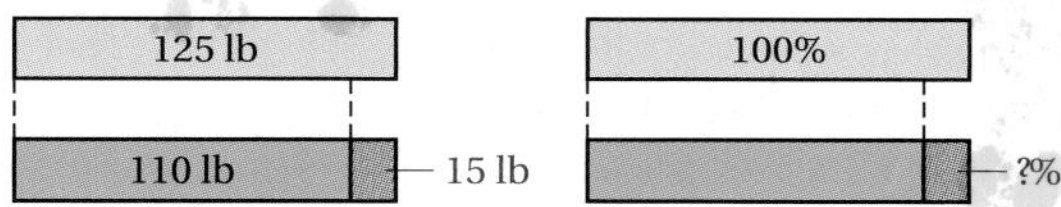

25. *Population Increase.* The population of the state of Nevada increased from 1,201,833 in 1990 to 2,241,154 in 2003. What was the percent of increase?

Source: U.S. Bureau of the Census

26. *Population Increase.* The population of the state of Utah increased from 1,722,850 in 1990 to 2,351,467 in 2003. What was the percent of increase?

Source: U.S. Bureau of the Census

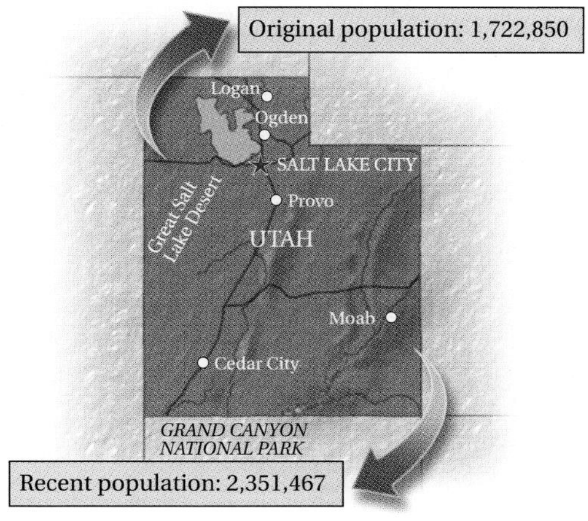

27. A person earns \$28,600 one year and receives a 5% raise in salary. What is the new salary?

28. A person earns \$43,200 one year and receives an 8% raise in salary. What is the new salary?

29. *Car Depreciation.* Irwin buys a car for \$21,566. It depreciates 25% each year that he owns it. What is the depreciated value of the car after 1 yr? after 2 yr?

30. *Car Depreciation.* Janice buys a car for \$22,688. It depreciates 25% each year that she owns it. What is the depreciated value of the car after 1 yr? after 2 yr?

31. *Doormat.* A bordered coir doormat that measures 42 in. × 24 in. × $1\frac{1}{2}$ in. has a retail price of \$89.95. Over the holidays it is on sale for \$65.49. What is the percent of decrease?

32. *Business Tote.* A leather business tote retails for \$239.99. An insurance company bought 30 totes for its sales staff at a reduced price of \$184.95 each. What was the percent of decrease in the price?

33. *Two-by-Four.* A cross-section of a standard or nominal "two-by-four" board actually measures $1\frac{1}{2}$ in. by $3\frac{1}{2}$ in. The rough board is 2 in. by 4 in. but is planed and dried to the finished size. What percent of the wood is removed in planing and drying?

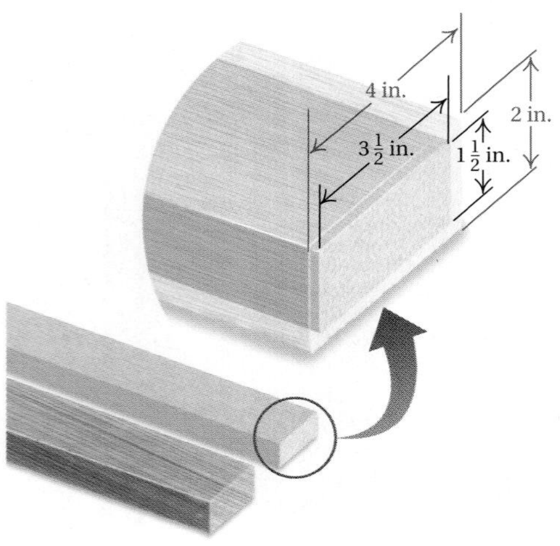

34. *Tipping.* Diners frequently add a 15% tip when charging a meal to a credit card. What is the total amount charged if the cost of the meal, without tip, is \$18? \$34? \$49?

35. *Population Decrease.* Between 1990 and 2000, the population of Detroit, Michigan, decreased from 1,028,000 to 951,000.

a) What is the percent of decrease?
b) If this percent of decrease over a 10-yr period repeated itself in the following decade, what would the population be in 2010?

Source: U.S. Bureau of the Census

36. *World Population.* World population is increasing by 1.14% each year. In 2004, it was 6.39 billion. How much will it be in 2010? 2015? 2018?

Source: *The World Factbook,* 2004

Life Insurance Rates for Smokers and Nonsmokers. The following table provides data showing how yearly rates (premiums) for a $500,000 term life insurance policy are increased for smokers. Complete the missing numbers in the table.

TYPICAL INSURANCE PREMIUMS (DOLLARS)

	AGE	RATE FOR NONSMOKER	RATE FOR SMOKER	PERCENT INCREASE FOR SMOKER
	35	$ 345	$ 630	83%
37.	40	$ 430	$ 735	
38.	45	$ 565		84%
39.	50	$ 780		100%
40.	55	$ 985		117%
41.	60	$1645	$2955	
42.	65	$2943	$5445	

Source: Pacific Life PL Protector Term Life Portfolio, OYT Rates

Population Increase. The following table provides data showing how the populations of various states increased from 1990 to 2003. Complete the missing numbers in the table.

	STATE	POPULATION IN 1990	POPULATION IN 2003	CHANGE	PERCENT CHANGE
43.	Alaska	550,043	648,818		
44.	Connecticut	3,287,116		196,256	
45.	Montana		917,621	118,556	
46.	Texas		22,118,509	5,131,999	
47.	Colorado	3,294,394		1,256,294	
48.	Pennsylvania	11,881,643	12,365,455		

Source: U.S. Bureau of the Census

49. *Car Depreciation.* A car generally depreciates 25% of its original value in the first year. A car is worth \$27,300 after the first year. What was its original cost?

50. *Car Depreciation.* Given normal use, an American-made car will depreciate 25% of its original cost the first year and 14% of its remaining value the second year. What is the value of a car at the end of the second year if its original cost was \$36,400? \$28,400? \$26,800?

51. *Strike Zone.* In baseball, the *strike zone* is normally a 17-in. by 30-in. rectangle. Some batters give the pitcher an advantage by swinging at pitches thrown out of the strike zone. By what percent is the area of the strike zone increased if a 2-in. border is added to the outside?

Source: Major League Baseball

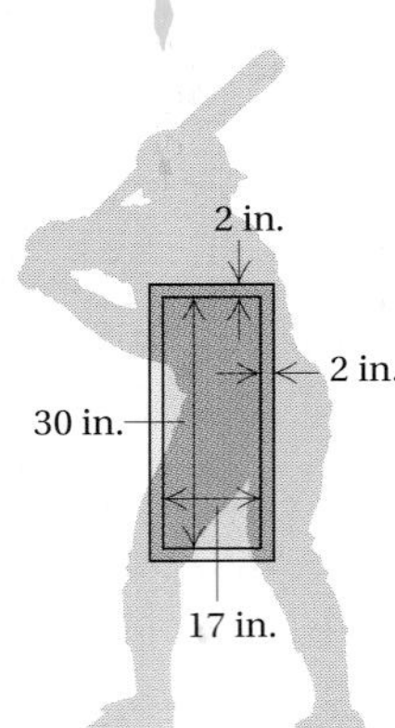

52. Tony has planted grass seed on a 24-ft by 36-ft area in his back yard. He has also installed a 6-ft by 8-ft garden. By what percent does the garden reduce the area he will have to mow?

6 ft 8 ft 36 ft 24 ft

53. $\mathbf{D}_\mathbf{W}$ Which is better for a wage earner, and why: a 10% raise followed by a 5% raise a year later, or a 5% raise followed by a 10% raise a year later?

54. $\mathbf{D}_\mathbf{W}$ A worker receives raises of 3%, 6%, and then 9%. By what percent has the original salary increased? Explain.

SKILL MAINTENANCE

Convert to decimal notation. [3.1b], [3.5a]

55. $\frac{25}{11}$ **56.** $\frac{11}{25}$ **57.** $\frac{27}{8}$ **58.** $\frac{43}{9}$ **59.** $\frac{23}{25}$

60. $\frac{20}{24}$ **61.** $\frac{14}{32}$ **62.** $\frac{2317}{1000}$ **63.** $\frac{34{,}809}{10{,}000}$ **64.** $\frac{27}{40}$

SYNTHESIS

65. *Adult Height.* It has been determined that at the age of 10, a girl has reached 84.4% of her final adult growth. Cynthia is 4 ft 8 in. at the age of 10. What will be her final adult height?

Source: *Dunlop Illustrated Encyclopedia of Facts.* New York: Sterling Publishing, 1970.

66. *Adult Height.* It has been determined that at the age of 15, a boy has reached 96.1% of his final adult height. Claude is 6 ft 4 in. at the age of 15. What will be his final adult height?

Source: *Dunlop Illustrated Encyclopedia of Facts.* New York: Sterling Publishing, 1970.

67. If p is 120% of q, then q is what percent of p?

68. A coupon allows a couple to have dinner and then have \$10 subtracted from the bill. Before subtracting \$10, however, the restaurant adds a tip of 15%. If the couple is presented with a bill for \$44.05, how much would the dinner (without tip) have cost without the coupon?

4.7 SALES TAX, COMMISSION, DISCOUNT, AND INTEREST

Objectives

a Solve applied problems involving sales tax and percent.

b Solve applied problems involving commission and percent.

c Solve applied problems involving discount and percent.

d Solve applied problems involving simple interest.

e Solve applied problems involving compound interest.

a Sales Tax

Sales tax computations represent a special type of percent of increase problem. The sales tax rate in Maryland is 5%. This means that the tax is 5% of the purchase price. Suppose the purchase price of a coat is \$124.95. The sales tax is then 5% of \$124.95, or 0.05×124.95, or 6.2475, or about \$6.25.

BILL:		
Purchase price	=	\$124.95
Sales tax (5% of \$124.95)	=	+ 6.25
Total price		\$131.20

The total that you pay is the price plus the sales tax:

$$\$124.95 + \$6.25, \quad \text{or} \quad \$131.20.$$

SALES TAX

Sales tax = Sales tax rate × Purchase price

Total price = Purchase price + Sales tax

EXAMPLE 1 *Florida Sales Tax.* The sales tax rate in Florida is 6%. How much tax is charged on the purchase of 4 inflatable rafts at \$89.99 each? What is the total price?

$89.99 plus 6% each

a) We first find the cost of the rafts. It is

$$4 \times \$89.99 = \$359.96.$$

b) The sales tax on items that cost \$359.96 is

Sales tax rate × Purchase price
↓ ↓
6% × \$359.96

or 0.06×359.96, or 21.5976. Thus the tax is \$21.60 (rounded to the nearest cent).

c) The total price is given by the purchase price plus the sales tax:

$$\$359.96 + \$21.60, \quad \text{or} \quad \$381.56.$$

To check, note that the total price is the purchase price plus 6% of the purchase price. Thus the total price is 106% of the purchase price. Since $1.06 \times 359.96 \approx 381.56$, we have a check. The sales tax is \$21.60 and the total price is \$381.56.

Do Exercises 1 and 2.

1. **California Sales Tax.** The sales tax rate in California is 7.25%. How much tax is charged on the purchase of a refrigerator that sells for \$668.95? What is the total price?

2. **Louisiana Sales Tax.** Sam buys 5 hardcover copies of Dean Koontz's novel *From the Corner of His Eye* for \$26.95 each. The sales tax rate in Louisiana is 4%. How much sales tax will be charged? What is the total price?

Answers on page A-11

3. The sales tax on the purchase of a night table that costs $849 is $50.94. What is the sales tax rate?

4. The sales tax on the purchase of a portable navigation system is $59.94 and the sales tax rate is 6%. Find the purchase price (the price before taxes are added).

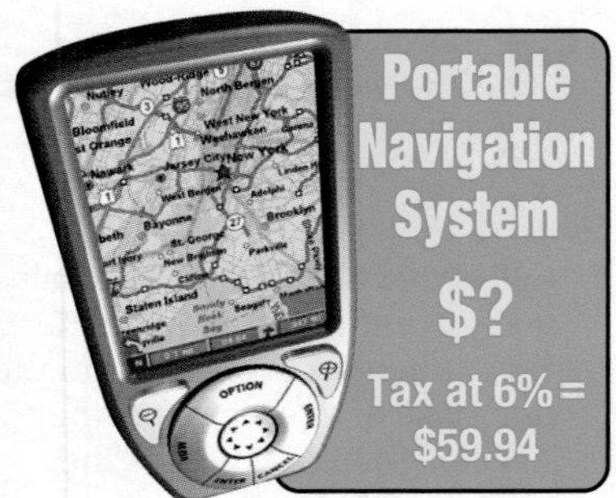

Answers on page A-11

EXAMPLE 2 The sales tax on the purchase of this wooden play center, which costs $1099, is $43.96. What is the sales tax rate?

Rephrase:	Sales tax	is	what percent	of	purchase price?
	↓	↓	↓	↓	↓
Translate:	$43.96	=	r	×	$1099

To solve the equation, we divide by 1099 on both sides:

$$\frac{43.96}{1099} = \frac{r \times 1099}{1099}$$

$$\frac{43.96}{1099} = r$$

$$0.04 = r$$

$$4\% = r.$$

The sales tax rate is 4%.

Do Exercise 3.

EXAMPLE 3 The sales tax on the purchase of a Tiffany torchiere lamp is $8.93 and the sales tax rate is 5%. Find the purchase price (the price before taxes are added).

Rephrase:	Sales tax	is	5%	of	what?	
	↓	↓	↓	↓	↓	
Translate:	$8.93	=	5%	×	b,	or $8.93 = 0.05 \times b$.

To solve, we divide by 0.05 on both sides:

$$\frac{8.93}{0.05} = \frac{0.05 \times b}{0.05}$$

$$\frac{8.93}{0.05} = b$$

$$\$178.60 = b.$$

The purchase price is $178.60.

Do Exercise 4.

b Commission

When you work for a **salary,** you receive the same amount of money each week or month. When you work for a **commission,** you are paid a percentage of the total sales for which you are responsible.

COMMISSION

Commission = Commission rate × Sales

EXAMPLE 4 *Exercise Equipment Sales.* A salesperson's commission rate is 16%. What is the commission from the sale of \$9700 worth of exercise equipment?

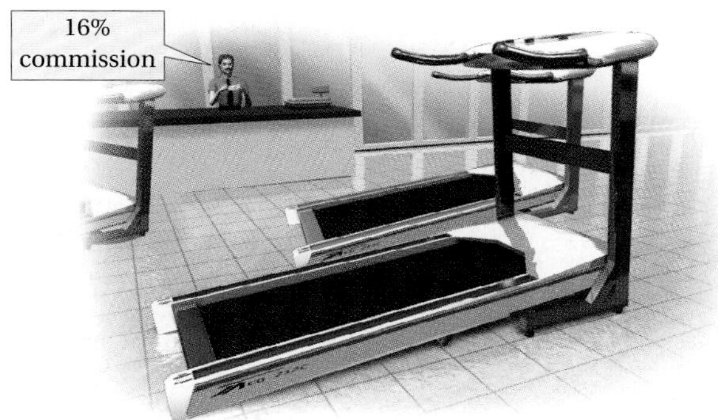

Commission	=	*Commission rate*	×	*Sales*
C	=	16%	×	9700
C	=	0.16	×	9700
C	=	1552		

The commission is \$1552.

Do Exercise 5.

EXAMPLE 5 *Farm Machinery Sales.* Dawn earns a commission of \$30,000 selling \$600,000 worth of farm machinery. What is the commission rate?

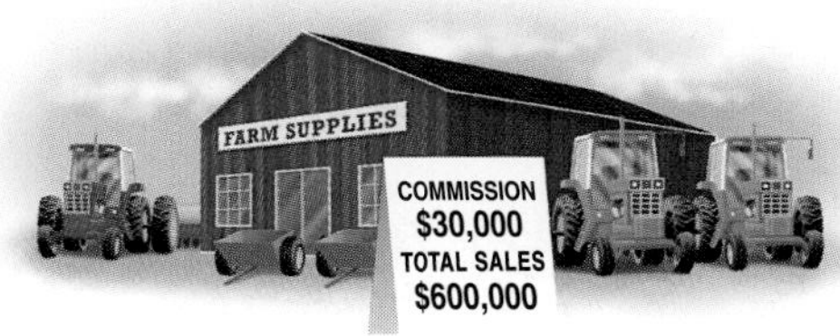

Commission	=	*Commission rate*	×	*Sales*
30,000	=	r	×	600,000

5. Raul's commission rate is 30%. What is the commission from the sale of \$18,760 worth of air conditioners?

Answer on page A-11

6. Liz earns a commission of $3000 selling $24,000 worth of U2 concert tickets. What is the commission rate?

7. Ben's commission rate is 16%. He receives a commission of $268 from sales of clothing. How many dollars worth of clothing were sold?

Answers on page A-11

To solve this equation, we divide by 600,000 on both sides:

$$\frac{30{,}000}{600{,}000} = \frac{r \times 600{,}000}{600{,}000}$$

$$\frac{1}{20} = r$$

$$0.05 = r$$

$$5\% = r.$$

The commission rate is 5%.

Do Exercise 6.

EXAMPLE 6 *Motorcycle Sales.* Joyce's commission rate is 12%. She receives a commission of $936 from the sale of a motorcycle. How much did the motorcycle cost?

Commission = *Commission rate* × *Sales*

936 = 12% × S, or $936 = 0.12 \times S$

To solve this equation, we divide by 0.12 on both sides:

$$\frac{936}{0.12} = \frac{0.12 \times S}{0.12}$$

$$\frac{936}{0.12} = S$$

$$7800 = S.$$

The motorcycle cost $7800.

Do Exercise 7.

C Discount

Suppose that the regular price of a rug is $60, and the rug is on sale at 25% off. Since 25% of $60 is $15, the sale price is $60 − $15, or $45. We call $60 the **original,** or **marked price,** 25% the **rate of discount,** $15 the **discount,** and $45 the **sale price.** Note that discount problems are a type of percent of decrease problem.

DISCOUNT AND SALE PRICE

Discount = Rate of discount × Original price

Sale price = Original price − Discount

EXAMPLE 7 A masonite door marked \$1389 is on sale at $33\frac{1}{3}\%$ off. What is the discount? the sale price?

a) *Discount* = *Rate of discount* × *Original price*

$$D = 33\frac{1}{3}\% \times 1389$$

$$D = \frac{1}{3} \times 1389$$

$$D = \frac{1389}{3} = 463$$

b) *Sale price* = *Original price* − *Discount*

$$S = 1389 - 463$$

$$S = 926$$

The discount is \$463 and the sale price is \$926.

Do Exercise 8.

EXAMPLE 8 *Antique Pricing.* The price of an antique table is marked down from \$620 to \$527. What is the rate of discount?

We first find the discount by subtracting the sale price from the original price:

$$620 - 527 = 93.$$

The discount is \$93.

Next, we use the equation for discount:

Discount = *Rate of discount* × *Original price*

$$93 = r \times 620.$$

To solve, we divide by 620 on both sides:

$$\frac{93}{620} = \frac{r \times 620}{620}$$

$$\frac{93}{620} = r$$

$$0.15 = r$$

$$15\% = r.$$

The discount rate is 15%.

> To check, note that a 15% discount rate means that 85% of the original price is paid:
> $0.85 \times 620 = 527.$

Do Exercise 9.

8. A suit marked \$540 is on sale at $33\frac{1}{3}\%$ off. What is the discount? the sale price?

9. The price of a pair of hiking boots is reduced from \$75 to \$60. Find the rate of discount.

Answers on page A-11

10. What is the simple interest on $4300 invested at an interest rate of 7% for 1 year?

11. What is the simple interest on a principal of $4300 invested at an interest rate of 7% for $\frac{3}{4}$ year?

Answers on page A-11

d Simple Interest

Suppose you put $1000 into an investment for 1 year. The $1000 is called the **principal.** If the **interest rate** is 8%, in addition to the principal, you get back 8% of the principal, which is

8% of $1000, or 0.08×1000, or $80.00.

The $80.00 is called the **simple interest.** It is, in effect, the price that a financial institution pays for the use of the money over time.

SIMPLE INTEREST FORMULA

The **simple interest I** on principal P, invested for t years at interest rate r, is given by

$$I = P \cdot r \cdot t.$$

EXAMPLE 9 What is the simple interest on $2500 invested at an interest rate of 6% for 1 year?

We use the formula $I = P \cdot r \cdot t$:

$$\begin{aligned} I = P \cdot r \cdot t &= \$2500 \times 6\% \times 1 \\ &= \$2500 \times 0.06 \\ &= \$150. \end{aligned}$$

The simple interest for 1 year is $150.

Do Exercise 10.

EXAMPLE 10 What is the simple interest on a principal of $2500 invested at an interest rate of 6% for $\frac{1}{4}$ year?

We use the formula $I = P \cdot r \cdot t$:

$$\begin{aligned} I = P \cdot r \cdot t &= \$2500 \times 6\% \times \frac{1}{4} \\ &= \frac{\$2500 \times 0.06}{4} \\ &= \$37.50. \end{aligned}$$

We could instead have found $\frac{1}{4}$ of 6% and then multiplied by 2500.

The simple interest for $\frac{1}{4}$ year is $37.50.

Do Exercise 11.

When time is given in days, we generally divide it by 365 to express the time as a fractional part of a year.

EXAMPLE 11 To pay for a shipment of tee shirts, New Wave Designs borrows \$8000 at $9\frac{3}{4}\%$ for 60 days. Find **(a)** the amount of simple interest that is due and **(b)** the total amount that must be paid after 60 days.

a) We express 60 days as a fractional part of a year:

$$I = P \cdot r \cdot t = \$8000 \times 9\frac{3}{4}\% \times \frac{60}{365}$$
$$= \$8000 \times 0.0975 \times \frac{60}{365}$$
$$\approx \$128.22.$$

The interest due for 60 days is \$128.22.

b) The total amount to be paid after 60 days is the principal plus the interest:

$$\$8000 + \$128.22 = \$8128.22.$$

The total amount due is \$8128.22.

Do Exercise 12.

12. The Glass Nook borrows \$4800 at $8\frac{1}{2}\%$ for 30 days. Find **(a)** the amount of simple interest due and **(b)** the total amount that must be paid after 30 days.

e Compound Interest

When interest is paid *on interest,* we call it **compound interest.** This is the type of interest usually paid on investments. Suppose you have \$5000 in a savings account at 6%. In 1 year, the account will contain the original \$5000 plus 6% of \$5000. Thus the total in the account after 1 year will be

106% of \$5000, or $1.06 \times \$5000$, or \$5300.

Now suppose that the total of \$5300 remains in the account for another year. At the end of this second year, the account will contain the \$5300 plus 6% of \$5300. The total in the account would thus be

106% of \$5300, or $1.06 \times \$5300$, or \$5618.

Note that in the second year, interest is earned on the first year's interest. When this happens, we say that interest is **compounded annually.**

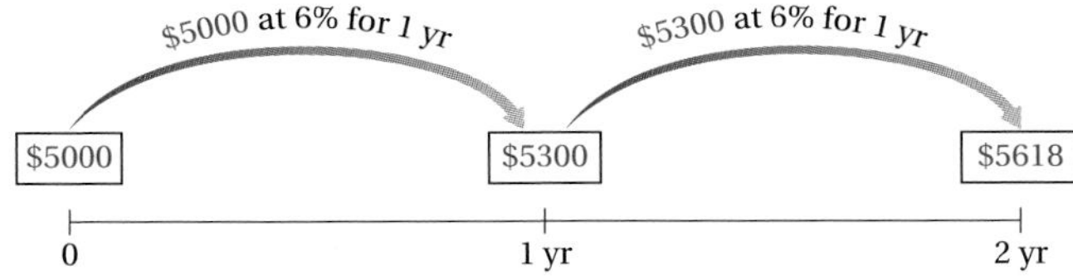

Answer on page A-11

13. Find the amount in an account if $2000 is invested at 9%, compounded annually, for 2 years.

EXAMPLE 12 Find the amount in an account if $2000 is invested at 8%, compounded annually, for 2 years.

a) After 1 year, the account will contain 108% of $2000:

$$1.08 \times \$2000 = \$2160.$$

b) At the end of the second year, the account will contain 108% of $2160:

$$1.08 \times \$2160 = \$2332.80.$$

The amount in the account after 2 years is $2332.80.

Do Exercise 13.

Suppose that the interest in Example 12 were **compounded semiannually**—that is, every half year. Interest would then be calculated twice a year at a rate of 8% ÷ 2, or 4% each time. The approach used in Example 12 can then be adapted, as follows.

After the first $\frac{1}{2}$ year, the account will contain 104% of $2000:

$$1.04 \times \$2000 = \$2080.$$

After a second $\frac{1}{2}$ year (1 full year), the account will contain 104% of $2080:

$$1.04 \times \$2080 = \$2163.20.$$

After a third $\frac{1}{2}$ year $\left(1\frac{1}{2}\text{ full years}\right)$, the account will contain 104% of $2163.20:

$$1.04 \times \$2163.20 = \$2249.728$$
$$\approx \$2249.73. \qquad \text{Rounding to the nearest cent}$$

Finally, after a fourth $\frac{1}{2}$ year (2 full years), the account will contain 104% of $2249.73:

$$1.04 \times \$2249.73 = \$2339.7192$$
$$\approx \$2339.72. \qquad \text{Rounding to the nearest cent}$$

Let's summarize our results and look at them another way:

End of 1st $\frac{1}{2}$ year $\rightarrow 1.04 \times 2000 = 2000 \times (1.04)^1$;
End of 2nd $\frac{1}{2}$ year $\rightarrow 1.04 \times (1.04 \times 2000) = 2000 \times (1.04)^2$;
End of 3rd $\frac{1}{2}$ year $\rightarrow 1.04 \times (1.04 \times 1.04 \times 2000) = 2000 \times (1.04)^3$;
End of 4th $\frac{1}{2}$ year $\rightarrow 1.04 \times (1.04 \times 1.04 \times 1.04 \times 2000) = 2000 \times (1.04)^4$.

Note that each multiplication was by 1.04 and that

$$\$2000 \times 1.04^4 \approx \$2339.72. \qquad \text{Using a calculator and rounding to the nearest cent}$$

We have illustrated the following result.

COMPOUND INTEREST FORMULA

If a principal P has been invested at interest rate r, compounded n times a year, in t years it will grow to an amount A given by

$$A = P \cdot \left(1 + \frac{r}{n}\right)^{n \cdot t}.$$

Answer on page A-11

Let's apply this formula to confirm our preceding discussion, where the amount invested is $P = \$2000$, the number of years is $t = 2$, and the number of compounding periods each year is $n = 2$. Substituting into the compound interest formula, we have

$$A = P \cdot \left(1 + \frac{r}{n}\right)^{n \cdot t} = 2000 \cdot \left(1 + \frac{8\%}{2}\right)^{2 \cdot 2}$$
$$= 2000 \cdot \left(1 + \frac{0.08}{2}\right)^{4} = 2000(1.04)^4$$
$$= 2000 \times 1.16985856 \approx \$2339.72.$$

If you are using a calculator, you could perform this computation in one step.

EXAMPLE 13 The Ibsens invest \$4000 in an account paying $5\frac{5}{8}\%$, compounded quarterly. Find the amount in the account after $2\frac{1}{2}$ years.

The compounding is quarterly, so n is 4. We substitute \$4000 for P, $5\frac{5}{8}\%$, or 0.05625, for r, 4 for n, and $2\frac{1}{2}$, or $\frac{5}{2}$, for t and compute A:

$$A = P \cdot \left(1 + \frac{r}{n}\right)^{n \cdot t} = \$4000 \cdot \left(1 + \frac{5\frac{5}{8}\%}{4}\right)^{4 \cdot 5/2}$$
$$= \$4000 \cdot \left(1 + \frac{0.05625}{4}\right)^{10}$$
$$= \$4000(1.0140625)^{10}$$
$$\approx \$4599.46.$$

The amount in the account after $2\frac{1}{2}$ years is \$4599.46.

Do Exercise 14.

14. A couple invests \$7000 in an account paying $6\frac{3}{8}\%$, compounded semiannually. Find the amount in the account after $1\frac{1}{2}$ years.

Answer on page A-11

CALCULATOR CORNER

Compound Interest A calculator is useful in computing compound interest. Not only does it do computations quickly but it also eliminates the need to round until the computation is completed. This minimizes "round-off errors" that occur when rounding is done at each stage of the computation. We must keep order of operations in mind when computing compound interest.

To find the amount due on a \$20,000 loan made for 25 days at 11% interest, compounded daily, we would compute $20{,}000\left(1 + \frac{0.11}{365}\right)^{25}$. To do this on a calculator, we press [2] [0] [0] [0] [0] [×] [(] [1] [+] [.] [1] [1] [÷] [3] [6] [5] [)] [y^x] (or [x^y]) [2] [5] [=]. Without parentheses, we would first find $1 + \frac{0.11}{365}$, raise this result to the 25th power, and then multiply by 20,000. To do this, we press [1] [+] [.] [1] [1] [÷] [3] [6] [5] [=] [y^x] (or [x^y]) [2] [5] [=] [×] [2] [0] [0] [0] [0]. In either case, the result is 20,151.23, rounded to the nearest cent.

Some calculators have business keys that allow such computations to be done more quickly.

Exercises:

1. Find the amount due on a \$16,000 loan made for 62 days at 13% interest, compounded daily.
2. An investment of \$12,500 is made for 90 days at 8.5% interest, compounded daily. How much is the investment worth after 90 days?

4.7 EXERCISE SET

For Extra Help

MathXL | MyMathLab | InterAct Math | Math Tutor Center | Video Lectures on CD Disc 2 | Student's Solutions Manual

a Solve.

1. *Kansas Sales Tax.* The sales tax rate in Kansas is 5.3%. How much sales tax would be charged on the purchase of a video game, Quest of the Planets, which sells for $49.99?

2. *New Jersey Sales Tax.* The sales tax rate in New Jersey is 6%. How much sales tax would be charged on the purchase of a copy of John Grisham's novel *A Painted House,* which sells for $27.95?

   Sources: Borders Bookstore; Andrea Sutcliffe, *Numbers*

3. *Tennessee Sales Tax.* The sales tax rate in Tennessee is 7%. How much sales tax would be charged on the purchase of a lawn mower that costs $279?

4. *Arizona Sales Tax.* The sales tax rate in Arizona is 5.6%. How much sales tax would be charged on the purchase of a lawn mower that costs $279?

5. *Utah Sales Tax.* The sales tax rate in Utah is 4.75%. How much tax is charged on the purchase of 5 telephones at $69 apiece? What is the total price?

6. *New York Sales Tax.* The sales tax rate in New York is 4.25%. How much tax is charged on the purchase of 5 teapots at $37.99 apiece? What is the total price?

7. The sales tax is $48 on the purchase of a dining room set that sells for $960. What is the sales tax rate?

8. The sales tax is $9.12 on the purchase of a patio set that sells for $456. What is the sales tax rate?

9. The sales tax on the purchase of a used car is $100 and the sales tax rate is 5%. Find the purchase price (the price before taxes are added).

10. The sales tax on the purchase of a new boat is $112 and the sales tax rate is 2%. Find the purchase price.

11. The sales tax on the purchase of a dining room set is $28 and the sales tax rate is 3.5%. Find the purchase price.

12. The sales tax on the purchase of a portable DVD player is $24.75 and the sales tax rate is 5.5%. Find the purchase price.

13. The sales tax rate in Austin is 2% for the city and county and 6.25% for the state. Find the total amount paid for 2 shower units at $332.50 apiece.

14. The sales tax rate in Omaha is 1.5% for the city and 5% for the state. Find the total amount paid for 3 air conditioners at $260 apiece.

15. The sales tax on an automobile purchase of $18,400 is $1030.40. What is the sales tax rate?

16. The sales tax on an automobile purchase of $15,800 is $979.60. What is the sales tax rate?

b Solve.

17. Katrina's commission rate is 6%. What is the commission from the sale of $45,000 worth of furnaces?

18. Jose's commission rate is 32%. What is the commission from the sale of $12,500 worth of sailboards?

19. Mitchell earns $120 selling $2400 worth of television sets in a consignment shop. What is the commission rate?

20. Donna earns $408 selling $3400 worth of shoes. What is the commission rate?

21. An art gallery's commission rate is 40%. They receive a commission of $392. How many dollars worth of artwork were sold?

22. A real estate agent's commission rate is 7%. She receives a commission of $5600 from the sale of a home. How much did the home sell for?

23. A real estate commission is 6%. What is the commission from the sale of a $98,000 home?

24. Chuck earns $1147.50 selling $7650 worth of ski passes. What is the commission rate?

25. Miguel's commission is increased according to how much he sells. He receives a commission of 5% for the first $2000 and 8% on the amount over $2000. What is the total commission on sales of $6000?

26. Lucinda earns a salary of $500 a month, plus a 2% commission on sales. One month, she sold $990 worth of encyclopedias. What were her wages that month?

c Find what is missing.

	MARKED PRICE	RATE OF DISCOUNT	DISCOUNT	SALE PRICE
27.	$300	10%		
28.	$2000	40%		
29.	$17	15%		
30.	$20	25%		
31.		10%	$12.50	
32.		15%	$65.70	
33.	$600		$240	
34.	$12,800		$1920	

35. Find the discount and the rate of discount for the car seat in this ad.

Sale
Car Seat
$149.99
Was $179.99

36. Find the marked price and the rate of discount for the basketball system in this ad.

d Find the simple interest.

	PRINCIPAL	RATE OF INTEREST	TIME	SIMPLE INTEREST
37.	$200	4%	1 year	
38.	$450	2%	1 year	
39.	$2000	8.4%	$\frac{1}{2}$ year	
40.	$200	7.7%	$\frac{1}{2}$ year	
41.	$4300	10.56%	$\frac{1}{4}$ year	
42.	$8000	9.42%	$\frac{1}{6}$ year	
43.	$20,000	$4\frac{5}{8}\%$	1 year	
44.	$100,000	$3\frac{7}{8}\%$	1 year	
45.	$50,000	$5\frac{3}{8}\%$	$\frac{1}{4}$ year	
46.	$80,000	$6\frac{3}{4}\%$	$\frac{1}{12}$ year	

Solve. Assume that simple interest is being calculated in each case.

47. Animal Instinct, a pet supply shop, borrows $6500 at 5% for 90 days. Find **(a)** the amount of interest due and **(b)** the total amount that must be paid after 90 days.

48. Andante's Cafe borrows $4500 at 12% for 60 days. Find **(a)** the amount of interest due and **(b)** the total amount that must be paid after 60 days.

49. CopiPix, Inc., borrows $10,000 at 9% for 60 days. Find **(a)** the amount of interest due and **(b)** the total amount that must be paid after 60 days.

50. Sal's Laundry borrows $8000 at 10% for 90 days. Find **(a)** the amount of interest due and **(b)** the total amount that must be paid after 90 days.

51. Jean's Garage borrows $5600 at 10% for 30 days. Find **(a)** the amount of interest due and **(b)** the total amount that must be paid after 30 days.

52. Shear Delights, a hair salon, borrows $3600 at 4% for 30 days. Find **(a)** the amount of interest due and **(b)** the total amount that must be paid after 30 days.

e Interest is compounded annually. Find the amount in the account after the given length of time. Round to the nearest cent.

	PRINCIPAL	RATE OF INTEREST	TIME	AMOUNT IN THE ACCOUNT
53.	$400	5%	2 years	
54.	$450	4%	2 years	
55.	$2000	8.8%	4 years	
56.	$4000	7.7%	4 years	
57.	$4300	10.56%	6 years	
58.	$8000	9.42%	6 years	
59.	$20,000	$6\frac{5}{8}\%$	25 years	
60.	$100,000	$5\frac{7}{8}\%$	30 years	

Interest is compounded semiannually. Find the amount in the account after the given length of time. Round to the nearest cent.

	PRINCIPAL	RATE OF INTEREST	TIME	AMOUNT IN THE ACCOUNT
61.	$4000	6%	1 year	
62.	$1000	5%	1 year	
63.	$20,000	8.8%	4 years	
64.	$40,000	7.7%	4 years	
65.	$5000	10.56%	6 years	
66.	$8000	9.42%	8 years	
67.	$20,000	$7\frac{5}{8}\%$	25 years	
68.	$100,000	$4\frac{7}{8}\%$	30 years	

Solve.

69. A family invests $4000 in an account paying 6%, compounded monthly. How much is in the account after 5 months?

70. The O'Hares invest $6000 in an account paying 8%, compounded quarterly. How much is in the account after 18 months?

71. **DW** Which is a better investment and why: $1000 invested at $7\frac{3}{4}$% simple interest for 1 year, or $1000 invested at 7% compounded monthly for 1 year?

72. **DW** A firm must choose between borrowing $5000 at 10% for 30 days and borrowing $10,000 at 8% for 60 days. Give arguments in favor of and against each option.

SKILL MAINTENANCE

VOCABULARY REINFORCEMENT

In each of Exercises 73–80, fill in the blank with the correct term from the given list. Some of the choices may not be used.

73. If the product of two numbers is 1, they are ______________ of each other. [2.2b]

74. A number is ______________ if its ones digit is even and the sum of its digits is divisible by 3. [1.8a]

75. The number 0 is the ______________ identity. [1.2a]

76. A ratio is the ______________ of two quantities. [4.1a]

77. The distance around an object is its ______________. [1.2b]

78. A number is ______________ if the sum of its digits is divisible by 3. [1.8a]

79. A natural number that has exactly two different factors, only itself and 1, is called a ______________ number. [1.7c]

80. When two pairs of numbers have the same ratio, they are ______________. [4.1c]

divisible by 3	quotient
divisible by 4	reciprocals
divisible by 6	proportional
divisible by 9	composite
perimeter	prime
area	additive
average	multiplicative

SYNTHESIS

81. Herb collects baseball memorabilia. He bought two autographed plaques, but became short of funds and had to sell them quickly for $200 each. On one, he made a 20% profit and on the other, he lost 20%. Did he make or lose money on the sale?

82. 🖩 Gordon receives a 10% commission on the first $5000 in sales and 15% on all sales beyond $5000. If Gordon receives a commission of $2405, how much did he sell? Use a calculator and trial and error if you wish.

Effective Yield. The *effective yield* is the yearly rate of simple interest that corresponds to a rate for which interest is compounded two or more times a year. For example, if P is invested at 12%, compounded quarterly, we would multiply P by $(1 + 0.12/4)^4$, or 1.03^4. Since $1.03^4 \approx 1.126$, the 12% compounded quarterly corresponds to an effective yield of approximately 12.6%. In Exercises 83 and 84, find the effective yield for the indicated account.

83. 🖩 The account pays 9% compounded monthly.

84. 🖩 The account pays 10% compounded daily.

4.8 INTEREST RATES ON CREDIT CARDS AND LOANS

Objective

a Solve applied problems involving interest rates on credit cards and loans.

a Credit Cards and Loans

Look at the following graphs. They offer good reason for a study of the real-world applications of percent, interest, loans, and credit cards.

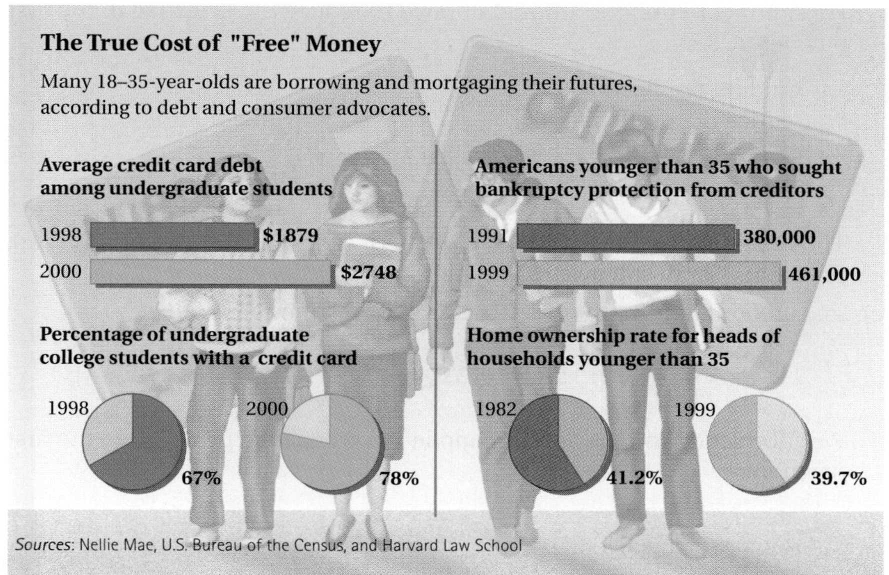

Comparing interest rates is essential if one is to become financially responsible. A small change in an interest rate can make a *large* difference in the cost of a loan. When you make a payment on a loan, do you know how much of that payment is interest and how much is applied to reducing the principal?

We begin with an example involving credit cards. A balance carried on a credit card is a type of loan. In a recent year in the United States, 100,000 young adults declared bankruptcy because of excessive credit card debt. The money you obtain through the use of a credit card is not "free" money. There is a price (interest) to be paid for the privilege.

EXAMPLE 1 *Credit Cards.* After the holidays, Sarah has a balance of $3216.28 on a credit card with an annual percentage rate (APR) of 19.7%. She decides not to make additional purchases with this card until she has paid off the balance.

a) Many credit cards require a minimum monthly payment of 2% of the balance. What is Sarah's minimum payment on a balance of $3216.28? Round the answer to the nearest dollar.

b) Find the amount of interest and the amount applied to reduce the principal in the minimum payment found in part (a).

c) If Sarah had transferred her balance to a card with an APR of 12.5%, how much of her first payment would be interest and how much would be applied to reduce the principal?

d) Compare the amounts for 12.5% from part (c) with the amounts for 19.7% from part (b).

1. Credit Cards. After the holidays, Jamal has a balance of \$4867.59 on a credit card with an annual percentage rate (APR) of 21.3%. He decides not to make additional purchases with this card until he has paid off the balance.

a) Many credit cards require a minimum monthly payment of 2% of the balance. What is Jamal's minimum payment on a balance of \$4867.59? Round the answer to the nearest dollar.

b) Find the amount of interest and the amount applied to reduce the principal in the minimum payment found in part (a).

c) If Jamal had transferred his balance to a card with an APR of 13.6%, how much of his first payment would be interest and how much would be applied to reduce the principal?

d) Compare the amounts for 13.6% from part (c) with the amounts for 21.3% from part (b).

Answers on page A-11

We solve as follows.

a) We multiply the balance of \$3216.28 by 2%:

$0.02 \times \$3216.28 = \$64.3256.$ Sarah's minimum payment, rounded to the nearest dollar, is \$64.

b) The amount of interest on \$3216.28 at 19.7% for one month* is given by

$$I = P \cdot r \cdot t = \$3216.28 \times 0.197 \times \frac{1}{12} \approx \$52.80.$$

We subtract to find the amount applied to reduce the principal in the first payment:

$$\begin{aligned}\text{Amount applied to reduce the principal} &= \text{Minimum payment} - \text{Interest for the month}\\ &= \$64 - \$52.80\\ &= \$11.20.\end{aligned}$$

Thus the principal of \$3216.28 is decreased by only \$11.20 with the first payment. (Sarah still owes \$3205.08.)

c) The amount of interest on \$3216.28 at 12.5% for one month is

$$I = P \cdot r \cdot t = \$3216.28 \times 0.125 \times \frac{1}{12} \approx \$33.50.$$

We subtract to find the amount applied to reduce the principal in the first payment:

$$\begin{aligned}\text{Amount applied to reduce the principal} &= \text{Minimum payment} - \text{Interest for the month}\\ &= \$64 - \$33.50\\ &= \$30.50.\end{aligned}$$

Thus the principal of \$3216.28 is decreased by \$30.50 with the first payment. (Sarah still owes \$3185.78.)

d) Let's organize the information for both rates in the following table.

BALANCE BEFORE FIRST PAYMENT	FIRST MONTH'S PAYMENT	% APR	AMOUNT OF INTEREST	AMOUNT APPLIED TO PRINCIPAL	BALANCE AFTER FIRST PAYMENT
\$3216.28	\$64	19.7%	\$52.80	\$11.20	\$3205.08
3216.28	64	12.5	33.50	30.50	3185.78

Difference in balance after first payment → \$19.30

At 19.7%, the interest is \$52.80 and the principal is decreased by \$11.20. At 12.5%, the interest is \$33.50 and the principal is decreased by \$30.50. Thus the principal is decreased by \$30.50 − \$11.20, or \$19.30 more with the 12.5% rate than with the 19.7% rate. Thus the interest at 19.7% is \$19.30 greater than the interest at 12.5%.

*Actually, the interest on a credit card is computed daily with a rate called a daily percentage rate (DPR). The DPR for Example 1 would be 19.7%/365 ≈ 0.054%. When no payments or additional purchases are made during the month, the difference in total interest for the month is minimal and we will not deal with it here.

Do Exercise 1 on the preceding page.

Even though the mathematics of the information in the chart below is beyond the scope of this text, it is interesting to compare how long it takes to pay off the balance in Example 1 if Sarah continues to pay \$64 for each payment with how long it takes if she pays double that amount, \$128, for each payment. Financial consultants frequently tell clients that if they want to take control of their debt, they should pay double the minimum payment.

RATE	PAYMENT PER MONTH	NUMBER OF PAYMENTS TO PAY OFF DEBT	TOTAL PAID BACK	ADDITIONAL COST OF PURCHASES
19.7%	\$ 64	107, or 8 yr 11 mo	\$6848	\$3631.72
19.7	128	33, or 2 yr 9 mo	4224	1007.72
12.5	64	72, or 6 yr	4608	1391.72
12.5	128	29, or 2 yr 5 mo	3712	495.72

As with most loans, if you pay an extra amount toward the principal with each payment, the length of the loan can be greatly reduced. Note that at the rate of 19.7%, it will take Sarah almost 9 yr to pay off her debt if she pays only \$64 per month and does not make additional purchases. If she transfers her balance to a card with a 12.5% rate and pays \$128 per month, she could eliminate her debt in approximately $2\frac{1}{2}$ yr. You can see how debt can get out of control if you continue to make purchases and pay only the minimum payment. The debt will never be eliminated.

The Federal Stafford Loan program provides educational loans to students at interest rates that are much lower than those on credit cards. Payments on a loan do not begin until 6 months after graduation. At that time, the student has 10 yr, or 120 monthly payments, to pay off the loan.

EXAMPLE 2 *Federal Stafford Loans.* After graduation, the balance on Taylor's Stafford loan is \$28,650. If the rate on his loan is 3.37%, he will make 120 payments of approximately \$282 each to pay off the loan.

a) Find the amount of interest and the amount of principal in the first payment.

b) If the interest rate were 5.25%, he would make 120 monthly payments of approximately \$307 each. How much more of the first payment is interest if the loan is 5.25% rather than 3.37%?

c) Compare the total amount of interest on the loan at 3.37% with the amount on the loan at 5.25%. How much more would Taylor pay in interest on the 5.25% loan than on the 3.37% loan?

We solve as follows.

a) We use the formula $I = P \cdot r \cdot t$, substituting \$28,650 for P, 0.0337 for r, and 1/12 for t:

$$I = \$28{,}650 \times 0.0337 \times \frac{1}{12}$$

$$\approx \$80.46.$$

The amount of interest in the first payment is \$80.46. The payment is \$282. We subtract to determine the amount applied to the principal:

$$\$282 - \$80.46 = \$201.54.$$

With the first payment, the principal will be reduced by \$201.54.

2. **Federal Stafford Loans.** After graduation, the balance on Maggie's Stafford loan is $32,680. To pay off the loan at 3.37%, she will make 120 payments of approximately $321 each.

   a) Find the amount of interest and the amount of principal in the first payment.

   b) If the interest rate were 5.5%, she would make 120 payments of approximately $355 each. How much more of the first payment is interest if the loan is 5.5% rather than 3.37%?

   c) Compare the total amount of interest on the loan at 3.37% with the amount of interest on the loan at 5.5%. How much more would Maggie pay in interest on the 5.5% loan than on the 3.37% loan?

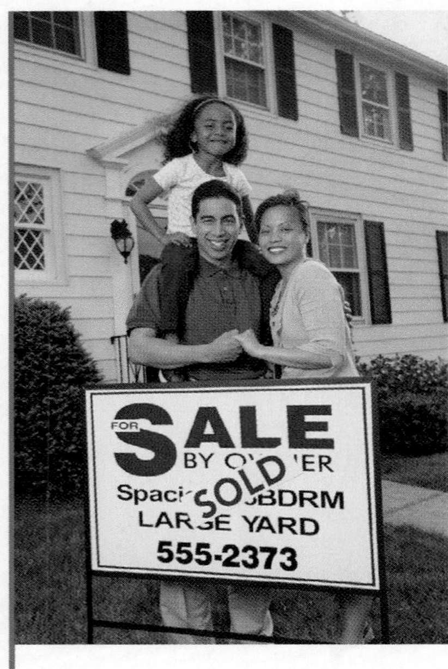

Answers on page A-11

b) The interest at 5.25% would be

$$I = \$28{,}650 \times 0.0525 \times \frac{1}{12}$$

$$\approx \$125.34.$$

At the rate of 5.25%, the additional interest in the first payment is

$\$125.34 - \$80.46 = \$44.88.$

The higher interest rate results in an additional $44.88 in interest in the first payment.

c) For the 3.37% loan, there will be 120 payments of $282 each:

$120 \times \$282 = \$33{,}840.$

The total amount of interest at this rate is

$\$33{,}840 - \$28{,}650 = \$5190.$

For the 5.25% loan, there will be 120 payments of $307 each:

$120 \times \$307 = \$36{,}840.$

The total amount of interest at this rate is

$\$36{,}840 - \$28{,}650 = \$8190.$

At the rate of 5.25%, Taylor would pay

$\$8190 - \$5190 = \$3000$

more in interest than at the rate of 3.37%.

Do Exercise 2.

EXAMPLE 3 *Home Loans.* The Sawyers recently purchased their first home. They borrowed $153,000 at $6\frac{5}{8}\%$ for 30 yr (360 payments). Their monthly payment (excluding insurance and taxes) is $979.68.

a) How much of the first payment is interest and how much is applied to reduce the principal?

b) If the Sawyers pay the entire 360 payments, how much interest will be paid on the loan?

We solve as follows.

a) To find the amount of interest paid in the first payment, we use the formula $I = P \cdot r \cdot t$:

$$I = P \cdot r \cdot t = \$153{,}000 \times 0.06625 \times \frac{1}{12} \approx \$844.69.$$

The amount applied to the principal is

$\$979.68 - \$844.69$, or $134.99.

b) Over the 30-yr period, the total paid will be

$360 \times \$979.68$, or $352,684.80.

The total amount of interest paid over the lifetime of the loan is

$\$352{,}684.80 - \$153{,}000$, or $199,684.80.

Do Exercises 3 and 4 on the following page.

AMORTIZATION TABLES

If we make 360 calculations as in Example 3(a) and continue with a decreased principal as in Margin Exercise 3, we can create an *amortization table,* part of which is shown below. Such tables are also found in reference books. The beginning, middle, and last part of the loan described are shown. Look over the table and note how small a portion of the payment reduces the principal at the beginning of the loan and how that portion increases throughout the lifetime of the loan. Do you see again why a loan is not "free"?

MORTGAGE AMORTIZATION PROGRAM

MORTGAGE AMOUNT: $153,000
INTEREST RATE: 6.625%
NUMBER OF YEARS: 30
MONTHLY PAYMENTS ARE: $979.68

PAYMENT	PRINCIPAL	INTEREST	BALANCE	
1	$134.99	$844.69	$152,865.01	← Example 3
2	$135.73	$843.94	$152,729.28	← Margin Exercise 3
3	$136.48	$843.19	$152,592.80	
4	$137.24	$842.44	$152,455.56	
5	$137.99	$841.68	$152,317.56	
6	$138.76	$840.92	$152,178.81	
7	$139.52	$840.15	$152,039.29	
8	$140.29	$839.38	$151,898.99	
9	$141.07	$838.61	$151,757.93	
10	$141.85	$837.83	$151,616.08	
11	$142.63	$837.05	$151,473.45	
12	$143.42	$836.26	$151,330.04	
⋮	⋮	⋮ ↳ Interest for 12 periods =		$10,086.14
174	$349.91	$629.77	$113,721.56	
175	$351.84	$627.84	$113,369.72	
176	$353.78	$625.90	$113,015.94	
177	$355.73	$623.94	$112,660.20	
178	$357.70	$621.98	$112,302.51	
179	$359.67	$620.00	$111,942.83	
180	$361.66	$618.02	$111,581.18	
181	$363.65	$616.02	$111,217.52	
182	$365.66	$614.01	$110,851.86	
183	$367.68	$611.99	$110,484.18	
184	$369.71	$609.96	$110,114.47	
185	$371.75	$607.92	$109,742.71	
186	$373.80	$605.87	$109,368.91	
⋮	⋮	⋮ ↳ Interest for 12 periods =		$8033.22
349	$917.04	$62.63	$10,427.84	
350	$922.11	$57.57	$9,505.74	
351	$927.20	$52.48	$8,578.54	
352	$932.32	$47.36	$7,646.23	
353	$937.46	$42.21	$6,708.76	
354	$942.64	$37.04	$5,766.13	
355	$947.84	$31.83	$4,818.28	
356	$953.07	$26.60	$3,865.21	
357	$958.34	$21.34	$2,906.87	
358	$963.63	$16.05	$1,943.24	
359	$968.95	$10.73	$974.30	
360	$974.30	$0.00	$0.00	
		↳ Interest for 12 periods =		$405.84

Total interest for 360 periods = $199,684.80

Refer to Example 3 for Margin Exercises 3 and 4.

3. **Home Loans.** Since the principal has been reduced by the first payment, at the time of the second payment of the Sawyers' 30-yr loan, the new principal is the decreased principal

$153,000 − $134.99,

or

$152,865.01.

Use $152,865.01 as the principal, and determine how much of the second payment is interest and how much is applied to reduce the principal. (In effect, repeat Example 3(a) using the new principal.)

4. **Home Loans.** The Sawyers decide to change the period of their home loan from 30 yr to 15 yr. Their monthly payment increases to $1343.33.

   a) How much of the first payment is interest and how much is applied to reduce the principal?

   b) If the Sawyers pay the entire 180 payments, how much interest will be paid on this loan?

   c) Compare the amount of interest to pay off the 15-yr loan with the amount of interest to pay off the 30-yr loan.

Answers on page A-11

5. Refinancing a Home Loan. Consider Example 4 for a 15-yr loan. The new monthly payment is $1250.14.

a) How much of the first payment is interest and how much is applied to reduce the principal?

b) If the Sawyers pay the entire 180 payments, how much interest will be paid on this loan?

c) Compare the amount of interest to pay off the 15-yr loan at $5\frac{1}{2}\%$ with the amount of interest to pay off the 15-yr loan at $6\frac{5}{8}\%$ in Margin Exercise 4.

Answers on page A-11

EXAMPLE 4 *Refinancing a Home Loan.* Refer to Example 3. Ten months after the Sawyers buy their home financed at a rate of $6\frac{5}{8}\%$, the rates drop to $5\frac{1}{2}\%$. After much consideration, they decide to refinance even though the new loan will cost them $1200 in refinance charges. They have reduced the principal a small amount in the 10 payments they have made, but they decide to again borrow $153,000 for 30 yr at the new rate. Their new monthly payment is $868.72.

a) How much of the first payment is interest and how much is applied to the principal?

b) Compare the amounts at $5\frac{1}{2}\%$ found in part (a) with the amounts at $6\frac{5}{8}\%$ found in Example 3(a).

c) With the lower house payment, how long will it take the Sawyers to recoup the refinance charge of $1200?

d) If the Sawyers pay the entire 360 payments, how much interest will be paid on this loan? How much less is the total interest at $5\frac{1}{2}\%$ than at $6\frac{5}{8}\%$?

We solve as follows.

a) To find the interest paid in the first payment, we use the formula $I = P \cdot r \cdot t$:

$$I = P \cdot r \cdot t = \$153{,}000 \times 0.055 \times \frac{1}{12} = \$701.25.$$

The amount applied to the principal is

$868.72 − $701.25, or $167.47.

b) We compare the amounts found in part (a) with the amounts found in Example 3(a):

Rate	*Monthly payment*	*Interest in first payment*	*Amount applied to principal*
$6\frac{5}{8}\%$	$979.68	$844.69	$134.99
$5\frac{1}{2}\%$	$868.72	$701.25	$167.47

At $5\frac{1}{2}\%$, the amount of interest in the first payment is $844.69 − $701.25, or $143.44, less than at $6\frac{5}{8}\%$. The amount applied to the principal is $167.47 − $134.99, or $32.48, more.

c) The monthly payment at $5\frac{1}{2}\%$ is $979.68 − $868.72, or $110.96, less than the payment at $6\frac{5}{8}\%$. The total savings each month is approximately $111. We can divide the cost of the refinancing by this monthly savings to determine the number of months it will take to recoup the $1200 refinancing charge: $1200 ÷ $111 ≈ 11. It will take the Sawyers approximately 11 months to break even.

d) Over the 30-yr period, the total paid will be

360 × $868.72, or $312,739.20.

The total amount of interest paid over the lifetime of the loan is

$312,739.20 − $153,000, or $159,739.20.

The total interest paid at $5\frac{1}{2}\%$ is

$199,684.80 (see Example 3) − $159,739.20, or $39,945.60,

less than the total interest paid at $6\frac{5}{8}\%$. Thus the $5\frac{1}{2}\%$ loan saves the Sawyers approximately $40,000 in interest charges over the 30 yr.

Do Exercise 5.

4.8 EXERCISE SET

For Extra Help

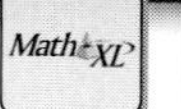
MathXL

MyMathLab

InterAct Math

Math Tutor Center

Video Lectures on CD Disc 2

Student's Solutions Manual

a Solve.

1. *Credit Cards.* At the end of his freshman year of college, Antonio has a balance of \$4876.54 on a credit card with an annual percentage rate (APR) of 21.3%. He decides not to make additional purchases with his card until he has paid off the balance.

a) Many credit cards require a minimum monthly payment of 2% of the balance. What is Antonio's minimum payment on a balance of \$4876.54? Round the answer to the nearest dollar.
b) Find the amount of interest and the amount applied to reduce the principal in the minimum payment found in part (a).
c) If Antonio had transferred his balance to a card with an APR of 12.6%, how much of his first payment would be interest and how much would be applied to reduce the principal?
d) Compare the amounts for 12.6% from part (c) with the amounts for 21.3% from part (b).

2. *Credit Cards.* At the end of her junior year of college, Becky had a balance of \$5328.88 on a credit card with an annual percentage rate (APR) of 18.7%. She decides not to make additional purchases with this card until she has paid off the balance.

a) Many credit cards require a minimum monthly payment of 2% of the balance. What is Becky's minimum payment on a balance of \$5328.88? Round the answer to the nearest dollar.
b) Find the amount of interest and the amount applied to reduce the principal in the minimum payment found in part (a).
c) If Becky had transferred her balance to a card with an APR of 13.2%, how much of her first payment would be interest and how much would be applied to reduce the principal?
d) Compare the amounts for 13.2% from part (c) with the amounts for 18.7% from part (b).

3. *Federal Stafford Loans.* After graduation, the balance on Grace's Stafford loan is \$44,560. To pay off the loan at 3.37%, she will make 120 payments of approximately \$437.93 each.

a) Find the amount of interest and the amount applied to reduce the principal in the first payment.
b) If the interest rate were 4.75%, she would make 120 monthly payments of approximately \$467.20 each. How much more of the first payment is interest if the loan is 4.75% rather than 3.37%?
c) Compare the total amount of interest on a loan at 3.37% with the amount on the loan at 4.75%. How much more would Grace pay on the 4.75% loan than on the 3.37% loan?

4. *Federal Stafford Loans.* After graduation, the balance on Ricky's Stafford loan is \$38,970. To pay off the loan at 3.37%, he will make 120 payments of approximately \$382.99 each.

a) Find the amount of interest and the amount applied to reduce the principal in the first payment.
b) If the interest rate were 5.4%, he would make 120 monthly payments of approximately \$421 each. How much more of the first payment is interest if the loan is 5.4% rather than 3.37%?
c) Compare the total amount of interest on the loan at 3.37% with the amount on the loan at 5.4%. How much more would Ricky pay on the 5.4% loan than on the 3.37% loan?

5. *Home Loan.* The Martinez family recently purchased a home. They borrowed \$164,000 at $6\frac{1}{4}\%$ for 30 yr (360 payments). Their monthly payment (excluding insurance and taxes) is \$1009.78.

a) How much of the first payment is interest and how much is applied to reduce the principal?
b) If this family pays the entire 360 payments, how much interest will be paid on the loan?
c) Determine the new principal after the first payment. Use that new principal to determine how much of the second payment is interest and how much is applied to reduce the principal.

6. *Home Loan.* The Kaufmans recently purchased a home. They borrowed \$136,000 at 5.75% for 30 yr (360 payments). Their monthly payment (excluding insurance and taxes) is \$793.66.

a) How much of the first payment is interest and how much is applied to reduce the principal?
b) If the Kaufmans pay the entire 360 payments, how much interest will be paid on the loan?
c) Determine the new principal after the first payment. Use that new principal to determine how much of the second payment is interest and how much is applied to reduce the principal.

7. *Refinancing a Home Loan.* Refer to Exercise 5. The Martinez family decides to change the period of their home loan to 15 yr. Their monthly payment increases to $1406.17.

   a) How much of the first payment is interest and how much is applied to reduce the principal?
   b) If the Martinez family pays the entire 180 payments, how much interest will be paid on the loan?
   c) Compare the amount of interest to pay off the 15-yr loan with the amount of interest to pay off the 30-yr loan.

8. *Refinancing a Home Loan.* Refer to Exercise 6. The Kaufmans decide to change the period of their home loan to 15 yr. Their monthly payment increases to $1129.36.

   a) How much of the first payment is interest and how much is applied to reduce the principal?
   b) If the Kaufmans pay the entire 180 payments, how much interest will be paid on the loan?
   c) Compare the amount of interest to pay off the 15-yr loan with the amount of interest to pay off the 30-yr loan.

Complete the following table, assuming monthly payments as given.

	INTEREST RATE	HOME MORTGAGE	TIME OF LOAN	MONTHLY PAYMENT	PRINCIPAL AFTER FIRST PAYMENT	PRINCIPAL AFTER SECOND PAYMENT
9.	6.98%	$100,000	360 mos	$663.96		
10.	6.98%	$100,000	180 mos	$897.71		
11.	8.04%	$100,000	180 mos	$957.96		
12.	8.04%	$100,000	360 mos	$736.55		
13.	7.24%	$150,000	360 mos	$1022.25		
14.	7.24%	$75,000	180 mos	$684.22		
15.	7.24%	$200,000	180 mos	$1824.60		
16.	7.24%	$180,000	360 mos	$1226.70		

17. *Dealership's Car Loan Offer.* For a trip to Colorado, Michael and Rebecca buy a 2005 Toyota Sienna van whose selling price is $23,950. For financing, they accept the promotion from the manufacturer that offers a 48-month loan at 2.9% with 10% down. Their monthly payment is $454.06.

   a) What is the down payment? the amount borrowed?
   b) How much of the first payment is interest and how much is applied to reduce the principal?
   c) What is the total interest cost of the loan if they pay all of the 48 payments?

18. *New-Car Loan.* After working at her first job for 2 yr, Janice buys a new Saturn for $16,385. She makes a down payment of $1385 and finances $15,000 for 4 yr at a new-car loan rate of 8.99%. Her monthly payment is $373.20.

   a) How much of her first payment is interest and how much is applied to reduce the principal?
   b) Find the principal balance at the beginning of the second month and determine how much less interest she will pay in the second payment than in the first.
   c) What is the total interest cost of the loan if she pays all of the 48 payments?

19. *Used-Car Loan.* Twin brothers, Jerry and Terry, each take a job at the college cafeteria in order to have the money to make payments on the purchase of a 2002 Chrysler PT Cruiser for $11,900. They make a down payment of 5% and finance the remainder at 9.3% for 3 yr. (Used-car loan rates are generally higher than new-car loan rates.) Their monthly payment is $361.08.

a) What is the down payment? the amount borrowed?
b) How much of the first payment is interest and how much is applied to reduce the principal?
c) If they pay all 36 payments, how much interest will they pay for the loan?

20. *Used-Car Loan.* For his construction job, Clint buys a 2003 Dodge Ram 1500 truck for $13,800. He makes a down payment of $1380 and finances the remainder for 4 yr at 8.8%. The monthly payment is $307.89.

a) How much is financed?
b) How much of the first payment is interest and how much is applied to reduce the principal?
c) If he pays all 48 payments, how much interest will he pay for the loan?

21. **DW** On the basis of the skills of mathematics you have obtained in this section, discuss the significant new ideas you now have about interest rates and credit cards that you didn't have before.

22. **DW** Examine the information in the graphs at the beginning of the section. Discuss how a knowledge of this section might have been of help to some of these students.

23. **DW** Compare the following two purchases and describe a situation in which each purchase is the best choice.

Purchase A: A new car for $15,145. The loan is for $14,500 at 6.9% for 4 yr. The monthly payment is $346.55.

Purchase B: A used car for $10,600. The loan is for $9300 at 12.5% for 3 yr. The monthly payment is $311.12.

24. **DW** Look over the examples and exercises in this section. What seems to happen to the monthly payment on a loan if the time of payment changes from 30 yr to 15 yr, assuming the interest rate stays the same? Discuss the pros and cons of both time periods.

SKILL MAINTENANCE

Solve. [4.1d]

25. $\frac{x}{12} = \frac{24}{16}$

26. $\frac{7}{2} = \frac{11}{x}$

Solve. [3.4b]

27. $0.64 \cdot x = 170$

28. $28.5 = 25.6 \times y$

Find decimal notation. [3.5a]

29. $\frac{5}{9}$

30. $\frac{23}{11}$

31. $\frac{11}{12}$

32. $\frac{13}{7}$

33. $\frac{15}{7}$

34. $\frac{19}{12}$

Convert to standard notation. [3.3b]

35. $4.03 trillion

36. 5.8 million

37. 42.7 million

38. 6.09 trillion

4 Summary and Review

The review that follows is meant to prepare you for a chapter exam. It consists of three parts. The first part, Concept Reinforcement, is designed to increase understanding through true/false exercises. The second part is a list of important properties and formulas. The third part is the Review Exercises. These provide practice exercises for the exam, together with references to section objectives so you can go back and review. Before beginning, stop and look back over the skills you have obtained. What skills in mathematics do you have now that you did not have before studying this chapter?

CONCEPT REINFORCEMENT

Determine whether the statement is true or false. Answers are given at the back of the book.

______ **1.** A fixed principal invested for 4 yr will earn more interest when interest is compounded quarterly than when interest is compounded semi-annually.

______ **2.** Of the numbers 0.5%, $\frac{5}{1000}$%, $\frac{1}{2}$%, $\frac{1}{5}$, and $0.\overline{1}$, the largest number is $0.\overline{1}$.

______ **3.** If principal A equals principal B and principal A is invested for 2 yr at 4% compounded quarterly while principal B is invested for 4 yr at 2% compounded semi-annually, the interest earned from each investment is the same.

______ **4.** The symbol % is equivalent to $\times \frac{1}{10}$.

IMPORTANT PROPERTIES AND FORMULAS

Commission = Commission rate × Sales
Discount = Rate of discount × Original price
Sale price = Original price − Discount

Simple Interest: $I = P \cdot r \cdot t$

Compound Interest: $A = P \cdot \left(1 + \frac{r}{n}\right)^{n \cdot t}$

Review Exercises

Write fraction notation for the ratio. Do not simplify. [4.1a]

1. 47 to 84

2. 46 to 1.27

3. 83 to 100

4. 0.72 to 197

Find the ratio of the first number to the second number and simplify. [4.1a]

5. 9 to 12

6. 3.6 to 6.4

7. *Gas Mileage.* The Chrysler PT Cruiser will travel 377 mi on 14.5 gal of gasoline in highway driving. What is the rate in miles per gallon? [4.1b]

Source: DaimlerChrysler Corporation

8. *CD-ROM Spin Rate.* A 12x CD-ROM on a computer will spin 472,500 revolutions if left running for 75 min. What is the rate of its spin in revolutions per minute (rpm)? [4.1b]

Source: *Electronic Engineering Times,* June 1997

9. A lawn requires 319 gal of water for every 500 ft^2. What is the rate in gallons per square foot? [4.1b]

10. *Turkey Servings.* A 25-lb turkey serves 18 people. Find the rate in servings per pound. [4.1b]

Determine whether the two pairs of numbers are proportional. [4.1c]

11. 9, 15 and 36, 59

12. 24, 37 and 42, 64.75

Solve. [4.1d]

13. $\frac{8}{9} = \frac{x}{36}$

14. $\frac{6}{x} = \frac{48}{56}$

15. $\frac{120}{\frac{3}{7}} = \frac{7}{x}$

16. $\frac{4.5}{120} = \frac{0.9}{x}$

Solve. [4.1e]

17. If 3 dozen eggs cost \$2.67, how much will 5 dozen eggs cost?

18. *Quality Control.* A factory manufacturing computer circuits found 3 defective circuits in a lot of 65 circuits. At this rate, how many defective circuits can be expected in a lot of 585 circuits?

19. A train travels 448 mi in 7 hr. At this rate, how far will it travel in 13 hr?

20. *Garbage Production.* It is known that 5 people produce 13 kg of garbage in one day. San Diego, California, has 1,266,753 people. How many kilograms of garbage are produced in San Diego in one day?

21. *Exchanging Money.* On 29 January 2006, 1 U.S. dollar was worth about 0.8244 European Monetary Unit (Euro).

a) How much would 250 U.S. dollars be worth in Euros?
b) Jamal was traveling in France and saw a sweatshirt that cost 50 Euros. How much would it cost in U.S. dollars?

22. Fifteen acres are required to produce 54 bushels of tomatoes. At this rate, how many acres are required to produce 97.2 bushels of tomatoes?

23. *Snow to Water.* Under typical conditions, $1\frac{1}{2}$ ft of snow will melt to 2 in. of water. To how many inches of water will $4\frac{1}{2}$ ft of snow melt?

24. *Lawyers in Michigan.* In Michigan, there are 2.3 lawyers for every 1000 people. The population of Detroit is 911,402. How many lawyers would you expect there to be in Detroit?

Source: U.S. Bureau of the Census

Find percent notation for the decimal notation in the sentence in Exercises 25 and 26. [4.2b]

25. Of all the snacks eaten on Super Bowl Sunday, 0.56 of them are chips and salsa.

Source: Korbel Research and Pace Foods

26. Of all the vehicles in Mexico City, 0.017 of them are taxis.

Source: *The Handy Geography Answer Book*

Find percent notation. [4.3a]

27. $\frac{3}{8}$

28. $\frac{1}{3}$

Find decimal notation. [4.2b]

29. 73.5%

30. $6\frac{1}{2}\%$

Find fraction notation. [4.3b]

31. 24%

32. 6.3%

Translate to a percent equation. Then solve. [4.4a, b]

33. 30.6 is what percent of 90?

34. 63 is 84 percent of what?

35. What is $38\frac{1}{2}\%$ of 168?

Translate to a proportion. Then solve. [4.5a, b]

36. 24 percent of what is 16.8?

37. 42 is what percent of 30?

38. What is 10.5% of 84?

Solve. [4.6a, b]

39. *Favorite Ice Creams.* According to a recent survey, 8.9% of those interviewed chose chocolate as their favorite ice cream flavor and 4.2% chose butter pecan. Of the 2500 students in a freshman class, how many would you expect to choose chocolate as their favorite ice cream? butter pecan?

Source: International Ice Cream Association

40. *Prescriptions.* Of the 295 million people in the United States, 123.64 million take at least one kind of prescription drug per day. What percent take at least one kind of prescription drug per day?

Source: American Society of Health-System Pharmacies

41. *Water Output.* The average person expels 200 mL of water per day by sweating. This is 8% of the total output of water from the body. How much is the total output of water?

Source: Elaine N. Marieb, *Essentials of Human Anatomy and Physiology*, 6th ed. Boston: Addison Wesley Longman, Inc., 2000

42. *Test Scores.* Jason got a 75 on a math test. He was allowed to go to the math lab and take a retest. He increased his score to 84. What was the percent of increase?

43. *Test Scores.* Jenny got an 80 on a math test. By taking a retest in the math lab, she increased her score by 15%. What was her new score?

Solve. [4.7a, b, c]

44. A state charges a meals tax of $4\frac{1}{2}\%$. What is the meals tax charged on a dinner party costing $320?

45. In a certain state, a sales tax of $378 is collected on the purchase of a used car for $7560. What is the sales tax rate?

46. Kim earns $753.50 selling $6850 worth of televisions. What is the commission rate?

47. An air conditioner has a marked price of $350. It is placed on sale at 12% off. What are the discount and the sale price?

48. The price of a fax machine is marked down from $305 to $262.30. What is the rate of discount?

49. An insurance salesperson receives a 7% commission. If \$42,000 worth of life insurance is sold, what is the commission?

50. Find the rate of discount.

Our lowest price
of the season!
18-ft metal frame
399^{69}
Reg. 489.99
Family-size
metal-frame pool

Solve. [4.7d, e]

51. What is the simple interest on \$1800 at 6% for $\frac{1}{3}$ year?

52. The Dress Shack borrows \$24,000 at 10% simple interest for 60 days. Find **(a)** the amount of interest due and **(b)** the total amount that must be paid after 60 days.

53. What is the simple interest on \$2200 principal at the interest rate of 5.5% for 1 year?

54. The Kleins invest \$7500 in an investment account paying an annual interest rate of 12%, compounded monthly. How much is in the account after 3 months?

55. Find the amount in an investment account if \$8000 is invested at 9%, compounded annually, for 2 years.

Solve. [4.8a]

56. *Credit Cards.* At the end of her junior year of college, Judy has a balance of \$6428.74 on a credit card with an annual percentage rate (APR) of 18.7%. She decides not to make additional purchases with this card until she has paid off the balance.

a) Many credit cards require a minimum payment of 2% of the balance. What is Judy's minimum payment on a balance of \$6428.74? Round the answer to the nearest dollar.

b) Find the amount of interest and the amount applied to reduce the principal in the minimum payment found in part (a).

c) If Judy had transferred her balance to a card with an APR of 13.2%, how much of her first payment would be interest and how much would be applied to reduce the principal?

d) Compare the amounts for 13.2% from part (c) with the amounts for 18.7% from part (b).

57. $^{D}{}_{W}$ Ollie buys a microwave oven during a 10%-off sale. The sale price that Ollie paid was \$162. To find the original price, Ollie calculates 10% of \$162 and adds that to \$162. Is this correct? Why or why not? [4.7c]

58. $^{D}{}_{W}$ Which is the better deal for a consumer and why: a discount of 40% or a discount of 20% followed by another of 22%? [4.7c]

SYNTHESIS

59. Shine-and-Glo Painters uses 2 gal of finishing paint for every 3 gal of primer. Each gallon of finishing paint covers 450 ft^2. If a surface of 4950 ft^2 needs both primer and finishing paint, how many gallons of each should be purchased? [4.1e]

60. Tara's Dress Shop reduces the price of a dress by 40% during a sale. By what percent must the store increase the sale price, after the sale, to get back to the original price? [4.7c]

4 Chapter Test

For Extra Help

Work It Out! Chapter Test Video Solutions on CD

Write fraction notation for the ratio. Do not simplify.

1. 85 to 97

2. 0.34 to 124

Find the ratio of the first number to the second number and simplify.

3. 18 to 20

4. 0.75 to 0.96

5. *Gas Mileage.* The 2005 Chevrolet Malibu Maxx will travel 319 mi on 14.5 gal of gasoline in city driving. What is the rate in miles per gallon?

Source: General Motors Corporation

6. *Ham Servings.* A 12-lb shankless ham contains 16 servings. What is the rate in servings per pound?

Determine whether the two pairs of numbers are proportional.

7. 7, 8 and 63, 72

8. 1.3, 3.4 and 5.6, 15.2

Solve.

9. $\frac{9}{4} = \frac{27}{x}$

10. $\frac{150}{2.5} = \frac{x}{6}$

Solve.

11. *Time Loss.* A watch loses 2 min in 10 hr. At this rate, how much will it lose in 24 hr?

12. *Map Scaling.* On a map, 3 in. represents 225 mi. If two cities are 7 in. apart on the map, how far are they apart in reality?

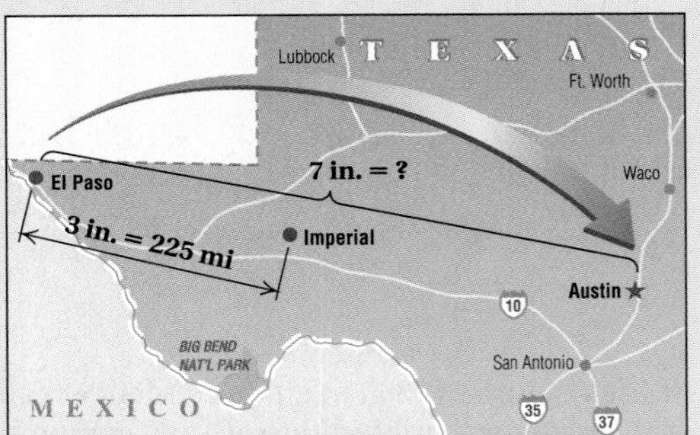

13. *Exchanging Money.* On 29 January 2006, 1 U.S. dollar was worth about 7.7565 Hong Kong dollars.

a) How much would 450 U.S. dollars be worth in Hong Kong dollars?

b) Mitchell was traveling in Hong Kong and saw a DVD player that cost 795 Hong Kong dollars. How much would it cost in U.S. dollars?

14. *Automobile Violations.* In a recent year, the Indianapolis Police Department employed 1088 officers and made 37,493 arrests. At this rate, how many arrests could be made if the number of officers were increased to 2500?

Source: *Indianapolis Star,* 12-31-00

15. *Bookmobiles.* Since 1991, the number of bookmobiles has decreased by approximately 6.4%. Find decimal notation for 6.4%.

Source: American Library Association

16. *Gravity.* The gravity of Mars is 0.38 as strong as Earth's. Find percent notation for 0.38.

Source: www.marsinstitute.info/epo/mermarsfacts.html

17. Find percent notation for $\frac{11}{8}$.

18. Find fraction notation for 65%.

19. Translate to a percent equation. Then solve.

What is 40% of 55?

20. Translate to a proportion. Then solve.

What percent of 80 is 65?

Solve.

21. *Cruise Ship Passengers.* Of the passengers on a typical cruise ship, 16% are in the 25–34 age group and 23% are in the 35–44 age group. A cruise ship has 2500 passengers. How many are in the 25–34 age group? the 35–44 age group?

Source: Polk

22. Derek Jeter of the New York Yankees got 202 hits during the 2005 baseball season. This was about 30.9% of his at-bats. How many at-bats did he have?

Source: Major League Baseball

23. *Airline Profits.* Profits of the entire U.S. airline industry decreased from \$5.5 billion in 1999 to \$2.7 billion in 2000. Find the percent of decrease.

Source: Air Transport Association

24. There are 6.6 billion people living in the world today. It is estimated that the total number who have ever lived is about 120 billion. What percent of people who have ever lived are alive today?

Source: *The Handy Geography Answer Book*

25. *Maine Sales Tax.* The sales tax rate in Maine is 5%. How much tax is charged on a purchase of \$324? What is the total price?

26. Gwen's commission rate is 15%. What is the commission from the sale of \$4200 worth of merchandise?

27. The marked price of a DVD player is \$200 and the item is on sale at 20% off. What are the discount and the sale price?

28. What is the simple interest on a principal of \$120 at the interest rate of 7.1% for 1 year?

29. The Burnham Parents–Teachers Association invests \$5200 at 6% simple interest. How much is in the account after $\frac{1}{2}$ year?

30. Find the amount in an account if \$1000 is invested at $5\frac{3}{8}\%$, compounded annually, for 2 years.

31. The Suarez family invests \$10,000 at an annual interest rate of 4.9%, compounded monthly. How much is in the account after 3 years?

32. *Job Opportunities.* The table below lists job opportunities, in millions, in 2002 and projected increases to 2012. Find the missing numbers.

OCCUPATION	NUMBER OF JOBS IN 2002 (in millions)	NUMBER OF JOBS IN 2012 (in millions)	CHANGE	PERCENT OF INCREASE
Retail salespersons	4.1	4.8	0.7	17.1%
Registered nurses	2.3		0.6	
Post-secondary teachers	1.6	2.2		
Food preparation and service workers		2.4	0.4	
Restaurant servers		2.5		19.0%

Source: Department of Labor

33. Find the discount and the discount rate of the washer–dryer duet in this ad.

34. *Home Loan.* Complete the following table, assuming the monthly payment as given.

Interest Rate	7.4%
Mortgage	\$120,000
Time of Loan	360 mos
Monthly Payment	\$830.86
Principal after First Payment	
Principal after Second Payment	

SYNTHESIS

35. By selling a home without using a realtor, Juan and Marie can avoid paying a 7.5% commission. They receive an offer of \$180,000 from a potential buyer. In order to give a comparable offer, for what price would a realtor need to sell the house? Round to the nearest hundred.

36. Karen's commission rate is 16%. She invests her commission from the sale of \$15,000 worth of merchandise at an interest rate of 12%, compounded quarterly. How much is Karen's investment worth after 6 months?

Data, Graphs, and Statistics

5

Real-World Application

The rhinoceros is considered one of the world's most endangered animals. The worldwide total number of rhinoceroses is approximately 17,800. Using the pictograph shown in Example 3 on page 351, determine how many more black rhinos there are than Indian/Nepalese rhinos.

This problem appears as Example 3 in Section 5.2.

5.1 AVERAGES, MEDIANS, AND MODES

Objectives

a Find the average of a set of numbers and solve applied problems involving averages.

b Find the median of a set of numbers and solve applied problems involving medians.

c Find the mode of a set of numbers and solve applied problems involving modes.

d Compare two sets of data using their means.

In real-world applications, we can use tables and graphs of various kinds to show information about data and to extract information from data that can lead us to make analyses and predictions. Graphs allow us to communicate a message from the data.

For example, the following two pages show data regarding the number of alcohol-related traffic deaths in recent years. Examine each method of presentation. Which method, if any, do you like the best and why?

Paragraph

The National Highway Traffic Safety Administration has released data regarding the number of alcohol-related traffic deaths for various years. In 1982, there were 26,173 deaths; in 1984, there were 24,762 deaths; in 1986, there were 25,017 deaths; in 1988, there were 23,833 deaths; in 1990, there were 22,587 deaths; in 1992, there were 18,290 deaths; in 1994, there were 17,308 deaths; in 1996, there were 17,749 deaths; in 1998, there were 16,673 deaths; in 2000, there were 17,380 deaths; and finally, in 2002, there were 17,419 deaths.

Table

YEAR	ALCOHOL-RELATED TRAFFIC DEATHS
1982	26,173
1984	24,762
1986	25,017
1988	23,833
1990	22,587
1992	18,290
1994	17,308
1996	17,749
1998	16,673
2000	17,380
2002	17,419

Source: *2002 FARS Annual Report File*, FHWA's Highway Statistics Annual Series, from the National Center for Statistics and Analysis

Pictograph

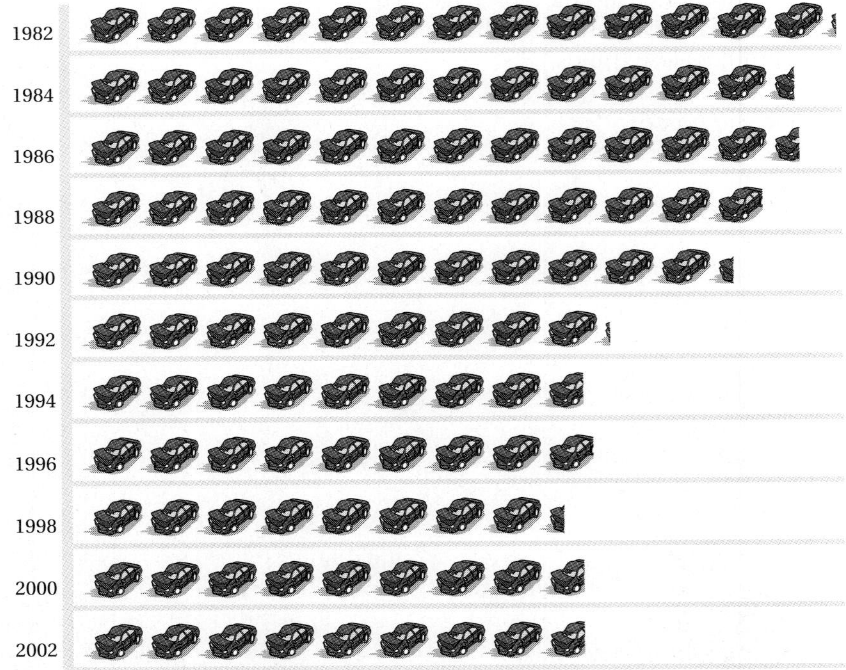

Bar Graph

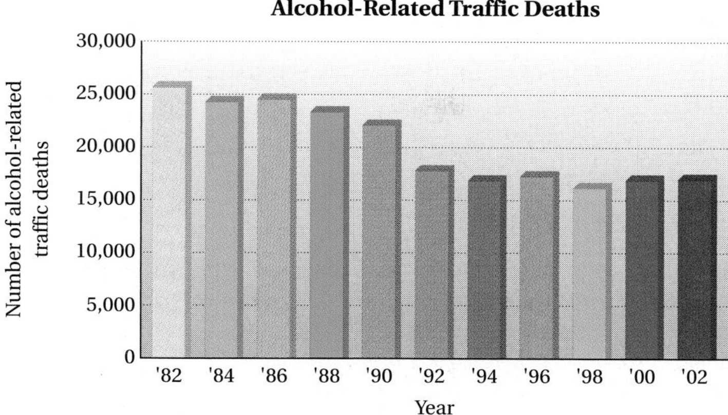

Line Graph

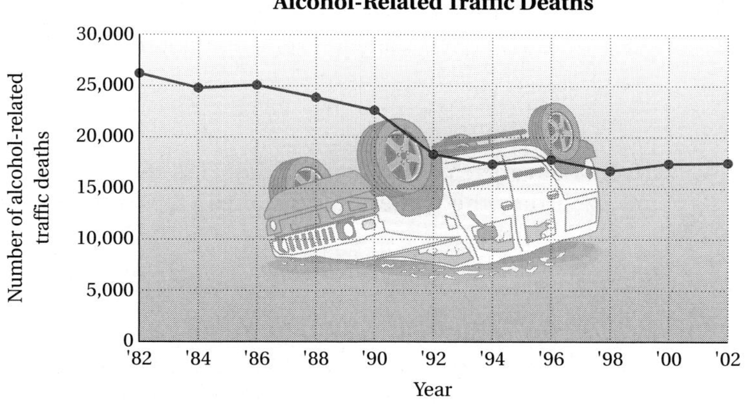

Most people would not find the paragraph method for displaying the data most useful. It takes time to read, and it is hard to look for a trend. The bar and line graphs are more helpful if we want to see the overall trend of decreased number of deaths and to make a prediction about years to come. The exact data values are most easily read in tables.

In this chapter, we will learn not only how to extract information from various kinds of tables and graphs, but also how to create various kinds of graphs, such as bar, line, and circle.

a Averages

A **statistic** is a number describing a set of data. One statistic is a *center point*, or *measure of central tendency*, that characterizes the data. The most common kind of center point is the *mean*, or *average*, of a set of numbers. We first considered averages in Section 1.6 and extended the coverage in Sections 2.6 and 3.4.

Let's consider the data of the number of alcohol-related traffic deaths:

26,173, 24,762, 25,017, 23,833, 22,587, 18,290, 17,308, 17,749, 16,673, 17,380, 17,419.

What is the average of this set of numbers? First, we add the numbers:

$$26{,}173 + 24{,}762 + 25{,}017 + 23{,}833 + 22{,}587 + 18{,}290 + 17{,}308 + 17{,}749 + 16{,}673 + 17{,}380 + 17{,}419 = 227{,}191.$$

Study Tips

THE AW MATH TUTOR CENTER

www.aw-bc.com/tutorcenter

The AW Math Tutor Center is staffed by highly qualified mathematics instructors who provide students with tutoring on text examples and odd-numbered exercises. Tutoring is provided free to students who have bought a new textbook with a special access card bound with the book. Tutoring is available by toll-free telephone, toll-free fax, e-mail, and the internet. White-board technology allows tutors and students to actually see problems worked while they "talk" live in real time during the tutoring sessions. If you purchased a book without this card, you can purchase an access code through your bookstore using ISBN# 0-201-72170-8. (This is also discussed in the Preface.)

Find the average.

1. 14, 175, 36

2. 75, 36.8, 95.7, 12.1

3. A student scored the following on five tests: 68, 85, 82, 74, 96. What was the average score?

4. In the first five games, a basketball player scored points as follows: 26, 21, 13, 14, 23. Find the average number of points scored per game.

5. Home-Run Batting Average. At the end of the 2004 baseball season, Barry Bonds had the most career home runs of any active player in the major leagues, 703 in 19 seasons. What was his average number of home runs per season?

Source: Major League Baseball

Answers on page A-12

Next, we divide by the number of data items, 11:

$$\frac{227{,}191}{11} \approx 20{,}654. \qquad \text{Rounding to the nearest one}$$

Note that if the number of traffic deaths had been the average (same) for 11 yr, we would have

$$20{,}654 + 20{,}654 + 20{,}654 + 20{,}654 + 20{,}654 + 20{,}654 + 20{,}654 + 20{,}654 + 20{,}654 + 20{,}654 + 20{,}654$$
$$= 227{,}194 \approx 227{,}191.$$

The number 20,654 is called the *average* of the set of numbers. It is also called the **arithmetic** (pronounced ăr´ ĭth-mĕt´-ĭk) **mean** or simply the **mean.**

AVERAGE

To find the **average** of a set of numbers, add the numbers and then divide by the number of data items.

EXAMPLE 1 On a 4-day trip, a car was driven the following number of miles each day: 240, 302, 280, 320. What was the average number of miles per day?

$$\frac{240 + 302 + 280 + 320}{4} = \frac{1142}{4}, \quad \text{or} \quad 285.5$$

The car was driven an average of 285.5 mi per day. Had the car been driven exactly 285.5 mi each day, the same total distance (1142 mi) would have been traveled.

Do Exercises 1–4.

EXAMPLE 2 *Scoring Average.* Kareem Abdul-Jabbar is the all-time leading scorer in the history of the National Basketball Association. He scored 38,387 points in 1560 games. What was the average number of points scored per game? Round to the nearest tenth.

Source: National Basketball Association

We already know the total number of points, 38,387, and the number of games, 1560. We divide and round to the nearest tenth:

$$\frac{38{,}387}{1560} = 24.60705\ldots \approx 24.6.$$

Abdul-Jabbar's average was 24.6 points per game.

Do Exercise 5.

EXAMPLE 3 *Gas Mileage.* The 2005 Chevrolet Cobalt LS, shown here, is estimated to get 450 mi of city driving on 18 gal of gasoline. What is the expected average number of miles per gallon—that is, what is its gas mileage for city driving?

Source: *Motor Trend,* July 2005

We divide the total number of miles, 450, by the total number of gallons, 18:

$$\frac{450}{18} = 25 \text{ mpg}.$$

The Chevrolet Cobalt's expected average is 25 miles per gallon for city driving.

Do Exercise 6.

6. Gas Mileage. The 2005 Chevrolet Cobalt LS gets 612 mi of highway driving on 18 gal of gasoline. What is the average number of miles per gallon—that is, what is its gas mileage for highway driving?

Source: *Motor Trend,* July 2005

EXAMPLE 4 *Grade Point Average.* In most colleges, students are assigned grade point values for grades obtained. The **grade point average,** or **GPA,** is the average of the grade point values for each credit hour taken. At most colleges, grade point values are assigned as follows:

A: 4.0 B: 3.0 C: 2.0 D: 1.0 F: 0.0

Meg earned the following grades for one semester. What was her grade point average?

COURSE	GRADE	NUMBER OF CREDIT HOURS IN COURSE
Colonial History	B	3
Basic Mathematics	A	4
English Literature	A	3
French	C	4
Time Management	D	1

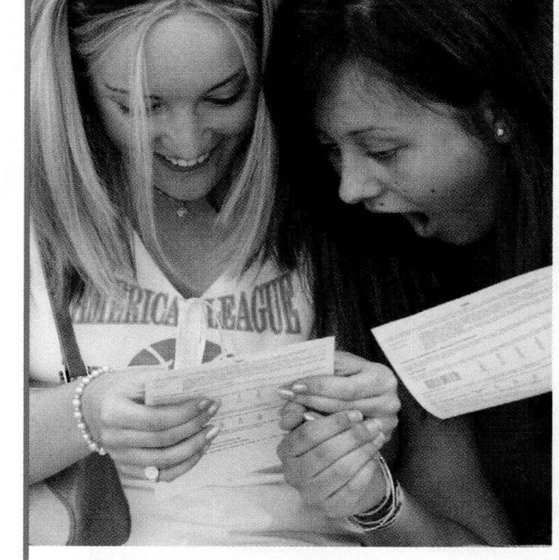

To find the GPA, we first multiply the grade point value (in color below) by the number of credit hours in the course to determine the number of *quality points,* and then add, as follows:

Colonial History	$3.0 \cdot 3 =$	9
Basic Mathematics	$4.0 \cdot 4 =$	16
English Literature	$4.0 \cdot 3 =$	12
French	$2.0 \cdot 4 =$	8
Time Management	$1.0 \cdot 1 =$	1
		46 (Total)

Answer on page A-12

7. Grade Point Average. Alex earned the following grades one semester.

GRADE	NUMBER OF CREDIT HOURS IN COURSE
B	3
C	4
C	4
A	2

What was Alex's grade point average? Assume that the grade point values are 4.0 for an A, 3.0 for a B, and so on. Round to the nearest tenth.

8. Grading. To get an A in math, Rosa must score an average of 90 on the tests. On the first three tests, her scores were 80, 100, and 86. What is the lowest score that Rosa can get on the last test and still get an A?

Answers on page A-12

The total number of credit hours taken is 3 + 4 + 3 + 4 + 1, or 15. We divide the number of quality points, 46, by the number of credits, 15, and round to the nearest tenth:

$$\text{GPA} = \frac{46}{15} \approx 3.1.$$

Meg's grade point average was 3.1.

Do Exercise 7.

EXAMPLE 5 *Grading.* To get a B in math, Geraldo must score an average of 80 on the tests. On the first four tests, his scores were 79, 88, 64, and 78. What is the lowest score that Geraldo can get on the last test and still get a B?

We can find the total of the five scores needed as follows:

$$80 + 80 + 80 + 80 + 80 = 5 \cdot 80, \quad \text{or} \quad 400.$$

The total of the scores on the first four tests is

$$79 + 88 + 64 + 78 = 309.$$

Thus Geraldo needs to get at least

$$400 - 309, \quad \text{or} \quad 91$$

in order to get a B. We can check this as follows:

$$\frac{79 + 88 + 64 + 78 + 91}{5} = \frac{400}{5}, \quad \text{or} \quad 80.$$

Do Exercise 8.

CALCULATOR CORNER

Computing Averages Averages can be computed easily on a calculator. We must keep the rules for order of operations in mind when doing this. For example, to calculate

$$\frac{84 + 92 + 79}{3}$$

on a calculator with parenthesis keys, we press [(] [8] [4] [+] [9] [2] [+] [7] [9] [)] [÷] [3] [=]. On a calculator without parenthesis keys, we first add the numbers in the numerator and then divide that result by 3. To do this, we press [8] [4] [+] [9] [2] [+] [7] [9] [=] [÷] [3] [=]. In either case, the result is 85.

Exercises:

1. Use a calculator to perform the computation in Example 1.
2. Use a calculator to perform the computations in Margin Exercises 1–4.
3. What would the result have been if we had not used parentheses in the first set of keystrokes above? (Keep the rules for order of operations in mind.)

b Medians

Another type of center-point statistic is the *median.* Medians are useful when we wish to de-emphasize unusually extreme scores. For example, suppose a small class scored as follows on an exam.

Phil:	78	Pat:	56
Jill:	81	Olga:	84
Matt:	82		

Let's first list the scores in order from smallest to largest:

56, 78, 81, 82, 84.

↑ Middle score (81)

The middle score—in this case, 81—is called the **median.** Note that because of the extremely low score of 56, the average of the scores is 76.2. In this example, the median may be a more appropriate center-point statistic.

EXAMPLE 6 What is the median of this set of numbers?

99, 870, 91, 98, 106, 90, 98

We first rearrange the numbers in order from smallest to largest. Then we locate the middle number, 98.

90, 91, 98, 98, 99, 106, 870

↑ Middle number (98)

The median is 98.

Do Exercises 9–11.

MEDIAN

Once a set of data is listed in order, from smallest to largest, the **median** is the middle number if there is an odd number of data items. If there is an even number of items, the median is the number that is the average of the two middle numbers.

EXAMPLE 7 What is the median of this set of numbers?

69, 80, 61, 63, 62, 65

We first rearrange the numbers in order from smallest to largest. There is an even number of numbers. We look for the middle two, which are 63 and 65. The median is halfway between 63 and 65, the number 64.

61, 62, 63, 65, 69, 80

The average of the middle numbers is $\frac{63 + 65}{2}$, or 64.

The median is 64.

Find the median.

9. 17, 13, 18, 14, 19

10. 20, 14, 13, 19, 16, 18, 17

11. 78, 81, 83, 91, 103, 102, 122, 119, 88

Answers on page A-12

Find the median.

12. Salaries of Part-Time Typists.
$3300, $4000, $3900, $3600, $3800, $3400

13. 68, 34, 67, 69, 34, 70

Find the modes of these data.

14. 23, 45, 45, 45, 78

15. 34, 34, 67, 67, 68, 70

16. 13, 24, 27, 28, 67, 89

17. In a lab, Gina determined the mass, in grams, of each of five eggs:

15 g, 19 g, 19 g, 14 g, 18 g.

a) What is the mean?

b) What is the median?

c) What is the mode?

Answers on page A-12

EXAMPLE 8 *Salaries.* The following are the salaries of the top four employees of Verducci's Dress Company. What is the median of the salaries?

$85,000, $100,000, $78,000, $84,000

We rearrange the numbers in order from smallest to largest. The two middle numbers are $84,000 and $85,000. Thus the median is halfway between $84,000 and $85,000 (the average of $84,000 and $85,000):

$78,000, $\underbrace{\$84{,}000, \quad \$85{,}000,}$ $100,000

$$\text{Median} = \frac{\$84{,}000 + \$85{,}000}{2} = \frac{\$169{,}000}{2} = \$84{,}500.$$

Do Exercises 12 and 13.

C Modes

The final type of center-point statistic is the *mode.*

MODE

The **mode** of a set of data is the number or numbers that occur most often. If each number occurs the same number of times, there is *no* mode.

EXAMPLE 9 Find the mode of these data.

13, 14, 17, 17, 18, 19

The number that occurs most often is 17. Thus the mode is 17.

A set of data has just one average (mean) and just one median, but it can have more than one mode. It is also possible for a set of data to have no mode—when all numbers are equally represented. For example, the set of data 5, 7, 11, 13, 19 has no mode.

EXAMPLE 10 Find the modes of these data.

33, 34, 34, 34, 35, 36, 37, 37, 37, 38, 39, 40

There are two numbers that occur most often, 34 and 37. Thus the modes are 34 and 37.

Do Exercises 14–17.

Which statistic is best for a particular situation? If someone is bowling, the *average* from several games is a good indicator of that person's ability. If someone is applying for a job, the *median* salary at that business is often most indicative of what people are earning there because although executives tend to make a lot more money, there are fewer of them. Finally, if someone is reordering for a clothing store, the *mode* of the sizes sold is probably the most important statistic.

d Comparing Two Sets of Data

We have seen how to calculate averages, medians, and modes from data. A way to analyze two sets of data is to make a determination about which of two groups is "better." One way to do so is by comparing the averages.

EXAMPLE 11 *Battery Testing.* An experiment is performed to compare battery quality. Two kinds of battery were tested to see how long, in hours, they kept a portable CD player running. On the basis of this test, which battery is better?

BATTERY A: ETERNREADY TIMES (in hours)			BATTERY B: STURDYCELL TIMES (in hours)		
27.9	28.3	27.4	28.3	27.6	27.8
27.6	27.9	28.0	27.4	27.6	27.9
26.8	27.7	28.1	26.9	27.8	28.1
28.2	26.9	27.4	27.9	28.7	27.6

Note that it is difficult to analyze the data at a glance because the numbers are close together. We need a way to compare the two groups. Let's compute the average of each set of data.

Battery A: Average

$$= \frac{27.9 + 28.3 + 27.4 + 27.6 + 27.9 + 28.0 + 26.8 + 27.7 + 28.1 + 28.2 + 26.9 + 27.4}{12}$$

$$= \frac{332.2}{12} \approx 27.68$$

Battery B: Average

$$= \frac{28.3 + 27.6 + 27.8 + 27.4 + 27.6 + 27.9 + 26.9 + 27.8 + 28.1 + 27.9 + 28.7 + 27.6}{12}$$

$$= \frac{333.6}{12} = 27.8$$

We see that the average time of battery B is higher than that of battery A and thus conclude that battery B is "better." (It should be noted that statisticians might question whether these differences are what they call "significant." The answer to that question belongs to a later math course.)

Do Exercise 18.

18. **Growth of Wheat.** Rudy experiments to see which of two kinds of wheat is better. (In this situation, the shorter wheat is considered "better.") He grows both kinds under similar conditions and measures stalk heights, in inches, as follows. Which kind is better?

WHEAT A STALK HEIGHTS (in inches)			
16.2	42.3	19.5	25.7
25.6	18.0	15.6	41.7
22.6	26.4	18.4	12.6
41.5	13.7	42.0	21.6

WHEAT B STALK HEIGHTS (in inches)			
19.7	18.4	19.7	17.2
19.7	14.6	32.0	25.7
14.0	21.6	42.5	32.6
22.6	10.9	26.7	22.8

Answer on page A-12

5.1 EXERCISE SET

For Extra Help

MathXL | MyMathLab | InterAct Math | Math Tutor Center | Video Lectures on CD Disc 3 | Student's Solutions Manual

a, b, c For each set of numbers, find the average, the median, and any modes that exist.

1. 17, 19, 29, 18, 14, 29

2. 72, 83, 85, 88, 92

3. 5, 37, 20, 20, 35, 5, 25

4. 13, 32, 25, 27, 13

5. 4.3, 7.4, 1.2, 5.7, 7.4

6. 13.4, 13.4, 12.6, 42.9

7. 234, 228, 234, 229, 234, 278

8. $29.95, $28.79, $30.95, $29.95

9. *Atlantic Storms and Hurricanes.* The following bar graph shows the number of Atlantic storms or hurricanes that formed in various months from 1980 to 2000. What is the average number for the 9 months given? the median? the mode?

Source: Colorado State University

10. *Tornadoes.* The following bar graph shows the average number of tornado deaths by month for years 1950–1999. What is the average number of tornado deaths for the 12 months? the median? the mode?

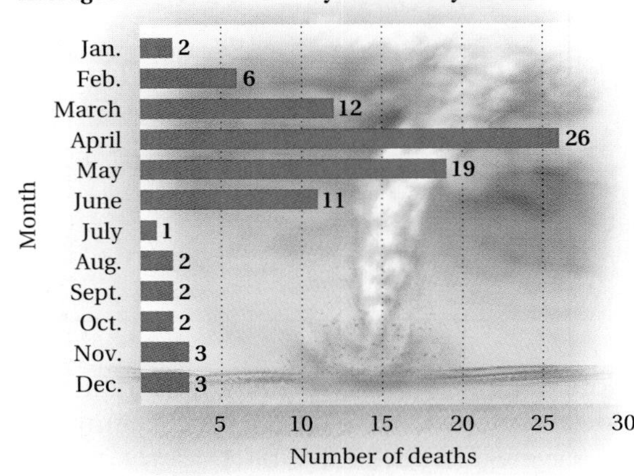

Source: National Weather Service's Storm Prediction Center

11. *Gas Mileage.* The Acura RSX Type-S gets 279 mi of highway driving on 9 gal of gasoline. What is the average number of miles expected per gallon—that is, what is its gas mileage?
Sources: EPA; *Car and Driver,* September 2005

12. *Gas Mileage.* The Chevrolet Cobalt SS Supercharged gets 322 mi of city driving on 14 gal of gasoline. What is the average number of miles expected per gallon—that is, what is its gas mileage?
Sources: EPA; *Car and Driver,* September 2005

Grade Point Average. The tables in Exercises 13 and 14 show the grades of a student for one semester. In each case, find the grade point average. Assume that the grade point values are 4.0 for an A, 3.0 for a B, and so on. Round to the nearest tenth.

13.

GRADE	NUMBER OF CREDIT HOURS IN COURSE
B	4
A	5
D	3
C	4

14.

GRADE	NUMBER OF CREDIT HOURS IN COURSE
A	5
C	4
F	3
B	5

15. *Salmon Prices.* The following prices per pound of Atlantic salmon were found at five fish markets:

$6.99, $8.49, $8.99, $6.99, $9.49.

What was the average price per pound? the median price? the mode?

16. *Cheddar Cheese Prices.* The following prices per pound of sharp cheddar cheese were found at five supermarkets:

$5.99, $6.79, $5.99, $6.99, $6.79.

What was the average price per pound? the median price? the mode?

17. *Grading.* To get a B in math, Rich must score an average of 80 on five tests. Scores on the first four tests were 80, 74, 81, and 75. What is the lowest score that Rich can get on the last test and still receive a B?

18. *Grading.* To get an A in math, Cybil must score an average of 90 on five tests. Scores on the first four tests were 90, 91, 81, and 92. What is the lowest score that Cybil can get on the last test and still receive an A?

19. *Length of Pregnancy.* Marta's first three pregnancies lasted for 270 days, 259 days, and 272 days. In order for Marta's average pregnancy to equal the worldwide average of 266 days, how long must her fourth pregnancy last?

Source: David Crystal (ed.), *The Cambridge Factfinder.* Cambridge CB2 1RP: Cambridge University Press, 1993, p. 84.

20. *Male Height.* Jason's brothers are 174 cm, 180 cm, 179 cm, and 172 cm tall. The average male is 176.5 cm tall. How tall is Jason if he and his brothers have an average height of 176.5 cm?

Solve.

21. *Light-Bulb Testing.* An experiment is performed to compare the lives of two types of light bulb. Several bulbs of each type were tested and the results are listed in the following table. On the basis of this test, which bulb is better?

BULB A: HOTLIGHT TIMES (in hours)			BULB B: BRIGHTBULB TIMES (in hours)		
983	964	1214	979	1083	1344
1417	1211	1521	984	1445	975
1084	1075	892	1492	1325	1283
1423	949	1322	1325	1352	1432

22. *Cola Testing.* An experiment is conducted to determine which of two colas tastes better. Students drank each cola and gave it a rating from 1 to 10. The results are given in the following table. On the basis of this test, which cola tastes better?

COLA A: VERVCOLA				COLA B: COLA-COLA			
6	8	10	7	10	9	9	6
7	9	9	8	8	8	10	7
5	10	9	10	8	7	4	3
9	4	7	6	7	8	10	9

23. DW You are applying for an entry-level job at a large firm. You can be informed of the mean, median, or mode salary. Which of the three figures would you request? Why?

24. DW Is it possible for a driver to average 20 mph on a 30-mi trip and still receive a ticket for driving 75 mph? Why or why not?

SKILL MAINTENANCE

Multiply.

25. $14 \cdot 14$ [1.3a]

26. $\frac{2}{3} \cdot \frac{2}{3}$ [2.2a]

27. 1.4×1.4 [3.3a]

28. 1.414×1.414 [3.3a]

29. 12.86×17.5 [3.3a]

30. 222×0.5678 [3.3a]

31. $\frac{4}{5} \cdot \frac{3}{28}$ [2.2a]

32. $\frac{28}{45} \cdot \frac{3}{2}$ [2.2a]

Convert to percent notation. [4.3a]

33. $\frac{19}{16}$

34. $\frac{11}{16}$

35. $\frac{64}{125}$

36. $\frac{9781}{10,000}$

SYNTHESIS

Bowling Averages. Bowling averages are always computed by rounding down to the nearest integer. For example, suppose a bowler gets a total of 599 for 3 games. To find the average, we divide 599 by 3 and drop the amount to the right of the decimal point:

$$\frac{599}{3} \approx 199.67. \qquad \text{The bowler's average is 199.}$$

In each case, find the bowling average.

37. 547 in 3 games

38. 4621 in 27 games

39. *Hank Aaron.* Hank Aaron averaged $34\frac{7}{22}$ home runs per year over a 22-yr career. After 21 yr, Aaron had averaged $35\frac{10}{21}$ home runs per year. How many home runs did Aaron hit in his final year?

40. The ordered set of data 18, 21, 24, a, 36, 37, b has a median of 30 and an average of 32. Find a and b.

41. *Grades.* Because of a poor grade on the fifth of five tests, Chris's average test grade fell from 90.5 to 84.0. What did Chris score on the fifth test? Assume that all tests are equally important.

42. *Price Negotiations.* Amy offers \$3200 for a used Ford Taurus advertised at \$4000. The first offer from Jim, the car's owner, is to "split the difference" and sell the car for $(3200 + 4000) \div 2$, or \$3600. Amy's second offer is to split the difference between Jim's offer and her first offer. Jim's second offer is to split the difference between Amy's second offer and his first offer. If this pattern continues and Amy accepts Jim's third (and final) offer, how much will she pay for the car?

5.2 TABLES AND PICTOGRAPHS

Objectives

a Extract and interpret data from tables.

Extract and interpret data from pictographs.

a Reading and Interpreting Tables

A **table** is often used to present data in rows and columns.

EXAMPLE 1 *Nutrition Information.* The following table lists daily nutrition information for eleven diets.

Diet	Average daily calories	Grams of fat	Grams of saturated fat	Grams of carbohydrates	Grams of protein	Grams of fiber*	Fruits and vegetables†
Atkins (Induction)	1640	61	19	8	31	8	6
Atkins (Ongoing Weight Loss)	1520	60	20	11	29	12	6
eDiets	1450	23	5	53	24	19	12
Jenny Craig	1520	18	7	62	20	16	6
Ornish (Eat More, Weigh Less)	1520	6	1	77	16	31	17
Slim-Fast	1540	22	6	57	21	21	12
South Beach (Phase One)	1530	51	14	15	34	9	12
South Beach (Phase Two)	1340	39	9	38	22	19	13
Volumetrics	1500	23	7	55	22	20	14
Weight Watchers	1450	24	7	56	20	20	11
Zone (Men's Menu)	1660	27	7	42	30	21	17

Source: Consumer Reports *Per 1000 calories †Daily servings

a) Which diet has the least number of average daily calories?

b) Which diet has the greatest number of grams of carbohydrates?

c) Which diet has the least number of grams of fiber?

d) Find the average number of grams of fat in the diets listed.

Careful examination of the table will give the answers.

a) To determine which diet has the least number of average daily calories, look down the column headed "Average Daily Calories" and find the smallest number. That number is 1340. Then look across that row to find the name of the diet, South Beach (Phase Two).

b) To determine which diet has the greatest amount of carbohydrates, look down the column headed "Grams of Carbohydrates" and find the largest number. That number is 77 g. Then look across that row to find the diet, Ornish (Eat More, Weigh Less).

c) To determine which diet has the least amount of fiber, look down the column headed "Grams of Fiber" and find the smallest number. That number is 8 g. Then look across that row to find the diet, Atkins (Induction).

d) Find the average of all the numbers in the column headed "Grams of Fat":

$$\frac{61 + 60 + 23 + 18 + 6 + 22 + 51 + 39 + 23 + 24 + 27}{11} = \frac{354}{11} \approx 32.2 \text{ g.}$$

Use the table in Example 1 to answer Margin Exercises 1–7.

1. Which diet has the greatest number of average daily calories?

2. Which diet has the greatest number of servings of fruit and vegetables?

3. Which diet has the least number of grams of fat?

4. Find the average number of grams of carbohydrates in the diets listed.

Answers on page A-12

The average daily number of grams of fat in these eleven diets is about 32.2 g.

Do Exercises 1–7. (Exercises 1–4 are on the preceding page.)

EXAMPLE 2 *Wheaties Nutrition Facts.* Most foods are required by law to provide factual information regarding nutrition, as shown in the following table of Nutrition Facts from a box of Wheaties cereal. Although this can be very helpful to the consumer, one must be careful in interpreting the data. The % Daily Value figures shown here are based on a 2000-calorie diet. Your daily values may be higher or lower, depending on your calorie needs or intake.

Nutrition Facts

Serving Size 1 cup (30g)
Servings Per Container 17

Amount Per Serving	Wheaties	with ½ cup skim milk
Calories	110	150
Calories from Fat	10	10
	% Daily value**	
Total Fat 1g*	**1**%	**2**%
Saturated Fat 0g	**0**%	**0**%
Cholesterol 0mg	**0**%	**1**%
Sodium 220mg	**9**%	**12**%
Potassium 110mg	**3**%	**9**%
Total Carbohydrate 24g	**8**%	**10**%
Dietary Fiber 3g	**12**%	**12**%
Sugars 4g		
Other Carbohydrate 17g		
Protein 3g		

Vitamin A	10%	15%
Vitamin C	10%	10%
Calcium	0%	15%
Iron	45%	45%
Vitamin D	10%	25%
Thiamin	50%	50%
Riboflavin	50%	60%
Niacin	50%	50%
Vitamin B_6	50%	50%
Folic Acid	50%	50%
Phosphorus	10%	20%
Magnesium	8%	10%
Zinc	50%	50%
Copper	4%	4%

*Amount in Cereal. A serving of cereal plus skim milk provides 1g fat, less than 5mg cholesterol, 280mg sodium, 310mg potassium, 30g carbohydrate (10g sugars), and 7g protein.

**Percent Daily Values are based on a 2,000 calorie diet. Your daily values may be higher or lower depending on your calorie needs.

Source: General Mills

Suppose your morning bowl of cereal consists of 2 cups of Wheaties with 1 cup of skim milk.

a) How many calories have you consumed?

b) What percent of the daily value of total fat have you consumed?

c) A nutritionist recommends that you look for foods that provide 10% or more of the daily value for vitamin C. Do you get that with your bowl of Wheaties?

d) Suppose you are trying to limit your daily caloric intake to 2500 calories. How many bowls of cereal would it take to exceed the 2500 calories?

Careful examination of the table of nutrition facts will give the answers.

a) Look at the column marked "with 1/2 cup skim milk" and note that 1 cup of cereal with 1/2 cup skim milk contains 150 calories. Since you are having twice that amount, you are consuming

$$2 \times 150, \quad \text{or} \quad 300 \text{ calories.}$$

b) Read across from "Total Fat" and note that in 1 cup of cereal with 1/2 cup skim milk, you get 2% of the daily value of fat. Since you are doubling that, you get 4% of the daily value of fat.

c) Find the row labeled "Vitamin C" on the left and look under the column labeled "with 1/2 cup skim milk." Note that you get 10% of the daily value for "1 cup with 1/2 cup of skim milk," and since you are doubling that, you are more than satisfying the 10% requirement.

d) From part (a), we know that you are consuming 300 calories per bowl. Dividing 2500 by 300 gives $\frac{2500}{300} \approx 8.33$. Thus if you eat 9 bowls of cereal in this manner, you will exceed the 2500 calories.

Do Exercises 8–12.

5. Find the average amount of protein in the diets listed.

6. Find the median of the amount of protein in the diets listed.

7. Find the average, the median, and the mode of the amount of saturated fat in the diets listed.

Use the Nutrition Facts data from the Wheaties box and the bowl of cereal described in Example 2 to answer Margin Exercises 8–12.

8. How many calories from fat are in your bowl of cereal?

9. A nutritionist recommends that you look for foods that provide 10% or more of the daily value for iron. Do you get that with your bowl of Wheaties?

10. How much sodium have you consumed? (*Hint*: Use the data listed in the first footnote below the table of nutrition facts.)

11. What daily value of sodium have you consumed?

12. How much protein have you consumed? (*Hint*: Use the data listed in the first footnote below the table of nutrition facts.)

Answers on page A-12

b Reading and Interpreting Pictographs

Pictographs (or *picture graphs*) are another way to show information. Instead of actually listing the amounts to be considered, a **pictograph** uses symbols to represent the amounts. In addition, a *key* is given telling what each symbol represents.

EXAMPLE 3 *Rhino Population.* The rhinoceros is considered one of the world's most endangered animals. The worldwide total number of rhinoceroses is approximately 17,800. The following pictograph shows the population of the five remaining rhino species. Located in the graph is a key that tells you that each symbol represents 300 rhinos.

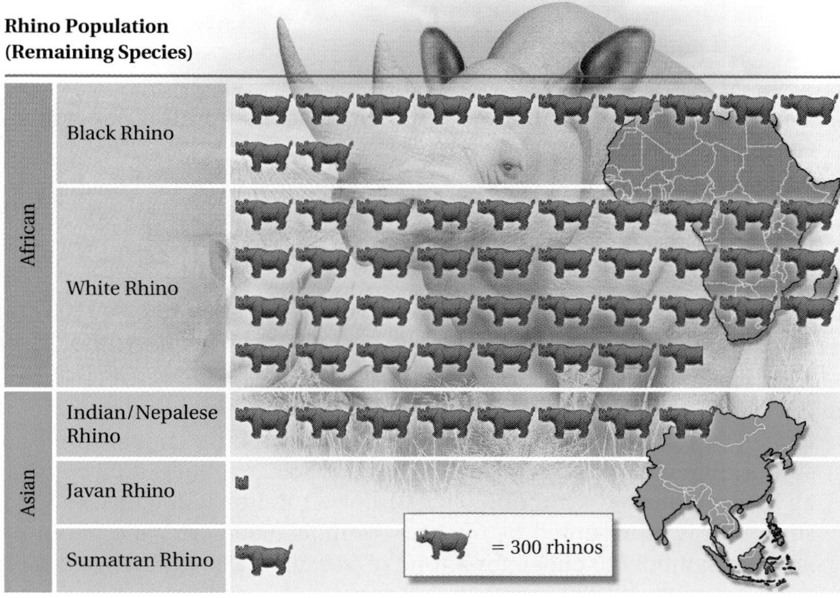

Source: International Rhino Foundation, 2005

a) Which species has the greatest number of rhinos?

b) Which species has the least number of rhinos?

c) How many more black rhinos are there than Indian/Nepalese rhinos?

We can compute the answers by first reading the pictograph.

a) The species with the most symbols has the greatest number of rhinos: white rhino, with $37\frac{3}{4}$ (symbols) $\times$ 300, or 11,325 rhinos.

b) The species with the fewest symbols has the least number of rhinos: Javan rhino, with about $\frac{1}{5}$ (of a symbol) $\times$ 300, or 60 rhinos.

c) From the graph we see that there are 12×300, or 3600, black rhinos, and 8×300, or 2400, Indian/Nepalese rhinos. Thus, there are $3600 - 2400$, or 1200, more black rhinos than Indian/Nepalese rhinos.

Do Exercises 13–15.

You have probably noticed that, although they seem to be very easy to read, pictographs are difficult to draw accurately because whole symbols reflect loose approximations due to significant rounding. In pictographs, you also need to use some mathematics to find the actual amounts.

On July 5, 2004, at the Cincinnati Zoo and Botanical Gardens, Emi, a critically endangered Sumatran rhinoceros, became the first rhino to produce two calves in captivity.

Use the pictograph in Example 3 to answer Margin Exercises 13–15.

13. How many more Indian/ Nepalese rhinos are there than Sumatran rhinos?

14. How does the black rhino population compare with the Sumatran rhino population?

15. What is the average number of rhinos in these five species?

Answers on page A-12

Use the pictograph in Example 4 to answer Margin Exercises 16–18.

16. Determine the approximate coffee consumption per capita of France.

17. Determine the approximate coffee consumption per capita of Italy.

18. The approximate coffee consumption of Finland is about the same as the combined coffee consumptions of Switzerland and the United States. What is the approximate coffee consumption of Finland?

Answers on page A-12

EXAMPLE 4 *Coffee Consumption.* For selected countries, the following pictograph shows approximately how many cups of coffee each person (per capita) drinks annually.

Coffee Consumption

Source: Beverage Marketing Corporation

a) Determine the approximate annual coffee consumption per capita of Germany.

b) Which two countries have the greatest difference in coffee consumption? Estimate that difference.

We use the data from the pictograph as follows.

a) Germany's consumption is represented by 11 whole symbols (1100 cups) and, though it is visually debatable, about $\frac{1}{8}$ of another symbol (about 13 cups), for a total of 1113 cups.

b) Visually, we see that Switzerland has the most consumption and that the United States has the least consumption. Switzerland's annual coffee consumption per capita is represented by 12 whole symbols (1200 cups) and about $\frac{1}{5}$ of another symbol (20 cups), for a total of 1220 cups. U.S. consumption is represented by 6 whole symbols (600 cups) and about $\frac{1}{10}$ of another symbol (10 cups), for a total of 610 cups. The difference between these amounts is $1220 - 610$, or 610 cups.

One advantage of pictographs is that the appropriate choice of a symbol will tell you, at a glance, the kind of measurement being made. Another advantage is that the comparison of amounts represented in the graph can be expressed more easily by just counting symbols. For instance, in Example 3, the ratio of Indian/Nepalese rhinoceroses to black rhinoceroses is 8:12.

There are at least three disadvantages of pictographs:

1. To make a pictograph easy to read, the amounts must be rounded significantly to the unit that a symbol represents. This makes it difficult to accurately represent an amount.
2. It is difficult to determine very accurately how much a partial symbol represents.
3. Some mathematics is required to finally compute the amount represented, since there is usually no explicit statement of the amount.

Do Exercises 16–18.

5.2 EXERCISE SET

For Extra Help

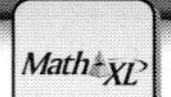
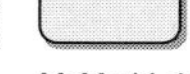

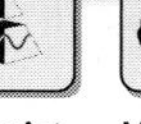

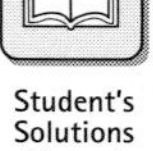

MathXL | MyMathLab | InterAct Math | Math Tutor Center | Video Lectures on CD Disc 3 | Student's Solutions Manual

a *Planets.* Use the following table, which lists information about the planets, for Exercises 1–10.

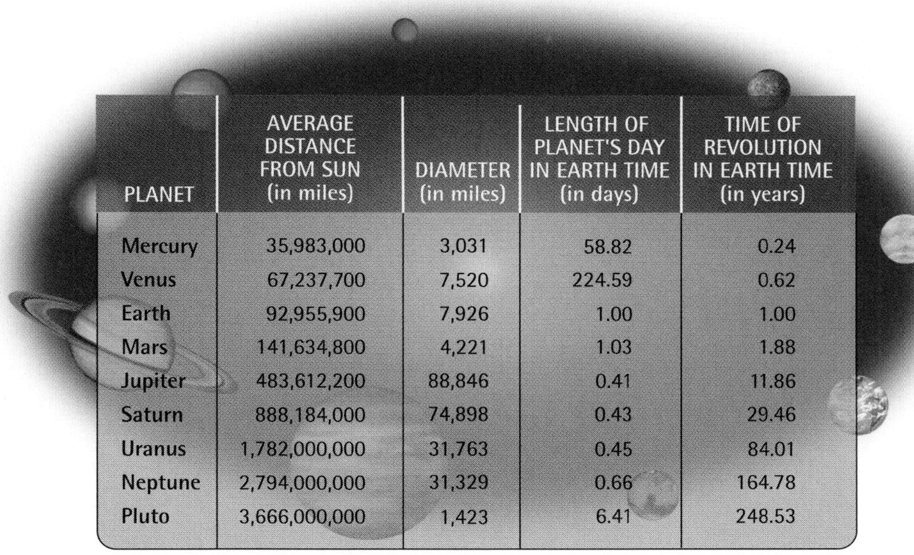

PLANET	AVERAGE DISTANCE FROM SUN (in miles)	DIAMETER (in miles)	LENGTH OF PLANET'S DAY IN EARTH TIME (in days)	TIME OF REVOLUTION IN EARTH TIME (in years)
Mercury	35,983,000	3,031	58.82	0.24
Venus	67,237,700	7,520	224.59	0.62
Earth	92,955,900	7,926	1.00	1.00
Mars	141,634,800	4,221	1.03	1.88
Jupiter	483,612,200	88,846	0.41	11.86
Saturn	888,184,000	74,898	0.43	29.46
Uranus	1,782,000,000	31,763	0.45	84.01
Neptune	2,794,000,000	31,329	0.66	164.78
Pluto	3,666,000,000	1,423	6.41	248.53

Source: *The Handy Science Answer Book*, Gale Research, Inc.

1. Find the average distance from the sun to Jupiter.

2. How long is a day on Venus?

3. Which planet has a time of revolution of 164.78 yr?

4. Which planet has a diameter of 4221 mi?

5. Which planets have an average distance from the sun that is greater than 1,000,000 mi?

6. Which planets have a diameter that is less than 100,000 mi?

7. About how many Earth diameters would it take to equal one Jupiter diameter?

8. How much longer is the longest time of revolution than the shortest?

9. What are the average, the median, and the mode of the diameters of the planets?

10. What are the average, the median, and the mode of the average distances from the sun of the planets?

Heat Index. In warm weather, a person can feel hotter due to reduced heat loss from the skin caused by higher humidity. The **temperature–humidity index,** or **apparent temperature,** is what the temperature would have to be with no humidity in order to give the same heat effect. The following table lists the apparent temperatures for various actual temperatures and relative humidities. Use this table for Exercises 11–22.

ACTUAL TEMPERATURE (°F)	RELATIVE HUMIDITY									
	10%	20%	30%	40%	50%	60%	70%	80%	90%	100%
	APPARENT TEMPERATURE (°F)									
75°	75	77	79	80	82	84	86	88	90	92
80°	80	82	85	87	90	92	94	97	99	102
85°	85	88	91	94	97	100	103	106	108	111
90°	90	93	97	100	104	107	111	114	118	121
95°	95	99	103	107	111	115	119	123	127	131
100°	100	105	109	114	118	123	127	132	137	141
105°	105	110	115	120	125	131	136	141	146	151

In Exercises 11–14, find the apparent temperature for the given actual temperature and humidity combinations.

11. 80°, 60%

12. 90°, 70%

13. 85°, 90%

14. 95°, 80%

15. How many listed temperature–humidity combinations give an apparent temperature of 100°?

16. How many listed temperature–humidity combinations give an apparent temperature of 111°?

17. At a relative humidity of 50%, what actual temperatures give an apparent temperature above 100°?

18. At a relative humidity of 90%, what actual temperatures give an apparent temperature above 100°?

19. At an actual temperature of 95°, what relative humidities give an apparent temperature above 100°?

20. At an actual temperature of 85°, what relative humidities give an apparent temperature above 100°?

21. At an actual temperature of 85°, by how much would the humidity have to increase in order to raise the apparent temperature from 94° to 108°?

22. At an actual temperature of 80°, by how much would the humidity have to increase in order to raise the apparent temperature from 87° to 102°?

Cigarette Consumption. The May 2004 U.S. Surgeon General's report, *The Health Consequences of Smoking,* reports that smoking causes many diseases and reduces the health of smokers. The per capita cigarette consumption in the United States for specific years from 1920 to 2003 is listed in the table below. Use these data for Exercises 23–26.

U.S. Cigarette Consumption (1920 – 2003) (Per capita among persons 18 and older)

YEAR	1920	1930	1940	1950	1960	1970	1980	1990	2000	2003
Cigarette Consumption	665	1485	1976	3552	4171	3985	3849	2817	2092	1903

Sources: U.S. Department of Health and Human Services, Centers for Disease Control and Prevention; USDA, Economic Research Service

23. Find the cigarette consumption in 1940 and in 1980. What was the percent of increase in the consumption from 1940 to 1980?

24. Find the cigarette consumption in 1960 and in 2000. What was the percent of decrease in the consumption from 1960 to 2000?

25. Find the average of the cigarette consumption for the years 1920 to 1950. Find the average of the cigarette consumption for the years 1970 to 2000. By how many cigarettes per capita does the latter average exceed the former?

26. Find the average of the cigarette consumption for the years 2000 and 2003. Find the average of the cigarette consumption for the years 1920 to 2000. By how many cigarettes per capita does the average for 1920 to 2000 exceed the average for the years 2000 and 2003?

b *World Population Growth.* The following pictograph shows world population in various years. Use the pictograph for Exercises 27–34.

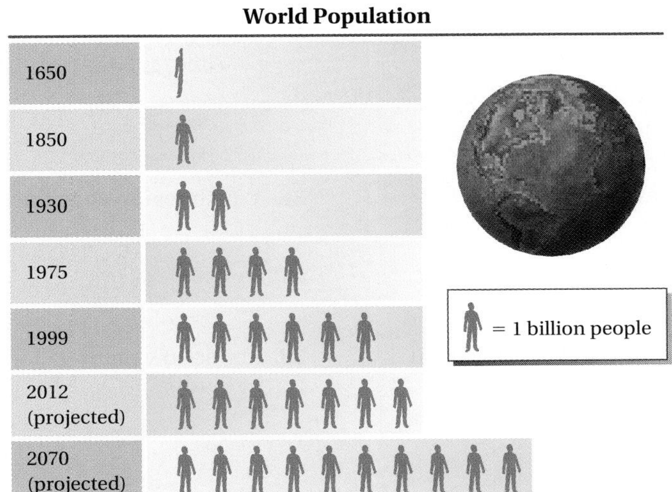

Source: U.S. Bureau of the Census, International Data Base

27. What was the world population in 1850?

28. What was the world population in 1975?

29. In which year will the population be the greatest?

30. In which year was the population the least?

31. Between which two years was the amount of growth the least?

32. Between which two years was the amount of growth the greatest?

33. How much greater will the world population in 2012 be than in 1975? What is the percent of increase?

34. How much greater was the world population in 1999 than in 1930? What is the percent of increase?

Water Consumption. The following pictograph shows water consumption, per person, in different regions of the world in a recent year. Use the pictograph for Exercises 35–40.

Region	Water Consumption
North America	
Central America	
Europe	
Oceania	
Asia	
South America	
Africa	

= 100,000 gallons

Sources: World Resources Institute; U.S. Energy Information Administration

35. Which region consumes the least water?

36. Which region consumes the most water?

37. About how many gallons are consumed per person in North America?

38. About how many gallons are consumed per person in Europe?

39. Approximately how many more gallons are consumed per person in North America than in Asia?

40. Approximately how many more gallons are consumed per person in Central America than in Africa?

41. D_W Loreena is drawing a pictograph in which dollar bills are used as symbols to represent the tuition at various private colleges. Should each dollar bill represent $10,000, $1000, or $100? Why?

42. D_W What advantage(s) does a table have over a pictograph?

SKILL MAINTENANCE

Solve. [4.6a]

43. *Kitchen Costs.* The average cost of a kitchen is $26,888. Some of the cost percentages are as follows.

Cabinets:	50%
Countertops:	15%
Appliances:	8%
Fixtures:	3%

Find the costs for each part of a kitchen.

Source: National Kitchen and Bath Association

44. *Bathroom Costs.* The average cost of a bathroom is $11,605. Some of the cost percentages are as follows.

Cabinets:	31%
Countertops:	11%
Labor:	25%
Flooring:	6%

Find the costs for each part of a bathroom.

Source: National Kitchen and Bath Association

Convert to fraction notation and simplify. [4.3b]

45. 24%

46. 45%

47. 4.8%

48. 6.4%

49. 53.1%

50. 87.3%

51. 100%

52. 2%

SYNTHESIS

53. Redraw the pictograph appearing in Example 4 as one in which each symbol represents 150 cups of coffee.

5.3 BAR GRAPHS AND LINE GRAPHS

Objectives

a Extract and interpret data from bar graphs.

b Draw bar graphs.

c Extract and interpret data from line graphs.

d Draw line graphs.

A **bar graph** is convenient for showing comparisons because you can tell at a glance which amount represents the largest or smallest quantity. Of course, since a bar graph is a more abstract form of pictograph, this is true of pictographs as well. However, with bar graphs, a *second scale* is usually included so that a more accurate determination of the amount can be made.

a Reading and Interpreting Bar Graphs

EXAMPLE 1 *Major League Sports Fans.* As NASCAR racing increases in popularity, its fan base is growing rapidly. The following horizontal bar graph shows the number of fans of various major league sports.

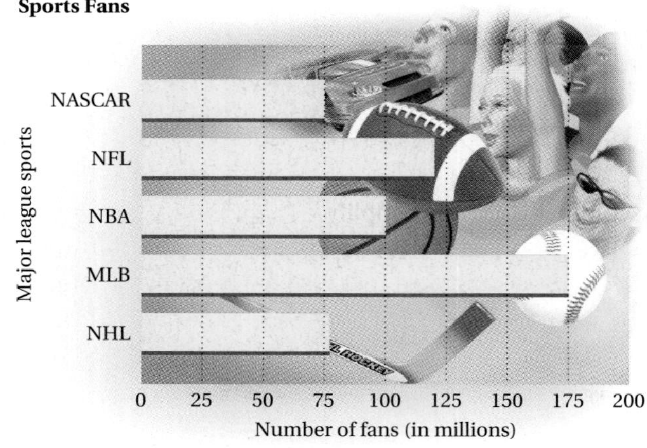

Sources: NASCAR; the individual leagues; ESPN

a) Which of the major league sports shown in the graph above has the greatest number of fans?

b) About how many fans does the NFL (National Football League) have?

c) Which of the major league sports has about 100 million fans?

We look at the graph to answer the questions.

a) The longest bar is for MLB (Major League Baseball). Thus that sport has the greatest number of fans.

b) We move to the right along the bar representing the NFL. We can read, fairly accurately, that this sport has approximately 120 million fans.

c) We locate the line representing 100 million and then go up until we reach a bar that ends at 100 million. We then go to the left and read the name of the sport, NBA (National Basketball Association).

Do Exercises 1–3.

Use the bar graph in Example 1 to answer Margin Exercises 1–3.

1. About how many fans does the NHL (National Hockey League) have?

2. Which sport has the fewest number of fans?

3. Which sports have 100 million or more fans?

Answers on page A-13

Bar graphs are often drawn vertically and sometimes a double bar graph is used to make comparisons.

EXAMPLE 2 *Breast Cancer.* The following graph indicates the incidence and mortality rates of breast cancer for women of various age groups.

When Breast Cancer Strikes

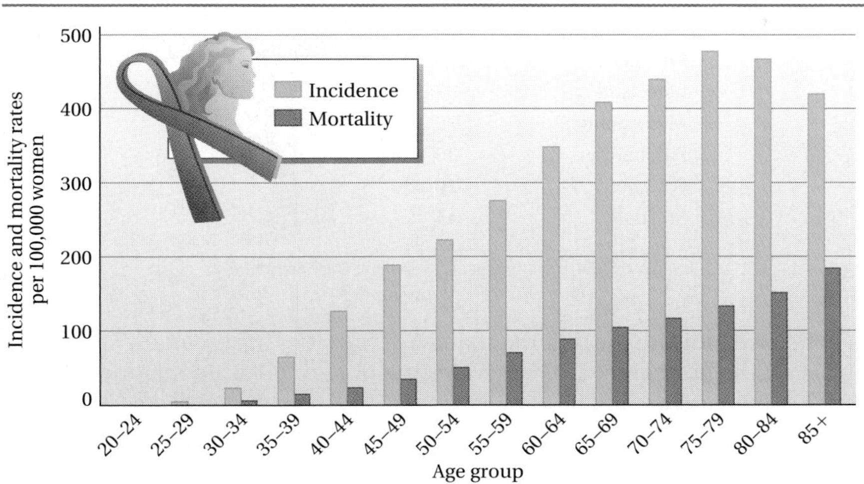

Source: National Cancer Institute

a) Approximately how many women, per 100,000, develop breast cancer between the ages of 40 and 44?

b) In what age range is the mortality rate for breast cancer approximately 100 for every 100,000 women?

c) In what age range is the incidence of breast cancer the highest?

d) Does the incidence of breast cancer seem to increase from the youngest to the oldest age group?

We look at the graph to answer the questions.

a) We go to the right, across the bottom, to the green bar above the age group 40–44. Next, we go up to the top of that bar and, from there, back to the left to read approximately 130 on the vertical scale. About 130 of every 100,000 women develop breast cancer between the ages of 40 and 44.

b) We read up the vertical scale to the number 100. From there we move to the right until we come to the top of a red bar. Moving down that bar, we find that in the 65–69 age group, about 100 of every 100,000 women die of breast cancer.

c) We look for the tallest green bar and read the age range below it. The incidence of breast cancer is highest for women in the 75–79 age group.

d) Looking at the heights of the bars, we see that the incidence of breast cancer increases to a high point in the 75–79 age group and then decreases.

Do Exercises 4–7.

Use the bar graph in Example 2 to answer Margin Exercises 4–7.

4. Approximately how many women, per 100,000, develop breast cancer between the ages of 35 and 39?

5. In what age group is the mortality rate the highest?

6. In what age group do about 350 of every 100,000 women develop breast cancer?

7. Does the breast cancer mortality rate seem to increase from the youngest to the oldest age group?

Answers on page A-13

8. Planetary Moons. Make a horizontal bar graph to show the number of moons orbiting the various planets.

PLANET	MOONS
Earth	1
Mars	2
Jupiter	17
Saturn	28
Uranus	21
Neptune	8
Pluto	1

Source: National Aeronautics and Space Administration

b Drawing Bar Graphs

EXAMPLE 3 *Police Officers.* Listed below are the numbers of police officers per 10,000 people in various cities. Make a vertical graph of the data.

CITY	POLICE OFFICERS, PER 10,000 PEOPLE
Cincinnati	31
Cleveland	37
Minneapolis	26
Pittsburgh	30
Columbus	26
St. Louis	44

Sources: U.S. Bureau of the Census; FBI Uniform Crime Reports

First, we indicate the different names of the cities in six equally spaced intervals on the horizontal scale and give the horizontal scale the title "Cities." (See the figure on the left below.)

Next, we scale the vertical axis. To do so, we look over the data and note that it ranges from 26 to 44. We start the vertical scaling at 0, labeling the marks by 5's from 0 to 50. We give the vertical scale the title "Number of Officers (per 10,000 people)."

Finally, we draw vertical bars to show the various numbers, as shown in the figure on the right. We give the graph an overall title, "Police Officers."

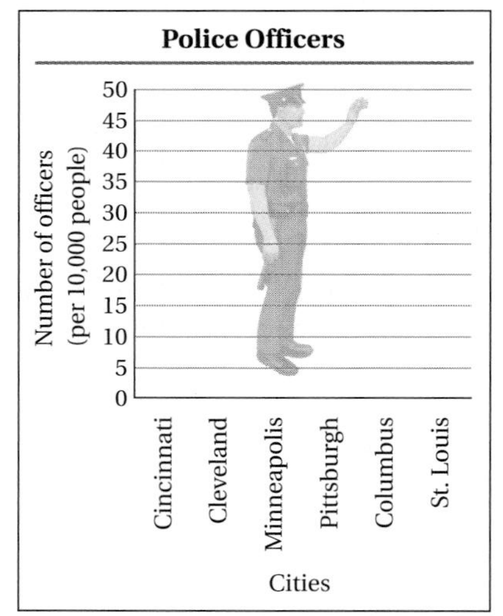

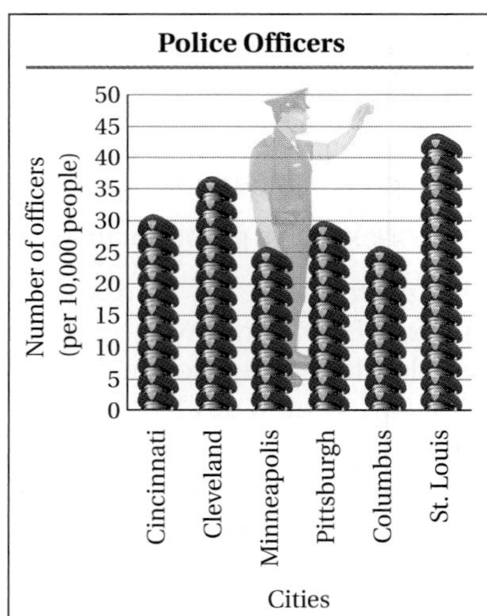

Sources: U.S. Bureau of the Census; FBI Uniform Crime Reports

Do Exercise 8.

c Reading and Interpreting Line Graphs

Line graphs are often used to show a change over time as well as to indicate patterns or trends.

Answer on page A-13

EXAMPLE 4 *New Home Sales.* The following line graph shows the number of new home sales, in thousands, over a 12-month period. The jagged line at the base of the vertical scale indicates an unnecessary portion of the scale. Note that the vertical scale differs from the horizontal scale so that the data can be shown reasonably.

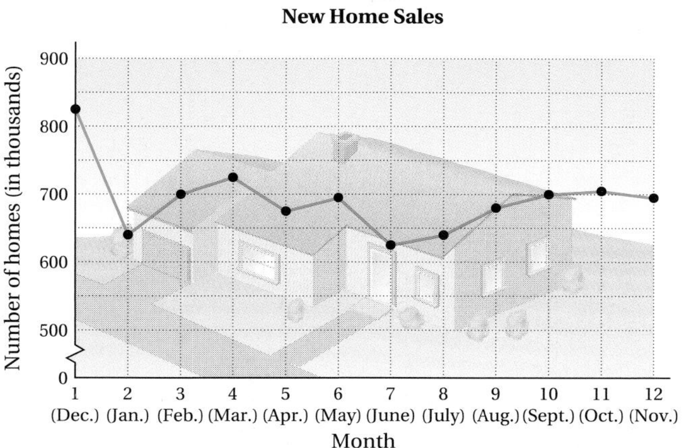

Source: U.S. Department of Commerce

a) For which month were new home sales the greatest?

b) Between which months did new home sales increase?

c) For which months were new home sales about 700 thousand?

We look at the graph to answer the questions.

a) The greatest number of new home sales was about 825 thousand in month 1, December.

b) Reading the graph from left to right, we see that new home sales increased from month 2 to month 3, from month 3 to month 4, from month 5 to month 6, from month 7 to month 8, from month 8 to month 9, from month 9 to month 10, and from month 10 to month 11.

c) We look from left to right along the line at 700.

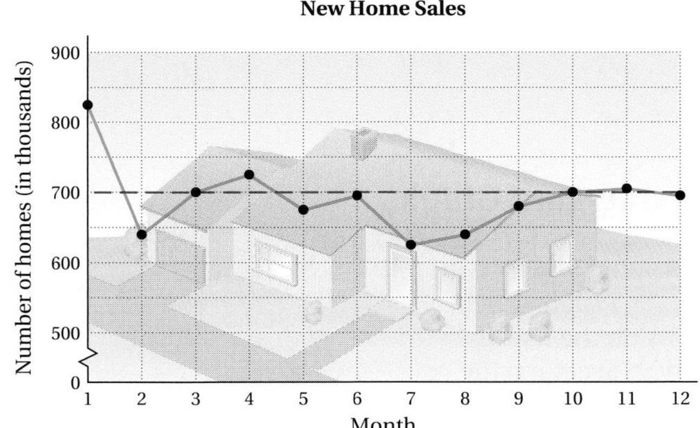

We see that points are close to 700 thousand at months 3, 6, 10, 11, and 12.

Do Exercises 9–11.

Use the line graph in Example 4 to answer Margin Exercises 9–11.

9. For which month were new home sales lowest?

10. Between which months did new home sales decrease?

11. For which months were new home sales about 650 thousand?

Answers on page A-13

Use the line graph in Example 5 to answer Margin Exercises 12–14.

12. Estimate the monthly payment for a loan of 25 yr.

13. What time period corresponds to a monthly payment of about $850?

14. By how much does the monthly payment decrease when the loan period is increased from 5 yr to 20 yr?

Answers on page A-13

EXAMPLE 5 *Monthly Loan Payment.* Suppose that you borrow $110,000 at an interest rate of 9% to buy a home. The following graph shows the monthly payment required to pay off the loan, depending on the length of the loan. (*Caution:* A low monthly payment means that you will pay more interest over the duration of the loan.)

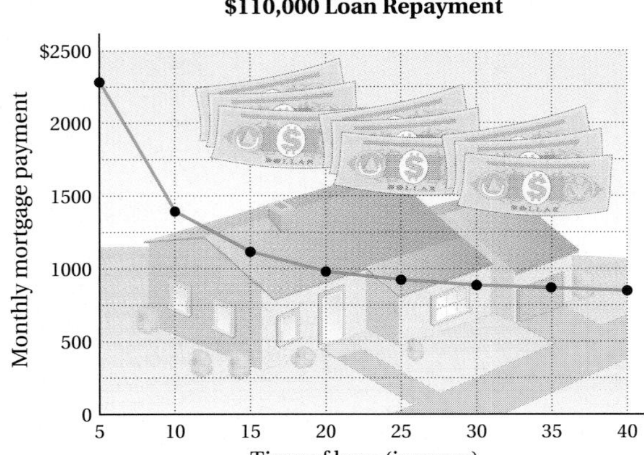

a) Estimate the monthly payment for a loan of 15 yr.

b) What time period corresponds to a monthly payment of about $1400?

c) By how much does the monthly payment decrease when the loan period is increased from 10 yr to 20 yr?

We look at the graph to answer the questions.

a) We find the time period labeled "15" on the bottom scale and move up from that point to the line. We then go straight across to the left and find that the monthly payment is about $1100.

b) We locate $1400 on the vertical axis. Then we move to the right until we hit the line. The point $1400 is on the line at the 10-yr time period.

c) The graph shows that the monthly payment for 10 yr is about $1400; for 20 yr, it is about $990. Thus the monthly payment is decreased by $1400 − $990, or $410. (It should be noted that you will pay back $990 · 20 · 12 − $1400 · 10 · 12, or $69,600, more in interest for a 20-yr loan.)

Do Exercises 12–14.

d Drawing Line Graphs

EXAMPLE 6 *Cell Phones with Internet Access.* Listed at right are projections on the use of cell phones with access to the Internet. Make a line graph of the data.

YEAR	CELL PHONES WITH WEB ACCESS (in millions)
2001	29.4
2002	69.6
2003	120.1
2004	152.4
2005	171.1

Source: Forrester Research

First, we indicate the different years on the horizontal scale and give the horizontal scale the title "Year." (See the figure on the left below.) Next, we scale the vertical axis by 25's to show the number of phones, in millions, and give the vertical scale the title "Number of cell phones (in millions)." We also give the graph the overall title "Cell Phones with Web Access."

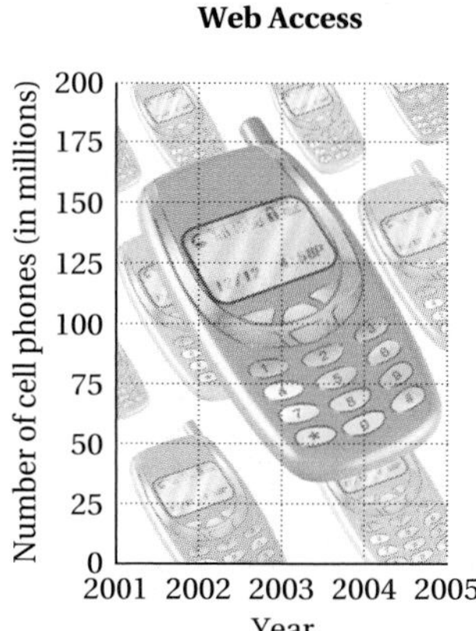

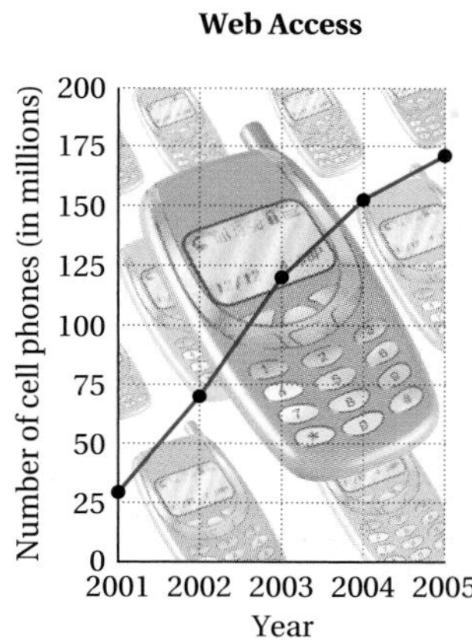

Next, we mark the number of phones at the appropriate level above each year. Then we draw line segments connecting the points. The dramatic change over time can now be observed easily from the graph.

Do Exercise 15.

15. Cell Phones. Listed below are projections on the use of cell phones with or without access to the Internet. Make a line graph of the data.

YEAR	NUMBER OF CELL PHONES (in millions)
2001	119.8
2002	135.3
2003	150.2
2004	163.8
2005	176.9

Source: Forrester Research

Answer on page A-13

Study Tips

ATTITUDE AND THE POWER OF YOUR CHOICES

Making the right choices can give you the power to succeed in learning mathematics.

You can choose to improve your attitude and raise the academic goals that you have set for yourself. Projecting a positive attitude toward your study of mathematics and expecting a positive outcome can make it easier for you to learn and to perform well in this course.

Here are some positive choices you can make:

- Choose to allocate the proper amount of time to learn.
- Choose to place the primary responsibility for learning on yourself.
- Choose to make a strong commitment to learning.

Well-known American psychologist William James once said, "The one thing that will guarantee the successful conclusion of a doubtful undertaking is faith in the beginning that you can do it." Having a positive attitude and making the right choices will add to your confidence in this course and multiply your successes.

5.3 EXERCISE SET

For Extra Help

a *Chocolate Desserts.* The following horizontal bar graph shows the average caloric content of various kinds of chocolate desserts. Use the bar graph for Exercises 1–12.

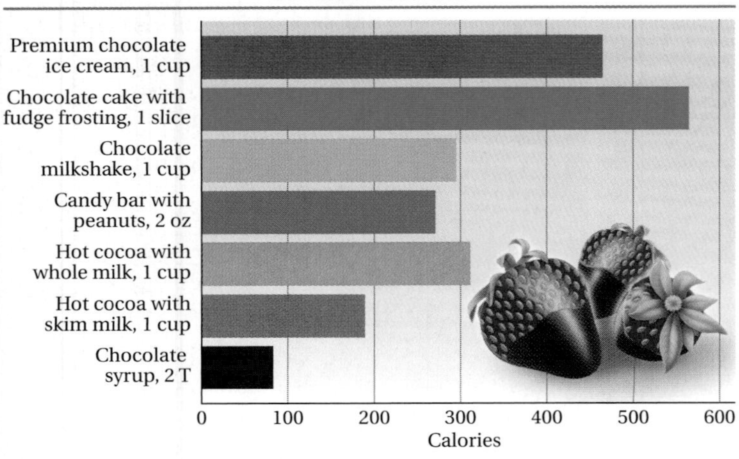

Source: Better Homes and Gardens, December 1996

1. Estimate how many calories there are in 1 cup of hot cocoa with skim milk.

2. Estimate how many calories there are in 1 cup of premium chocolate ice cream.

3. Which dessert has the highest caloric content?

4. Which dessert has the lowest caloric content?

5. Which dessert contains about 460 calories?

6. Which desserts contain about 300 calories?

7. How many more calories are there in 1 cup of hot cocoa made with whole milk than in 1 cup of hot cocoa made with skim milk?

8. Fred generally drinks a 4-cup chocolate milkshake. How many calories does he consume?

9. Kristin likes to eat 2 cups of premium chocolate ice cream at bedtime. How many calories does she consume?

10. Barney likes to eat a 6-oz chocolate bar with peanuts for lunch. How many calories does he consume?

11. Paul adds a 2-oz chocolate bar with peanuts to his diet each day for 1 yr (365 days) and makes no other changes in his eating or exercise habits. Consumption of 3500 extra calories will add about 1 lb to his body weight. How many pounds will he have gained at the end of the year?

12. Tricia adds one slice of chocolate cake with fudge frosting to her diet each day for 1 yr (365 days) and makes no other changes in her eating or exercise habits. The consumption of 3500 extra calories will add about 1 lb to her body weight. How many pounds will she have gained at the end of the year?

Education and Earnings. Side-by-side bar graphs allow for comparisons. The one shown at right provides data on the effect of education on earning power for men and women from 1970 to 2002. Use the bar graph for Exercises 13–20.

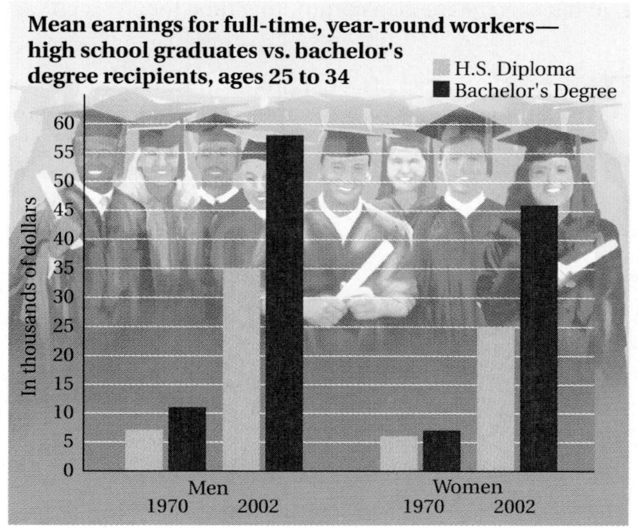

Source: USA Group Foundation

13. How much were the mean earnings for men with bachelor's degrees in 1970? in 2002? How much had they increased? What was the percent of increase?

14. How much were the mean earnings for women with bachelor's degrees in 1970? in 2002? How much had they increased? What was the percent of increase?

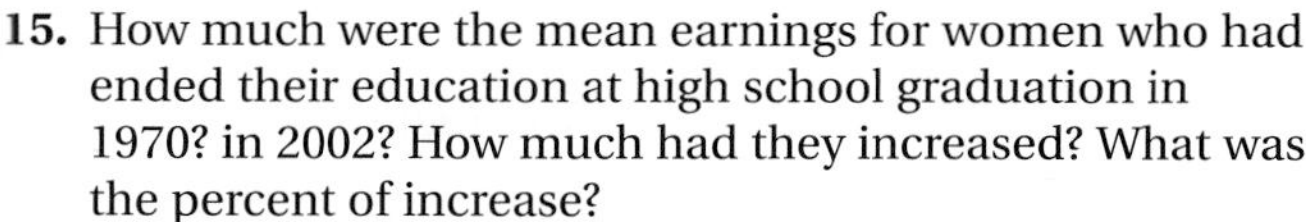

15. How much were the mean earnings for women who had ended their education at high school graduation in 1970? in 2002? How much had they increased? What was the percent of increase?

16. How much were the mean earnings for men who had ended their education at high school graduation in 1970? in 2002? How much had they increased? What was the percent of increase?

17. In 1970, how much more did men with bachelor's degrees earn than men who ended their education at high school graduation?

18. In 2002, how much more did men with bachelor's degrees earn than men who ended their education at high school graduation?

19. In 2002, how much more did women with bachelor's degrees earn than men who ended their education at high school graduation?

20. In 1970, how much more did men with bachelor's degrees earn than women who ended their education at high school graduation?

b

21. *Commuting Time.* The following table lists the average commuting time in six metropolitan areas with more than 1 million people. Make a vertical bar graph to illustrate the data.

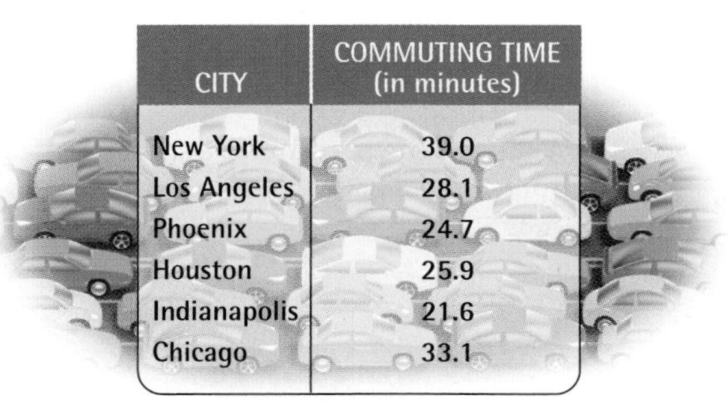

CITY	COMMUTING TIME (in minutes)
New York	39.0
Los Angeles	28.1
Phoenix	24.7
Houston	25.9
Indianapolis	21.6
Chicago	33.1

Source: U.S. Bureau of the Census

Use the data and the bar graph in Exercise 21 to do Exercises 22–25.

22. Which city has the greatest commuting time?

23. Which city has the least commuting time?

24. What was the median commuting time for all six cities?

25. What was the average commuting time for all six cities?

26. *Airline Net Profits.* The net profits (the amount remaining after all deductions like expenses have been made) of U.S. airlines in various years are listed in the table below. Make a horizontal bar graph illustrating the data.

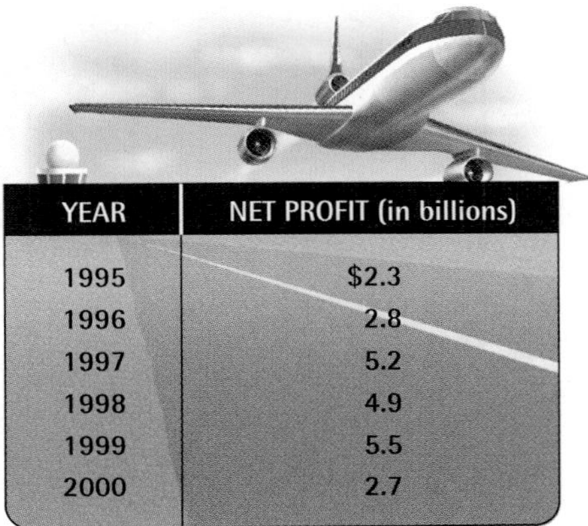

YEAR	NET PROFIT (in billions)
1995	$2.3
1996	2.8
1997	5.2
1998	4.9
1999	5.5
2000	2.7

Source: Air Transportation Association

Use the data and the bar graph in Exercise 26 to do Exercises 27–32.

27. Between which pairs of years was there an increase in profit?

28. Between which pairs of years was there a decrease in profit?

29. What was the percent of decrease between 1999 and 2000?

30. What was the percent of increase between 1996 and 1997?

31. What was the average net profit for all 6 yr?

32. What was the median net profit for all 6 yr?

C *Golf Distances.* In recent years, new equipment and technology have had a tremendous impact on the distance that a golfer can hit a golf ball. The line graph below shows the average driving distances for years from 1980 to 2004. Use the graph for Exercises 33–36.

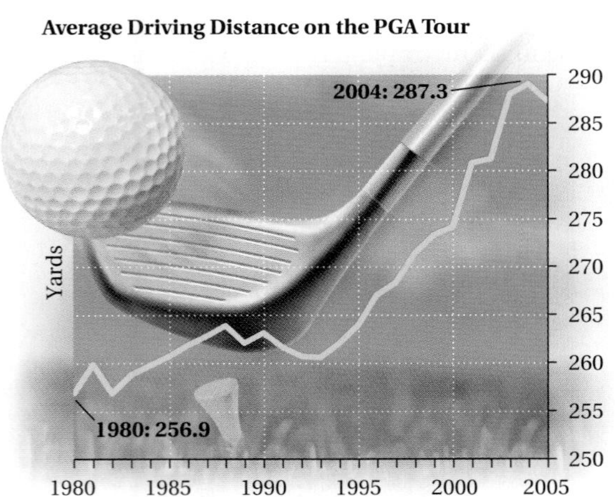

Source: PGA of America

33. How much farther was the driving distance in 2004 than in 1980?

34. What was the percent of increase in the driving distance from 1980 to 2004?

35. In what years was the average driving distance about 264 yd?

36. In what year was the average driving distance about 270 yd?

d Make a line graph of the data in the following tables (Exercises 37 and 42), using the horizontal axis to scale "Year."

37. *Longevity Beyond Age 65.* These data tell us the number of years that a 65-year-old male in the given year can expect to live. Draw a line graph.

YEAR	AVERAGE NUMBER OF YEARS MEN ARE ESTIMATED TO LIVE BEYOND AGE 65
1980	14
1990	15
2000	15.9
2010	16.4
2020	16.9
2030	17.5

Source: 2000 Social Security Report

38. What was the percent of increase in longevity (years beyond 65) between 1980 and 2000?

39. What is the expected percent of increase in longevity between 1980 and 2030?

40. What will be the percent of increase in longevity between 2020 and 2030?

41. What will be the percent of increase in longevity between 2000 and 2030?

42. *Homicide Rate in Baltimore.* These data indicate the number of homicides in Baltimore, Maryland, for several years. Draw a line graph.

YEAR	NUMBER OF MURDERS COMMITTED
1995	325
1996	331
1997	311
1998	313
1999	305
2000	262

Source: Baltimore Police Department

43. Between which two years was the increase in the number of murders the greatest?

44. Between which two years was the increase in the number of murders the least?

45. What is the average number of murders committed per year from 1995 to 2000?

46. What is the median number of murders committed?

47. What was the percent of decrease in the number of murders from 1999 to 2000?

48. What was the percent of increase in the number of murders from 1995 to 1996?

49. DW Can bar graphs always, sometimes, or never be converted to line graphs? Why?

50. DW Compare bar graphs and line graphs. Discuss why you might use one rather than the other to graph a particular set of data.

SKILL MAINTENANCE

VOCABULARY REINFORCEMENT

In each of Exercises 51–58, fill in the blank with the correct term from the given list. Some of the choices may not be used.

51. The set of numbers 1, 2, 3, 4, 5, . . . is called the set of ______________ numbers. [1.1b]

52. To find the average of a set of numbers, ______________ the numbers and then ______________ by the number of items of data. [5.1a]

53. The ______________ interest I on principal P, invested for t years at interest rate r, is given by $I = P \cdot r \cdot t$. [4.7d]

54. If an item has a regular price of \$100 and is on sale for 25% off, \$100 is called the ______________, 25% the ______________, 25% of \$100, or \$25, the ______________, and \$100 − \$25, or \$75, the ______________. [4.7c]

55. When interest is paid on interest, it is called ______________ interest. [4.7e]

56. The decimal $0.\overline{1518}$ is an example of a(n) ______________ decimal. [3.5a]

57. In the equation 103 − 13 = 90, the ______________ is 13. [1.2c]

58. The decimal 0.125 is an example of a(n) ______________ decimal. [3.5a]

add
subtract
multiply
divide
simple
compound
discount
rate of discount
sale price
marked price
repeating
terminating
natural
whole
difference
minuend
subtrahend

SYNTHESIS

59. *Movie Theater Screens.* Draw a line graph of the data presented in the bar graph at right. Use the line graph and the data to make the best estimate you can of the number of screens in 2000, in 2001, and in 2003. Explain.

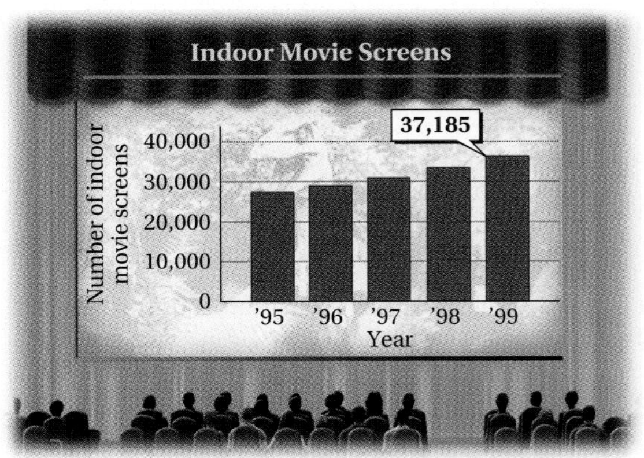

Source: Motion Picture Association of America

5.4 CIRCLE GRAPHS

Objectives

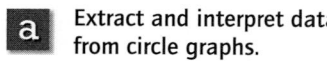

a Extract and interpret data from circle graphs.

b Draw circle graphs.

We often use **circle graphs,** also called **pie charts,** to show the percent of a quantity used in different categories. Circle graphs can also be used very effectively to show visually the *ratio* of one category to another. In either case, it is quite often necessary to use mathematics to find the actual amounts represented for each specific category.

a Reading and Interpreting Circle Graphs

EXAMPLE 1 *Costs of Owning a Dog.* The following circle graph shows the relative costs of raising a dog from birth to death.

Costs of Owning a Dog

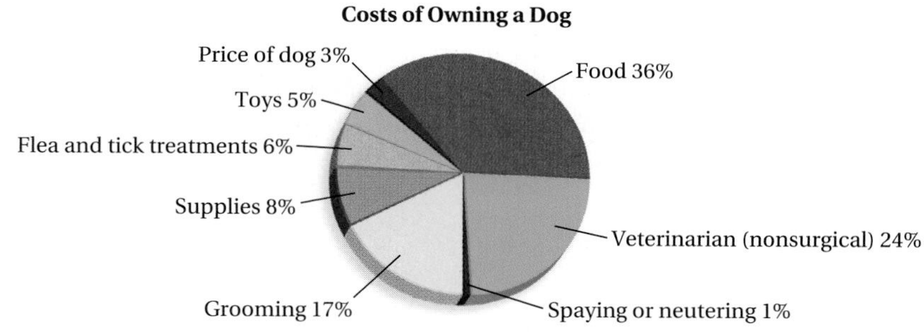

Source: The American Pet Products Manufacturers Association

a) Which item costs the most?

b) What percent of the total cost is spent on grooming?

c) Which item involves 24% of the cost?

d) The American Pet Products Manufacturers Association estimates that the total cost of owning a dog for its lifetime is \$6600. How much of that amount is spent for food?

e) What percent of the expense is for grooming and flea and tick treatments?

We look at the sections of the graph to find the answers.

a) The largest section (or sector) of the graph, 36%, is for food.

b) We see that grooming is 17% of the cost.

c) Nonsurgical veterinarian bills account for 24% of the cost.

d) The section of the graph representing food costs is 36%; 36% of \$6600 is \$2376.

e) We add the percents corresponding to grooming and flea and tick treatments. We have

$$17\% \text{ (grooming)} + 6\% \text{ (flea and tick treatments)} = 23\%.$$

Do Exercises 1–4.

Use the circle graph in Example 1 to answer Margin Exercises 1–4.

1. Which item costs the least?

2. What percent is not spent on either toys or supplies?

3. How much of the \$6600 lifetime cost of owning a dog is for grooming?

4. What part of the expense is for supplies and for buying the dog?

Answers on page A-13

5. Lengths of Engagement of Married Couples. The data below relate the percent of married couples who were engaged for a certain time period before marriage. Use this information to draw a circle graph.

ENGAGEMENT PERIOD	PERCENT
Less than 1 yr	24
1–2 yr	21
More than 2 yr	35
Never engaged	20

Source: Bruskin Goldring Research

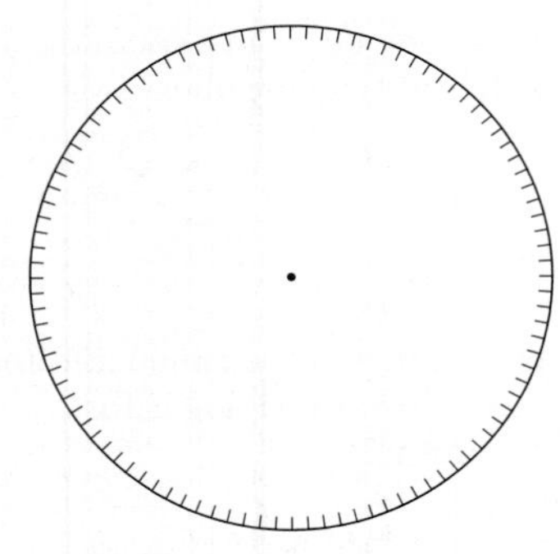

Answer on page A-13

Study Tips

TUNE OUT DISTRACTIONS

Are the places you generally study noisy? If there is constant noise in your home, dorm, or sorority/fraternity house, consider finding a quiet place in the library—maybe a place where the flow of people is minimized and you are not distracted by people-watching!

b Drawing Circle Graphs

To draw a circle graph, or pie chart, like the one in Example 1, think of a pie cut into 100 equally sized pieces. We would then shade in a wedge equal in size to 36 of these pieces to represent 36% for food. We shade a wedge equal in size to 5 of these pieces to represent 5% for toys, and so on.

EXAMPLE 2 *Fruit Juice Sales.* The percents of various kinds of fruit juice sold are given in the list at right. Use this information to draw a circle graph.

Source: Beverage Marketing Corporation

Apple:	14%
Orange:	56%
Blends:	6%
Grape:	5%
Grapefruit:	4%
Prune:	1%
Other:	14%

Using a circle with 100 equally spaced tick marks, we start with the 14% given for apple juice. We draw a line from the center to any tick mark. Then we count off 14 ticks and draw another line. We shade the wedge with a color—in this case, red—and label the wedge as shown in the figure on the left below.

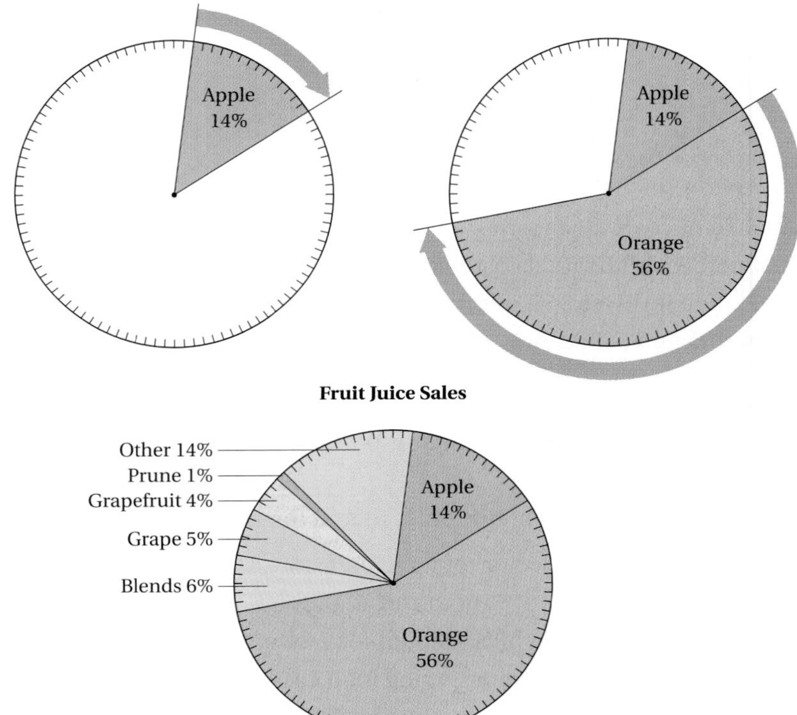

To shade a wedge for orange juice, at 56%, we start at one side of the apple wedge, count off 56 ticks, and draw another line. We shade the wedge with a different color—in this case, orange—and label the wedge as shown in the figure on the right above. Continuing in this manner and choosing different colors, we obtain the graph shown above. Finally, we give the graph the overall title "Fruit Juice Sales."

Do Exercise 5.

Translating for Success

1. *Vacation Miles.* The Saenz family drove their new van 13,640.8 mi in the first year. Of this total, 2018.2 mi were driven while on vacation. How many nonvacation miles did they drive?

2. *Rail Miles.* Of the recent $15\frac{1}{2}$ million passenger miles on a rail passenger line, 80% were transportation-to-work miles. How many rail miles, in millions, were to and from work?

3. *Sales Tax Rate.* The sales tax on the purchase of 10 bath towels that cost \$129.50 is \$8.42. What is the sales tax rate?

4. *Water Level.* During heavy rains in early spring, the water level in a pond rose 0.5 in. every 35 min. How much did the water rise in 90 min?

5. *Marathon Training.* When training for a marathon, Yuriy ran $15\frac{1}{2}$ mi per day 6 days a week. This distance is 80% of the distance Yuriy ran per day. How far did Yuriy run each day?

The goal of these matching questions is to practice step (2), *Translate,* of the five-step problem-solving process. Translate each word problem to an equation and select a correct translation from equations A–O.

A. $8.42 \cdot x = 129.50$

B. $x = 80\% \cdot 15\frac{1}{2}$

C. $x = \dfrac{84 - 68}{84}$

D. $2018.2 + x = 13{,}640.8$

E. $\dfrac{5}{100} = \dfrac{x}{3875}$

F. $2018.2 = x \cdot 13{,}640.8$

G. $4\frac{1}{6} \cdot 73 = x$

H. $\dfrac{x}{5} = \dfrac{100}{3875}$

I. $15\frac{1}{2} = 80\% \cdot x$

J. $8.42 = x \cdot 129.50$

K. $\dfrac{0.5}{35} = \dfrac{x}{90}$

L. $x \cdot 4\frac{1}{6} = 73$

M. $x = \dfrac{84 - 68}{68}$

N. $x = 8.42\% \cdot 129.50$

O. $0.5 \times 35 = 90 \cdot x$

Answers on page A-13

6. *Vacation Miles.* The Ning family drove 2018.2 mi on their summer vacation. If they put a total of 13,640.8 mi on their new van during that year, what percent were vacation miles?

7. *Sales Tax.* The sales tax rate is 8.42%. Salena purchased 10 pillows at \$12.95 each. How much tax is charged on this purchase?

8. *Charity Donations.* Rachel donated \$5 to her favorite charity for each \$100 she earned. One month, she earned \$3875. How much did she donate that month?

9. *Tuxedos.* Emil Tailoring Company purchased 73 yd of fabric for a new line of tuxedos. How many tuxedos can be produced if it takes $4\frac{1}{6}$ yd of fabric for each tuxedo?

10. *Percent of Increase.* In a calculus-based physics course, Mime got 68% on the first exam and 84% on the second. What was the percent of increase in her score?

5.4 EXERCISE SET

For Extra Help

MathXL | MyMathLab | InterAct Math | Math Tutor Center | Video Lectures on CD Disc 3 | Student's Solutions Manual

a *Popular Majors.* This circle graph shows the most popular majors among incoming freshmen. Use this graph for Exercises 1–6.

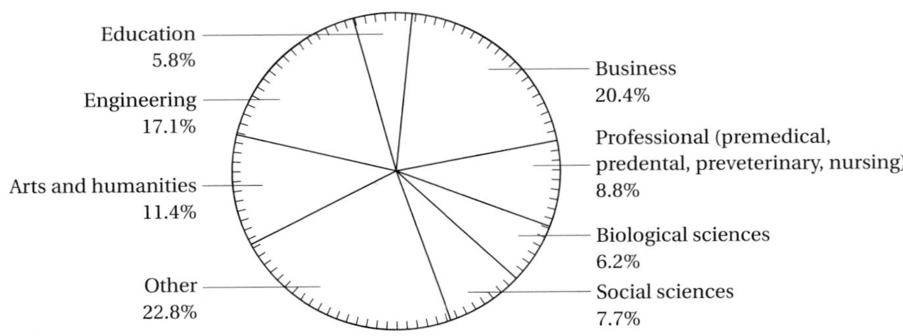

1. What percent of freshmen major in engineering?

2. Together, what percent of freshmen have biological science or professional majors?

3. In the fall of 2004, the University of Minnesota had approximately 10,562 freshmen. According to the data in the graph above, how many would be expected to major in education?

4. Texas Tech had 5766 freshmen in the fall of 2004. How many freshmen would be expected to major in the arts and humanities?

5. What percent of all freshmen do not major in biological or social sciences?

6. What percent of freshmen do not major in business or education?

Family Expenses. This circle graph shows expenses as a percent of income for a family of four. Use the graph for Exercises 7–10.

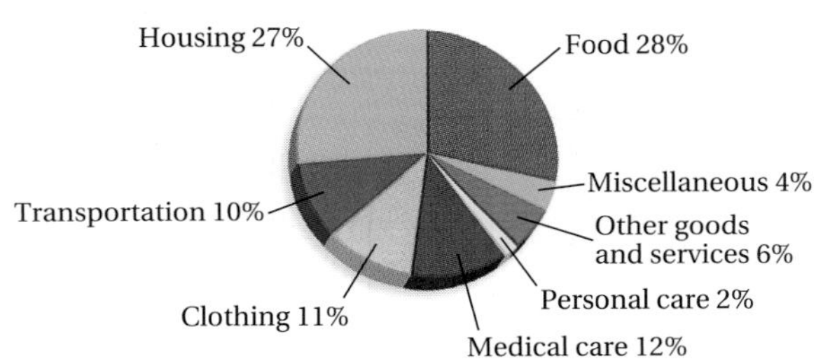

Source: U.S. Bureau of Labor Statistics

7. Which item accounts for the greatest expense?

8. In a family with a $4000 monthly income, how much is spent for transportation?

9. Some surveys combine medical care with personal care. What percent would be spent on those two items combined?

10. In a family with a $2000 monthly income, what is the ratio of the amount spent on medical care to the amount spent on personal care?

b Use the given information to complete a circle graph. Note that each circle is divided into 100 sections.

11. *Holiday Baking.* The table below lists the percentages of the times when people do their holiday baking.

PREFERENCE	PERCENT
Pre-dawn, before 6 A.M.	28.7%
Late night, 9 P.M. to midnight	11.7%
Overnight, midnight to 5 A.M.	11.5%
Marathon style, all day/night	10.5%
Take day off work	3.4%
Don't bake for the holidays	23.4%
Don't know	10.8%

Source: Land O'Lakes Holiday Bakeline

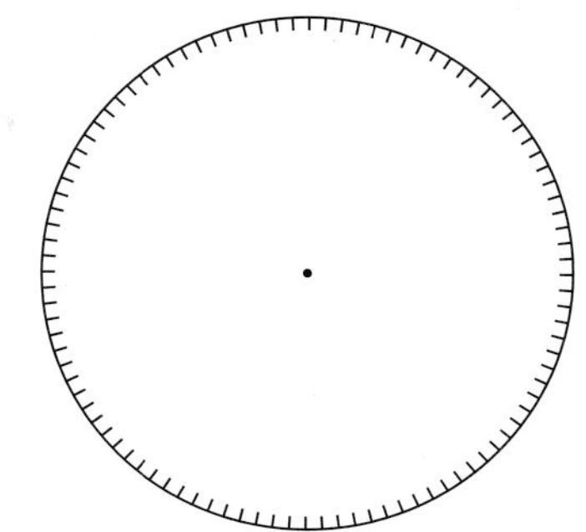

12. *Wealthy Givers.* The table below lists the amounts that people with a net worth of more than $1 million make in charitable contributions.

DONATIONS	PERCENT
Under $1000	18%
$1000–$2499	21%
$2500–$4999	20%
$5000–$9999	17%
$10,000 or more	18%
Don't contribute/no answer	6%

Source: Yankelovich Partners

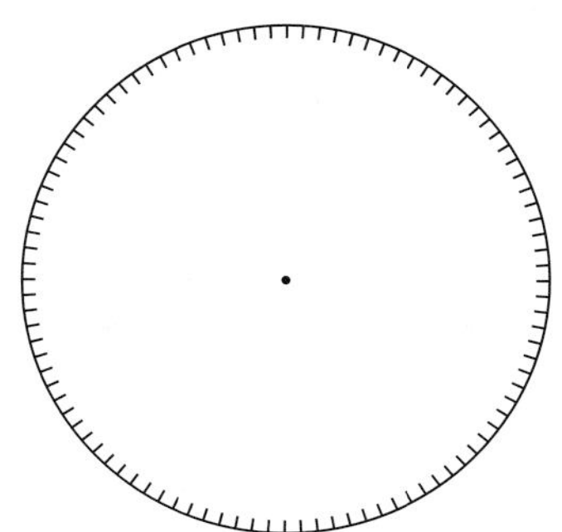

13. *Pregnancy Weight Gain.* The table below lists the amounts of weight gain during pregnancy.

WEIGHT GAIN (in pounds)	PERCENT
Less than 20	22%
21–30	32%
31–40	27%
41 or more	19%

Source: National Vital Statistics Report

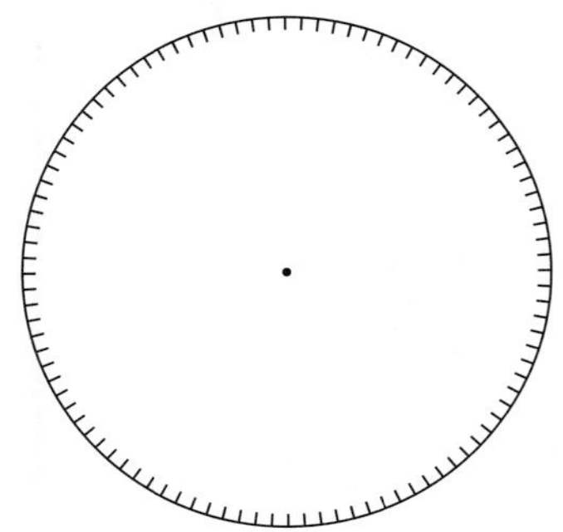

14. *e-mail.* The table below lists the numbers of e-mails people receive per day at work.

NUMBER OF E-MAILS PER DAY	PERCENT
Less than 1	28%
1–5	20%
6–10	12%
11–20	9%
21 or more	31%

Source: John J. Heldrich Center for Workforce Development

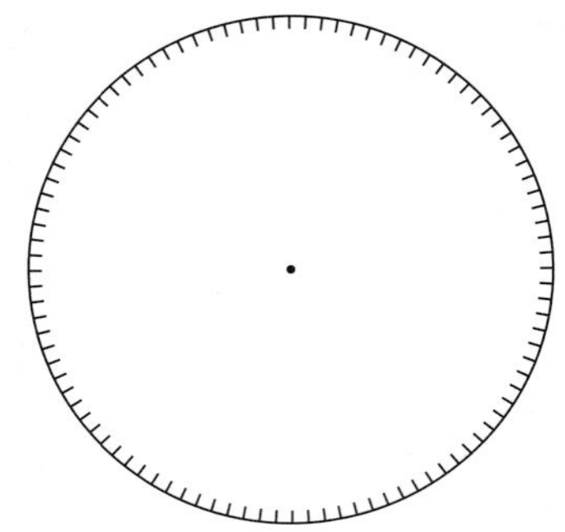

15. *Causes of Spinal Cord Injuries.* The table below lists the causes of spinal cord injury.

CAUSES	PERCENT
Motor vehicle accidents	44%
Acts of violence	24%
Falls	22%
Sports	8%
Other	2%

Source: National Spinal Cord Injury Association

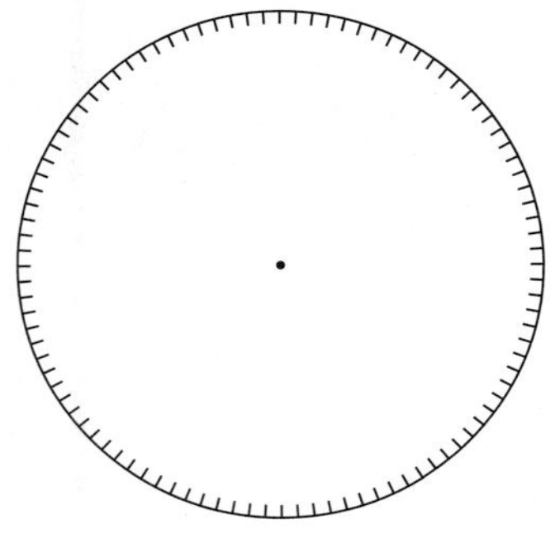

16. *Kids in Foster Care.* There are approximately one-half million children in foster care in the United States. Most of these children are under the age of 10. The table below lists the percentages by ages of children in foster care.

AGE GROUP	PERCENT
Under 1	3%
1–5	25%
6–10	27%
11–15	27%
16+	18%

Source: The Administration for Children and Families

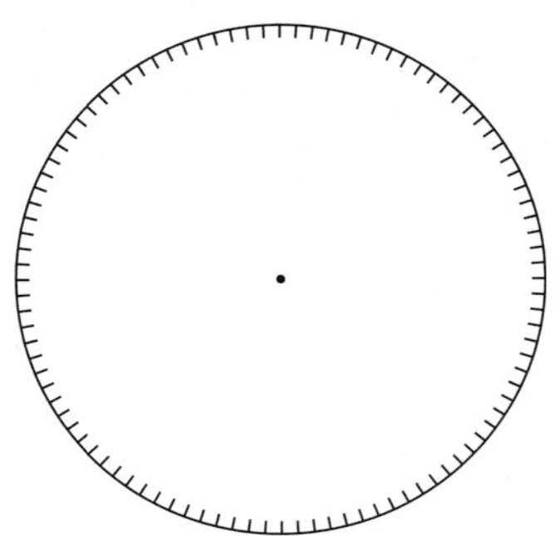

17. **D**$_W$ Discuss the advantages of being able to read a circle graph.

18. **D**$_W$ Compare circle graphs to bar graphs.

5 Summary and Review

The review that follows is meant to prepare you for a chapter exam. It consists of two parts. The first part, Concept Reinforcement, is designed to increase understanding of the concepts through true/false exercises. The second part is the Review Exercises. These provide practice exercises for the exam, together with references to section objectives so you can go back and review. Before beginning, stop and look back over the skills you have obtained. What skills in mathematics do you have now that you did not have before studying this chapter?

CONCEPT REINFORCEMENT

Determine whether the statement is true or false. Answers are given at the back of the book.

_______ **1.** A set of data has just one average and just one median, but it can have more than one mode.

_______ **2.** To find the average of a set of numbers, add the numbers and then multiply by the number of items of data.

_______ **3.** It is possible for the average, median, and mode of a set of data to be the same number.

_______ **4.** If each number in a set of data occurs the same number of times, there is no mode.

Review Exercises

FedEx Mailing Costs. There are three types of Federal Express delivery service for packages of various weights within Zone 5 (shipments from 601 to 1000 mi from origin), as shown in the table at right. Use this table for Exercises 1–6. [5.2a]

1. Find the cost of a 3-lb FedEx Priority Overnight delivery.

2. Find the cost of a 10-lb FedEx Standard Overnight delivery.

3. How much would you save by sending the package listed in Exercise 1 by FedEx 2Day delivery?

4. How much would you save by sending the package in Exercise 2 by FedEx 2Day delivery?

	Delivery by 10:00 a.m. next business day	Delivery by 3:00 p.m. next business day	Delivery by 4:30 p.m. second business day
FEDEX LETTER	FEDEX PRIORITY OVERNIGHT®	FEDEX STANDARD OVERNIGHT®	FEDEX 2DAY®
up to 8 oz.	$ 17.55	$ 15.95	$ n/a
All other packaging / weight in lbs. 1 lb.	$ 26.50	$ 23.25	$ 9.40
2 lbs.	29.75	26.25	10.10
3	32.25	28.50	11.30
4	36.00	31.50	13.10
5	38.75	33.25	14.50
6	41.50	36.25	15.80
7	45.25	39.75	17.30
8	47.50	41.75	18.80
9	51.00	45.00	20.20
10	53.50	47.00	21.60
11	56.50	49.50	23.00

Source: Federal Express Corporation

5. Is there any difference in price between sending a 5-oz package FedEx Priority Overnight and sending an 8-oz package in the same way?

6. An author has a 4-lb manuscript to send by FedEx Standard Overnight delivery to her publisher. She calls and the package is picked up. Later that day she completes work on another part of her manuscript that weighs 5 lb. She calls and sends it by FedEx Standard Overnight delivery to the same address. How much could she have saved if she had waited and sent both packages as one?

U.S. Police Forces. This pictograph shows the number of officers in the largest U.S. police forces. Use the graph for Exercises 7–10.

America's Largest Police Forces

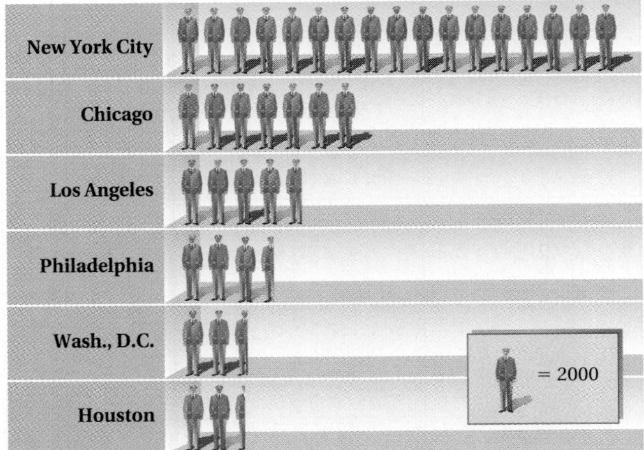

Source: International Association of Chiefs of Police

7. About how many officers are in the Chicago police force? [5.2b]

8. Which city has about 9000 officers on its force? [5.2b]

9. Of the cities listed, which has the smallest police force? [5.2b]

10. Estimate the average size of these six police forces. [5.1a], [5.2b]

Find the mode. [5.1c]

11. 26, 34, 43, 26, 51

12. 17, 7, 11, 11, 14, 17, 18

13. 0.2, 0.2, 1.7, 1.9, 2.4, 0.2

14. 700, 700, 800, 2700, 800

15. $14, $17, $21, $29, $17, $2

16. 20, 20, 20, 20, 20, 500

17. One summer, a student earned the following amounts over a four-week period: $102, $112, $130, and $98. What was the average amount earned per week? the median? [5.1a, b]

18. *Gas Mileage.* A 2001 Ford Focus gets 528 mi of highway driving on 16 gal of gasoline. What is the gas mileage? [5.1a]

Source: Ford Motor Company

19. To get an A in math, Marcus must score an average of 90 on four tests. His scores on the first three tests were 94, 78, and 92. What is the lowest score that he can make on the last test and still get an A? [5.1a]

Calorie Content in Fast Foods. Wendy's Hamburgers is a national food franchise. The following bar graph shows the caloric content of various sandwiches sold by Wendy's. Use the graph for Exercises 20–27. [5.3a]

Wendy's Sandwiches

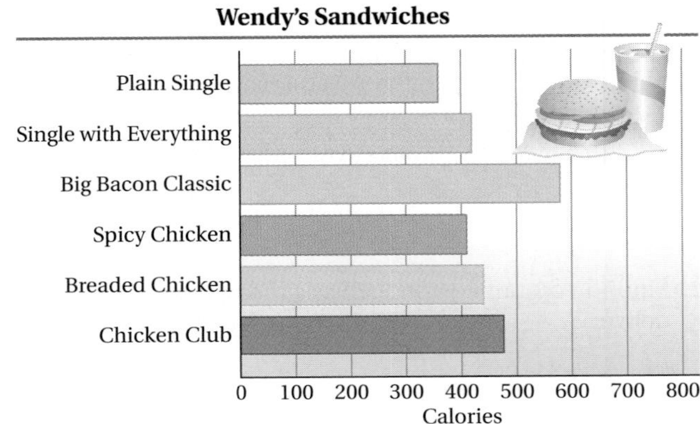

Source: Wendy's International

20. How many calories are in a Single with Everything?

21. How many calories are in a Breaded Chicken sandwich?

22. Which sandwich has the highest caloric content?

23. Which sandwich has the lowest caloric content?

24. Which sandwich contains about 360 calories?

25. Which sandwich contains about 470 calories?

26. How many more calories are in a Chicken Club than in a Single with Everything?

27. How many more calories are in a Big Bacon Classic than in a Plain Single?

Accidents by Driver Age. The following line graph shows the number of accidents per 100 drivers, by age. Use the graph for Exercises 28–33. [5.3c]

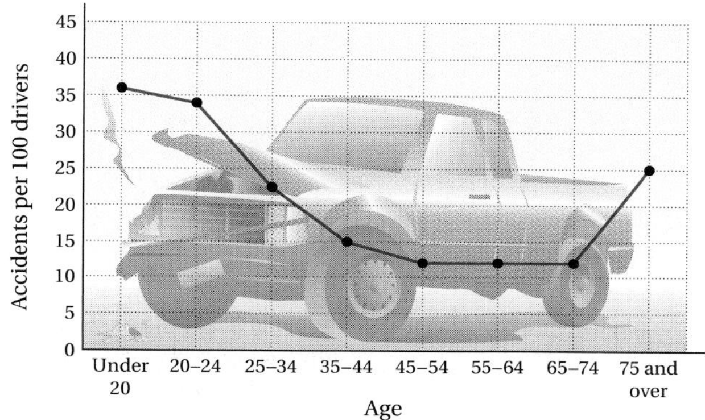

28. Which age group has the greatest number of accidents per 100 drivers?

29. What is the fewest number of accidents per 100 in any age group?

30. How many more accidents do people 75 yr of age and older have than those in the age range of 65–74?

31. Between what ages does the number of accidents stay basically the same?

32. How many fewer accidents do people 25–34 yr of age have than those 20–24 yr of age?

33. Which age group has accidents just about three times as often as people 55–64 yr of age?

Hotel Preferences. This circle graph shows hotel preferences for travelers. Use the graph for Exercises 34–37. [5.4a]

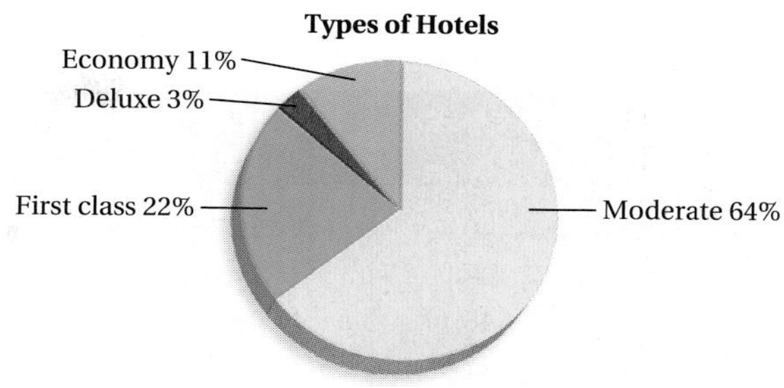

34. What percent of travelers prefer a first-class hotel?

35. What percent of travelers prefer an economy hotel?

36. Suppose 2500 travelers arrive in a city one day. How many of them might seek a moderate room?

37. What percent of travelers prefer either a first-class or a deluxe hotel?

First-Class Postage. The following table shows the cost of first-class postage in various years. Use the table for Exercises 38 and 39.

YEAR	FIRST-CLASS POSTAGE
1989	25¢
1991	29¢
1995	32¢
1999	33¢
2001	34¢
2002	37¢
2006	39¢

Source: U.S. Postal Service

38. Make a vertical bar graph of the data. [5.3b]

39. Make a line graph of the data. [5.3d]

40. *Battery Testing.* An experiment is performed to compare battery quality. Two kinds of battery were tested to see how long, in hours, they kept a hand radio running. On the basis of this test, which battery is better? [5.1d]

BATTERY A: TIMES (in hours)			BATTERY B: TIMES (in hours)		
38.9	39.3	40.4	39.3	38.6	38.8
53.1	41.7	38.0	37.4	47.6	37.9
36.8	47.7	48.1	46.9	37.8	38.1
38.2	46.9	47.4	47.9	50.1	38.2

Find the average. [5.1a]

41. 26, 34, 43, 51

42. 11, 14, 17, 18, 7

43. 0.2, 1.7, 1.9, 2.4

44. 700, 2700, 3000, 900, 1900

45. \$2, \$14, \$17, \$17, \$21, \$29

46. 20, 190, 280, 470, 470, 500

Find the median. [5.1b]

47. 26, 34, 43, 51

48. 7, 11, 14, 17, 18

49. 0.2, 1.7, 1.9, 2.4

50. 700, 900, 1900, 2700, 3000

51. \$2, \$17, \$21, \$29, \$14, \$17

52. 470, 20, 190, 280, 470, 500

53. *Grade Point Average.* Find the grade point average for one semester given the following grades. Assume the grade point values are 4.0 for A, 3.0 for B, and so on. Round to the nearest tenth. [5.1a]

COURSE	GRADE	NUMBER OF CREDIT HOURS IN COURSE
Basic math	A	5
English	B	3
Computer applications	C	4
Russian	B	3
College skills	B	1

54. DW Compare and contrast averages, medians, and modes. Discuss why you might use one over the others to analyze a set of data. [5.1a, b, c]

55. DW Find a real-world situation that fits this equation: [5.1a]

$$T = \frac{(20{,}500 + 22{,}800 + 23{,}400 + 26{,}000)}{4}.$$

SYNTHESIS

56. The ordered set of data 298, 301, 305, a, 323, b, 390 has a median of 316 and an average of 326. Find a and b. [5.1a, b]

5 Chapter Test

For Extra Help

Work It Out!
Chapter Test Video
Solutions on CD

Desirable Body Weights. The following tables list the desirable body weights for men and women over age 25. Use the tables for Exercises 1–4.

DESIRABLE WEIGHT OF MEN			
HEIGHT	SMALL FRAME (in pounds)	MEDIUM FRAME (in pounds)	LARGE FRAME (in pounds)
5 ft 7 in.	138	152	166
5 ft 9 in.	146	160	174
5 ft 11 in.	154	169	184
6 ft 1 in.	163	179	194
6 ft 3 in.	172	188	204

DESIRABLE WEIGHT OF WOMEN			
HEIGHT	SMALL FRAME (in pounds)	MEDIUM FRAME (in pounds)	LARGE FRAME (in pounds)
5 ft 1 in.	105	113	122
5 ft 3 in.	111	120	130
5 ft 5 in.	118	128	139
5 ft 7 in.	126	137	147
5 ft 9 in.	134	144	155

Source: U.S. Department of Agriculture

1. What is the desirable weight for a 6 ft 1 in. man with a medium frame?

2. What is the desirable weight for a 5 ft 3 in. woman with a small frame?

3. What size woman has a desirable weight of 120 lb?

4. What size man has a desirable weight of 169 lb?

Waste Generated. The number of pounds of waste generated per person per year varies greatly among countries around the world. In the pictograph at right, each symbol represents approximately 100 lb of waste. Use the pictograph for Exercises 5–8.

5. In which country does each person generate 600 lb of waste per year?

6. In which country does each person generate 1000 lb of waste per year?

7. How many pounds of waste per person per year are generated in France?

8. How many pounds of waste per person per year are generated in Finland?

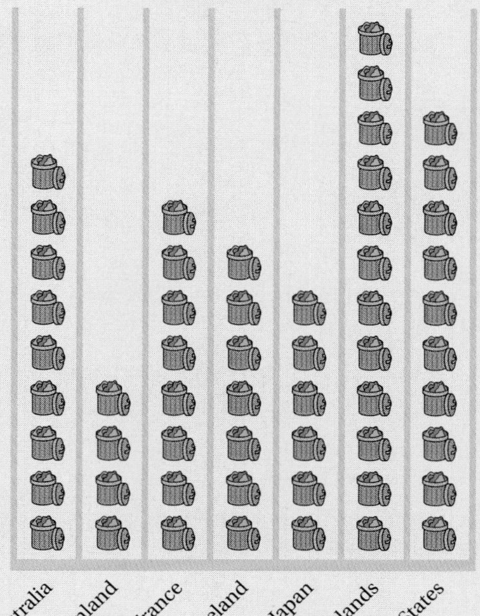

Source: Data from OECD Environmental Data Compendium: 2002

Find the average.

9. 45, 49, 52, 52

10. 1, 1, 3, 5, 3

11. 3, 17, 17, 18, 18, 20

Find the median and the mode.

12. 45, 49, 52, 52

13. 1, 1, 3, 5, 3

14. 3, 17, 17, 18, 18, 20

15. *Gas Mileage.* A 2006 Mitsubishi Eclipse GT V-6 gets 432 mi of highway driving on 16 gal of gasoline. What is the gas mileage?

Source: Mitsubishi Motors

16. *Grades.* To get a C in chemistry, a student must score an average of 70 on four tests. Scores on the first three tests were 68, 71, and 65. What is the lowest score that the student can make on the last test and still get a C?

Food Dollars Spent Away from Home. The line graph below shows the percentage of food dollars spent away from home for various years, and projected to 2010. Use the graph for Exercises 17–20.

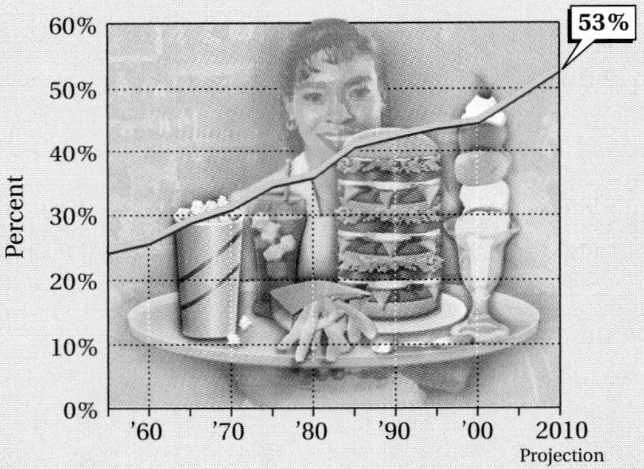

Sources: U.S. Bureau of Labor Statistics; National Restaurant Association

17. What percent of meals will be eaten away from home in 2010?

18. What percent of meals were eaten away from home in 1985?

19. In what year was the percent of meals eaten away from home about 30%?

20. In what year will the percent of meals eaten away from home be about 50%?

21. *Animal Speeds.* The following table lists maximum speeds of movement for various animals, in miles per hour, compared to the speed of the fastest human. Make a vertical bar graph of the data.

ANIMAL	SPEED (in miles per hour)
Antelope	61
Peregrine falcon	225
Cheetah	70
Fastest human	28
Greyhound	42
Golden eagle	150
Grant's gazelle	47

Source: Barbara Ann Kipfer, *The Order of Things.* New York: Random House, 1998.

Refer to the table and the graph in Exercise 21 for Exercises 22–25.

22. By how much does the fastest speed exceed the slowest speed?

23. Does a human have a chance of outrunning a greyhound? Explain.

24. Find the average of all the speeds.

25. Find the median of all the speeds.

26. *Retailing Losses.* The following table lists ways in which American retailers lost money recently. Construct a circle graph representing these data.

TYPE OF LOSS	PERCENT
Employee theft	44
Shoplifting	32.7
Administrative error	17.5
Vendor fraud	5.1
Other	0.7

Source: University of Florida, Department of Sociology for Sensormatic Electronics Corporation

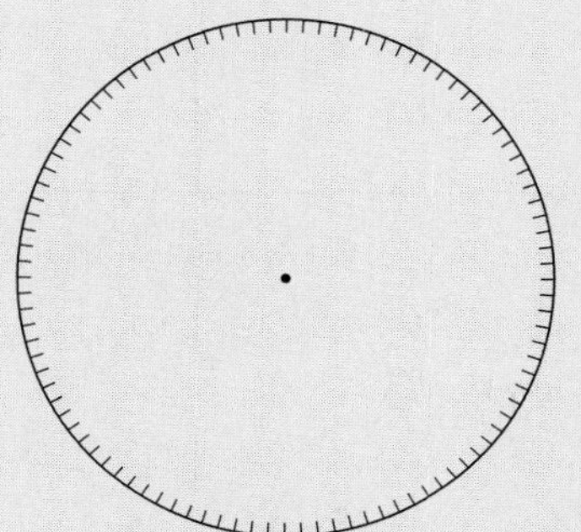

27. In reference to Exercise 26, it is known that retailers lost \$23 billion. Using the percents from the table and the circle graph, find the amount of money lost from each type of loss.

Porsche Sales. The table below lists the number of Porsche sales in the United States for various years. Use the table for Exercises 28 and 29.

YEAR	U.S. PORSCHE SALES
1996	7,152
1997	12,980
1998	17,239
1999	20,877
2000	23,000

Sources: Autodata, Bridge Information Systems

28. Make a bar graph of the data.

29. Make a line graph of the data.

30. *Chocolate Bars.* An experiment is performed to compare the quality of new Swiss chocolate bars being introduced in the United States. People were asked to taste the candies and rate them on a scale of 1 to 10. On the basis of this test, which chocolate bar is better?

BAR A: SWISS PECAN			BAR B: SWISS HAZELNUT		
9	10	8	10	6	8
10	9	7	9	10	10
6	9	10	8	7	6
7	8	8	9	10	8

31. *Grade Point Average.* Find the grade point average for one semester given the following grades. Assume the grade point values are 4.0 for A, 3.0 for B, and so on. Round to the nearest tenth.

COURSE	GRADE	NUMBER OF CREDIT HOURS IN COURSE
Introductory algebra	B	3
English	A	3
Business	C	4
Spanish	B	3
Typing	B	2

SYNTHESIS

32. The ordered set of data 69, 71, 73, a, 78, 98, b has a median of 74 and a mean of 82. Find a and b.

6

Geometry

Real-World Application

The average American eats 54 qt of popcorn annually. This quantity of popcorn would fill a bag measuring 10.5 in. by 10.5 in. by 28 in. Find the volume of such a bag.

Source: The Popcorn Board

This problem appears as Margin Exercise 2 in Section 6.5.

6.1 BASIC GEOMETRIC FIGURES

Objectives

a Draw and name segments, rays, and lines. Also, identify endpoints, if they exist.

b Name an angle in five different ways, and given an angle, measure it with a protractor.

c Classify an angle as right, straight, acute, or obtuse.

d Identify perpendicular lines.

e Classify a triangle as equilateral, isosceles, or scalene and as right, obtuse, or acute. Given a polygon of twelve, ten, or fewer sides, classify it as a dodecagon, decagon, and so on.

f Given a polygon of n sides, find the sum of its angle measures using the formula $(n - 2) \cdot 180°$.

In geometry we study sets of points. A **geometric figure** (or *figure*) is simply a set of points. Thus a figure can be a set with one point, a set with two points, or sets that look like those below.

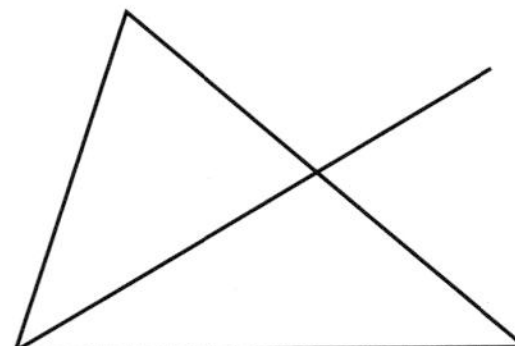

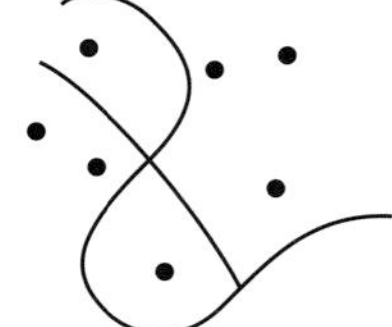

a Segments, Rays, and Lines

A **segment** is a geometric figure consisting of two points, called *endpoints*, and all points between them. The segment whose endpoints are A and B is shown below. It can be named $\overline{AB}$ or $\overline{BA}$.

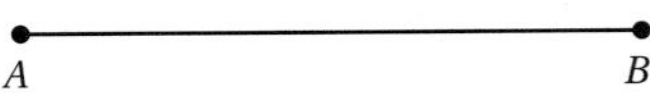

Do Exercise 1.

We get an idea of a geometric figure called a ray by thinking of a ray of light. A **ray** consists of a segment, say $\overline{AB}$, and all points X such that B is between A and X, that is, $\overline{AB}$ and all points "beyond" B.

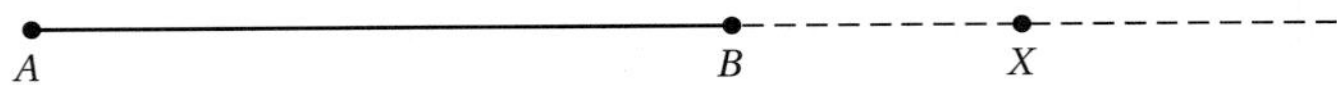

A ray is usually drawn as shown below. It has just one endpoint. The arrow indicates that it extends forever in one direction.

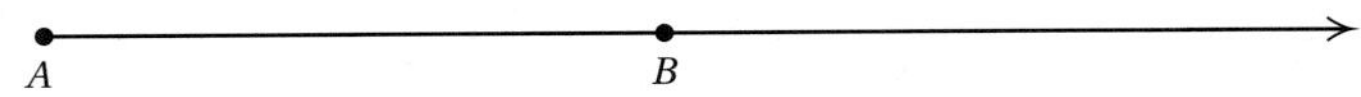

A ray is named $\overrightarrow{AB}$, where B is some point on the ray other than A. The endpoint is always listed first. Thus rays $\overrightarrow{AB}$ and $\overrightarrow{BA}$ are different.

Do Exercises 2–5.

Two rays such as $\overrightarrow{PQ}$ and $\overrightarrow{QP}$ make up what is known as a **line.** A line can be named with a smaller letter m, as shown below, or it can be named by two points P and Q on the line as $\overleftrightarrow{PQ}$.

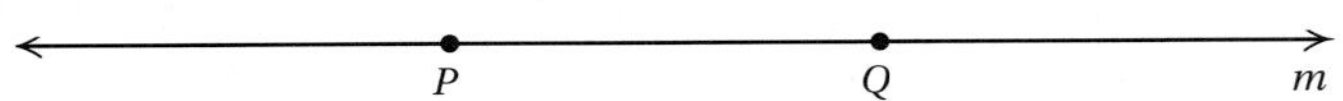

Do Exercises 6–11 on the following page.

1. a) Draw a segment.

b) Label its endpoints E and F.

c) Name this segment in two ways.

2. Draw two points P and Q.

3. Draw $\overline{PQ}$.

4. Draw $\overrightarrow{PQ}$. What is its endpoint?

5. Use a colored pencil to draw $\overrightarrow{QP}$. What is its endpoint?

Answers on page A-15

Lines in the same plane are called **coplanar.** Coplanar lines that do not intersect are called **parallel.** For example, lines l and m below are *parallel* ($l \parallel m$).

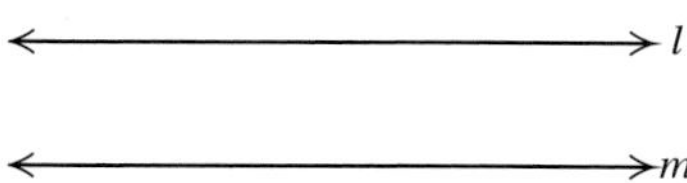

The figure below shows two lines that cross. Their *intersection* is D. They are also called **intersecting lines.**

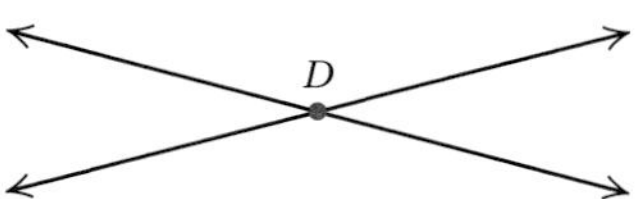

b Measuring Angles

We see a real-world application of *angles* of various types in the spokes of these bicycles and the different back postures of the riders.

Style of Biking Determines Cycling Posture

Road About 180° flat	**Mountain** About 45°	**Comfort** About 90°

Riders prefer a more aerodynamic flat-back position.	Riders prefer a semi-upright position to help lift the front wheel over obstacles.	Riders prefer an upright position that lessens stress on the lower back and neck.

Source: USA TODAY research

An **angle** is a set of points consisting of two **rays,** or half-lines, with a common endpoint. The endpoint is called the **vertex.**

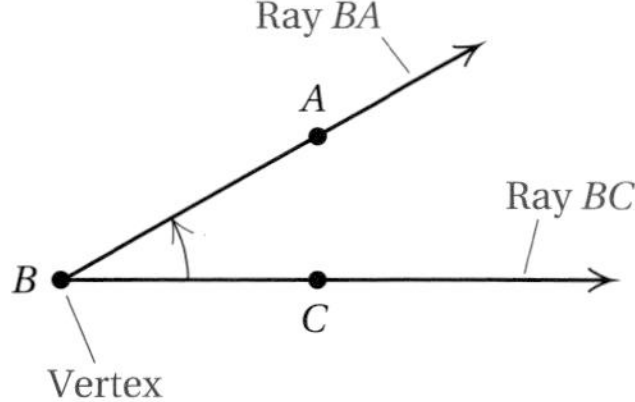

The rays are called the *sides*. The angle above can be named

angle ABC, angle CBA, $\angle ABC$, $\angle CBA$, or $\angle B$.

Note that the name of the vertex is either in the middle or, if no confusion results, listed by itself.

Do Exercises 12 and 13.

6. Draw two points R and S.

7. Draw $\overline{RS}$. What are its endpoints?

8. Draw $\overrightarrow{RS}$. What is its endpoint?

9. Draw $\overrightarrow{SR}$. What is its endpoint?

10. Draw $\overleftrightarrow{RS}$. What are its endpoints?

11. Name this line in seven different ways.

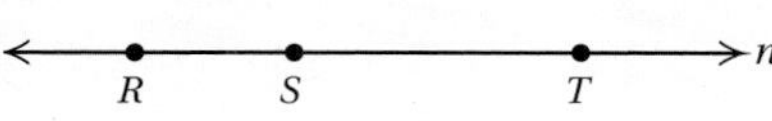

Name the angle in five different ways.

12.

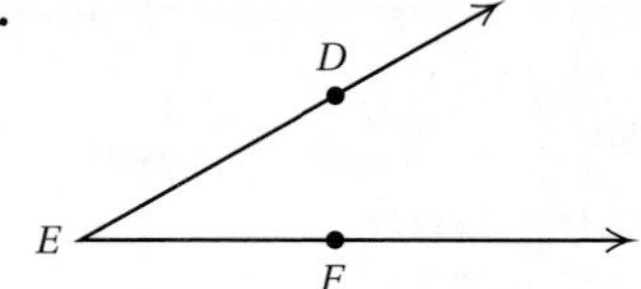

13.

P Q R

Answers on page A-15

14. Use a protractor to measure this angle.

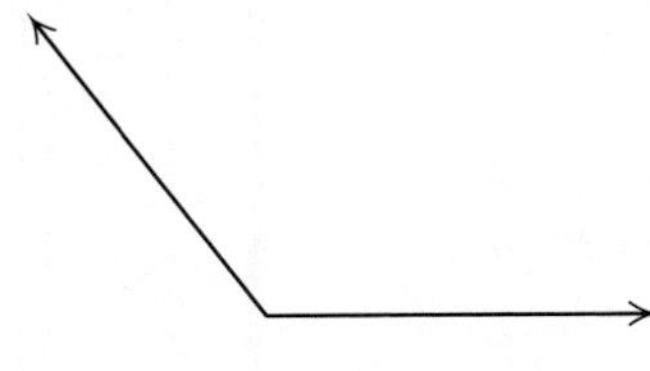

Answer on page A-15

Measuring angles is similar to measuring segments. To measure angles, we start with some arbitrary angle and assign to it a measure of 1. We call it a *unit angle*. Suppose that $\angle U$, shown below, is a unit angle. Let's measure $\angle DEF$. If we made 3 copies of $\angle U$, they would "fill up" $\angle DEF$. Thus the measure of $\angle DEF$ would be 3.

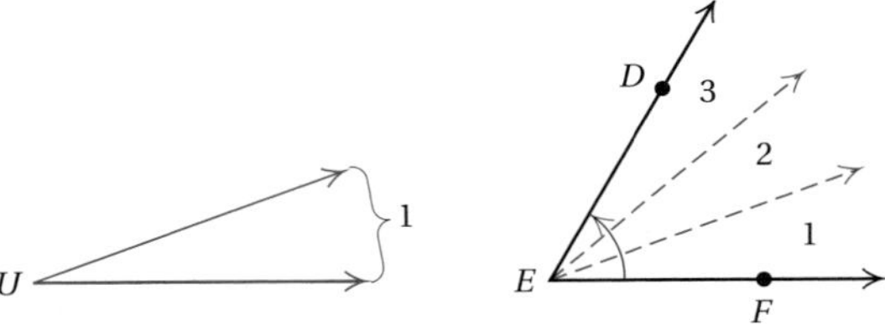

The unit most commonly used for angle measure is the degree. Below is such a unit. Its measure is 1 degree, or 1°.

A 1° angle:

Here are some other angles with their degree measures.

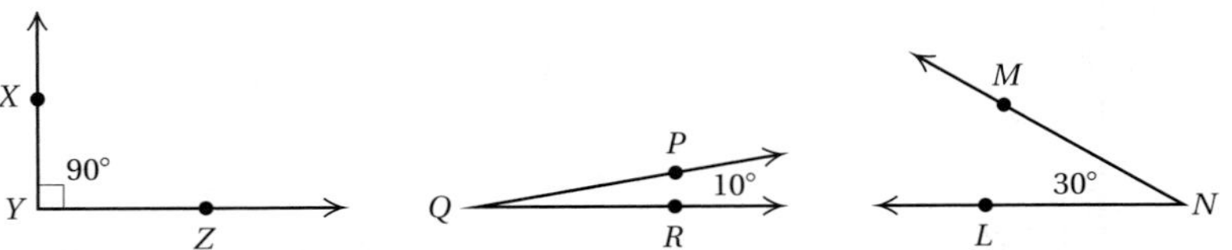

To indicate the *measure* of $\angle XYZ$, we write $m\angle XYZ = 90°$. The symbol ⌝ is sometimes drawn on a figure to indicate a 90° angle.

A device called a **protractor** is used to measure angles. Protractors have two scales. To measure an angle like $\angle Q$ below, we place the protractor's ▲ at the vertex and line up one of the angle's sides at 0°. Then we check where the angle's other side crosses the scale. In the figure below, 0° is on the inside scale, so we check where the angle's other side crosses the inside scale. We see that $m\angle Q = 145°$. The notation $m\angle Q$ is read "the measure of angle Q."

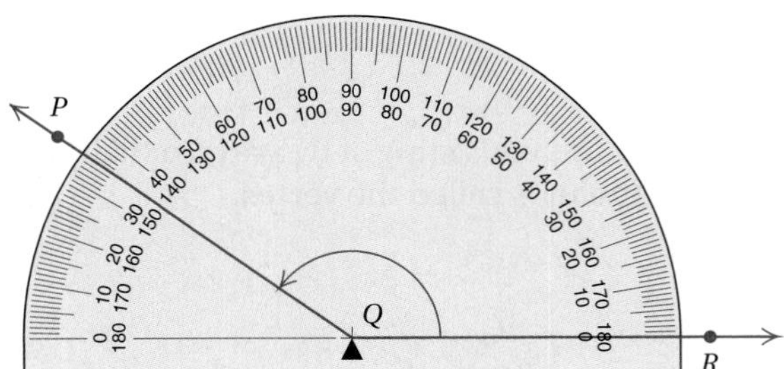

Do Exercise 14.

Study Tips

WORKING WITH A CLASSMATE

If you are finding it difficult to master a particular topic or concept, try talking about it with a classmate. Verbalizing your questions about the material might help clarify it. If your classmate is also finding the material difficult, it is possible that the majority of the people in your class are confused and you can ask your instructor to explain the concept again.

Let's find the measure of $\angle ABC$. This time we will use the 0° on the outside scale. We see that $m\,\angle ABC = 42°$.

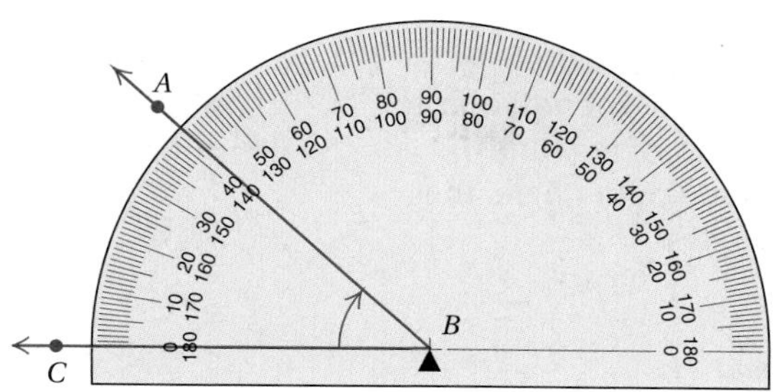

Do Exercise 15.

15. Use a protractor to measure this angle.

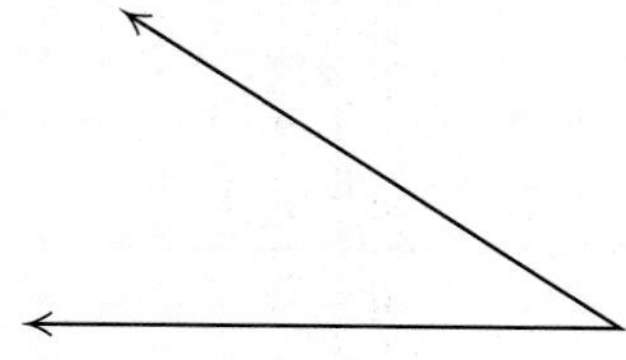

C Classifying Angles

The following are ways in which we classify angles.

TYPES OF ANGLES

Right angle: An angle whose measure is 90°.

Straight angle: An angle whose measure is 180°.

Acute angle: An angle whose measure is greater than 0° and less than 90°.

Obtuse angle: An angle whose measure is greater than 90° and less than 180°.

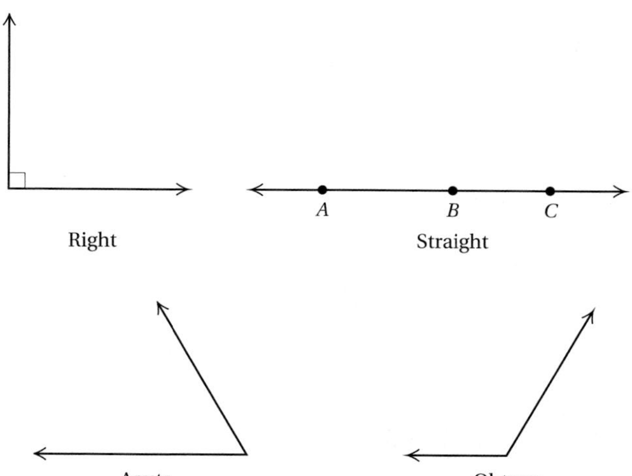

Do Exercises 16–19.

Classify the angle as right, straight, acute, or obtuse. Use a protractor if necessary.

16.

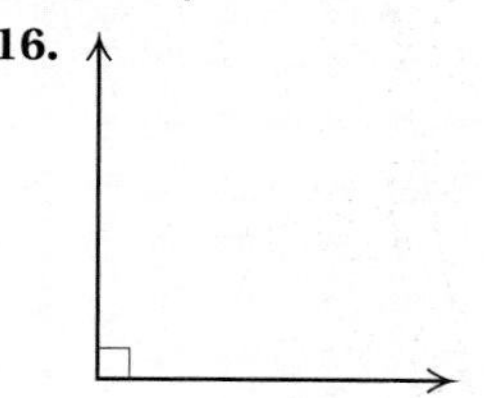

17.

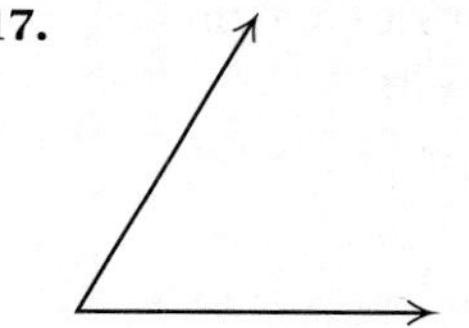

18.

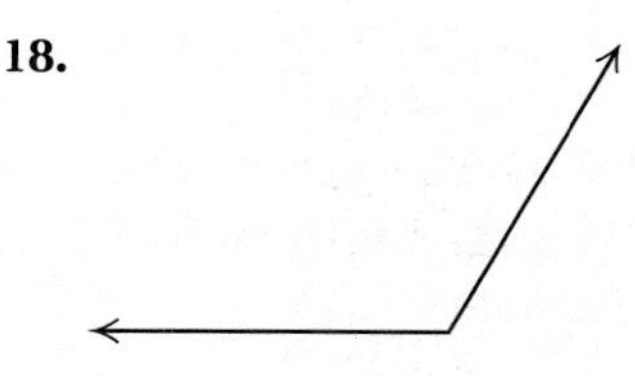

19.

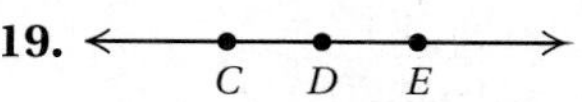

Answers on page A-15

d Perpendicular Lines

Two lines are **perpendicular** if they intersect to form a right angle.

To say that $\overleftrightarrow{AB}$ is perpendicular to $\overleftrightarrow{RS}$, we write $\overleftrightarrow{AB} \perp \overleftrightarrow{RS}$. If two lines intersect to form one right angle, they form four right angles.

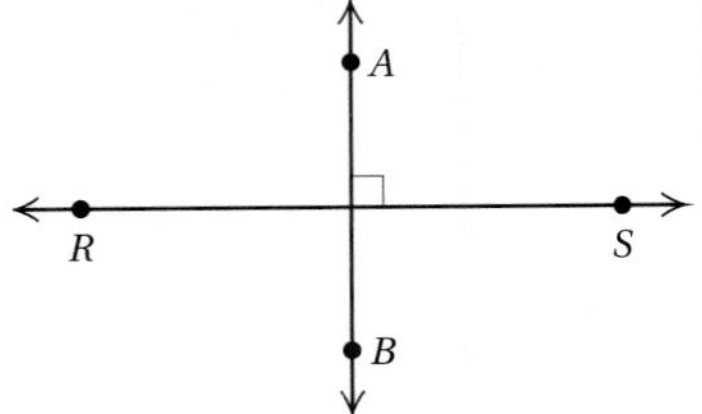

Do Exercises 20 and 21.

Determine whether the pair of lines is perpendicular. Use a protractor.

20.

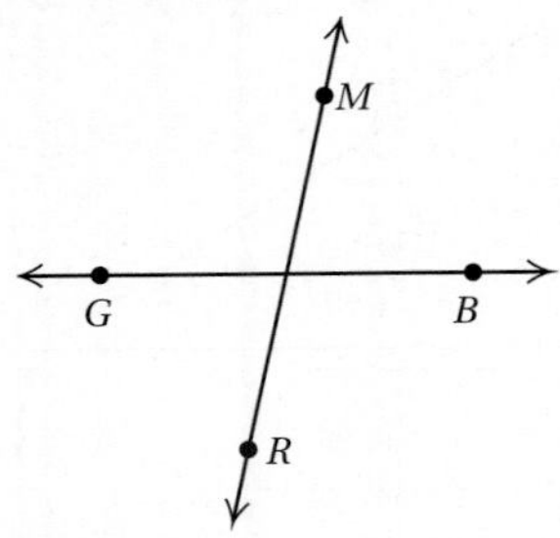

21.

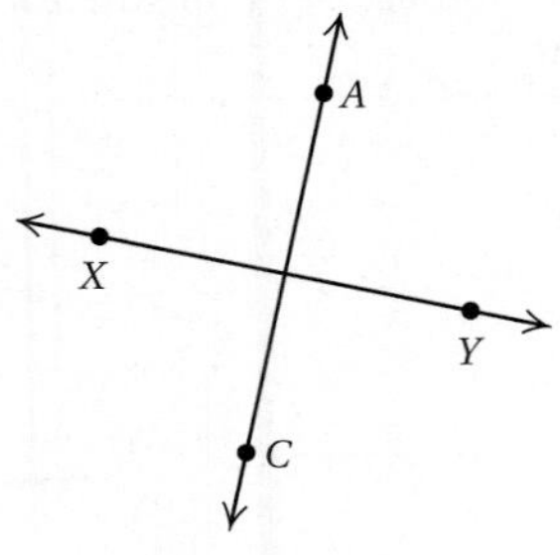

e Polygons

The figures below are examples of **polygons.**

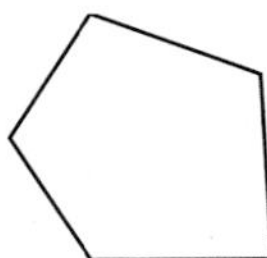

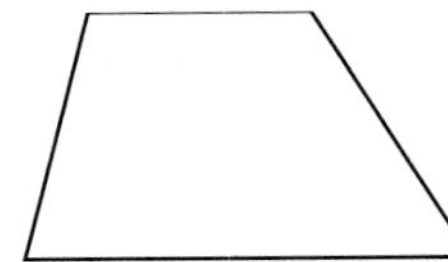

A **triangle** is a polygon made up of three segments, or sides. Consider these triangles. The triangle with vertices A, B, and C can be named $\triangle ABC$.

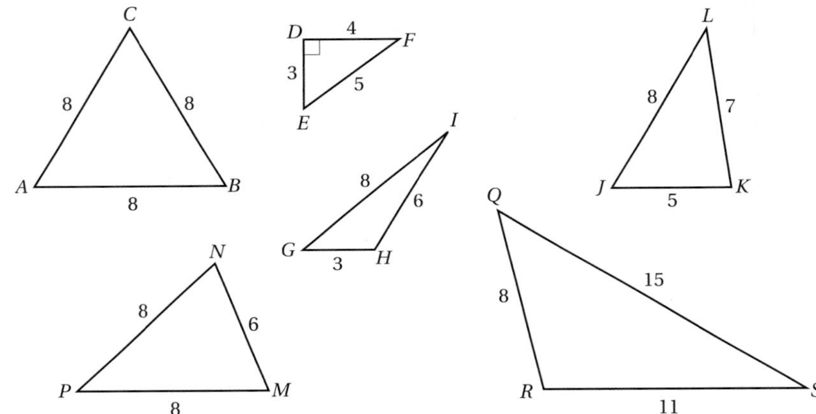

We can classify triangles according to sides and according to angles.

TYPES OF TRIANGLES

Equilateral triangle: All sides are the same length.
Isosceles triangle: Two or more sides are the same length.
Scalene triangle: All sides are of different lengths.
Right triangle: One angle is a right angle.
Obtuse triangle: One angle is an obtuse angle.
Acute triangle: All three angles are acute.

Do Exercises 22–25.

22. Which triangles at right are:
a) equilateral?
b) isosceles?
c) scalene?

23. Are all equilateral triangles isosceles?

24. Are all isosceles triangles equilateral?

25. Which triangles at right are:
a) right triangles?
b) obtuse triangles?
c) acute triangles?

Answers on page A-15

We can further classify polygons as follows.

NUMBER OF SIDES	POLYGON	NUMBER OF SIDES	POLYGON
4	Quadrilateral	8	Octagon
5	Pentagon	9	Nonagon
6	Hexagon	10	Decagon
7	Heptagon	12	Dodecagon

Do Exercises 26–31.

f Sum of the Angle Measures of a Polygon

The sum of the angle measures of a triangle is 180°. To see this, note that we can think of cutting apart a triangle as shown on the left below. If we reassemble the pieces, we see that a straight angle is formed.

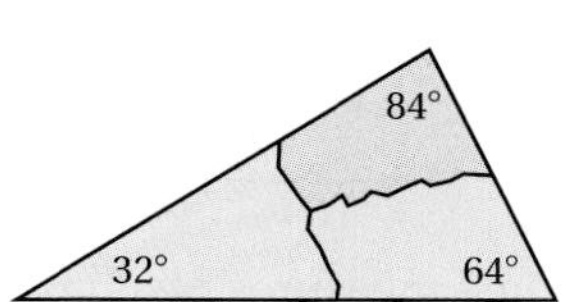

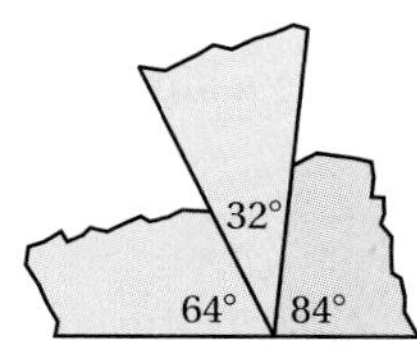

$64° + 32° + 84° = 180°$

SUM OF THE ANGLE MEASURES OF A TRIANGLE

In any triangle *ABC*, the sum of the measures of the angles is 180°:

$$m(\angle A) + m(\angle B) + m(\angle C) = 180°.$$

Do Exercise 32.

If we know the measures of two angles of a triangle, we can calculate the third.

EXAMPLE 1 Find the missing angle measure.

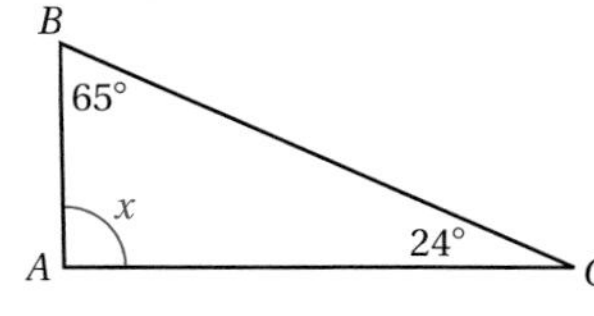

$$m(\angle A) + m(\angle B) + m(\angle C) = 180°$$
$$x + 65° + 24° = 180°$$
$$x + 89° = 180°$$
$$x = 180° - 89°$$
$$x = 91°$$

Do Exercise 33 on the following page.

Classify the polygon by name.

26.

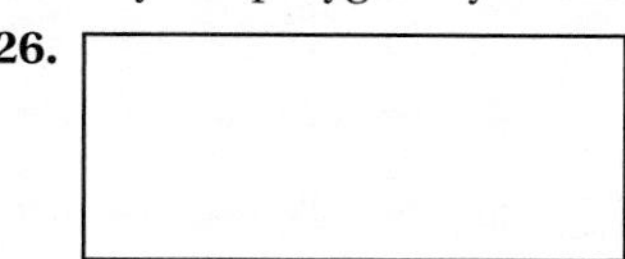

27.

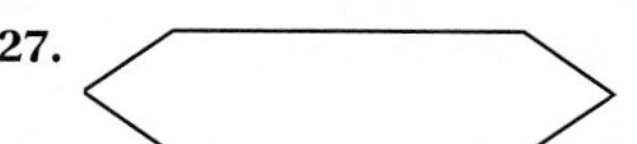

28.

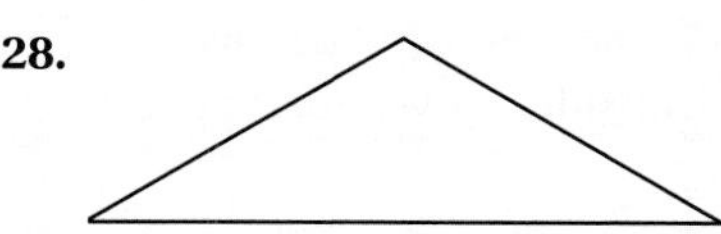

29.

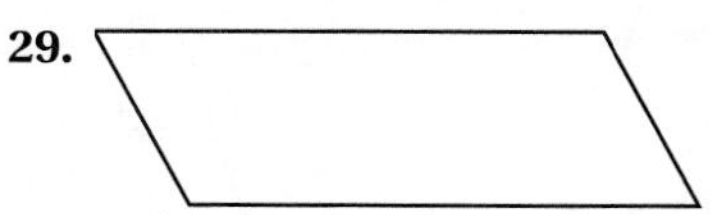

30.

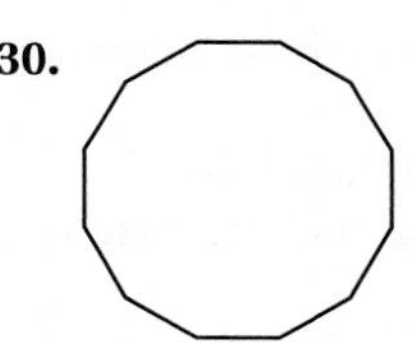

31.

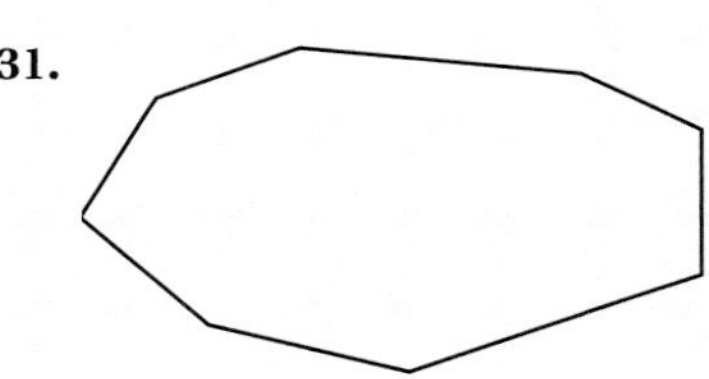

32. Find $m(\angle P) + m(\angle Q) + m(\angle R)$.

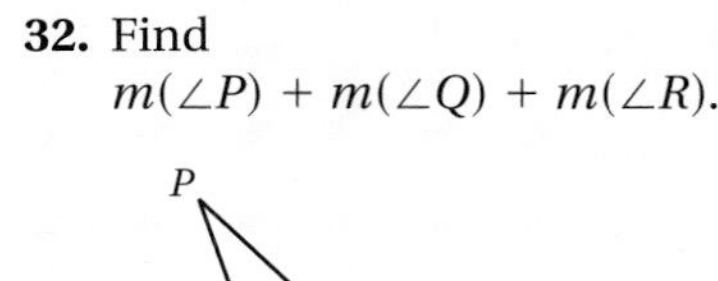

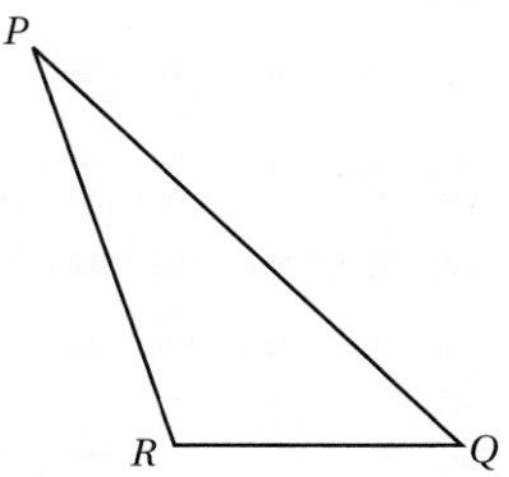

Answers on page A-15

33. Find the missing angle measure.

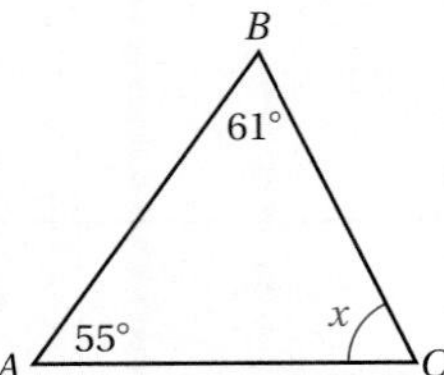

34. Consider a five-sided figure:

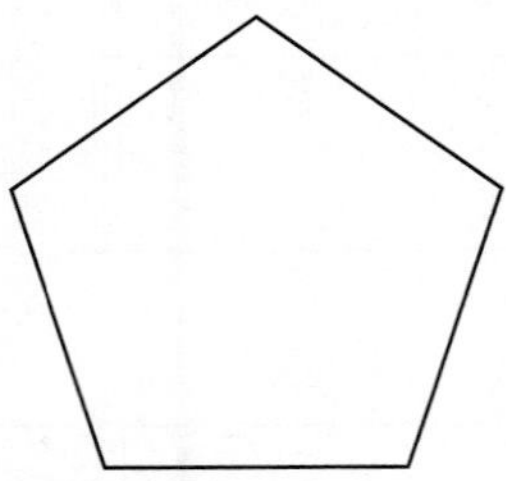

Complete.

a) The figure can be divided into _____ triangles.

b) The sum of the angle measures of each triangle is _____.

c) The sum of the angle measures of the polygon is _____ · 180°, or _____.

35. What is the sum of the angle measures of an octagon?

36. What is the sum of the angle measures of a 25-sided figure?

Answers on page A-15

Now let's use this idea to find the sum of the measures of the angles of a polygon of n sides. First let's consider a four-sided figure:

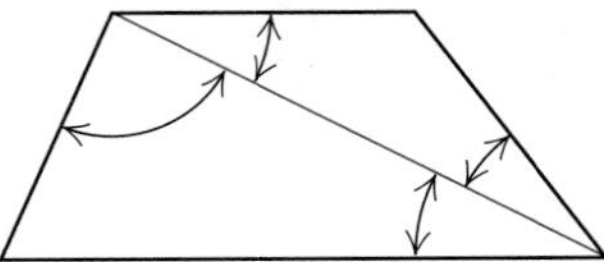

We can divide the figure into two triangles. The sum of the angle measures of each triangle is 180°. We have two triangles, so the sum of the angle measures of the figure is $2 \cdot 180°$, or 360°.

Do Exercise 34.

If a polygon has n sides, it can be divided into $n - 2$ triangles, each having 180° as the sum of its angle measures. Thus the sum of the angle measures of the polygon is $(n - 2) \cdot 180°$.

SUM OF ANGLE MEASURES

If a polygon has n sides, then the sum of its angle measures is $(n - 2) \cdot 180°$.

EXAMPLE 2 What is the sum of the angle measures of a hexagon?

A hexagon has 6 sides. We use the formula $(n - 2) \cdot 180°$:

$$\begin{aligned}(n - 2) \cdot 180° &= (6 - 2) \cdot 180° \\ &= 4 \cdot 180° \\ &= 720°.\end{aligned}$$

Do Exercises 35 and 36.

6.1 EXERCISE SET

For Extra Help

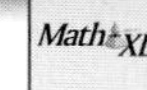
MathXL

MyMathLab

InterAct Math

Math Tutor Center

Video Lectures on CD Disc 3

Student's Solutions Manual

a

1. Draw the segment whose endpoints are G and H. Name the segment in two ways.

G $\qquad$ H

2. Draw the segment whose endpoints are C and D. Name the segment in two ways.

C $\qquad$ D

3. Draw the ray with endpoint Q. Name the ray.

Q $\qquad$ D

4. Draw the ray with endpoint D. Name the ray.

Q $\qquad$ D

Name the line in seven different ways.

5. l $\quad D \quad E \quad F$

6. m $\quad J \quad K \quad T$

b Name the angle in five different ways.

7.

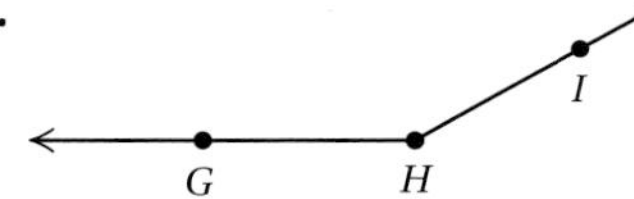

8.

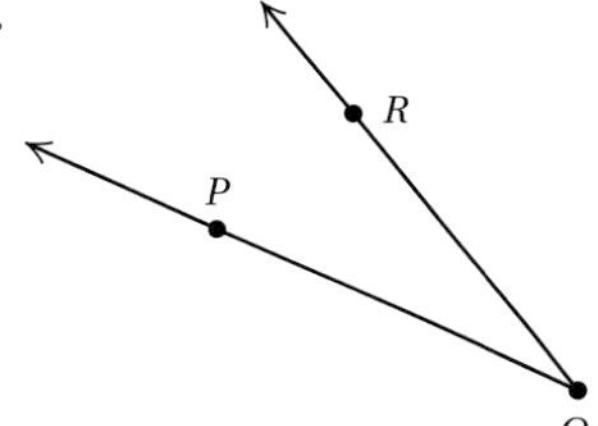

Use a protractor to measure the angle.

9.

10.

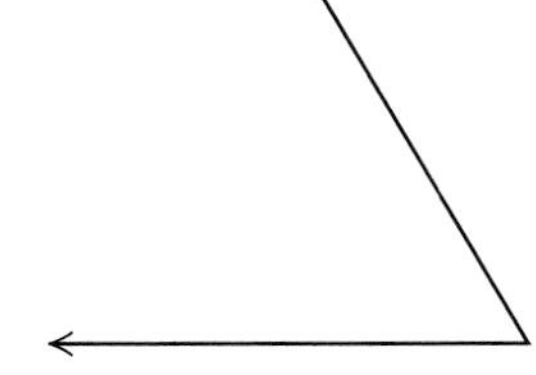

11.

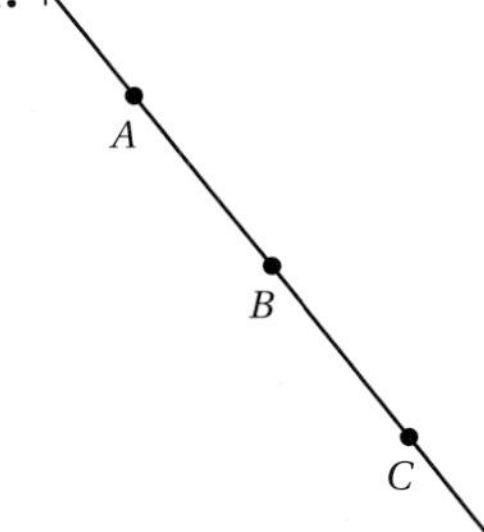

12. 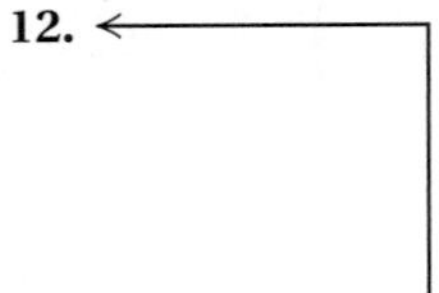

13. 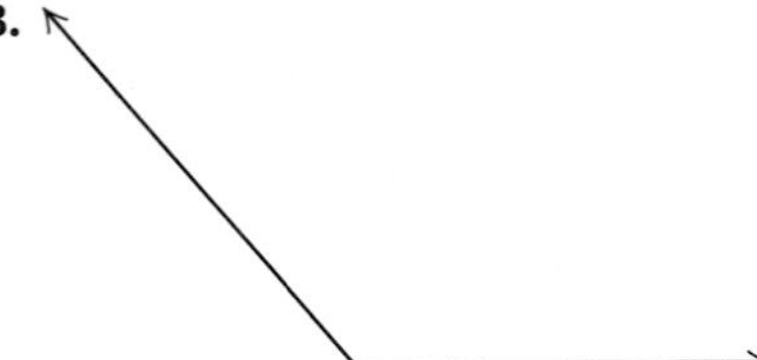

14. 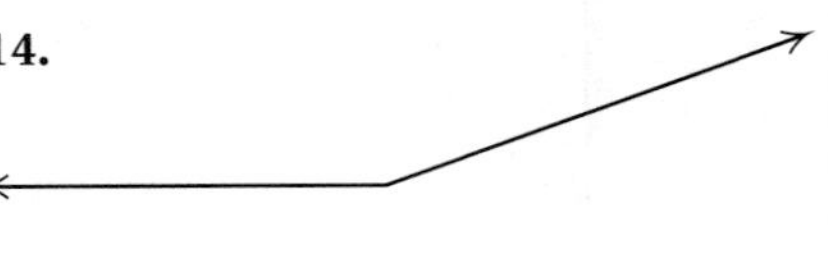

15.–22. Classify each of the angles in Exercises 7–14 as right, straight, acute, or obtuse.

23.–26. Classify each of the angles in Margin Exercises 12–15 as right, straight, acute, or obtuse.

d Determine whether the pair of lines is perpendicular. Use a protractor.

27.

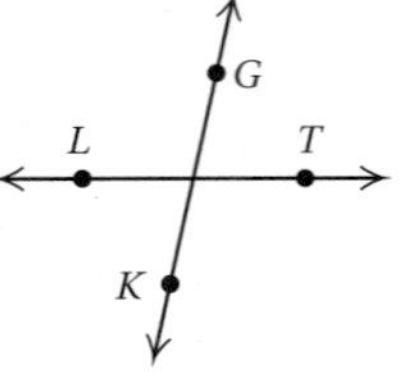

28.

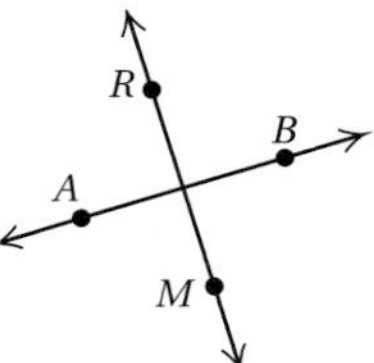

29.

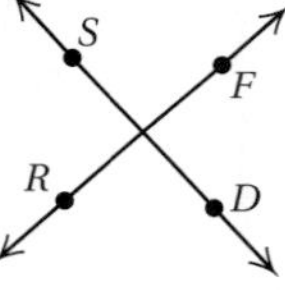

30.

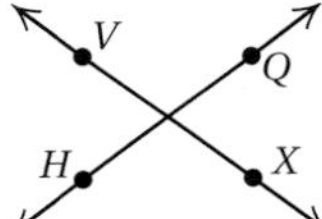

e Classify the triangle as equilateral, isosceles, or scalene. Then classify it as right, obtuse, or acute.

31.

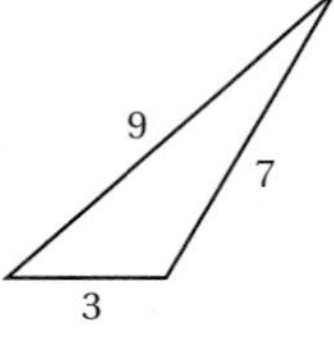

32.

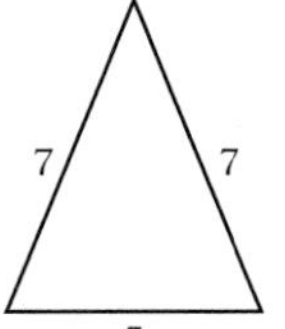

33.

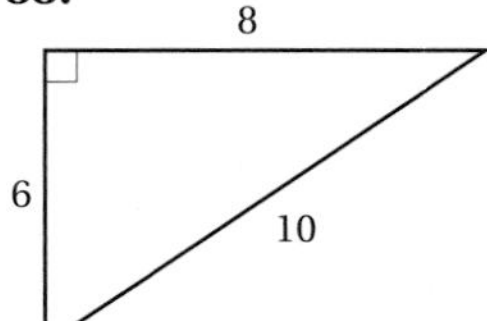

34.

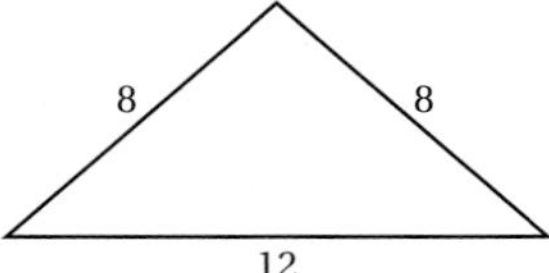

35.

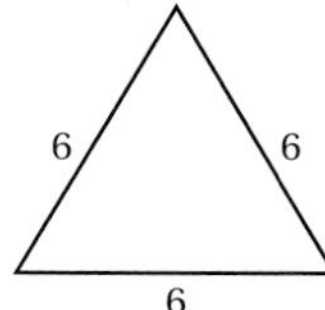

36.

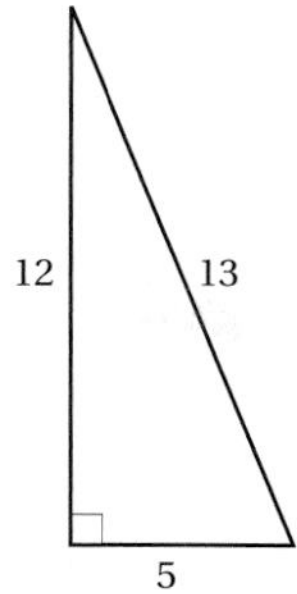

37.

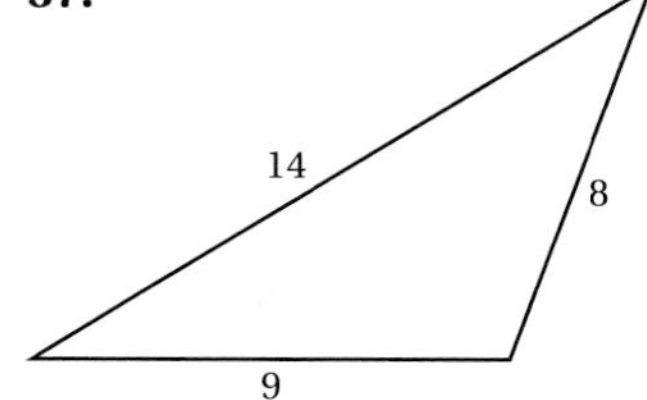

38.

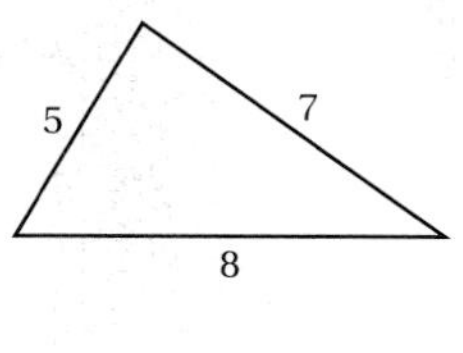

Classify the polygon by name.

39.

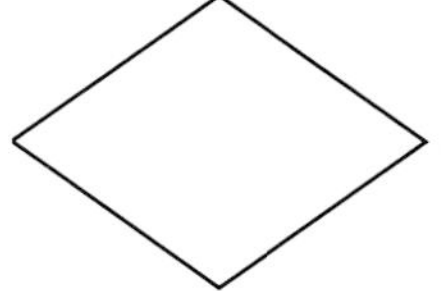

40.

41.

42.

43.

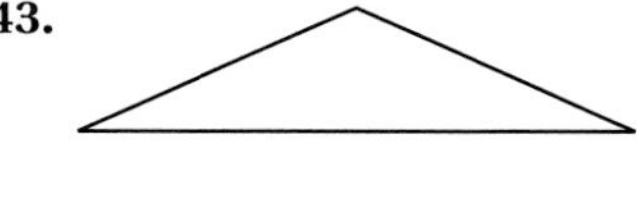

44.

45.

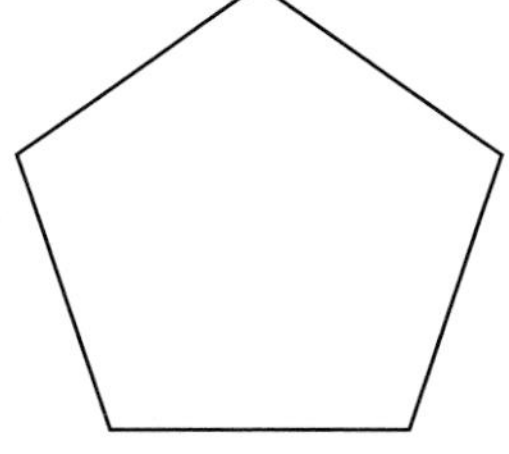

46.

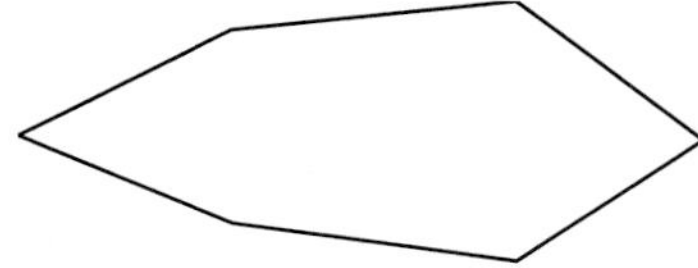

47.

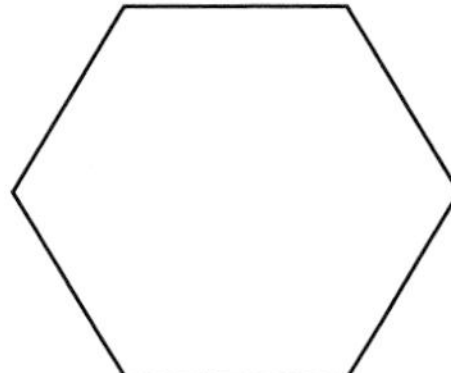

48.

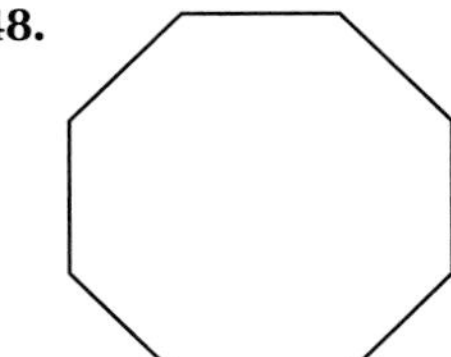

f Find the sum of the angle measures of each of the following.

49. A decagon

50. A quadrilateral

51. A heptagon

52. A nonagon

53. A 14-sided polygon

54. A 17-sided polygon

55. A 20-sided polygon

56. A 32-sided polygon

Find the missing angle measure.

57.

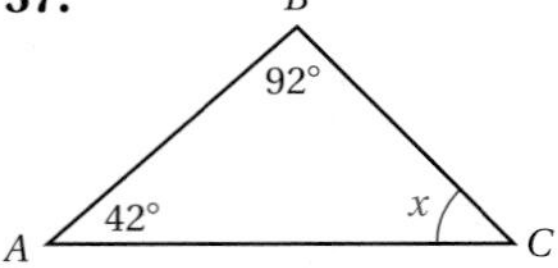

58.

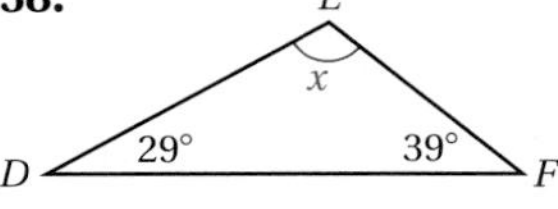

59.

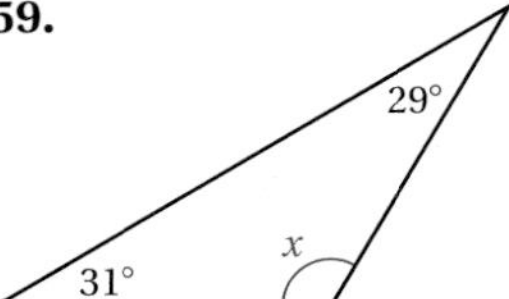

60.

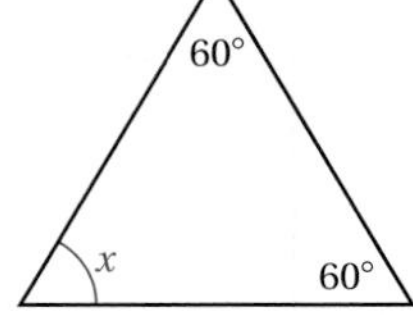

61. In $\triangle RST$, $m(\angle S) = 58°$ and $m(\angle T) = 79°$. Find $m(\angle R)$.

62. In $\triangle KNP$, $m(\angle K) = 137°$ and $m(\angle P) = 12°$. Find $m(\angle N)$.

63. DW Explain how you might use triangles to find the sum of the angle measures of this figure.

64. DW Explain a procedure that could be used to determine the measure of an angle's supplement from the measure of the angle's complement.

SKILL MAINTENANCE

Find the simple interest. [4.7d]

	PRINCIPAL	RATE OF INTEREST	TIME	SIMPLE INTEREST
65.	\$2000	8%	1 year	
66.	\$750	6%	$\frac{1}{2}$ year	
67.	\$4000	7.4%	$\frac{1}{2}$ year	
68.	\$200,000	6.7%	$\frac{1}{12}$ year	

Interest is compounded semiannually. Find the amount in the account after the given length of time. Round to the nearest cent. [4.7e]

	PRINCIPAL	RATE OF INTEREST	TIME	AMOUNT IN THE ACCOUNT
69.	\$25,000	6%	5 years	
70.	\$150,000	$6\frac{7}{8}\%$	15 years	
71.	\$150,000	7.4%	20 years	
72.	\$160,000	7.4%	20 years	

SYNTHESIS

73. In the figure, $m\angle 1 = 79.8°$ and $m\angle 6 = 33.07°$. Find $m\angle 2$, $m\angle 3$, $m\angle 4$, and $m\angle 5$.

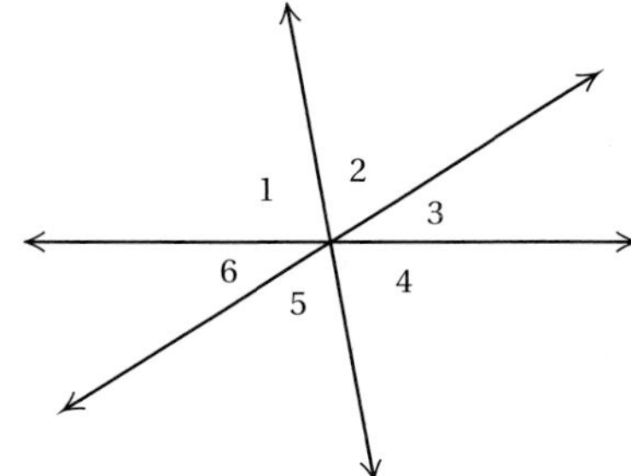

74. In the figure, $m\angle 2 = 42.17°$ and $m\angle 3 = 81.9°$. Find $m\angle 1$, $m\angle 4$, $m\angle 5$, and $m\angle 6$.

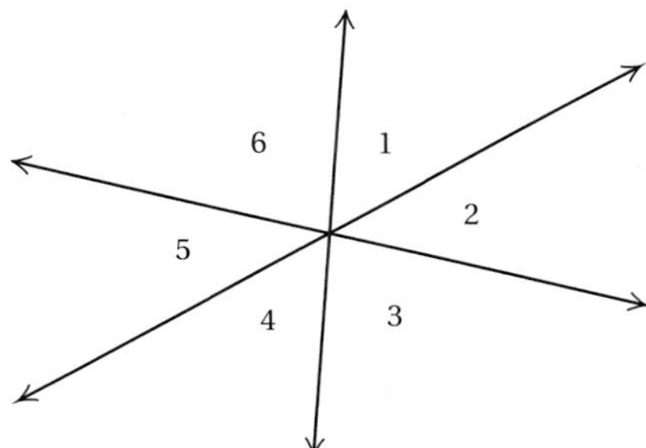

75. Find $m\angle ACB$, $m\angle CAB$, $m\angle EBC$, $m\angle EBA$, $m\angle AEB$, and $m\angle ADB$ in the rectangle shown below.

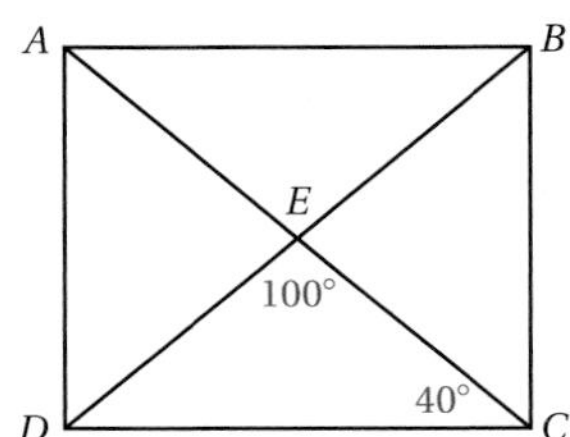

6.2 PERIMETER

Objectives

a Find the perimeter of a polygon.

b Solve applied problems involving perimeter.

Find the perimeter of the polygon.

1.

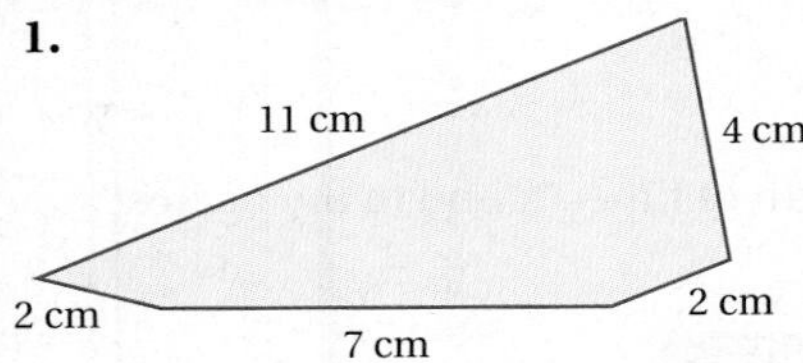

2.

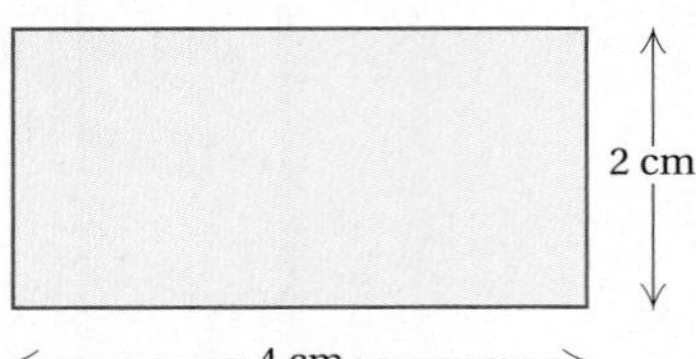

3. Find the perimeter of a rectangle that is 2 cm by 4 cm.

2 cm

4 cm

Answers on page A-15

a Finding Perimeters

PERIMETER OF A POLYGON

A **polygon** is a geometric figure with three or more sides. The **perimeter** of a **polygon** is the distance around it, or the sum of the lengths of its sides.

EXAMPLE 1 Find the perimeter of this polygon.

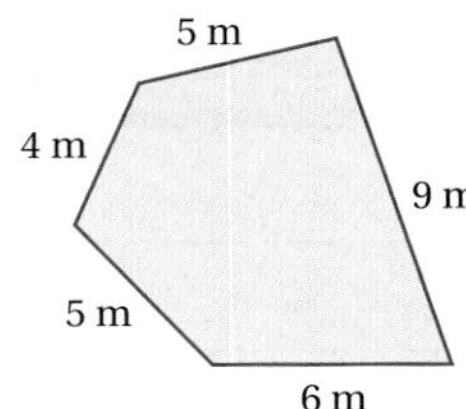

We add the lengths of the sides. Since all units are the same, we add the numbers, keeping meters (m) as the unit.

$$\begin{aligned}\text{Perimeter} &= 6\text{ m} + 5\text{ m} + 4\text{ m} + 5\text{ m} + 9\text{ m}\\ &= (6 + 5 + 4 + 5 + 9)\text{ m}\\ &= 29\text{ m}\end{aligned}$$

Do Exercises 1 and 2.

A **rectangle** is a figure with four sides and four 90° angles, like the one shown in Example 2.

EXAMPLE 2 Find the perimeter of a rectangle that is 3 cm by 4 cm.

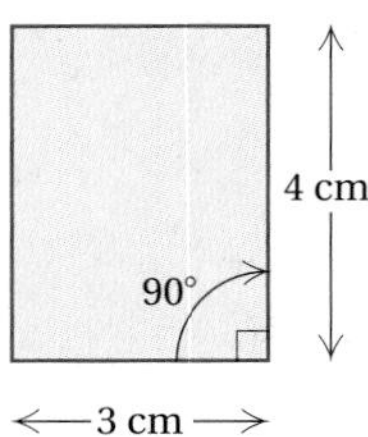

$$\begin{aligned}\text{Perimeter} &= 3\text{ cm} + 3\text{ cm} + 4\text{ cm} + 4\text{ cm}\\ &= (3 + 3 + 4 + 4)\text{ cm}\\ &= 14\text{ cm}\end{aligned}$$

Do Exercise 3.

PERIMETER OF A RECTANGLE

The **perimeter of a rectangle** is twice the sum of the length and the width, or 2 times the length plus 2 times the width:

$$\mathbf{P = 2 \cdot (l + w)}, \quad \text{or} \quad \mathbf{P = 2 \cdot l + 2 \cdot w}.$$

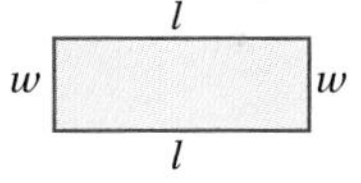

EXAMPLE 3 Find the perimeter of a rectangle that is 4.3 ft by 7.8 ft.

$$\begin{aligned} P &= 2 \cdot (l + w) \\ &= 2 \cdot (4.3 \text{ ft} + 7.8 \text{ ft}) \\ &= 2 \cdot (12.1 \text{ ft}) \\ &= 24.2 \text{ ft} \end{aligned}$$

Do Exercises 4 and 5.

A square is a rectangle with all sides the same length.

EXAMPLE 4 Find the perimeter of a square whose sides are 9 mm long.

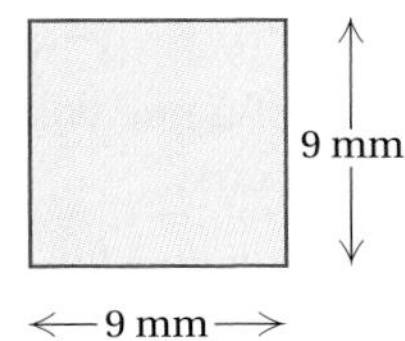

$$\begin{aligned} P &= 9 \text{ mm} + 9 \text{ mm} + 9 \text{ mm} + 9 \text{ mm} \\ &= (9 + 9 + 9 + 9) \text{ mm} \\ &= 36 \text{ mm} \end{aligned}$$

Do Exercise 6.

PERIMETER OF A SQUARE

The **perimeter of a square** is four times the length of a side:

$$\mathbf{P = 4 \cdot s}.$$

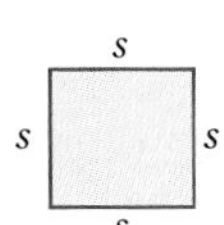

4. Find the perimeter of a rectangle that is 5.25 yd by 3.5 yd.

5. Find the perimeter of a rectangle that is $8\frac{1}{4}$ in. by $5\frac{2}{3}$ in.

6. Find the perimeter of a square with sides of length 10 km.

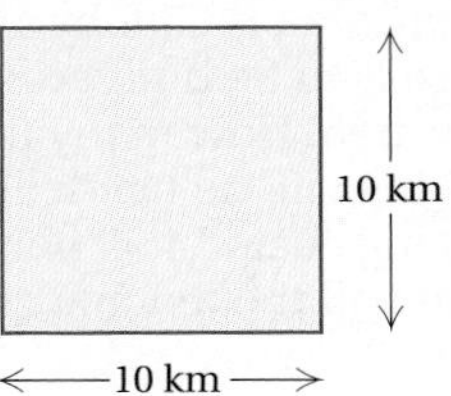

Answers on page A-15

7. Find the perimeter of a square with sides of length $5\frac{1}{4}$ yd.

8. Find the perimeter of a square with sides of length 7.8 km.

9. A play area is 25 ft by 10 ft. A fence is to be built around the play area. How many feet of fencing will be needed? If fencing costs \$4.95 per foot, what will the fencing cost?

Answers on page A-15

EXAMPLE 5 Find the perimeter of a square whose sides are $20\frac{1}{8}$ in. long.

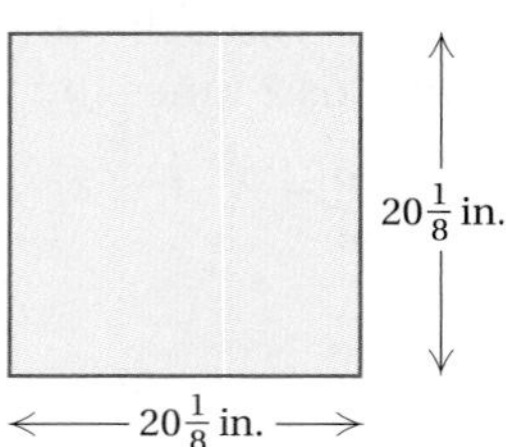

$$P = 4 \cdot s = 4 \cdot 20\frac{1}{8}\text{ in.}$$
$$= 4 \cdot \frac{161}{8}\text{ in.} = \frac{4 \cdot 161}{4 \cdot 2}\text{ in.}$$
$$= \frac{161}{2} \cdot \frac{4}{4}\text{ in.} = 80\frac{1}{2}\text{ in.}$$

Do Exercises 7 and 8.

b Solving Applied Problems

EXAMPLE 6 A vegetable garden is 20 ft by 15 ft. A fence is to be built around the garden. How many feet of fence will be needed? If fencing sells for \$2.95 per foot, what will the fencing cost?

1. **Familiarize.** We make a drawing and let $P =$ the perimeter.

2. **Translate.** The perimeter of the garden is given by

$$P = 2 \cdot (l + w) = 2 \cdot (20\text{ ft} + 15\text{ ft}).$$

3. **Solve.** We calculate the perimeter as follows:

$$P = 2 \cdot (20\text{ ft} + 15\text{ ft}) = 2 \cdot (35\text{ ft}) = 70\text{ ft}.$$

Then we multiply by \$2.95 to find the cost of the fencing:

$$\text{Cost} = \$2.95 \times \text{Perimeter} = \$2.95 \times 70\text{ ft} = \$206.50.$$

4. **Check.** The check is left to the student.
5. **State.** The 70 ft of fencing that is needed will cost \$206.50.

Do Exercise 9.

6.2 EXERCISE SET

For Extra Help

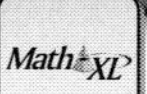

MathXL | MyMathLab | InterAct Math | Math Tutor Center | Video Lectures on CD Disc 3 | Student's Solutions Manual

a Find the perimeter of the polygon.

1.

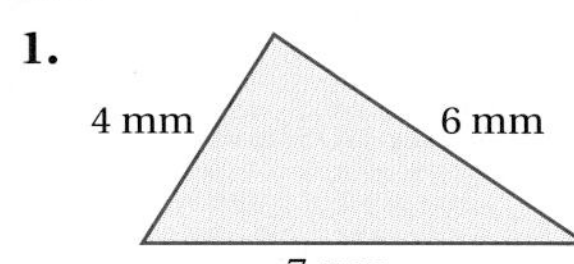

2.

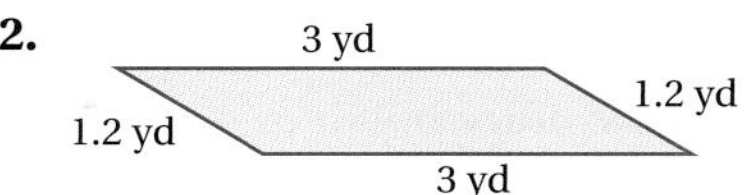

3.

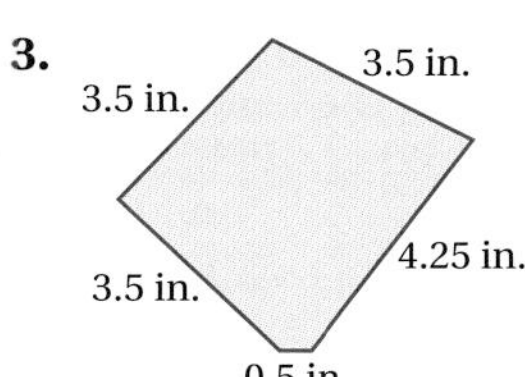

4.

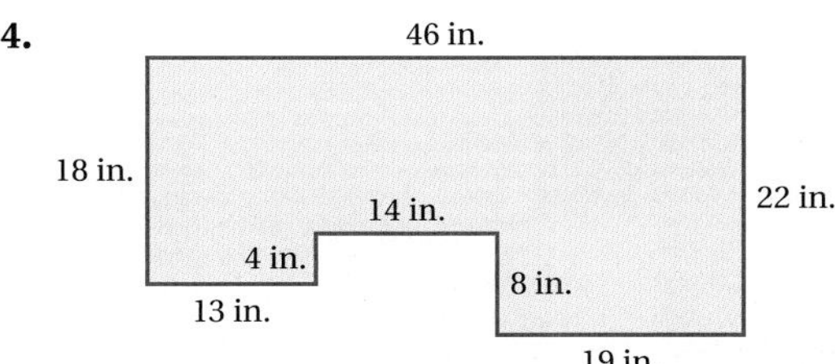

5.

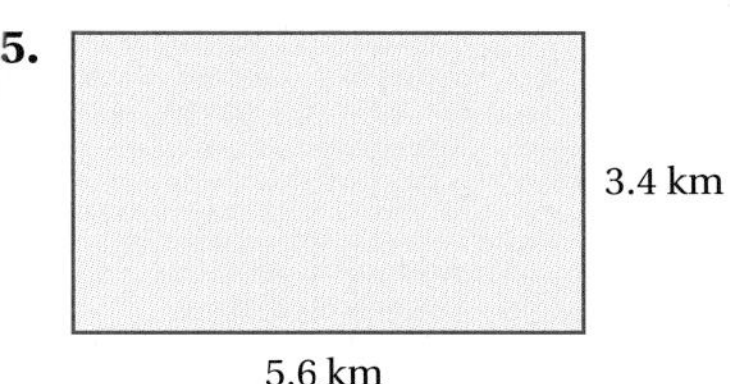

6.

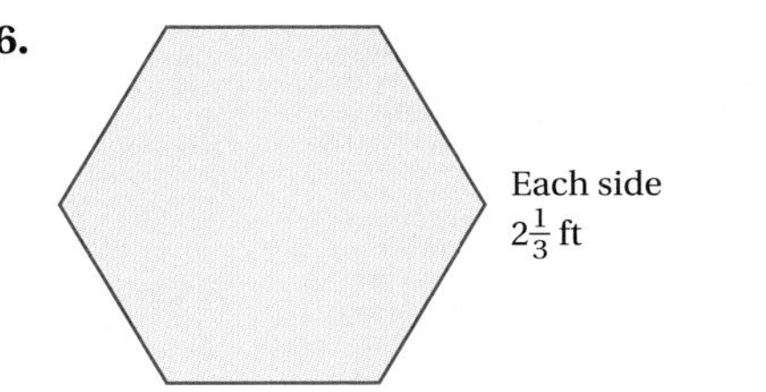

Find the perimeter of the rectangle.

7. 5 ft by 10 ft

8. 2.5 m by 100 m

9. 34.67 cm by 4.9 cm

10. $3\frac{1}{2}$ yd by $4\frac{1}{2}$ yd

Find the perimeter of the square.

11. 22 ft on a side

12. 56.9 km on a side

13. 45.5 mm on a side

14. $3\frac{1}{8}$ yd on a side

b Solve.

15. A security fence is to be built around a 173-m by 240-m rectangular field. What is the perimeter of the field? If 5-ft–high galvanized fence wire costs $7.29 per meter, what will the fencing cost?

16. *Softball Diamond.* A standard-sized slow-pitch softball diamond is a square with sides of length 65 ft. What is the perimeter of this softball diamond? (This is the distance you would have to run if you hit a home run.)

Source: American Softball Association

17. A piece of flooring tile is a square with sides of length 30.5 cm. What is the perimeter of a piece of tile?

18. A rectangular posterboard is 61.8 cm by 87.9 cm. What is the perimeter of the board?

19. A rain gutter is to be installed around the office building shown in the figure.

a) Find the perimeter of the office building.
b) If the gutter costs \$4.59 per foot, what is the total cost of the gutter?

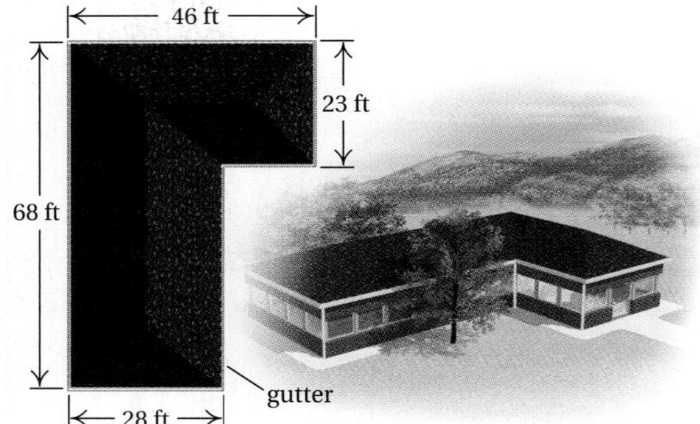

20. A carpenter is to build a fence around a 9-m by 12-m garden.

a) The posts are 3 m apart. How many posts will be needed?
b) The posts cost \$2.40 each. How much will the posts cost?
c) The fence will surround all but 3 m of the garden, which will be a gate. How long will the fence be?
d) The fence costs \$2.85 per meter. What will the cost of the fence be?
e) The gate costs \$9.95. What is the total cost of the materials?

21. $\mathbf{D}_\mathbf{W}$ Create for a fellow student a development of the formula

$$P = 2 \cdot (l + w) = 2 \cdot l + 2 \cdot w$$

for the perimeter of a rectangle.

22. $\mathbf{D}_\mathbf{W}$ Create for a fellow student a development of the formula

$$P = 4 \cdot s$$

for the perimeter of a square.

SKILL MAINTENANCE

23. Find the simple interest on \$600 at 6.4% for $\frac{1}{2}$ yr. [4.7d]

24. Find the simple interest on \$600 at 8% for 2 yr. [4.7d]

Evaluate. [1.6b]

25. 10^3

26. 11^3

27. 15^2

28. 22^2

29. 7^2

30. 4^3

Solve.

31. *Sales Tax.* In a certain state, a sales tax of \$878 is collected on the purchase of a car for \$17,560. What is the sales tax rate? [4.7a]

32. *Commission Rate.* Rich earns \$1854.60 selling \$16,860 worth of cell phones. What is the commission rate? [4.7b]

SYNTHESIS

Find the perimeter, in feet, of the figure.

33.

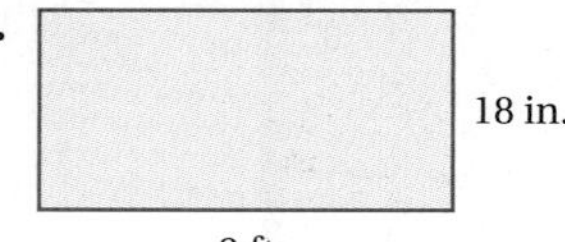

34.

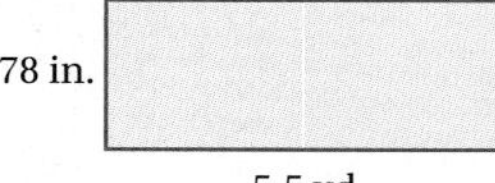

6.3 AREA

Objectives

a Find the area of a rectangle and a square.

b Find the area of a parallelogram, a triangle, and a trapezoid.

c Solve applied problems involving areas of rectangles, squares, parallelograms, triangles, and trapezoids.

a Rectangles and Squares

A polygon and its interior form a plane region. We can find the area of a *rectangular region,* or *rectangle,* by filling it in with square units. Two such units, a *square inch* and a *square centimeter,* are shown below.

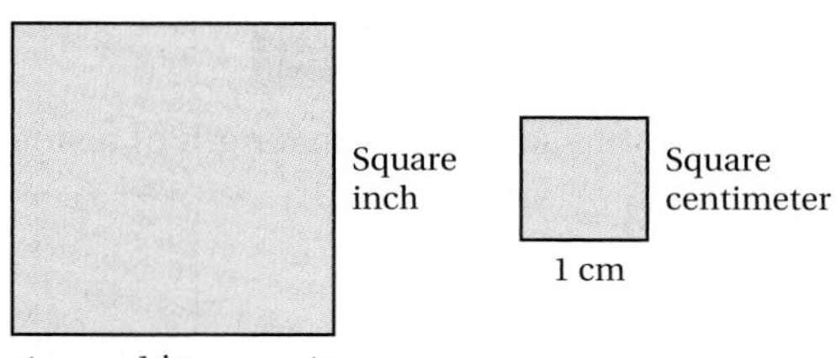

EXAMPLE 1 What is the area of this region?

We have a rectangular array. Since the region is filled with 12 square centimeters, its area is 12 square centimeters (sq cm), or 12 cm^2. The number of units is 3×4, or 12.

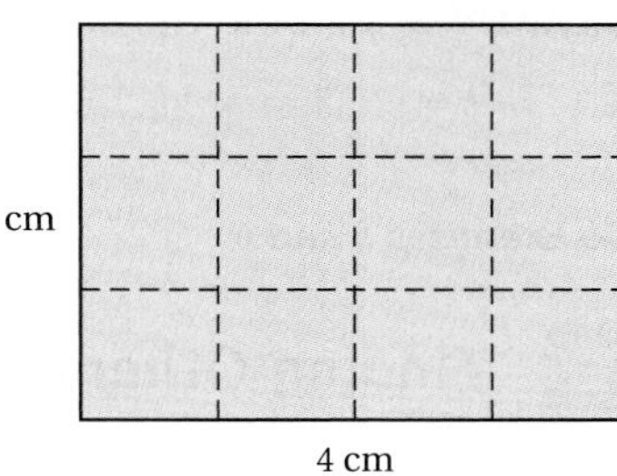

Do Exercise 1.

1. What is the area of this region? Count the number of square centimeters.

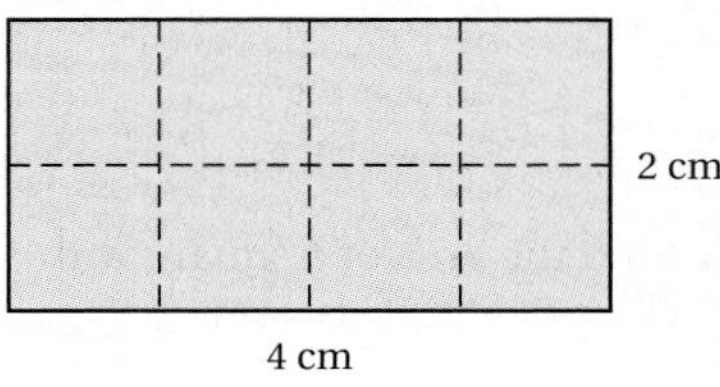

AREA OF A RECTANGLE

The **area of a rectangle** is the product of the length l and the width w:

$$A = l \cdot w.$$

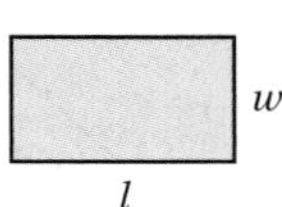

2. Find the area of a rectangle that is 7 km by 8 km.

EXAMPLE 2 Find the area of a rectangle that is 7 yd by 4 yd.

$$A = l \cdot w = 7 \text{ yd} \cdot 4 \text{ yd}$$
$$= 7 \cdot 4 \cdot \text{yd} \cdot \text{yd} = 28 \text{ yd}^2$$

We think of yd · yd as $(\text{yd})^2$ and denote it yd^2. Thus we read "28 yd^2" as "28 square yards."

Do Exercises 2 and 3.

3. Find the area of a rectangle that is $5\frac{1}{4}$ yd by $3\frac{1}{2}$ yd.

Answers on page A-15

4. Find the area of a square with sides of length 12 km.

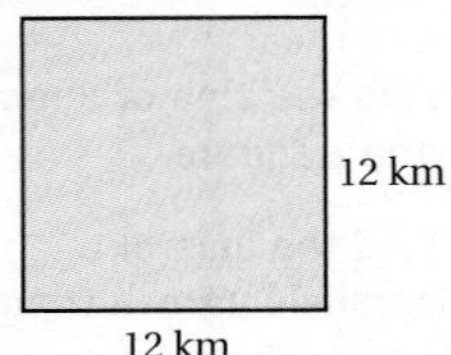

5. Find the area of a square with sides of length 10.9 m.

6. Find the area of a square with sides of length $3\frac{1}{2}$ yd.

Answers on page A-15

EXAMPLE 3 Find the area of a square with sides of length 9 mm.

$$A = (9\text{ mm}) \cdot (9\text{ mm})$$
$$= 9 \cdot 9 \cdot \text{mm} \cdot \text{mm}$$
$$= 81\text{ mm}^2$$

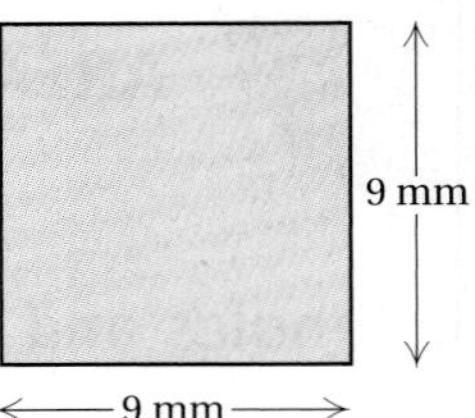

Do Exercise 4.

AREA OF A SQUARE

The **area of a square** is the square of the length of a side:

$A = s \cdot s$, or $A = s^2$.

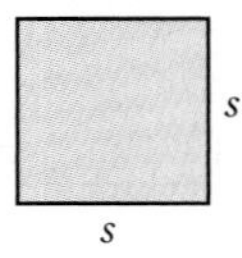

EXAMPLE 4 Find the area of a square with sides of length 20.3 m.

$A = s \cdot s = 20.3\text{ m} \times 20.3\text{ m} = 20.3 \times 20.3 \times \text{m} \times \text{m} = 412.09\text{ m}^2$

Do Exercises 5 and 6.

b Finding Other Areas

PARALLELOGRAMS

A **parallelogram** is a four-sided figure with two pairs of parallel sides, as shown below.

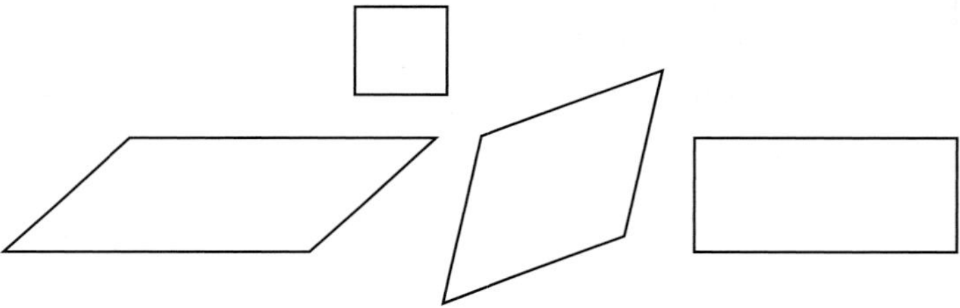

To find the area of a parallelogram, consider the one below.

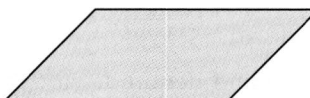

If we cut off a piece and move it to the other end, we get a rectangle.

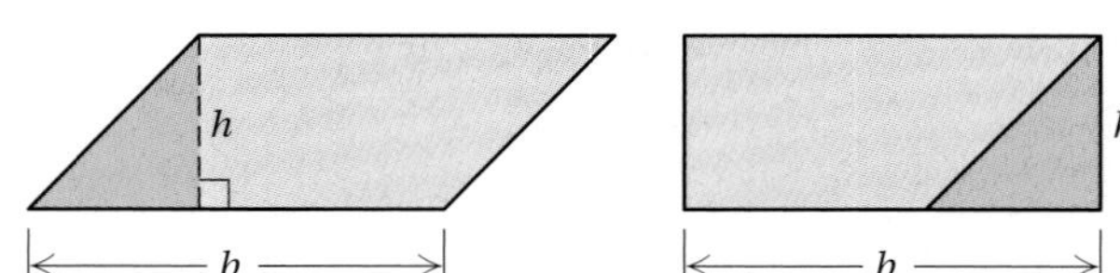

We can find the area by multiplying the length b, called a **base,** by h, called the **height.**

AREA OF A PARALLELOGRAM

The **area of a parallelogram** is the product of the length of a base b and the height h:

$$A = b \cdot h.$$

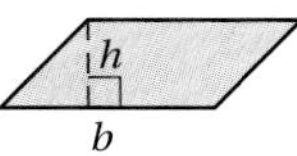

EXAMPLE 5 Find the area of this parallelogram.

$$\begin{aligned} A &= b \cdot h \\ &= 7\text{ km} \cdot 5\text{ km} \\ &= 35\text{ km}^2 \end{aligned}$$

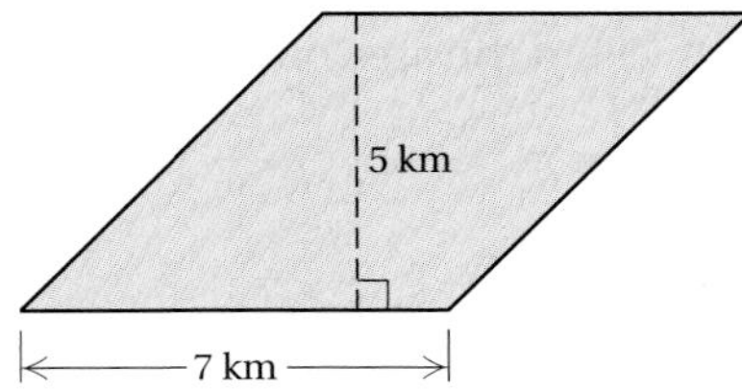

EXAMPLE 6 Find the area of this parallelogram.

$$\begin{aligned} A &= b \cdot h \\ &= 1.2\text{ m} \times 6\text{ m} \\ &= 7.2\text{ m}^2 \end{aligned}$$

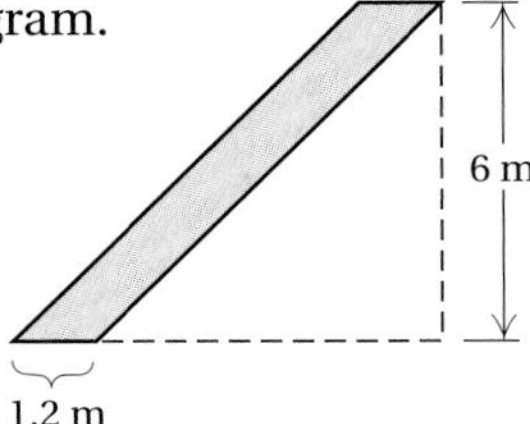

Do Exercises 7 and 8.

TRIANGLES

To find the area of a triangle like the one shown on the left below, think of cutting out another just like it and placing it as shown on the right below.

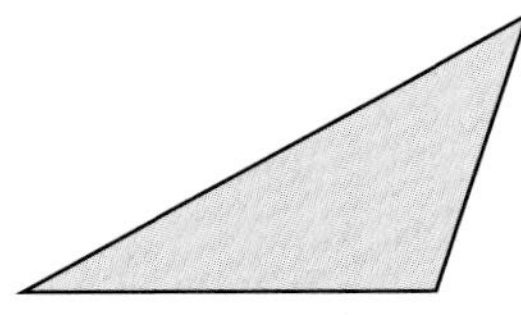

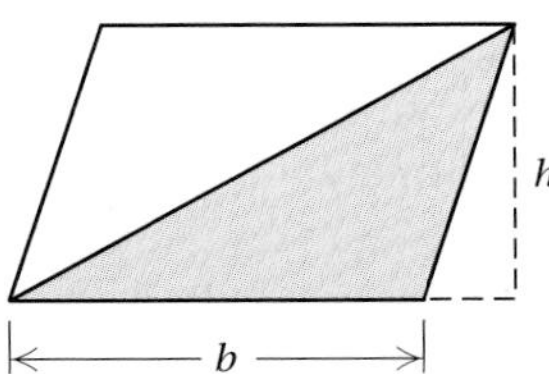

The resulting figure is a parallelogram whose area is

$$b \cdot h.$$

The triangle we started with has half the area of the parallelogram, or

$$\frac{1}{2} \cdot b \cdot h.$$

AREA OF A TRIANGLE

The **area of a triangle** is half the length of the base times the height:

$$A = \frac{1}{2} \cdot b \cdot h.$$

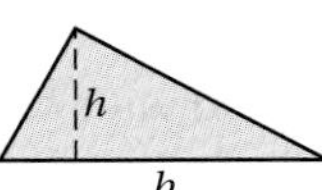

Find the area.

7.

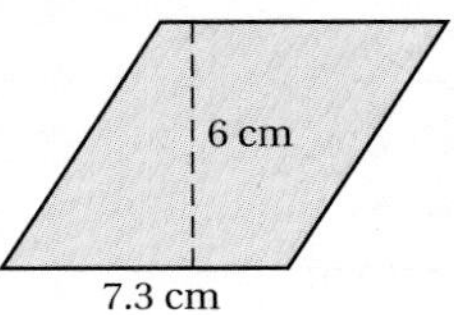

8.

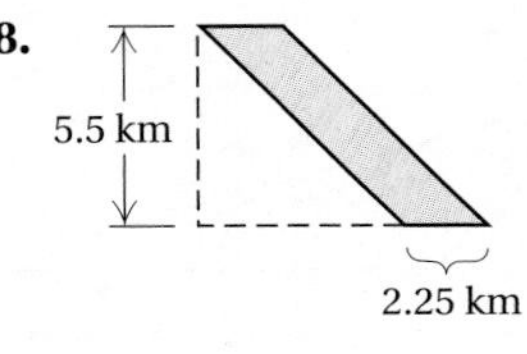

Answers on page A-15

Find the area.

9.

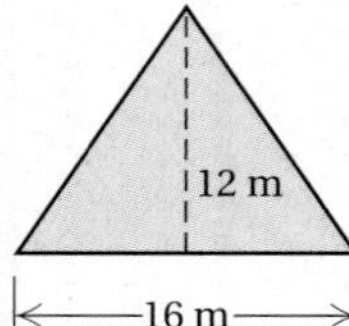

10.

3.4 cm

11 cm

Answers on page A-15

EXAMPLE 7 Find the area of this triangle.

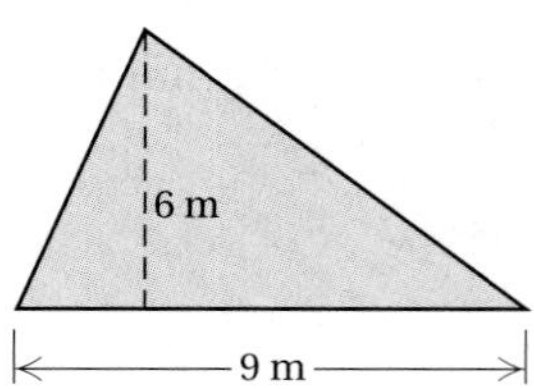

$$A = \frac{1}{2} \cdot b \cdot h$$
$$= \frac{1}{2} \cdot 9\text{ m} \cdot 6\text{ m}$$
$$= \frac{9 \cdot 6}{2}\text{ m}^2$$
$$= 27\text{ m}^2$$

EXAMPLE 8 Find the area of this triangle.

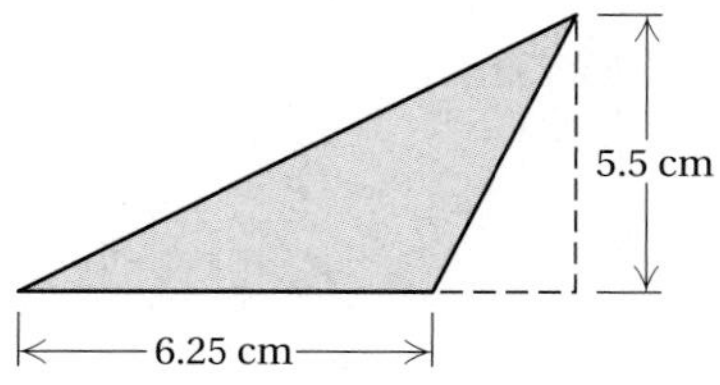

$$A = \frac{1}{2} \cdot b \cdot h$$
$$= \frac{1}{2} \times 6.25\text{ cm} \times 5.5\text{ cm}$$
$$= 0.5 \times 6.25 \times 5.5\text{ cm}^2$$
$$= 17.1875\text{ cm}^2$$

Do Exercises 9 and 10.

TRAPEZOIDS

A **trapezoid** is a polygon with four sides, two of which, the **bases,** are parallel to each other.

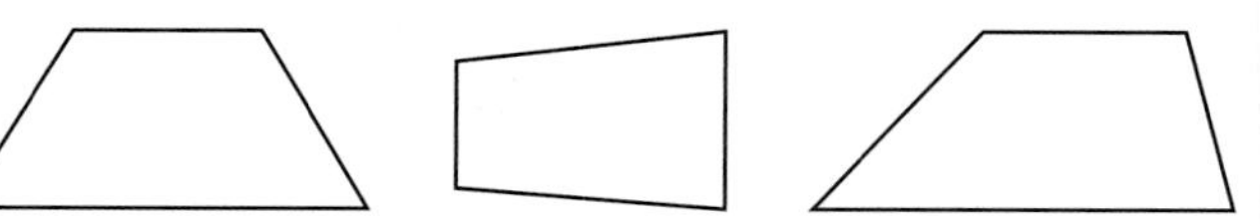

To find the area of a trapezoid, think of cutting out another just like it.

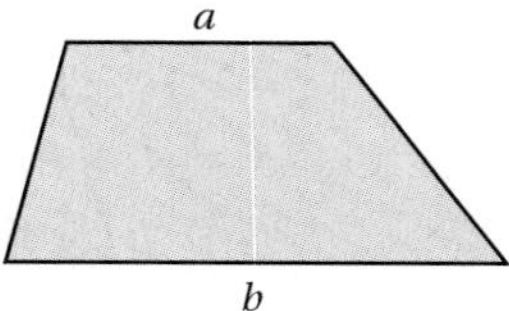

Then place the second one like this.

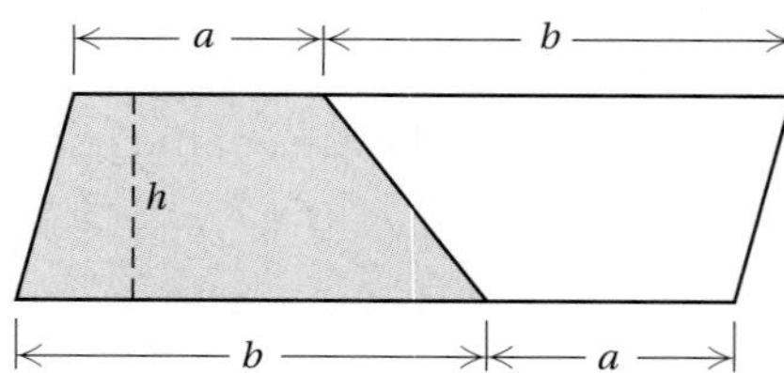

The resulting figure is a parallelogram whose area is

$h \cdot (a + b)$. The base is $a + b$.

The trapezoid we started with has half the area of the parallelogram, or

$$\frac{1}{2} \cdot h \cdot (a + b).$$

AREA OF A TRAPEZOID

The **area of a trapezoid** is half the product of the height and the sum of the lengths of the parallel sides (bases):

$$\mathbf{A = \frac{1}{2} \cdot h \cdot (a + b),} \quad \text{or} \quad \mathbf{A = \frac{a + b}{2} \cdot h.}$$

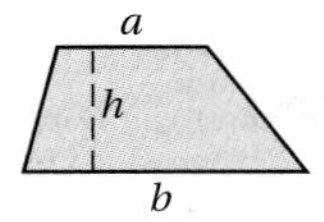

EXAMPLE 9 Find the area of this trapezoid.

$$A = \frac{1}{2} \cdot h \cdot (a + b)$$

$$= \frac{1}{2} \cdot 7 \text{ cm} \cdot (12 + 18) \text{ cm}$$

$$= \frac{7 \cdot 30}{2} \cdot \text{cm}^2 = \frac{7 \cdot 15 \cdot 2}{1 \cdot 2} \text{cm}^2$$

$$= \frac{7 \cdot 15}{1} \cdot \frac{2}{2} \text{cm}^2 = 105 \text{ cm}^2$$

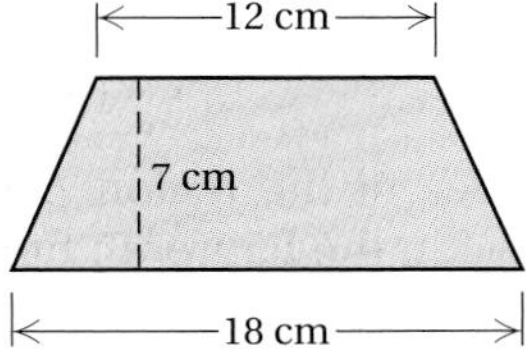

Do Exercises 11 and 12.

C Solving Applied Problems

EXAMPLE 10 *Baseball's Strike Zones.* Major League Baseball underwent a rule change between the 2000 and 2001 seasons. Over the years before 2001, the strike zone evolved to something other than what was defined in the rule book. The zone in the rule book is described by rectangle *ABCD* below. The zone used before 2001 is described by the region *AQRST*. The figure shown here represents the zones for a normal-sized player, but they vary depending on the height of the player.

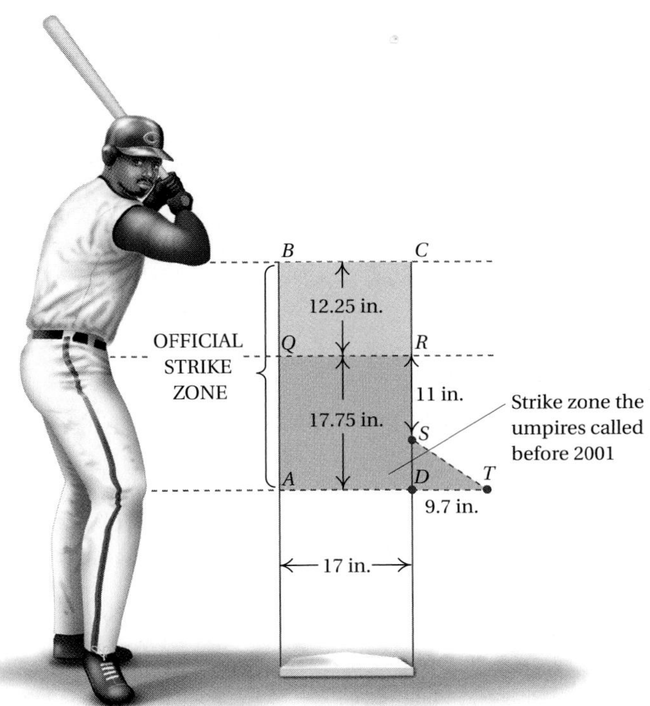

Find the area.

11.

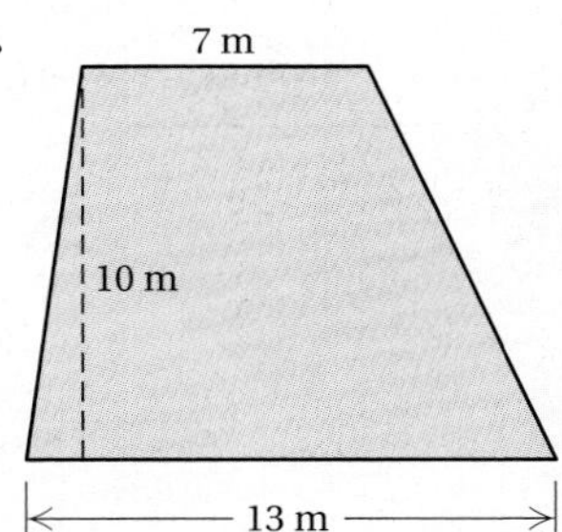

12.

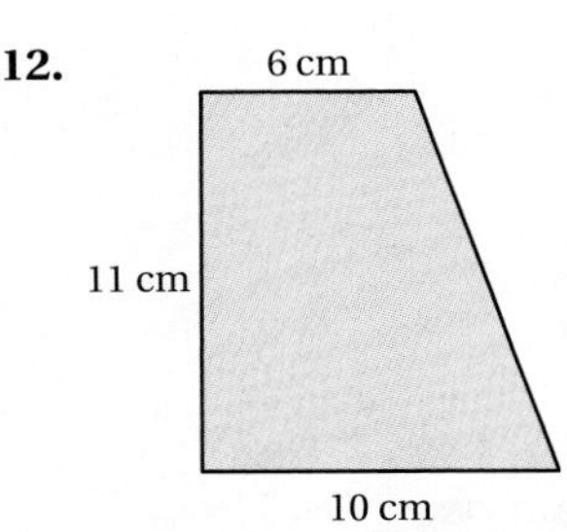

Answers on page A-15

13. Find the area of this kite.

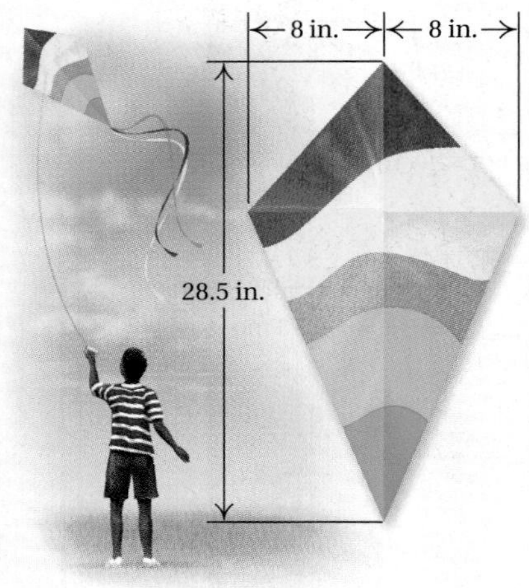

Answers on page A-15

Study Tips

STUDYING THE ART PIECES

When you study a section of a mathematics text, read it slowly, observing all the details of the corresponding art pieces that are discussed in the paragraphs. Also note the precise color markings in the art that enhances the learning process. These tips apply especially to this chapter because geometry, by its nature, is quite visual.

a) Find the area of the new zone.

b) Find the area of the former zone.

c) How much larger is the rule-book zone than the former zone?

d) By what percent has the area of the strike zone been increased by the change?

Sources: *The Cincinnati Inquirer;* Major League Baseball; Gannett News Service; *The Sporting News Official Baseball Rules Book*

This is a multistep problem that makes use of many of the skills we have learned in this book.

a) The area of rectangle *ABCD* (the rule-book zone), denoted A_1, is the length times width:

$$A_1 = l \cdot w = (17.75 \text{ in.} + 12.25 \text{ in.}) \cdot (17 \text{ in.}) = (30 \text{ in.}) \cdot (17 \text{ in.}) = 510 \text{ in}^2.$$

b) The area of the region *AQRST* (the former zone), denoted A_2, is shown shaded in the figure. To find that area, we add the area of rectangle *AQRD* to the area of triangle *SDT*. We first determine

$$\text{Area of } AQRD = l \cdot w = (17.75 \text{ in.}) \cdot (17 \text{ in.}) = 301.75 \text{ in}^2.$$

To find the area of triangle *SDT*, we first note that the length of the base is given as 9.7 in. To find the height that is the length of segment *SD*, we subtract 11 in. from 17.75 in.: 17.75 in. − 11 in. = 6.75 in. Then

$$\text{Area of triangle } SDT = \frac{1}{2} \cdot b \cdot h = \frac{1}{2}(9.7 \text{ in.}) \cdot (6.75 \text{ in.}) = 32.7375 \text{ in}^2.$$

The area of the former zone A_2 is the sum of the areas of the triangle *SDT* and the rectangle *AQRD*:

$$A_2 = 301.75 \text{ in}^2 + 32.7375 \text{ in}^2 = 334.4875 \text{ in}^2 \approx 334.5 \text{ in}^2.$$

c) To find the increase in the area, we subtract A_2 from A_1:

$$510 \text{ in}^2 - 334.5 \text{ in}^2 = 175.5 \text{ in}^2.$$

d) To determine the percent of increase, note that we are asking "What percent of the former area is the increase?" We translate this to an equation as follows:

What percent of 334.5 is 175.5?

$$p \cdot 334.5 = 175.5$$

We solve the equation:

$$p \cdot 334.5 = 175.5$$

$$\frac{p \cdot 334.5}{334.5} = \frac{175.5}{334.5}$$

$$p = \frac{175.5}{334.5} \approx 0.5247 \approx 52\%.$$

There was an increase of approximately 52% in the strike zone.

Do Exercise 13.

6.3 EXERCISE SET

For Extra Help

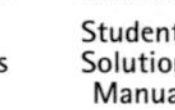

MathXL | MyMathLab | InterAct Math | Math Tutor Center | Video Lectures on CD Disc 3 | Student's Solutions Manual

a Find the area.

1.

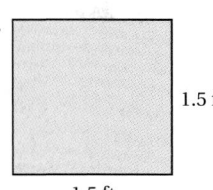

2.

1.5 ft

1.5 ft

3.

2 in.

0.7 in.

4.

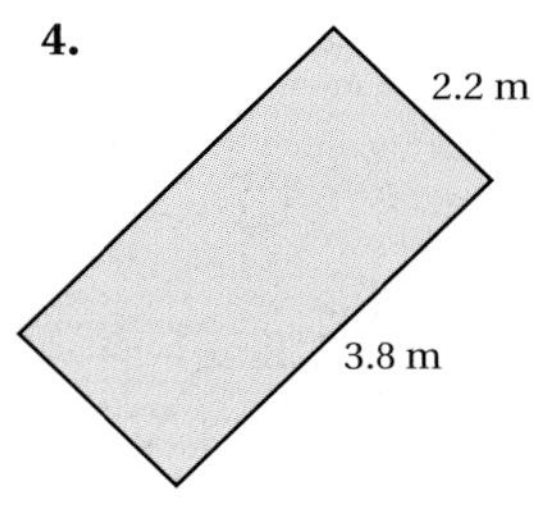

5.

$2\frac{1}{2}$ yd

$2\frac{1}{2}$ yd

6.

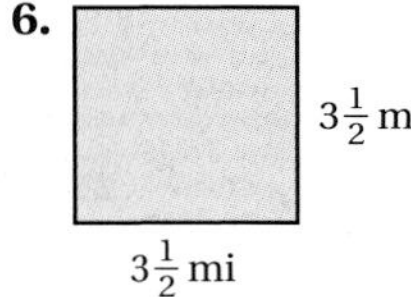

7.

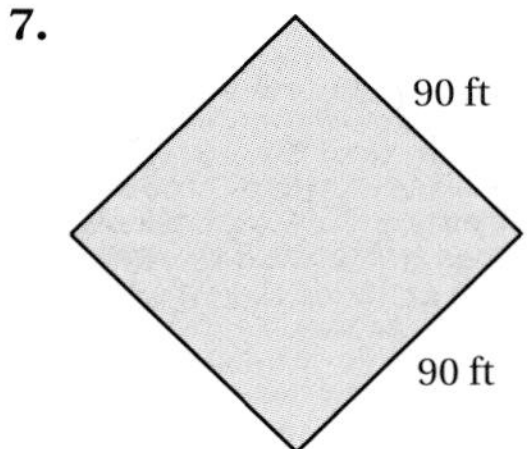

8.

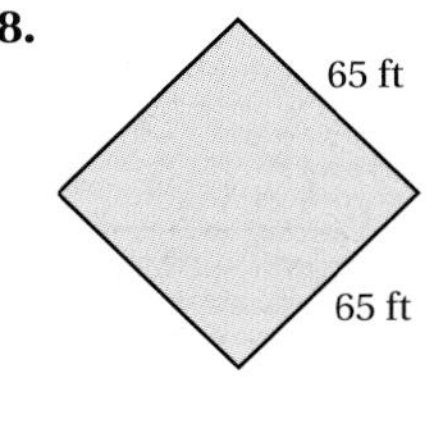

Find the area of the rectangle.

9. 5 ft by 10 ft

10. 14 yd by 8 yd

11. 34.67 cm by 4.9 cm

12. 2.45 km by 100 km

13. $4\frac{2}{3}$ in. by $8\frac{5}{6}$ in.

14. $10\frac{1}{3}$ mi by $20\frac{2}{3}$ mi

Find the area of the square.

15. 22 ft on a side

16. 18 yd on a side

17. 56.9 km on a side

18. 45.5 m on a side

19. $5\frac{3}{8}$ yd on a side

20. $7\frac{2}{3}$ ft on a side

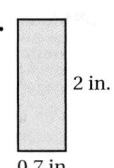

Find the area.

21.

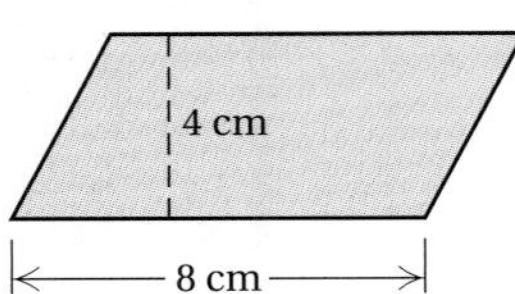

22.

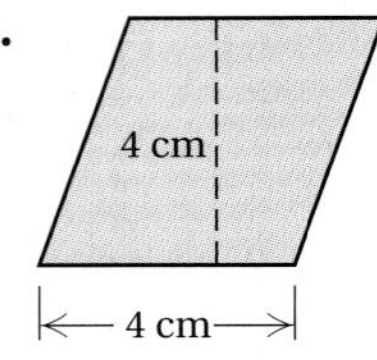

23.

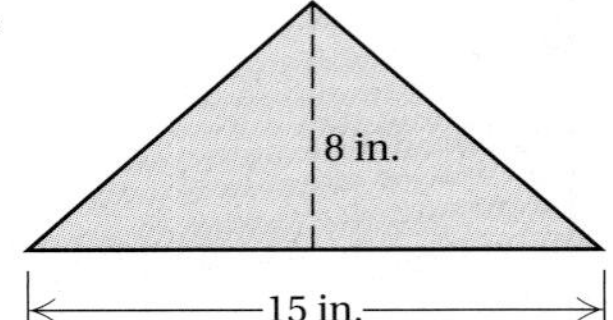

24.

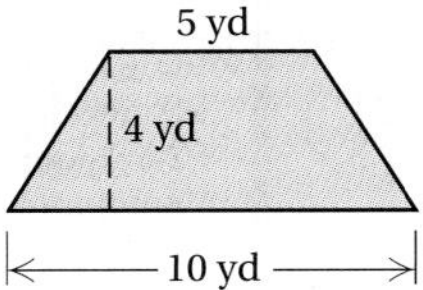

25.

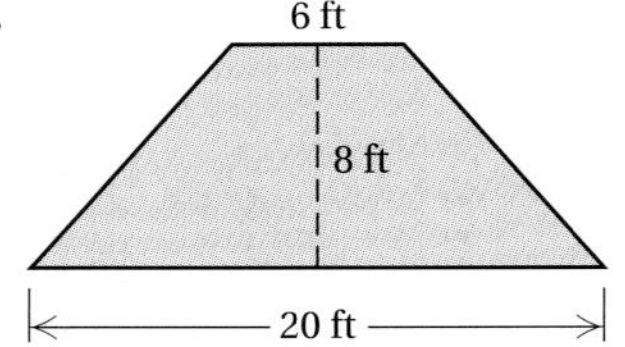

26.

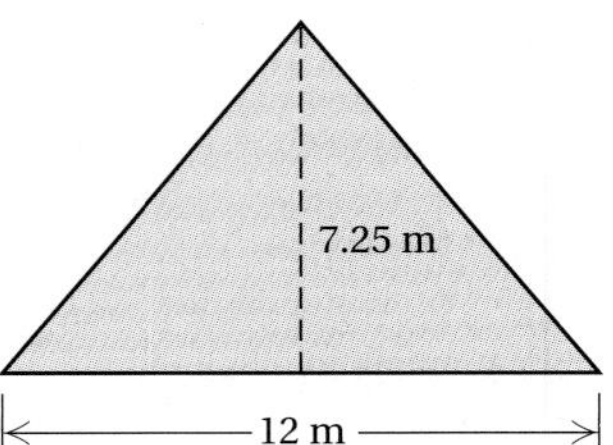

27.

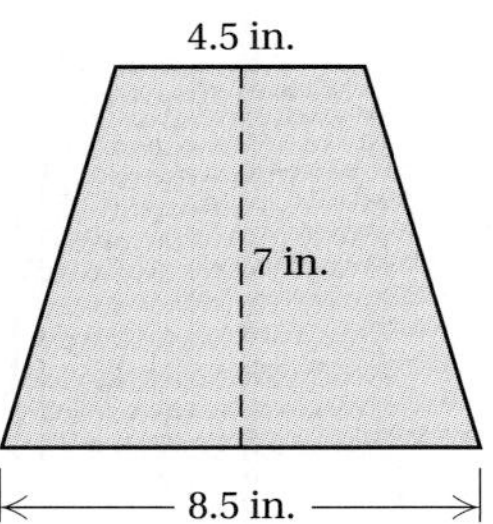

28.

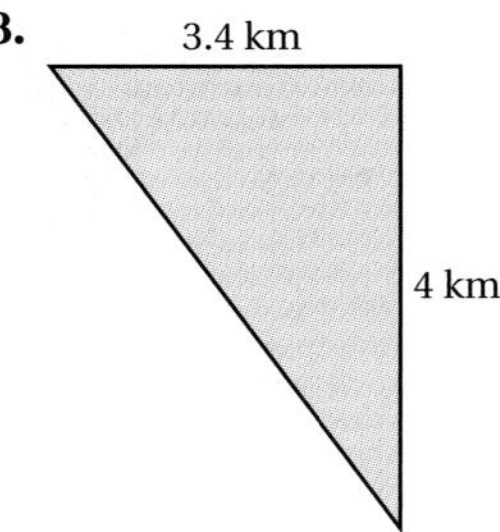

29.

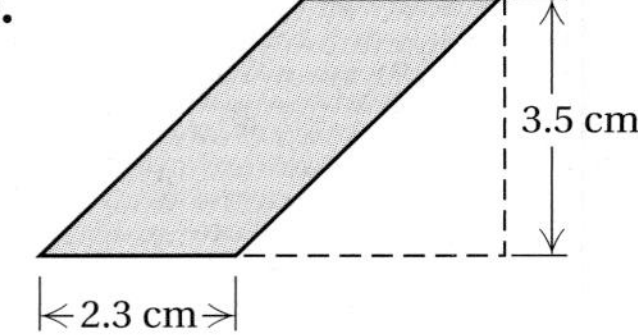

30.

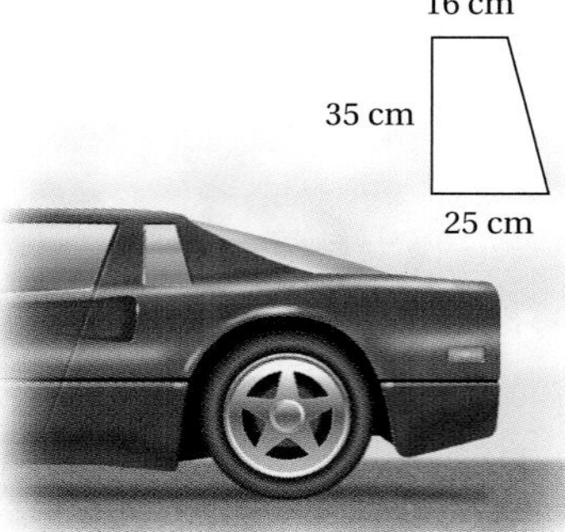

31.

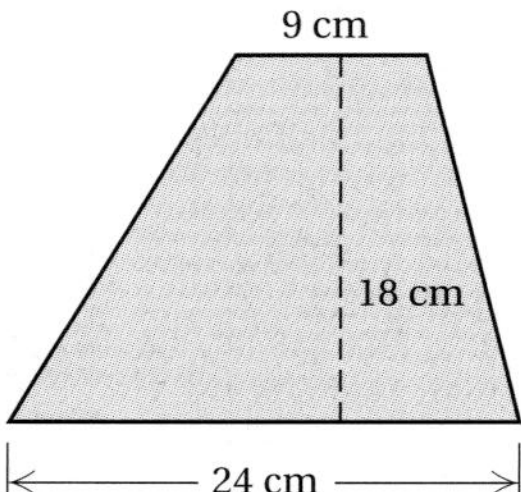

32.

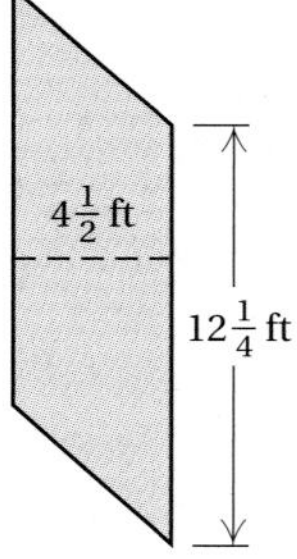

33.

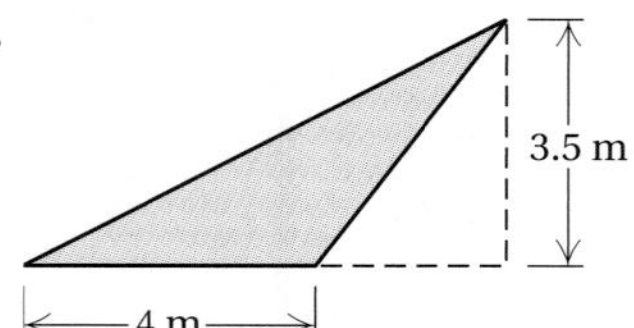

34.

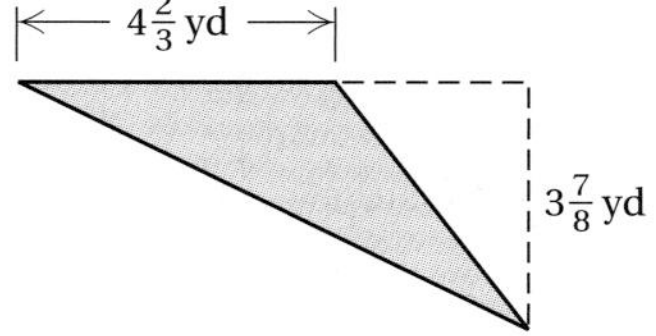

C Solve.

35. *Area of a Lawn.* A lot is 40 m by 36 m. A house 27 m by 9 m is built on the lot. How much area is left over for a lawn?

36. *Area of a Field.* A field is 240.8 m by 450.2 m. Part of the field, 160.4 m by 90.6 m, is paved for a parking lot. How much area is unpaved?

37. *Mowing Expense.* A square basketball court $19\frac{1}{2}$ ft on a side is placed on an 80-ft by $110\frac{2}{3}$-ft lawn.
    a) Find the area of the lawn.
    b) It costs \$0.012 per square foot to have the lawn mowed. What is the total cost of the mowing?

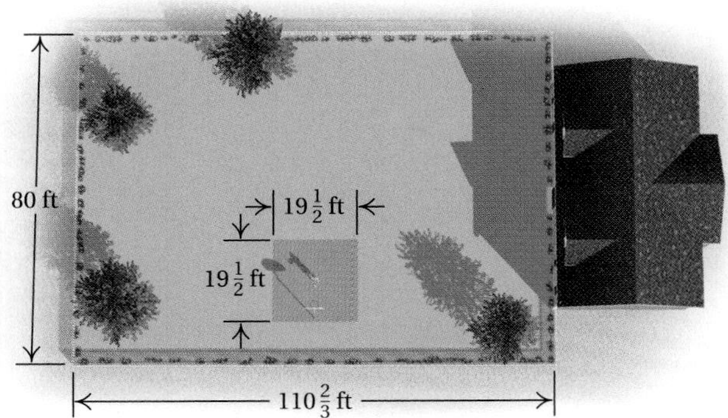

38. *Mowing Expense.* A square flower bed 10.5 ft on a side is dug on a 90-ft by $67\frac{1}{4}$-ft lawn.
    a) Find the area of the lawn.
    b) It costs \$0.03 per square foot to have the lawn mowed. What is the total cost of the mowing?

39. *Area of a Sidewalk.* Franklin Construction Company builds a sidewalk around two sides of the Municipal Trust Bank building, as shown in the figure. What is the area of the sidewalk?

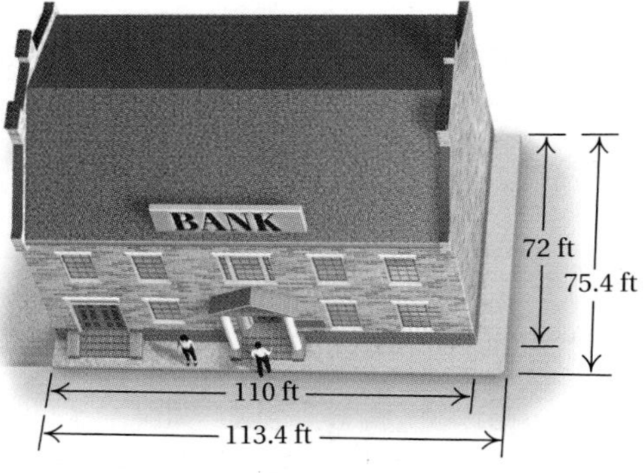

40. *Margin Area.* A standard sheet of typewriter paper is $8\frac{1}{2}$ in. by 11 in. We generally type on a $7\frac{1}{2}$-in. by 9-in. area of the paper. What is the area of the margin?

41. *Painting Costs.* A room is 15 ft by 20 ft. The ceiling is 8 ft above the floor. There are two windows in the room, each 3 ft by 4 ft. The door is $2\frac{1}{2}$ ft by $6\frac{1}{2}$ ft.
    a) What is the total area of the walls and the ceiling?
    b) A gallon of paint will cover 86.625 ft^2. How many gallons of paint are needed for the room, including the ceiling?
    c) Paint costs \$17.95 a gallon. How much will it cost to paint the room?

42. *Carpeting Costs.* A restaurant owner wants to carpet a 15-yd by 20-yd room.
    a) How many square yards of carpeting are needed?
    b) The carpeting she wants is \$18.50 per square yard. How much will it cost to carpet the room?

Find the area of the shaded region.

43.

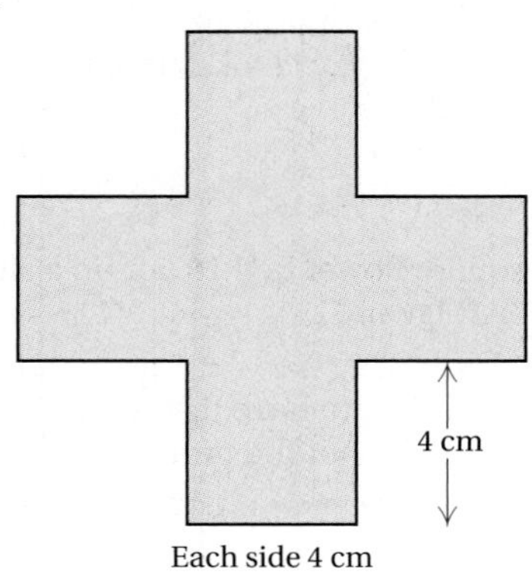

44.

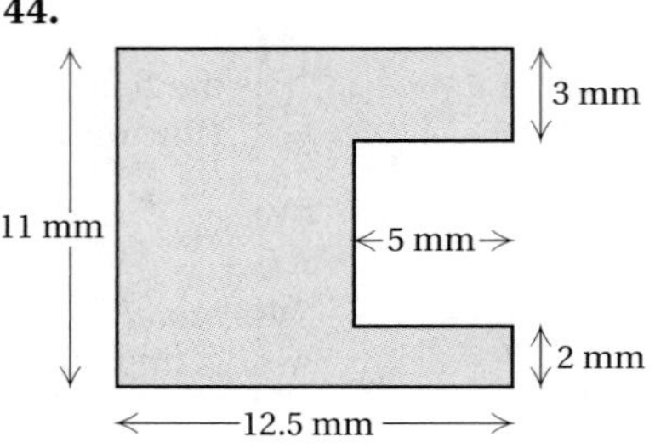

45.

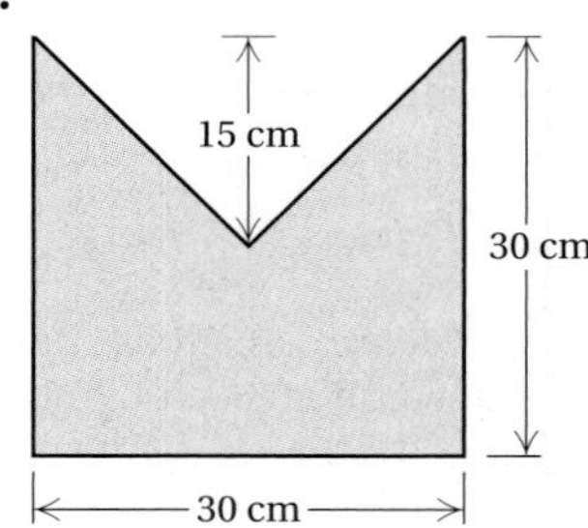

46.

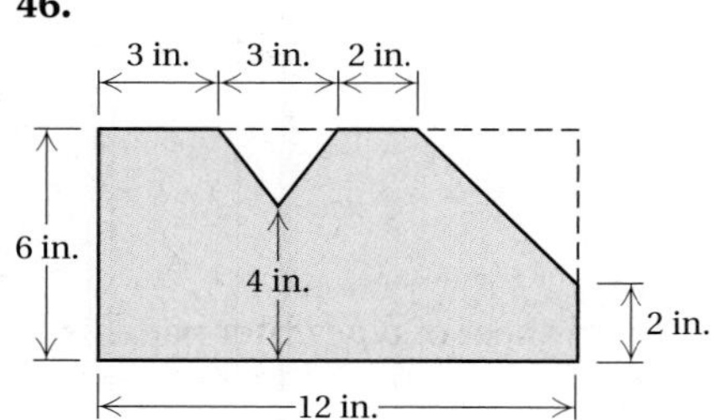

47.

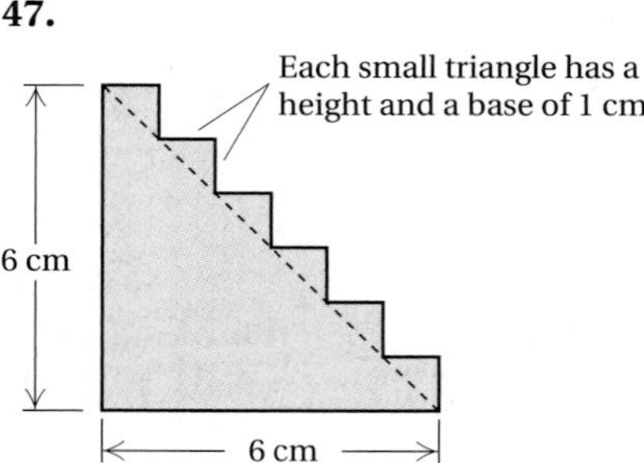

48.

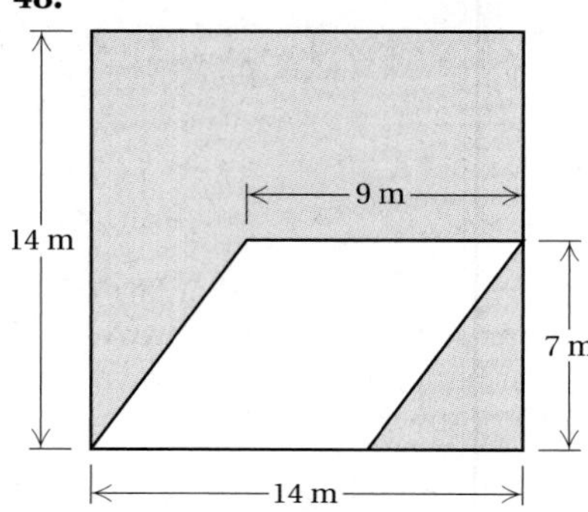

49. *Triangular Sail.* A rectangular piece of sailcloth is 36 ft by 24 ft. A triangular area with a height of 4.6 ft and a base of 5.2 ft is cut from the sailcloth. How much area is left over?

50. *Building Area.* Find the total area of the sides and ends of the building shown below.

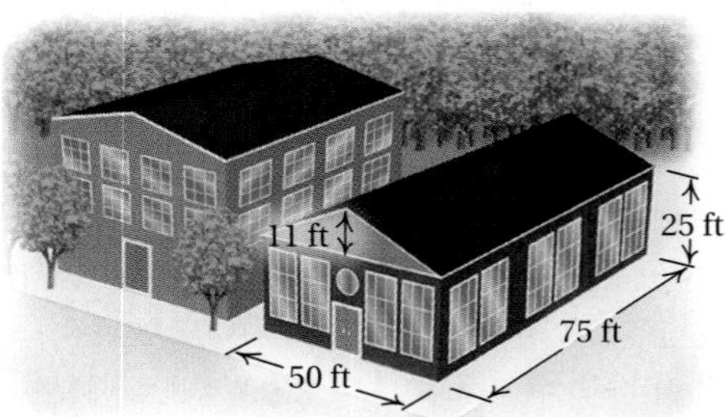

51. D_W The length and the width of one rectangle are each three times the length and the width of another rectangle. Is the area of the first rectangle three times the area of the other rectangle? Why or why not?

52. D_W Explain how the area of a triangle can be found by considering the area of a parallelogram.

SKILL MAINTENANCE

VOCABULARY REINFORCEMENT

In each of Exercises 53–60, fill in the blank with the correct term from the given list. Some of the choices may not be used.

53. A number is divisible by 8 if the number named by its last _______________ digits is divisible by 8. [1.8a]

54. A parallelogram is a four-sided figure with two pairs of _______________ sides. [6.3b]

55. Two lines are _______________ if they intersect to form a right angle. [6.1d]

56. A natural number, other than 1, that is not _______________ is composite. [1.7c]

57. A(n) _______________ is a set of points consisting of two rays. [6.1b]

58. To convert from _______________ to _______________, move the decimal point two places to the left and change the ¢ sign at the end to the $ sign in front. [3.3b]

59. The _______________ of a polygon is the sum of the lengths of its sides. [6.2a]

60. The number 1 is known as the _______________ identity, and the number 0 is known as the _______________ identity. [1.2a], [1.3a]

prime
composite
two
three
dollars
cents
perimeter
parallel
perpendicular
additive
multiplicative
parallel
perpendicular
vertex
angle
area

SYNTHESIS

61. Find the area, in square inches, of the shaded region.

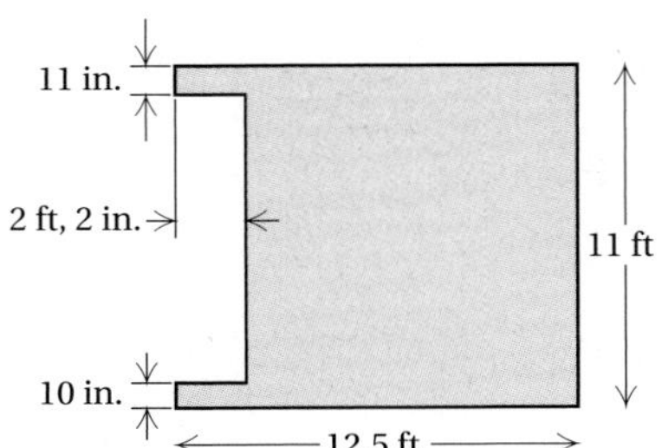

62. Find the area, in square feet, of the shaded region.

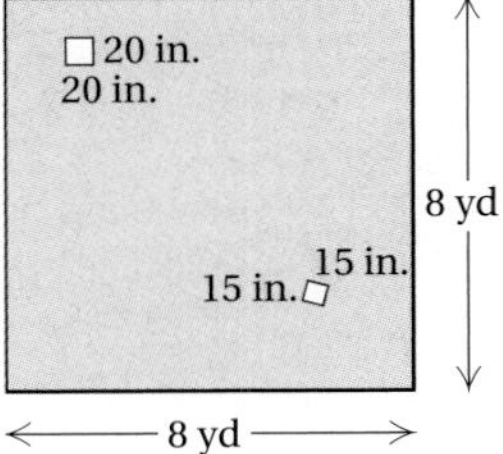

6.4 CIRCLES

Objectives

a Find the length of a radius of a circle given the length of a diameter, and find the length of a diameter given the length of a radius.

b Find the circumference of a circle given the length of a diameter or a radius.

c Find the area of a circle given the length of a radius.

d Solve applied problems involving circles.

a Radius and Diameter

Shown below is a circle with center O. Segment $\overline{AC}$ is a *diameter*. A **diameter** is a segment that passes through the center of the circle and has endpoints on the circle. Segment $\overline{OB}$ is called a *radius*. A **radius** is a segment with one endpoint on the center and the other endpoint on the circle.

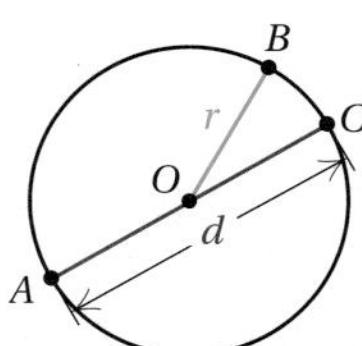

DIAMETER AND RADIUS

Suppose that d is the length of the diameter of a circle and r is the length of the radius. Then

$$\boldsymbol{d = 2 \cdot r} \quad \text{and} \quad \boldsymbol{r = \frac{d}{2}}.$$

EXAMPLE 1 Find the length of a radius of this circle.

$$r = \frac{d}{2}$$
$$= \frac{12\text{ m}}{2}$$
$$= 6\text{ m}$$

The radius is 6 m.

12 m

EXAMPLE 2 Find the length of a diameter of this circle.

$$d = 2 \cdot r$$
$$= 2 \cdot \frac{1}{4}\text{ ft}$$
$$= \frac{1}{2}\text{ ft}$$

The diameter is $\frac{1}{2}$ ft.

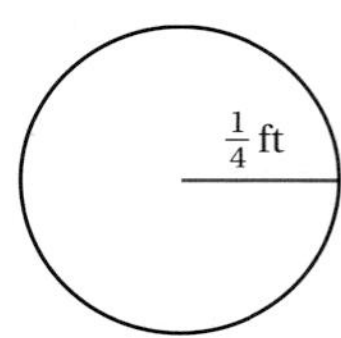

Do Exercises 1 and 2.

1. Find the length of a radius.

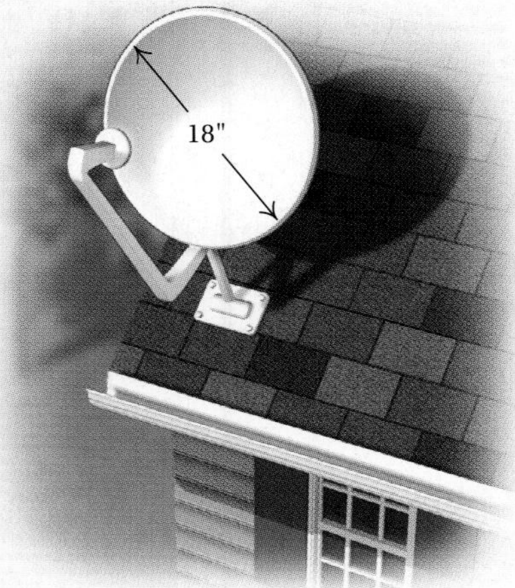

2. Find the length of a diameter.

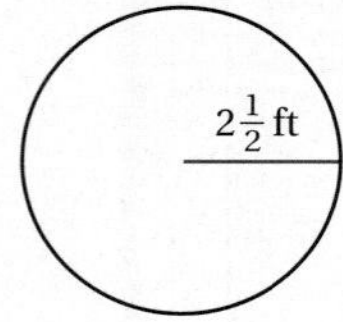

Answers on page A-15

b Circumference

The **circumference** of a circle is the distance around it. Calculating circumference is similar to finding the perimeter of a polygon.

To find a formula for the circumference of any circle given its diameter, we first need to consider the ratio C/d. Take a 12-oz soda can and measure the circumference C with a tape measure. Also measure the diameter d. The results are shown in the figure. Then

$$\frac{C}{d} = \frac{7.8 \text{ in.}}{2.5 \text{ in.}} \approx 3.1.$$

Suppose we did this with cans and circles of several sizes. We would get a number close to 3.1. For any circle, if we divide the circumference C by the diameter d, we get the same number. We call this number π (pi). The *exact* value of the ratio C/d is π; 3.14 and 22/7 are approximations of π.

CIRCUMFERENCE AND DIAMETER

The circumference C of a circle of diameter d is given by

$$\boldsymbol{C = \pi \cdot d.}$$

The number π is about 3.14, or about $\frac{22}{7}$.

EXAMPLE 3 Find the circumference of this circle. Use 3.14 for π.

$$\begin{aligned} C &= \pi \cdot d \\ &\approx 3.14 \times 6 \text{ cm} \\ &= 18.84 \text{ cm} \end{aligned}$$

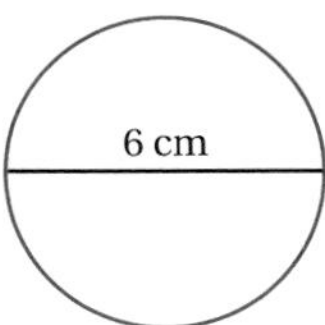

The circumference is about 18.84 cm.

Do Exercise 3.

3. Find the circumference of this circle. Use 3.14 for π.

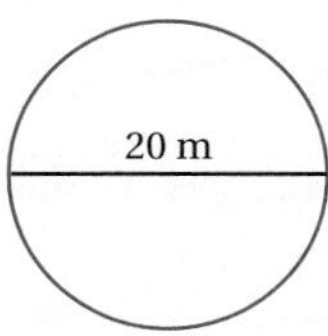

Answer on page A-15

4. Find the circumference of this circle. Use $\frac{22}{7}$ for π.

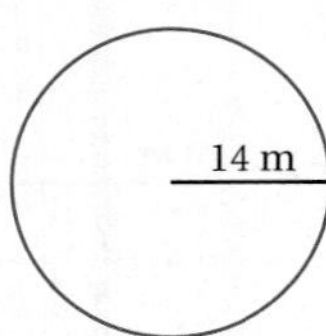

5. Find the perimeter of this figure. Use 3.14 for π.

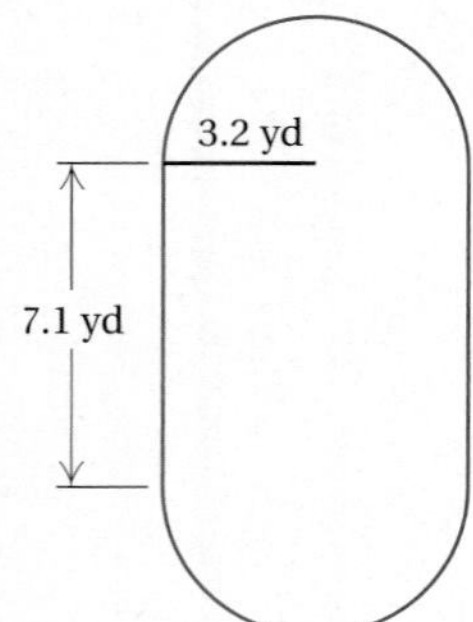

Answers on page A-15

Since $d = 2 \cdot r$, where r is the length of a radius, it follows that

$$C = \pi \cdot d = \pi \cdot (2 \cdot r).$$

CIRCUMFERENCE AND RADIUS

The circumference C of a circle of radius r is given by

$$\mathbf{C = 2 \cdot \pi \cdot r.}$$

EXAMPLE 4 Find the circumference of this circle. Use $\frac{22}{7}$ for π.

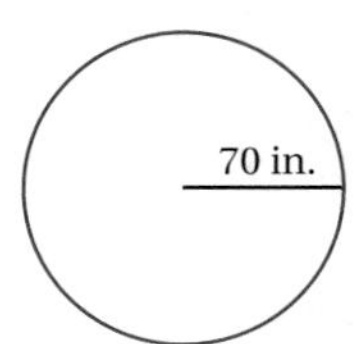

$$\begin{aligned} C &= 2 \cdot \pi \cdot r \\ &\approx 2 \cdot \frac{22}{7} \cdot 70 \text{ in.} \\ &= 2 \cdot 22 \cdot \frac{70}{7} \text{ in.} \\ &= 44 \cdot 10 \text{ in.} \\ &= 440 \text{ in.} \end{aligned}$$

The circumference is about 440 in.

EXAMPLE 5 Find the perimeter of this figure. Use 3.14 for π.

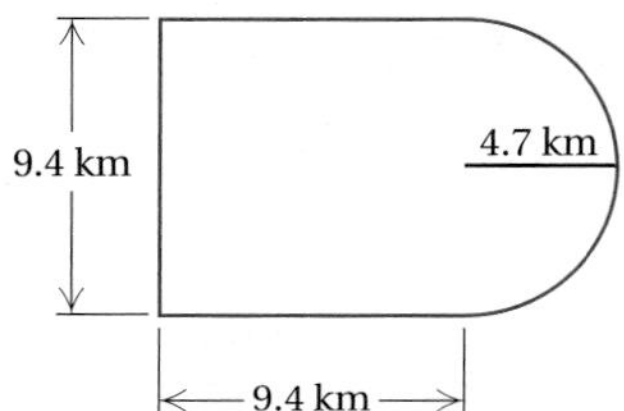

We let $P =$ the perimeter. We see that we have half a circle attached to a square. Thus we add half the circumference of the circle to the lengths of the three line segments.

$$\begin{aligned} P &= \begin{matrix}\text{Length of}\\ \text{three sides}\\ \text{of the square}\end{matrix} + \begin{matrix}\text{Half of the}\\ \text{circumference}\\ \text{of the circle}\end{matrix} \\ &= 3 \times 9.4 \text{ km} + \frac{1}{2} \times 2 \times \pi \times 4.7 \text{ km} \\ &= 28.2 \text{ km} + 3.14 \times 4.7 \text{ km} \\ &= 28.2 \text{ km} + 14.758 \text{ km} \\ &= 42.958 \text{ km} \end{aligned}$$

The perimeter is about 42.958 km.

Do Exercises 4 and 5.

C Area

Below is a circle of radius r.

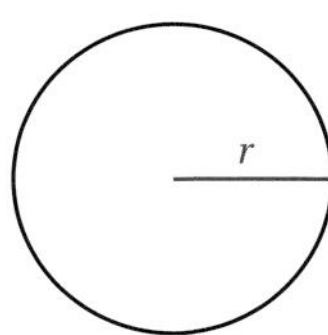

Think of cutting half the circular region into small pieces and arranging them as shown below.

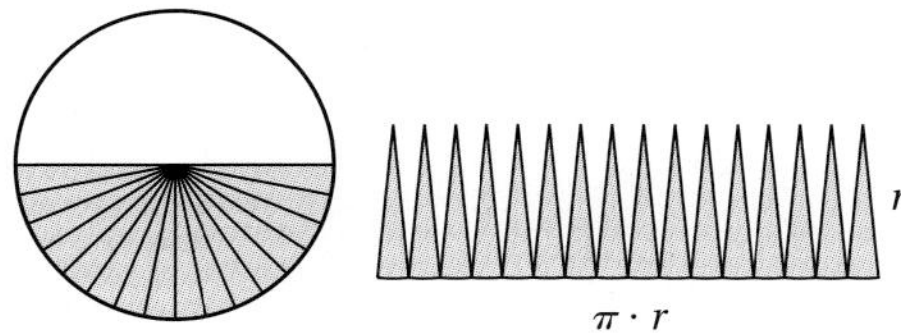

Then imagine cutting the other half of the circular region and arranging the pieces in with the others as shown below.

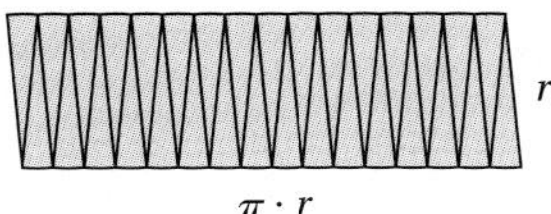

This is almost a parallelogram. The base has length $\frac{1}{2} \cdot 2 \cdot \pi \cdot r$, or $\pi \cdot r$ (half the circumference) and the height is r. Thus the area is

$$(\pi \cdot r) \cdot r.$$

This is the area of a circle.

AREA OF A CIRCLE

The **area of a circle** with radius of length r is given by

$$\boldsymbol{A = \pi \cdot r \cdot r}, \quad \text{or} \quad \boldsymbol{A = \pi \cdot r^2}.$$

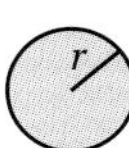

EXAMPLE 6 Find the area of this circle. Use $\frac{22}{7}$ for π.

$$\begin{aligned} A &= \pi \cdot r \cdot r \\ &\approx \frac{22}{7} \cdot 14\text{ cm} \cdot 14\text{ cm} \\ &= \frac{22}{7} \cdot 196\text{ cm}^2 \\ &= 616\text{ cm}^2 \end{aligned}$$

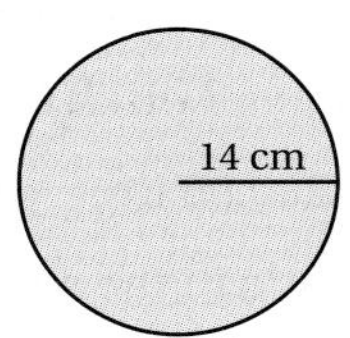

The area is about 616 cm^2.

Do Exercise 6.

6. Find the area of this circle. Use $\frac{22}{7}$ for π.

5 km

Answer on page A-15

7. Find the area of this circle. Use 3.14 for π. Round to the nearest hundredth.

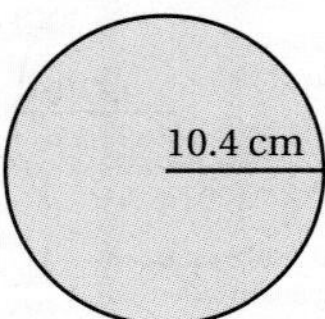

8. Which is larger and by how much: a 10-ft–square flower bed or a 12-ft–diameter flower bed?

Answers on page A-15

EXAMPLE 7 Find the area of this circle. Use 3.14 for π. Round to the nearest hundredth.

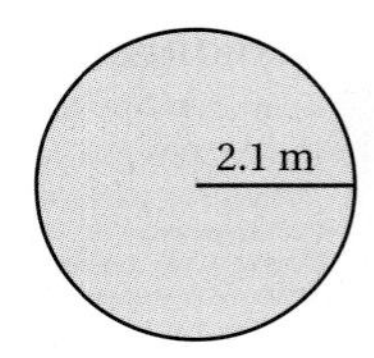

$$\begin{aligned} A &= \pi \cdot r \cdot r \\ &\approx 3.14 \times 2.1 \text{ m} \times 2.1 \text{ m} \\ &= 3.14 \times 4.41 \text{ m}^2 \\ &= 13.8474 \text{ m}^2 \\ &\approx 13.85 \text{ m}^2 \end{aligned}$$

The area is about 13.85 m^2.

Caution!

Remember that circumference is always measured in linear units like ft, m, cm, yd, and so on. But area is measured in square units like ft^2, m^2, cm^2, yd^2, and so on.

Do Exercise 7.

d Solving Applied Problems

EXAMPLE 8 *Area of Pizza Pans.* Which is larger and by how much: a pizza made in a 16-in.–square pizza pan or a pizza made in a 16-in.–diameter circular pan?

First, we make a drawing of each.

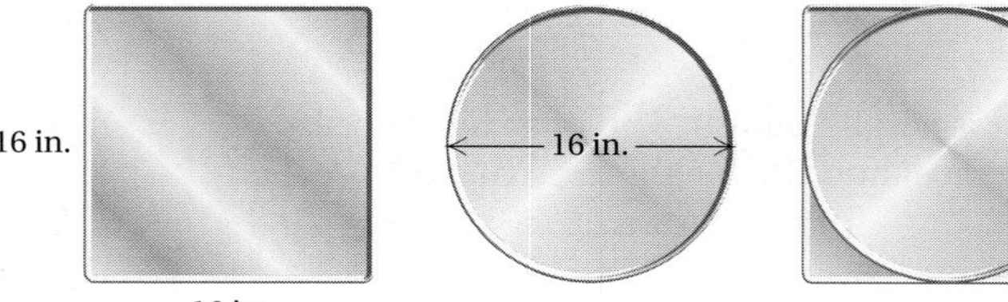

Then we compute areas.

The area of the square is

$$\begin{aligned} A &= s \cdot s \\ &= 16 \text{ in.} \times 16 \text{ in.} = 256 \text{ in}^2. \end{aligned}$$

The diameter of the circle is 16 in., so the radius is 16 in./2, or 8 in. The area of the circle is

$$\begin{aligned} A &= \pi \cdot r \cdot r \\ &\approx 3.14 \times 8 \text{ in.} \times 8 \text{ in.} = 200.96 \text{ in}^2. \end{aligned}$$

We see that the square pizza is larger by about

$$256 \text{ in}^2 - 200.96 \text{ in}^2, \quad \text{or} \quad 55.04 \text{ in}^2.$$

Thus the pizza made in the square pan is larger, by about 55.04 in^2.

Do Exercise 8.

For Extra Help

MathXL | MyMathLab | InterAct Math | Math Tutor Center | Video Lectures on CD Disc 3 | Student's Solutions Manual

a, b, c For each circle, find the length of a diameter, the circumference, and the area. Use $\frac{22}{7}$ for π.

1.

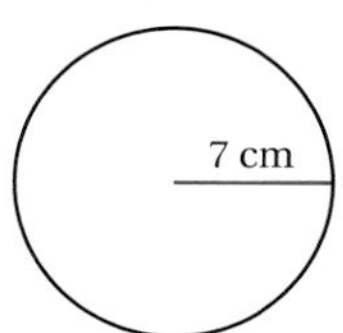

2.

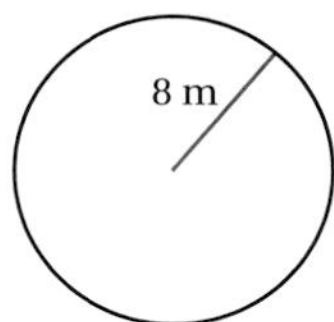

3.

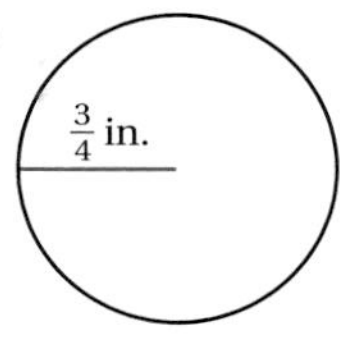

4.

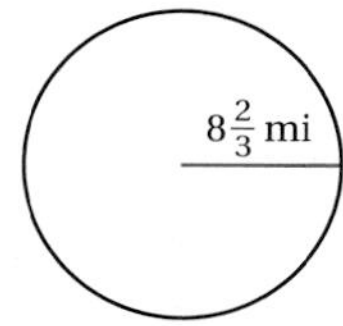

For each circle, find the length of a radius, the circumference, and the area. Use 3.14 for π.

5.

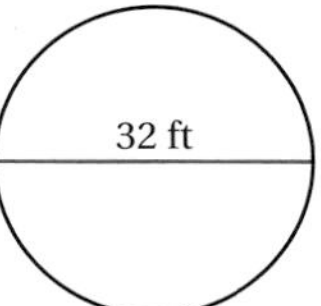

6.

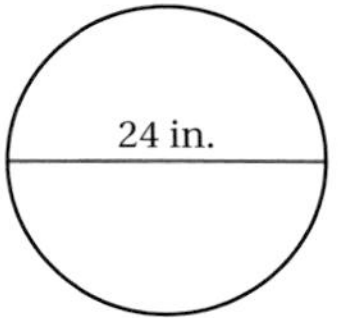

7.

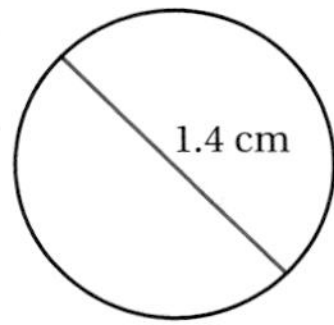

8.

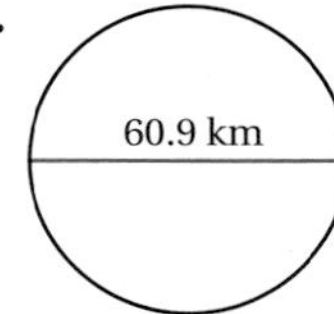

d Solve. Use 3.14 for π.

9. *Soda-Can Top.* The top of a soda can has a 6-cm diameter. What is the radius? the circumference? the area?

10. *Penny.* A penny has a 1-cm radius. What is the diameter? the circumference? the area?

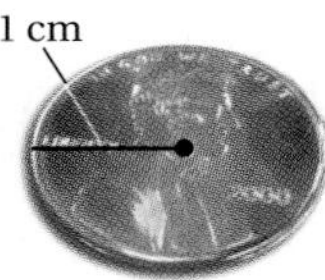

11. *Trampoline.* The standard backyard trampoline has a diameter of 14 ft. What is the area?

Source: International Trampoline Industry Association, Inc.

12. *Area of Pizza Pans.* Which is larger and by how much: a pizza made in a 12-in.–square pizza pan or a pizza made in a 12-in.–diameter circular pan?

13. *Dimensions of a Quarter.* The circumference of a quarter is 7.85 cm. What is the diameter? the radius? the area?

14. *Dimensions of a Dime.* The circumference of a dime is 2.23 in. What is the diameter? the radius? the area?

15. *Gypsy-Moth Tape.* To protect an elm tree in your backyard, you need to attach gypsy moth caterpillar tape around the trunk. The tree has a 1.1-ft diameter. What length of tape is needed?

16. *Earth.* The diameter of the earth at the equator is 7926.41 mi. What is the circumference of the earth at the equator?

Source: *The Handy Geography Answer Book*

17. *Swimming-Pool Walk.* You want to install a 1-yd–wide walk around a circular swimming pool. The diameter of the pool is 20 yd. What is the area of the walk?

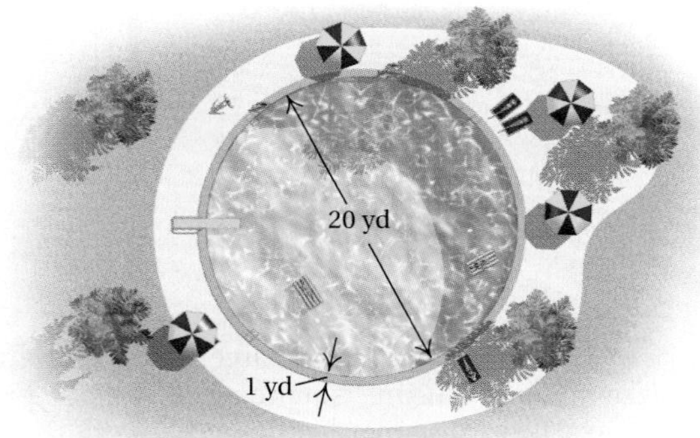

18. *Roller-Rink Floor.* A roller-rink floor is shown below. Each end is a semicircle. What is its area? If hardwood flooring costs $32.50 per square meter, how much will the flooring cost?

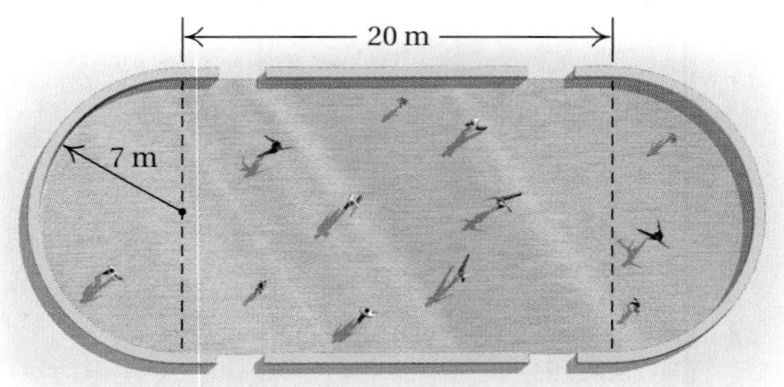

Find the perimeter. Use 3.14 for π.

19.

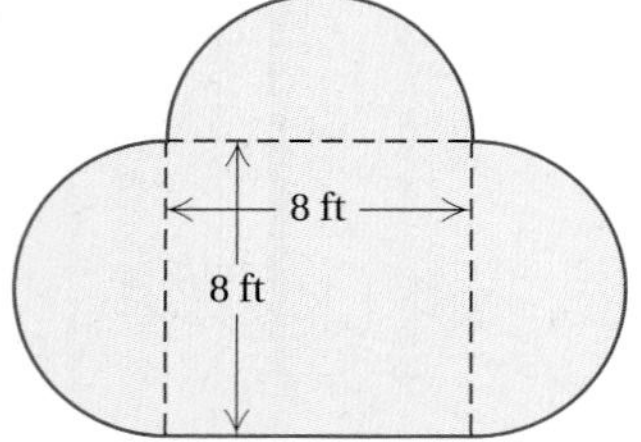

20.

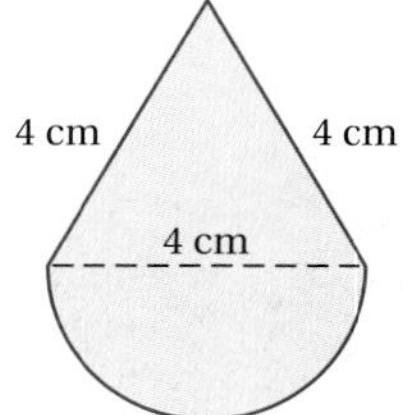

21.

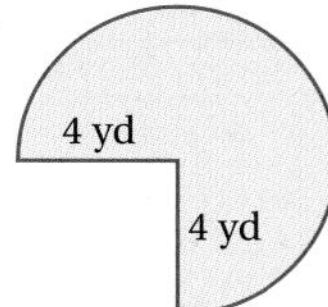

22.

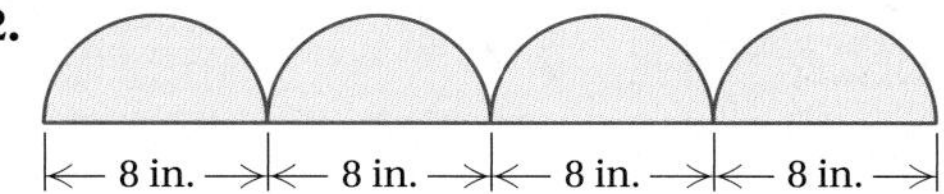

23.

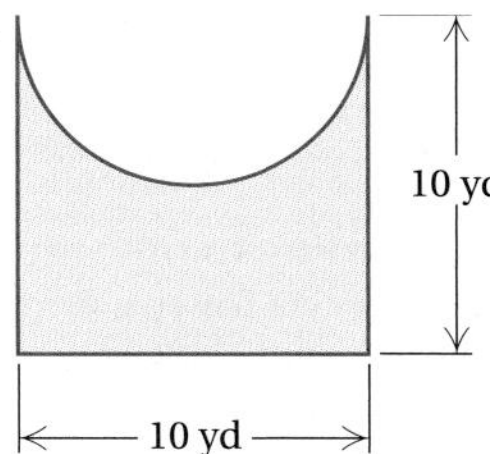

24.

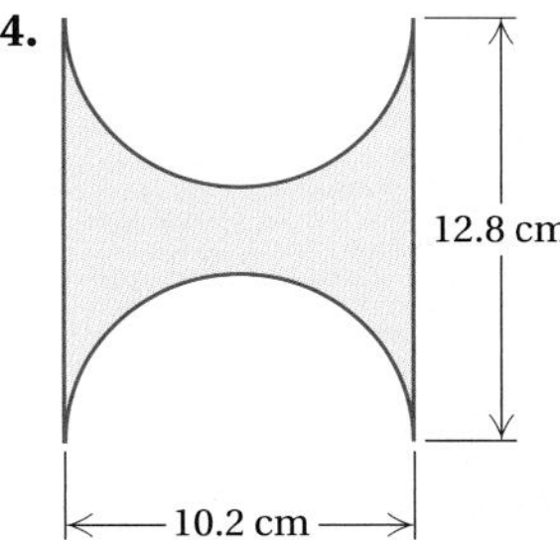

Find the area of the shaded region. Use 3.14 for π.

25.

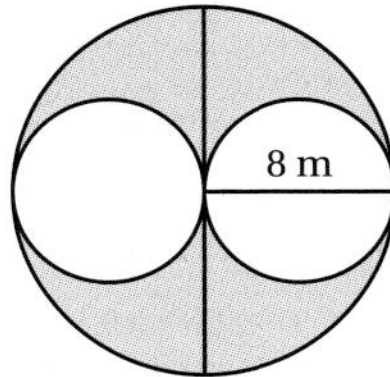

26.

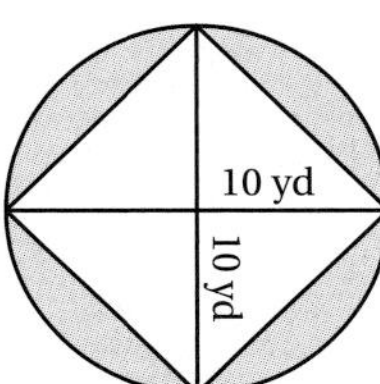

27.

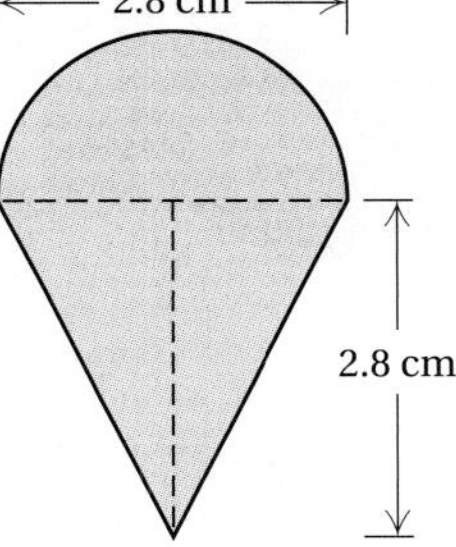

28.

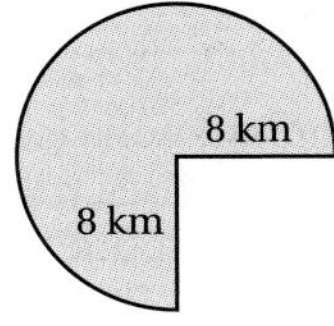

29.

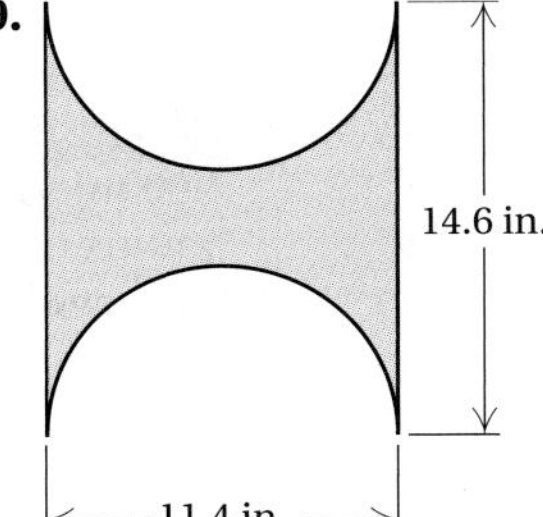

30.

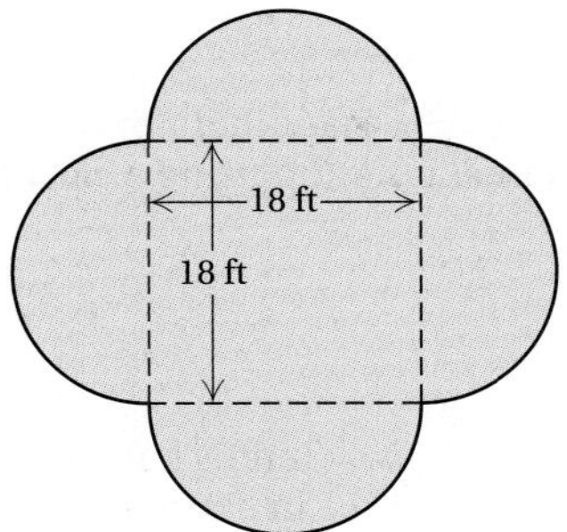

31. **D_W** Explain why a 16-in.–diameter pizza that costs \$16.25 is a better buy than a 10-in.–diameter pizza that costs \$7.85.

32. **D_W** The radius of one circle is twice the length of another circle's radius. Is the area of the first circle twice the area of the other circle? Why or why not?

SKILL MAINTENANCE

Evaluate. [1.6b]

33. 2^4

34. 17^2

35. 5^3

36. 8^2

Convert to percent notation. [4.3a]

37. $\frac{3}{8}$

38. $\frac{5}{8}$

39. $\frac{2}{3}$

40. $\frac{1}{5}$

41. Find the rate of discount. [4.7c]

42. Jack's commission is increased according to how much he sells. He receives a commission of 6% for the first \$3000 and 10% on the amount over \$3000. What is the total commission on sales of \$8500? [4.7b]

43. If 2 cans of tomato paste cost \$1.49, how many cans of tomato paste can you buy for \$7.45? [4.1e]

44. The weight of a human brain is 2.5% of total body weight. A person weighs 200 lb. What does the brain weigh? [4.6a]

SYNTHESIS

45. 🖩 $\pi \approx \frac{3927}{1250}$ is another approximation for π. Find decimal notation using a calculator.

46. 🖩 The distance from Kansas City to Indianapolis is 500 mi. A car was driven this distance using tires with a radius of 14 in. How many revolutions of each tire occurred on the trip? Use $\frac{22}{7}$ for π.

47. *Tennis Balls.* Tennis balls are generally packed vertically three in a can, one on top of another. Suppose the diameter of a tennis ball is d. Find the height of the stack of balls. Find the circumference of one ball. Which is greater? Explain.

6.5 VOLUME AND SURFACE AREA

Objectives

a Find the volume and the surface area of a rectangular solid.

b Given the radius and the height, find the volume of a circular cylinder.

c Given the radius, find the volume of a sphere.

d Given the radius and the height, find the volume of a circular cone.

e Solve applied problems involving volume of rectangular solids, circular cylinders, spheres, and cones.

a Rectangular Solids

The **volume** of a **rectangular solid** is the number of unit cubes needed to fill it.

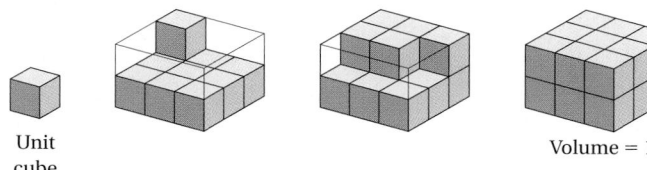

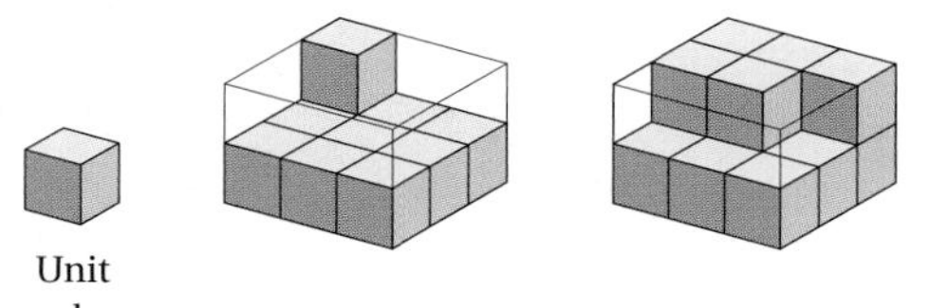

Two other units are shown below.

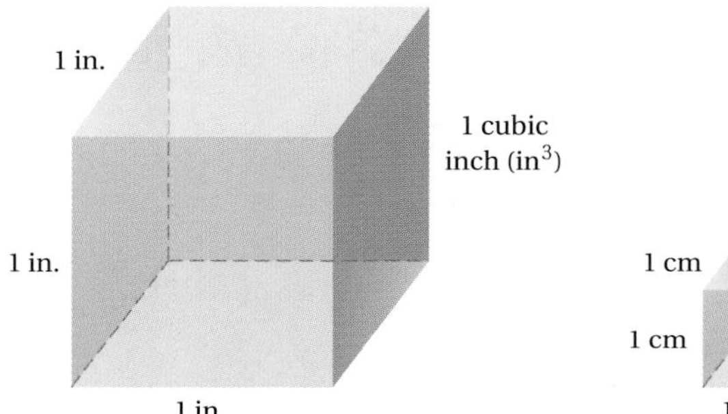

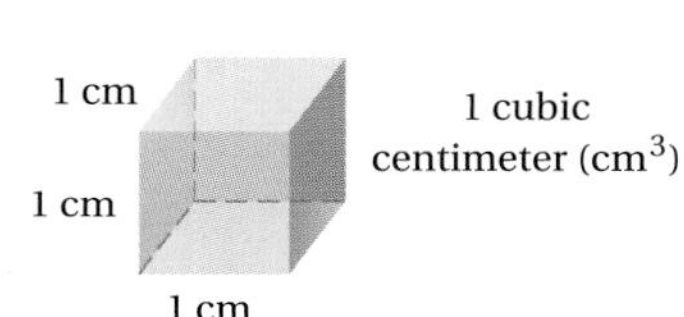

EXAMPLE 1 Find the volume.

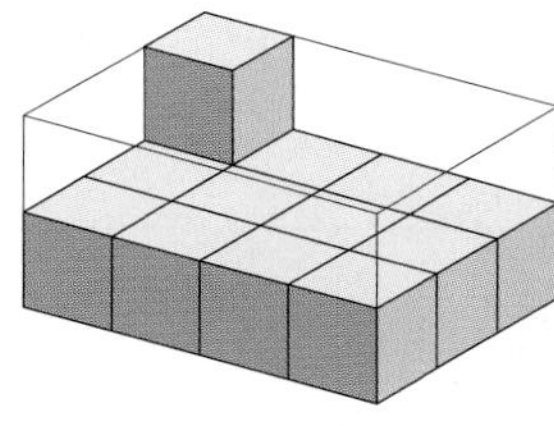

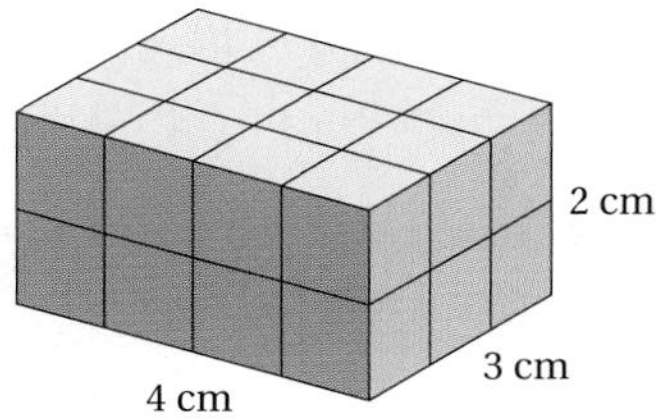

The figure is made up of 2 layers of 12 cubes each, so its volume is 24 cubic centimeters (cm^3).

Do Exercise 1.

VOLUME OF A RECTANGULAR SOLID

The **volume of a rectangular solid** is found by multiplying length by width by height:

$$V = l \cdot w \cdot h.$$

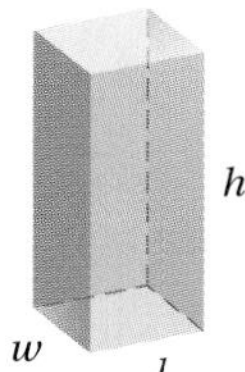

1. Find the volume.

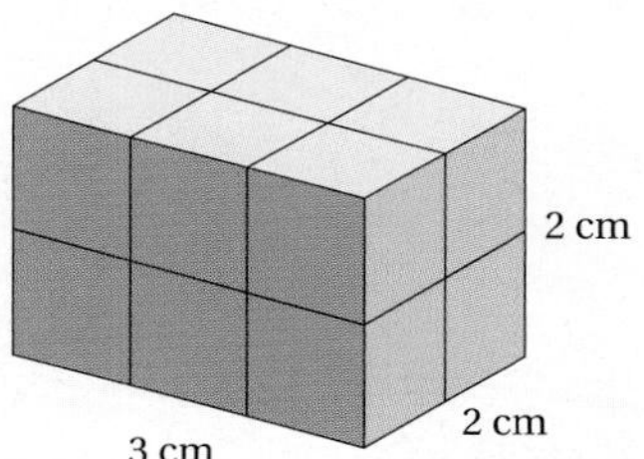

Answer on page A-15

2. Popcorn. The average American eats 54 qt of popcorn annually. This quantity of popcorn would fill a bag measuring 10.5 in. by 10.5 in. by 28 in. Find the volume of such a bag.

Source: The Popcorn Board

EXAMPLE 2 *Carry-on Luggage.* The largest piece of luggage that you can carry on an airplane measures 23 in. by 10 in. by 13 in. Find the volume of this solid.

$$V = l \cdot w \cdot h$$
$$= 23 \text{ in.} \cdot 10 \text{ in.} \cdot 13 \text{ in.}$$
$$= 230 \cdot 13 \text{ in}^3$$
$$= 2990 \text{ in}^3$$

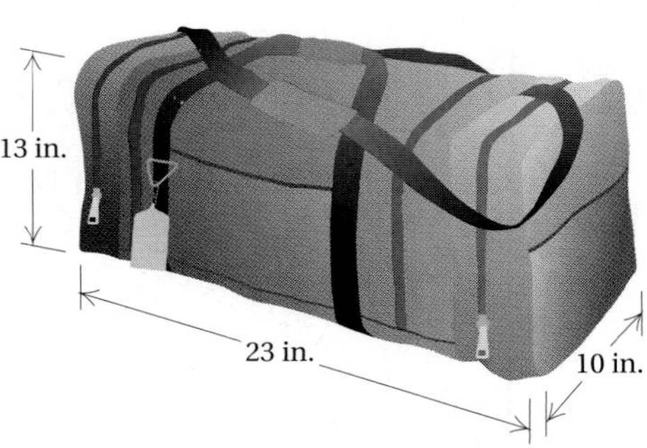

Do Exercises 2 and 3.

The **surface area** of a rectangular solid is the total area of the six rectangles that form the surface of the solid. For the rectangular solid below, we can show the six rectangles with a diagram.

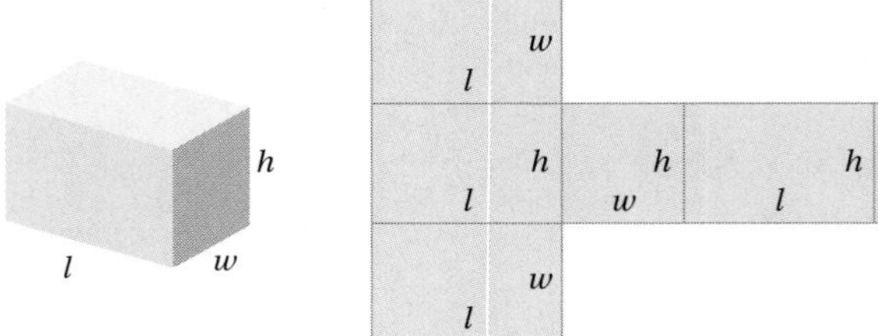

$$SA = l \cdot w + l \cdot w + l \cdot h + w \cdot h + l \cdot h + w \cdot h$$
$$= lw + lw + lh + wh + lh + wh$$
$$= 2lw + 2lh + 2wh, \quad \text{or} \quad 2(lw + lh + wh)$$

3. Cord of Wood. A cord of wood measures 4 ft by 4 ft by 8 ft. What is the volume of a cord of wood?

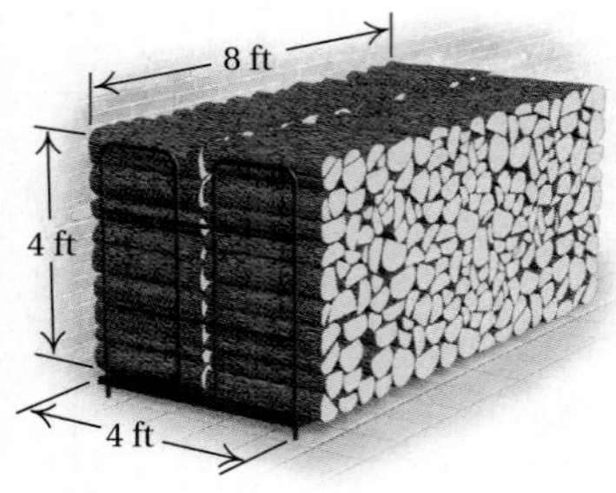

SURFACE AREA OF A RECTANGULAR SOLID

The surface area of a rectangular solid with length l, width w, and height h is given by the formula

$$\text{SA} = 2lw + 2lh + 2wh, \quad \text{or} \quad 2(lw + lh + wh).$$

EXAMPLE 3 Find the surface area of this rectangular solid.

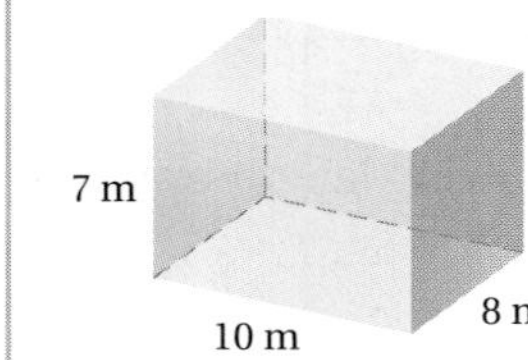

$$\text{SA} = 2lw + 2lh + 2wh$$
$$= 2 \cdot 10 \text{ m} \cdot 8 \text{ m} + 2 \cdot 10 \text{ m} \cdot 7 \text{ m} + 2 \cdot 8 \text{ m} \cdot 7 \text{ m}$$
$$= 160 \text{ m}^2 + 140 \text{ m}^2 + 112 \text{ m}^2$$
$$= 412 \text{ m}^2$$

The units used for area are square units.
The units used for volume are cubic units.

Answers on page A-15

Do Exercises 4 and 5.

b Cylinders

A rectangular solid is shown below. Note that we can think of the volume as the product of the area of the base times the height:

$$\begin{aligned} V &= l \cdot w \cdot h \\ &= (l \cdot w) \cdot h \\ &= (\text{Area of the base}) \cdot h \\ &= B \cdot h, \end{aligned}$$

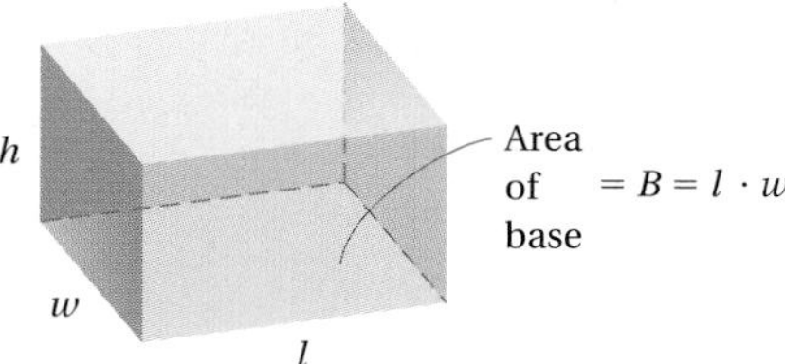

where B represents the area of the base.

Like rectangular solids, **circular cylinders** have bases of equal area that lie in parallel planes. The bases of circular cylinders are circular regions.

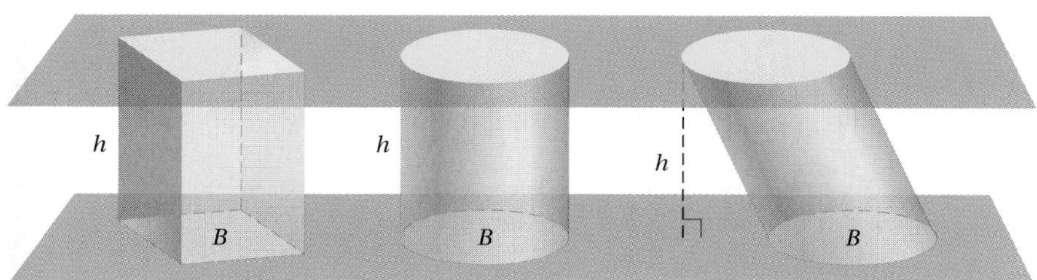

The volume of a circular cylinder is found in a manner similar to finding the volume of a rectangular solid. The volume is the product of the area of the base times the height. The height is always measured perpendicular to the base.

VOLUME OF A CIRCULAR CYLINDER

The **volume of a circular cylinder** is the product of the area of the base B and the height h:

$$V = B \cdot h, \quad \text{or} \quad V = \pi \cdot r^2 \cdot h.$$

Find the volume and the surface area of the rectangular solid.

4.

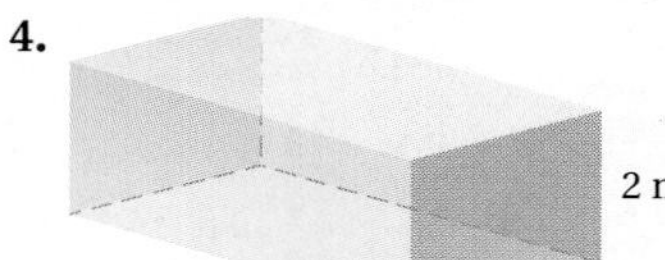

5.

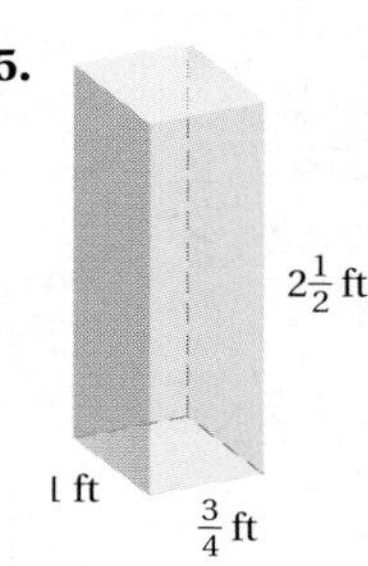

Answers on page A-15

Study Tips

TAKING NOTES

Effective note-taking can greatly enhance your learning of mathematics. Here are some suggestions.

- This textbook has been written and designed so that it represents a quality set of notes at the same time that it teaches. Thus you might not need to take as many notes in class. Just watch, listen, and ask questions as the class moves along, rather than racing to keep up your note-taking.
- If you have any difficulty keeping up with the instructor, use abbreviations to speed up your note-taking. Consider standard abbreviations like "Ex" for "Example," "≈" for "is approximately equal to," or "∴" for "therefore." Create your own abbreviations as well.

6. Find the volume of this cylinder. Use 3.14 for π.

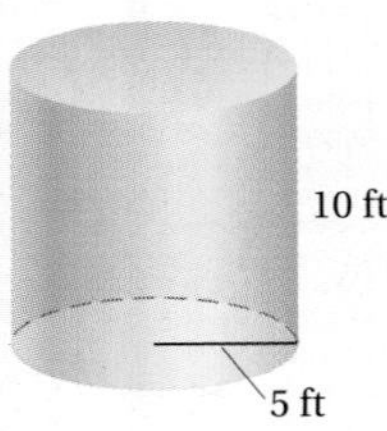

7. Find the volume of this cylinder. Use $\frac{22}{7}$ for π.

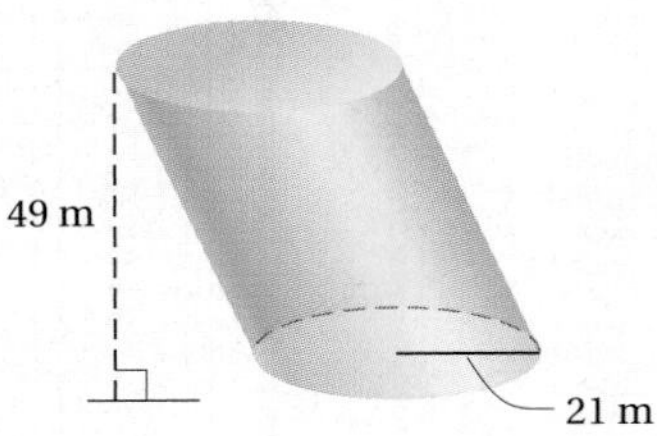

8. Find the volume of this sphere. Use $\frac{22}{7}$ for π.

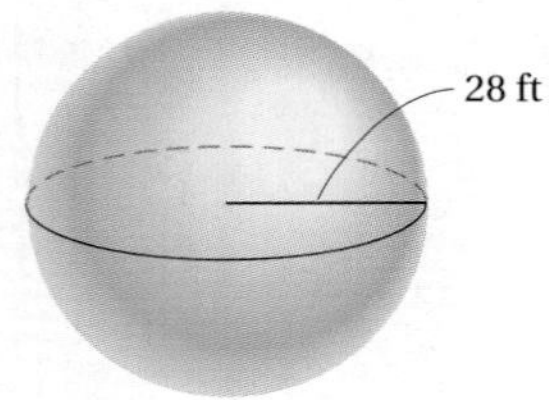

9. The radius of a standard-sized golf ball is 2.1 cm. Find the volume of a standard-sized golf ball. Use 3.14 for π.

Answers on page A-15

EXAMPLE 4 Find the volume of this circular cylinder. Use 3.14 for π.

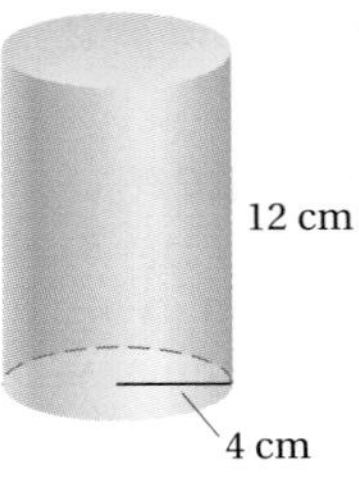

$$\begin{aligned} V = Bh &= \pi \cdot r^2 \cdot h \\ &\approx 3.14 \times 4 \text{ cm} \times 4 \text{ cm} \times 12 \text{ cm} \\ &= 602.88 \text{ cm}^3 \end{aligned}$$

Do Exercises 6 and 7.

C Spheres

A **sphere** is the three-dimensional counterpart of a circle. It is the set of all points in space that are a given distance (the radius) from a given point (the center).

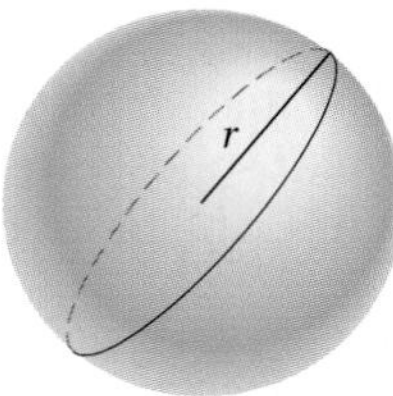

We find the volume of a sphere as follows.

VOLUME OF A SPHERE

The **volume of a sphere** of radius r is given by

$$V = \frac{4}{3} \cdot \pi \cdot r^3.$$

EXAMPLE 5 *Bowling Ball.* The radius of a standard-sized bowling ball is 4.2915 in. Find the volume of a standard-sized bowling ball. Round to the nearest hundredth of a cubic inch. Use 3.14 for π.

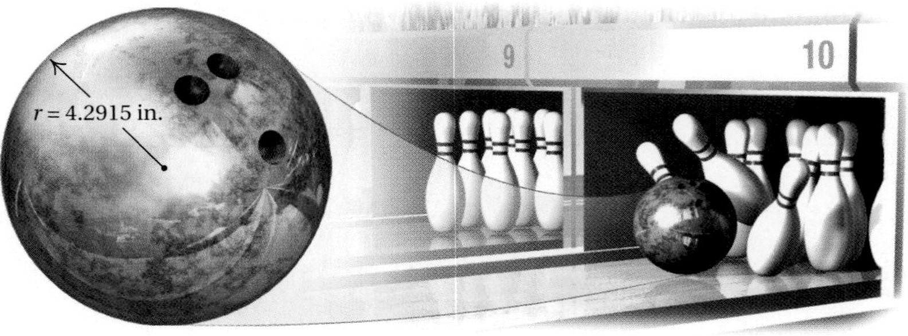

We have

$$\begin{aligned} V &= \tfrac{4}{3} \cdot \pi \cdot r^3 \\ &\approx \tfrac{4}{3} \times 3.14 \times (4.2915 \text{ in.})^3 \\ &\approx 330.90 \text{ in}^3. \qquad \text{Using a calculator} \end{aligned}$$

Do Exercises 8 and 9.

d Cones

Consider a circle in a plane and choose any point P not in the plane. The circular region, together with the set of all segments connecting P to a point on the circle, is called a **circular cone.**

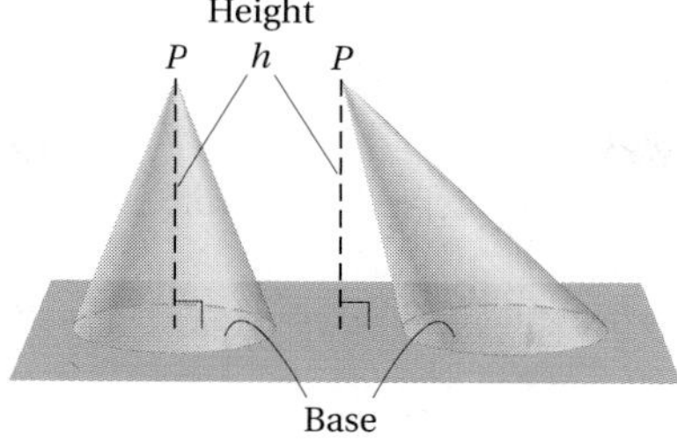

We find the volume of a cone as follows.

VOLUME OF A CIRCULAR CONE

The **volume of a circular cone** with base radius r is one-third the product of the base area and the height:

$$V = \frac{1}{3} \cdot B \cdot h = \frac{1}{3} \pi \cdot r^2 \cdot h.$$

EXAMPLE 6 Find the volume of this circular cone. Use 3.14 for π.

$$V = \tfrac{1}{3}\pi \cdot r^2 \cdot h$$
$$\approx \tfrac{1}{3} \times 3.14 \times 3\text{ cm} \times 3\text{ cm} \times 7\text{ cm}$$
$$= 65.94\text{ cm}^3$$

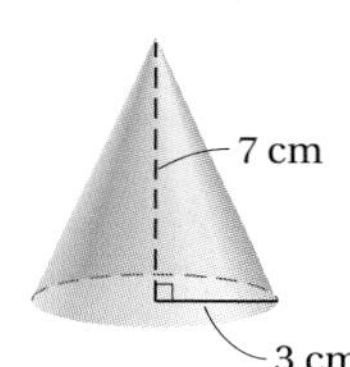

Do Exercises 10 and 11.

10. Find the volume of this cone. Use 3.14 for π.

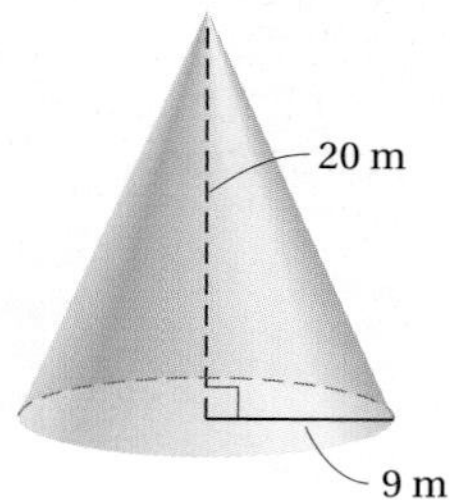

11. Find the volume of this cone. Use $\frac{22}{7}$ for π.

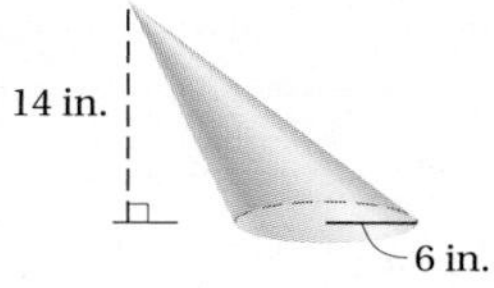

Answers on page A-15

CALCULATOR CORNER

The [π] Key Many calculators have a [π] key that can be used to enter the value of π in a computation. It might be necessary to press a [2nd] or [SHIFT] key before pressing the [π] key on some calculators. Since 3.14 is a rounded value for π, results obtained using the [π] key might be slightly different from those obtained when 3.14 is used for the value of π in a computation.

To find the volume of the circular cylinder in Example 4, we press [2nd] [π] [×] [4] [×] [4] [×] [1] [2] [=] or [SHIFT] [π] [×] [4] [×] [4] [×] [1] [2] [=]. The result is approximately 603.19. Note that this is slightly different from the result found using 3.14 for the value of π.

Exercises:

1. Use a calculator with a [π] key to perform the computations in Examples 4–6.
2. Use a calculator with a [π] key to perform the computations in Margin Exercises 6–11.

12. Medicine Capsule. A cold capsule is 8 mm long and 4 mm in diameter. Find the volume of the capsule. Use 3.14 for π. (*Hint*: First find the length of the cylindrical section.)

Answer on page A-15

e Solving Applied Problems

EXAMPLE 7 *Propane Gas Tank.* A propane gas tank is shaped like a circular cylinder with half of a sphere at each end. Find the volume of the tank if the cylindrical section is 5 ft long with a 4-ft diameter. Use 3.14 for π.

1. Familiarize. We first make a drawing.

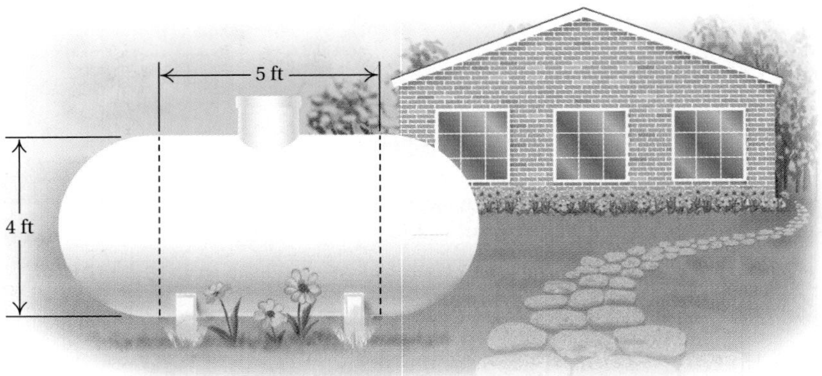

2. Translate. This is a two-step problem. We first find the volume of the cylindrical portion. Then we find the volume of the two ends and add. Note that the radius is 2 ft and that together the two ends make a sphere. We let

$$V = \pi \cdot r^2 \cdot h + \frac{4}{3} \cdot \pi \cdot r^3,$$

where V is the total volume. Then

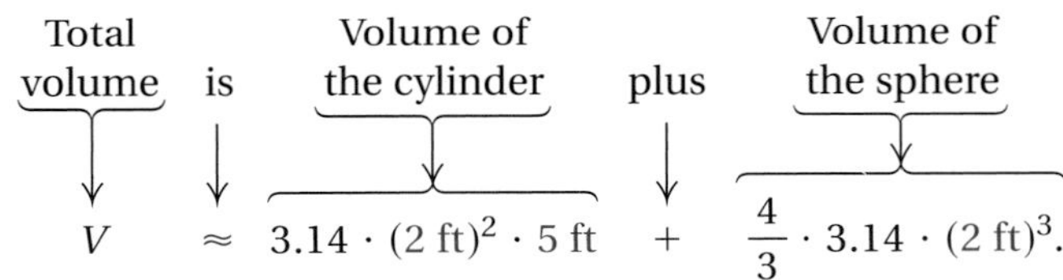

3. Solve. The volume of the cylinder is approximately

$$3.14 \cdot (2 \text{ ft})^2 \cdot 5 \text{ ft} = 3.14 \cdot 2 \text{ ft} \cdot 2 \text{ ft} \cdot 5 \text{ ft}$$
$$= 62.8 \text{ ft}^3.$$

The volume of the two ends is approximately

$$\frac{4}{3} \cdot 3.14 \cdot (2 \text{ ft})^3 = \frac{4}{3} \cdot 3.14 \cdot 2 \text{ ft} \cdot 2 \text{ ft} \cdot 2 \text{ ft}$$
$$\approx 33.5 \text{ ft}^3.$$

The total volume is about

$$62.8 \text{ ft}^3 + 33.5 \text{ ft}^3 = 96.3 \text{ ft}^3.$$

4. Check. The check is left to the student.

5. State. The volume of the tank is about 96.3 ft^3.

Do Exercise 12.

6.5 EXERCISE SET

For Extra Help

MathXL | MyMathLab | InterAct Math | Math Tutor Center | Video Lectures on CD Disc 3 | Student's Solutions Manual

a Find the volume and the surface area of the rectangular solid.

1.

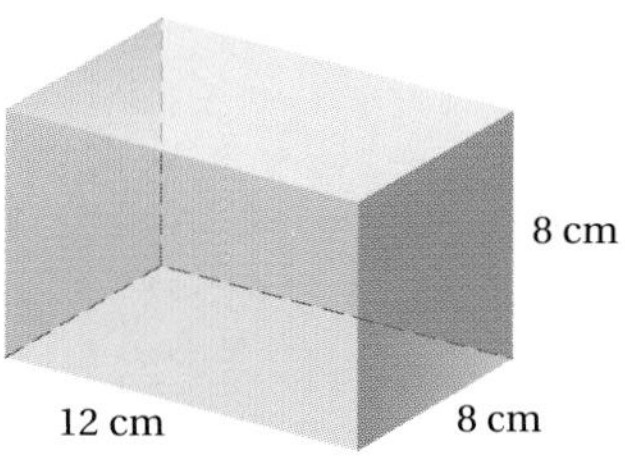

2.

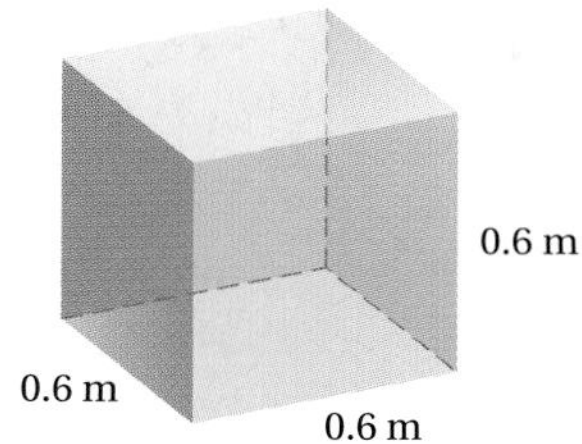

3.

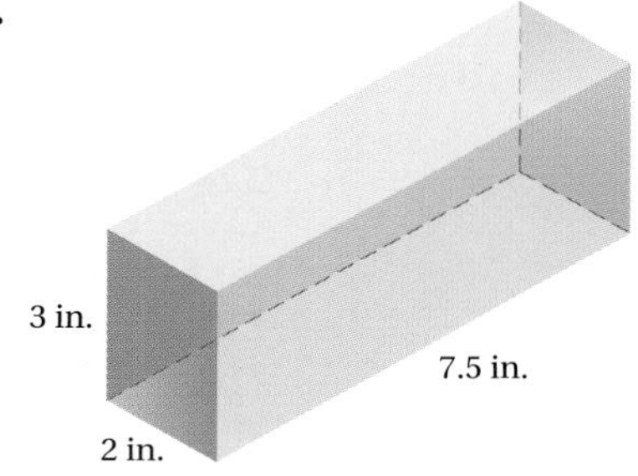

4.

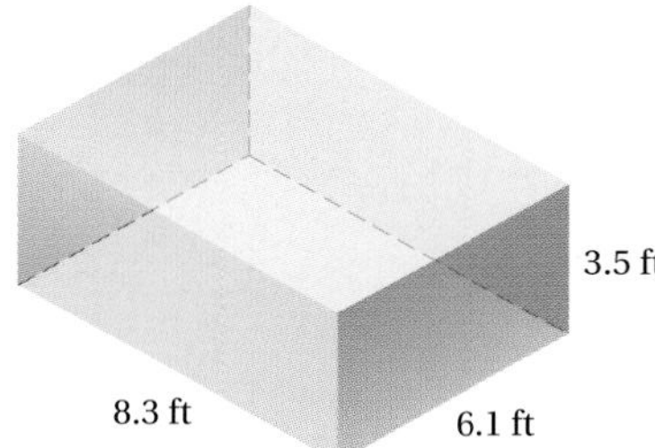

5.

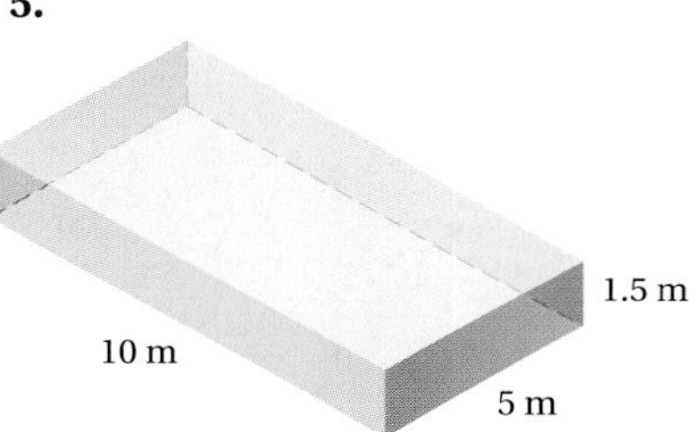

6.

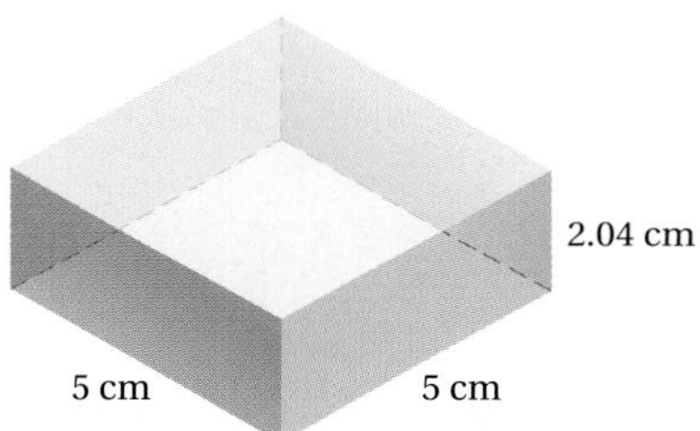

7.

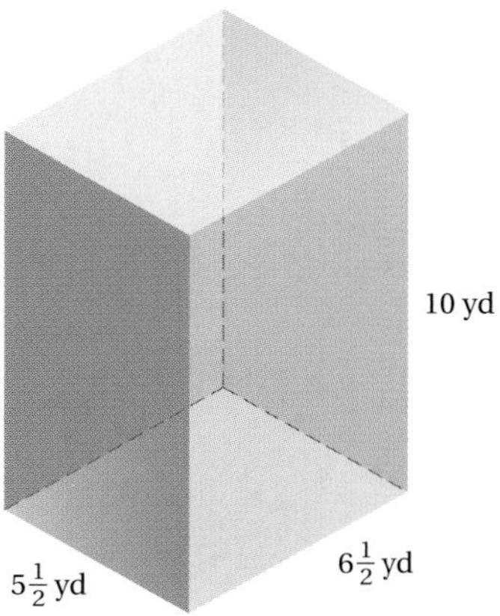

8.

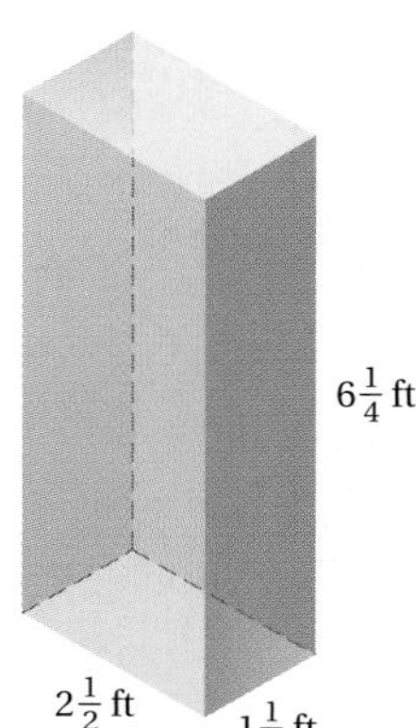

b Find the volume of the circular cylinder. Use 3.14 for π in Exercises 9–12. Use $\frac{22}{7}$ for π in Exercises 13 and 14.

9.

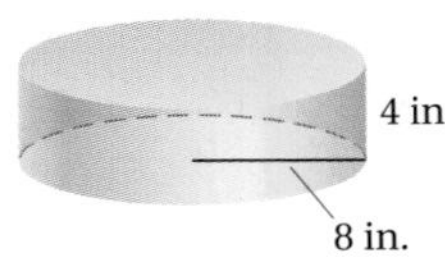

10.

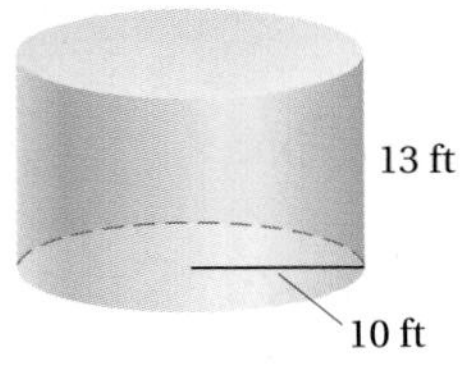

11.

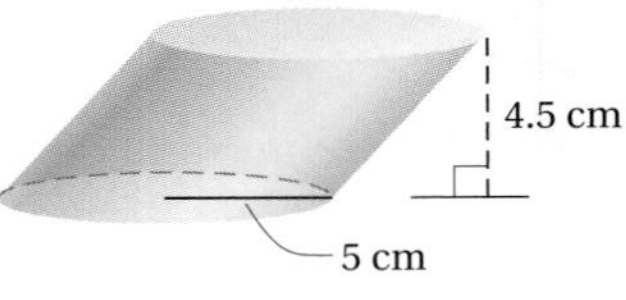

12.

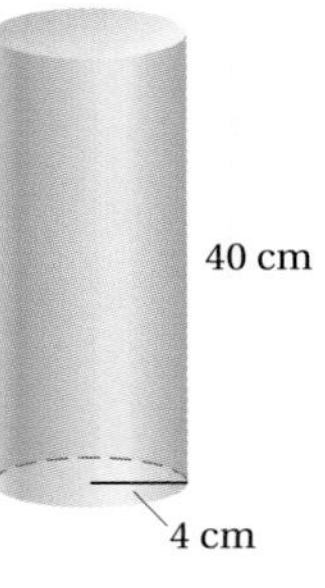

13.

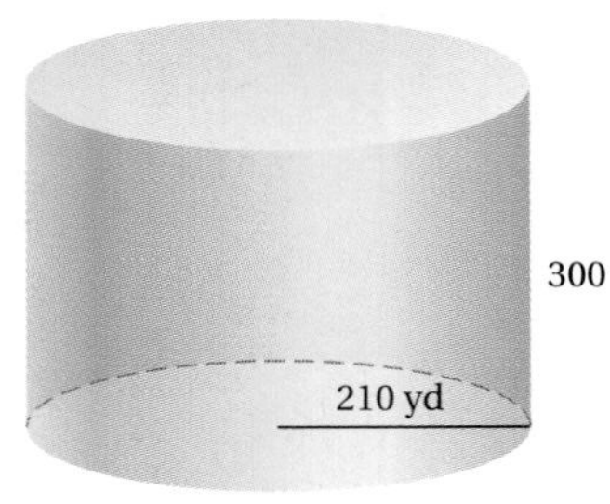

14.

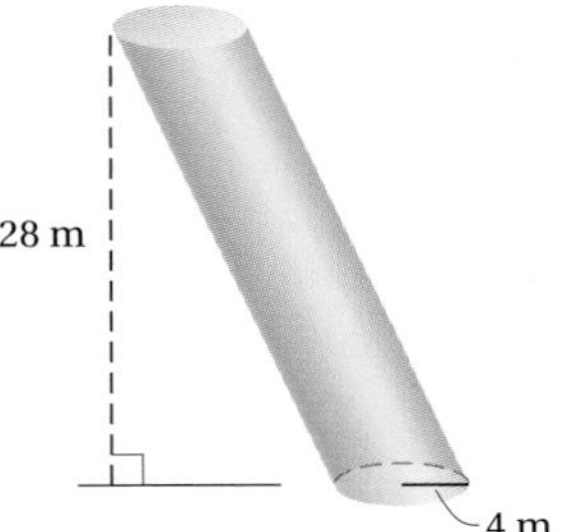

c Find the volume of the sphere. Use 3.14 for π in Exercises 15–18. Use $\frac{22}{7}$ for π in Exercises 19 and 20.

15.

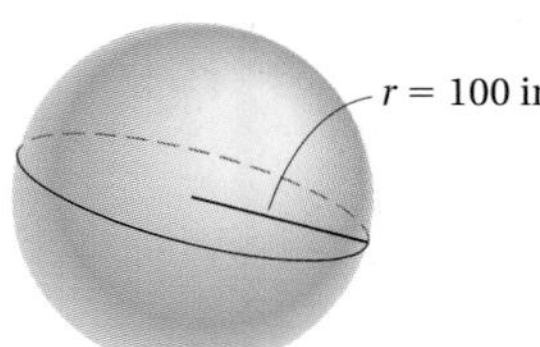

16.

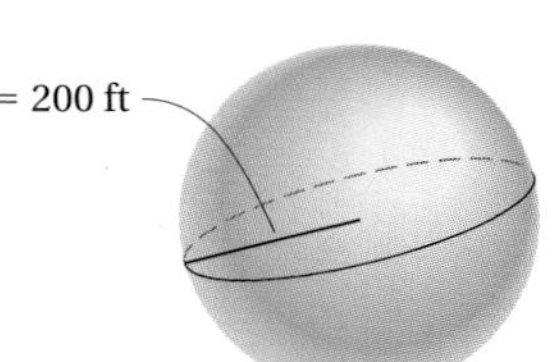

17.

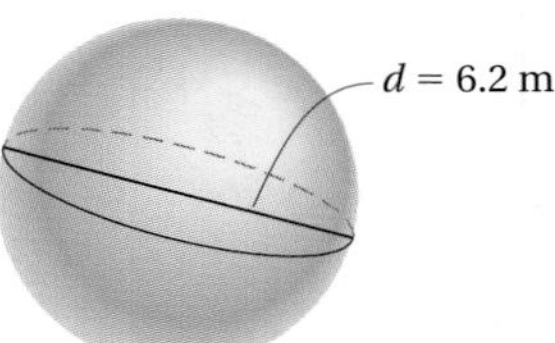

18.

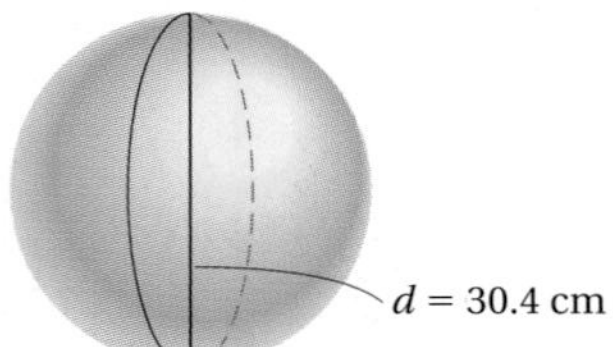

19.

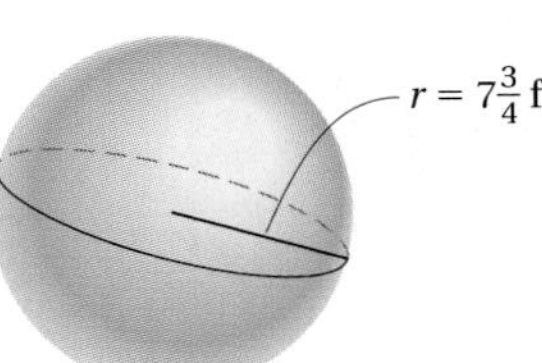

20.

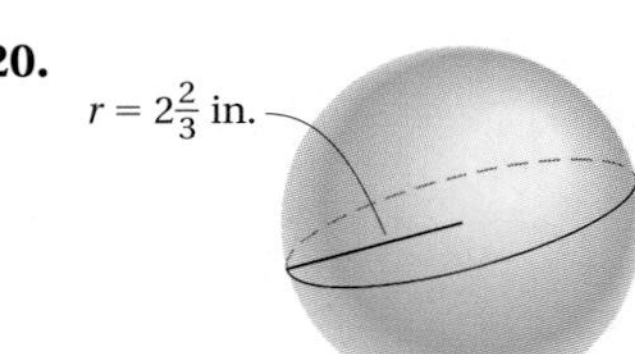

d Find the volume of the circular cone. Use 3.14 for π in Exercises 21 and 22. Use $\frac{22}{7}$  for π in Exercises 23 and 24.

21.

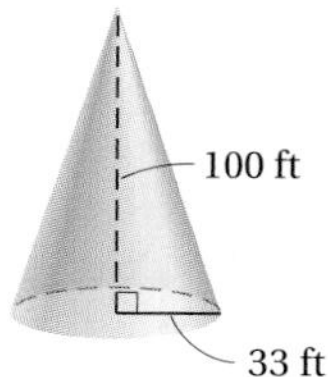

22.

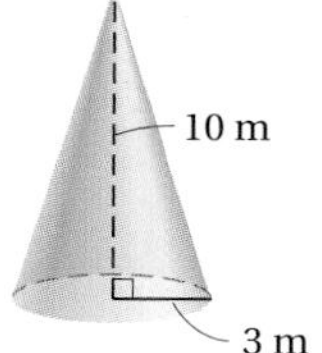

23.

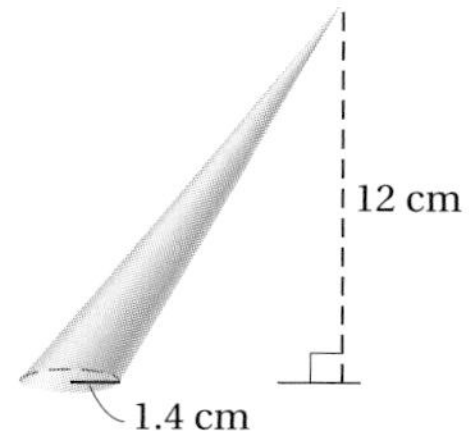

24.

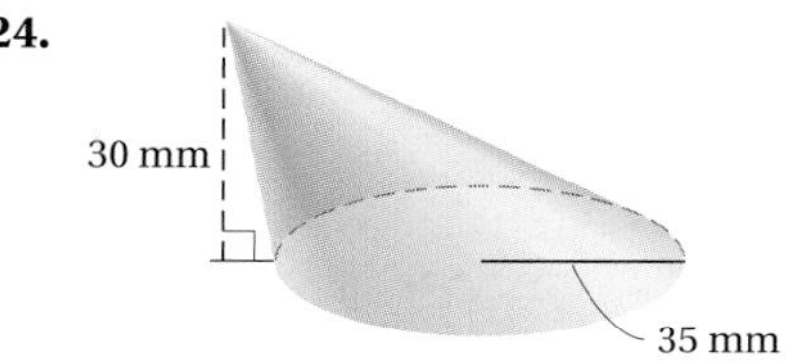

e Solve.

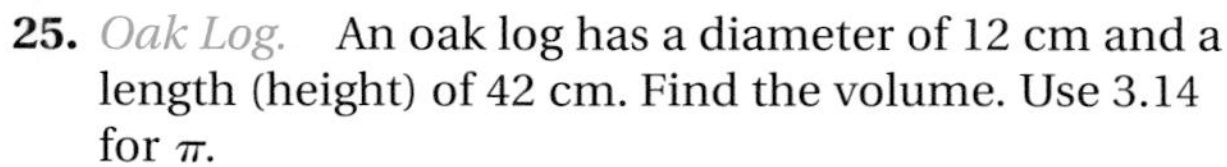

25. *Oak Log.* An oak log has a diameter of 12 cm and a length (height) of 42 cm. Find the volume. Use 3.14 for π.

26. *Ladder Rung.* A rung of a ladder is 2 in. in diameter and 16 in. long. Find the volume. Use 3.14 for π.

27. *Barn Silo.* A barn silo, excluding the top, is a circular cylinder. The silo is 6 m in diameter and the height is 13 m. Find the volume of the silo excluding the top. Use 3.14 for π.

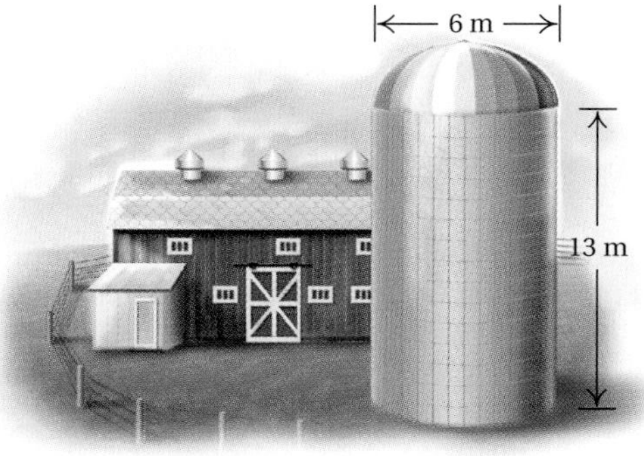

28. *Volume of a Trash Can.* The diameter of the base of a cylindrical trash can is 0.7 yd. The height is 1.1 yd. Find the volume. Use 3.14 for π.

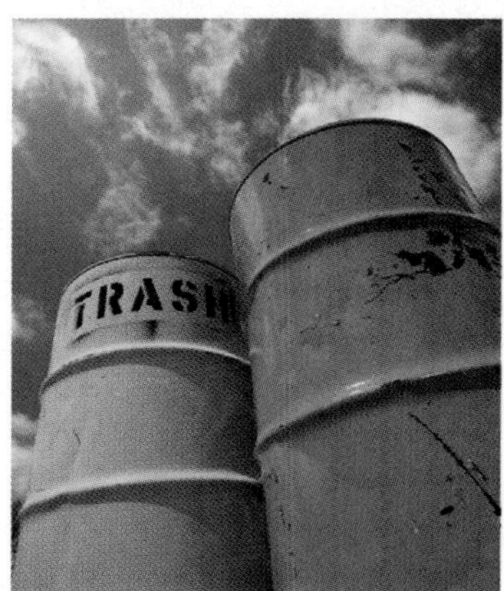

29. *Tennis Ball.* The diameter of a tennis ball is 6.5 cm. Find the volume. Use 3.14 for π.

30. *Spherical Gas Tank.* The diameter of a spherical gas tank is 6 m. Find the volume. Use 3.14 for π.

31. *Volume of Earth.* The diameter of the earth is about 3980 mi. Find the volume of the earth. Use 3.14 for π. Round to the nearest ten cubic miles.

32. *Astronomy.* The radius of Pluto's moon is about 500 km. Find the volume of this satellite. Use $\frac{22}{7}$ for π.

33. *Tennis-Ball Packaging.* Tennis balls are generally packaged in circular cylinders that hold 3 balls each. The diameter of a tennis ball is 6.5 cm. Find the volume of a can of tennis balls. Use 3.14 for π.

34. *Golf-Ball Packaging.* The box shown is just big enough to hold 3 golf balls. If the radius of a golf ball is 2.1 cm, how much air surrounds the three balls? Use 3.14 for π.

35. *Water Storage.* A water storage tank is a right circular cylinder with a radius of 14 cm and a height of 100 cm. What is the tank's volume? Use $\frac{22}{7}$ for π.

36. *Oceanography.* A research submarine is capsule-shaped. Find the volume of the submarine if it has a length of 10 m and a diameter of 8 m. Use 3.14 for π. (*Hint*: First find the length of the cylindrical section.)

37. *Metallurgy.* If all the gold in the world could be gathered together, it would form a cube 18 yd on a side. Find the volume of the world's gold.

38. The volume of a ball is 36π cm^3. Find the dimensions of a rectangular box that is just large enough to hold the ball.

39. **DW** How could you use the volume formulas given in this section to help estimate the volume of an egg?

40. **DW** The design of a modern home includes a cylindrical tower that will be capped with either a 10-ft–high dome or a 10-ft–high cone. Which type of cap will be more energy-efficient and why?

SKILL MAINTENANCE

Convert to fraction notation. [4.3b]

41. 35%

42. 85.5%

43. $37\frac{1}{2}\%$

44. $66.\overline{6}\%$

45. $83.\overline{3}\%$

46. $16\frac{2}{3}\%$

Solve. [1.5a]

47. A ream of paper contains 500 sheets. How many sheets are there in 15 reams?

48. A lab technician separates a vial containing 140 cc of blood into test tubes, each of which contains 3 cc of blood. How many test tubes can be filled? How much blood is left over?

SYNTHESIS

49. 🖩 The width of a dollar bill is 2.3125 in., the length is 6.0625 in., and the thickness is 0.0041 in. Find the volume occupied by one million one-dollar bills.

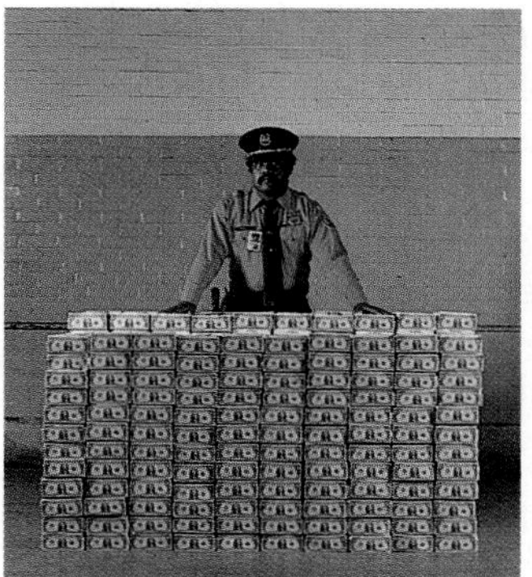

50. 🖩 Audio-cassette cases are typically 7 cm by 10.75 cm by 1.5 cm and contain 90 min of music. Compact-disc cases are typically 12.4 cm by 14.1 cm by 1 cm and contain 50 min of music. Which container holds the most music per cubic centimeter?

51. 🖩 A 2-cm–wide stream of water passes through a 30-m–long garden hose. At the instant that the water is turned off, how many liters of water are in the hose? Use 3.141593 for π.

52. 🖩 The volume of a basketball is 2304π cm^3. Find the volume of a cube-shaped box that is just large enough to hold a ball.

53. *Circumference of Earth.* The circumference of the earth at the equator is about 24,901.55 mi. Due to the irregular shape of the earth, the circumference of a circle of longitude wrapped around the earth between the north and south poles is about 24,859.82 mi. Describe and carry out a procedure for estimating the volume of the earth.

Source: *The Handy Geography Answer Book*

54. 🖩 A sphere with diameter 1 m is circumscribed by a cube. How much more volume is in the cube?

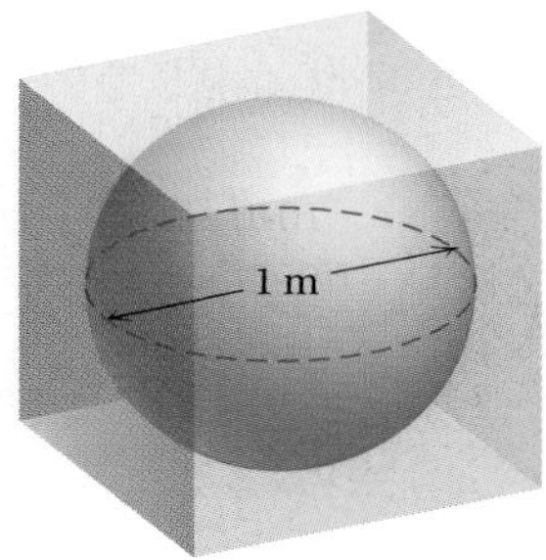

55. 🖩 A cube is circumscribed by a sphere with a 1-m diameter. How much more volume is in the sphere?

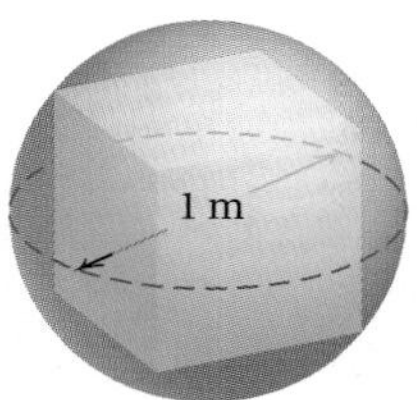

6.6 RELATIONSHIPS BETWEEN ANGLE MEASURES

Objectives

a Identify complementary and supplementary angles and find the measure of a complement or a supplement of a given angle.

b Determine whether segments are congruent and whether angles are congruent.

c Use the Vertical Angle Property to find measures of angles.

d Identify pairs of corresponding angles, interior angles, and alternate interior angles and apply properties of transversals and parallel lines to find measures of angles.

1. Identify each pair of complementary angles.

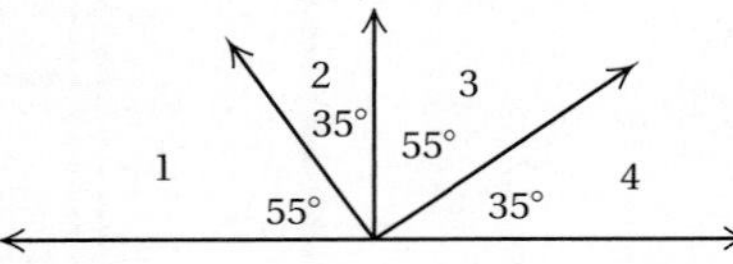

Find the measure of a complement of the angle.

2.

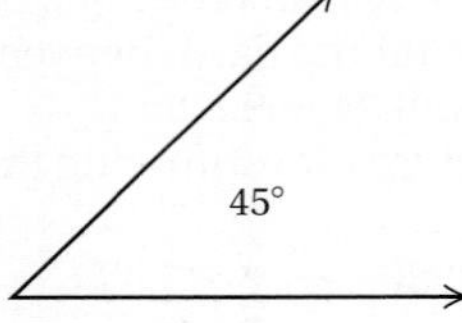

3.
18°

4.

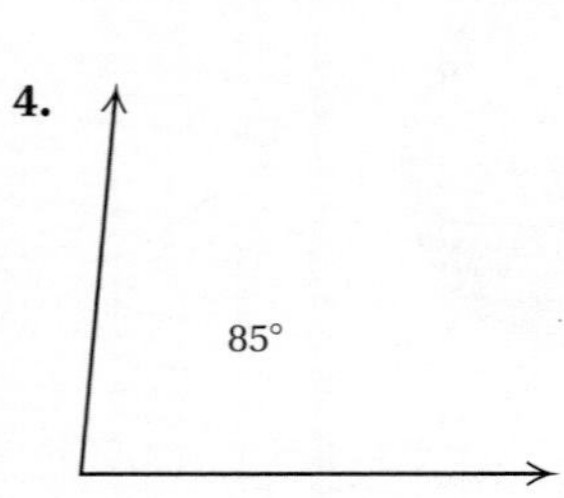

Answers on page A-16

a Complementary and Supplementary Angles

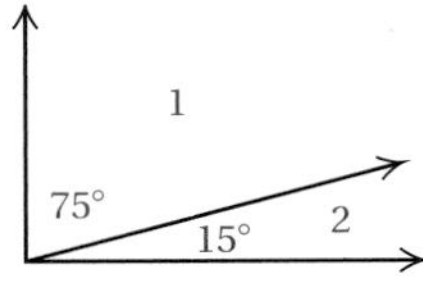

$\angle 1$ and $\angle 2$ above are **complementary** angles.

$$m\angle 1 + m\angle 2 = 90°$$
$$75° + 15° = 90°$$

COMPLEMENTARY ANGLES

Two angles are **complementary** if the sum of their measures is 90°. Each angle is called a **complement** of the other.

If two angles are complementary, each is an acute angle. When complementary angles are adjacent to each other, they form a right angle.

EXAMPLE 1 Identify each pair of complementary angles.

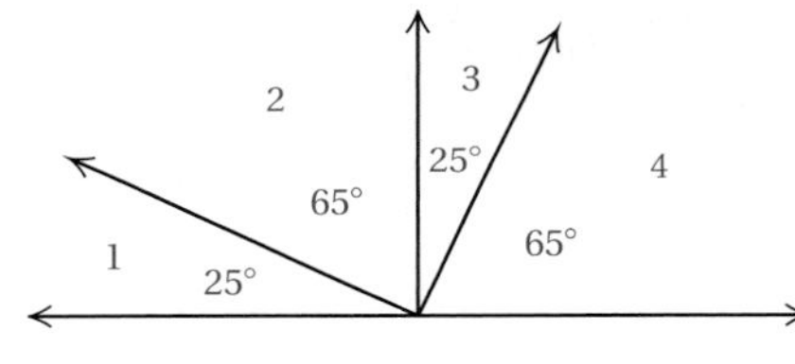

$\angle 1$ and $\angle 2$ $\quad 25° + 65° = 90° \quad$ $\angle 2$ and $\angle 3$

$\angle 1$ and $\angle 4$ $\quad\quad$ $\angle 3$ and $\angle 4$

Do Exercise 1.

EXAMPLE 2 Find the measure of a complement of an angle of 39°.

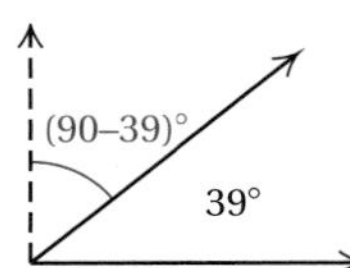

$$90° - 39° = 51°$$

The measure of a complement is 51°.

Do Exercises 2–4.

Next, consider $\angle 1$ and $\angle 2$ as shown below. Because the sum of their measures is 180°, $\angle 1$ and $\angle 2$ are said to be **supplementary.** Note that when supplementary angles are adjacent, they form a straight angle.

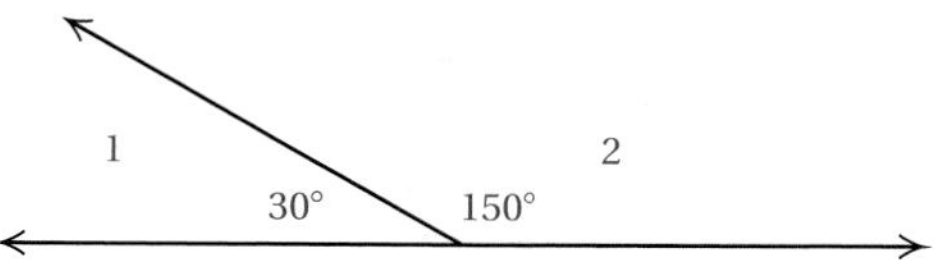

$$m\angle 1 + m\angle 2 = 180°$$
$$30° + 150° = 180°$$

5. Identify each pair of supplementary angles.

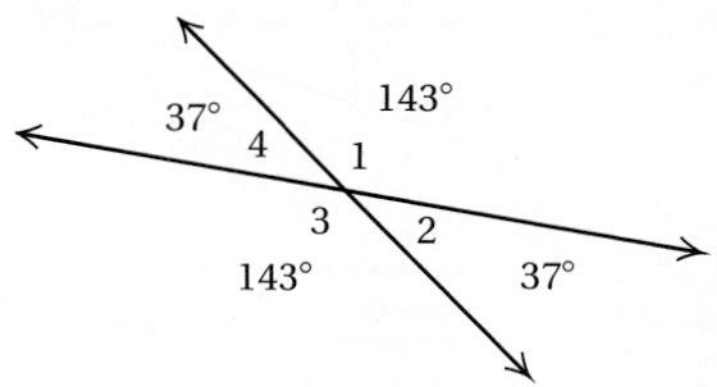

SUPPLEMENTARY ANGLES

Two angles are **supplementary** if the sum of their measures is 180°. Each angle is called a **supplement** of the other.

EXAMPLE 3 Identify each pair of supplementary angles.

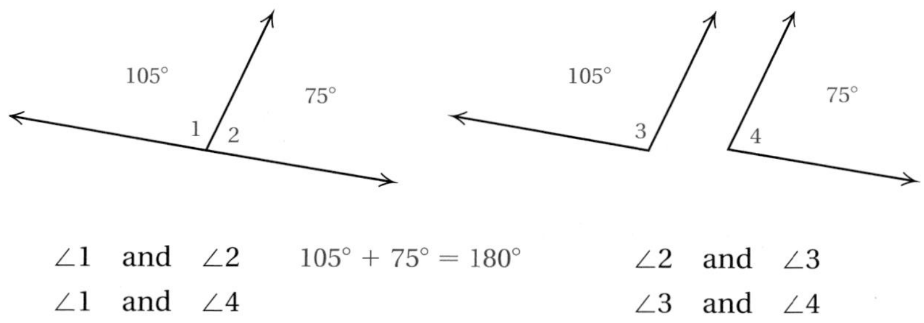

$\angle 1$ and $\angle 2$ $\quad 105° + 75° = 180°$ $\quad \angle 2$ and $\angle 3$

$\angle 1$ and $\angle 4$ $\quad \angle 3$ and $\angle 4$

Do Exercise 5.

EXAMPLE 4 Find the measure of a supplement of an angle of 112°.

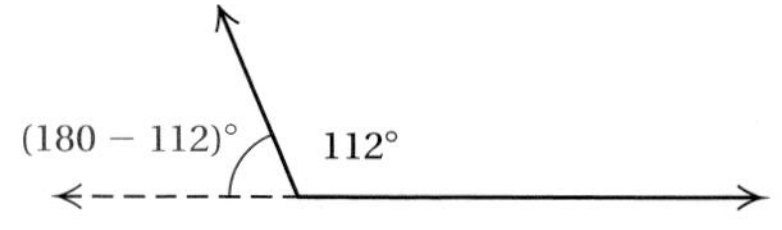

$$180° - 112° = 68°$$

The measure of a supplement is 68°.

Do Exercises 6–8.

Find the measure of a supplement of an angle with the given measure.

6. 38°

7. 157°

8. 90°

Answers on page A-16

Determine whether the given segments are congruent. Use a ruler.

9.

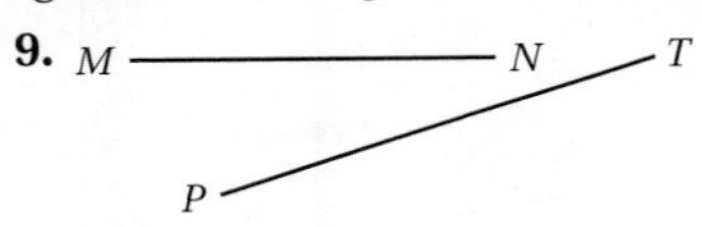

10. F G H K

Answers on page A-16

b Congruent Segments and Angles

Congruent figures have the same size and shape. They fit together exactly.

CONGRUENT SEGMENTS

Two segments are **congruent** if and only if they have the same length.

EXAMPLE 5 Use a ruler to show that $\overline{PQ}$ and $\overline{RS}$ are congruent.

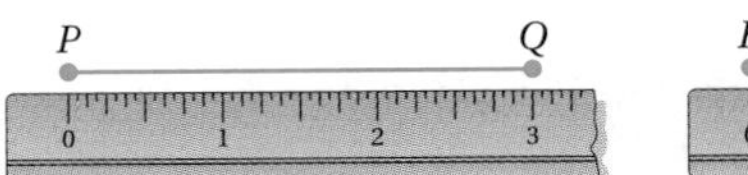

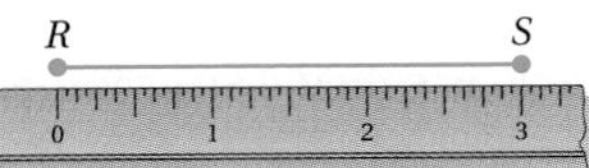

Since both segments have the same length, $\overline{PQ}$ and $\overline{RS}$ are congruent. To say that $\overline{PQ}$ and $\overline{RS}$ are congruent, we write

$$\overline{PQ} \cong \overline{RS}.$$

EXAMPLE 6 Which pairs of segments are congruent? Use a ruler.

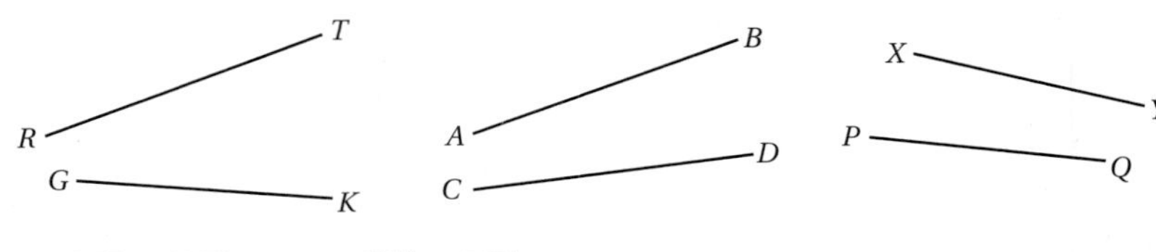

$\overline{AB} \cong \overline{CD}$ and $\overline{PQ} \cong \overline{XY}$.

Do Exercises 9 and 10.

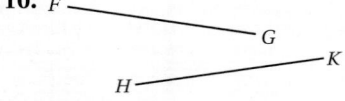

CONGRUENT ANGLES

Two angles are **congruent** if and only if they have the same measure.

EXAMPLE 7 Use a protractor to show that $\angle P$ and $\angle Q$ are congruent.

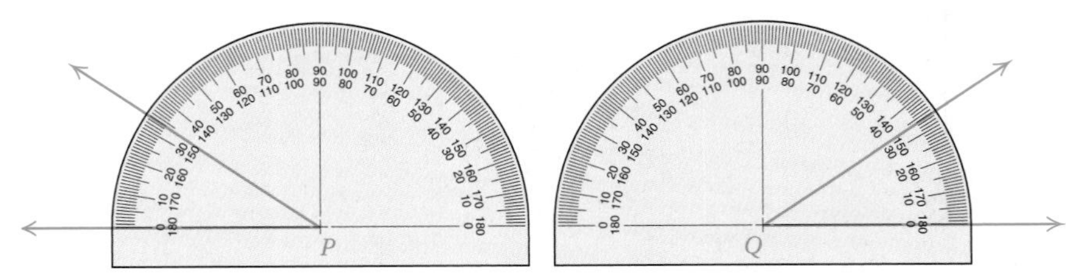

Since $m\angle P = m\angle Q = 34°$, $\angle P$ and $\angle Q$ are congruent. To say that $\angle P$ and $\angle Q$ are congruent, we write

$$\angle P \cong \angle Q.$$

EXAMPLE 8 Which pairs of angles are congruent? Use a protractor.

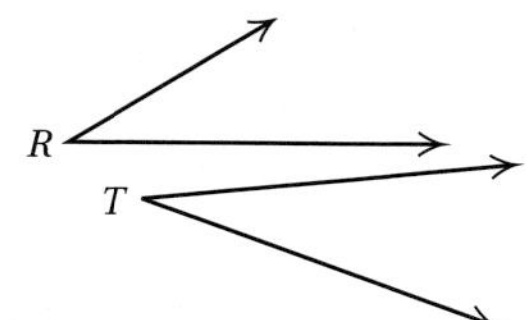

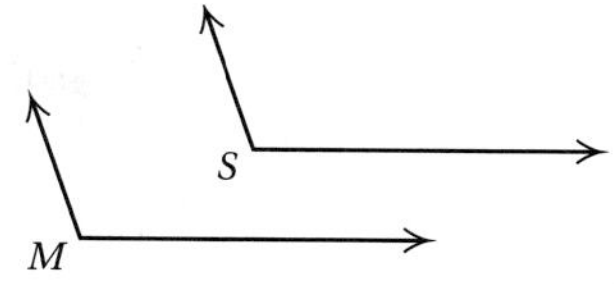

$\angle M \cong \angle S$ since $m\angle M = m\angle S = 108°$.

Do Exercises 11 and 12.

If two angles are congruent, then their supplements are congruent and their complements are congruent.

C Vertical Angles

When $\overleftrightarrow{RT}$ intersects $\overleftrightarrow{SQ}$ at P, four angles are formed:

$\angle SPT$

$\angle RPQ$

$\angle SPR$

$\angle QPT$

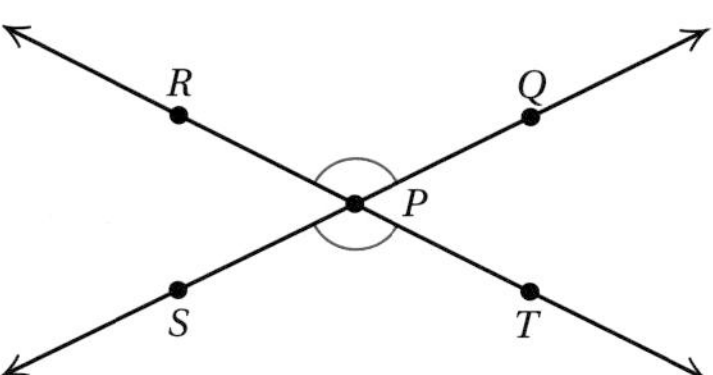

Pairs of angles such as $\angle RPQ$ and $\angle SPT$ are called **vertical angles.**

VERTICAL ANGLES

Two nonstraight angles are **vertical angles** if and only if their sides form two pairs of opposite rays.

Vertical angles are supplements of the same angle. Thus they are congruent.

THE VERTICAL ANGLE PROPERTY

Vertical angles are congruent.

Determine whether the pair of angles is congruent. Use a protractor.

11.

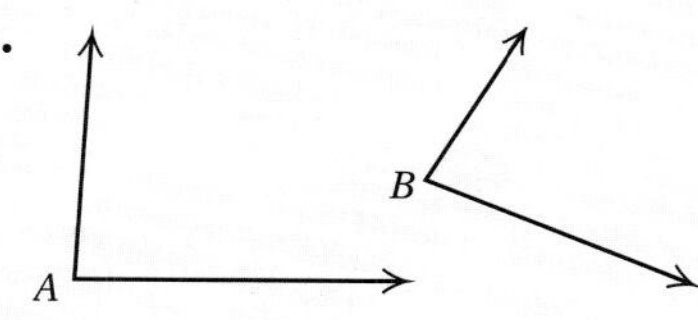

12.

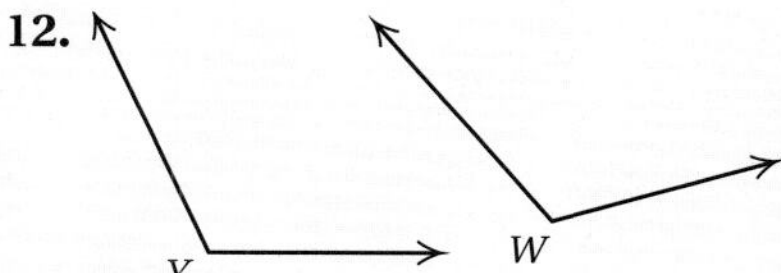

Answers on page A-16

13. In the figure below, $m\angle 2 = 41°$ and $m\angle 4 = 10°$. Find $m\angle 1$, $m\angle 3$, $m\angle 5$, and $m\angle 6$.

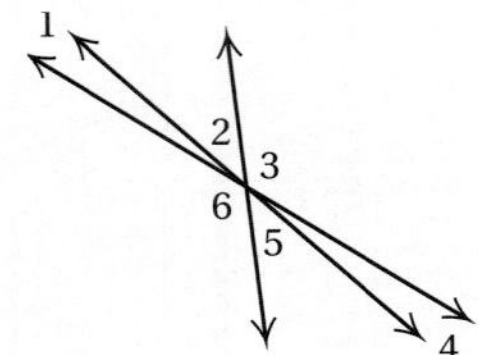

EXAMPLE 9 In the figure below, $m\angle 1 = 23°$ and $m\angle 3 = 34°$. Find $m\angle 2$, $m\angle 4$, $m\angle 5$, and $m\angle 6$.

Since $\angle 1$ and $\angle 4$ are vertical angles, $m\angle 4 = 23°$. Likewise, $\angle 3$ and $\angle 6$ are vertical angles, so $m\angle 6 = 34°$.

$$m\angle 1 + m\angle 2 + m\angle 3 = 180$$

$$23 + m\angle 2 + 34 = 180 \quad \text{Substituting}$$

$$m\angle 2 = 180 - 57$$

$$m\angle 2 = 123°$$

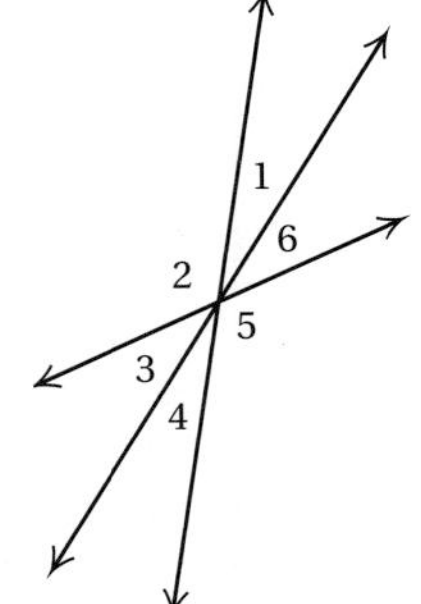

Since $\angle 2$ and $\angle 5$ are vertical angles, $m\angle 5 = 123°$.

Do Exercise 13.

d Transversals and Angles

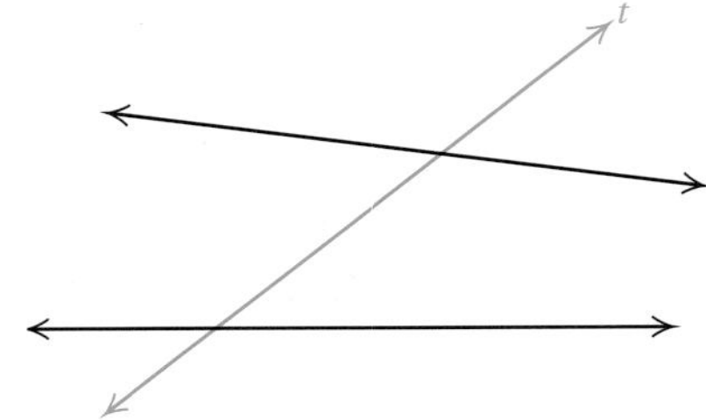

TRANSVERSAL

A **transversal** is a line that intersects two or more coplanar lines in different points.

When a transversal intersects a pair of lines, eight angles are formed. Certain pairs of these angles have special names.

CORRESPONDING ANGLES

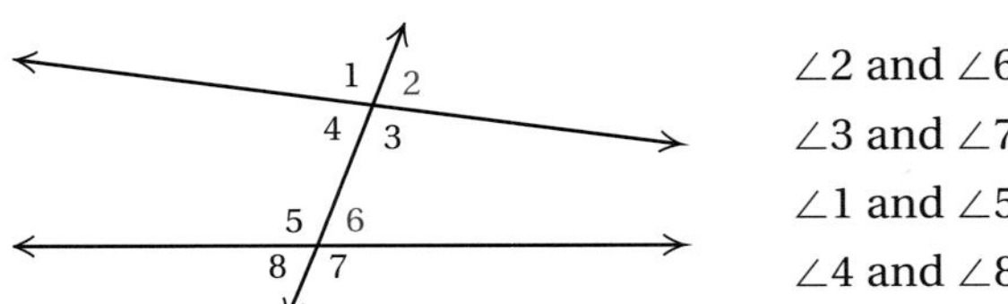

$\angle 2$ and $\angle 6$

$\angle 3$ and $\angle 7$

$\angle 1$ and $\angle 5$

$\angle 4$ and $\angle 8$

Answers on page A-16

INTERIOR ANGLES

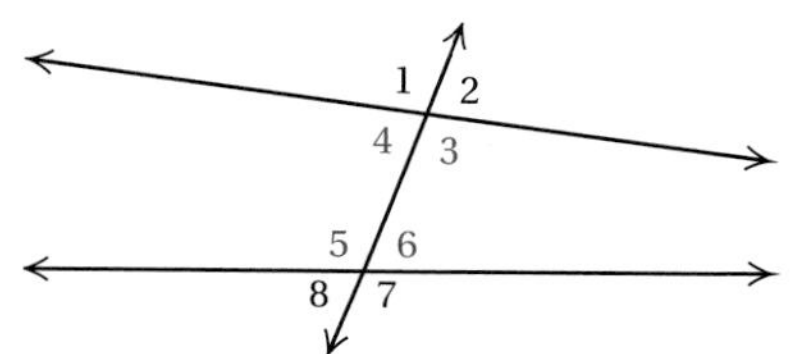

$\angle 3$, $\angle 4$, $\angle 5$, and $\angle 6$

ALTERNATE INTERIOR ANGLES

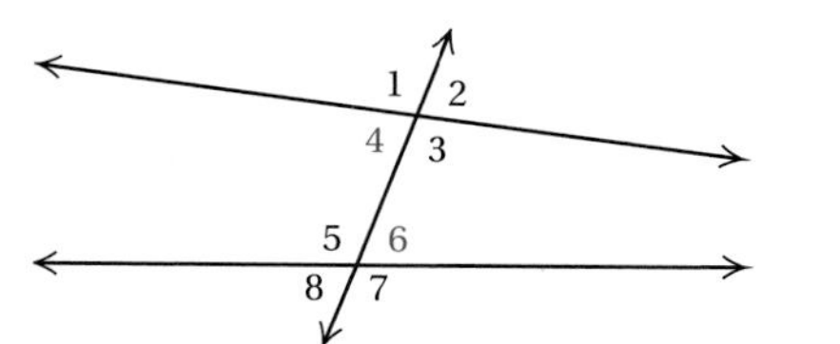

$\angle 4$ and $\angle 6$

$\angle 3$ and $\angle 5$

EXAMPLE 10 Identify all pairs of corresponding angles, all interior angles, and all pairs of alternate interior angles.

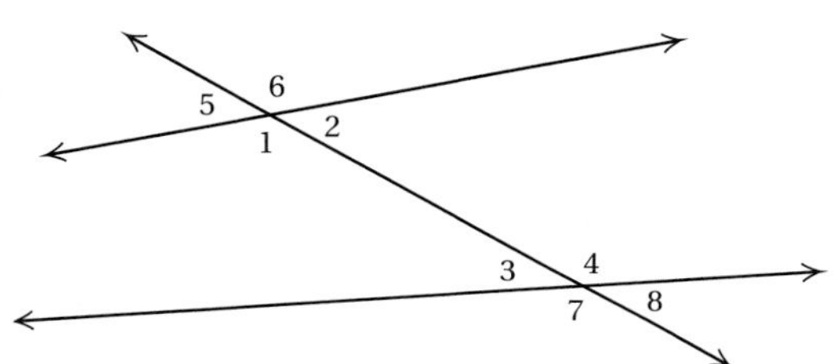

Corresponding angles:	$\angle 6$ and $\angle 4$, $\angle 2$ and $\angle 8$, $\angle 5$ and $\angle 3$, $\angle 1$ and $\angle 7$
Interior angles:	$\angle 1$, $\angle 2$, $\angle 3$, $\angle 4$
Alternate interior angles:	$\angle 1$ and $\angle 4$, $\angle 2$ and $\angle 3$

Do Exercises 14–16.

Given a line l and a point P not on l, there is at most one line that contains P and is parallel to l.

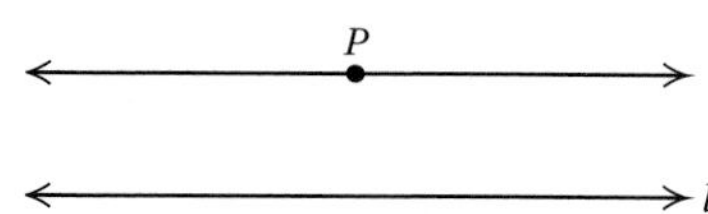

If two lines are parallel, the following relations hold.

Use the following figure to answer Margin Exercises 14–16.

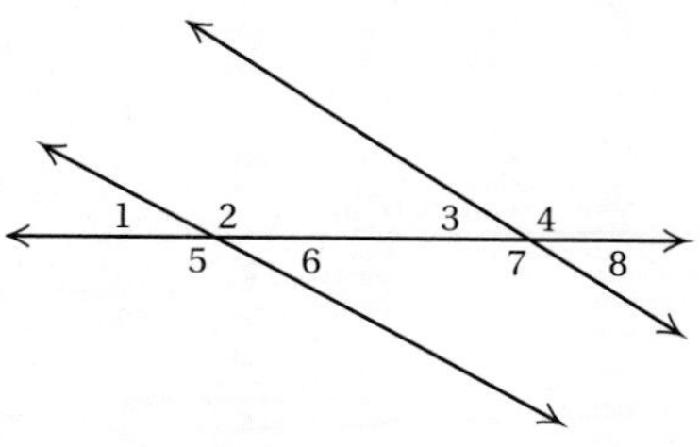

14. Identify all pairs of corresponding angles.

15. Identify all interior angles.

16. Identify all pairs of alternate interior angles.

Answers on page A-16

17. If $l \parallel m$ and $m\angle 3 = 51°$, what are the measures of the other angles?

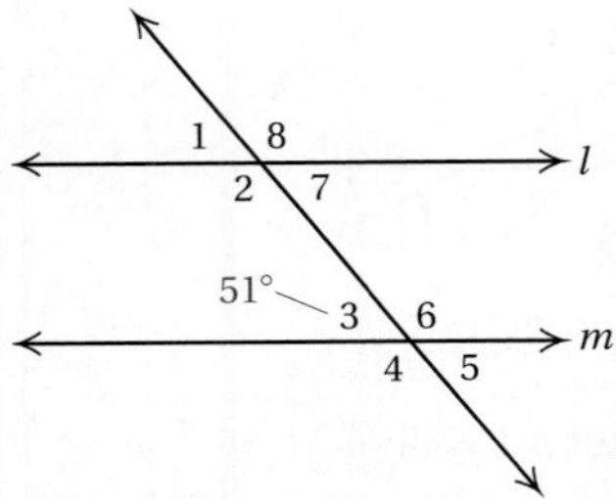

Answers on page A-16

PROPERTIES OF PARALLEL LINES

1. If a transversal intersects two parallel lines, then the corresponding angles are congruent.

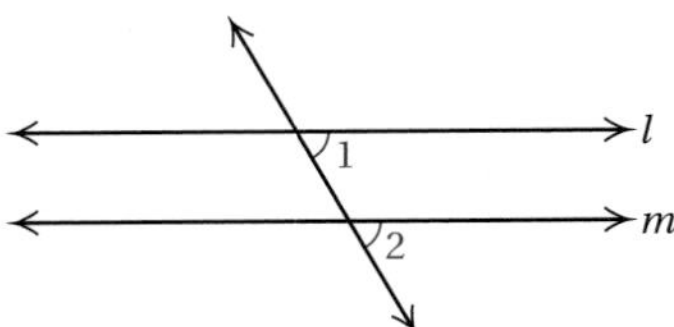

If $l \parallel m$, then $\angle 1 \cong \angle 2$.

2. If a transversal intersects two parallel lines, then the alternate interior angles are congruent.

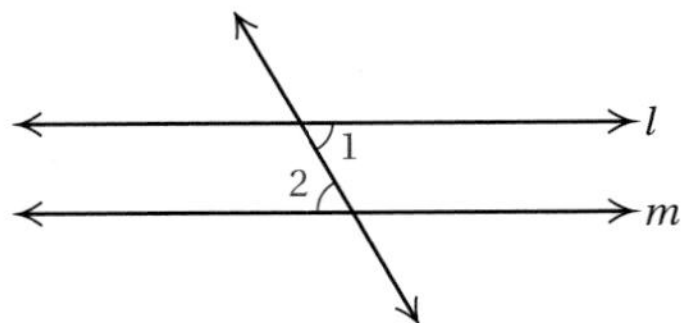

If $l \parallel m$, then $\angle 1 \cong \angle 2$.

3. In a plane, if two lines are parallel to a third line, then the two lines are parallel to each other.

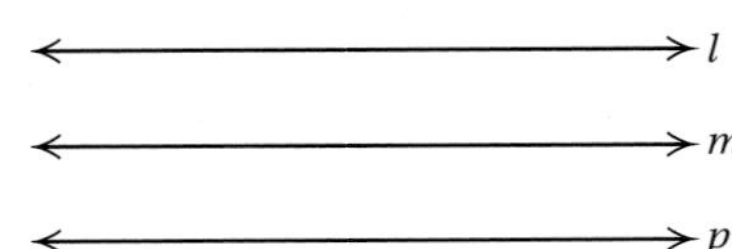

If $l \parallel p$ and $m \parallel p$, then $l \parallel m$.

4. If a transversal intersects two parallel lines, then the interior angles on the same side of the transversal are supplementary.

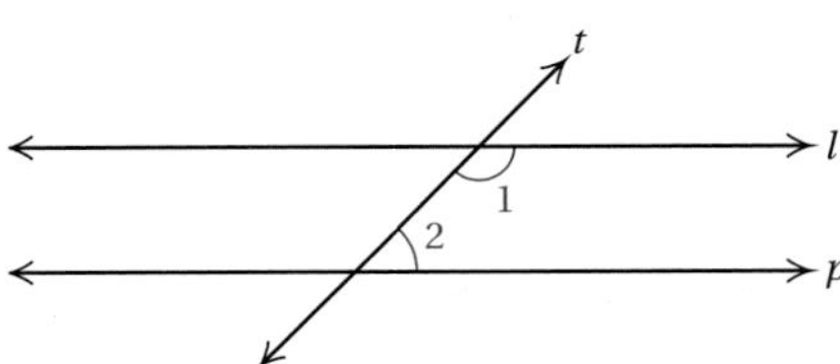

If $l \parallel p$, then $m\angle 1 + m\angle 2 = 180°$.

5. If a transversal is perpendicular to one of two parallel lines, then it is perpendicular to the other.

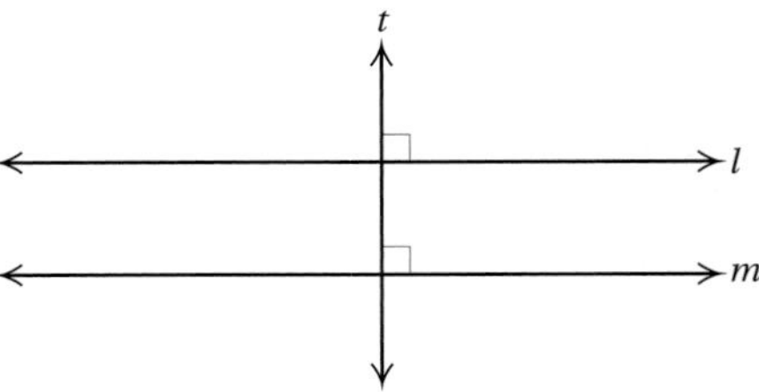

EXAMPLE 11 If $l \parallel m$ and $m\angle 1 = 40°$, what are the measures of the other angles?

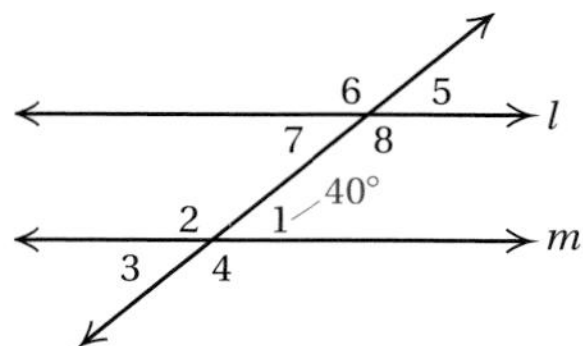

$m\angle 7 = 40°$	Using Property 2
$m\angle 5 = 40°$	Using Property 1
$m\angle 8 = 140°$	Using Property 4
$m\angle 3 = 40°$	$\angle 1$ and $\angle 3$ are vertical angles
$m\angle 4 = 140°$	Using Property 1 and $m\angle 8 = 140°$
$m\angle 2 = 140°$	$\angle 2$ and $\angle 4$ are vertical angles and $m\angle 4 = 140°$
$m\angle 6 = 140°$	$\angle 6$ and $\angle 8$ are vertical angles and $m\angle 8 = 140°$

Do Exercise 17 on the preceding page.

EXAMPLE 12 If $\overline{PT} \parallel \overline{SR}$, which pairs of angles are congruent?

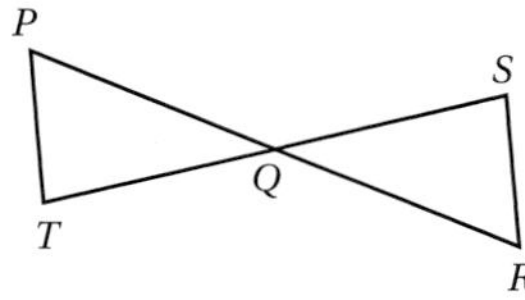

$\angle TPQ \cong \angle SRQ$ and $\angle PTQ \cong \angle RSQ$ Using Property 2

$\angle PQT \cong \angle RQS$ and $\angle PQS \cong \angle RQT$ Vertical angles

Do Exercise 18.

EXAMPLE 13 If $\overline{DE} \parallel \overline{BC}$, which pairs of angles are congruent?

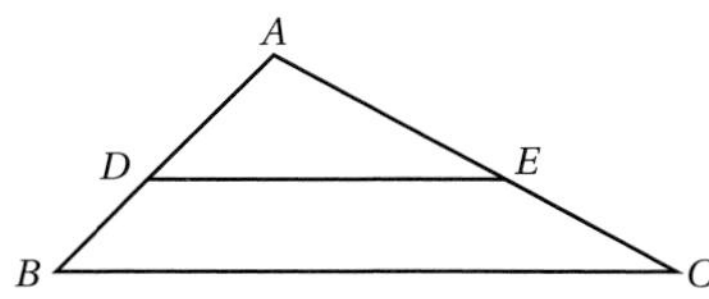

$\angle ADE \cong \angle ABC$ and $\angle AED \cong \angle ACB$ Using Property 1

Do Exercise 19.

18. If $\overline{AB} \parallel \overline{CD}$, which pairs of angles are congruent?

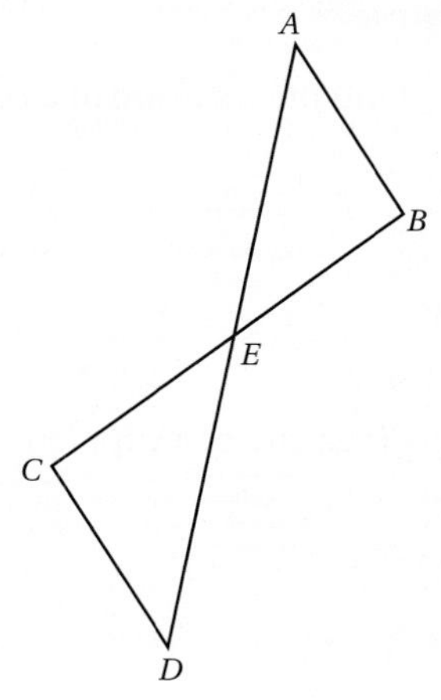

19. If $\overline{PQ} \parallel \overline{RS}$, which pairs of angles are congruent?

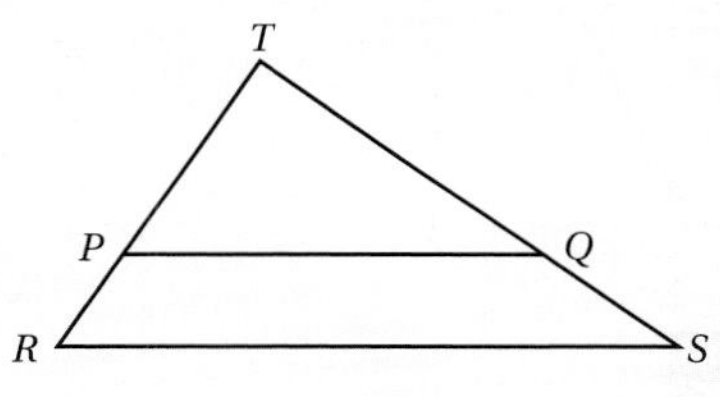

Answers on page A-16

6.6 EXERCISE SET

For Extra Help

MathXL | MyMathLab | InterAct Math | Math Tutor Center | Video Lectures on CD Disc 3 | Student's Solutions Manual

a Find the measure of a complement of an angle with the given measure.

1. 11°

2. 83°

3. 67°

4. 5°

5. 58°

6. 32°

7. 29°

8. 54°

Find the measure of a supplement of an angle with the given measure.

9. 3°

10. 54°

11. 139°

12. 13°

13. 85°

14. 129°

15. 102°

16. 45°

b Determine whether the pair of segments is congruent. Use a ruler.

17.

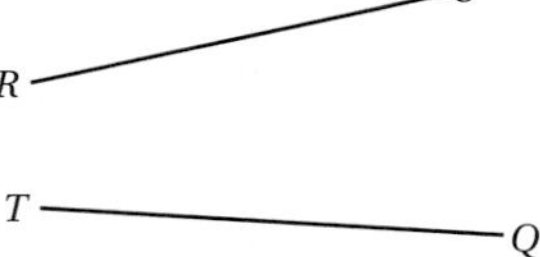

18.

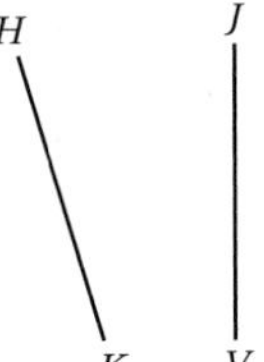

Determine whether the pair of angles is congruent. Use a protractor.

19.

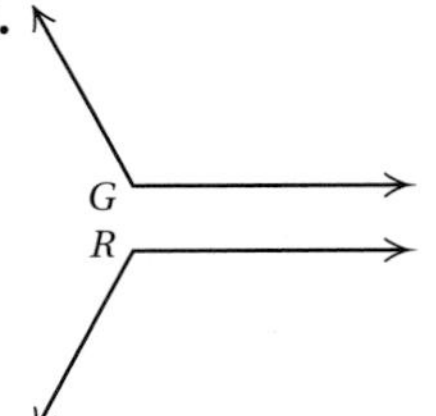

20.

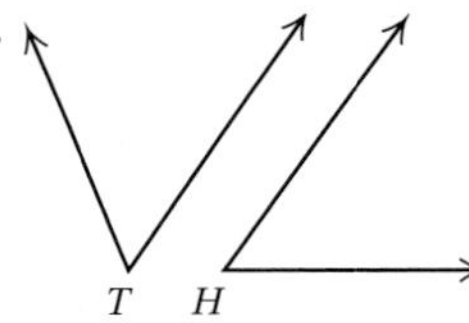

c

21. In the figure, $m\angle 1 = 80°$ and $m\angle 5 = 67°$. Find $m\angle 2$, $m\angle 3$, $m\angle 4$, and $m\angle 6$.

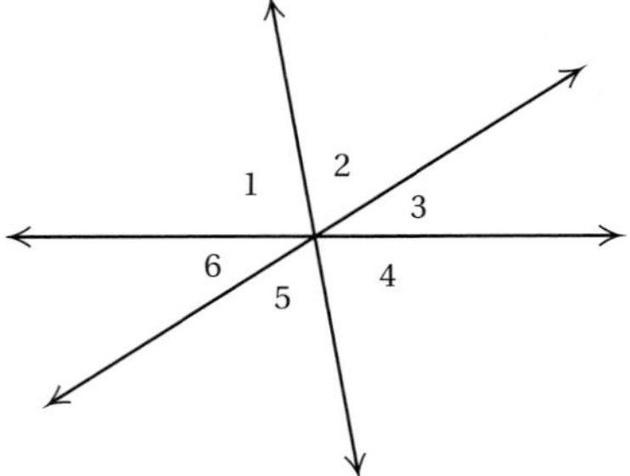

22. In the figure, $m\angle 2 = 42°$ and $m\angle 4 = 56°$. Find $m\angle 1$, $m\angle 3$, $m\angle 5$, and $m\angle 6$.

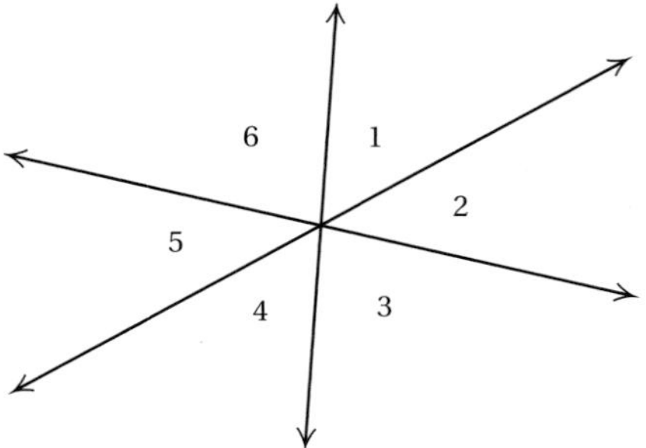

d In Exercises 23 and 24, **(a)** identify all pairs of corresponding angles, **(b)** identify all interior angles, and **(c)** identify all pairs of alternate interior angles.

23.

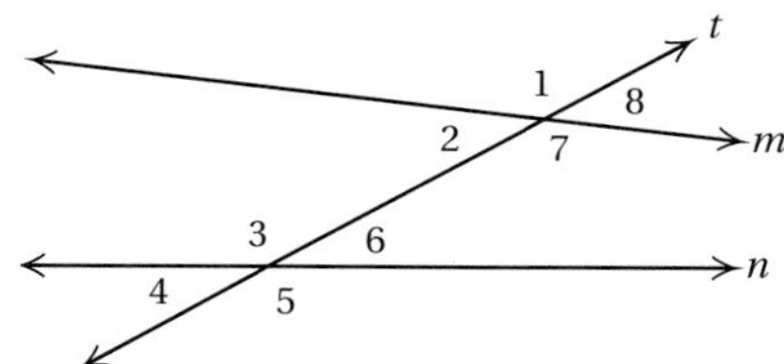

Lines m and n
Transversal t

24.

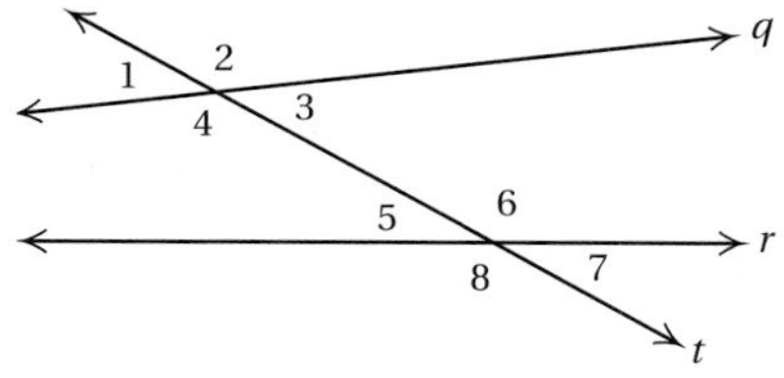

Lines q and r
Transversal t

25. If $m \parallel n$ and $m\angle 4 = 125°$, what are the measures of the other angles?

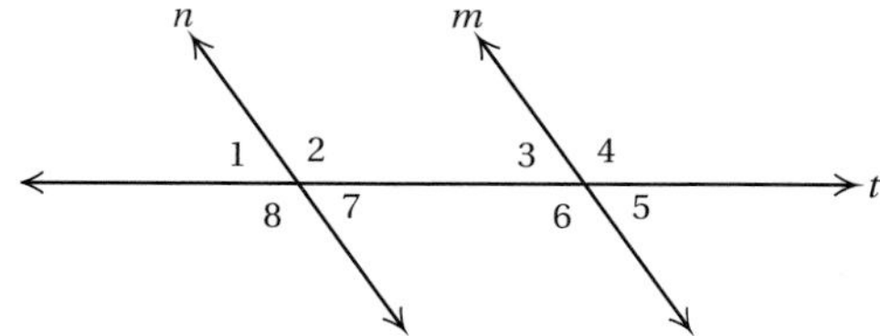

26. If $m \parallel n$ and $m\angle 8 = 34°$, what are the measures of the other angles?

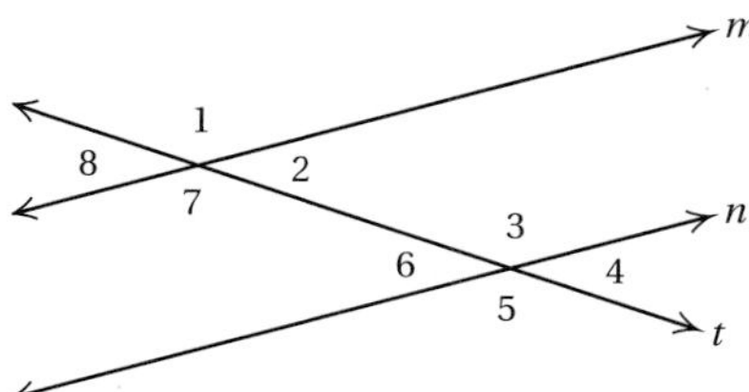

In each figure, $\overline{AB} \parallel \overline{CD}$. Identify pairs of congruent angles. Where possible, give the measures of the angles.

27.

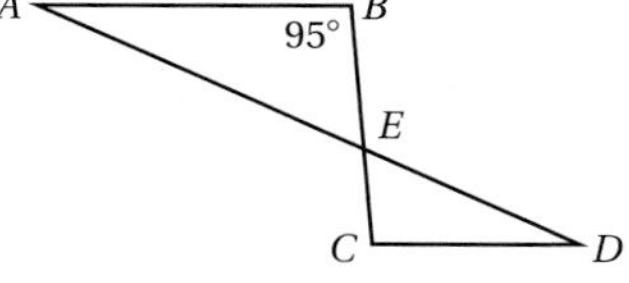

28.

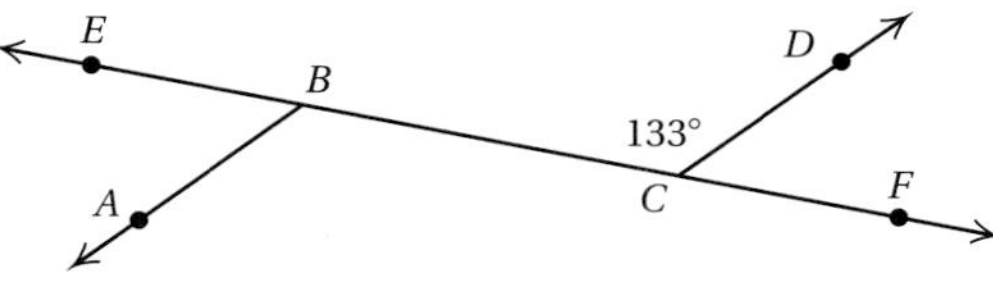

29.

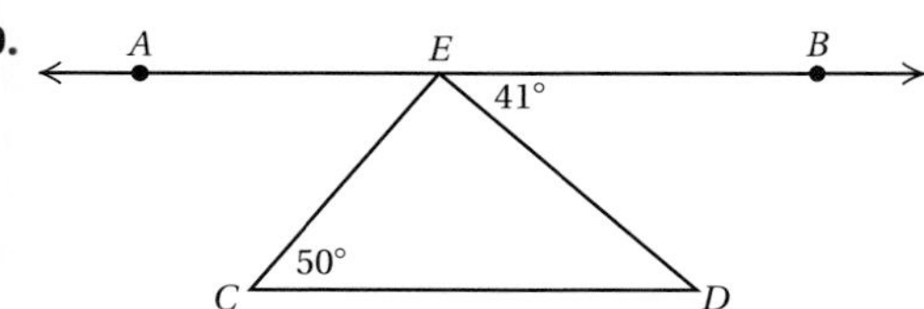

30.

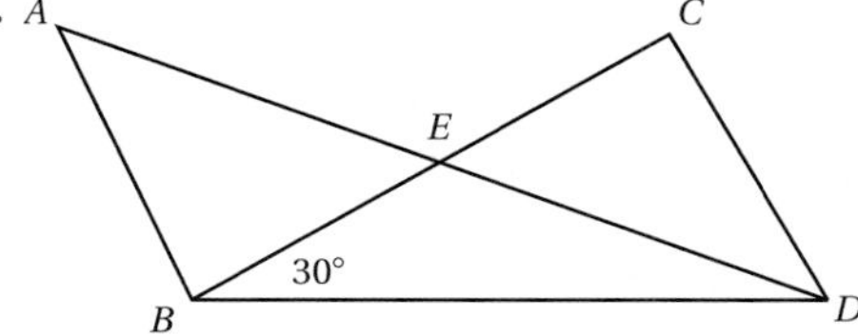

SKILL MAINTENANCE

Multiply. [2.4d]

31. $6 \times 1\frac{7}{8}$

32. $10\frac{3}{4} \times 1\frac{1}{2}$

33. $8\frac{3}{7} \times 14$

34. $2\frac{2}{3} \times 5\frac{1}{2}$

6.7 CONGRUENT TRIANGLES AND PROPERTIES OF PARALLELOGRAMS

Objectives

a Identify the corresponding parts of congruent triangles and show why triangles are congruent using SAS, SSS, and ASA.

b Use properties of parallelograms to find lengths of sides and measures of angles of parallelograms.

a Congruent Triangles

Triangles can be classified by their angles.

Acute: All angles acute
Right: One right angle
Obtuse: One obtuse angle
Equiangular: All angles congruent

Triangles can also be classified by their sides.

Equilateral: All sides congruent
Isosceles: At least two sides congruent
Scalene: No sides congruent

We know that congruent figures fit together exactly.

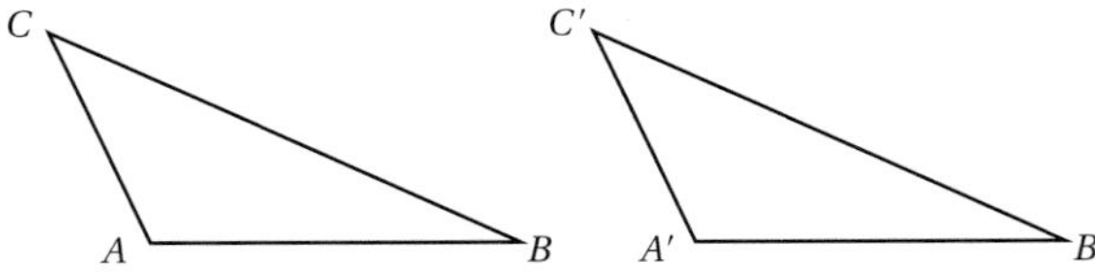

B' is read "B prime."

These triangles will fit together exactly if we match A with A', B with B', and C with C'. On the other hand, if we match A with B', B with C', and C with A', the triangles will not fit together exactly. The matching of vertices determines corresponding sides and angles.

EXAMPLES Consider $\triangle ABC$ and $\triangle A'B'C'$ above.

1. If we match A with A', B with B', and C with C', what are the corresponding sides?

$\overline{AB} \leftrightarrow \overline{A'B'}$
$\overline{BC} \leftrightarrow \overline{B'C'}$
$\overline{AC} \leftrightarrow \overline{A'C'}$

$\leftrightarrow$ means "corresponds to."

2. If we match A with B', B with C', and C with A', what are the corresponding angles?

$\angle A \leftrightarrow \angle B'$ $\angle B \leftrightarrow \angle C'$ $\angle C \leftrightarrow \angle A'$

If $A \leftrightarrow A'$, $B \leftrightarrow B'$, and $C \leftrightarrow C'$, then we write $ABC \leftrightarrow A'B'C'$.

CONGRUENT TRIANGLES

Two triangles are **congruent** if and only if their vertices can be matched so that the corresponding angles and sides are congruent.

The corresponding sides and angles of two congruent triangles are called *corresponding parts* of congruent triangles. Corresponding parts of congruent triangles are always congruent.

We write $\triangle ABC \cong \triangle A'B'C'$ to say that $\triangle ABC$ and $\triangle A'B'C'$ are congruent. We agree that this symbol also tells us the way in which the vertices are matched.

$$\triangle ABC \cong \triangle A'B'C'$$

$\triangle ABC \cong \triangle A'B'C'$ means that

$\angle A \cong \angle A'$ and $\overline{AB} \cong \overline{A'B'}$

$\angle B \cong \angle B'$ $\quad \overline{AC} \cong \overline{A'C'}$

$\angle C \cong \angle C'$ $\quad \overline{BC} \cong \overline{B'C'}$.

1. Suppose that $\triangle ABC \cong \triangle DEF$. What are the congruent corresponding parts?

EXAMPLE 3 Suppose that $\triangle PQR \cong \triangle STV$. What are the congruent corresponding parts?

Angles	*Sides*
$\angle P \cong \angle S$	$\overline{PQ} \cong \overline{ST}$
$\angle Q \cong \angle T$	$\overline{PR} \cong \overline{SV}$
$\angle R \cong \angle V$	$\overline{QR} \cong \overline{TV}$

Do Exercise 1.

EXAMPLE 4 Name the corresponding parts of these congruent triangles.

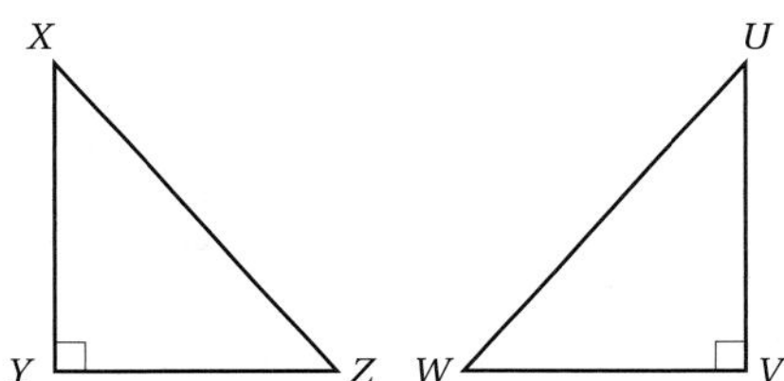

Angles	*Sides*
$\angle X \cong \angle U$	$\overline{XY} \cong \overline{UV}$
$\angle Y \cong \angle V$	$\overline{YZ} \cong \overline{VW}$
$\angle Z \cong \angle W$	$\overline{ZX} \cong \overline{WU}$

Do Exercise 2.

2. Name the corresponding parts of these congruent triangles.

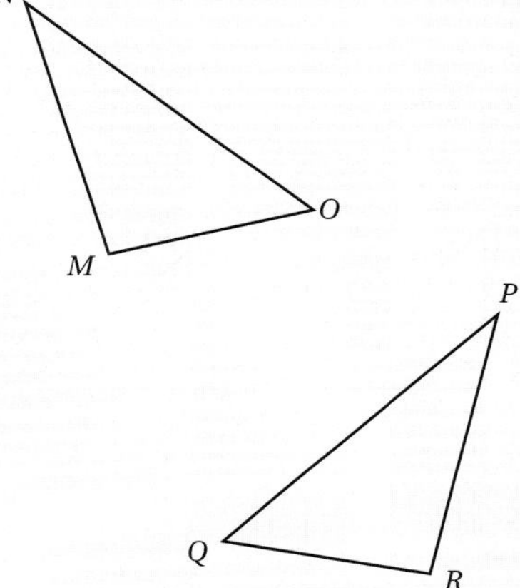

Sometimes we can show that triangles are congruent without already knowing that all six corresponding parts are congruent.

On a full sheet of paper, draw $\triangle ABC$. On another sheet of paper, make a copy of $\angle A$. Label the copy $\angle D$. On the sides of $\angle D$, copy $\overline{AB}$ and $\overline{AC}$. Label the copy $\overline{DE}$ and $\overline{DF}$. Draw $\overline{EF}$. Cut out $\triangle DEF$ and $\triangle ABC$ and place them together. What do you conclude?

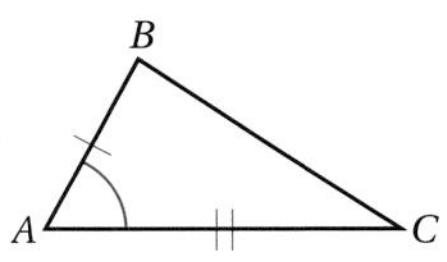

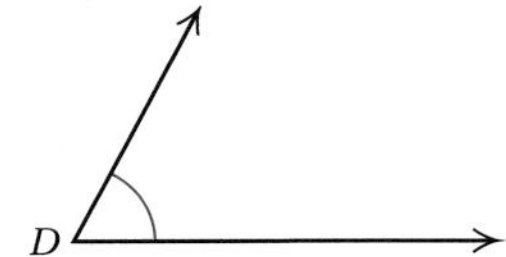

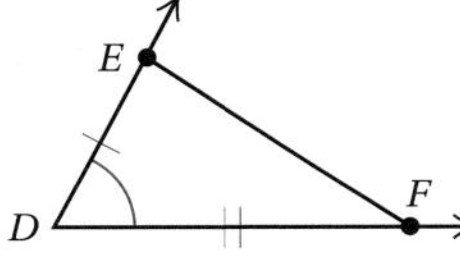

Answers on page A-16

3. Which pairs of triangles are congruent by the SAS property?

a)

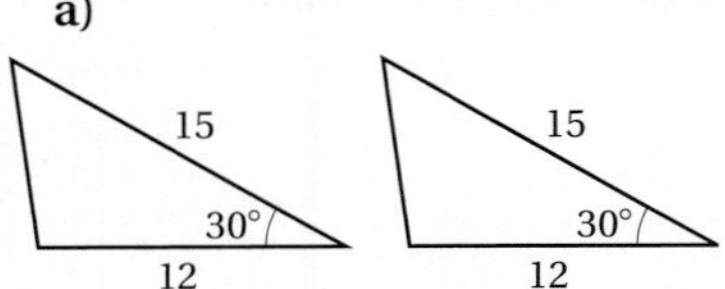

b)

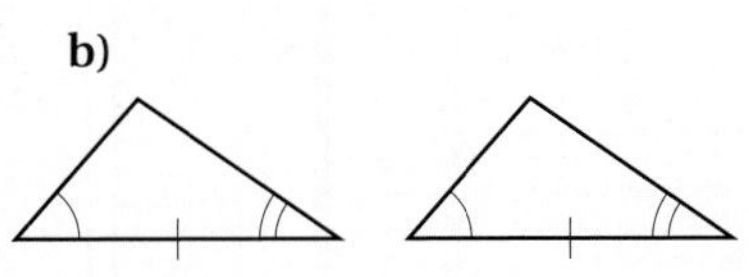

c)

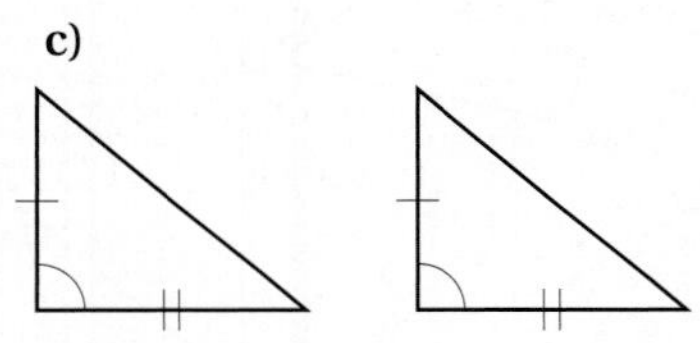

d)

Answers on page A-16

THE SIDE–ANGLE–SIDE (SAS) PROPERTY

Two triangles are congruent if two sides and the included angle of one triangle are congruent to two sides and the included angle of the other triangle.

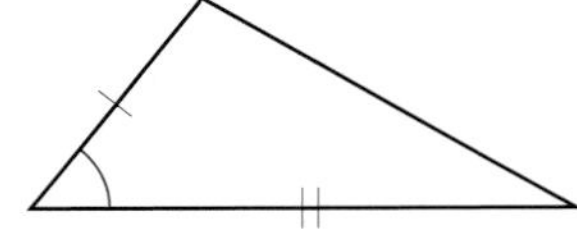

EXAMPLE 5 Which pairs of triangles are congruent by the SAS property?

a)

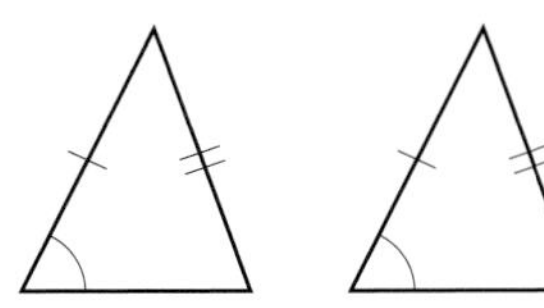

b)

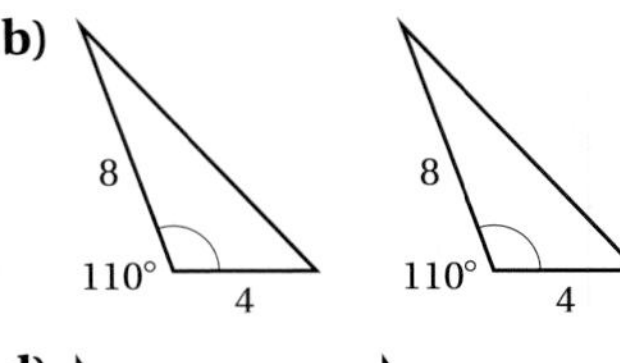

c)

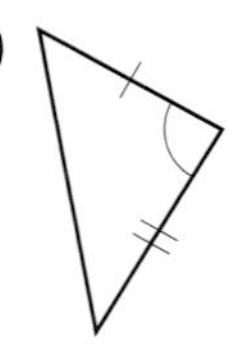

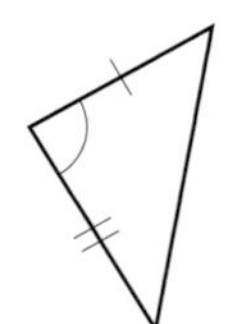

d)

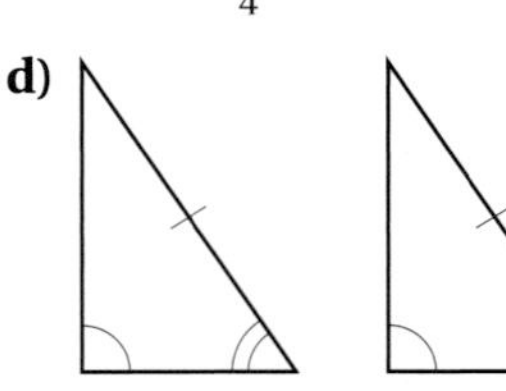

Pairs (b) and (c) are congruent by the SAS property.

Do Exercise 3.

On a sheet of paper, draw a triangle. Then copy this triangle by copying each of its sides. Cut both triangles out and place them together. This suggests the following property.

THE SIDE–SIDE–SIDE (SSS) PROPERTY

If three sides of one triangle are congruent to three sides of another triangle, then the triangles are congruent.

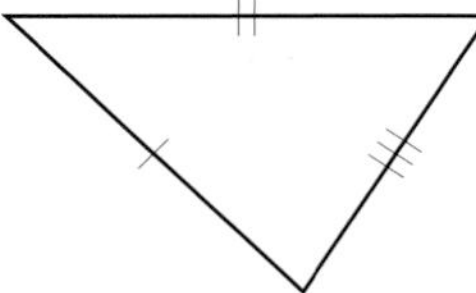

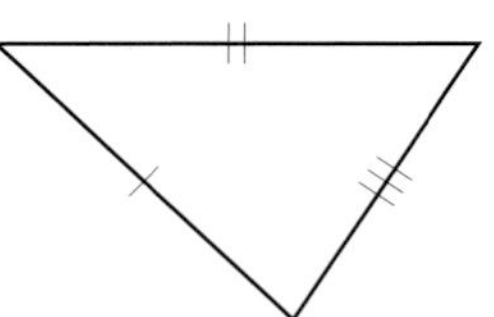

EXAMPLE 6 Which pairs of triangles are congruent by the SSS property?

a)

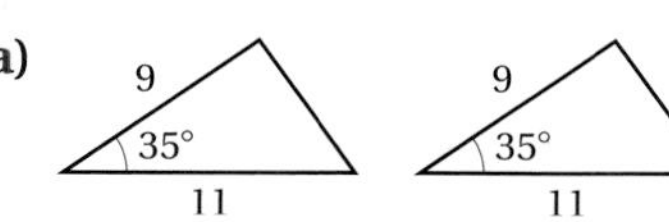

b)

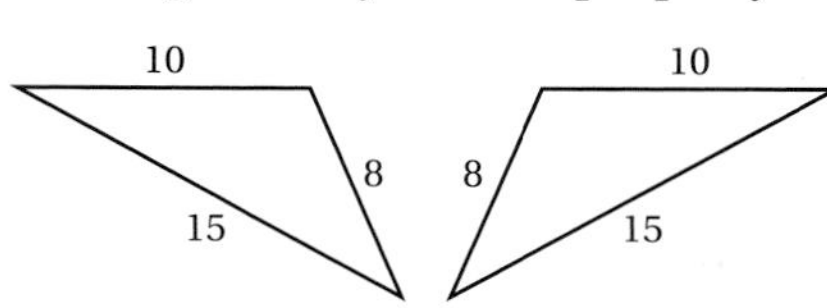

c)

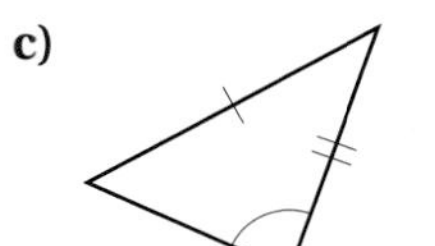

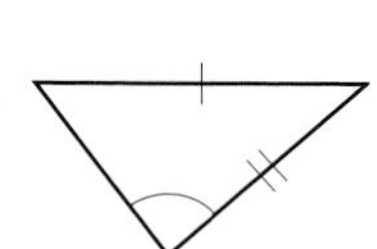

d)

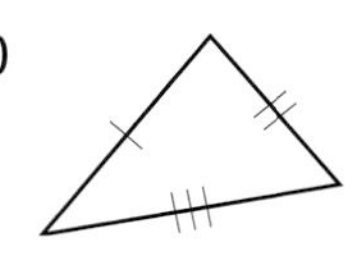

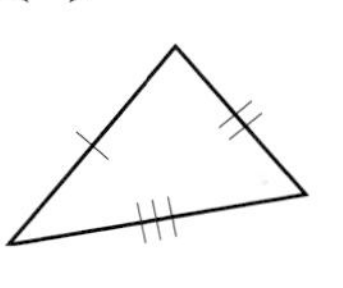

Pairs (b) and (d) are congruent by the SSS property.

Do Exercise 4.

We have shown triangles to be congruent using SAS and SSS. A third way to show congruence is shown below.

On a full sheet of paper, draw a triangle, $\triangle ABC$. On another sheet of paper, draw a segment $\overline{DE}$ so that $DE = AB$*. At D, make a copy of $\angle A$. At E, make a copy of $\angle B$. Label the third vertex of the copy F. Cut out $\triangle ABC$ and $\triangle DEF$ and place them together. What do you conclude?

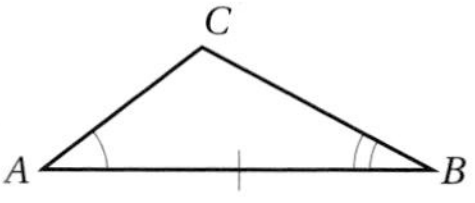

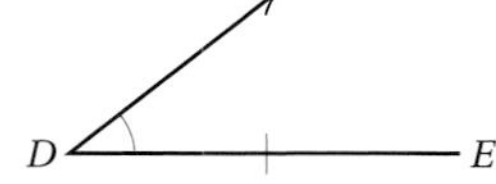

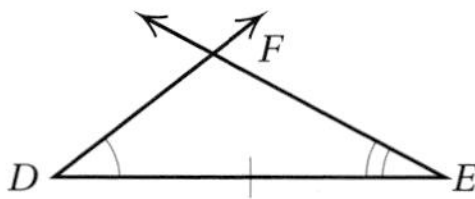

THE ANGLE–SIDE–ANGLE (ASA) PROPERTY

If two angles and the included side of a triangle are congruent to two angles and the included side of another triangle, then the triangles are congruent.

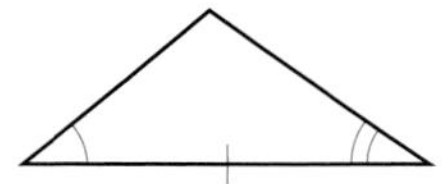

EXAMPLE 7 Which pairs of triangles are congruent by the ASA property?

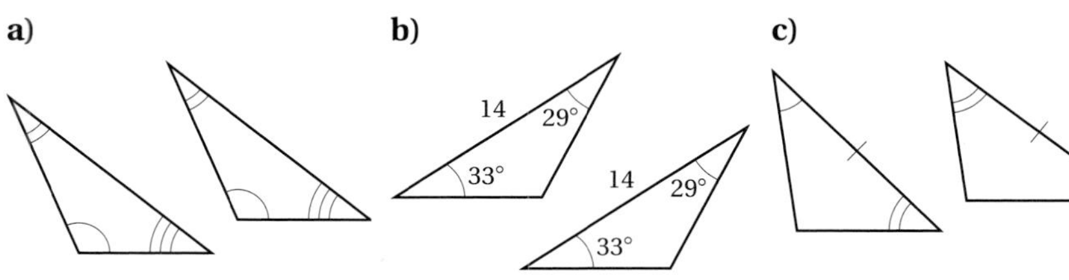

Pairs (b) and (c) are congruent by the ASA property.

Do Exercise 5.

4. Which pairs of triangles are congruent by the SSS property?

a)

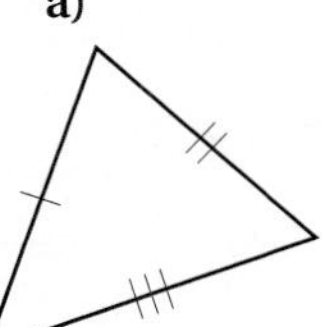

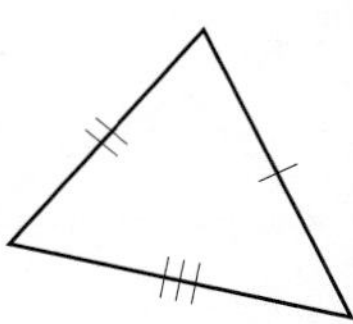

b)

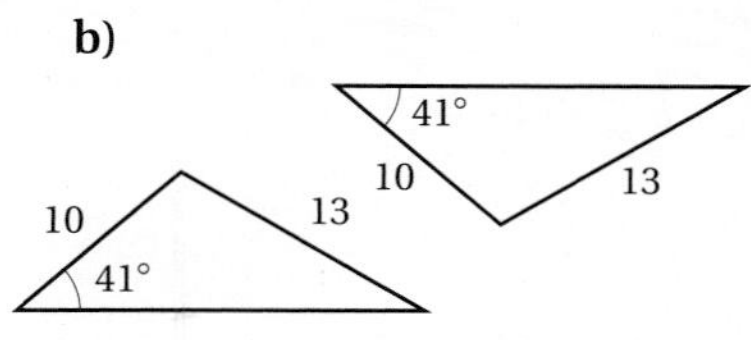

5. Which pairs of triangles are congruent by the ASA property?

a)

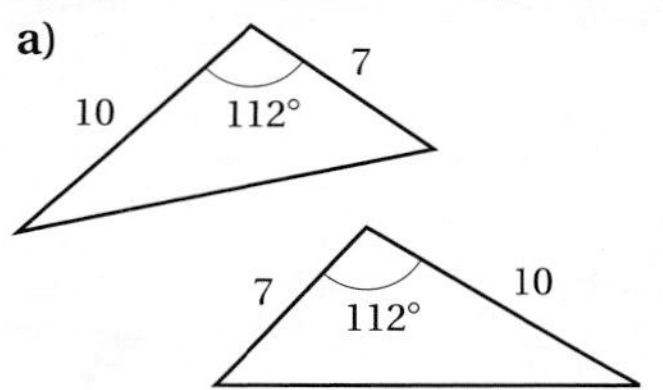

b)

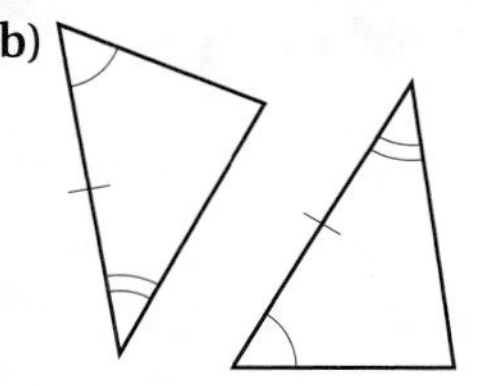

c)

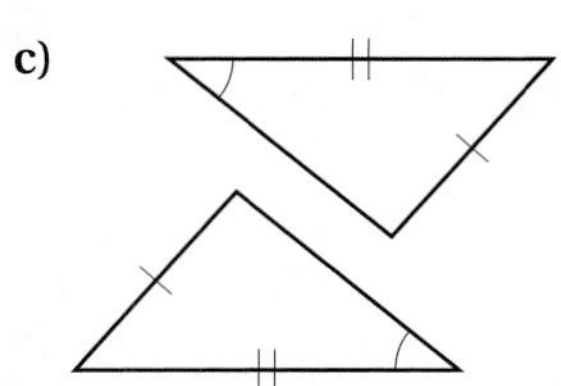

---

*$\overline{DE}$ denotes the segment with endpoints D and E. DE denotes the length of $\overline{DE}$.

Answers on page A-16

Which property (if any) should be used to show that the pair of triangles is congruent?

6.

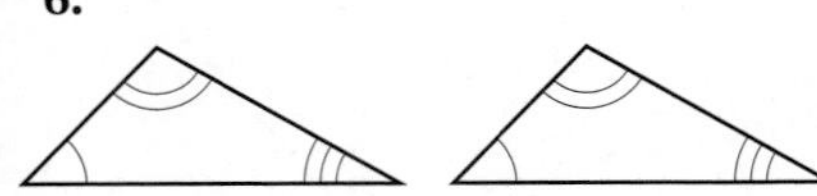

7.

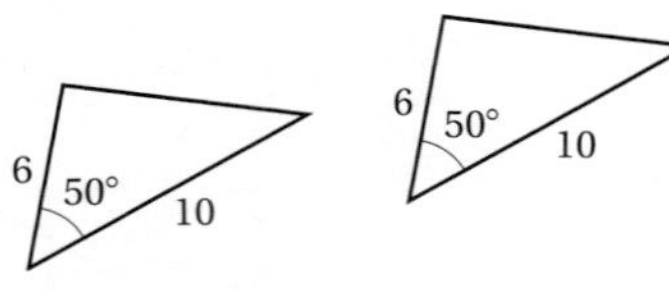

8.

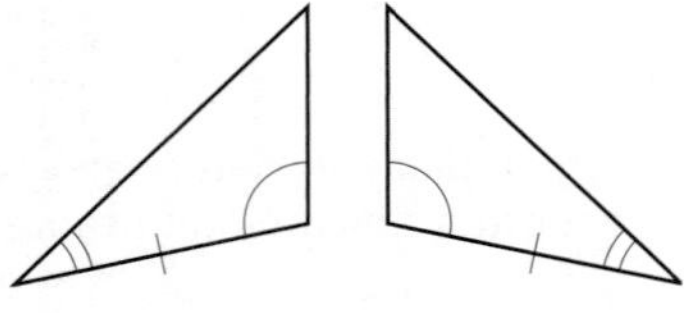

9.

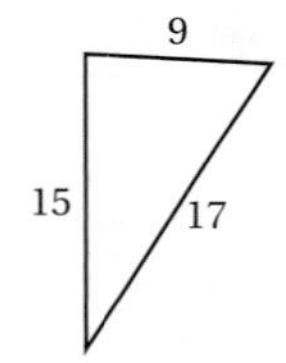

10. In this figure, $\overline{AB} \perp \overline{ED}$ and B is the midpoint of $\overline{ED}$. Explain why $\triangle ABD \cong \triangle ABE$.

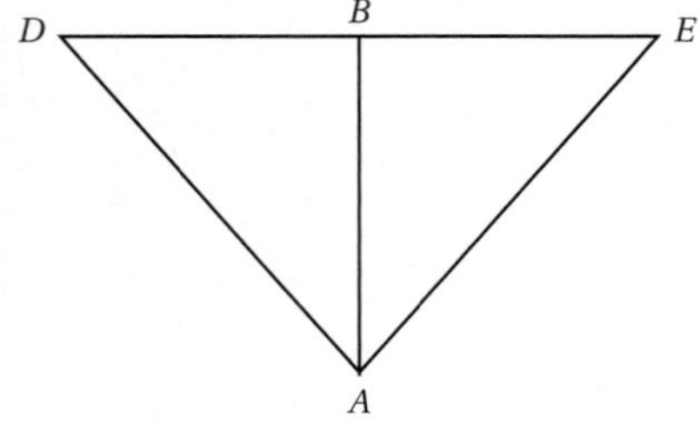

Answers on page A-16

EXAMPLES Which property (if any) should be used to show that these pairs of triangles are congruent?

8.

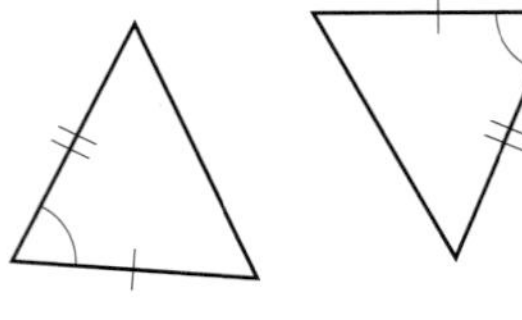

Use SAS.

9.

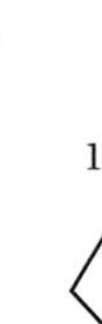
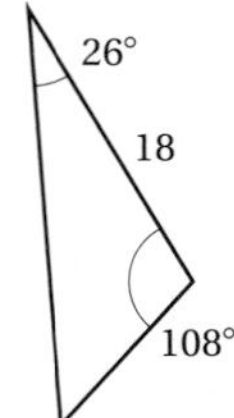

Use ASA.

10.

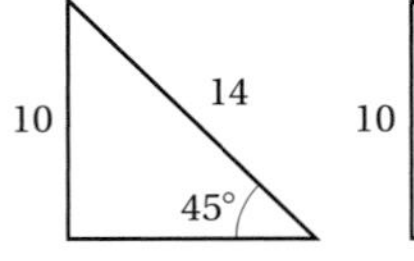

None.

11.

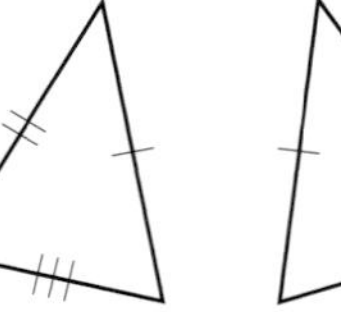

Use SSS.

Do Exercises 6–9.

It is important to be able to explain why triangles are congruent.

EXAMPLE 12 In $\triangle ABC$ and $\triangle DEF$, $\overline{AB} \cong \overline{DE}$, $\overline{AC} \cong \overline{DF}$, and $\angle A \cong \angle D$. Explain why the triangles are congruent.

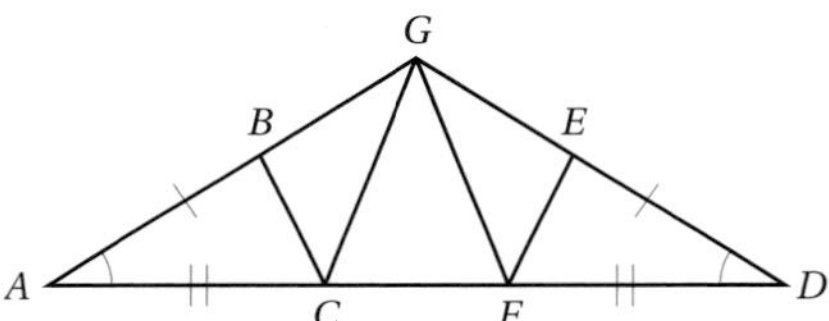

We have two sides and an included angle of $\triangle ABC$ congruent to the corresponding parts of $\triangle DEF$. Thus, $\triangle ABC \cong \triangle DEF$ by SAS.

EXAMPLE 13 In $\triangle CPD$ and $\triangle EQD$, $\overline{CP} \perp \overline{QP}$ and $\overline{EQ} \perp \overline{QP}$. Also, $\angle QDE \cong \angle PDC$ and D is the midpoint of $\overline{QP}$. Explain why $\triangle CPD \cong \triangle EQD$.

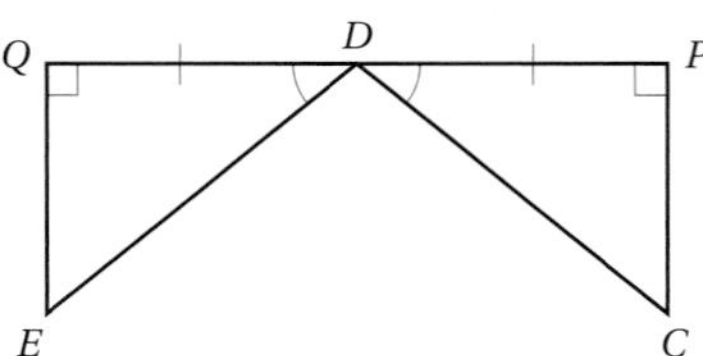

The perpendicular sides form right angles, which are congruent. Since D is the midpoint of $\overline{QP}$, we know that $\overline{QD} \cong \overline{PD}$. With $\angle QDE \cong \angle PDC$, we have $\triangle CPD \cong \triangle EQD$ by ASA.

Do Exercise 10.

Sometimes we can conclude that angles and segments are congruent by first showing that triangles are congruent.

EXAMPLE 14 $\overline{AB} \cong \overline{BC}$ and $\overline{EB} \cong \overline{DB}$. What can you conclude?

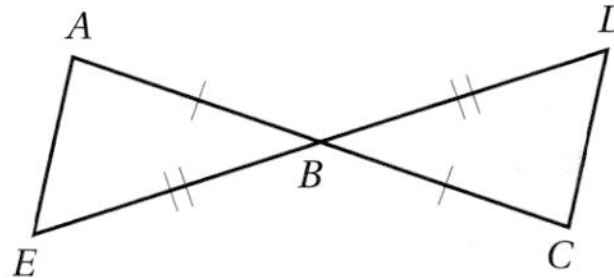

Since $\angle ABE$ and $\angle CBD$ are vertical angles, $\angle ABE \cong \angle CBD$. Thus, $\triangle ABE \cong \triangle CBD$ by SAS. As corresponding parts, $\overline{AE} \cong \overline{CD}$, $\angle A \cong \angle C$, and $\angle E \cong \angle D$.

EXAMPLE 15 Explain how you can use congruent triangles to find the distance across a marsh.

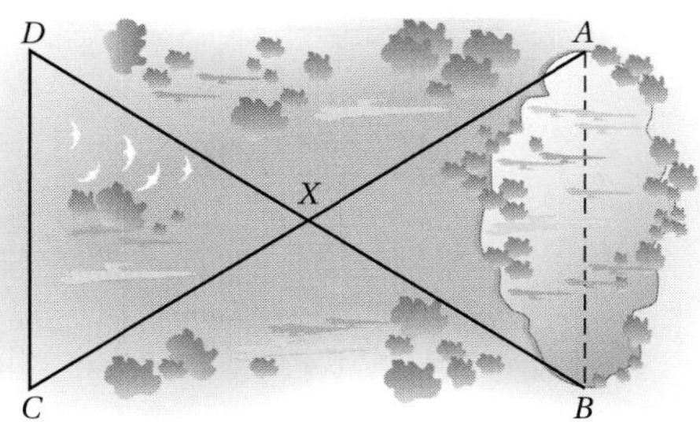

We mark off distances AX and BX and then extend $\overline{AX}$ and $\overline{BX}$ so that point X becomes the midpoint of $\overline{AC}$ and $\overline{BD}$. Then $\triangle ABX \cong \triangle CDX$ by SAS. Thus, $\overline{DC} \cong \overline{AB}$ as corresponding parts. Then we can measure $\overline{DC}$ knowing that $DC = AB$.

Do Exercises 11 and 12.

b Properties of Parallelograms

A quadrilateral is a polygon with four sides. A **diagonal** of a quadrilateral is a segment that joins two opposite vertices.

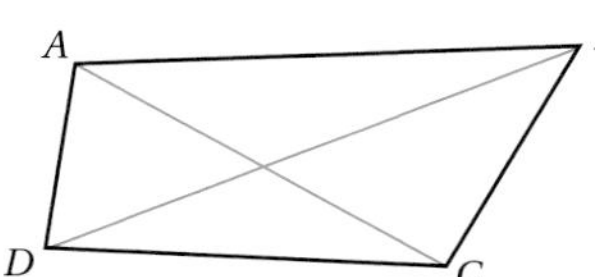

$\overline{AC}$ and $\overline{BD}$ are diagonals.

The sum of the measures of the angles of a quadrilateral is 360°.

A parallelogram is a quadrilateral with two pairs of parallel sides.

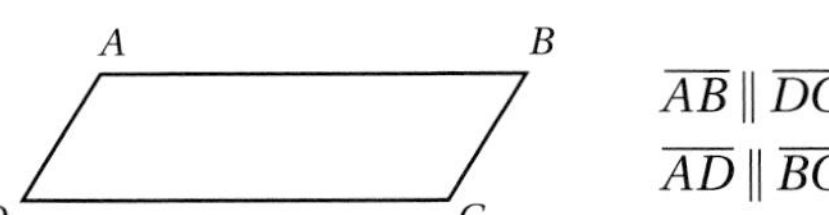

$\overline{AB} \parallel \overline{DC}$

$\overline{AD} \parallel \overline{BC}$

Draw two pairs of parallel lines to form parallelogram $ABCD$. Compare the lengths of opposite sides. Compare the measures of opposite angles. Compare the measures of consecutive angles. Draw diagonal $\overline{AC}$. How are $\triangle ADC$ and $\triangle CBA$ related? Draw diagonal $\overline{BD}$, intersecting $\overline{AC}$ at point E. What is special about point E?

11. $\angle R \cong \angle T$, $\angle W \cong \angle V$, and $\overline{RW} \cong \overline{TV}$. What can you conclude about this figure?

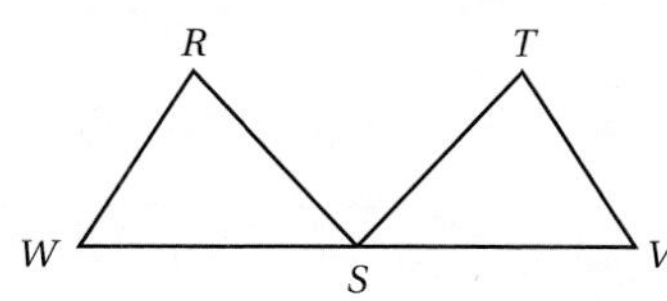

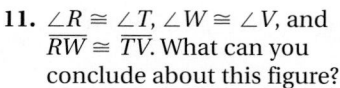

12. On a pair of pinking shears, the indicated angles and sides are congruent. How do you know that P is the midpoint of $\overline{GR}$?

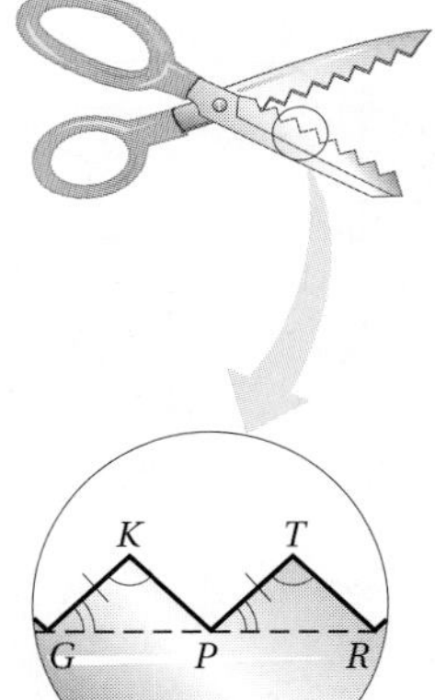

Answers on page A-16

Using the comparisons and the fact that corresponding parts of congruent triangles are congruent, we can list the following properties of parallelograms.

PROPERTIES OF PARALLELOGRAMS

1. A diagonal of a parallelogram determines two congruent triangles.
2. The opposite angles of a parallelogram are congruent.
3. The opposite sides of a parallelogram are congruent.
4. Consecutive angles of a parallelogram are supplementary.
5. The diagonals of a parallelogram bisect each other.

EXAMPLE 16 If $m\angle A = 120°$, find the measures of the other angles of parallelogram $ABCD$.

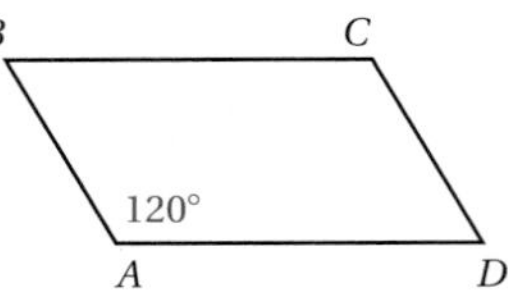

$m\angle C = 120°$ Using Property 2

$m\angle B = 60°$ Using Property 4

$m\angle D = 60°$ Using Property 2

EXAMPLE 17 Find AB and BC.

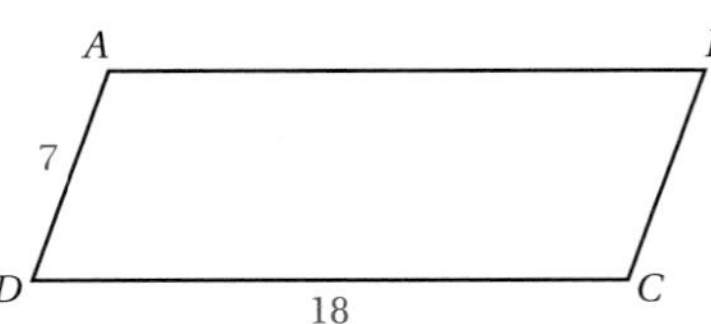

$AB = 18$ and $BC = 7$ Using Property 3

Do Exercises 13–16.

Find the measure of each angle.

13.

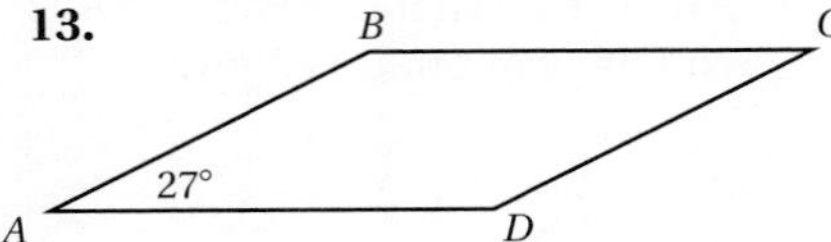

14. P Q 114° S R

Find the length of each side.

15.

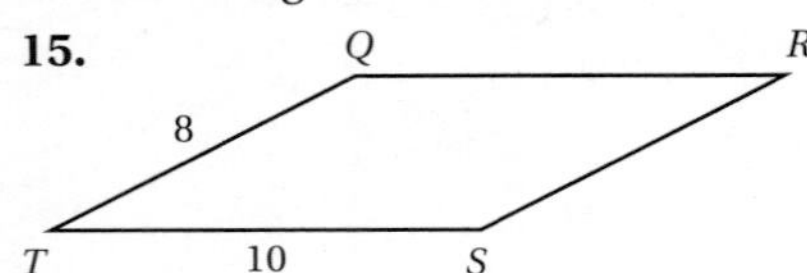

16. The perimeter of ▱$DEFG$ is 68.

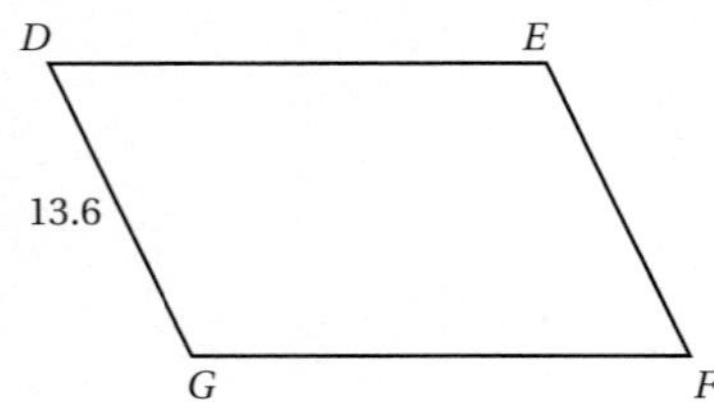

Answers on page A-16

6.7 EXERCISE SET

For Extra Help

MathXL | MyMathLab | InterAct Math | Math Tutor Center | Video Lectures on CD Disc 3 | Student's Solutions Manual

a Name the corresponding parts of the congruent triangles.

1. $\triangle ABC \cong \triangle RST$

2. $\triangle MNQ \cong \triangle HJK$

3. $\triangle DEF \cong \triangle GHK$

4. $\triangle ABC \cong \triangle ABC$

5. $\triangle XYZ \cong \triangle UVW$

6. $\triangle ABC \cong \triangle ACB$

Name the corresponding parts of the congruent triangles.

7.

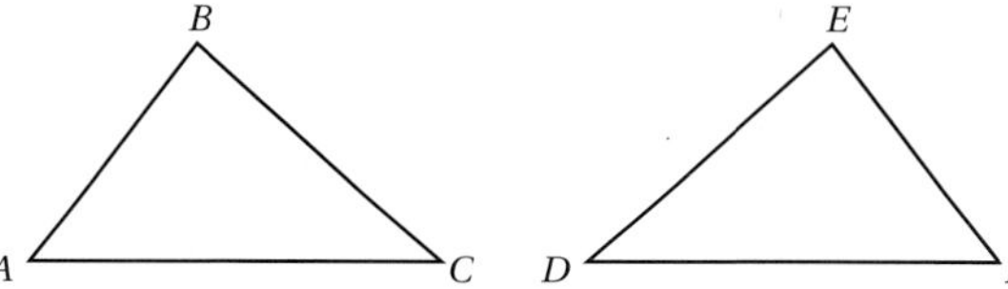

8.

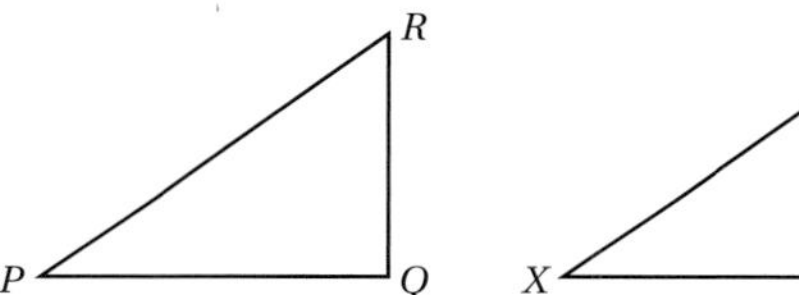

9.

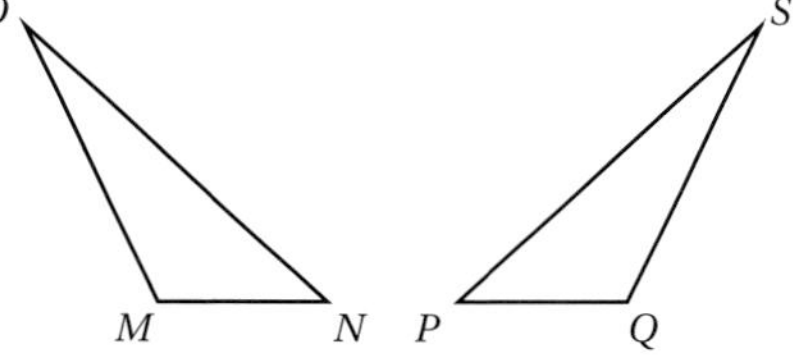

10.

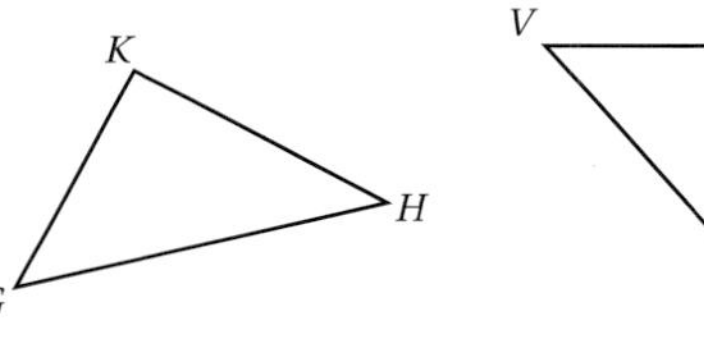

Determine whether the pair of triangles is congruent by the SAS property.

11.

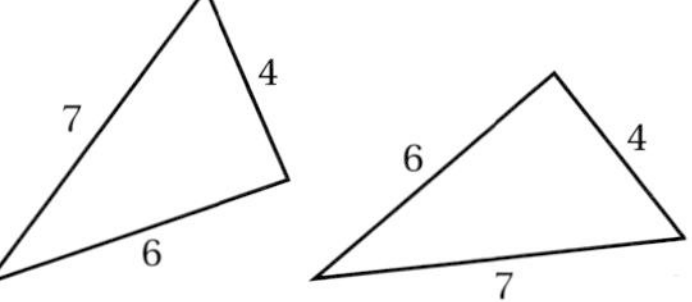

12.

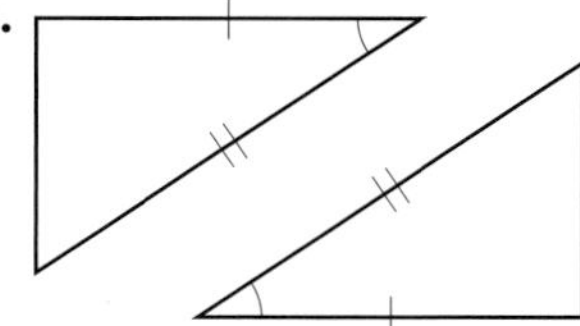

13.

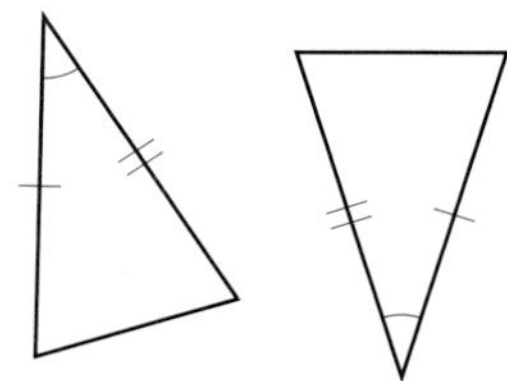

14.

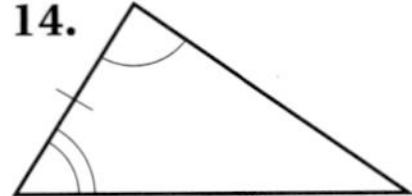

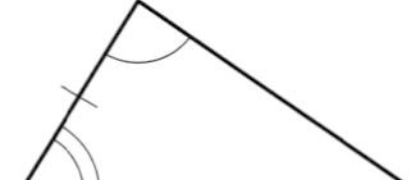

15.

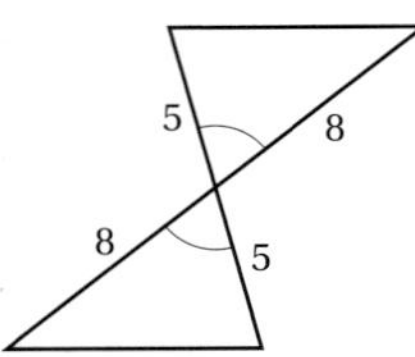

16.

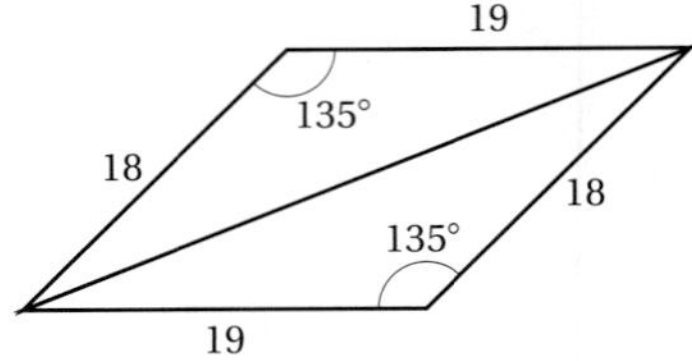

Determine whether the pair of triangles is congruent by the SSS property.

17.

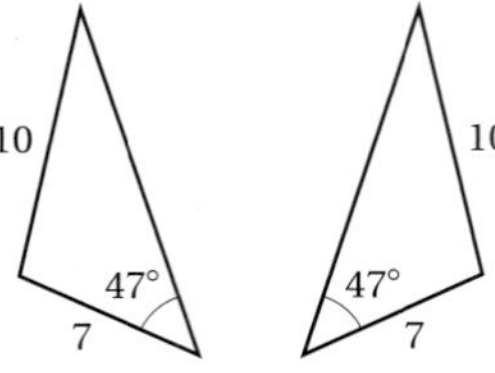

18.

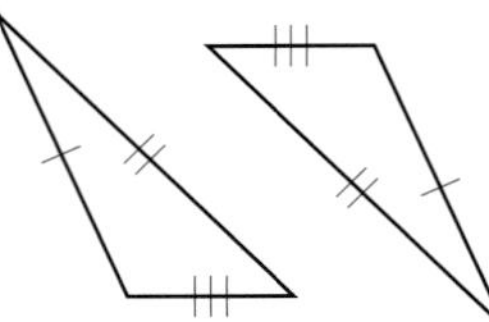

19.

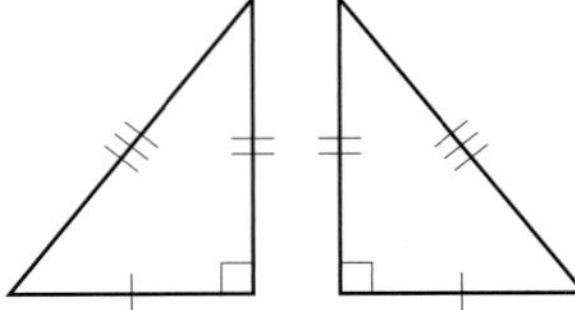

20.

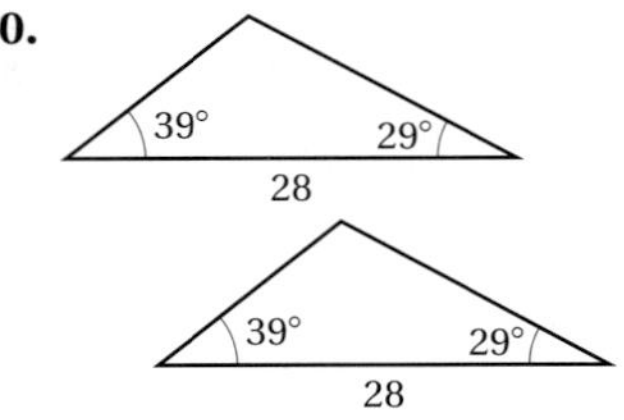

21.

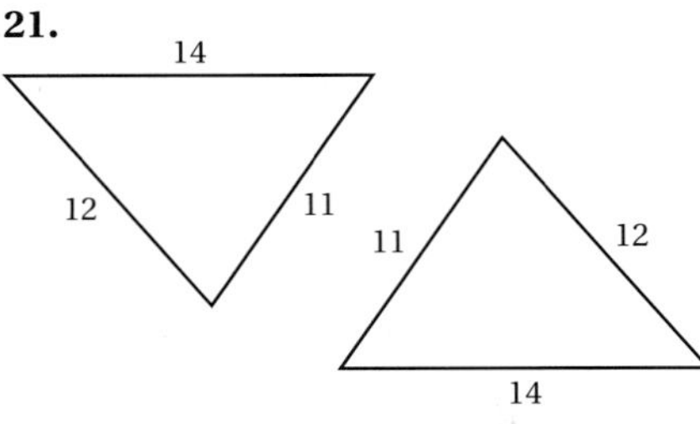

22.

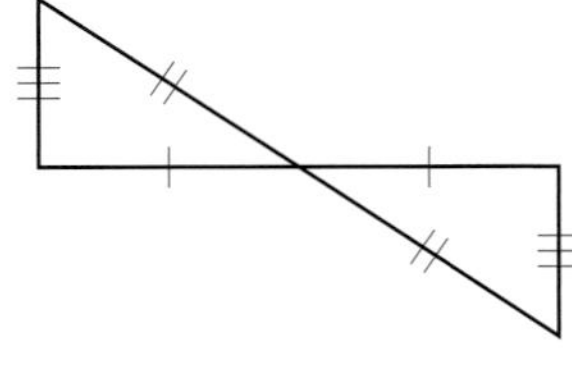

Determine whether the pair of triangles is congruent by the ASA property.

23.

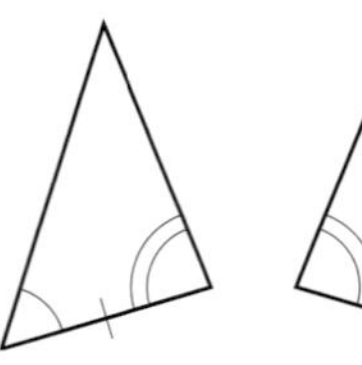

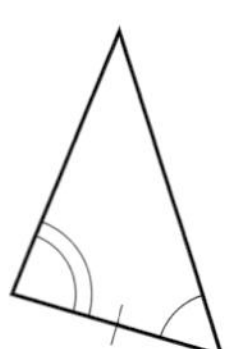

24.

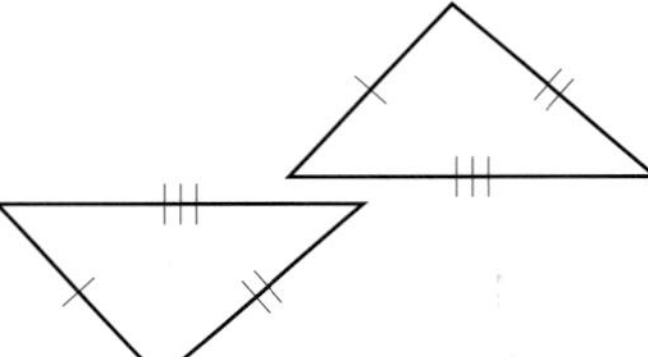

25.

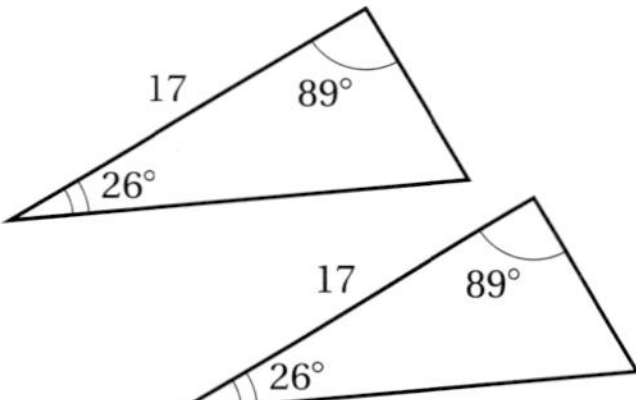

26.

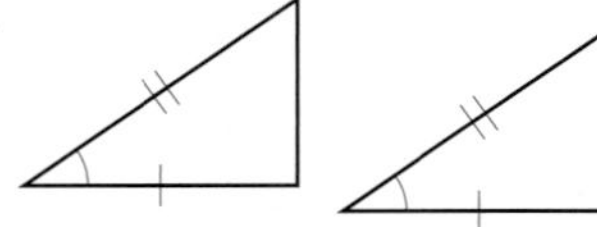

27.

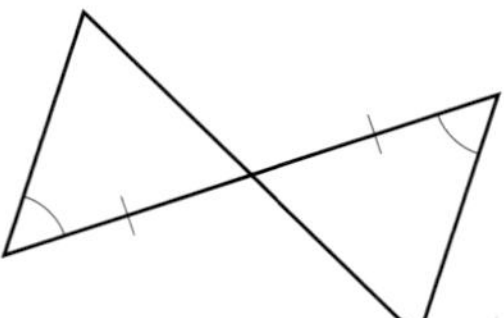

28.

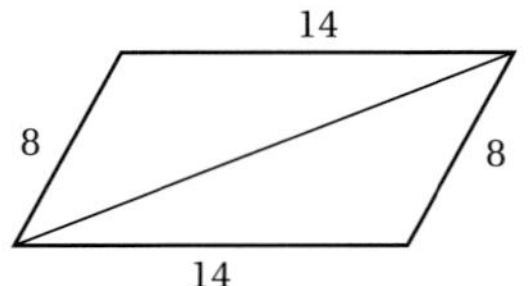

Which property (if any) should be used to show that the pair of triangles is congruent?

29.

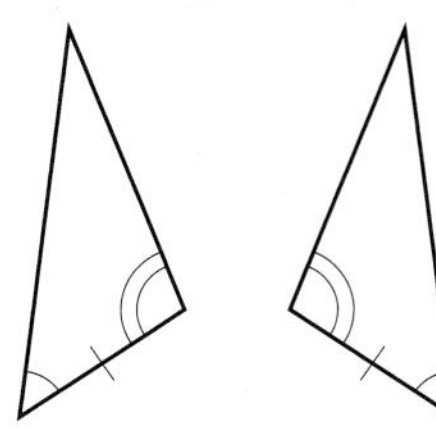

30.

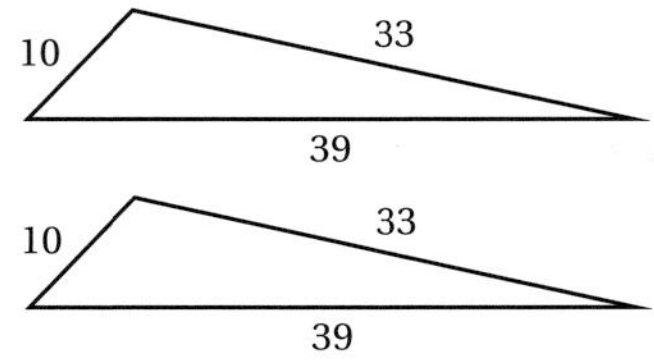

31.

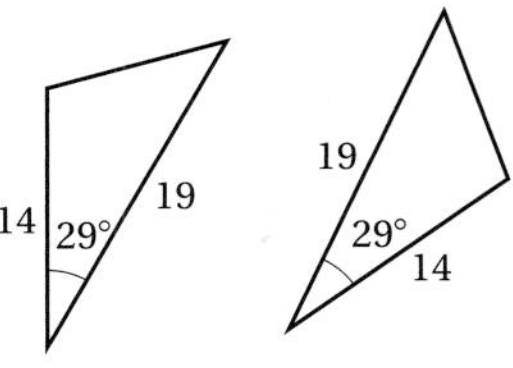

32.

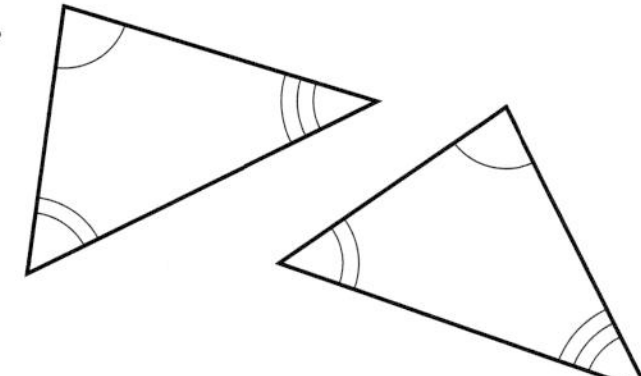

33.

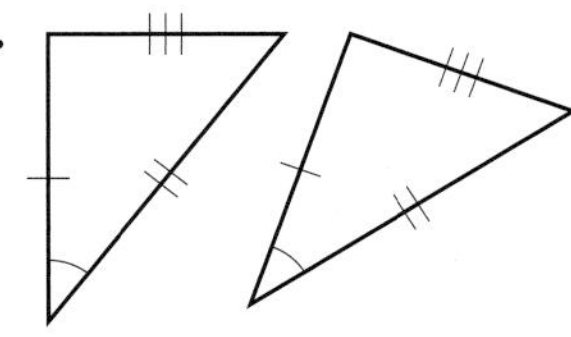

34.

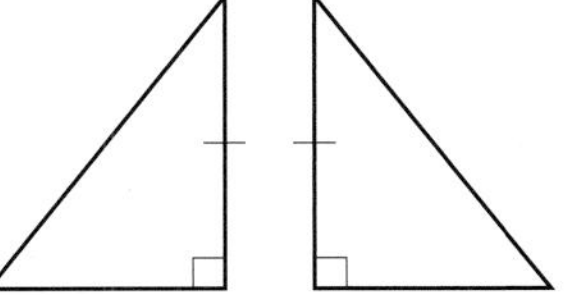

Explain why the triangles indicated in parentheses are congruent.

35. R is the midpoint of both $\overline{PT}$ and $\overline{QS}$. ($\triangle PRQ \cong \triangle TRS$)

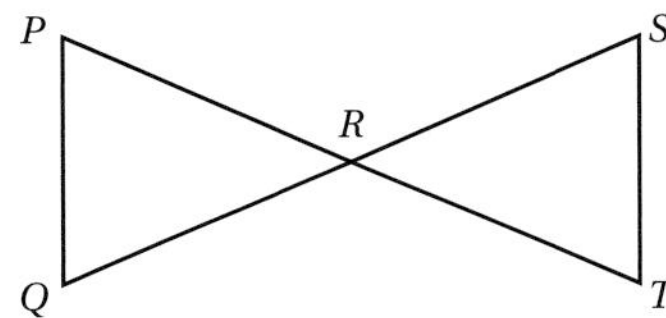

36. $\angle 1$ and $\angle 2$ are right angles, X is the midpoint of $\overline{AY}$, and $\overline{XB} \cong \overline{YZ}$. ($\triangle ABX \cong \triangle XZY$)

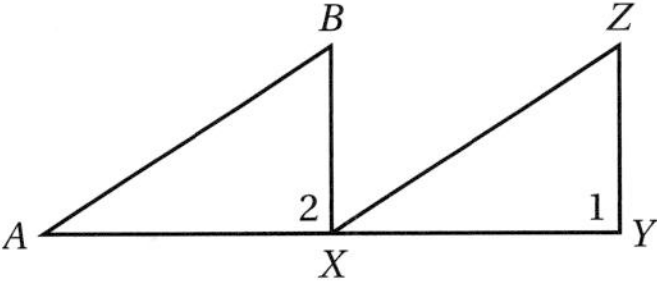

37. L is the midpoint of $\overline{KM}$ and $\overline{GL} \perp \overline{KM}$. ($\triangle KLG \cong \triangle MLG$)

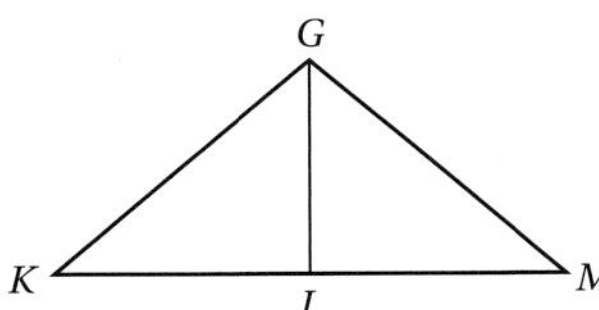

38. X is the midpoint of $\overline{QS}$ and $\overline{RP}$ with $RQ = SP$. ($\triangle RQX \cong \triangle PSX$)

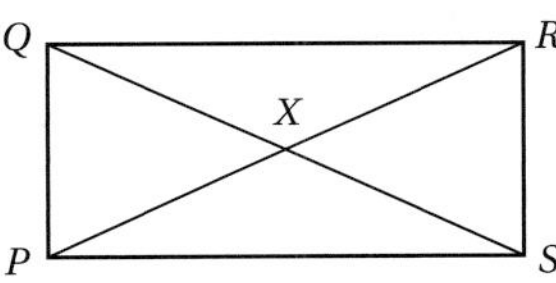

39. $\triangle AEB$ and $\triangle CDB$ are isosceles with $\overline{AE} \cong \overline{AB} \cong \overline{CB} \cong \overline{CD}$. Also, B is the midpoint of $\overline{ED}$. ($\triangle AEB \cong \triangle CDB$)

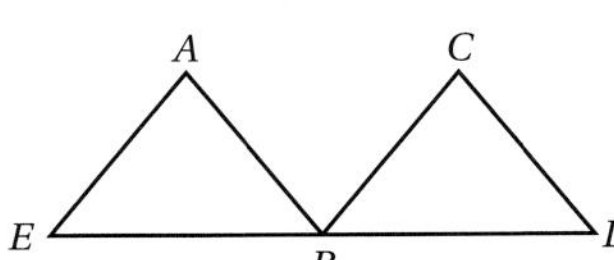

40. $\overline{AB} \perp \overline{BE}$ and $\overline{DE} \perp \overline{BE}$. $\overline{AB} \cong \overline{DE}$ and $\angle BAC \cong \angle EDC$. ($\triangle ABC \cong \triangle DEC$)

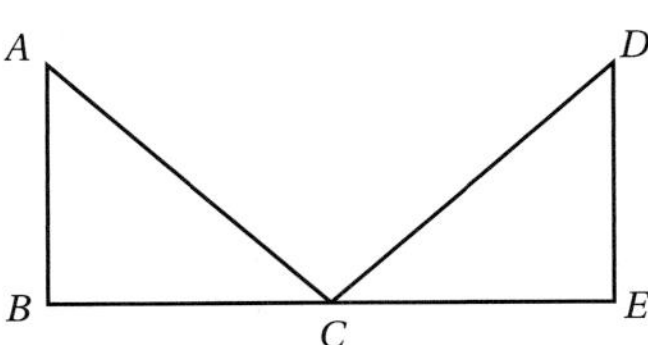

What can you conclude about each figure using the given information?

41. $\overline{GK} \perp \overline{LJ}$, $\overline{HK} \cong \overline{KJ}$, and $\overline{GK} \cong \overline{LK}$

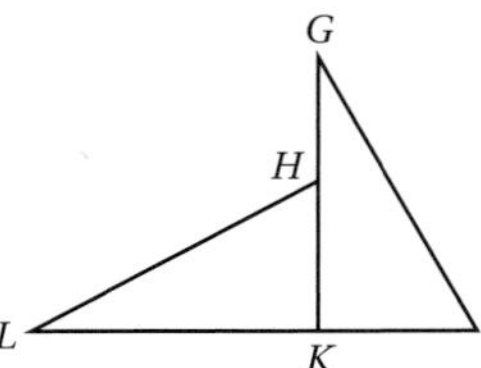

42. $\overline{AB} \cong \overline{DC}$ and $\angle BAC \cong \angle DCA$

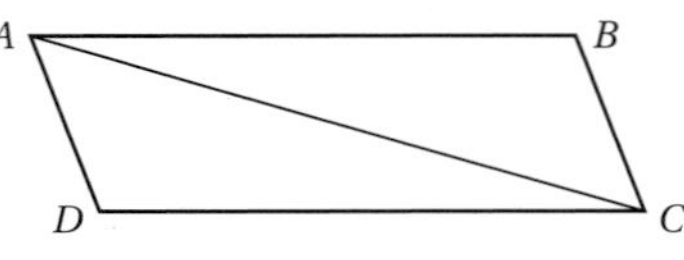

Use corresponding parts to solve Exercises 43 and 44.

43. On this national flag, the indicated segments and angles are congruent. Explain why P is the midpoint of $\overline{EF}$.

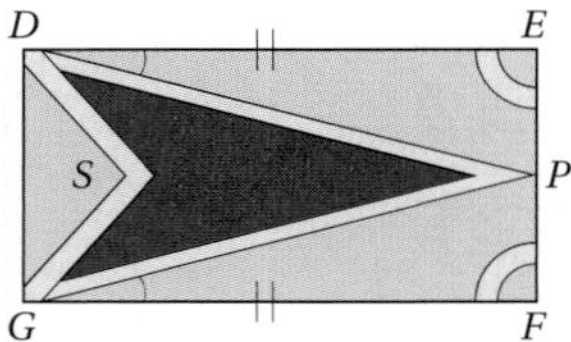

44. The indicated sides of a kite are congruent. Explain how you know that $\angle 1 \cong \angle 2$.

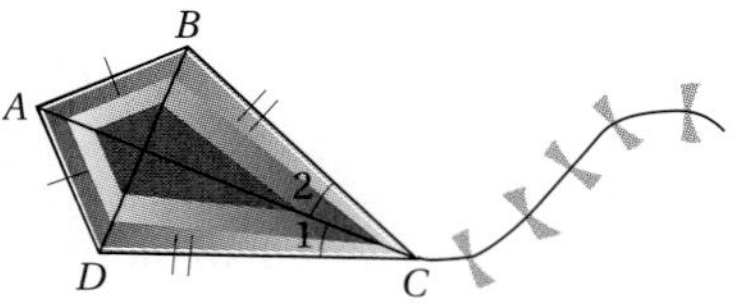

b Find the measures of the angles of the parallelogram.

45.

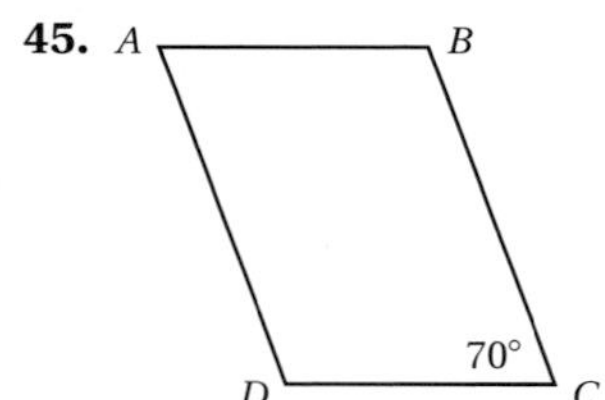

46. E F 110° H G

47. J K 71° M L

48. P S 105° Q R

Find the lengths of the sides of the parallelogram.

49.

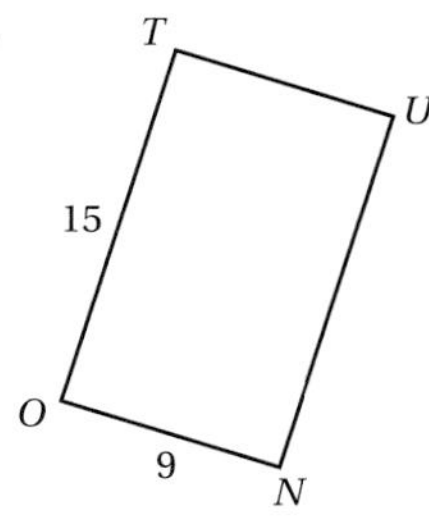

50.

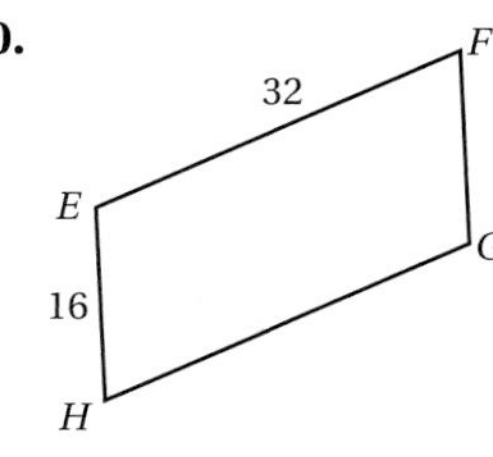

51. The perimeter of ▱$JKLM$ is 22.

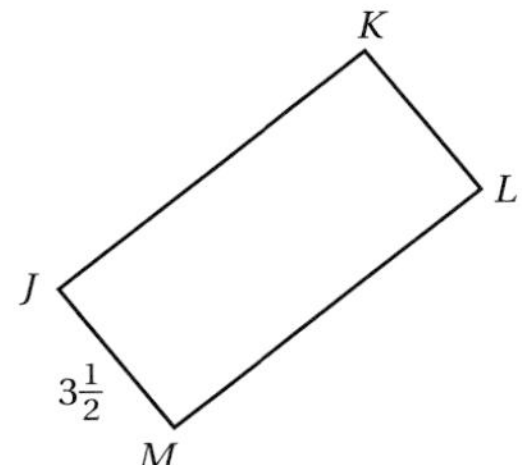

52. The perimeter of ▱$WXYZ$ is 248.

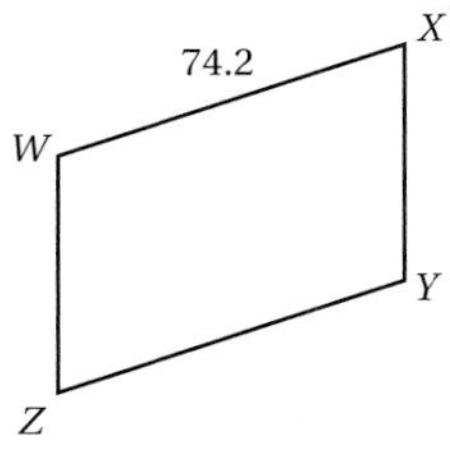

53. $AB = 14$ and $BD = 19$. Find the length of each diagonal.

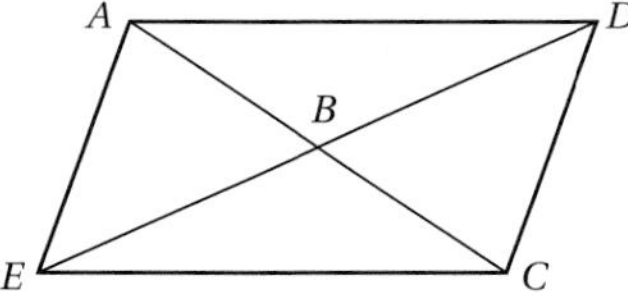

54. $EJ = 23$ and $GJ = 13$. Find the length of each diagonal.

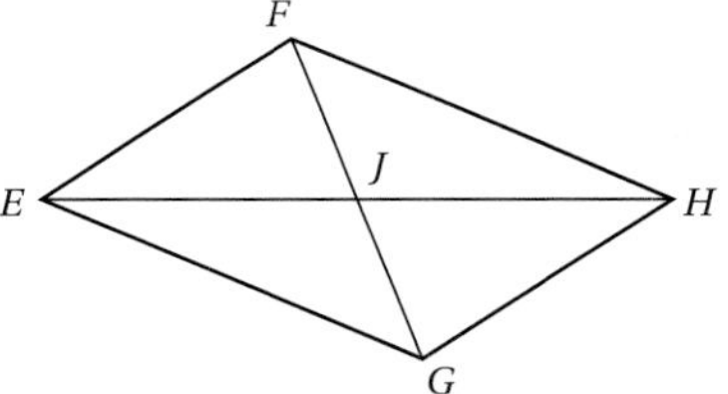

SKILL MAINTENANCE

Convert to percent notation. [4.2b], [4.3a]

55. 0.452

56. $\frac{1}{3}$

57. $\frac{11}{20}$

58. $\frac{22}{25}$

59. *Tourist Spending.* Foreign tourists spend $13.1 billion in this country annually. The most money, $2.7 billion, is spent in Florida. What is the ratio of amount spent in Florida to total amount spent? What is the ratio of total amount spent to amount spent in Florida? [4.1a]

60. One person in four plays a musical instrument. In a given group of people, what is the ratio of those who play an instrument to total number of people? What is the ratio of those who do not play an instrument to total number of people? [4.1a]

Divide. Find decimal notation for the answer. [3.4a]

61. $21 \div 12$

62. $23.4 \div 10$

63. $23.4 \div 100$

64. $23.4 \div 1000$

65. Multiply 3.14×4.41. Round to the nearest hundredth. [3.3a], [3.1d]

Objectives

a Identify the corresponding parts of similar triangles and determine which sides of a given pair of triangles have lengths that are proportional.

b Find lengths of sides of similar triangles using proportions.

6.8 SIMILAR TRIANGLES

a Proportions and Similar Triangles

We know that congruent figures have the same shape and size. *Similar figures* have the same shape, but are not necessarily the same size.

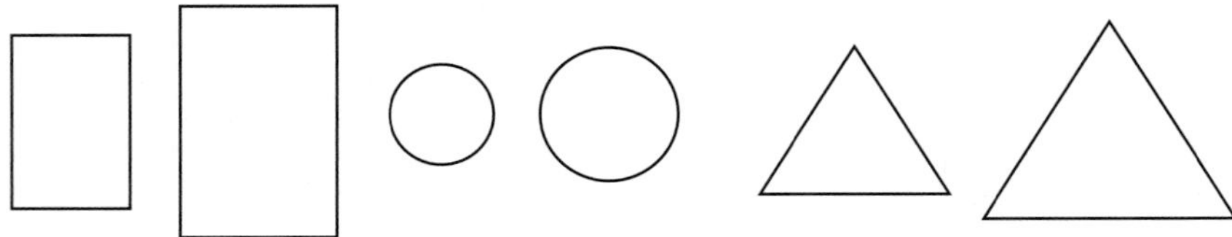

Similar figures

EXAMPLE 1 Which pairs of triangles appear to be similar?

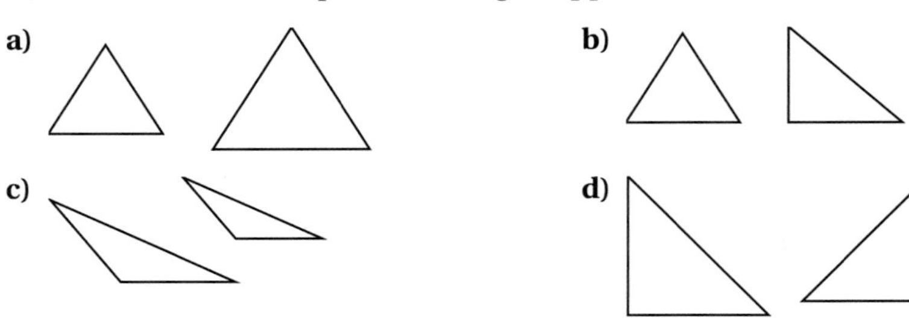

Pairs (a), (c), and (d) appear to be similar.

Do Exercise 1.

Similar triangles have corresponding sides and angles.

EXAMPLE 2 $\triangle ABC$ and $\triangle DEF$ are similar. Name their corresponding sides and angles.

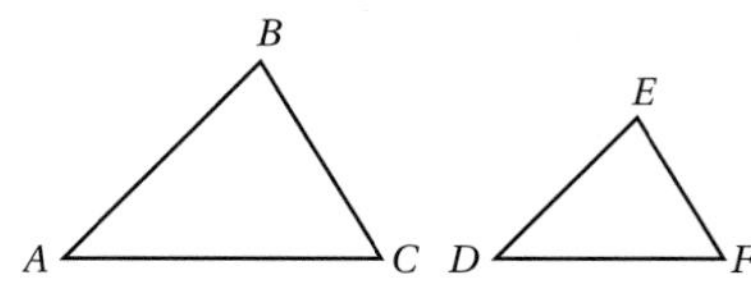

$\overline{AB} \leftrightarrow \overline{DE}$ $\quad \angle A \leftrightarrow \angle D$

$\overline{AC} \leftrightarrow \overline{DF}$ $\quad \angle B \leftrightarrow \angle E$

$\overline{BC} \leftrightarrow \overline{EF}$ $\quad \angle C \leftrightarrow \angle F$

Do Exercise 2.

1. Which pairs of triangles appear to be similar?

a)

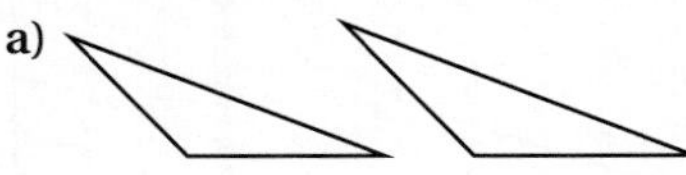

b)

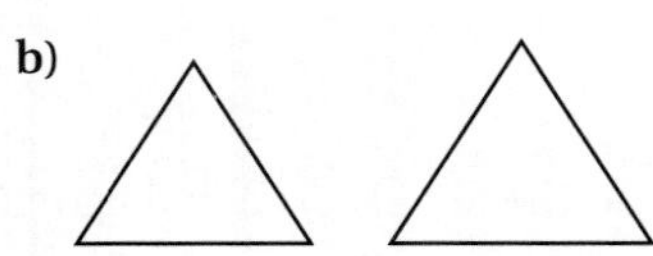

c)

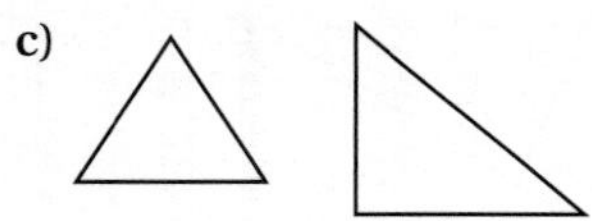

d)

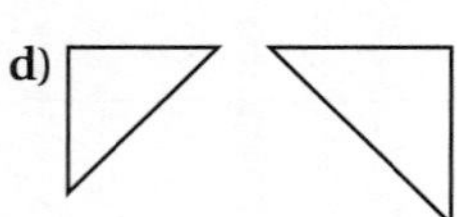

2. $\triangle PQR$ and $\triangle GHK$ are similar. Name their corresponding sides and angles.

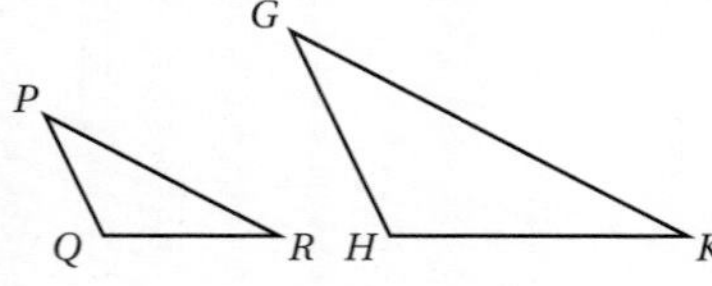

Answers on page A-16

SIMILAR TRIANGLES

Two triangles are **similar** if and only if their vertices can be matched so that the corresponding angles are congruent and the lengths of corresponding sides are proportional.

To say that $\triangle ABC$ and $\triangle DEF$ are similar, we write "$\triangle ABC \sim \triangle DEF$." We will agree that this symbol also tells us the way in which the vertices are matched.

$$\triangle ABC \sim \triangle DEF$$

Thus, $\triangle ABC \sim \triangle DEF$ means that

$$\angle A \cong \angle D$$
$$\angle B \cong \angle E \quad \text{and} \quad \frac{AB}{DE} = \frac{AC}{DF} = \frac{BC}{EF}.$$
$$\angle C \cong \angle F$$

EXAMPLE 3 Suppose that $\triangle PQR \sim \triangle STV$. Which angles are congruent? Which sides are proportional?

$$\angle P \cong \angle S$$
$$\angle Q \cong \angle T \quad \text{and} \quad \frac{PQ}{ST} = \frac{PR}{SV} = \frac{QR}{TV}.$$
$$\angle R \cong \angle V$$

Do Exercise 3.

EXAMPLE 4 These triangles are similar. Which sides are proportional?

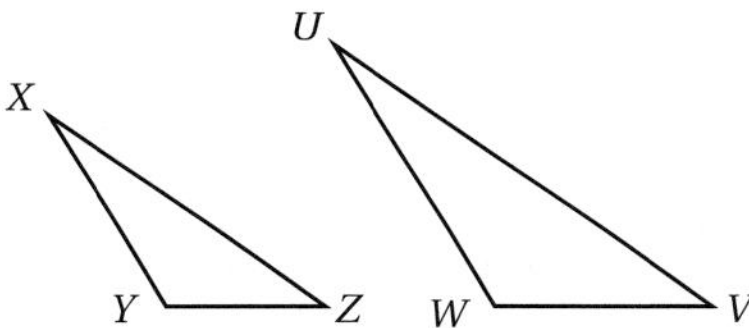

It appears that if we match X with U, Y with W, and Z with V, the corresponding angles will be congruent. Thus,

$$\frac{XY}{UW} = \frac{XZ}{UV} = \frac{YZ}{WV}.$$

Do Exercise 4.

3. Suppose that $\triangle JKL \sim \triangle ABC$. Which angles are congruent? Which sides are proportional?

4. These triangles are similar. Which sides are proportional?

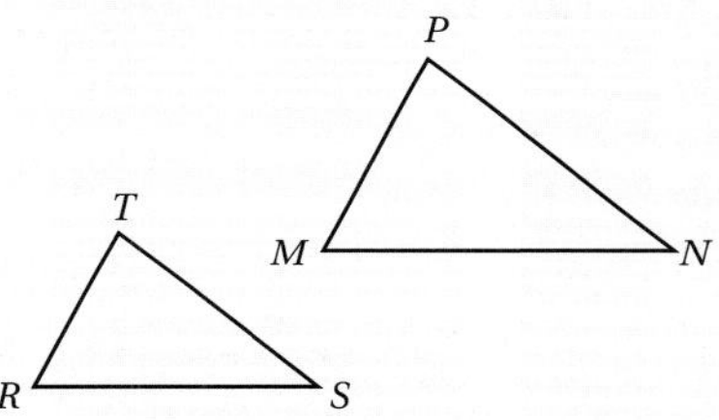

Answers on page A-16

5. If $\triangle WNE \sim \triangle CBT$, find BT and CT.

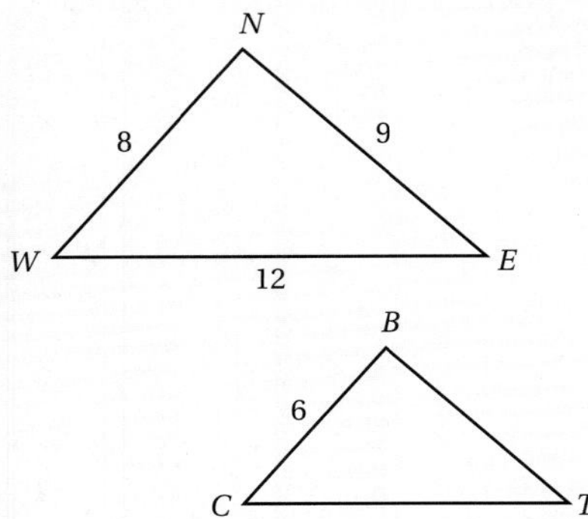

Answers on page A-16

b Proportions and Similar Triangles

We can find lengths of sides in similar triangles.

EXAMPLE 5 If $\triangle RAE \sim \triangle GQL$, find QL and GL.

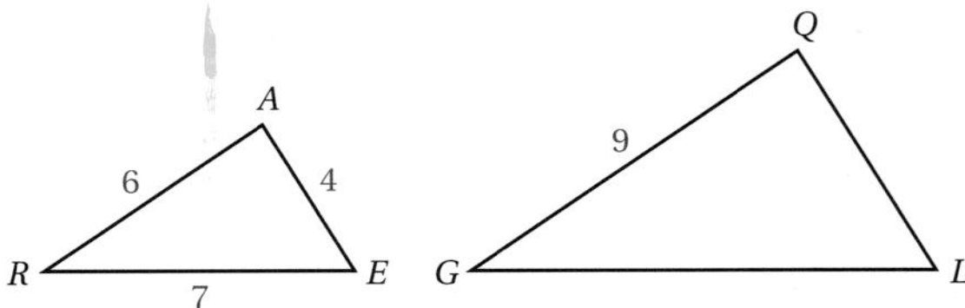

Since $\triangle RAE \sim \triangle GQL$, the corresponding sides are proportional. Thus,

$$\frac{6}{9} = \frac{4}{QL}$$

$$6(QL) = 9 \cdot 4 \quad \text{Equating cross products}$$

$$6(QL) = 36$$

$$QL = 6 \quad \text{Dividing by 6 on both sides}$$

and

$$\frac{6}{9} = \frac{7}{GL}$$

$$6(GL) = 9 \cdot 7$$

$$6(GL) = 63$$

$$GL = 10\tfrac{1}{2}.$$

Do Exercise 5.

EXAMPLE 6 If $\overline{AB} \parallel \overline{CD}$, find CD.

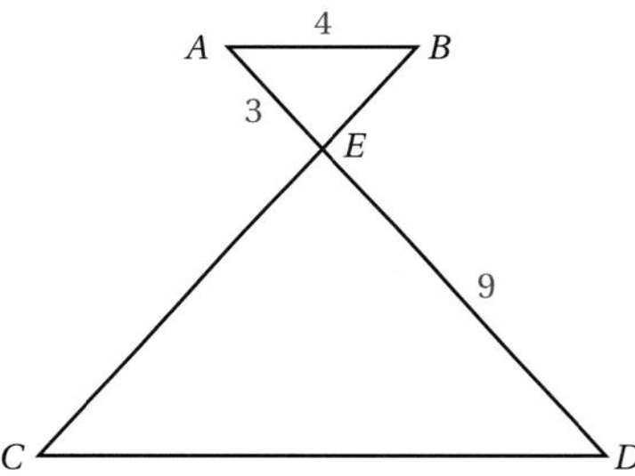

Recall that if a transversal intersects two parallel lines, then the alternate interior angles are congruent (Section 6.6). Thus,

$$\angle A \cong \angle D \quad \text{and} \quad \angle C \cong \angle B,$$

because they are pairs of alternate interior angles. Since $\angle AEB$ and $\angle DEC$ are vertical angles, they are congruent. Thus by definition

$$\triangle AEB \sim \triangle DEC$$

and the lengths of the corresponding sides are proportional. Thus,

$$\frac{AE}{DE} = \frac{AB}{CD}.$$

Solve:

$$\frac{3}{9} = \frac{4}{CD} \quad \text{Substituting}$$
$$3(CD) = 9 \cdot 4 \quad \text{Equating cross products}$$
$$3(CD) = 36$$
$$CD = 12. \quad \text{Dividing by 3 on both sides}$$

Do Exercise 6.

Similar triangles and proportions can often be used to find lengths that would ordinarily be difficult to measure. For example, we could find the height of a flagpole without climbing it or the distance across a river without crossing it.

EXAMPLE 7 How high is a flagpole that casts a 56-ft shadow at the same time that a 6-ft man casts a 5-ft shadow?

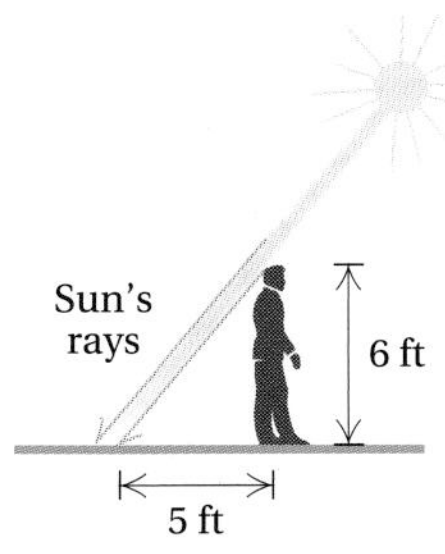

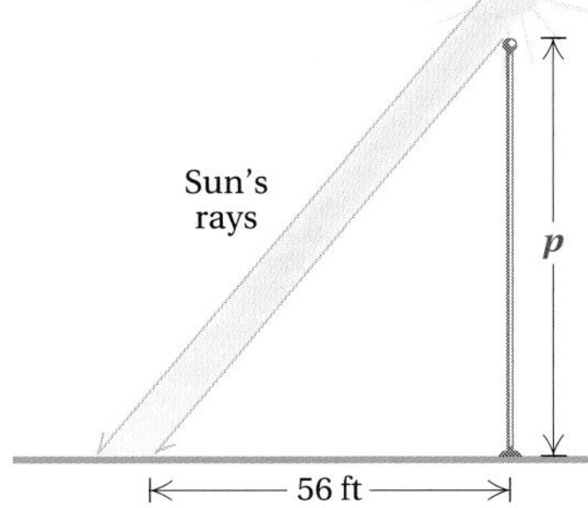

If we use the sun's rays to represent the third side of the triangle in our drawing of the situation, we see that we have similar triangles. Let p = the height of the flagpole. The ratio of 6 to p is the same as the ratio of 5 to 56. Thus we have the proportion

$$\begin{array}{l}\text{Height of man} \rightarrow \\ \text{Height of pole} \rightarrow\end{array} \frac{6}{p} = \frac{5}{56}. \begin{array}{l}\leftarrow \text{Length of shadow of man} \\ \leftarrow \text{Length of shadow of pole}\end{array}$$

Solve:

$$6 \cdot 56 = 5 \cdot p \quad \text{Equating cross products}$$
$$\frac{6 \cdot 56}{5} = \frac{5 \cdot p}{5} \quad \text{Dividing by 5 on both sides}$$
$$\frac{6 \cdot 56}{5} = p \quad \text{Simplifying}$$
$$67.2 = p$$

The height of the flagpole is 67.2 ft.

Do Exercise 7.

6. If $\overline{QR} \parallel \overline{ST}$, find QR.

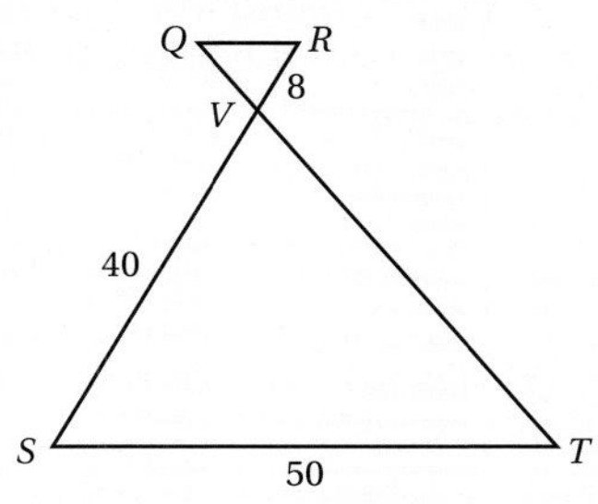

7. How high is a flagpole that casts a 45-ft shadow at the same time that a 5.5-ft woman casts a 10-ft shadow?

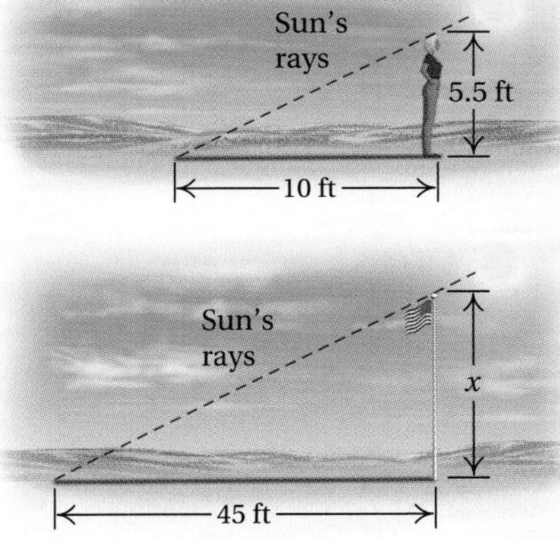

Answers on page A-16

8. **Rafters of a House.** Referring to Example 8, find the length y of the rafter of the house to the nearest tenth of a foot.

EXAMPLE 8 *Rafters of a House.* Carpenters use similar triangles to determine the lengths of rafters for a house. They first choose the pitch of the roof, or the ratio of the rise over the run. Then using a triangle with that ratio, they calculate the length of the rafter needed for the house. Loren is constructing rafters for a roof with a 6/12 pitch on a house that is 30 ft wide. Using a rafter guide, Loren knows that the rafter length corresponding to the 6/12 pitch is 13.4. Find the length x of the rafter of the house to the nearest tenth of a foot.

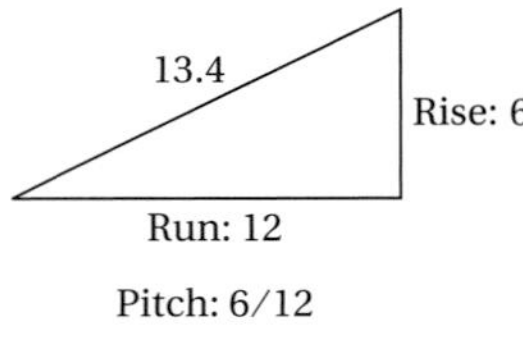

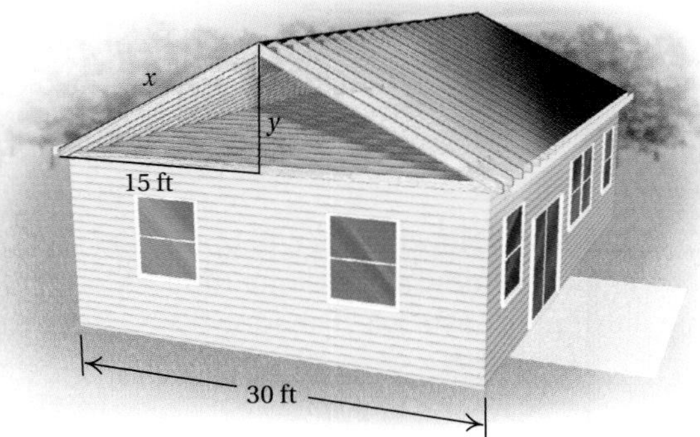

We see from the drawing that we have the proportion

Length of rafter in 6/12 triangle → $\frac{13.4}{x} = \frac{12}{15}$ ← Run in 6/12 triangle
Length of rafter on the house → x; 15 ← Run in similar triangle on the house

Solve: $13.4 \cdot 15 = x \cdot 12$ Equating cross products

$$\frac{13.4 \cdot 15}{12} = \frac{x \cdot 12}{12}$$ Dividing by 12 on both sides

$$\frac{13.4 \cdot 15}{12} = x$$

$16.8 \text{ ft} \approx x$ Rounding to the nearest tenth of a foot

The length of the rafter x of the house is about 16.8 ft.

Do Exercise 8.

Answers on page A-16

Study Tips

TIME MANAGEMENT (PART 3)

Here are some additional tips to help you with time management. (See also the Study Tips on time management in Sections 1.3 and 4.4.)

- **Are you a morning or an evening person?** If you are an evening person, it might be best to avoid scheduling early-morning classes. If you are a morning person, do the opposite. Nothing can drain your study time and effectiveness like fatigue.
- **Keep on schedule.** Your course syllabus provides the semester's schedule. Use a daily planner or personal digital assistant to outline your time for the semester. Be sure to note deadlines involving term papers and exams so you can begin a task early.
- **Balance your class schedule.** You may be someone who prefers large blocks of time for study on the off days. In that case, it might be advantageous for you to take courses that meet only three days a week.

Translating for Success

The goal of these matching questions is to practice step (2), *Translate,* of the five-step problem-solving process. Translate each word problem to an equation and select a correct translation from equations A–O.

1. *Servings of Pork.* An 8-lb pork roast contains 37 servings of meat. How many pounds of pork would be needed for 55 servings?

2. *Planting Corn.* Each year, Prairie State Farm plants 64 acres of corn. With good weather, $\frac{3}{4}$ of the planting can be completed by April 20. How many acres can be planted with good weather?

3. *Cruise Cost.* A group of 6 college students pays $4608 for a spring break cruise. What is each person's share?

4. *Sales Tax Rate.* The sales tax is $14.95 on the purchase of a new ladder that costs $299. What is the sales tax rate?

5. *Volume of a Sphere.* Find the volume of a sphere whose radius is 7.2 cm.

6. *Inheritance.* Each of 6 children inherits $4608 from their mother's estate. What is the total inheritance?

7. *Sales Tax.* Erica buys 5 pairs of earrings at $14.95 each. The sales tax rate in Indiana is 6%. How much sales tax will be charged?

8. *Volume of a Cone.* Find the volume of a circular cone with a 6-cm base radius and a height of 7.2 cm.

9. *Volume of a Storage Tank.* The diameter of a cylindrical grain-storage tank is $5\frac{1}{2}$ yd. Its height is 7 yd. Find its volume.

10. *Perimeter of a Photo.* A rectangular photo is 14.5 cm by 9.4 cm. What is the perimeter of the photo?

A. $x = \frac{1}{3}\pi \cdot 6^2 \cdot (7.2)$

B. $6 \cdot x = \$4608$

C. $x = \frac{4}{3} \cdot \pi \cdot 6^2 \cdot (7.2)$

D. $x = \pi \cdot \left(5\frac{1}{2} \div 2\right)^2 \cdot 7$

E. $x = 6\% \times 5 \times \$14.95$

F. $x = \frac{1}{3}\pi\left(5\frac{1}{2}\right)^2$

G. $\frac{3}{4} \cdot 64 = x$

H. $\$14.95 = x \cdot \$299$

I. $x = 2(14.5 + 9.4)$

J. $\frac{3}{4} \cdot x = 64$

K. $\frac{8}{37} = \frac{x}{55}$

L. $x = 4(14.5 + 9.4)$

M. $x = 6 \cdot \$4608$

N. $8 \cdot 37 = 55 \cdot x$

O. $x = \frac{4}{3}\pi(7.2)^3$

Answers on page A-16

6.8 EXERCISE SET

For Extra Help

MathXL | MyMathLab | InterAct Math | Math Tutor Center | Video Lectures on CD Disc 3 | Student's Solutions Manual

a For each pair of similar triangles, name the corresponding sides and angles.

1.

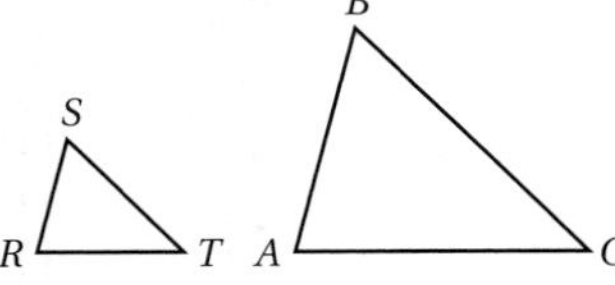

2.

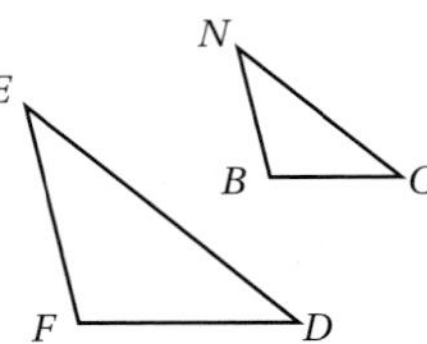

3.

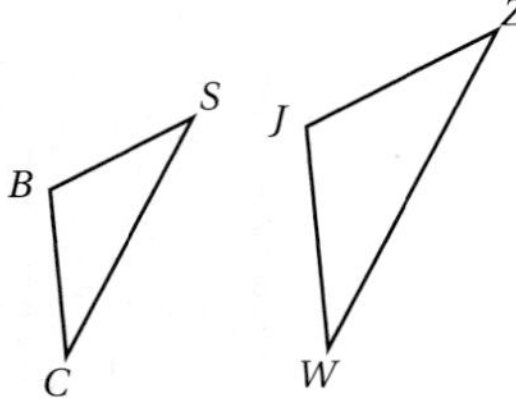

4.

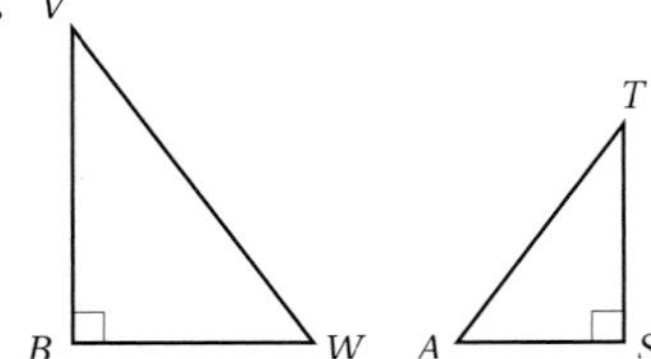

For each pair of similar triangles, name the congruent angles and proportional sides.

5. $\triangle ABC \sim \triangle RST$

6. $\triangle PQR \sim \triangle STV$

7. $\triangle MES \sim \triangle CLF$

8. $\triangle SMH \sim \triangle WLK$

Name the proportional sides in these similar triangles.

9.

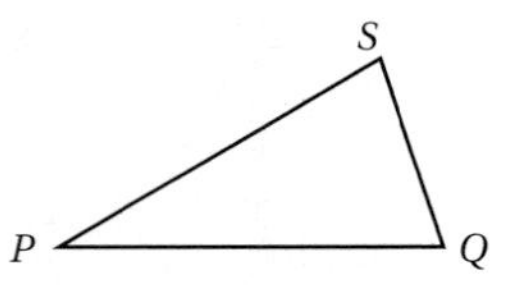

D

M N

10.

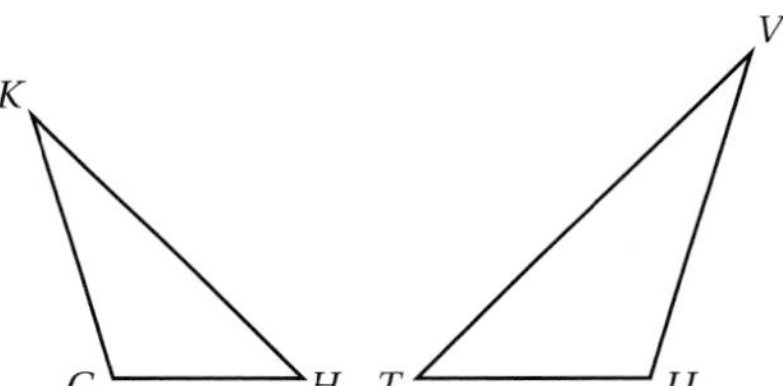

11.

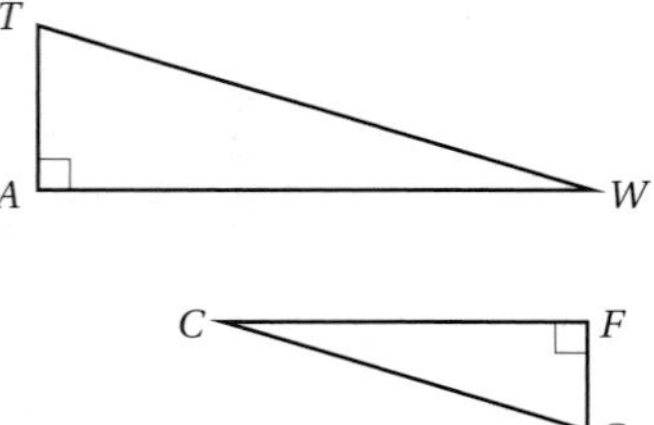

12.

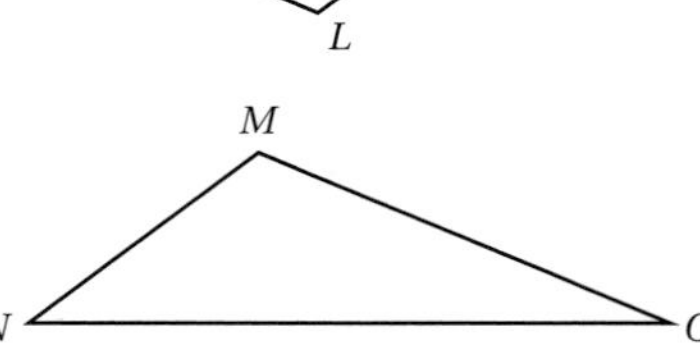

b Find the missing lengths.

13. If $\triangle ABC \sim \triangle PQR$, find QR and PR.

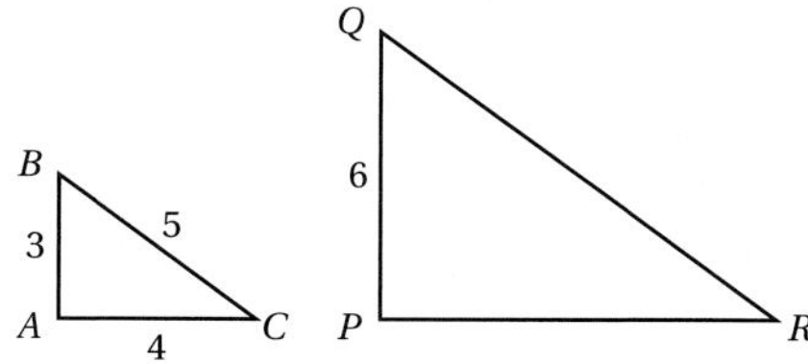

14. If $\triangle MAC \sim \triangle GET$, find AM and GT.

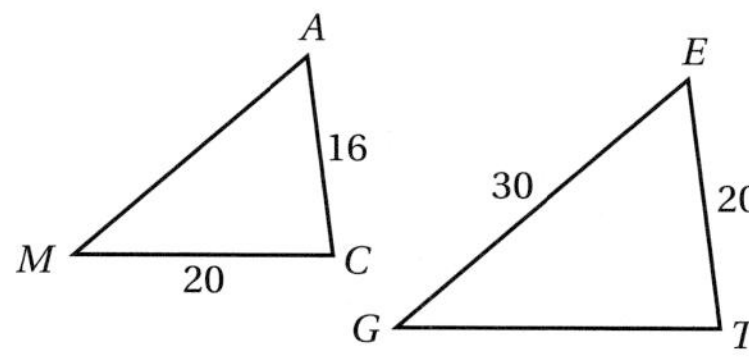

15. If $\overline{AD} \parallel \overline{CB}$, find EC.

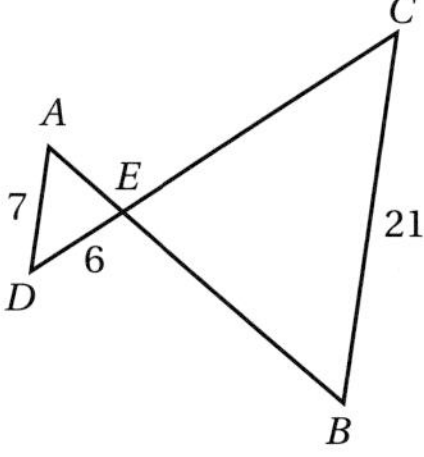

16. If $\overline{LN} \parallel \overline{PM}$, find QM.

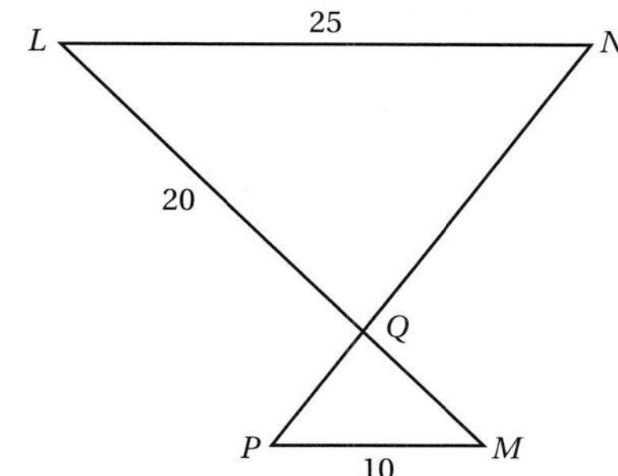

17. How high is a tree that casts a 27-ft shadow at the same time that a 4-ft fence post casts a 3-ft shadow?

18. How high is a flagpole that casts a 42-ft shadow at the same time that a $5\frac{1}{2}$-ft woman casts a 7-ft shadow?

19. Find the distance across the river. Assume that the ratio of d to 25 ft is the same as the ratio of 40 ft to 10 ft.

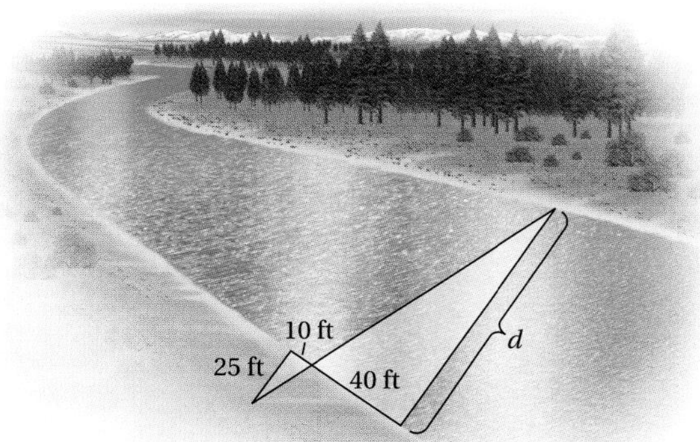

20. To measure the height of a hill, a string is drawn tight from level ground to the top of the hill. A 3-ft stick is placed under the string, touching it at point P, a distance of 5 ft from point G, where the string touches the ground. The string is then detached and found to be 120 ft long. How high is the hill?

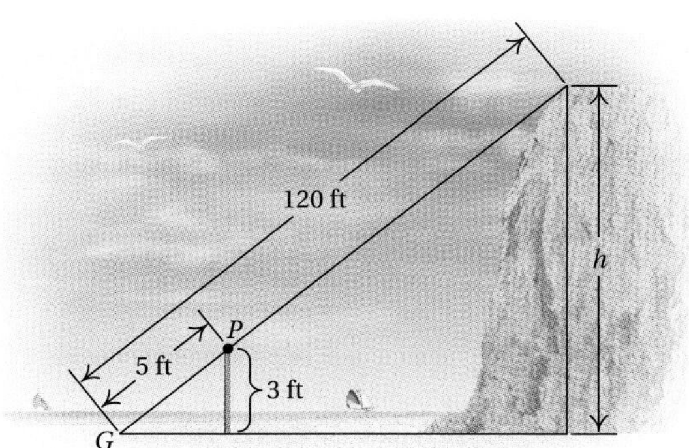

21. D_W Is it possible for two triangles to have two pairs of sides that are proportional without the triangles being similar? Why or why not?

22. D_W Design for a classmate a problem involving similar triangles for which

$$\frac{18}{128.95} = \frac{x}{789.89}.$$

SKILL MAINTENANCE

Multiply. [2.4d], [3.3a]

23. $2\frac{4}{5} \times 10\frac{1}{2}$

24. 3.05×0.08

25. $8 \times 9\frac{3}{4}$

26. 10.01×6.11

6 Summary and Review

The review that follows is meant to prepare you for a chapter exam. It consists of three parts. The first part, Concept Reinforcement, is designed to increase understanding of the concepts through true/false exercises. The second part is a list of important properties and formulas. The third part is the Review Exercises. These provide practice exercises for the exam, together with references to section objectives so you can go back and review. Before beginning, stop and look back over the skills you have obtained. What skills in mathematics do you have now that you did not have before studying this chapter?

CONCEPT REINFORCEMENT

Determine whether the statement is true or false. Answers are given at the back of the book.

_______ **1.** The area of a square is four times the length of a side.

_______ **2.** The area of a parallelogram with base 8 cm and height 5 cm is the same as the area of a rectangle with length 8 cm and width 5 cm.

_______ **3.** If $\triangle ACF \sim \triangle RTY$, then $\triangle ACF \cong \triangle RTY$.

_______ **4.** The measure of any obtuse angle is larger than the measure of any acute angle.

_______ **5.** Some right triangles are equilateral triangles.

IMPORTANT PROPERTIES AND FORMULAS

Perimeter of a Rectangle:	$P = 2 \cdot (l + w)$, or $P = 2 \cdot l + 2 \cdot w$
Perimeter of a Square:	$P = 4 \cdot s$
Area of a Rectangle:	$A = l \cdot w$
Area of a Square:	$A = s \cdot s$, or $A = s^2$
Area of a Parallelogram:	$A = b \cdot h$
Area of a Triangle:	$A = \frac{1}{2} \cdot b \cdot h$
Area of a Trapezoid:	$A = \frac{1}{2} \cdot h \cdot (a + b)$
Radius and Diameter of a Circle:	$d = 2 \cdot r$, or $r = \frac{d}{2}$
Circumference of a Circle:	$C = \pi \cdot d$, or $C = 2 \cdot \pi \cdot r$
Area of a Circle:	$A = \pi \cdot r \cdot r$, or $A = \pi \cdot r^2$
Volume of a Rectangular Solid:	$V = l \cdot w \cdot h$
Surface Area of a Rectangular Solid:	$SA = 2lw + 2lh + 2wh$, or $2(lw + lh + wh)$
Volume of a Circular Cylinder:	$V = \pi \cdot r^2 \cdot h$
Volume of a Sphere:	$V = \frac{4}{3} \cdot \pi \cdot r^3$
Volume of a Cone:	$V = \frac{1}{3} \cdot \pi \cdot r^2 \cdot h$
Sum of Angle Measures of a Triangle:	$m(\angle A) + m(\angle B) + m(\angle C) = 180°$

Review Exercises

Use a protractor to measure the angle. [6.1b]

1.

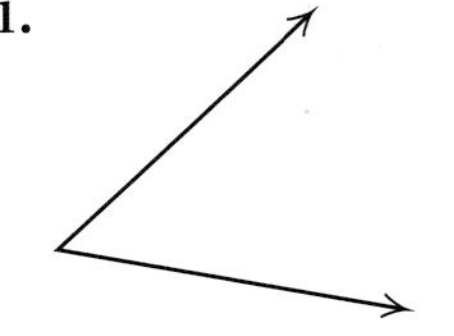

2.

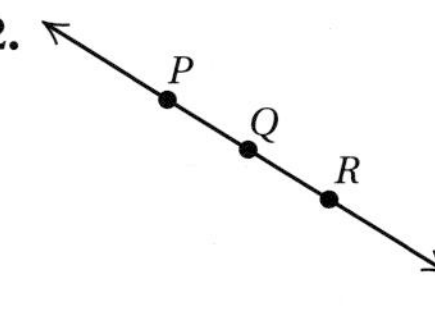

3.

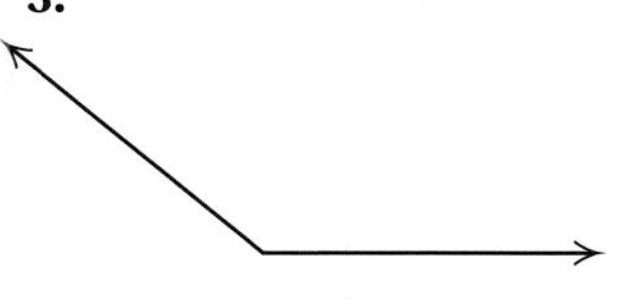

4.

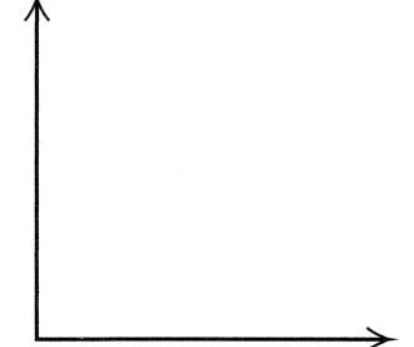

5.–8. Classify each of the angles in Exercises 1–4 as right, straight, acute, or obtuse. [6.1c]

Use the following triangle for Exercises 9–11.

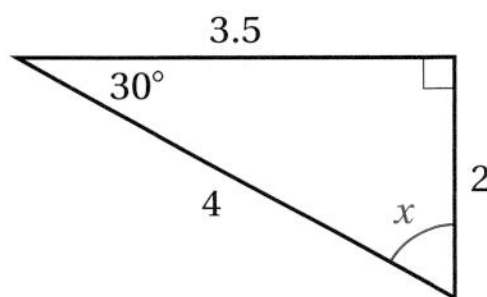

9. Find the missing angle measure. [6.1f]

10. Classify the triangle as equilateral, isosceles, or scalene. [6.1e]

11. Classify the triangle as right, obtuse, or acute. [6.1e]

12. Find the sum of the angle measures of a hexagon. [6.1f]

Find the perimeter. [6.2a]

13.

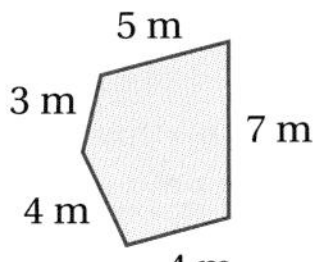

14.

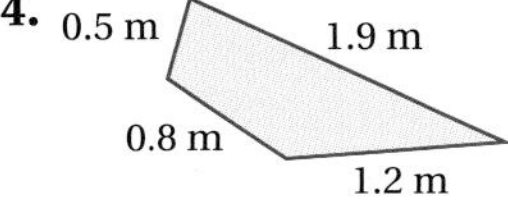

15. *Tennis Court.* The dimensions of a standard-sized tennis court are 78 ft by 36 ft. Find the perimeter and the area of the tennis court. [6.2b], [6.3c]

Find the perimeter and the area. [6.2a], [6.3a]

16.

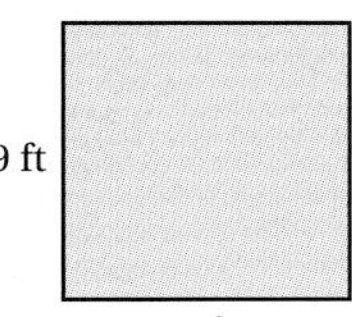

17.

Find the area. [6.3b]

18.

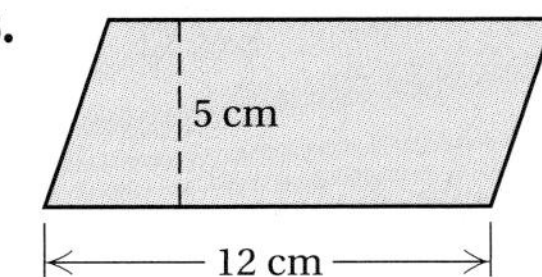

19.

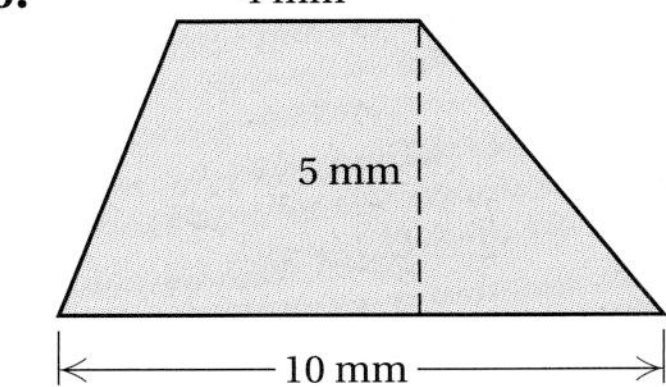

20.

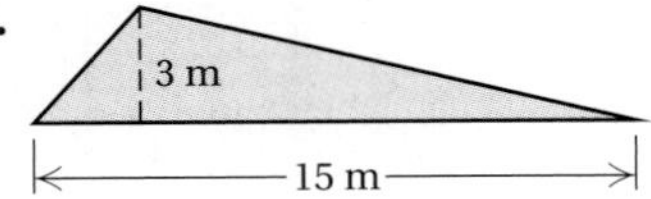

21.

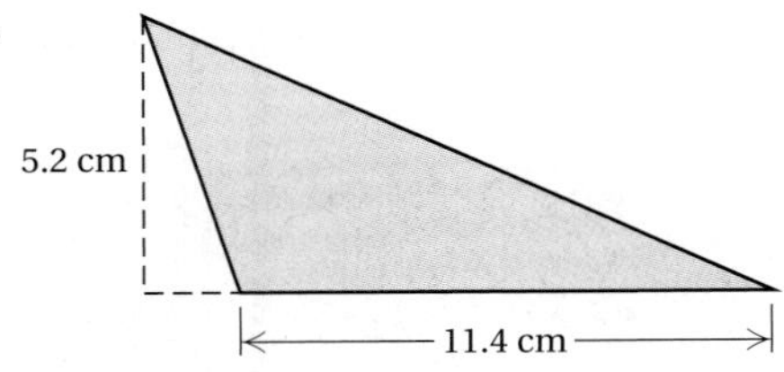

22.

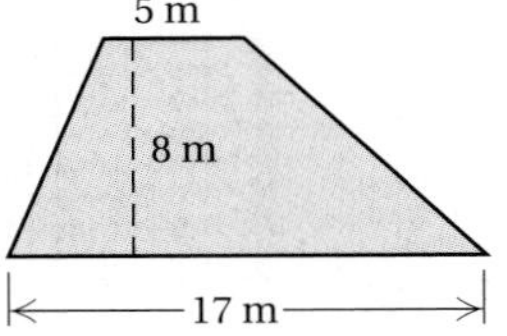

23.

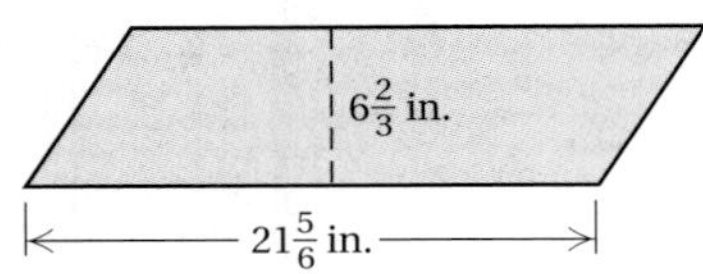

24. *Seeded Area.* A grassy area is to be seeded around three sides of a building and has equal width on the three sides, as shown below. What is the seeded area? [6.3c]

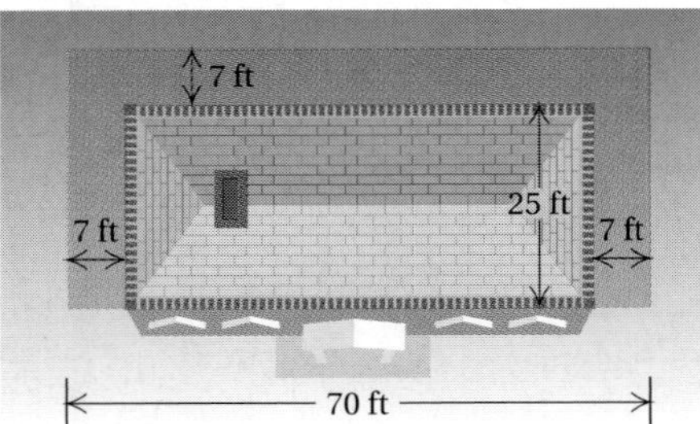

Find the length of a radius of the circle. [6.4a]

25.

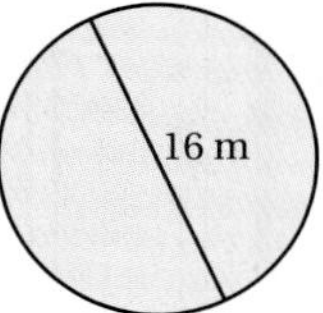

26.

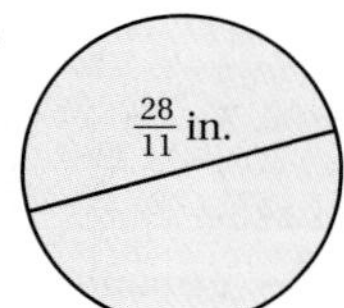

Find the length of a diameter of the circle. [6.4a]

27.

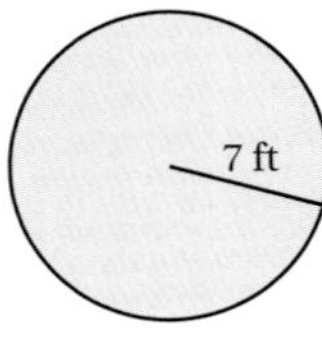

28.

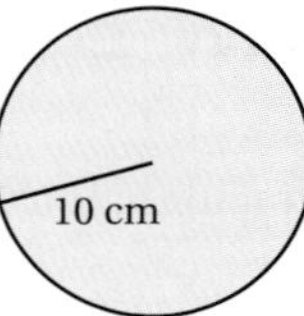

29. Find the circumference of the circle in Exercise 25. Use 3.14 for π. [6.4b]

30. Find the circumference of the circle in Exercise 26. Use $\frac{22}{7}$ for π. [6.4b]

31. Find the area of the circle in Exercise 25. Use 3.14 for π. [6.4c]

32. Find the area of the circle in Exercise 26. Use $\frac{22}{7}$ for π. [6.4c]

33. Find the area of the shaded region. Use 3.14 for π. [6.4d]

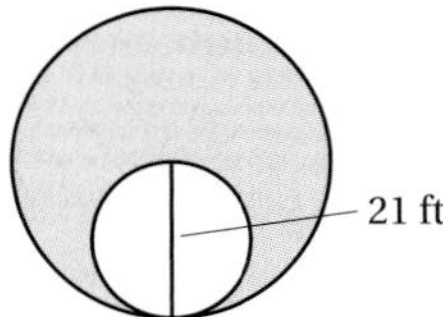

Find the volume and the surface area. [6.5a]

34.

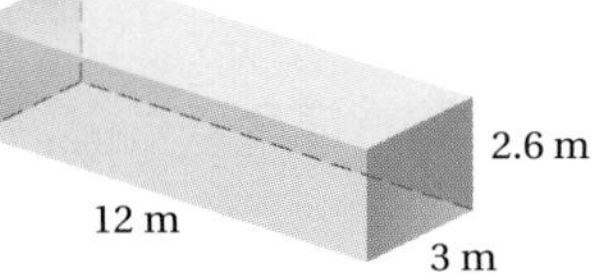

35.

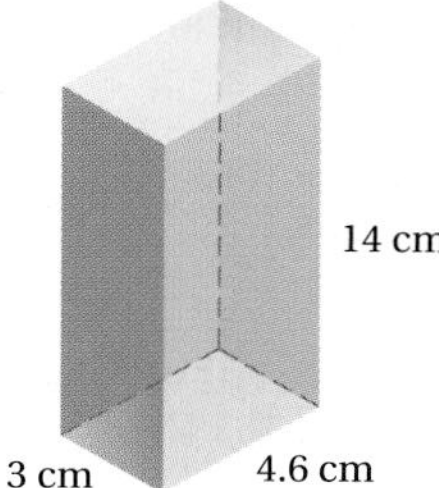

Find the volume. Use 3.14 for π.

36. [6.5b] **37.** [6.5d]

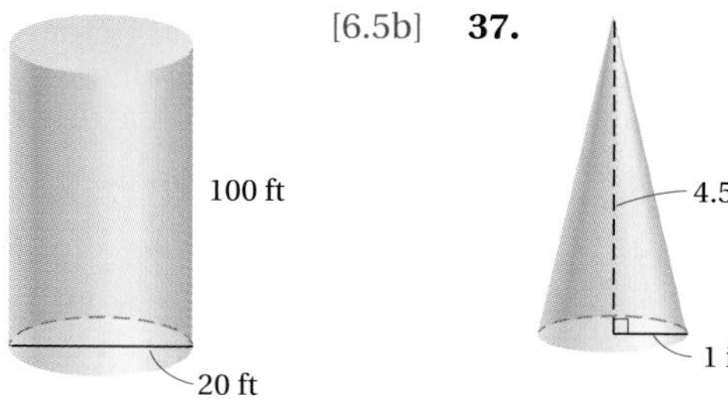

38. [6.5c]

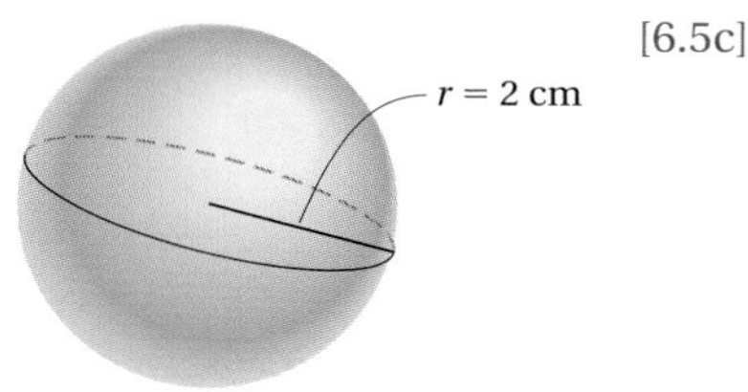

39. [6.5b]

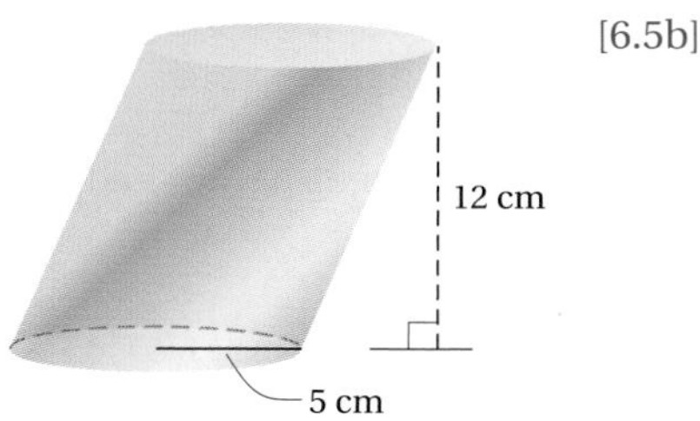

Find the measure of a complement of an angle with the given measure. [6.6a]

40. 82°

41. 5°

Find the measure of a supplement of an angle with the given measure. [6.6a]

42. 33°

43. 133°

44. In this figure, $m\angle 1 = 38°$ and $m\angle 5 = 105°$. Find $m\angle 2$, $m\angle 3$, $m\angle 4$, and $m\angle 6$. [6.6c]

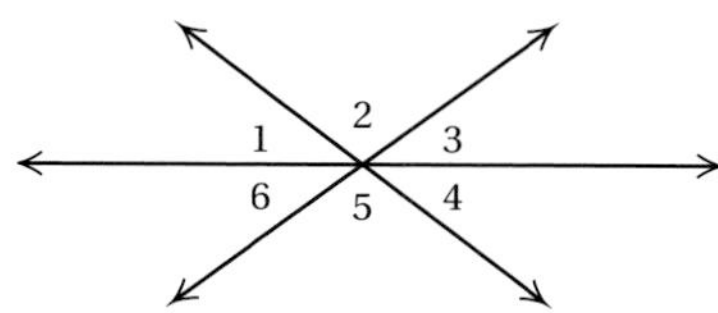

45. In this figure, identify **(a)** all pairs of corresponding angles, **(b)** all interior angles, and **(c)** all pairs of alternate interior angles. [6.6d]

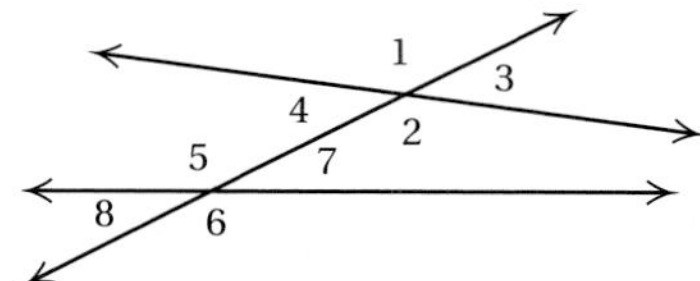

46. If $m \parallel n$ and $m\angle 4 = 135°$, what are the measures of the other angles? [6.6d]

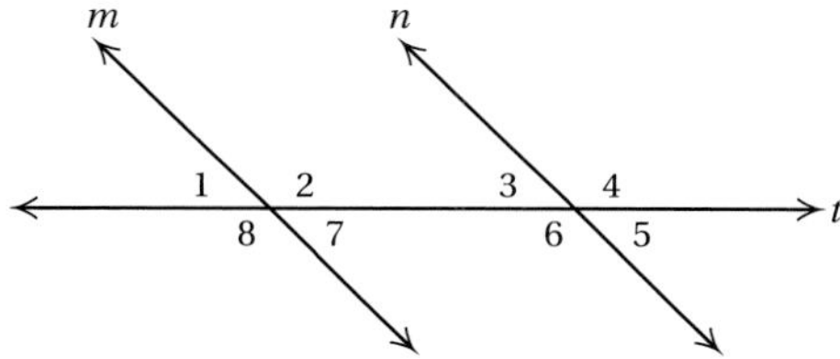

Name the corresponding parts of these congruent triangles. [6.7a]

47. $\triangle DHJ \cong \triangle RZK$

48.

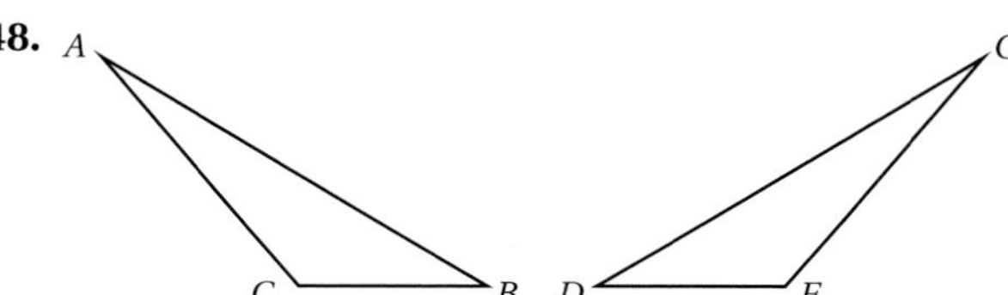

Which property (if any) should be used to show that the pair of triangles is congruent? [6.7a]

49.

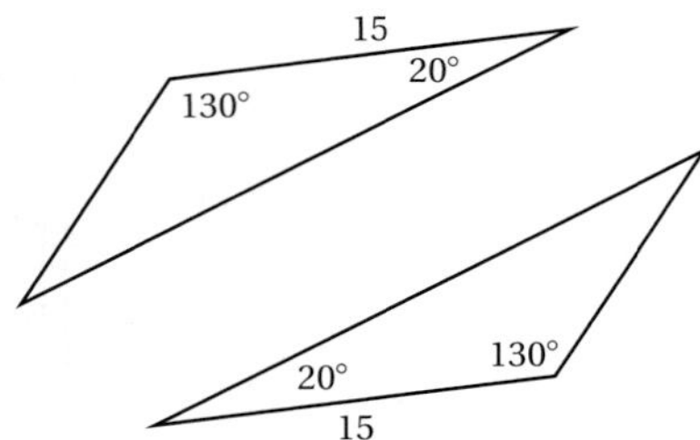

50.

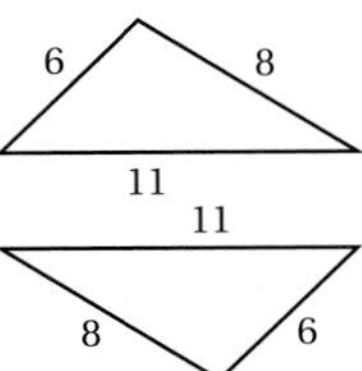

51.

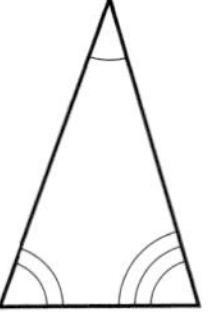

52. J is the midpoint of $\overline{IK}$ and $\overline{HI} \parallel \overline{KL}$. Explain why $\triangle JIH \cong \triangle JKL$. [6.7a]

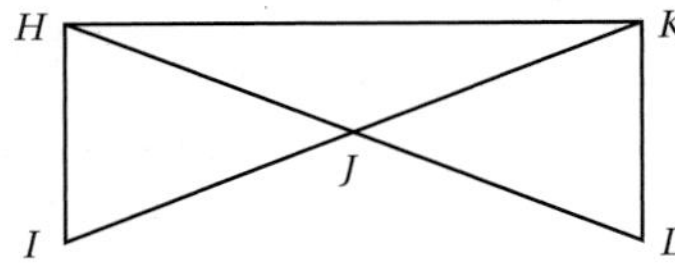

53. Find the measures of the angles and the lengths of the sides of this parallelogram. [6.7b]

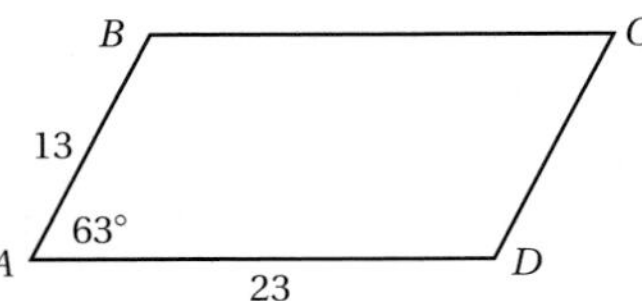

54. If $\triangle CQW \sim \triangle FAS$, name the congruent angles and the proportional sides. [6.8a]

55. If $\triangle NMO \sim \triangle STR$, find MO. [6.8b]

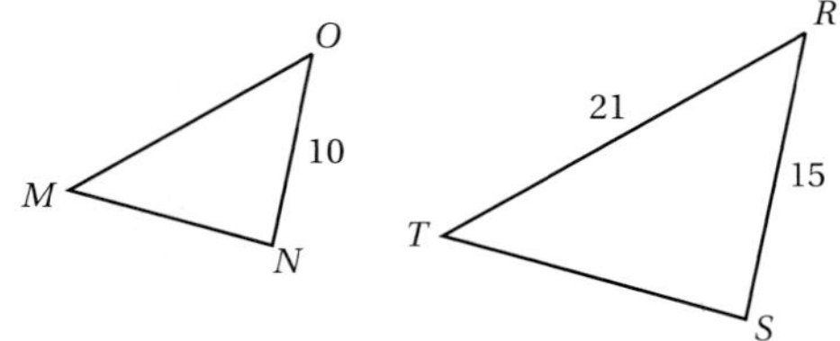

56. ^{D}W Describe the difference among linear, area, and volume units of measure. [6.2a], [6.3a], [6.5a]

57. ^{D}W Which occupies more volume: two spheres, each with radius r, or one sphere with radius $2r$? Explain why. [6.5c]

SYNTHESIS

58. A square is cut in half so that the perimeter of the resulting rectangle is 30 ft. Find the area of the original square. [6.2a], [6.3a]

59. Find the area, in square meters, of the shaded region. [6.3c]

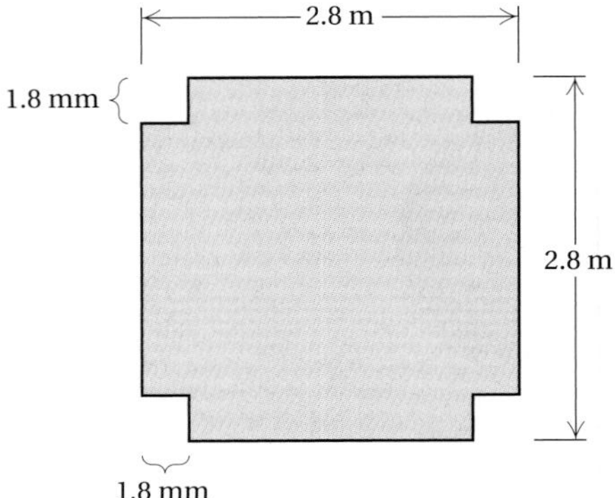

60. Find the area, in square centimeters, of the shaded region. [6.3c]

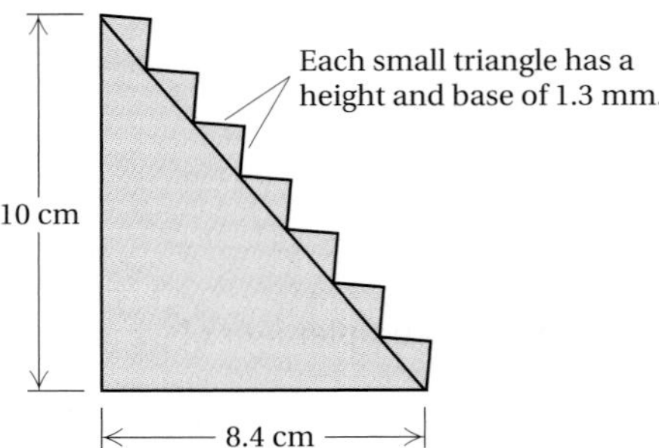

6 Chapter Test

For Extra Help

Work It Out! Chapter Test Video Solutions on CD

Use a protractor to measure the angle.

1.

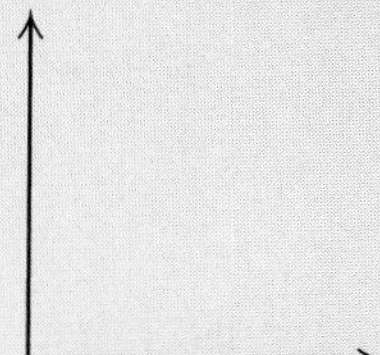

2.

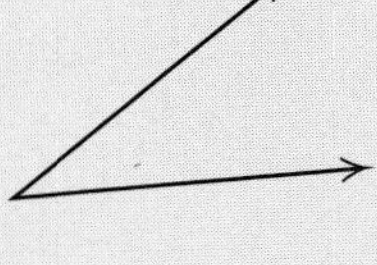

3.

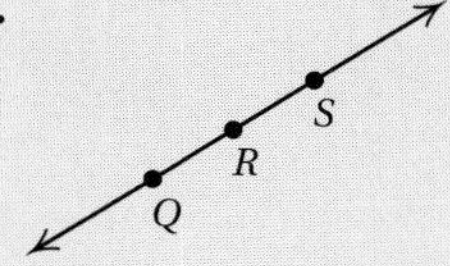

4.

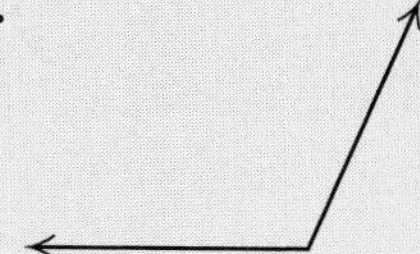

5.–8. Classify each of the angles in Exercises 1–4 as right, straight, acute, or obtuse.

Use the following triangle for Exercises 9–11.

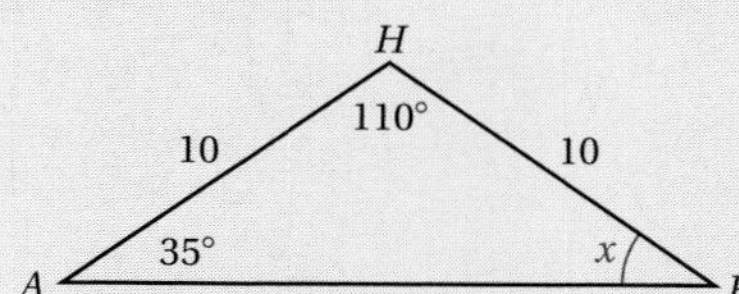

9. Find the missing angle measure.

10. Classify the triangle as equilateral, isosceles, or scalene.

11. Classify the triangle as right, obtuse, or acute.

12. Find the sum of the angle measures of a pentagon.

Find the perimeter and the area.

13.

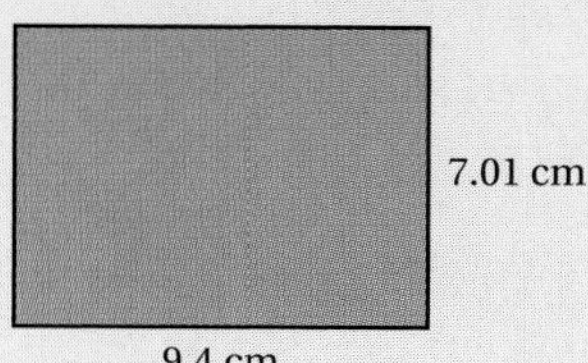

14.

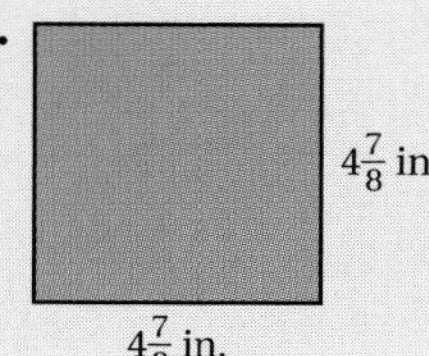

Find the area.

15.

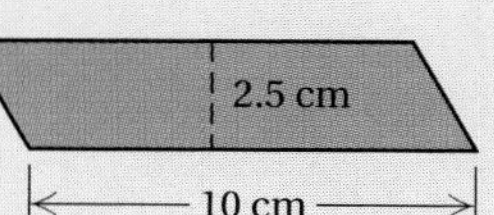

16.

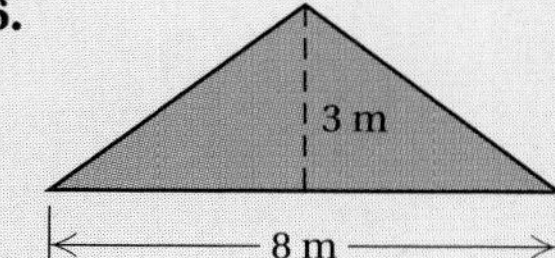

17.

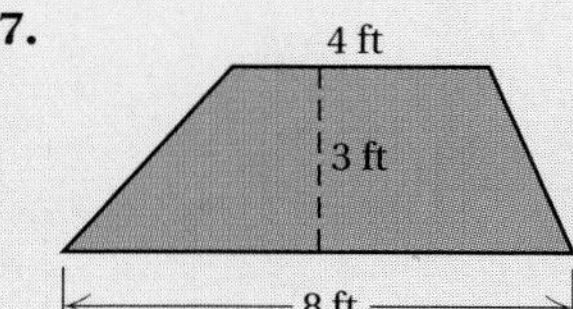

18. Find the length of a diameter of this circle.

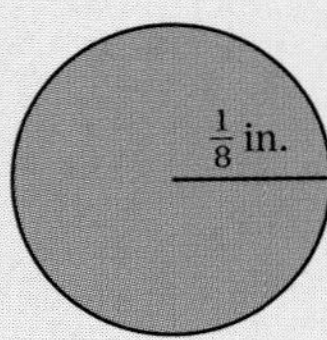

19. Find the length of a radius of this circle.

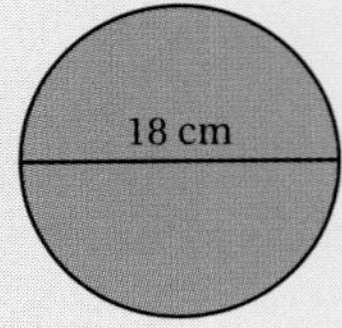

20. Find the circumference of the circle in Exercise 18. Use $\frac{22}{7}$ for π.

21. Find the area of the circle in Exercise 19. Use 3.14 for π.

22. Find the perimeter and the area of the shaded region. Use 3.14 for π.

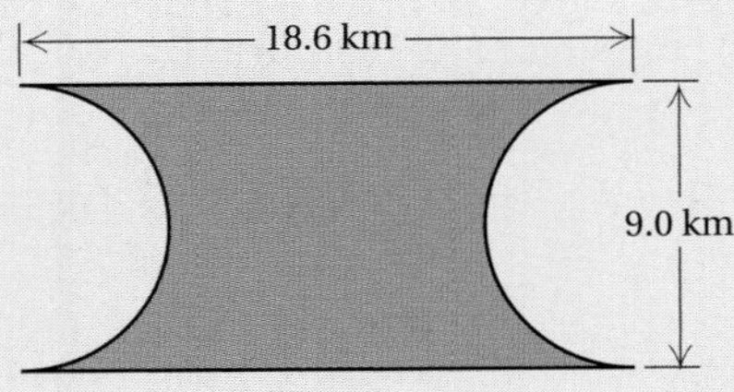

23. Find the volume and the surface area.

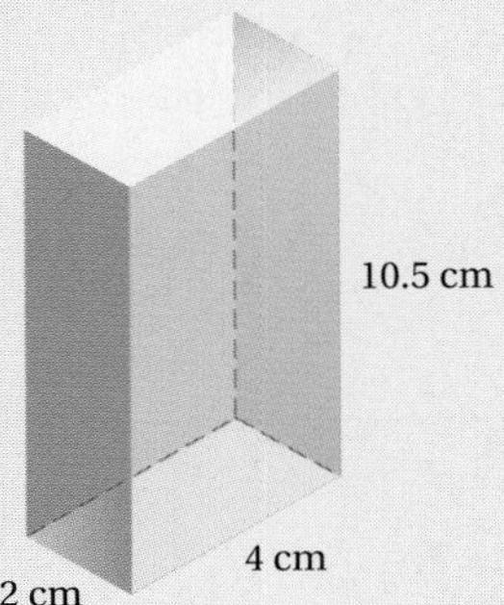

24. A twelve-box carton of 12-oz juice boxes comes in a rectangular box $10\frac{1}{2}$ in. by 8 in. by 5 in. What is the volume of the carton?

Find the volume. Use 3.14 for π.

25.

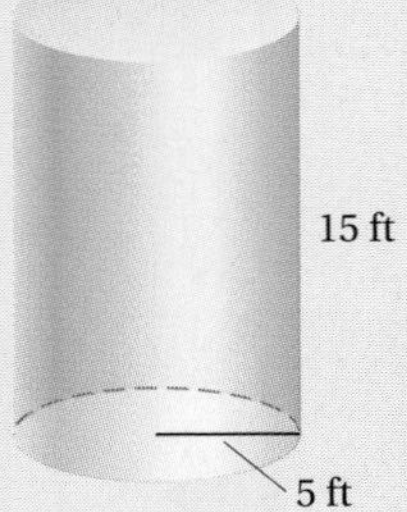

26.

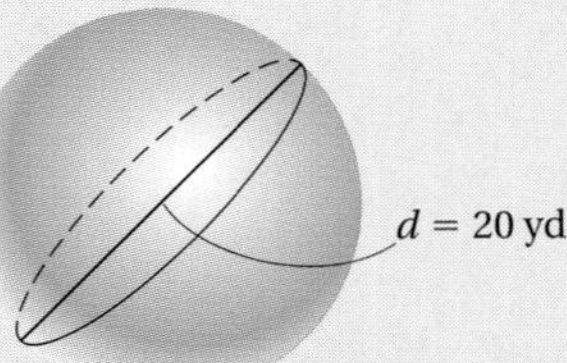

27.

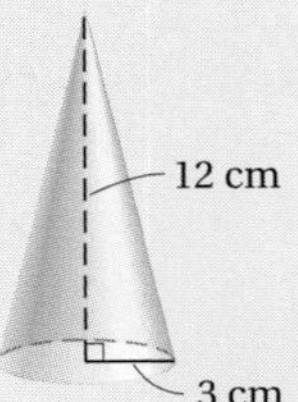

28. Find the measure of a supplement of an angle of 31°.

29. Find the measure of a complement of an angle of 79°.

30. In the figure, $m\angle 1 = 62°$ and $m\angle 5 = 110°$. Find $m\angle 2$, $m\angle 3$, $m\angle 4$, and $m\angle 6$.

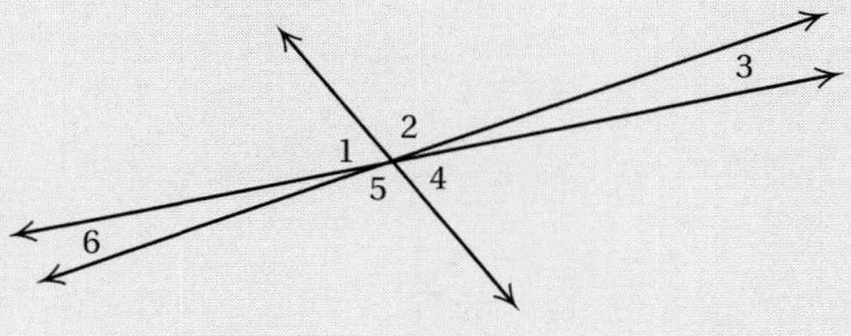

31. If $m \parallel n$ and $m\angle 4 = 120°$, what are the measures of the other angles?

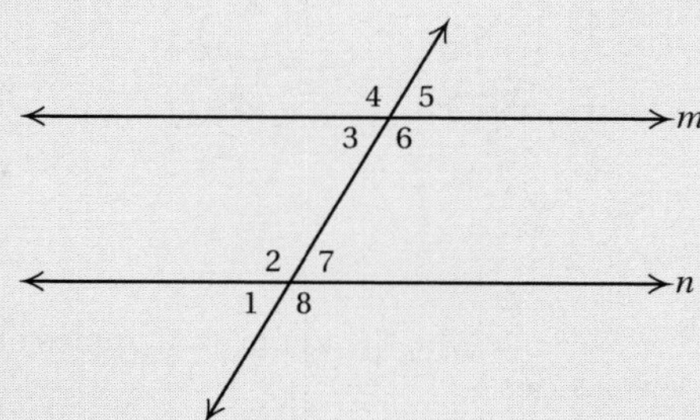

32. Name the corresponding parts of these congruent triangles: $\triangle CWS \cong \triangle ATZ$.

Which property (if any) would you use to show that $\triangle RST \cong \triangle DEF$ with the given information?

33. $\overline{RS} \cong \overline{DE}$, $\overline{RT} \cong \overline{DF}$, and $\angle R \cong \angle D$

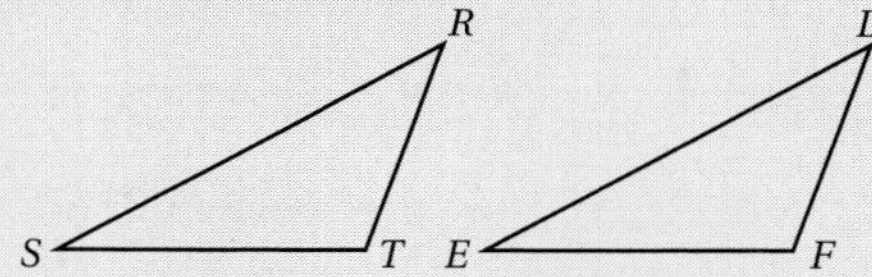

34. $\angle R \cong \angle D$, $\angle S \cong \angle E$, and $\angle T \cong \angle F$

35. $\overline{RS} \cong \overline{DE}$, $\angle R \cong \angle D$, and $\angle S \cong \angle E$

36. $\angle R \cong \angle D$, $\overline{RT} \cong \overline{DF}$, and $\overline{ST} \cong \overline{EF}$

37. The perimeter of $\square DEFG$ is 62. Find the measures of the angles and the lengths of the sides.

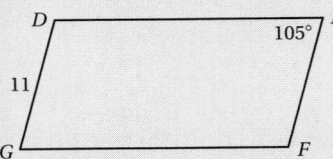

38. In $\square JKLM$, $JN = 3.2$ and $KN = 3$. Find the lengths of the diagonals, $\overline{LJ}$ and $\overline{KM}$.

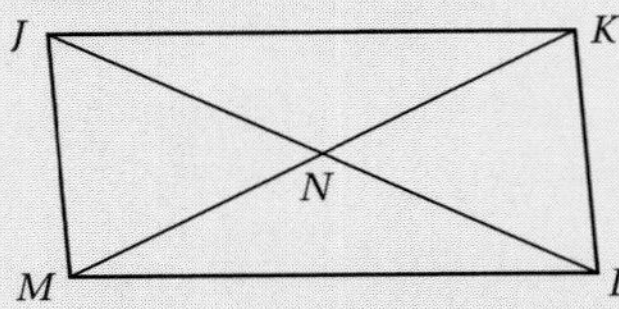

39. If $\triangle ERS \sim \triangle TGF$, name the congruent angles and the proportional sides.

40. If $\triangle GTR \sim \triangle ZEK$, find EK and ZK.

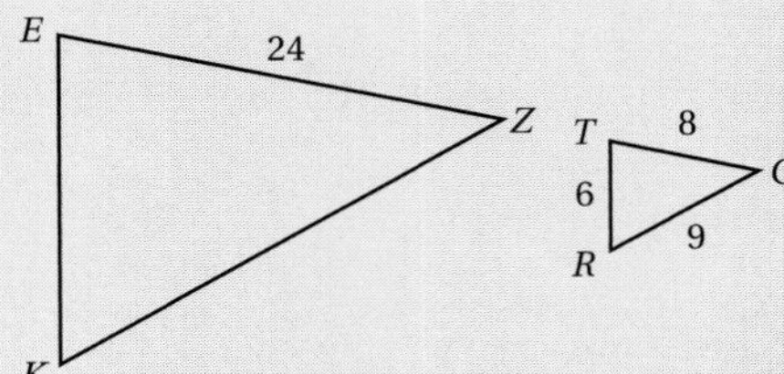

SYNTHESIS

Find the area of the shaded region. (Note that the figures are not drawn in perfect proportion.) Give the answer in square feet.

41.

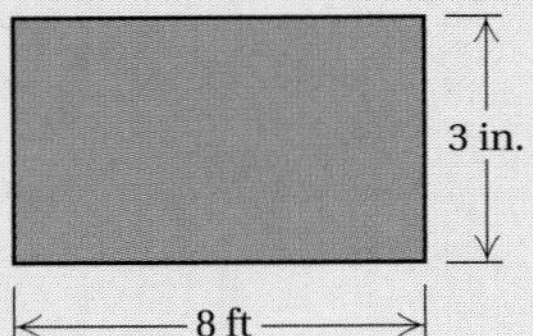

42.

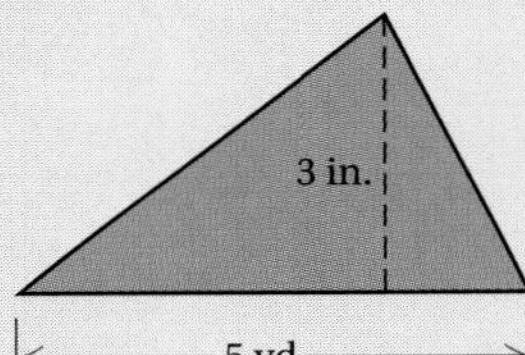

Find the volume of the solid. (Note that the solids are not drawn in perfect proportion.) Give the answer in cubic feet. Use 3.14 for π.

43.

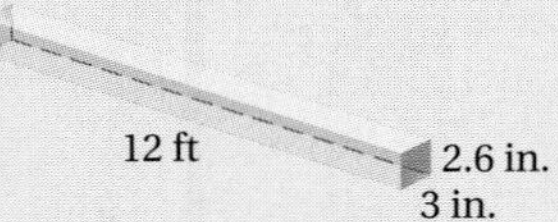

44.

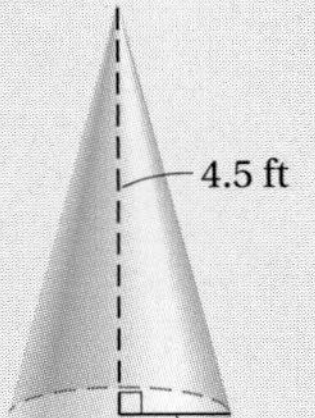

45.

Introduction to Real Numbers and Algebraic Expressions

Real-World Application

Surface temperatures on Mars vary from $-128°C$ during polar night to $27°C$ at the equator during midday at the closest point in orbit to the sun. Find the difference between the highest value and the lowest value in this temperature range.

Source: Mars Institute

This problem appears as Example 13 in Section 7.4.

Objectives

a Evaluate algebraic expressions by substitution.

b Translate phrases to algebraic expressions.

1. Translate this problem to an equation. Use the graph below.

Mountain Peaks. There are 92 mountain peaks in the United States higher than 14,000 ft. The bar graph below shows data for six of these. How much higher is Mt. Fairweather than Mt. Rainer?

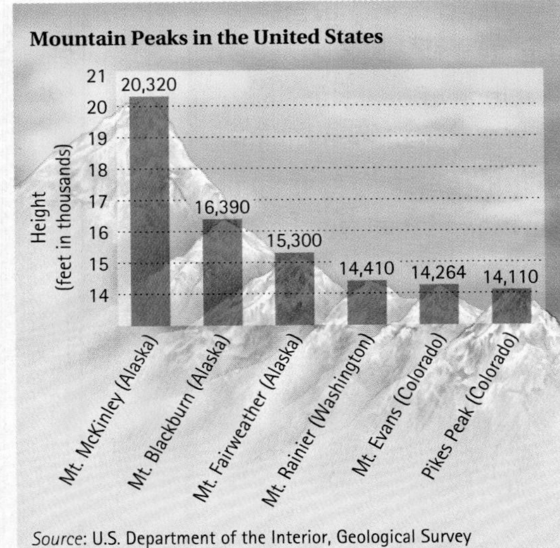

Answer on page A-17

7.1 INTRODUCTION TO ALGEBRA

The study of algebra involves the use of equations to solve problems. Equations are constructed from algebraic expressions. The purpose of this section is to introduce you to the types of expressions encountered in algebra.

a Evaluating Algebraic Expressions

In arithmetic, you have worked with expressions such as

$$49 + 75, \quad 8 \times 6.07, \quad 29 - 14, \quad \text{and} \quad \frac{5}{6}.$$

In algebra, we use certain letters for numbers and work with *algebraic expressions* such as

$$x + 75, \quad 8 \times y, \quad 29 - t, \quad \text{and} \quad \frac{a}{b}.$$

Sometimes a letter can represent various numbers. In that case, we call the letter a **variable.** Let a = your age. Then a is a variable since a changes from year to year. Sometimes a letter can stand for just one number. In that case, we call the letter a **constant.** Let b = your date of birth. Then b is a constant.

Where do algebraic expressions occur? Most often we encounter them when we are solving applied problems. For example, consider the bar graph shown at left, one that we might find in a book or magazine. Suppose we want to know how much higher Mt. McKinley is than Mt. Evans. Using arithmetic, we might simply subtract. But let's see how we can find this out using algebra. We translate the problem into a statement of equality, an equation. It could be done as follows:

Height of Mt. Evans	plus	How much more	is	Height of Mt. McKinley
↓	↓	↓	↓	↓
14,264	+	x	=	20,320.

Note that we have an algebraic expression, $14{,}264 + x$, on the left of the equals sign. To find the number x, we can subtract 14,264 on both sides of the equation:

$$14{,}264 + x = 20{,}320$$
$$14{,}264 + x - 14{,}264 = 20{,}320 - 14{,}264$$
$$x = 6056.$$

This value of x gives the answer, 6056 ft.

We call $14{,}264 + x$ an *algebraic expression* and $14{,}264 + x = 20{,}320$ an *algebraic equation.* Note that there is no equals sign, =, in an algebraic expression.

In arithmetic, you probably would do this subtraction without ever considering an equation. *In algebra, more complex problems are difficult to solve without first writing an equation.*

Do Exercise 1.

An **algebraic expression** consists of variables, constants, numerals, and operation signs. When we replace a variable with a number, we say that we are **substituting** for the variable. This process is called **evaluating the expression.**

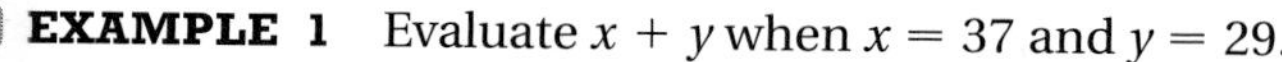

EXAMPLE 1 Evaluate $x + y$ when $x = 37$ and $y = 29$.

We substitute 37 for x and 29 for y and carry out the addition:

$$x + y = 37 + 29 = 66.$$

The number 66 is called the **value** of the expression.

Algebraic expressions involving multiplication can be written in several ways. For example, "8 times a" can be written as

$$8 \times a, \quad 8 \cdot a, \quad (8a), \quad \text{or simply} \quad 8a.$$

Two letters written together without an operation symbol, such as ab, also indicate a multiplication.

EXAMPLE 2 Evaluate $3y$ when $y = 14$.

$$3y = 3(14) = 42$$

Do Exercises 2–4.

EXAMPLE 3 *Area of a Rectangle.* The area A of a rectangle of length l and width w is given by the formula $A = lw$. Find the area when l is 24.5 in. and w is 16 in.

We substitute 24.5 in. for l and 16 in. for w and carry out the multiplication:

$$\begin{aligned} A = lw &= (24.5 \text{ in.})(16 \text{ in.}) \\ &= (24.5)(16)(\text{in.})(\text{in.}) \\ &= 392 \text{ in}^2, \text{ or } 392 \text{ square inches.} \end{aligned}$$

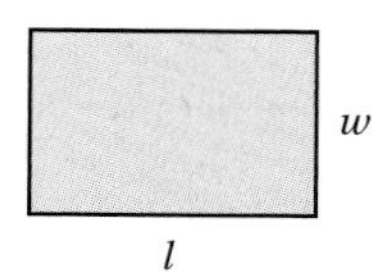

Do Exercise 5.

Algebraic expressions involving division can also be written in several ways. For example, "8 divided by t" can be written as

$$8 \div t, \quad \frac{8}{t}, \quad 8/t, \quad \text{or} \quad 8 \cdot \frac{1}{t},$$

where the fraction bar is a division symbol.

EXAMPLE 4 Evaluate $\frac{a}{b}$ when $a = 63$ and $b = 9$.

We substitute 63 for a and 9 for b and carry out the division:

$$\frac{a}{b} = \frac{63}{9} = 7.$$

2. Evaluate $a + b$ when $a = 38$ and $b = 26$.

3. Evaluate $x - y$ when $x = 57$ and $y = 29$.

4. Evaluate $4t$ when $t = 15$.

5. Find the area of a rectangle when l is 24 ft and w is 8 ft.

6. Evaluate a/b when $a = 200$ and $b = 8$.

7. Evaluate $10p/q$ when $p = 40$ and $q = 25$.

Answers on page A-17

8. **Motorcycle Travel.** Find the time it takes to travel 660 mi if the speed is 55 mph.

EXAMPLE 5 Evaluate $\frac{12m}{n}$ when $m = 8$ and $n = 16$.

$$\frac{12m}{n} = \frac{12 \cdot 8}{16} = \frac{96}{16} = 6$$

Do Exercises 6 and 7 on the preceding page.

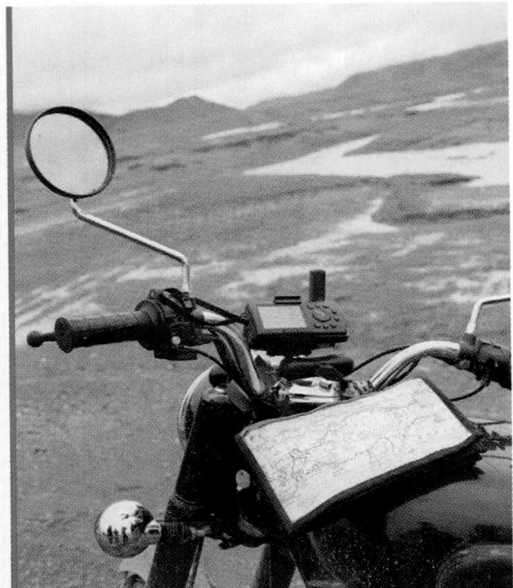

EXAMPLE 6 *Motorcycle Travel.* Ed takes a trip on his motorcycle. He wants to travel 660 mi on a particular day. The time t, in hours, that it takes to travel 660 mi is given by

$$t = \frac{660}{r},$$

where r is the speed of Ed's motorcycle. Find the time of travel if the speed r is 60 mph.

We substitute 60 for r and carry out the division:

$$t = \frac{660}{r} = \frac{660}{60} = 11 \text{ hr.}$$

Do Exercise 8.

CALCULATOR CORNER

Evaluating Algebraic Expressions *To the student and the instructor:* This book contains a series of *optional* discussions on using a calculator. A calculator is *not* a requirement for this textbook. There are many kinds of calculators and different instructions for their usage. We have included instructions here for the scientific keys on a graphing calculator such as a TI-84 Plus. Be sure to consult your user's manual as well. Also, check with your instructor about whether you are allowed to use a calculator in the course.

We can evaluate algebraic expressions on a calculator by making the appropriate substitutions, keeping in mind the rules for order of operations, and then carrying out the resulting calculations. To evaluate $12m/n$ when $m = 8$ and $n = 16$, as in Example 5, we enter $12 \cdot 8/16$ by pressing 1 2 × 8 ÷ 1 6 ENTER. The result is 6.

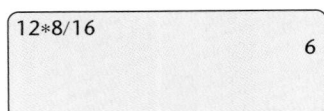

Exercises: Evaluate.

1. $\frac{12m}{n}$, when $m = 42$ and $n = 9$
2. $a + b$, when $a = 8.2$ and $b = 3.7$
3. $b - a$, when $a = 7.6$ and $b = 9.4$
4. $27xy$, when $x = 12.7$ and $y = 100.4$
5. $3a + 2b$, when $a = 2.9$ and $b = 5.7$
6. $2a + 3b$, when $a = 7.3$ and $b = 5.1$

Answer on page A-17

b Translating to Algebraic Expressions

In algebra, we translate problems to equations. The different parts of an equation are translations of word phrases to algebraic expressions. It is easier to translate if we know that certain words often translate to certain operation symbols.

To the student:

In the preface, at the front of the text, you will find a Student Organizer card. This pullout card will help you keep track of important dates and useful contact information. You can also use it to plan time for class, study, work, and relaxation. By managing your time wisely, you will provide yourself the best possible opportunity to be successful in this course.

KEY WORDS, PHRASES, AND CONCEPTS

ADDITION (+)	SUBTRACTION (−)	MULTIPLICATION (·)	DIVISION (÷)
add	subtract	multiply	divide
added to	subtracted from	multiplied by	divided by
sum	difference	product	quotient
total	minus	times	
plus	less than	of	
more than	decreased by		
increased by	take away		

EXAMPLE 7 Translate to an algebraic expression:

Twice (or two times) some number.

Think of some number, say, 8. We can write 2 times 8 as 2×8, or $2 \cdot 8$. We multiplied by 2. Do the same thing using a variable. We can use any variable we wish, such as x, y, m, or n. Let's use y to stand for some number. If we multiply by 2, we get an expression

$y \times 2$, $2 \times y$, $2 \cdot y$, or $2y$.

In algebra, $2y$ is the expression generally used.

EXAMPLE 8 Translate to an algebraic expression:

Thirty-eight percent of some number.

Let $n =$ the number. The word "of" translates to a multiplication symbol, so we get the following expressions as a translation:

$38\% \cdot n$, $0.38 \times n$, or $0.38n$.

EXAMPLE 9 Translate to an algebraic expression:

Seven less than some number.

We let

x represent the number.

Now if the number were 23, then 7 less than 23 is 16, that is, $(23 - 7)$, not $(7 - 23)$. If we knew the number to be 345, then the translation would be $345 - 7$. If the number is x, then the translation is

$x - 7$.

Caution!

Note that $7 - x$ is *not* a correct translation of the expression in Example 9. The expression $7 - x$ is a translation of "seven minus some number" or "some number less than seven."

Translate to an algebraic expression.

9. Eight less than some number

10. Eight more than some number

11. Four less than some number

12. Half of a number

13. Six more than eight times some number

14. The difference of two numbers

15. Fifty-nine percent of some number

16. Two hundred less than the product of two numbers

17. The sum of two numbers

Answers on page A-17

EXAMPLE 10 Translate to an algebraic expression:

Eighteen more than a number.

We let

$t =$ the number.

Now if the number were 26, then the translation would be $26 + 18$, or $18 + 26$. If we knew the number to be 174, then the translation would be $174 + 18$, or $18 + 174$. If the number is t, then the translation is

$$t + 18, \quad \text{or} \quad 18 + t.$$

EXAMPLE 11 Translate to an algebraic expression:

A number divided by 5.

We let

$m =$ the number.

Now if the number were 76, then the translation would be $76 \div 5$, or $76/5$, or $\frac{76}{5}$. If the number were 213, then the translation would be $213 \div 5$, or $213/5$, or $\frac{213}{5}$. If the number is m, then the translation is

$$m \div 5, \quad m/5, \quad \text{or} \quad \frac{m}{5}.$$

EXAMPLE 12 Translate each phrase to an algebraic expression.

PHRASE	ALGEBRAIC EXPRESSION
Five more than some number	$n + 5$, or $5 + n$
Half of a number	$\frac{1}{2}t$, $\frac{t}{2}$, or $t/2$
Five more than three times some number	$3p + 5$, or $5 + 3p$
The difference of two numbers	$x - y$
Six less than the product of two numbers	$mn - 6$
Seventy-six percent of some number	$76\%z$, or $0.76z$
Four less than twice some number	$2x - 4$

Do Exercises 9–17.

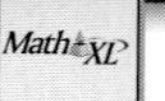

7.1 EXERCISE SET

For Extra Help

MathXL | MyMathLab | InterAct Math | Math Tutor Center | Video Lectures on CD Disc 4 | Student's Solutions Manual

a Substitute to find values of the expressions in each of the following applied problems.

1. *Commuting Time.* It takes Erin 24 min less time to commute to work than it does George. Suppose that the variable x stands for the time it takes George to get to work. Then $x - 24$ stands for the time it takes Erin to get to work. How long does it take Erin to get to work if it takes George 56 min? 93 min? 105 min?

2. *Enrollment Costs.* At Emmett Community College, it costs \$600 to enroll in the 8 A.M. section of Elementary Algebra. Suppose that the variable n stands for the number of students who enroll. Then $600n$ stands for the total amount of money collected for this course. How much is collected if 34 students enroll? 78 students? 250 students?

3. *Area of a Triangle.* The area A of a triangle with base b and height h is given by $A = \frac{1}{2}bh$. Find the area when $b = 45$ m (meters) and $h = 86$ m.

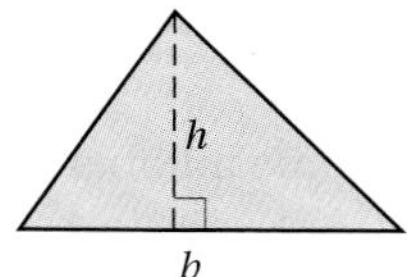

4. *Area of a Parallelogram.* The area A of a parallelogram with base b and height h is given by $A = bh$. Find the area of the parallelogram when the height is 15.4 cm (centimeters) and the base is 6.5 cm.

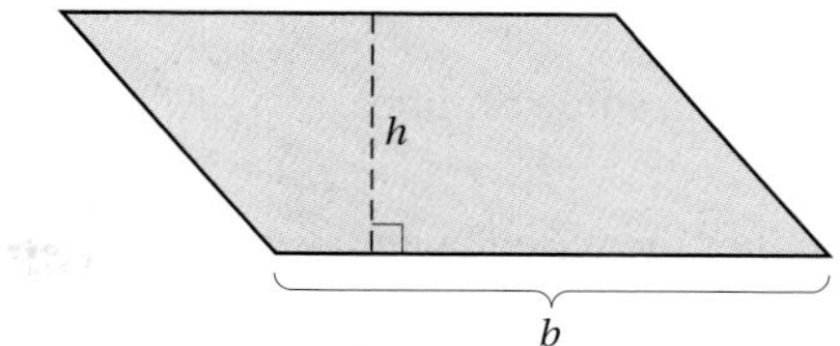

5. *Distance Traveled.* A driver who drives at a constant speed of r mph for t hr will travel a distance d mi given by $d = rt$ mi. How far will a driver travel at a speed of 65 mph for 4 hr?

6. *Simple Interest.* The simple interest I on a principal of P dollars at interest rate r for time t, in years, is given by $I = Prt$. Find the simple interest on a principal of \$4800 at 9% for 2 yr. (*Hint*: 9% = 0.09.)

7. *Hockey Goal.* The front of a regulation hockey goal is a rectangle that is 6 ft wide and 4 ft high. Find its area.

Source: National Hockey League

8. *Zoology.* A great white shark has triangular teeth. Each tooth measures about 5 cm across the base and has a height of 6 cm. Find the surface area of one side of one tooth. (See Exercise 3.)

Evaluate.

9. $8x$, when $x = 7$

10. $6y$, when $y = 7$

11. $\frac{c}{d}$, when $c = 24$ and $d = 3$

12. $\frac{p}{q}$, when $p = 16$ and $q = 2$

13. $\frac{3p}{q}$, when $p = 2$ and $q = 6$

14. $\frac{5y}{z}$, when $y = 15$ and $z = 25$

15. $\frac{x + y}{5}$, when $x = 10$ and $y = 20$

16. $\frac{p + q}{2}$, when $p = 2$ and $q = 16$

17. $\frac{x - y}{8}$, when $x = 20$ and $y = 4$

18. $\frac{m - n}{5}$, when $m = 16$ and $n = 6$

b Translate each phrase to an algebraic expression. Use any letter for the variable unless directed otherwise.

19. Seven more than some number

20. Nine more than some number

21. Twelve less than some number

22. Fourteen less than some number

23. Some number increased by four

24. Some number increased by thirteen

25. b more than a

26. c more than d

27. x divided by y

28. c divided by h

29. x plus w

30. s added to t

31. m subtracted from n

32. p subtracted from q

33. The sum of two numbers

34. The sum of nine and some number

35. Twice some number

36. Three times some number

37. Three multiplied by some number

38. The product of eight and some number

39. Six more than four times some number

40. Two more than six times some number

41. Eight less than the product of two numbers

42. The product of two numbers minus seven

43. Five less than twice some number

44. Six less than seven times some number

45. Three times some number plus eleven

46. Some number times 8 plus 5

47. The sum of four times a number plus three times another number

48. Five times a number minus eight times another number

49. The product of 89% and your salary

50. 67% of the women attending

51. Your salary after a 5% salary increase if your salary before the increase was s

52. The price of a blouse after a 30% reduction if the price before the reduction was P

53. Danielle drove at a speed of 65 mph for t hours. How far did Danielle travel?

54. Juan has d dollars before spending $29.95 on a DVD of the movie *Chicago*. How much did Juan have after the purchase?

55. Lisa had $50 before spending x dollars on pizza. How much money remains?

56. Dino drove his pickup truck at 55 mph for t hours. How far did he travel?

57. D_W If the length of a rectangle is doubled, does the area double? Why or why not?

58. D_W If the height and the base of a triangle are doubled, what happens to the area? Explain.

SKILL MAINTENANCE

Find the prime factorization. [1.7d]

59. 54 **60.** 32 **61.** 108 **62.** 192 **63.** 1023

Find the LCM. [1.9a]

64. 6, 18 **65.** 6, 24, 32 **66.** 10, 20, 30 **67.** 16, 24, 32 **68.** 18, 36, 44

SYNTHESIS

Evaluate.

69. $\dfrac{a - 2b + c}{4b - a}$, when $a = 20$, $b = 10$, and $c = 5$

70. $\dfrac{x}{y} - \dfrac{5}{x} + \dfrac{2}{y}$, when $x = 30$ and $y = 6$

71. $\dfrac{12 - c}{c + 12b}$, when $b = 1$ and $c = 12$

72. $\dfrac{2w - 3z}{7y}$, when $w = 5$, $y = 6$, and $z = 1$

7.2 THE REAL NUMBERS

Objectives

a State the integer that corresponds to a real-world situation.

b Graph rational numbers on a number line.

c Convert from fraction notation to decimal notation for a rational number.

d Determine which of two real numbers is greater and indicate which, using < or >; given an inequality like $a > b$, write another inequality with the same meaning. Determine whether an inequality like $-3 \leq 5$ is true or false.

e Find the absolute value of a real number.

Study Tips

THE AW MATH TUTOR CENTER

www.aw-bc.com/tutorcenter

The AW Math Tutor Center is staffed by highly qualified mathematics instructors who provide students with tutoring on text examples and odd-numbered exercises. Tutoring is provided free to students who have bought a new textbook with a special access card bound with the book. Tutoring is available by toll-free telephone, toll-free fax, e-mail, and the Internet. White-board technology allows tutors and students to actually see problems worked while they "talk" live in real time during the tutoring sessions. If you purchased a book without this card, you can purchase an access code through your bookstore using ISBN 0-201-72170-8. (This is also discussed in the Preface.)

A **set** is a collection of objects. (See Appendix E for more on sets.) For our purposes, we will most often be considering sets of numbers. One way to name a set uses what is called **roster notation.** For example, roster notation for the set containing the numbers 0, 2, and 5 is $\{0, 2, 5\}$.

Sets that are part of other sets are called **subsets.** In this section, we become acquainted with the set of *real numbers* and its various subsets.

Two important subsets of the real numbers are listed below using roster notation.

NATURAL NUMBERS

The set of **natural numbers** $= \{1, 2, 3, \ldots\}$. These are the numbers used for counting.

WHOLE NUMBERS

The set of **whole numbers** $= \{0, 1, 2, 3, \ldots\}$. This is the set of natural numbers with 0 included.

We can represent these sets on a number line. The natural numbers are those to the right of zero. The whole numbers are the natural numbers and zero.

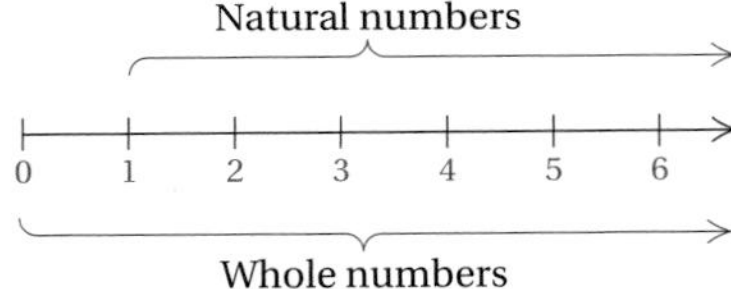

We create a new set, called the *integers,* by starting with the whole numbers, 0, 1, 2, 3, and so on. For each natural number 1, 2, 3, and so on, we obtain a new number to the left of zero on the number line:

For the number 1, there will be an *opposite* number -1 (negative 1).

For the number 2, there will be an *opposite* number -2 (negative 2).

For the number 3, there will be an *opposite* number -3 (negative 3), and so on.

The **integers** consist of the whole numbers and these new numbers.

INTEGERS

The set of **integers** $= \{\ldots, -5, -4, -3, -2, -1, 0, 1, 2, 3, 4, 5, \ldots\}$.

We picture the integers on a number line as follows.

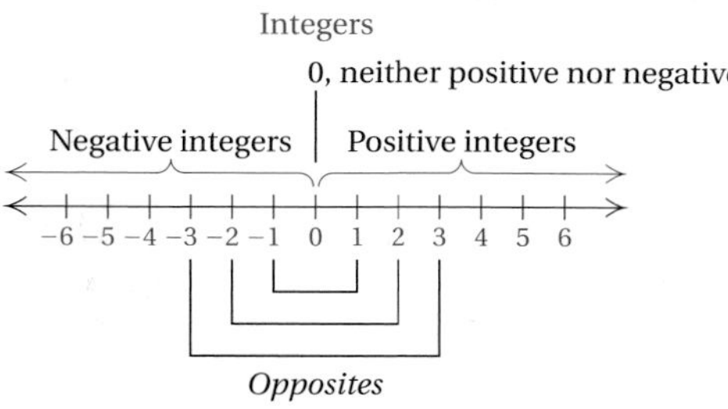

We call these new numbers to the left of 0 **negative integers.** The natural numbers are also called **positive integers.** Zero is neither positive nor negative. We call -1 and 1 **opposites** of each other. Similarly, -2 and 2 are opposites, -3 and 3 are opposites, -100 and 100 are opposites, and 0 is its own opposite. Pairs of opposite numbers like -3 and 3 are the same distance from 0. The integers extend infinitely on the number line to the left and right of zero.

a Integers and the Real World

Integers correspond to many real-world problems and situations. The following examples will help you get ready to translate problem situations that involve integers to mathematical language.

EXAMPLE 1 Tell which integer corresponds to this situation: The temperature is 4 degrees below zero.

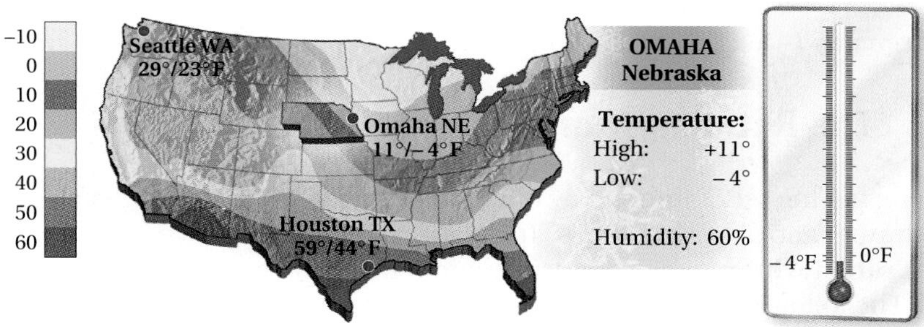

The integer -4 corresponds to the situation. The temperature is $-4°$.

EXAMPLE 2 *"Jeopardy."* Tell which integer corresponds to this situation: A contestant missed a $600 question on the television game show "Jeopardy."

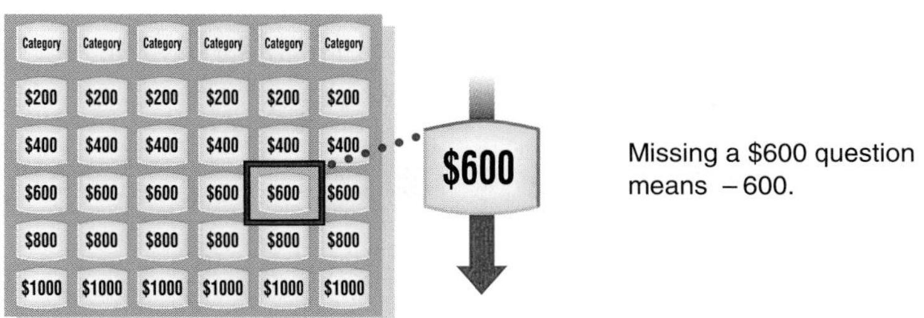

Missing a $600 question causes a $600 loss on the score—that is, the contestant earns -600 dollars.

State the integers that correspond to the given situation.

1. The halfback gained 8 yd on the first down. The quarterback was sacked for a 5-yd loss on the second down.

2. **Temperature High and Low.** The highest recorded temperature in Nevada is 125°F on June 29, 1994, in Laughlin. The lowest recorded temperature in Nevada is 50°F below zero on June 8, 1937, in San Jacinto.
   Sources: National Climatic Data Center, Asheville, NC, and Storm Phillips, STORMFAX, INC.

3. **Stock Decrease.** The price of Wendy's stock decreased from \$41 per share to \$38 per share over a recent time period.
   Source: The New York Stock Exchange

4. At 10 sec before liftoff, ignition occurs. At 156 sec after liftoff, the first stage is detached from the rocket.

5. A submarine dove 120 ft, rose 50 ft, and then dove 80 ft.

Answers on page A-17

EXAMPLE 3 *Elevation.* Tell which integer corresponds to this situation: The shores of California's largest lake, the Salton Sea, are 227 ft below sea level.

Source: *National Geographic*, February 2005, p. 88. Salton Sea, by Joel K. Bourne, Jr., Senior Writer.

The integer -227 corresponds to the situation. The elevation is -227 ft.

EXAMPLE 4 *Stock Price Change.* Tell which integers correspond to this situation: The price of Pearson Education stock decreased from \$27 per share to \$11 per share over a recent time period. The price of Safeway stock increased from \$20 per share to \$22 per share over a recent time period.

Source: The New York Stock Exchange

The integer -16 corresponds to the decrease in the stock value. The integer 2 represents the increase in stock value.

Do Exercises 1–5.

b The Rational Numbers

We created the set of integers by obtaining a negative number for each natural number and also including 0. To create a larger number system, called the set of **rational numbers,** we consider quotients of integers with nonzero divisors. The following are some examples of rational numbers:

$$\frac{2}{3},\quad -\frac{2}{3},\quad \frac{7}{1},\quad 4,\quad -3,\quad 0,\quad \frac{23}{-8},\quad 2.4,\quad -0.17,\quad 10\frac{1}{2}.$$

The number $-\frac{2}{3}$ (read "negative two-thirds") can also be named $\frac{-2}{3}$ or $\frac{2}{-3}$; that is,

$$-\frac{a}{b} = \frac{-a}{b} = \frac{a}{-b}.$$

The number 2.4 can be named $\frac{24}{10}$ or $\frac{12}{5}$, and -0.17 can be named $-\frac{17}{100}$. We can describe the set of rational numbers as follows.

RATIONAL NUMBERS

The set of **rational numbers** = the set of numbers $\frac{a}{b}$, where a and b are integers and b is not equal to 0 ($b \neq 0$).

Note that this new set of numbers, the rational numbers, contains the whole numbers, the integers, the arithmetic numbers (also called the nonnegative rational numbers), and the negative rational numbers.

We picture the rational numbers on a number line as follows.

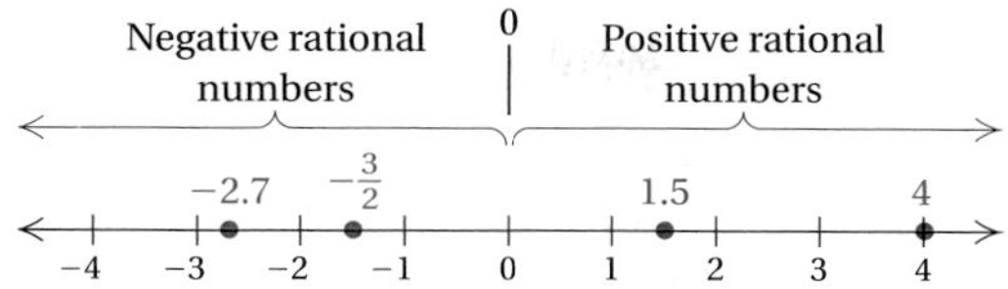

To **graph** a number means to find and mark its point on the number line. Some rational numbers are graphed in the preceding figure.

EXAMPLE 5 Graph: $\frac{5}{2}$.

The number $\frac{5}{2}$ can be named $2\frac{1}{2}$, or 2.5. Its graph is halfway between 2 and 3.

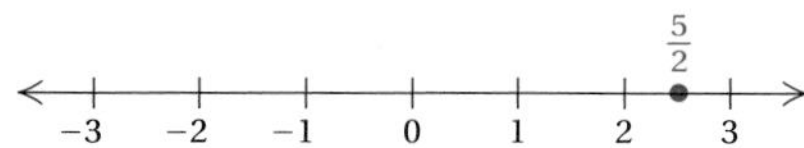

EXAMPLE 6 Graph: −3.2.

The graph of −3.2 is $\frac{2}{10}$ of the way from −3 to −4.

−3.2

−4 −3 −2 −1 0 1 2

EXAMPLE 7 Graph: $\frac{13}{8}$.

The number $\frac{13}{8}$ can be named $1\frac{5}{8}$, or 1.625. The graph is $\frac{5}{8}$ of the way from 1 to 2.

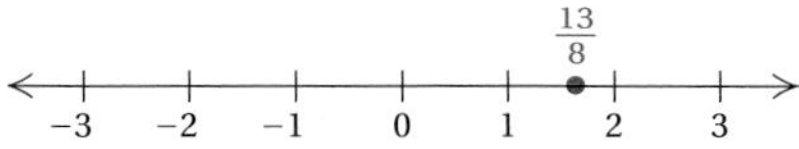

Do Exercises 6–8.

C Notation for Rational Numbers

Each rational number can be named using fraction or decimal notation.

EXAMPLE 8 Convert to decimal notation: $-\frac{5}{8}$.

We first find decimal notation for $\frac{5}{8}$. Since $\frac{5}{8}$ means 5 ÷ 8, we divide.

$$\begin{array}{r} 0.625 \\ 8\,\overline{)\,5.000} \\ \underline{4\,8} \\ 20 \\ \underline{16} \\ 40 \\ \underline{40} \\ 0 \end{array}$$

Thus, $\frac{5}{8} = 0.625$, so $-\frac{5}{8} = -0.625$.

Graph on a number line.

6. $-\frac{7}{2}$

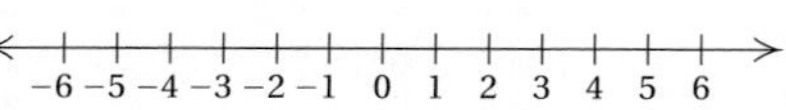

7. −1.4

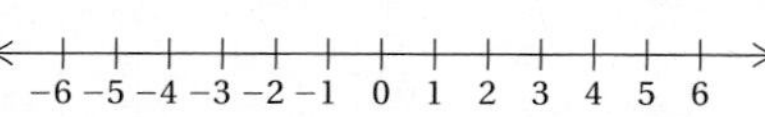

8. $\frac{11}{4}$

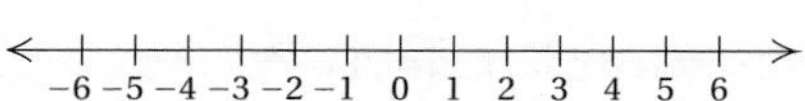

Answers on page A-17

CALCULATOR CORNER

Negative Numbers on a Calculator; Converting to Decimal Notation We use the opposite key (−) to enter negative numbers on a graphing calculator. Note that this is different from the − key, which is used for the operation of subtraction. To convert $-\frac{5}{8}$ to decimal notation, as in Example 8, we press (−) 5 ÷ 8 ENTER. The result is −0.625.

Exercises: Convert each of the following negative numbers to decimal notation.

1. $-\frac{3}{4}$
2. $-\frac{9}{20}$
3. $-\frac{1}{8}$
4. $-\frac{9}{5}$
5. $-\frac{27}{40}$
6. $-\frac{11}{16}$
7. $-\frac{7}{2}$
8. $-\frac{19}{25}$

Convert to decimal notation.

9. $-\frac{3}{8}$

10. $-\frac{6}{11}$

11. $\frac{4}{3}$

Answers on page A-17

CALCULATOR CORNER

Approximating Square Roots and π Square roots are found by pressing 2ND $\sqrt{}$. ($\sqrt{\ }$ is the second operation associated with the x^2 key.)

To find an approximation for $\sqrt{48}$, we press 2ND $\sqrt{}$ 4 8 ENTER. The approximation 6.92820323 is displayed.

To find $8 \cdot \sqrt{13}$, we press

The approximation 28.8444102 is displayed. The number π is used widely enough to have its own key. (π is the second operation associated with the ^ key.)

To approximate π, we press 2ND π ENTER. The approximation 3.141592654 is displayed.

Exercises: Approximate.

1. $\sqrt{76}$
2. $\sqrt{317}$
3. $15 \cdot \sqrt{20}$
4. $29 + \sqrt{42}$
5. π
6. $29 \cdot \pi$
7. $\pi \cdot 13^2$
8. $5 \cdot \pi + 8 \cdot \sqrt{237}$

Decimal notation for $-\frac{5}{8}$ is -0.625. We consider -0.625 to be a **terminating decimal.** Decimal notation for some numbers repeats.

EXAMPLE 9 Convert to decimal notation: $\frac{7}{11}$.

$$\begin{array}{r} 0.6363\ldots \\ 11\overline{)7.0000} \\ \underline{66} \\ 40 \\ \underline{33} \\ 70 \\ \underline{66} \\ 40 \\ \underline{33} \\ 7 \end{array}$$

Dividing

We can abbreviate repeating decimal notation by writing a bar over the repeating part—in this case, $0.\overline{63}$. Thus, $\frac{7}{11} = 0.\overline{63}$.

Each rational number can be expressed in either terminating or repeating decimal notation.

The following are other examples to show how each rational number can be named using fraction or decimal notation:

$$0 = \frac{0}{8}, \quad \frac{27}{100} = 0.27, \quad -8\frac{3}{4} = -8.75, \quad -\frac{13}{6} = -2.1\overline{6}.$$

Do Exercises 9–11.

d The Real Numbers and Order

Every rational number has a point on the number line. However, there are some points on the line for which there is no rational number. These points correspond to what are called **irrational numbers.**

What kinds of numbers are irrational? One example is the number π, which is used in finding the area and the circumference of a circle: $A = \pi r^2$ and $C = 2\pi r$.

Another example of an irrational number is the square root of 2, named $\sqrt{2}$. It is the length of the diagonal of a square with sides of length 1. It is also the number that when multiplied by itself gives 2—that is, $\sqrt{2} \cdot \sqrt{2} = 2$. There is no rational number that can be multiplied by itself to get 2. But the following are rational *approximations*:

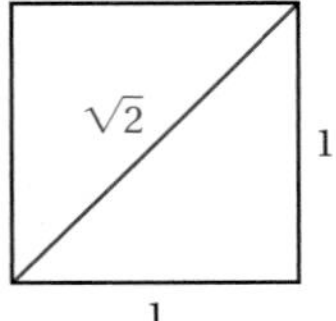

1.4 is an approximation of $\sqrt{2}$ because $(1.4)^2 = 1.96$;

1.41 is a better approximation because $(1.41)^2 = 1.9881$;

1.4142 is an even better approximation because $(1.4142)^2 = 1.99996164$.

We can find rational approximations for square roots using a calculator.

Decimal notation for rational numbers *either* terminates *or* repeats. Decimal notation for irrational numbers *neither* terminates *nor* repeats.

Some other examples of irrational numbers are $\sqrt{3}$, $-\sqrt{8}$, $\sqrt{11}$, and 0.121221222122221 . . . . Whenever we take the square root of a number that is not a perfect square, we will get an irrational number.

The rational numbers and the irrational numbers together correspond to all the points on a number line and make up what is called the **real-number system.**

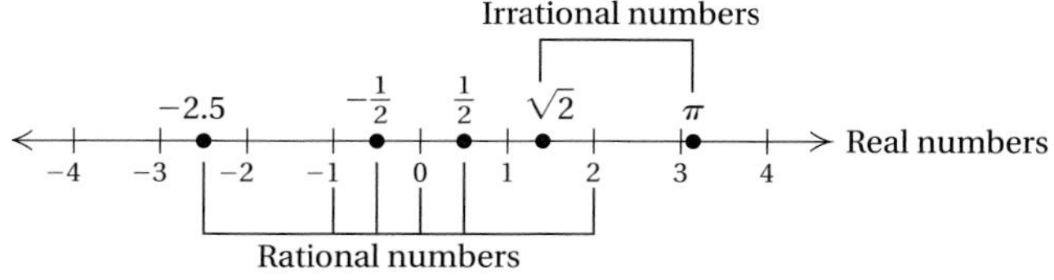

REAL NUMBERS

The set of **real numbers** = The set of all numbers corresponding to points on the number line.

The real numbers consist of the rational numbers and the irrational numbers. The following figure shows the relationships among various kinds of numbers.

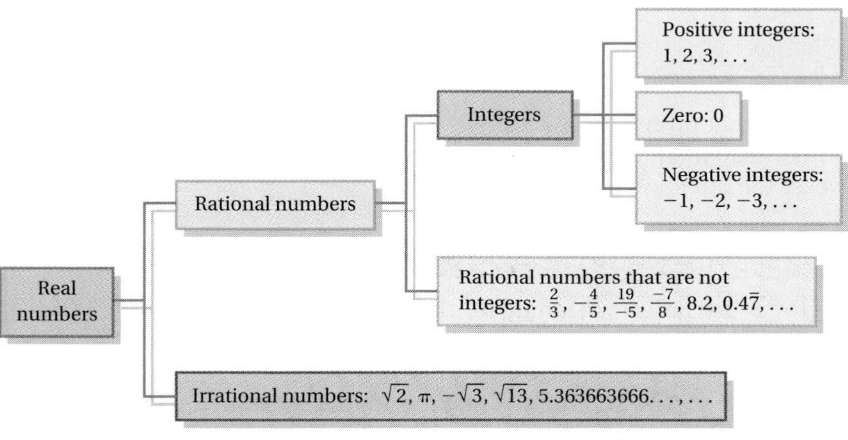

ORDER

Real numbers are named in order on the number line, with larger numbers named farther to the right. For any two numbers on the line, the one to the left is less than the one to the right.

$-8 < 6$

−9 −8 −7 −6 −5 −4 −3 −2 −1 0 1 2 3 4 5 6 7 8 9

$-3 > -7$

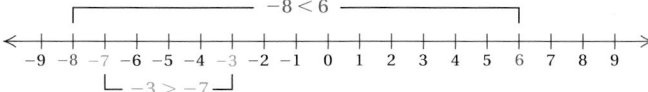

We use the symbol **<** to mean "**is less than.**" The sentence $-8 < 6$ means "-8 is less than 6." The symbol **>** means "**is greater than.**" The sentence $-3 > -7$ means "-3 is greater than -7." The sentences $-8 < 6$ and $-3 > -7$ are **inequalities.**

Use either < or > for □ to write a true sentence.

12. -3 □ 7

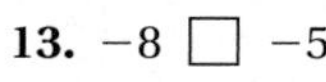

13. -8 □ -5

14. 7 □ -10

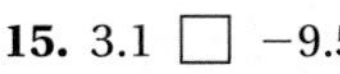

15. 3.1 □ -9.5

16. $-\frac{2}{3}$ □ -1

17. $-\frac{11}{8}$ □ $\frac{23}{15}$

18. $-\frac{2}{3}$ □ $-\frac{5}{9}$

19. -4.78 □ -5.01

Answers on page A-17

Write another inequality with the same meaning.

20. $-5 < 7$

21. $x > 4$

Write true or false.

22. $-4 \leq -6$

23. $7.8 \geq 7.8$

24. $-2 \leq \frac{3}{8}$

Answers on page A-17

EXAMPLES Use either < or > for □ to write a true sentence.

10. 2 □ 9 — Since 2 is to the left of 9, 2 is less than 9, so $2 < 9$.

11. -7 □ 3 — Since -7 is to the left of 3, we have $-7 < 3$.

12. 6 □ -12 — Since 6 is to the right of -12, then $6 > -12$.

13. -18 □ -5 — Since -18 is to the left of -5, we have $-18 < -5$.

14. -2.7 □ $-\frac{3}{2}$ — The answer is $-2.7 < -\frac{3}{2}$.

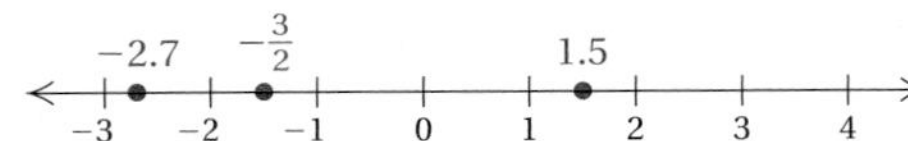

15. 1.5 □ -2.7 — The answer is $1.5 > -2.7$.

16. 1.38 □ 1.83 — The answer is $1.38 < 1.83$.

17. -3.45 □ 1.32 — The answer is $-3.45 < 1.32$.

18. -4 □ 0 — The answer is $-4 < 0$.

19. 5.8 □ 0 — The answer is $5.8 > 0$.

20. $\frac{5}{8}$ □ $\frac{7}{11}$ — We convert to decimal notation: $\frac{5}{8} = 0.625$ and $\frac{7}{11} = 0.6363\ldots$. Thus, $\frac{5}{8} < \frac{7}{11}$.

21. $-\frac{1}{2}$ □ $-\frac{1}{3}$ — The answer is $-\frac{1}{2} < -\frac{1}{3}$.

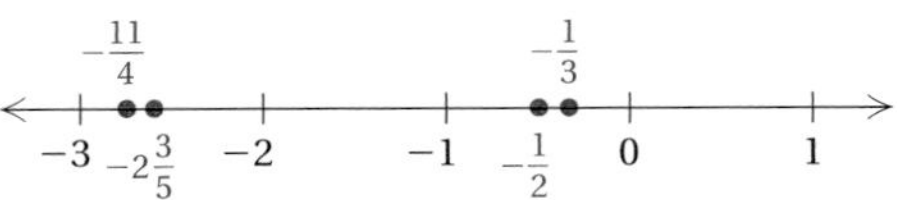

22. $-2\frac{3}{5}$ □ $-\frac{11}{4}$ — The answer is $-2\frac{3}{5} > -\frac{11}{4}$.

Do Exercises 12–19 on the preceding page.

Note that both $-8 < 6$ and $6 > -8$ are true. Every true inequality yields another true inequality when we interchange the numbers or variables and reverse the direction of the inequality sign.

ORDER; >, <

$a < b$ also has the meaning $b > a$.

EXAMPLES Write another inequality with the same meaning.

23. $-3 > -8$ — The inequality $-8 < -3$ has the same meaning.

24. $a < -5$ — The inequality $-5 > a$ has the same meaning.

A helpful mental device is to think of an inequality sign as an "arrow" with the arrow pointing to the smaller number.

Do Exercises 20 and 21.

Note that all positive real numbers are greater than zero and all negative real numbers are less than zero.

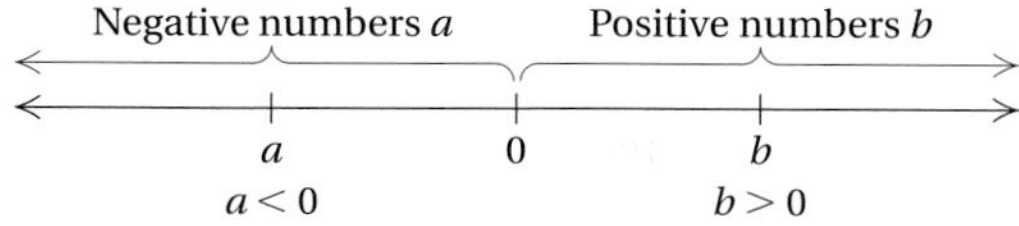

If b is a positive real number, then $b > 0$.
If a is a negative real number, then $a < 0$.

Expressions like $a \leq b$ and $b \geq a$ are also inequalities. We read $\boldsymbol{a \leq b}$ as "***a* is less than or equal to *b*.**" We read $\boldsymbol{a \geq b}$ as "***a* is greater than or equal to *b*.**"

EXAMPLES Write true or false for the statement.

25. $-3 \leq 5.4$ True since $-3 < 5.4$ is true

26. $-3 \leq -3$ True since $-3 = -3$ is true

27. $-5 \geq 1\frac{2}{3}$ False since neither $-5 > 1\frac{2}{3}$ nor $-5 = 1\frac{2}{3}$ is true

Do Exercises 22–24 on the preceding page.

e Absolute Value

From the number line, we see that numbers like 4 and -4 are the same distance from zero. Distance is always a nonnegative number. We call the distance of a number from zero on a number line the **absolute value** of the number.

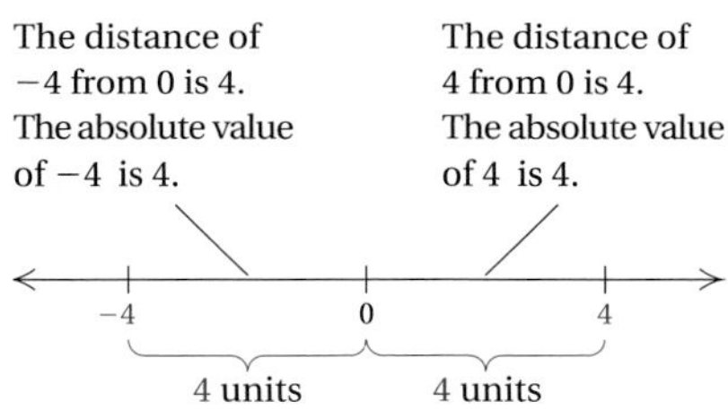

ABSOLUTE VALUE

The **absolute value** of a number is its distance from zero on a number line. We use the symbol $|x|$ to represent the absolute value of a number x.

CALCULATOR CORNER

Absolute Value The absolute-value operation is the first item in the Catalog on the TI-84 Plus graphing calculator. To find $|-7|$, as in Example 28, we first press 2ND CATALOG ENTER to copy "abs(" to the home screen. (CATALOG is the second operation associated with the 0 numeric key.) Then we press (−) 7) ENTER. The result is 7. To find $\left|-\frac{1}{2}\right|$ and express the result as a fraction, we press 2ND CATALOG ENTER (−) 1 ÷ 2) MATH 1 ENTER. The result is $\frac{1}{2}$.

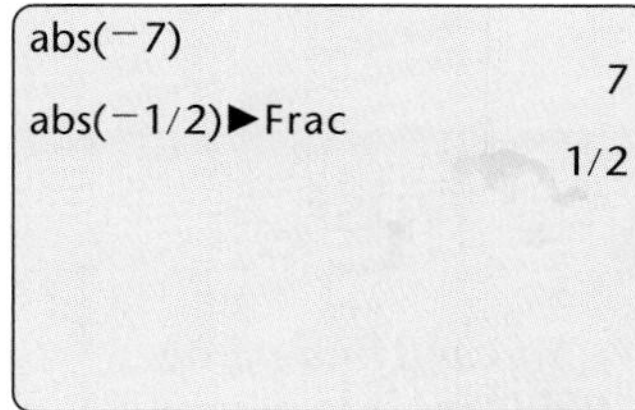

Exercises: Find the absolute value.

1. $|-5|$
2. $|17|$
3. $|0|$
4. $|6.48|$
5. $|-12.7|$
6. $|-0.9|$
7. $\left|-\frac{5}{7}\right|$
8. $\left|\frac{4}{3}\right|$

Find the absolute value.

25. $|8|$

26. $|-9|$

27. $\left|-\frac{2}{3}\right|$

28. $|5.6|$

Answers on page A-17

FINDING ABSOLUTE VALUE

a) If a number is negative, its absolute value is positive.
b) If a number is positive or zero, its absolute value is the same as the number.

EXAMPLES Find the absolute value.

28. $|-7|$ The distance of -7 from 0 is 7, so $|-7| = 7$.

29. $|12|$ The distance of 12 from 0 is 12, so $|12| = 12$.

30. $|0|$ The distance of 0 from 0 is 0, so $|0| = 0$.

31. $\left|\frac{3}{2}\right| = \frac{3}{2}$

32. $|-2.73| = 2.73$

Do Exercises 25–28.

Study Tips

USING THIS TEXTBOOK

You will find many Study Tips throughout the book. An index of all Study Tips can be found at the front of the book. One of the most important ways to improve your math study skills is to learn the proper use of the textbook. Here we highlight a few points that we consider most helpful.

- **Be sure to note the special symbols a, b, c, and so on, that correspond to the objectives you are to be able to perform.** The first time you see them is in the margin at the beginning of each section; the second time is in the subheadings of each section; and the third time is in the exercise set for the section. You will also find them next to the skill maintenance exercises in each exercise set and in the review exercises at the end of the chapter, as well as in the answers to the chapter tests and the cumulative reviews. These objective symbols allow you to refer to the appropriate place in the text whenever you need to review a topic.
- **Read and study each step of each example.** The examples include important side comments that explain each step. These carefully chosen examples and notes prepare you for success in the exercise set.
- **Stop and do the margin exercises as you study a section.** Doing the margin exercises is one of the most effective ways to enhance your ability to learn mathematics from this text. Don't deprive yourself of its benefits!
- **Note the icons listed at the top of each exercise set.** These refer to the many distinctive multimedia study aids that accompany the book.
- **Odd-numbered exercises.** Usually an instructor assigns some odd-numbered exercises. When you complete these, you can check your answers at the back of the book. If you miss any, check your work in the *Student's Solutions Manual* or ask your instructor for guidance.
- **Even-numbered exercises.** Whether or not your instructor assigns the even-numbered exercises, always do some on your own. Remember, there are no answers given for the class tests, so you need to practice doing exercises without answers. Check your answers later with a friend or your instructor.

7.2 EXERCISE SET

For Extra Help

MathXL | MyMathLab | InterAct Math | Math Tutor Center | Video Lectures on CD Disc 4 | Student's Solutions Manual

a State the integers that correspond to the situation.

1. *Pollution Fine.* In 2003, The Colonial Pipeline Company was fined a record $34 million for pollution.
Source: greenconsumerguide.com

2. *Lake Powell.* The water level of Lake Powell, a desert reservoir behind Glen Canyon Dam in northern Arizona and southeastern Utah, has dropped 130 ft since 2000.

3. On Wednesday, the temperature was 24° above zero. On Thursday, it was 2° below zero.

4. A student deposited her tax refund of $750 in a savings account. Two weeks later, she withdrew $125 to pay sorority fees.

5. *Temperature Extremes.* The highest temperature ever created on Earth was 950,000,000°F. The lowest temperature ever created was approximately 460°F below zero.
Source: *Guinness Book of World Records*

6. *Extreme Climate.* Verkhoyansk, a river port in northeast Siberia, has the most extreme climate on the planet. Its average monthly winter temperature is 58.5°F below zero, and its average monthly summer temperature is 56.5°F.
Source: *Guinness Book of World Records*

7. In bowling, the Alley Cats are 34 pins behind the Strikers going into the last frame. Describe the situation of each team.

8. During a video game, Maggie intercepted a missile worth 20 points, lost a starship worth 150 points, and captured a landing base worth 300 points.

b Graph the number on the number line.

9. $\frac{10}{3}$

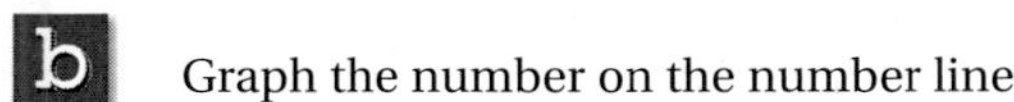

10. $-\frac{17}{4}$

−6 −5 −4 −3 −2 −1 0 1 2 3 4 5 6

11. −5.2

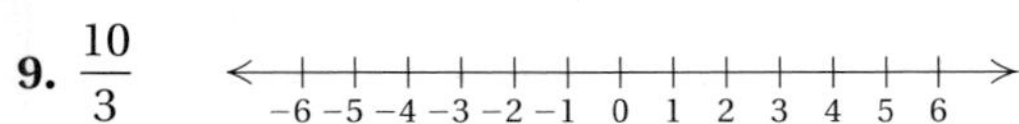

12. 4.78

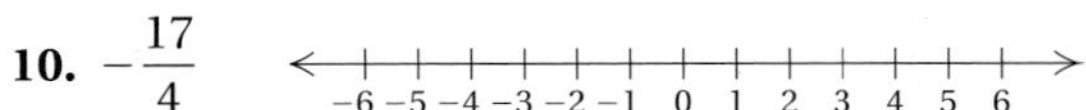

13. $-4\frac{2}{5}$

−6 −5 −4 −3 −2 −1 0 1 2 3 4 5 6

14. $2\frac{6}{11}$

−6 −5 −4 −3 −2 −1 0 1 2 3 4 5 6

c Convert to decimal notation.

15. $-\frac{7}{8}$

16. $-\frac{3}{16}$

17. $\frac{5}{6}$

18. $\frac{5}{3}$

19. $-\frac{7}{6}$

20. $-\frac{5}{12}$

21. $\frac{2}{3}$

22. $-\frac{11}{9}$

23. $\frac{1}{10}$

24. $\frac{1}{4}$

25. $-\frac{1}{2}$

26. $\frac{9}{8}$

27. $\frac{4}{25}$

28. $-\frac{7}{20}$

d Use either < or > for □ to write a true sentence.

29. 8 □ 0

30. 3 □ 0

31. −8 □ 3

32. 6 □ −6

33. −8 □ 8

34. 0 □ −9

35. −8 □ −5

36. −4 □ −3

37. −5 □ −11

38. −3 □ −4

39. −6 □ −5

40. −10 □ −14

41. 2.14 □ 1.24

42. −3.3 □ −2.2

43. −14.5 □ 0.011

44. 17.2 □ −1.67

45. −12.88 □ −6.45

46. −14.34 □ −17.88

47. $-\frac{1}{2}$ □ $-\frac{2}{3}$

48. $-\frac{5}{4}$ □ $-\frac{3}{4}$

49. $-\frac{2}{3}$ □ $\frac{1}{3}$

50. $\frac{3}{4}$ □ $-\frac{5}{4}$

51. $\frac{5}{12}$ □ $\frac{11}{25}$

52. $-\frac{13}{16}$ □ $-\frac{5}{9}$

Write true or false.

53. $-3 \geq -11$

54. $5 \leq -5$

55. $0 \geq 8$

56. $-5 \leq 7$

Write an inequality with the same meaning.

57. $-6 > x$

58. $x < 8$

59. $-10 \leq y$

60. $12 \geq t$

Find the absolute value.

61. $|-3|$

62. $|-7|$

63. $|10|$

64. $|11|$

65. $|0|$

66. $|-2.7|$

67. $|-30.4|$

68. $|325|$

69. $\left|-\frac{2}{3}\right|$

70. $\left|-\frac{10}{7}\right|$

71. $\left|\frac{0}{4}\right|$

72. $|14.8|$

73. $\left|-3\frac{5}{8}\right|$

74. $\left|-7\frac{4}{5}\right|$

75. D_W When Jennifer's calculator gives a decimal approximation for $\sqrt{2}$ and that approximation is promptly squared, the result is 2. Yet, when that same approximation is entered by hand and then squared, the result is not exactly 2. Why do you suppose this happens?

76. D_W How many rational numbers are there between 0 and 1? Why?

SKILL MAINTENANCE

Convert to decimal notation. [4.2b]

77. 63%

78. $23\frac{4}{5}\%$

79. 110%

80. 22.76%

Convert to percent notation. [4.3a]

81. $\frac{13}{25}$

82. $\frac{5}{4}$

83. $\frac{5}{6}$

84. $\frac{19}{32}$

SYNTHESIS

List in order from the least to the greatest.

85. $-\frac{2}{3}, \frac{1}{2}, -\frac{3}{4}, -\frac{5}{6}, \frac{3}{8}, \frac{1}{6}$

86. $\frac{2}{3}, -\frac{1}{7}, \frac{1}{3}, -\frac{2}{7}, -\frac{2}{3}, \frac{2}{5}, -\frac{1}{3}, -\frac{2}{5}, \frac{9}{8}$

87. $-5.16, -4.24, -8.76, 5.23, 1.85, -2.13$

88. $-8\frac{7}{8}, 7^1, -5, |-6|, 4, |3|, -8\frac{5}{8}, -100, 0, 1^7, \frac{14}{4}, -\frac{67}{8}$

Given that $0.\overline{3} = \frac{1}{3}$ and $0.\overline{6} = \frac{2}{3}$, express each of the following as a quotient or ratio of two integers.

89. $0.\overline{1}$

90. $0.\overline{9}$

91. $5.\overline{5}$

7.3 ADDITION OF REAL NUMBERS

Objectives

a Add real numbers without using a number line.

b Find the opposite, or additive inverse, of a real number.

c Solve applied problems involving addition of real numbers.

Add using a number line.

1. $0 + (-3)$

2. $1 + (-4)$

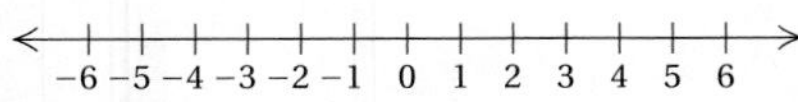

3. $-3 + (-2)$

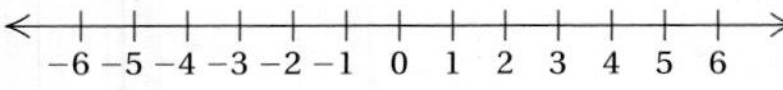

4. $-3 + 7$

5. $-2.4 + 2.4$

6. $-\frac{5}{2} + \frac{1}{2}$

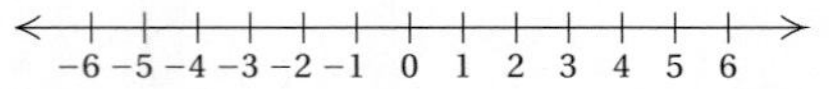

Answers on page A-18

In this section, we consider addition of real numbers. First, to gain an understanding, we add using a number line. Then we consider rules for addition.

ADDITION ON A NUMBER LINE

To do the addition $a + b$ on a number line, we start at 0. Then we move to a and then move according to b.

a) If b is positive, we move from a to the right.
b) If b is negative, we move from a to the left.
c) If b is 0, we stay at a.

EXAMPLE 1 Add: $3 + (-5)$.

We start at 0 and move to 3. Then we move 5 units left since -5 is negative.

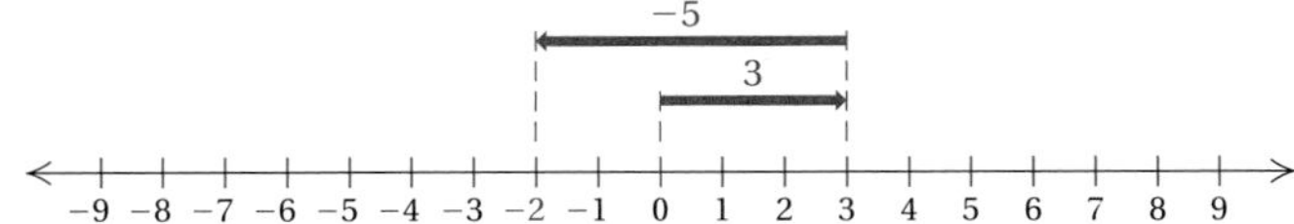

$3 + (-5) = -2$

EXAMPLE 2 Add: $-4 + (-3)$.

We start at 0 and move to -4. Then we move 3 units left since -3 is negative.

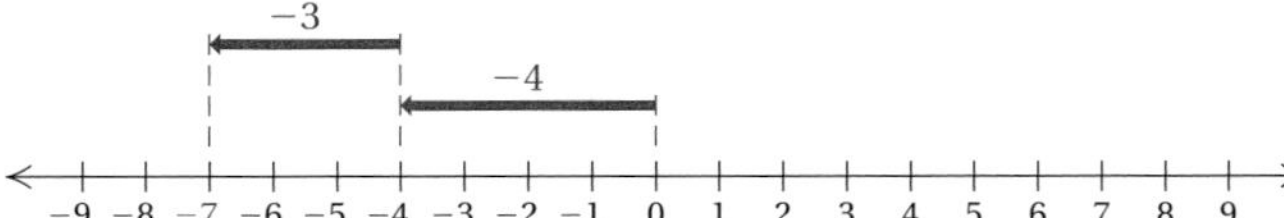

$-4 + (-3) = -7$

EXAMPLE 3 Add: $-4 + 9$.

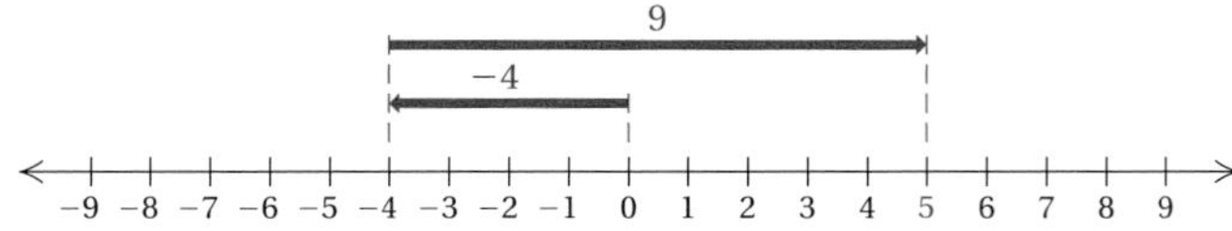

$-4 + 9 = 5$

EXAMPLE 4 Add: $-5.2 + 0$.

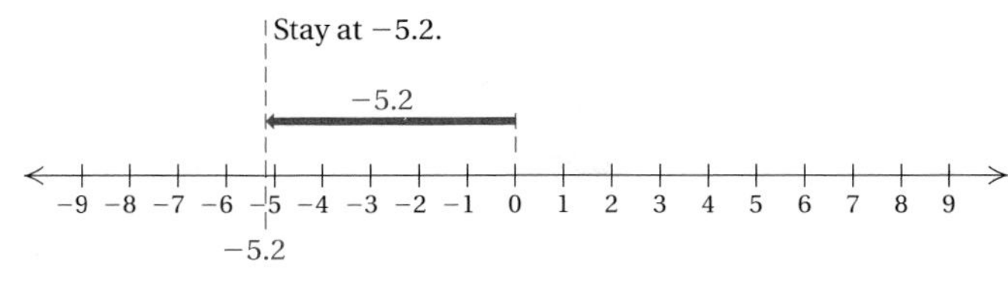

$-5.2 + 0 = -5.2$

Do Exercises 1–6 on the preceding page.

a Adding Without a Number Line

You may have noticed some patterns in the preceding examples. These lead us to rules for adding without using a number line that are more efficient for adding larger numbers.

RULES FOR ADDITION OF REAL NUMBERS

1. *Positive numbers*: Add the same as arithmetic numbers. The answer is positive.
2. *Negative numbers*: Add absolute values. The answer is negative.
3. *A positive and a negative number*: Subtract the smaller absolute value from the larger. Then:
   - **a)** If the positive number has the greater absolute value, the answer is positive.
   - **b)** If the negative number has the greater absolute value, the answer is negative.
   - **c)** If the numbers have the same absolute value, the answer is 0.
4. *One number is zero*: The sum is the other number.

Rule 4 is known as the **identity property of 0.** It says that for any real number a, $a + 0 = a$.

EXAMPLES Add without using a number line.

5. $-12 + (-7) = -19$ Two negatives. Add the absolute values: $|-12| + |-7| = 12 + 7 = 19$. Make the answer *negative*: -19.

6. $-1.4 + 8.5 = 7.1$ One negative, one positive. Find the absolute values: $|-1.4| = 1.4$; $|8.5| = 8.5$. Subtract the smaller absolute value from the larger: $8.5 - 1.4 = 7.1$. The *positive* number, 8.5, has the larger absolute value, so the answer is *positive*: 7.1.

7. $-36 + 21 = -15$ One negative, one positive. Find the absolute values: $|-36| = 36$, $|21| = 21$. Subtract the smaller absolute value from the larger: $36 - 21 = 15$. The *negative* number, -36, has the larger absolute value, so the answer is *negative*: -15.

Add without using a number line.

7. $-5 + (-6)$

8. $-9 + (-3)$

9. $-4 + 6$

10. $-7 + 3$

11. $5 + (-7)$

12. $-20 + 20$

13. $-11 + (-11)$

14. $10 + (-7)$

15. $-0.17 + 0.7$

16. $-6.4 + 8.7$

17. $-4.5 + (-3.2)$

18. $-8.6 + 2.4$

19. $\frac{5}{9} + \left(-\frac{7}{9}\right)$

20. $-\frac{1}{5} + \left(-\frac{3}{4}\right)$

Answers on page A-18

Add.

21. $(-15) + (-37) + 25 + 42 + (-59) + (-14)$

22. $42 + (-81) + (-28) + 24 + 18 + (-31)$

23. $-2.5 + (-10) + 6 + (-7.5)$

24. $-35 + 17 + 14 + (-27) + 31 + (-12)$

Find the opposite, or additive inverse, of each of the following.

25. -4

26. 8.7

27. -7.74

28. $-\frac{8}{9}$

29. 0

30. 12

Answers on page A-18

8. $1.5 + (-1.5) = 0$ The numbers have the same absolute value. The sum is 0.

9. $-\frac{7}{8} + 0 = -\frac{7}{8}$ One number is zero. The sum is $-\frac{7}{8}$.

10. $-9.2 + 3.1 = -6.1$

11. $-\frac{3}{2} + \frac{9}{2} = \frac{6}{2} = 3$

12. $-\frac{2}{3} + \frac{5}{8} = -\frac{16}{24} + \frac{15}{24} = -\frac{1}{24}$

Do Exercises 7–20 on the preceding page.

Suppose we want to add several numbers, some positive and some negative, as follows. How can we proceed?

$$15 + (-2) + 7 + 14 + (-5) + (-12)$$

We can change grouping and order as we please when adding. For instance, we can group the positive numbers together and the negative numbers together and add them separately. Then we add the two results.

EXAMPLE 13 Add: $15 + (-2) + 7 + 14 + (-5) + (-12)$.

a) $15 + 7 + 14 = 36$ Adding the positive numbers

b) $-2 + (-5) + (-12) = -19$ Adding the negative numbers

$36 + (-19) = 17$ Adding (a) and (b)

We can also add the numbers in any other order we wish, say, from left to right as follows:

$$\begin{aligned} 15 + (-2) + 7 + 14 + (-5) + (-12) &= 13 + 7 + 14 + (-5) + (-12) \\ &= 20 + 14 + (-5) + (-12) \\ &= 34 + (-5) + (-12) \\ &= 29 + (-12) \\ &= 17 \end{aligned}$$

Do Exercises 21–24.

b Opposites, or Additive Inverses

Suppose we add two numbers that are **opposites,** such as 6 and -6. The result is 0. When opposites are added, the result is always 0. Such numbers are also called **additive inverses.** Every real number has an opposite, or additive inverse.

OPPOSITES, OR ADDITIVE INVERSES

Two numbers whose sum is 0 are called **opposites,** or **additive inverses,** of each other.

EXAMPLES Find the opposite, or additive inverse, of each number.

14. 34 The opposite of 34 is -34 because $34 + (-34) = 0$.

15. -8 The opposite of -8 is 8 because $-8 + 8 = 0$.

16. 0 The opposite of 0 is 0 because $0 + 0 = 0$.

17. $-\frac{7}{8}$ The opposite of $-\frac{7}{8}$ is $\frac{7}{8}$ because $-\frac{7}{8} + \frac{7}{8} = 0$.

Do Exercises 25–30 on the preceding page.

To name the opposite, we use the symbol $-$, as follows.

SYMBOLIZING OPPOSITES

The opposite, or additive inverse, of a number a can be named $-a$ (read "the opposite of a," or "the additive inverse of a").

Note that if we take a number, say, 8, and find its opposite, -8, and then find the opposite of the result, we will have the original number, 8, again.

THE OPPOSITE OF AN OPPOSITE

The **opposite of the opposite** of a number is the number itself. (The additive inverse of the additive inverse of a number is the number itself.) That is, for any number a,

$$-(-a) = a.$$

EXAMPLE 18 Evaluate $-x$ and $-(-x)$ when $x = 16$.

If $x = 16$, then $-x = -16$. The opposite of 16 is -16.

If $x = 16$, then $-(-x) = -(-16) = 16$. The opposite of the opposite of 16 is 16.

EXAMPLE 19 Evaluate $-x$ and $-(-x)$ when $x = -3$.

If $x = -3$, then $-x = -(-3) = 3$.

If $x = -3$, then $-(-x) = -(-(-3)) = -(3) = -3$.

Note that in Example 19 we used a second set of parentheses to show that we are substituting the negative number -3 for x. Symbolism like $--x$ is not considered meaningful.

Do Exercises 31–36.

A symbol such as -8 is usually read "negative 8." It could be read "the additive inverse of 8," because the additive inverse of 8 is negative 8. It could also be read "the opposite of 8," because the opposite of 8 is -8. Thus a symbol like -8 can be read in more than one way. It is never correct to read -8 as "minus 8."

Caution!

A symbol like $-x$, which has a variable, should be read "the opposite of x" or "the additive inverse of x" and *not* "negative x," because we do not know whether x represents a positive number, a negative number, or 0. You can check this in Examples 18 and 19.

Evaluate $-x$ and $-(-x)$ when:

31. $x = 14$.

32. $x = 1$.

33. $x = -19$.

34. $x = -1.6$.

35. $x = \frac{2}{3}$.

36. $x = -\frac{9}{8}$.

Answers on page A-18

Find the opposite. (Change the sign.)

37. -4

38. -13.4

39. 0

40. $\frac{1}{4}$

41. Change in Class Size. During the first two weeks of the semester in Jim's algebra class, 4 students withdrew, 8 students enrolled in the class, and 6 students were dropped as "no shows." By how many students had the class size changed at the end of the first two weeks?

Answers on page A-18

Study Tips

SMALL STEPS LEAD TO GREAT SUCCESS (PART 1)

What is your long-term goal for getting an education? How does math help you to attain that goal? As you begin this course, approach each short-term task, such as going to class, asking questions, using your time wisely, and doing your homework, as part of the framework of your long-term goal.

"What man actually needs is not a tensionless state but rather the struggling and striving for a worthwhile goal, a freely chosen task."

Victor Frankl

We can use the symbolism $-a$ to restate the definition of opposite, or additive inverse.

THE SUM OF OPPOSITES

For any real number a, the **opposite,** or **additive inverse,** of a, expressed as $-a$, is such that

$$a + (-a) = (-a) + a = 0.$$

SIGNS OF NUMBERS

A negative number is sometimes said to have a "negative sign." A positive number is said to have a "positive sign." When we replace a number with its opposite, we can say that we have "changed its sign."

EXAMPLES Find the opposite. (Change the sign.)

20. -3 $\quad -(-3) = 3$

21. $-\frac{2}{13}$ $\quad -\left(-\frac{2}{13}\right) = \frac{2}{13}$

22. 0 $\quad -(0) = 0$

23. 14 $\quad -(14) = -14$

Do Exercises 37–40.

C Applications and Problem Solving

Addition of real numbers occurs in many real-world situations.

EXAMPLE 24 *Lake Level.* In the course of one four-month period, the water level of Lake Champlain went down 2 ft, up 1 ft, down 5 ft, and up 3 ft. How much had the lake level changed at the end of the four months?

We let $T =$ the total change in the level of the lake. Then the problem translates to a sum:

$$\underbrace{\text{Total change}}_{T} \ \underbrace{\text{is}}_{=} \ \underbrace{\text{1st change}}_{-2} \ \underbrace{\text{plus}}_{+} \ \underbrace{\text{2nd change}}_{1} \ \underbrace{\text{plus}}_{+} \ \underbrace{\text{3rd change}}_{(-5)} \ \underbrace{\text{plus}}_{+} \ \underbrace{\text{4th change.}}_{3}$$

Adding from left to right, we have

$$\begin{aligned} T = -2 + 1 + (-5) + 3 &= -1 + (-5) + 3 \\ &= -6 + 3 \\ &= -3. \end{aligned}$$

The lake level has dropped 3 ft at the end of the four-month period.

Do Exercise 41.

7.3 EXERCISE SET

For Extra Help

MathXL
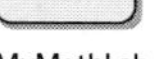
MyMathLab

InterAct Math

Math Tutor Center

Video Lectures on CD Disc 4

Student's Solutions Manual

a Add. Do not use a number line except as a check.

1. $2 + (-9)$

2. $-5 + 2$

3. $-11 + 5$

4. $4 + (-3)$

5. $-6 + 6$

6. $8 + (-8)$

7. $-3 + (-5)$

8. $-4 + (-6)$

9. $-7 + 0$

10. $-13 + 0$

11. $0 + (-27)$

12. $0 + (-35)$

13. $17 + (-17)$

14. $-15 + 15$

15. $-17 + (-25)$

16. $-24 + (-17)$

17. $18 + (-18)$

18. $-13 + 13$

19. $-28 + 28$

20. $11 + (-11)$

21. $8 + (-5)$

22. $-7 + 8$

23. $-4 + (-5)$

24. $10 + (-12)$

25. $13 + (-6)$

26. $-3 + 14$

27. $-25 + 25$

28. $50 + (-50)$

29. $53 + (-18)$

30. $75 + (-45)$

31. $-8.5 + 4.7$

32. $-4.6 + 1.9$

33. $-2.8 + (-5.3)$

34. $-7.9 + (-6.5)$

35. $-\frac{3}{5} + \frac{2}{5}$

36. $-\frac{4}{3} + \frac{2}{3}$

37. $-\frac{2}{9} + \left(-\frac{5}{9}\right)$

38. $-\frac{4}{7} + \left(-\frac{6}{7}\right)$

39. $-\frac{5}{8} + \frac{1}{4}$

40. $-\frac{5}{6} + \frac{2}{3}$

41. $-\frac{5}{8} + \left(-\frac{1}{6}\right)$

42. $-\frac{5}{6} + \left(-\frac{2}{9}\right)$

43. $-\frac{3}{8} + \frac{5}{12}$

44. $-\frac{7}{16} + \frac{7}{8}$

45. $-\frac{1}{6} + \frac{7}{10}$

46. $-\frac{11}{18} + \left(-\frac{3}{4}\right)$

47. $\frac{7}{15} + \left(-\frac{1}{9}\right)$

48. $-\frac{4}{21} + \frac{3}{14}$

49. $76 + (-15) + (-18) + (-6)$

50. $29 + (-45) + 18 + 32 + (-96)$

51. $-44 + \left(-\frac{3}{8}\right) + 95 + \left(-\frac{5}{8}\right)$

52. $24 + 3.1 + (-44) + (-8.2) + 63$

53. $98 + (-54) + 113 + (-998) + 44 + (-612)$

54. $-458 + (-124) + 1025 + (-917) + 218$

b Find the opposite, or additive inverse.

55. 24

56. -64

57. -26.9

58. 48.2

Evaluate $-x$ when:

59. $x = 8$.

60. $x = -27$.

61. $x = -\frac{13}{8}$.

62. $x = \frac{1}{236}$.

Evaluate $-(-x)$ when:

63. $x = -43$.

64. $x = 39$.

65. $x = \frac{4}{3}$.

66. $x = -7.1$.

Find the opposite. (Change the sign.)

67. -24

68. -12.3

69. $-\frac{3}{8}$

70. 10

 Solve.

71. *Tallest Mountain.* The tallest mountain in the world, when measured from base to peak, is Mauna Kea (White Mountain) in Hawaii. From its base 19,684 ft below sea level in the Hawaiian Trough, it rises 33,480 ft. What is the elevation of the peak above sea level?

Source: *The Guinness Book of Records*

72. *Telephone Bills.* Erika's cell-phone bill for July was \$82. She sent a check for \$50 and then made \$37 worth of calls in August. How much did she then owe on her cell-phone bill?

73. *Temperature Changes.* One day the temperature in Lawrence, Kansas, is 32°F at 6:00 A.M. It rises 15° by noon, but falls 50° by midnight when a cold front moves in. What is the final temperature?

74. *Stock Changes.* On a recent day, the price of a stock opened at a value of \$61.38. During the day, it rose \$4.75, dropped \$7.38, and rose \$5.13. Find the value of the stock at the end of the day.

75. *Profits and Losses.* A business expresses a profit as a positive number and refers to it as operating "in the black." A loss is expressed as a negative number and is referred to as operating "in the red." The profits and losses of Xponent Corporation over various years are shown in the bar graph below. Find the sum of the profits and losses.

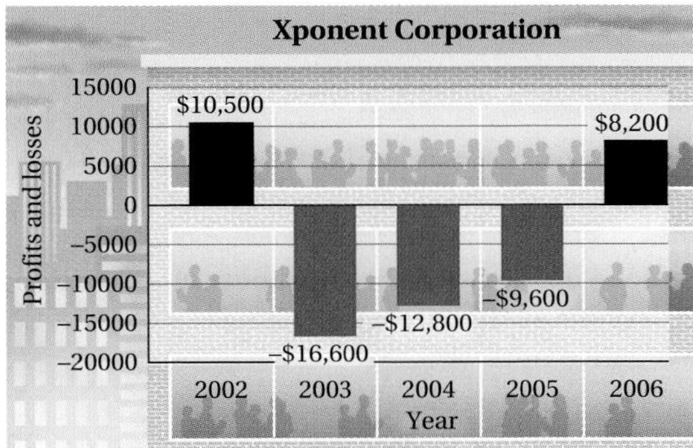

76. *Football Yardage.* In a college football game, the quarterback attempted passes with the following results. Find the total gain or loss.

TRY	GAIN OR LOSS
1st	13-yd gain
2nd	12-yd loss
3rd	21-yd gain

77. *Credit Card Bills.* On August 1, Lyle's credit card bill shows that he owes \$470. During the month of August, Lyle sends a check for \$45 to the credit card company, charges another \$160 in merchandise, and then pays off another \$500 of his bill. What is the new balance of Lyle's account at the end of August?

78. *Account Balance.* Leah has \$460 in a checking account. She writes a check for \$530, makes a deposit of \$75, and then writes a check for \$90. What is the balance in her account?

79. D_W Without actually performing the addition, explain why the sum of all integers from -50 to 50 is 0.

80. D_W Explain in your own words why the sum of two negative numbers is always negative.

SKILL MAINTENANCE

Convert to decimal notation. [4.2b]

81. 57%

82. 71.3%

83. $23\frac{4}{5}\%$

84. $92\frac{7}{8}\%$

Convert to percent notation. [4.3a]

85. $\frac{5}{4}$

86. $\frac{1}{8}$

87. $\frac{13}{25}$

88. $\frac{13}{32}$

SYNTHESIS

89. For what numbers x is $-x$ negative?

90. For what numbers x is $-x$ positive?

For each of Exercises 91 and 92, choose the correct answer from the selections given.

91. If a is positive and b is negative, then $-a + b$ is:

a) Positive.
b) Negative.
c) 0.
d) Cannot be determined without more information

92. If $a = b$ and a and b are negative, then $-a + (-b)$ is:

a) Positive.
b) Negative.
c) 0.
d) Cannot be determined without more information

7.4 SUBTRACTION OF REAL NUMBERS

Objectives

a Subtract real numbers and simplify combinations of additions and subtractions.

b Solve applied problems involving subtraction of real numbers.

a Subtraction

We now consider subtraction of real numbers.

SUBTRACTION

The difference $a - b$ is the number c for which $a = b + c$.

Consider, for example, $45 - 17$. *Think*: What number can we add to 17 to get 45? Since $45 = 17 + 28$, we know that $45 - 17 = 28$. Let's consider an example whose answer is a negative number.

EXAMPLE 1 Subtract: $3 - 7$.

Think: What number can we add to 7 to get 3? The number must be negative. Since $7 + (-4) = 3$, we know the number is -4: $3 - 7 = -4$. That is, $3 - 7 = -4$ because $7 + (-4) = 3$.

Do Exercises 1–3.

The definition above does not provide the most efficient way to do subtraction. We can develop a faster way to subtract. As a rationale for the faster way, let's compare $3 + 7$ and $3 - 7$ on a number line.

To find $3 + 7$ on a number line, we move 3 units to the right from 0 since 3 is positive. Then we move 7 units farther to the right since 7 is positive.

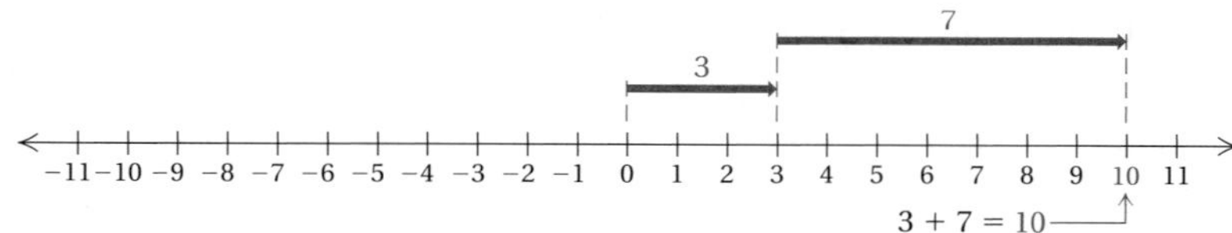

To find $3 - 7$, we do the "opposite" of adding 7: We move 7 units to the *left* to do the subtracting. This is the same as *adding* the opposite of 7, -7, to 3.

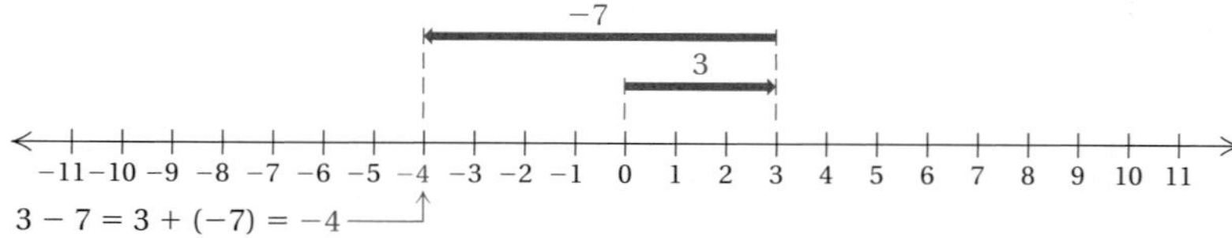

Do Exercises 4–6.

Look for a pattern in the examples shown at right.

SUBTRACTING	ADDING AN OPPOSITE
$5 - 8 = -3$	$5 + (-8) = -3$
$-6 - 4 = -10$	$-6 + (-4) = -10$
$-7 - (-2) = -5$	$-7 + 2 = -5$

Do Exercises 7–10 on the following page.

Subtract.

1. $-6 - 4$

Think: What number can be added to 4 to get -6:

$\square + 4 = -6$?

2. $-7 - (-10)$

Think: What number can be added to -10 to get -7:

$\square + (-10) = -7$?

3. $-7 - (-2)$

Think: What number can be added to -2 to get -7:

$\square + (-2) = -7$?

Subtract. Use a number line, doing the "opposite" of addition.

4. $-4 - (-3)$

−6 −5 −4 −3 −2 −1 0 1 2 3 4 5 6

5. $-4 - (-6)$

−6 −5 −4 −3 −2 −1 0 1 2 3 4 5 6

6. $5 - 9$

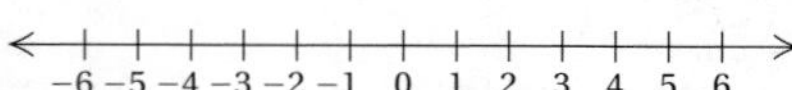

Answers on page A-18

Perhaps you have noticed that we can subtract by adding the opposite of the number being subtracted. This can always be done.

SUBTRACTING BY ADDING THE OPPOSITE

For any real numbers a and b,

$$a - b = a + (-b).$$

(To subtract, add the opposite, or additive inverse, of the number being subtracted.)

This is the method generally used for quick subtraction of real numbers.

EXAMPLES Subtract.

2. $2 - 6 = 2 + (-6) = -4$

The opposite of 6 is -6. We change the subtraction to addition and add the opposite. *Check*: $-4 + 6 = 2$.

3. $4 - (-9) = 4 + 9 = 13$

The opposite of -9 is 9. We change the subtraction to addition and add the opposite. *Check*: $13 + (-9) = 4$.

4. $-4.2 - (-3.6) = -4.2 + 3.6 = -0.6$

Adding the opposite. *Check*: $-0.6 + (-3.6) = -4.2$.

5. $-\frac{1}{2} - \left(-\frac{3}{4}\right) = -\frac{1}{2} + \frac{3}{4}$

$= -\frac{2}{4} + \frac{3}{4} = \frac{1}{4}$

Adding the opposite. *Check*: $\frac{1}{4} + \left(-\frac{3}{4}\right) = -\frac{1}{2}$.

Do Exercises 11–16.

EXAMPLES Read each of the following. Then subtract by adding the opposite of the number being subtracted.

6. $3 - 5$ Read "three minus five is three plus the opposite of five"

$3 - 5 = 3 + (-5) = -2$

7. $\frac{1}{8} - \frac{7}{8}$ Read "one-eighth minus seven-eighths is one-eighth plus the opposite of seven-eighths"

$\frac{1}{8} - \frac{7}{8} = \frac{1}{8} + \left(-\frac{7}{8}\right) = -\frac{6}{8}$, or $-\frac{3}{4}$

8. $-4.6 - (-9.8)$ Read "negative four point six minus negative nine point eight is negative four point six plus the opposite of negative nine point eight"

$-4.6 - (-9.8) = -4.6 + 9.8 = 5.2$

9. $-\frac{3}{4} - \frac{7}{5}$ Read "negative three-fourths minus seven-fifths is negative three-fourths plus the opposite of seven-fifths"

$-\frac{3}{4} - \frac{7}{5} = -\frac{3}{4} + \left(-\frac{7}{5}\right) = -\frac{15}{20} + \left(-\frac{28}{20}\right) = -\frac{43}{20}$

Do Exercises 17–21 on the following page.

Complete the addition and compare with the subtraction.

7. $4 - 6 = -2$;
$4 + (-6) =$ ______

8. $-3 - 8 = -11$;
$-3 + (-8) =$ ______

9. $-5 - (-9) = 4$;
$-5 + 9 =$ ______

10. $-5 - (-3) = -2$;
$-5 + 3 =$ ______

Subtract.

11. $2 - 8$

12. $-6 - 10$

13. $12.4 - 5.3$

14. $-8 - (-11)$

15. $-8 - (-8)$

16. $\frac{2}{3} - \left(-\frac{5}{6}\right)$

Answers on page A-18

Read each of the following. Then subtract by adding the opposite of the number being subtracted.

17. $3 - 11$

18. $12 - 5$

19. $-12 - (-9)$

20. $-12.4 - 10.9$

21. $-\frac{4}{5} - \left(-\frac{4}{5}\right)$

Simplify.

22. $-6 - (-2) - (-4) - 12 + 3$

23. $\frac{2}{3} - \frac{4}{5} - \left(-\frac{11}{15}\right) + \frac{7}{10} - \frac{5}{2}$

24. $-9.6 + 7.4 - (-3.9) - (-11)$

25. Temperature Extremes. The highest temperature ever recorded in the United States is 134°F in Greenland Ranch, California, on July 10, 1913. The lowest temperature ever recorded is −80°F in Prospect Creek, Alaska, on January 23, 1971. How much higher was the temperature in Greenland Ranch than that in Prospect Creek?

Source: National Oceanographic and Atmospheric Administration

Answers on page A-18

When several additions and subtractions occur together, we can make them all additions.

EXAMPLES Simplify.

10. $8 - (-4) - 2 - (-4) + 2 = 8 + 4 + (-2) + 4 + 2$ Adding the opposite

$= 16$

11. $8.2 - (-6.1) + 2.3 - (-4) = 8.2 + 6.1 + 2.3 + 4 = 20.6$

12.
$$\frac{3}{4} - \left(-\frac{1}{12}\right) - \frac{5}{6} - \frac{2}{3} = \frac{9}{12} + \frac{1}{12} + \left(-\frac{10}{12}\right) + \left(-\frac{8}{12}\right)$$
$$= \frac{9 + 1 + (-10) + (-8)}{12}$$
$$= \frac{-8}{12} = -\frac{8}{12} = -\frac{2}{3}$$

Do Exercises 22–24.

b Applications and Problem Solving

Let's now see how we can use subtraction of real numbers to solve applied problems.

EXAMPLE 13 *Surface Temperatures on Mars.* Surface temperatures on Mars vary from −128°C during polar night to 27°C at the equator during midday at the closest point in orbit to the sun. Find the difference between the highest value and the lowest value in this temperature range.

Source: Mars Institute

We let D = the difference in the temperatures. Then the problem translates to the following subtraction:

Difference in temperature	is	Highest temperature	minus	Lowest temperature
↓	↓	↓	↓	↓
D	$=$	27	$-$	(-128)
D	$= 27 + 128 = 155$			

The difference in the temperatures is 155°C.

Do Exercise 25.

7.4 EXERCISE SET

For Extra Help

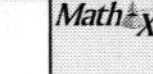
MathXL

MyMathLab

InterAct Math
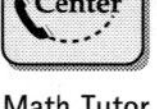
Math Tutor Center
Video Lectures on CD Disc 4

Student's Solutions Manual

a Subtract.

1. $2 - 9$

2. $3 - 8$

3. $-8 - (-2)$

4. $-6 - (-8)$

5. $-11 - (-11)$

6. $-6 - (-6)$

7. $12 - 16$

8. $14 - 19$

9. $20 - 27$

10. $30 - 4$

11. $-9 - (-3)$

12. $-7 - (-9)$

13. $-40 - (-40)$

14. $-9 - (-9)$

15. $7 - (-7)$

16. $4 - (-4)$

17. $8 - (-3)$

18. $-7 - 4$

19. $-6 - 8$

20. $6 - (-10)$

21. $-4 - (-9)$

22. $-14 - 2$

23. $-6 - (-5)$

24. $-4 - (-3)$

25. $8 - (-10)$

26. $5 - (-6)$

27. $-5 - (-2)$

28. $-3 - (-1)$

29. $-7 - 14$

30. $-9 - 16$

31. $0 - (-5)$

32. $0 - (-1)$

33. $-8 - 0$

34. $-9 - 0$

35. $7 - (-5)$

36. $7 - (-4)$

37. $2 - 25$

38. $18 - 63$

39. $-42 - 26$

40. $-18 - 63$

41. $-71 - 2$

42. $-49 - 3$

43. $24 - (-92)$

44. $48 - (-73)$

45. $-50 - (-50)$

46. $-70 - (-70)$

47. $-\frac{3}{8} - \frac{5}{8}$

48. $\frac{3}{9} - \frac{9}{9}$

49. $\frac{3}{4} - \frac{2}{3}$

50. $\frac{5}{8} - \frac{3}{4}$

51. $-\frac{3}{4} - \frac{2}{3}$

52. $-\frac{5}{8} - \frac{3}{4}$

53. $-\frac{5}{8} - \left(-\frac{3}{4}\right)$

54. $-\frac{3}{4} - \left(-\frac{2}{3}\right)$

55. $6.1 - (-13.8)$

56. $1.5 - (-3.5)$

57. $-2.7 - 5.9$

58. $-3.2 - 5.8$

59. $0.99 - 1$

60. $0.87 - 1$

61. $-79 - 114$

62. $-197 - 216$

63. $0 - (-500)$

64. $500 - (-1000)$

65. $-2.8 - 0$

66. $6.04 - 1.1$

67. $7 - 10.53$

68. $8 - (-9.3)$

69. $\frac{1}{6} - \frac{2}{3}$

70. $-\frac{3}{8} - \left(-\frac{1}{2}\right)$

71. $-\frac{4}{7} - \left(-\frac{10}{7}\right)$

72. $\frac{12}{5} - \frac{12}{5}$

73. $-\frac{7}{10} - \frac{10}{15}$

74. $-\frac{4}{18} - \left(-\frac{2}{9}\right)$

75. $\frac{1}{5} - \frac{1}{3}$

76. $-\frac{1}{7} - \left(-\frac{1}{6}\right)$

77. $\frac{5}{12} - \frac{7}{16}$

78. $-\frac{1}{35} - \left(-\frac{9}{40}\right)$

79. $-\frac{2}{15} - \frac{7}{12}$

80. $\frac{2}{21} - \frac{9}{14}$

Simplify.

81. $18 - (-15) - 3 - (-5) + 2$

82. $22 - (-18) + 7 + (-42) - 27$

83. $-31 + (-28) - (-14) - 17$

84. $-43 - (-19) - (-21) + 25$

85. $-34 - 28 + (-33) - 44$

86. $39 + (-88) - 29 - (-83)$

87. $-93 - (-84) - 41 - (-56)$

88. $84 + (-99) + 44 - (-18) - 43$

89. $-5.4 - (-30.9) + 30.8 + 40.2 - (-12)$

90. $14.9 - (-50.7) + 20 - (-32.8)$

91. $-\frac{7}{12} + \frac{3}{4} - \left(-\frac{5}{8}\right) - \frac{13}{24}$

92. $-\frac{11}{16} + \frac{5}{32} - \left(-\frac{1}{4}\right) + \frac{7}{8}$

b Solve.

93. *Ocean Depth.* The deepest point in the Pacific Ocean is the Marianas Trench, with a depth of 10,924 m. The deepest point in the Atlantic Ocean is the Puerto Rico Trench, with a depth of 8605 m. What is the difference in the elevation of the two trenches?

Source: *The World Almanac and Book of Facts*

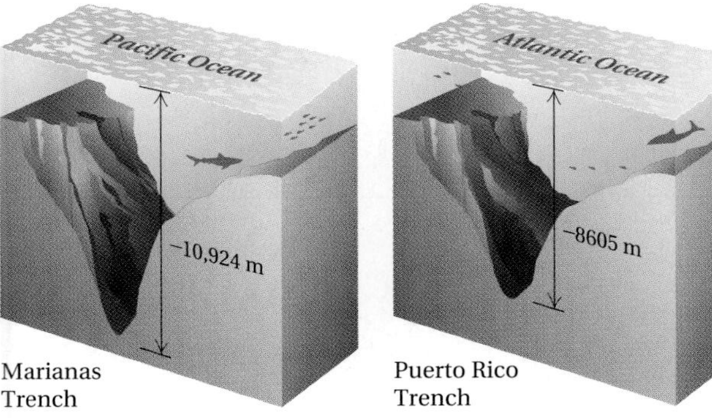

94. *Elevations in Africa.* The elevation of the highest point in Africa, Mt. Kilimanjaro, Tanzania, is 19,340 ft. The lowest elevation, at Lake Assal, Djibouti, is −512 ft. What is the difference in the elevations of the two locations?

95. Claire has a charge of $476.89 on her credit card, but she then returns a sweater that cost $128.95. How much does she now owe on her credit card?

96. Chris has $720 in a checking account. He writes a check for $970 to pay for a sound system. What is the balance in his checking account?

97. *Home-Run Differential.* In baseball, the difference between the number of home runs hit by a team's players and the number of home runs allowed by its pitchers is called the *home-run differential,* that is,

$$\text{Home run differential} = \frac{\text{Number of}}{\text{home runs hit}} - \frac{\text{Number of home}}{\text{runs allowed}}.$$

Teams strive for a positive home-run differential.

a) In a recent year, Atlanta hit 197 home runs and allowed 120. Find its home-run differential.

b) In a recent year, San Francisco hit 153 home runs and allowed 194. Find its home-run differential.

Source: Major League Baseball

98. *Temperature Records.* The greatest recorded temperature change in one 24-hour period occurred between January 23 and January 24, 1916, in Browning, Montana, where the temperature fell from 44°F to −56°F. By how much did the temperature drop?

Source: *The Guinness Book of Records*, 2004

99. *Low Points on Continents.* The lowest point in Africa is Lake Assal, which is 512 ft below sea level. The lowest point in South America is the Valdes Peninsula, which is 131 ft below sea level. How much lower is Lake Assal than the Valdes Peninsula?

Source: National Geographic Society

100. *Elevation Changes.* The lowest elevation in North America, Death Valley, California, is 282 ft below sea level. The highest elevation in North America, Mount McKinley, Alaska, is 20,320 ft. Find the difference in elevation between the highest point and the lowest.

Source: National Geographic Society

101. D_W If a negative number is subtracted from a positive number, will the result always be positive? Why or why not?

102. D_W Write a problem for a classmate to solve. Design the problem so that the solution is "The temperature dropped to −9°."

SKILL MAINTENANCE

Simplify. [1.6c]

103. $256 \div 64 \div 2^3 + 100$

104. $5 \cdot 6 + (7 \cdot 2)^2$

105. $2^5 \div 4 + 20 \div 2^2$

106. $65 - 5^2 \div 5 - 5 \cdot 2$

107. Add and simplify: $\frac{1}{8} + \frac{7}{12} + \frac{5}{24}$. [2.3a]

108. Simplify: $\frac{164}{256}$. [2.1e]

SYNTHESIS

Tell whether the statement is true or false for all integers a and b. If false, give an example to show why.

109. $a - 0 = 0 - a$

110. $0 - a = a$

111. If $a \neq b$, then $a - b \neq 0$.

112. If $a = -b$, then $a + b = 0$.

113. If $a + b = 0$, then a and b are opposites.

114. If $a - b = 0$, then $a = -b$.

7.5 MULTIPLICATION OF REAL NUMBERS

Objectives

a Multiply real numbers.

b Solve applied problems involving multiplication of real numbers.

a Multiplication

Multiplication of real numbers is very much like multiplication of arithmetic numbers. The only difference is that we must determine whether the answer is positive or negative.

MULTIPLICATION OF A POSITIVE NUMBER AND A NEGATIVE NUMBER

To see how to multiply a positive number and a negative number, consider the pattern of the following.

This number decreases by 1 each time. This number decreases by 5 each time.

$$\begin{aligned} 4 \cdot 5 &= 20 \\ 3 \cdot 5 &= 15 \\ 2 \cdot 5 &= 10 \\ 1 \cdot 5 &= 5 \\ 0 \cdot 5 &= 0 \\ -1 \cdot 5 &= -5 \\ -2 \cdot 5 &= -10 \\ -3 \cdot 5 &= -15 \end{aligned}$$

Do Exercise 1.

According to this pattern, it looks as though the product of a negative number and a positive number is negative. That is the case, and we have the first part of the rule for multiplying numbers.

THE PRODUCT OF A POSITIVE AND A NEGATIVE NUMBER

To multiply a positive number and a negative number, multiply their absolute values. The answer is negative.

EXAMPLES Multiply.

1. $8(-5) = -40$
2. $-\frac{1}{3} \cdot \frac{5}{7} = -\frac{5}{21}$
3. $(-7.2)5 = -36$

Do Exercises 2–7.

MULTIPLICATION OF TWO NEGATIVE NUMBERS

How do we multiply two negative numbers? Again, we look for a pattern.

This number decreases by 1 each time. This number increases by 5 each time.

$$\begin{aligned} 4 \cdot (-5) &= -20 \\ 3 \cdot (-5) &= -15 \\ 2 \cdot (-5) &= -10 \\ 1 \cdot (-5) &= -5 \\ 0 \cdot (-5) &= 0 \\ -1 \cdot (-5) &= 5 \\ -2 \cdot (-5) &= 10 \\ -3 \cdot (-5) &= 15 \end{aligned}$$

1. Complete, as in the example.

$$\begin{aligned} 4 \cdot 10 &= 40 \\ 3 \cdot 10 &= 30 \\ 2 \cdot 10 &= \\ 1 \cdot 10 &= \\ 0 \cdot 10 &= \\ -1 \cdot 10 &= \\ -2 \cdot 10 &= \\ -3 \cdot 10 &= \end{aligned}$$

Multiply.

2. $-3 \cdot 6$

3. $20 \cdot (-5)$

4. $4 \cdot (-20)$

5. $-\frac{2}{3} \cdot \frac{5}{6}$

6. $-4.23(7.1)$

7. $\frac{7}{8}\left(-\frac{4}{5}\right)$

8. Complete, as in the example.

$$\begin{aligned} 3 \cdot (-10) &= -30 \\ 2 \cdot (-10) &= -20 \\ 1 \cdot (-10) &= \\ 0 \cdot (-10) &= \\ -1 \cdot (-10) &= \\ -2 \cdot (-10) &= \\ -3 \cdot (-10) &= \end{aligned}$$

Answers on page A-18

Multiply.

9. $-9 \cdot (-3)$

10. $-16 \cdot (-2)$

11. $-7 \cdot (-5)$

12. $-\frac{4}{7}\left(-\frac{5}{9}\right)$

13. $-\frac{3}{2}\left(-\frac{4}{9}\right)$

14. $-3.25(-4.14)$

Multiply.

15. $5(-6)$

16. $(-5)(-6)$

17. $(-3.2) \cdot 0$

18. $\left(-\frac{4}{5}\right)\left(\frac{10}{3}\right)$

19. $0 \cdot (-34.2)$

20. $23 \cdot 0 \cdot \left(-4\frac{2}{3}\right)$

Answers on page A-18

Do Exercise 8 on the preceding page.

According to the pattern, it appears that the product of two negative numbers is positive. That is actually so, and we have the second part of the rule for multiplying real numbers.

THE PRODUCT OF TWO NEGATIVE NUMBERS

To multiply two negative numbers, multiply their absolute values. The answer is positive.

Do Exercises 9–14.

The following is another way to consider the rules we have for multiplication.

To multiply two nonzero real numbers:

a) Multiply the absolute values.
b) If the signs are the same, the answer is positive.
c) If the signs are different, the answer is negative.

MULTIPLICATION BY ZERO

The only case that we have not considered is multiplying by zero. As with other numbers, the product of any real number and 0 is 0.

THE MULTIPLICATION PROPERTY OF ZERO

For any real number a,

$$a \cdot 0 = 0 \cdot a = 0.$$

(The product of 0 and any real number is 0.)

EXAMPLES Multiply.

4. $(-3)(-4) = 12$
5. $-1.6(2) = -3.2$
6. $-19 \cdot 0 = 0$
7. $\left(-\frac{5}{6}\right)\left(-\frac{1}{9}\right) = \frac{5}{54}$
8. $0 \cdot (-452) = 0$
9. $23 \cdot 0 \cdot \left(-8\frac{2}{3}\right) = 0$

Do Exercises 15–20.

MULTIPLYING MORE THAN TWO NUMBERS

When multiplying more than two real numbers, we can choose order and grouping as we please.

EXAMPLES Multiply.

10. $-8 \cdot 2(-3) = -16(-3)$ Multiplying the first two numbers
$= 48$

11. $-8 \cdot 2(-3) = 24 \cdot 2$ Multiplying the negatives. Every pair of negative numbers gives a positive product.
$= 48$

12. $-3(-2)(-5)(4) = 6(-5)(4)$ Multiplying the first two numbers
$= (-30)4$
$= -120$

13. $\left(-\frac{1}{2}\right)(8)\left(-\frac{2}{3}\right)(-6) = (-4)4$ Multiplying the first two numbers and the last two numbers
$= -16$

14. $-5 \cdot (-2) \cdot (-3) \cdot (-6) = 10 \cdot 18 = 180$

15. $(-3)(-5)(-2)(-3)(-6) = (-30)(18) = -540$

Considering that the product of a pair of negative numbers is positive, we see the following pattern.

The product of an even number of negative numbers is positive.
The product of an odd number of negative numbers is negative.

Do Exercises 21–26.

EXAMPLE 16 Evaluate $2x^2$ when $x = 3$ and when $x = -3$.

$2x^2 = 2(3)^2 = 2(9) = 18;$
$2x^2 = 2(-3)^2 = 2(9) = 18$

Let's compare the expressions $(-x)^2$ and $-x^2$.

EXAMPLE 17 Evaluate $(-x)^2$ and $-x^2$ when $x = 5$.

$(-x)^2 = (-5)^2 = (-5)(-5) = 25;$ Substitute 5 for x. Then evaluate the power.

$-x^2 = -(5)^2 = -25$ Substitute 5 for x. Evaluate the power. Then find the opposite.

EXAMPLE 18 Evaluate $(-a)^2$ and $-a^2$ when $a = -4$.

To make sense of the substitutions and computations, we introduce extra brackets into the expressions.

$(-a)^2 = [-(-4)]^2 = [4]^2 = 16;$
$-a^2 = -(-4)^2 = -(16) = -16$

Multiply.

21. $5 \cdot (-3) \cdot 2$

22. $-3 \times (-4.1) \times (-2.5)$

23. $-\frac{1}{2} \cdot \left(-\frac{4}{3}\right) \cdot \left(-\frac{5}{2}\right)$

24. $-2 \cdot (-5) \cdot (-4) \cdot (-3)$

25. $(-4)(-5)(-2)(-3)(-1)$

26. $(-1)(-1)(-2)(-3)(-1)(-1)$

27. Evaluate $(-x)^2$ and $-x^2$ when $x = 2$.

28. Evaluate $(-x)^2$ and $-x^2$ when $x = -3$.

29. Evaluate $3x^2$ when $x = 4$ and when $x = -4$.

Answers on page A-18

30. Chemical Reaction. During a chemical reaction, the temperature in the beaker increased by 3°C every minute until 1:34 P.M. If the temperature was −17°C at 1:10 P.M., when the reaction began, what was the temperature at 1:34 P.M.?

Answer on page A-18

The expressions $(-x)^2$ and $-x^2$ are *not* equivalent. That is, they do not have the same value for every allowable replacement of the variable by a real number. To find $(-x)^2$, we take the opposite and then square. To find $-x^2$, we find the square and then take the opposite.

Do Exercises 27–29 on the preceding page.

b Applications and Problem Solving

We now consider multiplication of real numbers in real-world applications.

EXAMPLE 19 *Chemical Reaction.* During a chemical reaction, the temperature in the beaker decreased by 2°C every minute until 10:23 A.M. If the temperature was 17°C at 10:00 A.M., when the reaction began, what was the temperature at 10:23 A.M.?

This is a multistep problem. We first find the total number of degrees that the temperature dropped, using −2° for each minute. Since it dropped 2° for each of the 23 minutes, we know that the total drop d is given by

$$d = 23 \cdot (-2) = -46.$$

To determine the temperature after this time period, we find the sum of 17 and −46, or

$$T = 17 + (-46) = -29.$$

Thus the temperature at 10:23 A.M. was −29°C.

Do Exercise 30.

Study Tips

ATTITUDE AND THE POWER OF YOUR CHOICES

Making the right choices can give you the power to succeed in learning mathematics.

You can choose to improve your attitude and raise the academic goals that you have set for yourself. Projecting a positive attitude toward your study of mathematics and expecting a positive outcome can make it easier for you to learn and to perform well in this course.

Here are some positive choices you can make:

- Choose to allocate the proper amount of time to learn.
- Choose to place the primary responsibility for learning on yourself.
- Choose to make a strong commitment to learning.

Well-known American psychologist William James once said, "The one thing that will guarantee the successful conclusion of a doubtful undertaking is faith in the beginning that you can do it."

7.5 EXERCISE SET

For Extra Help

MathXL | MyMathLab | InterAct Math | Math Tutor Center | Video Lectures on CD Disc 4 | Student's Solutions Manual

a Multiply.

1. $-4 \cdot 2$

2. $-3 \cdot 5$

3. $-8 \cdot 6$

4. $-5 \cdot 2$

5. $8 \cdot (-3)$

6. $9 \cdot (-5)$

7. $-9 \cdot 8$

8. $-10 \cdot 3$

9. $-8 \cdot (-2)$

10. $-2 \cdot (-5)$

11. $-7 \cdot (-6)$

12. $-9 \cdot (-2)$

13. $15 \cdot (-8)$

14. $-12 \cdot (-10)$

15. $-14 \cdot 17$

16. $-13 \cdot (-15)$

17. $-25 \cdot (-48)$

18. $39 \cdot (-43)$

19. $-3.5 \cdot (-28)$

20. $97 \cdot (-2.1)$

21. $9 \cdot (-8)$

22. $7 \cdot (-9)$

23. $4 \cdot (-3.1)$

24. $3 \cdot (-2.2)$

25. $-5 \cdot (-6)$

26. $-6 \cdot (-4)$

27. $-7 \cdot (-3.1)$

28. $-4 \cdot (-3.2)$

29. $\frac{2}{3} \cdot \left(-\frac{3}{5}\right)$

30. $\frac{5}{7} \cdot \left(-\frac{2}{3}\right)$

31. $-\frac{3}{8} \cdot \left(-\frac{2}{9}\right)$

32. $-\frac{5}{8} \cdot \left(-\frac{2}{5}\right)$

33. -6.3×2.7

34. -4.1×9.5

35. $-\frac{5}{9} \cdot \frac{3}{4}$

36. $-\frac{8}{3} \cdot \frac{9}{4}$

37. $7 \cdot (-4) \cdot (-3) \cdot 5$

38. $9 \cdot (-2) \cdot (-6) \cdot 7$

39. $-\frac{2}{3} \cdot \frac{1}{2} \cdot \left(-\frac{6}{7}\right)$

40. $-\frac{1}{8} \cdot \left(-\frac{1}{4}\right) \cdot \left(-\frac{3}{5}\right)$

41. $-3 \cdot (-4) \cdot (-5)$

42. $-2 \cdot (-5) \cdot (-7)$

43. $-2 \cdot (-5) \cdot (-3) \cdot (-5)$

44. $-3 \cdot (-5) \cdot (-2) \cdot (-1)$

45. $\frac{1}{5}\left(-\frac{2}{9}\right)$

46. $-\frac{3}{5}\left(-\frac{2}{7}\right)$

47. $-7 \cdot (-21) \cdot 13$

48. $-14 \cdot (34) \cdot 12$

49. $-4 \cdot (-1.8) \cdot 7$

50. $-8 \cdot (-1.3) \cdot (-5)$

51. $-\frac{1}{9}\left(-\frac{2}{3}\right)\left(\frac{5}{7}\right)$

52. $-\frac{7}{2}\left(-\frac{5}{7}\right)\left(-\frac{2}{5}\right)$

53. $4 \cdot (-4) \cdot (-5) \cdot (-12)$

54. $-2 \cdot (-3) \cdot (-4) \cdot (-5)$

55. $0.07 \cdot (-7) \cdot 6 \cdot (-6)$

56. $80 \cdot (-0.8) \cdot (-90) \cdot (-0.09)$

57. $\left(-\frac{5}{6}\right)\left(\frac{1}{8}\right)\left(-\frac{3}{7}\right)\left(-\frac{1}{7}\right)$

58. $\left(\frac{4}{5}\right)\left(-\frac{2}{3}\right)\left(-\frac{15}{7}\right)\left(\frac{1}{2}\right)$

59. $(-14) \cdot (-27) \cdot 0$

60. $7 \cdot (-6) \cdot 5 \cdot (-4) \cdot 3 \cdot (-2) \cdot 1 \cdot 0$

61. $(-8)(-9)(-10)$

62. $(-7)(-8)(-9)(-10)$

63. $(-6)(-7)(-8)(-9)(-10)$

64. $(-5)(-6)(-7)(-8)(-9)(-10)$

65. $(-1)^{12}$

66. $(-1)^{9}$

67. Evaluate $(-x)^2$ and $-x^2$ when $x = 4$ and when $x = -4$.

68. Evaluate $(-x)^2$ and $-x^2$ when $x = 10$ and when $x = -10$.

69. Evaluate $(-3x)^2$ and $-3x^2$ when $x = 7$.

70. Evaluate $(-2x)^2$ and $-2x^2$ when $x = 3$.

71. Evaluate $5x^2$ when $x = 2$ and when $x = -2$.

72. Evaluate $2x^2$ when $x = 5$ and when $x = -5$.

73. Evaluate $-2x^3$ when $x = 1$ and when $x = -1$.

74. Evaluate $-3x^3$ when $x = 2$ and when $x = -2$.

b Solve.

75. *Lost Weight.* Dave lost 2 lb each week for a period of 10 weeks. Express his total weight change as an integer.

76. *Stock Loss.* Emma lost \$3 each day for a period of 5 days in the value of a stock she owned. Express her total loss as an integer.

77. *Chemical Reaction.* The temperature of a chemical compound was 0°C at 11:00 A.M. During a reaction, it dropped 3°C per minute until 11:18 A.M. What was the temperature at 11:18 A.M.?

78. *Chemical Reaction.* The temperature in a chemical compound was −5°C at 3:20 P.M. During a reaction, it increased 2°C per minute until 3:52 P.M. What was the temperature at 3:52 P.M.?

79. *Stock Price.* The price of ePDQ.com began the day at \$23.75 per share and dropped \$1.38 per hour for 8 hr. What was the price of the stock after 8 hr?

80. *Population Decrease.* The population of a rural town was 12,500. It decreased 380 each year for 4 yr. What was the population of the town after 4 yr?

81. *Diver's Position.* After diving 95 m below the sea level, a diver rises at a rate of 7 meters per minute for 9 min. Where is the diver in relation to the surface?

82. *Checking Account Balance.* Karen had \$68 in her checking account. After she had written checks to make seven purchases at \$13 each, what was the balance in her checking account?

83. ^D_W Multiplication can be thought of as repeated addition. Using this concept and a number line, explain why $3 \cdot (-5) = -15$.

84. ^D_W What rule have we developed that would tell you the sign of $(-7)^8$ and $(-7)^{11}$ without doing the computations? Explain.

SKILL MAINTENANCE

85. Find the LCM of 36 and 60. [1.9a]

86. Find the prime factorization of 4608. [1.7d]

Simplify. [2.1e]

87. $\frac{26}{39}$

88. $\frac{48}{54}$

89. $\frac{264}{484}$

90. $\frac{1025}{6625}$

91. $\frac{275}{800}$

92. $\frac{111}{201}$

93. $\frac{11}{264}$

94. $\frac{78}{13}$

SYNTHESIS

For each of Exercises 95 and 96, choose the correct answer from the selections given.

95. If a is positive and b is negative, then $-ab$ is:

a) Positive.
b) Negative.
c) 0.
d) Cannot be determined without more information

96. If a is positive and b is negative, then $(-a)(-b)$ is:

a) Positive.
b) Negative.
c) 0.
d) Cannot be determined without more information

97. Below is a number line showing 0 and two positive numbers x and y. Use a compass or ruler to locate as best you can the following:

$2x,\quad 3x,\quad 2y,\quad -x,\quad -y,\quad x + y,\quad x - y,\quad x - 2y.$

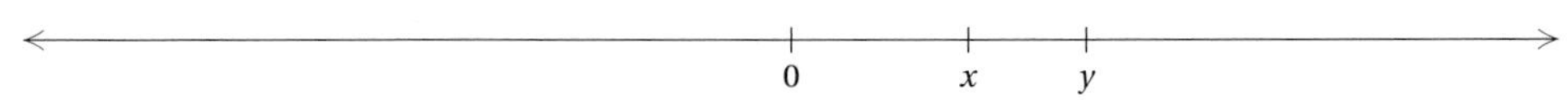

7.6 DIVISION OF REAL NUMBERS

Objectives

a Divide integers.

b Find the reciprocal of a real number.

c Divide real numbers.

d Solve applied problems involving division of real numbers.

We now consider division of real numbers. The definition of division results in rules for division that are the same as those for multiplication.

a Division of Integers

DIVISION

The quotient $a \div b$, or $\frac{a}{b}$, where $b \neq 0$, is that unique real number c for which $a = b \cdot c$.

Let's use the definition to divide integers.

EXAMPLES Divide, if possible. Check your answer.

1. $14 \div (-7) = -2$ — *Think*: What number multiplied by -7 gives 14? That number is -2. *Check*: $(-2)(-7) = 14$.

2. $\frac{-32}{-4} = 8$ — *Think*: What number multiplied by -4 gives -32? That number is 8. *Check*: $8(-4) = -32$.

3. $\frac{-10}{7} = -\frac{10}{7}$ — *Think*: What number multiplied by 7 gives -10? That number is $-\frac{10}{7}$. *Check*: $-\frac{10}{7} \cdot 7 = -10$.

4. $\frac{-17}{0}$ **is not defined.** — *Think*: What number multiplied by 0 gives -17? There is no such number because the product of 0 and any number is 0.

The rules for division are the same as those for multiplication.

To multiply or divide two real numbers (where the divisor is nonzero):

a) Multiply or divide the absolute values.
b) If the signs are the same, the answer is positive.
c) If the signs are different, the answer is negative.

Do Exercises 1–6.

EXCLUDING DIVISION BY 0

Example 4 shows why we cannot divide -17 by 0. We can use the same argument to show why we cannot divide any nonzero number b by 0. Consider $b \div 0$. We look for a number that when multiplied by 0 gives b. There is no such number because the product of 0 and any number is 0. Thus we cannot divide a nonzero number b by 0.

On the other hand, if we divide 0 by 0, we look for a number c such that $0 \cdot c = 0$. But $0 \cdot c = 0$ for any number c. Thus it appears that $0 \div 0$ could be any number we choose. Getting any answer we want when we divide 0 by 0 would be very confusing. Thus we agree that division by zero is not defined.

Divide.

1. $6 \div (-3)$
Think: What number multiplied by -3 gives 6?

2. $\frac{-15}{-3}$
Think: What number multiplied by -3 gives -15?

3. $-24 \div 8$
Think: What number multiplied by 8 gives -24?

4. $\frac{-48}{-6}$

5. $\frac{30}{-5}$

6. $\frac{30}{-7}$

Answers on page A-19

EXCLUDING DIVISION BY 0

Division by 0 is not defined.

$a \div 0$, or $\frac{a}{0}$, is not defined for all real numbers a.

DIVIDING 0 BY OTHER NUMBERS

Note that

$$0 \div 8 = 0 \text{ because } 0 = 0 \cdot 8; \qquad \frac{0}{-5} = 0 \text{ because } 0 = 0 \cdot (-5).$$

DIVIDENDS OF 0

Zero divided by any nonzero real number is 0:

$$\frac{0}{a} = 0; \qquad a \neq 0.$$

EXAMPLES Divide.

5. $0 \div (-6) = 0$

6. $\frac{0}{12} = 0$

7. $\frac{-3}{0}$ is not defined.

Do Exercises 7 and 8.

b Reciprocals

When two numbers like $\frac{1}{2}$ and 2 are multiplied, the result is 1. Such numbers are called **reciprocals** of each other. Every nonzero real number has a reciprocal, also called a **multiplicative inverse.**

RECIPROCALS

Two numbers whose product is 1 are called **reciprocals,** or **multiplicative inverses,** of each other.

EXAMPLES Find the reciprocal.

8. $\frac{7}{8}$ The reciprocal of $\frac{7}{8}$ is $\frac{8}{7}$ because $\frac{7}{8} \cdot \frac{8}{7} = 1$.

9. -5 The reciprocal of -5 is $-\frac{1}{5}$ because $-5\left(-\frac{1}{5}\right) = 1$.

10. 3.9 The reciprocal of 3.9 is $\frac{1}{3.9}$ because $3.9\left(\frac{1}{3.9}\right) = 1$.

11. $-\frac{1}{2}$ The reciprocal of $-\frac{1}{2}$ is -2 because $\left(-\frac{1}{2}\right)(-2) = 1$.

12. $-\frac{2}{3}$ The reciprocal of $-\frac{2}{3}$ is $-\frac{3}{2}$ because $\left(-\frac{2}{3}\right)\left(-\frac{3}{2}\right) = 1$.

13. $\frac{1}{3/4}$ The reciprocal of $\frac{1}{3/4}$ is $\frac{3}{4}$ because $\left(\frac{1}{3/4}\right)\left(\frac{3}{4}\right) = 1$.

Divide, if possible.

7. $\frac{-5}{0}$

8. $\frac{0}{-3}$

Find the reciprocal.

9. $\frac{2}{3}$

10. $-\frac{5}{4}$

11. -3

12. $-\frac{1}{5}$

13. 1.6

14. $\frac{1}{2/3}$

Answers on page A-19

15. Complete the following table.

NUMBER	OPPOSITE	RECIPROCAL
$\frac{2}{3}$		
$-\frac{5}{4}$		
0		
1		
-8		
-4.5		

Answers on page A-19

Study Tips

TAKE THE TIME!

The foundation of all your study skills is *time*! If you invest your time, we will help you achieve success.

"Nine-tenths of wisdom is being wise in time."

Theodore Roosevelt

RECIPROCAL PROPERTIES

For $a \neq 0$, the reciprocal of a can be named $\frac{1}{a}$ and the reciprocal of $\frac{1}{a}$ is a.

The reciprocal of a nonzero number $\frac{a}{b}$ can be named $\frac{b}{a}$.

The number 0 has no reciprocal.

Do Exercises 9–14 on the preceding page.

The reciprocal of a positive number is also a positive number, because their product must be the positive number 1. The reciprocal of a negative number is also a negative number, because their product must be the positive number 1.

THE SIGN OF A RECIPROCAL

The reciprocal of a number has the same sign as the number itself.

Caution!

It is important *not* to confuse *opposite* with *reciprocal*. Keep in mind that the opposite, or additive inverse, of a number is what we add to the number to get 0. The reciprocal, or multiplicative inverse, is what we multiply the number by to get 1.

Compare the following.

NUMBER	OPPOSITE (Change the sign.)	RECIPROCAL (Invert but do not change the sign.)
$-\frac{3}{8}$	$\frac{3}{8}$	$-\frac{8}{3}$
19	-19	$\frac{1}{19}$
$\frac{18}{7}$	$-\frac{18}{7}$	$\frac{7}{18}$
-7.9	7.9	$-\frac{1}{7.9}$, or $-\frac{10}{79}$
0	0	Not defined

$\left(-\frac{3}{8}\right)\left(-\frac{8}{3}\right) = 1$

$-\frac{3}{8} + \frac{3}{8} = 0$

Do Exercise 15.

Division of Real Numbers

We know that we can subtract by adding an opposite. Similarly, we can divide by multiplying by a reciprocal.

RECIPROCALS AND DIVISION

For any real numbers a and b, $b \neq 0$,

$$a \div b = \frac{a}{b} = a \cdot \frac{1}{b}.$$

(To divide, multiply by the reciprocal of the divisor.)

EXAMPLES Rewrite the division as a multiplication.

14. $-4 \div 3$ — $-4 \div 3$ is the same as $-4 \cdot \frac{1}{3}$

15. $\frac{6}{-7}$ — $\frac{6}{-7} = 6\left(-\frac{1}{7}\right)$

16. $\frac{x+2}{5}$ — $\frac{x+2}{5} = (x+2)\frac{1}{5}$ Parentheses are necessary here.

17. $\frac{-17}{1/b}$ — $\frac{-17}{1/b} = -17 \cdot b$

18. $\frac{3}{5} \div \left(-\frac{9}{7}\right)$ — $\frac{3}{5} \div \left(-\frac{9}{7}\right) = \frac{3}{5}\left(-\frac{7}{9}\right)$

Do Exercises 16–20.

When actually doing division calculations, we sometimes multiply by a reciprocal and we sometimes divide directly. With fraction notation, it is usually better to multiply by a reciprocal. With decimal notation, it is usually better to divide directly.

EXAMPLES Divide by multiplying by the reciprocal of the divisor.

19. $\frac{2}{3} \div \left(-\frac{5}{4}\right) = \frac{2}{3} \cdot \left(-\frac{4}{5}\right) = -\frac{8}{15}$

20. $-\frac{5}{6} \div \left(-\frac{3}{4}\right) = -\frac{5}{6} \cdot \left(-\frac{4}{3}\right) = \frac{20}{18} = \frac{10 \cdot 2}{9 \cdot 2} = \frac{10}{9} \cdot \frac{2}{2} = \frac{10}{9}$

Caution!

Be careful not to change the sign when taking a reciprocal!

21. $-\frac{3}{4} \div \frac{3}{10} = -\frac{3}{4} \cdot \left(\frac{10}{3}\right) = -\frac{30}{12} = -\frac{5}{2} \cdot \frac{6}{6} = -\frac{5}{2}$

Rewrite the division as a multiplication.

16. $\frac{4}{7} \div \left(-\frac{3}{5}\right)$

17. $\frac{5}{-8}$

18. $\frac{a-b}{7}$

19. $\frac{-23}{1/a}$

20. $-5 \div 7$

Divide by multiplying by the reciprocal of the divisor.

21. $\frac{4}{7} \div \left(-\frac{3}{5}\right)$

22. $-\frac{8}{5} \div \frac{2}{3}$

23. $-\frac{12}{7} \div \left(-\frac{3}{4}\right)$

24. Divide: $21.7 \div (-3.1)$.

Answers on page A-19

Find two equal expressions for the number with negative signs in different places.

25. $\dfrac{-5}{6}$

26. $-\dfrac{8}{7}$

27. $\dfrac{10}{-3}$

Answers on page A-19

Study Tips

VIDEO LECTURES ON CD

(ISBN-13 978-0-321-34832-6)
(ISBN-10 0-321-34832-X)

Developed and produced especially for this text, this complete set of digitized videos on CD-ROM features an engaging team of instructors, who present material and concepts by using examples and exercises from every section of the text.

With decimal notation, it is easier to carry out long division than to multiply by the reciprocal.

EXAMPLES Divide.

22. $-27.9 \div (-3) = \dfrac{-27.9}{-3} = 9.3$ Do the long division $3\overline{)27.9}$ (quotient 9.3). The answer is positive.

23. $-6.3 \div 2.1 = -3$ Do the long division $2.1.\overline{)6.3_{\wedge}}$ (quotient 3.). The answer is negative.

Do Exercises 21–24 on the preceding page.

Consider the following:

1. $\dfrac{2}{3} = \dfrac{2}{3} \cdot 1 = \dfrac{2}{3} \cdot \dfrac{-1}{-1} = \dfrac{2(-1)}{3(-1)} = \dfrac{-2}{-3}.$ Thus, $\dfrac{2}{3} = \dfrac{-2}{-3}.$

   (A negative number divided by a negative number is positive.)

2. $-\dfrac{2}{3} = -1 \cdot \dfrac{2}{3} = \dfrac{-1}{1} \cdot \dfrac{2}{3} = \dfrac{-1 \cdot 2}{1 \cdot 3} = \dfrac{-2}{3}.$ Thus, $-\dfrac{2}{3} = \dfrac{-2}{3}.$

   (A negative number divided by a positive number is negative.)

3. $\dfrac{-2}{3} = \dfrac{-2}{3} \cdot 1 = \dfrac{-2}{3} \cdot \dfrac{-1}{-1} = \dfrac{-2(-1)}{3(-1)} = \dfrac{2}{-3}.$ Thus, $-\dfrac{2}{3} = \dfrac{2}{-3}.$

   (A positive number divided by a negative number is negative.)

We can use the following properties to make sign changes in fraction notation.

SIGN CHANGES IN FRACTION NOTATION

For any numbers a and b, $b \neq 0$:

1. $\dfrac{-a}{-b} = \dfrac{a}{b}$

(The opposite of a number a divided by the opposite of another number b is the same as the quotient of the two numbers a and b.)

2. $\dfrac{-a}{b} = \dfrac{a}{-b} = -\dfrac{a}{b}$

(The opposite of a number a divided by another number b is the same as the number a divided by the opposite of the number b, and both are the same as the opposite of *a divided by b*.)

Do Exercises 25–27.

d Applications and Problem Solving

EXAMPLE 24 *Chemical Reaction.* During a chemical reaction, the temperature in the beaker decreased every minute by the same number of degrees. The temperature was 56°F at 10:10 A.M. By 10:42 A.M., the temperature had dropped to −12°F. By how many degrees did it change each minute?

We first determine by how many degrees d the temperature changed altogether. We subtract −12 from 56:

$$d = 56 - (-12) = 56 + 12 = 68.$$

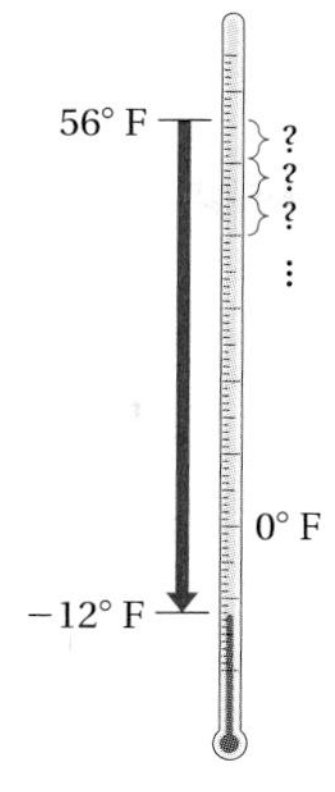

The temperature changed a total of 68°. We can express this as −68° since the temperature dropped.

The amount of time t that passed was 42 − 10, or 32 min. Thus the number of degrees T that the temperature dropped each minute is given by

$$T = \frac{d}{t} = \frac{-68}{32} = -2.125.$$

The change was −2.125°F per minute.

Do Exercise 28.

28. Chemical Reaction. During a chemical reaction, the temperature in the beaker decreased every minute by the same number of degrees. The temperature was 71°F at 2:12 P.M. By 2:37 P.M., the temperature had changed to −14°F. By how many degrees did it change each minute?

Answer on page A-19

CALCULATOR CORNER

Operations on the Real Numbers We can perform operations on the real numbers on a graphing calculator. Recall that negative numbers are entered using the opposite key, (−), rather than the subtraction operation key, −. Consider the sum −5 + (−3.8). We use parentheses when we write this sum in order to separate the addition symbol and the "opposite of" symbol and thus make the expression more easily read. When we enter this calculation on a graphing calculator, however, the parentheses are not necessary. We can press (−) 5 + (−) 3 . 8 ENTER. The result is −8.8. Note that it is not incorrect to enter the parentheses. The result will be the same if this is done.

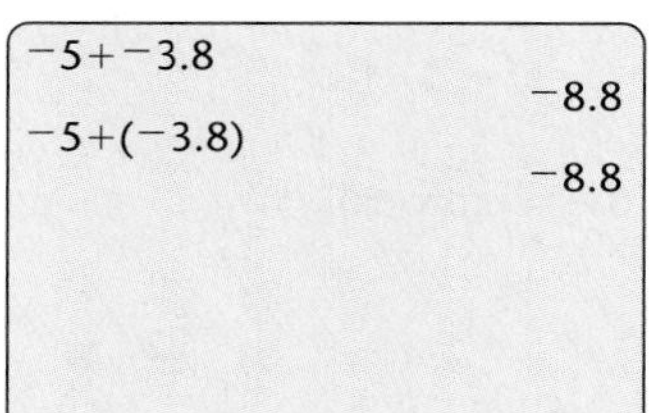

To find the difference 10 − (−17), we press 1 0 − (−) 1 7 ENTER. The result is 27. We can also multiply and divide real numbers. To find −5 · (−7), we press (−) 5 × (−) 7 ENTER, and to find 45 ÷ (−9), we press 4 5 ÷ (−) 9 ENTER. Note that it is not necessary to use parentheses in any of these calculations.

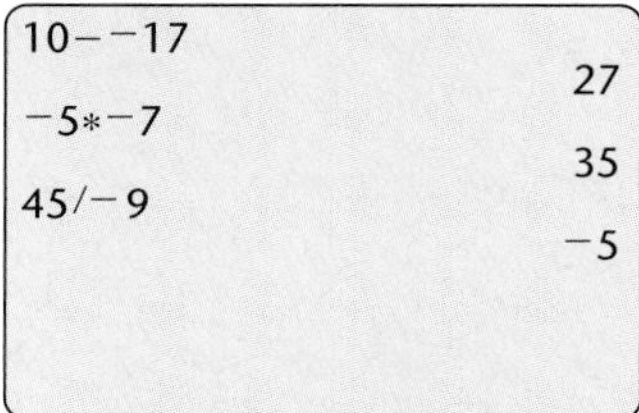

Exercises: Use a calculator to perform the operation.

1. $-8 + 4$
2. $1.2 + (-1.5)$
3. $-7 + (-5)$
4. $-7.6 + (-1.9)$
5. $-8 - 4$
6. $1.2 - (-1.5)$
7. $-7 - (-5)$
8. $-7.6 - (-1.9)$
9. $-8 \cdot 4$
10. $1.2 \cdot (-1.5)$
11. $-7 \cdot (-5)$
12. $-7.6 \cdot (-1.9)$
13. $-8 \div 4$
14. $1.2 \div (-1.5)$
15. $-7 \div (-5)$
16. $-7.6 \div (-1.9)$

7.6 EXERCISE SET

For Extra Help

MathXL | MyMathLab | InterAct Math | Math Tutor Center | Video Lectures on CD Disc 4 | Student's Solutions Manual

a Divide, if possible. Check each answer.

1. $48 \div (-6)$

2. $\frac{42}{-7}$

3. $\frac{28}{-2}$

4. $24 \div (-12)$

5. $\frac{-24}{8}$

6. $-18 \div (-2)$

7. $\frac{-36}{-12}$

8. $-72 \div (-9)$

9. $\frac{-72}{9}$

10. $\frac{-50}{25}$

11. $-100 \div (-50)$

12. $\frac{-200}{8}$

13. $-108 \div 9$

14. $\frac{-63}{-7}$

15. $\frac{200}{-25}$

16. $-300 \div (-16)$

17. $\frac{75}{0}$

18. $\frac{0}{-5}$

19. $\frac{0}{-2.6}$

20. $\frac{-23}{0}$

b Find the reciprocal.

21. $\frac{15}{7}$

22. $\frac{3}{8}$

23. $-\frac{47}{13}$

24. $-\frac{31}{12}$

25. 13

26. -10

27. 4.3

28. -8.5

29. $\frac{1}{-7.1}$

30. $\frac{1}{-4.9}$

31. $\frac{p}{q}$

32. $\frac{s}{t}$

33. $\frac{1}{4y}$

34. $\frac{-1}{8a}$

35. $\frac{2a}{3b}$

36. $\frac{-4y}{3x}$

C Rewrite the division as a multiplication.

37. $4 \div 17$

38. $5 \div (-8)$

39. $\dfrac{8}{-13}$

40. $-\dfrac{13}{47}$

41. $\dfrac{13.9}{-1.5}$

42. $-\dfrac{47.3}{21.4}$

43. $\dfrac{x}{\frac{1}{y}}$

44. $\dfrac{13}{x}$

45. $\dfrac{3x + 4}{5}$

46. $\dfrac{4y - 8}{-7}$

47. $\dfrac{5a - b}{5a + b}$

48. $\dfrac{2x + x^2}{x - 5}$

Divide.

49. $\frac{3}{4} \div \left(-\frac{2}{3}\right)$

50. $\frac{7}{8} \div \left(-\frac{1}{2}\right)$

51. $-\frac{5}{4} \div \left(-\frac{3}{4}\right)$

52. $-\frac{5}{9} \div \left(-\frac{5}{6}\right)$

53. $-\frac{2}{7} \div \left(-\frac{4}{9}\right)$

54. $-\frac{3}{5} \div \left(-\frac{5}{8}\right)$

55. $-\frac{3}{8} \div \left(-\frac{8}{3}\right)$

56. $-\frac{5}{8} \div \left(-\frac{6}{5}\right)$

57. $-6.6 \div 3.3$

58. $-44.1 \div (-6.3)$

59. $\dfrac{-11}{-13}$

60. $\dfrac{-1.9}{20}$

61. $\dfrac{48.6}{-3}$

62. $\dfrac{-17.8}{3.2}$

63. $\dfrac{-9}{17 - 17}$

64. $\dfrac{-8}{-5 + 5}$

d *Percent of Increase or Decrease in Employment.* A percent of increase is generally positive and a percent of decrease is generally negative. The following table lists estimates of the number of job opportunities for various occupations in 2002 and 2012. In Exercises 65–68, find the missing numbers.

	OCCUPATION	NUMBER OF JOBS IN 2002 (in thousands)	NUMBER OF JOBS IN 2012 (in thousands)	CHANGE	PERCENT OF INCREASE OR DECREASE
	Electrician	659	814	155	23.5%
	Travel agent	118	102	−16	−13.6%
65.	Fitness trainer/ aerobic instructor	183	264	81	
66.	Child-care worker	1211	1353	142	
67.	Telemarketer	428	406	−22	
68.	Aerospace engineer	78	74	−4	

Source: U.S. Bureau of Labor Statistics

69. $\mathbf{D_W}$ Explain how multiplication can be used to justify why a negative number divided by a positive number is negative.

70. $\mathbf{D_W}$ Explain how multiplication can be used to justify why a negative number divided by a negative number is positive.

SKILL MAINTENANCE

Simplify. [1.6c]

71. $2^3 - 5 \cdot 3 + 8 \cdot 10 \div 2$

72. $16 \cdot 2^3 - 5 \cdot 3 + 80 \div 10 \cdot 2$

73. $1000 \div 100 \div 10$

74. $216 \cdot 6^3 \div 6^2$

75. Simplify: $\frac{264}{468}$. [2.1e]

76. Convert to decimal notation: 47.7%. [4.2b]

77. Convert to percent notation: $\frac{7}{8}$. [4.3a]

78. Simplify: $\frac{40}{60}$. [2.1e]

79. Divide and simplify: $\frac{12}{25} \div \frac{32}{75}$. [2.2c]

80. Multiply and simplify: $\frac{12}{25} \cdot \frac{32}{75}$. [2.2a]

SYNTHESIS

81. Find the reciprocal of -10.5. What happens if you take the reciprocal of the result?

82. Determine those real numbers a for which the opposite of a is the same as the reciprocal of a.

Tell whether the expression represents a positive number or a negative number when a and b are negative.

83. $\frac{-a}{b}$

84. $\frac{-a}{-b}$

85. $-\left(\frac{a}{-b}\right)$

86. $-\left(\frac{-a}{b}\right)$

87. $-\left(\frac{-a}{-b}\right)$

88. Of all possible quotients of the numbers 10, $-\frac{1}{2}$, -5, and $\frac{1}{5}$, which two produce the largest quotient? Which two produce the smallest quotient?

7.7 PROPERTIES OF REAL NUMBERS

a Equivalent Expressions

In solving equations and doing other kinds of work in algebra, we manipulate expressions in various ways. For example, instead of $x + x$, we might write $2x$, knowing that the two expressions represent the same number for any allowable replacement of x. In that sense, the expressions $x + x$ and $2x$ are **equivalent**, as are $\frac{3}{x}$ and $\frac{3x}{x^2}$, even though 0 is not an allowable replacement because division by 0 is not defined.

EQUIVALENT EXPRESSIONS

Two expressions that have the same value for all allowable replacements are called **equivalent.**

The expressions $x + 3x$ and $5x$ are *not* equivalent.

Do Exercises 1 and 2.

In this section, we will consider several laws of real numbers that will allow us to find equivalent expressions. The first two laws are the *identity properties of 0 and 1.*

THE IDENTITY PROPERTY OF 0

For any real number a,

$$a + 0 = 0 + a = a.$$

(The number 0 is the *additive identity.*)

THE IDENTITY PROPERTY OF 1

For any real number a,

$$a \cdot 1 = 1 \cdot a = a.$$

(The number 1 is the *multiplicative identity.*)

We often refer to the use of the identity property of 1 as "multiplying by 1." We can use this method to find equivalent fraction expressions. Recall from arithmetic that to multiply with fraction notation, we multiply numerators and denominators. (See also Section 2.2.)

EXAMPLE 1 Write a fraction expression equivalent to $\frac{2}{3}$ with a denominator of $3x$:

$$\frac{2}{3} = \frac{\square}{3x}.$$

Objectives

a Find equivalent fraction expressions and simplify fraction expressions.

b Use the commutative and associative laws to find equivalent expressions.

c Use the distributive laws to multiply expressions like 8 and $x - y$.

d Use the distributive laws to factor expressions like $4x - 12 + 24y$.

e Collect like terms.

Complete the table by evaluating each expression for the given values.

1.

Value	$x + x$	$2x$
$x = 3$		
$x = -6$		
$x = 4.8$		

2.

Value	$x + 3x$	$5x$
$x = 2$		
$x = -6$		
$x = 4.8$		

Answers on page A-19

3. Write a fraction expression equivalent to $\frac{3}{4}$ with a denominator of 8:

$$\frac{3}{4} = \frac{\square}{8}.$$

4. Write a fraction expression equivalent to $\frac{3}{4}$ with a denominator of $4t$:

$$\frac{3}{4} = \frac{\square}{4t}.$$

Simplify.

5. $\frac{3y}{4y}$

6. $-\frac{16m}{12m}$

7. $\frac{5xy}{40y}$

8. $\frac{18p}{24pq}$

9. Evaluate $x + y$ and $y + x$ when $x = -2$ and $y = 3$.

10. Evaluate xy and yx when $x = -2$ and $y = 5$.

Answers on page A-19

Note that $3x = 3 \cdot x$. We want fraction notation for $\frac{2}{3}$ that has a denominator of $3x$, but the denominator 3 is missing a factor of x. Thus we multiply by 1, using x/x as an equivalent expression for 1:

$$\frac{2}{3} = \frac{2}{3} \cdot 1 = \frac{2}{3} \cdot \frac{x}{x} = \frac{2x}{3x}.$$

The expressions $2/3$ and $2x/3x$ are equivalent. They have the same value for any allowable replacement. Note that $2x/3x$ is not defined for a replacement of 0, but for all nonzero real numbers, the expressions $2/3$ and $2x/3x$ have the same value.

Do Exercises 3 and 4.

In algebra, we consider an expression like $2/3$ to be "simplified" from $2x/3x$. To find such simplified expressions, we use the identity property of 1 to remove a factor of 1. (See also Section 2.2.)

EXAMPLE 2 Simplify: $-\frac{20x}{12x}$.

$-\frac{20x}{12x} = -\frac{5 \cdot 4x}{3 \cdot 4x}$ We look for the largest factor common to both the numerator and the denominator and factor each.

$= -\frac{5}{3} \cdot \frac{4x}{4x}$ Factoring the fraction expression

$= -\frac{5}{3} \cdot 1$ $\frac{4x}{4x} = 1$

$= -\frac{5}{3}$ Removing a factor of 1 using the identity property of 1

EXAMPLE 3 Simplify: $\frac{14ab}{56a}$.

$$\frac{14ab}{56a} = \frac{14a \cdot b}{14a \cdot 4} = \frac{14a}{14a} \cdot \frac{b}{4} = 1 \cdot \frac{b}{4} = \frac{b}{4}$$

Do Exercises 5–8.

b The Commutative and Associative Laws

THE COMMUTATIVE LAWS

Let's examine the expressions $x + y$ and $y + x$, as well as xy and yx.

EXAMPLE 4 Evaluate $x + y$ and $y + x$ when $x = 4$ and $y = 3$.

We substitute 4 for x and 3 for y in both expressions:

$$x + y = 4 + 3 = 7; \qquad y + x = 3 + 4 = 7.$$

EXAMPLE 5 Evaluate xy and yx when $x = 23$ and $y = -12$.

We substitute 23 for x and -12 for y in both expressions:

$$xy = 23 \cdot (-12) = -276; \qquad yx = (-12) \cdot 23 = -276.$$

Do Exercises 9 and 10 on the preceding page.

Note that the expressions $x + y$ and $y + x$ have the same values no matter what the variables stand for. Thus they are equivalent. Therefore, when we add two numbers, the order in which we add does not matter. Similarly, the expressions xy and yx are equivalent. They also have the same values, no matter what the variables stand for. Therefore, when we multiply two numbers, the order in which we multiply does not matter.

The following are examples of general patterns or laws.

THE COMMUTATIVE LAWS

Addition. For any numbers a and b,

$$a + b = b + a.$$

(We can change the order when adding without affecting the answer.)

Multiplication. For any numbers a and b,

$$ab = ba.$$

(We can change the order when multiplying without affecting the answer.)

Using a commutative law, we know that $x + 2$ and $2 + x$ are equivalent. Similarly, $3x$ and $x(3)$ are equivalent. Thus, in an algebraic expression, we can replace one with the other and the result will be equivalent to the original expression.

EXAMPLE 6 Use the commutative laws to write an expression equivalent to $y + 5$, ab, and $7 + xy$.

An expression equivalent to $y + 5$ is $5 + y$ by the commutative law of addition.

An expression equivalent to ab is ba by the commutative law of multiplication.

An expression equivalent to $7 + xy$ is $xy + 7$ by the commutative law of addition. Another expression equivalent to $7 + xy$ is $7 + yx$ by the commutative law of multiplication. Another equivalent expression is $yx + 7$.

Do Exercises 11–13.

THE ASSOCIATIVE LAWS

Now let's examine the expressions $a + (b + c)$ and $(a + b) + c$. Note that these expressions involve the use of parentheses as *grouping* symbols, and they also involve three numbers. Calculations within parentheses are to be done first.

EXAMPLE 7 Calculate and compare: $3 + (8 + 5)$ and $(3 + 8) + 5$.

$3 + (8 + 5) = 3 + 13$ Calculating within parentheses first; adding the 8 and 5

$= 16;$

$(3 + 8) + 5 = 11 + 5$ Calculating within parentheses first; adding the 3 and 8

$= 16$

Use a commutative law to write an equivalent expression.

11. $x + 9$

12. pq

13. $xy + t$

Answers on page A-19

14. Calculate and compare:

$8 + (9 + 2)$ and $(8 + 9) + 2$.

15. Calculate and compare:

$10 \cdot (5 \cdot 3)$ and $(10 \cdot 5) \cdot 3$.

Use an associative law to write an equivalent expression.

16. $r + (s + 7)$

17. $9(ab)$

Answers on page A-19

The two expressions in Example 7 name the same number. Moving the parentheses to group the additions differently does not affect the value of the expression.

EXAMPLE 8 Calculate and compare: $3 \cdot (4 \cdot 2)$ and $(3 \cdot 4) \cdot 2$.

$$3 \cdot (4 \cdot 2) = 3 \cdot 8 = 24; \qquad (3 \cdot 4) \cdot 2 = 12 \cdot 2 = 24$$

Do Exercises 14 and 15.

You may have noted that when only addition is involved, parentheses can be placed any way we please without affecting the answer. When only multiplication is involved, parentheses also can be placed any way we please without affecting the answer.

THE ASSOCIATIVE LAWS

Addition. For any numbers a, b, and c,

$$a + (b + c) = (a + b) + c.$$

(Numbers can be grouped in any manner for addition.)

Multiplication. For any numbers a, b, and c,

$$a \cdot (b \cdot c) = (a \cdot b) \cdot c.$$

(Numbers can be grouped in any manner for multiplication.)

EXAMPLE 9 Use an associative law to write an expression equivalent to $(y + z) + 3$ and $8(xy)$.

An expression equivalent to $(y + z) + 3$ is $y + (z + 3)$ by the associative law of addition.

An expression equivalent to $8(xy)$ is $(8x)y$ by the associative law of multiplication.

Do Exercises 16 and 17.

The associative laws say parentheses can be placed any way we please when only additions or only multiplications are involved. Thus we often omit them. For example,

$$x + (y + 2) \text{ means } x + y + 2, \quad \text{and} \quad (lw)h \text{ means } lwh.$$

USING THE COMMUTATIVE AND ASSOCIATIVE LAWS TOGETHER

EXAMPLE 10 Use the commutative and associative laws to write at least three expressions equivalent to $(x + 5) + y$.

a) $(x + 5) + y = x + (5 + y)$
$= x + (y + 5)$
Using the associative law first and then using the commutative law

b) $(x + 5) + y = y + (x + 5)$
$= y + (5 + x)$
Using the commutative law first and then the commutative law again

c) $(x + 5) + y = (5 + x) + y$
$= 5 + (x + y)$
Using the commutative law first and then the associative law

EXAMPLE 11 Use the commutative and associative laws to write at least three expressions equivalent to $(3x)y$.

a) $(3x)y = 3(xy)$ — Using the associative law first and then using the commutative law

$= 3(yx)$

b) $(3x)y = y(3x)$ — Using the commutative law twice

$= y(x \cdot 3)$

c) $(3x)y = (x \cdot 3)y$ — Using the commutative law, and then the associative law, and then the commutative law again

$= x(3y)$

$= x(y \cdot 3)$

Do Exercises 18 and 19.

Use the commutative and associative laws to write at least three equivalent expressions.

18. $4(tu)$

19. $r + (2 + s)$

C The Distributive Laws

The *distributive laws* are the basis of many procedures in both arithmetic and algebra. They are probably the most important laws that we use to manipulate algebraic expressions. The distributive law of multiplication over addition involves two operations: addition and multiplication.

Let's begin by considering a multiplication problem from arithmetic:

$$\begin{array}{r l} 45 & \\ \times\ \ 7 & \\ \hline 35 & \leftarrow \text{This is } 7 \cdot 5. \\ 280 & \leftarrow \text{This is } 7 \cdot 40. \\ \hline 315 & \leftarrow \text{This is the sum } 7 \cdot 40 + 7 \cdot 5. \end{array}$$

To carry out the multiplication, we actually added two products. That is,

$$7 \cdot 45 = 7(40 + 5) = 7 \cdot 40 + 7 \cdot 5.$$

Let's examine this further. If we wish to multiply a sum of several numbers by a factor, we can either add and then multiply, or multiply and then add.

EXAMPLE 12 Compute in two ways: $5 \cdot (4 + 8)$.

a) $5 \cdot (4 + 8)$ — Adding within parentheses first, and then multiplying

$= 5 \cdot 12$

$= 60$

b) $(5 \cdot 4) + (5 \cdot 8)$ — Distributing the multiplication to terms within parentheses first and then adding

$= 20 + 40$

$= 60$

Do Exercises 20–22.

Compute.

20. a) $7 \cdot (3 + 6)$

b) $(7 \cdot 3) + (7 \cdot 6)$

21. a) $2 \cdot (10 + 30)$

b) $(2 \cdot 10) + (2 \cdot 30)$

22. a) $(2 + 5) \cdot 4$

b) $(2 \cdot 4) + (5 \cdot 4)$

THE DISTRIBUTIVE LAW OF MULTIPLICATION OVER ADDITION

For any numbers a, b, and c,

$$a(b + c) = ab + ac.$$

Answers on page A-19

Calculate.

23. **a)** $4(5 - 3)$

b) $4 \cdot 5 - 4 \cdot 3$

24. **a)** $-2 \cdot (5 - 3)$

b) $-2 \cdot 5 - (-2) \cdot 3$

25. **a)** $5 \cdot (2 - 7)$

b) $5 \cdot 2 - 5 \cdot 7$

What are the terms of the expression?

26. $5x - 8y + 3$

27. $-4y - 2x + 3z$

Multiply.

28. $3(x - 5)$

29. $5(x + 1)$

30. $\frac{3}{5}(p + q - t)$

Answers on page A-19

In the statement of the distributive law, we know that in an expression such as $ab + ac$, the multiplications are to be done first according to the rules for order of operations. (See Section 1.6.) So, instead of writing $(4 \cdot 5) + (4 \cdot 7)$, we can write $4 \cdot 5 + 4 \cdot 7$. However, in $a(b + c)$, we cannot omit the parentheses. If we did, we would have $ab + c$, which means $(ab) + c$. For example, $3(4 + 2) = 18$, but $3 \cdot 4 + 2 = 14$.

There is another distributive law that relates multiplication and subtraction. This law says that to multiply by a difference, we can either subtract and then multiply, or multiply and then subtract.

THE DISTRIBUTIVE LAW OF MULTIPLICATION OVER SUBTRACTION

For any numbers a, b, and c,

$$a(b - c) = ab - ac.$$

We often refer to "*the* distributive law" when we mean *either* or *both* of these laws.

Do Exercises 23–25.

What do we mean by the *terms* of an expression? **Terms** are separated by addition signs. If there are subtraction signs, we can find an equivalent expression that uses addition signs.

EXAMPLE 13 What are the terms of $3x - 4y + 2z$?

We have

$3x - 4y + 2z = 3x + (-4y) + 2z.$ Separating parts with + signs

The terms are $3x$, $-4y$, and $2z$.

Do Exercises 26 and 27.

The distributive laws are a basis for a procedure in algebra called **multiplying.** In an expression like $8(a + 2b - 7)$, we multiply each term inside the parentheses by 8:

$$8(a + 2b - 7) = 8 \cdot a + 8 \cdot 2b - 8 \cdot 7 = 8a + 16b - 56.$$

EXAMPLES Multiply.

14. $9(x - 5) = 9x - 9(5)$ Using the distributive law of multiplication over subtraction

$= 9x - 45$

15. $\frac{2}{3}(w + 1) = \frac{2}{3} \cdot w + \frac{2}{3} \cdot 1$ Using the distributive law of multiplication over addition

$= \frac{2}{3}w + \frac{2}{3}$

16. $\frac{4}{3}(s - t + w) = \frac{4}{3}s - \frac{4}{3}t + \frac{4}{3}w$ Using both distributive laws

Do Exercises 28–30.

EXAMPLE 17 Multiply: $-4(x - 2y + 3z)$.

$$-4(x - 2y + 3z) = -4 \cdot x - (-4)(2y) + (-4)(3z) \quad \text{Using both distributive laws}$$
$$= -4x - (-8y) + (-12z) \quad \text{Multiplying}$$
$$= -4x + 8y - 12z$$

We can also do this problem by first finding an equivalent expression with all plus signs and then multiplying:

$$-4(x - 2y + 3z) = -4[x + (-2y) + 3z]$$
$$= -4 \cdot x + (-4)(-2y) + (-4)(3z)$$
$$= -4x + 8y - 12z.$$

Do Exercises 31–33.

EXAMPLES Name the property illustrated by the equation.

Equation	*Property*
18. $5x = x(5)$	Commutative law of multiplication
19. $a + (8.5 + b) = (a + 8.5) + b$	Associative law of addition
20. $0 + 11 = 11$	Identity property of 0
21. $(-5s)t = -5(st)$	Associative law of multiplication
22. $\frac{3}{4} \cdot 1 = \frac{3}{4}$	Identity property of 1
23. $12.5(w - 3) = 12.5w - 12.5(3)$	Distributive law of multiplication over subtraction
24. $y + \frac{1}{2} = \frac{1}{2} + y$	Commutative law of addition

Do Exercises 34–40.

d Factoring

Factoring is the reverse of multiplying. To factor, we can use the distributive laws in reverse:

$$ab + ac = a(b + c) \quad \text{and} \quad ab - ac = a(b - c).$$

FACTORING

To **factor** an expression is to find an equivalent expression that is a product.

Look at Example 14. To *factor* $9x - 45$, we find an equivalent expression that is a product, $9(x - 5)$. When all the terms of an expression have a factor in common, we can "factor it out" using the distributive laws. Note the following.

$9x$ has the factors $9, -9, 3, -3, 1, -1, x, -x, 3x, -3x, 9x, -9x$;

-45 has the factors $1, -1, 3, -3, 5, -5, 9, -9, 15, -15, 45, -45$

Multiply.

31. $-2(x - 3)$

32. $5(x - 2y + 4z)$

33. $-5(x - 2y + 4z)$

Name the property illustrated by the equation.

34. $(-8a)b = -8(ab)$

35. $p \cdot 1 = p$

36. $m + 34 = 34 + m$

37. $2(t + 5) = 2t + 2(5)$

38. $0 + k = k$

39. $-8x = x(-8)$

40. $x + (4.3 + b) = (x + 4.3) + b$

Answers on page A-20

Factor.

41. $6x - 12$

42. $3x - 6y + 9$

43. $bx + by - bz$

44. $16a - 36b + 42$

45. $\frac{3}{8}x - \frac{5}{8}y + \frac{7}{8}$

46. $-12x + 32y - 16z$

Answers on page A-20

We generally remove the largest common factor. In this case, that factor is 9. Thus,

$$\begin{aligned} 9x - 45 &= 9 \cdot x - 9 \cdot 5 \\ &= 9(x - 5). \end{aligned}$$

Remember that an expression has been factored when we have found an equivalent expression that is a product. Above, we note that $9x - 45$ and $9(x - 5)$ are equivalent expressions. The expression $9x - 45$ is the difference of $9x$ and 45; the expression $9(x - 5)$ is the product of 9 and $(x - 5)$.

EXAMPLES Factor.

25. $5x - 10 = 5 \cdot x - 5 \cdot 2$ Try to do this step mentally.
$= 5(x - 2)$ You can check by multiplying.

26. $ax - ay + az = a(x - y + z)$

27. $9x + 27y - 9 = 9 \cdot x + 9 \cdot 3y - 9 \cdot 1 = 9(x + 3y - 1)$

Note in Example 27 that you might, at first, just factor out a 3, as follows:

$$\begin{aligned} 9x + 27y - 9 &= 3 \cdot 3x + 3 \cdot 9y - 3 \cdot 3 \\ &= 3(3x + 9y - 3). \end{aligned}$$

At this point, the mathematics is correct, but the answer is not because there is another factor of 3 that can be factored out, as follows:

$$\begin{aligned} 3 \cdot 3x + 3 \cdot 9y - 3 \cdot 3 &= 3(3x + 9y - 3) \\ &= 3(3 \cdot x + 3 \cdot 3y - 3 \cdot 1) \\ &= 3 \cdot 3(x + 3y - 1) \\ &= 9(x + 3y - 1). \end{aligned}$$

We now have a correct answer, but it took more work than we did in Example 27. Thus it is better to look for the greatest common factor at the outset.

EXAMPLES Factor. Try to write just the answer, if you can.

28. $5x - 5y = 5(x - y)$

29. $-3x + 6y - 9z = -3(x - 2y + 3z)$

We usually factor out a negative factor when the first term is negative. The way we factor can depend on the situation in which we are working. We might also factor the expression in Example 29 as follows:

$$-3x + 6y - 9z = 3(-x + 2y - 3z).$$

30. $18z - 12x - 24 = 6(3z - 2x - 4)$

31. $\frac{1}{2}x + \frac{3}{2}y - \frac{1}{2} = \frac{1}{2}(x + 3y - 1)$

Remember that you can always check factoring by multiplying. Keep in mind that an expression is factored when it is written as a product.

Do Exercises 41–46.

e Collecting Like Terms

Terms such as $5x$ and $-4x$, whose variable factors are exactly the same, are called **like terms.** Similarly, numbers, such as -7 and 13, are like terms. Also, $3y^2$ and $9y^2$ are like terms because the variables are raised to the same power. Terms such as $4y$ and $5y^2$ are not like terms, and $7x$ and $2y$ are not like terms.

The process of **collecting like terms** is also based on the distributive laws. We can apply the distributive law when a factor is on the right because of the commutative law of multiplication.

Later in this text, terminology like "collecting like terms" and "combining like terms" will also be referred to as "simplifying."

EXAMPLES Collect like terms. Try to write just the answer, if you can.

32. $4x + 2x = (4 + 2)x = 6x$ Factoring out the x using a distributive law

33. $2x + 3y - 5x - 2y = 2x - 5x + 3y - 2y$
$= (2 - 5)x + (3 - 2)y = -3x + y$

34. $3x - x = 3x - 1x = (3 - 1)x = 2x$

35. $x - 0.24x = 1 \cdot x - 0.24x = (1 - 0.24)x = 0.76x$

36. $x - 6x = 1 \cdot x - 6 \cdot x = (1 - 6)x = -5x$

37. $4x - 7y + 9x - 5 + 3y - 8 = 13x - 4y - 13$

38. $\frac{2}{3}a - b + \frac{4}{5}a + \frac{1}{4}b - 10 = \frac{2}{3}a - 1 \cdot b + \frac{4}{5}a + \frac{1}{4}b - 10$
$= \left(\frac{2}{3} + \frac{4}{5}\right)a + \left(-1 + \frac{1}{4}\right)b - 10$
$= \left(\frac{10}{15} + \frac{12}{15}\right)a + \left(-\frac{4}{4} + \frac{1}{4}\right)b - 10$
$= \frac{22}{15}a - \frac{3}{4}b - 10$

Do Exercises 47–53.

Collect like terms.

47. $6x - 3x$

48. $7x - x$

49. $x - 9x$

50. $x - 0.41x$

51. $5x + 4y - 2x - y$

52. $3x - 7x - 11 + 8y + 4 - 13y$

53. $-\frac{2}{3} - \frac{3}{5}x + y + \frac{7}{10}x - \frac{2}{9}y$

Answers on page A-20

Study Tips LEARNING RESOURCES

Are you aware of all the learning resources that exist for this textbook? Many details are given in the Preface.

- The *Student's Solutions Manual* contains fully worked-out solutions to the odd-numbered exercises in the exercise sets, with the exception of the discussion and writing exercises, as well as solutions to all exercises in Chapter Reviews, Chapter Tests, and Cumulative Reviews. You can order this through the bookstore or by calling 1-800-282-0693.
- An extensive set of *videos* supplements this text. These are available on CD-ROM by calling 1-800-282-0693.
- *Tutorial software* called InterAct Math also accompanies this text. If it is not available in the campus learning center, you can order it by calling 1-800-282-0693.
- The Addison-Wesley *Math Tutor Center* is available for help with the odd-numbered exercises. You can order this service by calling 1-800-824-7799.
- Extensive help is available online via MyMathLab and/or MathXL. Ask your instructor for information about these or visit MyMathLab.com and MathXL.com.

7.7 EXERCISE SET

For Extra Help

MathXL | MyMathLab | InterAct Math | Math Tutor Center | Video Lectures on CD Disc 4 | Student's Solutions Manual

a Find an equivalent expression with the given denominator.

1. $\frac{3}{5} = \frac{\square}{5y}$ **2.** $\frac{5}{8} = \frac{\square}{8t}$ **3.** $\frac{2}{3} = \frac{\square}{15x}$ **4.** $\frac{6}{7} = \frac{\square}{14y}$ **5.** $\frac{2}{x} = \frac{\square}{x^2}$ **6.** $\frac{4}{9x} = \frac{\square}{9xy}$

Simplify.

7. $-\frac{24a}{16a}$ **8.** $-\frac{42t}{18t}$ **9.** $-\frac{42ab}{36ab}$ **10.** $-\frac{64pq}{48pq}$ **11.** $\frac{20st}{15t}$ **12.** $\frac{21w}{7wz}$

b Write an equivalent expression. Use a commutative law.

13. $y + 8$ **14.** $x + 3$ **15.** mn **16.** ab

17. $9 + xy$ **18.** $11 + ab$ **19.** $ab + c$ **20.** $rs + t$

Write an equivalent expression. Use an associative law.

21. $a + (b + 2)$ **22.** $3(vw)$ **23.** $(8x)y$ **24.** $(y + z) + 7$

25. $(a + b) + 3$ **26.** $(5 + x) + y$ **27.** $3(ab)$ **28.** $(6x)y$

Use the commutative and associative laws to write three equivalent expressions.

29. $(a + b) + 2$ **30.** $(3 + x) + y$ **31.** $5 + (v + w)$ **32.** $6 + (x + y)$

33. $(xy)3$ **34.** $(ab)5$ **35.** $7(ab)$ **36.** $5(xy)$

c Multiply.

37. $2(b + 5)$ **38.** $4(x + 3)$ **39.** $7(1 + t)$ **40.** $4(1 + y)$

41. $6(5x + 2)$ **42.** $9(6m + 7)$ **43.** $7(x + 4 + 6y)$ **44.** $4(5x + 8 + 3p)$

45. $7(x - 3)$

46. $15(y - 6)$

47. $-3(x - 7)$

48. $1.2(x - 2.1)$

49. $\frac{2}{3}(b - 6)$

50. $\frac{5}{8}(y + 16)$

51. $7.3(x - 2)$

52. $5.6(x - 8)$

53. $-\frac{3}{5}(x - y + 10)$

54. $-\frac{2}{3}(a + b - 12)$

55. $-9(-5x - 6y + 8)$

56. $-7(-2x - 5y + 9)$

57. $-4(x - 3y - 2z)$

58. $8(2x - 5y - 8z)$

59. $3.1(-1.2x + 3.2y - 1.1)$

60. $-2.1(-4.2x - 4.3y - 2.2)$

List the terms of the expression.

61. $4x + 3z$

62. $8x - 1.4y$

63. $7x + 8y - 9z$

64. $8a + 10b - 18c$

d Factor. Check by multiplying.

65. $2x + 4$

66. $5y + 20$

67. $30 + 5y$

68. $7x + 28$

69. $14x + 21y$

70. $18a + 24b$

71. $5x + 10 + 15y$

72. $9a + 27b + 81$

73. $8x - 24$

74. $10x - 50$

75. $-4y + 32$

76. $-6m + 24$

77. $8x + 10y - 22$

78. $9a + 6b - 15$

79. $ax - a$

80. $by - 9b$

81. $ax - ay - az$

82. $cx + cy - cz$

83. $-18x + 12y + 6$

84. $-14x + 21y + 7$

85. $\frac{2}{3}x - \frac{5}{3}y + \frac{1}{3}$

86. $\frac{3}{5}a + \frac{4}{5}b - \frac{1}{5}$

e Collect like terms.

87. $9a + 10a$

88. $12x + 2x$

89. $10a - a$

90. $-16x + x$

91. $2x + 9z + 6x$

92. $3a - 5b + 7a$

93. $7x + 6y^2 + 9y^2$

94. $12m^2 + 6q + 9m^2$

95. $41a + 90 - 60a - 2$

96. $42x - 6 - 4x + 2$

97. $23 + 5t + 7y - t - y - 27$

98. $45 - 90d - 87 - 9d + 3 + 7d$

99. $\frac{1}{2}b + \frac{1}{2}b$

100. $\frac{2}{3}x + \frac{1}{3}x$

101. $2y + \frac{1}{4}y + y$

102. $\frac{1}{2}a + a + 5a$

103. $11x - 3x$

104. $9t - 17t$

105. $6n - n$

106. $100t - t$

107. $y - 17y$

108. $3m - 9m + 4$

109. $-8 + 11a - 5b + 6a - 7b + 7$

110. $8x - 5x + 6 + 3y - 2y - 4$

111. $9x + 2y - 5x$

112. $8y - 3z + 4y$

113. $11x + 2y - 4x - y$

114. $13a + 9b - 2a - 4b$

115. $2.7x + 2.3y - 1.9x - 1.8y$

116. $6.7a + 4.3b - 4.1a - 2.9b$

117. $\frac{13}{2}a + \frac{9}{5}b - \frac{2}{3}a - \frac{3}{10}b - 42$

118. $\frac{11}{4}x + \frac{2}{3}y - \frac{4}{5}x - \frac{1}{6}y + 12$

119. $\mathbf{D_W}$ The distributive law was introduced before the discussion on collecting like terms. Why do you think this was done?

120. $\mathbf{D_W}$ Find two algebraic expressions for the total area of this figure. Explain the equivalence of the expressions in terms of the distributive law.

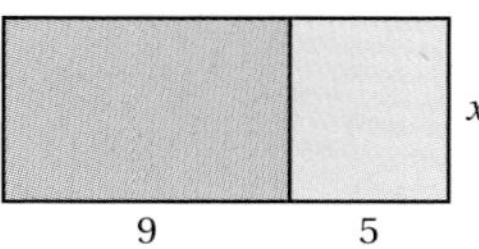

SKILL MAINTENANCE

Find the LCM. [1.9a]

121. 16, 18

122. 18, 24

123. 16, 18, 24

124. 12, 15, 20

125. 16, 32

126. 24, 72

127. 15, 45, 90

128. 18, 54, 108

129. Add and simplify: $\frac{11}{12} + \frac{15}{16}$. [2.3a]

130. Subtract and simplify: $\frac{7}{8} - \frac{2}{3}$. [2.3b]

131. Subtract and simplify: $\frac{1}{8} - \frac{1}{3}$. [2.3b]

132. Convert to percent notation: $\frac{3}{10}$. [4.3a]

SYNTHESIS

Tell whether the expressions are equivalent. Give an example if they are not.

133. $3t + 5$ and $3 \cdot 5 + t$

134. $4x$ and $x + 4$

135. $5m + 6$ and $6 + 5m$

136. $(x + y) + z$ and $z + (x + y)$

137. Factor: $q + qr + qrs + qrst$.

138. Collect like terms:
$21x + 44xy + 15y - 16x - 8y - 38xy + 2y + xy$.

7.8 SIMPLIFYING EXPRESSIONS; ORDER OF OPERATIONS

Objectives

a Find an equivalent expression for an opposite without parentheses, where an expression has several terms.

b Simplify expressions by removing parentheses and collecting like terms.

c Simplify expressions with parentheses inside parentheses.

d Simplify expressions using rules for order of operations.

Find an equivalent expression without parentheses.

1. $-(x + 2)$

2. $-(5x + 2y + 8)$

Answers on page A-20

We now expand our ability to manipulate expressions by first considering opposites of sums and differences. Then we simplify expressions involving parentheses.

a Opposites of Sums

What happens when we multiply a real number by -1? Consider the following products:

$$-1(7) = -7, \qquad -1(-5) = 5, \qquad -1(0) = 0.$$

From these examples, it appears that when we multiply a number by -1, we get the opposite, or additive inverse, of that number.

THE PROPERTY OF −1

For any real number a,

$$-1 \cdot a = -a.$$

(Negative one times a is the opposite, or additive inverse, of a.)

The property of -1 enables us to find certain expressions equivalent to opposites of sums.

EXAMPLES Find an equivalent expression without parentheses.

1. $-(3 + x) = -1(3 + x)$ Using the property of -1

$= -1 \cdot 3 + (-1)x$ Using a distributive law, multiplying each term by -1

$= -3 + (-x)$ Using the property of -1

$= -3 - x$

2. $-(3x + 2y + 4) = -1(3x + 2y + 4)$ Using the property of -1

$= -1(3x) + (-1)(2y) + (-1)4$ Using a distributive law

$= -3x - 2y - 4$ Using the property of -1

Do Exercises 1 and 2.

Suppose we want to remove parentheses in an expression like

$$-(x - 2y + 5).$$

We can first rewrite any subtractions inside the parentheses as additions. Then we take the opposite of each term:

$$-(x - 2y + 5) = -[x + (-2y) + 5]$$
$$= -x + 2y - 5.$$

The most efficient method for removing parentheses is to replace each term in the parentheses with its opposite ("change the sign of every term"). Doing so for $-(x - 2y + 5)$, we obtain $-x + 2y - 5$ as an equivalent expression.

EXAMPLES Find an equivalent expression without parentheses.

3. $-(5 - y) = -5 + y = y + (-5) = y - 5$ Changing the sign of each term

4. $-(2a - 7b - 6) = -2a + 7b + 6$

5. $-(-3x + 4y + z - 7w - 23) = 3x - 4y - z + 7w + 23$

Do Exercises 3–6.

Find an equivalent expression without parentheses. Try to do this in one step.

3. $-(6 - t)$

4. $-(x - y)$

5. $-(-4a + 3t - 10)$

6. $-(18 - m - 2n + 4z)$

b Removing Parentheses and Simplifying

When a sum is added, as in $5x + (2x + 3)$, we can simply remove, or drop, the parentheses and collect like terms because of the associative law of addition:

$$5x + (2x + 3) = 5x + 2x + 3 = 7x + 3.$$

On the other hand, when a sum is subtracted, as in $3x - (4x + 2)$, no "associative" law applies. However, we can subtract by adding an opposite. We then remove parentheses by changing the sign of each term inside the parentheses and collecting like terms.

EXAMPLE 6 Remove parentheses and simplify.

$3x - (4x + 2) = 3x + [-(4x + 2)]$ Adding the opposite of $(4x + 2)$

$= 3x + (-4x - 2)$ Changing the sign of each term inside the parentheses

$= 3x - 4x - 2$

$= -x - 2$ Collecting like terms

Caution!

Note that $3x - (4x + 2) \neq 3x - 4x + 2$. That is, $3x - (4x + 2)$ is *not* equivalent to $3x - 4x + 2$. You cannot simply drop the parentheses.

Do Exercises 7 and 8.

Remove parentheses and simplify.

7. $5x - (3x + 9)$

8. $5y - 2 - (2y - 4)$

In practice, the first three steps of Example 6 are usually combined by changing the sign of each term in parentheses and then collecting like terms.

EXAMPLES Remove parentheses and simplify.

7. $5y - (3y + 4) = 5y - 3y - 4$ Removing parentheses by changing the sign of every term inside the parentheses

$= 2y - 4$ Collecting like terms

8. $3x - 2 - (5x - 8) = 3x - 2 - 5x + 8$

$= -2x + 6$, or $6 - 2x$

9. $(3a + 4b - 5) - (2a - 7b + 4c - 8)$

$= 3a + 4b - 5 - 2a + 7b - 4c + 8$

$= a + 11b - 4c + 3$

Do Exercises 9–11.

Remove parentheses and simplify.

9. $6x - (4x + 7)$

10. $8y - 3 - (5y - 6)$

11. $(2a + 3b - c) - (4a - 5b + 2c)$

Answers on page A-20

Next, consider subtracting an expression consisting of several terms multiplied by a number other than 1 or -1.

EXAMPLE 10 Remove parentheses and simplify.

$$\begin{aligned} x - 3(x + y) &= x + [-3(x + y)] && \text{Adding the opposite of } 3(x + y) \\ &= x + [-3x - 3y] && \text{Multiplying } x + y \text{ by } -3 \\ &= x - 3x - 3y \\ &= -2x - 3y && \text{Collecting like terms} \end{aligned}$$

EXAMPLES Remove parentheses and simplify.

11. $3y - 2(4y - 5) = 3y - 8y + 10$ Multiplying each term in parentheses by -2

$= -5y + 10$

12. $(2a + 3b - 7) - 4(-5a - 6b + 12)$

$= 2a + 3b - 7 + 20a + 24b - 48 = 22a + 27b - 55$

13. $2y - \frac{1}{3}(9y - 12) = 2y - 3y + 4 = -y + 4$

14. $6.4(5x - 3y) - 2.5(8x + y) = 32x - 19.2y - 20x - 2.5y = 12x - 21.7y$

Do Exercises 12–16.

C Parentheses Within Parentheses

In addition to parentheses, some expressions contain other grouping symbols such as brackets [ ] and braces { }.

> When more than one kind of grouping symbol occurs, do the computations in the innermost ones first. Then work from the inside out.

EXAMPLES Simplify.

15. $[3 - (7 + 3)] = [3 - 10] = -7$

16. $\{8 - [9 - (12 + 5)]\} = \{8 - [9 - 17]\}$ Computing $12 + 5$

$= \{8 - [-8]\}$ Computing $9 - 17$

$= 8 + 8 = 16$

17. $\left[(-4) \div \left(-\frac{1}{4}\right)\right] \div \frac{1}{4} = [(-4) \cdot (-4)] \div \frac{1}{4}$ Working within the brackets; computing $(-4) \div \left(-\frac{1}{4}\right)$

$= 16 \div \frac{1}{4}$

$= 16 \cdot 4 = 64$

18. $4(2 + 3) - \{7 - [4 - (8 + 5)]\}$

$= 4 \cdot 5 - \{7 - [4 - 13]\}$ Working with the innermost parentheses first

$= 20 - \{7 - [-9]\}$ Computing $4 \cdot 5$ and $4 - 13$

$= 20 - 16$ Computing $7 - [-9]$

$= 4$

Do Exercises 17–20.

Remove parentheses and simplify.

12. $y - 9(x + y)$

13. $5a - 3(7a - 6)$

14. $4a - b - 6(5a - 7b + 8c)$

15. $5x - \dfrac{1}{4}(8x + 28)$

16. $4.6(5x - 3y) - 5.2(8x + y)$

Simplify.

17. $12 - (8 + 2)$

18. $\{9 - [10 - (13 + 6)]\}$

19. $[24 \div (-2)] \div (-2)$

20. $5(3 + 4) - \{8 - [5 - (9 + 6)]\}$

Answers on page A-20

EXAMPLE 19 Simplify.

$[5(x + 2) - 3x] - [3(y + 2) - 7(y - 3)]$

$= [5x + 10 - 3x] - [3y + 6 - 7y + 21]$ Working with the innermost parentheses first

$= [2x + 10] - [-4y + 27]$ Collecting like terms within brackets

$= 2x + 10 + 4y - 27$ Removing brackets

$= 2x + 4y - 17$ Collecting like terms

Do Exercise 21.

21. Simplify:

$[3(x + 2) + 2x] -$
$[4(y + 2) - 3(y - 2)]$.

d Order of Operations

When several operations are to be done in a calculation or a problem, we apply the same rules that we did in Section 1.6. We repeat them here for review. (If you did not study that section earlier, you should do so now.)

RULES FOR ORDER OF OPERATIONS

1. Do all calculations within grouping symbols before operations outside.
2. Evaluate all exponential expressions.
3. Do all multiplications and divisions in order from left to right.
4. Do all additions and subtractions in order from left to right.

These rules are consistent with the way in which most computers and scientific calculators perform calculations.

EXAMPLE 20 Simplify: $-34 \cdot 56 - 17$.

There are no parentheses or powers, so we start with the third step.

$-34 \cdot 56 - 17 = -1904 - 17$ Doing all multiplications and divisions in order from left to right

$= -1921$ Doing all additions and subtractions in order from left to right

EXAMPLE 21 Simplify: $25 \div (-5) + 50 \div (-2)$.

There are no calculations inside parentheses or powers. The parentheses with (-5) and (-2) are used only to represent the negative numbers. We begin by doing all multiplications and divisions.

$25 \div (-5) + 50 \div (-2)$

$= -5 + (-25)$ Doing all multiplications and divisions in order from left to right

$= -30$ Doing all additions and subtractions in order from left to right.

Do Exercises 22–24.

Simplify.

22. $23 - 42 \cdot 30$

23. $32 \div 8 \cdot 2$

24. $-24 \div 3 - 48 \div (-4)$

Answers on page A-20

Simplify.

25. $-4^3 + 52 \cdot 5 + 5^3 - (4^2 - 48 \div 4)$

26. $\dfrac{5 - 10 - 5 \cdot 23}{2^3 + 3^2 - 7}$

Answers on page A-20

EXAMPLE 22 Simplify: $-2^4 + 51 \cdot 4 - (37 + 23 \cdot 2)$.

$-2^4 + 51 \cdot 4 - (37 + 23 \cdot 2)$

$= -2^4 + 51 \cdot 4 - (37 + 46)$ Working within the parentheses

$= -2^4 + 51 \cdot 4 - 83$

$= -16 + 51 \cdot 4 - 83$ Evaluating exponential expressions. Note that $-2^4 \neq (-2)^4$.

$= -16 + 204 - 83$ Doing all multiplications

$= 188 - 83$ Doing all additions and subtractions in order from left to right

$= 105$

A fraction bar can play the role of a grouping symbol.

EXAMPLE 23 Simplify: $\dfrac{-64 \div (-16) \div (-2)}{2^3 - 3^2}$.

An equivalent expression with brackets as grouping symbols is

$$[-64 \div (-16) \div (-2)] \div [2^3 - 3^2].$$

This shows, in effect, that we do the calculations in the numerator and then in the denominator, and divide the results:

$$\frac{-64 \div (-16) \div (-2)}{2^3 - 3^2} = \frac{4 \div (-2)}{8 - 9} = \frac{-2}{-1} = 2.$$

Do Exercises 25 and 26.

CALCULATOR CORNER

Order of Operations and Grouping Symbols Parentheses are necessary in some calculations in order to ensure that operations are performed in the desired order. To simplify $-5(3 - 6) - 12$, we press (−) 5 (3 − 6) − 1 2 ENTER. The result is 3. Without parentheses, the computation is $-5 \cdot 3 - 6 - 12$, and the result is -33.

−5(3−6)−12
3
−5∗3−6−12
−33

When a negative number is raised to an even power, parentheses must also be used. To find $(-3)^4$, we press ((−) 3) ^ 4 ENTER. The result is 81. Without parentheses, the computation is $-3^4 = -1 \cdot 3^4 = -1 \cdot 81 = -81$.

To simplify an expression like $\dfrac{49 - 104}{7 + 4}$, we must enter it as $(49 - 104) \div (7 + 4)$. We press (4 9 − 1 0 4) ÷ (7 + 4) ENTER. The result is -5.

(49−104)/(7+4)
−5

Exercises: Calculate.

1. $-8 + 4(7 - 9) + 5$
2. $-3[2 + (-5)]$
3. $7[4 - (-3)] + 5[3^2 - (-4)]$
4. $(-7)^6$
5. $(-17)^5$
6. $(-104)^3$
7. -7^6
8. -17^5
9. -104^3
10. $\dfrac{38 - 178}{5 + 30}$
11. $\dfrac{311 - 17^2}{2 - 13}$
12. $785 - \dfrac{285 - 5^4}{17 + 3 \cdot 51}$

Translating for Success

1. *Calories in Cereal.* There are 140 calories in a $1\frac{1}{2}$-cup serving of Brand A cereal. How many calories are there in 6 cups of the cereal?

2. *Calories in Cereal.* There are 140 calories in 6 cups of Brand B cereal. How many calories are there in a $1\frac{1}{2}$-cup serving of the cereal?

3. *Gallons of Gasoline.* Jared's SUV traveled 310 mi on 15.5 gal of gasoline. At this rate, how many gallons would be needed to travel 465 mi?

4. *Gallons of Gasoline.* Elizabeth's new fuel-efficient car traveled 465 mi on 15.5 gal of gasoline. At this rate, how many gallons will be needed to travel 310 mi?

5. *Perimeter.* Find the perimeter of a rectangular field that measures 83.7 m by 62.4 m.

The goal of these matching questions is to practice step (2), *Translate,* of the five-step problem-solving process. Translate each word problem to an equation and select a correct translation from equations A–O.

A. $\dfrac{310}{15.5} = \dfrac{465}{x}$

B. $180 = 1\frac{1}{2} \cdot x$

C. $x = 71\frac{1}{8} - 76\frac{1}{2}$

D. $71\frac{1}{8} \cdot x = 74$

E. $74 \cdot 71\frac{1}{8} = x$

F. $x = 83.7 + 62.4$

G. $71\frac{1}{8} + x = 76\frac{1}{2}$

H. $x = 1\frac{2}{3} \cdot 180$

I. $\dfrac{140}{6} = \dfrac{x}{1\frac{1}{2}}$

J. $x = 2(83.7 + 62.4)$

K. $\dfrac{465}{15.5} = \dfrac{310}{x}$

L. $x = 83.7 \cdot 62.4$

M. $x = 180 \div 1\frac{2}{3}$

N. $\dfrac{140}{1\frac{1}{2}} = \dfrac{x}{6}$

O. $x = 1\frac{2}{3} \div 180$

Answers on page A-20

6. *Electric Bill.* Last month Todd's electric bills for his two rentals were \$83.70 and \$62.40. What was the total electric bill for the two properties?

7. *Package Tape.* A postal service center uses rolls of package tape that each contain 180 ft of tape. If it takes an average of $1\frac{2}{3}$ ft per package, how many packages can be taped with one roll?

8. *Online Price.* Jane spent \$180 for an area rug in a department store. Later she saw the same rug for sale online and realized she had paid $1\frac{1}{2}$ times the online price. What was the online price?

9. *Heights of Sons.* Henry's three sons play basketball on three different college teams. Jeff, Jason, and Jared's heights are 74 in., $71\frac{1}{8}$ in., and $76\frac{1}{2}$ in., respectively. How much taller is Jared than Jason?

10. *Total Investment.* An investor bought 74 shares of stock at \$$71\frac{1}{8}$ per share. What was the total investment?

7.8 EXERCISE SET

For Extra Help

MathXL | MyMathLab | InterAct Math | Math Tutor Center | Video Lectures on CD Disc 4 | Student's Solutions Manual

a Find an equivalent expression without parentheses.

1. $-(2x + 7)$

2. $-(8x + 4)$

3. $-(8 - x)$

4. $-(a - b)$

5. $-(4a - 3b + 7c)$

6. $-(x - 4y - 3z)$

7. $-(6x - 8y + 5)$

8. $-(4x + 9y + 7)$

9. $-(3x - 5y - 6)$

10. $-(6a - 4b - 7)$

11. $-(-8x - 6y - 43)$

12. $-(-2a + 9b - 5c)$

b Remove parentheses and simplify.

13. $9x - (4x + 3)$

14. $4y - (2y + 5)$

15. $2a - (5a - 9)$

16. $12m - (4m - 6)$

17. $2x + 7x - (4x + 6)$

18. $3a + 2a - (4a + 7)$

19. $2x - 4y - 3(7x - 2y)$

20. $3a - 9b - 1(4a - 8b)$

21. $15x - y - 5(3x - 2y + 5z)$

22. $4a - b - 4(5a - 7b + 8c)$

23. $(3x + 2y) - 2(5x - 4y)$

24. $(-6a - b) - 5(2b + a)$

25. $(12a - 3b + 5c) - 5(-5a + 4b - 6c)$

26. $(-8x + 5y - 12) - 6(2x - 4y - 10)$

c Simplify.

27. $[9 - 2(5 - 4)]$

28. $[6 - 5(8 - 4)]$

29. $8[7 - 6(4 - 2)]$

30. $10[7 - 4(7 - 5)]$

31. $[4(9 - 6) + 11] - [14 - (6 + 4)]$

32. $[7(8 - 4) + 16] - [15 - (7 + 8)]$

33. $[10(x + 3) - 4] + [2(x - 1) + 6]$

34. $[9(x + 5) - 7] + [4(x - 12) + 9]$

35. $[7(x + 5) - 19] - [4(x - 6) + 10]$

36. $[6(x + 4) - 12] - [5(x - 8) + 14]$

37. $3\{[7(x - 2) + 4] - [2(2x - 5) + 6]\}$

38. $4\{[8(x - 3) + 9] - [4(3x - 2) + 6]\}$

39. $4\{[5(x - 3) + 2] - 3[2(x + 5) - 9]\}$

40. $3\{[6(x - 4) + 5] - 2[5(x + 8) - 3]\}$

d Simplify.

41. $8 - 2 \cdot 3 - 9$

42. $8 - (2 \cdot 3 - 9)$

43. $(8 - 2 \cdot 3) - 9$

44. $(8 - 2)(3 - 9)$

45. $[(-24) \div (-3)] \div \left(-\frac{1}{2}\right)$

46. $[32 \div (-2)] \div \left(-\frac{1}{4}\right)$

47. $16 \cdot (-24) + 50$

48. $10 \cdot 20 - 15 \cdot 24$

49. $2^4 + 2^3 - 10$

50. $40 - 3^2 - 2^3$

51. $5^3 + 26 \cdot 71 - (16 + 25 \cdot 3)$

52. $4^3 + 10 \cdot 20 + 8^2 - 23$

53. $4 \cdot 5 - 2 \cdot 6 + 4$

54. $4 \cdot (6 + 8)/(4 + 3)$

55. $4^3/8$

56. $5^3 - 7^2$

57. $8(-7) + 6(-5)$

58. $10(-5) + 1(-1)$

59. $19 - 5(-3) + 3$

60. $14 - 2(-6) + 7$

61. $9 \div (-3) + 16 \div 8$

62. $-32 - 8 \div 4 - (-2)$

63. $-4^2 + 6$

64. $-5^2 + 7$

65. $-8^2 - 3$

66. $-9^2 - 11$

67. $12 - 20^3$

68. $20 + 4^3 \div (-8)$

69. $2 \cdot 10^3 - 5000$

70. $-7(3^4) + 18$

71. $6[9 - (3 - 4)]$

72. $8[(6 - 13) - 11]$

73. $-1000 \div (-100) \div 10$

74. $256 \div (-32) \div (-4)$

75. $8 - (7 - 9)$

76. $(8 - 7) - 9$

77. $\dfrac{10 - 6^2}{9^2 + 3^2}$

78. $\dfrac{5^2 - 4^3 - 3}{9^2 - 2^2 - 1^5}$

79. $\dfrac{3(6 - 7) - 5 \cdot 4}{6 \cdot 7 - 8(4 - 1)}$

80. $\dfrac{20(8 - 3) - 4(10 - 3)}{10(2 - 6) - 2(5 + 2)}$

81. $\dfrac{|2^3 - 3^2| + |12 \cdot 5|}{-32 \div (-16) \div (-4)}$

82. $\dfrac{|3 - 5|^2 - |7 - 13|}{|12 - 9| + |11 - 14|}$

83. $\mathbf{D_W}$ Jake keys in 18/2 · 3 on his calculator and expects the result to be 3. What mistake is he making?

84. $\mathbf{D_W}$ Determine whether $|-x|$ and $|x|$ are equivalent. Explain.

SKILL MAINTENANCE

VOCABULARY REINFORCEMENT

In each of Exercises 85–92, fill in the blank with the correct term from the given list. Some of the choices may not be used and some may be used more than once.

85. The set of ______________ is $\{\ldots, -5, -4, -3, -2, -1, 0, 1, 2, 3, \ldots\}$. [7.2a]

86. Two numbers whose sum is 0 are called ______________ of each other. [7.3b]

87. The ______________ of addition says that $a + b = b + a$ for any real numbers a and b. [7.7b]

88. The ______________ states that for any real number a, $a \cdot 1 = 1 \cdot a = a$. [7.7a]

89. The ______________ of addition says that $a + (b + c) = (a + b) + c$ for any real numbers a, b, and c. [7.7b]

90. The ______________ of multiplication says that $a(bc) = (ab)c$ for any real numbers a, b, and c. [7.7b]

91. Two numbers whose product is 1 are called ______________ of each other. [7.6b]

92. The equation $y + 0 = y$ illustrates the ______________. [7.7a]

natural numbers
multiplicative inverses
distributive law
associative law
whole numbers
additive inverses
identity property of 0
property of -1
integers
commutative law
identity property of 1
real numbers

SYNTHESIS

Find an equivalent expression by enclosing the last three terms in parentheses preceded by a minus sign.

93. $6y + 2x - 3a + c$

94. $x - y - a - b$

95. $6m + 3n - 5m + 4b$

Simplify.

96. $z - \{2z - [3z - (4z - 5z) - 6z] - 7z\} - 8z$

97. $\{x - [f - (f - x)] + [x - f]\} - 3x$

98. $x - \{x - 1 - [x - 2 - (x - 3 - \{x - 4 - [x - 5 - (x - 6)]\})]\}$

99. Use your calculator to do the following.

a) Evaluate $x^2 + 3$ when $x = 7$, when $x = -7$, and when $x = -5.013$.

b) Evaluate $1 - x^2$ when $x = 5$, when $x = -5$, and when $x = -10.455$.

100. Express $3^3 + 3^3 + 3^3$ as a power of 3.

Find the average.

101. $-15, 20, 50, -82, -7, -2$

102. $-1, 1, 2, -2, 3, -8, -10$

7 Summary and Review

The review that follows is meant to prepare you for a chapter exam. It consists of three parts. The first part, Concept Reinforcement, is designed to increase understanding of the concepts through true/false exercises. The second part is a list of important properties and formulas. The third part is the Review Exercises. These provide practice exercises for the exam, together with references to section objectives so you can go back and review. Before beginning, stop and look back over the skills you have obtained. What skills in mathematics do you have now that you did not have before studying this chapter?

CONCEPT REINFORCEMENT

Determine whether the statement is true or false. Answers are given at the back of the book.

________ **1.** The set of whole numbers is a subset of the set of integers.

________ **2.** All rational numbers can be named using fraction or decimal notation.

________ **3.** The product of an even number of negative numbers is negative.

________ **4.** The operation of subtraction is not commutative.

________ **5.** The product of a number and its multiplicative inverse is -1.

________ **6.** Decimal notation for irrational numbers neither repeats nor terminates.

________ **7.** $a < b$ also has the meaning $b \geq a$.

IMPORTANT PROPERTIES AND FORMULAS

Properties of the Real-Number System

The Commutative Laws: $a + b = b + a, \quad ab = ba$

The Associative Laws: $a + (b + c) = (a + b) + c, \quad a(bc) = (ab)c$

The Identity Properties: $a + 0 = 0 + a = a, \quad a \cdot 1 = 1 \cdot a = a$

The Inverse Properties: For any real number a, there is an opposite $-a$ such that $a + (-a) = (-a) + a = 0$.

For any nonzero real number a, there is a reciprocal $\frac{1}{a}$ such that $a \cdot \frac{1}{a} = \frac{1}{a} \cdot a = 1$.

The Distributive Laws: $a(b + c) = ab + ac, \quad a(b - c) = ab - ac$

Review Exercises

The review exercises that follow are for practice. Answers are at the back of the book. If you miss an exercise, restudy the objective indicated in red after the exercise or the direction line that precedes it.

1. Evaluate $\frac{x - y}{3}$ when $x = 17$ and $y = 5$. [7.1a]

2. Translate to an algebraic expression: [7.1b]
Nineteen percent of some number.

3. Tell which integers correspond to this situation: [7.2a]
David has a debt of $45 and Joe has $72 in his savings account.

4. Find: $|-38|$. [7.2e]

Graph the number on a number line. [7.2b]

5. -2.5

6. $\frac{8}{9}$

Use either $<$ or $>$ for $\square$ to write a true sentence. [7.2d]

7. $-3 \square 10$

8. $-1 \square -6$

9. $0.126 \square -12.6$

10. $-\frac{2}{3} \square -\frac{1}{10}$

Find the opposite. [7.3b]

11. 3.8

12. $-\frac{3}{4}$

Find the reciprocal. [7.6b]

13. $\frac{3}{8}$

14. -7

15. Evaluate $-x$ when $x = -34$. [7.3b]

16. Evaluate $-(-x)$ when $x = 5$. [7.3b]

Compute and simplify.

17. $4 + (-7)$ [7.3a]

18. $6 + (-9) + (-8) + 7$ [7.3a]

19. $-3.8 + 5.1 + (-12) + (-4.3) + 10$ [7.3a]

20. $-3 - (-7) + 7 - 10$ [7.4a]

21. $-\frac{9}{10} - \frac{1}{2}$ [7.4a]

22. $-3.8 - 4.1$ [7.4a]

23. $-9 \cdot (-6)$ [7.5a]

24. $-2.7(3.4)$ [7.5a]

25. $\frac{2}{3} \cdot \left(-\frac{3}{7}\right)$ [7.5a]

26. $3 \cdot (-7) \cdot (-2) \cdot (-5)$ [7.5a]

27. $35 \div (-5)$ [7.6a]

28. $-5.1 \div 1.7$ [7.6c]

29. $-\frac{3}{11} \div \left(-\frac{4}{11}\right)$ [7.6c]

Simplify. [7.8d]

30. $(-3.4 - 12.2) - 8(-7)$

31. $\frac{-12(-3) - 2^3 - (-9)(-10)}{3 \cdot 10 + 1}$

32. $-16 \div 4 - 30 \div (-5)$

33. $\frac{9[(7 - 14) - 13]}{|-2(8) - 4|}$

Solve.

34. On the first, second, and third downs, a football team had these gains and losses: 5-yd gain, 12-yd loss, and 15-yd gain, respectively. Find the total gain (or loss). [7.3c]

35. Kaleb's total assets are \$170. He borrows \$300. What are his total assets now? [7.4b]

36. *Stock Price.* The value of EFX Corp. stock began the day at \$17.68 per share and dropped \$1.63 per hour for 8 hr. What was the price of the stock after 8 hr? [7.5b]

37. *Checking Account Balance.* Yuri had \$68 in his checking account. After writing checks to make seven purchases of DVDs at the same price for each, the balance in his account was −\$64.65 What was the price of each DVD? [7.6d]

Multiply. [7.7c]

38. $5(3x - 7)$

39. $-2(4x - 5)$

40. $10(0.4x + 1.5)$

41. $-8(3 - 6x)$

Factor. [7.7d]

42. $2x - 14$

43. $-6x + 6$

44. $5x + 10$

45. $-3x + 12y - 12$

Collect like terms. [7.7e]

46. $11a + 2b - 4a - 5b$

47. $7x - 3y - 9x + 8y$

48. $6x + 3y - x - 4y$

49. $-3a + 9b + 2a - b$

Remove parentheses and simplify.

50. $2a - (5a - 9)$ [7.8b]

51. $3(b + 7) - 5b$ [7.8b]

52. $3[11 - 3(4 - 1)]$ [7.8c]

53. $2[6(y - 4) + 7]$ [7.8c]

54. $[8(x + 4) - 10] - [3(x - 2) + 4]$ [7.8c]

55. $5\{[6(x - 1) + 7] - [3(3x - 4) + 8]\}$ [7.8c]

Answer True or False. [7.2d]

56. $-9 \leq 11$

57. $-11 \geq -3$

58. Write another inequality with the same meaning as $-3 < x$. [7.2d]

59. **DW** Explain the notion of the opposite of a number in as many ways as possible. [7.3b]

60. **DW** Is the absolute value of a number always positive? Why or why not? [7.2e]

SYNTHESIS

Simplify. [7.2e], [7.4a], [7.6a], [7.8d]

61. $-\left|\frac{7}{8} - \left(-\frac{1}{2}\right) - \frac{3}{4}\right|$

62. $(|2.7 - 3| + 3^2 - |-3|) \div (-3)$

63. $2000 - 1990 + 1980 - 1970 + \cdots + 20 - 10$

64. Find a formula for the perimeter of the following figure. [7.7e]

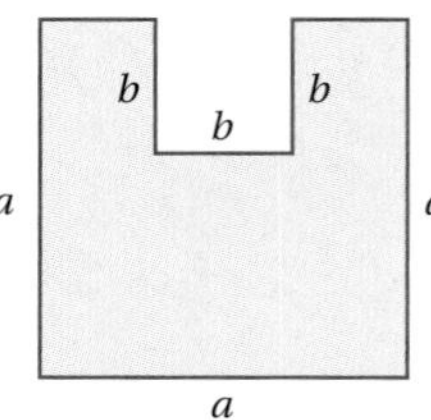

7 Chapter Test

For Extra Help

Work It Out! Chapter Test Video Solutions on CD

1. Evaluate $\frac{3x}{y}$ when $x = 10$ and $y = 5$.

2. Write an algebraic expression: Nine less than some number.

3. Find the area of a triangle when the height h is 30 ft and the base b is 16 ft.

Use either $<$ or $>$ for $\square$ to write a true sentence.

4. $-4 \square 0$

5. $-3 \square -8$

6. $-0.78 \square -0.87$

7. $-\frac{1}{8} \square \frac{1}{2}$

Find the absolute value.

8. $|-7|$

9. $\left|\frac{9}{4}\right|$

10. $|-2.7|$

Find the opposite.

11. $\frac{2}{3}$

12. -1.4

13. Evaluate $-x$ when $x = -8$.

Find the reciprocal.

14. -2

15. $\frac{4}{7}$

Compute and simplify.

16. $3.1 - (-4.7)$

17. $-8 + 4 + (-7) + 3$

18. $-\frac{1}{5} + \frac{3}{8}$

19. $2 - (-8)$

20. $3.2 - 5.7$

21. $\frac{1}{8} - \left(-\frac{3}{4}\right)$

22. $4 \cdot (-12)$

23. $-\frac{1}{2} \cdot \left(-\frac{3}{8}\right)$

24. $-45 \div 5$

25. $-\frac{3}{5} \div \left(-\frac{4}{5}\right)$

26. $4.864 \div (-0.5)$

27. $-2(16) - |2(-8) - 5^3|$

28. $-20 \div (-5) + 36 \div (-4)$

29. *Antarctica Highs and Lows.* The continent of Antarctica, which lies in the southern hemisphere, experiences winter in July. The average high temperature is $-67°F$ and the average low temperature is $-81°F$. How much higher is the average high than the average low?

Source: National Climatic Data Center

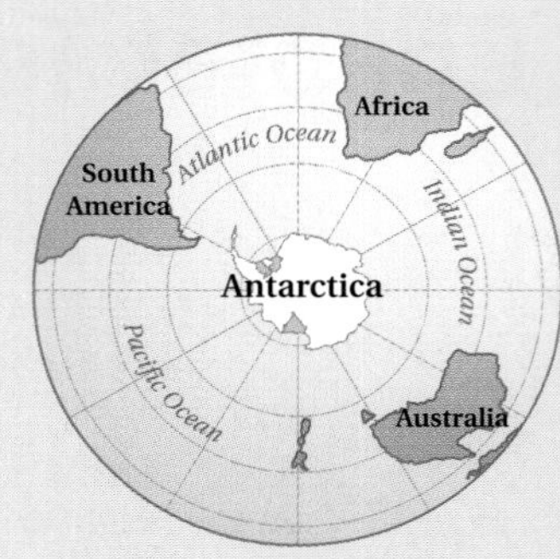

30. Maureen is a stockbroker. She kept track of the changes in the stock market over a period of 5 weeks. By how many points had the market risen or fallen over this time?

WEEK 1	WEEK 2	WEEK 3	WEEK 4	WEEK 5
Down 13 pts	Down 16 pts	Up 36 pts	Down 11 pts	Up 19 pts

31. *Population Decrease.* The population of a city was 18,600. It dropped 420 each year for 6 yr. What was the population of the city after 6 yr?

32. *Chemical Experiment.* During a chemical reaction, the temperature in the beaker decreased every minute by the same number of degrees. The temperature was 16°C at 11:08 A.M. By 11:43 A.M., the temperature had dropped to $-17°C$. By how many degrees did it drop each minute?

Multiply.

33. $3(6 - x)$

34. $-5(y - 1)$

Factor.

35. $12 - 22x$

36. $7x + 21 + 14y$

Simplify.

37. $6 + 7 - 4 - (-3)$

38. $5x - (3x - 7)$

39. $4(2a - 3b) + a - 7$

40. $4\{3[5(y - 3) + 9] + 2(y + 8)\}$

41. $256 \div (-16) \div 4$

42. $2^3 - 10[4 - (-2 + 18)3]$

43. Write an inequality with the same meaning as $x \le -2$.

SYNTHESIS

Simplify.

44. $|-27 - 3(4)| - |-36| + |-12|$

45. $a - \{3a - [4a - (2a - 4a)]\}$

46. Find a formula for the perimeter of the figure shown here.

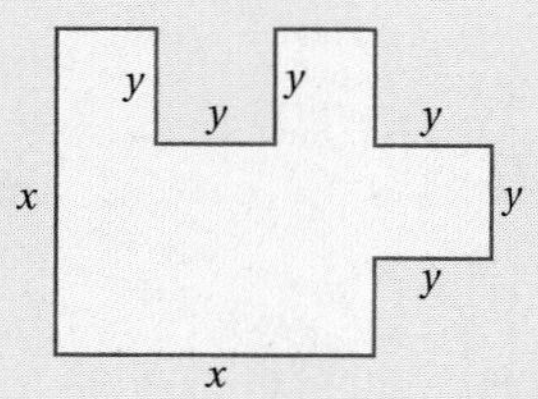

Solving Equations and Inequalities

8

Real-World Application

The average top speed of the three fastest roller coasters in the world is 109 mph. The third fastest roller coaster, Superman the Escape (at Six Flags Magic Mountain, Los Angeles, CA) reaches a top speed of 20 mph less than the fastest roller coaster, Top Thrill Dragster (in Cedar Point, Sandusky, OH). The second fastest roller coaster, Dodonpa (in Fujikyu Highlands, Japan), has a top speed of 107 mph. What is the top speed of the fastest roller coaster?

Source: *Fortune Small Business*, June 2004, p. 48

This problem appears as Example 7 in Section 8.6.

8.1 SOLVING EQUATIONS: THE ADDITION PRINCIPLE

Objectives

a Determine whether a given number is a solution of a given equation.

b Solve equations using the addition principle.

a Equations and Solutions

In order to solve problems, we must learn to solve equations.

EQUATION

An **equation** is a number sentence that says that the expressions on either side of the equals sign, =, represent the same number.

Here are some examples:

$$3 + 2 = 5, \quad 14 - 10 = 1 + 3, \quad x + 6 = 13, \quad 3x - 2 = 7 - x.$$

Equations have expressions on each side of the equals sign. The sentence "$14 - 10 = 1 + 3$" asserts that the expressions $14 - 10$ and $1 + 3$ name the same number.

Some equations are true. Some are false. Some are neither true nor false.

EXAMPLES Determine whether the equation is true, false, or neither.

1. $3 + 2 = 5$ — The equation is *true.*
2. $7 - 2 = 4$ — The equation is *false.*
3. $x + 6 = 13$ — The equation is *neither* true nor false, because we do not know what number x represents.

Do Exercises 1–3.

Determine whether the equation is true, false, or neither.

1. $5 - 8 = -4$

2. $12 + 6 = 18$

3. $x + 6 = 7 - x$

SOLUTION OF AN EQUATION

Any replacement for the variable that makes an equation true is called a **solution** of the equation. To solve an equation means to find *all* of its solutions.

One way to determine whether a number is a solution of an equation is to evaluate the expression on each side of the equals sign by substitution. If the values are the same, then the number is a solution.

EXAMPLE 4 Determine whether 7 is a solution of $x + 6 = 13$.

We have

$x + 6 = 13$		Writing the equation
$7 + 6 \ ? \ 13$		Substituting 7 for x
$13 \mid 13$	TRUE	

Since the left-hand and the right-hand sides are the same, we have a solution. No other number makes the equation true, so the only solution is the number 7.

Answers on page A-21

EXAMPLE 5 Determine whether 19 is a solution of $7x = 141$.

$7x = 141$	Writing the equation
$7(19) \;?\; 141$	Substituting 19 for x
$133 \mid$	FALSE

Since the left-hand and the right-hand sides are not the same, we do not have a solution.

Do Exercises 4–7.

Determine whether the given number is a solution of the given equation.

4. 8; $x + 4 = 12$

5. 0; $x + 4 = 12$

6. -3; $7 + x = -4$

7. $-\frac{3}{5}$; $-5x = 3$

Answers on page A-21

b Using the Addition Principle

Consider the equation

$$x = 7.$$

We can easily see that the solution of this equation is 7. If we replace x with 7, we get

$$7 = 7, \quad \text{which is true.}$$

Now consider the equation of Example 4:

$$x + 6 = 13.$$

In Example 4, we discovered that the solution of this equation is also 7, but the fact that 7 is the solution is not as obvious. We now begin to consider principles that allow us to start with an equation like $x + 6 = 13$ and end up with an *equivalent equation,* like $x = 7$, in which the variable is alone on one side and for which the solution is easier to find.

EQUIVALENT EQUATIONS

Equations with the same solutions are called **equivalent equations.**

One of the principles that we use in solving equations involves adding. An equation $a = b$ says that a and b stand for the same number. Suppose this is true, and we add a number c to the number a. We get the same answer if we add c to b, because a and b are the same number.

THE ADDITION PRINCIPLE FOR EQUATIONS

For any real numbers a, b, and c,

$$a = b \quad \text{is equivalent to} \quad a + c = b + c.$$

Let's again solve the equation $x + 6 = 13$ using the addition principle. We want to get x alone on one side. To do so, we use the addition principle, choosing to add -6 because $6 + (-6) = 0$:

$x + 6 = 13$	
$x + 6 + (-6) = 13 + (-6)$	Using the addition principle: adding -6 on both sides
$x + 0 = 7$	Simplifying
$x = 7.$	Identity property of 0: $x + 0 = x$

The solution of $x + 6 = 13$ is 7.

8. Solve using the addition principle:

$$x + 2 = 11.$$

Do Exercise 8.

When we use the addition principle, we sometimes say that we "add the same number on both sides of the equation." This is also true for subtraction, since we can express every subtraction as an addition. That is, since

$$a - c = b - c \quad \text{is equivalent to} \quad a + (-c) = b + (-c),$$

the addition principle tells us that we can "subtract the same number on both sides of the equation."

EXAMPLE 6 Solve: $x + 5 = -7$.

We have

$x + 5 = -7$	
$x + 5 - 5 = -7 - 5$	Using the addition principle: adding -5 on both sides or subtracting 5 on both sides
$x + 0 = -12$	Simplifying
$x = -12.$	Identity property of 0

9. Solve using the addition principle, subtracting 5 on both sides:

$$x + 5 = -8.$$

To check the answer, we substitute -12 in the original equation.

Check:

$$\begin{array}{r|l} x + 5 & -7 \\ \hline -12 + 5 \ ? & -7 \\ -7 & \end{array} \quad \text{TRUE}$$

The solution of the original equation is -12.

In Example 6, to get x alone, we used the addition principle and subtracted 5 on both sides. This eliminated the 5 on the left. We started with $x + 5 = -7$, and, using the addition principle, we found a simpler equation $x = -12$ for which it was easy to *"see"* the solution. The equations $x + 5 = -7$ and $x = -12$ are *equivalent.*

Do Exercise 9.

Now we use the addition principle to solve an equation that involves a subtraction.

10. Solve: $t - 3 = 19$.

EXAMPLE 7 Solve: $a - 4 = 10$.

We have

$a - 4 = 10$	
$a - 4 + 4 = 10 + 4$	Using the addition principle: adding 4 on both sides
$a + 0 = 14$	Simplifying
$a = 14.$	Identity property of 0

Check:

$$\begin{array}{r|l} a - 4 & 10 \\ \hline 14 - 4 \ ? & 10 \\ 10 & \end{array} \quad \text{TRUE}$$

The solution is 14.

Do Exercise 10.

Answers on page A-21

EXAMPLE 8 Solve: $-6.5 = y - 8.4$.

We have

$$-6.5 = y - 8.4$$
$$-6.5 + 8.4 = y - 8.4 + 8.4 \quad \text{Using the addition principle: adding 8.4 on both sides to eliminate } -8.4 \text{ on the right}$$
$$1.9 = y.$$

Check:
$$\begin{array}{c|l} -6.5 = y - 8.4 & \\ \hline -6.5 \; ? \; 1.9 - 8.4 & \\ & -6.5 \quad \text{TRUE} \end{array}$$

The solution is 1.9.

Note that equations are reversible. That is, if $a = b$ is true, then $b = a$ is true. Thus when we solve $-6.5 = y - 8.4$, we can reverse it and solve $y - 8.4 = -6.5$ if we wish.

Do Exercises 11 and 12.

EXAMPLE 9 Solve: $-\frac{2}{3} + x = \frac{5}{2}$.

We have

$$-\frac{2}{3} + x = \frac{5}{2}$$
$$\frac{2}{3} - \frac{2}{3} + x = \frac{2}{3} + \frac{5}{2} \quad \text{Adding } \tfrac{2}{3} \text{ on both sides}$$
$$x = \frac{2}{3} + \frac{5}{2}$$
$$x = \frac{2}{3} \cdot \frac{2}{2} + \frac{5}{2} \cdot \frac{3}{3} \quad \text{Multiplying by 1 to obtain equivalent fraction expressions with the least common denominator 6}$$
$$x = \frac{4}{6} + \frac{15}{6}$$
$$x = \frac{19}{6}.$$

Check:
$$\begin{array}{c|c} -\frac{2}{3} + x = \frac{5}{2} & \\ \hline -\frac{2}{3} + \frac{19}{6} \; ? & \frac{5}{2} \\ -\frac{4}{6} + \frac{19}{6} & \\ \frac{15}{6} & \\ \frac{5}{2} & \text{TRUE} \end{array}$$

The solution is $\frac{19}{6}$.

Do Exercises 13 and 14.

Solve.

11. $8.7 = n - 4.5$

12. $y + 17.4 = 10.9$

Solve.

13. $x + \frac{1}{2} = -\frac{3}{2}$

14. $t - \frac{13}{4} = \frac{5}{8}$

Answers on page A-21

8.1 EXERCISE SET

For Extra Help

MathXL | MyMathLab | InterAct Math | Math Tutor Center | Video Lectures on CD Disc 4 | Student's Solutions Manual

a Determine whether the given number is a solution of the given equation.

1. 15; $x + 17 = 32$

2. 35; $t + 17 = 53$

3. 21; $x - 7 = 12$

4. 36; $a - 19 = 17$

5. -7; $6x = 54$

6. -9; $8y = -72$

7. 30; $\frac{x}{6} = 5$

8. 49; $\frac{y}{8} = 6$

9. 19; $5x + 7 = 107$

10. 9; $9x + 5 = 86$

11. -11; $7(y - 1) = 63$

12. -18; $x + 3 = 3 + x$

b Solve using the addition principle. Don't forget to check!

13. $x + 2 = 6$

Check: $x + 2 = 6$?

14. $y + 4 = 11$

Check: $y + 4 = 11$?

15. $x + 15 = -5$

Check: $x + 15 = -5$?

16. $t + 10 = 44$

Check: $t + 10 = 44$?

17. $x + 6 = -8$

Check: $x + 6 = -8$?

18. $z + 9 = -14$

19. $x + 16 = -2$

20. $m + 18 = -13$

21. $x - 9 = 6$

22. $x - 11 = 12$

23. $x - 7 = -21$

24. $x - 3 = -14$

25. $5 + t = 7$

26. $8 + y = 12$

27. $-7 + y = 13$

28. $-8 + y = 17$

29. $-3 + t = -9$

30. $-8 + t = -24$

31. $x + \frac{1}{2} = 7$

32. $24 = -\frac{7}{10} + r$

33. $12 = a - 7.9$

34. $2.8 + y = 11$

35. $r + \frac{1}{3} = \frac{8}{3}$

36. $t + \frac{3}{8} = \frac{5}{8}$

37. $m + \frac{5}{6} = -\frac{11}{12}$

38. $x + \frac{2}{3} = -\frac{5}{6}$

39. $x - \frac{5}{6} = \frac{7}{8}$

40. $y - \frac{3}{4} = \frac{5}{6}$

41. $-\frac{1}{5} + z = -\frac{1}{4}$

42. $-\frac{1}{8} + y = -\frac{3}{4}$

43. $7.4 = x + 2.3$

44. $8.4 = 5.7 + y$

45. $7.6 = x - 4.8$

46. $8.6 = x - 7.4$

47. $-9.7 = -4.7 + y$

48. $-7.8 = 2.8 + x$

49. $5\frac{1}{6} + x = 7$

50. $5\frac{1}{4} = 4\frac{2}{3} + x$

51. $q + \frac{1}{3} = -\frac{1}{7}$

52. $52\frac{3}{8} = -84 + x$

53. **D$_W$** Explain the difference between equivalent expressions and equivalent equations.

54. **D$_W$** When solving an equation using the addition principle, how do you determine which number to add or subtract on both sides of the equation?

SKILL MAINTENANCE

55. Add: $-3 + (-8)$. [7.3a]

56. Subtract: $-3 - (-8)$. [7.4a]

57. Multiply: $-\frac{2}{3} \cdot \frac{5}{8}$. [7.5a]

58. Divide: $-\frac{3}{7} \div \left(-\frac{9}{7}\right)$. [7.6c]

59. Divide: $\frac{2}{3} \div \left(-\frac{4}{9}\right)$. [7.6c]

60. Add: $-8.6 + 3.4$. [7.3a]

61. Subtract: $-\frac{2}{3} - \left(-\frac{5}{8}\right)$. [7.4a]

62. Multiply: $(-25.4)(-6.8)$. [7.5a]

Translate to an algebraic expression. [7.1b]

63. Jane had \$83 before paying x dollars for a pair of tennis shoes. How much does she have left?

64. Justin drove his S-10 pickup truck 65 mph for t hours. How far did he drive?

SYNTHESIS

Solve.

65. $-356.788 = -699.034 + t$

66. $-\frac{4}{5} + \frac{7}{10} = x - \frac{3}{4}$

67. $x + \frac{4}{5} = -\frac{2}{3} - \frac{4}{15}$

68. $8 - 25 = 8 + x - 21$

69. $16 + x - 22 = -16$

70. $x + x = x$

71. $x + 3 = 3 + x$

72. $x + 4 = 5 + x$

73. $-\frac{3}{2} + x = -\frac{5}{17} - \frac{3}{2}$

74. $|x| = 5$

75. $|x| + 6 = 19$

8.2 SOLVING EQUATIONS: THE MULTIPLICATION PRINCIPLE

Objective

 Solve equations using the multiplication principle.

a Using the Multiplication Principle

Suppose that $a = b$ is true, and we multiply a by some number c. We get the same number if we multiply b by c, because a and b are the same number.

1. Solve. Multiply on both sides.

$$6x = 90$$

THE MULTIPLICATION PRINCIPLE FOR EQUATIONS

For any real numbers a, b, and c, $c \neq 0$,

$$a = b \quad \text{is equivalent to} \quad a \cdot c = b \cdot c.$$

When using the multiplication principle, we sometimes say that we "multiply on both sides of the equation by the same number."

EXAMPLE 1 Solve: $5x = 70$.

To get x alone, we multiply by the *multiplicative inverse,* or *reciprocal,* of 5. Then we get the *multiplicative identity* 1 times x, or $1 \cdot x$, which simplifies to x. This allows us to eliminate 5 on the left.

$5x = 70$	The reciprocal of 5 is $\frac{1}{5}$.
$\frac{1}{5} \cdot 5x = \frac{1}{5} \cdot 70$	Multiplying by $\frac{1}{5}$ to get $1 \cdot x$ and eliminate 5 on the left
$1 \cdot x = 14$	Simplifying
$x = 14$	Identity property of 1: $1 \cdot x = x$

2. Solve. Divide on both sides.

$$4x = -7$$

Check:

$$\begin{array}{c|c} 5x & 70 \\ \hline 5 \cdot 14 \; ? & 70 \\ 70 & \end{array}$$

TRUE

The solution is 14.

The multiplication principle also tells us that we can "divide on both sides of the equation by a nonzero number." This is because division is the same as multiplying by a reciprocal. That is,

$$\frac{a}{c} = \frac{b}{c} \quad \text{is equivalent to} \quad a \cdot \frac{1}{c} = b \cdot \frac{1}{c}, \quad \text{when } c \neq 0.$$

In an expression like $5x$ in Example 1, the number 5 is called the **coefficient.** Example 1 could be done as follows, dividing on both sides by 5, the coefficient of x.

EXAMPLE 2 Solve: $5x = 70$.

$5x = 70$	
$\frac{5x}{5} = \frac{70}{5}$	Dividing by 5 on both sides
$1 \cdot x = 14$	Simplifying
$x = 14$	Identity property of 1

Answers on page A-21

Do Exercises 1 and 2 on the preceding page.

3. Solve: $-6x = 108$.

EXAMPLE 3 Solve: $-4x = 92$.

We have

$$-4x = 92$$

$$\frac{-4x}{-4} = \frac{92}{-4}$$ Using the multiplication principle. Dividing by -4 on both sides is the same as multiplying by $-\frac{1}{4}$.

$$1 \cdot x = -23$$ Simplifying

$$x = -23.$$ Identity property of 1

Check:

$$\begin{array}{c|l} -4x & = 92 \\ \hline -4(-23) & ?\ 92 \\ 92 & \end{array}$$ TRUE

The solution is -23.

Do Exercise 3.

4. Solve: $-x = -10$.

EXAMPLE 4 Solve: $-x = 9$.

We have

$$-x = 9$$

$$-1 \cdot x = 9$$ Using the property of -1: $-x = -1 \cdot x$

$$\frac{-1 \cdot x}{-1} = \frac{9}{-1}$$ Dividing by -1 on both sides

$$1 \cdot x = -9$$

$$x = -9.$$

Check:

$$\begin{array}{c|l} -x & = 9 \\ \hline -(-9) & ?\ 9 \\ 9 & \end{array}$$ TRUE

The solution is -9.

5. Solve: $-x = -10$.

Do Exercise 4.

We can also solve the equation $-x = 9$ by multiplying as follows.

EXAMPLE 5 Solve: $-x = 9$.

We have

$$-x = 9$$

$$-1(-x) = -1 \cdot 9$$ Multiplying by -1 on both sides

$$-1 \cdot (-1) \cdot x = -9$$

$$1 \cdot x = -9$$

$$x = -9.$$

The solution is -9.

Do Exercise 5.

Answers on page A-21

6. Solve: $\frac{2}{3} = -\frac{5}{6}y$.

Solve.

7. $1.12x = 8736$

8. $6.3 = -2.1y$

Answers on page A-21

In practice, it is generally more convenient to divide on both sides of the equation if the coefficient of the variable is in decimal notation or is an integer. If the coefficient is in fraction notation, it is more convenient to multiply by a reciprocal.

EXAMPLE 6 Solve: $\frac{3}{8} = -\frac{5}{4}x$.

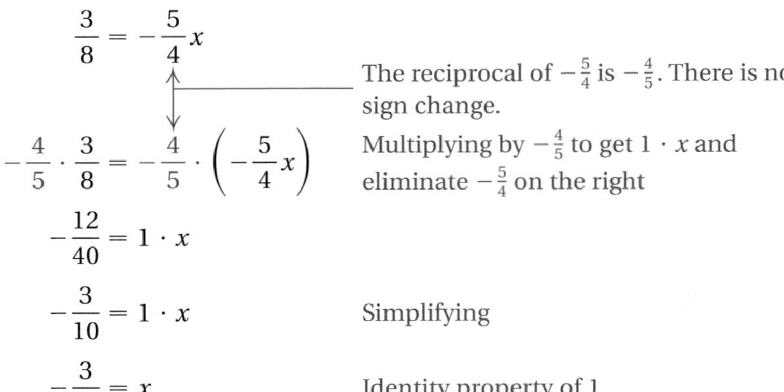

$$\frac{3}{8} = -\frac{5}{4}x$$

The reciprocal of $-\frac{5}{4}$ is $-\frac{4}{5}$. There is no sign change.

$$-\frac{4}{5} \cdot \frac{3}{8} = -\frac{4}{5} \cdot \left(-\frac{5}{4}x\right)$$ Multiplying by $-\frac{4}{5}$ to get $1 \cdot x$ and eliminate $-\frac{5}{4}$ on the right

$$-\frac{12}{40} = 1 \cdot x$$

$$-\frac{3}{10} = 1 \cdot x$$ Simplifying

$$-\frac{3}{10} = x$$ Identity property of 1

Check:

$$\frac{3}{8} = -\frac{5}{4}x$$

$$\frac{3}{8} \;?\; -\frac{5}{4}\left(-\frac{3}{10}\right)$$

$$\frac{3}{8}$$ TRUE

The solution is $-\frac{3}{10}$.

Note that equations are reversible. That is, if $a = b$ is true, then $b = a$ is true. Thus when we solve $\frac{3}{8} = -\frac{5}{4}x$, we can reverse it and solve $-\frac{5}{4}x = \frac{3}{8}$ if we wish.

Do Exercise 6.

EXAMPLE 7 Solve: $1.16y = 9744$.

$$1.16y = 9744$$

$$\frac{1.16y}{1.16} = \frac{9744}{1.16}$$ Dividing by 1.16 on both sides

$$y = \frac{9744}{1.16}$$

$$y = 8400$$ Using a calculator to divide

Check:

$$1.16y = 9744$$

$$1.16(8400) \;?\; 9744$$

$$9744$$ TRUE

The solution is 8400.

Do Exercises 7 and 8.

Now we use the multiplication principle to solve an equation that involves division.

9. Solve: $-14 = \frac{-y}{2}$.

EXAMPLE 8 Solve: $\frac{-y}{9} = 14$.

$$\frac{-y}{9} = 14$$

$$9 \cdot \frac{-y}{9} = 9 \cdot 14 \quad \text{Multiplying by 9 on both sides}$$

$$-y = 126$$

$$-1 \cdot (-y) = -1 \cdot 126 \quad \text{Multiplying by } -1 \text{ on both sides}$$

$$y = -126$$

Check:

$$\frac{-y}{9} = 14$$

$$\frac{-(-126)}{9} \;?\; 14$$

$$\frac{126}{9}$$

$$14 \qquad \text{TRUE}$$

The solution is -126.

There are other ways to solve the equation in Example 8. One is by multiplying by -9 on both sides as follows:

$$-9 \cdot \frac{-y}{9} = -9 \cdot 14$$

$$\frac{(-9)(-y)}{9} = -126$$

$$\frac{9y}{9} = -126$$

$$y = -126.$$

Do Exercise 9.

Answer on page A-21

Study Tips

TIME MANAGEMENT (PART 1)

Time is the most critical factor in your success in learning mathematics. Have reasonable expectations about the time you need to study math. (See also the Study Tips on time management in Sections 11.2 and 11.6.)

- **Juggling time.** Working 40 hours per week and taking 12 credit hours is equivalent to working two full-time jobs. Can you handle such a load? Your ratio of number of work hours to number of credit hours should be about 40/3, 30/6, 20/9, 10/12, or 5/14.
- **A rule of thumb on study time.** Budget about 2–3 hours for homework and study per week for every hour of class time.

"You cannot increase the quality or quantity of your achievement or performance except to the degree in which you increase your ability to use time effectively."

Brian Tracy, motivational speaker

8.2 EXERCISE SET

For Extra Help

MathXL

MyMathLab

InterAct Math

Math Tutor Center

Video Lectures on CD Disc 4

Student's Solutions Manual

a Solve using the multiplication principle. Don't forget to check!

1. $6x = 36$

Check: $6x = 36$ | ?

2. $3x = 51$

Check: $3x = 51$ | ?

3. $5x = 45$

Check: $5x = 45$ | ?

4. $8x = 72$

Check: $8x = 72$ | ?

5. $84 = 7x$

6. $63 = 9x$

7. $-x = 40$

8. $53 = -x$

9. $-x = -1$

10. $-47 = -t$

11. $7x = -49$

12. $8x = -56$

13. $-12x = 72$

14. $-15x = 105$

15. $-21x = -126$

16. $-13x = -104$

17. $\frac{t}{7} = -9$

18. $\frac{y}{-8} = 11$

19. $\frac{3}{4}x = 27$

20. $\frac{4}{5}x = 16$

21. $\frac{-t}{3} = 7$

22. $\frac{-x}{6} = 9$

23. $-\frac{m}{3} = \frac{1}{5}$

24. $\frac{1}{8} = -\frac{y}{5}$

25. $-\frac{3}{5}r = \frac{9}{10}$

26. $\frac{2}{5}y = -\frac{4}{15}$

27. $-\frac{3}{2}r = -\frac{27}{4}$

28. $-\frac{3}{8}x = -\frac{15}{16}$

29. $6.3x = 44.1$

30. $2.7y = 54$

31. $-3.1y = 21.7$

32. $-3.3y = 6.6$

33. $38.7m = 309.6$

34. $29.4m = 235.2$

35. $-\frac{2}{3}y = -10.6$

36. $-\frac{9}{7}y = 12.06$

37. $\frac{-x}{5} = 10$

38. $\frac{-x}{8} = -16$

39. $-\frac{t}{2} = 7$

40. $\frac{m}{-3} = 10$

41. $\mathbf{D_W}$ When solving an equation using the multiplication principle, how do you determine by what number to multiply or divide on both sides of the equation?

42. $\mathbf{D_W}$ Are the equations $x = 5$ and $x^2 = 25$ equivalent? Why or why not?

SKILL MAINTENANCE

Collect like terms. [7.7e]

43. $3x + 4x$

44. $6x + 5 - 7x$

45. $-4x + 11 - 6x + 18x$

46. $8y - 16y - 24y$

Remove parentheses and simplify. [7.8b]

47. $3x - (4 + 2x)$

48. $2 - 5(x + 5)$

49. $8y - 6(3y + 7)$

50. $-2a - 4(5a - 1)$

Translate to an algebraic expression. [7.1b]

51. Patty drives her van for 8 hr at a speed of r mph. How far does she drive?

52. A triangle has a height of 10 meters and a base of b meters. What is the area of the triangle?

SYNTHESIS

Solve.

53. $-0.2344m = 2028.732$

54. $0 \cdot x = 0$

55. $0 \cdot x = 9$

56. $4|x| = 48$

57. $2|x| = -12$

Solve for x.

58. $ax = 5a$

59. $3x = \frac{b}{a}$

60. $cx = a^2 + 1$

61. $\frac{a}{b}x = 4$

62. A student makes a calculation and gets an answer of 22.5. On the last step, she multiplies by 0.3 when she should have divided by 0.3. What is the correct answer?

8.3 USING THE PRINCIPLES TOGETHER

Objectives

a Solve equations using both the addition and the multiplication principles.

b Solve equations in which like terms may need to be collected.

c Solve equations by first removing parentheses and collecting like terms; solve equations with no solutions and equations with an infinite number of solutions.

1. Solve: $9x + 6 = 51$.

Solve.

2. $8x - 4 = 28$

3. $-\frac{1}{2}x + 3 = 1$

Answers on page A-21

a Applying Both Principles

Consider the equation $3x + 4 = 13$. It is more complicated than those we discussed in the preceding two sections. In order to solve such an equation, we first isolate the x-term, $3x$, using the addition principle. Then we apply the multiplication principle to get x by itself.

EXAMPLE 1 Solve: $3x + 4 = 13$.

$$3x + 4 = 13$$

$$3x + 4 - 4 = 13 - 4 \qquad \text{Using the addition principle: subtracting 4 on both sides}$$

First isolate the x-term. → $3x = 9$ Simplifying

$$\frac{3x}{3} = \frac{9}{3} \qquad \text{Using the multiplication principle: dividing by 3 on both sides}$$

Then isolate x. → $x = 3$ Simplifying

Check:

$3x + 4 = 13$	
$3 \cdot 3 + 4$?	13
$9 + 4$	
13	TRUE

We use the rules for order of operations to carry out the check. We find the product $3 \cdot 3$. Then we add 4.

The solution is 3.

Do Exercise 1.

EXAMPLE 2 Solve: $-5x - 6 = 16$.

$$-5x - 6 = 16$$

$$-5x - 6 + 6 = 16 + 6 \qquad \text{Adding 6 on both sides}$$

$$-5x = 22$$

$$\frac{-5x}{-5} = \frac{22}{-5} \qquad \text{Dividing by } -5 \text{ on both sides}$$

$$x = -\frac{22}{5}, \text{ or } -4\frac{2}{5} \qquad \text{Simplifying}$$

Check:

$-5x - 6 = 16$	
$-5\left(-\frac{22}{5}\right) - 6$?	16
$22 - 6$	
16	TRUE

The solution is $-\frac{22}{5}$.

Do Exercises 2 and 3.

EXAMPLE 3 Solve: $45 - t = 13$.

$$45 - t = 13$$
$$-45 + 45 - t = -45 + 13 \quad \text{Adding } -45 \text{ on both sides}$$
$$-t = -32$$
$$-1(-t) = -1(-32) \quad \text{Multiplying by } -1 \text{ on both sides}$$
$$t = 32$$

The number 32 checks and is the solution.

Do Exercise 4.

EXAMPLE 4 Solve: $16.3 - 7.2y = -8.18$.

$$16.3 - 7.2y = -8.18$$
$$-16.3 + 16.3 - 7.2y = -16.3 + (-8.18) \quad \text{Adding } -16.3 \text{ on both sides}$$
$$-7.2y = -24.48$$
$$\frac{-7.2y}{-7.2} = \frac{-24.48}{-7.2} \quad \text{Dividing by } -7.2 \text{ on both sides}$$
$$y = 3.4$$

Check:

$16.3 - 7.2y$	$= -8.18$
$16.3 - 7.2(3.4)$	? -8.18
$16.3 - 24.48$	
-8.18	TRUE

The solution is 3.4.

Do Exercises 5 and 6.

b Collecting Like Terms

If there are like terms on one side of the equation, we collect them before using the addition or the multiplication principle.

EXAMPLE 5 Solve: $3x + 4x = -14$.

$$3x + 4x = -14$$
$$7x = -14 \quad \text{Collecting like terms}$$
$$\frac{7x}{7} = \frac{-14}{7} \quad \text{Dividing by 7 on both sides}$$
$$x = -2$$

The number -2 checks, so the solution is -2.

Do Exercises 7 and 8.

If there are like terms on opposite sides of the equation, we get them on the same side by using the addition principle. Then we collect them. In other words, we get all terms with a variable on one side and all numbers on the other.

4. Solve: $-18 - m = -57$.

Solve.

5. $-4 - 8x = 8$

6. $41.68 = 4.7 - 8.6y$

Solve.

7. $4x + 3x = -21$

8. $x - 0.09x = 728$

Answers on page A-21

Solve.

9. $7y + 5 = 2y + 10$

10. $5 - 2y = 3y - 5$

EXAMPLE 6 Solve: $2x - 2 = -3x + 3$.

$$
\begin{aligned}
2x - 2 &= -3x + 3 \\
2x - 2 + 2 &= -3x + 3 + 2 && \text{Adding 2} \\
2x &= -3x + 5 && \text{Collecting like terms} \\
2x + 3x &= -3x + 3x + 5 && \text{Adding } 3x \\
5x &= 5 && \text{Simplifying} \\
\frac{5x}{5} &= \frac{5}{5} && \text{Dividing by 5} \\
x &= 1 && \text{Simplifying}
\end{aligned}
$$

Check:

$$
\begin{array}{c|c}
\multicolumn{2}{c}{2x - 2 = -3x + 3} \\
\hline
2 \cdot 1 - 2 \ ? & -3 \cdot 1 + 3 \\
2 - 2 & -3 + 3 \\
0 & 0
\end{array}
$$

Substituting in the original equation

TRUE

The solution is 1.

Do Exercises 9 and 10.

In Example 6, we used the addition principle to get all terms with a variable on one side and all numbers on the other side. Then we collected like terms and proceeded as before. If there are like terms on one side at the outset, they should be collected before proceeding.

Solve.

11. $7x - 17 + 2x = 2 - 8x + 15$

12. $3x - 15 = 5x + 2 - 4x$

EXAMPLE 7 Solve: $6x + 5 - 7x = 10 - 4x + 3$.

$$
\begin{aligned}
6x + 5 - 7x &= 10 - 4x + 3 \\
-x + 5 &= 13 - 4x && \text{Collecting like terms} \\
4x - x + 5 &= 13 - 4x + 4x && \text{Adding } 4x \text{ to get all terms with a variable on one side} \\
3x + 5 &= 13 && \text{Simplifying; that is, collecting like terms} \\
3x + 5 - 5 &= 13 - 5 && \text{Subtracting 5} \\
3x &= 8 && \text{Simplifying} \\
\frac{3x}{3} &= \frac{8}{3} && \text{Dividing by 3} \\
x &= \frac{8}{3} && \text{Simplifying}
\end{aligned}
$$

The number $\frac{8}{3}$ checks, so it is the solution.

Do Exercises 11 and 12.

CLEARING FRACTIONS AND DECIMALS

In general, equations are easier to solve if they do not contain fractions or decimals. Consider, for example,

$$\frac{1}{2}x + 5 = \frac{3}{4} \quad \text{and} \quad 2.3x + 7 = 5.4.$$

Answers on page A-21

If we multiply by 4 on both sides of the first equation and by 10 on both sides of the second equation, we have

$$4\left(\frac{1}{2}x + 5\right) = 4 \cdot \frac{3}{4} \quad \text{and} \quad 10(2.3x + 7) = 10 \cdot 5.4$$

$$4 \cdot \frac{1}{2}x + 4 \cdot 5 = 4 \cdot \frac{3}{4} \quad \text{and} \quad 10 \cdot 2.3x + 10 \cdot 7 = 10 \cdot 5.4$$

$$2x + 20 = 3 \quad \text{and} \quad 23x + 70 = 54.$$

The first equation has been "cleared of fractions" and the second equation has been "cleared of decimals." Both resulting equations are equivalent to the original equations and are easier to solve. *It is your choice* whether to clear fractions or decimals, but doing so often eases computations.

The easiest way to clear an equation of fractions is to multiply *every term on both sides* by the **least common multiple of all the denominators.**

EXAMPLE 8 Solve: $\frac{2}{3}x - \frac{1}{6} + \frac{1}{2}x = \frac{7}{6} + 2x$.

The number 6 is the least common multiple of all the denominators. We multiply by 6 on both sides.

$$6\left(\frac{2}{3}x - \frac{1}{6} + \frac{1}{2}x\right) = 6\left(\frac{7}{6} + 2x\right)$$ Multiplying by 6 on both sides

$$6 \cdot \frac{2}{3}x - 6 \cdot \frac{1}{6} + 6 \cdot \frac{1}{2}x = 6 \cdot \frac{7}{6} + 6 \cdot 2x$$ Using the distributive law (*Caution*! Be sure to multiply *all* the terms by 6.)

$$4x - 1 + 3x = 7 + 12x$$ Simplifying. Note that the fractions are cleared.

$$7x - 1 = 7 + 12x$$ Collecting like terms

$$7x - 1 - 12x = 7 + 12x - 12x$$ Subtracting $12x$

$$-5x - 1 = 7$$ Collecting like terms

$$-5x - 1 + 1 = 7 + 1$$ Adding 1

$$-5x = 8$$ Collecting like terms

$$\frac{-5x}{-5} = \frac{8}{-5}$$ Dividing by -5

$$x = -\frac{8}{5}$$

Check: $\frac{2}{3}x - \frac{1}{6} + \frac{1}{2}x = \frac{7}{6} + 2x$

$\frac{2}{3}\left(-\frac{8}{5}\right) - \frac{1}{6} + \frac{1}{2}\left(-\frac{8}{5}\right)$	? $\frac{7}{6} + 2\left(-\frac{8}{5}\right)$	
$-\frac{16}{15} - \frac{1}{6} - \frac{8}{10}$	$\frac{7}{6} - \frac{16}{5}$	
$-\frac{32}{30} - \frac{5}{30} - \frac{24}{30}$	$\frac{35}{30} - \frac{96}{30}$	
$\frac{-32 - 5 - 24}{30}$	$-\frac{61}{30}$	
$-\frac{61}{30}$		TRUE

13. Solve: $\frac{7}{8}x - \frac{1}{4} + \frac{1}{2}x = \frac{3}{4} + x$.

14. Solve: $41.68 = 4.7 - 8.6y$.

Solve.

15. $2(2y + 3) = 14$

16. $5(3x - 2) = 35$

Answers on page A-21

The solution is $-\frac{8}{5}$.

Do Exercise 13.

To illustrate clearing decimals, we repeat Example 4, but this time we clear the equation of decimals first. Compare both methods.

To clear an equation of decimals, we count the greatest number of decimal places in any one number. If the greatest number of decimal places is 1, we multiply every term on both sides by 10; if it is 2, we multiply by 100; and so on.

EXAMPLE 9 Solve: $16.3 - 7.2y = -8.18$.

The greatest number of decimal places in any one number is *two*. Multiplying by 100, which has *two* 0's, will clear all decimals.

$100(16.3 - 7.2y) = 100(-8.18)$	Multiplying by 100 on both sides
$100(16.3) - 100(7.2y) = 100(-8.18)$	Using the distributive law
$1630 - 720y = -818$	Simplifying
$1630 - 720y - 1630 = -818 - 1630$	Subtracting 1630
$-720y = -2448$	Collecting like terms
$\frac{-720y}{-720} = \frac{-2448}{-720}$	Dividing by -720
$y = \frac{17}{5}$, or 3.4	

The number $\frac{17}{5}$, or 3.4, checks, as shown in Example 4, so it is the solution.

Do Exercise 14.

C Equations Containing Parentheses

To solve certain kinds of equations that contain parentheses, we first use the distributive laws to remove the parentheses. Then we proceed as before.

EXAMPLE 10 Solve: $8x = 2(12 - 2x)$.

$8x = 2(12 - 2x)$	
$8x = 24 - 4x$	Using the distributive laws to multiply and remove parentheses
$8x + 4x = 24 - 4x + 4x$	Adding $4x$ to get all the x-terms on one side
$12x = 24$	Collecting like terms
$\frac{12x}{12} = \frac{24}{12}$	Dividing by 12
$x = 2$	

The number 2 checks, so the solution is 2.

Do Exercises 15 and 16.

Here is a procedure for solving the types of equation discussed in this section.

AN EQUATION-SOLVING PROCEDURE

1. Multiply on both sides to clear the equation of fractions or decimals. (This is optional, but it can ease computations.)
2. If parentheses occur, multiply to remove them using the *distributive laws.*
3. Collect like terms on each side, if necessary.
4. Get all terms with variables on one side and all numbers (constant terms) on the other side, using the *addition principle.*
5. Collect like terms again, if necessary.
6. Multiply or divide to solve for the variable, using the *multiplication principle.*
7. Check all possible solutions in the original equation.

EXAMPLE 11 Solve: $2 - 5(x + 5) = 3(x - 2) - 1$.

$$2 - 5(x + 5) = 3(x - 2) - 1$$

$$2 - 5x - 25 = 3x - 6 - 1 \quad \text{Using the distributive laws to multiply and remove parentheses}$$

$$-5x - 23 = 3x - 7 \quad \text{Collecting like terms}$$

$$-5x - 23 + 5x = 3x - 7 + 5x \quad \text{Adding } 5x$$

$$-23 = 8x - 7 \quad \text{Collecting like terms}$$

$$-23 + 7 = 8x - 7 + 7 \quad \text{Adding 7}$$

$$-16 = 8x \quad \text{Collecting like terms}$$

$$\frac{-16}{8} = \frac{8x}{8} \quad \text{Dividing by 8}$$

$$-2 = x$$

Check:

$2 - 5(x + 5) = 3(x - 2) - 1$	
$2 - 5(-2 + 5)$?	$3(-2 - 2) - 1$
$2 - 5(3)$	$3(-4) - 1$
$2 - 15$	$-12 - 1$
-13	-13 TRUE

The solution is -2.

Do Exercises 17 and 18.

EQUATIONS WITH INFINITELY MANY SOLUTIONS

The types of equations we have considered thus far in Sections 8.1–8.3 have all had exactly one solution. We now look at two other possibilities.

Consider

$$3 + x = x + 3.$$

Let's explore the solutions in Margin Exercises 19–22.

Do Exercises 19–22.

Solve.

17. $3(7 + 2x) = 30 + 7(x - 1)$

18. $4(3 + 5x) - 4 = 3 + 2(x - 2)$

Determine whether the given number is a solution of the given equation.

19. 10; $3 + x = x + 3$

20. -7; $3 + x = x + 3$

21. $\frac{1}{2}$; $3 + x = x + 3$

22. 0; $3 + x = x + 3$

Answers on page A-21

Determine whether the given number is a solution of the given equation.

23. 10; $3 + x = x + 8$

24. -7; $3 + x = x + 8$

25. $\frac{1}{2}$; $3 + x = x + 8$

26. 0; $3 + x = x + 8$

Solve.

27. $30 + 5(x + 3) = -3 + 5x + 48$

28. $2x + 7(x - 4) = 13 + 9x$

Answers on page A-21

We know by the commutative law of addition that this equation holds for any replacement of x with a real number. (See Section 7.7.) We have confirmed some of these solutions in Margin Exercises 19–22. Suppose we try to solve this equation using the addition principle:

$$\begin{aligned} 3 + x &= x + 3 \\ -x + 3 + x &= -x + x + 3 \quad \text{Adding } -x \\ 3 &= 3. \quad \text{TRUE} \end{aligned}$$

We end with a true equation. The original equation holds for all real-number replacements. Every real number is a solution. Thus the number of solutions is **infinite.**

EXAMPLE 12 Solve: $7x - 17 = 4 + 7(x - 3)$.

$$\begin{aligned} 7x - 17 &= 4 + 7(x - 3) \\ 7x - 17 &= 4 + 7x - 21 \quad \text{Using the distributive law to multiply and remove parentheses} \\ 7x - 17 &= 7x - 17 \quad \text{Collecting like terms} \\ -7x + 7x - 17 &= -7x + 7x - 17 \quad \text{Adding } -7x \\ -17 &= -17 \quad \text{TRUE} \end{aligned}$$

Every real number is a solution. There are infinitely many solutions.

EQUATIONS WITH NO SOLUTION

Now consider

$$3 + x = x + 8.$$

Let's explore the solutions in Margin Exercises 23–26.

Do Exercises 23–26.

None of the replacements in Margin Exercises 23–26 is a solution of the given equation. In fact, there are no solutions. Let's try to solve this equation using the addition principle:

$$\begin{aligned} 3 + x &= x + 8 \\ -x + 3 + x &= -x + x + 8 \quad \text{Adding } -x \\ 3 &= 8. \quad \text{FALSE} \end{aligned}$$

We end with a false equation. The original equation is false for all real-number replacements. Thus it has **no** solutions.

EXAMPLE 13 Solve: $3x + 4(x + 2) = 11 + 7x$.

$$\begin{aligned} 3x + 4(x + 2) &= 11 + 7x \\ 3x + 4x + 8 &= 11 + 7x \quad \text{Using the distributive law to multiply and remove parentheses} \\ 7x + 8 &= 11 + 7x \quad \text{Collecting like terms} \\ 7x + 8 - 7x &= 11 + 7x - 7x \quad \text{Subtracting } 7x \\ 8 &= 11 \quad \text{FALSE} \end{aligned}$$

There are no solutions.

Do Exercises 27 and 28.

The following is a guideline for solving linear equations of the types that we have considered in Sections 8.1–8.3.

RESULTING EQUATION	NUMBER OF SOLUTIONS	SOLUTION(S)
$x = a$, where a is a real number	One	The number a
A true equation such as $3 = 3$, $-11 = -11$, or $0 = 0$	Infinitely many	Every real number is a solution.
A false equation such as $3 = 8$, $-4 = 5$, or $0 = -5$	Zero	There are no solutions.

CALCULATOR CORNER

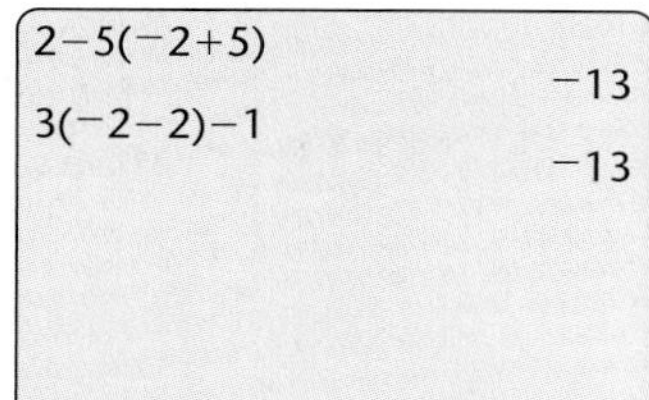

Checking Possible Solutions To check the possible solutions of an equation on a calculator, we can substitute and carry out the calculations on each side of the equation just as we do when we check by hand. To check the possible solution −2 in Example 11, for instance, we first substitute −2 for x in the expression on the left side of the equation. We press 2 − 5 ((−) 2 + 5) ENTER. We get −13. Next, we substitute −2 for x in the expression on the right side of the equation. We then press 3 ((−) 2 − 2) − 1 ENTER. Again we get −13. Since the two sides of the equation have the same value when x is −2, we know that −2 is the solution of the equation.

A table can also be used to check the possible solutions of an equation. First, we enter the left side and the right side of the equation on the Y = or equation editor screen. To do this, we first press Y=. If an expression for Y1 is currently entered, we place the cursor on it and press CLEAR to delete it. We do the same for any other entries that are present.

Next, we position the cursor to the right of Y1 = and enter the left side of the equation by pressing 2 − 5 (X,T,Θ,n + 5). Then we position the cursor beside Y2 = and enter the right side of the equation by pressing 3 (X,T,Θ,n − 2) − 1. Now we press 2ND TBLSET to display the Table Setup screen. (TBLSET is the second operation associated with the WINDOW key.) On the Indpnt line, we position the cursor on "Ask" and press ENTER to set up a table in ASK mode. (The settings for TblStart and ΔTbl are irrelevant in ASK mode.)

Now we press 2ND TABLE to display the table. (TABLE is the second operation associated with the GRAPH key.) We then enter the possible solution, −2, by pressing (−) 2 ENTER. We see that Y1 = −13 = Y2 for this value of x. This confirms that the left and right sides of the equation have the same value for $x = -2$, so −2 is the solution of the equation.

Plot1 Plot2 Plot3
\Y1 = 2−5(X+5)
\Y2 = 3(X−2)−1
\Y3 =
\Y4 =
\Y5 =
\Y6 =
\Y7 =

TABLE SETUP
TblStart=1
ΔTbl=1
Indpnt: Auto **Ask**
Depend: **Auto** Ask

X	Y1	Y2
−2	−13	−13

X =

Exercises:

1. Use substitution to check the solutions found in Examples 6, 7, and 10.
2. Use a table set in ASK mode to check the solutions found in Examples 6, 7, and 10.

8.3 EXERCISE SET

For Extra Help

MathXL | MyMathLab | InterAct Math | Math Tutor Center | Video Lectures on CD Disc 4 | Student's Solutions Manual

a Solve. Don't forget to check!

1. $5x + 6 = 31$

Check: $5x + 6 = 31$?

2. $7x + 6 = 13$

Check: $7x + 6 = 13$?

3. $8x + 4 = 68$

Check: $8x + 4 = 68$?

4. $4y + 10 = 46$

Check: $4y + 10 = 46$?

5. $4x - 6 = 34$

6. $5y - 2 = 53$

7. $3x - 9 = 33$

8. $4x - 19 = 5$

9. $7x + 2 = -54$

10. $5x + 4 = -41$

11. $-45 = 3 + 6y$

12. $-91 = 9t + 8$

13. $-4x + 7 = 35$

14. $-5x - 7 = 108$

15. $-8x - 24 = -29\frac{1}{3}$

16. $\frac{3}{2}x - 24 = -36$

Solve.

17. $5x + 7x = 72$

Check: $5x + 7x = 72$?

18. $8x + 3x = 55$

Check: $8x + 3x = 55$?

19. $8x + 7x = 60$

Check: $8x + 7x = 60$?

20. $8x + 5x = 104$

Check: $8x + 5x = 104$?

21. $4x + 3x = 42$

22. $7x + 18x = 125$

23. $-6y - 3y = 27$

24. $-5y - 7y = 144$

25. $-7y - 8y = -15$

26. $-10y - 3y = -39$

27. $x + \frac{1}{3}x = 8$

28. $x + \frac{1}{4}x = 10$

29. $10.2y - 7.3y = -58$

30. $6.8y - 2.4y = -88$

31. $8y - 35 = 3y$

32. $4x - 6 = 6x$

33. $8x - 1 = 23 - 4x$

34. $5y - 2 = 28 - y$

35. $2x - 1 = 4 + x$

36. $4 - 3x = 6 - 7x$

37. $6x + 3 = 2x + 11$

38. $14 - 6a = -2a + 3$

39. $5 - 2x = 3x - 7x + 25$

40. $-7z + 2z - 3z - 7 = 17$

41. $4 + 3x - 6 = 3x + 2 - x$

42. $5 + 4x - 7 = 4x - 2 - x$

43. $4y - 4 + y + 24 = 6y + 20 - 4y$

44. $5y - 7 + y = 7y + 21 - 5y$

Solve. Clear fractions or decimals first.

45. $\frac{7}{2}x + \frac{1}{2}x = 3x + \frac{3}{2} + \frac{5}{2}x$

46. $\frac{7}{8}x - \frac{1}{4} + \frac{3}{4}x = \frac{1}{16} + x$

47. $\frac{2}{3} + \frac{1}{4}t = \frac{1}{3}$

48. $-\frac{3}{2} + x = -\frac{5}{6} - \frac{4}{3}$

49. $\frac{2}{3} + 3y = 5y - \frac{2}{15}$

50. $\frac{1}{2} + 4m = 3m - \frac{5}{2}$

51. $\frac{5}{3} + \frac{2}{3}x = \frac{25}{12} + \frac{5}{4}x + \frac{3}{4}$

52. $1 - \frac{2}{3}y = \frac{9}{5} - \frac{y}{5} + \frac{3}{5}$

53. $2.1x + 45.2 = 3.2 - 8.4x$

54. $0.96y - 0.79 = 0.21y + 0.46$

55. $1.03 - 0.62x = 0.71 - 0.22x$

56. $1.7t + 8 - 1.62t = 0.4t - 0.32 + 8$

57. $\frac{2}{7}x - \frac{1}{2}x = \frac{3}{4}x + 1$

58. $\frac{5}{16}y + \frac{3}{8}y = 2 + \frac{1}{4}y$

C Solve.

59. $3(2y - 3) = 27$

60. $8(3x + 2) = 30$

61. $40 = 5(3x + 2)$

62. $9 = 3(5x - 2)$

63. $-23 + y = y + 25$

64. $17 - t = -t + 68$

65. $-23 + x = x - 23$

66. $y - \frac{2}{3} = -\frac{2}{3} + y$

67. $2(3 + 4m) - 9 = 45$

68. $5x + 5(4x - 1) = 20$

69. $5r - (2r + 8) = 16$

70. $6b - (3b + 8) = 16$

71. $6 - 2(3x - 1) = 2$

72. $10 - 3(2x - 1) = 1$

73. $5x + 5 - 7x = 15 - 12x + 10x - 10$

74. $3 - 7x + 10x - 14 = 9 - 6x + 9x - 20$

75. $22x - 5 - 15x + 3 = 10x - 4 - 3x + 11$

76. $11x - 6 - 4x + 1 = 9x - 8 - 2x + 12$

77. $5(d + 4) = 7(d - 2)$

78. $3(t - 2) = 9(t + 2)$

79. $8(2t + 1) = 4(7t + 7)$

80. $7(5x - 2) = 6(6x - 1)$

81. $3(r - 6) + 2 = 4(r + 2) - 21$

82. $5(t + 3) + 9 = 3(t - 2) + 6$

83. $19 - (2x + 3) = 2(x + 3) + x$

84. $13 - (2c + 2) = 2(c + 2) + 3c$

85. $2[4 - 2(3 - x)] - 1 = 4[2(4x - 3) + 7] - 25$

86. $5[3(7 - t) - 4(8 + 2t)] - 20 = -6[2(6 + 3t) - 4]$

87. $11 - 4(x + 1) - 3 = 11 + 2(4 - 2x) - 16$

88. $6(2x - 1) - 12 = 7 + 12(x - 1)$

89. $22x - 1 - 12x = 5(2x - 1) + 4$

90. $2 + 14x - 9 = 7(2x + 1) - 14$

91. $0.7(3x + 6) = 1.1 - (x + 2)$

92. $0.9(2x + 8) = 20 - (x + 5)$

93. $^{D}_{W}$ What procedure would you follow to solve an equation like $0.23x + \frac{17}{3} = -0.8 + \frac{3}{4}x$? Could your procedure be streamlined? If so, how?

94. $^{D}_{W}$ You are trying to explain to a classmate how equations can arise with infinitely many solutions and with no solutions. Give such an explanation. Does having no solution mean that 0 is a solution? Explain.

SKILL MAINTENANCE

95. Divide: $-22.1 \div 3.4$. [7.6c]

96. Multiply: $-22.1(3.4)$. [7.5a]

97. Factor: $7x - 21 - 14y$. [7.7d]

98. Factor: $8y - 88x + 8$. [7.7d]

Simplify.

99. $-3 + 2(-5)^2(-3) - 7$ [7.8d]

100. $3x + 2[4 - 5(2x - 1)]$ [7.8c]

101. $23(2x - 4) - 15(10 - 3x)$ [7.8b]

102. $256 \div 64 \div 4^2$ [7.8d]

SYNTHESIS

Solve.

103. $\frac{2}{3}\left(\frac{7}{8} - 4x\right) - \frac{5}{8} = \frac{3}{8}$

104. $\frac{1}{4}(8y + 4) - 17 = -\frac{1}{2}(4y - 8)$

105. $\frac{4 - 3x}{7} = \frac{2 + 5x}{49} - \frac{x}{14}$

106. The width of a rectangle is 5 ft, its length is $(3x + 2)$ ft, and its area is 75 ft^2. Find x.

8.4 FORMULAS

Objectives

a Evaluate a formula.

b Solve a formula for a specified letter.

1. **Storm Distance.** Suppose that it takes the sound of thunder 14 sec to reach you. How far away is the storm?

2. **Socks from Cotton.** Referring to Example 2, find the number of socks that can be made from 65 bales of cotton.

Answers on page A-21

a Evaluating Formulas

A **formula** is a "recipe" for doing a certain type of calculation. Formulas are often given as equations. When we replace the variables in an equation with numbers and calculate the result, we are **evaluating** the formula. We did some evaluating in Section 7.1.

Let's consider another example. A formula that has to do with weather is $M = \frac{1}{5}t$. You see a flash of lightning. After a few seconds you hear the thunder associated with that flash. How far away was the lightning?

Your distance from the storm is M miles. You can find that distance by counting the number of seconds t that it takes the sound of the thunder to reach you and then multiplying by $\frac{1}{5}$.

EXAMPLE 1 *Storm Distance.* Consider the formula $M = \frac{1}{5}t$. It takes 10 sec for the sound of thunder to reach you after you have seen a flash of lightning. How far away is the storm?

We substitute 10 for t and calculate M:

$$M = \frac{1}{5}t = \frac{1}{5}(10) = 2.$$

The storm is 2 mi away.

EXAMPLE 2 *Socks from Cotton.* Consider the formula $S = 4321x$, where S is the number of socks of normal size that can be produced from x bales of cotton. You see a shipment of 300 bales of cotton taken off a ship. How many socks can be made from the cotton?

Source: *Country Woman Magazine*

We substitute 300 for x and calculate S:

$$S = 4321x = 4321(300) = 1{,}296{,}300.$$

Thus, 1,296,300 socks can be made from 300 bales of cotton.

Do Exercises 1 and 2.

EXAMPLE 3 *Distance, Rate, and Time.* The distance d that a car will travel at a rate, or speed, r in time t is given by

$$d = rt.$$

A car travels at 75 miles per hour (mph) for 4.5 hr. How far will it travel?

We substitute 75 for r and 4.5 for t and calculate d:

$$d = rt = (75)(4.5) = 337.5 \text{ mi}.$$

The car will travel 337.5 mi.

Do Exercise 3.

3. Distance, Rate, and Time. A car travels at 55 mph for 6.2 hr. How far will it travel?

b Solving Formulas

Refer to Example 2. Suppose a clothing company wants to produce S socks and needs to know how many bales of cotton to order. If this calculation is to be repeated many times, it might be helpful to first solve the formula for x:

$$S = 4321x$$

$$\frac{S}{4321} = x. \qquad \text{Dividing by 4321}$$

Then we can substitute a number for S and calculate x. For example, if the number of socks S to be produced is 432,100, then

$$x = \frac{S}{4321} = \frac{432{,}100}{4321} = 100.$$

The company would need to order 100 bales of cotton.

EXAMPLE 4 Solve for t: $M = \frac{1}{5}t$.

$$M = \frac{1}{5}t \qquad \text{We want this letter alone.}$$

$$5 \cdot M = 5 \cdot \frac{1}{5}t \qquad \text{Multiplying by 5 on both sides}$$

$$5M = t$$

For $M = 2$ in Example 4, $t = 5M = 5(2)$, or 10.

EXAMPLE 5 *Distance, Rate, and Time.* Solve for t: $d = rt$.

$$d = rt \qquad \text{We want this letter alone.}$$

$$\frac{d}{r} = \frac{rt}{r} \qquad \text{Dividing by } r$$

$$\frac{d}{r} = \frac{r}{r} \cdot t$$

$$\frac{d}{r} = t \qquad \text{Simplifying}$$

Do Exercises 4–6.

4. Solve for q: $B = \frac{1}{3}q$.

5. Distance, Rate, and Time. Solve for r: $d = rt$.

6. Electricity. Solve for I: $E = IR$. (This formula relates voltage E, current I, and resistance R.)

Answers on page A-21

Solve for x.

7. $y = x + 5$

8. $y = x - 7$

9. $y = x - b$

10. Solve for y: $9y = 5x$.

11. Solve for p: $ap = bq$.

12. Solve for x: $y = mx + b$.

13. Solve for Q: $tQ - p = a$.

Answers on page A-21

EXAMPLE 6 Solve for x: $y = x + 3$.

$y = x + 3$	We want this letter alone.
$y - 3 = x + 3 - 3$	Subtracting 3
$y - 3 = x$	Simplifying

EXAMPLE 7 Solve for x: $y = x - a$.

$y = x - a$	We want this letter alone.
$y + a = x - a + a$	Adding a
$y + a = x$	Simplifying

Do Exercises 7–9.

EXAMPLE 8 Solve for y: $6y = 3x$.

$6y = 3x$	We want this letter alone.
$\frac{6y}{6} = \frac{3x}{6}$	Dividing by 6
$y = \frac{1}{2}x$	Simplifying

EXAMPLE 9 Solve for y: $by = ax$.

$by = ax$	We want this letter alone.
$\frac{by}{b} = \frac{ax}{b}$	Dividing by b
$y = \frac{ax}{b}$	Simplifying

Do Exercises 10 and 11.

EXAMPLE 10 Solve for x: $ax + b = c$.

$ax + b = c$	We want this letter alone.
$ax + b - b = c - b$	Subtracting b
$ax = c - b$	Simplifying
$\frac{ax}{a} = \frac{c - b}{a}$	Dividing by a
$x = \frac{c - b}{a}$	Simplifying

Do Exercises 12 and 13.

To solve a formula for a given letter, identify the letter and:

1. Multiply on both sides to clear fractions or decimals, if that is needed.
2. Collect like terms on each side, if necessary.
3. Get all terms with the letter to be solved for on one side of the equation and all other terms on the other side.
4. Collect like terms again, if necessary.
5. Solve for the letter in question.

EXAMPLE 11 *Circumference.* Solve for r: $C = 2\pi r$. This is a formula for the circumference C of a circle of radius r.

$C = 2\pi r$ We want this letter alone.

$\frac{C}{2\pi} = \frac{2\pi r}{2\pi}$ Dividing by 2π

$\frac{C}{2\pi} = r$

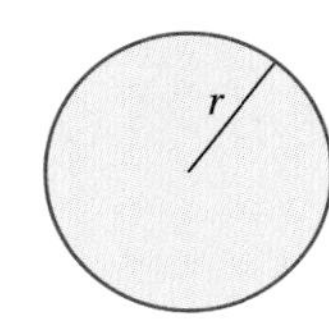

EXAMPLE 12 *Averages.* Solve for a: $A = \frac{a + b + c}{3}$. This is a formula for the average A of three numbers a, b, and c.

$A = \frac{a + b + c}{3}$ We want the letter a alone.

$3 \cdot A = 3 \cdot \frac{a + b + c}{3}$ Multiplying by 3 on both sides

$3A = a + b + c$ Simplifying

$3A - b - c = a$ Subtracting b and c

Do Exercises 14 and 15.

14. Circumference. Solve for D:

$$C = \pi D.$$

(This is a formula for the circumference C of a circle of diameter D.)

15. Averages. Solve for c:

$$A = \frac{a + b + c + d}{4}.$$

Answers on page A-22

Study Tips

HIGHLIGHTING

Reading and highlighting a section before your instructor lectures on it allows you to maximize your learning and understanding during the lecture.

- **Try to keep one section ahead of your syllabus.** If you study ahead of your lectures, you can concentrate on what is being explained in them, rather than trying to write everything down. You can then take notes only of special points or of questions related to what is happening in class.
- **Highlight important points.** You are probably used to highlighting key points as you study. If that works for you, continue to do so. But you will notice many design features throughout this book that already highlight important points. Thus you may not need to highlight as much as you generally do.
- **Highlight points that you do not understand.** Use a unique mark to indicate trouble spots that can lead to questions to be asked during class, in a tutoring session, or when calling or contacting the AW Math Tutor Center.

8.4 EXERCISE SET

For Extra Help

MathXL | MyMathLab | InterAct Math | Math Tutor Center | Video Lectures on CD Disc 4 | Student's Solutions Manual

a, **b** Solve.

1. *Furnace Output.* The formula

$$B = 30a$$

is used in New England to estimate the minimum furnace output B, in Btu's, for a modern house with a square feet of flooring.

Source: U.S. Department of Energy

a) Determine the minimum furnace output for a 1900-ft^2 modern house.
b) Solve for a. That is, solve $B = 30a$ for a.

2. *Furnace Output.* The formula

$$B = 50a$$

is used in New England to estimate the minimum furnace output B, in Btu's, for an old, poorly insulated house with a square feet of flooring.

Source: U.S. Department of Energy

a) Determine the minimum furnace output for a 3200-ft^2 old, poorly insulated house.
b) Solve for a. That is, solve $B = 50a$ for a.

3. *Distance from a Storm.* The formula

$$M = \frac{1}{5}t$$

can be used to determine how far M, in miles, you are from lightning when its thunder takes t seconds to reach your ears.

a) It takes 8 sec for the sound of thunder to reach you after you have seen the lightning. How far away is the storm?
b) Solve for t.

4. *Electrical Power.* The power rating P, in watts, of an electrical appliance is determined by

$$P = I \cdot V,$$

where I is the current, in amperes, and V is measured in volts.

a) A kitchen requires 30 amps of current and the voltage in the house is 115 volts. What is the wattage of the kitchen?
b) Solve for I; for V.

5. *College Enrollment.* At many colleges, the number of "full-time-equivalent" students f is given by

$$f = \frac{n}{15},$$

where n is the total number of credits for which students have enrolled in a given semester.

a) Determine the number of full-time-equivalent students on a campus in which students registered for a total of 21,345 credits.
b) Solve for n.

6. *Surface Area of a Cube.* The surface area A of a cube with side s is given by

$$A = 6s^2.$$

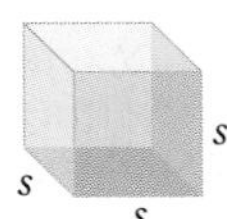

a) Find the surface area of a cube with sides of 3 in.
b) Solve for s^2.

7. *Calorie Density.* The calorie density D, in calories per ounce, of a food that contains c calories and weighs w ounces is given by

$$D = \frac{c}{w}.$$

Eight ounces of fat-free milk contains 84 calories. Find the calorie density of fat-free milk.

Source: *Nutrition Action Healthletter*, March 2000, p. 9. Center for Science in the Public Interest, Suite 300; 1875 Connecticut Ave NW, Washington, D.C. 20008.

8. *Wavelength of a Musical Note.* The wavelength w, in meters per cycle, of a musical note is given by

$$w = \frac{r}{f},$$

where r is the speed of the sound, in meters per second, and f is the frequency, in cycles per second. The speed of sound in air is 344 m/sec. What is the wavelength of a note whose frequency in air is 24 cycles per second?

9. *Size of a League Schedule.* When all n teams in a league play every other team twice, a total of N games are played, where

$$N = n^2 - n.$$

A soccer league has 7 teams and all teams play each other twice. How many games are played?

10. *Size of a League Schedule.* When all n teams in a league play every other team twice, a total of N games are played, where

$$N = n^2 - n.$$

A basketball league has 11 teams and all teams play each other twice. How many games are played?

b Solve for the indicated letter.

11. $y = 5x$, for x

12. $d = 55t$, for t

13. $a = bc$, for c

14. $y = mx$, for x

15. $y = 13 + x$, for x

16. $y = x - \frac{2}{3}$, for x

17. $y = x + b$, for x

18. $y = x - A$, for x

19. $y = 5 - x$, for x

20. $y = 10 - x$, for x

21. $y = a - x$, for x

22. $y = q - x$, for x

23. $8y = 5x$, for y

24. $10y = -5x$, for y

25. $By = Ax$, for x

26. $By = Ax$, for y

27. $W = mt + b$, for t

28. $W = mt - b$, for t

29. $y = bx + c$, for x

30. $y = bx - c$, for x

31. $A = \dfrac{a + b + c}{3}$, for b

32. $A = \dfrac{a + b + c}{3}$, for c

33. $A = at + b$, for t

34. $S = rx + s$, for x

35. *Area of a Parallelogram:*

$A = bh$, for h

(Area A, base b, height h)

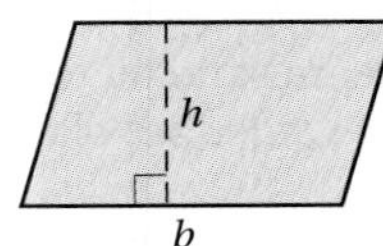

36. *Distance, Rate, Time:*

$d = rt$, for r

(Distance d, speed r, time t)

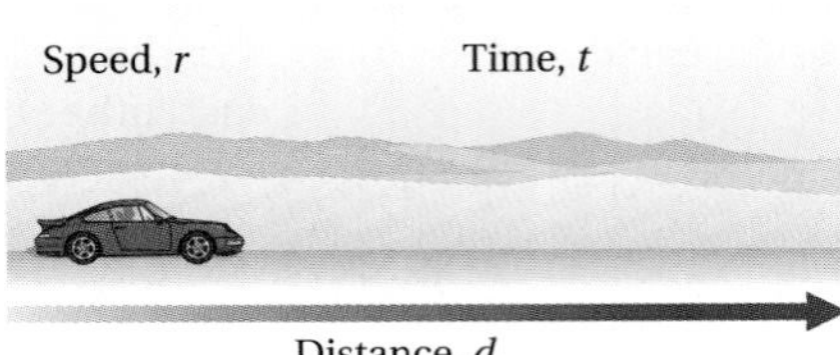

37. *Perimeter of a Rectangle:*

$P = 2l + 2w$, for w

(Perimeter P, length l, width w)

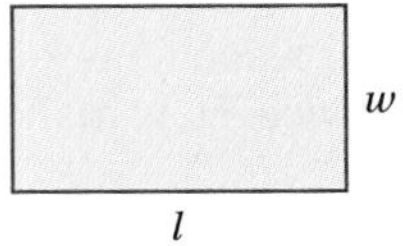

38. *Area of a Circle:*

$A = \pi r^2$, for r^2

(Area A, radius r)

39. *Average of Two Numbers:*

$A = \dfrac{a + b}{2}$, for a

a $\quad A = \dfrac{a+b}{2} \quad$ b

40. *Area of a Triangle:*

$A = \dfrac{1}{2}bh$, for b

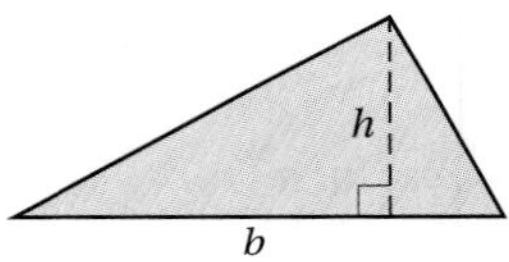

41. *Force:*

$F = ma$, for a

(Force F, mass m, acceleration a)

42. *Simple Interest:*

$I = Prt$, for P

(Interest I, principal P, interest rate r, time t)

43. *Relativity:*

$E = mc^2$, for c^2

(Energy E, mass m, speed of light c)

44. $Q = \dfrac{p - q}{2}$, for p

45. $Ax + By = c$, for x

46. $Ax + By = c$, for y

47. $v = \dfrac{3k}{t}$, for t

48. $P = \dfrac{ab}{c}$, for c

49. $\mathbf{D_W}$ Devise an application in which it would be useful to solve the equation $d = rt$ for r. (See Exercise 36.)

50. $\mathbf{D_W}$ The equations

$$P = 2l + 2w \quad \text{and} \quad w = \frac{P}{2} - l$$

are equivalent formulas involving the perimeter P, the length l, and the width w of a rectangle. (See Exercise 37.) Devise a problem for which the second of the two formulas would be more useful.

SKILL MAINTENANCE

51. Convert to decimal notation: $\frac{23}{25}$. [3.1b]

52. Add: $-23 + (-67)$. [7.3a]

53. Add: $0.082 + (-9.407)$. [7.3a]

54. Subtract: $-23 - (-67)$. [7.4a]

55. Subtract: $-45.8 - (-32.6)$. [7.4a]

56. Remove parentheses and simplify: [7.8b]

$$4a - 8b - 5(5a - 4b).$$

Convert to decimal notation. [4.2b]

57. 3.1%

58. 67.1%

59. Add: $-\frac{2}{3} + \frac{5}{6}$. [7.3a]

60. Subtract: $-\frac{2}{3} - \frac{5}{6}$. [7.4a]

SYNTHESIS

61. *Female Caloric Needs.* The number of calories K needed each day by a moderately active woman who weighs w pounds, is h inches tall, and is a years old can be estimated by the formula

$$K = 917 + 6(w + h - a).$$

Source: Parker, M., *She Does Math.* Mathematical Association of America, p. 96

a) Elaine is moderately active, weighs 120 lb, is 67 in. tall, and is 23 yr old. What are her caloric needs?

b) Solve the formula for a; for h; for w.

62. *Male Caloric Needs.* The number of calories K needed each day by a moderately active man who weighs w kilograms, is h centimeters tall, and is a years old can be estimated by the formula

$$K = 19.18w + 7h - 9.52a + 92.4.$$

Source: Parker, M., *She Does Math.* Mathematical Association of America, p. 96

a) Marv is moderately active, weighs 97 kg, is 185 cm tall, and is 55 yr old. What are his caloric needs?

b) Solve the formula for a; for h; for w.

Solve.

63. $H = \frac{2}{a - b}$, for b; for a

64. $P = 4m + 7mn$, for m

65. In $A = lw$, l and w both double. What is the effect on A?

66. In $P = 2a + 2b$, P doubles. Do a and b necessarily both double?

67. In $A = \frac{1}{2}bh$, b increases by 4 units and h does not change. What happens to A?

68. Solve for F: $D = \frac{1}{E + F}$.

8.5 APPLICATIONS OF PERCENT

Objective

Solve applied problems involving percent.

a Translating and Solving

Many applied problems involve percent. Here we begin to see how equation solving can enhance our problem-solving skills. For background on the manipulative skills of percent notation, see Sections 4.2–4.5.

In solving percent problems, we first *translate* the problem to an equation. Then we *solve* the equation using the techniques discussed in Sections 8.1–8.3. The key words in the translation are as follows.

KEY WORDS IN PERCENT TRANSLATIONS

"**Of**" translates to "·" or "×".

"**Is**" translates to "=".

"**What number**" translates to any letter.

% translates to "$\times \frac{1}{100}$" or "× 0.01".

EXAMPLE 1 Translate:

28% of 5 is what number?

$$28\% \cdot 5 = a$$

This is a percent equation.

EXAMPLE 2 Translate:

45% of what number is 28?

$$45\% \times b = 28$$

EXAMPLE 3 Translate:

What percent of 90 is 7?

$$n \cdot 90 = 7$$

Do Exercises 1–6.

Translate to an equation. Do not solve.

1. 13% of 80 is what number?

2. What number is 60% of 70?

3. 43 is 20% of what number?

4. 110% of what number is 30?

5. 16 is what percent of 80?

6. What percent of 94 is 10.5?

Answers on page A-22

Percent problems are actually of three different types. Although the method we present does *not* require that you be able to identify which type we are studying, it is helpful to know them.

We know that

15 is 25% of 60, or

$15 = 25\% \times 60$.

We can think of this as:

Amount = Percent number × Base.

Each of the three types of percent problems depends on which of the three pieces of information is missing.

1. Finding the ***amount*** (the result of taking the percent)

   Example: What number is 25% of 60?

   Translation: $y = 25\% \cdot 60$

2. Finding the ***base*** (the number you are taking the percent of)

   Example: 15 is 25% of what number?

   Translation: $15 = 25\% \cdot y$

3. Finding the ***percent number*** (the percent itself)

   Example: 15 is what percent of 60?

   Translation: $15 = y \cdot 60$

FINDING THE AMOUNT

EXAMPLE 4 What number is 11% of 49?

What number is 11% of 49?

Translate: $a = 11\% \times 49$

Solve: The letter is by itself. To solve the equation, we need only convert 11% to decimal notation and multiply:

$$a = 11\% \times 49 = 0.11 \times 49 = 5.39.$$

Thus, 5.39 is 11% of 49. The answer is 5.39.

Do Exercise 7.

FINDING THE BASE

EXAMPLE 5 3 is 16% of what number?

3 is 16% of what number?

Translate: $3 = 16\% \times b$

$3 = 0.16 \times b$ Converting 16% to decimal notation

Solve: In this case, the letter is not by itself. To solve the equation, we divide by 0.16 on both sides:

$$3 = 0.16 \times b$$

$$\frac{3}{0.16} = \frac{0.16 \times b}{0.16} \quad \text{Dividing by 0.16}$$

$$18.75 = b. \quad \text{Simplifying}$$

The answer is 18.75.

Do Exercise 8.

7. What number is 2.4% of 80?

8. 25.3 is 22% of what number?

Answers on page A-22

Study Tips

USING THE SUPPLEMENTS

The new mathematical skills and concepts presented in the lectures will be of increased value to you if you begin the homework assignment as soon as possible after the lecture. Then if you still have difficulty with any of the exercises, you have time to access supplementary resources such as:

- *Student's Solutions Manual*
- Video Lectures on CD
- MathXL Tutorials on CD
- AW Math Tutor Center
- MyMathLab
- MathXL
- Work It Out! Chapter Test Video Solutions on CD

9. What percent of \$50 is \$18?

10. Areas of Alaska and Arizona. The area of Arizona is 19% of the area of Alaska. The area of Alaska is 586,400 mi^2. What is the area of Arizona?

Answers on page A-22

FINDING THE PERCENT NUMBER

In solving these problems, you *must* remember to convert to percent notation after you have solved the equation.

EXAMPLE 6 \$32 is what percent of \$50?

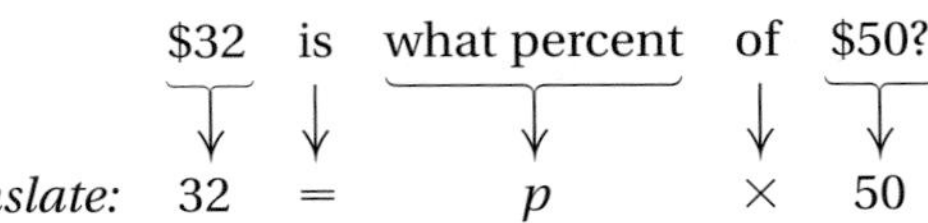

Translate: $32 = p \times 50$

Solve: To solve the equation, we divide by 50 on both sides and convert the answer to percent notation:

$$32 = p \times 50$$

$$\frac{32}{50} = \frac{p \times 50}{50} \qquad \text{Dividing by 50}$$

$$0.64 = p$$

$$64\% = p. \qquad \text{Converting to percent notation}$$

Thus, \$32 is 64% of \$50. The answer is 64%.

Do Exercise 9.

EXAMPLE 7 *Coronary Heart Disease.* In 2001, there were about 206 million people age 20 or older in the United States. About 6.4% of them had coronary heart disease. How many had coronary heart disease?

Source: American Heart Association

To solve the problem, we first reword and then translate. We let $a =$ the number of people age 20 or older in the United States with coronary heart disease.

Rewording: What number is 6.4% of 206?

Translate: $a = 6.4\% \times 206$

Solve: The letter is by itself. To solve the equation, we need only convert 6.4% to decimal notation and multiply:

$$a = 6.4\% \times 206 = 0.064 \times 206 = 13.184.$$

Thus, 13.184 million is 6.4% of 206 million, so in 2001 about 13.184 million people age 20 or older in the United States had coronary heart disease.

Do Exercise 10.

EXAMPLE 8 *Digital Camera.* At one time, Best Buy had a Canon PowerShot digital camera on sale for \$349.99. This was 87.5% of the list price. What was the list price?

To solve the problem, we first reword and then translate. We let $L =$ the list price.

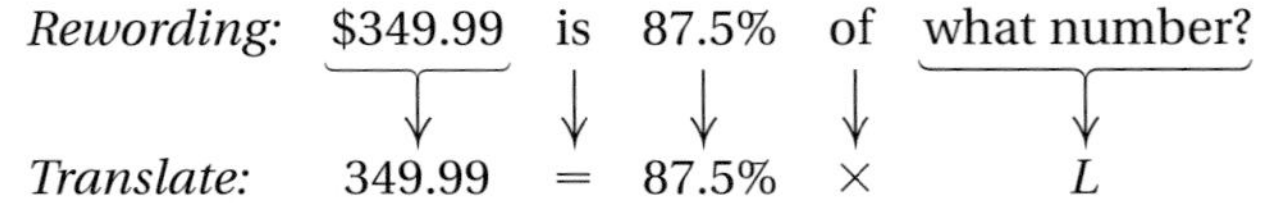

Translate: $349.99 = 87.5\% \times L$

Solve: To solve the equation, we convert 87.5% to decimal notation and divide by 0.875 on both sides:

$$349.99 = 87.5\% \times L$$

$$349.99 = 0.875 \times L \quad \text{Converting to decimal notation}$$

$$\frac{349.99}{0.875} = \frac{0.875 \times L}{0.875} \quad \text{Dividing by 0.875}$$

$$399.99 \approx L. \quad \text{Simplifying using a calculator and rounding to the nearest cent}$$

The list price was about $399.99.

Do Exercise 11.

EXAMPLE 9 *Apple iPod.* An Apple iPod 20.0 GB digital audio player was on sale at Best Buy for $249.99, decreased from a normal list price of $299.99. What was the percent of decrease?

To solve the problem, we must first determine the amount of decrease from the original price:

Original price	minus	Sale price	=	Decrease
↓	↓	↓	↓	↓
$299.99	−	$249.99	=	$50.00.

Using the $50 decrease, we reword and translate. We let p = the percent of decrease. We want to know, "What percent of the *original* price is $50?"

Rewording:	$50	is	what percent	of	$299.99?
	↓	↓	↓	↓	↓
Translate:	50	=	p	×	299.99

Solve: To solve the equation, we divide by 299.99 on both sides and convert the answer to percent notation:

$$50 = p \times 299.99$$

$$\frac{50}{299.99} = \frac{p \times 299.99}{299.99} \quad \text{Dividing by 299.99}$$

$$0.167 \approx p \quad \text{Simplifying and converting to percent notation}$$

$$16.7\% = p.$$

Thus the percent of decrease was about 16.7%.

Do Exercise 12.

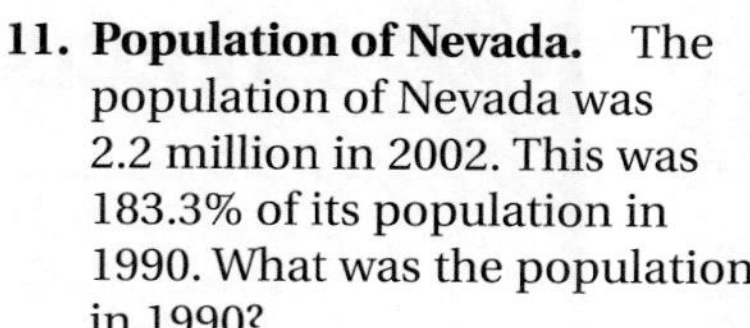

11. Population of Nevada. The population of Nevada was 2.2 million in 2002. This was 183.3% of its population in 1990. What was the population in 1990?

Source: U.S. Bureau of the Census

12. Job Opportunities. There were 252 thousand medical assistants in 1998. Job opportunities are expected to grow to 398 thousand by 2008. What is the percent of increase?

Source: *Handbook of U.S. Labor Statistics*

Answers on page A-22

8.5 EXERCISE SET

For Extra Help

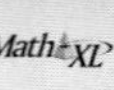

MathXL | MyMathLab | InterAct Math | Math Tutor Center | Video Lectures on CD Disc 4 | Student's Solutions Manual

a Solve.

1. What percent of 180 is 36?

2. What percent of 76 is 19?

3. 45 is 30% of what number?

4. 20.4 is 24% of what number?

5. What number is 65% of 840?

6. What number is 50% of 50? (This was a $500.00 question on the "Who Wants To Be a Millionaire?" television quiz show.)

7. 30 is what percent of 125?

8. 57 is what percent of 300?

9. 12% of what number is 0.3?

10. 7 is 175% of what number?

11. 2 is what percent of 40?

12. 40 is 2% of what number?

13. What percent of 68 is 17?

14. What percent of 150 is 39?

15. What number is 35% of 240?

16. What number is 1% of one million?

17. What percent of 125 is 30?

18. What percent of 60 is 75?

19. What percent of 300 is 48?

20. What percent of 70 is 70?

21. 14 is 30% of what number?

22. 54 is 24% of what number?

23. What number is 2% of 40?

24. What number is 40% of 2?

25. 0.8 is 16% of what number?

26. 25 is what percent of 50?

27. 54 is 135% of what number?

28. 8 is 2% of what number?

Costs of Owning a Dog. The American Pet Products Manufacturers Association estimates that the total cost of owning a dog for its lifetime is $6600. The following circle graph shows the relative costs of raising a dog from birth to death.

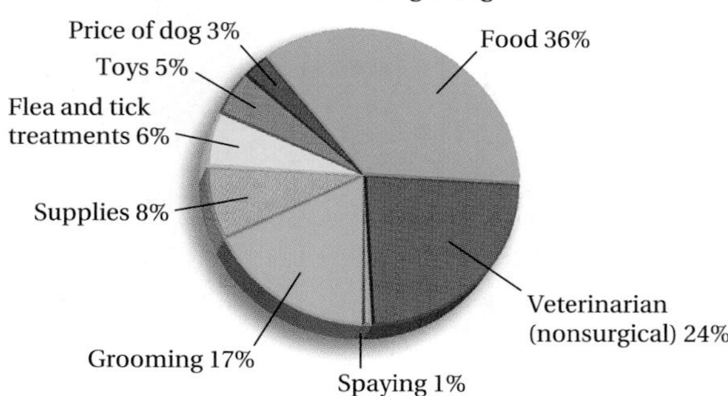

Source: The American Pet Products Manufacturers Association

Complete the following table of costs of owning a dog for its lifetime.

	EXPENSE ITEM	COST		EXPENSE ITEM	COST
29.	Price of dog		**30.**	Food	
31.	Veterinarian		**32.**	Grooming	
33.	Supplies		**34.**	Flea and tick treatments	

35. *Car Sales.* In 2002, there were 8,317,954 retail sales of cars in the United States. Of that total, 2,268,093 were imported. Japan manufactured 1,003,745 of those imports, and Germany manufactured 564,910 of them. What percent of the imported cars were manufactured in Japan? in Germany?

Source: Ward's Communications, *World Almanac and Book of Facts* 2004

36. *Truck Sales.* In 2002, 17,118,000 new motor vehicles were sold in the United States. Of these, 9,036,000 were trucks. Imported trucks made up 1,061,000 of those sales. What percent of the motor vehicles sold were trucks? What percent of the truck sales were imported trucks?

Source: U.S. Bureau of Economic Analysis

37. *Batting Average.* At one point in a recent season, Sammy Sosa of the Chicago Cubs had 193 hits. His batting average was 0.320, or 32%. That is, of the total number of at-bats, 32% were hits. How many at-bats did he have?

Source: Major League Baseball

38. *Pass Completions.* At one point in a recent season, Peyton Manning of the Indianapolis Colts had completed 357 passes. This was 62.5% of his attempts. How many attempts did he make?

Source: National Football League

39. *Student Loans.* To finance her community college education, Sarah takes out a Stafford loan for $6500. After a year, Sarah decides to pay off the interest, which is 3% of $6500. How much will she pay?

40. *Student Loans.* Paul takes out a federal Stafford loan for $5400. After a year, Paul decides to pay off the interest, which is 4.5% of $5400. How much will he pay?

41. *Tipping.* Leon left a $4 tip for a meal that cost $25.

a) What percent of the cost of the meal was the tip?
b) What was the total cost of the meal including the tip?

42. *Tipping.* Selena left a $12.76 tip for a meal that cost $58.

a) What percent of the cost of the meal was the tip?
b) What was the total cost of the meal including the tip?

43. *Tipping.* Leon left a 15% tip for a meal that cost $25.

a) How much was the tip?
b) What was the total cost of the meal including the tip?

44. *Tipping.* Sam, Selena, Rachel, and Clement left a 15% tip for a meal that cost $58.

a) How much was the tip?
b) What was the total cost of the meal including the tip?

45. *Tipping.* Leon left a 15% tip of $4.32 for a meal.

a) What was the cost of the meal before the tip?
b) What was the total cost of the meal including the tip?

46. *Tipping.* Selena left a 15% tip of $8.40 for a meal.

a) What was the cost of the meal before the tip?
b) What was the total cost of the meal including the tip?

47. In a medical study of a group of pregnant women with "poor" diets, 16 of the women, or 8%, had babies who were in good or excellent health. How many women were in the original study?

48. In a medical study of a group of pregnant women with "good-to-excellent" diets, 285 of the women, or 95%, had babies who were in good or excellent health. How many women were in the original study?

49. *Body Fat.* The author of this text exercises regularly at a local YMCA that recently offered a body-fat percentage test to its members. The device used measures the passage of a very low voltage of electricity through the body. The author's body-fat percentage was found to be 16.5% and he weighs 191 lb. What part, in pounds, of his body weight is fat?

50. *Junk Mail.* The U.S. Postal Service reports that we open and read 78% of the junk mail that we receive. A sports instructional videotape company sends out 10,500 advertising brochures.

Source: U.S. Postal Service

a) How many of the brochures can it expect to be opened and read?
b) The company sells videos to 189 of the people who receive the brochure. What percent of the 10,500 people who receive the brochure buy the video?

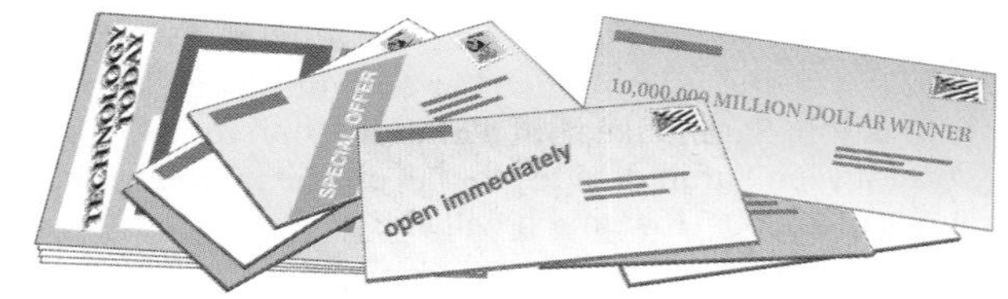

Life Insurance Rates for Smokers and Nonsmokers. The data in the following table illustrate how yearly rates (premiums) for a $500,000 term life insurance policy are increased for smokers. Complete the table by finding the missing numbers. Round to the nearest percent and dollar.

TYPICAL INSURANCE PREMIUMS (IN DOLLARS)

	AGE	RATE FOR NONSMOKER	RATE FOR SMOKER	RATE INCREASE	PERCENT OF INCREASE FOR SMOKER
	35	$255	$680	$425	167%
51.	40	$335	$990		
52.	45	$485			208%
53.	50	$735			198%
54.	55	$945	$3330		
55.	60	$1510	$5445		
56.	65	$2545			242%

Source: Faith Financial Planners, Inc.

57. $^{\mathrm{D}}_{\mathrm{W}}$ The 80/20 rule is commonly quoted in the field of business. It asserts that 80% of your results will come from 20% of your activities. Discuss how this might affect you as a student and as an employee.

58. $^{\mathrm{D}}_{\mathrm{W}}$ Comment on the following quote by Yogi Berra, a famous Major League Hall of Fame baseball player: "Ninety percent of hitting is mental. The other half is physical."

SKILL MAINTENANCE

Compute.

59. $9.076 \div 0.05$ [3.4a]

60. 9.076×0.05 [3.3a]

61. $1.089 + 10.89 + 0.1089$ [3.2a]

62. $1000.23 - 156.0893$ [3.2b]

Remove parentheses and simplify. [7.8b]

63. $-5a + 3c - 2(c - 3a)$

64. $4(x - 2y) - (y - 3x)$

Add. [7.3a]

65. $-6.5 + 2.6$

66. $-\frac{3}{8} + (-5) + \frac{1}{4} + (-1)$

Fill in the blank with a word that makes the statement true. [1.6c]

67. To simplify the calculation $18 - 24 \div 3 - 48 \div (-4)$, do all the ______________ calculations first, and then the ______________ calculations.

68. To simplify the calculation $18 - 24^3 \div 48 \div (-4)^2$, do all the ______________ calculations first, and then the ______________ calculations, and finally the ______________ calculation.

SYNTHESIS

69. It has been determined that at the age of 15, a boy has reached 96.1% of his final adult height. Jaraan is 6 ft 4 in. at the age of 15. What will his final adult height be?

70. It has been determined that at the age of 10, a girl has reached 84.4% of her final adult height. Dana is 4 ft 8 in. at the age of 10. What will her final adult height be?

8.6 APPLICATIONS AND PROBLEM SOLVING

Objective

a Solve applied problems by translating to equations.

a Five Steps for Solving Problems

We have discussed many new equation-solving tools in this chapter and used them for applications and problem solving. Here we consider a five-step strategy that can be very helpful in solving problems.

FIVE STEPS FOR PROBLEM SOLVING IN ALGEBRA

1. *Familiarize* yourself with the problem situation.
2. *Translate* the problem to an equation.
3. *Solve* the equation.
4. *Check* the answer in the original problem.
5. *State* the answer to the problem clearly.

Of the five steps, the most important is probably the first one: becoming familiar with the problem situation. The table below lists some hints for familiarization.

TO FAMILIARIZE YOURSELF WITH A PROBLEM

- If a problem is given in words, read it carefully. Reread the problem, perhaps aloud. Try to verbalize the problem as if you were explaining it to someone else.
- Choose a variable (or variables) to represent the unknown and clearly state what the variable represents. Be descriptive! For example, let $L =$ the length, $d =$ the distance, and so on.
- Make a drawing and label it with known information, using specific units if given. Also, indicate unknown information.
- Find further information. Look up formulas or definitions with which you are not familiar. (Geometric formulas appear on the inside back cover of this text.) Consult a reference librarian or the Internet.
- Create a table that lists all the information you have available. Look for patterns that may help in the translation to an equation.
- Think of a possible answer and check the guess. Note the manner in which the guess is checked.

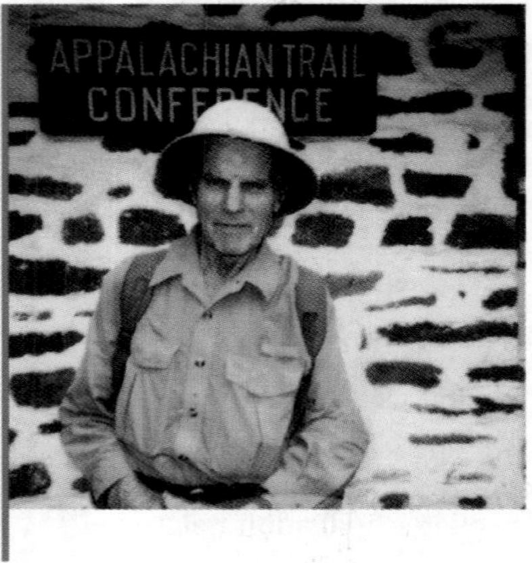

EXAMPLE 1 *Hiking.* In 1998, at age 79, Earl Shaffer became the oldest person to through-hike all 2100 miles of the Appalachian Trail—from Springer Mountain, Georgia, to Mount Katahdin, Maine. Shaffer through-hiked the trail three times, in 1948 (Georgia to Maine), in 1965 (Maine to Georgia), and in 1998 (Georgia to Maine) near the 50th anniversary of his first hike. At one point in 1998, Shaffer stood atop Big Walker Mountain, Virginia, which is three times as far from the northern end as from the southern end. How far was Shaffer from each end of the trail?

Source: Appalachian Trail Conference; Earl Shaffer Foundation

1. **Familiarize.** Let's consider a drawing.

To become familiar with the problem, let's guess a possible distance that Shaffer stood from Springer Mountain—say, 600 mi. Three times 600 mi is 1800 mi. Since 600 mi + 1800 mi = 2400 mi and 2400 mi is greater than 2100 mi, we see that our guess is too large. Rather than guess again, let's use the skills we have obtained in the ability to solve equations. We let

d = the distance, in miles, to the southern end, and

$3d$ = the distance, in miles, to the northern end.

(We could also let x = the distance to the northern end and $\frac{1}{3}x$ = the distance to the southern end.)

2. **Translate.** From the drawing, we see that the lengths of the two parts of the trail must add up to 2100 mi. This leads to our translation.

Distance to southern end	plus	Distance to northern end	is	2100 mi
↓	↓	↓	↓	↓
d	$+$	$3d$	$=$	2100

3. **Solve.** We solve the equation:

$$d + 3d = 2100$$
$$4d = 2100 \quad \text{Collecting like terms}$$
$$\frac{4d}{4} = \frac{2100}{4} \quad \text{Dividing by 4}$$
$$d = 525.$$

4. **Check.** As expected, d is less than 600 mi. If $d = 525$ mi, then $3d = 1575$ mi. Since 525 mi + 1575 mi = 2100 mi, we have a check.

5. **State.** Atop Big Walker Mountain, Shaffer stood 525 mi from Springer Mountain and 1575 mi from Mount Katahdin.

Do Exercise 1.

1. **Running.** In 1997, Yiannis Kouros of Australia set the record for the greatest distance run in 24 hr by running 188 mi. After 8 hr, he was approximately twice as far from the finish line as he was from the start. How far had he run?

Source: *Guinness World Records*, 2004

Answer on page A-22

EXAMPLE 2 *Rocket Sections.* A rocket is divided into three sections: the payload and navigation section in the top, the fuel section in the middle, and the rocket engine section in the bottom. The top section is one-sixth the length of the bottom section. The middle section is one-half the length of the bottom section. The total length of the rocket is 240 ft. Find the length of each section.

1. **Familiarize.** We first make a drawing.

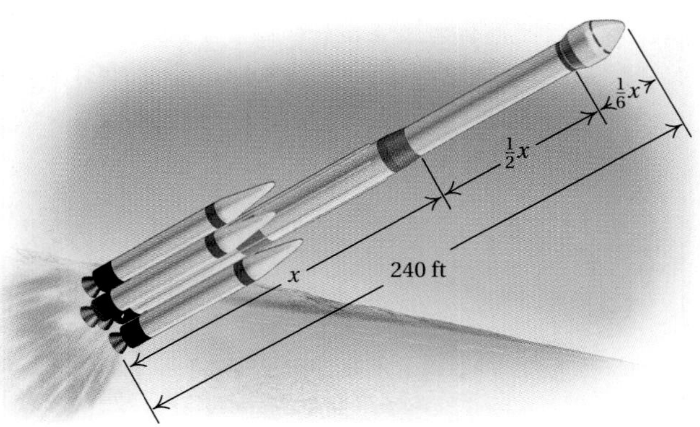

Because the lengths of the top and the middle sections are expressed in terms of the length of the bottom section, we let

$$x = \text{the length of the bottom section.}$$

Then $\frac{1}{6}x = $ the length of the top section

and $\frac{1}{2}x = $ the length of the middle section.

2. **Translate.** From the statement of the problem and the drawing, we see that the lengths add up to 240 ft. That gives us our translation:

Length of bottom section	plus	Length of middle section	plus	Length of top section	is	Total length
↓	↓	↓	↓	↓	↓	↓
x	$+$	$\frac{1}{2}x$	$+$	$\frac{1}{6}x$	$=$	$240.$

3. **Solve.** We begin by clearing fractions and then solving as follows:

$$x + \frac{1}{2}x + \frac{1}{6}x = 240 \quad \text{The LCM of all the denominators is 6.}$$

$$6\left(x + \frac{1}{2}x + \frac{1}{6}x\right) = 6 \cdot 240 \quad \text{Multiplying by the LCM, 6}$$

$$6 \cdot x + 6 \cdot \frac{1}{2}x + 6 \cdot \frac{1}{6}x = 6 \cdot 240 \quad \text{Using the distributive law}$$

$$6x + 3x + x = 1440 \quad \text{Simplifying}$$

$$10x = 1440 \quad \text{Collecting like terms}$$

$$\frac{10x}{10} = \frac{1440}{10} \quad \text{Dividing by 10}$$

$$x = 144.$$

4. **Check.** Do we have an answer to the *problem*? If the length of the bottom section is 144 ft, then the length of the middle section is $\frac{1}{2} \cdot 144$ ft, or 72 ft, and the length of the top section is $\frac{1}{6} \cdot 144$ ft, or 24 ft. These lengths add up to 240 ft. Our answer checks.
5. **State.** The length of the bottom section is 144 ft, the length of the middle section is 72 ft, and the length of the top section is 24 ft. (Note the importance of including the unit, feet, in the answer.)

Do Exercise 2.

Recall that the set of integers $= \{\ldots, -5, -4, -3, -2, -1, 0, 1, 2, 3, 4, 5, \ldots\}$. Before we solve the next problem, we need to learn some additional terminology regarding integers.

The following are examples of **consecutive integers:** 16, 17, 18, 19, 20; and $-31, -30, -29, -28$. Note that consecutive integers can be represented in the form x, $x + 1$, $x + 2$, and so on.

The following are examples of **consecutive even integers:** 16, 18, 20, 22, 24; and $-52, -50, -48, -46$. Note that consecutive even integers can be represented in the form x, $x + 2$, $x + 4$, and so on.

The following are examples of **consecutive odd integers:** 21, 23, 25, 27, 29; and $-71, -69, -67, -65$. Note that consecutive odd integers can be represented in the form x, $x + 2$, $x + 4$, and so on.

EXAMPLE 3 *Interstate Mile Markers.* U.S. interstate highways post numbered markers every mile to indicate location in case of an accident or breakdown. In many states, the numbers on the markers increase from west to east. The sum of two consecutive mile markers on I-70 in Kansas is 559. Find the numbers on the markers.

Source: Federal Highway Administration, Ed Rotalewski

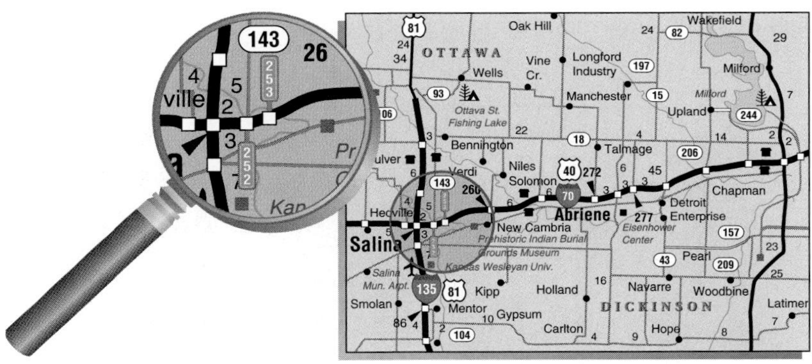

1. **Familiarize.** The numbers on the mile markers are consecutive positive integers. Thus if we let $x =$ the smaller number, then $x + 1 =$ the larger number.

 To become familiar with the problem, we can make a table. First, we guess a value for x; then we find $x + 1$. Finally, we add the two numbers and check the sum.

x	$x + 1$	Sum of x and $x + 1$
114	115	229
252	253	505
302	303	605

2. **Gourmet Sandwiches.** A gourmet sandwich shop located near a college campus specializes in sandwiches prepared in buns of length 18 in. Suppose Jenny, Emma, and Sarah buy one of these sandwiches and take it back to their apartment. Since they have different appetites, Jenny cuts the sandwich in such a way that Emma gets half of what Jenny gets and Sarah gets three-fourths of what Jenny gets. Find the length of each person's sandwich.

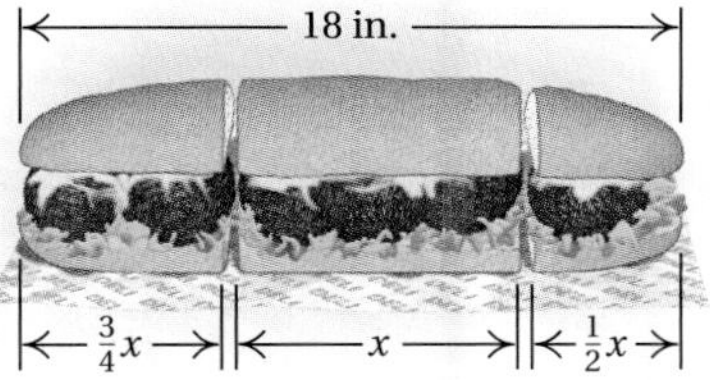

Answer on page A-22

3. Interstate Mile Markers. The sum of two consecutive mile markers on I-90 in upstate New York is 627. (On I-90 in New York, the marker numbers increase from east to west.) Find the numbers on the markers.

Source: New York State Department of Transportation

From the table, we see that the first marker will be between 252 and 302. We could continue guessing and solve the problem this way, but let's work on developing our algebra skills.

2. **Translate.** We reword the problem and translate as follows.

First integer	plus	Second integer	is	559	Rewording
↓	↓	↓	↓	↓	
x	$+$	$(x + 1)$	$=$	559	Translating

3. **Solve.** We solve the equation:

$$x + (x + 1) = 559$$
$$2x + 1 = 559 \quad \text{Collecting like terms}$$
$$2x + 1 - 1 = 559 - 1 \quad \text{Subtracting 1}$$
$$2x = 558$$
$$\frac{2x}{2} = \frac{558}{2} \quad \text{Dividing by 2}$$
$$x = 279.$$

If x is 279, then $x + 1$ is 280.

4. **Check.** Our possible answers are 279 and 280. These are consecutive positive integers and $279 + 280 = 559$, so the answers check.
5. **State.** The mile markers are 279 and 280.

Do Exercise 3.

EXAMPLE 4 *Copy Machine Rental.* It costs \$225 per month plus 1.2¢ per copy to rent a copy machine. A law firm needs to lease a copy machine for use during a special case that they anticipate will take 3 months. If they allot a budget of \$1100, how many copies can they make?

1. **Familiarize.** Suppose that the law firm makes 20,000 copies. Then the cost is monthly charges plus copy charges, or

3(\$225)	plus	Cost per copy	times	Number of copies
↓	↓	↓	↓	↓
\$675	$+$	\$0.012	$\cdot$	20,000,

Answer on page A-22

which is \$915. This process familiarizes us with the way in which a calculation is made. Note that we convert 1.2¢ to \$0.012 so that all information is in the same unit, dollars. Otherwise, we will not get the correct answer.

We let $c =$ the number of copies that can be made for the budget of \$1100.

2. **Translate.** We reword the problem and translate as follows.

$$\underbrace{\text{Monthly cost}}_{3(\$225)} \;\; \underbrace{\text{plus}}_{+} \;\; \underbrace{\text{Cost per copy}}_{\$0.012} \;\; \underbrace{\text{times}}_{\cdot} \;\; \underbrace{\text{Number of copies}}_{c} \;\; \underbrace{\text{is}}_{=} \;\; \underbrace{\text{Budget}}_{\$1100}$$

3. **Solve.** We solve the equation:

$$3(225) + 0.012c = 1100$$
$$675 + 0.012c = 1100$$
$$0.012c = 425 \qquad \text{Subtracting 675}$$
$$\frac{0.012c}{0.012} = \frac{425}{0.012} \qquad \text{Dividing by 0.012}$$
$$c \approx 35{,}417. \qquad \text{Rounding to the nearest one}$$

4. **Check.** We check in the original problem. The cost for 35,417 pages is 35,417(\$0.012) = \$425.004. The rental for 3 months is 3(\$225) = \$675. The total cost is then \$425.004 + \$675 ≈ \$1100, which is the \$1100 that was allotted.

5. **State.** The law firm can make 35,417 copies on the copy rental allotment of \$1100.

Do Exercise 4.

4. Copy Machine Rental. The law firm in Example 4 decides to increase its budget to \$1400 for the 3-month period. How many copies can they make for \$1400?

EXAMPLE 5 *Perimeter of NBA Court.* The perimeter of an NBA basketball court is 288 ft. The length is 44 ft longer than the width. Find the dimensions of the court.

Source: National Basketball Association

1. **Familiarize.** We first make a drawing.

We let $w =$ the width of the rectangle. Then $w + 44 =$ the length. The perimeter P of a rectangle is the distance around the rectangle and is given by the formula $2l + 2w = P$, where

$l =$ the length and $w =$ the width.

Answer on page A-22

5. Perimeter of High School Basketball Court. The perimeter of a standard high school basketball court is 268 ft. The length is 34 ft longer than the width. Find the dimensions of the court.

Source: Indiana High School Athletic Association

Answer on page A-22

2. Translate. To translate the problem, we substitute $w + 44$ for l and 288 for P:

$$2l + 2w = P$$
$$2(w + 44) + 2w = 288.$$

Caution! Parentheses are important here.

3. Solve. We solve the equation:

$$2(w + 44) + 2w = 288$$
$$2 \cdot w + 2 \cdot 44 + 2w = 288 \quad \text{Using the distributive law}$$
$$4w + 88 = 288 \quad \text{Collecting like terms}$$
$$4w + 88 - 88 = 288 - 88 \quad \text{Subtracting 88}$$
$$4w = 200$$
$$\frac{4w}{4} = \frac{200}{4} \quad \text{Dividing by 4}$$
$$w = 50.$$

Thus possible dimensions are

$w = 50$ ft and $l = w + 44 = 50 + 44$, or 94 ft.

4. Check. If the width is 50 ft and the length is 94 ft, then the perimeter is $2(50 \text{ ft}) + 2(94 \text{ ft})$, or 288 ft. This checks.

5. State. The width is 50 ft and the length is 94 ft.

Do Exercise 5.

EXAMPLE 6 *Roof Gable.* In a triangular gable end of a roof, the angle of the peak is twice as large as the angle of the back side of the house. The measure of the angle on the front side is 20° greater than the angle on the back side. How large are the angles?

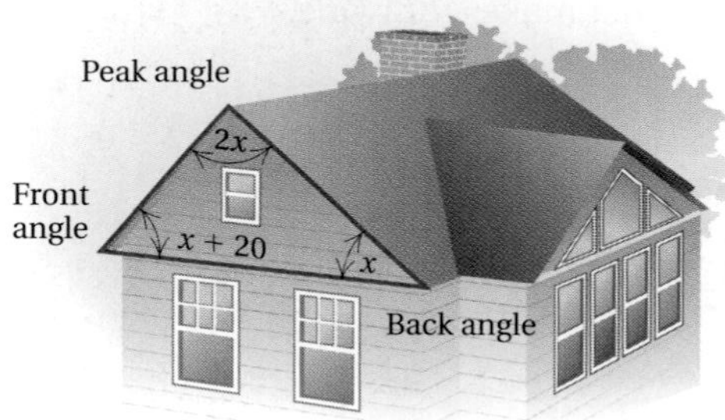

1. Familiarize. We first make a drawing as shown above. We let

measure of back angle $= x$.

Then measure of peak angle $= 2x$

and measure of front angle $= x + 20$.

2. **Translate.** To translate, we need to recall a geometric fact. (You might, as part of step 1, look it up in a geometry book or in the list of formulas on the inside back cover.) Remember, the measures of the angles of a triangle total 180°.

Measure of back angle	plus	Measure of peak angle	plus	Measure of front angle	is	180°
↓	↓	↓	↓	↓	↓	↓
x	$+$	$2x$	$+$	$(x + 20)$	$=$	$180°$

3. **Solve.** We solve the equation:

$$\begin{aligned} x + 2x + (x + 20) &= 180 \\ 4x + 20 &= 180 \\ 4x + 20 - 20 &= 180 - 20 \\ 4x &= 160 \\ \frac{4x}{4} &= \frac{160}{4} \\ x &= 40. \end{aligned}$$

Possible measures for the angles are as follows:

Back angle: $x = 40°$;
Peak angle: $2x = 2(40) = 80°$;
Front angle: $x + 20 = 40 + 20 = 60°$.

4. **Check.** Consider our answers: 40°, 80°, and 60°. The peak is twice the back and the front is 20° greater than the back. The sum is 180°. The angles check.

5. **State.** The measures of the angles are 40°, 80°, and 60°.

Caution!

Units are important in answers. Remember to include them, where appropriate.

Do Exercise 6.

Caution!

Always be sure to answer the original problem completely. For instance, in Example 1, we need to find *two* numbers: the distances from *each* end of the trail to the hiker. Similarly, in Example 3, we need to find two mile markers, and in Example 5, we need to find two dimensions, not just the width.

EXAMPLE 7 *Top Speeds of Roller Coasters.* The average top speed of the three fastest roller coasters in the world is 109 mph. The third fastest roller coaster, Superman the Escape (at Six Flags Magic Mountain, Los Angeles, CA) reaches a top speed of 20 mph less than the fastest roller coaster, Top Thrill Dragster (in Cedar Point, Sandusky, OH). The second fastest roller coaster, Dodonpa (in Fujikyu Highlands, Japan) has a top speed of 107 mph. What is the top speed of the fastest roller coaster?

Source: *Fortune Small Business*, June 2004, p. 48

6. The second angle of a triangle is three times as large as the first. The third angle measures 30° more than the first angle. Find the measures of the angles.

Answer on page A-22

7. Average Test Score. Sam's average score on his first three math tests is 77. He scored 62 on the first test. On the third test, he scored 9 more than he scored on his second test. What did he score on the second and third tests?

1. **Familiarize.** The **average** of a set of numbers is the sum of the numbers divided by the number of addends. (For more on average, see Section 5.1.)

   We are given that the second fastest speed is 107 mph. Suppose the three top speeds are 90, 107, and 112. The average is then

   $$\frac{90 + 107 + 112}{3} = \frac{309}{3} = 103,$$

   which is too low. Instead of continuing to guess, let's use the equation-solving skills we have learned in this chapter. We let x represent the top speed of the fastest roller coaster. Then $x - 20$ is the top speed of the third fastest roller coaster.

2. **Translate.** We reword the problem and translate as follows:

   $$\frac{\text{Speed of fastest coaster} + \text{Speed of second fastest coaster} + \text{Speed of third fastest coaster}}{\text{Number of roller coasters}} = \text{Average speed of three fastest roller coasters}$$

   $$\frac{x + 107 + (x - 20)}{3} = 109.$$

3. **Solve.** We solve as follows:

   $$\frac{x + 107 + (x - 20)}{3} = 109$$

   $$3 \cdot \frac{x + 107 + (x - 20)}{3} = 3 \cdot 109$$ Multiplying by 3 on both sides to clear the fraction

   $$x + 107 + (x - 20) = 327$$

   $$2x + 87 = 327$$ Collecting like terms

   $$2x = 240$$ Subtracting 87

   $$x = 120.$$ Dividing by 2

4. **Check.** If the top speed of the fastest roller coaster is 120 mph, then the top speed of the third fastest is $120 - 20$, or 100 mph. The average of the top speeds of the three fastest is $(120 + 107 + 100) \div 3 = 327 \div 3$, or 109 mph. The answer checks.

5. **State.** The speed of the fastest roller coaster in the world is 120 mph.

Do Exercise 7.

Answer on page A-22

EXAMPLE 8 *Simple Interest.* An investment is made at 6% simple interest for 1 year. It grows to \$768.50. How much was originally invested (the principal)?

1. **Familiarize.** Suppose that \$100 was invested. Recalling the formula for simple interest, $I = Prt$, we know that the interest for 1 year on \$100 at 6% simple interest is given by $I = \$100 \cdot 0.06 \cdot 1 = \$6$. Then, at the end of the year, the amount in the account is found by adding the principal and the interest:

$$\begin{array}{ccccc} \text{Principal} & + & \text{Interest} & = & \text{Amount} \\ \downarrow & & \downarrow & & \downarrow \\ \$100 & + & \$6 & = & \$106. \end{array}$$

In this problem, we are working backward. We are trying to find the principal, which is the original investment. We let $x =$ the principal. Then the interest earned is $6\%x$.

2. **Translate.** We reword the problem and then translate.

$$\begin{array}{ccccc} \text{Principal} & + & \text{Interest} & = & \text{Amount} \\ \downarrow & & \downarrow & & \downarrow \\ x & + & 6\%x & = & 768.50 \end{array}$$

Interest is 6% of the principal.

3. **Solve.** We solve the equation:

$$x + 6\%x = 768.50$$

$$x + 0.06x = 768.50 \quad \text{Converting to decimal notation}$$

$$1x + 0.06x = 768.50 \quad \text{Identity property of 1}$$

$$1.06x = 768.50 \quad \text{Collecting like terms}$$

$$\frac{1.06x}{1.06} = \frac{768.50}{1.06} \quad \text{Dividing by 1.06}$$

$$x = 725.$$

4. **Check.** We check by taking 6% of \$725 and adding it to \$725:

$$6\% \times \$725 = 0.06 \times 725 = \$43.50.$$

Then $\$725 + \$43.50 = \$768.50$, so \$725 checks.

5. **State.** The original investment was \$725.

Do Exercise 8.

8. Simple Interest. An investment is made at 7% simple interest for 1 year. It grows to \$8988. How much was originally invested (the principal)?

EXAMPLE 9 *Selling a Home.* The Landers are planning to sell their home. If they want to be left with \$117,500 after paying 6% of the selling price to a realtor as a commission, for how much must they sell the house?

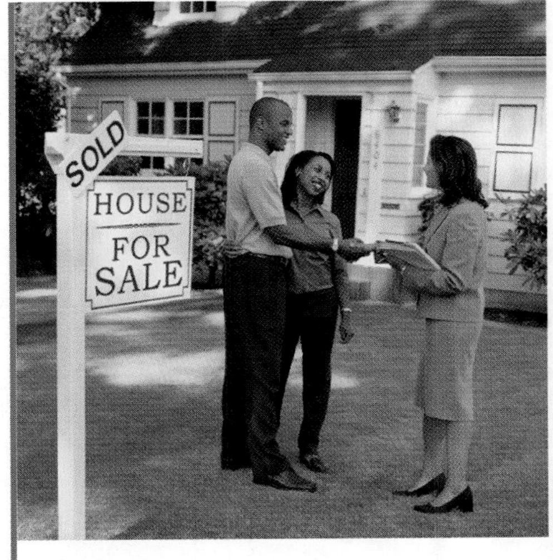

1. **Familiarize.** Suppose the Landers sell the house for \$120,000. A 6% commission can be determined by finding 6% of \$120,000:

$$6\% \text{ of } \$120{,}000 = 0.06(\$120{,}000) = \$7200.$$

Subtracting this commission from \$120,000 would leave the Landers with

$$\$120{,}000 - \$7200 = \$112{,}800.$$

This shows that in order for the Landers to clear \$117,500, the house must sell for more than \$120,000. To determine what the sale price must be, we could check more guesses. Instead, we let $x =$ the selling price, in dollars. With a 6% commission, the realtor would receive $0.06x$.

Answer on page A-22

9. Price Before Sale. The price of a suit was decreased to a sale price of \$526.40. This was a 20% reduction. What was the former price?

Answer on page A-22

2. Translate. We reword the problem and translate as follows.

Selling price	less	Commission	is	Amount remaining.
x	$-$	$0.06x$	$=$	$117{,}500$

3. Solve. We solve the equation:

$$x - 0.06x = 117{,}500$$
$$1x - 0.06x = 117{,}500$$
$$0.94x = 117{,}500 \quad \text{Collecting like terms. Had we noted that after the commission has been paid, 94\% remains, we could have begun with this equation.}$$
$$\frac{094x}{0.94} = \frac{117{,}500}{0.94} \quad \text{Dividing by 0.94}$$
$$x = 125{,}000.$$

4. Check. To check, we first find 6% of \$125,000:

6% of \$125,000 = 0.06(\$125,000) = \$7500. This is the commission.

Next, we subtract the commission to find the remaining amount:

\$125,000 − \$7500 = \$117,500.

Since, after the commission, the Landers are left with \$117,500, our answer checks. Note that the \$125,000 selling price is greater than \$120,000, as predicted in the *Familiarize* step.

5. State. To be left with \$117,500, the Landers must sell the house for \$125,000.

Do Exercise 9.

Caution!

The problem in Example 9 is easy to solve with algebra. Without algebra, it is not. A common error in such a problem is to take 6% of the price after commission and then subtract or add. Note that 6% of the selling price (6% · \$125,000 = \$7500) is not equal to 6% of the amount that the Landers want to be left with (6% · \$117,500 = \$7050).

Study Tips

PROBLEM-SOLVING TIPS

The more problems you solve, the more your skills will improve.

1. Look for patterns when solving problems. Each time you study an example in a text, you may observe a pattern for problems that you will encounter later in the exercise sets or in other practical situations.
2. When translating in mathematics, consider the dimensions of the variables and constants in the equation. The variables that represent length should all be in the same unit, those that represent money should all be in dollars or all in cents, and so on.
3. Make sure that units appear in the answer whenever appropriate and that you have completely answered the original problem.

Translating for Success

The goal of these matching questions is to practice step (2), *Translate,* of the five-step problem-solving process. Translate each word problem to an equation and select a correct translation from equations A–O.

1. *Angle Measures.* The measure of the second angle of a triangle is 51° more than that of the first angle. The measure of the third angle is 3° less than twice the first angle. Find the measures of the angles.

2. *Sales Tax.* Tina paid $3976 for a used car. This amount included 5% for sales tax. How much did the car cost before tax?

3. *Perimeter.* The perimeter of a rectangle is 2347 ft. The length is 28 ft greater than the width. Find the length and the width.

4. *Fraternity or Sorority Membership.* At Arches Tech University, 3976 students belong to a fraternity or a sorority. This is 35% of the total enrollment. What is the total enrollment at Arches Tech?

5. *Fraternity or Sorority Membership.* At Moab Tech University, thirty-five percent of the students belong to a fraternity or a sorority. The total enrollment of the university is 11,360 students. How many students belong to either a fraternity or a sorority?

6. *Island Population.* There are 180 thousand people living on a small Caribbean island. The women outnumber the men by 96 thousand. How many men live on the island?

7. *Wire Cutting.* A 384-m wire is cut into three pieces. The second piece is 3 m longer than the first. The third is four-fifths as long as the first. How long is each piece?

8. *Locker Numbers.* The numbers on three adjoining lockers are consecutive integers whose sum is 384. Find the integers.

9. *Fraternity or Sorority Membership.* The total enrollment at Canyonlands Tech University is 11,360 students. Of these, 3976 students belong to a fraternity or a sorority. What percent of the students belong to a fraternity or sorority?

10. *Width of a Rectangle.* The length of a rectangle is 96 ft. The perimeter of the rectangle is 3976 ft. Find the width.

A. $x + (x - 3) + \frac{4}{5}x = 384$

B. $x + (x + 51) + (2x - 3) = 180$

C. $x + (x + 96) = 180$

D. $2 \cdot 96 + 2x = 3976$

E. $x + (x + 1) + (x + 2) = 384$

F. $3976 = x \cdot 11{,}360$

G. $2x + 2(x + 28) = 2347$

H. $3976 = x + 5\%x$

I. $x + (x + 28) = 2347$

J. $x = 35\% \cdot 11{,}360$

K. $x + 96 = 3976$

L. $x + (x + 3) + \frac{4}{5}x = 384$

M. $x + (x + 2) + (x + 4) = 384$

N. $35\% \cdot x = 3976$

O. $x + (x + 28) = 2347$

Answers on page A-22

8.6 EXERCISE SET

For Extra Help

MathXL | MyMathLab | InterAct Math | Math Tutor Center | Video Lectures on CD Disc 4 | Student's Solutions Manual

a Solve. *Even though you might find the answer quickly in some other way, practice using the five-step problem-solving strategy.*

1. *Pipe Cutting.* A 240-in. pipe is cut into two pieces. One piece is three times the length of the other. Find the lengths of the pieces.

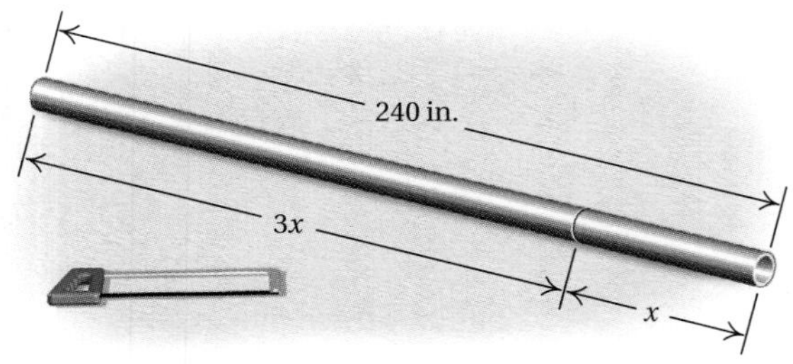

2. *Board Cutting.* A 72-in. board is cut into two pieces. One piece is 2 in. longer than the other. Find the lengths of the pieces.

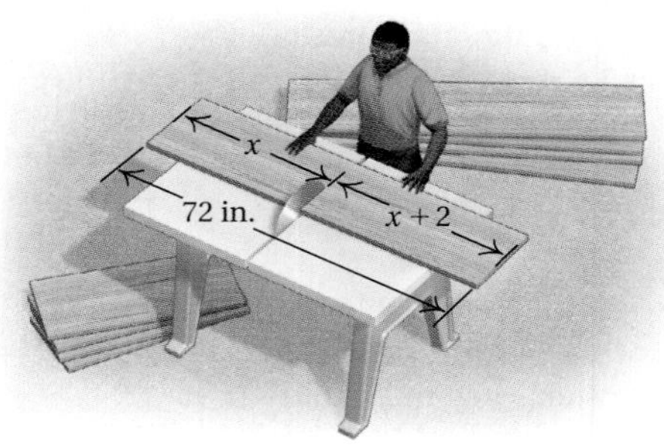

3. *Cinnamon Life.* Recently, the cost of four 21-oz boxes of Cinnamon Life cereal was $17.16. What was the cost of one box?

4. *Area of Lake Ontario.* The area of Lake Superior is about four times the area of Lake Ontario. The area of Lake Superior is 30,172 mi^2. What is the area of Lake Ontario?

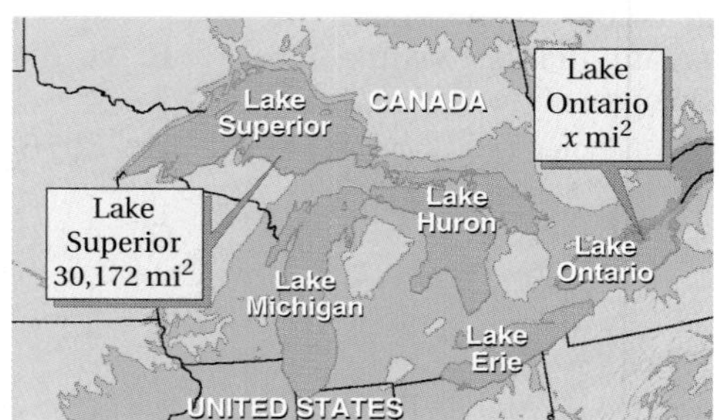

5. *Women's Dresses.* In a recent year, the total amount spent on women's blouses was $6.5 billion. This was $0.2 billion more than what was spent on women's dresses. How much was spent on women's dresses?

6. *Statue of Liberty.* The height of the Eiffel Tower is 974 ft, which is about 669 ft higher than the Statue of Liberty. What is the height of the Statue of Liberty?

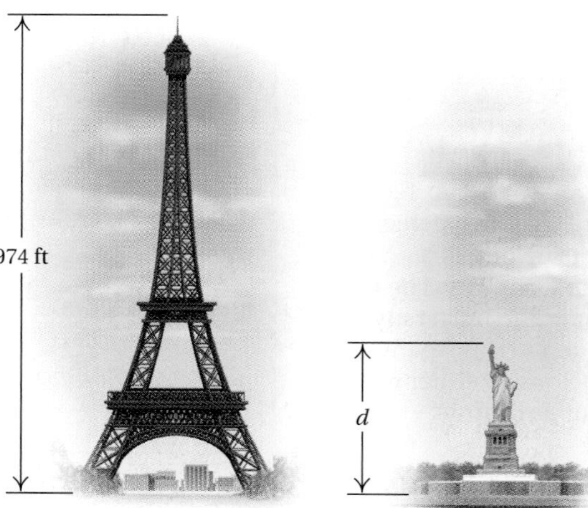

7. *Iditarod Race.* The Iditarod sled dog race extends for 1049 mi from Anchorage to Nome. If a musher is twice as far from Anchorage as from Nome, how many miles of the race has the musher completed?

Source: Iditarod Trail Commission

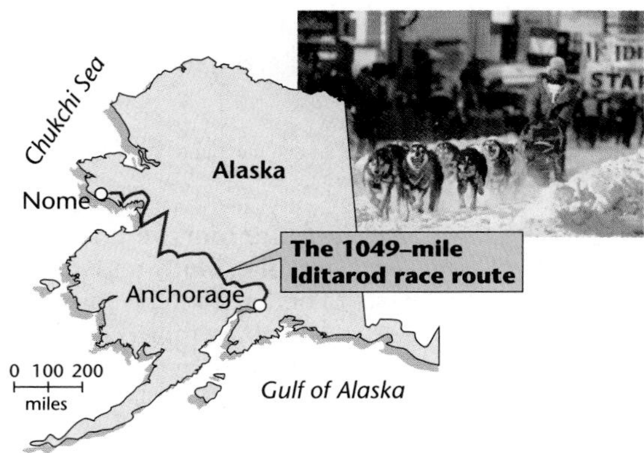

8. *Home Remodeling.* In a recent year, Americans spent a total of $35 billion to remodel bathrooms and kitchens. Twice as much was spent on kitchens as bathrooms. How much was spent on each?

9. *Consecutive Apartment Numbers.* The apartments in Vincent's apartment house are numbered consecutively on each floor. The sum of his number and his next-door neighbor's number is 2409. What are the two numbers?

10. *Consecutive Post Office Box Numbers.* The sum of the numbers on two consecutive post office boxes is 547. What are the numbers?

11. *Consecutive Ticket Numbers.* The numbers on Sam's three raffle tickets are consecutive integers. The sum of the numbers is 126. What are the numbers?

12. *Consecutive Ages.* The ages of Whitney, Wesley, and Wanda are consecutive integers. The sum of their ages is 108. What are their ages?

13. *Consecutive Odd Integers.* The sum of three consecutive odd integers is 189. What are the integers?

14. *Consecutive Integers.* Three consecutive integers are such that the first plus one-half the second plus seven less than twice the third is 2101. What are the integers?

15. *Standard Billboard Sign.* A standard rectangular highway billboard sign has a perimeter of 124 ft. The length is 6 ft more than three times the width. Find the dimensions of the sign.

16. *Two-by-Four.* The perimeter of a cross section or end of a "two-by-four" piece of lumber is 10 in. The length is 2 in. more than the width. Find the actual dimensions of the cross section of a two-by-four.

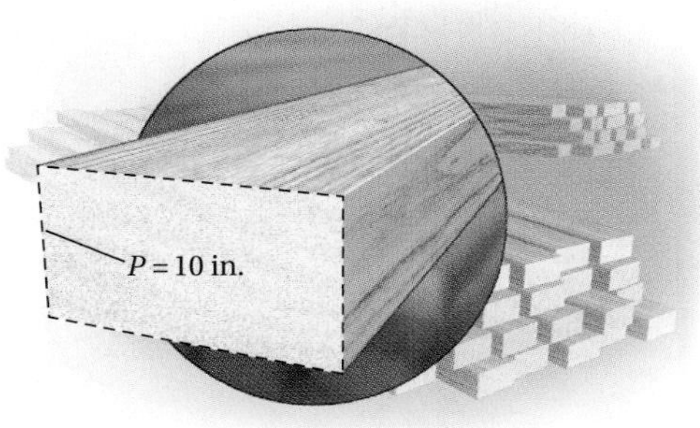

17. *Price of Sneakers.* Amy paid \$63.75 for a pair of New Balance 903 running shoes during a 15%-off sale. What was the regular price?

18. *Price of a CD Player.* Doug paid \$72 for a shockproof portable CD player during a 20%-off sale. What was the regular price?

19. *Price of a Textbook.* Evelyn paid \$89.25, including 5% tax, for her biology textbook. How much did the book itself cost?

20. *Price of a Printer.* Jake paid \$100.70, including 6% tax, for a color printer. How much did the printer itself cost?

21. *Parking Costs.* A hospital parking lot charges \$1.50 for the first hour or part thereof, and \$1.00 for each additional hour or part thereof. A weekly pass costs \$27.00 and allows unlimited parking for 7 days. Suppose that each visit Ed makes to the hospital lasts $1\frac{1}{2}$ hr. What is the minimum number of times that Ed would have to visit per week to make it worthwhile for him to buy the pass?

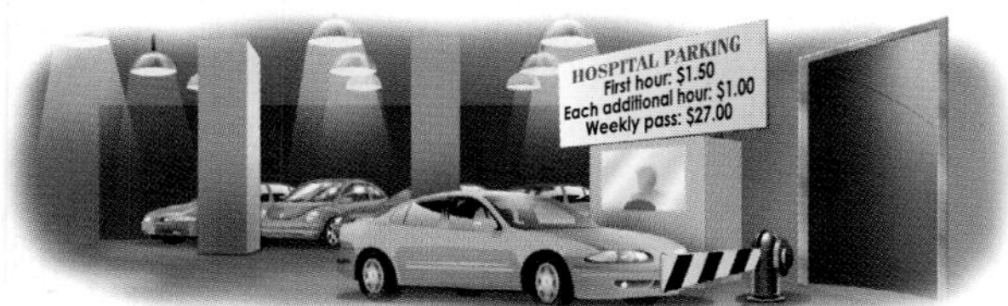

22. *Van Rental.* Value Rent-A-Car rents vans at a daily rate of \$84.95 plus 60 cents per mile. Molly rents a van to deliver electrical parts to her customers. She is allotted a daily budget of \$320. How many miles can she drive for \$320? (*Hint*: 60¢ = \$0.60.)

23. *Triangular Field.* The second angle of a triangular field is three times as large as the first angle. The third angle is 40° greater than the first angle. How large are the angles?

24. *Triangular Parking Lot.* The second angle of a triangular parking lot is four times as large as the first angle. The third angle is 45° less than the sum of the other two angles. How large are the angles?

25. *Triangular Backyard.* A home has a triangular backyard. The second angle of the triangle is 5° more than the first angle. The third angle is 10° more than three times the first angle. Find the angles of the triangular yard.

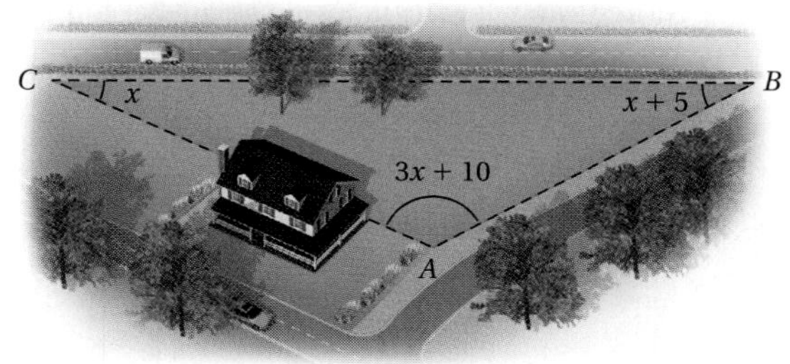

26. *Boarding Stable.* A rancher needs to form a triangular horse pen using ropes next to a stable. The second angle is three times the first angle. The third angle is 15° less than the first angle. Find the angles of the triangular pen.

27. *Stock Prices.* Sarah's investment in AOL/Time Warner stock grew 28% to \$448. How much did she invest?

28. *Savings Interest.* Sharon invested money in a savings account at a rate of 6% simple interest. After 1 yr, she has \$6996 in the account. How much did Sharon originally invest?

29. *Credit Cards.* The balance in Will's Mastercard® account grew 2%, to \$870, in one month. What was his balance at the beginning of the month?

30. *Loan Interest.* Alvin borrowed money from a cousin at a rate of 10% simple interest. After 1 yr, \$7194 paid off the loan. How much did Alvin borrow?

31. *Taxi Fares.* In Beniford, taxis charge \$3 plus 75¢ per mile for an airport pickup. How far from the airport can Courtney travel for \$12?

32. *Taxi Fares.* In Cranston, taxis charge \$4 plus 90¢ per mile for an airport pickup. How far from the airport can Ralph travel for \$17.50?

33. *Tipping.* Leon left a 15% tip for a meal. The total cost of the meal, including the tip, was \$41.40. What was the cost of the meal before the tip was added?

34. *Tipping.* Selena left an 18% tip for a meal. The total cost of the meal, including the tip, was \$40.71. What was the cost of the meal before the tip was added?

35. *Average Price.* Tom paid an average of \$34 per tie for a recent purchase of three ties. The price of one tie was twice as much as another, and the remaining tie cost \$27. What were the prices of the other two ties?

36. *Average Test Score.* Jaci averaged 84 on her first three history exams. The first score was 67. The second score was 7 less than the third score. What did she score on the second and third exams?

37. $\mathbf{D_W}$ Write a problem for a classmate to solve so that it can be translated to the equation

$$\frac{2}{3}x + (x + 5) + x = 375.$$

38. $\mathbf{D_W}$ Erin returns a tent that she bought during a storewide 35%-off sale that has ended. She is offered store credit for 125% of what she paid (not to be used on sale items). Is this fair to Erin? Why or why not?

SKILL MAINTENANCE

Calculate.

39. $-\frac{4}{5} - \frac{3}{8}$ [7.4a]

40. $-\frac{4}{5} + \frac{3}{8}$ [7.3a]

41. $-\frac{4}{5} \cdot \frac{3}{8}$ [7.5a]

42. $-\frac{4}{5} \div \frac{3}{8}$ [7.6c]

43. $\frac{1}{10} \div \left(-\frac{1}{100}\right)$ [7.6c]

44. $-25.6 \div (-16)$ [7.6c]

45. $-25.6(-16)$ [7.5a]

46. $-25.6 - (-16)$ [7.4a]

47. $-25.6 + (-16)$ [7.3a]

48. $(-0.02) \div (-0.2)$ [7.6c]

SYNTHESIS

49. Apples are collected in a basket for six people. One-third, one-fourth, one-eighth, and one-fifth are given to four people, respectively. The fifth person gets ten apples with one apple remaining for the sixth person. Find the original number of apples in the basket.

50. *Test Questions.* A student scored 78 on a test that had 4 seven-point fill-ins and 24 three-point multiple-choice questions. The student had one fill-in wrong. How many multiple-choice questions did the student answer correctly?

51. The area of this triangle is 2.9047 in^2. Find x.

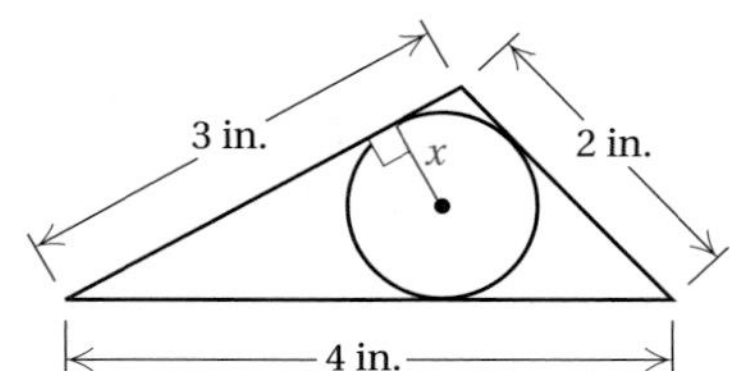

52. A storekeeper goes to the bank to get $10 worth of change. She requests twice as many quarters as half dollars, twice as many dimes as quarters, three times as many nickels as dimes, and no pennies or dollars. How many of each coin did the storekeeper get?

53. In one city, a sales tax of 9% was added to the price of gasoline as registered on the pump. Suppose a driver asked for $10 worth of gas. The attendant filled the tank until the pump read $9.10 and charged the driver $10. Something was wrong. Use algebra to correct the error.

8.7 SOLVING INEQUALITIES

Objectives

a Determine whether a given number is a solution of an inequality.

b Graph an inequality on a number line.

c Solve inequalities using the addition principle.

d Solve inequalities using the multiplication principle.

e Solve inequalities using the addition and multiplication principles together.

We now extend our equation-solving principles to the solving of inequalities.

a Solutions of Inequalities

In Section 7.2, we defined the symbols $>$ (is greater than), $<$ (is less than), $\geq$ (is greater than or equal to), and $\leq$ (is less than or equal to). For example, $3 \leq 4$ and $3 \leq 3$ are both true, but $-3 \leq -4$ and $0 \geq 2$ are both false.

An **inequality** is a number sentence with $>$, $<$, $\geq$, or $\leq$ as its verb—for example,

$$-4 > t, \quad x < 3, \quad 2x + 5 \geq 0, \quad \text{and} \quad -3y + 7 \leq -8.$$

Some replacements for a variable in an inequality make it true and some make it false.

SOLUTION

A replacement that makes an inequality true is called a **solution.** The set of all solutions is called the **solution set.** When we have found the set of all solutions of an inequality, we say that we have **solved** the inequality.

EXAMPLES Determine whether the number is a solution of $x < 2$.

1. -2.7 Since $-2.7 < 2$ is true, -2.7 is a solution.
2. 2 Since $2 < 2$ is false, 2 is not a solution.

EXAMPLES Determine whether the number is a solution of $y \geq 6$.

3. 6 Since $6 \geq 6$ is true, 6 is a solution.
4. $-\frac{4}{3}$ Since $-\frac{4}{3} \geq 6$ is false, $-\frac{4}{3}$ is not a solution.

Do Exercises 1 and 2.

Determine whether each number is a solution of the inequality.

1. $x > 3$

a) 2 b) 0

c) -5 d) 15.4

e) 3 f) $-\frac{2}{5}$

2. $x \leq 6$

a) 6 b) 0

c) -4.3 d) 25

e) -6 f) $\frac{5}{8}$

Answers on page A-22

b Graphs of Inequalities

Some solutions of $x < 2$ are $-3, 0, 1, 0.45, -8.9, -\pi, \frac{5}{8}$, and so on. In fact, there are infinitely many real numbers that are solutions. Because we cannot list them all individually, it is helpful to make a drawing that represents all the solutions.

A **graph** of an inequality is a drawing that represents its solutions. An inequality in one variable can be graphed on a number line. An inequality in two variables can be graphed on a coordinate plane; we will study such graphs in Chapter 9.

Graph.

3. $x \leq 4$

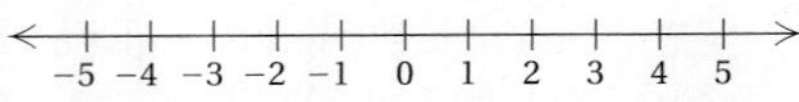

4. $x > -2$

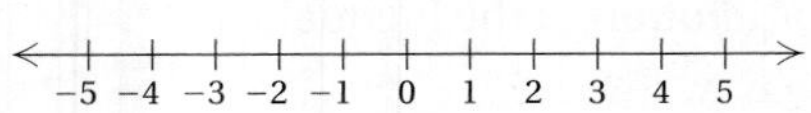

5. $-2 < x \leq 4$

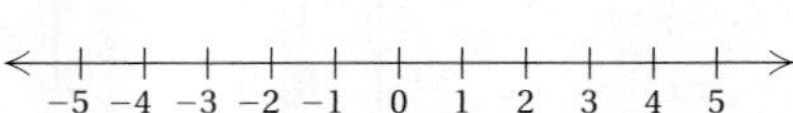

Answers on page A-22

EXAMPLE 5 Graph: $x < 2$.

The solutions of $x < 2$ are all those numbers less than 2. They are shown on the graph by shading all points to the left of 2. The open circle at 2 indicates that 2 is *not* part of the graph.

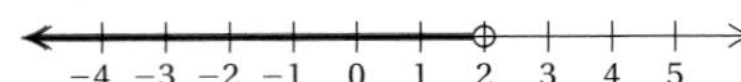

EXAMPLE 6 Graph: $x \geq -3$.

The solutions of $x \geq -3$ are shown on the number line by shading the point for -3 and all points to the right of -3. The closed circle at -3 indicates that -3 *is* part of the graph.

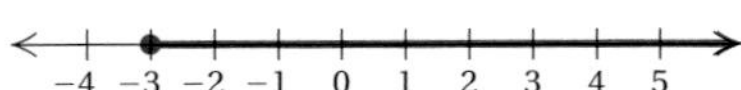

EXAMPLE 7 Graph: $-3 \leq x < 2$.

The inequality $-3 \leq x < 2$ is read "-3 is less than or equal to x *and* x is less than 2," or "x is greater than or equal to -3 *and* x is less than 2." In order to be a solution of this inequality, a number must be a solution of both $-3 \leq x$ and $x < 2$. The number 1 is a solution, as are -1.7, 0, 1.5, and $\frac{3}{8}$. We can see from the graphs below that the solution set consists of the numbers that overlap in the two solution sets in Examples 5 and 6:

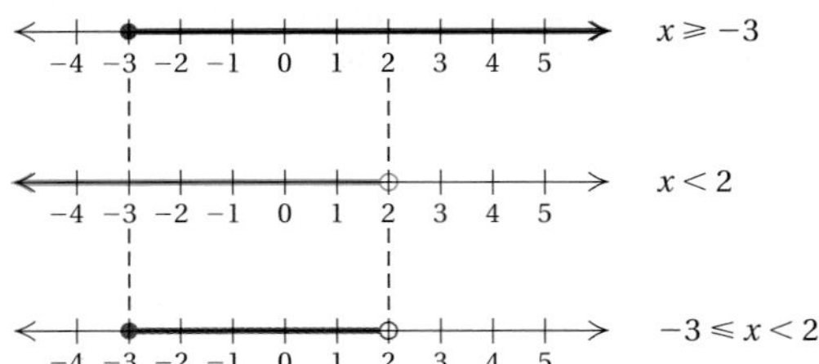

The open circle at 2 means that 2 is *not* part of the graph. The closed circle at -3 means that -3 *is* part of the graph. The other solutions are shaded.

Do Exercises 3–5.

C Solving Inequalities Using the Addition Principle

Consider the true inequality $3 < 7$. If we add 2 on both sides, we get another true inequality:

$$3 + 2 < 7 + 2, \quad \text{or} \quad 5 < 9.$$

Similarly, if we add -4 on both sides of $x + 4 < 10$, we get an *equivalent* inequality:

$$x + 4 + (-4) < 10 + (-4),$$

or $$x < 6.$$

To say that $x + 4 < 10$ and $x < 6$ are **equivalent** is to say that they have the same solution set. For example, the number 3 is a solution of $x + 4 < 10$. It is also a solution of $x < 6$. The number -2 is a solution of $x < 6$. It is also a solution of $x + 4 < 10$. Any solution of one is a solution of the other—they are equivalent.

THE ADDITION PRINCIPLE FOR INEQUALITIES

For any real numbers a, b, and c:

$a < b$ is equivalent to $a + c < b + c$;

$a > b$ is equivalent to $a + c > b + c$;

$a \le b$ is equivalent to $a + c \le b + c$;

$a \ge b$ is equivalent to $a + c \ge b + c$.

In other words, when we add or subtract the same number on both sides of an inequality, the direction of the inequality symbol is not changed.

As with equation solving, when solving inequalities, our goal is to isolate the variable on one side. Then it is easier to determine the solution set.

EXAMPLE 8 Solve: $x + 2 > 8$. Then graph.

We use the addition principle, subtracting 2 on both sides:

$$x + 2 - 2 > 8 - 2$$
$$x > 6.$$

From the inequality $x > 6$, we can determine the solutions directly. Any number greater than 6 makes the last sentence true and is a solution of that sentence. Any such number is also a solution of the original sentence. Thus the inequality is solved. The graph is as follows:

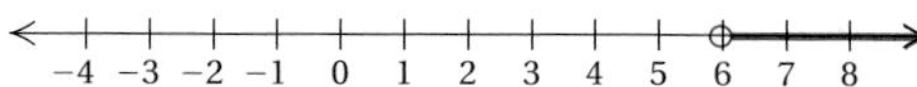

We cannot check all the solutions of an inequality by substitution, as we can check solutions of equations, because there are too many of them. A partial check can be done by substituting a number greater than 6—say, 7—into the original inequality:

$x + 2 > 8$	
$7 + 2$	8
9	TRUE

Since $9 > 8$ is true, 7 is a solution. Any number greater than 6 is a solution.

EXAMPLE 9 Solve: $3x + 1 \le 2x - 3$. Then graph.

We have

$$3x + 1 \le 2x - 3$$
$$3x + 1 - 1 \le 2x - 3 - 1 \quad \text{Subtracting 1}$$
$$3x \le 2x - 4 \quad \text{Simplifying}$$
$$3x - 2x \le 2x - 4 - 2x \quad \text{Subtracting } 2x$$
$$x \le -4. \quad \text{Simplifying}$$

The graph is as follows:

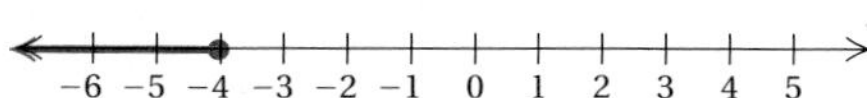

Remember that the graph is a drawing that represents the solutions of the original inequality.

Solve. Then graph.

6. $x + 3 > 5$

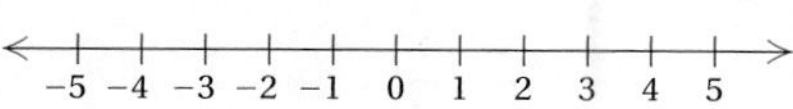

7. $x - 1 \le 2$

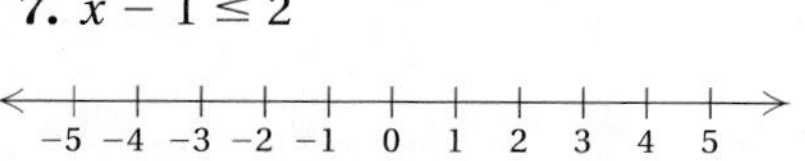

8. $5x + 1 < 4x - 2$

−5 −4 −3 −2 −1 0 1 2 3 4 5

Answers on page A-22

Solve.

9. $x + \frac{2}{3} \geq \frac{4}{5}$

10. $5y + 2 \leq -1 + 4y$

Answers on page A-22

In Example 9, any number less than or equal to -4 is a solution. The following are some solutions:

$$-4, \quad -5, \quad -6, \quad -\frac{13}{3}, \quad -204.5, \quad \text{and} \quad -18\pi.$$

Besides drawing a graph, we can also describe all the solutions of an inequality using **set notation.** We could just begin to list them in a set using roster notation (see p. 480), as follows:

$$\left\{-4, -5, -6, -\frac{13}{3}, -204.5, -18\pi, \ldots\right\}.$$

We can never list them all this way, however. Seeing this set without knowing the inequality makes it difficult for us to know what real numbers we are considering. There is, however, another kind of notation that we can use. It is

$$\{x \mid x \leq -4\},$$

which is read

"The set of all x such that x is less than or equal to -4."

This shorter notation for sets is called **set-builder notation.**

From now on, we will use this notation when solving inequalities.

Do Exercises 6–8 on the preceding page.

EXAMPLE 10 Solve: $x + \frac{1}{3} > \frac{5}{4}$.

We have

$$x + \frac{1}{3} > \frac{5}{4}$$

$$x + \frac{1}{3} - \frac{1}{3} > \frac{5}{4} - \frac{1}{3} \qquad \text{Subtracting } \tfrac{1}{3}$$

$$x > \frac{5}{4} \cdot \frac{3}{3} - \frac{1}{3} \cdot \frac{4}{4} \qquad \text{Multiplying by 1 to obtain a common denominator}$$

$$x > \frac{15}{12} - \frac{4}{12}$$

$$x > \frac{11}{12}.$$

Any number greater than $\frac{11}{12}$ is a solution. The solution set is

$$\left\{x \mid x > \tfrac{11}{12}\right\},$$

which is read

"The set of all x such that x is greater than $\frac{11}{12}$."

When solving inequalities, you may obtain an answer like $\frac{11}{12} < x$. Recall from Chapter 7 that this has the same meaning as $x > \frac{11}{12}$. Thus the solution set in Example 10 can be described as $\left\{x \mid \frac{11}{12} < x\right\}$ or as $\left\{x \mid x > \frac{11}{12}\right\}$. The latter is used most often.

Do Exercises 9 and 10.

d Solving Inequalities Using the Multiplication Principle

There is a multiplication principle for inequalities that is similar to that for equations, but it must be modified. When we are multiplying on both sides by a negative number, the direction of the inequality symbol must be changed.

Let's see what happens. Consider the true inequality $3 < 7$. If we multiply on both sides by a *positive* number, like 2, we get another true inequality:

$3 \cdot 2 < 7 \cdot 2$, or $6 < 14$. True

If we multiply on both sides by a *negative* number, like -2, and we do not change the direction of the inequality symbol, we get a *false* inequality:

$3 \cdot (-2) < 7 \cdot (-2)$, or $-6 < -14$. False

The fact that $6 < 14$ is true but $-6 < -14$ is false stems from the fact that the negative numbers, in a sense, mirror the positive numbers. That is, whereas 14 is to the *right* of 6 on a number line, the number -14 is to the *left* of -6. Thus, if we reverse (change the direction of) the inequality symbol, we get a *true* inequality: $-6 > -14$.

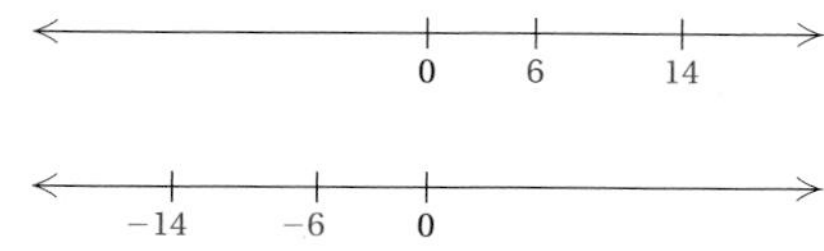

THE MULTIPLICATION PRINCIPLE FOR INEQUALITIES

For any real numbers a and b, and any *positive* number c:

$a < b$ is equivalent to $ac < bc$;

$a > b$ is equivalent to $ac > bc$.

For any real numbers a and b, and any *negative* number c:

$a < b$ is equivalent to $ac > bc$;

$a > b$ is equivalent to $ac < bc$.

Similar statements hold for $\le$ and $\ge$.

In other words, when we multiply or divide by a positive number on both sides of an inequality, the direction of the inequality symbol stays the same. When we multiply or divide by a negative number on both sides of an inequality, the direction of the inequality symbol is reversed.

EXAMPLE 11 Solve: $4x < 28$. Then graph.

We have

$4x < 28$

$\frac{4x}{4} < \frac{28}{4}$ Dividing by 4

The symbol stays the same.

$x < 7$. Simplifying

The solution set is $\{x \mid x < 7\}$. The graph is as follows:

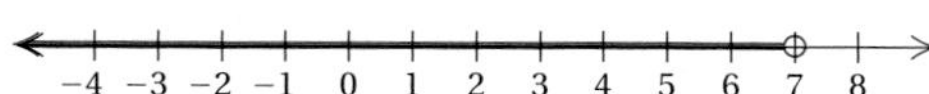

Do Exercises 11 and 12.

Solve. Then graph.

11. $8x < 64$

12. $5y \ge 160$

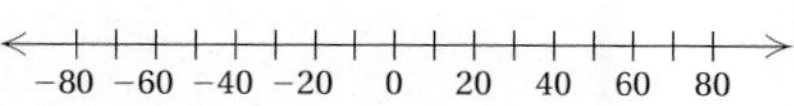

Answers on page A-22

Solve.

13. $-4x \leq 24$

14. $-5y > 13$

15. Solve: $7 - 4x < 8$.

Answers on page A-22

EXAMPLE 12 Solve: $-2y < 18$. Then graph.

$$-2y < 18$$

$$\frac{-2y}{-2} > \frac{18}{-2} \quad \text{Dividing by } -2$$

The symbol must be reversed!

$$y > -9. \quad \text{Simplifying}$$

The solution set is $\{y \mid y > -9\}$. The graph is as follows:

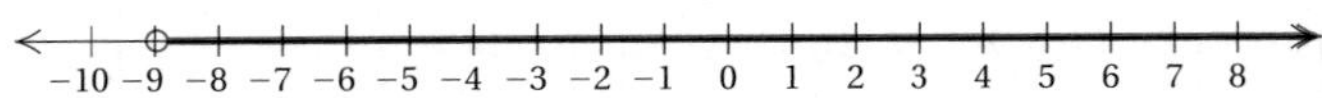

Do Exercises 13 and 14.

e Using the Principles Together

All of the equation-solving techniques used in Sections 8.1–8.3 can be used with inequalities, provided we remember to reverse the inequality symbol when multiplying or dividing on both sides by a negative number.

EXAMPLE 13 Solve: $6 - 5y > 7$.

$$6 - 5y > 7$$

$$-6 + 6 - 5y > -6 + 7 \quad \text{Adding } -6\text{. The symbol stays the same.}$$

$$-5y > 1 \quad \text{Simplifying}$$

$$\frac{-5y}{-5} < \frac{1}{-5} \quad \text{Dividing by } -5$$

The symbol must be reversed because we are dividing by a *negative* number, -5.

$$y < -\frac{1}{5}. \quad \text{Simplifying}$$

The solution set is $\{y \mid y < -\frac{1}{5}\}$.

Do Exercise 15.

EXAMPLE 14 Solve: $8y - 5 > 17 - 5y$.

$$-17 + 8y - 5 > -17 + 17 - 5y \quad \text{Adding } -17\text{. The symbol stays the same.}$$

$$8y - 22 > -5y \quad \text{Simplifying}$$

$$-8y + 8y - 22 > -8y - 5y \quad \text{Adding } -8y$$

$$-22 > -13y \quad \text{Simplifying}$$

$$\frac{-22}{-13} < \frac{-13y}{-13} \quad \text{Dividing by } -13$$

The symbol must be reversed because we are dividing by a *negative* number, -13.

$$\frac{22}{13} < y.$$

The solution set is $\{y \mid \frac{22}{13} < y\}$, or $\{y \mid y > \frac{22}{13}\}$.

Do Exercise 16.

It is typical to try to solve an inequality by isolating the variable on the left side of the inequality. Although this is not necessary, it does prevent having to reverse the inequality symbol at the end. Let's solve the inequality in Example 14 again, but this time we will isolate the variable on the left.

EXAMPLE 15 Solve: $8y - 5 > 17 - 5y$.

Note that if we add $5y$ on both sides, the coefficient of the y-term will be positive after like terms have been collected.

$8y - 5 + 5y > 17 - 5y + 5y$	Adding $5y$
$13y - 5 > 17$	Simplifying
$13y - 5 + 5 > 17 + 5$	Adding 5
$13y > 22$	Simplifying
$\frac{13y}{13} > \frac{22}{13}$	Dividing by 13. We leave the inequality symbol the same because we are dividing by a positive number.
$y > \frac{22}{13}$	

The solution set is $\{y \mid y > \frac{22}{13}\}$.

Do Exercise 17.

EXAMPLE 16 Solve: $3(x - 2) - 1 < 2 - 5(x + 6)$.

We have

$3(x - 2) - 1 < 2 - 5(x + 6)$	
$3x - 6 - 1 < 2 - 5x - 30$	Using the distributive law to multiply and remove parentheses
$3x - 7 < -5x - 28$	Simplifying
$3x + 5x < -28 + 7$	Adding $5x$ and 7 to get all x-terms on one side and all other terms on the other side
$8x < -21$	Simplifying
$x < \frac{-21}{8}$, or $-\frac{21}{8}$.	Dividing by 8

The solution set is $\{x \mid x < -\frac{21}{8}\}$.

Do Exercise 18.

16. Solve. Use a method like the one used in Example 14.

$24 - 7y \leq 11y - 14$

17. Solve. Use a method like the one used in Example 15.

$24 - 7y \leq 11y - 14$

18. Solve:

$3(7 + 2x) \leq 30 + 7(x - 1)$.

Answers on page A-22

19. Solve:

$2.1x + 43.2 \geq 1.2 - 8.4x.$

20. Solve:

$\frac{3}{4} + x < \frac{7}{8}x - \frac{1}{4} + \frac{1}{2}x.$

Answers on page A-22

EXAMPLE 17 Solve: $16.3 - 7.2p \leq -8.18$.

The greatest number of decimal places in any one number is *two*. Multiplying by 100, which has two 0's, will clear decimals. Then we proceed as before.

$$16.3 - 7.2p \leq -8.18$$

$$100(16.3 - 7.2p) \leq 100(-8.18) \quad \text{Multiplying by 100}$$

$$100(16.3) - 100(7.2p) \leq 100(-8.18) \quad \text{Using the distributive law}$$

$$1630 - 720p \leq -818 \quad \text{Simplifying}$$

$$1630 - 720p - 1630 \leq -818 - 1630 \quad \text{Subtracting 1630}$$

$$-720p \leq -2448 \quad \text{Simplifying}$$

$$\frac{-720p}{-720} \geq \frac{-2448}{-720} \quad \text{Dividing by } -720$$

The symbol must be reversed.

$$p \geq 3.4$$

The solution set is $\{p \mid p \geq 3.4\}$.

Do Exercise 19.

EXAMPLE 18 Solve: $\frac{2}{3}x - \frac{1}{6} + \frac{1}{2}x > \frac{7}{6} + 2x$.

The number 6 is the least common multiple of all the denominators. Thus we multiply by 6 on both sides.

$$\frac{2}{3}x - \frac{1}{6} + \frac{1}{2}x > \frac{7}{6} + 2x$$

$$6\left(\frac{2}{3}x - \frac{1}{6} + \frac{1}{2}x\right) > 6\left(\frac{7}{6} + 2x\right) \quad \text{Multiplying by 6 on both sides}$$

$$6 \cdot \frac{2}{3}x - 6 \cdot \frac{1}{6} + 6 \cdot \frac{1}{2}x > 6 \cdot \frac{7}{6} + 6 \cdot 2x \quad \text{Using the distributive law}$$

$$4x - 1 + 3x > 7 + 12x \quad \text{Simplifying}$$

$$7x - 1 > 7 + 12x \quad \text{Collecting like terms}$$

$$7x - 1 - 12x > 7 + 12x - 12x \quad \text{Subtracting } 12x$$

$$-5x - 1 > 7 \quad \text{Collecting like terms}$$

$$-5x - 1 + 1 > 7 + 1 \quad \text{Adding 1}$$

$$-5x > 8 \quad \text{Simplifying}$$

$$\frac{-5x}{-5} < \frac{8}{-5} \quad \text{Dividing by } -5$$

The symbol must be reversed.

$$x < -\frac{8}{5}$$

The solution set is $\{x \mid x < -\frac{8}{5}\}$.

Do Exercise 20.

8.7 EXERCISE SET

For Extra Help

MathXL | MyMathLab | InterAct Math | Math Tutor Center | Video Lectures on CD Disc 4 | Student's Solutions Manual

a

Determine whether each number is a solution of the given inequality.

1. $x > -4$

a) 4
b) 0
c) -4
d) 6
e) 5.6

2. $x \leq 5$

a) 0
b) 5
c) -1
d) -5
e) $7\frac{1}{4}$

3. $x \geq 6.8$

a) -6
b) 0
c) 6
d) 8
e) $-3\frac{1}{2}$

4. $x < 8$

a) 8
b) -10
c) 0
d) 11
e) -4.7

b

Graph on a number line.

5. $x > 4$

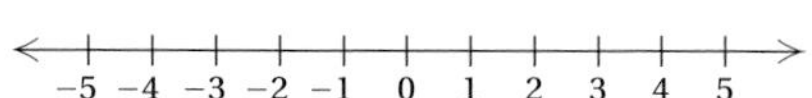

6. $x < 0$

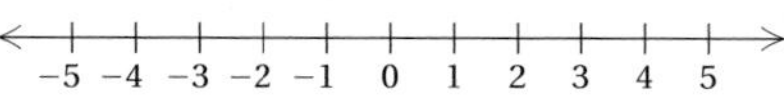

7. $t < -3$

−5 −4 −3 −2 −1 0 1 2 3 4 5

8. $y > 5$

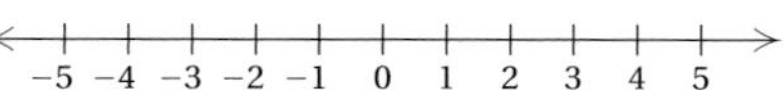

9. $m \geq -1$

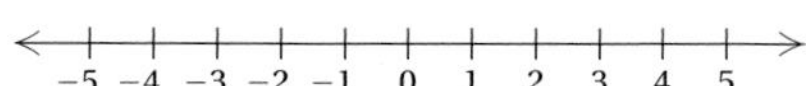

10. $x \leq -2$

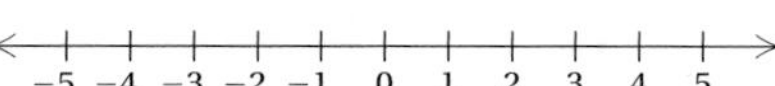

11. $-3 < x \leq 4$

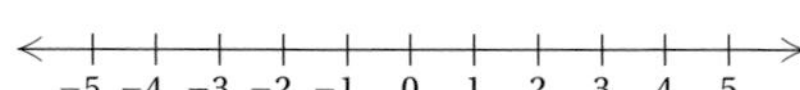

12. $-5 \leq x < 2$

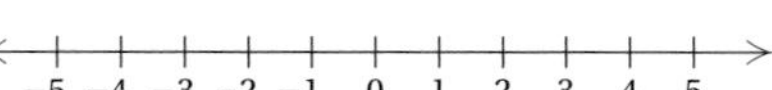

13. $0 < x < 3$

−5 −4 −3 −2 −1 0 1 2 3 4 5

14. $-5 \leq x \leq 0$

−5 −4 −3 −2 −1 0 1 2 3 4 5

c

Solve using the addition principle. Then graph.

15. $x + 7 > 2$

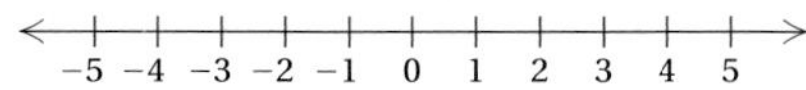

16. $x + 5 > 2$

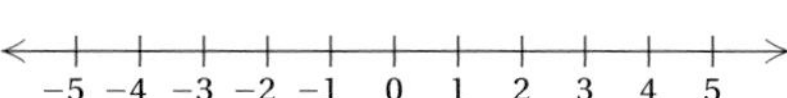

17. $x + 8 \leq -10$

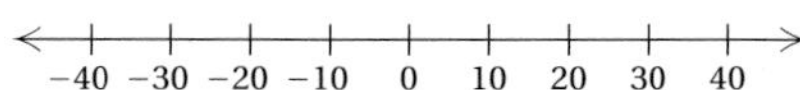

18. $x + 8 \leq -11$

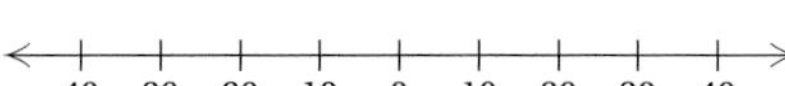

Solve using the addition principle.

19. $y - 7 > -12$

20. $y - 9 > -15$

21. $2x + 3 > x + 5$

22. $2x + 4 > x + 7$

23. $3x + 9 \leq 2x + 6$

24. $3x + 18 \leq 2x + 16$

25. $5x - 6 < 4x - 2$

26. $9x - 8 < 8x - 9$

27. $-9 + t > 5$

28. $-8 + p > 10$

29. $y + \frac{1}{4} \leq \frac{1}{2}$

30. $x - \frac{1}{3} \leq \frac{5}{6}$

31. $x - \frac{1}{3} > \frac{1}{4}$

32. $x + \frac{1}{8} > \frac{1}{2}$

d Solve using the multiplication principle. Then graph.

33. $5x < 35$

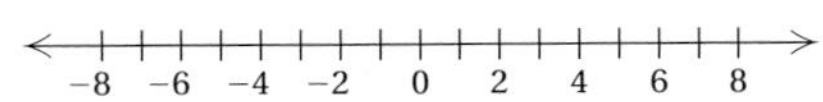

34. $8x \geq 32$

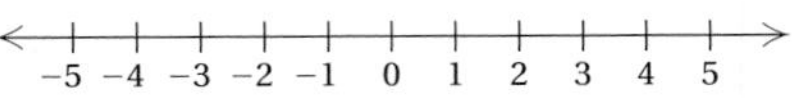

35. $-12x > -36$

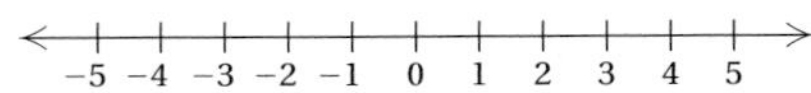

36. $-16x > -64$

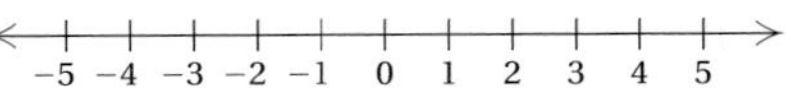

Solve using the multiplication principle.

37. $5y \geq -2$

38. $3x < -4$

39. $-2x \leq 12$

40. $-3x \leq 15$

41. $-4y \geq -16$

42. $-7x < -21$

43. $-3x < -17$

44. $-5y > -23$

45. $-2y > \frac{1}{7}$

46. $-4x \leq \frac{1}{9}$

47. $-\frac{6}{5} \leq -4x$

48. $-\frac{7}{9} > 63x$

e Solve using the addition and multiplication principles.

49. $4 + 3x < 28$

50. $3 + 4y < 35$

51. $3x - 5 \leq 13$

52. $5y - 9 \leq 21$

53. $13x - 7 < -46$

54. $8y - 6 < -54$

55. $30 > 3 - 9x$

56. $48 > 13 - 7y$

57. $4x + 2 - 3x \leq 9$

58. $15x + 5 - 14x \leq 9$

59. $-3 < 8x + 7 - 7x$

60. $-8 < 9x + 8 - 8x - 3$

61. $6 - 4y > 4 - 3y$

62. $9 - 8y > 5 - 7y + 2$

63. $5 - 9y \leq 2 - 8y$

64. $6 - 18x \leq 4 - 12x - 5x$

65. $19 - 7y - 3y < 39$

66. $18 - 6y - 4y < 63 + 5y$

67. $2.1x + 45.2 > 3.2 - 8.4x$

68. $0.96y - 0.79 \leq 0.21y + 0.46$

69. $\frac{x}{3} - 2 \leq 1$

70. $\frac{2}{3} + \frac{x}{5} < \frac{4}{15}$

71. $\frac{y}{5} + 1 \leq \frac{2}{5}$

72. $\frac{3x}{4} - \frac{7}{8} \geq -15$

73. $3(2y - 3) < 27$

74. $4(2y - 3) > 28$

75. $2(3 + 4m) - 9 \geq 45$

76. $3(5 + 3m) - 8 \leq 88$

77. $8(2t + 1) > 4(7t + 7)$

78. $7(5y - 2) > 6(6y - 1)$

79. $3(r - 6) + 2 < 4(r + 2) - 21$

80. $5(x + 3) + 9 \leq 3(x - 2) + 6$

81. $0.8(3x + 6) \geq 1.1 - (x + 2)$

82. $0.4(2x + 8) \geq 20 - (x + 5)$

83. $\frac{5}{3} + \frac{2}{3}x < \frac{25}{12} + \frac{5}{4}x + \frac{3}{4}$

84. $1 - \frac{2}{3}y \geq \frac{9}{5} - \frac{y}{5} + \frac{3}{5}$

85. **DW** Are the inequalities $3x - 4 < 10 - 4x$ and $2(x - 5) > 3(2x - 6)$ equivalent? Why or why not?

86. **DW** Explain in your own words why it is necessary to reverse the inequality symbol when multiplying on both sides of an inequality by a negative number.

SKILL MAINTENANCE

Add or subtract. [7.3a], [7.4a]

87. $-56 + (-18)$

88. $-2.3 + 7.1$

89. $-\frac{3}{4} + \frac{1}{8}$

90. $8.12 - 9.23$

91. $-56 - (-18)$

92. $-\frac{3}{4} - \frac{1}{8}$

93. $-2.3 - 7.1$

94. $-8.12 + 9.23$

Simplify.

95. $5 - 3^2 + (8 - 2)^2 \cdot 4$ [7.8d]

96. $10 \div 2 \cdot 5 - 3^2 + (-5)^2$ [7.8d]

97. $5(2x - 4) - 3(4x + 1)$ [7.8b]

98. $9(3 + 5x) - 4(7 + 2x)$ [7.8b]

SYNTHESIS

99. Determine whether each number is a solution of the inequality $|x| < 3$.

a) 0
b) -2
c) -3
d) 4
e) 3
f) 1.7
g) -2.8

100. Graph $|x| < 3$ on a number line.

$-5\ -4\ -3\ -2\ -1\ 0\ 1\ 2\ 3\ 4\ 5$

Solve.

101. $x + 3 < 3 + x$

102. $x + 4 > 3 + x$

8.8 APPLICATIONS AND PROBLEM SOLVING WITH INEQUALITIES

Objectives

a Translate number sentences to inequalities.

b Solve applied problems using inequalities.

The five steps for problem solving can be used for problems involving inequalities.

a Translating to Inequalities

Before solving problems that involve inequalities, we list some important phrases to look for. Sample translations are listed as well.

IMPORTANT WORDS	SAMPLE SENTENCE	TRANSLATION
is at least	Bill is at least 21 years old.	$b \geq 21$
is at most	At most 5 students dropped the course.	$n \leq 5$
cannot exceed	To qualify, earnings cannot exceed $12,000.	$r \leq 12{,}000$
must exceed	The speed must exceed 15 mph.	$s > 15$
is less than	Tucker's weight is less than 50 lb.	$w < 50$
is more than	Boston is more than 200 miles away.	$d > 200$
is between	The film was between 90 and 100 minutes long.	$90 < t < 100$
no more than	Bing weighs no more than 90 lb.	$w \leq 90$
no less than	Valerie scored no less than 8.3.	$s \geq 8.3$

The following phrases deserve special attention.

TRANSLATING "AT LEAST" AND "AT MOST"

A quantity x is at least some amount q: $x \geq q$.
(If x is at least q, it cannot be less than q.)

A quantity x is at most some amount q: $x \leq q$.
(If x is at most q, it cannot be more than q.)

Do Exercises 1–10.

Translate.

1. Maggie scored no less than 92 on her English exam.
2. The average credit card holder is at least $4000 in debt.
3. The price of that PT Cruiser is at most $21,900.
4. The time of the test was between 45 and 55 min.
5. Normandale Community College is more than 15 mi away.
6. Tania's weight is less than 110 lb.
7. That number is greater than -2.
8. The costs of production of that CD-ROM cannot exceed $12,500.
9. At most, 11.4% of all deaths in Arizona are from cancer.
10. Yesterday, at least 23 people got tickets for speeding.

Answers on page A-23

b Solving Problems

EXAMPLE 1 *Catering Costs.* To cater a party, Curtis' Barbeque charges a $50 setup fee plus $15 per person. The cost of Hotel Pharmacy's end-of-season softball party cannot exceed $450. How many people can attend the party?

Source: Curtis' All American Barbeque, Putney, Vermont

1. **Familiarize.** Suppose that 20 people were to attend the party. The cost would then be $\$50 + \$15 \cdot 20$, or $350. This shows that more than 20 people could attend without exceeding $450. Instead of making another guess, we let n = the number of people in attendance.

11. Butter Temperatures. Butter stays solid at Fahrenheit temperatures below 88°. The formula

$$F = \frac{9}{5}C + 32$$

can be used to convert Celsius temperatures C to Fahrenheit temperatures F. Determine (in terms of an inequality) those Celsius temperatures for which butter stays solid.

Answer on page A-23

2. **Translate.** The cost of the party will be \$50 for the setup fee plus \$15 times the number of people attending. We can reword and translate to an inequality as follows:

Rewording:	The setup fee	plus	the cost of the meals	cannot exceed	\$450.
	↓	↓	↓	↓	↓
Translating:	50	+	$15 \cdot n$	$\leq$	450

3. **Solve.** We solve the inequality for n:

$$50 + 15n \leq 450$$
$$50 + 15n - 50 \leq 450 - 50 \quad \text{Subtracting 50}$$
$$15n \leq 400 \quad \text{Simplifying}$$
$$\frac{15n}{15} \leq \frac{400}{15} \quad \text{Dividing by 15}$$
$$n \leq \frac{400}{15}$$
$$n \leq 26\tfrac{2}{3}. \quad \text{Simplifying}$$

4. **Check.** Although the solution set of the inequality is all numbers less than or equal to $26\frac{2}{3}$, since $n =$ the number of people in attendance, we round *down* to 26. If 26 people attend, the cost will be $\$50 + \$15 \cdot 26$, or \$440, and if 27 attend, the cost will exceed \$450.

5. **State.** At most 26 people can attend the party.

Do Exercise 11.

Caution!

Solutions of problems should always be checked using the original wording of the problem. In some cases, answers might need to be whole numbers or integers or rounded off in a particular direction.

EXAMPLE 2 *Nutrition.* The U.S. Department of Agriculture recommends that for a typical 2000-calorie daily diet, no more than 20 g of saturated fat be consumed. In the first three days of a four-day vacation, Anthony consumed 26 g, 17 g, and 22 g of saturated fat. Determine (in terms of an inequality) how many grams of saturated fat Anthony can consume on the fourth day if he is to average no more than 20 g of saturated fat per day.

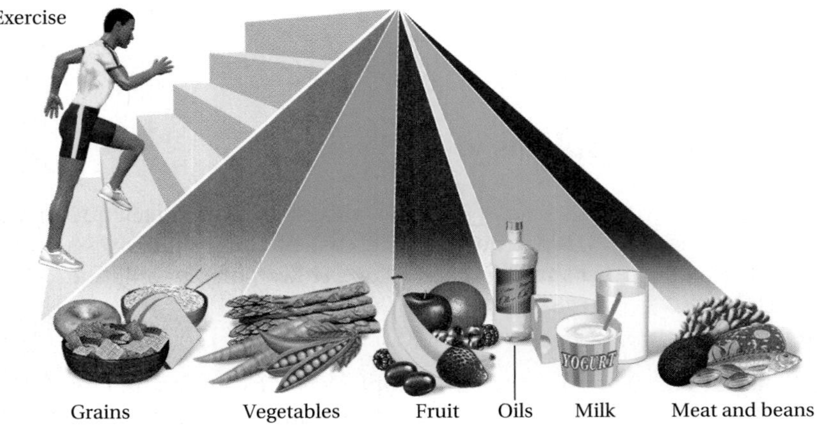

Sources: U.S. Department of Health and Human Services and Department of Agriculture

1. **Familiarize.** Suppose Anthony consumed 19 g of saturated fat on the fourth day. His daily average for the vacation would then be

$$\frac{26\text{ g} + 17\text{ g} + 22\text{ g} + 19\text{ g}}{4} = 21\text{ g}.$$

This shows that Anthony cannot consume 19 g of saturated fat on the fourth day, if he is to average no more than 20 g of fat per day. We let $x =$ the number of grams of fat that Anthony consumes on the fourth day.

2. **Translate.** We reword the problem and translate to an inequality as follows:

Rewording:	The average consumption of saturated fat	should be no more than	20 g.
Translating:	$\dfrac{26 + 17 + 22 + x}{4}$	$\leq$	20

3. **Solve.** Because of the fraction expression, it is convenient to use the multiplication principle first to solve the inequality:

$$\frac{26 + 17 + 22 + x}{4} \leq 20$$

$$4\left(\frac{26 + 17 + 22 + x}{4}\right) \leq 4 \cdot 20 \qquad \text{Multiplying by 4}$$

$$26 + 17 + 22 + x \leq 80$$

$$65 + x \leq 80 \qquad \text{Simplifying}$$

$$x \leq 15. \qquad \text{Subtracting 65}$$

4. **Check.** As a partial check, we show that Anthony can consume 15 g of saturated fat on the fourth day and not exceed a 20-g average for the four days:

$$\frac{26 + 17 + 22 + 15}{4} = \frac{80}{4} = 20.$$

5. **State.** Anthony's average intake of saturated fat for the vacation will not exceed 20 g per day if he consumes no more than 15 g of saturated fat on the fourth day.

Do Exercise 12.

Translate to an inequality and solve.

12. **Test Scores.** A pre-med student is taking a chemistry course in which four tests are given. To get an A, she must average at least 90 on the four tests. The student got scores of 91, 86, and 89 on the first three tests. Determine (in terms of an inequality) what scores on the last test will allow her to get an A.

Answer on page A-23

Study Tips

CHECKLIST

The foundation of all your study skills is TIME!

- ☐ Are you approaching your study of mathematics with an assertive, positive attitude?
- ☐ Are you making use of the textbook supplements, such as the AW Math Tutor Center, the *Student's Solutions Manual*, and the videos?
- ☐ Have you determined the location of the learning resource centers on your campus, such as a math lab, tutor center, and your instructor's office?
- ☐ Are you stopping to work the margin exercises when directed to do so?
- ☐ Are you keeping one section ahead in your syllabus?

8.8 EXERCISE SET

For Extra Help

MathXL | MyMathLab | InterAct Math | Math Tutor Center | Video Lectures on CD Disc 4 | Student's Solutions Manual

a Translate to an inequality.

1. A number is at least 7.

2. A number is greater than or equal to 5.

3. The baby weighs more than 2 kilograms (kg).

4. Between 75 and 100 people attended the concert.

5. The speed of the train was between 90 and 110 mph.

6. At least 400,000 people attended the Million Man March.

7. At most 1,200,000 people attended the Million Man March.

8. The amount of acid is not to exceed 40 liters (L).

9. The cost of gasoline is no less than \$1.50 per gallon.

10. The temperature is at most $-2°$.

11. A number is greater than 8.

12. A number is less than 5.

13. A number is less than or equal to -4.

14. A number is greater than or equal to 18.

15. The number of people is at least 1300.

16. The cost is at most \$4857.95.

17. The amount of acid is not to exceed 500 liters.

18. The cost of gasoline is no less than 94 cents per gallon.

19. Two more than three times a number is less than 13.

20. Five less than one-half a number is greater than 17.

b Solve.

21. *Test Scores.* A student is taking a literature course in which four tests are given. To get a B, he must average at least 80 on the four tests. The student got scores of 82, 76, and 78 on the first three tests. Determine (in terms of an inequality) what scores on the last test will allow him to get at least a B.

22. *Test Scores.* Your quiz grades are 73, 75, 89, and 91. Determine (in terms of an inequality) what scores on the last quiz will allow you to get an average quiz grade of at least 85.

23. *Gold Temperatures.* Gold stays solid at Fahrenheit temperatures below 1945.4°. Determine (in terms of an inequality) those Celsius temperatures for which gold stays solid. Use the formula given in Margin Exercise 11.

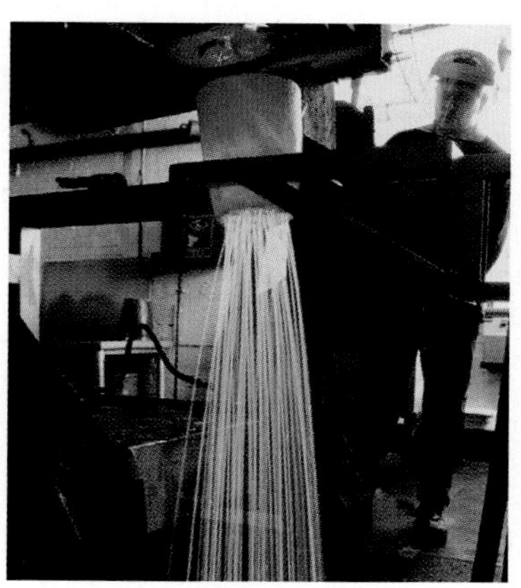

24. *Body Temperatures.* The human body is considered to be fevered when its temperature is higher than 98.6°F. Using the formula given in Margin Exercise 11, determine (in terms of an inequality) those Celsius temperatures for which the body is fevered.

25. *World Records in the 1500-m Run.* The formula

$$R = -0.075t + 3.85$$

can be used to predict the world record in the 1500-m run t years after 1930. Determine (in terms of an inequality) those years for which the world record will be less than 3.5 min.

26. *World Records in the 200-m Dash.* The formula

$$R = -0.028t + 20.8$$

can be used to predict the world record in the 200-m dash t years after 1920. Determine (in terms of an inequality) those years for which the world record will be less than 19.0 sec.

27. *Sizes of Envelopes.* Rhetoric Advertising is a direct-mail company. It determines that for a particular campaign, it can use any envelope with a fixed width of $3\frac{1}{2}$ in. and an area of at least $17\frac{1}{2}$ in^2. Determine (in terms of an inequality) those lengths that will satisfy the company constraints.

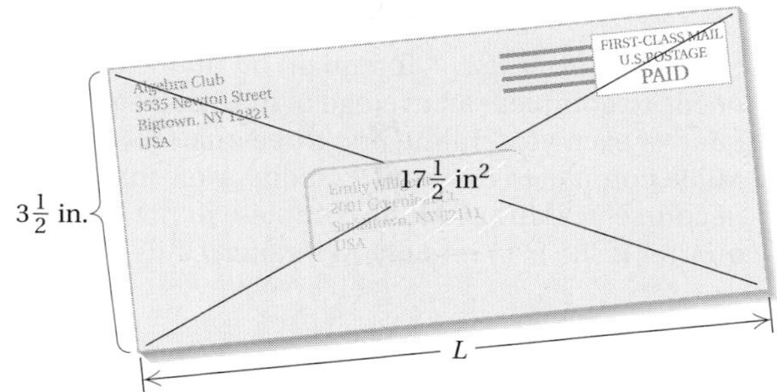

28. *Sizes of Packages.* An overnight delivery service accepts packages of up to 165 in. in length and girth combined. (Girth is the distance around the package.) A package has a fixed girth of 53 in. Determine (in terms of an inequality) those lengths for which a package is acceptable.

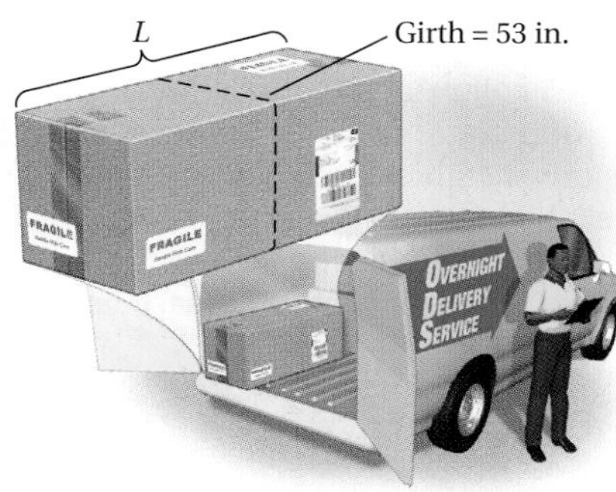

29. *Blueprints.* To make copies of blueprints, Vantage Reprographics charges a $5 setup fee plus $4 per copy. Myra can spend no more than $65 for the copying. What numbers of copies will allow her to stay within budget?

30. *Banquet Costs.* The women's volleyball team can spend at most $450 for its awards banquet at a local restaurant. If the restaurant charges a $40 setup fee plus $16 per person, at most how many can attend?

31. *Phone Costs.* Simon claims that it costs him at least $3.00 every time he calls an overseas customer. If his typical call costs 75¢ plus 45¢ for each minute, how long do his calls typically last? (*Hint*: 75¢ = $0.75.)

32. *Parking Costs.* Laura is certain that every time she parks in the municipal garage it costs her at least $2.20. If the garage charges 45¢ plus 25¢ for each half hour, for how long is Laura's car generally parked?

33. *College Tuition.* Angelica's financial aid stipulates that her tuition not exceed $1000. If her local community college charges a $35 registration fee plus $375 per course, what is the greatest number of courses for which Angelica can register?

34. *Furnace Repairs.* RJ's Plumbing and Heating charges $25 plus $30 per hour for emergency service. Gary remembers being billed over $100 for an emergency call. How long was RJ's there?

Tuition, Fees, and Refunds
Fees and Refunds Policy
Fees are due at the time of registration.
Tuition $375 per course
Registration $35
Late Registration $25
Additional Drop/Add $25
Late Payment 1% of past due balance
Tuition Refunds
Prior to start of class 100%
Before the second class meeting 90%
Before the fourth class meeting 25%
After the fourth class meeting 0%

35. *Nutrition.* Following the guidelines of the Food and Drug Administration, Dale tries to eat at least 5 servings of fruits or vegetables each day. For the first six days of one week, he had 4, 6, 7, 4, 6, and 4 servings. How many servings of fruits or vegetables should Dale eat on Saturday, in order to average at least 5 servings per day for the week?

36. *College Course Load.* To remain on financial aid, Millie needs to complete an average of at least 7 credits per quarter each year. In the first three quarters of 2005, Millie completed 5, 7, and 8 credits. How many credits of course work must Millie complete in the fourth quarter if she is to remain on financial aid?

37. *Perimeter of a Rectangle.* The width of a rectangle is fixed at 8 ft. What lengths will make the perimeter at least 200 ft? at most 200 ft?

38. *Perimeter of a Triangle.* One side of a triangle is 2 cm shorter than the base. The other side is 3 cm longer than the base. What lengths of the base will allow the perimeter to be greater than 19 cm?

39. *Area of a Rectangle.* The width of a rectangle is fixed at 4 cm. For what lengths will the area be less than 86 cm^2?

L

4 cm $A < 86\ cm^2$ 4 cm

L

40. *Area of a Rectangle.* The width of a rectangle is fixed at 16 yd. For what lengths will the area be at least 264 yd^2?

41. *Insurance-covered Repairs.* Most insurance companies will replace a vehicle if an estimated repair exceeds 80% of the "blue-book" value of the vehicle. Michelle's insurance company paid $8500 for repairs to her Subaru after an accident. What can be concluded about the blue-book value of the car?

42. *Insurance-covered Repairs.* Following an accident, Jeff's Ford pickup was replaced by his insurance company because the damage was so extensive. Before the damage, the blue-book value of the truck was $21,000. How much would it have cost to repair the truck? (See Exercise 41.)

43. *Fat Content in Foods.* Reduced Fat Skippy® peanut butter contains 12 g of fat per serving. In order for a food to be labeled "reduced fat," it must have at least 25% less fat than the regular item. What can you conclude about the number of grams of fat in a serving of the regular Skippy peanut butter?

Source: Best Foods

44. *Fat Content in Foods.* Reduced Fat Chips Ahoy!® cookies contain 5 g of fat per serving. What can you conclude about the number of grams of fat in regular Chips Ahoy! cookies (see Exercise 43)?

Source: Nabisco Brands, Inc.

45. *Pond Depth.* On July 1, Garrett's Pond was 25 ft deep. Since that date, the water level has dropped $\frac{2}{3}$ ft per week. For what dates will the water level not exceed 21 ft?

46. *Weight Gain.* A 3-lb puppy is gaining weight at a rate of $\frac{3}{4}$ lb per week. When will the puppy's weight exceed $22\frac{1}{2}$ lb?

47. *Area of a Triangular Flag.* As part of an outdoor education course, Wanda needs to make a bright-colored triangular flag with an area of at least 3 ft^2. What heights can the triangle be if the base is $1\frac{1}{2}$ ft?

48. *Area of a Triangular Sign.* Zoning laws in Harrington prohibit displaying signs with areas exceeding 12 ft^2. If Flo's Marina is ordering a triangular sign with an 8-ft base, how tall can the sign be?

49. *Electrician Visits.* Dot's Electric made 17 customer calls last week and 22 calls this week. How many calls must be made next week in order to maintain an average of at least 20 for the three-week period?

50. *Volunteer Work.* George and Joan do volunteer work at a hospital. Joan worked 3 more hr than George, and together they worked more than 27 hr. What possible numbers of hours did each work?

51. D_W If f represents Fran's age and t represents Todd's age, write a sentence that would translate to $t + 3 < f$.

52. D_W Explain how the meanings of "Five more than a number" and "Five is more than a number" differ.

SKILL MAINTENANCE

VOCABULARY REINFORCEMENT

In each of Exercises 53–60, fill in the blank with the correct term from the given list. Some of the choices may not be used.

53. The product of a(n) ______________ number of negative numbers is always positive. [7.5a]

54. The product of a(n) ______________ number of negative numbers is always negative. [7.5a]

55. The ______________ inverse of a negative number is always positive. [7.3b]

56. The ______________ inverse of a negative number is always negative. [7.6b]

57. Equations with the same solutions are called ______________ equations. [8.1b]

58. The ______________ for equations asserts that when we add the same number to the expressions on each side of the equation, we get equivalent equations. [8.1b]

59. The ______________ for inequalities asserts that when we multiply or divide by a negative number on both sides of an inequality, the direction of the inequality symbol ______________. [8.7d]

60. Any replacement for the variable that makes an equation true is called a(n) ______________ of the equation. [8.1a]

addition principle
multiplication principle
solution
replacement
variable
is reversed
stays the same
even
odd
multiplicative
additive
equivalent

SYNTHESIS

61. *Ski Wax.* Green ski wax works best between 5° and 15° Fahrenheit. Determine those Celsius temperatures for which green ski wax works best.

62. *Parking Fees.* Mack's Parking Garage charges $4.00 for the first hour and $2.50 for each additional hour. For how long has a car been parked when the charge exceeds $16.50?

63. *Nutritional Standards.* In order for a food to be labeled "lowfat," it must have fewer than 3 g of fat per serving. Reduced fat Tortilla Pops® contain 60% less fat than regular nacho cheese tortilla chips, but still cannot be labeled lowfat. What can you conclude about the fat content of a serving of nacho cheese tortilla chips?

64. *Parking Fees.* When asked how much the parking charge is for a certain car (see Exercise 62), Mack replies "between 14 and 24 dollars." For how long has the car been parked?

8 Summary and Review

The review that follows is meant to prepare you for a chapter exam. It consists of three parts. The first part, Concept Reinforcement, is designed to increase understanding of the concepts through true/false exercises. The second part is a list of important properties and formulas. The third part is the Review Exercises. These provide practice exercises for the exam, together with references to section objectives so you can go back and review. Before beginning, stop and look back over the skills you have obtained. What skills in mathematics do you have now that you did not have before studying this chapter?

CONCEPT REINFORCEMENT

Determine whether the statement is true or false. Answers are given at the back of the book.

________ **1.** If $x > y$, then $-x < -y$.

________ **2.** Consecutive odd integers are 2 units apart.

________ **3.** For any number n, $n \geq n$.

________ **4.** $3 - x = 4x$ and $5x = -3$ are equivalent equations.

________ **5.** Some equations have no solution.

________ **6.** $2x - 7 < 11$ and $x < -9$ are equivalent inequalities.

IMPORTANT PROPERTIES AND FORMULAS

The Addition Principle for Equations: For any real numbers a, b, and c: $a = b$ is equivalent to $a + c = b + c$.

The Multiplication Principle for Equations: For any real numbers a, b, and c, $c \neq 0$: $a = b$ is equivalent to $a \cdot c = b \cdot c$.

The Addition Principle for Inequalities: For any real numbers a, b, and c:
$a < b$ is equivalent to $a + c < b + c$;
$a > b$ is equivalent to $a + c > b + c$;
$a \leq b$ is equivalent to $a + c \leq b + c$;
$a \geq b$ is equivalent to $a + c \geq b + c$.

The Multiplication Principle for Inequalities: For any real numbers a and b, and any *positive* number c:
$a < b$ is equivalent to $ac < bc$; $a > b$ is equivalent to $ac > bc$.
For any real numbers a and b, and any *negative* number c:
$a < b$ is equivalent to $ac > bc$; $a > b$ is equivalent to $ac < bc$.

Review Exercises

Solve. [8.1b]

1. $x + 5 = -17$

2. $n - 7 = -6$

3. $x - 11 = 14$

4. $y - 0.9 = 9.09$

Solve. [8.2a]

5. $-\frac{2}{3}x = -\frac{1}{6}$

6. $-8x = -56$

7. $-\frac{x}{4} = 48$

8. $15x = -35$

9. $\frac{4}{5}y = -\frac{3}{16}$

Solve. [8.3a]

10. $5 - x = 13$

11. $\frac{1}{4}x - \frac{5}{8} = \frac{3}{8}$

Solve. [8.3b, c]

12. $5t + 9 = 3t - 1$

13. $7x - 6 = 25x$

14. $14y = 23y - 17 - 10$

15. $0.22y - 0.6 = 0.12y + 3 - 0.8y$

16. $\frac{1}{4}x - \frac{1}{8}x = 3 - \frac{1}{16}x$

17. $14y + 17 + 7y = 9 + 21y + 8$

Solve. [8.3c]

18. $4(x + 3) = 36$

19. $3(5x - 7) = -66$

20. $8(x - 2) - 5(x + 4) = 20 + x$

21. $-5x + 3(x + 8) = 16$

22. $6(x - 2) - 16 = 3(2x - 5) + 11$

Determine whether the given number is a solution of the inequality $x \leq 4$. [8.7a]

23. -3

24. 7

25. 4

Solve. Write set notation for the answers. [8.7c, d, e]

26. $y + \frac{2}{3} \geq \frac{1}{6}$

27. $9x \geq 63$

28. $2 + 6y > 14$

29. $7 - 3y \geq 27 + 2y$

30. $3x + 5 < 2x - 6$

31. $-4y < 28$

32. $4 - 8x < 13 + 3x$

33. $-4x \leq \frac{1}{3}$

Graph on a number line. [8.7b, e]

34. $4x - 6 < x + 3$

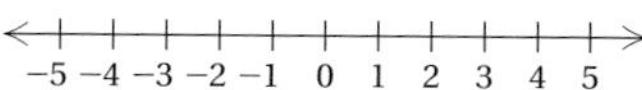

35. $-2 < x \leq 5$

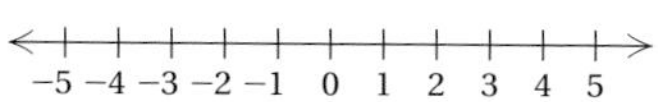

36. $y > 0$

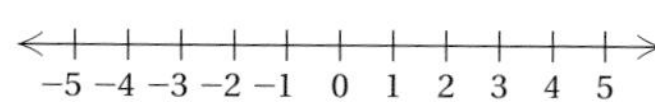

Solve. [8.4b]

37. $C = \pi d$, for d

38. $V = \frac{1}{3}Bh$, for B

39. $A = \frac{a + b}{2}$, for a

40. $y = mx + b$, for x

Solve. [8.6a]

41. *Dimensions of Wyoming.* The state of Wyoming is roughly in the shape of a rectangle whose perimeter is 1280 mi. The length is 90 mi more than the width. Find the dimensions.

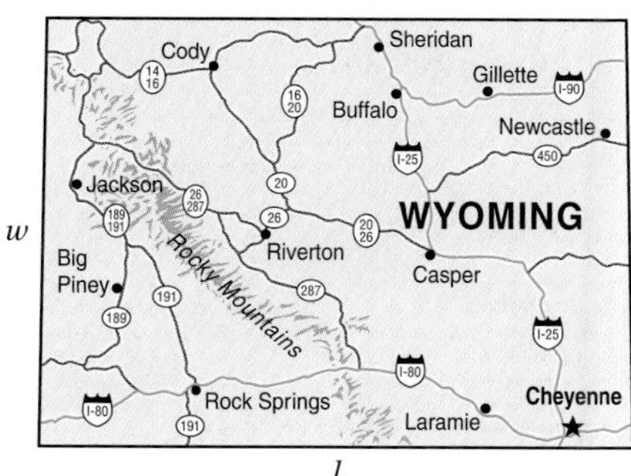

42. *Interstate Mile Markers.* The sum of two consecutive mile markers on I-5 in California is 691. Find the numbers on the markers.

43. An entertainment center sold for $2449 in June. This was $332 more than the cost in February. Find the cost in February.

44. Ty is paid a commission of $4 for each appliance he sells. One week, he received $108 in commissions. How many appliances did he sell?

45. The measure of the second angle of a triangle is 50° more than that of the first angle. The measure of the third angle is 10° less than twice the first angle. Find the measures of the angles.

Solve. [8.5a]

46. What number is 20% of 75?

47. Fifteen is what percent of 80?

48. 18 is 3% of what number?

49. *Job Opportunities.* There were 905 thousand child-care workers in 1998. Job opportunities are expected to grow to 1141 thousand by 2008. What is the percent of increase?

Source: *Handbook of U.S. Labor Statistics*

Solve. [8.6a]

50. After a 30% reduction, a bread maker is on sale for $154. What was the marked price (the price before the reduction)?

51. A hotel manager's salary is $61,410, which is a 15% increase over the previous year's salary. What was the previous salary?

52. A tax-exempt charity received a bill of $145.90 for a sump pump. The bill incorrectly included sales tax of 5%. How much does the charity actually owe?

Solve. [8.8b]

53. *Test Scores.* Your test grades are 71, 75, 82, and 86. What is the lowest grade that you can get on the next test and still have an average test score of at least 80?

54. The length of a rectangle is 43 cm. What widths will make the perimeter greater than 120 cm?

55. $\mathbf{D}_\mathbf{W}$ Would it be better to receive a 5% raise and then an 8% raise or the other way around? Why? [8.5a]

56. $\mathbf{D}_\mathbf{W}$ Are the inequalities $x > -5$ and $-x < 5$ equivalent? Why or why not? [8.7d]

SYNTHESIS

Solve.

57. $2|x| + 4 = 50$ [7.2e], [8.3a]

58. $|3x| = 60$ [7.2e], [8.2a]

59. $y = 2a - ab + 3$, for a [8.4b]

8 Chapter Test

For Extra Help

Work It Out! Chapter Test Video Solutions on CD

Solve.

1. $x + 7 = 15$

2. $t - 9 = 17$

3. $3x = -18$

4. $-\frac{4}{7}x = -28$

5. $3t + 7 = 2t - 5$

6. $\frac{1}{2}x - \frac{3}{5} = \frac{2}{5}$

7. $8 - y = 16$

8. $-\frac{2}{5} + x = -\frac{3}{4}$

9. $3(x + 2) = 27$

10. $-3x - 6(x - 4) = 9$

11. $0.4p + 0.2 = 4.2p - 7.8 - 0.6p$

12. $4(3x - 1) + 11 = 2(6x + 5) - 8$

13. $-2 + 7x + 6 = 5x + 4 + 2x$

Solve. Write set notation for the answers.

14. $x + 6 \leq 2$

15. $14x + 9 > 13x - 4$

16. $12x \leq 60$

17. $-2y \geq 26$

18. $-4y \leq -32$

19. $-5x \geq \frac{1}{4}$

20. $4 - 6x > 40$

21. $5 - 9x \geq 19 + 5x$

Graph on a number line.

22. $y \leq 9$

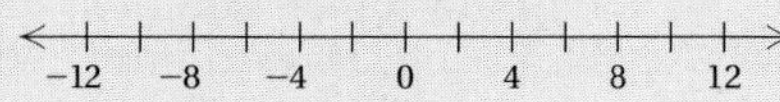

23. $6x - 3 < x + 2$

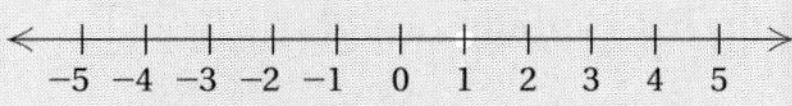

24. $-2 \leq x \leq 2$

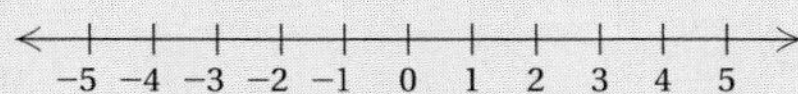

Solve.

25. What number is 24% of 75?

26. 15.84 is what percent of 96?

27. 800 is 2% of what number?

28. *Job Opportunities.* Job opportunities for physician's assistants are expected to increase from 58,000 in 2000 to 89,000 in 2010. What is the percent of increase?

Source: *Monthly Labor Review*, November 2001

29. *Perimeter of a Photograph.* The perimeter of a rectangular photograph is 36 cm. The length is 4 cm greater than the width. Find the width and the length.

30. *Charitable Contributions.* About 35.9% of all charitable contributions are made to religious organizations. In 2003, about \$86.4 billion was given to religious organizations. How much was given to charities in general?

Source: AAFRC Trust for Philanthropy/Giving USA 2004

31. *Raffle Tickets.* The numbers on three raffle tickets are consecutive integers whose sum is 7530. Find the integers.

32. *Savings Account.* Money is invested in a savings account at 5% simple interest. After 1 year, there is \$924 in the account. How much was originally invested?

33. *Board Cutting.* An 8-m board is cut into two pieces. One piece is 2 m longer than the other. How long are the pieces?

34. *Lengths of a Rectangle.* The width of a rectangle is 96 yd. Find all possible lengths such that the perimeter of the rectangle will be at least 540 yd.

35. *Budgeting.* Jason has budgeted an average of \$95 a month for entertainment. For the first five months of the year, he has spent \$98, \$89, \$110, \$85, and \$83. How much can Jason spend in the sixth month without exceeding his average budget?

36. *Copy Machine Rental.* It costs \$225 per month plus 1.2¢ per copy to rent a copy machine. A catalog publisher needs to lease a copy machine for use during a special project that they anticipate will take 3 months. They decide to rent the copier, but must stay within a budget of \$2400 for copies. Determine (in terms of an inequality) the number of copies they can make and still remain within budget.

37. Solve $A = 2\pi rh$ for r.

38. Solve $y = 8x + b$ for x.

SYNTHESIS

39. Solve $c = \dfrac{1}{a - d}$ for d.

40. Solve: $3|w| - 8 = 37$.

41. A movie theater had a certain number of tickets to give away. Five people got the tickets. The first got one-third of the tickets, the second got one-fourth of the tickets, and the third got one-fifth of the tickets. The fourth person got eight tickets, and there were five tickets left for the fifth person. Find the total number of tickets given away.

Graphs of Linear Equations

Real-World Application

The online retail sales of jewelry y, in billions of dollars, is predicted by

$$y = 0.81x + 2,$$

where x is the number of years since 2003. Determine the online retail sales of jewelry in 2003, in 2008, and in 2015.

Graph the equation and use the graph to estimate online sales in 2010. Then determine the year in which sales will be \$9.29 billion.

Source: Forrester Research

This problem appears as Example 11 in Section 9.1.

9.1 GRAPHS AND APPLICATIONS OF LINEAR EQUATIONS

Objectives

a Plot points associated with ordered pairs of numbers; determine the quadrant in which a point lies.

b Find the coordinates of a point on a graph.

c Determine whether an ordered pair is a solution of an equation with two variables.

d Graph linear equations of the type $y = mx + b$ and $Ax + By = C$, identifying the y-intercept.

e Solve applied problems involving graphs of linear equations.

Plot these points on the graph below.

1. $(4, 5)$ **2.** $(5, 4)$

3. $(-2, 5)$ **4.** $(-3, -4)$

5. $(5, -3)$ **6.** $(-2, -1)$

7. $(0, -3)$ **8.** $(2, 0)$

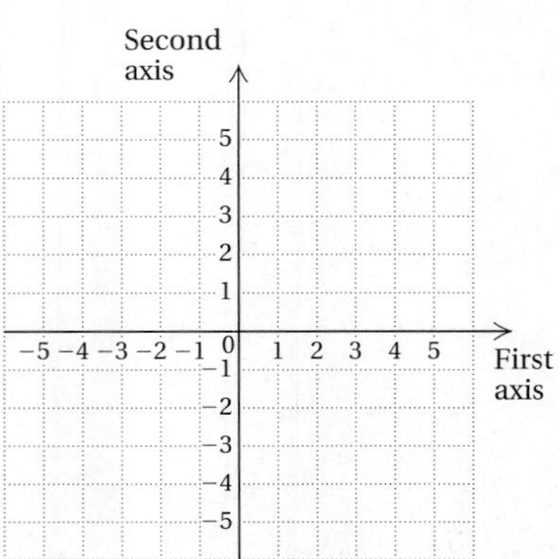

Answers on page A-24

You probably have seen bar graphs like the following in newspapers and magazines. Note that a straight line can be drawn along the tops of the bars. Such a line is a *graph of a linear equation.* In this chapter, we study how to graph linear equations and consider properties such as slope and intercepts. Many applications of these topics will also be considered.

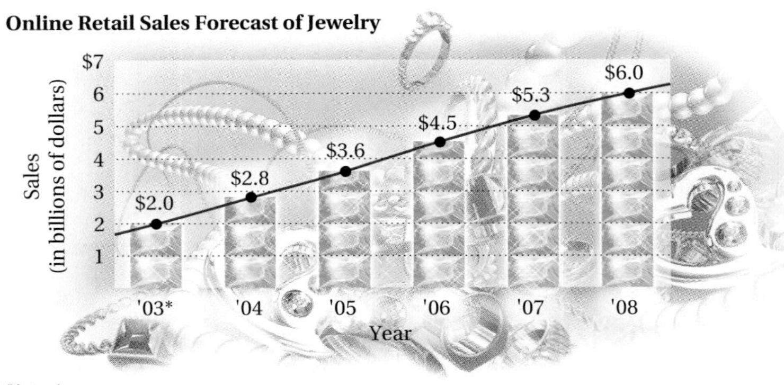

*Actual
Source: Forrester Research

a Plotting Ordered Pairs

In Chapter 8, we graphed numbers and inequalities in one variable on a line. To enable us to graph an equation that contains two variables, we now learn to graph number pairs on a plane.

On a number line, each point is the graph of a number. On a plane, each point is the graph of a number pair. We use two perpendicular number lines called **axes.** They cross at a point called the **origin.** The arrows show the positive directions.

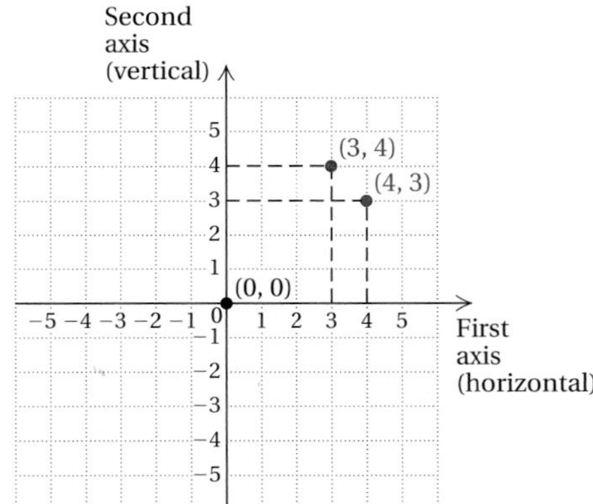

Consider the ordered pair $(3, 4)$. The numbers in an ordered pair are called **coordinates.** In $(3, 4)$, the **first coordinate (abscissa)** is 3 and the **second coordinate (ordinate)** is 4. To plot $(3, 4)$, we start at the origin and move horizontally to the 3. Then we move up vertically 4 units and make a "dot."

The point $(4, 3)$ is also plotted. Note that $(3, 4)$ and $(4, 3)$ give different points. The order of the numbers in the pair is indeed important. They are called **ordered pairs** because it makes a difference which number comes first. The coordinates of the origin are $(0, 0)$.

EXAMPLE 1 Plot the point $(-5, 2)$.

The first number, -5, is negative. Starting at the origin, we move -5 units in the horizontal direction (5 units to the left). The second number, 2, is positive. We move 2 units in the vertical direction (up).

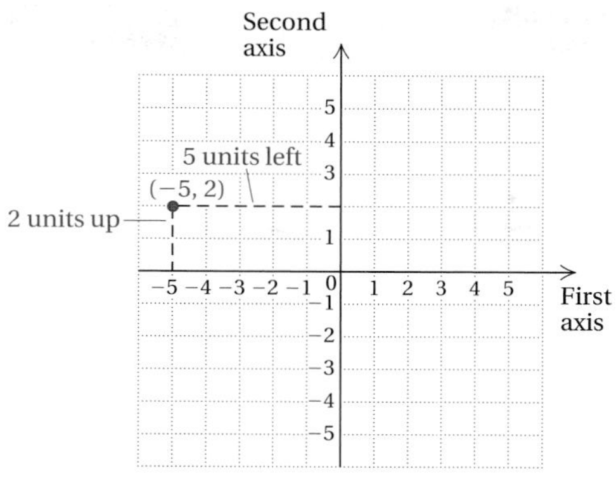

Caution!

The *first* coordinate of an ordered pair is always graphed in a *horizontal* direction and the *second* coordinate is always graphed in a *vertical* direction.

Do Exercises 1–8 on the preceding page.

The figure below shows some points and their coordinates. In region I (the *first quadrant*), both coordinates of any point are positive. In region II (the *second quadrant*), the first coordinate is negative and the second positive. In region III (the *third quadrant*), both coordinates are negative. In region IV (the *fourth quadrant*), the first coordinate is positive and the second is negative.

EXAMPLE 2 In which quadrant, if any, are the points $(-4, 5)$, $(5, -5)$, $(2, 4)$, $(-2, -5)$, and $(-5, 0)$ located?

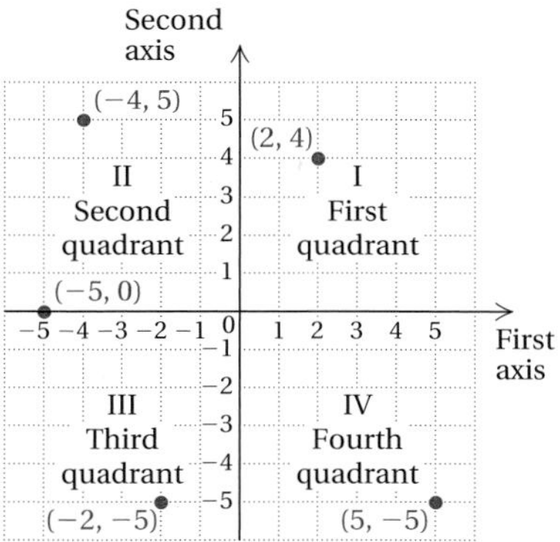

The point $(-4, 5)$ is in the second quadrant. The point $(5, -5)$ is in the fourth quadrant. The point $(2, 4)$ is in the first quadrant. The point $(-2, -5)$ is in the third quadrant. The point $(-5, 0)$ is on an axis and is *not* in any quadrant.

Do Exercises 9–15.

b Finding Coordinates

To find the coordinates of a point, we see how far to the right or left of zero it is located and how far up or down from zero.

9. What can you say about the coordinates of a point in the third quadrant?

10. What can you say about the coordinates of a point in the fourth quadrant?

In which quadrant, if any, is the point located?

11. $(5, 3)$

12. $(-6, -4)$

13. $(10, -14)$

14. $(-13, 9)$

15. $(0, -3)$

16. Find the coordinates of points A, B, C, D, E, F, and G on the graph below.

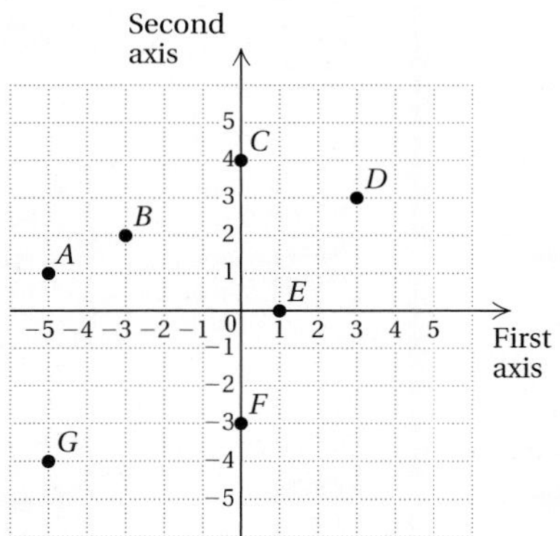

Answers on page A-24

17. Determine whether $(2, -4)$ is a solution of $4q - 3p = 22$.

18. Determine whether $(2, -4)$ is a solution of $7a + 5b = -6$.

Answers on page A-24

EXAMPLE 3 Find the coordinates of points A, B, C, D, E, F, and G.

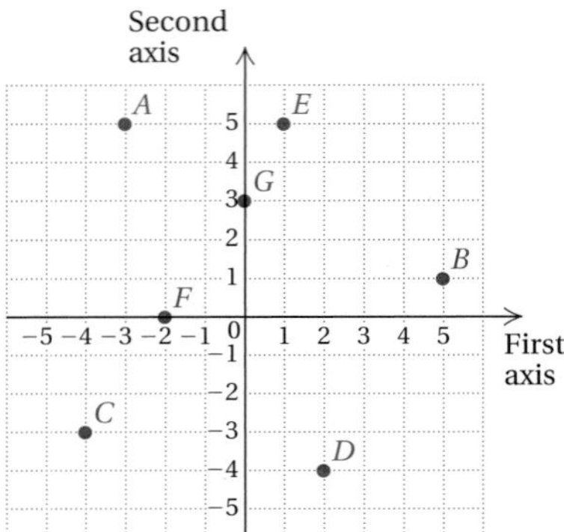

Point A is 3 units to the left (horizontal direction) and 5 units up (vertical direction). Its coordinates are $(-3, 5)$. Point D is 2 units to the right and 4 units down. Its coordinates are $(2, -4)$. The coordinates of the other points are as follows:

B: $(5, 1)$; C: $(-4, -3)$;

E: $(1, 5)$; F: $(-2, 0)$; G: $(0, 3)$.

Do Exercise 16 on the preceding page.

C Solutions of Equations

Now we begin to learn how graphs can be used to represent solutions of equations. When an equation contains two variables, the solutions of the equation are *ordered pairs* in which each number in the pair corresponds to a letter in the equation. Unless stated otherwise, to determine whether a pair is a solution, we use the first number in each pair to replace the variable that occurs first *alphabetically*.

EXAMPLE 4 Determine whether each of the following pairs is a solution of $4q - 3p = 22$: $(2, 7)$ and $(-1, 6)$.

For $(2, 7)$, we substitute 2 for p and 7 for q (using alphabetical order of variables):

$$\begin{array}{r|l} 4q - 3p & 22 \\ \hline 4 \cdot 7 - 3 \cdot 2 \ ? & 22 \\ 28 - 6 & \\ 22 & \end{array} \quad \text{TRUE}$$

Thus, $(2, 7)$ is a solution of the equation.

For $(-1, 6)$, we substitute -1 for p and 6 for q:

$$\begin{array}{r|l} 4q - 3p & 22 \\ \hline 4 \cdot 6 - 3 \cdot (-1) \ ? & 22 \\ 24 + 3 & \\ 27 & \end{array} \quad \text{FALSE}$$

Thus, $(-1, 6)$ is *not* a solution of the equation.

Do Exercises 17 and 18.

EXAMPLE 5 Show that the pairs $(3, 7)$, $(0, 1)$, and $(-3, -5)$ are solutions of $y = 2x + 1$. Then graph the three points and use the graph to determine another pair that is a solution.

To show that a pair is a solution, we substitute, replacing x with the first coordinate and y with the second coordinate of each pair:

$$\begin{array}{c|l} y & = 2x + 1 \\ \hline 7 & ?\ 2 \cdot 3 + 1 \\ & 6 + 1 \\ & 7 \qquad \text{TRUE} \end{array} \qquad \begin{array}{c|l} y & = 2x + 1 \\ \hline 1 & ?\ 2 \cdot 0 + 1 \\ & 0 + 1 \\ & 1 \qquad \text{TRUE} \end{array}$$

$$\begin{array}{c|l} y & = 2x + 1 \\ \hline -5 & ?\ 2(-3) + 1 \\ & -6 + 1 \\ & -5 \qquad \text{TRUE} \end{array}$$

In each of the three cases, the substitution results in a true equation. Thus the pairs are all solutions.

We plot the points as shown at right. The order of the points follows the alphabetical order of the variables. That is, x comes before y, so x-values are first coordinates and y-values are second coordinates. Similarly, we also label the horizontal axis as the x-axis and the vertical axis as the y-axis.

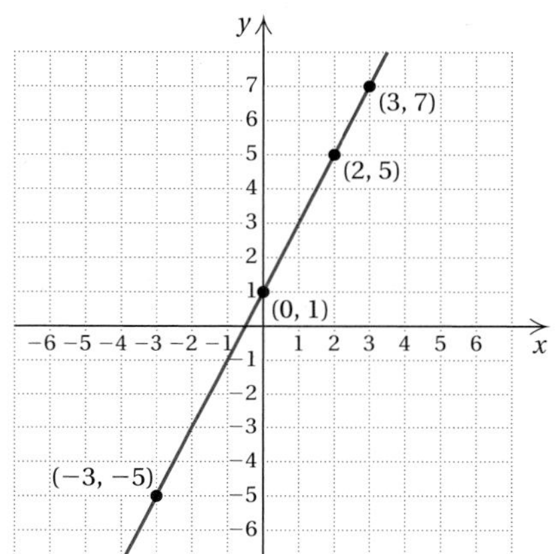

Note that the three points appear to "line up." That is, they appear to be on a straight line. Will other points that line up with these points also represent solutions of $y = 2x + 1$? To find out, we use a straightedge and lightly sketch a line passing through $(3, 7)$, $(0, 1)$, and $(-3, -5)$.

The line appears to pass through $(2, 5)$ as well. Let's see if this pair is a solution of $y = 2x + 1$:

$$\begin{array}{c|l} y & = 2x + 1 \\ \hline 5 & ?\ 2 \cdot 2 + 1 \\ & 4 + 1 \\ & 5 \qquad \text{TRUE} \end{array}$$

Thus, $(2, 5)$ is a solution.

Do Exercise 19.

Example 5 leads us to suspect that any point on the line that passes through $(3, 7)$, $(0, 1)$, and $(-3, -5)$ represents a solution of $y = 2x + 1$. In fact, every solution of $y = 2x + 1$ is represented by a point on that line and every point on that line represents a solution. The line is the *graph* of the equation.

19. Use the graph in Example 5 to find at least two more points that are solutions of $y = 2x + 1$.

Answer on page A-24

Complete the table and graph.

20. $y = -2x$

x	y	(x, y)
-3		
-1		
0		
1		
3		

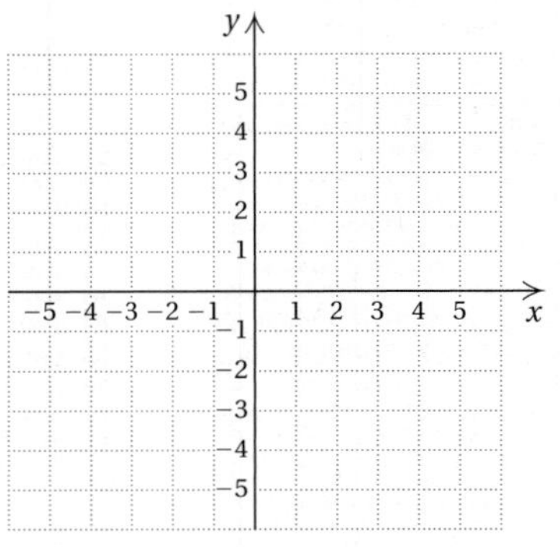

21. $y = \frac{1}{2}x$

x	y	(x, y)
4		
2		
0		
-2		
-4		
-1		

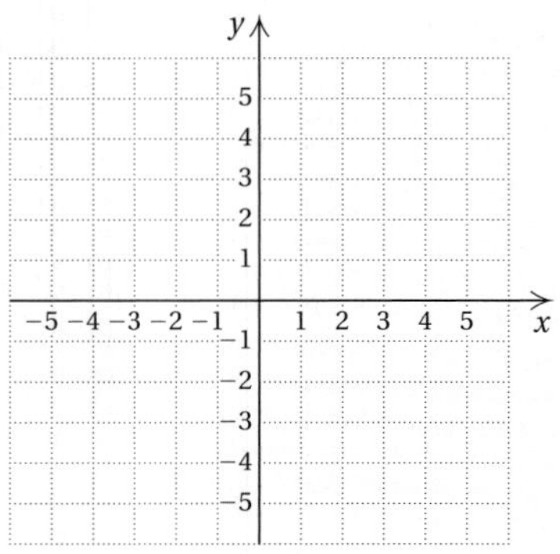

Answers on page A-24

GRAPH OF AN EQUATION

The **graph** of an equation is a drawing that represents all its solutions.

d Graphs of Linear Equations

Equations like $y = 2x + 1$ and $4q - 3p = 22$ are said to be **linear** because the graph of each equation is a straight line. In general, any equation equivalent to one of the form $y = mx + b$ or $Ax + By = C$, where m, b, A, B, and C are constants (not variables) and A and B are not both 0, is linear.

To graph a linear equation:

1. Select a value for one variable and calculate the corresponding value of the other variable. Form an ordered pair using alphabetical order as indicated by the variables.
2. Repeat step (1) to obtain at least two other ordered pairs. Two points are essential to determine a straight line. A third point serves as a check.
3. Plot the ordered pairs and draw a straight line passing through the points.

In general, calculating three (or more) ordered pairs is not difficult for equations of the form $y = mx + b$. We simply substitute values for x and calculate the corresponding values for y.

EXAMPLE 6 Graph: $y = 2x$.

First, we find some ordered pairs that are solutions. We choose *any* number for x and then determine y by substitution. Since $y = 2x$, we find y by doubling x. Suppose that we choose 3 for x. Then

$$y = 2x = 2 \cdot 3 = 6.$$

We get a solution: the ordered pair $(3, 6)$.

Suppose that we choose 0 for x. Then

$$y = 2x = 2 \cdot 0 = 0.$$

We get another solution: the ordered pair $(0, 0)$.

For a third point, we make a negative choice for x. If x is -3, we have

$$y = 2x = 2 \cdot (-3) = -6.$$

We now have enough points to plot the line, but if we wish, we can compute more. If a number takes us off the graph paper, we either do not use it or we use larger paper or rescale the axes. Continuing in this manner, we create a table like the one shown below.

Now we plot these points. We draw the line, or graph, with a straightedge and label it $y = 2x$.

x	y $y = 2x$	(x, y)
3	6	(3, 6)
1	2	(1, 2)
0	0	(0, 0)
−2	−4	(−2, −4)
−3	−6	(−3, −6)

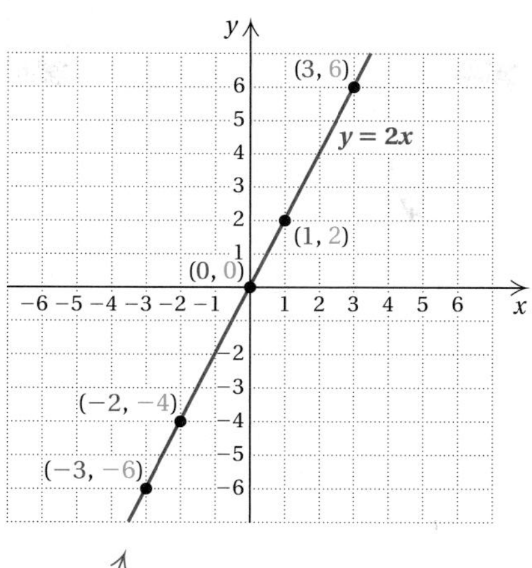

(1) Choose x.
(2) Compute y.
(3) Form the pair (x, y).
(4) Plot the points.

Caution!

Keep in mind that you can choose *any* number for x and then compute y. Our choice of certain numbers in the examples does not dictate the ones you can choose.

Do Exercises 20 and 21 on the preceding page.

EXAMPLE 7 Graph: $y = -3x + 1$.

We select a value for x, compute y, and form an ordered pair. Then we repeat the process for other choices of x.

If $x = 2$, then $y = -3 \cdot 2 + 1 = -5$, and $(2, -5)$ is a solution.
If $x = 0$, then $y = -3 \cdot 0 + 1 = 1$, and $(0, 1)$ is a solution.
If $x = -1$, then $y = -3 \cdot (-1) + 1 = 4$, and $(-1, 4)$ is a solution.

Results are often listed in a table, as shown below. The points corresponding to each pair are then plotted.

x	y $y = -3x + 1$	(x, y)
2	−5	(2, −5)
0	1	(0, 1)
−1	4	(−1, 4)

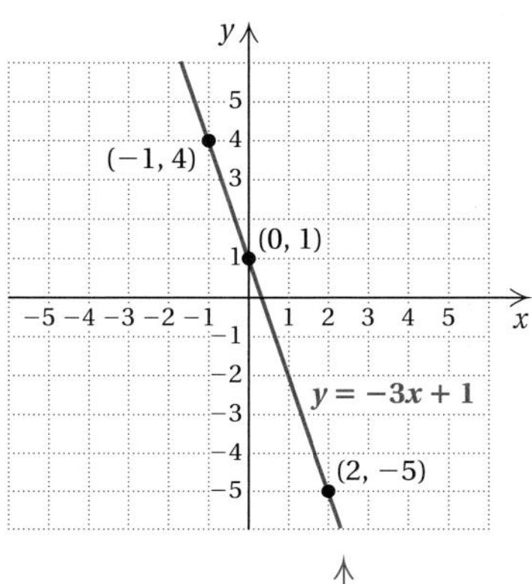

(1) Choose x.
(2) Compute y.
(3) Form the pair (x, y).
(4) Plot the points.

Graph.

22. $y = 2x + 3$

x	y	(x, y)

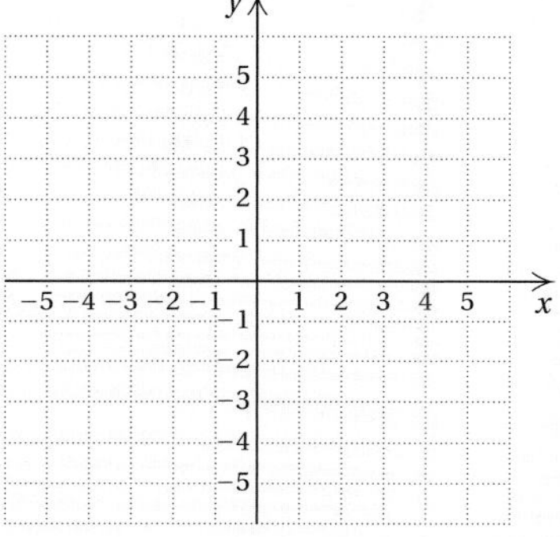

23. $y = -\frac{1}{2}x - 3$

x	y	(x, y)

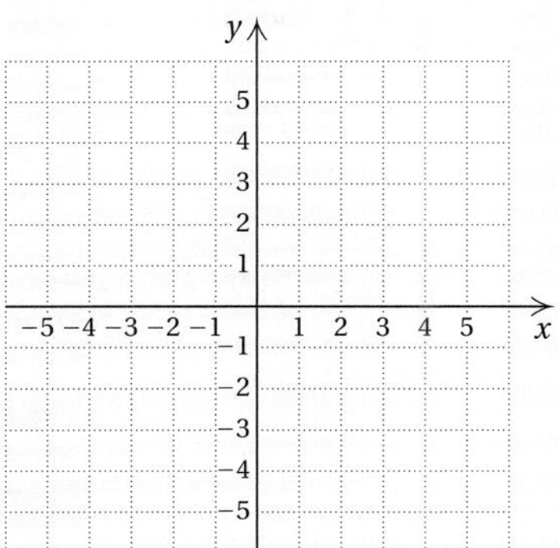

Answers on page A-24

Note that all three points line up. If they did not, we would know that we had made a mistake. When only two points are plotted, a mistake is harder to detect. We use a ruler or other straightedge to draw a line through the points. Every point on the line represents a solution of $y = -3x + 1$.

Do Exercises 22 and 23 on the preceding page.

In Example 6, we saw that $(0, 0)$ is a solution of $y = 2x$. It is also the point at which the graph crosses the y-axis. Similarly, in Example 7, we saw that $(0, 1)$ is a solution of $y = -3x + 1$. It is also the point at which the graph crosses the y-axis. A generalization can be made: If x is replaced with 0 in the equation $y = mx + b$, then the corresponding y-value is $m \cdot 0 + b$, or b. Thus any equation of the form $y = mx + b$ has a graph that passes through the point $(0, b)$. Since $(0, b)$ is the point at which the graph crosses the y-axis, it is called the **y-intercept.** Sometimes, for convenience, we simply refer to b as the y-intercept.

y-INTERCEPT

The graph of the equation $y = mx + b$ passes through the **y-intercept** $(0, b)$.

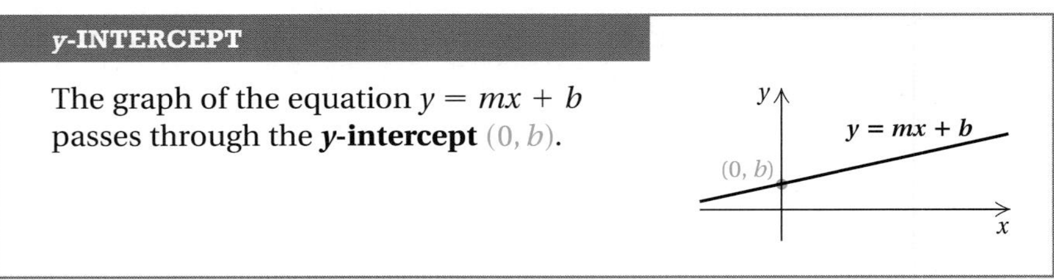

CALCULATOR CORNER

Finding Solutions of Equations A table of values representing ordered pairs that are solutions of an equation can be displayed on a graphing calculator. To do this for the equation in Example 7, $y = -3x + 1$, we first press Y= to access the equation-editor screen. Then we clear any equations that are present. (See the Calculator Corner in Section 8.3 for the procedure for doing this.) Next, we enter the equation by positioning the cursor beside "Y1 =" and pressing (−) 3 X,T,Θ,n + 1. Now we press 2ND TBLSET to display the table set-up screen. (TBLSET is the second function associated with the WINDOW key.) You can choose to supply the x-values yourself or you can set the calculator to supply them. To supply them yourself, follow the procedure for selecting ASK mode on p. 571. To have the calculator supply the x-values, set "Indpnt" to "Auto" by positioning the cursor over "Auto" and pressing ENTER. "Depend" should also be set to "Auto."

When "Indpnt" is set to "Auto," the graphing calculator will supply values of x, beginning with the value specified as TBLSTART and continuing by adding the value of △TBL to the preceding value for x. Below, we show a table of values that starts with $x = -2$ and adds 1 to the preceding x-value. We press (−) 2 ▽ 1 or (−) 2 ENTER 1 to select a minimum x-value of -2 and an increment of 1. To display the table, we press 2ND TABLE. (TABLE is the second operation associated with the GRAPH key.) We can use the △ and ▽ keys to scroll up and down through the table to see other solutions of the equation.

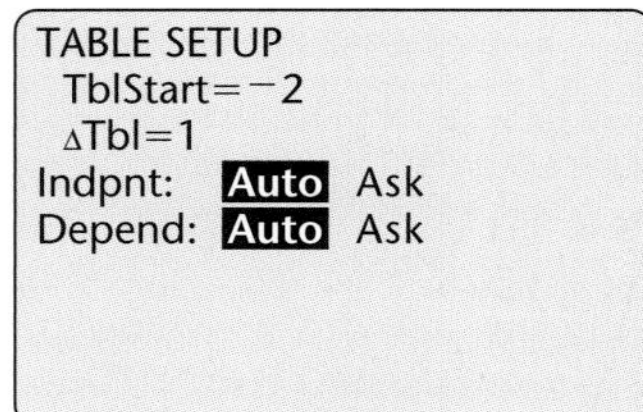

X	Y1
−2	7
−1	4
0	1
1	−2
2	−5
3	−8
4	−11

X = −2

Exercise:

1. Create a table of ordered pairs that are solutions of the equations in Examples 6 and 8.

EXAMPLE 8 Graph $y = \frac{2}{5}x + 4$ and identify the y-intercept.

We select a value for x, compute y, and form an ordered pair. Then we repeat the process for other choices of x. In this case, using multiples of 5 avoids fractions. We try to avoid graphing ordered pairs with fractions because they are difficult to graph accurately.

If $x = 0$, then $y = \frac{2}{5} \cdot 0 + 4 = 4$, and $(0, 4)$ is a solution.

If $x = 5$, then $y = \frac{2}{5} \cdot 5 + 4 = 6$, and $(5, 6)$ is a solution.

If $x = -5$, then $y = \frac{2}{5} \cdot (-5) + 4 = 2$, and $(-5, 2)$ is a solution.

The following table lists these solutions. Next, we plot the points and see that they form a line. Finally, we draw and label the line.

x	y $y = \frac{2}{5}x + 4$	(x, y)
0	4	$(0, 4)$
5	6	$(5, 6)$
-5	2	$(-5, 2)$

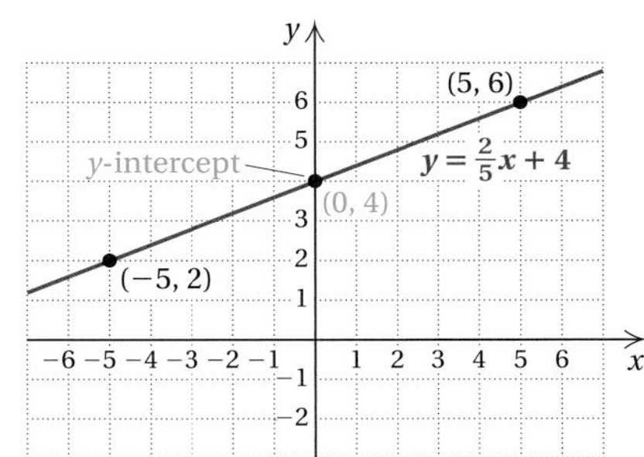

We see that $(0, 4)$ is a solution of $y = \frac{2}{5}x + 4$. It is the y-intercept. Because the equation is in the form $y = mx + b$, we can read the y-intercept directly from the equation as follows:

$$y = \frac{2}{5}x + 4 \qquad (0, 4) \text{ is the } y\text{-intercept.}$$

Do Exercises 24 and 25.

Calculating ordered pairs is generally easiest when y is isolated on one side of the equation, as in $y = mx + b$. To graph an equation in which y is not isolated, we can use the addition and multiplication principles to solve for y (see Section 8.3).

EXAMPLE 9 Graph $3y + 5x = 0$ and identify the y-intercept.

To find an equivalent equation in the form $y = mx + b$, we solve for y:

$$3y + 5x = 0$$

$$3y + 5x - 5x = 0 - 5x \qquad \text{Subtracting } 5x$$

$$3y = -5x \qquad \text{Collecting like terms}$$

$$\frac{3y}{3} = \frac{-5x}{3} \qquad \text{Dividing by 3}$$

$$y = -\frac{5}{3}x.$$

Graph the equation and identify the y-intercept.

24. $y = \frac{3}{5}x + 2$

x	y	(x, y)

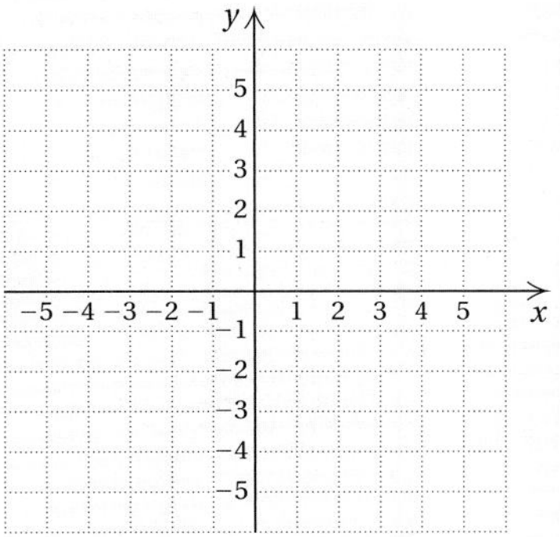

25. $y = -\frac{3}{5}x - 1$

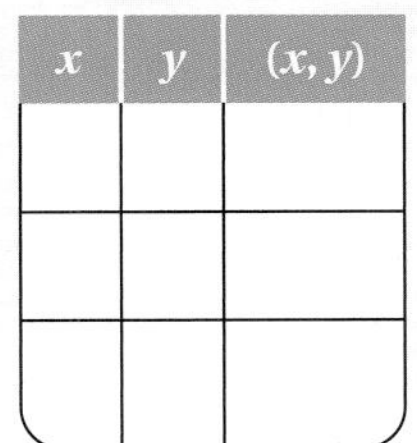

x	y	(x, y)

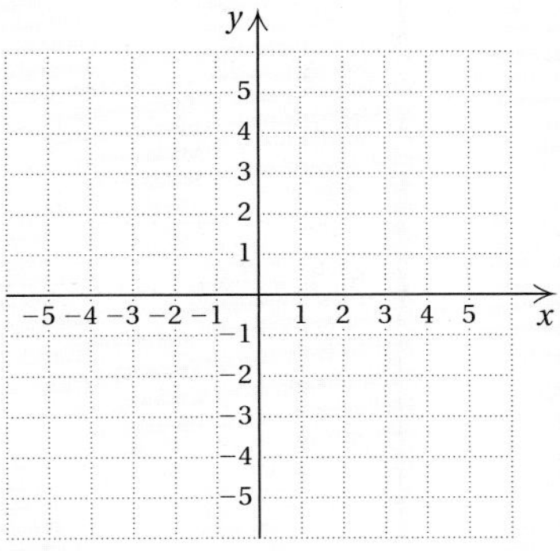

Answers on page A-24

Graph the equation and identify the y-intercept.

26. $5y + 4x = 0$

x	y	
0		←y-intercept

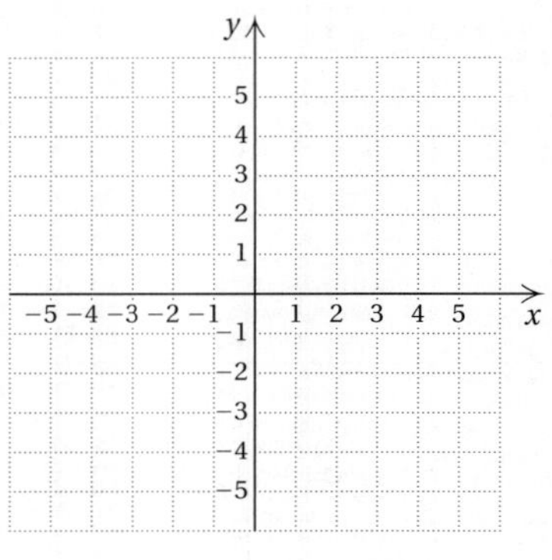

Because all the equations above are equivalent, we can use $y = -\frac{5}{3}x$ to draw the graph of $3y + 5x = 0$. To graph $y = -\frac{5}{3}x$, we select x-values and compute y-values. In this case, if we select multiples of 3, we can avoid fractions.

If $x = 0$, then $y = -\frac{5}{3} \cdot 0 = 0$.

If $x = 3$, then $y = -\frac{5}{3} \cdot 3 = -5$.

If $x = -3$, then $y = -\frac{5}{3} \cdot (-3) = 5$.

We list these solutions in a table. Next, we plot the points and see that they form a line. Finally, we draw and label the line. The y-intercept is $(0, 0)$.

x	y	
0	0	←y-intercept
3	−5	
−3	5	

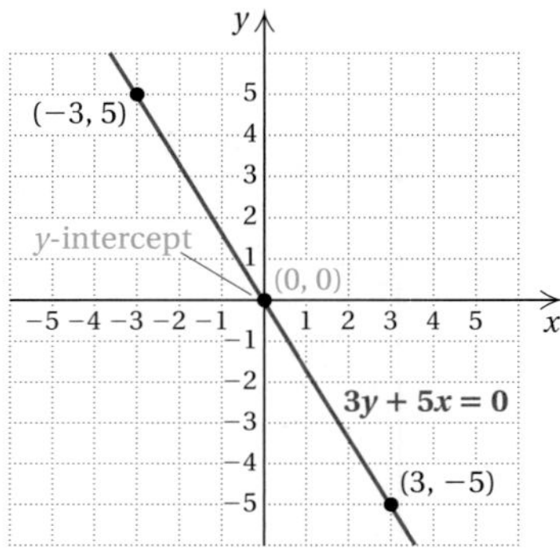

Do Exercises 26 and 27.

27. $4y = 3x$

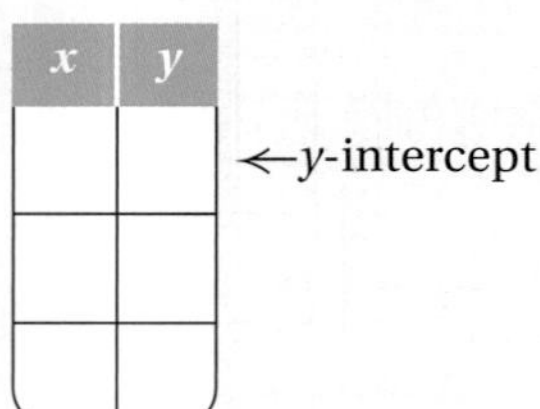

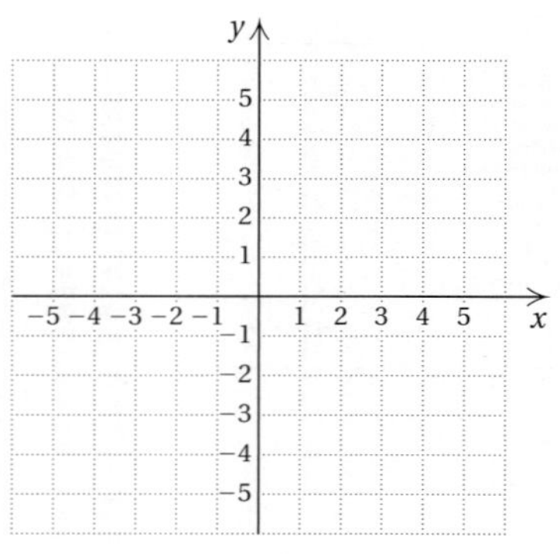

EXAMPLE 10 Graph $4y + 3x = -8$ and identify the y-intercept.

To find an equivalent equation in the form $y = mx + b$, we solve for y:

$$4y + 3x = -8$$

$$4y + 3x - 3x = -8 - 3x \quad \text{Subtracting } 3x$$

$$4y = -3x - 8 \quad \text{Simplifying}$$

$$\frac{1}{4} \cdot 4y = \frac{1}{4} \cdot (-3x - 8) \quad \text{Multiplying by } \tfrac{1}{4} \text{ or dividing by 4}$$

$$y = \frac{1}{4} \cdot (-3x) - \frac{1}{4} \cdot 8 \quad \text{Using the distributive law}$$

$$y = -\frac{3}{4}x - 2. \quad \text{Simplifying}$$

Thus, $4y + 3x = -8$ is equivalent to $y = -\frac{3}{4}x - 2$. The y-intercept is $(0, -2)$. We find two other pairs using multiples of 4 for x to avoid fractions. We then complete and label the graph as shown.

x	y	
0	−2	←y-intercept
4	−5	
−4	1	

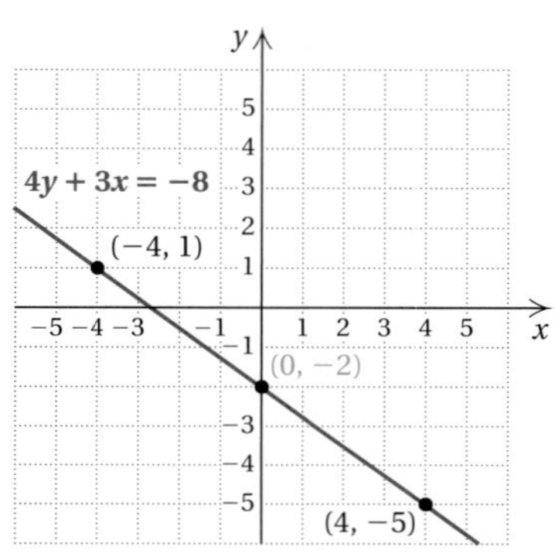

Answers on page A-24

Do Exercises 28 and 29.

e Applications of Linear Equations

Mathematical concepts become more understandable through visualization. Throughout this text, you will occasionally see the heading **AG** Algebraic–Graphical Connection, as in Example 11, which follows. In this feature, the algebraic approach is enhanced and expanded with a graphical connection. Relating a solution of an equation to a graph can often give added meaning to the algebraic solution.

EXAMPLE 11 *Online Retail Sales of Jewelry.* The online retail sales of jewelry y, in billions of dollars, is predicted by

$$y = 0.81x + 2,$$

where x is the number of years since 2003—that is, $x = 0$ corresponds to 2003, $x = 5$ corresponds to 2008, and so on.

Source: Forrester Research

a) Determine the online retail sales of jewelry in 2003, 2008, and 2015.

b) Graph the equation and then use the graph to estimate online retail sales in 2010.

c) In what year would online sales be $9.29 billion?

a) The years 2003, 2008, and 2015 correspond to $x = 0$, $x = 5$, and $x = 12$, respectively. We substitute 0, 5, and 12 for x and then calculate y:

$$y = 0.81(0) + 2 = 0 + 2 = 2;$$
$$y = 0.81(5) + 2 = 4.05 + 2 = 6.05;$$
$$y = 0.81(12) + 2 = 9.72 + 2 = 11.72.$$

Online jewelry sales in 2003, 2008, and 2015 are estimated to be $2.0 billion, $6.05 billion, and $11.72 billion, respectively.

AG ALGEBRAIC–GRAPHICAL CONNECTION

b) We have three ordered pairs from part (a). We plot these points and see that they line up. Thus our calculations are probably correct. Since we are considering only the year 2003 and the number of years since 2003 ($x \geq 0$) and since the sales, in billions of dollars, for those years will be positive ($y > 0$), we need only the first quadrant for the graph. Then we use these points to draw a straight line through them. See Figure 1 on the following page.

Graph the equation and identify the y-intercept.

28. $5y - 3x = -10$

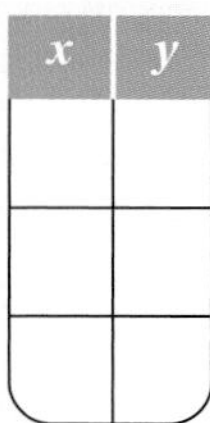

x	y

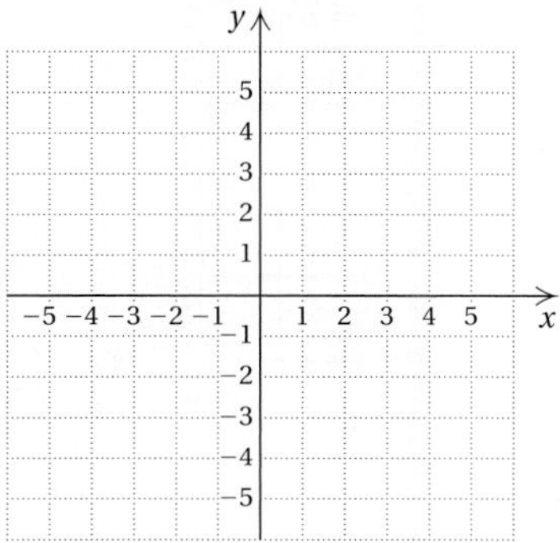

29. $5y + 3x = 20$

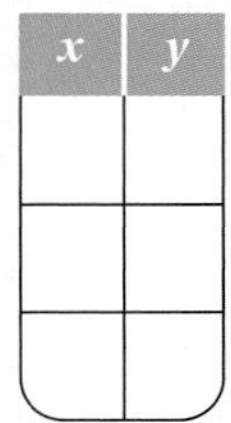

x	y

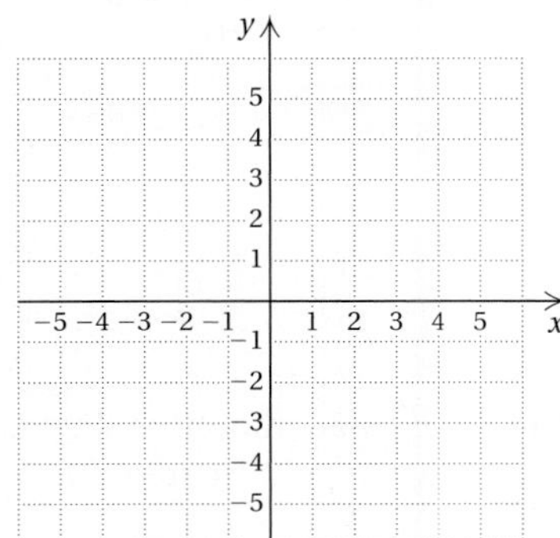

Answers on page A-24

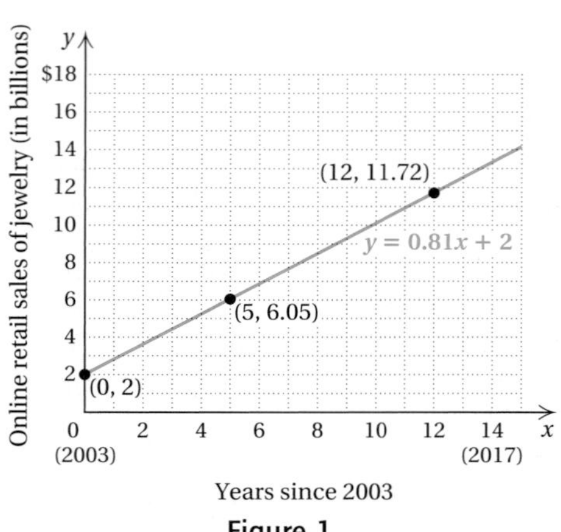

Figure 1

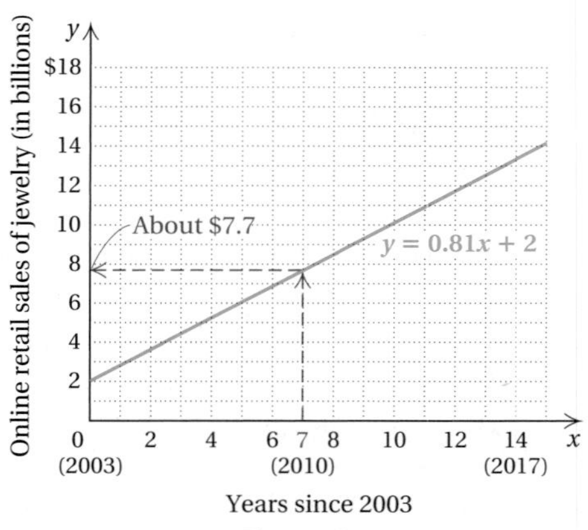

Figure 2

To use the graph to estimate sales in 2010, we first note in Figure 2 that this year corresponds to $x = 7$. We need to determine which y-value is paired with $x = 7$. We locate the point on the graph by moving up vertically from $x = 7$, and then find the value on the y-axis that corresponds to that point. It appears that online jewelry sales in 2010 will be about \$7.7 billion.

To find a more accurate value, we can simply substitute into the equation:

$$y = 0.81(7) + 2 = 5.67 + 2 = \$7.67.$$

c) We substitute 9.29 for y and solve for x:

$$y = 0.81x + 2$$
$$9.29 = 0.81x + 2 \quad \text{Substituting}$$
$$7.29 = 0.81x \quad \text{Subtracting 2}$$
$$x = 9. \quad \text{Dividing by 0.81}$$

In 9 years, or in 2012, online jewelry sales will be approximately \$9.29 billion.

Do Exercise 30.

Many equations in two variables have graphs that are not straight lines. Three such nonlinear graphs are shown below. We will cover some such graphs in the optional Calculator Corners throughout the text and in Chapter 15.

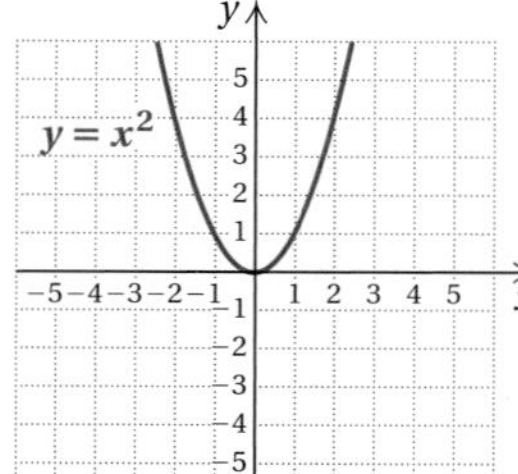

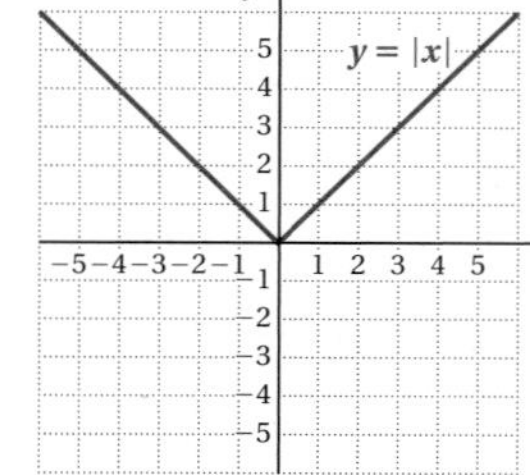

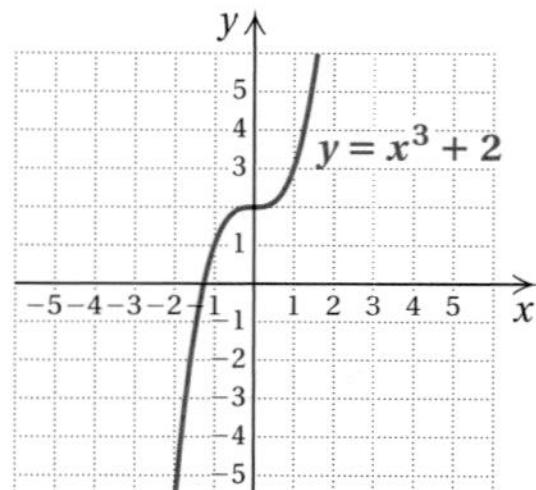

30. Value of a Color Copier. The value of Dupliographic's color copier is given by

$$v = -0.68t + 3.4,$$

where v is the value, in thousands of dollars, t years from the date of purchase.

a) Find the value after 1 yr, 2 yr, 4 yr, and 5 yr.

t	v
1	
2	
4	
5	

b) Graph the equation and use the graph to estimate the value of the copier after $2\frac{1}{2}$ yr.

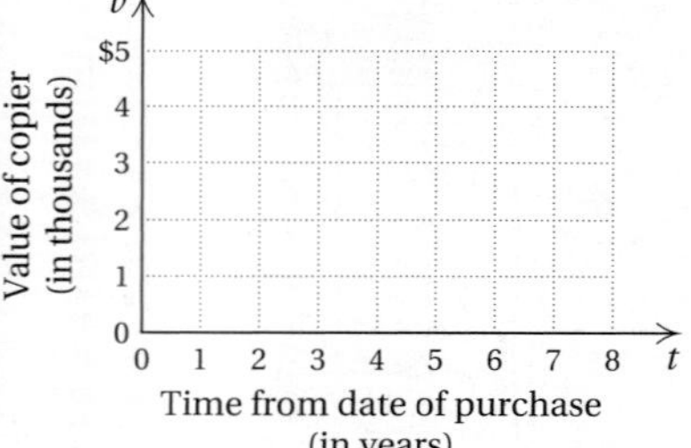

c) After what amount of time is the value of the copier \$1500?

Answers on page A-24

CALCULATOR CORNER

Graphing Equations Graphs of equations are displayed in the **viewing window** of a graphing calculator. The viewing window is the portion of the coordinate plane that appears on the calculator's screen. It is defined by the minimum and maximum values of x and y: Xmin, Xmax, Ymin, and Ymax. The notation [Xmin, Xmax, Ymin, Ymax] is used to represent these window settings or dimensions. For example, $[-12, 12, -8, 8]$ denotes a window that displays the portion of the x-axis from -12 to 12 and the portion of the y-axis from -8 to 8. In addition, the distance between tick marks on the axes is defined by the settings Xscl and Yscl. The Xres setting indicates the pixel resolution. We usually select Xres = 1. The window corresponding to the settings $[-20, 30, -12, 20]$, Xscl = 5, Yscl = 2, Xres = 1, is shown on the left below. Press (WINDOW) on the top row of the keypad of your calculator to display the current window settings. The settings for the **standard viewing window** are shown on the right below.

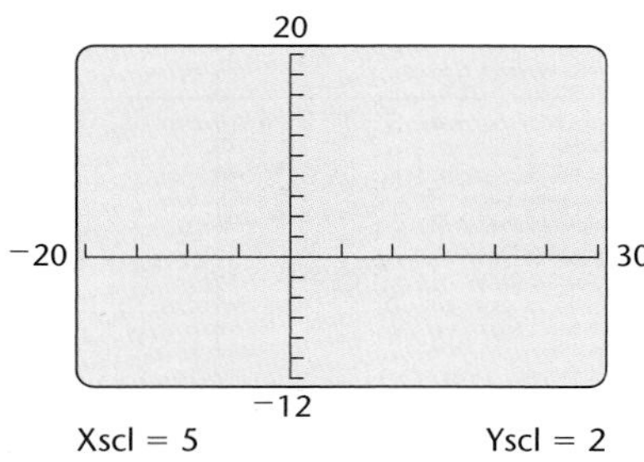

Xscl = 5 Yscl = 2

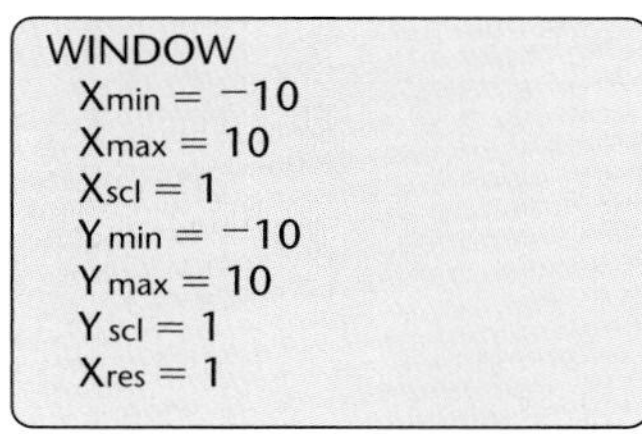

To change a setting, we position the cursor beside the setting we wish to change and enter the new value. For example, to change from the standard settings to $[-20, 30, -12, 20]$, Xscl = 5, Yscl = 2, on the WINDOW screen, we press (−) 2 0 ENTER 3 0 ENTER 5 ENTER (−) 1 2 ENTER 2 0 ENTER 2 ENTER. The ▽ key can be used instead of ENTER after typing each window setting. To see the window, we press (GRAPH) on the top row of the keypad. To return quickly to the standard window setting $[-10, 10, -10, 10]$, Xscl = 1, Yscl = 1, we press (ZOOM) (6).

Equations must be solved for y before they can be graphed on the TI-84 Plus. Consider the equation $3x + 2y = 6$. Solving for y, we get $y = \dfrac{6 - 3x}{2}$. We enter this equation as $y_1 = (6 - 3x)/2$ on the equation-editor screen as described in the Calculator Corner in Section 8.3 (see p. 571). Then we press (ZOOM) (6) to select the standard viewing window and display the graph.

Plot1 Plot2 Plot3
\Y1 = (6−3X)/2
\Y2 =
\Y3 =
\Y4 =
\Y5 =
\Y6 =
\Y7 =

$y = (6 - 3x)/2$

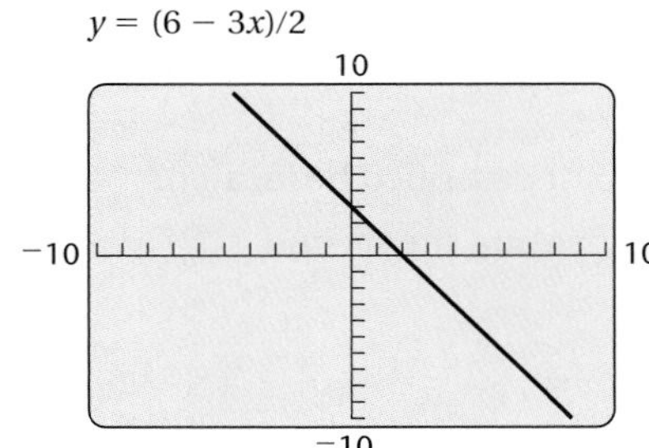

Exercises: Graph each equation in the standard viewing window $[-10, 10, -10, 10]$, Xscl = 1, Yscl = 1.

1. $y = 2x + 1$
2. $y = -3x + 1$
3. $y = -5x + 3$
4. $y = 4x - 5$
5. $4x - 5y = -10$
6. $5y + 5 = -3x$
7. $y = 2.085x + 5.08$
8. $y = -3.45x - 1.68$

9.1 EXERCISE SET

For Extra Help

MathXL | MyMathLab | InterAct Math | Math Tutor Center | Video Lectures on CD Disc 5 | Student's Solutions Manual

a

1. Plot these points.

$(2, 5)$ $(-1, 3)$ $(3, -2)$ $(-2, -4)$
$(0, 4)$ $(0, -5)$ $(5, 0)$ $(-5, 0)$

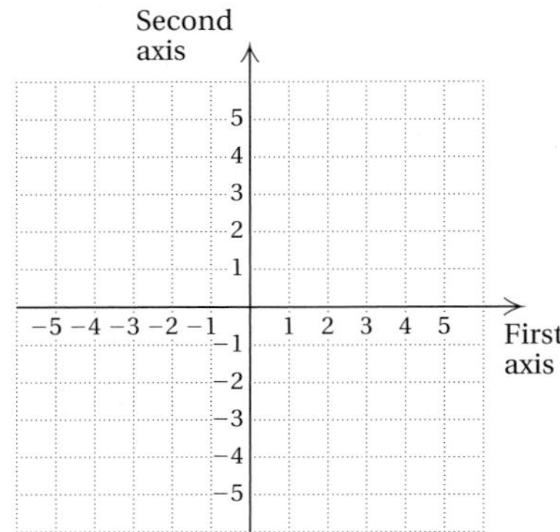

2. Plot these points.

$(4, 4)$ $(-2, 4)$ $(5, -3)$ $(-5, -5)$
$(0, 4)$ $(0, -4)$ $(3, 0)$ $(-4, 0)$

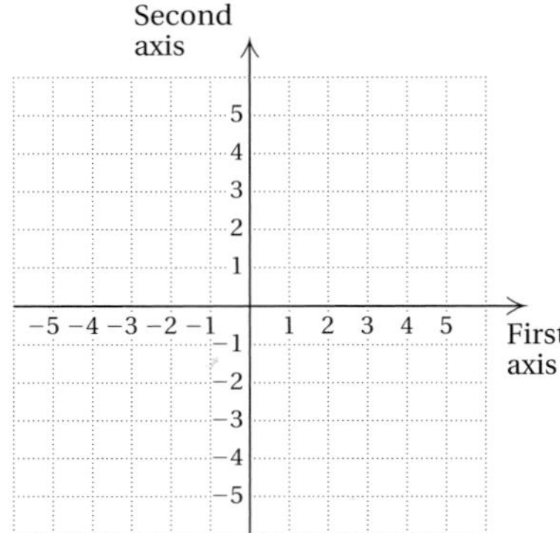

In which quadrant, if any, is the point located?

3. $(-5, 3)$

4. $(1, -12)$

5. $(100, -1)$

6. $(-2.5, 35.6)$

7. $(-6, -29)$

8. $(3.6, 105.9)$

9. $(3.8, 0)$

10. $(0, -492)$

11. $\left(-\frac{1}{3}, \frac{15}{7}\right)$

12. $\left(-\frac{2}{3}, -\frac{9}{8}\right)$

13. $\left(12\frac{7}{8}, -1\frac{1}{2}\right)$

14. $\left(23\frac{5}{8}, 81.74\right)$

In which quadrant(s) can the point described be located?

15. The first coordinate is negative and the second coordinate is positive.

16. The first and second coordinates are positive.

17. The first coordinate is positive.

18. The second coordinate is negative.

19. The first and second coordinates are equal.

20. The first coordinate is the additive inverse of the second coordinate.

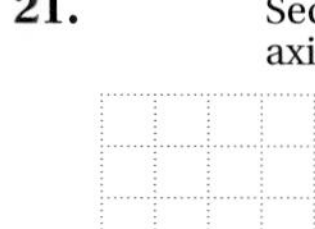

Find the coordinates of points A, B, C, D, and E.

21.

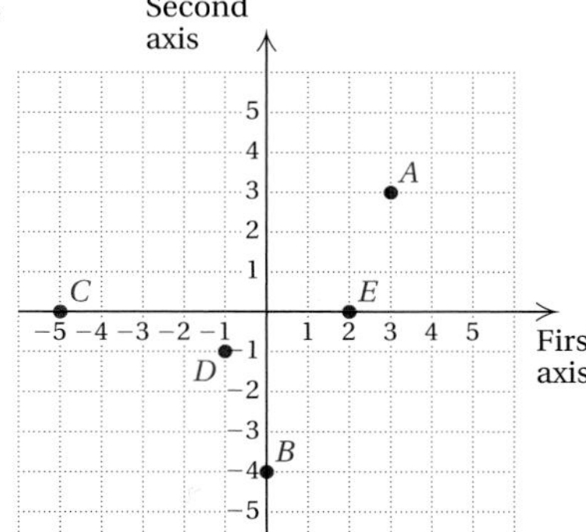

22.

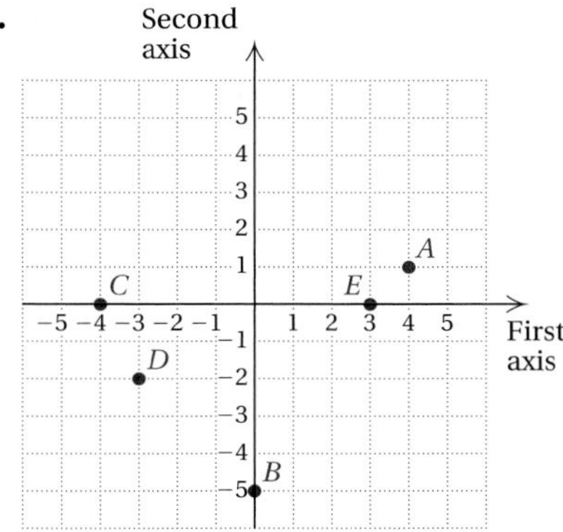

c Determine whether the given ordered pair is a solution of the equation.

23. $(2, 9)$; $y = 3x - 1$

24. $(1, 7)$; $y = 2x + 5$

25. $(4, 2)$; $2x + 3y = 12$

26. $(0, 5)$; $5x - 3y = 15$

27. $(3, -1)$; $3a - 4b = 13$

28. $(-5, 1)$; $2p - 3q = -13$

In Exercises 29–34, an equation and two ordered pairs are given. Show that each pair is a solution. Then use the graph of the two points to determine another solution. Answers may vary.

29. $y = x - 5$; $(4, -1)$ and $(1, -4)$

30. $y = x + 3$; $(-1, 2)$ and $(3, 6)$

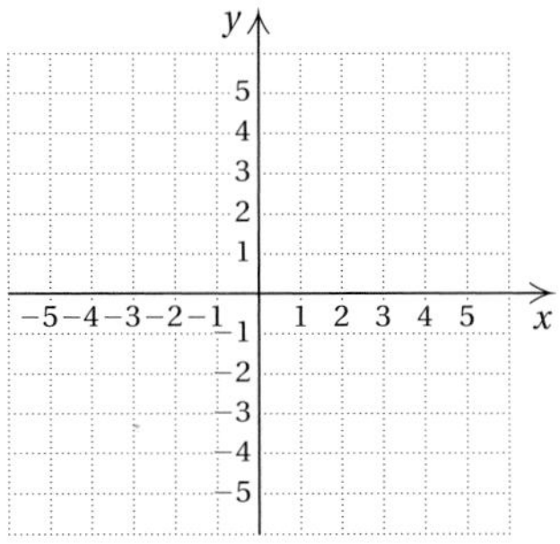

31. $y = \frac{1}{2}x + 3$; $(4, 5)$ and $(-2, 2)$

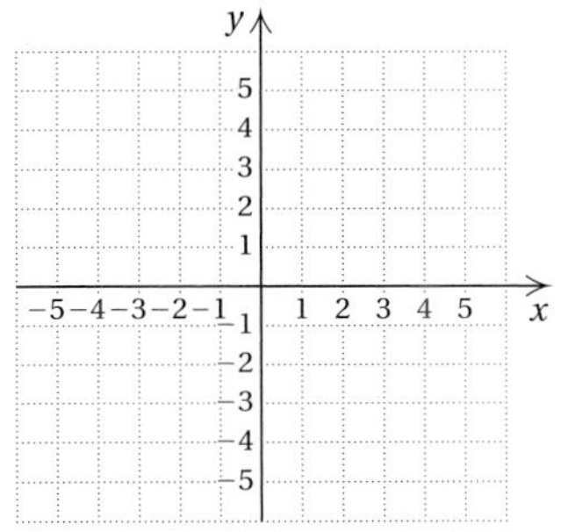

32. $3x + y = 7$; $(2, 1)$ and $(4, -5)$

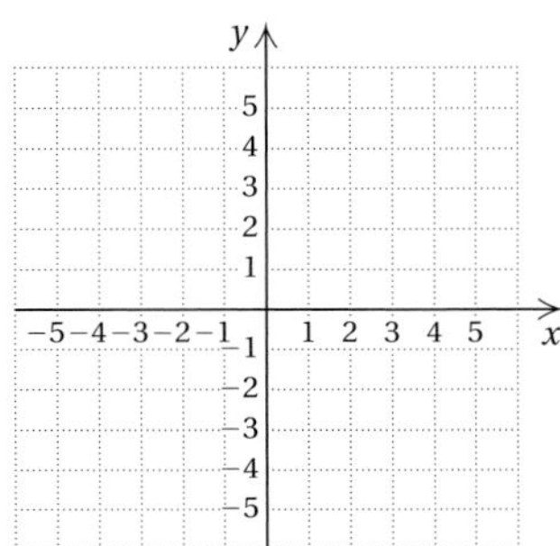

33. $4x - 2y = 10$; $(0, -5)$ and $(4, 3)$

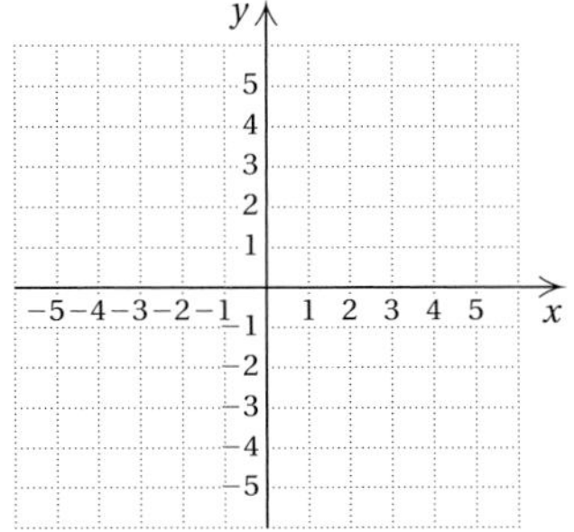

34. $6x - 3y = 3$; $(1, 1)$ and $(-1, -3)$

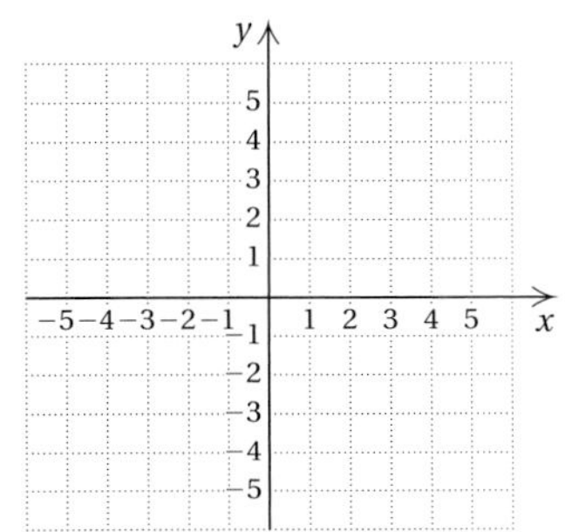

d Graph the equation and identify the y-intercept.

35. $y = x + 1$

x	y
-2	
-1	
0	
1	
2	
3	

36. $y = x - 1$

x	y
-2	
-1	
0	
1	
2	
3	

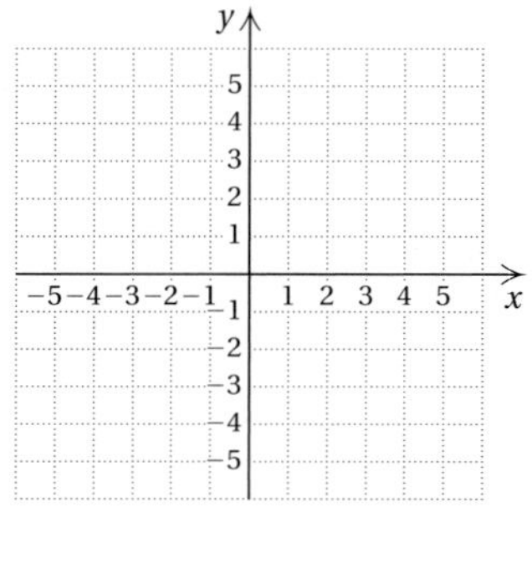

37. $y = x$

x	y
-2	
-1	
0	
1	
2	
3	

38. $y = -x$

x	y
-2	
-1	
0	
1	
2	
3	

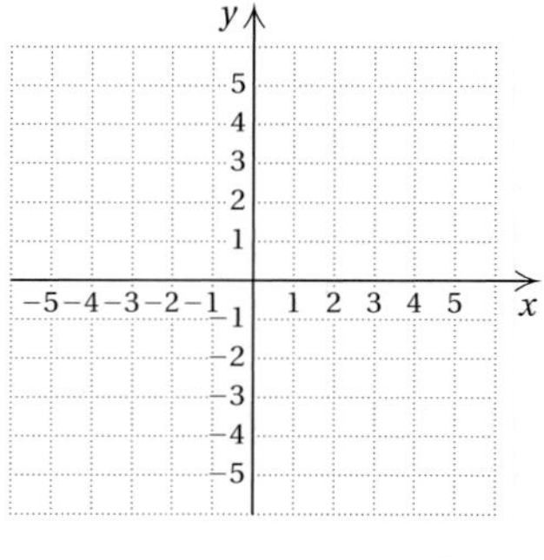

39. $y = \frac{1}{2}x$

x	y
-2	
0	
4	

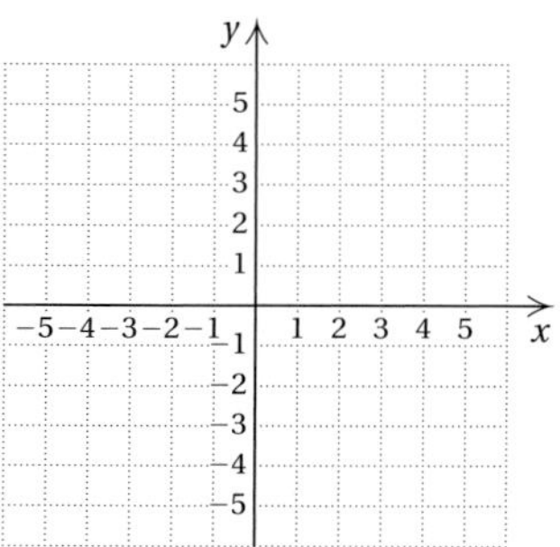

40. $y = \frac{1}{3}x$

x	y
-6	
0	
3	

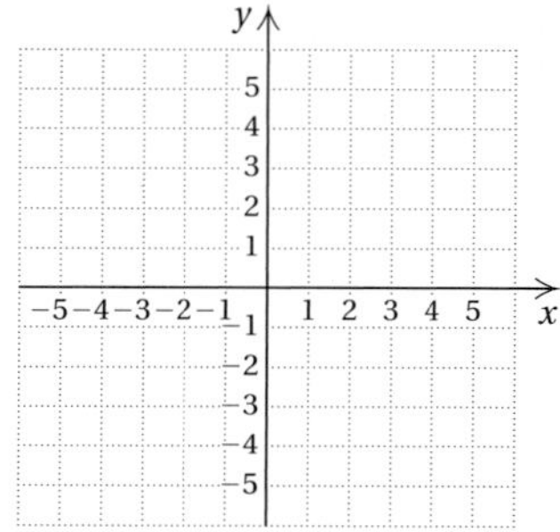

41. $y = x - 3$

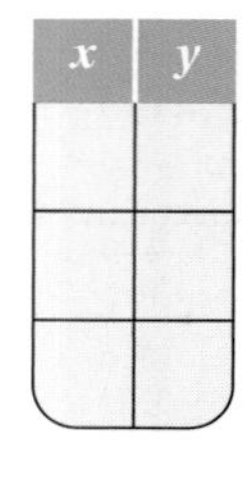

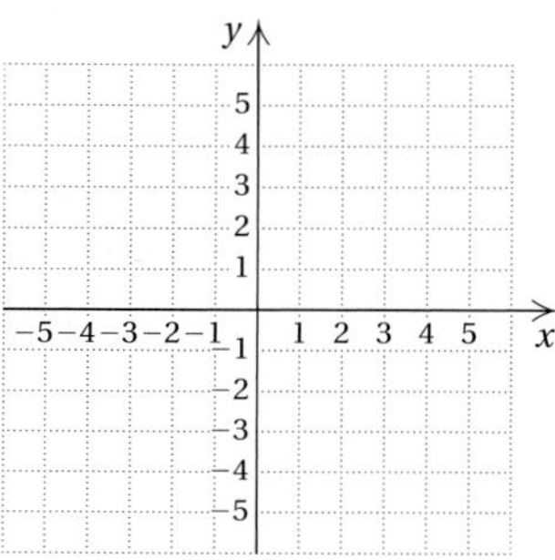

42. $y = x + 3$

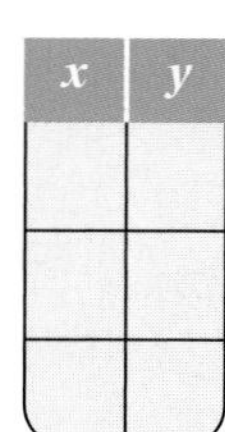

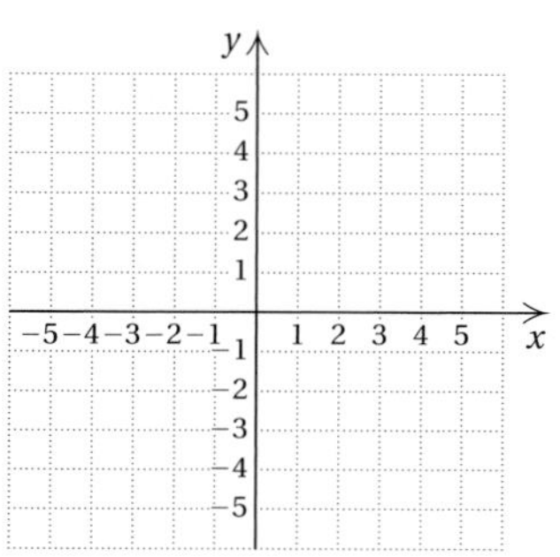

43. $y = 3x - 2$

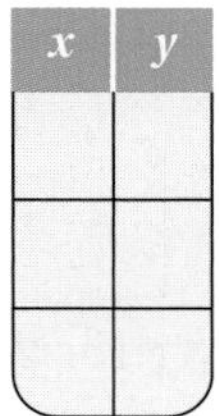

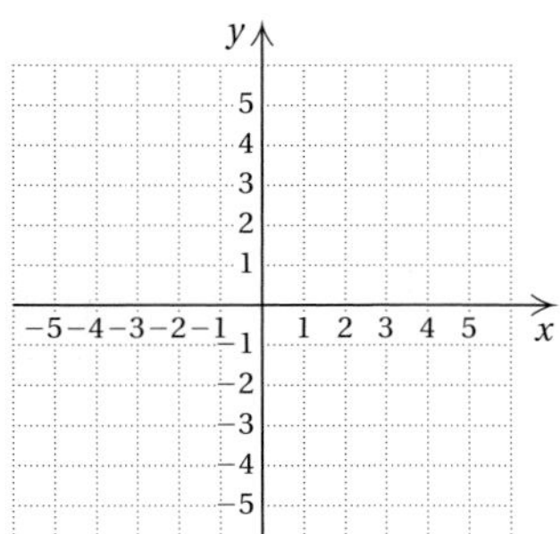

44. $y = 2x + 2$

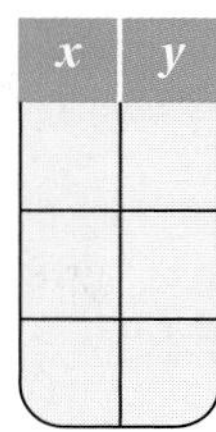

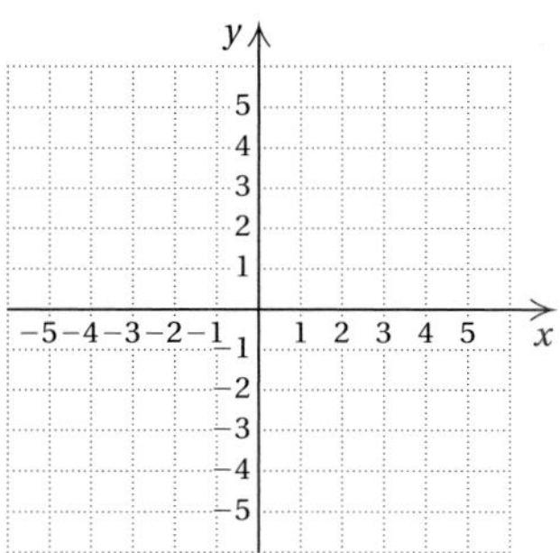

45. $y = \frac{1}{2}x + 1$

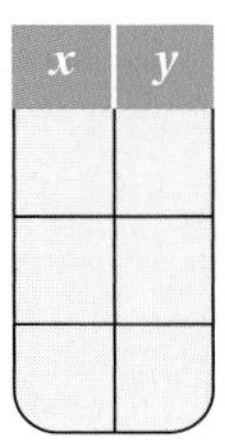

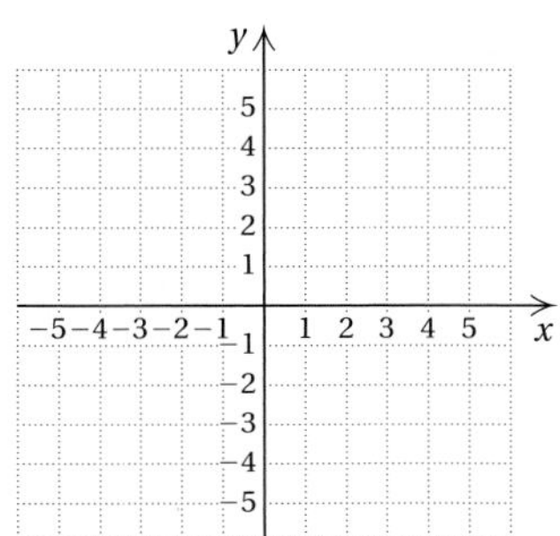

46. $y = \frac{1}{3}x - 4$

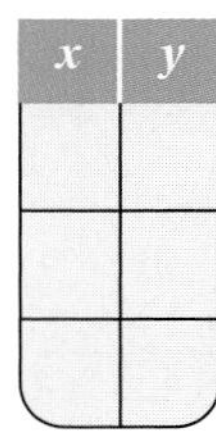

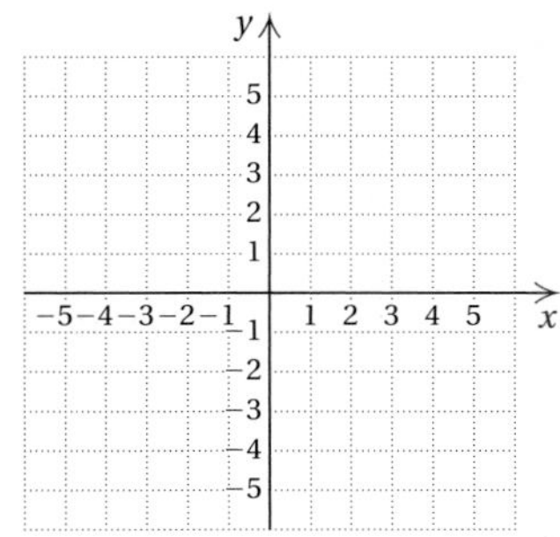

47. $x + y = -5$

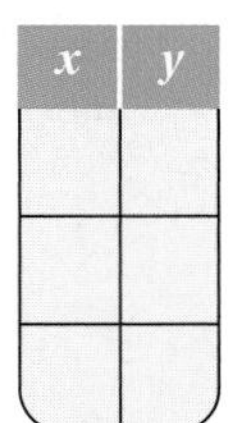

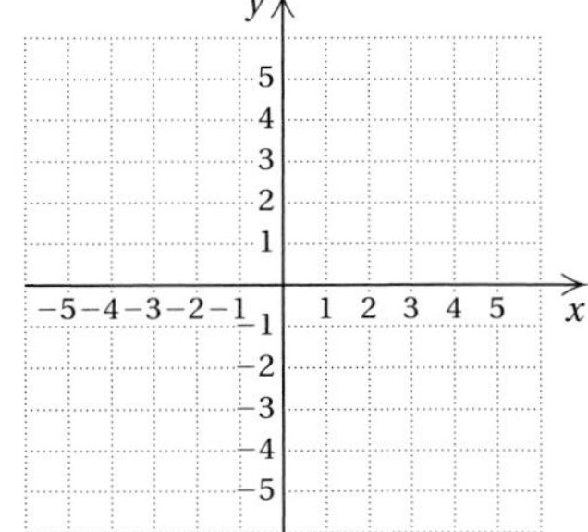

48. $x + y = 4$

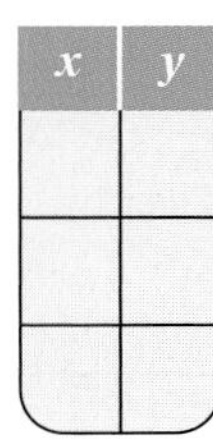

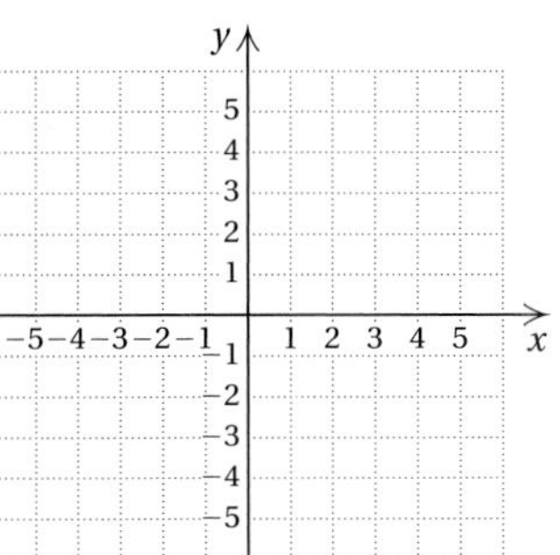

49. $y = \frac{5}{3}x - 2$

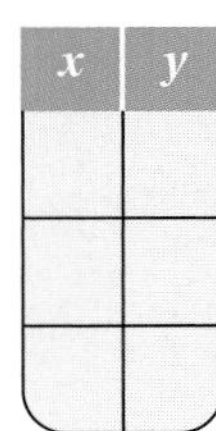

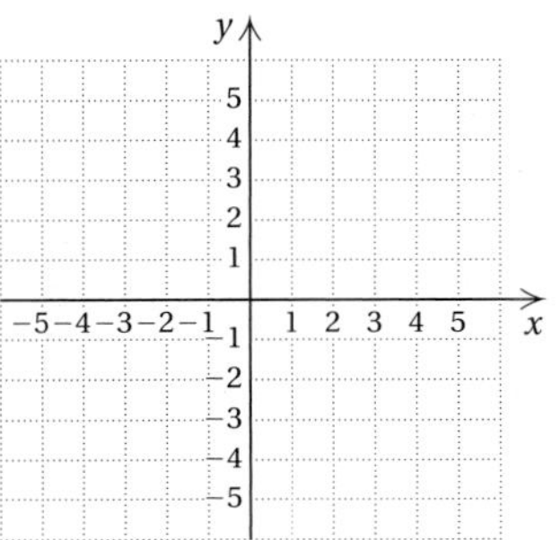

50. $y = \frac{5}{2}x + 3$

51. $x + 2y = 8$

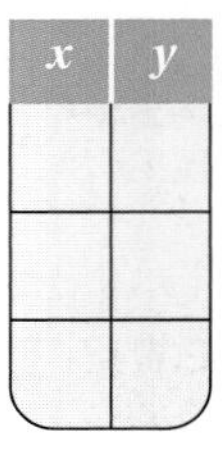

52. $x + 2y = -6$

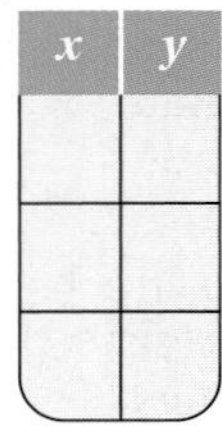

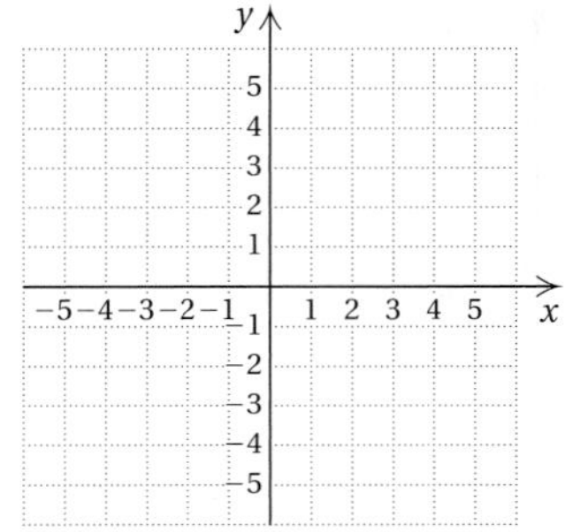

53. $y = \frac{3}{2}x + 1$

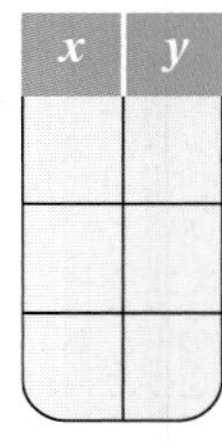

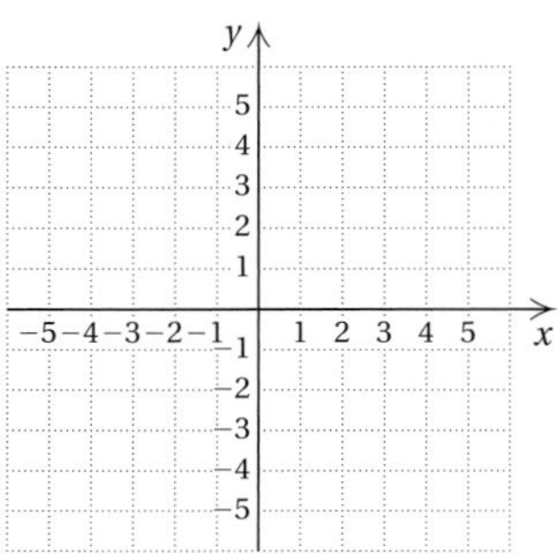

54. $y = -\frac{1}{2}x - 3$

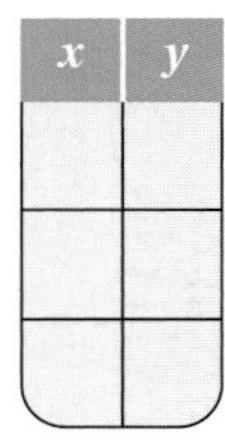

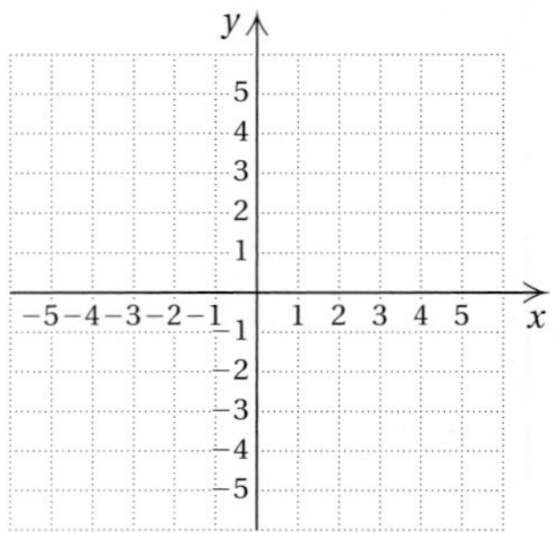

55. $8x - 2y = -10$

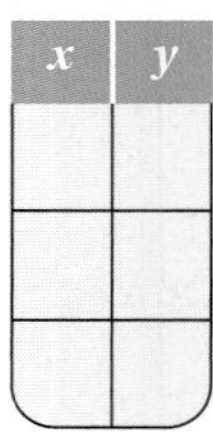

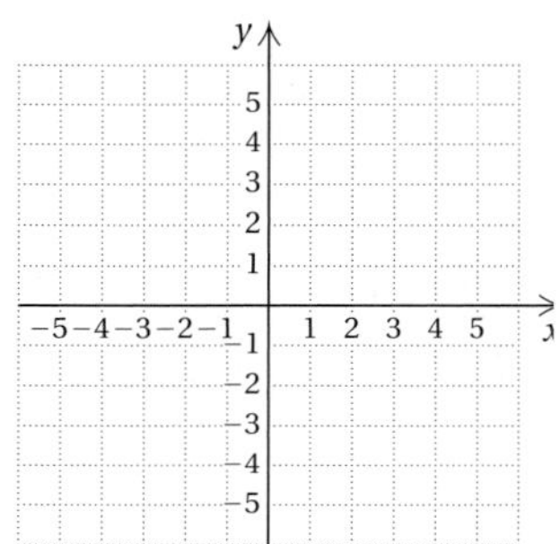

56. $6x - 3y = 9$

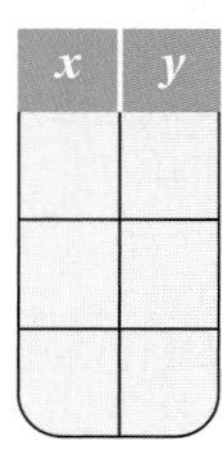

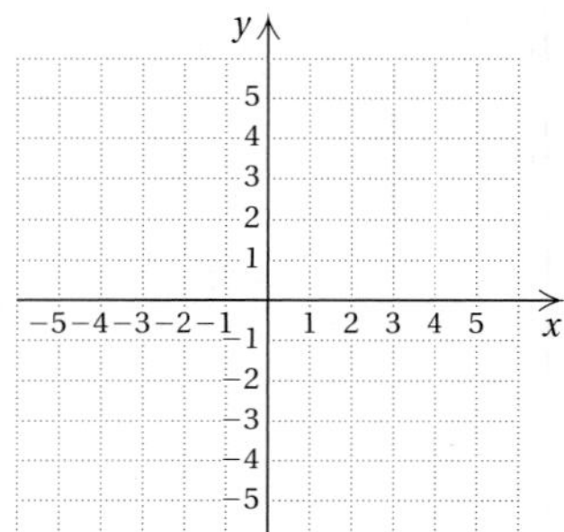

57. $8y + 2x = -4$

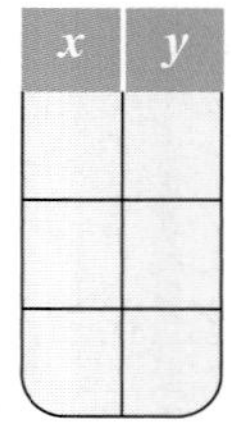

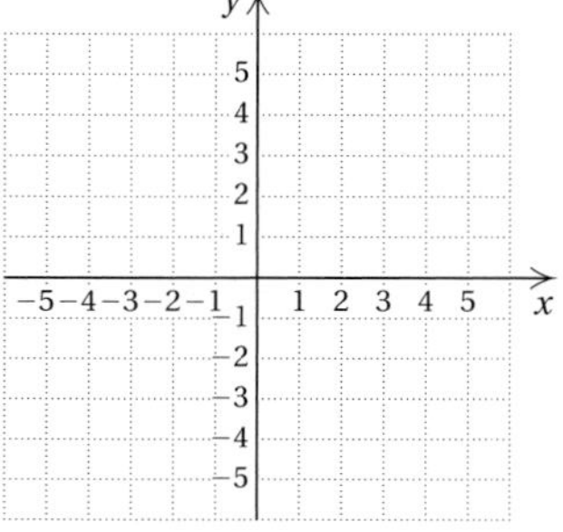

58. $6y + 2x = 8$

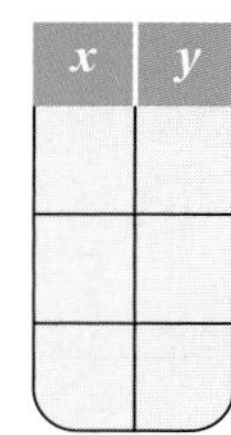

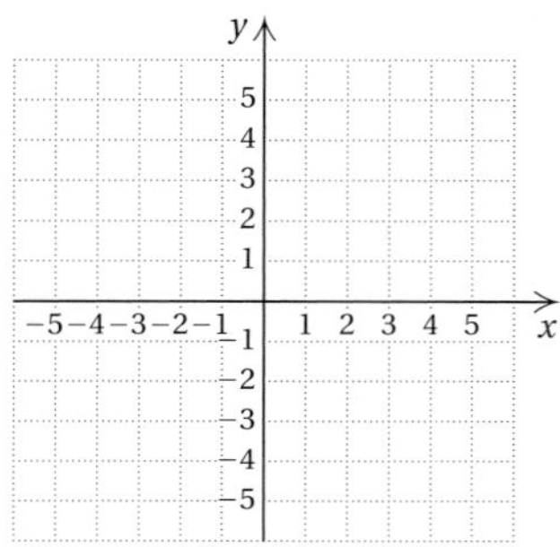

e Solve.

59. *Value of Computer Software.* The value V, in dollars, of a shopkeeper's inventory software program is given by $V = -50t + 300$, where t is the number of years since the shopkeeper first bought the program.

a) Find the value of the software after 0 yr, 4 yr, and 6 yr.

b) Graph the equation and then use the graph to estimate the value of the software after 5 yr.

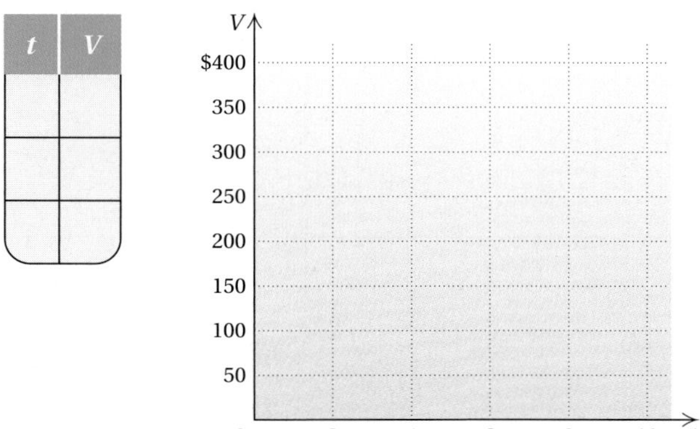

c) After how many years is the value of the software $150?

60. *SAT Math Scores.* The average SAT math scores M of potential college students can be approximated by $M = 1.4t + 518$, where t is the number of years since 2004.

Source: The College Board

a) Find the average SAT math score in 2004 ($t = 0$), 2005 ($t = 1$), and 2010.

b) Graph the equation and then use the graph to estimate the average SAT math score in 2008.

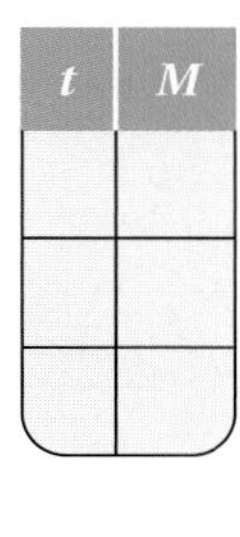

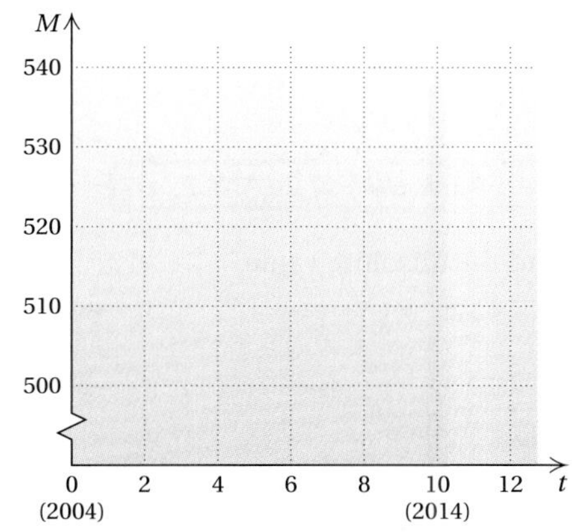

c) In what year will the average SAT math score be 539?

61. *Coffee Consumption.* The number of gallons N of coffee consumed each year by the average U.S. consumer can be approximated by $N = 0.8d + 21.2$, where d is the number of years since 1995.

Source: *Statistical Abstract of the United States,* 2003

a) Find the number of gallons of coffee consumed in 1996 ($d = 1$), 2000, 2006, and 2010.

b) Graph the equation and use the graph to estimate what the coffee consumption was in 2002.

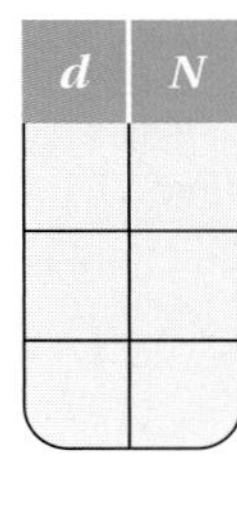

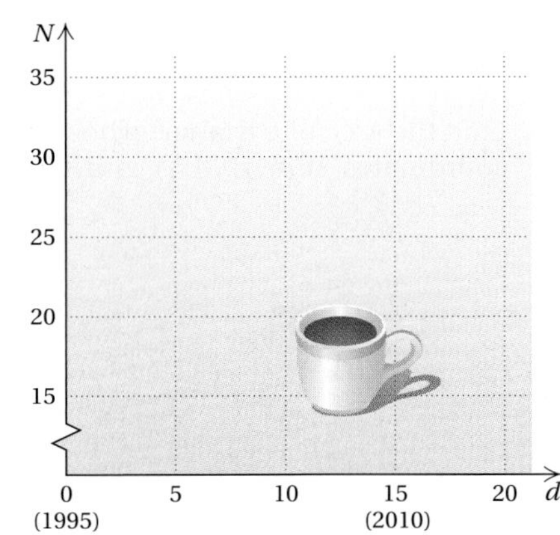

c) In what year will coffee consumption be about 31.6 gal?

62. *Record Temperature Drop.* On 22 January 1943, the temperature T, in degrees Fahrenheit, in Spearfish, South Dakota, could be approximated by $T = -2.15m + 54$, where m is the number of minutes since 9:00 that morning.

Source: *Information Please Almanac*

a) Find the temperature at 9:01 A.M., 9:08 A.M., and 9:20 A.M.

b) Graph the equation and use the graph to estimate the temperature at 9:15 A.M.

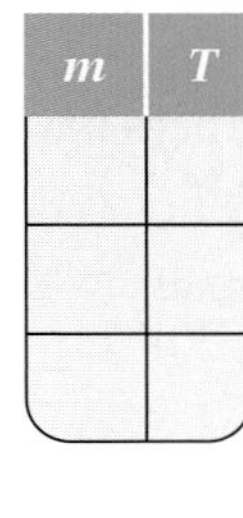

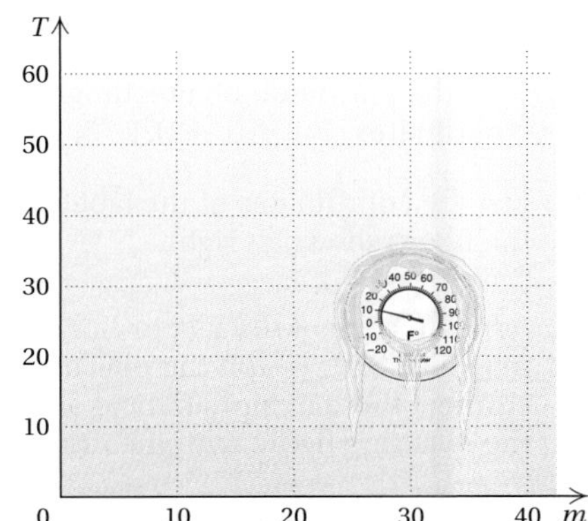

c) The temperature stopped dropping when it reached −4°F. At what time did this occur?

63. DW The equations $3x + 4y = 8$ and $y = -\frac{3}{4}x + 2$ are equivalent. Which equation is easier to graph and why?

64. DW Referring to Exercise 62, discuss why the linear equation no longer described the temperature after the temperature reached $-4°$.

SKILL MAINTENANCE

Find the absolute value. [7.2e]

65. $|-12|$

66. $|4.89|$

67. $|0|$

68. $\left|-\frac{4}{5}\right|$

69. $|-3.4|$

70. $|\sqrt{2}|$

71. $\left|\frac{2}{3}\right|$

72. $\left|-\frac{7}{8}\right|$

Solve. [8.5a]

73. *Baseball Ticket Prices.* In 2004, the average price of a ticket to a major-league baseball game was $19.82. This price was an increase of 17.9% over the price in 2000. What was the average price in 2000?

Source: Major League Baseball

74. *Tipping.* Erin left a 15% tip for a meal. The total cost of the meal, including the tip, was $21.16. What was the cost of the meal before the tip was added?

SYNTHESIS

75. The points $(-1, 1)$, $(4, 1)$, and $(4, -5)$ are three vertices of a rectangle. Find the coordinates of the fourth vertex.

76. Three parallelograms share the vertices $(-2, -3)$, $(-1, 2)$, and $(4, -3)$. Find the fourth vertex of each parallelogram.

77. Graph eight points such that the sum of the coordinates in each pair is 6.

78. Graph eight points such that the first coordinate minus the second coordinate is 1.

79. Find the perimeter of a rectangle whose vertices have coordinates $(5, 3)$, $(5, -2)$, $(-3, -2)$, and $(-3, 3)$.

80. Find the area of a triangle whose vertices have coordinates $(0, 9)$, $(0, -4)$, and $(5, -4)$.

81. List the coordinates of the labeled ordered pairs A–K in the figure shown at right.

82. Add 2 to each of the x-coordinates of the ordered pairs in Exercise 81. Graph the new ordered pairs and connect them in alphabetical order with lines. Compare the resulting figure with the original.

83. Subtract 3 from each of the y-coordinates of the ordered pairs in Exercise 81. Graph the new ordered pairs and connect them in alphabetical order with lines. Compare the resulting figure with the original.

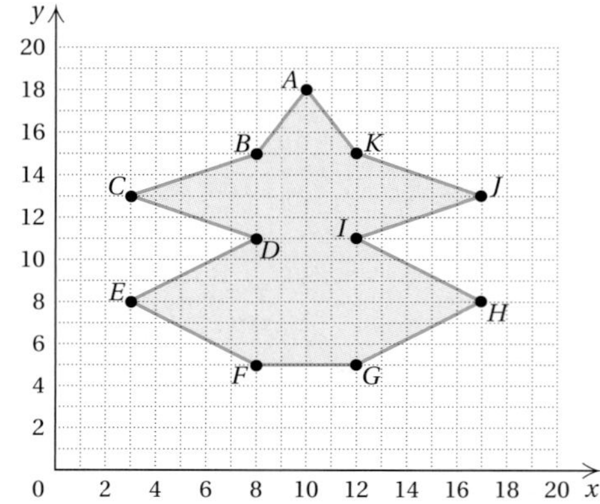

9.2 MORE WITH GRAPHING AND INTERCEPTS

a Graphing Using Intercepts

In Section 9.1, we graphed linear equations of the form $Ax + By = C$ by first solving for y to find an equivalent equation in the form $y = mx + b$. We did so because it is then easier to calculate the y-value that corresponds to a given x-value. Another convenient way to graph $Ax + By = C$ is to use **intercepts.** Look at the graph of $-2x + y = 4$ shown below.

The y-intercept is $(0, 4)$. It occurs where the line crosses the y-axis and thus will always have 0 as the first coordinate. The x-intercept is $(-2, 0)$. It occurs where the line crosses the x-axis and thus will always have 0 as the second coordinate.

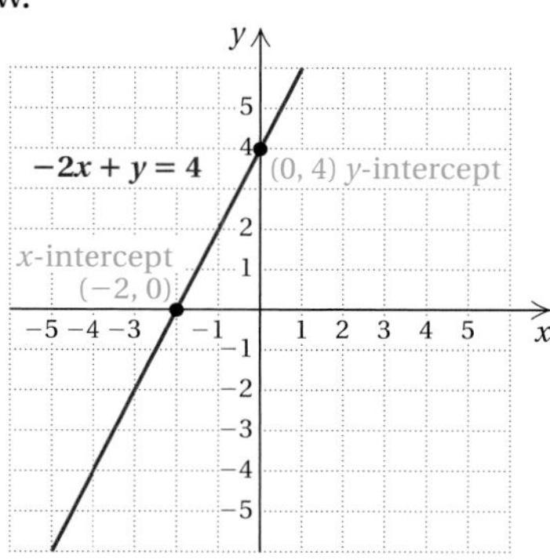

Do Exercise 1.

We find intercepts as follows.

INTERCEPTS

The **y-intercept** is $(0, b)$. To find b, let $x = 0$ and solve the original equation for y.

The **x-intercept** is $(a, 0)$. To find a, let $y = 0$ and solve the original equation for x.

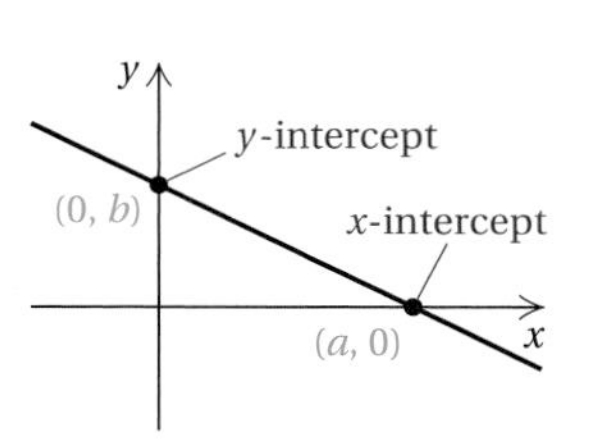

Now let's draw a graph using intercepts.

EXAMPLE 1 Consider $4x + 3y = 12$. Find the intercepts. Then graph the equation using the intercepts.

To find the y-intercept, we let $x = 0$. Then we solve for y:

$$4 \cdot 0 + 3y = 12$$
$$3y = 12$$
$$y = 4.$$

Thus, $(0, 4)$ is the y-intercept. Note that finding this intercept amounts to covering up the x-term and solving the rest of the equation.

To find the x-intercept, we let $y = 0$. Then we solve for x:

$$4x + 3 \cdot 0 = 12$$
$$4x = 12$$
$$x = 3.$$

Objectives

a Find the intercepts of a linear equation, and graph using intercepts.

b Graph equations equivalent to those of the type $x = a$ and $y = b$.

1. Look at the graph shown below.

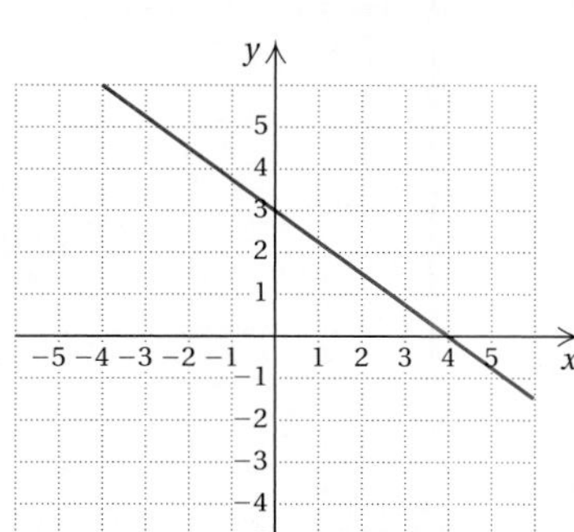

a) Find the coordinates of the y-intercept.

b) Find the coordinates of the x-intercept.

Answers on page A-26

Thus, (3, 0) is the x-intercept. Note that finding this intercept amounts to covering up the y-term and solving the rest of the equation.

We plot these points and draw the line, or graph.

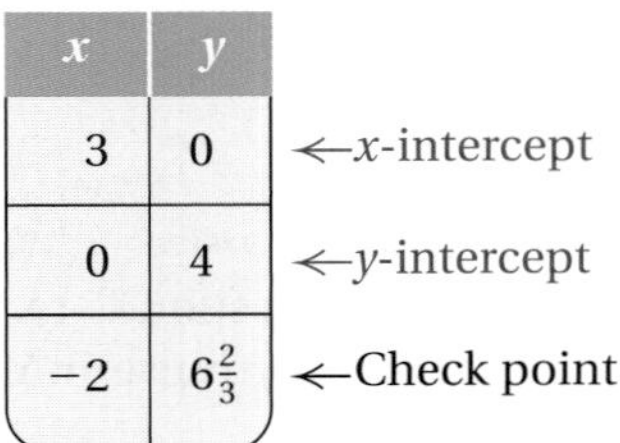

x	y	
3	0	← x-intercept
0	4	← y-intercept
−2	$6\frac{2}{3}$	← Check point

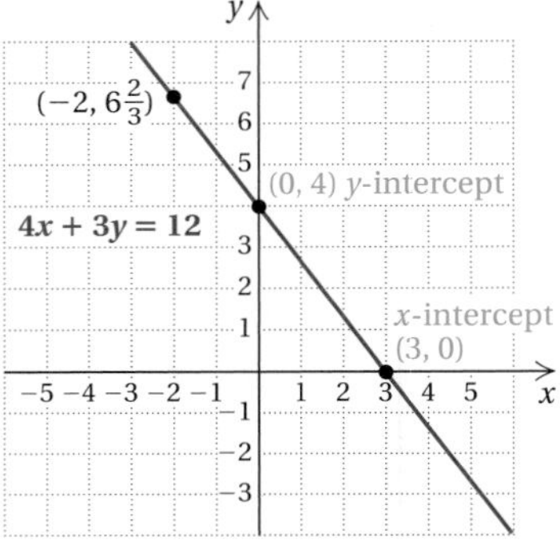

A third point should be used as a check. We substitute any convenient value for x and solve for y. In this case, we choose $x = -2$. Then

$$4(-2) + 3y = 12 \qquad \text{Substituting } -2 \text{ for } x$$
$$-8 + 3y = 12$$
$$3y = 12 + 8 = 20$$
$$y = \tfrac{20}{3}, \text{ or } 6\tfrac{2}{3}. \qquad \text{Solving for } y$$

It appears that the point $\left(-2, 6\frac{2}{3}\right)$ is on the graph, though graphing fraction values can be inexact. The graph is probably correct.

Do Exercises 2 and 3.

Graphs of equations of the type $y = mx$ pass through the origin. Thus the x-intercept and the y-intercept are the same, (0, 0). In such cases, we must calculate another point in order to complete the graph. Another point would also have to be calculated if a check is desired.

EXAMPLE 2 Graph: $y = 3x$.

We know that (0, 0) is both the x-intercept and the y-intercept. We calculate values at two other points and complete the graph, knowing that it passes through the origin (0, 0).

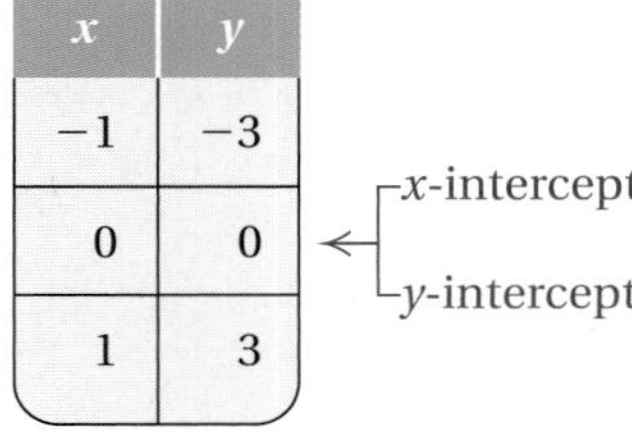

x	y	
−1	−3	
0	0	← x-intercept, y-intercept
1	3	

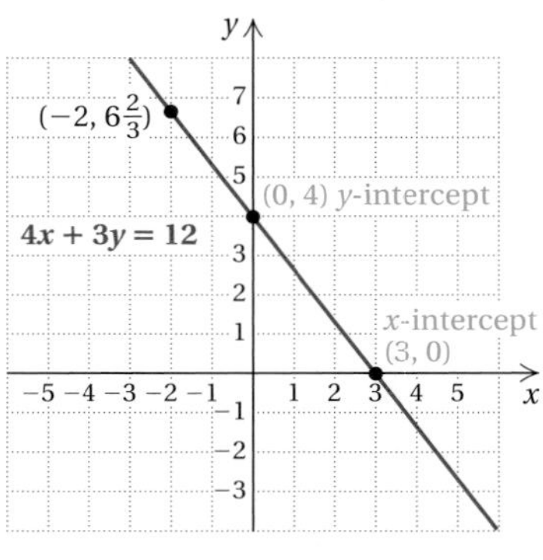

Do Exercises 4 and 5 on the following page.

For each equation, find the intercepts. Then graph the equation using the intercepts.

2. $2x + 3y = 6$

x	y	
	0	← x-intercept
0		← y-intercept
		← Check point

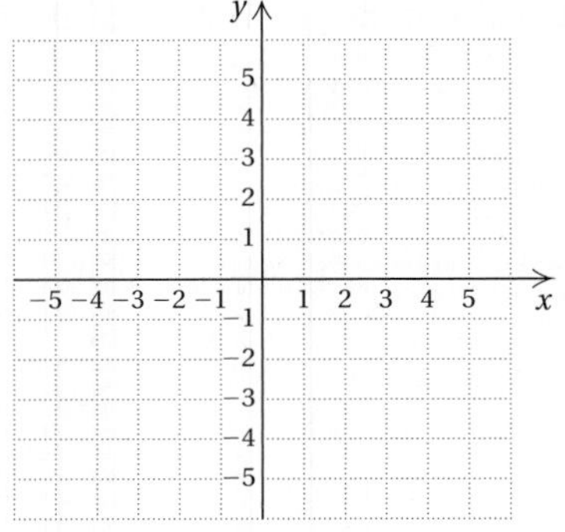

3. $3y - 4x = 12$

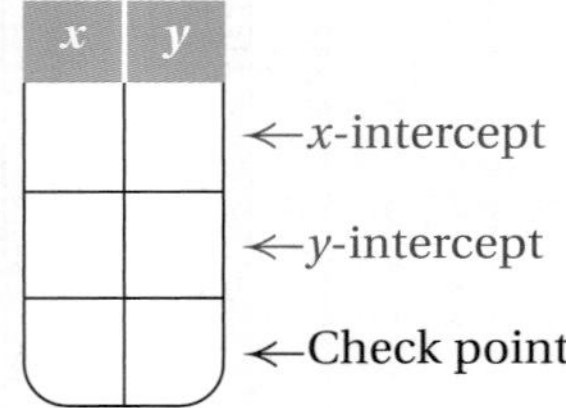

x	y	
		← x-intercept
		← y-intercept
		← Check point

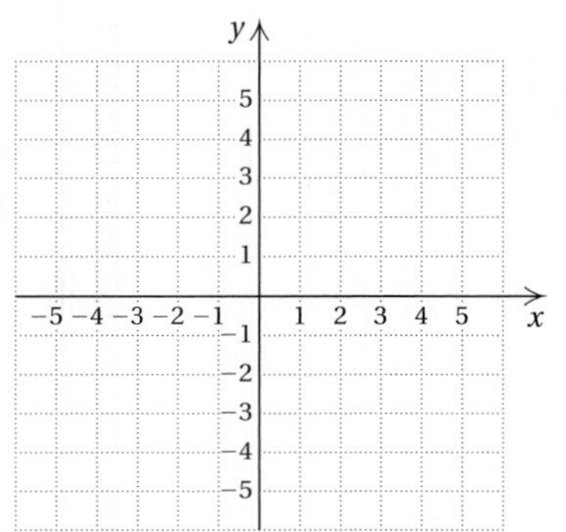

Answers on page A-26

CALCULATOR CORNER

Viewing the Intercepts Knowing the intercepts of a linear equation helps us to determine a good viewing window for the graph of the equation. For example, when we graph the equation $y = -x + 15$ in the standard window, we see only a small portion of the graph in the upper righthand corner of the screen, as shown on the left below.

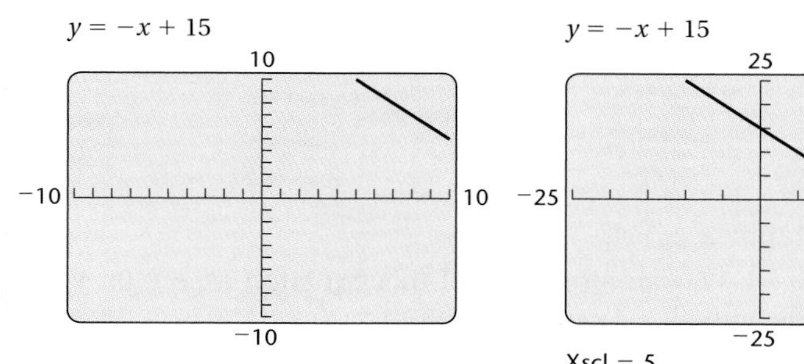

Using algebra, as we did in Example 1, we find that the intercepts of the graph of this equation are (0, 15) and (15, 0). This tells us that, if we are to see more of the graph than is shown on the left above, both Xmax and Ymax should be greater than 15. We can try different window settings until we find one that suits us. One good choice is [−25, 25, −25, 25], Xscl = 5, Yscl = 5, shown on the right above.

Exercises: Find the intercepts of each equation algebraically. Then graph the equation on a graphing calculator, choosing window settings that allow the intercepts to be seen clearly. (Settings may vary.)

1. $y = -7.5x - 15$
2. $y - 2.15x = 43$
3. $6x - 5y = 150$
4. $y = 0.2x - 4$
5. $y = 1.5x - 15$
6. $5x - 4y = 2$

b Equations Whose Graphs Are Horizontal or Vertical Lines

EXAMPLE 3 Graph: $y = 3$.

Consider $y = 3$. We can also think of this equation as $0 \cdot x + y = 3$. No matter what number we choose for x, we find that y is 3. We make up a table with all 3's in the y-column.

x	y
	3
	3
	3

y must be 3.

Choose any number for x. →

x	y
−2	3
0	3
4	3

← y-intercept (row 0, 3)

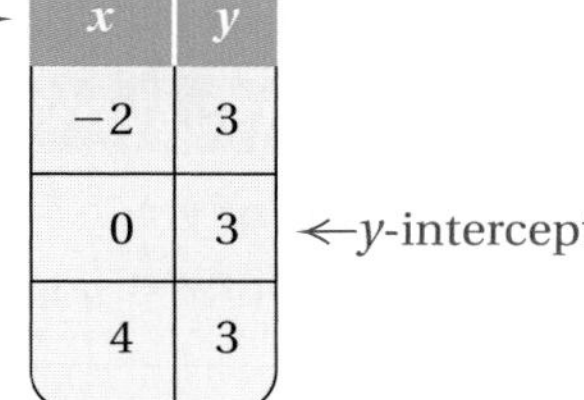

Graph.

4. $y = 2x$

x	y
−1	
0	
1	

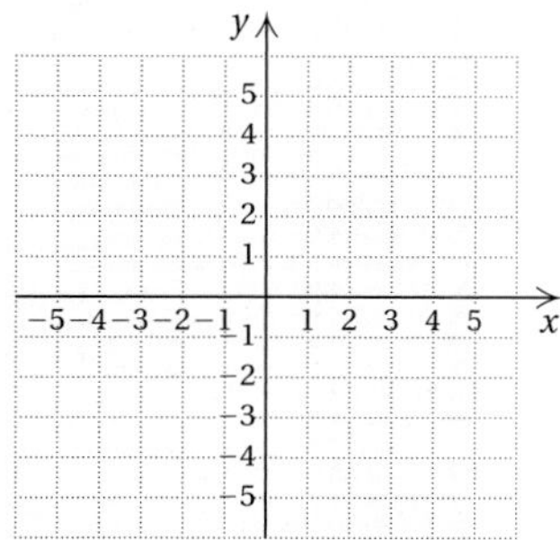

5. $y = -\frac{2}{3}x$

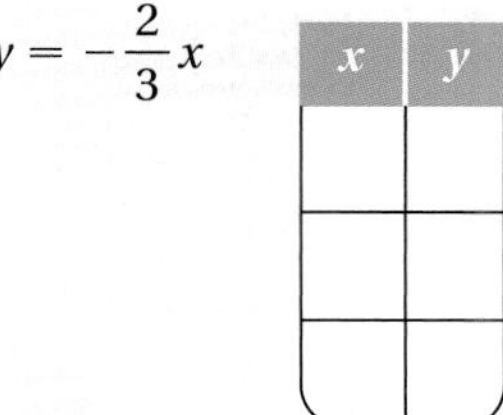

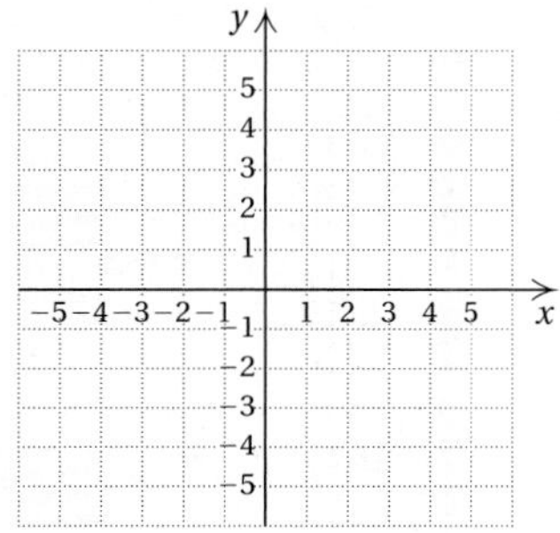

Answers on page A-26

Graph.

6. $x = 5$

x	y
5	
5	
5	

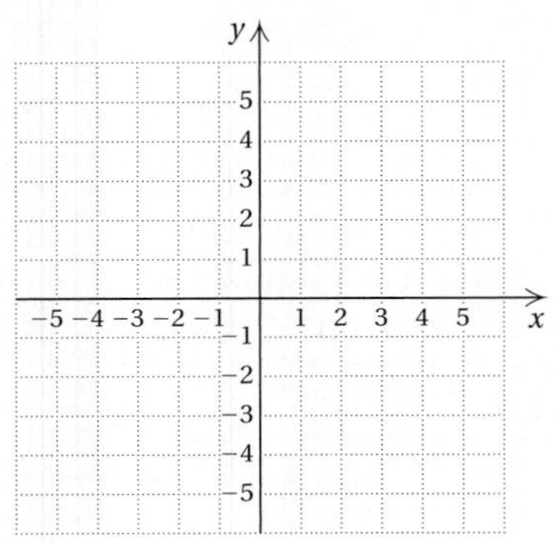

7. $y = -2$

x	y
	-2
	-2
	-2

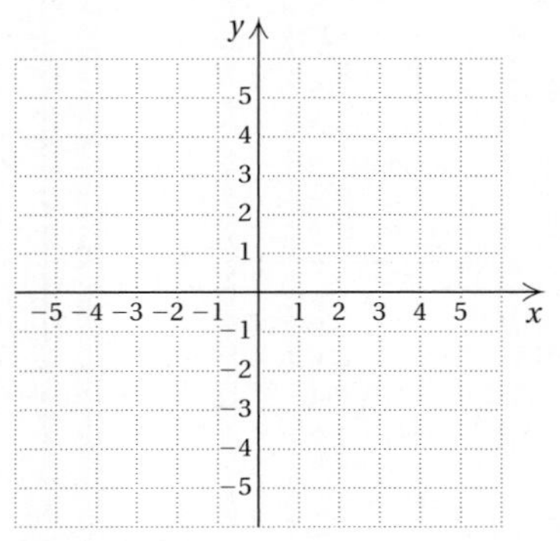

Answers on page A-26

When we plot the ordered pairs $(-2, 3)$, $(0, 3)$, and $(4, 3)$ and connect the points, we obtain a horizontal line. Any ordered pair $(x, 3)$ is a solution. So the line is parallel to the x-axis with y-intercept $(0, 3)$.

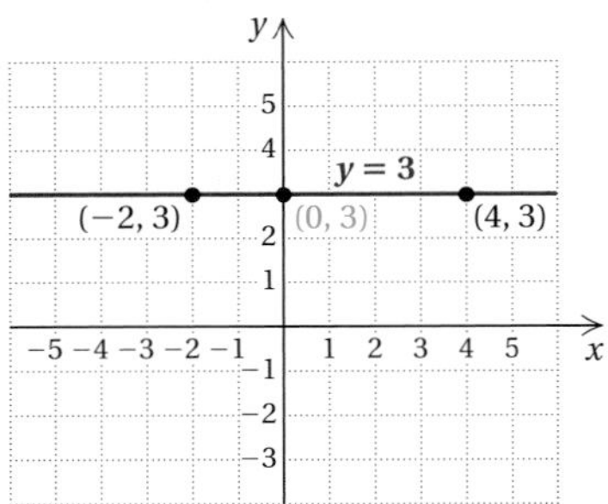

EXAMPLE 4 Graph: $x = -4$.

Consider $x = -4$. We can also think of this equation as $x + 0 \cdot y = -4$. We make up a table with all -4's in the x-column.

x	y
-4	
-4	
-4	
-4	

x must be -4.

Choose any number for y. →

x	y
-4	-5
-4	1
-4	3
-4	0

x-intercept → (last row)

When we plot the ordered pairs $(-4, -5)$, $(-4, 1)$, $(-4, 3)$, and $(-4, 0)$ and connect the points, we obtain a vertical line. Any ordered pair $(-4, y)$ is a solution. So the line is parallel to the y-axis with x-intercept $(-4, 0)$.

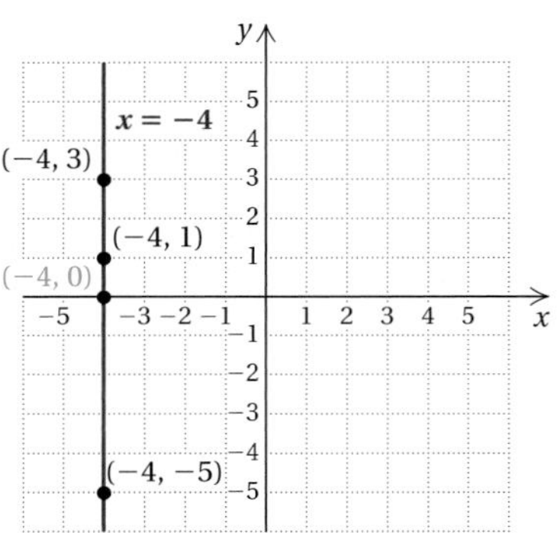

HORIZONTAL AND VERTICAL LINES

The graph of $y = b$ is a **horizontal line.** The y-intercept is $(0, b)$.
The graph of $x = a$ is a **vertical line.** The x-intercept is $(a, 0)$.

Do Exercises 6–9. (Exercises 8 and 9 are on the following page.)

The following is a general procedure for graphing linear equations.

GRAPHING LINEAR EQUATIONS

1. If the equation is of the type $x = a$ or $y = b$, the graph will be a line parallel to an axis; $x = a$ is vertical and $y = b$ is horizontal.
   Examples.

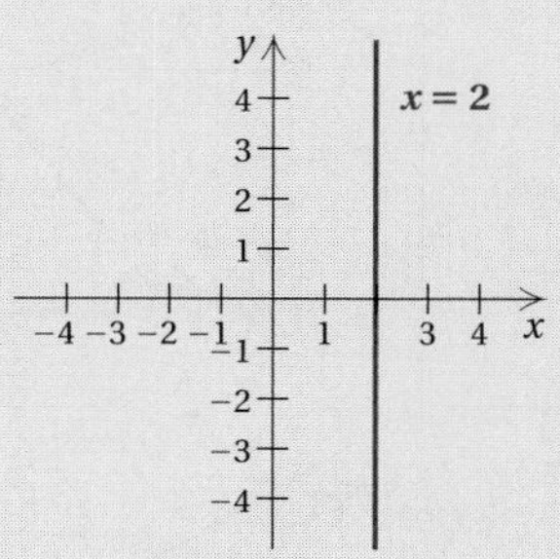

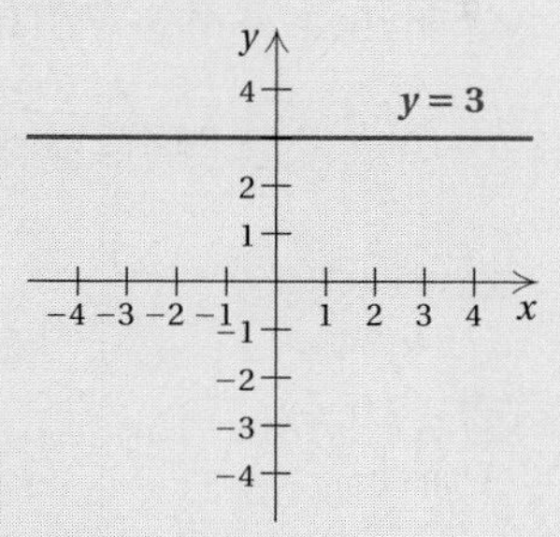

2. If the equation is of the type $y = mx$, both intercepts are the origin, $(0, 0)$. Plot $(0, 0)$ and two other points.
   Example.

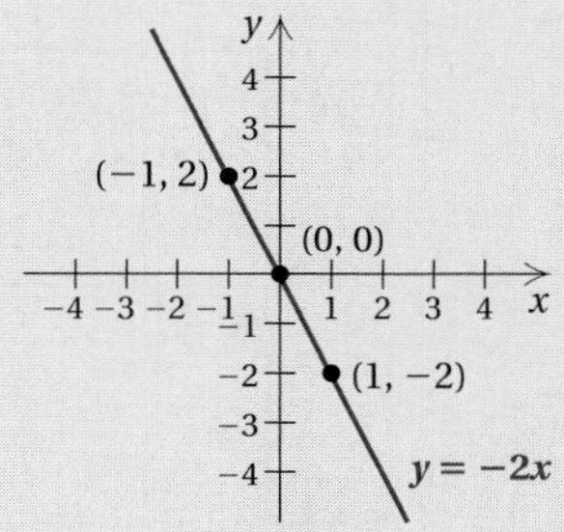

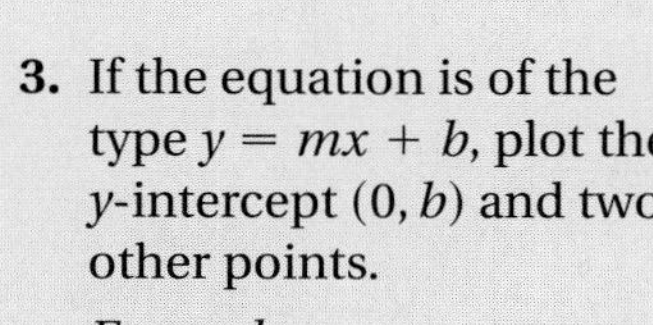

3. If the equation is of the type $y = mx + b$, plot the y-intercept $(0, b)$ and two other points.
   Example.

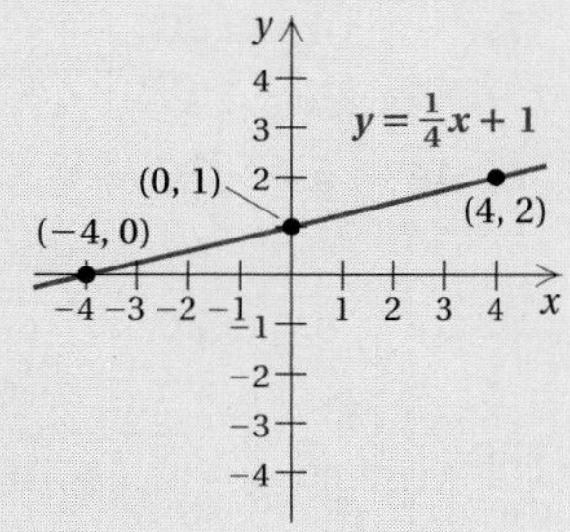

4. If the equation is of the type $Ax + By = C$, but not of the type $x = a$, $y = b$, $y = mx$, or $y = mx + b$, then either solve for y and proceed as with the equation $y = mx + b$, or graph using intercepts. If the intercepts are too close together, choose another point or points farther from the origin.
   Examples.

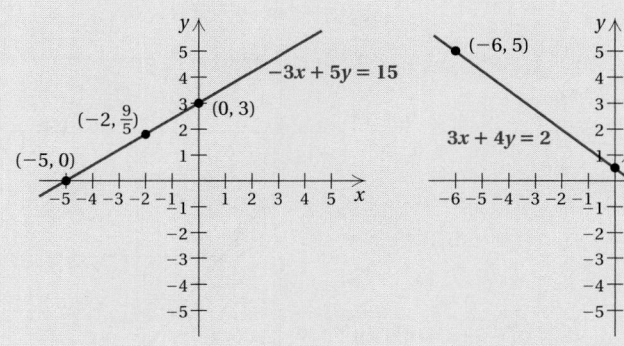

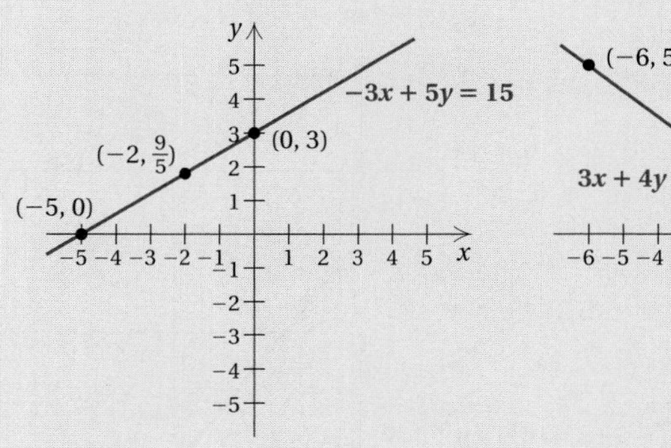

8. $x = 0$

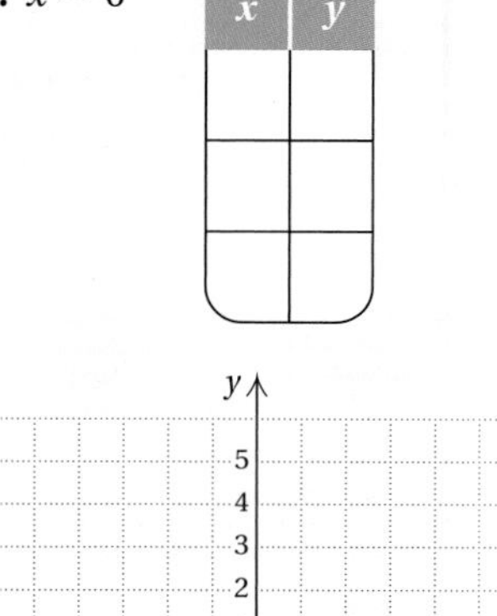

x	y

9. $x = -3$

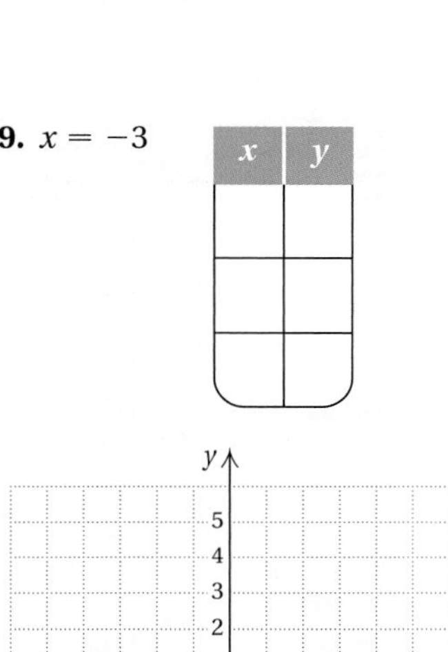

x	y

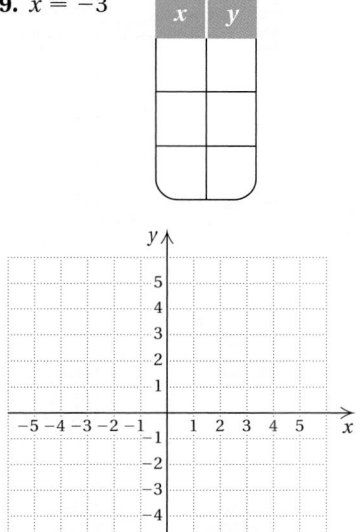

Answers on page A-27

Visualizing for Success

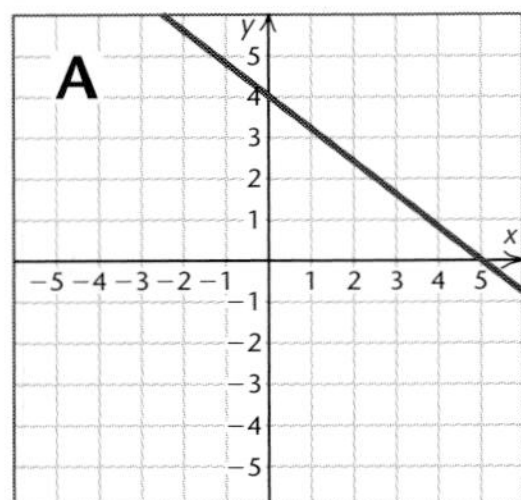

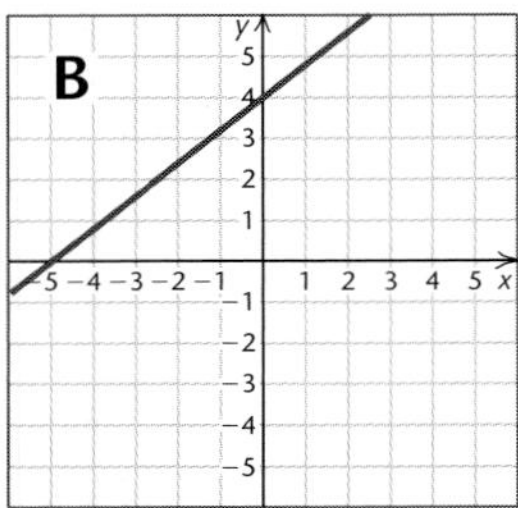

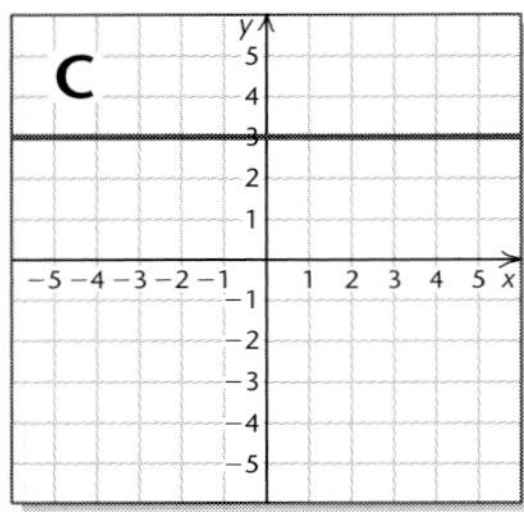

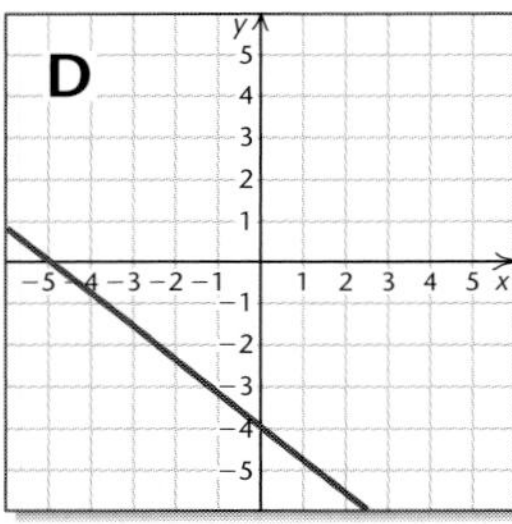

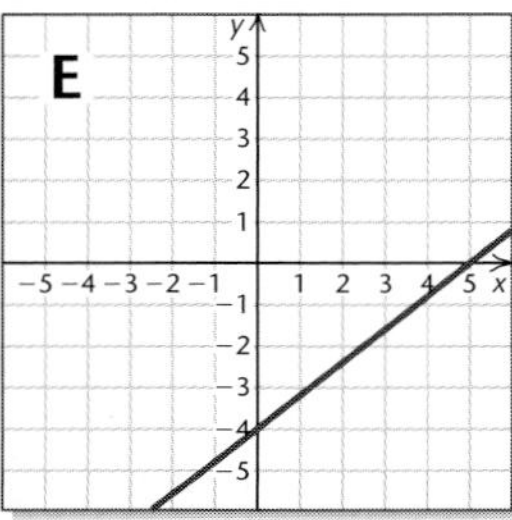

Match the equation with its graph.

1. $5y + 20 = 4x$
2. $y = 3$
3. $3x + 5y = 15$
4. $5y + 4x = 20$
5. $5y = 10 - 2x$
6. $4x + 5y + 20 = 0$
7. $5x - 4y = 20$
8. $4y + 5x + 20 = 0$
9. $5y - 4x = 20$
10. $x = -3$

Answers on page A-27

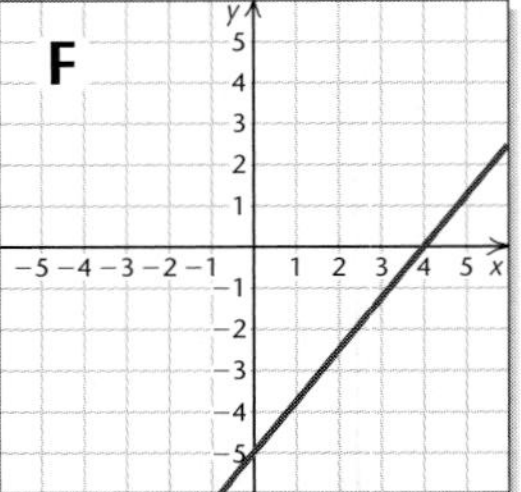

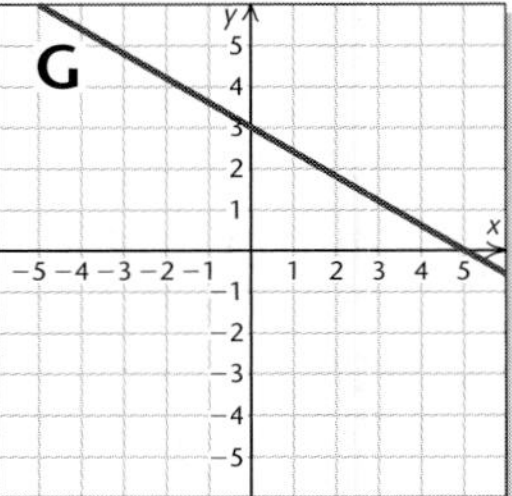

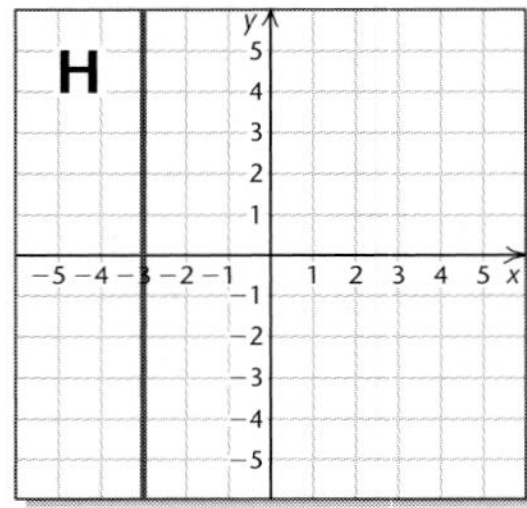

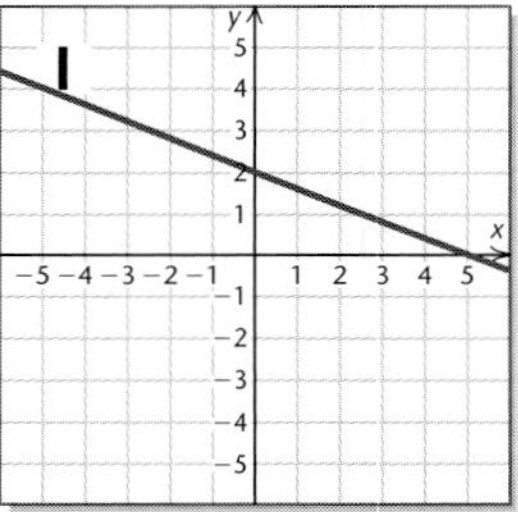

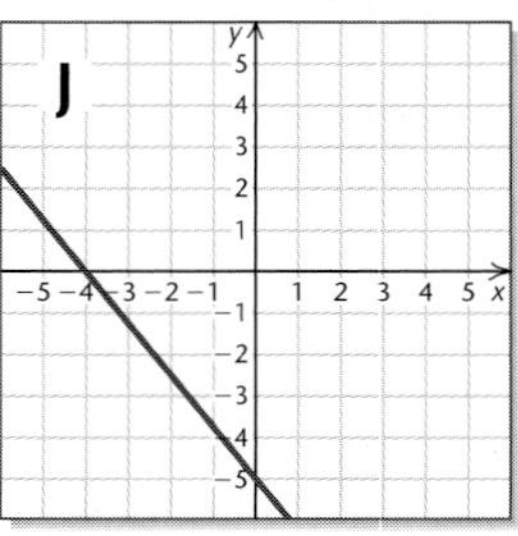

9.2 EXERCISE SET

For Extra Help

MathXL | MyMathLab | InterAct Math | Math Tutor Center | Video Lectures on CD Disc 5 | Student's Solutions Manual

a For Exercises 1–4, find **(a)** the coordinates of the y-intercept and **(b)** the coordinates of the x-intercept.

1.

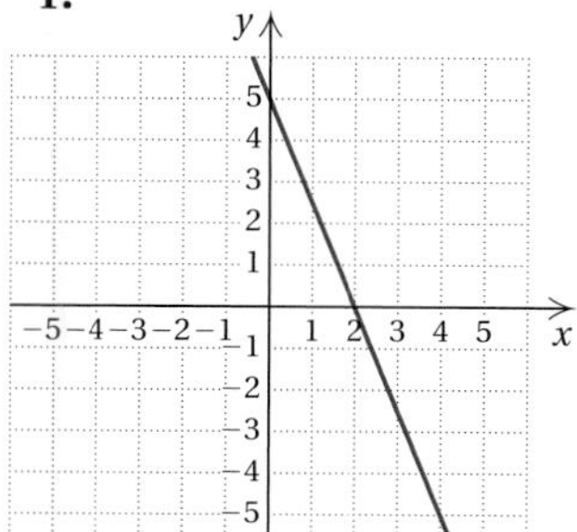

2.

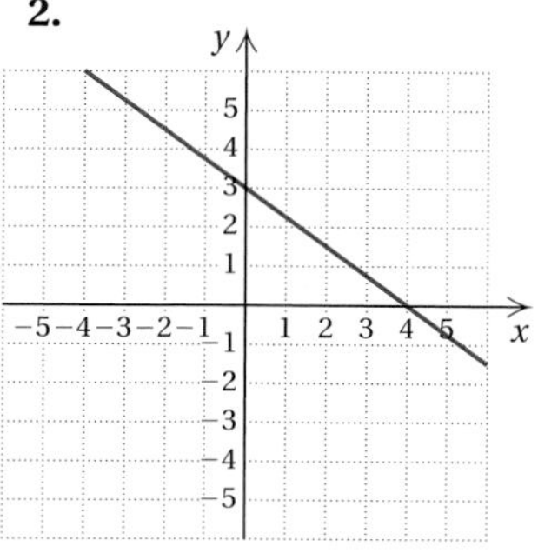

3.

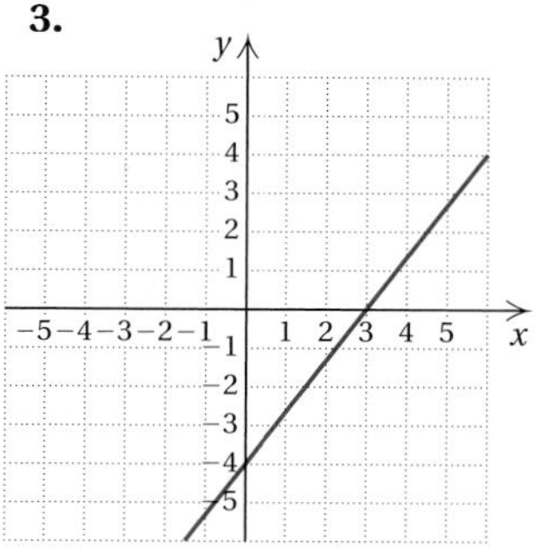

4.

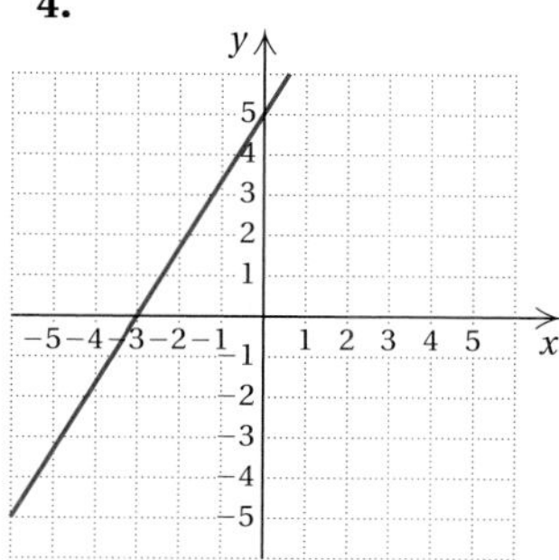

For Exercises 5–12, find **(a)** the coordinates of the y-intercept and **(b)** the coordinates of the x-intercept. Do not graph.

5. $3x + 5y = 15$

6. $5x + 2y = 20$

7. $7x - 2y = 28$

8. $3x - 4y = 24$

9. $-4x + 3y = 10$

10. $-2x + 3y = 7$

11. $6x - 3 = 9y$

12. $4y - 2 = 6x$

For each equation, find the intercepts. Then use the intercepts to graph the equation.

13. $x + 3y = 6$

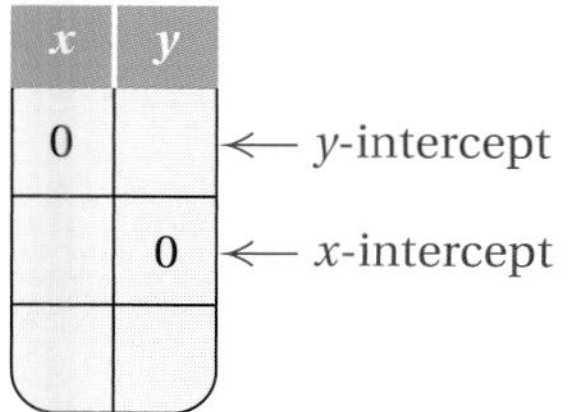

x	y	
0		← y-intercept
	0	← x-intercept

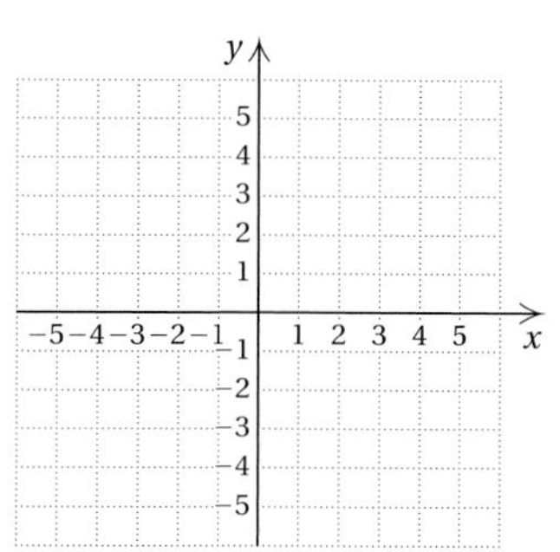

14. $x + 2y = 2$

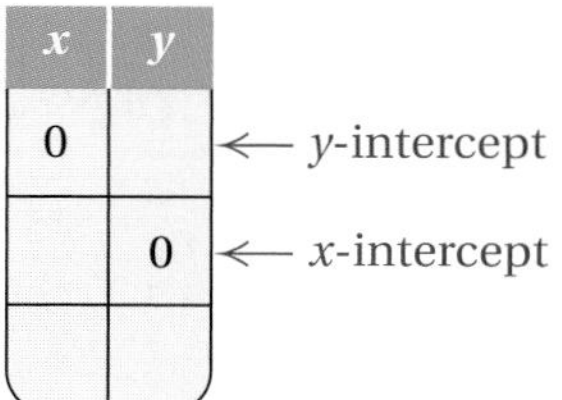

x	y	
0		← y-intercept
	0	← x-intercept

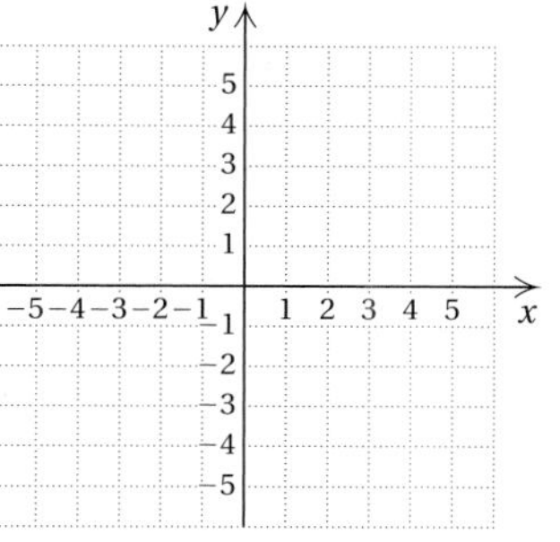

15. $-x + 2y = 4$

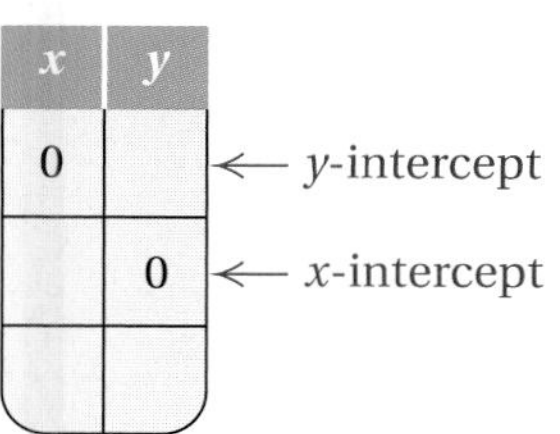

x	y	
0		← y-intercept
	0	← x-intercept

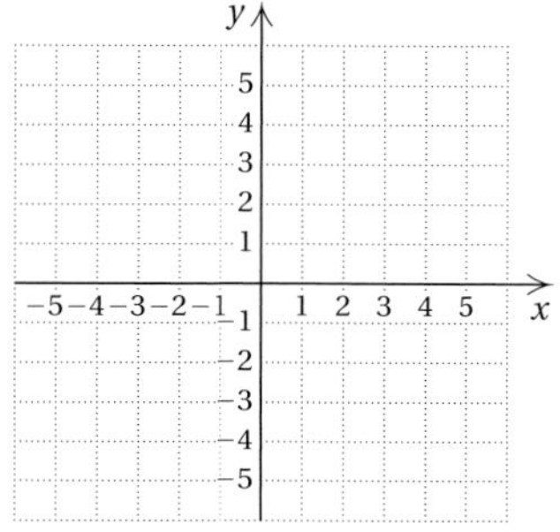

16. $-x + y = 5$

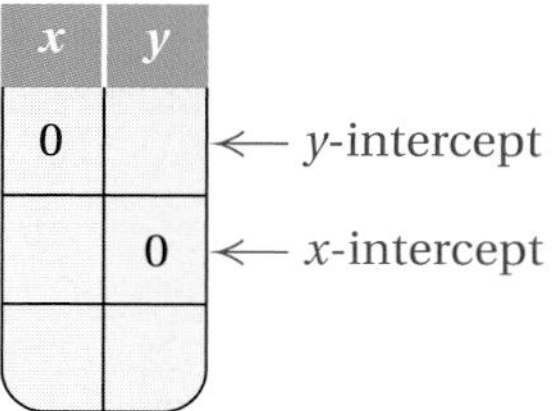

x	y	
0		← y-intercept
	0	← x-intercept

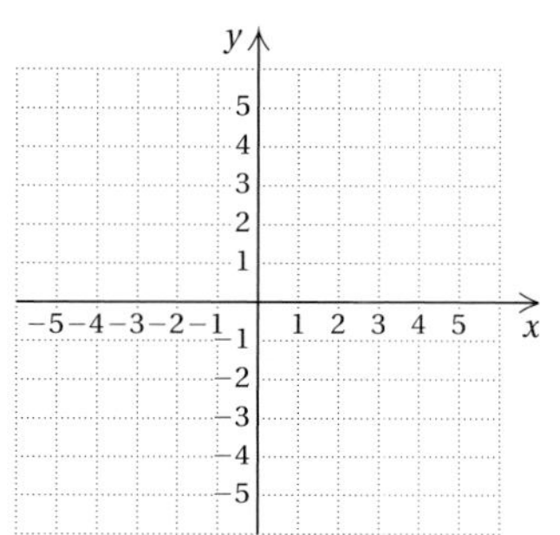

17. $3x + y = 6$

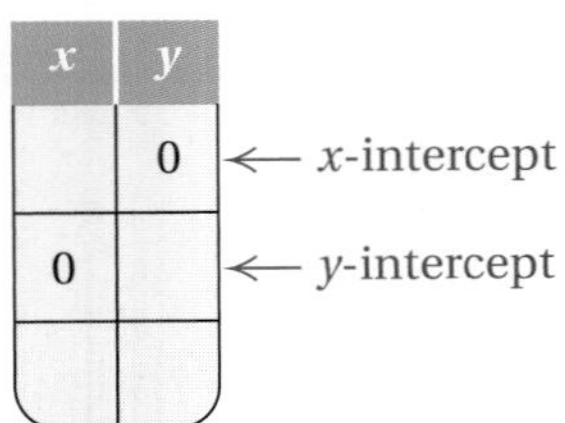

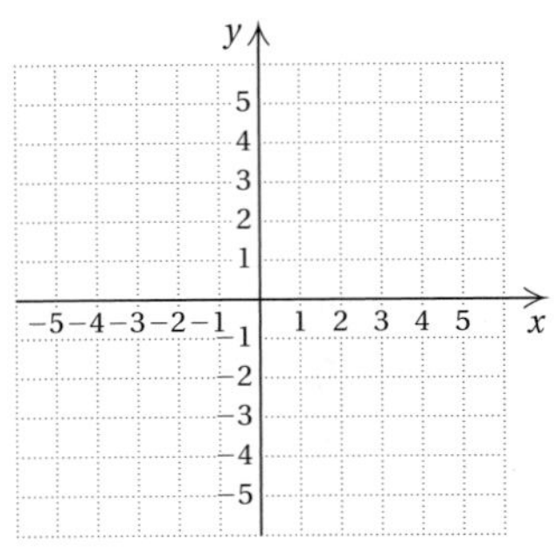

18. $2x + y = 6$

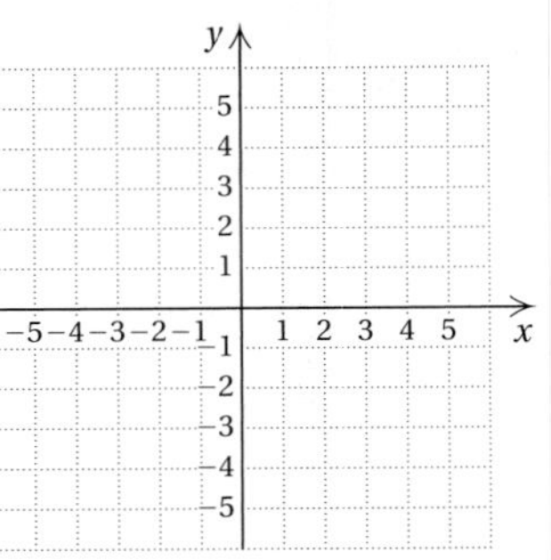

19. $2y - 2 = 6x$

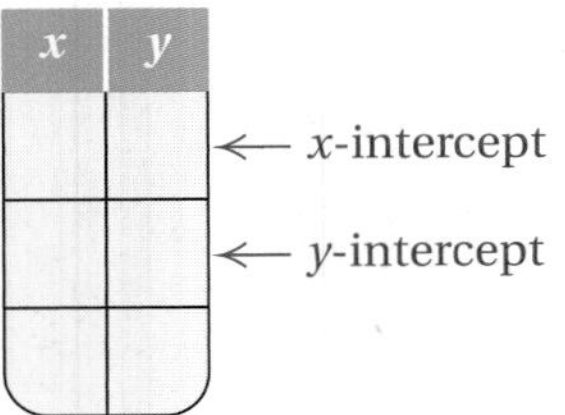

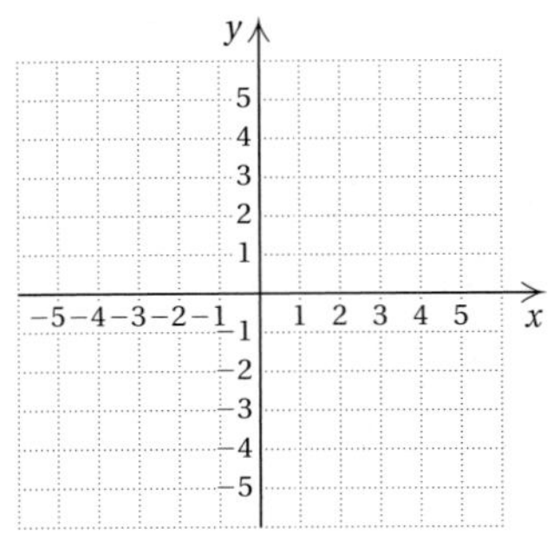

20. $3y - 6 = 9x$

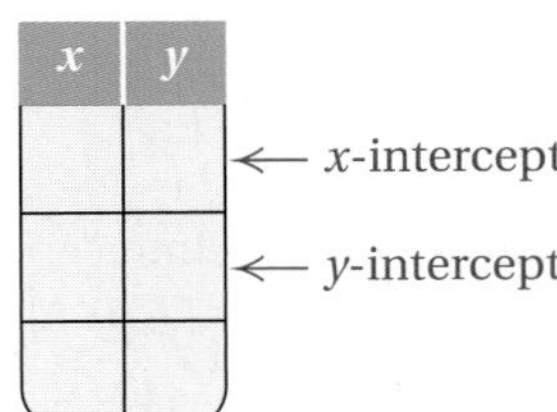

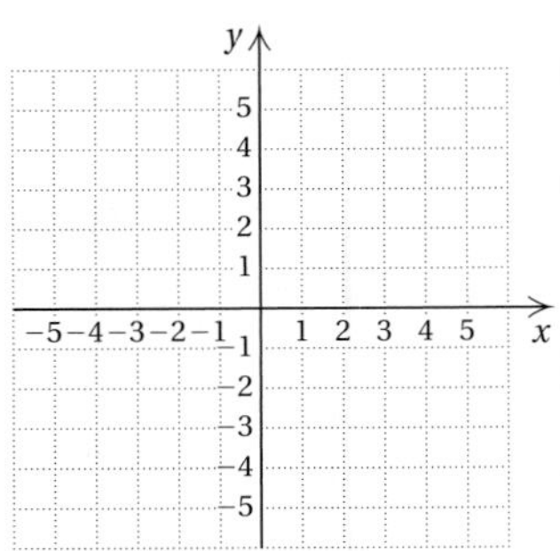

21. $3x - 9 = 3y$

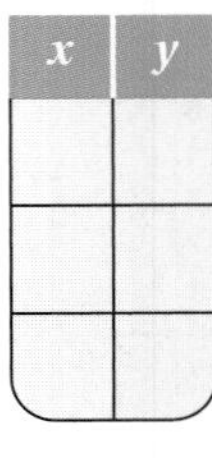

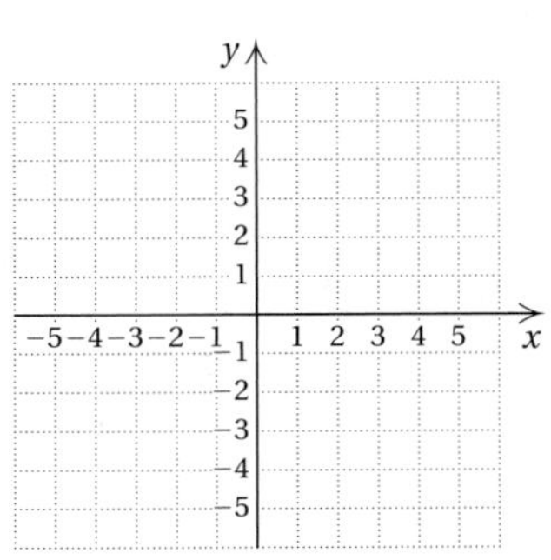

22. $5x - 10 = 5y$

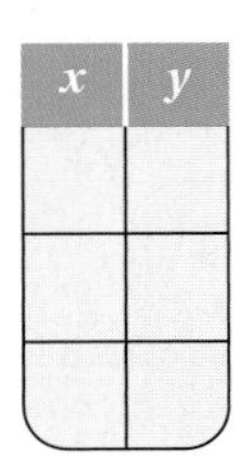

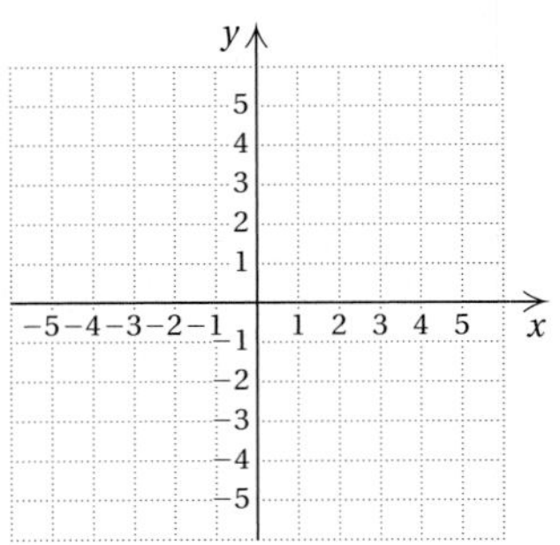

23. $2x - 3y = 6$

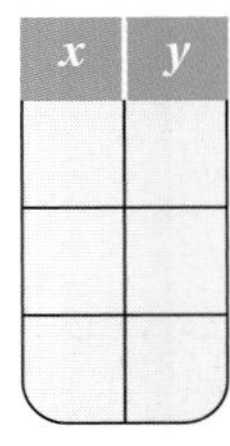

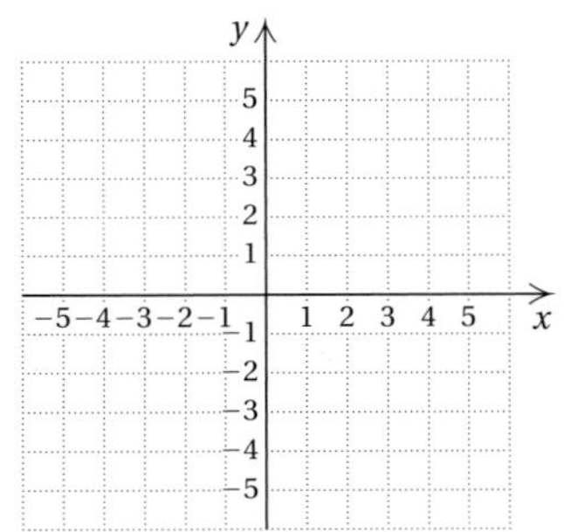

24. $2x - 5y = 10$

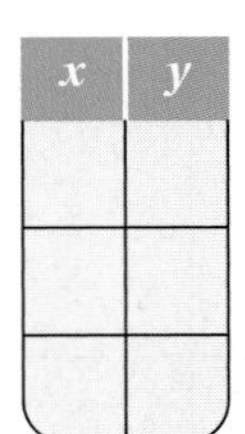

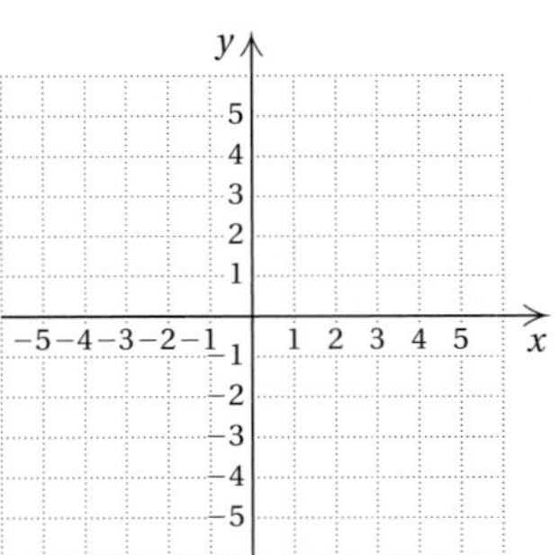

25. $4x + 5y = 20$

x	y

26. $2x + 6y = 12$

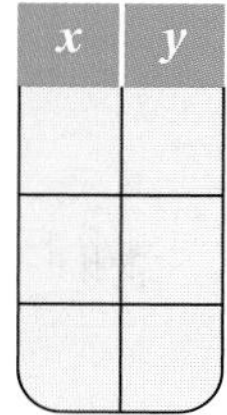

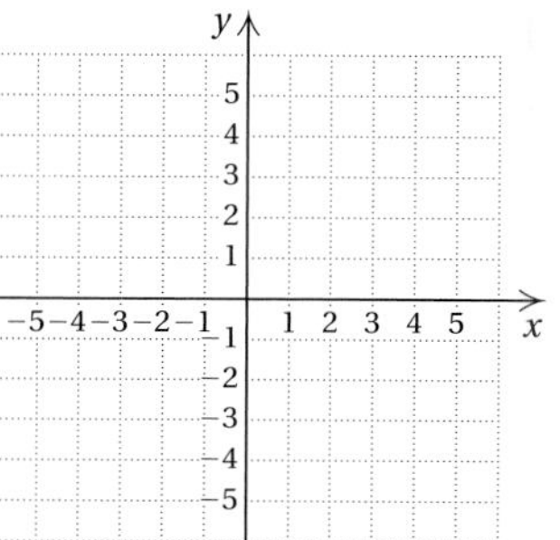

27. $2x + 3y = 8$

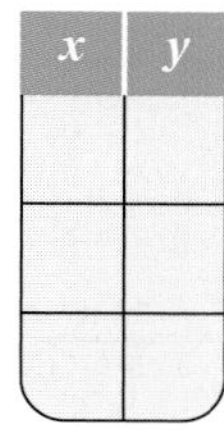

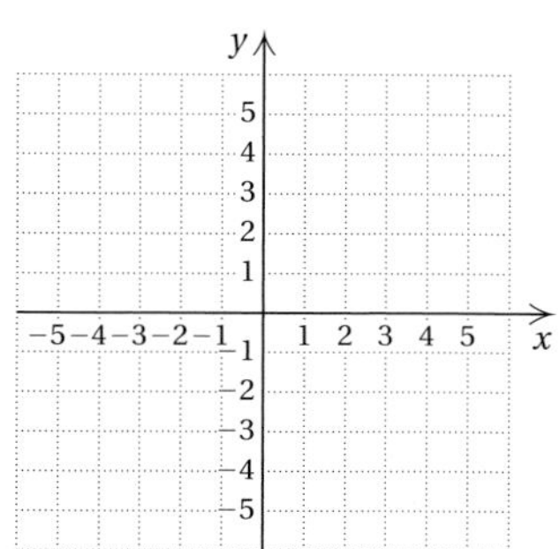

28. $x - 1 = y$

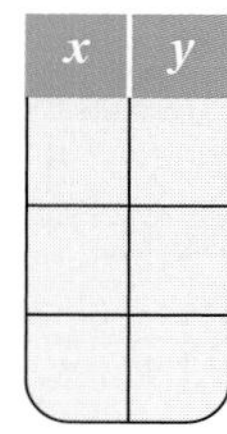

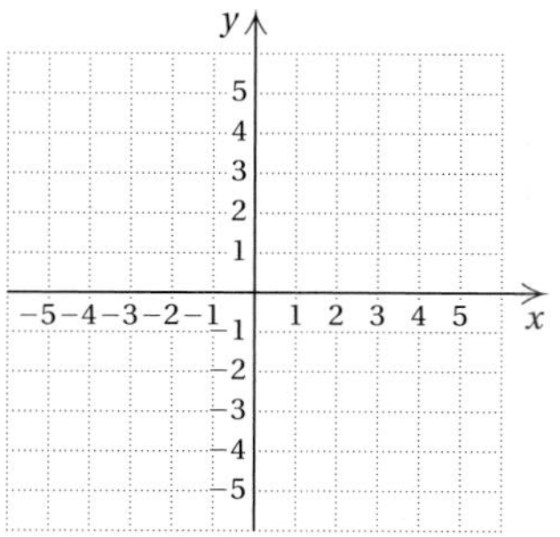

29. $x - 3 = y$

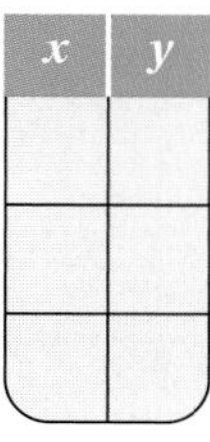

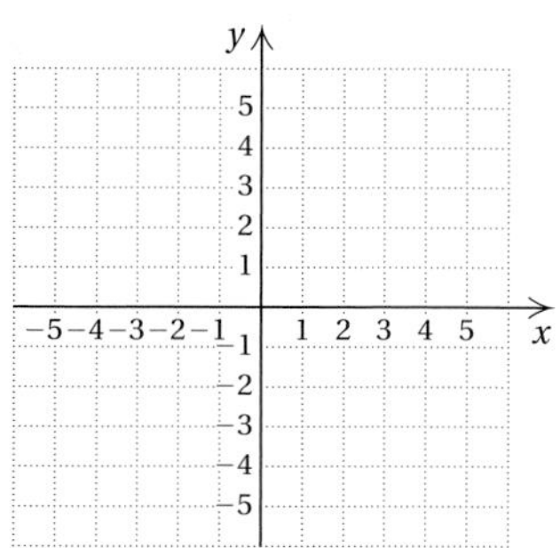

30. $2x - 1 = y$

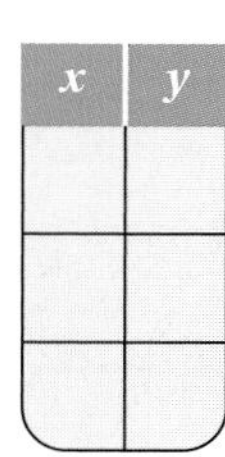

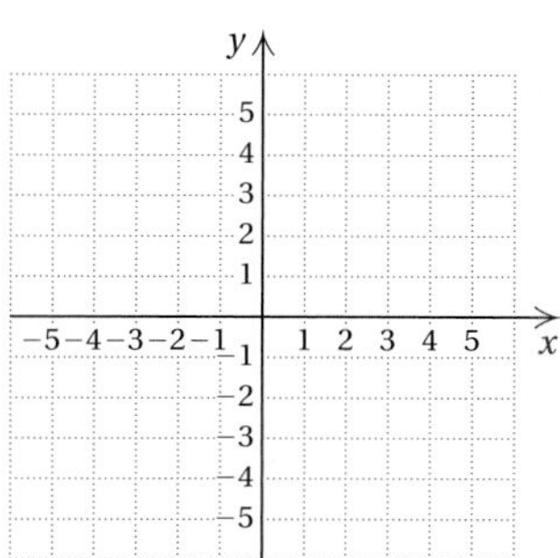

31. $3x - 2 = y$

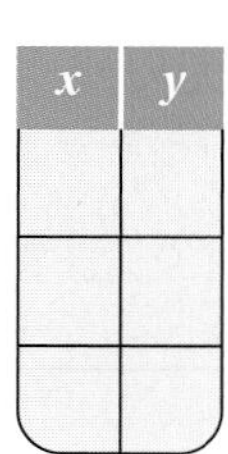

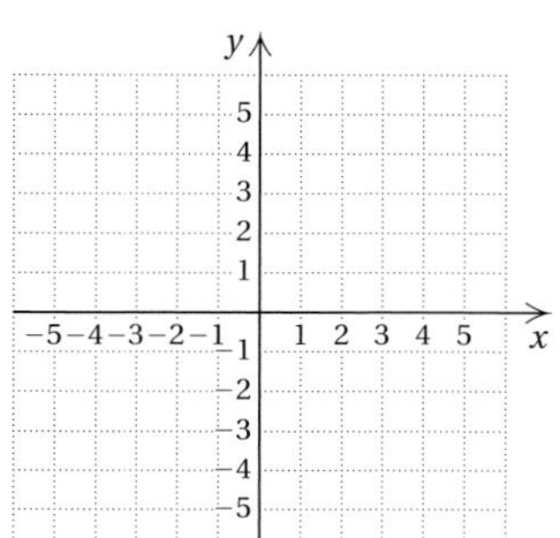

32. $4x - 3y = 12$

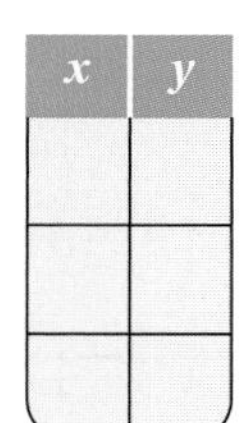

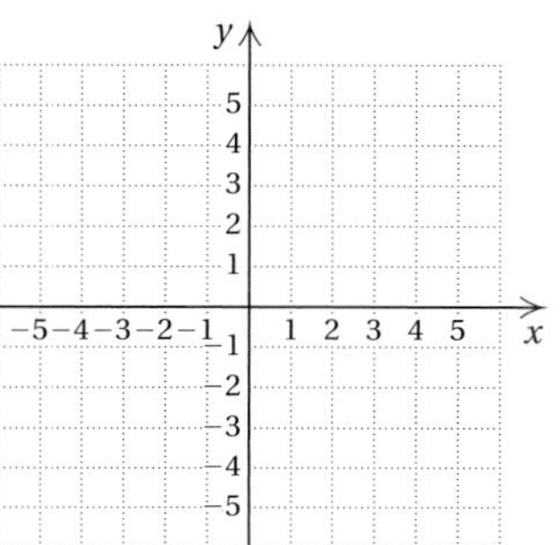

33. $6x - 2y = 12$

x	y

34. $7x + 2y = 6$

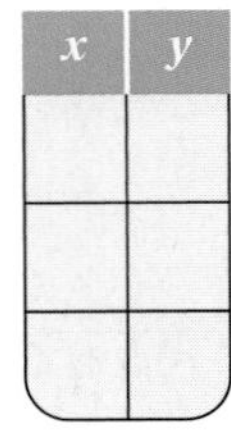

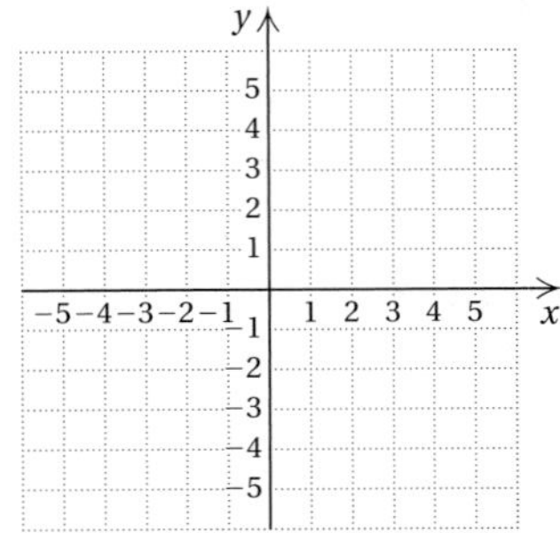

35. $3x + 4y = 5$

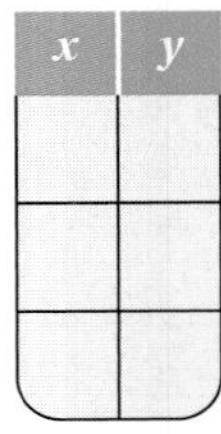

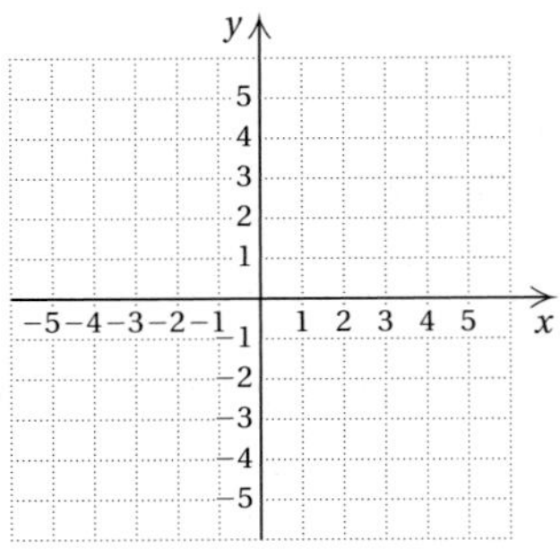

36. $y = -4 - 4x$

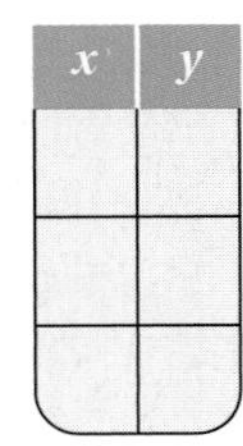

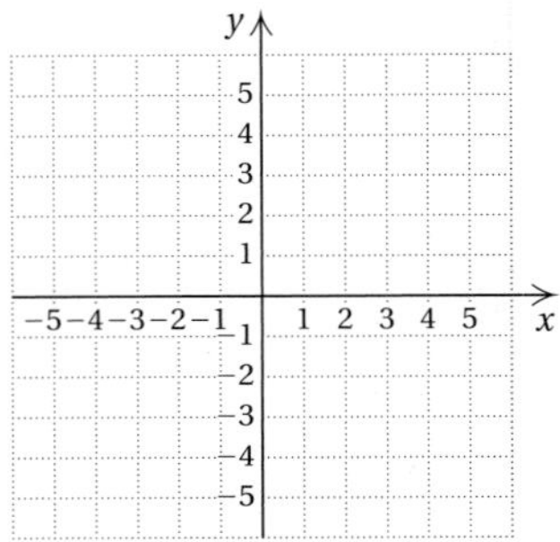

37. $y = -3 - 3x$

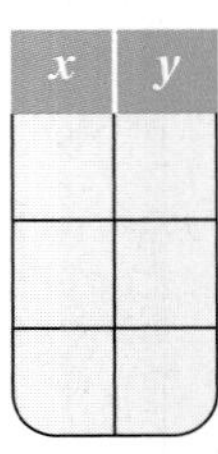

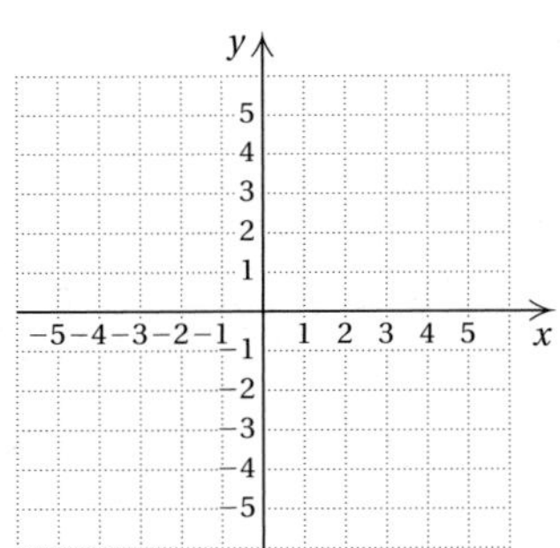

38. $-3x = 6y - 2$

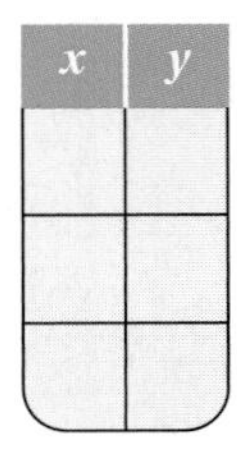

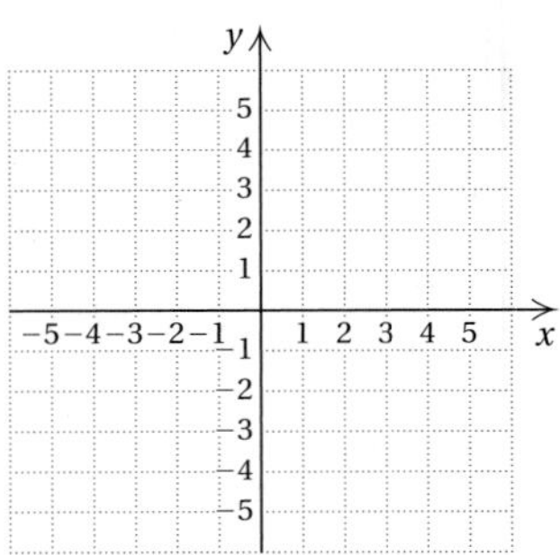

39. $y - 3x = 0$

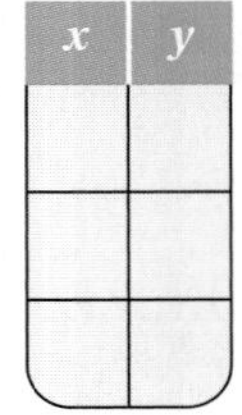

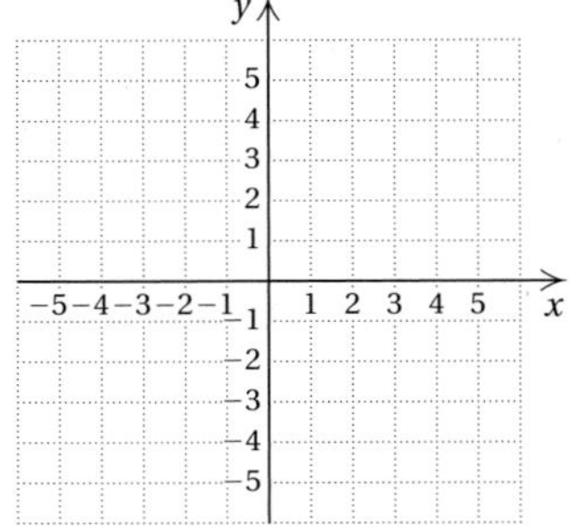

40. $x + 2y = 0$

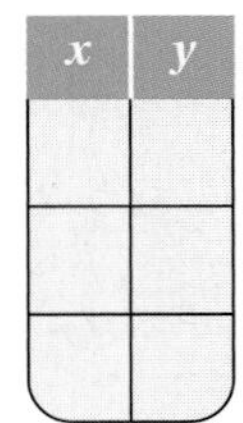

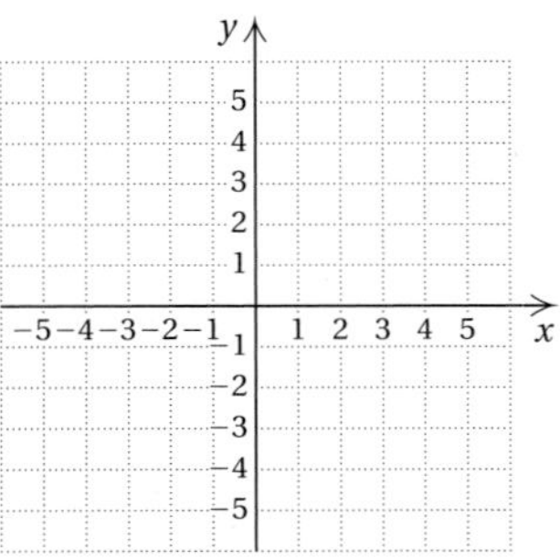

b Graph.

41. $x = -2$

x	y
-2	
-2	
-2	

42. $x = 1$

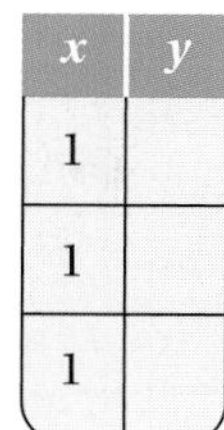

x	y
1	
1	
1	

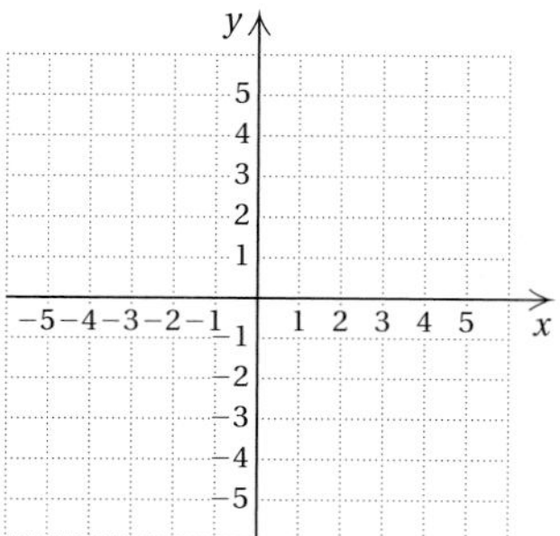

43. $y = 2$

x	y
	2
	2
	2

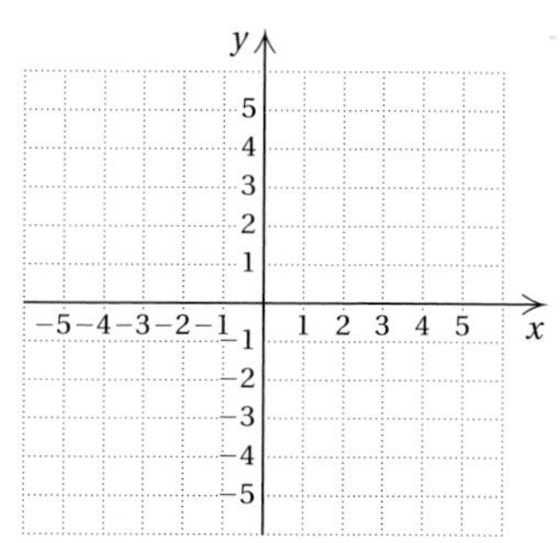

44. $y = -4$

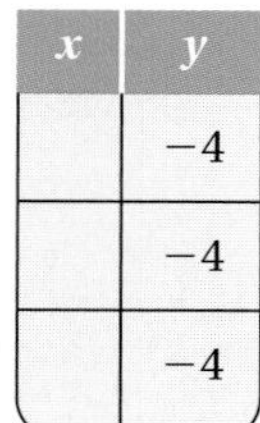

x	y
	-4
	-4
	-4

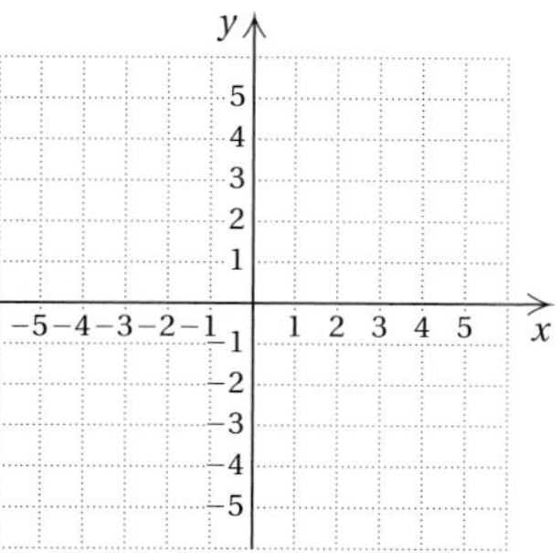

45. $x = 2$

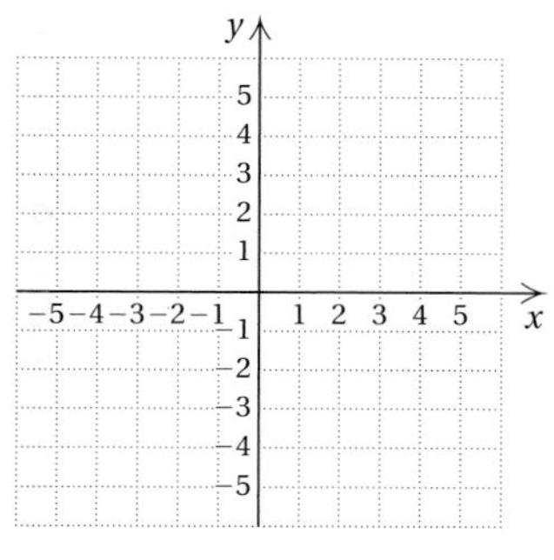

46. $x = 3$

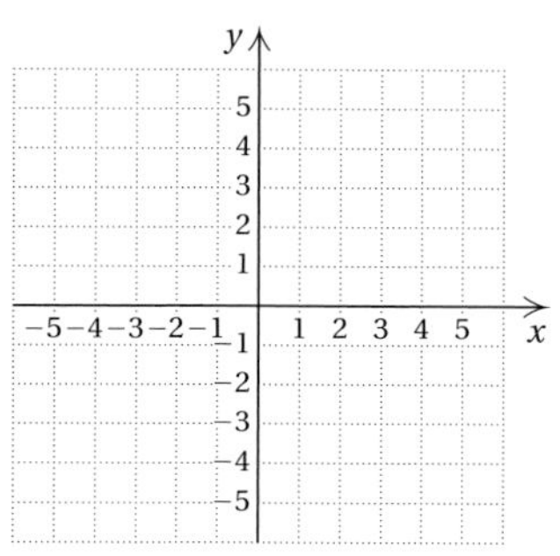

47. $y = 0$

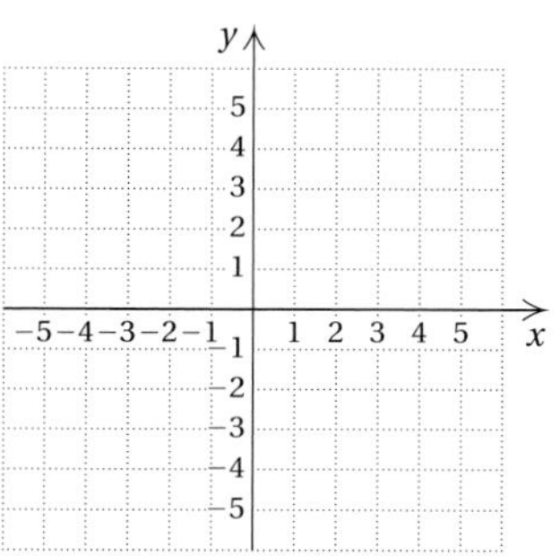

48. $y = -1$

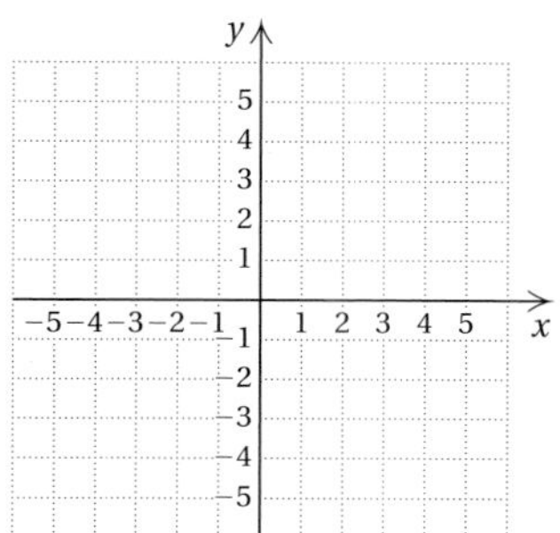

49. $x = \frac{3}{2}$

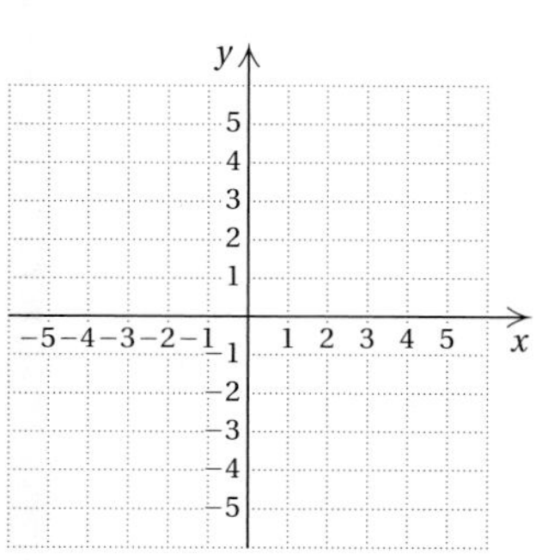

50. $x = -\frac{5}{2}$

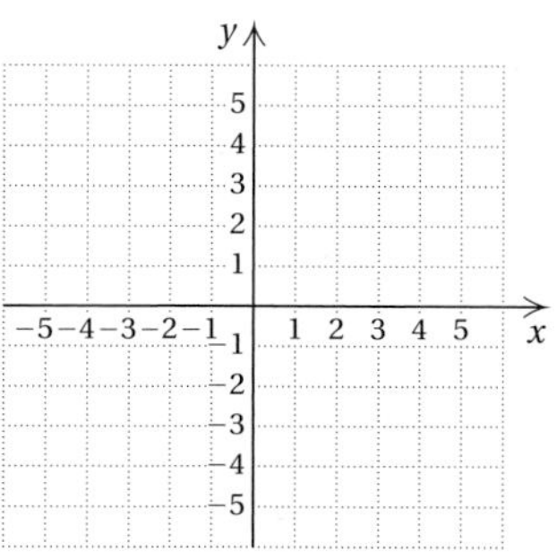

51. $3y = -5$

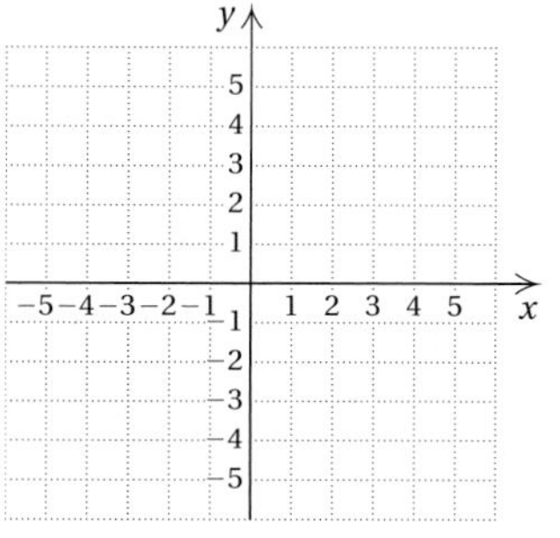

52. $12y = 45$

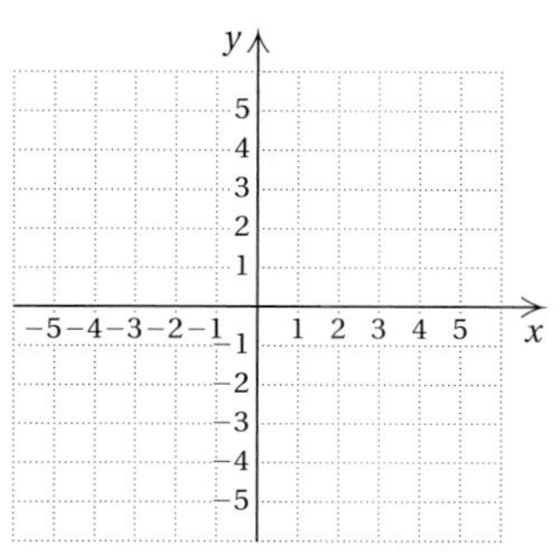

53. $4x + 3 = 0$

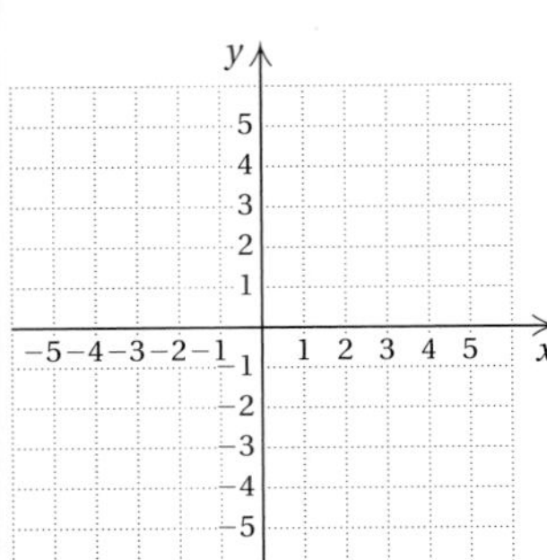

54. $-3x + 12 = 0$

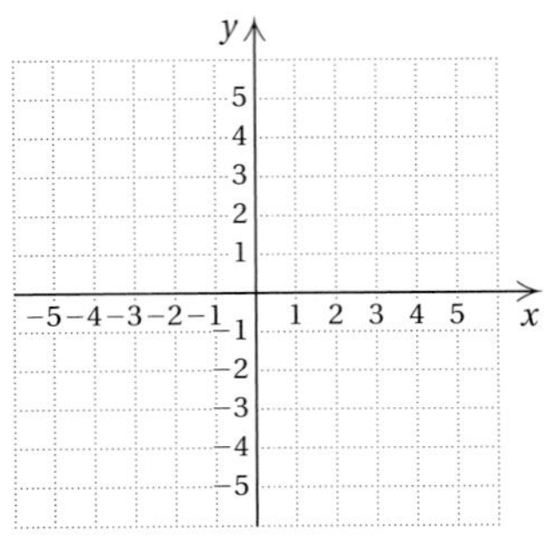

55. $48 - 3y = 0$

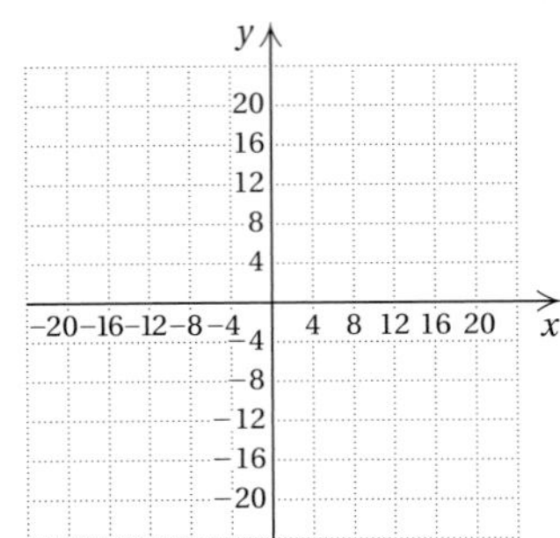

56. $63 + 7y = 0$

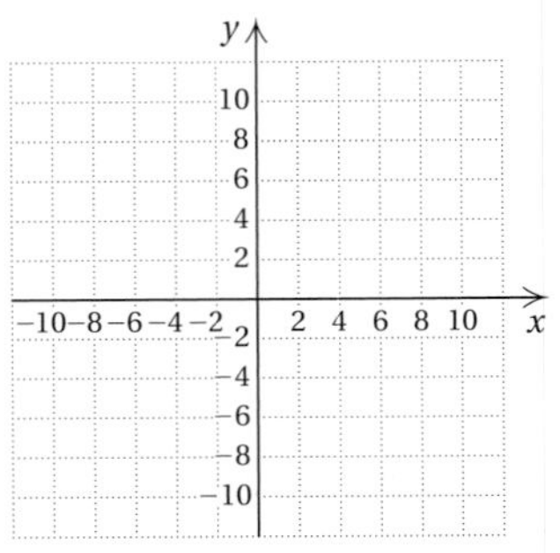

Write an equation for the graph shown.

57.

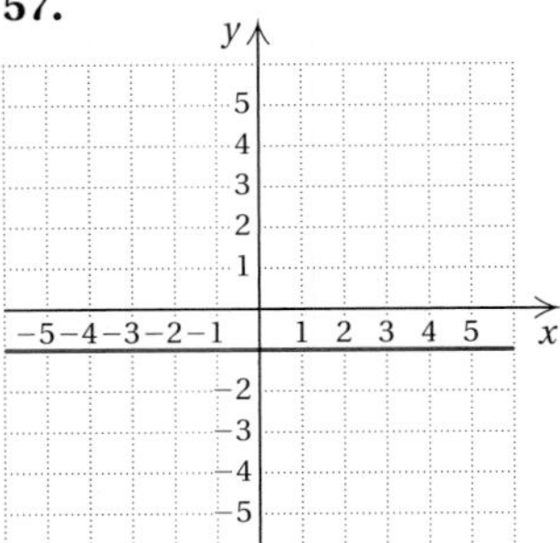

58.

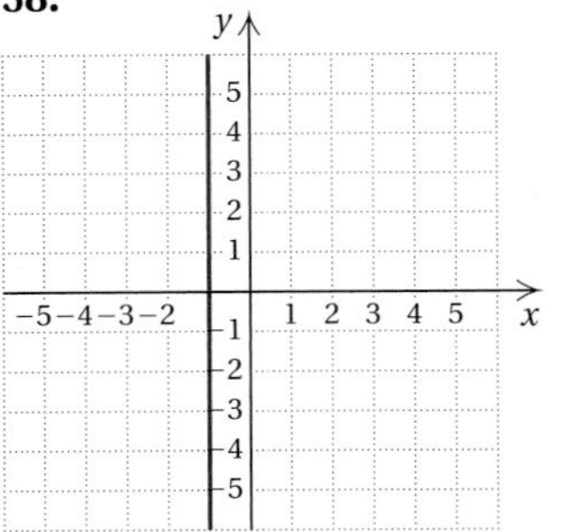

59.

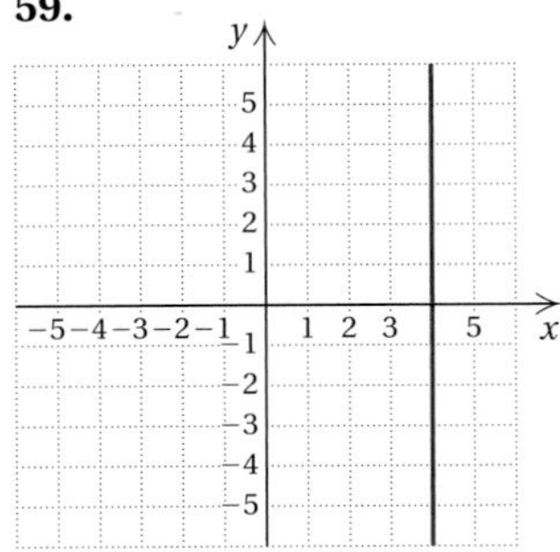

60.

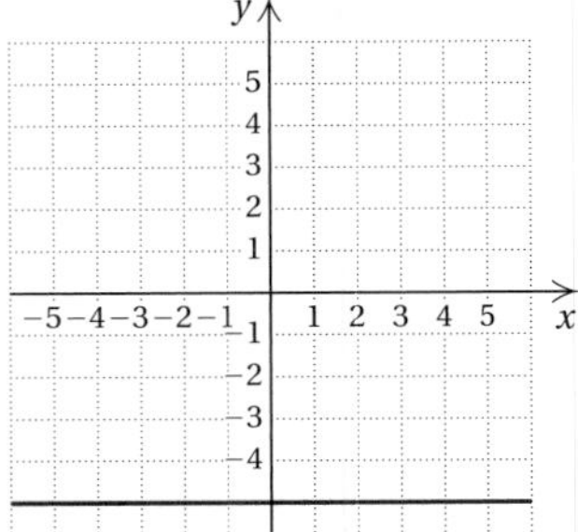

61. $\mathbf{D_W}$ If the graph of the equation $Ax + By = C$ is a horizontal line, what can you conclude about A? Why?

62. $\mathbf{D_W}$ Explain in your own words why the graph of $x = 7$ is a vertical line.

SKILL MAINTENANCE

Solve. [8.5a]

63. *Desserts.* If a restaurant sells 250 desserts in an evening, it is typical that 40 of them will be pie. What percent of the desserts sold will be pie?

64. *Tipping.* Harry left a 20% tip of \$6.50 for a meal. What was the cost of the meal before the tip?

Solve. [8.7e]

65. $-1.6x < 64$

66. $-12x - 71 \geq 13$

67. $x + (x - 1) < (x + 2) - (x + 1)$

68. $6 - 18x \leq 4 - 12x - 5x$

SYNTHESIS

69. Write an equation of a line parallel to the x-axis and passing through $(-3, -4)$.

70. Find the value of m such that the graph of $y = mx + 6$ has an x-intercept of $(2, 0)$.

71. Find the value of k such that the graph of $3x + k = 5y$ has an x-intercept of $(-4, 0)$.

72. Find the value of k such that the graph of $4x = k - 3y$ has a y-intercept of $(0, -8)$.

9.3 SLOPE AND APPLICATIONS

Objectives

a Given the coordinates of two points on a line, find the slope of the line, if it exists.

b Find the slope, or rate of change, in an applied problem involving slope.

c Find the slope of a line from an equation.

a Slope

We have considered two forms of a linear equation,

$$Ax + By = C \quad \text{and} \quad y = mx + b.$$

We found that from the form of the equation $y = mx + b$, we know certain information—namely, that the y-intercept of the line is $(0, b)$.

$$y = mx + b$$

? ← (from m) The y-intercept is $(0, b)$. (from b)

What about the constant m? Does it give us certain information about the line? Look at the following graphs and see if you can make any connection between the constant m and the "slant" of the line.

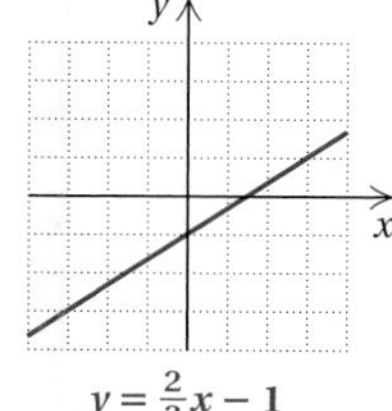

$y = \frac{2}{3}x - 1$

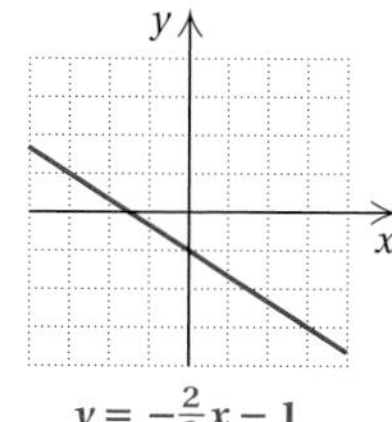

$y = -\frac{2}{3}x - 1$

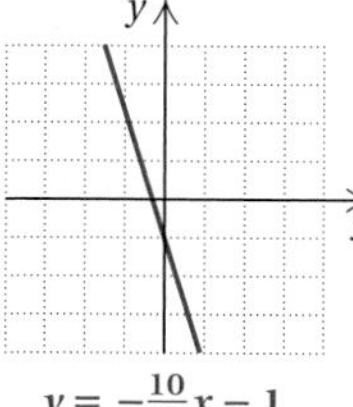

$y = -\frac{10}{3}x - 1$

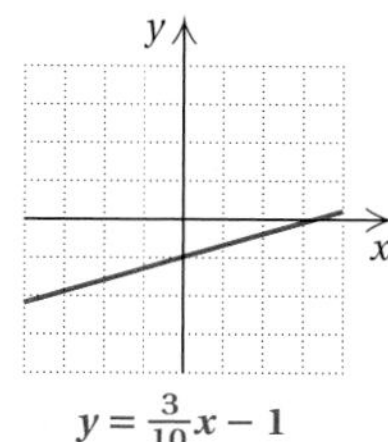

$y = \frac{3}{10}x - 1$

The graphs of some linear equations slant upward from left to right. Others slant downward. Some are vertical and some are horizontal. Some slant more steeply than others. We now look for a way to describe such possibilities with numbers.

Consider a line with two points marked P and Q. As we move from P to Q, the y-coordinate changes from 1 to 3 and the x-coordinate changes from 2 to 6. The change in y is $3 - 1$, or 2. The change in x is $6 - 2$, or 4.

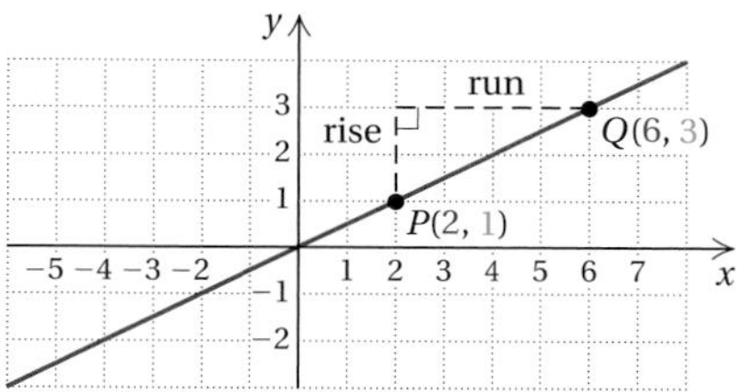

We call the change in y the **rise** and the change in x the **run.** The ratio rise/run is the same for any two points on a line. We call this ratio the **slope.** Slope describes the slant of a line. The slope of the line in the graph above is given by

$$\frac{\text{rise}}{\text{run}} = \frac{\text{the change in } y}{\text{the change in } x}, \text{ or } \frac{2}{4}, \text{ or } \frac{1}{2}.$$

SLOPE

The **slope** of a line containing points (x_1, y_1) and (x_2, y_2) is given by

$$m = \frac{\text{rise}}{\text{run}} = \frac{\text{the change in } y}{\text{the change in } x} = \frac{y_2 - y_1}{x_2 - x_1}.$$

Graph the line containing the points and find the slope in two different ways.

1. $(-2, 3)$ and $(3, 5)$

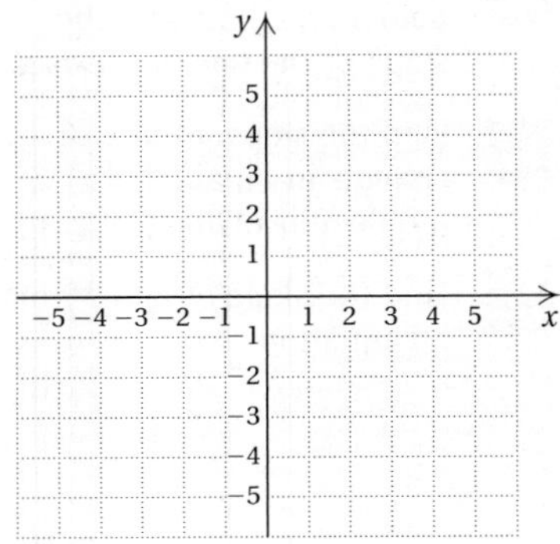

In the preceding definition, (x_1, y_1) and (x_2, y_2)—read "x sub-one, y sub-one and x sub-two, y sub-two"—represent two different points on a line. It does not matter which point is considered (x_1, y_1) and which is considered (x_2, y_2) so long as coordinates are subtracted in the same order in both the numerator and the denominator—for example,

$$\frac{y_2 - y_1}{x_2 - x_1} = \frac{y_1 - y_2}{x_1 - x_2}.$$

EXAMPLE 1 Graph the line containing the points $(-4, 3)$ and $(2, -6)$ and find the slope.

The graph is shown below. We consider (x_1, y_1) to be $(-4, 3)$ and (x_2, y_2) to be $(2, -6)$. From $(-4, 3)$ and $(2, -6)$, we see that the change in y, or the rise, is $-6 - 3$, or -9. The change in x, or the run, is $2 - (-4)$, or 6.

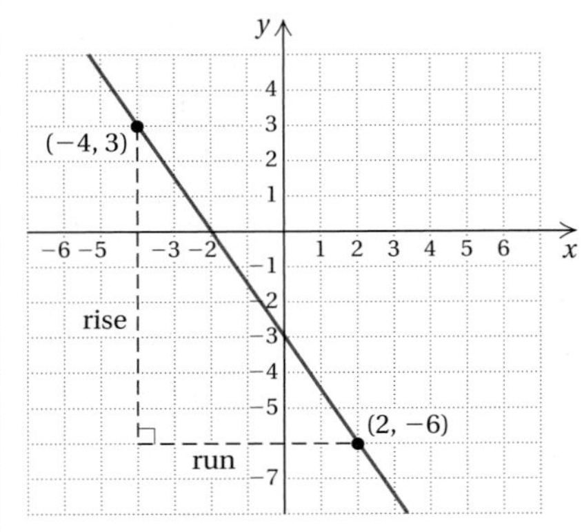

$$\begin{aligned}\text{Slope} &= \frac{\text{rise}}{\text{run}} = \frac{\text{change in } y}{\text{change in } x}\\ &= \frac{y_2 - y_1}{x_2 - x_1}\\ &= \frac{-6 - 3}{2 - (-4)}\\ &= \frac{-9}{6} = -\frac{9}{6}, \text{ or } -\frac{3}{2}.\end{aligned}$$

2. $(0, -3)$ and $(-3, 2)$

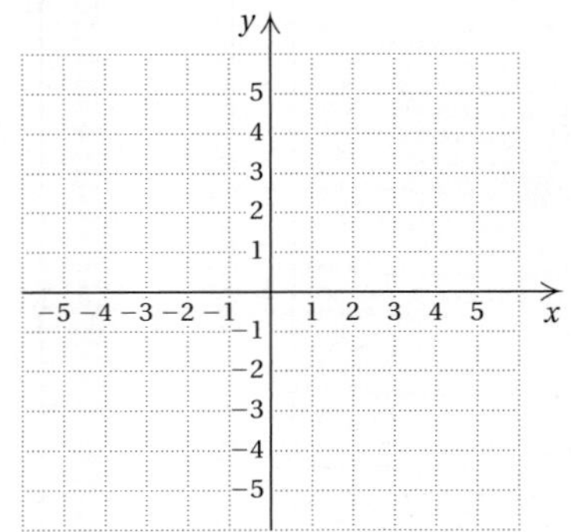

When we use the formula

$$m = \frac{y_2 - y_1}{x_2 - x_1},$$

we must remember to subtract the y-coordinates in the same order that we subtract the x-coordinates. Let's redo Example 1, where we consider (x_1, y_1) to be $(2, -6)$ and (x_2, y_2) to be $(-4, 3)$:

$$\text{Slope} = \frac{\text{change in } y}{\text{change in } x} = \frac{3 - (-6)}{-4 - 2} = \frac{9}{-6} = -\frac{3}{2}.$$

Do Exercises 1 and 2.

The slope of a line tells how it slants. A line with positive slope slants up from left to right. The larger the slope, the steeper the slant. A line with negative slope slants downward from left to right.

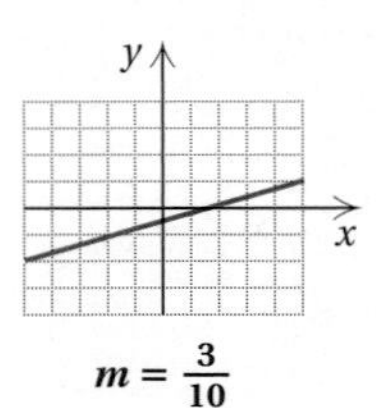

$m = \frac{3}{10}$

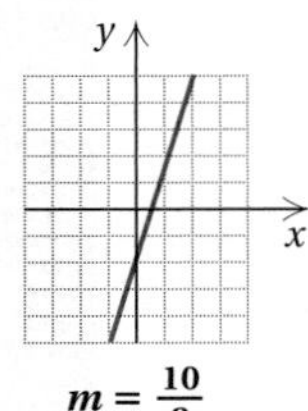

$m = \frac{10}{3}$

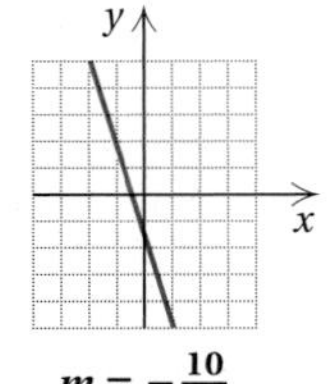

$m = -\frac{10}{3}$

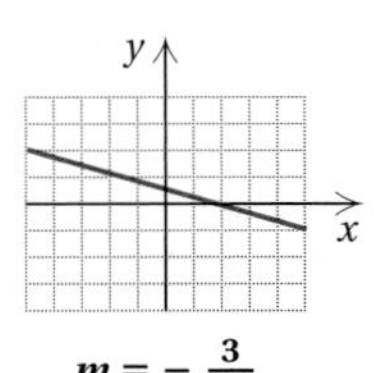

$m = -\frac{3}{10}$

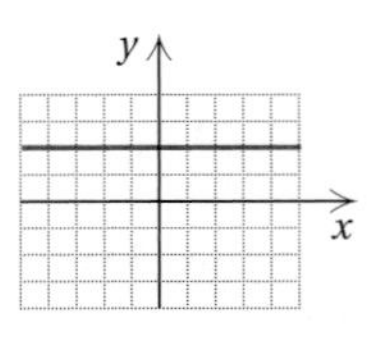

$m = 0$

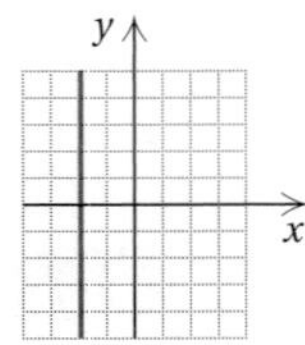

m is not defined.

Later in this section, in Examples 10 and 11, we will discuss the slope of a horizontal line and of a vertical line. The slope of a horizontal line is 0. The slope of a vertical line is not defined.

Answers on page A-28

b Applications of Slope; Rates of Change

Slope has many real-world applications. For example, numbers like 2%, 3%, and 6% are often used to represent the *grade* of a road, a measure of how steep a road on a hill or mountain is. For example, a 3% grade $\left(3\% = \frac{3}{100}\right)$ means that for every horizontal distance of 100 ft, the road rises 3 ft, and a −3% grade means that for every horizontal distance of 100 ft, the road drops 3 ft. (Road signs do not include negative signs. It's usually obvious whether you are climbing or descending.) The concept of grade also occurs in skiing or snowboarding, where a 7% grade is considered very tame, but a 70% grade is considered extremely steep. And in cardiology, a physician may change the grade of a treadmill to measure its effect on heartbeat.

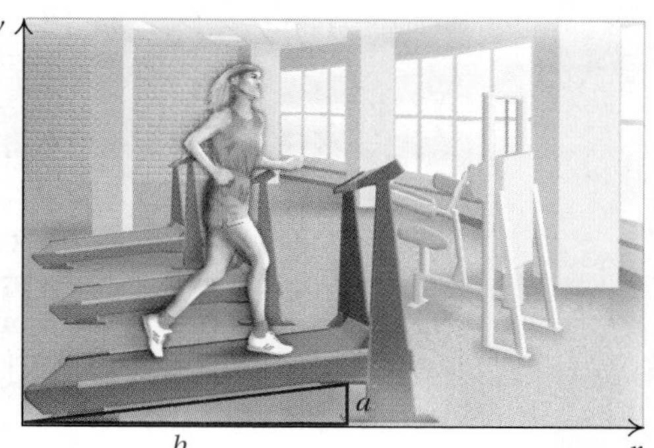

Architects and carpenters use slope when designing and building stairs, ramps, or roof pitches. Another application occurs in hydrology. When a river flows, the strength or force of the river depends on how far the river falls vertically compared to how far it flows horizontally.

EXAMPLE 2 *Skiing.* Among the steepest skiable terrain in North America, the Headwall on Mount Washington, in New Hampshire, drops 720 ft over a horizontal distance of 900 ft. Find the grade of the Headwall.

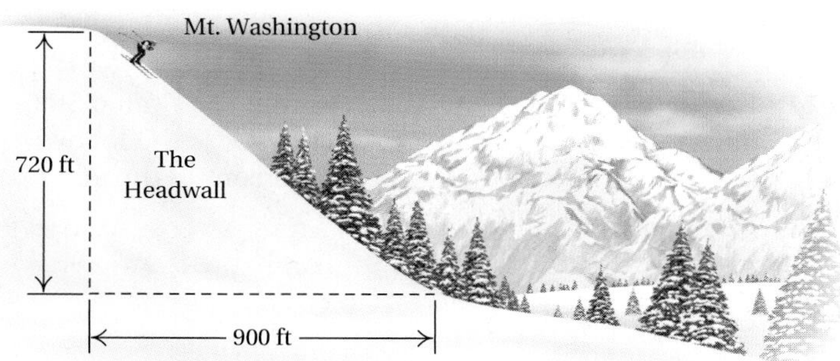

The grade of the Headwall is its slope, expressed as a percent:

$$m = \frac{720}{900} \begin{matrix} \longleftarrow \text{Vertical change} \\ \longleftarrow \text{Horizontal change} \end{matrix}$$

$$= \frac{8}{10} = 80\%.$$

Do Exercise 3.

3. **Construction.** Public buildings regularly include steps with 7-in. risers and 11-in. treads. Find the grade of such a stairway.

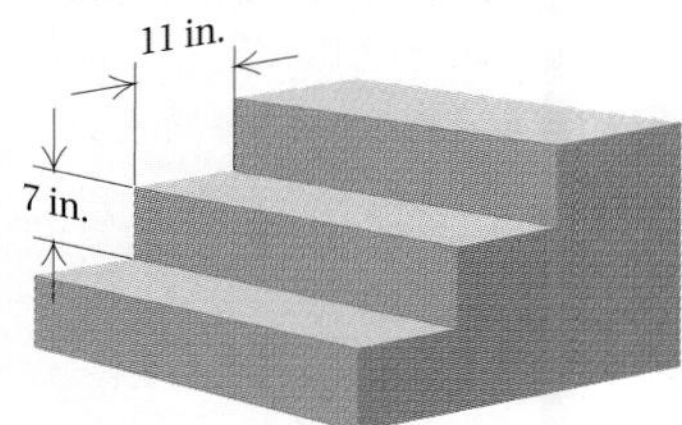

Answer on page A-28

4. **Cost of a Telephone Call.** The following graph shows data of interstate long-distance calling offered by AmeriCom in the Simplicity Plan. At what rate is the customer billed?

Source: AmeriCom

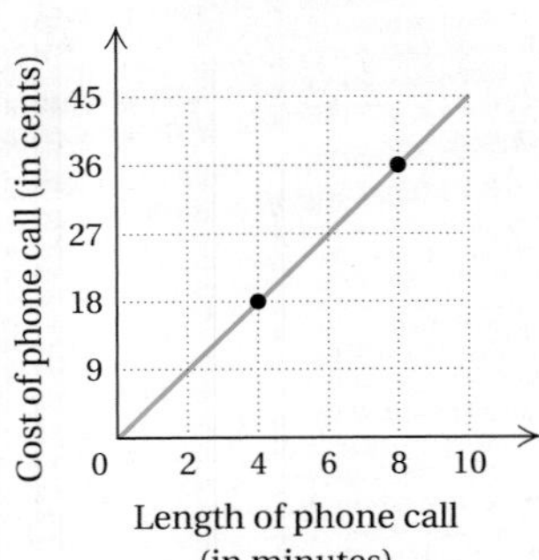

Answer on page A-28

Slope can also be considered as a **rate of change.**

EXAMPLE 3 *Haircutting.* Kiddie Kutters has a graph displaying data from a recent day's work. Use the graph to determine the slope, or the rate of change, of the number of haircuts with respect to time.

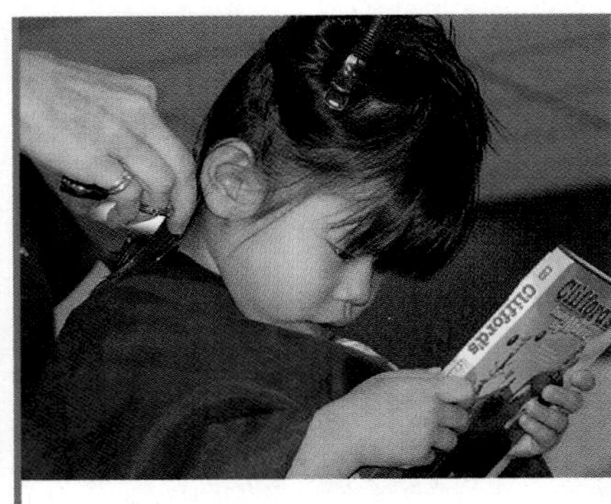

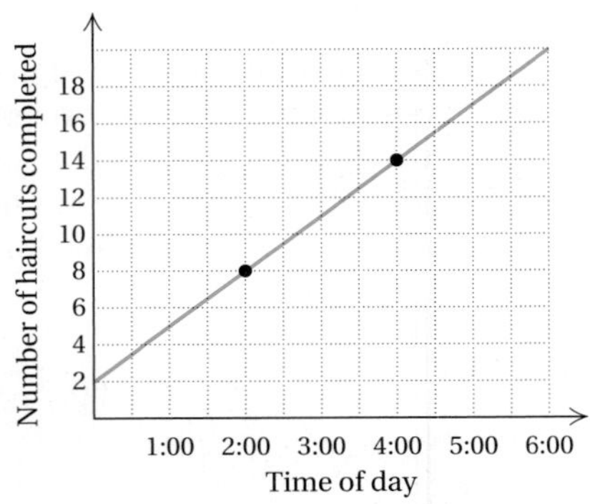

The vertical axis of the graph shows the number of haircuts and the horizontal axis the time, in units of one hour. We can describe the rate of change in the number of haircuts with respect to time as

$$\frac{\text{Haircuts}}{\text{Hour}}, \quad \text{or} \quad \textit{number of haircuts per hour.}$$

This value is the slope of the line. We determine two ordered pairs on the graph—in this case,

(2:00, 8 haircuts) and (4:00, 14 haircuts).

This tells us that in the 2 hr between 2:00 and 4:00, 6 haircuts were completed. Thus,

$$\text{Rate of change} = \frac{14 \text{ haircuts} - 8 \text{ haircuts}}{4{:}00 - 2{:}00} = \frac{6 \text{ haircuts}}{2 \text{ hours}} = 3 \text{ haircuts per hour.}$$

Do Exercise 4.

EXAMPLE 4 *Decreased Smoking.* Each year in the United States, the percent of the adult population who smoke declines. Use the following graph to determine the slope, or rate of change of the percent of the adult population who smoke with respect to time.

Source: U.S. Centers for Disease Control and Prevention

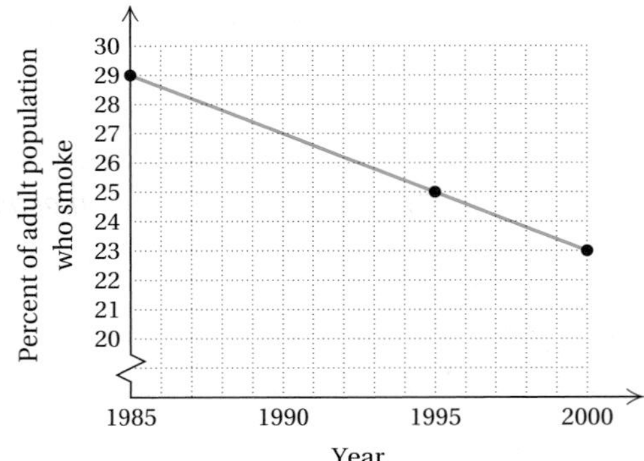

The vertical axis of the graph shows the percent of the adult population who smoke and the horizontal axis shows the years. We can describe the rate of change in the percent who smoke with respect to time as

$$\frac{\text{Percent who smoke}}{\text{Years}}, \quad \text{or} \quad \text{percent who smoke per year.}$$

This value is the slope of the line. We determine two ordered pairs on the graph—in this case,

(1995, 25) and (2000, 23).

This tells us that in the 5 yr from 1995 to 2000, the percent dropped from 25% to 23%. Thus

$$\text{Rate of change} = \frac{23\% - 25\%}{2000 - 1995} = \frac{-2\%}{5 \text{ yr}} = -\frac{2}{5}\% \text{ per year.}$$

Do Exercise 5.

5. **Death by Firearms.** Find the rate of change in the number of deaths by firearms.

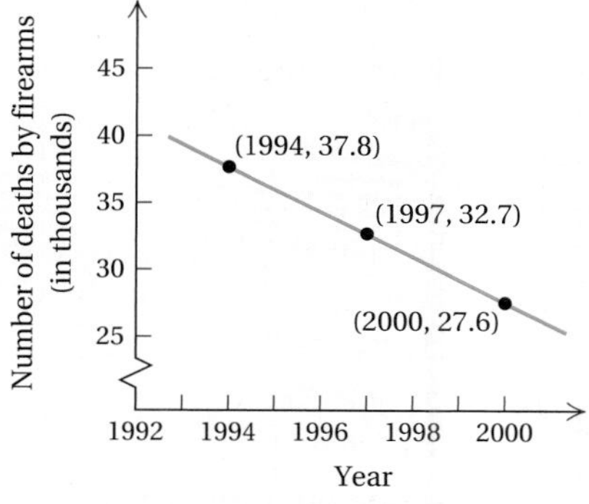

Source: National Vital Statistics Report

Answer on page A-28

Study Tips

SMALL STEPS LEAD TO GREAT SUCCESS (PART 2)

Chris Widener is a popular motivational speaker and writer. In his article "A Little Equation That Creates Big Results," he proposes the following equation: "Your Short-Term Actions Multiplied By Time = Your Long-Term Accomplishments."

Think of the major or career toward which you are working as a long-term accomplishment. We (your authors and instructors) are at a point in life where we realize the long-term benefits of learning mathematics. For you as students, it may be more difficult to see those long-term results. But make an effort to do so.

Widener goes on to say, "We need to take action on our dreams and beliefs every day." Think of the long-term goal as you do the short-term tasks of homework in math, studying for tests, and completing this course so you can move on to what it prepares you for.

Who writes best-selling novels? The person who only dreams of becoming a best-selling author or the one who also spends 4 hours a day doing research and working at a computer?

Who loses weight? The person who thinks about being thin or the one who also plans a healthy diet and runs 3 miles a day?

Who is successful at math? The person who only knows all the benefits of math or the one who also spends at least 2 hours studying outside of class for every hour spent inside class?

"The purpose of man is in action, not thought."

Thomas Carlyle, British historian/essayist

"Prepare for your success in little ways and you will eventually see results in big ways. It's almost magical."

Tom Morris, public philosopher/speaker

CALCULATOR CORNER

Visualizing Slope

Exercises: Graph each of the following sets of equations using the window settings $[-6, 6, -4, 4]$, Xscl = 1, Yscl = 1.

1. $y = x,\ y = 2x,$ $y = 5x,\ y = 10x$
   What do you think the graph of $y = 123x$ will look like?
2. $y = x,\ y = \frac{3}{4}x,$ $y = 0.38x,\ y = \frac{5}{32}x$
   What do you think the graph of $y = 0.000043x$ will look like?

Find the slope of the line.

6. $y = 4x + 11$

7. $y = -17x + 8$

8. $y = -x + \dfrac{1}{2}$

9. $y = \dfrac{2}{3}x - 1$

Find the slope of the line.

10. $4x + 4y = 7$

11. $5x - 4y = 8$

Answers on page A-28

C Finding the Slope from an Equation

It is possible to find the slope of a line from its equation. Let's consider the equation $y = 2x + 3$, which is in the form $y = mx + b$. We can find two points by choosing convenient values for x—say, 0 and 1—and substituting to find the corresponding y-values. We find the two points on the line to be $(0, 3)$ and $(1, 5)$. The slope of the line is found using the definition of slope:

$$m = \frac{\text{change in } y}{\text{change in } x} = \frac{5 - 3}{1 - 0} = \frac{2}{1} = 2.$$

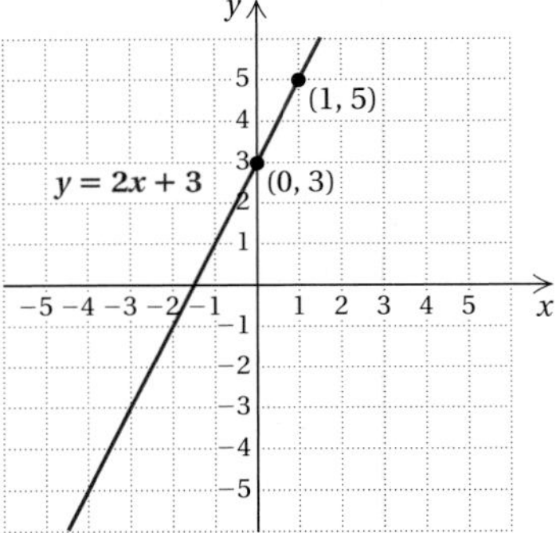

The slope is 2. Note that this is also the coefficient of the x-term in the equation $y = 2x + 3$.

DETERMINING SLOPE FROM THE EQUATION $y = mx + b$

The slope of the line $y = mx + b$ is m. To find the slope of a nonvertical line, solve the linear equation in x and y for y and get the resulting equation in the form $y = mx + b$. The coefficient of the x-term, m, is the slope of the line.

EXAMPLES Find the slope of the line.

5. $y = -3x + \dfrac{2}{9}$ → $m = -3 = \text{Slope}$

6. $y = \dfrac{4}{5}x$ → $m = \dfrac{4}{5} = \text{Slope}$

7. $y = x + 6$ → $m = 1 = \text{Slope}$

8. $y = -0.6x - 3.5$ → $m = -0.6 = \text{Slope}$

Do Exercises 6–9.

To find slope from an equation, we may have to first find an equivalent form of the equation.

EXAMPLE 9 Find the slope of the line $2x + 3y = 7$.

We solve for y to get the equation in the form $y = mx + b$:

$$2x + 3y = 7$$
$$3y = -2x + 7$$
$$y = \frac{-2x + 7}{3}$$
$$y = -\frac{2}{3}x + \frac{7}{3}. \qquad \text{This is } y = mx + b.$$

The slope is $-\frac{2}{3}$.

Do Exercises 10 and 11 on the preceding page.

What about the slope of a horizontal or a vertical line?

EXAMPLE 10 Find the slope of the line $y = 5$.

We can think of $y = 5$ as $y = 0x + 5$. Then from this equation, we see that $m = 0$. Consider the points $(-3, 5)$ and $(4, 5)$, which are on the line. The change in $y = 5 - 5$, or 0. The change in $x = -3 - 4$, or -7. We have

$$m = \frac{5 - 5}{-3 - 4} = \frac{0}{-7} = 0.$$

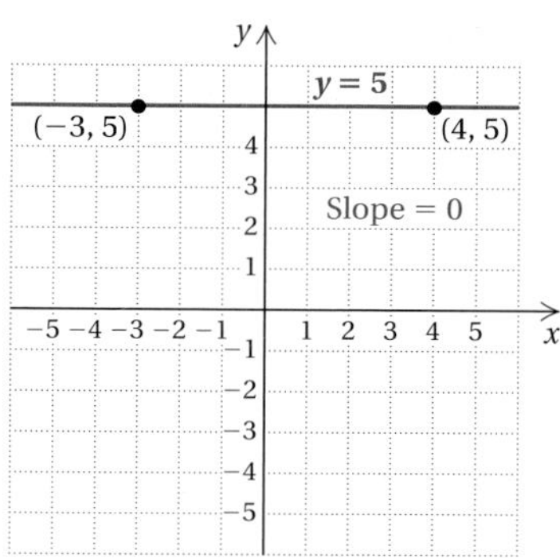

Any two points on a horizontal line have the same y-coordinate. The change in y is 0. Thus the slope of a horizontal line is 0.

EXAMPLE 11 Find the slope of the line $x = -4$.

Consider the points $(-4, 3)$ and $(-4, -2)$, which are on the line. The change in $y = 3 - (-2)$, or 5. The change in $x = -4 - (-4)$, or 0. We have

$$m = \frac{3 - (-2)}{-4 - (-4)} = \frac{5}{0}. \quad \text{Not defined}$$

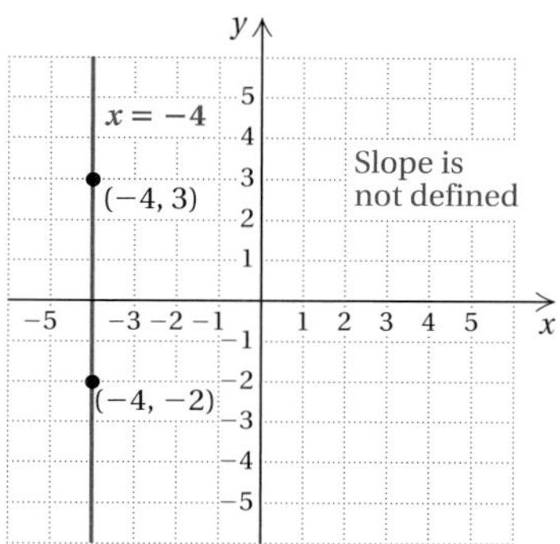

Since division by 0 is not defined, the slope of this line is not defined. The answer in this example is "The slope of this line is not defined."

SLOPE 0; SLOPE NOT DEFINED

The slope of a horizontal line is 0.

The slope of a vertical line is not defined.

Do Exercises 12 and 13.

CALCULATOR CORNER

Visualizing Slope

Exercises: Graph each of the following sets of equations using the window settings $[-6, 6, -4, 4]$, Xscl = 1, Yscl = 1.

1. $y = -x$, $y = -2x$, $y = -5x$, $y = -10x$

   What do you think the graph of $y = -123x$ will look like?

2. $y = -x$, $y = -\frac{3}{4}x$, $y = -0.38x$, $y = -\frac{5}{32}x$

   What do you think the graph of $y = -0.000043x$ will look like?

Find the slope, if it exists, of the line.

12. $x = 7$

13. $y = -5$

Answers on page A-28

9.3 EXERCISE SET

For Extra Help

MathXL | MyMathLab | InterAct Math | Math Tutor Center | Video Lectures on CD Disc 5 | Student's Solutions Manual

a Find the slope, if it exists, of the line.

1.

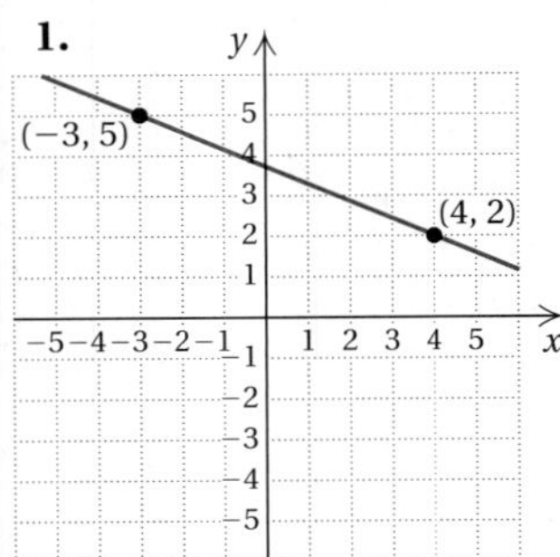

2.

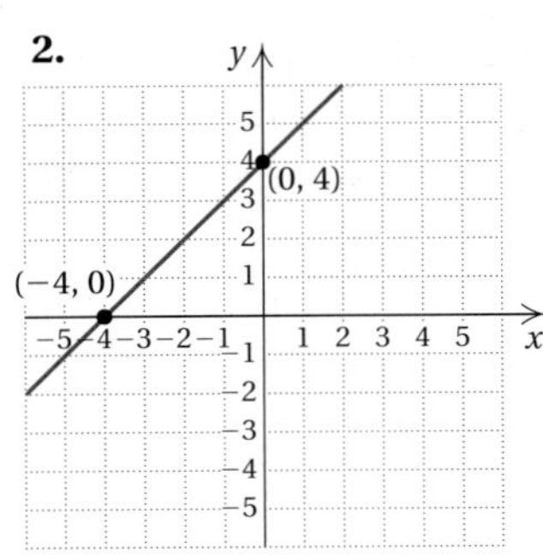

3.

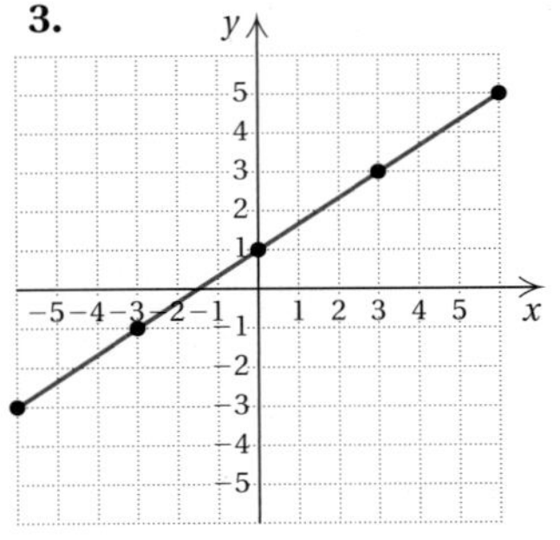

4.

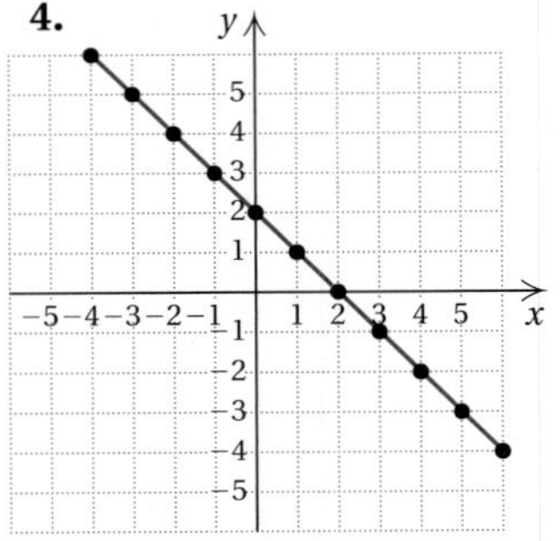

5.

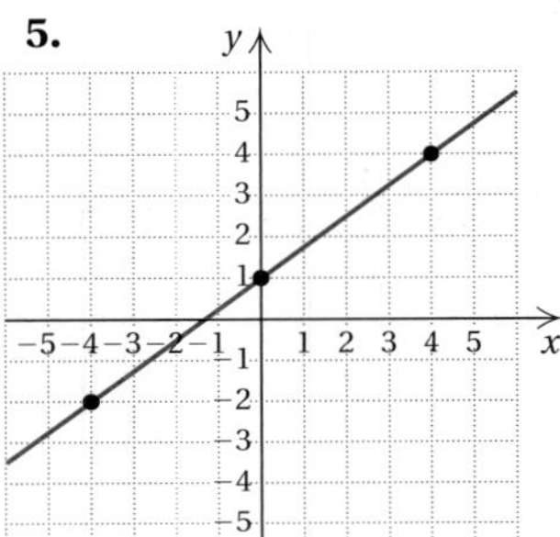

6.

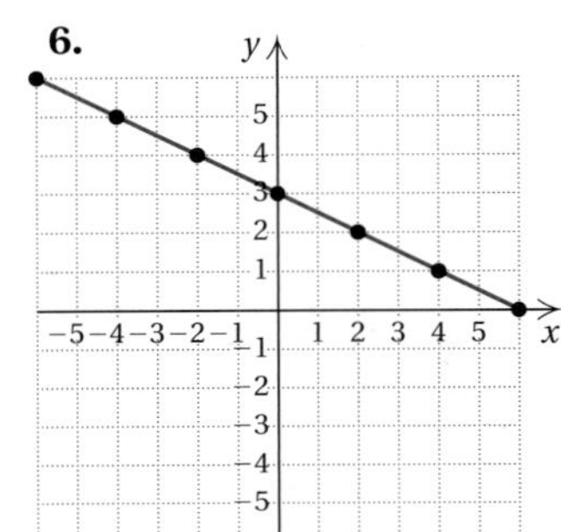

7.

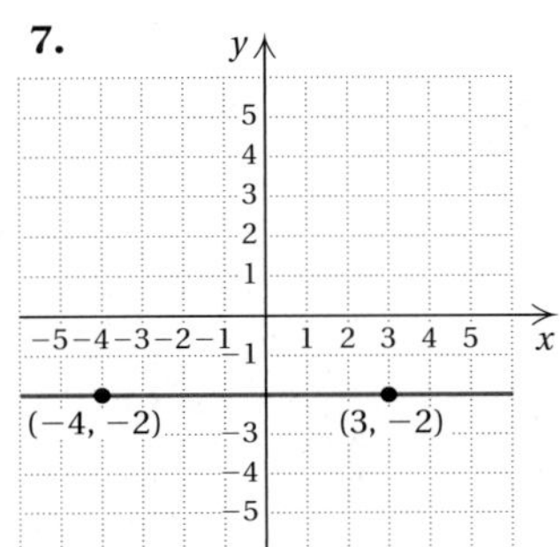

8.

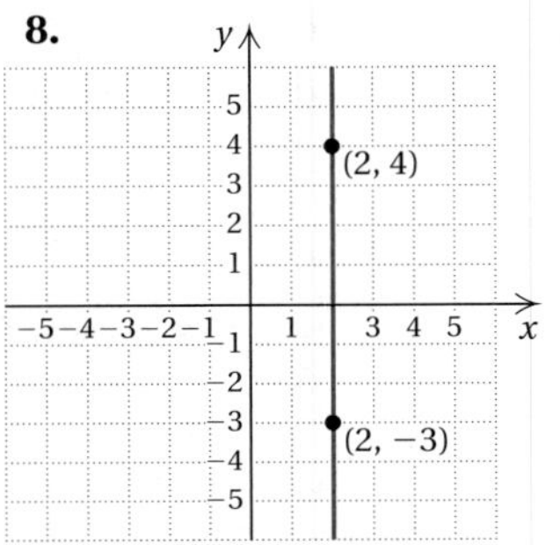

Graph the line containing the given pair of points and find the slope.

9. $(-2, 4), (3, 0)$

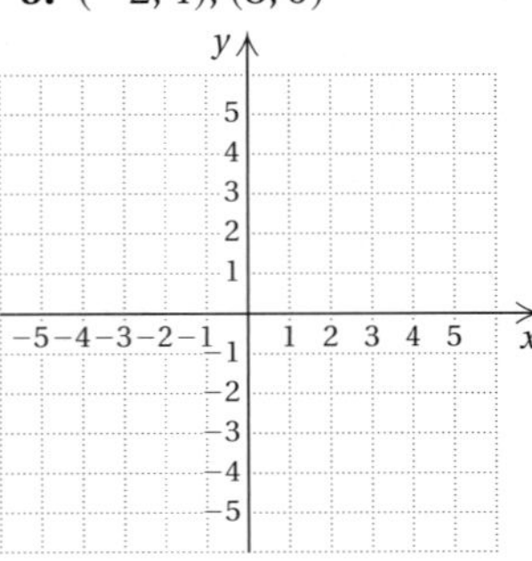

10. $(2, -4), (-3, 2)$

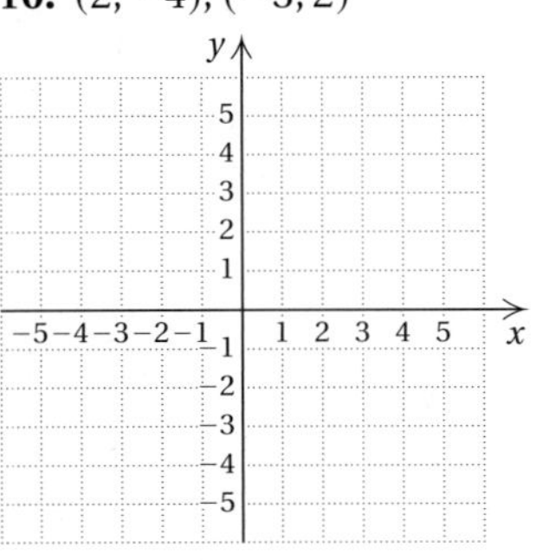

11. $(-4, 0), (-5, -3)$

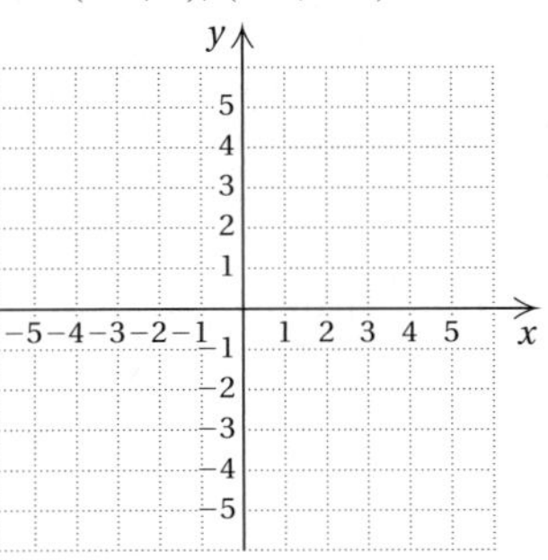

12. $(-3, 0), (-5, -2)$

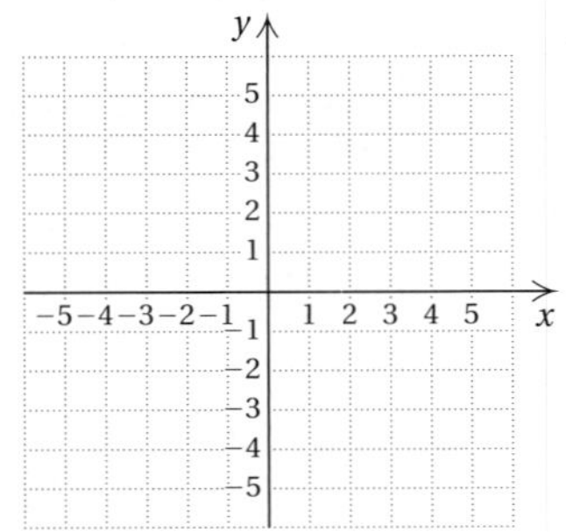

13. $(-4, 2), (2, -3)$

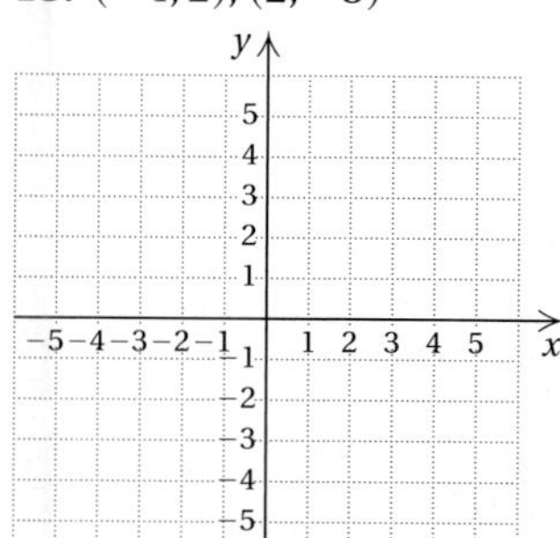

14. $(-3, 5), (4, -3)$

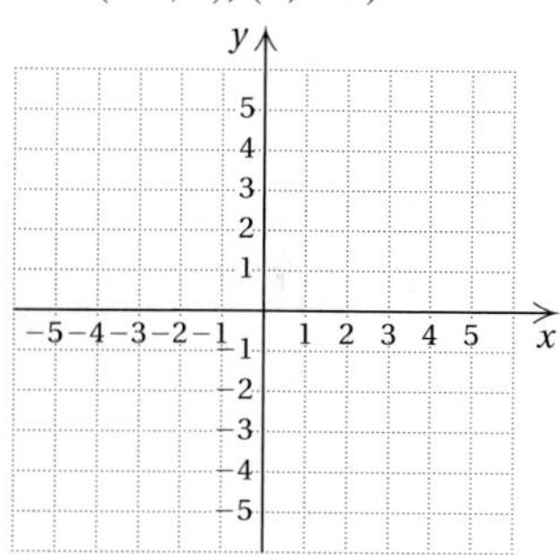

15. $(5, 3), (-3, -4)$

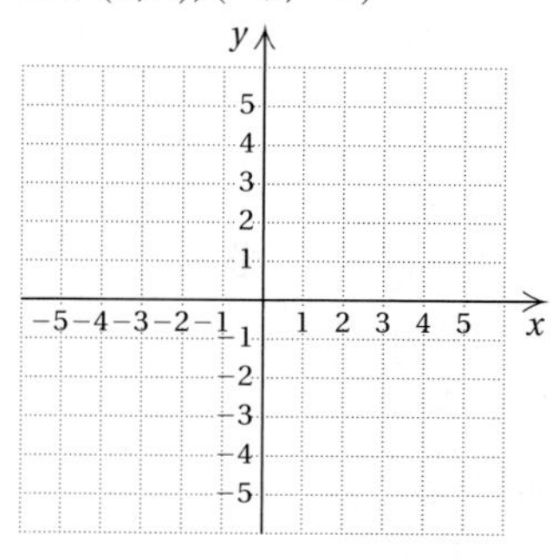

16. $(-4, -3), (2, 5)$

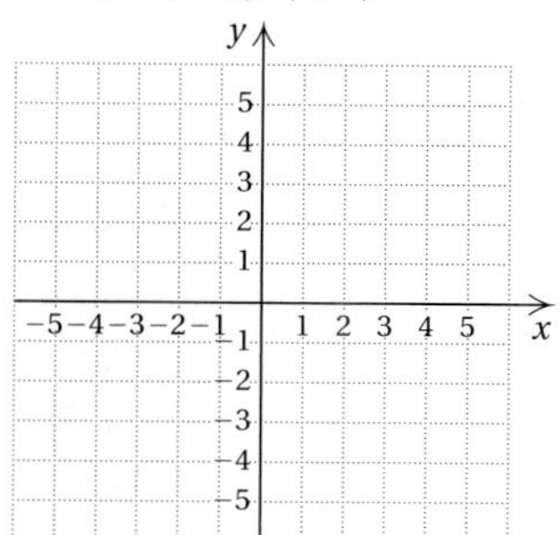

Find the slope, if it exists, of the line containing the given pair of points.

17. $\left(2, -\frac{1}{2}\right), \left(5, \frac{3}{2}\right)$

18. $\left(\frac{2}{3}, -1\right), \left(\frac{5}{3}, 2\right)$

19. $(4, -2), (4, 3)$

20. $(4, -3), (-2, -3)$

21. $(-11, 7), (15, -3)$

22. $(-13, 22), (8, -17)$

23. $\left(-\frac{1}{2}, \frac{3}{11}\right), \left(\frac{5}{4}, \frac{3}{11}\right)$

24. $(0.2, 4), (0.2, -0.04)$

b In Exercises 25–28, find the slope (or rate of change).

25. Find the slope (or pitch) of the roof.

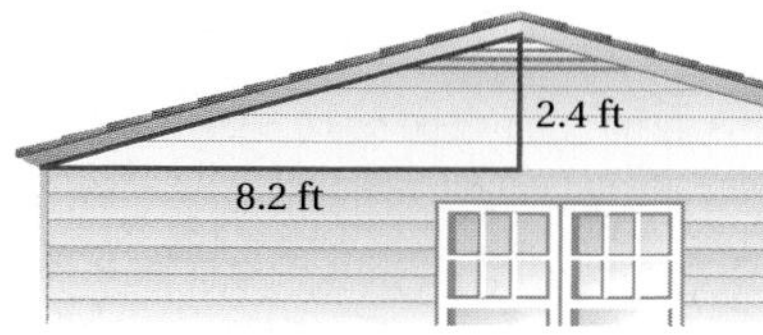

26. Find the slope (or grade) of the road.

27. Find the slope of the river.

28. Find the slope of the treadmill.

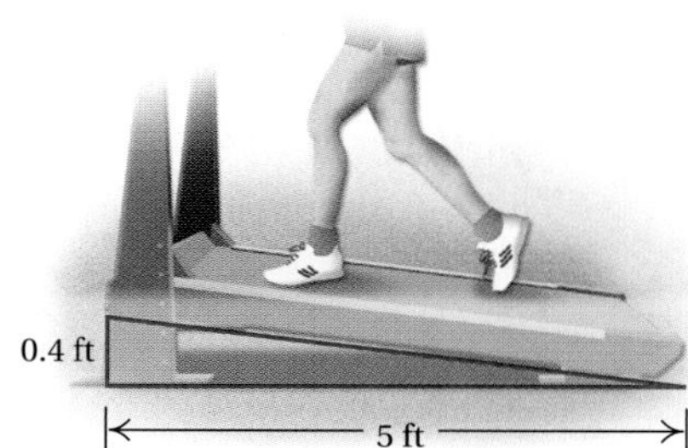

29. *Slope of Long's Peak.* From a base elevation of 9600 ft, Long's Peak in Colorado rises to a summit elevation of 14,255 ft over a horizontal distance of 15,840 ft. Find the grade of Long's Peak.

30. *Ramps for the Disabled.* In order to meet federal standards, a wheelchair ramp must not rise more than 1 ft over a horizontal distance of 12 ft. Express this slope as a grade.

In Exercises 31–34, use the graph to calculate a rate of change in which the units of the horizontal axis are used in the denominator.

31. *Gas Mileage.* The following graph shows data for a Honda Odyssey Minivan driven on interstate highways. Find the rate of change in miles per gallon, that is, the gas mileage.

Source: American Honda Motor Company, Inc.

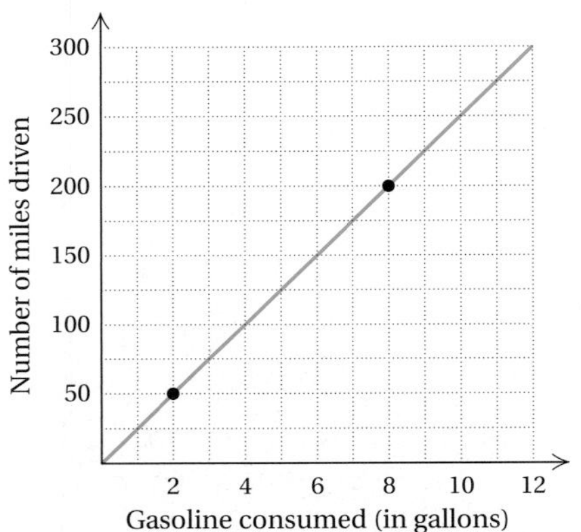

32. *Hairdresser.* Eve's Custom Cuts has a graph displaying data from a recent day of work. Find the rate of change of the number of haircuts with respect to time.

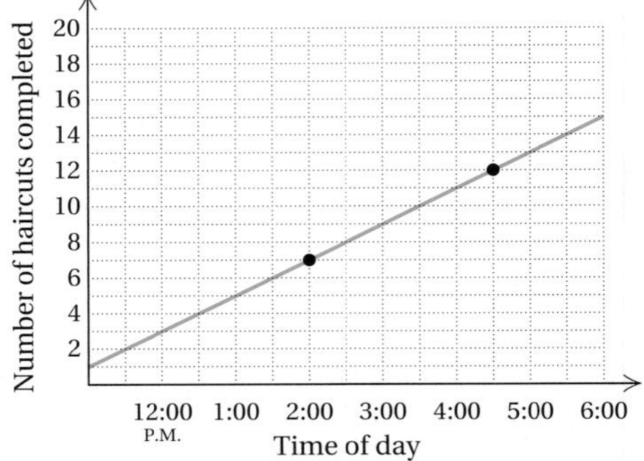

33. *Depreciation of an Office Machine.* The value of a particular color copier is represented in the following graph. Find the rate of change of the value with respect to time, in dollars per year.

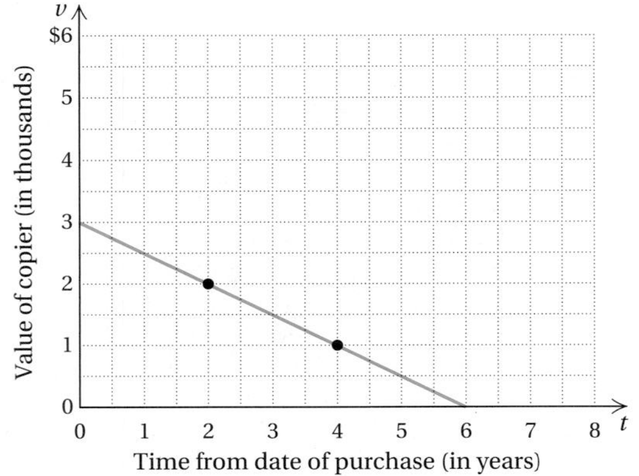

34. *Farmland.* The amount of farmland in the United States, in millions of acres, is represented in the following graph. Find the rate of change, rounded to the nearest hundred thousand, of the number of acres with respect to time, in number of acres per year.

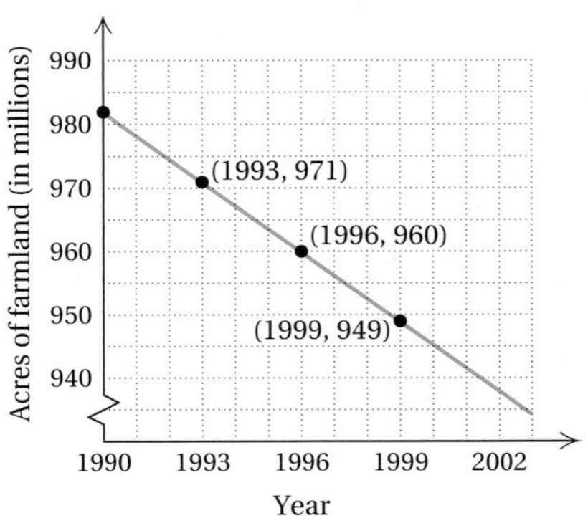

Source: U.S. Department of Agriculture

35. *Population Growth of Alaska.* The population of Alaska is illustrated in the following graph. Find the rate of change, to the nearest hundred, of the population with respect to time, in number of people per year.

Population Growth of Alaska

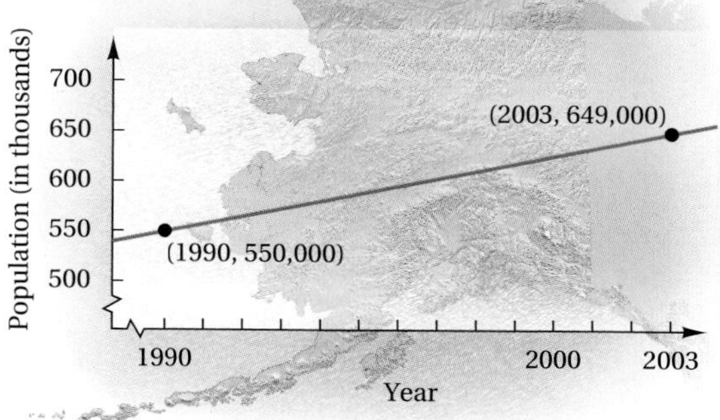

Source: U.S. Bureau of the Census

36. *Population Growth of Florida.* The population of Florida is illustrated in the following graph. Find the rate of change of the population with respect to time, in number of people per year.

Population Growth of Florida

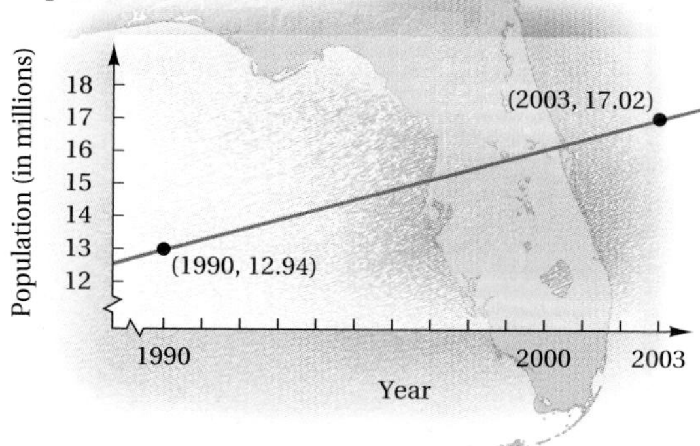

Source: U.S. Bureau of the Census

C Find the slope, if it exists, of the line.

37. $y = -10x + 7$

38. $y = \frac{10}{3}x - \frac{5}{7}$

39. $y = 3.78x - 4$

40. $y = -\frac{3}{5}x + 28$

41. $3x - y = 4$

42. $-2x + y = 8$

43. $x + 5y = 10$

44. $x - 4y = 8$

45. $3x + 2y = 6$

46. $2x - 4y = 8$

47. $x = \frac{2}{15}$

48. $y = -\frac{1}{3}$

49. $y = -2.74x$

50. $y = \frac{219}{298}x - 6.7$

51. $9x = 3y + 5$

52. $4y = 9x - 7$

53. $5x - 4y + 12 = 0$

54. $16 + 2x - 8y = 0$

55. $y = 4$

56. $x = -3$

Assuming that the scales on each axis of each graph are the same, explain how you can estimate the slope of the line that contains segment PQ without knowing the coordinates of the points P and Q.

57. DW

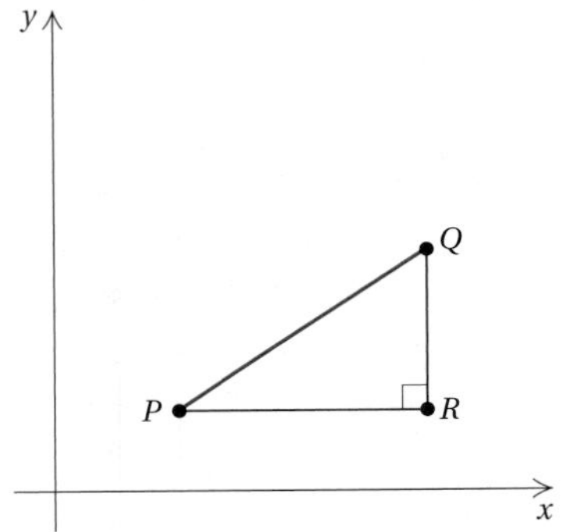

58. DW

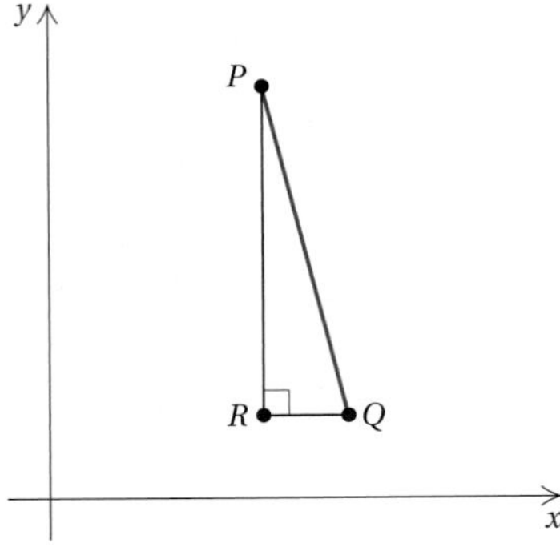

SKILL MAINTENANCE

Convert to fraction notation. [4.3b]

59. 16%

60. $33\frac{1}{3}\%$

61. 37.5%

62. 75%

Solve. [8.5a]

63. What is 15% of $23.80?

64. $7.29 is 15% of what number?

65. Jennifer left an $8.50 tip for a meal that cost $42.50. What percent of the cost of the meal was the tip?

66. Kristen left an 18% tip of $3.24 for a meal. What was the cost of the meal before the tip?

67. Juan left a 15% tip for a meal. The total cost of the meal, including the tip, was $51.92. What was the cost of the meal before the tip was added?

68. After a 25% reduction, a sweater is on sale for $41.25. What was the original price?

SYNTHESIS

In Exercises 69–72, find an equation for the graph shown.

69.

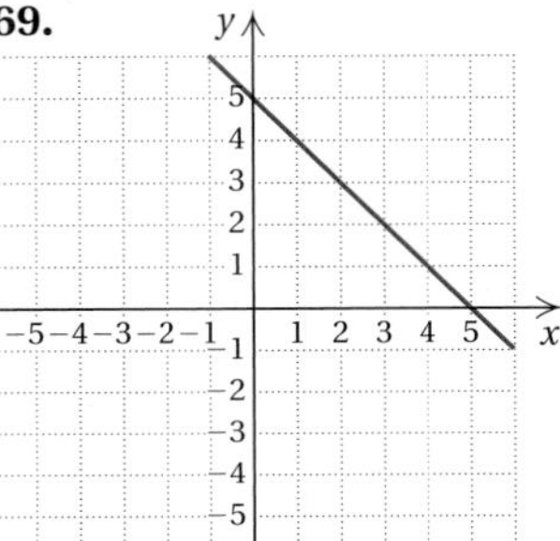

70.

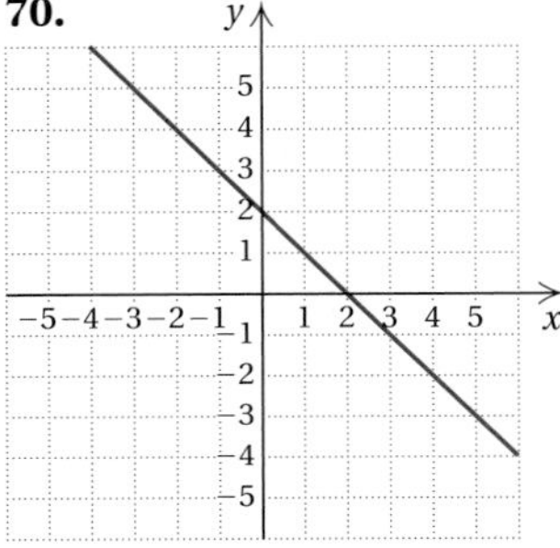

71.

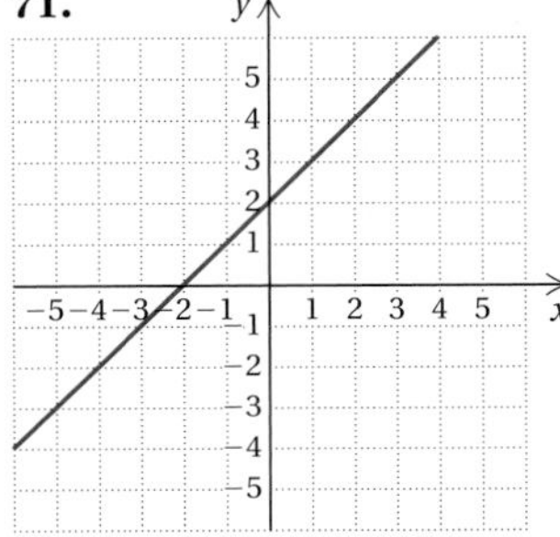

72.

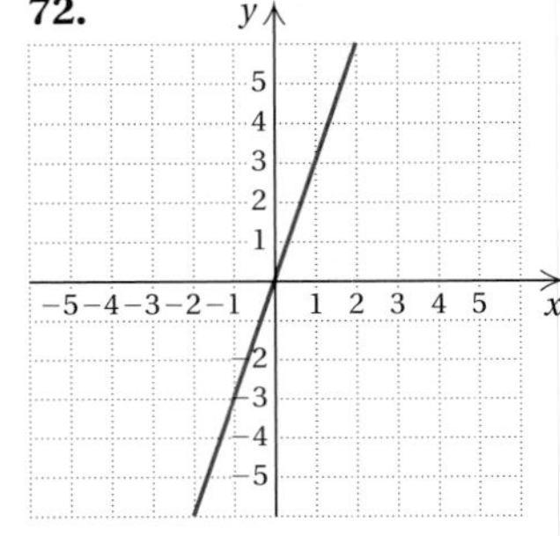

Graph the equation using the standard viewing window. Then construct a table of y-values for x-values starting at $x = -10$ with $\Delta\text{Tbl} = 0.1$.

73. $y = 0.35x - 7$

74. $y = 5.6 - x^2$

75. $y = x^3 - 5$

76. $y = 4 + 3x - x^2$

9.4 EQUATIONS OF LINES

Objectives

a Given an equation in the form $y = mx + b$, find the slope and the y-intercept; find an equation of a line when the slope and the y-intercept are given.

b Find an equation of a line when the slope and a point on the line are given.

c Find an equation of a line when two points on the line are given.

We have learned that the slope of a line and the y-intercept of the graph of the line can be read directly from the equation if it is in the form $y = mx + b$. The **slope** is m and the **y-intercept** is $(0, b)$. Here we use slope and y-intercept in order to examine linear equations in more detail.

a Finding an Equation of a Line When the Slope and the y-Intercept Are Given

We know from Sections 9.1 and 9.3 that in the equation $y = mx + b$, the slope is m and the y-intercept is $(0, b)$. Thus we call the equation $y = mx + b$ the **slope–intercept equation.***

THE SLOPE–INTERCEPT EQUATION: $y = mx + b$

The equation $y = mx + b$ is called the **slope–intercept equation.** The slope is m and the y-intercept is $(0, b)$.

EXAMPLE 1 Find the slope and the y-intercept of $2x - 3y = 8$.

We first solve for y:

$$2x - 3y = 8$$
$$-3y = -2x + 8 \quad \text{Subtracting } 2x$$
$$\frac{-3y}{-3} = \frac{-2x + 8}{-3} \quad \text{Dividing by } -3$$
$$y = \frac{-2x}{-3} + \frac{8}{-3}$$
$$y = \frac{2}{3}x - \frac{8}{3}$$

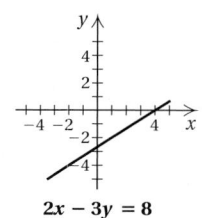

The slope is $\frac{2}{3}$. The y-intercept is $\left(0, -\frac{8}{3}\right)$.

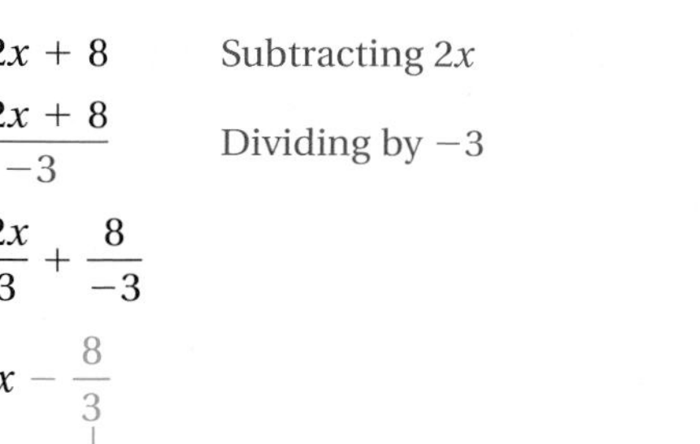

$2x - 3y = 8$

Do Exercises 1–5.

Find the slope and the y-intercept.

1. $y = 5x$

2. $y = -\frac{3}{2}x - 6$

3. $3x + 4y = 15$

4. $2y = 4x - 17$

5. $-7x - 5y = 22$

EXAMPLE 2 A line has slope -2.4 and y-intercept $(0, 11)$. Find an equation of the line.

We use the slope–intercept equation and substitute -2.4 for m and 11 for b:

$$y = mx + b$$
$$y = -2.4x + 11. \quad \text{Substituting}$$

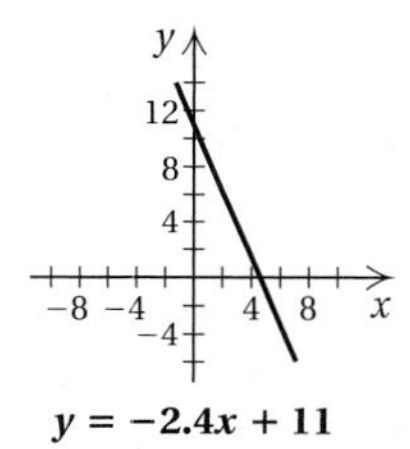

$y = -2.4x + 11$

Answers on page A-29

*For coverage of the point–slope equation form, see Appendix G.

6. A line has slope 3.5 and y-intercept $(0, -23)$. Find an equation of the line.

EXAMPLE 3 A line has slope 0 and y-intercept $(0, -6)$. Find an equation of the line.

We use the slope–intercept equation and substitute 0 for m and -6 for b:

$$y = mx + b$$
$$y = 0x + (-6) \qquad \text{Substituting}$$
$$y = -6.$$

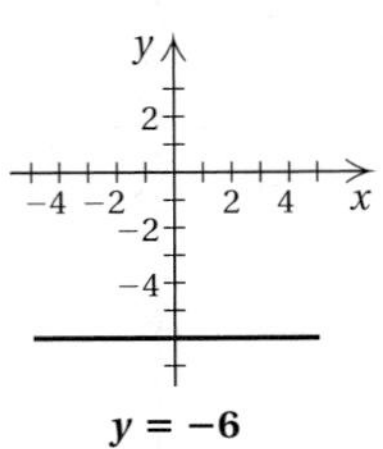

7. A line has slope 0 and y-intercept $(0, 13)$. Find an equation of the line.

EXAMPLE 4 A line has slope $-\frac{5}{3}$ and y-intercept $(0, 0)$. Find an equation of the line.

We use the slope–intercept equation and substitute $-\frac{5}{3}$ for m and 0 for b:

$$y = mx + b$$
$$y = -\tfrac{5}{3}x + 0 \qquad \text{Substituting}$$
$$y = -\tfrac{5}{3}x.$$

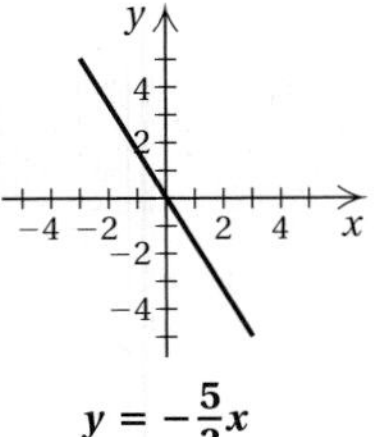

8. A line has slope -7.29 and y-intercept $(0, 0)$. Find an equation of the line.

Do Exercises 6–8.

b Finding an Equation of a Line When the Slope and a Point Are Given

Suppose we know the slope of a line and a certain point on that line. We can use the slope–intercept equation $y = mx + b$ to find an equation of the line. To write an equation in this form, we need to know the slope m and the y-intercept $(0, b)$.

EXAMPLE 5 Find an equation of the line with slope 3 that contains the point $(4, 1)$.

We know that the slope is 3, so the equation is $y = 3x + b$. This equation is true for $(4, 1)$. Using the point $(4, 1)$, we substitute 4 for x and 1 for y in $y = 3x + b$. Then we solve for b:

$$y = 3x + b \qquad \text{Substituting 3 for } m \text{ in } y = mx + b$$
$$1 = 3(4) + b \qquad \text{Substituting 4 for } x \text{ and 1 for } y$$
$$1 = 12 + b$$
$$-11 = b. \qquad \text{Solving for } b \text{, we find that the } y\text{-intercept is } (0, -11).$$

We use the equation $y = mx + b$ and substitute 3 for m and -11 for b:

$$y = 3x - 11.$$

Answers on page A-29

This is the equation of the line with slope 3 and y-intercept $(0, -11)$.

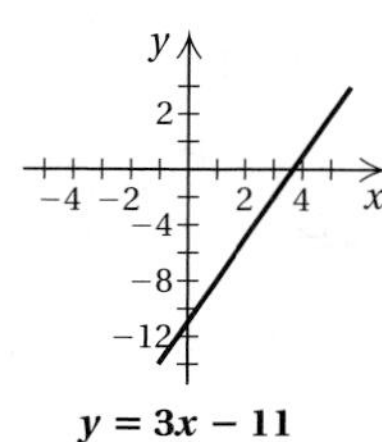

$y = 3x - 11$

EXAMPLE 6 Find an equation of the line with slope -5 that contains the point $(-2, 3)$.

We know that the slope is -5, so the equation is $y = -5x + b$. Using the point $(-2, 3)$, we substitute -2 for x and 3 for y in $y = -5x + b$. Then we solve for b:

$$y = -5x + b \qquad \text{Substituting } -5 \text{ for } m \text{ in } y = mx + b$$

$$3 = -5(-2) + b \qquad \text{Substituting } -2 \text{ for } x \text{ and } 3 \text{ for } y$$

$$3 = 10 + b$$

$$-7 = b. \qquad \text{Solving for } b$$

We use the equation $y = mx + b$ and substitute -5 for m and -7 for b:

$$y = -5x - 7.$$

This is the equation of the line with slope -5 and y-intercept $(0, -7)$.

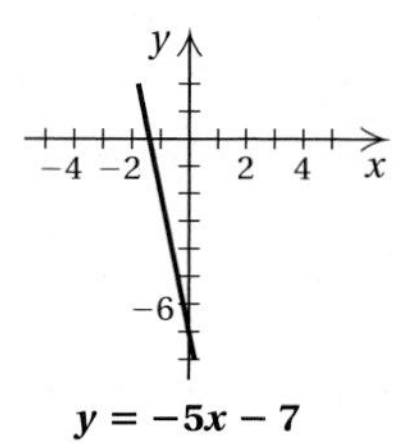

$y = -5x - 7$

Do Exercises 9–12.

Find an equation of the line that contains the given point and has the given slope.

9. $(4, 2), \quad m = 5$

10. $(-2, 1), \quad m = -3$

11. $(3, 5), \quad m = 6$

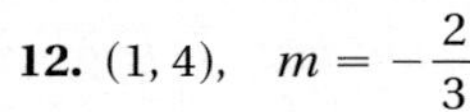

12. $(1, 4), \quad m = -\frac{2}{3}$

C Finding an Equation of a Line When Two Points Are Given

We can also use the slope–intercept equation to find an equation of a line when two points are given.

EXAMPLE 7 Find an equation of the line containing the points $(2, 3)$ and $(-6, 1)$.

First, we find the slope:

$$m = \frac{3 - 1}{2 - (-6)} = \frac{2}{8}, \text{ or } \frac{1}{4}.$$

Thus, $y = \frac{1}{4}x + b$. We then proceed as we did in Example 6, using either point to find b.

Answers on page A-29

Find an equation of the line containing the given points.

13. (2, 4) and (3, 5)

14. (−1, 2) and (−3, −2)

Answers on page A-29

We choose (2, 3) and substitute 2 for x and 3 for y:

$$y = \frac{1}{4}x + b \qquad \text{Substituting } \frac{1}{4} \text{ for } m \text{ in } y = mx + b$$

$$3 = \frac{1}{4} \cdot 2 + b \qquad \text{Substituting 2 for } x \text{ and 3 for } y$$

$$3 = \frac{1}{2} + b$$

$$\frac{5}{2} = b. \qquad \text{Solving for } b$$

We use the equation $y = mx + b$ and substitute $\frac{1}{4}$ for m and $\frac{5}{2}$ for b:

$$y = \frac{1}{4}x + \frac{5}{2}.$$

This is the equation of the line with slope $\frac{1}{4}$ and y-intercept $\left(0, \frac{5}{2}\right)$.

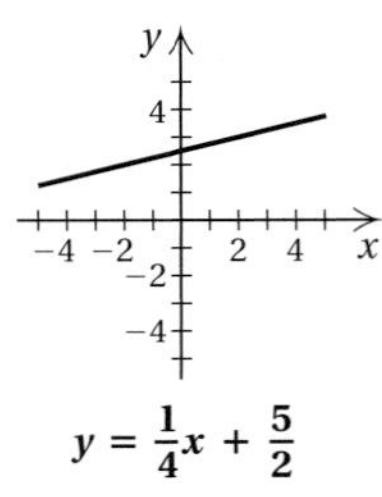

$y = \frac{1}{4}x + \frac{5}{2}$

Do Exercises 13 and 14.

Study Tips

BETTER TEST TAKING

How often do you make the following statement after taking a test: "I was able to do the homework, but I froze during the test"? Here are two tips to help you with this difficulty. Both are intended to make test taking less stressful by getting you to practice good test-taking habits on a daily basis.

- **Treat every homework exercise as if it were a test question.** If you had to work a problem at your job with no backup answer provided, what would you do? You would probably work it very deliberately, checking and rechecking every step. You might work it more than one time, or you might try to work it another way to check the result. Try to use this approach when doing your homework. Treat every exercise as though it were a test question with no answer at the back of the book.
- **Be sure that you do questions without answers as part of every homework assignment whether or not the instructor has assigned them!** One reason a test may seem such a different task is that questions on a test lack answers. That is the reason for taking a test: to see if you can do the questions without assistance. As part of your test preparation, be sure you do some exercises for which you do not have the answers. Thus when you take a test, you are doing a more familiar task.

The purpose of doing your homework using these approaches is to give you more test-taking practice beforehand. Let's use a sports analogy: At a basketball game, the players take lots of practice shots before the game. They play the first half, go to the locker room, and come out for the second half. What do they do before the second half, even though they have just played 20 minutes of basketball? They shoot baskets again! We suggest the same approach here. Create more and more situations in which you practice taking test questions by treating each homework exercise like a test question and by doing exercises for which you have no answers. Good luck!

"He who does not venture has no luck."

9.4 EXERCISE SET

For Extra Help

MathXL | MyMathLab | InterAct Math | Math Tutor Center | Video Lectures on CD Disc 5 | Student's Solutions Manual

a Find the slope and the y-intercept.

1. $y = -4x - 9$

2. $y = -2x + 3$

3. $y = 1.8x$

4. $y = -27.4x$

5. $-8x - 7y = 21$

6. $-2x - 8y = 16$

7. $4x = 9y + 7$

8. $5x + 4y = 12$

9. $-6x = 4y + 2$

10. $4.8x - 1.2y = 36$

11. $y = -17$

12. $y = 28$

Find an equation of the line with the given slope and y-intercept.

13. Slope $= -7$, y-intercept $= (0, -13)$

14. Slope $= 73$, y-intercept $= (0, 54)$

15. Slope $= 1.01$, y-intercept $= (0, -2.6)$

16. Slope $= -\frac{3}{8}$, y-intercept $= \left(0, \frac{7}{11}\right)$

b Find an equation of the line containing the given point and having the given slope.

17. $(-3, 0)$, $m = -2$

18. $(2, 5)$, $m = 5$

19. $(2, 4)$, $m = \frac{3}{4}$

20. $\left(\frac{1}{2}, 2\right)$, $m = -1$

21. $(2, -6)$, $m = 1$

22. $(4, -2)$, $m = 6$

23. $(0, 3)$, $m = -3$

24. $(-2, -4)$, $m = 0$

c Find an equation of the line that contains the given pair of points.

25. $(12, 16)$ and $(1, 5)$

26. $(-6, 1)$ and $(2, 3)$

27. $(0, 4)$ and $(4, 2)$

28. $(0, 0)$ and $(4, 2)$

29. $(3, 2)$ and $(1, 5)$

30. $(-4, 1)$ and $(-1, 4)$

31. $(-4, 5)$ and $(-2, -3)$

32. $(-2, -4)$ and $(2, -1)$

33. *Aerobic Exercise.* The line graph below describes the *target heart rate T,* in beats per minute, of a person of age a, who is exercising. The goal is to get the number of beats per minute to this target level.

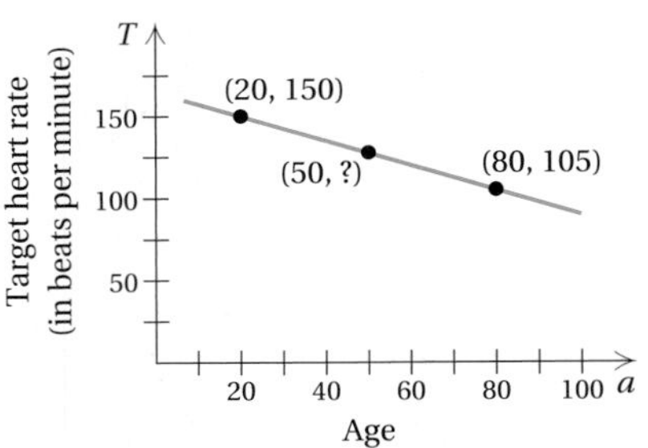

a) Find an equation of the line.

b) What is the rate of change of target heart rate with respect to time?

c) Use the equation to calculate the target heart rate of a person of age 50.

34. *Diabetes Cases.* The line graph below describes the number N, in millions, of persons diagnosed with diabetes in the United States in years x since 1992.

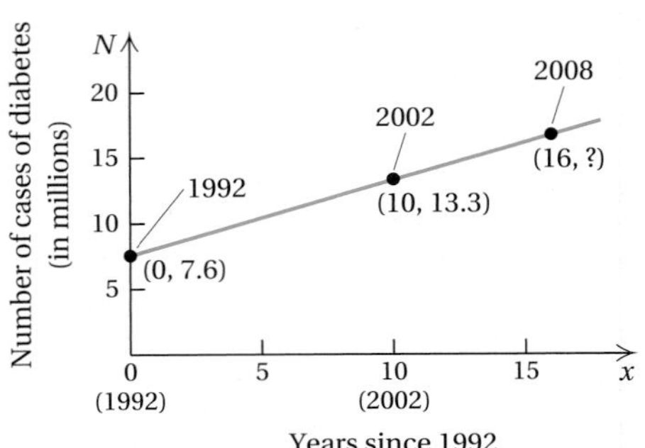

Source: U.S. National Center for Health Statistics

a) Find an equation of the line.

b) What is the rate of change of the number of cases of diabetes with respect to time?

c) Use the equation to predict the number of cases of diabetes in 2008.

35. $\mathbf{D_W}$ Do all graphs of linear equations have y-intercepts? Why or why not?

36. $\mathbf{D_W}$ Do all graphs of linear equations have x-intercepts? Why or why not?

SKILL MAINTENANCE

Solve. [8.3b, c]

37. $3x - 4(9 - x) = 17$

38. $2(5 + 2y) + 4y = 13$

39. $40(2x - 7) = 50(4 - 6x)$

40. $\frac{2}{3}(x - 5) = \frac{3}{8}(x + 5)$

41. $3x - 9x + 21x - 15x = 6x - 12 - 24x + 18$

42. $3x - (9x + 21x) - 15x = 6x - (12 - 24x) + 18$

43. $3(x - 9x) + 21(x - 15x) = 6(x - 12) - 24(x + 18)$

44. $3x - (9x + 21x - 15x) = 6x - (12 - 24x + 18)$

SYNTHESIS

45. Find an equation of the line that contains the point $(2, -3)$ and has the same slope as the line $3x - y + 4 = 0$.

46. Find an equation of the line that has the same y-intercept as the line $x - 3y = 6$ and contains the point $(5, -1)$.

47. Find an equation of the line with the same slope as the line $3x - 2y = 8$ and the same y-intercept as the line $2y + 3x = -4$.

9.5 GRAPHING USING THE SLOPE AND THE y-INTERCEPT

Objective

a Use the slope and the y-intercept to graph a line.

a Graphs Using the Slope and the y-Intercept

We can graph a line if we know the coordinates of two points on that line. We can also graph a line if we know the slope and the y-intercept.

EXAMPLE 1 Draw a line that has slope $\frac{1}{4}$ and y-intercept $(0, 2)$.

We plot $(0, 2)$ and from there move *up* 1 unit (since the numerator is *positive* and corresponds to the change in y) and *to the right* 4 units (since the denominator is *positive* and corresponds to the change in x). This locates the point $(4, 3)$. We plot $(4, 3)$ and draw a line passing through $(0, 2)$ and $(4, 3)$, as shown on the right below.

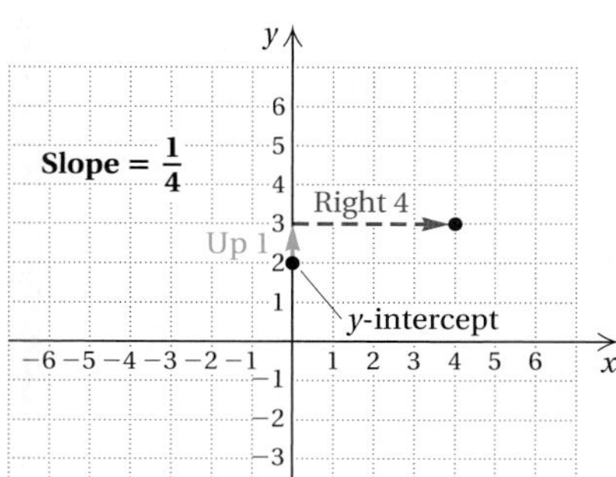

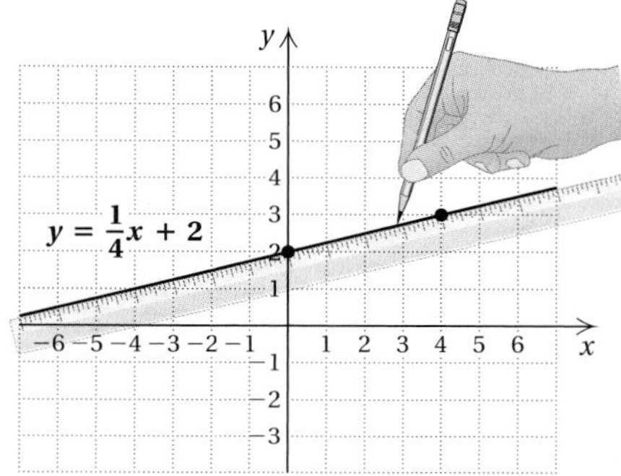

We are actually graphing the equation $y = \frac{1}{4}x + 2$.

EXAMPLE 2 Draw a line that has slope $-\frac{2}{3}$ and y-intercept $(0, 4)$.

We can think of $-\frac{2}{3}$ as $\frac{-2}{3}$. We plot $(0, 4)$ and from there move *down* 2 units (since the numerator is *negative*) and *to the right* 3 units (since the denominator is *positive*). We plot the point $(3, 2)$ and draw a line passing through $(0, 4)$ and $(3, 2)$.

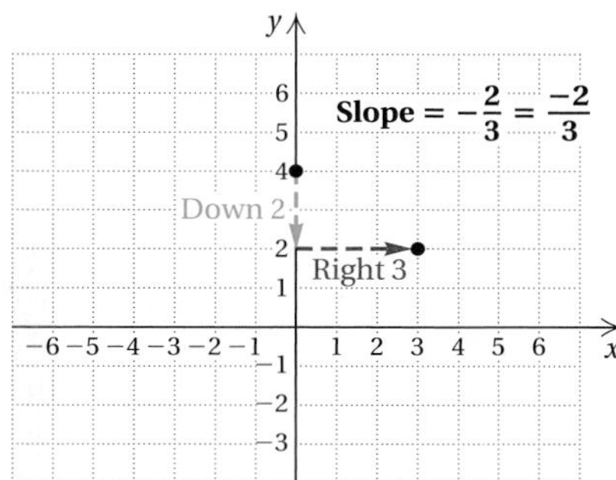

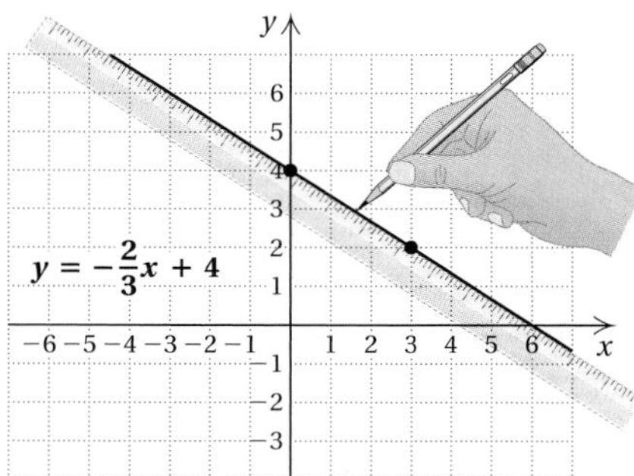

We are actually graphing the equation $y = -\frac{2}{3}x + 4$.

Do Exercises 1–3.

1. Draw a line that has slope $\frac{2}{5}$ and y-intercept $(0, -3)$. What equation is graphed?

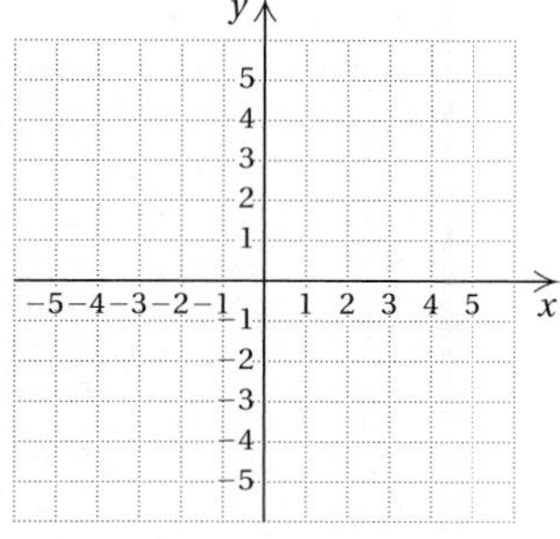

2. Draw a line that has slope $-\frac{2}{5}$ and y-intercept $(0, -3)$. What equation is graphed?

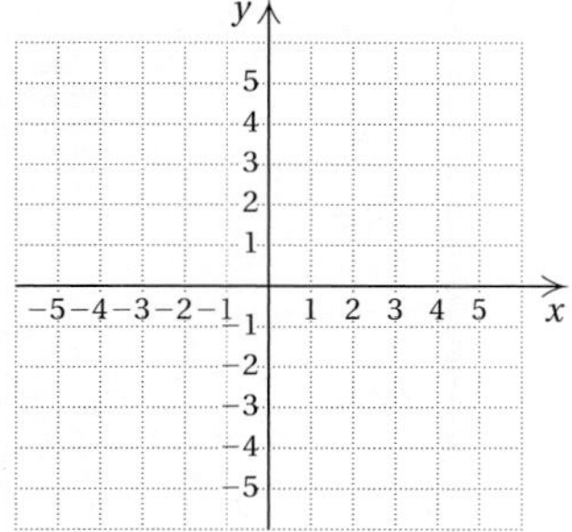

3. Draw a line that has slope 6 and y-intercept $(0, -3)$. Think of 6 as $\frac{6}{1}$. What equation is graphed?

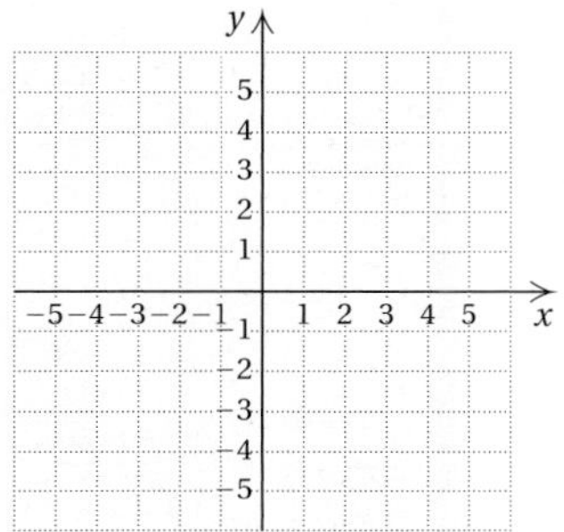

Answers on page A-29

4. Graph $y = \frac{3}{5}x - 4$ using the slope and the y-intercept.

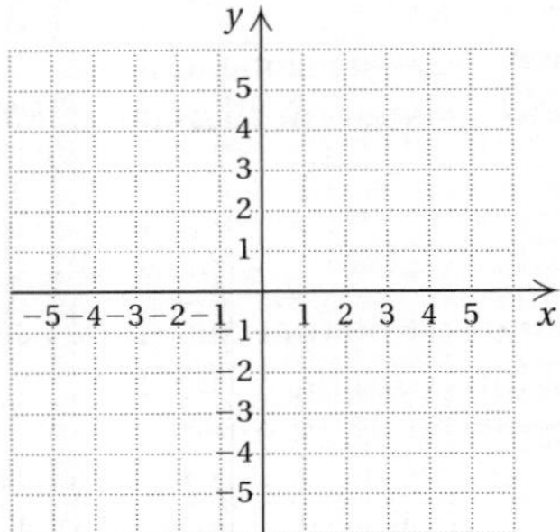

Answer on page A-29

We now use our knowledge of the slope–intercept equation to graph linear equations.

EXAMPLE 3 Graph $y = \frac{3}{4}x + 5$ using the slope and the y-intercept.

From the equation $y = \frac{3}{4}x + 5$, we see that the slope of the graph is $\frac{3}{4}$ and the y-intercept is $(0, 5)$. We plot $(0, 5)$ and then consider the slope, $\frac{3}{4}$. Starting at $(0, 5)$, we plot a second point by moving *up* 3 units (since the numerator is *positive*) and *to the right* 4 units (since the denominator is *positive*). We reach a new point, $(4, 8)$.

We can also rewrite the slope as $\frac{-3}{-4}$. We again start at the y-intercept, $(0, 5)$, but move *down* 3 units (since the numerator is *negative* and corresponds to the change in y) and *to the left* 4 units (since the denominator is *negative* and corresponds to the change in x). We reach another point, $(-4, 2)$. Once two or three points have been plotted, the line representing all solutions of $y = \frac{3}{4}x + 5$ can be drawn.

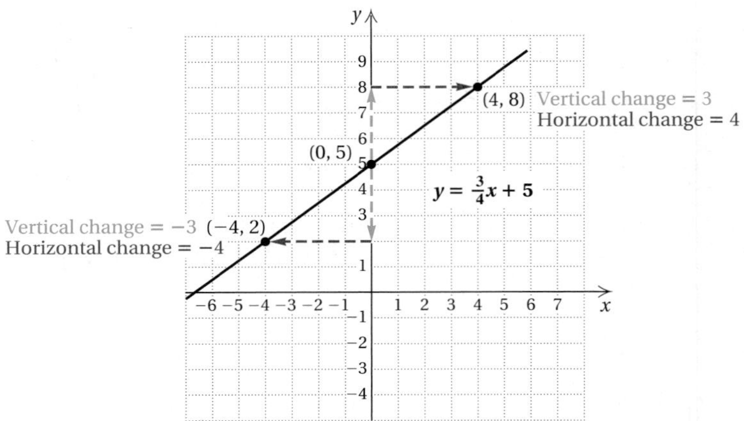

Do Exercise 4.

EXAMPLE 4 Graph $2x + 3y = 3$ using the slope and the y-intercept.

To graph $2x + 3y = 3$, we first rewrite the equation in slope–intercept form:

$$2x + 3y = 3$$
$$3y = -2x + 3 \qquad \text{Adding } -2x$$
$$\frac{1}{3} \cdot 3y = \frac{1}{3}(-2x + 3) \qquad \text{Multiplying by } \frac{1}{3}$$
$$y = -\frac{2}{3}x + 1. \qquad \text{Simplifying}$$

To graph $y = -\frac{2}{3}x + 1$, we first plot the y-intercept, $(0, 1)$. We can think of the slope as $\frac{-2}{3}$. Starting at $(0, 1)$ and using the slope, we find a second point by moving *down* 2 units (since the numerator is *negative*) and *to the right* 3 units (since the denominator is *positive*). We plot the new point, $(3, -1)$. In a similar manner, we can move from the point $(3, -1)$ to locate a third point, $(6, -3)$. The line can then be drawn.

Since $-\frac{2}{3} = \frac{2}{-3}$, an alternative approach is to again plot (0, 1), but this time move *up* 2 units (since the numerator is *positive*) and *to the left* 3 units (since the denominator is *negative*). This leads to another point on the graph, (−3, 3).

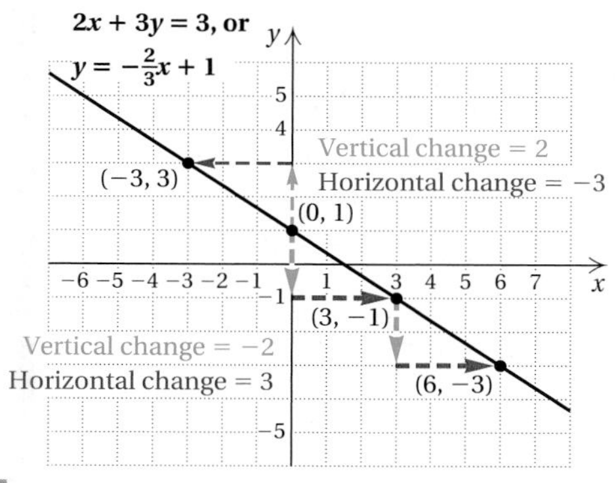

It helps to use both $\frac{2}{-3}$ and $\frac{-2}{3}$ to draw the graph.

Do Exercise 5.

5. Graph: $3x + 4y = 12$.

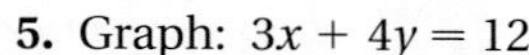

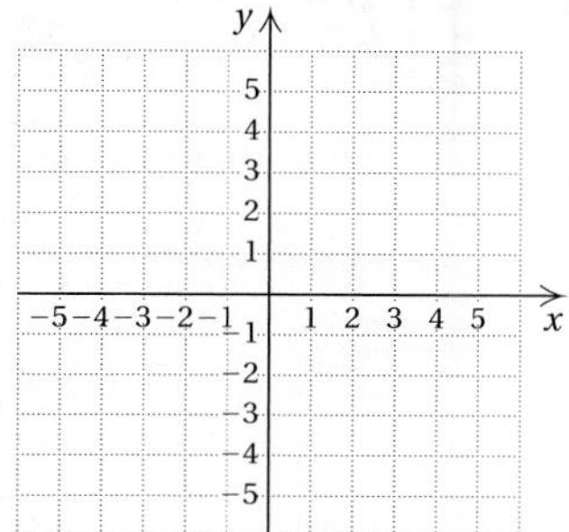

Answer on page A-29

Study Tips

TURNING NEGATIVES INTO POSITIVES

B. C. Forbes said, "History has demonstrated that notable winners *usually encountered heartbreaking obstacles before they triumphed. They won because they refused to become discouraged by their defeats."*

Here are some anecdotes about well-known people who turned what could have been a negative experience into a positive outcome.

- *Richard Bach* sold more than 7 million copies of his story about a "soaring" seagull, Jonathan Livingston Seagull. His work was turned down by 18 publishers before Macmillan finally published it in 1970.
- *Walt Disney* was once fired by a newspaper for what they said was his "lack of ideas." He went bankrupt several times before he built an entertainment empire that now includes Disneyland and Disney World.
- *Erik Weihenmayer,* a blind man, has climbed the tallest mountains in Africa and North and South America and recently climbed the tallest mountain in the world, Mt. Everest.
- *Hank Aaron* holds the all-time Major League home run record with a total of 755, topping the former record holder, Babe Ruth, who had 714. But Aaron also held the all-time record for many years for striking out 1383 times, also topping Babe Ruth, who struck out 1330 times!
- At the age of 15, *Michael Jordan* was cut from his school basketball team. He was told he was too small to play. Yet in 2000, he was selected by ESPN as the top athlete of the 20th century.
- *Albert Einstein* didn't speak until he was 4 years old and was not able to read until he was 7. He is now recognized as one of the greatest physicists of all time, having developed the famous theory of relativity.

In an article entitled "*Mistakes–Important Teacher,*" Josh Hinds writes, "Another approach (to negative experiences) is to remind ourselves that failures are not always failures, rather they are lessons. I would challenge you to find one occurrence in your own life where you have learned from a past mistake. While it can be true that we don't gain direct rewards from them, we still gain something of great importance. Therefore we need to explore our failures and take the time to use them as our teachers ..."

"Whoever makes no mistakes is doing nothing."

Dutch proverb

EXERCISE SET

For Extra Help

MathXL

MyMathLab

InterAct Math

Math Tutor Center

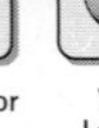
Video Lectures on CD Disc 5

Student's Solutions Manual

a Draw a line that has the given slope and y-intercept.

1. Slope $\frac{2}{5}$; y-intercept $(0, 1)$

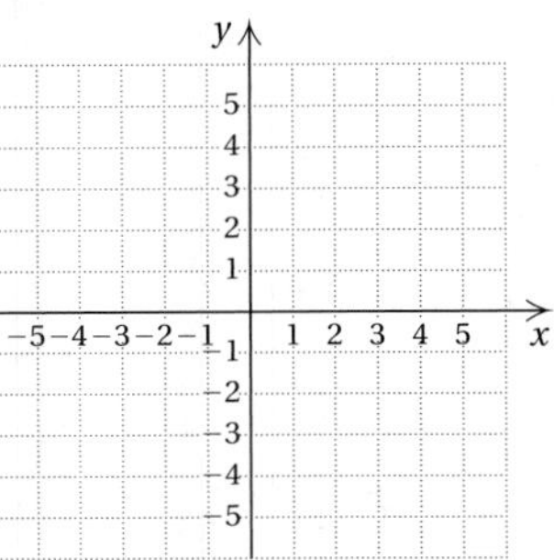

2. Slope $\frac{3}{5}$; y-intercept $(0, -1)$

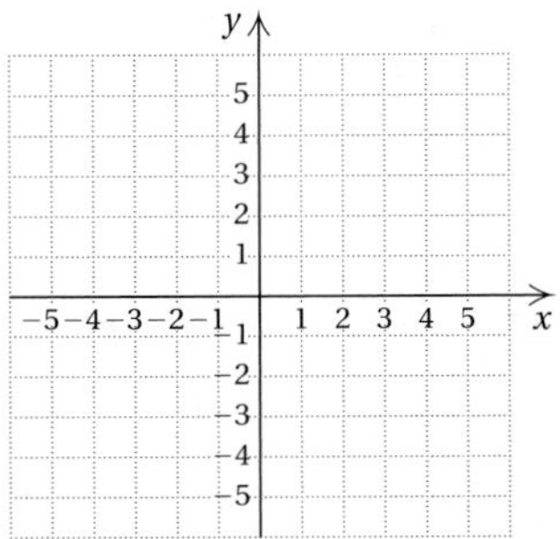

3. Slope $\frac{5}{3}$; y-intercept $(0, -2)$

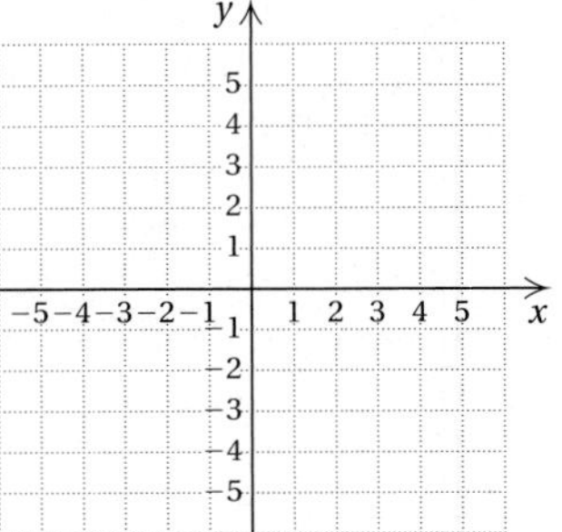

4. Slope $\frac{5}{2}$; y-intercept $(0, 1)$

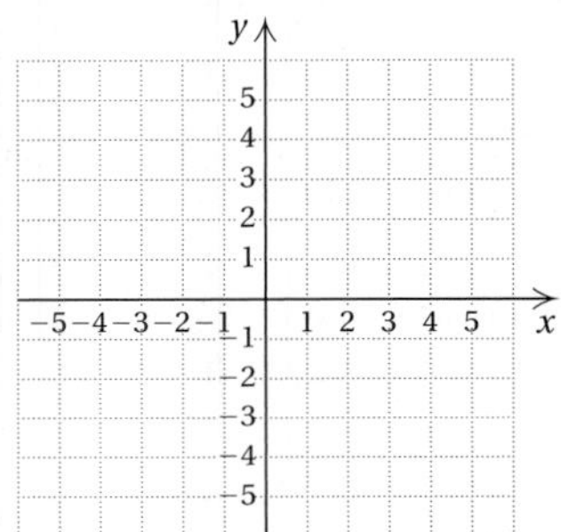

5. Slope $-\frac{3}{4}$; y-intercept $(0, 5)$

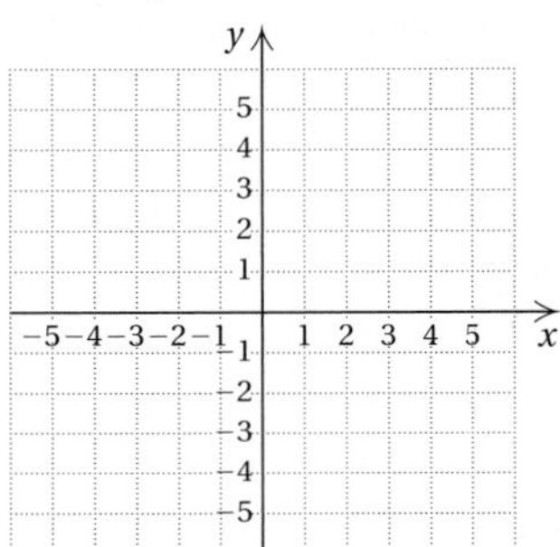

6. Slope $-\frac{4}{5}$; y-intercept $(0, 6)$

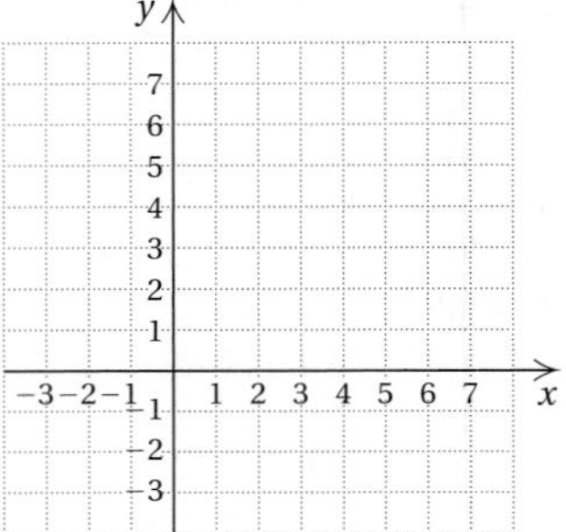

7. Slope $-\frac{1}{2}$; y-intercept $(0, 3)$

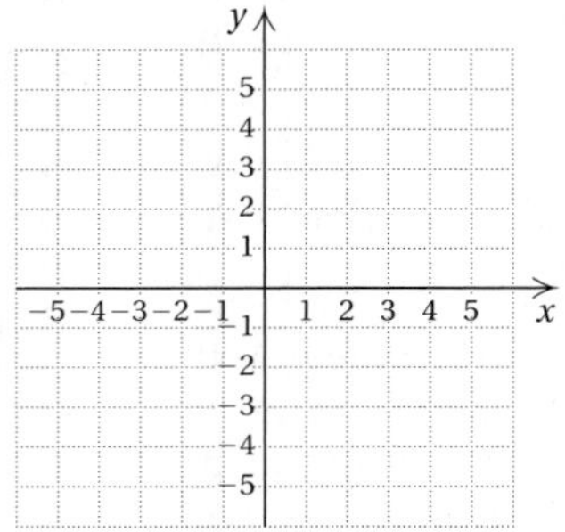

8. Slope $\frac{1}{3}$; y-intercept $(0, -4)$

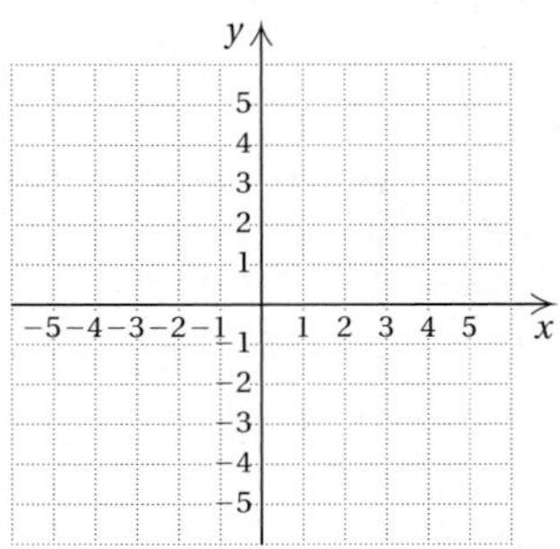

9. Slope 2; y-intercept $(0, -4)$

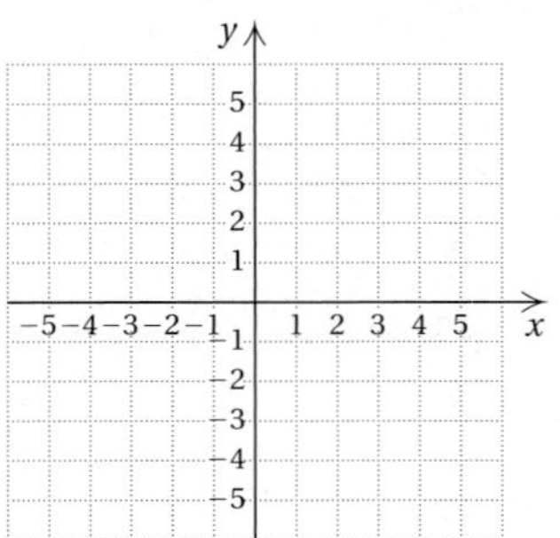

10. Slope -2; y-intercept $(0, -3)$

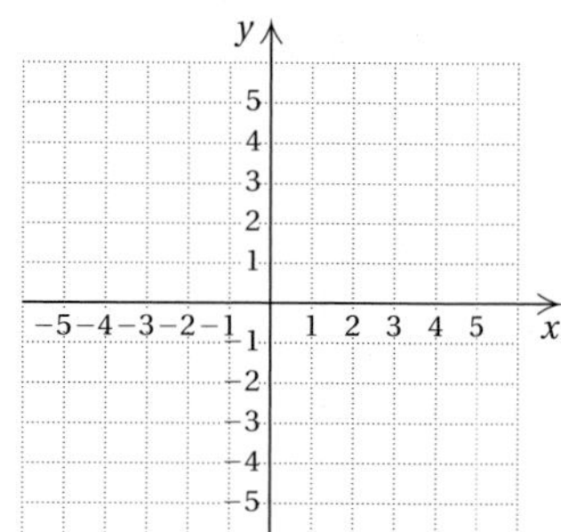

11. Slope -3; y-intercept $(0, 2)$

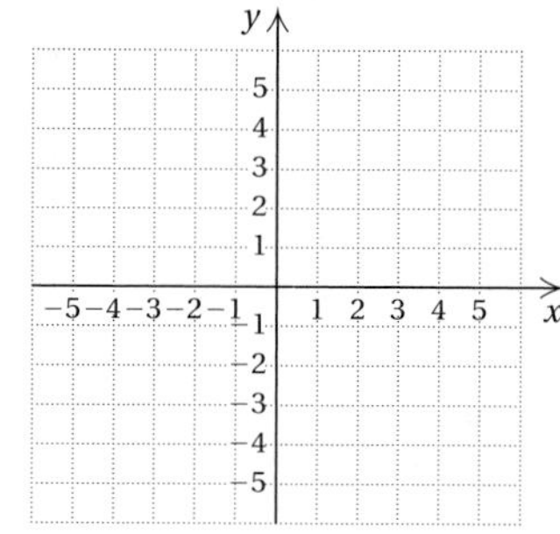

12. Slope 3; y-intercept $(0, 4)$

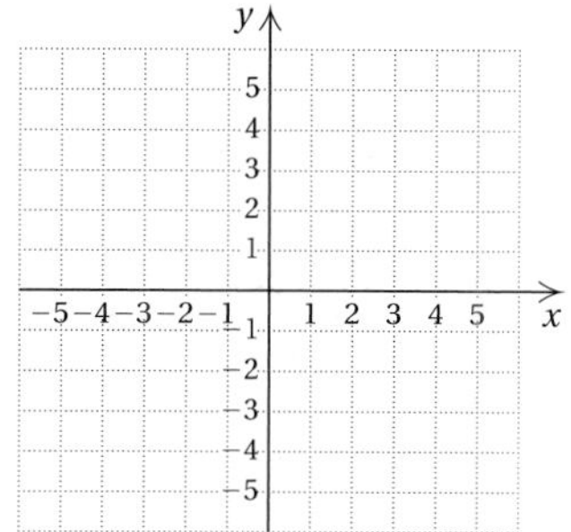

Graph using the slope and the y-intercept.

13. $y = \frac{3}{5}x + 2$

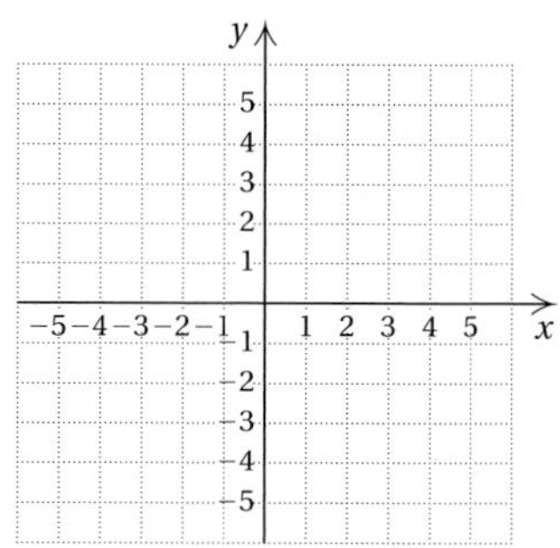

14. $y = -\frac{3}{5}x - 1$

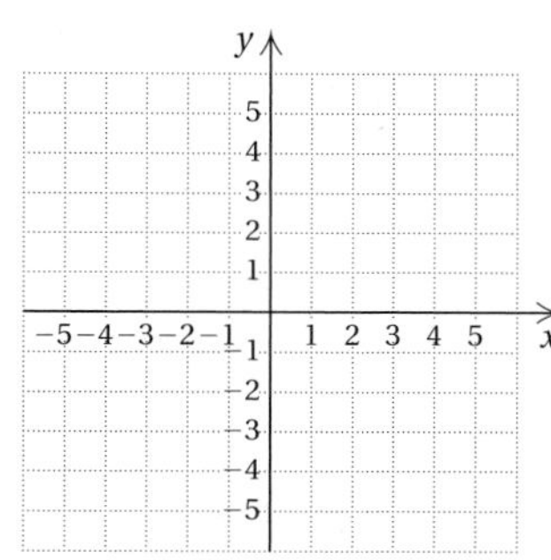

15. $y = -\frac{3}{5}x + 1$

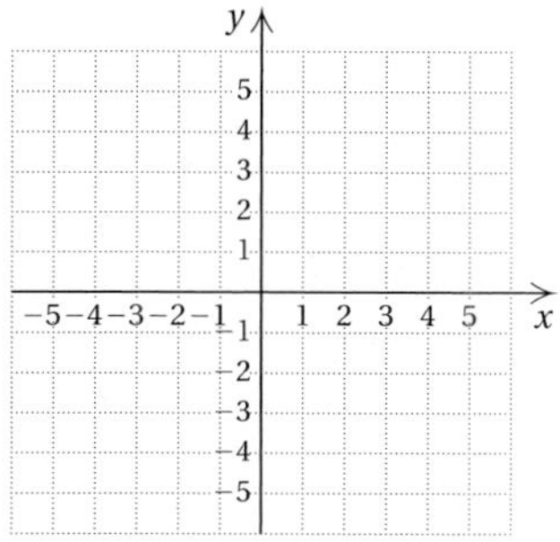

16. $y = \frac{3}{5}x - 2$

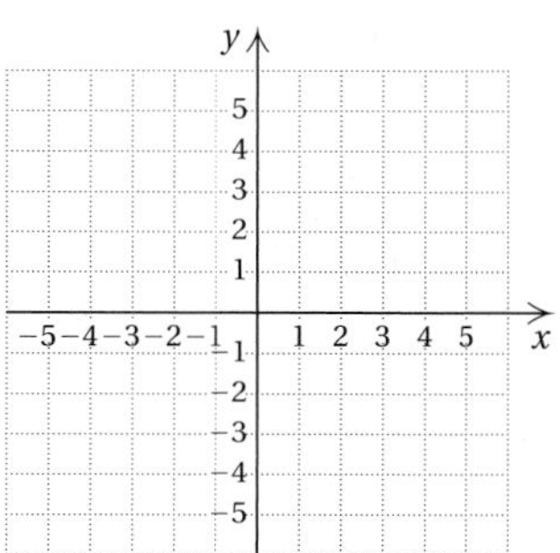

17. $y = \frac{5}{3}x + 3$

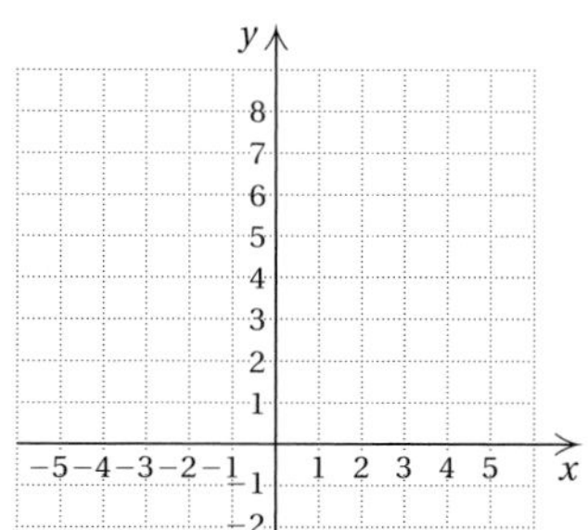

18. $y = \frac{5}{3}x - 2$

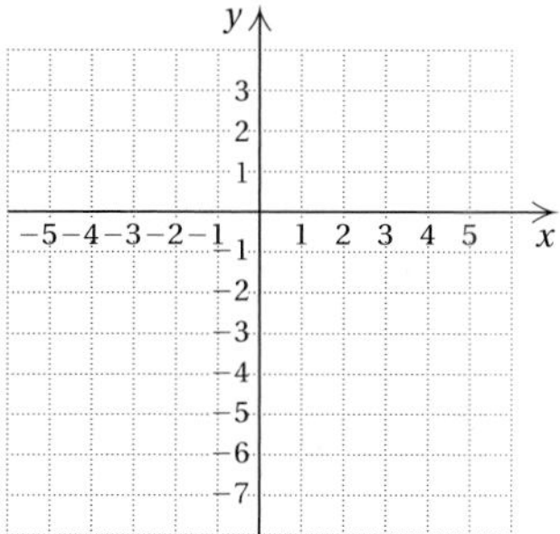

19. $y = -\frac{3}{2}x - 2$

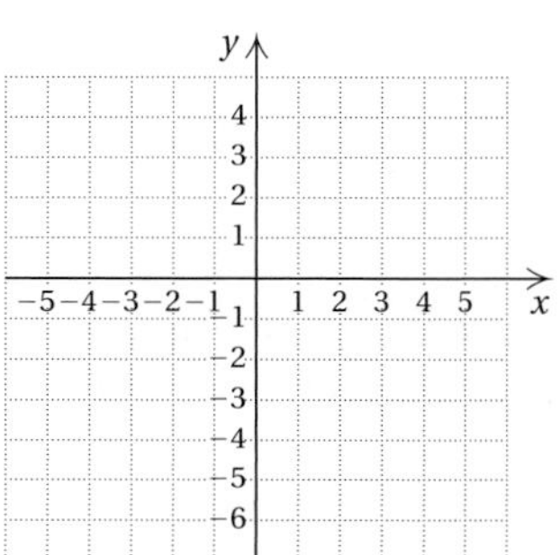

20. $y = -\frac{4}{3}x + 3$

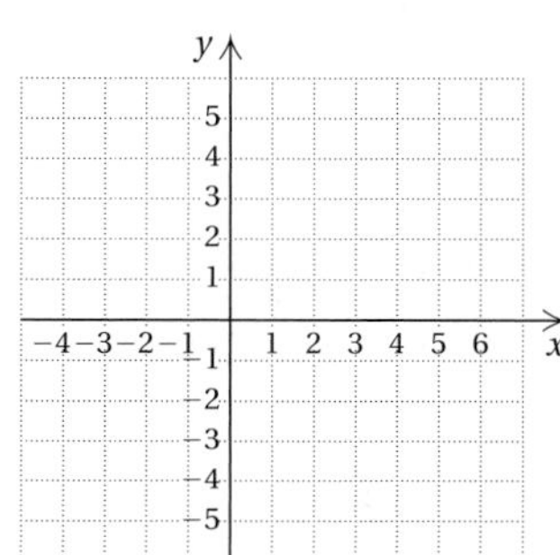

21. $2x + y = 1$

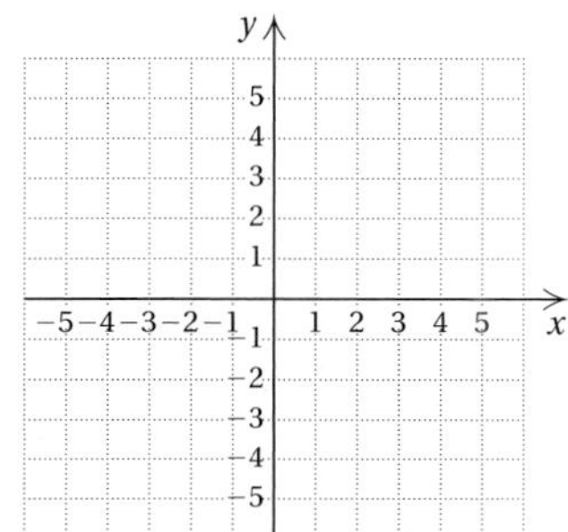

22. $3x + y = 2$

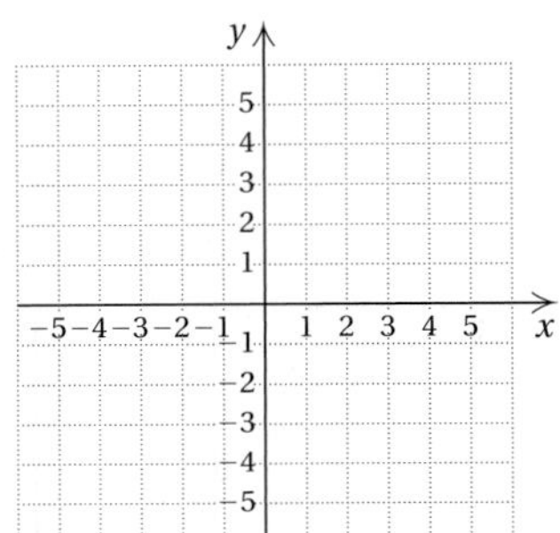

23. $3x - y = 4$

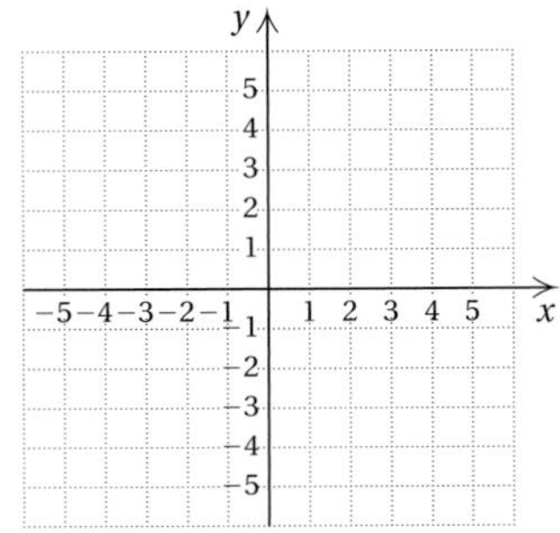

24. $2x - y = 5$

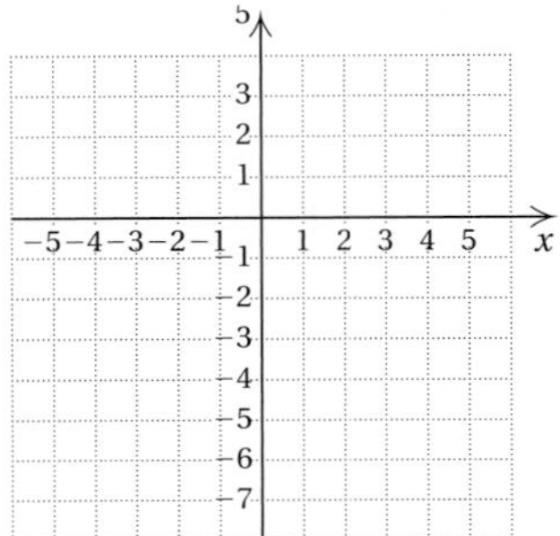

25. $2x + 3y = 9$

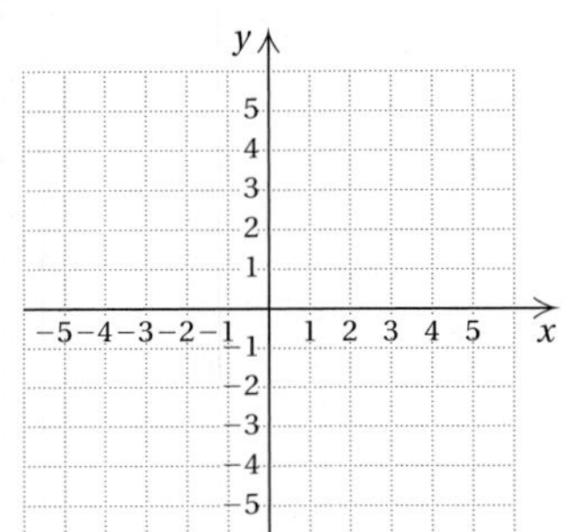

26. $4x + 5y = 15$

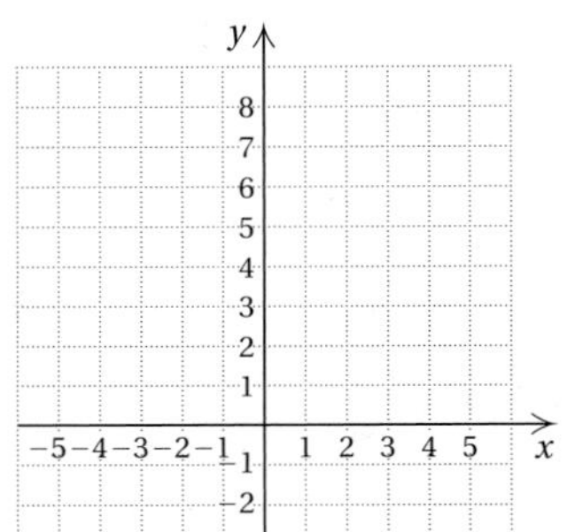

27. $x - 4y = 12$

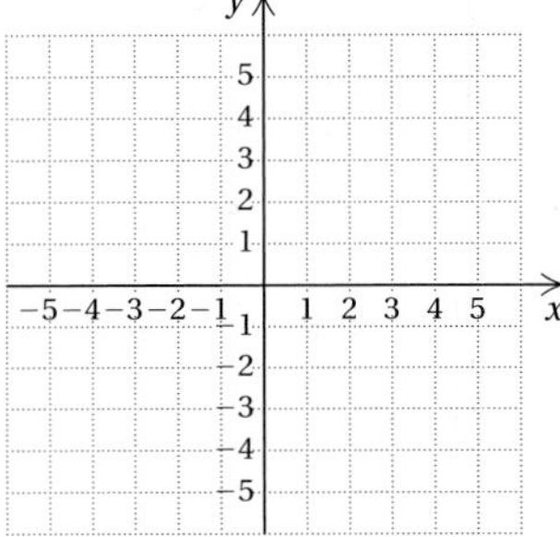

28. $x + 5y = 20$

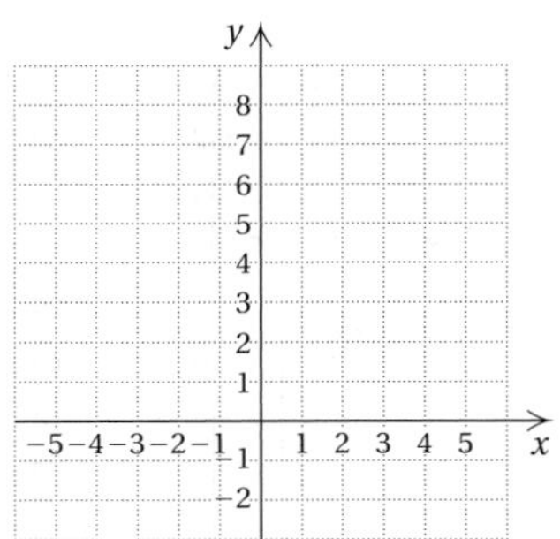

29. $x + 2y = 6$

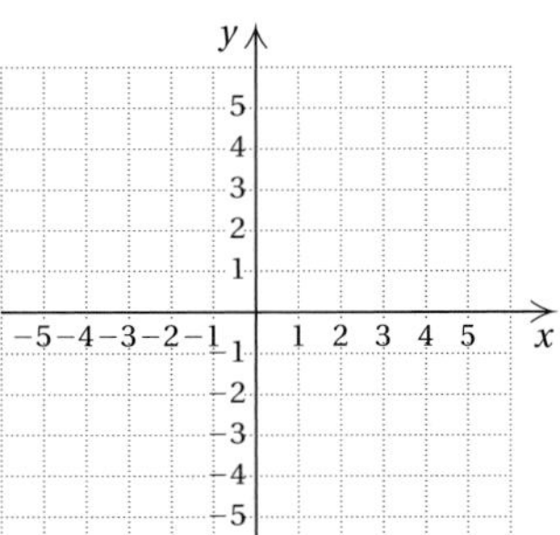

30. $x - 3y = 9$

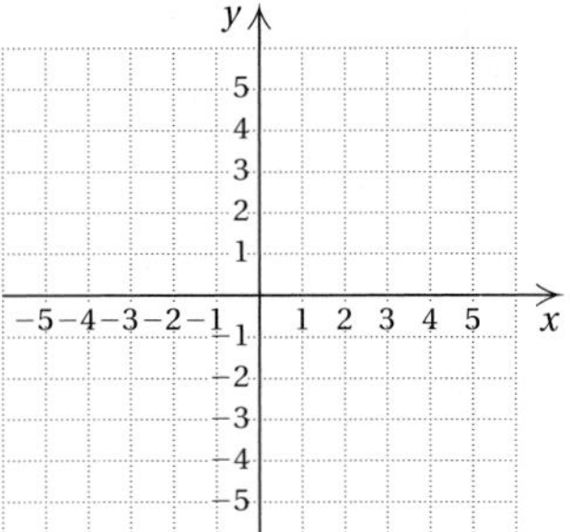

31. D_W Can a horizontal line be graphed using the method of Examples 3 and 4? Why or why not?

32. D_W Can a vertical line be graphed using the method of Examples 3 and 4? Why or why not?

SKILL MAINTENANCE

Find the slope of the line containing the given pair of points. [9.3a]

33. $(-2, -6), (8, 7)$

34. $(2, -6), (8, -7)$

35. $(4.5, -2.3), (14.5, 4.6)$

36. $(-0.8, -2.3), (-4.8, 0.1)$

37. $(-2, -6), (8, -6)$

38. $(-2, -6), (-2, 7)$

39. $(11, -1), (11, -4)$

40. $(-3, 5), (8, 5)$

41. *Kidney Transplants.* The number of kidney transplants in the United States has increased in recent years, as shown in the following graph. Find the rate of change in the number of kidney transplants with respect to time. Find the slope of the graph. [9.3b]

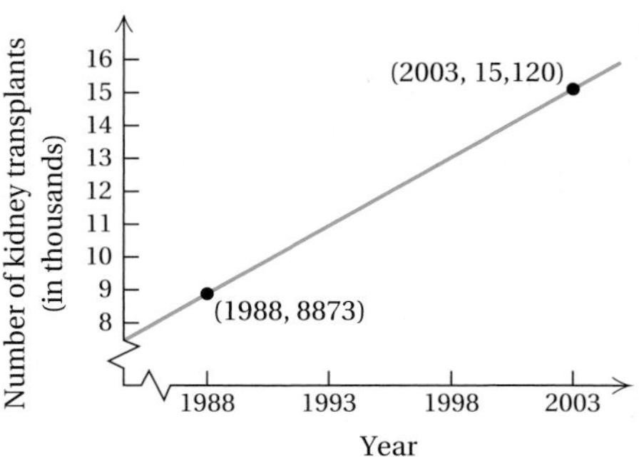

Source: National Organ Procurement and Transplantation Network

42. *Liver Transplants.* The number of liver transplants in the United States has increased in recent years, as shown in the following graph. Find the rate of change in the number of liver transplants with respect to time. Find the slope of the graph. [9.3b]

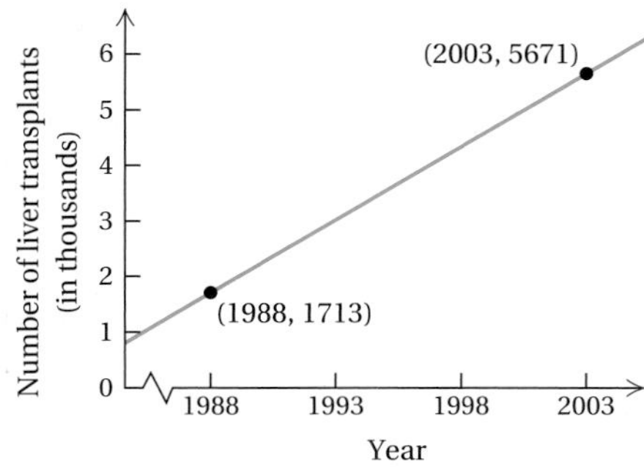

Source: National Organ Procurement and Transplantation Network

SYNTHESIS

43. *Refrigerator Size.* Kitchen designers recommend that a refrigerator be selected on the basis of the number of people in the household. For 1–2 people, a 16 ft^3 model is suggested. For each additional person, an additional 1.5 ft^3 is recommended. If x is the number of residents in excess of 2, find the slope–intercept equation for the recommended size of a refrigerator.

44. *Telephone Service.* In a recent promotion, AT&T charged a monthly fee of \$3.95 plus 7¢ for each minute of long-distance phone calls. If x is the number of minutes of long-distance calls, find the slope–intercept equation for the monthly bill.

45. Graph the line with slope 2 that passes through the point $(-3, 1)$.

9.6 PARALLEL AND PERPENDICULAR LINES

Objectives

a Determine whether the graphs of two linear equations are parallel.

b Determine whether the graphs of two linear equations are perpendicular.

When we graph a pair of linear equations, there are three possibilities:

1. The graphs are the same.
2. The graphs intersect at exactly one point.
3. The graphs are parallel (they do not intersect).

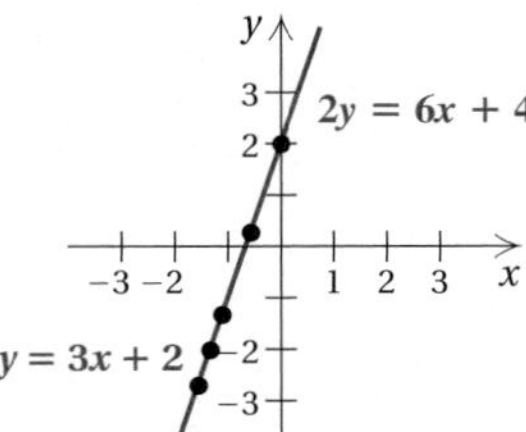

Equations have the same graph.

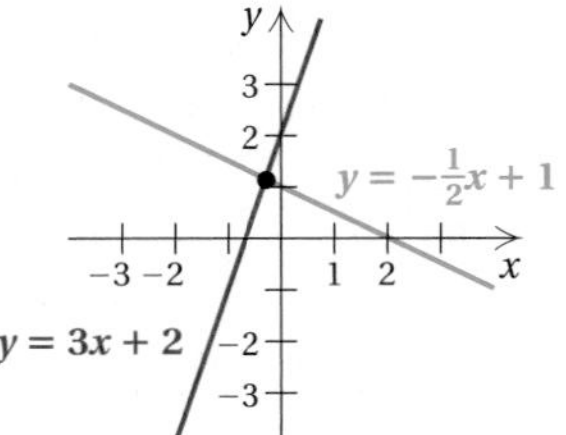

Graphs intersect at exactly one point.

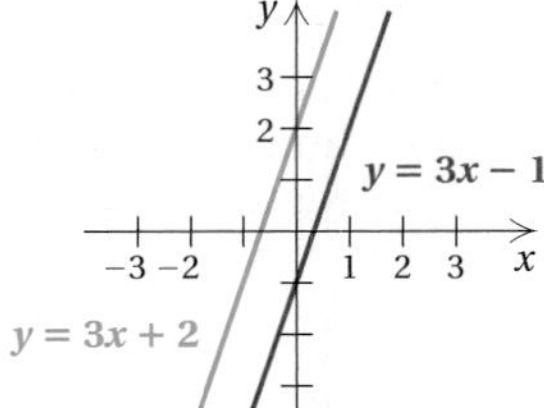

Graphs are parallel.

a Parallel Lines

The graphs shown below are of the linear equations

$$y = 2x + 5 \quad \text{and} \quad y = 2x - 3.$$

The slope of each line is 2. The y-intercepts are $(0, 5)$ and $(0, -3)$ and are different. The lines do not intersect and are parallel.

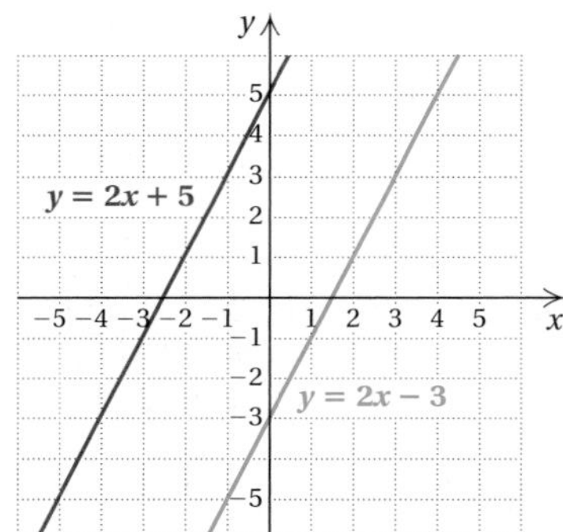

PARALLEL LINES

- Parallel nonvertical lines have the *same* slope, $m_1 = m_2$, and *different* y-intercepts, $b_1 \neq b_2$.
- Parallel horizontal lines have equations $y = p$ and $y = q$, where $p \neq q$.
- Parallel vertical lines have equations $x = p$ and $x = q$, where $p \neq q$.

By simply graphing, we may find it difficult to determine whether lines are parallel. Sometimes they may intersect only very far from the origin. We can use the preceding statements about slopes, y-intercepts, and parallel lines to determine for certain whether lines are parallel.

EXAMPLE 1 Determine whether the graphs of the lines $y = -3x + 4$ and $6x + 2y = -10$ are parallel.

The graphs of these equations are shown below, but they are not necessary in order to determine whether the lines are parallel.

We first solve each equation for y. In this case, the first equation is already solved for y.

a) $y = -3x + 4$

b) $6x + 2y = -10$

$$2y = -6x - 10$$

$$y = \frac{1}{2}(-6x - 10)$$

$$y = -3x - 5$$

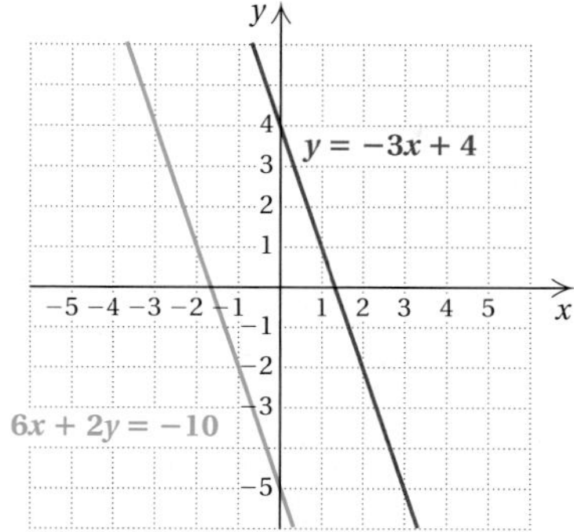

The slope of each line is -3. The y-intercepts are $(0, 4)$ and $(0, -5)$ and are different. The lines are parallel.

Do Exercises 1 and 2.

b Perpendicular Lines

Perpendicular lines in a plane are lines that intersect at a right angle. The measure of a right angle is 90°. The lines whose graphs are shown below are perpendicular. You can check this approximately by using a protractor or placing a rectangular piece of paper at the intersection.

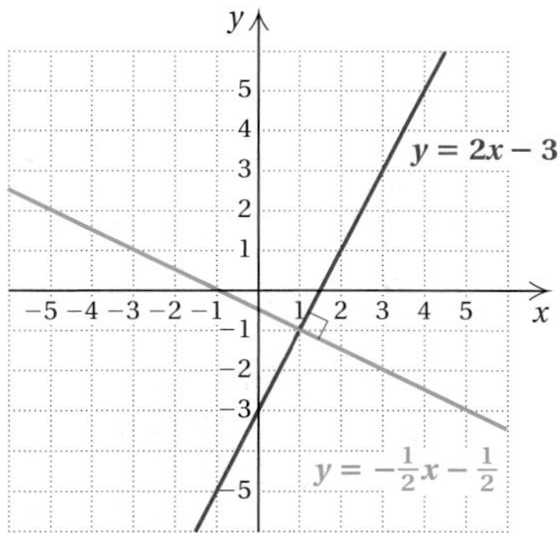

The slopes of the lines are 2 and $-\frac{1}{2}$. Note that $2\left(-\frac{1}{2}\right) = -1$. That is, the product of the slopes is -1.

PERPENDICULAR LINES

- Two nonvertical lines are perpendicular if the product of their slopes is -1, $m_1 \cdot m_2 = -1$. (If one line has slope m, the slope of the line perpendicular to it is $-1/m$.)
- If one equation in a pair of perpendicular lines is vertical, then the other is horizontal. These equations are of the form $x = a$ and $y = b$.

Determine whether the graphs of the pair of equations are parallel.

1. $y - 3x = 1$,
$-2y = 3x + 2$

2. $3x - y = -5$,
$y - 3x = -2$

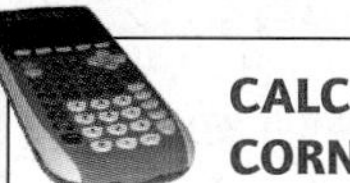

CALCULATOR CORNER

Parallel Lines Graph each pair of equations in Margin Exercises 1 and 2 in the standard viewing window, $[-10, 10, -10, 10]$. (Note that each equation must be solved for y so that it can be entered in "$y =$" form on the graphing calculator.) Determine whether the lines appear to be parallel.

Determine whether the graphs of the pair of equations are perpendicular.

3. $y = -\frac{3}{4}x + 7$,
$y = \frac{4}{3}x - 9$

4. $4x - 5y = 8$,
$6x + 9y = -12$

Answers on page A-30

CALCULATOR CORNER

Perpendicular Lines Graph each pair of equations in Margin Exercises 3 and 4 in the window $[-9, 9, -6, 6]$. (Note that the equations in Margin Exercise 4 must be solved for y so that they can be entered in "$y =$" form on the graphing calculator.) Determine whether the lines appear to be perpendicular. Note (in the viewing window) that more of the x-axis is shown than the y-axis. The dimensions were chosen to more accurately reflect the slopes of the lines.

EXAMPLE 2 Determine whether the graphs of the lines $3y = 9x + 3$ and $6y + 2x = 6$ are perpendicular.

The graphs are shown below, but they are not necessary in order to determine whether the lines are perpendicular.

We first solve each equation for y in order to determine the slopes:

a) $3y = 9x + 3$

$y = \frac{1}{3}(9x + 3)$

$y = 3x + 1;$

b) $6y + 2x = 6$

$6y = -2x + 6$

$y = \frac{1}{6}(-2x + 6)$

$y = -\frac{1}{3}x + 1.$

The slopes are 3 and $-\frac{1}{3}$. The product of the slopes is $3\left(-\frac{1}{3}\right) = -1$. The lines are perpendicular.

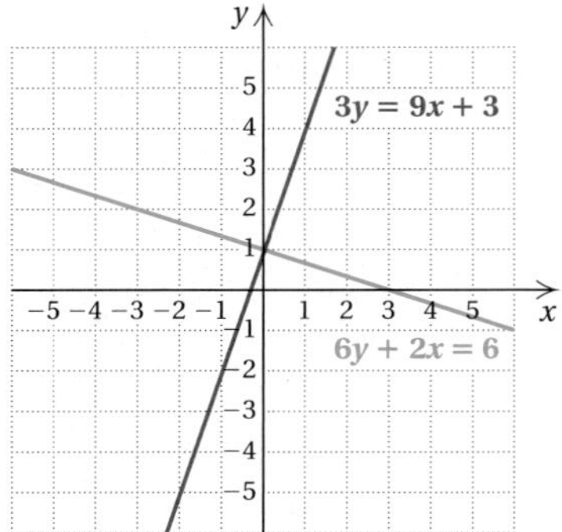

Do Exercises 3 and 4 on the preceding page.

Study Tips

SPECIAL VIDEOTAPES

In addition to the video lectures on CD for the book, there is a special CD-ROM, Work It Out! Chapter Test Video. Check with your instructor to see whether this CD is available on your campus.

There is also a special Math Study Skills for Students Video on CD that is designed to help you make better use of your math study time and improve your retention of concepts and procedures taught in classes from basic mathematics through intermediate algebra. (See the Preface for more information.)

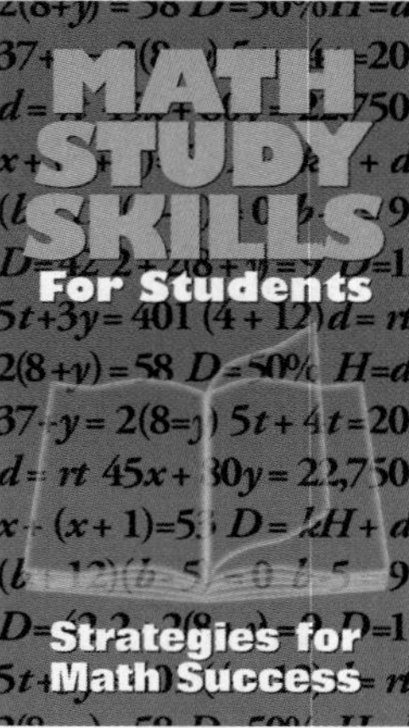

9.6 EXERCISE SET

For Extra Help

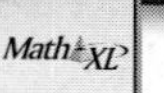
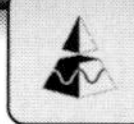
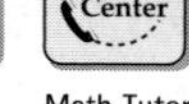

MathXL | MyMathLab | InterAct Math | Math Tutor Center | Video Lectures on CD Disc 5 | Student's Solutions Manual

a Determine whether the graphs of the equations are parallel lines.

1. $x + 4 = y,$
$y - x = -3$

2. $3x - 4 = y,$
$y - 3x = 8$

3. $y + 3 = 6x,$
$-6x - y = 2$

4. $y = -4x + 2,$
$-5 = -2y + 8x$

5. $10y + 32x = 16.4,$
$y + 3.5 = 0.3125x$

6. $y = 6.4x + 8.9,$
$5y - 32x = 5$

7. $y = 2x + 7,$
$5y + 10x = 20$

8. $y + 5x = -6,$
$3y + 5x = -15$

9. $3x - y = -9,$
$2y - 6x = -2$

10. $y - 6 = -6x,$
$-2x + y = 5$

11. $x = 3,$
$x = 4$

12. $y = 1,$
$y = -2$

b Determine whether the graphs of the equations are perpendicular lines.

13. $y = -4x + 3,$
$4y + x = -1$

14. $y = -\frac{2}{3}x + 4,$
$3x + 2y = 1$

15. $x + y = 6,$
$4y - 4x = 12$

16. $2x - 5y = -3,$
$5x + 2y = 6$

17. $y = -0.3125x + 11,$
$y - 3.2x = -14$

18. $y = -6.4x - 7,$
$64y - 5x = 32$

19. $y = -x + 8,$
$x - y = -1$

20. $2x + 6y = -3,$
$12y = 4x + 20$

21. $\frac{3}{8}x - \frac{y}{2} = 1,$
$\frac{4}{3}x - y + 1 = 0$

22. $\frac{1}{2}x + \frac{3}{4}y = 6,$
$-\frac{3}{2}x + y = 4$

23. $x = 0,$
$y = -2$

24. $x = -3,$
$y = 5$

a, b Determine whether the graphs of the equations are parallel, perpendicular, or neither.

25. $3y + 21 = 2x,$
$3y = 2x + 24$

26. $3y + 21 = 2x,$
$2y = 16 - 3x$

27. $3y = 2x - 21,$
$2y - 16 = 3x$

28. $3y + 2x + 7 = 0,$
$3y = 2x + 24$

29. D_W Consider two equations of the type $Ax + By = C$. Explain how you would go about showing that their graphs are perpendicular.

30. D_W Consider two equations of the type $Ax + By = C$. Explain how you would go about showing that their graphs are parallel.

SKILL MAINTENANCE

VOCABULARY REINFORCEMENT

In each of Exercises 31–38, fill in the blank with the correct term from the given list. Some of the choices may not be used.

31. Equations with the same solutions are called ________________. [8.1b]

32. The ________________ for equations asserts that when we subtract the same number on both sides of an equation, we get equivalent equations. [8.1b]

33. The ________________ for equations asserts that when we multiply or divide by the same nonzero number on both sides of an equation, we get equivalent equations. [8.2a]

34. ________________ lines are graphs of equations of the type $y = b$. [9.2b]

35. ________________ lines are graphs of equations of the type $x = a$. [9.2b]

36. The ________________ of a line is a number that indicates how the line slants. [9.3a]

37. The ________________ of a line, if it exists, indicates where the line crosses the x-axis. [9.3a]

38. The ________________ of a line, if it exists, indicates where the line crosses the y-axis. [9.2a]

vertical
horizontal
variable
addition principle
multiplication principle
coefficient
equivalent equations
slope
x-intercept
y-intercept
parallel
perpendicular

SYNTHESIS

39. Find an equation of a line that contains the point $(0, 6)$ and is parallel to $y - 3x = 4$.

40. Find an equation of the line that contains the point $(-2, 4)$ and is parallel to $y = 2x - 3$.

41. Find an equation of the line that contains the point $(0, 2)$ and is perpendicular to $3y - x = 0$.

42. Find an equation of the line that contains the point $(1, 0)$ and is perpendicular to $2x + y = -4$.

43. Find an equation of the line that has x-intercept $(-2, 0)$ and is parallel to $4x - 8y = 12$.

44. Find the value of k such that $4y = kx - 6$ and $5x + 20y = 12$ are parallel.

45. Find the value of k such that $4y = kx - 6$ and $5x + 20y = 12$ are perpendicular.

The lines in the graphs in Exercises 46 and 47 are perpendicular and the lines in the graph in Exercise 48 are parallel. Find an equation of each line.

46.

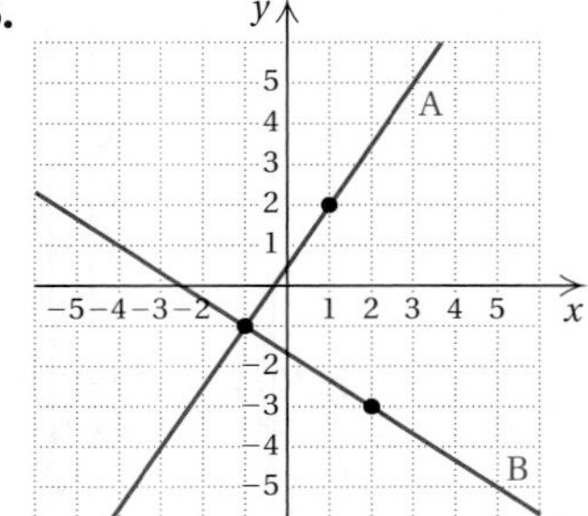

47.

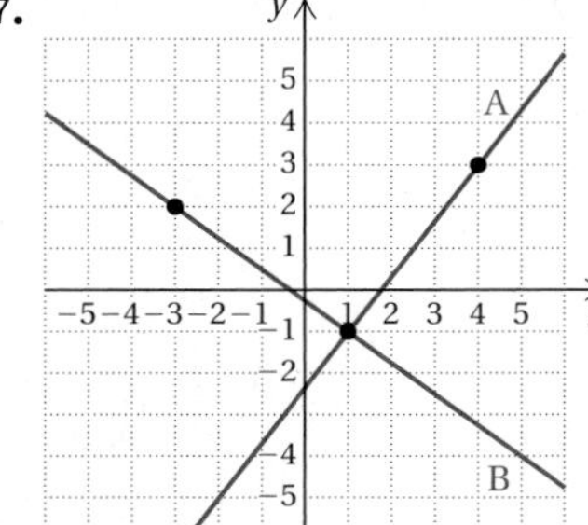

48.

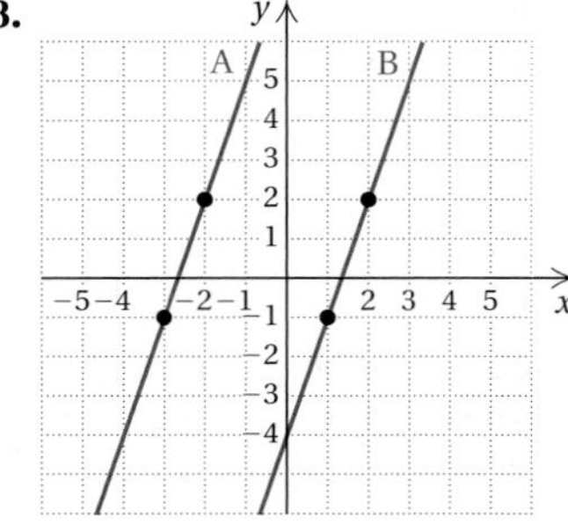

9.7 GRAPHING INEQUALITIES IN TWO VARIABLES

Objectives

a Determine whether an ordered pair of numbers is a solution of an inequality in two variables.

b Graph linear inequalities.

A graph of an inequality is a drawing that represents its solutions. An inequality in one variable can be graphed on a number line. An inequality in two variables can be graphed on a coordinate plane.

a Solutions of Inequalities in Two Variables

The solutions of inequalities in two variables are ordered pairs.

EXAMPLE 1 Determine whether $(-3, 2)$ is a solution of $5x + 4y < 13$.

We use alphabetical order to replace x with -3 and y with 2.

$$\begin{array}{r|l} 5x + 4y & < 13 \\ \hline 5(-3) + 4 \cdot 2 & ?\ 13 \\ -15 + 8 & \\ -7 & \quad \text{TRUE} \end{array}$$

Since $-7 < 13$ is true, $(-3, 2)$ is a solution.

EXAMPLE 2 Determine whether $(6, 8)$ is a solution of $5x + 4y < 13$.

We use alphabetical order to replace x with 6 and y with 8.

$$\begin{array}{r|l} 5x + 4y & < 13 \\ \hline 5(6) + 4(8) & ?\ 13 \\ 30 + 32 & \\ 62 & \quad \text{FALSE} \end{array}$$

Since $62 < 13$ is false, $(6, 8)$ is not a solution.

Do Exercises 1 and 2.

1. Determine whether $(4, 3)$ is a solution of $3x - 2y < 1$.

2. Determine whether $(2, -5)$ is a solution of $4x + 7y \geq 12$.

b Graphing Inequalities in Two Variables

EXAMPLE 3 Graph: $y > x$.

We first graph the line $y = x$. Every solution of $y = x$ is an ordered pair like $(3, 3)$. The first and second coordinates are the same. We draw the line $y = x$ dashed because its points (as shown on the left below) are *not* solutions of $y > x$.

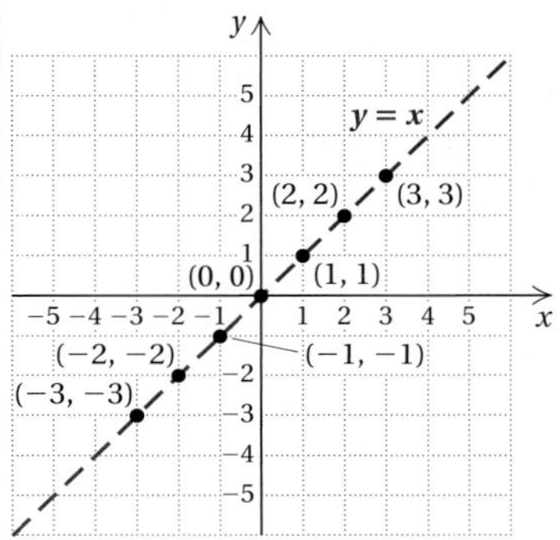

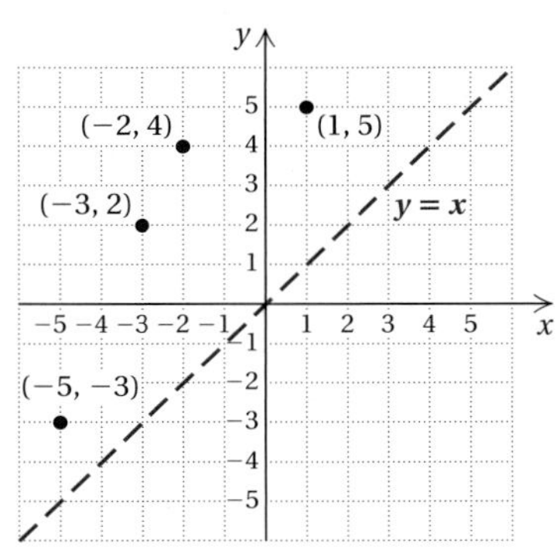

Answers on page A-30

3. Graph: $y < x$.

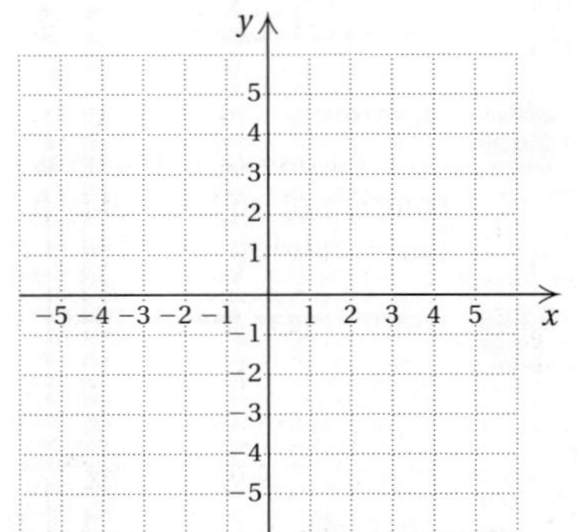

Now look at the graph on the right on the preceding page. Several ordered pairs are plotted in the half-plane above the line $y = x$. Each is a solution of $y > x$.

We can check a pair such as $(-2, 4)$ as follows:

$$\begin{array}{c|c} y > x \\ \hline 4 \ ? \ -2 & \text{TRUE} \end{array}$$

It turns out that any point on the same side of $y = x$ as $(-2, 4)$ is also a solution. *If we know that one point in a half-plane is a solution, then all points in that half-plane are solutions.* We could have chosen other points to check. The graph of $y > x$ is shown below. (Solutions are indicated by color shading throughout.) We shade the half-plane above $y = x$.

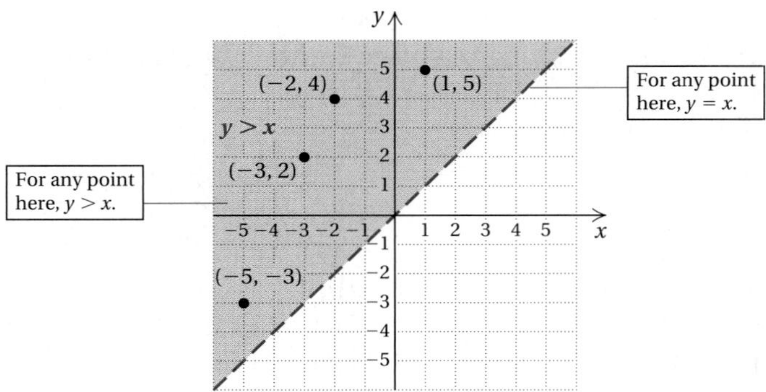

Do Exercise 3.

A **linear inequality** is one that we can get from a linear equation by changing the equals symbol to an inequality symbol. Every linear equation has a graph that is a straight line. The graph of a linear inequality is a half-plane, sometimes including the line along the edge.

To graph an inequality in two variables:

1. Replace the inequality symbol with an equals sign and graph this related linear equation.
2. If the inequality symbol is $<$ or $>$, draw the line dashed. If the inequality symbol is $\leq$ or $\geq$, draw the line solid.
3. The graph consists of a half-plane, either above or below or left or right of the line, and, if the line is solid, the line as well. To determine which half-plane to shade, choose a point *not on the line* as a test point. Substitute to find whether that point is a solution of the *inequality*. If it is, shade the half-plane containing that point. If it is not, shade the half-plane on the opposite side of the line.

Answer on page A-30

EXAMPLE 4 Graph: $5x - 2y < 10$.

1. We first graph the line $5x - 2y = 10$. The intercepts are $(0, -5)$ and $(2, 0)$. This line forms the boundary of the solutions of the inequality.
2. Since the inequality contains the $<$ symbol, points on the line are not solutions of the inequality, so we draw a dashed line.
3. To determine which half-plane to shade, we consider a test point *not* on the line. We try $(3, -2)$ and substitute:

$$\begin{array}{c|c} 5x - 2y & < 10 \\ \hline 5(3) - 2(-2) & ?\ 10 \\ 15 + 4 & \\ 19 & \end{array} \quad \text{FALSE}$$

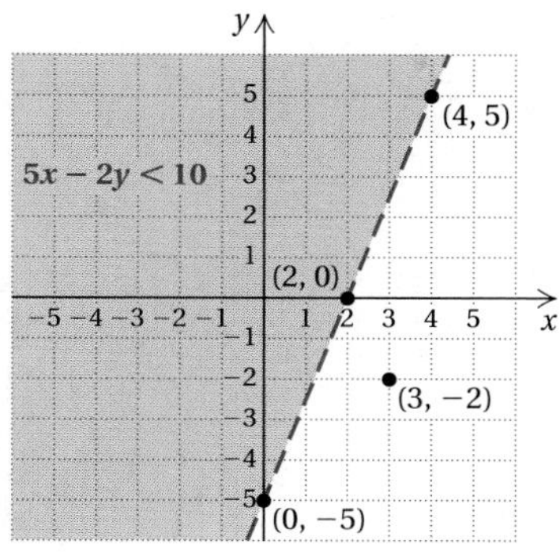

Since this inequality is false, the point $(3, -2)$ is *not* a solution; no point in the half-plane containing $(3, -2)$ is a solution. Thus the points in the opposite half-plane are solutions. The graph is shown above.

Do Exercise 4.

EXAMPLE 5 Graph: $2x + 3y \leq 6$.

1. First, we graph the line $2x + 3y = 6$. The intercepts are $(0, 2)$ and $(3, 0)$.
2. Since the inequality contains the $\leq$ symbol, we draw the line solid to indicate that any pair on the line is a solution.
3. Next, we choose a test point that does not belong to the line. We substitute to determine whether this point is a solution. The origin $(0, 0)$ is generally an easy one to use:

$$\begin{array}{c|c} 2x + 3y & \leq 6 \\ \hline 2 \cdot 0 + 3 \cdot 0 & ?\ 6 \\ 0 & \end{array} \quad \text{TRUE}$$

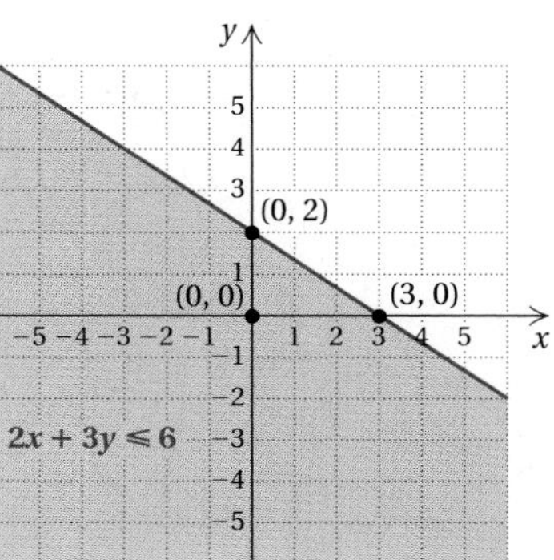

We see that $(0, 0)$ is a solution, so we shade the lower half-plane. Had the substitution given us a false inequality, we would have shaded the other half-plane.

Do Exercises 5 and 6.

4. Graph: $2x + 4y < 8$.

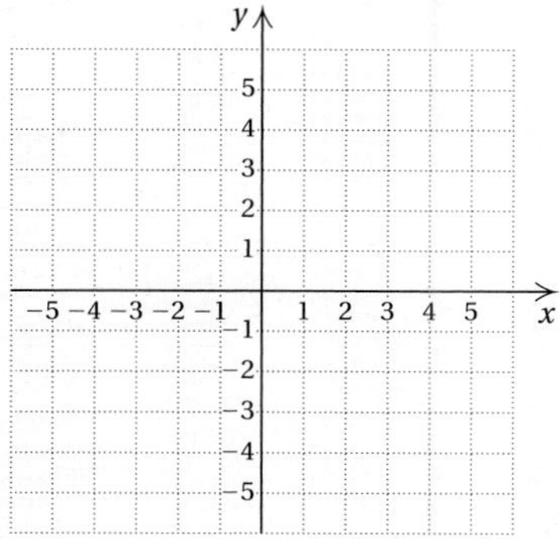

Graph.

5. $3x - 5y < 15$

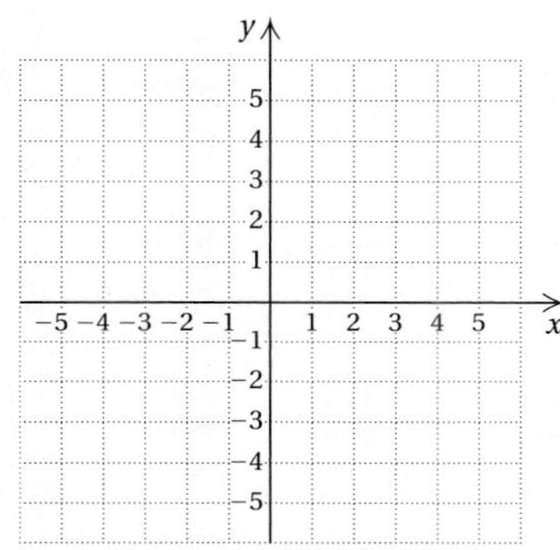

6. $2x + 3y \geq 12$

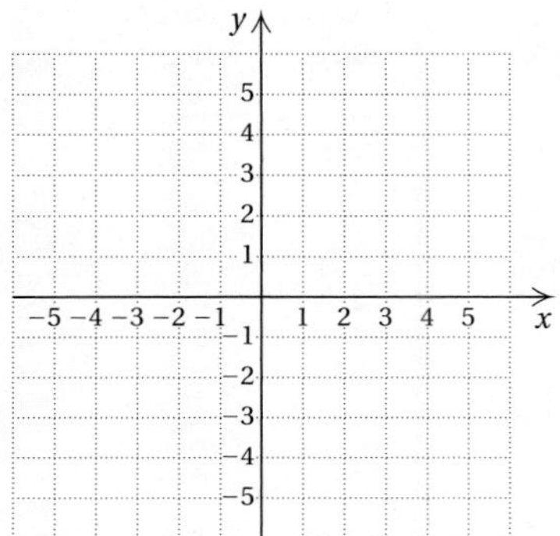

Answers on page A-30

Graph.

7. $x > -3$

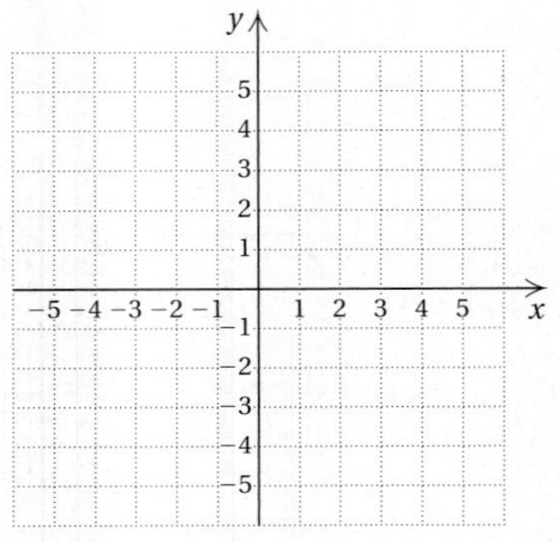

8. $y \leq 4$

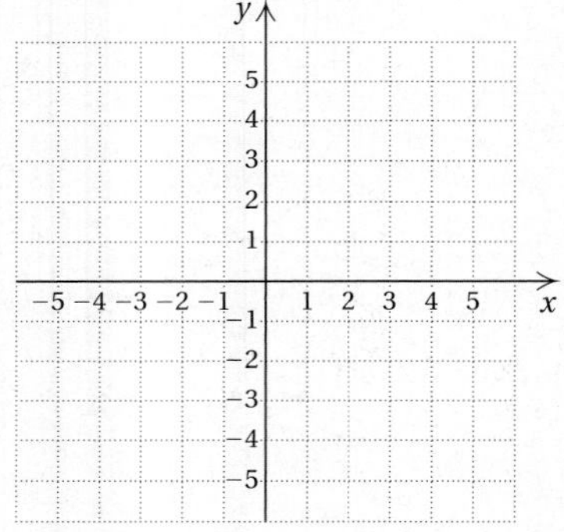

Answers on page A-31

EXAMPLE 6 Graph $x < 3$ on a plane.

There is no y-term in this inequality, but we can rewrite this inequality as $x + 0y < 3$. We use the same technique that we have used with the other examples.

1. We graph the related equation $x = 3$ on the plane.
2. Since the inequality symbol is $<$, we use a dashed line.
3. The graph is a half-plane either to the left or to the right of the line $x = 3$. To determine which, we consider a test point, $(-4, 5)$:

$$\begin{array}{r|l} x + 0y & < 3 \\ \hline -4 + 0(5) & ?\ 3 \\ -4 & \end{array} \quad \text{TRUE}$$

We see that $(-4, 5)$ is a solution, so all the pairs in the half-plane containing $(-4, 5)$ are solutions. We shade that half-plane.

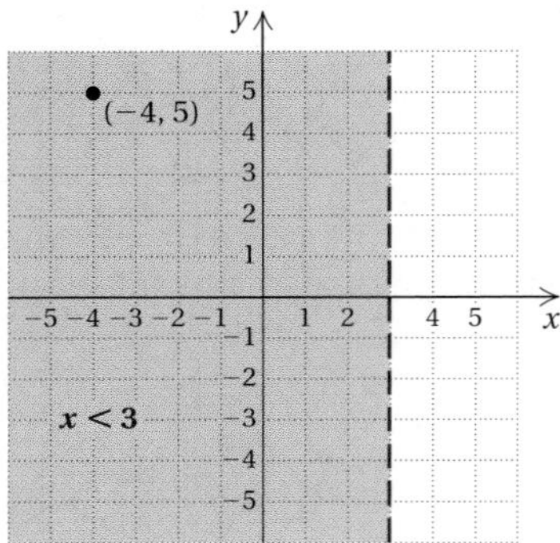

We see from the graph that the solutions of $x < 3$ are all those ordered pairs whose first coordinates are less than 3.

If we graph the inequality in Example 6 on a line rather than on a plane, its graph is as follows:

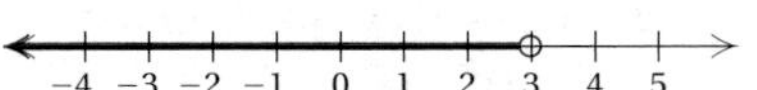

EXAMPLE 7 Graph: $y \geq -4$.

1. We first graph $y = -4$.
2. We use a solid line to indicate that all points on the line are solutions.
3. We then use $(2, 3)$ as a test point and substitute:

$$\begin{array}{c|c} 0x + y \geq -4 & \\ \hline 0(2) + 3 \ ? & -4 \\ 3 & \end{array} \quad \text{TRUE}$$

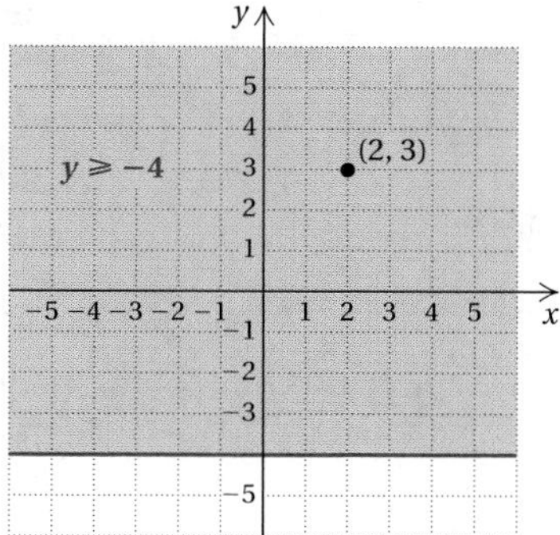

Since $(2, 3)$ is a solution, all points in the half-plane containing $(2, 3)$ are solutions. Note that this half-plane consists of all ordered pairs whose second coordinate is greater than or equal to -4.

Do Exercises 7 and 8 on the preceding page.

CALCULATOR CORNER

Graphs of Inequalities We can graph inequalities on a graphing calculator, shading the region of the solution set. To graph the inequality in Example 5,

$$2x + 3y \leq 6,$$

we first graph the line $2x + 3y = 6$. Solving for y, we get $y = \frac{6 - 2x}{3}$, or $y = -\frac{2}{3}x + 2$. We enter this equation on the equation-editor screen.

After determining algebraically that the solution set consists of all points below the line, we use the graphing calculator's "shade below" graph style to shade this region. On the equation-editor screen, we position the cursor over the graphstyle icon to the left of the equation and press ENTER repeatedly until the "shade below" icon appears. Then we press GRAPH to display the graph of the inequality.

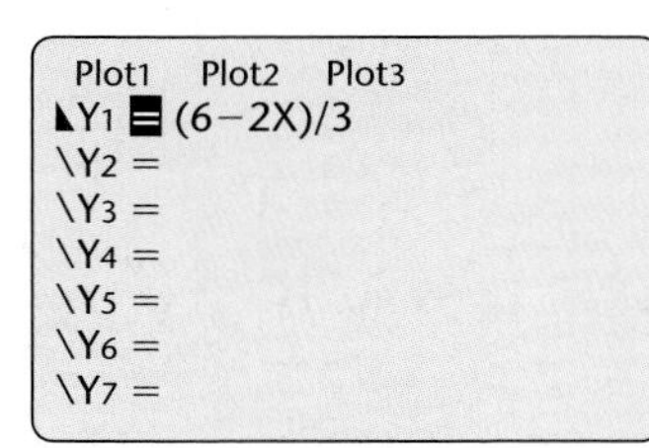

$y = \frac{6 - 2x}{3}$

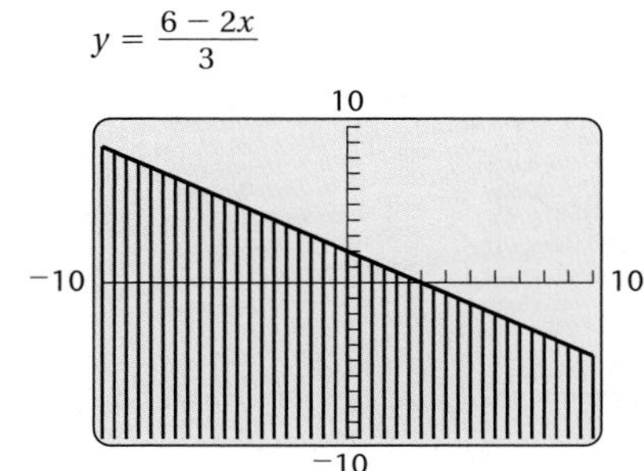

Note that we cannot graph an inequality like the one in Example 6, $x < 3$, on a graphing calculator because the related equation has no y-term and thus cannot be entered in "$y =$" form.

Exercises:

1. Use a graphing calculator to graph the inequalities in Margin Exercises 6 and 8.
2. Use a graphing calculator to graph the inequality in Example 7.

Visualizing for Success

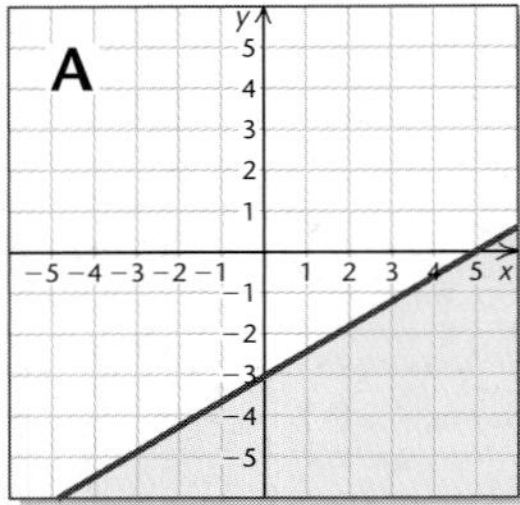

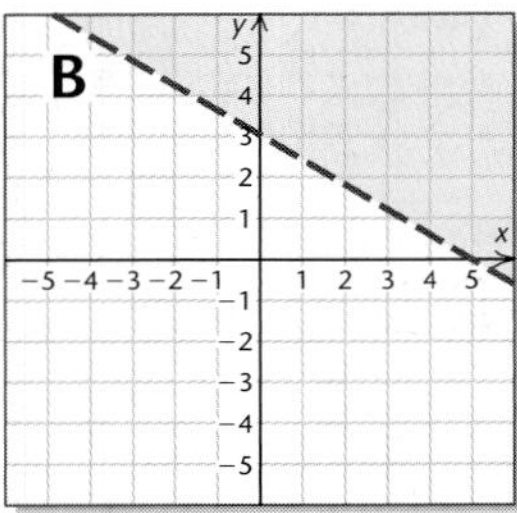

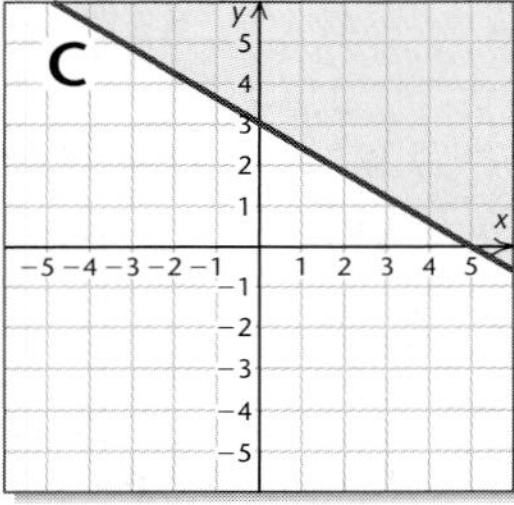

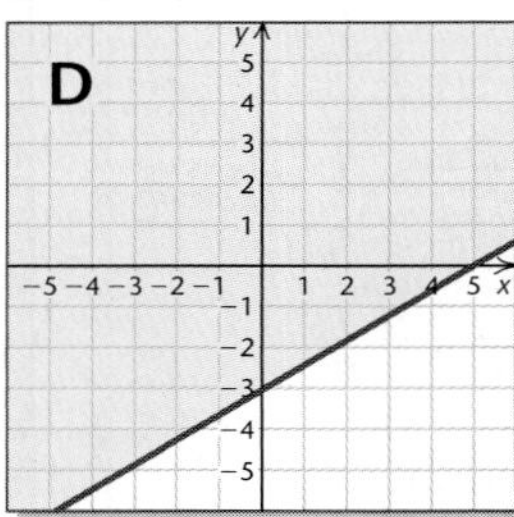

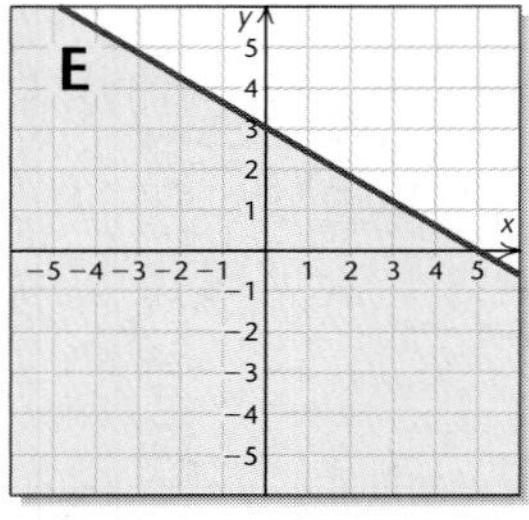

Match the equation or inequality with its graph.

1. $3x - 5y \leq 15$
2. $3x + 5y = 15$
3. $3x + 5y \leq 15$
4. $3x - 5y \geq 15$
5. $3x - 5y = 15$
6. $3x - 5y < 15$
7. $3x + 5y \geq 15$
8. $3x + 5y > 15$
9. $3x - 5y > 15$
10. $3x + 5y < 15$

Answers on page A-31

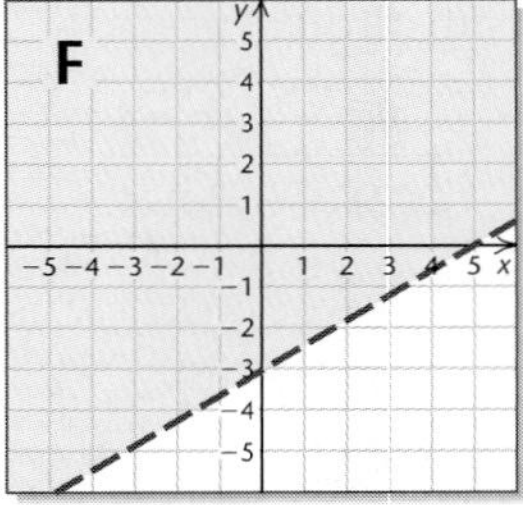

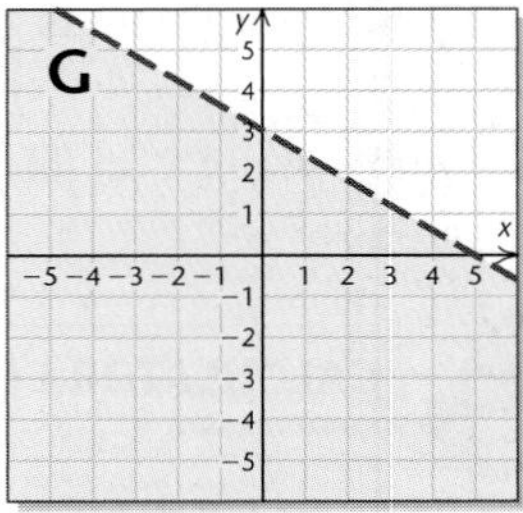

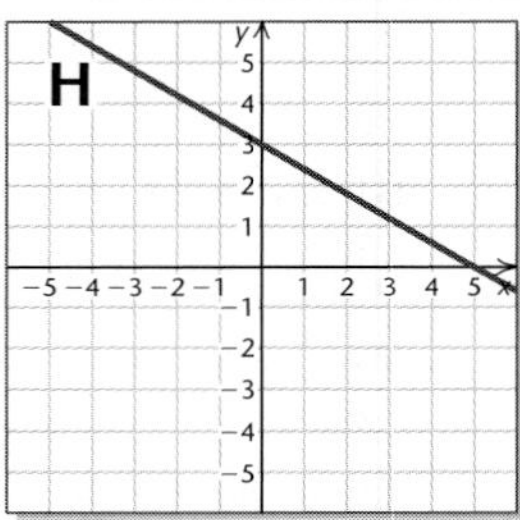

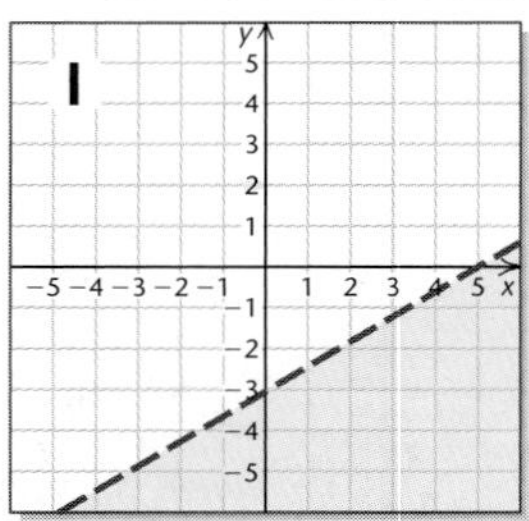

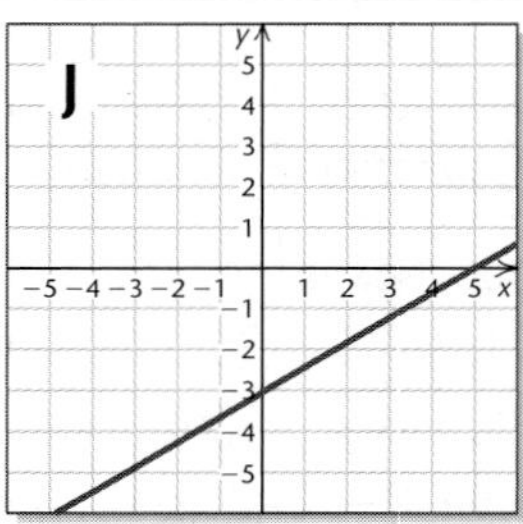

9.7 EXERCISE SET

For Extra Help

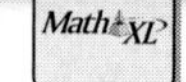
MathXL

MyMathLab

InterAct Math

Math Tutor Center

Video Lectures on CD Disc 5

Student's Solutions Manual

1. Determine whether $(-3, -5)$ is a solution of
$-x - 3y < 18$.

2. Determine whether $(2, -3)$ is a solution of
$5x - 4y \geq 1$.

3. Determine whether $\left(\frac{1}{2}, -\frac{1}{4}\right)$ is a solution of
$7y - 9x \leq -3$.

4. Determine whether $(-8, 5)$ is a solution of
$x + 0 \cdot y > 4$.

b Graph on a plane.

5. $x > 2y$

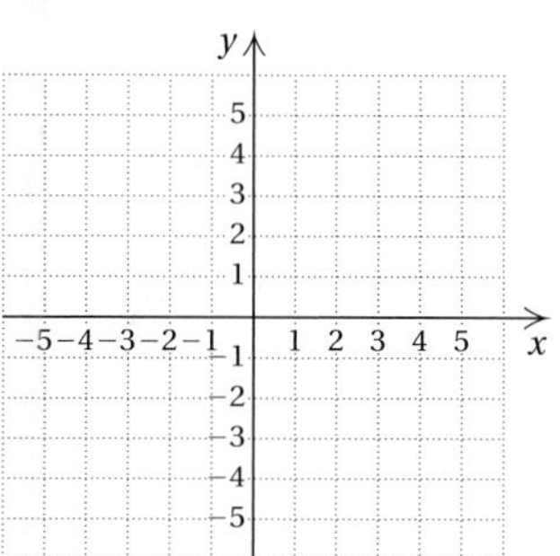

6. $x > 3y$

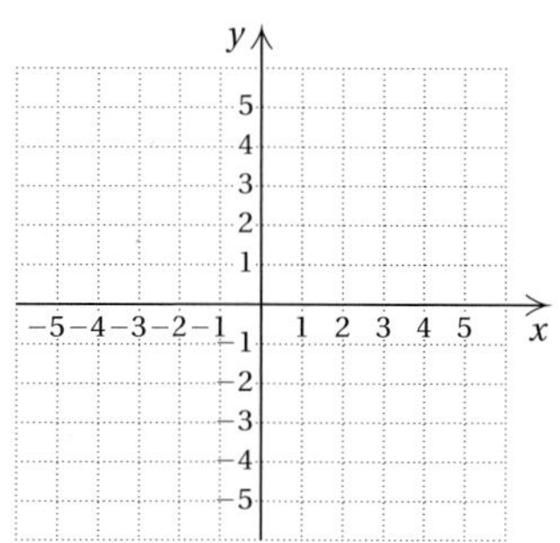

7. $y \leq x - 3$

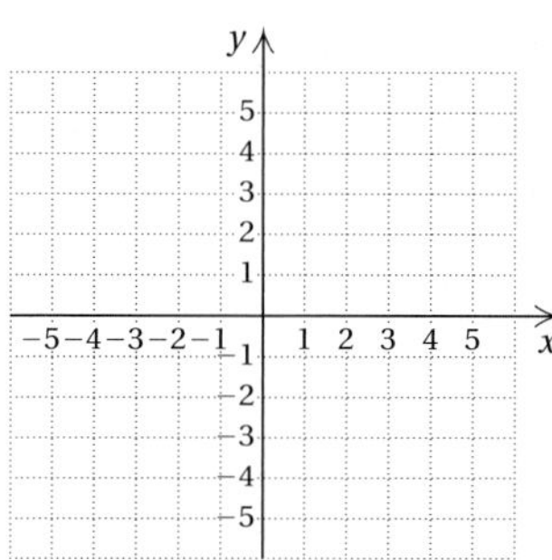

8. $y \leq x - 5$

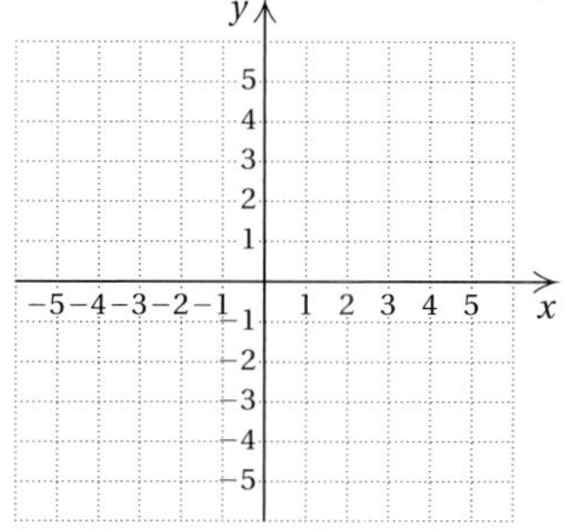

9. $y < x + 1$

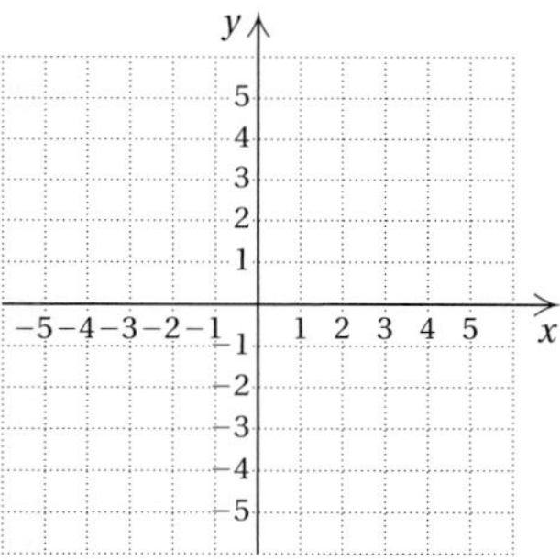

10. $y < x + 4$

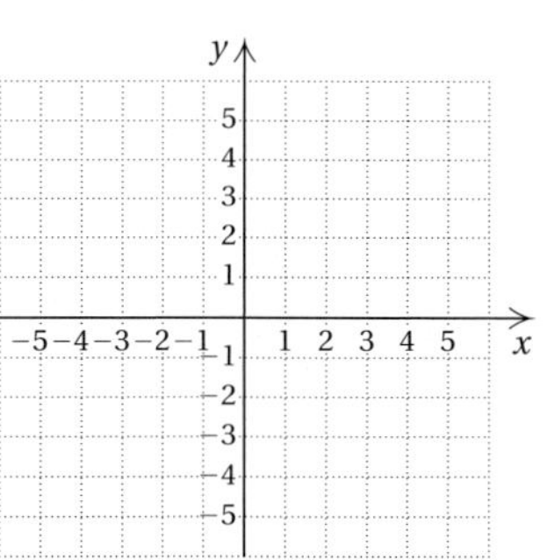

11. $y \geq x - 2$

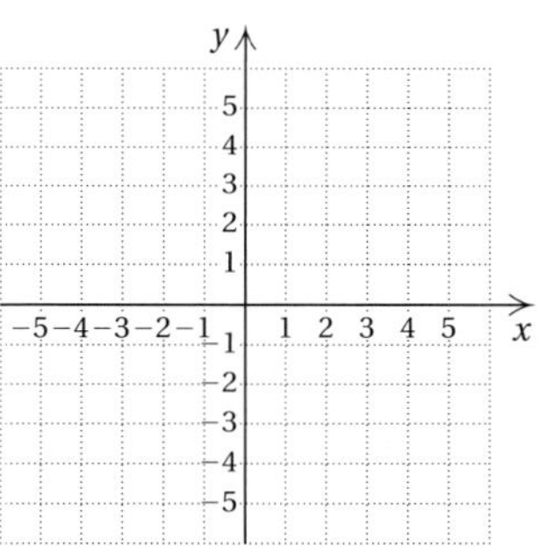

12. $y \geq x - 1$

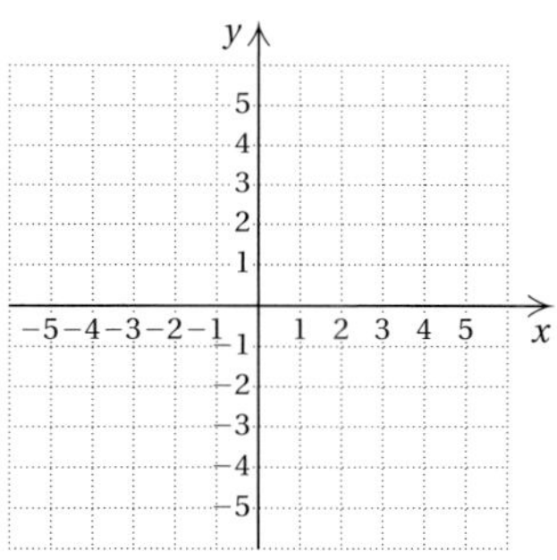

13. $y \leq 2x - 1$

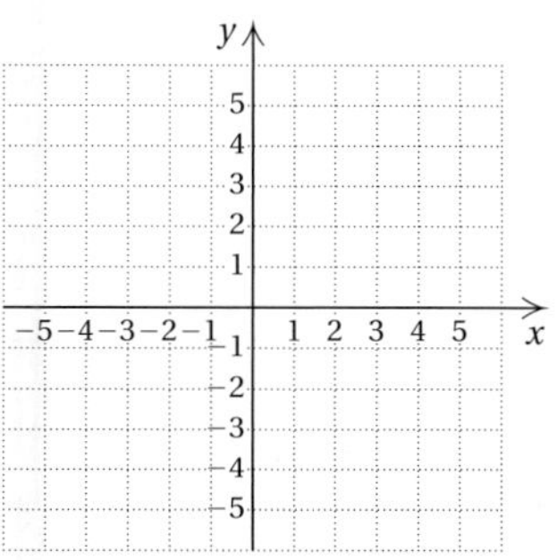

14. $y \leq 3x + 2$

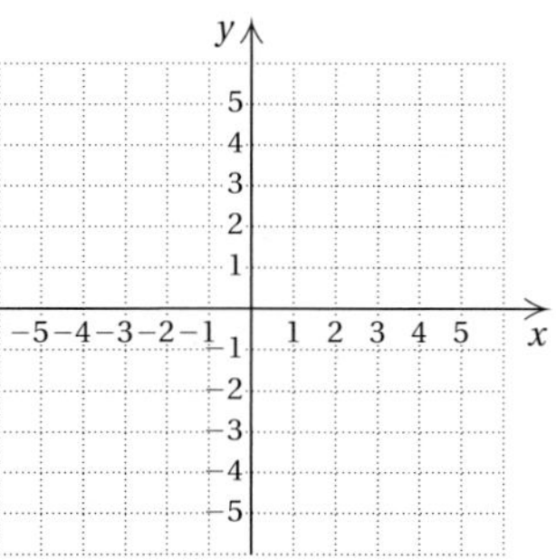

15. $x + y \leq 3$

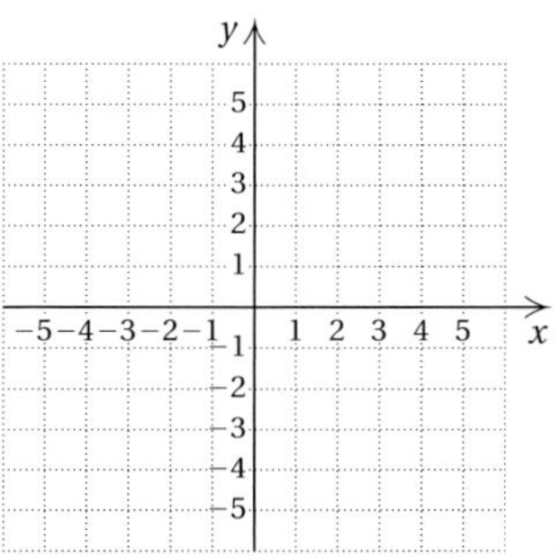

16. $x + y \leq 4$

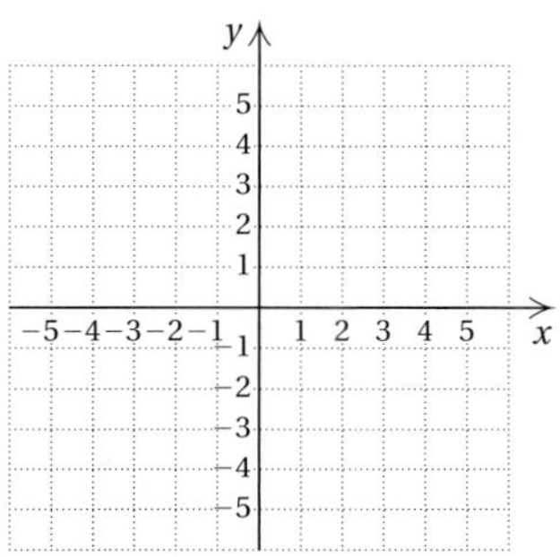

17. $x - y > 7$

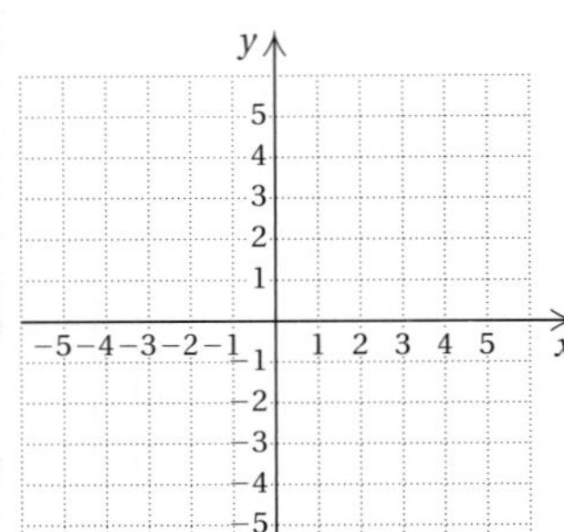

18. $x - y > -2$

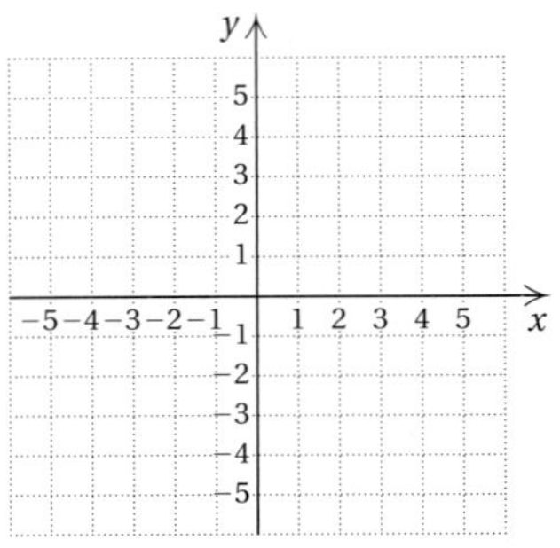

19. $2x + 3y \leq 12$

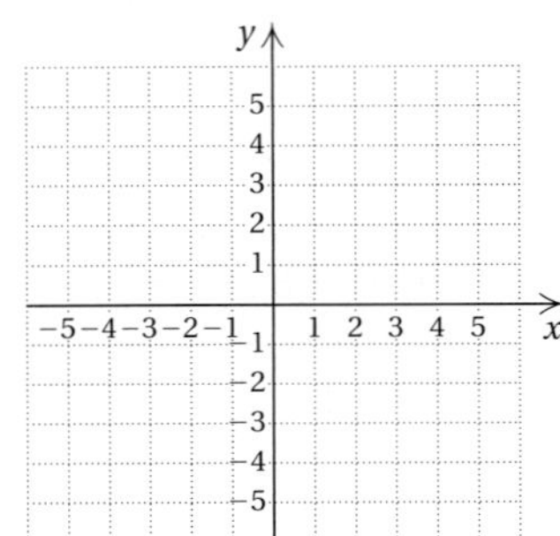

20. $5x + 4y \geq 20$

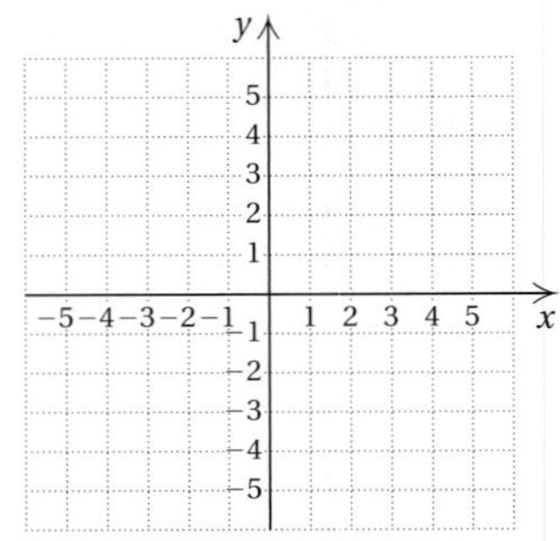

21. $y \geq 1 - 2x$

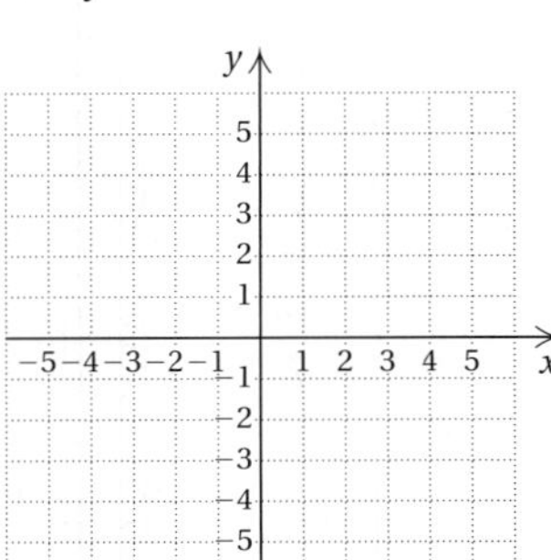

22. $y - 2x \leq -1$

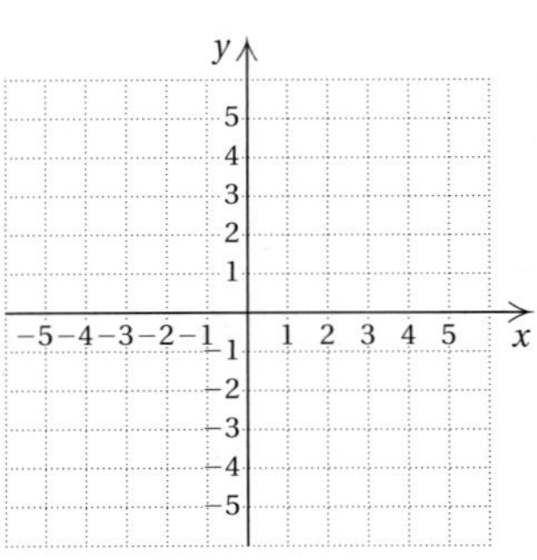

23. $2x - 3y > 6$

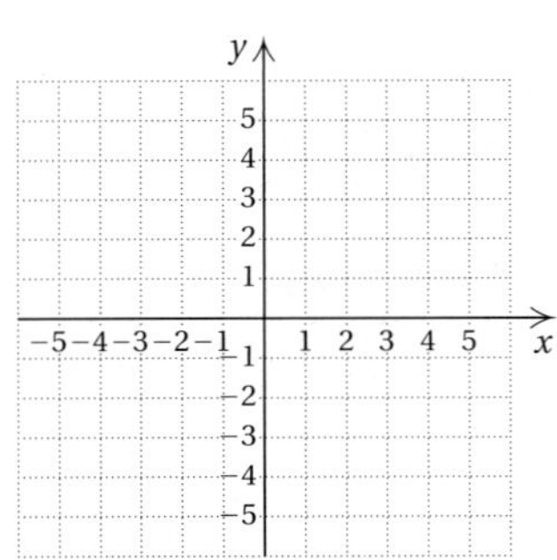

24. $5y - 2x \leq 10$

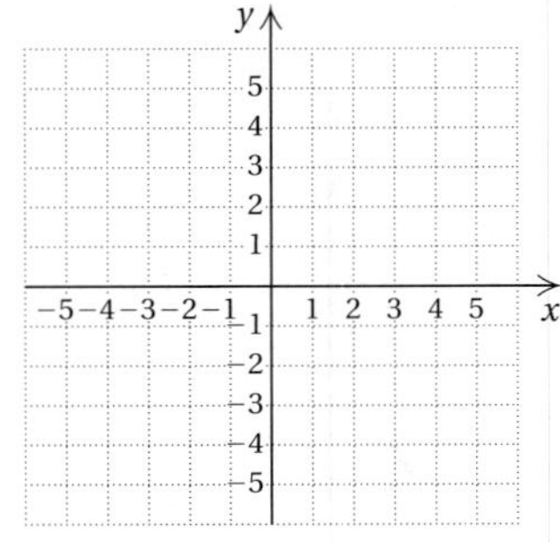

25. $y \leq 3$

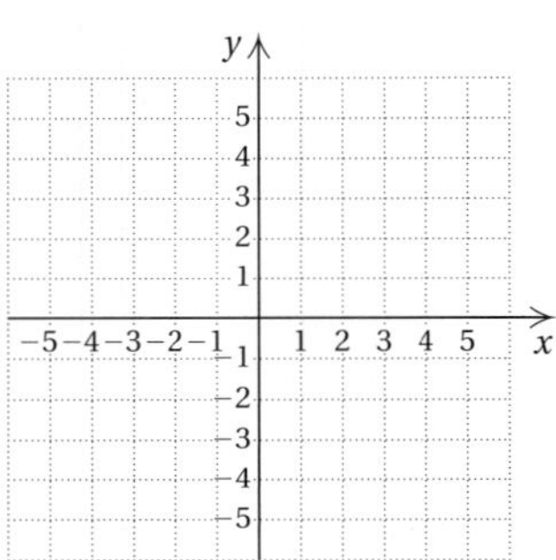

26. $y > -1$

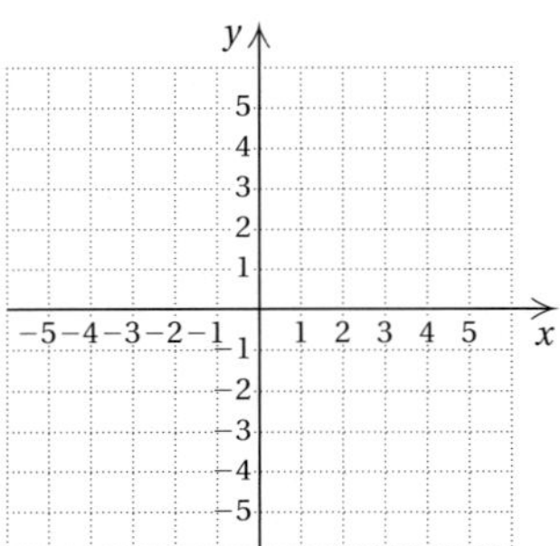

27. $x \geq -1$

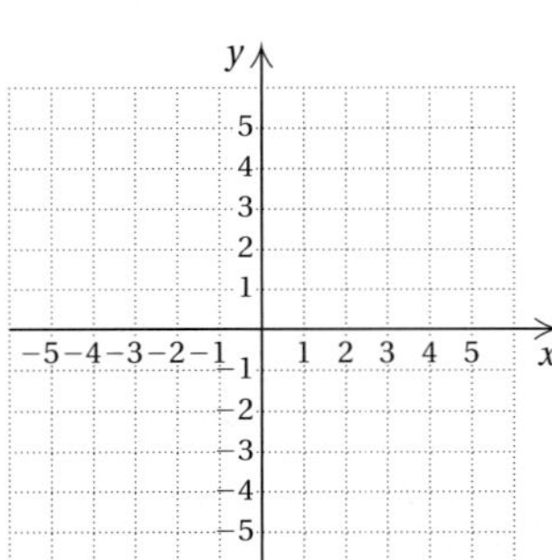

28. $x < 0$

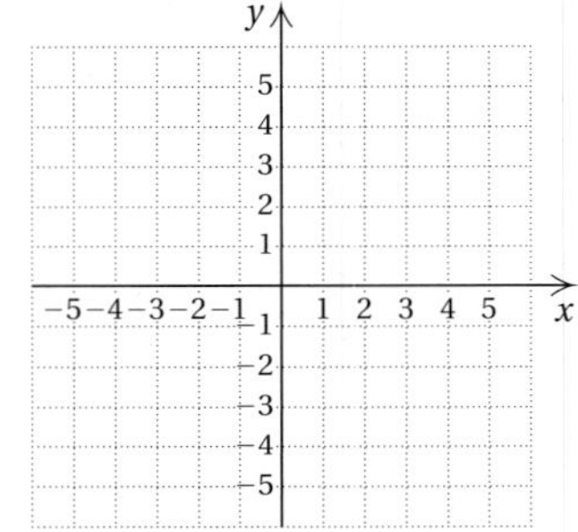

29. $^{\text{D}}\!{}_{\text{W}}$ Why is (0, 0) such a "convenient" test point?

30. $^{\text{D}}\!{}_{\text{W}}$ Is the graph of any inequality in the form $y > mx + b$ shaded *above* the line $y = mx + b$? Why or why not?

SKILL MAINTENANCE

Determine whether the graphs of the equations are parallel, perpendicular, or neither. [9.6a, b]

31. $5y + 50 = 4x$,
$5y = 4x + 15$

32. $5x + 4y = 12$,
$5y + 50 = 4x$

33. $5y + 50 = 4x$,
$4y = 5x + 12$

34. $4x + 5y + 35 = 0$,
$5y = 4x + 40$

SYNTHESIS

35. *Elevators.* Many elevators have a capacity of 1 metric ton (1000 kg). Suppose c children, each weighing 35 kg, and a adults, each weighing 75 kg, are on an elevator. Find and graph an inequality that asserts that the elevator is overloaded.

36. *Hockey Wins and Losses.* A hockey team determines that it needs at least 60 points for the season in order to make the playoffs. A win w is worth 2 points and a tie t is worth 1 point. Find and graph an inequality that describes the situation.

9 Summary and Review

The review that follows is meant to prepare you for a chapter exam. It consists of three parts. The first part, Concept Reinforcement, is designed to increase understanding of the concepts through true/false exercises. The second part is a list of important properties and formulas. The third part is the Review Exercises. These provide practice exercises for the exam, together with references to section objectives so you can go back and review. Before beginning, stop and look back over the skills you have obtained. What skills in mathematics do you have now that you did not have before studying this chapter?

CONCEPT REINFORCEMENT

Determine whether the statement is true or false. Answers are given at the back of the book.

_______ **1.** A slope of $-\frac{3}{4}$ is steeper than a slope of $-\frac{5}{2}$.

_______ **2.** The slope of the line that passes through (a, b) and (c, d) is $\frac{(d-b)}{(c-a)}$.

_______ **3.** The y-intercept of $Ax + By = C$ is $\left(0, \frac{C}{B}\right)$.

_______ **4.** The x- and y-intercepts of $y = mx$ are both $(0, 0)$.

_______ **5.** Parallel lines have the same y-intercept.

_______ **6.** Lines $y = mx + s$ and $y = -\frac{1}{m}x + t$, $m \neq 0$, are perpendicular.

IMPORTANT PROPERTIES AND FORMULAS

Slope $= m = \frac{y_2 - y_1}{x_2 - x_1}$

Slope–Intercept Equation: $y = mx + b$

Parallel Lines: Slopes equal, y-intercepts different

Perpendicular Lines: Product of slopes $= -1$

Review Exercises

Find the coordinates of the point. [9.1b]

1. A **2.** B **3.** C

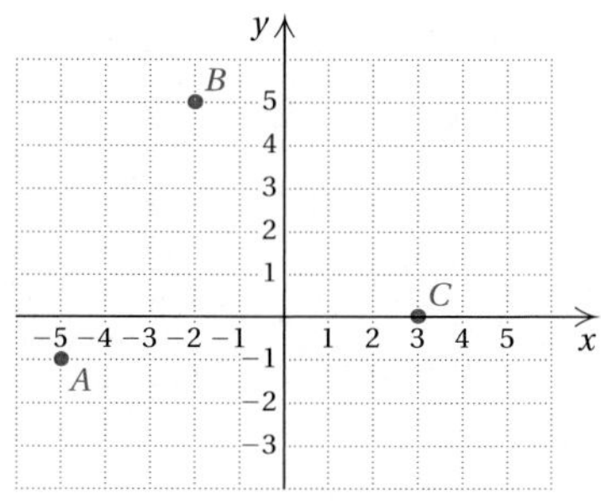

Plot the point. [9.1a]

4. $(2, 5)$ **5.** $(0, -3)$ **6.** $(-4, -2)$

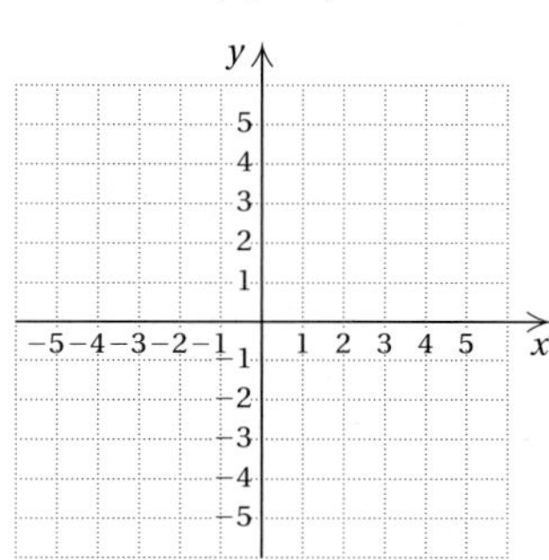

In which quadrant is the point located? [9.1a]

7. $(3, -8)$ **8.** $(-20, -14)$ **9.** $(4.9, 1.3)$

Determine whether the ordered pair is a solution of $2y - x = 10$. [9.1c]

10. $(2, -6)$ **11.** $(0, 5)$

12. Show that the ordered pairs $(0, -3)$ and $(2, 1)$ are solutions of the equation $2x - y = 3$. Then use the graph of the two points to determine another solution. Answers may vary. [9.1c]

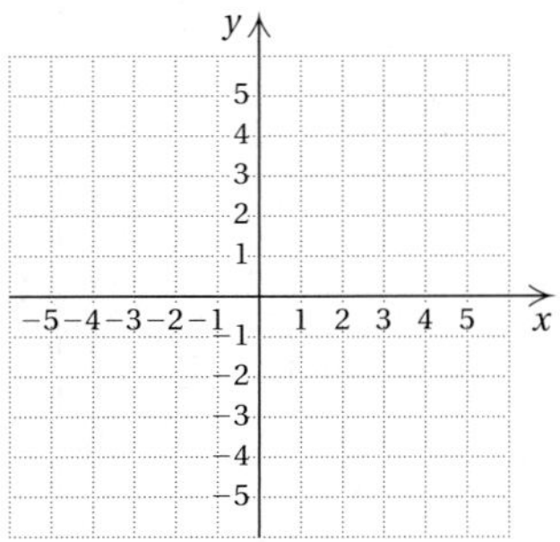

Graph the equation, identifying the y-intercept. [9.1d]

13. $y = 2x - 5$

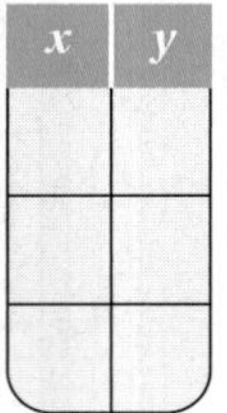

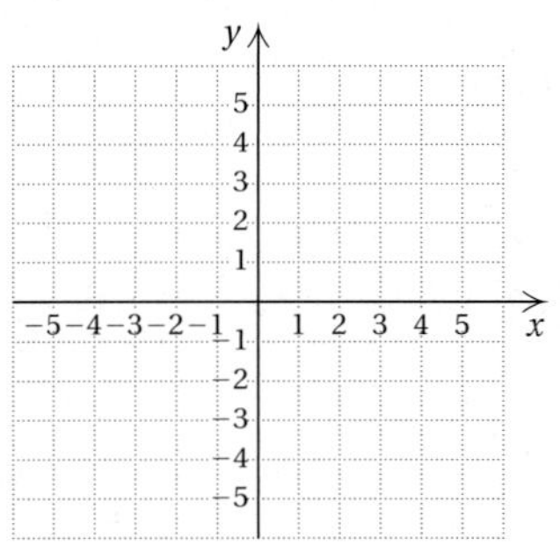

14. $y = -\frac{3}{4}x$

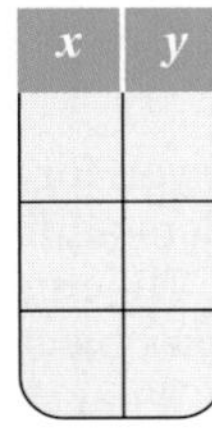

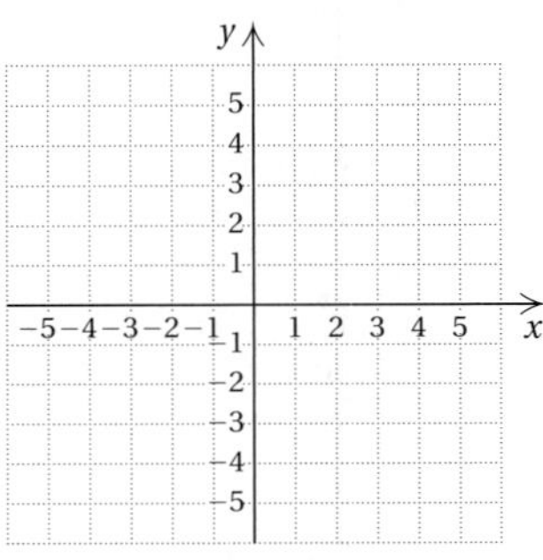

15. $y = -x + 4$

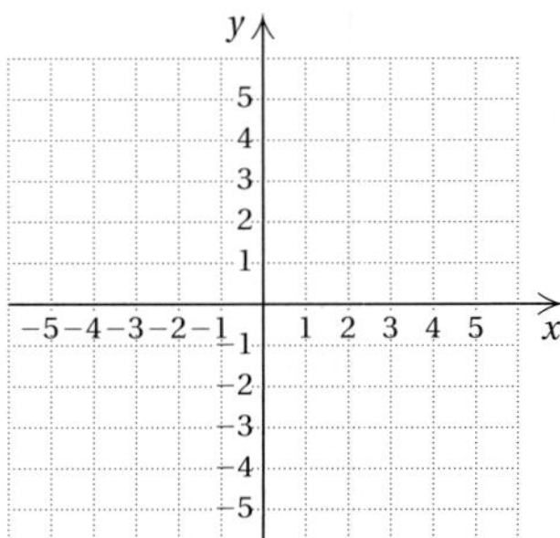

16. $y = 3 - 4x$

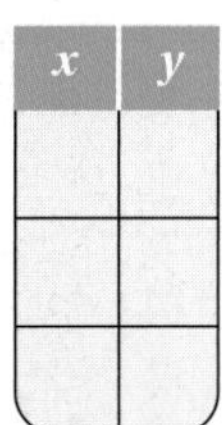

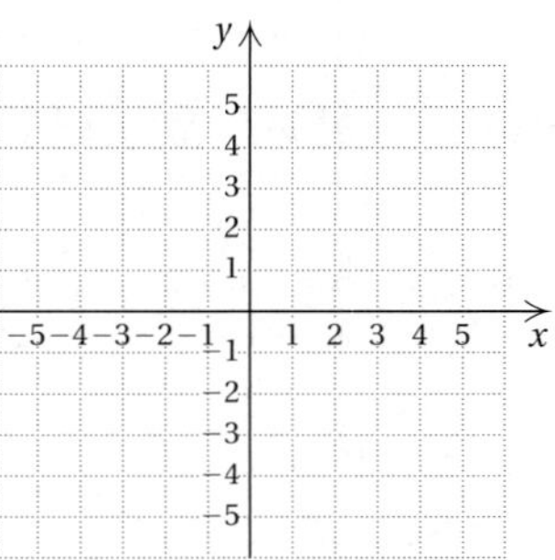

Graph the equation. [9.2b]

17. $y = 3$

x	y

18. $5x - 4 = 0$

x	y

Find the intercepts of the equation. Then graph the equation. [9.2a]

19. $x - 2y = 6$

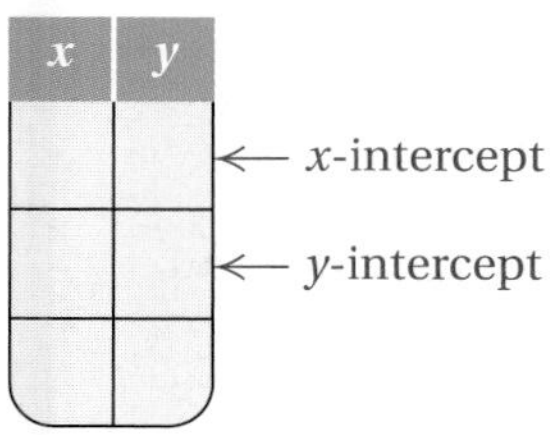

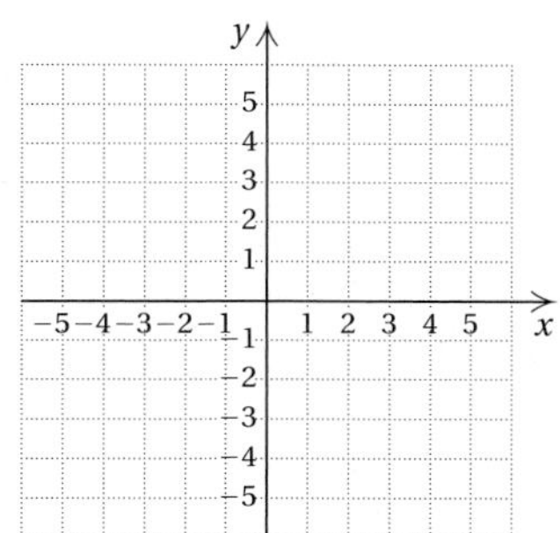

20. $5x - 2y = 10$

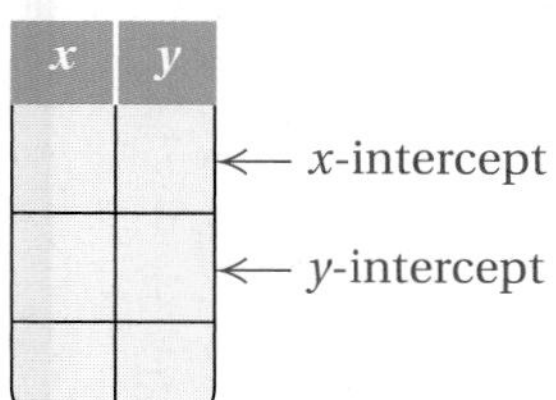

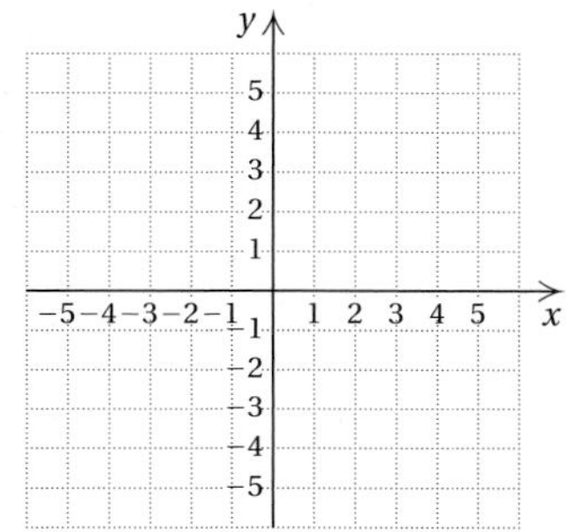

Solve. [9.1e]

21. *Kitchen Design.* Kitchen designers recommend that a refrigerator be selected on the basis of the number of people n in the household. The appropriate size S, in cubic feet, is given by

$$S = \frac{3}{2}n + 13.$$

a) Determine the recommended size of a refrigerator if the number of people is 1, 2, 5, and 10.

b) Graph the equation and use the graph to estimate the recommended size of a refrigerator for 3 people sharing an apartment.

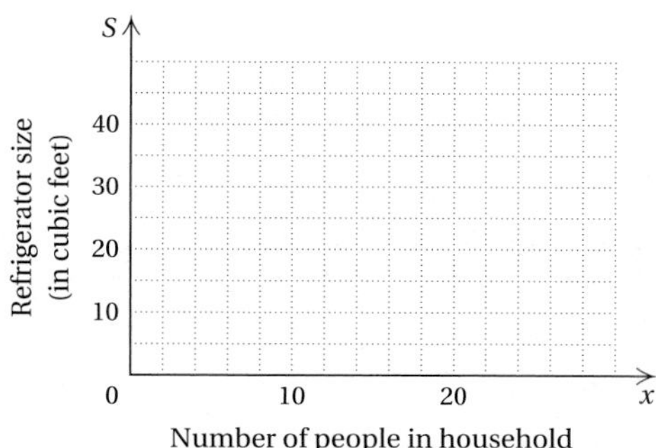

c) A refrigerator is 22 ft^3. For how many residents is it the recommended size?

22. *Snow Removal.* By 3:00 P.M., Erin had plowed 7 driveways and by 5:30 P.M. she had completed 13.

a) Find Erin's plowing rate in number of driveways per hour. [9.3b]

b) Find Erin's plowing rate in minutes per driveway. [9.3b]

23. *Manicures.* The following graph shows data from a recent day's work at the O'Hara School of Cosmetology. What is the rate of change, in number of manicures per hour? [9.3b]

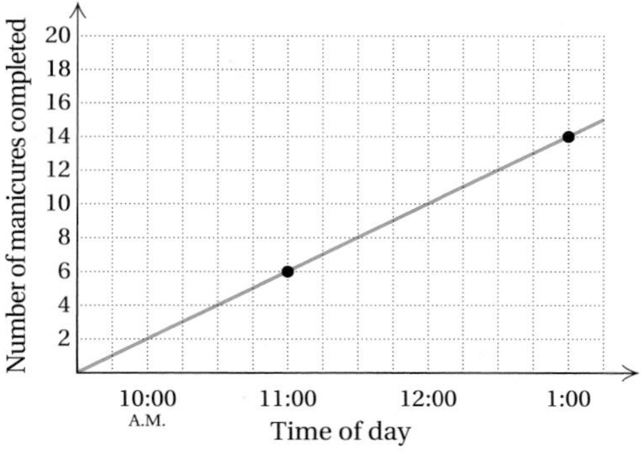

Find the slope. [9.3a]

24.

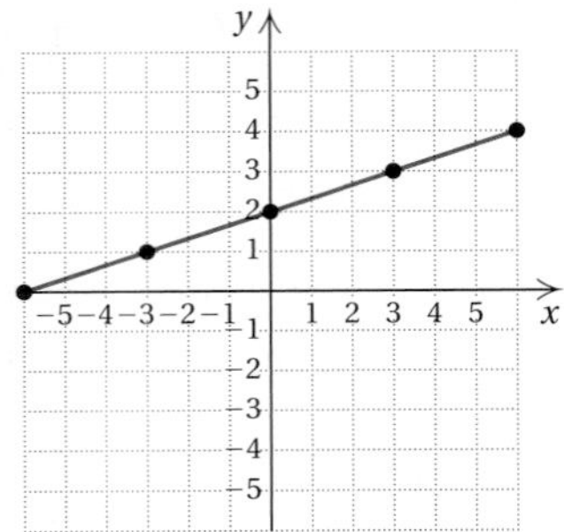

25.

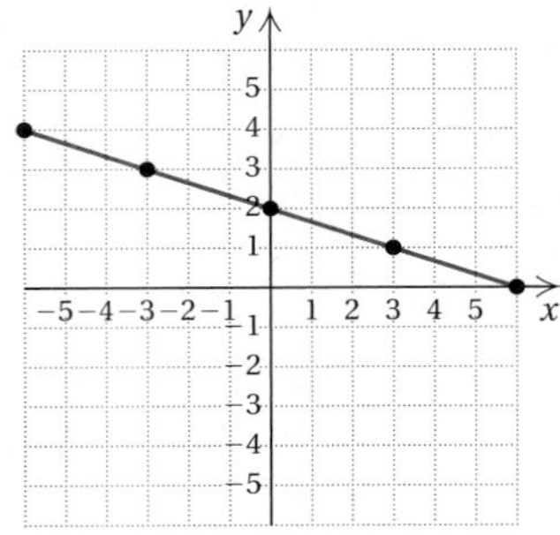

Graph the line containing the given pair of points and find the slope. [9.3a]

26. $(-5, -2), (5, 4)$

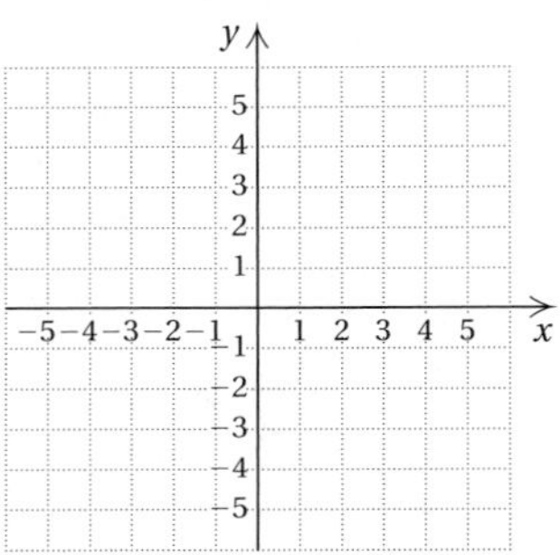

27. $(-5, 5), (4, -4)$

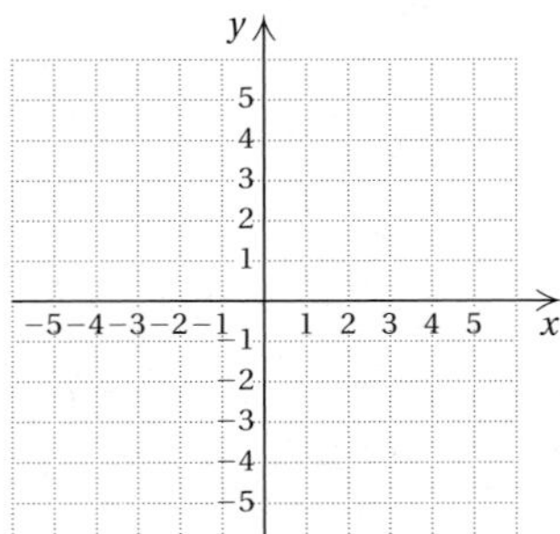

28. *Road Grade.* At one point, Beartooth Highway in Yellowstone National Park rises 315 ft over a horizontal distance of 4500 ft. Find the slope, or grade, of the road. [9.3b]

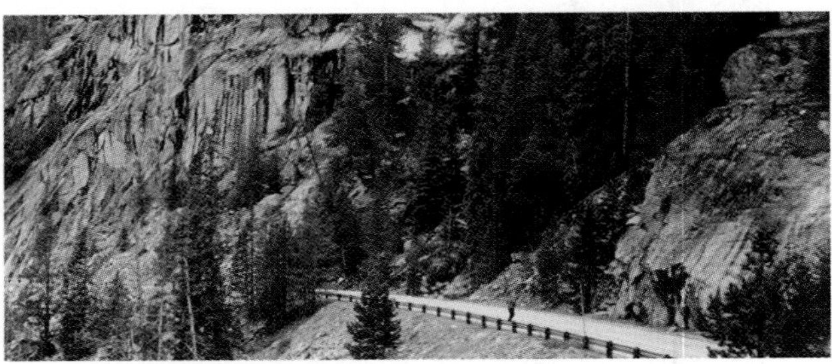

Find the slope, if it exists. [9.3c]

29. $y = -\frac{5}{8}x - 3$

30. $2x - 4y = 8$

31. $x = -2$

32. $y = 9$

Find the slope and the y-intercept. [9.4a]

33. $y = -9x + 46$

34. $x + y = 9$

35. $3x - 5y = 4$

Find an equation of the line with the given slope and y-intercept. [9.4a]

36. Slope $= -2.8$; y-intercept: $(0, 19)$

37. Slope $= \frac{5}{8}$; y-intercept: $\left(0, -\frac{7}{8}\right)$

Find an equation of the line containing the given point and with the given slope. [9.4b]

38. $(1, 2)$, $m = 3$

39. $(-2, -5)$, $m = \frac{2}{3}$

40. $(0, -4)$, $m = -2$

Find an equation of the line containing the given pair of points. [9.4c]

41. $(5, 7)$ and $(-1, 1)$

42. $(2, 0)$ and $(-4, -3)$

Solve. [9.4c]

43. *Women in the Labor Force.* The line graph below illustrates the percent of the female population age 16 and older in the work force for years since 1950.

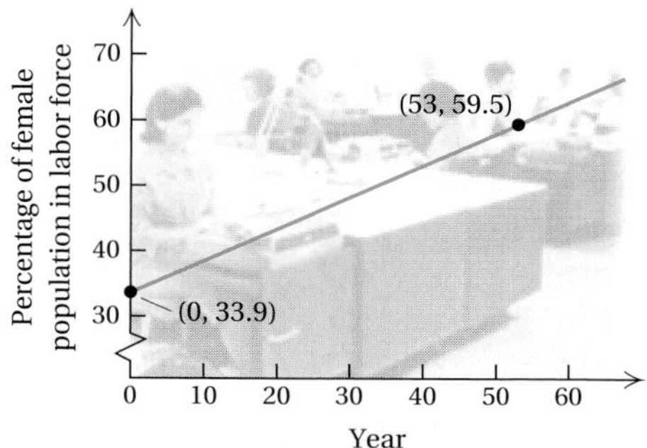

Source: U.S. Department of Labor

a) Find an equation of the line.
b) What is the rate of change in the percent of female workers in the labor force with respect to time?
c) Use the equation to find the percent of female workers in the labor force in 2005.

44. Draw a line that has slope -1 and y-intercept $(0, 4)$. [9.5a]

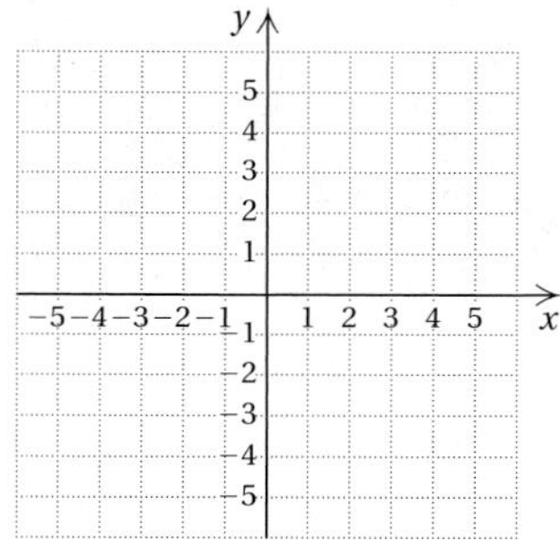

45. Draw a line that has slope $\frac{5}{3}$ and y-intercept $(0, -3)$. [9.5a]

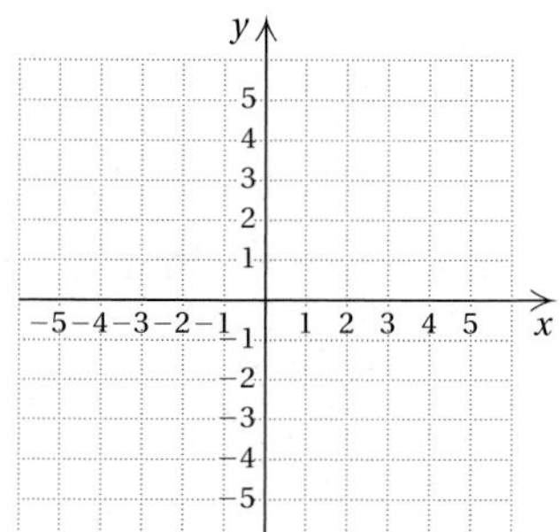

46. Graph $y = -\frac{3}{5}x + 2$ using the slope and the y-intercept. [9.5a]

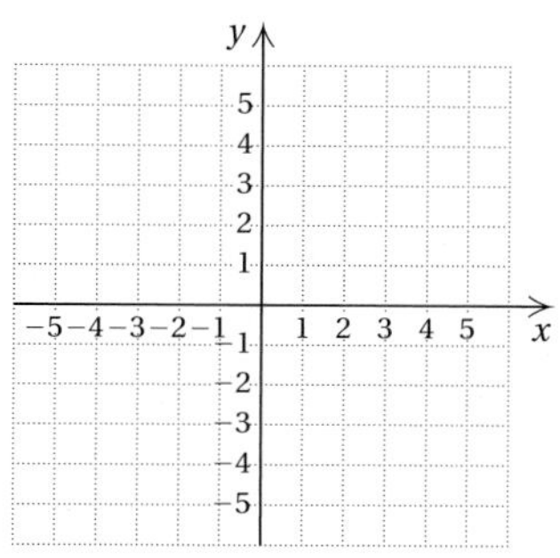

47. Graph $2y - 3x = 6$ using the slope and the y-intercept. [9.5a]

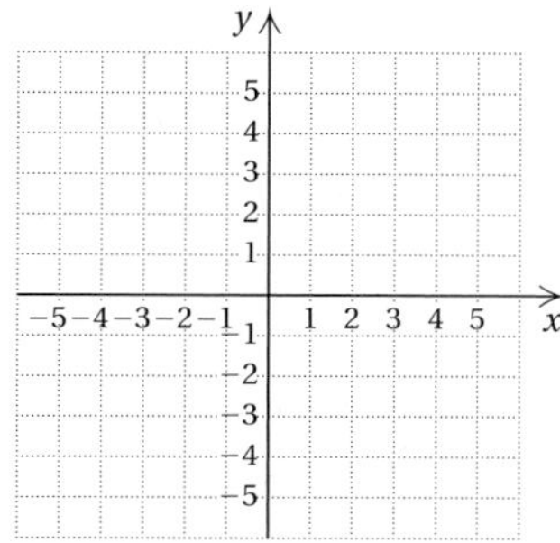

Determine whether the graphs of the equations are parallel, perpendicular, or neither. [9.6a, b]

48. $4x + y = 6,$
$4x + y = 8$

49. $2x + y = 10,$
$y = \frac{1}{2}x - 4$

50. $x + 4y = 8,$
$x = -4y - 10$

51. $3x - y = 6,$
$3x + y = 8$

Determine whether the given point is a solution of the inequality $x - 2y > 1$. [9.7a]

52. $(0, 0)$

53. $(1, 3)$

54. $(4, -1)$

Graph on a plane. [9.7b]

55. $x < y$

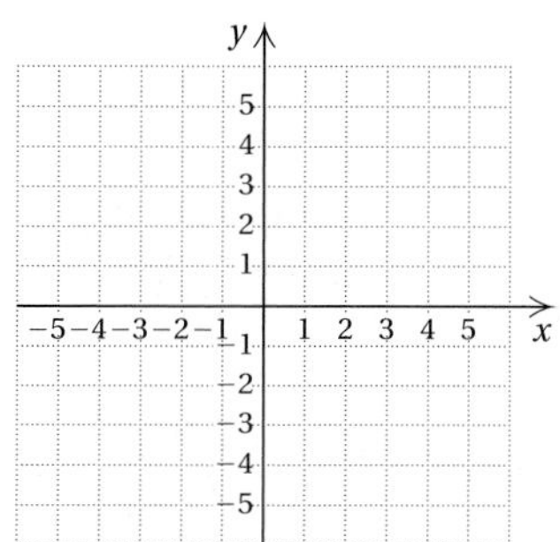

56. $x + 2y \geq 4$

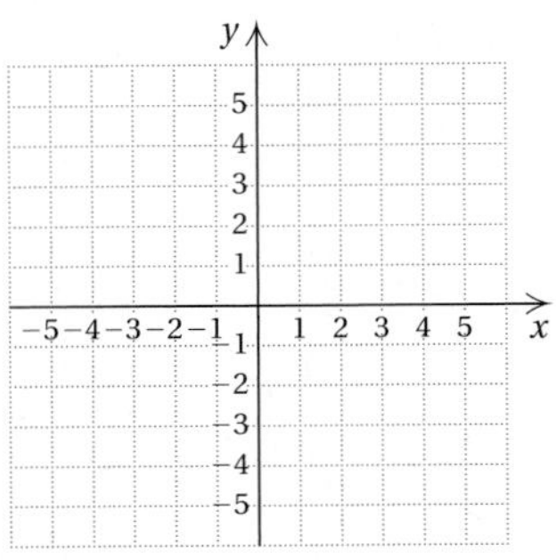

57. $x > -2$

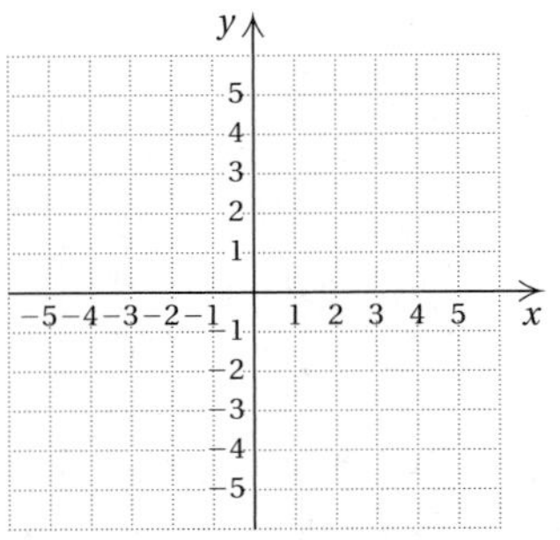

58. **D$_W$** Explain why the first coordinate of the y-intercept is always 0. [9.1d]

59. **D$_W$** Graph $x < 1$ on both a number line and a plane, and explain the difference between the graphs. [9.7b]

60. **D$_W$** Describe how you would graph $y = 0.37x + 2458$ using the slope and the y-intercept. You need not actually draw the graph. [9.5a]

SYNTHESIS

61. Find the value of m in $y = mx + 3$ such that $(-2, 5)$ is on the graph. [9.2a]

62. Find the area and the perimeter of a rectangle for which $(-2, 2)$, $(7, 2)$, and $(7, -3)$ are three of the vertices. [9.1a]

63. *Mountaineering.* As part of an ill-fated expedition to climb Mount Everest in 1996, author Jon Krakauer departed "The Balcony," elevation 27,600 ft, at 7:00 A.M. and reached the summit, elevation 29,028 ft, at 1:25 P.M. [9.3c]

Source: Jon Krakauer, *Into Thin Air*. New York: Villard, 1998.

a) Find Krakauer's rate of ascent in feet per minute.
b) Find Krakauer's rate of ascent in minutes per foot.

64. In chess, the knight can move to any of the eight squares shown by a, b, c, d, e, f, g, and h below. If lines are drawn from the beginning to the end of the move, what slopes are possible for these lines? [9.3a]

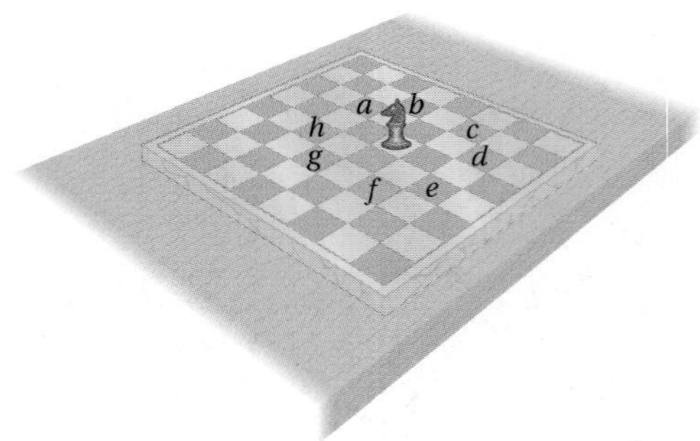

9 Chapter Test

For Extra Help

Work It Out! Chapter Test Video Solutions on CD

In which quadrant is the point located?

1. $\left(-\frac{1}{2}, 7\right)$

2. $(-5, -6)$

Find the coordinates of the point.

3. A

4. B

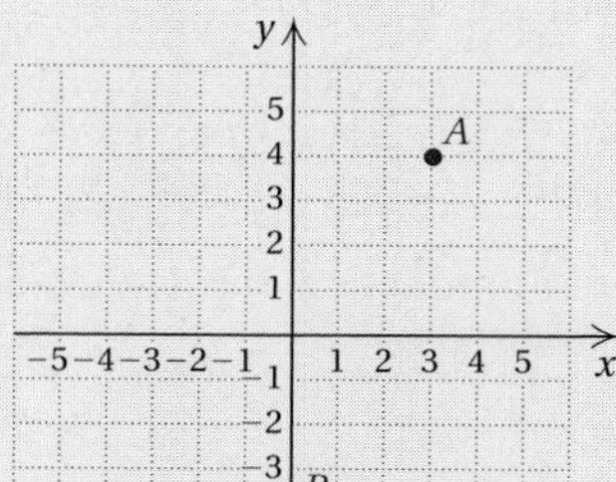

5. Show that the ordered pairs $(-4, -3)$ and $(-1, 3)$ are solutions of the equation $y - 2x = 5$. Then use the graph of the straight line containing the two points to determine another solution. Answers may vary.

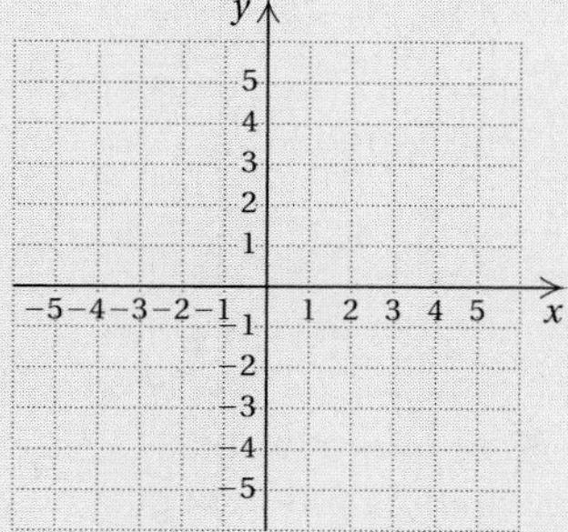

Graph the equation. Identify the y-intercept.

6. $y = 2x - 1$

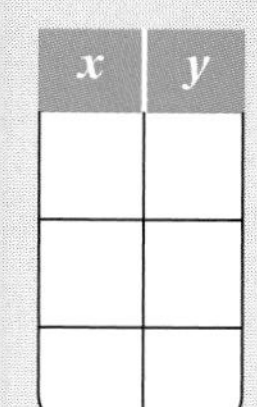

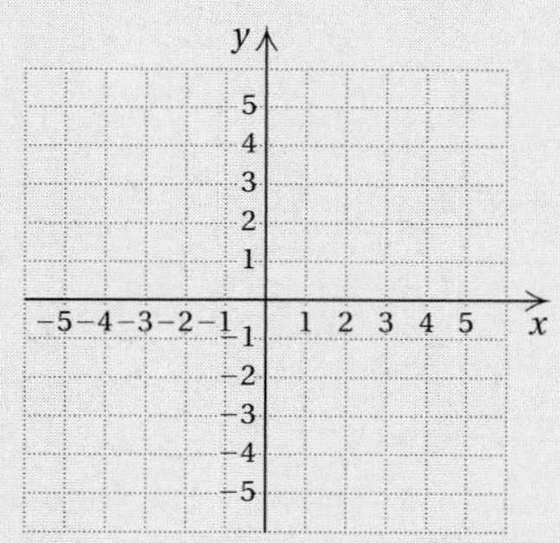

7. $y = -\frac{3}{2}x$

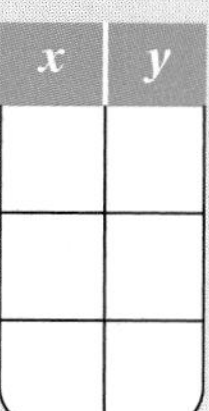

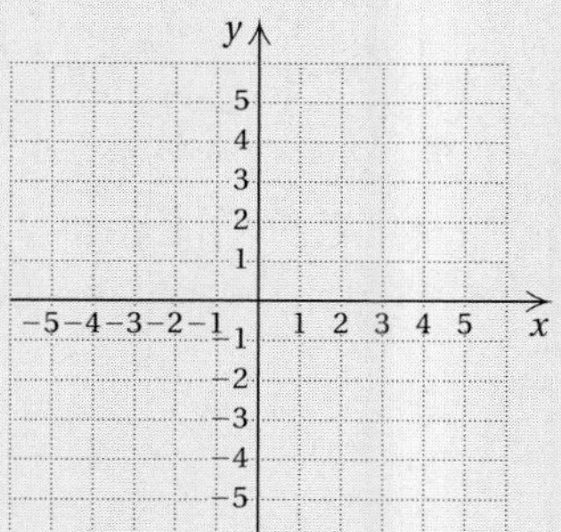

Graph the equation.

8. $2x + 8 = 0$

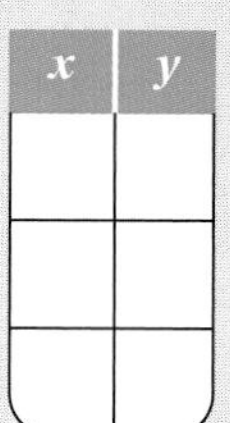

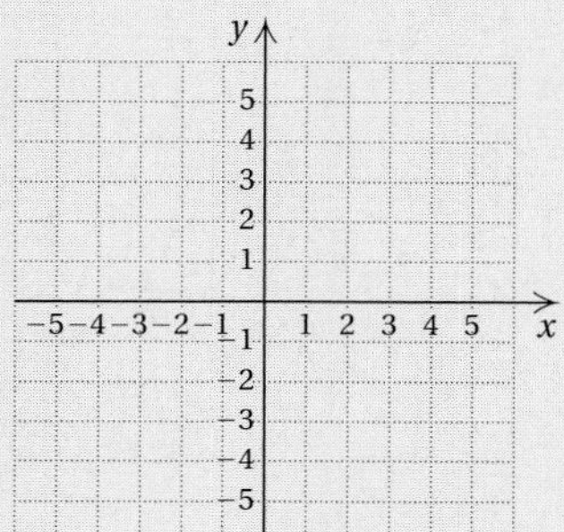

9. $y = 5$

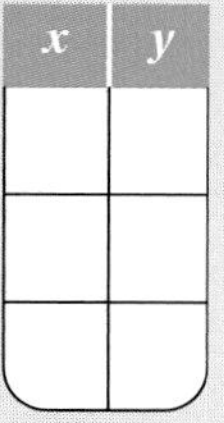

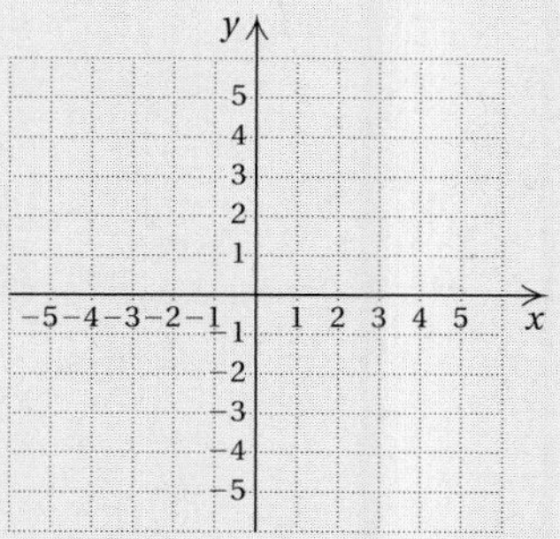

Find the intercepts of the equation. Then graph the equation.

10. $2x - 4y = -8$

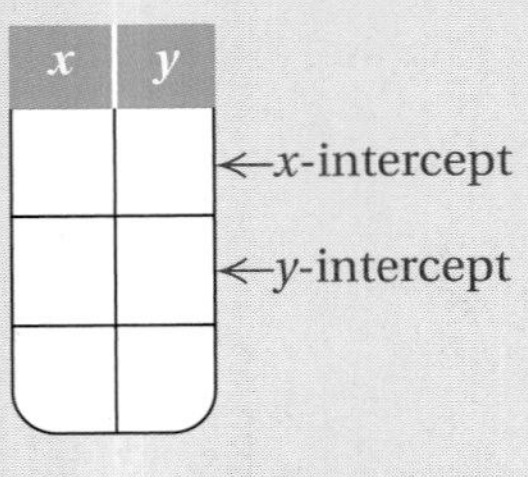

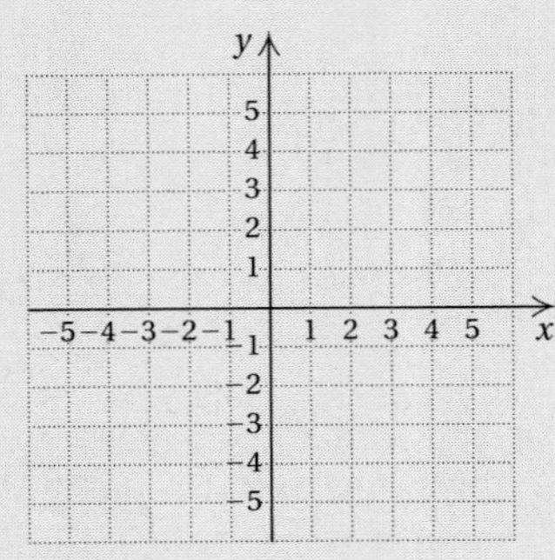

11. $2x - y = 3$

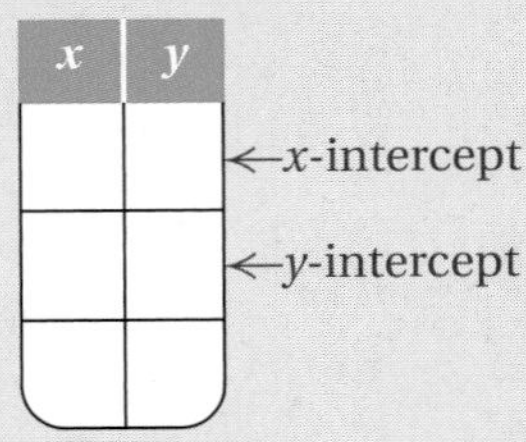

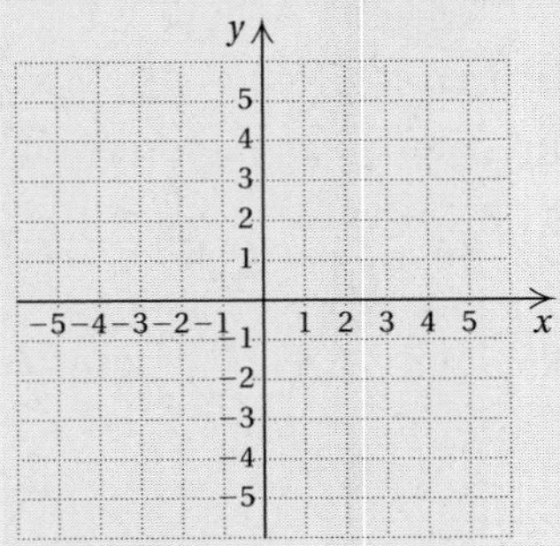

12. *Private-College Costs.* The yearly cost T, in thousands of dollars, of tuition and required fees at a private college (includes two- and four-year schools and does not include room and board) can be approximated by

$$T = \frac{3}{5}n + 5,$$

where n is the number of years since 1985. That is, $n = 0$ corresponds to 1985, $n = 5$ corresponds to 1990, and so on.

Source: *Statistical Abstract of the United States,* 2003

a) Find the cost of tuition in 1985, 1996, 2000, and 2004.

b) Graph the equation and then use the graph to estimate the cost of tuition in 2005.

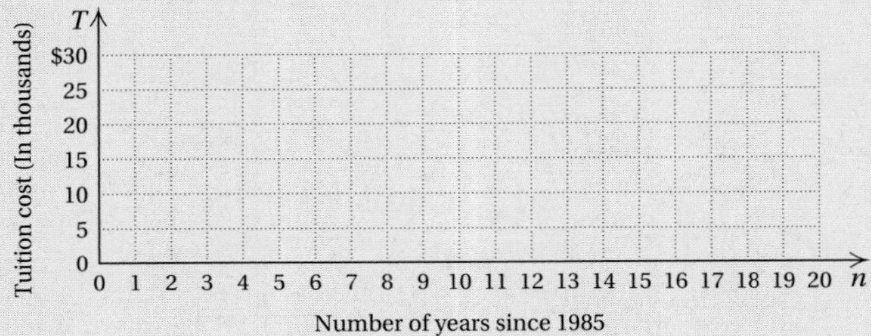

c) Predict the year in which the cost of tuition will be $23,000.

13. *Elevators.* At 2:38, Serge entered an elevator on the 34th floor of the Regency Hotel. At 2:40, he stepped off at the 5th floor.

a) Find the elevator's average rate of travel in number of floors per minute.

b) Find the elevator's average rate of travel in seconds per floor.

14. *Train Travel.* The following graph shows data concerning a recent train ride from Denver to Kansas City. At what rate did the train travel?

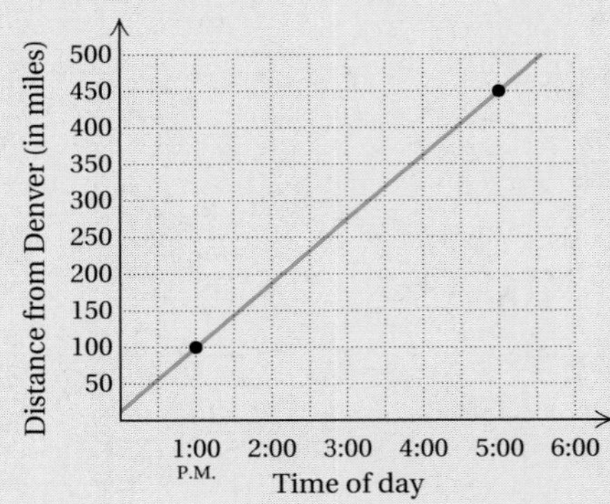

15. Find the slope.

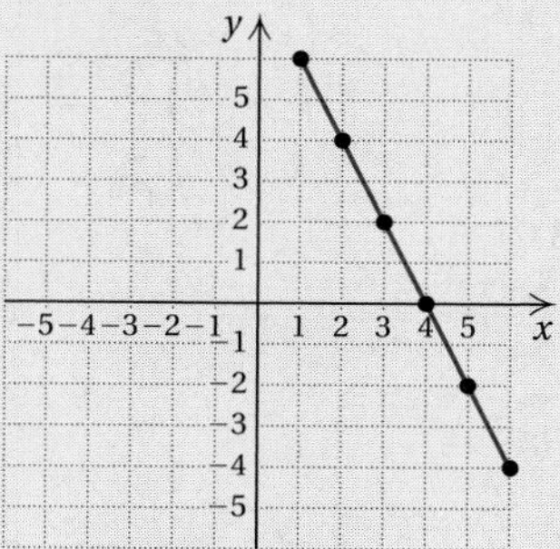

16. Graph the line containing $(-3, 1)$ and $(5, 4)$ and find the slope.

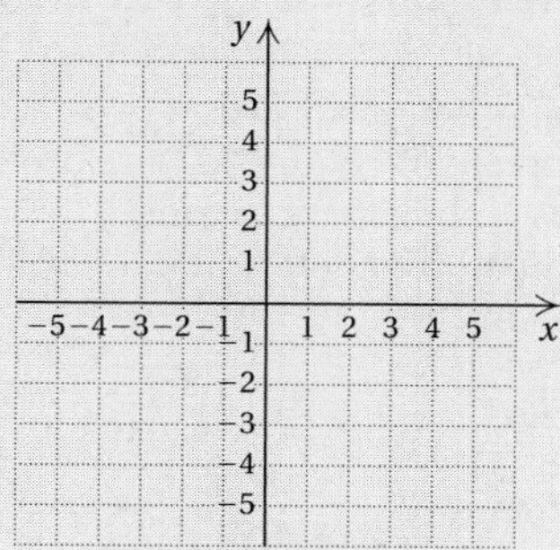

17. *Navigation.* Capital Rapids drops 54 ft vertically over a horizontal distance of 1080 ft. What is the slope of the rapids?

18. Find the slope, if it exists.

a) $2x - 5y = 10$
b) $x = -2$

19. Draw a graph of the line with slope $-\frac{3}{2}$ and y-intercept $(0, 1)$.

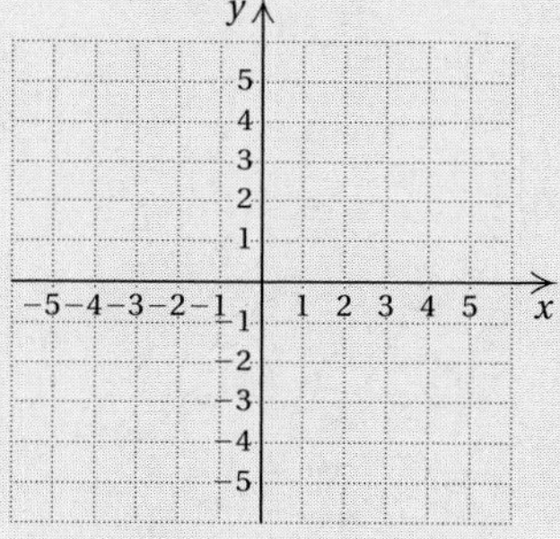

20. Graph $y = 2x - 3$ using the slope and the y-intercept.

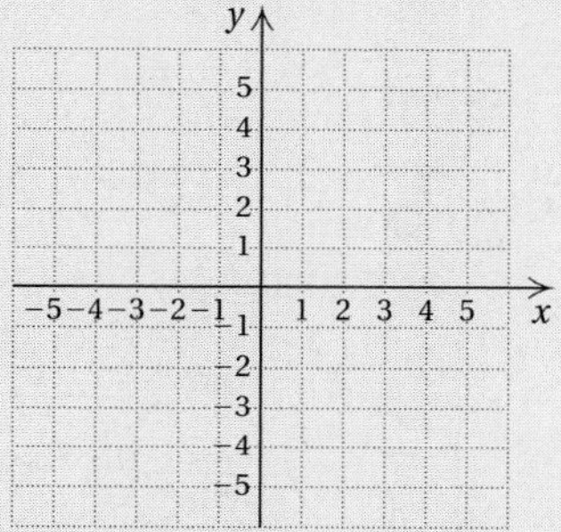

Find the slope and the y-intercept.

21. $y = 2x - \frac{1}{4}$

22. $-4x + 3y = -6$

Find an equation of the line with the given slope and y-intercept.

23. Slope $= 1.8$; y-intercept: $(0, -7)$

24. Slope $= -\frac{3}{8}$; y-intercept: $\left(0, -\frac{1}{8}\right)$

Find an equation of the line containing the given point and with the given slope.

25. $(3, 5)$, $m = 1$

26. $(-2, 0)$, $m = -3$

Find an equation of the line containing the given pair of points.

27. $(1, 1)$ and $(2, -2)$

28. $(4, -1)$ and $(-4, -3)$

29. *Lung Transplants.* The number of lung transplants in the United States has increased in recent years. The line graph at right describes the increase in the number of lung transplants for years since 1988.

a) Find an equation of the line.

b) What is the rate of change in the number of lung transplants with respect to time?

c) Use the equation to determine the number of lung transplants in 2005.

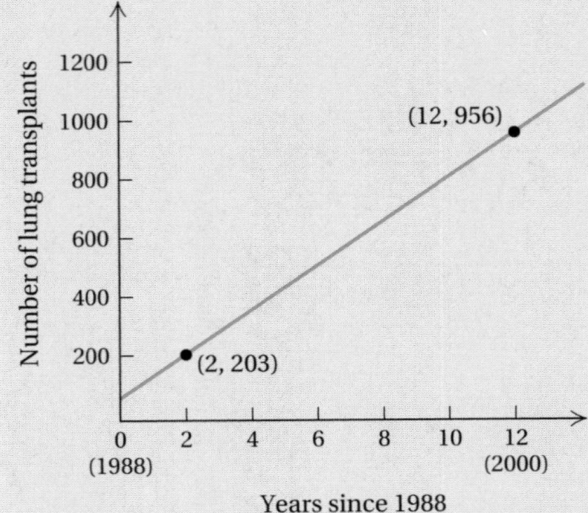

Source: National Organ Procurement and Transplantation Network

Determine whether the graphs of the equations are parallel, perpendicular, or neither.

30. $2x + y = 8,$
$2x + y = 4$

31. $2x + 5y = 2,$
$y = 2x + 4$

32. $x + 2y = 8,$
$-2x + y = 8$

Determine whether the given point is a solution of the inequality $3y - 2x < -2$.

33. $(0, 0)$

34. $(-4, -10)$

Graph on a plane.

35. $y > x - 1$

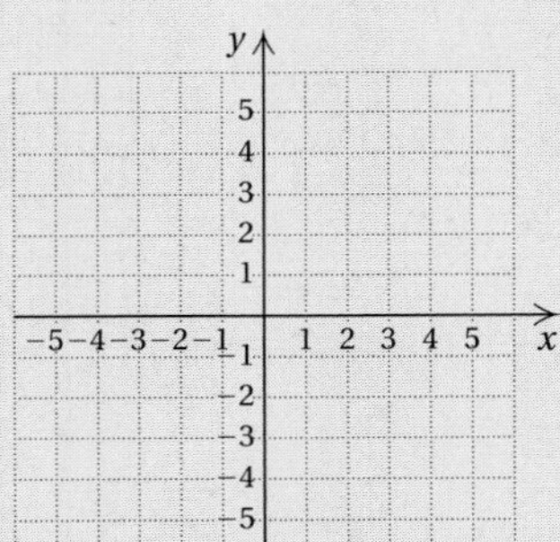

36. $2x - y \leq 4$

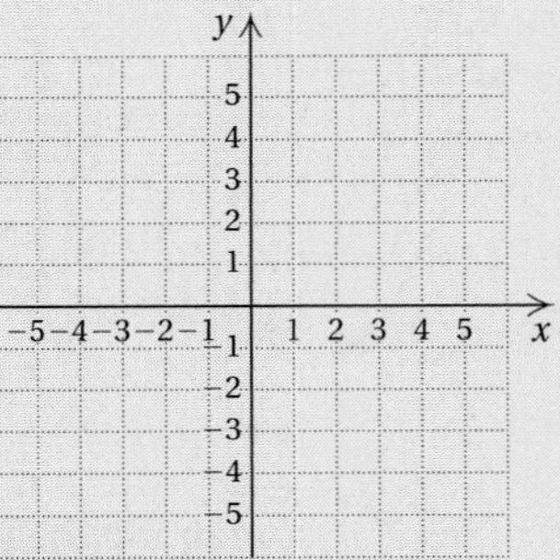

SYNTHESIS

37. A diagonal of a square connects the points $(-3, -1)$ and $(2, 4)$. Find the area and the perimeter of the square.

38. Find the value of k such that $3x + 7y = 14$ and $ky - 7x = -3$ are perpendicular.

Polynomials: Operations

10

Real World-Application

Total revenue of NASCAR (National Association of Stock Car Automobile Racing) is expected to be $3423 million by 2006. Convert this number to scientific notation.

Source: NASCAR

This problem appears as Exercise 64 in Exercise Set 10.2.

10.1 INTEGERS AS EXPONENTS

Objectives

a Tell the meaning of exponential notation.

b Evaluate exponential expressions with exponents of 0 and 1.

c Evaluate algebraic expressions containing exponents.

d Use the product rule to multiply exponential expressions with like bases.

e Use the quotient rule to divide exponential expressions with like bases.

f Express an exponential expression involving negative exponents with positive exponents.

What is the meaning of each of the following?

1. 5^4

2. x^5

3. $(3t)^2$

4. $3t^2$

5. $(-x)^4$

6. $-y^3$

Answers on page A-34

We introduced integer exponents of 2 or higher in Section R.5. Here we consider 0 and 1, as well as negative integers, as exponents.

a Exponential Notation

An exponent of 2 or greater tells how many times the base is used as a factor. For example,

$$a \cdot a \cdot a \cdot a = a^4.$$

In this case, the **exponent** is 4 and the **base** is a. An expression for a power is called **exponential notation.**

$a^n \leftarrow$ This is the exponent.
$\uparrow$
This is the base.

EXAMPLE 1 What is the meaning of 3^5? of n^4? of $(2n)^3$? of $50x^2$? of $(-n)^3$? of $-n^3$?

3^5 means $3 \cdot 3 \cdot 3 \cdot 3 \cdot 3$; $\quad n^4$ means $n \cdot n \cdot n \cdot n$;
$(2n)^3$ means $2n \cdot 2n \cdot 2n$; $\quad 50x^2$ means $50 \cdot x \cdot x$;
$(-n)^3$ means $(-n) \cdot (-n) \cdot (-n)$; $\quad -n^3$ means $-1 \cdot n \cdot n \cdot n$

Do Exercises 1–6.

We read exponential notation as follows: a^n is read the ***n*th power of *a*,** or simply ***a* to the *n*th,** or ***a* to the *n*.** We often read x^2 as "***x*-squared.**" The reason for this is that the area of a square of side x is $x \cdot x$, or x^2. We often read x^3 as "***x*-cubed.**" The reason for this is that the volume of a cube with length, width, and height x is $x \cdot x \cdot x$, or x^3.

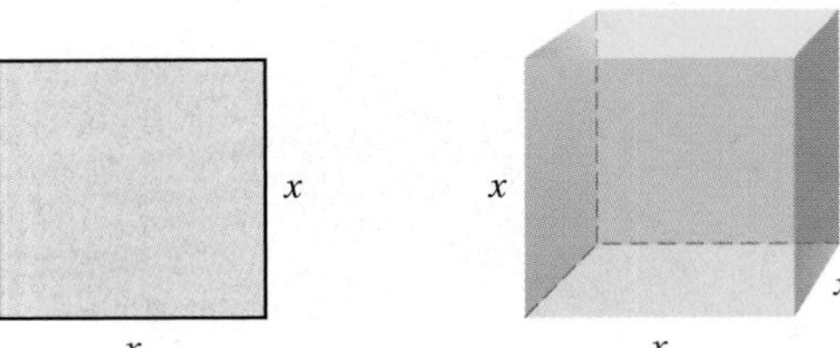

b One and Zero as Exponents

Look for a pattern in the following:

On each side, we **divide** by 8 at each step.

$8 \cdot 8 \cdot 8 \cdot 8 = 8^4$
$8 \cdot 8 \cdot 8 = 8^3$
$8 \cdot 8 = 8^2$
$8 = 8^?$
$1 = 8^?.$

On this side, the exponents **decrease** by 1.

To continue the pattern, we would say that

$$8 = 8^1 \quad \text{and} \quad 1 = 8^0.$$

We make the following definition.

EXPONENTS OF 0 AND 1

$a^1 = a$, for any number a;
$a^0 = 1$, for any nonzero number a

We consider 0^0 to be not defined. We will explain why later in this section.

EXAMPLE 2 Evaluate 5^1, $(-8)^1$, 3^0, $(-7.3)^0$, and $(186{,}892{,}046)^0$.

$5^1 = 5; \quad (-8)^1 = -8; \quad 3^0 = 1;$
$(-7.3)^0 = 1; \quad (186{,}892{,}046)^0 = 1$

Do Exercises 7–12.

Evaluate.

7. 6^1 **8.** 7^0

9. $(8.4)^1$ **10.** 8654^0

11. $(-1.4)^1$ **12.** 0^1

C Evaluating Algebraic Expressions

Algebraic expressions can involve exponential notation. For example, the following are algebraic expressions:

$$x^4, \qquad (3x)^3 - 2, \qquad a^2 + 2ab + b^2.$$

We evaluate algebraic expressions by replacing variables with numbers and following the rules for order of operations.

EXAMPLE 3 Evaluate $1000 - x^4$ when $x = 5$.

$$\begin{aligned} 1000 - x^4 &= 1000 - 5^4 && \text{Substituting} \\ &= 1000 - 5 \cdot 5 \cdot 5 \cdot 5 \\ &= 1000 - 625 \\ &= 375 \end{aligned}$$

EXAMPLE 4 *Area of a Compact Disc.* The standard compact disc used for software and music has a radius of 6 cm. Find the area of such a CD (ignoring the hole in the middle).

$$\begin{aligned} A &= \pi r^2 \\ &= \pi \cdot (6\text{ cm})^2 \\ &= \pi \cdot 6\text{ cm} \cdot 6\text{ cm} \\ &\approx 3.14 \times 36\text{ cm}^2 \\ &= 113.04\text{ cm}^2 \end{aligned}$$

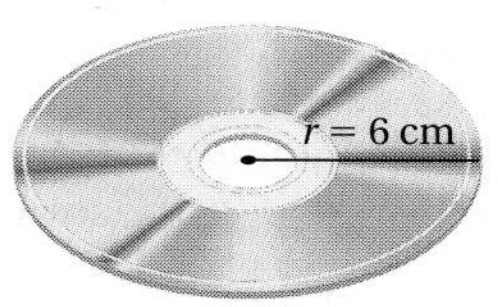

In Example 4, "cm²" means "square centimeters" and "≈" means "is approximately equal to."

EXAMPLE 5 Evaluate $(5x)^3$ when $x = -2$.

When we evaluate with a negative number, we often use extra parentheses to show the substitution.

$$\begin{aligned} (5x)^3 &= [5 \cdot (-2)]^3 && \text{Substituting} \\ &= [-10]^3 && \text{Multiplying within brackets first} \\ &= [-10] \cdot [-10] \cdot [-10] \\ &= -1000 && \text{Evaluating the power} \end{aligned}$$

Answers on page A-34

13. Evaluate t^3 when $t = 5$.

14. Evaluate $-5x^5$ when $x = -2$.

15. Find the area of a circle when $r = 32$ cm. Use 3.14 for π.

16. Evaluate $200 - a^4$ when $a = 3$.

17. Evaluate $t^1 - 4$ and $t^0 - 4$ when $t = 7$.

18. a) Evaluate $(4t)^2$ when $t = -3$.

b) Evaluate $4t^2$ when $t = -3$.

c) Determine whether $(4t)^2$ and $4t^2$ are equivalent.

Multiply and simplify.

19. $3^5 \cdot 3^5$

20. $x^4 \cdot x^6$

21. $p^4p^{12}p^8$

22. $x \cdot x^4$

23. $(a^2b^3)(a^7b^5)$

Answers on page A-34

EXAMPLE 6 Evaluate $5x^3$ when $x = -2$.

$5x^3 = 5 \cdot (-2)^3$ Substituting

$= 5 \cdot (-2) \cdot (-2) \cdot (-2)$ Evaluating the power first

$= 5(-8)$ $(-2)(-2)(-2) = -8$

$= -40$

Recall that two expressions are equivalent if they have the same value for all meaningful replacements. Note that Examples 5 and 6 show that $(5x)^3$ and $5x^3$ are *not* equivalent—that is, $(5x)^3 \neq 5x^3$.

Do Exercises 13–18.

d Multiplying Powers with Like Bases

There are several rules for manipulating exponential notation to obtain equivalent expressions. We first consider multiplying powers with like bases:

$$a^3 \cdot a^2 = \underbrace{(a \cdot a \cdot a)}_{\text{3 factors}}\underbrace{(a \cdot a)}_{\text{2 factors}} = \underbrace{a \cdot a \cdot a \cdot a \cdot a}_{\text{5 factors}} = a^5.$$

Since an integer exponent greater than 1 tells how many times we use a base as a factor, then $(a \cdot a \cdot a)(a \cdot a) = a \cdot a \cdot a \cdot a \cdot a = a^5$ by the associative law. Note that the exponent in a^5 is the sum of those in $a^3 \cdot a^2$, that is, $3 + 2 = 5$. Likewise,

$$b^4 \cdot b^3 = (b \cdot b \cdot b \cdot b)(b \cdot b \cdot b) = b^7, \quad \text{where} \quad 4 + 3 = 7.$$

Adding the exponents gives the correct result.

THE PRODUCT RULE

For any number a and any positive integers m and n,

$$a^m \cdot a^n = a^{m+n}.$$

(When multiplying with exponential notation, if the bases are the same, keep the base and add the exponents.)

EXAMPLES Multiply and simplify. By simplify, we mean write the expression as one number to a nonnegative power.

7. $5^6 \cdot 5^2 = 5^{6+2}$ Adding exponents: $a^m \cdot a^n = a^{m+n}$

$= 5^8$

8. $x^3 \cdot x^9 = x^{3+9} = x^{12}$

9. $m^5m^{10}m^3 = m^{5+10+3} = m^{18}$

10. $x \cdot x^8 = x^1 \cdot x^8$ Writing x as x^1

$= x^{1+8}$

$= x^9$

11. $(a^3b^2)(a^3b^5) = (a^3a^3)(b^2b^5)$

$= a^6b^7$

Do Exercises 19–23.

e Dividing Powers with Like Bases

The following suggests a rule for dividing powers with like bases, such as a^5/a^2:

$$\frac{a^5}{a^2} = \frac{a \cdot a \cdot a \cdot a \cdot a}{a \cdot a} = \frac{a \cdot a \cdot a \cdot a \cdot a}{1 \cdot a \cdot a} = \frac{a \cdot a \cdot a}{1} \cdot \frac{a \cdot a}{a \cdot a} = \frac{a \cdot a \cdot a}{1} \cdot 1$$
$$= a \cdot a \cdot a = a^3.$$

Note that the exponent in a^3 is the difference of those in $a^5 \div a^2$, that is, $5 - 2 = 3$. In a similar way, we have

$$\frac{t^9}{t^4} = \frac{t \cdot t \cdot t \cdot t \cdot t \cdot t \cdot t \cdot t \cdot t}{t \cdot t \cdot t \cdot t} = t^5, \quad \text{where } 9 - 4 = 5.$$

Subtracting exponents gives the correct answer.

THE QUOTIENT RULE

For any nonzero number a and any positive integers m and n,

$$\frac{a^m}{a^n} = a^{m-n}.$$

(When dividing with exponential notation, if the bases are the same, keep the base and subtract the exponent of the denominator from the exponent of the numerator.)

EXAMPLES Divide and simplify. By simplify, we mean write the expression as one number to a nonnegative power.

12. $\frac{6^5}{6^3} = 6^{5-3}$ Subtracting exponents
$= 6^2$

13. $\frac{x^8}{x^2} = x^{8-2}$
$= x^6$

14. $\frac{t^{12}}{t} = \frac{t^{12}}{t^1} = t^{12-1}$
$= t^{11}$

15. $\frac{p^5q^7}{p^2q^5} = \frac{p^5}{p^2} \cdot \frac{q^7}{q^5} = p^{5-2}q^{7-5}$
$= p^3q^2$

The quotient rule can also be used to explain the definition of 0 as an exponent. Consider the expression a^4/a^4, where a is nonzero:

$$\frac{a^4}{a^4} = \frac{a \cdot a \cdot a \cdot a}{a \cdot a \cdot a \cdot a} = 1.$$

This is true because the numerator and the denominator are the same. Now suppose we apply the rule for dividing powers with the same base:

$$\frac{a^4}{a^4} = a^{4-4} = a^0.$$

Since $a^4/a^4 = 1$ and $a^4/a^4 = a^0$, it follows that $a^0 = 1$, when $a \neq 0$.

We can explain why we do not define 0^0 using the quotient rule. We know that 0^0 is 0^{1-1}. But 0^{1-1} is also equal to $0^1/0^1$, or $0/0$. We have already seen that division by 0 is not defined, so 0^0 is also not defined.

Do Exercises 24–27.

Divide and simplify.

24. $\frac{4^5}{4^2}$

25. $\frac{y^6}{y^2}$

26. $\frac{p^{10}}{p}$

27. $\frac{a^7b^6}{a^3b^4}$

Answers on page A-34

Express with positive exponents. Then simplify.

28. 4^{-3}

29. 5^{-2}

30. 2^{-4}

31. $(-2)^{-3}$

32. $4p^{-3}$

33. $\frac{1}{x^{-2}}$

Answers on page A-34

f Negative Integers as Exponents

We can use the rule for dividing powers with like bases to lead us to a definition of exponential notation when the exponent is a negative integer. Consider $5^3/5^7$ and first simplify it using procedures we have learned for working with fractions:

$$\frac{5^3}{5^7} = \frac{5 \cdot 5 \cdot 5}{5 \cdot 5 \cdot 5 \cdot 5 \cdot 5 \cdot 5 \cdot 5} = \frac{5 \cdot 5 \cdot 5 \cdot 1}{5 \cdot 5 \cdot 5 \cdot 5 \cdot 5 \cdot 5 \cdot 5}$$

$$= \frac{5 \cdot 5 \cdot 5}{5 \cdot 5 \cdot 5} \cdot \frac{1}{5 \cdot 5 \cdot 5 \cdot 5} = \frac{1}{5^4}.$$

Now we apply the rule for dividing exponential expressions with the same bases. Then

$$\frac{5^3}{5^7} = 5^{3-7} = 5^{-4}.$$

From these two expressions for $5^3/5^7$, it follows that

$$5^{-4} = \frac{1}{5^4}.$$

This leads to our definition of negative exponents.

NEGATIVE EXPONENT

For any real number a that is nonzero and any integer n,

$$a^{-n} = \frac{1}{a^n}.$$

In fact, the numbers a^n and a^{-n} are reciprocals of each other because

$$a^n \cdot a^{-n} = a^n \cdot \frac{1}{a^n} = \frac{a^n}{a^n} = 1.$$

The following is another way to arrive at the definition of negative exponents.

On each side, we **divide** by 5 at each step.

$$5 \cdot 5 \cdot 5 \cdot 5 = 5^4$$
$$5 \cdot 5 \cdot 5 = 5^3$$
$$5 \cdot 5 = 5^2$$
$$5 = 5^1$$
$$1 = 5^0$$
$$\frac{1}{5} = 5^?$$
$$\frac{1}{25} = 5^?$$

On this side, the exponents **decrease** by 1.

To continue the pattern, it should follow that

$$\frac{1}{5} = \frac{1}{5^1} = 5^{-1} \quad \text{and} \quad \frac{1}{25} = \frac{1}{5^2} = 5^{-2}.$$

EXAMPLES Express using positive exponents. Then simplify.

16. $4^{-2} = \frac{1}{4^2} = \frac{1}{16}$

17. $(-3)^{-2} = \frac{1}{(-3)^2} = \frac{1}{(-3)(-3)} = \frac{1}{9}$

18. $m^{-3} = \frac{1}{m^3}$

19. $ab^{-1} = a\left(\frac{1}{b^1}\right) = a\left(\frac{1}{b}\right) = \frac{a}{b}$

20. $\frac{1}{x^{-3}} = x^{-(-3)} = x^3$

21. $3c^{-5} = 3\left(\frac{1}{c^5}\right) = \frac{3}{c^5}$

Example 20 might also be done as follows:

$$\frac{1}{x^{-3}} = \frac{1}{\frac{1}{x^3}} = 1 \cdot \frac{x^3}{1} = x^3.$$

Caution!

As shown in Examples 16 and 17, a negative exponent does not necessarily mean that an expression is negative.

Do Exercises 28–33 on the preceding page.

The rules for multiplying and dividing powers with like bases still hold when exponents are 0 or negative. We state them in a summary below.

EXAMPLES Simplify. By simplify, we generally mean write the expression as one number or variable to a nonnegative power.

22. $7^{-3} \cdot 7^6 = 7^{-3+6}$ Adding exponents
$= 7^3$

23. $x^4 \cdot x^{-3} = x^{4+(-3)} = x^1 = x$

24. $\frac{5^4}{5^{-2}} = 5^{4-(-2)}$ Subtracting exponents
$= 5^{4+2} = 5^6$

25. $\frac{x}{x^7} = x^{1-7} = x^{-6} = \frac{1}{x^6}$

26. $\frac{b^{-4}}{b^{-5}} = b^{-4-(-5)}$
$= b^{-4+5} = b^1 = b$

27. $y^{-4} \cdot y^{-8} = y^{-4+(-8)}$
$= y^{-12} = \frac{1}{y^{12}}$

Do Exercises 34–38.

The following is a summary of the definitions and rules for exponents that we have considered in this section.

DEFINITIONS AND RULES FOR EXPONENTS

1 as an exponent:	$a^1 = a$
0 as an exponent:	$a^0 = 1, a \neq 0$
Negative integers as exponents:	$a^{-n} = \frac{1}{a^n}, \frac{1}{a^{-n}} = a^n; a \neq 0$
Product Rule:	$a^m \cdot a^n = a^{m+n}$
Quotient Rule:	$\frac{a^m}{a^n} = a^{m-n}, a \neq 0$

Simplify.

34. $5^{-2} \cdot 5^4$

35. $x^{-3} \cdot x^{-4}$

36. $\frac{7^{-2}}{7^3}$

37. $\frac{b^{-2}}{b^{-3}}$

38. $\frac{t}{t^{-5}}$

Answers on page A-34

10.1 EXERCISE SET

For Extra Help

MathXL | MyMathLab | InterAct Math | Math Tutor Center | Video Lectures on CD Disc 5 | Student's Solutions Manual

a What is the meaning of each of the following?

1. 3^4

2. 4^3

3. $(-1.1)^5$

4. $(87.2)^6$

5. $\left(\frac{2}{3}\right)^4$

6. $\left(-\frac{5}{8}\right)^3$

7. $(7p)^2$

8. $(11c)^3$

9. $8k^3$

10. $17x^2$

11. $-6y^4$

12. $-q^5$

b Evaluate.

13. $a^0, a \neq 0$

14. $t^0, t \neq 0$

15. b^1

16. c^1

17. $\left(\frac{2}{3}\right)^0$

18. $\left(-\frac{5}{8}\right)^0$

19. $(-7.03)^1$

20. $\left(\frac{4}{5}\right)^1$

21. 8.38^0

22. 8.38^1

23. $(ab)^1$

24. $(ab)^0, a, b \neq 0$

25. ab^0

26. ab^1

c Evaluate.

27. m^3, when $m = 3$

28. x^6, when $x = 2$

29. p^1, when $p = 19$

30. x^{19}, when $x = 0$

31. $-x^4$, when $x = -3$

32. $-2y^7$, when $x = 2$

33. x^4, when $x = 4$

34. y^{15}, when $y = 1$

35. $y^2 - 7$, when $y = -10$

36. $z^5 + 5$, when $z = -2$

37. $161 - b^2$, when $b = 5$

38. $325 - v^3$, when $v = -3$

39. $x^1 + 3$ and $x^0 + 3$, when $x = 7$

40. $y^0 - 8$ and $y^1 - 8$, when $y = -3$

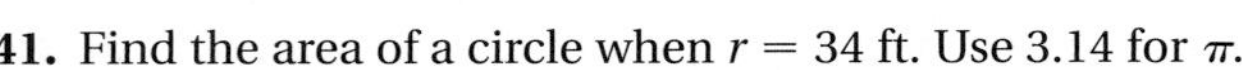

41. Find the area of a circle when $r = 34$ ft. Use 3.14 for π.

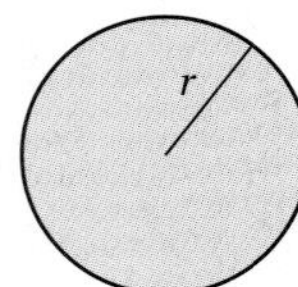

42. The area A of a square with sides of length s is given by $A = s^2$. Find the area of a square with sides of length 24 m.

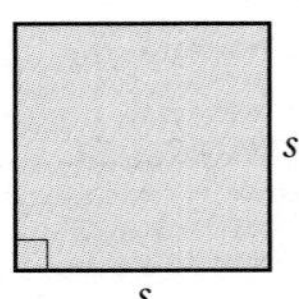

f Express using positive exponents. Then simplify.

43. 3^{-2}

44. 2^{-3}

45. 10^{-3}

46. 5^{-4}

47. 7^{-3}

48. 5^{-2}

49. a^{-3}

50. x^{-2}

51. $\frac{1}{8^{-2}}$

52. $\frac{1}{2^{-5}}$

53. $\frac{1}{y^{-4}}$

54. $\frac{1}{t^{-7}}$

55. $\frac{1}{z^{-n}}$

56. $\frac{1}{h^{-n}}$

Express using negative exponents.

57. $\frac{1}{4^3}$

58. $\frac{1}{5^2}$

59. $\frac{1}{x^3}$

60. $\frac{1}{y^2}$

61. $\frac{1}{a^5}$

62. $\frac{1}{b^7}$

d, **f** Multiply and simplify.

63. $2^4 \cdot 2^3$

64. $3^5 \cdot 3^2$

65. $8^5 \cdot 8^9$

66. $n^3 \cdot n^{20}$

67. $x^4 \cdot x^3$

68. $y^7 \cdot y^9$

69. $9^{17} \cdot 9^{21}$

70. $t^0 \cdot t^{16}$

71. $(3y)^4(3y)^8$

72. $(2t)^8(2t)^{17}$

73. $(7y)^1(7y)^{16}$

74. $(8x)^0(8x)^1$

75. $3^{-5} \cdot 3^8$

76. $5^{-8} \cdot 5^9$

77. $x^{-2} \cdot x$

78. $x \cdot x^{-1}$

79. $x^{14} \cdot x^3$

80. $x^9 \cdot x^4$

81. $x^{-7} \cdot x^{-6}$

82. $y^{-5} \cdot y^{-8}$

83. $a^{11} \cdot a^{-3} \cdot a^{-18}$

84. $a^{-11} \cdot a^{-3} \cdot a^{-7}$

85. $t^8 \cdot t^{-8}$

86. $m^{10} \cdot m^{-10}$

e, f Divide and simplify.

87. $\frac{7^5}{7^2}$

88. $\frac{5^8}{5^6}$

89. $\frac{8^{12}}{8^6}$

90. $\frac{8^{13}}{8^2}$

91. $\frac{y^9}{y^5}$

92. $\frac{x^{11}}{x^9}$

93. $\frac{16^2}{16^8}$

94. $\frac{7^2}{7^9}$

95. $\frac{m^6}{m^{12}}$

96. $\frac{a^3}{a^4}$

97. $\frac{(8x)^6}{(8x)^{10}}$

98. $\frac{(8t)^4}{(8t)^{11}}$

99. $\frac{(2y)^9}{(2y)^9}$

100. $\frac{(6y)^7}{(6y)^7}$

101. $\frac{x}{x^{-1}}$

102. $\frac{y^8}{y}$

103. $\frac{x^7}{x^{-2}}$

104. $\frac{t^8}{t^{-3}}$

105. $\frac{z^{-6}}{z^{-2}}$

106. $\frac{x^{-9}}{x^{-3}}$

107. $\frac{x^{-5}}{x^{-8}}$

108. $\frac{y^{-2}}{y^{-9}}$

109. $\frac{m^{-9}}{m^{-9}}$

110. $\frac{x^{-7}}{x^{-7}}$

Matching. In Exercises 111 and 112, match each item in the first column with the appropriate item in the second column by drawing connecting lines.

111.

5^2	$-\frac{1}{10}$
5^{-2}	$\frac{1}{10}$
$\left(\frac{1}{5}\right)^2$	$-\frac{1}{25}$
$\left(\frac{1}{5}\right)^{-2}$	10
-5^2	25
$(-5)^2$	-25
$-\left(-\frac{1}{5}\right)^2$	$\frac{1}{25}$
$\left(-\frac{1}{5}\right)^{-2}$	-10

112.

$-\left(\frac{1}{8}\right)^2$	16
$\left(\frac{1}{8}\right)^{-2}$	-16
8^{-2}	64
8^2	-64
-8^2	$\frac{1}{64}$
$(-8)^2$	$-\frac{1}{64}$
$\left(-\frac{1}{8}\right)^{-2}$	$-\frac{1}{16}$
$\left(-\frac{1}{8}\right)^2$	$\frac{1}{16}$

113. DW Suppose that the width of a square is three times the width of a second square. How do the areas of the squares compare? Why?

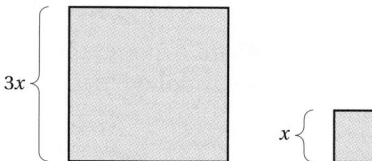

114. DW Suppose that the width of a cube is twice the width of a second cube. How do the volumes of the cubes compare? Why?

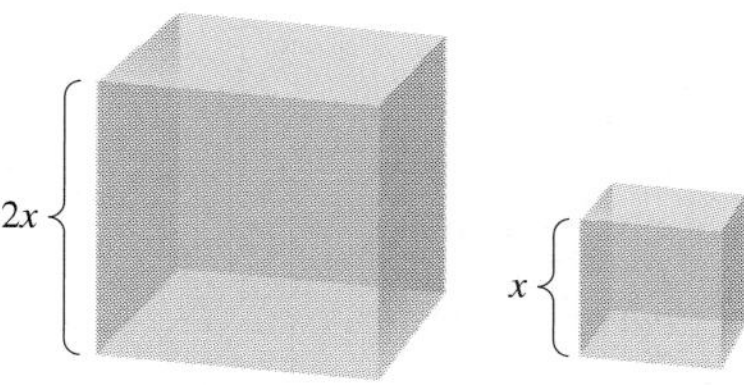

SKILL MAINTENANCE

Solve. [8.6a]

115. *Cutting a Submarine Sandwich.* A 12-in. submarine sandwich is cut into two pieces. One piece is twice as long as the other. How long are the pieces?

116. *Book Pages.* A book is opened. The sum of the page numbers on the facing pages is 457. Find the page numbers.

117. The perimeter of a rectangle is 640 ft. The length is 15 ft more than the width. Find the area of the rectangle.

118. The first angle of a triangle is 24° more than the second. The third angle is twice the first. Find the measures of the angles of the triangle.

Solve. [8.3c]

119. $-6(2 - x) + 10(5x - 7) = 10$

120. $-10(x - 4) = 5(2x + 5) - 7$

Factor. [7.7d]

121. $4x - 12 + 24y$

122. $256 - 2a - 4b$

SYNTHESIS

Determine whether each of the following is correct.

123. $(x + 1)^2 = x^2 + 1$

124. $(x - 1)^2 = x^2 - 2x + 1$

125. $(5x)^0 = 5x^0$

126. $\dfrac{x^3}{x^5} = x^2$

Simplify.

127. $(y^{2x})(y^{3x})$

128. $a^{5k} \div a^{3k}$

129. $\dfrac{a^{6t}(a^{7t})}{a^{9t}}$

130. $\dfrac{\left(\frac{1}{2}\right)^4}{\left(\frac{1}{2}\right)^5}$

131. $\dfrac{(0.8)^5}{(0.8)^3(0.8)^2}$

132. Determine whether $(a + b)^2$ and $a^2 + b^2$ are equivalent. (*Hint*: Choose values for a and b and evaluate.)

Use $>$, $<$, or $=$ for $\square$ to write a true sentence.

133. $3^5 \square 3^4$

134. $4^2 \square 4^3$

135. $4^3 \square 5^3$

136. $4^3 \square 3^4$

Evaluate.

137. $\dfrac{1}{-z^4}$, when $z = -10$

138. $\dfrac{1}{-z^5}$, when $z = -0.1$

10.2 EXPONENTS AND SCIENTIFIC NOTATION

Objectives

a Use the power rule to raise powers to powers.

b Raise a product to a power and a quotient to a power.

c Convert between scientific notation and decimal notation.

d Multiply and divide using scientific notation.

e Solve applied problems using scientific notation.

Simplify. Express the answers using positive exponents.

1. $(3^4)^5$

2. $(x^{-3})^4$

3. $(y^{-5})^{-3}$

4. $(x^4)^{-8}$

Answers on page A-34

We now enhance our ability to manipulate exponential expressions by considering three more rules. The rules are also applied to a new way to name numbers called *scientific notation*.

a Raising Powers to Powers

Consider an expression like $(3^2)^4$. We are raising 3^2 to the fourth power:

$$\begin{aligned}(3^2)^4 &= (3^2)(3^2)(3^2)(3^2)\\ &= (3\cdot 3)(3\cdot 3)(3\cdot 3)(3\cdot 3)\\ &= 3\cdot 3\cdot 3\cdot 3\cdot 3\cdot 3\cdot 3\cdot 3\\ &= 3^8.\end{aligned}$$

Note that in this case we could have multiplied the exponents:

$$(3^2)^4 = 3^{2\cdot 4} = 3^8.$$

Likewise, $(y^8)^3 = (y^8)(y^8)(y^8) = y^{24}$. Once again, we get the same result if we multiply the exponents:

$$(y^8)^3 = y^{8\cdot 3} = y^{24}.$$

THE POWER RULE

For any real number a and any integers m and n,

$$(a^m)^n = a^{mn}.$$

(To raise a power to a power, multiply the exponents.)

EXAMPLES Simplify. Express the answers using positive exponents.

1. $(3^5)^4 = 3^{5\cdot 4}$ Multiplying exponents
$= 3^{20}$

2. $(2^2)^5 = 2^{2\cdot 5} = 2^{10}$

3. $(y^{-5})^7 = y^{-5\cdot 7} = y^{-35} = \dfrac{1}{y^{35}}$

4. $(x^4)^{-2} = x^{4(-2)} = x^{-8} = \dfrac{1}{x^8}$

5. $(a^{-4})^{-6} = a^{(-4)(-6)} = a^{24}$

Do Exercises 1–4.

b Raising a Product or a Quotient to a Power

When an expression inside parentheses is raised to a power, the inside expression is the base. Let's compare $2a^3$ and $(2a)^3$:

$2a^3 = 2\cdot a\cdot a\cdot a;$ The base is a.

$(2a)^3 = (2a)(2a)(2a)$ The base is $2a$.

$= (2\cdot 2\cdot 2)(a\cdot a\cdot a)$ Using the associative and commutative laws of multiplication to regroup the factors

$= 2^3a^3$

$= 8a^3.$

We see that $2a^3$ and $(2a)^3$ are *not* equivalent. We also see that we can evaluate the power $(2a)^3$ by raising each factor to the power 3. This leads us to the following rule for raising a product to a power.

RAISING A PRODUCT TO A POWER

For any real numbers a and b and any integer n,

$$(ab)^n = a^n b^n.$$

(To raise a product to the nth power, raise each factor to the nth power.)

EXAMPLES Simplify.

6. $(4x^2)^3 = (4^1x^2)^3$ Since $4 = 4^1$

$= (4^1)^3 \cdot (x^2)^3$ Raising *each* factor to the third power

$= 4^3 \cdot x^6 = 64x^6$

7. $(5x^3y^5z^2)^4 = 5^4(x^3)^4(y^5)^4(z^2)^4$ Raising *each* factor to the fourth power

$= 625x^{12}y^{20}z^8$

8. $(-5x^4y^3)^3 = (-5)^3(x^4)^3(y^3)^3$

$= -125x^{12}y^9$

9. $[(-x)^{25}]^2 = (-x)^{50}$ Using the power rule

$= (-1 \cdot x)^{50}$ Using the property of -1 (Section 7.8)

$= (-1)^{50}x^{50}$

$= 1 \cdot x^{50}$ The product of an even number of negative factors is positive.

$= x^{50}$

10. $(5x^2y^{-2})^3 = 5^3(x^2)^3(y^{-2})^3 = 125x^6y^{-6}$ Be sure to raise *each* factor to the third power.

$= \dfrac{125x^6}{y^6}$

11. $(3x^3y^{-5}z^2)^4 = 3^4(x^3)^4(y^{-5})^4(z^2)^4 = 81x^{12}y^{-20}z^8 = \dfrac{81x^{12}z^8}{y^{20}}$

12. $(-x^4)^{-3} = (-1 \cdot x^4)^{-3} = (-1)^{-3} \cdot x^{4(-3)} = (-1)^{-3} \cdot x^{-12}$

$= \dfrac{1}{(-1)^3} \cdot \dfrac{1}{x^{12}} = \dfrac{1}{-1} \cdot \dfrac{1}{x^{12}} = -\dfrac{1}{x^{12}}$

13. $(-2x^{-5}y^4)^{-3} = (-2)^{-3}(x^{-5})^{-3}(y^4)^{-3} = \dfrac{1}{(-2)^3} \cdot x^{15} \cdot y^{-12}$

$= \dfrac{1}{-8} \cdot x^{15} \cdot \dfrac{1}{y^{12}} = -\dfrac{x^{15}}{8y^{12}}$

Do Exercises 5–11.

Simplify.

5. $(2x^5y^{-3})^4$

6. $(5x^5y^{-6}z^{-3})^2$

7. $[(-x)^{37}]^2$

8. $(3y^{-2}x^{-5}z^8)^3$

9. $(-y^8)^{-3}$

10. $(-2x^4)^{-3}$

11. $(-3x^2y^{-5})^{-3}$

Answers on page A-34

Simplify.

12. $\left(\frac{x^6}{5}\right)^2$

13. $\left(\frac{2t^5}{w^4}\right)^3$

14. $\left(\frac{x^4}{3}\right)^{-2}$

Do this two ways.

Answers on page A-34

There is a similar rule for raising a quotient to a power.

RAISING A QUOTIENT TO A POWER

For any real numbers a and b, $b \neq 0$, and any integer n,

$$\left(\frac{a}{b}\right)^n = \frac{a^n}{b^n}.$$

(To raise a quotient to the nth power, raise both the numerator and the denominator to the nth power.) Also,

$$\left(\frac{a}{b}\right)^{-n} = \left(\frac{b}{a}\right)^n = \frac{b^n}{a^n},\quad a \neq 0.$$

EXAMPLES Simplify.

14. $\left(\frac{x^2}{4}\right)^3 = \frac{(x^2)^3}{4^3} = \frac{x^6}{64}$

15. $\left(\frac{3a^4}{b^3}\right)^2 = \frac{(3a^4)^2}{(b^3)^2} = \frac{3^2(a^4)^2}{b^{3\cdot 2}} = \frac{9a^8}{b^6}$

16. $\left(\frac{y^3}{5}\right)^{-2} = \frac{(y^3)^{-2}}{5^{-2}} = \frac{y^{-6}}{5^{-2}} = \frac{\frac{1}{y^6}}{\frac{1}{5^2}} = \frac{1}{y^6} \div \frac{1}{5^2} = \frac{1}{y^6} \cdot \frac{5^2}{1} = \frac{25}{y^6}$

Example 16 might also be done as follows:

$$\left(\frac{y^3}{5}\right)^{-2} = \left(\frac{5}{y^3}\right)^2 \qquad \left(\frac{a}{b}\right)^{-n} = \left(\frac{b}{a}\right)^n$$

$$= \frac{5^2}{(y^3)^2} = \frac{25}{y^6}.$$

Do Exercises 12–14.

C Scientific Notation

There are many kinds of symbols, or notation, for numbers. You are already familiar with fraction notation, decimal notation, and percent notation. Now we study another, **scientific notation,** which makes use of exponential notation. Scientific notation is especially useful when calculations involve very large or very small numbers. The following are examples of scientific notation.

①

②

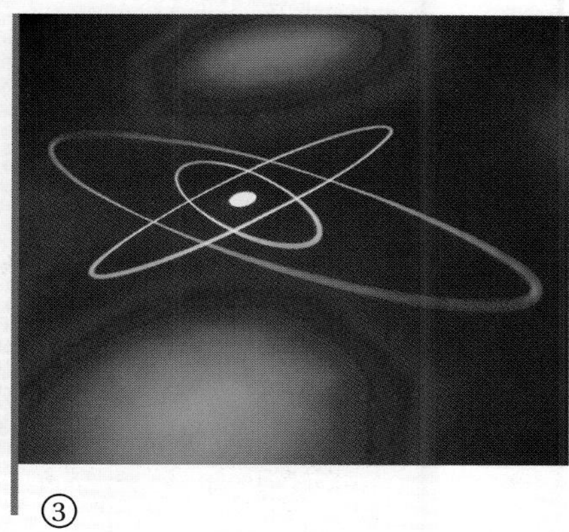
③

① *Niagara Falls*: On the Canadian side, during the summer the amount of water that spills over the falls in 1 day is about

4.9793×10^{10} gal = 49,793,000,000 gal.

② *The mass of the earth*: 6.615×10^{21} = 6,615,000,000,000,000,000,000 tons.

③ *The mass of a hydrogen atom*:

1.7×10^{-24} g = 0.0000000000000000000000017 g.

SCIENTIFIC NOTATION

Scientific notation for a number is an expression of the type

$$M \times 10^n,$$

where n is an integer, M is greater than or equal to 1 and less than 10 ($1 \le M < 10$), and M is expressed in decimal notation. 10^n is also considered to be scientific notation when $M = 1$.

You should try to make conversions to scientific notation mentally as much as possible. Here is a handy mental device.

A positive exponent in scientific notation indicates a large number (greater than or equal to 10) and a negative exponent indicates a small number (between 0 and 1).

EXAMPLES Convert to scientific notation.

17. $78{,}000 = 7.8 \times 10^4$

7.8,000.
4 places

Large number, so the exponent is positive.

18. $0.0000057 = 5.7 \times 10^{-6}$

0.000005.7
6 places

Small number, so the exponent is negative.

Caution!

Each of the following is *not* scientific notation.

$\underbrace{12.46} \times 10^7$

This number is greater than 10.

$\underbrace{0.347} \times 10^{-5}$

This number is less than 1.

Do Exercises 15 and 16 on the following page.

Convert to scientific notation.

15. 0.000517

16. 523,000,000

Convert to decimal notation.

17. 6.893×10^{11}

18. 5.67×10^{-5}

Multiply and write scientific notation for the result.

19. $(1.12 \times 10^{-8})(5 \times 10^{-7})$

20. $(9.1 \times 10^{-17})(8.2 \times 10^{3})$

Answers on page A-34

EXAMPLES Convert mentally to decimal notation.

19. $7.893 \times 10^5 = 789{,}300$

7.89300.

5 places

Positive exponent, so the answer is a large number.

20. $4.7 \times 10^{-8} = 0.000000047$

.00000004.7

8 places

Negative exponent, so the answer is a small number.

Do Exercises 17 and 18.

d Multiplying and Dividing Using Scientific Notation

MULTIPLYING

Consider the product

$$400 \cdot 2000 = 800{,}000.$$

In scientific notation, this is

$$(4 \times 10^2) \cdot (2 \times 10^3) = (4 \cdot 2)(10^2 \cdot 10^3) = 8 \times 10^5.$$

By applying the commutative and associative laws, we can find this product by multiplying $4 \cdot 2$, to get 8, and $10^2 \cdot 10^3$, to get 10^5 (we do this by adding the exponents).

EXAMPLE 21 Multiply: $(1.8 \times 10^6) \cdot (2.3 \times 10^{-4})$.

We apply the commutative and associative laws to get

$$\begin{aligned}(1.8 \times 10^6) \cdot (2.3 \times 10^{-4}) &= (1.8 \cdot 2.3) \times (10^6 \cdot 10^{-4}) \\ &= 4.14 \times 10^{6+(-4)} \\ &= 4.14 \times 10^2.\end{aligned}$$

We get 4.14 by multiplying 1.8 and 2.3. We get 10^2 by adding the exponents 6 and -4.

EXAMPLE 22 Multiply: $(3.1 \times 10^5) \cdot (4.5 \times 10^{-3})$.

We have

$$\begin{aligned}(3.1 \times 10^5) \cdot (4.5 \times 10^{-3}) &= (3.1 \times 4.5)(10^5 \cdot 10^{-3}) \\ &= 13.95 \times 10^2 && \text{Not scientific notation. 13.95 is greater than 10.} \\ &= (1.395 \times 10^1) \times 10^2 && \text{Substituting } 1.395 \times 10^1 \text{ for 13.95} \\ &= 1.395 \times (10^1 \times 10^2) && \text{Associative law} \\ &= 1.395 \times 10^3. && \text{Adding exponents. The answer is now in scientific notation.}\end{aligned}$$

Do Exercises 19 and 20.

DIVIDING

Consider the quotient

$$800{,}000 \div 400 = 2000.$$

In scientific notation, this is

$$(8 \times 10^5) \div (4 \times 10^2) = \frac{8 \times 10^5}{4 \times 10^2} = \frac{8}{4} \times \frac{10^5}{10^2} = 2 \times 10^3.$$

We can find this product by dividing 8 by 4, to get 2, and 10^5 by 10^2, to get 10^3 (we do this by subtracting the exponents.)

EXAMPLE 23 Divide: $(3.41 \times 10^5) \div (1.1 \times 10^{-3})$.

We have

$$(3.41 \times 10^5) \div (1.1 \times 10^{-3}) = \frac{3.41 \times 10^5}{1.1 \times 10^{-3}} = \frac{3.41}{1.1} \times \frac{10^5}{10^{-3}}$$
$$= 3.1 \times 10^{5-(-3)}$$
$$= 3.1 \times 10^8.$$

Divide and write scientific notation for the result.

21. $\dfrac{4.2 \times 10^5}{2.1 \times 10^2}$

22. $\dfrac{1.1 \times 10^{-4}}{2.0 \times 10^{-7}}$

Answers on page A-34

CALCULATOR CORNER

1.789E−11
1.789E−11

NORMAL **SCI** ENG
FLOAT 0123456789
RADIAN DEGREE
FUNC PAR POL SEQ
CONNECTED DOT
SEQUENTIAL SIMUL
REAL a+bi re^θi
FULL HORIZ G−T

1.8E6*2.3E−4
4.14E2

On a scientific calculator, the answer may appear as follows.

4.14 02

Scientific Notation To enter a number in scientific notation on a graphing calculator, we first type the decimal portion of the number and then press 2ND EE. (EE is the second operation associated with the , key.) Finally, we type the exponent, which can be at most two digits. For example, to enter 1.789×10^{-11} in scientific notation, we press 1 . 7 8 9 2ND EE (−) 1 1 ENTER. The decimal portion of the number appears before a small E and the exponent follows the E.

The graphing calculator can be used to perform computations using scientific notation. To find the product in Example 21 and express the result in scientific notation, we first set the calculator in Scientific mode by pressing MODE, positioning the cursor over Sci on the first line, and pressing ENTER. Then we press 2ND QUIT to go to the home screen and enter the computation by pressing 1 . 8 2ND EE 6 × 2 . 3 2ND EE (−) 4 ENTER.

Exercises: Multiply or divide and express the answer in scientific notation.

1. $(3.15 \times 10^7)(4.3 \times 10^{-12})$
2. $(4.76 \times 10^{-5})(1.9 \times 10^{10})$
3. $(8 \times 10^9)(4 \times 10^{-5})$
4. $(4 \times 10^4)(9 \times 10^7)$
5. $\dfrac{4.5 \times 10^6}{1.5 \times 10^{12}}$
6. $\dfrac{6.4 \times 10^{-5}}{1.6 \times 10^{-10}}$
7. $\dfrac{4 \times 10^{-9}}{5 \times 10^{16}}$
8. $\dfrac{9 \times 10^{11}}{3 \times 10^{-2}}$

23. Niagara Falls Water Flow. On the Canadian side, during the summer the amount of water that spills over the falls in 1 min is about

$$1.3088 \times 10^8 \text{ L}.$$

How much water spills over the falls in one day? Express the answer in scientific notation.

Answer on page A-34

Study Tips

WRITING ALL THE STEPS

Take the time to include all the steps when working your homework problems. Doing so will help you organize your thinking and avoid computational errors. If you find a wrong answer, having all the steps allows easier checking of your work. It will also give you complete, step-by-step solutions of the exercises that can be used to study for an exam.

Writing down all the steps and keeping your work organized may also give you a better chance of getting partial credit.

"Success comes before work only in the dictionary."

Anonymous

EXAMPLE 24 Divide: $(6.4 \times 10^{-7}) \div (8.0 \times 10^6)$.

We have

$$\begin{aligned}
(6.4 \times 10^{-7}) \div (8.0 \times 10^6) &= \frac{6.4 \times 10^{-7}}{8.0 \times 10^6} \\
&= \frac{6.4}{8.0} \times \frac{10^{-7}}{10^6} \\
&= 0.8 \times 10^{-7-6} \\
&= 0.8 \times 10^{-13} && \text{Not scientific notation. 0.8 is less than 1.} \\
&= (8.0 \times 10^{-1}) \times 10^{-13} && \text{Substituting } 8.0 \times 10^{-1} \text{ for } 0.8 \\
&= 8.0 \times (10^{-1} \times 10^{-13}) && \text{Associative law} \\
&= 8.0 \times 10^{-14}. && \text{Adding exponents}
\end{aligned}$$

Do Exercises 21 and 22 on the preceding page.

e Applications with Scientific Notation

EXAMPLE 25 *Distance from the Sun to Earth.* Light from the sun traveling at a rate of 300,000 kilometers per second (km/s) reaches Earth in 499 sec. Find the distance, expressed in scientific notation, from the sun to Earth.

The time t that it takes for light to reach Earth from the sun is 4.99×10^2 sec (s). The speed is 3.0×10^5 km/s. Recall that distance can be expressed in terms of speed and time as

$$\text{Distance} = \text{Speed} \cdot \text{Time}$$
$$d = rt.$$

We substitute 3.0×10^5 for r and 4.99×10^2 for t:

$$\begin{aligned}
d &= rt \\
&= (3.0 \times 10^5)(4.99 \times 10^2) && \text{Substituting} \\
&= 14.97 \times 10^7 \\
&= (1.497 \times 10^1) \times 10^7 \\
&= 1.497 \times (10^1 \times 10^7) \\
&= 1.497 \times 10^8 \text{ km}. && \text{Converting to scientific notation}
\end{aligned}$$

Thus the distance from the sun to Earth is 1.497×10^8 km.

Do Exercise 23.

EXAMPLE 26 *DNA.* A strand of DNA (deoxyribonucleic acid) is about 150 cm long and 1.3×10^{-10} cm wide. How many times longer is DNA than it is wide?

Source: Human Genome Project Information

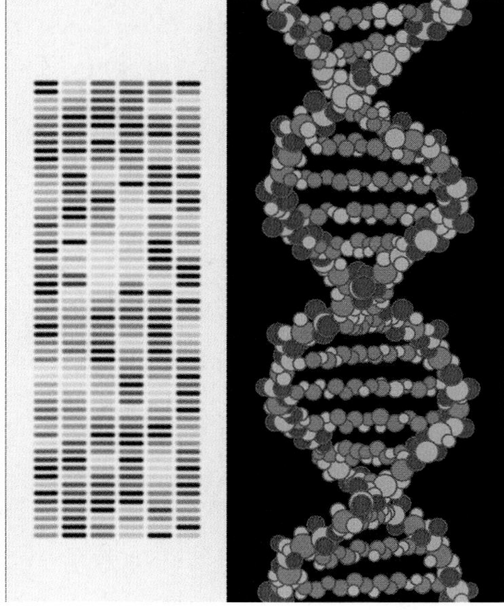

To determine how many times longer (N) DNA is than it is wide, we divide the length by the width:

$$N = \frac{150}{1.3 \times 10^{-10}} = \frac{150}{1.3} \times \frac{1}{10^{-10}}$$
$$\approx 115.385 \times 10^{10}$$
$$= (1.15385 \times 10^{2}) \times 10^{10}$$
$$= 1.15385 \times 10^{12}.$$

Thus the length of DNA is about 1.15385×10^{12} times its width.

Do Exercise 24.

The following is a summary of the definitions and rules for exponents that we have considered in this section and the preceding one.

DEFINITIONS AND RULES FOR EXPONENTS

Exponent of 1:	$a^1 = a$
Exponent of 0:	$a^0 = 1, a \neq 0$
Negative exponents:	$a^{-n} = \frac{1}{a^n}, \frac{1}{a^{-n}} = a^n, a \neq 0$
Product Rule:	$a^m \cdot a^n = a^{m+n}$
Quotient Rule:	$\frac{a^m}{a^n} = a^{m-n}, a \neq 0$
Power Rule:	$(a^m)^n = a^{mn}$
Raising a product to a power:	$(ab)^n = a^n b^n$
Raising a quotient to a power:	$\left(\frac{a}{b}\right)^n = \frac{a^n}{b^n}, b \neq 0;$
	$\left(\frac{a}{b}\right)^{-n} = \frac{b^n}{a^n}, b \neq 0, a \neq 0$
Scientific notation:	$M \times 10^n$, or 10^n, where $1 \leq M < 10$

24. Earth vs. Saturn. The mass of Earth is about 6×10^{21} metric tons. The mass of Saturn is about 5.7×10^{23} metric tons. About how many times the mass of Earth is the mass of Saturn? Express the answer in scientific notation.

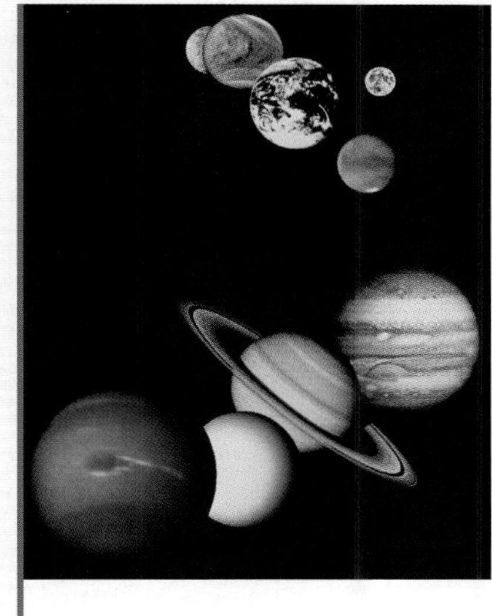

Answer on page A-34

10.2 EXERCISE SET

For Extra Help

MathXL | MyMathLab | InterAct Math | Math Tutor Center | Video Lectures on CD Disc 5 | Student's Solutions Manual

a, **b** Simplify.

1. $(2^3)^2$

2. $(5^2)^4$

3. $(5^2)^{-3}$

4. $(7^{-3})^5$

5. $(x^{-3})^{-4}$

6. $(a^{-5})^{-6}$

7. $(a^{-2})^9$

8. $(x^{-5})^6$

9. $(t^{-3})^{-6}$

10. $(a^{-4})^{-7}$

11. $(t^4)^{-3}$

12. $(t^5)^{-2}$

13. $(x^{-2})^{-4}$

14. $(t^{-6})^{-5}$

15. $(ab)^3$

16. $(xy)^2$

17. $(ab)^{-3}$

18. $(xy)^{-6}$

19. $(mn^2)^{-3}$

20. $(x^3y)^{-2}$

21. $(4x^3)^2$

22. $4(x^3)^2$

23. $(3x^{-4})^2$

24. $(2a^{-5})^3$

25. $(x^4y^5)^{-3}$

26. $(t^5x^3)^{-4}$

27. $(x^{-6}y^{-2})^{-4}$

28. $(x^{-2}y^{-7})^{-5}$

29. $(a^{-2}b^7)^{-5}$

30. $(q^5r^{-1})^{-3}$

31. $(5r^{-4}t^3)^2$

32. $(4x^5y^{-6})^3$

33. $(a^{-5}b^7c^{-2})^3$

34. $(x^{-4}y^{-2}z^9)^2$

35. $(3x^3y^{-8}z^{-3})^2$

36. $(2a^2y^{-4}z^{-5})^3$

37. $(-4x^3y^{-2})^2$

38. $(-8x^3y^{-2})^3$

39. $(-a^{-3}b^{-2})^{-4}$

40. $(-p^{-4}q^{-3})^{-2}$

41. $\left(\frac{y^3}{2}\right)^2$

42. $\left(\frac{a^5}{3}\right)^3$

43. $\left(\frac{a^2}{b^3}\right)^4$

44. $\left(\frac{x^3}{y^4}\right)^5$

45. $\left(\frac{y^2}{2}\right)^{-3}$

46. $\left(\frac{a^4}{3}\right)^{-2}$

47. $\left(\frac{7}{x^{-3}}\right)^2$

48. $\left(\frac{3}{a^{-2}}\right)^3$

49. $\left(\frac{x^2y}{z}\right)^3$

50. $\left(\frac{m}{n^4p}\right)^3$

51. $\left(\frac{a^2b}{cd^3}\right)^{-2}$

52. $\left(\frac{2a^2}{3b^4}\right)^{-3}$

c Convert to scientific notation.

53. 28,000,000,000

54. 4,900,000,000,000

55. 907,000,000,000,000,000

56. 168,000,000,000,000

57. 0.00000304

58. 0.000000000865

59. 0.000000018

60. 0.00000000002

61. 100,000,000,000

62. 0.0000001

Convert the number in the sentence to scientific notation.

63. *Population of the United States.* As of July 2005, the population of the United States was about 296 million (1 million $= 10^6$).
Source: U.S. Bureau of the Census

64. *NASCAR.* Total revenue of NASCAR (National Association of Stock Car Automobile Racing) is expected to be $3423 million by 2006.
Source: NASCAR

65. *State Lottery.* Typically, the probability of winning a state lottery is about 1/10,000,000.

66. *Cancer Death Rate.* In Michigan, the death rate due to cancer is about 127.1/1000.
Source: AARP

Convert to decimal notation.

67. 8.74×10^7

68. 1.85×10^8

69. 5.704×10^{-8}

70. 8.043×10^{-4}

71. 10^7

72. 10^6

73. 10^{-5}

74. 10^{-8}

d Multiply or divide and write scientific notation for the result.

75. $(3 \times 10^4)(2 \times 10^5)$

76. $(3.9 \times 10^8)(8.4 \times 10^{-3})$

77. $(5.2 \times 10^5)(6.5 \times 10^{-2})$

78. $(7.1 \times 10^{-7})(8.6 \times 10^{-5})$

79. $(9.9 \times 10^{-6})(8.23 \times 10^{-8})$

80. $(1.123 \times 10^4) \times 10^{-9}$

81. $\dfrac{8.5 \times 10^8}{3.4 \times 10^{-5}}$

82. $\dfrac{5.6 \times 10^{-2}}{2.5 \times 10^5}$

83. $(3.0 \times 10^6) \div (6.0 \times 10^9)$

84. $(1.5 \times 10^{-3}) \div (1.6 \times 10^{-6})$

85. $\dfrac{7.5 \times 10^{-9}}{2.5 \times 10^{12}}$

86. $\dfrac{4.0 \times 10^{-3}}{8.0 \times 10^{20}}$

Solve.

87. *River Discharge.* The average discharge at the mouths of the Amazon River is 4,200,000 cubic feet per second. How much water is discharged from the Amazon River in 1 yr? Express the answer in scientific notation.

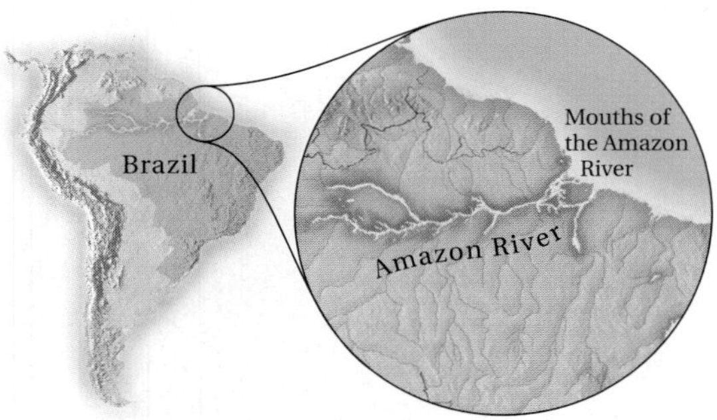

88. *Computers.* A gigabyte is a measure of a computer's storage capacity. One gigabyte holds about one billion bytes of information. If a firm's computer network contains 2500 gigabytes of memory, how many bytes are in the network? Express the answer in scientific notation.

89. *Earth vs. Jupiter.* The mass of Earth is about 6×10^{21} metric tons. The mass of Jupiter is about 1.908×10^{24} metric tons. About how many times the mass of Earth is the mass of Jupiter? Express the answer in scientific notation.

90. *Water Contamination.* In the United States, 200 million gal of used motor oil is improperly disposed of each year. One gallon of used oil can contaminate one million gallons of drinking water. How many gallons of drinking water can 200 million gallons of oil contaminate? Express the answer in scientific notation.

Source: *The Macmillan Visual Almanac*

91. *Stars.* It is estimated that there are 10 billion trillion stars in the known universe. Express the number of stars in scientific notation (1 billion $= 10^9$; 1 trillion $= 10^{12}$).

92. *Closest Star.* Excluding the sun, the closest star to Earth is Proxima Centauri, which is 4.3 light-years away (one light-year $= 5.88 \times 10^{12}$ mi). How far, in miles, is Proxima Centauri from Earth? Express the answer in scientific notation.

93. *Earth vs. Sun.* The mass of Earth is about 6×10^{21} metric tons. The mass of the sun is about 1.998×10^{27} metric tons. About how many times the mass of Earth is the mass of the sun? Express the answer in scientific notation.

94. *Red Light.* The wavelength of light is given by the velocity divided by the frequency. The velocity of red light is 300,000,000 m/sec, and its frequency is 400,000,000,000,000 cycles per second. What is the wavelength of red light? Express the answer in scientific notation.

Space Travel. Use the following information for Exercises 95 and 96.

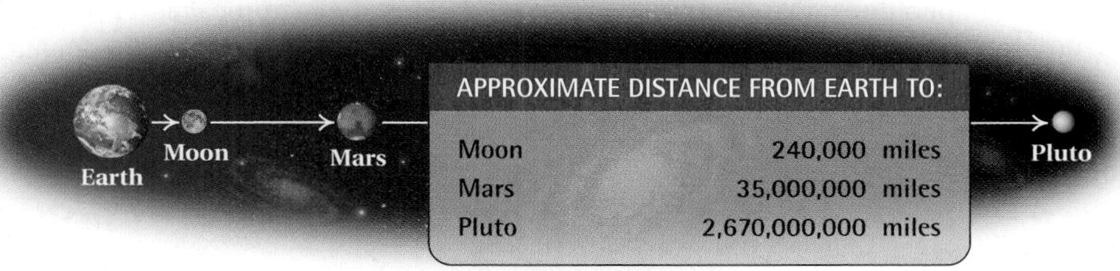

APPROXIMATE DISTANCE FROM EARTH TO:	
Moon	240,000 miles
Mars	35,000,000 miles
Pluto	2,670,000,000 miles

95. *Time to Reach Mars.* Suppose that it takes about 3 days for a space vehicle to travel from Earth to the moon. About how long would it take the same space vehicle traveling at the same speed to reach Mars? Express the answer in scientific notation.

96. *Time to Reach Pluto.* Suppose that it takes about 3 days for a space vehicle to travel from Earth to the moon. About how long would it take the same space vehicle traveling at the same speed to reach Pluto? Express the answer in scientific notation.

97. D_W Explain in your own words when exponents should be added and when they should be multiplied.

98. D_W Without performing actual computations, explain why 3^{-29} is smaller than 2^{-29}.

SKILL MAINTENANCE

Factor. [7.7d]

99. $9x - 36$

100. $4x - 2y + 16$

101. $3s + 3t + 24$

102. $-7x - 14$

Solve. [8.3b]

103. $2x - 4 - 5x + 8 = x - 3$

104. $8x + 7 - 9x = 12 - 6x + 5$

Solve. [8.3c]

105. $8(2x + 3) - 2(x - 5) = 10$

106. $4(x - 3) + 5 = 6(x + 2) - 8$

Graph. [9.1d], [9.2a]

107. $y = x - 5$

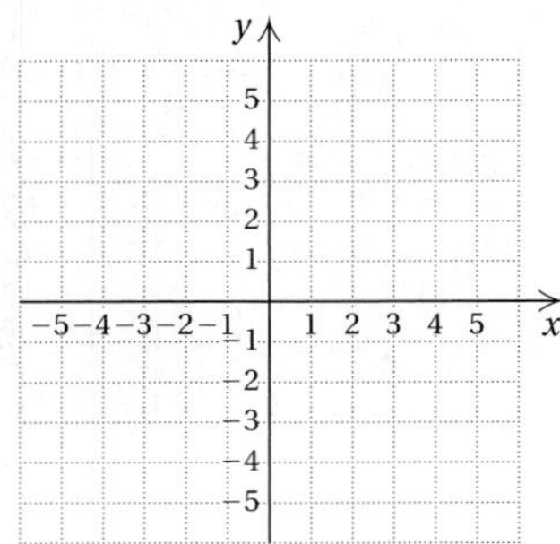

108. $2x + y = 8$

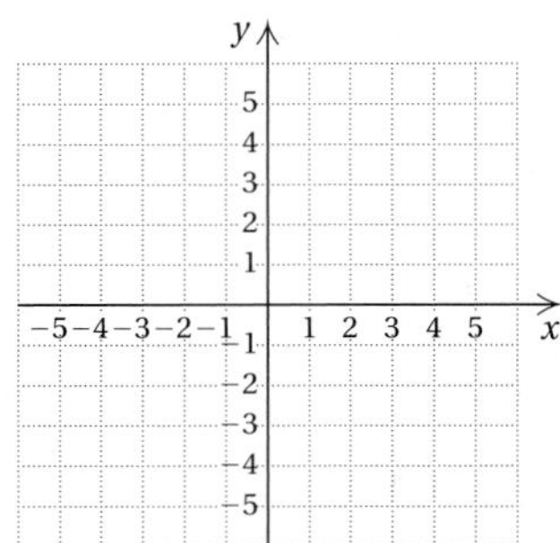

SYNTHESIS

109. Carry out the indicated operations. Express the result in scientific notation.

$$\frac{(5.2 \times 10^6)(6.1 \times 10^{-11})}{1.28 \times 10^{-3}}$$

110. Find the reciprocal and express it in scientific notation.

$$6.25 \times 10^{-3}$$

Simplify.

111. $\frac{(5^{12})^2}{5^{25}}$

112. $\frac{a^{22}}{(a^2)^{11}}$

113. $\frac{(3^5)^4}{3^5 \cdot 3^4}$

114. $\left(\frac{5x^{-2}}{3y^{-2}z}\right)^0$

115. $\frac{49^{18}}{7^{35}}$

116. $\left(\frac{1}{a}\right)^{-n}$

117. $\frac{(0.4)^5}{[(0.4^3)]^2}$

118. $\left(\frac{4a^3b^{-2}}{5c^{-3}}\right)^1$

Determine whether each of the following is true for all pairs of integers m and n and all positive numbers x and y.

119. $x^m \cdot y^n = (xy)^{mn}$

120. $x^m \cdot y^m = (xy)^{2m}$

121. $(x - y)^m = x^m - y^m$

122. $-x^m = (-x)^m$

123. $(-x)^{2m} = x^{2m}$

124. $x^{-m} = \frac{-1}{x^m}$

10.3 INTRODUCTION TO POLYNOMIALS

Objectives

a. Evaluate a polynomial for a given value of the variable.

b. Identify the terms of a polynomial.

c. Identify the like terms of a polynomial.

d. Identify the coefficients of a polynomial.

e. Collect the like terms of a polynomial.

f. Arrange a polynomial in descending order, or collect the like terms and then arrange in descending order.

g. Identify the degree of each term of a polynomial and the degree of the polynomial.

h. Identify the missing terms of a polynomial.

i. Classify a polynomial as a monomial, binomial, trinomial, or none of these.

We have already learned to evaluate and to manipulate certain kinds of algebraic expressions. We will now consider algebraic expressions called *polynomials.*

The following are examples of *monomials in one variable*:

$$3x^2, \quad 2x, \quad -5, \quad 37p^4, \quad 0.$$

Each expression is a constant or a constant times some variable to a nonnegative integer power.

MONOMIAL

A **monomial** is an expression of the type ax^n, where a is a real-number constant and n is a nonnegative integer.

Algebraic expressions like the following are **polynomials:**

$$\tfrac{3}{4}y^5, \quad -2, \quad 5y + 3, \quad 3x^2 + 2x - 5, \quad -7a^3 + \tfrac{1}{2}a, \quad 6x, \quad 37p^4, \quad x, \quad 0.$$

POLYNOMIAL

A **polynomial** is a monomial or a combination of sums and/or differences of monomials.

The following algebraic expressions are *not* polynomials:

$$\textbf{(1)}\ \frac{x+3}{x-4}, \qquad \textbf{(2)}\ 5x^3 - 2x^2 + \frac{1}{x}, \qquad \textbf{(3)}\ \frac{1}{x^3 - 2}.$$

Expressions (1) and (3) are not polynomials because they represent quotients, not sums or differences. Expression (2) is not a polynomial because

$$\frac{1}{x} = x^{-1},$$

and this is not a monomial because the exponent is negative.

Do Exercise 1.

1. Write three polynomials.

Answer on page A-35

a Evaluating Polynomials and Applications

When we replace the variable in a polynomial with a number, the polynomial then represents a number called a **value** of the polynomial. Finding that number, or value, is called **evaluating the polynomial.** We evaluate a polynomial using the rules for order of operations (Section 7.8).

EXAMPLE 1 Evaluate the polynomial when $x = 2$.

a) $3x + 5 = 3 \cdot 2 + 5$
$= 6 + 5$
$= 11$

b) $2x^2 - 7x + 3 = 2 \cdot 2^2 - 7 \cdot 2 + 3$
$= 2 \cdot 4 - 7 \cdot 2 + 3$
$= 8 - 14 + 3$
$= -3$

Evaluate the polynomial when $x = 3$.

2. $-4x - 7$

3. $-5x^3 + 7x + 10$

Evaluate the polynomial when $x = -5$.

4. $5x + 7$

5. $2x^2 + 5x - 4$

6. Use *only* the graph shown in Example 3 to evaluate the polynomial $2x - 2$ when $x = 4$ and when $x = -1$.

7. Referring to Example 4, determine the total number of games to be played in a league of 12 teams.

8. Perimeter of a Baseball Diamond. The perimeter P of a square of side x is given by the polynomial equation $P = 4x$.

A baseball diamond is a square 90 ft on a side. Find the perimeter of a baseball diamond.

Answers on page A-35

EXAMPLE 2 Evaluate the polynomial when $x = -4$.

a) $2 - x^3 = 2 - (-4)^3 = 2 - (-64)$
$= 2 + 64$
$= 66$

b) $-x^2 - 3x + 1 = -(-4)^2 - 3(-4) + 1$
$= -16 + 12 + 1$
$= -3$

Do Exercises 2–5.

AG *ALGEBRAIC–GRAPHICAL CONNECTION*

An equation like $y = 2x - 2$, which has a polynomial on one side and y on the other, is called a **polynomial equation.** Here and in many places throughout the book, we will connect graphs to related concepts.

Recall from Chapter 9 that in order to plot points before graphing an equation, we choose values for x and compute the corresponding y-values. If the equation has y on one side and a polynomial involving x on the other, then determining y is the same as evaluating the polynomial. Once the graph of such an equation has been drawn, we can evaluate the polynomial for a given x-value by finding the y-value that is paired with it on the graph.

EXAMPLE 3 Use *only* the given graph of $y = 2x - 2$ to evaluate the polynomial $2x - 2$ when $x = 3$.

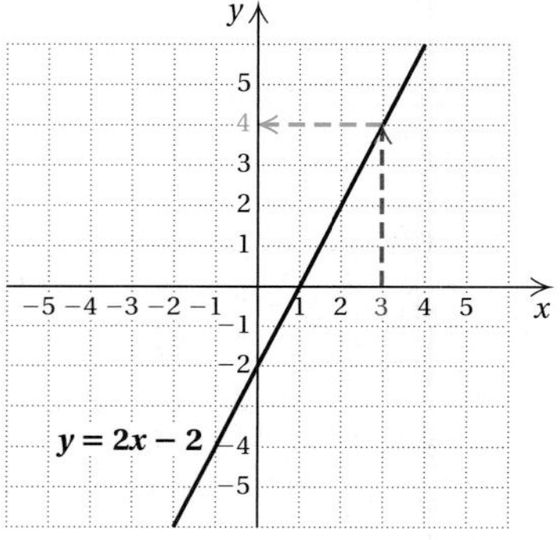

First, we locate 3 on the x-axis. From there we move vertically to the graph of the equation and then horizontally to the y-axis. There we locate the y-value that is paired with 3. Although our drawing may not be precise, it appears that the y-value 4 is paired with 3. Thus the value of $2x - 2$ is 4 when $x = 3$.

Do Exercise 6.

Polynomial equations can be used to model many real-world situations.

EXAMPLE 4 *Games in a Sports League.* In a sports league of x teams in which each team plays every other team twice, the total number of games N to be played is given by the polynomial equation

$$N = x^2 - x.$$

A women's slow-pitch softball league has 10 teams. What is the total number of games to be played?

We evaluate the polynomial when $x = 10$:

$$N = x^2 - x = 10^2 - 10 = 100 - 10 = 90.$$

The league plays 90 games.

Do Exercises 7 and 8.

EXAMPLE 5 *Medical Dosage.* The concentration C, in parts per million, of a certain antibiotic in the bloodstream after t hours is given by the polynomial equation

$$C = -0.05t^2 + 2t + 2.$$

Find the concentration after 2 hr.

To find the concentration after 2 hr, we evaluate the polynomial when $t = 2$:

$$\begin{aligned} C &= -0.05t^2 + 2t + 2 \\ &= -0.05(2)^2 + 2(2) + 2 \\ &= -0.05(4) + 2(2) + 2 \\ &= -0.2 + 4 + 2 \\ &= -0.2 + 6 \\ &= 5.8. \end{aligned}$$

Carrying out the calculation using the rules for order of operations

The concentration after 2 hr is 5.8 parts per million.

9. Medical Dosage.

a) Referring to Example 5, determine the concentration after 3 hr by evaluating the polynomial when $t = 3$.

b) Use *only* the graph showing medical dosage to check the value found in part (a).

AG *ALGEBRAIC–GRAPHICAL CONNECTION*

The polynomial equation in Example 5 can be graphed if we evaluate the polynomial for several values of t. We list the values in a table and show the graph below. Note that the concentration peaks at the 20-hr mark and after slightly more than 40 hr, the concentration is 0. Since neither time nor concentration can be negative, our graph uses only the first quadrant.

	C
t	$C = -0.05t^2 + 2t + 2$
0	2
2	5.8 ← Example 5
10	17
20	22
30	17

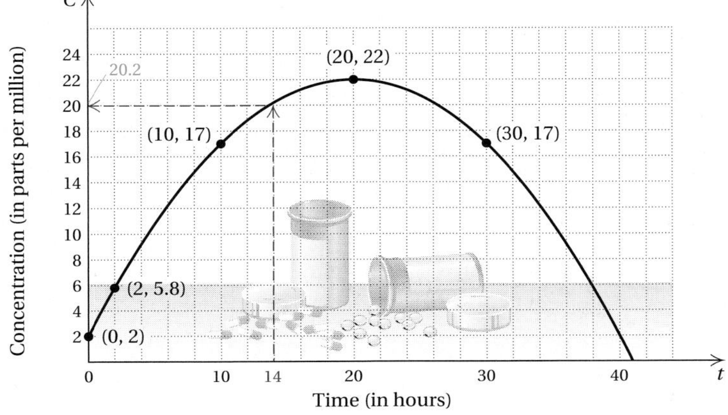

10. Medical Dosage. Referring to Example 5, use *only* the graph showing medical dosage to estimate the value of the polynomial when $t = 26$.

Do Exercises 9 and 10.

Answers on page A-35

CALCULATOR CORNER

Evaluating Polynomials (*Note*: If you set your graphing calculator in Sci (scientific) mode to do the exercises in Section 10.2, return it to Normal mode now.)

There are several ways to evaluate polynomials on a graphing calculator. One method uses a table. To evaluate the polynomial in Example 2(b), $-x^2 - 3x + 1$, when $x = -4$, we first enter $y_1 = -x^2 - 3x + 1$ on the equation-editor screen. Then we set up a table in ASK mode (see p. 571) and enter the value -4 for x. We see that when $x = -4$, the value of Y1 is -3. This is the value of the polynomial when $x = -4$.

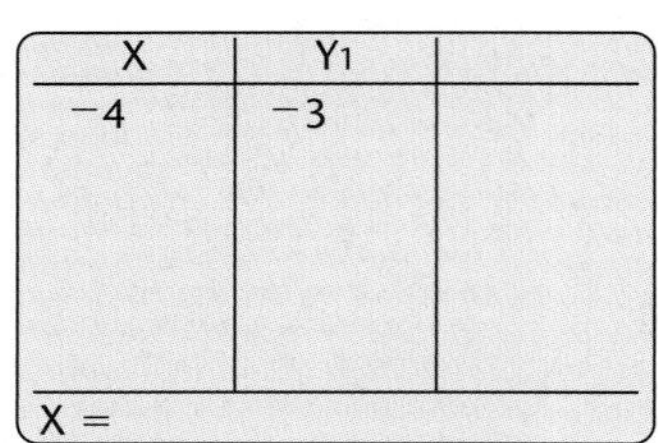

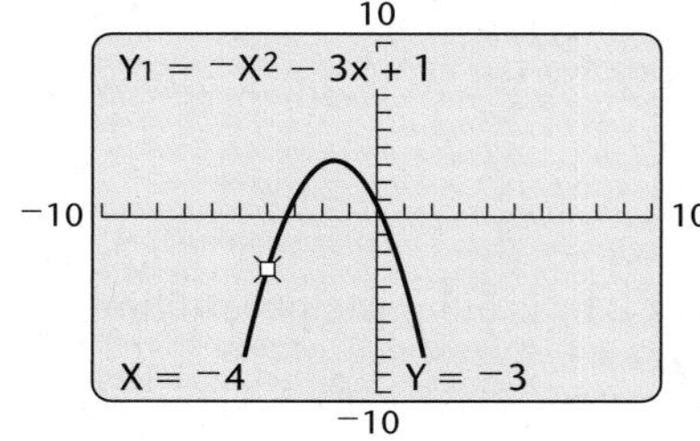

We can also use the Value feature from the CALC menu to evaluate this polynomial. First, we graph $y_1 = -x^2 - 3x + 1$ in a window that includes the x-value -4. We will use the standard window (see p. 647). Then we press 2ND CALC 1 or 2ND CALC ENTER to access the CALC menu and select item 1, Value. Now we supply the desired x-value by pressing (−) 4. We then press ENTER to see X = −4, Y = −3 at the bottom of the screen. Thus, when $x = -4$, the value of $-x^2 - 3x + 1$ is -3.

Exercises: Use the Value feature to evaluate the polynomial for the given values of x.

1. $-x^2 - 3x + 1$, when $x = -2$, $x = -0.5$, and $x = 4$
2. $3x^2 - 5x + 2$, when $x = -3$, $x = 1$, and $x = 2.6$
3. $2x^2 - x - 8$, when $x = -3$, $x = 1.8$, and $x = 3$
4. $-5x^2 + 3x + 7$, when $x = -1$, $x = 2$, and $x = 3.4$

Find an equivalent polynomial using only additions.

11. $-9x^3 - 4x^5$

12. $-2y^3 + 3y^7 - 7y - 9$

Answers on page A-35

b Identifying Terms

As we saw in Section 7.4, subtractions can be rewritten as additions. For any polynomial that has some subtractions, we can find an equivalent polynomial using only additions.

EXAMPLES Find an equivalent polynomial using only additions.

6. $-5x^2 - x = -5x^2 + (-x)$

7. $4x^5 - 2x^6 + 4x - 7 = 4x^5 + (-2x^6) + 4x + (-7)$

Do Exercises 11 and 12.

When a polynomial has only additions, the monomials being added are called **terms.** In Example 6, the terms are $-5x^2$ and $-x$. In Example 7, the terms are $4x^5$, $-2x^6$, $4x$, and -7.

EXAMPLE 8 Identify the terms of the polynomial

$$4x^7 + 3x + 12 + 8x^3 + 5x.$$

Terms: $4x^7$, $3x$, 12, $8x^3$, and $5x$.

If there are subtractions, you can *think* of them as additions without rewriting.

EXAMPLE 9 Identify the terms of the polynomial

$$3t^4 - 5t^6 - 4t + 2.$$

Terms: $3t^4$, $-5t^6$, $-4t$, and 2.

Do Exercises 13 and 14.

c Like Terms

When terms have the same variable and the variable is raised to the same power, we say that they are **like terms.**

EXAMPLES Identify the like terms in the polynomials.

10. $4x^3 + 5x - 4x^2 + 2x^3 + x^2$

Like terms: $4x^3$ and $2x^3$ — Same variable and exponent

Like terms: $-4x^2$ and x^2 — Same variable and exponent

11. $6 - 3a^2 - 8 - a - 5a$

Like terms: 6 and -8 — Constant terms are like terms because $6 = 6x^0$ and $-8 = -8x^0$.

Like terms: $-a$ and $-5a$

Do Exercises 15–17.

d Coefficients

The coefficient of the term $5x^3$ is 5. In the following polynomial, the color numbers are the **coefficients,** 3, -2, 5, and 4:

$$3x^5 - 2x^3 + 5x + 4.$$

EXAMPLE 12 Identify the coefficient of each term in the polynomial

$$3x^4 - 4x^3 + \frac{1}{2}x^2 + x - 8.$$

The coefficient of the first term is 3.

The coefficient of the second term is -4.

The coefficient of the third term is $\frac{1}{2}$.

The coefficient of the fourth term is 1.

The coefficient of the fifth term is -8.

Do Exercise 18.

Identify the terms of the polynomial.

13. $3x^2 + 6x + \frac{1}{2}$

14. $-4y^5 + 7y^2 - 3y - 2$

Identify the like terms in the polynomial.

15. $4x^3 - x^3 + 2$

16. $4t^4 - 9t^3 - 7t^4 + 10t^3$

17. $5x^2 + 3x - 10 + 7x^2 - 8x + 11$

18. Identify the coefficient of each term in the polynomial

$$2x^4 - 7x^3 - 8.5x^2 + 10x - 4.$$

Answers on page A-35

Collect like terms.

19. $3x^2 + 5x^2$

20. $4x^3 - 2x^3 + 2 + 5$

21. $\frac{1}{2}x^5 - \frac{3}{4}x^5 + 4x^2 - 2x^2$

22. $24 - 4x^3 - 24$

23. $5x^3 - 8x^5 + 8x^5$

24. $-2x^4 + 16 + 2x^4 + 9 - 3x^5$

Collect like terms.

25. $7x - x$

26. $5x^3 - x^3 + 4$

27. $\frac{3}{4}x^3 + 4x^2 - x^3 + 7$

28. $\frac{4}{5}x^4 - x^4 + x^5 - \frac{1}{5} - \frac{1}{4}x^4 + 10$

Answers on page A-35

e Collecting Like Terms

We can often simplify polynomials by **collecting like terms,** or **combining like terms.** To do this, we use the distributive laws. We factor out the variable expression and add or subtract the coefficients. We try to do this mentally as much as possible.

EXAMPLES Collect like terms.

13. $2x^3 - 6x^3 = (2 - 6)x^3$ Using a distributive law
$= -4x^3$

14. $5x^2 + 7 + 4x^4 + 2x^2 - 11 - 2x^4 = (5 + 2)x^2 + (4 - 2)x^4 + (7 - 11)$
$= 7x^2 + 2x^4 - 4$

Note that using the distributive laws in this manner allows us to collect like terms by adding or subtracting the coefficients. Often the middle step is omitted and we add or subtract mentally, writing just the answer. In collecting like terms, we may get 0.

EXAMPLE 15 Collect like terms: $3x^5 + 2x^2 - 3x^5 + 8$.

$$\begin{aligned} 3x^5 + 2x^2 - 3x^5 + 8 &= (3 - 3)x^5 + 2x^2 + 8 \\ &= 0x^5 + 2x^2 + 8 \\ &= 2x^2 + 8 \end{aligned}$$

Do Exercises 19–24.

Expressing a term like x^2 by showing 1 as a factor, $1 \cdot x^2$, may make it easier to understand how to factor or collect like terms.

EXAMPLES Collect like terms.

16. $5x^2 + x^2 = 5x^2 + 1x^2$ Replacing x^2 with $1x^2$
$= (5 + 1)x^2$ Using a distributive law
$= 6x^2$

17. $5x^8 - 6x^5 - x^8 = 5x^8 - 6x^5 - 1x^8$ $x^8 = 1x^8$
$= (5 - 1)x^8 - 6x^5$
$= 4x^8 - 6x^5$

18. $\frac{2}{3}x^4 - x^3 - \frac{1}{6}x^4 + \frac{2}{5}x^3 - \frac{3}{10}x^3 = \left(\frac{2}{3} - \frac{1}{6}\right)x^4 + \left(-1 + \frac{2}{5} - \frac{3}{10}\right)x^3$ $-x^3 = -1 \cdot x^3$
$= \left(\frac{4}{6} - \frac{1}{6}\right)x^4 + \left(-\frac{10}{10} + \frac{4}{10} - \frac{3}{10}\right)x^3$
$= \frac{3}{6}x^4 - \frac{9}{10}x^3$
$= \frac{1}{2}x^4 - \frac{9}{10}x^3$

Do Exercises 25–28.

f Descending and Ascending Order

Note in the following polynomial that the exponents decrease from left to right. We say that the polynomial is arranged in **descending order:**

$$2x^4 - 8x^3 + 5x^2 - x + 3.$$

The term with the largest exponent is first. The term with the next largest exponent is second, and so on. The associative and commutative laws allow us to arrange the terms of a polynomial in descending order.

EXAMPLES Arrange the polynomial in descending order.

19. $6x^5 + 4x^7 + x^2 + 2x^3 = 4x^7 + 6x^5 + 2x^3 + x^2$

20. $\frac{2}{3} + 4x^5 - 8x^2 + 5x - 3x^3 = 4x^5 - 3x^3 - 8x^2 + 5x + \frac{2}{3}$

Do Exercises 29–31.

EXAMPLE 21 Collect like terms and then arrange in descending order:

$2x^2 - 4x^3 + 3 - x^2 - 2x^3.$

$2x^2 - 4x^3 + 3 - x^2 - 2x^3 = x^2 - 6x^3 + 3$ Collecting like terms

$= -6x^3 + x^2 + 3$ Arranging in descending order

Do Exercises 32 and 33.

We usually arrange polynomials in descending order, but not always. The opposite order is called **ascending order.** Generally, if an exercise is written in a certain order, we give the answer in that same order.

g Degrees

The **degree** of a term is the exponent of the variable. The degree of the term $-5x^3$ is 3.

EXAMPLE 22 Identify the degree of each term of $8x^4 - 3x + 7$.

The degree of $8x^4$ is 4.

The degree of $-3x$ is 1. Recall that $x = x^1$.

The degree of 7 is 0. Think of 7 as $7x^0$. Recall that $x^0 = 1$.

The **degree of a polynomial** is the largest of the degrees of the terms, unless it is the polynomial 0. The polynomial 0 is a special case. We agree that it has *no* degree either as a term or as a polynomial. This is because we can express 0 as $0 = 0x^5 = 0x^7$, and so on, using any exponent we wish.

EXAMPLE 23 Identify the degree of the polynomial $5x^3 - 6x^4 + 7$.

$5x^3 - 6x^4 + 7.$ The largest exponent is 4.

The degree of the polynomial is 4.

Do Exercises 34 and 35.

Let's summarize the terminology that we have learned, using the polynomial $3x^4 - 8x^3 + 5x^2 + 7x - 6$.

TERM	COEFFICIENT	DEGREE OF THE TERM	DEGREE OF THE POLYNOMIAL
$3x^4$	3	4	
$-8x^3$	-8	3	
$5x^2$	5	2	4
$7x$	7	1	
-6	-6	0	

Arrange the polynomial in descending order.

29. $x + 3x^5 + 4x^3 + 5x^2 + 6x^7 - 2x^4$

30. $4x^2 - 3 + 7x^5 + 2x^3 - 5x^4$

31. $-14 + 7t^2 - 10t^5 + 14t^7$

Collect like terms and then arrange in descending order.

32. $3x^2 - 2x + 3 - 5x^2 - 1 - x$

33. $-x + \frac{1}{2} + 14x^4 - 7x - 1 - 4x^4$

Identify the degree of each term and the degree of the polynomial.

34. $-6x^4 + 8x^2 - 2x + 9$

35. $4 - x^3 + \frac{1}{2}x^6 - x^5$

Answers on page A-35

h Missing Terms

If a coefficient is 0, we generally do not write the term. We say that we have a **missing term.**

EXAMPLE 24 Identify the missing terms in the polynomial

$$8x^5 - 2x^3 + 5x^2 + 7x + 8.$$

There is no term with x^4. We say that the x^4-term is missing.

Do Exercises 36–39.

For certain skills or manipulations, we can write missing terms with zero coefficients or leave space.

EXAMPLE 25 Write the polynomial $x^4 - 6x^3 + 2x - 1$ in two ways: with its missing terms and by leaving space for them.

a) $x^4 - 6x^3 + 2x - 1 = x^4 - 6x^3 + 0x^2 + 2x - 1$ Writing with the missing x^2-term

b) $x^4 - 6x^3 + 2x - 1 = x^4 - 6x^3 \qquad + 2x - 1$ Leaving space for the missing x^2-term

EXAMPLE 26 Write the polynomial $y^5 - 1$ in two ways: with its missing terms and by leaving space for them.

a) $y^5 - 1 = y^5 + 0y^4 + 0y^3 + 0y^2 + 0y - 1$

b) $y^5 - 1 = y^5 \qquad\qquad\qquad\qquad - 1$

Do Exercises 40 and 41.

i Classifying Polynomials

Polynomials with just one term are called **monomials.** Polynomials with just two terms are called **binomials.** Those with just three terms are called **trinomials.** Those with more than three terms are generally not specified with a name.

EXAMPLE 27

MONOMIALS	BINOMIALS	TRINOMIALS	NONE OF THESE
$4x^2$	$2x + 4$	$3x^3 + 4x + 7$	$4x^3 - 5x^2 + x - 8$
9	$3x^5 + 6x$	$6x^7 - 7x^2 + 4$	$z^5 + 2z^4 - z^3 + 7z + 3$
$-23x^{19}$	$-9x^7 - 6$	$4x^2 - 6x - \frac{1}{2}$	$4x^6 - 3x^5 + x^4 - x^3 + 2x - 1$

Do Exercises 42–45.

Identify the missing terms in the polynomial.

36. $2x^3 + 4x^2 - 2$

37. $-3x^4$

38. $x^3 + 1$

39. $x^4 - x^2 + 3x + 0.25$

Write the polynomial in two ways: with its missing terms and by leaving space for them.

40. $2x^3 + 4x^2 - 2$

41. $a^4 + 10$

Classify the polynomial as a monomial, binomial, trinomial, or none of these.

42. $5x^4$

43. $4x^3 - 3x^2 + 4x + 2$

44. $3x^2 + x$

45. $3x^2 + 2x - 4$

Answers on page A-35

10.3 EXERCISE SET

For Extra Help

MathXL | MyMathLab | InterAct Math | Math Tutor Center | Video Lectures on CD Disc 5 | Student's Solutions Manual

a Evaluate the polynomial when $x = 4$ and when $x = -1$.

1. $-5x + 2$

2. $-8x + 1$

3. $2x^2 - 5x + 7$

4. $3x^2 + x - 7$

5. $x^3 - 5x^2 + x$

6. $7 - x + 3x^2$

Evaluate the polynomial when $x = -2$ and when $x = 0$.

7. $\frac{1}{3}x + 5$

8. $8 - \frac{1}{4}x$

9. $x^2 - 2x + 1$

10. $5x + 6 - x^2$

11. $-3x^3 + 7x^2 - 3x - 2$

12. $-2x^3 + 5x^2 - 4x + 3$

13. *Skydiving.* During the first 13 sec of a jump, the number of feet S that a skydiver falls in t seconds can be approximated by the polynomial equation

$$S = 11.12t^2.$$

Approximately how far has a skydiver fallen 10 sec after having jumped from a plane?

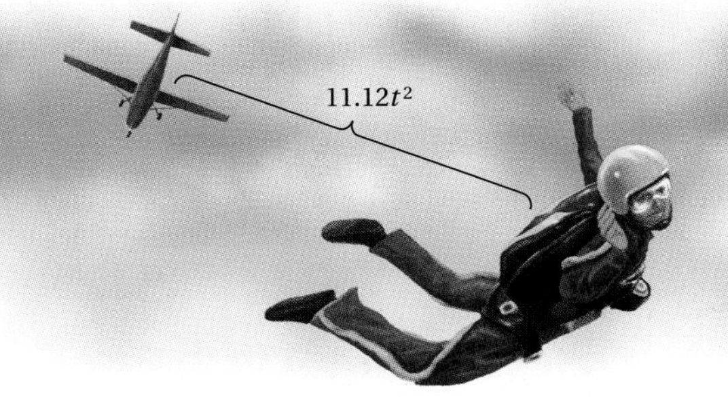

14. *Skydiving.* For jumps that exceed 13 sec, the polynomial equation

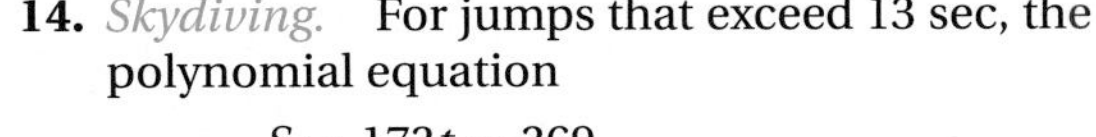

$$S = 173t - 369$$

can be used to approximate the distance S, in feet, that a skydiver has fallen in t seconds. Approximately how far has a skydiver fallen 20 sec after having jumped from a plane?

15. *Electricity Consumption.* The net consumption of electricity in China can be approximated by the polynomial equation

$$E = 90.28t + 1138.34,$$

where E is the consumption of electricity, in billions of kilowatt-hours, and t is the number of years since 2001—that is, $t = 0$ corresponds to 2001, $t = 9$ corresponds to 2010, and so on.

Source: Energy Information Administration (EIA), International Energy Outlook 2004

a) Approximate the consumption of electricity, in billions of kilowatt-hours, in 2001, 2005, 2010, 2015, and 2025.

b) Check the results of part (a) using the graph below.

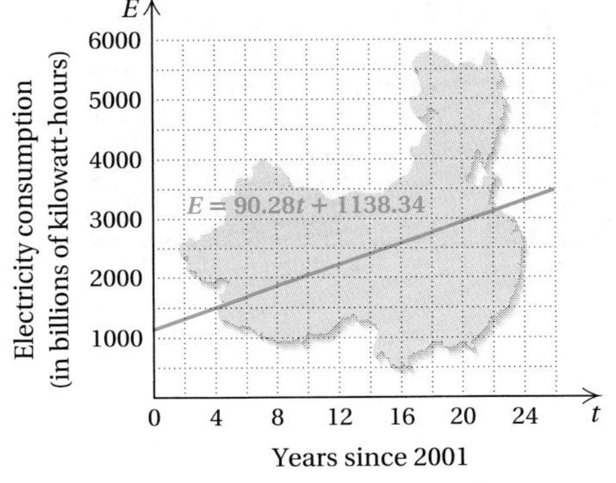

16. *Electricity Consumption.* The net consumption of electricity in the United States can be approximated by the polynomial equation

$$E = 75.72t + 3378.11,$$

where E is the consumption of electricity, in billions of kilowatt-hours, and t is the number of years since 2001—that is, $t = 0$ corresponds to 2001, $t = 9$ corresponds to 2010, and so on.

Source: Energy Information Administration (EIA), International Energy Outlook 2004

a) Approximate the consumption of electricity, in billions of kilowatt-hours, in 2001, 2005, 2010, 2015, and 2025.

b) Check the results of part (a) using the graph below.

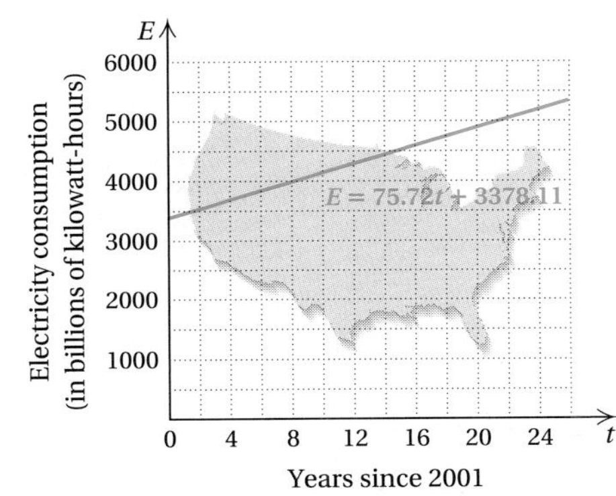

17. *Total Revenue.* Hadley Electronics is marketing a new kind of plasma TV. The firm determines that when it sells x TVs, its total revenue R (the total amount of money taken in) will be

$$R = 280x - 0.4x^2 \text{ dollars.}$$

What is the total revenue from the sale of 75 TVs? 100 TVs?

18. *Total Cost.* Hadley Electronics determines that the total cost C of producing x plasma TVs is given by

$$C = 5000 + 0.6x^2 \text{ dollars.}$$

What is the total cost of producing 500 TVs? 650 TVs?

19. The graph of the polynomial equation $y = 5 - x^2$ is shown below. Use *only* the graph to estimate the value of the polynomial when $x = -3$, $x = -1$, $x = 0$, $x = 1.5$, and $x = 2$.

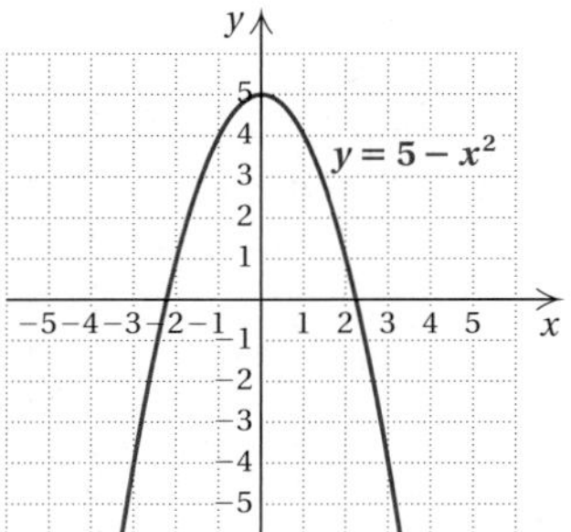

20. The graph of the polynomial equation $y = 6x^3 - 6x$ is shown below. Use *only* the graph to estimate the value of the polynomial when $x = -1$, $x = -0.5$, $x = 0.5$, $x = 1$, and $x = 1.1$.

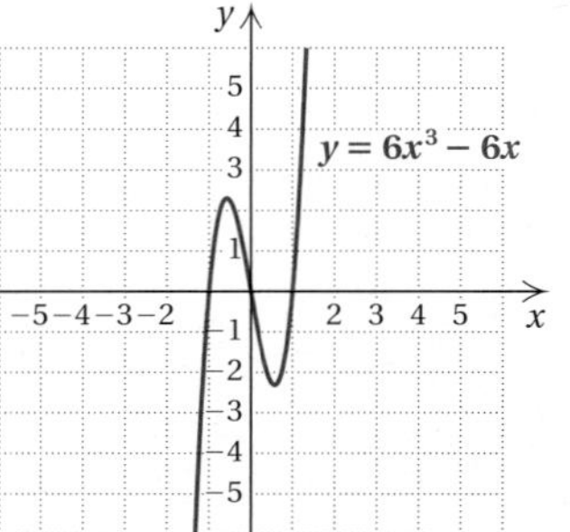

Hearing-Impaired Americans. The number N, in millions, of hearing-impaired Americans of age x can be approximated by the polynomial equation

$$N = -0.00006x^3 + 0.006x^2 - 0.1x + 1.9.$$

The graph of this equation is shown at right. Use either the graph or the polynomial equation for Exercises 21 and 22.

Source: American Speech-Language Hearing Association

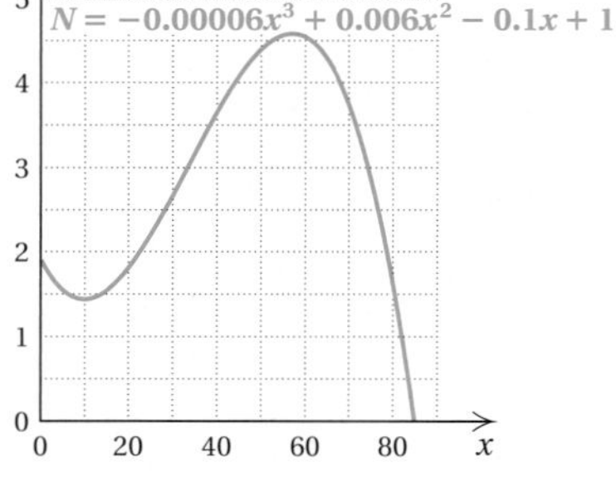

21. Approximate the number of hearing-impaired Americans of ages 20 and 40.

22. Approximate the number of hearing-impaired Americans of ages 50 and 60.

Memorizing words. Participants in a psychology experiment were able to memorize an average of M words in t minutes, where $M = -0.001t^3 + 0.1t^2$. Use the graph below for Exercises 23–28.

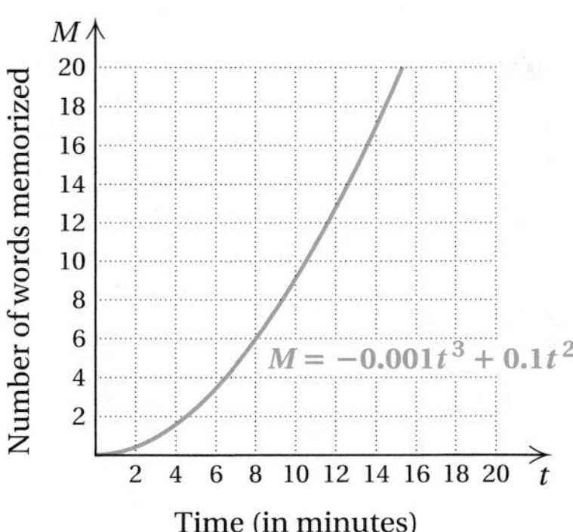

23. Estimate the number of words memorized after 10 min.

24. Estimate the number of words memorized after 14 min.

25. Find the approximate value of M for $t = 8$.

26. Find the approximate value of M for $t = 12$.

27. Estimate the value of M when t is 13.

28. Estimate the value of M when t is 7.

b Identify the terms of the polynomial.

29. $2 - 3x + x^2$

30. $2x^2 + 3x - 4$

31. $-2x^4 + \frac{1}{3}x^3 - x + 3$

32. $-\frac{2}{5}x^5 - x^3 + 6$

c Identify the like terms in the polynomial.

33. $5x^3 + 6x^2 - 3x^2$

34. $3x^2 + 4x^3 - 2x^2$

35. $2x^4 + 5x - 7x - 3x^4$

36. $-3t + t^3 - 2t - 5t^3$

37. $3x^5 - 7x + 8 + 14x^5 - 2x - 9$

38. $8x^3 + 7x^2 - 11 - 4x^3 - 8x^2 - 29$

d Identify the coefficient of each term of the polynomial.

39. $-3x + 6$

40. $2x - 4$

41. $5x^2 + \frac{3}{4}x + 3$

42. $\frac{2}{3}x^2 - 5x + 2$

43. $-5x^4 + 6x^3 - 2.7x^2 + 8x - 2$

44. $7x^3 - 4x^2 - 4.2x + 5$

e Collect like terms.

45. $2x - 5x$

46. $2x^2 + 8x^2$

47. $x - 9x$

48. $x - 5x$

49. $5x^3 + 6x^3 + 4$

50. $6x^4 - 2x^4 + 5$

51. $5x^3 + 6x - 4x^3 - 7x$

52. $3a^4 - 2a + 2a + a^4$

53. $6b^5 + 3b^2 - 2b^5 - 3b^2$

54. $2x^2 - 6x + 3x + 4x^2$

55. $\frac{1}{4}x^5 - 5 + \frac{1}{2}x^5 - 2x - 37$

56. $\frac{1}{3}x^3 + 2x - \frac{1}{6}x^3 + 4 - 16$

57. $6x^2 + 2x^4 - 2x^2 - x^4 - 4x^2$

58. $8x^2 + 2x^3 - 3x^3 - 4x^2 - 4x^2$

59. $\frac{1}{4}x^3 - x^2 - \frac{1}{6}x^2 + \frac{3}{8}x^3 + \frac{5}{16}x^3$

60. $\frac{1}{5}x^4 + \frac{1}{5} - 2x^2 + \frac{1}{10} - \frac{3}{15}x^4 + 2x^2 - \frac{3}{10}$

f Arrange the polynomial in descending order.

61. $x^5 + x + 6x^3 + 1 + 2x^2$

62. $3 + 2x^2 - 5x^6 - 2x^3 + 3x$

63. $5y^3 + 15y^9 + y - y^2 + 7y^8$

64. $9p - 5 + 6p^3 - 5p^4 + p^5$

Collect like terms and then arrange in descending order.

65. $3x^4 - 5x^6 - 2x^4 + 6x^6$

66. $-1 + 5x^3 - 3 - 7x^3 + x^4 + 5$

67. $-2x + 4x^3 - 7x + 9x^3 + 8$

68. $-6x^2 + x - 5x + 7x^2 + 1$

69. $3x + 3x + 3x - x^2 - 4x^2$

70. $-2x - 2x - 2x + x^3 - 5x^3$

71. $-x + \frac{3}{4} + 15x^4 - x - \frac{1}{2} - 3x^4$

72. $2x - \frac{5}{6} + 4x^3 + x + \frac{1}{3} - 2x$

g Identify the degree of each term of the polynomial and the degree of the polynomial.

73. $2x - 4$

74. $6 - 3x$

75. $3x^2 - 5x + 2$

76. $5x^3 - 2x^2 + 3$

77. $-7x^3 + 6x^2 + \frac{3}{5}x + 7$

78. $5x^4 + \frac{1}{4}x^2 - x + 2$

79. $x^2 - 3x + x^6 - 9x^4$

80. $8x - 3x^2 + 9 - 8x^3$

81. Complete the following table for the polynomial $-7x^4 + 6x^3 - 3x^2 + 8x - 2$.

TERM	COEFFICIENT	DEGREE OF THE TERM	DEGREE OF THE POLYNOMIAL
$-7x^4$			
$6x^3$	6		
		2	
$8x$		1	
	-2		

82. Complete the following table for the polynomial $3x^2 + 8x^5 - 46x^3 + 6x - 2.4 - \frac{1}{2}x^4$.

TERM	COEFFICIENT	DEGREE OF THE TERM	DEGREE OF THE POLYNOMIAL
		5	
$-\frac{1}{2}x^4$		4	
	-46		
$3x^2$		2	
	6		
-2.4			

h Identify the missing terms in the polynomial.

83. $x^3 - 27$

84. $x^5 + x$

85. $x^4 - x$

86. $5x^4 - 7x + 2$

87. $2x^3 - 5x^2 + x - 3$

88. $-6x^3$

Write the polynomial in two ways: with its missing terms and by leaving space for them.

89. $x^3 - 27$

90. $x^5 + x$

91. $x^4 - x$

92. $5x^4 - 7x + 2$

93. $2x^3 - 5x^2 + x - 3$

94. $-6x^3$

i Classify the polynomial as a monomial, binomial, trinomial, or none of these.

95. $x^2 - 10x + 25$

96. $-6x^4$

97. $x^3 - 7x^2 + 2x - 4$

98. $x^2 - 9$

99. $4x^2 - 25$

100. $2x^4 - 7x^3 + x^2 + x - 6$

101. $40x$

102. $4x^2 + 12x + 9$

103. $\mathbf{D_W}$ Is it better to evaluate a polynomial before or after like terms have been collected? Why?

104. $\mathbf{D_W}$ Explain why an understanding of the rules for order of operations is essential when evaluating polynomials.

SKILL MAINTENANCE

105. Three tired campers stopped for the night. All they had to eat was a bag of apples. During the night, one awoke and ate one-third of the apples. Later, a second camper awoke and ate one-third of the apples that remained. Much later, the third camper awoke and ate one-third of those apples yet remaining after the other two had eaten. When they got up the next morning, 8 apples were left. How many apples did they begin with? [8.6a]

Subtract. [7.4a]

106. $1 - 20$

107. $\frac{1}{8} - \frac{5}{6}$

108. $\frac{3}{8} - \left(-\frac{1}{4}\right)$

109. $5.6 - 8.2$

110. Solve: $3(x + 2) = 5x - 9$. [8.3c]

111. Solve $C = ab - r$ for b. [8.4b]

112. A nut dealer has 1800 lb of peanuts, 1500 lb of cashews, and 700 lb of almonds. What percent of the total is peanuts? cashews? almonds? [8.5a]

113. Factor: $3x - 15y + 63$. [7.7d]

SYNTHESIS

Collect like terms.

114. $6x^3 \cdot 7x^2 - (4x^3)^2 + (-3x^3)^2 - (-4x^2)(5x^3) - 10x^5 + 17x^6$

115. $(3x^2)^3 + 4x^2 \cdot 4x^4 - x^4(2x)^2 + ((2x)^2)^3 - 100x^2(x^2)^2$

116. Construct a polynomial in x (meaning that x is the variable) of degree 5 with four terms and coefficients that are integers.

117. What is the degree of $(5m^5)^2$?

118. A polynomial in x has degree 3. The coefficient of x^2 is 3 less than the coefficient of x^3. The coefficient of x is three times the coefficient of x^2. The remaining coefficient is 2 more than the coefficient of x^3. The sum of the coefficients is -4. Find the polynomial.

Use the CALC feature and choose VALUE on your graphing calculator to find the values in each of the following.

119. Exercise 19

120. Exercise 20

121. Exercise 21

122. Exercise 22

10.4 ADDITION AND SUBTRACTION OF POLYNOMIALS

a Addition of Polynomials

To add two polynomials, we can write a plus sign between them and then collect like terms. Depending on the situation, you may see polynomials written in descending order, ascending order, or neither. Generally, if an exercise is written in a particular order, we write the answer in that same order.

EXAMPLE 1 Add: $(-3x^3 + 2x - 4) + (4x^3 + 3x^2 + 2)$.

$$\begin{aligned}(-3x^3 + 2x - 4) &+ (4x^3 + 3x^2 + 2)\\ &= (-3 + 4)x^3 + 3x^2 + 2x + (-4 + 2) \quad \text{Collecting like terms}\\ &= x^3 + 3x^2 + 2x - 2\end{aligned}$$

EXAMPLE 2 Add:

$$\left(\tfrac{2}{3}x^4 + 3x^2 - 2x + \tfrac{1}{2}\right) + \left(-\tfrac{1}{3}x^4 + 5x^3 - 3x^2 + 3x - \tfrac{1}{2}\right).$$

We have

$$\begin{aligned}&\left(\tfrac{2}{3}x^4 + 3x^2 - 2x + \tfrac{1}{2}\right) + \left(-\tfrac{1}{3}x^4 + 5x^3 - 3x^2 + 3x - \tfrac{1}{2}\right)\\ &\quad= \left(\tfrac{2}{3} - \tfrac{1}{3}\right)x^4 + 5x^3 + (3 - 3)x^2 + (-2 + 3)x + \left(\tfrac{1}{2} - \tfrac{1}{2}\right) \quad \text{Collecting like terms}\\ &\quad= \tfrac{1}{3}x^4 + 5x^3 + x.\end{aligned}$$

We can add polynomials as we do because they represent numbers. After some practice, you will be able to add mentally.

Do Exercises 1–4.

EXAMPLE 3 Add: $(3x^2 - 2x + 2) + (5x^3 - 2x^2 + 3x - 4)$.

$$\begin{aligned}(3x^2 - 2x + 2) &+ (5x^3 - 2x^2 + 3x - 4)\\ &= 5x^3 + (3 - 2)x^2 + (-2 + 3)x + (2 - 4) \quad \text{You might do this step mentally.}\\ &= 5x^3 + x^2 + x - 2 \quad \text{Then you would write only this.}\end{aligned}$$

Do Exercises 5 and 6.

We can also add polynomials by writing like terms in columns.

EXAMPLE 4 Add: $9x^5 - 2x^3 + 6x^2 + 3$ and $5x^4 - 7x^2 + 6$ and $3x^6 - 5x^5 + x^2 + 5$.

We arrange the polynomials with the like terms in columns.

$$\begin{array}{rrrrrrl} & 9x^5 & & -\ 2x^3 & +\ 6x^2 & +\ 3 & \\ & & 5x^4 & & -\ 7x^2 & +\ 6 & \text{We leave spaces for missing terms.}\\ 3x^6 & -\ 5x^5 & & & +\ \ x^2 & +\ 5 & \\ \hline 3x^6 & +\ 4x^5 & +\ 5x^4 & -\ 2x^3 & & +\ 14 & \text{Adding}\end{array}$$

We write the answer as $3x^6 + 4x^5 + 5x^4 - 2x^3 + 14$ without the space.

Objectives

a Add polynomials.

b Simplify the opposite of a polynomial.

c Subtract polynomials.

d Use polynomials to represent perimeter and area.

Add.

1. $(3x^2 + 2x - 2) + (-2x^2 + 5x + 5)$

2. $(-4x^5 + x^3 + 4) + (7x^4 + 2x^2)$

3. $(31x^4 + x^2 + 2x - 1) + (-7x^4 + 5x^3 - 2x + 2)$

4. $(17x^3 - x^2 + 3x + 4) + \left(-15x^3 + x^2 - 3x - \dfrac{2}{3}\right)$

Add mentally. Try to write just the answer.

5. $(4x^2 - 5x + 3) + (-2x^2 + 2x - 4)$

6. $(3x^3 - 4x^2 - 5x + 3) + \left(5x^3 + 2x^2 - 3x - \dfrac{1}{2}\right)$

Answers on page A-35

Add.

7. $\begin{array}{r} -2x^3 + 5x^2 - 2x + 4 \\ x^4 \qquad + 6x^2 + 7x - 10 \\ \underline{-9x^4 + 6x^3 + x^2 \qquad - 2} \end{array}$

8. $-3x^3 + 5x + 2$ and $x^3 + x^2 + 5$ and $x^3 - 2x - 4$

Simplify.

9. $-(4x^3 - 6x + 3)$

10. $-(5x^4 + 3x^2 + 7x - 5)$

11. $-\left(14x^{10} - \frac{1}{2}x^5 + 5x^3 - x^2 + 3x\right)$

Subtract.

12. $(7x^3 + 2x + 4) - (5x^3 - 4)$

13. $(-3x^2 + 5x - 4) - (-4x^2 + 11x - 2)$

Answers on page A-35

Do Exercises 7 and 8.

b Opposites of Polynomials

In Section 7.8, we used the property of -1 to show that we can find the opposite of an expression. For example, the opposite of $x - 2y + 5$ of can be written as

$$-(x - 2y + 5)$$

by changing the sign of every term:

$$-(x - 2y + 5) = -x + 2y - 5.$$

This applies to polynomials as well.

OPPOSITES OF POLYNOMIALS

To find an equivalent polynomial for the **opposite,** or **additive inverse,** of a polynomial, change the sign of every term. This is the same as multiplying by -1.

EXAMPLE 5 Simplify: $-(x^2 - 3x + 4)$.

$$-(x^2 - 3x + 4) = -x^2 + 3x - 4$$

EXAMPLE 6 Simplify: $-(-t^3 - 6t^2 - t + 4)$.

$$-(-t^3 - 6t^2 - t + 4) = t^3 + 6t^2 + t - 4$$

EXAMPLE 7 Simplify: $-\left(-7x^4 - \frac{5}{9}x^3 + 8x^2 - x + 67\right)$.

$$-\left(-7x^4 - \frac{5}{9}x^3 + 8x^2 - x + 67\right) = 7x^4 + \frac{5}{9}x^3 - 8x^2 + x - 67$$

Do Exercises 9–11.

c Subtraction of Polynomials

Recall that we can subtract a real number by adding its opposite, or additive inverse: $a - b = a + (-b)$. This allows us to subtract polynomials.

EXAMPLE 8 Subtract:

$$(9x^5 + x^3 - 2x^2 + 4) - (2x^5 + x^4 - 4x^3 - 3x^2).$$

We have

$(9x^5 + x^3 - 2x^2 + 4) - (2x^5 + x^4 - 4x^3 - 3x^2)$

$= 9x^5 + x^3 - 2x^2 + 4 + [-(2x^5 + x^4 - 4x^3 - 3x^2)]$ Adding the opposite

$= 9x^5 + x^3 - 2x^2 + 4 - 2x^5 - x^4 + 4x^3 + 3x^2$ Finding the opposite by changing the sign of *each* term

$= 7x^5 - x^4 + 5x^3 + x^2 + 4.$ Adding (collecting like terms)

Do Exercises 12 and 13.

As with similar work in Section 7.8, we combine steps by changing the sign of each term of the polynomial being subtracted and collecting like terms. Try to do this mentally as much as possible.

EXAMPLE 9 Subtract: $(9x^5 + x^3 - 2x) - (-2x^5 + 5x^3 + 6)$.

$(9x^5 + x^3 - 2x) - (-2x^5 + 5x^3 + 6)$

$= 9x^5 + x^3 - 2x + 2x^5 - 5x^3 - 6$ Finding the opposite by changing the sign of each term

$= 11x^5 - 4x^3 - 2x - 6$ Adding (collecting like terms)

Do Exercises 14 and 15.

We can use columns to subtract. We replace coefficients with their opposites, as shown in Example 8.

EXAMPLE 10 Write in columns and subtract:

$(5x^2 - 3x + 6) - (9x^2 - 5x - 3)$.

a)
$$\begin{array}{r} 5x^2 - 3x + 6 \\ \underline{-(9x^2 - 5x - 3)} \end{array}$$
Writing like terms in columns

b)
$$\begin{array}{r} 5x^2 - 3x + 6 \\ \underline{-9x^2 + 5x + 3} \end{array}$$
Changing signs

c)
$$\begin{array}{r} 5x^2 - 3x + 6 \\ \underline{-9x^2 + 5x + 3} \\ -4x^2 + 2x + 9 \end{array}$$
Adding

If you can do so without error, you can arrange the polynomials in columns and write just the answer, remembering to change the signs and add.

EXAMPLE 11 Write in columns and subtract:

$(x^3 + x^2 + 2x - 12) - (-2x^3 + x^2 - 3x)$.

$$\begin{array}{r} x^3 + x^2 + 2x - 12 \\ \underline{-(-2x^3 + x^2 - 3x \qquad)} \\ 3x^3 \qquad + 5x - 12. \end{array}$$
Leaving space for the missing term

Changing the signs and adding

Do Exercises 16 and 17.

d Polynomials and Geometry

EXAMPLE 12 Find a polynomial for the sum of the areas of these rectangles.

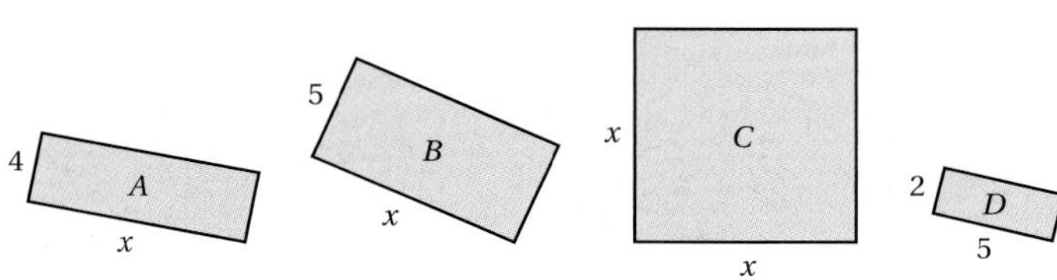

Recall that the area of a rectangle is the product of the length and the width. The sum of the areas is a sum of products. We find these products and then collect like terms.

Subtract.

14. $(-6x^4 + 3x^2 + 6) - (2x^4 + 5x^3 - 5x^2 + 7)$

15. $\left(\frac{3}{2}x^3 - \frac{1}{2}x^2 + 0.3\right) - \left(\frac{1}{2}x^3 + \frac{1}{2}x^2 + \frac{4}{3}x + 1.2\right)$

Write in columns and subtract.

16. $(4x^3 + 2x^2 - 2x - 3) - (2x^3 - 3x^2 + 2)$

17. $(2x^3 + x^2 - 6x + 2) - (x^5 + 4x^3 - 2x^2 - 4x)$

Answers on page A-35

18. Find a polynomial for the sum of the perimeters and the areas of the rectangles.

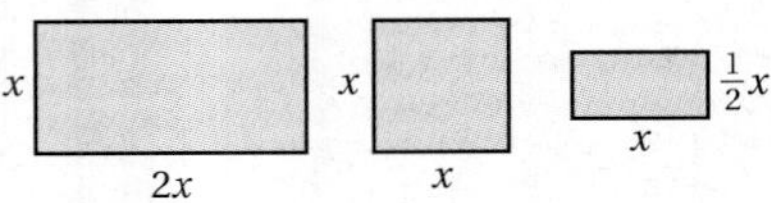

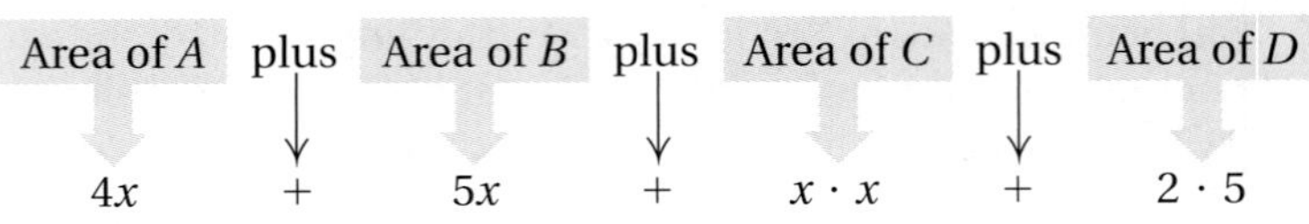

Area of A plus Area of B plus Area of C plus Area of D
4x + 5x + x · x + 2 · 5

We collect like terms:

$$4x + 5x + x^2 + 10 = x^2 + 9x + 10.$$

Do Exercise 18.

19. Lawn Area. An 8-ft by 8-ft shed is placed on a lawn x ft on a side. Find a polynomial for the remaining area.

EXAMPLE 13 *Lawn Area.* A water fountain with a 4-ft by 4-ft square base is placed on a square grassy park area that is x ft on a side. To determine the amount of grass seed needed for the lawn, find a polynomial for the grassy area.

We make a drawing of the situation as shown here. We then reword the problem and write the polynomial as follows.

Area of park	−	Area of base of fountain	=	Area left over
$x \cdot x$	−	$4 \cdot 4$	=	Area left over

Then $(x^2 - 16)\text{ ft}^2$ = Area left over.

Do Exercise 19.

Answers on page A-35

CALCULATOR CORNER

Checking Addition and Subtraction of Polynomials A table set in AUTO mode can be used to perform a partial check that polynomials have been added or subtracted correctly. To check Example 3, we enter $y_1 = (3x^2 - 2x + 2) + (5x^3 - 2x^2 + 3x - 4)$ and $y_2 = 5x^3 + x^2 + x - 2$. If the addition has been done correctly, the values of y_1 and y_2 will be the same regardless of the table settings used.

X	Y1	Y2
−2	−40	−40
−1	−7	−7
0	−2	−2
1	5	5
2	44	44
3	145	145
4	338	338
X = −2		

A graph can also be used to check addition and subtraction. See the Calculator Corner on p. 764 for the procedure.

Exercises: Use a table to determine whether the sum or difference is correct.

1. $(-3x^3 + 2x - 4) + (4x^3 + 3x^2 + 2) = x^3 + 3x^2 + 2x - 2$
2. $(x^3 - 2x^2 + 3x - 7) + (3x^2 - 4x + 5) = x^3 + x^2 - x - 2$
3. $(5x^2 - 7x + 4) + (2x^2 + 3x - 6) = 7x^2 + 4x - 2$
4. $(9x^5 + x^3 - 2x) - (-2x^5 + 5x^3 + 6) = 11x^5 - 4x^3 - 2x - 6$
5. $(3x^4 - 2x^2 - 1) - (2x^4 - 3x^2 - 4) = x^4 + x^2 - 5$
6. $(-2x^3 + 3x^2 - 4x + 5) - (3x^2 + 2x + 8) = -2x^3 - 6x - 3$

10.4 EXERCISE SET

For Extra Help

MathXL | MyMathLab | InterAct Math | Math Tutor Center  | Video Lectures on CD Disc 5  | Student's Solutions Manual

a Add.

1. $(3x + 2) + (-4x + 3)$

2. $(6x + 1) + (-7x + 2)$

3. $(-6x + 2) + (x^2 + \frac{1}{2}x - 3)$

4. $(x^2 - \frac{5}{3}x + 4) + (8x - 9)$

5. $(x^2 - 9) + (x^2 + 9)$

6. $(x^3 + x^2) + (2x^3 - 5x^2)$

7. $(3x^2 - 5x + 10) + (2x^2 + 8x - 40)$

8. $(6x^4 + 3x^3 - 1) + (4x^2 - 3x + 3)$

9. $(1.2x^3 + 4.5x^2 - 3.8x) + (-3.4x^3 - 4.7x^2 + 23)$

10. $(0.5x^4 - 0.6x^2 + 0.7) + (2.3x^4 + 1.8x - 3.9)$

11. $(1 + 4x + 6x^2 + 7x^3) + (5 - 4x + 6x^2 - 7x^3)$

12. $(3x^4 - 6x - 5x^2 + 5) + (6x^2 - 4x^3 - 1 + 7x)$

13. $\left(\frac{1}{4}x^4 + \frac{2}{3}x^3 + \frac{5}{8}x^2 + 7\right) + \left(-\frac{3}{4}x^4 + \frac{3}{8}x^2 - 7\right)$

14. $\left(\frac{1}{3}x^9 + \frac{1}{5}x^5 - \frac{1}{2}x^2 + 7\right) + \left(-\frac{1}{5}x^9 + \frac{1}{4}x^4 - \frac{3}{5}x^5 + \frac{3}{4}x^2 + \frac{1}{2}\right)$

15. $(0.02x^5 - 0.2x^3 + x + 0.08) + (-0.01x^5 + x^4 - 0.8x - 0.02)$

16. $(0.03x^6 + 0.05x^3 + 0.22x + 0.05) + \left(\frac{7}{100}x^6 - \frac{3}{100}x^3 + 0.5\right)$

17. $(9x^8 - 7x^4 + 2x^2 + 5) + (8x^7 + 4x^4 - 2x) + (-3x^4 + 6x^2 + 2x - 1)$

18. $(4x^5 - 6x^3 - 9x + 1) + (6x^3 + 9x^2 + 9x) + (-4x^3 + 8x^2 + 3x - 2)$

19.
$$\begin{array}{rrrrr} 0.15x^4 & + 0.10x^3 & - 0.9x^2 & & \\ & - 0.01x^3 & + 0.01x^2 & + x & \\ 1.25x^4 & & + 0.11x^2 & & + 0.01 \\ & 0.27x^3 & & & + 0.99 \\ \underline{-0.35x^4} & & \underline{+ 15x^2} & & \underline{- 0.03} \end{array}$$

20.
$$\begin{array}{rrrrr} 0.05x^4 & + 0.12x^3 & - 0.5x^2 & & \\ & - 0.02x^3 & + 0.02x^2 & + 2x & \\ 1.5x^4 & & + 0.01x^2 & & + 0.15 \\ & 0.25x^3 & & & + 0.85 \\ \underline{-0.25x^4} & & \underline{+ 10x^2} & & \underline{- 0.04} \end{array}$$

b Simplify.

21. $-(-5x)$

22. $-(x^2 - 3x)$

23. $-(-x^2 + \frac{3}{2}x - 2)$

24. $-(-4x^3 - x^2 - \frac{1}{4}x)$

25. $-(12x^4 - 3x^3 + 3)$

26. $-(4x^3 - 6x^2 - 8x + 1)$

27. $-(3x - 7)$

28. $-(-2x + 4)$

29. $-(4x^2 - 3x + 2)$

30. $-(-6a^3 + 2a^2 - 9a + 1)$

31. $-\left(-4x^4 + 6x^2 + \frac{3}{4}x - 8\right)$

32. $-(-5x^4 + 4x^3 - x^2 + 0.9)$

c Subtract.

33. $(3x + 2) - (-4x + 3)$

34. $(6x + 1) - (-7x + 2)$

35. $(-6x + 2) - (x^2 + x - 3)$

36. $(x^2 - 5x + 4) - (8x - 9)$

37. $(x^2 - 9) - (x^2 + 9)$

38. $(x^3 + x^2) - (2x^3 - 5x^2)$

39. $(6x^4 + 3x^3 - 1) - (4x^2 - 3x + 3)$

40. $(-4x^2 + 2x) - (3x^3 - 5x^2 + 3)$

41. $(1.2x^3 + 4.5x^2 - 3.8x) - (-3.4x^3 - 4.7x^2 + 23)$

42. $(0.5x^4 - 0.6x^2 + 0.7) - (2.3x^4 + 1.8x - 3.9)$

43. $\left(\frac{5}{8}x^3 - \frac{1}{4}x - \frac{1}{3}\right) - \left(-\frac{1}{8}x^3 + \frac{1}{4}x - \frac{1}{3}\right)$

44. $\left(\frac{1}{5}x^3 + 2x^2 - 0.1\right) - \left(-\frac{2}{5}x^3 + 2x^2 + 0.01\right)$

45. $(0.08x^3 - 0.02x^2 + 0.01x) - (0.02x^3 + 0.03x^2 - 1)$

46. $(0.8x^4 + 0.2x - 1) - \left(\frac{7}{10}x^4 + \frac{1}{5}x - 0.1\right)$

Subtract.

47. $\begin{array}{l} x^2 + 5x + 6 \\ \underline{-(x^2 + 2x)} \end{array}$

48. $\begin{array}{l} x^3 \qquad\quad + 1 \\ \underline{-(x^3 + x^2 \qquad)} \end{array}$

49. $\begin{array}{l} 5x^4 + 6x^3 - 9x^2 \\ \underline{-(-6x^4 - 6x^3 \qquad\qquad + 8x + 9)} \end{array}$

50. $\begin{array}{l} 5x^4 \qquad + 6x^2 - 3x + 6 \\ \underline{-(\quad 6x^3 + 7x^2 - 8x - 9)} \end{array}$

51. $\begin{array}{l} x^5 \qquad\qquad\qquad\qquad\qquad - 1 \\ \underline{-(x^5 - x^4 + x^3 - x^2 + x - 1)} \end{array}$

52. $\begin{array}{l} x^5 + x^4 - x^3 + x^2 - x + 2 \\ \underline{-(x^5 - x^4 + x^3 - x^2 - x + 2)} \end{array}$

Solve.

Find a polynomial for the perimeter of the figure.

53.

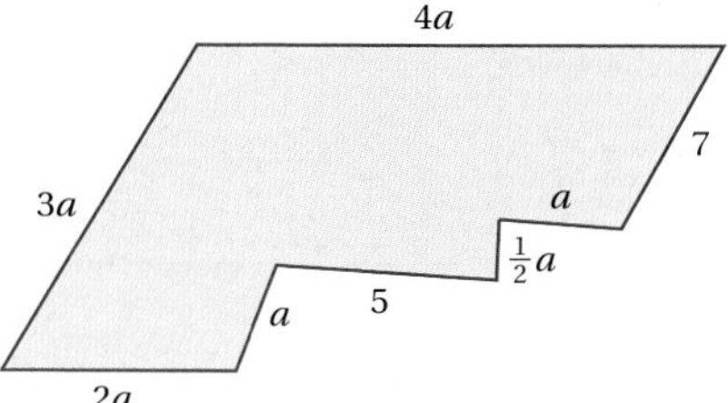

54.

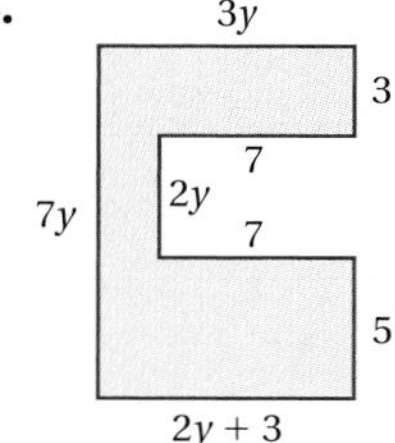

55. Find a polynomial for the sum of the areas of these rectangles.

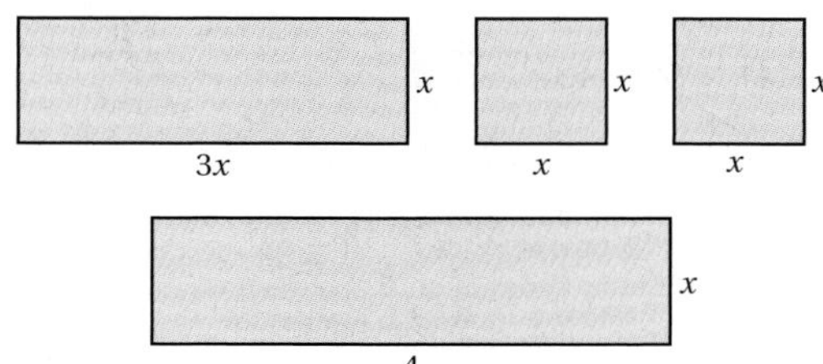

56. Find a polynomial for the sum of the areas of these circles.

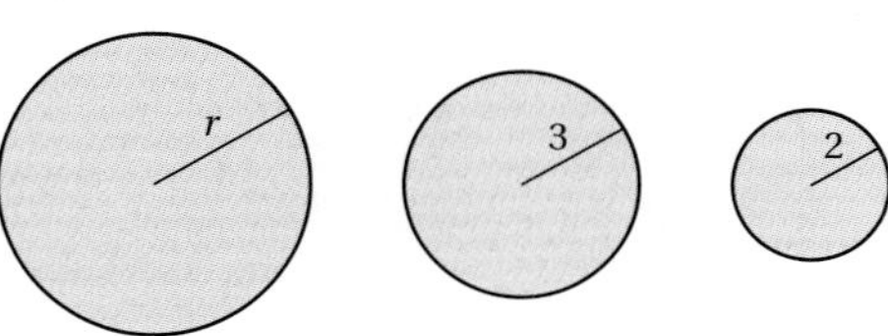

Find two algebraic expressions for the area of each figure. First, regard the figure as one large rectangle, and then regard the figure as a sum of four smaller rectangles.

57.

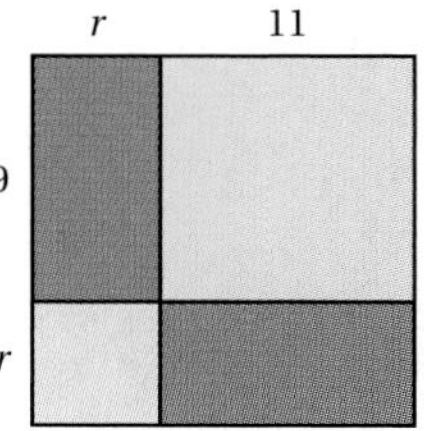

58.

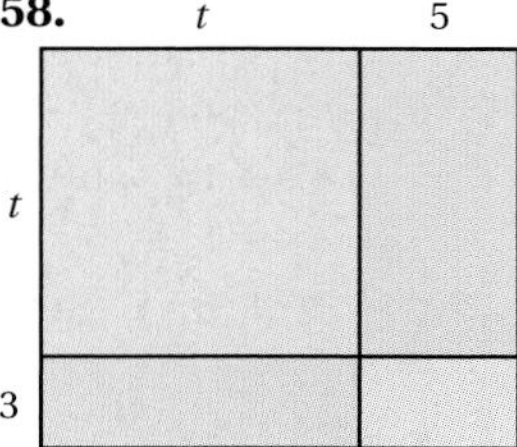

59.

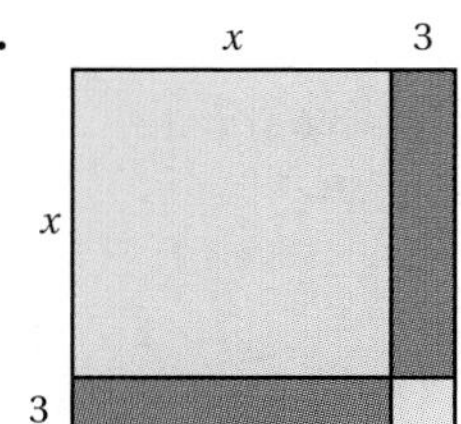

60.

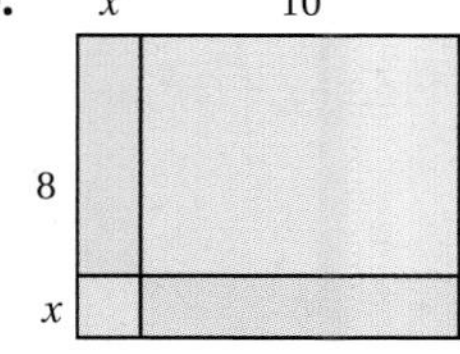

Find a polynomial for the shaded area of the figure.

61. 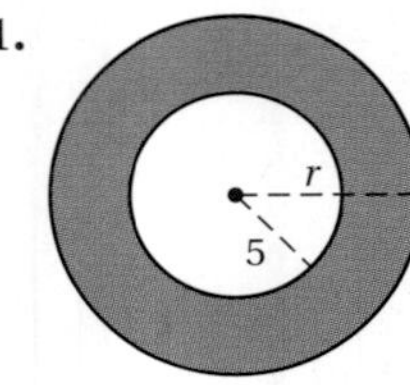

62. 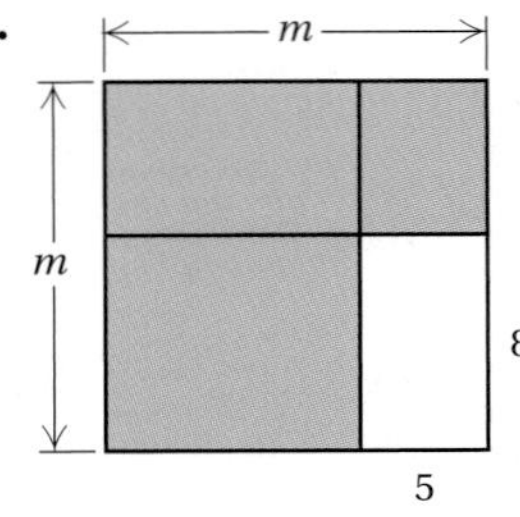

63. 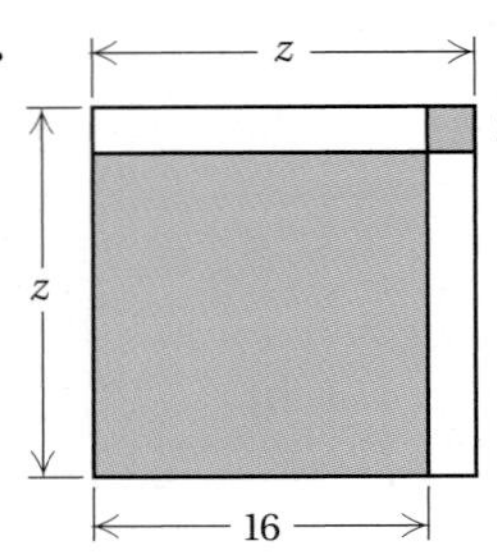

64. 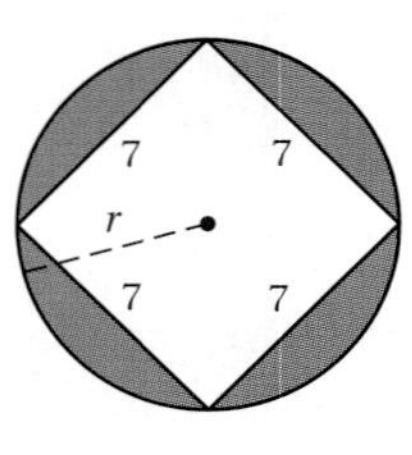

65. $\mathrm{D_W}$ Is the sum of two binomials ever a trinomial? Why or why not?

66. $\mathrm{D_W}$ Which, if any, of the commutative, associative, and distributive laws are needed for adding polynomials? Why?

SKILL MAINTENANCE

Solve. [8.3b]

67. $8x + 3x = 66$

68. $5x - 7x = 38$

69. $\frac{3}{8}x + \frac{1}{4} - \frac{3}{4}x = \frac{11}{16} + x$

70. $5x - 4 = 26 - x$

71. $1.5x - 2.7x = 22 - 5.6x$

72. $3x - 3 = -4x + 4$

Solve. [8.3c]

73. $6(y - 3) - 8 = 4(y + 2) + 5$

74. $8(5x + 2) = 7(6x - 3)$

Solve. [8.7e]

75. $3x - 7 \leq 5x + 13$

76. $2(x - 4) > 5(x - 3) + 7$

SYNTHESIS

Find a polynomial for the surface area of the right rectangular solid.

77. 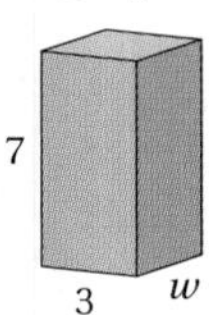

78. 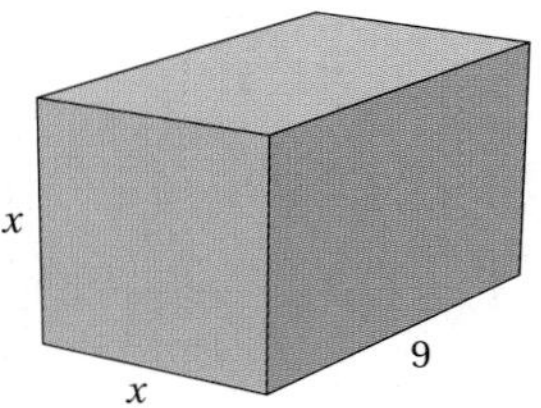

79. 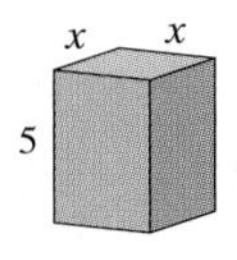

80. 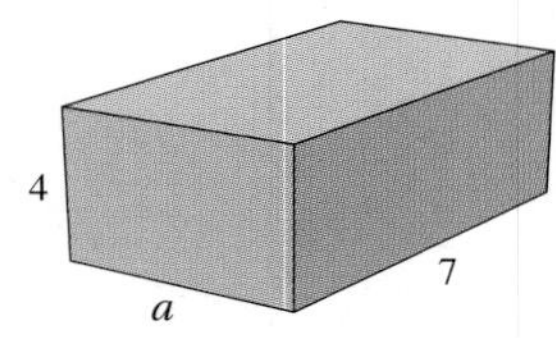

81. Find $(y - 2)^2$ using the four parts of this square.

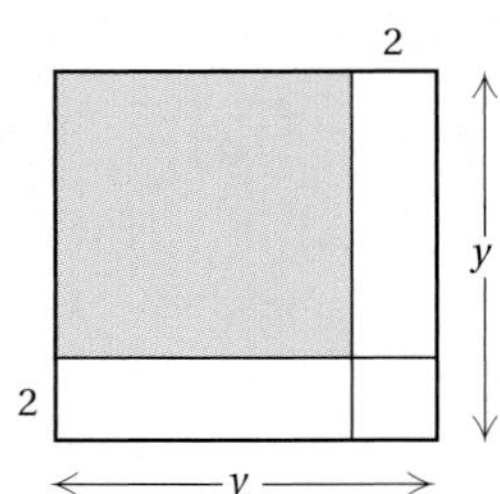

Simplify.

82. $(3x^2 - 4x + 6) - (-2x^2 + 4) + (-5x - 3)$

83. $(7y^2 - 5y + 6) - (3y^2 + 8y - 12) + (8y^2 - 10y + 3)$

84. $(-4 + x^2 + 2x^3) - (-6 - x + 3x^3) - (-x^2 - 5x^3)$

85. $(-y^4 - 7y^3 + y^2) + (-2y^4 + 5y - 2) - (-6y^3 + y^2)$

10.5 MULTIPLICATION OF POLYNOMIALS

Objectives

a Multiply monomials.

b Multiply a monomial and any polynomial.

c Multiply two binomials.

d Multiply any two polynomials.

We now multiply polynomials using techniques based, for the most part, on the distributive laws, but also on the associative and commutative laws. As we proceed in this chapter, we will develop special ways to find certain products.

a Multiplying Monomials

Consider $(3x)(4x)$. We multiply as follows:

$(3x)(4x) = 3 \cdot x \cdot 4 \cdot x$ By the associative law of multiplication

$= 3 \cdot 4 \cdot x \cdot x$ By the commutative law of multiplication

$= (3 \cdot 4)(x \cdot x)$ By the associative law

$= 12x^2.$ Using the product rule for exponents

MULTIPLYING MONOMIALS

To find an equivalent expression for the product of two monomials, multiply the coefficients and then multiply the variables using the product rule for exponents.

EXAMPLES Multiply.

1. $5x \cdot 6x = (5 \cdot 6)(x \cdot x)$ By the associative and commutative laws

$= 30x^2$ Multiplying the coefficients and multiplying the variables

2. $(3x)(-x) = (3x)(-1x)$

$= (3)(-1)(x \cdot x) = -3x^2$

3. $(-7x^5)(4x^3) = (-7 \cdot 4)(x^5 \cdot x^3)$

$= -28x^{5+3}$ Adding the exponents

$= -28x^8$ Simplifying

After some practice, you can do this mentally. Multiply the coefficients and then the variables by keeping the base and adding the exponents. Write only the answer.

Do Exercises 1–8.

Multiply.

1. $(3x)(-5)$

2. $(-x) \cdot x$

3. $(-x)(-x)$

4. $(-x^2)(x^3)$

5. $3x^5 \cdot 4x^2$

6. $(4y^5)(-2y^6)$

7. $(-7y^4)(-y)$

8. $7x^5 \cdot 0$

Answers on page A-36

b Multiplying a Monomial and Any Polynomial

To find an equivalent expression for the product of a monomial, such as $2x$, and a binomial, such as $5x + 3$, we use a distributive law and multiply each term of $5x + 3$ by $2x$.

EXAMPLE 4 Multiply: $2x(5x + 3)$.

$2x(5x + 3) = (2x)(5x) + (2x)(3)$ Using a distributive law

$= 10x^2 + 6x$ Multiplying the monomials

EXAMPLE 5 Multiply: $5x(2x^2 - 3x + 4)$.

$$5x(2x^2 - 3x + 4) = (5x)(2x^2) - (5x)(3x) + (5x)(4)$$
$$= 10x^3 - 15x^2 + 20x$$

MULTIPLYING A MONOMIAL AND A POLYNOMIAL

To multiply a monomial and a polynomial, multiply each term of the polynomial by the monomial.

EXAMPLE 6 Multiply: $-2x^2(x^3 - 7x^2 + 10x - 4)$.

$$-2x^2(x^3 - 7x^2 + 10x - 4)$$
$$= (-2x^2)(x^3) - (-2x^2)(7x^2) + (-2x^2)(10x) - (-2x^2)(4)$$
$$= -2x^5 + 14x^4 - 20x^3 + 8x^2$$

Do Exercises 9–11.

C Multiplying Two Binomials

To find an equivalent expression for the product of two binomials, we use the distributive laws more than once. In Example 7, we use a distributive law three times.

EXAMPLE 7 Multiply: $(x + 5)(x + 4)$.

$(x + 5)(x + 4) = x(x + 4) + 5(x + 4)$	Using a distributive law
$= x \cdot x + x \cdot 4 + 5 \cdot x + 5 \cdot 4$	Using a distributive law on each part
$= x^2 + 4x + 5x + 20$	Multiplying the monomials
$= x^2 + 9x + 20$	Collecting like terms

To visualize the product in Example 7, consider a rectangle of length $x + 5$ and width $x + 4$.

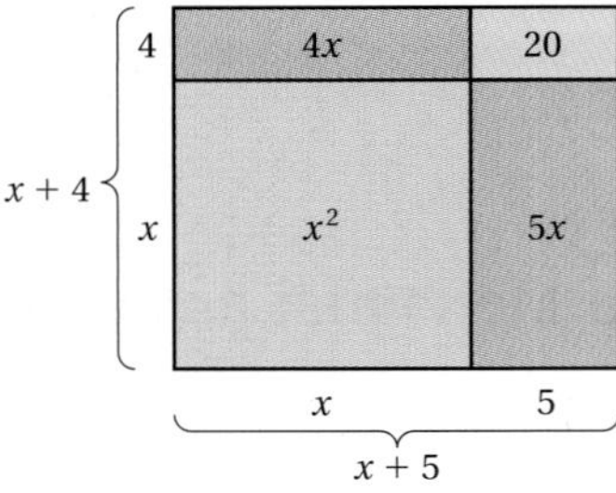

The total area can be expressed as $(x + 5)(x + 4)$ or, by adding the four smaller areas, $x^2 + 4x + 5x + 20$, or $x^2 + 9x + 20$.

Do Exercises 12–14.

Multiply.

9. $4x(2x + 4)$

10. $3t^2(-5t + 2)$

11. $-5x^3(x^3 + 5x^2 - 6x + 8)$

12. Multiply: $(y + 2)(y + 7)$.

a) Fill in the blanks in the steps of the solution below.

$(y + 2)(y + 7)$

$= y \cdot$ ______ $+ 2 \cdot$ ______

$= y \cdot$ ______ $+ y \cdot$ ______

$+ 2 \cdot$ ______ $+ 2 \cdot$ ______

$=$ ______ $+$ ______

$+$ ______ $+$ ______

$= y^2 +$ ______ $+ 14$

b) Write an algebraic expression that represents the total area of the four smaller rectangles in the figure shown here.

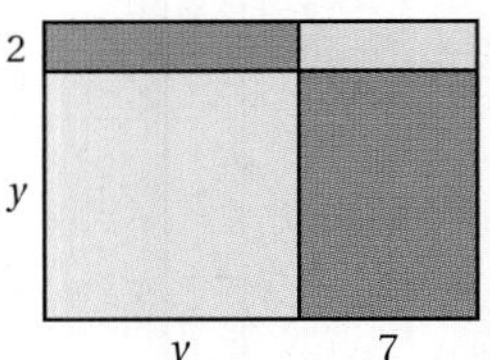

Multiply.

13. $(x + 8)(x + 5)$

14. $(x + 5)(x - 4)$

Answers on page A-36

EXAMPLE 8 Multiply: $(4x + 3)(x - 2)$.

$$\begin{aligned}(4x+3)(x-2) &= 4x(x-2) + 3(x-2) && \text{Using a distributive law}\\ &= 4x \cdot x - 4x \cdot 2 + 3 \cdot x - 3 \cdot 2 && \text{Using a distributive law on each part}\\ &= 4x^2 - 8x + 3x - 6 && \text{Multiplying the monomials}\\ &= 4x^2 - 5x - 6 && \text{Collecting like terms}\end{aligned}$$

Do Exercises 15 and 16.

Multiply.

15. $(5x + 3)(x - 4)$

16. $(2x - 3)(3x - 5)$

d Multiplying Any Two Polynomials

Let's consider the product of a binomial and a trinomial. We use a distributive law four times. You may see ways to skip some steps and do the work mentally.

EXAMPLE 9 Multiply: $(x^2 + 2x - 3)(x^2 + 4)$.

$$\begin{aligned}(x^2+2x-3)(x^2+4) &= x^2(x^2+4) + 2x(x^2+4) - 3(x^2+4)\\ &= x^2 \cdot x^2 + x^2 \cdot 4 + 2x \cdot x^2 + 2x \cdot 4 - 3 \cdot x^2 - 3 \cdot 4\\ &= x^4 + 4x^2 + 2x^3 + 8x - 3x^2 - 12\\ &= x^4 + 2x^3 + x^2 + 8x - 12\end{aligned}$$

Do Exercises 17 and 18.

Multiply.

17. $(x^2 + 3x - 4)(x^2 + 5)$

18. $(3y^2 - 7)(2y^3 - 2y + 5)$

PRODUCT OF TWO POLYNOMIALS

To multiply two polynomials P and Q, select one of the polynomials—say, P. Then multiply each term of P by every term of Q and collect like terms.

To use columns for long multiplication, multiply each term in the top row by every term in the bottom row. We write like terms in columns, and then add the results. Such multiplication is like multiplying with whole numbers.

$$\begin{array}{r} 3\,2\,1\\ \times\quad 1\,2\\ \hline 6\,4\,2\\ 3\,2\,1\\ \hline 3\,8\,5\,2\end{array}\qquad \begin{array}{r} 300 + 20 + 1\\ \times\qquad\qquad 10 + 2\\ \hline 600 + 40 + 2\\ 3000 + 200 + 10\\ \hline 3000 + 800 + 50 + 2\end{array}\qquad \begin{array}{l} \\ \\ \text{Multiplying the top row by 2}\\ \text{Multiplying the top row by 10}\\ \text{Adding}\end{array}$$

EXAMPLE 10 Multiply: $(4x^3 - 2x^2 + 3x)(x^2 + 2x)$.

$$\begin{array}{rl} 4x^3 - 2x^2 + 3x & \\ x^2 + 2x & \\ \hline 8x^4 - 4x^3 + 6x^2 & \text{Multiplying the top row by } 2x\\ 4x^5 - 2x^4 + 3x^3 & \text{Multiplying the top row by } x^2\\ \hline 4x^5 + 6x^4 - x^3 + 6x^2 & \text{Collecting like terms}\\ & \text{Line up like terms in columns.}\end{array}$$

Answers on page A-36

EXAMPLE 11 Multiply: $(5x^3 - 3x + 4)(-2x^2 - 3)$.

When missing terms occur, it helps to leave spaces for them and align like terms as we multiply.

$$
\begin{array}{rrrrrl}
 & 5x^3 & & -3x & +\ 4 & \\
 & & -2x^2 & & -\ 3 & \\
\hline
 & -15x^3 & & +9x & -\ 12 & \quad \text{Multiplying by } -3 \\
-10x^5 & +\ 6x^3 & -\ 8x^2 & & & \quad \text{Multiplying by } -2x^2 \\
\hline
-10x^5 & -\ 9x^3 & -\ 8x^2 & +\ 9x & -\ 12 & \quad \text{Collecting like terms}
\end{array}
$$

Do Exercises 19 and 20.

EXAMPLE 12 Multiply: $(2x^2 + 3x - 4)(2x^2 - x + 3)$.

$$
\begin{array}{rrrrrl}
 & & 2x^2 & +\ 3x & -\ 4 & \\
 & & 2x^2 & -\ \ x & +\ 3 & \\
\hline
 & & 6x^2 & +\ 9x & -\ 12 & \quad \text{Multiplying by } 3 \\
 & -2x^3 & -\ 3x^2 & +\ 4x & & \quad \text{Multiplying by } -x \\
4x^4 & +\ 6x^3 & -\ 8x^2 & & & \quad \text{Multiplying by } 2x^2 \\
\hline
4x^4 & +\ 4x^3 & -\ 5x^2 & +\ 13x & -\ 12 & \quad \text{Collecting like terms}
\end{array}
$$

Do Exercise 21.

Multiply.

19.
$$
\begin{array}{r}
3x^2 - 2x + 4 \\
x + 5 \\
\hline
\end{array}
$$

20.
$$
\begin{array}{r}
-5x^2 + 4x + 2 \\
-4x^2 - 8 \\
\hline
\end{array}
$$

21. Multiply.
$$
\begin{array}{r}
3x^2 - 2x - 5 \\
2x^2 + x - 2 \\
\hline
\end{array}
$$

Answers on page A-36

CALCULATOR CORNER

Checking Multiplication of Polynomials A partial check of multiplication of polynomials can be performed graphically on the TI-84 Plus graphing calculator. Consider the product $(x + 3)(x - 2) = x^2 + x - 6$. We will use two graph styles to determine whether this product is correct. First, we press MODE to determine whether SEQUENTIAL mode is selected. If it is not, we position the blinking cursor over SEQUENTIAL and then press ENTER. Next, on the Y = screen, we enter $y_1 = (x + 3)(x - 2)$ and $y_2 = x^2 + x - 6$. We will select the line-graph style for y_1 and the path style for y_2. To select these graph styles, we use ◀ to position the cursor over the icon to the left of the equation and press ENTER repeatedly until the desired style of icon appears, as shown below.

```
NORMAL SCI ENG
FLOAT 0123456789
RADIAN DEGREE
FUNC PAR POL SEQ
CONNECTED DOT
SEQUENTIAL SIMUL
REAL a+bi re^θi
FULL HORIZ G-T
```

```
Plot1  Plot2  Plot3
\Y1 ≡ (X+3)(X−2)
-0Y2 ≡ X²+X−6
\Y3 =
\Y4 =
\Y5 =
\Y6 =
\Y7 =
```

$y_1 = (x + 3)(x - 2),\ y_2 = x^2 + x - 6$

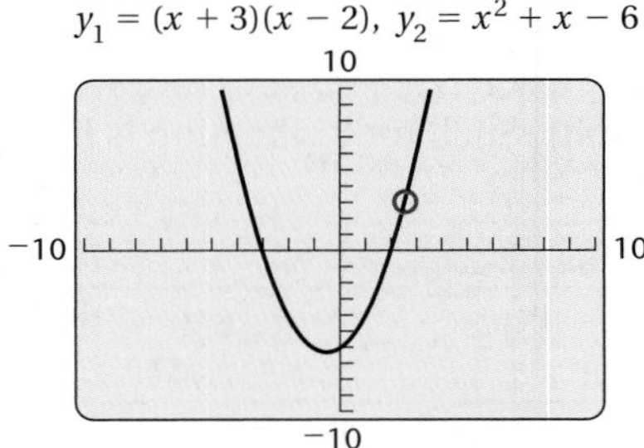

The graphing calculator will graph y_1 first as a solid line. Then it will graph y_2 as the circular cursor traces the leading edge of the graph, allowing us to determine visually whether the graphs coincide. In this case, the graphs appear to coincide, so the factorization is probably correct.

A table can also be used to perform a partial check of a product. See the Calculator Corner on p. 756 for the procedure.

Exercises: Determine graphically whether the product is correct.

1. $(x + 5)(x + 4) = x^2 + 9x + 20$
2. $(4x + 3)(x - 2) = 4x^2 - 5x - 6$
3. $(5x + 3)(x - 4) = 5x^2 + 17x - 12$
4. $(2x - 3)(3x - 5) = 6x^2 - 19x - 15$

10.5 EXERCISE SET

For Extra Help

MathXL | MyMathLab | InterAct Math | Math Tutor Center | Video Lectures on CD Disc 5 | Student's Solutions Manual

a Multiply.

1. $(8x^2)(5)$
2. $(4x^2)(-2)$
3. $(-x^2)(-x)$
4. $(-x^3)(x^2)$
5. $(8x^5)(4x^3)$
6. $(10a^2)(2a^2)$
7. $(0.1x^6)(0.3x^5)$
8. $(0.3x^4)(-0.8x^6)$
9. $\left(-\frac{1}{5}x^3\right)\left(-\frac{1}{3}x\right)$
10. $\left(-\frac{1}{4}x^4\right)\left(\frac{1}{5}x^8\right)$
11. $(-4x^2)(0)$
12. $(-4m^5)(-1)$
13. $(3x^2)(-4x^3)(2x^6)$
14. $(-2y^5)(10y^4)(-3y^3)$

b Multiply.

15. $2x(-x + 5)$
16. $3x(4x - 6)$
17. $-5x(x - 1)$
18. $-3x(-x - 1)$
19. $x^2(x^3 + 1)$
20. $-2x^3(x^2 - 1)$
21. $3x(2x^2 - 6x + 1)$
22. $-4x(2x^3 - 6x^2 - 5x + 1)$
23. $(-6x^2)(x^2 + x)$
24. $(-4x^2)(x^2 - x)$
25. $(3y^2)(6y^4 + 8y^3)$
26. $(4y^4)(y^3 - 6y^2)$

c Multiply.

27. $(x + 6)(x + 3)$
28. $(x + 5)(x + 2)$
29. $(x + 5)(x - 2)$
30. $(x + 6)(x - 2)$
31. $(x - 4)(x - 3)$
32. $(x - 7)(x - 3)$
33. $(x + 3)(x - 3)$
34. $(x + 6)(x - 6)$
35. $(5 - x)(5 - 2x)$
36. $(3 + x)(6 + 2x)$
37. $(2x + 5)(2x + 5)$
38. $(3x - 4)(3x - 4)$
39. $\left(x - \frac{5}{2}\right)\left(x + \frac{2}{5}\right)$
40. $\left(x + \frac{4}{3}\right)\left(x + \frac{3}{2}\right)$
41. $(x - 2.3)(x + 4.7)$
42. $(2x + 0.13)(2x - 0.13)$

Write an algebraic expression that represents the total area of the four smaller rectangles.

43.

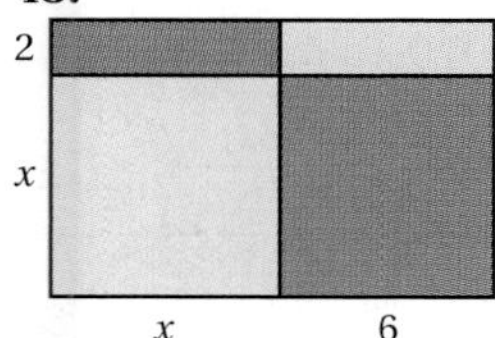

44.

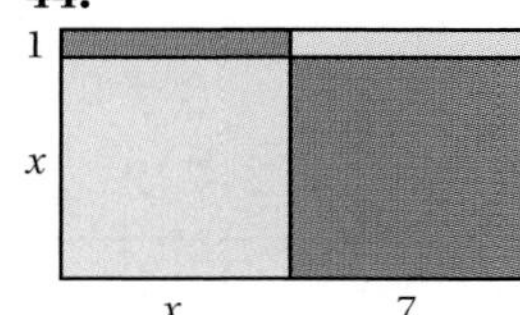

45.

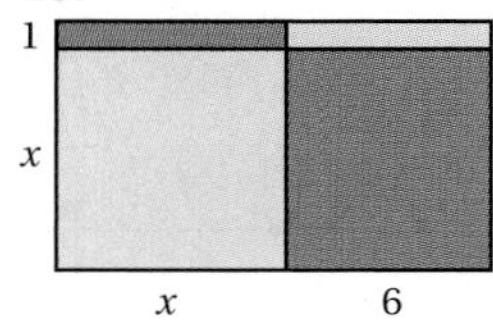

46.

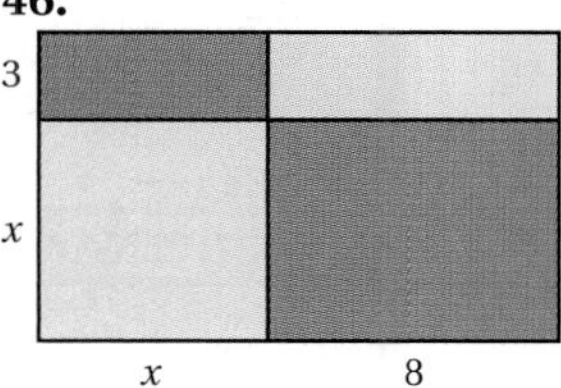

Draw and label rectangles similar to the one following Example 7 to illustrate each product.

47. $x(x + 5)$
48. $x(x + 2)$
49. $(x + 1)(x + 2)$
50. $(x + 3)(x + 1)$
51. $(x + 5)(x + 3)$
52. $(x + 4)(x + 6)$

d Multiply.

53. $(x^2 + x + 1)(x - 1)$

54. $(x^2 + x - 2)(x + 2)$

55. $(2x + 1)(2x^2 + 6x + 1)$

56. $(3x - 1)(4x^2 - 2x - 1)$

57. $(y^2 - 3)(3y^2 - 6y + 2)$

58. $(3y^2 - 3)(y^2 + 6y + 1)$

59. $(x^3 + x^2)(x^3 + x^2 - x)$

60. $(x^3 - x^2)(x^3 - x^2 + x)$

61. $(-5x^3 - 7x^2 + 1)(2x^2 - x)$

62. $(-4x^3 + 5x^2 - 2)(5x^2 + 1)$

63. $(1 + x + x^2)(-1 - x + x^2)$

64. $(1 - x + x^2)(1 - x + x^2)$

65. $(2t^2 - t - 4)(3t^2 + 2t - 1)$

66. $(3a^2 - 5a + 2)(2a^2 - 3a + 4)$

67. $(x - x^3 + x^5)(x^2 - 1 + x^4)$

68. $(x - x^3 + x^5)(3x^2 + 3x^6 + 3x^4)$

69. $(x^3 + x^2 + x + 1)(x - 1)$

70. $(x + 2)(x^3 - x^2 + x - 2)$

71. $(x + 1)(x^3 + 7x^2 + 5x + 4)$

72. $(x + 2)(x^3 + 5x^2 + 9x + 3)$

73. $\left(x - \frac{1}{2}\right)\left(2x^3 - 4x^2 + 3x - \frac{2}{5}\right)$

74. $\left(x + \frac{1}{3}\right)\left(6x^3 - 12x^2 - 5x + \frac{1}{2}\right)$

75. $\mathbf{D_W}$ Under what conditions will the product of two binomials be a trinomial?

76. $\mathbf{D_W}$ How can the following figure be used to show that $(x + 3)^2 \neq x^2 + 9$?

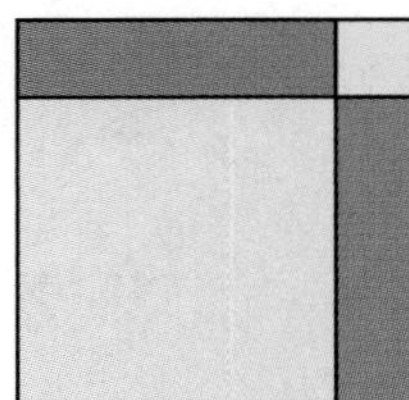

SKILL MAINTENANCE

Simplify.

77. $-\frac{1}{4} - \frac{1}{2}$ [7.4a]

78. $-3.8 - (-10.2)$ [7.4a]

79. $(10 - 2)(10 + 2)$ [7.8d]

80. $10 - 2 + (-6)^2 \div 3 \cdot 2$ [7.8d]

Factor. [7.7d]

81. $15x - 18y + 12$

82. $16x - 24y + 36$

83. $-9x - 45y + 15$

84. $100x - 100y + 1000a$

85. Graph: $y = \frac{1}{2}x - 3$. [9.5a]

86. Solve: $4(x - 3) = 5(2 - 3x) + 1$. [8.3c]

SYNTHESIS

Find a polynomial for the shaded area of the figure.

87.

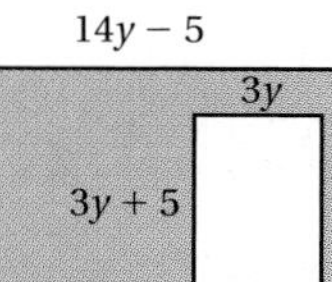

88.

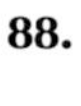

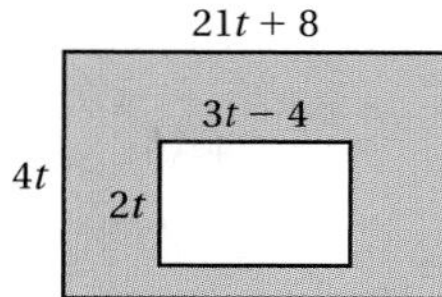

89. A box with a square bottom is to be made from a 12-in.-square piece of cardboard. Squares with side x are cut out of the corners and the sides are folded up. Find the polynomials for the volume and the outside surface area of the box.

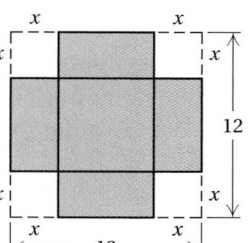

For each figure, determine what the missing number must be in order for the figure to have the given area.

90. Area $= x^2 + 7x + 10$

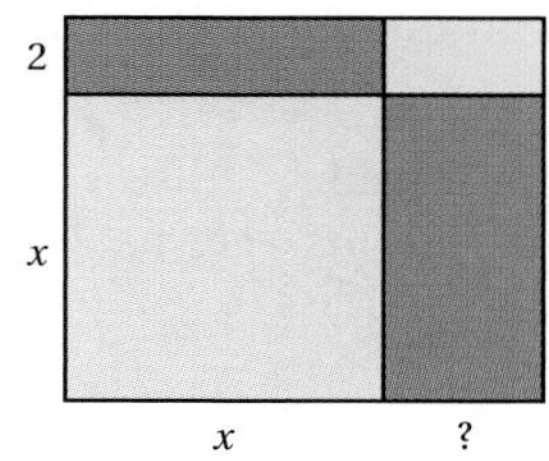

91. Area $= x^2 + 8x + 15$

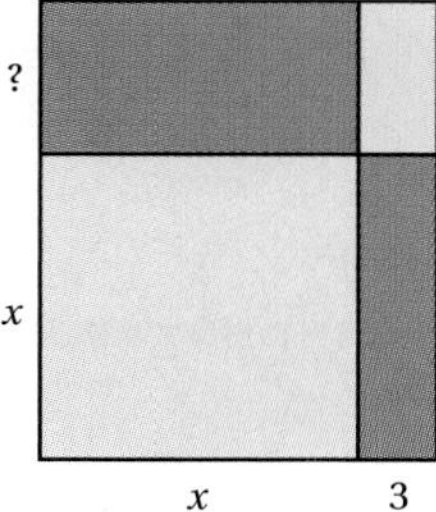

92. An open wooden box is a cube with side x cm. The box, including its bottom, is made of wood that is 1 cm thick. Find a polynomial for the interior volume of the cube.

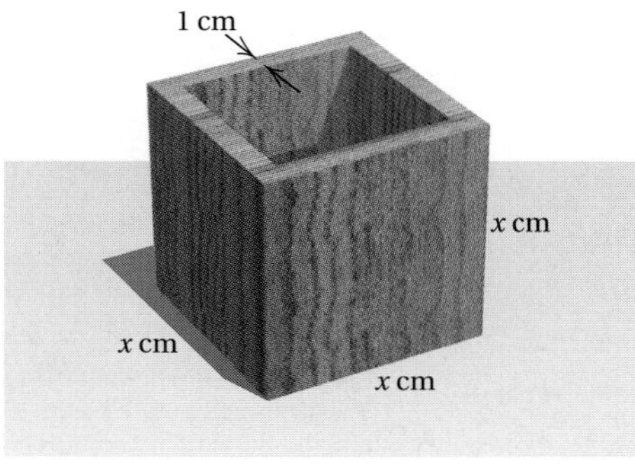

93. Find a polynomial for the volume of the solid shown below.

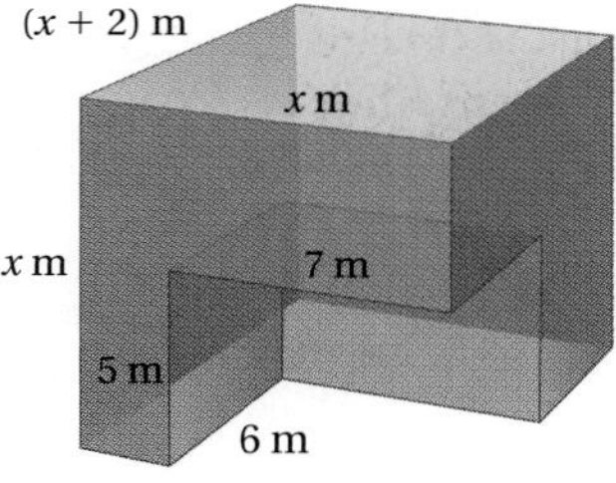

Compute and simplify.

94. $(x + 3)(x + 6) + (x + 3)(x + 6)$

95. $(x - 2)(x - 7) - (x - 7)(x - 2)$

96. $(x + 5)^2 - (x - 3)^2$

97. Extend the pattern and simplify:

$$(x - a)(x - b)(x - c)(x - d) \cdots (x - z).$$

98. Use a graphing calculator to check your answers to Exercises 15, 29, and 53. Use graphs, tables, or both, as directed by your instructor.

10.6 SPECIAL PRODUCTS

Objectives

- **a** Multiply two binomials mentally using the FOIL method.
- **b** Multiply the sum and the difference of two terms mentally.
- **c** Square a binomial mentally.
- **d** Find special products when polynomial products are mixed together.

We encounter certain products so often that it is helpful to have faster methods of computing. Such techniques are called *special products*. We now consider special ways of multiplying any two binomials.

a Products of Two Binomials Using FOIL

To multiply two binomials, we can select one binomial and multiply each term of that binomial by every term of the other. Then we collect like terms. Consider the product $(x + 3)(x + 7)$:

$$\begin{aligned}(x + 3)(x + 7) &= x(x + 7) + 3(x + 7)\\ &= x \cdot x + x \cdot 7 + 3 \cdot x + 3 \cdot 7\\ &= x^2 + 7x + 3x + 21\\ &= x^2 + 10x + 21.\end{aligned}$$

This example illustrates a special technique for finding the product of two binomials:

$$(x + 3)(x + 7) = \underbrace{x \cdot x}_{\text{First terms}} + \underbrace{7 \cdot x}_{\text{Outside terms}} + \underbrace{3 \cdot x}_{\text{Inside terms}} + \underbrace{3 \cdot 7}_{\text{Last terms}}.$$

To remember this method of multiplying, we use the initials **FOIL.**

THE FOIL METHOD

To multiply two binomials, $A + B$ and $C + D$, multiply the First terms AC, the Outside terms AD, the Inside terms BC, and then the Last terms BD. Then collect like terms, if possible.

$$(A + B)(C + D) = AC + AD + BC + BD$$

1. Multiply First terms: AC.
2. Multiply Outside terms: AD.
3. Multiply Inside terms: BC.
4. Multiply Last terms: BD.

↓

FOIL

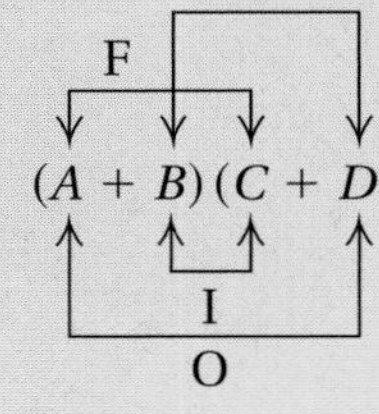

Study Tips

CHECKLIST

The foundation of all your study skills is TIME!

- ☐ Are you taking the time to include all the steps when working your homework and the tests?
- ☐ Are you using the time-management suggestions we have given so you have the proper amount of time to study mathematics?
- ☐ Have you been using the supplements for the text such as the *Student's Solutions Manual* and the AW Math Tutor Center?
- ☐ Have you memorized the rules for special products of polynomials and for manipulating expressions with exponents?

EXAMPLE 1 Multiply: $(x + 8)(x^2 - 5)$.

We have

$$\begin{aligned}(x + 8)(x^2 - 5) &= \overset{F}{x \cdot x^2} + \overset{O}{x \cdot (-5)} + \overset{I}{8 \cdot x^2} + \overset{L}{8(-5)}\\ &= x^3 - 5x + 8x^2 - 40\\ &= x^3 + 8x^2 - 5x - 40.\end{aligned}$$

Since each of the original binomials is in descending order, we write the product in descending order, as is customary, but this is not a "must."

Often we can collect like terms after we have multiplied.

EXAMPLES Multiply.

2. $(x + 6)(x - 6) = x^2 - 6x + 6x - 36$ Using FOIL
$= x^2 - 36$ Collecting like terms

3. $(x + 7)(x + 4) = x^2 + 4x + 7x + 28$
$= x^2 + 11x + 28$

4. $(y - 3)(y - 2) = y^2 - 2y - 3y + 6$
$= y^2 - 5y + 6$

5. $(x^3 - 5)(x^3 + 5) = x^6 + 5x^3 - 5x^3 - 25$
$= x^6 - 25$

6. $(4t^3 + 5)(3t^2 - 2) = 12t^5 - 8t^3 + 15t^2 - 10$

Do Exercises 1–8.

EXAMPLES Multiply.

7. $\left(x - \frac{2}{3}\right)\left(x + \frac{2}{3}\right) = x^2 + \frac{2}{3}x - \frac{2}{3}x - \frac{4}{9}$
$= x^2 - \frac{4}{9}$

8. $(x^2 - 0.3)(x^2 - 0.3) = x^4 - 0.3x^2 - 0.3x^2 + 0.09$
$= x^4 - 0.6x^2 + 0.09$

9. $(3 - 4x)(7 - 5x^3) = 21 - 15x^3 - 28x + 20x^4$
$= 21 - 28x - 15x^3 + 20x^4$

(*Note*: If the original polynomials are in ascending order, it is natural to write the product in ascending order, but this is not a "must.")

10. $(5x^4 + 2x^3)(3x^2 - 7x) = 15x^6 - 35x^5 + 6x^5 - 14x^4$
$= 15x^6 - 29x^5 - 14x^4$

Do Exercises 9–12.

We can show the FOIL method geometrically as follows.

The area of the large rectangle is $(A + B)(C + D)$.

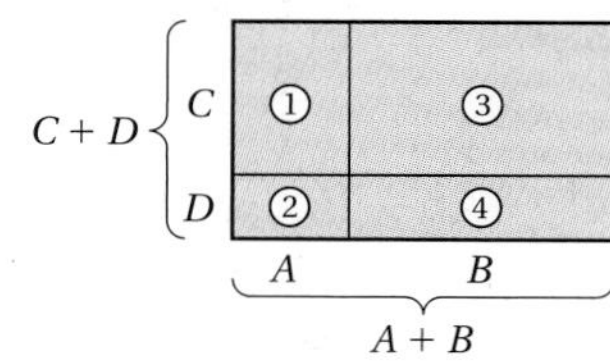

The area of rectangle ① is AC.
The area of rectangle ② is AD.
The area of rectangle ③ is BC.
The area of rectangle ④ is BD.

The area of the large rectangle is the sum of the areas of the smaller rectangles. Thus,

$$(A + B)(C + D) = AC + AD + BC + BD.$$

b Multiplying Sums and Differences of Two Terms

Consider the product of the sum and the difference of the same two terms, such as

$$(x + 2)(x - 2).$$

Multiply mentally, if possible. If you need extra steps, be sure to use them.

1. $(x + 3)(x + 4)$

2. $(x + 3)(x - 5)$

3. $(2x - 1)(x - 4)$

4. $(2x^2 - 3)(x - 2)$

5. $(6x^2 + 5)(2x^3 + 1)$

6. $(y^3 + 7)(y^3 - 7)$

7. $(t + 5)(t + 3)$

8. $(2x^4 + x^2)(-x^3 + x)$

Multiply.

9. $\left(x + \frac{4}{5}\right)\left(x - \frac{4}{5}\right)$

10. $(x^3 - 0.5)(x^2 + 0.5)$

11. $(2 + 3x^2)(4 - 5x^2)$

12. $(6x^3 - 3x^2)(5x^2 - 2x)$

Answers on page A-36

Multiply.

13. $(x + 5)(x - 5)$

14. $(2x - 3)(2x + 3)$

Since this is the product of two binomials, we can use FOIL. This type of product occurs so often, however, that it would be valuable if we could use an even faster method. To find a faster way to compute such a product, look for a pattern in the following:

a) $(x + 2)(x - 2) = x^2 - 2x + 2x - 4$ Using FOIL
$= x^2 - 4;$

b) $(3x - 5)(3x + 5) = 9x^2 + 15x - 15x - 25$
$= 9x^2 - 25.$

Do Exercises 13 and 14.

Perhaps you discovered in each case that when you multiply the two binomials, two terms are opposites, or additive inverses, which add to 0 and "drop out."

Multiply.

15. $(x + 2)(x - 2)$

16. $(x - 7)(x + 7)$

17. $(6 - 4y)(6 + 4y)$

18. $(2x^3 - 1)(2x^3 + 1)$

19. $\left(x - \frac{2}{5}\right)\left(x + \frac{2}{5}\right)$

Answers on page A-36

PRODUCT OF THE SUM AND THE DIFFERENCE

The product of the sum and the difference of the same two terms is the square of the first term minus the square of the second term:

$$(A + B)(A - B) = A^2 - B^2.$$

It is helpful to memorize this rule in both words and symbols. (If you do forget it, you can, of course, use FOIL.)

EXAMPLES Multiply. (Carry out the rule and say the words as you go.)

$(A + B)(A - B) = A^2 - B^2$
↓ ↓ ↓ ↓ ↓ ↓

11. $(x + 4)(x - 4) = x^2 - 4^2$ "The square of the first term, x^2, minus the square of the second, 4^2"
$= x^2 - 16$ Simplifying

12. $(5 + 2w)(5 - 2w) = 5^2 - (2w)^2$
$= 25 - 4w^2$

13. $(3x^2 - 7)(3x^2 + 7) = (3x^2)^2 - 7^2$
$= 9x^4 - 49$

14. $(-4x - 10)(-4x + 10) = (-4x)^2 - 10^2$
$= 16x^2 - 100$

15. $\left(x + \frac{3}{8}\right)\left(x - \frac{3}{8}\right) = x^2 - \left(\frac{3}{8}\right)^2 = x^2 - \frac{9}{64}$

Do Exercises 15–19.

C Squaring Binomials

Consider the square of a binomial, such as $(x + 3)^2$. This can be expressed as $(x + 3)(x + 3)$. Since this is the product of two binomials, we can again use FOIL. But again, this type of product occurs so often that we would like to use an even faster method. Look for a pattern in the following:

a) $(x + 3)^2 = (x + 3)(x + 3)$
$= x^2 + 3x + 3x + 9$
$= x^2 + 6x + 9;$

b) $(x - 3)^2 = (x - 3)(x - 3)$
$= x^2 - 3x - 3x + 9$
$= x^2 - 6x + 9;$

c) $(5 + 3p)^2 = (5 + 3p)(5 + 3p)$
$= 25 + 15p + 15p + 9p^2$
$= 25 + 30p + 9p^2;$

d) $(3x - 5)^2 = (3x - 5)(3x - 5)$
$= 9x^2 - 15x - 15x + 25$
$= 9x^2 - 30x + 25.$

Do Exercises 20 and 21.

When squaring a binomial, we multiply a binomial by itself. Perhaps you noticed that two terms are the same and when added give twice the product of the terms in the binomial. The other two terms are squares.

SQUARE OF A BINOMIAL

The square of a sum or a difference of two terms is the square of the first term, plus or minus twice the product of the two terms, plus the square of the last term:

$$(A + B)^2 = A^2 + 2AB + B^2; \qquad (A - B)^2 = A^2 - 2AB + B^2.$$

It is helpful to memorize this rule in both words and symbols.

EXAMPLES Multiply. (Carry out the rule and say the words as you go.)

$(A + B)^2 = A^2 + 2 \cdot A \cdot B + B^2$

16. $(x + 3)^2 = x^2 + 2 \cdot x \cdot 3 + 3^2$ "x^2 plus 2 times x times 3 plus 3^2"
$= x^2 + 6x + 9$

$(A - B)^2 = A^2 - 2 \cdot A \cdot B + B^2$

17. $(t - 5)^2 = t^2 - 2 \cdot t \cdot 5 + 5^2$ "t^2 minus 2 times t times 5 plus 5^2"
$= t^2 - 10t + 25$

18. $(2x + 7)^2 = (2x)^2 + 2 \cdot 2x \cdot 7 + 7^2 = 4x^2 + 28x + 49$

19. $(5x - 3x^2)^2 = (5x)^2 - 2 \cdot 5x \cdot 3x^2 + (3x^2)^2 = 25x^2 - 30x^3 + 9x^4$

20. $(2.3 - 5.4m)^2 = 2.3^2 - 2(2.3)(5.4m) + (5.4m)^2$
$= 5.29 - 24.84m + 29.16m^2$

Do Exercises 22–27.

Caution!

Although the square of a product is the product of the squares, the square of a sum is *not* the sum of the squares. That is, $(AB)^2 = A^2B^2$, but

The term $2AB$ is missing.

$$(A + B)^2 \neq A^2 + B^2.$$

To confirm this inequality, note, using the rules for order of operations, that

$$(7 + 5)^2 = 12^2 = 144,$$

whereas

$$7^2 + 5^2 = 49 + 25 = 74, \quad \text{and} \quad 74 \neq 144.$$

Multiply.

20. $(x + 8)(x + 8)$

21. $(x - 5)(x - 5)$

Multiply.

22. $(x + 2)^2$

23. $(a - 4)^2$

24. $(2x + 5)^2$

25. $(4x^2 - 3x)^2$

26. $(7.8 + 1.2y)(7.8 + 1.2y)$

27. $(3x^2 - 5)(3x^2 - 5)$

Answers on page A-36

28. In the figure at right, describe in terms of area the sum $A^2 + B^2$. How can the figure be used to verify that $(A + B)^2 \neq A^2 + B^2$?

Answer on page A-36

Study Tips

MEMORIZING FORMULAS

Memorizing can be a very helpful tool in the study of mathematics. Don't underestimate its power as you consider the special products. Consider putting the rules, in words and in math symbols, on index cards and go over them many times.

We can look at the rule for finding $(A + B)^2$ geometrically as follows. The area of the large square is

$$(A + B)(A + B) = (A + B)^2.$$

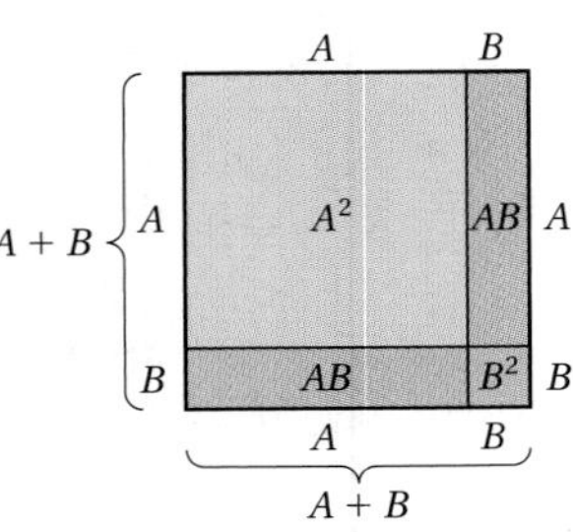

This is equal to the sum of the areas of the smaller rectangles:

$$A^2 + AB + AB + B^2 = A^2 + 2AB + B^2.$$

Thus, $(A + B)^2 = A^2 + 2AB + B^2$.

Do Exercise 28.

d Multiplication of Various Types

Let's now try several types of multiplications mixed together so that we can learn to sort them out. When you multiply, first see what kind of multiplication you have. Then use the best method.

MULTIPLYING TWO POLYNOMIALS

1. Is it the product of a monomial and a polynomial? If so, multiply each term of the polynomial by the monomial.
   Example: $5x(x + 7) = 5x \cdot x + 5x \cdot 7 = 5x^2 + 35x$
2. Is it the product of the sum and the difference of the *same* two terms? If so, use the following:
   $$(A + B)(A - B) = A^2 - B^2.$$
   The product of the sum and the difference of the same two terms is the difference of the squares. [The answer has 2 terms.]
   Example: $(x + 7)(x - 7) = x^2 - 7^2 = x^2 - 49$
3. Is the product the square of a binomial? If so, use the following:
   $$(A + B)(A + B) = (A + B)^2 = A^2 + 2AB + B^2,$$
   or $$(A - B)(A - B) = (A - B)^2 = A^2 - 2AB + B^2.$$
   The square of a binomial is the square of the first term, plus or minus *twice* the product of the two terms, plus the square of the last term. [The answer has 3 terms.]
   Example: $(x + 7)(x + 7) = (x + 7)^2$
   $= x^2 + 2 \cdot x \cdot 7 + 7^2 = x^2 + 14x + 49$
4. Is it the product of two binomials other than those above? If so, use FOIL. [The answer will have 3 or 4 terms.]
   Example: $(x + 7)(x - 4) = x^2 - 4x + 7x - 28 = x^2 + 3x - 28$
5. Is it the product of two polynomials other than those above? If so, multiply each term of one by every term of the other. Use columns if you wish. [The answer will have 2 or more terms, usually more than 2 terms.]
   Example:
   $$\begin{aligned}(x^2 - 3x + 2)(x + 7) &= x^2(x + 7) - 3x(x + 7) + 2(x + 7)\\ &= x^2 \cdot x + x^2 \cdot 7 - 3x \cdot x - 3x \cdot 7\\ &\quad + 2 \cdot x + 2 \cdot 7\\ &= x^3 + 7x^2 - 3x^2 - 21x + 2x + 14\\ &= x^3 + 4x^2 - 19x + 14\end{aligned}$$

Remember that FOIL will *always* work for two binomials. You can use it instead of either of rules 2 and 3, but those rules will make your work go faster.

EXAMPLE 21 Multiply: $(x + 3)(x - 3)$.

$(x + 3)(x - 3) = x^2 - 9$ Using method 2 (the product of the sum and the difference of two terms)

EXAMPLE 22 Multiply: $(t + 7)(t - 5)$.

$(t + 7)(t - 5) = t^2 + 2t - 35$ Using method 4 (the product of two binomials, but neither the square of a binomial nor the product of the sum and the difference of two terms)

EXAMPLE 23 Multiply: $(x + 6)(x + 6)$.

$$(x + 6)(x + 6) = x^2 + 2(6)x + 36$$ Using method 3 (the square of a binomial sum)

$$= x^2 + 12x + 36$$

EXAMPLE 24 Multiply: $2x^3(9x^2 + x - 7)$.

$2x^3(9x^2 + x - 7) = 18x^5 + 2x^4 - 14x^3$ Using method 1 (the product of a monomial and a trinomial; multiplying each term of the trinomial by the monomial)

EXAMPLE 25 Multiply: $(5x^3 - 7x)^2$.

$$(5x^3 - 7x)^2 = 25x^6 - 2(5x^3)(7x) + 49x^2$$ Using method 3 (the square of a binomial difference)

$$= 25x^6 - 70x^4 + 49x^2$$

EXAMPLE 26 Multiply: $\left(3x + \frac{1}{4}\right)^2$.

$$\left(3x + \tfrac{1}{4}\right)^2 = 9x^2 + 2(3x)\left(\tfrac{1}{4}\right) + \tfrac{1}{16}$$ Using method 3 (the square of a binomial sum. To get the middle term, we multiply $3x$ by $\frac{1}{4}$ and double.)

$$= 9x^2 + \tfrac{3}{2}x + \tfrac{1}{16}$$

EXAMPLE 27 Multiply: $\left(4x - \frac{3}{4}\right)^2$.

$$\left(4x - \tfrac{3}{4}\right)^2 = 16x^2 - 2(4x)\left(\tfrac{3}{4}\right) + \tfrac{9}{16}$$ Using method 3 (the square of a binomial difference)

$$= 16x^2 - 6x + \tfrac{9}{16}$$

EXAMPLE 28 Multiply: $(p + 3)(p^2 + 2p - 1)$.

$$\begin{array}{r} p^2 + 2p - 1 \\ p + 3 \\ \hline 3p^2 + 6p - 3 \\ p^3 + 2p^2 - p \phantom{{}-3} \\ \hline p^3 + 5p^2 + 5p - 3 \end{array}$$

Using method 5 (the product of two polynomials)

Multiplying by 3

Multiplying by p

Do Exercises 29–36.

Multiply.

29. $(x + 5)(x + 6)$

30. $(t - 4)(t + 4)$

31. $4x^2(-2x^3 + 5x^2 + 10)$

32. $(9x^2 + 1)^2$

33. $(2a - 5)(2a + 8)$

34. $\left(5x + \dfrac{1}{2}\right)^2$

35. $\left(2x - \dfrac{1}{2}\right)^2$

36. $(x^2 - x + 4)(x - 2)$

Answers on page A-36

Visualizing for Success

1

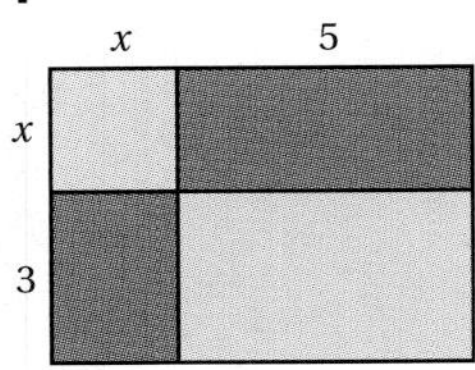

2

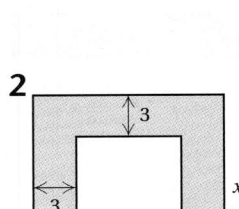

3

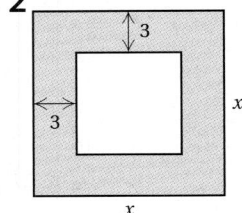

4

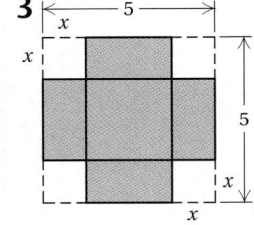

5

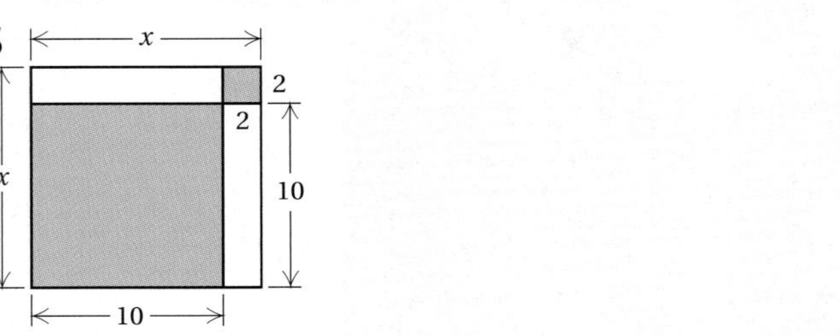

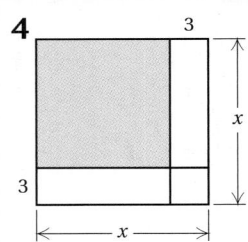

In each of Exercises 1–10, find two algebraic expressions for the shaded area of the figure from the list below.

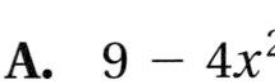

A. $9 - 4x^2$
B. $x^2 - (x - 6)^2$
C. $(x + 3)(x - 3)$
D. $10^2 + 2^2$
E. $x^2 + 8x + 15$
F. $(x + 5)(x + 3)$
G. $x^2 - 6x + 9$
H. $(3 - 2x)^2 + 4x(3 - 2x)$
I. $(x + 3)^2$
J. $(5x + 3)^2$
K. $(5 - 2x)^2 + 4x(5 - 2x)$
L. $x^2 - 9$
M. 104
N. $x^2 - 15$
O. $12x - 36$
P. $25x^2 + 30x + 9$
Q. $(x - 5)(x - 3)$
$\quad + 3(x - 5) + 5(x - 3)$
R. $(x - 3)^2$
S. $25 - 4x^2$
T. $x^2 + 6x + 9$

Answers on page A-37

6

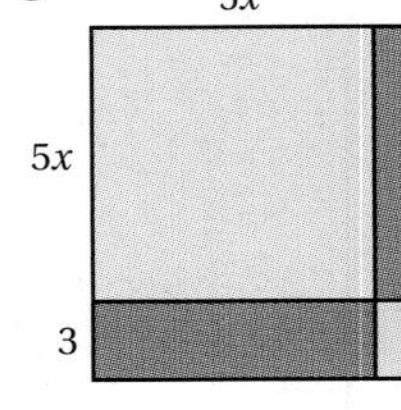

7

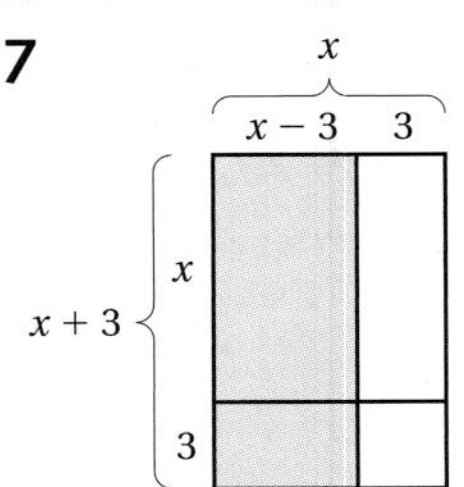

8

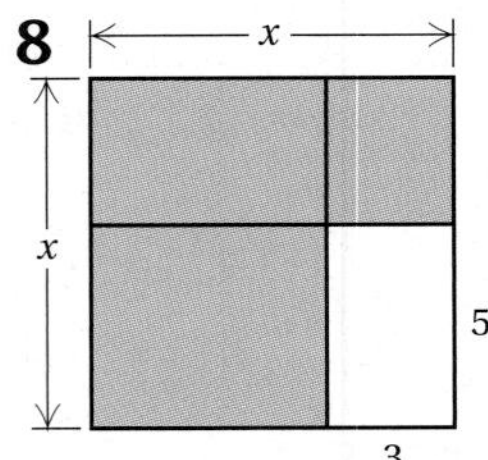

9

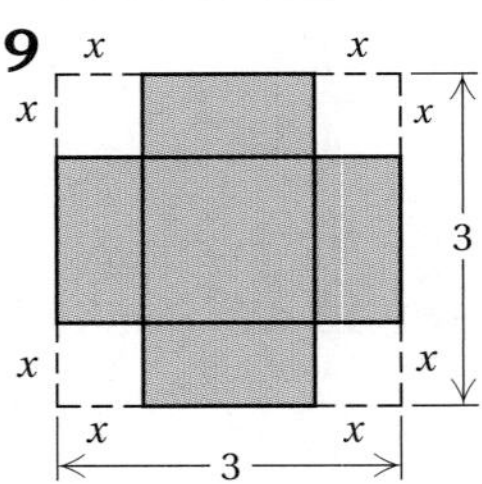

10

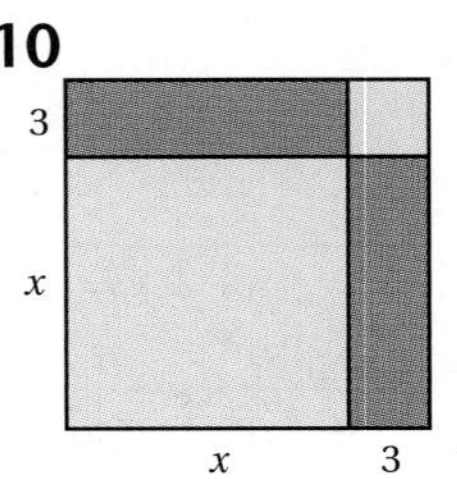

10.6 EXERCISE SET

For Extra Help

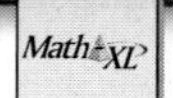
MathXL

MyMathLab

InterAct Math

Math Tutor Center
Video Lectures on CD Disc 5

Student's Solutions Manual

a Multiply. Try to write only the answer. If you need more steps, be sure to use them.

1. $(x + 1)(x^2 + 3)$

2. $(x^2 - 3)(x - 1)$

3. $(x^3 + 2)(x + 1)$

4. $(x^4 + 2)(x + 10)$

5. $(y + 2)(y - 3)$

6. $(a + 2)(a + 3)$

7. $(3x + 2)(3x + 2)$

8. $(4x + 1)(4x + 1)$

9. $(5x - 6)(x + 2)$

10. $(x - 8)(x + 8)$

11. $(3t - 1)(3t + 1)$

12. $(2m + 3)(2m + 3)$

13. $(4x - 2)(x - 1)$

14. $(2x - 1)(3x + 1)$

15. $\left(p - \frac{1}{4}\right)\left(p + \frac{1}{4}\right)$

16. $\left(q + \frac{3}{4}\right)\left(q + \frac{3}{4}\right)$

17. $(x - 0.1)(x + 0.1)$

18. $(x + 0.3)(x - 0.4)$

19. $(2x^2 + 6)(x + 1)$

20. $(2x^2 + 3)(2x - 1)$

21. $(-2x + 1)(x + 6)$

22. $(3x + 4)(2x - 4)$

23. $(a + 7)(a + 7)$

24. $(2y + 5)(2y + 5)$

25. $(1 + 2x)(1 - 3x)$

26. $(-3x - 2)(x + 1)$

27. $\left(\frac{3}{8}y - \frac{5}{6}\right)\left(\frac{3}{8}y - \frac{5}{6}\right)$

28. $\left(\frac{1}{5}x - \frac{2}{7}\right)\left(\frac{1}{5}x + \frac{2}{7}\right)$

29. $(x^2 + 3)(x^3 - 1)$

30. $(x^4 - 3)(2x + 1)$

31. $(3x^2 - 2)(x^4 - 2)$

32. $(x^{10} + 3)(x^{10} - 3)$

33. $(2.8x - 1.5)(4.7x + 9.3)$

34. $\left(x - \frac{3}{8}\right)\left(x + \frac{4}{7}\right)$

35. $(3x^5 + 2)(2x^2 + 6)$

36. $(1 - 2x)(1 + 3x^2)$

37. $(8x^3 + 1)(x^3 + 8)$

38. $(4 - 2x)(5 - 2x^2)$

39. $(4x^2 + 3)(x - 3)$

40. $(7x - 2)(2x - 7)$

41. $(4y^4 + y^2)(y^2 + y)$

42. $(5y^6 + 3y^3)(2y^6 + 2y^3)$

b Multiply mentally, if possible. If you need extra steps, be sure to use them.

43. $(x + 4)(x - 4)$

44. $(x + 1)(x - 1)$

45. $(2x + 1)(2x - 1)$

46. $(x^2 + 1)(x^2 - 1)$

47. $(5m - 2)(5m + 2)$

48. $(3x^4 + 2)(3x^4 - 2)$

49. $(2x^2 + 3)(2x^2 - 3)$

50. $(6x^5 - 5)(6x^5 + 5)$

51. $(3x^4 - 4)(3x^4 + 4)$

52. $(t^2 - 0.2)(t^2 + 0.2)$

53. $(x^6 - x^2)(x^6 + x^2)$

54. $(2x^3 - 0.3)(2x^3 + 0.3)$

55. $(x^4 + 3x)(x^4 - 3x)$

56. $\left(\frac{3}{4} + 2x^3\right)\left(\frac{3}{4} - 2x^3\right)$

57. $(x^{12} - 3)(x^{12} + 3)$

58. $(12 - 3x^2)(12 + 3x^2)$

59. $(2y^8 + 3)(2y^8 - 3)$

60. $\left(m - \frac{2}{3}\right)\left(m + \frac{2}{3}\right)$

61. $\left(\frac{5}{8}x - 4.3\right)\left(\frac{5}{8}x + 4.3\right)$

62. $(10.7 - x^3)(10.7 + x^3)$

c Multiply mentally, if possible. If you need extra steps, be sure to use them.

63. $(x + 2)^2$

64. $(2x - 1)^2$

65. $(3x^2 + 1)^2$

66. $\left(3x + \frac{3}{4}\right)^2$

67. $\left(a - \frac{1}{2}\right)^2$

68. $\left(2a - \frac{1}{5}\right)^2$

69. $(3 + x)^2$

70. $(x^3 - 1)^2$

71. $(x^2 + 1)^2$

72. $(8x - x^2)^2$

73. $(2 - 3x^4)^2$

74. $(6x^3 - 2)^2$

75. $(5 + 6t^2)^2$

76. $(3p^2 - p)^2$

77. $\left(x - \frac{5}{8}\right)^2$

78. $(0.3y + 2.4)^2$

d Multiply mentally, if possible.

79. $(3 - 2x^3)^2$

80. $(x - 4x^3)^2$

81. $4x(x^2 + 6x - 3)$

82. $8x(-x^5 + 6x^2 + 9)$

83. $\left(2x^2 - \frac{1}{2}\right)\left(2x^2 - \frac{1}{2}\right)$

84. $(-x^2 + 1)^2$

85. $(-1 + 3p)(1 + 3p)$

86. $(-3q + 2)(3q + 2)$

87. $3t^2(5t^3 - t^2 + t)$

88. $-6x^2(x^3 + 8x - 9)$

89. $(6x^4 + 4)^2$

90. $(8a + 5)^2$

91. $(3x + 2)(4x^2 + 5)$

92. $(2x^2 - 7)(3x^2 + 9)$

93. $(8 - 6x^4)^2$

94. $\left(\frac{1}{5}x^2 + 9\right)\left(\frac{3}{5}x^2 - 7\right)$

95. $(t - 1)(t^2 + t + 1)$

96. $(y + 5)(y^2 - 5y + 25)$

Compute each of the following and compare.

97. $3^2 + 4^2$; $(3 + 4)^2$

98. $6^2 + 7^2$; $(6 + 7)^2$

99. $9^2 - 5^2$; $(9 - 5)^2$

100. $11^2 - 4^2$; $(11 - 4)^2$

Find the total area of all the shaded rectangles.

101.

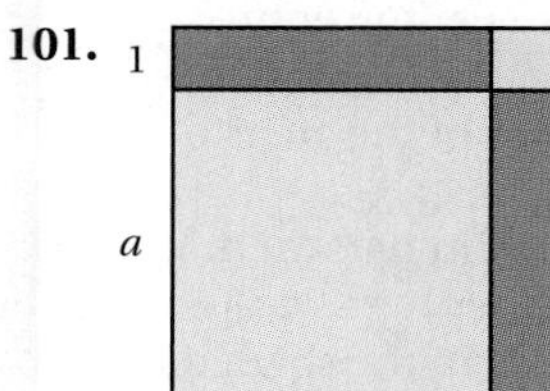

102.

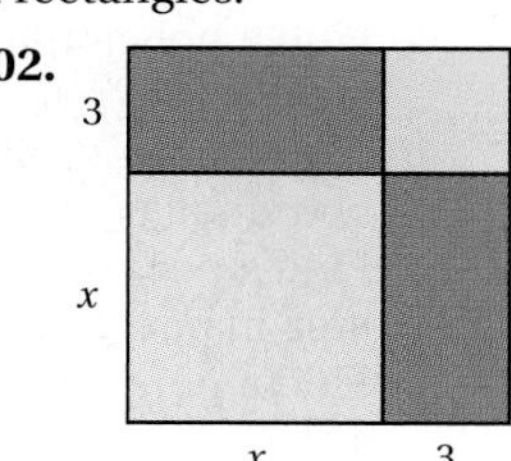

103.

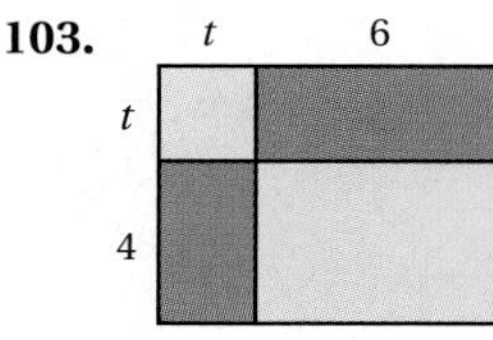

104.

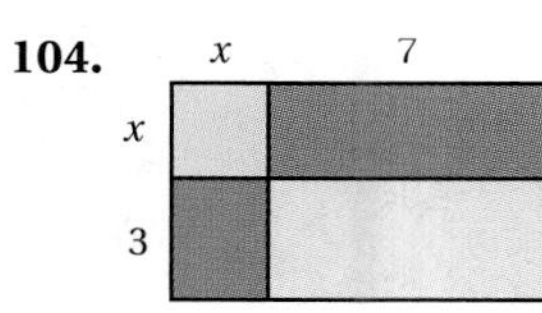

105. D_W Under what conditions is the product of two binomials a binomial?

106. D_W Brittney feels that since the FOIL method can be used to find the product of any two binomials, she needn't study the other special products. What advice would you give her?

SKILL MAINTENANCE

107. *Electricity Usage.* In apartment 3B, lamps, an air conditioner, and a television set are all operating at the same time. The lamps use 10 times as many watts of electricity as the television set, and the air conditioner uses 40 times as many watts as the television set. The total wattage used in the apartment is 2550. How many watts are used by each appliance? [8.6a]

Solve. [8.3c]

108. $3x - 8x = 4(7 - 8x)$

109. $3(x - 2) = 5(2x + 7)$

110. $5(2x - 3) - 2(3x - 4) = 20$

Solve. [8.4b]

111. $3x - 2y = 12$, for y

112. $3a - 5d = 4$, for a

SYNTHESIS

Multiply.

113. $5x(3x - 1)(2x + 3)$

114. $[(2x - 3)(2x + 3)](4x^2 + 9)$

115. $[(a - 5)(a + 5)]^2$

116. $(a - 3)^2(a + 3)^2$
(*Hint*: Examine Exercise 115.)

117. $(3t^4 - 2)^2(3t^4 + 2)^2$
(*Hint*: Examine Exercise 115.)

118. $[3a - (2a - 3)][3a + (2a - 3)]$

Solve.

119. $(x + 2)(x - 5) = (x + 1)(x - 3)$

120. $(2x + 5)(x - 4) = (x + 5)(2x - 4)$

121. *Factors and Sums.* To *factor* a number is to express it as a product. Since $12 = 4 \cdot 3$, we say that 12 is *factored* and that 4 and 3 are *factors* of 12. In the following table, the top number has been factored in such a way that the sum of the factors is the bottom number. For example, in the first column, 40 has been factored as $5 \cdot 8$, and $5 + 8 = 13$, the bottom number. Such thinking is important in algebra when we factor trinomials of the type $x^2 + bx + c$. Find the missing numbers in the table.

Product	40	63	36	72	−140	−96	48	168	110			
Factor	5									−9	−24	−3
Factor	8									−10	18	
Sum	13	16	−20	−38	−4	4	−14	−29	−21			18

122. A factored polynomial for the shaded area in this rectangle is $(A + B)(A - B)$.

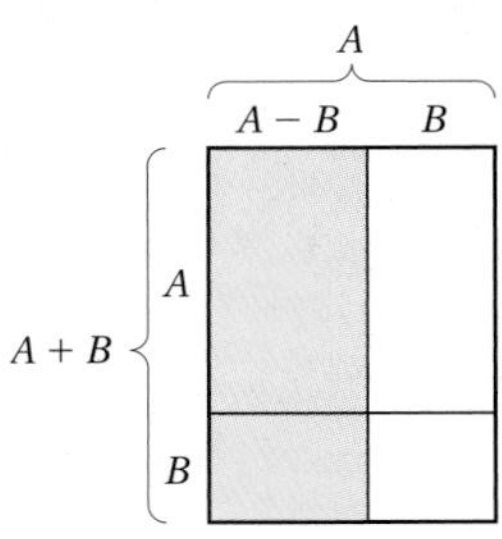

a) Find a polynomial for the area of the entire rectangle.
b) Find a polynomial for the sum of the areas of the two small unshaded rectangles.
c) Find a polynomial for the area in part (a) minus the area in part (b).
d) Find a polynomial for the area of the shaded region and compare this with the polynomial found in part (c).

Use the TABLE or GRAPH feature to check whether each of the following is correct.

123. $(x - 1)^2 = x^2 - 2x + 1$

124. $(x - 2)^2 = x^2 - 4x - 4$

125. $(x - 3)(x + 3) = x^2 - 6$

126. $(x - 3)(x + 2) = x^2 - x - 6$

10.7 OPERATIONS WITH POLYNOMIALS IN SEVERAL VARIABLES

Objectives

a Evaluate a polynomial in several variables for given values of the variables.

b Identify the coefficients and the degrees of the terms of a polynomial and the degree of a polynomial.

c Collect like terms of a polynomial.

d Add polynomials.

e Subtract polynomials.

f Multiply polynomials.

The polynomials that we have been studying have only one variable. A **polynomial in several variables** is an expression like those you have already seen, but with more than one variable. Here are two examples:

$$3x + xy^2 + 5y + 4, \qquad 8xy^2z - 2x^3z - 13x^4y^2 + 15.$$

a Evaluating Polynomials

EXAMPLE 1 Evaluate the polynomial $4 + 3x + xy^2 + 8x^3y^3$ when $x = -2$ and $y = 5$.

We replace x with -2 and y with 5:

$$\begin{aligned} 4 + 3x + xy^2 + 8x^3y^3 &= 4 + 3(-2) + (-2) \cdot 5^2 + 8(-2)^3 \cdot 5^3 \\ &= 4 - 6 - 50 - 8000 \\ &= -8052. \end{aligned}$$

EXAMPLE 2 *Male Caloric Needs.* The number of calories needed each day by a moderately active man who weighs w kilograms, is h centimeters tall, and is a years old can be estimated by the polynomial

$$19.18w + 7h - 9.52a + 92.4.$$

The author of this text is moderately active, weighs 87 kg, is 185 cm tall, and is 64 yr old. What are his daily caloric needs?

Source: Parker, M., *She Does Math.* Mathematical Association of America

We evaluate the polynomial for $w = 87$, $h = 185$, and $a = 64$:

$$\begin{aligned} &19.18w + 7h - 9.52a + 92.4 \\ &= 19.18(87) + 7(185) - 9.52(64) + 92.4 \qquad \text{Substituting} \\ &= 2446.78. \end{aligned}$$

His daily caloric need is about 2447 calories.

Do Exercises 1–3.

1. Evaluate the polynomial

$$4 + 3x + xy^2 + 8x^3y^3$$

when $x = 2$ and $y = -5$.

2. Evaluate the polynomial

$$8xy^2 - 2x^3z - 13x^4y^2 + 5$$

when $x = -1$, $y = 3$, and $z = 4$.

3. Female Caloric Needs. The number of calories needed each day by a moderately active woman who weighs w pounds, is h inches tall, and is a years old can be estimated by the polynomial

$$917 + 6w + 6h - 6a.$$

Christine is moderately active, weighs 125 lb, is 64 in. tall, and is 27 yr old. What are her daily caloric needs?

Source: Parker, M., *She Does Math.* Mathematical Association of America

Answers on page A-37

4. Identify the coefficient of each term:

$-3xy^2 + 3x^2y - 2y^3 + xy + 2.$

5. Identify the degree of each term and the degree of the polynomial

$4xy^2 + 7x^2y^3z^2 - 5x + 2y + 4.$

Collect like terms.

6. $4x^2y + 3xy - 2x^2y$

7. $-3pq - 5pqr^3 - 12 + 8pq + 5pqr^3 + 4$

Answers on page A-37

b Coefficients and Degrees

The **degree** of a term is the sum of the exponents of the variables. The **degree of a polynomial** is the degree of the term of highest degree.

EXAMPLE 3 Identify the coefficient and the degree of each term and the degree of the polynomial

$$9x^2y^3 - 14xy^2z^3 + xy + 4y + 5x^2 + 7.$$

TERM	COEFFICIENT	DEGREE	DEGREE OF THE POLYNOMIAL
$9x^2y^3$	9	5	
$-14xy^2z^3$	-14	6	6
xy	1	2	
$4y$	4	1	Think: $4y = 4y^1$.
$5x^2$	5	2	
7	7	0	Think: $7 = 7x^0$, or $7x^0y^0z^0$.

Do Exercises 4 and 5.

c Collecting Like Terms

Like terms have exactly the same variables with exactly the same exponents. For example,

$3x^2y^3$ and $-7x^2y^3$ are like terms;
$9x^4z^7$ and $12x^4z^7$ are like terms.

But

$13xy^5$ and $-2x^2y^5$ are *not* like terms, because the x-factors have different exponents;

and

$3xyz^2$ and $4xy$ are *not* like terms, because there is no factor of z^2 in the second expression.

Collecting like terms is based on the distributive laws.

EXAMPLES Collect like terms.

4. $5x^2y + 3xy^2 - 5x^2y - xy^2 = (5 - 5)x^2y + (3 - 1)xy^2 = 2xy^2$
5. $8a^2 - 2ab + 7b^2 + 4a^2 - 9ab - 17b^2 = 12a^2 - 11ab - 10b^2$
6. $7xy - 5xy^2 + 3xy^2 - 7 + 6x^3 + 9xy - 11x^3 + y - 1$
   $= -2xy^2 + 16xy - 5x^3 + y - 8$

Do Exercises 6 and 7.

d Addition

We can find the sum of two polynomials in several variables by writing a plus sign between them and then collecting like terms.

EXAMPLE 7 Add: $(-5x^3 + 3y - 5y^2) + (8x^3 + 4x^2 + 7y^2)$.

$$(-5x^3 + 3y - 5y^2) + (8x^3 + 4x^2 + 7y^2)$$
$$= (-5 + 8)x^3 + 4x^2 + 3y + (-5 + 7)y^2$$
$$= 3x^3 + 4x^2 + 3y + 2y^2$$

EXAMPLE 8 Add:

$$(5xy^2 - 4x^2y + 5x^3 + 2) + (3xy^2 - 2x^2y + 3x^3y - 5).$$

We first look for like terms. They are $5xy^2$ and $3xy^2$, $-4x^2y$ and $-2x^2y$, and 2 and -5. We collect these. Since there are no more like terms, the answer is

$$8xy^2 - 6x^2y + 5x^3 + 3x^3y - 3.$$

Do Exercises 8–10.

Add.

8. $(4x^3 + 4x^2 - 8y - 3) + (-8x^3 - 2x^2 + 4y + 5)$

9. $(13x^3y + 3x^2y - 5y) + (x^3y + 4x^2y - 3xy + 3y)$

10. $(-5p^2q^4 + 2p^2q^2 + 3q) + (6pq^2 + 3p^2q + 5)$

e Subtraction

We subtract a polynomial by adding its opposite, or additive inverse. The opposite of the polynomial $4x^2y - 6x^3y^2 + x^2y^2 - 5y$ is

$$-(4x^2y - 6x^3y^2 + x^2y^2 - 5y) = -4x^2y + 6x^3y^2 - x^2y^2 + 5y.$$

EXAMPLE 9 Subtract:

$$(4x^2y + x^3y^2 + 3x^2y^3 + 6y + 10) - (4x^2y - 6x^3y^2 + x^2y^2 - 5y - 8).$$

We have

$$(4x^2y + x^3y^2 + 3x^2y^3 + 6y + 10) - (4x^2y - 6x^3y^2 + x^2y^2 - 5y - 8)$$
$$= 4x^2y + x^3y^2 + 3x^2y^3 + 6y + 10 - 4x^2y + 6x^3y^2 - x^2y^2 + 5y + 8$$

Finding the opposite by changing the sign of each term

$$= 7x^3y^2 + 3x^2y^3 - x^2y^2 + 11y + 18.$$

Collecting like terms. (Try to write just the answer!)

Caution!

Do *not* add exponents when collecting like terms—that is,

$7x^3 + 8x^3 \neq 15x^6$; ← Wrong

$7x^3 + 8x^3 = 15x^3$. ← Correct

Do Exercises 11 and 12.

Subtract.

11. $(-4s^4t + s^3t^2 + 2s^2t^3) - (4s^4t - 5s^3t^2 + s^2t^2)$

12. $(-5p^4q + 5p^3q^2 - 3p^2q^3 - 7q^4 - 2) - (4p^4q - 4p^3q^2 + p^2q^3 + 2q^4 - 7)$

Answers on page A-37

Multiply.

13. $(x^2y^3 + 2x)(x^3y^2 + 3x)$

14. $(p^4q - 2p^3q^2 + 3q^3)(p + 2q)$

Multiply.

15. $(3xy + 2x)(x^2 + 2xy^2)$

16. $(x - 3y)(2x - 5y)$

17. $(4x + 5y)^2$

18. $(3x^2 - 2xy^2)^2$

19. $(2xy^2 + 3x)(2xy^2 - 3x)$

20. $(3xy^2 + 4y)(-3xy^2 + 4y)$

21. $(3y + 4 - 3x)(3y + 4 + 3x)$

22. $(2a + 5b + c)(2a - 5b - c)$

Answers on page A-37

f Multiplication

To multiply polynomials in several variables, we can multiply each term of one by every term of the other. We can use columns for long multiplications as with polynomials in one variable. We multiply each term at the top by every term at the bottom. We write like terms in columns, and then we add.

EXAMPLE 10 Multiply: $(3x^2y - 2xy + 3y)(xy + 2y)$.

$$\begin{array}{rrrrl} & 3x^2y & - 2xy & + 3y & \\ & & xy & + 2y & \\ \hline & 6x^2y^2 & - 4xy^2 & + 6y^2 & \text{Multiplying by } 2y \\ 3x^3y^2 & - 2x^2y^2 & + 3xy^2 & & \text{Multiplying by } xy \\ \hline 3x^3y^2 & + 4x^2y^2 & - xy^2 & + 6y^2 & \text{Adding} \end{array}$$

Do Exercises 13 and 14.

Where appropriate, we use the special products that we have learned.

EXAMPLES Multiply.

11. $(x^2y + 2x)(xy^2 + y^2) = x^3y^3 + x^2y^3 + 2x^2y^2 + 2xy^2$ Using FOIL

12. $(p + 5q)(2p - 3q) = 2p^2 - 3pq + 10pq - 15q^2$
$= 2p^2 + 7pq - 15q^2$

$(A + B)^2 = A^2 + 2 \cdot A \cdot B + B^2$

13. $(3x + 2y)^2 = (3x)^2 + 2(3x)(2y) + (2y)^2 = 9x^2 + 12xy + 4y^2$

$(A - B)^2 = A^2 - 2 \cdot A \cdot B + B^2$

14. $(2y^2 - 5x^2y)^2 = (2y^2)^2 - 2(2y^2)(5x^2y) + (5x^2y)^2$
$= 4y^4 - 20x^2y^3 + 25x^4y^2$

$(A + B)(A - B) = A^2 - B^2$

15. $(3x^2y + 2y)(3x^2y - 2y) = (3x^2y)^2 - (2y)^2 = 9x^4y^2 - 4y^2$

16. $(-2x^3y^2 + 5t)(2x^3y^2 + 5t) = (5t - 2x^3y^2)(5t + 2x^3y^2)$
$= (5t)^2 - (2x^3y^2)^2 = 25t^2 - 4x^6y^4$

$(A - B)(A + B) = A^2 - B^2$

17. $(2x + 3 - 2y)(2x + 3 + 2y) = (2x + 3)^2 - (2y)^2$
$= 4x^2 + 12x + 9 - 4y^2$

Remember that FOIL will always work when you are multiplying binomials. You can use it instead of the rules for special products, but those rules will make your work go faster.

Do Exercises 15–22.

For Extra Help

MathXL | MyMathLab | InterAct Math | Math Tutor Center | Video Lectures on CD Disc 5 | Student's Solutions Manual

a Evaluate the polynomial when $x = 3$, $y = -2$, and $z = -5$.

1. $x^2 - y^2 + xy$

2. $x^2 + y^2 - xy$

3. $x^2 - 3y^2 + 2xy$

4. $x^2 - 4xy + 5y^2$

5. $8xyz$

6. $-3xyz^2$

7. $xyz^2 - z$

8. $xy - xz + yz$

Lung Capacity. The polynomial equation

$$C = 0.041h - 0.018A - 2.69$$

can be used to estimate the lung capacity C, in liters, of a female of height h, in centimeters, and age A, in years. Use this formula for Exercises 9 and 10.

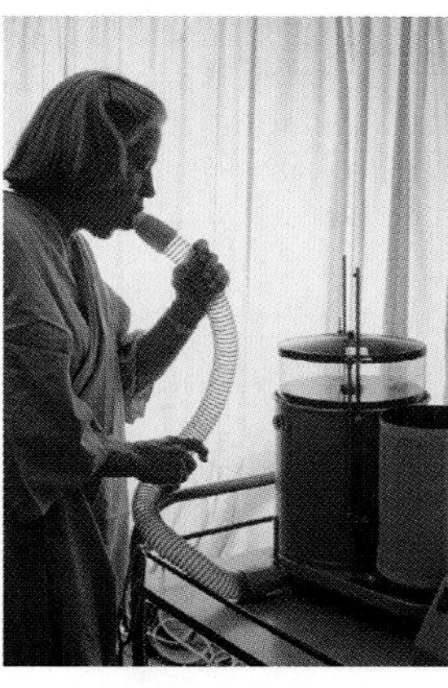

9. Find the lung capacity of a 20-year-old woman who is 165 cm tall.

10. Find the lung capacity of a 50-year-old woman who is 160 cm tall.

Altitude of a Launched Object. The altitude h, in meters, of a launched object is given by the polynomial equation

$$h = h_0 + vt - 4.9t^2,$$

where h_0 is the height, in meters, from which the launch occurs, v is the initial upward speed (or velocity), in meters per second (m/s), and t is the number of seconds for which the object is airborne. Use this formula for Exercises 11 and 12.

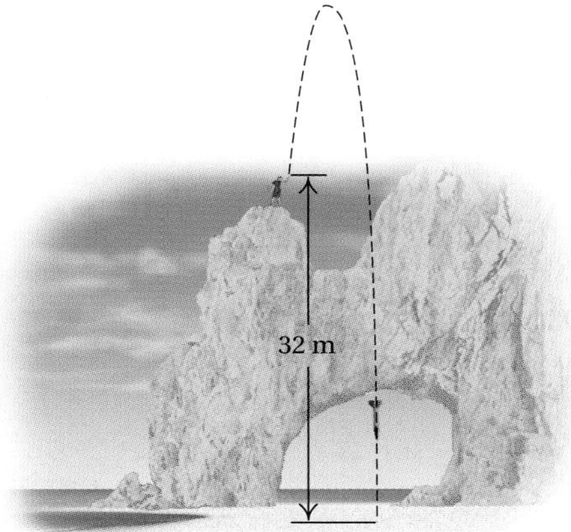

11. A model rocket is launched from the top of the Lands End Arch, near San Lucas, Baja, Mexico, 32 m above the ground. The upward speed is 40 m/s. How high will the rocket be 2 sec after the blastoff?

12. A golf ball is thrown upward with an initial speed of 30 m/s by a golfer atop the Washington Monument, which is 160 m above the ground. How high above the ground will the ball be after 3 sec?

Surface Area of a Right Circular Cylinder. The surface area S of a right circular cylinder is given by the polynomial equation

$$S = 2\pi rh + 2\pi r^2,$$

where h is the height and r is the radius of the base. Use this formula for Exercises 13 and 14.

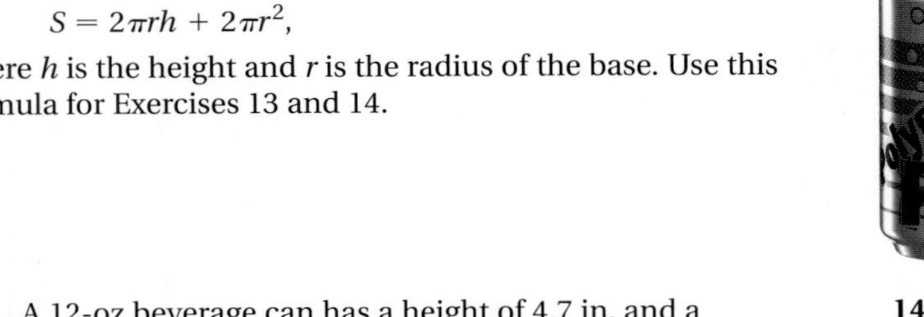

13. A 12-oz beverage can has a height of 4.7 in. and a radius of 1.2 in. Evaluate the polynomial when $h = 4.7$ and $r = 1.2$ to find the area of the can. Use 3.14 for π.

14. A 26-oz coffee can has a height of 6.5 in. and a radius of 2.5 in. Evaluate the polynomial when $h = 6.5$ and $r = 2.5$ to find the area of the can. Use 3.14 for π.

Surface Area of a Silo. A silo is a structure that is shaped like a right circular cylinder with a half sphere on top. The surface area S of a silo of height h and radius r (including the area of the base) is given by the polynomial equation $S = 2\pi rh + \pi r^2$. Note that h is the height of the entire silo.

15. A container of tennis balls is silo-shaped, with a height of $7\frac{1}{2}$ in. and a radius of $1\frac{1}{4}$ in. Find the surface area of the container. Use 3.14 for π.

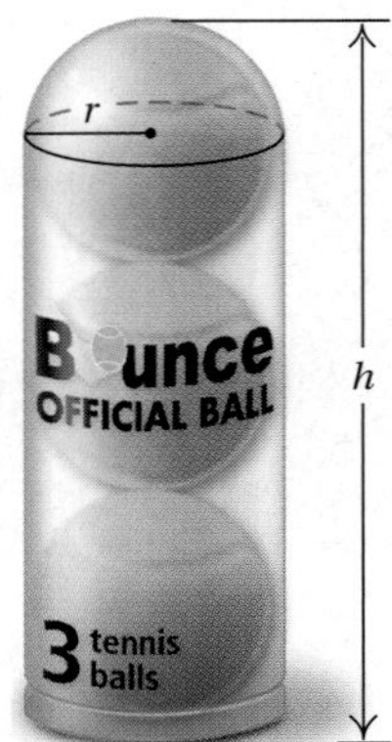

16. A $1\frac{1}{2}$-oz bottle of roll-on deodorant has a height of 4 in. and a radius of $\frac{3}{4}$ in. Find the surface area of the bottle if the bottle is shaped like a silo. Use 3.14 for π.

b Identify the coefficient and the degree of each term of the polynomial. Then find the degree of the polynomial.

17. $x^3y - 2xy + 3x^2 - 5$

18. $5y^3 - y^2 + 15y + 1$

19. $17x^2y^3 - 3x^3yz - 7$

20. $6 - xy + 8x^2y^2 - y^5$

c Collect like terms.

21. $a + b - 2a - 3b$

22. $y^2 - 1 + y - 6 - y^2$

23. $3x^2y - 2xy^2 + x^2$

24. $m^3 + 2m^2n - 3m^2 + 3mn^2$

25. $6au + 3av + 14au + 7av$

26. $3x^2y - 2z^2y + 3xy^2 + 5z^2y$

27. $2u^2v - 3uv^2 + 6u^2v - 2uv^2$

28. $3x^2 + 6xy + 3y^2 - 5x^2 - 10xy - 5y^2$

d Add.

29. $(2x^2 - xy + y^2) + (-x^2 - 3xy + 2y^2)$

30. $(2zt - z^2 + 5t^2) + (z^2 - 3zt + t^2)$

31. $(r - 2s + 3) + (2r + s) + (s + 4)$

32. $(ab - 2a + 3b) + (5a - 4b) + (3a + 7ab - 8b)$

33. $(b^3a^2 - 2b^2a^3 + 3ba + 4) + (b^2a^3 - 4b^3a^2 + 2ba - 1)$

34. $(2x^2 - 3xy + y^2) + (-4x^2 - 6xy - y^2) + (x^2 + xy - y^2)$

e Subtract.

35. $(a^3 + b^3) - (a^2b - ab^2 + b^3 + a^3)$

36. $(x^3 - y^3) - (-2x^3 + x^2y - xy^2 + 2y^3)$

37. $(xy - ab - 8) - (xy - 3ab - 6)$

38. $(3y^4x^2 + 2y^3x - 3y - 7) - (2y^4x^2 + 2y^3x - 4y - 2x + 5)$

39. $(-2a + 7b - c) - (-3b + 4c - 8d)$

40. Subtract $5a + 2b$ from the sum of $2a + b$ and $3a - b$.

f Multiply.

41. $(3z - u)(2z + 3u)$

42. $(a - b)(a^2 + b^2 + 2ab)$

43. $(a^2b - 2)(a^2b - 5)$

44. $(xy + 7)(xy - 4)$

45. $(a^3 + bc)(a^3 - bc)$

46. $(m^2 + n^2 - mn)(m^2 + mn + n^2)$

47. $(y^4x + y^2 + 1)(y^2 + 1)$

48. $(a - b)(a^2 + ab + b^2)$

49. $(3xy - 1)(4xy + 2)$

50. $(m^3n + 8)(m^3n - 6)$

51. $(3 - c^2d^2)(4 + c^2d^2)$

52. $(6x - 2y)(5x - 3y)$

53. $(m^2 - n^2)(m + n)$

54. $(pq + 0.2)(0.4pq - 0.1)$

55. $(xy + x^5y^5)(x^4y^4 - xy)$

56. $(x - y^3)(2y^3 + x)$

57. $(x + h)^2$

58. $(3a + 2b)^2$

59. $(r^3t^2 - 4)^2$

60. $(3a^2b - b^2)^2$

61. $(p^4 + m^2n^2)^2$

62. $(2ab - cd)^2$

63. $\left(2a^3 - \frac{1}{2}b^3\right)^2$

64. $-3x(x + 8y)^2$

65. $3a(a - 2b)^2$

66. $(a^2 + b + 2)^2$

67. $(2a - b)(2a + b)$

68. $(x - y)(x + y)$

69. $(c^2 - d)(c^2 + d)$

70. $(p^3 - 5q)(p^3 + 5q)$

71. $(ab + cd^2)(ab - cd^2)$

72. $(xy + pq)(xy - pq)$

73. $(x + y - 3)(x + y + 3)$

74. $(p + q + 4)(p + q - 4)$

75. $[x + y + z][x - (y + z)]$

76. $[a + b + c][a - (b + c)]$

77. $(a + b + c)(a - b - c)$

78. $(3x + 2 - 5y)(3x + 2 + 5y)$

79. $(x^2 - 4y + 2)(3x^2 + 5y - 3)$

80. $(2x^2 - 7y + 4)(x^2 + y - 3)$

81. **DW** Is it possible for a polynomial in four variables to have a degree less than 4? Why or why not?

82. **DW** Can the sum of two trinomials in several variables be a trinomial in one variable? Why or why not?

SKILL MAINTENANCE

In which quadrant is the point located? [9.1a]

83. $(2, -5)$

84. $(-8, -9)$

85. $(16, 23)$

86. $(-3, 2)$

Graph. [9.2b]

87. $2x = -10$

88. $y = -4$

89. $8y - 16 = 0$

90. $x = 4$

SYNTHESIS

Find a polynomial for the shaded area. (Leave results in terms of π where appropriate.)

91.

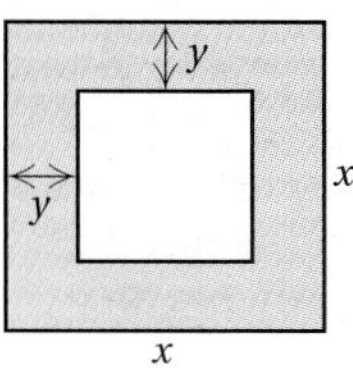

92.

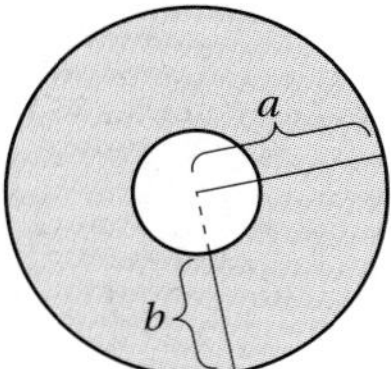

93.

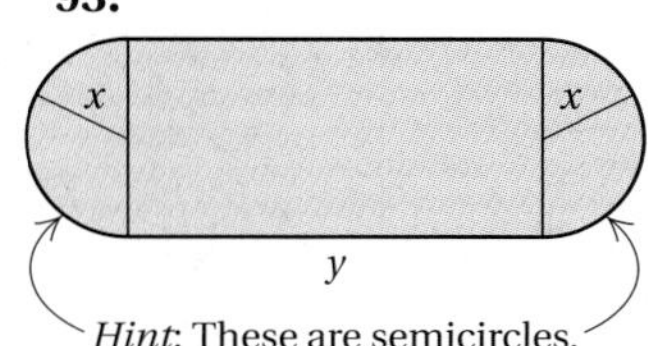

94.

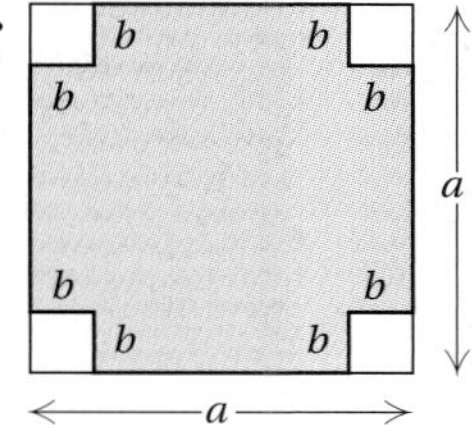

Find a formula for the surface area of the solid object. Leave results in terms of π.

95.

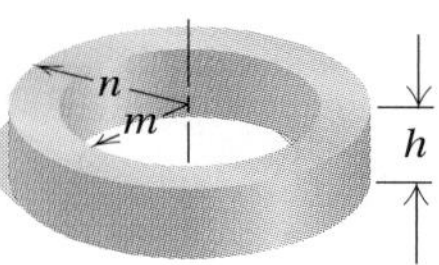

96.

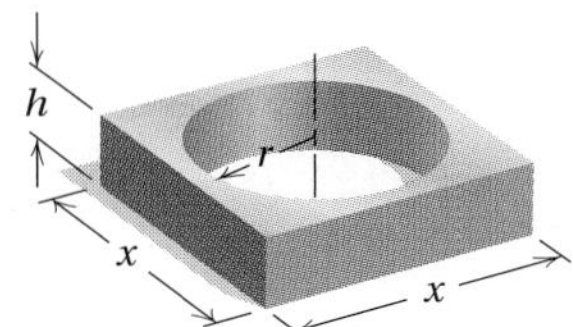

97. *Observatory Paint Costs.* The observatory at Danville University is shaped like a silo that is 40 ft high and 30 ft wide (see Exercise 15). The Heavenly Bodies Astronomy Club is to paint the exterior of the observatory using paint that covers 250 ft^2 per gallon. How many gallons should they purchase?

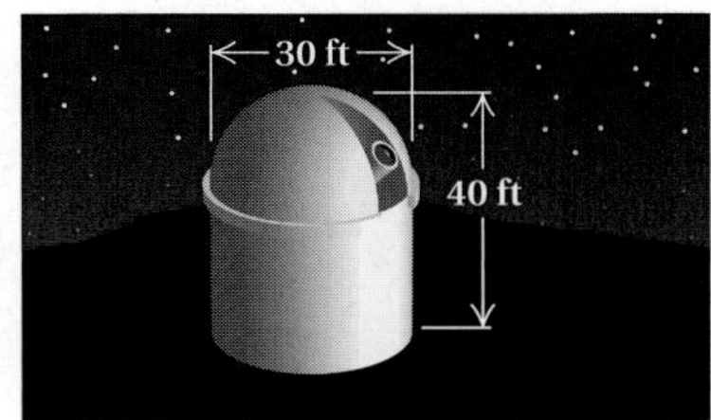

98. *Interest Compounded Annually.* An amount of money P that is invested at the yearly interest rate r grows to the amount

$$P(1 + r)^t$$

after t years. Find a polynomial that can be used to determine the amount to which P will grow after 2 yr.

99. Suppose that \$10,400 is invested at 8.5% compounded annually. How much is in the account at the end of 5 yr? (See Exercise 98.)

100. Multiply: $(x + a)(x - b)(x - a)(x + b)$.

10.8 DIVISION OF POLYNOMIALS

Objectives

a Divide a polynomial by a monomial.

b Divide a polynomial by a divisor that is a binomial.

Divide.

1. $\dfrac{20x^3}{5x}$

2. $\dfrac{-28x^{14}}{4x^3}$

3. $\dfrac{-56p^5q^7}{2p^2q^6}$

4. $\dfrac{x^5}{4x}$

Answers on page A-37

In this section, we consider division of polynomials. You will see that such division is similar to what is done in arithmetic.

a Dividing by a Monomial

We first consider division by a monomial. When dividing a monomial by a monomial, we use the quotient rule of Section 10.1 to subtract exponents when the bases are the same. We also divide the coefficients.

EXAMPLES Divide.

1. $\dfrac{10x^2}{2x} = \dfrac{10}{2} \cdot \dfrac{x^2}{x} = 5x^{2-1} = 5x$

2. $\dfrac{x^9}{3x^2} = \dfrac{1x^9}{3x^2} = \dfrac{1}{3} \cdot \dfrac{x^9}{x^2} = \dfrac{1}{3}x^{9-2} = \dfrac{1}{3}x^7$

3. $\dfrac{-18x^{10}}{3x^3} = \dfrac{-18}{3} \cdot \dfrac{x^{10}}{x^3} = -6x^{10-3} = -6x^7$

4. $\dfrac{42a^2b^5}{-3ab^2} = \dfrac{42}{-3} \cdot \dfrac{a^2}{a} \cdot \dfrac{b^5}{b^2} = -14a^{2-1}b^{5-2} = -14ab^3$

Caution!

The coefficients are divided but the exponents are subtracted.

Do Exercises 1–4.

To divide a polynomial by a monomial, we note that since

$$\frac{A}{C} + \frac{B}{C} = \frac{A+B}{C},$$

it follows that

$$\frac{A+B}{C} = \frac{A}{C} + \frac{B}{C}.$$ Switching the left and right sides of the equation

This is actually the procedure we use when performing divisions like $86 \div 2$. Although we might write

$$\frac{86}{2} = 43,$$

we could also calculate as follows:

$$\frac{86}{2} = \frac{80+6}{2} = \frac{80}{2} + \frac{6}{2} = 40 + 3 = 43.$$

Similarly, to divide a polynomial by a monomial, we divide each term by the monomial.

EXAMPLE 5 Divide: $(9x^8 + 12x^6) \div (3x^2)$.

We have

$$(9x^8 + 12x^6) \div (3x^2) = \frac{9x^8 + 12x^6}{3x^2}$$

$$= \frac{9x^8}{3x^2} + \frac{12x^6}{3x^2}.$$ To see this, add and get the original expression.

We now perform the separate divisions:

$$\frac{9x^8}{3x^2} + \frac{12x^6}{3x^2} = \frac{9}{3} \cdot \frac{x^8}{x^2} + \frac{12}{3} \cdot \frac{x^6}{x^2}$$
$$= 3x^{8-2} + 4x^{6-2}$$
$$= 3x^6 + 4x^4.$$

Caution!

The coefficients are *divided*, but the exponents are *subtracted*.

To check, we multiply the quotient $3x^6 + 4x^4$ by the divisor $3x^2$:

$$3x^2(3x^6 + 4x^4) = (3x^2)(3x^6) + (3x^2)(4x^4) = 9x^8 + 12x^6.$$

This is the polynomial that was being divided, so our answer is $3x^6 + 4x^4$.

Do Exercises 5–7.

EXAMPLE 6 Divide and check: $(10a^5b^4 - 2a^3b^2 + 6a^2b) \div (2a^2b)$.

$$\frac{10a^5b^4 - 2a^3b^2 + 6a^2b}{2a^2b} = \frac{10a^5b^4}{2a^2b} - \frac{2a^3b^2}{2a^2b} + \frac{6a^2b}{2a^2b}$$
$$= \frac{10}{2}a^{5-2}b^{4-1} - \frac{2}{2}a^{3-2}b^{2-1} + \frac{6}{2}$$
$$= 5a^3b^3 - ab + 3$$

Check:

$$2a^2b(5a^3b^3 - ab + 3) = 2a^2b \cdot 5a^3b^3 - 2a^2b \cdot ab + 2a^2b \cdot 3$$
$$= 10a^5b^4 - 2a^3b^2 + 6a^2b$$

Our answer, $5a^3b^3 - ab + 3$, checks.

To divide a polynomial by a monomial, divide each term by the monomial.

Do Exercises 8 and 9.

b Dividing by a Binomial

Let's first consider long division as it is performed in arithmetic. When we divide, we repeat the following procedure.

To carry out long division:

1. Divide,
2. Multiply,
3. Subtract, and
4. Bring down the next term.

We review this by considering the division $3711 \div 8$.

```
        4  <------  ① Divide: 37 ÷ 8 ≈ 4.
8 ) 3 7 1 1
    3 2  <--------  ② Multiply: 4 × 8 = 32.
      5 1           ③ Subtract: 37 − 32 = 5.
        ↑
        └---------  ④ Bring down the 1.
```

```
      4 6 3
8 ) 3 7 1 1
    3 2
      5 1
      4 8
        3 1
        2 4
          7
```

Divide. Check the result.

5. $(28x^7 + 32x^5) \div (4x^3)$

6. $(2x^3 + 6x^2 + 4x) \div (2x)$

7. $(6x^2 + 3x - 2) \div 3$

Divide and check.

8. $(8x^2 - 3x + 1) \div 2$

9. $\dfrac{2x^4y^6 - 3x^3y^4 + 5x^2y^3}{x^2y^2}$

Answers on page A-37

10. Divide and check:

$(x^2 + x - 6) \div (x + 3)$.

Answer on page A-37

Study Tips

FORMING A STUDY GROUP

Consider forming a study group with some of your fellow students. Exchange e-mail addresses, telephone numbers, and schedules so that you can coordinate study time for homework and tests.

Next, we repeat the process two more times. We obtain the complete division as shown on the right above. The quotient is 463. The remainder is 7, expressed as R = 7. We write the answer as

$$463 \text{ R } 7 \quad \text{or} \quad 463 + \frac{7}{8} = 463\frac{7}{8}.$$

We check by multiplying the quotient, 463, by the divisor, 8, and adding the remainder, 7:

$$8 \cdot 463 + 7 = 3704 + 7 = 3711.$$

Now let's look at long division with polynomials. We use this procedure when the divisor is not a monomial. We write polynomials in descending order and then write in missing terms.

EXAMPLE 7 Divide $x^2 + 5x + 6$ by $x + 2$.

$$\begin{array}{r} x \\ x + 2\overline{)x^2 + 5x + 6} \\ \underline{x^2 + 2x} \\ 3x \end{array}$$

- x ← Divide the first term by the first term: $x^2/x = x$. Ignore the term 2.
- $x^2 + 2x$ ← Multiply x above by the divisor, $x + 2$.
- $3x$ ← Subtract: $(x^2 + 5x) - (x^2 + 2x) = x^2 + 5x - x^2 - 2x = 3x$.

We now "bring down" the next term of the dividend—in this case, 6.

$$\begin{array}{r} x + 3 \\ x + 2\overline{)x^2 + 5x + 6} \\ \underline{x^2 + 2x} \\ 3x + 6 \\ \underline{3x + 6} \\ 0 \end{array}$$

- $x + 3$ ← Divide the first term by the first term: $3x/x = 3$.
- $3x + 6$ ← The 6 has been "brought down."
- $3x + 6$ ← Multiply 3 by the divisor, $x + 2$.
- 0 ← Subtract: $(3x + 6) - (3x + 6) = 3x + 6 - 3x - 6 = 0$.

The quotient is $x + 3$. The remainder is 0, expressed as R = 0. A remainder of 0 is generally not listed in an answer.

To check, we multiply the quotient by the divisor and add the remainder, if any, to see if we get the dividend:

$$\underbrace{(x + 2)}_{\text{Divisor}} \cdot \underbrace{(x + 3)}_{\text{Quotient}} + \underbrace{0}_{\text{Remainder}} = \underbrace{x^2 + 5x + 6}_{\text{Dividend}}. \qquad \text{The division checks.}$$

Do Exercise 10.

EXAMPLE 8 Divide and check: $(x^2 + 2x - 12) \div (x - 3)$.

$$\begin{array}{r} x \\ x - 3\overline{)x^2 + 2x - 12} \\ \underline{x^2 - 3x} \\ 5x \end{array}$$

- x ← Divide the first term by the first term: $x^2/x = x$.
- $x^2 - 3x$ ← Multiply x above by the divisor, $x - 3$.
- $5x$ ← Subtract: $(x^2 + 2x) - (x^2 - 3x) = x^2 + 2x - x^2 + 3x = 5x$.

We now "bring down" the next term of the dividend—in this case, -12.

$$\begin{array}{r} x + 5 \\ x - 3\overline{)x^2 + 2x - 12} \\ \underline{x^2 - 3x} \\ 5x - 12 \\ \underline{5x - 15} \\ 3 \end{array}$$

- $x + 5$ ← Divide the first term by the first term: $5x/x = 5$.
- $5x - 12$ ← Bring down the -12.
- $5x - 15$ ← Multiply 5 above by the divisor, $x - 3$.
- 3 ← Subtract: $(5x - 12) - (5x - 15) = 5x - 12 - 5x + 15 = 3$.

The answer is $x + 5$ with $R = 3$, or

$$\underbrace{x+5}_{\text{Quotient}} + \frac{3 \;\rightarrow\; \text{Remainder}}{\underbrace{x-3}_{\text{Divisor}}}.$$

(This is the way answers will be given at the back of the book.)

Check: We can check by multiplying the divisor by the quotient and adding the remainder, as follows:

$$\begin{aligned}(x-3)(x+5)+3 &= x^2+2x-15+3\\ &= x^2+2x-12.\end{aligned}$$

Divide and check.

11. $x - 2\overline{)x^2 + 2x - 8}$

When dividing, an answer may "come out even" (that is, have a remainder of 0, as in Example 7), or it may not (as in Example 8). **If a remainder is not 0, we continue dividing until the degree of the remainder is less than the degree of the divisor.** Check this in each of Examples 7 and 8.

12. $x + 3\overline{)x^2 + 7x + 10}$

Do Exercises 11 and 12.

EXAMPLE 9 Divide and check: $(x^3 + 1) \div (x + 1)$.

$$\begin{array}{rll}
 & x^2 - x + 1 & \\
x+1\,\big) & x^3 + 0x^2 + 0x + 1 & \leftarrow \text{Fill in the missing terms (see Section 10.3).}\\
 & \underline{x^3 + x^2} & \text{Subtract: } x^3 - (x^3 + x^2) = -x^2.\\
 & -x^2 + 0x & \\
 & \underline{-x^2 - x} & \text{Subtract: } -x^2 - (-x^2 - x) = x.\\
 & x + 1 & \\
 & \underline{x + 1} & \text{Subtract: } (x+1) - (x+1) = 0.\\
 & 0 &
\end{array}$$

The answer is $x^2 - x + 1$. The check is left to the student.

Divide and check.

13. $(x^3 - 1) \div (x - 1)$

EXAMPLE 10 Divide and check: $(9x^4 - 7x^2 - 4x + 13) \div (3x - 1)$.

$$\begin{array}{rll}
 & 3x^3 + x^2 - 2x - 2 & \\
3x-1\,\big) & 9x^4 + 0x^3 - 7x^2 - 4x + 13 & \leftarrow \text{Fill in the missing term.}\\
 & \underline{9x^4 - 3x^3} & \text{Subtract: } 9x^4 - (9x^4 - 3x^3) = 3x^3.\\
 & 3x^3 - 7x^2 & \\
 & \underline{3x^3 - x^2} & \text{Subtract: } (3x^3 - 7x^2) - (3x^3 - x^2) = -6x^2.\\
 & -6x^2 - 4x & \\
 & \underline{-6x^2 + 2x} & \text{Subtract: } (-6x^2 - 4x) - (-6x^2 + 2x) = -6x.\\
 & -6x + 13 & \\
 & \underline{-6x + 2} & \text{Subtract: } (-6x + 13) - (-6x + 2) = 11.\\
 & 11 &
\end{array}$$

14. $(8x^4 + 10x^2 + 2x + 9) \div (4x + 2)$

The answer is $3x^3 + x^2 - 2x - 2$ with $R = 11$; or

$$3x^3 + x^2 - 2x - 2 + \frac{11}{3x-1}.$$

Check: $(3x - 1)(3x^3 + x^2 - 2x - 2) + 11$

$$\begin{aligned}&= -3x^3 - x^2 + 2x + 2 + 9x^4 + 3x^3 - 6x^2 - 6x + 11\\ &= 9x^4 - 7x^2 - 4x + 13\end{aligned}$$

Do Exercises 13 and 14.

Answers on page A-37

10.8 EXERCISE SET

For Extra Help

MathXL | MyMathLab | InterAct Math | Math Tutor Center | Video Lectures on CD Disc 5 | Student's Solutions Manual

a Divide and check.

1. $\dfrac{24x^4}{8}$

2. $\dfrac{-2u^2}{u}$

3. $\dfrac{25x^3}{5x^2}$

4. $\dfrac{16x^7}{-2x^2}$

5. $\dfrac{-54x^{11}}{-3x^8}$

6. $\dfrac{-75a^{10}}{3a^2}$

7. $\dfrac{64a^5b^4}{16a^2b^3}$

8. $\dfrac{-34p^{10}q^{11}}{-17pq^9}$

9. $\dfrac{24x^4 - 4x^3 + x^2 - 16}{8}$

10. $\dfrac{12a^4 - 3a^2 + a - 6}{6}$

11. $\dfrac{u - 2u^2 - u^5}{u}$

12. $\dfrac{50x^5 - 7x^4 + x^2}{x}$

13. $(15t^3 + 24t^2 - 6t) \div (3t)$

14. $(25t^3 + 15t^2 - 30t) \div (5t)$

15. $(20x^6 - 20x^4 - 5x^2) \div (-5x^2)$

16. $(24x^6 + 32x^5 - 8x^2) \div (-8x^2)$

17. $(24x^5 - 40x^4 + 6x^3) \div (4x^3)$

18. $(18x^6 - 27x^5 - 3x^3) \div (9x^3)$

19. $\dfrac{18x^2 - 5x + 2}{2}$

20. $\dfrac{15x^2 - 30x + 6}{3}$

21. $\dfrac{12x^3 + 26x^2 + 8x}{2x}$

22. $\dfrac{2x^4 - 3x^3 + 5x^2}{x^2}$

23. $\dfrac{9r^2s^2 + 3r^2s - 6rs^2}{3rs}$

24. $\dfrac{4x^4y - 8x^6y^2 + 12x^8y^6}{4x^4y}$

b Divide.

25. $(x^2 + 4x + 4) \div (x + 2)$

26. $(x^2 - 6x + 9) \div (x - 3)$

27. $(x^2 - 10x - 25) \div (x - 5)$

28. $(x^2 + 8x - 16) \div (x + 4)$

29. $(x^2 + 4x - 14) \div (x + 6)$

30. $(x^2 + 5x - 9) \div (x - 2)$

31. $\dfrac{x^2 - 9}{x + 3}$

32. $\dfrac{x^2 - 25}{x - 5}$

33. $\dfrac{x^5 + 1}{x + 1}$

34. $\dfrac{x^4 - 81}{x - 3}$

35. $\dfrac{8x^3 - 22x^2 - 5x + 12}{4x + 3}$

36. $\dfrac{2x^3 - 9x^2 + 11x - 3}{2x - 3}$

37. $(x^6 - 13x^3 + 42) \div (x^3 - 7)$

38. $(x^6 + 5x^3 - 24) \div (x^3 - 3)$

39. $(15x^3 + 8x^2 + 11x + 12) \div (5x + 1)$

40. $(20x^4 - 2x^3 + 5x + 3) \div (2x - 3)$

41. $(t^3 - t^2 + t - 1) \div (t - 1)$

42. $(t^3 - t^2 + t - 1) \div (t + 1)$

43. $\mathbf{D_W}$ How is the distributive law used when dividing a polynomial by a binomial?

44. $\mathbf{D_W}$ On an assignment, Emma *incorrectly* writes

$$\frac{12x^3 - 6x}{3x} = 4x^2 - 6x.$$

What mistake do you think she is making and how might you convince her that a mistake has been made?

SKILL MAINTENANCE

VOCABULARY REINFORCEMENT

In each of Exercises 45–52, fill in the blank with the correct term from the given list. Some of the choices may not be used.

45. The ______________ rule asserts that when multiplying with exponential notation, if the bases are the same, keep the base and add the exponents. [10.1d]

46. A(n) ______________ is an expression of the type ax^n, where a is a real-number constant and n is a nonnegative integer. [10.3a, i]

47. The ______________ principle asserts that when we multiply or divide by the same nonzero number on each side of an equation, we get ______________ equations. [8.2a]

48. Vertical lines are graphs of equations of the type ______________. [9.2b]

49. A(n) ______________ is a polynomial with three terms, such as $5x^4 - 7x^2 + 4$. [10.3i]

50. The ______________ rule asserts that when dividing with exponential notation, if the bases are the same, keep the base and subtract the exponent of the denominator from the exponent of the numerator. [10.1e]

51. The ______________ of a number is its distance from zero on a number line. [7.2e]

52. The ______________ of the line $y = mx + b$ is m. [9.4a]

$x = a$

$y = b$

slope

positive

absolute value

equivalent

inverse

quotient

product

monomial

binomial

trinomial

addition

multiplication

SYNTHESIS

Divide.

53. $(x^4 + 9x^2 + 20) \div (x^2 + 4)$

54. $(y^4 + a^2) \div (y + a)$

55. $(5a^3 + 8a^2 - 23a - 1) \div (5a^2 - 7a - 2)$

56. $(15y^3 - 30y + 7 - 19y^2) \div (3y^2 - 2 - 5y)$

57. $(6x^5 - 13x^3 + 5x + 3 - 4x^2 + 3x^4) \div (3x^3 - 2x - 1)$

58. $(5x^7 - 3x^4 + 2x^2 - 10x + 2) \div (x^2 - x + 1)$

59. $(a^6 - b^6) \div (a - b)$

60. $(x^5 + y^5) \div (x + y)$

If the remainder is 0 when one polynomial is divided by another, the divisor is a *factor* of the dividend. Find the value(s) of c for which $x - 1$ is a factor of the polynomial.

61. $x^2 + 4x + c$

62. $2x^2 + 3cx - 8$

63. $c^2x^2 - 2cx + 1$

10 Summary and Review

The review that follows is meant to prepare you for a chapter exam. It consists of three parts. The first part, Concept Reinforcement, is designed to increase understanding of the concepts through true/false exercises. The second part is a list of important properties and formulas. The third part is the Review Exercises. These provide practice exercises for the exam, together with references to section objectives so you can go back and review. Before beginning, stop and look back over the skills you have obtained. What skills in mathematics do you have now that you did not have before studying this chapter?

CONCEPT REINFORCEMENT

Determine whether the statement is true or false. Answers are given at the back of the book.

_______ **1.** If one of the terms in an algebraic expression has a power in it, then the expression is a polynomial.

_______ **2.** All trinomials are polynomials.

_______ **3.** $x^2 \cdot x^3 = x^6$

_______ **4.** $(x + y)^2 \neq x^2 + y^2$

_______ **5.** The square of the difference of two expressions is the difference of the squares of the two expressions

_______ **6.** The product of the sum and the difference of two expressions is the difference of the squares of the expressions.

IMPORTANT PROPERTIES AND FORMULAS

FOIL:	$(A + B)(C + D) = AC + AD + BC + BD$
Square of a Sum:	$(A + B)(A + B) = (A + B)^2 = A^2 + 2AB + B^2$
Square of a Difference:	$(A - B)(A - B) = (A - B)^2 = A^2 - 2AB + B^2$
Product of a Sum and a Difference:	$(A + B)(A - B) = A^2 - B^2$
Definitions and Rules for Exponents:	See p. 721.

Review Exercises

Multiply and simplify. [10.1d, f]

1. $7^2 \cdot 7^{-4}$

2. $y^7 \cdot y^3 \cdot y$

3. $(3x)^5 \cdot (3x)^9$

4. $t^8 \cdot t^0$

Divide and simplify. [10.1e, f]

5. $\dfrac{4^5}{4^2}$

6. $\dfrac{a^5}{a^8}$

7. $\dfrac{(7x)^4}{(7x)^4}$

Simplify.

8. $(3t^4)^2$ [10.2a, b]

9. $(2x^3)^2(-3x)^2$ [10.1d], [10.2a, b]

10. $\left(\dfrac{2x}{y}\right)^{-3}$ [10.2b]

11. Express using a negative exponent: $\frac{1}{t^5}$. [10.1f]

12. Express using a positive exponent: y^{-4}. [10.1f]

13. Convert to scientific notation: 0.0000328. [10.2c]

14. Convert to decimal notation: 8.3×10^6. [10.2c]

Multiply or divide and write scientific notation for the result. [10.2d]

15. $(3.8 \times 10^4)(5.5 \times 10^{-1})$

16. $\frac{1.28 \times 10^{-8}}{2.5 \times 10^{-4}}$

17. *Diet-Drink Consumption.* It has been estimated that there will be 292 million people in the United States by 2005 and that on average, each of them will drink 15.3 gal of diet drinks that year. How many gallons of diet drinks will be consumed by the entire population in 2005? Express the answer in scientific notation. [10.2e]
Source: U.S. Department of Agriculture

18. Evaluate the polynomial $x^2 - 3x + 6$ when $x = -1$. [10.3a]

19. Identify the terms of the polynomial $-4y^5 + 7y^2 - 3y - 2$. [10.3b]

20. Identify the missing terms in $x^3 + x$. [10.3h]

21. Identify the degree of each term and the degree of the polynomial $4x^3 + 6x^2 - 5x + \frac{5}{3}$. [10.3g]

Classify the polynomial as a monomial, a binomial, a trinomial, or none of these. [10.3i]

22. $4x^3 - 1$

23. $4 - 9t^3 - 7t^4 + 10t^2$

24. $7y^2$

Collect like terms and then arrange in descending order. [10.3f]

25. $3x^2 - 2x + 3 - 5x^2 - 1 - x$

26. $-x + \frac{1}{2} + 14x^4 - 7x^2 - 1 - 4x^4$

Add. [10.4a]

27. $(3x^4 - x^3 + x - 4) + (x^5 + 7x^3 - 3x^2 - 5) + (-5x^4 + 6x^2 - x)$

28. $(3x^5 - 4x^4 + x^3 - 3) + (3x^4 - 5x^3 + 3x^2) + (-5x^5 - 5x^2) + (-5x^4 + 2x^3 + 5)$

Subtract. [10.4c]

29. $(5x^2 - 4x + 1) - (3x^2 + 1)$

30. $(3x^5 - 4x^4 + 3x^2 + 3) - (2x^5 - 4x^4 + 3x^3 + 4x^2 - 5)$

31. Find a polynomial for the perimeter and for the area. [10.4d], [10.5b]

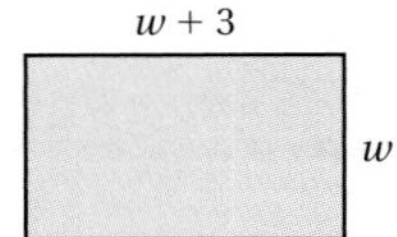

32. Find two algebraic expressions for the area of this figure. First, regard the figure as one large rectangle, and then regard the figure as a sum of four smaller rectangles. [10.4d]

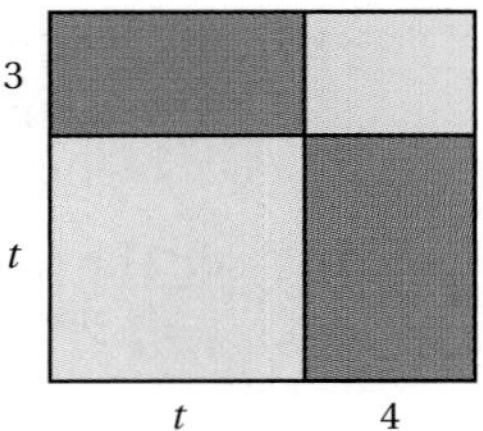

Multiply.

33. $\left(x + \frac{2}{3}\right)\left(x + \frac{1}{2}\right)$ [10.6a]

34. $(7x + 1)^2$ [10.6c]

35. $(4x^2 - 5x + 1)(3x - 2)$ [10.5d]

36. $(3x^2 + 4)(3x^2 - 4)$ [10.6b]

37. $5x^4(3x^3 - 8x^2 + 10x + 2)$ [10.5b]

38. $(x + 4)(x - 7)$ [10.6a]

39. $(3y^2 - 2y)^2$ [10.6c]

40. $(2t^2 + 3)(t^2 - 7)$ [10.6a]

41. Evaluate the polynomial

$$2 - 5xy + y^2 - 4xy^3 + x^6$$

when $x = -1$ and $y = 2$. [10.7a]

42. Identify the coefficient and the degree of each term of the polynomial

$$x^5y - 7xy + 9x^2 - 8.$$

Then find the degree of the polynomial. [10.7b]

Collect like terms. [10.7c]

43. $y + w - 2y + 8w - 5$

44. $m^6 - 2m^2n + m^2n^2 + n^2m - 6m^3 + m^2n^2 + 7n^2m$

45. Add: [10.7d]
$(5x^2 - 7xy + y^2) + (-6x^2 - 3xy - y^2) + (x^2 + xy - 2y^2).$

46. Subtract: [10.7e]
$(6x^3y^2 - 4x^2y - 6x) - (-5x^3y^2 + 4x^2y + 6x^2 - 6).$

Multiply. [10.7f]

47. $(p - q)(p^2 + pq + q^2)$ **48.** $\left(3a^4 - \frac{1}{3}b^3\right)^2$

Divide.

49. $(10x^3 - x^2 + 6x) \div (2x)$ [10.8a]

50. $(6x^3 - 5x^2 - 13x + 13) \div (2x + 3)$ [10.8b]

51. The graph of the polynomial equation $y = 10x^3 - 10x$ is shown below. Use *only* the graph to estimate the value of the polynomial when $x = -1$, $x = -0.5$, $x = 0.5$, and $x = 1$. [10.3a]

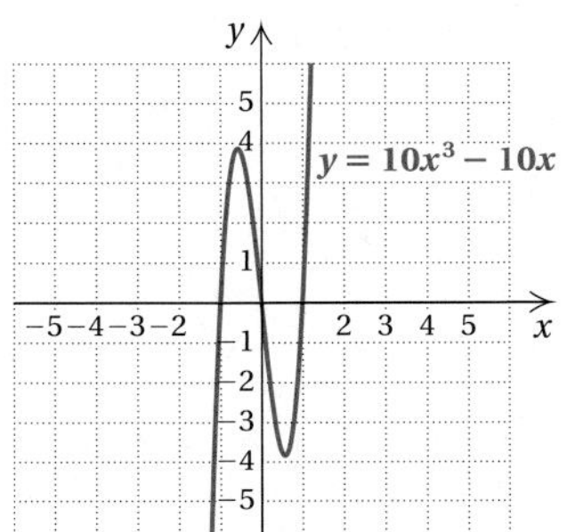

52. D_W Explain why the expression 578.6×10^{-7} is not in scientific notation. [10.2c]

53. D_W Write a short explanation of the difference between a monomial, a binomial, a trinomial, and a general polynomial. [10.3i]

SYNTHESIS

Find a polynomial for the shaded area. [10.4d], [10.6b]

54.

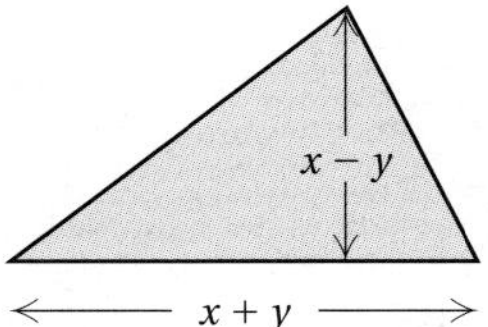

55.

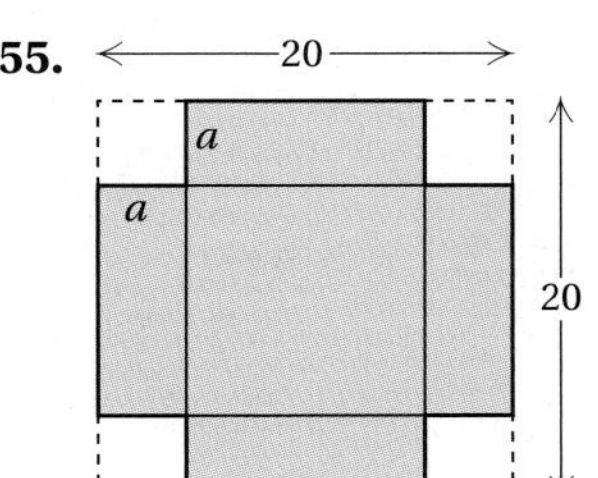

56. Collect like terms: [10.1d], [10.2a], [10.3e]
$-3x^5 \cdot 3x^3 - x^6(2x)^2 + (3x^4)^2 + (2x^2)^4 - 40x^2(x^3)^2.$

57. Solve: [8.3b], [10.6a]

$$(x - 7)(x + 10) = (x - 4)(x - 6).$$

58. The product of two polynomials is $x^5 - 1$. One of the polynomials is $x - 1$. Find the other. [10.8b]

59. A rectangular garden is twice as long as it is wide and is surrounded by a sidewalk that is 4 ft wide (see the figure below). The area of the sidewalk is 256 ft^2. Find the dimensions of the garden. [8.3b], [10.4d], [10.5a], [10.6a]

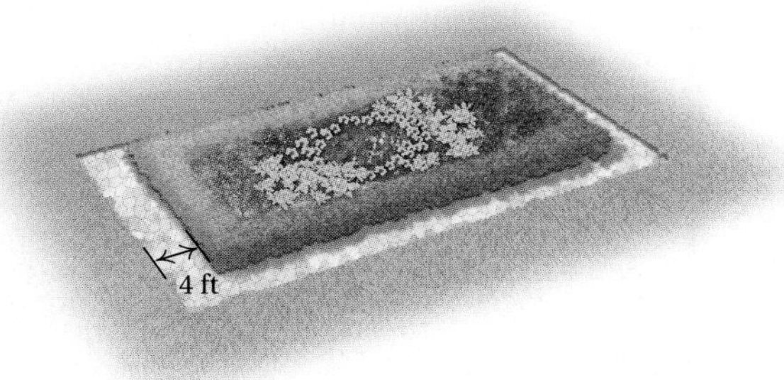

10 Chapter Test

For Extra Help

Work It Out!
Chapter Test Video
Solutions on CD

Multiply and simplify.

1. $6^{-2} \cdot 6^{-3}$

2. $x^6 \cdot x^2 \cdot x$

3. $(4a)^3 \cdot (4a)^8$

Divide and simplify.

4. $\frac{3^5}{3^2}$

5. $\frac{x^3}{x^8}$

6. $\frac{(2x)^5}{(2x)^5}$

Simplify.

7. $(x^3)^2$

8. $(-3y^2)^3$

9. $(2a^3b)^4$

10. $\left(\frac{ab}{c}\right)^3$

11. $(3x^2)^3(-2x^5)^3$

12. $3(x^2)^3(-2x^5)^3$

13. $2x^2(-3x^2)^4$

14. $(2x)^2(-3x^2)^4$

15. Express using a positive exponent:
5^{-3}.

16. Express using a negative exponent:
$\frac{1}{y^8}$.

17. Convert to scientific notation:
3,900,000,000.

18. Convert to decimal notation:
5×10^{-8}.

Multiply or divide and write scientific notation for the answer.

19. $\frac{5.6 \times 10^6}{3.2 \times 10^{-11}}$

20. $(2.4 \times 10^5)(5.4 \times 10^{16})$

21. *CD-ROM Memory.* A CD-ROM can contain about 600 million pieces of information (bytes). How many sound files, each containing 40,000 bytes, can a CD-ROM hold? Express the answer in scientific notation.

22. Evaluate the polynomial $x^5 + 5x - 1$ when $x = -2$.

23. Identify the coefficient of each term of the polynomial $\frac{1}{3}x^5 - x + 7$.

24. Identify the degree of each term and the degree of the polynomial $2x^3 - 4 + 5x + 3x^6$.

25. Classify the polynomial $7 - x$ as a monomial, a binomial, a trinomial, or none of these.

Collect like terms.

26. $4a^2 - 6 + a^2$

27. $y^2 - 3y - y + \frac{3}{4}y^2$

28. Collect like terms and then arrange in descending order:

$$3 - x^2 + 2x^3 + 5x^2 - 6x - 2x + x^5.$$

Add.

29. $(3x^5 + 5x^3 - 5x^2 - 3) +$
$(x^5 + x^4 - 3x^3 - 3x^2 + 2x - 4)$

30. $\left(x^4 + \frac{2}{3}x + 5\right) + \left(4x^4 + 5x^2 + \frac{1}{3}x\right)$

Subtract.

31. $(2x^4 + x^3 - 8x^2 - 6x - 3) - (6x^4 - 8x^2 + 2x)$

32. $(x^3 - 0.4x^2 - 12) - (x^5 + 0.3x^3 + 0.4x^2 + 9)$

Multiply.

33. $-3x^2(4x^2 - 3x - 5)$

34. $\left(x - \frac{1}{3}\right)^2$

35. $(3x + 10)(3x - 10)$

36. $(3b + 5)(b - 3)$

37. $(x^6 - 4)(x^8 + 4)$

38. $(8 - y)(6 + 5y)$

39. $(2x + 1)(3x^2 - 5x - 3)$

40. $(5t + 2)^2$

41. Collect like terms:

$$x^3y - y^3 + xy^3 + 8 - 6x^3y - x^2y^2 + 11.$$

42. Subtract:

$$(8a^2b^2 - ab + b^3) - (-6ab^2 - 7ab - ab^3 + 5b^3).$$

43. Multiply: $(3x^5 - 4y^5)(3x^5 + 4y^5)$.

Divide.

44. $(12x^4 + 9x^3 - 15x^2) \div (3x^2)$

45. $(6x^3 - 8x^2 - 14x + 13) \div (3x + 2)$

46. The graph of the polynomial equation $y = x^3 - 5x - 1$ is shown at right. Use *only* the graph to estimate the value of the polynomial when $x = -1$, $x = -0.5$, $x = 0.5$, $x = 1$, and $x = 1.1$.

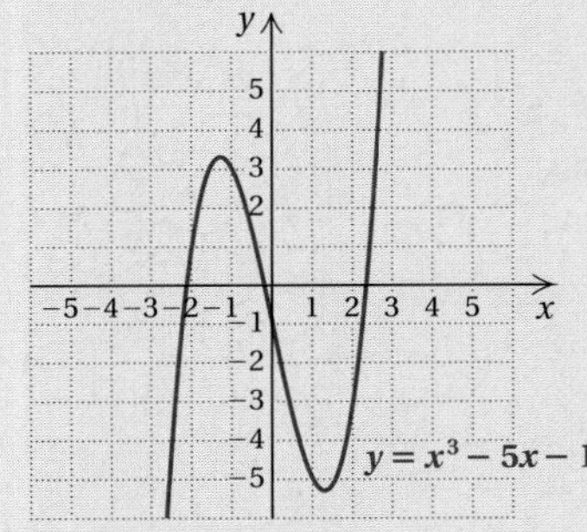

47. Find a polynomial for the surface area of this right rectangular solid.

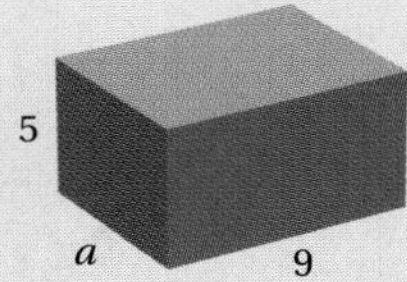

48. Find two algebraic expressions for the area of this figure. First, regard the figure as one large rectangle, and then regard the figure as a sum of four smaller rectangles.

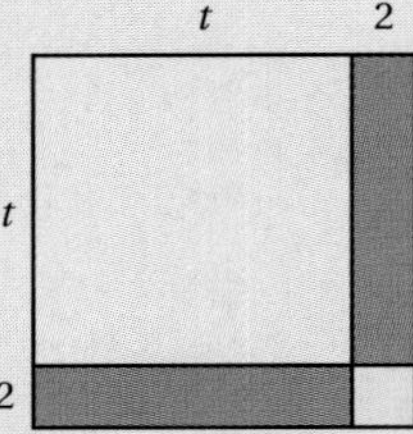

SYNTHESIS

49. The height of a box is 1 less than its length, and the length is 2 more than its width. Find the volume in terms of the length.

50. Solve: $(x - 5)(x + 5) = (x + 6)^2$.

Polynomials: Factoring

Real-World Application

An outdoor-education ropes course includes a 25-ft cable that slopes downward from a height of 37 ft to a height of 30 ft. How far is it between the trees that the cable connects?

This problem appears as Example 5 in Section 11.8.

11.1 INTRODUCTION TO FACTORING

Objectives

a Find the greatest common factor, the GCF, of monomials.

b Factor polynomials when the terms have a common factor, factoring out the greatest common factor.

c Factor certain expressions with four terms using factoring by grouping.

We introduce factoring with a review of factoring natural numbers. Consider the product $15 = 3 \cdot 5$. We say that 3 and 5 are **factors** of 15 and that $3 \cdot 5$ is a **factorization** of 15. Since $15 = 15 \cdot 1$, we also know that 15 and 1 are factors of 15 and that $15 \cdot 1$ is a factorization of 15.

a Finding the Greatest Common Factor

The numbers 20 and 30 have several factors in common, among them 2 and 5. The greatest of the common factors is called the **greatest common factor, GCF.** One way to find the GCF is by making a list of factors of each number.

List all the factors of 20: $\underline{1}, \underline{2}, 4, \underline{5}, \underline{10}$, and 20.

List all the factors of 30: $\underline{1}, \underline{2}, 3, \underline{5}, 6, \underline{10}$, 15, and 30.

Now list the numbers common to both lists, the common factors:

1, 2, 5, and 10.

Then the greatest common factor, the GCF, is 10, the largest number in the common list.

The preceding procedure gives meaning to the notion of a GCF, but the following method, using prime factorizations, is generally faster.

EXAMPLE 1 Find the GCF of 20 and 30.

We find the prime factorization of each number. Then we draw lines between the common factors.

$$20 = 2 \cdot 2 \cdot 5$$

$$30 = 2 \cdot 3 \cdot 5$$

The GCF $= 2 \cdot 5 = 10$.

EXAMPLE 2 Find the GCF of 180 and 420.

We find the prime factorization of each number. Then we draw lines between the common factors.

$$180 = 2 \cdot 2 \cdot 3 \cdot 3 \cdot 5 = 2^2 \cdot 3^2 \cdot 5^1$$

$$420 = 2 \cdot 2 \cdot 3 \cdot 5 \cdot 7 = 2^2 \cdot 3^1 \cdot 5^1 \cdot 7^1$$

The GCF $= 2 \cdot 2 \cdot 3 \cdot 5 = 2^2 \cdot 3^1 \cdot 5^1 = 60$. Note how we can use the exponents to determine the GCF. There are 2 lines for the 2's, 1 line for the 3, 1 line for the 5, and no line for the 7.

EXAMPLE 3 Find the GCF of 30 and 77.

We find the prime factorization of each number. Then we draw lines between the common factors, if any exist.

$$30 = 2 \cdot 3 \cdot 5 = 2^1 \cdot 3^1 \cdot 5^1$$

$$77 = 7 \cdot 11 = 7^1 \cdot 11^1$$

Since there is no common prime factor, the GCF is 1.

EXAMPLE 4 Find the GCF of 54, 90, and 252.

We find the prime factorization of each number. Then we draw lines between the common factors.

$$54 = 2 \cdot 3 \cdot 3 \cdot 3 = 2^1 \cdot 3^3,$$

$$90 = 2 \cdot 3 \cdot 3 \cdot 5 = 2^1 \cdot 3^2 \cdot 5^1,$$

$$252 = 2 \cdot 2 \cdot 3 \cdot 3 \cdot 7 = 2^2 \cdot 3^2 \cdot 7^1$$

The GCF $= 2^1 \cdot 3^2 = 18$.

Do Exercises 1–4.

Find the GCF.

1. 40, 100
2. 7, 21
3. 72, 360, 432
4. 3, 5, 22

Consider the product

$$12x^3(x^2 - 6x + 2) = 12x^5 - 72x^4 + 24x^3.$$

To factor the polynomial on the right, we reverse the process of multiplication:

$$12x^5 - 72x^4 + 24x^3 = 12x^3(x^2 - 6x + 2).$$

FACTOR; FACTORIZATION

To **factor** a polynomial is to express it as a product.

A **factor** of a polynomial P is a polynomial that can be used to express P as a product.

A **factorization** of a polynomial is an expression that names that polynomial as a product.

In the factorization

$$12x^5 - 72x^4 + 24x^3 = 12x^3(x^2 - 6x + 2),$$

the monomial $12x^3$ is called the GCF of the terms, $12x^5$, $-72x^4$, and $24x^3$. The first step in factoring polynomials is to find the GCF of the terms.

Consider the monomials

$$x^3,\ x^4,\ x^6, \text{ and } x^7.$$

The GCF of these monomials is x^3, found by noting that the smallest exponent of x is 3.

Consider

$$20x^2 \text{ and } 30x^5.$$

The GCF of 20 and 30 is 10. The GCF of x^2 and x^5 is x^2. Then the GCF of $20x^2$ and $30x^5$ is the product of the individual GCFs, $10x^2$.

Answers on page A-38

Find the GCF.

5. $12x^2, \ -16x^3$

6. $3y^6, \ -5y^3, \ 2y^2$

7. $-24m^5n^6, \ 12mn^3, \ -16m^2n^2, \ 8m^4n^4$

8. $-35x^7, \ -49x^6, \ -14x^5, \ -63x^3$

Answers on page A-38

EXAMPLE 5 Find the GCF of $15x^5$, $-12x^4$, $27x^3$, and $-3x^2$.

First, we find a prime factorization of the coefficients, attaching a factor of -1 for the negative coefficients.

$$15x^5 = 3 \cdot 5 \cdot x^5,$$
$$-12x^4 = -1 \cdot 2 \cdot 2 \cdot 3 \cdot x^4,$$
$$27x^3 = 3 \cdot 3 \cdot 3 \cdot x^3,$$
$$-3x^2 = -1 \cdot 3 \cdot x^2$$

The greatest *positive* common factor of the coefficients is 3.

Then we find the GCF of the powers of x. That GCF is x^2, because 2 is the smallest exponent of x. Thus the GCF of the set of monomials is $3x^2$.

What about the -1 factors in Example 5? Strictly speaking, both 1 and -1 are factors of any number or expression. We see this as follows:

$$3x^2 = 1 \cdot 3x^2 = (-1)(-3x^2).$$

Because the coefficient -3 is less than the coefficient 3, we consider $3x^2$, and not $-3x^2$, the GCF.

EXAMPLE 6 Find the GCF of $14p^2y^3$, $-8py^2$, $2py$, and $4p^3$.

We have

$$14p^2y^3 = 2 \cdot 7 \cdot p^2 \cdot y^3,$$
$$-8py^2 = -1 \cdot 2 \cdot 2 \cdot 2 \cdot p \cdot y^2,$$
$$2py = 2 \cdot p \cdot y,$$
$$4p^3 = 2 \cdot 2 \cdot p^3.$$

The greatest positive common factor of the coefficients is 2, the GCF of the powers of p is p, and the GCF of the powers of y is 1 since there is no y-factor in the last monomial. Thus the GCF is $2p$.

TO FIND THE GCF OF TWO OR MORE MONOMIALS

1. Find the prime factorization of the coefficients, including -1 as a factor if any coefficient is negative.
2. Determine any common prime factors of the coefficients. For each one that occurs, include it as a factor of the GCF. If none occurs, use 1 as a factor.
3. Examine each of the variables as factors. If any appear as a factor of all the monomials, include it as a factor, using the smallest exponent of the variable. If none occurs in all the monomials, use 1 as a factor.
4. The GCF is the product of the results of steps (2) and (3).

Do Exercises 5–8.

b Factoring When Terms Have a Common Factor

The polynomials we consider most when factoring are those with more than one term. To multiply a monomial and a polynomial with more than one term, we multiply each term of the polynomial by the monomial using the distributive laws:

$$a(b + c) = ab + ac \quad \text{and} \quad a(b - c) = ab - ac.$$

To factor, we do the reverse. We express a polynomial as a product using the distributive laws in reverse:

$$ab + ac = a(b + c) \quad \text{and} \quad ab - ac = a(b - c).$$

Compare.

Multiply	*Factor*
$3x(x^2 + 2x - 4)$	$3x^3 + 6x^2 - 12x$
$= 3x \cdot x^2 + 3x \cdot 2x - 3x \cdot 4$	$= 3x \cdot x^2 + 3x \cdot 2x - 3x \cdot 4$
$= 3x^3 + 6x^2 - 12x$	$= 3x(x^2 + 2x - 4)$

Caution!

Consider the following:

$$3x^3 + 6x^2 - 12x = 3 \cdot x \cdot x \cdot x + 2 \cdot 3 \cdot x \cdot x - 2 \cdot 2 \cdot 3 \cdot x.$$

The terms of the polynomial, $3x^3$, $6x^2$, and $-12x$, have been factored but the polynomial itself has not been factored. This is not what we mean by a factorization of the polynomial. The *factorization* is

$$3x(x^2 + 2x - 4). \leftarrow \text{A product}$$

The expressions $3x$ and $x^2 + 2x - 4$ are *factors* of $3x^3 + 6x^2 - 12x$.

Do Exercises 9 and 10.

To factor, we first find the GCF of all terms. It may be 1.

EXAMPLE 7 Factor: $7x^2 + 14$.

We have

$$\begin{aligned} 7x^2 + 14 &= 7 \cdot x^2 + 7 \cdot 2 && \text{Factoring each term} \\ &= 7(x^2 + 2). && \text{Factoring out the GCF, 7} \end{aligned}$$

Check: We multiply to check:

$$7(x^2 + 2) = 7 \cdot x^2 + 7 \cdot 2 = 7x^2 + 14.$$

9. a) Multiply: $3(x + 2)$.

b) Factor: $3x + 6$.

10. a) Multiply: $2x(x^2 + 5x + 4)$.

b) Factor: $2x^3 + 10x^2 + 8x$.

Answers on page A-38

Factor. Check by multiplying.

11. $x^2 + 3x$

12. $3y^6 - 5y^3 + 2y^2$

13. $9x^4y^2 - 15x^3y + 3x^2y$

14. $\frac{3}{4}t^3 + \frac{5}{4}t^2 + \frac{7}{4}t + \frac{1}{4}$

15. $35x^7 - 49x^6 + 14x^5 - 63x^3$

16. $84x^2 - 56x + 28$

Factor.

17. $x^2(x + 7) + 3(x + 7)$

18. $x^2(a + b) + 2(a + b)$

Answers on page A-38

EXAMPLE 8 Factor: $16x^3 + 20x^2$.

$$16x^3 + 20x^2 = (4x^2)(4x) + (4x^2)(5) \quad \text{Factoring each term}$$
$$= 4x^2(4x + 5) \quad \text{Factoring out the GCF, } 4x^2$$

Suppose in Example 8 that you had not recognized the GCF and removed only part of it, as follows:

$$16x^3 + 20x^2 = (2x^2)(8x) + (2x^2)(10)$$
$$= 2x^2(8x + 10).$$

Note that $8x + 10$ still has a common factor of 2. You need not begin again. Just continue factoring out common factors, as follows, until finished:

$$= 2x^2(2 \cdot 4x + 2 \cdot 5)$$
$$= 2x^2[2(4x + 5)]$$
$$= 4x^2(4x + 5).$$

EXAMPLE 9 Factor: $15x^5 - 12x^4 + 27x^3 - 3x^2$.

$$15x^5 - 12x^4 + 27x^3 - 3x^2 = (3x^2)(5x^3) - (3x^2)(4x^2) + (3x^2)(9x) - (3x^2)(1)$$
$$= 3x^2(5x^3 - 4x^2 + 9x - 1) \quad \text{Factoring out the GCF, } 3x^2$$

Caution!

Don't forget the term -1.

Check: We multiply to check:

$$3x^2(5x^3 - 4x^2 + 9x - 1)$$
$$= (3x^2)(5x^3) - (3x^2)(4x^2) + (3x^2)(9x) - (3x^2)(1)$$
$$= 15x^5 - 12x^4 + 27x^3 - 3x^2.$$

As you become more familiar with factoring, you will be able to spot the GCF without factoring each term. Then you can write just the answer.

EXAMPLES Factor.

10. $24x^2 + 12x - 36 = 12(2x^2 + x - 3)$

11. $8m^3 - 16m = 8m(m^2 - 2)$

12. $14p^2y^3 - 8py^2 + 2py = 2py(7py^2 - 4y + 1)$

13. $\frac{4}{5}x^2 + \frac{1}{5}x + \frac{2}{5} = \frac{1}{5}(4x^2 + x + 2)$

Do Exercises 11–16.

There are two important points to keep in mind as we study this chapter.

TIPS FOR FACTORING

- Before doing any other kind of factoring, first try to factor out the GCF.
- Always check the result of factoring by multiplying.

C Factoring by Grouping: Four Terms

Certain polynomials with four terms can be factored using a method called *factoring by grouping.*

EXAMPLE 14 Factor: $x^2(x + 1) + 2(x + 1)$.

The binomial $x + 1$ is common to both terms:

$$x^2(x + 1) + 2(x + 1) = (x^2 + 2)(x + 1).$$

The factorization is $(x^2 + 2)(x + 1)$.

Do Exercises 17 and 18 on the preceding page.

Consider the four-term polynomial

$$x^3 + x^2 + 2x + 2.$$

There is no factor other than 1 that is common to all the terms. We can, however, factor $x^3 + x^2$ and $2x + 2$ separately:

$$x^3 + x^2 = x^2(x + 1); \qquad \text{Factoring } x^3 + x^2$$
$$2x + 2 = 2(x + 1). \qquad \text{Factoring } 2x + 2$$

We have grouped certain terms and factored each polynomial separately:

$$\begin{aligned} x^3 + x^2 + 2x + 2 &= (x^3 + x^2) + (2x + 2) \\ &= x^2(x + 1) + 2(x + 1) \\ &= (x^2 + 2)(x + 1), \end{aligned}$$

as in Example 14. This method is called **factoring by grouping.** We began with a polynomial with four terms. After grouping and removing common factors, we obtained a polynomial with two parts, each having a common factor $x + 1$. Not all polynomials with four terms can be factored by this procedure, but it does give us a method to try.

Study Tips

CHECKLIST

The foundation of all your study skills is TIME!

- ☐ Are you staying on schedule and on time for class and adapting your study time and class schedule to your personality?
- ☐ Did you study the examples in this chapter carefully?
- ☐ Are you asking questions at appropriate times in class and with your tutors?
- ☐ Are you doing exercises without answers as part of every homework assignment to prepare you for tests?

Factor by grouping.

19. $x^3 + 7x^2 + 3x + 21$

20. $8t^3 + 2t^2 + 12t + 3$

21. $3m^5 - 15m^3 + 2m^2 - 10$

22. $3x^3 - 6x^2 - x + 2$

23. $4x^3 - 6x^2 - 6x + 9$

24. $y^4 - 2y^3 - 2y - 10$

Answers on page A-38

EXAMPLES Factor by grouping.

15. $6x^3 - 9x^2 + 4x - 6$

$= (6x^3 - 9x^2) + (4x - 6)$

$= 3x^2(2x - 3) + 2(2x - 3)$ Factoring each binomial

$= (3x^2 + 2)(2x - 3)$ Factoring out the common factor $2x - 3$

We think through this process as follows:

$$6x^3 - 9x^2 + 4x - 6 = 3x^2(2x - 3) \square (2x - 3)$$

(1) Factor the first two terms.

(2) The factor $2x - 3$ gives us a hint to the factorization of the last two terms.

(3) Now we ask ourselves, "What times $2x - 3$ is $4x - 6$?" The answer is $+2$.

16. $x^3 + x^2 + x + 1 = (x^3 + x^2) + (x + 1)$

Caution! Don't forget the 1.

$= x^2(x + 1) + 1(x + 1)$ Factoring each binomial

$= (x^2 + 1)(x + 1)$ Factoring out the common factor $x + 1$

17. $2x^3 - 6x^2 - x + 3$

$= (2x^3 - 6x^2) + (-x + 3)$ Separating into two binomials

$= 2x^2(x - 3) - 1(x - 3)$ *Check*: $-1(x - 3) = -x + 3$.

$= (2x^2 - 1)(x - 3)$ Factoring out the common factor $x - 3$

We can think through this process as follows.

(1) Factor the first two terms: $2x^3 - 6x^2 = 2x^2(x - 3)$.

(2) The factor $x - 3$ gives us a hint for factoring the last two terms:

$$2x^3 - 6x^2 - x + 3 = 2x^2(x - 3) \square (x - 3).$$

(3) Now we ask ourselves, "What times $x - 3$ is $-x + 3$?" The answer is -1.

18. $12x^5 + 20x^2 - 21x^3 - 35 = 4x^2(3x^3 + 5) - 7(3x^3 + 5)$

$= (4x^2 - 7)(3x^3 + 5)$

19. $x^3 + x^2 + 2x - 2 = x^2(x + 1) + 2(x - 1)$

This polynomial is not factorable using factoring by grouping. It may be factorable, but not by methods that we will consider in this text.

Do Exercises 19–24.

11.1 EXERCISE SET

For Extra Help

MathXL | MyMathLab | InterAct Math | Math Tutor Center | Video Lectures on CD Disc 6 | Student's Solutions Manual

a Find the GCF.

1. $x^2,\ -6x$
2. $x^2,\ 5x$
3. $3x^4,\ x^2$
4. $8x^4,\ -24x^2$
5. $2x^2,\ 2x,\ -8$
6. $8x^2,\ -4x,\ -20$
7. $-17x^5y^3,\ 34x^3y^2,\ 51xy$
8. $16p^6q^4,\ 32p^3q^3,\ -48pq^2$
9. $-x^2,\ -5x,\ -20x^3$
10. $-x^2,\ -6x,\ -24x^5$
11. $x^5y^5,\ x^4y^3,\ x^3y^3,\ -x^2y^2$
12. $-x^9y^6,\ -x^7y^5,\ x^4y^4,\ x^3y^3$

b Factor. Check by multiplying.

13. $x^2 - 6x$
14. $x^2 + 5x$
15. $2x^2 + 6x$
16. $8y^2 - 8y$
17. $x^3 + 6x^2$
18. $3x^4 - x^2$
19. $8x^4 - 24x^2$
20. $5x^5 + 10x^3$
21. $2x^2 + 2x - 8$
22. $8x^2 - 4x - 20$
23. $17x^5y^3 + 34x^3y^2 + 51xy$
24. $16p^6q^4 + 32p^5q^3 - 48pq^2$
25. $6x^4 - 10x^3 + 3x^2$
26. $5x^5 + 10x^2 - 8x$
27. $x^5y^5 + x^4y^3 + x^3y^3 - x^2y^2$
28. $x^9y^6 - x^7y^5 + x^4y^4 + x^3y^3$
29. $2x^7 - 2x^6 - 64x^5 + 4x^3$
30. $8y^3 - 20y^2 + 12y - 16$
31. $1.6x^4 - 2.4x^3 + 3.2x^2 + 6.4x$
32. $2.5x^6 - 0.5x^4 + 5x^3 + 10x^2$
33. $\frac{5}{3}x^6 + \frac{4}{3}x^5 + \frac{1}{3}x^4 + \frac{1}{3}x^3$
34. $\frac{5}{9}x^7 + \frac{2}{9}x^5 - \frac{4}{9}x^3 - \frac{1}{9}x$

c Factor.

35. $x^2(x + 3) + 2(x + 3)$
36. $3z^2(2z + 1) + (2z + 1)$
37. $5a^3(2a - 7) - (2a - 7)$
38. $m^4(8 - 3m) - 7(8 - 3m)$

Factor by grouping.

39. $x^3 + 3x^2 + 2x + 6$

40. $6z^3 + 3z^2 + 2z + 1$

41. $2x^3 + 6x^2 + x + 3$

42. $3x^3 + 2x^2 + 3x + 2$

43. $8x^3 - 12x^2 + 6x - 9$

44. $10x^3 - 25x^2 + 4x - 10$

45. $12p^3 - 16p^2 + 3p - 4$

46. $18x^3 - 21x^2 + 30x - 35$

47. $5x^3 - 5x^2 - x + 1$

48. $7x^3 - 14x^2 - x + 2$

49. $x^3 + 8x^2 - 3x - 24$

50. $2x^3 + 12x^2 - 5x - 30$

51. $2x^3 - 8x^2 - 9x + 36$

52. $20g^3 - 4g^2 - 25g + 5$

53. **D_W** Josh says that there is no need to print answers for Exercises 13–52 at the back of the book. Is he correct in saying this? Why or why not?

54. **D_W** Explain how one could construct a polynomial with four terms that can be factored by grouping.

SKILL MAINTENANCE

Solve.

55. $-2x < 48$ [8.7d]

56. $4x - 8x + 16 \geq 6(x - 2)$ [8.7e]

57. Divide: $\dfrac{-108}{-4}$. [7.6a]

58. Solve $A = \dfrac{p + q}{2}$ for p. [8.4b]

Multiply. [10.6d]

59. $(y + 5)(y + 7)$

60. $(y + 7)^2$

61. $(y + 7)(y - 7)$

62. $(y - 7)^2$

Find the intercepts of the equation. Then graph the equation. [9.2a]

63. $x + y = 4$

64. $x - y = 3$

65. $5x - 3y = 15$

66. $y - 3x = 6$

SYNTHESIS

Factor.

67. $4x^5 + 6x^3 + 6x^2 + 9$

68. $x^6 + x^4 + x^2 + 1$

69. $x^{12} + x^7 + x^5 + 1$

70. $x^3 - x^2 - 2x + 5$

71. $p^3 + p^2 - 3p + 10$

11.2 FACTORING TRINOMIALS OF THE TYPE $x^2 + bx + c$

Objective

Factor trinomials of the type $x^2 + bx + c$ by examining the constant term c.

a Factoring $x^2 + bx + c$

We now begin a study of the factoring of trinomials. We first factor trinomials like

$$x^2 + 5x + 6 \quad \text{and} \quad x^2 + 3x - 10$$

by a refined *trial-and-error process.* In this section, we restrict our attention to trinomials of the type $ax^2 + bx + c$, where $a = 1$. The coefficient a is often called the **leading coefficient.**

To understand the factoring that follows, compare the following multiplications:

$$\begin{aligned}
& \qquad\quad\ \ \text{F} \quad\ \ \text{O} \quad\ \ \text{I} \quad\ \ \text{L} \\
& \qquad\quad\ \downarrow \quad\ \downarrow \quad\ \downarrow \quad\ \downarrow \\
(x + 2)(x + 5) &= x^2 + 5x + 2x + 2 \cdot 5 \\
&= x^2 + 7x + 10; \\
(x - 2)(x - 5) &= x^2 - 5x - 2x + 2 \cdot 5 \\
&= x^2 - 7x + 10; \\
(x + 3)(x - 7) &= x^2 - 7x + 3x + 3(-7) \\
&= x^2 - 4x - 21; \\
(x - 3)(x + 7) &= x^2 + 7x - 3x + (-3)7 \\
&= x^2 + 4x - 21.
\end{aligned}$$

Note that for all four products:

- The product of the two binomials is a trinomial.
- The coefficient of x in the trinomial is the sum of the constant terms in the binomials.
- The constant term in the trinomial is the product of the constant terms in the binomials.

These observations lead to a method for factoring certain trinomials. The first type we consider has a positive constant term, just as in the first two multiplications above.

CONSTANT TERM POSITIVE

To factor $x^2 + 7x + 10$, we think of FOIL in reverse. We multiplied x times x to get the first term of the trinomial, so we know that the first term of each binomial factor is x. Next, we look for numbers p and q such that

$$x^2 + 7x + 10 = (x + p)(x + q).$$

To get the middle term and the last term of the trinomial, we look for two numbers p and q whose product is 10 and whose sum is 7. Those numbers are 2 and 5. Thus the factorization is

$$(x + 2)(x + 5).$$

Check: $(x + 2)(x + 5) = x^2 + 5x + 2x + 10$
$\qquad\qquad\qquad\qquad\quad = x^2 + 7x + 10.$

---

1. Consider the trinomial $x^2 + 7x + 12$.

a) Complete the following table.

PAIRS OF FACTORS	SUMS OF FACTORS
1, 12	13
−1, −12	
2, 6	
−2, −6	
3, 4	
−3, −4	

b) Explain why you need to consider only positive factors, as in the following table.

PAIRS OF FACTORS	SUMS OF FACTORS
1, 12	
2, 6	
3, 4	

c) Factor: $x^2 + 7x + 12$.

2. Factor: $x^2 + 13x + 36$.

Answers on page A-39

3. Explain why you would *not* consider the pairs of factors listed below in factoring $y^2 - 8y + 12$.

PAIRS OF FACTORS	SUMS OF FACTORS
1, 12	
2, 6	
3, 4	

Factor.

4. $x^2 - 8x + 15$

5. $t^2 - 9t + 20$

Answers on page A-39

EXAMPLE 1 Factor: $x^2 + 5x + 6$.

Think of FOIL in reverse. The first term of each factor is x: $(x + \quad)(x + \quad)$. Next, we look for two numbers whose product is 6 and whose sum is 5. All the pairs of factors of 6 are shown in the table on the left below. Since both the product, 6, and the sum, 5, of the pair of numbers must be positive, we need consider only the positive factors, listed in the table on the right.

PAIRS OF FACTORS	SUMS OF FACTORS
1, 6	7
−1, −6	−7
2, 3	5
−2, −3	−5

PAIRS OF FACTORS	SUMS OF FACTORS
1, 6	7
2, 3	**5**

The numbers we need are 2 and 3.

The factorization is $(x + 2)(x + 3)$. We can check by multiplying to see whether we get the original trinomial.

Check: $(x + 2)(x + 3) = x^2 + 3x + 2x + 6 = x^2 + 5x + 6$.

Do Exercises 1 and 2 on the preceding page.

Compare these multiplications:

$$(x - 2)(x - 5) = x^2 - 5x - 2x + 10 = x^2 - 7x + 10;$$
$$(x + 2)(x + 5) = x^2 + 5x + 2x + 10 = x^2 + 7x + 10.$$

TO FACTOR $x^2 + bx + c$ WHEN c IS POSITIVE

When the constant term of a trinomial is positive, look for two numbers with the same sign. The sign is that of the middle term:

$$x^2 - 7x + 10 = (x - 2)(x - 5);$$

$$x^2 + 7x + 10 = (x + 2)(x + 5).$$

EXAMPLE 2 Factor: $y^2 - 8y + 12$.

Since the constant term, 12, is positive and the coefficient of the middle term, −8, is negative, we look for a factorization of 12 in which both factors are negative. Their sum must be −8.

PAIRS OF FACTORS	SUMS OF FACTORS
−1, −12	−13
−2, −6	**−8**
−3, −4	−7

The numbers we need are −2 and −6.

The factorization is $(y - 2)(y - 6)$. The student should check by multiplying.

Do Exercises 3–5.

CONSTANT TERM NEGATIVE

As we saw in two of the multiplications earlier in this section, the product of two binomials can have a negative constant term:

$$(x + 3)(x - 7) = x^2 - 4x - 21$$

and

$$(x - 3)(x + 7) = x^2 + 4x - 21.$$

Note that when the signs of the constants in the binomials are reversed, only the sign of the middle term in the product changes.

EXAMPLE 3 Factor: $x^2 - 8x - 20$.

The constant term, -20, must be expressed as the product of a negative number and a positive number. Since the sum of these two numbers must be negative (specifically, -8), the negative number must have the greater absolute value.

PAIRS OF FACTORS	SUMS OF FACTORS
1, −20	−19
2, −10	**−8**
4, −5	−1
5, −4	1
10, −2	8
20, −1	19

The numbers we need are 2 and −10.

Because these sums are all positive, for this problem all of the corresponding pairs can be disregarded. Note that in all three pairs, the positive number has the greater absolute value.

The numbers that we are looking for are 2 and −10. The factorization is $(x + 2)(x - 10)$.

Check: $(x + 2)(x - 10) = x^2 - 10x + 2x - 20$
$= x^2 - 8x - 20.$

TO FACTOR $x^2 + bx + c$ WHEN c IS NEGATIVE

When the constant term of a trinomial is negative, look for two numbers whose product is negative. One must be positive and the other negative:

$$x^2 - 4x - 21 = (x + 3)(x - 7);$$

$$x^2 + 4x - 21 = (x - 3)(x + 7).$$

Consider pairs of numbers for which the number with the larger absolute value has the same sign as b, the coefficient of the middle term.

Do Exercises 6 and 7. (Exercise 7 is on the following page.)

6. Consider $x^2 - 5x - 24$.

a) Explain why you would *not* consider the pairs of factors listed below in factoring $x^2 - 5x - 24$.

PAIRS OF FACTORS	SUMS OF FACTORS
−1, 24	
−2, 12	
−3, 8	
−4, 6	

b) Explain why you *would* consider the pairs of factors listed below in factoring $x^2 - 5x - 24$.

PAIRS OF FACTORS	SUMS OF FACTORS
1, −24	
2, −12	
3, −8	
4, −6	

c) Factor: $x^2 - 5x - 24$.

Answers on page A-39

EXAMPLE 4 Factor: $t^2 - 24 + 5t$.

It helps to first write the trinomial in descending order: $t^2 + 5t - 24$. Since the constant term, -24, is negative, we look for a factorization of -24 in which one factor is positive and one factor is negative. Their sum must be 5, so the positive factor must have the larger absolute value. Thus we consider only pairs of factors in which the positive term has the larger absolute value.

PAIRS OF FACTORS	SUMS OF FACTORS
$-1, 24$	23
$-2, 12$	10
$-3, 8$	**5** ← The numbers we need are -3 and 8.
$-4, 6$	2

The factorization is $(t - 3)(t + 8)$. The check is left to the student.

Do Exercises 8 and 9.

EXAMPLE 5 Factor: $x^4 - x^2 - 110$.

Consider this trinomial as $(x^2)^2 - x^2 - 110$. We look for numbers p and q such that

$$x^4 - x^2 - 110 = (x^2 + p)(x^2 + q).$$

Since the constant term, -110, is negative, we look for a factorization of -110 in which one factor is positive and one factor is negative. Their sum must be -1. The middle-term coefficient, -1, is small compared to -110. This tells us that the desired factors are close to each other in absolute value. The numbers we want are 10 and -11. The factorization is

$$(x^2 + 10)(x^2 - 11).$$

EXAMPLE 6 Factor: $a^2 + 4ab - 21b^2$.

We consider the trinomial in the equivalent form

$$a^2 + 4ba - 21b^2.$$

This way we think of $-21b^2$ as the "constant" term and $4b$ as the "coefficient" of the middle term. Then we try to express $-21b^2$ as a product of two factors whose sum is $4b$. Those factors are $-3b$ and $7b$. The factorization is $(a - 3b)(a + 7b)$.

Check: $(a - 3b)(a + 7b) = a^2 + 7ab - 3ba - 21b^2$
$= a^2 + 4ab - 21b^2.$

There are polynomials that are not factorable.

EXAMPLE 7 Factor: $x^2 - x + 5$.

Since 5 has very few factors, we can easily check all possibilities.

PAIRS OF FACTORS	SUMS OF FACTORS
$5, 1$	6
$-5, -1$	-6

7. Consider $x^2 + 10x - 24$.

a) Explain why you would *not* consider the pairs of factors listed below in factoring $x^2 + 10x - 24$.

PAIRS OF FACTORS	SUMS OF FACTORS
$1, -24$	
$2, -12$	
$3, -8$	
$4, -6$	

b) Explain why you *would* consider the pairs of factors listed below in factoring $x^2 + 10x - 24$.

PAIRS OF FACTORS	SUMS OF FACTORS
$-1, 24$	
$-2, 12$	
$-3, 8$	
$-4, 6$	

c) Factor: $x^2 + 10x - 24$.

Factor.

8. $a^2 - 24 + 10a$

9. $-24 - 10t + t^2$

Answers on page A-39

There are no factors whose sum is -1. Thus the polynomial is *not* factorable into factors that are polynomials.

In this text, a polynomial like $x^2 - x + 5$ that cannot be factored further is said to be **prime.** In more advanced courses, polynomials like $x^2 - x + 5$ can be factored and are not considered prime.

Do Exercises 10–12.

Often factoring requires two or more steps. In general, when told to factor, we should *factor completely*. This means that the final factorization should not contain any factors that can be factored further.

EXAMPLE 8 Factor: $2x^3 - 20x^2 + 50x$.

Always look first for a common factor. This time there is one, $2x$, which we factor out first:

$$2x^3 - 20x^2 + 50x = 2x(x^2 - 10x + 25).$$

Now consider $x^2 - 10x + 25$. Since the constant term is positive and the coefficient of the middle term is negative, we look for a factorization of 25 in which both factors are negative. Their sum must be -10.

PAIRS OF FACTORS	SUMS OF FACTORS
$-25, -1$	-26
$\mathbf{-5, -5,}$	$\mathbf{-10}$ ← The numbers we need are -5 and -5.

The factorization of $x^2 - 10x + 25$ is $(x - 5)(x - 5)$, or $(x - 5)^2$. The final factorization is $2x(x - 5)^2$. We check by multiplying:

$$\begin{aligned} 2x(x - 5)^2 &= 2x(x^2 - 10x + 25) \\ &= (2x)(x^2) - (2x)(10x) + (2x)(25) \\ &= 2x^3 - 20x^2 + 50x. \end{aligned}$$

Do Exercises 13–15.

Once any common factors have been factored out, the following summary can be used to factor $x^2 + bx + c$.

TO FACTOR $x^2 + bx + c$

1. First arrange in descending order.
2. Use a trial-and-error process that looks for factors of c whose sum is b.
3. If c is positive, the signs of the factors are the same as the sign of b.
4. If c is negative, one factor is positive and the other is negative. If the sum of two factors is the opposite of b, changing the sign of each factor will give the desired factors whose sum is b.
5. Check by multiplying.

Factor.

10. $y^2 - 12 - 4y$

11. $t^4 + 5t^2 - 14$

12. $x^2 + 2x + 7$

Factor.

13. $x^3 + 4x^2 - 12x$

14. $p^2 - pq - 3pq^2$

15. $3x^3 + 24x^2 + 48x$

Answers on page A-39

Factor.

16. $14 + 5x - x^2$

17. $-x^2 + 3x + 18$

Answers on page A-39

LEADING COEFFICIENT -1

EXAMPLE 9 Factor: $10 - 3x - x^2$.

Note that the polynomial is written in ascending order. When we write it in descending order, we get

$$-x^2 - 3x + 10,$$

which has a leading coefficient of -1. Before factoring in such a case, we can factor out a -1, as follows:

$$-x^2 - 3x + 10 = -1(x^2 + 3x - 10).$$

Then we proceed to factor $x^2 + 3x - 10$. We get

$$-x^2 - 3x + 10 = -1(x^2 + 3x - 10) = -1(x + 5)(x - 2).$$

We can also express this answer in two other ways by multiplying either binomial by -1. Thus each of the following is a correct answer:

$$-x^2 - 3x + 10 = -1(x + 5)(x - 2)$$
$$= (-x - 5)(x - 2) \quad \text{Multiplying } x + 5 \text{ by } -1$$
$$= (x + 5)(-x + 2). \quad \text{Multiplying } x - 2 \text{ by } -1$$

Do Exercises 16 and 17.

Study Tips

TIME MANAGEMENT (PART 2)

Here are some additional tips to help you with time management. (See also the Study Tips on time management in Sections 8.2 and 11.6.)

- **Are you a morning or an evening person?** If you are an evening person, it might be best to avoid scheduling early-morning classes. If you are a morning person, do the opposite, but go to bed earlier to compensate. Nothing can drain your study time and effectiveness like fatigue.
- **Keep on schedule.** Your course syllabus provides a plan for the semester's schedule. Use a write-on calendar, daily planner, Palm Pilot or other PDA, or laptop computer to outline your time for the semester. Be sure to note deadlines involving term papers and exams so you can begin a task early, breaking it down into smaller segments that can be accomplished more easily.
- **Balance your class schedule.** You may be someone who prefers large blocks of time for study on the off days. In that case, it might be advantageous for you to take courses that meet only three days a week. Keep in mind, however, that this might be a problem when tests in more than one course are scheduled for the same day.

"Time is our most important asset, yet we tend to waste it, kill it, and spend it rather than invest it."

Jim Rohn, motivational speaker

11.2 EXERCISE SET

For Extra Help

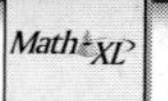
MathXL

MyMathLab

InterAct Math

Math Tutor Center

Video Lectures on CD Disc 6

Student's Solutions Manual

a Factor. Remember that you can check by multiplying.

1. $x^2 + 8x + 15$

PAIRS OF FACTORS	SUMS OF FACTORS

2. $x^2 + 5x + 6$

PAIRS OF FACTORS	SUMS OF FACTORS

3. $x^2 + 7x + 12$

PAIRS OF FACTORS	SUMS OF FACTORS

4. $x^2 + 9x + 8$

PAIRS OF FACTORS	SUMS OF FACTORS

5. $x^2 - 6x + 9$

PAIRS OF FACTORS	SUMS OF FACTORS

6. $y^2 - 11y + 28$

PAIRS OF FACTORS	SUMS OF FACTORS

7. $x^2 - 5x - 14$

PAIRS OF FACTORS	SUMS OF FACTORS

8. $a^2 + 7a - 30$

PAIRS OF FACTORS	SUMS OF FACTORS

9. $b^2 + 5b + 4$

PAIRS OF FACTORS	SUMS OF FACTORS

10. $z^2 - 8z + 7$

PAIRS OF FACTORS	SUMS OF FACTORS

11. $x^2 + \frac{2}{3}x + \frac{1}{9}$

PAIRS OF FACTORS	SUMS OF FACTORS

12. $x^2 - \frac{2}{5}x + \frac{1}{25}$

PAIRS OF FACTORS	SUMS OF FACTORS

13. $d^2 - 7d + 10$

14. $t^2 - 12t + 35$

15. $y^2 - 11y + 10$

16. $x^2 - 4x - 21$

17. $x^2 + x + 1$

18. $x^2 + 5x + 3$

19. $x^2 - 7x - 18$

20. $y^2 - 3y - 28$

21. $x^3 - 6x^2 - 16x$

22. $x^3 - x^2 - 42x$

23. $y^3 - 4y^2 - 45y$

24. $x^3 - 7x^2 - 60x$

25. $-2x - 99 + x^2$

26. $x^2 - 72 + 6x$

27. $c^4 + c^2 - 56$

28. $b^4 + 5b^2 - 24$

29. $a^4 + 2a^2 - 35$

30. $x^4 - x^2 - 6$

31. $x^2 + x - 42$

32. $x^2 + 2x - 15$

33. $7 - 2p + p^2$

34. $11 - 3w + w^2$

35. $x^2 + 20x + 100$

36. $a^2 + 19a + 88$

37. $30 + 7x - x^2$

38. $45 + 4x - x^2$

39. $24 - a^2 - 10a$

40. $-z^2 + 36 - 9z$

41. $x^4 - 21x^3 - 100x^2$

42. $x^4 - 20x^3 + 96x^2$

43. $x^2 - 21x - 72$

44. $4x^2 + 40x + 100$

45. $x^2 - 25x + 144$

46. $y^2 - 21y + 108$

47. $a^2 + a - 132$

48. $a^2 + 9a - 90$

49. $120 - 23x + x^2$

50. $96 + 22d + d^2$

51. $108 - 3x - x^2$

52. $112 + 9y - y^2$

53. $y^2 - 0.2y - 0.08$

54. $t^2 - 0.3t - 0.10$

55. $p^2 + 3pq - 10q^2$

56. $a^2 + 2ab - 3b^2$

57. $84 - 8t - t^2$

58. $72 - 6m - m^2$

59. $m^2 + 5mn + 4n^2$

60. $x^2 + 11xy + 24y^2$

61. $s^2 - 2st - 15t^2$

62. $p^2 + 5pq - 24q^2$

63. $6a^{10} - 30a^9 - 84a^8$

64. $7x^9 - 28x^8 - 35x^7$

65. **DW** Gwyneth factors $x^3 - 8x^2 + 15x$ as $(x^2 - 5x)(x - 3)$. Is she wrong? Why or why not? What advice would you offer?

66. **DW** When searching for a factorization, why do we list pairs of numbers with the correct *product* instead of pairs of numbers with the correct *sum*?

67. **DW** Without multiplying $(x - 17)(x - 18)$, explain why it cannot possibly be a factorization of $x^2 + 35x + 306$.

68. **DW** What is the advantage of writing out the prime factorization of c when factoring $x^2 + bx + c$ with a large value of c?

SKILL MAINTENANCE

Multiply. [10.6d]

69. $8x(2x^2 - 6x + 1)$

70. $(7w + 6)(4w - 11)$

71. $(7w + 6)^2$

72. $(4w - 11)^2$

73. $(4w - 11)(4w + 11)$

74. $-y(-y^2 + 3y - 5)$

75. $(3x - 5y)(2x + 7y)$

76. Simplify: $(3x^4)^3$. [10.2a, b]

Solve. [8.3a]

77. $3x - 8 = 0$

78. $2x + 7 = 0$

Solve.

79. *Arrests for Counterfeiting.* In a recent year, 29,200 people were arrested for counterfeiting. This number was down 1.2% from the preceding year. How many people were arrested the preceding year? [8.5a]

80. The first angle of a triangle is four times as large as the second. The measure of the third angle is 30° greater than that of the second. Find the angle measures. [8.6a]

SYNTHESIS

81. Find all integers m for which $y^2 + my + 50$ can be factored.

82. Find all integers b for which $a^2 + ba - 50$ can be factored.

Factor completely.

83. $x^2 - \frac{1}{2}x - \frac{3}{16}$

84. $x^2 - \frac{1}{4}x - \frac{1}{8}$

85. $x^2 + \frac{30}{7}x - \frac{25}{7}$

86. $\frac{1}{3}x^3 + \frac{1}{3}x^2 - 2x$

87. $b^{2n} + 7b^n + 10$

88. $a^{2m} - 11a^m + 28$

Find a polynomial in factored form for the shaded area. (Leave answers in terms of π.)

89.

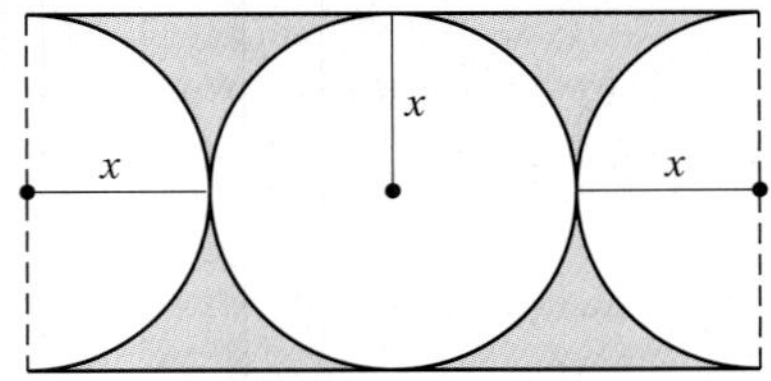

90.

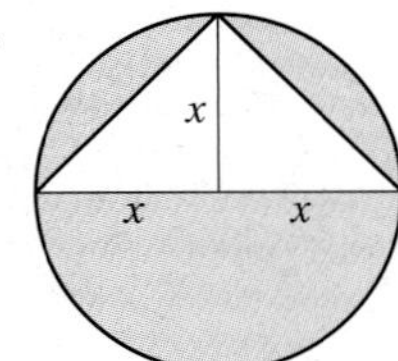

11.3 FACTORING $ax^2 + bx + c$, $a \neq 1$: THE FOIL METHOD

Objective

a Factor trinomials of the type $ax^2 + bx + c$, $a \neq 1$, using the FOIL method.

In Section 11.2, we learned a trial-and-error method to factor trinomials of the type $x^2 + bx + c$. In this section, we factor trinomials in which the coefficient of the leading term x^2 is not 1. The procedure we learn is a refined trial-and-error method.

a The FOIL Method

We want to factor trinomials of the type $ax^2 + bx + c$. Consider the following multiplication:

$$\begin{aligned}(2x + 5)(3x + 4) &= \overset{\text{F}}{6x^2} + \overset{\text{O}}{8x} + \overset{\text{I}}{15x} + \overset{\text{L}}{20} \\ &= 6x^2 + 23x + 20\end{aligned}$$

F	O + I		L
$2 \cdot 3$	$2 \cdot 4$	$5 \cdot 3$	$5 \cdot 4$

To factor $6x^2 + 23x + 20$, we reverse the above multiplication, using what we might call an "unFOIL" process. We look for two binomials $rx + p$ and $sx + q$ whose product is $(rx + p)(sx + q) = 6x^2 + 23x + 20$. The product of the First terms must be $6x^2$. The product of the Outside terms plus the product of the Inside terms must be $23x$. The product of the Last terms must be 20. We know from the preceding discussion that the answer is $(2x + 5)(3x + 4)$. Generally, however, finding such an answer is a refined trial-and-error process. It turns out that $(-2x - 5)(-3x - 4)$ is also a correct answer, but we generally choose an answer in which the first coefficients are positive.

We will use the following trial-and-error method.

THE FOIL METHOD

To factor $ax^2 + bx + c$, $a \neq 1$, using the FOIL method:

1. Factor out the largest common factor, if one exists.
2. Find two First terms whose product is ax^2.

$$(\square x + \quad)(\square x + \quad) = ax^2 + bx + c.$$

FOIL

3. Find two Last terms whose product is c:

$$(\quad x + \square)(\quad x + \square) = ax^2 + bx + c.$$

FOIL

4. Look for Outer and Inner products resulting from steps (2) and (3) for which the sum is bx:

$$(\square x + \square)(\square x + \square) = ax^2 + bx + c.$$

I O FOIL

5. Always check by multiplying.

The ac-method in Section 11.4

To the student: In Section 11.4, we will consider an alternative method for the same kind of factoring. It involves factoring by grouping and is called the *ac*-method.

To the instructor: We present two ways to factor general trinomials in Sections 11.3 and 11.4: the FOIL method in Section 11.3 and the *ac*-method in Section 11.4. You can teach both methods and let the student use the one that he or she prefers or you can select just one.

Factor.

1. $2x^2 - x - 15$

2. $12x^2 - 17x - 5$

Answers on page A-40

Study Tips

READING EXAMPLES

A careful study of the examples in these sections on factoring is critical. *Read them carefully* to ensure success!

EXAMPLE 1 Factor: $3x^2 - 10x - 8$.

1) First, we check for a common factor. Here there is none (other than 1 or -1).

2) Find two **F**irst terms whose product is $3x^2$.

The only possibilities for the **F**irst terms are $3x$ and x, so any factorization must be of the form

$$(3x + \quad)(x + \quad).$$

3) Find two **L**ast terms whose product is -8.

Possible factorizations of -8 are

$$(-8) \cdot 1, \quad 8 \cdot (-1), \quad (-2) \cdot 4, \quad \text{and} \quad 2 \cdot (-4).$$

Since the First terms are not identical, we must also consider

$$1 \cdot (-8), \quad (-1) \cdot 8, \quad 4 \cdot (-2), \quad \text{and} \quad (-4) \cdot 2.$$

4) Inspect the **O**uter and **I**nner products resulting from steps (2) and (3). Look for a combination in which the sum of the products is the middle term, $-10x$:

Trial	*Product*	
$(3x - 8)(x + 1)$	$3x^2 + 3x - 8x - 8$ $= 3x^2 - 5x - 8$	← Wrong middle term
$(3x + 8)(x - 1)$	$3x^2 - 3x + 8x - 8$ $= 3x^2 + 5x - 8$	← Wrong middle term
$(3x - 2)(x + 4)$	$3x^2 + 12x - 2x - 8$ $= 3x^2 + 10x - 8$	← Wrong middle term
$(3x + 2)(x - 4)$	$3x^2 - 12x + 2x - 8$ $= 3x^2 - 10x - 8$	← **Correct middle term!**
$(3x + 1)(x - 8)$	$3x^2 - 24x + x - 8$ $= 3x^2 - 23x - 8$	← Wrong middle term
$(3x - 1)(x + 8)$	$3x^2 + 24x - x - 8$ $= 3x^2 + 23x - 8$	← Wrong middle term
$(3x + 4)(x - 2)$	$3x^2 - 6x + 4x - 8$ $= 3x^2 - 2x - 8$	← Wrong middle term
$(3x - 4)(x + 2)$	$3x^2 + 6x - 4x - 8$ $= 3x^2 + 2x - 8$	← Wrong middle term

The correct factorization is $(3x + 2)(x - 4)$.

5) Check: $(3x + 2)(x - 4) = 3x^2 - 10x - 8$.

Two observations can be made from Example 1. First, we listed all possible trials even though we could have stopped after having found the correct factorization. We did this to show that each trial differs only in the middle term of the product. **Second, note that as in Section 11.2, only the sign of the middle term changes when the signs in the binomials are reversed:**

Plus ↓, Minus ↓

$$(3x + 4)(x - 2) = 3x^2 - 2x - 8$$

Minus ↓, Plus ↓ ← Middle term changes sign

$$(3x - 4)(x + 2) = 3x^2 + 2x - 8.$$

Do Exercises 1 and 2.

EXAMPLE 2 Factor: $24x^2 - 76x + 40$.

1) First, we factor out the largest common factor, 4:

$$4(6x^2 - 19x + 10).$$

Now we factor the trinomial $6x^2 - 19x + 10$.

2) Because $6x^2$ can be factored as $3x \cdot 2x$ or $6x \cdot x$, we have these possibilities for factorizations:

$$(3x + \quad)(2x + \quad) \quad \text{or} \quad (6x + \quad)(x + \quad).$$

3) There are four pairs of factors of 10 and they each can be listed in two ways:

$10, 1 \qquad -10, -1 \qquad 5, 2 \qquad -5, -2$

and

$1, 10 \qquad -1, -10 \qquad 2, 5 \qquad -2, -5.$

4) The two possibilities from step (2) and the eight possibilities from step (3) give $2 \cdot 8$, or 16 possibilities for factorizations. We look for **O**uter and **I**nner products resulting from steps (2) and (3) for which the sum is the middle term, $-19x$. Since the sign of the middle term is negative, but the sign of the last term, 10, is positive, the two factors of 10 must both be negative. This means only four pairings from step (3) need be considered. We first try these factors with

$$(3x + \quad)(2x + \quad).$$

If none gives the correct factorization, we will consider

$$(6x + \quad)(x + \quad).$$

Trial	*Product*	
$(3x - 10)(2x - 1)$	$6x^2 - 3x - 20x + 10$	
	$= 6x^2 - 23x + 10$	← Wrong middle term
$(3x - 1)(2x - 10)$	$6x^2 - 30x - 2x + 10$	
	$= 6x^2 - 32x + 10$	← Wrong middle term
$(3x - 5)(2x - 2)$	$6x^2 - 6x - 10x + 10$	
	$= 6x^2 - 16x + 10$	← Wrong middle term
$(3x - 2)(2x - 5)$	$6x^2 - 15x - 4x + 10$	
	$= 6x^2 - 19x + 10$	← **Correct middle term!**

Since we have a correct factorization, we need not consider

$$(6x + \quad)(x + \quad).$$

The factorization of $6x^2 - 19x + 10$ is $(3x - 2)(2x - 5)$, but *do not forget the common factor*! We must include it in order to factor the original trinomial:

$$24x^2 - 76x + 40 = 4(6x^2 - 19x + 10)$$
$$= 4(3x - 2)(2x - 5).$$

5) Check: $4(3x - 2)(2x - 5) = 4(6x^2 - 19x + 10) = 24x^2 - 76x + 40.$

Caution!

When factoring any polynomial, always look for a common factor. Failure to do so is such a common error that this caution bears repeating.

Study Tips

SKILL MAINTENANCE EXERCISES

It is never too soon to begin reviewing for the final examination. The Skill Maintenance exercises found in each exercise set review and reinforce skills taught in earlier sections. Include all of these exercises in your weekly preparation. Answers to both odd-numbered and even-numbered exercises appear at the back of the book.

Factor.

3. $3x^2 - 19x + 20$

4. $20x^2 - 46x + 24$

5. Factor: $6x^2 + 7x + 2$.

Answers on page A-40

In Example 2, look again at the possibility $(3x - 5)(2x - 2)$. Without multiplying, we can reject such a possibility. To see why, consider the following:

$$(3x - 5)(2x - 2) = 2(3x - 5)(x - 1).$$

The expression $2x - 2$ has a common factor, 2. But we removed the *largest* common factor in the first step. If $2x - 2$ were one of the factors, then 2 would have to be a common factor in addition to the original 4. Thus, $(2x - 2)$ cannot be part of the factorization of the original trinomial.

Given that the largest common factor is factored out at the outset, we need not consider factorizations that have a common factor.

Do Exercises 3 and 4.

EXAMPLE 3 Factor: $10x^2 + 37x + 7$.

1) There is no common factor (other than 1 or -1).

2) Because $10x^2$ factors as $10x \cdot x$ or $5x \cdot 2x$, we have these possibilities for factorizations:

$$(10x + \quad)(x + \quad) \quad \text{or} \quad (5x + \quad)(2x + \quad).$$

3) There are two pairs of factors of 7 and they each can be listed in two ways:

$$1, 7 \qquad -1, -7 \qquad \text{and} \qquad 7, 1 \qquad -7, -1.$$

4) From steps (2) and (3), we see that there are 8 possibilities for factorizations. Look for **O**uter and **I**nner products for which the sum is the middle term. Because all coefficients in $10x^2 + 37x + 7$ are positive, we need consider only positive factors of 7. The possibilities are

$$(10x + 1)(x + 7) = 10x^2 + 71x + 7,$$
$$(10x + 7)(x + 1) = 10x^2 + 17x + 7,$$
$$(5x + 7)(2x + 1) = 10x^2 + 19x + 7,$$
$$(5x + 1)(2x + 7) = 10x^2 + 37x + 7. \qquad \leftarrow \text{Correct middle term}$$

The factorization is $(5x + 1)(2x + 7)$.

5) Check: $(5x + 1)(2x + 7) = 10x^2 + 37x + 7$.

Do Exercise 5.

TIPS FOR FACTORING $ax^2 + bx + c$, $a \neq 1$

- Always factor out the largest common factor, if one exists.
- Once the common factor has been factored out of the original trinomial, no binomial factor can contain a common factor (other than 1 or -1).
- If c is positive, then the signs in both binomial factors must match the sign of b. (This assumes that $a > 0$.)
- Reversing the signs in the binomials reverses the sign of the middle term of their product.
- Organize your work so that you can keep track of which possibilities have or have not been checked.
- Always check by multiplying.

EXAMPLE 4 Factor: $10x + 8 - 3x^2$.

An important problem-solving strategy is to find a way to make new problems look like problems we already know how to solve. (See Example 9 in Section 11.2.) The factoring tips above apply only to trinomials of the form $ax^2 + bx + c$, with $a > 0$. This leads us to rewrite $10x + 8 - 3x^2$ in descending order:

$$10x + 8 - 3x^2 = -3x^2 + 10x + 8. \quad \text{Writing in descending order}$$

Although $-3x^2 + 10x + 8$ looks similar to the trinomials we have factored, the tips above require a positive leading coefficient. This can be attained by factoring out -1:

$$-3x^2 + 10x + 8 = -1(3x^2 - 10x - 8) \quad \text{Factoring out } -1 \text{ changes the signs of the coefficients.}$$
$$= -1(3x + 2)(x - 4). \quad \text{Using the result from Example 1}$$

The factorization of $10x + 8 - 3x^2$ is $-1(3x + 2)(x - 4)$. Other correct answers are

$$10x + 8 - 3x^2 = (3x + 2)(-x + 4) \quad \text{Multiplying } x - 4 \text{ by } -1$$
$$= (-3x - 2)(x - 4). \quad \text{Multiplying } 3x + 2 \text{ by } -1$$

Do Exercises 6 and 7.

Factor.

6. $2 - x - 6x^2$

7. $2x + 8 - 6x^2$

EXAMPLE 5 Factor: $6p^2 - 13pq - 28q^2$.

1) Factor out a common factor, if any.
   There is none (other than 1 or -1).
2) Factor the first term, $6p^2$.
   Possibilities are $2p$, $3p$ and $6p$, p. We have these as possibilities for factorizations:
   $$(2p + \quad)(3p + \quad) \quad \text{or} \quad (6p + \quad)(p + \quad).$$
3) Factor the last term, $-28q^2$, which has a negative coefficient.
   There are six pairs of factors and each can be listed in two ways:
   $$-28q, q \qquad 28q, -q \qquad -14q, 2q \qquad 14q, -2q \qquad -7q, 4q \qquad 7q, -4q$$
   and
   $$q, -28q \qquad -q, 28q \qquad 2q, -14q \qquad -2q, 14q \qquad 4q, -7q \qquad -4q, 7q.$$
4) The coefficient of the middle term is negative, so we look for combinations of factors from steps (2) and (3) such that the sum of their products has a negative coefficient. We try some possibilities:
   $$(2p + q)(3p - 28q) = 6p^2 - 53pq - 28q^2,$$
   $$(2p - 7q)(3p + 4q) = 6p^2 - 13pq - 28q^2. \quad \leftarrow \textbf{Correct middle term}$$
   The factorization of $6p^2 - 13pq - 28q^2$ is $(2p - 7q)(3p + 4q)$.
5) The check is left to the student.

Do Exercises 8 and 9.

Factor.

8. $6a^2 - 5ab + b^2$

9. $6x^2 + 15xy + 9y^2$

Answers on page A-40

CALCULATOR CORNER

Checking Factorizations A partial check of a factorization can be performed using a table or a graph. To check the factorization $6x^3 - 9x^2 + 4x - 6 = (3x^2 + 2)(2x - 3)$, for example, we enter $y_1 = 6x^3 - 9x^2 + 4x - 6$ and $y_2 = (3x^2 + 2)(2x - 3)$ on the equation-editor screen (see page 642). Then we set up a table in AUTO mode (see page 647). If the factorization is correct, the values of y_1 and y_2 will be the same regardless of the table settings used.

X	Y1	Y2
−3	−261	−261
−2	−98	−98
−1	−25	−25
0	−6	−6
1	−5	−5
2	14	14
3	87	87
X = −3		

We can also graph $y_1 = 6x^3 - 9x^2 + 4x - 6$ and $y_2 = (3x^2 + 2)(2x - 3)$. If the graphs appear to coincide, the factorization is probably correct.

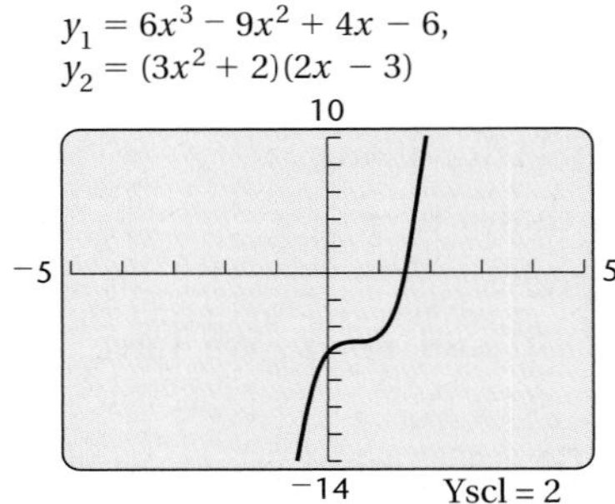

Keep in mind that these procedures provide only a partial check since we cannot view all possible values of x in a table nor can we see the entire graph.

Exercises: Use a table or a graph to determine whether the factorization is correct.

1. $24x^2 - 76x + 40 = 4(3x - 2)(2x - 5)$
2. $4x^2 - 5x - 6 = (4x + 3)(x - 2)$
3. $5x^2 + 17x - 12 = (5x + 3)(x - 4)$
4. $10x^2 + 37x + 7 = (5x - 1)(2x + 7)$
5. $12x^2 - 17x - 5 = (6x + 1)(2x - 5)$
6. $12x^2 - 17x - 5 = (4x + 1)(3x - 5)$
7. $x^2 - 4 = (x - 2)(x - 2)$
8. $x^2 - 4 = (x + 2)(x - 2)$

11.3 EXERCISE SET

For Extra Help

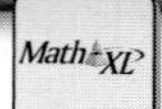
MathXL

MyMathLab

InterAct Math

Math Tutor Center

Video Lectures on CD Disc 6

Student's Solutions Manual

a Factor.

1. $2x^2 - 7x - 4$

2. $3x^2 - x - 4$

3. $5x^2 - x - 18$

4. $4x^2 - 17x + 15$

5. $6x^2 + 23x + 7$

6. $6x^2 - 23x + 7$

7. $3x^2 + 4x + 1$

8. $7x^2 + 15x + 2$

9. $4x^2 + 4x - 15$

10. $9x^2 + 6x - 8$

11. $2x^2 - x - 1$

12. $15x^2 - 19x - 10$

13. $9x^2 + 18x - 16$

14. $2x^2 + 5x + 2$

15. $3x^2 - 5x - 2$

16. $18x^2 - 3x - 10$

17. $12x^2 + 31x + 20$

18. $15x^2 + 19x - 10$

19. $14x^2 + 19x - 3$

20. $35x^2 + 34x + 8$

21. $9x^2 + 18x + 8$

22. $6 - 13x + 6x^2$

23. $49 - 42x + 9x^2$

24. $16 + 36x^2 + 48x$

25. $24x^2 + 47x - 2$

26. $16p^2 - 78p + 27$

27. $35x^2 - 57x - 44$

28. $9a^2 + 12a - 5$

29. $20 + 6x - 2x^2$

30. $15 + x - 2x^2$

31. $12x^2 + 28x - 24$

32. $6x^2 + 33x + 15$

33. $30x^2 - 24x - 54$

34. $18t^2 - 24t + 6$

35. $4y + 6y^2 - 10$

36. $-9 + 18x^2 - 21x$

37. $3x^2 - 4x + 1$

38. $6t^2 + 13t + 6$

39. $12x^2 - 28x - 24$

40. $6x^2 - 33x + 15$

41. $-1 + 2x^2 - x$

42. $-19x + 15x^2 + 6$

43. $9x^2 - 18x - 16$

44. $14y^2 + 35y + 14$

45. $15x^2 - 25x - 10$

46. $18x^2 + 3x - 10$

47. $12p^3 + 31p^2 + 20p$

48. $15x^3 + 19x^2 - 10x$

49. $16 + 18x - 9x^2$

50. $33t - 15 - 6t^2$

51. $-15x^2 + 19x - 6$

52. $1 + p - 2p^2$

53. $14x^4 + 19x^3 - 3x^2$

54. $70x^4 + 68x^3 + 16x^2$

55. $168x^3 - 45x^2 + 3x$

56. $144x^5 + 168x^4 + 48x^3$

57. $15x^4 - 19x^2 + 6$

58. $9x^4 + 18x^2 + 8$

59. $25t^2 + 80t + 64$

60. $9x^2 - 42x + 49$

61. $6x^3 + 4x^2 - 10x$

62. $18x^3 - 21x^2 - 9x$

63. $25x^2 + 79x + 64$

64. $9y^2 + 42y + 47$

65. $6x^2 - 19x - 5$

66. $2x^2 + 11x - 9$

67. $12m^2 - mn - 20n^2$

68. $12a^2 - 17ab + 6b^2$

69. $6a^2 - ab - 15b^2$

70. $3p^2 - 16pq - 12q^2$

71. $9a^2 + 18ab + 8b^2$

72. $10s^2 + 4st - 6t^2$

73. $35p^2 + 34pq + 8q^2$

74. $30a^2 + 87ab + 30b^2$

75. $18x^2 - 6xy - 24y^2$

76. $15a^2 - 5ab - 20b^2$

77. D_W Explain how the factoring in Exercise 21 can be used to aid the factoring in Exercise 71.

78. D_W A student presents the following work:

$$\begin{aligned} 4x^2 + 28x + 48 &= (2x + 6)(2x + 8) \\ &= 2(x + 3)(x + 4). \end{aligned}$$

Is it correct? Explain.

SKILL MAINTENANCE

Solve. [8.4b]

79. $A = pq - 7$, for q

80. $y = mx + b$, for x

81. $3x + 2y = 6$, for y

82. $p - q + r = 2$, for q

Solve. [8.7e]

83. $5 - 4x < -11$

84. $2x - 4(x + 3x) \geq 6x - 8 - 9x$

85. Graph: $y = \dfrac{2}{5}x - 1$. [9.1d]

86. Divide: $\dfrac{y^{12}}{y^4}$. [10.1e]

Find the intercepts of the equation. [9.2a]

87. $4x - 16y = 64$

88. $4x + 16y = 64$

89. $x - 1.3y = 6.5$

90. $\frac{2}{3}x + \frac{5}{8}y = \frac{5}{12}$

91. $y = 4 - 5x$

92. $y = 2x - 5$

SYNTHESIS

Factor.

93. $20x^{2n} + 16x^n + 3$

94. $-15x^{2m} + 26x^m - 8$

95. $3x^{6a} - 2x^{3a} - 1$

96. $x^{2n+1} - 2x^{n+1} + x$

97.–106. Use the TABLE feature to check the factoring in Exercises 15–24.

11.4 FACTORING $ax^2 + bx + c, a \neq 1$: THE *ac*-METHOD

Objective

a Factor trinomials of the type $ax^2 + bx + c, a \neq 1$, using the *ac*-method.

a The *ac*-Method

Another method for factoring trinomials of the type $ax^2 + bx + c, a \neq 1$, involves the product, ac, of the leading coefficient a and the last term c. It is called the ***ac*-method.** Because it uses factoring by grouping, it is also referred to as the **grouping method.**

We know how to factor the trinomial $x^2 + 5x + 6$. We look for factors of the constant term, 6, whose sum is the coefficient of the middle term, 5. What happens when the leading coefficient is not 1? To factor a trinomial like $3x^2 - 10x - 8$, we can use a method similar to what we used for $x^2 + 5x + 6$. That method is outlined as follows.

THE *ac*-METHOD

To factor $ax^2 + bx + c, a \neq 1$, using the *ac*-method:

1. Factor out a common factor, if any.
2. Multiply the leading coefficient a and the constant c.
3. Try to factor the product ac so that the sum of the factors is b. That is, find integers p and q such that $pq = ac$ and $p + q = b$.
4. Split the middle term. That is, write it as a sum using the factors found in step (3).
5. Factor by grouping.
6. Check by multiplying.

EXAMPLE 1 Factor: $3x^2 - 10x - 8$.

1) First, we factor out a common factor, if any. There is none (other than 1 or -1).

2) We multiply the leading coefficient, 3, and the constant, -8:

$$3(-8) = -24.$$

3) Then we look for a factorization of -24 in which the sum of the factors is the coefficient of the middle term, -10.

PAIRS OF FACTORS	SUMS OF FACTORS	
$-1, \ 24$	23	
$1, -24$	-23	
$-2, \ 12$	10	
$\mathbf{2, -12}$	$\mathbf{-10}$	← $2 + (-12) = -10$
$-3, \ 8$	5	
$3, \ -8$	-5	
$-4, \ 6$	2	
$4, \ -6$	-2	

4) Next, we split the middle term as a sum or a difference using the factors found in step (3): $-10x = 2x - 12x$.

5) Finally, we factor by grouping, as follows:

$$3x^2 - 10x - 8 = 3x^2 + 2x - 12x - 8$$ Substituting $2x - 12x$ for $-10x$

$$= (3x^2 + 2x) + (-12x - 8)$$

$$= x(3x + 2) - 4(3x + 2)$$ Factoring by grouping

$$= (x - 4)(3x + 2).$$

We can also split the middle term as $-12x + 2x$. We still get the same factorization, although the factors may be in a different order. Note the following:

$$3x^2 - 10x - 8 = 3x^2 - 12x + 2x - 8$$ Substituting $-12x + 2x$ for $-10x$

$$= (3x^2 - 12x) + (2x - 8)$$

$$= 3x(x - 4) + 2(x - 4)$$ Factoring by grouping

$$= (3x + 2)(x - 4).$$

6) Check: $(3x + 2)(x - 4) = 3x^2 - 10x - 8.$

Do Exercises 1 and 2.

Factor.

1. $6x^2 + 7x + 2$

2. $12x^2 - 17x - 5$

EXAMPLE 2 Factor: $8x^2 + 8x - 6$.

1) First, we factor out a common factor, if any. The number 2 is common to all three terms, so we factor it out: $2(4x^2 + 4x - 3)$.
2) Next, we factor the trinomial $4x^2 + 4x - 3$. We multiply the leading coefficient and the constant, 4 and -3: $4(-3) = -12$.
3) We try to factor -12 so that the sum of the factors is 4.

PAIRS OF FACTORS	SUMS OF FACTORS
-1, 12	11
1, -12	-11
-2, 6	4 ← $-2 + 6 = 4$
2, -6	-4
-3, 4	1
3, -4	-1

4) Then we split the middle term, $4x$, as follows: $4x = -2x + 6x$.
5) Finally, we factor by grouping:

$$4x^2 + 4x - 3 = 4x^2 - 2x + 6x - 3$$ Substituting $-2x + 6x$ for $4x$

$$= (4x^2 - 2x) + (6x - 3)$$

$$= 2x(2x - 1) + 3(2x - 1)$$ Factoring by grouping

$$= (2x + 3)(2x - 1).$$

The factorization of $4x^2 + 4x - 3$ is $(2x + 3)(2x - 1)$. But don't forget the common factor! We must include it to get a factorization of the original trinomial: $8x^2 + 8x - 6 = 2(2x + 3)(2x - 1)$.

6) Check: $2(2x + 3)(2x - 1) = 2(4x^2 + 4x - 3) = 8x^2 + 8x - 6.$

Do Exercises 3 and 4.

Factor.

3. $6x^2 + 15x + 9$

4. $20x^2 - 46x + 24$

Answers on page A-40

11.4 EXERCISE SET

For Extra Help

MathXL | MyMathLab | InterAct Math | Math Tutor Center | Video Lectures on CD Disc 6 | Student's Solutions Manual

a Factor. Note that the middle term has already been split.

1. $x^2 + 2x + 7x + 14$

2. $x^2 + 3x + x + 3$

3. $x^2 - 4x - x + 4$

4. $a^2 + 5a - 2a - 10$

5. $6x^2 + 4x + 9x + 6$

6. $3x^2 - 2x + 3x - 2$

7. $3x^2 - 4x - 12x + 16$

8. $24 - 18y - 20y + 15y^2$

9. $35x^2 - 40x + 21x - 24$

10. $8x^2 - 6x - 28x + 21$

11. $4x^2 + 6x - 6x - 9$

12. $2x^4 - 6x^2 - 5x^2 + 15$

13. $2x^4 + 6x^2 + 5x^2 + 15$

14. $9x^4 - 6x^2 - 6x^2 + 4$

Factor by grouping.

15. $2x^2 + 7x - 4$

16. $5x^2 + x - 18$

17. $3x^2 - 4x - 15$

18. $3x^2 + x - 4$

19. $6x^2 + 23x + 7$

20. $6x^2 + 13x + 6$

21. $3x^2 - 4x + 1$

22. $7x^2 - 15x + 2$

23. $4x^2 - 4x - 15$

24. $9x^2 - 6x - 8$

25. $2x^2 + x - 1$

26. $15x^2 + 19x - 10$

27. $9x^2 - 18x - 16$

28. $2x^2 - 5x + 2$

29. $3x^2 + 5x - 2$

30. $18x^2 + 3x - 10$

31. $12x^2 - 31x + 20$

32. $15x^2 - 19x - 10$

33. $14x^2 - 19x - 3$

34. $35x^2 - 34x + 8$

35. $9x^2 + 18x + 8$

36. $6 - 13x + 6x^2$

37. $49 - 42x + 9x^2$

38. $25x^2 + 40x + 16$

39. $24x^2 - 47x - 2$

40. $16a^2 + 78a + 27$

41. $5 - 9a^2 - 12a$

42. $17x - 4x^2 + 15$

43. $20 + 6x - 2x^2$

44. $15 + x - 2x^2$

45. $12x^2 + 28x - 24$

46. $6x^2 + 33x + 15$

47. $30x^2 - 24x - 54$

48. $18t^2 - 24t + 6$

49. $4y + 6y^2 - 10$

50. $-9 + 18x^2 - 21x$

51. $3x^2 - 4x + 1$

52. $6t^2 + t - 15$

53. $12x^2 - 28x - 24$

54. $6x^2 - 33x + 15$

55. $-1 + 2x^2 - x$

56. $-19x + 15x^2 + 6$

57. $9x^2 + 18x - 16$

58. $14y^2 + 35y + 14$

59. $15x^2 - 25x - 10$

60. $18x^2 + 3x - 10$

61. $12p^3 + 31p^2 + 20p$

62. $15x^3 + 19x^2 - 10x$

63. $4 - x - 5x^2$

64. $1 - p - 2p^2$

65. $33t - 15 - 6t^2$

66. $-15x^2 - 19x - 6$

67. $14x^4 + 19x^3 - 3x^2$

68. $70x^4 + 68x^3 + 16x^2$

69. $168x^3 - 45x^2 + 3x$

70. $144x^5 + 168x^4 + 48x^3$

71. $15x^4 - 19x^2 + 6$

72. $9x^4 + 18x^2 + 8$

73. $25t^2 + 80t + 64$

74. $9x^2 - 42x + 49$

75. $6x^3 + 4x^2 - 10x$

76. $18x^3 - 21x^2 - 9x$

77. $25x^2 + 79x + 64$

78. $9y^2 + 42y + 47$

79. $6x^2 - 19x - 5$

80. $2x^2 + 11x - 9$

81. $12m^2 - mn - 20n^2$

82. $12a^2 - 17ab + 6b^2$

83. $6a^2 - ab - 15b^2$

84. $3p^2 - 16pq - 12q^2$

85. $9a^2 - 18ab + 8b^2$

86. $10s^2 + 4st - 6t^2$

87. $35p^2 + 34pq + 8q^2$

88. $30a^2 + 87ab + 30b^2$

89. $18x^2 - 6xy - 24y^2$

90. $15a^2 - 5ab - 20b^2$

91. $60x + 18x^2 - 6x^3$

92. $60x + 4x^2 - 8x^3$

93. $35x^5 - 57x^4 - 44x^3$

94. $15x^3 + 33x^4 + 6x^5$

95. D$_W$ If you have studied both the FOIL and the *ac*-methods of factoring $ax^2 + bx + c, a \neq 1$, decide which method you think is better and explain why.

96. D$_W$ Explain factoring $ax^2 + bx + c, a \neq 1$, using the *ac*-method as though you were teaching a fellow student.

SKILL MAINTENANCE

Solve. [8.7d, e]

97. $-10x > 1000$

98. $-3.8x \leq -824.6$

99. $6 - 3x \geq -18$

100. $3 - 2x - 4x > -9$

101. $\frac{1}{2}x - 6x + 10 \leq x - 5x$

102. $-2(x + 7) > -4(x - 5)$

103. $3x - 6x + 2(x - 4) > 2(9 - 4x)$

104. $-6(x - 4) + 8(4 - x) \leq 3(x - 7)$

Solve. [8.6a]

105. The earth is a sphere (or ball) that is about 40,000 km in circumference. Find the radius of the earth, in kilometers and in miles. Use 3.14 for π. (*Hint*: 1 km $\approx$ 0.62 mi.)

106. The second angle of a triangle is 10° less than twice the first. The third angle is 15° more than four times the first. Find the measure of the second angle.

SYNTHESIS

Factor.

107. $9x^{10} - 12x^5 + 4$

108. $24x^{2n} + 22x^n + 3$

109. $16x^{10} + 8x^5 + 1$

110. $(a + 4)^2 - 2(a + 4) + 1$

111.–120. Use the TABLE feature to check the factoring in Exercises 15–24.

11.5 FACTORING TRINOMIAL SQUARES AND DIFFERENCES OF SQUARES

Objectives

a Recognize trinomial squares.

b Factor trinomial squares.

c Recognize differences of squares.

d Factor differences of squares, being careful to factor completely.

It would be helpful to memorize this table of perfect squares.

NUMBER, N	PERFECT SQUARE, N^2
1	1
2	4
3	9
4	16
5	25
6	36
7	49
8	64
9	81
10	100
11	121
12	144
13	169
14	196
15	225
16	256
17	289
18	324
19	361
20	400
21	441
22	484
23	529
24	576
25	625

In this section, we first learn to factor trinomials that are squares of binomials. Then we factor binomials that are differences of squares.

a Recognizing Trinomial Squares

Some trinomials are squares of binomials. For example, the trinomial $x^2 + 10x + 25$ is the square of the binomial $x + 5$. To see this, we can calculate $(x + 5)^2$. It is $x^2 + 2 \cdot x \cdot 5 + 5^2$, or $x^2 + 10x + 25$. A trinomial that is the square of a binomial is called a **trinomial square,** or a **perfect-square trinomial.**

In Chapter 10, we considered squaring binomials as special-product rules:

$$(A + B)^2 = A^2 + 2AB + B^2;$$
$$(A - B)^2 = A^2 - 2AB + B^2.$$

We can use these equations in reverse to factor trinomial squares.

TRINOMIAL SQUARES

$$A^2 + 2AB + B^2 = (A + B)^2;$$
$$A^2 - 2AB + B^2 = (A - B)^2$$

How can we recognize when an expression to be factored is a trinomial square? Look at $A^2 + 2AB + B^2$ and $A^2 - 2AB + B^2$. In order for an expression to be a trinomial square:

a) The two expressions A^2 and B^2 must be squares, such as

$$4, \quad x^2, \quad 25x^4, \quad 16t^2.$$

When the coefficient is a perfect square and the power(s) of the variable(s) is (are) even, then the expression is a perfect square.

b) There must be no minus sign before A^2 or B^2.

c) If we multiply A and B and double the result, $2 \cdot AB$, we get either the remaining term or its opposite.

EXAMPLE 1 Determine whether $x^2 + 6x + 9$ is a trinomial square.

a) We know that x^2 and 9 are squares.

b) There is no minus sign before x^2 or 9.

c) If we multiply the square roots, x and 3, and double the product, we get the remaining term: $2 \cdot x \cdot 3 = 6x$.

Thus, $x^2 + 6x + 9$ is the square of a binomial. In fact, $x^2 + 6x + 9 = (x + 3)^2$.

EXAMPLE 2 Determine whether $x^2 + 6x + 11$ is a trinomial square.

The answer is no, because only one term, x^2, is a square.

EXAMPLE 3 Determine whether $16x^2 + 49 - 56x$ is a trinomial square.

It helps to first write the trinomial in descending order:

$$16x^2 - 56x + 49.$$

a) We know that $16x^2$ and 49 are squares.
b) There is no minus sign before $16x^2$ or 49.
c) If we multiply the square roots, $4x$ and 7, and double the product, we get the opposite of the remaining term: $2 \cdot 4x \cdot 7 = 56x$; $56x$ is the opposite of $-56x$.

Thus, $16x^2 + 49 - 56x$ is a trinomial square. In fact, $16x^2 - 56x + 49 = (4x - 7)^2$.

Do Exercises 1–8.

Determine whether each is a trinomial square. Write "yes" or "no."

1. $x^2 + 8x + 16$
2. $25 - x^2 + 10x$
3. $t^2 - 12t + 4$
4. $25 + 20y + 4y^2$
5. $5x^2 + 16 - 14x$
6. $16x^2 + 40x + 25$
7. $p^2 + 6p - 9$
8. $25a^2 + 9 - 30a$

b Factoring Trinomial Squares

We can use the factoring methods from Sections 11.2–11.4 to factor trinomial squares, but there is a faster method using the following equations.

FACTORING TRINOMIAL SQUARES

$$A^2 + 2AB + B^2 = (A + B)^2;$$
$$A^2 - 2AB + B^2 = (A - B)^2$$

We consider 3 to be a square root of 9 because $3^2 = 9$. Similarly, A is a square root of A^2. We use square roots of the squared terms and the sign of the remaining term to factor a trinomial square.

EXAMPLE 4 Factor: $x^2 + 6x + 9$.

$$x^2 + 6x + 9 = x^2 + 2 \cdot x \cdot 3 + 3^2 = (x + 3)^2$$
$$A^2 + 2\ A\ B + B^2 = (A + B)^2$$

The sign of the middle term is positive.

EXAMPLE 5 Factor: $x^2 + 49 - 14x$.

$$x^2 + 49 - 14x = x^2 - 14x + 49 \quad \text{Changing to descending order}$$
$$= x^2 - 2 \cdot x \cdot 7 + 7^2 \quad \text{The sign of the middle term is negative.}$$
$$= (x - 7)^2$$

EXAMPLE 6 Factor: $16x^2 - 40x + 25$.

$$16x^2 - 40x + 25 = (4x)^2 - 2 \cdot 4x \cdot 5 + 5^2 = (4x - 5)^2$$
$$A^2 - 2\ A\ B + B^2 = (A - B)^2$$

Do Exercises 9–13.

Factor.

9. $x^2 + 2x + 1$
10. $1 - 2x + x^2$
11. $4 + t^2 + 4t$
12. $25x^2 - 70x + 49$
13. $49 - 56y + 16y^2$

Answers on page A-41

Factor.

14. $48m^2 + 75 + 120m$

15. $p^4 + 18p^2 + 81$

16. $4z^5 - 20z^4 + 25z^3$

17. $9a^2 + 30ab + 25b^2$

Answers on page A-41

EXAMPLE 7 Factor: $t^4 + 20t^2 + 100$.

$$t^4 + 20t^2 + 100 = (t^2)^2 + 2(t^2)(10) + 10^2$$
$$= (t^2 + 10)^2$$

EXAMPLE 8 Factor: $75m^3 + 210m^2 + 147m$.

Always look first for a common factor. This time there is one, $3m$:

$$75m^3 + 210m^2 + 147m = 3m[25m^2 + 70m + 49]$$
$$= 3m[(5m)^2 + 2(5m)(7) + 7^2]$$
$$= 3m(5m + 7)^2.$$

EXAMPLE 9 Factor: $4p^2 - 12pq + 9q^2$.

$$4p^2 - 12pq + 9q^2 = (2p)^2 - 2(2p)(3q) + (3q)^2$$
$$= (2p - 3q)^2$$

Do Exercises 14–17.

C Recognizing Differences of Squares

The following polynomials are *differences of squares*:

$$x^2 - 9, \qquad 4t^2 - 49, \qquad a^2 - 25b^2.$$

To factor a difference of squares such as $x^2 - 9$, think about the formula we used in Chapter 10:

$$(A + B)(A - B) = A^2 - B^2.$$

Equations are reversible, so we also know the following.

DIFFERENCE OF SQUARES

$$A^2 - B^2 = (A + B)(A - B)$$

Thus,

$$x^2 - 9 = (x + 3)(x - 3).$$

To use this formula, we must be able to recognize when it applies. A **difference of squares** is an expression like the following:

$$A^2 - B^2.$$

How can we recognize such expressions? Look at $A^2 - B^2$. In order for a binomial to be a difference of squares:

a) There must be two expressions, both squares, such as

$$4x^2, \quad 9, \quad 25t^4, \quad 1, \quad x^6, \quad 49y^8.$$

b) The terms must have different signs.

EXAMPLE 10 Is $9x^2 - 64$ a difference of squares?

a) The first expression is a square: $9x^2 = (3x)^2$.
The second expression is a square: $64 = 8^2$.

b) The terms have different signs, $+9x^2$ and -64.

Thus we have a difference of squares, $(3x)^2 - 8^2$.

EXAMPLE 11 Is $25 - t^3$ a difference of squares?

a) The expression t^3 is not a square.

The expression is not a difference of squares.

EXAMPLE 12 Is $-4x^2 + 16$ a difference of squares?

a) The expressions $4x^2$ and 16 are squares: $4x^2 = (2x)^2$ and $16 = 4^2$.

b) The terms have different signs, $-4x^2$ and $+16$.

Thus we have a difference of squares. We can also see this by rewriting in the equivalent form: $16 - 4x^2$.

Do Exercises 18–24.

d Factoring Differences of Squares

To factor a difference of squares, we use the following equation.

FACTORING A DIFFERENCE OF SQUARES

$$A^2 - B^2 = (A + B)(A - B)$$

To factor a difference of squares $A^2 - B^2$, we find A and B, which are square roots of the expressions A^2 and B^2. We then use A and B to form two factors. One is the sum $A + B$, and the other is the difference $A - B$.

EXAMPLE 13 Factor: $x^2 - 4$.

$$x^2 - 4 = x^2 - 2^2 = (x + 2)(x - 2)$$
$$\uparrow \quad \uparrow \quad \uparrow \quad \uparrow \quad \uparrow \quad \uparrow$$
$$A^2 - B^2 = (A + B)(A - B)$$

EXAMPLE 14 Factor: $9 - 16t^4$.

$$9 - 16t^4 = 3^2 - (4t^2)^2 = (3 + 4t^2)(3 - 4t^2)$$
$$\uparrow \quad \uparrow \quad \uparrow \quad \uparrow \quad \uparrow \quad \uparrow$$
$$A^2 - B^2 = (A + B)(A - B)$$

Determine whether each is a difference of squares. Write "yes" or "no."

18. $x^2 - 25$

19. $t^2 - 24$

20. $y^2 + 36$

21. $4x^2 - 15$

22. $16x^4 - 49$

23. $9w^6 - 1$

24. $-49 + 25t^2$

Answers on page A-41

Factor.

25. $x^2 - 9$

26. $4t^2 - 64$

27. $a^2 - 25b^2$

28. $64x^4 - 25x^6$

29. $5 - 20t^6$
[*Hint*: $1 = 1^2$, $t^6 = (t^3)^2$.]

Answers on page A-41

EXAMPLE 15 Factor: $m^2 - 4p^2$.

$$m^2 - 4p^2 = m^2 - (2p)^2 = (m + 2p)(m - 2p)$$

EXAMPLE 16 Factor: $x^2 - \frac{1}{9}$.

$$x^2 - \frac{1}{9} = x^2 - \left(\frac{1}{3}\right)^2 = \left(x + \frac{1}{3}\right)\left(x - \frac{1}{3}\right)$$

EXAMPLE 17 Factor: $18x^2 - 50x^6$.

Always look first for a factor common to all terms. This time there is one, $2x^2$.

$$\begin{aligned} 18x^2 - 50x^6 &= 2x^2(9 - 25x^4) \\ &= 2x^2[3^2 - (5x^2)^2] \\ &= 2x^2(3 + 5x^2)(3 - 5x^2) \end{aligned}$$

EXAMPLE 18 Factor: $49x^4 - 9x^6$.

$$\begin{aligned} 49x^4 - 9x^6 &= x^4(49 - 9x^2) \\ &= x^4(7 + 3x)(7 - 3x) \end{aligned}$$

Do Exercises 25–29.

Caution!

Note carefully in these examples that a difference of squares is *not* the square of the difference; that is,

$$A^2 - B^2 \neq (A - B)^2.$$

For example,

$$(45 - 5)^2 = 40^2 = 1600,$$

but

$$45^2 - 5^2 = 2025 - 25 = 2000.$$

Similarly,

$$A^2 - 2AB + B^2 \neq (A - B)(A + B).$$

For example,

$$(10 - 3)(10 + 3) = 7 \cdot 13 = 91,$$

but

$$\begin{aligned} 10^2 - 2 \cdot 10 \cdot 3 + 3^2 &= 100 - 2 \cdot 10 \cdot 3 + 9 \\ &= 100 - 60 + 9 \\ &= 49. \end{aligned}$$

FACTORING COMPLETELY

If a factor with more than one term can still be factored, you should do so. When no factor can be factored further, you have **factored completely.** Always factor completely whenever told to factor.

EXAMPLE 19 Factor: $p^4 - 16$.

$$p^4 - 16 = (p^2)^2 - 4^2$$
$$= (p^2 + 4)(p^2 - 4) \quad \text{Factoring a difference of squares}$$
$$= (p^2 + 4)(p + 2)(p - 2) \quad \text{Factoring further; } p^2 - 4 \text{ is a difference of squares.}$$

The polynomial $p^2 + 4$ cannot be factored further into polynomials with real coefficients.

Caution!

Apart from possibly removing a common factor, you cannot factor a sum of squares. In particular,

$$A^2 + B^2 \neq (A + B)^2.$$

Consider $25x^2 + 100$. Here a sum of squares has a common factor, 25. Factoring, we get $25(x^2 + 4)$, where $x^2 + 4$ is prime. For example,

$$x^2 + 4 \neq (x + 2)^2.$$

EXAMPLE 20 Factor: $\frac{1}{16}x^8 - 81$.

$$\frac{1}{16}x^8 - 81 = \left(\frac{1}{4}x^4 + 9\right)\left(\frac{1}{4}x^4 - 9\right) \quad \text{Factoring a difference of squares}$$
$$= \left(\frac{1}{4}x^4 + 9\right)\left(\frac{1}{2}x^2 + 3\right)\left(\frac{1}{2}x^2 - 3\right) \quad \text{Factoring further. The factor } \tfrac{1}{4}x^4 - 9 \text{ is a difference of squares.}$$

The polynomial $\frac{1}{4}x^4 + 9$ cannot be factored further into polynomials with real coefficients.

EXAMPLE 21 Factor: $y^4 - 16x^{12}$.

$$y^4 - 16x^{12} = (y^2 + 4x^6)(y^2 - 4x^6) \quad \text{Factoring a difference of squares}$$
$$= (y^2 + 4x^6)(y + 2x^3)(y - 2x^3) \quad \text{Factoring further. The factor } y^2 - 4x^6 \text{ is a difference of squares.}$$

TIPS FOR FACTORING

- Always look first for a common factor! If there is one, factor it out.
- Be alert for trinomial squares and differences of squares. Once recognized, they can be factored without trial and error.
- Always factor completely.
- Check by multiplying.

Do Exercises 30–32.

Factor completely.

30. $81x^4 - 1$

31. $16 - \frac{1}{81}y^8$

32. $49p^4 - 25q^6$

Answers on page A-41

11.5 EXERCISE SET

For Extra Help

MathXL | MyMathLab | InterAct Math | Math Tutor Center | Video Lectures on CD Disc 6 | Student's Solutions Manual

a Determine whether each of the following is a trinomial square.

1. $x^2 - 14x + 49$

2. $x^2 - 16x + 64$

3. $x^2 + 16x - 64$

4. $x^2 - 14x - 49$

5. $x^2 - 2x + 4$

6. $x^2 + 3x + 9$

7. $9x^2 - 36x + 24$

8. $36x^2 - 24x + 16$

b Factor completely. Remember to look first for a common factor and to check by multiplying.

9. $x^2 - 14x + 49$

10. $x^2 - 20x + 100$

11. $x^2 + 16x + 64$

12. $x^2 + 20x + 100$

13. $x^2 - 2x + 1$

14. $x^2 + 2x + 1$

15. $4 + 4x + x^2$

16. $4 + x^2 - 4x$

17. $q^4 - 6q^2 + 9$

18. $64 + 16a^2 + a^4$

19. $49 + 56y + 16y^2$

20. $75 + 48a^2 - 120a$

21. $2x^2 - 4x + 2$

22. $2x^2 - 40x + 200$

23. $x^3 - 18x^2 + 81x$

24. $x^3 + 24x^2 + 144x$

25. $12q^2 - 36q + 27$

26. $20p^2 + 100p + 125$

27. $49 - 42x + 9x^2$

28. $64 - 112x + 49x^2$

29. $5y^4 + 10y^2 + 5$

30. $a^4 + 14a^2 + 49$

31. $1 + 4x^4 + 4x^2$

32. $1 - 2a^5 + a^{10}$

33. $4p^2 + 12pq + 9q^2$

34. $25m^2 + 20mn + 4n^2$

35. $a^2 - 6ab + 9b^2$

36. $x^2 - 14xy + 49y^2$

37. $81a^2 - 18ab + b^2$

38. $64p^2 + 16pq + q^2$

39. $36a^2 + 96ab + 64b^2$

40. $16m^2 - 40mn + 25n^2$

c Determine whether each of the following is a difference of squares.

41. $x^2 - 4$

42. $x^2 - 36$

43. $x^2 + 25$

44. $x^2 + 9$

45. $x^2 - 45$

46. $x^2 - 80y^2$

47. $16x^2 - 25y^2$

48. $-1 + 36x^2$

d Factor completely. Remember to look first for a common factor.

49. $y^2 - 4$

50. $q^2 - 1$

51. $p^2 - 9$

52. $x^2 - 36$

53. $-49 + t^2$

54. $-64 + m^2$

55. $a^2 - b^2$

56. $p^2 - q^2$

57. $25t^2 - m^2$

58. $w^2 - 49z^2$

59. $100 - k^2$

60. $81 - w^2$

61. $16a^2 - 9$

62. $25x^2 - 4$

63. $4x^2 - 25y^2$

64. $9a^2 - 16b^2$

65. $8x^2 - 98$

66. $24x^2 - 54$

67. $36x - 49x^3$

68. $16x - 81x^3$

69. $\left(\frac{1}{16} - 49x^8\right)$

70. $\left(\frac{1}{625}x^8 - 49\right)$

71. $(0.09y^2 - 0.0004)$

72. $(0.16p^2 - 0.0025)$

73. $49a^4 - 81$

74. $25a^4 - 9$

75. $a^4 - 16$

76. $y^4 - 1$

77. $5x^4 - 405$

78. $4x^4 - 64$

79. $1 - y^8$

80. $x^8 - 1$

81. $x^{12} - 16$

82. $x^8 - 81$

83. $y^2 - \frac{1}{16}$

84. $x^2 - \frac{1}{25}$

85. $25 - \frac{1}{49}x^2$

86. $\frac{1}{4} - 9q^2$

87. $16m^4 - t^4$

88. $p^4q^4 - 1$

89. **D$_W$** Explain in your own words how to determine whether a polynomial is a trinomial square.

90. **D$_W$** Spiro concludes that since $x^2 - 9 = (x - 3)(x + 3)$, it must follow that $x^2 + 9 = (x + 3)(x + 3)$. What mistake is he making? How would you go about correcting the misunderstanding?

SKILL MAINTENANCE

Divide. [7.6a, c]

91. $(-110) \div 10$

92. $-1000 \div (-2.5)$

93. $\left(-\frac{2}{3}\right) \div \frac{4}{5}$

94. $8.1 \div (-9)$

95. $-64 \div (-32)$

96. $-256 \div 1.6$

Find a polynomial for the shaded area. (Leave results in terms of π where appropriate.) [10.4d]

97.

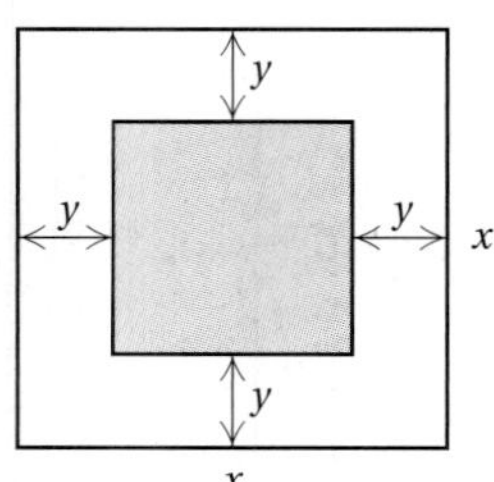

98.

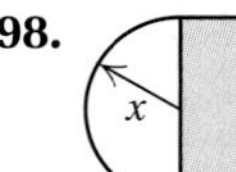
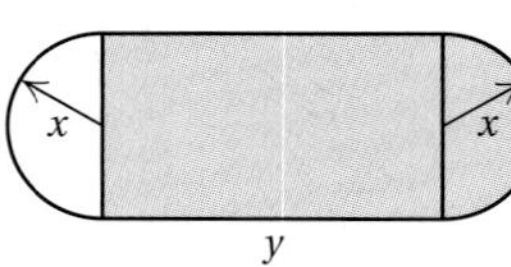

Simplify.

99. $y^5 \cdot y^7$ [10.1d]

100. $(5a^2b^3)^2$ [10.2a, b]

Find the intercepts. Then graph the equation. [9.2a]

101. $y - 6x = 6$

102. $3x - 5y = 15$

SYNTHESIS

Factor completely, if possible.

103. $49x^2 - 216$

104. $27x^3 - 13x$

105. $x^2 + 22x + 121$

106. $x^2 - 5x + 25$

107. $18x^3 + 12x^2 + 2x$

108. $162x^2 - 82$

109. $x^8 - 2^8$

110. $4x^4 - 4x^2$

111. $3x^5 - 12x^3$

112. $3x^2 - \frac{1}{3}$

113. $18x^3 - \frac{8}{25}x$

114. $x^2 - 2.25$

115. $0.49p - p^3$

116. $3.24x^2 - 0.81$

117. $0.64x^2 - 1.21$

118. $1.28x^2 - 2$

119. $(x + 3)^2 - 9$

120. $(y - 5)^2 - 36q^2$

121. $x^2 - \left(\frac{1}{x}\right)^2$

122. $a^{2n} - 49b^{2n}$

123. $81 - b^{4k}$

124. $9x^{18} + 48x^9 + 64$

125. $9b^{2n} + 12b^n + 4$

126. $(x + 7)^2 - 4x - 24$

127. $(y + 3)^2 + 2(y + 3) + 1$

128. $49(x + 1)^2 - 42(x + 1) + 9$

Find c such that the polynomial is the square of a binomial.

129. $cy^2 + 6y + 1$

130. $cy^2 - 24y + 9$

Use the TABLE feature to determine whether the factorization is correct.

131. $x^2 + 9 = (x + 3)(x + 3)$

132. $x^2 - 49 = (x - 7)(x + 7)$

133. $x^2 + 9 = (x + 3)^2$

134. $x^2 - 49 = (x - 7)^2$

Objective

a Factor polynomials completely using any of the methods considered in this chapter.

11.6 FACTORING: A GENERAL STRATEGY

a We now combine all of our factoring techniques and consider a general strategy for factoring polynomials. Here we will encounter polynomials of all the types we have considered, in random order, so you will have the opportunity to determine which method to use.

FACTORING STRATEGY

To factor a polynomial:

a) Always look first for a common factor. If there is one, factor out the largest common factor.

b) Then look at the number of terms.

Two terms: Determine whether you have a difference of squares, $A^2 - B^2$. Do not try to factor a sum of squares: $A^2 + B^2$.

Three terms: Determine whether the trinomial is a square. If it is, you know how to factor. If not, try trial and error, using FOIL or the *ac*-method.

Four terms: Try factoring by grouping.

c) *Always factor completely.* If a factor with more than one term can still be factored, you should factor it. When no factor can be factored further, you have finished.

d) Check by multiplying.

EXAMPLE 1 Factor: $5t^4 - 80$.

a) We look for a common factor:

$$5t^4 - 80 = 5(t^4 - 16).$$

b) The factor $t^4 - 16$ has only two terms. It is a difference of squares: $(t^2)^2 - 4^2$. We factor $t^4 - 16$ and then include the common factor:

$$5(t^2 + 4)(t^2 - 4).$$

c) We see that one of the factors, $t^2 - 4$, is again a difference of squares. We factor it:

$$5(t^2 + 4)(t + 2)(t - 2).$$

↑ This is a sum of squares. It cannot be factored!

We have factored completely because no factor with more than one term can be factored further.

d) Check:
$$\begin{aligned} 5(t^2 + 4)(t + 2)(t - 2) &= 5(t^2 + 4)(t^2 - 4) \\ &= 5(t^4 - 16) \\ &= 5t^4 - 80. \end{aligned}$$

EXAMPLE 2 Factor: $2x^3 + 10x^2 + x + 5$.

a) We look for a common factor. There isn't one.

b) There are four terms. We try factoring by grouping:

$$\begin{aligned} &2x^3 + 10x^2 + x + 5 \\ &\quad = (2x^3 + 10x^2) + (x + 5) && \text{Separating into two binomials} \\ &\quad = 2x^2(x + 5) + 1(x + 5) && \text{Factoring each binomial} \\ &\quad = (2x^2 + 1)(x + 5). && \text{Factoring out the common factor } x + 5 \end{aligned}$$

c) None of these factors can be factored further, so we have factored completely.

d) Check: $(2x^2 + 1)(x + 5) = 2x^2 \cdot x + 2x^2 \cdot 5 + 1 \cdot x + 1 \cdot 5$
$= 2x^3 + 10x^2 + x + 5.$

EXAMPLE 3 Factor: $x^5 - 2x^4 - 35x^3$.

a) We look first for a common factor. This time there is one, x^3:

$$x^5 - 2x^4 - 35x^3 = x^3(x^2 - 2x - 35).$$

b) The factor $x^2 - 2x - 35$ has three terms, but it is not a trinomial square. We factor it using trial and error:

$$x^5 - 2x^4 - 35x^3 = x^3(x^2 - 2x - 35) = x^3(x - 7)(x + 5).$$

Don't forget to include the common factor in the final answer!

c) No factor with more than one term can be factored further, so we have factored completely.

d) Check: $x^3(x - 7)(x + 5) = x^3(x^2 - 2x - 35) = x^5 - 2x^4 - 35x^3.$

EXAMPLE 4 Factor: $x^4 - 10x^2 + 25$.

a) We look first for a common factor. There isn't one.

b) There are three terms. We see that this polynomial is a trinomial square. We factor it:

$$x^4 - 10x^2 + 25 = (x^2)^2 - 2 \cdot x^2 \cdot 5 + 5^2 = (x^2 - 5)^2.$$

We could use trial and error if we have not recognized that we have a trinomial square.

c) Since $x^2 - 5$ cannot be factored further, we have factored completely.

d) Check: $(x^2 - 5)^2 = (x^2)^2 - 2(x^2)(5) + 5^2 = x^4 - 10x^2 + 25.$

Do Exercises 1–5.

EXAMPLE 5 Factor: $6x^2y^4 - 21x^3y^5 + 3x^2y^6$.

a) We look first for a common factor:

$$6x^2y^4 - 21x^3y^5 + 3x^2y^6 = 3x^2y^4(2 - 7xy + y^2).$$

Factor.

1. $3m^4 - 3$

2. $x^6 + 8x^3 + 16$

3. $2x^4 + 8x^3 + 6x^2$

4. $3x^3 + 12x^2 - 2x - 8$

5. $8x^3 - 200x$

Answers on page A-41

b) There are three terms in $2 - 7xy + y^2$. We determine whether the trinomial is a square. Since only y^2 is a square, we do not have a trinomial square. Can the trinomial be factored by trial and error? A key to the answer is that x is only in the term $-7xy$. The polynomial might be in a form like $(1 - y)(2 + y)$, but there would be no x in the middle term. Thus, $2 - 7xy + y^2$ cannot be factored.

c) Have we factored completely? Yes, because no factor with more than one term can be factored further.

d) The check is left to the student.

Study Tips

TIME MANAGEMENT (PART 3)

Here are some additional tips to help you with time management. (See also the Study Tips on time management in Sections 8.2 and 11.2.)

- **Avoid "time killers."** We live in a media age, and the Internet, e-mail, television, and movies are all time killers. Keep track of the time you spend on such activities and compare it to the time you spend studying.
- **Prioritize your tasks.** Be careful about taking on too many college activities that fall outside of academics. These activities are important but keep them to a minimum to be sure that you have enough time for your studies.
- **Be aggressive about your study tasks.** Instead of worrying over your math homework or test preparation, do something to get yourself started. If the task is large, break it down into smaller parts, and do one at a time. You will be surprised at how quickly the large task can then be completed.

"Time is more valuable than money. You can get more money, but you can't get more time."

Jim Rohn, motivational speaker

EXAMPLE 6 Factor: $(p + q)(x + 2) + (p + q)(x + y)$.

a) We look for a common factor:

$$(p + q)(x + 2) + (p + q)(x + y) = (p + q)[(x + 2) + (x + y)]$$
$$= (p + q)(2x + y + 2).$$

b) There are three terms in $2x + y + 2$, but this trinomial cannot be factored further.

c) Neither factor can be factored further, so we have factored completely.

d) The check is left to the student.

EXAMPLE 7 Factor: $px + py + qx + qy$.

a) We look first for a common factor. There isn't one.

b) There are four terms. We try factoring by grouping:

$$px + py + qx + qy = p(x + y) + q(x + y)$$
$$= (p + q)(x + y).$$

c) Have we factored completely? Since neither factor can be factored further, we have factored completely.

d) Check: $(p + q)(x + y) = px + py + qx + qy$.

EXAMPLE 8 Factor: $25x^2 + 20xy + 4y^2$.

a) We look first for a common factor. There isn't one.

b) There are three terms. We determine whether the trinomial is a square. The first term and the last term are squares:

$$25x^2 = (5x)^2 \quad \text{and} \quad 4y^2 = (2y)^2.$$

Since twice the product of $5x$ and $2y$ is the other term,

$$2 \cdot 5x \cdot 2y = 20xy,$$

the trinomial is a perfect square.

We factor by writing the square roots of the square terms and the sign of the middle term:

$$25x^2 + 20xy + 4y^2 = (5x + 2y)^2.$$

c) Since $5x + 2y$ cannot be factored further, we have factored completely.

d) Check: $(5x + 2y)^2 = (5x)^2 + 2(5x)(2y) + (2y)^2$
$= 25x^2 + 20xy + 4y^2.$

EXAMPLE 9 Factor: $p^2q^2 + 7pq + 12$.

a) We look first for a common factor. There isn't one.

b) There are three terms. We determine whether the trinomial is a square. The first term is a square, but neither of the other terms is a square, so we do not have a trinomial square. We factor, thinking of the product pq as a single variable. We consider this possibility for factorization:

$$(pq + \quad)(pq + \quad).$$

We factor the last term, 12. All the signs are positive, so we consider only positive factors. Possibilities are 1, 12 and 2, 6 and 3, 4. The pair 3, 4 gives a sum of 7 for the coefficient of the middle term. Thus,

$$p^2q^2 + 7pq + 12 = (pq + 3)(pq + 4).$$

c) No factor with more than one term can be factored further, so we have factored completely.

d) Check: $(pq + 3)(pq + 4) = (pq)(pq) + 4 \cdot pq + 3 \cdot pq + 3 \cdot 4$
$= p^2q^2 + 7pq + 12.$

Factor.

6. $15x^4 + 5x^2y - 10y^2$

7. $10p^6q^2 + 4p^5q^3 + 2p^4q^4$

8. $(a - b)(x + 5) + (a - b)(x + y^2)$

EXAMPLE 10 Factor: $8x^4 - 20x^2y - 12y^2$.

a) We look first for a common factor:

$$8x^4 - 20x^2y - 12y^2 = 4(2x^4 - 5x^2y - 3y^2).$$

b) There are three terms in $2x^4 - 5x^2y - 3y^2$. We determine whether the trinomial is a square. Since none of the terms is a square, we do not have a trinomial square. We factor $2x^4$. Possibilities are $2x^2$, x^2 and $2x$, x^3 and others. We also factor the last term, $-3y^2$. Possibilities are $3y$, $-y$ and $-3y$, y and others. We look for factors such that the sum of their products is the middle term. The x^2 in the middle term, $-5x^2y$, should lead us to try $(2x^2)(x^2)$. We try some possibilities:

$$(2x^2 - y)(x^2 + 3y) = 2x^4 + 5x^2y - 3y^2,$$
$$(2x^2 + y)(x^2 - 3y) = 2x^4 - 5x^2y - 3y^2.$$

c) No factor with more than one term can be factored further, so we have factored completely. The factorization, including the common factor, is

$$4(2x^2 + y)(x^2 - 3y).$$

d) Check: $4(2x^2 + y)(x^2 - 3y) = 4[(2x^2)(x^2) + 2x^2(-3y) + yx^2 + y(-3y)]$
$= 4[2x^4 - 6x^2y + x^2y - 3y^2]$
$= 4(2x^4 - 5x^2y - 3y^2)$
$= 8x^4 - 20x^2y - 12y^2.$

9. $ax^2 + ay + bx^2 + by$

10. $x^4 + 2x^2y^2 + y^4$

11. $x^2y^2 + 5xy + 4$

EXAMPLE 11 Factor: $a^4 - 16b^4$.

a) We look first for a common factor. There isn't one.

b) There are two terms. Since $a^4 = (a^2)^2$ and $16b^4 = (4b^2)^2$, we see that we do have a difference of squares. Thus,

$$a^4 - 16b^4 = (a^2 + 4b^2)(a^2 - 4b^2)$$

c) The last factor can be factored further. It is also a difference of squares.

$$a^4 - 16b^4 = (a^2 + 4b^2)(a + 2b)(a - 2b)$$

d) Check: $(a^2 + 4b^2)(a + 2b)(a - 2b) = (a^2 + 4b^2)(a^2 - 4b^2) = a^4 - 16b^4$.

12. $p^4 - 81q^4$

Do Exercises 6–12.

Answers on page A-41

11.6 EXERCISE SET

For Extra Help

MathXL | MyMathLab | InterAct Math | Math Tutor Center | Video Lectures on CD Disc 6 | Student's Solutions Manual

a Factor completely.

1. $3x^2 - 192$

2. $2t^2 - 18$

3. $a^2 + 25 - 10a$

4. $y^2 + 49 + 14y$

5. $2x^2 - 11x + 12$

6. $8y^2 - 18y - 5$

7. $x^3 + 24x^2 + 144x$

8. $x^3 - 18x^2 + 81x$

9. $x^3 + 3x^2 - 4x - 12$

10. $x^3 - 5x^2 - 25x + 125$

11. $48x^2 - 3$

12. $50x^2 - 32$

13. $9x^3 + 12x^2 - 45x$

14. $20x^3 - 4x^2 - 72x$

15. $x^2 + 4$

16. $t^2 + 25$

17. $x^4 + 7x^2 - 3x^3 - 21x$

18. $m^4 + 8m^3 + 8m^2 + 64m$

19. $x^5 - 14x^4 + 49x^3$

20. $2x^6 + 8x^5 + 8x^4$

21. $20 - 6x - 2x^2$

22. $45 - 3x - 6x^2$

23. $x^2 - 6x + 1$

24. $x^2 + 8x + 5$

25. $4x^4 - 64$

26. $5x^5 - 80x$

27. $1 - y^8$

28. $t^8 - 1$

29. $x^5 - 4x^4 + 3x^3$

30. $x^6 - 2x^5 + 7x^4$

31. $\frac{1}{81}x^6 - \frac{8}{27}x^3 + \frac{16}{9}$

32. $36a^2 - 15a + \frac{25}{16}$

33. $mx^2 + my^2$

34. $12p^2 + 24q^3$

35. $9x^2y^2 - 36xy$

36. $x^2y - xy^2$

37. $2\pi rh + 2\pi r^2$

38. $10p^4q^4 + 35p^3q^3 + 10p^2q^2$

39. $(a + b)(x - 3) + (a + b)(x + 4)$

40. $5c(a^3 + b) - (a^3 + b)$

41. $(x - 1)(x + 1) - y(x + 1)$

42. $3(p - q) - q^2(p - q)$

43. $n^2 + 2n + np + 2p$

44. $a^2 - 3a + ay - 3y$

45. $6q^2 - 3q + 2pq - p$

46. $2x^2 - 4x + xy - 2y$

47. $4b^2 + a^2 - 4ab$

48. $x^2 + y^2 - 2xy$

49. $16x^2 + 24xy + 9y^2$

50. $9c^2 + 6cd + d^2$

51. $49m^4 - 112m^2n + 64n^2$

52. $4x^2y^2 + 12xyz + 9z^2$

53. $y^4 + 10y^2z^2 + 25z^4$

54. $0.01x^4 - 0.1x^2y^2 + 0.25y^4$

55. $\frac{1}{4}a^2 + \frac{1}{3}ab + \frac{1}{9}b^2$

56. $4p^2q + pq^2 + 4p^3$

57. $a^2 - ab - 2b^2$

58. $3b^2 - 17ab - 6a^2$

59. $2mn - 360n^2 + m^2$

60. $15 + x^2y^2 + 8xy$

61. $m^2n^2 - 4mn - 32$

62. $p^2q^2 + 7pq + 6$

63. $r^5s^2 - 10r^4s + 16r^3$

64. $p^5q^2 + 3p^4q - 10p^3$

65. $a^5 + 4a^4b - 5a^3b^2$

66. $2s^6t^2 + 10s^3t^3 + 12t^4$

67. $a^2 - \frac{1}{25}b^2$

68. $p^2 - \frac{1}{49}b^2$

69. $x^2 - y^2$

70. $p^2q^2 - r^2$

71. $16 - p^4q^4$

72. $15a^4 - 15b^4$

73. $1 - 16x^{12}y^{12}$

74. $81a^4 - b^4$

75. $q^3 + 8q^2 - q - 8$

76. $m^3 - 7m^2 - 4m + 28$

77. $112xy + 49x^2 + 64y^2$

78. $4ab^5 - 32b^4 + a^2b^6$

79. **D**W Kelly factored $16 - 8x + x^2$ as $(x - 4)^2$, while Tony factored it as $(4 - x)^2$. Evaluate each expression for several values of x. Then explain why both answers are correct.

80. **D**W Describe in your own words a strategy that can be used to factor polynomials.

SKILL MAINTENANCE

Find an equation of a line containing the given point and having the given slope. [9.4b]

81. $(-4, 0);\ m = -3$

82. $(0, -4);\ m = 8$

83. $(-4, 5);\ m = -\frac{2}{3}$

84. $(3, -2);\ m = -0.28$

85. Divide: $\frac{7}{5} \div \left(-\frac{11}{10}\right)$. [7.6c]

86. Multiply: $(5x - t)^2$. [10.6c]

87. Solve $A = aX + bX - 7$ for X. [8.4b]

88. Solve: $4(x - 9) - 2(x + 7) < 14$. [8.7e]

SYNTHESIS

Factor completely.

89. $a^4 - 2a^2 + 1$

90. $x^4 + 9$

91. $12.25x^2 - 7x + 1$

92. $\frac{1}{5}x^2 - x + \frac{4}{5}$

93. $5x^2 + 13x + 7.2$

94. $x^3 - (x - 3x^2) - 3$

95. $18 + y^3 - 9y - 2y^2$

96. $-(x^4 - 7x^2 - 18)$

97. $a^3 + 4a^2 + a + 4$

98. $x^3 + x^2 - (4x + 4)$

99. $x^3 - x^2 - 4x + 4$

100. $3x^4 - 15x^2 + 12$

101. $y^2(y - 1) - 2y(y - 1) + (y - 1)$

102. $y^2(y + 1) - 4y(y + 1) - 21(y + 1)$

103. $(y + 4)^2 + 2x(y + 4) + x^2$

104. $6(x - 1)^2 + 7y(x - 1) - 3y^2$

Objectives

a Solve equations (already factored) using the principle of zero products.

b Solve quadratic equations by factoring and then using the principle of zero products.

11.7 SOLVING QUADRATIC EQUATIONS BY FACTORING

Second-degree equations like $x^2 + x - 156 = 0$ and $9 - x^2 = 0$ are examples of *quadratic equations*.

QUADRATIC EQUATION

A **quadratic equation** is an equation equivalent to an equation of the type

$$ax^2 + bx + c = 0, \quad a \neq 0.$$

In order to solve quadratic equations, we need a new equation-solving principle.

a The Principle of Zero Products

The product of two numbers is 0 if one or both of the numbers is 0. Furthermore, *if any product is* 0, *then a factor must be* 0. For example:

If $7x = 0$, then we know that $x = 0$.

If $x(2x - 9) = 0$, then we know that $x = 0$ or $2x - 9 = 0$.

If $(x + 3)(x - 2) = 0$, then we know that $x + 3 = 0$ or $x - 2 = 0$.

Caution!

In a product such as $ab = 24$, we cannot conclude with certainty that a is 24 or that b is 24, but if $ab = 0$, we can conclude that $a = 0$ or $b = 0$.

EXAMPLE 1 Solve: $(x + 3)(x - 2) = 0$.

We have a product of 0. This equation will be true when either factor is 0. Thus it is true when

$$x + 3 = 0 \quad \text{or} \quad x - 2 = 0.$$

Here we have two simple equations that we know how to solve:

$$x = -3 \quad \text{or} \quad x = 2.$$

Each of the numbers -3 and 2 is a solution of the original equation, as we can see in the following checks.

Check: For -3:

$$\begin{array}{c|c} (x + 3)(x - 2) & 0 \\ \hline (-3 + 3)(-3 - 2) \stackrel{?}{} & 0 \\ 0(-5) & \\ 0 & \end{array} \quad \text{TRUE}$$

For 2:

$$\begin{array}{c|c} (x + 3)(x - 2) & 0 \\ \hline (2 + 3)(2 - 2) \stackrel{?}{} & 0 \\ 5(0) & \\ 0 & \end{array} \quad \text{TRUE}$$

Study Tips

WORKING WITH A CLASSMATE

If you are finding it difficult to master a particular topic or concept, try talking about it with a classmate. Verbalizing your questions about the material might help clarify it. If your classmate is also finding the material difficult, it is possible that the majority of the people in your class are confused and you can ask your instructor to explain the concept again.

We now have a principle to help in solving quadratic equations.

THE PRINCIPLE OF ZERO PRODUCTS

An equation $ab = 0$ is true if and only if $a = 0$ is true or $b = 0$ is true, or both are true. (A product is 0 if and only if one or both of the factors is 0.)

EXAMPLE 2 Solve: $(5x + 1)(x - 7) = 0$.

We have

$$(5x + 1)(x - 7) = 0$$

$5x + 1 = 0$ *or* $x - 7 = 0$ Using the principle of zero products

$5x = -1$ *or* $x = 7$ Solving the two equations separately

$x = -\frac{1}{5}$ *or* $x = 7$.

Check: For $-\frac{1}{5}$:

$$\begin{array}{c|l} (5x + 1)(x - 7) = 0 & \\ \hline (5(-\frac{1}{5}) + 1)(-\frac{1}{5} - 7) \;?\; 0 & \\ (-1 + 1)(-7\frac{1}{5}) & \\ 0(-7\frac{1}{5}) & \\ 0 & \text{TRUE} \end{array}$$

For 7:

$$\begin{array}{c|l} (5x + 1)(x - 7) = 0 & \\ \hline (5(7) + 1)(7 - 7) \;?\; 0 & \\ (35 + 1) \cdot 0 & \\ 36 \cdot 0 & \\ 0 & \text{TRUE} \end{array}$$

The solutions are $-\frac{1}{5}$ and 7.

When you solve an equation using the principle of zero products, a check by substitution, as in Examples 1 and 2, will detect errors in solving.

Do Exercises 1–3.

When some factors have only one term, you can still use the principle of zero products.

EXAMPLE 3 Solve: $x(2x - 9) = 0$.

We have

$$x(2x - 9) = 0$$

$x = 0$ *or* $2x - 9 = 0$ Using the principle of zero products

$x = 0$ *or* $2x = 9$

$x = 0$ *or* $x = \frac{9}{2}$.

Check: For 0:

$$\begin{array}{c|l} x(2x - 9) = 0 & \\ \hline 0 \cdot (2 \cdot 0 - 9) \;?\; 0 & \\ 0 \cdot (-9) & \\ 0 & \text{TRUE} \end{array}$$

For $\frac{9}{2}$:

$$\begin{array}{c|l} x(2x - 9) = 0 & \\ \hline \frac{9}{2} \cdot (2 \cdot \frac{9}{2} - 9) \;?\; 0 & \\ \frac{9}{2} \cdot (9 - 9) & \\ \frac{9}{2} \cdot 0 & \\ 0 & \text{TRUE} \end{array}$$

Do Exercise 4.

Solve using the principle of zero products.

1. $(x - 3)(x + 4) = 0$

2. $(x - 7)(x - 3) = 0$

3. $(4t + 1)(3t - 2) = 0$

4. Solve: $y(3y - 17) = 0$.

Answers on page A-42

5. Solve: $x^2 - x - 6 = 0$.

Solve.

6. $x^2 - 3x = 28$

7. $x^2 = 6x - 9$

Solve.

8. $x^2 - 4x = 0$

9. $9x^2 = 16$

Answers on page A-42

b Using Factoring to Solve Equations

Using factoring and the principle of zero products, we can solve some new kinds of equations. Thus we have extended our equation-solving abilities.

EXAMPLE 4 Solve: $x^2 + 5x + 6 = 0$.

There are no like terms to collect, and we have a squared term. We first factor the polynomial. Then we use the principle of zero products.

$$x^2 + 5x + 6 = 0$$

$$(x + 2)(x + 3) = 0 \quad \text{Factoring}$$

$$x + 2 = 0 \quad or \quad x + 3 = 0 \quad \text{Using the principle of zero products}$$

$$x = -2 \quad or \quad x = -3$$

Check: For -2:

$$\begin{array}{c|c} x^2 + 5x + 6 & 0 \\ \hline (-2)^2 + 5(-2) + 6 \ ? & 0 \\ 4 - 10 + 6 & \\ -6 + 6 & \\ 0 & \end{array} \quad \text{TRUE}$$

For -3:

$$\begin{array}{c|c} x^2 + 5x + 6 & 0 \\ \hline (-3)^2 + 5(-3) + 6 \ ? & 0 \\ 9 - 15 + 6 & \\ -6 + 6 & \\ 0 & \end{array} \quad \text{TRUE}$$

The solutions are -2 and -3.

Caution!

Keep in mind that you *must* have 0 on one side of the equation before you can use the principle of zero products. Get all nonzero terms on one side and 0 on the other.

Do Exercise 5.

EXAMPLE 5 Solve: $x^2 - 8x = -16$.

We first add 16 to get a 0 on one side:

$$x^2 - 8x = -16$$

$$x^2 - 8x + 16 = 0 \quad \text{Adding 16}$$

$$(x - 4)(x - 4) = 0 \quad \text{Factoring}$$

$$x - 4 = 0 \quad or \quad x - 4 = 0 \quad \text{Using the principle of zero products}$$

$$x = 4 \quad or \quad x = 4. \quad \text{Solving each equation}$$

There is only one solution, 4. The check is left to the student.

Do Exercises 6 and 7.

EXAMPLE 6 Solve: $x^2 + 5x = 0$.

$$x^2 + 5x = 0$$

$$x(x + 5) = 0 \quad \text{Factoring out a common factor}$$

$$x = 0 \quad or \quad x + 5 = 0 \quad \text{Using the principle of zero products}$$

$$x = 0 \quad or \quad x = -5$$

The solutions are 0 and -5. The check is left to the student.

EXAMPLE 7 Solve: $4x^2 = 25$.

$$4x^2 = 25$$

$$4x^2 - 25 = 0 \quad \text{Subtracting 25 on both sides to get 0 on one side}$$

$$(2x - 5)(2x + 5) = 0 \quad \text{Factoring a difference of squares}$$

$$2x - 5 = 0 \quad or \quad 2x + 5 = 0$$

$$2x = 5 \quad or \quad 2x = -5 \quad \text{Solving each equation}$$

$$x = \frac{5}{2} \quad or \quad x = -\frac{5}{2}$$

The solutions are $\frac{5}{2}$ and $-\frac{5}{2}$. The check is left to the student.

Do Exercises 8 and 9 on the preceding page.

Solve.

10. $-2x^2 + 13x - 21 = 0$

EXAMPLE 8 Solve: $-5x^2 + 2x + 3 = 0$.

In this case, the leading coefficient of the trinomial is negative. Thus we first multiply by -1 and then proceed as we have in Examples 1–7.

$$-5x^2 + 2x + 3 = 0$$

$$-1(-5x^2 + 2x + 3) = -1 \cdot 0 \quad \text{Multiplying by } -1$$

$$5x^2 - 2x - 3 = 0 \quad \text{Simplifying}$$

$$(5x + 3)(x - 1) = 0 \quad \text{Factoring}$$

$$5x + 3 = 0 \quad or \quad x - 1 = 0 \quad \text{Using the principle of zero products}$$

$$5x = -3 \quad or \quad x = 1$$

$$x = -\frac{3}{5} \quad or \quad x = 1$$

The solutions are $-\frac{3}{5}$ and 1. The check is left to the student.

Do Exercises 10 and 11.

11. $10 - 3x - x^2 = 0$

EXAMPLE 9 Solve: $(x + 2)(x - 2) = 5$.

Be careful with an equation like this one! It might be tempting to set each factor equal to 5. **Remember: We must have a 0 on one side.** We first carry out the product on the left. Then we subtract 5 on both sides to get 0 on one side. Then we proceed with the principle of zero products.

$$(x + 2)(x - 2) = 5$$

$$x^2 - 4 = 5 \quad \text{Multiplying on the left}$$

$$x^2 - 4 - 5 = 5 - 5 \quad \text{Subtracting 5}$$

$$x^2 - 9 = 0 \quad \text{Simplifying}$$

$$(x + 3)(x - 3) = 0 \quad \text{Factoring}$$

$$x + 3 = 0 \quad or \quad x - 3 = 0 \quad \text{Using the principle of zero products}$$

$$x = -3 \quad or \quad x = 3$$

The solutions are -3 and 3. The check is left to the student.

Do Exercise 12.

12. Solve: $(x + 1)(x - 1) = 8$.

Answers on page A-42

13. Find the x-intercepts of the graph shown below.

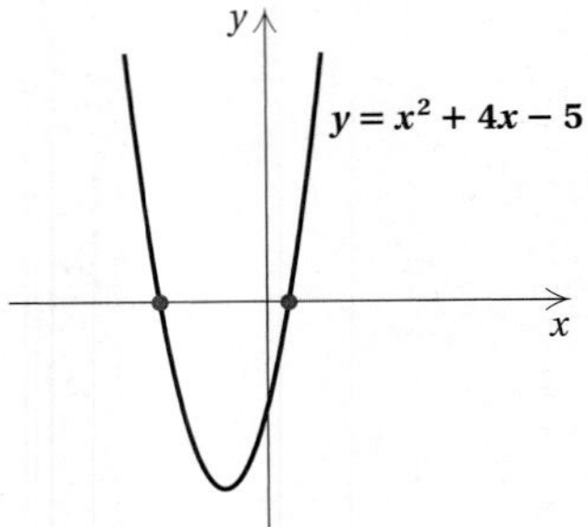

AG ALGEBRAIC–GRAPHICAL CONNECTION

In Chapter 9, we graphed linear equations of the type $y = mx + b$ and $Ax + By = C$. Recall that to find the x-intercept, we replaced y with 0 and solved for x. This procedure can also be used to find the x-intercepts when an equation of the form $y = ax^2 + bx + c$, $a \neq 0$, is to be graphed. Although the details of creating such graphs will be left to Chapter 15, we consider them briefly here from the standpoint of finding the x-intercepts. The graphs are shaped like the following curves. Note that each x-intercept represents a solution of $ax^2 + bx + c = 0$.

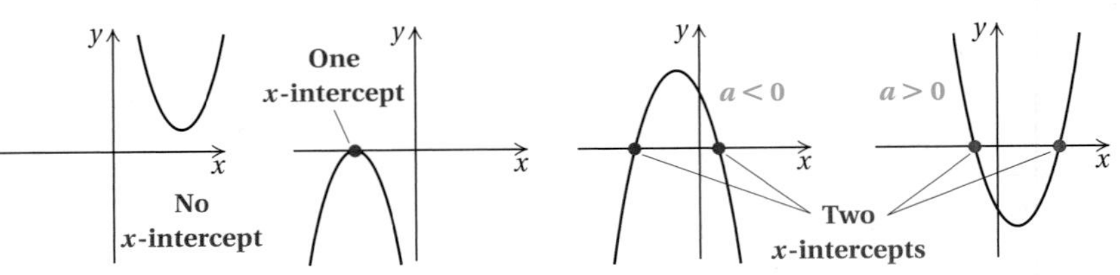

EXAMPLE 10 Find the x-intercepts of the graph of $y = x^2 - 4x - 5$ shown at right. (The grid is intentionally not included.)

To find the x-intercepts, we let $y = 0$ and solve for x:

$$y = x^2 - 4x - 5$$

$$0 = x^2 - 4x - 5 \quad \text{Substituting 0 for } y$$

14. Use *only* the graph shown below to solve $3x - x^2 = 0$.

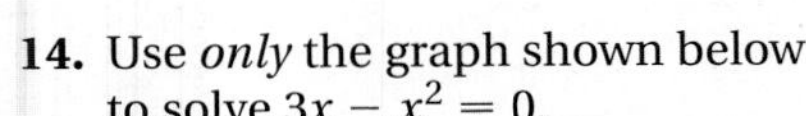

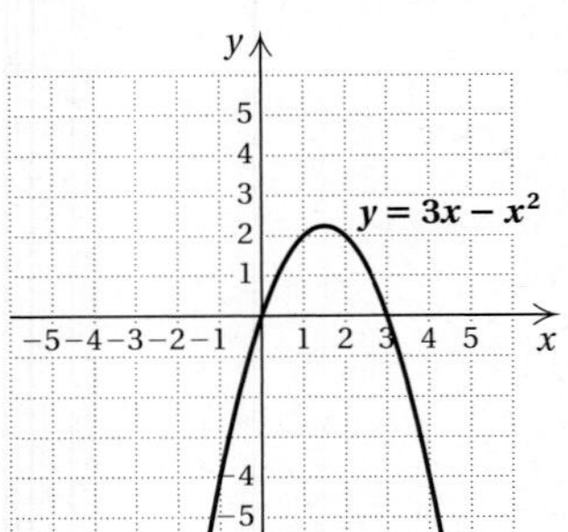

$$0 = (x - 5)(x + 1) \quad \text{Factoring}$$

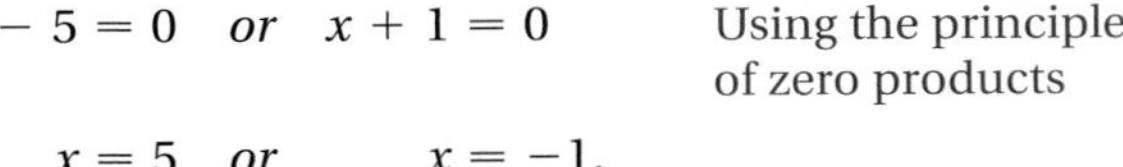

$$x - 5 = 0 \quad or \quad x + 1 = 0 \quad \text{Using the principle of zero products}$$

$$x = 5 \quad or \quad x = -1.$$

The solutions of the equation $0 = x^2 - 4x - 5$ are 5 and -1. The x-intercepts of the graph of $y = x^2 - 4x - 5$ are $(5, 0)$ and $(-1, 0)$. We can now label them on the graph.

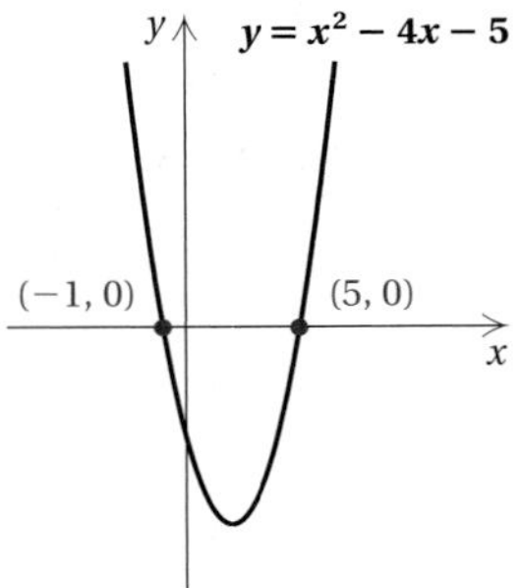

Do Exercises 13 and 14.

Answers on page A-42

CALCULATOR CORNER

Solving Quadratic Equations We can solve quadratic equations graphically. Consider the equation $x^2 + 2x = 8$. First, we must write the equation with 0 on one side. To do this, we subtract 8 on both sides of the equation; we get $x^2 + 2x - 8 = 0$. Next, we graph $y = x^2 + 2x - 8$ in a window that shows the x-intercepts. The standard window works well in this case.

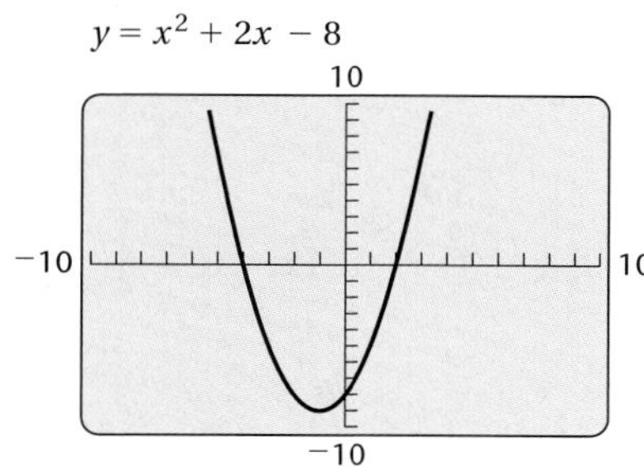

The solutions of the equation are the values of x for which $x^2 + 2x - 8 = 0$. These are also the first coordinates of the x-intercepts of the graph. We use the ZERO feature from the CALC menu to find these numbers. To find the solution corresponding to the leftmost x-intercept, we first press [2ND] [CALC] [2] to select the ZERO feature. The prompt "Left Bound?" appears. Next, we use the [◁] or the [▷] key to move the cursor to the left of the intercept and press [ENTER]. Now the prompt "Right Bound?" appears. Then we move the cursor to the right of the intercept and press [ENTER]. The prompt "Guess?" appears. We move the cursor close to the intercept and press [ENTER] again. We now see the cursor positioned at the leftmost x-intercept and the coordinates of that point, $x = -4$, $y = 0$, are displayed. Thus, $x^2 + 2x - 8 = 0$ when $x = -4$. This is one solution of the equation.

We can repeat this procedure to find the first coordinate of the other x-intercept. We see that $x = 2$ at that point. Thus the solutions of the equation $x^2 + 2x - 8 = 0$ are -4 and 2. Note that the x-intercepts of the graph of $y = x^2 + 2x - 8$ are $(-4, 0)$ and $(2, 0)$.

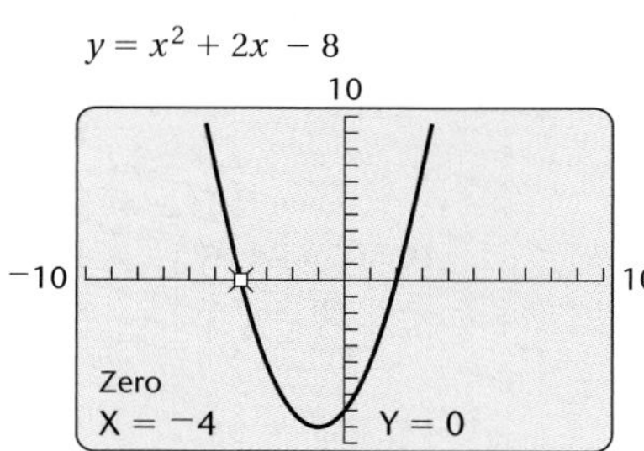

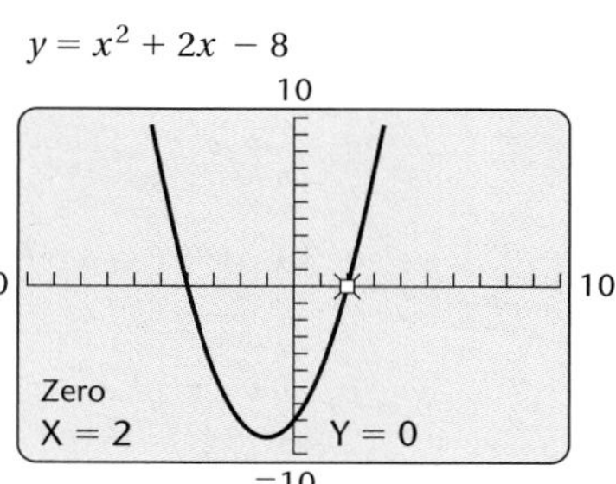

Exercises:

1. Solve each of the equations in Examples 4–8 graphically.

11.7 EXERCISE SET

For Extra Help

MathXL | MyMathLab | InterAct Math | Math Tutor Center | Video Lectures on CD Disc 6 | Student's Solutions Manual

a Solve using the principle of zero products.

1. $(x+4)(x+9)=0$

2. $(x+2)(x-7)=0$

3. $(x+3)(x-8)=0$

4. $(x+6)(x-8)=0$

5. $(x+12)(x-11)=0$

6. $(x-13)(x+53)=0$

7. $x(x+3)=0$

8. $y(y+5)=0$

9. $0=y(y+18)$

10. $0=x(x-19)$

11. $(2x+5)(x+4)=0$

12. $(2x+9)(x+8)=0$

13. $(5x+1)(4x-12)=0$

14. $(4x+9)(14x-7)=0$

15. $(7x-28)(28x-7)=0$

16. $(13x+14)(6x-5)=0$

17. $2x(3x-2)=0$

18. $55x(8x-9)=0$

19. $\left(\frac{1}{5}+2x\right)\left(\frac{1}{9}-3x\right)=0$

20. $\left(\frac{7}{4}x-\frac{1}{16}\right)\left(\frac{2}{3}x-\frac{16}{15}\right)=0$

21. $(0.3x-0.1)(0.05x+1)=0$

22. $(0.1x+0.3)(0.4x-20)=0$

23. $9x(3x-2)(2x-1)=0$

24. $(x+5)(x-75)(5x-1)=0$

b Solve by factoring and using the principle of zero products. Remember to check.

25. $x^2+6x+5=0$

26. $x^2+7x+6=0$

27. $x^2+7x-18=0$

28. $x^2+4x-21=0$

29. $x^2-8x+15=0$

30. $x^2-9x+14=0$

31. $x^2-8x=0$

32. $x^2-3x=0$

33. $x^2+18x=0$

34. $x^2+16x=0$

35. $x^2=16$

36. $100=x^2$

37. $9x^2-4=0$

38. $4x^2-9=0$

39. $0=6x+x^2+9$

40. $0=25+x^2+10x$

41. $x^2+16=8x$

42. $1+x^2=2x$

43. $5x^2=6x$

44. $7x^2=8x$

45. $6x^2-4x=10$

46. $3x^2-7x=20$

47. $12y^2-5y=2$

48. $2y^2+12y=-10$

49. $t(3t + 1) = 2$

50. $x(x - 5) = 14$

51. $100y^2 = 49$

52. $64a^2 = 81$

53. $x^2 - 5x = 18 + 2x$

54. $3x^2 + 8x = 9 + 2x$

55. $10x^2 - 23x + 12 = 0$

56. $12x^2 + 17x - 5 = 0$

Find the x-intercepts for the graph of the equation. (The grids are intentionally not included.)

57.

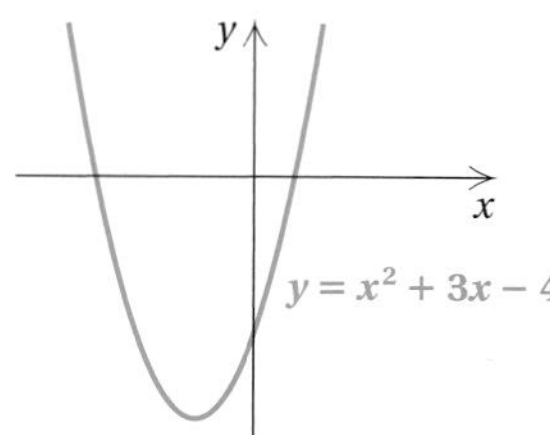

58.

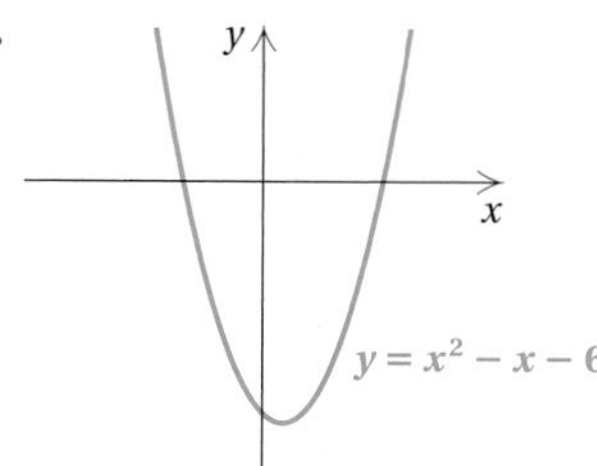

59.

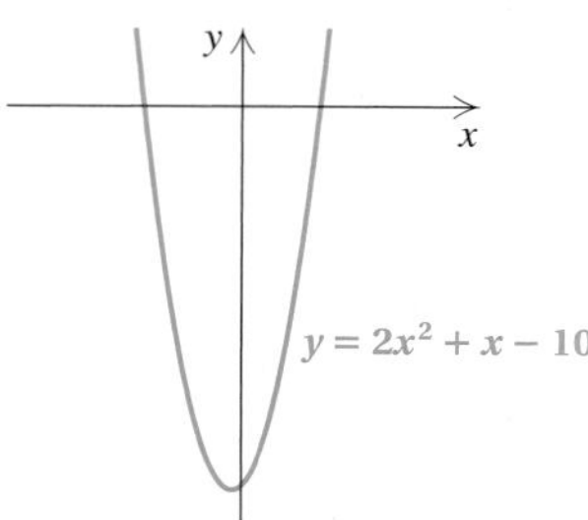

60.

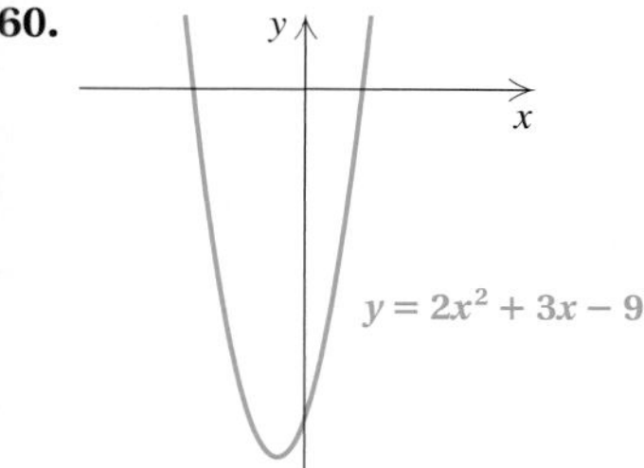

61.

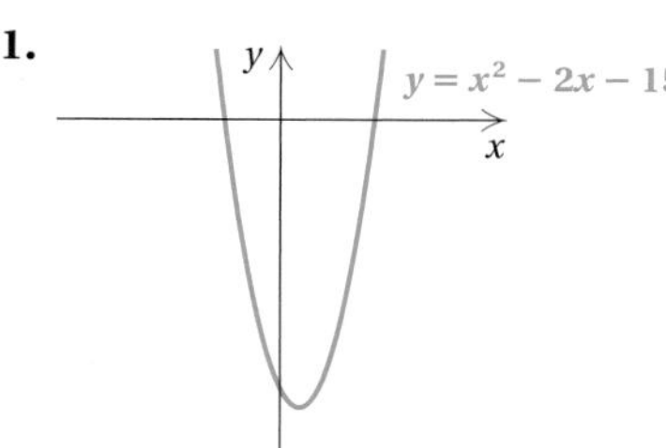

62.

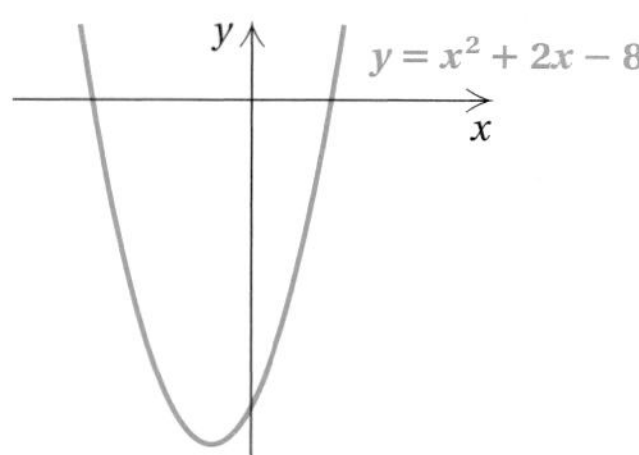

63. Use the following graph to solve $x^2 - 3x - 4 = 0$.

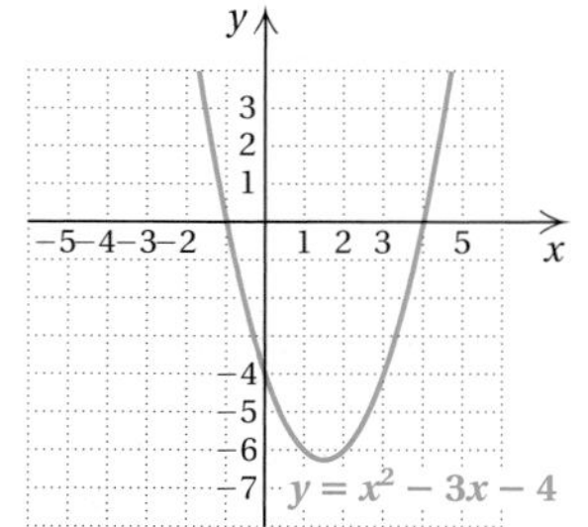

64. Use the following graph to solve $x^2 + x - 6 = 0$.

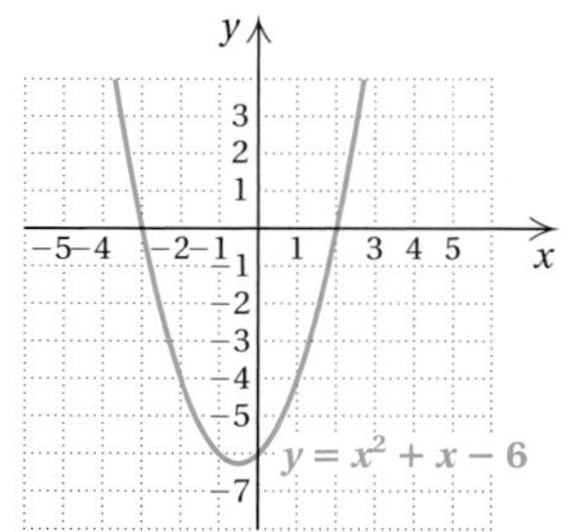

65. Use the following graph to solve $-x^2 + 2x + 3 = 0$.

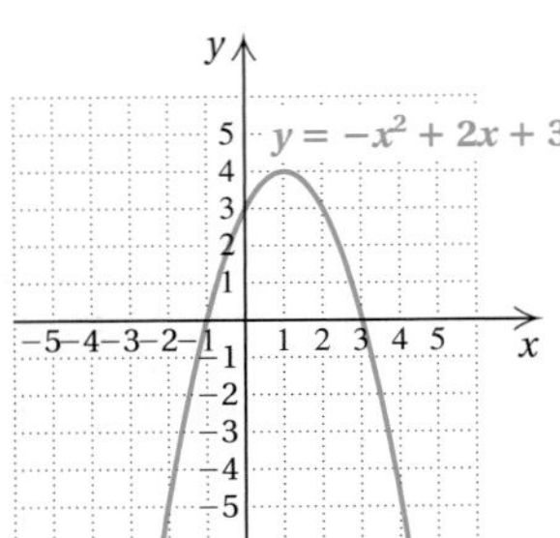

66. Use the following graph to solve $-x^2 - x + 6 = 0$.

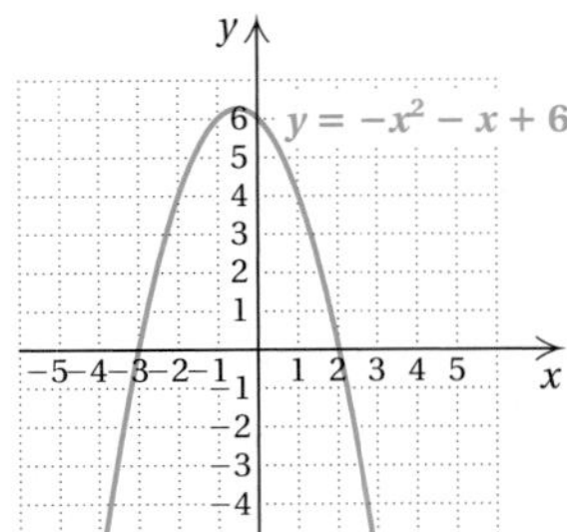

67. **DW** What is wrong with the following? Explain the correct method of solution.

$$(x - 3)(x + 4) = 8$$
$$x - 3 = 8 \quad or \quad x + 4 = 8$$
$$x = 11 \quad or \quad x = 4$$

68. **DW** What is incorrect about solving $x^2 = 3x$ by dividing by x on both sides?

SKILL MAINTENANCE

Translate to an algebraic expression. [7.1b]

69. The square of the sum of a and b

70. The sum of the squares of a and b

Divide. [7.6c]

71. $144 \div (-9)$

72. $-24.3 \div 5.4$

73. $-\frac{5}{8} \div \frac{3}{16}$

74. $-\frac{3}{16} \div \left(-\frac{5}{8}\right)$

SYNTHESIS

Solve.

75. $b(b + 9) = 4(5 + 2b)$

76. $y(y + 8) = 16(y - 1)$

77. $(t - 3)^2 = 36$

78. $(t - 5)^2 = 2(5 - t)$

79. $x^2 - \frac{1}{64} = 0$

80. $x^2 - \frac{25}{36} = 0$

81. $\frac{5}{16}x^2 = 5$

82. $\frac{27}{25}x^2 = \frac{1}{3}$

83. Find an equation that has the given numbers as solutions. For example, 3 and -2 are solutions of $x^2 - x - 6 = 0$.

a) $-3, 4$
b) $-3, -4$
c) $\frac{1}{2}, \frac{1}{2}$
d) $5, -5$
e) $0, 0.1, \frac{1}{4}$

84. *Matching.* Match each equation in the first column with the equivalent equation in the second column.

$x^2 + 10x - 2 = 0$	$4x^2 + 8x + 36 = 0$
$(x - 6)(x + 3) = 0$	$(2x + 8)(2x - 5) = 0$
$5x^2 - 5 = 0$	$9x^2 - 12x + 24 = 0$
$(2x - 5)(x + 4) = 0$	$(x + 1)(5x - 5) = 0$
$x^2 + 2x + 9 = 0$	$x^2 - 3x - 18 = 0$
$3x^2 - 4x + 8 = 0$	$2x^2 + 20x - 4 = 0$

Use a graphing calculator to find the solutions of the equation. Round solutions to the nearest hundredth.

85. $x^2 - 9.10x + 15.77 = 0$

86. $-x^2 + 0.63x + 0.22 = 0$

87. $0.84x^2 - 2.30x = 0$

88. $6.4x^2 - 8.45x - 94.06 = 0$

11.8 APPLICATIONS OF QUADRATIC EQUATIONS

Objectives

a Solve applied problems involving quadratic equations that can be solved by factoring.

b Solve applied problems involving the Pythagorean theorem and quadratic equations that can be solved by factoring.

a Applied Problems, Quadratic Equations, and Factoring

We can now use our new method for solving quadratic equations and the five steps for solving problems.

EXAMPLE 1 *Manufacturing Marble Slabs.* Marble Supreme sells rectangular marble slabs used for tempering fudge in candy shops. The most popular slab that Marble Supreme sells is twice as long as it is wide and has an area of 7200 in^2. What are the dimensions of the slab?

1. **Familiarize.** We first make a drawing. Recall that the area of any rectangle is Length · Width. We let x = the width of the slab, in inches. The length is then $2x$.

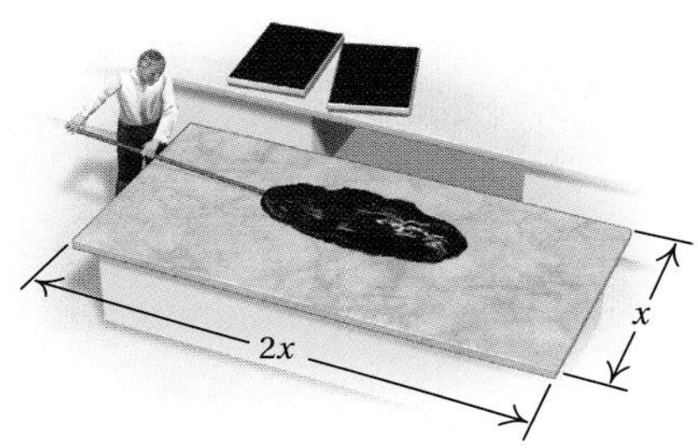

2. **Translate.** We reword and translate as follows:

Rewording: The area of the rectangle is 7200 cm^2.

Translating: $2x \cdot x = 7200$

3. **Solve.** We solve the equation as follows:

$$2x \cdot x = 7200$$
$$2x^2 = 7200$$
$$2x^2 - 7200 = 0 \quad \text{Subtracting 7200 to get 0 on one side}$$
$$2(x^2 - 3600) = 0 \quad \text{Removing a common factor of 2}$$
$$2(x - 60)(x + 60) = 0 \quad \text{Factoring a difference of squares}$$
$$(x - 60)(x + 60) = 0 \quad \text{Dividing by 2}$$
$$x - 60 = 0 \quad or \quad x + 60 = 0 \quad \text{Using the principle of zero products}$$
$$x = 60 \quad or \quad x = -60. \quad \text{Solving each equation}$$

4. **Check.** The solutions of the equation are 60 and -60. Since the width must be positive, -60 cannot be a solution. To check 60 in., we note that if the width is 60 in., then the length is $2 \cdot 60$ in. = 120 in., and the area is 60 in. · 120 in. = 7200 in^2. Thus the solution 60 checks.

5. **State.** The slab is 60 in. wide and 120 in. long.

Do Exercise 1.

1. **Dimensions of Picture.** A rectangular picture is twice as long as it is wide. If the area of the picture is 288 in^2, find its dimensions.

Answer on page A-42

2. Dimensions of a Sail. The mainsail of Stacey's lightning-styled sailboat has an area of 125 ft^2. The sail is 15 ft taller than it is wide. Find the height and the width of the sail.

Answer on page A-42

EXAMPLE 2 *Racing Sailboat.* The height of a triangular sail on a racing sailboat is 9 ft more than the base. The area of the triangle is 110 ft^2. Find the height and the base of the sail.

Source: Whitney Gladstone, North Graphics, San Diego, CA

1. **Familiarize.** We first make a drawing. If you don't remember the formula for the area of a triangle, look it up in the list of formulas at the back of this book or in a geometry book. The area is $\frac{1}{2}$(base)(height).

   We let b = the base of the triangle. Then $b + 9$ = the height.

2. **Translate.** It helps to reword this problem before translating:

$\frac{1}{2}$	times	Base	times	Height	is	110.	Rewording
↓	↓	↓	↓	↓	↓	↓	
$\frac{1}{2}$	$\cdot$	b	$\cdot$	$(b+9)$	$=$	110	Translating

3. **Solve.** We solve the equation as follows:

$$\frac{1}{2} \cdot b \cdot (b + 9) = 110$$

$$\frac{1}{2}(b^2 + 9b) = 110 \qquad \text{Multiplying}$$

$$2 \cdot \frac{1}{2}(b^2 + 9b) = 2 \cdot 110 \qquad \text{Multiplying by 2}$$

$$b^2 + 9b = 220 \qquad \text{Simplifying}$$

$$b^2 + 9b - 220 = 220 - 220 \qquad \text{Subtracting 220 to get 0 on one side}$$

$$b^2 + 9b - 220 = 0$$

$$(b - 11)(b + 20) = 0 \qquad \text{Factoring}$$

$$b - 11 = 0 \quad \text{or} \quad b + 20 = 0 \qquad \text{Using the principle of zero products}$$

$$b = 11 \quad \text{or} \quad b = -20.$$

4. **Check.** The base of a triangle cannot have a negative length, so -20 cannot be a solution. Suppose the base is 11 ft. The height is 9 ft more than the base, so the height is 11 ft + 9 ft, or 20 ft, and the area is $\frac{1}{2}(11)(20)$, or 110 ft^2. These numbers check in the original problem.

5. **State.** The height is 20 ft and the base is 11 ft.

Do Exercise 2.

EXAMPLE 3 *Games in a Sports League.* In a sports league of x teams in which each team plays every other team twice, the total number N of games to be played is given by

$$x^2 - x = N.$$

Maggie's basketball league plays a total of 240 games. How many teams are in the league?

1., 2. Familiarize and **Translate.** We are given that x is the number of teams in a league and N is the number of games. To familiarize yourself with this problem, reread Example 4 in Section 10.3 where we first considered it. To find the number of teams x in a league in which 240 games are played, we substitute 240 for N in the equation:

$x^2 - x = 240.$ Substituting 240 for N

3. Solve. We solve the equation as follows:

$$x^2 - x = 240$$

$x^2 - x - 240 = 240 - 240$ Subtracting 240 to get 0 on one side

$$x^2 - x - 240 = 0$$

$(x - 16)(x + 15) = 0$ Factoring

$x - 16 = 0$ *or* $x + 15 = 0$ Using the principle of zero products

$x = 16$ *or* $x = -15.$

4. Check. The solutions of the equation are 16 and -15. Since the number of teams cannot be negative, -15 cannot be a solution. But 16 checks, since $16^2 - 16 = 256 - 16 = 240$.

5. State. There are 16 teams in the league.

Do Exercise 3.

3. Use $N = x^2 - x$ for the following.

a) **Volleyball League.** Amy's volleyball league has 19 teams. What is the total number of games to be played?

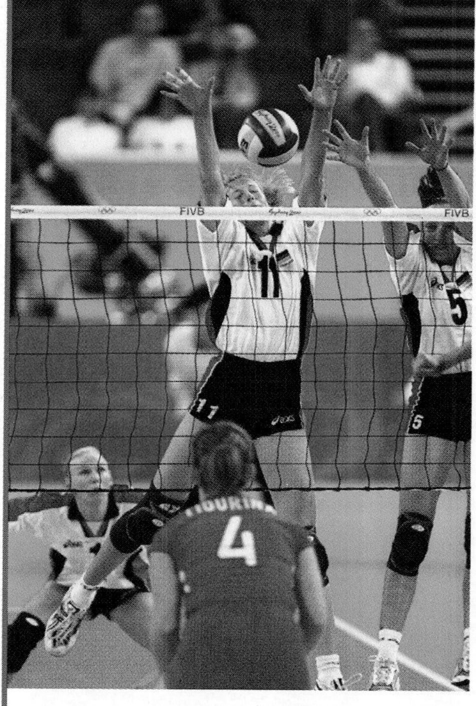

b) **Softball League.** Barry's slow-pitch softball league plays a total of 72 games. How many teams are in the league?

Answers on page A-42

Study Tips — FIVE STEPS FOR PROBLEM SOLVING

Are you remembering to use the five steps for problem solving that were developed in Section 8.6?

1. **Familiarize** yourself with the situation.
   a) Carefully read and reread until you understand *what* you are being asked to find.
   b) Draw a diagram or see if there is a formula that applies.
   c) Assign a letter, or *variable,* to the unknown.
2. **Translate** the problem to an equation using the letter or variable.
3. **Solve** the equation.
4. **Check** the answer in the original wording of the problem.
5. **State** the answer to the problem clearly with appropriate units.

"Most worthwhile achievements are the result of many little things done in a simple direction."

Nido Quebin, speaker/entrepreneur

4. Page Numbers. The product of the page numbers on two facing pages of a book is 506. Find the page numbers.

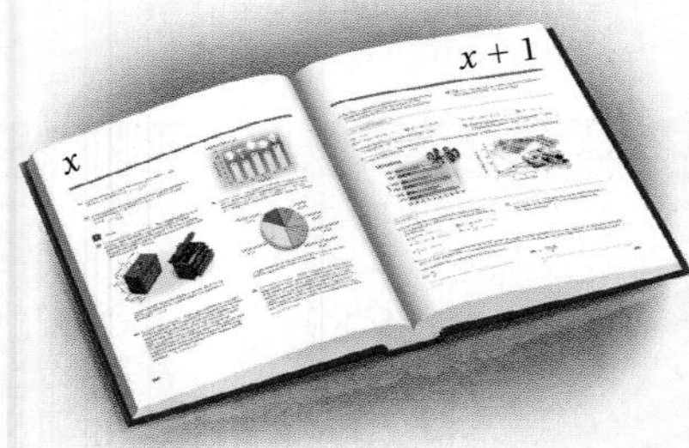

Answer on page A-42

Study Tips

ASKING QUESTIONS

Don't be afraid to ask questions in class. Most instructors welcome this and encourage students to ask them. Other students probably have the same questions you do.

"Better to ask twice than lose your way once."

Danish proverb

EXAMPLE 4 *Athletic Numbers.* The product of the numbers of two consecutive entrants in a marathon race is 156. Find the numbers.

1. **Familiarize.** The numbers are consecutive integers. Recall that consecutive integers are next to each other, such as 49 and 50, or -6 and -5. Let $x =$ the smaller integer; then $x + 1 =$ the larger integer.
2. **Translate.** It helps to reword the problem before translating:

First integer	times	Second integer	is	156.	Rewording
x	$\cdot$	$(x + 1)$	$=$	156	Translating

3. **Solve.** We solve the equation as follows:

$$x(x + 1) = 156$$

$$x^2 + x = 156 \quad \text{Multiplying}$$

$$x^2 + x - 156 = 156 - 156 \quad \text{Subtracting 156 to get 0 on one side}$$

$$x^2 + x - 156 = 0 \quad \text{Simplifying}$$

$$(x - 12)(x + 13) = 0 \quad \text{Factoring}$$

$$x - 12 = 0 \quad or \quad x + 13 = 0 \quad \text{Using the principle of zero products}$$

$$x = 12 \quad or \quad x = -13.$$

4. **Check.** The solutions of the equation are 12 and -13. When x is 12, then $x + 1$ is 13, and $12 \cdot 13 = 156$. The numbers 12 and 13 are consecutive integers that are solutions to the problem. When x is -13, then $x + 1$ is -12, and $(-13)(-12) = 156$. The numbers -13 and -12 are consecutive integers, but they are not solutions of the problem because negative numbers are not used as entry numbers.
5. **State.** The entry numbers are 12 and 13.

Do Exercise 4.

b The Pythagorean Theorem

The following problems involve the Pythagorean theorem, which relates the lengths of the sides of a *right* triangle. A triangle is a **right triangle** if it has a 90°, or *right*, angle. The side opposite the 90° angle is called the **hypotenuse.** The other sides are called **legs.**

THE PYTHAGOREAN THEOREM

In any right triangle, if a and b are the lengths of the legs and c is the length of the hypotenuse, then

$$a^2 + b^2 = c^2.$$

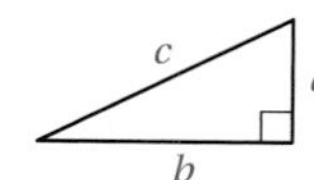

The symbol ⌐ denotes a 90° angle.

EXAMPLE 5 *Physical Education.* An outdoor-education ropes course includes a 25-ft cable that slopes downward from a height of 37 ft to a height of 30 ft. How far is it between the trees that the cable connects?

1. **Familiarize.** We make a drawing as above, noting that when we subtract 30 ft from 37 ft, we get the height of the right triangle that is formed. We let b = the distance between the trees.

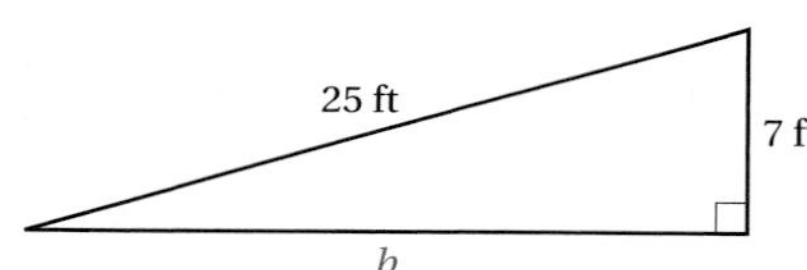

2. **Translate.** A right triangle is formed, so we can use the Pythagorean theorem:

$$a^2 + b^2 = c^2$$

$$7^2 + b^2 = 25^2.$$ Substituting 7 for the length of a leg and 25 for the length of the hypotenuse

3. **Solve.** We solve the equation as follows:

$$7^2 + b^2 = 25^2$$

$$49 + b^2 = 625$$ Squaring 7 and 25

$$b^2 - 576 = 0$$ Subtracting 625

$$(b - 24)(b + 24) = 0$$ Factoring

$$b - 24 = 0 \quad or \quad b + 24 = 0$$ Using the principle of zero products

$$b = 24 \quad or \quad b = -24.$$

4. **Check.** Since the distance between the trees cannot be negative, -24 cannot be a solution. If the distance is 24 ft, we have $7^2 + 24^2 = 49 + 576 = 625$, which is 25^2. Thus, 24 checks and is a solution.

5. **State.** The distance between the trees is 24 ft.

Do Exercise 5.

5. **Reach of a Ladder.** Twila has a 26-ft ladder leaning against her house. If the bottom of the ladder is 10 ft from the base of the house, how high does the ladder reach?

Answer on page A-42

EXAMPLE 6 *Ladder Settings.* A ladder of length 13 ft is placed against a building in such a way that the distance from the top of the ladder to the ground is 7 ft more than the distance from the bottom of the ladder to the building. Find both distances.

1. **Familiarize.** We first make a drawing. The ladder and the missing distances form the hypotenuse and legs of a right triangle. We let $x =$ the length of the side (leg) across the bottom. Then $x + 7 =$ the length of the other side (leg). The hypotenuse has length 13 ft.

2. **Translate.** Since a right triangle is formed, we can use the Pythagorean theorem:

$$a^2 + b^2 = c^2$$

$$x^2 + (x + 7)^2 = 13^2. \qquad \text{Substituting}$$

3. **Solve.** We solve the equation as follows:

$x^2 + (x^2 + 14x + 49) = 169$	Squaring the binomial and 13
$2x^2 + 14x + 49 = 169$	Collecting like terms
$2x^2 + 14x + 49 - 169 = 169 - 169$	Subtracting 169 to get 0 on one side
$2x^2 + 14x - 120 = 0$	Simplifying
$2(x^2 + 7x - 60) = 0$	Factoring out a common factor
$x^2 + 7x - 60 = 0$	Dividing by 2
$(x + 12)(x - 5) = 0$	Factoring
$x + 12 = 0 \quad or \quad x - 5 = 0$	Using the principle of zero products
$x = -12 \quad or \quad x = 5.$	

4. **Check.** The negative integer -12 cannot be the length of a side. When $x = 5$, $x + 7 = 12$, and $5^2 + 12^2 = 13^2$. So 5 and 12 check.

5. **State.** The distance from the top of the ladder to the ground is 12 ft. The distance from the bottom of the ladder to the building is 5 ft.

Do Exercise 6.

6. **Right-Triangle Geometry.** The length of one leg of a right triangle is 1 m longer than the other. The length of the hypotenuse is 5 m. Find the lengths of the legs.

Answer on page A-42

Translating for Success

1. *Angle Measures.* The measures of the angles of a triangle are three consecutive integers. Find the measures of the angles.

2. *Rectangle Dimensions.* The area of a rectangle is 3599 ft^2. The length is 2 ft longer than the width. Find the dimensions of the rectangle.

3. *Sales Tax.* Claire paid $40,704 for a new SUV. This included 6% for sales tax. How much did the SUV cost before tax?

4. *Wire Cutting.* A 180-m wire is cut into three pieces. The third piece is 2 m longer than the first. The second is two-thirds as long as the first. How long is each piece?

5. *Perimeter.* The perimeter of a rectangle is 240 ft. The length is 2 ft greater than the width. Find the length and the width.

The goal of these matching questions is to practice step (2), *Translate,* of the five-step problem-solving process. Translate each word problem to an equation and select a correct translation from equations A–O.

A. $2x \cdot x = 288$

B. $x(x + 60) = 7021$

C. $59 = x \cdot 60$

D. $x^2 + (x + 2)^2 = 3599$

E. $x^2 + (x + 70)^2 = 130^2$

F. $6\% \cdot x = 40{,}704$

G. $2(x + 2) + 2x = 240$

H. $\frac{1}{2}x(x - 1) = 1770$

I. $x + \frac{2}{3}x + (x + 2) = 180$

J. $59\% \cdot x = 60$

K. $x + 6\% \cdot x = 40{,}704$

L. $2x^2 + x = 288$

M. $x(x + 2) = 3599$

N. $x^2 + 60 = 7021$

O. $x + (x + 1) + (x + 2) = 180$

Answers on page A-42

6. *Cell-Phone Tower.* A guy wire on a cell-phone tower is 130 ft long and is attached to the top of the tower. The height of the tower is 70 ft longer than the distance from the point on the ground where the wire is attached to the bottom of the tower. Find the height of the tower.

7. *Sales Meeting Attendance.* PTQ Corporation holds a sales meeting in Tucson. Of the 60 employees, 59 of them attend the meeting. What percent attend the meeting?

8. *Dimensions of a Pool.* A rectangular swimming pool is twice as long as it is wide. The area of the surface is 288 ft^2. Find the dimensions of the pool.

9. *Dimensions of a Triangle.* The height of a triangle is 1 cm less than the length of the base. The area of the triangle is 1770 cm^2. Find the height and the length of the base.

10. *Width of a Rectangle.* The length of a rectangle is 60 ft longer than the width. Find the width if the area of the rectangle is 7021 ft^2.

11.8 EXERCISE SET

For Extra Help

MathXL | MyMathLab | InterAct Math | Math Tutor Center | Video Lectures on CD Disc 6 | Student's Solutions Manual

a, **b** Solve.

1. *Design.* The screen of the TI-84 Plus graphing calculator is nearly rectangular. The length of the rectangle is 2 cm more than the width. If the area of the rectangle is 24 cm^2, find the length and the width.

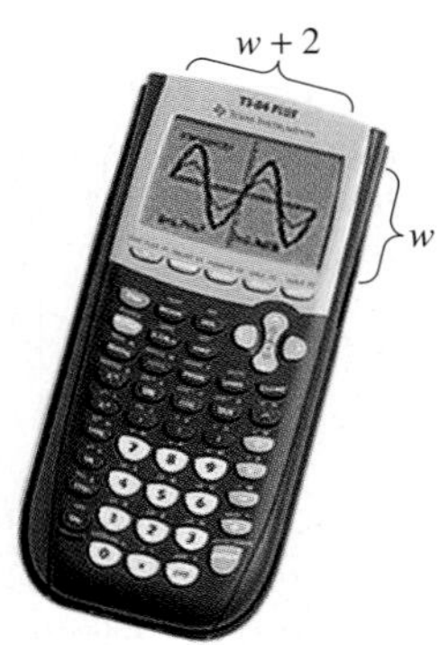

2. *Area of a Garden.* The length of a rectangular garden is 4 m greater than the width. The area of the garden is 96 m^2. Find the length and the width.

3. *Furnishings.* A rectangular table in Arlo's House of Tunes is six times as long as it is wide. The area of the table is 24 ft^2. Find the length and the width of the table.

4. *Dimensions of Picture.* A rectangular picture is three times as long as it is wide. The area of the picture is 588 in^2. Find the dimensions of the picture.

5. *Dimensions of a Triangle.* A triangle is 10 cm wider than it is tall. The area is 28 cm^2. Find the height and the base.

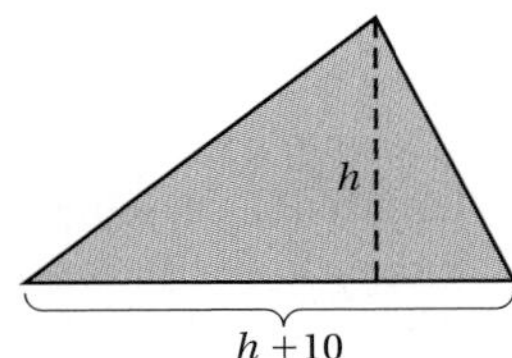

6. *Dimensions of a Triangle.* The height of a triangle is 3 cm less than the length of the base. The area of the triangle is 35 cm^2. Find the height and the length of the base.

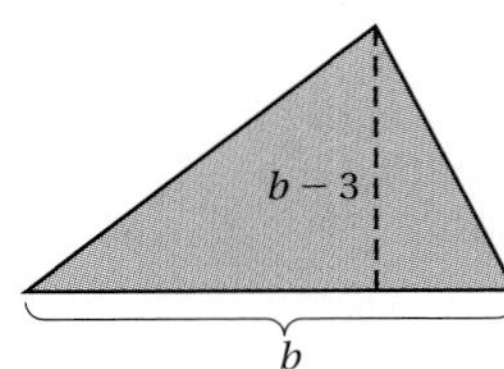

7. *Road Design.* A triangular traffic island has a base half as long as its height. The island has an area of 64 m^2. Find the base and the height.

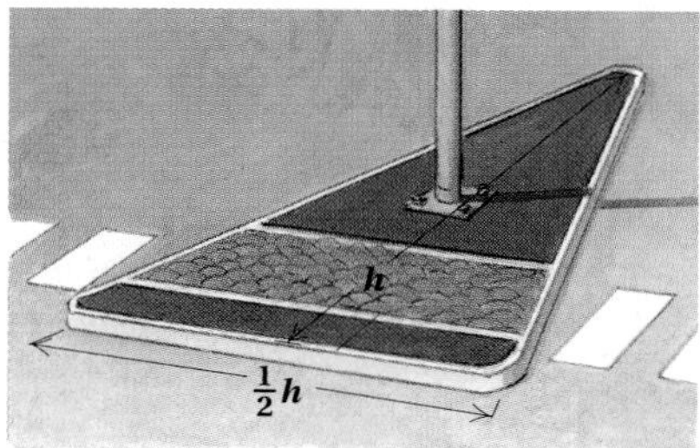

8. *Dimensions of a Sail.* The height of the jib sail on a Lightning sailboat is 5 ft greater than the length of its "foot." The area of the sail is 42 ft^2. Find the length of the foot and the height of the sail.

Games in a League. Use $x^2 - x = N$ for Exercises 9–12.

9. A chess league has 14 teams. What is the total number of games to be played if each team plays every other team twice?

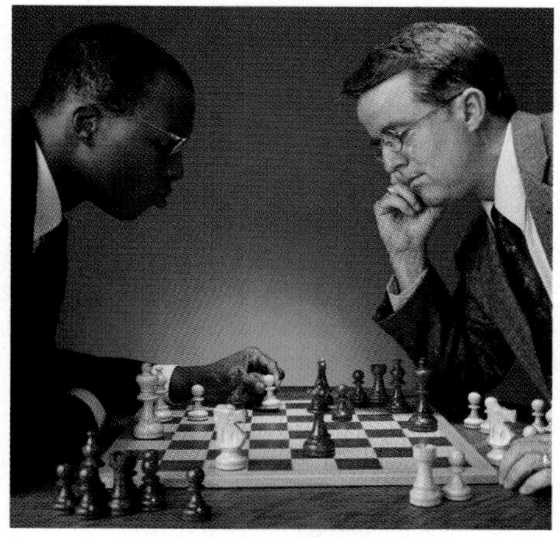

10. A women's volleyball league has 23 teams. What is the total number of games to be played if each team plays every other team twice?

11. A slow-pitch softball league plays a total of 132 games. How many teams are in the league if each team plays every other team twice?

12. A basketball league plays a total of 90 games. How many teams are in the league if each team plays every other team twice?

Handshakes. A researcher wants to investigate the potential spread of germs by contact. She knows that the number of possible handshakes within a group of x people, assuming each person shakes every other person's hand only once, is given by

$$N = \frac{1}{2}(x^2 - x).$$

Use this formula for Exercises 13–16.

13. There are 100 people at a party. How many handshakes are possible?

14. There are 40 people at a meeting. How many handshakes are possible?

15. Everyone at a meeting shook hands with each other. There were 300 handshakes in all. How many people were at the meeting?

16. Everyone at a party shook hands with each other. There were 153 handshakes in all. How many people were at the party?

17. *Toasting.* During a toast at a party, there were 190 "clicks" of glasses. How many people took part in the toast?

18. *High-fives.* After winning the championship, all Detroit Pistons teammates exchanged "high-fives." Altogether there were 66 high-fives. How many players were there?

19. *Consecutive Page Numbers.* The product of the page numbers on two facing pages of a book is 210. Find the page numbers.

20. *Consecutive Page Numbers.* The product of the page numbers on two facing pages of a book is 420. Find the page numbers.

21. The product of two consecutive even integers is 168. Find the integers. (See Section 2.6.)

22. The product of two consecutive even integers is 224. Find the integers. (See Section 2.6.)

23. The product of two consecutive odd integers is 255. Find the integers.

24. The product of two consecutive odd integers is 143. Find the integers.

25. *Right-Triangle Geometry.* The length of one leg of a right triangle is 8 ft. The length of the hypotenuse is 2 ft longer than the other leg. Find the length of the hypotenuse and the other leg.

26. *Right-Triangle Geometry.* The length of one leg of a right triangle is 24 ft. The length of the other leg is 16 ft shorter than the hypotenuse. Find the length of the hypotenuse and the other leg.

27. *Roadway Design.* Elliott Street is 24 ft wide when it ends at Main Street in Brattleboro, Vermont. A 40-ft long diagonal crosswalk allows pedestrians to cross Main Street to or from either corner of Elliott Street (see the figure). Determine the width of Main Street.

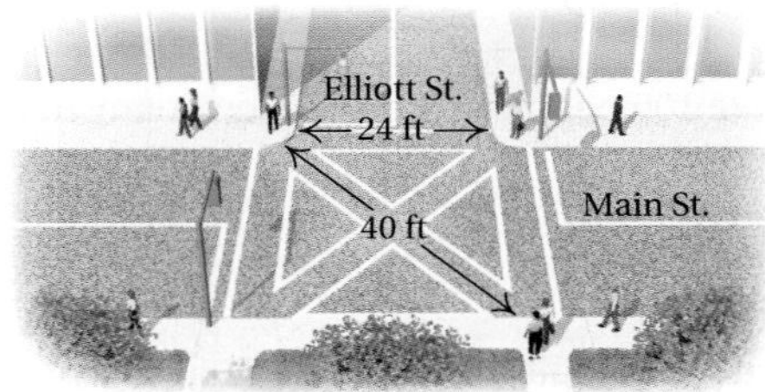

28. *Sailing.* The mainsail of a Lightning sailboat is a right triangle in which the hypotenuse is called the leech. If a 24-ft tall mainsail has a leech length of 26 ft and if Dacron® sailcloth costs $10 per square foot, find the cost of a new mainsail.

29. *Lookout Tower.* The diagonal braces in a lookout tower are 15 ft long and span a distance of 12 ft. How high does each brace reach vertically?

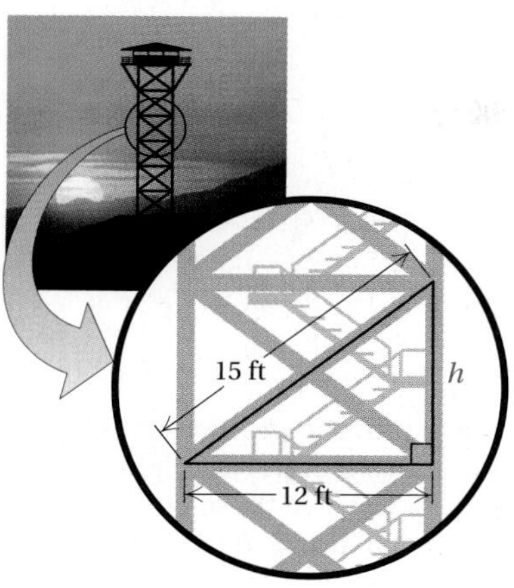

30. *Aviation.* Engine failure forced Geraldine to pilot her Cessna 150 to an emergency landing. To land, Geraldine's plane glided 17,000 ft over a 15,000-ft stretch of deserted highway. From what altitude did the descent begin?

31. *Architecture.* An architect has allocated a rectangular space of 264 ft^2 for a square dining room and a 10-ft wide kitchen, as shown in the figure. Find the dimensions of each room.

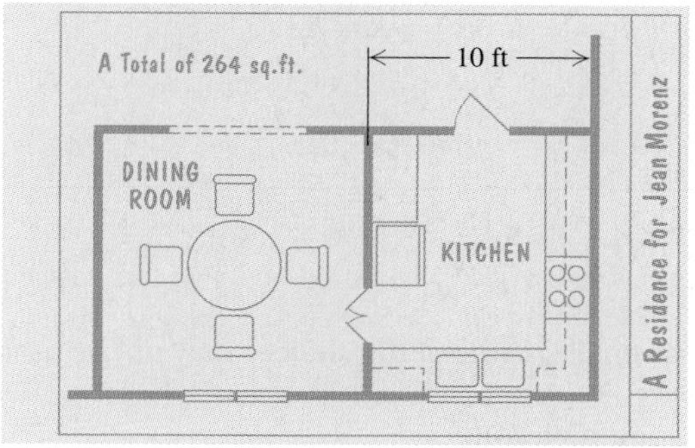

32. *Guy Wire.* The guy wire on a TV antenna is 1 m longer than the height of the antenna. If the guy wire is anchored 3 m from the foot of the antenna, how tall is the antenna?

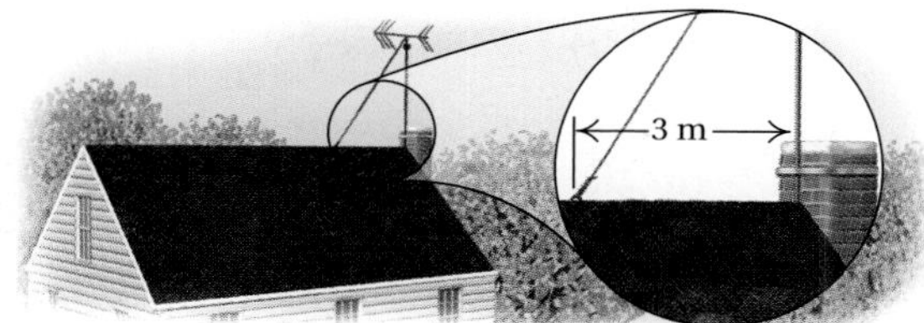

Rocket Launch. A model rocket is launched with an initial velocity of 180 ft/sec. Its height h, in feet, after t seconds is given by the formula

$$h = 180t - 16t^2.$$

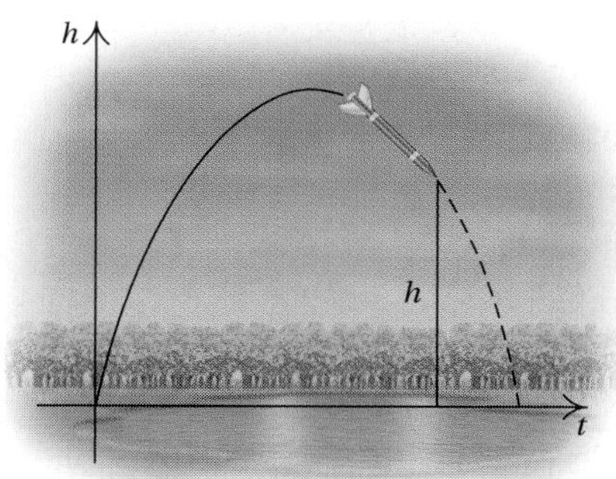

33. After how many seconds will the rocket first reach a height of 464 ft?

34. After how many seconds from launching will the rocket again be at that same height of 464 ft? (See Exercise 33.)

35. The sum of the squares of two consecutive odd positive integers is 74. Find the integers.

36. The sum of the squares of two consecutive odd positive integers is 130. Find the integers.

37. **DW** An archaeologist has measuring sticks of 3 ft, 4 ft, and 5 ft. Explain how she could draw a 7-ft by 9-ft rectangle on a piece of land being excavated.

38. **DW** Look closely at the problem-solving techniques developed in this chapter. What kinds of equations do we use? In order to solve these equations, what additional new skill do we need? Compare the skills learned in this chapter with those of Chapter 2.

SKILL MAINTENANCE

VOCABULARY REINFORCEMENT

In each of Exercises 39–46, fill in the blank with the correct term from the given list. Some of the choices may not be used and some may be used more than once.

39. To ____________ a polynomial is to express it as a product. [11.1b]

40. A(n) ____________ of a polynomial P is a polynomial that can be used to express P as a product. [11.1b]

41. A factorization of a polynomial is an expression that names that polynomial as a(n) ____________. [11.1b]

42. When factoring, always look first for the ____________. [11.1b]

43. The expression $-5x^2 + 8x - 7$ is an example of a ____________. [10.3a]

44. The ____________ asserts that when dividing with exponential notation, if the bases are the same, keep the base and subtract the exponent of the denominator from the exponent of the numerator. [10.1e]

45. For the graph of the equation $4x - 3y = 12$, the pair $(0, -4)$ is known as the ____________. [9.2a]

46. For the graph of the equation $4x - 3y = 12$, the number $\frac{4}{3}$ is known as the ____________. [9.3c]

quotient rule
product rule
slope
common factor
common multiple
factor
x-intercept
y-intercept
binomial
trinomial
quotient
product

SYNTHESIS

47. *Telephone Service.* Use the information in the figure below to determine the height of the telephone pole.

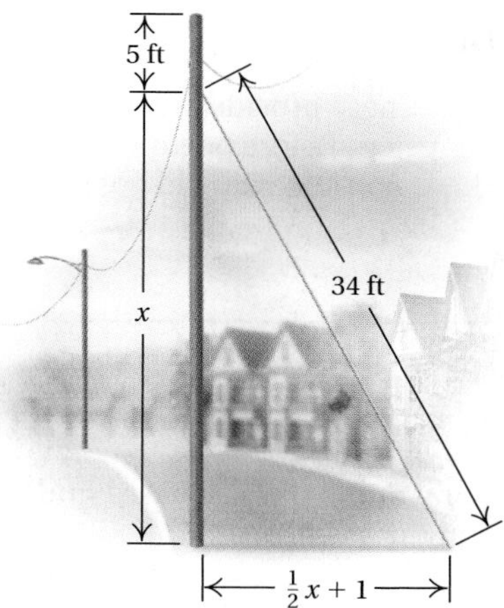

48. *Roofing.* A *square* of shingles covers 100 ft^2 of surface area. How many squares will be needed to reshingle the house shown?

49. *Pool Sidewalk.* A cement walk of constant width is built around a 20-ft by 40-ft rectangular pool. The total area of the pool and the walk is 1500 ft^2. Find the width of the walk.

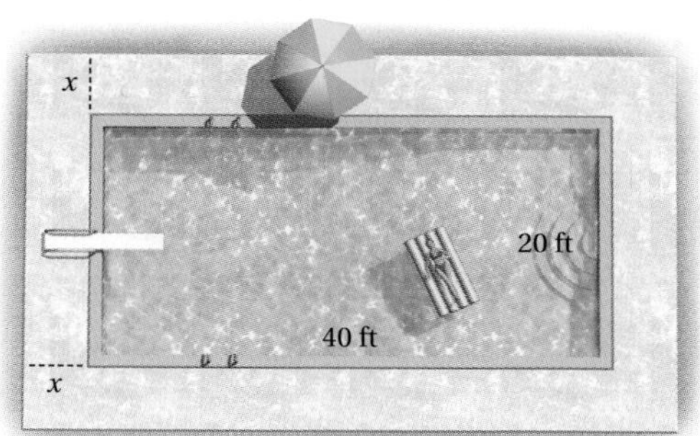

50. *Rain-Gutter Design.* An open rectangular gutter is made by turning up the sides of a piece of metal 20 in. wide. The area of the cross-section of the gutter is 50 in^2. Find the depth of the gutter.

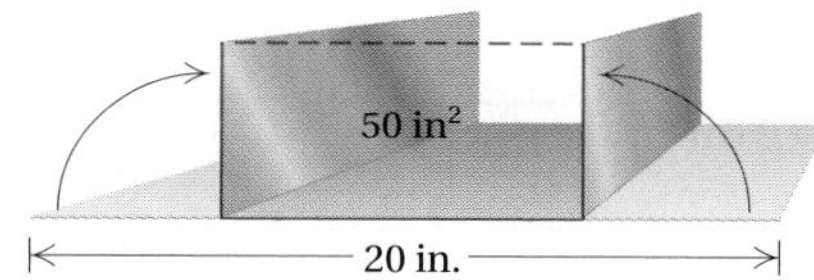

51. *Dimensions of an Open Box.* A rectangular piece of cardboard is twice as long as it is wide. A 4-cm square is cut out of each corner, and the sides are turned up to make a box with an open top. The volume of the box is 616 cm^3. Find the original dimensions of the cardboard.

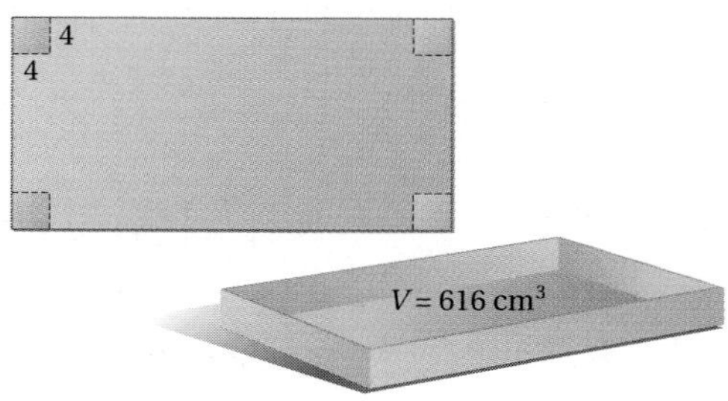

52. Solve for x.

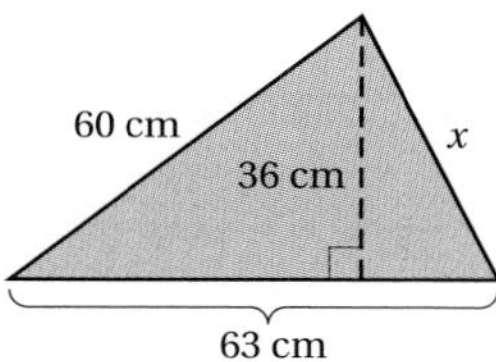

53. *Dimensions of a Closed Box.* The total surface area of a closed box is 350 ft^2. The box is 9 ft high and has a square base and lid. Find the length of a side of the base.

54. The ones digit of a number less than 100 is 4 greater than the tens digit. The sum of the number and the product of the digits is 58. Find the number.

11 Summary and Review

The review that follows is meant to prepare you for a chapter exam. It consists of three parts. The first part, Concept Reinforcement, is designed to increase understanding of the concepts through true/false exercises. The second part is a list of important properties and formulas. The third part is the Review Exercises. These provide practice exercises for the exam, together with references to section objectives so you can go back and review. Before beginning, stop and look back over the skills you have obtained. What skills in mathematics do you have now that you did not have before studying this chapter?

CONCEPT REINFORCEMENT

Determine whether the statement is true or false. Answers are given at the back of the book.

_______ **1.** The greatest common factor (GCF) of a set of natural numbers is at least 1 and always less than or equal to the smallest number in the set.

_______ **2.** To factor $x^2 + bx + c$, use a trial-and-error process that looks for factors of b whose sum is c.

_______ **3.** $(x - a)^2 \neq -(a - x)^2$

_______ **4.** A product is 0 if and only if all factors are 0.

_______ **5.** In order that the principle of zero products can be used, one side of the equation must be 0.

IMPORTANT PROPERTIES AND FORMULAS

Factoring Formulas: $A^2 - B^2 = (A + B)(A - B)$, $A^2 + 2AB + B^2 = (A + B)^2$, $A^2 - 2AB + B^2 = (A - B)^2$

The Principle of Zero Products: An equation $ab = 0$ is true if and only if $a = 0$ is true or $b = 0$ is true, or both are true.

The Pythagorean Theorem: $a^2 + b^2 = c^2$

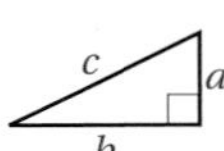

Review Exercises

Find the GCF. [11.1a]

1. $-15y^2,\ 25y^6$

2. $12x^3,\ -60x^2y,\ 36xy$

Factor completely. [11.6a]

3. $5 - 20x^6$

4. $x^2 - 3x$

5. $9x^2 - 4$

6. $x^2 + 4x - 12$

7. $x^2 + 14x + 49$

8. $6x^3 + 12x^2 + 3x$

9. $x^3 + x^2 + 3x + 3$

10. $6x^2 - 5x + 1$

11. $x^4 - 81$

12. $9x^3 + 12x^2 - 45x$

13. $2x^2 - 50$

14. $x^4 + 4x^3 - 2x - 8$

15. $16x^4 - 1$

16. $8x^6 - 32x^5 + 4x^4$

17. $75 + 12x^2 + 60x$

18. $x^2 + 9$

19. $x^3 - x^2 - 30x$

20. $4x^2 - 25$

21. $9x^2 + 25 - 30x$

22. $6x^2 - 28x - 48$

23. $x^2 - 6x + 9$

24. $2x^2 - 7x - 4$

25. $18x^2 - 12x + 2$

26. $3x^2 - 27$

27. $15 - 8x + x^2$

28. $25x^2 - 20x + 4$

29. $49b^{10} + 4a^8 - 28a^4b^5$

30. $x^2y^2 + xy - 12$

31. $12a^2 + 84ab + 147b^2$

32. $m^2 + 5m + mt + 5t$

33. $32x^4 - 128y^4z^4$

Solve. [11.7a, b]

34. $(x - 1)(x + 3) = 0$

35. $x^2 + 2x - 35 = 0$

36. $x^2 + x - 12 = 0$

37. $3x^2 + 2 = 5x$

38. $2x^2 + 5x = 12$

39. $16 = x(x - 6)$

Solve. [11.8a]

40. *Sharks' Teeth.* Sharks' teeth are shaped like triangles. The height of a tooth of a great white shark is 1 cm longer than the base. The area is 15 cm^2. Find the height and the base.

41. The product of two consecutive even integers is 288. Find the integers.

42. The product of two consecutive odd integers is 323. Find the integers.

43. *Tree Supports.* A duckbill-anchor system is used to support a newly planted Bradford pear tree. The cable is 2 ft longer than the distance from the base of the tree to where the cable is attached to the tree. The cables are anchored 4 ft from the tree. How far from the ground are the cables attached to the tree?

44. If the sides of a square are lengthened by 3 km, the area increases to 81 km^2. Find the length of a side of the original square.

Find the x-intercepts for the graph of the equation. [11.7b]

45. $y = x^2 + 9x + 20$

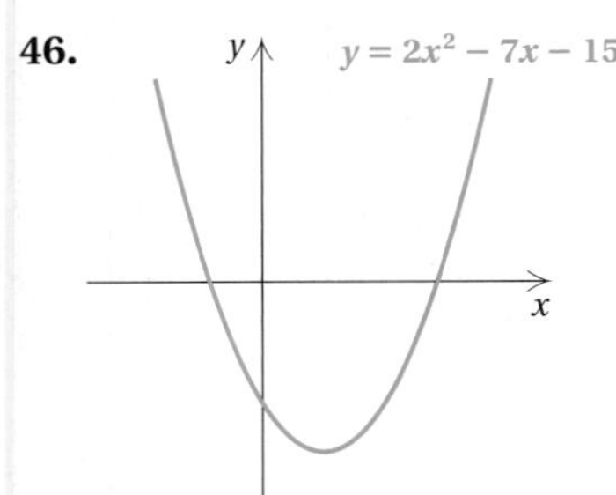

46. $y = 2x^2 - 7x - 15$

47. DW Write a problem for a classmate to solve such that only one of the two solutions of a quadratic equation can be used as an answer. [11.8a]

48. DW On a quiz, Sheri writes the factorization of $4x^2 - 100$ as $(2x - 10)(2x + 10)$. Explain Sheri's mistake. [11.5d]

SYNTHESIS

Solve. [11.8a]

49. The pages of a book measure 15 cm by 20 cm. Margins of equal width surround the printing on each page and constitute one-half of the area of the page. Find the width of the margins.

50. The cube of a number is the same as twice the square of the number. Find all such numbers.

51. The length of a rectangle is two times its width. When the length is increased by 20 and the width decreased by 1, the area is 160. Find the original length and width.

Solve. [11.7b]

52. $x^2 + 25 = 0$

53. $(x - 2)(x + 3)(2x - 5) = 0$

54. $(x - 3)4x^2 + 3x(x - 3) - (x - 3)10 = 0$

55. Find a polynomial for the shaded area in the figure below.

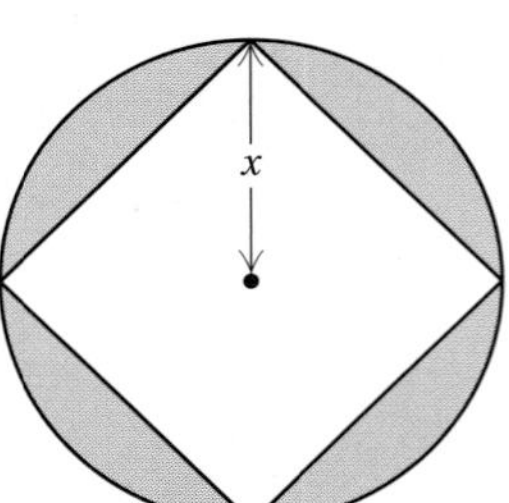

11 Chapter Test

For Extra Help

Work It Out! Chapter Test Video Solutions on CD

1. Find the GCF: $28x^3, 48x^7$.

Factor completely.

2. $x^2 - 7x + 10$

3. $x^2 + 25 - 10x$

4. $6y^2 - 8y^3 + 4y^4$

5. $x^3 + x^2 + 2x + 2$

6. $x^2 - 5x$

7. $x^3 + 2x^2 - 3x$

8. $28x - 48 + 10x^2$

9. $4x^2 - 9$

10. $x^2 - x - 12$

11. $6m^3 + 9m^2 + 3m$

12. $3w^2 - 75$

13. $60x + 45x^2 + 20$

14. $3x^4 - 48$

15. $49x^2 - 84x + 36$

16. $5x^2 - 26x + 5$

17. $x^4 + 2x^3 - 3x - 6$

18. $80 - 5x^4$

19. $4x^2 - 4x - 15$

20. $6t^3 + 9t^2 - 15t$

21. $3m^2 - 9mn - 30n^2$

Solve.

22. $x^2 - x - 20 = 0$

23. $2x^2 + 7x = 15$

24. $x(x - 3) = 28$

Solve.

25. The length of a rectangle is 2 m more than the width. The area of the rectangle is 48 m^2. Find the length and the width.

26. The base of a triangle is 6 cm greater than twice the height. The area is 28 cm^2. Find the height and the base.

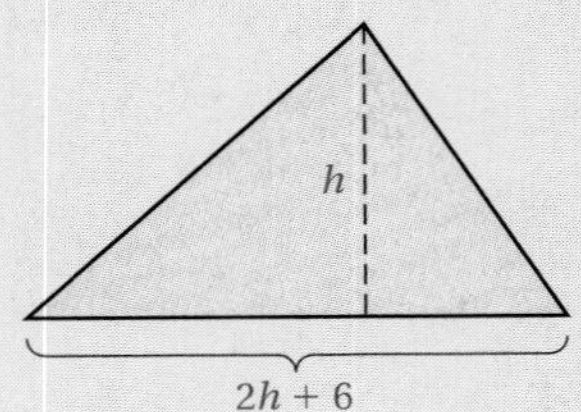

27. *Masonry Corner.* A mason wants to be sure he has a right corner in a building's foundation. He marks a point 3 ft from the corner along one wall and another point 4 ft from the corner along the other wall. If the corner is a right angle, what should the distance be between the two marked points?

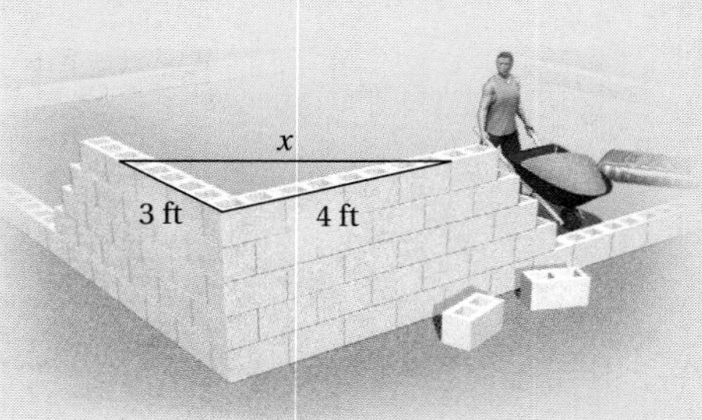

Find the x-intercepts for the graph of the equation.

28.

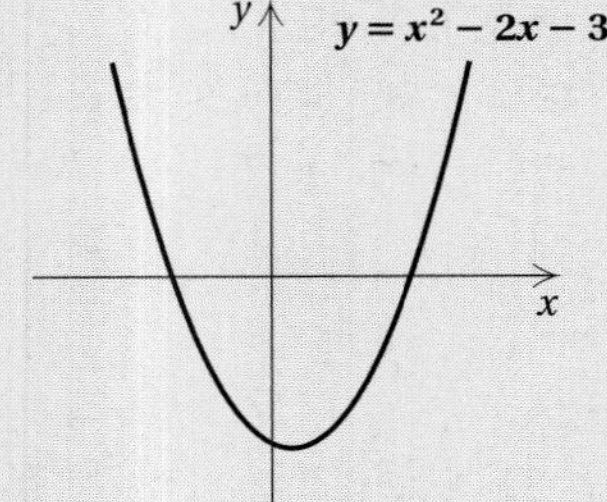

29.

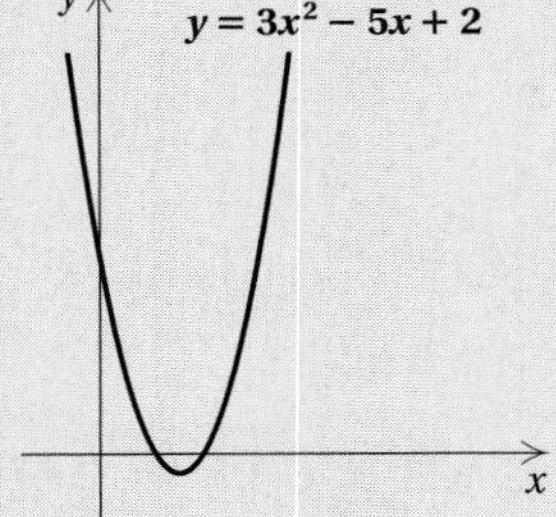

SYNTHESIS

30. The length of a rectangle is five times its width. When the length is decreased by 3 and the width is increased by 2, the area of the new rectangle is 60. Find the original length and width.

31. Factor: $(a + 3)^2 - 2(a + 3) - 35$.

32. Solve: $20x(x + 2)(x - 1) = 5x^3 - 24x - 14x^2$.

33. If $x + y = 4$ and $x - y = 6$, then $x^2 - y^2 = ?$

a) 2
b) 10
c) 34
d) 24

Rational Expressions and Equations

12

Real-World Application

To estimate how many frogs there are in a rain forest, a research team tags 600 frogs and then releases them. Later, the team catches 300 frogs and notes that 25 of them have been tagged. Estimate the total frog population in the rain forest.

This problem appears as Exercise 46 in Section 12.7.

12.1 MULTIPLYING AND SIMPLIFYING RATIONAL EXPRESSIONS

Objectives

a Find all numbers for which a rational expression is not defined.

b Multiply a rational expression by 1, using an expression such as A/A.

c Simplify rational expressions by factoring the numerator and the denominator and removing factors of 1.

d Multiply rational expressions and simplify.

Find all numbers for which the rational expression is not defined.

1. $\dfrac{16}{x-3}$

2. $\dfrac{2x-7}{x^2+5x-24}$

3. $\dfrac{x+5}{8}$

Answers on page A-42

a Rational Expressions and Replacements

Rational numbers are quotients of integers. Some examples are

$$\frac{2}{3},\quad \frac{4}{-5},\quad \frac{-8}{17},\quad \frac{563}{1}.$$

The following are called **rational expressions** or **fraction expressions.** They are quotients, or ratios, of polynomials:

$$\frac{3}{4},\quad \frac{z}{6},\quad \frac{5}{x+2},\quad \frac{t^2+3t-10}{7t^2-4}.$$

A rational expression is also a division. For example,

$$\frac{3}{4} \text{ means } 3 \div 4 \text{ and } \frac{x-8}{x+2} \text{ means } (x-8) \div (x+2).$$

Because rational expressions indicate division, we must be careful to avoid denominators of zero. When a variable is replaced with a number that produces a denominator equal to zero, the rational expression is not defined. For example, in the expression

$$\frac{x-8}{x+2},$$

when x is replaced with -2, the denominator is 0, and the expression is *not* defined:

$$\frac{x-8}{x+2} = \frac{-2-8}{-2+2} = \frac{-10}{0}. \longleftarrow \text{Division by 0 is not defined.}$$

When x is replaced with a number other than -2, such as 3, the expression *is* defined because the denominator is nonzero:

$$\frac{x-8}{x+2} = \frac{3-8}{3+2} = \frac{-5}{5} = -1.$$

EXAMPLE 1 Find all numbers for which the rational expression

$$\frac{x+4}{x^2-3x-10}$$

is not defined.

The value of the numerator has no bearing on whether or not a rational expression is defined. To determine which numbers make the rational expression not defined, we set the *denominator* equal to 0 and solve:

$$x^2-3x-10=0$$

$$(x-5)(x+2)=0 \qquad \text{Factoring}$$

$$x-5=0 \quad or \quad x+2=0 \qquad \text{Using the principle of zero products (see Section 11.7)}$$

$$x=5 \quad or \quad x=-2.$$

The rational expression is not defined for the replacement numbers 5 and -2.

Do Exercises 1–3.

b Multiplying by 1

We multiply rational expressions in the same way that we multiply fraction notation in arithmetic. For a review, see Section 2.1. We saw that

$$\frac{3}{7} \cdot \frac{2}{5} = \frac{3 \cdot 2}{7 \cdot 5} = \frac{6}{35}.$$

MULTIPLYING RATIONAL EXPRESSIONS

To multiply rational expressions, multiply numerators and multiply denominators:

$$\frac{A}{B} \cdot \frac{C}{D} = \frac{AC}{BD}.$$

For example,

$$\frac{x-2}{3} \cdot \frac{x+2}{x+7} = \frac{(x-2)(x+2)}{3(x+7)}.$$ Multiplying the numerators and the denominators

Note that we leave the numerator, $(x - 2)(x + 2)$, and the denominator, $3(x + 7)$, in factored form because it is easier to simplify if we do not multiply. In order to learn to simplify, we first need to consider multiplying the rational expression by 1.

Any rational expression with the same numerator and denominator is a symbol for 1:

$$\frac{19}{19} = 1, \quad \frac{x+8}{x+8} = 1, \quad \frac{3x^2-4}{3x^2-4} = 1, \quad \frac{-1}{-1} = 1.$$

EQUIVALENT EXPRESSIONS

Expressions that have the same value for all allowable (or meaningful) replacements are called **equivalent expressions.**

We can multiply by 1 to obtain an *equivalent expression.* At this point, we select expressions for 1 arbitrarily. Later, we will have a system for our choices when we add and subtract.

EXAMPLES Multiply.

2. $\frac{3x+2}{x+1} \cdot 1 = \frac{3x+2}{x+1} \cdot \frac{2x}{2x} = \frac{(3x+2)2x}{(x+1)2x}$ Using the identity property of 1. We arbitrarily choose $2x/2x$ as a symbol for 1.

3. $\frac{x+2}{x-7} \cdot \frac{x+3}{x+3} = \frac{(x+2)(x+3)}{(x-7)(x+3)}$ We arbitrarily choose $(x + 3)/(x + 3)$ as a symbol for 1.

4. $\frac{2+x}{2-x} \cdot \frac{-1}{-1} = \frac{(2+x)(-1)}{(2-x)(-1)}$ Using $(-1)/(-1)$ as a symbol for 1

Do Exercises 4–6.

Multiply.

4. $\frac{2x+1}{3x-2} \cdot \frac{x}{x}$

5. $\frac{x+1}{x-2} \cdot \frac{x+2}{x+2}$

6. $\frac{x-8}{x-y} \cdot \frac{-1}{-1}$

Answers on page A-42

Simplify.

7. $\frac{5y}{y}$

8. $\frac{9x^2}{36x}$

Answers on page A-43

Study Tips

TUNE OUT DISTRACTIONS

Do you generally study in noisy places? If there is constant noise in your home, dorm, or other study area, consider finding a quiet place in the library—maybe an uncrowded place so you are not distracted with people-watching!

"The ability to concentrate and use your time well is everything."

Lee Iacocca, former CEO of Chrysler Corporation

C Simplifying Rational Expressions

Simplifying rational expressions is similar to simplifying fraction expressions in arithmetic. For a review, see Section 2.1. We saw there, for example, that an expression like $\frac{15}{40}$ can be simplified as follows:

$$\frac{15}{40} = \frac{3 \cdot 5}{8 \cdot 5}$$ Factoring the numerator and the denominator. Note the common factor, 5.

$$= \frac{3}{8} \cdot \frac{5}{5}$$ Factoring the fraction expression

$$= \frac{3}{8} \cdot 1$$ $\frac{5}{5} = 1$

$$= \frac{3}{8}.$$ Using the identity property of 1, or "removing a factor of 1"

Similar steps are followed when simplifying rational expressions: We factor and remove a factor of 1, using the fact that

$$\frac{ab}{cb} = \frac{a}{c} \cdot \frac{b}{b} = \frac{a}{c} \cdot 1 = \frac{a}{c}.$$

In algebra, instead of simplifying

$$\frac{15}{40},$$

we may need to simplify an expression like

$$\frac{x^2 - 16}{x + 4}.$$

Just as factoring is important in simplifying in arithmetic, so too is it important in simplifying rational expressions. The factoring we use most is the factoring of polynomials, which we studied in Chapter 11.

To simplify, we can do the reverse of multiplying. We factor the numerator and the denominator and "remove" a factor of 1.

EXAMPLE 5 Simplify: $\frac{8x^2}{24x}$.

$$\frac{8x^2}{24x} = \frac{8 \cdot x \cdot x}{3 \cdot 8 \cdot x}$$ Factoring the numerator and the denominator. Note the common factor, $8x$.

$$= \frac{8x}{8x} \cdot \frac{x}{3}$$ Factoring the rational expression

$$= 1 \cdot \frac{x}{3}$$ $\frac{8x}{8x} = 1$

$$= \frac{x}{3}$$ We removed a factor of 1.

Do Exercises 7 and 8.

EXAMPLES Simplify.

6. $\dfrac{5a + 15}{10} = \dfrac{5(a + 3)}{5 \cdot 2}$ Factoring the numerator and the denominator

$= \dfrac{5}{5} \cdot \dfrac{a + 3}{2}$ Factoring the rational expression

$= 1 \cdot \dfrac{a + 3}{2}$ $\dfrac{5}{5} = 1$

$= \dfrac{a + 3}{2}$ Removing a factor of 1

7. $\dfrac{6a + 12}{7a + 14} = \dfrac{6(a + 2)}{7(a + 2)}$ Factoring the numerator and the denominator

$= \dfrac{6}{7} \cdot \dfrac{a + 2}{a + 2}$ Factoring the rational expression

$= \dfrac{6}{7} \cdot 1$ $\dfrac{a + 2}{a + 2} = 1$

$= \dfrac{6}{7}$ Removing a factor of 1

8. $\dfrac{6x^2 + 4x}{2x^2 + 2x} = \dfrac{2x(3x + 2)}{2x(x + 1)}$ Factoring the numerator and the denominator

$= \dfrac{2x}{2x} \cdot \dfrac{3x + 2}{x + 1}$ Factoring the rational expression

$= 1 \cdot \dfrac{3x + 2}{x + 1}$ $\dfrac{2x}{2x} = 1$

$= \dfrac{3x + 2}{x + 1}$ Removing a factor of 1

Caution!

Note in this step that you *cannot* remove the x's because x is not a factor of the entire numerator, $3x + 2$, and the entire denominator, $x + 1$.

9. $\dfrac{x^2 + 3x + 2}{x^2 - 1} = \dfrac{(x + 2)(x + 1)}{(x + 1)(x - 1)}$ Factoring the numerator and the denominator

$= \dfrac{x + 1}{x + 1} \cdot \dfrac{x + 2}{x - 1}$ Factoring the rational expression

$= 1 \cdot \dfrac{x + 2}{x - 1}$ $\dfrac{x + 1}{x + 1} = 1$

$= \dfrac{x + 2}{x - 1}$ Removing a factor of 1

Simplify.

9. $\dfrac{2x^2 + x}{3x^2 + 2x}$

10. $\dfrac{x^2 - 1}{2x^2 - x - 1}$

11. $\dfrac{7x + 14}{7}$

12. $\dfrac{12y + 24}{48}$

Answers on page A-43

Simplify.

13. $\frac{x-8}{8-x}$

14. $\frac{c-d}{d-c}$

15. $\frac{-x-7}{x+7}$

Answers on page A-43

CANCELING

You may have encountered canceling when working with rational expressions. With great concern, we mention it as a possible way to speed up your work. Our concern is that canceling be done with care and understanding. Example 9 might have been done faster as follows:

$$\frac{x^2+3x+2}{x^2-1} = \frac{(x+2)(x+1)}{(x+1)(x-1)}$$ Factoring the numerator and the denominator

$$= \frac{(x+2)\cancel{(x+1)}}{\cancel{(x+1)}(x-1)}$$ When a factor of 1 is noted, it is canceled, as shown: $\frac{x+1}{x+1} = 1$.

$$= \frac{x+2}{x-1}.$$ Simplifying

Caution!

The difficulty with canceling is that it is often applied incorrectly, as in the following situations:

$$\frac{\cancel{x}+3}{\cancel{x}} = 3; \qquad \frac{\cancel{4}+1}{\cancel{4}+2} = \frac{1}{2}; \qquad \frac{1\cancel{5}}{\cancel{5}4} = \frac{1}{4}.$$

Wrong! Wrong! Wrong!

In each of these situations, the expressions canceled were *not* factors of 1. Factors are parts of products. For example, in $2 \cdot 3$, 2 and 3 are factors, but in $2 + 3$, 2 and 3 are *not* factors. If you can't factor, you can't cancel. If in doubt, don't cancel!

Do Exercises 9–12 on the preceding page.

OPPOSITES IN RATIONAL EXPRESSIONS

Expressions of the form $a - b$ and $b - a$ are opposites of each other. When either of these binomials is multiplied by -1, the result is the other binomial:

$$-1(a-b) = -a+b = b+(-a) = b-a;$$
$$-1(b-a) = -b+a = a+(-b) = a-b.$$

Multiplication by -1 reverses the order in which subtraction occurs.

Consider, for example,

$$\frac{x-4}{4-x}.$$

At first glance, it appears as though the numerator and the denominator do not have any common factors other than 1. But $x - 4$ and $4 - x$ are opposites, or additive inverses, of each other. Thus we can rewrite one as the opposite of the other by factoring out a -1.

EXAMPLE 10 Simplify: $\frac{x-4}{4-x}$.

$$\frac{x-4}{4-x} = \frac{x-4}{-(x-4)} = \frac{x-4}{-1(x-4)}$$ $4 - x = -(x - 4)$; $4 - x$ and $x - 4$ are opposites.

$$= -1 \cdot \frac{x-4}{x-4}$$
$$= -1 \cdot 1$$
$$= -1$$

Do Exercises 13–15.

d Multiplying and Simplifying

We try to simplify after we multiply. That is why we leave the numerator and the denominator in factored form.

EXAMPLE 11 Multiply and simplify: $\frac{5a^3}{4} \cdot \frac{2}{5a}$.

$$\frac{5a^3}{4} \cdot \frac{2}{5a} = \frac{5a^3(2)}{4(5a)}$$ Multiplying the numerators and the denominators

$$= \frac{2 \cdot 5 \cdot a \cdot a \cdot a}{2 \cdot 2 \cdot 5 \cdot a}$$ Factoring the numerator and the denominator

$$= \frac{\cancel{2} \cdot \cancel{5} \cdot \cancel{a} \cdot a \cdot a}{\cancel{2} \cdot 2 \cdot \cancel{5} \cdot \cancel{a}}$$ Removing a factor of 1: $\frac{2 \cdot 5 \cdot a}{2 \cdot 5 \cdot a} = 1$

$$= \frac{a^2}{2}$$ Simplifying

EXAMPLE 12 Multiply and simplify: $\frac{x^2 + 6x + 9}{x^2 - 4} \cdot \frac{x - 2}{x + 3}$.

$$\frac{x^2 + 6x + 9}{x^2 - 4} \cdot \frac{x - 2}{x + 3} = \frac{(x^2 + 6x + 9)(x - 2)}{(x^2 - 4)(x + 3)}$$ Multiplying the numerators and the denominators

$$= \frac{(x + 3)(x + 3)(x - 2)}{(x + 2)(x - 2)(x + 3)}$$ Factoring the numerator and the denominator

$$= \frac{\cancel{(x + 3)}(x + 3)\cancel{(x - 2)}}{(x + 2)\cancel{(x - 2)}\cancel{(x + 3)}}$$ Removing a factor of 1: $\frac{(x + 3)(x - 2)}{(x + 3)(x - 2)} = 1$

$$= \frac{x + 3}{x + 2}$$ Simplifying

Do Exercise 16.

EXAMPLE 13 Multiply and simplify: $\frac{x^2 + x - 2}{15} \cdot \frac{5}{2x^2 - 3x + 1}$.

$$\frac{x^2 + x - 2}{15} \cdot \frac{5}{2x^2 - 3x + 1} = \frac{(x^2 + x - 2)5}{15(2x^2 - 3x + 1)}$$ Multiplying the numerators and the denominators

$$= \frac{(x + 2)(x - 1)5}{5(3)(x - 1)(2x - 1)}$$ Factoring the numerator and the denominator

$$= \frac{(x + 2)\cancel{(x - 1)}\cancel{5}}{\cancel{5}(3)\cancel{(x - 1)}(2x - 1)}$$ Removing a factor of 1: $\frac{(x - 1)5}{(x - 1)5} = 1$

$$= \frac{x + 2}{3(2x - 1)}$$ Simplifying

You need not carry out this multiplication.

Do Exercise 17.

16. Multiply and simplify:

$$\frac{a^2 - 4a + 4}{a^2 - 9} \cdot \frac{a + 3}{a - 2}.$$

17. Multiply and simplify:

$$\frac{x^2 - 25}{6} \cdot \frac{3}{x + 5}.$$

Answers on page A-43

CALCULATOR CORNER

Checking Multiplication and Simplification We can use the TABLE feature as a partial check that rational expressions have been multiplied and/or simplified correctly. To check the simplification in Example 9,

$$\frac{x^2 + 3x + 2}{x^2 - 1} = \frac{x + 2}{x - 1},$$

we first enter $y_1 = (x^2 + 3x + 2)/(x^2 - 1)$ and $y_2 = (x + 2)/(x - 1)$. Then, using AUTO mode, we look at a table of values of y_1 and y_2. If the simplification is correct, the values should be the same for all allowable replacements.

$y_1 = (x^2 + 3x + 2)/(x^2 - 1)$,
$y_2 = (x + 2)/(x - 1)$

X	Y1	Y2
−4	.4	.4
−3	.25	.25
−2	0	0
−1	ERROR	−.5
0	−2	−2
1	ERROR	ERROR
2	4	4
X = −4		

The ERROR messages indicate that −1 and 1 are not allowable replacements in the first expression, and 1 is not an allowable replacement in the second. For all other numbers, we see that y_1 and y_2 are the same, so the simplification appears to be correct. Remember, this is only a partial check since we cannot check all possible values.

Exercises: Use the TABLE feature to determine whether each of the following appears to be correct.

1. $\dfrac{8x^2}{24x} = \dfrac{x}{3}$
2. $\dfrac{5x + 15}{10} = \dfrac{x + 3}{2}$
3. $\dfrac{x + 3}{x} = 3$
4. $\dfrac{x^2 + 3x - 4}{x^2 - 16} = \dfrac{x - 1}{x + 4}$
5. $\dfrac{x^2 + 2x - 3}{x^2 - 4} \cdot \dfrac{4x - 8}{x + 3} = \dfrac{4x - 1}{x + 2}$
6. $\dfrac{x^2 - 25}{6} \cdot \dfrac{3}{x + 5} = \dfrac{x - 5}{3}$
7. $\dfrac{x^2 + 6x + 9}{x^2 - 4} \cdot \dfrac{x - 2}{x + 3} = \dfrac{x + 3}{x + 2}$
8. $\dfrac{x^2}{x^2 - 3x} \cdot \dfrac{x^2 - 9}{3} = \dfrac{x(x + 3)}{3}$

12.1 EXERCISE SET

For Extra Help

MathXL | MyMathLab | InterAct Math | Math Tutor Center | Video Lectures on CD Disc 6 | Student's Solutions Manual

a Find all numbers for which the rational expression is not defined.

1. $\dfrac{-3}{2x}$

2. $\dfrac{24}{-8y}$

3. $\dfrac{5}{x - 8}$

4. $\dfrac{y - 4}{y + 6}$

5. $\dfrac{3}{2y + 5}$

6. $\dfrac{x^2 - 9}{4x - 12}$

7. $\dfrac{x^2 + 11}{x^2 - 3x - 28}$

8. $\dfrac{p^2 - 9}{p^2 - 7p + 10}$

9. $\dfrac{m^3 - 2m}{m^2 - 25}$

10. $\dfrac{7 - 3x + x^2}{49 - x^2}$

11. $\dfrac{x - 4}{3}$

12. $\dfrac{x^2 - 25}{14}$

b Multiply. Do not simplify. Note that in each case you are multiplying by 1.

13. $\dfrac{4x}{4x} \cdot \dfrac{3x^2}{5y}$

14. $\dfrac{5x^2}{5x^2} \cdot \dfrac{6y^3}{3z^4}$

15. $\dfrac{2x}{2x} \cdot \dfrac{x - 1}{x + 4}$

16. $\dfrac{2a - 3}{5a + 2} \cdot \dfrac{a}{a}$

17. $\dfrac{3 - x}{4 - x} \cdot \dfrac{-1}{-1}$

18. $\dfrac{x - 5}{5 - x} \cdot \dfrac{-1}{-1}$

19. $\dfrac{y + 6}{y + 6} \cdot \dfrac{y - 7}{y + 2}$

20. $\dfrac{x^2 + 1}{x^3 - 2} \cdot \dfrac{x - 4}{x - 4}$

c Simplify.

21. $\dfrac{8x^3}{32x}$

22. $\dfrac{4x^2}{20x}$

23. $\dfrac{48p^7q^5}{18p^5q^4}$

24. $\dfrac{-76x^8y^3}{-24x^4y^3}$

25. $\dfrac{4x - 12}{4x}$

26. $\dfrac{5a - 40}{5}$

27. $\dfrac{3m^2 + 3m}{6m^2 + 9m}$

28. $\dfrac{4y^2 - 2y}{5y^2 - 5y}$

29. $\dfrac{a^2 - 9}{a^2 + 5a + 6}$

30. $\dfrac{t^2 - 25}{t^2 + t - 20}$

31. $\dfrac{a^2 - 10a + 21}{a^2 - 11a + 28}$

32. $\dfrac{x^2 - 2x - 8}{x^2 - x - 6}$

33. $\dfrac{x^2 - 25}{x^2 - 10x + 25}$

34. $\dfrac{x^2 + 8x + 16}{x^2 - 16}$

35. $\dfrac{a^2 - 1}{a - 1}$

36. $\dfrac{t^2 - 1}{t + 1}$

37. $\dfrac{x^2 + 1}{x + 1}$

38. $\dfrac{m^2 + 9}{m + 3}$

39. $\dfrac{6x^2 - 54}{4x^2 - 36}$

40. $\dfrac{8x^2 - 32}{4x^2 - 16}$

41. $\dfrac{6t + 12}{t^2 - t - 6}$

42. $\dfrac{4x + 32}{x^2 + 9x + 8}$

43. $\dfrac{2t^2 + 6t + 4}{4t^2 - 12t - 16}$

44. $\dfrac{3a^2 - 9a - 12}{6a^2 + 30a + 24}$

45. $\dfrac{t^2 - 4}{(t + 2)^2}$

46. $\dfrac{m^2 - 10m + 25}{m^2 - 25}$

47. $\frac{6 - x}{x - 6}$

48. $\frac{t - 3}{3 - t}$

49. $\frac{a - b}{b - a}$

50. $\frac{y - x}{-x + y}$

51. $\frac{6t - 12}{2 - t}$

52. $\frac{5a - 15}{3 - a}$

53. $\frac{x^2 - 1}{1 - x}$

54. $\frac{a^2 - b^2}{b^2 - a^2}$

d Multiply and simplify.

55. $\frac{4x^3}{3x} \cdot \frac{14}{x}$

56. $\frac{18}{x^3} \cdot \frac{5x^2}{6}$

57. $\frac{3c}{d^2} \cdot \frac{4d}{6c^3}$

58. $\frac{3x^2y}{2} \cdot \frac{4}{xy^3}$

59. $\frac{x^2 - 3x - 10}{x^2 - 4x + 4} \cdot \frac{x - 2}{x - 5}$

60. $\frac{t^2}{t^2 - 4} \cdot \frac{t^2 - 5t + 6}{t^2 - 3t}$

61. $\frac{a^2 - 9}{a^2} \cdot \frac{a^2 - 3a}{a^2 + a - 12}$

62. $\frac{x^2 + 10x - 11}{x^2 - 1} \cdot \frac{x + 1}{x + 11}$

63. $\frac{4a^2}{3a^2 - 12a + 12} \cdot \frac{3a - 6}{2a}$

64. $\frac{5v + 5}{v - 2} \cdot \frac{v^2 - 4v + 4}{v^2 - 1}$

65. $\frac{t^4 - 16}{t^4 - 1} \cdot \frac{t^2 + 1}{t^2 + 4}$

66. $\frac{x^4 - 1}{x^4 - 81} \cdot \frac{x^2 + 9}{x^2 + 1}$

67. $\dfrac{(x+4)^3}{(x+2)^3} \cdot \dfrac{x^2+4x+4}{x^2+8x+16}$

68. $\dfrac{(t-2)^3}{(t-1)^3} \cdot \dfrac{t^2-2t+1}{t^2-4t+4}$

69. $\dfrac{5a^2-180}{10a^2-10} \cdot \dfrac{20a+20}{2a-12}$

70. $\dfrac{2t^2-98}{4t^2-4} \cdot \dfrac{8t+8}{16t-112}$

71. $^{\mathbf{D}}\mathbf{W}$ How is the process of canceling related to the identity property of 1?

72. $^{\mathbf{D}}\mathbf{W}$ Explain how a rational expression can be formed for which -3 and 4 are not allowable replacements.

SKILL MAINTENANCE

Solve.

73. *Consecutive Even Integers.* The product of two consecutive even integers is 360. Find the integers. [11.8a]

74. *Chemistry.* About 5 L of oxygen can be dissolved in 100 L of water at 0°C. This is 1.6 times the amount that can be dissolved in the same volume of water at 20°C. How much oxygen can be dissolved in 100 L at 20°C? [8.6a]

Factor. [11.6a]

75. $x^2 - x - 56$

76. $a^2 - 16a + 64$

77. $x^5 - 2x^4 - 35x^3$

78. $2y^3 - 10y^2 + y - 5$

79. $16 - t^4$

80. $10x^2 + 80x + 70$

81. $x^2 - 9x + 14$

82. $x^2 + x + 7$

83. $16x^2 - 40xy + 25y^2$

84. $a^2 - 9ab + 14b^2$

SYNTHESIS

Simplify.

85. $\dfrac{x^4-16y^4}{(x^2+4y^2)(x-2y)}$

86. $\dfrac{(a-b)^2}{b^2-a^2}$

87. $\dfrac{t^4-1}{t^4-81} \cdot \dfrac{t^2-9}{t^2+1} \cdot \dfrac{(t-9)^2}{(t+1)^2}$

88. $\dfrac{(t+2)^3}{(t+1)^3} \cdot \dfrac{t^2+2t+1}{t^2+4t+4} \cdot \dfrac{t+1}{t+2}$

89. $\dfrac{x^2-y^2}{(x-y)^2} \cdot \dfrac{x^2-2xy+y^2}{x^2-4xy-5y^2}$

90. $\dfrac{x-1}{x^2+1} \cdot \dfrac{x^4-1}{(x-1)^2} \cdot \dfrac{x^2-1}{x^4-2x^2+1}$

91. Select any number x, multiply by 2, add 5, multiply by 5, subtract 25, and divide by 10. What do you get? Explain how this procedure can be used for a number trick.

12.2 DIVISION AND RECIPROCALS

Objectives

a Find the reciprocal of a rational expression.

b Divide rational expressions and simplify.

There is a similarity between what we do with rational expressions and what we do with rational numbers. In fact, after variables have been replaced with rational numbers, a rational expression represents a rational number.

a Finding Reciprocals

Two expressions are reciprocals of each other if their product is 1. The reciprocal of a rational expression is found by interchanging the numerator and the denominator.

EXAMPLES

1. The reciprocal of $\frac{2}{5}$ is $\frac{5}{2}$. $\left(\text{This is because } \frac{2}{5} \cdot \frac{5}{2} = \frac{10}{10} = 1.\right)$
2. The reciprocal of $\frac{2x^2 - 3}{x + 4}$ is $\frac{x + 4}{2x^2 - 3}$.
3. The reciprocal of $x + 2$ is $\frac{1}{x + 2}$. $\left(\text{Think of } x + 2 \text{ as } \frac{x + 2}{1}.\right)$

Do Exercises 1–4.

Find the reciprocal.

1. $\frac{7}{2}$

2. $\frac{x^2 + 5}{2x^3 - 1}$

3. $x - 5$

4. $\frac{1}{x^2 - 3}$

b Division

We divide rational expressions in the same way that we divide fraction notation in arithmetic. For a review, see Section 2.2.

DIVIDING RATIONAL EXPRESSIONS

To divide by a rational expression, multiply by its reciprocal:

$$\frac{A}{B} \div \frac{C}{D} = \frac{A}{B} \cdot \frac{D}{C} = \frac{AD}{BC}.$$

Then factor and, if possible, simplify.

EXAMPLE 4 Divide: $\frac{3}{4} \div \frac{9}{5}$.

$$\frac{3}{4} \div \frac{9}{5} = \frac{3}{4} \cdot \frac{5}{9}$$ Multiplying by the reciprocal of the divisor

$$= \frac{3 \cdot 5}{4 \cdot 9} = \frac{3 \cdot 5}{2 \cdot 2 \cdot 3 \cdot 3}$$ Factoring

$$= \frac{\cancel{3} \cdot 5}{2 \cdot 2 \cdot \cancel{3} \cdot 3}$$ Removing a factor of 1: $\frac{3}{3} = 1$

$$= \frac{5}{12}$$ Simplifying

Do Exercise 5.

5. Divide: $\frac{3}{5} \div \frac{7}{10}$.

Answers on page A-43

6. Divide: $\frac{x}{8} \div \frac{x}{5}$.

EXAMPLE 5 Divide: $\frac{2}{x} \div \frac{3}{x}$.

$$\frac{2}{x} \div \frac{3}{x} = \frac{2}{x} \cdot \frac{x}{3}$$ Multiplying by the reciprocal of the divisor

$$= \frac{2 \cdot x}{x \cdot 3} = \frac{2 \cdot \cancel{x}}{\cancel{x} \cdot 3}$$ Removing a factor of 1: $\frac{x}{x} = 1$

$$= \frac{2}{3}$$

Do Exercise 6.

7. Divide:

$$\frac{x-3}{x+5} \div \frac{x+5}{x-2}.$$

EXAMPLE 6 Divide: $\frac{x+1}{x+2} \div \frac{x-1}{x+3}$.

$$\frac{x+1}{x+2} \div \frac{x-1}{x+3} = \frac{x+1}{x+2} \cdot \frac{x+3}{x-1}$$ Multiplying by the reciprocal of the divisor

$$= \frac{(x+1)(x+3)}{(x+2)(x-1)}$$

We usually do not carry out the multiplication in the numerator or the denominator. It is not wrong to do so, but the factored form is often more useful.

Do Exercise 7.

8. Divide:

$$\frac{a^2+5a}{6} \div \frac{a^2-25}{18a}.$$

EXAMPLE 7 Divide: $\frac{4}{x^2-7x} \div \frac{28x}{x^2-49}$.

$$\frac{4}{x^2-7x} \div \frac{28x}{x^2-49} = \frac{4}{x^2-7x} \cdot \frac{x^2-49}{28x}$$ Multiplying by the reciprocal

$$= \frac{4(x^2-49)}{(x^2-7x)(28x)}$$

$$= \frac{2 \cdot 2 \cdot (x-7)(x+7)}{x(x-7) \cdot 2 \cdot 2 \cdot 7 \cdot x}$$ Factoring the numerator and the denominator

$$= \frac{\cancel{2} \cdot \cancel{2} \cdot \cancel{(x-7)}(x+7)}{x\cancel{(x-7)} \cdot \cancel{2} \cdot \cancel{2} \cdot 7 \cdot x}$$ Removing a factor of 1: $\frac{2 \cdot 2 \cdot (x-7)}{2 \cdot 2 \cdot (x-7)} = 1$

$$= \frac{x+7}{7x^2}$$

Do Exercise 8.

Answers on page A-43

EXAMPLE 8 Divide and simplify: $\dfrac{x+1}{x^2-1} \div \dfrac{x+1}{x^2-2x+1}$.

$$\frac{x+1}{x^2-1} \div \frac{x+1}{x^2-2x+1}$$

$$= \frac{x+1}{x^2-1} \cdot \frac{x^2-2x+1}{x+1}$$ Multiplying by the reciprocal

$$= \frac{(x+1)(x^2-2x+1)}{(x^2-1)(x+1)}$$

$$= \frac{(x+1)(x-1)(x-1)}{(x-1)(x+1)(x+1)}$$ Factoring the numerator and the denominator

$$= \frac{\cancel{(x+1)}\cancel{(x-1)}(x-1)}{\cancel{(x-1)}\cancel{(x+1)}(x+1)}$$ Removing a factor of 1: $\dfrac{(x+1)(x-1)}{(x+1)(x-1)} = 1$

$$= \frac{x-1}{x+1}$$

EXAMPLE 9 Divide and simplify: $\dfrac{x^2-2x-3}{x^2-4} \div \dfrac{x+1}{x+5}$.

$$\frac{x^2-2x-3}{x^2-4} \div \frac{x+1}{x+5}$$

$$= \frac{x^2-2x-3}{x^2-4} \cdot \frac{x+5}{x+1}$$ Multiplying by the reciprocal

$$= \frac{(x^2-2x-3)(x+5)}{(x^2-4)(x+1)}$$

$$= \frac{(x-3)(x+1)(x+5)}{(x-2)(x+2)(x+1)}$$ Factoring the numerator and the denominator

$$= \frac{(x-3)\cancel{(x+1)}(x+5)}{(x-2)(x+2)\cancel{(x+1)}}$$ Removing a factor of 1: $\dfrac{x+1}{x+1} = 1$

$$= \frac{(x-3)(x+5)}{(x-2)(x+2)}.$$ ← You need not carry out the multiplications in the numerator and the denominator.

Do Exercises 9–11.

Divide and simplify.

9. $\dfrac{x-3}{x+5} \div \dfrac{x+2}{x+5}$

10. $\dfrac{x^2-5x+6}{x+5} \div \dfrac{x+2}{x+5}$

11. $\dfrac{y^2-1}{y+1} \div \dfrac{y^2-2y+1}{y+1}$

Answers on page A-43

12.2 EXERCISE SET

For Extra Help

MathXL | MyMathLab | InterAct Math | Math Tutor Center | Video Lectures on CD Disc 6 | Student's Solutions Manual

a Find the reciprocal.

1. $\frac{4}{x}$

2. $\frac{a+3}{a-1}$

3. $x^2 - y^2$

4. $x^2 - 5x + 7$

5. $\frac{1}{a+b}$

6. $\frac{x^2}{x^2-3}$

7. $\frac{x^2+2x-5}{x^2-4x+7}$

8. $\frac{(a-b)(a+b)}{(a+4)(a-5)}$

b Divide and simplify.

9. $\frac{2}{5} \div \frac{4}{3}$

10. $\frac{3}{10} \div \frac{3}{2}$

11. $\frac{2}{x} \div \frac{8}{x}$

12. $\frac{t}{3} \div \frac{t}{15}$

13. $\frac{a}{b^2} \div \frac{a^2}{b^3}$

14. $\frac{x^2}{y} \div \frac{x^3}{y^3}$

15. $\frac{a+2}{a-3} \div \frac{a-1}{a+3}$

16. $\frac{x-8}{x+9} \div \frac{x+2}{x-1}$

17. $\frac{x^2-1}{x} \div \frac{x+1}{x-1}$

18. $\frac{4y-8}{y+2} \div \frac{y-2}{y^2-4}$

19. $\frac{x+1}{6} \div \frac{x+1}{3}$

20. $\frac{a}{a-b} \div \frac{b}{a-b}$

21. $\frac{5x-5}{16} \div \frac{x-1}{6}$

22. $\frac{4y-12}{12} \div \frac{y-3}{3}$

23. $\frac{-6+3x}{5} \div \frac{4x-8}{25}$

24. $\frac{-12+4x}{4} \div \frac{-6+2x}{6}$

25. $\frac{a+2}{a-1} \div \frac{3a+6}{a-5}$

26. $\frac{t-3}{t+2} \div \frac{4t-12}{t+1}$

27. $\frac{x^2-4}{x} \div \frac{x-2}{x+2}$

28. $\frac{x+y}{x-y} \div \frac{x^2+y}{x^2-y^2}$

29. $\frac{x^2-9}{4x+12} \div \frac{x-3}{6}$

30. $\frac{a-b}{2a} \div \frac{a^2-b^2}{8a^3}$

31. $\frac{c^2+3c}{c^2+2c-3} \div \frac{c}{c+1}$

32. $\frac{y+5}{2y} \div \frac{y^2-25}{4y^2}$

33. $\dfrac{2y^2 - 7y + 3}{2y^2 + 3y - 2} \div \dfrac{6y^2 - 5y + 1}{3y^2 + 5y - 2}$

34. $\dfrac{x^2 + x - 20}{x^2 - 7x + 12} \div \dfrac{x^2 + 10x + 25}{x^2 - 6x + 9}$

35. $\dfrac{x^2 - 1}{4x + 4} \div \dfrac{2x^2 - 4x + 2}{8x + 8}$

36. $\dfrac{5t^2 + 5t - 30}{10t + 30} \div \dfrac{2t^2 - 8}{6t^2 + 36t + 54}$

37. **DW** Is the reciprocal of a product the product of the reciprocals? Why or why not?

38. **DW** Explain why 5, -1, and 7 are *not* allowable replacements in the division

$$\frac{x + 3}{x - 5} \div \frac{x - 7}{x + 1}.$$

SKILL MAINTENANCE

Solve.

39. Bonnie is taking an astronomy course. In order to receive an A, she must average at least 90 after four exams. Bonnie scored 96, 98, and 89 on the first three tests. Determine (in terms of an inequality) what scores on the last test will earn her an A. [8.8b]

40. *Triangle Dimensions.* The base of a triangle is 4 in. less than twice the height. The area is 35 in^2. Find the height and the base. [11.8a]

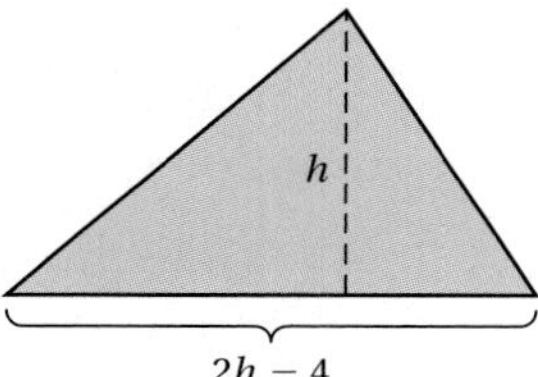

Subtract. [10.4c]

41. $(8x^3 - 3x^2 + 7) - (8x^2 + 3x - 5)$

42. $(3p^2 - 6pq + 7q^2) - (5p^2 - 10pq + 11q^2)$

Simplify. [10.2a, b]

43. $(2x^{-3}y^4)^2$

44. $(5x^6y^{-4})^3$

45. $\left(\dfrac{2x^3}{y^5}\right)^2$

46. $\left(\dfrac{a^{-3}}{b^4}\right)^5$

SYNTHESIS

Simplify.

47. $\dfrac{3a^2 - 5ab - 12b^2}{3ab + 4b^2} \div (3b^2 - ab)$

48. $\dfrac{3x + 3y + 3}{9x} \div \dfrac{x^2 + 2xy + y^2 - 1}{x^4 + x^2}$

49. $\dfrac{a^2b^2 + 3ab^2 + 2b^2}{a^2b^4 + 4b^4} \div (5a^2 + 10a)$

50. The volume of this rectangular solid is $x - 3$. What is its height?

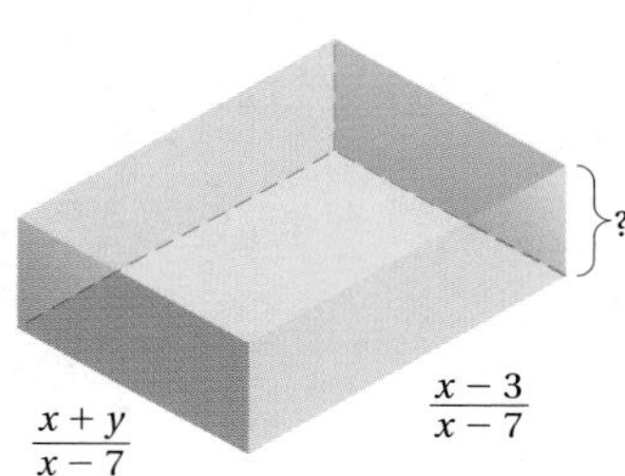

12.3 LEAST COMMON MULTIPLES AND DENOMINATORS

Objectives

a Find the LCM of several numbers by factoring.

b Add fractions, first finding the LCD.

c Find the LCM of algebraic expressions by factoring.

Find the LCM by factoring.

1. 16, 18

2. 6, 12

3. 2, 5

4. 24, 30, 20

Answers on page A-43

a Least Common Multiples

To add when denominators are different, we first find a common denominator. For a review, see Section 2.3. We saw there, for example, that to add $\frac{5}{12}$ and $\frac{7}{30}$, we first look for the **least common multiple, LCM,** of both 12 and 30. That number becomes the **least common denominator, LCD.** To find the LCM of 12 and 30, we factor:

$$12 = 2 \cdot 2 \cdot 3;$$

$$30 = 2 \cdot 3 \cdot 5.$$

The LCM is the number that has 2 as a factor twice, 3 as a factor once, and 5 as a factor once:

$$\text{LCM} = 2 \cdot 2 \cdot 3 \cdot 5 = 60.$$

12 is a factor of the LCM. (first three factors: $2 \cdot 2 \cdot 3$)

30 is a factor of the LCM. (last three factors: $2 \cdot 3 \cdot 5$)

FINDING LCMS

To find the LCM, use each factor the greatest number of times that it appears in any one factorization.

EXAMPLE 1 Find the LCM of 24 and 36.

$$\left.\begin{aligned} 24 &= 2 \cdot 2 \cdot 2 \cdot 3 \\ 36 &= 2 \cdot 2 \cdot 3 \cdot 3 \end{aligned}\right\} \quad \text{LCM} = 2 \cdot 2 \cdot 2 \cdot 3 \cdot 3, \text{ or } 72$$

Do Exercises 1–4.

b Adding Using the LCD

Let's finish adding $\frac{5}{12}$ and $\frac{7}{30}$:

$$\frac{5}{12} + \frac{7}{30} = \frac{5}{2 \cdot 2 \cdot 3} + \frac{7}{2 \cdot 3 \cdot 5}.$$

The least common denominator, LCD, is $2 \cdot 2 \cdot 3 \cdot 5$. To get the LCD in the first denominator, we need a 5. To get the LCD in the second denominator, we need another 2. We get these numbers by multiplying by forms of 1:

$$\frac{5}{12} + \frac{7}{30} = \frac{5}{2 \cdot 2 \cdot 3} \cdot \frac{5}{5} + \frac{7}{2 \cdot 3 \cdot 5} \cdot \frac{2}{2} \quad \text{Multiplying by 1}$$

$$= \frac{25}{2 \cdot 2 \cdot 3 \cdot 5} + \frac{14}{2 \cdot 3 \cdot 5 \cdot 2} \quad \text{Each denominator is now the LCD.}$$

$$= \frac{39}{2 \cdot 2 \cdot 3 \cdot 5} \quad \text{Adding the numerators and keeping the LCD}$$

$$= \frac{\cancel{3} \cdot 13}{2 \cdot 2 \cdot \cancel{3} \cdot 5} \quad \text{Factoring the numerator and removing a factor of 1: } \frac{3}{3} = 1$$

$$= \frac{13}{20}. \quad \text{Simplifying}$$

EXAMPLE 2 Add: $\frac{5}{12} + \frac{11}{18}$.

$$\left.\begin{aligned} 12 &= 2 \cdot 2 \cdot 3 \\ 18 &= 2 \cdot 3 \cdot 3 \end{aligned}\right\} \quad \text{LCD} = 2 \cdot 2 \cdot 3 \cdot 3, \text{ or } 36$$

$$\frac{5}{12} + \frac{11}{18} = \frac{5}{2 \cdot 2 \cdot 3} \cdot \frac{3}{3} + \frac{11}{2 \cdot 3 \cdot 3} \cdot \frac{2}{2} = \frac{15 + 22}{2 \cdot 2 \cdot 3 \cdot 3} = \frac{37}{36}$$

Do Exercises 5–8.

Add, first finding the LCD. Simplify if possible.

5. $\frac{3}{16} + \frac{1}{18}$

6. $\frac{1}{6} + \frac{1}{12}$

7. $\frac{1}{2} + \frac{3}{5}$

8. $\frac{1}{24} + \frac{1}{30} + \frac{3}{20}$

C LCMs of Algebraic Expressions

To find the LCM of two or more algebraic expressions, we factor them. Then we use each factor the greatest number of times that it occurs in any one expression. In Section 12.4, each LCM will become an LCD used to add rational expressions.

EXAMPLE 3 Find the LCM of $12x$, $16y$, and $8xyz$.

$$\left.\begin{aligned} 12x &= 2 \cdot 2 \cdot 3 \cdot x \\ 16y &= 2 \cdot 2 \cdot 2 \cdot 2 \cdot y \\ 8xyz &= 2 \cdot 2 \cdot 2 \cdot x \cdot y \cdot z \end{aligned}\right\} \quad \begin{aligned} \text{LCM} &= 2 \cdot 2 \cdot 2 \cdot 2 \cdot 3 \cdot x \cdot y \cdot z \\ &= 48xyz \end{aligned}$$

EXAMPLE 4 Find the LCM of $x^2 + 5x - 6$ and $x^2 - 1$.

$$\left.\begin{aligned} x^2 + 5x - 6 &= (x + 6)(x - 1) \\ x^2 - 1 &= (x + 1)(x - 1) \end{aligned}\right\} \quad \text{LCM} = (x + 6)(x - 1)(x + 1)$$

EXAMPLE 5 Find the LCM of $x^2 + 4$, $x + 1$, and 5.

These expressions do not share a common factor other than 1, so the LCM is their product:

$$5(x^2 + 4)(x + 1).$$

EXAMPLE 6 Find the LCM of $x^2 - 25$ and $2x - 10$.

$$\left.\begin{aligned} x^2 - 25 &= (x + 5)(x - 5) \\ 2x - 10 &= 2(x - 5) \end{aligned}\right\} \quad \text{LCM} = 2(x + 5)(x - 5)$$

EXAMPLE 7 Find the LCM of $x^2 - 4y^2$, $x^2 - 4xy + 4y^2$, and $x - 2y$.

$$\left.\begin{aligned} x^2 - 4y^2 &= (x - 2y)(x + 2y) \\ x^2 - 4xy + 4y^2 &= (x - 2y)(x - 2y) \\ x - 2y &= x - 2y \end{aligned}\right\} \quad \begin{aligned} \text{LCM} &= (x + 2y)(x - 2y)(x - 2y) \\ &= (x + 2y)(x - 2y)^2 \end{aligned}$$

Do Exercises 9–12.

Find the LCM.

9. $12xy^2$, $15x^3y$

10. $y^2 + 5y + 4$, $y^2 + 2y + 1$

11. $t^2 + 16$, $t - 2$, 7

12. $x^2 + 2x + 1$, $3x^2 - 3x$, $x^2 - 1$

Answers on page A-43

12.3 EXERCISE SET

For Extra Help

MathXL | MyMathLab | InterAct Math | Math Tutor Center | Video Lectures on CD Disc 6 | Student's Solutions Manual

a Find the LCM.

1. 12, 27

2. 10, 15

3. 8, 9

4. 12, 18

5. 6, 9, 21

6. 8, 36, 40

7. 24, 36, 40

8. 4, 5, 20

9. 10, 100, 500

10. 28, 42, 60

b Add, first finding the LCD. Simplify if possible.

11. $\frac{7}{24} + \frac{11}{18}$

12. $\frac{7}{60} + \frac{2}{25}$

13. $\frac{1}{6} + \frac{3}{40}$

14. $\frac{5}{24} + \frac{3}{20}$

15. $\frac{1}{20} + \frac{1}{30} + \frac{2}{45}$

16. $\frac{2}{15} + \frac{5}{9} + \frac{3}{20}$

c Find the LCM.

17. $6x^2,\ 12x^3$

18. $2a^2b,\ 8ab^3$

19. $2x^2,\ 6xy,\ 18y^2$

20. $p^3q,\ p^2q,\ pq^2$

21. $2(y - 3),\ 6(y - 3)$

22. $5(m + 2),\ 15(m + 2)$

23. $t,\ t + 2,\ t - 2$

24. $y,\ y - 5,\ y + 5$

25. $x^2 - 4,\ x^2 + 5x + 6$

26. $x^2 - 4,\ x^2 - x - 2$

27. $t^3 + 4t^2 + 4t,\ t^2 - 4t$

28. $m^4 - m^2,\ m^3 - m^2$

29. $a + 1,\ (a - 1)^2,\ a^2 - 1$

30. $a^2 - 2ab + b^2,\ a^2 - b^2,\ 3a + 3b$

31. $m^2 - 5m + 6,\ m^2 - 4m + 4$

32. $2x^2 + 5x + 2,\ 2x^2 - x - 1$

33. $2 + 3x,\ 4 - 9x^2,\ 2 - 3x$

34. $9 - 4x^2,\ 3 + 2x,\ 3 - 2x$

35. $10v^2 + 30v,\ 5v^2 + 35v + 60$

36. $12a^2 + 24a,\ 4a^2 + 20a + 24$

37. $9x^3 - 9x^2 - 18x,\ 6x^5 - 24x^4 + 24x^3$

38. $x^5 - 4x^3,\ x^3 + 4x^2 + 4x$

39. $x^5 + 4x^4 + 4x^3,\ 3x^2 - 12,\ 2x + 4$

40. $x^5 + 2x^4 + x^3,\ 2x^3 - 2x,\ 5x - 5$

41. D_W If the LCM of a binomial and a trinomial is the trinomial, what relationship exists between the two expressions?

42. D_W Explain why the product of two numbers is not always their least common multiple.

SKILL MAINTENANCE

Factor. [11.6a]

43. $x^2 - 6x + 9$

44. $6x^2 + 4x$

45. $x^2 - 9$

46. $x^2 + 4x - 21$

47. $x^2 + 6x + 9$

48. $x^2 - 4x - 21$

Complete the table finding the LCM, the GCF, and the product of each pair of expressions. [10.5a], [11.1a], [12.3a]

	EXPRESSIONS	LCM	GCF	PRODUCT
	$12x^3,\ 8x^2$	$24x^3$	$4x^2$	$96x^5$
49.	$40x^3,\ 24x^4$			
50.	$16x^5,\ 48x^6$			
51.	$20x^2,\ 10x$			
52.	$12ab,\ 16ab^3$			
53.	$10x^2,\ 24x^3$			
54.	$a^5,\ a^{15}$			

SYNTHESIS

55. *Running.* Pedro and Maria leave the starting point of a fitness loop at the same time. Pedro jogs a lap in 6 min and Maria jogs one in 8 min. Assuming they continue to run at the same pace, when will they next meet at the starting place?

56. Look for a pattern in Exercises 49–54. See if you can discover a formula connecting LCM and GCF.

12.4 ADDING RATIONAL EXPRESSIONS

Objective

a Add rational expressions.

Add.

1. $\dfrac{5}{9} + \dfrac{2}{9}$

2. $\dfrac{3}{x-2} + \dfrac{x}{x-2}$

3. $\dfrac{4x+5}{x-1} + \dfrac{2x-1}{x-1}$

Answers on page A-43

a Adding Rational Expressions

We add rational expressions as we do rational numbers.

ADDING RATIONAL EXPRESSIONS WITH LIKE DENOMINATORS

To add when the denominators are the same, add the numerators and keep the same denominator. Then simplify if possible.

EXAMPLES Add.

1. $\dfrac{x}{x+1} + \dfrac{2}{x+1} = \dfrac{x+2}{x+1}$

2. $\dfrac{2x^2+3x-7}{2x+1} + \dfrac{x^2+x-8}{2x+1} = \dfrac{(2x^2+3x-7)+(x^2+x-8)}{2x+1}$

$= \dfrac{3x^2+4x-15}{2x+1}$

3. $\dfrac{x-5}{x^2-9} + \dfrac{2}{x^2-9} = \dfrac{(x-5)+2}{x^2-9} = \dfrac{x-3}{x^2-9}$

$= \dfrac{x-3}{(x-3)(x+3)}$ Factoring

$= \dfrac{\cancel{x-3}}{\cancel{(x-3)}(x+3)}$ Removing a factor of 1: $\dfrac{x-3}{x-3} = 1$

$= \dfrac{1}{x+3}$ Simplifying

As in Example 3, simplifying should be done if possible after adding.

Do Exercises 1–3.

When denominators are different, we find the least common denominator, LCD. The procedure we use is as follows.

ADDING RATIONAL EXPRESSIONS WITH DIFFERENT DENOMINATORS

To add rational expressions with different denominators:

1. Find the LCM of the denominators. This is the least common denominator (LCD).
2. For each rational expression, find an equivalent expression with the LCD. To do so, multiply by 1 using an expression for 1 made up of factors of the LCD that are missing from the original denominator.
3. Add the numerators. Write the sum over the LCD.
4. Simplify if possible.

EXAMPLE 4 Add: $\frac{5x^2}{8} + \frac{7x}{12}$.

First, we find the LCD:

$$\left.\begin{aligned} 8 &= 2 \cdot 2 \cdot 2 \\ 12 &= 2 \cdot 2 \cdot 3 \end{aligned}\right\} \quad \text{LCD} = 2 \cdot 2 \cdot 2 \cdot 3, \text{ or } 24.$$

Compare the factorization $8 = 2 \cdot 2 \cdot 2$ with the factorization of the LCD, $24 = 2 \cdot 2 \cdot 2 \cdot 3$. The factor of 24 that is missing from 8 is 3. Compare $12 = 2 \cdot 2 \cdot 3$ and $24 = 2 \cdot 2 \cdot 2 \cdot 3$. The factor of 24 that is missing from 12 is 2.

We multiply each term by a symbol for 1 to get the LCD in each expression, and then add and, if possible, simplify:

$$\frac{5x^2}{8} + \frac{7x}{12} = \frac{5x^2}{2 \cdot 2 \cdot 2} + \frac{7x}{2 \cdot 2 \cdot 3}$$

$$= \frac{5x^2}{2 \cdot 2 \cdot 2} \cdot \frac{3}{3} + \frac{7x}{2 \cdot 2 \cdot 3} \cdot \frac{2}{2} \qquad \text{Multiplying by 1 to get the same denominators}$$

$$= \frac{15x^2}{24} + \frac{14x}{24} = \frac{15x^2 + 14x}{24}.$$

Add.

4. $\frac{3x}{16} + \frac{5x^2}{24}$

5. $\frac{3}{16x} + \frac{5}{24x^2}$

EXAMPLE 5 Add: $\frac{3}{8x} + \frac{5}{12x^2}$.

First, we find the LCD:

$$\left.\begin{aligned} 8x &= 2 \cdot 2 \cdot 2 \cdot x \\ 12x^2 &= 2 \cdot 2 \cdot 3 \cdot x \cdot x \end{aligned}\right\} \quad \text{LCD} = 2 \cdot 2 \cdot 2 \cdot 3 \cdot x \cdot x, \text{ or } 24x^2.$$

The factors of the LCD missing from $8x$ are 3 and x. The factor of the LCD missing from $12x^2$ is 2. We multiply each term by 1 to get the LCD in each expression, and then add and, if possible, simplify:

$$\frac{3}{8x} + \frac{5}{12x^2} = \frac{3}{8x} \cdot \frac{3 \cdot x}{3 \cdot x} + \frac{5}{12x^2} \cdot \frac{2}{2}$$

$$= \frac{9x}{24x^2} + \frac{10}{24x^2} = \frac{9x + 10}{24x^2}.$$

Do Exercises 4 and 5.

6. Add:

$$\frac{3}{x^3 - x} + \frac{4}{x^2 + 2x + 1}.$$

EXAMPLE 6 Add: $\frac{2a}{a^2 - 1} + \frac{1}{a^2 + a}$.

First, we find the LCD:

$$\left.\begin{aligned} a^2 - 1 &= (a - 1)(a + 1) \\ a^2 + a &= a(a + 1) \end{aligned}\right\} \quad \text{LCD} = a(a - 1)(a + 1).$$

We multiply each term by 1 to get the LCD in each expression, and then add and simplify:

$$\frac{2a}{(a - 1)(a + 1)} \cdot \frac{a}{a} + \frac{1}{a(a + 1)} \cdot \frac{a - 1}{a - 1}$$

$$= \frac{2a^2}{a(a - 1)(a + 1)} + \frac{a - 1}{a(a - 1)(a + 1)}$$

$$= \frac{2a^2 + a - 1}{a(a - 1)(a + 1)}$$

$$= \frac{(a + 1)(2a - 1)}{a(a - 1)(a + 1)}. \qquad \text{Factoring the numerator in order to simplify}$$

Answers on page A-43

7. Add:

$$\frac{x-2}{x+3}+\frac{x+7}{x+8}.$$

8. Add:

$$\frac{5}{x^2+17x+16}+\frac{3}{x^2+9x+8}.$$

Answers on page A-43

Then

$$=\frac{(a+1)(2a-1)}{a(a-1)(a+1)} \quad \text{Removing a factor of 1: } \frac{a+1}{a+1}=1$$

$$=\frac{2a-1}{a(a-1)}.$$

Do Exercise 6 on the preceding page.

EXAMPLE 7 Add: $\dfrac{x+4}{x-2}+\dfrac{x-7}{x+5}$.

First, we find the LCD. It is just the product of the denominators:

$$\text{LCD} = (x-2)(x+5).$$

We multiply by 1 to get the LCD in each expression, and then add and simplify:

$$\frac{x+4}{x-2}\cdot\frac{x+5}{x+5}+\frac{x-7}{x+5}\cdot\frac{x-2}{x-2}$$

$$=\frac{(x+4)(x+5)}{(x-2)(x+5)}+\frac{(x-7)(x-2)}{(x-2)(x+5)}$$

$$=\frac{x^2+9x+20}{(x-2)(x+5)}+\frac{x^2-9x+14}{(x-2)(x+5)}$$

$$=\frac{x^2+9x+20+x^2-9x+14}{(x-2)(x+5)}=\frac{2x^2+34}{(x-2)(x+5)}.$$

Do Exercise 7.

EXAMPLE 8 Add: $\dfrac{x}{x^2+11x+30}+\dfrac{-5}{x^2+9x+20}$.

$$\frac{x}{x^2+11x+30}+\frac{-5}{x^2+9x+20}$$

$$=\frac{x}{(x+5)(x+6)}+\frac{-5}{(x+5)(x+4)}$$

Factoring the denominators in order to find the LCD. The LCD is $(x+4)(x+5)(x+6)$.

$$=\frac{x}{(x+5)(x+6)}\cdot\frac{x+4}{x+4}+\frac{-5}{(x+5)(x+4)}\cdot\frac{x+6}{x+6}$$

Multiplying by 1

$$=\frac{x(x+4)+(-5)(x+6)}{(x+4)(x+5)(x+6)}=\frac{x^2+4x-5x-30}{(x+4)(x+5)(x+6)}$$

$$=\frac{x^2-x-30}{(x+4)(x+5)(x+6)}$$

$$=\frac{(x-6)(x+5)}{(x+4)(x+5)(x+6)}$$

$$=\frac{(x-6)}{(x+4)(x+6)}$$

Always simplify at the end if possible: $\dfrac{x+5}{x+5}=1$.

Do Exercise 8.

DENOMINATORS THAT ARE OPPOSITES

When one denominator is the opposite of the other, we can first multiply either expression by 1 using $-1/-1$.

EXAMPLES

9. $\frac{x}{2} + \frac{3}{-2} = \frac{x}{2} + \frac{3}{-2} \cdot \frac{-1}{-1}$ Multiplying by 1 using $\frac{-1}{-1}$

$= \frac{x}{2} + \frac{-3}{2}$ The denominators are now the same.

$= \frac{x + (-3)}{2} = \frac{x - 3}{2}$

10. $\frac{3x + 4}{x - 2} + \frac{x - 7}{2 - x} = \frac{3x + 4}{x - 2} + \frac{x - 7}{2 - x} \cdot \frac{-1}{-1}$

We could have chosen to multiply this expression by $-1/-1$. We multiply only one expression, *not* both.

$= \frac{3x + 4}{x - 2} + \frac{-x + 7}{x - 2}$ *Note:* $(2 - x)(-1) = -2 + x = x - 2$.

$= \frac{(3x + 4) + (-x + 7)}{x - 2} = \frac{2x + 11}{x - 2}$

Do Exercises 9 and 10.

FACTORS THAT ARE OPPOSITES

Suppose that when we factor to find the LCD, we find factors that are opposites. The easiest way to handle this is to first go back and multiply by $-1/-1$ appropriately to change factors so that they are not opposites.

EXAMPLE 11 Add: $\frac{x}{x^2 - 25} + \frac{3}{10 - 2x}$.

First, we factor to find the LCD:

$x^2 - 25 = (x - 5)(x + 5);$

$10 - 2x = 2(5 - x).$

We note that there is an $x - 5$ as one factor of $x^2 - 25$ and a $5 - x$ as one factor of $10 - 2x$. If the denominator of the second expression were $2x - 10$, this situation would not occur. To rewrite the second expression with a denominator of $2x - 10$, we multiply by 1 using $-1/-1$, and then continue as before:

$$\frac{x}{x^2 - 25} + \frac{3}{10 - 2x} = \frac{x}{(x - 5)(x + 5)} + \frac{3}{10 - 2x} \cdot \frac{-1}{-1}$$

$$= \frac{x}{(x - 5)(x + 5)} + \frac{-3}{2x - 10}$$

$$= \frac{x}{(x - 5)(x + 5)} + \frac{-3}{2(x - 5)} \qquad \text{LCD} = 2(x - 5)(x + 5)$$

$$= \frac{x}{(x - 5)(x + 5)} \cdot \frac{2}{2} + \frac{-3}{2(x - 5)} \cdot \frac{x + 5}{x + 5}$$

$$= \frac{2x}{2(x - 5)(x + 5)} + \frac{-3(x + 5)}{2(x - 5)(x + 5)}$$

$$= \frac{2x - 3(x + 5)}{2(x - 5)(x + 5)} = \frac{2x - 3x - 15}{2(x - 5)(x + 5)}$$

$$= \frac{-x - 15}{2(x - 5)(x + 5)}. \qquad \text{Collecting like terms}$$

Do Exercise 11.

Add.

9. $\frac{x}{4} + \frac{5}{-4}$

10. $\frac{2x + 1}{x - 3} + \frac{x + 2}{3 - x}$

11. Add:
$\frac{x + 3}{x^2 - 16} + \frac{5}{12 - 3x}$.

Answers on page A-43

12.4 EXERCISE SET

For Extra Help

MathXL | MyMathLab | InterAct Math | Math Tutor Center | Video Lectures on CD Disc 6 | Student's Solutions Manual

a Add. Simplify if possible.

1. $\frac{5}{8}+\frac{3}{8}$

2. $\frac{3}{16}+\frac{5}{16}$

3. $\frac{1}{3+x}+\frac{5}{3+x}$

4. $\frac{4x+6}{2x-1}+\frac{5-8x}{-1+2x}$

5. $\frac{x^2+7x}{x^2-5x}+\frac{x^2-4x}{x^2-5x}$

6. $\frac{4}{x+y}+\frac{9}{y+x}$

7. $\frac{2}{x}+\frac{5}{x^2}$

8. $\frac{3}{y^2}+\frac{6}{y}$

9. $\frac{5}{6r}+\frac{7}{8r}$

10. $\frac{13}{18x}+\frac{7}{24x}$

11. $\frac{4}{xy^2}+\frac{6}{x^2y}$

12. $\frac{8}{ab^3}+\frac{3}{a^2b}$

13. $\frac{2}{9t^3}+\frac{1}{6t^2}$

14. $\frac{5}{c^2d^3}+\frac{-4}{7cd^2}$

15. $\frac{x+y}{xy^2}+\frac{3x+y}{x^2y}$

16. $\frac{2c-d}{c^2d}+\frac{c+d}{cd^2}$

17. $\frac{3}{x-2}+\frac{3}{x+2}$

18. $\frac{2}{y+1}+\frac{2}{y-1}$

19. $\frac{3}{x+1}+\frac{2}{3x}$

20. $\frac{4}{5y}+\frac{7}{y-2}$

21. $\frac{2x}{x^2-16}+\frac{x}{x-4}$

22. $\dfrac{4x}{x^2 - 25} + \dfrac{x}{x + 5}$

23. $\dfrac{5}{z + 4} + \dfrac{3}{3z + 12}$

24. $\dfrac{t}{t - 3} + \dfrac{5}{4t - 12}$

25. $\dfrac{3}{x - 1} + \dfrac{2}{(x - 1)^2}$

26. $\dfrac{8}{(y + 3)^2} + \dfrac{5}{y + 3}$

27. $\dfrac{4a}{5a - 10} + \dfrac{3a}{10a - 20}$

28. $\dfrac{9x}{6x - 30} + \dfrac{3x}{4x - 20}$

29. $\dfrac{x + 4}{x} + \dfrac{x}{x + 4}$

30. $\dfrac{a}{a - 3} + \dfrac{a - 3}{a}$

31. $\dfrac{4}{a^2 - a - 2} + \dfrac{3}{a^2 + 4a + 3}$

32. $\dfrac{a}{a^2 - 2a + 1} + \dfrac{1}{a^2 - 5a + 4}$

33. $\dfrac{x + 3}{x - 5} + \dfrac{x - 5}{x + 3}$

34. $\dfrac{3x}{2y - 3} + \dfrac{2x}{3y - 2}$

35. $\dfrac{a}{a^2 - 1} + \dfrac{2a}{a^2 - a}$

36. $\dfrac{3x + 2}{3x + 6} + \dfrac{x - 2}{x^2 - 4}$

37. $\dfrac{7}{8} + \dfrac{5}{-8}$

38. $\dfrac{5}{-3} + \dfrac{11}{3}$

39. $\dfrac{3}{t} + \dfrac{4}{-t}$

40. $\dfrac{5}{-a} + \dfrac{8}{a}$

41. $\dfrac{2x + 7}{x - 6} + \dfrac{3x}{6 - x}$

42. $\dfrac{2x - 7}{5x - 8} + \dfrac{6 + 10x}{8 - 5x}$

43. $\dfrac{y^2}{y-3} + \dfrac{9}{3-y}$

44. $\dfrac{t^2}{t-2} + \dfrac{4}{2-t}$

45. $\dfrac{b-7}{b^2-16} + \dfrac{7-b}{16-b^2}$

46. $\dfrac{a-3}{a^2-25} + \dfrac{a-3}{25-a^2}$

47. $\dfrac{a^2}{a-b} + \dfrac{b^2}{b-a}$

48. $\dfrac{x^2}{x-7} + \dfrac{49}{7-x}$

49. $\dfrac{x+3}{x-5} + \dfrac{2x-1}{5-x} + \dfrac{2(3x-1)}{x-5}$

50. $\dfrac{3(x-2)}{2x-3} + \dfrac{5(2x+1)}{2x-3} + \dfrac{3(x+1)}{3-2x}$

51. $\dfrac{2(4x+1)}{5x-7} + \dfrac{3(x-2)}{7-5x} + \dfrac{-10x-1}{5x-7}$

52. $\dfrac{5(x-2)}{3x-4} + \dfrac{2(x-3)}{4-3x} + \dfrac{3(5x+1)}{4-3x}$

53. $\dfrac{x+1}{(x+3)(x-3)} + \dfrac{4(x-3)}{(x-3)(x+3)} + \dfrac{(x-1)(x-3)}{(3-x)(x+3)}$

54. $\dfrac{2(x+5)}{(2x-3)(x-1)} + \dfrac{3x+4}{(2x-3)(1-x)} + \dfrac{x-5}{(3-2x)(x-1)}$

55. $\dfrac{6}{x-y} + \dfrac{4x}{y^2-x^2}$

56. $\dfrac{a-2}{3-a} + \dfrac{4-a^2}{a^2-9}$

57. $\dfrac{4-a}{25-a^2} + \dfrac{a+1}{a-5}$

58. $\dfrac{x+2}{x-7} + \dfrac{3-x}{49-x^2}$

59. $\dfrac{2}{t^2+t-6} + \dfrac{3}{t^2-9}$

60. $\dfrac{10}{a^2-a-6} + \dfrac{3a}{a^2+4a+4}$

61. **D$_W$** Explain why the expressions

$$\frac{1}{3-x} \text{ and } \frac{1}{x-3}$$

are opposites.

62. **D$_W$** A student insists on finding a common denominator by always multiplying the denominators of the expressions being added. How could this approach be improved?

SKILL MAINTENANCE

Subtract. [10.4c]

63. $(x^2 + x) - (x + 1)$

64. $(4y^3 - 5y^2 + 7y - 24) - (-9y^3 + 9y^2 - 5y + 49)$

Simplify. [10.2a, b]

65. $(2x^4y^3)^{-3}$

66. $\left(\dfrac{x^3}{5y}\right)^2$

67. $\left(\dfrac{x^{-4}}{y^7}\right)^3$

68. $(5x^{-2}y^{-3})^2$

Graph.

69. $y = \dfrac{1}{2}x - 5$ [9.1d], [9.2a]

70. $2y + x + 10 = 0$ [9.1d], [9.2a]

71. $y = 3$ [9.2b]

72. $x = -5$ [9.2b]

Solve.

73. $3x - 7 = 5x + 9$ [8.3b]

74. $2a + 8 = 13 - 4a$ [8.3b]

75. $x^2 - 8x + 15 = 0$ [11.7b]

76. $x^2 - 7x = 18$ [11.7b]

SYNTHESIS

Find the perimeter and the area of the figure.

77.

$\dfrac{y+4}{3}$

$\dfrac{y-2}{5}$

78.

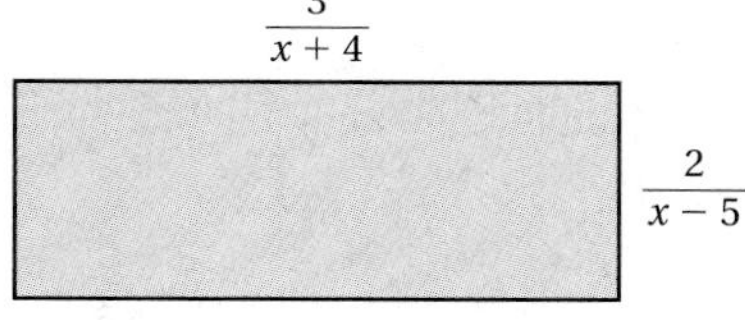

$\dfrac{3}{x+4}$

$\dfrac{2}{x-5}$

Add. Simplify if possible.

79. $\dfrac{5}{z+2} + \dfrac{4z}{z^2-4} + 2$

80. $\dfrac{-2}{y^2-9} + \dfrac{4y}{(y-3)^2} + \dfrac{6}{3-y}$

81. $\dfrac{3z^2}{z^4-4} + \dfrac{5z^2-3}{2z^4+z^2-6}$

82. Find an expression equivalent to

$$\frac{a-3b}{a-b}$$

that is a sum of two rational expressions. Answers may vary.

83.–86. Use the TABLE feature to check the additions in Exercises 29–32.

12.5 SUBTRACTING RATIONAL EXPRESSIONS

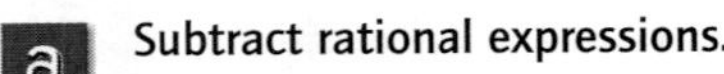

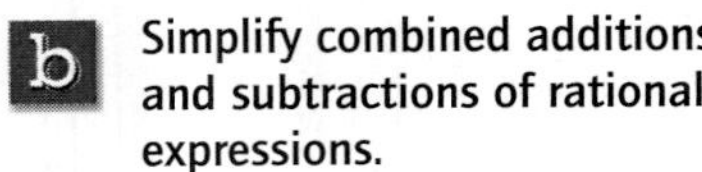

Objectives

a Subtract rational expressions.

b Simplify combined additions and subtractions of rational expressions.

Subtract.

1. $\dfrac{7}{11} - \dfrac{3}{11}$

2. $\dfrac{7}{y} - \dfrac{2}{y}$

3. $\dfrac{2x^2 + 3x - 7}{2x + 1} - \dfrac{x^2 + x - 8}{2x + 1}$

Answers on page A-44

a Subtracting Rational Expressions

We subtract rational expressions as we do rational numbers.

SUBTRACTING RATIONAL EXPRESSIONS WITH LIKE DENOMINATORS

To subtract when the denominators are the same, subtract the numerators and keep the same denominator. Then simplify if possible.

EXAMPLE 1 Subtract: $\dfrac{8}{x} - \dfrac{3}{x}$.

$$\frac{8}{x} - \frac{3}{x} = \frac{8 - 3}{x} = \frac{5}{x}$$

EXAMPLE 2 Subtract: $\dfrac{3x}{x + 2} - \dfrac{x - 2}{x + 2}$.

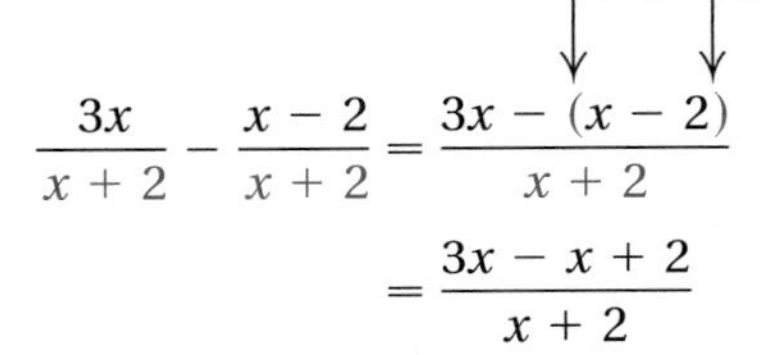

$$\frac{3x}{x + 2} - \frac{x - 2}{x + 2} = \frac{3x - (x - 2)}{x + 2}$$

Caution!

The parentheses are important to make sure that you subtract the entire numerator.

$$= \frac{3x - x + 2}{x + 2} \qquad \text{Removing parentheses}$$

$$= \frac{2x + 2}{x + 2}$$

Do Exercises 1–3.

To subtract rational expressions with different denominators, we use a procedure similar to what we used for addition, except that we subtract numerators and write the difference over the LCD.

SUBTRACTING RATIONAL EXPRESSIONS WITH DIFFERENT DENOMINATORS

To subtract rational expressions with different denominators:

1. Find the LCM of the denominators. This is the least common denominator (LCD).
2. For each rational expression, find an equivalent expression with the LCD. To do so, multiply by 1 using a symbol for 1 made up of factors of the LCD that are missing from the original denominator.
3. Subtract the numerators. Write the difference over the LCD.
4. Simplify if possible.

EXAMPLE 3 Subtract: $\dfrac{x+2}{x-4} - \dfrac{x+1}{x+4}$.

The LCD $= (x-4)(x+4)$.

$$\frac{x+2}{x-4}\cdot\frac{x+4}{x+4} - \frac{x+1}{x+4}\cdot\frac{x-4}{x-4} \quad \text{Multiplying by 1}$$

$$= \frac{(x+2)(x+4)}{(x-4)(x+4)} - \frac{(x+1)(x-4)}{(x-4)(x+4)}$$

$$= \frac{x^2+6x+8}{(x-4)(x+4)} - \frac{x^2-3x-4}{(x-4)(x+4)}$$

Subtracting this numerator. Don't forget the parentheses.

$$= \frac{x^2+6x+8-(x^2-3x-4)}{(x-4)(x+4)}$$

$$= \frac{x^2+6x+8-x^2+3x+4}{(x-4)(x+4)} \quad \text{Removing parentheses}$$

$$= \frac{9x+12}{(x-4)(x+4)}$$

Do Exercise 4.

4. Subtract:

$$\frac{x-2}{3x} - \frac{2x-1}{5x}.$$

EXAMPLE 4 Subtract: $\dfrac{x}{x^2+5x+6} - \dfrac{2}{x^2+3x+2}$.

$$\frac{x}{x^2+5x+6} - \frac{2}{x^2+3x+2}$$

$$= \frac{x}{(x+2)(x+3)} - \frac{2}{(x+2)(x+1)} \quad \text{LCD} = (x+1)(x+2)(x+3)$$

$$= \frac{x}{(x+2)(x+3)}\cdot\frac{x+1}{x+1} - \frac{2}{(x+2)(x+1)}\cdot\frac{x+3}{x+3}$$

$$= \frac{x^2+x}{(x+1)(x+2)(x+3)} - \frac{2x+6}{(x+1)(x+2)(x+3)}$$

Subtracting this numerator. Don't forget the parentheses.

$$= \frac{x^2+x-(2x+6)}{(x+1)(x+2)(x+3)}$$

$$= \frac{x^2+x-2x-6}{(x+1)(x+2)(x+3)}$$

$$= \frac{x^2-x-6}{(x+1)(x+2)(x+3)}$$

$$= \frac{(x+2)(x-3)}{(x+1)(x+2)(x+3)}$$

$$= \frac{\cancel{(x+2)}(x-3)}{(x+1)\cancel{(x+2)}(x+3)} \quad \text{Simplifying by removing a factor of 1: } \frac{x+2}{x+2} = 1$$

$$= \frac{x-3}{(x+1)(x+3)}.$$

Do Exercise 5.

5. Subtract:

$$\frac{x}{x^2+15x+56} - \frac{6}{x^2+13x+42}.$$

DENOMINATORS THAT ARE OPPOSITES

When one denominator is the opposite of the other, we can first multiply one expression by $-1/-1$ to obtain a common denominator.

Answers on page A-44

Subtract.

6. $\frac{x}{3} - \frac{2x - 1}{-3}$

7. $\frac{3x}{x - 2} - \frac{x - 3}{2 - x}$

Answers on page A-44

EXAMPLE 5 Subtract: $\frac{x}{5} - \frac{3x - 4}{-5}$.

$$\frac{x}{5} - \frac{3x - 4}{-5} = \frac{x}{5} - \frac{3x - 4}{-5} \cdot \frac{-1}{-1}$$ Multiplying by 1 using $\frac{-1}{-1}$

This is equal to 1 (not -1).

$$= \frac{x}{5} - \frac{(3x - 4)(-1)}{(-5)(-1)}$$

$$= \frac{x}{5} - \frac{4 - 3x}{5}$$

$$= \frac{x - (4 - 3x)}{5}$$ Remember the parentheses!

$$= \frac{x - 4 + 3x}{5} = \frac{4x - 4}{5}$$

EXAMPLE 6 Subtract: $\frac{5y}{y - 5} - \frac{2y - 3}{5 - y}$.

$$\frac{5y}{y - 5} - \frac{2y - 3}{5 - y} = \frac{5y}{y - 5} - \frac{2y - 3}{5 - y} \cdot \frac{-1}{-1}$$

$$= \frac{5y}{y - 5} - \frac{(2y - 3)(-1)}{(5 - y)(-1)}$$

$$= \frac{5y}{y - 5} - \frac{3 - 2y}{y - 5}$$

$$= \frac{5y - (3 - 2y)}{y - 5}$$ Remember the parentheses!

$$= \frac{5y - 3 + 2y}{y - 5} = \frac{7y - 3}{y - 5}$$

Do Exercises 6 and 7.

FACTORS THAT ARE OPPOSITES

Suppose that when we factor to find the LCD, we find factors that are opposites. Then we multiply by $-1/-1$ appropriately to change factors so that they are not opposites.

EXAMPLE 7 Subtract: $\frac{p}{64 - p^2} - \frac{5}{p - 8}$.

Factoring $64 - p^2$, we get $(8 - p)(8 + p)$. Note that the factors $8 - p$ in the first denominator and $p - 8$ in the second denominator are opposites. We multiply the first expression by $-1/-1$ to avoid this situation. Then we proceed as before.

$$\frac{p}{64 - p^2} - \frac{5}{p - 8} = \frac{p}{64 - p^2} \cdot \frac{-1}{-1} - \frac{5}{p - 8}$$

$$= \frac{-p}{p^2 - 64} - \frac{5}{p - 8}$$

$$= \frac{-p}{(p - 8)(p + 8)} - \frac{5}{p - 8}$$ LCD $= (p - 8)(p + 8)$

$$= \frac{-p}{(p - 8)(p + 8)} - \frac{5}{p - 8} \cdot \frac{p + 8}{p + 8}$$

Multiplying, we have

$$\frac{-p}{(p-8)(p+8)} - \frac{5p+40}{(p-8)(p+8)}$$

Subtracting this numerator. Don't forget the parentheses.

$$= \frac{-p-(5p+40)}{(p-8)(p+8)}$$

$$= \frac{-p-5p-40}{(p-8)(p+8)} = \frac{-6p-40}{(p-8)(p+8)}.$$

Do Exercise 8.

b Combined Additions and Subtractions

Now let's look at some combined additions and subtractions.

EXAMPLE 8 Perform the indicated operations and simplify:

$$\frac{x+9}{x^2-4} + \frac{5-x}{4-x^2} - \frac{2+x}{x^2-4}.$$

$$\frac{x+9}{x^2-4} + \frac{5-x}{4-x^2} - \frac{2+x}{x^2-4}$$

$$= \frac{x+9}{x^2-4} + \frac{5-x}{4-x^2} \cdot \frac{-1}{-1} - \frac{2+x}{x^2-4}$$

$$= \frac{x+9}{x^2-4} + \frac{x-5}{x^2-4} - \frac{2+x}{x^2-4} = \frac{(x+9)+(x-5)-(2+x)}{x^2-4}$$

$$= \frac{x+9+x-5-2-x}{x^2-4} = \frac{x+2}{x^2-4}$$

$$= \frac{\cancel{(x+2)} \cdot 1}{\cancel{(x+2)}(x-2)} = \frac{1}{x-2}.$$

Do Exercise 9.

EXAMPLE 9 Perform the indicated operations and simplify:

$$\frac{1}{x} - \frac{1}{x^2} + \frac{2}{x+1}.$$

The LCD $= x \cdot x(x+1)$, or $x^2(x+1)$.

$$\frac{1}{x} \cdot \frac{x(x+1)}{x(x+1)} - \frac{1}{x^2} \cdot \frac{(x+1)}{(x+1)} + \frac{2}{x+1} \cdot \frac{x^2}{x^2}$$

$$= \frac{x(x+1)}{x^2(x+1)} - \frac{x+1}{x^2(x+1)} + \frac{2x^2}{x^2(x+1)}$$

Subtracting this numerator. Don't forget the parentheses.

$$= \frac{x(x+1)-(x+1)+2x^2}{x^2(x+1)}$$

$$= \frac{x^2+x-x-1+2x^2}{x^2(x+1)} \qquad \text{Removing parentheses}$$

$$= \frac{3x^2-1}{x^2(x+1)}$$

Do Exercise 10.

8. Subtract:

$$\frac{y}{16-y^2} - \frac{7}{y-4}.$$

9. Perform the indicated operations and simplify:

$$\frac{x+2}{x^2-9} - \frac{x-7}{9-x^2} + \frac{-8-x}{x^2-9}.$$

10. Perform the indicated operations and simplify:

$$\frac{1}{x} - \frac{5}{3x} + \frac{2x}{x+1}.$$

Answers on page A-44

12.5 EXERCISE SET

For Extra Help

MathXL | MyMathLab | InterAct Math | Math Tutor Center | Video Lectures on CD Disc 6 | Student's Solutions Manual

a Subtract. Simplify if possible.

1. $\frac{7}{x} - \frac{3}{x}$

2. $\frac{5}{a} - \frac{8}{a}$

3. $\frac{y}{y-4} - \frac{4}{y-4}$

4. $\frac{t^2}{t+5} - \frac{25}{t+5}$

5. $\frac{2x-3}{x^2+3x-4} - \frac{x-7}{x^2+3x-4}$

6. $\frac{x+1}{x^2-2x+1} - \frac{5-3x}{x^2-2x+1}$

7. $\frac{a-2}{10} - \frac{a+1}{5}$

8. $\frac{y+3}{2} - \frac{y-4}{4}$

9. $\frac{4z-9}{3z} - \frac{3z-8}{4z}$

10. $\frac{a-1}{4a} - \frac{2a+3}{a}$

11. $\frac{4x+2t}{3xt^2} - \frac{5x-3t}{x^2t}$

12. $\frac{5x+3y}{2x^2y} - \frac{3x+4y}{xy^2}$

13. $\frac{5}{x+5} - \frac{3}{x-5}$

14. $\frac{3t}{t-1} - \frac{8t}{t+1}$

15. $\frac{3}{2t^2-2t} - \frac{5}{2t-2}$

16. $\frac{11}{x^2-4} - \frac{8}{x+2}$

17. $\frac{2s}{t^2-s^2} - \frac{s}{t-s}$

18. $\frac{3}{12+x-x^2} - \frac{2}{x^2-9}$

19. $\frac{y-5}{y} - \frac{3y-1}{4y}$

20. $\frac{3x-2}{4x} - \frac{3x+1}{6x}$

21. $\frac{a}{x+a} - \frac{a}{x-a}$

22. $\frac{a}{a-b} - \frac{a}{a+b}$

23. $\frac{11}{6} - \frac{5}{-6}$

24. $\frac{5}{9} - \frac{7}{-9}$

25. $\frac{5}{a} - \frac{8}{-a}$

26. $\frac{8}{x} - \frac{3}{-x}$

27. $\frac{4}{y-1} - \frac{4}{1-y}$

28. $\frac{5}{a-2} - \frac{3}{2-a}$

29. $\frac{3-x}{x-7} - \frac{2x-5}{7-x}$

30. $\frac{t^2}{t-2} - \frac{4}{2-t}$

31. $\frac{a-2}{a^2-25} - \frac{6-a}{25-a^2}$

32. $\frac{x-8}{x^2-16} - \frac{x-8}{16-x^2}$

33. $\frac{4-x}{x-9} - \frac{3x-8}{9-x}$

34. $\frac{4x-6}{x-5} - \frac{7-2x}{5-x}$

35. $\frac{5x}{x^2-9} - \frac{4}{3-x}$

36. $\frac{8x}{16-x^2} - \frac{5}{x-4}$

37. $\dfrac{t^2}{2t^2 - 2t} - \dfrac{1}{2t - 2}$

38. $\dfrac{4}{5a^2 - 5a} - \dfrac{2}{5a - 5}$

39. $\dfrac{x}{x^2 + 5x + 6} - \dfrac{2}{x^2 + 3x + 2}$

40. $\dfrac{a}{a^2 + 11a + 30} - \dfrac{5}{a^2 + 9a + 20}$

b Perform the indicated operations and simplify.

41. $\dfrac{3(2x + 5)}{x - 1} - \dfrac{3(2x - 3)}{1 - x} + \dfrac{6x - 1}{x - 1}$

42. $\dfrac{a - 2b}{b - a} - \dfrac{3a - 3b}{a - b} + \dfrac{2a - b}{a - b}$

43. $\dfrac{x - y}{x^2 - y^2} + \dfrac{x + y}{x^2 - y^2} - \dfrac{2x}{x^2 - y^2}$

44. $\dfrac{x - 3y}{2(y - x)} + \dfrac{x + y}{2(x - y)} - \dfrac{2x - 2y}{2(x - y)}$

45. $\dfrac{2(x - 1)}{2x - 3} - \dfrac{3(x + 2)}{2x - 3} - \dfrac{x - 1}{3 - 2x}$

46. $\dfrac{5(2y + 1)}{2y - 3} - \dfrac{3(y - 1)}{3 - 2y} - \dfrac{3(y - 2)}{2y - 3}$

47. $\dfrac{10}{2y - 1} - \dfrac{6}{1 - 2y} + \dfrac{y}{2y - 1} + \dfrac{y - 4}{1 - 2y}$

48. $\dfrac{(x + 1)(2x - 1)}{(2x - 3)(x - 3)} - \dfrac{(x - 3)(x + 1)}{(3 - x)(3 - 2x)} + \dfrac{(2x + 1)(x + 3)}{(3 - 2x)(x - 3)}$

49. $\dfrac{a + 6}{4 - a^2} - \dfrac{a + 3}{a + 2} + \dfrac{a - 3}{2 - a}$

50. $\dfrac{4t}{t^2 - 1} - \dfrac{2}{t} - \dfrac{2}{t + 1}$

51. $\dfrac{2z}{1 - 2z} + \dfrac{3z}{2z + 1} - \dfrac{3}{4z^2 - 1}$

52. $\dfrac{1}{x - y} - \dfrac{2x}{x^2 - y^2} + \dfrac{1}{x + y}$

53. $\dfrac{1}{x + y} - \dfrac{1}{x - y} + \dfrac{2x}{x^2 - y^2}$

54. $\dfrac{2b}{a^2 - b^2} - \dfrac{1}{a + b} + \dfrac{1}{a - b}$

55. **DW** Are parentheses as important when adding rational expressions as they are when subtracting? Why or why not?

56. **DW** Is it possible to add or subtract rational expressions without knowing how to factor? Why or why not?

SKILL MAINTENANCE

Simplify.

57. $\frac{x^8}{x^3}$ [10.1e]

58. $3x^4 \cdot 10x^8$ [10.1d]

59. $(a^2b^{-5})^{-4}$ [10.2a, b]

60. $\frac{54x^{10}}{3x^7}$ [10.1e]

61. $\frac{66x^2}{11x^5}$ [10.1e]

62. $5x^{-7} \cdot 2x^4$ [10.1d]

Find a polynomial for the shaded area of the figure. [10.4d]

63.

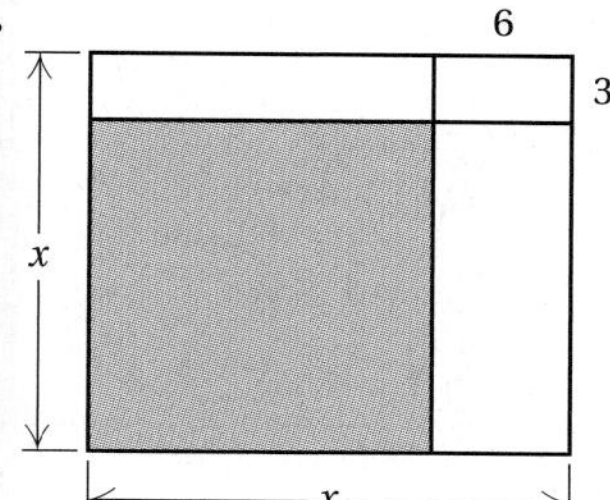

64.

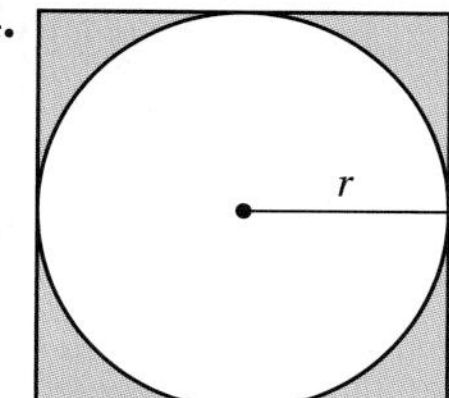

SYNTHESIS

Perform the indicated operations and simplify.

65. $\frac{2x+11}{x-3} \cdot \frac{3}{x+4} + \frac{2x+1}{4+x} \cdot \frac{3}{3-x}$

66. $\frac{x^2}{3x^2-5x-2} - \frac{2x}{3x+1} \cdot \frac{1}{x-2}$

67. $\frac{x}{x^4-y^4} - \left(\frac{1}{x+y}\right)^2$

68. $\left(\frac{a}{a-b} + \frac{b}{a+b}\right)\left(\frac{1}{3a+b} + \frac{2a+6b}{9a^2-b^2}\right)$

69. The perimeter of the following right triangle is $2a + 5$. Find the length of the missing side and the area.

$\frac{a^2-5a-9}{a-6}$

$\frac{a^2-6}{a-6}$

70.–73. Use the TABLE feature to check the subtractions in Exercises 15, 16, 19, and 20.

12.6 SOLVING RATIONAL EQUATIONS

Objective

a Solve rational equations.

1. Solve: $\frac{3}{4} + \frac{5}{8} = \frac{x}{12}$.

Answer on page A-44

Study Tips

RECORDING YOUR LECTURES

Consider recording your lectures and playing them back when convenient (for example, while commuting to campus). It can even be advantageous to record math lectures. (Be sure to get permission from your instructor before doing so, however.) Important points can be emphasized verbally.

a Rational Equations

In Sections 12.1–12.5, we studied operations with *rational expressions.* These expressions have no equals signs. We can add, subtract, multiply, or divide and simplify expressions, but we cannot solve if there are no equals signs—as, for example, in

$$\frac{x^2 + 6x + 9}{x^2 - 4} \cdot \frac{x - 2}{x + 3}, \quad \frac{x + y}{x - y} \div \frac{x^2 + y}{x^2 - y^2}, \quad \text{and} \quad \frac{a + 3}{a^2 - 16} + \frac{5}{12 - 3a}.$$

Operation signs occur. There are no equals signs!

Most often, the result of our calculation is another rational expression that has not been cleared of fractions.

Equations *do have* equals signs, and we can clear them of fractions as we did in Section 8.3. A **rational,** or **fraction, equation,** is an equation containing one or more rational expressions. Here are some examples:

$$\frac{2}{3} + \frac{5}{6} = \frac{x}{9}, \quad x + \frac{6}{x} = -5, \quad \text{and} \quad \frac{x^2}{x - 1} = \frac{1}{x - 1}.$$

There are equals signs as well as operation signs.

SOLVING RATIONAL EQUATIONS

To solve a rational equation, the first step is to clear the equation of fractions. To do this, multiply all terms on both sides of the equation by the LCM of all the denominators. Then carry out the equation-solving process as we learned it in Chapter 8.

When clearing an equation of fractions, we use the terminology LCM instead of LCD because we are *not* adding or subtracting rational expressions.

EXAMPLE 1 Solve: $\frac{2}{3} + \frac{5}{6} = \frac{x}{9}$.

The LCM of all denominators is $2 \cdot 3 \cdot 3$, or 18. We multiply all terms on both sides by 18:

$$18\left(\frac{2}{3} + \frac{5}{6}\right) = 18 \cdot \frac{x}{9}$$ Multiplying by the LCM on both sides

$$18 \cdot \frac{2}{3} + 18 \cdot \frac{5}{6} = 18 \cdot \frac{x}{9}$$ Multiplying each term by the LCM to remove parentheses

$$12 + 15 = 2x$$ Simplifying. Note that we have now cleared fractions.

$$27 = 2x$$

$$\frac{27}{2} = x.$$

The solution is $\frac{27}{2}$.

Do Exercise 1.

EXAMPLE 2 Solve: $\frac{x}{6} - \frac{x}{8} = \frac{1}{12}$.

The LCM is 24. We multiply all terms on both sides by 24:

$$\frac{x}{6} - \frac{x}{8} = \frac{1}{12}$$

$$24\left(\frac{x}{6} - \frac{x}{8}\right) = 24 \cdot \frac{1}{12}$$ Multiplying by the LCM on both sides

$$24 \cdot \frac{x}{6} - 24 \cdot \frac{x}{8} = 24 \cdot \frac{1}{12}$$ Multiplying to remove parentheses

Be sure to multiply each term by the LCM.

$$4x - 3x = 2$$ Simplifying

$$x = 2.$$

Check:

$$\frac{x}{6} - \frac{x}{8} = \frac{1}{12}$$

$\frac{2}{6} - \frac{2}{8}$	$\frac{1}{12}$
$\frac{1}{3} - \frac{1}{4}$	
$\frac{4}{12} - \frac{3}{12}$	
$\frac{1}{12}$	TRUE

This checks, so the solution is 2.

Do Exercise 2.

EXAMPLE 3 Solve: $\frac{1}{x} = \frac{1}{4 - x}$.

The LCM is $x(4 - x)$. We multiply all terms on both sides by $x(4 - x)$:

$$\frac{1}{x} = \frac{1}{4 - x}$$

$$x(4 - x) \cdot \frac{1}{x} = x(4 - x) \cdot \frac{1}{4 - x}$$ Multiplying by the LCM on both sides

$$4 - x = x$$ Simplifying

$$4 = 2x$$

$$x = 2.$$

Check:

$$\frac{1}{x} = \frac{1}{4 - x}$$

$\frac{1}{2}$	$\frac{1}{4 - 2}$
	$\frac{1}{2}$ TRUE

This checks, so the solution is 2.

Do Exercise 3.

2. Solve: $\frac{x}{4} - \frac{x}{6} = \frac{1}{8}$.

3. Solve: $\frac{1}{x} = \frac{1}{6 - x}$.

Answers on page A-44

4. Solve: $\frac{1}{2x} + \frac{1}{x} = -12.$

AC ALGEBRAIC–GRAPHICAL CONNECTION

We can obtain a visual check of the solutions of a rational equation by graphing. For example, consider the equation

$$\frac{x}{4} + \frac{x}{2} = 6.$$

We can examine the solution by graphing the equations

$$y = \frac{x}{4} + \frac{x}{2} \quad \text{and} \quad y = 6$$

using the same set of axes.

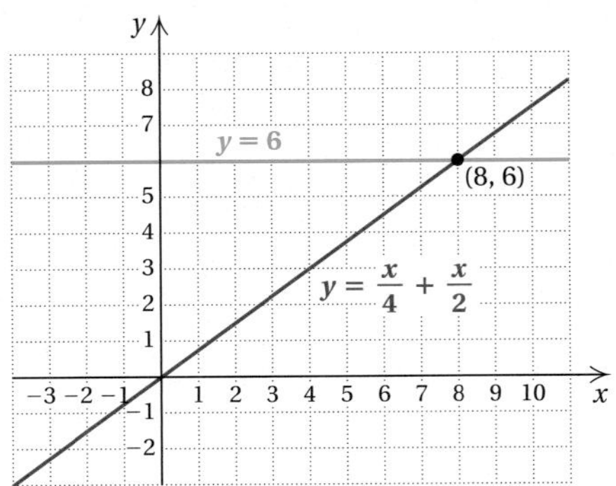

The y-values for each equation will be the same where the graphs intersect. The x-value of that point will yield that value, so it will be the solution of the equation. It appears from the graph that when $x = 8$, the value of $x/4 + x/2$ is 6. We can check by substitution:

$$\frac{x}{4} + \frac{x}{2} = \frac{8}{4} + \frac{8}{2} = 2 + 4 = 6.$$

Thus the solution is 8.

Caution!

We have introduced a new use of the LCM in this section. We previously used the LCM in adding or subtracting rational expressions. *Now* we have equations with equals signs. We clear fractions by multiplying both sides of the equation by the LCM. This eliminates the denominators. Do *not* make the mistake of trying to clear fractions when you do not have an equation.

EXAMPLE 4 Solve: $\frac{2}{3x} + \frac{1}{x} = 10.$

The LCM is $3x$. We multiply all terms on both sides by $3x$:

$$\frac{2}{3x} + \frac{1}{x} = 10$$

$$3x\left(\frac{2}{3x} + \frac{1}{x}\right) = 3x \cdot 10 \quad \text{Multiplying by the LCM on both sides}$$

$$3x \cdot \frac{2}{3x} + 3x \cdot \frac{1}{x} = 3x \cdot 10 \quad \text{Multiplying to remove parentheses}$$

$$2 + 3 = 30x \quad \text{Simplifying}$$

$$5 = 30x$$

$$\frac{5}{30} = x$$

$$\frac{1}{6} = x.$$

The check is left to the student. The solution is $\frac{1}{6}$.

Answer on page A-44

Do Exercise 4.

EXAMPLE 5 Solve: $x + \frac{6}{x} = -5$.

The LCM is x. We multiply all terms on both sides by x:

$$x + \frac{6}{x} = -5$$

$$x\left(x + \frac{6}{x}\right) = -5x \qquad \text{Multiplying by } x \text{ on both sides}$$

$$x \cdot x + x \cdot \frac{6}{x} = -5x \qquad \text{Note that each rational expression on the left is now multiplied by } x.$$

$$x^2 + 6 = -5x \qquad \text{Simplifying}$$

$$x^2 + 5x + 6 = 0 \qquad \text{Adding } 5x \text{ to get a 0 on one side}$$

$$(x + 3)(x + 2) = 0 \qquad \text{Factoring}$$

$$x + 3 = 0 \quad or \quad x + 2 = 0 \qquad \text{Using the principle of zero products}$$

$$x = -3 \quad or \quad x = -2.$$

Check: For -3:

$$x + \frac{6}{x} = -5$$

$-3 + \frac{6}{-3}$	-5
$-3 - 2$	
-5	TRUE

For -2:

$$x + \frac{6}{x} = -5$$

$-2 + \frac{6}{-2}$	-5
$-2 - 3$	
-5	TRUE

Both of these check, so there are two solutions, -1 and -2.

Do Exercise 5.

CHECKING POSSIBLE SOLUTIONS

When we multiply both sides of an equation by the LCM, the resulting equation might have solutions that are *not* solutions of the original equation. Thus we must *always* check possible solutions in the original equation.

1. If you have carried out all algebraic procedures correctly, you need only check if a number makes a denominator 0 in the original equation. If it does make a denominator 0, it is *not* a solution.
2. To be sure that no computational errors have been made and that you indeed have a solution, a complete check is necessary, as we did in Chapter 2.

5. Solve: $x + \frac{1}{x} = 2$.

6. Solve: $\frac{x^2}{x + 2} = \frac{4}{x + 2}$.

7. Solve: $\frac{4}{x - 2} + \frac{1}{x + 2} = \frac{26}{x^2 - 4}$.

Answers on page A-44

Example 6 illustrates the importance of checking all possible solutions.

CALCULATOR CORNER

Checking Solutions of Rational Equations We can use a table to check possible solutions of rational equations. Consider the equation in Example 6 and the possible solutions that were found, 1 and -1. To check these solutions, we enter $y_1 = x^2/(x - 1)$ and $y_2 = 1/(x - 1)$ on the equation-editor screen. Then, with a table set in ASK mode, we enter $x = 1$. The ERROR messages indicate that 1 is not a solution because it is not an allowable replacement for x in the equation. Next, we enter $x = -1$. Since y_1 and y_2 have the same value, we know that the equation is true when $x = -1$, and thus -1 is a solution.

$y_1 = x^2/(x - 1)$,
$y_2 = 1/(x - 1)$

X	Y1	Y2
1	ERROR	ERROR
−1	−.5	−.5

X =

Exercises: Use a graphing calculator to check the possible solutions in each of the following.

1. Examples 1, 3, 5, and 7
2. Margin Exercises 1, 3, 6, and 7

EXAMPLE 6 Solve: $\dfrac{x^2}{x-1} = \dfrac{1}{x-1}$.

The LCM is $x - 1$. We multiply all terms on both sides by $x - 1$:

$$\frac{x^2}{x-1} = \frac{1}{x-1}$$

$$(x-1)\cdot\frac{x^2}{x-1} = (x-1)\cdot\frac{1}{x-1} \quad \text{Multiplying by } x-1 \text{ on both sides}$$

$$x^2 = 1 \quad \text{Simplifying}$$

$$x^2 - 1 = 0 \quad \text{Subtracting 1 to get a 0 on one side}$$

$$(x-1)(x+1) = 0 \quad \text{Factoring}$$

$$x - 1 = 0 \quad or \quad x + 1 = 0 \quad \text{Using the principle of zero products}$$

$$x = 1 \quad or \quad x = -1.$$

The numbers 1 and -1 are possible solutions.

Check:

For 1:

$$\frac{x^2}{x-1} = \frac{1}{x-1}$$

$$\frac{1^2}{1-1} \;\Big|\; \frac{1}{1-1}$$

$$\frac{1}{0} \;\Big|\; \frac{1}{0} \quad \text{NOT DEFINED}$$

For -1:

$$\frac{x^2}{x-1} = \frac{1}{x-1}$$

$$\frac{(-1)^2}{(-1)-1} \;\Big|\; \frac{1}{(-1)-1}$$

$$\frac{1}{-2} \;\Big|\; \frac{1}{-2}$$

$$-\frac{1}{2} \;\Big|\; -\frac{1}{2} \quad \text{TRUE}$$

We look at the original equation and see that 1 makes a denominator 0 and is therefore not a solution. The number -1 checks and is a solution.

Do Exercise 6 on the preceding page.

EXAMPLE 7 Solve: $\dfrac{3}{x-5} + \dfrac{1}{x+5} = \dfrac{2}{x^2-25}$.

The LCM is $(x - 5)(x + 5)$. We multiply all terms on both sides by $(x - 5)(x + 5)$:

$$(x-5)(x+5)\left(\frac{3}{x-5} + \frac{1}{x+5}\right) = (x-5)(x+5)\left(\frac{2}{x^2-25}\right)$$

Multiplying by the LCM on both sides

$$(x-5)(x+5)\cdot\frac{3}{x-5} + (x-5)(x+5)\cdot\frac{1}{x+5} = (x-5)(x+5)\cdot\frac{2}{x^2-25}$$

$$3(x+5) + (x-5) = 2 \quad \text{Simplifying}$$

$$3x + 15 + x - 5 = 2 \quad \text{Removing parentheses}$$

$$4x + 10 = 2$$

$$4x = -8$$

$$x = -2.$$

The check is left to the student. The number -2 checks and is the solution.

Do Exercise 7 on the preceding page.

Study Tips

ARE YOU CALCULATING OR SOLVING?

One of the common difficulties with this chapter is knowing for sure the task at hand. Are you combining expressions using operations to get another *rational expression*, or are you solving equations for which the results are numbers that are *solutions* of an equation? To learn to make these decisions, complete the following list by writing in the blank the type of answer you should get: "Rational expression" or "Solutions." You need not complete the mathematical operations.

TASK	ANSWER (Just write "Rational expression" or "Solutions.")
1. Add: $\frac{4}{x-2}+\frac{1}{x+2}$.	
2. Solve: $\frac{4}{x-2}=\frac{1}{x+2}$.	
3. Subtract: $\frac{4}{x-2}-\frac{1}{x+2}$.	
4. Multiply: $\frac{4}{x-2}\cdot\frac{1}{x+2}$.	
5. Divide: $\frac{4}{x-2}\div\frac{1}{x+2}$.	
6. Solve: $\frac{4}{x-2}+\frac{1}{x+2}=\frac{26}{x^2-4}$.	
7. Perform the indicated operations and simplify: $\frac{4}{x-2}+\frac{1}{x+2}-\frac{26}{x^2-4}$.	
8. Solve: $\frac{x^2}{x-1}=\frac{1}{x-1}$.	
9. Solve: $\frac{2}{y^2-25}=\frac{3}{y-5}+\frac{1}{y-5}$.	
10. Solve: $\frac{x}{x+4}-\frac{4}{x-4}=\frac{x^2+16}{x^2-16}$.	
11. Perform the indicated operations and simplify: $\frac{x}{x+4}-\frac{4}{x-4}-\frac{x^2+16}{x^2-16}$.	
12. Solve: $\frac{5}{y-3}-\frac{30}{y^2-9}=1$.	
13. Add: $\frac{5}{y-3}+\frac{30}{y^2-9}+1$.	

12.6 EXERCISE SET

For Extra Help

MathXL | MyMathLab | InterAct Math | Math Tutor Center | Video Lectures on CD Disc 6 | Student's Solutions Manual

a Solve. Don't forget to check!

1. $\frac{4}{5} - \frac{2}{3} = \frac{x}{9}$

2. $\frac{x}{20} = \frac{3}{8} - \frac{4}{5}$

3. $\frac{3}{5} + \frac{1}{8} = \frac{1}{x}$

4. $\frac{2}{3} + \frac{5}{6} = \frac{1}{x}$

5. $\frac{3}{8} + \frac{4}{5} = \frac{x}{20}$

6. $\frac{3}{5} + \frac{2}{3} = \frac{x}{9}$

7. $\frac{1}{x} = \frac{2}{3} - \frac{5}{6}$

8. $\frac{1}{x} = \frac{1}{8} - \frac{3}{5}$

9. $\frac{1}{6} + \frac{1}{8} = \frac{1}{t}$

10. $\frac{1}{8} + \frac{1}{12} = \frac{1}{t}$

11. $x + \frac{4}{x} = -5$

12. $\frac{10}{x} - x = 3$

13. $\frac{x}{4} - \frac{4}{x} = 0$

14. $\frac{x}{5} - \frac{5}{x} = 0$

15. $\frac{5}{x} = \frac{6}{x} - \frac{1}{3}$

16. $\frac{4}{x} = \frac{5}{x} - \frac{1}{2}$

17. $\frac{5}{3x} + \frac{3}{x} = 1$

18. $\frac{5}{2y} + \frac{8}{y} = 1$

19. $\frac{t-2}{t+3} = \frac{3}{8}$

20. $\frac{x-7}{x+2} = \frac{1}{4}$

21. $\frac{2}{x+1} = \frac{1}{x-2}$

22. $\frac{8}{y-3} = \frac{6}{y+4}$

23. $\frac{x}{6} - \frac{x}{10} = \frac{1}{6}$

24. $\frac{x}{8} - \frac{x}{12} = \frac{1}{8}$

25. $\frac{t+2}{5} - \frac{t-2}{4} = 1$

26. $\frac{x+1}{3} - \frac{x-1}{2} = 1$

27. $\frac{5}{x-1} = \frac{3}{x+2}$

28. $\frac{x-7}{x-9} = \frac{2}{x-9}$

29. $\frac{a-3}{3a+2} = \frac{1}{5}$

30. $\frac{x+7}{8x-5} = \frac{2}{3}$

31. $\dfrac{x-1}{x-5} = \dfrac{4}{x-5}$

32. $\dfrac{y+11}{y+8} = \dfrac{3}{y+8}$

33. $\dfrac{2}{x+3} = \dfrac{5}{x}$

34. $\dfrac{6}{y} = \dfrac{5}{y-8}$

35. $\dfrac{x-2}{x-3} = \dfrac{x-1}{x+1}$

36. $\dfrac{t+5}{t-2} = \dfrac{t-2}{t+4}$

37. $\dfrac{1}{x+3} + \dfrac{1}{x-3} = \dfrac{1}{x^2-9}$

38. $\dfrac{4}{x-3} + \dfrac{2x}{x^2-9} = \dfrac{1}{x+3}$

39. $\dfrac{x}{x+4} - \dfrac{4}{x-4} = \dfrac{x^2+16}{x^2-16}$

40. $\dfrac{5}{y-3} - \dfrac{30}{y^2-9} = 1$

41. $\dfrac{4-a}{8-a} = \dfrac{4}{a-8}$

42. $\dfrac{3}{x-7} = \dfrac{x+10}{x-7}$

43. $2 - \dfrac{a-2}{a+3} = \dfrac{a^2-4}{a+3}$

44. $\dfrac{5}{x-1} + x + 1 = \dfrac{5x+4}{x-1}$

Solve.

45. $\frac{x+1}{x+2} = \frac{x+3}{x+4}$

46. $\frac{x^2}{x^2-4} = \frac{x}{x+2} - \frac{2x}{2-x}$

47. $4a - 3 = \frac{a+13}{a+1}$

48. $\frac{3x-9}{x-3} = \frac{5x-4}{2}$

49. $\frac{4}{y-2} - \frac{2y-3}{y^2-4} = \frac{5}{y+2}$

50. $\frac{y^2-4}{y+3} = 2 - \frac{y-2}{y+3}$

51. ^{D}W Why is it especially important to check the possible solutions to a rational equation?

52. ^{D}W How can a graph be used to determine how many solutions an equation has?

SKILL MAINTENANCE

VOCABULARY REINFORCEMENT

In each of Exercises 53–60, fill in the blank with the correct term from the given list. Some of the choices may not be used.

53. A rational expression is a(n) ______________ of two polynomials. [12.1a]

54. A factor of a polynomial P is a polynomial that can be used to express P as a(n) ______________. [11.1a]

55. Two expressions are ______________ of each other if their product is 1. [7.6b]

56. When ______________, always remember to look first for the greatest common factor. [11.1b]

57. To find the LCM, use each factor the ______________ number of times that it appears in any one factorization. [12.3a]

58. When solving rational equations, always check a possible solution to see if it makes a denominator 0. If it does, it is ______________ a solution. [12.6a]

59. The quotient rule asserts that when dividing with exponential notation, if the bases are the same, keep the base and ______________ the exponent of the denominator from the exponent of the numerator. [10.1e]

60. Two expressions are ______________ of each other if their sum is 0. [7.3b]

not
always
factor
add
subtract
sum
product
smallest
greatest
factoring
quotient
reciprocals
additive inverses
exponents

SYNTHESIS

Solve.

61. $\frac{x}{x^2+3x-4} + \frac{x+1}{x^2+6x+8} = \frac{2x}{x^2+x-2}$

62. $\frac{3a-5}{a^2+4a+3} + \frac{2a+2}{a+3} = \frac{a-3}{a+1}$

63. Use a graphing calculator to check the solutions to Exercises 1–4.

64. Use a graphing calculator to check the solutions to Exercises 13, 15, and 25.

12.7 APPLICATIONS USING RATIONAL EQUATIONS AND PROPORTIONS

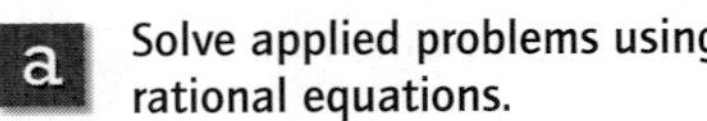

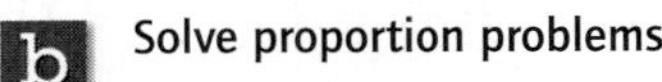

Objectives

a Solve applied problems using rational equations.

b Solve proportion problems.

In many areas of study, applications involving rates, proportions, or reciprocals translate to rational equations. By using the five steps for problem solving and the skills of Sections 12.1–12.6, we can now solve such problems.

a Solving Applied Problems

PROBLEMS INVOLVING WORK

EXAMPLE 1 *Recyclable Work.* Erin and Tara work as volunteers at a community recycling depot. Erin can sort a morning's accumulation of recyclables in 4 hr, while Tara requires 6 hr to do the same job. How long would it take them, working together, to sort the recyclables?

1. **Familiarize.** We familiarize ourselves with the problem by considering two *incorrect* ways of translating the problem to mathematical language.

   a) A common *incorrect* way to translate the problem is to add the two times: 4 hr + 6 hr = 10 hr. Let's think about this. Erin can do the job alone in 4 hr. If Erin and Tara work together, whatever time it takes them should be *less* than 4 hr. Thus we reject 10 hr as a solution, but we do have a partial check on any answer we get. The answer should be less than 4 hr.

   b) Another *incorrect* way to translate the problem is as follows. Suppose the two people split up the sorting job in such a way that Erin does half the sorting and Tara does the other half. Then

   Erin sorts $\frac{1}{2}$ the recyclables in $\frac{1}{2}(4\text{ hr})$, or 2 hr,

   and Tara sorts $\frac{1}{2}$ the recyclables in $\frac{1}{2}(6\text{ hr})$, or 3 hr.

   But time is wasted since Erin would finish 1 hr earlier than Tara. In effect, they have not worked together to get the job done as fast as possible. If Erin helps Tara after completing her half, the entire job could be done in a time somewhere between 2 hr and 3 hr.

   We proceed to a translation by considering how much of the job is finished in 1 hr, 2 hr, 3 hr, and so on. It takes Erin 4 hr to do the sorting job alone. Then, in 1 hr, she can do $\frac{1}{4}$ of the job. It takes Tara 6 hr to do the job alone. Then, in 1 hr, he can do $\frac{1}{6}$ of the job. Working together, they can do

   $$\frac{1}{4} + \frac{1}{6}, \text{ or } \frac{5}{12} \text{ of the job in 1 hr.}$$

   In 2 hr, Erin can do $2\left(\frac{1}{4}\right)$ of the job and Tara can do $2\left(\frac{1}{6}\right)$ of the job. Working together, they can do

   $$2\left(\frac{1}{4}\right) + 2\left(\frac{1}{6}\right), \text{ or } \frac{5}{6} \text{ of the job in 2 hr.}$$

Study Tips

BEING A TUTOR

Try being a tutor for a fellow student. You can maximize your understanding and retention of concepts if you explain the material to someone else.

Continuing this reasoning, we can create a table like the following one.

TIME	FRACTION OF THE JOB COMPLETED		
	ERIN	TARA	TOGETHER
1 hr	$\frac{1}{4}$	$\frac{1}{6}$	$\frac{1}{4} + \frac{1}{6}$, or $\frac{5}{12}$
2 hr	$2\left(\frac{1}{4}\right)$	$2\left(\frac{1}{6}\right)$	$2\left(\frac{1}{4}\right) + 2\left(\frac{1}{6}\right)$, or $\frac{5}{6}$
3 hr	$3\left(\frac{1}{4}\right)$	$3\left(\frac{1}{6}\right)$	$3\left(\frac{1}{4}\right) + 3\left(\frac{1}{6}\right)$, or $1\frac{1}{4}$
t hr	$t\left(\frac{1}{4}\right)$	$t\left(\frac{1}{6}\right)$	$t\left(\frac{1}{4}\right) + t\left(\frac{1}{6}\right)$

From the table, we see that if they work 3 hr, the fraction of the job completed is $1\frac{1}{4}$, which is more of the job than needs to be done. We see again that the answer is somewhere between 2 hr and 3 hr. What we want is a number t such that the fraction of the job that gets completed is 1; that is, the job is just completed.

2. **Translate.** From the table, we see that the time we want is some number t for which

$$t\left(\frac{1}{4}\right) + t\left(\frac{1}{6}\right) = 1, \quad \text{or} \quad \frac{t}{4} + \frac{t}{6} = 1,$$

where 1 represents the idea that the entire job is completed in time t.

3. **Solve.** We solve the equation:

$$12\left(\frac{t}{4} + \frac{t}{6}\right) = 12 \cdot 1 \qquad \text{Multiplying by the LCM, which is } 2 \cdot 2 \cdot 3, \text{ or } 12$$

$$12 \cdot \frac{t}{4} + 12 \cdot \frac{t}{6} = 12$$

$$3t + 2t = 12$$

$$5t = 12$$

$$t = \frac{12}{5}, \text{ or } 2\frac{2}{5} \text{ hr.}$$

4. **Check.** The check can be done by recalculating:

$$\frac{12}{5}\left(\frac{1}{4}\right) + \frac{12}{5}\left(\frac{1}{6}\right) = \frac{3}{5} + \frac{2}{5} = \frac{5}{5} = 1.$$

We also have another check in what we learned from the *Familiarize* step. The answer, $2\frac{2}{5}$ hr, is between 2 hr and 3 hr (see the table), and it is less than 4 hr, the time it takes Erin working alone.

5. **State.** It takes $2\frac{2}{5}$ hr for them to do the sorting, working together.

1. Wall Construction. By checking work records, a contractor finds that it takes Eduardo 6 hr to construct a wall of a certain size. It takes Yolanda 8 hr to construct the same wall. How long would it take if they worked together?

Answer on page A-44

Study Tips

TEST TAKING: MORE ON DOING EVEN-NUMBERED EXERCISES

In an earlier study tip (p. 682), as a way to improve your test-taking skills, we encouraged you to build some even-numbered exercises into your homework. Here we explore this issue further.

Working a test is different from working your homework, when the answers are provided. When taking the test, you are "on your own," so to speak. Keep the following tips in mind when taking your next test or quiz.

1. Work a bit slower and deliberately, taking a fresh piece of paper to redo the problem. Check your work against your previous work to see if there is a difference and why. This is especially helpful if you finish the test early and have extra time.
2. Use estimation techniques to solve the problem as a check.
3. Do the checks to applied problems that we so often discuss in the book.

THE WORK PRINCIPLE

Suppose $a =$ the time it takes A to do a job, $b =$ the time it takes B to do the same job, and $t =$ the time it takes them to do the same job working together. Then

$$\frac{t}{a} + \frac{t}{b} = 1, \quad \text{or} \quad \frac{1}{a} + \frac{1}{b} = \frac{1}{t}.$$

Do Exercise 1.

PROBLEMS INVOLVING MOTION

Problems that deal with distance, speed (or rate), and time are called **motion problems.** Translation of these problems involves the distance formula, $d = r \cdot t$, and/or the equivalent formulas $r = d/t$ and $t = d/r$.

MOTION FORMULAS

The following are the formulas for motion problems:

$d = rt$; Distance = Rate · Time (basic formula)

$r = \dfrac{d}{t}$; Rate = Distance/Time

$t = \dfrac{d}{r}$. Time = Distance/Rate

EXAMPLE 2 *Animal Speeds.* A zebra can run 15 mph faster than an elephant. A zebra can run 8 mi in the same time that an elephant can run 5 mi. Find the speed of each animal.

Source: *World Almanac*, 2005, p. 179

1. **Familiarize.** We first make a drawing. Let $r =$ the speed of the elephant. Then $r + 15 =$ the speed of the zebra.

Recall that sometimes we need to find a formula in order to solve an application. A formula that relates the notions of distance, speed, and time is $d = rt$, or

Distance = Speed · Time.

(Indeed, you may need to look up such a formula.)

Since each animal travels for the same length of time, we can use just t for time. We organize the information in a chart, as follows.

	d = DISTANCE	r · SPEED	t TIME	
Elephant	5	r	t	$\rightarrow 5 = rt$
Zebra	8	$r + 15$	t	$\rightarrow 8 = (r + 15)t$

2. **Translate.** We can apply the formula $d = rt$ along the rows of the table to obtain two equations:

$$5 = rt, \qquad (1)$$
$$8 = (r + 15)t. \qquad (2)$$

We know that the animals travel for the same length of time. Thus if we solve each equation for t and set the results equal to each other, we get an equation in terms of r.

Solving $5 = rt$ for t: $\quad t = \frac{5}{r}$

Solving $8 = (r + 15)t$ for t: $\quad t = \frac{8}{r + 15}$

Since the times are the same, we have the following equation:

$$\frac{5}{r} = \frac{8}{r + 15}.$$

3. **Solve.** To solve the equation, we first multiply both sides by the LCM, which is $r(r + 15)$:

$$r(r + 15) \cdot \frac{5}{r} = r(r + 15) \cdot \frac{8}{r + 15} \quad \text{Multiplying both sides by the LCM, which is } r(r + 15)$$
$$5(r + 15) = 8r \quad \text{Simplifying}$$
$$5r + 75 = 8r \quad \text{Removing parentheses}$$
$$75 = 3r$$
$$25 = r.$$

We now have a possible solution. The speed of the elephant is 25 mph, and the speed of the zebra is $r + 15 = 25 + 15$, or 40 mph.

4. **Check.** We first reread the problem to see what we were to find. We check the speeds of 25 for the elephant and 40 for the zebra. The zebra does travel 15 mph faster than the elephant and will travel farther than the elephant, which runs at a slower speed. If the zebra runs 8 mi at 40 mph, the time it has traveled is $\frac{8}{40}$, or $\frac{1}{5}$ hr. If the elephant runs 5 mi at 25 mph, the time it has traveled is $\frac{5}{25}$, or $\frac{1}{5}$ hr. Since the times are the same, the speeds check.

5. **State.** The speed of the elephant is 25 mph and the speed of the zebra is 40 mph.

Do Exercise 2.

2. **Driving speed.** Nancy drives 20 mph faster than her father, Greg. In the same time that Nancy travels 180 mi, her father travels 120 mi. Find their speeds.

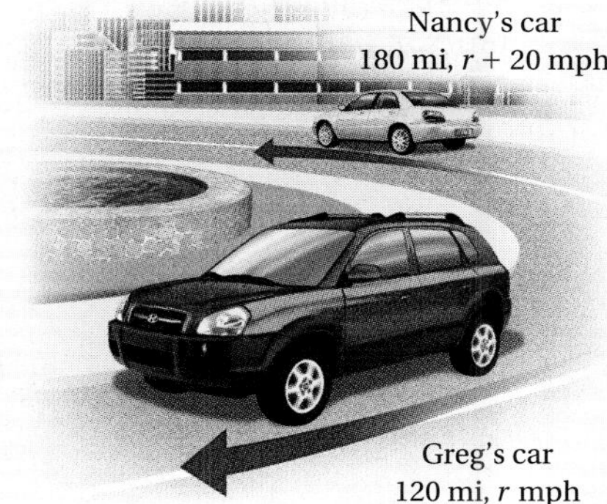

Answer on page A-44

3. Find the ratio of 145 km to 2.5 liters (L).

4. **Batting Average.** Recently, a baseball player got 7 hits in 25 times at bat. What was the rate, or batting average, in hits per times at bat?

5. Impulses in nerve fibers travel 310 km in 2.5 hr. What is the rate, or speed, in kilometers per hour?

6. A lake of area 550 yd^2 contains 1320 fish. What is the population density of the lake in number of fish per square yard?

7. **Automotive Mileage.** In highway driving, a Chrysler PT Cruiser will travel 377 mi on 14.5 gal of gasoline. How much gas will be required for a 900-mi trip?

Source: DaimlerChrysler Corporation

Answers on page A-44

b Applications Involving Proportions

We now consider applications with proportions. A **proportion** involves ratios. A **ratio** of two quantities is their quotient. For example, 73% is the ratio of 73 to 100, $\frac{73}{100}$. The ratio of two different kinds of measure is called a **rate.** Suppose an animal travels 720 ft in 2.5 hr. Its **rate,** or **speed,** is then

$$\frac{720 \text{ ft}}{2.5 \text{ hr}} = 288 \frac{\text{ft}}{\text{hr}}.$$

Do Exercises 3–6.

PROPORTION

An equality of ratios, $A/B = C/D$, is called a **proportion.** The numbers within a proportion are said to be **proportional** to each other.

EXAMPLE 3 *Mileage.* A 2004 Toyota Prius is a gasoline–electric car that travels 240 mi in city driving on 4 gal of gas. Find the amount of gas required for 360 mi of city driving.

Source: *Motor Trend*, May 2004, p. 89

1. **Familiarize.** We know that the Toyota can travel 240 mi on 4 gal of gas. Thus we can set up a ratio, letting x = the amount of gas required to drive 360 mi.
2. **Translate.** We assume that the car uses gas at the same rate in all city driving. Thus the ratios are the same and we can write a proportion. Note that the units of *mileage* are in the numerators and the units of *gasoline* are in the denominators.

$$\begin{matrix}\text{Miles}\rightarrow \\ \text{Gas}\rightarrow\end{matrix} \frac{240}{4} = \frac{360}{x} \begin{matrix}\leftarrow\text{Miles} \\ \leftarrow\text{Gas}\end{matrix}$$

3. **Solve.** To solve for x, we multiply both sides by the LCM, which is $4x$:

$$4x \cdot \frac{240}{4} = 4x \cdot \frac{360}{x} \quad \text{Multiplying by } 4x$$
$$240x = 1440 \quad \text{Simplifying}$$
$$\frac{240x}{240} = \frac{1440}{240} \quad \text{Dividing by 240}$$
$$x = 6. \quad \text{Simplifying}$$

We can also use cross products to solve the proportion:

$$\frac{240}{4} = \frac{360}{x} \qquad 240x \text{ and } 4 \cdot 360 \text{ are cross products.}$$

$$240x = 4 \cdot 360 \quad \text{Equating cross products}$$
$$\frac{240x}{240} = \frac{4 \cdot 360}{240} \quad \text{Dividing by 240}$$
$$x = 6.$$

4. **Check.** The check is left to the student.
5. **State.** The Toyota Prius will require 6 gal of gas for 360 mi of city driving.

Do Exercise 7 on the preceding page.

EXAMPLE 4 *Environmental Science.* To determine the number of fish in a lake, a park ranger catches 225 fish, tags them, and throws them back into the lake. Later, 108 fish are caught, and 15 of them are found to be tagged. Estimate how many fish are in the lake.

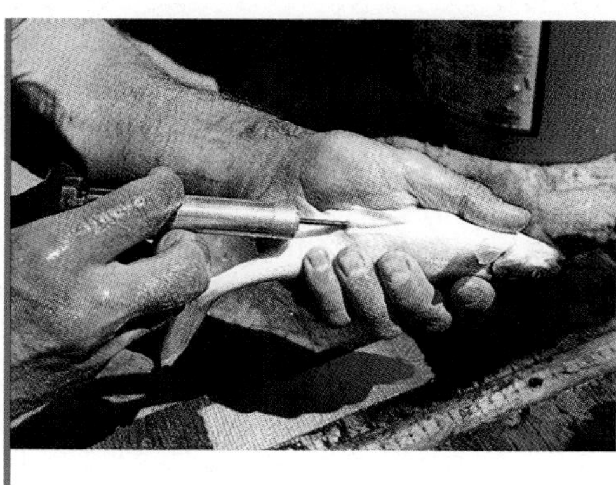

1. **Familiarize.** The ratio of the number of fish tagged to the total number of fish in the lake, F, is $\frac{225}{F}$. Of the 108 fish caught later, 15 fish were tagged. The ratio of fish tagged to fish caught is $\frac{15}{108}$.
2. **Translate.** Assuming that the two ratios are the same, we can translate to a proportion.

$$\text{Fish tagged originally} \rightarrow \frac{225}{F} = \frac{15}{108} \leftarrow \text{Tagged fish caught later}$$
$$\text{Fish in lake} \rightarrow F \qquad 108 \leftarrow \text{Fish caught later}$$

3. **Solve.** We solve the proportion. We multiply by the LCM, which is $108F$:

$$108F \cdot \frac{225}{F} = 108F \cdot \frac{15}{108} \quad \text{Multiplying by } 108F$$
$$108 \cdot 225 = F \cdot 15$$
$$\frac{108 \cdot 225}{15} = F \quad \text{Dividing by 15}$$
$$1620 = F.$$

4. **Check.** The check is left to the student.
5. **State.** We estimate that there are about 1620 fish in the lake.

Do Exercise 8.

In the following example, we predict whether an important home-run record can be broken.

EXAMPLE 5 *Home Runs by the Chicago Cubs.* After 14 games of the 2004 Major League Baseball season, which consists of 162 games, the Chicago Cubs had hit 30 home runs. The club record for a season (1998) was 212 home runs. At the pace they were hitting home runs in 2004, would the Cubs break their club record?

Source: Major League Baseball

8. Environmental Science. To determine the number of humpback whales in a pod, a marine biologist, using tail markings, identifies 27 members of the pod. Several weeks later, 40 whales from the pod are randomly sighted. Of the 40 sighted, 12 are from the 27 originally identified. Estimate the number of whales in the pod.

Answer on page A-45

9. **Barry Bonds' Walks.** Barry Bonds of the San Francisco Giants set a major league record in 2002 by getting 198 walks in 143 games. After playing in 122 games in 2004, he had 191 walks. Assuming he played in 143 games in 2004, would he break his record? (Barry Bonds actually got 232 walks in 147 games in 2004.)

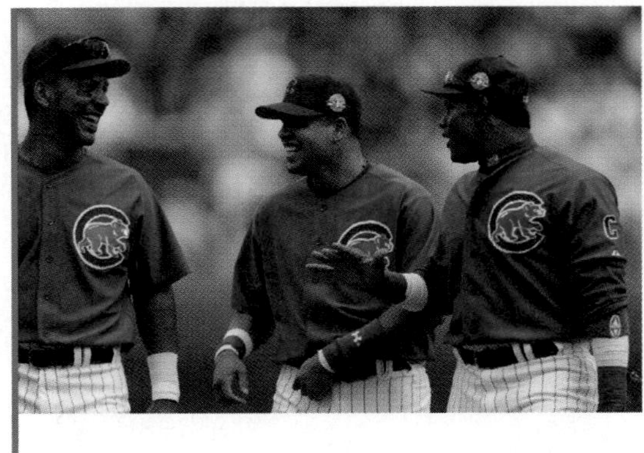

1. **Familiarize.** Let's assume that the pace of hitting 30 home runs in 14 games continues for the entire 162-game season. We let $H =$ the number of home runs that the Cubs hit in 162 games.
2. **Translate.** Assuming the rate continues, the ratios are the same and we have the proportion

$$\frac{\text{Number of home runs} \rightarrow H}{\text{Number of games} \rightarrow 162} = \frac{30 \leftarrow \text{Number of home runs}}{14 \leftarrow \text{Number of games}}.$$

3. **Solve.** We solve the proportion:

$$\frac{H}{162} = \frac{30}{14}$$

$$14H = 162 \cdot 30 \quad \text{Equating cross products}$$

$$\frac{14H}{14} = \frac{162 \cdot 30}{14} \quad \text{Dividing by 14}$$

$$H \approx 347.$$

4. **Check.** The check is left to the student.
5. **State.** Since $347 > 212$, we could predict that the Cubs would break their team record of 212. (The Cubs actually got 235 home runs in 2004.)

Do Exercise 9.

SIMILAR TRIANGLES

Proportions arise in geometry when we are studying *similar triangles.* If two triangles are **similar,** then their corresponding angles have the same measure and their corresponding sides are proportional. To illustrate, if triangle ABC is similar to triangle RST, then angles A and R have the same measure, angles B and S have the same measure, angles C and T have the same measure, and

$$\frac{a}{r} = \frac{b}{s} = \frac{c}{t}.$$

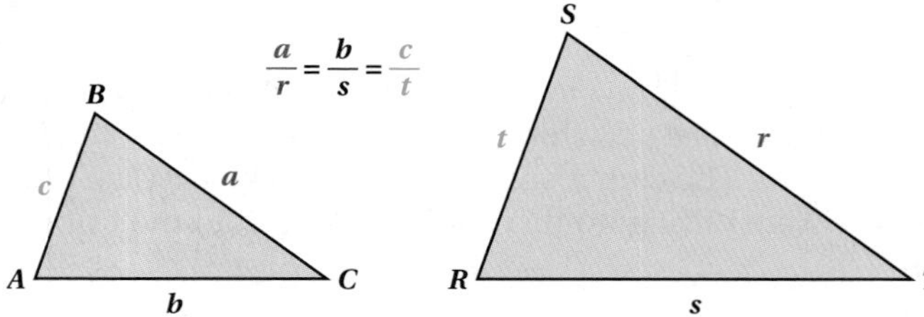

Answer on page A-45

SIMILAR TRIANGLES

In **similar triangles,** corresponding angles have the same measure and the lengths of corresponding sides are proportional.

EXAMPLE 6 *Similar Triangles.* Triangles ABC and XYZ below are similar triangles. Solve for z if $x = 10$, $a = 8$, and $c = 5$.

We make a sketch, write a proportion, and then solve. Note that side a is always opposite angle A, side x is always opposite angle X, and so on.

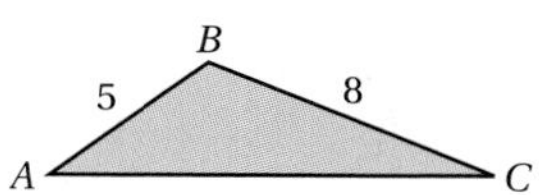

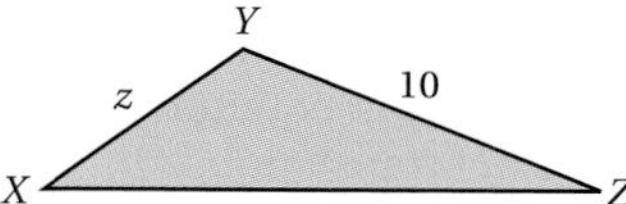

We have

$$\frac{z}{5} = \frac{10}{8}$$ The proportion $\frac{5}{z} = \frac{8}{10}$ could also be used.

$$40 \cdot \frac{z}{5} = 40 \cdot \frac{10}{8}$$ Multiplying by 40

$$8z = 50$$

$$z = \frac{50}{8}, \text{ or } 6.25.$$ Dividing by 8

EXAMPLE 7 *F-106 Blueprint.* A blueprint for an F-106 Delta Dart fighter plane is a scale drawing, as shown below. Each wing has a triangular shape. The blueprint shows similar triangles. Find the length of side a of the wing.

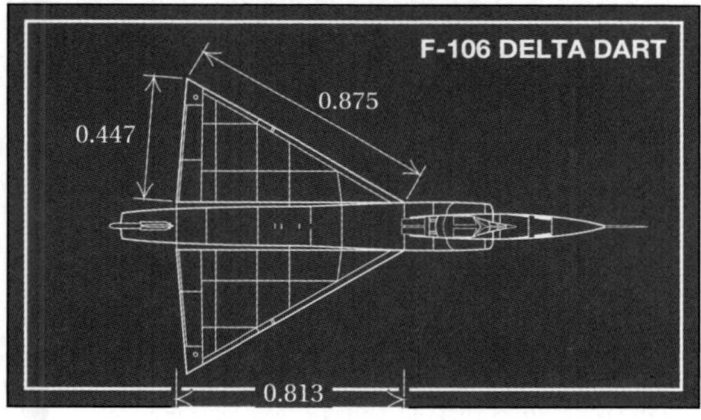

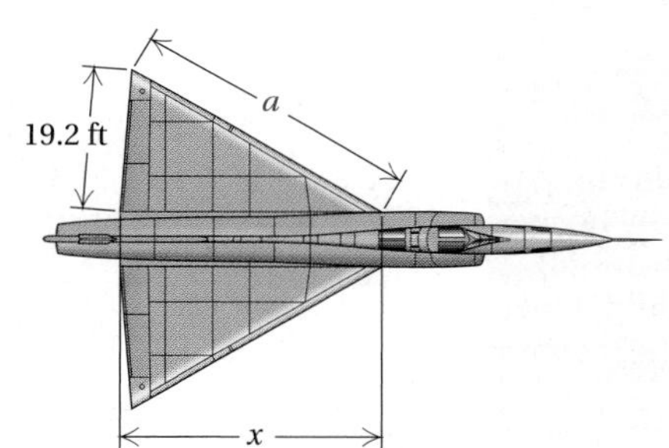

We let a = the length of the wing. Thus we have the proportion

Length on the blueprint → 0.447 / Length of the wing → 19.2 = 0.875 ← Length on the blueprint / a ← Length of the wing.

$$\frac{0.447}{19.2} = \frac{0.875}{a}.$$

Solve: $0.447 \cdot a = 19.2 \cdot 0.875$ Equating cross products

$$a = \frac{19.2 \cdot 0.875}{0.447}$$ Dividing by 0.447

$$a \approx 37.6 \text{ ft}$$

The length of side a of the wing is about 37.6 ft.

Do Exercises 10 and 11.

10. Height of a Flagpole. How high is a flagpole that casts a 45-ft shadow at the same time that a 5.5-ft woman casts a 10-ft shadow?

Sun's rays
5.5 ft
10 ft
Sun's rays
x
45 ft

11. F-106 Blueprint. Referring to Example 7, find the length x on the plane.

Answers on page A-45

Translating for Success

1. *Search Engine Ads.* In 2005, it was estimated that \$3.6 billion was spent in advertising on Internet search engines such as Google® and Yahoo®. This was a 25% increase over the amount spent in 2004. How much was spent in 2004?

   Source: Forrester Research

2. *Cycling Distance.* A bicyclist traveled 197 mi in 7 days. At this rate, how many miles could the cyclist travel in 30 days?

3. *Bicycling.* The speed of one bicyclist is 2 km/h faster than the speed of another bicyclist. The first bicyclist travels 60 km in the same amount of time that it takes the second to travel 50 km. Find the speed of each bicyclist.

4. *Filling Time.* A swimming pool can be filled in 5 hr by hose A alone and in 6 hr by hose B alone. How long would it take to fill the tank if both hoses were working?

5. *Office Budget.* Emma has \$36 budgeted for office stationery. Engraved stationery costs \$20 for the first 25 sheets and \$0.08 for each additional sheet. How many engraved sheets of stationery can Emma order and still stay within her budget?

The goal of these matching questions is to practice step (2), *Translate,* of the five-step problem-solving process. Translate each word problem to an equation and select a correct translation from equations A–O.

A. $2x + 2(x + 1) = 613$

B. $x^2 + (x + 1)^2 = 613$

C. $\dfrac{60}{x + 2} = \dfrac{50}{x}$

D. $20 + 0.08(x - 25) = 36$

E. $\dfrac{197}{7} = \dfrac{x}{30}$

F. $x + (x + 1) = 613$

G. $\dfrac{7}{197} = \dfrac{x}{30}$

H. $x^2 + (x + 2)^2 = 612$

I. $x^2 + (x + 1)^2 = 612$

J. $\dfrac{50}{x + 2} = \dfrac{60}{x}$

K. $x + 25\% \cdot x = 3.6$

L. $t + 5 = 7$

M. $x^2 + (x + 1)^2 = 452$

N. $\dfrac{1}{5} + \dfrac{1}{6} = \dfrac{1}{t}$

O. $x^2 + (x + 2)^2 = 452$

Answers on page A-45

6. *Sides of a Square.* If the sides of a square are increased by 2 ft, the area of the original square plus the area of the enlarged square is 452 ft^2. Find the length of a side of the original square.

7. *Consecutive Integers.* The sum of two consecutive integers is 613. Find the integers.

8. *Sums of Squares.* The sum of the squares of two consecutive odd integers is 612. Find the integers.

9. *Sums of Squares.* The sum of the squares of two consecutive integers is 613. Find the integers.

10. *Rectangle Dimensions.* The length of a rectangle is 1 ft longer than its width. Find the dimensions of the rectangle such that the perimeter of the rectangle is 613 ft.

12.7 EXERCISE SET

For Extra Help

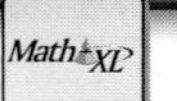

MathXL | MyMathLab | InterAct Math | Math Tutor Center | Video Lectures on CD Disc 6 | Student's Solutions Manual

a Solve.

1. *Construction.* It takes Mandy 4 hr to put up paneling in a room. Omar takes 5 hr to do the same job. How long would it take them, working together, to panel the room?

2. *Carpentry.* By checking work records, a carpenter finds that Juanita can build a small shed in 12 hr. Anton can do the same job in 16 hr. How long would it take if they worked together?

3. *Shoveling.* Vern can shovel the snow from his driveway in 45 min. Nina can do the same job in 60 min. How long would it take Nina and Vern to shovel the driveway if they worked together?

4. *Raking.* Zoë can rake her yard in 4 hr. Steffi does the same job in 3 hr. How long would it take the two of them, working together, to rake the yard?

5. *Wiring.* By checking work records, a contractor finds that Kenny Dewitt can wire a room addition in 9 hr. It takes Betty Wohnt 7 hr to wire the same room. How long would it take if they worked together?

6. *Plumbing.* By checking work records, a plumber finds that Raul can plumb a house in 48 hr. Mira can do the same job in 36 hr. How long would it take if they worked together?

7. *Gardening.* Nicole can weed her vegetable garden in 50 min. Glen can weed the same garden in 40 min. How long would it take if they worked together?

8. *Harvesting.* Bobbi can pick a quart of raspberries in 20 min. Blanche can pick a quart in 25 min. How long would it take if Bobbi and Blanche worked together?

9. *Office Printers.* The HP Officejet 4215 All-In-One printer, fax, scanner, and copier can print in black one copy of a company's year-end report in 10 min. The HP Officejet 7410 All-In-One can print the same report in 6 min. How long would it take the two printers, working together, to print one copy of the report?

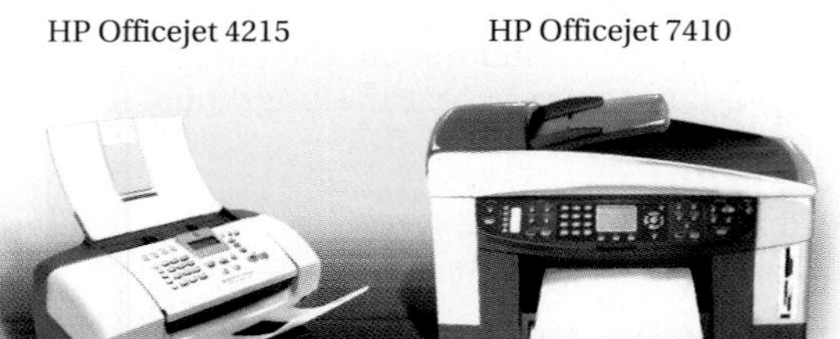

10. *Office Copiers.* The HP Officejet 7410 All-In-One printer, fax, scanner, and copier can copy in color a staff training manual in 9 min. The HP Officejet 4215 All-In-One can copy the same report in 15 min. How long would it take the two copiers, working together, to make one copy of the manual?

11. *Car Speed.* Rick drives his four-wheel-drive truck 40 km/h faster than Sarah drives her Saturn. While Sarah travels 150 km, Rick travels 350 km. Find their speeds.

Complete this table and the equations as part of the *Familiarize* step.

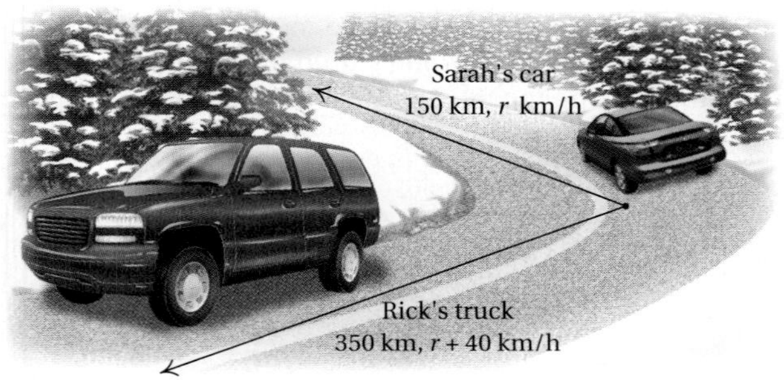

$d = r \cdot t$

	DISTANCE	SPEED	TIME	
Car	150	r		→ $150 = r(\ \ )$
Truck	350		t	→ $350 = (\ \ )t$

12. *Car Speed.* A passenger car travels 30 km/h faster than a delivery truck. While the car goes 400 km, the truck goes 250 km. Find their speeds.

13. *Train Speed.* The speed of a B & M freight train is 14 mph slower than the speed of an Amtrak passenger train. The freight train travels 330 mi in the same time that it takes the passenger train to travel 400 mi. Find the speed of each train.

Complete this table and the equations as part of the *Familiarize* step.

$d = r \cdot t$

	DISTANCE	SPEED	TIME	
B & M	330		t	→ $330 = (\ \ )t$
Amtrak	400	r		→ $400 = r(\ \ )$

14. *Train Speed.* The speed of a freight train is 15 mph slower than the speed of a passenger train. The freight train travels 390 mi in the same time that it takes the passenger train to travel 480 mi. Find the speed of each train.

15. *Trucking Speed.* A long-distance trucker traveled 120 mi in one direction during a snowstorm. The return trip in rainy weather was accomplished at double the speed and took 3 hr less time. Find the speed going.

16. *Car Speed.* After driving 126 mi, a person found that the drive would have taken 1 hr less time by increasing the speed by 8 mph. What was the actual speed?

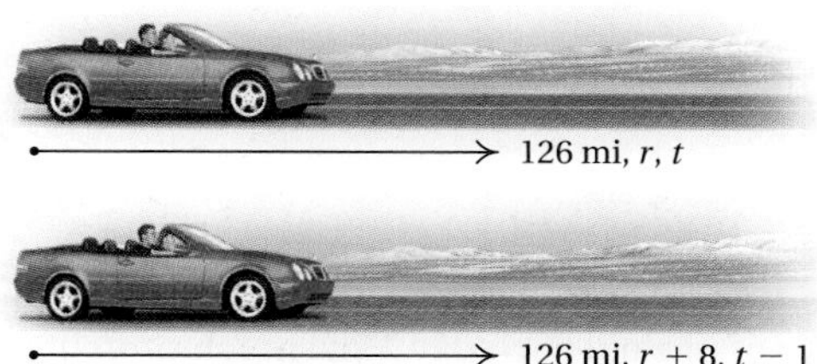

17. *Bicycle Speed.* Hank bicycles 5 km/h slower than Kelly. In the time that it takes Hank to bicycle 42 km, Kelly can bicycle 57 km. How fast does each bicyclist travel?

18. *Driving Speed.* Hillary's Lexus travels 30 mph faster than Bill's Harley. In the same time that Bill travels 75 mi, Hillary travels 120 mi. Find their speeds.

19. *Walking Speed.* Bonnie power walks 3 km/h faster than Ralph. In the time that it takes Ralph to walk 7.5 km, Bonnie walks 12 km. Find their speeds.

20. *Cross-Country Skiing.* Gerard cross-country skis 4 km/h faster than Sally. In the time that it takes Sally to ski 18 km, Gerard skis 24 km. Find their speeds.

21. *Tractor Speed.* Manley's tractor is just as fast as Caledonia's. It takes Manley 1 hr more than it takes Caledonia to drive to town. If Manley is 20 mi from town and Caledonia is 15 mi from town, how long does it take Caledonia to drive to town?

22. *Boat Speed.* Tory and Emilio's motorboats both travel at the same speed. Tory pilots her boat 40 km before docking. Emilio continues for another 2 hr, traveling a total of 100 km before docking. How long did it take Tory to navigate the 40 km?

b Find the ratio of the following. Simplify if possible.

23. 10 divorces, 18 marriages

24. 800 mi, 50 gal

25. *Speed of Black Racer.* A black racer snake travels 4.6 km in 2 hr. What is the speed in kilometers per hour?

26. *Speed of Light.* Light travels 558,000 mi in 3 sec. What is the speed in miles per second?

Solve.

27. *Protein Needs.* A 120-lb person should eat a minimum of 44 g of protein each day. How much protein should a 180-lb person eat each day?

28. *Coffee Beans.* The coffee beans from 14 trees are required to produce 7.7 kg of coffee (this is the average amount that each person in the United States drinks each year). How many trees are required to produce 320 kg of coffee?

29. *Hemoglobin.* A normal 10-cc specimen of human blood contains 1.2 g of hemoglobin. How much hemoglobin would 16 cc of the same blood contain?

30. *Walking Speed.* Wanda walked 234 km in 14 days. At this rate, how far would she walk in 42 days?

31. *Honey Bees.* Making 1 lb of honey requires 20,000 trips by bees to flowers to gather nectar. How many pounds of honey would 35,000 trips produce?

Source: Tom Turpin, Professor of Entomology, Purdue University

32. *Cockroaches and Horses.* A cockroach can run about 2 mi/hr (mph). The average body length of a cockroach is 1 in. The average body length of a horse is 8 ft (96 in.). If we assume that a horse's speed-to-length ratio is the same as that of a cockroach, how fast can a horse run?

Source: Tom Turpin, Professor of Entomology, Purdue University

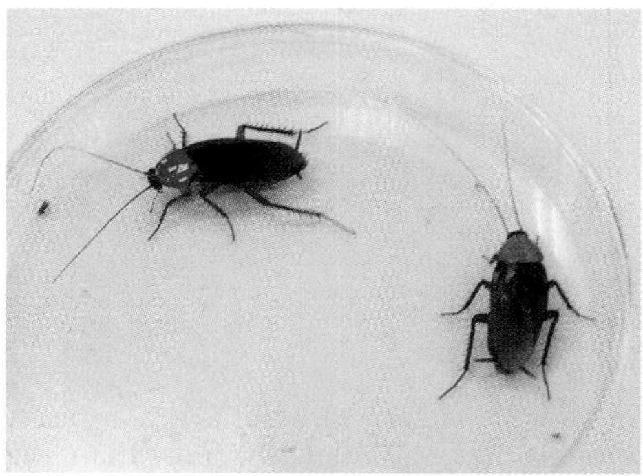

Professor Turpin founded the annual cockroach race at Purdue University.

33. *Money.* The ratio of the weight of copper to the weight of zinc in a U.S. penny is $\frac{1}{39}$. If 50 kg of zinc is being turned into pennies, how much copper is needed?

34. *Baking.* In a potato bread recipe, the ratio of milk to flour is $\frac{3}{13}$. If 5 cups of milk are used, how many cups of flour are used?

35. *Ichiro Suzuki.* In the 2004 Major League Baseball season, Ichiro Suzuki, playing for the Seattle Mariners of the American League, collected 72 hits in 217 at-bats in his first 48 games.

a) The ratio of number of hits to number of at-bats, rounded to the nearest thousandth, is a player's *batting average.* What was Suzuki's batting average in his first 48 games?

b) Based on the ratio of number of hits to number of games, how many hits would he get in the 162-game season?

c) Based on the ratio of number of hits to number of at-bats and assuming he bats 700 times in 2004, how many hits would he get?*

36. *Barry Bonds.* In the 2004 Major League Baseball season, Barry Bonds, playing for the San Francisco Giants of the National League, collected 44 hits in 117 at-bats in his first 49 games.

a) The ratio of number of hits to number of at-bats, rounded to the nearest thousandth, is a player's *batting average.* What was Bonds' batting average in his first 49 games?

b) Based on the ratio of number of hits to number of games, how many hits would he get in the 162-game season?

c) Based on the ratio of number of hits to number of at-bats and assuming he bats 400 times in 2004, how many hits would he get?†

Hat Sizes. Hat sizes are determined by measuring the circumference of one's head in either inches or centimeters. Use ratio and proportion to complete the missing parts of the following table.

	HAT SIZE	HEAD CIRCUMFERENCE (in inches)	HEAD CIRCUMFERENCE (in centimeters)
	$6\frac{3}{4}$	$21\frac{1}{5}$ in.	53.8 cm
37.	7		
38.			56.8 cm
39.		$22\frac{4}{5}$ in.	
40.	$7\frac{3}{8}$		
41.			59.8 cm
42.		24 in.	

*Ichiro Suzuki finished the 2004 American League baseball season with 262 hits in 704 at-bats and the highest 2004 American League season batting average of 0.372.
†Barry Bonds finished the 2004 National League baseball season with 135 hits in 373 at-bats and the highest 2004 National League batting average of 0.362.

43. *Estimating Trout Population.* To determine the number of trout in a lake, a conservationist catches 112 trout, tags them, and throws them back into the lake. Later, 82 trout are caught; 32 of them are tagged. Estimate the number of trout in the lake.

44. *Grass Seed.* It takes 60 oz of grass seed to seed 3000 ft^2 of lawn. At this rate, how much would be needed to seed 5000 ft^2 of lawn?

45. *Quality Control.* A sample of 144 firecrackers contained 9 "duds." How many duds would you expect in a sample of 3200 firecrackers?

46. *Frog Population.* To estimate how many frogs there are in a rain forest, a research team tags 600 frogs and then releases them. Later the team catches 300 frogs and notes that 25 of them have been tagged. Estimate the total frog population in the rain forest.

47. *Weight on Mars.* The ratio of the weight of an object on Mars to the weight of the same object on Earth is 0.4 to 1.

a) How much would a 12-ton rocket weigh on Mars?
b) How much would a 120-lb astronaut weigh on Mars?

48. *Weight on Moon.* The ratio of the weight of an object on the moon to the weight of the same object on Earth is 0.16 to 1.

a) How much would a 12-ton rocket weigh on the moon?
b) How much would a 180-lb astronaut weigh on the moon?

Geometry. For each pair of similar triangles, find the length of the indicated side.

49. b:

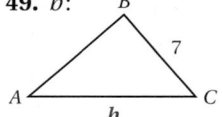

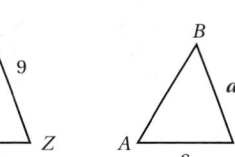

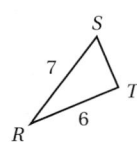

B 7 A C b

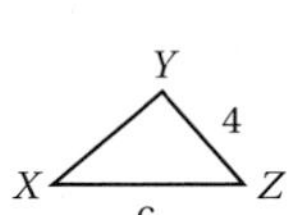

50. a:

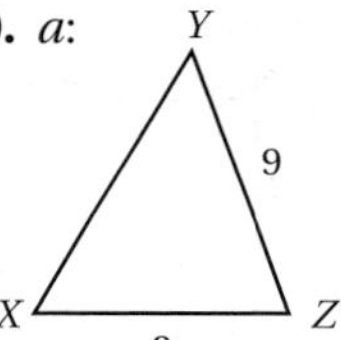

B a A C 6

51. f:

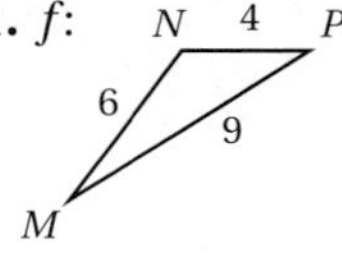

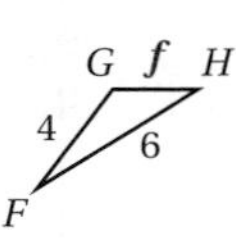

52. r:

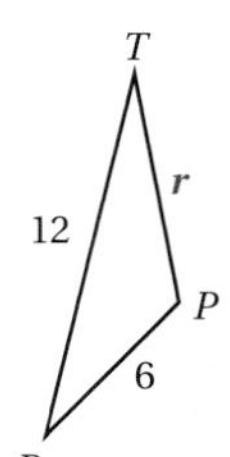

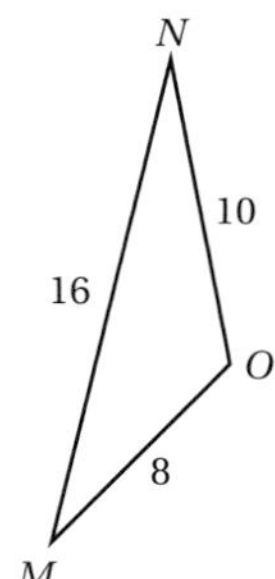

53. h:

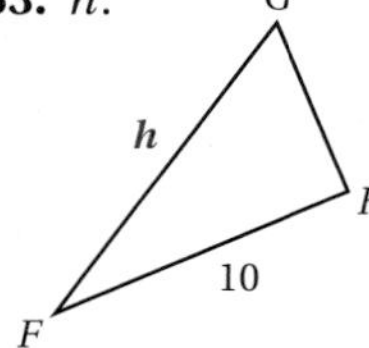

S 7 T R 6

54. n:

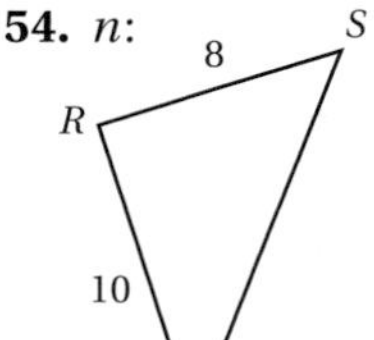

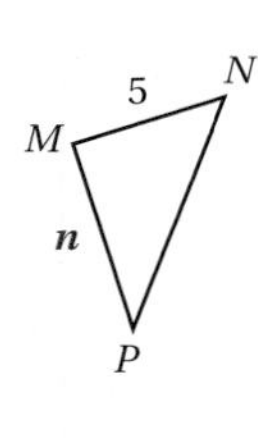

55. l:

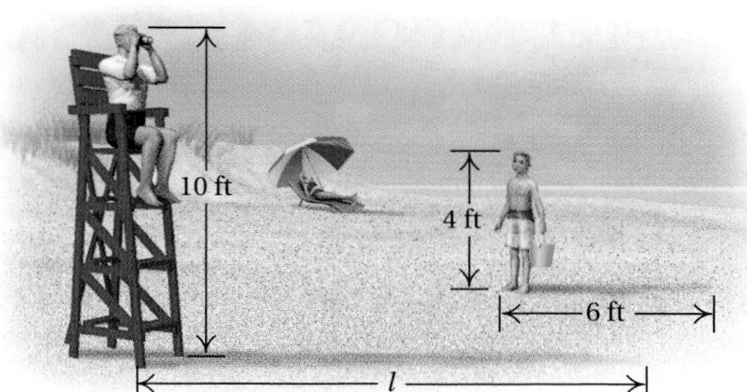

56. h:

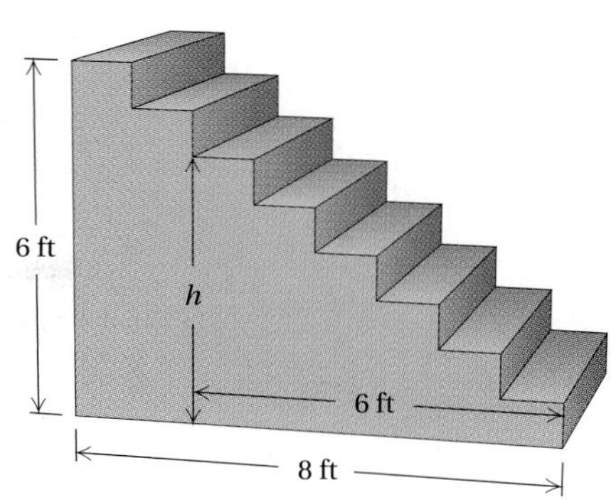

57. D_W Look over Example 2 of Section 12.7. What new equation-solving skill have we developed in this chapter that enables us to solve this problem? How do these equations differ from others in the preceding parts of this book?

58. D_W Look over Example 2 of Section 12.7. In order to solve the equation, what other skills of this chapter are required?

SKILL MAINTENANCE

Simplify. [10.1d]

59. $x^5 \cdot x^6$

60. $x^{-5} \cdot x^6$

61. $x^{-5} \cdot x^{-6}$

62. $x^5 \cdot x^{-6}$

Graph. [9.1d]

63. $y = 2x - 6$

64. $y = -2x + 6$

65. $3x + 2y = 12$

66. $x - 3y = 6$

67. $y = -\frac{3}{4}x + 2$

68. $y = \frac{2}{5}x - 4$

SYNTHESIS

69. Ann and Betty work together and complete a sales report in 4 hr. It would take Betty 6 hr longer, working alone, to do the job than it would Ann. How long would it take each of them to do the job working alone?

70. Express 100 as the sum of two numbers for which the ratio of one number, increased by 5, to the other number, decreased by 5, is 4.

71. How soon, in minutes, after 5 o'clock will the hands on a clock first be together?

72. Rachel allows herself 1 hr to reach a sales appointment 50 mi away. After she has driven 30 mi, she realizes that she must increase her speed by 15 mph in order to get there on time. What was her speed for the first 30 mi?

73. Solve $\frac{t}{a} + \frac{t}{b} = 1$ for t.

12.8 COMPLEX RATIONAL EXPRESSIONS

Objective

a Simplify complex rational expressions.

a Simplifying Complex Rational Expressions

A **complex rational expression,** or **complex fraction expression,** is a rational expression that has one or more rational expressions within its numerator or denominator. Here are some examples:

$$\frac{1+\frac{2}{x}}{3}, \quad \frac{\frac{x+y}{2}}{\frac{2x}{x+1}}, \quad \frac{\frac{1}{3}+\frac{1}{5}}{\frac{2}{x}-\frac{x}{y}}.$$

These are rational expressions within the complex rational expression.

There are two methods to simplify complex rational expressions. We will consider them both.

To the instructor and the student: *Students can be instructed to either try both methods and then choose the one that works best for them, or use the method chosen by the instructor.*

METHOD 1

MULTIPLYING BY THE LCM OF ALL THE DENOMINATORS

To simplify a complex rational expression:

1. First, find the LCM of all the denominators of all the rational expressions occurring *within* both the numerator and the denominator of the complex rational expression.
2. Then multiply by 1 using LCM/LCM.
3. If possible, simplify by removing a factor of 1.

EXAMPLE 1 Simplify: $\dfrac{\frac{1}{2}+\frac{3}{4}}{\frac{5}{6}-\frac{3}{8}}$.

We have

$$\frac{\frac{1}{2}+\frac{3}{4}}{\frac{5}{6}-\frac{3}{8}}$$

The denominators *within* the complex rational expression are 2, 4, 6, and 8. The LCM of these denominators is 24. We multiply by 1 using $\frac{24}{24}$. This amounts to multiplying both the numerator *and* the denominator by 24.

$$= \frac{\frac{1}{2}+\frac{3}{4}}{\frac{5}{6}-\frac{3}{8}}\cdot\frac{24}{24} \qquad \text{Multiplying by 1}$$

$$= \frac{\left(\frac{1}{2}+\frac{3}{4}\right)24}{\left(\frac{5}{6}-\frac{3}{8}\right)24}$$

← Multiplying the numerator by 24

← Multiplying the denominator by 24

Using the distributive laws, we carry out the multiplications:

$$= \frac{\frac{1}{2}(24) + \frac{3}{4}(24)}{\frac{5}{6}(24) - \frac{3}{8}(24)}$$

$$= \frac{12 + 18}{20 - 9} \quad \text{Simplifying}$$

$$= \frac{30}{11}.$$

Multiplying in this manner has the effect of clearing fractions in both the top and the bottom of the complex rational expression.

Do Exercise 1.

1. Simplify. Use method 1.

$$\frac{\frac{1}{3} + \frac{4}{5}}{\frac{7}{8} - \frac{5}{6}}$$

EXAMPLE 2 Simplify: $\dfrac{\frac{3}{x} + \frac{1}{2x}}{\frac{1}{3x} - \frac{3}{4x}}$.

The denominators within the complex expression are x, $2x$, $3x$, and $4x$. The LCM of these denominators is $12x$. We multiply by 1 using $12x/12x$.

$$\frac{\frac{3}{x} + \frac{1}{2x}}{\frac{1}{3x} - \frac{3}{4x}} \cdot \frac{12x}{12x} = \frac{\left(\frac{3}{x} + \frac{1}{2x}\right)12x}{\left(\frac{1}{3x} - \frac{3}{4x}\right)12x} = \frac{\frac{3}{x}(12x) + \frac{1}{2x}(12x)}{\frac{1}{3x}(12x) - \frac{3}{4x}(12x)}$$

$$= \frac{36 + 6}{4 - 9} = -\frac{42}{5}$$

Do Exercise 2.

2. Simplify. Use method 1.

$$\frac{\frac{x}{2} + \frac{2x}{3}}{\frac{1}{x} - \frac{x}{2}}$$

EXAMPLE 3 Simplify: $\dfrac{1 - \frac{1}{x}}{1 - \frac{1}{x^2}}$.

The denominators within the complex expression are x and x^2. The LCM of these denominators is x^2. We multiply by 1 using x^2/x^2. Then, after obtaining a single rational expression, we simplify:

$$\frac{1 - \frac{1}{x}}{1 - \frac{1}{x^2}} \cdot \frac{x^2}{x^2} = \frac{\left(1 - \frac{1}{x}\right)x^2}{\left(1 - \frac{1}{x^2}\right)x^2} = \frac{1(x^2) - \frac{1}{x}(x^2)}{1(x^2) - \frac{1}{x^2}(x^2)} = \frac{x^2 - x}{x^2 - 1}$$

$$= \frac{x\cancel{(x - 1)}}{(x + 1)\cancel{(x - 1)}} = \frac{x}{x + 1}.$$

Do Exercise 3.

3. Simplify. Use method 1.

$$\frac{1 + \frac{1}{x}}{1 - \frac{1}{x^2}}$$

Answers on page A-45

4. Simplify. Use method 2.

$$\frac{\frac{1}{3}+\frac{4}{5}}{\frac{7}{8}-\frac{5}{6}}$$

Answer on page A-45

METHOD 2

ADDING IN THE NUMERATOR AND THE DENOMINATOR

To simplify a complex rational expression:

1. Add or subtract, as necessary, to get a single rational expression in the numerator.
2. Add or subtract, as necessary, to get a single rational expression in the denominator.
3. Divide the numerator by the denominator.
4. If possible, simplify by removing a factor of 1.

We will redo Examples 1–3 using this method.

EXAMPLE 4 Simplify: $\dfrac{\frac{1}{2}+\frac{3}{4}}{\frac{5}{6}-\frac{3}{8}}$.

The LCM of 2 and 4 in the numerator is 4. The LCM of 6 and 8 in the denominator is 24. We have

$$\frac{\frac{1}{2}+\frac{3}{4}}{\frac{5}{6}-\frac{3}{8}}=\frac{\frac{1}{2}\cdot\frac{2}{2}+\frac{3}{4}}{\frac{5}{6}\cdot\frac{4}{4}-\frac{3}{8}\cdot\frac{3}{3}}$$

Multiplying the $\frac{1}{2}$ by 1 to get the common denominator, 4

Multiplying the $\frac{5}{6}$ and the $\frac{3}{8}$ by 1 to get the common denominator, 24

$$=\frac{\frac{2}{4}+\frac{3}{4}}{\frac{20}{24}-\frac{9}{24}}$$

$$=\frac{\frac{5}{4}}{\frac{11}{24}}$$ Adding in the numerator; subtracting in the denominator

$$=\frac{5}{4}\cdot\frac{24}{11}$$ Multiplying by the reciprocal of the divisor

$$=\frac{5\cdot 3\cdot 2\cdot 2\cdot 2}{2\cdot 2\cdot 11}$$ Factoring

$$=\frac{5\cdot 3\cdot 2\cdot \not{2}\cdot \not{2}}{\not{2}\cdot \not{2}\cdot 11}$$ Removing a factor of 1: $\frac{2\cdot 2}{2\cdot 2}=1$

$$=\frac{30}{11}.$$

Do Exercise 4.

EXAMPLE 5 Simplify: $\dfrac{\dfrac{3}{x}+\dfrac{1}{2x}}{\dfrac{1}{3x}-\dfrac{3}{4x}}$.

We have

$$\frac{\frac{3}{x}+\frac{1}{2x}}{\frac{1}{3x}-\frac{3}{4x}} = \frac{\frac{3}{x}\cdot\frac{2}{2}+\frac{1}{2x}}{\frac{1}{3x}\cdot\frac{4}{4}-\frac{3}{4x}\cdot\frac{3}{3}}$$

Finding the LCD, $2x$, and multiplying by 1 in the numerator

Finding the LCD, $12x$, and multiplying by 1 in the denominator

$$= \frac{\frac{6}{2x}+\frac{1}{2x}}{\frac{4}{12x}-\frac{9}{12x}} = \frac{\frac{7}{2x}}{\frac{-5}{12x}}$$

Adding in the numerator and subtracting in the denominator

$$= \frac{7}{2x}\cdot\frac{12x}{-5}$$

Multiplying by the reciprocal of the divisor

$$= \frac{7}{2x}\cdot\frac{6(2x)}{-5}$$

Factoring

$$= \frac{7}{\cancel{2x}}\cdot\frac{6(\cancel{2x})}{-5}$$

Removing a factor of 1: $\frac{2x}{2x}=1$

$$= \frac{42}{-5} = -\frac{42}{5}.$$

Do Exercise 5.

5. Simplify. Use method 2.

$$\frac{\frac{x}{2}+\frac{2x}{3}}{\frac{1}{x}-\frac{x}{2}}$$

EXAMPLE 6 Simplify: $\dfrac{1-\dfrac{1}{x}}{1-\dfrac{1}{x^2}}$.

We have

$$\frac{1-\frac{1}{x}}{1-\frac{1}{x^2}} = \frac{1\cdot\frac{x}{x}-\frac{1}{x}}{1\cdot\frac{x^2}{x^2}-\frac{1}{x^2}}$$

Finding the LCD, x, and multiplying by 1 in the numerator

Finding the LCD, x^2, and multiplying by 1 in the denominator

$$= \frac{\frac{x-1}{x}}{\frac{x^2-1}{x^2}}$$

Subtracting in the numerator and subtracting in the denominator

$$= \frac{x-1}{x}\cdot\frac{x^2}{x^2-1}$$

Multiplying by the reciprocal of the divisor

$$= \frac{(x-1)x\cdot x}{x(x-1)(x+1)}$$

Factoring

$$= \frac{\cancel{(x-1)}\cancel{x}\cdot x}{\cancel{x}\cancel{(x-1)}(x+1)}$$

Removing a factor of 1: $\frac{x(x-1)}{x(x-1)}=1$

$$= \frac{x}{x+1}.$$

Do Exercise 6.

6. Simplify. Use method 2.

$$\frac{1+\frac{1}{x}}{1-\frac{1}{x^2}}$$

Answers on page A-45

EXERCISE SET

For Extra Help

 MathXL
 MyMathLab
 InterAct Math
 Math Tutor Center
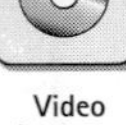 Video Lectures on CD Disc 6
Student's Solutions Manual

a Simplify.

1. $\dfrac{1 + \frac{9}{16}}{1 - \frac{3}{4}}$

2. $\dfrac{6 - \frac{3}{8}}{4 + \frac{5}{6}}$

3. $\dfrac{1 - \frac{3}{5}}{1 + \frac{1}{5}}$

4. $\dfrac{2 + \frac{2}{3}}{2 - \frac{2}{3}}$

5. $\dfrac{\frac{1}{2} + \frac{3}{4}}{\frac{5}{8} - \frac{5}{6}}$

6. $\dfrac{\frac{3}{4} + \frac{7}{8}}{\frac{2}{3} - \frac{5}{6}}$

7. $\dfrac{\frac{1}{x} + 3}{\frac{1}{x} - 5}$

8. $\dfrac{2 - \frac{1}{a}}{4 + \frac{1}{a}}$

9. $\dfrac{4 - \frac{1}{x^2}}{2 - \frac{1}{x}}$

10. $\dfrac{\frac{2}{y} + \frac{1}{2y}}{y + \frac{y}{2}}$

11. $\dfrac{8 + \frac{8}{d}}{1 + \frac{1}{d}}$

12. $\dfrac{3 + \frac{2}{t}}{3 - \frac{2}{t}}$

13. $\dfrac{\frac{x}{8} - \frac{8}{x}}{\frac{1}{8} + \frac{1}{x}}$

14. $\dfrac{\frac{2}{m} + \frac{m}{2}}{\frac{m}{3} - \frac{3}{m}}$

15. $\dfrac{1 + \frac{1}{y}}{1 - \frac{1}{y^2}}$

16. $\dfrac{\frac{1}{q^2} - 1}{\frac{1}{q} + 1}$

17. $\dfrac{\frac{1}{5} - \frac{1}{a}}{\frac{5 - a}{5}}$

18. $\dfrac{\frac{4}{t}}{4 + \frac{1}{t}}$

19. $\dfrac{\frac{1}{a} + \frac{1}{b}}{\frac{1}{a^2} - \frac{1}{b^2}}$

20. $\dfrac{\frac{1}{x^2} - \frac{1}{y^2}}{\frac{2}{x} - \frac{2}{y}}$

21. $\dfrac{\frac{p}{q} + \frac{q}{p}}{\frac{1}{p} + \frac{1}{q}}$

22. $\dfrac{x - 3 + \frac{2}{x}}{x - 4 + \frac{3}{x}}$

23. $\dfrac{\frac{2}{a} + \frac{4}{a^2}}{\frac{5}{a^3} - \frac{3}{a}}$

24. $\dfrac{\frac{5}{x^3} - \frac{1}{x^2}}{\frac{2}{x} + \frac{3}{x^2}}$

25. $\dfrac{\dfrac{2}{7a^4} - \dfrac{1}{14a}}{\dfrac{3}{5a^2} + \dfrac{2}{15a}}$

26. $\dfrac{\dfrac{5}{4x^3} - \dfrac{3}{8x}}{\dfrac{3}{2x} + \dfrac{3}{4x^3}}$

27. $\dfrac{\dfrac{a}{b} + \dfrac{c}{d}}{\dfrac{b}{a} + \dfrac{d}{c}}$

28. $\dfrac{\dfrac{a}{b} - \dfrac{c}{d}}{\dfrac{b}{a} - \dfrac{d}{c}}$

29. $\dfrac{\dfrac{x}{5y^3} + \dfrac{3}{10y}}{\dfrac{3}{10y} + \dfrac{x}{5y^3}}$

30. $\dfrac{\dfrac{a}{6b^3} + \dfrac{4}{9b^2}}{\dfrac{5}{6b} - \dfrac{1}{9b^3}}$

31. $\dfrac{\dfrac{3}{x+1} + \dfrac{1}{x}}{\dfrac{2}{x+1} + \dfrac{3}{x}}$

32. $\dfrac{x - 7 + \dfrac{5}{x-1}}{x - 3 + \dfrac{1}{x-1}}$

33. **D**$_\text{W}$ Why is factoring an important skill when simplifying rational expressions?

34. **D**$_\text{W}$ Why is the distributive law especially important when using method 1 of this section?

SKILL MAINTENANCE

Add. [10.4a]

35. $(2x^3 - 4x^2 + x - 7) + (4x^4 + x^3 + 4x^2 + x)$

36. $(2x^3 - 4x^2 + x - 7) + (-2x^3 + 4x^2 - x + 7)$

Factor. [11.6a]

37. $p^2 - 10p + 25$

38. $p^2 + 10p + 25$

39. $50p^2 - 100$

40. $5p^2 - 40p - 100$

Solve. [11.8a]

41. *Perimeter of a Rectangle.* The length of a rectangle is 3 yd greater than the width. The area of the rectangle is 10 yd^2. Find the perimeter.

42. *Ladder Distances.* A ladder of length 13 ft is placed against a building in such a way that the distance from the top of the ladder to the ground is 7 ft more than the distance from the bottom of the ladder to the building. Find these distances.

SYNTHESIS

43. Find the reciprocal of $\dfrac{2}{x-1} - \dfrac{1}{3x-2}$.

Simplify.

44. $\left[\dfrac{\dfrac{x+1}{x-1} + 1}{\dfrac{x+1}{x-1} - 1}\right]^5$

45. $1 + \dfrac{1}{1 + \dfrac{1}{1 + \dfrac{1}{1 + \dfrac{1}{x}}}}$

46. $\dfrac{\dfrac{z}{1 - \dfrac{z}{2 + 2z}} - 2z}{\dfrac{2z}{5z - 2} - 3}$

12.9 DIRECT AND INVERSE VARIATION

Objectives

a Find an equation of direct variation given a pair of values of the variables.

b Solve applied problems involving direct variation.

c Find an equation of inverse variation given a pair of values of the variables.

d Solve applied problems involving inverse variation.

a Equations of Direct Variation

A bicycle is traveling at a speed of 15 km/h. In 1 hr, it goes 15 km; in 2 hr, it goes 30 km; in 3 hr, it goes 45 km; and so on. We can form a set of ordered pairs using the number of hours as the first coordinate and the number of kilometers traveled as the second coordinate. These determine a set of ordered pairs:

$$(1, 15),\ (2, 30),\ (3, 45),\ (4, 60),\quad \text{and so on.}$$

Note that the second coordinate is always 15 times the first.

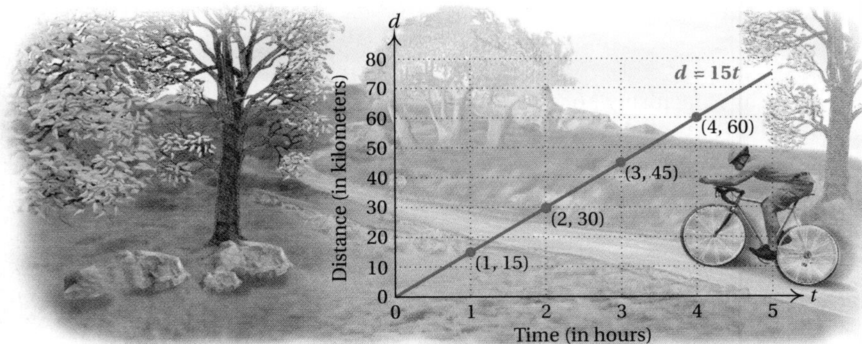

In this example, distance is a constant multiple of time, so we say that there is *direct variation* and that distance *varies directly* as time. The *equation of variation* is $d = 15t$.

DIRECT VARIATION

When a situation translates to an equation described by $y = kx$, with k a positive constant, we say that **y varies directly as x.** The equation $y = kx$ is called an **equation of direct variation.**

In direct variation, as one variable increases, the other variable increases as well. This is shown in the graph above.

The terminologies

"y varies as x,"

"y is directly proportional to x," and

"y is proportional to x"

also imply direct variation and are used in many situations. The constant k is called the **constant of proportionality** or the **variation constant.** It can be found if one pair of values of x and y is known. Once k is known, other pairs can be determined.

Study Tips

MAKING APPLICATIONS REAL

Newspapers and magazines are full of mathematical applications. Find such an application and share it with your class. As you develop more skills in mathematics, you will find yourself observing the world from a different perspective, seeing mathematics everywhere. Math courses become more interesting when we connect the concepts to the real world.

EXAMPLE 1 Find an equation of variation in which y varies directly as x and $y = 7$ when $x = 25$.

We first substitute to find k:

$y = kx$

$7 = k \cdot 25$ Substituting 25 for x and 7 for y

$\frac{7}{25} = k$, or $k = 0.28$. Solving for k, the variation constant

Then the equation of variation is

$y = 0.28x$.

The answer is the equation $y = 0.28x$, *not* $k = 0.28$. We can visualize the example by looking at the graph.

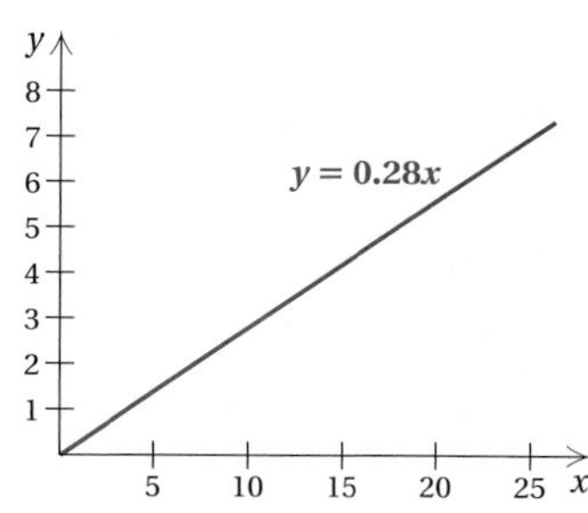

We see that when y varies directly as x, the constant of proportionality is also the slope of the associated graph—the rate at which y changes with respect to x.

EXAMPLE 2 Find an equation in which s varies directly as t and $s = 10$ when $t = 15$. Then find the value of s when $t = 32$.

We have

$s = kt$ We know that s varies directly as t.

$10 = k \cdot 15$ Substituting 10 for s and 15 for t

$\frac{10}{15} = k$, or $k = \frac{2}{3}$. Solving for k

Thus the equation of variation is $s = \frac{2}{3}t$.

$s = \frac{2}{3}t$

$s = \frac{2}{3} \cdot 32$ Substituting 32 for t in the equation of variation

$s = \frac{64}{3}$, or $21\frac{1}{3}$.

The value of s is $21\frac{1}{3}$ when $t = 32$.

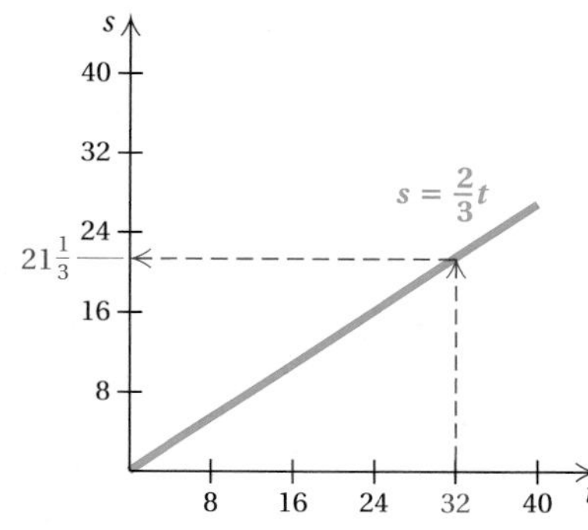

Do Exercises 1 and 2.

1. Find an equation of variation in which y varies directly as x and $y = 84$ when $x = 12$. Then find the value of y when $x = 41$.

2. Find an equation of variation in which y varies directly as x and $y = 50$ when $x = 80$. Then find the value of y when $x = 20$.

Answers on page A-45

3. **Electricity Costs.** The cost C of operating a television varies directly as the number n of hours that it is in operation. It costs \$14.00 to operate a standard-size color TV continuously for 30 days.

   a) Find an equation of variation.

   b) At this rate, how much would it cost to operate the TV for 1 day? for 1 hour?

4. **Weight on Venus.** The weight V of an object on Venus varies directly as its weight E on Earth. A person weighing 165 lb on Earth would weigh 145.2 lb on Venus.

   a) Find an equation of variation.

   b) How much would a person weighing 198 lb on Earth weigh on Venus?

Answers on page A-45

b Applications of Direct Variation

EXAMPLE 3 *Karat Ratings of Gold Objects.* The beauty of gold has inspired artisans throughout the centuries, as shown by this magnificent sarcophagus, which contained Tutankhamen's mummy. Today we know that the karat rating K of a gold object varies directly as the actual percentage P of gold in the object. A 14-karat gold ring is 58.25% gold. What is the percentage of gold in a 10-karat chain?

Source: Barbara Ann Kipfer, *The Order of Things.* New York: Random House, 1998.

1., 2. Familiarize and **Translate.** The problem states that we have direct variation between the variables K and P. Thus an equation $K = kP$, $k > 0$, applies. As the percentage of gold increases, the karat rating increases. The letters K and k represent different quantities.

3. Solve. The mathematical manipulation has two parts. First, we determine the equation of variation by substituting known values for K and P to find the variation constant k. Second, we compute the percentage of gold in a 10-karat chain.

a) First, we find an equation of variation:

$$K = kP$$

$$14 = k(0.5825) \quad \text{Substituting 14 for } K \text{ and 58.25\%, or 0.5825, for } P$$

$$\frac{14}{0.5825} = k$$

$$24.03 \approx k. \quad \text{Dividing and rounding to the nearest hundredth}$$

The equation of variation is $K = 24.03P$.

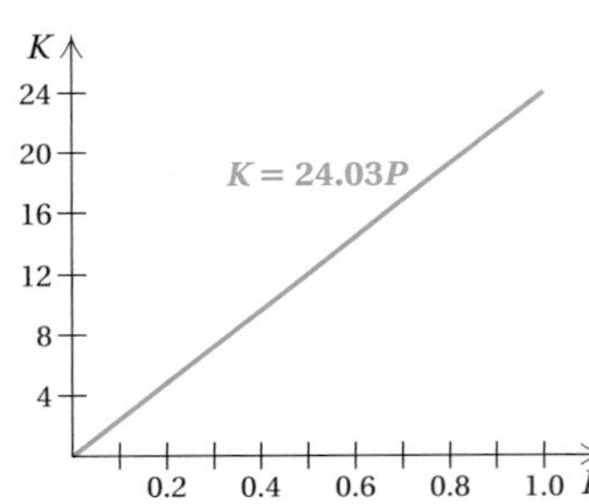

b) We then use the equation to find the percentage of gold in a 10-karat chain:

$$K = 24.03P$$

$$10 = 24.03P \quad \text{Substituting 10 for } K$$

$$\frac{10}{24.03} = P$$

$$0.416 \approx P$$

$$41.6\% \approx P.$$

4. Check. The check might be done by repeating the computations. You might also do some reasoning about the answer. The karat rating decreased from 14 to 10. Similarly, the percentage decreased from 58.25% to 41.6%.

5. State. A 10-karat chain is 41.6% gold.

Do Exercises 3 and 4.

Let's consider direct variation from the standpoint of a graph. The graph of $y = kx$, $k > 0$, always goes through the origin and rises from left to right. Note that as x increases, y increases; and as x decreases, y decreases. This is why the terminology "direct" is used. What one variable does, the other does as well.

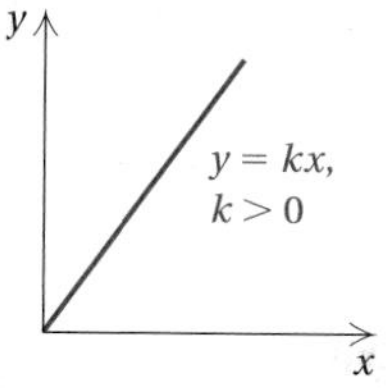

C Equations of Inverse Variation

A car is traveling a distance of 20 mi. At a speed of 5 mph, it will take 4 hr; at 20 mph, it will take 1 hr; at 40 mph, it will take $\frac{1}{2}$ hr; and so on. We use speed as the first coordinate and the time as the second coordinate. These determine a set of ordered pairs:

$$(5, 4),\ (20, 1),\ \left(40, \tfrac{1}{2}\right),\ \left(60, \tfrac{1}{3}\right),\quad \text{and so on.}$$

Note that the product of speed and time for each of these pairs is 20. Note too that as the speed *increases,* the time *decreases.*

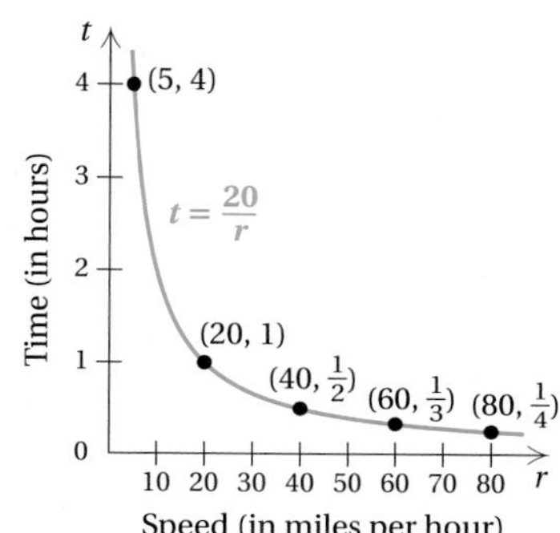

In this case, the product of speed and time is constant so we say that there is *inverse variation* and that time *varies inversely* as speed. The equation of variation is

$$rt = 20 \text{ (a constant)}, \quad \text{or} \quad t = \frac{20}{r}.$$

INVERSE VARIATION

When a situation translates to an equation described by $y = k/x$, with k a positive constant, we say that **y varies inversely as x.** The equation $y = k/x$ is called an **equation of inverse variation.**

In inverse variation, as one variable increases, the other variable decreases.

The terminology

"y is inversely proportional to x"

also implies inverse variation and is used in some situations. The constant k is again called the **constant of proportionality** or the **variation constant.**

5. Find an equation of variation in which y varies inversely as x and $y = 105$ when $x = 0.6$. Then find the value of y when $x = 20$.

6. Find an equation of variation in which y varies inversely as x and $y = 45$ when $x = 20$. Then find the value of y when $x = 1.6$.

Answers on page A-45

EXAMPLE 4 Find an equation of variation in which y varies inversely as x and $y = 145$ when $x = 0.8$. Then find the value of y when $x = 25$.

We first substitute to find k:

$$y = \frac{k}{x}$$
$$145 = \frac{k}{0.8}$$
$$(0.8)145 = k$$
$$116 = k.$$

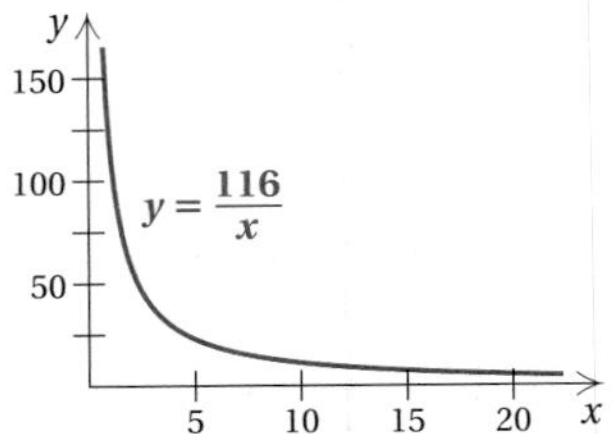

The equation of variation is $y = 116/x$. The answer is the equation $y = 116/x$, *not* $k = 116$.

When $x = 25$, we have

$$y = \frac{116}{x}$$
$$y = \frac{116}{25} \quad \text{Substituting 25 for } x$$
$$y = 4.64.$$

The value of y is 4.64 when $x = 25$.

Do Exercises 5 and 6.

The graph of $y = k/x$, $k > 0$, is shaped like the figure at right for positive values of x. (You need not know how to graph such equations at this time.) Note that as x increases, y decreases; and as x decreases, y increases. This is why the terminology "inverse" is used. One variable does the opposite of what the other does.

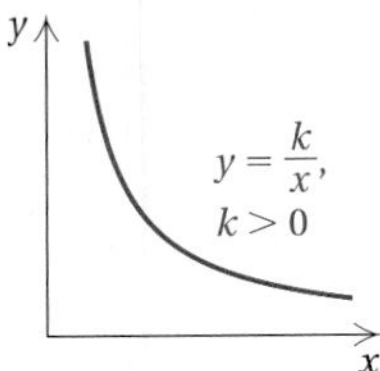

d Applications of Inverse Variation

Often in an applied situation we must decide which kind of variation, if any, might apply to the problem.

EXAMPLE 5 *Work Time.* Molly is a maintenance supervisor. She notes that it takes 4 hr for 20 people to wash and wax the floors in a building. How long would it then take 25 people to do the job?

1. **Familiarize.** Think about the problem situation. What kind of variation would be used? It seems reasonable that the more people there are working on the job, the less time it will take to finish. (One might argue that too many people in a crowded area would be counterproductive, but we will disregard that possibility.) Thus inverse variation might apply. We let $T =$ the time to do the job, in hours, and $N =$ the number of people. Assuming inverse variation, we know that an equation $T = k/N$, $k > 0$, applies. As the number of people increases, the time it takes to do the job decreases.

2. **Translate.** We write an equation of variation:

$$T = \frac{k}{N}.$$

Time varies inversely as the number of people involved.

3. **Solve.** The mathematical manipulation has two parts. First, we find the equation of variation by substituting known values for T and N to find k. Second, we compute the amount of time it would take 25 people to do the job.

a) First, we find an equation of variation:

$$T = \frac{k}{N}$$

$$4 = \frac{k}{20} \quad \text{Substituting 4 for } T \text{ and 20 for } N$$

$$20 \cdot 4 = k.$$

$$80 = k.$$

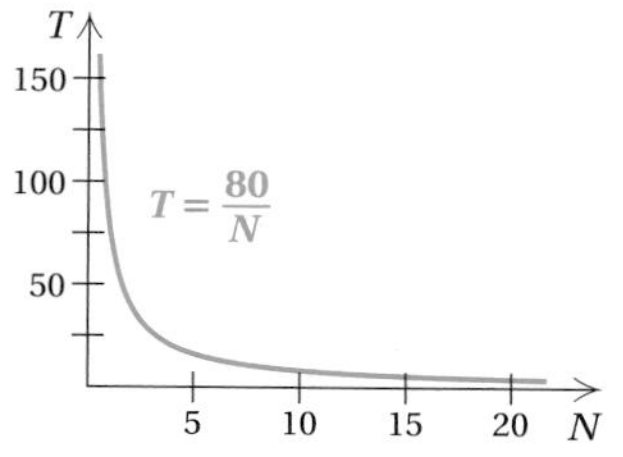

The equation of variation is $T = \frac{80}{N}$.

b) We then use the equation to find the amount of time that it takes 25 people to do the job:

$$T = \frac{80}{N}$$

$$= \frac{80}{25} \quad \text{Substituting 25 for } N$$

$$= 3.2.$$

4. **Check.** The check might be done by repeating the computations. We might also analyze the results. The number of people increased from 20 to 25. Did the time decrease? It did, and this confirms what we expect with inverse variation.

5. **State.** It should take 3.2 hr for 25 people to complete the job.

Do Exercises 7 and 8.

7. Referring to Example 5, determine how long it would take 10 people to do the job.

8. **Time of Travel.** The time t required to drive a fixed distance varies inversely as the speed r. It takes 5 hr at 60 km/h to drive a fixed distance.

a) Find an equation of variation.

b) How long would it take at 40 km/h?

Answers on page A-45

12.9 EXERCISE SET

For Extra Help

MathXL

MyMathLab

InterAct Math

Math Tutor Center

Video Lectures on CD Disc 6

Student's Solutions Manual

a Find an equation of variation in which y varies directly as x and the following are true. Then find the value of y when $x = 20$.

1. $y = 36$ when $x = 9$

2. $y = 60$ when $x = 16$

3. $y = 0.8$ when $x = 0.5$

4. $y = 0.7$ when $x = 0.4$

5. $y = 630$ when $x = 175$

6. $y = 400$ when $x = 125$

7. $y = 500$ when $x = 60$

8. $y = 200$ when $x = 300$

b Solve.

9. *Wages and Work Time.* A person's paycheck P varies directly as the number H of hours worked. For working 15 hr, the pay is $84.

a) Find an equation of variation.
b) Find the pay for 35 hr of work.

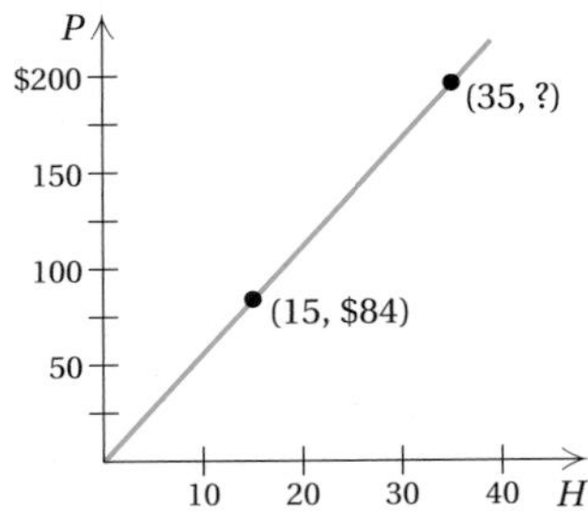

10. *Interest and Interest Rate.* The interest I earned in 1 yr on a fixed principal varies directly as the interest rate r. An investment earns $53.55 at an interest rate of 4.25%.

a) Find an equation of variation.
b) How much will the investment earn at a rate of 5.75%?

11. *Cost of Sand.* The cost C, in dollars, to fill a sandbox varies directly as the depth S, in inches, of the sand. Ella, the director of Creekside Daycare, checks at her local hardware store and finds that it would cost $67.50 to fill the box with 6 in. of sand. She decides to fill the sandbox to a depth of 9 in.

a) Find an equation of variation.
b) How much will the sand cost Ella?

12. *Cost of Cement.* The cost C, in dollars, of cement needed to pave a driveway varies directly as the depth D, in inches, of the driveway. John checks at his local building materials store and finds that it costs $500 to install his driveway with a depth of 8 in. He decides to build a stronger driveway at a depth of 12 in.

a) Find an equation of variation.
b) How much will it cost for the cement?

13. *Lunar Weight.* The weight M of an object on the moon varies directly as its weight E on Earth. Your author, Marv Bittinger, weighs 192 lb, but would weigh only 32 lb on the moon.

a) Find an equation of variation.
b) Marv's wife, Elaine, weighs 110 lb on Earth. How much would she weigh on the moon?
c) Marv's granddaughter, Maggie, would weigh only 5 lb on the moon. How much does Maggie weigh on Earth?

14. *Mars Weight.* The weight M of an object on Mars varies directly as its weight E on Earth. In 1999, Chen Yanqing, who weighs 128 lb, set a world record for her weight class with a lift (snatch) of 231 lb. On Mars, this lift would be only 88 lb.

Source: *The Guinness Book of Records*, 2001

a) Find an equation of variation.
b) How much would Yanqing weigh on Mars?

15. *Computer Megahertz.* The number of computer instructions N per second varies directly as the speed S of its internal processor. A processor with a speed of 25 megahertz can perform 2,000,000 instructions per second.

a) Find an equation of variation.
b) How many instructions will the same processor perform if it is running at a speed of 200 megahertz?

16. *Water in Human Body.* The number of kilograms W of water in a human body varies directly as the total body weight B. A person who weighs 75 kg contains 54 kg of water.

a) Find an equation of variation.
b) How many kilograms of water are in a person who weighs 95 kg?

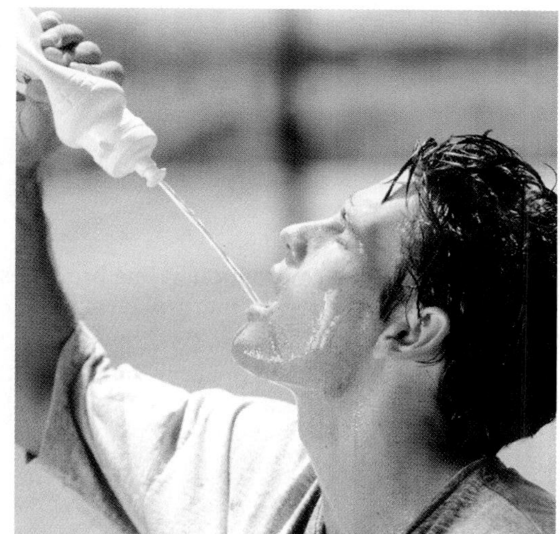

17. *Steak Servings.* The number of servings S of meat that can be obtained from round steak varies directly as the weight W. From 9 kg of round steak, one can get 70 servings of meat. How many servings can one get from 12 kg of round steak?

18. *Turkey Servings.* A chef is planning meals in a refreshment tent at a golf tournament. The number of servings S of meat that can be obtained from a turkey varies directly as its weight W. From a turkey weighing 30.8 lb, one can get 40 servings of meat. How many servings can be obtained from a 19.8-lb turkey?

c Find an equation of variation in which y varies inversely as x and the following are true. Then find the value of y when $x = 10$.

19. $y = 3$ when $x = 25$

20. $y = 2$ when $x = 45$

21. $y = 10$ when $x = 8$

22. $y = 10$ when $x = 7$

23. $y = 6.25$ when $x = 0.16$

24. $y = 0.125$ when $x = 8$

25. $y = 50$ when $x = 42$

26. $y = 25$ when $x = 42$

27. $y = 0.2$ when $x = 0.3$

28. $y = 0.4$ when $x = 0.6$

b, **d** Solve.

29. *Production and Time.* A production line produces 15 CD players every 8 hr. How many players can it produce in 37 hr?

a) What kind of variation might apply to this situation?
b) Solve the problem.

30. *Wages and Work Time.* A person works for 15 hr and makes $93.75. How much will the person make by working 35 hr?

a) What kind of variation might apply to this situation?
b) Solve the problem.

31. *Cooking Time.* It takes 4 hr for 9 cooks to prepare the food for a wedding rehearsal dinner. How long will it take 8 cooks to prepare the dinner?

a) What kind of variation might apply to this situation?
b) Solve the problem.

32. *Work Time.* It takes 16 hr for 2 people to resurface a tennis court. How long will it take 6 people to do the job?

a) What kind of variation might apply to this situation?
b) Solve the problem.

33. *Miles per Gallon.* To travel a fixed distance, the number of gallons N of gasoline needed is inversely proportional to the miles-per-gallon rating P of the car. A car that gets 20 miles per gallon (mpg) needs 14 gal to travel the distance.

a) Find an equation of variation.

b) How much gas will be needed for a car that gets 28 mpg?

34. *Miles per Gallon.* To travel a fixed distance, the number of gallons N of gasoline needed is inversely proportional to the miles-per-gallon rating P of the car. A car that gets 25 miles per gallon (mpg) needs 12 gal to travel the distance.

a) Find an equation of variation.

b) How much gas will be needed for a car that gets 20 mpg?

35. *Electrical Current.* The current I in an electrical conductor varies inversely as the resistance R of the conductor. The current is 96 amperes when the resistance is 20 ohms.

a) Find an equation of variation.

b) What is the current when the resistance is 60 ohms?

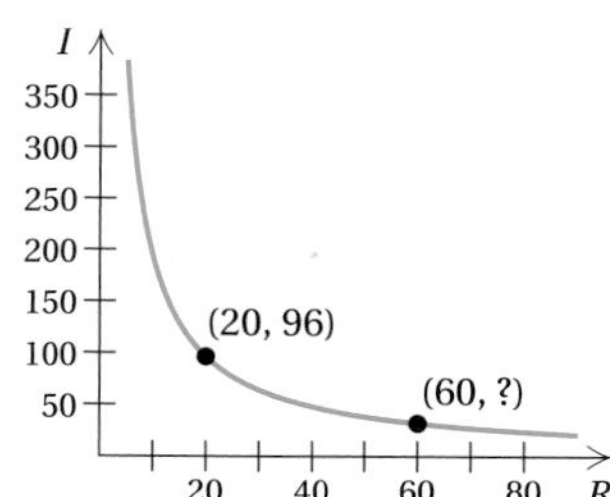

36. *Gas Volume.* The volume V of a gas varies inversely as the pressure P on it. The volume of a gas is 200 cm^3 under a pressure of 32 kg/cm^2.

a) Find an equation of variation.

b) What will be its volume under a pressure of 20 kg/cm^2?

37. *Answering Questions.* For a fixed time limit for a quiz, the number of minutes m that a student should allow for each question on a quiz is inversely proportional to the number of questions n on the quiz. A 16-question quiz means that students have 2.5 min per question.

a) Find an equation of variation.

b) How many questions would appear on a quiz in which students have 4 min per question?

38. *Pumping Time.* The time t required to empty a tank varies inversely as the rate r of pumping. A pump can empty a tank in 90 min at a rate of 1200 L/min.

a) Find an equation of variation.

b) How long will it take the pump to empty the tank at a rate of 2000 L/min?

39. *Apparent Size.* The apparent size A of an object varies inversely as the distance d of the object from the eye. A flagpole 30 ft from an observer appears to be 27.5 ft tall. How tall will the same flagpole appear to be if it is 100 ft from the eye?

40. *Driving Time.* The time t required to drive a fixed distance varies inversely as the speed r. It takes 5 hr at 55 mph to drive a fixed distance. How long would it take at 40 mph?

DW In Exercises 41–44, determine whether the situation represents direct variation, inverse variation, or neither. Give a reason for your answer.

41. The cost of mailing a package in the United States and the weight of the package

42. The number of hours that a student watches TV per week and the student's grade point average

43. The weight of a turkey and the cooking time

44. The number of plays that it takes to go 80 yd for a touchdown and the average gain per play

SKILL MAINTENANCE

Solve. [12.6a]

45. $\frac{x+2}{x+5} = \frac{x-4}{x-6}$

46. $\frac{x-3}{x-5} = \frac{x+5}{x+1}$

Solve. [11.7b]

47. $x^2 - 25x + 144 = 0$

48. $t^2 + 21t + 108 = 0$

49. $35x^2 + 8 = 34x$

50. $14x^2 - 19x - 3 = 0$

Calculate. [7.8d]

51. $3^7 \div 3^4 \div 3^3 \div 3$

52. $\frac{37 - 5(4-6)}{2 \cdot 6 + 8}$

53. $-5^2 + 4 \cdot 6$

54. $(-5)^2 + 4 \cdot 6$

SYNTHESIS

55. Graph the equation that corresponds to Exercise 12. Then use the TABLE feature to create a table with TblStart = 1 and ΔTbl = 1. What happens to the y-values as the x-values become larger?

56. Graph the equation that corresponds to Exercise 17. Then use the TABLE feature to create a table with TblStart = 1 and ΔTbl = 1. What happens to the y-values as the x-values become larger?

Write an equation of variation for the situation.

57. The square of the pitch P of a vibrating string varies directly as the tension t on the string.

58. In a stream, the amount S of salt carried varies directly as the sixth power of the speed V of the stream.

59. The power P in a windmill varies directly as the cube of the wind speed V.

60. The volume V of a sphere varies directly as the cube of the radius r.

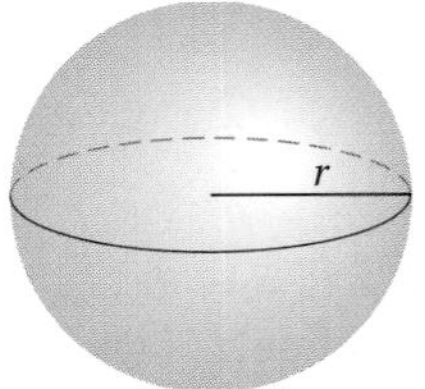

12 Summary and Review

The review that follows is meant to prepare you for a chapter exam. It consists of two parts. The first part, Concept Reinforcement, is designed to increase understanding of the concepts through true/false exercises. The second part is the Review Exercises. These provide practice exercises for the exam, together with references to section objectives so you can go back and review. Before beginning, stop and look back over the skills you have obtained. What skills in mathematics do you have now that you did not have before studying this chapter?

CONCEPT REINFORCEMENT

Determine whether the statement is true or false. Answers are given at the back of the book.

_______ **1.** To find the LCM, use each factor the least number of times that it appears in any one factorization.

_______ **2.** The value of the numerator has no bearing on whether or not a rational expression is defined.

_______ **3.** To add or subtract rational expressions when the denominators are the same, add or subtract the numerator and keep the same denominator.

_______ **4.** When clearing an equation of fractions, multiply both sides of the equation by the LCM.

_______ **5.** When a situation translates to an equation described by $y = k/x$, with k a positive constant, y varies directly as x.

Review Exercises

Find all numbers for which the rational expression is not defined. [12.1a]

1. $\dfrac{3}{x}$

2. $\dfrac{4}{x-6}$

3. $\dfrac{x+5}{x^2-36}$

4. $\dfrac{x^2-3x+2}{x^2+x-30}$

5. $\dfrac{-4}{(x+2)^2}$

6. $\dfrac{x-5}{5}$

Simplify. [12.1c]

7. $\dfrac{4x^2-8x}{4x^2+4x}$

8. $\dfrac{14x^2-x-3}{2x^2-7x+3}$

9. $\dfrac{(y-5)^2}{y^2-25}$

Multiply and simplify. [12.1d]

10. $\dfrac{a^2-36}{10a} \cdot \dfrac{2a}{a+6}$

11. $\dfrac{6t-6}{2t^2+t-1} \cdot \dfrac{t^2-1}{t^2-2t+1}$

Divide and simplify. [12.2b]

12. $\dfrac{10-5t}{3} \div \dfrac{t-2}{12t}$

13. $\dfrac{4x^4}{x^2-1} \div \dfrac{2x^3}{x^2-2x+1}$

Find the LCM. [12.3c]

14. $3x^2,\ 10xy,\ 15y^2$

15. $a - 2,\ 4a - 8$

16. $y^2 - y - 2,\ y^2 - 4$

Add and simplify. [12.4a]

17. $\dfrac{x+8}{x+7} + \dfrac{10-4x}{x+7}$

18. $\dfrac{3}{3x-9} + \dfrac{x-2}{3-x}$

19. $\dfrac{2a}{a+1} + \dfrac{4a}{a^2-1}$

20. $\dfrac{d^2}{d-c} + \dfrac{c^2}{c-d}$

Subtract and simplify. [12.5a]

21. $\dfrac{6x-3}{x^2-x-12} - \dfrac{2x-15}{x^2-x-12}$

22. $\dfrac{3x-1}{2x} - \dfrac{x-3}{x}$

23. $\dfrac{x+3}{x-2} - \dfrac{x}{2-x}$

24. $\dfrac{1}{x^2-25} - \dfrac{x-5}{x^2-4x-5}$

25. Perform the indicated operations and simplify: [12.5b]

$$\frac{3x}{x+2} - \frac{x}{x-2} + \frac{8}{x^2-4}.$$

Simplify. [12.8a]

26. $\dfrac{\dfrac{1}{z}+1}{\dfrac{1}{z^2}-1}$

27. $\dfrac{\dfrac{c}{d}-\dfrac{d}{c}}{\dfrac{1}{c}+\dfrac{1}{d}}$

Solve. [12.6a]

28. $\dfrac{3}{y} - \dfrac{1}{4} = \dfrac{1}{y}$

29. $\dfrac{15}{x} - \dfrac{15}{x+2} = 2$

Solve. [12.7a]

30. *Highway Work.* In checking records, a contractor finds that crew A can pave a certain length of highway in 9 hr, while crew B can do the same job in 12 hr. How long would it take if they worked together?

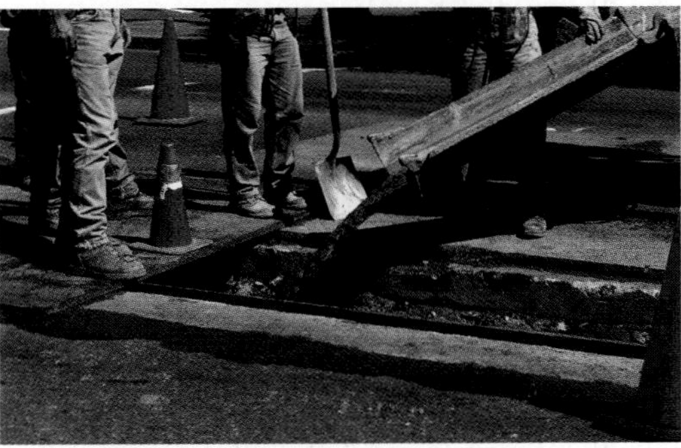

31. *Train Speed.* A manufacturer is testing two high-speed trains. One train travels 40 km/h faster than the other. While one train travels 70 km, the other travels 60 km. Find the speed of each train.

32. *Airplane Speed.* One plane travels 80 mph faster than another. While one travels 1750 mi, the other travels 950 mi. Find the speed of each plane.

Solve. [12.7b]

33. *Quality Control.* A sample of 250 calculators contained 8 defective calculators. How many defective calculators would you expect to find in a sample of 5000?

34. *Pizza Proportions.* A certain kind of pizza at Finnelli's Pizzeria uses the following ratio: 5 parts sausage to 7 parts cheese, 6 parts onion to 13 parts green pepper, and 9 parts pepperoni to 14 parts cheese.

a) Finnelli's makes several pizzas with green pepper and onion. They use 2 cups of green pepper. How much onion would they use?

b) Finnelli's makes several pizzas with sausage and cheese. They use 3 cups of sausage. How much cheese would they use?

c) Finnelli's makes several pizzas with pepperoni and cheese. They use 6 cups of pepperoni. How much cheese would they use?

35. *Estimating Whale Population.* To determine the number of blue whales in the world's oceans, marine biologists tag 500 blue whales in various parts of the world. Later, 400 blue whales are checked, and it is found that 20 of them are tagged. Estimate the blue whale population.

36. Triangles *ABC* and *XYZ* below are similar. Find the value of x.

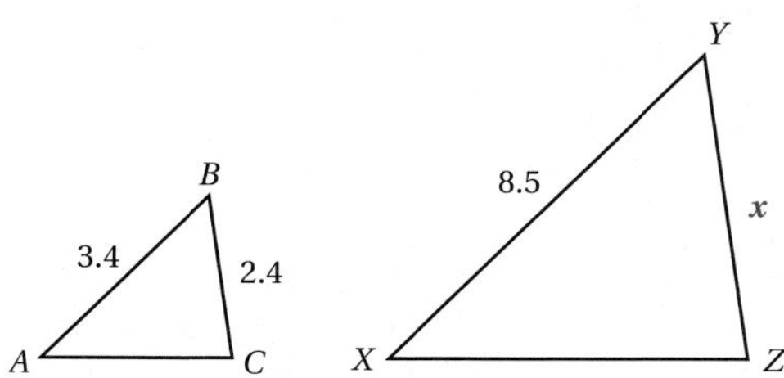

Find an equation of variation in which y varies directly as x and the following are true. Then find the value of y when $x = 20$. [12.9a]

37. $y = 12$ when $x = 4$

38. $y = 4$ when $x = 8$

39. $y = 0.4$ when $x = 0.5$

Find an equation of variation in which y varies inversely as x and the following are true. Then find the value of y when $x = 5$. [12.9c]

40. $y = 5$ when $x = 6$

41. $y = 0.5$ when $x = 2$

42. $y = 1.3$ when $x = 0.5$

Solve.

43. *Wages.* A person's paycheck P varies directly as the number H of hours worked. The pay is $165.00 for working 20 hr. Find the pay for 35 hr of work. [12.9b]

44. *Washing Time.* It takes 5 hr for 2 washing machines to wash a fixed amount of laundry. How long would it take 10 washing machines to do the same job? (The number of hours varies inversely as the number of washing machines.) [12.9d]

DW Carry out the direction for each of the following. Explain the use of the LCM in each case.

45. Add: $\frac{4}{x-2} + \frac{1}{x+2}$. [12.4a]

46. Subtract: $\frac{4}{x-2} - \frac{1}{x+2}$. [12.5a]

47. Solve: $\frac{4}{x-2} + \frac{1}{x+2} = \frac{26}{x^2-4}$. [12.6a]

48. Simplify: $\dfrac{1-\frac{2}{x}}{1+\frac{x}{4}}$. [12.8a]

SYNTHESIS

Simplify.

49. $\frac{2a^2+5a-3}{a^2} \cdot \frac{5a^3+30a^2}{2a^2+7a-4} \div \frac{a^2+6a}{a^2+7a+12}$
[12.1d], [12.2b]

50. $\frac{12a}{(a-b)(b-c)} - \frac{2a}{(b-a)(c-b)}$ [12.5a]

51. Compare

$$\frac{A+B}{B} = \frac{C+D}{D}$$

with the proportion

$$\frac{A}{B} = \frac{C}{D}.$$

[12.7b]

12 Chapter Test

For Extra Help

Work It Out! Chapter Test Video Solutions on CD

Find all numbers for which the rational expression is not defined.

1. $\dfrac{8}{2x}$

2. $\dfrac{5}{x + 8}$

3. $\dfrac{x - 7}{x^2 - 49}$

4. $\dfrac{x^2 + x - 30}{x^2 - 3x + 2}$

5. $\dfrac{11}{(x - 1)^2}$

6. $\dfrac{x + 2}{2}$

7. Simplify:

$$\frac{6x^2 + 17x + 7}{2x^2 + 7x + 3}.$$

8. Multiply and simplify:

$$\frac{a^2 - 25}{6a} \cdot \frac{3a}{a - 5}.$$

9. Divide and simplify:

$$\frac{25x^2 - 1}{9x^2 - 6x} \div \frac{5x^2 + 9x - 2}{3x^2 + x - 2}.$$

10. Find the LCM:

$$y^2 - 9,\ y^2 + 10y + 21,\ y^2 + 4y - 21.$$

Add or subtract. Simplify if possible.

11. $\dfrac{16 + x}{x^3} + \dfrac{7 - 4x}{x^3}$

12. $\dfrac{5 - t}{t^2 + 1} - \dfrac{t - 3}{t^2 + 1}$

13. $\dfrac{x - 4}{x - 3} + \dfrac{x - 1}{3 - x}$

14. $\dfrac{x - 4}{x - 3} - \dfrac{x - 1}{3 - x}$

15. $\dfrac{5}{t-1} + \dfrac{3}{t}$

16. $\dfrac{1}{x^2 - 16} - \dfrac{x+4}{x^2 - 3x - 4}$

17. $\dfrac{1}{x-1} + \dfrac{4}{x^2-1} - \dfrac{2}{x^2 - 2x + 1}$

18. Simplify: $\dfrac{9 - \dfrac{1}{y^2}}{3 - \dfrac{1}{y}}$.

Solve.

19. $\dfrac{7}{y} - \dfrac{1}{3} = \dfrac{1}{4}$

20. $\dfrac{15}{x} - \dfrac{15}{x-2} = -2$

Find an equation of variation in which y varies directly as x and the following are true. Then find the value of y when $x = 25$.

21. $y = 6$ when $x = 3$

22. $y = 1.5$ when $x = 3$

Find an equation of variation in which y varies inversely as x and the following are true. Then find the value of y when $x = 100$.

23. $y = 6$ when $x = 3$

24. $y = 11$ when $x = 2$

Solve.

25. *Train Travel.* The distance d traveled by a train varies directly as the time t that it travels. The train travels 60 km in $\frac{1}{2}$ hr. How far will it travel in 2 hr?

26. *Concrete Work.* It takes 3 hr for 2 concrete mixers to mix a fixed amount of concrete. The number of hours varies inversely as the number of concrete mixers used. How long would it take 5 concrete mixers to do the same job?

27. *Quality Control.* A sample of 125 spark plugs contained 4 defective spark plugs. How many defective spark plugs would you expect to find in a sample of 500?

28. *Zebra Population.* A game warden catches, tags, and then releases 15 zebras. A month later, a sample of 20 zebras is collected and 6 of them have tags. Use this information to estimate the size of the zebra population in that area.

29. *Copying Time.* Kopy Kwik has 2 copiers. One can copy a year-end report in 20 min. The other can copy the same document in 30 min. How long would it take both machines, working together, to copy the report?

30. *Driving Speed.* Craig drives 20 km/h faster than Marilyn. In the same time that Marilyn drives 225 km, Craig drives 325 km. Find the speed of each car.

31. This pair of triangles is similar. Find the missing length x.

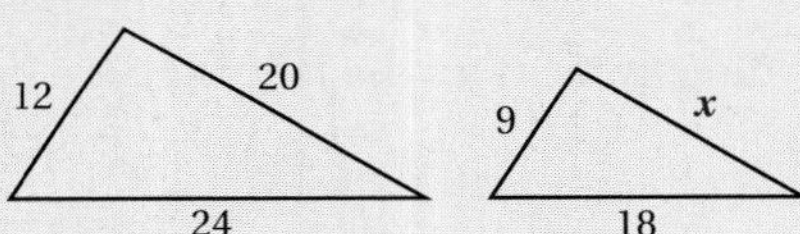

SYNTHESIS

32. Reggie and Rema work together to mulch the flower beds around an office complex in $2\frac{6}{7}$ hr. Working alone, it would take Reggie 6 hr more than it would take Rema. How long would it take each of them to complete the landscaping working alone?

33. Simplify: $1 + \cfrac{1}{1 + \cfrac{1}{1 + \cfrac{1}{a}}}$.

Systems of Equations

13

Real-World Application

At a local “paint swap,” Kari found large supplies of Skylite Pink (12.5% red pigment) and MacIntosh Red (20% red pigment). How many gallons of each color should Kari pick up in order to mix a gallon of Summer Rose (17% red pigment)?

This problem appears as Exercise 6 in Section 13.4.

Objectives

a Determine whether an ordered pair is a solution of a system of equations.

b Solve systems of two linear equations in two variables by graphing.

13.1 SYSTEMS OF EQUATIONS IN TWO VARIABLES

a Systems of Equations and Solutions

Many problems can be solved more easily by translating to two equations in two variables. The following is such a **system of equations:**

$$x + y = 8,$$
$$2x - y = 1.$$

SOLUTION OF A SYSTEM OF EQUATIONS

A **solution** of a system of two equations is an ordered pair that makes both equations true.

Look at the graphs shown below. Recall that a graph of an equation is a drawing that represents its solution set. Each point on the graph corresponds to a solution of that equation. Which points (ordered pairs) are solutions of *both* equations?

The graph shows that there is only one. It is the point P where the graphs cross. This point looks as if its coordinates are $(3, 5)$. We check to see if $(3, 5)$ is a solution of *both* equations, substituting 3 for x and 5 for y.

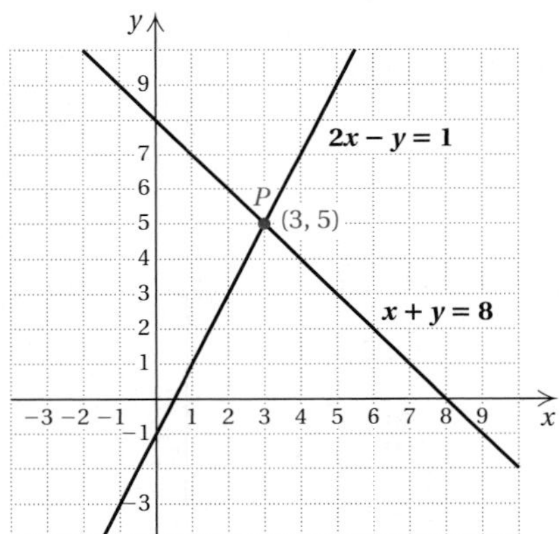

Check:

$x + y = 8$		
$3 + 5$	?	8
8		TRUE

$2x - y = 1$		
$2 \cdot 3 - 5$	?	1
$6 - 5$		
1		TRUE

There is just one solution of the system of equations. It is $(3, 5)$. In other words, $x = 3$ and $y = 5$.

EXAMPLE 1 Determine whether $(1, 2)$ is a solution of the system

$$y = x + 1,$$
$$2x + y = 4.$$

We check by substituting alphabetically 1 for x and 2 for y.

Check:

$y = x + 1$		
2	?	$1 + 1$
		2 TRUE

$2x + y = 4$		
$2 \cdot 1 + 2$	?	4
$2 + 2$		
4		TRUE

This checks, so $(1, 2)$ is a solution of the system.

EXAMPLE 2 Determine whether $(-3, 2)$ is a solution of the system

$$p + q = -1,$$
$$q + 3p = 4.$$

We check by substituting alphabetically -3 for p and 2 for q.

Check:

$p + q = -1$	
$-3 + 2 \;?\; -1$	
-1	TRUE

$q + 3p = 4$	
$2 + 3(-3) \;?\; 4$	
$2 - 9$	
-7	FALSE

The point $(-3, 2)$ is not a solution of $q + 3p = 4$. Thus it is not a solution of the system.

Example 2 illustrates that an ordered pair may be a solution of one equation while *not* a solution of *both* equations. If that is the case, it is *not* a solution of the system.

Do Exercises 1 and 2.

b Graphing Systems of Equations

Recall that the **graph** of an equation is a drawing that represents its solution set. If the graph of an equation is a line, then every point on the line corresponds to an ordered pair that is a solution of the equation. If we graph a **system** of two linear equations, we graph both equations and find the coordinates of the points of intersection, if any exist.

EXAMPLE 3 Solve this system of equations by graphing:

$$x + y = 6,$$
$$x = y + 2.$$

We graph the equations using any of the methods studied in Chapter 9. Point P with coordinates $(4, 2)$ looks as if it is the solution.

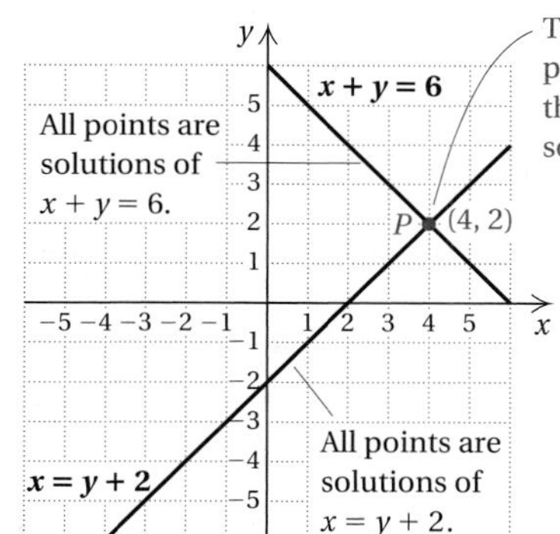

We check the pair as follows.

Check:

$x + y = 6$	
$4 + 2 \;?\; 6$	
6	TRUE

$x = y + 2$	
$4 \;?\; 2 + 2$	
4	TRUE

The solution is $(4, 2)$.

Do Exercise 3.

Determine whether the given ordered pair is a solution of the system of equations.

1. $(2, -3)$; $x = 2y + 8,$
$2x + y = 1$

Check:

$x = 2y + 8$
?

$2x + y = 1$
?

2. $(20, 40)$; $a = \frac{1}{2}b,$
$b - a = 60$

Check:

$a = \frac{1}{2}b$
?

$b - a = 60$
?

3. Solve this system by graphing:

$$2x + y = 1,$$
$$x = 2y + 8.$$

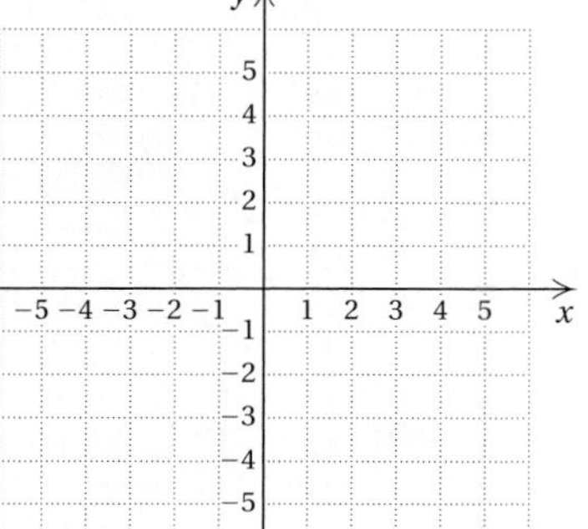

Answers on page A-46

4. Solve this system by graphing:

$$x = -4,$$
$$y = 3.$$

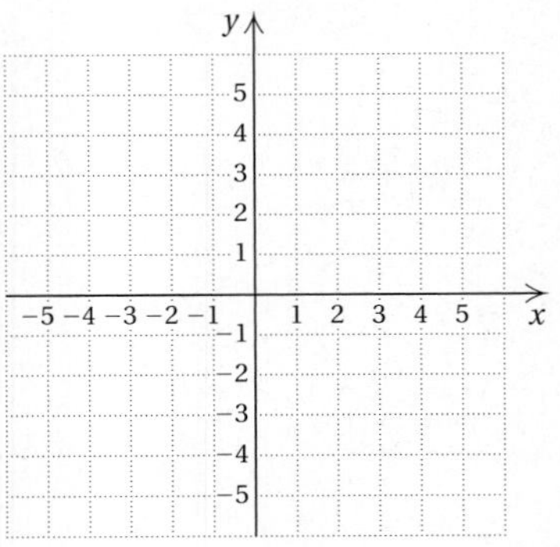

EXAMPLE 4 Solve this system of equations by graphing:

$$x = 2,$$
$$y = -3.$$

The graph of $x = 2$ is a vertical line, and the graph of $y = -3$ is a horizontal line. They intersect at the point $(2, -3)$.

The solution is $(2, -3)$.

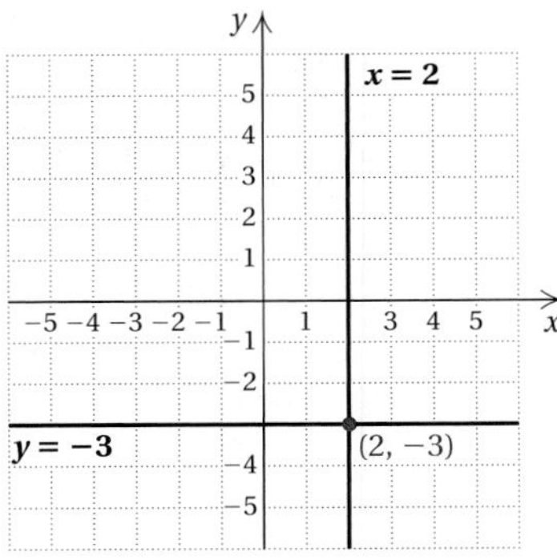

Do Exercise 4.

5. Solve this system by graphing:

$$y + 4 = x,$$
$$x - y = -2.$$

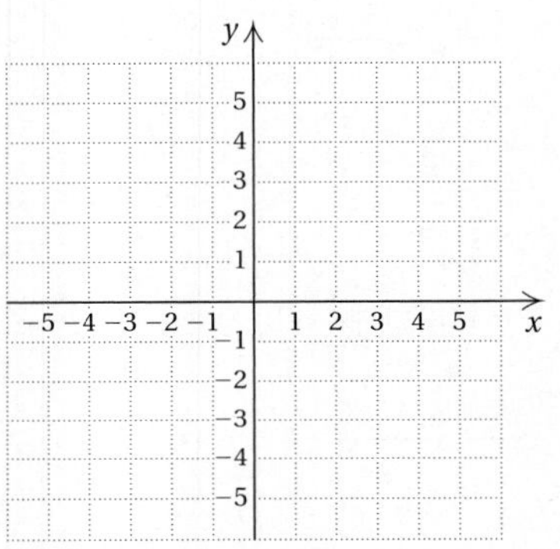

Sometimes the equations in a system have graphs that are parallel lines.

EXAMPLE 5 Solve this system of equations by graphing:

$$y = 3x + 4,$$
$$y = 3x - 3.$$

We graph the equations, again using any of the methods studied in Chapter 3. The lines have the same slope, 3, and different y-intercepts, $(0, 4)$ and $(0, -3)$, so they are parallel.

There is no point at which the lines cross, so the system has no solution. The solution set is the empty set, denoted $\varnothing$, or $\{\ \}$.

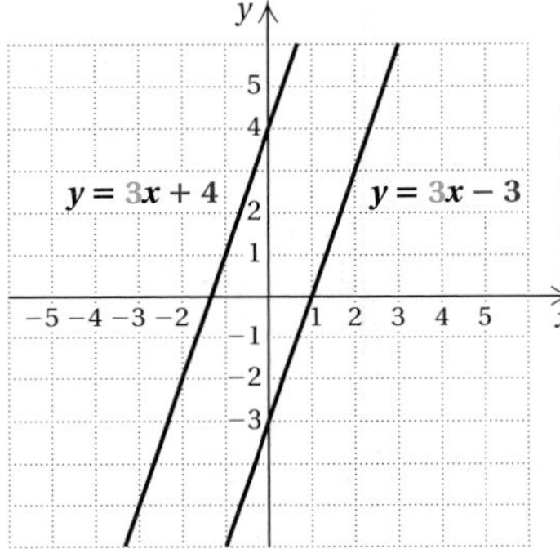

Do Exercise 5.

6. Solve this system by graphing:

$$2x + y = 4,$$
$$-6x - 3y = -12.$$

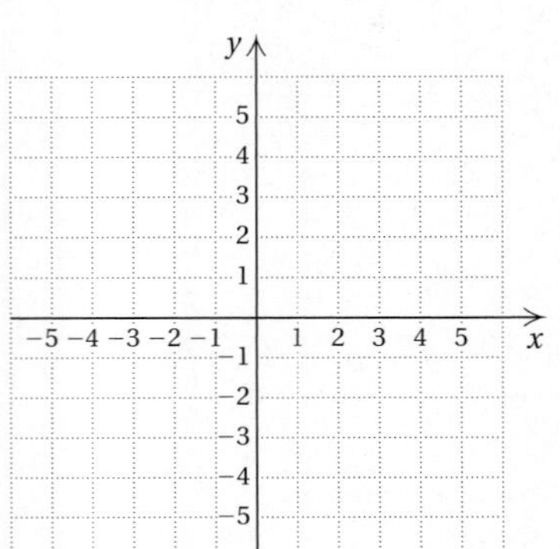

Sometimes the equations in a system have the same graph.

EXAMPLE 6 Solve this system of equations by graphing:

$$2x + 3y = 6,$$
$$-8x - 12y = -24.$$

We graph the equations and see that the graphs are the same. Thus any solution of one of the equations is a solution of the other. Each equation has an infinite number of solutions, some of which are indicated on the graph.

On the following page, we check one such solution, $(0, 2)$: the y-intercept of each equation.

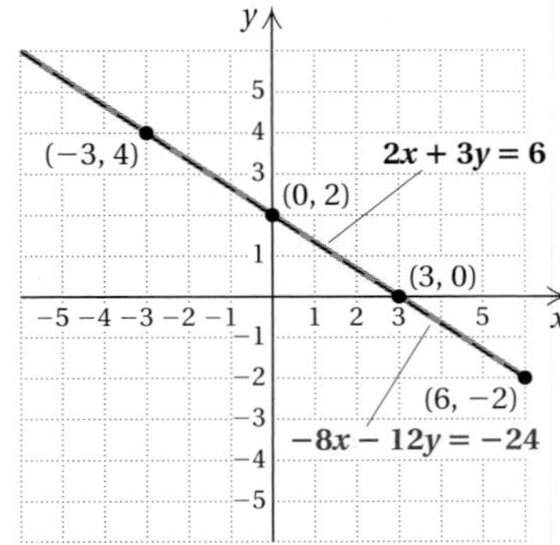

Answers on page A-46

Check:

$$\begin{array}{c|l} 2x + 3y = 6 & \\ \hline 2(0) + 3(2) \ ?\ 6 & \\ 0 + 6 & \\ 6 & \text{TRUE} \end{array} \qquad \begin{array}{c|l} -8x - 12y = -24 & \\ \hline -8(0) - 12(2) \ ?\ -24 & \\ 0 - 24 & \\ -24 & \text{TRUE} \end{array}$$

We leave it to the student to check that $(-3, 4)$ is also a solution of the system. If $(0, 2)$ and $(-3, 4)$ are solutions, then all points on the line containing them are solutions. The system has an infinite number of solutions.

Do Exercise 6 on the preceding page.

When we graph a system of two equations in two variables, we obtain one of the following three results.

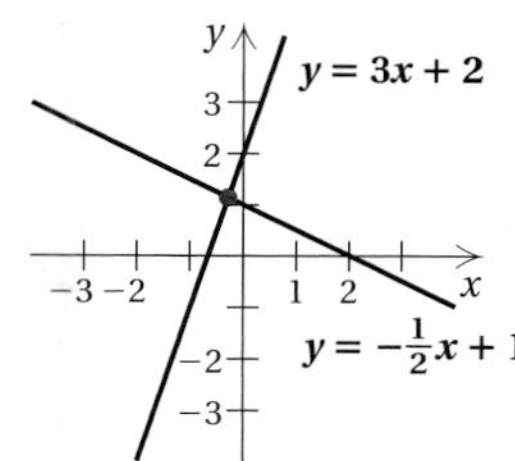

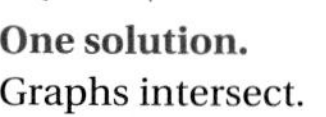

One solution.
Graphs intersect.

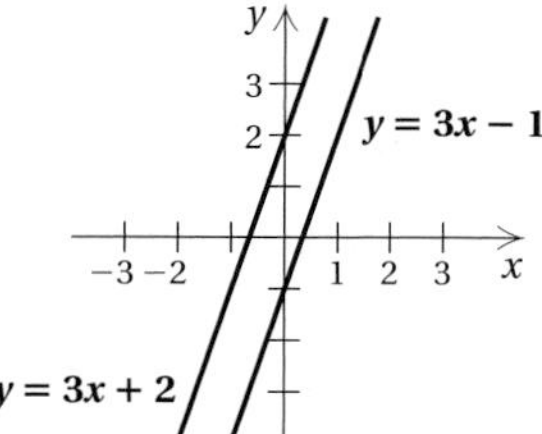

No solution.
Graphs are parallel.

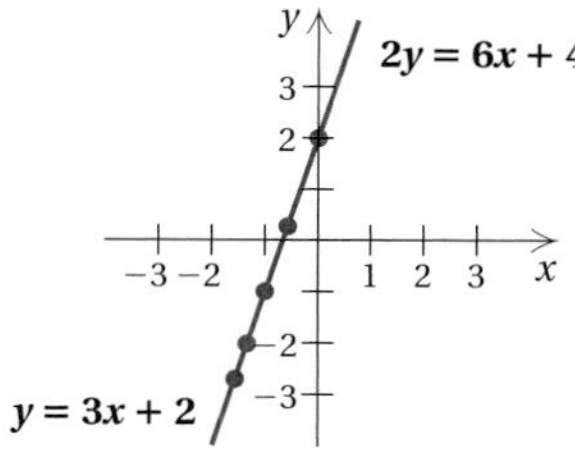

Infinitely many solutions.
Equations have the same graph.

AC ALGEBRAIC–GRAPHICAL CONNECTION

To bring together the concepts of Chapters 6–13, let's take an algebraic–graphical look at equation solving. Such interpretation is useful when using a graphing calculator.

Consider the equation $6 - x = x - 2$. Let's solve it algebraically as we did in Chapter 8:

$$\begin{aligned} 6 - x &= x - 2 & \\ 6 &= 2x - 2 && \text{Adding } x \\ 8 &= 2x && \text{Adding 2} \\ 4 &= x. && \text{Dividing by 2} \end{aligned}$$

Can we also solve the equation graphically? We can, as we see in the following two methods.

METHOD 1 Solve $6 - x = x - 2$ graphically.

We let $y = 6 - x$ and $y = x - 2$. Graphing the system of equations gives us the graph at right. The point of intersection is $(4, 2)$. Note that the x-coordinate of the intersection is 4. This value for x is also the *solution* of the equation $6 - x = x - 2$.

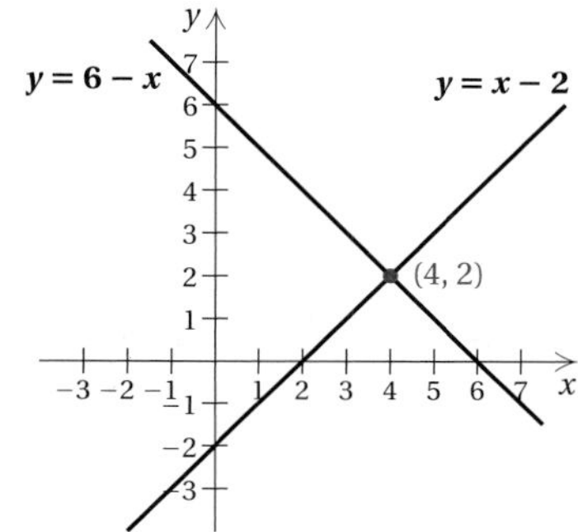

Do Exercise 7.

7. a) Solve $2x - 1 = 8 - x$ algebraically.

b) Solve $2x - 1 = 8 - x$ graphically using method 1.

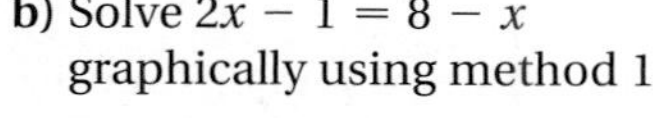

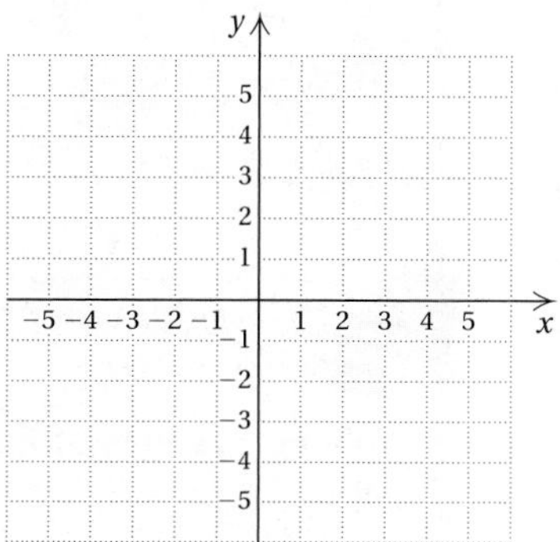

c) Compare your answers to parts (a) and (b).

Answers on page A-46

8. a) Solve $2x - 1 = 8 - x$ graphically using method 2.

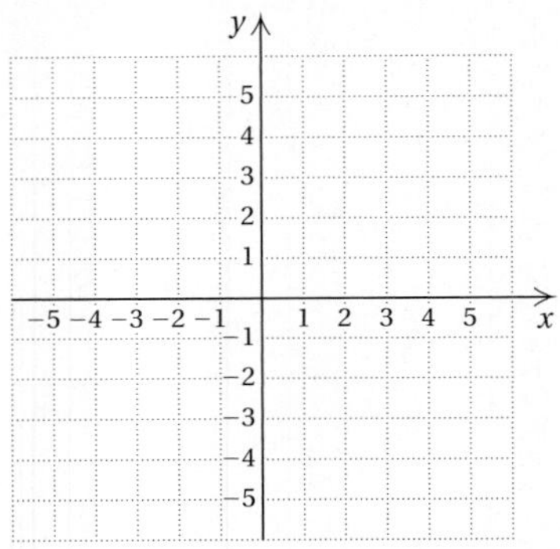

b) Compare your answers to Margin Exercises 7(a), 7(b), and 8(a).

METHOD 2 Solve $6 - x = x - 2$ graphically.

Adding x and -6 on both sides, we obtain the form $0 = 2x - 8$. In this case, we let $y = 0$ and $y = 2x - 8$. Since $y = 0$ is the x-axis, we need graph only $y = 2x - 8$ and see where it crosses the x-axis. Note that the x-intercept of $y = 2x - 8$ is $(4, 0)$. The x-coordinate of this ordered pair is also the *solution* of the equation $6 - x = x - 2$.

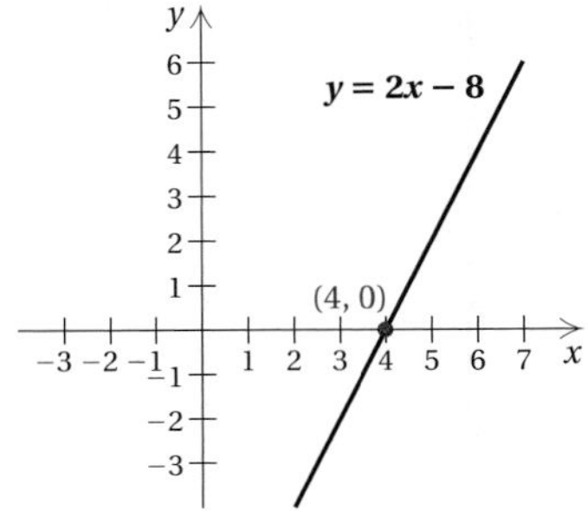

Do Exercise 8.

Let's compare the two methods. Using method 1, we graph two equations. The solution of the original equation is the x-coordinate of the point of intersection. Using method 2, we find that the solution of the original equation is the x-coordinate of the x-intercept of the graph.

Answers on page A-46

CALCULATOR CORNER

Solving Systems of Equations We can solve a system of two equations in two variables on a graphing calculator. Consider the system of equations in Example 3,

$$x + y = 6,$$
$$x = y + 2.$$

First, we solve the equations for y, obtaining $y = -x + 6$ and $y = x - 2$. Then we enter $y_1 = -x + 6$ and $y_2 = x - 2$ on the equation-editor screen and graph the equations. We can use the standard viewing window, $[-10, 10, -10, 10]$.

We will use the INTERSECT feature to find the coordinates of the point of intersection of the lines. To access this feature, we press 2ND CALC 5. (CALC is the second operation associated with the TRACE key.) The query "First curve?" appears on the graph screen. The blinking cursor is positioned on the graph of y_1. We press ENTER to indicate that this is the first curve involved in the intersection. Next, the query "Second curve?" appears and the blinking cursor is positioned on the graph of y_2. We press ENTER to indicate that this is the second curve. Now the query "Guess?" appears, so we use the ▷ and ◁ keys to move the cursor close to the point of intersection and press ENTER. The coordinates of the point of intersection of the graphs, $x = 4$, $y = 2$, appear at the bottom of the screen. Thus the solution of the system of equations is $(4, 2)$.

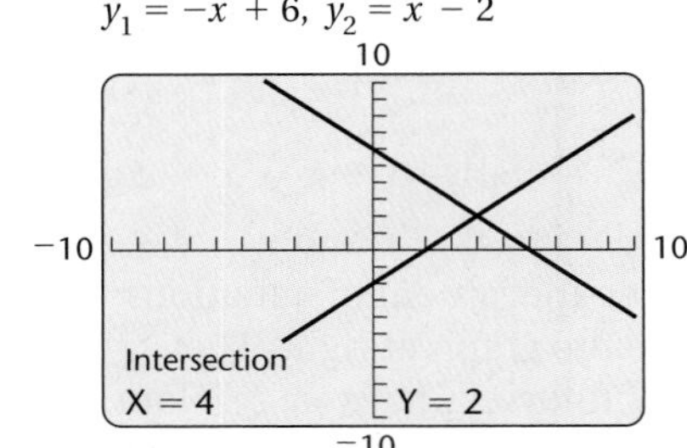

Exercises: Use a graphing calculator to solve the system of equations.

1. $x + y = 2,$
   $y = x + 4$
2. $y = x + 5,$
   $2x + y = 5$
3. $x - y = 5,$
   $y = 2x - 7$
4. $x + 3y = -1,$
   $x - y = -5$
5. $3x + 5y = 19,$
   $4x = 10 + y$
6. $3x = y + 6,$
   $1 = 2x + y$

13.1 EXERCISE SET

For Extra Help

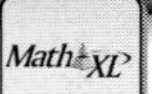

MathXL | MyMathLab | InterAct Math | Math Tutor Center | Video Lectures on CD Disc 7 | Student's Solutions Manual

a Determine whether the given ordered pair is a solution of the system of equations. Use alphabetical order of the variables.

1. $(1, 5)$; $5x - 2y = -5,$
$3x - 7y = -32$

2. $(3, 2)$; $2x + 3y = 12,$
$x - 4y = -5$

3. $(4, 2)$; $3b - 2a = -2,$
$b + 2a = 8$

4. $(6, -6)$; $t + 2s = 6,$
$t - s = -12$

5. $(15, 20)$; $3x - 2y = 5,$
$6x - 5y = -10$

6. $(-1, -5)$; $4r + s = -9,$
$3r = 2 + s$

7. $(-1, 1)$; $x = -1,$
$x - y = -2$

8. $(-3, 4)$; $2x = -y - 2,$
$y = -4$

9. $(18, 3)$; $y = \frac{1}{6}x,$
$2x - y = 33$

10. $(-3, 1)$; $y = -\frac{1}{3}x,$
$3y = -5x - 12$

b Solve the system of equations by graphing.

11. $x - y = 2,$
$x + y = 6$

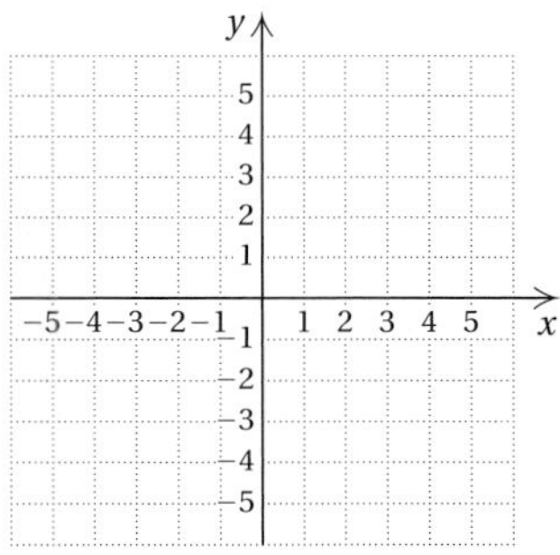

12. $x + y = 3,$
$x - y = 1$

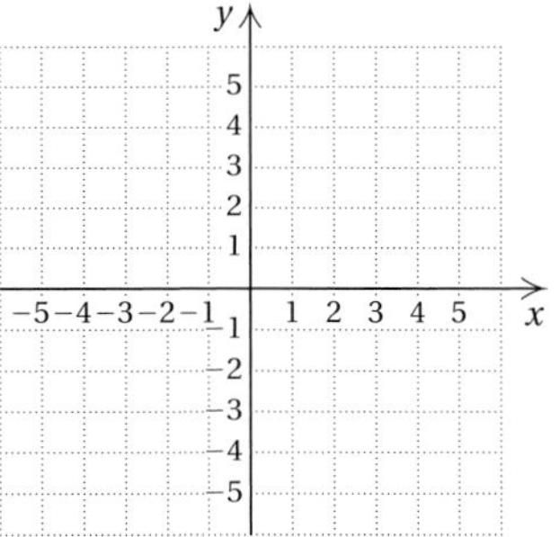

13. $8x - y = 29,$
$2x + y = 11$

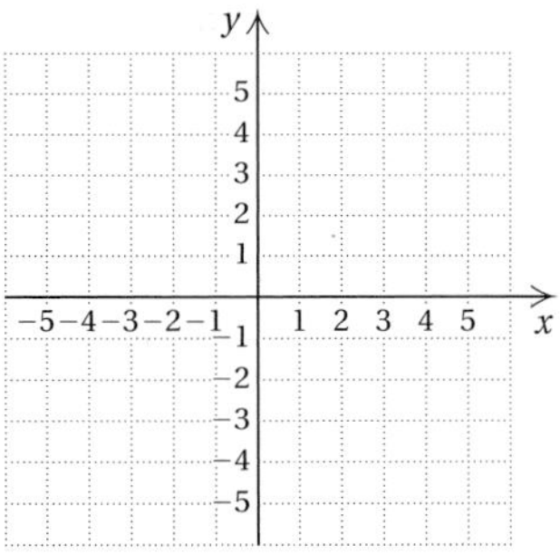

14. $4x - y = 10,$
$3x + 5y = 19$

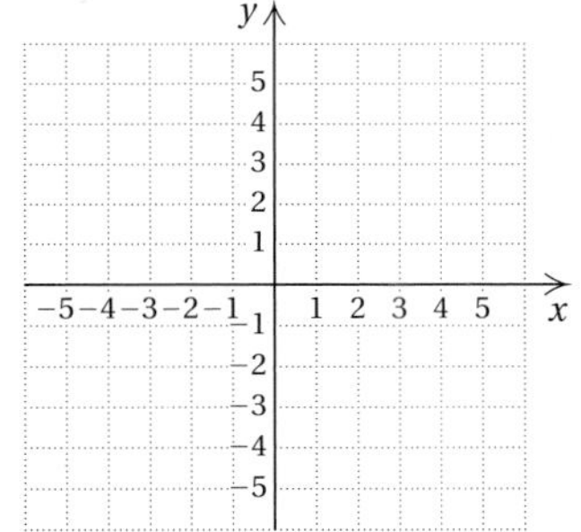

15. $u = v,$
$4u = 2v - 6$

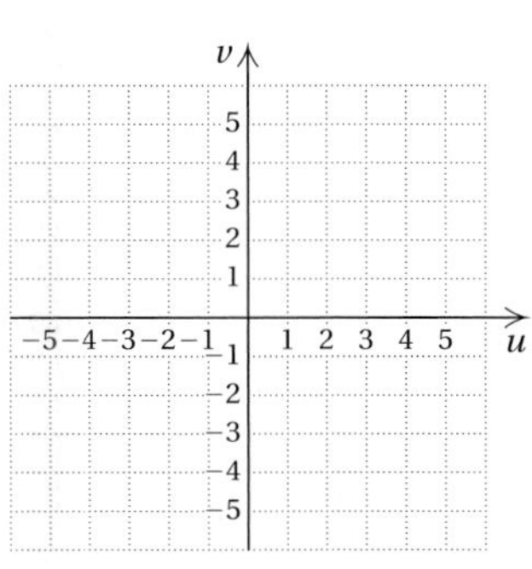

16. $x = 3y,$
$3y - 6 = 2x$

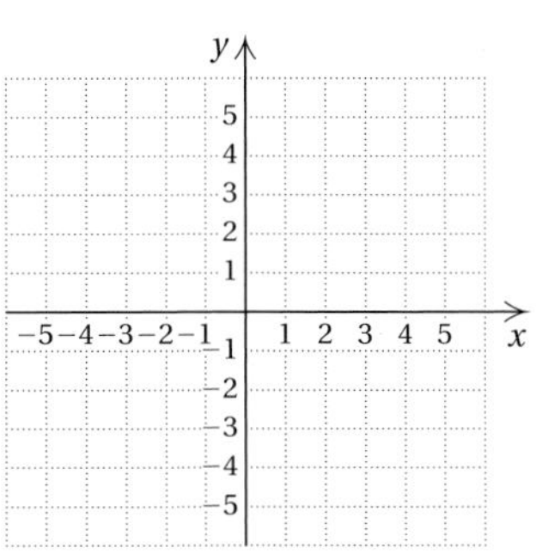

17. $x = -y,$
$x + y = 4$

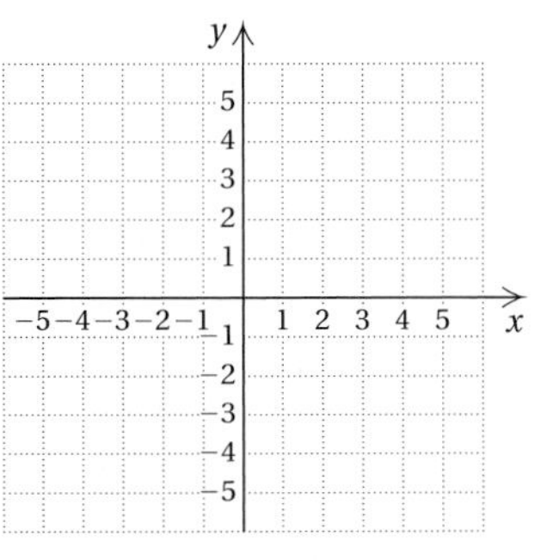

18. $-3x = 5 - y,$
$2y = 6x + 10$

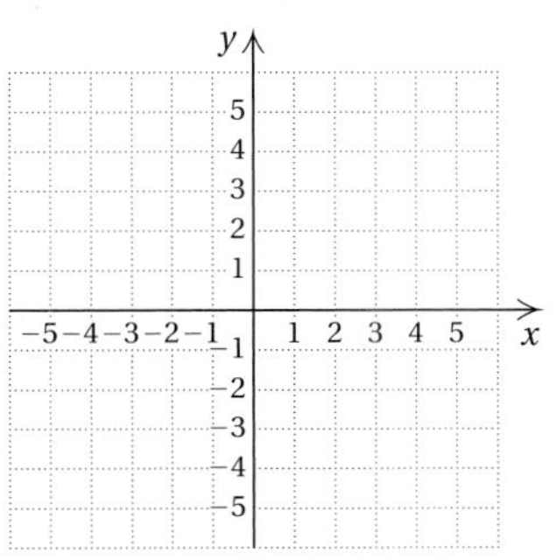

19. $a = \frac{1}{2}b + 1,$
$a - 2b = -2$

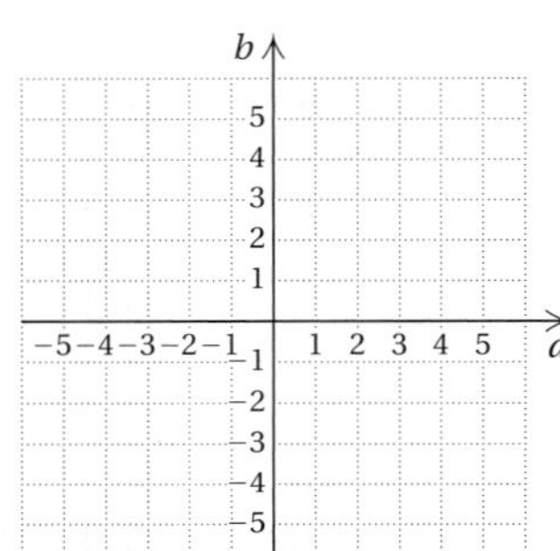

20. $x = \frac{1}{3}y + 2,$
$-2x - y = 1$

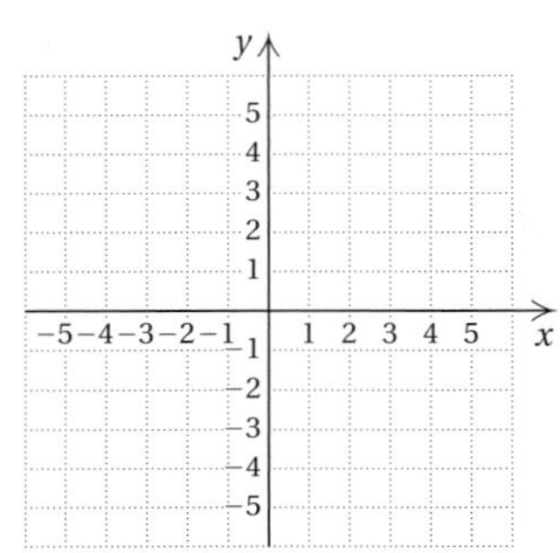

21. $y - 2x = 0,$
$y = 6x - 2$

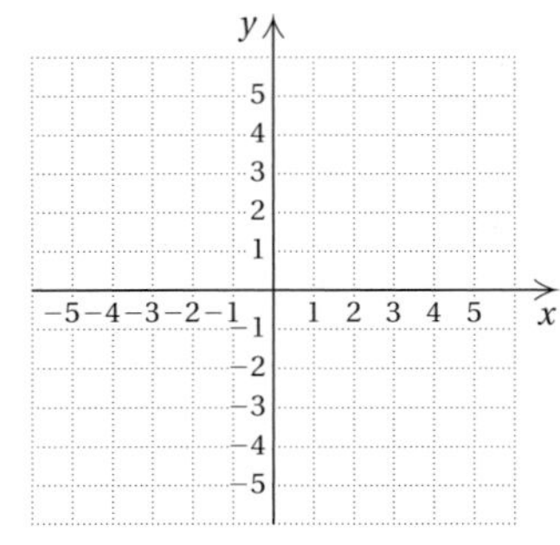

22. $y = 3x,$
$y = -3x + 2$

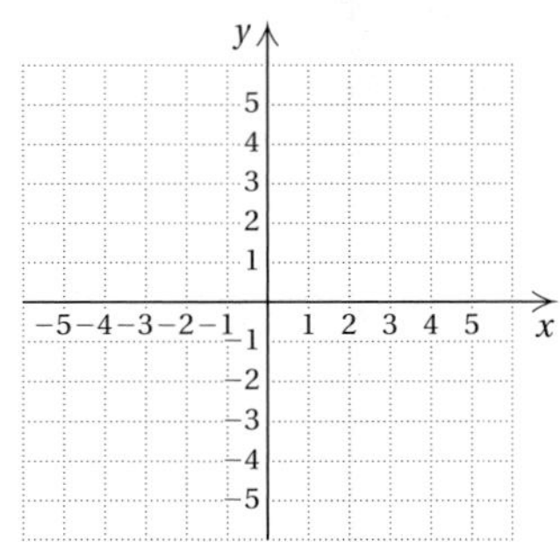

23. $x + y = 9,$
$3x + 3y = 27$

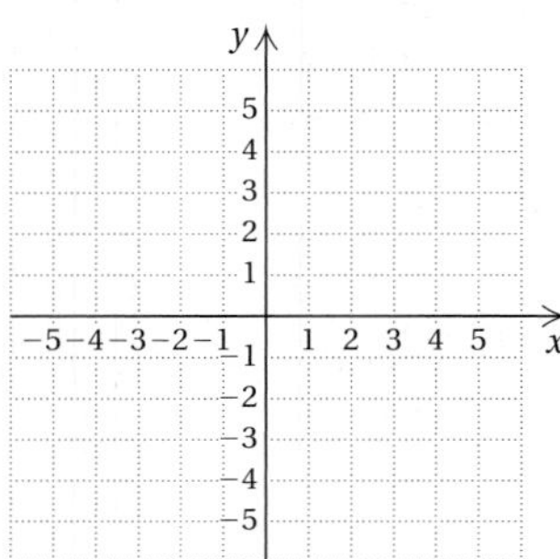

24. 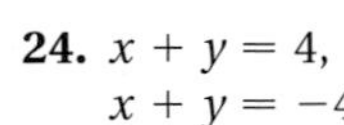
$x + y = 4,$
$x + y = -4$

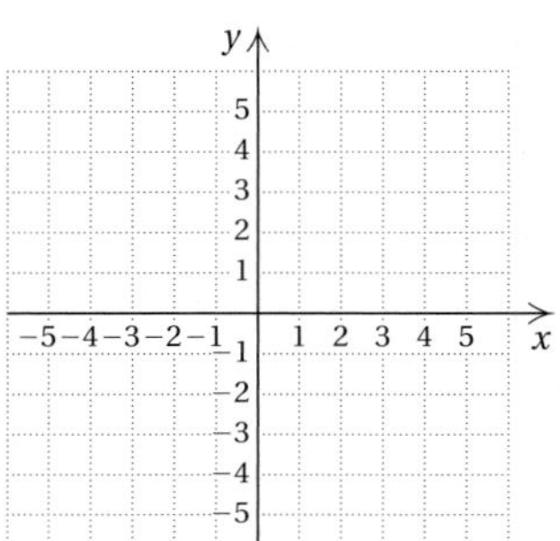

25. $x = 5,$
$y = -3$

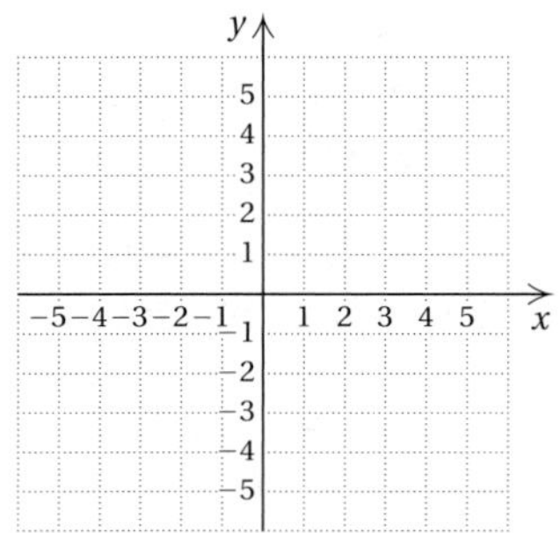

26. $y = 2,$
$y = -4$

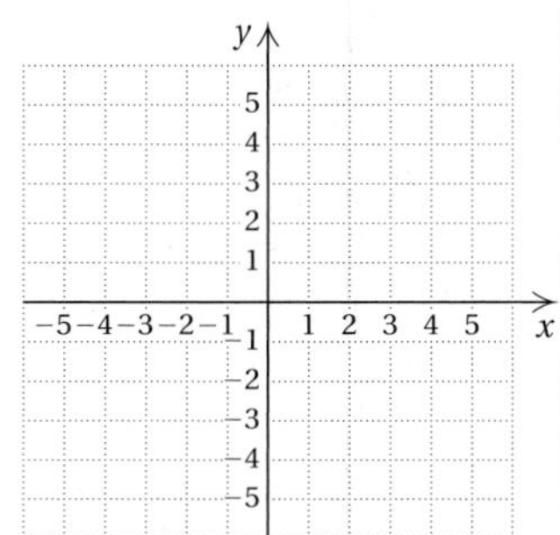

27. **D**W Suppose you have shown that the solution of the equation $3x - 1 = 9 - 2x$ is 2. How can this result be used to determine where the graphs of $y = 3x - 1$ and $y = 9 - 2x$ intersect?

28. **D**W Graph this system of equations. What happens when you try to determine a solution from the graph?

$x - 2y = 6,$
$3x + 2y = 4$

SKILL MAINTENANCE

Simplify.

29. $\frac{1}{x} - \frac{1}{x^2} + \frac{1}{x + 1}$ [12.5b]

30. $\frac{3 - x}{x - 2} - \frac{x - 7}{2 - x}$ [12.5a]

31. $\frac{x + 2}{x - 4} - \frac{x + 1}{x + 4}$ [12.5a]

32. $\frac{2x^2 - x - 15}{x^2 - 9}$ [12.1c]

Classify the polynomial as a monomial, a binomial, a trinomial, or none of these. [10.3i]

33. $5x^2 - 3x + 7$

34. $4x^3 - 2x^2$

35. $1.8x^5$

36. $x^3 + 2x^2 - 3x + 1$

SYNTHESIS

37. The solution of the following system is $(2, -3)$. Find A and B.

$Ax - 3y = 13,$
$x - By = 8$

38. Find an equation to go with $5x + 2y = 11$ such that the solution of the system is $(3, -2)$. Answers may vary.

39. Find a system of equations with $(6, -2)$ as a solution. Answers may vary.

40.–47. Use the TABLE feature on a graphing calculator to check your answers to Exercises 11–18.

13.2 THE SUBSTITUTION METHOD

Objectives

a Solve a system of two equations in two variables by the substitution method when one of the equations has a variable alone on one side.

b Solve a system of two equations in two variables by the substitution method when neither equation has a variable alone on one side.

c Solve applied problems by translating to a system of two equations and then solving using the substitution method.

Consider the following system of equations:

$$3x + 7y = 5,$$
$$6x - 7y = 1.$$

Suppose we try to solve this system graphically. We obtain the graph shown at right.

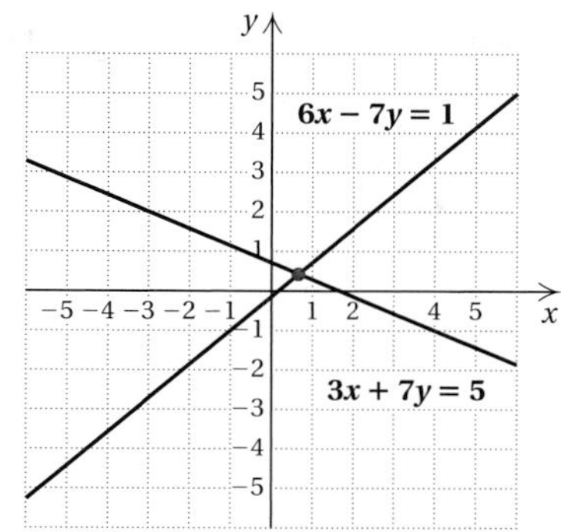

What is the solution? It is rather difficult to tell exactly. It would appear that the coordinates of the point are not integers. It turns out that the solution is $\left(\frac{2}{3}, \frac{3}{7}\right)$. We need techniques involving algebra to determine the solution exactly. Graphing helps us picture the solution of a system of equations, but solving by graphing, though practical in many applications, is not always fast or accurate in cases where solutions are not integers. We now learn other methods using algebra. Because they use algebra, they are called **algebraic.**

a Solving by the Substitution Method

One nongraphical method for solving systems is known as the **substitution method.** In Example 1, we use the substitution method to solve a system we graphed in Example 3 of Section 13.1.

EXAMPLE 1 Solve the system

$$x + y = 6, \quad (1)$$
$$x = y + 2. \quad (2)$$

Equation (2) says that x and $y + 2$ name the same thing. Thus in equation (1), we can substitute $y + 2$ for x:

$x + y = 6$	Equation (1)
$(y + 2) + y = 6.$	Substituting $y + 2$ for x

This last equation has only one variable. We solve it:

$y + 2 + y = 6$	Removing parentheses
$2y + 2 = 6$	Collecting like terms
$2y + 2 - 2 = 6 - 2$	Subtracting 2 on both sides
$2y = 4$	Simplifying
$\frac{2y}{2} = \frac{4}{2}$	Dividing by 2
$y = 2.$	Simplifying

We have found the y-value of the solution. To find the x-value, we return to the original pair of equations. Substituting into either equation will give us the x-value.

Study Tips

BEGINNING TO STUDY FOR THE FINAL EXAM (PART 1)

It is never too soon to begin to study for the final examination. Take a few minutes each week to review the highlighted information, such as formulas, properties, and procedures. Make special use of the Summary and Reviews, Chapter Tests, and Cumulative Reviews, as well as the supplements such as the Work It Out! Chapter Test Video Solutions on CD, the Interact Math Tutorial Web site, MyMathLab, and MathXL. The Cumulative Review/Final Examination for Chapters 1–9 is a sample final exam.

"Practice does not make perfect; practice makes permanent."

Dr. Richard Chase, former president, Wheaton College

1. Solve by the substitution method. Do not graph.

$$x + y = 5,$$
$$x = y + 1$$

2. Solve by the substitution method:

$$a - b = 4,$$
$$b = 2 - a.$$

Answers on page A-46

We choose equation (2) because it has x alone on one side:

$x = y + 2$ Equation (2)
$\quad = 2 + 2$ Substituting 2 for y
$\quad = 4.$

The ordered pair $(4, 2)$ may be a solution. Note that we are using alphabetical order in listing the coordinates in an ordered pair. That is, since x precedes y, we list 4 before 2 in the pair $(4, 2)$.

We check as follows.

Check:

$x + y = 6$		$x = y + 2$	
$4 + 2 \;?\; 6$		$4 \;?\; 2 + 2$	
6	TRUE	4	TRUE

Since $(4, 2)$ checks, we have the solution. We could also express the answer as $x = 4, y = 2$.

Note in Example 1 that substituting 2 for y in equation (1) will also give us the x-value of the solution:

$$x + y = 6$$
$$x + 2 = 6$$
$$x = 4.$$

Do Exercise 1.

EXAMPLE 2 Solve the system

$t = 1 - 3s,$ **(1)**
$s - t = 11.$ **(2)**

We substitute $1 - 3s$ for t in equation (2):

$s - t = 11$ Equation (2)
$s - (1 - 3s) = 11.$ Substituting $1 - 3s$ for t

Remember to use parentheses when you substitute.

Now we solve for s:

$s - 1 + 3s = 11$ Removing parentheses
$4s - 1 = 11$ Collecting like terms
$4s = 12$ Adding 1
$s = 3.$ Dividing by 4

Next, we substitute 3 for s in equation (1) of the original system:

$t = 1 - 3s$ Equation (1)
$\quad = 1 - 3 \cdot 3$ Substituting 3 for s
$\quad = -8.$

The pair $(3, -8)$ checks and is the solution. Remember: We list the answer in alphabetical order, (s, t). That is, since s comes before t in the alphabet, 3 is listed first and -8 second.

Do Exercise 2.

b Solving for the Variable First

Sometimes neither equation of a pair has a variable alone on one side. Then we solve one equation for one of the variables and proceed as before, substituting into the *other* equation. If possible, we solve in either equation for a variable that has a coefficient of 1.

EXAMPLE 3 Solve the system

$$x - 2y = 6, \qquad (1)$$
$$3x + 2y = 4. \qquad (2)$$

We solve one equation for one variable. Since the coefficient of x is 1 in equation (1), it is easier to solve that equation for x:

$$x - 2y = 6 \qquad \text{Equation (1)}$$
$$x = 6 + 2y. \qquad \text{Adding } 2y \qquad (3)$$

We substitute $6 + 2y$ for x in equation (2) of the original pair and solve for y:

$$3x + 2y = 4 \qquad \text{Equation (2)}$$
$$3(6 + 2y) + 2y = 4 \qquad \text{Substituting } 6 + 2y \text{ for } x$$
$$18 + 6y + 2y = 4 \qquad \text{Removing parentheses}$$
$$18 + 8y = 4 \qquad \text{Collecting like terms}$$
$$8y = -14 \qquad \text{Subtracting 18}$$
$$y = \frac{-14}{8}, \text{ or } -\frac{7}{4}. \qquad \text{Dividing by 8}$$

To find x, we go back to either of the original equations (1) or (2) or to equation (3), which we solved for x. It is generally easier to use an equation like equation (3) where we have solved for a specific variable. We substitute $-\frac{7}{4}$ for y in equation (3) and compute x:

$$x = 6 + 2y \qquad \text{Equation (3)}$$
$$= 6 + 2\left(-\tfrac{7}{4}\right) \qquad \text{Substituting } -\tfrac{7}{4} \text{ for } y$$
$$= 6 - \tfrac{7}{2} = \tfrac{5}{2}.$$

We check the ordered pair $\left(\frac{5}{2}, -\frac{7}{4}\right)$.

Check:

$$\begin{array}{c|c} x - 2y & 6 \\ \hline \frac{5}{2} - 2\left(-\frac{7}{4}\right) & ?\ 6 \\ \frac{5}{2} + \frac{7}{2} & \\ \frac{12}{2} & \\ 6 & \text{TRUE} \end{array} \qquad \begin{array}{c|c} 3x + 2y & 4 \\ \hline 3 \cdot \frac{5}{2} + 2\left(-\frac{7}{4}\right) & ?\ 4 \\ \frac{15}{2} - \frac{7}{2} & \\ \frac{8}{2} & \\ 4 & \text{TRUE} \end{array}$$

Since $\left(\frac{5}{2}, -\frac{7}{4}\right)$ checks, it is the solution. This solution would have been difficult to find graphically because it involves fractions.

Do Exercise 3.

c Solving Applied Problems

Now let's solve an applied problem using systems of equations and the substitution method.

3. Solve:

$$x - 2y = 8,$$
$$2x + y = 8.$$

Caution!

A solution of a system of equations in two variables is an ordered *pair* of numbers. Once you have solved for one variable, don't forget the other. A common mistake is to solve for only one variable.

Answer on page A-46

4. Community Garden. A rectangular community garden is to be enclosed with 92 m of fencing. In order to allow for compost storage, the garden must be 4 m longer than it is wide. Determine the dimensions of the garden.

Answer on page A-46

EXAMPLE 4 *Standard Billboard.* A standard rectangular highway billboard has a perimeter of 124 ft. The length is 34 ft more than the width. Find the length and the width.

Source: Eller Sign Company

1. **Familiarize.** We make a drawing and label it. We let $l =$ the length and $w =$ the width.

2. **Translate.** The perimeter of the rectangle is given by the formula $2l + 2w$. We translate each statement, as follows.

The perimeter is 124 ft.
$$2l + 2w = 124$$

The length is 34 ft longer than the width.
$$l = 34 + w$$

We now have a system of equations:

$$2l + 2w = 124, \quad (1)$$
$$l = 34 + w. \quad (2)$$

3. **Solve.** We solve the system. To begin, we substitute $34 + w$ for l in the first equation and solve:

$$2(34 + w) + 2w = 124 \quad \text{Substituting for } 34 + w\ l \text{ in equation (1)}$$
$$2 \cdot 34 + 2 \cdot w + 2w = 124 \quad \text{Removing parentheses}$$
$$4w + 68 = 124 \quad \text{Collecting like terms}$$
$$4w = 56 \quad \text{Subtracting 68}$$
$$w = 14. \quad \text{Dividing by 4}$$

We go back to one of the original equations and substitute 14 for w:

$$l = 34 + w = 34 + 14 = 48. \quad \text{Substituting in equation (2)}$$

4. **Check.** If the length is 48 ft and the width is 14 ft, then the length is 34 ft more than the width ($48 - 14 = 34$), and the perimeter is $2(48 \text{ ft}) + 2(14 \text{ ft})$, or 124 ft. Thus these dimensions check.
5. **State.** The width is 14 ft and the length is 48 ft.

The problem in Example 4 illustrates that many problems that can be solved by translating to *one* equation in *one* variable may actually be easier to solve by translating to *two* equations in *two* variables.

Do Exercise 4.

13.2 EXERCISE SET

For Extra Help

MathXL | MyMathLab | InterAct Math | Math Tutor Center | Video Lectures on CD Disc 7 | Student's Solutions Manual

a Solve using the substitution method.

1. $x + y = 10,$
$y = x + 8$

2. $x + y = 4,$
$y = 2x + 1$

3. $y = x - 6,$
$x + y = -2$

4. $y = x + 1,$
$2x + y = 4$

5. $y = 2x - 5,$
$3y - x = 5$

6. $y = 2x + 1,$
$x + y = -2$

7. $x = -2y,$
$x + 4y = 2$

8. $r = -3s,$
$r + 4s = 10$

b Solve using the substitution method. First, solve one equation for one variable.

9. $x - y = 6,$
$x + y = -2$

10. $s + t = -4,$
$s - t = 2$

11. $y - 2x = -6,$
$2y - x = 5$

12. $x - y = 5,$
$x + 2y = 7$

13. $2x + 3y = -2,$
$2x - y = 9$

14. $x + 2y = 10,$
$3x + 4y = 8$

15. $x - y = -3,$
$2x + 3y = -6$

16. $3b + 2a = 2,$
$-2b + a = 8$

17. $r - 2s = 0,$
$4r - 3s = 15$

18. $y - 2x = 0,$
$3x + 7y = 17$

c Solve.

19. *Perimeter of NBA Court.* The perimeter of an NBA-sized basketball court is 288 ft. The length is 44 ft longer than the width. Find the dimensions of the court.

Source: National Basketball Association

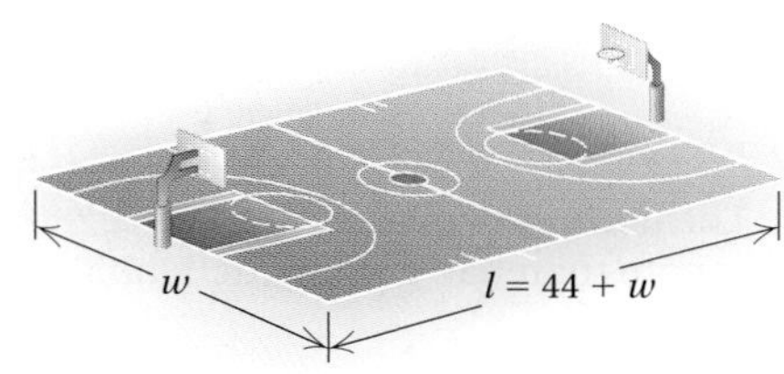

20. *Perimeter of High School Court.* The perimeter of a standard high school basketball court is 268 ft. The length is 34 ft longer than the width. Find the dimensions of the court.

Source: Indiana High School Athletic Association

21. *Two-by-Four.* The perimeter of a cross section of a "two-by-four" piece of lumber is 10 in. The length is 2 in. more than the width. Find the actual dimensions of a cross section of a two-by-four.

22. *Rose Garden.* The perimeter of a rectangular rose garden is 400 m. The length is 3 m more than twice the width. Find the length and the width.

23. *Dimensions of Wyoming.* The state of Wyoming is a rectangle with a perimeter of 1280 mi. The width is 90 mi less than the length. Find the length and the width.

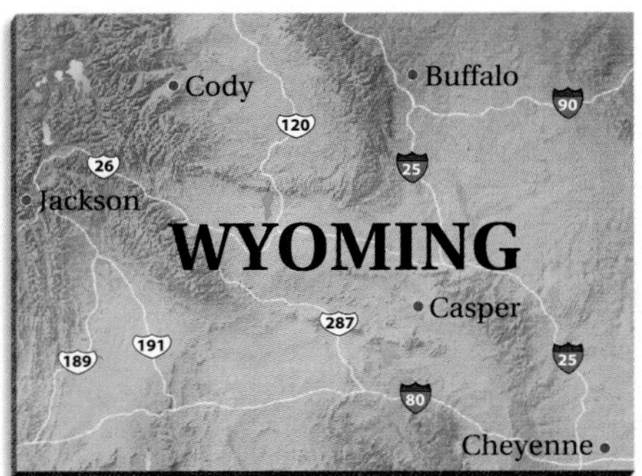

24. *Dimensions of Colorado.* The state of Colorado is roughly in the shape of a rectangle whose perimeter is 1300 mi. The width is 110 mi less than the length. Find the length and the width.

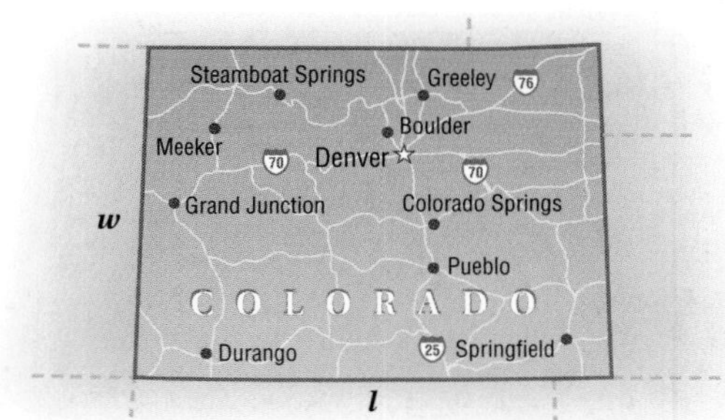

25. *Racquetball.* A regulation racquetball court should have a perimeter of 120 ft, with a length that is twice the width. Find the length and the width of a court.

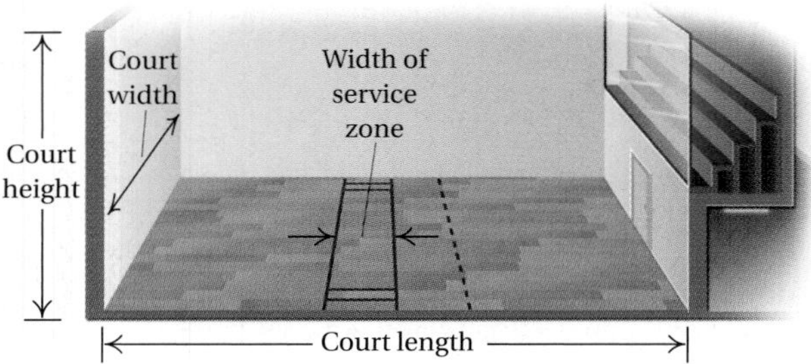

26. *Racquetball.* The height of the front wall of a standard racquetball court is four times the width of the service zone (see the figure). Together, these measurements total 25 ft. Find the height and the width.

27. *Lacrosse.* The perimeter of a lacrosse field is 340 yd. The length is 10 yd less than twice the width. Find the length and the width.

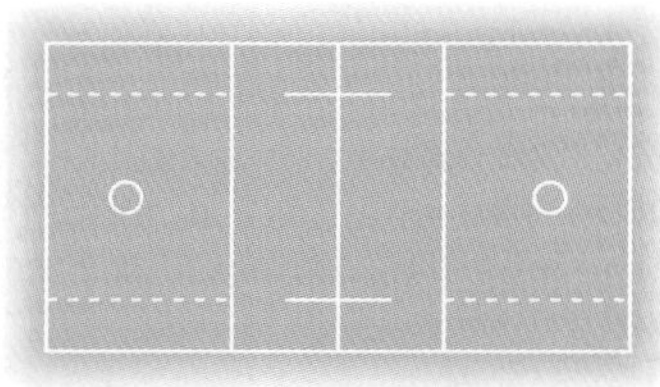

28. *Soccer.* The perimeter of a soccer field is 280 yd. The width is 5 more than half the length. Find the length and the width.

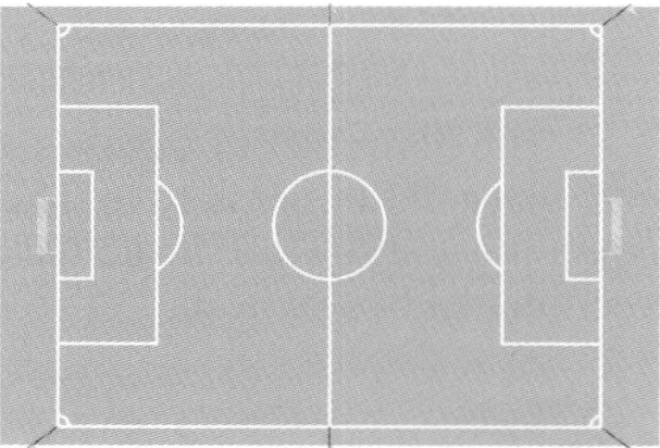

29. The sum of two numbers is 37. One number is 5 more than the other. Find the numbers.

30. The sum of two numbers is 26. One number is 12 more than the other. Find the numbers.

31. Find two numbers whose sum is 52 and whose difference is 28.

32. Find two numbers whose sum is 63 and whose difference is 5.

33. The difference between two numbers is 12. Two times the larger is five times the smaller. What are the numbers?

34. The difference between two numbers is 18. Twice the smaller number plus three times the larger is 74. What are the numbers?

35. DW Janine can tell by inspection that the system

$x = 2y - 1,$

$x = 2y + 3$

has no solution. How can she tell?

36. DW Joel solves every system of two equations (in x and y) by first solving for y in the first equation and then substituting into the second equation. Is he using the best approach? Why or why not?

SKILL MAINTENANCE

Graph. [9.1d], [9.2a, b]

37. $2x - 3y = 6$

38. $2x + 3y = 6$

39. $y = 2x - 5$

40. $y = 4$

Factor completely. [11.6a]

41. $6x^2 - 13x + 6$

42. $4p^2 - p - 3$

43. $4x^2 + 3x + 2$

44. $9a^2 - 25$

Simplify. [10.1d, e, f]

45. $\frac{x^{-2}}{x^{-5}}$

46. $x^2 \cdot x^5$

47. $x^{-2} \cdot x^{-5}$

48. $\frac{a^2b^{-3}}{a^5b^{-6}}$

SYNTHESIS

Solve using a graphing calculator and its CALC-INTERSECT feature. Then solve algebraically and decide which method you prefer to use.

49. $x - y = 5,$
$x + 2y = 7$

50. $y - 2x = -6,$
$2y - x = 5$

51. $y - 2.35x = -5.97,$
$2.14y - x = 4.88$

52. $y = 1.2x - 32.7,$
$y = -0.7x + 46.15$

53. *Softball.* The perimeter of a softball diamond is two-thirds of the perimeter of a baseball diamond. Together, the two perimeters measure 200 yd. Find the distance between the bases in each sport.

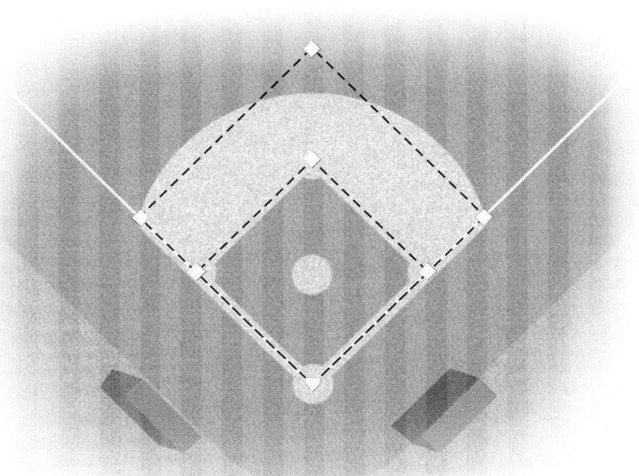

54. Write a system of two linear equations that can be solved more quickly—but still precisely—by a graphing calculator than by substitution. Time yourself using both methods to solve the system.

13.3 THE ELIMINATION METHOD

Objectives

a Solve a system of two equations in two variables using the elimination method when no multiplication is necessary.

b Solve a system of two equations in two variables using the elimination method when multiplication is necessary.

a Solving by the Elimination Method

The **elimination method** for solving systems of equations makes use of the *addition principle*. Some systems are much easier to solve using this method. For example, to solve the system

$$2x + 3y = 13, \quad (1)$$
$$4x - 3y = 17 \quad (2)$$

by substitution, we would need to first solve for a variable in one of the equations. Were we to solve equation (1) for y, we would find (after several steps) that $y = \frac{13}{3} - \frac{2}{3}x$. We could then use the expression $\frac{13}{3} - \frac{2}{3}x$ in equation (2) as a replacement for y:

$$4x - 3\left(\tfrac{13}{3} - \tfrac{2}{3}x\right) = 17.$$

As you can see, although substitution could be used to solve this system, doing so is not easy. Fortunately, another method, elimination, can be used to solve systems and, on problems like this, is simpler to use.

EXAMPLE 1 Solve the system

$$2x + 3y = 13, \quad (1)$$
$$4x - 3y = 17. \quad (2)$$

The key to the advantage of the elimination method for solving this system involves the $3y$ in one equation and the $-3y$ in the other. The terms are opposites. If we add the terms on the sides of the equations, the y-terms will add to 0, and in effect, the variable y will be eliminated.

We will use the addition principle for equations. According to equation (2), $4x - 3y$ and 17 are the same number. Thus we can use a vertical form and add $4x - 3y$ to the left side of equation (1) and 17 to the right side—in effect, adding the same number on both sides of equation (1):

$$\begin{aligned} 2x + 3y &= 13 && (1) \\ 4x - 3y &= 17 && (2) \\ \hline 6x + 0y &= 30, \text{ or} && \text{Adding} \\ 6x &= 30. \end{aligned}$$

We have "eliminated" one variable. This is why we call this the **elimination method.** We now have an equation with just one variable that can be solved for x:

$$\begin{aligned} 6x &= 30 \\ x &= 5. \end{aligned}$$

Next, we substitute 5 for x in either of the original equations:

$$\begin{aligned} 2x + 3y &= 13 && \text{Equation (1)} \\ 2(5) + 3y &= 13 && \text{Substituting 5 for } x \\ 10 + 3y &= 13 \\ 3y &= 3 \\ y &= 1. && \text{Solving for } y \end{aligned}$$

We check the ordered pair (5, 1).

Check:

$$\begin{array}{c|l} 2x + 3y = 13 & \\ \hline 2(5) + 3(1) \ ?\ 13 & \\ 10 + 3 & \\ 13 & \text{TRUE} \end{array} \qquad \begin{array}{c|l} 4x - 3y = 17 & \\ \hline 4(5) - 3(1) \ ?\ 17 & \\ 20 - 3 & \\ 17 & \text{TRUE} \end{array}$$

Since (5, 1) checks, it is the solution. We can see the solution in the graph shown below.

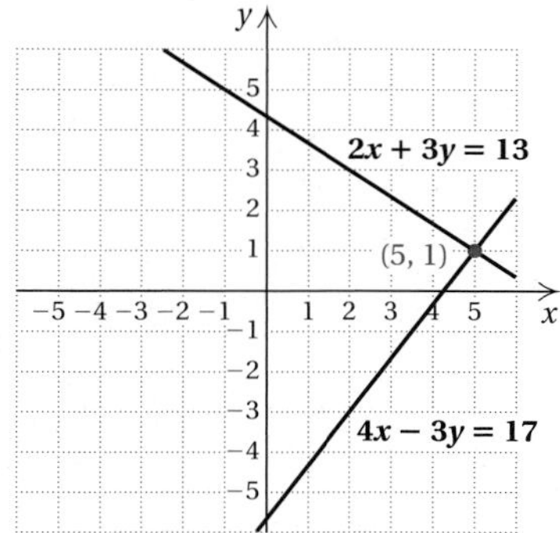

Do Exercises 1 and 2.

Solve using the elimination method.

1. $x + y = 5,$
$2x - y = 4$

2. $-2x + y = -4,$
$2x - 5y = 12$

b Using the Multiplication Principle First

The elimination method allows us to eliminate a variable. We may need to multiply by certain numbers first, however, so that terms become opposites.

EXAMPLE 2 Solve the system

$$2x + 3y = 8, \quad (1)$$
$$x + 3y = 7. \quad (2)$$

If we add, we will not eliminate a variable. However, if the $3y$ were $-3y$ in one equation, we could eliminate y. Thus we multiply by -1 on both sides of equation (2) and then add, using a vertical form:

$$\begin{array}{rll} 2x + 3y = & 8 & \text{Equation (1)} \\ -x - 3y = & -7 & \text{Multiplying equation (2) by } -1 \\ \hline x = & 1. & \text{Adding} \end{array}$$

Next, we substitute 1 for x in one of the original equations:

$$\begin{array}{rl} x + 3y = 7 & \text{Equation (2)} \\ 1 + 3y = 7 & \text{Substituting 1 for } x \\ 3y = 6 & \\ y = 2. & \text{Solving for } y \end{array}$$

Answers on page A-47

3. Solve. Multiply one equation by -1 first.

$$5x + 3y = 17,$$
$$5x - 2y = -3$$

4. Solve the system

$$3x - 2y = -30,$$
$$5x - 2y = -46.$$

Answers on page A-47

We check the ordered pair $(1, 2)$.

Check:

$2x + 3y = 8$		
$2 \cdot 1 + 3 \cdot 2$	? 8	
$2 + 6$		
8		TRUE

$x + 3y = 7$		
$1 + 3 \cdot 2$	? 7	
$1 + 6$		
7		TRUE

Since $(1, 2)$ checks, it is the solution.

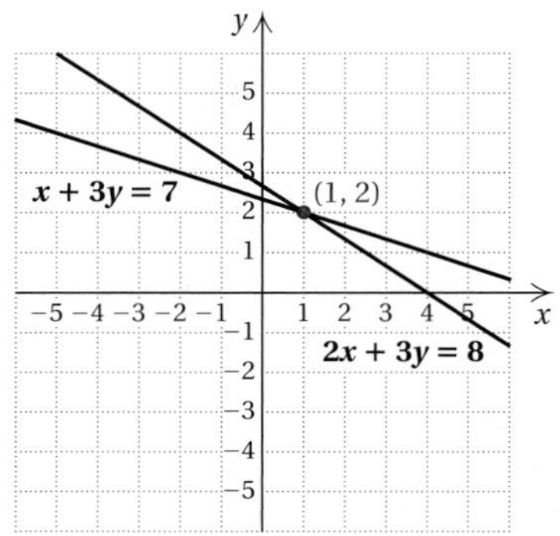

Do Exercises 3 and 4.

In Example 2, we used the multiplication principle, multiplying by -1. However, we often need to multiply by something other than -1.

EXAMPLE 3 Solve the system

$$3x + 6y = -6, \quad (1)$$
$$5x - 2y = 14. \quad (2)$$

Looking at the terms with variables, we see that if $-2y$ were $-6y$, we would have terms that are opposites. We can achieve this by multiplying by 3 on both sides of equation (2). Then we add and solve for x:

$3x + 6y = -6$	Equation (1)
$15x - 6y = 42$	Multiplying by 3 on both sides of equation (2)
$18x \quad\quad = 36$	Adding
$x = 2.$	Solving for x

Next, we substitute 2 for x in either of the original equations. We choose the first:

$3x + 6y = -6$	Equation (1)
$3 \cdot 2 + 6y = -6$	Substituting 2 for x
$6 + 6y = -6$	
$6y = -12$	
$y = -2.$	Solving for y

We check the ordered pair $(2, -2)$.

Check:

$3x + 6y = -6$		
$3 \cdot 2 + 6 \cdot (-2)$	? -6	
$6 + (-12)$		
-6		TRUE

$5x - 2y = 14$		
$5 \cdot 2 - 2 \cdot (-2)$	? 14	
$10 - (-4)$		
14		TRUE

Since (2, −2) checks, it is the solution.

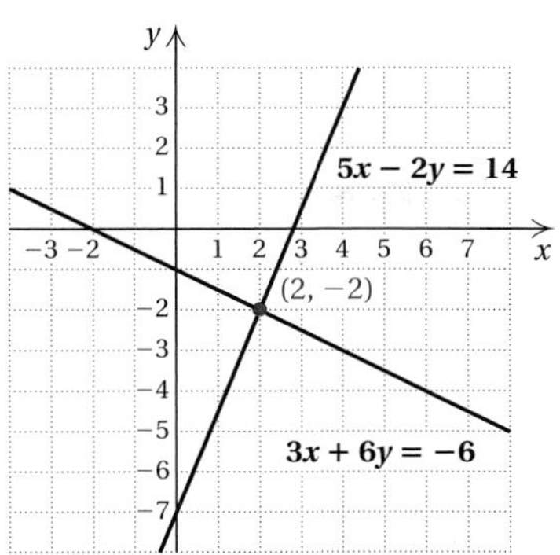

Solve the system.

5. $4a + 7b = 11,$
$2a + 3b = 5$

6. $3x - 8y = 2,$
$5x + 2y = -12$

Do Exercises 5 and 6.

Part of the strategy in using the elimination method is making a decision about which variable to eliminate. So long as the algebra has been carried out correctly, the solution can be found by eliminating *either* variable. We multiply so that terms involving the variable to be eliminated are opposites. It is helpful to first get each equation in a form equivalent to $Ax + By = C$.

EXAMPLE 4 Solve the system

$3y + 1 + 2x = 0,$ (1)

$5x = 7 - 4y.$ (2)

We first rewrite each equation in a form equivalent to $Ax + By = C$:

$2x + 3y = -1,$ (1) Subtracting 1 on both sides and rearranging terms

$5x + 4y = 7.$ (2) Adding $4y$ on both sides

We decide to eliminate the x-term. We do this by multiplying by 5 on both sides of equation (1) and by −2 on both sides of equation (2). Then we add and solve for y:

$10x + 15y = -5$ Multiplying by 5 on both sides of equation (1)

$-10x - 8y = -14$ Multiplying by −2 on both sides of equation (2)

$7y = -19$ Adding

$y = \frac{-19}{7}$, or $-\frac{19}{7}$. Solving for y

Next, we substitute $-\frac{19}{7}$ for y in one of the original equations:

$2x + 3y = -1$ Equation (1)

$2x + 3\left(-\frac{19}{7}\right) = -1$ Substituting $-\frac{19}{7}$ for y

$2x - \frac{57}{7} = -1$

$2x = -1 + \frac{57}{7}$

$2x = -\frac{7}{7} + \frac{57}{7}$

$2x = \frac{50}{7}$

$\frac{1}{2} \cdot 2x = \frac{1}{2} \cdot \frac{50}{7}$ Multiplying by $\frac{1}{2}$ on both sides of the equation

$x = \frac{50}{14}$

$x = \frac{25}{7}.$ Simplifying

CALCULATOR CORNER

Solving Systems of Equations Use the INTERSECT feature to solve the systems of equations in Margin Exercises 1–6. (See the Calculator Corner on p. 972 for the procedure.)

Answers on page A-47

Caution!

Solving a *system* of equations in two variables requires finding an ordered *pair* of numbers. Once you have solved for one variable, don't forget the other, and remember to list the ordered-pair solution using alphabetical order.

7. Solve the system

$$3x = 5 + 2y,$$
$$2x + 3y - 1 = 0.$$

We check the ordered pair $\left(\frac{25}{7}, -\frac{19}{7}\right)$.

Check:

$$\begin{array}{c|c} 3y + 1 + 2x & 0 \\ \hline 3\left(-\frac{19}{7}\right) + 1 + 2\left(\frac{25}{7}\right) \ ? & 0 \\ -\frac{57}{7} + \frac{7}{7} + \frac{50}{7} & \\ 0 & \end{array} \quad \text{TRUE}$$

$$\begin{array}{c|c} 5x & 7 - 4y \\ \hline 5\left(\frac{25}{7}\right) \ ? & 7 - 4\left(-\frac{19}{7}\right) \\ \frac{125}{7} & \frac{49}{7} + \frac{76}{7} \\ & \frac{125}{7} \end{array} \quad \text{TRUE}$$

The solution is $\left(\frac{25}{7}, -\frac{19}{7}\right)$.

Do Exercise 7.

Let's consider a system with no solution and see what happens when we apply the elimination method.

EXAMPLE 5 Solve the system

$$y - 3x = 2, \quad \textbf{(1)}$$
$$y - 3x = 1. \quad \textbf{(2)}$$

We multiply by -1 on both sides of equation (2) and then add:

$$\begin{aligned} y - 3x &= 2 && \text{Equation (1)} \\ -y + 3x &= -1 && \text{Multiplying by } -1 \text{ on both sides of equation (2)} \\ \hline 0 &= 1. && \text{Adding} \end{aligned}$$

We obtain a false equation, $0 = 1$, so there is *no solution*. (See Section 8.3c.) The slope–intercept forms of these equations are

$$y = 3x + 2,$$
$$y = 3x + 1.$$

The slopes, 3, are the same and the y-intercepts, $(0, 2)$ and $(0, 1)$, are different. Thus the lines are parallel. They do not intersect.

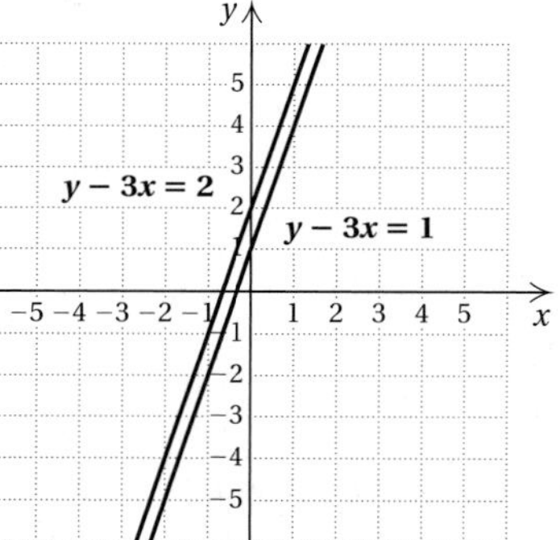

Do Exercise 8.

8. Solve the system

$$2x + y = 15,$$
$$4x + 2y = 23.$$

Answers on page A-47

Sometimes there is an infinite number of solutions. Let's look at a system that we graphed in Example 6 of Section 13.1.

EXAMPLE 6 Solve the system

$$2x + 3y = 6, \quad (1)$$
$$-8x - 12y = -24. \quad (2)$$

We multiply by 4 on both sides of equation (1) and then add the two equations:

$$\begin{aligned} 8x + 12y &= 24 && \text{Multiplying by 4 on both sides of equation (1)} \\ -8x - 12y &= -24 \\ \hline 0 &= 0. && \text{Adding} \end{aligned}$$

We have eliminated both variables, and what remains, $0 = 0$, is an equation easily seen to be true. If this happens when we use the elimination method, we have an infinite number of solutions. (See Section 8.3c.)

Do Exercise 9.

When decimals or fractions appear, we first multiply to clear them. Then we proceed as before.

EXAMPLE 7 Solve the system

$$\frac{1}{3}x + \frac{1}{2}y = -\frac{1}{6}, \quad (1)$$
$$\frac{1}{2}x + \frac{2}{5}y = \frac{7}{10}. \quad (2)$$

The number 6 is a multiple of all the denominators of equation (1). The number 10 is a multiple of all the denominators of equation (2). We multiply by 6 on both sides of equation (1) and by 10 on both sides of equation (2):

$$6\left(\frac{1}{3}x + \frac{1}{2}y\right) = 6\left(-\frac{1}{6}\right) \qquad 10\left(\frac{1}{2}x + \frac{2}{5}y\right) = 10\left(\frac{7}{10}\right)$$
$$6 \cdot \frac{1}{3}x + 6 \cdot \frac{1}{2}y = -1 \qquad 10 \cdot \frac{1}{2}x + 10 \cdot \frac{2}{5}y = 7$$
$$2x + 3y = -1; \qquad 5x + 4y = 7.$$

The resulting system is

$$2x + 3y = -1,$$
$$5x + 4y = 7.$$

As we saw in Example 4, the solution of this system is $\left(\frac{25}{7}, -\frac{19}{7}\right)$.

Do Exercises 10 and 11 on the following page.

9. Solve the system

$$5x - 2y = 3,$$
$$-15x + 6y = -9.$$

CALCULATOR CORNER

Solving Systems of Equations

1. Consider the system of equations in Example 5. In order to enter these equations on a graphing calculator, we must first solve each for y. What happens when we do this? What does this indicate about the nature of the solutions of the system of equations?
2. Consider the system of equations in Example 6. In order to enter these equations on a graphing calculator, we must first solve each for y. What happens when we do this? What does this indicate about the nature of the solutions of the system of equations?

Answer on page A-47

Solve the system.

10. $\frac{1}{2}x + \frac{3}{10}y = \frac{1}{5}$,
$\frac{3}{5}x + \quad y = -\frac{2}{5}$

11. $3.3x + \ 6.6y = -6.6$,
$0.1x - 0.04y = 0.28$

Answers on page A-47

The following is a summary that compares the graphical, substitution, and elimination methods for solving systems of equations.

METHOD	STRENGTHS	WEAKNESSES
Graphical	Can "see" solution.	Inexact when solution involves numbers that are not integers or are very large and off the graph.
Substitution	Works well when solutions are not integers. Easy to use when a variable is alone on one side.	Introduces extensive computations with fractions for more complicated systems where coefficients are not 1 or −1. Cannot "see" solution.
Elimination	Works well when solutions are not integers, when coefficients are not 1 or −1, and when coefficients involve decimals or fractions.	Cannot "see" solution.

Study Tips

Although this does not qualify as a Study Tip, you can use the information to motivate your study of mathematics. The book Best Jobs for the 21st Century, *3rd ed., developed by Michael Farr with database work by Laurence Shatkin, Ph.D., ranks 500 jobs on the basis of overall scores for pay, growth rate through 2010, and number of annual openings. Note that the use of mathematics is significant in most of the 20 top jobs.*

THE 20 BEST JOBS

JOB	ANNUAL EARNINGS	GROWTH RATE THROUGH 2010	ANNUAL OPENINGS
1. Computer software engineers, applications	$70,210	100.0%	28,000
2. Computer systems analysts	61,990	59.7	34,000
3. Computer and information systems managers	82,480	47.9	28,000
4. Teachers, postsecondary	52,115	23.5	184,000
5. Management analysts	57,970	28.9	50,000
6. Registered nurses	46,670	25.6	140,000
7. Computer software engineers, systems software	73,280	89.7	23,000
8. Medical and health services managers	59,220	32.3	27,000
9. Sales agents, financial services	59,690	22.3	55,000
10. Sales agents, securities and commodities	59,690	22.3	55,000
11. Securities, commodities, and financial services sales agents	59,690	22.3	55,000
12. Computer support specialists	38,560	97.0	40,000
13. Sales managers	71,620	32.8	21,000
14. Computer security specialists	53,770	81.9	18,000
15. Network and computer systems administrators	53,770	81.9	18,000
16. Financial managers	70,210	18.5	53,000
17. Financial managers, branch or department	70,210	18.5	53,000
18. Treasurers, comptrollers, and chief financial officers	70,210	18.5	53,000
19. Accountants	45,380	18.5	100,000
20. Accountants and auditors	45,380	18.5	100,000

Source: Based on "The 20 Best Jobs," from *Best Jobs for the 21st Century,* 3rd ed. Indianapolis: Jist Works, 2004, p. 16.

13.3 EXERCISE SET

For Extra Help

MathXL | MyMathLab | InterAct Math | Math Tutor Center | Video Lectures on CD Disc 7 | Student's Solutions Manual

a Solve using the elimination method.

1. $x - y = 7,$
$x + y = 5$

2. $x + y = 11,$
$x - y = 7$

3. $x + y = 8,$
$-x + 2y = 7$

4. $x + y = 6,$
$-x + 3y = -2$

5. $5x - y = 5,$
$3x + y = 11$

6. $2x - y = 8,$
$3x + y = 12$

7. $4a + 3b = 7,$
$-4a + b = 5$

8. $7c + 5d = 18,$
$c - 5d = -2$

9. $8x - 5y = -9,$
$3x + 5y = -2$

10. $3a - 3b = -15,$
$-3a - 3b = -3$

11. $4x - 5y = 7,$
$-4x + 5y = 7$

12. $2x + 3y = 4,$
$-2x - 3y = -4$

b Solve using the multiplication principle first. Then add.

13. $x + y = -7,$
$3x + y = -9$

14. $-x - y = 8,$
$2x - y = -1$

15. $3x - y = 8,$
$x + 2y = 5$

16. $x + 3y = 19,$
$x - y = -1$

17. $x - y = 5,$
$4x - 5y = 17$

18. $x + y = 4,$
$5x - 3y = 12$

19. $2w - 3z = -1,$
$3w + 4z = 24$

20. $7p + 5q = 2,$
$8p - 9q = 17$

21. $2a + 3b = -1,$
$3a + 5b = -2$

22. $3x - 4y = 16,$
$5x + 6y = 14$

23. $x = 3y,$
$5x + 14 = y$

24. $5a = 2b,$
$2a + 11 = 3b$

25. $2x + 5y = 16,$
$3x - 2y = 5$

26. $3p - 2q = 8,$
$5p + 3q = 7$

27. $p = 32 + q,$
$3p = 8q + 6$

28. $3x = 8y + 11,$
$x + 6y - 8 = 0$

29. $3x - 2y = 10,$
$-6x + 4y = -20$

30. $2x + y = 13,$
$4x + 2y = 23$

31. $0.06x + 0.05y = 0.07,$
$0.4x - 0.3y = 1.1$

32. $1.8x - 2y = 0.9,$
$0.04x + 0.18y = 0.15$

33. $\frac{1}{3}x + \frac{3}{2}y = \frac{5}{4},$
$\frac{3}{4}x - \frac{5}{6}y = \frac{3}{8}$

34. $x - \frac{3}{2}y = 13,$
$\frac{3}{2}x - y = 17$

35. $-4.5x + 7.5y = 6,$
$-x + 1.5y = 5$

36. $0.75x + 0.6y = -0.3,$
$3.9x + 5.2y = 96.2$

37. **D$_W$** The following lists the steps a student uses to solve a system of equations, but an error has been made. Find and describe the error and correct the answer.

$$\begin{aligned} 3x - y &= 4 \\ 2x + y &= 16 \\ \hline 5x \quad &= 20 \\ x &= 4 \end{aligned}$$

$$\begin{aligned} 3x - y &= 4 \\ 3(4) - y &= 4 \\ y &= 4 - 12 \\ y &= -8 \end{aligned}$$

The solution is $(4, -8)$.

38. **D$_W$** Explain how the addition and multiplication principles are used in this section. Then count the number of times that these principles are used in Example 4.

SKILL MAINTENANCE

VOCABULARY REINFORCEMENT

In each of Exercises 39–46, fill in the blank with the correct term from the given list. Some of the choices may not be used.

39. Parallel lines have the same ______________ and different ______________. [9.6a]

40. Two nonvertical lines are ______________ if the product of their slopes is -1. [9.6b]

41. A(n) ______________ of a system of two equations is an ordered pair that makes both equations true. [13.1a]

42. If a situation gives rise to an equation $y = kx$, where k is a positive constant, we say that we have ______________ variation. [12.9a]

43. The graph of $y = b$ is a(n) ______________ line. [9.2b]

44. If a situation gives rise to an equation $y = k/x$, where k is a positive constant, we say that we have ______________ variation. [12.9c]

45. The equation $y = mx + b$ is called the ______________ equation. [9.4a]

46. The ______________ of an equation is a drawing that represents its solution set. [9.1c]

horizontal
vertical
nonvertical
direct
inverse
parallel
perpendicular
x-intercepts
y-intercepts
solution
slope
slope–intercept
graph

SYNTHESIS

47.–56. Use the TABLE feature to check the possible solutions to Exercises 1–10.

57.–66. Use a graphing calculator and the CALC-INTERSECT feature to solve the systems in Exercises 21–30.

Solve using the substitution method, the elimination method, or the graphing method.

67. $3(x - y) = 9,$
$x + y = 7$

68. $2(x - y) = 3 + x,$
$x = 3y + 4$

69. $2(5a - 5b) = 10,$
$-5(6a + 2b) = 10$

70. $\frac{x}{3} + \frac{y}{2} = 1\frac{1}{3},$
$x + 0.05y = 4$

71. $y = -\frac{2}{7}x + 3,$
$y = \frac{4}{5}x + 3$

72. $y = \frac{2}{5}x - 7,$
$y = \frac{2}{5}x + 4$

Solve for x and y.

73. $y = ax + b,$
$y = x + c$

74. $ax + by + c = 0,$
$ax + cy + b = 0$

13.4 APPLICATIONS AND PROBLEM SOLVING

Objective

a Solve applied problems by translating to a system of two equations in two variables.

1. Chicken and Hamburger Prices. Fast Rick's Burger restaurant decides to include chicken on its menu. It offers a special two-and-one promotion. The price of one hamburger and two pieces of chicken is \$5.39, and the price of two hamburgers and one piece of chicken is \$5.68. Find the price of one hamburger and the price of one piece of chicken.

Answer on page A-47

a We now use systems of equations to solve applied problems that involve two equations in two variables.

EXAMPLE 1 *Pizza and Soda Prices.* A campus vendor charges \$3.50 for one slice of pizza and one medium soda and \$9.15 for three slices of pizza and two medium sodas. Determine the price of one medium soda and the price of one slice of pizza.

1. **Familiarize.** We let $p =$ the price of one slice of pizza and $s =$ the price of one medium soda.
2. **Translate.** The price of one slice of pizza and one medium soda is \$3.50. This gives us one equation:

   $$p + s = 3.50.$$

   The price of three slices of pizza and two medium sodas is \$9.15. This gives us another equation:

   $$3p + 2s = 9.15.$$

3. **Solve.** We solve the system of equations

   $$p + s = 3.50, \qquad \textbf{(1)}$$
   $$3p + 2s = 9.15. \qquad \textbf{(2)}$$

   Which method should we use? As we discussed in Section 13.3, any method can be used. Each has its advantages and disadvantages. We decide to proceed with the elimination method, because we see that if we multiply each side of equation (1) by -2 and add, the s-terms can be eliminated. (We could also multiply equation (1) by -3 and eliminate p.)

   $$\begin{aligned} -2p - 2s &= -7.00 && \text{Multiplying equation (1) by } -2 \\ 3p + 2s &= 9.15 && \text{Equation (2)} \\ \hline p &= 2.15. && \text{Adding} \end{aligned}$$

   Next, we substitute 2.15 for p in equation (1) and solve for s:

   $$\begin{aligned} p + s &= 3.50 \\ 2.15 + s &= 3.50 \\ s &= 1.35. \end{aligned}$$

4. **Check.** The sum of the prices for one slice of pizza and one medium soda is

   $$\$2.15 + \$1.35, \quad \text{or} \quad \$3.50.$$

   Three times the price of one slice of pizza plus twice the price of a medium soda is

   $$3(\$2.15) + 2(\$1.35), \quad \text{or} \quad \$9.15.$$

   The prices check.
5. **State.** The price of one slice of pizza is \$2.15, and the price of one medium soda is \$1.35.

Do Exercise 1.

EXAMPLE 2 *IMAX Movie Prices.* There were 322 people at a recent showing of the IMAX 3D movie *Nascar 3D: The IMAX Experience.* Admission was \$8.75 each for adults and \$6.00 each for children, and receipts totaled \$2531.50. How many adults and how many children attended?

1. **Familiarize.** There are many ways in which to familiarize ourselves with a problem situation. This time, let's make a guess and do some calculations. The total number of people at the movie was 322, so we choose numbers that total 322. Let's try

   242 adults and
   80 children.

   How much money was taken in? The problem says that adults paid \$8.75 each, so the total amount of money collected from the adults was

   242(\$8.75), or \$2117.50.

   Children paid \$6.00 each, so the total amount of money collected from the children was

   80(\$6.00), or \$480.

   This makes the total receipts \$2117.50 + \$480, or \$2597.50.

   Our guess is not the answer to the problem because the total taken in, according to the problem, was \$2531.50. If we were to continue guessing, we would need to add more children and fewer adults, since our first guess gave us an amount of total receipts that was higher than \$2351.50. The steps we have used to see if our guesses are correct help us to understand the actual steps involved in solving the problem.

   Let's list the information in a table. That usually helps in the familiarization process. We let a = the number of adults and c = the number of children.

	ADULTS	CHILDREN	TOTAL	
Admission	\$8.75	\$6.00		
Number Attending	a	c	322	→ $a + c = 322$
Money Taken In	$8.75a$	$6.00c$	\$2531.50	→ $8.75a + 6.00c = 2531.50$

2. **Translate.** The total number of people attending was 322, so

   $a + c = 322.$

   The amount taken in from the adults was $8.75a$, and the amount taken in from the children was $6.00c$. These amounts are in dollars. The total was \$2531.50, so we have

   $8.75a + 6.00c = 2531.50.$

   We can multiply by 100 on both sides to clear decimals. Thus we have a translation to a system of equations:

   $a + c = 322,$ **(1)**
   $875a + 600c = 253{,}150.$ **(2)** Multiplying by 100

2. **Game Admissions.** There were 166 paid admissions to a game. The price was \$3.10 each for adults and \$1.75 each for children. The amount taken in was \$459.25. How many adults and how many children attended?

Complete the following table to aid with the familiarization.

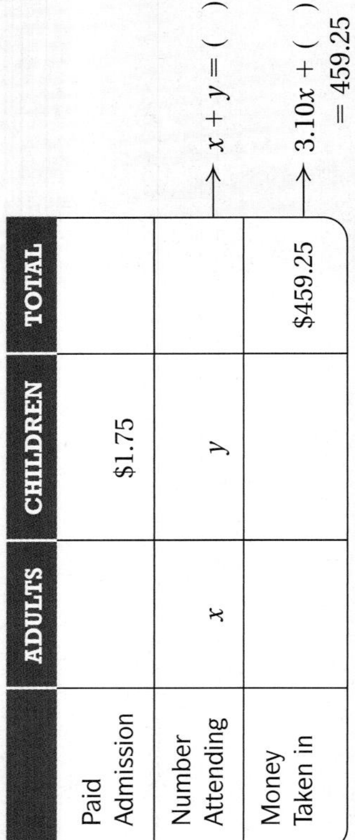

	ADULTS	CHILDREN	TOTAL	
Paid Admission		\$1.75		
Number Attending	x	y		→ $x + y = (\ \ )$
Money Taken in			\$459.25	→ $3.10x + (\ \ ) = 459.25$

3. **Solve.** We solve the system. We use the elimination method since the equations are both in the form $Ax + By = C$. (A case can certainly be made for using the substitution method since we can solve for one of the variables quite easily in the first equation. Very often a decision is just a matter of choice.) We multiply by -600 on both sides of equation (1) and then add and solve for a:

$$\begin{aligned} -600a - 600c &= -193{,}200 && \text{Multiplying by } -600 \\ 875a + 600c &= 253{,}150 && \\ \hline 275a &= 59{,}950 && \text{Adding} \\ a &= \frac{59{,}950}{275} && \text{Dividing by 275} \\ a &= 218. && \end{aligned}$$

Next, we go back to equation (1), substituting 218 for a, and solve for c:

$$\begin{aligned} a + c &= 322 \\ 218 + c &= 322 \\ c &= 104. \end{aligned}$$

4. **Check.** The check is left to the student. It is similar to what we did in the *Familiarize* step.

5. **State.** Attending the showing were 218 adults and 104 children.

Do Exercise 2.

EXAMPLE 3 *Mixture of Solutions.* A chemist has one solution that is 80% acid (that is, 8 parts are acid and 2 parts are water) and another solution that is 30% acid. What is needed is 200 L of a solution that is 62% acid. The chemist will prepare it by mixing the two solutions. How much of each should be used?

1. **Familiarize.** We can make a drawing of the situation. The chemist uses x liters of the first solution and y liters of the second solution. We can also arrange the information in a table.

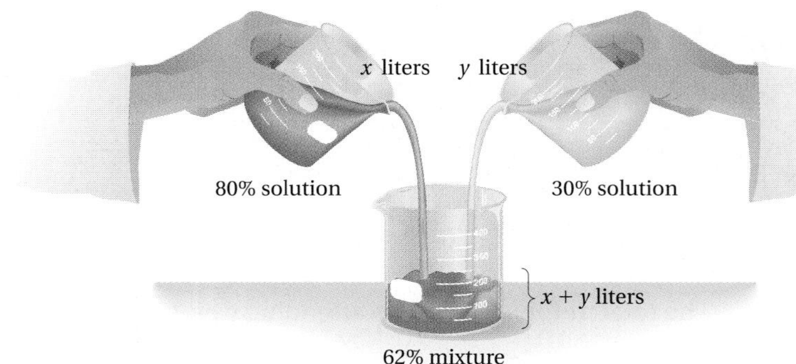

	FIRST SOLUTION	SECOND SOLUTION	MIXTURE	
Amount of Solution	x	y	200 L	→ $x + y = 200$
Percent of Acid	80%	30%	62%	
Amount of Acid In Solution	$80\%x$	$30\%y$	$62\% \times 200$, or 124 L	→ $80\%x + 30\%y = 124$

Answer on page A-47

2. **Translate.** The chemist uses x liters of the first solution and y liters of the second. Since the total is to be 200 L, we have

Total amount of solution: $x + y = 200.$

The amount of acid in the new mixture is to be 62% of 200 L, or 124 L. The amounts of acid from the two solutions are $80\%x$ and $30\%y$. Thus,

Total amount of acid: $80\%x + 30\%y = 124$

or $0.8x + 0.3y = 124.$

We clear decimals by multiplying by 10 on both sides:

$$10(0.8x + 0.3y) = 10 \cdot 124$$
$$8x + 3y = 1240.$$

Thus we have a translation to a system of equations:

$$x + y = 200, \quad (1)$$
$$8x + 3y = 1240. \quad (2)$$

3. **Solve.** We solve the system. We use the elimination method, again because equations are in the form $Ax + By = C$ and a multiplication in one equation will allow us to eliminate a variable, but substitution would also work. We multiply by -3 on both sides of equation (1) and then add and solve for x:

$$-3x - 3y = -600 \quad \text{Multiplying by } -3$$
$$8x + 3y = 1240$$
$$5x = 640 \quad \text{Adding}$$
$$x = \frac{640}{5} \quad \text{Dividing by 5}$$
$$x = 128.$$

Next, we go back to equation (1) and substitute 128 for x:

$$x + y = 200$$
$$128 + y = 200$$
$$y = 72.$$

The solution is $x = 128$ and $y = 72$.

4. **Check.** The sum of 128 and 72 is 200. Also, 80% of 128 is 102.4 and 30% of 72 is 21.6. These add up to 124.

5. **State.** The chemist should use 128 L of the 80%-acid solution and 72 L of the 30%-acid solution.

Do Exercise 3.

EXAMPLE 4 *Candy Mixtures.* A bulk wholesaler wishes to mix some candy worth 45 cents per pound and some worth 80 cents per pound to make 350 lb of a mixture worth 65 cents per pound. How much of each type of candy should be used?

1. **Familiarize.** Arranging the information in a table will help. We let $x =$ the amount of 45-cents candy and $y =$ the amount of 80-cents candy.

3. Mixture of Solutions. One solution is 50% alcohol and a second is 70% alcohol. How much of each should be mixed in order to make 30 L of a solution that is 55% alcohol?

Complete the following table to aid in the familiarization.

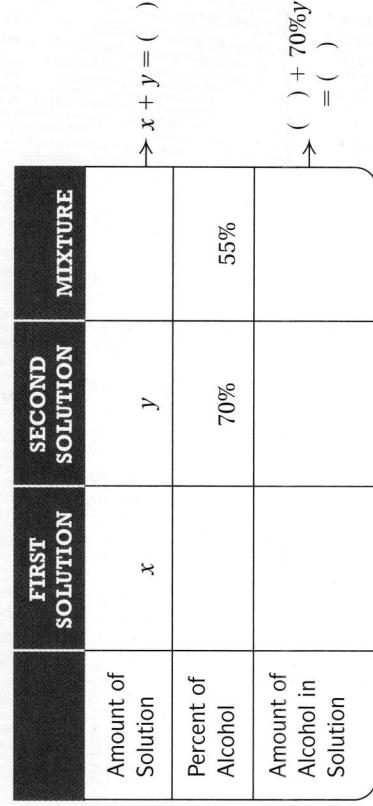

	FIRST SOLUTION	SECOND SOLUTION	MIXTURE	
Amount of Solution	x	y		→ $x + y = (\ \ )$
Percent of Alcohol		70%	55%	
Amount of Alcohol in Solution				→ $(\ \ ) + 70\%y = (\ \ )$

Answer on page A-47

4. Mixture of Grass Seeds. Grass seed A is worth \$1.40 per pound and seed B is worth \$1.75 per pound. How much of each should be mixed in order to make 50 lb of a mixture worth \$1.54 per pound?

Complete the following table to aid in the familiarization.

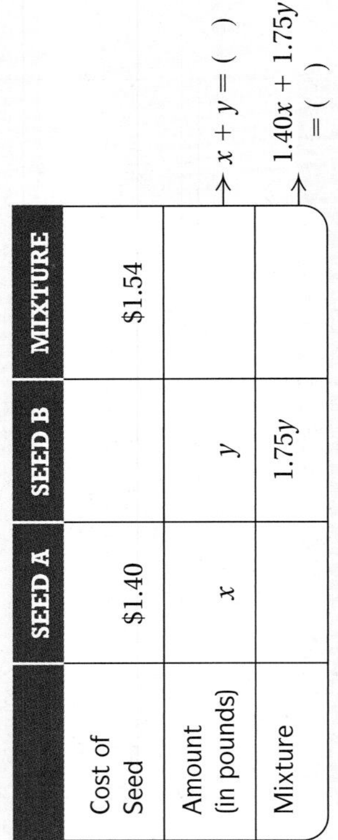

	SEED A	SEED B	MIXTURE	
Cost of Seed	\$1.40		\$1.54	
Amount (in pounds)	x	y		$\rightarrow x + y = (\ \ )$
Mixture		$1.75y$		$\rightarrow 1.40x + 1.75y = (\ \ )$

	INEXPENSIVE CANDY	EXPENSIVE CANDY	MIXTURE	
Cost of Candy	45 cents	80 cents	65 cents	
Amount (in pounds)	x	y	350	$\rightarrow x + y = 350$
Total Cost	$45x$	$80y$	65 cents · (350), or 22,750 cents	$\rightarrow 45x + 80y = 22{,}750$

Note the similarity of this problem to Example 2. Here we consider types of candy instead of groups of people.

2. **Translate.** We translate as follows. From the second row of the table, we find that

Total amount of candy: $x + y = 350.$

Our second equation will come from the costs. The value of the inexpensive candy, in cents, is $45x$ (x pounds at 45 cents per pound). The value of the expensive candy is $80y$, and the value of the mixture is 65×350, or 22,750 cents. Thus we have

Total cost of mixture: $45x + 80y = 22{,}750.$

Remember the problem-solving tip about dimension symbols. In this last equation, all expressions are given in cents. We could have expressed them all in dollars, but we do not want some in cents and some in dollars. Thus we have a translation to a system of equations:

$$x + y = 350, \quad (1)$$
$$45x + 80y = 22{,}750. \quad (2)$$

3. **Solve.** We solve the system using the elimination method again. We multiply by -45 on both sides of equation (1) and then add and solve for y:

$$\begin{aligned} -45x - 45y &= -15{,}750 \quad &&\text{Multiplying by } -45 \\ 45x + 80y &= 22{,}750 \\ \hline 35y &= 7{,}000 \quad &&\text{Adding} \\ y &= \frac{7{,}000}{35} \\ y &= 200. \end{aligned}$$

Next, we go back to equation (1), substituting 200 for y, and solve for x:

$$\begin{aligned} x + y &= 350 \\ x + 200 &= 350 \\ x &= 150. \end{aligned}$$

4. **Check.** We consider $x = 150$ lb and $y = 200$ lb. The sum is 350 lb. The value of the candy is $45(150) + 80(200)$, or 22,750 cents and each pound of the mixture is worth $22{,}750 \div 350$, or 65 cents. These values check.

5. **State.** The grocer should mix 150 lb of the 45-cents candy with 200 lb of the 80-cents candy.

Answer on page A-47

Do Exercise 4.

EXAMPLE 5 *Coin Value.* A student assistant at the university copy center has some nickels and dimes to use for change when students make copies. The value of the coins is \$7.40. There are 26 more dimes than nickels. How many of each kind of coin are there?

1. **Familiarize.** We let $d =$ the number of dimes and $n =$ the number of nickels.
2. **Translate.** We have one equation at once:

   $d = n + 26.$

   The value of the nickels, in cents, is $5n$, since each coin is worth 5 cents. The value of the dimes, in cents, is $10d$, since each coin is worth 10 cents. The total value is given as \$7.40. Since we have the values of the nickels and dimes *in cents,* we must use cents for the total value. This is 740. This gives us another equation:

   $10d + 5n = 740.$

   We now have a system of equations:

   $d = n + 26,$ **(1)**

   $10d + 5n = 740.$ **(2)**

3. **Solve.** Since we have d alone on one side of one equation, we use the substitution method. We substitute $n + 26$ for d in equation (2):

$10d + 5n = 740$	Equation (2)
$10(n + 26) + 5n = 740$	Substituting $n + 26$ for d
$10n + 260 + 5n = 740$	Removing parentheses
$15n + 260 = 740$	Collecting like terms
$15n = 480$	Subtracting 260
$n = \frac{480}{15}$, or 32.	Dividing by 15

   Next, we substitute 32 for n in either of the original equations to find d. We use equation (1):

   $d = n + 26 = 32 + 26 = 58.$

4. **Check.** We have 58 dimes and 32 nickels. There are 26 more dimes than nickels. The value of the coins is 58(\$0.10) + 32(\$0.05), which is \$7.40. This checks.
5. **State.** The student assistant has 58 dimes and 32 nickels.

5. **Coin Value.** On a table are 20 coins, quarters and dimes. Their value is \$3.05. How many of each kind of coin are there?

Answer on page A-47

Do Exercise 5 on the preceding page.

Look back over Examples 2–5. The problems are quite similar in their structure. Compare them and try to see the similarities. The problems in Examples 2–5 are often called *mixture problems*. These problems provide a pattern, or model, for many related problems.

PROBLEM-SOLVING TIP

When solving problems, see if they are patterned or modeled after other problems that you have studied.

Study Tips

TROUBLE SPOTS

By now you have probably encountered certain topics that gave you more difficulty than others. It is important to know that this happens to every person who studies mathematics. Unfortunately, frustration is often part of the learning process and it is important not to give up when difficulty arises.

One source of frustration for many students is not being able to set aside sufficient time for studying. Family commitments, work schedules, and extracurricular activities are just a few of the time demands that many students face. Couple these demands with a math lesson that seems to require a greater than usual amount of study time, and it is no wonder that many students often feel frustrated. Below are some study tips that might be useful if and when troubles arise.

- **Realize that everyone—even your instructor—has been stumped at times when studying math.** You are not the first person, nor will you be the last, to encounter a "roadblock."
- **Whether working alone or with a classmate, try to allow enough study time so that you won't need to constantly glance at a clock.** Difficult material is best mastered when your mind is completely focused on the subject matter. Thus, if you are tired, it is usually best to study early the next morning or to take a ten-minute "power-nap" in order to make the most productive use of your time. Consider redoing the weekly planner on the Student Organizer in the Preface. You may need to adjust your schedule frequently. PLAN FOR SUCCESS with extra study time!
- **Talk about your trouble spot with a classmate.** It is possible that she or he is also having difficulty with the same material. If that is the case, perhaps the majority of your class is confused and your instructor's coverage of the topic is not yet finished. If your classmate *does* understand the topic that is troubling you, patiently allow him or her to explain it to you. By verbalizing the math in question, your classmate may help clarify the material for both of you. Perhaps you will be able to return the favor for your classmate when he or she is struggling with a topic that you understand.
- **Try to study in a "controlled" environment.** What we mean by this is that you can often put yourself in a setting that will enable you to maximize your powers of concentration. For example, some students may succeed in studying at home or in a dorm room, but for many these settings are filled with distractions. Consider a trip to a library, classroom building, or perhaps the attic or basement if such a setting is more conducive to studying. If you plan on working with a classmate, try to find a location in which conversation will not be bothersome to others.
- **When working on difficult material, it is often helpful to first "back up" and review the most recent material that did make sense.** This can build your confidence and create a momentum that can often carry you through the roadblock. Sometimes a small piece of information that appeared in a previous section is all that is needed for your problem spot to disappear. When the difficult material is finally mastered, try to make use of what is fresh in your mind by taking a "sneak preview" of what your next topic for study will be.

For Extra Help

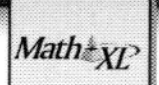
MathXL

MyMathLab

InterAct Math

Math Tutor Center

Video Lectures on CD Disc 7

Student's Solutions Manual

a Solve.

1. *Basketball Scoring.* In a recent game of the 2004–2005 basketball season, the San Antonio Spurs scored 85 of their points on a combination of 39 two- and three-point baskets. How many of each type of shot were made?
Source: National Basketball Association

2. *Basketball Scoring.* Shaquille O'Neill of the Miami Heat once scored 36 points on 22 shots in an NBA game, shooting only two-pointers and foul shots (one point). How many of each type of shot did he make?
Source: National Basketball Association

3. *Film Processing.* Cord Camera charges $9.00 for processing a 24-exposure roll of film and $12.60 for processing a 36-exposure roll of film. After Jack's field trip to observe a variety of architectural styles, he took 17 rolls of film to Cord Camera and paid $171 for processing. How many rolls of each type were processed?
Source: Cord Camera

4. *Paid Admissions.* Following the baseball season, the players on a junior college team decided to go to a major-league baseball game. Ticket prices for the game are shown in the table below. They bought 29 tickets of two types, Upper Box and Lower Reserved. The cost of all the tickets was $318. How many of each kind of ticket did they buy?

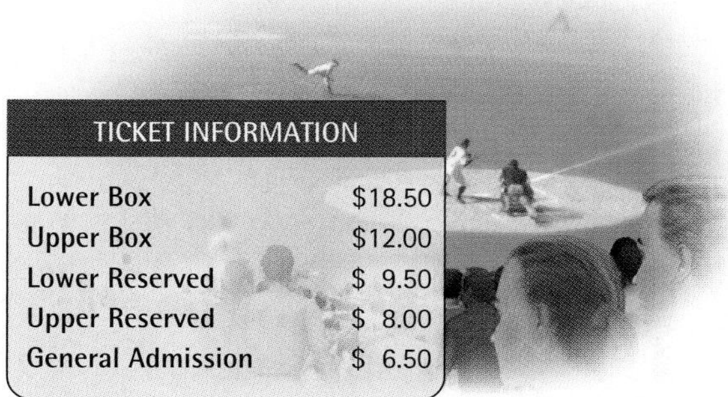

TICKET INFORMATION	
Lower Box	$18.50
Upper Box	$12.00
Lower Reserved	$ 9.50
Upper Reserved	$ 8.00
General Admission	$ 6.50

5. *Grain Mixtures for Horses.* Brianna is a barn manager at a horse stable. She needs to calculate the correct mix of grain and hay to feed her horse. On the basis of her horse's age, weight, and workload, she determines that he needs to eat 15 lb of feed per day, with an average protein content of 8%. Hay contains 6% protein, whereas grain has a 12% protein content. How many pounds of hay and grain should she feed her horse each day?
Source: *Michael Plumb's Horse Journal*, February 1996, pp. 26–29

6. *Paint Mixtures.* At a local "paint swap," Kari found large supplies of Skylite Pink (12.5% red pigment) and MacIntosh Red (20% red pigment). How many gallons of each color should Kari pick up in order to mix a gallon of Summer Rose (17% red pigment)?

7. *Investments.* Cassandra has a number of \$50 and \$100 savings bonds to use for part of her college expenses. The total value of the bonds is \$1250. There are 7 more \$50 bonds than \$100 bonds. How many of each type of bond does she have?

8. *Food Prices.* Mr. Chooeydough's Pizza Parlor charges \$3.70 for a slice of pizza and a soda and \$9.65 for three slices of pizza and two sodas. Determine the cost of one soda and the cost of one slice of pizza.

9. *Ticket Sales.* There were 203 tickets sold for a volleyball game. For activity-card holders, the price was \$2.25 each, and for non-cardholders, the price was \$3 each. The total amount of money collected was \$513. How many of each type of ticket were sold?

10. *Paid Admissions.* There were 429 people at a play. Admission was \$8 each for adults and \$4.50 each for children. The total receipts were \$2641. How many adults and how many children attended?

11. *The Butterfly Exhibit.* On one day during a weekend, 1630 people visited The Butterfly Exhibit at White River Gardens in Indianapolis, Indiana. Admission was \$7 each for adults and \$6 each for children. The receipts totaled \$11,080. How many adults and how many children visited that day?

12. *Zoo Admissions.* During the summer months, the Bronx Zoo charges \$12 each for adults and \$9 each for children and seniors. One July day, a total of \$9780 was collected from 960 admissions. How many adult admissions were there?

Source: Bronx Zoo

13. *Mixture of Solutions.* Solution A is 50% acid and solution B is 80% acid. How many liters of each should be used in order to make 100 L of a solution that is 68% acid? Complete the following table to aid in the familiarization.

	SOLUTION A	SOLUTION B	MIXTURE	
Amount of Solution	x	y	L	$\rightarrow x + y = (\quad)$
Percent of Acid	50%		68%	
Amount of Acid in Solution		80%y	68% × 100 or L	$\rightarrow$ 50%x + ( ) = ( )

14. *Mixture of Solutions.* Solution A is 30% alcohol and solution B is 75% alcohol. How much of each should be used in order to make 100 L of a solution that is 50% alcohol?

15. *Coin Value.* A parking meter contains dimes and quarters worth \$15.25. There are 103 coins in all. How many of each type of coin are there?

16. *Coin Value.* A vending machine contains nickels and dimes worth \$14.50. There are 95 more nickels than dimes. How many of each type of coin are there?

17. *Coffee Blends.* Cafebucks coffee shop mixes Brazilian coffee worth \$19 per pound with Turkish coffee worth \$22 per pound. The mixture is to sell for \$20 per pound. How much of each type of coffee should be used in order to make a 300-lb mixture? Complete the following table to aid in the familiarization.

	BRAZILIAN COFFEE	TURKISH COFFEE	MIXTURE	
Cost of Coffee	\$19		\$20	
Amount (in pounds)	x	y	300	→ $x + y = (\quad)$
Mixture		$22y$	20(300), or \$6000	→ $19x + (\quad) = 6000$

18. *Coffee Blends.* The Java Joint wishes to mix organic Kenyan coffee beans that sell for \$7.25 per pound with organic Venezuelan beans that sell for \$8.50 per pound in order to form a 50-lb batch of Morning Blend that sells for \$8.00 per pound. How many pounds of each type of bean should be used to make the blend?

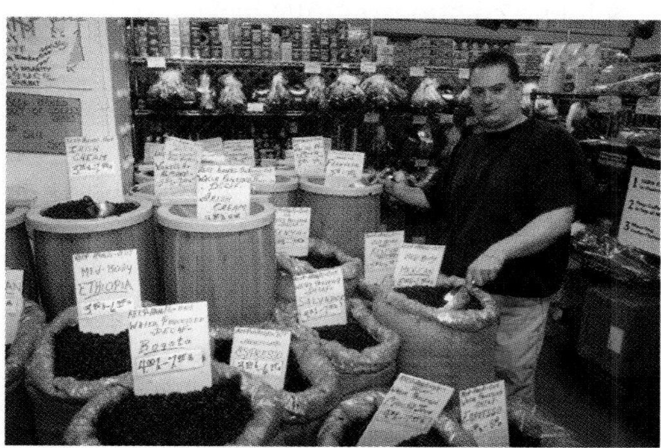

19. *Horticulture.* A solution containing 28% fungicide is to be mixed with a solution containing 40% fungicide to make 300 L of a solution containing 36% fungicide. How much of each solution should be used?

20. *Production.* Clear Shine window cleaner is 12% alcohol and Sunstream window cleaner is 30% alcohol. How much of each should be used to make 90 oz of a cleaner that is 20% alcohol?

21. *Printing.* A printer knows that a page of print contains 830 words if large type is used and 1050 words if small type is used. A document containing 11,720 words fills exactly 12 pages. How many pages are in the large type? in the small type?

22. *Paint Mixture.* A merchant has two kinds of paint. If 9 gal of the inexpensive paint is mixed with 7 gal of the expensive paint, the mixture will be worth \$19.70 per gallon. If 3 gal of the inexpensive paint is mixed with 5 gal of the expensive paint, the mixture will be worth \$19.825 per gallon. What is the price per gallon of each type of paint?

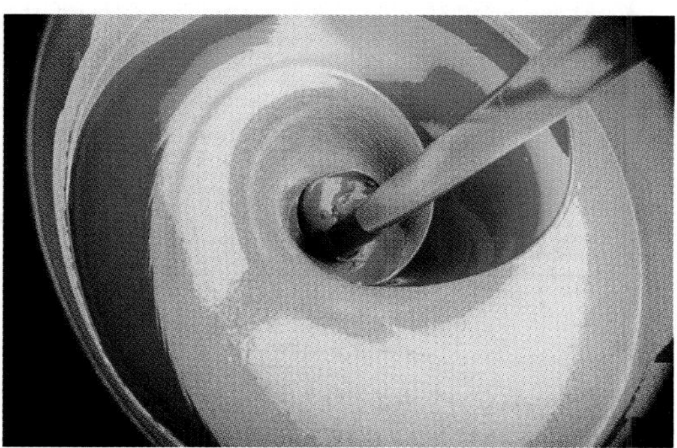

23. *Mixed Nuts.* A customer has asked a caterer to provide 60 lb of nuts, 60% of which are to be cashews. The caterer has available mixtures of 70% cashews and 45% cashews. How many pounds of each mixture should be used?

24. *Mixture of Grass Seeds.* Grass seed A is worth \$2.50 per pound and seed B is worth \$1.75 per pound. How much of each would you use in order to make 75 lb of a mixture worth \$2.14 per pound?

25. *Test Scores.* You are taking a test in which items of type A are worth 10 points and items of type B are worth 15 points. It takes 3 min to complete each item of type A and 6 min to complete each item of type B. The total time allowed is 60 min and you do exactly 16 questions. How many questions of each type did you complete? Assuming that all your answers were correct, what was your score?

26. *Gold Alloys.* A goldsmith has two alloys that are different purities of gold. The first is three-fourths pure gold and the second is five-twelfths pure gold. How many ounces of each should be melted and mixed in order to obtain a 6-oz mixture that is two-thirds pure gold?

27. *Ages.* The Kuyatts' house is twice as old as the Marconis' house. Eight years ago, the Kuyatts' house was three times as old as the Marconis' house. How old is each house?

28. *Ages.* David is twice as old as his daughter. In 4 yr, David's age will be three times what his daughter's age was 6 yr ago. How old are they now?

29. *Ages.* Randy is four times as old as Mandy. In 12 yr, Mandy's age will be half of Randy's. How old are they now?

30. *Ages.* Jennifer is twice as old as Ramon. The sum of their ages 7 yr ago was 13. How old are they now?

31. *Supplementary Angles.* **Supplementary angles** are angles whose sum is 180°. Two supplementary angles are such that one is 30° more than two times the other. Find the angles.

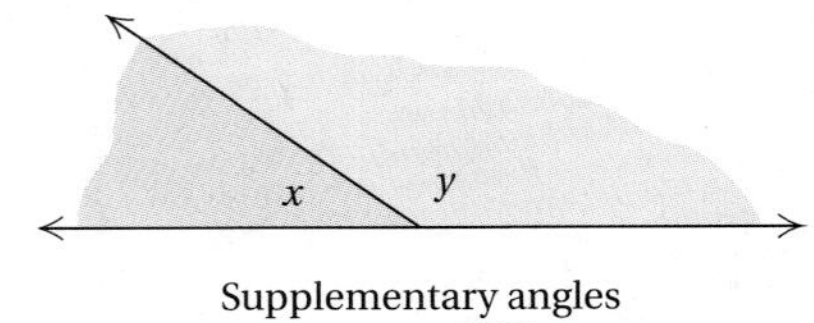

Supplementary angles
$x + y = 180°$

32. *Supplementary Angles.* Two supplementary angles are such that one is 8° less than three times the other. Find the angles.

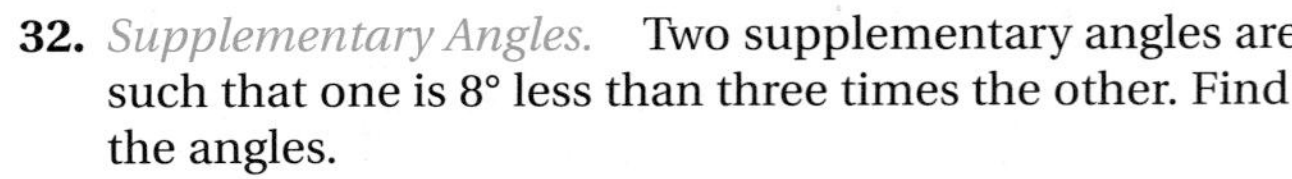

33. *Complementary Angles.* **Complementary angles** are angles whose sum is 90°. Two complementary angles are such that their difference is 34°. Find the angles.

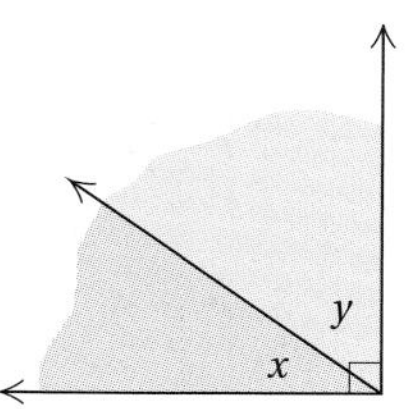

Complementary angles
$x + y = 90°$

34. *Complementary Angles.* Two angles are complementary. One angle is 42° more than one-half the other. Find the angles.

35. *Octane Ratings.* In most areas of the United States, gas stations offer three grades of gasoline, indicated by octane ratings on the pumps, such as 87, 89, and 93. When a tanker delivers gas, it brings only two grades of gasoline, the highest and the lowest, filling two large underground tanks. If you purchase the middle grade, the pump's computer mixes the other two grades appropriately. How much 87-octane gas and 93-octane gas should be blended in order to make 18 gal of 89-octane gas?

Source: Exxon

36. *Octane Ratings.* Referring to Exercise 35, suppose the pump grades offered are 85, 87, and 91. How much 85-octane gas and 91-octane gas should be blended in order to make 12 gal of 87-octane gas?

Source: Exxon

37. *Cough Syrup.* Dr. Zeke's cough syrup is 2% alcohol. Vitabrite cough syrup is 5% alcohol. How much of each type should be used in order to prepare an 80-oz batch of cough syrup that is 3% alcohol?

38. *Suntan Lotion.* Lisa has a tube of Kinney's suntan lotion that is rated 15 spf and a second tube of Coppertone that is 30 spf. How many fluid ounces of each type of lotion should be mixed in order to create 50 fluid ounces of sunblock that is rated 20 spf?

39. $\mathbf{D_W}$ Look over Example 4 of this section. What new equation-solving skills have we developed in this chapter that enable us to solve this problem? How do these equations differ from others in the preceding parts of this book?

40. $\mathbf{D_W}$ Look over Example 4 of this section. What equation-solving skills learned elsewhere in this book are most important in solving systems of equations?

SKILL MAINTENANCE

Factor. [11.6a]

41. $25x^2 - 81$

42. $36 - a^2$

43. $4x^2 + 100$

44. $4x^2 - 100$

Find the intercepts. Then graph the equation. [9.2a]

45. $y = -2x - 3$

46. $y = -0.1x + 0.4$

47. $5x - 2y = -10$

48. $2.5x + 4y = 10$

Simplify. [12.1c]

49. $\dfrac{x^2 - 5x + 6}{x^2 - 4}$

50. $\dfrac{x^2 - 25}{x^2 - 10x + 25}$

Subtract. [12.5a]

51. $\dfrac{x - 2}{x + 3} - \dfrac{2x - 5}{x - 4}$

52. $\dfrac{x + 7}{x^2 - 1} - \dfrac{3}{x + 1}$

SYNTHESIS

53. *Milk Mixture.* A farmer has 100 L of milk that is 4.6% butterfat. How much skim milk (no butterfat) should be mixed with it in order to make milk that is 3.2% butterfat?

54. One year, Shannon made $288 from two investments: $1100 was invested at one yearly rate and $1800 at a rate that was 1.5% higher. Find the two rates of interest.

55. *Automobile Maintenance.* An automobile radiator contains 16 L of antifreeze and water. This mixture is 30% antifreeze. How much of this mixture should be drained and replaced with pure antifreeze so that the mixture will be 50% antifreeze?

56. *Employer Payroll.* An employer has a daily payroll of $1225 when employing some workers at $80 per day and others at $85 per day. When the number of $80 workers is increased by 50% and the number of $85 workers is decreased by $\frac{1}{5}$, the new daily payroll is $1540. How many were originally employed at each rate?

57. A two-digit number is six times the sum of its digits. The tens digit is 1 more than the units digit. Find the number.

13.5 APPLICATIONS WITH MOTION

Objective

a Solve motion problems using the formula $d = rt$.

a We first studied problems involving motion in Chapter 13. Here we extend our problem-solving skills by solving certain motion problems whose solutions can be found using systems of equations. Recall the motion formula.

THE MOTION FORMULA

Distance = Rate (or speed) · Time

$$d = rt$$

We have five steps for problem solving. The tips in the margin at right are also helpful when solving motion problems.

As we saw in Chapter 13, there are motion problems that can be solved with just one equation. Let's start with another such problem.

TIPS FOR SOLVING MOTION PROBLEMS

1. Draw a diagram using an arrow or arrows to represent distance and the direction of each object in motion.
2. Organize the information in a chart.
3. Look for as many things as you can that are the same so that you can write equations.

EXAMPLE 1 *Car Travel.* Two cars leave Ashland at the same time traveling in opposite directions. One travels at 60 mph and the other at 30 mph. In how many hours will they be 150 mi apart?

1. **Familiarize.** We first make a drawing.

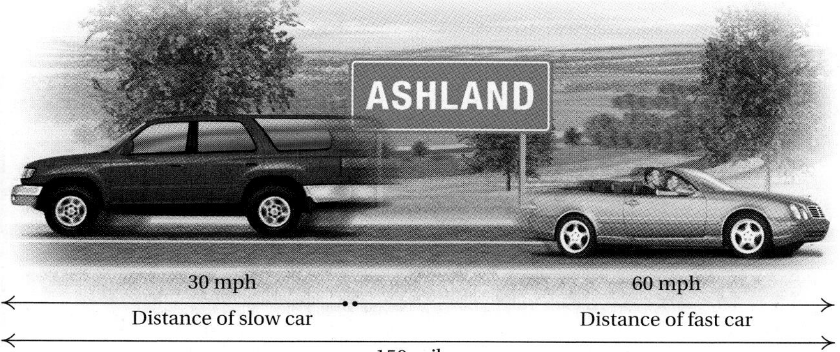

From the wording of the problem and the drawing, we see that the distances may *not* be the same. But the times that the cars travel are the same, so we can use just t for time. We can organize the information in a chart.

	d = DISTANCE	r · SPEED	t TIME
Fast Car	Distance of fast car	60	t
Slow Car	Distance of slow car	30	t
Total	150		

Study Tips

VARYING ORDER IN TEST PREPARATION

When studying for a test, it is common for students to study material in the order in which it appears in the book. Educational research has shown that when students do so, they tend to remember material at the beginning and at the end the best and the material in the middle the least. So when you're preparing for a test, especially when you allow yourself two or three days to study (as you should), vary what you study at the beginning and at the end of your study sessions.

1. **Car Travel.** Two cars leave town at the same time traveling in opposite directions. One travels at 48 mph and the other at 60 mph. How far apart will they be 3 hr later? (*Hint*: The times are the same. Be *sure* to make a drawing.)

2. **Car Travel.** Two cars leave town at the same time traveling in the same direction. One travels at 35 mph and the other at 40 mph. In how many hours will they be 15 mi apart? (*Hint*: The times are the same. Be *sure* to make a drawing.)

Answers on page A-47

2. **Translate.** From the drawing, we see that

   (Distance of fast car) + (Distance of slow car) = 150.

   Then using $d = rt$ in each row of the table, we get

   $60t + 30t = 150.$

3. **Solve.** We solve the equation:

$$60t + 30t = 150$$
$$90t = 150 \qquad \text{Collecting like terms}$$
$$t = \frac{150}{90}, \text{ or } \frac{5}{3}, \text{ or } 1\frac{2}{3} \text{ hr.} \qquad \text{Dividing by 90}$$

4. **Check.** When $t = \frac{5}{3}$ hr,

$$(\text{Distance of fast car}) + (\text{Distance of slow car}) = 60\left(\frac{5}{3}\right) + 30\left(\frac{5}{3}\right)$$
$$= 100 + 50, \text{ or } 150 \text{ mi.}$$

   Thus the time of $\frac{5}{3}$ hr, or $1\frac{2}{3}$ hr, checks.

5. **State.** In $1\frac{2}{3}$ hr, the cars will be 150 mi apart.

Do Exercises 1 and 2.

Now let's solve some motion problems using systems of equations.

EXAMPLE 2 *Train Travel.* A train leaves Stanton traveling east at 35 miles per hour (mph). An hour later, another train leaves Stanton on a parallel track at 40 mph. How far from Stanton will the second (or faster) train catch up with the first (or slower) train?

1. **Familiarize.** We first make a drawing.

From the drawing, we see that the distances are the same. Let's call the distance d. We don't know the times. We let t = the time for the faster train. Then the time for the slower train = $t + 1$, since it left 1 hr earlier. We can organize the information in a chart.

$d = r \cdot t$

	DISTANCE	SPEED	TIME	
Slow Train	d	35	$t + 1$	→ $d = 35(t + 1)$
Fast Train	d	40	t	→ $d = 40t$

2. **Translate.** In motion problems, we look for quantities that are the same so that we can write equations. From each row of the chart, we get an equation, $d = rt$. Thus we have two equations:

$$d = 35(t + 1), \quad (1)$$
$$d = 40t. \quad (2)$$

3. **Solve.** Since we have a variable alone on one side, we solve the system using the substitution method:

$35(t + 1) = 40t$ Using the substitution method (substituting $35(t + 1)$ for d in equation 2)

$35t + 35 = 40t$ Removing parentheses

$35 = 5t$ Subtracting $35t$

$\frac{35}{5} = t$ Dividing by 5

$7 = t.$

The problem asks us to find how far from Stanton the fast train catches up with the other. Thus we need to find d. We can do this by substituting 7 for t in the equation $d = 40t$:

$$d = 40(7) = 280.$$

4. **Check.** If the time is 7 hr, then the distance that the slow train travels is $35(7 + 1)$, or 280 mi. The fast train travels $40(7)$, or 280 mi. Since the distances are the same, we know how far from Stanton the trains will be when the fast train catches up with the other.

5. **State.** The fast train will catch up with the slow train 280 mi from Stanton.

Do Exercise 3.

EXAMPLE 3 *Boat Travel.* A motorboat took 3 hr to make a downstream trip with a 6-km/h current. The return trip against the same current took 5 hr. Find the speed of the boat in still water.

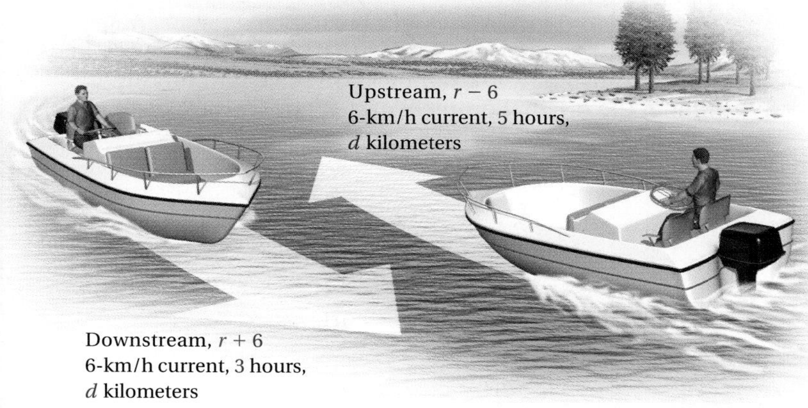

1. **Familiarize.** We first make a drawing. From the drawing, we see that the distances are the same. Let's call the distance d. We let r = the speed of the boat in still water. Then, when the boat is traveling downstream, its speed is $r + 6$ (the current helps the boat along). When it is traveling upstream, its speed is $r - 6$ (the current holds the boat back).

3. **Car Travel.** A car leaves Spokane traveling north at 56 km/h. Another car leaves Spokane 1 hr later traveling north at 84 km/h. How far from Spokane will the second car catch up with the first? (*Hint*: The cars travel the same distance.)

Answer on page A-47

4. **Air Travel.** An airplane flew for 5 hr with a 25-km/h tail wind. The return flight against the same wind took 6 hr. Find the speed of the airplane in still air. (*Hint*: The distance is the same both ways. The speeds are $r + 25$ and $r - 25$, where r is the speed in still air.)

We can organize the information in a chart. In this case, the distances are the same, so we use the formula $d = rt$.

	DISTANCE (d)	SPEED (r)	TIME (t)	
Downstream	d	$r + 6$	3	$\to d = (r + 6)3$
Upstream	d	$r - 6$	5	$\to d = (r - 6)5$

2. **Translate.** From each row of the chart, we get an equation, $d = rt$:

$$d = (r + 6)3, \quad \textbf{(1)}$$
$$d = (r - 6)5. \quad \textbf{(2)}$$

3. **Solve.** Since there is a variable alone on one side of an equation, we solve the system using substitution:

$$(r + 6)3 = (r - 6)5 \quad \text{Substituting } (r + 6)3 \text{ for } d \text{ in equation (2)}$$
$$3r + 18 = 5r - 30 \quad \text{Removing parentheses}$$
$$-2r + 18 = -30 \quad \text{Subtracting } 5r$$
$$-2r = -48 \quad \text{Subtracting 18}$$
$$r = \frac{-48}{-2}, \text{ or } 24. \quad \text{Dividing by } -2$$

4. **Check.** When $r = 24$, $r + 6 = 30$, and $30 \cdot 3 = 90$, the distance downstream. When $r = 24$, $r - 6 = 18$, and $18 \cdot 5 = 90$, the distance upstream. In both cases, we get the same distance.

5. **State.** The speed in still water is 24 km/h.

MORE TIPS FOR SOLVING MOTION PROBLEMS

1. Translating to a system of equations eases the solution of many motion problems.
2. At the end of the problem, always ask yourself, "Have I found what the problem asked for?" You might have solved for a certain variable but still not have answered the question of the original problem. For instance, in Example 2 we solve for t but the question of the original problem asks for d. Thus we need to continue the *Solve* step.

Do Exercise 4.

Answer on page A-47

Translating for Success

The goal of these matching questions is to practice step (2), *Translate,* of the five-step problem-solving process. Translate each word problem to an equation or a system of equations and select a correct translation from A–O.

1. *Car Travel.* Two cars leave town at the same time traveling in different directions. One travels 50 mph and the other travels 55 mph. In how many hours will they be 50 mi apart?

2. *Mixture of Solutions.* Solution A is 20% alcohol and solution B is 60% alcohol. How much of each should be used in order to make 10 L of a solution that is 50% alcohol?

3. *Triangle Dimensions.* The height of a triangle is 3 cm less than the base. The area is 27 cm^2. Find the height and the base.

4. *Fish Population.* To determine the number of fish in a lake, a conservationist catches 85 fish, tags them, and throws them back into the lake. Later, 60 fish are caught, 25 of which are tagged. How many fish are in the lake?

5. *Supplementary Angles.* Two angles are supplementary. One angle measures 36° more than three times the measure of the other. Find the measure of each angle.

A. $20\%x + 60\%y = 50\% \cdot 10,$
$x + y = 10$

B. $18 + 0.35x = 100$

C. $55x + 50x = 50$

D. $x + (3x + 36) + (x + 7) = 90$

E. $\dfrac{85}{x} = \dfrac{25}{60}$

F. $\dfrac{1}{15} + \dfrac{1}{9} = \dfrac{1}{x}$

G. $\dfrac{1}{2}x(x - 3) = 27$

H. $x^2 + (x + 4)^2 = 8^2$

I. $8^2 + x^2 = (x + 4)^2$

J. $x + (3x + 36) = 180$

K. $20x + 60y = 5,$
$x + y = 10$

L. $x + (3x + 36) + (x - 7) = 180$

M. $18 + 35x = 100$

N. $\dfrac{x}{85} = \dfrac{25}{60}$

O. $x + (3x + 36) = 90$

Answers on page A-47

6. *Triangle Dimensions.* The length of one leg of a right triangle is 8 m. The length of the hypotenuse is 4 m longer than the length of the other leg. Find the lengths of the hypotenuse and the other leg.

7. *Costs of Promotional Buttons.* The vice-president of a fraternity has \$100 to spend on promotional buttons for pledge week. There is a setup fee of \$18 and a cost of 35¢ per button. How many buttons can he purchase?

8. *Triangle Measures.* The second angle of a triangle measures 36° more than three times the measure of the first. The measure of the third angle is 7° less than the first. Find the measure of each angle of the triangle.

9. *Complementary Angles.* Two angles are complementary. One angle measures 36° more than three times the measure of the other. Find the measure of each angle.

10. *Work Time.* It takes Maggie 15 hr to put a roof on a house. It takes Claire 9 hr to put a roof on the same type of house. How long would it take to complete the house if they worked together?

EXERCISE SET

For Extra Help

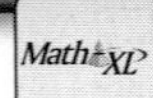
MathXL

MyMathLab

InterAct Math

Math Tutor Center

Video Lectures on CD Disc 7

Student's Solutions Manual

a Solve. In Exercises 1–6, complete the table to aid the translation.

1. *Car Travel.* Two cars leave town at the same time going in the same direction. One travels at 30 mph and the other travels at 46 mph. In how many hours will they be 72 mi apart?

$d = r \cdot t$

	DISTANCE	SPEED	TIME
Slow Car	Distance of slow car		t
Fast Car	Distance of fast car	46	

2. *Car and Truck Travel.* A truck and a car leave a service station at the same time and travel in the same direction. The truck travels at 55 mph and the car at 40 mph. They can maintain CB radio contact within a range of 10 mi. When will they lose contact?

$d = r \cdot t$

	DISTANCE	SPEED	TIME
Truck	Distance of truck	55	
Car	Distance of car		t

3. *Train Travel.* A train leaves a station and travels east at 72 mph. Three hours later, a second train leaves on a parallel track and travels east at 120 mph. When will it overtake the first train?

$d = r \cdot t$

	DISTANCE	SPEED	TIME	
Slow Train	d		$t + 3$	→ $d = 72(\quad)$
Fast Train	d	120		→ $d = (\quad)t$

4. *Airplane Travel.* A private airplane leaves an airport and flies due south at 192 mph. Two hours later, a jet leaves the same airport and flies due south at 960 mph. When will the jet overtake the plane?

$d = r \cdot t$

	DISTANCE	SPEED	TIME	
Private Plane	d	192		→ $d = 192(\quad)$
Jet	d		t	→ $d = (\quad)(t)$

5. *Canoeing.* A canoeist paddled for 4 hr with a 6-km/h current to reach a campsite. The return trip against the same current took 10 hr. Find the speed of the canoe in still water.

$d = r \cdot t$

	DISTANCE	SPEED	TIME	
Down-stream	d	$r + 6$		→ $d = (\quad)4$
Upstream	d		10	→ $\quad = (r - 6)10$

6. *Airplane Travel.* An airplane flew for 4 hr with a 20-km/h tail wind. The return flight against the same wind took 5 hr. Find the speed of the plane in still air.

$d = r \cdot t$

	DISTANCE	SPEED	TIME	
With Wind	d		4	→ $d = (\quad)4$
Against Wind	d	$r - 20$		→ $d = (\quad)5$

7. *Train Travel.* It takes a passenger train 2 hr less time than it takes a freight train to make the trip from Central City to Clear Creek. The passenger train averages 96 km/h, while the freight train averages 64 km/h. How far is it from Central City to Clear Creek?

8. *Airplane Travel.* It takes a small jet 4 hr less time than it takes a propeller-driven plane to travel from Glen Rock to Oakville. The jet averages 637 km/h, while the propeller plane averages 273 km/h. How far is it from Glen Rock to Oakville?

9. *Motorboat Travel.* On a weekend outing, Antoine rents a motorboat for 8 hr to travel down the river and back. The rental operator tells him to go for 3 hr downstream, leaving him 5 hr to return upstream.

a) If the river current flows at a speed of 6 mph, how fast must Antoine travel in order to return in 8 hr?

b) How far downstream did Antoine travel before he turned back?

10. *Airplane Travel.* For spring break some students flew to Cancun. From Mexico City, the airplane took 2 hr to fly 600 mi against a head wind. The return trip with the wind took $1\frac{2}{3}$ hr. Find the speed of the plane in still air.

11. *Running.* A toddler takes off running down the sidewalk at 230 ft/min. One minute later, a worried mother runs after the child at 660 ft/min. When will the mother overtake the toddler?

12. *Airplane Travel.* Two airplanes start at the same time and fly toward each other from points 1000 km apart at rates of 420 km/h and 330 km/h. When will they meet?

13. *Motorcycle Travel.* A motorcycle breaks down and the rider must walk the rest of the way to work. The motorcycle was being driven at 45 mph, and the rider walks at a speed of 6 mph. The distance from home to work is 25 mi, and the total time for the trip was 2 hr. How far did the motorcycle go before it broke down?

14. *Walking and Jogging.* A student walks and jogs to college each day. She averages 5 km/h walking and 9 km/h jogging. The distance from home to college is 8 km, and she makes the trip in 1 hr. How far does the student jog?

15. D_W Discuss the advantages of using a table to organize information when solving a motion problem.

16. D_W From the formula $d = rt$, derive two other formulas, one for r and one for t. Discuss the kinds of problems for which each formula might be useful.

SKILL MAINTENANCE

Simplify. [12.1c]

17. $\dfrac{8x^2}{24x}$

18. $\dfrac{5x^8y^4}{10x^3y}$

19. $\dfrac{5a + 15}{10}$

20. $\dfrac{12x - 24}{48}$

21. $\dfrac{2x^2 - 50}{x^2 - 25}$

22. $\dfrac{x^2 - 1}{x^4 - 1}$

23. $\dfrac{x^2 - 3x - 10}{x^2 - 2x - 15}$

24. $\dfrac{6x^2 + 15x - 36}{2x^2 - 5x + 3}$

25. $\dfrac{(x^2 + 6x + 9)(x - 2)}{(x^2 - 4)(x + 3)}$

26. $\dfrac{x^2 + 25}{x^2 - 25}$

27. $\dfrac{6x^2 + 18x + 12}{6x^2 - 6}$

28. $\dfrac{x^3 + 3x^2 + 2x + 6}{2x^3 + 6x^2 + x + 3}$

SYNTHESIS

29. *Lindbergh's Flight.* Charles Lindbergh flew the Spirit of St. Louis in 1927 from New York to Paris at an average speed of 107.4 mph. Eleven years later, Howard Hughes flew the same route, averaged 217.1 mph, and took 16 hr and 57 min less time. Find the length of their route.

30. *Car Travel.* A car travels from one town to another at a speed of 32 mph. If it had gone 4 mph faster, it could have made the trip in $\frac{1}{2}$ hr less time. How far apart are the towns?

31. *River Cruising.* An afternoon sightseeing cruise up river and back down river is scheduled to last 1 hr. The speed of the current is 4 mph, and the speed of the riverboat in still water is 12 mph. How far upstream should the pilot travel before turning around?

13 Summary and Review

The review that follows is meant to prepare you for a chapter exam. It consists of three parts. The first part, Concept Reinforcement, is designed to increase understanding of the concepts through true/false exercises. The second part is a list of important properties and formulas. The third part is the Review Exercises. These provide practice exercises for the exam, together with references to section objectives so you can go back and review. Before beginning, stop and look back over the skills you have obtained. What skills in mathematics do you have now that you did not have before studying this chapter?

CONCEPT REINFORCEMENT

Determine whether the statement is true or false. Answers are given at the back of the book.

_______ **1.** Every system of two equations has one and only one ordered pair as a solution.

_______ **2.** The system of equations $y = \frac{a}{c}x + b$ and $y = \frac{a}{c}x - b$, $b \neq 0$, has no solution.

_______ **3.** The solution of the system of equations $x = a$ and $y = b$ is (a, b).

_______ **4.** The solution of the system of equations $y = \frac{a}{c}x + b$ and $y = -\frac{c}{a}x + b$ is $(b, 0)$.

_______ **5.** A solution of a system of two equations is an ordered pair that makes at least one equation true.

IMPORTANT PROPERTIES AND FORMULAS

Motion Formula: $d = rt$

Review Exercises

Determine whether the given ordered pair is a solution of the system of equations. [13.1a]

1. $(6, -1)$; $x - y = 3$, $2x + 5y = 6$

2. $(2, -3)$; $2x + y = 1$, $x - y = 5$

3. $(-2, 1)$; $x + 3y = 1$, $2x - y = -5$

4. $(-4, -1)$; $x - y = 3$, $x + y = -5$

Solve the system by graphing. [13.1b]

5. $x + y = 4$, $x - y = 8$

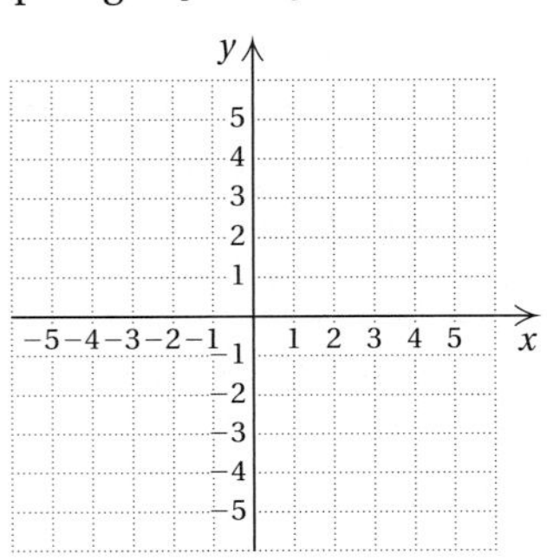

6. $x + 3y = 12$, $2x - 4y = 4$

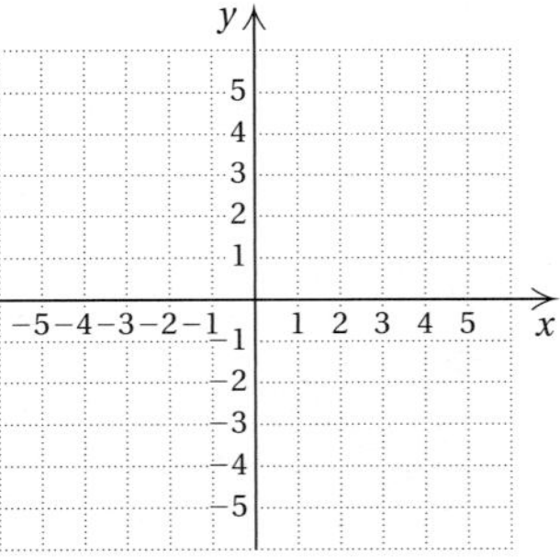

7. $y = 5 - x$, $3x - 4y = -20$

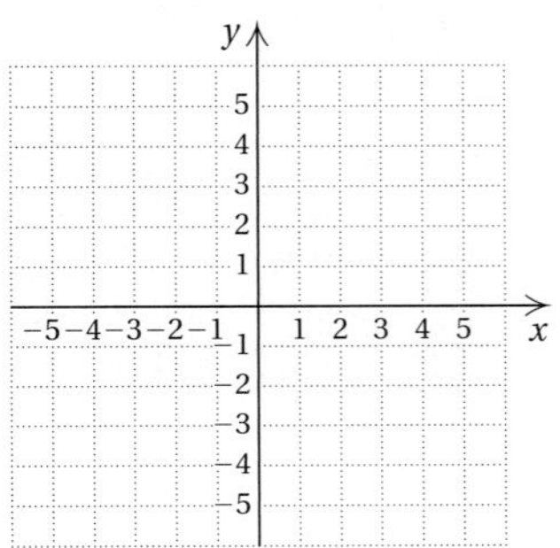

8. $3x - 2y = -4,$
$2y - 3x = -2$

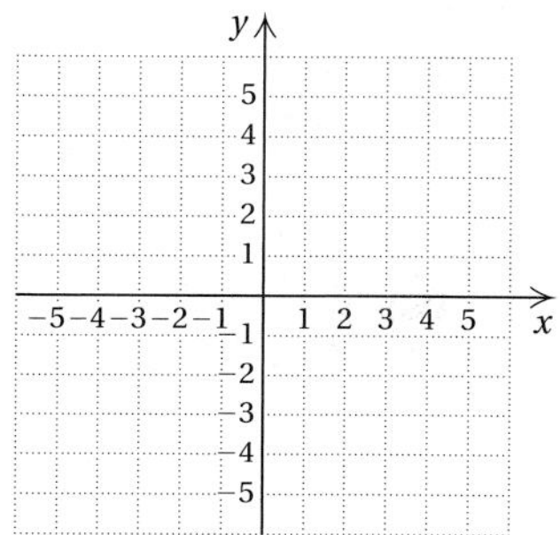

Solve the system using the substitution method. [13.2a]

9. $y = 5 - x,$
$3x - 4y = -20$

10. $x + y = 6,$
$y = 3 - 2x$

11. $x - y = 4,$
$y = 2 - x$

12. $s + t = 5,$
$s = 13 - 3t$

Solve the system using the substitution method. [13.2b]

13. $x + 2y = 6,$
$2x + 3y = 8$

14. $3x + y = 1,$
$x - 2y = 5$

Solve the system using the elimination method. [13.3a]

15. $x + y = 4,$
$2x - y = 5$

16. $x + 2y = 9,$
$3x - 2y = -5$

17. $x - y = 8,$
$2x + y = 7$

Solve the system using the elimination method. [13.3b]

18. $2x + 3y = 8,$
$5x + 2y = -2$

19. $5x - 2y = 2,$
$3x - 7y = 36$

20. $-x - y = -5,$
$2x - y = 4$

21. $6x + 2y = 4,$
$10x + 7y = -8$

22. $-6x - 2y = 5,$
$12x + 4y = -10$

23. $\frac{2}{3}x + y = -\frac{5}{3},$
$x - \frac{1}{3}y = -\frac{13}{3}$

Solve. [13.2c], [13.4a]

24. *Rectangle Dimensions.* The perimeter of a rectangle is 96 cm. The length is 27 cm more than the width. Find the length and the width.

25. *Paid Admissions.* There were 508 people at a rock concert. Orchestra seats cost \$25 each and balcony seats cost \$18 each. The total receipts were \$11,223. Find the number of orchestra seats and the number of balcony seats sold for the concert.

26. *Window Cleaner.* Clear Shine window cleaner is 30% alcohol, whereas Sunstream window cleaner is 60% alcohol. How much of each is needed to make 80 L of a cleaner that is 45% alcohol?

27. *Weights of Elephants.* A zoo has both an Asian and an African elephant. The African elephant weighs 2400 kg more than the Asian elephant. Together, they weigh 12,000 kg. How much does each elephant weigh?

28. *Mixed Nuts.* Sandy's Catering needs to provide 13 lb of mixed nuts for a wedding reception. The wedding couple has allocated \$71 for nuts. Peanuts cost \$4.50 per pound and fancy nuts cost \$7.00 per pound. How many pounds of each type should be mixed?

29. *Phone Rates.* Recently, AT&T offered an unlimited long-distance calling plan to anyone in the United States, 24 hours a day, 7 days a week, for \$29.95 a month. Another plan charges 7¢ a minute all day every day, but costs an additional \$3.95 per month. For what number of minutes will the two plans cost the same?

30. *Octane Ratings.* The octane rating of a gasoline is a measure of the amount of isooctane in the gas. How much 87-octane gas and 95-octane gas should be blended in order to end up with a 10-gal batch of 93-octane gas?

Source: Champlain Electric and Petroleum Equipment

31. *Age.* Jeff is three times as old as his son. In 13 yr, Jeff will be twice as old as his son. How old is each now?

32. *Complementary Angles.* Two angles are complementary. Their difference is 26°. Find the measure of each angle.

33. *Supplementary Angles.* Two angles are supplementary. Their difference is 26°. Find the measure of each angle.

Solve. [13.5a]

34. *Air Travel.* An airplane flew for 4 hr with a 15-km/h tail wind. The return flight against the wind took 5 hr. Find the speed of the airplane in still air.

$d = r \cdot t$

	DISTANCE	SPEED	TIME
Going			
Returning			

35. *Car Travel.* One car leaves Phoenix, Arizona, on Interstate highway I-10 traveling at a speed of 55 mph. Two hours later, another car leaves Phoenix on the same highway, but travels at a speed limit of 75 mph. How far from Phoenix will the second car catch up to the other?

$d = r \cdot t$

	DISTANCE	SPEED	TIME
Slow Car			
Fast Car			

36. DW Janine can tell by inspection that the system

$$y = 2x - 1,$$
$$y = 2x + 3$$

has no solution. How did she determine this? [13.1b]

37. DW Which of the five problem-solving steps have you found the most challenging? Why? [13.4a], [13.5b]

SYNTHESIS

38. *Value of a Horse.* Stephanie agreed to work as a stablehand for 1 yr. At the end of that time, she was to receive $2400 and one horse. After 7 months, she quit the job, but still received the horse and $1000. What was the value of the horse? [13.4a]

39. The solution of the following system is (6, 2). Find C and D. [13.1a]

$$2x - Dy = 6,$$
$$Cx + 4y = 14$$

40. Solve: [13.2a]

$$3(x - y) = 4 + x,$$
$$x = 5y + 2.$$

Each of the following shows the graph of a system of equations. Find the equations. [9.4c], [13.1b]

41.

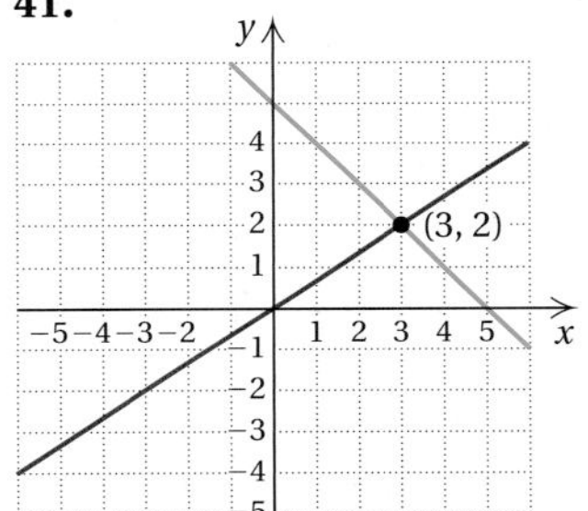

42.

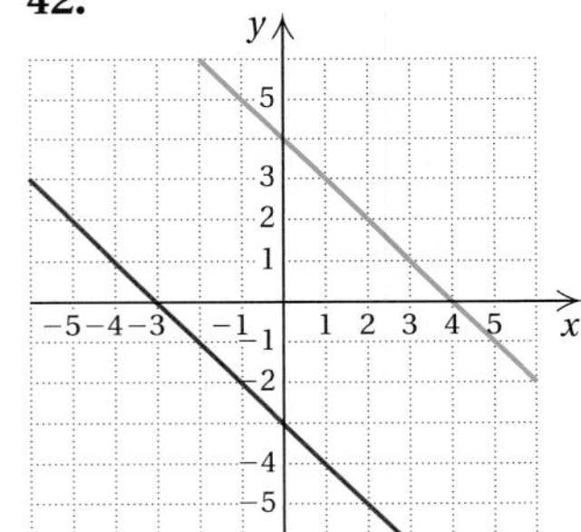

43. *Ancient Chinese Math Problem.* Several ancient Chinese books included problems that can be solved by translating to systems of equations. *Arithmetical Rules in Nine Sections* is a book of 246 problems compiled by a Chinese mathematician, Chang Tsang, who died in 152 B.C. One of the problems is: Suppose there are a number of rabbits and pheasants confined in a cage. In all, there are 35 heads and 94 feet. How many rabbits and how many pheasants are there? Solve the problem. [13.4a]

13 Chapter Test

For Extra Help

Work It Out!
Chapter Test Video
Solutions on CD

1. Determine whether the given ordered pair is a solution of the system of equations.

$$(-2, -1); \quad x = 4 + 2y, \quad 2y - 3x = 4$$

2. Solve this system by graphing. Show your work.

$$x - y = 3, \quad x - 2y = 4.$$

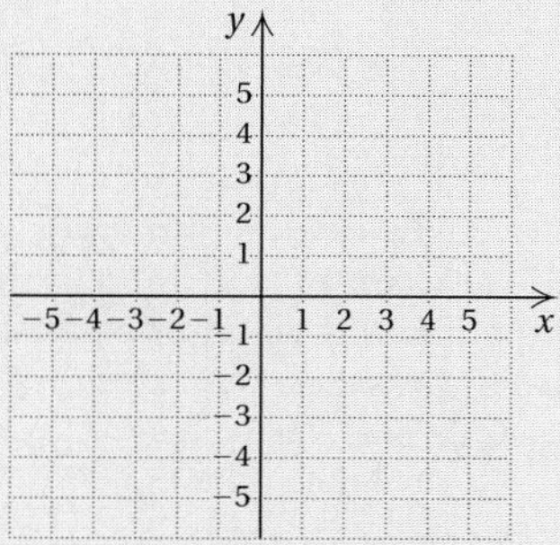

Solve the system using the substitution method.

3. $y = 6 - x,$ $2x - 3y = 22$

4. $x + 2y = 5,$ $x + y = 2$

5. $y = 5x - 2,$ $y - 2 = 5x$

Solve the system using the elimination method.

6. $x - y = 6,$ $3x + y = -2$

7. $\frac{1}{2}x - \frac{1}{3}y = 8,$ $\frac{2}{3}x + \frac{1}{2}y = 5$

8. $-4x - 9y = 4,$ $6x + 3y = 1$

9. $2x + 3y = 13,$ $3x - 5y = 10$

Solve.

10. *Rectangle Dimensions.* The perimeter of a rectangular field is 8266 yd. The length is 84 yd more than the width. Find the length and the width.

11. *Mixture of Solutions.* Solution A is 25% acid, and solution B is 40% acid. How much of each is needed to make 60 L of a solution that is 30% acid?

12. *Motorboat Travel.* A motorboat traveled for 2 hr with an 8-km/h current. The return trip against the same current took 3 hr. Find the speed of the motorboat in still water.

13. *Carnival Prices.* A carnival comes to town and makes an income of $4275 one day. Twice as much was made on concessions as on the rides. How much did the concessions bring in? How much did the rides bring in?

14. *Farm Acreage.* The Rolling Velvet Horse Farm allots 650 acres to plant hay and oats. The owners know that their needs are best met if they plant 180 acres more of hay than of oats. How many acres of each should they plant?

15. *Supplementary Angles.* Two angles are supplementary. One angle measures 45° more than twice the measure of the other. Find the measure of each angle.

16. *Octane Ratings.* The octane rating of a gasoline is a measure of the amount of isooctane in the gas. How much 87-octane gas and 93-octane gas should be blended in order to end up with 12 gal of 91-octane gas?

Source: Champlain Electric and Petroleum Equipment

17. *Phone Rates.* Recently, SBC offered a domestic calling plan for \$2.95 per month plus 10¢ per minute. Another plan charges \$1.95 per month plus 15¢ per minute. For what number of minutes will the two plans cost the same?

18. *Ski Trip.* A group of students drive both a car and an SUV on a ski trip. The car left first and traveled at 55 mph. The SUV left 2 hr later and traveled at 65 mph. How long will it take the SUV to catch up to the car?

SYNTHESIS

19. Find the numbers C and D such that $(-2, 3)$ is a solution of the system

$$Cx - 4y = 7,$$
$$3x + Dy = 8.$$

20. *Ticket Line.* You are in line at a ticket window. There are two more people ahead of you than there are behind you. In the entire line, there are three times as many people as there are behind you. How many are ahead of you in line?

Each of the following shows the graph of a system of equations. Find the equations.

21.

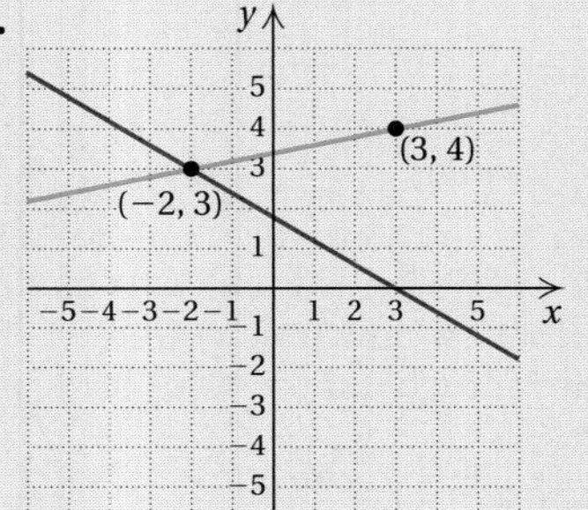

22.

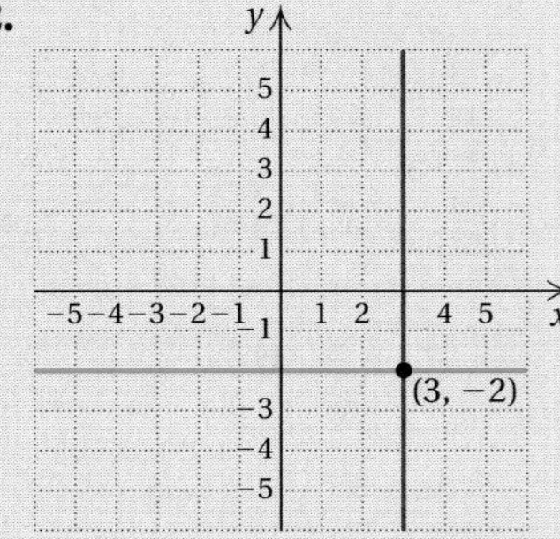

Radical Expressions and Equations

Real-World Application

A women's fast-pitch softball diamond is actually a square 60 ft on a side. How far is it from home plate to second base? Give an exact answer and an approximation to three decimal places.

This problem appears as Example 5 in Section 14.6.

14.1 INTRODUCTION TO RADICAL EXPRESSIONS

Objectives

- **a** Find the principal square roots and their opposites of the whole numbers from 0^2 to 25^2.
- **b** Approximate square roots of real numbers using a calculator.
- **c** Solve applied problems involving square roots.
- **d** Identify radicands of radical expressions.
- **e** Identify whether a radical expression represents a real number.
- **f** Simplify a radical expression with a perfect-square radicand.

Find the square roots.

1. 36
2. 64
3. 121
4. 144

Find the following.

5. $\sqrt{16}$
6. $\sqrt{49}$
7. $\sqrt{100}$
8. $\sqrt{441}$
9. $-\sqrt{49}$
10. $-\sqrt{169}$

Use a calculator to approximate each of the following square roots to three decimal places.

11. $\sqrt{15}$
12. $\sqrt{30}$
13. $\sqrt{980}$
14. $-\sqrt{667.8}$
15. $\sqrt{\frac{2}{3}}$
16. $-\sqrt{\frac{203.4}{67.82}}$

Answers on page A-48

a Square Roots

When we raise a number to the second power, we have squared the number. Sometimes we may need to find the number that was squared. We call this process finding a square root of a number.

SQUARE ROOT

The number c is a **square root** of a if $c^2 = a$.

Every positive number has two square roots. For example, the square roots of 25 are 5 and -5 because $5^2 = 25$ and $(-5)^2 = 25$. The positive square root is also called the **principal square root.** The symbol $\sqrt{\ }$ is called a **radical*** (or **square root**) symbol. The radical symbol represents only the principal square root. Thus, $\sqrt{25} = 5$. To name the negative square root of a number, we use $-\sqrt{\ }$. The number 0 has only one square root, 0.

EXAMPLE 1 Find the square roots of 81.

The square roots are 9 and -9.

EXAMPLE 2 Find $\sqrt{225}$.

There are two square roots of 225, 15 and -15. We want the principal, or positive, square root since this is what $\sqrt{\ }$ represents. Thus, $\sqrt{225} = 15$.

EXAMPLE 3 Find $-\sqrt{64}$.

The symbol $\sqrt{64}$ represents the positive square root. Then $-\sqrt{64}$ represents the negative square root. That is, $\sqrt{64} = 8$, so $-\sqrt{64} = -8$.

Do Exercises 1–10.

We can think of the processes of "squaring" and "finding square roots" as inverses of each other. We square a number and get one answer. When we find the square roots of the answer, we get the original number *and* its opposite.

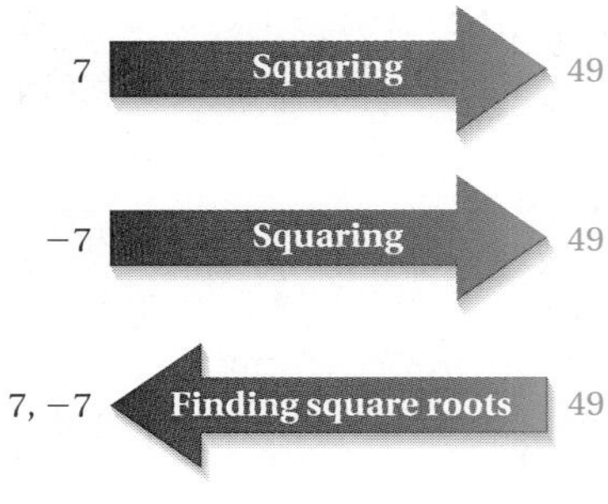

*Radicals can be other than square roots, but we will consider only square-root radicals in Chapter 14. See Appendix C for other types of radicals.

b Approximating Square Roots

We often need to use rational numbers to *approximate* square roots that are irrational. Such approximations can be found using a calculator with a square-root key $\sqrt{\ }$.

EXAMPLES Use a calculator to approximate each of the following.

	Number	*Using a calculator with a 10-digit readout*	*Rounded to three decimal places*
4.	$\sqrt{10}$	3.162277660	3.162
5.	$-\sqrt{583.8}$	−24.16195356	−24.162
6.	$\sqrt{\frac{48}{55}}$	0.934198733	0.934

Do Exercises 11–16 on the preceding page.

c Applications of Square Roots

We now consider an application involving a formula with a radical expression.

EXAMPLE 7 *Speed of a Skidding Car.* After an accident, how do police determine the speed at which the car had been traveling? The formula $r = 2\sqrt{5L}$ can be used to approximate the speed r, in miles per hour, of a car that has left a skid mark of length L, in feet. What was the speed of a car that left skid marks of length **(a)** 30 ft? **(b)** 150 ft?

a) We substitute 30 for L and find an approximation:

$$r = 2\sqrt{5L} = 2\sqrt{5 \cdot 30} = 2\sqrt{150} \approx 24.495.$$

The speed of the car was about 24.5 mph.

CALCULATOR CORNER

Approximating Square Roots We can use a graphing calculator to approximate square roots. To approximate $\sqrt{18}$, for example, we press 2ND $\sqrt{\ }$ 1 8) ENTER. ($\sqrt{\ }$ is the second operation associated with the x^2 key.) Although it is not necessary for this example to include the left parenthesis, we do so here in order to close the set of parentheses that are opened when the graphing calculator displays "$\sqrt{\ }$(." To approximate $-\sqrt{8.65}$, we press (−) 2ND $\sqrt{\ }$ 8 . 6 5) ENTER. We see that $\sqrt{18} \approx 4.243$ and $-\sqrt{8.65} \approx -2.941$.

```
√(18)
            4.242640687
-√(8.65)
           -2.941088234
```

Exercises: Use a graphing calculator to approximate each of the following to three decimal places.

1. $\sqrt{43}$
2. $\sqrt{101}$
3. $\sqrt{10{,}467}$
4. $\sqrt{\frac{2}{5}}$
5. $-\sqrt{9406}$
6. $-\sqrt{\frac{11}{17}}$

17. Speed of a Skidding Car. Referring to Example 7, determine the speed of a car that left skid marks of length **(a)** 40 ft; **(b)** 123 ft.

Identify the radicand.

18. $\sqrt{227}$

19. $-\sqrt{45 + x}$

20. $\sqrt{\frac{x}{x + 2}}$

21. $8\sqrt{x^2 + 4}$

Determine whether the expression represents a real number. Write "yes" or "no."

22. $-\sqrt{25}$

23. $\sqrt{-25}$

24. $-\sqrt{-36}$

25. $-\sqrt{36}$

Answers on page A-48

b) We substitute 150 for L and find an approximation:

$$r = 2\sqrt{5L} = 2\sqrt{5 \cdot 150} \approx 54.772.$$

The speed of the car was about 54.8 mph.

Do Exercise 17.

d Radicands and Radical Expressions

When an expression is written under a radical, we have a **radical expression.** Here are some examples:

$$\sqrt{14}, \quad \sqrt{x}, \quad 8\sqrt{x^2 + 4}, \quad \sqrt{\frac{x^2 - 5}{2}}.$$

The expression written under the radical is called the **radicand.**

EXAMPLES Identify the radicand in each expression.

8. $-\sqrt{105}$ The radicand is 105.

9. $\sqrt{x} + 2$ The radicand is x.

10. $\sqrt{x + 2}$ The radicand is $x + 2$.

11. $6\sqrt{y^2 - 5}$ The radicand is $y^2 - 5$.

12. $\sqrt{\frac{a - b}{a + b}}$ The radicand is $\frac{a - b}{a + b}$.

Do Exercises 18–21.

e Expressions That Are Meaningful as Real Numbers

The square of any nonzero number is always positive. For example, $8^2 = 64$ and $(-11)^2 = 121$. There are no real numbers that when squared yield negative numbers. For example, $\sqrt{-100}$ does not represent a real number because there are no real numbers that when squared yield -100. We can try to square 10 and -10, but we know that $10^2 = 100$ and $(-10)^2 = 100$. Neither square is -100. Thus the following expressions do not represent real numbers (they are meaningless as real numbers):

$$\sqrt{-100}, \quad \sqrt{-49}, \quad -\sqrt{-3}.$$

EXCLUDING NEGATIVE RADICANDS

Radical expressions with negative radicands do not represent real numbers.

Later in your study of mathematics, you may encounter a number system called the **complex numbers** in which negative numbers have square roots.

Do Exercises 22–25.

f Perfect-Square Radicands

The expression $\sqrt{x^2}$, with a perfect-square radicand, x^2, can be troublesome to simplify. Recall that $\sqrt{}$ denotes the principal square root. That is, the answer is nonnegative (either positive or zero). If x represents a nonnegative number, $\sqrt{x^2}$ simplifies to x. If x represents a negative number, $\sqrt{x^2}$ simplifies to $-x$ (the opposite of x), which is positive.

Suppose that $x = 3$. Then

$$\sqrt{x^2} = \sqrt{3^2} = \sqrt{9} = 3.$$

Suppose that $x = -3$. Then

$$\sqrt{x^2} = \sqrt{(-3)^2} = \sqrt{9} = 3, \quad \text{the } \textit{opposite} \text{ of } -3.$$

Note that 3 is the *absolute value* of both 3 and -3. In general, when replacements for x are considered to be *any* real numbers, it follows that

$$\sqrt{x^2} = |x|,$$

and when $x = 3$ or $x = -3$,

$$\sqrt{x^2} = \sqrt{3^2} = |3| = 3 \quad \text{and} \quad \sqrt{x^2} = \sqrt{(-3)^2} = |-3| = 3.$$

PRINCIPAL SQUARE ROOT OF A^2

For any real number A,

$$\sqrt{A^2} = |A|.$$

(That is, for any real number A, the principal square root of A^2 is the absolute value of A.)

EXAMPLES Simplify. Assume that expressions under radicals represent any real number.

13. $\sqrt{10^2} = |10| = 10$

14. $\sqrt{(-7)^2} = |-7| = 7$

15. $\sqrt{(3x)^2} = |3x|$ Absolute-value notation is necessary.

16. $\sqrt{a^2b^2} = \sqrt{(ab)^2} = |ab|$

17. $\sqrt{x^2 + 2x + 1} = \sqrt{(x + 1)^2} = |x + 1|$

Do Exercises 26–31.

Fortunately, in many cases, it can be assumed that radicands that are variable expressions do not represent the square of a negative number. When this assumption is made, the need for absolute-value symbols disappears. Then

$$\text{for } x \geq 0, \quad \sqrt{x^2} = x,$$

since x is nonnegative.

PRINCIPAL SQUARE ROOT OF A^2

For any nonnegative real number A,

$$\sqrt{A^2} = A.$$

(That is, for any nonnegative real number A, the principal square root of A^2 is A.)

Simplify. Assume that expressions under radicals represent any real number.

26. $\sqrt{(-13)^2}$

27. $\sqrt{(7w)^2}$

28. $\sqrt{(xy)^2}$

29. $\sqrt{x^2y^2}$

30. $\sqrt{(x - 11)^2}$

31. $\sqrt{x^2 + 8x + 16}$

Answers on page A-48

Simplify. Assume that radicands do not represent the square of a negative number.

32. $\sqrt{(xy)^2}$

33. $\sqrt{x^2y^2}$

34. $\sqrt{(x-11)^2}$

35. $\sqrt{x^2+8x+16}$

36. $\sqrt{25y^2}$

37. $\sqrt{\frac{1}{4}t^2}$

Answers on page A-48

EXAMPLES Simplify. Assume that radicands do not represent the square of a negative number.

18. $\sqrt{(3x)^2} = 3x$ — Since $3x$ is assumed to be nonnegative, $|3x| = 3x$.

19. $\sqrt{a^2b^2} = \sqrt{(ab)^2} = ab$ — Since ab is assumed to be nonnegative, $|ab| = ab$.

20. $\sqrt{x^2+2x+1} = \sqrt{(x+1)^2} = x+1$ — Since $x+1$ is assumed to be nonnegative

Do Exercises 32–37.

RADICALS AND ABSOLUTE VALUE

Henceforth, in this text we will assume that no radicands are formed by raising negative quantities to even powers.

We make this assumption in order to eliminate some confusion and because it is valid in many applications. As you study further in mathematics, however, you will frequently have to make a determination about expressions under radicals being nonnegative or positive. This will often be necessary in calculus.

Study Tips

BEGINNING TO STUDY FOR THE FINAL EXAM (PART 2)

The best scenario for preparing for a final exam is to do so over a period of at least two weeks. Work in a diligent, disciplined manner, doing some final-exam preparation each day. Here is a detailed plan that many find useful.

1. **Begin by browsing through each chapter, reviewing the highlighted or boxed information regarding important formulas in both the text and the Summary and Review.** There may be some formulas that you will need to memorize.
2. **Retake each chapter test that you took in class, assuming your instructor has returned it. Otherwise, use the chapter test in the book.** Restudy the objectives in the text that correspond to each question you missed.
3. **Then work any Cumulative Review that covers chapters up to that point.** Be careful to avoid any questions corresponding to objectives not covered. Again, restudy the objectives in the text that correspond to each question you missed.
4. **If you are still missing questions, use the supplements for extra review.** For example, you might check out the videos, the *Student's Solutions Manual*, the InterAct Math Tutorial Web site, MathXL, or MyMathLab.
5. **For remaining difficulties, see your instructor, go to a tutoring session, or participate in a study group.**
6. **Check for former final exams that may be on file in the math department or a study center, or with students who have already taken the course.** Use them for practice, being alert to trouble spots.
7. **Take the Final Examination in the text during the last couple of days before the final.** Set aside the same amount of time that you will have for the final. See how much of the final exam you can complete under test-like conditions.

"The door of opportunity won't open unless you do some pushing."

Anonymous

14.1 EXERCISE SET

For Extra Help

MathXL | MyMathLab | InterAct Math | Math Tutor Center | Video Lectures on CD Disc 7 | Student's Solutions Manual

a Find the square roots.

1. 4

2. 1

3. 9

4. 16

5. 100

6. 121

7. 169

8. 144

9. 256

10. 625

Simplify.

11. $\sqrt{4}$

12. $\sqrt{1}$

13. $-\sqrt{9}$

14. $-\sqrt{25}$

15. $-\sqrt{36}$

16. $-\sqrt{81}$

17. $-\sqrt{225}$

18. $\sqrt{400}$

19. $\sqrt{361}$

20. $-\sqrt{441}$

b Use a calculator to approximate the square roots. Round to three decimal places.

21. $\sqrt{5}$

22. $\sqrt{8}$

23. $\sqrt{432}$

24. $-\sqrt{8196}$

25. $-\sqrt{347.7}$

26. $-\sqrt{204.788}$

27. $\sqrt{\frac{278}{36}}$

28. $-\sqrt{\frac{567}{788}}$

29. $-\sqrt{8 \cdot 9 \cdot 200}$

30. $\sqrt{\frac{47 \cdot 83}{947.03}}$

c *Parking-Lot Arrival Spaces.* The attendants at a parking lot park cars in temporary spaces before the cars are taken to permanent parking stalls. The number N of such spaces needed is approximated by the formula $N = 2.5\sqrt{A}$, where A is the average number of arrivals during peak hours.

31. Find the number of spaces needed when the average number of arrivals is **(a)** 25; **(b)** 89.

32. Find the number of spaces needed when the average number of arrivals is **(a)** 62; **(b)** 100.

Hang Time. An athlete's *hang time* (time airborne for a jump) T, in seconds, is given by $T = 0.144\sqrt{V}$, where V is the athlete's vertical leap, in inches.

Source: Peter Brancazio, "The Mechanics of a Slam Dunk," *Popular Mechanics*, November 1991. Courtesy of Peter Brancazio, Brooklyn College.

33. Carl Landry of the Purdue University basketball team can jump 33.5 in. vertically. Find his hang time.

34. Brandon McKnight of the Purdue University basketball team can jump 33 in. vertically. Find his hang time.

35. Jeff Trepagnier of the USC basketball team can jump 40 in. vertically. Find his hang time.

36. Keston Roberts of the Binghampton University basketball team can jump 38 in. vertically. Find his hang time.

d Identify the radicand.

37. $\sqrt{200}$

38. $\sqrt{16z}$

39. $-\sqrt{a-4}$

40. $\sqrt{3t+10}$

41. $5\sqrt{t^2+1}$

42. $-9\sqrt{x^2+16}$

43. $x^2y\sqrt{\frac{3}{x+2}}$

44. $ab^2\sqrt{\frac{a}{a+b}}$

e Determine whether the expression represents a real number. Write "yes" or "no."

45. $\sqrt{-16}$

46. $\sqrt{-81}$

47. $-\sqrt{81}$

48. $-\sqrt{64}$

49. $-\sqrt{-25}$

50. $\sqrt{-(-49)}$

f Simplify. Remember that we have assumed that radicands do not represent the square of a negative number.

51. $\sqrt{c^2}$

52. $\sqrt{x^2}$

53. $\sqrt{9x^2}$

54. $\sqrt{16y^2}$

55. $\sqrt{(8p)^2}$

56. $\sqrt{(7pq)^2}$

57. $\sqrt{(ab)^2}$

58. $\sqrt{(6y)^2}$

59. $\sqrt{(34d)^2}$

60. $\sqrt{(53b)^2}$

61. $\sqrt{(x + 3)^2}$

62. $\sqrt{(d - 3)^2}$

63. $\sqrt{a^2 - 10a + 25}$

64. $\sqrt{x^2 + 2x + 1}$

65. $\sqrt{4a^2 - 20a + 25}$

66. $\sqrt{9p^2 + 12p + 4}$

67. D_W What is the difference between "**the** square root of 10" and "**a** square root of 10"?

68. D_W Explain the difference between the two descriptions of the principal square root of A^2 given on p. 1021.

SKILL MAINTENANCE

Solve. [13.4a]

69. *Supplementary Angles.* Two angles are supplementary. One angle is 3° less than twice the other. Find the measures of the angles.

70. *Complementary Angles.* Two angles are complementary. The sum of the measure of the first angle and half the measure of the second is 64°. Find the measures of the angles.

71. *Food Expenses.* The amount F that a family spends on food varies directly as its income I. A family making \$39,200 a year will spend \$10,192 on food. At this rate, how much would a family making \$41,000 spend on food? [12.9b]

Divide and simplify. [12.2b]

72. $\dfrac{x - 3}{x + 4} \div \dfrac{x^2 - 9}{x + 4}$

73. $\dfrac{x^2 + 10x - 11}{x^2 - 1} \div \dfrac{x + 11}{x + 1}$

74. $\dfrac{x^4 - 16}{x^4 - 1} \div \dfrac{x^2 + 4}{x^2 + 1}$

SYNTHESIS

75. Use only the graph of $y = \sqrt{x}$, shown below, to approximate $\sqrt{3}$, $\sqrt{5}$, and $\sqrt{7}$. Answers may vary.

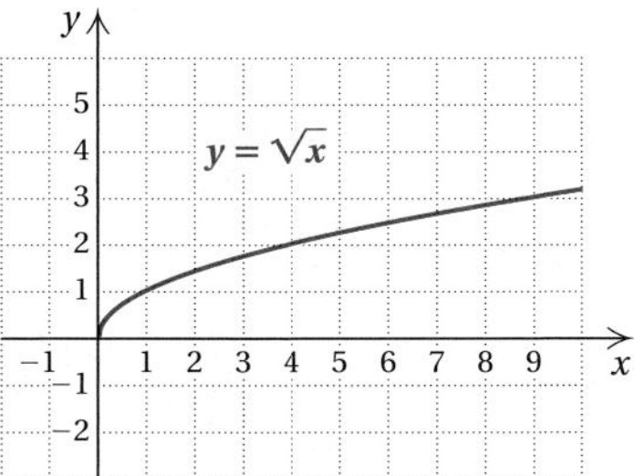

76. Between what two consecutive integers is $\sqrt{78}$?

Solve.

77. $\sqrt{x^2} = 16$

78. $\sqrt{y^2} = -7$

79. $t^2 = 49$

80. Suppose that the area of a square is 3. Find the length of a side.

14.2 MULTIPLYING AND SIMPLIFYING WITH RADICAL EXPRESSIONS

Objectives

a Simplify radical expressions.

b Simplify radical expressions where radicands are powers.

c Multiply radical expressions and simplify if possible.

1. Simplify.

a) $\sqrt{25} \cdot \sqrt{16}$

b) $\sqrt{25 \cdot 16}$

Multiply.

2. $\sqrt{3}\sqrt{11}$

3. $\sqrt{5}\sqrt{5}$

4. $\sqrt{\frac{5}{11}}\sqrt{\frac{6}{7}}$

5. $\sqrt{x}\sqrt{x+1}$

6. $\sqrt{x+2}\sqrt{x-2}$

Answers on page A-49

a Simplifying by Factoring

To see how to multiply with radical notation, consider the following.

a) $\sqrt{9} \cdot \sqrt{4} = 3 \cdot 2 = 6$ This is a product of square roots.

b) $\sqrt{9 \cdot 4} = \sqrt{36} = 6$ This is the square root of a product.

Note that

$$\sqrt{9} \cdot \sqrt{4} = \sqrt{9 \cdot 4}.$$

Do Exercise 1.

We can multiply radical expressions by multiplying the radicands.

THE PRODUCT RULE FOR RADICALS

For any nonnegative radicands A and B,

$$\sqrt{A} \cdot \sqrt{B} = \sqrt{A \cdot B}.$$

(The product of square roots is the square root of the product of the radicands.)

EXAMPLES Multiply.

1. $\sqrt{5}\sqrt{7} = \sqrt{5 \cdot 7} = \sqrt{35}$

2. $\sqrt{8}\sqrt{8} = \sqrt{8 \cdot 8} = \sqrt{64} = 8$

3. $\sqrt{\frac{2}{3}}\sqrt{\frac{4}{5}} = \sqrt{\frac{2}{3} \cdot \frac{4}{5}} = \sqrt{\frac{8}{15}}$

4. $\sqrt{2x}\sqrt{3x-1} = \sqrt{2x(3x-1)} = \sqrt{6x^2 - 2x}$

Do Exercises 2–6.

To factor radical expressions, we can use the product rule for radicals in reverse.

FACTORING RADICAL EXPRESSIONS

$$\sqrt{AB} = \sqrt{A}\sqrt{B}.$$

In some cases, we can simplify after factoring.

A square-root radical expression is simplified when its radicand has no factors that are perfect squares.

When simplifying a square-root radical expression, we first determine whether the radicand is a perfect square. Then we determine whether it has perfect-square factors. The radicand is then factored and the radical expression simplified using the preceding rule.

Compare the following:

$$\sqrt{50} = \sqrt{10 \cdot 5} = \sqrt{10}\,\sqrt{5};$$
$$\sqrt{50} = \sqrt{25 \cdot 2} = \sqrt{25}\,\sqrt{2} = 5\sqrt{2}.$$

In the second case, the radicand has the perfect-square factor 25. If you do not recognize perfect-square factors, try factoring the radicand into its prime factors. For example,

$$\sqrt{50} = \sqrt{2 \cdot \underbrace{5 \cdot 5}} = 5\sqrt{2}.$$

↑

Perfect square (a pair of the same factors)

Square-root radical expressions in which the radicand has no perfect-square factors, such as $5\sqrt{2}$, are considered to be in simplest form.

EXAMPLES Simplify by factoring.

5. $\sqrt{18} = \sqrt{9 \cdot 2}$ — Identifying a perfect-square factor and factoring the radicand. The factor 9 is a perfect square.
$= \sqrt{9} \cdot \sqrt{2}$ — Factoring into a product of radicals
$= 3\sqrt{2}$ — The radicand has no factors that are perfect squares.

6. $\sqrt{48t} = \sqrt{16 \cdot 3 \cdot t}$ — Identifying a perfect-square factor and factoring the radicand. The factor 16 is a perfect square.
$= \sqrt{16}\,\sqrt{3t}$ — Factoring into a product of radicals
$= 4\sqrt{3t}$ — Taking a square root

7. $\sqrt{20t^2} = \sqrt{4 \cdot 5 \cdot t^2}$ — Identifying perfect-square factors and factoring the radicand. The factors 4 and t^2 are perfect squares.
$= \sqrt{4}\,\sqrt{t^2}\,\sqrt{5}$ — Factoring into a product of several radicals
$= 2t\sqrt{5}$ — Taking square roots. No absolute-value signs are necessary since we have assumed that expressions under radicals are nonnegative.

8. $\sqrt{x^2 - 6x + 9} = \sqrt{(x - 3)^2} = x - 3$ — No absolute-value signs are necessary since we have assumed that expressions under radicals are nonnegative.

9. $\sqrt{36x^2} = \sqrt{36}\,\sqrt{x^2} = 6x$, or $\sqrt{36x^2} = \sqrt{(6x)^2} = 6x$

10. $\sqrt{3x^2 + 6x + 3} = \sqrt{3(x^2 + 2x + 1)}$ — Factoring the radicand
$= \sqrt{3(x + 1)^2}$ — Factoring further
$= \sqrt{3}\,\sqrt{(x + 1)^2}$ — Factoring into a product of radicals
$= \sqrt{3}(x + 1)$ — Taking the square root

Do Exercises 7–14.

Simplify by factoring.

7. $\sqrt{32}$

8. $\sqrt{92}$

9. $\sqrt{128t}$

10. $\sqrt{363q}$

11. $\sqrt{63x^2}$

12. $\sqrt{x^2 + 14x + 49}$

13. $\sqrt{81m^2}$

14. $\sqrt{3x^2 - 60x + 300}$

Answers on page A-49

Simplify.

15. $\sqrt{t^4}$

16. $\sqrt{t^{20}}$

17. $\sqrt{h^{46}}$

Simplify by factoring.

18. $\sqrt{x^7}$

19. $\sqrt{24x^{11}}$

Multiply and simplify.

20. $\sqrt{3}\sqrt{6}$

21. $\sqrt{2}\sqrt{50}$

Answers on page A-49

b Simplifying Square Roots of Powers

To take the square root of an even power such as x^{10}, we note that $x^{10} = (x^5)^2$. Then

$$\sqrt{x^{10}} = \sqrt{(x^5)^2} = x^5.$$

We can find the answer by taking half the exponent. That is,

$$\sqrt{x^{10}} = x^5. \leftarrow \tfrac{1}{2}(10) = 5$$

EXAMPLES Simplify.

11. $\sqrt{x^6} = \sqrt{(x^3)^2} = x^3$ $\quad \tfrac{1}{2}(6) = 3$

12. $\sqrt{x^8} = x^4$

13. $\sqrt{t^{22}} = t^{11}$

Do Exercises 15–17.

If an odd power occurs, we express the power in terms of the largest even power. Then we simplify the even power as in Examples 11–13.

EXAMPLE 14 Simplify by factoring: $\sqrt{x^9}$.

$$\sqrt{x^9} = \sqrt{x^8 \cdot x}$$
$$= \sqrt{x^8}\sqrt{x}$$
$$= x^4\sqrt{x}$$

Caution!

Note that $\sqrt{x^9} \neq x^3$.

EXAMPLE 15 Simplify by factoring: $\sqrt{32x^{15}}$.

$\sqrt{32x^{15}} = \sqrt{16 \cdot 2 \cdot x^{14} \cdot x}$ We factor the radicand, looking for perfect-square factors. The largest even power is 14.

$= \sqrt{16}\sqrt{x^{14}}\sqrt{2x}$ Factoring into a product of radicals. Perfect-square factors are usually listed first.

$= 4x^7\sqrt{2x}$ Simplifying

Do Exercises 18 and 19.

c Multiplying and Simplifying

Sometimes we can simplify after multiplying. We leave the radicand in factored form and factor further to determine perfect-square factors. Then we simplify the perfect-square factors.

EXAMPLE 16 Multiply and then simplify by factoring: $\sqrt{2}\sqrt{14}$.

$\sqrt{2}\sqrt{14} = \sqrt{2 \cdot 14}$ Multiplying

$= \sqrt{2 \cdot 2 \cdot 7}$ Factoring

$= \sqrt{2 \cdot 2}\sqrt{7}$ Looking for perfect-square factors, pairs of factors

$= 2\sqrt{7}$

Do Exercises 20 and 21.

EXAMPLE 17 Multiply and then simplify by factoring: $\sqrt{3x^2}\sqrt{9x^3}$.

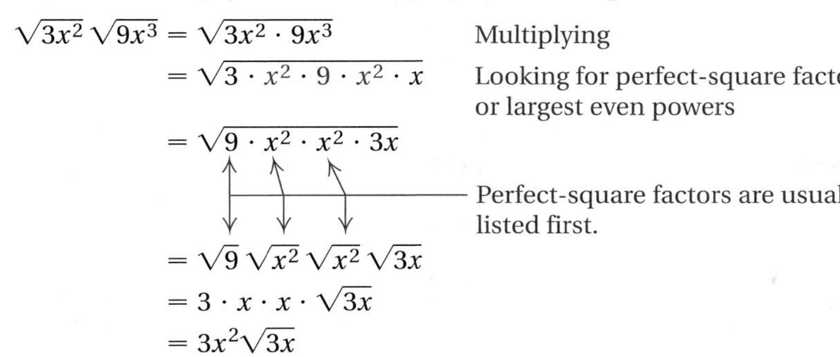

$$\sqrt{3x^2}\sqrt{9x^3} = \sqrt{3x^2 \cdot 9x^3}$$ Multiplying

$$= \sqrt{3 \cdot x^2 \cdot 9 \cdot x^2 \cdot x}$$ Looking for perfect-square factors or largest even powers

$$= \sqrt{9 \cdot x^2 \cdot x^2 \cdot 3x}$$ Perfect-square factors are usually listed first.

$$= \sqrt{9}\sqrt{x^2}\sqrt{x^2}\sqrt{3x}$$

$$= 3 \cdot x \cdot x \cdot \sqrt{3x}$$

$$= 3x^2\sqrt{3x}$$

In doing an example like the preceding one, it might be helpful to do more factoring, as follows:

$$\sqrt{3x^2} \cdot \sqrt{9x^3} = \sqrt{3 \cdot \underline{x \cdot x} \cdot \underline{3 \cdot 3} \cdot \underline{x \cdot x} \cdot x}.$$

Then we look for pairs of factors, as shown, and simplify perfect-square factors:

$$= 3 \cdot x \cdot x\sqrt{3x}$$

$$= 3x^2\sqrt{3x}.$$

EXAMPLE 18 Simplify: $\sqrt{20cd^2}\sqrt{35cd^5}$.

$$\sqrt{20cd^2}\sqrt{35cd^5}$$

$$= \sqrt{20cd^2 \cdot 35cd^5}$$ Multiplying

$$= \sqrt{\underline{2 \cdot 2} \cdot 5 \cdot c \cdot \underline{d \cdot d} \cdot 5 \cdot 7 \cdot c \cdot \underline{d \cdot d} \cdot \underline{d \cdot d} \cdot d}$$ Looking for pairs of factors

$$= \sqrt{\underline{2 \cdot 2} \cdot \underline{5 \cdot 5} \cdot \underline{c \cdot c} \cdot \underline{d \cdot d} \cdot \underline{d \cdot d} \cdot \underline{d \cdot d} \cdot 7d}$$

$$= 2 \cdot 5 \cdot c \cdot d \cdot d \cdot d\sqrt{7d}$$

$$= 10cd^3\sqrt{7d}$$

Do Exercises 22–24.

Multiply and simplify.

22. $\sqrt{2x^3}\sqrt{8x^3y^4}$

23. $\sqrt{10xy^2}\sqrt{5x^2y^3}$

24. $\sqrt{28q^2r} \cdot \sqrt{21q^3r^7}$

Answers on page A-49

We know that $\sqrt{AB} = \sqrt{A}\sqrt{B}$. That is, the square root of a product is the product of the square roots. What about the square root of a sum? That is, is the square root of a sum equal to the sum of the square roots? To check, consider $\sqrt{A+B}$ and $\sqrt{A} + \sqrt{B}$ when $A = 16$ and $B = 9$:

$$\sqrt{A+B} = \sqrt{16+9} = \sqrt{25} = 5;$$

and

$$\sqrt{A} + \sqrt{B} = \sqrt{16} + \sqrt{9} = 4 + 3 = 7.$$

Thus we see the following.

Caution!

The square root of a sum is not the sum of the square roots.

$$\sqrt{A+B} \neq \sqrt{A} + \sqrt{B}$$

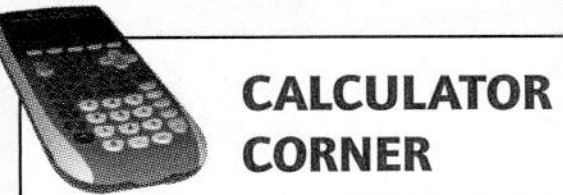

CALCULATOR CORNER

Simplifying Radical Expressions

Exercises: Use a table or a graph to determine whether each of the following is true.

1. $\sqrt{x+4} = \sqrt{x} + 2$
2. $\sqrt{3+x} = \sqrt{3} + x$
3. $\sqrt{x-2} = \sqrt{x} - \sqrt{2}$
4. $\sqrt{9x} = 3\sqrt{x}$

14.2 EXERCISE SET

For Extra Help

MathXL | MyMathLab | InterAct Math | Math Tutor Center  | Video Lectures on CD Disc 7  | Student's Solutions Manual

a Simplify by factoring.

1. $\sqrt{12}$ **2.** $\sqrt{8}$ **3.** $\sqrt{75}$ **4.** $\sqrt{50}$ **5.** $\sqrt{20}$

6. $\sqrt{45}$ **7.** $\sqrt{600}$ **8.** $\sqrt{300}$ **9.** $\sqrt{486}$ **10.** $\sqrt{567}$

11. $\sqrt{9x}$ **12.** $\sqrt{4y}$ **13.** $\sqrt{48x}$ **14.** $\sqrt{40m}$

15. $\sqrt{16a}$ **16.** $\sqrt{49b}$ **17.** $\sqrt{64y^2}$ **18.** $\sqrt{9x^2}$

19. $\sqrt{13x^2}$ **20.** $\sqrt{23s^2}$ **21.** $\sqrt{8t^2}$ **22.** $\sqrt{125a^2}$

23. $\sqrt{180}$ **24.** $\sqrt{320}$ **25.** $\sqrt{288y}$ **26.** $\sqrt{363p}$

27. $\sqrt{28x^2}$ **28.** $\sqrt{20x^2}$ **29.** $\sqrt{x^2 - 6x + 9}$ **30.** $\sqrt{t^2 + 22t + 121}$

31. $\sqrt{8x^2 + 8x + 2}$ **32.** $\sqrt{20x^2 - 20x + 5}$ **33.** $\sqrt{36y + 12y^2 + y^3}$ **34.** $\sqrt{x - 2x^2 + x^3}$

b Simplify by factoring.

35. $\sqrt{t^6}$

36. $\sqrt{x^{18}}$

37. $\sqrt{x^{12}}$

38. $\sqrt{x^{16}}$

39. $\sqrt{x^5}$

40. $\sqrt{x^3}$

41. $\sqrt{t^{19}}$

42. $\sqrt{p^{17}}$

43. $\sqrt{(y-2)^8}$

44. $\sqrt{(x+3)^6}$

45. $\sqrt{4(x+5)^{10}}$

46. $\sqrt{16(a-7)^4}$

47. $\sqrt{36m^3}$

48. $\sqrt{250y^3}$

49. $\sqrt{8a^5}$

50. $\sqrt{12b^7}$

51. $\sqrt{104p^{17}}$

52. $\sqrt{284m^{23}}$

53. $\sqrt{448x^6y^3}$

54. $\sqrt{243x^5y^4}$

c Multiply and then, if possible, simplify by factoring.

55. $\sqrt{3}\sqrt{18}$

56. $\sqrt{5}\sqrt{10}$

57. $\sqrt{15}\sqrt{6}$

58. $\sqrt{3}\sqrt{27}$

59. $\sqrt{18}\sqrt{14x}$

60. $\sqrt{12}\sqrt{18x}$

61. $\sqrt{3x}\sqrt{12y}$

62. $\sqrt{7x}\sqrt{21y}$

63. $\sqrt{13}\sqrt{13}$

64. $\sqrt{11}\sqrt{11x}$

65. $\sqrt{5b}\sqrt{15b}$

66. $\sqrt{6a}\sqrt{18a}$

67. $\sqrt{2t}\sqrt{2t}$

68. $\sqrt{7a}\sqrt{7a}$

69. $\sqrt{ab}\sqrt{ac}$

70. $\sqrt{xy}\sqrt{xz}$

71. $\sqrt{2x^2y}\sqrt{4xy^2}$

72. $\sqrt{15mn^2}\sqrt{5m^2n}$

73. $\sqrt{18}\sqrt{18}$

74. $\sqrt{16}\sqrt{16}$

75. $\sqrt{5}\sqrt{2x-1}$

76. $\sqrt{3}\sqrt{4x+2}$

77. $\sqrt{x+2}\sqrt{x+2}$

78. $\sqrt{x-9}\sqrt{x-9}$

79. $\sqrt{18x^2y^3}\sqrt{6xy^4}$

80. $\sqrt{12x^3y^2}\sqrt{8xy}$

81. $\sqrt{50x^4y^6}\sqrt{10xy}$

82. $\sqrt{10xy^2}\sqrt{5x^2y^3}$

83. $\sqrt{99p^4q^3}\sqrt{22p^5q^2}$

84. $\sqrt{75m^8n^9}\sqrt{50m^5n^7}$

85. $\sqrt{24a^2b^3c^4}\sqrt{32a^5b^4c^7}$

86. $\sqrt{18p^5q^2r^{11}}\sqrt{108p^3q^6r^9}$

87. $\mathbf{D_W}$ Are the rules for manipulating expressions with exponents important when simplifying radical expressions? Why or why not?

88. $\mathbf{D_W}$ Explain the error(s) in the following:
$\sqrt{x^2 - 25} = \sqrt{x^2} - \sqrt{25} = x - 5.$

SKILL MAINTENANCE

Solve. [13.3a, b]

89. $x - y = -6,$
$x + y = 2$

90. $3x + 5y = 6,$
$5x + 3y = 4$

91. $3x - 2y = 4,$
$2x + 5y = 9$

92. $4a - 5b = 25,$
$a - b = 7$

Solve.

93. *Storage Area Dimensions.* The perimeter of a rectangular storage area is 84 ft. The length is 18 ft greater than the width. Find the area of the rectangle. [13.4a]

94. *Movie Revenue.* There were 411 people at a movie. Admission was \$7.00 each for adults and \$3.75 each for children, and receipts totaled \$2678.75. How many adults and how many children attended? [13.4a]

95. *Insecticide Mixtures.* A solution containing 30% insecticide is to be mixed with a solution containing 50% insecticide in order to make 200 L of a solution containing 42% insecticide. How much of each solution should be used? [13.4a]

96. *Canoe Travel.* Greg and Beth paddled to a picnic spot downriver in 2 hr. It took them 3 hr to return against the current. If the speed of the current was 2 mph, at what speed were they paddling the canoe? [13.5a]

SYNTHESIS

Factor.

97. $\sqrt{5x - 5}$

98. $\sqrt{x^2 - x - 2}$

99. $\sqrt{x^2 - 36}$

100. $\sqrt{2x^2 - 5x - 12}$

101. $\sqrt{x^3 - 2x^2}$

102. $\sqrt{a^2 - b^2}$

Simplify.

103. $\sqrt{0.25}$

104. $\sqrt{0.01}$

Multiply and then simplify by factoring.

105. $(\sqrt{2y})(\sqrt{3})(\sqrt{8y})$

106. $\sqrt{18(x - 2)}\sqrt{20(x - 2)^3}$

107. $\sqrt{27(x + 1)}\sqrt{12y(x + 1)^2}$

108. $\sqrt{2^{109}}\sqrt{x^{306}}\sqrt{x^{11}}$

109. $\sqrt{x}\sqrt{2x}\sqrt{10x^5}$

110. $\sqrt{a}(\sqrt{a^3} - 5)$

14.3 QUOTIENTS INVOLVING RADICAL EXPRESSIONS

Objectives

a Divide radical expressions.

b Simplify square roots of quotients.

c Rationalize the denominator of a radical expression.

a Dividing Radical Expressions

Consider the expressions

$$\frac{\sqrt{25}}{\sqrt{16}} \quad \text{and} \quad \sqrt{\frac{25}{16}}.$$

Let's evaluate them separately:

a) $\dfrac{\sqrt{25}}{\sqrt{16}} = \dfrac{5}{4}$ because $\sqrt{25} = 5$ and $\sqrt{16} = 4$;

b) $\sqrt{\dfrac{25}{16}} = \dfrac{5}{4}$ because $\dfrac{5}{4} \cdot \dfrac{5}{4} = \dfrac{25}{16}$.

We see that both expressions represent the same number. This suggests that the quotient of two square roots is the square root of the quotient of the radicands.

THE QUOTIENT RULE FOR RADICALS

For any nonnegative number A and any positive number B,

$$\frac{\sqrt{A}}{\sqrt{B}} = \sqrt{\frac{A}{B}}.$$

(The quotient of two square roots is the square root of the quotient of the radicands.)

EXAMPLES Divide and simplify.

1. $\dfrac{\sqrt{27}}{\sqrt{3}} = \sqrt{\dfrac{27}{3}} = \sqrt{9} = 3$

2. $\dfrac{\sqrt{30a^5}}{\sqrt{6a^2}} = \sqrt{\dfrac{30a^5}{6a^2}} = \sqrt{5a^3} = \sqrt{5 \cdot a^2 \cdot a} = \sqrt{a^2} \cdot \sqrt{5a} = a\sqrt{5a}$

Do Exercises 1–3.

Divide and simplify.

1. $\dfrac{\sqrt{96}}{\sqrt{6}}$

2. $\dfrac{\sqrt{75}}{\sqrt{3}}$

3. $\dfrac{\sqrt{42x^5}}{\sqrt{7x^2}}$

Answers on page A-49

b Square Roots of Quotients

To find the square root of certain quotients, we can reverse the quotient rule for radicals. We can take the square root of a quotient by taking the square roots of the numerator and the denominator separately.

SQUARE ROOTS OF QUOTIENTS

For any nonnegative number A and any positive number B,

$$\sqrt{\frac{A}{B}} = \frac{\sqrt{A}}{\sqrt{B}}.$$

(We can take the square roots of the numerator and the denominator separately.)

EXAMPLES Simplify by taking the square roots of the numerator and the denominator separately.

3. $\sqrt{\frac{25}{9}} = \frac{\sqrt{25}}{\sqrt{9}} = \frac{5}{3}$ Taking the square roots of the numerator and the denominator

4. $\sqrt{\frac{1}{16}} = \frac{\sqrt{1}}{\sqrt{16}} = \frac{1}{4}$ Taking the square roots of the numerator and the denominator

5. $\sqrt{\frac{49}{t^2}} = \frac{\sqrt{49}}{\sqrt{t^2}} = \frac{7}{t}$

Do Exercises 4–6.

We are assuming that expressions for numerators are nonnegative and expressions for denominators are positive. Thus we need not be concerned about absolute-value signs or zero denominators.

Sometimes a rational expression can be simplified to one that has a perfect-square numerator and a perfect-square denominator.

EXAMPLES Simplify.

6. $\sqrt{\frac{18}{50}} = \sqrt{\frac{9 \cdot 2}{25 \cdot 2}} = \sqrt{\frac{9}{25} \cdot \frac{2}{2}} = \sqrt{\frac{9}{25} \cdot 1}$

$= \sqrt{\frac{9}{25}} = \frac{\sqrt{9}}{\sqrt{25}} = \frac{3}{5}$

7. $\sqrt{\frac{2560}{2890}} = \sqrt{\frac{256 \cdot 10}{289 \cdot 10}} = \sqrt{\frac{256}{289} \cdot \frac{10}{10}} = \sqrt{\frac{256}{289} \cdot 1}$

$= \sqrt{\frac{256}{289}} = \frac{\sqrt{256}}{\sqrt{289}} = \frac{16}{17}$

8. $\frac{\sqrt{48x^3}}{\sqrt{3x^7}} = \sqrt{\frac{48x^3}{3x^7}} = \sqrt{\frac{16}{x^4}}$ Simplifying the radicand

$= \frac{\sqrt{16}}{\sqrt{x^4}} = \frac{4}{x^2}$

Do Exercises 7–9.

C Rationalizing Denominators

Sometimes in mathematics it is useful to find an equivalent expression without a radical in the denominator. This provides a standard notation for expressing results. The procedure for finding such an expression is called **rationalizing the denominator.** We carry this out by multiplying by 1 in either of two ways.

> To rationalize a denominator:
>
> **Method 1.** Multiply by 1 under the radical to make the denominator of the radicand a perfect square.
>
> **Method 2.** Multiply by 1 outside the radical to make the radicand in the denominator a perfect square.

Simplify.

4. $\sqrt{\frac{16}{9}}$

5. $\sqrt{\frac{1}{25}}$

6. $\sqrt{\frac{36}{x^2}}$

Simplify.

7. $\sqrt{\frac{18}{32}}$

8. $\sqrt{\frac{2250}{2560}}$

9. $\frac{\sqrt{98y}}{\sqrt{2y^{11}}}$

Answers on page A-49

10. Rationalize the denominator:

$$\sqrt{\frac{3}{5}}.$$

a) Use method 1.

b) Use method 2.

Rationalize the denominator.

11. $\sqrt{\frac{5}{8}}$

(*Hint*: Multiply the radicand by $\frac{2}{2}$.)

12. $\frac{10}{\sqrt{3}}$

Answers on page A-49

EXAMPLE 9 Rationalize the denominator: $\sqrt{\frac{2}{3}}$.

METHOD 1 We multiply by 1, choosing $\frac{3}{3}$ for 1. This makes the denominator of the radicand a perfect square:

$$\sqrt{\frac{2}{3}} = \sqrt{\frac{2}{3} \cdot \frac{3}{3}} \qquad \text{Multiplying by 1}$$

$$= \sqrt{\frac{6}{9}} = \frac{\sqrt{6}}{\sqrt{9}} \qquad \text{The radicand in the denominator, 9, is a perfect square.}$$

$$= \frac{\sqrt{6}}{3}.$$

METHOD 2 We can also rationalize by first taking the square roots of the numerator and the denominator. Then we multiply by 1, using $\sqrt{3}/\sqrt{3}$:

$$\sqrt{\frac{2}{3}} = \frac{\sqrt{2}}{\sqrt{3}}$$

$$= \frac{\sqrt{2}}{\sqrt{3}} \cdot \frac{\sqrt{3}}{\sqrt{3}} \qquad \text{Multiplying by 1}$$

$$= \frac{\sqrt{2} \cdot \sqrt{3}}{\sqrt{3} \cdot \sqrt{3}} = \frac{\sqrt{6}}{\sqrt{9}} \qquad \text{The radicand, 9, in the denominator is a perfect square.}$$

$$= \frac{\sqrt{6}}{3}.$$

Do Exercise 10.

We can always multiply by 1 to make a denominator a perfect square. Then we can take the square root of the denominator.

EXAMPLE 10 Rationalize the denominator: $\sqrt{\frac{5}{18}}$.

The denominator, 18, is not a perfect square. Factoring, we get $18 = 3 \cdot 3 \cdot 2$. If we had another factor of 2, however, we would have a perfect square, 36. Thus we multiply by 1, choosing $\frac{2}{2}$. This makes the denominator a perfect square.

$$\sqrt{\frac{5}{18}} = \sqrt{\frac{5}{3 \cdot 3 \cdot 2}} = \sqrt{\frac{5}{3 \cdot 3 \cdot 2} \cdot \frac{2}{2}} = \sqrt{\frac{10}{36}} = \frac{\sqrt{10}}{\sqrt{36}} = \frac{\sqrt{10}}{6}$$

EXAMPLE 11 Rationalize the denominator: $\frac{8}{\sqrt{7}}$.

This time we obtain an expression without a radical in the denominator by multiplying by 1, choosing $\sqrt{7}/\sqrt{7}$:

$$\frac{8}{\sqrt{7}} = \frac{8}{\sqrt{7}} \cdot \frac{\sqrt{7}}{\sqrt{7}} = \frac{8\sqrt{7}}{\sqrt{49}} = \frac{8\sqrt{7}}{7}.$$

Caution!

$8\sqrt{7} \neq \sqrt{56}$.

Do Exercises 11 and 12.

EXAMPLE 12 Rationalize the denominator: $\frac{\sqrt{3}}{\sqrt{2}}$.

We look at the denominator. It is $\sqrt{2}$. We multiply by 1, choosing $\sqrt{2}/\sqrt{2}$:

$$\frac{\sqrt{3}}{\sqrt{2}} = \frac{\sqrt{3}}{\sqrt{2}} \cdot \frac{\sqrt{2}}{\sqrt{2}} = \frac{\sqrt{3} \cdot \sqrt{2}}{\sqrt{2} \cdot \sqrt{2}} = \frac{\sqrt{6}}{\sqrt{4}} = \frac{\sqrt{6}}{2}, \text{ or } \frac{1}{2}\sqrt{6}.$$

EXAMPLES Rationalize the denominator.

13. $\frac{\sqrt{5}}{\sqrt{x}} = \frac{\sqrt{5}}{\sqrt{x}} \cdot \frac{\sqrt{x}}{\sqrt{x}}$ Multiplying by 1

$= \frac{\sqrt{5}\sqrt{x}}{\sqrt{x}\sqrt{x}}$

$= \frac{\sqrt{5x}}{x}$ $\sqrt{x} \cdot \sqrt{x} = x$ by the definition of square root

14. $\frac{\sqrt{49a^5}}{\sqrt{12}} = \frac{\sqrt{49a^5}}{\sqrt{12}} \cdot \frac{\sqrt{3}}{\sqrt{3}}$ Factoring 12, we get $2 \cdot 2 \cdot 3$, so we need another factor of 3 in order for the radicand in the denominator to be a perfect square. We multiply by $\sqrt{3}/\sqrt{3}$.

$= \frac{\sqrt{49a^5}\sqrt{3}}{\sqrt{12}\sqrt{3}}$

$= \frac{\sqrt{49 \cdot a^4 \cdot a \cdot 3}}{\sqrt{36}} = \frac{\sqrt{49}\sqrt{a^4}\sqrt{3a}}{\sqrt{36}}$

$= \frac{7a^2\sqrt{3a}}{6}$

Do Exercises 13–15.

Rationalize the denominator.

13. $\frac{\sqrt{3}}{\sqrt{7}}$

14. $\frac{\sqrt{5}}{\sqrt{r}}$

15. $\frac{\sqrt{64y^2}}{\sqrt{7}}$

Answers on page A-49

Study Tips

BEGINNING TO STUDY FOR THE FINAL EXAM (PART 3): THREE DAYS TO TWO WEEKS OF STUDY TIME

1. **Begin by browsing through each chapter, reviewing the highlighted or boxed information regarding important formulas in both the text and the Summary and Review.** There may be some formulas that you will need to memorize.
2. **Retake each chapter test that you took in class, assuming your instructor has returned it. Otherwise, use the chapter test in the book.** Restudy the objectives in the text that correspond to each question you missed.
3. **Work the Cumulative Review/Final Examination during the last couple of days before the final.** Set aside the same amount of time that you will have for the final. See how much of the final exam you can complete under test-like conditions. Be careful to avoid any questions corresponding to objectives not covered. Again, restudy the objectives in the text that correspond to each question you missed.
4. **For remaining difficulties, see your instructor, go to a tutoring session, or participate in a study group.**

"It is a great piece of skill to know how to guide your luck, even while waiting for it."

Baltasar Gracian,
seventeenth-century Spanish philosopher and writer

14.3 EXERCISE SET

For Extra Help

MathXL | MyMathLab | InterAct Math | Math Tutor Center | Video Lectures on CD Disc 7 | Student's Solutions Manual

a Divide and simplify.

1. $\dfrac{\sqrt{18}}{\sqrt{2}}$

2. $\dfrac{\sqrt{20}}{\sqrt{5}}$

3. $\dfrac{\sqrt{108}}{\sqrt{3}}$

4. $\dfrac{\sqrt{60}}{\sqrt{15}}$

5. $\dfrac{\sqrt{65}}{\sqrt{13}}$

6. $\dfrac{\sqrt{45}}{\sqrt{15}}$

7. $\dfrac{\sqrt{3}}{\sqrt{75}}$

8. $\dfrac{\sqrt{3}}{\sqrt{48}}$

9. $\dfrac{\sqrt{12}}{\sqrt{75}}$

10. $\dfrac{\sqrt{18}}{\sqrt{32}}$

11. $\dfrac{\sqrt{8x}}{\sqrt{2x}}$

12. $\dfrac{\sqrt{18b}}{\sqrt{2b}}$

13. $\dfrac{\sqrt{63y^3}}{\sqrt{7y}}$

14. $\dfrac{\sqrt{48x^3}}{\sqrt{3x}}$

b Simplify.

15. $\sqrt{\dfrac{16}{49}}$

16. $\sqrt{\dfrac{9}{49}}$

17. $\sqrt{\dfrac{1}{36}}$

18. $\sqrt{\dfrac{1}{4}}$

19. $-\sqrt{\dfrac{16}{81}}$

20. $-\sqrt{\dfrac{25}{49}}$

21. $\sqrt{\dfrac{64}{289}}$

22. $\sqrt{\dfrac{81}{361}}$

23. $\sqrt{\dfrac{1690}{1960}}$

24. $\sqrt{\dfrac{1210}{6250}}$

25. $\sqrt{\dfrac{25}{x^2}}$

26. $\sqrt{\dfrac{36}{a^2}}$

27. $\sqrt{\dfrac{9a^2}{625}}$

28. $\sqrt{\dfrac{x^2y^2}{256}}$

29. $\dfrac{\sqrt{50y^{15}}}{\sqrt{2y^{25}}}$

30. $\dfrac{\sqrt{3t^{15}}}{\sqrt{12t}}$

31. $\dfrac{\sqrt{7x^{23}}}{\sqrt{343x^5}}$

32. $\dfrac{\sqrt{125q^3}}{\sqrt{5q^{19}}}$

C Rationalize the denominator.

33. $\sqrt{\dfrac{2}{5}}$

34. $\sqrt{\dfrac{2}{7}}$

35. $\sqrt{\dfrac{7}{8}}$

36. $\sqrt{\dfrac{3}{8}}$

37. $\sqrt{\dfrac{1}{12}}$

38. $\sqrt{\dfrac{7}{12}}$

39. $\sqrt{\dfrac{5}{18}}$

40. $\sqrt{\dfrac{1}{18}}$

41. $\dfrac{3}{\sqrt{5}}$

42. $\dfrac{4}{\sqrt{3}}$

43. $\sqrt{\dfrac{8}{3}}$

44. $\sqrt{\dfrac{12}{5}}$

45. $\sqrt{\dfrac{3}{x}}$

46. $\sqrt{\dfrac{2}{x}}$

47. $\sqrt{\dfrac{x}{y}}$

48. $\sqrt{\dfrac{a}{b}}$

49. $\sqrt{\dfrac{x^2}{20}}$

50. $\sqrt{\dfrac{x^2}{18}}$

51. $\dfrac{\sqrt{7}}{\sqrt{2}}$

52. $\dfrac{\sqrt{3}}{\sqrt{5}}$

53. $\dfrac{\sqrt{9}}{\sqrt{8}}$

54. $\dfrac{\sqrt{4}}{\sqrt{27}}$

55. $\dfrac{\sqrt{3}}{\sqrt{2}}$

56. $\dfrac{\sqrt{2}}{\sqrt{5}}$

57. $\dfrac{2}{\sqrt{2}}$

58. $\frac{3}{\sqrt{3}}$

59. $\frac{\sqrt{5}}{\sqrt{11}}$

60. $\frac{\sqrt{7}}{\sqrt{27}}$

61. $\frac{\sqrt{7}}{\sqrt{12}}$

62. $\frac{\sqrt{5}}{\sqrt{18}}$

63. $\frac{\sqrt{48}}{\sqrt{32}}$

64. $\frac{\sqrt{56}}{\sqrt{40}}$

65. $\frac{\sqrt{450}}{\sqrt{18}}$

66. $\frac{\sqrt{224}}{\sqrt{14}}$

67. $\frac{\sqrt{3}}{\sqrt{x}}$

68. $\frac{\sqrt{2}}{\sqrt{y}}$

69. $\frac{4y}{\sqrt{5}}$

70. $\frac{8x}{\sqrt{3}}$

71. $\frac{\sqrt{a^3}}{\sqrt{8}}$

72. $\frac{\sqrt{x^3}}{\sqrt{27}}$

73. $\frac{\sqrt{56}}{\sqrt{12x}}$

74. $\frac{\sqrt{45}}{\sqrt{8a}}$

75. $\frac{\sqrt{27c}}{\sqrt{32c^3}}$

76. $\frac{\sqrt{7x^3}}{\sqrt{12x}}$

77. $\frac{\sqrt{y^5}}{\sqrt{xy^2}}$

78. $\frac{\sqrt{x^3}}{\sqrt{xy}}$

79. $\frac{\sqrt{45mn^2}}{\sqrt{32m}}$

80. $\frac{\sqrt{16a^4b^6}}{\sqrt{128a^6b^6}}$

81. **DW** Why is it important to know how to multiply radical expressions before learning how to divide them?

82. **DW** Describe a method that could be used to rationalize the *numerator* of a radical expression.

SKILL MAINTENANCE

Solve. [13.3a, b]

83. $x = y + 2,$
$x + y = 6$

84. $4x - y = 10,$
$4x + y = 70$

85. $2x - 3y = 7,$
$2x - 3y = 9$

86. $2x - 3y = 7,$
$-4x + 6y = -14$

87. $x + y = -7,$
$x - y = 2$

88. $2x + 3y = 8,$
$5x - 4y = -2$

Divide and simplify. [12.2b]

89. $\dfrac{x^2 - 49}{x + 8} \div \dfrac{x^2 - 14x + 49}{x^2 + 15x + 56}$

90. $\dfrac{x - 2}{x - 3} \div \dfrac{x - 4}{x - 5}$

91. $\dfrac{a^2 - 25}{6} \div \dfrac{a + 5}{3}$

92. $\dfrac{x - 2}{x + 3} \div \dfrac{x^2 - 4x + 4}{x^2 - 9}$

Multiply. [10.6b]

93. $(3x - 7)(3x + 7)$

94. $(4a - 5b)(4a + 5b)$

Collect like terms. [7.7e]

95. $9x - 5y + 12x - 4y$

96. $17a + 9b - 3a - 15b$

SYNTHESIS

Periods of Pendulums. The period T of a pendulum is the time it takes the pendulum to move from one side to the other and back. A formula for the period is

$$T = 2\pi\sqrt{\frac{L}{32}},$$

where T is in seconds and L is in feet. Use 3.14 for π.

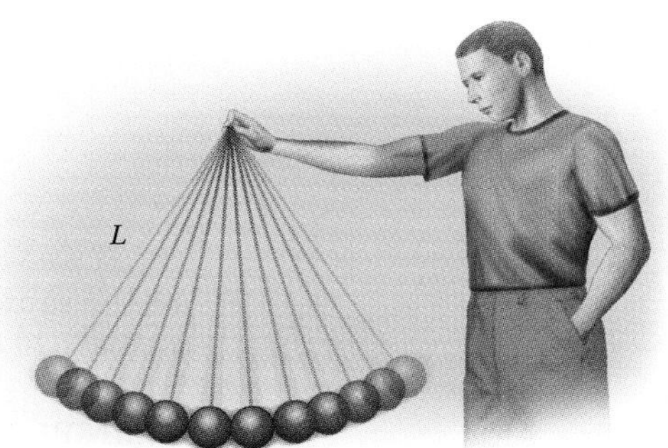

97. Find the periods of pendulums of lengths 2 ft, 8 ft, 64 ft, and 100 ft.

98. Find the period of a pendulum of length $\frac{2}{3}$ in.

99. The pendulum of a grandfather clock is $(32/\pi^2)$ ft long. How long does it take to swing from one side to the other?

100. The pendulum of a grandfather clock is $(45/\pi^2)$ ft long. How long does it take to swing from one side to the other?

Rationalize the denominator.

101. $\sqrt{\dfrac{5}{1600}}$

102. $\sqrt{\dfrac{3}{1000}}$

103. $\sqrt{\dfrac{1}{5x^3}}$

104. $\sqrt{\dfrac{3x^2y}{a^2x^5}}$

105. $\sqrt{\dfrac{3a}{b}}$

106. $\sqrt{\dfrac{1}{5zw^2}}$

107. $\sqrt{0.009}$

108. $\sqrt{0.012}$

Simplify.

109. $\sqrt{\dfrac{1}{x^2} - \dfrac{2}{xy} + \dfrac{1}{y^2}}$

110. $\sqrt{2 - \dfrac{4}{z^2} + \dfrac{2}{z^4}}$

14.4 ADDITION, SUBTRACTION, AND MORE MULTIPLICATION

Objectives

a Add or subtract with radical notation, using the distributive law to simplify.

b Multiply expressions involving radicals, where some of the expressions contain more than one term.

c Rationalize denominators having two terms.

a Addition and Subtraction

We can add any two real numbers. The sum of 5 and $\sqrt{2}$ can be expressed as

$$5 + \sqrt{2}.$$

We cannot simplify this unless we use rational approximations such as $5 + \sqrt{2} \approx 5 + 1.414 = 6.414$. However, when we have *like radicals,* a sum can be simplified using the distributive laws and collecting like terms. **Like radicals** have the same radicands.

EXAMPLE 1 Add: $3\sqrt{5} + 4\sqrt{5}$.

Suppose we were considering $3x + 4x$. Recall that to add, we use a distributive law as follows:

$$3x + 4x = (3 + 4)x = 7x.$$

The situation is similar in this example, but we let $x = \sqrt{5}$:

$$3\sqrt{5} + 4\sqrt{5} = (3 + 4)\sqrt{5} \quad \text{Using a distributive law to factor out } \sqrt{5}$$
$$= 7\sqrt{5}.$$

If we wish to add or subtract as we did in Example 1, the radicands must be the same. Sometimes after simplifying the radical terms, we discover that we have like radicals.

EXAMPLES Add or subtract. Simplify, if possible, by collecting like radical terms.

2. $5\sqrt{2} - \sqrt{18} = 5\sqrt{2} - \sqrt{9 \cdot 2}$ Factoring 18
$= 5\sqrt{2} - \sqrt{9}\sqrt{2}$
$= 5\sqrt{2} - 3\sqrt{2}$
$= (5 - 3)\sqrt{2}$ Using a distributive law to factor out the common factor, $\sqrt{2}$
$= 2\sqrt{2}$

3. $\sqrt{4x^3} + 7\sqrt{x} = \sqrt{4 \cdot x^2 \cdot x} + 7\sqrt{x}$
$= 2x\sqrt{x} + 7\sqrt{x}$
$= (2x + 7)\sqrt{x}$ Using a distributive law to factor out $\sqrt{x}$

Don't forget the parentheses!

4. $\sqrt{x^3 - x^2} + \sqrt{4x - 4} = \sqrt{x^2(x - 1)} + \sqrt{4(x - 1)}$ Factoring radicands
$= \sqrt{x^2}\sqrt{x - 1} + \sqrt{4}\sqrt{x - 1}$
$= x\sqrt{x - 1} + 2\sqrt{x - 1}$
$= (x + 2)\sqrt{x - 1}$ Using a distributive law to factor out the common factor, $\sqrt{x - 1}$. Don't forget the parentheses!

Add or subtract and simplify by collecting like radical terms, if possible.

1. $3\sqrt{2} + 9\sqrt{2}$

2. $8\sqrt{5} - 3\sqrt{5}$

3. $2\sqrt{10} - 7\sqrt{40}$

4. $\sqrt{24} + \sqrt{54}$

5. $\sqrt{9x + 9} - \sqrt{4x + 4}$

Answers on page A-49

Do Exercises 1–5 on the preceding page.

Sometimes rationalizing denominators enables us to combine like radicals.

EXAMPLE 5 Add: $\sqrt{3} + \sqrt{\frac{1}{3}}$.

$$\sqrt{3} + \sqrt{\frac{1}{3}} = \sqrt{3} + \sqrt{\frac{1}{3} \cdot \frac{3}{3}}$$ Multiplying by 1 in order to rationalize the denominator

$$= \sqrt{3} + \sqrt{\frac{3}{9}}$$

$$= \sqrt{3} + \frac{\sqrt{3}}{\sqrt{9}}$$

$$= \sqrt{3} + \frac{\sqrt{3}}{3}$$

$$= 1 \cdot \sqrt{3} + \frac{1}{3}\sqrt{3}$$

$$= \left(1 + \frac{1}{3}\right)\sqrt{3}$$ Factoring out the common factor, $\sqrt{3}$

$$= \frac{4}{3}\sqrt{3}$$

Do Exercises 6 and 7.

Add or subtract.

6. $\sqrt{2} + \sqrt{\frac{1}{2}}$

7. $\sqrt{\frac{5}{3}} + \sqrt{\frac{3}{5}}$

b Multiplication

Now let's multiply where some of the expressions may contain more than one term. To do this, we use procedures already studied in this chapter as well as the distributive laws and special products for multiplying with polynomials.

EXAMPLE 6 Multiply: $\sqrt{2}(\sqrt{3} + \sqrt{7})$.

$$\sqrt{2}(\sqrt{3} + \sqrt{7}) = \sqrt{2}\sqrt{3} + \sqrt{2}\sqrt{7}$$ Multiplying using a distributive law

$$= \sqrt{6} + \sqrt{14}$$ Using the rule for multiplying with radicals

EXAMPLE 7 Multiply: $(2 + \sqrt{3})(5 - 4\sqrt{3})$.

$$(2 + \sqrt{3})(5 - 4\sqrt{3}) = 2 \cdot 5 - 2 \cdot 4\sqrt{3} + \sqrt{3} \cdot 5 - \sqrt{3} \cdot 4\sqrt{3}$$ Using FOIL

$$= 10 - 8\sqrt{3} + 5\sqrt{3} - 4 \cdot 3$$

$$= 10 - 8\sqrt{3} + 5\sqrt{3} - 12$$

$$= -2 - 3\sqrt{3}$$

Answers on page A-49

Multiply.

8. $\sqrt{3}(\sqrt{5}+\sqrt{2})$

9. $(1-\sqrt{2})(4+3\sqrt{5})$

10. $(\sqrt{2}+\sqrt{a})(\sqrt{2}-\sqrt{a})$

11. $(5+\sqrt{x})^2$

12. $(3-\sqrt{7})(3+\sqrt{7})$

Find the conjugate of the expression.

13. $7+\sqrt{5}$

14. $\sqrt{5}-\sqrt{2}$

15. $1-\sqrt{x}$

Answers on page A-49

EXAMPLE 8 Multiply: $(\sqrt{3}-\sqrt{x})(\sqrt{3}+\sqrt{x})$.

$$(\sqrt{3}-\sqrt{x})(\sqrt{3}+\sqrt{x}) = (\sqrt{3})^2-(\sqrt{x})^2 \quad \text{Using } (A-B)(A+B)=A^2-B^2$$
$$= 3 - x$$

EXAMPLE 9 Multiply: $(3-\sqrt{p})^2$.

$$(3-\sqrt{p})^2 = 3^2 - 2\cdot 3\cdot\sqrt{p} + (\sqrt{p})^2 \quad \text{Using } (A-B)^2 = A^2-2AB+B^2$$
$$= 9 - 6\sqrt{p} + p$$

EXAMPLE 10 Multiply: $(2+\sqrt{5})^2$.

$$(2+\sqrt{5})^2 = 2^2 + 2\cdot 2\sqrt{5} + (\sqrt{5})^2 \quad \text{Using } (A+B)^2 = A^2+2AB+B^2$$
$$= 4 + 4\sqrt{5} + 5$$
$$= 9 + 4\sqrt{5}$$

Do Exercises 8–12.

C More on Rationalizing Denominators

Note in Example 8 that the result has no radicals. This will happen whenever we multiply expressions such as $\sqrt{a}-\sqrt{b}$ and $\sqrt{a}+\sqrt{b}$. We see this in the following:

$$(\sqrt{a}+\sqrt{b})(\sqrt{a}-\sqrt{b}) = (\sqrt{a})^2-(\sqrt{b})^2 = a-b.$$

Expressions such as $\sqrt{3}-\sqrt{x}$ and $\sqrt{3}+\sqrt{x}$ are known as **conjugates**; so too are $2+\sqrt{5}$ and $2-\sqrt{5}$. We can use conjugates to rationalize a denominator that involves a sum or difference of two terms, where one or both are radicals. To do so, we multiply by 1 using the conjugate to form the expression for 1.

Do Exercises 13–15.

EXAMPLE 11 Rationalize the denominator: $\dfrac{3}{2+\sqrt{5}}$.

We multiply by 1 using the conjugate of $2+\sqrt{5}$, which is $2-\sqrt{5}$, as the numerator and the denominator of the expression for 1:

$$\frac{3}{2+\sqrt{5}} = \frac{3}{2+\sqrt{5}}\cdot\frac{2-\sqrt{5}}{2-\sqrt{5}} \quad \text{Multiplying by 1}$$
$$= \frac{3(2-\sqrt{5})}{(2+\sqrt{5})(2-\sqrt{5})} \quad \text{Multiplying}$$
$$= \frac{6-3\sqrt{5}}{2^2-(\sqrt{5})^2} \quad \text{Using } (A+B)(A-B)=A^2-B^2$$
$$= \frac{6-3\sqrt{5}}{4-5}$$
$$= \frac{6-3\sqrt{5}}{-1}$$
$$= -6+3\sqrt{5}, \text{ or } 3\sqrt{5}-6.$$

EXAMPLE 12 Rationalize the denominator: $\dfrac{\sqrt{3}+\sqrt{5}}{\sqrt{3}-\sqrt{5}}$.

We multiply by 1 using the conjugate of $\sqrt{3}-\sqrt{5}$, which is $\sqrt{3}+\sqrt{5}$, as the numerator and the denominator of the expression for 1:

$$\frac{\sqrt{3}+\sqrt{5}}{\sqrt{3}-\sqrt{5}} = \frac{\sqrt{3}+\sqrt{5}}{\sqrt{3}-\sqrt{5}} \cdot \frac{\sqrt{3}+\sqrt{5}}{\sqrt{3}+\sqrt{5}} \qquad \text{Multiplying by 1}$$

$$= \frac{(\sqrt{3}+\sqrt{5})^2}{(\sqrt{3}-\sqrt{5})(\sqrt{3}+\sqrt{5})}$$

$$= \frac{(\sqrt{3})^2 + 2\sqrt{3}\sqrt{5} + (\sqrt{5})^2}{(\sqrt{3})^2 - (\sqrt{5})^2} \qquad \text{Using } (A+B)^2 = A^2 + 2AB + B^2 \text{ and } (A+B)(A-B) = A^2 - B^2$$

$$= \frac{3 + 2\sqrt{15} + 5}{3 - 5}$$

$$= \frac{8 + 2\sqrt{15}}{-2}$$

$$= \frac{2(4+\sqrt{15})}{2(-1)} \qquad \text{Factoring in order to simplify}$$

$$= \frac{2}{2} \cdot \frac{4+\sqrt{15}}{-1}$$

$$= \frac{4+\sqrt{15}}{-1}$$

$$= -4 - \sqrt{15}.$$

Do Exercises 16 and 17.

EXAMPLE 13 Rationalize the denominator: $\dfrac{5}{2+\sqrt{x}}$.

We multiply by 1 using the conjugate of $2+\sqrt{x}$, which is $2-\sqrt{x}$, as the numerator and the denominator of the expression for 1:

$$\frac{5}{2+\sqrt{x}} = \frac{5}{2+\sqrt{x}} \cdot \frac{2-\sqrt{x}}{2-\sqrt{x}} \qquad \text{Multiplying by 1}$$

$$= \frac{5(2-\sqrt{x})}{(2+\sqrt{x})(2-\sqrt{x})}$$

$$= \frac{5 \cdot 2 - 5 \cdot \sqrt{x}}{2^2 - (\sqrt{x})^2} \qquad \text{Using } (A+B)(A-B) = A^2 - B^2$$

$$= \frac{10 - 5\sqrt{x}}{4 - x}.$$

Do Exercise 18.

Rationalize the denominator.

16. $\dfrac{3}{7+\sqrt{5}}$

17. $\dfrac{\sqrt{5}+\sqrt{7}}{\sqrt{5}-\sqrt{7}}$

18. Rationalize the denominator:

$$\frac{7}{1-\sqrt{x}}.$$

Answers on page A-49

14.4 EXERCISE SET

For Extra Help

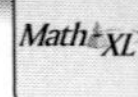

MathXL | MyMathLab | InterAct Math | Math Tutor Center | Video Lectures on CD Disc 7 | Student's Solutions Manual

a Add or subtract. Simplify by collecting like radical terms, if possible.

1. $7\sqrt{3} + 9\sqrt{3}$

2. $6\sqrt{2} + 8\sqrt{2}$

3. $7\sqrt{5} - 3\sqrt{5}$

4. $8\sqrt{2} - 5\sqrt{2}$

5. $6\sqrt{x} + 7\sqrt{x}$

6. $9\sqrt{y} + 3\sqrt{y}$

7. $4\sqrt{d} - 13\sqrt{d}$

8. $2\sqrt{a} - 17\sqrt{a}$

9. $5\sqrt{8} + 15\sqrt{2}$

10. $3\sqrt{12} + 2\sqrt{3}$

11. $\sqrt{27} - 2\sqrt{3}$

12. $7\sqrt{50} - 3\sqrt{2}$

13. $\sqrt{45} - \sqrt{20}$

14. $\sqrt{27} - \sqrt{12}$

15. $\sqrt{72} + \sqrt{98}$

16. $\sqrt{45} + \sqrt{80}$

17. $2\sqrt{12} + \sqrt{27} - \sqrt{48}$

18. $9\sqrt{8} - \sqrt{72} + \sqrt{98}$

19. $\sqrt{18} - 3\sqrt{8} + \sqrt{50}$

20. $3\sqrt{18} - 2\sqrt{32} - 5\sqrt{50}$

21. $2\sqrt{27} - 3\sqrt{48} + 3\sqrt{12}$

22. $3\sqrt{48} - 2\sqrt{27} - 3\sqrt{12}$

23. $\sqrt{4x} + \sqrt{81x^3}$

24. $\sqrt{12x^2} + \sqrt{27}$

25. $\sqrt{27} - \sqrt{12x^2}$

26. $\sqrt{81x^3} - \sqrt{4x}$

27. $\sqrt{8x + 8} + \sqrt{2x + 2}$

28. $\sqrt{12x + 12} + \sqrt{3x + 3}$

29. $\sqrt{x^5 - x^2} + \sqrt{9x^3 - 9}$

30. $\sqrt{16x - 16} + \sqrt{25x^3 - 25x^2}$

31. $4a\sqrt{a^2b} + a\sqrt{a^2b^3} - 5\sqrt{b^3}$

32. $3x\sqrt{y^3x} - x\sqrt{yx^3} + y\sqrt{y^3x}$

33. $\sqrt{3} - \sqrt{\frac{1}{3}}$

34. $\sqrt{2} - \sqrt{\frac{1}{2}}$

35. $5\sqrt{2} + 3\sqrt{\frac{1}{2}}$

36. $4\sqrt{3} + 2\sqrt{\frac{1}{3}}$

37. $\sqrt{\frac{2}{3}} - \sqrt{\frac{1}{6}}$

38. $\sqrt{\frac{1}{2}} - \sqrt{\frac{1}{8}}$

b Multiply.

39. $\sqrt{3}(\sqrt{5} - 1)$

40. $\sqrt{2}(\sqrt{2} + \sqrt{3})$

41. $(2 + \sqrt{3})(5 - \sqrt{7})$

42. $(\sqrt{5} + \sqrt{7})(2\sqrt{5} - 3\sqrt{7})$

43. $(2 - \sqrt{5})^2$

44. $(\sqrt{3} + \sqrt{10})^2$

45. $(\sqrt{2} + 8)(\sqrt{2} - 8)$

46. $(1 + \sqrt{7})(1 - \sqrt{7})$

47. $(\sqrt{6} - \sqrt{5})(\sqrt{6} + \sqrt{5})$

48. $(\sqrt{3} + \sqrt{10})(\sqrt{3} - \sqrt{10})$

49. $(3\sqrt{5} - 2)(\sqrt{5} + 1)$

50. $(\sqrt{5} - 2\sqrt{2})(\sqrt{10} - 1)$

51. $(\sqrt{x} - \sqrt{y})^2$

52. $(\sqrt{w} + 11)^2$

c Rationalize the denominator.

53. $\dfrac{2}{\sqrt{3} - \sqrt{5}}$

54. $\dfrac{5}{3 + \sqrt{7}}$

55. $\dfrac{\sqrt{3} - \sqrt{2}}{\sqrt{3} + \sqrt{2}}$

56. $\dfrac{2 - \sqrt{7}}{\sqrt{3} - \sqrt{2}}$

57. $\dfrac{4}{\sqrt{10} + 1}$

58. $\dfrac{6}{\sqrt{11} - 3}$

59. $\dfrac{1 - \sqrt{7}}{3 + \sqrt{7}}$

60. $\dfrac{2 + \sqrt{8}}{1 - \sqrt{5}}$

61. $\dfrac{3}{4 + \sqrt{x}}$

62. $\dfrac{8}{2 - \sqrt{x}}$

63. $\dfrac{3 + \sqrt{2}}{8 - \sqrt{x}}$

64. $\dfrac{4 - \sqrt{3}}{6 + \sqrt{y}}$

65. **DW** Explain why it is important for the signs within a pair of conjugates to differ.

66. **DW** Describe a method that could be used to rationalize a numerator that contains the sum of two radical expressions.

SKILL MAINTENANCE

Solve.

67. $3x + 5 + 2(x - 3) = 4 - 6x$ [8.3c]

68. $3(x - 4) - 2 = 8(2x + 3)$ [8.3c]

69. $x^2 - 5x = 6$ [11.7b]

70. $x^2 + 10 = 7x$ [11.7b]

Multiply and simplify. [12.1d]

71. $\dfrac{7x^9}{27} \cdot \dfrac{9}{7x^3}$

72. $\dfrac{3}{x^2 - 9} \cdot \dfrac{x^2 - 6x + 9}{12}$

Solve.

73. *Juice Mixtures.* Jolly Juice is 3% real fruit juice, and Real Squeeze is 6% real fruit juice. How many liters of each should be combined in order to make an 8-L mixture that is 5.4% real fruit juice? [13.4a]

74. *Travel Time.* The time t that it takes a bus to travel a fixed distance varies inversely as its speed r. At a speed of 40 mph, it takes $\frac{1}{2}$ hr to travel a fixed distance. How long will it take to travel the same distance at 60 mph? [12.9d]

75. The graph of the polynomial equation $y = x^3 - 5x^2 + x - 2$ is shown at right. Use either the graph or the equation to estimate or find the value of the polynomial when $x = -1$, $x = 0$, $x = 1$, $x = 3$, and $x = 4.85$. [10.3a]

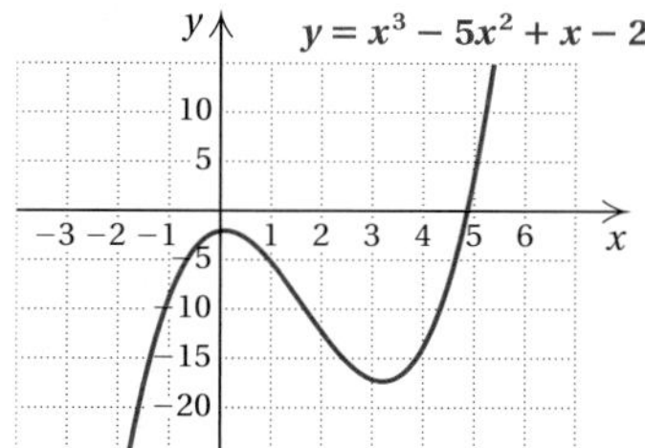

SYNTHESIS

76. Evaluate $\sqrt{a^2 + b^2}$ and $\sqrt{a^2} + \sqrt{b^2}$ when $a = 2$ and $b = 3$.

77. On the basis of Exercise 76, determine whether $\sqrt{a^2 + b^2}$ and $\sqrt{a^2} + \sqrt{b^2}$ are equivalent.

Use the GRAPH and TABLE features to determine whether each of the following is correct.

78. $\sqrt{9x^3} + \sqrt{x} = \sqrt{9x^3 + x}$

79. $\sqrt{x^2 + 4} = \sqrt{x} + 2$

Add or subtract as indicated.

80. $\frac{3}{5}\sqrt{24} + \frac{2}{5}\sqrt{150} - \sqrt{96}$

81. $\frac{1}{3}\sqrt{27} + \sqrt{8} + \sqrt{300} - \sqrt{18} - \sqrt{162}$

82. Three students were asked to simplify $\sqrt{10} + \sqrt{50}$. Their answers were $\sqrt{10}(1 + \sqrt{5})$, $\sqrt{10} + 5\sqrt{2}$, and $\sqrt{2}(5 + \sqrt{5})$. Which, if any, are correct?

Determine whether each of the following is true. Show why or why not.

83. $(3\sqrt{x + 2})^2 = 9(x + 2)$

84. $(\sqrt{x + 2})^2 = x + 2$

14.5 RADICAL EQUATIONS

Objectives

a Solve radical equations with one or two radical terms isolated, using the principle of squaring once.

b Solve radical equations with two radical terms, using the principle of squaring twice.

c Solve applied problems using radical equations.

1. Solve: $\sqrt{3x} - 5 = 3$.

a Solving Radical Equations

The following are examples of *radical equations*:

$$\sqrt{2x} - 4 = 7, \qquad \sqrt{x+1} = \sqrt{2x-5}.$$

A **radical equation** has variables in one or more radicands. To solve radical equations, we first convert them to equations without radicals. We do this for square-root radical equations by squaring both sides of the equation, using the following principle.

THE PRINCIPLE OF SQUARING

If an equation $a = b$ is true, then the equation $a^2 = b^2$ is true.

To solve square-root radical equations, we first try to get a radical by itself. That is, we try to isolate the radical. Then we use the principle of squaring. This allows us to eliminate one radical.

EXAMPLE 1 Solve: $\sqrt{2x} - 4 = 7$.

$$\sqrt{2x} - 4 = 7$$

$$\sqrt{2x} = 11 \quad \text{Adding 4 to isolate the radical}$$

$$(\sqrt{2x})^2 = 11^2 \quad \text{Squaring both sides}$$

$$2x = 121 \quad \sqrt{2x}\cdot\sqrt{2x} = 2x\text{, by the definition of square root}$$

$$x = \frac{121}{2} \quad \text{Dividing by 2}$$

Check:

$$\sqrt{2x} - 4 = 7$$

$$\sqrt{2\cdot\frac{121}{2}} - 4 \ ?\ 7$$

$$\sqrt{121} - 4$$

$$11 - 4$$

$$7 \quad \text{TRUE}$$

The solution is $\frac{121}{2}$.

Do Exercise 1.

EXAMPLE 2 Solve: $2\sqrt{x+2} = \sqrt{x+10}$.

Each radical is isolated. We proceed with the principle of squaring.

$$(2\sqrt{x+2})^2 = (\sqrt{x+10})^2 \quad \text{Squaring both sides}$$

$$2^2(\sqrt{x+2})^2 = (\sqrt{x+10})^2 \quad \text{Raising each factor of the product on the left to the second power}$$

$$4(x+2) = x + 10 \quad \text{Simplifying}$$

$$4x + 8 = x + 10 \quad \text{Removing parentheses}$$

$$3x = 2 \quad \text{Subtracting } x \text{ and } 8$$

$$x = \frac{2}{3} \quad \text{Dividing by 3}$$

Answer on page A-50

Check: $2\sqrt{x+2} = \sqrt{x+10}$

$$
\begin{array}{c|c}
2\sqrt{\frac{2}{3}+2} \ ? & \sqrt{\frac{2}{3}+10} \\
2\sqrt{\frac{8}{3}} & \sqrt{\frac{32}{3}} \\
2\sqrt{\frac{4\cdot 2}{3}} & \sqrt{\frac{16\cdot 2}{3}} \\
4\sqrt{\frac{2}{3}} & 4\sqrt{\frac{2}{3}}
\end{array}
\quad \text{TRUE}
$$

The number $\frac{2}{3}$ checks. The solution is $\frac{2}{3}$.

Do Exercises 2 and 3.

It is important to check when using the principle of squaring. This principle may not produce equivalent equations. When we square both sides of an equation, the new equation may have solutions that the first one does not. For example, the equation

$$x = 1 \qquad (1)$$

has just one solution, the number 1. When we square both sides, we get

$$x^2 = 1, \qquad (2)$$

which has two solutions, 1 and -1. The equations $x = 1$ and $x^2 = 1$ do not have the same solutions and thus are not equivalent. Whereas it is true that any solution of equation (1) is a solution of equation (2), it is *not* true that any solution of equation (2) is a solution of equation (1).

Caution!

When the principle of squaring is used to solve an equation, all possible solutions *must* be checked in the original equation!

Sometimes we may need to apply the principle of zero products after squaring. (See Section 11.7.)

EXAMPLE 3 Solve: $x - 5 = \sqrt{x+7}$.

$$
\begin{aligned}
x - 5 &= \sqrt{x+7} \\
(x-5)^2 &= (\sqrt{x+7})^2 && \text{Using the principle of squaring} \\
x^2 - 10x + 25 &= x + 7 \\
x^2 - 11x + 18 &= 0 && \text{Subtracting } x \text{ and } 7 \\
(x-9)(x-2) &= 0 && \text{Factoring} \\
x - 9 = 0 \quad or \quad x - 2 &= 0 && \text{Using the principle of zero products} \\
x = 9 \quad or \quad x &= 2
\end{aligned}
$$

Check: For 9:

$$
\begin{array}{c|c}
x-5 = & \sqrt{x+7} \\
\hline
9-5 \ ? & \sqrt{9+7} \\
4 & 4
\end{array}
\quad \text{TRUE}
$$

For 2:

$$
\begin{array}{c|c}
x-5 = & \sqrt{x+7} \\
\hline
2-5 \ ? & \sqrt{2+7} \\
-3 & 3
\end{array}
\quad \text{FALSE}
$$

The number 9 checks, but 2 does not. Thus the solution is 9.

Do Exercise 4.

Solve.

2. $\sqrt{3x+1} = \sqrt{2x+3}$

3. $3\sqrt{x+1} = \sqrt{x+12}$

4. Solve: $x - 1 = \sqrt{x+5}$.

Answers on page A-50

5. Solve: $1 + \sqrt{1 - x} = x$.

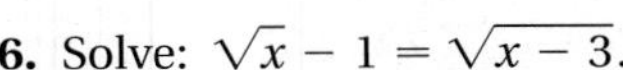

6. Solve: $\sqrt{x} - 1 = \sqrt{x - 3}$.

Answers on page A-50

AG ALGEBRAIC–GRAPHICAL CONNECTION

We can visualize or check the solutions of a radical equation graphically. Consider the equation of Example 3:

$$x - 5 = \sqrt{x + 7}.$$

We can examine the solutions by graphing the equations

$$y = x - 5 \quad \text{and} \quad y = \sqrt{x + 7}$$

using the same set of axes. A hand-drawn graph of $y = \sqrt{x + 7}$ would involve approximating square roots on a calculator.

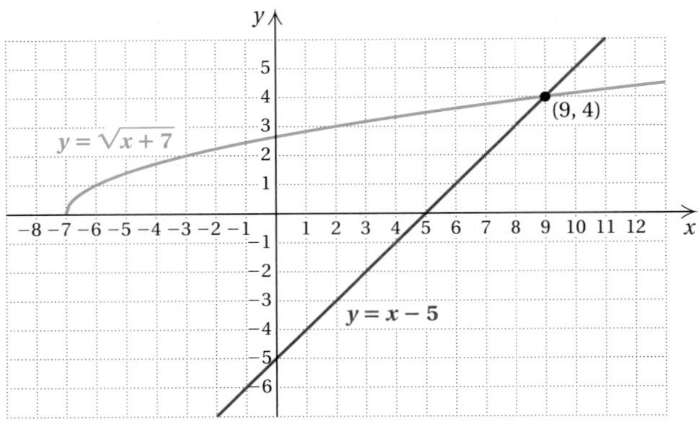

It appears that when $x = 9$, the values of $y = x - 5$ and $y = \sqrt{x + 7}$ are the same, 4. We can check this as we did in Example 3. Note also that the graphs *do not* intersect at $x = 2$.

EXAMPLE 4 Solve: $3 + \sqrt{27 - 3x} = x$.

In this case, we must first isolate the radical.

$3 + \sqrt{27 - 3x} = x$	
$\sqrt{27 - 3x} = x - 3$	Subtracting 3 to isolate the radical
$(\sqrt{27 - 3x})^2 = (x - 3)^2$	Using the principle of squaring
$27 - 3x = x^2 - 6x + 9$	
$0 = x^2 - 3x - 18$	Adding $3x$ and subtracting 27 to obtain 0 on the left
$0 = (x - 6)(x + 3)$	Factoring
$x - 6 = 0$ *or* $x + 3 = 0$	Using the principle of zero products
$x = 6$ *or* $x = -3$	

Check: For 6:

$$\begin{array}{r|l} 3 + \sqrt{27 - 3x} & x \\ \hline 3 + \sqrt{27 - 3 \cdot 6} & ?\ 6 \\ 3 + \sqrt{9} & \\ 3 + 3 & \\ 6 & \end{array}$$

TRUE

For −3:

$$\begin{array}{r|l} 3 + \sqrt{27 - 3x} & x \\ \hline 3 + \sqrt{27 - 3 \cdot (-3)} & ?\ -3 \\ 3 + \sqrt{27 + 9} & \\ 3 + \sqrt{36} & \\ 3 + 6 & \\ 9 & \end{array}$$

FALSE

The number 6 checks, but −3 does not. The solution is 6.

Do Exercise 5 on the preceding page.

b Using the Principle of Squaring More Than Once

Sometimes when we have two radical terms, we may need to apply the principle of squaring a second time.

EXAMPLE 5 Solve: $\sqrt{x} - 1 = \sqrt{x - 5}$.

We have

$\sqrt{x} - 1 = \sqrt{x - 5}$	
$(\sqrt{x} - 1)^2 = (\sqrt{x - 5})^2$	Using the principle of squaring
$(\sqrt{x})^2 - 2 \cdot \sqrt{x} \cdot 1 + 1^2 = x - 5$	Using $(A - B)^2 = A^2 - 2AB + B^2$ on the left side
$x - 2\sqrt{x} + 1 = x - 5$	Simplifying. Only one radical term remains.
$-2\sqrt{x} = -6$	Isolating the radical, subtracting x and 1
$\sqrt{x} = 3$	
$(\sqrt{x})^2 = 3^2$	Using the principle of squaring
$x = 9.$	

The check is left to the student. The number 9 checks and is the solution.

The following is a procedure for solving square-root radical equations.

SOLVING SQUARE-ROOT RADICAL EQUATIONS

To solve square-root radical equations:

1. Isolate one of the radical terms.
2. Use the principle of squaring.
3. If a radical term remains, perform steps (1) and (2) again.
4. Solve the equation and check possible solutions.

Do Exercise 6 on the preceding page.

CALCULATOR CORNER

Solving Radical Equations

We can solve radical equations on a graphing calculator. Consider the equation in Example 3, $x - 5 = \sqrt{x + 7}$. We first graph each side of the equation. We enter $y_1 = x - 5$ and $y_2 = \sqrt{x + 7}$ on the equation-editor screen and graph the equations, using the window $[-2, 12, -6, 6]$. Note that there is one point of intersection. Use the INTERSECT feature to find its coordinates. (See the Calculator Corner on p. 972 for the procedure.)

The first coordinate, 9, is the value of x for which $y_1 = y_2$, or $x - 5 = \sqrt{x + 7}$. It is the solution of the equation. Note that the graph shows a single solution whereas the algebraic solution in Example 3 yields two possible solutions, 9 and 2, that must be checked. The check shows that 9 is the only solution.

$y_1 = x - 5$, $y_2 = \sqrt{x + 7}$

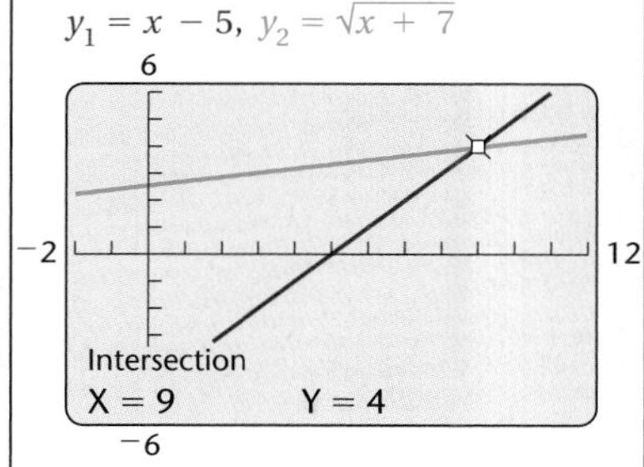

Exercises:

1. Solve the equations in Examples 4 and 5 graphically.
2. Solve the equations in Margin Exercises 1–6 graphically.

7. How far to the horizon can you see through an airplane window at a height, or altitude, of 38,000 ft?

8. A sailor climbs 40 ft up the mast of a ship to a crow's nest. How far can he see to the horizon?

9. How far above sea level must a sailor climb on the mast of a ship in order to see 10.2 mi out to an iceberg?

Answers on page A-50

C Applications

Sighting to the Horizon. How far can you see from a given height? The equation

$$D = \sqrt{2h}$$

can be used to approximate the distance D, in miles, that a person can see to the horizon from a height h, in feet.

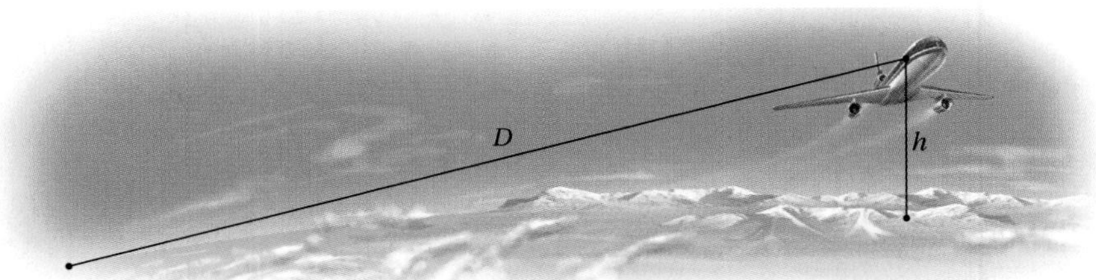

EXAMPLE 6 How far to the horizon can you see through an airplane window at a height, or altitude, of 30,000 ft?

We substitute 30,000 for h in $D = \sqrt{2h}$ and find an approximation using a calculator:

$$D = \sqrt{2 \cdot 30{,}000} \approx 245 \text{ mi}.$$

You can see for about 245 mi to the horizon.

Do Exercises 7 and 8.

EXAMPLE 7 *Height of a Ranger Station.* How high must a ranger station be in order for the ranger to see out to a fire on the horizon 15.4 mi away?

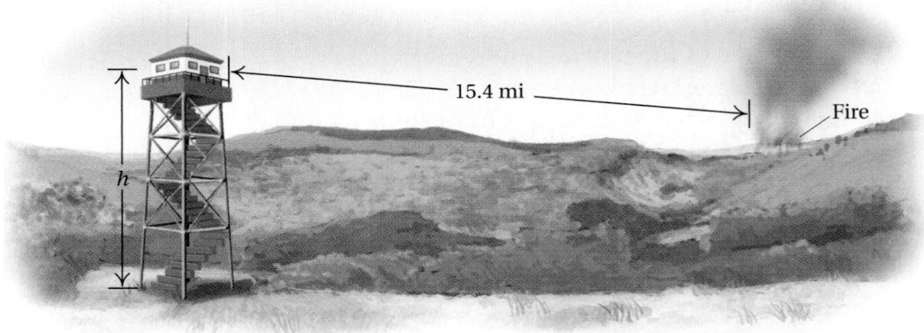

We substitute 15.4 for D in $D = \sqrt{2h}$ and solve:

$$15.4 = \sqrt{2h}$$
$$(15.4)^2 = (\sqrt{2h})^2 \quad \text{Using the principle of squaring}$$
$$237.16 = 2h$$
$$\frac{237.16}{2} = h$$
$$118.58 = h.$$

The height of the ranger tower must be about 119 ft in order for the ranger to see out to a fire 15.4 mi away.

Do Exercise 9.

14.5 EXERCISE SET

For Extra Help

MathXL | MyMathLab | InterAct Math | Math Tutor Center | Video Lectures on CD Disc 7 | Student's Solutions Manual

a Solve.

1. $\sqrt{x} = 6$

2. $\sqrt{x} = 1$

3. $\sqrt{x} = 4.3$

4. $\sqrt{x} = 6.2$

5. $\sqrt{y + 4} = 13$

6. $\sqrt{y - 5} = 21$

7. $\sqrt{2x + 4} = 25$

8. $\sqrt{2x + 1} = 13$

9. $3 + \sqrt{x - 1} = 5$

10. $4 + \sqrt{y - 3} = 11$

11. $6 - 2\sqrt{3n} = 0$

12. $8 - 4\sqrt{5n} = 0$

13. $\sqrt{5x - 7} = \sqrt{x + 10}$

14. $\sqrt{4x - 5} = \sqrt{x + 9}$

15. $\sqrt{x} = -7$

16. $\sqrt{x} = -5$

17. $\sqrt{2y + 6} = \sqrt{2y - 5}$

18. $2\sqrt{3x - 2} = \sqrt{2x - 3}$

19. $x - 7 = \sqrt{x - 5}$

20. $\sqrt{x + 7} = x - 5$

21. $x - 9 = \sqrt{x - 3}$

22. $\sqrt{x + 18} = x - 2$

23. $2\sqrt{x - 1} = x - 1$

24. $x + 4 = 4\sqrt{x + 1}$

25. $\sqrt{5x + 21} = x + 3$

26. $\sqrt{27 - 3x} = x - 3$

27. $\sqrt{2x - 1} + 2 = x$

28. $x = 1 + 6\sqrt{x - 9}$

29. $\sqrt{x^2 + 6} - x + 3 = 0$

30. $\sqrt{x^2 + 5} - x + 2 = 0$

31. $\sqrt{x^2 - 4} - x = 6$

32. $\sqrt{x^2 - 5x + 7} = x - 3$

33. $\sqrt{(p + 6)(p + 1)} - 2 = p + 1$

34. $\sqrt{(4x + 5)(x + 4)} = 2x + 5$

35. $\sqrt{4x - 10} = \sqrt{2 - x}$

36. $\sqrt{2 - x} = \sqrt{3x - 7}$

 Solve. Use the principle of squaring twice.

37. $\sqrt{x-5} = 5 - \sqrt{x}$

38. $\sqrt{x+9} = 1 + \sqrt{x}$

39. $\sqrt{y+8} - \sqrt{y} = 2$

40. $\sqrt{3x+1} = 1 - \sqrt{x+4}$

41. $\sqrt{x-4} + \sqrt{x+1} = 5$

42. $1 + \sqrt{x+7} = \sqrt{3x-2}$

43. $\sqrt{x} - 1 = \sqrt{x-31}$

44. $\sqrt{2x-5} - 1 = \sqrt{x-3}$

 Solve.

Use the formula $D = \sqrt{2h}$ for Exercises 45–48.

45. How far to the horizon can you see through an airplane window at a height, or altitude, of 27,000 ft?

46. How far to the horizon can you see through an airplane window at a height, or altitude, of 32,000 ft?

47. How far above sea level must a pilot fly in order to see to a horizon that is 180 mi away?

48. A person can see 220 mi to the horizon through an airplane window. How high above sea level is the airplane?

Speed of a Skidding Car. How do police determine how fast a car had been traveling after an accident has occurred? The formula

$$r = 2\sqrt{5L}$$

can be used to approximate the speed r, in miles per hour, of a car that has left a skid mark of length L, in feet. (See Example 7 in Section 14.1.) Use this formula for Exercises 49 and 50.

49. How far will a car skid at 65 mph? at 75 mph?

50. How far will a car skid at 55 mph? at 90 mph?

51. DW Explain why possible solutions of radical equations must be checked.

52. DW Determine whether the statement below is true or false and explain your answer.

The solution of $\sqrt{11 - 2x} = -3$ is 1.

SKILL MAINTENANCE

VOCABULARY REINFORCEMENT

In each of Exercises 53–60, fill in the blank with the correct term from the given list. Some of the choices may not be used and some may be used more than once.

53. Parallel lines have the same ____________ and different ____________. [9.6a]

54. The number c is a(n) ____________ of a if $c^2 = a$. [14.1a]

55. The number c is a(n) ____________ of a if $c^2 = a$ and c is either zero or positive. [14.1a]

56. If a situation gives rise to an equation $y = kx$, where k is a ____________ constant, we say that we have ____________ variation. [12.9a]

57. The ____________ rule asserts that when dividing with exponential notation, if the bases are the same, keep the base and subtract the exponent of the denominator from the exponent of the numerator. [10.1e]

58. If a situation gives rise to an equation $y = k/x$, where k is a ____________ constant, we say that we have ____________ variation. [12.9c]

59. The ____________ rule for radicals asserts that for any nonnegative number A and any positive number B, $\frac{\sqrt{A}}{\sqrt{B}} = \sqrt{\frac{A}{B}}$. [14.3a]

60. The ____________ rule for radicals asserts that for any nonnegative number A and any positive number B, $\sqrt{A}\sqrt{B} = \sqrt{AB}$. [14.2a]

y-intercepts
product
quotient
perpendicular
square root
slope
principal square root
slope–intercept
vertical
solution
direct
positive
inverse
nonvertical
graphs

SYNTHESIS

Solve.

61. $\sqrt{5x^2 + 5} = 5$

62. $\sqrt{x} = -x$

63. $4 + \sqrt{19 - x} = 6 + \sqrt{4 - x}$

64. $x = (x - 2)\sqrt{x}$

65. $\sqrt{x + 3} = \frac{8}{\sqrt{x - 9}}$

66. $\frac{12}{\sqrt{5x + 6}} = \sqrt{2x + 5}$

67.–70. Use a graphing calculator to check your answers to Exercises 11–14.

14.6 APPLICATIONS WITH RIGHT TRIANGLES

Objectives

a Given the lengths of any two sides of a right triangle, find the length of the third side.

b Solve applied problems involving right triangles.

a Right Triangles

A **right triangle** is a triangle with a 90° angle, as shown in the figure below. The small square in the corner indicates the 90° angle.

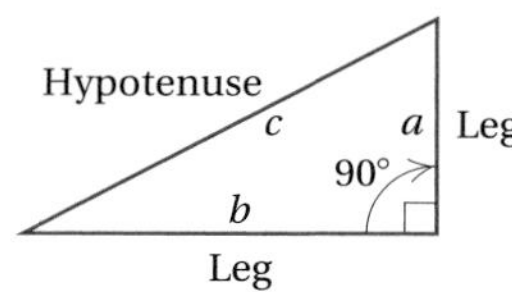

In a right triangle, the longest side is called the **hypotenuse.** It is also the side opposite the right angle. The other two sides are called **legs.** We generally use the letters a and b for the lengths of the legs and c for the length of the hypotenuse. They are related as follows.

THE PYTHAGOREAN THEOREM

In any right triangle, if a and b are the lengths of the legs and c is the length of the hypotenuse, then

$$a^2 + b^2 = c^2.$$

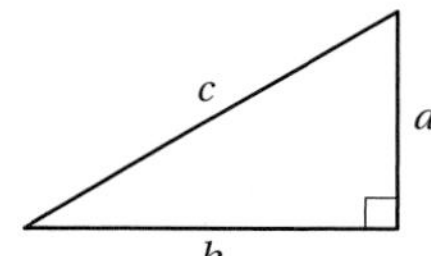

The equation $a^2 + b^2 = c^2$ is called the **Pythagorean equation.**

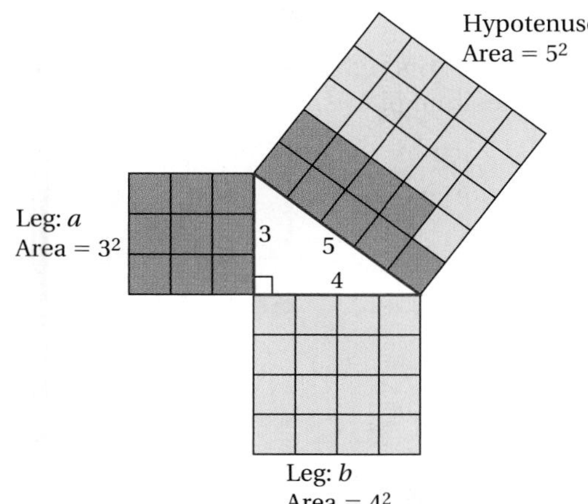

$$a^2 + b^2 = c^2$$
$$3^2 + 4^2 = 5^2$$
$$9 + 16 = 25$$

The Pythagorean theorem is named after the ancient Greek mathematician Pythagoras (569?–500? B.C.). It is uncertain who actually proved this result the first time. A proof can be found in most geometry books.

If we know the lengths of any two sides of a right triangle, we can find the length of the third side.

1. Find the length of the hypotenuse of this right triangle. Give an exact answer and an approximation to three decimal places.

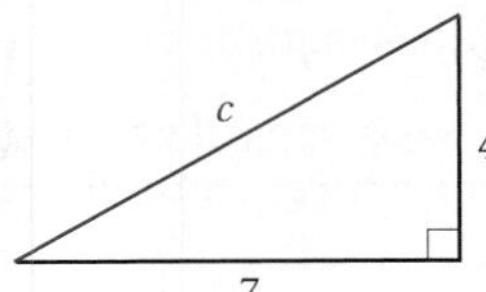

2. Find the length of the leg of this right triangle. Give an exact answer and an approximation to three decimal places.

14
11
a

EXAMPLE 1 Find the length of the hypotenuse of this right triangle. Give an exact answer and an approximation to three decimal places.

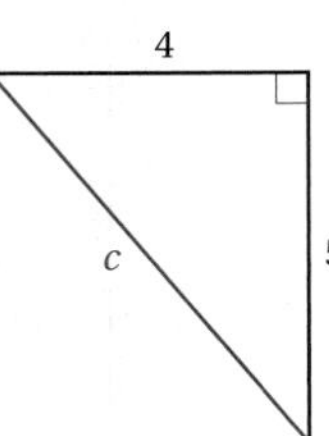

$$4^2 + 5^2 = c^2 \quad \text{Substituting in the Pythagorean equation}$$
$$16 + 25 = c^2$$
$$41 = c^2$$
$$c = \sqrt{41}$$
$$\approx 6.403 \quad \text{Using a calculator}$$

EXAMPLE 2 Find the length of the leg of this right triangle. Give an exact answer and an approximation to three decimal places.

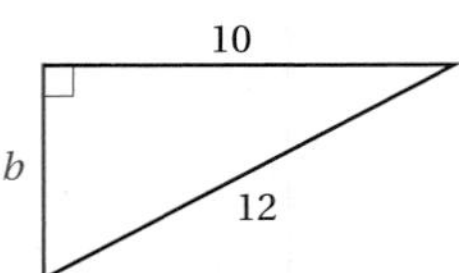

$$10^2 + b^2 = 12^2 \quad \text{Substituting in the Pythagorean equation}$$
$$100 + b^2 = 144$$
$$b^2 = 144 - 100$$
$$b^2 = 44$$
$$b = \sqrt{44}$$
$$\approx 6.633 \quad \text{Using a calculator}$$

Do Exercises 1 and 2.

EXAMPLE 3 Find the length of the leg of this right triangle. Give an exact answer and an approximation to three decimal places.

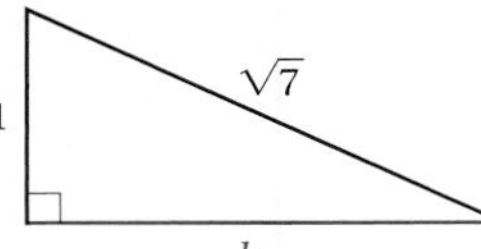

$$1^2 + b^2 = (\sqrt{7})^2 \quad \text{Substituting in the Pythagorean equation}$$
$$1 + b^2 = 7$$
$$b^2 = 7 - 1 = 6$$
$$b = \sqrt{6}$$
$$\approx 2.449 \quad \text{Using a calculator}$$

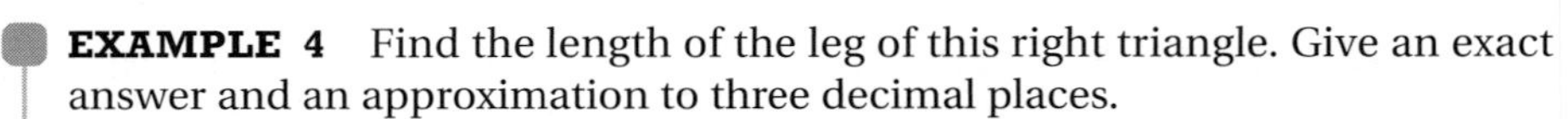

EXAMPLE 4 Find the length of the leg of this right triangle. Give an exact answer and an approximation to three decimal places.

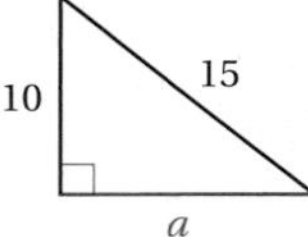

$$a^2 + 10^2 = 15^2$$
$$a^2 + 100 = 225$$
$$a^2 = 225 - 100$$
$$a^2 = 125$$
$$a = \sqrt{125}$$
$$\approx 11.180 \quad \text{Using a calculator}$$

Do Exercises 3 and 4 on the following page.

Answers on page A-50

b Applications

EXAMPLE 5 *Dimensions of a Softball Diamond.* A women's fast-pitch softball diamond is actually a square 60 ft on a side. How far is it from home plate to second base? (This can be helpful information when lining up the bases.) Give an exact answer and an approximation to three decimal places.

1. **Familiarize.** We first make a drawing. We note that the first and second base lines, together with a line from home to second, form a right triangle. We label the unknown distance d.

2. **Translate.** We use the Pythagorean equation:

$$a^2 + b^2 = c^2 \quad \text{Pythagorean equation}$$
$$60^2 + 60^2 = d^2. \quad \text{Substituting 60 for } a \text{ and } b \text{ and replacing } c \text{ with } d$$

3. **Solve.** We solve as follows:

$$60^2 + 60^2 = d^2$$
$$3600 + 3600 = d^2 \quad \text{Squaring}$$
$$7200 = d^2$$
$$\sqrt{7200} = d \quad \text{Exact answer}$$
$$84.853 \approx d. \quad \text{Approximate answer}$$

4. **Check.** We check the calculations using the Pythagorean equation: $60^2 + 60^2 = 7200$ and $(84.853)^2 \approx 7200$. The distance checks.

5. **State.** The distance from home plate to second base is $\sqrt{7200}$ ft, or about 84.853 ft.

Do Exercise 5.

Find the length of the leg of the right triangle. Give an exact answer and an approximation to three decimal places.

3.

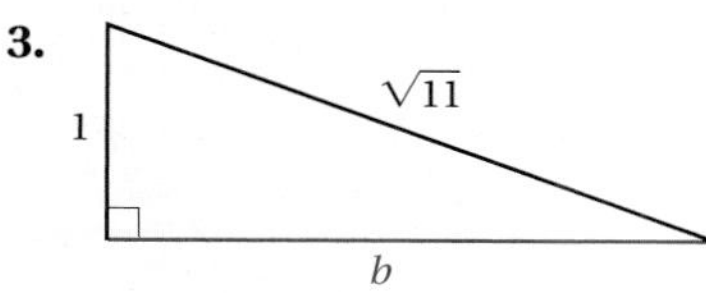

4.

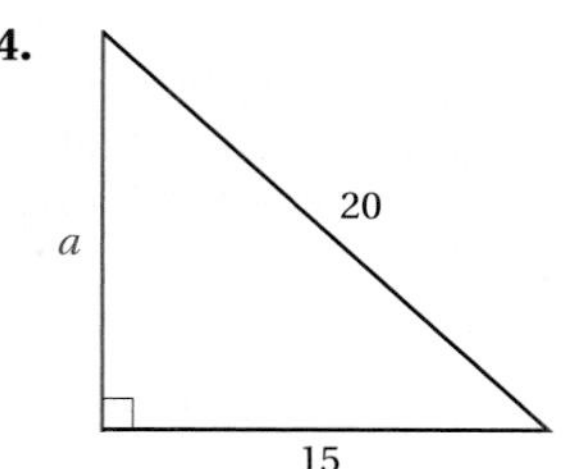

5. **Guy Wire.** How long is a guy wire reaching from the top of a 15-ft pole to a point on the ground 10 ft from the pole? Give an exact answer and an approximation to three decimal places.

Answers on page A-50

Translating for Success

1. *Coin Mixture.* A collection of nickels and quarters is worth \$9.35. There are 59 coins in all. How many of each coin are there?

2. *Diagonal of a Square.* Find the length of a diagonal of a square whose sides are 8 ft long.

3. *Shoveling Time.* It takes Mark 55 min to shovel 4 in. of snow from his driveway. It takes Eric 75 min to do the same job. How long would it take if they worked together?

4. *Angles of a Triangle.* The second angle of a triangle is three times as large as the first. The third is 17° less than the sum of the other angles. Find the measures of the angles.

5. *Perimeter.* The perimeter of a rectangle is 568 ft. The length is 26 ft greater than the width. Find the length and the width.

The goal of these matching questions is to practice step (2), *Translate,* of the five-step problem-solving process. Translate each word problem to an equation or a system of equations and select a correct translation from equations A–O.

A. $5x + 25y = 9.35,$
$x + y = 59$

B. $4^2 + x^2 = 8^2$

C. $x(x + 26) = 568$

D. $8 = x \cdot 24$

E. $\frac{75}{x} = \frac{105}{x + 5}$

F. $\frac{75}{x} = \frac{55}{x + 5}$

G. $2x + 2(x + 26) = 568$

H. $x + 3x + (x + 3x - 17) = 180$

I. $x + 3x + (3x - 17) = 180$

J. $0.05x + 0.25y = 9.35,$
$x + y = 59$

K. $8^2 + 8^2 = x^2$

L. $x^2 + (x + 26)^2 = 568$

M. $\frac{1}{4} + \frac{1}{x} = \frac{1}{55}$

N. $\frac{1}{55} + \frac{1}{75} = \frac{1}{x}$

O. $x + 5\% \cdot x = 8568$

Answers on page A-50

6. *Car Travel.* One horse travels 75 km in the same time that a horse traveling 5 km/h faster travels 105 km. Find the speed of each horse.

7. *Money Borrowed.* Emma borrows some money at 5% simple interest. After 1 yr, \$8568 pays off her loan. How much did she originally borrow?

8. *TV Time.* The average amount of time per day that TV sets in the United States are turned on is 8 hr. What percent of the time are our TV sets on?

Source: Nielsen Media Research

9. *Ladder Height.* An 8-ft plank is leaning against a shed. The bottom of the plank is 4 ft from the building. How high is the top of the plank?

10. *Lengths of a Rectangle.* The area of a rectangle is 568 ft^2. The length is 26 ft greater than the width. Find the length and the width.

14.6 EXERCISE SET

For Extra Help

MathXL | MyMathLab | InterAct Math | Math Tutor Center | Video Lectures on CD Disc 7 | Student's Solutions Manual

a Find the length of the third side of the right triangle. Where appropriate, give both an exact answer and an approximation to three decimal places.

1.

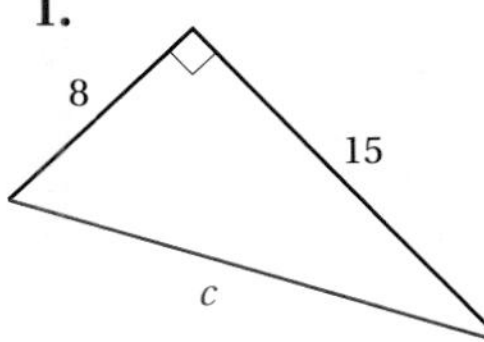

2.

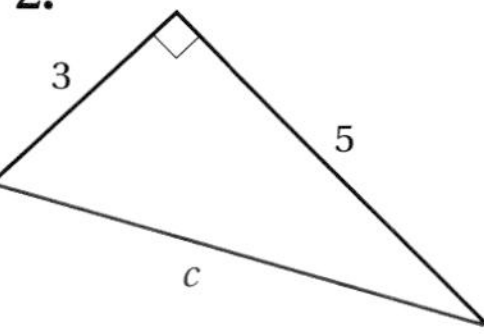

3.

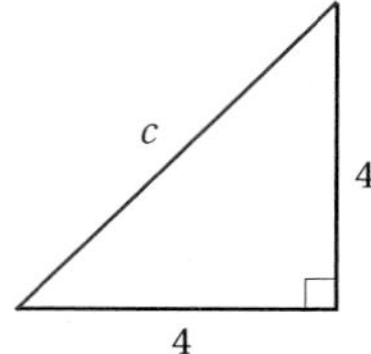

4.

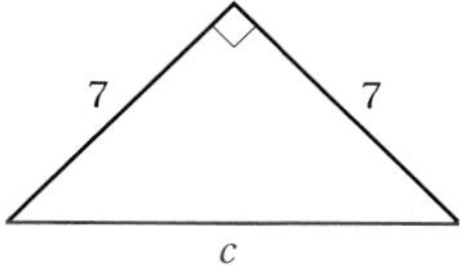

5.

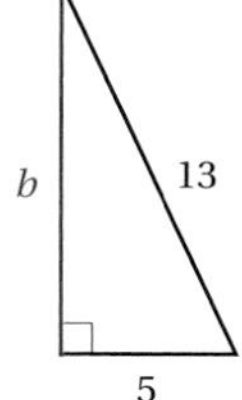

6.

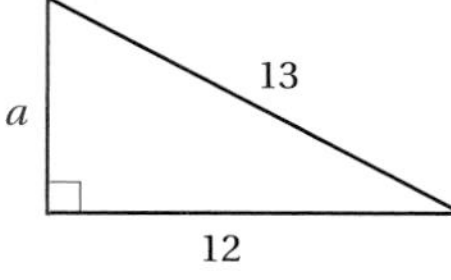

7.

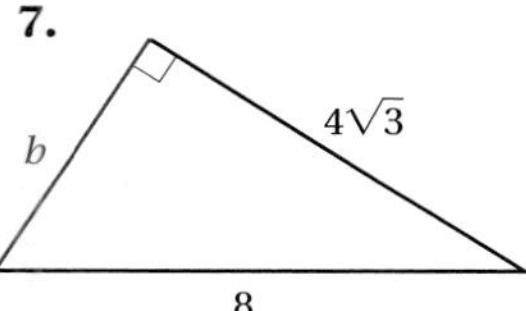

8.

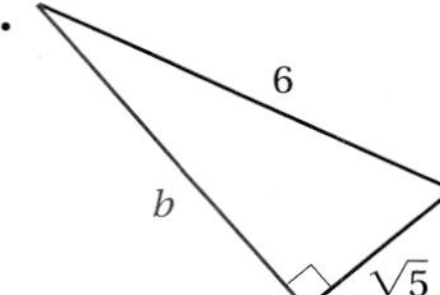

In a right triangle, find the length of the side not given. Where appropriate, give both an exact answer and an approximation to three decimal places.

9. $a = 10, \quad b = 24$

10. $a = 5, \quad b = 12$

11. $a = 9, \quad c = 15$

12. $a = 18, \quad c = 30$

13. $b = 1, \quad c = \sqrt{5}$

14. $b = 1, \quad c = \sqrt{2}$

15. $a = 1, \quad c = \sqrt{3}$

16. $a = \sqrt{3}, \quad b = \sqrt{5}$

17. $c = 10, \quad b = 5\sqrt{3}$

18. $a = 5, \quad b = 5$

19. $a = \sqrt{2}, \quad b = \sqrt{7}$

20. $c = \sqrt{7}, \quad a = \sqrt{2}$

b Solve. Don't forget to make a drawing. Give an exact answer and an approximation to three decimal places.

21. *Airport Distance.* An airplane is flying at an altitude of 4100 ft. The slanted distance directly to the airport is 15,100 ft. How far is the airplane horizontally from the airport?

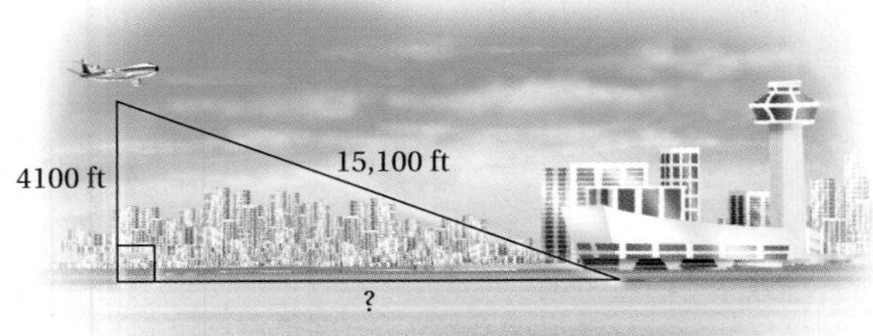

22. *Surveying Distance.* A surveyor had poles located at points P, Q, and R. The distances that the surveyor was able to measure are marked on the drawing. What is the approximate distance from P to R?

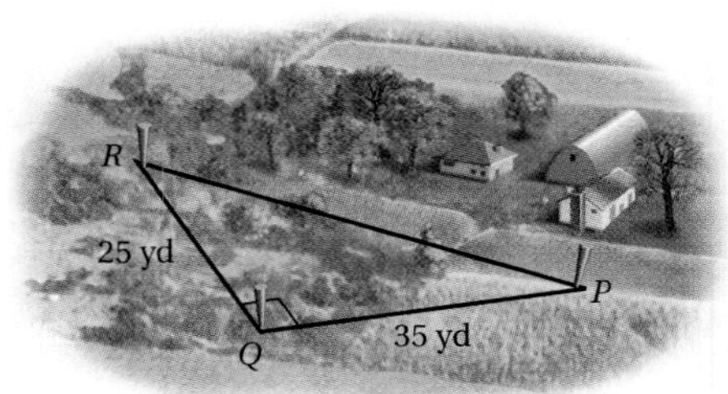

23. *Cordless Telephones.* Becky's new cordless telephone has clear reception up to 300 ft from its base. Her phone is located near a window in her apartment, 180 ft above ground level. How far into her backyard can Becky use her phone?

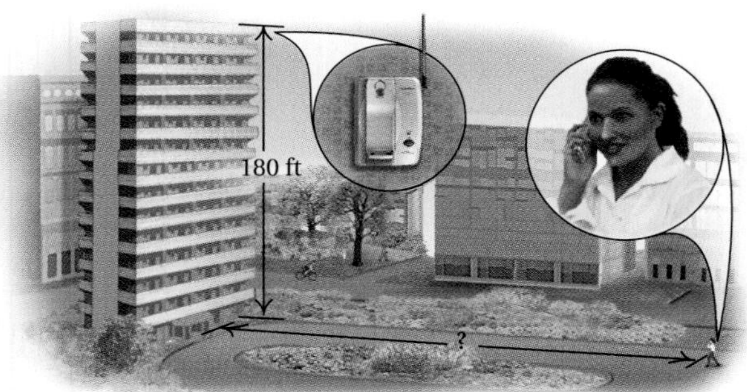

24. *Rope Course.* An outdoor rope course consists of a cable that slopes downward from a height of 37 ft to a resting place 30 ft above the ground. The trees that the cable connects are 24 ft apart. How long is the cable?

25. *Diagonal of a Square.* Find the length of a diagonal of a square whose sides are 3 cm long.

26. *Ladder Height.* A 10-m ladder is leaning against a building. The bottom of the ladder is 5 m from the building. How high is the top of the ladder?

27. *Guy Wire.* How long is a guy wire reaching from the top of a 12-ft pole to a point on the ground 8 ft from the base of the pole?

28. *Diagonal of a Soccer Field.* The largest regulation soccer field is 100 yd wide and 130 yd long. Find the length of a diagonal of such a field.

29. **DW** Can a carpenter use a 28-ft ladder to repair clapboard that is 28 ft above ground level? Why or why not?

30. **DW** In an **equilateral triangle,** all sides have the same length. Can a right triangle ever be equilateral? Why or why not?

SKILL MAINTENANCE

Solve. [13.3a, b]

31. $5x + 7 = 8y,$
$3x = 8y - 4$

32. $5x + y = 17,$
$-5x + 2y = 10$

33. $3x - 4y = -11,$
$5x + 6y = 12$

34. $x + y = -9,$
$x - y = -11$

35. Find the slope of the line $4 - x = 3y$. [9.4a]

36. Find the slope of the line containing the points $(8, -3)$ and $(0, -8)$. [9.3a]

SYNTHESIS

37. Find x.

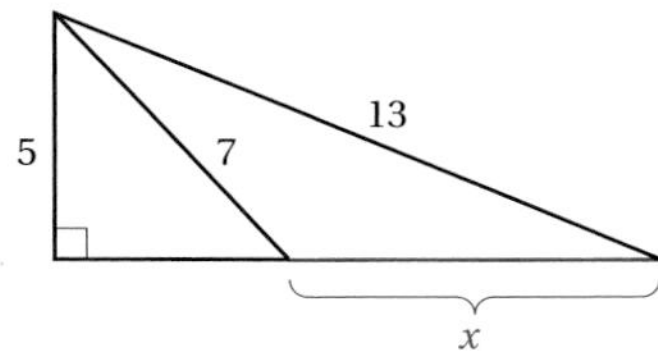

38. *Cordless Telephones.* Virginia's AT&T 9002 cordless phone has a range of 1000 ft. Her apartment is a corner unit, located as shown in the figure below. Will Virginia be able to use the phone at the community pool?

Source: AT&T

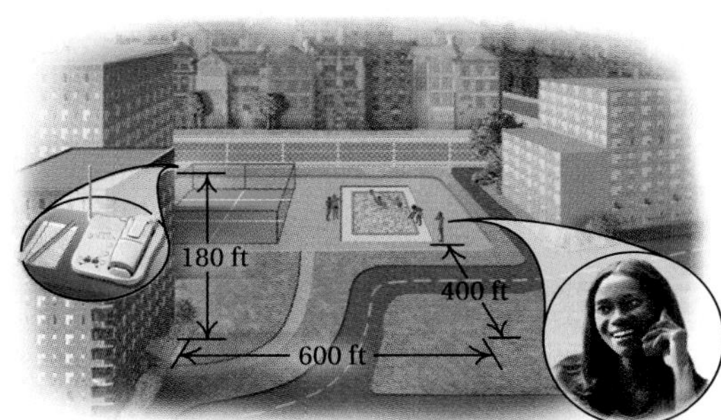

14 Summary and Review

The review that follows is meant to prepare you for a chapter exam. It consists of three parts. The first part, Concept Reinforcement, is designed to increase understanding of the concepts through true/false exercises. The second part is a list of important properties and formulas. The third part is the Review Exercises. These provide practice exercises for the exam, together with references to section objectives so you can go back and review. Before beginning, stop and look back over the skills you have obtained. What skills in mathematics do you have now that you did not have before studying this chapter?

CONCEPT REINFORCEMENT

Determine whether the statement is true or false. Answers are given at the back of the book.

_______ **1.** The radical symbol $\sqrt{}$ represents only the principal square root.

_______ **2.** The square root of a sum is not the sum of the square roots.

_______ **3.** If an equation $a = b$ is true, then the equation $a^2 = b^2$ is true.

_______ **4.** If an equation $a^2 = b^2$ is true, then the equation $a = b$ is true.

_______ **5.** For any nonnegative real number A, the principal square root of A^2 is $-A$.

_______ **6.** Every nonnegative number has two square roots.

_______ **7.** There are no real numbers that when squared yield negative numbers.

_______ **8.** When both sides of an equation are squared, the new equation may have solutions that the first equation does not.

IMPORTANT PROPERTIES AND FORMULAS

Product Rule for Radicals: $\sqrt{A}\,\sqrt{B} = \sqrt{AB}$

Quotient Rule for Radicals: $\dfrac{\sqrt{A}}{\sqrt{B}} = \sqrt{\dfrac{A}{B}}$

Principle of Squaring: If an equation $a = b$ is true, then the equation $a^2 = b^2$ is true.

Pythagorean Equation: $a^2 + b^2 = c^2$, where a and b are the lengths of the legs of a right triangle and c is the length of the hypotenuse.

Review Exercises

Find the square roots. [14.1a]

1. 64

2. 400

Simplify. [14.1a]

3. $\sqrt{36}$

4. $-\sqrt{169}$

Use a calculator to approximate each of the following square roots to three decimal places. [14.1b]

5. $\sqrt{3}$

6. $\sqrt{99}$

7. $-\sqrt{320.12}$

8. $\sqrt{\dfrac{11}{20}}$

9. $-\sqrt{\dfrac{47.3}{11.2}}$

10. $18\sqrt{11 \cdot 43.7}$

Identify the radicand. [14.1d]

11. $\sqrt{x^2 + 4}$

12. $\sqrt{5ab^3}$

Determine whether the expression represents a real number. Write "yes" or "no." [14.1e]

13. $\sqrt{-22}$

14. $-\sqrt{49}$

15. $\sqrt{-36}$

16. $\sqrt{-10.2}$

17. $-\sqrt{-4}$

18. $\sqrt{2(-3)}$

Simplify. [14.1f]

19. $\sqrt{m^2}$

20. $\sqrt{(x-4)^2}$

Multiply. [14.2c]

21. $\sqrt{3}\,\sqrt{7}$

22. $\sqrt{x-3}\,\sqrt{x+3}$

Simplify by factoring. [14.2a]

23. $-\sqrt{48}$

24. $\sqrt{32t^2}$

25. $\sqrt{t^2-49}$

26. $\sqrt{x^2+16x+64}$

Simplify by factoring. [14.2b]

27. $\sqrt{x^8}$

28. $\sqrt{m^{15}}$

Multiply and simplify. [14.2c]

29. $\sqrt{6}\,\sqrt{10}$

30. $\sqrt{5x}\,\sqrt{8x}$

31. $\sqrt{5x}\,\sqrt{10xy^2}$

32. $\sqrt{20a^3b}\,\sqrt{5a^2b^2}$

Simplify. [14.3b]

33. $\sqrt{\dfrac{25}{64}}$

34. $\sqrt{\dfrac{20}{45}}$

35. $\sqrt{\dfrac{49}{t^2}}$

Rationalize the denominator. [14.3c]

36. $\sqrt{\dfrac{1}{2}}$

37. $\sqrt{\dfrac{1}{8}}$

38. $\sqrt{\dfrac{5}{y}}$

39. $\dfrac{2}{\sqrt{3}}$

Divide and simplify. [14.3a, c]

40. $\dfrac{\sqrt{27}}{\sqrt{45}}$

41. $\dfrac{\sqrt{45x^2y}}{\sqrt{54y}}$

42. Rationalize the denominator: [14.4c]

$$\frac{4}{2+\sqrt{3}}.$$

Simplify. [14.4a]

43. $10\sqrt{5}+3\sqrt{5}$

44. $\sqrt{80}-\sqrt{45}$

45. $3\sqrt{2}-5\sqrt{\dfrac{1}{2}}$

Simplify. [14.4b]

46. $(2+\sqrt{3})^2$

47. $(2+\sqrt{3})(2-\sqrt{3})$

Solve. [14.5a]

48. $\sqrt{x-3}=7$

49. $\sqrt{5x+3}=\sqrt{2x-1}$

50. $1+x=\sqrt{1+5x}$

51. Solve: [14.5b]

$$\sqrt{x}=\sqrt{x-5}+1.$$

In a right triangle, find the length of the side not given. Give an exact answer and an approximation to three decimal places. [14.6a]

52. $a=15, \quad c=25$

53. $a=1, \quad b=\sqrt{2}$

Solve. [14.6b]

54. *Airplane Descent.* A pilot is instructed to descend from 30,000 ft to 20,000 ft over a horizontal distance of 50,000 ft. What distance will the plane travel during this descent?

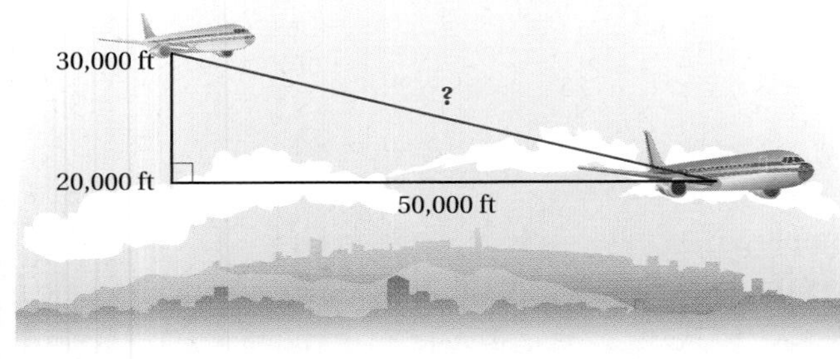

55. *Lookout Tower.* The diagonal braces in a lookout tower are 15 ft long and span a distance of 12 ft. How high does each brace reach vertically?

Solve. [14.1c], [14.5c]

56. *Speed of a Skidding Car.* The formula $r = 2\sqrt{5L}$ can be used to approximate the speed r, in miles per hour, of a car that has left a skid mark of length L, in feet.

a) What was the speed of a car that left skid marks of length 200 ft?

b) How far will a car skid at 90 mph?

57. D_W Explain why the following is incorrect: [14.3b]

$$\sqrt{\frac{9 + 100}{25}} = \frac{3 + 10}{5}.$$

58. D_W Determine whether each of the following is correct for all real numbers. Explain why or why not. [14.2a]

a) $\sqrt{5x^2} = |x|\sqrt{5}$

b) $\sqrt{b^2 - 4} = b - 2$

c) $\sqrt{x^2 + 16} = x + 4$

SYNTHESIS

59. *Distance Driven.* Two cars leave a service station at the same time. One car travels east at a speed of 50 mph, and the other travels south at a speed of 60 mph. After one-half hour, how far apart are they? [14.6b]

60. Simplify: $\sqrt{\sqrt{\sqrt{256}}}$. [14.2a]

61. Solve $A = \sqrt{a^2 + b^2}$ for b. [14.5a]

62. Find x. [14.6a]

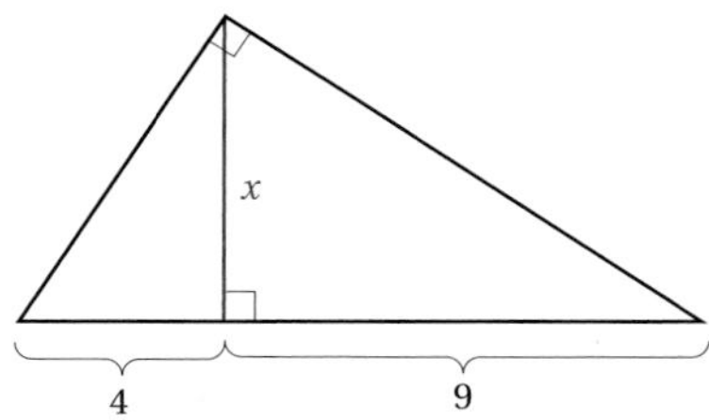

14 Chapter Test

For Extra Help

Work It Out!
Chapter Test Video
Solutions on CD

1. Find the square roots of 81.

Simplify.

2. $\sqrt{64}$

3. $-\sqrt{25}$

Approximate the expression involving square roots to three decimal places.

4. $\sqrt{116}$

5. $-\sqrt{87.4}$

6. $\sqrt{\dfrac{96 \cdot 38}{214.2}}$

7. Identify the radicand in $\sqrt{4 - y^3}$.

Determine whether the expression represents a real number. Write "yes" or "no."

8. $\sqrt{24}$

9. $\sqrt{-23}$

Simplify.

10. $\sqrt{a^2}$

11. $\sqrt{36y^2}$

Multiply.

12. $\sqrt{5}\,\sqrt{6}$

13. $\sqrt{x - 8}\,\sqrt{x + 8}$

Simplify by factoring.

14. $\sqrt{27}$

15. $\sqrt{25x - 25}$

16. $\sqrt{t^5}$

Multiply and simplify.

17. $\sqrt{5}\,\sqrt{10}$

18. $\sqrt{3ab}\,\sqrt{6ab^3}$

Simplify.

19. $\sqrt{\dfrac{27}{12}}$

20. $\sqrt{\dfrac{144}{a^2}}$

Rationalize the denominator.

21. $\sqrt{\dfrac{2}{5}}$

22. $\sqrt{\dfrac{2x}{y}}$

Divide and simplify.

23. $\dfrac{\sqrt{27}}{\sqrt{32}}$

24. $\dfrac{\sqrt{35x}}{\sqrt{80xy^2}}$

Add or subtract.

25. $3\sqrt{18} - 5\sqrt{18}$

26. $\sqrt{5} + \sqrt{\frac{1}{5}}$

Simplify.

27. $(4 - \sqrt{5})^2$

28. $(4 - \sqrt{5})(4 + \sqrt{5})$

29. Rationalize the denominator: $\frac{10}{4 - \sqrt{5}}$.

30. In a right triangle, $a = 8$ and $b = 4$. Find c. Give an exact answer and an approximation to three decimal places.

Solve.

31. $\sqrt{3x} + 2 = 14$

32. $\sqrt{6x + 13} = x + 3$

33. $\sqrt{1 - x} + 1 = \sqrt{6 - x}$

34. *Sighting to the Horizon.* The equation $D = \sqrt{2h}$ can be used to approximate the distance D, in miles, that a person can see to the horizon from a height h, in feet.

a) How far to the horizon can you see through an airplane window at a height of 28,000 ft?

b) Christina can see about 261 mi to the horizon through an airplane window. How high is the airplane?

35. *Lacrosse.* A regulation lacrosse field is 60 yd wide and 110 yd long. Find the length of a diagonal of such a field.

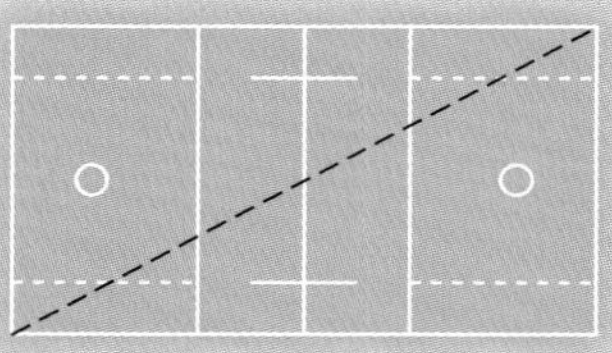

SYNTHESIS

Simplify.

36. $\sqrt{\sqrt{\sqrt{625}}}$

37. $\sqrt{y^{16n}}$

Quadratic Equations

Real-World Application

An airplane flies 1449 mi against the wind and 1539 mi with the wind in a total time of 5 hr. The speed of the airplane in still air is 600 mph. What is the speed of the wind?

This problem appears as Exercise 18 in Section 15.5.

15.1 INTRODUCTION TO QUADRATIC EQUATIONS

Objectives

a Write a quadratic equation in standard form $ax^2 + bx + c = 0$, $a > 0$, and determine the coefficients a, b, and c.

b Solve quadratic equations of the type $ax^2 + bx = 0$, where $b \neq 0$, by factoring.

c Solve quadratic equations of the type $ax^2 + bx + c = 0$, where $b \neq 0$ and $c \neq 0$, by factoring.

d Solve applied problems involving quadratic equations.

AC ALGEBRAIC–GRAPHICAL CONNECTION

Before we begin this chapter, let's look back at some algebraic–graphical equation-solving concepts and their interrelationships. In Chapter 9, we considered the graph of a *linear equation* $y = mx + b$. For example, the graph of the equation $y = \frac{5}{2}x - 4$ and its x-intercept are shown below.

If $y = 0$, then $x = \frac{8}{5}$. Thus the x-intercept is $\left(\frac{8}{5}, 0\right)$. This point is also the intersection of the graphs of $y = \frac{5}{2}x - 4$ and $y = 0$.

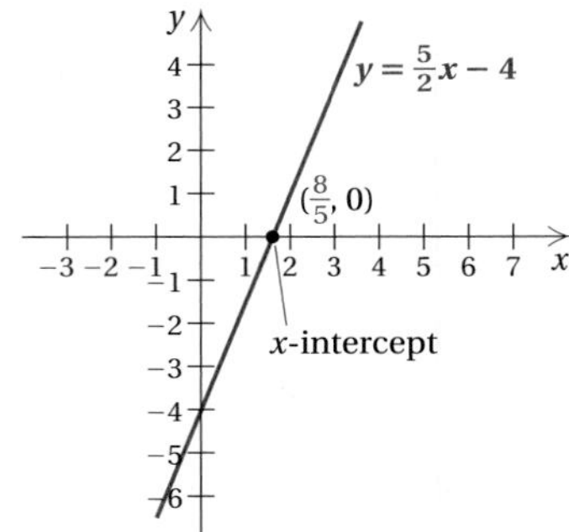

In Chapter 8, we learned how to solve linear equations like $0 = \frac{5}{2}x - 4$ algebraically (using algebra). We proceeded as follows:

$$0 = \frac{5}{2}x - 4$$

$$4 = \frac{5}{2}x \qquad \text{Adding 4}$$

$$8 = 5x \qquad \text{Multiplying by 2}$$

$$\frac{8}{5} = x. \qquad \text{Dividing by 5}$$

We see that $\frac{8}{5}$, the solution of $0 = \frac{5}{2}x - 4$, is the first coordinate of the x-intercept of the graph of $y = \frac{5}{2}x - 4$.

Do Exercise 1.

In this chapter, we build on these ideas by applying them to quadratic equations. In Section 11.7, we briefly considered the graph of a *quadratic equation*

$$y = ax^2 + bx + c, \quad a \neq 0.$$

For example, the graph of the equation $y = x^2 + 6x + 8$ and its x-intercepts are shown below.

The x-intercepts are $(-4, 0)$ and $(-2, 0)$. We will develop in detail the creation of such graphs in Section 15.6. The points $(-4, 0)$ and $(-2, 0)$ are the intersections of the graphs of $y = x^2 + 6x + 8$ and $y = 0$.

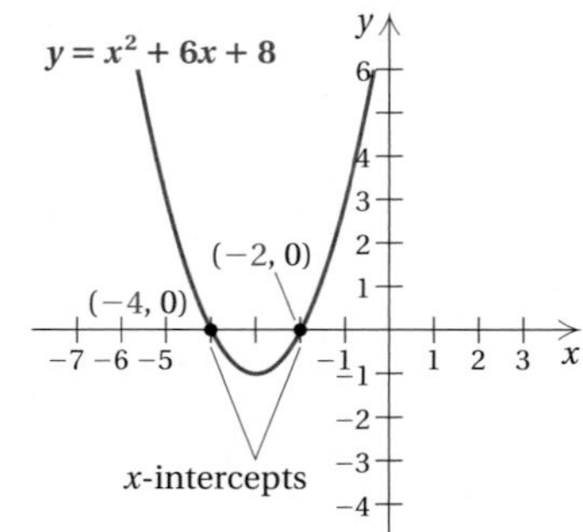

1. a) Consider the linear equation $y = -\frac{2}{3}x - 3$.Find the intercepts and graph the equation.

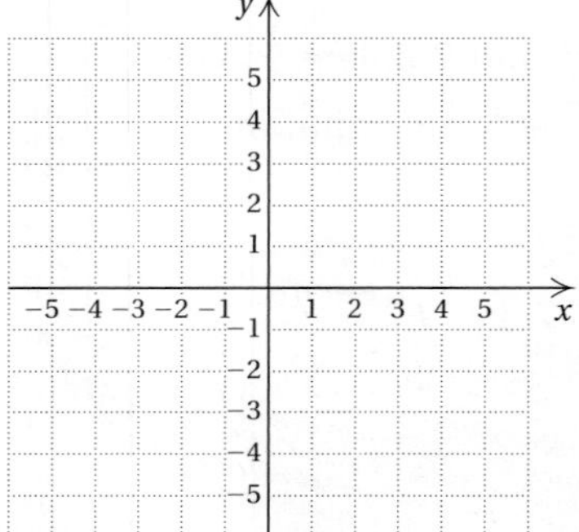

b) Solve the equation

$$0 = -\frac{2}{3}x - 3.$$

c) Complete: The solution of the equation $0 = -\frac{2}{3}x - 3$ is ________. This value is the ________ of the x-intercept, (____, ____), of the graph of $y = -\frac{2}{3}x - 3$.

Answers on page A-50

We began studying the solution of quadratic equations like

$$x^2 + 6x + 8 = 0$$

in Section 11.7. There we used factoring for such solutions:

$x^2 + 6x + 8 = 0$

$(x + 4)(x + 2) = 0$ Factoring

$x + 4 = 0$ *or* $x + 2 = 0$ Using the principle of zero products

$x = -4$ *or* $x = -2.$

We see that the solutions of $x^2 + 6x + 8 = 0$, -4 and -2, are the first coordinates of the x-intercepts, $(-4, 0)$ and $(-2, 0)$, of the graph of $y = x^2 + 6x + 8$.

Do Exercise 2.

We will enhance our ability to solve quadratic equations in Sections 15.1–15.3.

a Standard Form

The following are **quadratic equations.** They contain polynomials of second degree.

$4x^2 + 7x - 5 = 0,$

$3t^2 - \frac{1}{2}t = 9,$

$5y^2 = -6y,$

$5m^2 = 15$

The quadratic equation $4x^2 + 7x - 5 = 0$ is said to be in **standard form.** Although the quadratic equation $4x^2 = 5 - 7x$ is equivalent to the preceding equation, it is *not* in standard form.

QUADRATIC EQUATION

A **quadratic equation** is an equation equivalent to an equation of the type

$$ax^2 + bx + c = 0, \quad a > 0,$$

where a, b, and c are real-number constants. We say that the preceding is the **standard form of a quadratic equation.**

We define $a > 0$ to ease the proof of the quadratic formula, which we consider later, and to ease solving by factoring, which we review in this section. Suppose we are studying an equation like $-3x^2 + 8x - 2 = 0$. It is not in standard form. We can find an equivalent equation that is in standard form by multiplying by -1 on both sides:

$-1(-3x^2 + 8x - 2) = -1(0)$

$3x^2 - 8x + 2 = 0.$ Standard form

2. Consider the quadratic equation $y = x^2 - 2x - 3$ and its graph shown below.

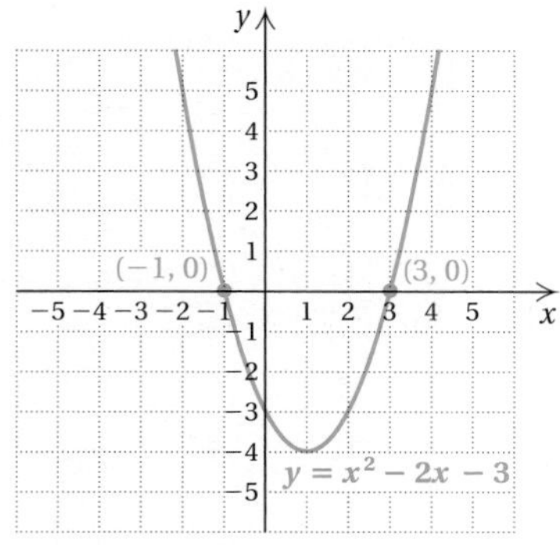

a) Solve the equation

$$x^2 - 2x - 3 = 0.$$

(*Hint*: Use the principle of zero products.)

b) Complete: The solutions of the equation $x^2 - 2x - 3 = 0$ are ____ and ____. These values are the ________ of the x-intercepts, (____, ____) and (____, ____), of the graph of $y = x^2 - 2x - 3$.

Answers on page A-51

Write in standard form and determine a, b, and c.

3. $6x^2 = 3 - 7x$

4. $y^2 = 8y$

5. $3 - x^2 = 9x$

6. $3x + 5x^2 = x^2 - 4 + 2x$

7. $5x^2 = 21$

Solve.

8. $2x^2 + 8x = 0$

9. $10x^2 - 6x = 0$

Answers on page A-51

EXAMPLES Write in standard form and determine a, b, and c.

1. $4x^2 + 7x - 5 = 0$ The equation is already in standard form.

$a = 4;\quad b = 7;\quad c = -5$

2. $3x^2 - 0.5x = 9$

$3x^2 - 0.5x - 9 = 0$ Subtracting 9. This is standard form.

$a = 3;\quad b = -0.5;\quad c = -9$

3. $-4y^2 = 5y$

$-4y^2 - 5y = 0$ Subtracting $5y$

Not positive!

$4y^2 + 5y = 0$ Multiplying by -1. This is standard form.

$a = 4;\quad b = 5;\quad c = 0$

Do Exercises 3–7.

b Solving Quadratic Equations of the Type $ax^2 + bx = 0$

Sometimes we can use factoring and the principle of zero products to solve quadratic equations. We are actually reviewing methods that we introduced in Section 11.7.

When $c = 0$ and $b \neq 0$, we can always factor and use the principle of zero products (see Section 11.7 for a review).

EXAMPLE 4 Solve: $7x^2 + 2x = 0$.

$$7x^2 + 2x = 0$$

$$x(7x + 2) = 0 \quad \text{Factoring}$$

$$x = 0 \quad or \quad 7x + 2 = 0 \quad \text{Using the principle of zero products}$$

$$x = 0 \quad or \quad 7x = -2$$

$$x = 0 \quad or \quad x = -\tfrac{2}{7}$$

Check: For 0:

$$\begin{array}{c|c} 7x^2 + 2x & 0 \\ \hline 7 \cdot 0^2 + 2 \cdot 0 & ?\ 0 \\ 0 & \end{array} \quad \text{TRUE}$$

For $-\frac{2}{7}$:

$$\begin{array}{c|c} 7x^2 + 2x & 0 \\ \hline 7(-\frac{2}{7})^2 + 2(-\frac{2}{7}) & ?\ 0 \\ 7(\frac{4}{49}) - \frac{4}{7} & \\ \frac{4}{7} - \frac{4}{7} & \\ 0 & \end{array} \quad \text{TRUE}$$

The solutions are 0 and $-\frac{2}{7}$.

Caution!

You may be tempted to divide each term in an equation like the one in Example 4 by x. This method would yield the equation

$$7x + 2 = 0,$$

whose only solution is $-\frac{2}{7}$. In effect, since 0 is also a solution of the original equation, we have divided by 0. The error of such division means the loss of one of the solutions.

EXAMPLE 5 Solve: $4x^2 - 8x = 0$.

We have

$$4x^2 - 8x = 0$$
$$4x(x - 2) = 0 \quad \text{Factoring}$$
$$4x = 0 \quad or \quad x - 2 = 0 \quad \text{Using the principle of zero products}$$
$$x = 0 \quad or \quad x = 2.$$

The solutions are 0 and 2.

A quadratic equation of the type $ax^2 + bx = 0$, where $c = 0$ and $b \neq 0$, will always have 0 as one solution and a nonzero number as the other solution.

Do Exercises 8 and 9 on the preceding page.

AG ***ALGEBRAIC–GRAPHICAL CONNECTION***

Let's visualize the solutions in Example 5.

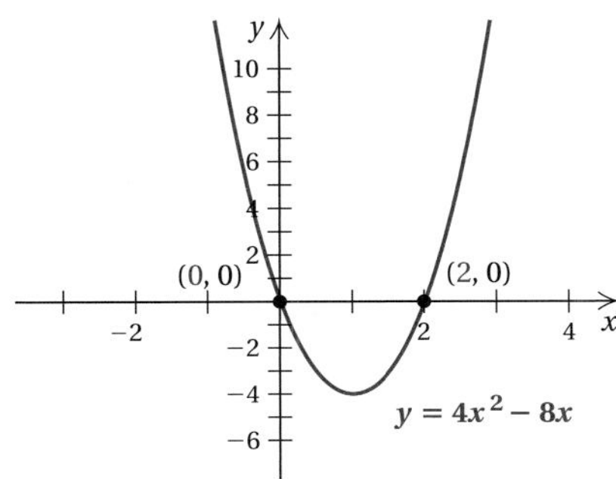

We see that the solutions of $4x^2 - 8x = 0$, 0 and 2, are the first coordinates of the x-intercepts, $(0, 0)$ and $(2, 0)$, of the graph of $y = 4x^2 - 8x$.

C Solving Quadratic Equations of the Type $ax^2 + bx + c = 0$

When neither b nor c is 0, we can sometimes solve by factoring.

EXAMPLE 6 Solve: $2x^2 - x - 21 = 0$.

We have

$$2x^2 - x - 21 = 0$$
$$(2x - 7)(x + 3) = 0 \quad \text{Factoring}$$
$$2x - 7 = 0 \quad or \quad x + 3 = 0 \quad \text{Using the principle of zero products}$$
$$2x = 7 \quad or \quad x = -3$$
$$x = \tfrac{7}{2} \quad or \quad x = -3.$$

The solutions are $\frac{7}{2}$ and -3.

EXAMPLE 7 Solve: $(y - 3)(y - 2) = 6(y - 3)$.

We write the equation in standard form and then try to factor:

$$y^2 - 5y + 6 = 6y - 18 \quad \text{Multiplying}$$
$$y^2 - 11y + 24 = 0 \quad \text{Standard form}$$
$$(y - 8)(y - 3) = 0 \quad \text{Factoring}$$
$$y - 8 = 0 \quad or \quad y - 3 = 0 \quad \text{Using the principle of zero products}$$
$$y = 8 \quad or \quad y = 3.$$

The solutions are 8 and 3.

Do Exercises 10 and 11 on the following page.

AG ***ALGEBRAIC–GRAPHICAL CONNECTION***

Let's visualize the solutions in Example 6.

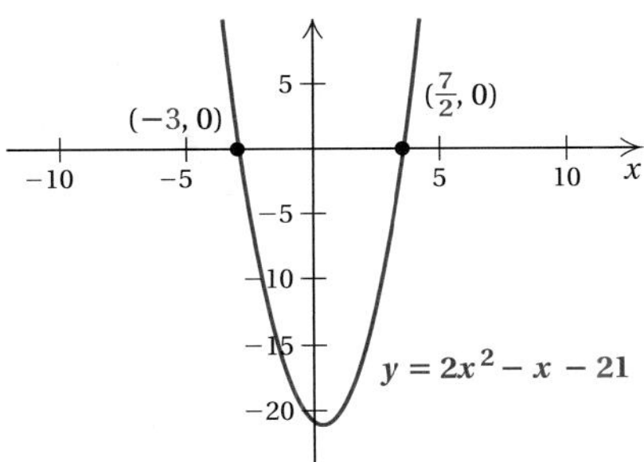

We see that the solutions of $2x^2 - x - 21 = 0$, -3 and $\frac{7}{2}$, are the first coordinates of the x-intercepts, $(-3, 0)$ and $\left(\frac{7}{2}, 0\right)$, of the graph of $y = 2x^2 - x - 21$.

Solve.

10. $4x^2 + 5x - 6 = 0$

11. $(x - 1)(x + 1) = 5(x - 1)$

12. Solve:

$$\frac{20}{x + 5} - \frac{1}{x - 4} = 1.$$

Answers on page A-51

Recall that to solve a rational equation, we multiply both sides by the LCM of all the denominators. We may obtain a quadratic equation after a few steps. When that happens, we know how to finish solving, but we must remember to check possible solutions because a replacement may result in division by 0. See Section 12.6.

EXAMPLE 8 Solve: $\dfrac{3}{x - 1} + \dfrac{5}{x + 1} = 2.$

We multiply by the LCM, which is $(x - 1)(x + 1)$:

$$(x - 1)(x + 1) \cdot \left(\frac{3}{x - 1} + \frac{5}{x + 1}\right) = 2 \cdot (x - 1)(x + 1).$$

We use the distributive law on the left:

$$(x - 1)(x + 1) \cdot \frac{3}{x - 1} + (x - 1)(x + 1) \cdot \frac{5}{x + 1} = 2(x - 1)(x + 1)$$

$$3(x + 1) + 5(x - 1) = 2(x - 1)(x + 1)$$

$$3x + 3 + 5x - 5 = 2(x^2 - 1)$$

$$8x - 2 = 2x^2 - 2$$

$$0 = 2x^2 - 8x$$

$$0 = 2x(x - 4) \quad \text{Factoring}$$

$$2x = 0 \quad or \quad x - 4 = 0$$

$$x = 0 \quad or \quad x = 4.$$

Check: For 0:

$$\frac{3}{x - 1} + \frac{5}{x + 1} = 2$$

$$\frac{3}{0 - 1} + \frac{5}{0 + 1} \stackrel{?}{=} 2$$

$$\frac{3}{-1} + \frac{5}{1}$$

$$-3 + 5$$

$$2 \quad \text{TRUE}$$

For 4:

$$\frac{3}{x - 1} + \frac{5}{x + 1} = 2$$

$$\frac{3}{4 - 1} + \frac{5}{4 + 1} \stackrel{?}{=} 2$$

$$\frac{3}{3} + \frac{5}{5}$$

$$1 + 1$$

$$2 \quad \text{TRUE}$$

The solutions are 0 and 4.

Do Exercise 12.

d Solving Applied Problems

EXAMPLE 9 *Diagonals of a Polygon.* The number of diagonals d of a polygon of n sides is given by the formula

$$d = \frac{n^2 - 3n}{2}.$$

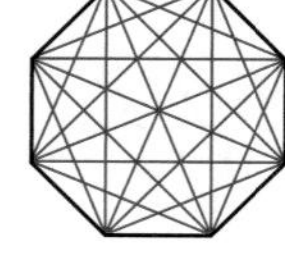

If a polygon has 27 diagonals, how many sides does it have?

1. **Familiarize.** We can make a drawing to familiarize ourselves with the problem. We draw an octagon (8 sides) and count the diagonals and see that there are 20. Let's check this in the formula. We evaluate the formula

for $n = 8$:

$$d = \frac{8^2 - 3(8)}{2} = \frac{64 - 24}{2} = \frac{40}{2} = 20.$$

2. **Translate.** We know that the number of diagonals is 27. We substitute 27 for d:

$$27 = \frac{n^2 - 3n}{2}.$$

3. **Solve.** We solve the equation for n, reversing the equation first for convenience:

$$\frac{n^2 - 3n}{2} = 27$$

$$n^2 - 3n = 54 \quad \text{Multiplying by 2 to clear fractions}$$

$$n^2 - 3n - 54 = 0 \quad \text{Subtracting 54}$$

$$(n - 9)(n + 6) = 0 \quad \text{Factoring}$$

$$n - 9 = 0 \quad or \quad n + 6 = 0$$

$$n = 9 \quad or \quad n = -6.$$

4. **Check.** Since the number of sides cannot be negative, -6 cannot be a solution. We leave it to the student to show by substitution that 9 checks.

5. **State.** The polygon has 9 sides (it is a nonagon).

Do Exercise 13 on the following page.

Study Tips

READ FOR SUCCESS

In his article "The Daily Dozen Disciplines for Massive Success in 2001 & Beyond," Jerry Clark comments, "Research has shown that 58% of high school graduates never read another book from cover to cover the rest of their adult lives, that 78% of the population have not been in a bookstore in the last 5 years, and that 97% of the population of the U.S. do not have library cards." Clark then suggests spending at least 15 minutes each day reading an empowering and uplifting book or article. The following books are some suggestions from your author. Their motivating words may empower you in your study of mathematics.

1. *Fish*, by Stephen C. Lundin, Harry Paul, and John Christensen (Hyperion). This was a Wall Street Journal Business Bestseller. Though it has a strange title, it discusses a remarkable way to boost morale and improve results.
2. *True Success: A New Philosophy of Excellence*, by Tom Morris (Grosset/Putnam). Morris was a well-loved philosophy professor at Notre Dame. Students, especially athletes, flocked to his classes.
3. *The Road Less Traveled, Abounding Grace*, by M. Scott Peck. Noted psychiatrist and author of many excellent books, Peck has amazing insights and wisdom about life.
4. *The Weight of Glory, The Great Divorce*, by C. S. Lewis. British author and scholar, noted for his philosophical exposition, Lewis also wrote science-fiction fantasies with moral overtones.

"Far worse than not reading books is not realizing that it matters!"

Jim Rohn, motivational speaker

13. Consider the following heptagon, that is, a polygon with 7 sides.

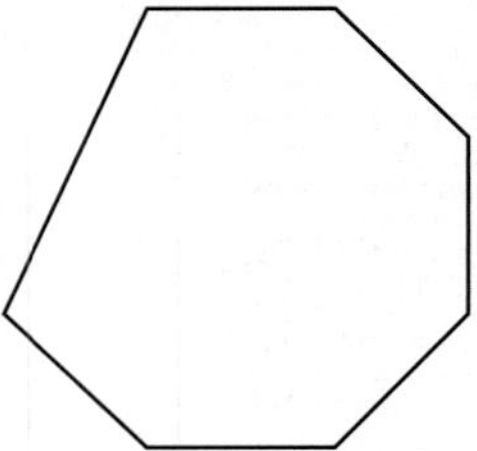

a) Draw all the diagonals and then count them.

b) Use the formula

$$d = \frac{n^2 - 3n}{2}$$

to check your answer to part (a) by evaluating the formula for $n = 7$.

c) A polygon has 44 diagonals. How many sides does it have?

Answers on page A-51

CALCULATOR CORNER

Solving Quadratic Equations A quadratic equation written with 0 on one side of the equals sign can be solved using the ZERO feature of a graphing calculator. (See the Calculator Corner on p. 859 for the procedure.)

We can also use the INTERSECT feature to solve a quadratic equation. Consider the equation in Margin Exercise 11,

$$(x - 1)(x + 1) = 5(x - 1).$$

First, we enter $y_1 = (x - 1)(x + 1)$ and $y_2 = 5(x - 1)$ on the equation-editor screen and graph the equations, using the window $[-5, 5, -5, 20]$, Yscl $= 2$. We see that there are two points of intersection, so the equation has two solutions.

Next, we use the INTERSECT feature to find the coordinates of the lefthand point of intersection. (See the Calculator Corner on p. 972 for the procedure.) The first coordinate of this point, 1, is one solution of the equation. We use the INTERSECT feature again to find the other solution, 4.

$y_1 = (x - 1)(x + 1),\ y_2 = 5(x - 1)$

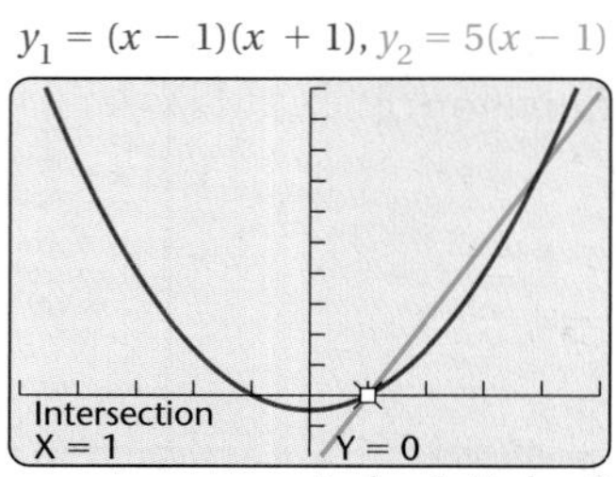

Xscl = 1, Yscl = 2

$y_1 = (x - 1)(x + 1),\ y_2 = 5(x - 1)$

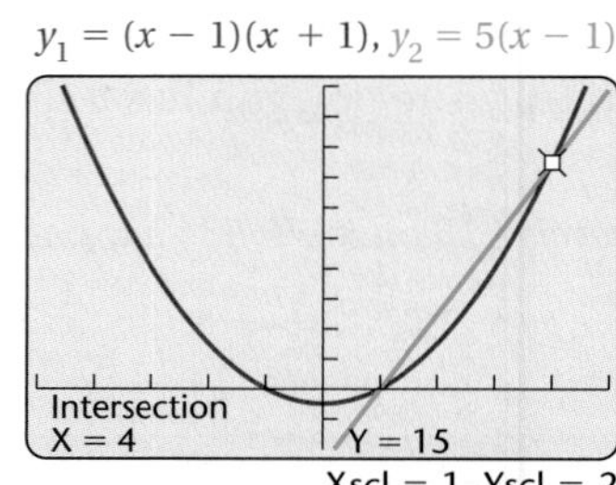

Xscl = 1, Yscl = 2

Note that we could use the ZERO feature to solve this equation if we first write it with 0 on one side, that is, $(x - 1)(x + 1) - 5(x - 1) = 0$.

Exercises: Solve.

1. $5x^2 - 8x + 3 = 0$
2. $2x^2 - 7x - 15 = 0$
3. $6(x - 3) = (x - 3)(x - 2)$
4. $(x + 1)(x - 4) = 3(x - 4)$

15.1 EXERCISE SET

For Extra Help

MathXL | MyMathLab | InterAct Math | Math Tutor Center | Video Lectures on CD Disc 8 | Student's Solutions Manual

a Write in standard form and determine a, b, and c.

1. $x^2 - 3x + 2 = 0$

2. $x^2 - 8x - 5 = 0$

3. $7x^2 = 4x - 3$

4. $9x^2 = x + 5$

5. $5 = -2x^2 + 3x$

6. $3x - 1 = 5x^2 + 9$

b Solve.

7. $x^2 + 5x = 0$

8. $x^2 + 7x = 0$

9. $3x^2 + 6x = 0$

10. $4x^2 + 8x = 0$

11. $5x^2 = 2x$

12. $11x = 3x^2$

13. $4x^2 + 4x = 0$

14. $8x^2 - 8x = 0$

15. $0 = 10x^2 - 30x$

16. $0 = 10x^2 - 50x$

17. $11x = 55x^2$

18. $33x^2 = -11x$

19. $14t^2 = 3t$

20. $6m = 19m^2$

21. $5y^2 - 3y^2 = 72y + 9y$

22. $63p - 16p^2 = 17p + 58p^2$

c Solve.

23. $x^2 + 8x - 48 = 0$

24. $x^2 - 16x + 48 = 0$

25. $5 + 6x + x^2 = 0$

26. $x^2 + 10 + 11x = 0$

27. $18 = 7p + p^2$

28. $t^2 + 14t = -24$

29. $-15 = -8y + y^2$

30. $q^2 + 14 = 9q$

31. $x^2 + 10x + 25 = 0$

32. $x^2 + 6x + 9 = 0$

33. $r^2 = 8r - 16$

34. $x^2 + 1 = 2x$

35. $6x^2 + x - 2 = 0$

36. $2x^2 - 11x + 15 = 0$

37. $3a^2 = 10a + 8$

38. $15b - 9b^2 = 4$

39. $6x^2 - 4x = 10$

40. $3x^2 - 7x = 20$

41. $2t^2 + 12t = -10$

42. $12w^2 - 5w = 2$

43. $t(t - 5) = 14$

44. $6z^2 + z - 1 = 0$

45. $t(9 + t) = 4(2t + 5)$

46. $3y^2 + 8y = 12y + 15$

47. $16(p - 1) = p(p + 8)$

48. $(2x - 3)(x + 1) = 4(2x - 3)$

49. $(t - 1)(t + 3) = t - 1$

50. $(x - 2)(x + 2) = x + 2$

Solve.

51. $\frac{24}{x-2}+\frac{24}{x+2}=5$

52. $\frac{8}{x+2}+\frac{8}{x-2}=3$

53. $\frac{1}{x}+\frac{1}{x+6}=\frac{1}{4}$

54. $\frac{1}{x}+\frac{1}{x+9}=\frac{1}{20}$

55. $1+\frac{12}{x^2-4}=\frac{3}{x-2}$

56. $\frac{5}{t-3}-\frac{30}{t^2-9}=1$

57. $\frac{r}{r-1}+\frac{2}{r^2-1}=\frac{8}{r+1}$

58. $\frac{x+2}{x^2-2}=\frac{2}{3-x}$

59. $\frac{x-1}{1-x}=-\frac{x+8}{x-8}$

60. $\frac{4-x}{x-4}+\frac{x+3}{x-3}=0$

61. $\frac{5}{y+4}-\frac{3}{y-2}=4$

62. $\frac{2z+11}{2z+8}=\frac{3z-1}{z-1}$

d Solve.

63. *Diagonals.* A decagon is a figure with 10 sides. How many diagonals does a decagon have?

64. *Diagonals.* A hexagon is a figure with 6 sides. How many diagonals does a hexagon have?

65. *Diagonals.* A polygon has 14 diagonals. How many sides does it have?

66. *Diagonals.* A polygon has 9 diagonals. How many sides does it have?

67. $^{\mathbf{D}}\mathbf{W}$ Explain how the graph of $y=(x-2)(x+3)$ is related to the solutions of the equation $(x-2)(x+3)=0$.

68. $^{\mathbf{D}}\mathbf{W}$ Explain how you might go about constructing a quadratic equation whose solutions are -5 and 7.

SKILL MAINTENANCE

Simplify. [14.1a], [14.2a]

69. $\sqrt{64}$

70. $-\sqrt{169}$

71. $\sqrt{8}$

72. $\sqrt{12}$

73. $\sqrt{20}$

74. $\sqrt{88}$

75. $\sqrt{405}$

76. $\sqrt{1020}$

Use a calculator to approximate the square roots. Round to three decimal places. [14.1b]

77. $\sqrt{7}$

78. $\sqrt{23}$

79. $\sqrt{\frac{7}{3}}$

80. $\sqrt{524.77}$

SYNTHESIS

Solve.

81. $4m^2-(m+1)^2=0$

82. $x^2+\sqrt{22}x=0$

83. $\sqrt{5}x^2-x=0$

84. $\sqrt{7}x^2+\sqrt{3}x=0$

Use a graphing calculator to solve the equation.

85. $3x^2-7x=20$

86. $x(x-5)=14$

87. $3x^2+8x=12x+15$

88. $(x-2)(x+2)=x+2$

89. $(x-2)^2+3(x-2)=4$

90. $(x+3)^2=4$

91. $16(x-1)=x(x+8)$

92. $x^2+2.5x+1.5625=9.61$

15.2 SOLVING QUADRATIC EQUATIONS BY COMPLETING THE SQUARE

Objectives

a Solve quadratic equations of the type $ax^2 = p$.

b Solve quadratic equations of the type $(x + c)^2 = d$.

c Solve quadratic equations by completing the square.

d Solve certain problems involving quadratic equations of the type $ax^2 = p$.

a Solving Quadratic Equations of the Type $ax^2 = p$

For equations of the type $ax^2 = p$, we first solve for x^2 and then apply the *principle of square roots,* which states that a positive number has two square roots. The number 0 has one square root, 0.

THE PRINCIPLE OF SQUARE ROOTS

- The equation $x^2 = d$ has two real solutions when $d > 0$. The solutions are $\sqrt{d}$ and $-\sqrt{d}$.
- The equation $x^2 = d$ has no real-number solution when $d < 0$.
- The equation $x^2 = 0$ has 0 as its only solution.

EXAMPLE 1 Solve: $x^2 = 3$.

$$x^2 = 3$$

$x = \sqrt{3}$ *or* $x = -\sqrt{3}$ Using the principle of square roots

Check: For $\sqrt{3}$:

$$x^2 = 3$$
$$(\sqrt{3})^2 \; ? \; 3$$
$$3 \quad \text{TRUE}$$

For $-\sqrt{3}$:

$$x^2 = 3$$
$$(-\sqrt{3})^2 \; ? \; 3$$
$$3 \quad \text{TRUE}$$

The solutions are $\sqrt{3}$ and $-\sqrt{3}$.

Do Exercise 1.

EXAMPLE 2 Solve: $\frac{1}{8}x^2 = 0$.

$\frac{1}{8}x^2 = 0$

$x^2 = 0$ Multiplying by 8

$x = 0$ Using the principle of square roots

The solution is 0.

Do Exercise 2.

EXAMPLE 3 Solve: $-3x^2 + 7 = 0$.

$-3x^2 + 7 = 0$

$-3x^2 = -7$ Subtracting 7

$x^2 = \frac{-7}{-3} = \frac{7}{3}$ Dividing by -3

$x = \sqrt{\frac{7}{3}}$ *or* $x = -\sqrt{\frac{7}{3}}$ Using the principle of square roots

$x = \sqrt{\frac{7}{3} \cdot \frac{3}{3}}$ *or* $x = -\sqrt{\frac{7}{3} \cdot \frac{3}{3}}$ Rationalizing the denominators

$x = \frac{\sqrt{21}}{3}$ *or* $x = -\frac{\sqrt{21}}{3}$

1. Solve: $x^2 = 10$.

2. Solve: $6x^2 = 0$.

Answers on page A-51

3. Solve: $2x^2 - 3 = 0$.

Check: For $\frac{\sqrt{21}}{3}$:

$$\begin{array}{c|l} -3x^2 + 7 = 0 & \\ \hline -3\left(\frac{\sqrt{21}}{3}\right)^2 + 7 \ ? \ 0 & \\ -3 \cdot \frac{21}{9} + 7 & \\ -7 + 7 & \\ 0 & \text{TRUE} \end{array}$$

For $-\frac{\sqrt{21}}{3}$:

$$\begin{array}{c|l} -3x^2 + 7 = 0 & \\ \hline -3\left(-\frac{\sqrt{21}}{3}\right)^2 + 7 \ ? \ 0 & \\ -3 \cdot \frac{21}{9} + 7 & \\ -7 + 7 & \\ 0 & \text{TRUE} \end{array}$$

The solutions are $\frac{\sqrt{21}}{3}$ and $-\frac{\sqrt{21}}{3}$.

Do Exercise 3.

b Solving Quadratic Equations of the Type $(x + c)^2 = d$

Solve.

4. $(x - 3)^2 = 16$

5. $(x + 4)^2 = 11$

In an equation of the type $(x + c)^2 = d$, we have the square of a binomial equal to a constant. We can use the principle of square roots to solve such an equation.

EXAMPLE 4 Solve: $(x - 5)^2 = 9$.

$$(x - 5)^2 = 9$$

$$x - 5 = 3 \quad or \quad x - 5 = -3 \qquad \text{Using the principle of square roots}$$

$$x = 8 \quad or \quad x = 2$$

The solutions are 8 and 2.

EXAMPLE 5 Solve: $(x + 2)^2 = 7$.

$$(x + 2)^2 = 7$$

$$x + 2 = \sqrt{7} \quad or \quad x + 2 = -\sqrt{7} \qquad \text{Using the principle of square roots}$$

$$x = -2 + \sqrt{7} \quad or \quad x = -2 - \sqrt{7}$$

The solutions are $-2 + \sqrt{7}$ and $-2 - \sqrt{7}$, or simply $-2 \pm \sqrt{7}$ (read "-2 plus or minus $\sqrt{7}$").

Do Exercises 4 and 5.

In Examples 4 and 5, the left sides of the equations are squares of binomials. If we can express an equation in such a form, we can proceed as we did in those examples.

EXAMPLE 6 Solve: $x^2 + 8x + 16 = 49$.

$$x^2 + 8x + 16 = 49 \qquad \text{The left side is the square of a binomial; } A^2 + 2AB + B^2 = (A + B)^2.$$

$$(x + 4)^2 = 49$$

$$x + 4 = 7 \quad or \quad x + 4 = -7 \qquad \text{Using the principle of square roots}$$

$$x = 3 \quad or \quad x = -11$$

The solutions are 3 and -11.

Answers on page A-51

Do Exercises 6 and 7.

Solve.

6. $x^2 - 6x + 9 = 64$

7. $x^2 - 2x + 1 = 5$

C Completing the Square

We have seen that a quadratic equation like $(x - 5)^2 = 9$ can be solved by using the principle of square roots. We also noted that an equation like $x^2 + 8x + 16 = 49$ can be solved in the same manner because the expression on the left side is the square of a binomial, $(x + 4)^2$. This second procedure is the basis for a method of solving quadratic equations called **completing the square.** *It can be used to solve any quadratic equation.*

Suppose we have the following quadratic equation:

$$x^2 + 10x = 4.$$

If we could add to both sides of the equation a constant that would make the expression on the left the square of a binomial, we could then solve the equation using the principle of square roots.

How can we determine what to add to $x^2 + 10x$ in order to construct the square of a binomial? We want to find a number a such that the following equation is satisfied:

$$x^2 + 10x + a^2 = (x + a)(x + a) = x^2 + 2ax + a^2.$$

Thus, a is such that $2a = 10$. Solving for a, we get $a = 5$; that is, a is half of the coefficient of x in $x^2 + 10x$. Since $a^2 = \left(\frac{10}{2}\right)^2 = 5^2 = 25$, we add 25 to our original expression:

$x^2 + 10x + 25$ is the square of $x + 5$;

that is,

$$x^2 + 10x + 25 = (x + 5)^2.$$

COMPLETING THE SQUARE

To **complete the square** of an expression like $x^2 + bx$, we take half of the coefficient of x and square it. Then we add that number, which is $(b/2)^2$.

Returning to solve our original equation, we first add 25 on *both* sides to complete the square. Then we solve as follows:

$$x^2 + 10x = 4 \quad \text{Original equation}$$

$$x^2 + 10x + 25 = 4 + 25 \quad \text{Adding 25: } \left(\tfrac{10}{2}\right)^2 = 5^2 = 25$$

$$(x + 5)^2 = 29$$

$$x + 5 = \sqrt{29} \quad or \quad x + 5 = -\sqrt{29} \quad \text{Using the principle of square roots}$$

$$x = -5 + \sqrt{29} \quad or \quad x = -5 - \sqrt{29}.$$

The solutions are $-5 \pm \sqrt{29}$.

We have seen that a quadratic equation $(x + c)^2 = d$ can be solved by using the principle of square roots. Any quadratic equation can be put in this form by completing the square. Then we can solve as before.

Answers on page A-51

Solve.

8. $x^2 - 6x + 8 = 0$

9. $x^2 + 8x - 20 = 0$

10. Solve: $x^2 - 12x + 23 = 0$.

11. Solve: $x^2 - 3x - 10 = 0$.

Answers on page A-51

EXAMPLE 7 Solve: $x^2 + 6x + 8 = 0$.

We have

$$x^2 + 6x + 8 = 0$$
$$x^2 + 6x = -8. \quad \text{Subtracting 8}$$

We take half of 6, $\frac{6}{2} = 3$, and square it, to get 3^2, or 9. Then we add 9 on *both* sides of the equation. This makes the left side the square of a binomial. We have now completed the square.

$$x^2 + 6x + 9 = -8 + 9 \quad \text{Adding 9. The left side is the square of a binomial.}$$
$$(x + 3)^2 = 1$$
$$x + 3 = 1 \quad or \quad x + 3 = -1 \quad \text{Using the principle of square roots}$$
$$x = -2 \quad or \quad x = -4$$

The solutions are -2 and -4.

Do Exercises 8 and 9.

EXAMPLE 8 Solve $x^2 - 4x - 7 = 0$ by completing the square.

We have

$$x^2 - 4x - 7 = 0$$
$$x^2 - 4x = 7 \quad \text{Adding 7}$$
$$x^2 - 4x + 4 = 7 + 4 \quad \text{Adding 4: } \left(\frac{-4}{2}\right)^2 = (-2)^2 = 4$$
$$(x - 2)^2 = 11$$
$$x - 2 = \sqrt{11} \quad or \quad x - 2 = -\sqrt{11} \quad \text{Using the principle of square roots}$$
$$x = 2 + \sqrt{11} \quad or \quad x = 2 - \sqrt{11}.$$

The solutions are $2 \pm \sqrt{11}$.

Do Exercise 10.

Example 7, as well as the following example, can be solved more easily by factoring. We solve them by completing the square only to illustrate that completing the square can be used to solve *any* quadratic equation.

EXAMPLE 9 Solve $x^2 + 3x - 10 = 0$ by completing the square.

We have

$$x^2 + 3x - 10 = 0$$
$$x^2 + 3x = 10$$
$$x^2 + 3x + \frac{9}{4} = 10 + \frac{9}{4} \quad \text{Adding } \frac{9}{4}\text{: } \left(\frac{3}{2}\right)^2 = \frac{9}{4}$$
$$\left(x + \frac{3}{2}\right)^2 = \frac{40}{4} + \frac{9}{4} = \frac{49}{4}$$
$$x + \frac{3}{2} = \frac{7}{2} \quad or \quad x + \frac{3}{2} = -\frac{7}{2} \quad \text{Using the principle of square roots}$$
$$x = \frac{4}{2} \quad or \quad x = -\frac{10}{2}$$
$$x = 2 \quad or \quad x = -5.$$

The solutions are 2 and -5.

Do Exercise 11.

When the coefficient of x^2 is not 1, we can make it 1, as shown in the following example.

12. Solve: $2x^2 + 3x - 3 = 0$.

EXAMPLE 10 Solve $2x^2 = 3x + 1$ by completing the square.

We first obtain standard form. Then we multiply by $\frac{1}{2}$ on both sides to make the x^2-coefficient 1.

$$2x^2 = 3x + 1$$

$$2x^2 - 3x - 1 = 0 \qquad \text{Finding standard form}$$

$$\frac{1}{2}(2x^2 - 3x - 1) = \frac{1}{2} \cdot 0 \qquad \text{Multiplying by } \frac{1}{2} \text{ to make the } x^2\text{-coefficient 1}$$

$$x^2 - \frac{3}{2}x - \frac{1}{2} = 0$$

$$x^2 - \frac{3}{2}x = \frac{1}{2} \qquad \text{Adding } \frac{1}{2}$$

$$x^2 - \frac{3}{2}x + \frac{9}{16} = \frac{1}{2} + \frac{9}{16} \qquad \text{Adding } \frac{9}{16}\text{: } \left[\frac{1}{2}\left(-\frac{3}{2}\right)\right]^2 = \left[-\frac{3}{4}\right]^2 = \frac{9}{16}$$

$$\left(x - \frac{3}{4}\right)^2 = \frac{8}{16} + \frac{9}{16}$$

$$\left(x - \frac{3}{4}\right)^2 = \frac{17}{16}$$

$$x - \frac{3}{4} = \frac{\sqrt{17}}{4} \quad \text{or} \quad x - \frac{3}{4} = -\frac{\sqrt{17}}{4} \qquad \text{Using the principle of square roots}$$

$$x = \frac{3}{4} + \frac{\sqrt{17}}{4} \quad \text{or} \quad x = \frac{3}{4} - \frac{\sqrt{17}}{4}$$

The solutions are $\frac{3 \pm \sqrt{17}}{4}$.

SOLVING BY COMPLETING THE SQUARE

To solve a quadratic equation $ax^2 + bx + c = 0$ by completing the square:

1. If $a \neq 1$, multiply by $1/a$ so that the x^2-coefficient is 1.
2. If the x^2-coefficient is 1, add so that the equation is in the form

$$x^2 + bx = -c, \quad \text{or} \quad x^2 + \frac{b}{a}x = -\frac{c}{a} \text{ if step (1) has been applied.}$$

3. Take half of the x-coefficient and square it. Add the result on both sides of the equation.
4. Express the side with the variables as the square of a binomial.
5. Use the principle of square roots and complete the solution.

Completing the square provides a base for the quadratic formula, which we will discuss in Section 15.3. It also has other uses in later mathematics courses.

Do Exercise 12.

Answer on page A-51

13. Falling Object. The Transco Tower in Houston is 901 ft tall. How long would it take an object to fall to the ground from the top?

Source: *The New York Times Almanac*

Answer on page A-51

d Applications

EXAMPLE 11 *Falling Object.* As of this writing, the CN Tower in Toronto is considered the world's tallest building and free-standing structure. It is about 1815 ft tall. How long would it take an object to fall to the ground from the top?

1. **Familiarize.** If we did not know anything about this problem, we might consider looking up a formula in a mathematics or physics book. A formula that fits this situation is

$$s = 16t^2,$$

where s is the distance, in feet, traveled by a body falling freely from rest in t seconds. This formula is actually an approximation in that it does not account for air resistance. In this problem, we know the distance s to be 1815 ft. We want to determine the time t for the object to reach the ground.

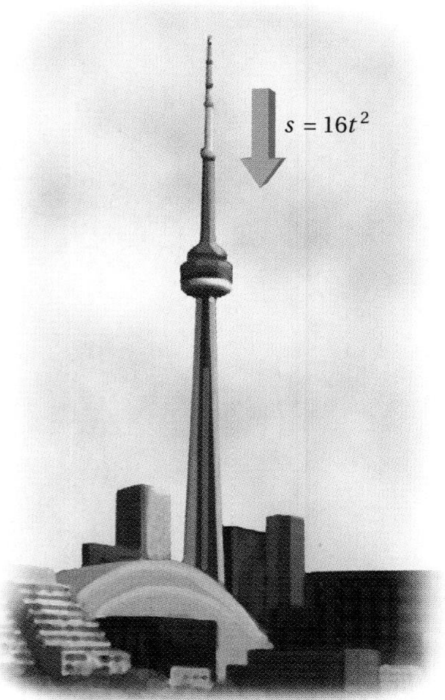

2. **Translate.** We know that the distance is 1815 ft and that we need to solve for t. We substitute 1815 for s: $1815 = 16t^2$. This gives us a translation.

3. **Solve.** We solve the equation:

$$1815 = 16t^2$$

$$\frac{1815}{16} = t^2 \qquad \text{Solving for } t^2$$

$$\sqrt{\frac{1815}{16}} = t \quad or \quad -\sqrt{\frac{1815}{16}} = t \qquad \text{Using the principle of square roots}$$

$$10.7 \approx t \quad or \quad -10.7 \approx t. \qquad \text{Using a calculator to find the square root and rounding to the nearest tenth}$$

4. **Check.** The number -10.7 cannot be a solution because time cannot be negative in this situation. We substitute 10.7 in the original equation:

$$s = 16(10.7)^2 = 16(114.49) = 1831.84.$$

This answer is close: $1831.84 \approx 1815$. Remember that we rounded to approximate our solution, $t \approx 10.7$. Thus we have a check.

5. **State.** It takes about 10.7 sec for an object to fall to the ground from the top of the CN Tower.

Do Exercise 13.

15.2 EXERCISE SET

For Extra Help

MathXL | MyMathLab | InterAct Math | Math Tutor Center | Video Lectures on CD Disc 8 | Student's Solutions Manual

a Solve.

1. $x^2 = 121$

2. $x^2 = 100$

3. $5x^2 = 35$

4. $5x^2 = 45$

5. $5x^2 = 3$

6. $2x^2 = 9$

7. $4x^2 - 25 = 0$

8. $9x^2 - 4 = 0$

9. $3x^2 - 49 = 0$

10. $5x^2 - 16 = 0$

11. $4y^2 - 3 = 9$

12. $36y^2 - 25 = 0$

13. $49y^2 - 64 = 0$

14. $8x^2 - 400 = 0$

b Solve.

15. $(x + 3)^2 = 16$

16. $(x - 4)^2 = 25$

17. $(x + 3)^2 = 21$

18. $(x - 3)^2 = 6$

19. $(x + 13)^2 = 8$

20. $(x - 13)^2 = 64$

21. $(x - 7)^2 = 12$

22. $(x + 1)^2 = 14$

23. $(x + 9)^2 = 34$

24. $(t + 5)^2 = 49$

25. $\left(x + \frac{3}{2}\right)^2 = \frac{7}{2}$

26. $\left(y - \frac{3}{4}\right)^2 = \frac{17}{16}$

27. $x^2 - 6x + 9 = 64$

28. $p^2 - 10p + 25 = 100$

29. $x^2 + 14x + 49 = 64$

30. $t^2 + 8t + 16 = 36$

c Solve by completing the square. Show your work.

31. $x^2 - 6x - 16 = 0$

32. $x^2 + 8x + 15 = 0$

33. $x^2 + 22x + 21 = 0$

34. $x^2 + 14x - 15 = 0$

35. $x^2 - 2x - 5 = 0$

36. $x^2 - 4x - 11 = 0$

37. $x^2 - 22x + 102 = 0$

38. $x^2 - 18x + 74 = 0$

39. $x^2 + 10x - 4 = 0$

40. $x^2 - 10x - 4 = 0$

41. $x^2 - 7x - 2 = 0$

42. $x^2 + 7x - 2 = 0$

43. $x^2 + 3x - 28 = 0$

44. $x^2 - 3x - 28 = 0$

45. $x^2 + \frac{3}{2}x - \frac{1}{2} = 0$

46. $x^2 - \frac{3}{2}x - 2 = 0$

47. $2x^2 + 3x - 17 = 0$

48. $2x^2 - 3x - 1 = 0$

49. $3x^2 + 4x - 1 = 0$

50. $3x^2 - 4x - 3 = 0$

51. $2x^2 = 9x + 5$

52. $2x^2 = 5x + 12$

53. $6x^2 + 11x = 10$

54. $4x^2 + 12x = 7$

Solve.

55. *Petronas Towers.* At a height of 1483 ft, the Petronas Towers in Kuala Lumpur is one of the tallest buildings in the world. How long would it take an object to fall from the top?

Source: *The New York Times Almanac*

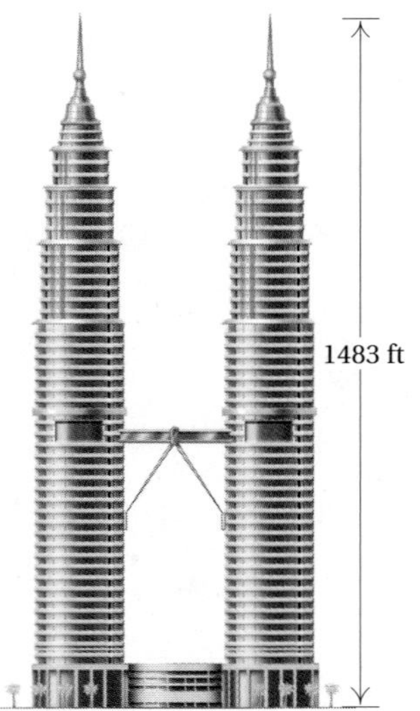

56. *Jin Mao Building.* At a height of 1381 ft, the Jin Mao Building in Shanghai is one of the tallest buildings in the world. How long would it take an object to fall from the top?

Source: *The New York Times Almanac*

57. *Free-Fall Record.* The world record for free-fall to the ground, by a man without a parachute, is 311 ft and is held by Dar Robinson. Approximately how long did the fall take?

Source: *Sports Illustrated*

58. *Free-Fall Record.* The world record for free-fall to the ground, by a woman without a parachute, into a cushioned landing area is 175 ft and is held by Kitty O'Neill. Approximately how long did the fall take?

59. $\mathrm{D_W}$ Corey asserts that the solution of a quadratic equation is $3 \pm \sqrt{14}$ and states that there is only one solution. What mistake is being made?

60. $\mathrm{D_W}$ If a quadratic equation can be solved by factoring, what type of number(s) will generally be solutions?

SKILL MAINTENANCE

VOCABULARY REINFORCEMENT

In each of Exercises 61–68, fill in the blank with the correct term from the given list. Some of the choices may not be used and some may be used more than once.

61. The ______________ rule asserts that when multiplying with exponential notation, if the bases are the same, we keep the base and add the exponents. [10.1d]

62. A(n) ______________ equation is an equation ______________ to an equation of the type $ax^2 + bx + c = 0$, where a, b, and c are real-number constants and $a > 0$. [15.1a]

63. The number -5 is not the ______________ of 25. [14.1a]

64. The number c is a(n) ______________ of a if $c^2 = a$. [14.1a]

65. The ______________ rule asserts that when dividing with exponential notation, if the bases are the same, we keep the base and subtract the exponent of the denominator from the exponent of the numerator. [10.1e]

66. The ______________ rule for radicals asserts that for any nonnegative radicands A and B, $\sqrt{A} \cdot \sqrt{B} = \sqrt{A \cdot B}$. [14.2a]

67. The ______________ rule for radicals asserts that for any nonnegative radicand A and positive number B, $\dfrac{\sqrt{A}}{\sqrt{B}} = \sqrt{\dfrac{A}{B}}$. [14.3a]

68. The ______________ rule asserts that to raise a power to a power, we ______________ the exponents. [10.2a]

square root
slope–intercept
quadratic
principal square root
add
rational
quotient
power
subtract
divide
product
equivalent
solution
multiply

SYNTHESIS

Find b such that the trinomial is a square.

69. $x^2 + bx + 36$

70. $x^2 + bx + 55$

71. $x^2 + bx + 128$

72. $4x^2 + bx + 16$

73. $x^2 + bx + c$

74. $ax^2 + bx + c$

Solve.

75. $4.82x^2 = 12{,}000$

76. $\dfrac{x}{2} = \dfrac{32}{x}$

77. $\dfrac{x}{9} = \dfrac{36}{4x}$

78. $\dfrac{4}{m^2 - 7} = 1$

15.3 THE QUADRATIC FORMULA

Objectives

a Solve quadratic equations using the quadratic formula.

b Find approximate solutions of quadratic equations using a calculator.

We learn to complete the square to prove a general formula that can be used to solve quadratic equations even when they cannot be solved by factoring.

a Solving Using the Quadratic Formula

Each time you solve by completing the square, you perform nearly the same steps. When we repeat the same kind of computation many times, we look for a formula so we can speed up our work. Consider

$$ax^2 + bx + c = 0, \quad a > 0.$$

Let's solve by completing the square. As we carry out the steps, compare them with Example 10 in the preceding section.

$$x^2 + \frac{b}{a}x + \frac{c}{a} = 0 \qquad \text{Multiplying by } \frac{1}{a}$$

$$x^2 + \frac{b}{a}x = -\frac{c}{a} \qquad \text{Adding } -\frac{c}{a}$$

Half of $\frac{b}{a}$ is $\frac{b}{2a}$. The square is $\frac{b^2}{4a^2}$. Thus we add $\frac{b^2}{4a^2}$ on both sides.

$$x^2 + \frac{b}{a}x + \frac{b^2}{4a^2} = -\frac{c}{a} + \frac{b^2}{4a^2} \qquad \text{Adding } \frac{b^2}{4a^2}$$

$$\left(x + \frac{b}{2a}\right)^2 = -\frac{4ac}{4a^2} + \frac{b^2}{4a^2} \qquad \text{Factoring the left side and finding a common denominator on the right}$$

$$\left(x + \frac{b}{2a}\right)^2 = \frac{b^2 - 4ac}{4a^2}$$

$$x + \frac{b}{2a} = \sqrt{\frac{b^2 - 4ac}{4a^2}} \quad or \quad x + \frac{b}{2a} = -\sqrt{\frac{b^2 - 4ac}{4a^2}} \qquad \text{Using the principle of square roots}$$

Since $a > 0$, $\sqrt{4a^2} = 2a$, so we can simplify as follows:

$$x + \frac{b}{2a} = \frac{\sqrt{b^2 - 4ac}}{2a} \quad or \quad x + \frac{b}{2a} = -\frac{\sqrt{b^2 - 4ac}}{2a}.$$

Thus,

$$x = -\frac{b}{2a} + \frac{\sqrt{b^2 - 4ac}}{2a} \quad or \quad x = -\frac{b}{2a} - \frac{\sqrt{b^2 - 4ac}}{2a},$$

so $$x = -\frac{b}{2a} \pm \frac{\sqrt{b^2 - 4ac}}{2a},$$

or $$x = \frac{-b \pm \sqrt{b^2 - 4ac}}{2a}.$$

We now have the following.

THE QUADRATIC FORMULA

The solutions of $ax^2 + bx + c = 0$ are given by

$$x = \frac{-b \pm \sqrt{b^2 - 4ac}}{2a}.$$

The formula also holds when $a < 0$. A similar proof would show this, but we will not consider it here.

EXAMPLE 1 Solve $5x^2 - 8x = -3$ using the quadratic formula.

We first find standard form and determine a, b, and c:

$$5x^2 - 8x + 3 = 0;$$

$$a = 5, \quad b = -8, \quad c = 3.$$

We then use the quadratic formula:

$$x = \frac{-b \pm \sqrt{b^2 - 4ac}}{2a}$$

$$x = \frac{-(-8) \pm \sqrt{(-8)^2 - 4 \cdot 5 \cdot 3}}{2 \cdot 5} \quad \text{Substituting}$$

Caution!

Be sure to write the fraction bar all the way across.

$$x = \frac{8 \pm \sqrt{64 - 60}}{10}$$

$$x = \frac{8 \pm \sqrt{4}}{10}$$

$$x = \frac{8 \pm 2}{10}$$

$$x = \frac{8 + 2}{10} \quad \text{or} \quad x = \frac{8 - 2}{10}$$

$$x = \frac{10}{10} \quad \text{or} \quad x = \frac{6}{10}$$

$$x = 1 \quad \text{or} \quad x = \frac{3}{5}.$$

The solutions are 1 and $\frac{3}{5}$.

Do Exercise 1.

It would have been easier to solve the equation in Example 1 by factoring. We used the quadratic formula only to illustrate that it can be used to solve any quadratic equation. The following is a general procedure for solving a quadratic equation.

SOLVING QUADRATIC EQUATIONS

To solve a quadratic equation:

1. Check to see if it is in the form $ax^2 = p$ or $(x + c)^2 = d$. If it is, use the principle of square roots as in Section 15.2.
2. If it is not in the form of (1), write it in standard form, $ax^2 + bx + c = 0$ with a and b nonzero.
3. Then try factoring.
4. If it is not possible to factor or if factoring seems difficult, use the quadratic formula.

The solutions of a quadratic equation can always be found using the quadratic formula. They cannot always be found by factoring. (When the radicand $b^2 - 4ac \geq 0$, the equation has real-number solutions. When $b^2 - 4ac < 0$, the equation has no real-number solutions.)

1. Solve using the quadratic formula:

$$2x^2 = 4 - 7x.$$

Answer on page A-51

AG ALGEBRAIC–GRAPHICAL CONNECTION

Let's visualize the solutions in Example 2.

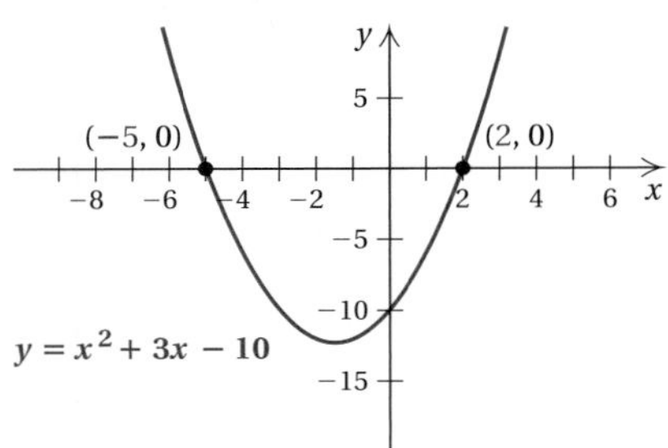

We see that the solutions of $x^2 + 3x - 10 = 0$, -5 and 2, are the first coordinates of the x-intercepts, $(-5, 0)$ and $(2, 0)$, of the graph of $y = x^2 + 3x - 10$.

EXAMPLE 2 Solve $x^2 + 3x - 10 = 0$ using the quadratic formula.

The equation is in standard form. So we determine a, b, and c:

$$x^2 + 3x - 10 = 0;$$
$$a = 1, \quad b = 3, \quad c = -10.$$

We then use the quadratic formula:

$$x = \frac{-b \pm \sqrt{b^2 - 4ac}}{2a}$$
$$= \frac{-3 \pm \sqrt{3^2 - 4 \cdot 1 \cdot (-10)}}{2 \cdot 1} \qquad \text{Substituting}$$
$$= \frac{-3 \pm \sqrt{9 + 40}}{2}$$
$$= \frac{-3 \pm \sqrt{49}}{2} = \frac{-3 \pm 7}{2}.$$

Thus,

$$x = \frac{-3 + 7}{2} = \frac{4}{2} = 2 \quad or \quad x = \frac{-3 - 7}{2} = \frac{-10}{2} = -5.$$

The solutions are 2 and -5.

Note that the radicand ($b^2 - 4ac = 49$) in the quadratic formula is a perfect square, so we could have used factoring to solve.

Do Exercise 2.

2. Solve using the quadratic formula:

$$x^2 - 3x - 10 = 0.$$

EXAMPLE 3 Solve $x^2 = 4x + 7$ using the quadratic formula. Compare using the quadratic formula here with completing the square as we did in Example 8 of Section 15.2.

We first find standard form and determine a, b, and c:

$$x^2 - 4x - 7 = 0;$$
$$a = 1, \quad b = -4, \quad c = -7.$$

We then use the quadratic formula:

$$x = \frac{-b \pm \sqrt{b^2 - 4ac}}{2a} = \frac{-(-4) \pm \sqrt{(-4)^2 - 4 \cdot 1 \cdot (-7)}}{2 \cdot 1} \qquad \text{Substituting}$$
$$= \frac{4 \pm \sqrt{16 + 28}}{2} = \frac{4 \pm \sqrt{44}}{2}$$
$$= \frac{4 \pm \sqrt{4 \cdot 11}}{2} = \frac{4 \pm \sqrt{4}\sqrt{11}}{2}$$
$$= \frac{4 \pm 2\sqrt{11}}{2} = \frac{2 \cdot 2 \pm 2\sqrt{11}}{2 \cdot 1}$$
$$= \frac{2(2 \pm \sqrt{11})}{2 \cdot 1} = \frac{2}{2} \cdot \frac{2 \pm \sqrt{11}}{1} \qquad \text{Factoring out 2 in the numerator and the denominator}$$
$$= 2 \pm \sqrt{11}.$$

The solutions are $2 + \sqrt{11}$ and $2 - \sqrt{11}$, or $2 \pm \sqrt{11}$.

Do Exercise 3.

3. Solve using the quadratic formula:

$$x^2 + 4x = 7.$$

4. Solve using the quadratic formula:

$$x^2 = x - 1.$$

Answers on page A-51

EXAMPLE 4 Solve $x^2 + x = -1$ using the quadratic formula.

We first find standard form and determine a, b, and c:

$$x^2 + x + 1 = 0;$$
$$a = 1, \quad b = 1, \quad c = 1.$$

We then use the quadratic formula:

$$x = \frac{-b \pm \sqrt{b^2 - 4ac}}{2a} = \frac{-1 \pm \sqrt{1^2 - 4 \cdot 1 \cdot 1}}{2 \cdot 1} = \frac{-1 \pm \sqrt{-3}}{2}.$$

Note that the radicand ($b^2 - 4ac = -3$) in the quadratic formula is negative. Thus there are no real-number solutions because square roots of negative numbers do not exist as real numbers.

Do Exercise 4 on the preceding page.

EXAMPLE 5 Solve $3x^2 = 7 - 2x$ using the quadratic formula.

We first find standard form and determine a, b, and c:

$$3x^2 + 2x - 7 = 0;$$
$$a = 3, \quad b = 2, \quad c = -7.$$

We then use the quadratic formula:

$$x = \frac{-b \pm \sqrt{b^2 - 4ac}}{2a} = \frac{-2 \pm \sqrt{2^2 - 4 \cdot 3 \cdot (-7)}}{2 \cdot 3} = \frac{-2 \pm \sqrt{4 + 84}}{2 \cdot 3}$$
$$= \frac{-2 \pm \sqrt{88}}{6} = \frac{-2 \pm \sqrt{4 \cdot 22}}{6} = \frac{-2 \pm \sqrt{4}\sqrt{22}}{6} = \frac{-2 \pm 2\sqrt{22}}{6}$$
$$= \frac{2(-1 \pm \sqrt{22})}{2 \cdot 3} = \frac{2}{2} \cdot \frac{-1 \pm \sqrt{22}}{3} = \frac{-1 \pm \sqrt{22}}{3}.$$

The solutions are $\frac{-1 + \sqrt{22}}{3}$ and $\frac{-1 - \sqrt{22}}{3}$, or $\frac{-1 \pm \sqrt{22}}{3}$.

Do Exercise 5.

b Approximate Solutions

A calculator can be used to approximate solutions.

EXAMPLE 6 Use a calculator to approximate to the nearest tenth the solutions to the equation in Example 5.

Using a calculator, we have

$$\frac{-1 + \sqrt{22}}{3} \approx 1.230138587 \approx 1.2 \text{ to the nearest tenth, and}$$
$$\frac{-1 - \sqrt{22}}{3} \approx -1.896805253 \approx -1.9 \text{ to the nearest tenth.}$$

The approximate solutions are 1.2 and -1.9.

Do Exercise 6 on the following page.

5. Solve using the quadratic formula:

$$5x^2 - 8x = 3.$$

Answer on page A-51

CALCULATOR CORNER

Visualizing Solutions of Quadratic Equations To see that there are no real-number solutions of the equation in Example 4,

$$x^2 + x = -1,$$

we graph $y_1 = x^2 + x$ and $y_2 = -1$.

$y_1 = x^2 + x, \; y_2 = -1$

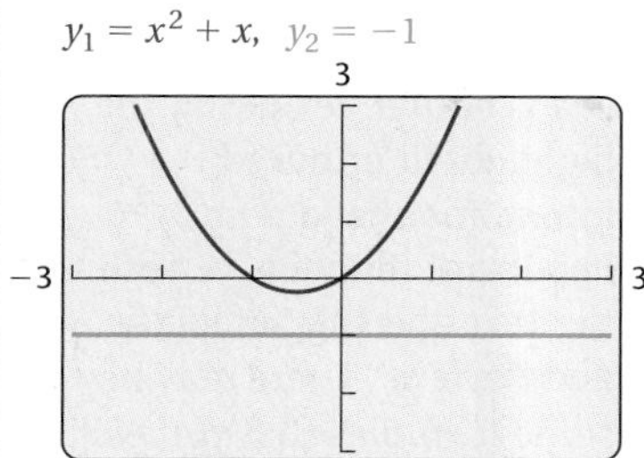

We see that the graphs do not intersect. Thus there is no real number for which $y_1 = y_2$, or $x^2 + x = -1$.

Exercises:

1. Explain how the graph of $y = x^2 + x + 1$ shows that the equation in Example 4, $x^2 + x = -1$, has no real-number solutions.
2. Use a graph to determine whether the equation $x^2 + x = 1$ has real-number solutions.
3. Use a graph to determine whether the equation $x^2 = x - 1$ has real-number solutions.

6. Approximate the solutions to the equation in Margin Exercise 5. Round to the nearest tenth.

Answer on page A-51

CALCULATOR CORNER

Approximating Solutions of Quadratic Equations In Example 5, we found that the solutions of the equation $3x^2 = 7 - 2x$ are $\frac{-1 + \sqrt{22}}{3}$ and $\frac{-1 - \sqrt{22}}{3}$. We can use a graphing calculator to approximate these solutions. To approximate $\frac{-1 + \sqrt{22}}{3}$, we press

((-) 1 + 2ND √ 2 2)) ÷ 3 ENTER. To approximate $\frac{-1 - \sqrt{22}}{3}$, we press ((-) 1 − 2ND √ 2 2)) ÷ 3 ENTER. We see that the solutions are approximately 1.2 and −1.9.

```
(−1+√(22))/3
              1.230138587
(−1−√(22))/3
             −1.896805253
```

Exercises: Use a graphing calculator to approximate the solutions to each of the following to the nearest tenth.

1. Example 3
2. Margin Exercise 3
3. Margin Exercise 5

COMPARING METHODS OF SOLVING QUADRATIC EQUATIONS

In Sections 15.1–15.3, we have studied three different methods of solving quadratic equations. Each of these methods has advantages and disadvantages, as outlined in the table below. Note that although the quadratic formula can be used to solve *any* quadratic equation, the other methods are sometimes faster and easier to use.

METHOD	ADVANTAGES	DISADVANTAGES
The quadratic formula	Can be used to solve *any* quadratic equation.	Can be slower than factoring or the principle of square roots.
The principle of square roots	Fastest way to solve equations of the form $ax^2 = p$, or $(x + k)^2 = p$. Can be used to solve *any* quadratic equation.	Can be slow when completing the square is required.
Factoring	Can be very fast.	Can be used only on certain equations. Many equations are difficult or impossible to solve by factoring.

Study Tips

BEGINNING TO STUDY FOR THE FINAL EXAM (PART 4): ONE OR TWO DAYS OF STUDY TIME

1. **Begin by browsing through each chapter, reviewing the highlighted or boxed information regarding important formulas in both the text and the Summary and Review.** There may be some formulas that you will need to memorize.
2. **Take the Final Examination in the text during the last couple of days before the final.** Set aside the same amount of time that you will have for the final. See how much of the final exam you can complete under test-like conditions. Be careful to avoid any questions corresponding to objectives not covered. Restudy the objectives in the text that correspond to each question you missed.
3. **Attend a final-exam review session if one is available.**

"Great is the art of beginning, but greater is the art of ending."

Henry Wadsworth Longfellow, nineteenth-century American poet

15.3 EXERCISE SET

For Extra Help

MathXL | MyMathLab | InterAct Math | Math Tutor Center | Video Lectures on CD Disc 8 | Student's Solutions Manual

a Solve. Try factoring first. If factoring is not possible or is difficult, use the quadratic formula.

1. $x^2 - 4x = 21$

2. $x^2 + 8x = 9$

3. $x^2 = 6x - 9$

4. $x^2 = 24x - 144$

5. $3y^2 - 2y - 8 = 0$

6. $3y^2 - 7y + 4 = 0$

7. $4x^2 + 4x = 15$

8. $4x^2 + 12x = 7$

9. $x^2 - 9 = 0$

10. $x^2 - 16 = 0$

11. $x^2 - 2x - 2 = 0$

12. $x^2 - 2x - 11 = 0$

13. $y^2 - 10y + 22 = 0$

14. $y^2 + 6y - 1 = 0$

15. $x^2 + 4x + 4 = 7$

16. $x^2 - 2x + 1 = 5$

17. $3x^2 + 8x + 2 = 0$

18. $3x^2 - 4x - 2 = 0$

19. $2x^2 - 5x = 1$

20. $4x^2 + 4x = 5$

21. $2y^2 - 2y - 1 = 0$

22. $4y^2 + 4y - 1 = 0$

23. $2t^2 + 6t + 5 = 0$

24. $4y^2 + 3y + 2 = 0$

25. $3x^2 = 5x + 4$

26. $2x^2 + 3x = 1$

27. $2y^2 - 6y = 10$

28. $5m^2 = 3 + 11m$

29. $\frac{x^2}{x+3} - \frac{5}{x+3} = 0$

30. $\frac{x^2}{x-4} - \frac{7}{x-4} = 0$

31. $x + 2 = \frac{3}{x+2}$

32. $x - 3 = \frac{5}{x-3}$

33. $\frac{1}{x} + \frac{1}{x+1} = \frac{1}{3}$

34. $\frac{1}{x} + \frac{1}{x+6} = \frac{1}{5}$

b Solve using the quadratic formula. Use a calculator to approximate the solutions to the nearest tenth.

35. $x^2 - 4x - 7 = 0$

36. $x^2 + 2x - 2 = 0$

37. $y^2 - 6y - 1 = 0$

38. $y^2 + 10y + 22 = 0$

39. $4x^2 + 4x = 1$

40. $4x^2 = 4x + 1$

41. $3x^2 - 8x + 2 = 0$

42. $3x^2 + 4x - 2 = 0$

43. **D_W** Write a quadratic equation in the form $y = ax^2 + bx + c$ that does not cross the x-axis.

44. **D_W** Under what condition(s) would using the quadratic formula *not* be the easiest way to solve a quadratic equation?

SKILL MAINTENANCE

Add or subtract. [14.4a]

45. $\sqrt{40} - 2\sqrt{10} + \sqrt{90}$

46. $\sqrt{54} - \sqrt{24}$

47. $\sqrt{18} + \sqrt{50} - 3\sqrt{8}$

48. $\sqrt{81x^3} - \sqrt{4x}$

49. Simplify: $\sqrt{80}$. [14.2a]

50. Multiply and simplify: $\sqrt{3x^2}\sqrt{9x^3}$. [14.2c]

51. Simplify: $\sqrt{9000x^{10}}$. [14.2b]

52. Rationalize the denominator: $\sqrt{\frac{7}{3}}$. [14.3c]

53. Find an equation of variation in which y varies inversely as x, and $y = 235$ when $x = 0.6$. [12.9c]

54. The time T to do a certain job varies inversely as the number N of people working. It takes 5 hr for 24 people to wash and wax the floors in a building. How long would it take 36 people to do the job? [12.9d]

SYNTHESIS

Solve.

55. $5x + x(x - 7) = 0$

56. $x(3x + 7) - 3x = 0$

57. $3 - x(x - 3) = 4$

58. $x(5x - 7) = 1$

59. $(y + 4)(y + 3) = 15$

60. $(y + 5)(y - 1) = 27$

61. $x^2 + (x + 2)^2 = 7$

62. $x^2 + (x + 1)^2 = 5$

63.–70. Use a graphing calculator to approximate the solutions of the equations in Exercises 35–42. Compare your answers with those found using the quadratic formula.

15.4 FORMULAS

Objective

a Solve a formula for a given letter.

a Solving Formulas

Formulas arise frequently in the natural and social sciences, business, engineering, and health care. In Section 8.4, we saw that the same steps that are used to solve linear equations can be used to solve a formula that appears in this form. Similarly, the steps that are used to solve a rational, radical, or quadratic equation can also be used to solve a formula that appears in one of these forms.

EXAMPLE 1 *Intelligence Quotient.* The formula $Q = \dfrac{100m}{c}$ is used to determine the intelligence quotient, Q, of a person of mental age m and chronological age c. Solve for c.

$$Q = \frac{100m}{c}$$

$$c \cdot Q = c \cdot \frac{100m}{c} \qquad \text{Multiplying by } c \text{ on both sides}$$

$$cQ = 100m \qquad \text{Simplifying}$$

$$c = \frac{100m}{Q} \qquad \text{Dividing by } Q \text{ on both sides}$$

This formula can be used to determine a person's chronological, or actual, age from his or her mental age and intelligence quotient.

Do Exercise 1.

1. a) Solve for I: $E = \dfrac{9R}{I}$.

b) Solve for R: $E = \dfrac{9R}{I}$.

EXAMPLE 2 Solve for x: $y = ax + bx - 4$.

$$y = ax + bx - 4 \qquad \text{We want this letter alone on one side.}$$

$$y + 4 = ax + bx \qquad \text{Adding 4}$$

$$y + 4 = (a + b)x \qquad \text{Collecting like terms}$$

$$\frac{y + 4}{(a + b)} = \frac{(a + b)x}{(a + b)} \qquad \text{Dividing by } a + b \text{ on both sides}$$

$$\frac{y + 4}{a + b} = x \qquad \text{Simplifying. The answer can also be written as } x = \frac{y + 4}{a + b}.$$

Do Exercise 2.

2. Solve for x: $y = ax - bx + 5$.

Answers on page A-51

3. Optics Formula. Solve for f:

$$\frac{1}{p} + \frac{1}{q} = \frac{1}{f}.$$

Answer on page A-51

Caution!

Had we performed the following steps in Example 2, we would *not* have solved for x:

$$y = ax + bx - 4$$

$$y - ax + 4 = bx \qquad \text{Subtracting } ax \text{ and adding 4}$$

x occurs on both sides of the equals sign.

$$\frac{y - ax + 4}{b} = x. \qquad \text{Dividing by } b$$

The mathematics of each step is correct, but since x occurs on both sides of the formula, *we have not solved the formula for x*. Remember that the letter being solved for should be **alone** on one side of the equation, with no occurrence of that letter on the other side!

EXAMPLE 3 Solve the following work formula for t:

$$\frac{t}{a} + \frac{t}{b} = 1.$$

We multiply by the LCM, which is ab:

$$ab \cdot \left(\frac{t}{a} + \frac{t}{b}\right) = ab \cdot 1 \qquad \text{Multiplying by } ab$$

$$ab \cdot \frac{t}{a} + ab \cdot \frac{t}{b} = ab \qquad \text{Using a distributive law to remove parentheses}$$

$$bt + at = ab \qquad \text{Simplifying}$$

$$(b + a)t = ab \qquad \text{Factoring out } t$$

$$t = \frac{ab}{b + a}. \qquad \text{Dividing by } b + a$$

Do Exercise 3.

EXAMPLE 4 *Distance to the Horizon.* Solve for h: $D = \sqrt{2h}$. (See Exercises 45–48 in Exercise Set 14.5.)

This is a radical equation. Recall that we first isolate the radical. Then we use the principle of squaring.

$$D = \sqrt{2h}$$

$$D^2 = (\sqrt{2h})^2 \qquad \text{Using the principle of squaring (Section 14.5)}$$

$$D^2 = 2h \qquad \text{Simplifying}$$

$$\frac{D^2}{2} = h \qquad \text{Dividing by 2}$$

EXAMPLE 5 Solve for g: $T = 2\pi\sqrt{\frac{L}{g}}$ (the period of a pendulum).

We have

$$\frac{T}{2\pi} = \sqrt{\frac{L}{g}} \qquad \text{Dividing by } 2\pi \text{ to isolate the radical}$$

$$\left(\frac{T}{2\pi}\right)^2 = \left(\sqrt{\frac{L}{g}}\right)^2. \qquad \text{Using the principle of squaring}$$

Then

$$\frac{T^2}{4\pi^2} = \frac{L}{g}$$

$$gT^2 = 4\pi^2 L \qquad \text{Multiplying by } 4\pi^2 g \text{ to clear fractions}$$

$$g = \frac{4\pi^2 L}{T^2}. \qquad \text{Dividing by } T^2 \text{ to get } g \text{ alone}$$

Do Exercises 4–6.

In most formulas, the letters represent nonnegative numbers, so we need not use absolute values when taking square roots.

EXAMPLE 6 *Torricelli's Theorem.* The speed v of a liquid leaving a water cooler from an opening is related to the height h of the top of the liquid above the opening by the formula

$$h = \frac{v^2}{2g}.$$

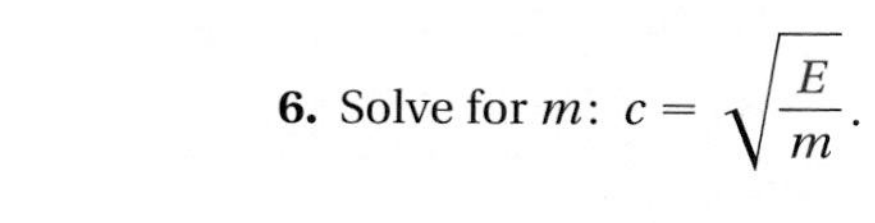

Solve for v.

Since v^2 appears by itself and there is no expression involving v, we first solve for v^2. Then we use the principle of square roots, taking only the nonnegative square root because v is nonnegative.

$2gh = v^2$ — Multiplying by $2g$ to clear fractions

$\sqrt{2gh} = v$ — Using the principle of square roots. Assume that v is nonnegative.

Do Exercise 7.

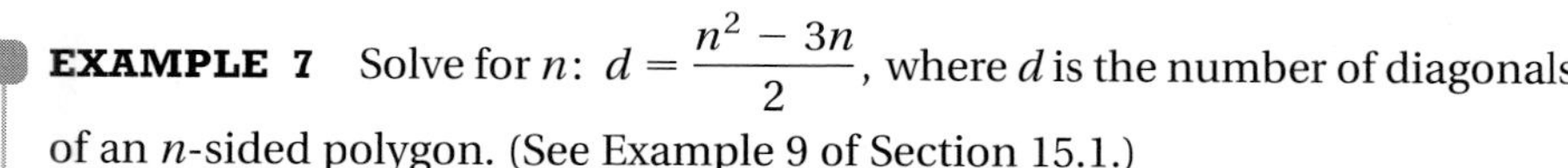

EXAMPLE 7 Solve for n: $d = \dfrac{n^2 - 3n}{2}$, where d is the number of diagonals of an n-sided polygon. (See Example 9 of Section 15.1.)

In this case, there is a term involving n as well as an n^2-term. Thus we must use the quadratic formula.

$$d = \frac{n^2 - 3n}{2}$$

$n^2 - 3n = 2d$ — Multiplying by 2 to clear fractions

$n^2 - 3n - 2d = 0$ — Finding standard form

$a = 1, \quad b = -3, \quad c = -2d$ — The variable is n; d represents a constant.

$n = \dfrac{-b \pm \sqrt{b^2 - 4ac}}{2a}$ — Quadratic formula

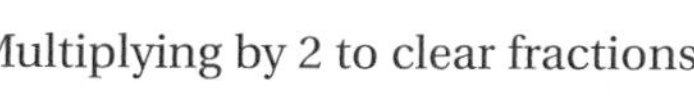

$= \dfrac{-(-3) \pm \sqrt{(-3)^2 - 4 \cdot 1 \cdot (-2d)}}{2 \cdot 1}$ — Substituting into the quadratic formula

$= \dfrac{3 + \sqrt{9 + 8d}}{2}$ — Using the positive root

Do Exercise 8.

4. Solve for L: $r = 2\sqrt{5L}$ (the speed of a skidding car).

5. Solve for L: $T = 2\pi\sqrt{\dfrac{L}{g}}$.

6. Solve for m: $c = \sqrt{\dfrac{E}{m}}$.

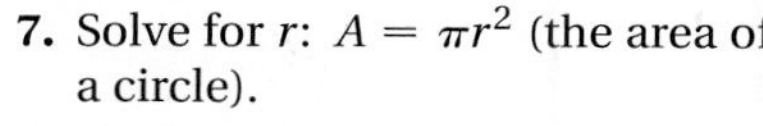

7. Solve for r: $A = \pi r^2$ (the area of a circle).

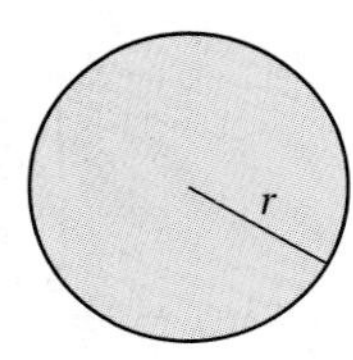

8. Solve for n: $N = n^2 - n$.

Answers on page A-51

15.4 EXERCISE SET

For Extra Help

MathXL | MyMathLab | InterAct Math | Math Tutor Center | Video Lectures on CD Disc 8 | Student's Solutions Manual

a Solve for the indicated letter.

1. $q = \frac{VQ}{I}$, for I
(An engineering formula)

2. $y = \frac{4A}{a}$, for a

3. $S = \frac{kmM}{d^2}$, for m

4. $S = \frac{kmM}{d^2}$, for M

5. $S = \frac{kmM}{d^2}$, for d^2

6. $T = \frac{10t}{W^2}$, for W^2

7. $T = \frac{10t}{W^2}$, for W

8. $S = \frac{kmM}{d^2}$, for d

9. $A = at + bt$, for t

10. $S = rx + sx$, for x

11. $y = ax + bx + c$, for x

12. $y = ax - bx - c$, for x

13. $\frac{t}{a} + \frac{t}{b} = 1$, for a
(A work formula)

14. $\frac{t}{a} + \frac{t}{b} = 1$, for b
(A work formula)

15. $\frac{1}{p} + \frac{1}{q} = \frac{1}{f}$, for p
(An optics formula)

16. $\frac{1}{p} + \frac{1}{q} = \frac{1}{f}$, for q
(An optics formula)

17. $A = \frac{1}{2}bh$, for b
(The area of a triangle)

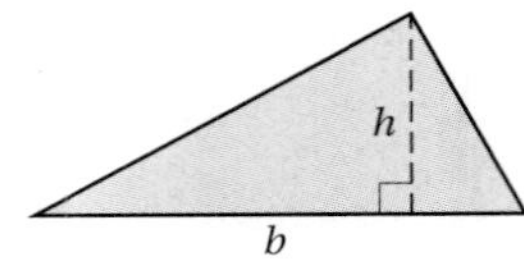

18. $s = \frac{1}{2}gt^2$, for g

19. $S = 2\pi r(r + h)$, for h
(The surface area of a right circular cylinder)

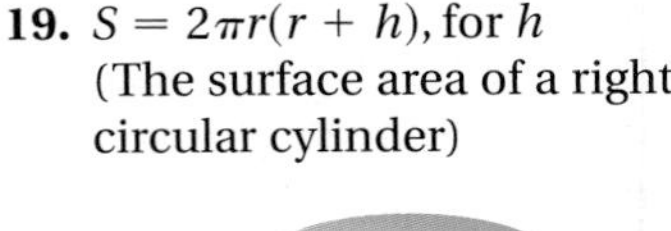
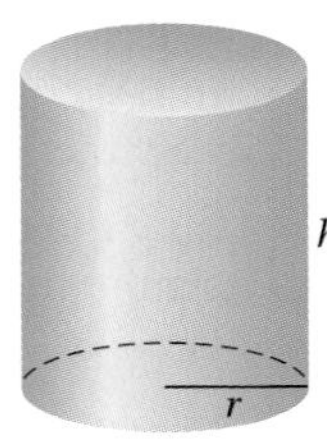

20. $S = 2\pi(r + h)$, for r

21. $\frac{1}{R} = \frac{1}{r_1} + \frac{1}{r_2}$, for R
(An electricity formula)

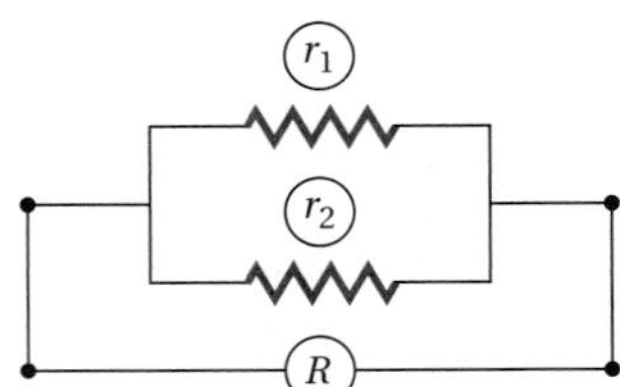

22. $\frac{1}{R} = \frac{1}{r_1} + \frac{1}{r_2}$, for r_1

23. $P = 17\sqrt{Q}$, for Q

24. $A = 1.4\sqrt{t}$, for t

25. $v = \sqrt{\frac{2gE}{m}}$, for E

26. $Q = \sqrt{\frac{aT}{c}}$, for T

27. $S = 4\pi r^2$, for r

28. $E = mc^2$, for c

29. $P = kA^2 + mA$, for A

30. $Q = ad^2 - cd$, for d

31. $c^2 = a^2 + b^2$, for a

32. $c = \sqrt{a^2 + b^2}$, for b

33. $s = 16t^2$, for t

34. $V = \pi r^2 h$, for r

35. $A = \pi r^2 + 2\pi rh$, for r

36. $A = 2\pi r^2 + 2\pi rh$, for r

37. $F = \frac{Av^2}{400}$, for v

38. $A = \frac{\pi r^2 S}{360}$, for r

39. $c = \sqrt{a^2 + b^2}$, for a

40. $c^2 = a^2 + b^2$, for b

41. $h = \frac{a}{2}\sqrt{3}$, for a
(The height of an equilateral triangle with sides of length a)

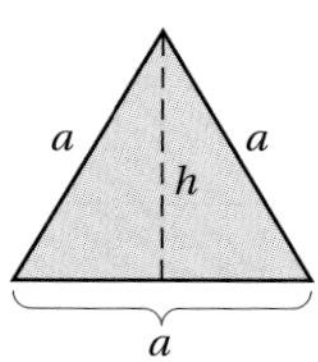

42. $d = s\sqrt{2}$, for s
(The hypotenuse of an isosceles right triangle with s the length of the legs)

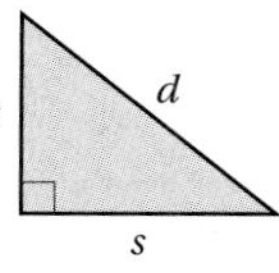

43. $n = aT^2 - 4T + m$, for T

44. $y = ax^2 + bx + c$, for x

45. $v = 2\sqrt{\frac{2kT}{\pi m}}$, for T

46. $E = \frac{1}{2}mv^2 + mgy$, for v

47. $3x^2 = d^2$, for x

48. $c = \sqrt{\frac{E}{m}}$, for E

49. $N = \frac{n^2 - n}{2}$, for n

50. $M = \frac{m}{\sqrt{1 - \left(\frac{v}{c}\right)^2}}$, for c

51. $S = \frac{a + b}{3b}$, for b

52. $Q = \frac{a - b}{2b}$, for b

53. $\frac{A - B}{AB} = Q$, for B

54. $L = \frac{Mt + g}{t}$, for t

55. $S = 180(n - 2)$, for n

56. $S = \frac{n}{2}(a + 1)$, for a

57. $A = P(1 + rt)$, for t
(An interest formula)

58. $A = P(1 + rt)$, for r
(An interest formula)

59. $\frac{A}{B} = \frac{C}{D}$, for D

60. $\frac{A}{B} = \frac{C}{D}$, for B

61. $C = \frac{Ka - b}{a}$, for a

62. $Q = \frac{Pt - h}{t}$, for t

63. **DW** Describe a situation in which the result of Example 3,

$$t = \frac{ab}{a + b},$$

would be especially useful.

64. **DW** Explain how you would solve the equation $0 = ax^2 + bx + c$ for x.

SKILL MAINTENANCE

In a right triangle, where a and b represent the legs and c represents the hypotenuse, find the length of the side not given. Give an exact answer and an approximation to three decimal places. [14.6a]

65. $a = 4, b = 7$

66. $b = 11, c = 14$

67. $a = 4, b = 5$

68. $a = 10, c = 12$

69. $c = 8\sqrt{17}, a = 2$

70. $a = \sqrt{2}, b = \sqrt{3}$

Solve. [14.6b]

71. *Guy Wire.* How long is a guy wire reaching from the top of an 18-ft pole to a point on the ground 10 ft from the pole? Give an exact answer and an approximation to three decimal places.

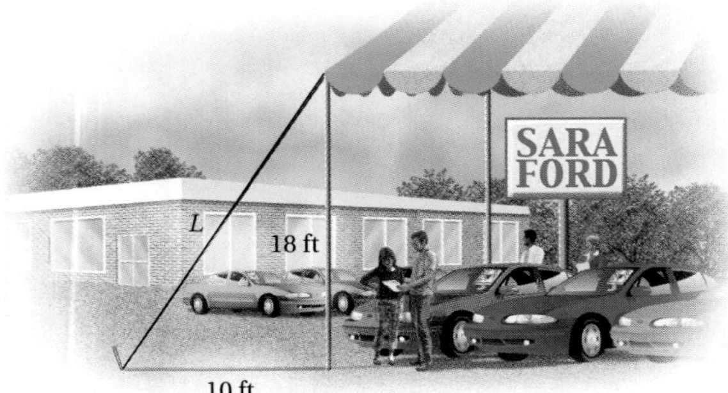

72. *Soccer Fields.* The smallest regulation soccer field is 50 yd wide and 100 yd long. Find the length of a diagonal of such a field.

Multiply and simplify. [14.2c]

73. $\sqrt{3x} \cdot \sqrt{6x}$

74. $\sqrt{8x^2} \cdot \sqrt{24x^3}$

75. $3\sqrt{t} \cdot \sqrt{t}$

76. $\sqrt{x^2} \cdot \sqrt{x^5}$

SYNTHESIS

77. The circumference C of a circle is given by $C = 2\pi r$.
a) Solve $C = 2\pi r$ for r.
b) The area is given by $A = \pi r^2$. Express the area in terms of the circumference C.

78. Referring to Exercise 77, express the circumference C in terms of the area A.

79. Solve $3ax^2 - x - 3ax + 1 = 0$ for x.

80. Solve $h = 16t^2 + vt + s$ for t.

15.5 APPLICATIONS AND PROBLEM SOLVING

Objective

a Solve applied problems using quadratic equations.

a Using Quadratic Equations to Solve Applied Problems

EXAMPLE 1 *Red Raspberry Patch.* The area of a rectangular red raspberry patch is 76 ft^2. The length is 7 ft longer than three times the width. Find the dimensions of the raspberry patch.

1. **Familiarize.** We first make a drawing and label it with both known and unknown information. We let $w =$ the width of the rectangle. The length of the rectangle is 7 ft longer than three times the width. Thus the length is $3w + 7$.

2. **Translate.** Recall that area is length $\times$ width. Thus we have two expressions for the area of the rectangle: $(3w + 7)(w)$ and 76. This gives us a translation:

$$(3w + 7)(w) = 76.$$

3. **Solve.** We solve the equation:

$$3w^2 + 7w = 76$$

$$3w^2 + 7w - 76 = 0$$

$$(3w + 19)(w - 4) = 0 \quad \text{Factoring (the quadratic formula could also be used)}$$

$$3w + 19 = 0 \quad or \quad w - 4 = 0 \quad \text{Using the principle of zero products}$$

$$3w = -19 \quad or \quad w = 4$$

$$w = -\tfrac{19}{3} \quad or \quad w = 4.$$

4. **Check.** We check in the original problem. We know that $-\frac{19}{3}$ is not a solution because width cannot be negative. When $w = 4$, $3w + 7 = 19$, and the area is 4(19), or 76. This checks.

5. **State.** The width of the rectangular raspberry patch is 4 ft, and the length is 19 ft.

Do Exercise 1.

1. **Pool Dimensions.** The area of a rectangular swimming pool is 68 yd^2. The length is 1 yd longer than three times the width. Find the dimensions of the rectangular swimming pool. Round to the nearest tenth.

Answer on page A-52

EXAMPLE 2 *Staircase.* A mason builds a staircase in such a way that the portion underneath the stairs forms a right triangle. The hypotenuse is 6 m long. The leg across the ground is 1 m longer than the leg next to the wall at the back. Find the lengths of the legs. Round to the nearest tenth.

1. **Familiarize.** We first make a drawing, letting $s =$ the length of the shorter leg. Then $s + 1 =$ the length of the other leg.
2. **Translate.** To translate, we use the Pythagorean equation:

$$s^2 + (s + 1)^2 = 6^2.$$

3. **Solve.** We solve the equation:

$$s^2 + (s + 1)^2 = 6^2$$
$$s^2 + s^2 + 2s + 1 = 36$$
$$2s^2 + 2s - 35 = 0.$$

Since we cannot factor, we use the quadratic formula:

$$a = 2, \quad b = 2, \quad c = -35$$
$$s = \frac{-b \pm \sqrt{b^2 - 4ac}}{2a} = \frac{-2 \pm \sqrt{2^2 - 4 \cdot 2(-35)}}{2 \cdot 2}$$
$$= \frac{-2 \pm \sqrt{4 + 280}}{4} = \frac{-2 \pm \sqrt{284}}{4}$$
$$= \frac{-2 \pm \sqrt{4 \cdot 71}}{4} = \frac{-2 \pm 2 \cdot \sqrt{71}}{2 \cdot 2}$$
$$= \frac{2(-1 \pm \sqrt{71})}{2 \cdot 2} = \frac{2}{2} \cdot \frac{-1 \pm \sqrt{71}}{2} = \frac{-1 \pm \sqrt{71}}{2}.$$

Using a calculator, we get approximations:

$$\frac{-1 + \sqrt{71}}{2} \approx 3.7 \quad or \quad \frac{-1 - \sqrt{71}}{2} \approx -4.7.$$

4. **Check.** Since the length of a leg cannot be negative, -4.7 does not check. But 3.7 does check. If the smaller leg s is 3.7, the other leg is $s + 1$, or 4.7. Then

$$(3.7)^2 + (4.7)^2 = 13.69 + 22.09 = 35.78.$$

Using a calculator, we get $\sqrt{35.78} \approx 5.98 \approx 6$. Note that our check is not exact because we are using an approximation for $\sqrt{71}$.

5. **State.** One leg is about 3.7 m long, and the other is about 4.7 m long.

Do Exercise 2.

2. Animal Pen. The hypotenuse of a right triangular animal pen at the zoo is 7 yd long. One leg is 2 yd longer than the other. Find the lengths of the legs. Round to the nearest tenth.

EXAMPLE 3 *Boat Speed.* The current in a stream moves at a speed of 2 km/h. A boat travels 24 km upstream and 24 km downstream in a total time of 5 hr. What is the speed of the boat in still water?

1. **Familiarize.** We first make a drawing. The distances are the same. We let $r =$ the speed of the boat in still water. Then when the boat is traveling upstream, its speed is $r - 2$. When it is traveling downstream, its speed is $r + 2$. We let t_1 represent the time it takes the boat to go upstream and t_2 the time it takes to go downstream. We summarize in a table.

Answer on page A-52

	d	r	t
Upstream	24	$r-2$	t_1
Downstream	24	$r+2$	t_2
Total time			5

$\longrightarrow t_1 = \dfrac{24}{r-2}$

$\longrightarrow t_2 = \dfrac{24}{r+2}$

2. **Translate.** Recall the basic formula for motion: $d = rt$. From it we can obtain an equation for time: $t = d/r$. Total time consists of the time to go upstream, t_1, plus the time to go downstream, t_2. Using $t = d/r$ and the rows of the table, we have

$$t_1 = \frac{24}{r-2} \quad \text{and} \quad t_2 = \frac{24}{r+2}.$$

Since the total time is 5 hr, $t_1 + t_2 = 5$, and we have

$$\frac{24}{r-2} + \frac{24}{r+2} = 5. \qquad \text{We have translated to an equation with one variable.}$$

3. **Solve.** We solve the equation. We multiply on both sides by the LCM, which is $(r-2)(r+2)$:

$$(r-2)(r+2)\cdot\left[\frac{24}{r-2} + \frac{24}{r+2}\right] = (r-2)(r+2)5$$

$$(r-2)(r+2)\cdot\frac{24}{r-2} + (r-2)(r+2)\cdot\frac{24}{r+2} = (r^2-4)5$$

$$24(r+2) + 24(r-2) = 5r^2 - 20$$

$$24r + 48 + 24r - 48 = 5r^2 - 20$$

$$-5r^2 + 48r + 20 = 0$$

$$5r^2 - 48r - 20 = 0 \qquad \text{Multiplying by } -1$$

$$(5r+2)(r-10) = 0 \qquad \text{Factoring}$$

$$5r + 2 = 0 \quad or \quad r - 10 = 0$$

Using the principle of zero products

$$5r = -2 \quad or \quad r = 10$$

$$r = -\tfrac{2}{5} \quad or \quad r = 10.$$

4. **Check.** Since speed cannot be negative, $-\frac{2}{5}$ cannot be a solution. But suppose the speed of the boat in still water is 10 km/h. The speed upstream is then $10 - 2$, or 8 km/h. The speed downstream is $10 + 2$, or 12 km/h. The time upstream, using $t = d/r$, is 24/8, or 3 hr. The time downstream is 24/12, or 2 hr. The total time is 5 hr. This checks.

5. **State.** The speed of the boat in still water is 10 km/h.

Do Exercise 3.

3. **Speed of a Stream.** The speed of a boat in still water is 12 km/h. The boat travels 45 km upstream and 45 km downstream in a total time of 8 hr. What is the speed of the stream? (*Hint*: Let s = the speed of the stream. Then $12 - s$ is the speed upstream and $12 + s$ is the speed downstream. Note also that $12 - s$ cannot be negative, because the boat must be going faster than the current if it is moving forward.)

Answer on page A-52

Translating for Success

1. *Guy Wire.* How long is a guy wire that reaches from the top of a 75-ft cellphone tower to a point on the ground 21 ft from the pole?

2. *Coin Mixture.* A collection of dimes and quarters is worth \$16.95. There are 90 coins in all. How many of each coin are there?

3. *Wire Cutting.* A 486-in. wire is cut into three pieces. The second piece is 5 in. longer than the first. The third is one-half as long as the first. How long is each piece?

4. *Amount Invested.* Money is invested at 6.2% simple interest. At the end of 1 yr, there is \$28,992.60 in the account. How much was originally invested?

5. *Stroke Victims.* In 2001, there were 163,538 deaths due to stroke. This was about 23.4% of the total number of people who had a stroke in 2001. How many people had a stroke in 2001?

The goal of these matching questions is to practice step (2), *Translate,* of the five-step problem-solving process. Translate each word problem to an equation or a system of equations and select a correct translation from A–O.

A. $x^2 + (x - 1)^2 = 7$

B. $\dfrac{600}{x} = \dfrac{600}{x+2} + 10$

C. $184{,}757 = x \cdot 865{,}000$

D. $163{,}538 = 23.4\% \cdot x$

E. $2x + 2(x - 1) = 49$

F. $x + (x + 5) + \dfrac{1}{2}x = 486$

G. $0.10x + 0.25y = 16.95,$
$x + y = 90$

H. $x + 25y = 16.95,$
$x + y = 90$

I. $x(x - 1) = 49$

J. $x^2 + (x - 1)^2 = 49$

K. $x^2 + 21^2 = 75^2$

L. $x + 6.2\%x = 28{,}992.60$

M. $75^2 + 21^2 = x^2$

N. $x + (x + 1) + (x + 2) = 894$

O. $\dfrac{600}{x} + \dfrac{600}{x-2} = 10$

Answers on page A-52

6. *Locker Numbers.* The numbers on three adjoining lockers are consecutive integers whose sum is 894. Find the integers.

7. *Triangle Dimensions.* The hypotenuse of a right triangle is 7 ft. The length of one leg is 1 ft shorter than the other. Find the lengths of the legs.

8. *Rectangle Dimensions.* The perimeter of a rectangle is 49 ft. The length is 1 ft shorter than the width. Find the length and the width.

9. *Car Travel.* Maggie drove her car 600 mi to see her friend. The return trip was 2 hr faster at a speed that was 10 mph greater. Find the time for the return trip.

10. *Mortality from Heart Attack.* Of the 865,000 people who had a heart attack in 2001, 184,757 died. What percent of those who had a heart attack died?

15.5 EXERCISE SET

For Extra Help

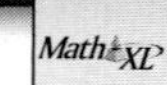

MathXL MyMathLab InterAct Math Math Tutor Center Video Lectures on CD Disc 8 Student's Solutions Manual

a Solve.

1. *HDTV Dimensions.* A high-definition television (HDTV) features a larger screen and greater clarity than a standard television. An HDTV might have a 70-in. diagonal screen with the width 27 in. greater than the height. Find the width and the height of a 70-in. HDTV screen.

2. The length of a rectangular pine forest is 2 mi greater than the width. The area is 80 mi^2. Find the length and the width.

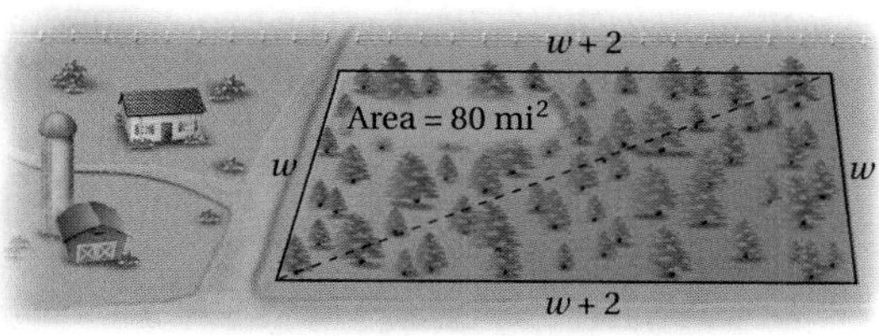

3. The length of a rectangular area rug is 3 ft greater than the width. The area is 70 ft^2. Find the length and the width.

4. *HDTV Dimensions.* When we say that a television is 42 in., we mean that the diagonal is 42 in. For a 42-in. television, the width is 15 in. more than the height. Find the dimensions of a 42-in. high-definition television.

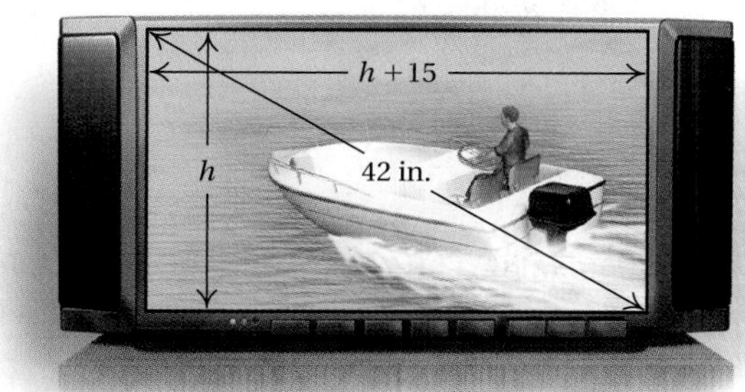

5. *Carpenter's Square.* A *square* is a carpenter's tool in the shape of a right triangle. One side, or leg, of a square is 8 in. longer than the other. The length of the hypotenuse is $8\sqrt{13}$ in. Find the lengths of the legs of the square.

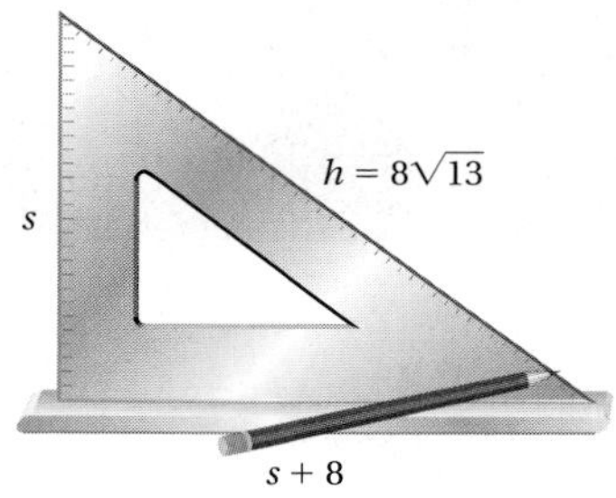

6. *Rectangle Dimensions.* The length of a rectangular lobby in a hotel is twice the width. The area is 50 m^2. Find the length and the width.

7. *Rectangle Dimensions.* The width of a rectangle is 4 cm less than the length. The area is 320 cm^2. Find the length and the width.

8. *Rectangle Dimensions.* The width of a rectangle is 3 cm less than the length. The area is 340 cm^2. Find the length and the width.

Find the approximate answers for Exercises 9–14. Round to the nearest tenth.

9. *Right-Triangle Dimensions.* The hypotenuse of a right triangle is 8 m long. One leg is 2 m longer than the other. Find the lengths of the legs.

10. *Right-Triangle Dimensions.* The hypotenuse of a right triangle is 5 cm long. One leg is 2 cm longer than the other. Find the lengths of the legs.

11. *Rectangle Dimensions.* The length of a rectangle is 2 in. greater than the width. The area is 20 in^2. Find the length and the width.

12. *Rectangle Dimensions.* The length of a rectangle is 3 ft greater than the width. The area is 15 ft^2. Find the length and the width.

13. *Rectangle Dimensions.* The length of a rectangle is twice the width. The area is 20 cm^2. Find the length and the width.

14. *Rectangle Dimensions.* The length of a rectangle is twice the width. The area is 10 m^2. Find the length and the width.

15. *Picture Frame.* A picture frame measures 25 cm by 20 cm. There is 266 cm^2 of picture showing. The frame is of uniform thickness. Find the thickness of the frame.

16. *Tablecloth.* A rectangular tablecloth measures 96 in. by 72 in. It is laid on a tabletop with an area of 5040 in^2, and hangs over the edge by the same amount on all sides. By how many inches does the cloth hang over the edge?

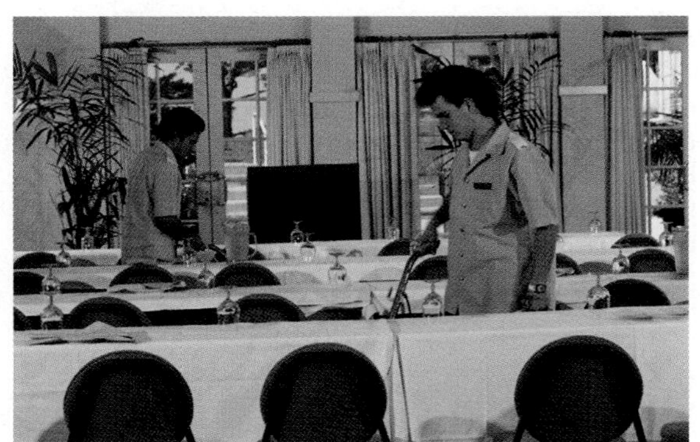

For Exercises 17–24, complete the table to help with the familiarization.

17. *Boat Speed.* The current in a stream moves at a speed of 3 km/h. A boat travels 40 km upstream and 40 km downstream in a total time of 14 hr. What is the speed of the boat in still water? Complete the following table to help with the familiarization.

	d	r	t
Upstream		$r - 3$	t_1
Downstream	40		t_2
Total Time			

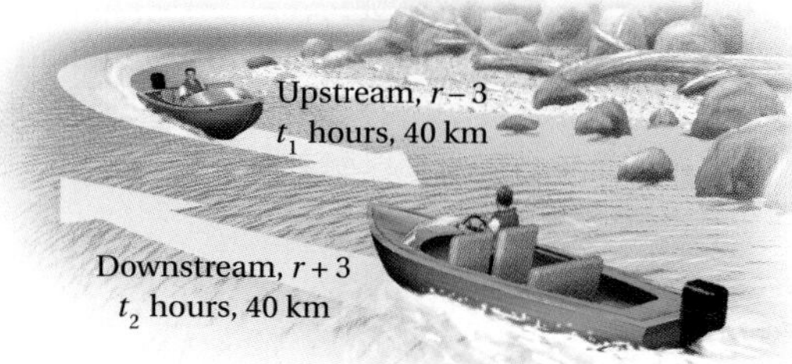

18. *Wind Speed.* An airplane flies 1449 mi against the wind and 1539 mi with the wind in a total time of 5 hr. The speed of the airplane in still air is 600 mph. What is the speed of the wind?

	d	r	t
With Wind	1539		
Against Wind		$600 - r$	
Total Time			5

19. *Wind Speed.* An airplane flies 520 km against the wind and 680 km with the wind in a total time of 4 hr. The speed of the airplane in still air is 300 km/h. What is the speed of the wind?

	d	r	t
With Wind		$300 + r$	
Against Wind	520		
Total Time			4

20. *Boat Speed.* The current in a stream moves at a speed of 4 mph. A boat travels 5 mi upstream and 13 mi downstream in a total time of 2 hr. What is the speed of the boat in still water?

	d	r	t
Upstream		$r - 4$	t_1
Downstream	13		t_2
Total Time			

21. *Speed of a Stream.* The speed of a boat in still water is 10 km/h. The boat travels 12 km upstream and 28 km downstream in a total time of 4 hr. What is the speed of the stream?

	d	r	t
Upstream			
Downstream			
Total Time			

22. *Speed of a Stream.* The speed of a boat in still water is 8 km/h. The boat travels 60 km upstream and 60 km downstream in a total time of 16 hr. What is the speed of the stream?

	d	r	t
Upstream			
Downstream			
Total Time			

23. *Boat Speed.* The current in a stream moves at a speed of 4 mph. A boat travels 4 mi upstream and 12 mi downstream in a total time of 2 hr. What is the speed of the boat in still water?

	d	r	t
Upstream			
Downstream			
Total Time			

24. *Boat Speed.* The current in a stream moves at a speed of 3 mph. A boat travels 45 mi upstream and 45 mi downstream in a total time of 8 hr. What is the speed of the boat in still water?

	d	r	t
Upstream			
Downstream			
Total Time			

25. *Speed of a Stream.* The speed of a boat in still water is 9 km/h. The boat travels 80 km upstream and 80 km downstream in a total time of 18 hr. What is the speed of the stream?

26. *Speed of a Stream.* The speed of a boat in still water is 10 km/h. The boat travels 48 km upstream and 48 km downstream in a total time of 10 hr. What is the speed of the stream?

D$_W$ Find and explain the error(s) in each of the following solutions of a quadratic equation.

27.
$$(x + 6)^2 = 16$$
$$x + 6 = \sqrt{16}$$
$$x + 6 = 4$$
$$x = -2$$

28.
$$x^2 + 2x - 8 = 0$$
$$(x + 4)(x - 2) = 0$$
$$x = 4 \quad or \quad x = -2$$

SKILL MAINTENANCE

Add or subtract. [14.4a]

29. $5\sqrt{2} + \sqrt{18}$

30. $7\sqrt{40} - 2\sqrt{10}$

31. $\sqrt{4x^3} - 7\sqrt{x}$

32. $\sqrt{24} - \sqrt{54}$

33. $\sqrt{2} + \sqrt{\frac{1}{2}}$

34. $\sqrt{3} - \sqrt{\frac{1}{3}}$

35. $\sqrt{24} + \sqrt{54} - \sqrt{48}$

36. $\sqrt{4x} + \sqrt{81x^3}$

SYNTHESIS

37. *Pizza.* What should the diameter d of a pizza be so that it has the same area as two 12-in. pizzas? Do you get more to eat with a 16-in. pizza or with two 12-in. pizzas?

38. *Golden Rectangle.* The so-called *golden rectangle* is said to be extremely pleasing visually and was used often by ancient Greek and Roman architects. The length of a golden rectangle is approximately 1.6 times the width. Find the dimensions of a golden rectangle if its area is 9000 m^2.

15.6 GRAPHS OF QUADRATIC EQUATIONS

Objectives

a Graph quadratic equations.

b Find the x-intercepts of a quadratic equation.

In this section, we will graph equations of the form

$$y = ax^2 + bx + c, \quad a \neq 0.$$

The polynomial on the right side of the equation is of second degree, or **quadratic.** Examples of the types of equations we are going to graph are

$$y = x^2, \qquad y = x^2 + 2x - 3, \qquad y = -2x^2 + 3.$$

a Graphing Quadratic Equations of the Type $y = ax^2 + bx + c$

Graphs of quadratic equations of the type $y = ax^2 + bx + c$ (where $a \neq 0$) are always cup-shaped. They have a **line of symmetry** like the dashed lines shown in the figures below. If we fold on this line, the two halves will match exactly. The curve goes on forever. The top or bottom point where the curve changes is called the **vertex.** The second coordinate is either the smallest value of y or the largest value of y. The vertex is also thought of as a turning point. Graphs of quadratic equations are called **parabolas.**

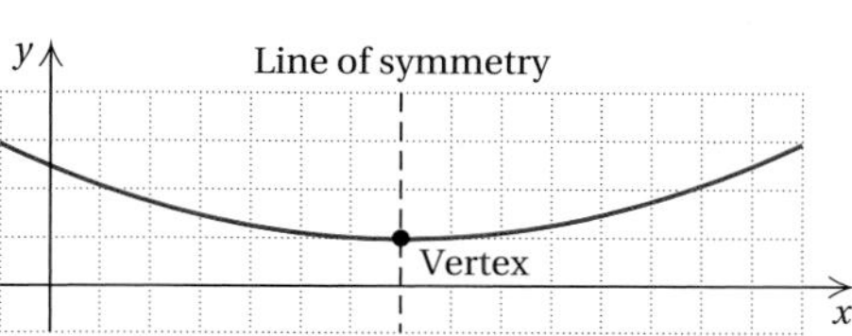

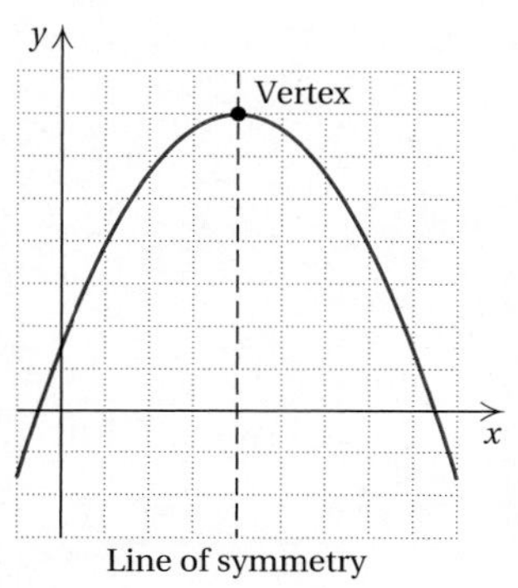

To graph a quadratic equation, we begin by choosing some numbers for x and computing the corresponding values of y.

EXAMPLE 1 Graph: $y = x^2$.

We choose numbers for x and find the corresponding values for y. Then we plot the ordered pairs (x, y) resulting from the computations and connect them with a smooth curve.

For $x = -3$, $y = x^2 = (-3)^2 = 9$.
For $x = -2$, $y = x^2 = (-2)^2 = 4$.
For $x = -1$, $y = x^2 = (-1)^2 = 1$.
For $x = 0$, $y = x^2 = (0)^2 = 0$.
For $x = 1$, $y = x^2 = (1)^2 = 1$.
For $x = 2$, $y = x^2 = (2)^2 = 4$.
For $x = 3$, $y = x^2 = (3)^2 = 9$.

x	y	(x, y)
-3	9	$(-3, 9)$
-2	4	$(-2, 4)$
-1	1	$(-1, 1)$
0	0	$(0, 0)$
1	1	$(1, 1)$
2	4	$(2, 4)$
3	9	$(3, 9)$

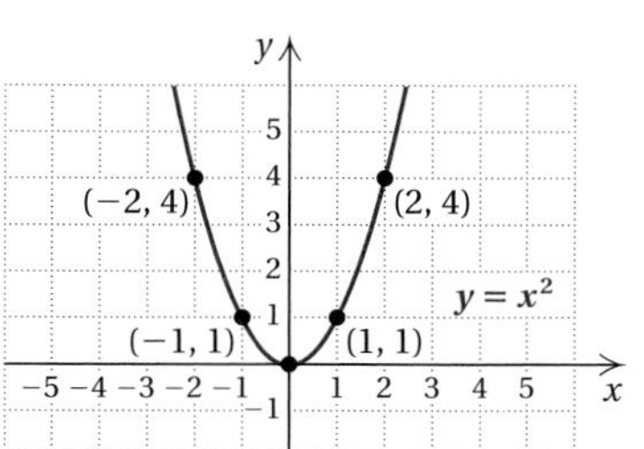

In Example 1, the vertex is the point $(0, 0)$. The second coordinate of the vertex, 0, is the smallest y-value. The y-axis $(x = 0)$ is the line of symmetry. Parabolas whose equations are $y = ax^2$ always have the origin $(0, 0)$ as the vertex and the y-axis as the line of symmetry.

How do we graph a quadratic equation? There are many methods, some of which you will study in your next mathematics course. Our goal here is to give you a basic graphing technique that is fairly easy to apply. A key in the graphing is knowing the vertex. By graphing it and then choosing x-values on both sides of the vertex, we can compute more points and complete the graph.

FINDING THE VERTEX

For a parabola given by the quadratic equation $y = ax^2 + bx + c$:

1. The x-coordinate of the vertex is $-\dfrac{b}{2a}$.

   The line of symmetry is $x = -b/(2a)$.
2. The second coordinate of the vertex is found by substituting the x-coordinate into the equation and computing y.

The proof that the vertex can be found in this way can be shown by completing the square in a manner similar to the proof of the quadratic formula, but it will not be considered here.

CALCULATOR CORNER

Graphing Quadratic Equations Use a graphing calculator to make a table of values for $y = x^2$. (See the Calculator Corner on p. 647 for the procedure.)

EXAMPLE 2 Graph: $y = -2x^2 + 3$.

We first find the vertex. The x-coordinate of the vertex is

$$-\frac{b}{2a} = -\frac{0}{2(-2)} = 0.$$

We substitute 0 for x into the equation to find the second coordinate of the vertex:

$$y = -2x^2 + 3 = -2(0)^2 + 3 = 3.$$

The vertex is $(0, 3)$. The line of symmetry is the y-axis $(x = 0)$. We choose some x-values on both sides of the vertex and graph the parabola.

For $x = 1$, $y = -2x^2 + 3 = -2(1)^2 + 3 = -2 + 3 = 1$.

For $x = -1$, $y = -2x^2 + 3 = -2(-1)^2 + 3 = -2 + 3 = 1$.

For $x = 2$, $y = -2x^2 + 3 = -2(2)^2 + 3 = -8 + 3 = -5$.

For $x = -2$, $y = -2x^2 + 3 = -2(-2)^2 + 3 = -8 + 3 = -5$.

x	y
0	3 ← This is the vertex.
1	1
−1	1
2	−5
−2	−5

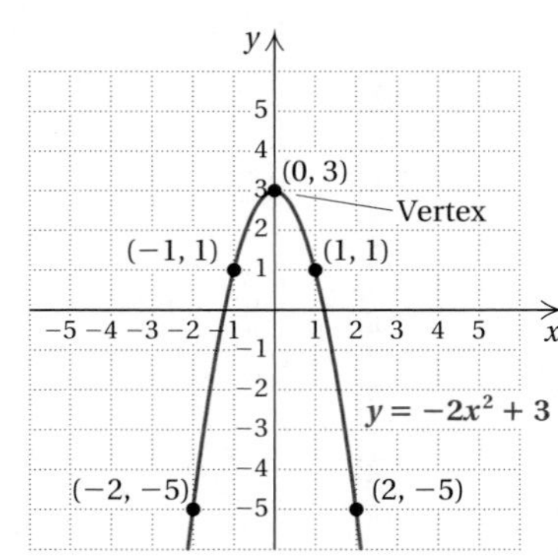

There are two other tips you might use when graphing quadratic equations. The first involves the coefficient of x^2. Note that a in $y = ax^2 + bx + c$ tells us whether the graph opens up or down. When a is positive, as in Example 1, the graph opens up; when a is negative, as in Example 2, the graph opens down. It is also helpful to plot the y-intercept. It occurs when $x = 0$.

TIPS FOR GRAPHING QUADRATIC EQUATIONS

1. Graphs of quadratic equations $y = ax^2 + bx + c$ are all parabolas. They are *smooth* cup-shaped symmetric curves, with no sharp points or kinks in them.
2. Find the vertex and the line of symmetry.
3. The graph of $y = ax^2 + bx + c$ opens up if $a > 0$. It opens down if $a < 0$.
4. Find the y-intercept. It occurs when $x = 0$, and it is easy to compute.

EXAMPLE 3 Graph: $y = x^2 + 2x - 3$.

We first find the vertex. The x-coordinate of the vertex is

$$-\frac{b}{2a} = -\frac{2}{2(1)} = -1.$$

We substitute -1 for x into the equation to find the second coordinate of the vertex:

$$\begin{aligned} y &= x^2 + 2x - 3 \\ &= (-1)^2 + 2(-1) - 3 \\ &= 1 - 2 - 3 \\ &= -4. \end{aligned}$$

The vertex is $(-1, -4)$. The line of symmetry is $x = -1$.

We choose some x-values on both sides of $x = -1$—say, -2, -3, -4 and 0, 1, 2—and graph the parabola. Since the coefficient of x^2 is 1, which is positive, we know that the graph opens up. Be sure to find y when $x = 0$. This gives the y-intercept.

x	y	
-1	-4	← Vertex
0	-3	← y-intercept
-2	-3	
1	0	
-3	0	
2	5	
-4	5	

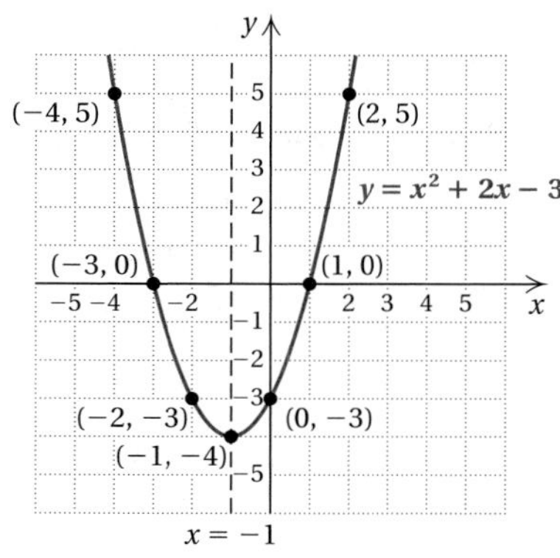

Do Exercises 1–3.

Graph. List the ordered pair for the vertex.

1. $y = x^2 - 3$

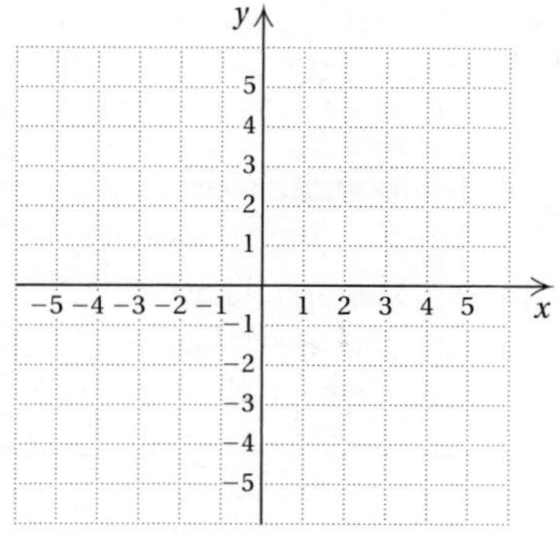

2. $y = -3x^2 + 6x$

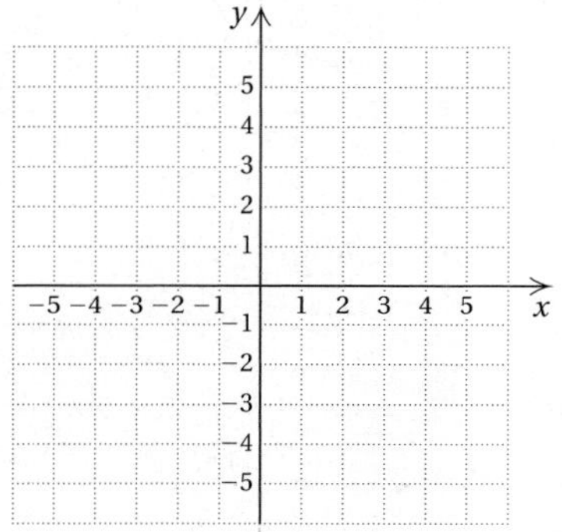

3. $y = x^2 - 4x + 4$

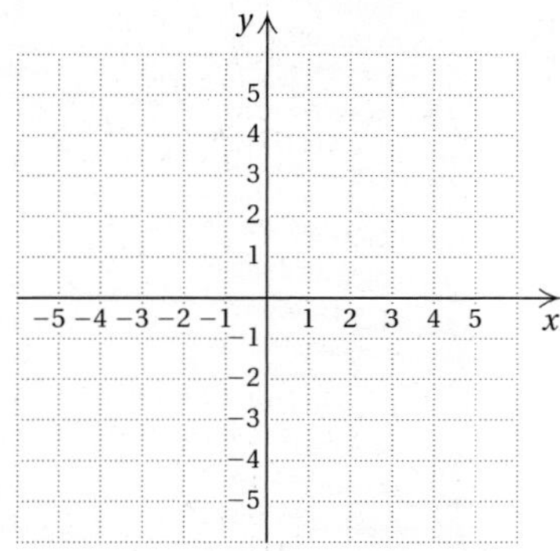

Answers on page A-52

b Finding the x-Intercepts of a Quadratic Equation

CALCULATOR CORNER

Graphing Quadratic Equations Use a graphing calculator to graph each of the equations in Margin Exercises 1–3.

The x-intercepts of $y = ax^2 + bx + c$ occur at those values of x for which $y = 0$. Thus the first coordinates of the x-intercepts are solutions of the equation

$$0 = ax^2 + bx + c.$$

We have been studying how to find such numbers in Sections 15.1–15.3.

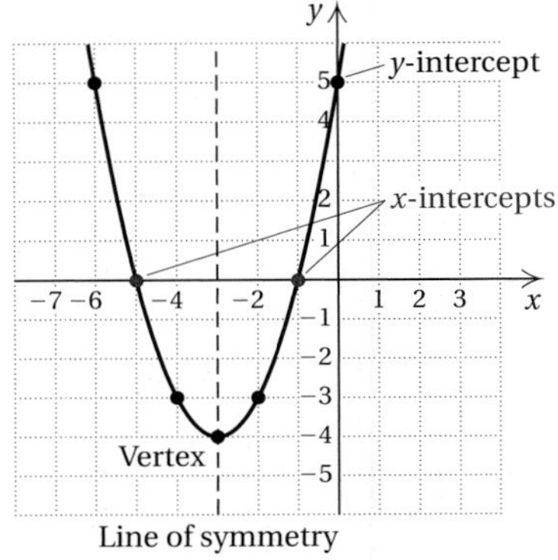

Find the x-intercepts.

4. $y = x^2 - 3$

5. $y = x^2 + 6x + 8$

6. $y = -2x^2 - 4x + 1$

7. $y = x^2 + 3$

Answers on page A-52

EXAMPLE 4 Find the x-intercepts of $y = x^2 - 4x + 1$.

We solve the equation

$$x^2 - 4x + 1 = 0.$$

Factoring is not convenient, so we use the quadratic formula:

$$a = 1, \quad b = -4, \quad c = 1$$

$$\begin{aligned} x &= \frac{-b \pm \sqrt{b^2 - 4ac}}{2a} \\ &= \frac{-(-4) \pm \sqrt{(-4)^2 - 4(1)(1)}}{2(1)} \\ &= \frac{4 \pm \sqrt{16 - 4}}{2} \\ &= \frac{4 \pm \sqrt{12}}{2} = \frac{4 \pm \sqrt{4 \cdot 3}}{2} \\ &= \frac{4 \pm 2\sqrt{3}}{2} = \frac{2 \cdot 2 \pm 2\sqrt{3}}{2 \cdot 1} \\ &= \frac{2}{2} \cdot \frac{2 \pm \sqrt{3}}{1} = 2 \pm \sqrt{3}. \end{aligned}$$

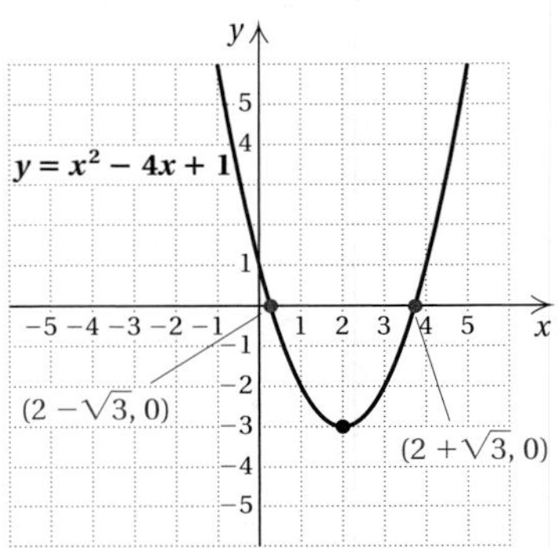

The x-intercepts are $(2 - \sqrt{3}, 0)$ and $(2 + \sqrt{3}, 0)$.

In the quadratic formula $x = \dfrac{-b \pm \sqrt{b^2 - 4ac}}{2a}$, the radicand $b^2 - 4ac$ is called the **discriminant.** The discriminant tells how many real-number solutions the equation $0 = ax^2 + bx + c$ has, so it also tells how many x-intercepts there are.

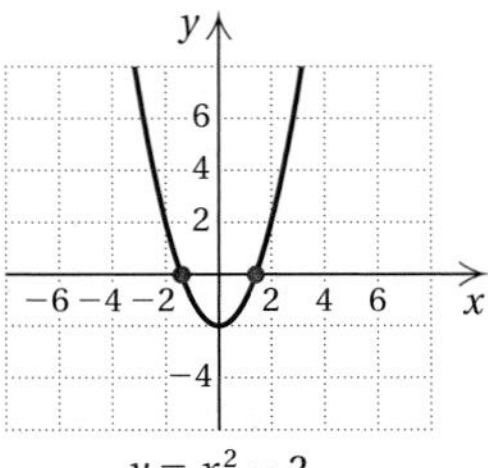

$y = x^2 - 2$
$b^2 - 4ac = 8 > 0$
Two real solutions
Two x-intercepts

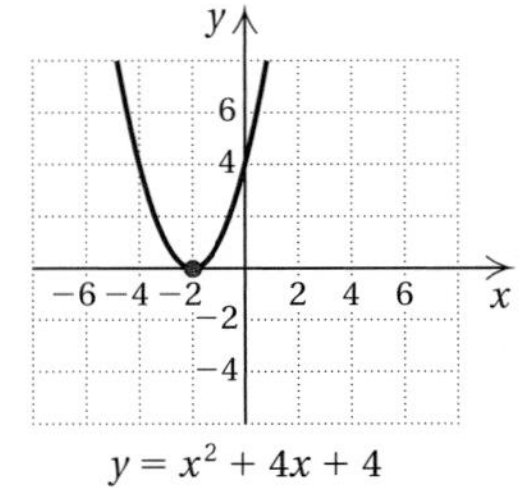

$y = x^2 + 4x + 4$
$b^2 - 4ac = 0$
One real solution
One x-intercept

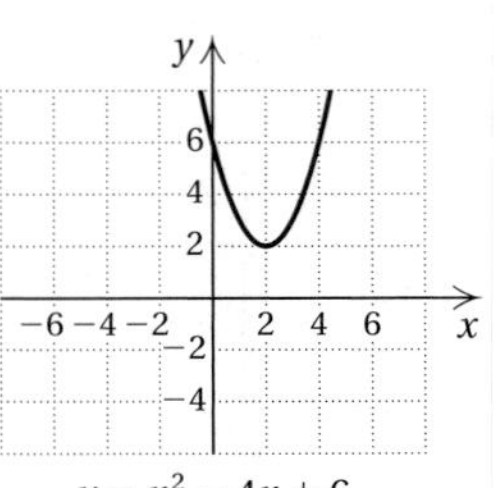

$y = x^2 - 4x + 6$
$b^2 - 4ac = -8 < 0$
No real solutions
No x-intercepts

Do Exercises 4–7.

Visualizing for Success

Match each equation or inequality with its graph.

1. $y = -4 + 4x - x^2$
2. $y = 5 - x^2$
3. $5x + 2y = -10$
4. $5x + 2y \leq 10$
5. $y < 5x$
6. $y = x^2 - 3x - 2$
7. $2x - 5y = 10$
8. $5x - 2y = 10$
9. $2x + 5y = 10$
10. $y = x^2 + 3x - 2$

Answers on page A-52

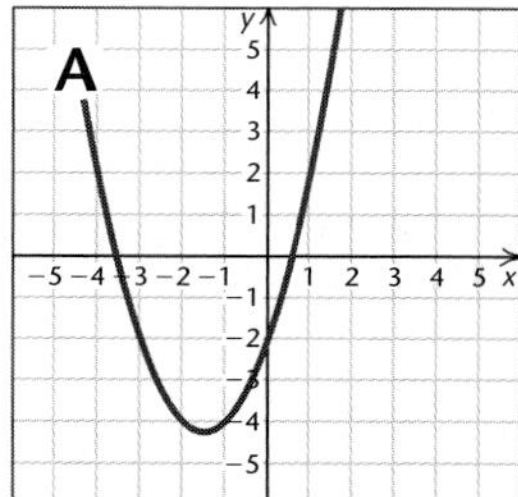

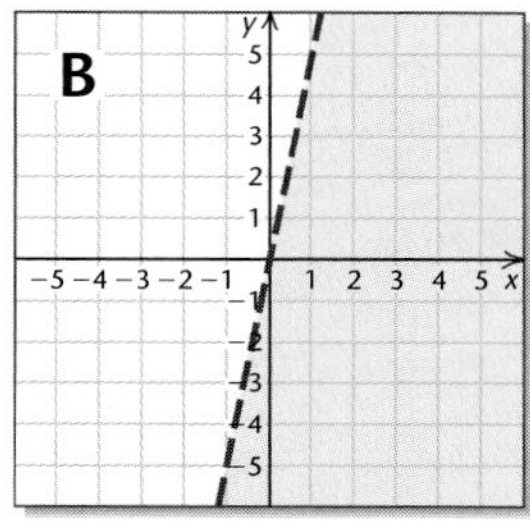

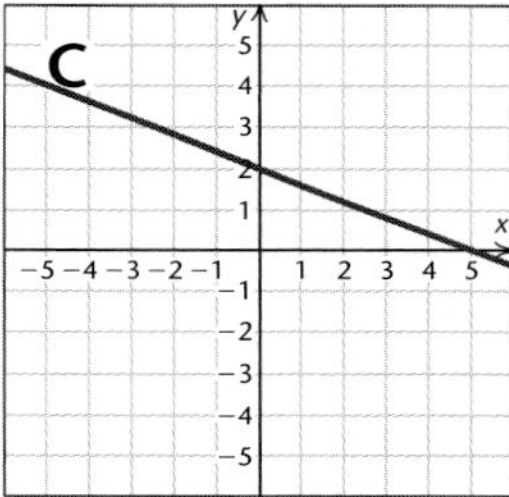

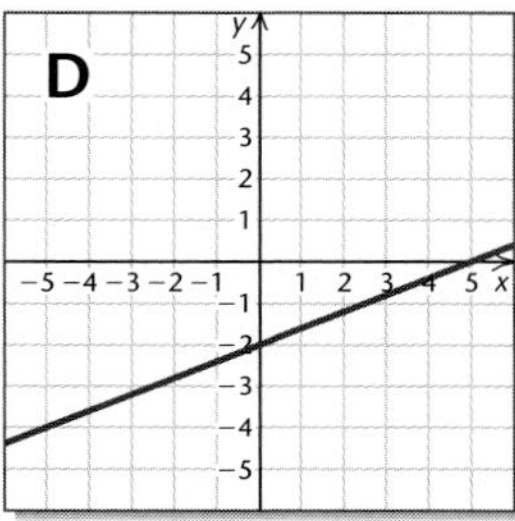

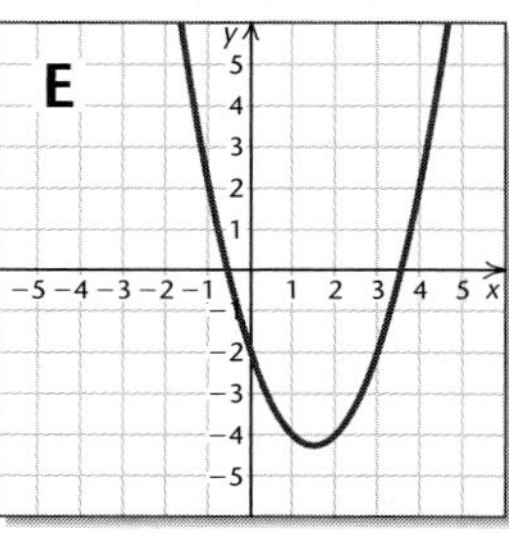

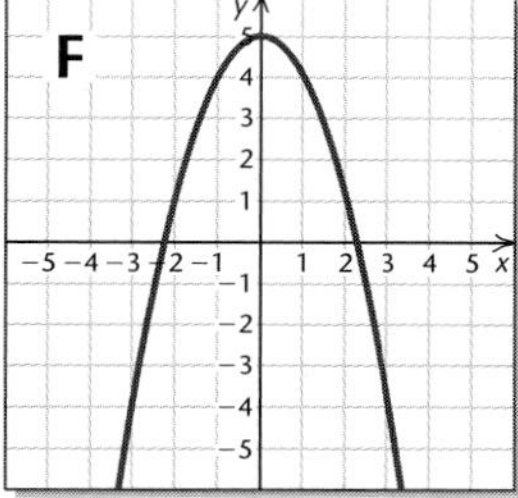

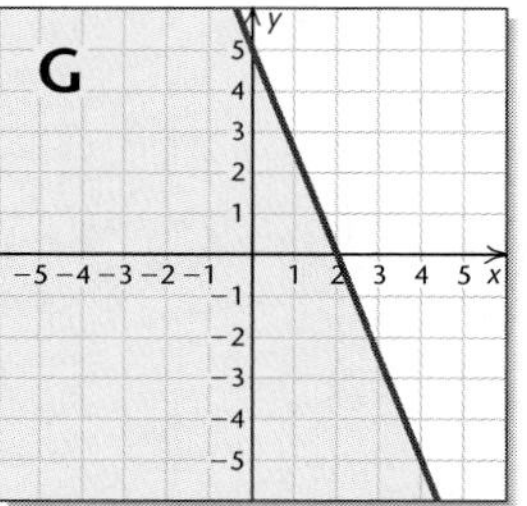

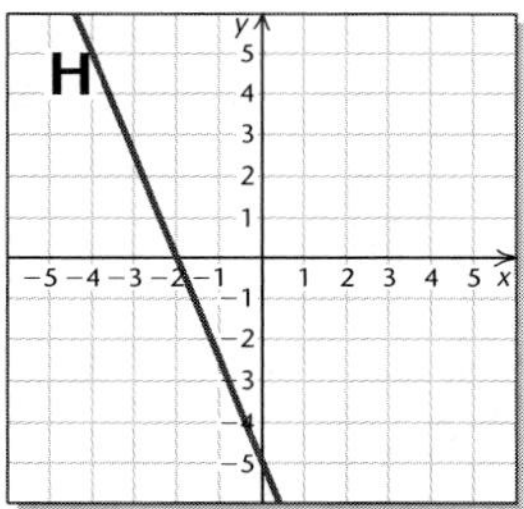

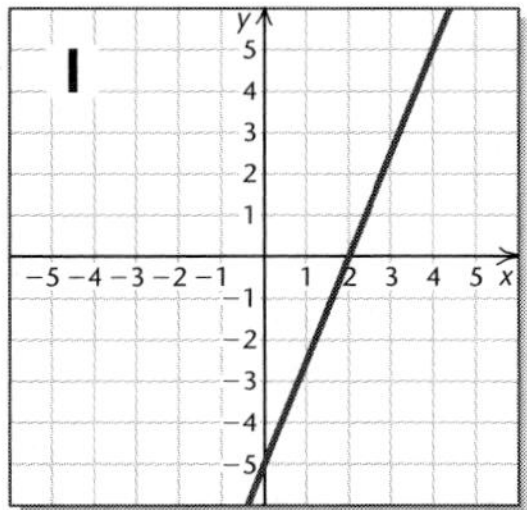

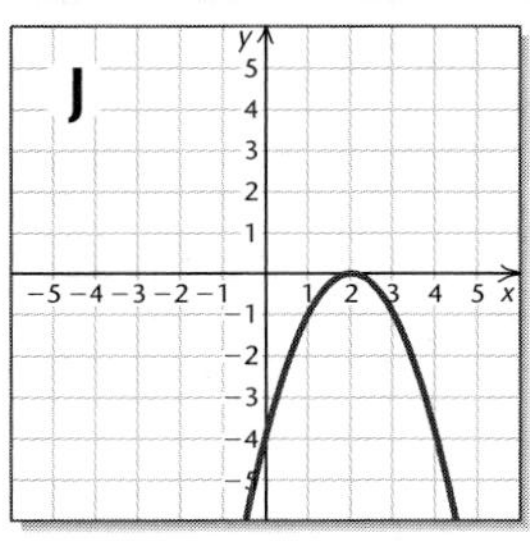

15.6 EXERCISE SET

For Extra Help

MathXL | MyMathLab | InterAct Math | Math Tutor Center | Video Lectures on CD Disc 8 | Student's Solutions Manual

a Graph the quadratic equation. In Exercises 1–8, label the ordered pairs for the vertex and the y-intercept.

1. $y = x^2 + 1$

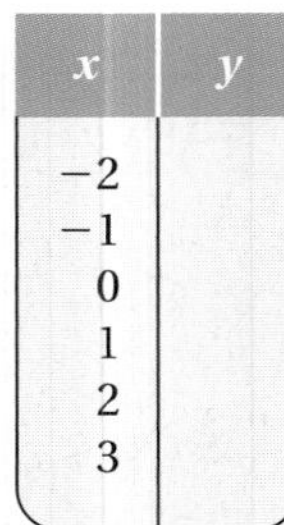

x	y
-2	
-1	
0	
1	
2	
3	

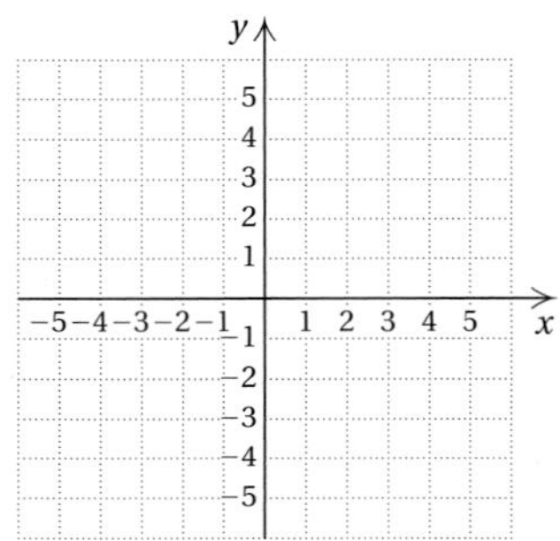

2. $y = 2x^2$

x	y
-2	
-1	
0	
1	
2	
3	

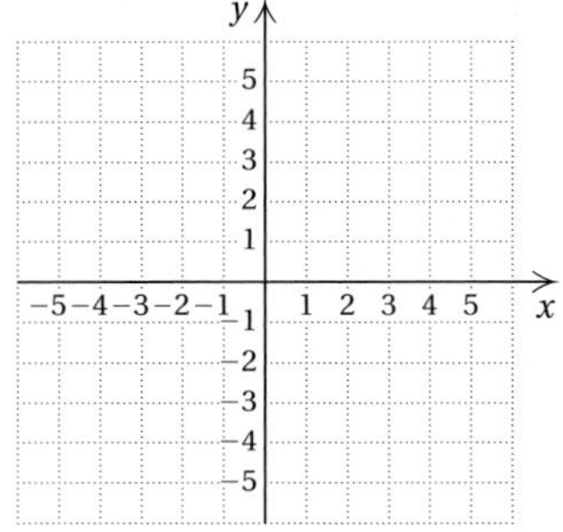

3. $y = -1 \cdot x^2$

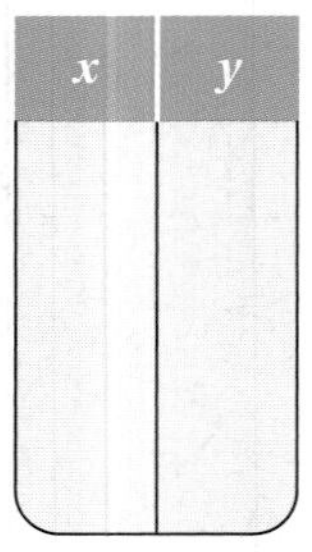

x	y

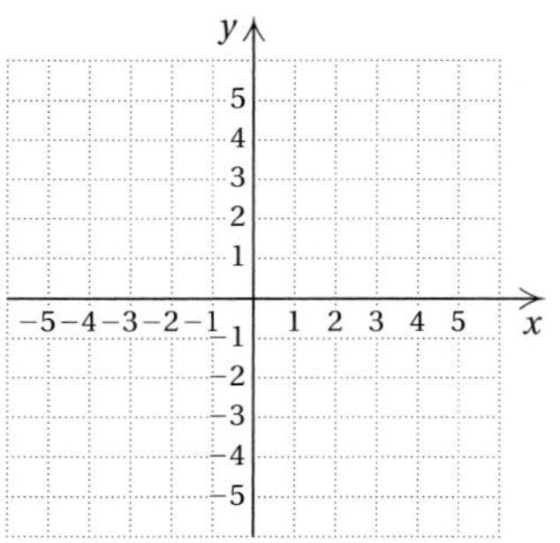

4. $y = x^2 - 1$

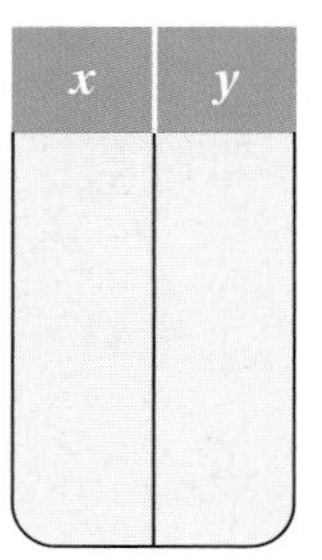

x	y

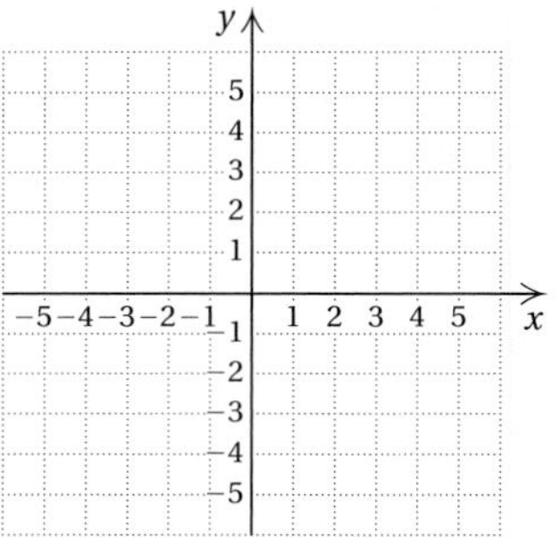

5. $y = -x^2 + 2x$

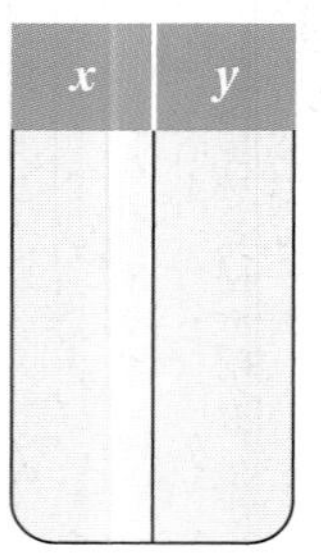

x	y

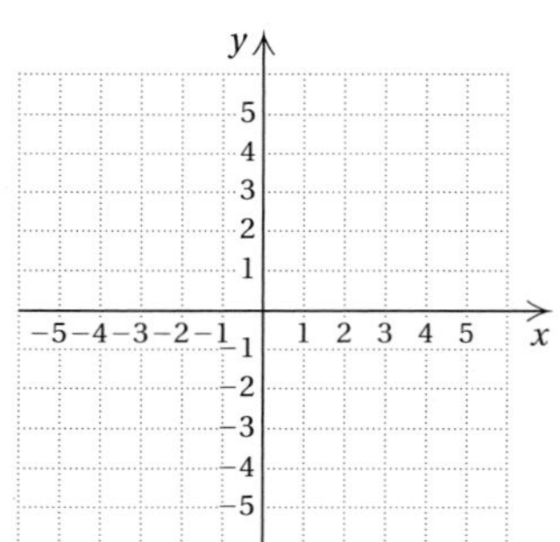

6. $y = x^2 + x - 2$

x	y

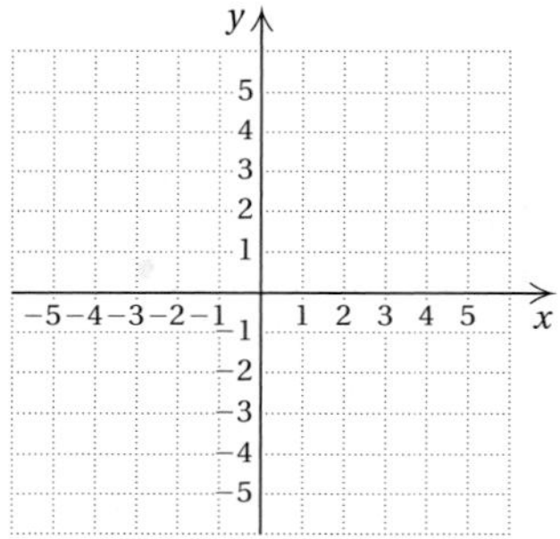

7. $y = 5 - x - x^2$

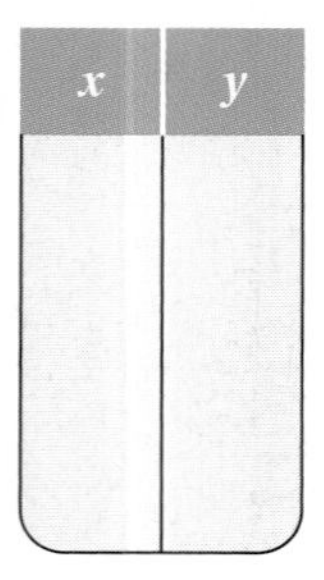

x	y

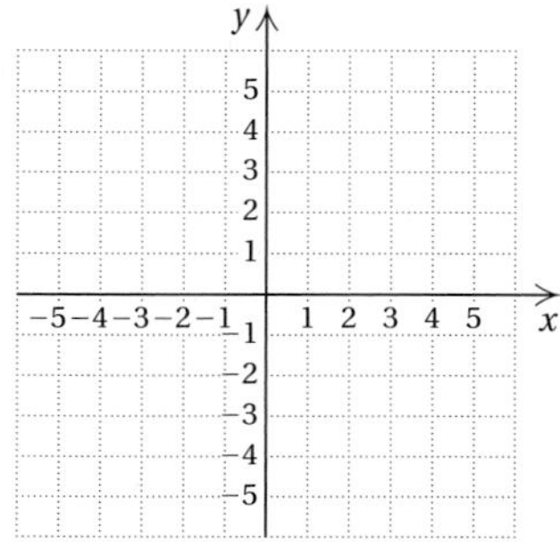

8. $y = x^2 + 2x + 1$

x	y

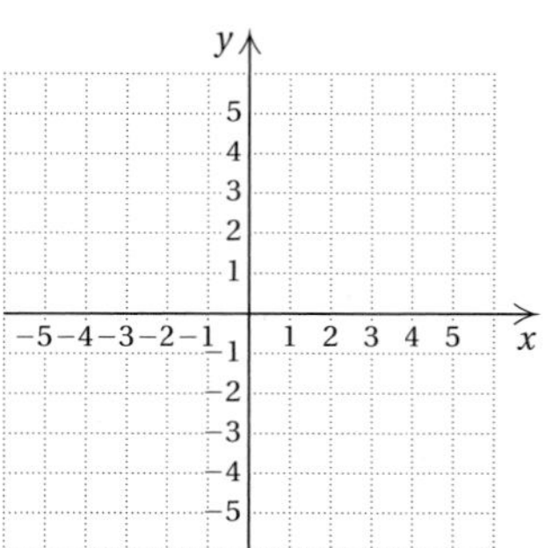

9. $y = x^2 - 2x + 1$

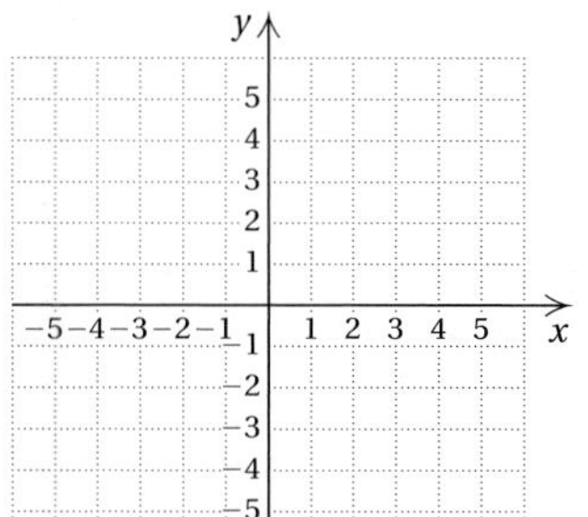

10. $y = -\frac{1}{2}x^2$

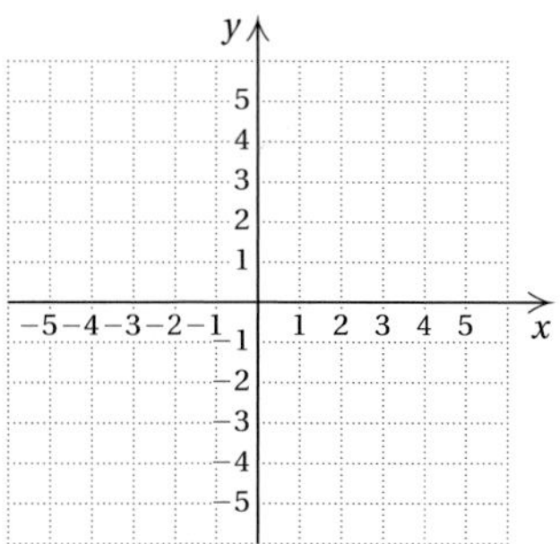

11. $y = -x^2 + 2x + 3$

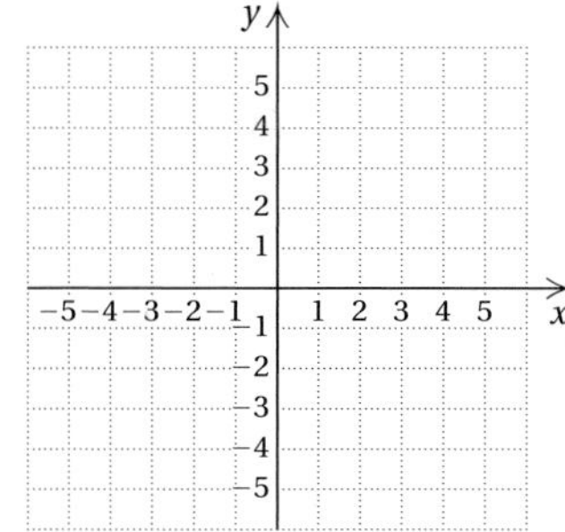

12. $y = -x^2 - 2x + 3$

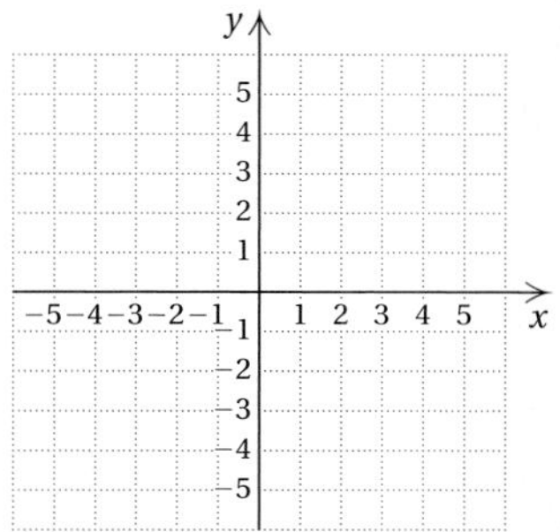

13. $y = -2x^2 - 4x + 1$

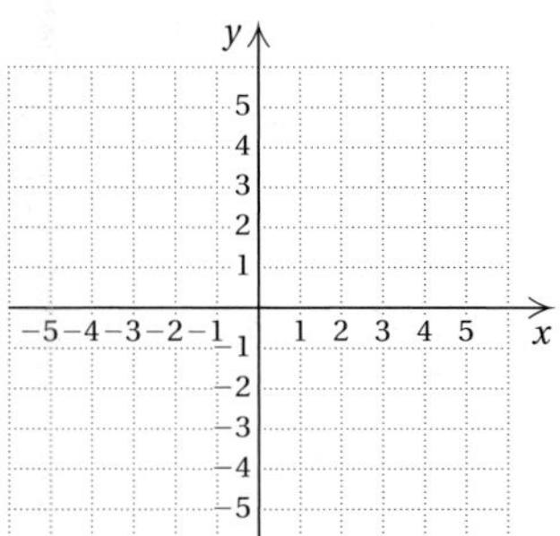

14. $y = 2x^2 + 4x - 1$

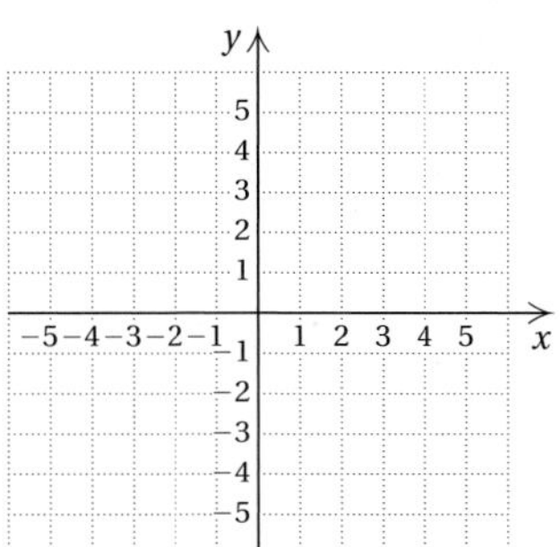

15. $y = 5 - x^2$

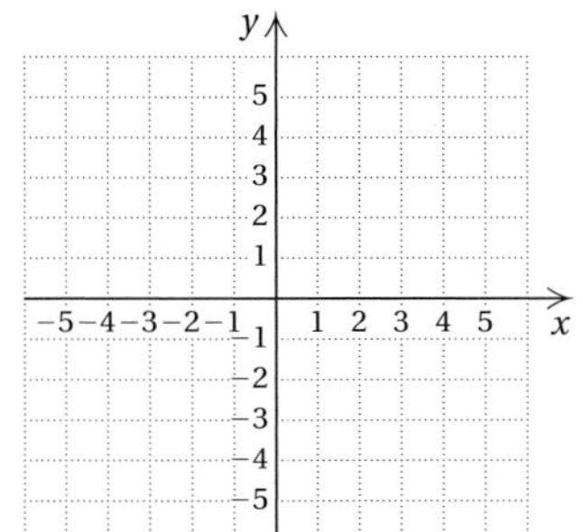

16. $y = 4 - x^2$

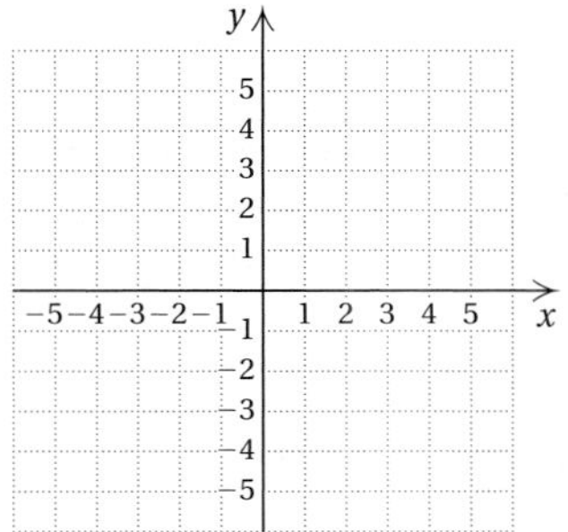

17. $y = \frac{1}{4}x^2$

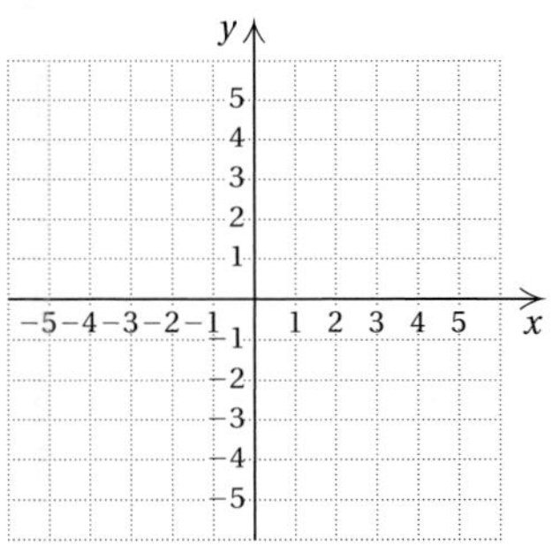

18. $y = -0.1x^2$

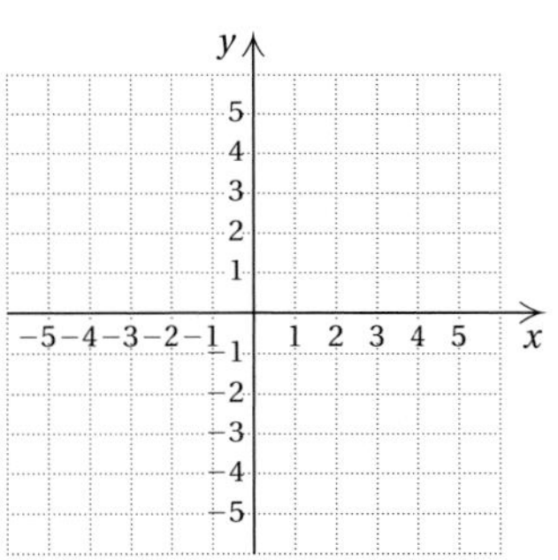

19. $y = -x^2 + x - 1$

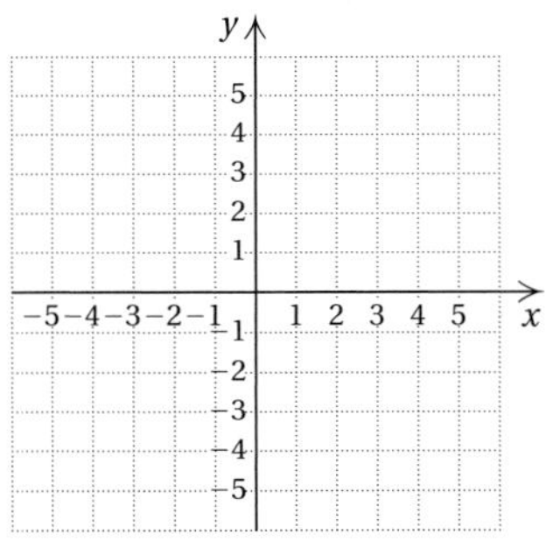

20. $y = x^2 + 2x$

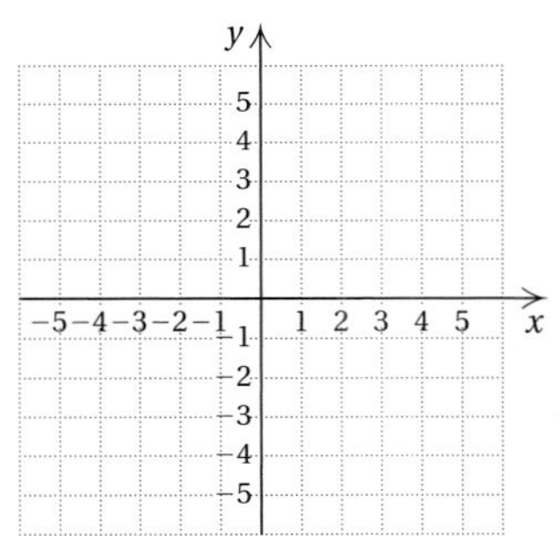

21. $y = -2x^2$

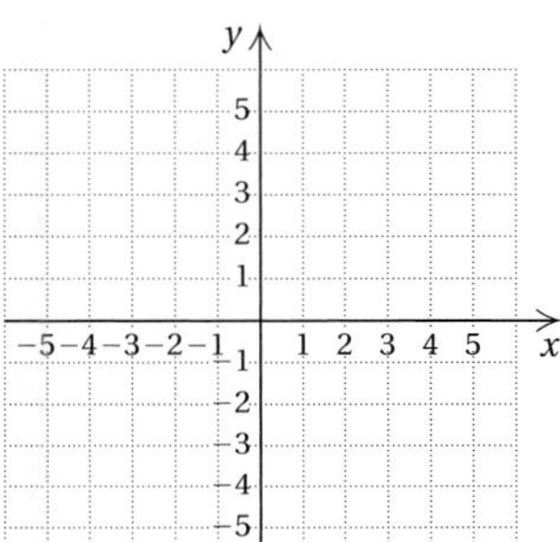

22. $y = -x^2 - 1$

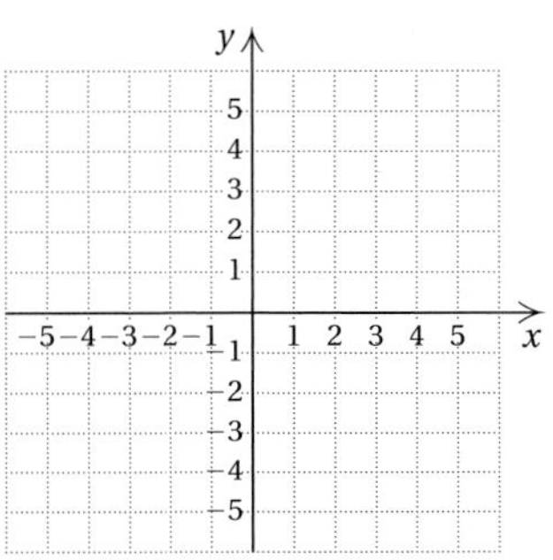

23. $y = x^2 - x - 6$

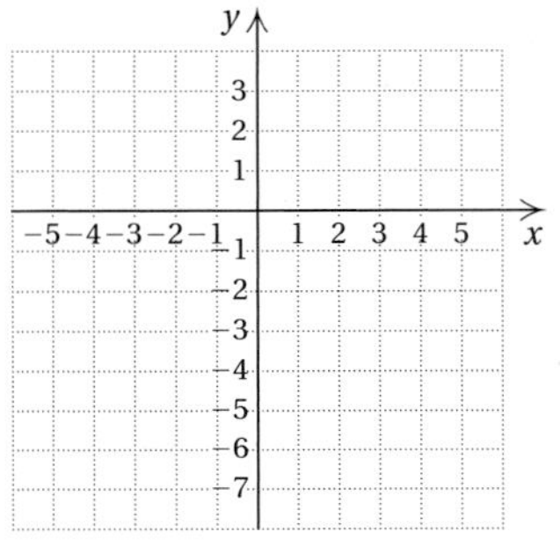

24. $y = 6 + x - x^2$

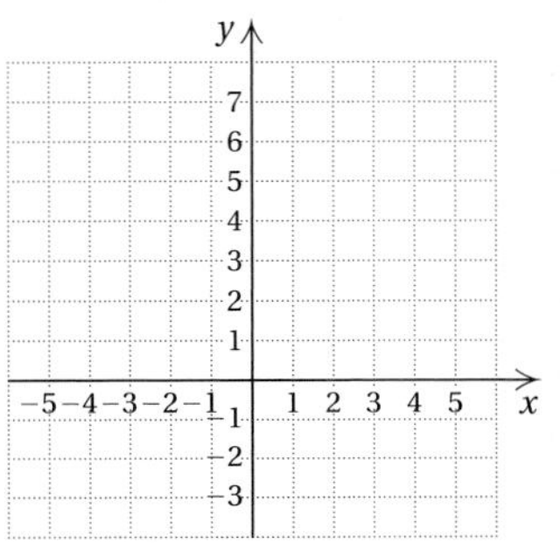

b Find the x-intercepts.

25. $y = x^2 - 2$

26. $y = x^2 - 7$

27. $y = x^2 + 5x$

28. $y = x^2 - 4x$

29. $y = 8 - x - x^2$

30. $y = 8 + x - x^2$

31. $y = x^2 - 6x + 9$

32. $y = x^2 + 10x + 25$

33. $y = -x^2 - 4x + 1$

34. $y = x^2 + 4x - 1$

35. $y = x^2 + 9$

36. $y = x^2 + 1$

37. D_W Suppose that the x-intercepts of a parabola are $(a_1, 0)$ and $(a_2, 0)$. What is the easiest way to find an equation for the line of symmetry? to find the coordinates of the vertex?

38. D_W Discuss the effect of the sign of a on the graph of $y = ax^2 + bx + c$.

SKILL MAINTENANCE

39. Add: $\sqrt{8} + \sqrt{50} + \sqrt{98} + \sqrt{128}$. [14.4a]

40. Multiply and simplify: $\sqrt{5y^4}\sqrt{125y}$. [14.2c]

41. Find an equation of variation in which y varies inversely as x and $y = 12.4$ when $x = 2.4$. [12.9c]

42. Evaluate $3x^4 + 3x - 7$ when $x = -2$. [10.3a]

SYNTHESIS

43. *Height of a Projectile.* The height H, in feet, of a projectile with an initial velocity of 96 ft/sec is given by the equation

$$H = -16t^2 + 96t,$$

where t is the time, in seconds. Use the graph of this equation, shown here, or any equation-solving technique to answer the following questions.

a) How many seconds after launch is the projectile 128 ft above ground?

b) When does the projectile reach its maximum height?

c) How many seconds after launch does the projectile return to the ground?

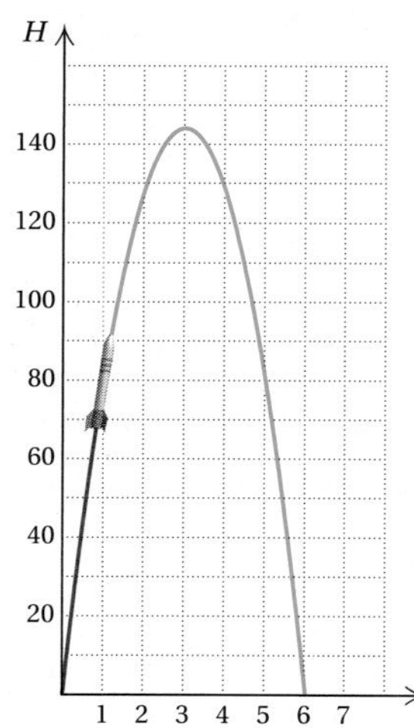

For each equation in Exercises 44–47, evaluate the discriminant $b^2 - 4ac$. Then use the answer to state how many real-number solutions exist for the equation.

44. $y = x^2 + 8x + 16$

45. $y = x^2 + 2x - 3$

46. $y = -2x^2 + 4x - 3$

47. $y = -0.02x^2 + 4.7x - 2300$

15.7 FUNCTIONS

Objectives

a Determine whether a correspondence is a function.

b Given a function described by an equation, find function values (outputs) for specified values (inputs).

c Draw a graph of a function.

d Determine whether a graph is that of a function.

e Solve applied problems involving functions and their graphs.

a Identifying Functions

We now develop one of the most important concepts in mathematics, **functions.** We have actually been studying functions all through this text; we just haven't identified them as such. Ordered pairs form a correspondence between first and second coordinates. A function is a special correspondence from one set of numbers to another. For example:

To each student in a college, there corresponds his or her student ID number.

To each item in a store, there corresponds its price.

To each real number, there corresponds the cube of that number.

In each case, the first set is called the **domain** and the second set is called the **range.** Given a member of the domain, there is *just one* member of the range to which it corresponds. This kind of correspondence is called a **function.**

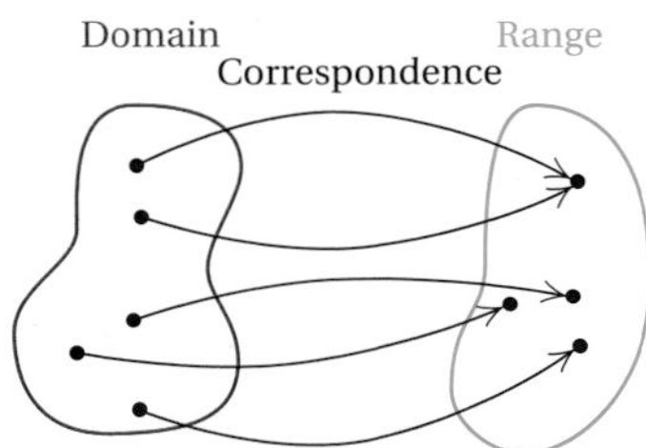

EXAMPLE 1 Determine whether the correspondence is a function.

f:

Domain		*Range*
1	→	\$107.4
2	→	\$ 34.1
3	→	\$ 29.6
4	→	\$ 19.6

g:

Domain		*Range*
3	→	5
4	→	9
5	→	−7
6	→	−7

h:

Domain		*Range*
New York	→	Mets
New York	→	Yankees
St. Louis	→	Cardinals
San Diego	→	Padres

p:

Domain		*Range*
Mets	→	New York
Yankees	→	New York
Cardinals	→	St. Louis
Padres	→	San Diego

The correspondence f *is* a function because each member of the domain is matched to only one member of the range.

The correspondence g *is* also a function because each member of the domain is matched to only one member of the range.

The correspondence h *is not* a function because one member of the domain, New York, is matched to more than one member of the range.

The correspondence p *is* a function because each member of the domain is paired with only one member of the range.

Determine whether the correspondence is a function.

1. *Domain* — *Range*

Cheetah → 70 mph
Human → 28 mph
Lion → 50 mph
Chicken → 9 mph

2. *Domain* — *Range*

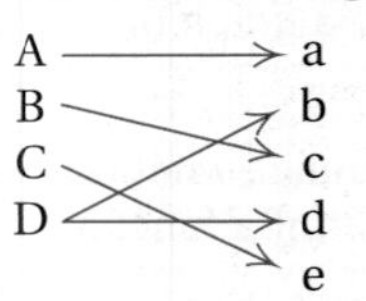

3. *Domain* — *Range*

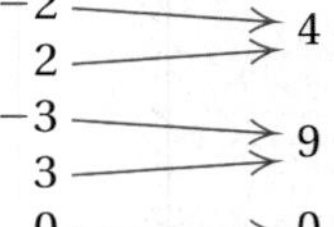

4. *Domain* — *Range*

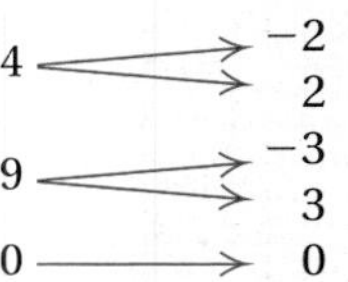

Determine whether each of the following is a function.

5. *Domain*
A set of numbers
Correspondence
Square each number and subtract 10.
Range
A set of numbers

6. *Domain*
A set of polygons
Correspondence
Find the perimeter of each polygon.
Range
A set of numbers

Answers on page A-53

FUNCTION, DOMAIN, AND RANGE

A **function** is a correspondence between a first set, called the **domain,** and a second set, called the **range,** such that each member of the domain corresponds to *exactly one* member of the range.

Do Exercises 1–4.

EXAMPLE 2 Determine whether the correspondence is a function.

	Domain	*Correspondence*	*Range*
a)	A family	Each person's weight	A set of positive numbers
b)	The natural numbers	Each number's square	A set of natural numbers
c)	The set of all states	Each state's members of the U.S. Senate	A set of U.S. Senators

a) The correspondence *is* a function because each person has *only one* weight.

b) The correspondence *is* a function because each natural number has *only one* square.

c) The correspondence *is not* a function because each state has two U.S. Senators.

Do Exercises 5 and 6.

When a correspondence between two sets is not a function, it may still be an example of a *relation.*

RELATION

A **relation** is a correspondence between a first set, called the **domain,** and a second set, called the **range,** such that each member of the domain corresponds to *at least one* member of the range.

Thus, although the correspondences of Examples 1 and 2 are not all functions, they *are* all relations. A function is a special type of relation—one in which each member of the domain is paired with *exactly one* member of the range.

b Finding Function Values

Most functions considered in mathematics are described by equations. A linear equation like $y = 2x + 3$, studied in Chapter 9, is called a **linear function.** A quadratic equation like $y = 4 - x^2$, studied in Chapter 15, is called a **quadratic function.**

Recall that when graphing $y = 2x + 3$, we chose x-values and then found corresponding y-values. For example, when $x = 4$,

$$y = 2x + 3 = 2 \cdot 4 + 3 = 11.$$

When thinking of functions, we call the number 4 an **input** and the number 11 an **output.**

It helps to think of a function as a machine; that is, think of putting a member of the domain (an input) into the machine. The machine knows the correspondence and gives out a member of the range (the output).

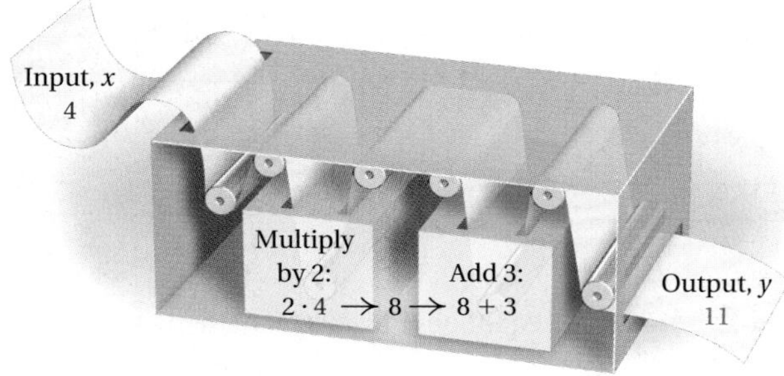

The function $y = 2x + 3$ can be named f and described by the equation $f(x) = 2x + 3$. We call the input x and the output $f(x)$. This is read "f of x," or "f at x," or "the value of f at x."

Caution!

The notation $f(x)$ *does not mean* "f times x" and should not be read that way.

The equation $f(x) = 2x + 3$ describes the function that takes an input x, multiplies it by 2, and then adds 3.

Input ↘ $f(x) = 2x \quad + 3$ ↗ **Multiply by 2**   **Add 3**

To find the output $f(4)$, we take the input 4, double it, and add 3 to get 11. That is, we substitute 4 into the formula for $f(x)$:

$$f(4) = 2 \cdot 4 + 3 = 11.$$

Outputs of functions are also called **function values.** For $f(x) = 2x + 3$, we know that $f(4) = 11$. We can say that "the function value at 4 is 11."

EXAMPLE 3 Find the indicated function value.

a) $f(5)$, for $f(x) = 3x + 2$ **b)** $g(3)$, for $g(z) = 5z^2 - 4$

c) $A(-2)$, for $A(r) = 3r^2 + 2r$ **d)** $f(-5)$, for $f(x) = x^2 + 3x - 4$

a) $f(5) = 3 \cdot 5 + 2 = 17$

b) $g(3) = 5(3)^2 - 4 = 41$

c) $A(-2) = 3(-2)^2 + 2(-2) = 8$

d) $f(-5) = (-5)^2 + 3(-5) - 4 = 25 - 15 - 4 = 6$

Do Exercises 7 and 8.

Find the function values.

7. $f(x) = 5x - 3$

a) $f(-6)$

b) $f(0)$

c) $f(1)$

d) $f(20)$

e) $f(-1.2)$

8. $g(x) = x^2 - 4x + 9$

a) $g(-2)$

b) $g(0)$

c) $g(5)$

d) $g(10)$

Answers on page A-53

CALCULATOR CORNER

Finding Function Values We can find function values on a graphing calculator. One method is to substitute inputs directly into the formula. Consider the function in Example 3(d), $f(x) = x^2 + 3x - 4$. To find $f(-5)$, we press 

. We find that $f(-5) = 6$.

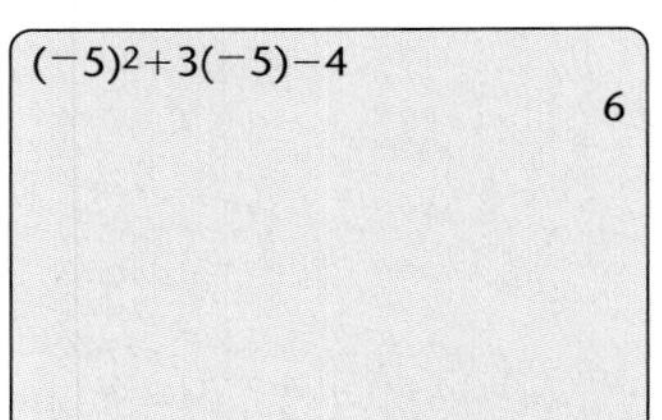

X	Y1	
−5	6	
X =		

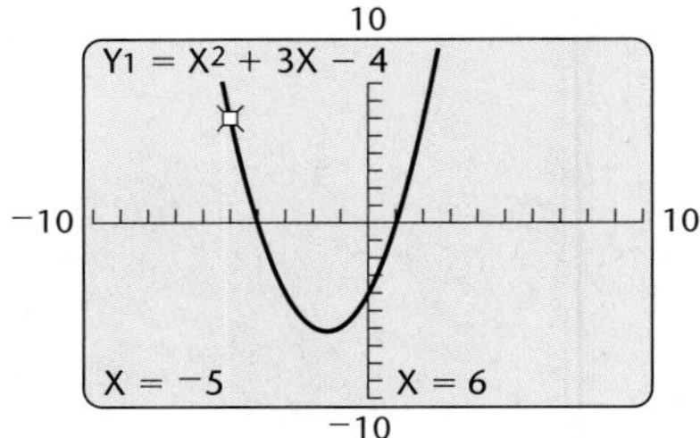

After we have entered the function as $y_1 = x^2 + 3x - 4$ on the equation-editor screen, there are several other methods that we can use to find function values. We can use a table set in ASK mode and enter $x = -5$. We see that the function value, y_1, is 6. We can also use the VALUE feature to evaluate the function. To do this, we first graph the function in a window that includes $x = -5$ and then press 2ND CALC 1 to access the VALUE feature. Next, we supply the desired x-value by pressing (−) 5. Finally, we press ENTER to see $x = -5$, $y = 6$ at the bottom of the screen. Again we see that the function value is 6.

There are other ways to find function values, but we will not discuss them here.

Exercises: Find the function values.

1. $f(-3.4)$, for $f(x) = 2x - 6$
2. $f(4)$, for $f(x) = -2.3x$
3. $f(-1)$, for $f(x) = x^2 - 3$
4. $f(3)$, for $f(x) = 2x^2 - x + 5$

C Graphs of Functions

To graph a function, we find ordered pairs (x, y) or $(x, f(x))$, plot them, and connect the points. Note that y and $f(x)$ are used interchangeably when we are working with functions and their graphs.

EXAMPLE 4 Graph: $f(x) = x + 2$.

A list of some function values is shown in this table. We plot the points and connect them. The graph is a straight line.

x	$f(x)$
−4	−2
−3	−1
−2	0
−1	1
0	2
1	3
2	4
3	5
4	6

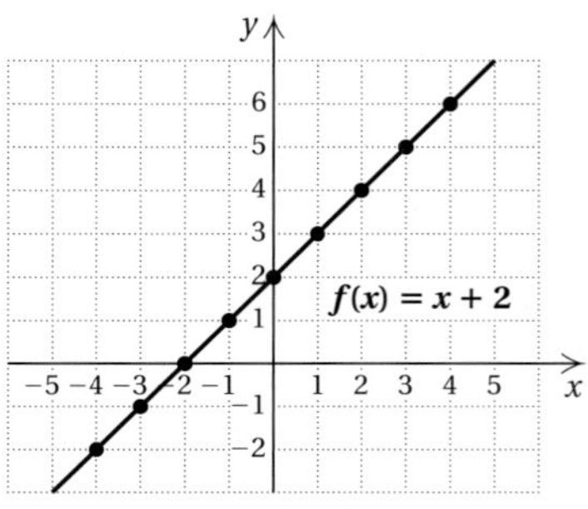

EXAMPLE 5 Graph: $g(x) = 4 - x^2$.

Recall from Section 9.6 that the graph is a parabola. We calculate some function values and draw the curve.

$$g(0) = 4 - 0^2 = 4 - 0 = 4,$$
$$g(-1) = 4 - (-1)^2 = 4 - 1 = 3,$$
$$g(2) = 4 - (2)^2 = 4 - 4 = 0,$$
$$g(-3) = 4 - (-3)^2 = 4 - 9 = -5$$

x	$g(x)$
−3	−5
−2	0
−1	3
0	4
1	3
2	0
3	−5

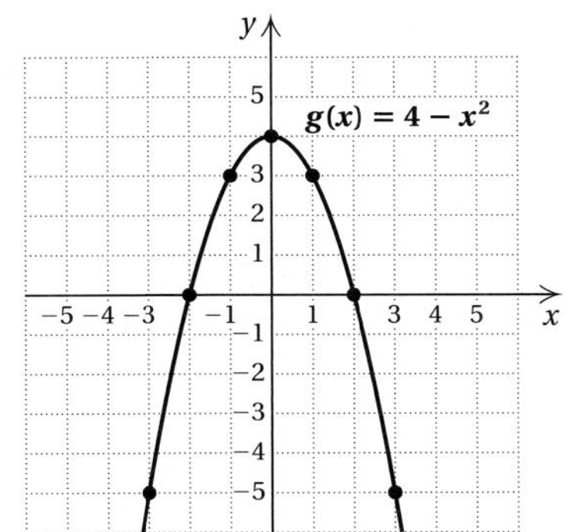

EXAMPLE 6 Graph: $h(x) = |x|$.

A list of some function values is shown in the following table. We plot the points and connect them. The graph is a V-shaped "curve" that rises on either side of the vertical axis.

x	$h(x)$
−3	3
−2	2
−1	1
0	0
1	1
2	2
3	3

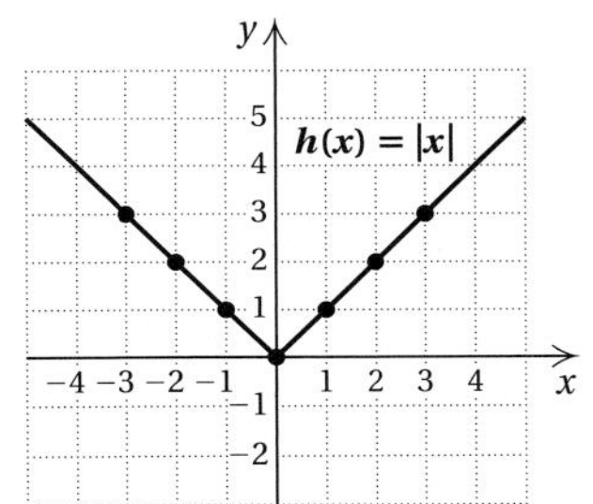

Do Exercises 9–11 on the following page.

d The Vertical-Line Test

Consider the function f described by $f(x) = x^2 - 5$. Its graph is shown at right. It is also the graph of the equation $y = x^2 - 5$.

To find a function value, like $f(3)$, from a graph, we locate the input on the horizontal axis, move vertically to the graph of the function, and then horizontally to find the output on the vertical axis, where members of the range can be found.

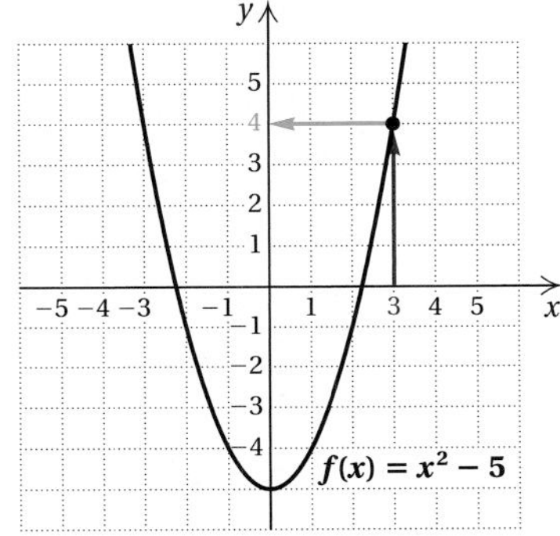

Study Tips

FINAL STUDY TIP

You are arriving at the end of your course in Introductory Algebra. If you have not begun to prepare for the final examination, be sure to read the comments in the Study Tips on pp. 975, 1022, 1037, and 1094.

"We make a living by what we get, but we make a life by what we give."

Winston Churchill

Graph.

9. $f(x) = x - 4$

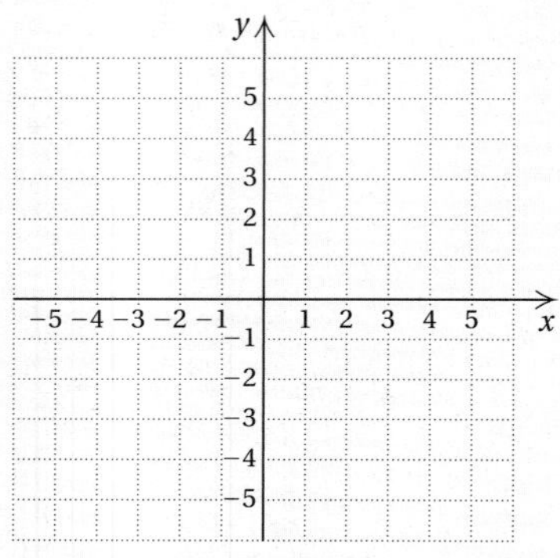

10. $g(x) = 5 - x^2$

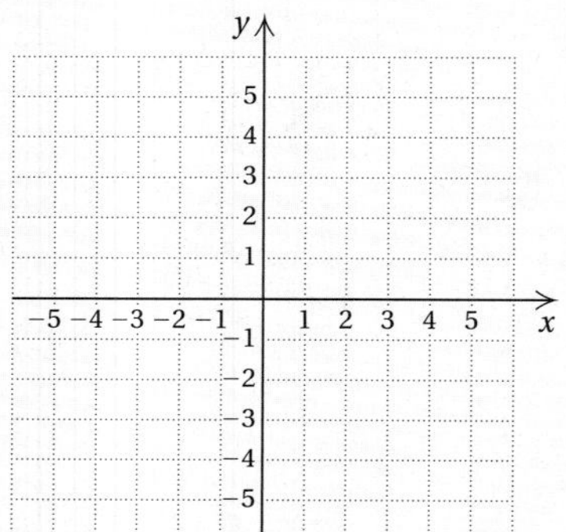

11. $t(x) = 3 - |x|$

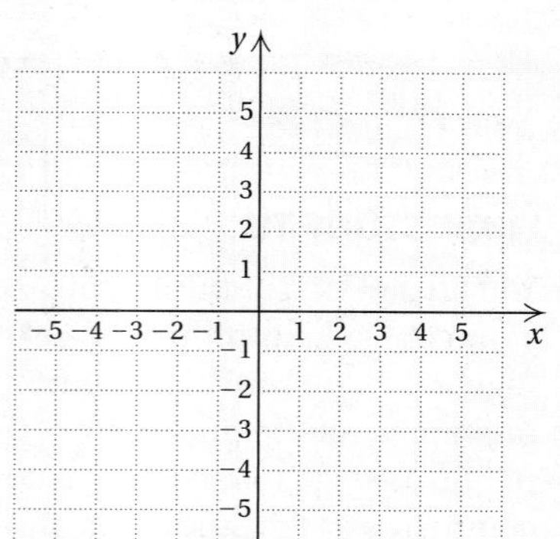

Answers on page A-53

Recall that when one member of the domain is paired with two or more different members of the range, the correspondence is *not* a function. Thus, when a graph contains two or more different points with the same first coordinate, the graph cannot represent a function. Points sharing a common first coordinate are vertically above or below each other (see the following graph).

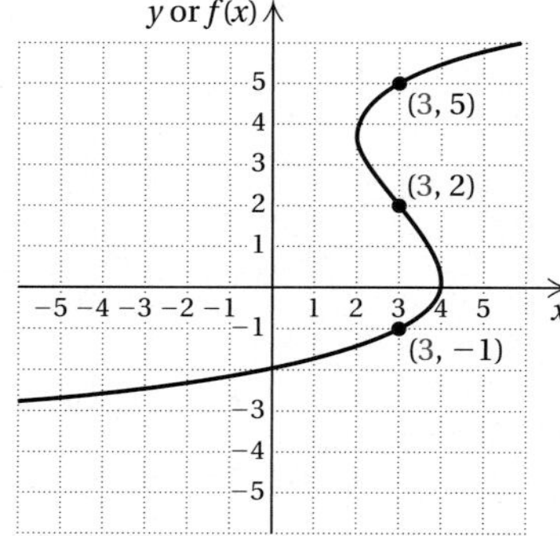

Since 3 is paired with more than one member of the range, the graph does not represent a function.

This observation leads to the *vertical-line test.*

THE VERTICAL-LINE TEST

A graph represents a function if it is impossible to draw a vertical line that intersects the graph more than once.

EXAMPLE 7 Determine whether each of the following is the graph of a function.

a) 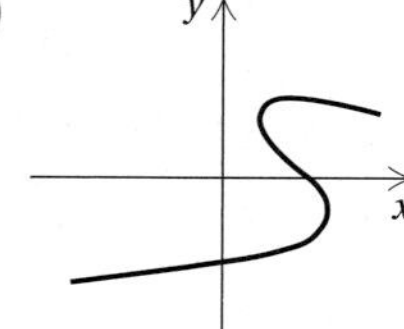

b) 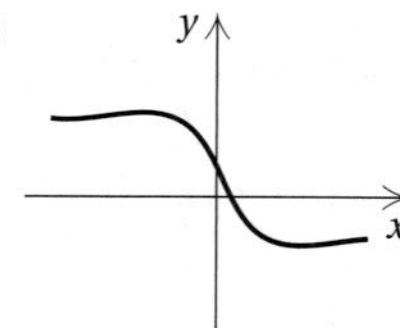

c) 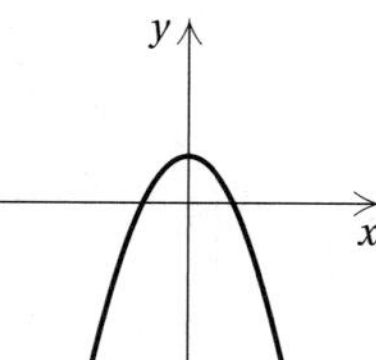

d) 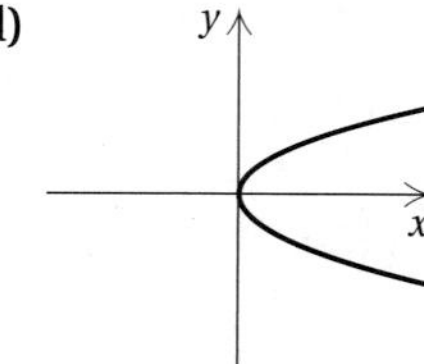

a) The graph *is not* that of a function because a vertical line crosses the graph at more than one point.

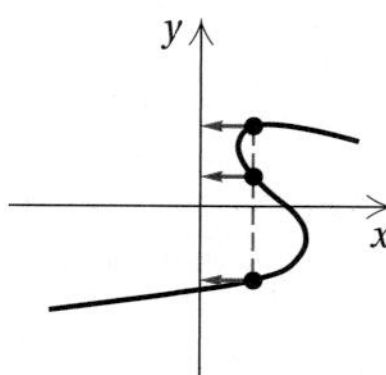

b) The graph *is* that of a function because no vertical line can cross the graph at more than one point. This can be confirmed with a ruler or straightedge.

c) The graph *is* that of a function.

d) The graph *is not* that of a function. There is a vertical line that crosses the graph more than once.

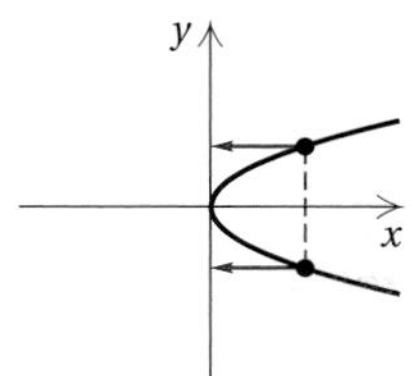

Do Exercises 12–15.

e Applications of Functions and Their Graphs

Functions are often described by graphs, whether or not an equation is given. To use a graph in an application, we note that each point on the graph represents a pair of values.

EXAMPLE 8 *Movie Revenue.* The following graph approximates the weekly revenue, in millions of dollars, from the movie *Math and the City*. The revenue is a function of the week, and no equation is given for the function.

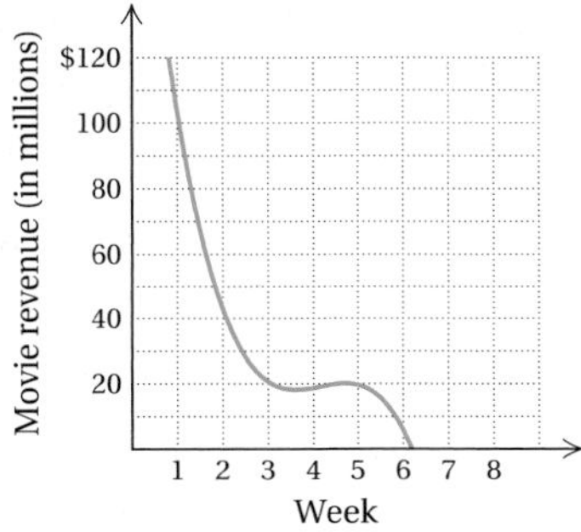

Use the graph to answer the following.

a) What was the movie revenue for week 1?

b) What was the movie revenue for week 5?

a) To estimate the revenue for week 1, we locate 1 on the horizontal axis and move directly up until we reach the graph. Then we move across to the vertical axis. We estimate that value to be about $105 million.

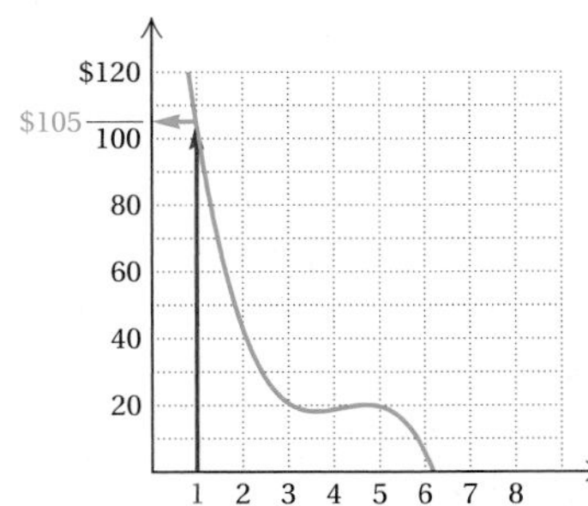

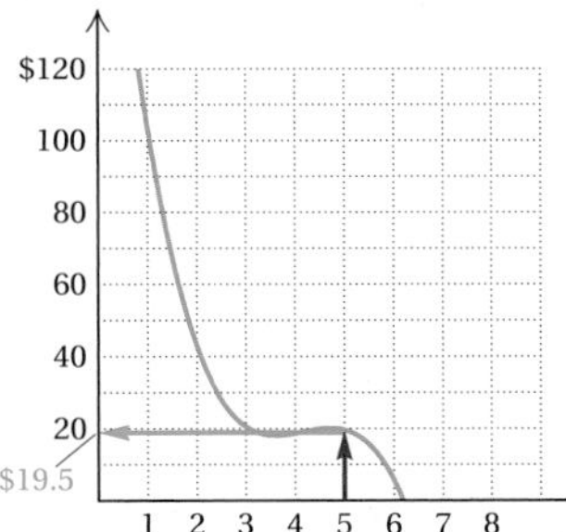

b) To estimate the revenue for week 5, we locate 5 on the horizontal axis and move directly up until we reach the graph. Then we move across to the vertical axis. We estimate that value to be about $19.5 million.

Do Exercises 16 and 17.

Determine whether each of the following is the graph of a function.

12.

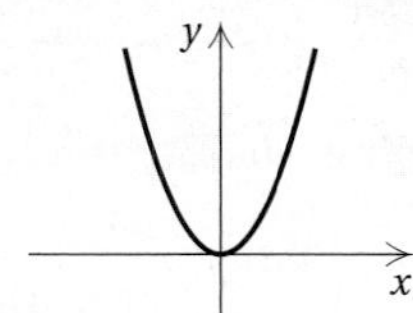

13.

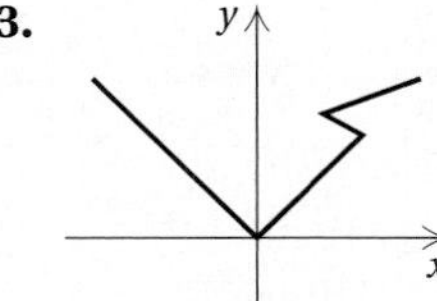

14.

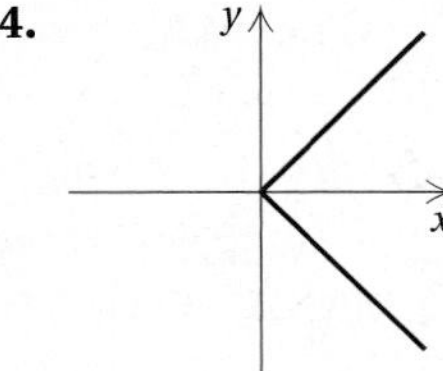

15.

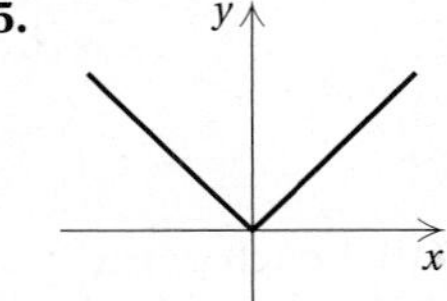

Referring to the graph in Example 8:

16. What was the movie revenue for week 2?

17. What was the movie revenue for week 6?

Answers on page A-53

15.7 EXERCISE SET

For Extra Help

MathXL | MyMathLab | InterAct Math | Math Tutor Center | Video Lectures on CD Disc 8 | Student's Solutions Manual

a Determine whether the correspondence is a function.

1. *Domain* → *Range*

2 → 9
5 → 8
19 → 9

2. *Domain* → *Range*

5 → 3
−3 → 3
7 → 7
−7 → 7

3. *Domain* → *Range*

−5 → 1
5 → 1
8 → 1

4. *Domain* → *Range*

6 → −6
6 → −7
7 → −7
3 → −3

5. *Domain* → *Range*

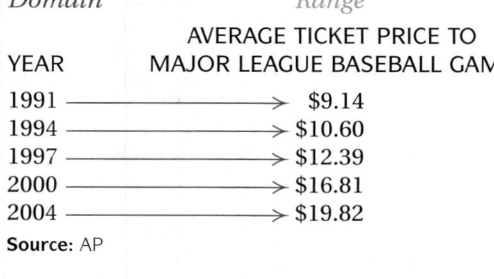

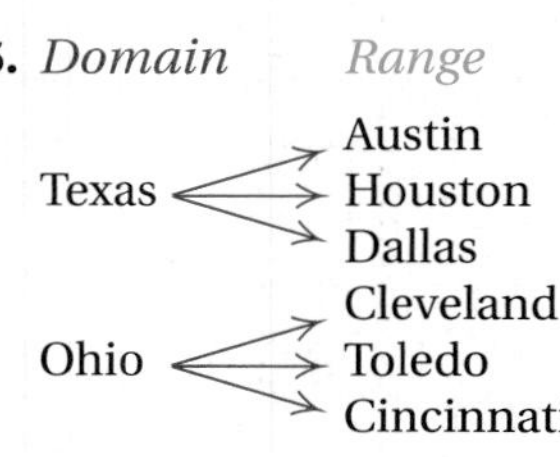

Texas → Austin
Texas → Houston
Texas → Dallas
Ohio → Cleveland
Ohio → Toledo
Ohio → Cincinnati

6. *Domain* → *Range*

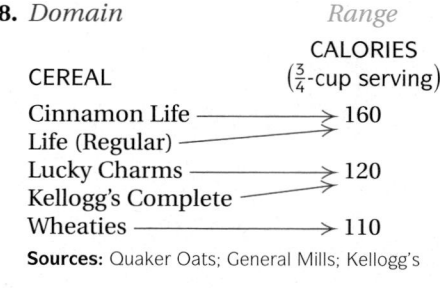

Austin → Texas
Houston → Texas
Dallas → Texas
Cleveland → Ohio
Toledo → Ohio
Cincinnati → Ohio

7. *Domain* → *Range*

YEAR	AVERAGE TICKET PRICE TO MAJOR LEAGUE BASEBALL GAME
1991	\$9.14
1994	\$10.60
1997	\$12.39
2000	\$16.81
2004	\$19.82

Source: AP

8. *Domain* → *Range*

CEREAL	CALORIES ($\frac{3}{4}$-cup serving)
Cinnamon Life	160
Life (Regular)	160
Lucky Charms	120
Kellogg's Complete	120
Wheaties	110

Sources: Quaker Oats; General Mills; Kellogg's

Determine whether each of the following is a function. Identify any relations that are not functions.

	Domain	*Correspondence*	*Range*
9.	A math class	Each person's seat number	A set of numbers
10.	A set of numbers	Square each number and then add 4.	A set of numbers
11.	A set of shapes	Find the area of each shape.	A set of numbers
12.	A family	Each person's eye color	A set of colors
13.	The people in a town	Each person's aunt	A set of females
14.	A set of avenues	Find an intersecting road.	A set of cross streets

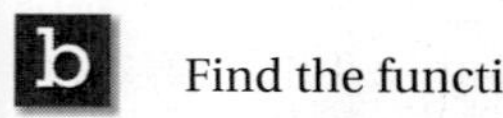

b Find the function values.

15. $f(x) = x + 5$

a) $f(4)$ **b)** $f(7)$
c) $f(-3)$ **d)** $f(0)$
e) $f(2.4)$ **f)** $f(\frac{2}{3})$

16. $g(t) = t - 6$

a) $g(0)$ **b)** $g(6)$
c) $g(13)$ **d)** $g(-1)$
e) $g(-1.08)$ **f)** $g(\frac{7}{8})$

17. $h(p) = 3p$

a) $h(-7)$ **b)** $h(5)$
c) $h(14)$ **d)** $h(0)$
e) $h(\frac{2}{3})$ **f)** $h(-54.2)$

18. $f(x) = -4x$

a) $f(6)$
b) $f\left(-\frac{1}{2}\right)$
c) $f(20)$
d) $f(11.8)$
e) $f(0)$
f) $f(-1)$

19. $g(s) = 3s + 4$

a) $g(1)$
b) $g(-7)$
c) $g(6.7)$
d) $g(0)$
e) $g(-10)$
f) $g\left(\frac{2}{3}\right)$

20. $h(x) = 19$, a constant function

a) $h(4)$
b) $h(-6)$
c) $h(12.5)$
d) $h(0)$
e) $h\left(\frac{2}{3}\right)$
f) $h(1234)$

21. $f(x) = 2x^2 - 3x$

a) $f(0)$
b) $f(-1)$
c) $f(2)$
d) $f(10)$
e) $f(-5)$
f) $f(-10)$

22. $f(x) = 3x^2 - 2x + 1$

a) $f(0)$
b) $f(1)$
c) $f(-1)$
d) $f(10)$
e) $f(2)$
f) $f(-3)$

23. $f(x) = |x| + 1$

a) $f(0)$
b) $f(-2)$
c) $f(2)$
d) $f(-3)$
e) $f(-10)$
f) $f(22)$

24. $g(t) = \sqrt{t}$

a) $g(4)$
b) $g(25)$
c) $g(16)$
d) $g(100)$
e) $g(50)$
f) $g(84)$

25. $f(x) = x^3$

a) $f(0)$
b) $f(-1)$
c) $f(2)$
d) $f(10)$
e) $f(-5)$
f) $f(-10)$

26. $f(x) = x^4 - 3$

a) $f(1)$
b) $f(-1)$
c) $f(0)$
d) $f(2)$
e) $f(-2)$
f) $f(10)$

27. *Estimating Heights.* An anthropologist can estimate the height of a male or a female, given the lengths of certain bones. A *humerus* is the bone from the elbow to the shoulder. The height, in centimeters, of a female with a humerus of x centimeters is given by the function

$$F(x) = 2.75x + 71.48.$$

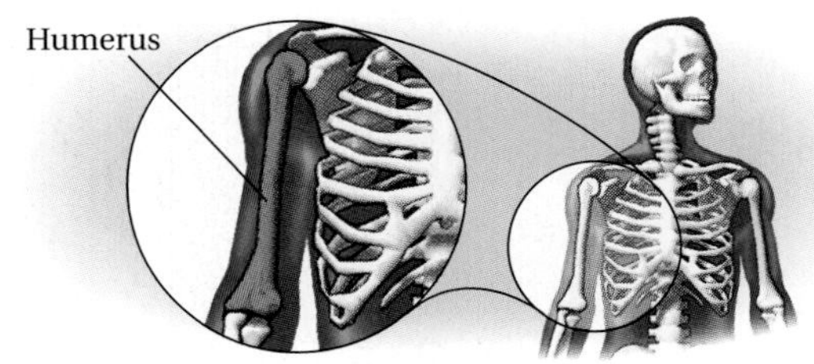

If a humerus is known to be from a female, how tall was she if the bone is **(a)** 32 cm long? **(b)** 30 cm long?

28. Refer to Exercise 27. When a humerus is from a male, the function

$$M(x) = 2.89x + 70.64$$

can be used to find the male's height, in centimeters. If a humerus is known to be from a male, how tall was he if the bone is **(a)** 30 cm long? **(b)** 35 cm long?

29. *Pressure at Sea Depth.* The function $P(d) = 1 + (d/33)$ gives the pressure, in *atmospheres* (atm), at a depth of d feet in the sea. Note that $P(0) = 1$ atm, $P(33) = 2$ atm, and so on. Find the pressure at 20 ft, 30 ft, and 100 ft.

30. *Temperature as a Function of Depth.* The function $T(d) = 10d + 20$ gives the temperature, in degrees Celsius, inside the earth as a function of the depth d, in kilometers. Find the temperature at 5 km, 20 km, and 1000 km.

31. *Melting Snow.* The function $W(d) = 0.112d$ approximates the amount, in centimeters, of water that results from d centimeters of snow melting. Find the amount of water that results from snow melting from depths of 16 cm, 25 cm, and 100 cm.

32. *Temperature Conversions.* The function $C(F) = \frac{5}{9}(F - 32)$ determines the Celsius temperature that corresponds to F degrees Fahrenheit. Find the Celsius temperature that corresponds to 62°F, 77°F, and 23°F.

Graph the function.

33. $f(x) = 3x - 1$

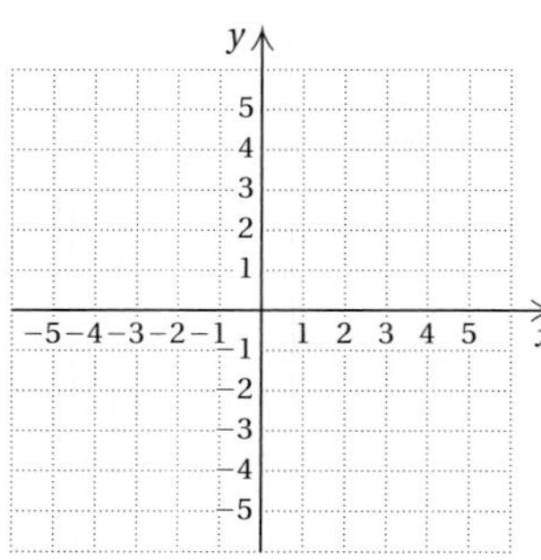

34. $g(x) = 2x + 5$

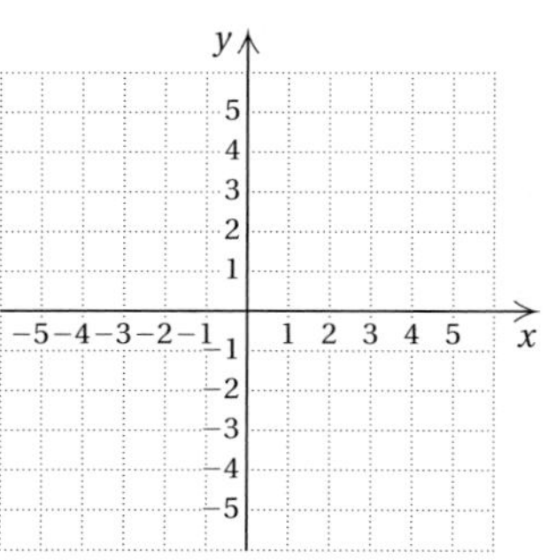

35. $g(x) = -2x + 3$

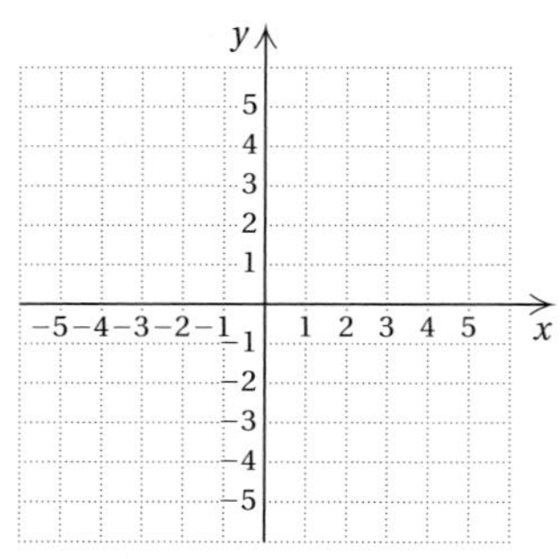

36. $f(x) = -\frac{1}{2}x + 2$

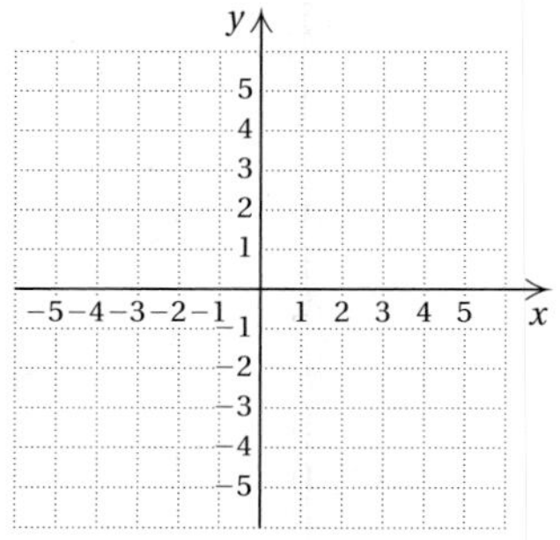

37. $f(x) = \frac{1}{2}x + 1$

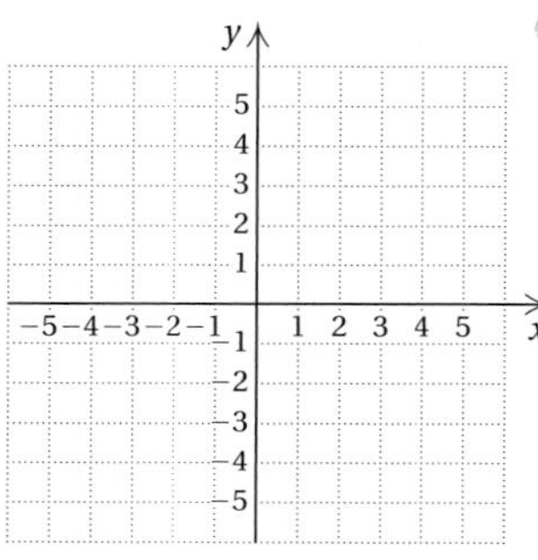

38. $f(x) = -\frac{3}{4}x - 2$

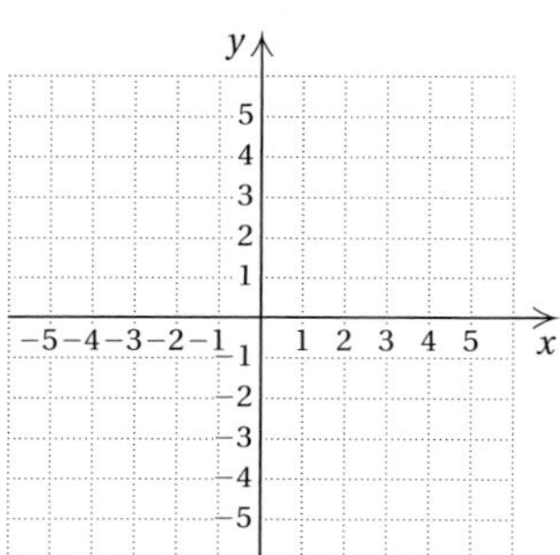

39. $f(x) = 2 - |x|$

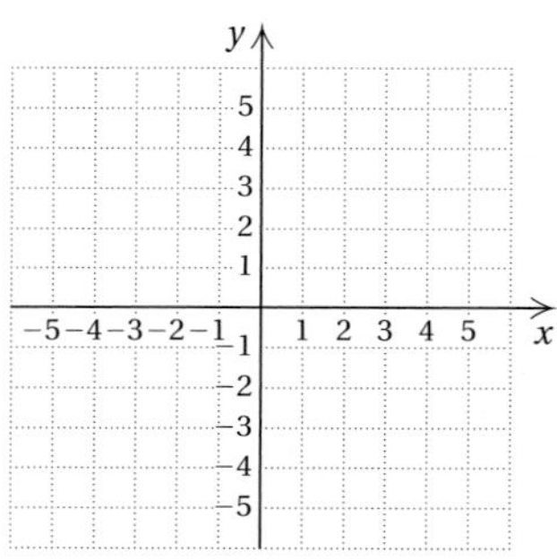

40. $f(x) = |x| - 4$

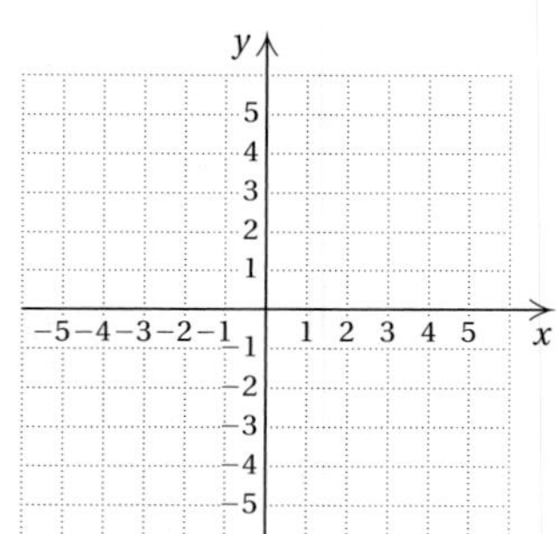

41. $f(x) = x^2$

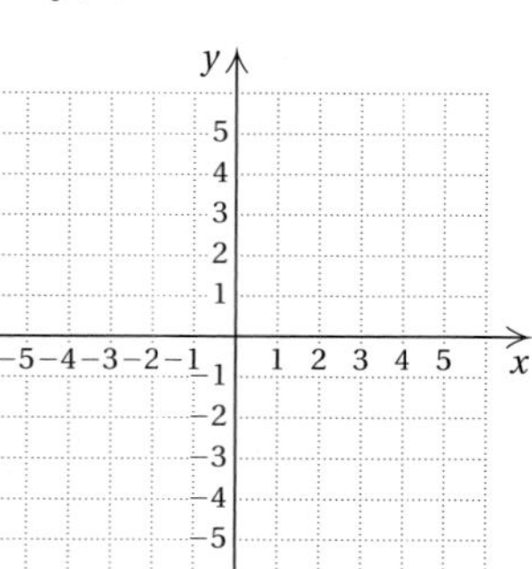

42. $f(x) = x^2 - 1$

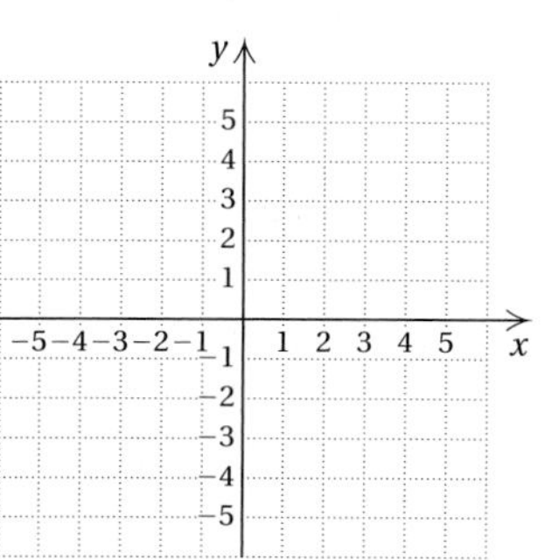

43. $f(x) = x^2 - x - 2$

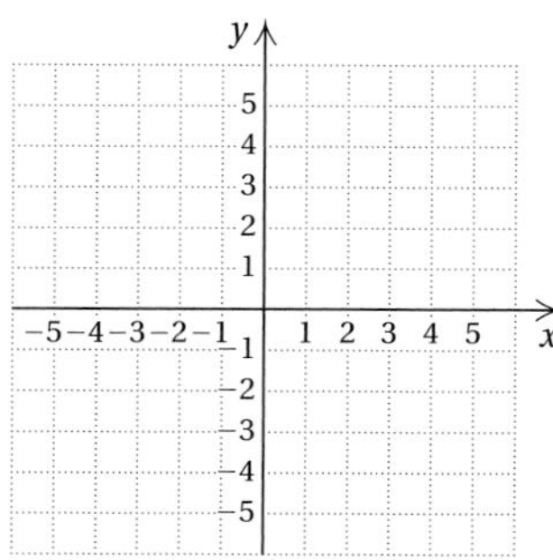

44. $f(x) = x^2 + 6x + 5$

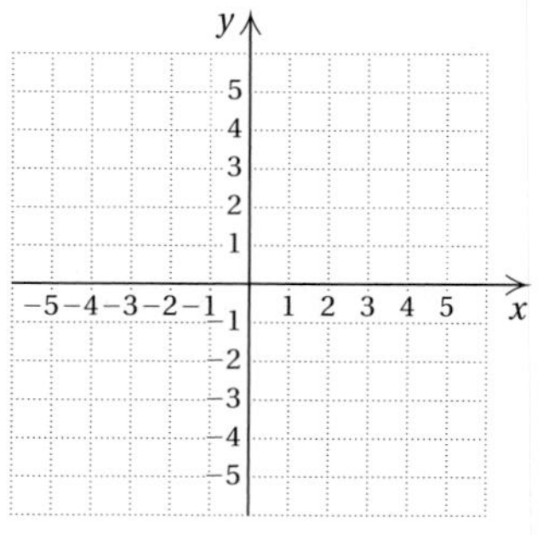

d Determine whether each of the following is the graph of a function.

45.

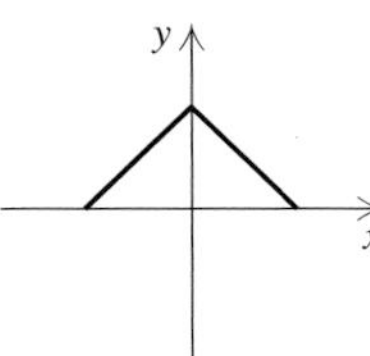

46.

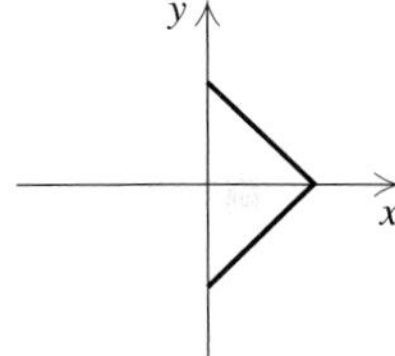

47.

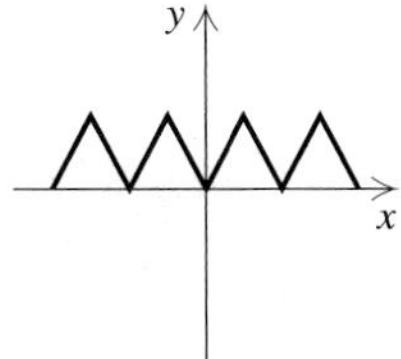

48.

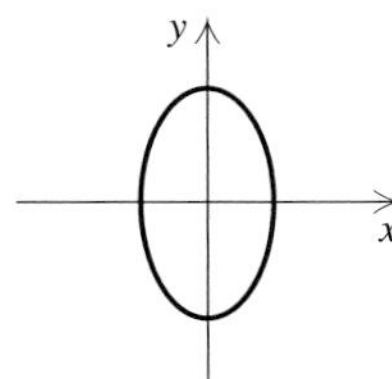

49.

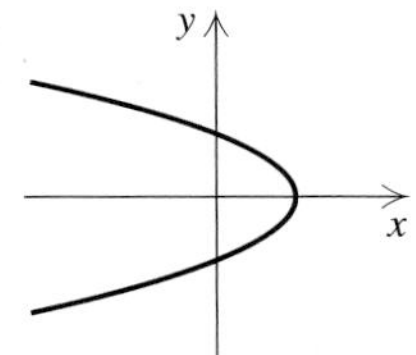

50.

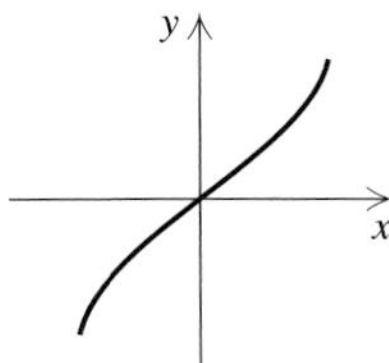

51.

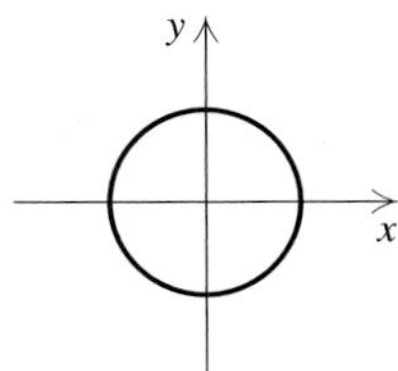

52.

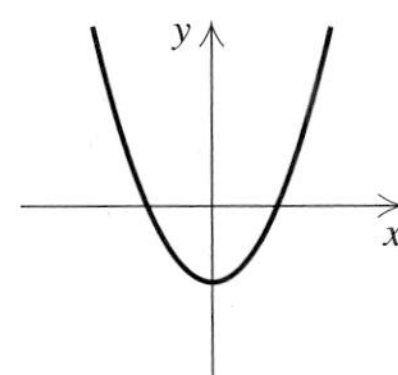

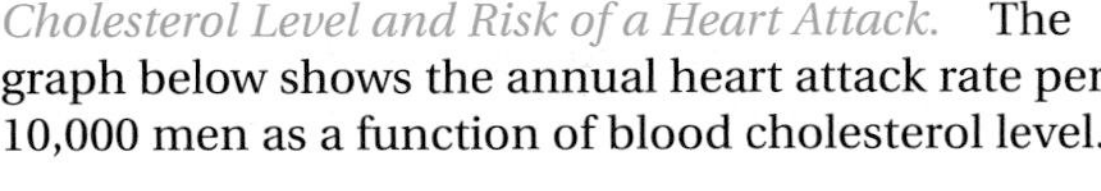

e *Cholesterol Level and Risk of a Heart Attack.* The graph below shows the annual heart attack rate per 10,000 men as a function of blood cholesterol level.

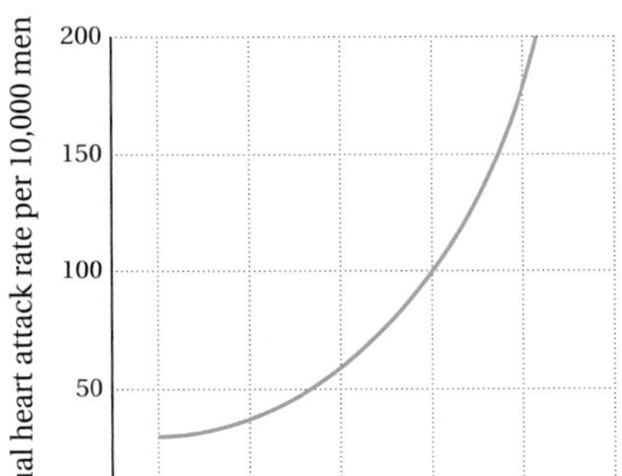

Source: Copyright 1989, CSPI. From *Nutrition Action Healthletter* (1875 Connecticut Avenue, N.W., Suite 300, Washington, DC 20009-5728)

53. Approximate the annual heart attack rate per 10,000 men for those whose blood cholesterol level is 225 mg/dl.

54. Approximate the annual heart attack rate per 10,000 men for those whose blood cholesterol level is 275 mg/dl.

55. D_W Is it possible for a function to have more numbers as outputs than as inputs? Why or why not?

56. D_W Look up the word "function" in a dictionary. Explain how that definition might be related to the mathematical one given in this section.

SKILL MAINTENANCE

Determine whether the pair of equations represents parallel lines. [9.6a]

57. $y = \frac{3}{4}x - 7,$
$3x + 4y = 7$

58. $y = \frac{3}{5},$
$y = -\frac{5}{3}$

Solve the system using the substitution method. [13.2b]

59. $2x - y = 6,$
$4x - 2y = 5$

60. $x - 3y = 2,$
$3x - 9y = 6$

SYNTHESIS

Graph.

61. $g(x) = x^3$

62. $f(x) = 2 + \sqrt{x}$

63. $f(x) = |x| + x$

64. $g(x) = |x| - x$

15 Summary and Review

The review that follows is meant to prepare you for a chapter exam. It consists of three parts. The first part, Concept Reinforcement, is designed to increase understanding of the concepts through true/false exercises. The second part is a list of important properties and formulas. The third part is the Review Exercises. These provide practice exercises for the exam, together with references to section objectives so you can go back and review. Before beginning, stop and look back over the skills you have obtained. What skills in mathematics do you have now that you did not have before studying this chapter?

CONCEPT REINFORCEMENT

Determine whether the statement is true or false. Answers are given at the back of the book.

_______ **1.** The equation $x^2 = -4$ has no real-number solutions.

_______ **2.** The solutions of $ax^2 + bx + c = 0$ are the first coordinates of the y-intercepts of the graph of $y = ax^2 + bx + c$.

_______ **3.** All graphs of quadratic equations, $y = ax^2 + bx + c$, have at least one y-intercept.

_______ **4.** If (p, q) is the vertex of the graph of $y = ax^2 + bx + c, a < 0$, then q is the largest value of y.

_______ **5.** If a quadratic equation $ax^2 + bx + c = 0$ has no real-number solutions, then the graph of $y = ax^2 + bx + c$ does not have an x-intercept.

_______ **6.** If (r, s) is the vertex of the graph of $y = ax^2 + bx + c$, then the axis of symmetry is $x = s$.

IMPORTANT PROPERTIES AND FORMULAS

Standard Form: $ax^2 + bx + c = 0, a > 0$

Principle of Square Roots: The equation $x^2 = d$, where $d > 0$, has two solutions, $\sqrt{d}$ and $-\sqrt{d}$. The solution of $x^2 = 0$ is 0.

Quadratic Formula: $x = \dfrac{-b \pm \sqrt{b^2 - 4ac}}{2a}$

Discriminant: $b^2 - 4ac$

The x-coordinate of the vertex of a parabola $= -\dfrac{b}{2a}$.

Review Exercises

Solve.

1. $8x^2 = 24$ [15.2a]

2. $40 = 5y^2$ [15.2a]

3. $5x^2 - 8x + 3 = 0$ [15.1c]

4. $3y^2 + 5y = 2$ [15.1c]

5. $(x + 8)^2 = 13$ [15.2b]

6. $9x^2 = 0$ [15.2a]

7. $5t^2 - 7t = 0$ [15.1b]

Solve. [15.3a]

8. $x^2 - 2x - 10 = 0$

9. $9x^2 - 6x - 9 = 0$

10. $x^2 + 6x = 9$

11. $1 + 4x^2 = 8x$

12. $6 + 3y = y^2$

13. $3m = 4 + 5m^2$

14. $3x^2 = 4x$

Solve. [15.1c]

15. $\frac{15}{x} - \frac{15}{x + 2} = 2$

16. $x + \frac{1}{x} = 2$

Solve by completing the square. Show your work. [15.2c]

17. $x^2 - 5x + 2 = 0$

18. $3x^2 - 2x - 5 = 0$

Approximate the solutions to the nearest tenth. [15.3b]

19. $x^2 - 5x + 2 = 0$

20. $4y^2 + 8y + 1 = 0$

21. Solve for T: $V = \frac{1}{2}\sqrt{1 + \frac{T}{L}}$. [15.4a]

Graph the quadratic equation. Label the ordered pairs for the vertex and the y-intercept. [15.6a]

22. $y = 2 - x^2$

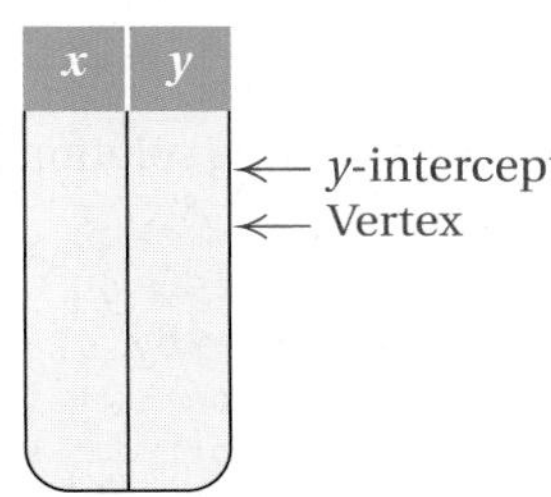

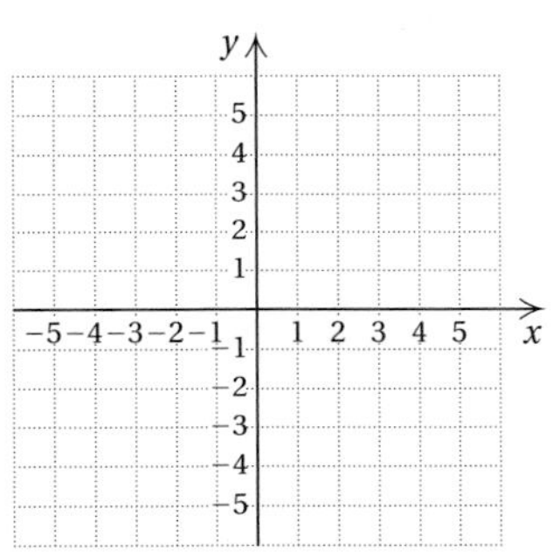

23. $y = x^2 - 4x - 2$

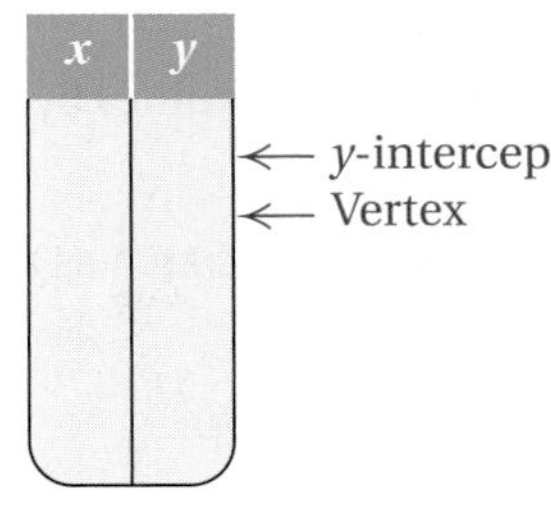

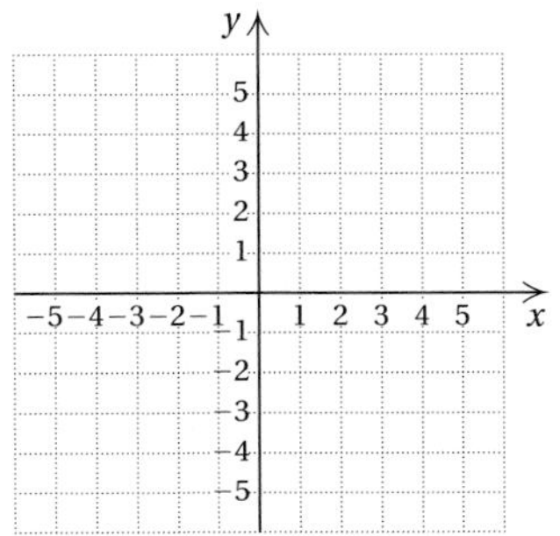

Find the x-intercepts. [15.6b]

24. $y = 2 - x^2$

25. $y = x^2 - 4x - 2$

Solve.

26. *Right-Triangle Dimensions.* The hypotenuse of a right triangle is 5 cm long. One leg is 3 cm longer than the other. Find the lengths of the legs. Round to the nearest tenth. [15.5a]

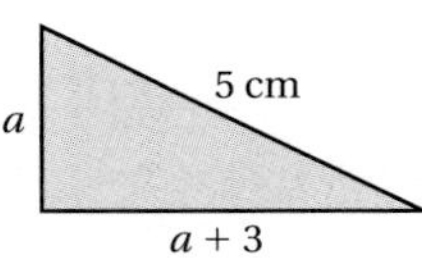

27. *Car-Loading Ramp.* The length of a loading ramp on a car hauler is 25 ft. This ramp and its height form the hypotenuse and one leg of a right triangle. The height of the ramp is 5 ft less than the length of the other leg. Find the height of the loading ramp. [15.5a]

28. *Lake Point Towers.* The height of Lake Point Towers in Chicago is 645 ft. How long would it take an object to fall to the ground from the top? [15.2d]

Find the function values. [15.7b]

29. If $f(x) = 2x - 5$, find $f(2)$, $f(-1)$, and $f(3.5)$.

30. If $g(x) = |x| - 1$, find $g(1)$, $g(-1)$, and $g(-20)$.

31. *Caloric Needs.* If you are moderately active, you need to consume each day about 15 calories per pound of body weight. The function $C(p) = 15p$ approximates the number of calories C that are needed to maintain body weight p, in pounds. How many calories are needed to maintain a body weight of 180 lb? [15.7e]

Graph the function. [15.7c]

32. $g(x) = 4 - x$

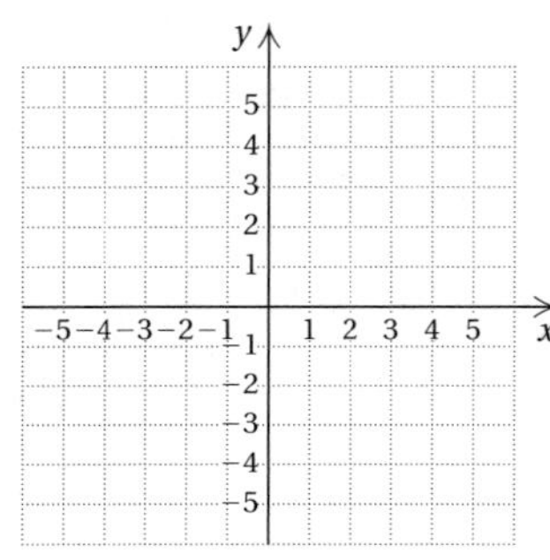

33. $f(x) = x^2 - 3$

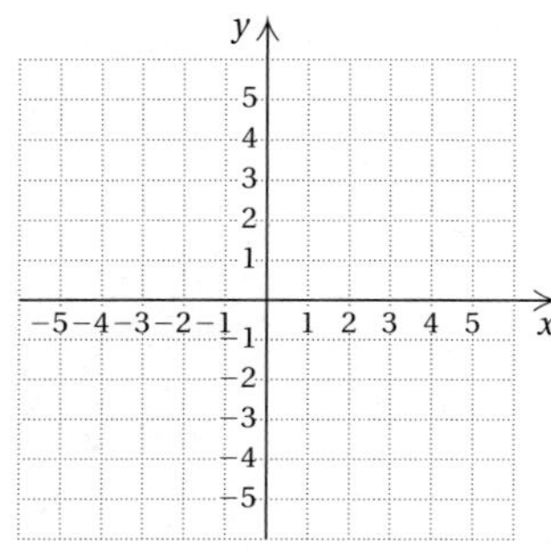

34. $h(x) = |x| - 5$

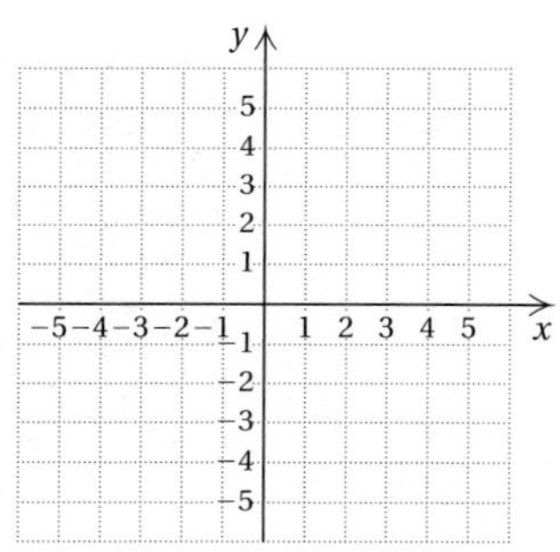

35. $f(x) = x^2 - 2x + 1$

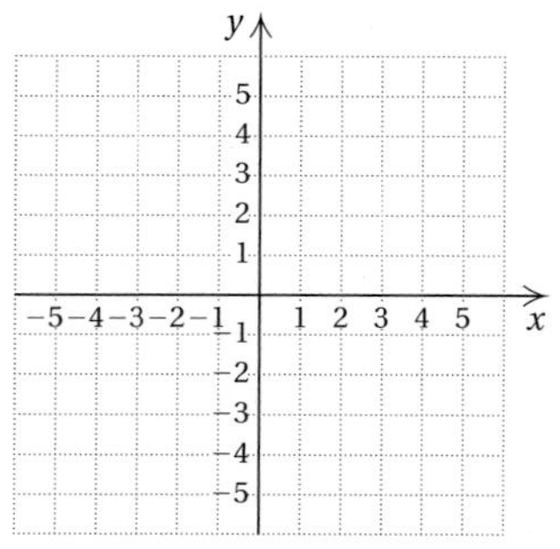

Determine whether each of the following is the graph of a function. [15.7d]

36.

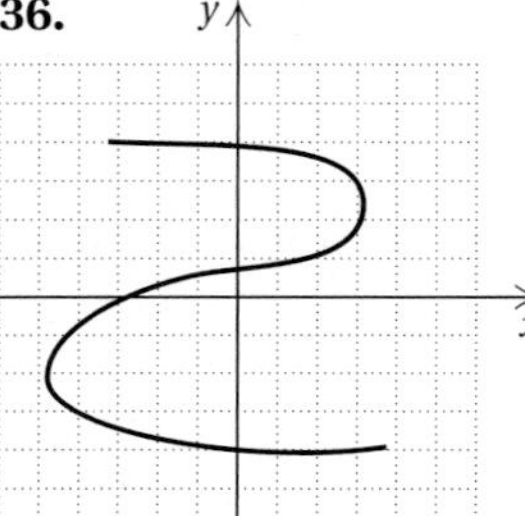

37.

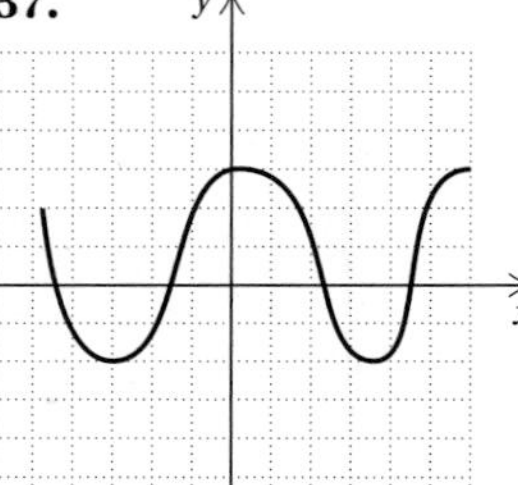

38. D_W List the names and give an example of as many types of equation as you can that you have learned to solve in this text. [8.1a], [12.6a], [13.1a], [14.5a], [15.1a]

39. D_W Find the errors in each of the following solutions of equations. [15.1b]

a)
$$\begin{aligned} x^2 + 20x &= 0 \\ x(x + 20) &= 0 \\ x + 20 &= 0 \\ x &= 20 \end{aligned}$$

b)
$$\begin{aligned} x^2 + x &= 6 \\ x(x + 1) &= 6 \\ x = 6 \quad &or \quad x + 1 = 6 \\ x = 6 \quad &or \quad x = 5 \end{aligned}$$

SYNTHESIS

40. Two consecutive integers have squares that differ by 63. Find the integers. [15.5a]

41. A square with sides of length s has the same area as a circle with a radius of 5 in. Find s. [15.5a]

42. Solve: $x - 4\sqrt{x} - 5 = 0$. [15.1c]

Use the graph of

$$y = (x + 3)^2$$

to solve each equation. [15.6b]

43. $(x + 3)^2 = 1$

44. $(x + 3)^2 = 4$

45. $(x + 3)^2 = 9$

46. $(x + 3)^2 = 0$

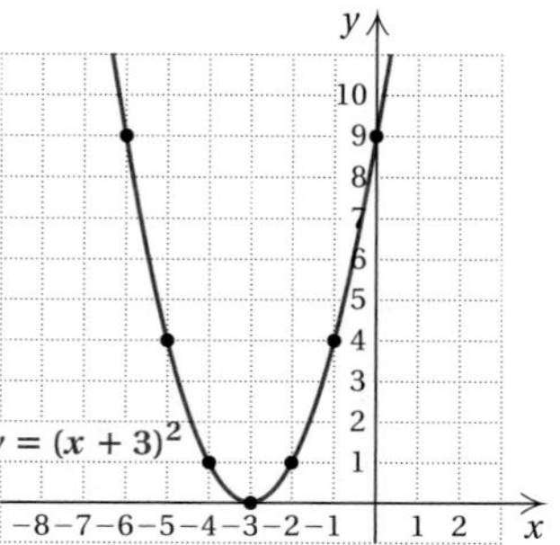

15 Chapter Test

For Extra Help

Work It Out!
Chapter Test Video
Solutions on CD

Solve.

1. $7x^2 = 35$

2. $7x^2 + 8x = 0$

3. $48 = t^2 + 2t$

4. $3y^2 - 5y = 2$

5. $(x - 8)^2 = 13$

6. $x^2 = x + 3$

7. $m^2 - 3m = 7$

8. $10 = 4x + x^2$

9. $3x^2 - 7x + 1 = 0$

10. $x - \frac{2}{x} = 1$

11. $\frac{4}{x} - \frac{4}{x + 2} = 1$

12. Solve $x^2 - 4x - 10 = 0$ by completing the square. Show your work.

13. Approximate the solutions to $x^2 - 4x - 10 = 0$ to the nearest tenth.

14. Solve for n: $d = an^2 + bn$.

15. Find the x-intercepts: $y = -x^2 + x + 5$.

Graph. Label the ordered pairs for the vertex and the y-intercept.

16. $y = 4 - x^2$

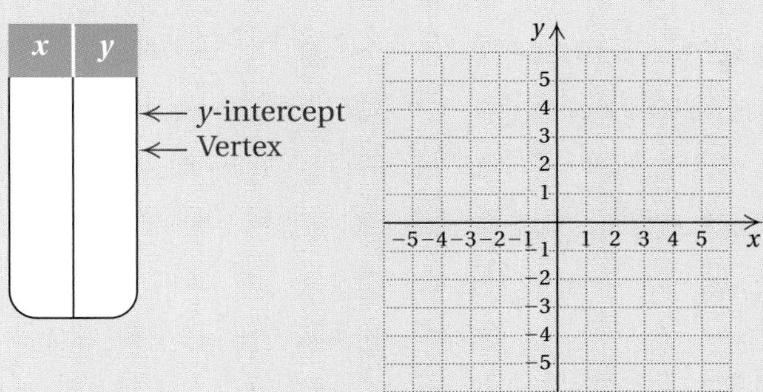

17. $y = -x^2 + x + 5$

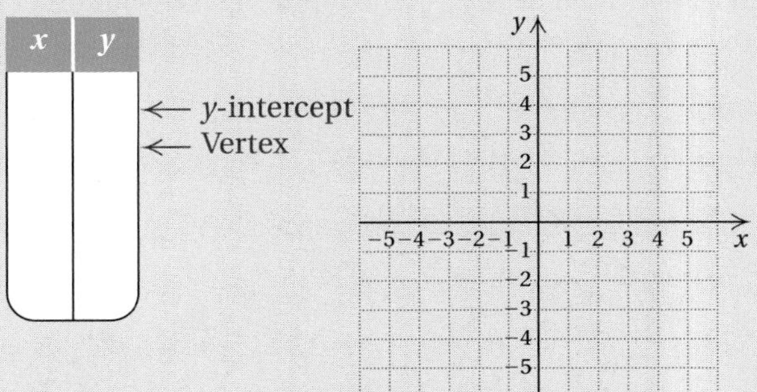

18. If $f(x) = \frac{1}{2}x + 1$, find $f(0)$, $f(1)$, and $f(2)$.

19. If $g(t) = -2|t| + 3$, find $g(-1)$, $g(0)$, and $g(3)$.

Solve.

20. *Rug Dimensions.* The width of a rectangular area rug is 4 m less than the length. The area is 16.25 m^2. Find the length and the width.

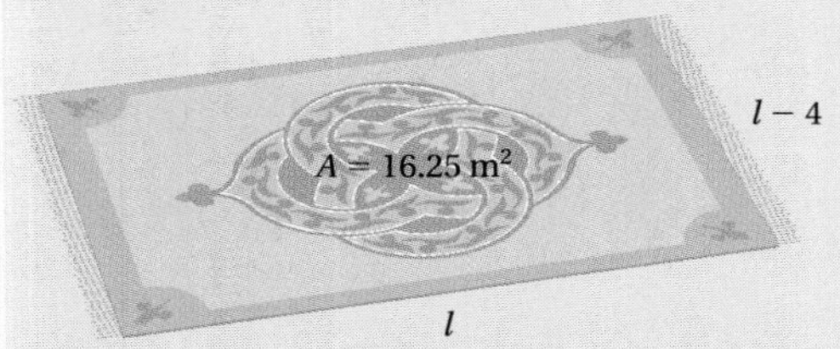

21. *Boat Speed.* The current in a stream moves at a speed of 2 km/h. A boat travels 44 km upstream and 52 km downstream in a total of 4 hr. What is the speed of the boat in still water?

22. *World Record for 10,000-m Run.* The world record for the 10,000-m run has been decreasing steadily since 1940. The record is approximately 30.18 min minus 0.06 times the number of years since 1940. The function $R(t) = 30.18 - 0.06t$ estimates the record R, in minutes, as a function of t, the time in years since 1940. Predict what the record will be in 2010.

Graph.

23. $h(x) = x - 4$

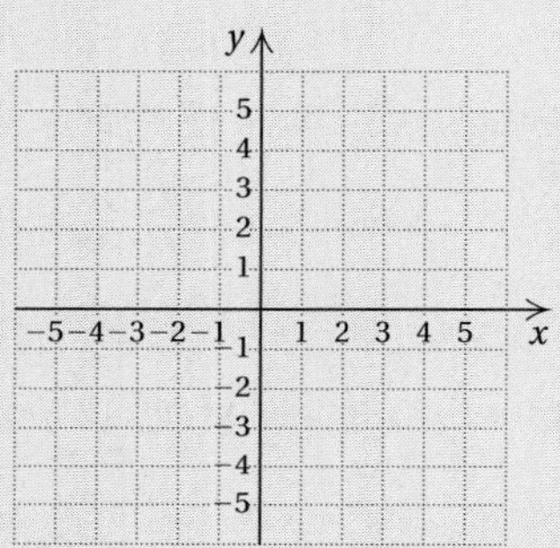

24. $g(x) = x^2 - 4$

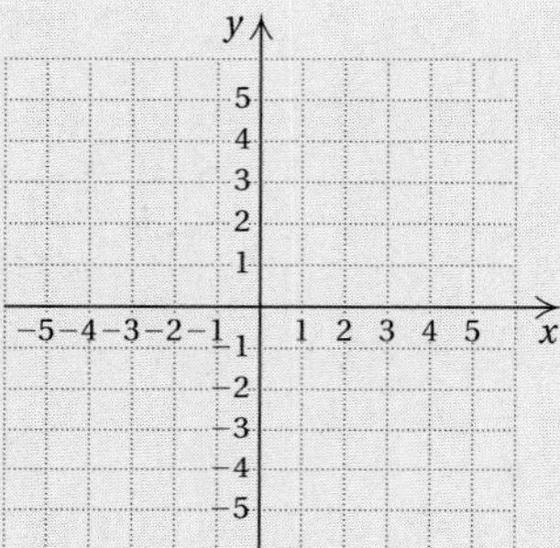

Determine whether each of the following is the graph of a function.

25.

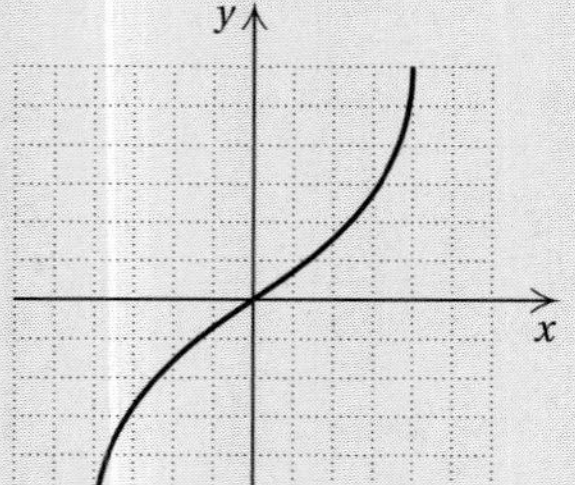

26.

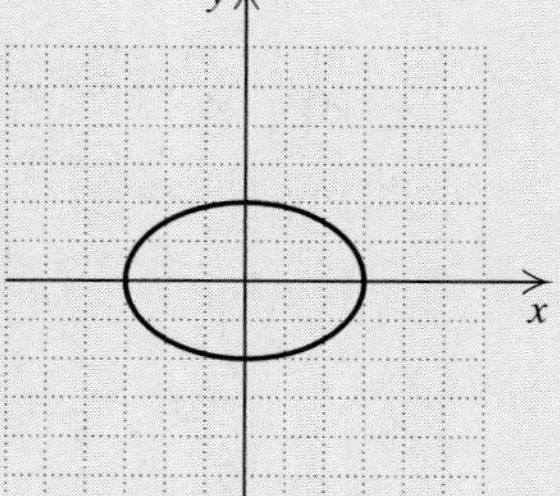

SYNTHESIS

27. Find the side of a square whose diagonal is 5 ft longer than a side.

28. Solve this system for x. Use the substitution method.

$$x - y = 2,$$
$$xy = 4$$

Appendixes

A LINEAR MEASURES: AMERICAN AND METRIC UNITS

Objectives

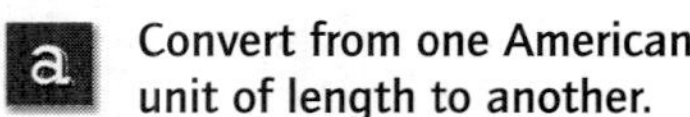
a Convert from one American unit of length to another.

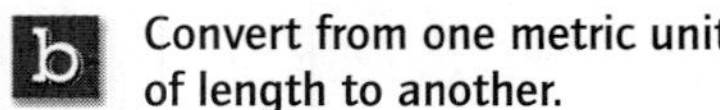
b Convert from one metric unit of length to another.

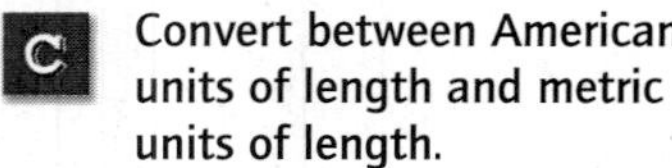
c Convert between American units of length and metric units of length.

Length, or distance, is one kind of measure. To find lengths, we start with some **unit segment** and assign to it a measure of 1. Suppose $\overline{AB}$ below is a unit segment.

Let's measure segment $\overline{CD}$ below, using $\overline{AB}$ as our unit segment.

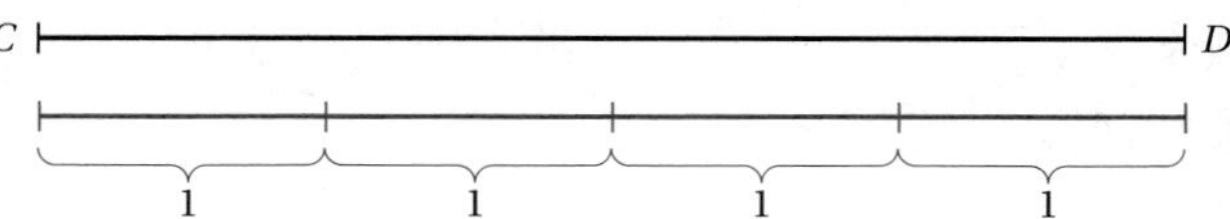

Since we can place 4 unit segments end to end along $\overline{CD}$, the measure of $\overline{CD}$ is 4.

Sometimes we have to use parts of units, called **subunits.** For example, the measure of the segment $\overline{MN}$ below is $1\frac{1}{2}$. We place one unit segment and one half-unit segment end to end.

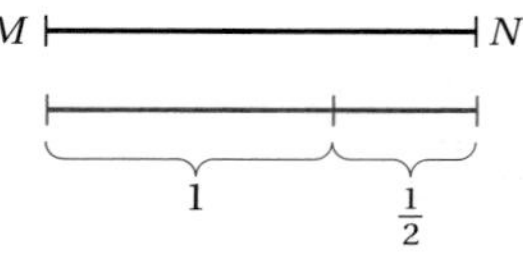

Do Exercises 1–4.

Use the unit below to measure the length of each segment or object.

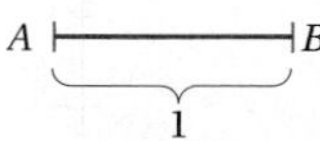

1.

2.

3.

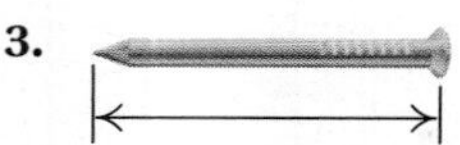

4.

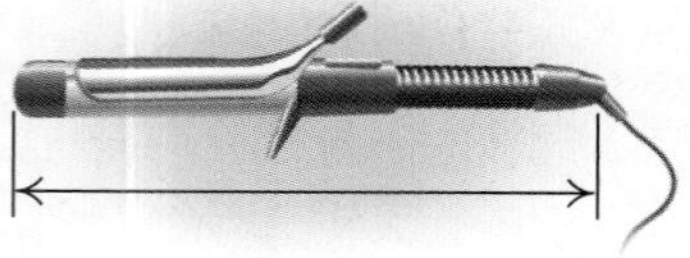

Answers on page A-55

a American Measures

American units of length are related as follows.

(Actual size, in inches)

AMERICAN UNITS OF LENGTH	
12 inches (in.) = 1 foot (ft)	3 feet = 1 yard (yd)
36 inches = 1 yard	5280 feet = 1 mile (mi)

We can visualize comparisons of the units as follows:

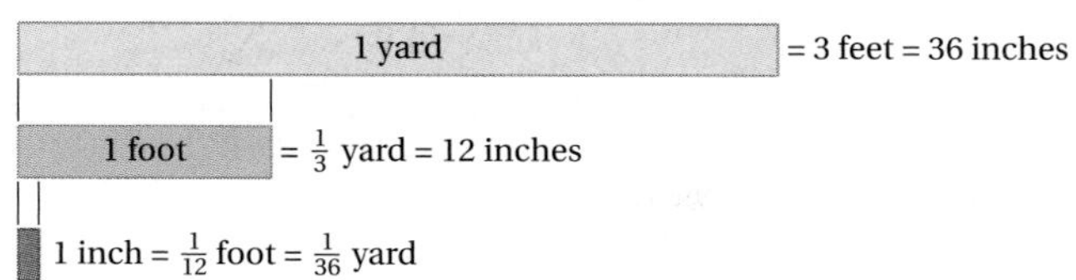

The symbolism 13 in. = 13″ and 27 ft = 27′ is also used for inches and feet. American units have also been called "English," or "British–American," because at one time they were used by both countries. Today, both Canada and England have officially converted to the metric system. However, if you travel in England, you will still see units such as "miles" on road signs.

To change from certain American units to others, we make substitutions. Such a substitution is usually helpful when we are converting from a *larger* unit to a *smaller* one.

EXAMPLE 1 Complete: $7\frac{1}{3}$ yd = ______ in.

$7\frac{1}{3}$ yd $= 7\frac{1}{3} \times 1$ yd We think of $7\frac{1}{3}$ yd as $7\frac{1}{3} \times$ yd, or $7\frac{1}{3} \times 1$ yd.

$= 7\frac{1}{3} \times 36$ in. Substituting 36 in. for 1 yd

$= \frac{22}{3} \times 36$ in.

$= 264$ in.

Do Exercises 5–7.

Sometimes it helps to use multiplying by 1 in making conversions. For example, 12 in. = 1 ft, so

$$\frac{12 \text{ in.}}{1 \text{ ft}} = 1 \quad \text{and} \quad \frac{1 \text{ ft}}{12 \text{ in.}} = 1.$$

If we divide 12 in. by 1 ft or 1 ft by 12 in., we get 1 because the lengths are the same. Let's first convert from *smaller* to *larger* units.

EXAMPLE 2 Complete: 48 in. = ______ ft.

We want to convert from "in." to "ft." We multiply by 1 using a symbol for 1 with "in." on the bottom and "ft" on the top to eliminate inches and to convert to feet:

$48 \text{ in.} = \frac{48 \text{ in.}}{1} \times \frac{1 \text{ ft}}{12 \text{ in.}}$ Multiplying by 1 using $\frac{1 \text{ ft}}{12 \text{ in.}}$ to eliminate in.

$= \frac{48 \text{ in.}}{12 \text{ in.}} \times 1 \text{ ft}$

$= \frac{48}{12} \times \frac{\text{in.}}{\text{in.}} \times 1 \text{ ft}$

$= 4 \times 1$ ft The $\frac{\text{in.}}{\text{in.}}$ acts like 1, so we can omit it.

$= 4$ ft.

Complete.

5. 8 yd = ______ in.

6. $2\frac{5}{6}$ yd = ______ ft

7. 3.8 mi = ______ in.

Answers on page A-55

Complete.

8. 72 in. = ________ ft

9. 17 in. = ________ ft

Complete.

10. 24 ft = ________ yd

11. 35 ft = ________ yd

Complete.

12. 26,400 ft = ________ mi

13. 2640 ft = ________ mi

Answers on page A-55

We can also look at this conversion as "canceling" units:

$$48 \text{ in.} = \frac{48 \cancel{\text{in.}}}{1} \times \frac{1 \text{ ft}}{12 \cancel{\text{in.}}} = \frac{48}{12} \times 1 \text{ ft} = 4 \text{ ft}.$$

This method is used not only in mathematics, as here, but also in fields such as medicine, chemistry, and physics.

Do Exercises 8 and 9.

EXAMPLE 3 Complete: 25 ft = ________ yd.

Since we are converting from "ft" to "yd," we choose a symbol for 1 with "yd" on the top and "ft" on the bottom:

$$25 \text{ ft} = 25 \text{ ft} \times \frac{1 \text{ yd}}{3 \text{ ft}}$$ 3 ft = 1 yd, so $\frac{3 \text{ ft}}{1 \text{ yd}} = 1$, and $\frac{1 \text{ yd}}{3 \text{ ft}} = 1$. We use $\frac{1 \text{ yd}}{3 \text{ ft}}$ to eliminate ft.

$$= \frac{25}{3} \times \frac{\text{ft}}{\text{ft}} \times 1 \text{ yd}$$

$$= 8\frac{1}{3} \times 1 \text{ yd}$$ The $\frac{\text{ft}}{\text{ft}}$ acts like 1, so we can omit it.

$$= 8\frac{1}{3} \text{ yd, or } 8.\overline{3} \text{ yd.}$$

Again, in this example, we can consider conversion from the point of view of canceling:

$$25 \text{ ft} = 25 \cancel{\text{ft}} \times \frac{1 \text{ yd}}{3 \cancel{\text{ft}}} = \frac{25}{3} \times 1 \text{ yd} = 8\frac{1}{3} \text{ yd, or } 8.\overline{3} \text{ yd.}$$

Do Exercises 10 and 11.

EXAMPLE 4 Complete: 23,760 ft = ________ mi.

We choose a symbol for 1 with "mi" on the top and "ft" on the bottom:

$$23{,}760 \text{ ft} = 23{,}760 \text{ ft} \times \frac{1 \text{ mi}}{5280 \text{ ft}}$$ 5280 ft = 1 mi, so $\frac{1 \text{ mi}}{5280 \text{ ft}} = 1$.

$$= \frac{23{,}760}{5280} \times \frac{\text{ft}}{\text{ft}} \times 1 \text{ mi}$$

$$= 4.5 \times 1 \text{ mi}$$ Dividing

$$= 4.5 \text{ mi.}$$

Let's also consider this example using canceling:

$$23{,}760 \text{ ft} = 23{,}760 \cancel{\text{ft}} \times \frac{1 \text{ mi}}{5280 \cancel{\text{ft}}}$$

$$= \frac{23{,}760}{5280} \times 1 \text{ mi}$$

$$= 4.5 \times 1 \text{ mi} = 4.5 \text{ mi.}$$

Do Exercises 12 and 13.

b Metric Measures

The **metric system** is used in most countries of the world, but very little in the United States. The metric system does not use inches, feet, pounds, and so on, although units for time and electricity are the same as those used now in the United States.

An advantage of the metric system is that it is easier to convert from one unit to another. That is because the metric system is based on the number 10.

The basic unit of length is the **meter.** It is just over a yard. In fact, 1 meter $\approx$ 1.1 yd.

(Comparative sizes are shown.)

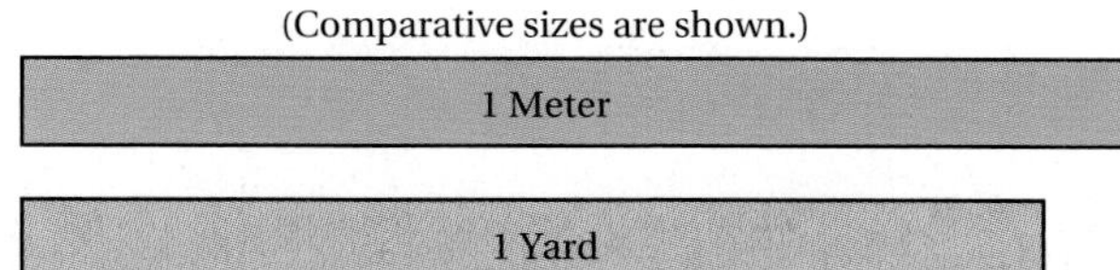

The other units of length are multiples of the length of a meter:

10 times a meter, 100 times a meter, 1000 times a meter, and so on,

or fractions of a meter:

$\frac{1}{10}$ of a meter, $\frac{1}{100}$ of a meter, $\frac{1}{1000}$ of a meter, and so on.

METRIC UNITS OF LENGTH

1 *kilo*meter (km) = 1000 meters (m)

1 *hecto*meter (hm) = 100 meters (m)

1 *deka*meter (dam) = 10 meters (m)

1 meter (m)

1 *deci*meter (dm) = $\frac{1}{10}$ meter (m)

1 *centi*meter (cm) = $\frac{1}{100}$ meter (m)

1 *milli*meter (mm) = $\frac{1}{1000}$ meter (m)

dam and *dm* are not used often.

You should memorize these names and abbreviations. Think of *kilo-* for 1000, *hecto-* for 100, *deka-* for 10, *deci-* for $\frac{1}{10}$, *centi-* for $\frac{1}{100}$, and *milli-* for $\frac{1}{1000}$. We will also use these prefixes when considering units of area, capacity, and mass.

Use a centimeter ruler. Measure each object.

14.

15.

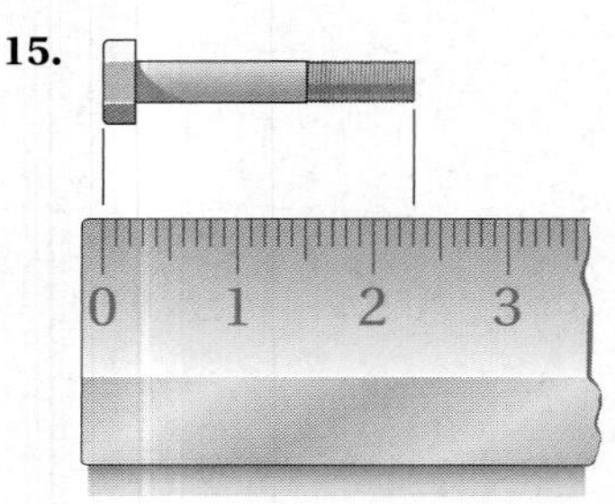

16.

Answers on page A-55

THINKING METRIC

To familiarize yourself with metric units, consider the following.

1 kilometer (1000 meters)	is slightly more than $\frac{1}{2}$ mile (0.6 mi).
1 meter	is just over a yard (1.1 yd).
1 centimeter (0.01 meter)	is a little more than the width of a paperclip (about 0.3937 inch).

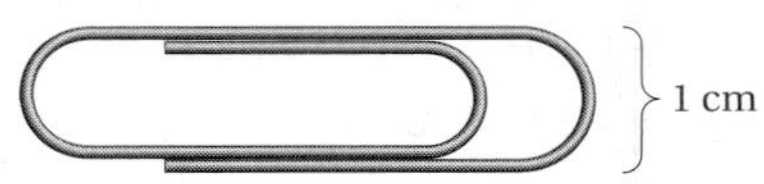

1 inch is about 2.54 centimeters.

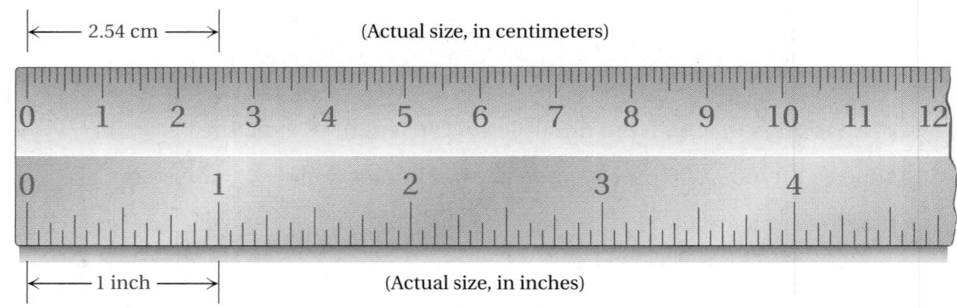

1 millimeter is about the diameter of a paperclip wire.

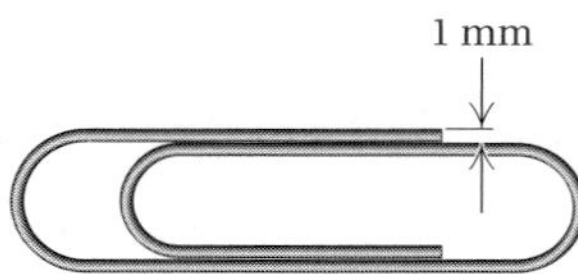

The millimeter (mm) is used to measure small distances, especially in industry.

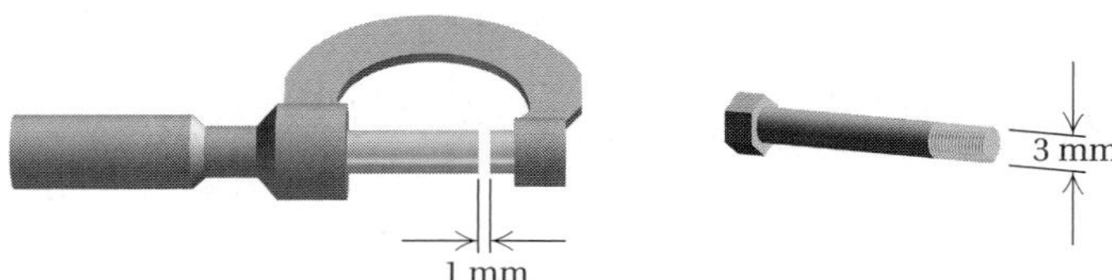

In many countries, the centimeter (cm) is used for body dimensions and clothing sizes.

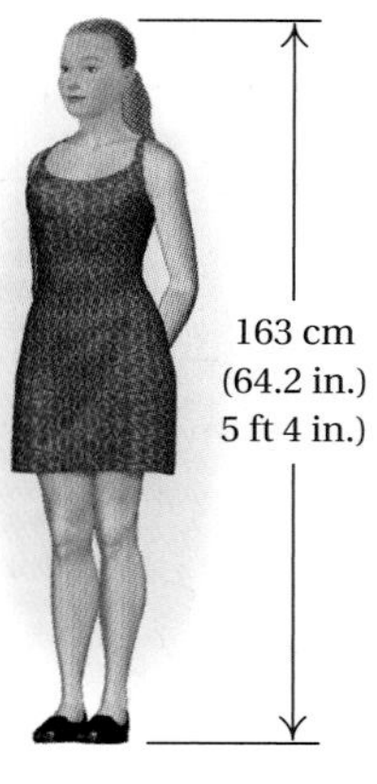

Do Exercises 14–16 on the preceding page.

The meter (m) is used for expressing dimensions of larger objects—say, the height of a diving board—and for shorter distances, like the length of a rug.

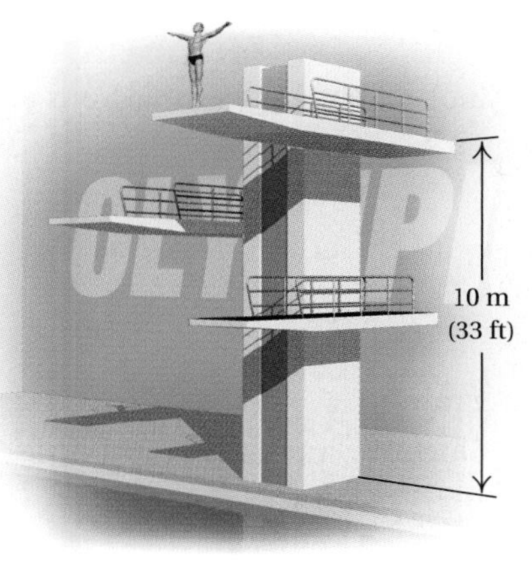

The kilometer (km) is used for longer distances, mostly in cases where miles are now being used.

1 mile is about 1.6 km.

1 km
1 mi

Do Exercises 17–22.

Complete with mm, cm, m, or km.

17. A stick of gum is 7 ________ long.

18. Minneapolis is 3213 ________ from San Francisco.

19. A penny is 1 ________ thick.

20. The halfback ran 7 ________ .

21. The book is 3 ________ thick.

22. The desk is 2 ________ long.

Answers on page A-55

Complete.

23. 23 km = ________ m

As with American units, when changing from a *larger* unit to a *smaller* unit, we usually make substitutions.

EXAMPLE 5 Complete: 4 km = ________ m.

We want to convert from "km" to "m." Since we are converting from a *larger* to a *smaller* unit, we use substitution.

$$\begin{aligned} 4 \text{ km} &= 4 \times 1 \text{ km} \\ &= 4 \times 1000 \text{ m} \quad \text{Substituting 1000 m for 1 km} \\ &= 4000 \text{ m} \end{aligned}$$

Do Exercises 23 and 24.

24. 4 hm = ________ m

Since

$$\frac{1}{10} \text{ m} = 1 \text{ dm}, \quad \frac{1}{100} \text{ m} = 1 \text{ cm}, \quad \text{and} \quad \frac{1}{1000} \text{ m} = 1 \text{ mm},$$

it follows that

$$1 \text{ m} = 10 \text{ dm}, \quad 1 \text{ m} = 100 \text{ cm}, \quad \text{and} \quad 1 \text{ m} = 1000 \text{ mm}.$$

EXAMPLE 6 Complete: 93.4 m = ________ cm.

We want to convert from "m" to "cm." Since we are converting from a *larger* to a *smaller* unit, we use substitution.

$$\begin{aligned} 93.4 \text{ m} &= 93.4 \times 1 \text{ m} \\ &= 93.4 \times 100 \text{ cm} \quad \text{Substituting 100 cm for 1 m} \\ &= 9340 \text{ cm} \end{aligned}$$

Complete.

25. 1.78 m = ________ cm

EXAMPLE 7 Complete: 0.248 m = ________ mm.

We want to convert from "m" to "mm." Since we are again converting from a *larger* to a *smaller* unit, we use substitution.

$$\begin{aligned} 0.248 \text{ m} &= 0.248 \times 1 \text{ m} \\ &= 0.248 \times 1000 \text{ mm} \quad \text{Substituting 1000 mm for 1 m} \\ &= 248 \text{ mm} \end{aligned}$$

Do Exercises 25 and 26.

26. 9.04 m = ________ mm

Answers on page A-55

We now convert from "m" to "km." Since we are converting from a *smaller* unit to a *larger* unit, we use multiplying by 1. We choose a symbol for 1 with "km" in the numerator and "m" in the denominator.

Complete.

27. 7814 m = ________ km

EXAMPLE 8 Complete: 2347 m = ________ km.

$$2347\text{ m} = 2347\text{ m} \times \frac{1\text{ km}}{1000\text{ m}}$$ Multiplying by 1 using $\frac{1\text{ km}}{1000\text{ m}}$

$$= \frac{2347}{1000} \times \frac{\text{m}}{\text{m}} \times 1\text{ km}$$ The $\frac{\text{m}}{\text{m}}$ acts like 1, so we omit it.

$$= 2.347\text{ km}$$ Dividing by 1000 moves the decimal point three places to the left.

Using canceling, we can work this example as follows:

$$2347\text{ m} = 2347\,\cancel{\text{m}} \times \frac{1\text{ km}}{1000\,\cancel{\text{m}}}$$

$$= \frac{2347}{1000} \times 1\text{ km} = 2.347\text{ km}.$$

28. 7814 m = ________ dam

Do Exercises 27 and 28.

Sometimes we multiply by 1 more than once.

EXAMPLE 9 Complete: 8.42 mm = ________ cm.

$$8.42\text{ mm} = 8.42\text{ mm} \times \frac{1\text{ m}}{1000\text{ mm}} \times \frac{100\text{ cm}}{1\text{ m}}$$ Multiplying by 1 using $\frac{1\text{ m}}{1000\text{ mm}}$ and $\frac{100\text{ cm}}{1\text{ m}}$

$$= \frac{8.42 \times 100}{1000} \times \frac{\text{mm}}{\text{mm}} \times \frac{\text{m}}{\text{m}} \times 1\text{ cm}$$

$$= \frac{842}{1000}\text{ cm} = 0.842\text{ cm}$$

Complete.

29. 9.67 mm = ________ cm

Do Exercises 29 and 30.

MENTAL CONVERSION

Look back over the examples and exercises done so far and you will see that changing from one unit to another in the metric system amounts to only the movement of a decimal point. That is because the metric system is based on 10. Let's find a faster way to convert. Look at the following table.

30. 89 km = ________ cm

1000 m	100 m	10 m	1 m	0.1 m	0.01 m	0.001 m
1 km	1 hm	1 dam	1 m	1 dm	1 cm	1 mm

Each place in the table has a value $\frac{1}{10}$ that to the left or 10 times that to the right. Thus moving one place in the table corresponds to moving one decimal place.

Answers on page A-55

Complete. Try to do this mentally using the table.

31. 6780 m = ________ km

32. 9.74 cm = ________ mm

33. 1 mm = ________ cm

34. 845.1 mm = ________ dm

Answers on page A-55

Let's convert mentally.

EXAMPLE 10 Complete: 8.42 mm = ________ cm.

Think: To go from mm to cm in the table is a move of one place to the left. Thus we move the decimal point one place to the left.

1000 m	100 m	10 m	1 m	0.1 m	0.01 m	0.001 m
1 km	1 hm	1 dam	1 m	1 dm	1 cm	1 mm

1 place to the left

8.42 0.8.42 8.42 mm = 0.842 cm

EXAMPLE 11 Complete: 1.886 km = ________ cm.

Think: To go from km to cm is a move of five places to the right. Thus we move the decimal point five places to the right.

1000 m	100 m	10 m	1 m	0.1 m	0.01 m	0.001 m
1 km	1 hm	1 dam	1 m	1 dm	1 cm	1 mm

5 places to the right

1.886 1.88600. 1.886 km = 188,600 cm

EXAMPLE 12 Complete: 3 m = ________ cm.

Think: To go from m to cm in the table is a move of two places to the right. Thus we move the decimal point two places to the right.

1000 m	100 m	10 m	1 m	0.1 m	0.01 m	0.001 m
1 km	1 hm	1 dam	1 m	1 dm	1 cm	1 mm

2 places to the right

3 3.00. 3 m = 300 cm

You should try to make metric conversions mentally as much as possible.

The fact that conversions can be done so easily is an important advantage of the metric system. The most commonly used metric units of length are km, m, cm, and mm. We have purposely used these more often than the others in the exercises.

Do Exercises 31–34.

C Converting Units

We can make conversions between American units and metric units by using the following table. These conversions are rounded approximations. Again, we either make a substitution or multiply by 1.

AMERICAN	METRIC
1 in.	2.540 cm
1 ft	0.305 m
1 yd	0.914 m
1 mi	1.609 km
0.621 mi	1 km
1.094 yd	1 m
3.281 ft	1 m
39.370 in.	1 m

This table is set up to enable us to make conversions by substitution.

EXAMPLE 13 Complete: 26.2 mi = ________ km.
(The length of the Olympic marathon)

$$\begin{aligned} 26.2 \text{ mi} &= 26.2 \times 1 \text{ mi} \\ &\approx 26.2 \times 1.609 \text{ km} \qquad \text{Substituting 1.609 km for 1 mi} \\ &= 42.1558 \text{ km} \end{aligned}$$

EXAMPLE 14 Complete: 2.16 m = ________ in.
(The height of Shaquille O'Neal of the Miami Heat)

$$\begin{aligned} 2.16 \text{ m} &= 2.16 \times 1 \text{ m} \\ &\approx 2.16 \times 39.37 \text{ in.} \qquad \text{Substituting 39.37 in. for 1 m} \\ &= 85.0392 \text{ in.} \end{aligned}$$

In an application like this one, the answer would probably be rounded to the nearest one, as 85 in.

EXAMPLE 15 Complete: 100 m = ________ ft.
(The length of the 100-meter dash)

$$\begin{aligned} 100 \text{ m} &= 100 \times 1 \text{ m} \\ &\approx 100 \times 3.281 \text{ ft} \qquad \text{Substituting 3.281 ft for 1 m} \\ &= 328.1 \text{ ft} \end{aligned}$$

EXAMPLE 16 Complete: 4544 km = ________ mi.
(The distance from New York to Los Angeles)

$$\begin{aligned} 4544 &= 4544 \times 1 \text{ km} \\ &\approx 4544 \times 0.621 \text{ mi} \qquad \text{Substituting 0.621 mi for 1 km} \\ &= 2821.824 \text{ mi} \end{aligned}$$

In practical situations, we would probably round this answer to 2822 mi.

Do Exercises 35–37.

Complete.

35. 100 yd = ________ m
(The length of a football field excluding the end zones)

36. 500 mi = ________ km
(The Indianapolis 500-mile race)

37. 2383 km = ________ mi
(The distance from St. Louis to Phoenix)

Answers on page A-55

Complete.

38. 568 mi = ________ km
(The distance from San Francisco to Las Vegas)

39. The height of the John Hancock Building in Chicago is 1127 ft. Find the height in meters.

40. Complete: 3.175 mm = ________ in.
(The thickness of a quarter)

Answers on page A-55

EXAMPLE 17 *Petronas Towers.* The height of the Petronas Towers in Kuala Lumpur, Malaysia, is 1483 ft. Find the height in meters.

Source: *The New York Times Almanac*

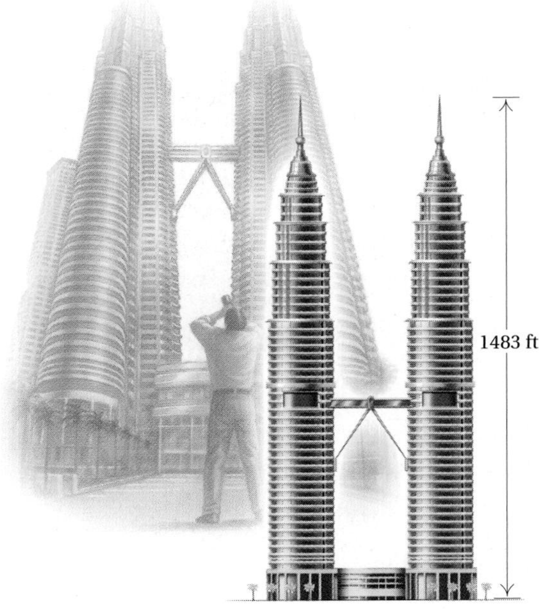

The height H, in meters, is given by

$$\begin{aligned} H &= 1483 \text{ ft} \\ &= 1483 \times 1 \text{ ft} \\ &\approx 1483 \times 0.305 \text{ m} \qquad \text{Substituting 0.305 m for 1 ft} \\ &= 452.315 \text{ m.} \end{aligned}$$

Do Exercises 38 and 39.

EXAMPLE 18 Complete: 0.10414 mm = ________ in.
(The thickness of a $1 bill)

In this case, we must make two substitutions or multiply by two forms of 1 since the chart on the preceding page does not provide an easy way to convert from millimeters to inches. Here we choose to multiply by forms of 1.

$$\begin{aligned} 0.10414 \text{ mm} &= 0.10414 \times 1 \cancel{\text{mm}} \times \frac{1 \text{ cm}}{10 \cancel{\text{mm}}} \\ &= 0.010414 \text{ cm} \\ &\approx 0.010414 \times 1 \cancel{\text{cm}} \times \frac{1 \text{ in.}}{2.54 \cancel{\text{cm}}} \\ &= 0.0041 \text{ in.} \end{aligned}$$

Do Exercise 40.

A

EXERCISE SET

a Complete.

1. 1 ft = ______ in.
2. 1 yd = ______ ft
3. 1 in. = ______ ft
4. 1 mi = ______ yd
5. 1 mi = ______ ft
6. 1 ft = ______ yd
7. 3 yd = ______ in.
8. 10 yd = ______ ft
9. 84 in. = ______ ft
10. 48 ft = ______ yd
11. 18 in. = ______ ft
12. 29 ft = ______ yd
13. 5 mi = ______ ft
14. 5 mi = ______ yd
15. 63 in. = ______ ft
16. 11,616 ft = ______ mi
17. 10 ft = ______ yd
18. 9.6 yd = ______ ft
19. 7.1 mi = ______ ft
20. 31,680 ft = ______ mi
21. $4\frac{1}{2}$ ft = ______ yd
22. 48 in. = ______ ft
23. 45 in. = ______ yd
24. $6\frac{1}{3}$ yd = ______ in.
25. 330 ft = ______ yd
26. 5280 yd = ______ mi
27. 3520 yd = ______ mi
28. 25 mi = ______ ft
29. 100 yd = ______ ft
30. 480 in. = ______ ft
31. 360 in. = ______ ft
32. 720 in. = ______ yd
33. 1 in. = ______ yd
34. 25 in. = ______ ft
35. 2 mi = ______ in.
36. 63,360 in. = ______ mi

b Complete. Do as much as possible mentally.

37. **a)** 1 km = ________ m
b) 1 m = ________ km

38. **a)** 1 hm = ________ m
b) 1 m = ________ hm

39. **a)** 1 dam = ________ m
b) 1 m = ________ dam

40. **a)** 1 dm = ________ m
b) 1 m = ________ dm

41. **a)** 1 cm = ________ m
b) 1 m = ________ cm

42. **a)** 1 mm = ________ m
b) 1 m = ________ mm

43. **a)** 6.7 km = ________ m
b) This conversion is from a larger unit to a smaller. Did you substitute or multiply by 1?

44. 27 km = ________ m

45. **a)** 98 cm = ________ m
b) This conversion is from a smaller unit to a larger. Did you substitute or multiply by 1?

46. 0.789 cm = ________ m

47. 8921 m = ________ km

48. 8664 m = ________ km

49. 56.66 m = ________ km

50. 4.733 m = ________ km

51. 5666 m = ________ cm

52. 869 m = ________ cm

53. 477 cm = ________ m

54. 6.27 mm = ________ m

55. 6.88 m = ________ cm

56. 6.88 m = ________ dm

57. 1 mm = ________ cm

58. 1 cm = ________ km

59. 1 km = ________ cm

60. 2 km = ________ cm

61. 14.2 cm = ________ mm

62. 25.3 cm = ________ mm

63. 8.2 mm = ________ cm

64. 9.7 mm = ________ cm

65. 4500 mm = ________ cm

66. 8,000,000 m = ________ km

67. 0.024 mm = ________ m

68. 60,000 mm = ________ dam

69. 6.88 m = ________ dam

70. 7.44 m = ________ hm

71. 2.3 dam = ________ dm

72. 9 km = ________ hm

73. 392 dam = ________ km

74. 0.056 mm = ________ dm

Complete the following table.

	OBJECT	MILLIMETERS (mm)	CENTIMETERS (cm)	METERS (m)
75.	Length of a calculator		18	
76.	Width of a calculator	85		
77.	Length of a piece of typing paper			0.278
78.	Length of a football field			109.09
79.	Width of a football field		4844	
80.	Width of a credit card	56		
81.	Length of 4 meter sticks			4
82.	Length of 3 meter sticks		300	
83.	Thickness of an index card	0.27		
84.	Thickness of a piece of cardboard		0.23	
85.	Height of the Sears Tower			442
86.	Height of the CN Tower (Toronto)	553,000		

C Complete.

87. 330 ft = ______ m
(The length of most baseball foul lines)

88. 12 in. = ______ cm
(The length of a common ruler)

89. 1171.4 km = ______ mi
(The distance from Cleveland to Atlanta)

90. 2 m = ______ ft
(The length of a desk)

91. 65 mph = ______ km/h
(The common speed limit in the United States)

92. 100 km/h = ______ mph
(A common speed limit in Canada)

93. 180 mi = ______ km
(The distance from Indianapolis to Chicago)

94. 141,600,000 mi = ______ km
(The farthest distance of Mars from the sun)

95. 70 mph = ______ km/h
(An interstate speed limit in Arizona)

96. 60 km/h = ______ mph
(A city speed limit in Canada)

97. 10 yd = ______ m
(The length needed for a first down in football)

98. 450 ft = ______ m
(The length of a long home run in baseball)

99. 2.08 m = ________ in.
(The height of Amare Stoudemire of the Phoenix Suns)

100. 76 in. = ________ m
(The height of Jason Kidd of the New Jersey Nets)

101. 381 m = ________ ft
(The height of the Empire State Building)

102. 1127 ft = ________ m
(The height of the John Hancock Center)

103. 7.5 in. = ________ cm
(The length of a pencil)

104. 15.7 cm = ________ in.
(The length of a $1 bill)

105. 2216 km = ________ mi
(The distance from Chicago to Miami)

106. 1862 mi = ________ km
(The distance from Seattle to Kansas City)

107. 13 mm = ________ in.
(The thickness of a plastic case for a DVD)

108. 0.25 in. = ________ mm
(The thickness of an eraser on a pencil)

Complete the following table.

	OBJECT	YARDS (yd)	CENTIMETERS (cm)	INCHES (in.)	METERS (m)	MILLIMETERS (mm)
109.	Length of a mousepad		23.8			
110.	Width of a mousepad		20.3			
111.	Width of a piece of typing paper			$8\frac{1}{2}$		
112.	Length of a football field	120				
113.	Width of a football field		4844			
114.	Width of a credit card					56
115.	Length of 4 yardsticks	4				
116.	Length of 3 meter sticks		300			
117.	Thickness of an index card				0.00027	
118.	Thickness of a piece of cardboard		0.23			
119.	Height of the Sears Tower				442	
120.	Height of Central Plaza, Hong Kong	409				

SYNTHESIS

121. Develop a formula to convert from inches to millimeters.

122. Develop a formula to convert from millimeters to inches. How does it relate to the answer for Exercise 121?

B WEIGHT AND MASS; MEDICAL APPLICATIONS

Objectives

a Convert from one American unit of weight to another.

b Convert from one metric unit of mass to another.

c Make conversions and solve applied problems concerning medical dosages.

There is a difference between **mass** and **weight,** but the terms are often used interchangeably. People sometimes use the word "weight" instead of "mass." Weight is related to the force of gravity. The farther you are from the center of the earth, the less you weigh. Your mass stays the same no matter where you are.

a Weight: The American System

AMERICAN UNITS OF WEIGHT	
1 ton (T) = 2000 pounds (lb)	1 lb = 16 ounces (oz)

The term "ounce" used here for weight is different from the "ounce" we will use for capacity in Appendix C. We convert units using the same techniques that we use with linear measure.

EXAMPLE 1 A well-known hamburger is called a "quarter-pounder." Find its name in ounces: a "______ ouncer."

$$\frac{1}{4}\text{ lb} = \frac{1}{4}\cdot 1\text{ lb} = \frac{1}{4}\cdot 16\text{ oz}$$

Substituting 16 oz for 1 lb. Since we are converting from a larger unit to a smaller unit, we use substitution.

$$= 4\text{ oz}$$

A "quarter-pounder" can also be called a "four-ouncer."

EXAMPLE 2 Complete: 15,360 lb = ______ T.

$$15{,}360\text{ lb} = 15{,}360\ \cancel{\text{lb}} \times \frac{1\text{ T}}{2000\ \cancel{\text{lb}}}$$

Multiplying by 1. Since we are converting from a smaller unit to a larger unit, we use multiplying by 1.

$$= \frac{15{,}360}{2000}\text{ T} = 7.68\text{ T}$$

Do Exercises 1–3.

Complete.

1. 5 lb = ______ oz

2. 8640 lb = ______ T

3. 1 T = ______ oz

b Mass: The Metric System

The basic unit of mass is the **gram** (g), which is the mass of 1 cubic centimeter (1 cm^3) of water. Since a cubic centimeter is small, a gram is a small unit of mass.

1 g = 1 gram = the mass of 1 cm^3 of water

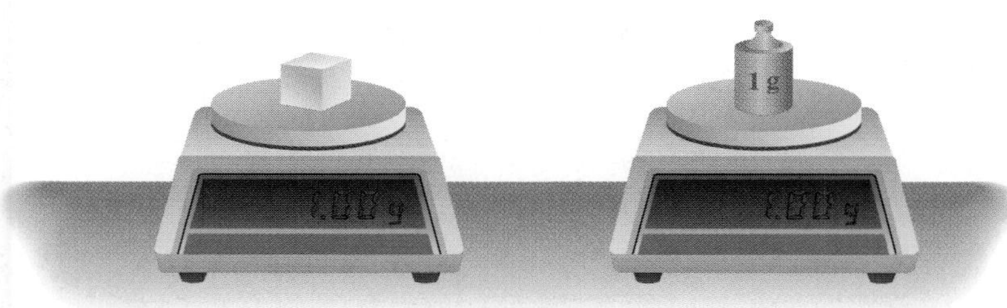

Answers on page A-56

The following table lists the metric units of mass. The prefixes are the same as those for length.

METRIC UNITS OF MASS

1 metric ton (t) = 1000 kilograms (kg)

1 *kilo*gram (kg) = 1000 grams (g)

1 *hecto*gram (hg) = 100 grams (g)

1 *deka*gram (dag) = 10 grams (g)

1 gram (g)

1 *deci*gram (dg) = $\frac{1}{10}$ gram (g)

1 *centi*gram (cg) = $\frac{1}{100}$ gram (g)

1 *milli*gram (mg) = $\frac{1}{1000}$ gram (g)

THINKING METRIC

One gram is about the mass of 1 raisin or 1 paperclip or 1 package of "Splenda" sweetener. Since 1 kg is about 2.2 lb, 1000 kg is about 2200 lb, or 1 metric ton (t), which is just a little more than 1 American ton (T), which is 2000 lb.

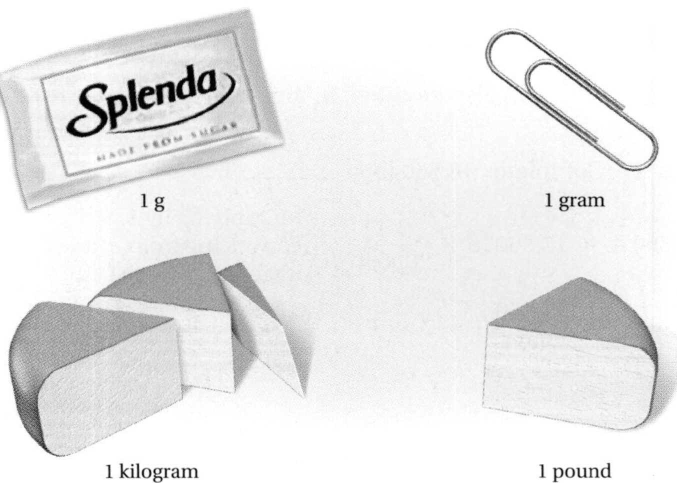

Small masses, such as dosages of medicine and vitamins, may be measured in milligrams (mg). The gram (g) is used for objects ordinarily measured in ounces, such as the mass of a letter, a piece of candy, a coin, or a small package of food.

Each 2.5 mg

15 g

2 g

2 kg

90 kg

The kilogram (kg) is used for larger food packages, such as fruit, or for human body mass. The metric ton (t) is used for very large masses, such as the mass of an automobile, a truckload of gravel, or an airplane.

Do Exercises 4–8.

CHANGING UNITS MENTALLY

As before, changing from one metric unit to another amounts to only the movement of a decimal point. We use this table.

1000 g	100 g	10 g	1 g	0.1 g	0.01 g	0.001 g
1 kg	1 hg	1 dag	1 g	1 dg	1 cg	1 mg

EXAMPLE 3 Complete: 8 kg = ______ g.

Think: To go from kg to g in the table is a move of three places to the right. Thus we move the decimal point three places to the right.

1000 g	100 g	10 g	1 g	0.1 g	0.01 g	0.001 g
1 kg	1 hg	1 dag	1 g	1 dg	1 cg	1 mg

3 places to the right

8.0   8.000.   8 kg = 8000 g

EXAMPLE 4 Complete: 4235 g = ______ kg.

Think: To go from g to kg in the table is a move of three places to the left. Thus we move the decimal point three places to the left.

Complete with mg, g, kg, or t.

4. A laptop computer has a mass of 6 ______ .

5. Eric has a body mass of 85.4 ______ .

6. This is a 3-______ vitamin.

7. A pen has a mass of 12 ______ .

8. A minivan has a mass of 3 ______ .

Complete.

9. 6.2 kg = ______ g

10. 304.8 cg = ______ g

Answers on page A-56

Complete.

11. 7.7 cg = ________ mg

12. 2344 mg = ________ cg

13. 67 dg = ________ mg

14. Complete:

1 mcg = ________ mg.

Answers on page A-56

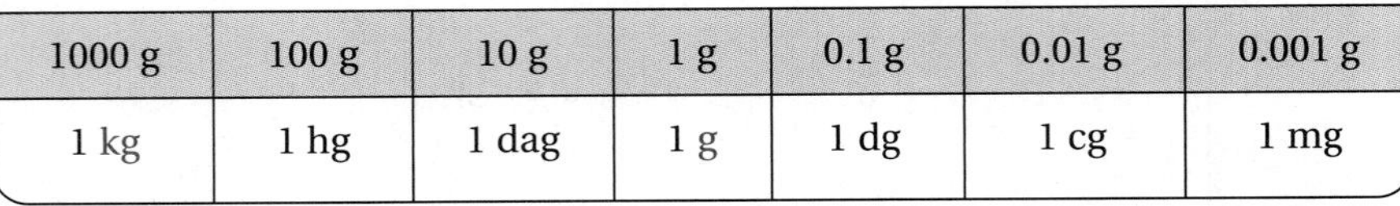

1000 g	100 g	10 g	1 g	0.1 g	0.01 g	0.001 g
1 kg	1 hg	1 dag	1 g	1 dg	1 cg	1 mg

3 places to the left

4235.0   4.235.0   4235 g = 4.235 kg

Do Exercises 9 and 10 on the preceding page.

EXAMPLE 5 Complete: 6.98 cg = ________ mg.

Think: To go from cg to mg is a move of one place to the right. Thus we move the decimal point one place to the right.

1000 g	100 g	10 g	1 g	0.1 g	0.01 g	0.001 g
1 kg	1 hg	1 dag	1 g	1 dg	1 cg	1 mg

1 place to the right

6.98   6.9.8   6.98 cg = 69.8 mg

The most commonly used metric units of mass are kg, g, cg, and mg. We have purposely used those more than the others in the exercises.

EXAMPLE 6 Complete: 89.21 mg = ________ g.

Think: To go from mg to g is a move of three places to the left. Thus we move the decimal point three places to the left.

1000 g	100 g	10 g	1 g	0.1 g	0.01 g	0.001 g
1 kg	1 hg	1 dag	1 g	1 dg	1 cg	1 mg

3 places to the left

89.21   0.089.21   89.21 mg = 0.08921 g

Do Exercises 11–13.

C Medical Applications

Another metric unit that is used in medicine is the microgram (mcg). It is defined as follows.

MICROGRAM

$$1 \text{ microgram} = 1 \text{ mcg} = \frac{1}{1{,}000{,}000}\text{ g} = 0.000001 \text{ g}$$

$$1{,}000{,}000 \text{ mcg} = 1 \text{ g}$$

EXAMPLE 7 Complete: 1 mg = _______ mcg.

We convert to grams and then to micrograms:

$$\begin{aligned} 1 \text{ mg} &= 0.001 \text{ g} \\ &= 0.001 \times 1 \text{ g} \\ &= 0.001 \times 1{,}000{,}000 \text{ mcg} \qquad \text{Substituting 1,000,000 mcg for 1 g} \\ &= 1000 \text{ mcg}. \end{aligned}$$

Do Exercise 14 on the preceding page.

EXAMPLE 8 *Medical Dosage.* Nitroglycerin sublingual tablets come in 0.4-mg tablets. How many micrograms are in each tablet?

Source: Steven R. Smith, M.D.

We are to complete: 0.4 mg = _______ mcg. Thus,

$$\begin{aligned} 0.4 \text{ mg} &= 0.4 \times 1 \text{ mg} \\ &= 0.4 \times 1000 \text{ mcg} \qquad \text{From Example 7, substituting 1000 mcg for 1 mg} \\ &= 400 \text{ mcg}. \end{aligned}$$

We can also do this problem in a manner similar to Example 7.

Do Exercise 15.

15. Medical Dosage. A physician prescribes 500 mcg of alprazolam, an antianxiety medication. How many milligrams is this dosage?

Source: Steven R. Smith, M.D.

Answer on page A-56

B EXERCISE SET

a Complete.

1. 1 T = ________ lb

2. 1 lb = ________ oz

3. 6000 lb = ________ T

4. 8 T = ________ lb

5. 4 lb = ________ oz

6. 10 lb = ________ oz

7. 6.32 T = ________ lb

8. 8.07 T = ________ lb

9. 3200 oz = ________ T

10. 6400 oz = ________ T

11. 80 oz = ________ lb

12. 960 oz = ________ lb

13. *Excelsior.* Western Excelsior is a company that makes a packing material called excelsior. Excelsior is produced from the wood of aspen trees. The largest grove of aspen trees on record contained 13,000,000 tons of aspen. How many pounds of aspen were there?

Source: Western Excelsior Corporation, Mancos CO

14. *Excelsior.* Western Excelsior buys 44,800 tons of aspen each year to make excelsior. How many pounds of aspen does it buy each year?

Source: Western Excelsior Corporation, Mancos CO

b Complete.

15. 1 kg = ________ g

16. 1 hg = ________ g

17. 1 dag = ________ g

18. 1 dg = ________ g

19. 1 cg = ________ g

20. 1 mg = ________ g

21. 1 g = ________ mg

22. 1 g = ________ cg

23. 1 g = ________ dg

24. 25 kg = ________ g

25. 234 kg = ________ g

26. 9403 g = ________ kg

27. 5200 g = ________ kg

28. 1.506 kg = ________ g

29. 67 hg = ________ kg

30. 45 cg = ______ g

31. 0.502 dg = ______ g

32. 0.0025 cg = ______ mg

33. 8492 g = ______ kg

34. 9466 g = ______ kg

35. 585 mg = ______ cg

36. 96.1 mg = ______ cg

37. 8 kg = ______ cg

38. 0.06 kg = ______ mg

39. 1 t = ______ kg

40. 2 t = ______ kg

41. 3.4 cg = ______ dag

42. 115 mg = ______ g

Complete.

43. 1 mg = ______ mcg

44. 1 mcg = ______ mg

45. 325 mcg = ______ mg

46. 0.45 mg = ______ mcg

Medical Dosage. Solve each of the following. (None of these medications should be taken without consulting your own physician.)

Source: Steven R. Smith, M.D.

47. Digoxin is a medication used to treat heart problems. A physician orders 0.125 mg of digoxin to be taken once daily. How many micrograms of digoxin are there in the daily dosage?

48. Digoxin is a medication used to treat heart problems. A physician orders 0.25 mg of digoxin to be taken once a day. How many micrograms of digoxin are there in the daily dosage?

49. Triazolam is a medication used for the short-term treatment of insomnia. A physician advises her patient to take one of the 0.125-mg tablets each night for 7 nights. How many milligrams of triazolam will the patient have ingested over that 7-day period? How many micrograms?

50. Clonidine is a medication used to treat high blood pressure. The usual starting dose of clonidine is one 0.1-mg tablet twice a day. If a patient is started on this dose by his physician, how many total milligrams of clonidine will the patient have taken before he returns to see his physician 14 days later? How many micrograms?

51. Cephalexin is an antibiotic that frequently is prescribed in a 500-mg tablet form. A physician prescribes 2 g of cephalexin per day for a patient with a skin abscess. How many 500-mg tablets would have to be taken in order to achieve this daily dosage?

52. Quinidine gluconate is a liquid mixture, part medicine and part water, which is administered intravenously. There is 80 mg of quinidine gluconate in each cubic centimeter (cc) of the liquid mixture. A physician orders 900 mg of quinidine gluconate to be administered daily to a patient with malaria. How much of the solution would have to be administered in order to achieve the recommended daily dosage?

53. Amoxicillin is a common antibiotic prescribed for children. It is a liquid suspension composed of part amoxicillin and part water. In one formulation of amoxicillin suspension, there is 250 mg of amoxicillin in 5 cc of the liquid suspension. A physician prescribes 400 mg per day for a 2-year-old child with an ear infection. How much of the amoxicillin liquid suspension would the child's parent need to administer in order to achieve the recommended daily dosage of amoxicillin?

54. Albuterol is a medication used for the treatment of asthma. It comes in an inhaler that contains 17 mg of albuterol mixed with a liquid. One actuation (inhalation) from the mouthpiece delivers a 90-mcg dose of albuterol.

a) A physician orders 2 inhalations 4 times per day. How many micrograms of albuterol does the patient inhale per day?

b) How many actuations/inhalations are contained in one inhaler?

c) Danielle is going away for 4 months of college and wants to take enough albuterol to last for that time. Her physician has prescribed 2 inhalations 4 times per day. Estimate how many inhalers Danielle will need to take with her for the 4-month period.

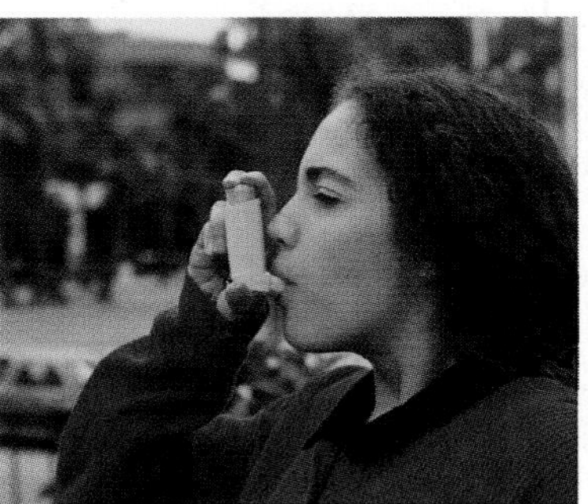

SYNTHESIS

55. A box of gelatin-mix packages weighs $15\frac{3}{4}$ lb. Each package weighs $1\frac{3}{4}$ oz. How many packages are in the box?

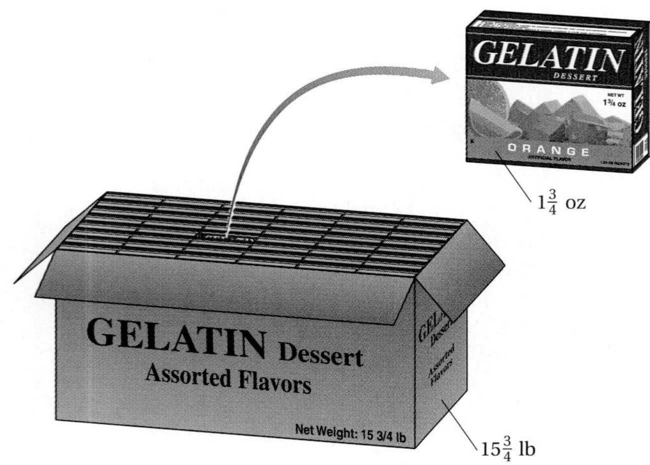

56. At \$0.90 a dozen, the cost of eggs is \$0.60 per pound. How much does an egg weigh?

C CAPACITY; MEDICAL APPLICATIONS

Objectives

a Convert from one unit of capacity to another.

b Solve applied problems concerning medical dosages.

a Capacity

AMERICAN UNITS

To answer a question like "How much soda is in the can?" we need measures of **capacity.** American units of capacity are fluid ounces, cups, pints, quarts, and gallons. These units are related as follows.

AMERICAN UNITS OF CAPACITY	
1 gallon (gal) = 4 quarts (qt)	1 pt = 2 cups = 16 fluid ounces (fl oz)
1 qt = 2 pints (pt)	1 cup = 8 fluid oz

Fluid ounces, abbreviated fl oz, are often referred to as ounces, or oz.

EXAMPLE 1 Complete: 9 gal = ______ oz.

Since we are converting from a *larger* unit to a *smaller* unit, we use substitution.

$$9 \text{ gal} = 9 \cdot 1 \text{ gal} = 9 \cdot 4 \text{ qt} \quad \text{Substituting 4 qt for 1 gal}$$
$$= 9 \cdot 4 \cdot 1 \text{ qt} = 9 \cdot 4 \cdot 2 \text{ pt} \quad \text{Substituting 2 pt for 1 qt}$$
$$= 9 \cdot 4 \cdot 2 \cdot 1 \text{ pt} = 9 \cdot 4 \cdot 2 \cdot 16 \text{ oz} \quad \text{Substituting 16 oz for 1 pt}$$
$$= 1152 \text{ oz}$$

EXAMPLE 2 Complete: 24 qt = ______ gal.

Since we are converting from a *smaller* unit to a *larger* unit, we multiply by 1 using 1 gal in the numerator and 4 qt in the denominator.

$$24 \text{ qt} = 24 \cancel{\text{qt}} \cdot \frac{1 \text{ gal}}{4 \cancel{\text{qt}}} = \frac{24}{4} \cdot 1 \text{ gal} = 6 \text{ gal}$$

Do Exercises 1 and 2.

Complete.

1. 5 gal = ______ pt

2. 80 qt = ______ gal

METRIC UNITS

One unit of capacity in the metric system is a **liter.** A liter is just a bit more than a quart. It is defined as follows.

Answers on page A-56

Complete with mL or L.

3. The patient received an injection of 2 ________ of penicillin.

4. There are 250 ________ in a coffee cup.

5. The gas tank holds 80 ________.

6. Bring home 8 ________ of milk.

Answers on page A-56

METRIC UNITS OF CAPACITY

1 liter (L) = 1000 cubic centimeters (1000 cm^3)

The script letter ℓ is also used for "liter."

The metric prefixes are also used with liters. The most common is **milli-.** The milliliter (mL) is, then, $\frac{1}{1000}$ liter. Thus,

$$1\text{ L} = 1000\text{ mL} = 1000\text{ cm}^3;$$
$$0.001\text{ L} = 1\text{ mL} = 1\text{ cm}^3.$$

Although the other metric prefixes are rarely used for capacity, we display them in the following table as we did for linear measure.

1000 L	100 L	10 L	1 L	0.1 L	0.01 L	0.001 L
1 kL	1 hL	1 daL	1 L	1 dL	1 cL	1 mL (cc)

A preferred unit for drug dosage is the milliliter (mL) or the cubic centimeter (cm^3). The notation "cc" is also used for cubic centimeter, especially in medicine. The milliliter and the cubic centimeter represent the same measure of capacity. A milliliter is about $\frac{1}{5}$ of a teaspoon.

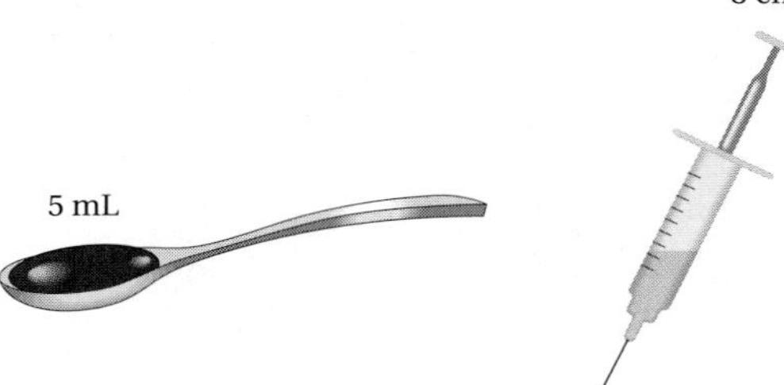

$$1\text{ mL} = 1\text{ cm}^3 = 1\text{ cc}$$

Volumes for which quarts and gallons are used are expressed in liters. Large volumes in business and industry are expressed using measures of cubic meters (m^3).

Do Exercises 3–6.

EXAMPLE 3 Complete: 4.5 L = ________ mL.

$4.5\text{ L} = 4.5 \times 1\text{ L} = 4.5 \times 1000\text{ mL}$ Substituting 1000 mL for 1 L

$= 4500\text{ mL}$

1000 L	100 L	10 L	1 L	0.1 L	0.01 L	0.001 L
1 kL	1 hL	1 daL	1 L	1 dL	1 cL	1 mL (cc)

3 places to the right

EXAMPLE 4 Complete: 280 mL = ______ L.

$280 \text{ mL} = 280 \times 1 \text{ mL}$

$= 280 \times 0.001 \text{ L}$ Substituting 0.001 L for 1 mL

$= 0.28 \text{ L}$

1000 L	100 L	10 L	1 L	0.1 L	0.01 L	0.001 L
1 kL	1 hL	1 daL	1 L	1 dL	1 cL	1 mL (cc)

3 places to the left

We do find metric units of capacity in frequent use in the United States—for example, in sizes of soda bottles and automobile engines.

Do Exercises 7 and 8.

b Medical Applications

The metric system is used extensively in medicine.

EXAMPLE 5 *Medical Dosage.* A physician orders 3.5 L of 5% dextrose in water (abbrev. D5W) to be administered over a 24-hr period. How many milliliters were ordered?

We convert 3.5 L to milliliters:

$$3.5 \text{ L} = 3.5 \times 1 \text{ L} = 3.5 \times 1000 \text{ mL} = 3500 \text{ mL}.$$

The physician had ordered 3500 mL of D5W.

Do Exercise 9.

EXAMPLE 6 *Medical Dosage.* Liquids at a pharmacy are often labeled in liters or milliliters. Thus if a physician's prescription is given in ounces, it must be converted. For conversion, a pharmacist knows that 1 oz ≈ 29.57 mL.* A prescription calls for 3 oz of theophylline. For how many milliliters is the prescription?

We convert as follows:

$$3 \text{ oz} = 3 \times 1 \text{ oz} \approx 3 \times 29.57 \text{ mL} = 88.71 \text{ mL}.$$

The prescription calls for 88.71 mL of theophylline.

Do Exercise 10.

*In practice, most physicians use 30 mL as an approximation to 1 oz.

Complete.

7. 0.97 L = ______ mL

8. 8990 mL = ______ L

9. Medical Dosage. A physician orders 2400 mL of 0.9% saline solution to be administered intravenously over a 24-hr period. How many liters were ordered?

10. Medical Dosage. A prescription calls for 2 oz of theophylline.

a) For how many milliliters is the prescription?

b) For how many liters is the prescription?

Answers on page A-56

C EXERCISE SET

a Complete.

1. 1 L = ______ mL = ______ cm^3

2. ______ L = 1 mL = ______ cm^3

3. 87 L = ______ mL

4. 806 L = ______ mL

5. 49 mL = ______ L

6. 19 mL = ______ L

7. 0.401 mL = ______ L

8. 0.816 mL = ______ L

9. 78.1 L = ______ cm^3

10. 99.6 L = ______ cm^3

11. 10 qt = ______ oz

12. 9.6 oz = ______ pt

13. 20 cups = ______ pt

14. 1 gal = ______ oz

15. 8 gal = ______ qt

16. 1 gal = ______ cups

17. 5 gal = ______ qt

18. 11 gal = ______ qt

19. 56 qt = ______ gal

20. 84 qt = ______ gal

21. 11 gal = ______ pt

22. 5 gal = ______ pt

Complete.

	OBJECT	GALLONS (gal)	QUARTS (qt)	PINTS (pt)	CUPS	OUNCES (oz)
23.	12-can package of 12-oz sodas					144
24.	6-bottle package of 16-oz sodas		3			
25.	Full tank of gasoline	16				
26.	Container of milk			8		
27.	Minute Maid Orange Juice				4	
28.	Scope Mouthwash					33
29.	Revlon Flex Shampoo					15
30.	Jhirmack Moisturizing Shampoo					20

Complete.

	OBJECT	LITERS (L)	MILLILITERS (mL)	CUBIC CENTIMETERS (cc)	CUBIC CENTIMETERS (cm^3)
31.	2-L bottle of soda	2			
32.	Heinz Vinegar		3755		
33.	Full tank of gasoline in Europe	64			
34.	Old Spice Aftershave				125
35.	Revlon Flex Shampoo			443	
36.	Jhirmack Moisturizing Shampoo		591		

b *Medical Dosage.* Solve each of the following.

Source: Steven R. Smith, M.D.

37. An emergency-room physician orders 2.0 L of Ringer's lactate to be administered over 2 hr for a patient in shock. How many milliliters is this?

38. An emergency-room physician orders 2.5 L of 0.9% saline solution over 4 hr for a patient suffering from dehydration. How many milliliters is this?

39. A physician orders 320 mL of 5% dextrose in water (D5W) solution to be administered intravenously over 4 hr. How many liters of D5W is this?

40. A physician orders 40 mL of 5% dextrose in water (D5W) solution to be administered intravenously over 2 hr to an elderly patient. How many liters of D5W is this?

41. A physician orders 0.5 oz of magnesia and alumina oral suspension antacid 4 times per day for a patient with indigestion. How many milliliters of the antacid is the patient to ingest in a day?

42. A physician orders 0.25 oz of magnesia and alumina oral suspension antacid 3 times per day for a child with upper abdominal discomfort. How many milliliters of the antacid is the child to ingest in a day?

43. A physician orders 0.5 L of normal saline solution. How many milliliters are ordered?

44. A physician has ordered that his patient receive 60 mL per hour of normal saline solution intravenously. How many liters of the saline solution is the patient to receive in a 24-hr period?

45. A physician wants her patient to receive 3.0 L of normal saline intravenously over a 24-hr period. How many milliliters per hour must the nurse administer?

46. A physician tells a patient to purchase 0.5 L of hydrogen peroxide. Commercially, hydrogen peroxide is found on the shelf in bottles that hold 4 oz, 8 oz, and 16 oz. Which bottle comes closest to filling the prescription?

Medical Dosage. Because patients do not always have a working knowledge of the metric system, physicians often prescribe dosages in teaspoons (t or tsp) and tablespoons (T or tbs). The units are related to each other and to the metric system as follows:

$$5 \text{ mL} \approx 1 \text{ tsp}, \qquad 3 \text{ tsp} = 1 \text{ T}.$$

Complete.

47. 45 mL = ______ tsp

48. 3 T = ______ tsp

49. 1 mL = ______ tsp

50. 18.5 mL = ______ tsp

51. 2 T = ______ tsp

52. 8.5 tsp = ______ T

53. 1 T = ______ mL

54. 18.5 mL = ______ T

SYNTHESIS

55. *Bees and Honey.* The average bee produces only $\frac{1}{8}$ tsp of honey in its lifetime. It takes 60,000 honeybees to produce 100 lb of honey. How much does a teaspoon of honey weigh?

D TIME AND TEMPERATURE

Objectives

a Convert from one unit of time to another.

b Convert between Celsius and Fahrenheit temperatures using the formulas

$$F = \frac{9}{5} \cdot C + 32$$

and

$$C = \frac{5}{9} \cdot (F - 32).$$

Complete.

1. 2 hr = ________ min

2. 4 yr = ________ days

3. 1 day = ________ min

4. 168 hr = ________ wk

Answers on page A-56

a Time

A table of units of time is shown below. The metric system sometimes uses "h" for hour and "s" for second, but we will use the more familiar "hr" and "sec."

UNITS OF TIME

1 day = 24 hours (hr)	1 year (yr) = $365\frac{1}{4}$ days
1 hr = 60 minutes (min)	1 week (wk) = 7 days
1 min = 60 seconds (sec)	

Since we cannot have $\frac{1}{4}$ day on the calendar, we give each year 365 days and every fourth year 366 days (a leap year), unless it is a year at the beginning of a century not divisible by 400.

EXAMPLE 1 Complete: 1 hr = ________ sec.

$$\begin{aligned} 1 \text{ hr} &= 60 \text{ min} \\ &= 60 \cdot 1 \text{ min} \\ &= 60 \cdot 60 \text{ sec} \qquad \text{Substituting 60 sec for 1 min} \\ &= 3600 \text{ sec} \end{aligned}$$

EXAMPLE 2 Complete: 5 yr = ________ days.

$$\begin{aligned} 5 \text{ yr} &= 5 \cdot 1 \text{ yr} \\ &= 5 \cdot 365\frac{1}{4} \text{ days} \qquad \text{Substituting } 365\frac{1}{4} \text{ days for 1 yr} \\ &= 1826\frac{1}{4} \text{ days} \end{aligned}$$

EXAMPLE 3 Complete: 4320 min = ________ days.

$$4320 \text{ min} = 4320 \cancel{\text{min}} \cdot \frac{1 \cancel{\text{hr}}}{60 \cancel{\text{min}}} \cdot \frac{1 \text{ day}}{24 \cancel{\text{hr}}} = \frac{4320}{60 \cdot 24} \text{ days} = 3 \text{ days}$$

Do Exercises 1–4.

b Temperature

Below are two temperature scales: **Fahrenheit** for American measure and **Celsius** for metric measure.

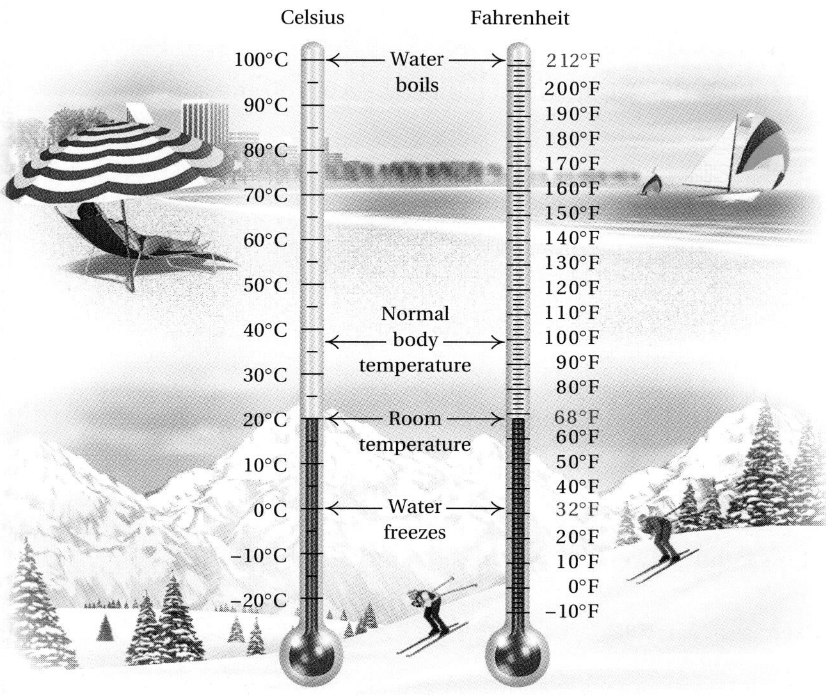

By laying a ruler or a piece of paper horizontally between the scales, we can make an approximate conversion from one measure of temperature to another and get an idea of how the temperature scales compare.

EXAMPLES Convert to Celsius (using the scales shown above). Approximate to the nearest ten degrees.

4. 212°F (Boiling point of water)	100°C	This is exact.
5. 32°F (Freezing point of water)	0°C	This is exact.
6. 105°F	40°C	This is approximate.

Do Exercises 5–7.

EXAMPLES Make an approximate conversion to Fahrenheit.

7. 44°C (Hot bath)	110°F	This is approximate.
8. 20°C (Room temperature)	68°F	This is exact.
9. 83°C	180°F	This is approximate.

Do Exercises 8–10.

Convert to Celsius. Approximate to the nearest ten degrees.

5. 180°F (Brewing coffee)

6. 25°F (Cold day)

7. −10°F (Miserably cold day)

Convert to Fahrenheit. Approximate to the nearest ten degrees.

8. 25°C (Warm day at the beach)

9. 40°C (Temperature of a patient with a high fever)

10. 10°C (A cold bath)

Answers on page A-56

Convert to Fahrenheit.

11. 80°C

12. 35°C

Convert to Celsius.

13. 95°F

14. 113°F

Answers on page A-56

The following formula allows us to make exact conversions from Celsius to Fahrenheit.

CELSIUS TO FAHRENHEIT

$$F = \frac{9}{5} \cdot C + 32, \quad \text{or} \quad F = 1.8 \cdot C + 32$$

$\left(\text{Multiply the Celsius temperature by } \frac{9}{5}, \text{ or 1.8, and add 32.}\right)$

EXAMPLES Convert to Fahrenheit.

10. 0°C (Freezing point of water)

$$F = \frac{9}{5} \cdot C + 32 = \frac{9}{5} \cdot 0 + 32 = 0 + 32 = 32°$$

Thus, 0°C = 32°F.

11. 37°C (Normal body temperature)

$$F = 1.8 \cdot C + 32 = 1.8 \cdot 37 + 32 = 66.6 + 32 = 98.6°$$

Thus, 37°C = 98.6°F.

Check the answers to Examples 10 and 11 using the scales on p. 1165.

Do Exercises 11 and 12.

The following formula allows us to make exact conversions from Fahrenheit to Celsius.

FAHRENHEIT TO CELSIUS

$$C = \frac{5}{9} \cdot (F - 32), \quad \text{or} \quad C = \frac{F - 32}{1.8}$$

$\left(\text{Subtract 32 from the Fahrenheit temperature and multiply by } \frac{5}{9} \text{ or divide by 1.8.}\right)$

EXAMPLES Convert to Celsius.

12. 212°F (Boiling point of water)

$$C = \frac{5}{9} \cdot (F - 32) = \frac{5}{9} \cdot (212 - 32) = \frac{5}{9} \cdot 180 = 100°$$

Thus, 212°F = 100°C.

13. 77°F

$$C = \frac{F - 32}{1.8} = \frac{77 - 32}{1.8} = \frac{45}{1.8} = 25°$$

Thus, 77°F = 25°C.

Check the answers to Examples 12 and 13 using the scales on p. 1165.

Do Exercises 13 and 14.

D EXERCISE SET

a Complete.

1. 1 day = ______ hr

2. 1 hr = ______ min

3. 1 min = ______ sec

4. 1 wk = ______ days

5. 1 yr = ______ days

6. 2 yr = ______ days

7. 180 sec = ______ hr

8. 60 sec = ______ hr

9. 492 sec = ______ min (The amount of time it takes for the rays of the sun to reach the earth)

10. 18,000 sec = ______ hr

11. 156 hr = ______ days

12. 444 hr = ______ days

13. 645 min = ______ hr

14. 375 min = ______ hr

15. 2 wk = ______ hr

16. 4 hr = ______ sec

17. 756 hr = ______ wk

18. 166,320 min = ______ wk

19. 2922 wk = ______ yr

20. 623 days = ______ wk

21. *Actual Time in a Day.* Although we round it to 24 hr, the actual length of a day is 23 hr 56 min 4.2 sec. How many seconds are there in an actual day?
Source: *The Handy Geography Answer Book*

22. *Time Length.* What length of time is 86,400 sec? Is it 1 hr, 1 day, 1 week, or 1 month? (This was a question on the TV game show "Who Wants to Be a Millionaire?".)

b Convert to Fahrenheit. Use the formula $F = \frac{9}{5} \cdot C + 32$ or $F = 1.8 \cdot C + 32$.

23. 25°C

24. 85°C

25. 40°C

26. 90°C

27. 86°C

28. 93°C

29. 58°C

30. 35°C

31. 2°C

32. 78°C

33. 5°C

34. 15°C

35. 3000°C
(The melting point of iron)

36. 1000°C
(The melting point of gold)

Convert to Celsius. Use the formula $C = \frac{5}{9} \cdot (F - 32)$ or $C = \frac{F - 32}{1.8}$.

37. 86°F

38. 59°F

39. 131°F

40. 140°F

41. 178°F

42. 195°F

43. 140°F

44. 107°F

45. 68°F

46. 50°F

47. 44°F

48. 120°F

49. 98.6°F
(Normal body temperature)

50. 104°F
(High-fevered body temperature)

51. *Highest Temperatures.* The highest temperature ever recorded in the world is 136°F in the desert of Libya in 1922. The highest temperature ever recorded in the United States is $56\frac{2}{3}$°C in California's Death Valley in 1913.

Source: *The Handy Geography Answer Book*

a) Convert each temperature to the other scale.
b) How much higher in degrees Fahrenheit was the world record than the U. S. record?

52. *Boiling Point and Altitude.* The boiling point of water actually changes with altitude. The boiling point is 212°F at sea level, but lowers about 1°F for every 500 ft that the altitude increases above sea level.

Sources: *The Handy Geography Answer Book; The New York Times Almanac*

a) What is the boiling point at an elevation of 1500 ft above sea level?
b) The elevation of Tucson is 2564 ft above sea level and that of Phoenix is 1117 ft. What is the boiling point in each city?
c) How much lower is the boiling point in Denver, whose elevation is 5280 ft, than in Tucson?
d) What is the boiling point at the top of Mt. McKinley in Alaska, the highest point in the United States, at 20,320 ft?

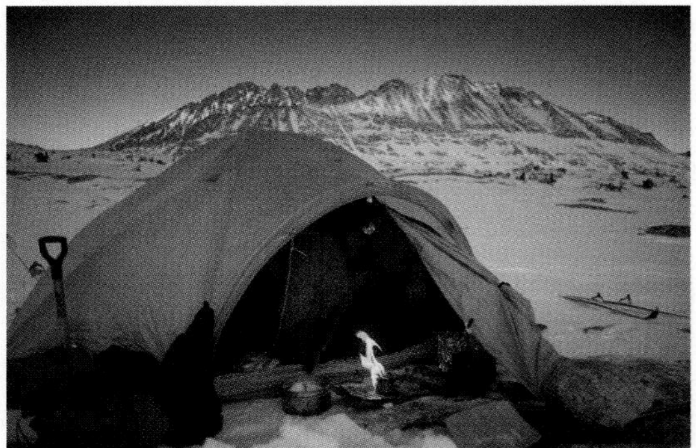

SYNTHESIS

53. Estimate the number of years in one million seconds.

54. Estimate the number of years in one billion seconds.

E SETS

a Naming Sets

To name the set of whole numbers less than 6, we can use the **roster method,** as follows: $\{0, 1, 2, 3, 4, 5\}$.

The set of real numbers x such that x is less than 6 cannot be named by listing all its members because there are infinitely many. We name such a set using **set-builder notation,** as follows: $\{x | x < 6\}$. This is read "The set of all x such that x is less than 6." See Section 8.7 for more on this notation.

Do Exercises 1 and 2.

b Set Membership and Subsets

The symbol $\in$ means **is a member of** or **belongs to,** or **is an element of.** Thus, $x \in A$ means x is a member of A or x belongs to A or x is an element of A.

EXAMPLE 1 Classify each of the following as true or false.

a) $1 \in \{1, 2, 3\}$
b) $1 \in \{2, 3\}$
c) $4 \in \{x | x \text{ is an even whole number}\}$
d) $5 \in \{x | x \text{ is an even whole number}\}$

a) Since 1 *is* listed as a member of the set, $1 \in \{1, 2, 3\}$ is true.
b) Since 1 *is not* a member of $\{2, 3\}$, the statement $1 \in \{2, 3\}$ is false.
c) Since 4 *is* an even whole number, $4 \in \{x | x \text{ is an even whole number}\}$ is a true statement.
d) Since 5 *is not* even, $5 \in \{x | x \text{ is an even whole number}\}$ is false.

Set membership can be illustrated with a diagram, as shown here.

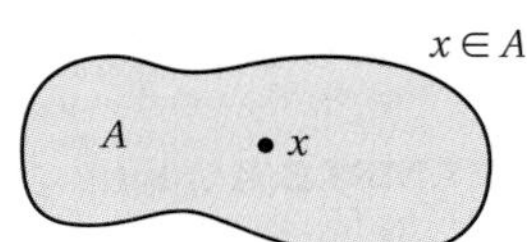

Do Exercises 3 and 4.

If every element of A is an element of B, then A is a **subset** of B. This is denoted $A \subseteq B$. The set of whole numbers is a subset of the set of integers. The set of rational numbers is a subset of the set of real numbers.

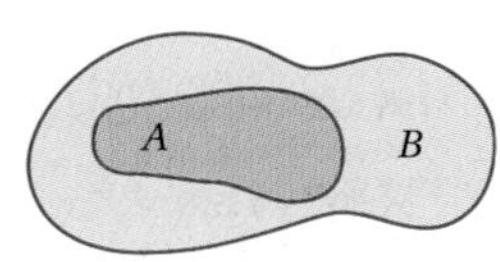

EXAMPLE 2 Classify each of the following as true or false.

a) $\{1, 2\} \subseteq \{1, 2, 3, 4\}$
b) $\{p, q, r, w\} \subseteq \{a, p, r, z\}$
c) $\{x | x < 6\} \subseteq \{x | x \leq 11\}$

a) Since every element of $\{1, 2\}$ is in the set $\{1, 2, 3, 4\}$, the statement $\{1, 2\} \subseteq \{1, 2, 3, 4\}$ is true.

Objectives

a Name sets using the roster method.

b Classify statements regarding set membership and subsets as true or false.

c Find the intersection and the union of sets.

Name the set using the roster method.

1. The set of whole numbers 0 through 7

2. $\{x |$ the square of x is $25\}$

Determine whether each of the following is true or false.

3. $8 \in \{x | x \text{ is an even whole number}\}$

4. $2 \in \{x | x \text{ is a prime number}\}$

Answers on page A-56

Determine whether each of the following is true or false.

5. $\{-2, -3, 4\} \subseteq \{-5, -4, -2, 7, -3, 5, 4\}$

6. $\{a, e, i, o, u\} \subseteq$ The set of all consonants

7. $\{x \mid x \le -8\} \subseteq \{x \mid x \le -7\}$

Find the intersection.

8. $\{-2, -3, 4, -4, 8\} \cap \{-5, -4, -2, 7, -3, 5, 4\}$

9. $\{a, e, i, o, u\} \cap \{m, a, r, v, i, n\}$

10. $\{a, e, i, o, u\} \cap$ The set of all consonants

Find the union.

11. $\{-2, -3, 4, -4, 8\} \cup \{-5, -4, -2, 7, -3, 5, 4\}$

12. $\{a, e, i, o, u\} \cup \{m, a, r, v, i, n\}$

13. $\{a, e, i, o, u\} \cup$ The set of all consonants

Answers on page A-56

b) Since $q \in \{p, q, r, w\}$, but $q \notin \{a, p, r, z\}$, the statement $\{p, q, r, w\} \subseteq \{a, p, r, z\}$ is false.

c) Since every number that is less than 6 is also less than 11, the statement $\{x \mid x < 6\} \subseteq \{x \mid x \le 11\}$ is true.

Do Exercises 5–7.

C Intersections and Unions

The **intersection** of sets A and B, denoted $A \cap B$, is the set of members that are common to both sets.

EXAMPLE 3 Find the intersection.

a) $\{0, 1, 3, 5, 25\} \cap \{2, 3, 4, 5, 6, 7, 9\}$ **b)** $\{a, p, q, w\} \cap \{p, q, t\}$

a) $\{0, 1, 3, 5, 25\} \cap \{2, 3, 4, 5, 6, 7, 9\} = \{3, 5\}$

b) $\{a, p, q, w\} \cap \{p, q, t\} = \{p, q\}$

Set intersection can be illustrated with a diagram, as shown here.

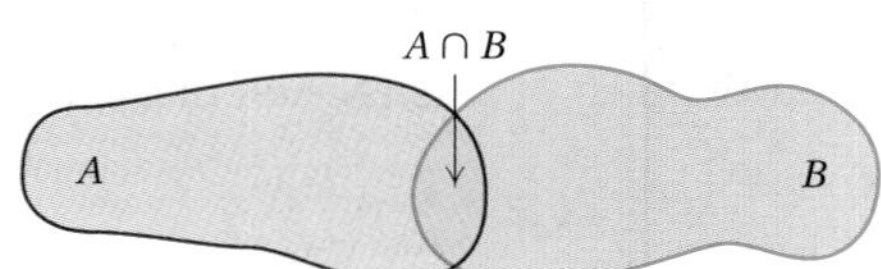

The set without members is known as the **empty set,** and is often named $\varnothing$, and sometimes $\{\ \}$. Each of the following is a description of the empty set:

$\{2, 3\} \cap \{5, 6, 7\}$;

$\{x \mid x$ is an even natural number$\} \cap \{x \mid x$ is an odd natural number$\}$.

Do Exercises 8–10.

Two sets A and B can be combined to form a set that contains the members of A as well as those of B. The new set is called the **union** of A and B, denoted $A \cup B$.

EXAMPLE 4 Find the union.

a) $\{0, 5, 7, 13, 27\} \cup \{0, 2, 3, 4, 5\}$ **b)** $\{a, c, e, g\} \cup \{b, d, f\}$

a) $\{0, 5, 7, 13, 27\} \cup \{0, 2, 3, 4, 5\} = \{0, 2, 3, 4, 5, 7, 13, 27\}$

Note that the 0 and the 5 are *not* listed twice in the solution.

b) $\{a, c, e, g\} \cup \{b, d, f\} = \{a, b, c, d, e, f, g\}$

Set union can be illustrated with a diagram, as shown here.

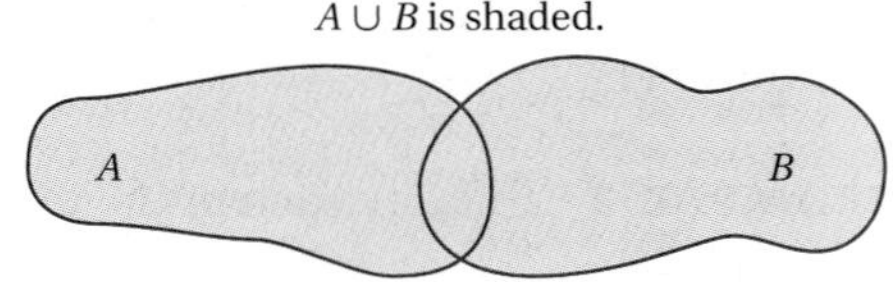

The solution set of the equation $(x - 3)(x + 2) = 0$ is $\{3, -2\}$. This set is the union of the solution sets of $x - 3 = 0$ and $x + 2 = 0$, which are $\{3\}$ and $\{-2\}$.

Do Exercises 11–13.

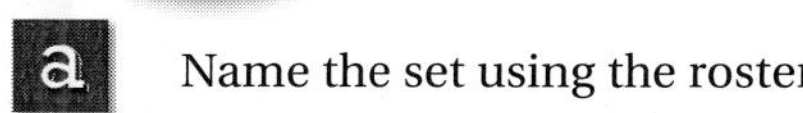

EXERCISE SET

a Name the set using the roster method.

1. The set of whole numbers 3 through 8

2. The set of whole numbers 101 through 107

3. The set of odd numbers between 40 and 50

4. The set of multiples of 5 between 11 and 39

5. $\{x \mid \text{the square of } x \text{ is } 9\}$

6. $\{x \mid x \text{ is the cube of } 0.2\}$

b Classify the statement as true or false.

7. $2 \in \{x \mid x \text{ is an odd number}\}$

8. $7 \in \{x \mid x \text{ is an odd number}\}$

9. Jeff Gordon $\in$ The set of all NASCAR drivers

10. Apple $\in$ The set of all fruit

11. $-3 \in \{-4, -3, 0, 1\}$

12. $0 \in \{-4, -3, 0, 1\}$

13. $\frac{2}{3} \in \{x \mid x \text{ is a rational number}\}$

14. Heads $\in$ The set of outcomes of flipping a penny

15. $\{4, 5, 8,\} \subseteq \{1, 3, 4, 5, 6, 7, 8, 9\}$

16. The set of vowels $\subseteq$ The set of consonants

17. $\{-1, -2, -3, -4, -5\} \subseteq \{-1, 2, 3, 4, 5\}$

18. The set of integers $\subseteq$ The set of rational numbers

c Find the intersection.

19. $\{a, b, c, d, e\} \cap \{c, d, e, f, g\}$

20. $\{a, e, i, o, u\} \cap \{q, u, i, c, k\}$

21. $\{1, 2, 5, 10\} \cap \{0, 1, 7, 10\}$

22. $\{0, 1, 7, 10\} \cap \{0, 1, 2, 5\}$

23. $\{1, 2, 5, 10\} \cap \{3, 4, 7, 8\}$

24. $\{a, e, i, o, u\} \cap \{m, n, f, g, h\}$

Find the union.

25. $\{a, e, i, o, u\} \cup \{q, u, i, c, k\}$

26. $\{a, b, c, d, e\} \cup \{c, d, e, f, g\}$

27. $\{0, 1, 7, 10\} \cup \{0, 1, 2, 5\}$

28. $\{1, 2, 5, 10\} \cup \{0, 1, 7, 10\}$

29. $\{a, e, i, o, u\} \cup \{m, n, f, g, h\}$

30. $\{1, 2, 5, 10\} \cup \{a, b\}$

SYNTHESIS

31. Find the union of the set of integers and the set of whole numbers.

32. Find the intersection of the set of odd integers and the set of even integers.

33. Find the union of the set of rational numbers and the set of irrational numbers.

34. Find the intersection of the set of even integers and the set of positive rational numbers.

35. Find the intersection of the set of rational numbers and the set of irrational numbers.

36. Find the union of the set of negative integers, the set of positive integers, and the set containing 0.

37. For a set A, find each of the following.

a) $A \cup \varnothing$ **b)** $A \cup A$
c) $A \cap A$ **d)** $A \cap \varnothing$

38. A set is *closed* under an operation if, when the operation is performed on its members, the result is in the set. For example, the set of real numbers is closed under the operation of addition since the sum of any two real numbers is a real number.

a) Is the set of even numbers closed under addition?
b) Is the set of odd numbers closed under addition?
c) Is the set $\{0, 1\}$ closed under addition?
d) Is the set $\{0, 1\}$ closed under multiplication?
e) Is the set of real numbers closed under multiplication?
f) Is the set of integers closed under division?

39. Experiment with sets of various types and determine whether the following distributive law for sets is true:

$$A \cap (B \cup C) = (A \cap B) \cup (A \cap C).$$

F FACTORING SUMS OR DIFFERENCES OF CUBES

Objective

a Factor sums and differences of two cubes.

a Factoring Sums or Differences of Cubes

We can factor the sum or the difference of two expressions that are cubes.

Consider the following products:

$$(A + B)(A^2 - AB + B^2) = A(A^2 - AB + B^2) + B(A^2 - AB + B^2)$$
$$= A^3 - A^2B + AB^2 + A^2B - AB^2 + B^3$$
$$= A^3 + B^3$$

and

$$(A - B)(A^2 + AB + B^2) = A(A^2 + AB + B^2) - B(A^2 + AB + B^2)$$
$$= A^3 + A^2B + AB^2 - A^2B - AB^2 - B^3$$
$$= A^3 - B^3.$$

The above equations (reversed) show how we can factor a sum or a difference of two cubes.

FACTORING SUMS OR DIFFERENCES OF CUBES

$A^3 + B^3 = (A + B)(A^2 - AB + B^2)$,
$A^3 - B^3 = (A - B)(A^2 + AB + B^2)$

N	N^3
0.2	0.008
0.1	0.001
0	0
1	1
2	8
3	27
4	64
5	125
6	216
7	343
8	512
9	729
10	1000

Note that what we are considering here is a sum or a difference of cubes. We are not cubing a binomial. For example, $(A + B)^3$ is *not* the same as $A^3 + B^3$. The table of cubes in the margin is helpful.

EXAMPLE 1 Factor: $x^3 - 8$.

We have

$$x^3 - 8 = x^3 - 2^3 = (x - 2)(x^2 + x \cdot 2 + 2^2).$$
$$\uparrow \quad \uparrow \quad \uparrow \; \uparrow \; \uparrow \quad \uparrow \; \uparrow \; \uparrow$$
$$\mathbf{A^3 - B^3 = (A - B)(A^2 + A\ \ B + B^2)}$$

This tells us that $x^3 - 8 = (x - 2)(x^2 + 2x + 4)$. Note that we cannot factor $x^2 + 2x + 4$. (It is not a trinomial square nor can it be factored by trial and error or the *ac*-method.) The check is left to the student.

Do Exercises 1 and 2.

Factor.

1. $x^3 - 27$

2. $64 - y^3$

EXAMPLE 2 Factor: $x^3 + 125$.

We have

$$x^3 + 125 = x^3 + 5^3 = (x + 5)(x^2 - x \cdot 5 + 5^2).$$
$$\uparrow \quad \uparrow \quad \uparrow \; \uparrow \; \uparrow \quad \uparrow \; \uparrow \; \uparrow$$
$$\mathbf{A^3 + B^3 = (A + B)(A^2 - A\ \ B + B^2)}$$

Thus, $x^3 + 125 = (x + 5)(x^2 - 5x + 25)$. The check is left to the student.

Do Exercises 3 and 4.

Factor.

3. $y^3 + 8$

4. $125 + t^3$

Answers on page A-56

Factor.

5. $27x^3 - y^3$

6. $8y^3 + z^3$

Factor.

7. $m^6 - n^6$

8. $16x^7y + 54xy^7$

9. $729x^6 - 64y^6$

10. $x^3 - 0.027$

Answers on page A-56

EXAMPLE 3 Factor: $x^3 - 27t^3$.

We have

$$x^3 - 27t^3 = x^3 - (3t)^3 = (x - 3t)(x^2 + x \cdot 3t + (3t)^2)$$

$$\uparrow \quad \uparrow \quad \uparrow \quad \uparrow \quad \uparrow \quad \uparrow \quad \uparrow \quad \uparrow$$

$$A^3 - B^3 = (A - B)(A^2 + A \quad B + B^2)$$

$$= (x - 3t)(x^2 + 3xt + 9t^2)$$

Do Exercises 5 and 6.

EXAMPLE 4 Factor: $128y^7 - 250x^6y$.

We first look for a common factor:

$$128y^7 - 250x^6y = 2y(64y^6 - 125x^6) = 2y[(4y^2)^3 - (5x^2)^3]$$
$$= 2y(4y^2 - 5x^2)(16y^4 + 20x^2y^2 + 25x^4).$$

EXAMPLE 5 Factor: $a^6 - b^6$.

We can express this polynomial as a difference of squares:

$$(a^3)^2 - (b^3)^2.$$

We factor as follows:

$$a^6 - b^6 = (a^3 + b^3)(a^3 - b^3).$$

One factor is a sum of two cubes, and the other factor is a difference of two cubes. We factor them:

$$(a + b)(a^2 - ab + b^2)(a - b)(a^2 + ab + b^2).$$

We have now factored completely.

In Example 5, had we thought of factoring first as a difference of two cubes, we would have had

$$(a^2)^3 - (b^2)^3 = (a^2 - b^2)(a^4 + a^2b^2 + b^4)$$
$$= (a + b)(a - b)(a^4 + a^2b^2 + b^4).$$

In this case, we might have missed some factors; $a^4 + a^2b^2 + b^4$ can be factored as $(a^2 - ab + b^2)(a^2 + ab + b^2)$, but we probably would not have known to do such factoring.

EXAMPLE 6 Factor: $64a^6 - 729b^6$.

$$64a^6 - 729b^6 = (8a^3 - 27b^3)(8a^3 + 27b^3)$$ Factoring a difference of squares

$$= [(2a)^3 - (3b)^3][(2a)^3 + (3b)^3].$$

Each factor is a sum or a difference of cubes. We factor each:

$$= (2a - 3b)(4a^2 + 6ab + 9b^2)(2a + 3b)(4a^2 - 6ab + 9b^2)$$

Sum of cubes:	$A^3 + B^3 = (A + B)(A^2 - AB + B^2)$;
Difference of cubes:	$A^3 - B^3 = (A - B)(A^2 + AB + B^2)$;
Difference of squares:	$A^2 - B^2 = (A + B)(A - B)$;
Sum of squares:	$A^2 + B^2$ cannot be factored using real numbers if the largest common factor has been removed.

Do Exercises 7–10.

F EXERCISE SET

a Factor.

1. $z^3 + 27$

2. $a^3 + 8$

3. $x^3 - 1$

4. $c^3 - 64$

5. $y^3 + 125$

6. $x^3 + 1$

7. $8a^3 + 1$

8. $27x^3 + 1$

9. $y^3 - 8$

10. $p^3 - 27$

11. $8 - 27b^3$

12. $64 - 125x^3$

13. $64y^3 + 1$

14. $125x^3 + 1$

15. $8x^3 + 27$

16. $27y^3 + 64$

17. $a^3 - b^3$

18. $x^3 - y^3$

19. $a^3 + \frac{1}{8}$

20. $b^3 + \frac{1}{27}$

21. $2y^3 - 128$

22. $3z^3 - 3$

23. $24a^3 + 3$

24. $54x^3 + 2$

25. $rs^3 + 64r$

26. $ab^3 + 125a$

27. $5x^3 - 40z^3$

28. $2y^3 - 54z^3$

29. $x^3 + 0.001$

30. $y^3 + 0.125$

31. $64x^6 - 8t^6$

32. $125c^6 - 8d^6$

33. $2y^4 - 128y$

34. $3z^5 - 3z^2$

35. $z^6 - 1$

36. $t^6 + 1$

37. $t^6 + 64y^6$

38. $p^6 - q^6$

SYNTHESIS

Consider these polynomials:

$(a + b)^3$; $a^3 + b^3$; $(a + b)(a^2 - ab + b^2)$;

$(a + b)(a^2 + ab + b^2)$; $(a + b)(a + b)(a + b)$.

39. Evaluate each polynomial when $a = -2$ and $b = 3$.

40. Evaluate each polynomial when $a = 4$ and $b = -1$.

Factor. Assume that variables in exponents represent natural numbers.

41. $x^{6a} + y^{3b}$

42. $a^3x^3 - b^3y^3$

43. $3x^{3a} + 24y^{3b}$

44. $\frac{8}{27}x^3 + \frac{1}{64}y^3$

45. $\frac{1}{24}x^3y^3 + \frac{1}{3}z^3$

46. $7x^3 - \frac{7}{8}$

47. $(x + y)^3 - x^3$

48. $(1 - x)^3 + (x - 1)^6$

49. $(a + 2)^3 - (a - 2)^3$

50. $y^4 - 8y^3 - y + 8$

G FINDING EQUATIONS OF LINES: POINT–SLOPE EQUATION

Objectives

a Find an equation of a line when the slope and a point are given.

b Find an equation of a line when two points are given.

In Section 9.4, we found equations of lines using the slope–intercept equation, $y = mx + b$. Here we introduce another form, the *point–slope equation,* and find equations of lines using both forms.

a Finding an Equation of a Line When the Slope and a Point Are Given

Suppose we know the slope of a line and the coordinates of one point on the line. We can use the slope–intercept equation to find an equation of the line. Or, we can use what is called a **point–slope equation.** We first develop a formula for such a line.

Suppose that a line of slope m passes through the point (x_1, y_1). For any other point (x, y) to lie on this line, we must have

$$\frac{y - y_1}{x - x_1} = m.$$

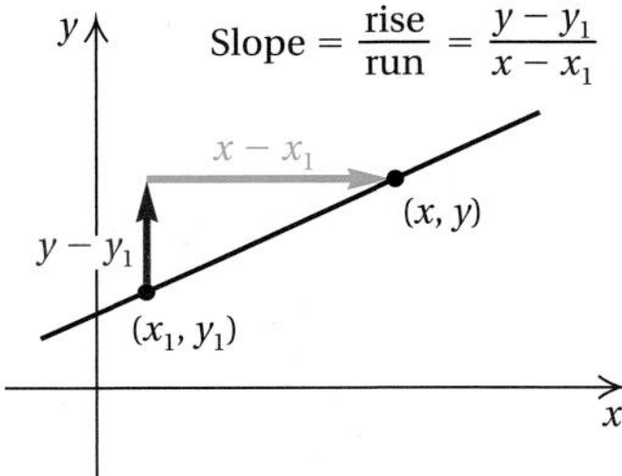

It is tempting to use this last equation as an equation of the line of slope m that passes through (x_1, y_1). The only problem with this form is that when x and y are replaced with x_1 and y_1, we have $\frac{0}{0} = m$, a false equation. To avoid this difficulty, we multiply by $x - x_1$ on both sides and simplify:

$$(x - x_1)\frac{y - y_1}{x - x_1} = m(x - x_1) \qquad \text{Multiplying by } x - x_1 \text{ on both sides}$$

$$y - y_1 = m(x - x_1). \qquad \text{Removing a factor of 1: } \frac{x - x_1}{x - x_1} = 1$$

This is the *point–slope* form of a linear equation.

POINT–SLOPE EQUATION

The **point–slope equation** of a line with slope m, passing through (x_1, y_1), is

$$y - y_1 = m(x - x_1).$$

Find an equation of the line with the given slope and containing the given point.

1. $m = -3,\ (-5, 4)$

2. $m = 5,\ (-2, 1)$

3. $m = 6,\ (3, -5)$

4. $m = -\frac{2}{3},\ (1, 2)$

Answers on page A-57

If we know the slope of a line and a certain point on the line, we can find an equation of the line using either the point–slope equation,

$$y - y_1 = m(x - x_1),$$

or the slope–intercept equation,

$$y = mx + b.$$

EXAMPLE 1 Find an equation of the line with slope -2 and containing the point $(-1, 3)$.

Using the Point–Slope Equation: We consider $(-1, 3)$ to be (x_1, y_1) and -2 to be the slope m, and substitute:

$$\begin{aligned} y - y_1 &= m(x - x_1) \\ y - 3 &= -2[x - (-1)] \quad \text{Substituting} \\ y - 3 &= -2(x + 1) \\ y - 3 &= -2x - 2 \\ y &= -2x - 2 + 3 \\ y &= -2x + 1. \end{aligned}$$

Using the Slope–Intercept Equation: The point $(-1, 3)$ is on the line, so it is a solution. Thus we can substitute -1 for x and 3 for y in $y = mx + b$. We also substitute -2 for m, the slope. Then we solve for b:

$$\begin{aligned} y &= mx + b \\ 3 &= -2 \cdot (-1) + b \quad \text{Substituting} \\ 3 &= 2 + b \\ 1 &= b. \quad \text{Solving for } b \end{aligned}$$

We then use the equation $y = mx + b$ and substitute -2 for m and 1 for b:

$$y = -2x + 1.$$

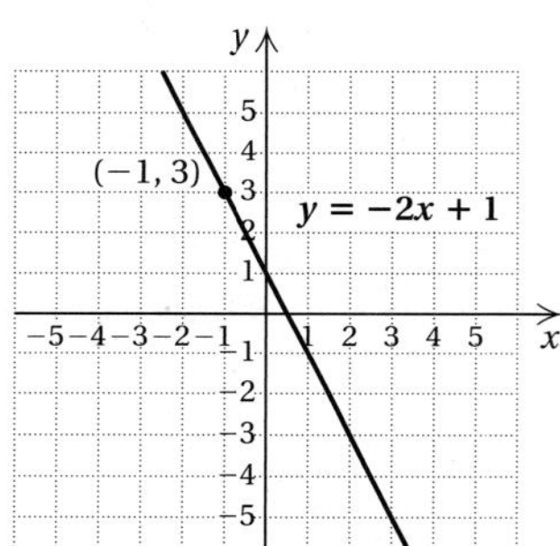

Do Exercises 1–4.

b Finding an Equation of a Line When Two Points Are Given

We can also use the point–slope equation or the slope–intercept equation to find an equation of a line when two points are given.

EXAMPLE 2 Find an equation of the line containing the points (3, 4) and (−5, 2).

First, we find the slope:

$$m = \frac{4-2}{3-(-5)} = \frac{2}{8}, \text{ or } \frac{1}{4}.$$

Now we have the slope and two points. We then proceed as we did in Example 1, using either point, and either the point–slope equation or the slope–intercept equation.

Using the Point–Slope Equation: We choose (3, 4) and substitute 3 for x_1, 4 for y_1, and $\frac{1}{4}$ for m:

$$y - y_1 = m(x - x_1)$$
$$y - 4 = \tfrac{1}{4}(x - 3) \quad \text{Substituting}$$
$$y - 4 = \tfrac{1}{4}x - \tfrac{3}{4}$$
$$y = \tfrac{1}{4}x - \tfrac{3}{4} + 4$$
$$y = \tfrac{1}{4}x - \tfrac{3}{4} + \tfrac{16}{4}$$
$$y = \tfrac{1}{4}x + \tfrac{13}{4}.$$

Using the Slope–Intercept Equation: We choose (3, 4) and substitute 3 for x, 4 for y, and $\frac{1}{4}$ for m and solve for b:

$$y = mx + b$$
$$4 = \tfrac{1}{4} \cdot 3 + b \quad \text{Substituting}$$
$$4 = \tfrac{3}{4} + b$$
$$4 - \tfrac{3}{4} = b$$
$$\tfrac{16}{4} - \tfrac{3}{4} = b$$
$$\tfrac{13}{4} = b. \quad \text{Solving for } b$$

Finally, we use the equation $y = mx + b$ and substitute $\frac{1}{4}$ for m and $\frac{13}{4}$ for b:

$$y = \tfrac{1}{4}x + \tfrac{13}{4}.$$

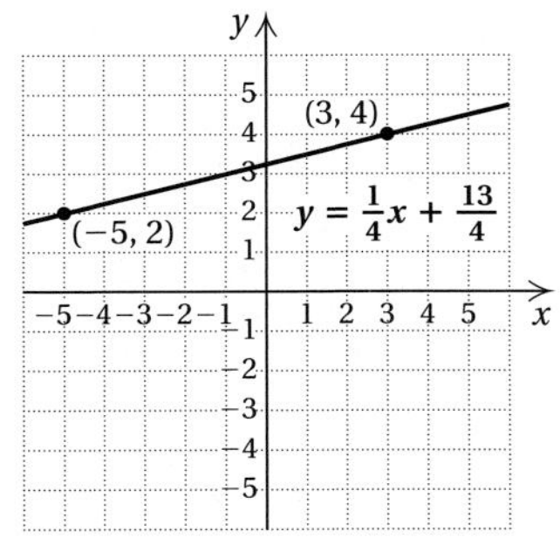

Do Exercises 5 and 6.

5. Find an equation of the line containing the points (3, −5) and (−1, 4).

6. Find an equation of the line containing the points (−3, 11) and (−4, 20).

Answers on page A-57

G EXERCISE SET

a Find an equation of the line having the given slope and containing the given point.

1. $m = 4,\ (5, 2)$

2. $m = 5,\ (4, 3)$

3. $m = -2,\ (2, 8)$

4. $m = -3,\ (9, 6)$

5. $m = 3,\ (-2, -2)$

6. $m = 1,\ (-1, -7)$

7. $m = -3,\ (-2, 0)$

8. $m = -2,\ (8, 0)$

9. $m = 0,\ (0, 4)$

10. $m = 0,\ (0, -7)$

11. $m = -\frac{4}{5},\ (2, 3)$

12. $m = \frac{2}{3},\ (1, -2)$

b Find an equation of the line containing the given pair of points.

13. $(2, 5)$ and $(4, 7)$

14. $(1, 4)$ and $(5, 6)$

15. $(-1, -1)$ and $(9, 9)$

16. $(-3, -3)$ and $(2, 2)$

17. $(0, -5)$ and $(3, 0)$

18. $(-4, 0)$ and $(0, 7)$

19. $(-4, -7)$ and $(-2, -1)$

20. $(-2, -3)$ and $(-4, -6)$

21. $(0, 0)$ and $(-4, 7)$

22. $(0, 0)$ and $(6, 1)$

23. $\left(\frac{2}{3}, \frac{3}{2}\right)$ and $\left(-3, \frac{5}{6}\right)$

24. $\left(\frac{1}{4}, -\frac{1}{2}\right)$ and $\left(\frac{3}{4}, 6\right)$

SYNTHESIS

25. Find an equation of the line that has the same y-intercept as the line $2x - y = -3$ and contains the point $(-1, -2)$.

26. Find an equation of the line with the same slope as the line $\frac{1}{2}x - \frac{1}{3}y = 10$ and the same y-intercept as the line $\frac{1}{4}x + 3y = -2$.

H EQUATIONS INVOLVING ABSOLUTE VALUE

Objective

a Solve equations with absolute-value expressions.

a Equations with Absolute Value

There are equations that have more than one solution. Examples are equations with absolute value. Remember, the absolute value of a number is its distance from 0 on the number line.

EXAMPLE 1 Solve: $|x| = 4$. Then graph on the number line.

Note that $|x| = |x - 0|$, so that $|x - 0|$ is the distance from x to 0. Thus solutions of the equation $|x| = 4$, or $|x - 0| = 4$, are those numbers x whose distance from 0 is 4. Those numbers are -4 and 4. The solution set is $\{-4, 4\}$. The graph consists of just two points, as shown.

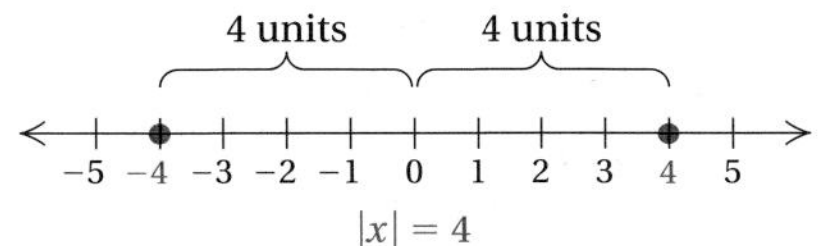

EXAMPLE 2 Solve: $|x| = 0$.

The only number whose absolute value is 0 is 0 itself. Thus the solution is 0. The solution set is $\{0\}$.

EXAMPLE 3 Solve: $|x| = -7$.

The absolute value of a number is always nonnegative. Thus there is no number whose absolute value is -7; consequently, the equation has no solution. The solution set is $\varnothing$.

Examples 1–3 lead us to the following principle for solving linear equations with absolute value.

THE ABSOLUTE-VALUE PRINCIPLE

For any positive number p and any algebraic expression X:

a) The solution of $|X| = p$ is those numbers that satisfy $X = -p$ or $X = p$.
b) The equation $|X| = 0$ is equivalent to the equation $X = 0$.
c) The equation $|X| = -p$ has no solution.

Do Exercises 1–3.

We can use the absolute-value principle with the addition and multiplication principles to solve equations with absolute value.

EXAMPLE 4 Solve: $2|x| + 5 = 9$.

We first use the addition and multiplication principles to get $|x|$ by itself. Then we use the absolute-value principle.

1. Solve: $|x| = 6$. Then graph on the number line.

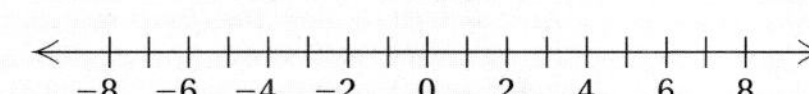

2. Solve: $|x| = -6$.

3. Solve: $|p| = 0$.

Answers on page A-57

Solve.

4. $|3x| = 6$

5. $4|x| + 10 = 27$

6. $3|x| - 2 = 10$

7. Solve: $|x - 4| = 1$. Use two methods as in Example 5.

8. Solve: $|3x - 4| = 17$.

9. Solve: $|6 + 2x| = -3$.

Answers on page A-57

We have

$$\begin{aligned} 2|x| + 5 &= 9 \\ 2|x| &= 4 && \text{Subtracting 5} \\ |x| &= 2 && \text{Dividing by 2} \\ x = -2 \quad &or \quad x = 2 && \text{Using the absolute-value principle} \end{aligned}$$

The solutions are -2 and 2. The solution set is $\{-2, 2\}$.

Do Exercises 4–6.

EXAMPLE 5 Solve: $|x - 2| = 3$.

We can consider solving this equation in two different ways.

METHOD 1 This allows us to see the meaning of the solutions graphically. The solution set consists of those numbers that are 3 units from 2 on the number line.

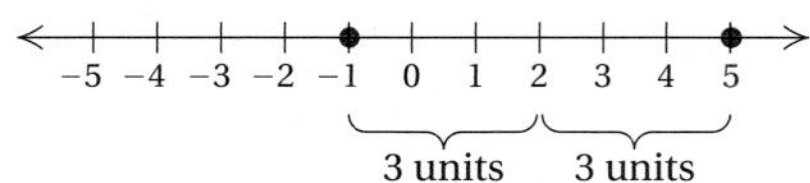

The solutions of $|x - 2| = 3$ are -1 and 5. The solution set is $\{-1, 5\}$.

METHOD 2 This method is more efficient. We use the absolute-value principle, replacing X with $x - 2$ and p with 3. Then we solve each equation separately.

$$\begin{aligned} |X| &= p \\ |x - 2| &= 3 \\ x - 2 = -3 \quad or \quad x - 2 &= 3 && \text{Absolute-value principle} \\ x = -1 \quad or \quad x &= 5 \end{aligned}$$

The solutions are -1 and 5. The solution set is $\{-1, 5\}$.

Do Exercise 7.

EXAMPLE 6 Solve: $|2x + 5| = 13$.

We use the absolute-value principle, replacing X with $2x + 5$ and p with 13:

$$\begin{aligned} |X| &= p \\ |2x + 5| &= 13 \\ 2x + 5 = -13 \quad or \quad 2x + 5 &= 13 && \text{Absolute-value principle} \\ 2x = -18 \quad or \quad 2x &= 8 \\ x = -9 \quad or \quad x &= 4. \end{aligned}$$

The solutions are -9 and 4. The solution set is $\{-9, 4\}$.

Do Exercise 8.

EXAMPLE 7 Solve: $|4 - 7x| = -8$.

Since absolute value is always nonnegative, this equation has no solution. The solution set is $\varnothing$.

Do Exercise 9.

H EXERCISE SET

a Solve.

1. $|x| = 3$

2. $|x| = 5$

3. $|x| = -3$

4. $|x| = -9$

5. $|q| = 0$

6. $|y| = 7.4$

7. $|x - 3| = 12$

8. $|3x - 2| = 6$

9. $|2x - 3| = 4$

10. $|5x + 2| = 3$

11. $|4x - 9| = 14$

12. $|9y - 2| = 17$

13. $|x| + 7 = 18$

14. $|x| - 2 = 6.3$

15. $574 = 283 + |t|$

16. $-562 = -2000 + |x|$

17. $|5x| = 40$

18. $|2y| = 18$

19. $|3x| - 4 = 17$

20. $|6x| + 8 = 32$

21. $7|w| - 3 = 11$

22. $5|x| + 10 = 26$

23. $\left|\frac{2x - 1}{3}\right| = 5$

24. $\left|\frac{4 - 5x}{6}\right| = 7$

25. $|m + 5| + 9 = 16$

26. $|t - 7| - 5 = 4$

27. $10 - |2x - 1| = 4$

28. $2|2x - 7| + 11 = 25$

29. $|3x - 4| = -2$

30. $|x - 6| = -8$

31. $\left|\frac{5}{9} + 3x\right| = \frac{1}{6}$

32. $\left|\frac{2}{3} - 4x\right| = \frac{4}{5}$

SYNTHESIS

Solve.

33. $|x + 5| = x + 5$

34. $1 - \left|\frac{1}{4}x + 8\right| = \frac{3}{4}$

35. $|7x - 2| = x + 4$

36. $|x - 1| = x - 1$

I THE DISTANCE FORMULA AND MIDPOINTS

Objectives

a Use the distance formula to find the distance between two points whose coordinates are known.

b Use the midpoint formula to find the midpoint of a segment when the coordinates of its endpoints are known.

Find the distance between the pair of points.

1. $(7, 12)$ and $(7, -2)$

2. $(6, 2)$ and $(-5, 2)$

Answers on page A-57

a The Distance Formula

Suppose that two points are on a horizontal line, and thus have the same second coordinate. We can find the distance between them by subtracting their first coordinates. This difference may be negative, depending on the order in which we subtract. So, to make sure we get a positive number, we take the absolute value of this difference. The distance between two points on a horizontal line (x_1, y_1) and (x_2, y_1) is thus $|x_2 - x_1|$. Similarly, the distance between two points on a vertical line (x_2, y_1) and (x_2, y_2) is $|y_2 - y_1|$.

EXAMPLES Find the distance between these points.

1. $(-5, 13)$ and $(-5, 2)$

We take the absolute value of the difference of the second coordinates, since the first coordinates are the same. The distance is

$$|13 - 2| = |11| = 11.$$

2. $(7, -3)$ and $(-5, -3)$

Since the second coordinates are the same, we take the absolute value of the difference of the first coordinates. The distance is

$$|-5 - 7| = |-12| = 12.$$

Do Exercises 1 and 2.

Now consider *any* two points (x_1, y_1) and (x_2, y_2). If $x_1 \neq x_2$ and $y_1 \neq y_2$, these points are vertices of a right triangle, as shown. The other vertex is then (x_2, y_1). The lengths of the legs are $|x_2 - x_1|$ and $|y_2 - y_1|$.

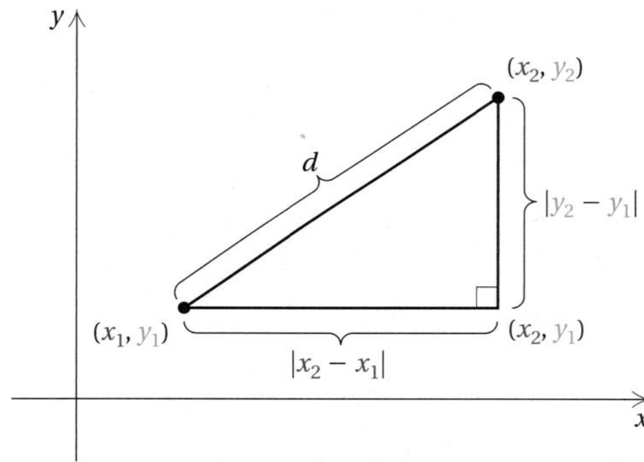

We find d, the length of the hypotenuse, by using the Pythagorean theorem (Sections 11.8 and 14.6):

$$d^2 = |x_2 - x_1|^2 + |y_2 - y_1|^2.$$

Since the square of a number is the same as the square of its opposite, we don't need these absolute-value signs. Thus,

$$d^2 = (x_2 - x_1)^2 + (y_2 - y_1)^2.$$

Taking the principal square root, we obtain the distance between two points.

THE DISTANCE FORMULA

The distance between any two points (x_1, y_1) and (x_2, y_2) is given by

$$d = \sqrt{(x_2 - x_1)^2 + (y_2 - y_1)^2}.$$

This formula holds even when the two points *are* on a vertical or a horizontal line.

EXAMPLE 3 Find the distance between $(4, -3)$ and $(-5, 4)$. Give an exact answer and an approximation to three decimal places.

We substitute into the distance formula:

$$\begin{aligned} d &= \sqrt{(-5 - 4)^2 + [4 - (-3)]^2} && \text{Substituting} \\ &= \sqrt{(-9)^2 + 7^2} \\ &= \sqrt{130} \approx 11.402. && \text{Using a calculator} \end{aligned}$$

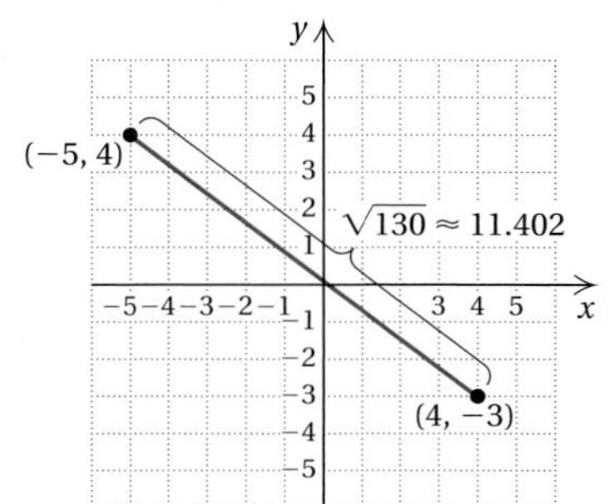

Do Exercises 3 and 4.

b Midpoints of Segments

The distance formula can be used to derive a formula for finding the midpoint of a segment when the coordinates of the endpoints are known.

THE MIDPOINT FORMULA

If the endpoints of a segment are (x_1, y_1) and (x_2, y_2), then the coordinates of the midpoints are

$$\left(\frac{x_1 + x_2}{2}, \frac{y_1 + y_2}{2}\right).$$

(To locate the midpoint, determine the average of the x-coordinates and the average of the y-coordinates.)

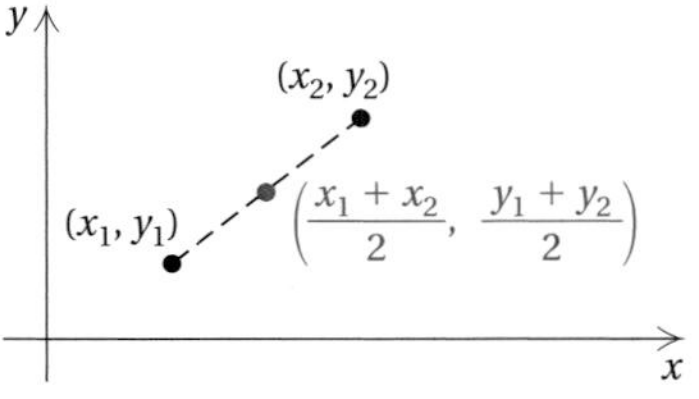

EXAMPLE 4 Find the midpoint of the segment with endpoints $(-2, 3)$ and $(4, -6)$.

Using the midpoint formula, we obtain

$$\left(\frac{-2 + 4}{2}, \frac{3 + (-6)}{2}\right), \quad \text{or} \quad \left(\frac{2}{2}, \frac{-3}{2}\right), \quad \text{or} \quad \left(1, -\frac{3}{2}\right).$$

Do Exercises 5 and 6.

Find the distance between the pair of points. Where appropriate, give an approximation to three decimal places.

3. $(2, 6)$ and $(-4, -2)$

4. $(-2, 1)$ and $(4, 2)$

Find the midpoint of the segment with the given endpoints.

5. $(-3, 1)$ and $(6, -7)$

6. $(10, -7)$ and $(8, -3)$

Answers on page A-57

I EXERCISE SET

a Find the distance between the pair of points. Where appropriate, give an approximation to three decimal places.

1. $(6, -4)$ and $(2, -7)$

2. $(1, 2)$ and $(-4, 14)$

3. $\left(-5, \frac{3}{4}\right)$ and $\left(-5, -\frac{3}{2}\right)$

4. $\left(-6, \frac{1}{2}\right)$ and $\left(-3, \frac{1}{2}\right)$

5. $(0, -4)$ and $(5, -6)$

6. $(8, 3)$ and $(8, -3)$

7. $(9, 9)$ and $(-9, -9)$

8. $(2, 22)$ and $(-8, 1)$

9. $(2.8, -3.5)$ and $(-4.3, -3.5)$

10. $(6.1, 2)$ and $(5.6, -4.4)$

11. $\left(\frac{5}{7}, \frac{1}{14}\right)$ and $\left(\frac{1}{7}, \frac{11}{14}\right)$

12. $(0, \sqrt{7})$ and $(\sqrt{6}, 0)$

13. $(-23, 10)$ and $(56, -17)$

14. $(34, -18)$ and $(-46, -38)$

15. (a, b) and $(0, 0)$

16. $(0, 0)$ and (p, q)

17. $(\sqrt{2}, -\sqrt{3})$ and $(-\sqrt{7}, \sqrt{5})$

18. $(\sqrt{8}, \sqrt{3})$ and $(-\sqrt{5}, -\sqrt{6})$

19. $(1000, -240)$ and $(-2000, 580)$

20. $(-3000, 560)$ and $(-430, -640)$

21. $\left(-\frac{2}{5}, 14\right)$ and $(0, 14)$

22. $(0, 7)$ and $(0, -7)$

b Find the midpoint of the segment with the given endpoints.

23. $(-1, 9)$ and $(4, -2)$

24. $(5, 10)$ and $(2, -4)$

25. $(3, 5)$ and $(-3, 6)$

26. $(7, -3)$ and $(4, 11)$

27. $(-10, -13)$ and $(8, -4)$

28. $(6, -2)$ and $(-5, 12)$

29. $(-3.4, 8.1)$ and $(2.9, -8.7)$

30. $(4.1, 6.9)$ and $(5.2, -6.9)$

31. $\left(\frac{1}{6}, -\frac{3}{4}\right)$ and $\left(-\frac{1}{3}, \frac{5}{6}\right)$

32. $\left(-\frac{4}{5}, -\frac{2}{3}\right)$ and $\left(\frac{1}{8}, \frac{3}{4}\right)$

33. $(\sqrt{2}, -1)$ and $(\sqrt{3}, 4)$

34. $(9, 2\sqrt{3})$ and $(-4, 5\sqrt{3})$

SYNTHESIS

Find the distance between the given points.

35. $(-1, 3k)$ and $(6, 2k)$

36. (a, b) and $(-a, -b)$

37. $(6m, -7n)$ and $(-2m, n)$

38. $(-3\sqrt{3}, 1 - \sqrt{6})$ and $(\sqrt{3}, 1 + \sqrt{6})$

If the sides of a triangle have lengths a, b, and c and $a^2 + b^2 = c^2$, then the triangle is a right triangle. Determine whether the given points are vertices of a right triangle.

39. $(-8, -5)$, $(6, 1)$, and $(-4, 5)$

40. $(9, 6)$, $(-1, 2)$, and $(1, -3)$

41. Find the midpoint of the segment with the endpoints $(2 - \sqrt{3}, 5\sqrt{2})$ and $(2 + \sqrt{3}, 3\sqrt{2})$.

42. Find the point on the y-axis that is equidistant from $(2, 10)$ and $(6, 2)$.

HIGHER ROOTS, AND RATIONAL NUMBERS AS EXPONENTS

Objectives

a Find principal square roots and their opposites.

b Simplify radical expressions with perfect-square radicands.

c Find cube roots, simplifying certain expressions.

d Simplify expressions involving odd and even roots.

e Write expressions with or without rational exponents, and simplify, if possible.

f Write expressions without negative exponents, and simplify, if possible.

g Use the laws of exponents with rational exponents.

Find the square roots.

1. 9

2. 36

3. 121

Simplify.

4. $\sqrt{1}$

5. $\sqrt{36}$

6. $\sqrt{\frac{81}{100}}$

7. $\sqrt{0.0064}$

Answers on page A-57

a Square Roots and Square-Root Functions

When we raise a number to the second power, we say that we have **squared** the number. Sometimes we may need to find the number that was squared. We call this process **finding a square root** of a number.

SQUARE ROOT

The number c is a **square root** of a if $c^2 = a$.

For example:

5 is a *square root* of 25 because $5^2 = 5 \cdot 5 = 25$;

-5 is a *square root* of 25 because $(-5)^2 = (-5)(-5) = 25$.

The number -4 does not have a real-number square root because there is no real number c such that $c^2 = -4$.

PROPERTIES OF SQUARE ROOTS

Every positive real number has two real-number square roots.

The number 0 has just one square root, 0 itself.

Negative numbers do not have real-number square roots.

EXAMPLE 1 Find the two square roots of 64.

The square roots of 64 are 8 and -8 because $8^2 = 64$ and $(-8)^2 = 64$.

Do Exercises 1–3.

PRINCIPAL SQUARE ROOT

The **principal square root** of a nonnegative number is its nonnegative square root. The symbol $\sqrt{a}$ represents the principal square root of a. To name the negative square root of a, we can write $-\sqrt{a}$.

EXAMPLES Simplify.

2. $\sqrt{25} = 5$

Remember: $\sqrt{}$ indicates the principal (nonnegative) square root.

3. $-\sqrt{25} = -5$

4. $\sqrt{\frac{81}{64}} = \frac{9}{8}$

5. $\sqrt{0.0049} = 0.07$

6. $-\sqrt{0.000001} = -0.001$

7. $\sqrt{0} = 0$

8. $\sqrt{-25}$ Does not exist as a real number. Negative numbers do not have real-number square roots.

Do Exercises 4–13. (Exercises 4–7 are on the preceding page.)

RADICAL; RADICAL EXPRESSION; RADICAND

The symbol $\sqrt{}$ is called a **radical.**

An expression written with a radical is called a **radical expression.**

The expression written under the radical is called the **radicand.**

These are radical expressions:

$$\sqrt{5}, \quad \sqrt{a}, \quad -\sqrt{5x}, \quad \sqrt{y^2 + 7}.$$

The radicands in these expressions are 5, a, $5x$, and $y^2 + 7$, respectively.

EXAMPLE 9 Identify the radicand in $\sqrt{x^2 - 9}$.

The radicand in $\sqrt{x^2 - 9}$ is $x^2 - 9$.

Do Exercises 14 and 15.

b Finding $\sqrt{a^2}$

In the expression $\sqrt{a^2}$, the radicand is a perfect square. It is tempting to think that $\sqrt{a^2} = a$, but we see below that this is not the case.

Suppose $a = 5$. Then we have $\sqrt{5^2}$, which is $\sqrt{25}$, or 5.

Suppose $a = -5$. Then we have $\sqrt{(-5)^2}$, which is $\sqrt{25}$, or 5.

Suppose $a = 0$. Then we have $\sqrt{0^2}$, which is $\sqrt{0}$, or 0.

The symbol $\sqrt{a^2}$ never represents a negative number. It represents the principal square root of a^2. Note the following.

SIMPLIFYING $\sqrt{a^2}$

$a \geq 0 \longrightarrow \sqrt{a^2} = a$

If a is positive or 0, the principal square root of a^2 is a.

$a < 0 \longrightarrow \sqrt{a^2} = -a$

If a is negative, the principal square root of a^2 is the opposite of a.

Find the following.

8. a) $\sqrt{16}$ b) $-\sqrt{16}$ c) $\sqrt{-16}$

9. a) $\sqrt{49}$ b) $-\sqrt{49}$ c) $\sqrt{-49}$

10. a) $\sqrt{144}$ b) $-\sqrt{144}$ c) $\sqrt{-144}$

11. $\sqrt{\dfrac{25}{64}}$

12. $-\sqrt{0.81}$

13. $\sqrt{1.44}$

Identify the radicand.

14. $\sqrt{28 + x}$

15. $\sqrt{\dfrac{y}{y + 3}}$

Answers on page A-57

Find the following. Assume that letters can represent *any* real number.

16. $\sqrt{y^2}$

17. $\sqrt{(-24)^2}$

18. $\sqrt{(5y)^2}$

19. $\sqrt{16y^2}$

20. $\sqrt{(x+7)^2}$

21. $\sqrt{4(x-2)^2}$

22. $\sqrt{49(y+5)^2}$

23. $\sqrt{x^2-6x+9}$

Find the following.

24. $\sqrt[3]{-64}$

25. $\sqrt[3]{27y^3}$

26. $\sqrt[3]{8(x+2)^3}$

27. $\sqrt[3]{-\frac{343}{64}}$

Answers on page A-57

In all cases, the radical expression represents the absolute value of a.

PRINCIPAL SQUARE ROOT OF a^2

For any real number a, $\sqrt{a^2} = |a|$. The principal (nonnegative) square root of a^2 is the absolute value of a.

The absolute value is used to ensure that the principal square root is nonnegative, which is as it is defined.

EXAMPLES Find the following. Assume that letters can represent any real number.

10. $\sqrt{(-16)^2} = |-16|$, or 16

11. $\sqrt{(3b)^2} = |3b| = |3| \cdot |b| = 3|b|$

$|3b|$ can be simplified to $3|b|$ because the absolute value of any product is the product of the absolute values. That is, $|a \cdot b| = |a| \cdot |b|$.

12. $\sqrt{(x-1)^2} = |x-1|$

13. $\sqrt{x^2+8x+16} = \sqrt{(x+4)^2}$
$= |x+4|$

Caution!

$|x+4|$ is *not* the same as $|x| + 4$.

Do Exercises 16–23.

C Cube Roots

CUBE ROOT

The number c is the **cube root** of a, written $\sqrt[3]{a}$, if the third power of c is a—that is, if $c^3 = a$, then $\sqrt[3]{a} = c$.

For example:

2 is the *cube root* of 8 because $2^3 = 2 \cdot 2 \cdot 2 = 8$;

-4 is the *cube root* of -64 because $(-4)^3 = (-4)(-4)(-4) = -64$.

We talk about *the* cube root of a number rather than *a* cube root because of the following.

Every real number has exactly one cube root in the system of real numbers. The symbol $\sqrt[3]{a}$ represents *the* cube root of a.

EXAMPLES Find the following.

14. $\sqrt[3]{8} = 2$ because $2^3 = 8$.

15. $\sqrt[3]{-27} = -3$

16. $\sqrt[3]{-\frac{216}{125}} = -\frac{6}{5}$

17. $\sqrt[3]{0.001} = 0.1$

18. $\sqrt[3]{x^3} = x$

19. $\sqrt[3]{-8} = -2$

20. $\sqrt[3]{0} = 0$

21. $\sqrt[3]{-8y^3} = \sqrt[3]{(-2y)^3} = -2y$

When we are determining a cube root, no absolute-value signs are needed because a real number has just one cube root. The real-number cube root of a positive number is positive. The real-number cube root of a negative number is negative. The cube root of 0 is 0. That is, $\sqrt[3]{a^3} = a$ whether $a > 0$, $a < 0$, or $a = 0$.

Do Exercises 24–27 on the preceding page.

d Odd and Even kth Roots

In the expression $\sqrt[k]{a}$, we call k the **index** and assume $k \geq 2$.

ODD ROOTS

The 5th root of a number a is the number c for which $c^5 = a$. There are also 7th roots, 9th roots, and so on. Whenever the number k in $\sqrt[k]{}$ is an odd number, we say that we are taking an **odd root.**

Every number has just one real-number odd root. For example, $\sqrt[3]{8} = 2$, $\sqrt[3]{-8} = -2$, and $\sqrt[3]{0} = 0$. If the number is positive, then the root is positive. If the number is negative, then the root is negative. If the number is 0, then the root is 0. Absolute-value signs are *not* needed when we are finding odd roots.

> If k is an *odd* natural number, then for any real number a,
>
> $\sqrt[k]{a^k} = a.$

EXAMPLES Find the following.

22. $\sqrt[5]{32} = 2$
23. $\sqrt[5]{-32} = -2$
24. $-\sqrt[5]{32} = -2$
25. $-\sqrt[5]{-32} = -(-2) = 2$
26. $\sqrt[7]{x^7} = x$
27. $\sqrt[7]{128} = 2$
28. $\sqrt[7]{-128} = -2$
29. $\sqrt[7]{0} = 0$
30. $\sqrt[5]{a^5} = a$
31. $\sqrt[9]{(x-1)^9} = x - 1$

Do Exercises 28–34.

EVEN ROOTS

When the index k in $\sqrt[k]{}$ is an even number, we say that we are taking an **even root.** When the index is 2, we do not write it. Every positive real number has two real-number kth roots when k is even. One of those roots is positive and one is negative. Negative real numbers do not have real-number kth roots when k is even. When we are finding even kth roots, absolute-value signs are sometimes necessary, as they are with square roots. For example,

$$\sqrt{64} = 8, \quad \sqrt[6]{64} = 2, \quad -\sqrt[6]{64} = -2, \quad \sqrt[6]{64x^6} = \sqrt[6]{(2x)^6} = |2x| = 2|x|.$$

Note that in $\sqrt[6]{64x^6}$, we need absolute-value signs because a variable is involved.

Find the following.

28. $\sqrt[5]{243}$
29. $\sqrt[5]{-243}$
30. $\sqrt[5]{x^5}$
31. $\sqrt[7]{y^7}$
32. $\sqrt[5]{0}$
33. $\sqrt[5]{-32x^5}$
34. $\sqrt[7]{(3x+2)^7}$

Find the following. Assume that letters can represent any real number.

35. $\sqrt[4]{81}$
36. $-\sqrt[4]{81}$
37. $\sqrt[4]{-81}$
38. $\sqrt[4]{0}$

Answers on page A-57

Find the following. Assume that letters can represent any real number.

39. $\sqrt[4]{16(x-2)^4}$

40. $\sqrt[6]{x^6}$

41. $\sqrt[8]{(x+3)^8}$

42. $\sqrt[7]{(x+3)^7}$

43. $\sqrt[5]{243x^5}$

Rewrite without rational exponents, and simplify, if possible.

44. $y^{1/4}$

45. $(3a)^{1/2}$

46. $16^{1/4}$

47. $(125)^{1/3}$

48. $(a^3b^2c)^{1/5}$

Answers on page A-57

EXAMPLES Find the following. Assume that letters can represent any real number.

32. $\sqrt[4]{16} = 2$

33. $-\sqrt[4]{16} = -2$

34. $\sqrt[4]{-16}$ Does not exist as a real number.

35. $\sqrt[4]{81x^4} = \sqrt[4]{(3x)^4} = 3|x|$

36. $\sqrt[6]{(y+7)^6} = |y+7|$

37. $\sqrt{81y^2} = \sqrt{(9y)^2} = 9|y|$

The following is a summary of how absolute value is used when we are taking even or odd roots.

SIMPLIFYING $\sqrt[k]{a^k}$

For any real number a:

a) $\sqrt[k]{a^k} = |a|$ when k is an *even* natural number. We use absolute value when k is even unless a is nonnegative.

b) $\sqrt[k]{a^k} = a$ when k is an *odd* natural number greater than 1. We do not use absolute value when k is odd.

Do Exercises 35–43. (Exercises 35–38 are on the preceding page.)

e Rational Exponents

Expressions like $a^{1/2}$, $5^{-1/4}$, and $(2y)^{4/5}$ have not yet been defined. We will define such expressions so that the general properties of exponents hold.

Consider $a^{1/2} \cdot a^{1/2}$. If we want to multiply by adding exponents, it must follow that $a^{1/2} \cdot a^{1/2} = a^{1/2+1/2}$, or a^1. Thus we should define $a^{1/2}$ to be a square root of a. Similarly, $a^{1/3} \cdot a^{1/3} \cdot a^{1/3} = a^{1/3+1/3+1/3}$, or a^1, so $a^{1/3}$ should be defined to mean $\sqrt[3]{a}$.

$a^{1/n}$

For any *nonnegative* real number a and any natural number index n $(n \neq 1)$,

$$a^{1/n} \quad \text{means} \quad \sqrt[n]{a} \text{ (the nonnegative } n\text{th root of } a\text{).}$$

Whenever we use rational exponents, we assume that the bases are nonnegative.

EXAMPLES Rewrite without rational exponents, and simplify, if possible.

38. $27^{1/3} = \sqrt[3]{27} = 3$

39. $(abc)^{1/5} = \sqrt[5]{abc}$

40. $x^{1/2} = \sqrt{x}$ An index of 2 is not written.

Do Exercises 44–48.

EXAMPLES Rewrite with rational exponents.

41. $\sqrt[5]{7xy} = (7xy)^{1/5}$

We need parentheses around the radicand here.

42. $8\sqrt[3]{xy} = 8(xy)^{1/3}$

43. $\sqrt[7]{\frac{x^3y}{9}} = \left(\frac{x^3y}{9}\right)^{1/7}$

Do Exercises 49–52.

How should we define $a^{2/3}$? If the general properties of exponents are to hold, we have $a^{2/3} = (a^{1/3})^2$, or $(a^2)^{1/3}$, or $(\sqrt[3]{a})^2$, or $\sqrt[3]{a^2}$. We define this accordingly.

$a^{m/n}$

For any natural numbers m and n ($n \neq 1$) and any nonnegative real number a,

$$a^{m/n} \quad \text{means} \quad \sqrt[n]{a^m}, \quad \text{or} \quad (\sqrt[n]{a})^m.$$

EXAMPLES Rewrite without rational exponents, and simplify, if possible.

44. $(27)^{2/3} = \sqrt[3]{27^2}$
$= (\sqrt[3]{27})^2$
$= 3^2$
$= 9$

45. $4^{3/2} = \sqrt[2]{4^3}$
$= (\sqrt[2]{4})^3$
$= 2^3$
$= 8$

Do Exercises 53–55.

EXAMPLES Rewrite with rational exponents.

The index becomes the denominator of the rational exponent.

46. $\sqrt[3]{9^4} = 9^{4/3}$

47. $(\sqrt[4]{7xy})^5 = (7xy)^{5/4}$

Do Exercises 56 and 57.

f Negative Rational Exponents

Negative rational exponents have a meaning similar to that of negative integer exponents.

$a^{-m/n}$

For any rational number m/n and any positive real number a,

$$a^{-m/n} \quad \text{means} \quad \frac{1}{a^{m/n}},$$

that is, $a^{m/n}$ and $a^{-m/n}$ are reciprocals.

Rewrite with rational exponents.

49. $\sqrt[3]{19ab}$ **50.** $19\sqrt[3]{ab}$

51. $\sqrt[5]{\frac{x^2y}{16}}$ **52.** $7\sqrt[4]{2ab}$

Rewrite without rational exponents, and simplify, if possible.

53. $x^{3/5}$ **54.** $8^{2/3}$

55. $4^{5/2}$

Rewrite with rational exponents.

56. $(\sqrt[3]{7abc})^4$ **57.** $\sqrt[5]{6^7}$

Answers on page A-57

Rewrite with positive exponents, and simplify, if possible.

58. $16^{-1/4}$

59. $(3xy)^{-7/8}$

60. $81^{-3/4}$

61. $7p^{3/4}q^{-6/5}$

62. $\left(\frac{11m}{7n}\right)^{-2/3}$

Use the laws of exponents to simplify.

63. $7^{1/3} \cdot 7^{3/5}$

64. $\frac{5^{7/6}}{5^{5/6}}$

65. $(9^{3/5})^{2/3}$

66. $(p^{-2/3}q^{1/4})^{1/2}$

Answers on page A-57

EXAMPLES Rewrite with positive exponents, and simplify, if possible.

48. $9^{-1/2} = \frac{1}{9^{1/2}} = \frac{1}{\sqrt{9}} = \frac{1}{3}$

49. $(5xy)^{-4/5} = \frac{1}{(5xy)^{4/5}}$

50. $64^{-2/3} = \frac{1}{64^{2/3}} = \frac{1}{(\sqrt[3]{64})^2} = \frac{1}{4^2} = \frac{1}{16}$

51. $4x^{-2/3}y^{1/5} = 4 \cdot \frac{1}{x^{2/3}} \cdot y^{1/5} = \frac{4y^{1/5}}{x^{2/3}}$

52. $\left(\frac{3r}{7s}\right)^{-5/2} = \left(\frac{7s}{3r}\right)^{5/2}$ Since $\left(\frac{a}{b}\right)^{-n} = \left(\frac{b}{a}\right)^{n}$

Do Exercises 58–62.

g Laws of Exponents

The same laws hold for rational-number exponents as for integer exponents. We list them for review.

For any real number a and any rational exponents m and n:

1. $a^m \cdot a^n = a^{m+n}$ In multiplying, we can add exponents if the bases are the same.
2. $\frac{a^m}{a^n} = a^{m-n}$ In dividing, we can subtract exponents if the bases are the same.
3. $(a^m)^n = a^{m \cdot n}$ To raise a power to a power, we can multiply the exponents.
4. $(ab)^m = a^m b^m$ To raise a product to a power, we can raise each factor to the power.
5. $\left(\frac{a}{b}\right)^n = \frac{a^n}{b^n}$ To raise a quotient to a power, we can raise both the numerator and the denominator to the power.

EXAMPLES Use the laws of exponents to simplify.

53. $3^{1/5} \cdot 3^{3/5} = 3^{1/5+3/5} = 3^{4/5}$ Adding exponents

54. $\frac{7^{1/4}}{7^{1/2}} = 7^{1/4-1/2} = 7^{1/4-2/4} = 7^{-1/4} = \frac{1}{7^{1/4}}$ Subtracting exponents

55. $(7.2^{2/3})^{3/4} = 7.2^{2/3 \cdot 3/4} = 7.2^{6/12} = 7.2^{1/2}$ Multiplying exponents

56. $(a^{-1/3}b^{2/5})^{1/2} = a^{-1/3 \cdot 1/2} \cdot b^{2/5 \cdot 1/2}$ Raising a product to a power and multiplying exponents

$= a^{-1/6}b^{1/5} = \frac{b^{1/5}}{a^{1/6}}$

Do Exercises 63–66.

J EXERCISE SET

a Find the square roots.

1. 16

2. 225

3. 144

4. 9

5. 400

6. 81

Simplify.

7. $-\sqrt{\frac{49}{36}}$

8. $-\sqrt{\frac{361}{9}}$

9. $\sqrt{196}$

10. $\sqrt{441}$

11. $\sqrt{0.0036}$

12. $\sqrt{0.04}$

Use a calculator to approximate to three decimal places.

13. $\sqrt{347}$

14. $-\sqrt{1839.2}$

15. $\sqrt{\frac{285}{74}}$

16. $\sqrt{\frac{839.4}{19.7}}$

Identify the radicand.

17. $9\sqrt{y^2 + 16}$

18. $-3\sqrt{p^2 - 10}$

19. $x^4y^5\sqrt{\frac{x}{y - 1}}$

20. $a^2b^2\sqrt{\frac{a^2 - b}{b}}$

b Find the following. Assume that letters can represent *any* real number.

21. $\sqrt{16x^2}$

22. $\sqrt{25t^2}$

23. $\sqrt{(-12c)^2}$

24. $\sqrt{(-9d)^2}$

25. $\sqrt{(p + 3)^2}$

26. $\sqrt{(2 - x)^2}$

27. $\sqrt{x^2 - 4x + 4}$

28. $\sqrt{9t^2 - 30t + 25}$

c Simplify.

29. $\sqrt[3]{27}$

30. $-\sqrt[3]{64}$

31. $\sqrt[3]{-64x^3}$

32. $\sqrt[3]{-125y^3}$

33. $\sqrt[3]{-216}$

34. $-\sqrt[3]{-1000}$

35. $\sqrt[3]{0.343(x + 1)^3}$

36. $\sqrt[3]{0.000008(y - 2)^3}$

d Find the following. Assume that letters can represent *any* real number.

37. $-\sqrt[4]{625}$

38. $-\sqrt[4]{256}$

39. $\sqrt[5]{-1}$

40. $\sqrt[5]{-32}$

41. $\sqrt[5]{-\frac{32}{243}}$

42. $\sqrt[5]{-\frac{1}{32}}$

43. $\sqrt[6]{x^6}$

44. $\sqrt[4]{(7b)^4}$

45. $\sqrt[10]{(-6)^{10}}$

46. $\sqrt[12]{(-10)^{12}}$

47. $\sqrt[7]{y^7}$

48. $\sqrt[3]{(-6)^3}$

49. $\sqrt[5]{(x-2)^5}$

50. $\sqrt[9]{(2xy)^9}$

e Rewrite without rational exponents, and simplify, if possible.

51. $y^{1/7}$

52. $x^{1/6}$

53. $8^{1/3}$

54. $16^{1/2}$

55. $(a^3b^3)^{1/5}$

56. $(x^2y^2)^{1/3}$

57. $16^{3/4}$

58. $4^{7/2}$

59. $49^{3/2}$

60. $27^{4/3}$

Rewrite with rational exponents.

61. $\sqrt{17}$

62. $\sqrt{x^3}$

63. $\sqrt[3]{18}$

64. $\sqrt[3]{23}$

65. $\sqrt[5]{xy^2z}$

66. $\sqrt[7]{x^3y^2z^2}$

67. $(\sqrt{3mn})^3$

68. $(\sqrt[3]{7xy})^4$

69. $(\sqrt[7]{8x^2y})^5$

70. $(\sqrt[6]{2a^5b})^7$

f Rewrite with positive exponents, and simplify, if possible.

71. $x^{-1/4}$

72. $\frac{1}{a^{-7/8}}$

73. $\left(\frac{7x}{8yz}\right)^{-3/5}$

74. $3^{-5/2}a^3b^{-7/3}$

g Use the laws of exponents to simplify. Write the answers with positive exponents.

75. $5^{3/4} \cdot 5^{1/8}$

76. $11^{2/3} \cdot 11^{1/2}$

77. $\frac{7^{5/8}}{7^{3/8}}$

78. $\frac{3^{5/8}}{3^{-1/8}}$

79. $\frac{4.9^{-1/6}}{4.9^{-2/3}}$

80. $\frac{2.3^{-3/10}}{2.3^{-1/5}}$

81. $(a^{2/3} \cdot b^{5/8})^4$

82. $(3^{2/9})^{3/5}$

83. $a^{2/3} \cdot a^{5/4}$

84. $(a^{-3/2})^{2/9}$

K NONLINEAR INEQUALITIES

Objectives

a Solve quadratic and other polynomial inequalities.

b Graph quadratic inequalities.

a Solving Polynomial Inequalities

Inequalities like the following are called **quadratic inequalities:**

$$x^2 + 3x - 10 < 0, \qquad 5x^2 - 3x + 2 \geq 0.$$

In each case, we have a polynomial of degree 2 on the left.

We will first consider solving a quadratic inequality, such as $ax^2 + bx + c > 0$, using the graph of a related equation, $y = ax^2 + bx + c$.

EXAMPLE 1 Solve: $x^2 + 3x - 10 > 0$.

Consider the equation $y = x^2 + 3x - 10$ and its graph. The graph opens up since the leading coefficient $(a = 1)$ is positive. We find the x-intercepts by setting the polynomial equal to 0 and solving:

$$x^2 + 3x - 10 = 0$$
$$(x + 5)(x - 2) = 0$$
$$x + 5 = 0 \quad or \quad x - 2 = 0$$
$$x = -5 \quad or \quad x = 2.$$

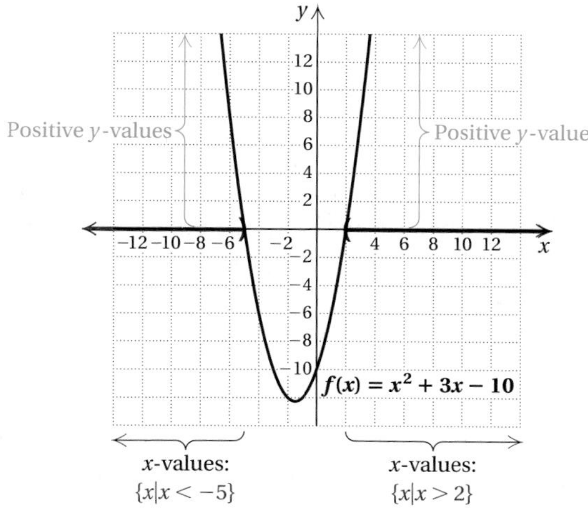

Values of y will be positive to the left and right of the intercepts, as shown. Thus the solution set of the inequality is

$$\{x \mid x < -5 \text{ or } x > 2\}.$$

Do Exercise 1.

1. Solve by graphing:

$$x^2 + 2x - 3 > 0.$$

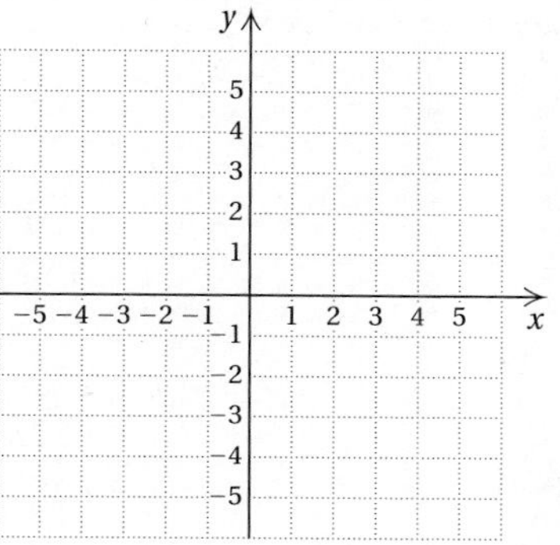

Answer on page A-57

2. Solve by graphing:

$$x^2 + 2x - 3 < 0.$$

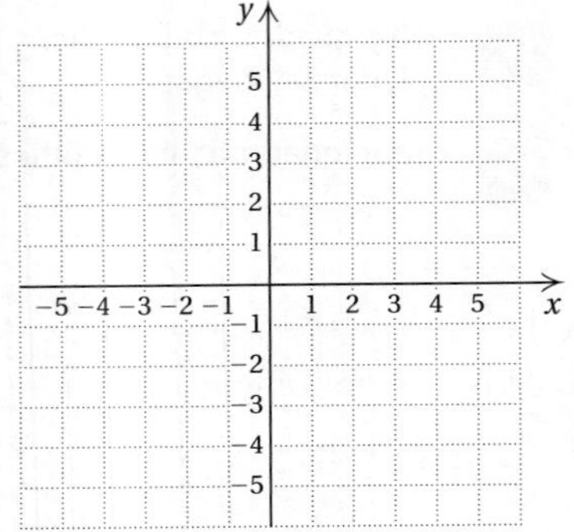

3. Solve by graphing:

$$x^2 + 2x - 3 \le 0.$$

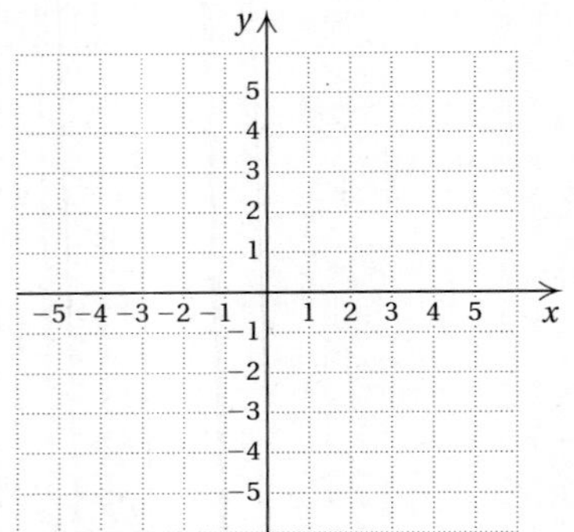

Answers on page A-57

EXAMPLE 2 Solve: $x^2 + 3x - 10 < 0$.

Looking again at the graph of $y = x^2 + 3x - 10$ or at least visualizing it tells us that y-values are negative for those x-values between -5 and 2.

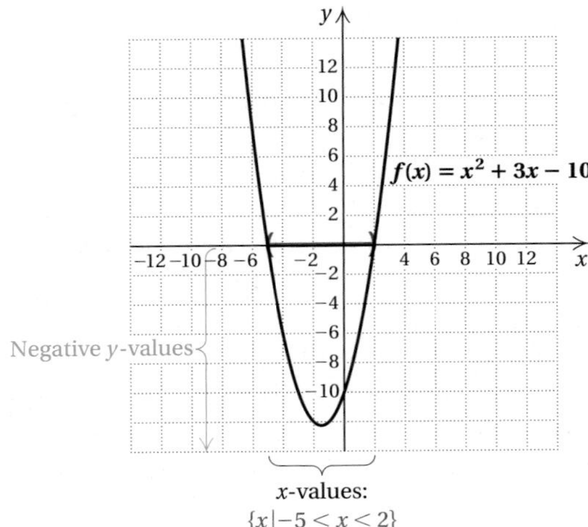

That is, the solution set is $\{x \mid -5 < x < 2\}$.

When an inequality contains $\le$ or $\ge$, the x-values of the x-intercepts must be included. Thus the solution set of the inequality $x^2 + 3x - 10 \le 0$ is $\{x \mid -5 \le x \le 2\}$.

Do Exercises 2 and 3.

In Examples 1 and 2, we see that the x-intercepts divide the number line into intervals. If a particular equation has a positive output for one number in an interval, it will be positive for all the numbers in the interval. The same is true for negative outputs. Thus we can merely make a test substitution in each interval to solve the inequality. This is very similar to our method of using test points to graph a linear inequality in a plane.

EXAMPLE 3 Solve: $x^2 + 3x - 10 < 0$.

We set the polynomial equal to 0 and solve. The solutions of $x^2 + 3x - 10 = 0$, or $(x + 5)(x - 2) = 0$, are -5 and 2. We then locate them on a number line as follows. Note that the numbers divide the number line into three intervals, which we will call A, B, and C.

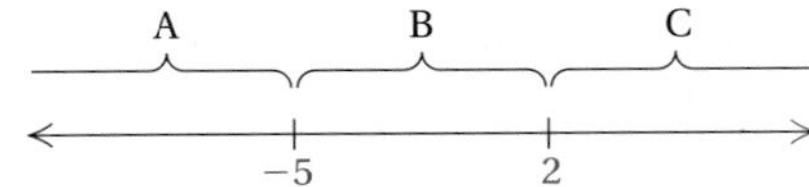

We choose a test number in interval A, say -7, and substitute -7 for x in $y = x^2 + 3x - 10$:

$$y = (-7)^2 + 3(-7) - 10 = 49 - 21 - 10 = 18.$$

Note that $18 > 0$, so the function values will be positive for any number in interval A.

Next, we try a test number in interval B, say 1, and find the corresponding function value:

$$y = 1^2 + 3(1) - 10 = 1 + 3 - 10 = -6.$$

Note that $-6 < 0$, so the function values will be negative for any number in interval B.

Next, we try a test number in interval C, say 4, and find the corresponding y-value:

$$f(4) = 4^2 + 3(4) - 10$$
$$= 16 + 12 - 10 = 18.$$

Note that $18 > 0$, so the y-values will be positive for any number in interval C.

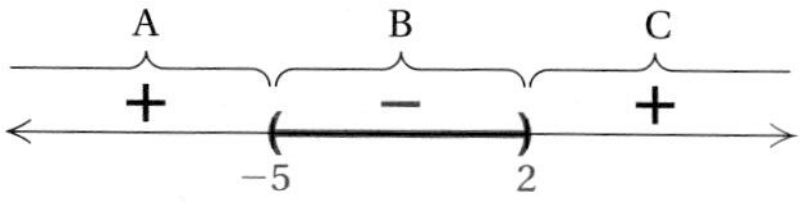

We are looking for numbers x for which $x^2 + 3x - 10 < 0$. Thus any number x in interval B is a solution. If the inequality had been $\leq$, it would have been necessary to include the intercepts -5 and 2 in the solution set as well. The solution set is $\{x \mid -5 < x < 2\}$.

Do Exercises 4 and 5.

EXAMPLE 4 Solve: $5x(x + 3)(x - 2) \geq 0$.

The solutions of $5x(x + 3)(x - 2) = 0$ are -3, 0, and 2. They divide the real-number line into four intervals, as shown below.

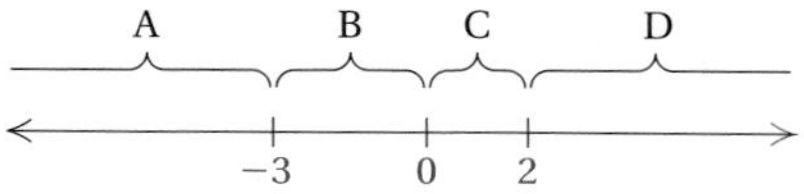

We try test numbers in each interval:

A: Test -5, $y = 5(-5)(-5 + 3)(-5 - 2) = -350 < 0$.

B: Test -2, $y = 5(-2)(-2 + 3)(-2 - 2) = 40 > 0$.

C: Test 1, $y = 5(1)(1 + 3)(1 - 2) = -20 < 0$.

D: Test 3, $y = 5(3)(3 + 3)(3 - 2) = 90 > 0$.

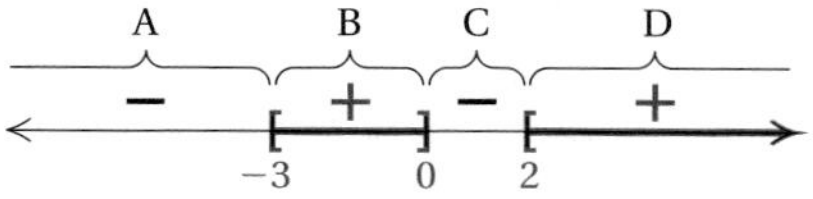

The expression is positive for values of x in intervals B and D. Since the inequality symbol is $\geq$, we need to include the x-intercepts. The solution set of the inequality is

$$\{x \mid -3 \leq x \leq 0 \text{ or } 2 \leq x\}.$$

Do Exercise 6.

b Graphing Quadratic Inequalities

Graphing quadratic inequalities involves the same procedure as graphing quadratic equations, but the graph will also include the interior region enclosed by the parabola or the exterior region outside the parabola.

Solve using the method of Example 3.

4. $x^2 + 3x > 4$

5. $x^2 + 3x \leq 4$

6. Solve: $6x(x + 1)(x - 1) < 0$.

Answers on page A-57

7. Graph: $y \geq x^2 - 2x - 3$.

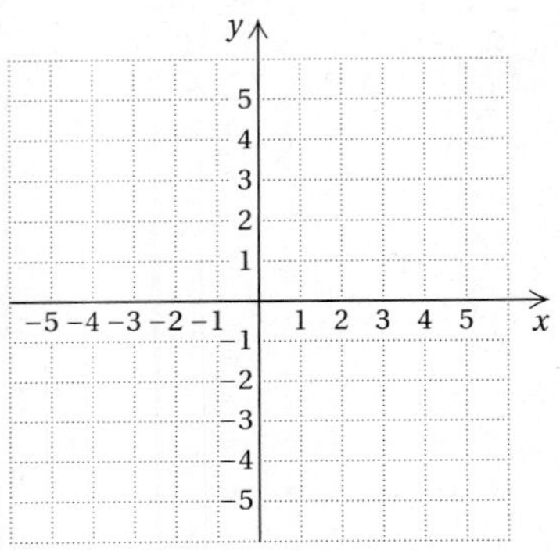

8. Graph: $y > -x^2 + 2x$.

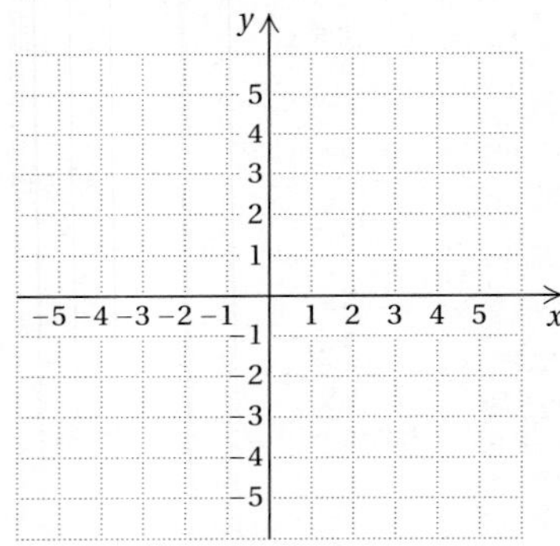

Answers on page A-58

EXAMPLE 5 Graph: $y \leq x^2 + 2x - 3$.

We first replace the inequality symbol with an equals sign and graph the equation:

$$y = x^2 + 2x - 3.$$

The x-coordinate of the vertex is

$$-\frac{b}{2a} = -\frac{2}{2 \cdot 1} = -1.$$

We substitute -1 for x in the equation to find the second coordinate of the vertex:

$$y = x^2 + 2x - 3 = (-1)^2 + 2(-1) - 3 = 1 - 2 - 3 = -4.$$

The vertex is $(-1, -4)$. The line of symmetry is $x = -1$. We choose some x-values on both sides of the vertex and graph the parabola.

For $x = 0$, $\quad y = x^2 + 2x - 3 = 0^2 + 2 \cdot 0 - 3 = -3$.
For $x = -2$, $\quad y = x^2 + 2x - 3 = (-2)^2 + 2(-2) - 3 = -3$.
For $x = 1$, $\quad y = x^2 + 2x - 3 = 1^2 + 2 \cdot 1 - 3 = 0$.
For $x = -3$, $\quad y = x^2 + 2x - 3 = (-3)^2 + 2(-3) - 3 = 0$.

x	y	
-1	-4	← This is the vertex.
0	-3	
-2	-3	
1	0	
-3	0	

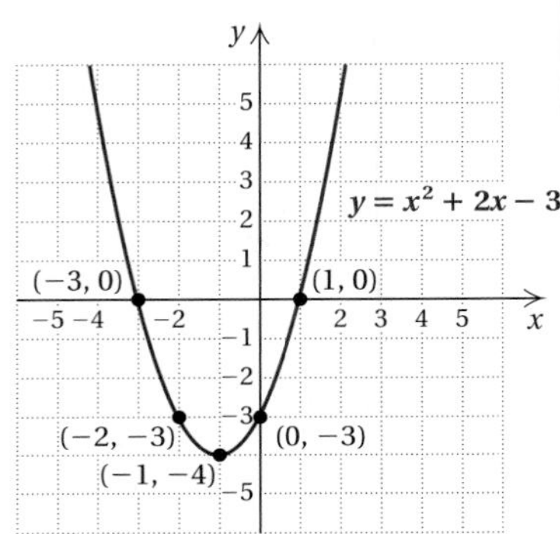

The inequality symbol is $\leq$, so we draw the curve solid.

To determine which region to shade, we choose a point not on the curve as a test point. The origin $(0, 0)$ is usually an easy one to use.

$$\begin{array}{c|l} y \leq x^2 + 2x - 3 & \\ \hline 0 & 0^2 + 2 \cdot 0 - 3 \\ & 0 + 0 - 3 \\ & -3 \qquad \text{FALSE} \end{array}$$

We see that $(0, 0)$ is *not* a solution, so we shade the exterior region. Had the substitution given us a true inequality, we would have shaded the interior.

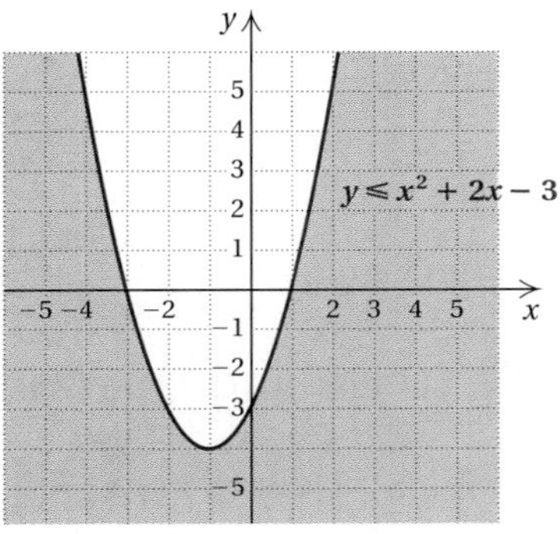

Do Exercises 7 and 8.

K EXERCISE SET

a Solve algebraically and verify results from the graph.

1. $(x - 6)(x + 2) > 0$

2. $(x - 5)(x + 1) > 0$

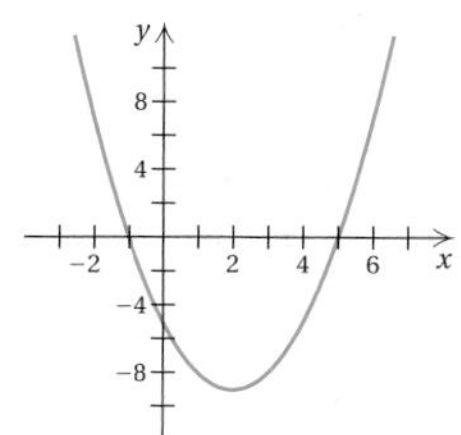

3. $4 - x^2 \geq 0$

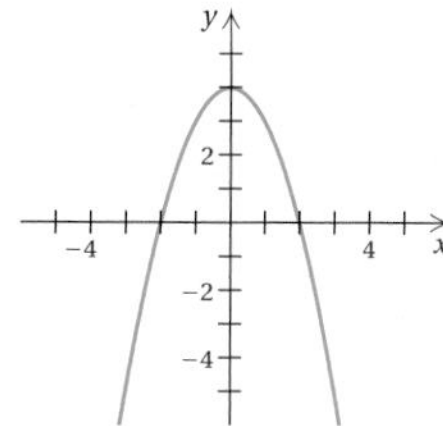

4. $9 - x^2 \leq 0$

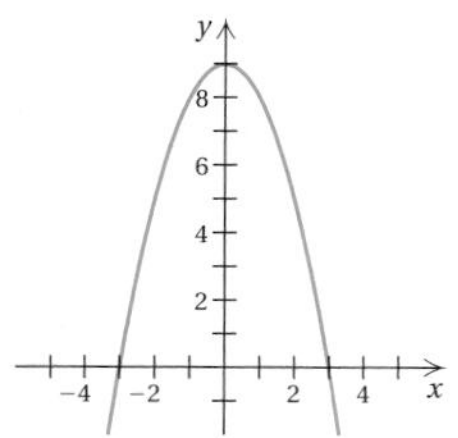

Solve.

5. $3(x + 1)(x - 4) \leq 0$

6. $(x - 7)(x + 3) \leq 0$

7. $x^2 - x - 2 < 0$

8. $x^2 + x - 2 < 0$

9. $x^2 - 2x + 1 \geq 0$

10. $x^2 + 6x + 9 < 0$

11. $x^2 + 8 < 6x$

12. $x^2 - 12 > 4x$

13. $3x(x + 2)(x - 2) < 0$

14. $5x(x + 1)(x - 1) > 0$

15. $(x + 9)(x - 4)(x + 1) > 0$

16. $(x - 1)(x + 8)(x - 2) < 0$

17. $(x + 3)(x + 2)(x - 1) < 0$

18. $(x - 2)(x - 3)(x + 1) < 0$

b Graph.

19. $y \leq 3x^2$

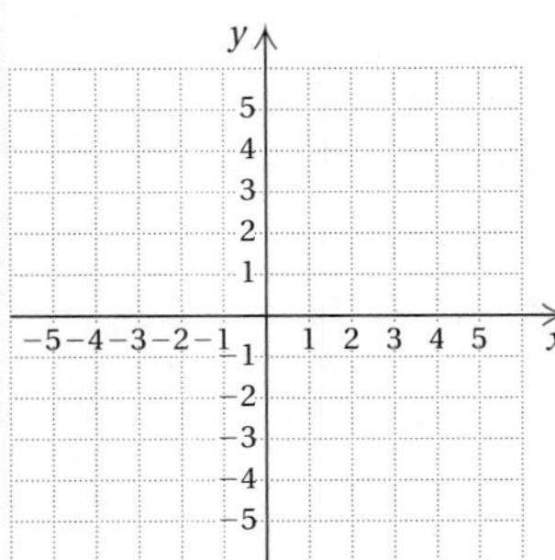

20. $y > -x^2$

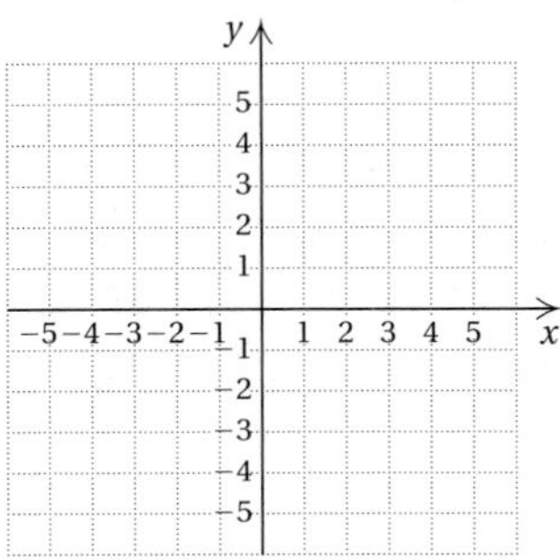

21. $y < 4 - x^2$

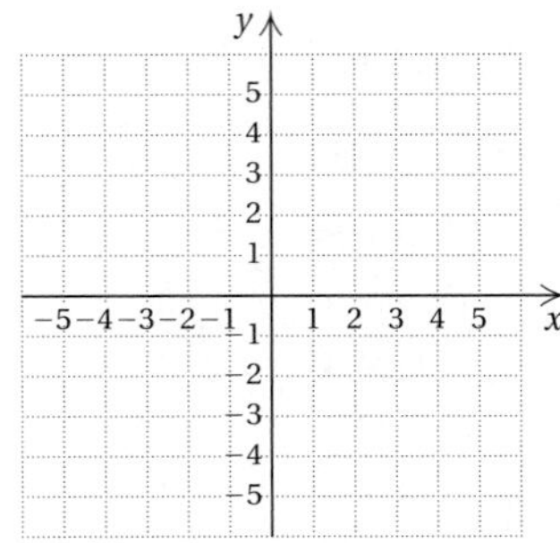

22. $y \leq 2 - x^2$

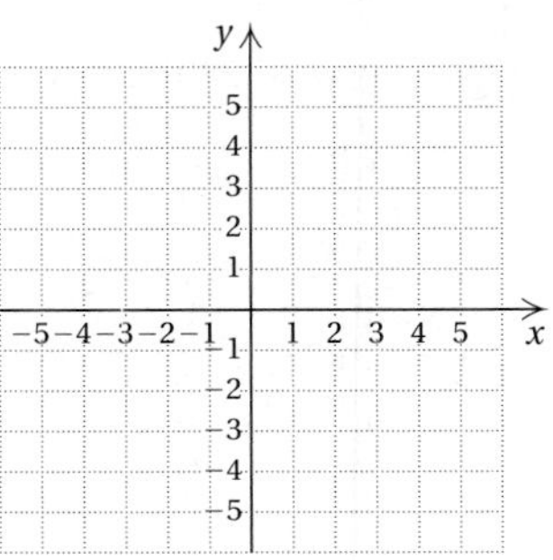

23. $y \leq \frac{1}{2}x^2 - 1$

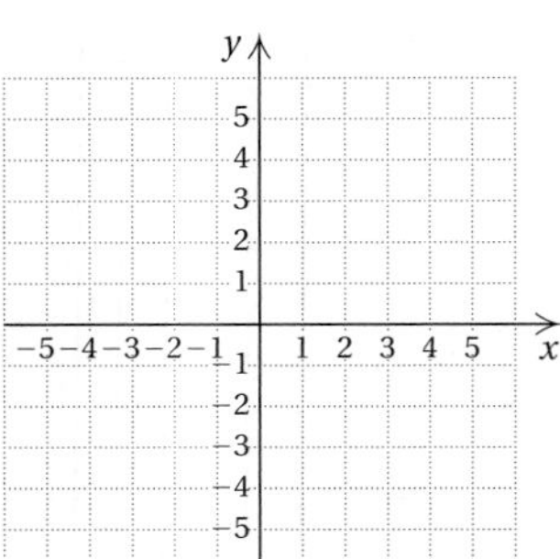

24. $y < -\frac{1}{2}x^2 + 2$

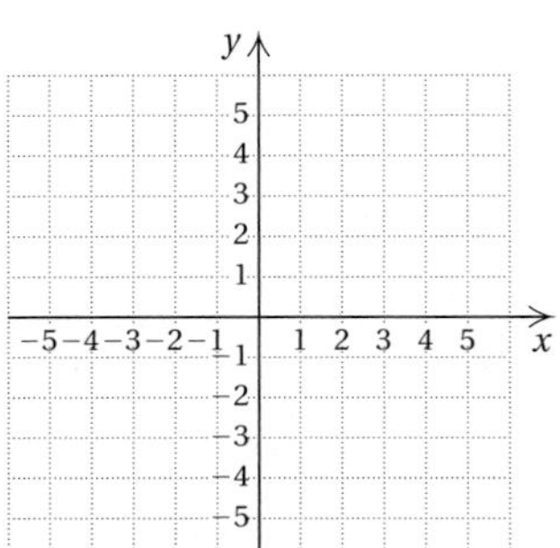

25. $y > x^2 + 4x - 1$

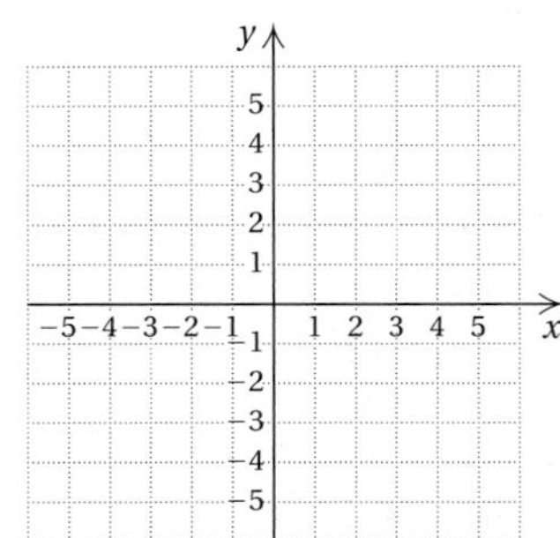

26. $y \leq x^2 - 2x - 3$

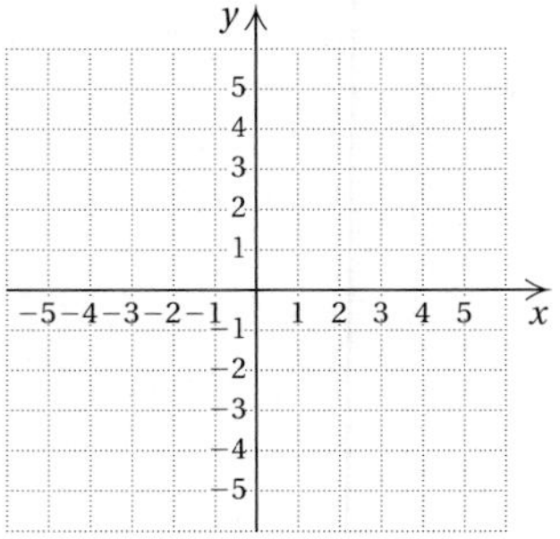

27. $y \geq x^2 + 2x + 1$

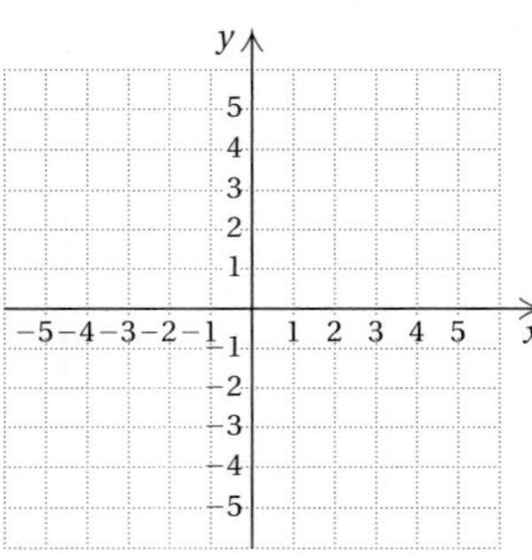

28. $y > x^2 + x - 6$

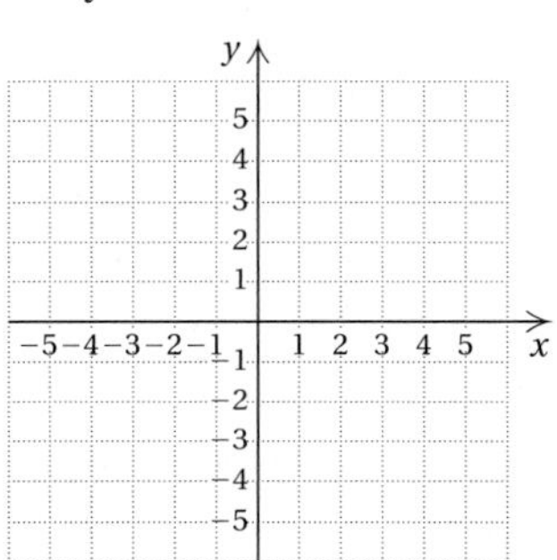

29. $y < 5 - x - x^2$

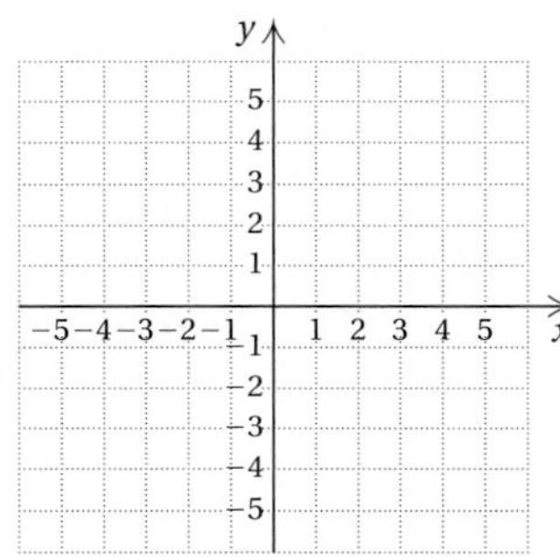

30. $y \geq -x^2 + 2x + 3$

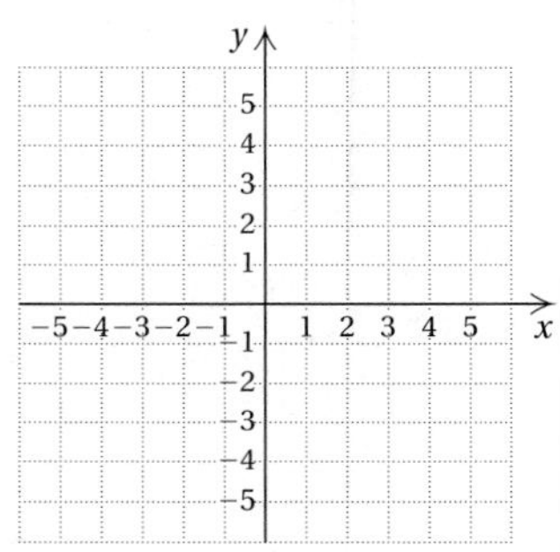

31. $y \leq -x^2 - 2x + 3$

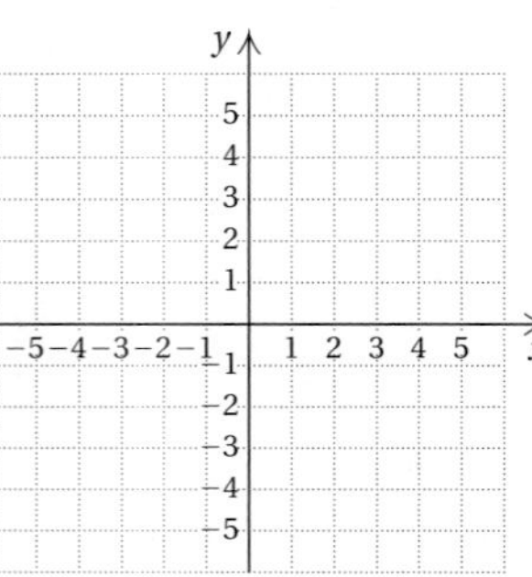

32. $y < -2x^2 - 4x + 1$

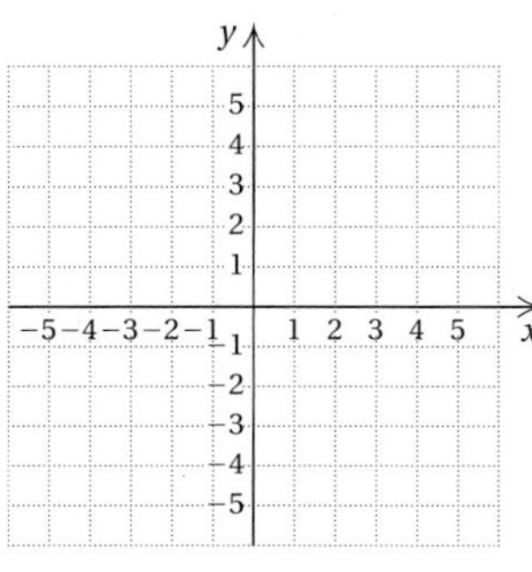

33. $y > 2x^2 + 4x - 1$

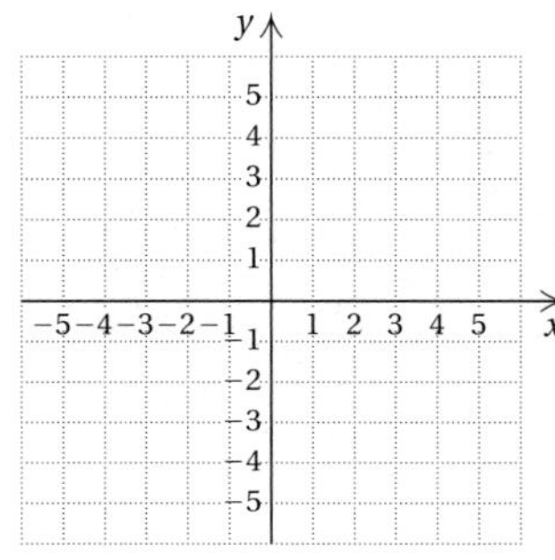

34. $y \leq x^2 + 2x - 5$

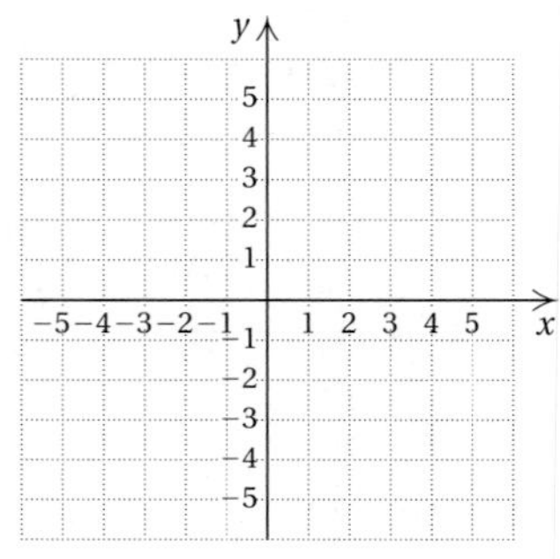

L SYSTEMS OF LINEAR INEQUALITIES

Objective

a Graph systems of linear inequalities and find coordinates of any vertices.

In Section 9.7, we studied the graphing of inequalities in two variables. Here we study *systems* of linear inequalities.

a Systems of Linear Inequalities

The following is an example of a system of two linear inequalities in two variables:

$$x + y \le 4,$$
$$x - y < 4.$$

A **solution** of a system of linear inequalities is an ordered pair that is a solution of *both* inequalities. We now graph solutions of systems of linear inequalities. To do so, we graph each inequality and determine where the graphs overlap, or intersect. That will be a region in which the ordered pairs are solutions of both inequalities.

EXAMPLE 1 Graph the solutions of the system

$$x + y \le 4,$$
$$x - y < 4.$$

We graph the inequality $x + y \le 4$ by first graphing the equation $x + y = 4$ using a solid red line. We consider $(0, 0)$ as a test point and find that it is a solution, so we shade all points on that side of the line using red shading. (See the graph on the left below.) The arrows at the ends of the line also indicate the half-plane, or region, that contains the solutions.

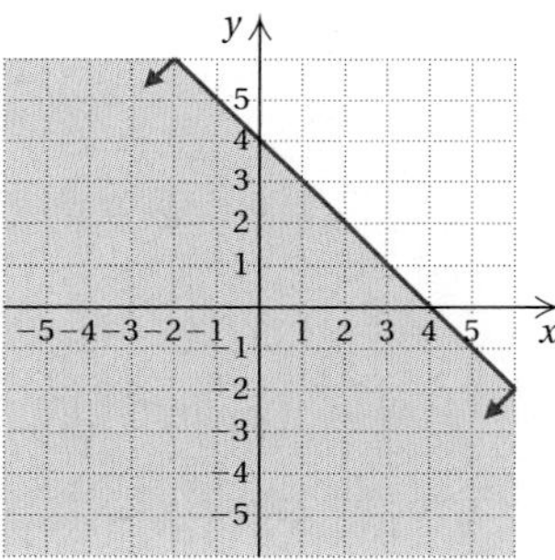

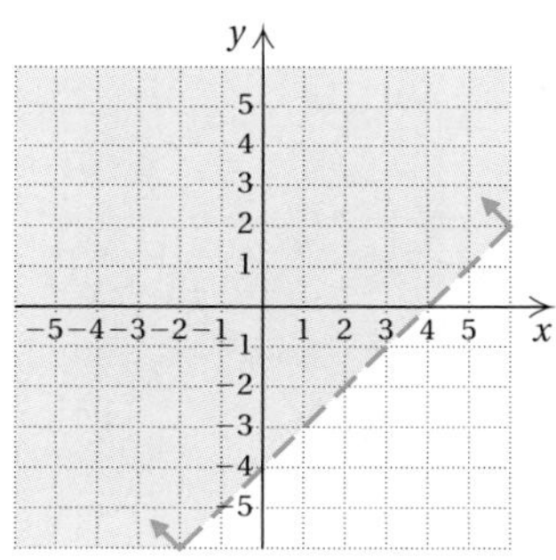

Next, we graph $x - y < 4$. We begin by graphing the equation $x - y = 4$ using a dashed blue line and consider $(0, 0)$ as a test point. Again, $(0, 0)$ is a solution so we shade that side of the line using blue shading. (See the graph on the right above.)

The solution set of the system is the region that is shaded both red and blue and part of the line $x + y = 4$. (See the graph below.)

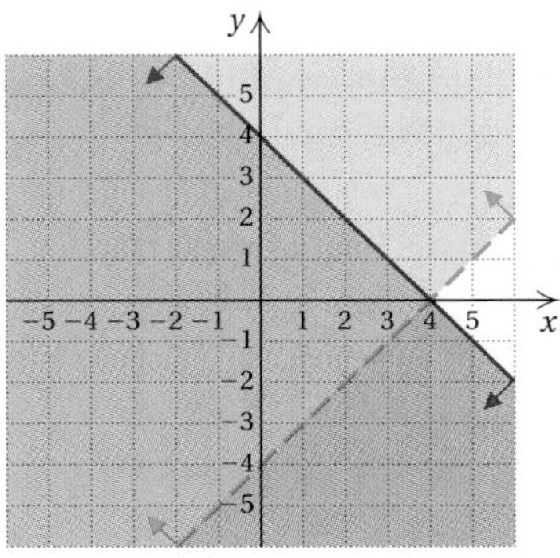

Do Exercise 1.

EXAMPLE 2 Graph: $-2 < x \leq 5$.

This is actually a system of inequalities:

$$-2 < x,$$
$$x \leq 5.$$

We graph the equation $-2 = x$ and see that the graph of the first inequality is the half-plane to the right of the line $-2 = x$. (See the graph on the left below.)

Next, we graph the second inequality, starting with the line $x = 5$, and find that its graph is the line and also the half-plane to the left of it. (See the graph on the right below.)

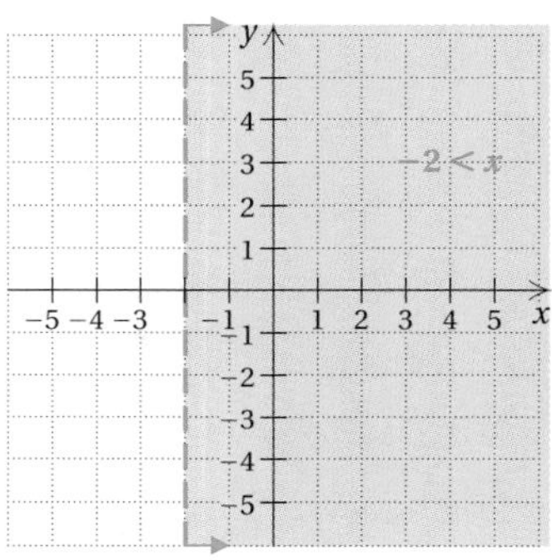

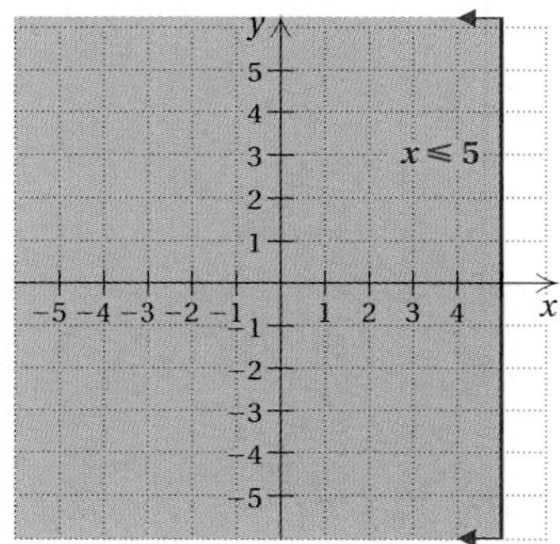

We shade the intersection of these graphs.

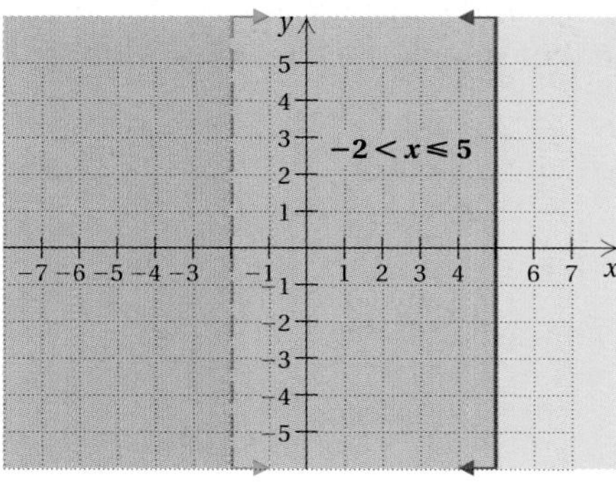

Do Exercise 2.

1. Graph:

$$x + y \geq 1,$$
$$y - x \geq 2.$$

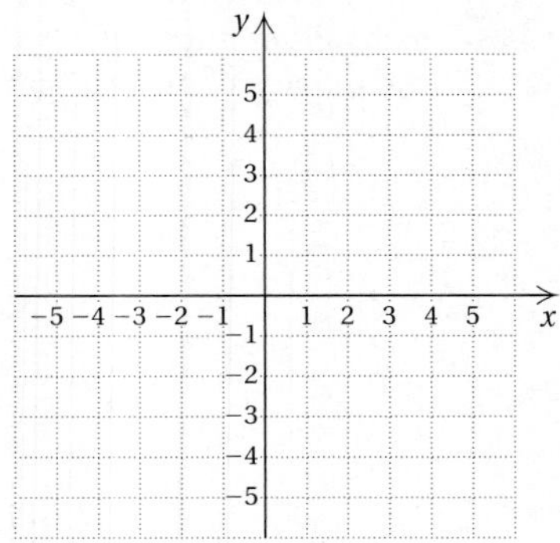

2. Graph: $-3 \leq y < 4$.

Answers on page A-58

A system of inequalities may have a graph that consists of a polygon and its interior. In *linear programming,* which is a topic rich in application that you may study in a later course, it is important to be able to find the vertices of such a polygon.

EXAMPLE 3 Graph the following system of inequalities. Find the coordinates of any vertices formed.

$$6x - 2y \leq 12, \quad (1)$$
$$y - 3 \leq 0, \quad (2)$$
$$x + y \geq 0 \quad (3)$$

We graph the lines $6x - 2y = 12$, $y - 3 = 0$, and $x + y = 0$ using solid lines. The regions for each inequality are indicated by the arrows at the ends of the lines. We then note where the regions overlap and shade the region of solutions using one color.

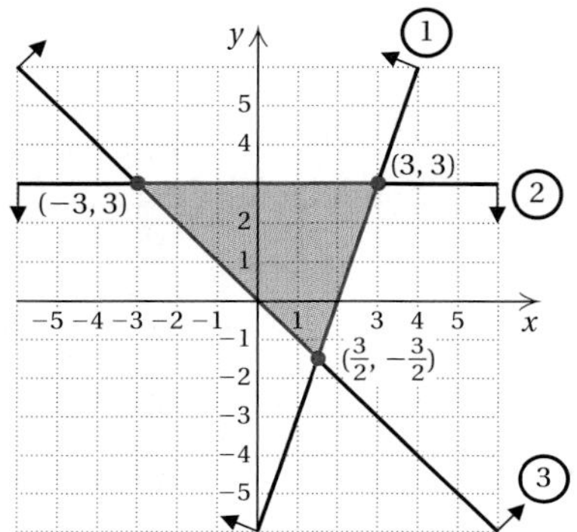

To find the vertices, we solve three different systems of equations. The system of equations from inequalities (1) and (2) is

$$6x - 2y = 12, \quad (1)$$
$$y - 3 = 0. \quad (2)$$

Solving, we obtain the vertex $(3, 3)$.

The system of equations from inequalities (1) and (3) is

$$6x - 2y = 12, \quad (1)$$
$$x + y = 0. \quad (3)$$

Solving, we obtain the vertex $\left(\frac{3}{2}, -\frac{3}{2}\right)$.

The system of equations from inequalities (2) and (3) is

$$y - 3 = 0, \quad (2)$$
$$x + y = 0. \quad (3)$$

Solving, we obtain the vertex $(-3, 3)$.

Do Exercise 3.

3. Graph the system of inequalities. Find the coordinates of any vertices formed.

$$5x + 6y \leq 30,$$
$$0 \leq y \leq 3,$$
$$0 \leq x \leq 4$$

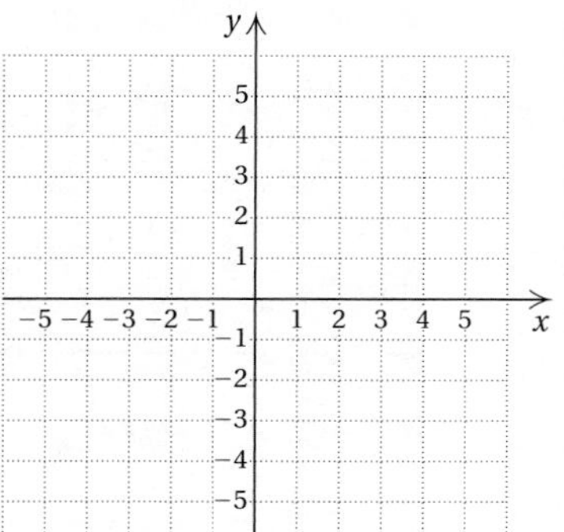

Answer on page A-58

4. Graph the system of inequalities. Find the coordinates of any vertices formed.

$$2x + 4y \le 8,$$
$$x + y \le 3,$$
$$x \ge 0,$$
$$y \ge 0$$

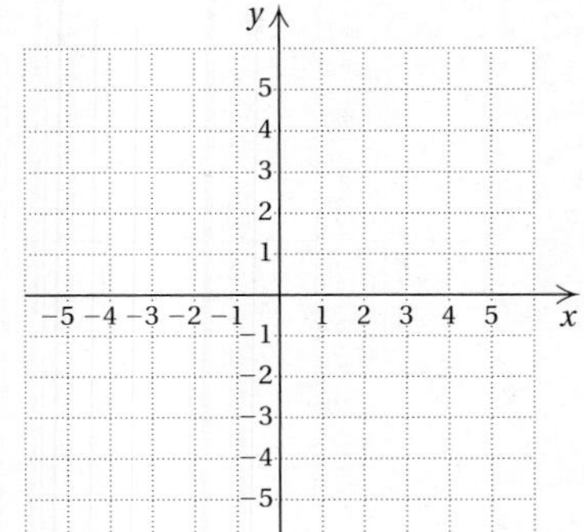

Answer on page A-58

EXAMPLE 4 Graph the following system of inequalities. Find the coordinates of any vertices formed.

$$x + y \le 16, \quad (1)$$
$$3x + 6y \le 60, \quad (2)$$
$$x \ge 0, \quad (3)$$
$$y \ge 0 \quad (4)$$

We graph each inequality using solid lines. The regions for each inequality are indicated by the arrows at the ends of the lines. We then note where the regions overlap and shade the region of solutions using one color.

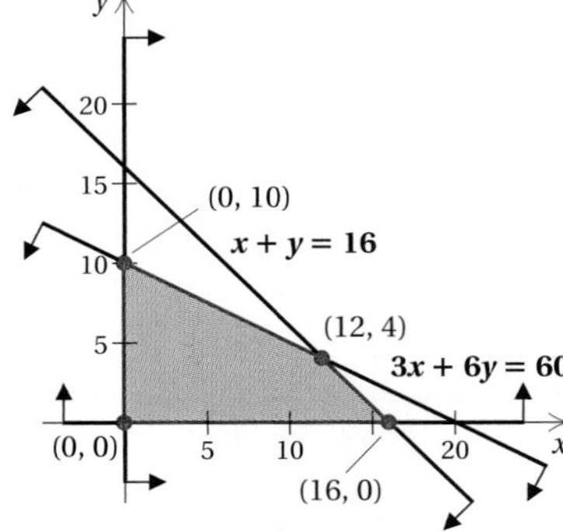

To find the vertices, we solve four different systems of equations. The system of equations from inequalities (1) and (2) is

$$x + y = 16, \quad (1)$$
$$3x + 6y = 60. \quad (2)$$

Solving, we obtain the vertex $(12, 4)$.

The system of equations from inequalities (1) and (4) is

$$x + y = 16, \quad (1)$$
$$y = 0. \quad (4)$$

Solving, we obtain the vertex $(16, 0)$.

The system of equations from inequalities (3) and (4) is

$$x = 0, \quad (3)$$
$$y = 0. \quad (4)$$

The vertex is $(0, 0)$.

The system of equations from inequalities (2) and (3) is

$$3x + 6y = 60, \quad (2)$$
$$x = 0. \quad (3)$$

Solving, we obtain the vertex $(0, 10)$.

Do Exercise 4.

L EXERCISE SET

a Graph the system of inequalities. Find the coordinates of any vertices formed.

1. $y \geq x,$
$y \leq -x + 2$

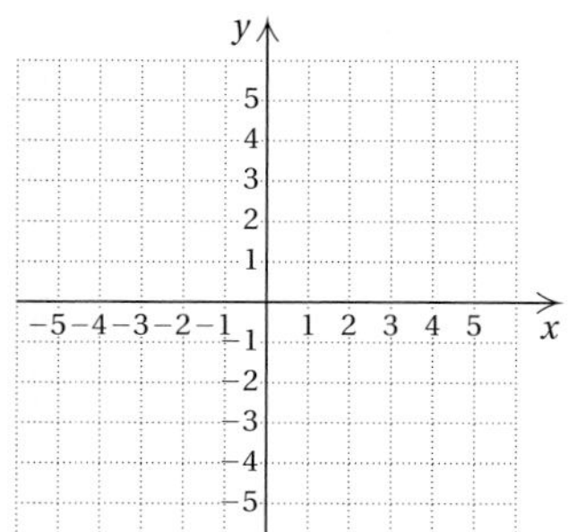

2. $y \geq x,$
$y \leq -x + 4$

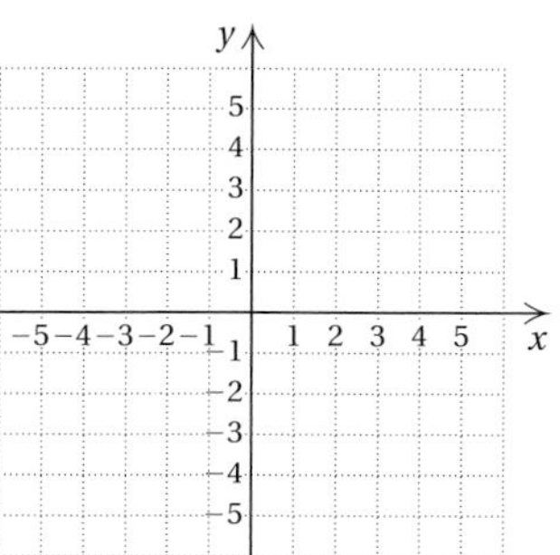

3. $y > x,$
$y < -x + 1$

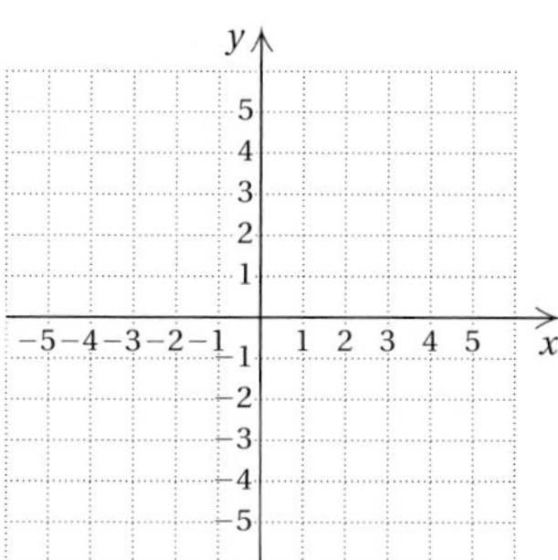

4. $y < x,$
$y > -x + 3$

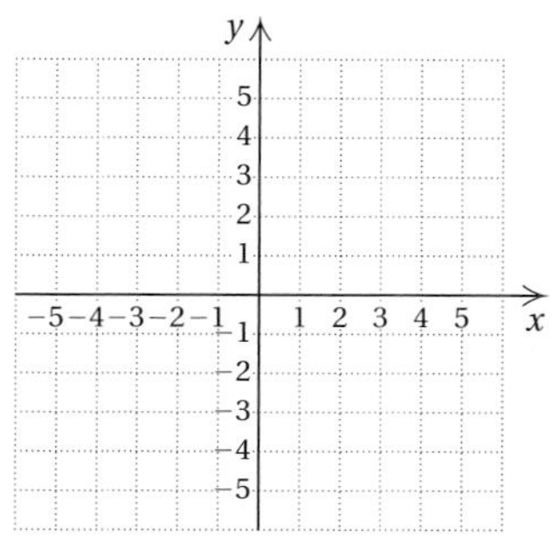

5. $y \geq -2,$
$x \geq 1$

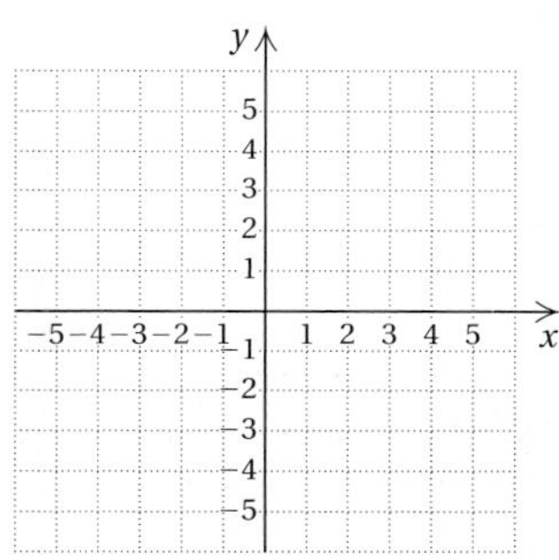

6. $y \leq -2,$
$x \geq 2$

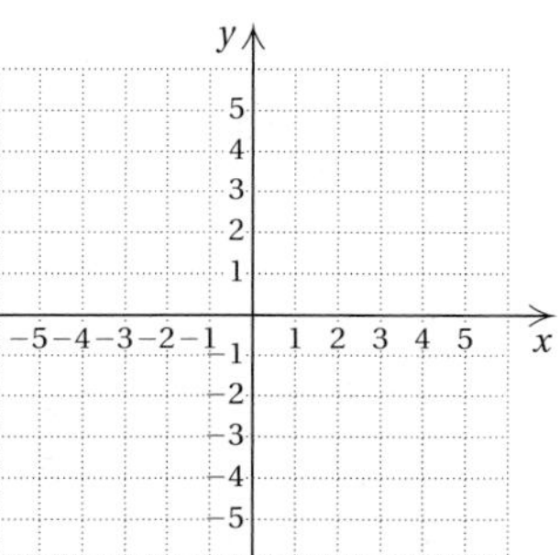

7. $x \leq 3,$
$y \geq -3x + 2$

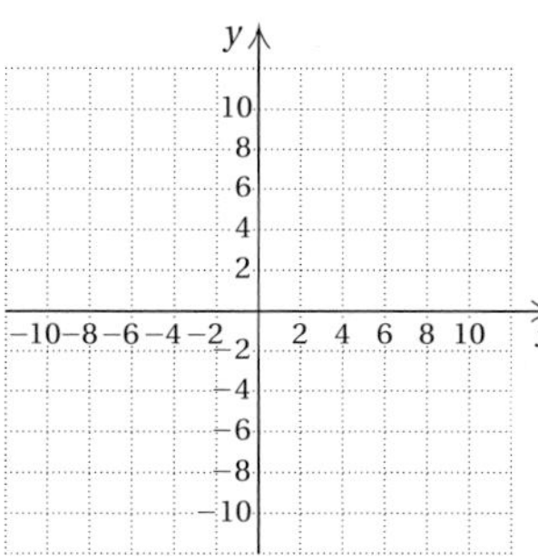

8. $x \geq -2,$
$y \leq -2x + 3$

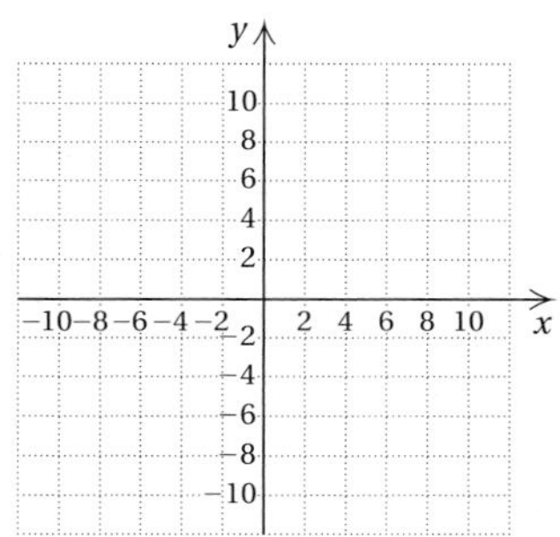

9. $y \leq 2x + 1,$
$y \geq -2x + 1,$
$x \leq 2$

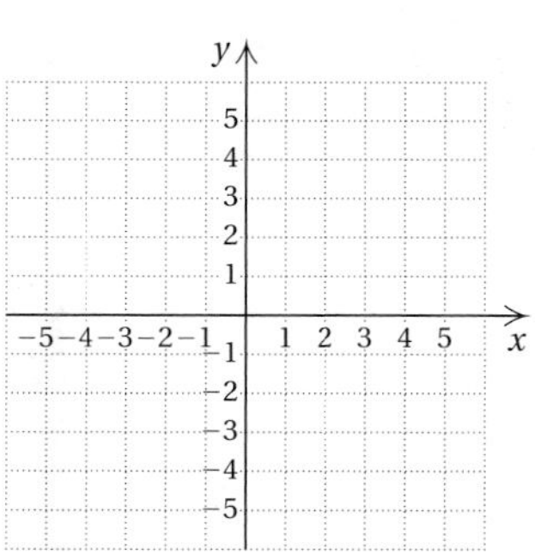

10. $x - y \leq 2,$
$x + 2y \geq 8,$
$y \leq 4$

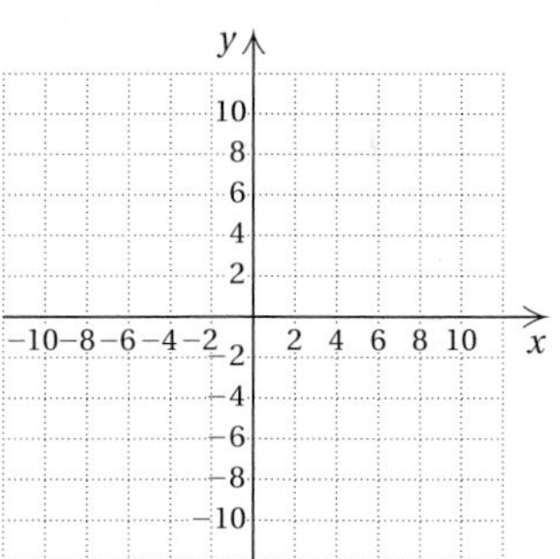

11. $x + y \leq 1,$
$x - y \leq 2$

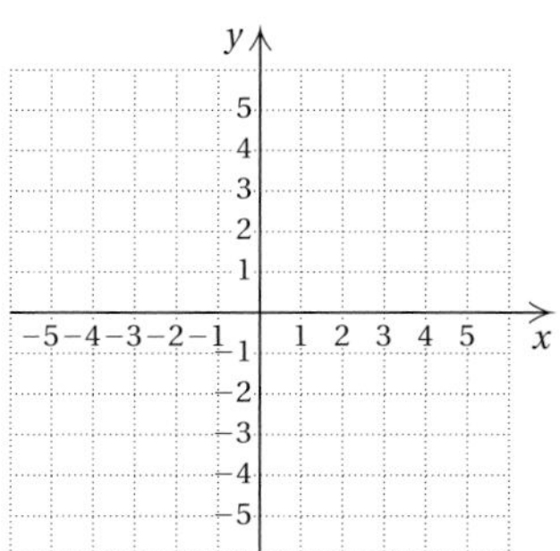

12. $x + y \leq 3,$
$x - y \leq 4$

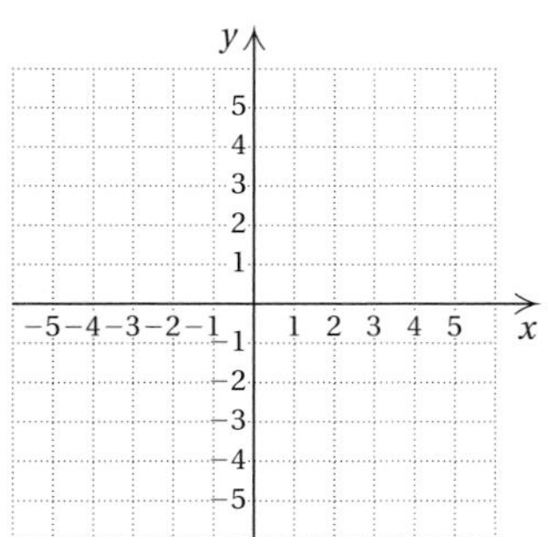

13. $x + 2y \le 12,$
$2x + y \le 12,$
$x \ge 0,$
$y \ge 0$

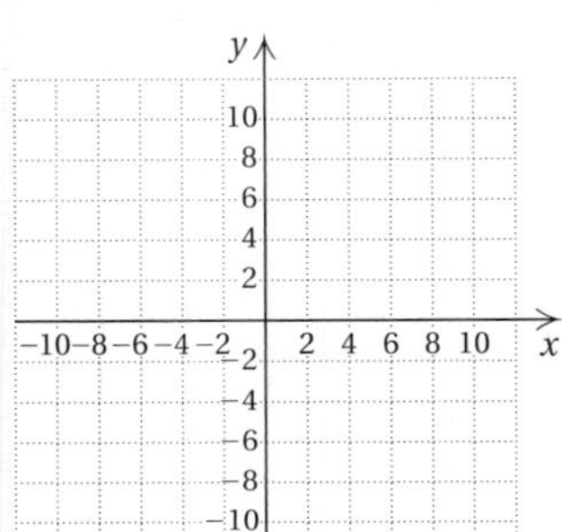

14. $4y - 3x \ge -12,$
$4y + 3x \ge -36,$
$y \le 0,$
$x \le 0$

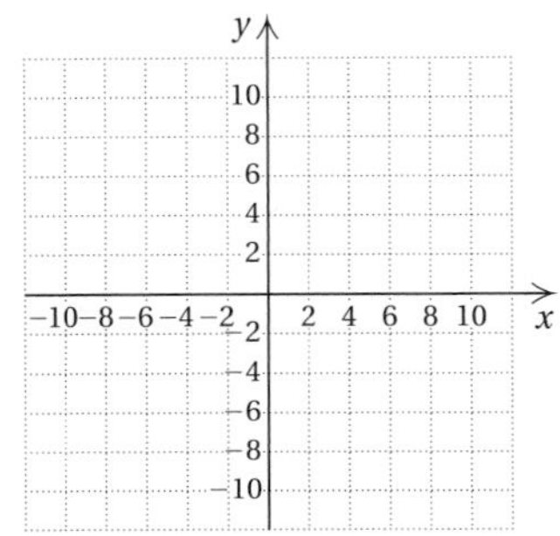

15. $8x + 5y \le 40,$
$x + 2y \le 8,$
$x \ge 0,$
$y \ge 0$

16. $y - x \ge 1,$
$y - x \le 3,$
$2 \le x \le 5$

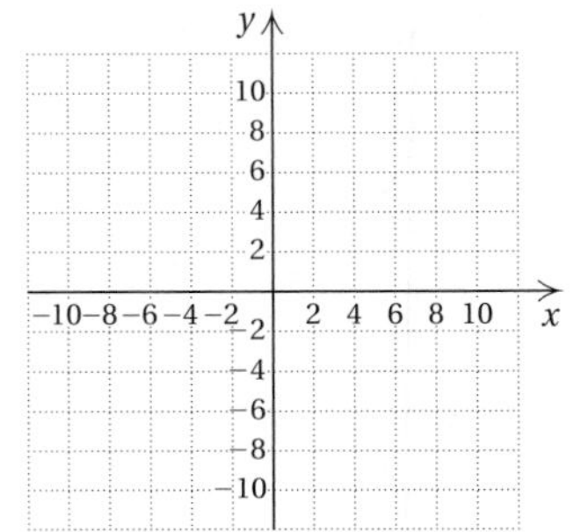

SYNTHESIS

17. *Luggage Size.* Unless an additional fee is paid, most major airlines will not check any luggage for which the sum of the item's length, width, and height exceeds 62 in. The U.S. Postal Service will ship a package only if the sum of the package's length and girth (distance around its midsection) does not exceed 130 in. Video Promotions is ordering several 30-in. long cases that will be both mailed and checked as luggage. Using w and h for width and height (in inches), respectively, write and graph an inequality that represents all acceptable combinations of width and height.

Source: U.S. Postal Service

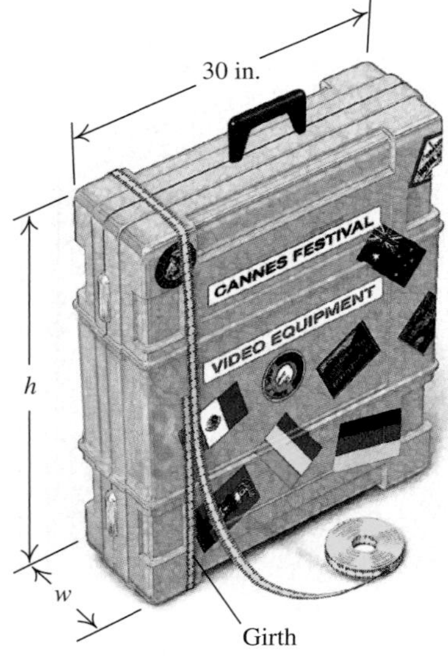

18. *Exercise Danger Zone.* It is dangerous to exercise when the weather is hot and humid. The solutions of the following system of inequalities give a "danger zone" for which it is dangerous to exercise intensely:

$$4H - 3F < 70,$$
$$F + H > 160,$$
$$2F + 3H > 390,$$

where F is the temperature, in degrees Fahrenheit, and H is the humidity.

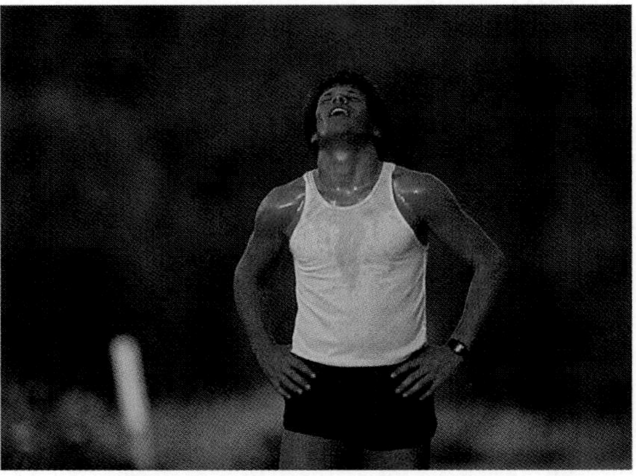

a) Draw the danger zone by graphing the system of inequalities.

b) Is it dangerous to exercise when $F = 80°$ and $H = 80\%$?

M PROBABILITY

Objective

a Compute the probability of a simple event.

We say that when a coin is tossed, the chances that it will fall heads are 1 out of 2, or the **probability** that it will fall heads is $\frac{1}{2}$. Of course, this does not mean that if a coin is tossed ten times, it will necessarily fall heads exactly five times. If the coin is tossed a great number of times, however, it will fall heads very nearly half of them.

Experimental and Theoretical Probability

If we toss a coin a great number of times, say 1000, and count the number of times it falls heads, we can determine the probability of its falling heads. If it falls heads 503 times, we would calculate the probability of the coin falling heads to be

$$\frac{503}{1000}, \quad \text{or} \quad 0.503.$$

This is an **experimental** determination of probability. Such a determination of probability is quite common.

If we consider a coin and reason that it is just as likely to fall heads as tails, we would calculate the probability to be $\frac{1}{2}$. This is a **theoretical** determination of probability. Experimentally, we can determine probabilities within certain limits. These may or may not agree with what we obtain theoretically.

a Computing Probabilities

EXPERIMENTAL PROBABILITIES

We first consider experimental determination of probability. The basic principle we use in computing such probabilities is as follows.

PRINCIPLE *P* (EXPERIMENTAL)

An experiment is performed in which n observations are made. If a situation E, or event, occurs m times out of the n observations, then we say that the **experimental probability** of that event is given by

$$P(E) = \frac{m}{n}.$$

EXAMPLE 1 *Sociological Survey.* An actual survey was conducted to determine the number of people who are left-handed, right-handed, or both. The results are shown in the graph.

a) Determine the probability that a person is left-handed.

b) Determine the probability that a person is ambidextrous (uses both hands equally well).

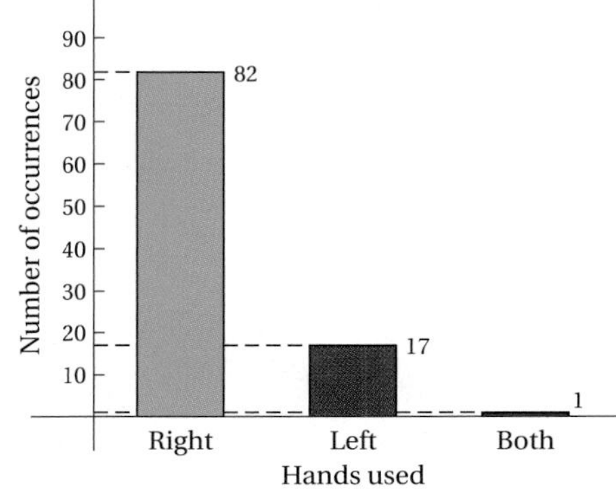

1. In reference to Example 1, what is the probability that a person is right-handed?

2. In Example 2, what is the probability that a home was tuned to NBC? What is the probability that a home was tuned to a network other than CBS, ABC, or NBC, or was not tuned in at all?

Answers on page A-59

a) The number of people who are right-handed was 82, the number who are left-handed was 17, and there was 1 person who is ambidextrous. The total number of observations was $82 + 17 + 1$, or 100. Thus the probability that a person is left-handed is P, where

$$P = \frac{17}{100}.$$

b) The probability that a person is ambidextrous is P, where

$$P = \frac{1}{100}.$$

Do Exercise 1.

EXAMPLE 2 *TV Ratings.* The major television networks and others such as cable TV are always concerned about the percentages of homes that have TVs and are watching their programs. It is too costly and unmanageable to contact every home in the country so a sample, or portion, of the homes are contacted. This is done by an electronic device attached to the TVs of about 1400 homes across the country. Viewing information is then fed into a computer. The following are the results of a recent survey.

Network	CBS	ABC	NBC	Other or not watching
Number of Homes Watching	258	231	206	705

What is the probability that a home was tuned to CBS during the time period considered? to ABC?

The probability that a home was tuned to CBS is P, where

$$P = \frac{258}{1400} \approx 0.184 = 18.4\%.$$

The probability that a home was tuned to ABC is P, where

$$P = \frac{231}{1400} = 0.165 = 16.5\%.$$

Do Exercise 2.

The numbers that we found in Example 2 and in Margin Exercise 2 (18.4 for CBS, 16.5 for ABC, and 14.7 for NBC) are called the *ratings*.

THEORETICAL PROBABILITIES

We need some terminology before we can continue. Suppose we perform an experiment such as flipping a coin, throwing a dart, drawing a card from a deck, or checking an item off an assembly line for quality. The results of an experiment are called **outcomes.** The set of all possible outcomes is called the **sample space.** An **event** is a set of outcomes, that is, the subset of the sample space. For example, for the experiment "throwing a dart," suppose the dartboard is as follows.

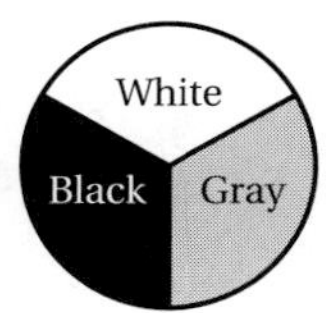

Then one event is

{black}, (the outcome is "hitting black")

which is a subset of the sample space

{black, white, gray}, (sample space)

assuming that the dart must hit the target somewhere.

We denote the probability that an event E occurs as $P(E)$. For example, "getting a head" may be denoted by H. Then $P(H)$ represents the probability of getting a head. When all the outcomes of an experiment have the same probability of occurring, we say that they are **equally likely.** A sample space that can be expressed as a union of equally likely events can allow us to calculate probabilities of other events.

PRINCIPLE *P* (THEORETICAL)

If an event E can occur m ways out of n possible equally likely outcomes of a sample space S, then the **theoretical probability** of that event is given by

$$P(E) = \frac{m}{n}.$$

A die (pl., dice) is a cube, with six faces, each containing a number of dots from 1 to 6.

EXAMPLE 3 What is the probability of rolling a 3 on a die?

On a fair die, there are 6 equally likely outcomes and there is 1 way to get a 3. By Principle P, $P(3) = \frac{1}{6}$.

EXAMPLE 4 What is the probability of rolling an even number on a die?

The event is getting an *even* number. It can occur in 3 ways (getting 2, 4, or 6). The number of equally likely outcomes is 6. By Principle P, $P(\text{even}) = \frac{3}{6}$, or $\frac{1}{2}$.

Do Exercise 3.

3. What is the probability of rolling a prime number on a die?

Answer on page A-59

We now use a number of examples related to a standard bridge deck of 52 cards. Such a deck is made up as shown in the following figure.

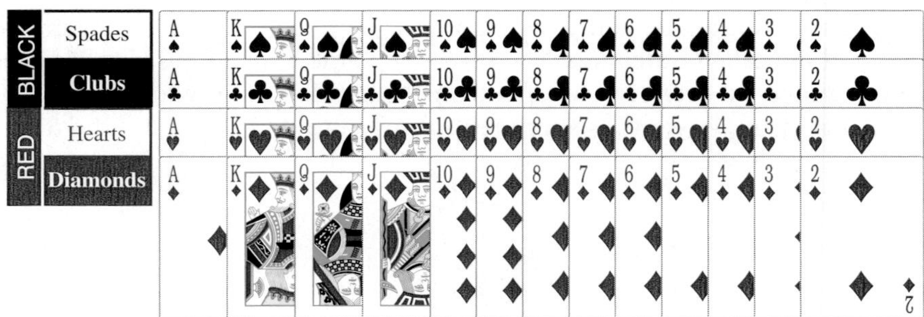

EXAMPLE 5 What is the probability of drawing an ace from a well-shuffled deck of 52 cards?

Since there are 52 outcomes (cards in the deck) and they are equally likely (from a well-shuffled deck) and there are 4 ways to obtain an ace, by Principle *P* we have

$$P(\text{drawing an ace}) = \frac{4}{52}, \quad \text{or} \quad \frac{1}{13}.$$

EXAMPLE 6 Suppose we select, without looking, one marble from a bag containing 3 red marbles and 4 green marbles. What is the probability of selecting a red marble?

There are 7 equally likely ways of selecting any marble, and since the number of ways of getting a red marble is 3,

$$P(\text{selecting a red marble}) = \frac{3}{7}.$$

Do Exercises 4 and 5.

If an event E cannot occur, then $P(E) = 0$. For example, in coin tossing, the event that a coin will land on its edge has probability 0. If an event E is certain to occur (that is, every trial is a success), then $P(E) = 1$. For example, in coin tossing, the event that a coin will fall either heads or tails has probability 1. In general, the probability that an event E will occur is a number from 0 to 1: $0 \le P(E) \le 1$.

Do Exercises 6 and 7.

4. Suppose we draw a card from a well-shuffled deck of 52 cards.

a) What is the probability of drawing a king?

b) What is the probability of drawing a spade?

c) What is the probability of drawing a black card?

d) What is the probability of drawing a jack or a queen?

5. Suppose we select, without looking, one marble from a bag containing 5 red marbles and 6 green marbles. What is the probability of selecting a green marble?

6. On a single roll of a die, what is the probability of getting a 7?

7. On a single roll of a die, what is the probability of getting a 1, 2, 3, 4, 5, or 6?

Answers on page A-59

EXERCISE SET

a

1. In an actual survey, 100 people were polled to determine the probability of a person wearing either glasses or contact lenses. Of those polled, 57 wore either glasses or contacts. What is the probability that a person wears either glasses or contacts? What is the probability that a person wears neither?

2. In another survey, 100 people were polled and asked to select a number from 1 to 5. The results are shown in the following table.

Number Choices	1	2	3	4	5
Number of People Who Selected That Number	18	24	23	23	12

What is the probability that the number selected is 1? 2? 3? 4? 5? What general conclusion might a psychologist make from this experiment?

Linguistics. An experiment was conducted to determine the relative occurrence of various letters of the English alphabet. A paragraph from a newspaper, one from a textbook, and one from a magazine were considered. In all, there was a total of 1044 letters. The number of occurrences of each letter of the alphabet is listed in the following table.

Letter	A	B	C	D	E	F	G	H	I	J	K	L	M
Number of Occurrences	78	22	33	33	140	24	22	63	60	2	9	35	30
Letter	N	O	P	Q	R	S	T	U	V	W	X	Y	Z
Number of Occurrences	74	74	27	4	67	67	95	31	10	22	8	13	1

Round answers to Exercises 3–6 to three decimal places.

3. What is the probability of the occurrence of the letter A? E? I? O? U?

4. What is the probability of a vowel occurring?

5. What is the probability of a consonant occurring?

6. What letter has the least probability of occurring? What is the probability of this letter not occurring?

Suppose we draw a card from a well-shuffled deck of 52 cards.

7. How many equally likely outcomes are there?

8. What is the probability of drawing a queen?

9. What is the probability of drawing a heart?

10. What is the probability of drawing a 4?

11. What is the probability of drawing a red card?

12. What is the probability of drawing a black card?

13. What is the probability of drawing an ace or a deuce?

14. What is the probability of drawing a 9 or a king?

Suppose we select, without looking, one marble from a bag containing 4 red marbles and 10 green marbles.

15. What is the probability of selecting a red marble?

16. What is the probability of selecting a green marble?

17. What is the probability of selecting a purple marble?

18. What is the probability of selecting a white marble?

SYNTHESIS

19. What is the probability of getting a total of 8 on a roll of a pair of dice? (Assume that the dice are different, say, one red and one black.)

20. What is the probability of getting a total of 7 on a roll of a pair of dice?

21. What is the probability of getting a total of 6 on a roll of a pair of dice?

22. What is the probability of getting a total of 3 on a roll of a pair of dice?

23. What is the probability of getting snake eyes (a total of 2) on a roll of a pair of dice?

24. What is the probability of getting box-cars (a total of 12) on a roll of a pair of dice?

APPLYING REASONING SKILLS

Objectives

a Draw conclusions using inductive reasoning.

b Draw conclusions using deductive reasoning.

The two basic types of reasoning commonly used to solve problems are inductive reasoning and deductive reasoning.

a Inductive Reasoning

Inductive reasoning is being used when conclusions are made on the basis of observations. This usually involves identifying a pattern exhibited by several items in a group and then drawing a general conclusion about the entire group. Four commonly occurring patterns are increasing, decreasing, alternating, and circular.

EXAMPLE 1 Find the next number in the sequence

2, 5, 8, 11, ...

The numbers increase in value in this sequence, so we have an increasing pattern. We observe that 3 is added to each number to get the next number. Thus the next number in the sequence is 11 + 3, or 14.

EXAMPLE 2 Find the next number in the sequence

50, 45, 40, 35, ...

The numbers decrease in value in this sequence, so we have a decreasing pattern. We observe that 5 is subtracted from each number to get the next number. Thus the next number in the sequence is 35 − 5, or 30.

EXAMPLE 3 Find the next number in the sequence

3, 4, 6, 9, 13, ...

The numbers increase in value in this sequence, so we have an increasing pattern. We observe that, first, 1 is added to 3 to get 4, then 2 is added to 4 to get 6, then 3 is added to 6 to get 9, and then 4 is added to 9 to get 13. To find the next number in the sequence, we must add 5 to 13. Thus the next number in the sequence is 13 + 5, or 18.

Do Exercises 1–3.

EXAMPLE 4 Find the next number in the sequence

4, 9, 6, 11, 8, ...

Note that the numbers go from smaller to larger to smaller to larger and so on. This is an alternating pattern. Observe that first 5 is added to 4 to get 9. Then 3 is subtracted from 9 to get 6. Next 5 is added to 6 to get 11, and then 3 is subtracted from 11 to get 8. The pattern is to add 5 and then subtract 3. To find the next number in the sequence, we add 5 to 8. Thus the next number in the sequence is 8 + 5, or 13.

Do Exercise 4.

Find the next number in the sequence.

1. 1, 6, 11, 16, ...

2. 35, 31, 27, 23, ...

3. 2, 4, 7, 11, 16, ...

4. Find the next number in the sequence

7, 13, 9, 15, 11, ...

Answers on page A-59

5. Find the next letter in the sequence

A, F, C, H, E, ...

6. Find the next number in the sequence

3, 13, 4, 12, 5, ...

Find the next figure in the sequence.

7\.

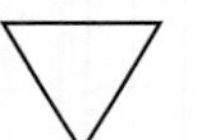
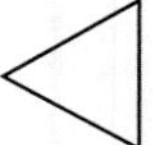

8\.

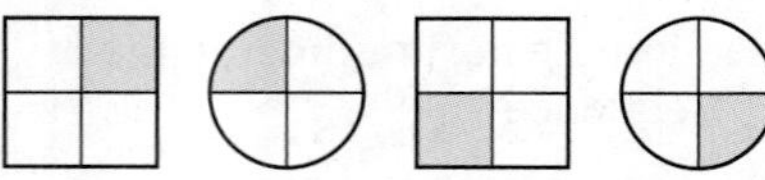

Answers on page A-59

EXAMPLE 5 Find the next letter in the sequence

A, Z, B, Y, C, ...

First, we convert this sequence of letters to a sequence of numbers by assigning a numerical value to each letter. We let A $= 1$, B $= 2$, C $= 3$, and so on, ending with Z $= 26$. Rewriting the sequence in numerical form, we have

1, 26, 2, 25, 3, ...

The numbers alternate between small and large values, so we have an alternating sequence. We can use a diagram to find a pattern:

$$1 \xrightarrow{+25} 26 \xrightarrow{-24} 2 \xrightarrow{+23} 25 \xrightarrow{-22} 3$$

The diagram indicates that the next number in the sequence is $3 + 21$, or 24. This corresponds to the letter X.

Do Exercises 5 and 6.

EXAMPLE 6 Find the next figure in the sequence.

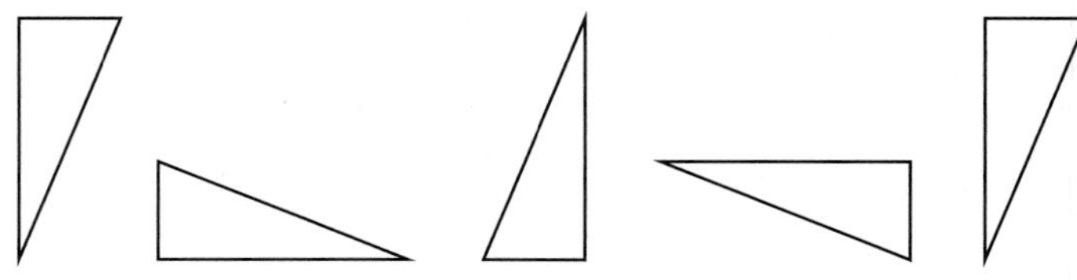

This pattern consists of a triangle that rotates one-fourth turn counterclockwise as the sequence moves from figure to figure. This is a circular pattern. To continue this pattern, we rotate the last triangle one-fourth turn counterclockwise to get the figure shown.

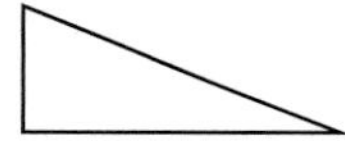

EXAMPLE 7 Find the next figure in the sequence.

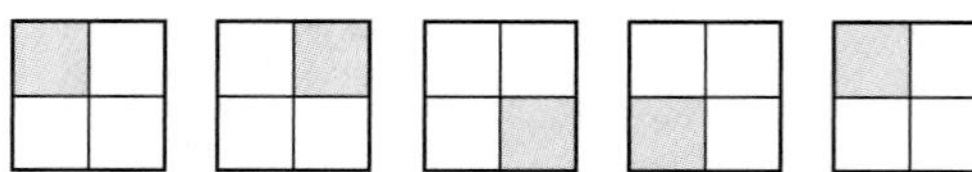

This sequence consists of a square divided into four regions, one of which is color. We observe that the color rotates clockwise as the sequence moves from figure to figure, so we have a circular pattern. To continue this pattern, the next figure in the sequence should be color in the upper righthand region, as shown here:

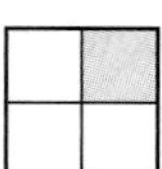

Do Exercises 7 and 8.

EXAMPLE 8 Find the next figure in the sequence.

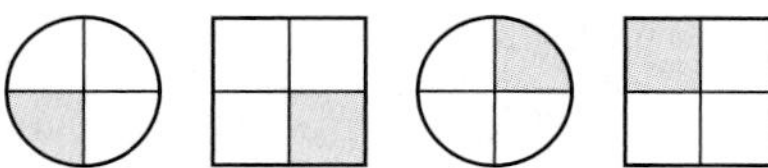

There are two patterns in this sequence. Circles and squares alternate to create one pattern. The second pattern is created by color that rotates counterclockwise as the sequence moves from figure to figure. The next figure in the shape pattern will be a circle. The next figure in the color sequence will be color in the lower lefthand region. Thus the next figure in the sequence is as shown below:

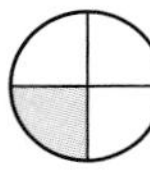

Do Exercises 9 and 10.

Find the next figure in the sequence.

9.

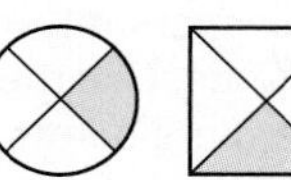

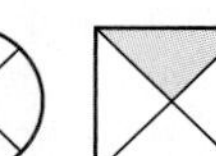

10.

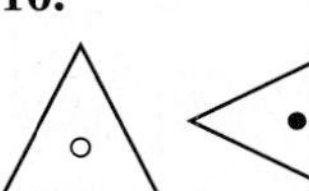
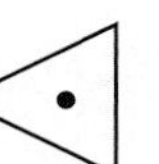
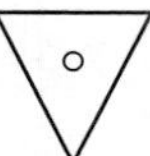
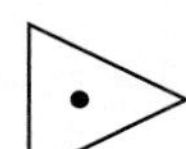

b Deductive Reasoning

Deductive reasoning is being used when conclusions are made on the basis of a set of facts. This usually involves deriving a series of new facts from those originally stated until a conclusion is reached.

EXAMPLE 9 A survey of 100 people shows that 65 own a dog, 25 own a cat, and 18 own both a dog and a cat. How many of those surveyed own neither a dog nor a cat?

First, we find that $65 + 25$, or 90 people, own a dog or a cat or both. From this number, we subtract the number of people who own both a dog and a cat, because they are counted twice—once as dog owners and once as cat owners. We have $90 - 18$, or 72. Finally, we subtract this number from 100, the number of people surveyed, to find how many own neither a dog nor a cat. We have $100 - 72$, or 28.

Do Exercises 11 and 12.

11. A poll of 125 people shows that 68 are on a low-fat diet, 39 are on a low-sodium diet, and 17 are on both a low-fat and a low-sodium diet. How many of those polled are on neither a low-fat nor a low-sodium diet?

12. A survey of 200 people shows that 125 of them subscribe to a daily newspaper, 70 subscribe to a weekly newsmagazine, and 65 subscribe to both. How many of those surveyed subscribe to neither a daily newspaper nor a weekly newsmagazine?

Answers on page A-59

EXAMPLE 10 Amy, Chloe, Marc, and Pete all work in the same building. One is a doctor, one is a lawyer, one is a computer analyst, and one is an architect. Use the statements below to determine who is the architect.

1. Chloe and Marc carpool with the computer analyst.
2. Amy and Pete work on the same floor of the building as the architect.
3. Chloe eats lunch with the lawyer and the architect.

We set up a table as shown below. We use the given statements to determine which jobs each person *cannot* hold. In statement 1, we learn that neither Chloe nor Marc is the computer analyst, so we put an X under "Analyst" in the rows labeled "Chloe" and "Marc." In statement 2, we learn that neither Amy nor Pete is the architect, so we put an X under "Architect" in the rows labeled "Amy" and "Pete." Statement 3 tells us that Chloe is neither the lawyer nor the architect, so we put an X under both "Lawyer" and "Architect" in the row labeled "Chloe." Now we see that the only empty cell of the table under "Architect" corresponds to the row labeled "Marc," so Marc is the architect.

	DOCTOR	LAWYER	ANALYST	ARCHITECT
Amy				X
Chloe		X	X	X
Marc			X	
Pete				X

Do Exercise 13.

13. Maria, Tonya, Carlos, and Frank all live in the same building. One is a student, one is a teacher, one is a salesperson, and one is a paramedic. Use the statements and the table below to determine who is the teacher.

1. Tonya lives on the same floor as the teacher and the paramedic.
2. Maria and Frank jog with the student.
3. Carlos and Frank eat dinner with the teacher.

	STUDENT	TEACHER	SALESPERSON	PARAMEDIC
Maria				
Tonya				
Carlos				
Frank				

Answers on page A-59

N EXERCISE SET

a Find the next letter or number in the sequence.

1. 7, 10, 13, 16, . . .

2. 25, 27, 21, 23, 17, . . .

3. B, G, L, Q, . . .

4. W, U, S, Q, . . .

5. 40, 41, 39, 42, 38, . . .

6. 1, 3, 4, 7, 11, . . .

Find the next figure in the sequence.

7. 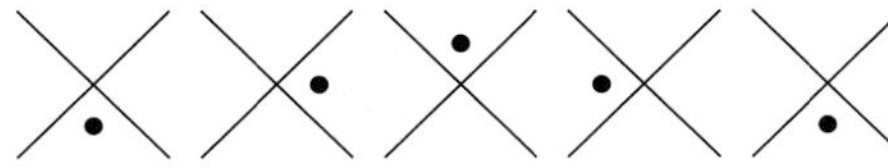

8. 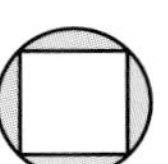 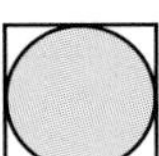 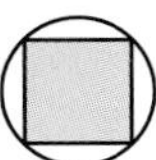

9. 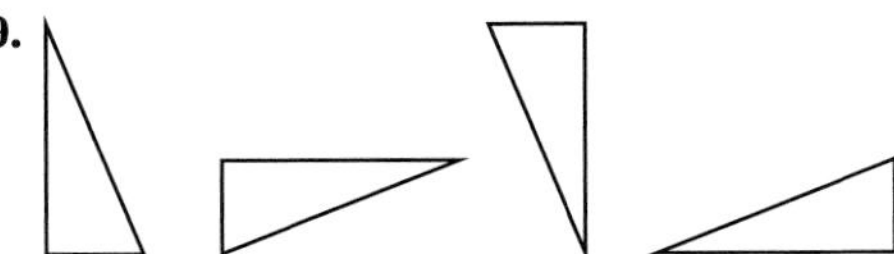

10. 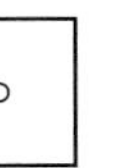  

b

11. A survey of 100 households shows that 78 of them have a VCR, 42 have a CD player, and 37 have both a VCR and a CD player. How many of those surveyed have neither a VCR nor a CD player?

12. A poll of 80 library patrons shows that, in the past month, 60 of them checked out at least one novel, 25 checked out at least one audiotape, and 13 checked out both a novel and an audiotape. How many of those polled checked out neither a novel nor an audiotape in the past month?

13. Max, Jay, Bev, and Lea live on the same street. For exercise, one jogs, one walks, one swims, and one rides a bicycle. Use the statements and the table below to determine who is the swimmer.

1. Max and Lea carpool with the swimmer.
2. Jay and Bev live across the street from the walker.
3. Jay's next-door neighbors are the bike rider and the swimmer.

	JOGGER	WALKER	SWIMMER	BIKER
Max				
Jay				
Bev				
Lea				

14. Antonio, Dale, Janet, and Lynn attend the same college. One majors in English, one in business, one in math, and one in Spanish. Use the statements and the table below to determine Lynn's major.

1. Antonio and Lynn eat lunch with the Spanish major.
2. Dale and Janet study with the business major.
3. Lynn is in the same speech class as the English and business majors.

	ENGLISH	BUSINESS	MATH	SPANISH
Antonio				
Dale				
Janet				
Lynn				

15. Pat, Chris, Casey, and Lee are in the same math class. One has brown hair, one has black hair, one has red hair, and one is a blond. Use the statements below to determine who has red hair.

1. Pat walks to class with the redhead and the blond.
2. Chris got a higher score on the last test than the students with black hair and blond hair.
3. Casey and Lee study with the redhead.

16. Joe, Sam, Fran, and Beth work in the same office. One is a supervisor, one is a typist, one is a receptionist, and one is a bookkeeper. Use the statements below to determine Joe's position.

1. Sam and Beth earn more than the typist.
2. Joe eats lunch with the receptionist and the bookkeeper.
3. The supervisor works later than Fran and Joe.

0 MULTISTEP APPLIED PROBLEMS

Objective

a Solve multistep applied problems.

a Solving Multistep Applied Problems

Often it is necessary to use a combination of mathematical skills in order to solve an applied problem. As you familiarize yourself with such a problem, you should break it down into small problems that can be solved in sequence in order to arrive at the desired result.

EXAMPLE 1 Four students plan to drive 2440 mi during a one-week spring break. They rent an SUV that gets 25 miles per gallon (mpg) on the highway for a weekly rate of \$395 (with unlimited mileage). The gasoline costs average \$1.35 per gallon. If they plan to split the total cost of transportation equally, what is the transportation cost per student?

1. **Familiarize.** First, we find the number of gallons of gas required. Next, we find the cost of the number of gallons of gasoline. Finally, we determine the total cost of the transportation and divide by 4 to determine the cost per student. We let $n =$ the number of gallons required, $c =$ the cost of the gasoline, and $t =$ the total cost of transportation.

2., 3. **Translate** and **Solve.** First, we translate to an equation that will enable us to find n.

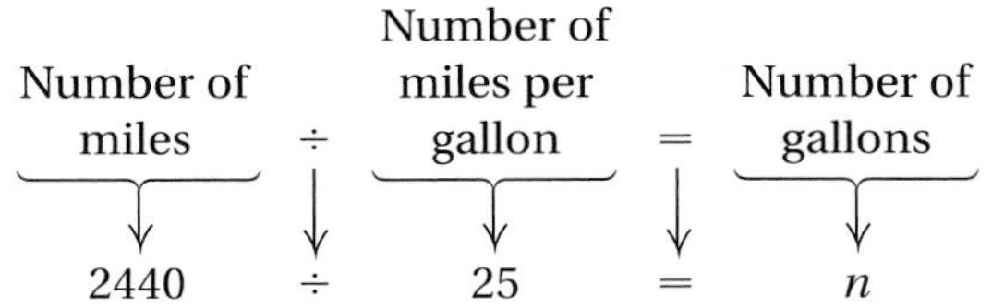

Number of miles ÷ Number of miles per gallon = Number of gallons
2440 ÷ 25 = n

To solve the equation, we carry out the division:

$2440 \div 25 = 97.6.$

Thus, $n = 97.6$ gallons.

Next, we translate to an equation that will enable us to find c.

Price per gallon × Number of gallons = Cost of gasoline

$\$1.35 \times 97.6 = c$

To solve the equation, we carry out the multiplication:

$1.35 \times 97.6 = 131.76.$

Thus, $c = \$131.76$.

We now find the total transportation cost, t.

Rental cost + Gasoline cost = Total cost of transportation

$\$395 + \$131.76 = t$

To solve the equation, we add:

$395 + 131.76 = 526.76.$

1. Three professors are planning to drive 810 mi to a conference. They rent a minivan for a week at a weekly rate of $285. The van gets 27 mpg and the average cost of gasoline is $1.53 per gallon. If they plan to split the total cost of transportation equally, what is the transportation cost per professor?

2. A homeowner hires a painter to paint the exterior of her home. She buys 7 gal of paint at $19.99 per gallon and two paintbrushes at $10.95 each. The sales tax rate is 5%. If the painter charges $20 an hour and it takes 32 hr to complete the job, what is the total cost of the job?

Answers on page A-59

The total cost of transportation is $526.76. Since there are four students, we divide the total cost by 4:

$$526.76 \div 4 = 131.69.$$

Thus the cost per student is $131.69.

4. **Check.** We multiply the cost per student by the number of students:

$$131.69 \times 4 = 526.76.$$

Then we subtract the rental cost of the SUV:

$$526.76 - 395 = 131.76.$$

Next, we divide the gasoline cost by the price per gallon to find the number of gallons used:

$$131.76 \div 1.35 = 97.6.$$

And finally, we multiply the number of gallons by the number of miles per gallon to obtain the total number of miles driven:

$$97.6 \times 25 = 2440.$$

The result checks.

5. **State.** The transportation cost per student is $131.69.

Do Exercise 1.

EXAMPLE 2 A law firm called A-1 Service to repair its air conditioner. It was charged $55 for the service call, $60 per hour for $1\frac{1}{2}$ hr of labor, and $42.79 for parts. What was the total cost of the repair?

1. **Familiarize.** First, we find the cost of labor. Then we add the individual costs to find the total cost of the repair. We let $x =$ the cost of the labor and $c =$ the total cost of the repair.

2., 3. **Translate** and **Solve.** First, we translate to an equation that will enable us to find x.

$$\underbrace{\text{Cost per hour}} \cdot \underbrace{\text{Number of hours}} = \underbrace{\text{Cost of labor}}$$

$$60 \cdot 1\tfrac{1}{2} = x$$

To solve the equation, we carry out the multiplication:

$$60 \cdot 1\frac{1}{2} = \frac{60}{1} \cdot \frac{3}{2} = \frac{180}{2} = 90.$$

Thus, $x = \$90$.

Now we add to find the total cost of the repair. We have

$$c = 55 + 90 + 42.79 = 187.79.$$

4. **Check.** We check by repeating the calculations. The result checks.

5. **State.** The total cost of the repair was $187.79.

Do Exercise 2.

0 EXERCISE SET

a Solve.

1. The Speedy Cab Company charges $3 for the first mile of a taxi ride and 75¢ for each additional mile. What is the cost of a 13-mi taxi ride?

2. A telephone company charges 25¢ for the first minute of a long-distance call and 10¢ for each additional minute. What is the cost of a 33-min long-distance call?

3. A driver plans to change her car's motor oil. She buys 4 qt of oil at $1.49 per quart and an oil filter for $4.95. What is the total cost of the oil change?

4. A consumer purchases 3 pairs of socks at $6.95 each and a belt for $36. What is the total cost of the purchases?

5. An employee at a sporting goods store is paid $8.50 per hour plus 2% commission on her sales. One week she worked 35 hr and had sales of $15,700. What was her total pay for the week?

6. A salesperson earns a commission of 5% on his first $20,000 of sales in a month and 8% on all sales over $20,000 for the month. One month he had sales of $45,000. How much did he earn in commissions for the month?

7. A family room measures 15 ft by 18 ft. Carpet costing $21 per square yard and pad costing $3.25 per square yard will be laid in the room. What is the total cost of the carpet and pad?

8. One pound of boneless honey-baked ham serves 6 people. A hostess is planning to serve this ham at a dinner party for 15 people. Ham costs $7.95 per pound. What is the cost of the ham for the dinner party?

9. A driver filled the gas tank when the odometer read 46,192.8. After the next fill-up, the odometer read 46,519.4. It took 11.5 gal to fill the tank. How many miles per gallon did the driver get?

10. A homeowner plans to paint the exterior of his home. He buys 9 gal of paint at $14.99 per gallon and a paint brush for $12.49. The sales tax rate is $5\frac{1}{2}$%. If he hires a painter who charges $15 per hour and it takes 45 hr to complete the job, what is the total cost of the job?

11. A cylindrical water tank has a height of 28 ft and a diameter of 10 ft. One cubic foot of water weighs 62.5 lb. How many pounds of water will the tank hold when it is full? (Use $\frac{22}{7}$ for π.)

12. A cylindrical tank with a height of 24 ft and a radius of 9 ft is filled with water. There is about 7.5 gal in a cubic foot of water. A pump can empty the tank at a rate of 120 gal per minute. How long will it take to empty the tank? (Use 3.14 for π.)

13. A lot measures 90 ft by 220 ft. A house measuring 30 ft by 85 ft sits on the lot. A bag of lawn fertilizer covers 5000 ft^2 and costs $7.95. How much will it cost to fertilize the lawn?

14. A caterer is preparing brunch for 32 people. She will serve 6 oz of orange juice to each person. A 64-oz carton of orange juice costs $3.49. What is the cost of the orange juice for the brunch?

15. A department store employee gets a 35% discount on shoes and a 30% discount on clothing. He buys a pair of shoes regularly priced at $69.95 and a shirt regularly priced at $54.95. What did the employee pay for these purchases?

16. A bedroom measures 4 m by 5 m. The ceiling is 3.5 m high. There are two windows in the room, each measuring 1 m by 4.5 m. The door to the room and the closet door each measure 1 m by 2.1 m. A liter of paint covers 20.2 m^2 and costs $3.98 per liter. How much will it cost to paint the walls of the room?

17. Sam, a paralegal who has just taken an entry-level job in state government, makes $26,300 a year. He can qualify for a home loan of up to $2\frac{1}{4}$ times his annual salary. With the current interest rate, his monthly payment on a mortgage is about 1% of the amount borrowed. If Sam borrows as much as he can, what will be his monthly payment?

18. John started a new job with an annual salary of $34,000. After three months, he received a cost-of-living adjustment that raised his salary by 2%. Then after one year, his salary was raised by 4%. What was his annual salary after both increases?

19. A decorative rug is placed in the center of a rectangular room. The distance between the carpet and the wall is the same all the way around the room. The room is 23 ft long and 18 ft wide. The rug is 150 ft^2. What is the distance between the carpet and the wall?

20. A wallpaper border is available in 15-ft rolls at $10.95 per roll. Sally is going to place this border around both the top and the bottom of a rectangular room that measures 22 ft by 12 ft. How many rolls will she need and what will be her total cost?

P THEA PRACTICE TEST

1. Green and White Landscaping employs 48 people during the summer. In August, $\frac{1}{3}$ of them return to school. Only $\frac{5}{8}$ of those remaining are employed for snow removal in the winter. How many people does Green and White employ during the winter?

   A. 10 **B.** 20 **C.** 46 **D.** 12

2. Debbie started a new job with an annual salary of \$24,000. In January, she received a cost-of-living adjustment that raised her salary by 3%. In April, her salary was raised by 5%. What was her annual salary after both increases?

   A. \$25,920 **B.** \$25,956 **C.** \$25,200 **D.** \$24,800

3. Lindsay paid \$4.60 for the use of 400 ft^3 of water. At that rate, how much would she have to pay for 700 ft^3 of water?

   A. \$32.20 **B.** \$2.63 **C.** \$8.80 **D.** \$8.05

4. Light travels at a speed of 3.0×10^5 km per second. Earth is, on average, 1.5×10^8 km from the sun. About how many minutes does it take for light to travel to Earth from the sun?

   A. 8.3 min **B.** 500 min **C.** 0.5 min **D.** 83 min

5. Jason began a hike at an elevation of 150 ft above sea level. He descended 265 ft, then climbed back up 75 ft. At what elevation was he at that point?

   A. 490 ft above sea level **B.** 340 ft above sea level **C.** 40 ft below sea level **D.** 190 ft below sea level

6. The following circle graph shows how college students, in general, use their spending money.

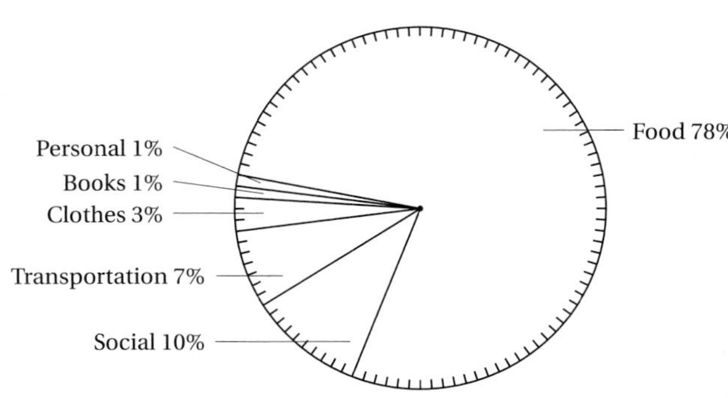

   After paying her fixed expenses, Carla has 25% of her income left for spending money. According to the circle graph, what percent of her income does she spend each month for social activities?

   A. 35% **B.** 2.5%
   C. 15% **D.** Not enough information given to determine the answer

7. Carol assembles 10 electronic parts per hour during the first half of her shift. She takes a half-hour lunch break. After lunch her rate drops to 8 parts per hour for the last half of her shift. Which of the following graphs represents the total number of parts assembled throughout the day?

A.

Parts assembled

Time worked

B.

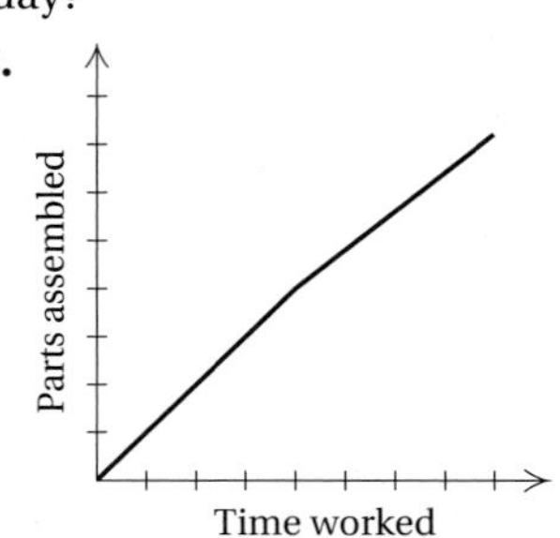

C.

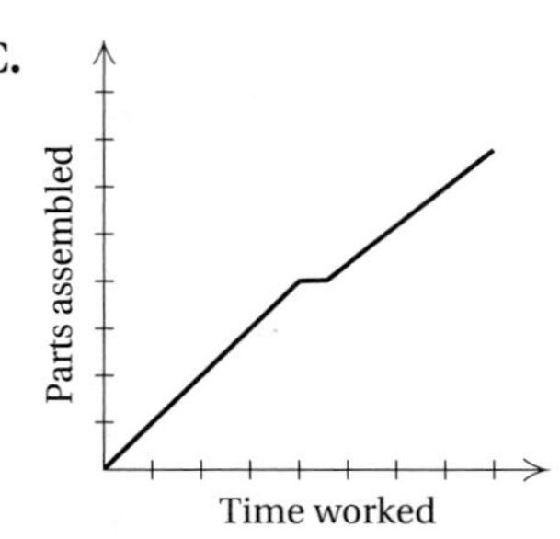

D.

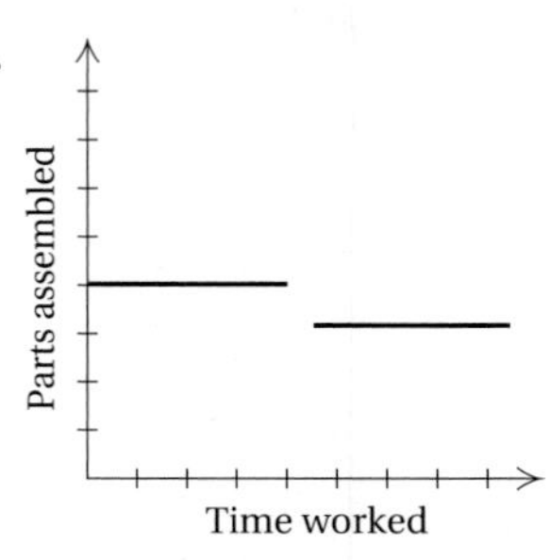

8. In order to estimate the market value of his home, Jordan needs to know the average selling price of other homes in his neighborhood. In the first half of the preceding year, three homes sold for an average selling price of $124,000. In the second half of the year, a home sold for $140,000 and another for $120,000. What was the average selling price of all the homes sold in the preceding year?

A. $124,000 **B.** $128,000 **C.** $130,200 **D.** $126,400

9. The curves below show the distribution of grades in two separate sections of a mathematics class.

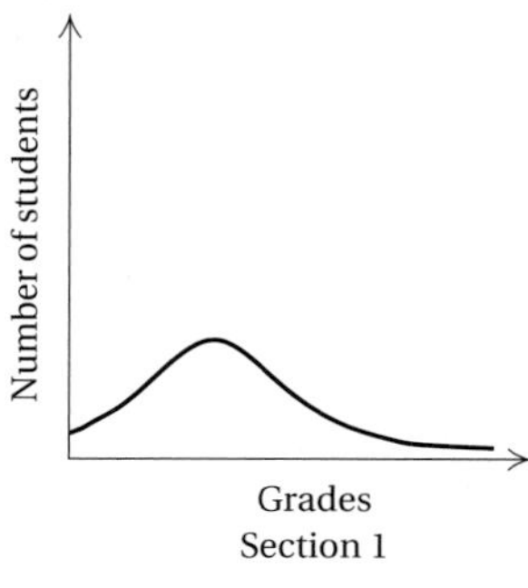

Section 1

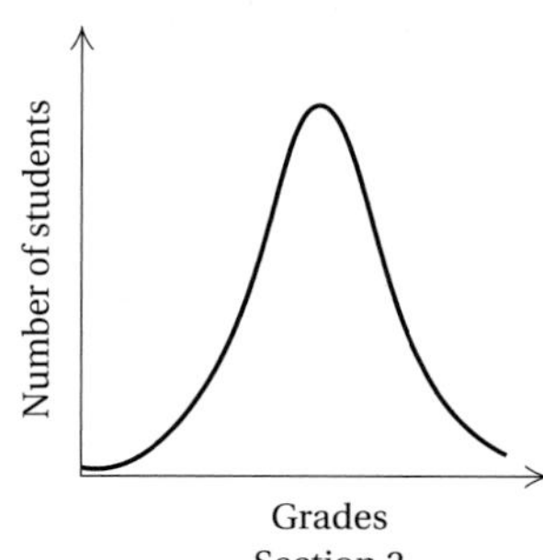

Section 2

Which of the following statements correctly analyzes the information presented in these distributions?

A. The mean grade for both sections was the same.
B. There were more students in section 2 than in section 1.
C. There was more variability in the grades for section 1 than in those for section 2.
D. Students in section 2 received poorer grades than students in section 1.

10. Use the graph of the line AB at right to answer the question that follows.

Which of the following equations represents the line AB?

A. $y = 3$ **B.** $x + y = 3$
C. $x = 3$ **D.** $y = x + 3$

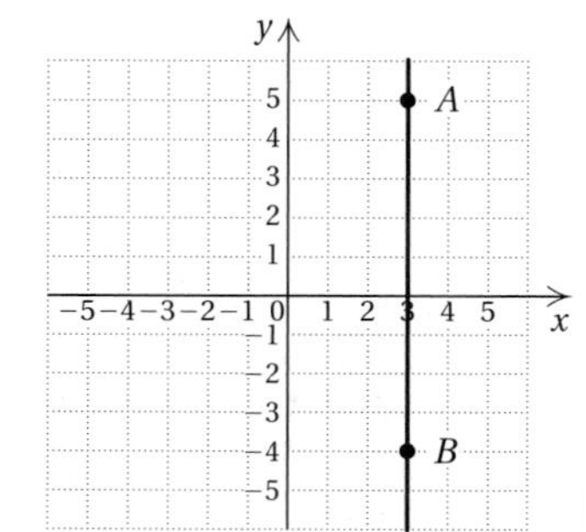

11. Find the slope, if it exists, of the line containing the points $(0, -4)$ and $(-2, -1)$.

A. $-\frac{3}{2}$ **B.** $-\frac{2}{3}$ **C.** $\frac{3}{2}$ **D.** Undefined

12. What are the coordinates of the y-intercept of the line whose equation is $2x - 5y = 10$?

A. $(0, -2)$ **B.** $(0, 5)$ **C.** $(5, 0)$ **D.** $(-2, 0)$

13. Which of the following is an equation of the line containing the points $(-5, 6)$ and $(-2, 4)$?

A. $y = -\frac{3}{2}x - \frac{3}{2}$ **B.** $y = -\frac{2}{3}x + 6$ **C.** $y = \frac{3}{2}x + \frac{27}{2}$ **D.** $y = -\frac{2}{3}x + \frac{8}{3}$

14. The graph at right shows how the amount of time t that it takes to polish the rotor blade of a helicopter depends on the number of machinists n working on it. Which of the following statements about the relationship is true?

A. The more machinists working, the longer it takes to polish the blade.
B. It takes 4 machinists 10 hr to polish the blade.
C. If the blade must be polished in 2 hr, 5 machinists can get it done.
D. In 10 hr, 2 machinists can polish the blade.

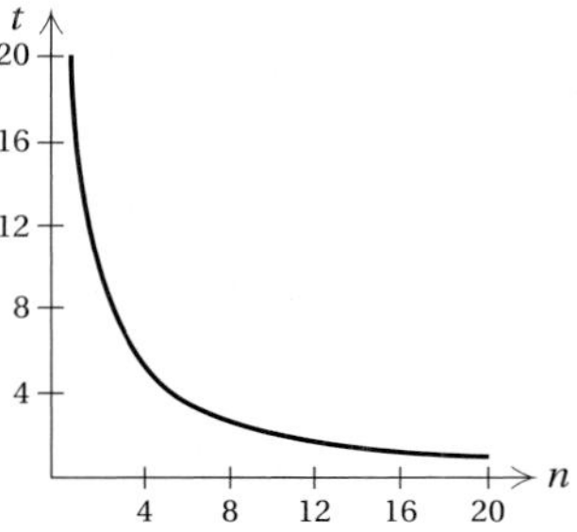

15. If $\frac{2}{5}x + 7 = 1$, what is the value of $10 - 3x$?

A. -15 **B.** 55 **C.** 17 **D.** -35

16. Solve $t = \frac{1}{5}(r + 5)$ for r.

A. $r = 5t - 5$ **B.** $r = 5t - 25$ **C.** $r = \frac{t - 1}{5}$ **D.** $r = \frac{1}{5}t - 1$

17. What is the solution of the system of equations $y = 2x^2 + 7x + 3$ and $4x + 3y = 9$?

A. $(-3, 7), \left(-\frac{1}{2}, \frac{11}{3}\right)$ **B.** $(0, 3), \left(-\frac{25}{6}, \frac{77}{9}\right)$

C. $\left(3\sqrt{3}, 3 - 4\sqrt{3}\right), \left(-3\sqrt{3}, 3 + 4\sqrt{3}\right)$ **D.** No solution

18. Which of the following equations correctly translates this statement? The product of the length l of a fish and the square of the girth g of the fish is 280 times the weight w of the fish.

A. $(l \times g)^2 = 280w$ **B.** $280(l \times g^2) = w$ **C.** $l \times g^2 = 280w$ **D.** $280(l + g) = w$

19. Which of the following graphs shows the solution of the system of equations $y - 5 + x = 2x$ and $y = x^2 - 2$?

A.

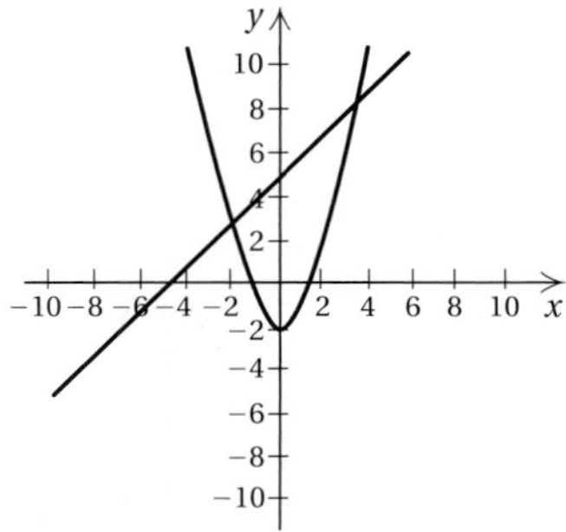

B.

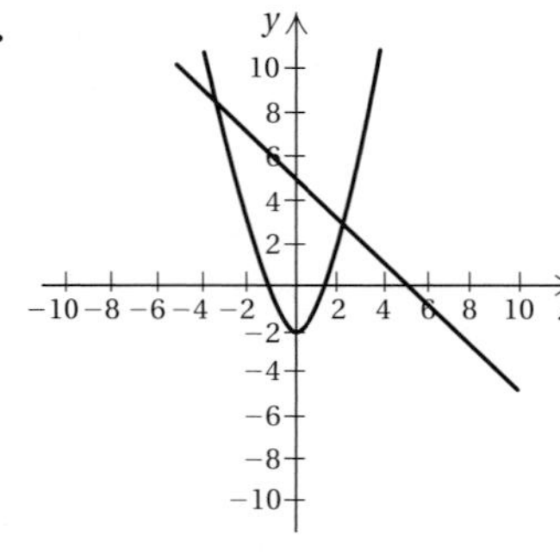

C.

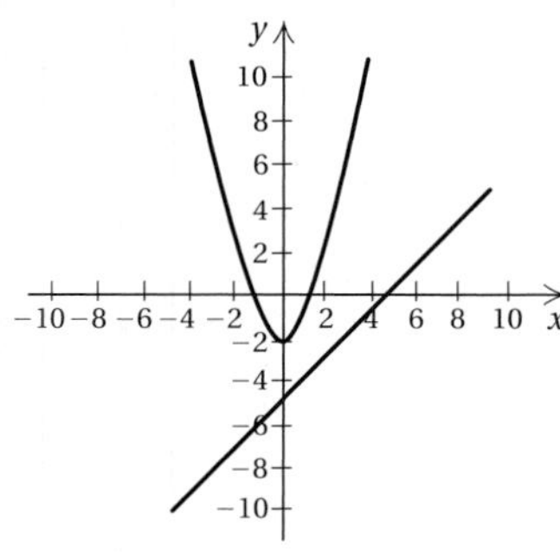

D.

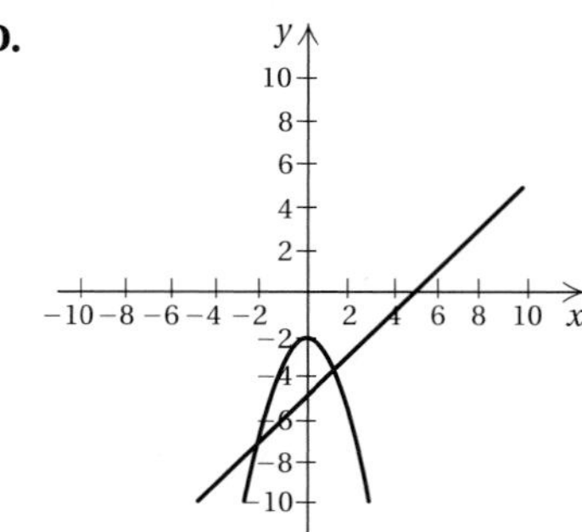

20. Karen earns 2% of the price of each home she sells as commission. Last week, she sold two houses and earned \$3500. The selling price of one house was $1\frac{1}{2}$ times the selling price of the other. What was the selling price of the more expensive house?

A. \$175,000 **B.** \$105,000 **C.** \$116,666 **D.** \$87,500

21. Party Time stores sell complete party packages. The Flyaway package contains paper products for 12 people plus 30 helium-filled balloons and sells for \$34.80. The Down-to-Earth package contains paper products for 15 people plus 1 balloon and sells for \$14.30. Party Time wants to begin selling a Party package that will contain paper products for 25 people plus 10 balloons. What should the selling price of the Party package be?

A. \$49.10 **B.** \$29.00 **C.** \$30.50 **D.** \$11.60

22. Which of the following is one factor of $3x^2 - 4x - 15$?

A. $(x - 5)$ **B.** $(3x - 5)$ **C.** $(x + 3)$ **D.** $(x - 3)$

23. Perform the multiplication: $(2x + 5)^2$.

A. $4x^2 + 25$ **B.** $4x^2 + 20x + 25$ **C.** $4x^2 + 10x + 25$ **D.** $2x^2 + 25$

24. Add and simplify, if possible: $\frac{x - 12}{x^2 + x - 6} + \frac{x}{x - 2}$.

A. $\frac{x + 6}{x + 3}$ **B.** $\frac{4x - 12}{x - 6}$ **C.** $\frac{2x - 12}{x^2 + 2x - 8}$ **D.** $\frac{2x - 12}{(x^2 + x - 6)(x - 2)}$

25. Add and simplify, if possible: $\sqrt{18x^3} - \sqrt{2x} + \sqrt{3x}$.

A. $\sqrt{18x^3 + x}$ **B.** $\sqrt{18x^3} + \sqrt{x}$ **C.** $(3x - 1)\sqrt{2x} + \sqrt{3x}$ **D.** $\sqrt{x}(2\sqrt{2} + \sqrt{3})$

26. If $f(x) = |2x - 7|$, find $f(2) \cdot f\left(\frac{1}{2}\right)$.

A. 18 **B.** 5 **C.** 25 **D.** 3

27. Shown at right is the graph of a quadratic function.

Which of the following equations is represented by this graph?

A. $y = x^2 - 5x - 6$ **B.** $y = x^2 + 5x - 6$
C. $y = x^2 + 7x + 6$ **D.** $y = x^2 - 7x + 6$

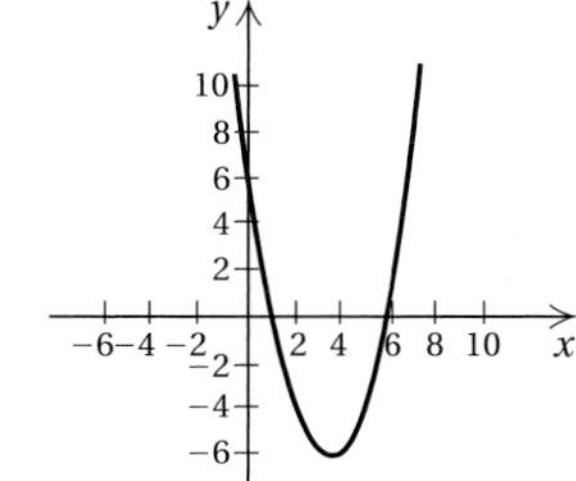

28. Shown at right is the graph of a quadratic inequality.

Which of the following inequalities describes the shaded region?

A. $y \leq x^2 + 2x - 8$ **B.** $y \geq x^2 + 2x - 8$
C. $y \leq (x - 1)^2 - 9$ **D.** $y \geq (x - 1)^2 - 9$

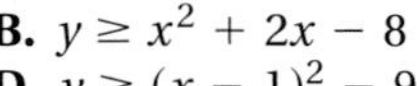

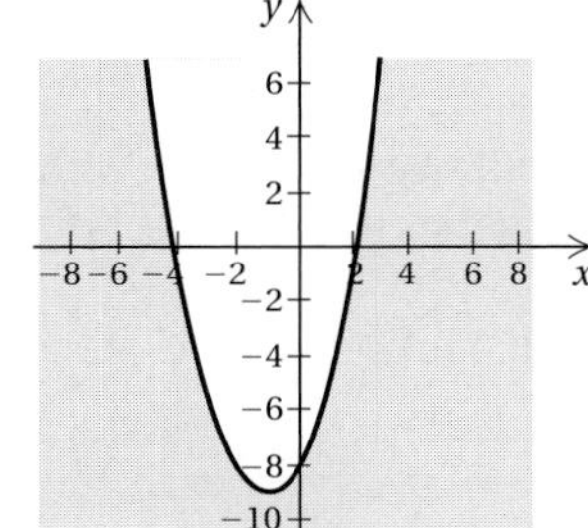

29. Which of the following numbers should be added to both sides of the equation $x^2 + 9x = 15$ in order to solve by completing the square?

A. 81 **B.** $\frac{81}{4}$ **C.** $\frac{9}{2}$ **D.** 125

30. Solve the equation $2x^2 - 5x = 8$ using the quadratic formula. In the solution, what is the number under the radical sign?

A. -39 **B.** 5 **C.** 41 **D.** 89

31. A rectangular piece of bound carpet lies in the center of a rectangular room. The distance between the carpet and the wall is the same all the way around the room. The room is 15 ft long and 12 ft wide. The carpet piece is 12 yd². What is the distance between the carpet and the wall?

A. 4 yd **B.** 4 ft **C.** $1\frac{1}{2}$ ft **D.** 3 ft

32. A window is square with a semicircular top section, as shown in the figure at right.

About how many square feet of glass will it take to make the window?

A. 16.1 ft^2 **B.** 12.5 ft^2
C. 23.1 ft^2 **D.** 28.3 ft^2

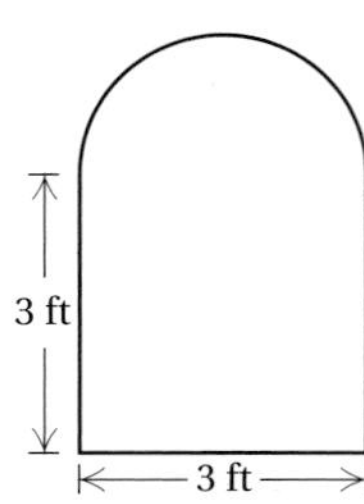

33. A decorative tin in the shape of a cylinder is full of candy. The tin is 8 in. high, and the diameter of the lid is 5 in. What is the volume of candy that the tin contains?

A. About 157 in^3 **B.** About 165 in^3 **C.** About 40 in^3 **D.** About 126 in^3

34. Metal Manufacturing is making a hand rest for a medical supply company in the shape shown at right.

How much metal is used to make the hand rest?

A. 1700 cm^2 **B.** 6500 cm^2
C. 2700 cm^2 **D.** 2550 cm^2

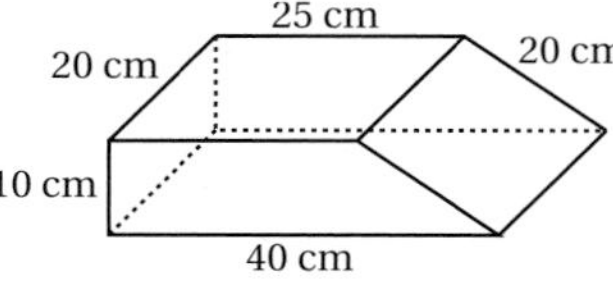

35. Buck Creek Township Fire Department has a fire truck with a ladder that extends straight to 60 ft. At a fire in an apartment building, the closest the truck could get to the burning building was 30 ft. About how high on the building did the ladder reach?

A. 52.0 ft **B.** 60.0 ft **C.** 30.0 ft **D.** 67.1 ft

36. Jessica is building a scale model of a house. The dining room is in the shape of a hexagon, as shown in the figure at right.

Polygon *ABCDEF*, the dining room of the house, is similar to polygon *PQRSTU*, the dining room of the model. *AB* measures 12 ft and *PQ* 4 in. How long should she make *UP* if *FA* is 5 ft?

A. 15 in. **B.** $\frac{3}{5}$ in.

C. $1\frac{2}{3}$ in. **D.** 3 in.

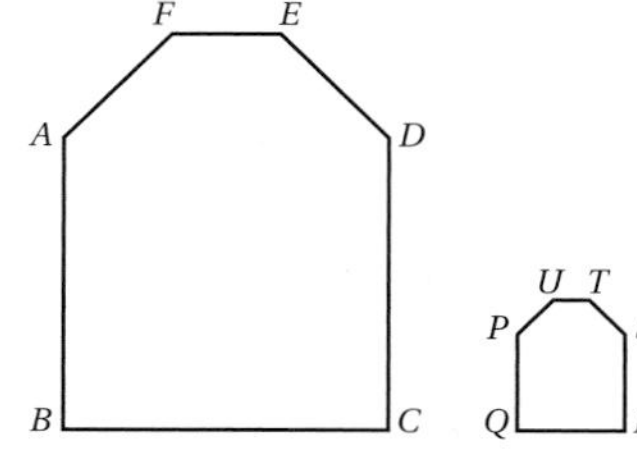

37. Shown at right are two triangles, $\triangle ABC$ and $\triangle DEF$.

If you know that $\overline{AC} \cong \overline{DF}$ and $\angle ABC \cong \angle DEF$, which of the following is a valid conclusion about the triangles?

A. $\triangle ABC \cong \triangle DEF$ by SAS.
B. $\triangle ABC \cong \triangle DEF$ by ASA.
C. $\triangle ABC \cong \triangle DEF$ by SSS.
D. None of these is a valid conclusion.

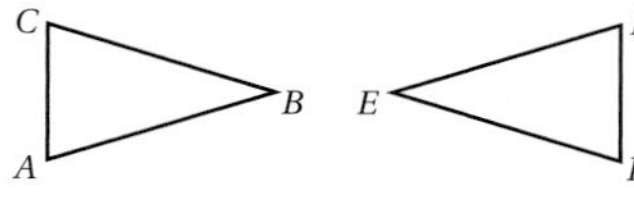

38. In the hexagon shown at right, $\overline{AB} \parallel \overline{DE}$. $\overline{BE}$ divides the figure into two quadrilaterals, *ABEF* and *DEBC*.

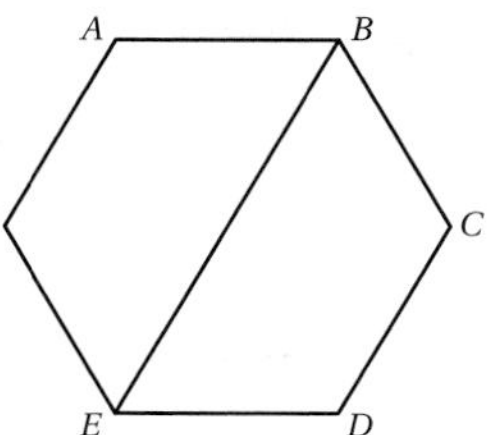

Which of the following is a valid conclusion about this figure?

A. $ABEF \cong DEBC$
B. $\angle CBE \cong \angle FEB$
C. $\angle ABE \cong \angle DEB$
D. None of these is a valid conclusion.

39. In the polygon shown at right, $\overline{BF}$ is perpendicular to both $\overline{BC}$ and $\overline{EG}$.

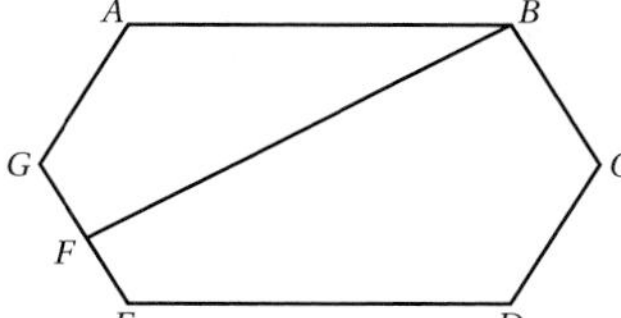

Which of the following is a valid conclusion?

A. $\overline{BC}$ is congruent to $\overline{EG}$.
B. $\overline{BC}$ is perpendicular to $\overline{EG}$.
C. $\overline{BC}$ is parallel to $\overline{EG}$.
D. None of these is a valid conclusion.

40. Use the statements below to answer the question that follows.

1. All employees of Poster Prints will get a pay raise this year.
2. Some employees of Poster Prints will be transferred to another location.
3. No management personnel will be transferred.
4. Jackie is an employee of Poster Prints.

Which of the following statements must be true?

A. Jackie will be transferred.
B. Jackie is in management.
C. Jackie will get a pay raise.
D. None of these statements must be true.

41. Mike, Linda, Jeff, and Paula are brothers and sisters. One is an engineer, one is an author, one is a consultant, and one is an accountant. Use the statements below to determine which one is an author.

1. Mike lives near the author and the consultant.
2. Linda is older than the engineer and the accountant.
3. Paula is younger than the author and the engineer.
4. Jeff does not live near any of his brothers or sisters.

Who is the author?

A. Mike **B.** Linda **C.** Jeff **D.** Paula

42. Use the pattern sequence below to answer the question that follows.

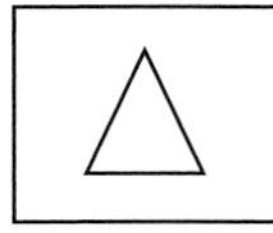
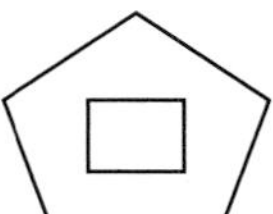
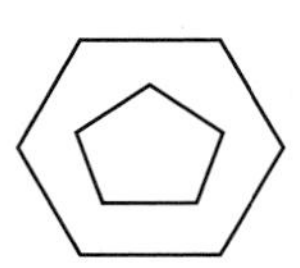

What figure comes next in the sequence?

A.
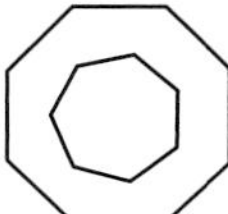

B.
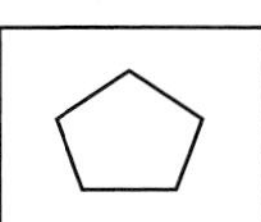

C.

D.

43. Use the pattern sequence below to answer the question that follows.

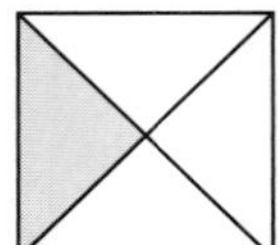 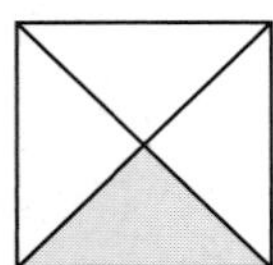 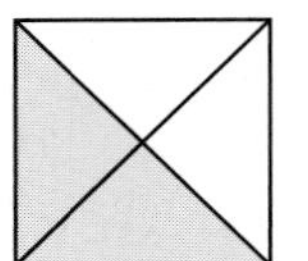 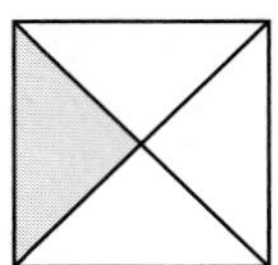 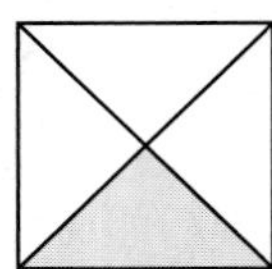 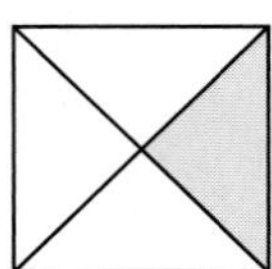 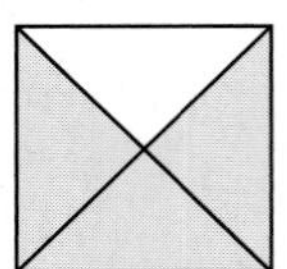 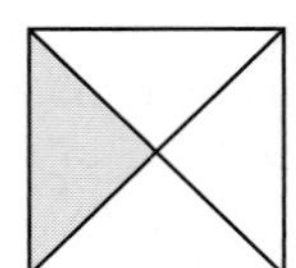 ?

What figure comes next in the sequence?

A. 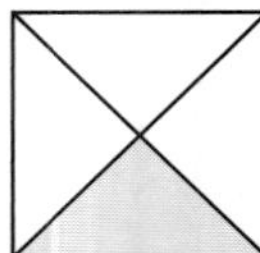
B. 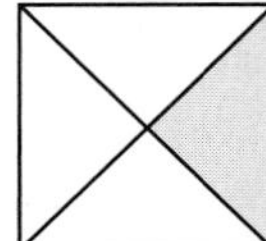
C. 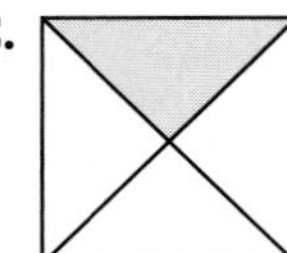
D. 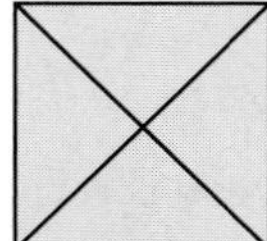

44. A flower garden is a square with a semicircle along each side, as shown in the figure at right.

A landscape designer wants to plant flowers along the perimeter of the garden. If there should be 4 plants for every meter, and each plant costs \$0.70, how much will the flowers cost?

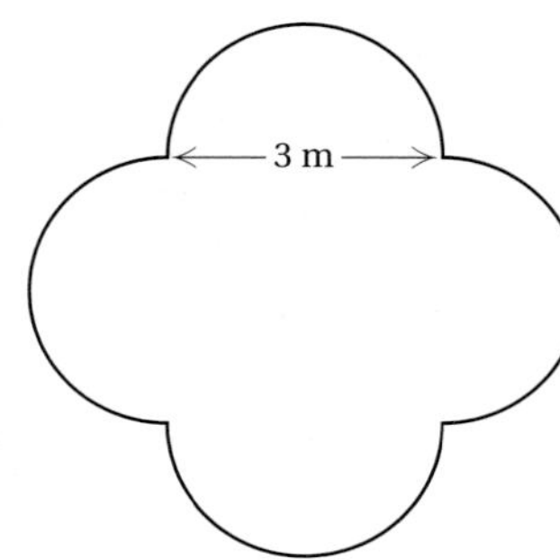

A. \$13.30 **B.** \$52.50
C. \$26.60 **D.** \$39.90

Use the information below for Questions 45 and 46.

Brenda is a graphics designer who has just accepted an entry-level job paying \$24,000 a year. She can qualify for a mortgage of up to $2\frac{1}{2}$ times her annual salary. With the current interest rate, the monthly payment on a mortgage is about 1% of the amount borrowed.

45. If Brenda borrows as much as she can, what will be her monthly mortgage payment?

A. \$600 **B.** \$240 **C.** \$480 **D.** \$120

46. Brenda has other fixed expenses. If she buys the house, she will have to pay \$100 a month for taxes and insurance. Income taxes take 20% of her earnings. Her car payment is \$175 a month, and auto insurance is \$100 a month. She spends 15% of her income on food. How much will she have left each month after paying her mortgage and other fixed expenses?

A. \$1675 **B.** \$325 **C.** \$925 **D.** \$1275

Q THEA TEST MATHEMATICS SKILLS

*SKILL**	*TEXT REFERENCE*
FUNDAMENTAL MATHEMATICS	
1. Solve word problems involving integers, fractions, decimals, and units of measurement.	
• Solve word problems involving integers.	Sections 1.5 and 7.2–7.6
• Solve word problems involving fractions.	Section 2.5
• Solve word problems involving decimals (including percents).	Sections 3.7, 4.6, 4.7, and 4.8
• Solve word problems involving ratio and proportions.	Sections 4.1 and 12.7
• Solve word problems involving units of measurement and conversions (including scientific notation).	Sections 6.2–6.5 and 10.2 Appendixes A, B, and C
2. Solve problems involving data interpretation and analysis.	
• Interpret information from line graphs, bar graphs, pictographs, and pie charts.	Sections 5.2–5.4
• Interpret data from tables.	Section 5.2
• Recognize appropriate graphic representations of various data.	Chapter 5
• Analyze and interpret data using measures of central tendency (mean, median, and mode). Analyze and interpret data using the concept of variability.	Section 5.1
ALGEBRA	
3. Graph numbers or number relationships.	
• Identify the graph of a given equation.	Sections 9.1, 9.2, 9.5, and 9.6
• Identify the graph of a given inequality.	Sections 8.7 and 9.7
• Find the slope and/or intercepts of a given line.	Sections 9.2, 9.3, and 9.4–9.6
• Find the equation of a line.	Section 9.4
• Recognize and interpret information from the graph of a function (including direct and inverse variation).	Sections 12.9 and 15.7
4. Solve one- and two-variable equations.	
• Find the value of the unknown in a given one-variable equation.	Sections 8.1–8.3
• Express one variable in terms of a second variable in two-variable equations.	Sections 8.4 and 13.2
• Solve systems of two equations in two variables (including graphical solutions).	Sections 13.1–13.3
5. Solve word problems involving one and two variables.	
• Identify the algebraic equivalent of a stated relationship.	Sections 7.1 and 8.6
• Solve word problems involving one and two unknowns.	Sections 8.5, 8.6, 8.8, 11.8, 12.7, 12.9, 13.4, 13.5, 14.6, and 15.5

*Many of these skills, particularly those in problem solving and interpreting information, are not confined to the sections or chapters specified here.

*SKILL**	*TEXT REFERENCE*
6. Understand operations with algebraic expressions and functional notation.	
• Factor quadratics and polynomials.	Sections 11.1–11.6
• Perform operations on and simplify polynomial expressions.	Sections 10.3–10.8
• Perform operations on and simplify rational expressions.	Sections 12.1–12.5
• Perform operations on and simplify radical expressions.	Sections 14.1–14.4
• Apply principles of functions and functional notation.	Section 15.7
7. Solve problems involving quadratic equations.	
• Graph quadratic functions.	Section 15.6
• Graph quadratic inequalities.	Appendix K
• Solve quadratic equations using factoring, completing the square, or the quadratic formula.	Sections 11.7 and 15.1–15.3
• Solve problems involving quadratic models.	Sections 11.8 and 15.5
GEOMETRY	
8. Solve problems involving geometric figures.	
• Solve problems involving two-dimensional geometric figures (e.g., perimeter and area problems).	Sections 6.2–6.4
• Solve problems involving three-dimensional geometric figures (e.g., volume and surface area problems).	Section 6.5
• Solve problems using the Pythagorean theorem.	Sections 11.8 and 14.6
9. Solve problems involving geometric concepts.	
• Solve problems using principles of similarity and congruence.	Sections 6.7 and 6.8
• Solve problems using principles of parallelism and perpendicularity.	Sections 6.6 and 9.6
PROBLEM SOLVING	
10. Apply reasoning skills.	
• Draw conclusions using inductive reasoning.	Appendix N
• Draw conclusions using deductive reasoning.	Appendix N
11. Solve applied problems involving a combination of mathematical skills.	
• Apply combinations of mathematical skills to solve problems.	Throughout
• Apply combinations of mathematical skills to solve a series of related problems.	Throughout

*Many of these skills, particularly those in problem solving and interpreting information, are not confined to the sections or chapters specified here.

Answers

CHAPTER 1

Margin Exercises, Section 1.1, pp. 2–6

1. 2 ten thousands **2.** 2 hundred thousands
3. 2 millions **4.** 2 ten millions
5. 2 tens **6.** 2 hundreds
7. 2 hundred thousands; 8 ten thousands; 0 thousands; 2 hundreds; 1 ten; 9 ones
8. 1 thousand + 8 hundreds + 9 tens + 5 ones
9. 2 ten thousands + 3 thousands + 4 hundreds + 1 ten + 6 ones
10. 3 thousands + 0 hundreds + 3 tens + 1 one, or 3 thousands + 3 tens + 1 one
11. 4 thousands + 1 hundred + 8 tens + 0 ones, or 4 thousands + 1 hundred + 8 tens
12. 1 hundred thousand + 5 ten thousands + 4 thousands + 6 hundreds + 1 ten + 6 ones
13. Forty-nine **14.** Sixteen **15.** Thirty-eight
16. Two hundred four
17. Forty-four thousand, one hundred fifty-five
18. One million, eight hundred seventy-nine thousand, two hundred four **19.** Six billion, four hundred forty-nine million **20.** 213,105,329 **21.** $<$ **22.** $>$ **23.** $>$
24. $<$ **25.** $<$ **26.** $>$

Exercise Set 1.1, p. 7

1. 5 thousands **3.** 5 hundreds **5.** 2 **7.** 1
9. 5 thousands + 7 hundreds + 0 tens + 2 ones, or 5 thousands + 7 hundreds + 2 ones
11. 9 ten thousands + 3 thousands + 9 hundreds + 8 tens + 6 ones
13. 2 thousands + 0 hundreds + 5 tens + 8 ones, or 2 thousands + 5 tens + 8 ones
15. 1 thousand + 2 hundreds + 6 tens + 8 ones
17. 4 hundred thousands + 0 ten thousands + 5 thousands + 6 hundreds + 9 tens + 8 ones, or 4 hundred thousands + 5 thousands + 6 hundreds + 9 tens + 8 ones
19. 2 hundred thousands + 7 ten thousands + 2 thousands + 1 hundred + 6 tens + 1 one
21. 1 million + 1 hundred thousand + 8 ten thousands + 0 thousands + 2 hundreds + 1 ten + 2 ones, or 1 million + 1 hundred thousand + 8 ten thousands + 2 hundreds + 1 ten + 2 ones **23.** Eighty-five
25. Eighty-eight thousand **27.** One hundred twenty-three thousand, seven hundred sixty-five **29.** Seven billion, seven hundred fifty-four million, two hundred eleven thousand, five hundred seventy-seven
31. 2,233,812 **33.** 8,000,000,000 **35.** Five hundred sixty-six thousand, two hundred eighty **37.** Seventy-six million, eighty-six thousand, seven hundred ninety-two
39. 9,460,000,000,000 **41.** 64,186,000 **43.** $<$
45. $>$ **47.** $<$ **49.** $>$ **51.** $>$ **53.** $>$
55. $1,800,607 < 2,136,068$, or $2,136,068 > 1,800,607$
57. $6482 > 4827$, or $4827 < 6482$
59. DW **61.** 138

Margin Exercises, Section 1.2, pp. 11–18

1. 13,465 **2.** 9745 **3.** 16,182 **4.** 27,474 **5.** 29 in.
6. 62 ft **7.** 16 in.; 26 in. **8.** $7 = 2 + 5$, or $7 = 5 + 2$
9. $17 = 9 + 8$, or $17 = 8 + 9$ **10.** $5 = 13 - 8$; $8 = 13 - 5$
11. $11 = 14 - 3$; $3 = 14 - 11$ **12.** 3801 **13.** 6328
14. 4747 **15.** 56 **16.** 205 **17.** 658 **18.** 2851
19. 1546

Calculator Corner, p. 12

1. 55 **2.** 121 **3.** 1602 **4.** 734 **5.** 1932 **6.** 864

Calculator Corner, p. 16

1. 67 **2.** 119 **3.** 2128 **4.** 2593

Exercise Set 1.2, p. 19

1. 387 **3.** 5198 **5.** 8503 **7.** 5266 **9.** 100
11. 8310 **13.** 6608 **15.** 34,432 **17.** 18,424
19. 2320 **21.** 114 mi **23.** 570 ft **25.** $7 = 3 + 4$, or $7 = 4 + 3$ **27.** $43 = 27 + 16$, or $43 = 16 + 27$
29. $6 = 15 - 9$; $9 = 15 - 6$ **31.** $23 = 32 - 9$; $9 = 32 - 23$
33. 44 **35.** 1126 **37.** 5382 **39.** 3831 **41.** 2168
43. 43,028 **45.** 84 **47.** 454 **49.** 771 **51.** 7019
53. 5745 **55.** 12,134 **57.** 4206 **59.** 10,305
61. DW **63.** 8 ten thousands
64. Five billion, two hundred ninety-four million, two hundred forty-seven thousand

65. $1 + 99 = 100, 2 + 98 = 100, \ldots, 49 + 51 = 100$. Then $49 \cdot 100 = 4900$ and $4900 + 50 + 100 = 5050$.

Margin Exercises, Section 1.3, pp. 23–36

1. 116 **2.** 148 **3.** 4938 **4.** 6740 **5.** 1035 **6.** 3024 **7.** 46,252 **8.** 205,065 **9.** 144,432 **10.** 287,232 **11.** 14,075,720 **12.** 391,760 **13.** 17,345,600 **14.** 56,200 **15.** 562,000 **16.** 45 sq ft **17.** $\frac{54}{6} = 9$; $6\overline{)54}$ with quotient 9 **18.** $15 = 5 \cdot 3$, or $15 = 3 \cdot 5$ **19.** $72 = 9 \cdot 8$, or $72 = 8 \cdot 9$ **20.** $6 = 12 \div 2$; $2 = 12 \div 6$ **21.** $6 = 42 \div 7$; $7 = 42 \div 6$ **22.** 6; $6 \cdot 9 = 54$ **23.** 6 R 7; $6 \cdot 9 = 54$, $54 + 7 = 61$ **24.** 4 R 5; $4 \cdot 12 = 48$, $48 + 5 = 53$ **25.** 6 R 13; $6 \cdot 24 = 144$, $144 + 13 = 157$ **26.** 59 R 3 **27.** 1475 R 5 **28.** 1015 **29.** 807 R 4 **30.** 1088 **31.** 360 R 4 **32.** 800 R 47 **33.** 40 **34.** 50 **35.** 70 **36.** 100 **37.** 40 **38.** 80 **39.** 90 **40.** 140 **41.** 470 **42.** 240 **43.** 290 **44.** 600 **45.** 800 **46.** 800 **47.** 9300 **48.** 8000 **49.** 8000 **50.** 19,000 **51.** 69,000 **52.** Eliminate the power sunroof and the power package. Answers may vary. **53.** **(a)** \$18,300; **(b)** yes **54.** $70 + 20 + 40 + 70 = 200$ **55.** $9300 - 6700 = 2600$ **56.** $23,000 - 12,000 = 11,000$ **57.** $840 \times 250 = 210,000$; $800 \times 200 = 160,000$

Calculator Corner, p. 25

1. 448 **2.** 21,970 **3.** 6380 **4.** 39,564 **5.** 180,480 **6.** 2,363,754

Calculator Corner, p. 33

1. 3 R 11 **2.** 28 **3.** 124 R 2 **4.** 131 R 18 **5.** 283 R 57 **6.** 843 R 187

Exercise Set 1.3, p. 37

1. 564 **3.** 2,340,000 **5.** 1527 **7.** 64,603 **9.** 4770 **11.** 3995 **13.** 46,080 **15.** 14,652 **17.** 207,672 **19.** 798,408 **21.** 20,723,872 **23.** 362,128 **25.** 20,064,048 **27.** 25,236,000 **29.** 529,984 sq mi **31.** 8100 sq ft **33.** $18 = 3 \cdot 6$, or $18 = 6 \cdot 3$ **35.** $22 = 22 \cdot 1$, or $22 = 1 \cdot 22$ **37.** $9 = 45 \div 5$; $5 = 45 \div 9$ **39.** $37 = 37 \div 1$; $1 = 37 \div 37$ **41.** 12 **43.** 1 **45.** 22 **47.** Not defined **49.** 55 R 2 **51.** 108 **53.** 307 **55.** 92 R 2 **57.** 1703 **59.** 127 **61.** 29 R 5 **63.** 40 R 12 **65.** 90 R 22 **67.** 29 **69.** 1609 R 2 **71.** 1007 R 1 **73.** 23 **75.** 107 R 1 **77.** 370 **79.** 609 R 15 **81.** 304 **83.** 3508 R 219 **85.** 8070 **87.** 50 **89.** 470 **91.** 730 **93.** 900 **95.** 100 **97.** 1000 **99.** 9100 **101.** 32,900 **103.** 6000 **105.** 8000 **107.** 45,000 **109.** 373,000 **111.** 180 **113.** 5720 **115.** 16,500 **117.** 5200 **119.** 31,000 **121.** 69,000 **123.** \$11,200 **125.** \$18,900; no **127.** Answers will vary depending on the options chosen. **129.** $50 \cdot 70 = 3500$ **131.** $30 \cdot 30 = 900$ **133.** $900 \cdot 300 = 270,000$ **135.** $400 \cdot 200 = 80,000$ **137.** **(a)** \$2,840,000; **(b)** \$2,850,000 **139.** DW **141.** Perimeter **142.** Equation; inequality **143.** Digits; periods **144.** Additive **145.** Dividend **146.** Factors; product **147.** Minuend **148.** Multiplicative **149.** 54, 122; 33, 2772; 4, 8 **151.** 30 buses **153.** 247,464 sq ft

Margin Exercises, Section 1.4, pp. 44–47

1. 7 **2.** 5 **3.** No **4.** Yes **5.** 5 **6.** 10 **7.** 5 **8.** 22 **9.** 22,490 **10.** 9022 **11.** 570 **12.** 3661 **13.** 8 **14.** 45 **15.** 77 **16.** 3311 **17.** 6114 **18.** 8 **19.** 16 **20.** 644 **21.** 96 **22.** 94

Exercise Set 1.4, p. 48

1. 14 **3.** 0 **5.** 29 **7.** 0 **9.** 8 **11.** 14 **13.** 1035 **15.** 25 **17.** 450 **19.** 90,900 **21.** 32 **23.** 143 **25.** 79 **27.** 45 **29.** 324 **31.** 743 **33.** 37 **35.** 66 **37.** 15 **39.** 48 **41.** 175 **43.** 335 **45.** 104 **47.** 45 **49.** 4056 **51.** 17,603 **53.** 18,252 **55.** 205 **57.** DW **59.** $7 = 15 - 8$; $8 = 15 - 7$ **60.** $6 = 48 \div 8$; $8 = 48 \div 6$ **61.** $<$ **62.** $>$ **63.** $>$ **64.** $<$ **65.** 142 R 5 **66.** 142 **67.** 334 **68.** 334 R 11 **69.** 347

Margin Exercises, Section 1.5, pp. 51–58

1. 11,277 adoptions **2.** 8441 adoptions **3.** 19,089 adoptions **4.** \$1874 **5.** \$171 **6.** \$5572 **7.** 9180 sq in. **8.** 378 cartons with 1 can left over **9.** 79 gal **10.** 181 seats

Translating for Success, p. 59

1. E **2.** M **3.** D **4.** G **5.** A **6.** O **7.** F **8.** K **9.** J **10.** H

Exercise Set 1.5, p. 60

1. 33,042 performances **3.** 1348 performances **5.** 2054 mi **7.** 18 rows **9.** 792,316 degrees; 1,291,900 degrees **11.** 192,268 degrees **13.** \$69,277 **15.** 240 mi **17.** 168 hr **19.** 225 squares **21.** \$24,456 **23.** 35 weeks; 2 episodes left over **25.** 236 gal **27.** 3616 mi **29.** 12,804 gal **31.** \$14,445 **33.** \$247 **35.** **(a)** 4200 sq ft; **(b)** 268 ft **37.** \$29,105,000,000 **39.** 151,500 **41.** 563 packages; 7 bars left over **43.** 384 mi; 27 in. **45.** 21 columns **47.** 56 full cartons; 11 books left over. If 1355 books are shipped, it will take 57 cartons. **49.** 32 \$10 bills **51.** \$400 **53.** 280 min, or 4 hr 40 min **55.** 525 min, or 8 hr 45 min **57.** 106 bones **59.** 3000 sq in. **61.** DW **63.** 234,600 **64.** 234,560 **65.** 235,000 **66.** 22,000 **67.** 16,000 **68.** 8000 **69.** 4000 **70.** 320,000 **71.** 720,000 **72.** 46,800,000 **73.** 792,000 mi; 1,386,000 mi

Margin Exercises, Section 1.6, pp. 67–72

1. 5^4 **2.** 5^5 **3.** 10^2 **4.** 10^4 **5.** 10,000 **6.** 100 **7.** 512 **8.** 32 **9.** 51 **10.** 30 **11.** 584 **12.** 84

13. 4; 1 **14.** 52; 52 **15.** 29 **16.** 1880 **17.** 253 **18.** 93 **19.** 1880 **20.** 305 **21.** 75 **22.** 4 **23.** 1496 ft **24.** 46 **25.** 4

Calculator Corner, p. 68

1. 243 **2.** 15,625 **3.** 20,736 **4.** 2048

Calculator Corner, p. 70

1. 49 **2.** 85 **3.** 36 **4.** 0 **5.** 73 **6.** 49

Exercise Set 1.6, p. 73

1. 3^4 **3.** 5^2 **5.** 7^5 **7.** 10^3 **9.** 49 **11.** 729 **13.** 20,736 **15.** 121 **17.** 22 **19.** 20 **21.** 100 **23.** 1 **25.** 49 **27.** 5 **29.** 434 **31.** 41 **33.** 88 **35.** 4 **37.** 303 **39.** 20 **41.** 70 **43.** 295 **45.** 32 **47.** 906 **49.** 62 **51.** 102 **53.** 32 **55.** $94 **57.** 401 **59.** 110 **61.** 7 **63.** 544 **65.** 708 **67.** 27 **69.** DW **71.** 452 **72.** 835 **73.** 13 **74.** 37 **75.** 2342 **76.** 4898 **77.** 25 **78.** 100 **79.** 104,286 mi^2 **80.** 98 gal **81.** 24; $1 + 5 \cdot (4 + 3) = 36$ **83.** 7; $12 \div (4 + 2) \cdot 3 - 2 = 4$

Margin Exercises, Section 1.7, pp. 77–81

1. Yes **2.** No **3.** 1, 2, 5, 10 **4.** 1, 3, 5, 9, 15, 45 **5.** 1, 2, 31, 62 **6.** 1, 2, 3, 4, 6, 8, 12, 24 **7.** $5 = 1 \cdot 5$; $45 = 9 \cdot 5$; $100 = 20 \cdot 5$ **8.** $10 = 1 \cdot 10$; $60 = 6 \cdot 10$; $110 = 11 \cdot 10$ **9.** 5, 10, 15, 20, 25, 30, 35, 40, 45, 50 **10.** Yes **11.** Yes **12.** No **13.** 2, 13, 19, 41, 73 are prime; 6, 12, 65, 99 are composite; 1 is neither **14.** $2 \cdot 3$ **15.** $2 \cdot 2 \cdot 3$ **16.** $3 \cdot 3 \cdot 5$ **17.** $2 \cdot 7 \cdot 7$ **18.** $2 \cdot 3 \cdot 3 \cdot 7$ **19.** $2 \cdot 2 \cdot 2 \cdot 2 \cdot 3 \cdot 3$ **20.** $2 \cdot 2 \cdot 2 \cdot 5 \cdot 7 \cdot 7$ **21.** $5 \cdot 5 \cdot 7 \cdot 11$

Calculator Corner, p. 78

1. Yes **2.** No **3.** No **4.** Yes **5.** No **6.** Yes **7.** Yes **8.** No **9.** Yes **10.** No

Exercise Set 1.7, p. 82

1. No **3.** Yes **5.** 1, 2, 3, 6, 9, 18 **7.** 1, 2, 3, 6, 9, 18, 27, 54 **9.** 1, 2, 4 **11.** 1 **13.** 1, 2, 7, 14, 49, 98 **15.** 1, 3, 5, 15, 17, 51, 85, 255 **17.** 4, 8, 12, 16, 20, 24, 28, 32, 36, 40 **19.** 20, 40, 60, 80, 100, 120, 140, 160, 180, 200 **21.** 3, 6, 9, 12, 15, 18, 21, 24, 27, 30 **23.** 12, 24, 36, 48, 60, 72, 84, 96, 108, 120 **25.** 10, 20, 30, 40, 50, 60, 70, 80, 90, 100 **27.** 9, 18, 27, 36, 45, 54, 63, 72, 81, 90 **29.** No **31.** Yes **33.** Yes **35.** No **37.** No **39.** Neither **41.** Composite **43.** Prime **45.** Prime **47.** $2 \cdot 2 \cdot 2$ **49.** $2 \cdot 7$ **51.** $2 \cdot 3 \cdot 7$ **53.** $5 \cdot 5$ **55.** $2 \cdot 5 \cdot 5$ **57.** $13 \cdot 13$ **59.** $2 \cdot 2 \cdot 5 \cdot 5$ **61.** $5 \cdot 7$ **63.** $2 \cdot 2 \cdot 2 \cdot 3 \cdot 3$ **65.** $7 \cdot 11$ **67.** $2 \cdot 2 \cdot 7 \cdot 103$ **69.** $3 \cdot 17$ **71.** $2 \cdot 2 \cdot 2 \cdot 2 \cdot 3 \cdot 5 \cdot 5$ **73.** $3 \cdot 7 \cdot 13$ **75.** $2 \cdot 3 \cdot 11 \cdot 17$ **77.** DW **79.** 26 **80.** 256 **81.** 425 **82.** 4200 **83.** 0 **84.** 22 **85.** 1 **86.** 3 **87.** $612 **88.** 201 min, or 3 hr 21 min **89.** Row 1: 48, 90, 432, 63; row 2: 7, 2, 2, 10, 8, 6, 21, 10; row 3: 9, 18, 36, 14, 12, 11, 21; row 4: 29, 19, 42

Margin Exercises, Section 1.8, pp. 84–87

1. Yes **2.** No **3.** Yes **4.** No **5.** Yes **6.** No **7.** Yes **8.** No **9.** Yes **10.** No **11.** No **12.** Yes **13.** No **14.** Yes **15.** No **16.** Yes **17.** No **18.** Yes **19.** No **20.** Yes **21.** Yes **22.** No **23.** No **24.** Yes **25.** Yes **26.** No **27.** No **28.** Yes **29.** No **30.** Yes **31.** Yes **32.** No

Exercise Set 1.8, p. 88

1. 46, 224, 300, 36, 45,270, 4444, 256, 8064, 21,568 **3.** 224, 300, 36, 4444, 256, 8064, 21,568 **5.** 300, 36, 45,270, 8064 **7.** 36, 45,270, 711, 8064 **9.** 324, 42, 501, 3009, 75, 2001, 402, 111,111, 1005 **11.** 55,555, 200, 75, 2345, 35, 1005 **13.** 324 **15.** 200 **17.** 313,332, 7624, 111,126, 876, 1110, 5128, 64,000, 9990 **19.** 313,332, 111,126, 876, 1110, 9990 **21.** 9990 **23.** 1110, 64,000, 9990 **25.** DW **27.** 138 **28.** 139 **29.** 874 **30.** 56 **31.** 26 **32.** 13 **33.** 234 **34.** 4003 **35.** 45 gal **36.** 4320 min **37.** $2 \cdot 2 \cdot 2 \cdot 3 \cdot 5 \cdot 5 \cdot 13$ **39.** $2 \cdot 2 \cdot 3 \cdot 3 \cdot 7 \cdot 11$ **41.** 95,238

Margin Exercises, Section 1.9, pp. 90–93

1. 45 **2.** 40 **3.** 30 **4.** 24 **5.** 10 **6.** 80 **7.** 40 **8.** 360 **9.** 864 **10.** 2520 **11.** 18 **12.** 24 **13.** 36 **14.** 210

Exercise Set 1.9, p. 94

1. 4 **3.** 50 **5.** 40 **7.** 54 **9.** 150 **11.** 120 **13.** 72 **15.** 420 **17.** 144 **19.** 288 **21.** 30 **23.** 105 **25.** 72 **27.** 60 **29.** 36 **31.** 900 **33.** 48 **35.** 50 **37.** 143 **39.** 420 **41.** 378 **43.** 810 **45.** 2160 **47.** 9828 **49.** Every 60 yr **51.** Every 420 yr **53.** DW **55.** $<$ **56.** 704 performances **57.** 33,135 **58.** 6939 **59.** 2 ten thousands + 4 thousands + 6 hundreds + 5 ones **60.** One hundred two thousand, nine hundred sixty **61.** 18,900 **63.** 5 in. by 24 in.

Concept Reinforcement, p. 96

1. False **2.** True **3.** True **4.** True **5.** False **6.** False

Summary and Review: Chapter 1, p. 96

1. 8 thousands **2.** 3 **3.** 2 thousands + 7 hundreds + 9 tens + 3 ones

4. 5 ten thousands + 6 thousands + 0 hundreds + 7 tens + 8 ones, or 5 ten thousands + 6 thousands + 7 tens + 8 ones **5.** 4 millions + 0 hundred thousands + 0 ten thousands + 7 thousands + 1 hundred + 0 tens + 1 one, or 4 millions + 7 thousands + 1 hundred + 1 one
6. Sixty-seven thousand, eight hundred nineteen
7. Two million, seven hundred eighty-one thousand, four hundred twenty-seven **8.** 1 billion, sixty-five million, seventy thousand, six hundred seven **9.** 476,588
10. 2,000,400,000 **11.** 14,272 **12.** 66,024 **13.** 21,788
14. 98,921 **15.** $10 = 6 + 4$, or $10 = 4 + 6$
16. $8 = 11 - 3$; $3 = 11 - 8$ **17.** 5148 **18.** 1689
19. 2274 **20.** 17,757 **21.** 345,800 **22.** 345,760
23. 346,000 **24.** 300,000 **25.** 41,300 + 19,700 = 61,000
26. 38,700 − 24,500 = 14,200 **27.** $400 \cdot 700 = 280{,}000$
28. $>$ **29.** $<$ **30.** 5,100,000 **31.** 6,276,800
32. 506,748 **33.** 27,589 **34.** 5,331,810
35. $56 = 8 \cdot 7$, or $56 = 7 \cdot 8$ **36.** $4 = 52 \div 13$; $13 = 52 \div 4$
37. 12 R 3 **38.** 5 **39.** 913 R 3 **40.** 384 R 1
41. 4 R 46 **42.** 54 **43.** 452 **44.** 4389 **45.** 8
46. 45 **47.** 4^3 **48.** 10,000 **49.** 36 **50.** 65
51. 233 **52.** 56 **53.** 32 **54.** 260 **55.** 165
56. \$502 **57.** \$484 **58.** 1982 **59.** 19 cartons
60. 14 beehives **61.** \$13,585 **62.** \$27,598
63. 137 beakers filled; 13 mL left over **64.** 98 ft^2; 42 ft
65. Prime **66.** Neither **67.** Composite **68.** $2 \cdot 5 \cdot 7$
69. $2 \cdot 3 \cdot 5$ **70.** $3 \cdot 3 \cdot 5$ **71.** $2 \cdot 3 \cdot 5 \cdot 5$
72. $2 \cdot 2 \cdot 2 \cdot 3 \cdot 3 \cdot 3 \cdot 3$ **73.** $2 \cdot 3 \cdot 5 \cdot 5 \cdot 5 \cdot 7$
74. Yes **75.** No **76.** No **77.** Yes **78.** 36
79. 90 **80.** 30 **81.** 1404
82. $\mathbf{D_W}$ $9432 = 9 \cdot 1000 + 4 \cdot 100 + 3 \cdot 10 + 2 \cdot 1 = 9(999 + 1) + 4(99 + 1) + 3(9 + 1) + 2 \cdot 1 = 9 \cdot 999 + 9 \cdot 1 + 4 \cdot 99 + 4 \cdot 1 + 3 \cdot 9 + 3 \cdot 1 + 2 \cdot 1$. Since 999, 99, and 9 are each a multiple of 9, $9 \cdot 999$, $4 \cdot 99$, and $3 \cdot 9$ are multiples of 9. This leaves $9 \cdot 1 + 4 \cdot 1 + 3 \cdot 1 + 2 \cdot 1$, or $9 + 4 + 3 + 2$. If $9 + 4 + 3 + 2$, the sum of the digits, is divisible by 9, then 9432 is divisible by 9.
83. $d = 8$ **84.** $a = 8, b = 4$ **85.** 6 days
86. 13, 11, 101, 37

Test: Chapter 1, p. 99

1. [1.1a] 5 **2.** [1.1b] 8 thousands + 8 hundreds + 4 tens + 3 ones **3.** [1.1c] Thirty-eight million, four hundred three thousand, two hundred seventy-seven
4. [1.2a] 9989 **5.** [1.2a] 63,791 **6.** [1.2a] 34
7. [1.2a] 10,515 **8.** [1.2d] 3630 **9.** [1.2d] 1039
10. [1.2d] 6848 **11.** [1.2d] 5175 **12.** [1.3a] 41,112
13. [1.3a] 5,325,600 **14.** [1.3a] 2405 **15.** [1.3a] 534,264
16. [1.3d] 3 R 3 **17.** [1.3d] 70 **18.** [1.3d] 97
19. [1.3d] 805 R 8 **20.** [1.5a] 1852 12-packs; 7 cakes left over **21.** [1.5a] 1,256,615 mi^2 **22.** **(a)** [1.2b], [1.3b] 300 in., 5000 in^2; 264 in., 3872 in^2; 228 in., 2888 in^2; **(b)** [1.5a] 2112 in^2 **23.** [1.5a] 206,330 voters
24. [1.5a] 1808 lb **25.** [1.5a] 20 staplers **26.** [1.4b] 46
27. [1.4b] 13 **28.** [1.4b] 14 **29.** [1.4b] 381 **30.** [1.3e] 35,000 **31.** [1.3e] 34,570 **32.** [1.3e] 34,600
33. [1.3f] 23,600 + 54,700 = 78,300
34. [1.3f] 54,800 − 23,600 = 31,200
35. [1.3f] $800 \cdot 500 = 400{,}000$ **36.** [1.6a] 12^4
37. [1.6b] 343 **38.** [1.6b] 100,000
39. [1.1d] $>$ **40.** [1.1d] $<$ **41.** [1.6c] 31 **42.** [1.6c] 98
43. [1.6c] 2 **44.** [1.6c] 18 **45.** [1.6d] 216
46. [1.6c] 92 **47.** [1.9a] 48 **48.** [1.9a] 600
49. [1.7c] Prime **50.** [1.7c] Composite
51. [1.7d] $2 \cdot 3 \cdot 3$ **52.** [1.7d] $2 \cdot 2 \cdot 3 \cdot 5$
53. [1.8a] Yes **54.** [1.8a] No **55.** [1.8a] No
56. [1.8a] Yes **57.** [1.3b], [1.5a], [1.6c] 336 in^2
58. [1.4b], [1.6c] 9

CHAPTER 2

Margin Exercises, Section 2.1, pp. 102–110

1. Numerator: 83; denominator: 100
2. Numerator: 27; denominator: 50
3. Numerator: 11; denominator: 25
4. Numerator: 21; denominator: 1000
5. $\frac{1}{2}$ **6.** $\frac{1}{3}$ **7.** $\frac{1}{3}$ **8.** $\frac{2}{3}$ **9.** $\frac{3}{4}$ **10.** $\frac{15}{16}$ **11.** $\frac{5}{4}$
12. $\frac{7}{4}$ **13.** 1 **14.** 1 **15.** 1 **16.** 1 **17.** 1 **18.** 1
19. 0 **20.** 0 **21.** 0 **22.** 0 **23.** Not defined
24. Not defined **25.** 8 **26.** 10 **27.** 346 **28.** 23
29. **30.** $\frac{15}{56}$ **31.** $\frac{32}{15}$ **32.** $\frac{3}{100}$ **33.** $\frac{14}{3}$
34. $\frac{8}{16}$ **35.** $\frac{30}{50}$ **36.** $\frac{52}{100}$ **37.** $\frac{200}{75}$ **38.** $\frac{12}{9}$ **39.** $\frac{18}{24}$
40. $\frac{90}{100}$ **41.** $\frac{9}{45}$ **42.** $\frac{56}{49}$ **43.** $\frac{1}{4}$ **44.** $\frac{5}{6}$ **45.** 5 **46.** $\frac{4}{3}$
47. $\frac{7}{8}$ **48.** $\frac{89}{78}$ **49.** $\frac{8}{7}$ **50.** $\frac{1}{4}$ **51.** $\frac{7}{24}$ **52.** $\frac{37}{80}$
53. $\frac{20}{100} = \frac{1}{5}$; $\frac{42}{100} = \frac{21}{50}$; $\frac{8}{100} = \frac{2}{25}$; $\frac{16}{100} = \frac{4}{25}$; $\frac{14}{100} = \frac{7}{50}$

Calculator Corner, p. 110

1. $\frac{14}{15}$ **2.** $\frac{7}{8}$ **3.** $\frac{138}{167}$ **4.** $\frac{7}{25}$

Exercise Set 2.1, p. 111

1. Numerator: 3; denominator: 4
3. Numerator: 11; denominator: 2
5. Numerator: 0; denominator: 7
7. $\frac{2}{4}$ **9.** $\frac{1}{8}$ **11.** $\frac{4}{3}$ **13.** $\frac{12}{16}$ **15.** $\frac{3}{4}$ **17.** $\frac{12}{12}$ **19.** $\frac{4}{8}$
21. $\frac{6}{12}$ **23.** **(a)** $\frac{2}{8}$; **(b)** $\frac{6}{8}$ **25.** **(a)** $\frac{3}{8}$; **(b)** $\frac{5}{8}$ **27.** Not defined
29. 1 **31.** 0 **33.** 403 **35.** Not defined **37.** 1
39. 0 **41.** $\frac{1}{6}$ **43.** $\frac{5}{8}$ **45.** $\frac{2}{15}$ **47.** $\frac{4}{15}$ **49.** $\frac{9}{16}$
51. $\frac{14}{39}$ **53.** $\frac{21}{4}$ **55.** $\frac{49}{64}$ **57.** $\frac{5}{10}$ **59.** $\frac{20}{32}$ **61.** $\frac{75}{45}$
63. $\frac{42}{132}$ **65.** $\frac{3}{4}$ **67.** $\frac{1}{5}$ **69.** 3 **71.** $\frac{3}{4}$ **73.** $\frac{7}{8}$ **75.** 6
77. $\frac{1}{3}$ **79.** $\mathbf{D_W}$ **81.** $\frac{1}{6}$ **83.** $\frac{2}{16}$, or $\frac{1}{8}$

Margin Exercises, Section 2.2, pp. 115–118

1. $\frac{7}{12}$ **2.** $\frac{1}{3}$ **3.** 6 **4.** $\frac{5}{2}$ **5.** $\frac{5}{2}$ **6.** $\frac{7}{10}$ **7.** $\frac{1}{9}$ **8.** 5
9. $\frac{8}{7}$ **10.** $\frac{8}{3}$ **11.** $\frac{1}{10}$ **12.** 100 **13.** 1 **14.** $\frac{14}{15}$ **15.** $\frac{4}{5}$
16. 32

Exercise Set 2.2, p. 119

1. $\frac{1}{3}$ **3.** $\frac{1}{6}$ **5.** $\frac{27}{10}$ **7.** $\frac{14}{9}$ **9.** 1 **11.** 1 **13.** 4 **15.** 9 **17.** $\frac{98}{5}$ **19.** 30 **21.** $\frac{1}{5}$ **23.** $\frac{9}{25}$ **25.** $\frac{11}{40}$ **27.** $\frac{5}{14}$ **29.** $\frac{6}{5}$ **31.** $\frac{1}{6}$ **33.** 6 **35.** $\frac{3}{10}$ **37.** $\frac{4}{5}$ **39.** $\frac{4}{15}$ **41.** 4 **43.** 2 **45.** $\frac{1}{8}$ **47.** $\frac{3}{7}$ **49.** 8 **51.** 35 **53.** 1 **55.** $\frac{2}{3}$ **57.** $\frac{9}{4}$ **59.** 144 **61.** 75 **63.** 2 **65.** $\frac{3}{5}$ **67.** 315 **69.** **DW** **71.** 67 **72.** 33 R 4 **73.** 285 R 2 **74.** 103 R 10 **75.** 67 **76.** 264 **77.** 8499 **78.** 4368 **79.** $\frac{3}{8}$

Margin Exercises, Section 2.3, pp. 121–126

1. $\frac{4}{5}$ **2.** 1 **3.** $\frac{1}{2}$ **4.** $\frac{3}{4}$ **5.** $\frac{5}{6}$ **6.** $\frac{29}{24}$ **7.** $\frac{5}{9}$ **8.** $\frac{413}{1000}$ **9.** $\frac{197}{210}$ **10.** $\frac{65}{72}$ **11.** $\frac{1}{2}$ **12.** $\frac{3}{8}$ **13.** $\frac{1}{2}$ **14.** $\frac{13}{18}$ **15.** $\frac{1}{2}$ **16.** $\frac{9}{112}$ **17.** $<$ **18.** $>$ **19.** $>$ **20.** $>$ **21.** $<$ **22.** $\frac{1}{6}$ **23.** $\frac{11}{40}$

Exercise Set 2.3, p. 127

1. 1 **3.** $\frac{3}{4}$ **5.** $\frac{3}{2}$ **7.** $\frac{7}{24}$ **9.** $\frac{3}{2}$ **11.** $\frac{19}{24}$ **13.** $\frac{9}{10}$ **15.** $\frac{29}{18}$ **17.** $\frac{31}{100}$ **19.** $\frac{41}{60}$ **21.** $\frac{189}{100}$ **23.** $\frac{7}{8}$ **25.** $\frac{13}{24}$ **27.** $\frac{17}{24}$ **29.** $\frac{3}{4}$ **31.** $\frac{437}{500}$ **33.** $\frac{53}{40}$ **35.** $\frac{391}{144}$ **37.** $\frac{2}{3}$ **39.** $\frac{3}{4}$ **41.** $\frac{5}{8}$ **43.** $\frac{1}{24}$ **45.** $\frac{1}{2}$ **47.** $\frac{9}{14}$ **49.** $\frac{3}{5}$ **51.** $\frac{7}{10}$ **53.** $\frac{17}{60}$ **55.** $\frac{53}{100}$ **57.** $\frac{26}{75}$ **59.** $\frac{9}{100}$ **61.** $\frac{13}{24}$ **63.** $\frac{1}{10}$ **65.** $\frac{1}{24}$ **67.** $\frac{13}{16}$ **69.** $\frac{31}{75}$ **71.** $\frac{13}{75}$ **73.** $<$ **75.** $>$ **77.** $<$ **79.** $<$ **81.** $>$ **83.** $>$ **85.** $<$ **87.** $\frac{1}{15}$ **89.** $\frac{2}{15}$ **91.** $\frac{1}{2}$ **93.** **DW** **95.** 537,179 popular votes **96.** 5 electoral votes **97.** 84 electoral votes **98.** 510,314 popular votes **99.** 21,468,755 votes **100.** 33,112,603 votes **101.** 1 **102.** Not defined **103.** Not defined **104.** 4 **105.** $\frac{19}{24}$ **107.** $\frac{145}{144}$

Margin Exercises, Section 2.4, pp. 130–136

1. $1\frac{2}{3}$ **2.** $2\frac{3}{4}$ **3.** $8\frac{3}{4}$ **4.** $12\frac{2}{3}$ **5.** $\frac{22}{5}$ **6.** $\frac{61}{10}$ **7.** $\frac{29}{6}$ **8.** $\frac{37}{4}$ **9.** $\frac{62}{3}$ **10.** $2\frac{1}{3}$ **11.** $1\frac{1}{10}$ **12.** $18\frac{1}{3}$ **13.** $7\frac{2}{5}$ **14.** $12\frac{1}{10}$ **15.** $13\frac{7}{12}$ **16.** $1\frac{1}{2}$ **17.** $3\frac{1}{6}$ **18.** $3\frac{5}{18}$ **19.** $3\frac{2}{3}$ **20.** 20 **21.** $1\frac{7}{8}$ **22.** $12\frac{4}{5}$ **23.** $8\frac{1}{3}$ **24.** 16 **25.** $7\frac{3}{7}$ **26.** $1\frac{7}{8}$ **27.** $\frac{7}{10}$

Calculator Corner, p. 136

1. $\frac{7}{12}$ **2.** $\frac{11}{10}$ **3.** $\frac{35}{16}$ **4.** $\frac{3}{10}$ **5.** $10\frac{2}{15}$ **6.** $1\frac{1}{28}$ **7.** $10\frac{11}{15}$ **8.** $2\frac{91}{115}$

Exercise Set 2.4, p. 137

1. $\frac{57}{4}, \frac{27}{4}, \frac{9}{4}$ **3.** $\frac{17}{3}$ **5.** $\frac{59}{6}$ **7.** $\frac{51}{4}$ **9.** $3\frac{3}{5}$ **11.** $5\frac{7}{10}$ **13.** $43\frac{1}{8}$ **15.** $6\frac{1}{2}$ **17.** $2\frac{11}{12}$ **19.** $14\frac{7}{12}$ **21.** $21\frac{1}{2}$ **23.** $27\frac{7}{8}$ **25.** $27\frac{13}{24}$ **27.** $1\frac{3}{5}$ **29.** $4\frac{1}{10}$ **31.** $15\frac{3}{8}$ **33.** $7\frac{5}{12}$ **35.** $13\frac{3}{8}$ **37.** $11\frac{5}{18}$ **39.** $22\frac{2}{3}$ **41.** $2\frac{5}{12}$ **43.** $8\frac{1}{6}$ **45.** $9\frac{31}{40}$ **47.** $24\frac{91}{100}$ **49.** $975\frac{4}{5}$ **51.** $6\frac{1}{4}$ **53.** $1\frac{1}{5}$ **55.** $3\frac{9}{16}$ **57.** $1\frac{1}{8}$ **59.** $1\frac{8}{43}$ **61.** $\frac{9}{40}$ **63.** **DW** **65.** 45,800 **66.** 45,770 **67.** Yes **68.** No **69.** No **70.** Yes **71.** No **72.** Yes **73.** Yes **74.** Yes **75.** 47 **76.** 889 **77.** 25 **78.** 477,978 **79.** $360\frac{60}{473}$ **81.** $35\frac{57}{64}$

Margin Exercises, Section 2.5, pp. 141–149

1. $\frac{3}{8}$ **2.** $\frac{8}{81}$ ft^2 **3.** 14 lb **4.** 320 loops **5.** 252 mi **6.** $\frac{9}{10}$ mi **7.** $\frac{5}{24}$ mi **8.** $232\frac{3}{20}$ mi **9.** $4\frac{5}{6}$ ft **10.** $354\frac{23}{24}$ gal **11.** $227\frac{1}{2}$ mi **12.** 20 mpg **13.** $240\frac{3}{4}$ ft^2

Translating for Success, p. 150

1. O **2.** K **3.** F **4.** D **5.** H **6.** G **7.** L **8.** E **9.** M **10.** J

Exercise Set 2.5, p. 151

1. $\frac{1}{12}$ **3.** $\frac{7}{16}$ L **5.** 24 bowls **7.** $\frac{5}{8}$ in. **9.** $\frac{21}{32}$ **11.** 30 mph **13.** \$115,500 **15.** Food: \$9000; housing: \$7200; clothing: \$3600; savings: \$4000; taxes: \$9000; other expenses: \$3200 **17.** $\frac{13}{12}$ lb **19.** 75 times **21.** 32 pairs **23.** 16 L **25.** 288 km; 108 km **27.** $\frac{5}{8}$" **29.** $\frac{1}{4}$ of the business **31.** $\frac{5}{6}$ lb **33.** $\frac{23}{12}$ mi **35.** 690 kg; $\frac{14}{23}$ cement; $\frac{5}{23}$ stone; $\frac{4}{23}$ sand; 1 **37.** $\frac{5}{12}$ hr **39.** $\frac{19}{24}$ cup **41.** $20\frac{1}{8}$ in. **43.** $5\frac{3}{8}$ yd **45.** $6\frac{5}{12}$ in. **47.** $28\frac{3}{4}$ yd **49.** $2\frac{7}{8}$ in., $4\frac{7}{8}$ in. **51.** $1\frac{9}{16}$ in. **53.** 15 mpg **55.** 68°F **57.** 45,000 beagles **59.** About 4,500,000 **61.** $62\frac{1}{2}$ ft^2 **63.** $13\frac{1}{3}$ tsp **65.** 4 cu ft **67.** $35\frac{115}{256}$ sq in. **69.** $59{,}538\frac{1}{8}$ sq ft

71. *Cake*: $1\frac{1}{8}$ cups all-purpose flour, $\frac{3}{8}$ cup sugar, $\frac{3}{8}$ cup cold butter, $\frac{3}{8}$ cup sour cream, $\frac{1}{4}$ teaspoon baking powder, $\frac{1}{4}$ teaspoon baking soda, $\frac{1}{2}$ egg, $\frac{1}{2}$ teaspoon almond extract; *filling*: $\frac{1}{2}$ package (4 ounces) cream cheese, $\frac{1}{8}$ cup sugar, $\frac{1}{2}$ egg, $\frac{3}{8}$ cup peach preserves, $\frac{1}{4}$ cup sliced almonds; *cake*: 9 cups all-purpose flour, 3 cups sugar, 3 cups cold butter, 3 cups sour cream, 2 teaspoons baking powder, 2 teaspoons baking soda, 4 eggs, 4 teaspoons almond extract; *filling*: 4 packages (8 ounces each) cream cheese, 1 cup sugar, 4 eggs, 3 cups peach preserves, 2 cups sliced almonds **73.** **DW** **75.** Divisor; quotient; dividend **76.** Common **77.** Composite **78.** Divisible; divisible **79.** Multiplications; divisions; additions; subtractions **80.** Addends **81.** Numerator **82.** Reciprocal **83.** $\frac{4}{15}$; \$320

Margin Exercises, Section 2.6, pp. 159–162

1. $\frac{1}{2}$ **2.** $\frac{3}{10}$ **3.** $20\frac{2}{3}$, or $\frac{62}{3}$ **4.** $9\frac{7}{8}$ in. **5.** $\frac{5}{9}$ **6.** $\frac{31}{40}$ **7.** $\frac{27}{56}$ **8.** 0 **9.** 1 **10.** $\frac{1}{2}$ **11.** 1 **12.** 12; answers may vary **13.** 32; answers may vary **14.** 27; answers may vary **15.** 15; answers may vary **16.** 1; answers may vary **17.** 1,000,000; answers may vary **18.** $22\frac{1}{2}$ **19.** 132 **20.** 37

Exercise Set 2.6, p. 163

1. $\frac{1}{24}$ **3.** $\frac{2}{5}$ **5.** $\frac{4}{7}$ **7.** $\frac{59}{30}$, or $1\frac{29}{30}$ **9.** $\frac{3}{20}$ **11.** $\frac{211}{8}$, or $26\frac{3}{8}$

13. $\frac{7}{16}$ **15.** $\frac{1}{36}$ **17.** $\frac{3}{8}$ **19.** $\frac{37}{48}$ **21.** $\frac{25}{72}$ **23.** $\frac{103}{16}$, or $6\frac{7}{16}$
25. $2\frac{41}{128}$ lb **27.** $7\frac{23}{50}$ sec **29.** $\frac{17}{6}$, or $2\frac{5}{6}$ **31.** $\frac{8395}{84}$, or $99\frac{79}{84}$
33. 0 **35.** 0 **37.** $\frac{1}{2}$ **39.** $\frac{1}{2}$ **41.** 0 **43.** 1 **45.** 6
47. 12 **49.** 19 **51.** 6 **53.** 12 **55.** 16 **57.** 3
59. 13 **61.** 2 **63.** $1\frac{1}{2}$ **65.** $\frac{1}{2}$ **67.** $271\frac{1}{2}$ **69.** 3
71. 100 **73.** $29\frac{1}{2}$ **75.** DW **77.** 3402 **78.** 1,038,180
79. 59 R 77 **80.** 348 **81.** 783 **82.** $\frac{8}{3}$ **83.** $\frac{3}{8}$
84. Prime: 5, 7, 23, 43; composite: 9, 14; neither: 1
85. 16 people **86.** 43 mg
87. **(a)** $13 \cdot 9\frac{1}{4} + 8\frac{1}{4} \cdot 7\frac{1}{4}$; **(b)** $\frac{2881}{16}$, or $180\frac{1}{16}$ in^2;
(c) Multiply before adding. **89.** $a = 2, b = 8$
91. The largest is $\frac{4}{3} + \frac{5}{2} = \frac{23}{6}$.

Concept Reinforcement, p. 167

1. True **2.** False **3.** True **4.** True **5.** False

Summary and Review: Chapter 2, p. 167

1. Numerator: 2; denominator: 7 **2.** $\frac{3}{5}$ **3.** $\frac{7}{6}$ **4.** 1
5. 0 **6.** $\frac{2}{5}$ **7.** 18 **8.** 4 **9.** $\frac{1}{3}$
10. Not defined **11.** Not defined **12.** $\frac{11}{23}$ **13.** $\frac{2}{7}$
14. $\frac{39}{40}$ **15.** $\frac{32}{225}$ **16.** $\frac{3}{100}$; $\frac{8}{100} = \frac{2}{25}$; $\frac{10}{100} = \frac{1}{10}$; $\frac{15}{100} = \frac{3}{20}$; $\frac{21}{100}$; $\frac{43}{100}$
17. $\frac{3}{2}$ **18.** 56 **19.** $\frac{5}{2}$ **20.** 24 **21.** $\frac{2}{3}$ **22.** $\frac{1}{14}$ **23.** $\frac{2}{3}$
24. $\frac{1}{22}$ **25.** $\frac{3}{20}$ **26.** $\frac{10}{7}$ **27.** $\frac{5}{4}$ **28.** $\frac{1}{3}$ **29.** 9 **30.** $\frac{36}{47}$
31. $\frac{9}{2}$ **32.** 2 **33.** $\frac{11}{6}$ **34.** $\frac{1}{4}$ **35.** $\frac{9}{4}$ **36.** 300
37. 1 **38.** $\frac{4}{9}$ **39.** $\frac{63}{40}$ **40.** $\frac{19}{48}$ **41.** $\frac{25}{12}$ **42.** $\frac{891}{1000}$
43. $\frac{1}{3}$ **44.** $\frac{1}{8}$ **45.** $\frac{5}{27}$ **46.** $\frac{11}{18}$ **47.** $>$ **48.** $>$ **49.** $\frac{15}{2}$
50. $\frac{67}{8}$ **51.** $\frac{13}{3}$ **52.** $\frac{75}{7}$ **53.** $2\frac{1}{3}$ **54.** $6\frac{3}{4}$ **55.** $12\frac{3}{5}$
56. $3\frac{1}{2}$ **57.** $\frac{3}{10}$ **58.** 240 **59.** $\frac{19}{40}$ **60.** $\frac{2}{5}$ **61.** $10\frac{2}{5}$
62. $11\frac{11}{15}$ **63.** $10\frac{2}{3}$ **64.** $8\frac{1}{4}$ **65.** $7\frac{7}{9}$ **66.** $4\frac{11}{15}$
67. $4\frac{3}{20}$ **68.** $13\frac{3}{8}$ **69.** 16 **70.** $3\frac{1}{2}$ **71.** $2\frac{21}{50}$ **72.** 6
73. 12 **74.** $1\frac{7}{17}$ **75.** $\frac{1}{8}$ **76.** $\frac{9}{10}$ **77.** 9 days
78. 1000 km **79.** $\frac{1}{3}$ cup; 2 cups **80.** $4\frac{1}{4}$ yd
81. 256,000,000 metric tons **82.** $177\frac{3}{4}$ in^2 **83.** $50\frac{1}{4}$ in^2
84. $1\frac{73}{100}$ in. **85.** 24 lb **86.** About $69\frac{3}{8}$ kg **87.** $8\frac{3}{8}$ cups
88. $63\frac{2}{3}$ pies; $19\frac{1}{3}$ pies **89.** 1 **90.** $\frac{7}{40}$ **91.** 3 **92.** $\frac{77}{240}$
93. $\frac{1}{2}$ **94.** 0 **95.** 1 **96.** 7 **97.** 10 **98.** $2\frac{1}{2}$
99. $28\frac{1}{2}$
100. DW Taking $\frac{1}{2}$ of a number is equivalent to multiplying the number by $\frac{1}{2}$. Dividing by $\frac{1}{2}$ is equivalent to multiplying by the reciprocal of $\frac{1}{2}$, or 2. Thus taking $\frac{1}{2}$ of a number is not the same as dividing by $\frac{1}{2}$.
101. DW It might be necessary to find the least common denominator before adding or subtracting. The least common denominator is the least common multiple of the denominators.
102. $a = 11{,}176$; $b = 9887$ **103.** $\frac{6}{3} + \frac{5}{4} = 3\frac{1}{4}$

Test: Chapter 2, p. 171

1. [2.1a] $\frac{3}{4}$ **2.** [2.3c] $>$ **3.** [2.2b] 1 **4.** [2.2b] 0
5. [2.2e] $\frac{1}{14}$ **6.** [2.2b] Not defined **7.** [2.2a] 32
8. [2.2a] $\frac{5}{2}$ **9.** [2.2a] $\frac{1}{10}$ **10.** [2.2a] $\frac{2}{9}$ **11.** [2.2b] $\frac{8}{5}$
12. [2.2b] 4 **13.** [2.2b] $\frac{1}{18}$ **14.** [2.2c] $\frac{3}{10}$ **15.** [2.2c] $\frac{8}{5}$
16. [2.2c] 18 **17.** [2.3a] 3 **18.** [2.3a] $\frac{37}{24}$ **19.** [2.3a] $\frac{921}{1000}$
20. [2.3b] $\frac{1}{3}$ **21.** [2.3b] $\frac{1}{12}$ **22.** [2.3b] $\frac{77}{120}$ **23.** [2.2d] 64
24. [2.3d] $\frac{1}{4}$ **25.** [2.4a] $\frac{7}{2}$ **26.** [2.4a] $8\frac{2}{9}$ **27.** [2.4b] $14\frac{1}{5}$
28. [2.4c] $4\frac{7}{24}$ **29.** [2.4d] $4\frac{1}{2}$ **30.** [2.4e] 2
31. [2.5a] 4375 students **32.** [2.5a] $\frac{3}{40}$ m
33. [2.5a] About 105 kg **34.** [2.5a] 80 books
35. [2.5a] **(a)** 3 in.; **(b)** $4\frac{1}{2}$ in. **36.** [2.5a] $6\frac{11}{36}$ ft
37. [2.6a] $\frac{3}{4}$ **38.** [2.6b] 0 **39.** [2.6b] 1 **40.** [2.6b] $18\frac{1}{2}$
41. [2.6b] 16 **42.** [2.5a] Rebecca walks $\frac{17}{56}$ mi farther.
43. [2.5a] $\frac{7}{48}$ acre

CHAPTER 3

Margin Exercises, Section 3.1, pp. 175–180

1. Eighty and thirty-one hundredths; seventy-seven and forty-three hundredths **2.** Two and six thousand seven hundred sixty-seven hundred-thousandths
3. Two hundred forty-five and eighty-nine hundredths
4. Thirty-four and sixty-four ten-thousandths
5. Thirty-one thousand, seventy-nine and seven hundred sixty-four thousandths **6.** $\frac{896}{1000}$ **7.** $\frac{2378}{100}$ **8.** $\frac{56{,}789}{10{,}000}$ **9.** $\frac{19}{10}$
10. 7.43 **11.** 0.406 **12.** 6.7089 **13.** 0.9 **14.** 0.057
15. 0.083 **16.** 4.3 **17.** 283.71 **18.** 456.013
19. 2.04 **20.** 0.06 **21.** 0.58 **22.** 1 **23.** 0.8989
24. 21.05 **25.** 2.8 **26.** 13.9 **27.** 234.4 **28.** 7.0
29. 0.64 **30.** 7.83 **31.** 34.68 **32.** 0.03 **33.** 0.943
34. 8.004 **35.** 43.112 **36.** 37.401 **37.** 7459.355
38. 7459.35 **39.** 7459.4 **40.** 7459 **41.** 7460
42. 7500 **43.** 7000

Exercise Set 3.1, p. 181

1. Sixty-three and five hundredths **3.** Twenty-six and fifty-nine hundredths **5.** Eight and thirty-five hundredths
7. Eighty-six and eighty-nine hundredths **9.** Thirty-four and eight hundred ninety-one thousandths
11. $\frac{83}{10}$ **13.** $\frac{356}{100}$ **15.** $\frac{4603}{100}$ **17.** $\frac{13}{100{,}000}$ **19.** $\frac{10{,}008}{10{,}000}$
21. $\frac{20{,}003}{1000}$ **23.** 0.8 **25.** 8.89 **27.** 3.798 **29.** 0.0078
31. 0.00019 **33.** 0.376193 **35.** 99.44 **37.** 3.798
39. 2.1739 **41.** 8.953073 **43.** 0.58 **45.** 0.91
47. 0.001 **49.** 235.07 **51.** $\frac{4}{100}$ **53.** 0.4325 **55.** 0.1
57. 0.5 **59.** 2.7 **61.** 123.7 **63.** 0.89 **65.** 0.67
67. 1.00 **69.** 0.09 **71.** 0.325 **73.** 17.002
75. 10.101 **77.** 9.999 **79.** 800 **81.** 809.473
83. 809 **85.** 34.5439 **87.** 34.54 **89.** 35 **91.** DW
93. 6170 **94.** 6200 **95.** 6000 **96.** 54 **97.** $6\frac{3}{5}$
98. $2 \cdot 2 \cdot 2 \cdot 2 \cdot 5 \cdot 5 \cdot 5$, or $2^4 \cdot 5^3$
99. $2 \cdot 3 \cdot 3 \cdot 5 \cdot 17$, or $2 \cdot 3^2 \cdot 5 \cdot 17$ **100.** $2 \cdot 7 \cdot 11 \cdot 13$
101. $2 \cdot 2 \cdot 2 \cdot 7 \cdot 7 \cdot 11$, or $2^3 \cdot 7^2 \cdot 11$
103. 2.000001, 2.0119, 2.018, 2.0302, 2.1, 2.108, 2.109
105. 6.78346 **107.** 0.03030

Margin Exercises, Section 3.2, pp. 184–188

1. 10.917 **2.** 34.2079 **3.** 4.969 **4.** 3.5617 **5.** 9.40544 **6.** 912.67 **7.** 2514.773 **8.** 10.754 **9.** 0.339 **10.** 0.5345 **11.** 0.5172 **12.** 7.36992 **13.** 1194.22 **14.** 4.9911 **15.** 38.534 **16.** 14.164 **17.** 2133.5
18. The "balance forward" column should read:
$3078.92
2738.23
2659.67
2890.47
2877.33
2829.33
2868.91
2766.04
2697.45
2597.45

Calculator Corner, p. 186

1. 317.645 **2.** 506.553 **3.** 17.15 **4.** 49.08 **5.** 4.4 **6.** 33.83 **7.** 454.74 **8.** 0.99

Exercise Set 3.2, p. 189

1. 334.37 **3.** 1576.215 **5.** 132.560 **7.** 50.0248 **9.** 40.007 **11.** 977.955 **13.** 771.967 **15.** 8754.8221 **17.** 49.02 **19.** 85.921 **21.** 2.4975 **23.** 3.397 **25.** 8.85 **27.** 3.37 **29.** 1.045 **31.** 3.703 **33.** 0.9902 **35.** 99.66 **37.** 4.88 **39.** 0.994 **41.** 17.802 **43.** 51.13 **45.** 32.7386 **47.** 4.0622 **49.** 11.65 **51.** 384.68 **53.** 582.97 **55.** 15,335.3
57. The balance forward should read:
$ 9704.56
9677.12
10,677.12
10,553.17
10,429.15
10,416.72
12,916.72
12,778.94
12,797.82
9997.82
59. D$_W$ **61.** 2720 **62.** $2 \cdot 2 \cdot 3 \cdot 19$ **63.** $\frac{1}{6}$ **64.** $\frac{34}{45}$ **65.** 6166 **66.** 5366 **67.** $16\frac{1}{2}$ servings **68.** $60\frac{1}{5}$ mi **69.** 345.8

Margin Exercises, Section 3.3, pp. 193–196

1. 529.48 **2.** 5.0594 **3.** 34.2906 **4.** 0.348 **5.** 0.0348 **6.** 0.00348 **7.** 0.000348 **8.** 34.8 **9.** 348 **10.** 3480 **11.** 34,800 **12.** 3,700,000 **13.** 6,100,000,000 **14.** 2,700,000,000 **15.** 1569¢ **16.** 17¢ **17.** $0.35 **18.** $5.77

Calculator Corner, p. 195

1. 48.6 **2.** 6930.5 **3.** 142.803 **4.** 0.5076 **5.** 7916.4 **6.** 20.4153

Exercise Set 3.3, p. 197

1. 60.2 **3.** 6.72 **5.** 0.252 **7.** 0.522 **9.** 237.6 **11.** 583,686.852 **13.** 780 **15.** 8.923 **17.** 0.09768 **19.** 0.782 **21.** 521.6 **23.** 3.2472 **25.** 897.6 **27.** 322.07 **29.** 55.68 **31.** 3487.5 **33.** 50.0004 **35.** 114.42902 **37.** 13.284 **39.** 90.72 **41.** 0.0028728 **43.** 0.72523 **45.** 1.872115 **47.** 45,678 **49.** 2888¢ **51.** 66¢ **53.** $0.34 **55.** $34.45 **57.** 258,700,000,000 **59.** 748,900,000 **61.** D$_W$ **63.** $11\frac{1}{5}$ **64.** $\frac{35}{72}$ **65.** $2\frac{7}{15}$ **66.** $7\frac{2}{15}$ **67.** 342 **68.** 87 **69.** 4566 **70.** 1257 **71.** 87 **72.** 1176 R 14 **73.** 10^{21} = 1 sextillion **75.** 10^{24} = 1 septillion

Margin Exercises, Section 3.4, pp. 200–206

1. 0.6 **2.** 1.5 **3.** 0.47 **4.** 0.32 **5.** 3.75 **6.** 0.25 **7.** **(a)** 375; **(b)** 15 **8.** 4.9 **9.** 12.8 **10.** 15.625 **11.** 12.78 **12.** 0.001278 **13.** 0.09847 **14.** 67.832 **15.** 0.78314 **16.** 1105.6 **17.** 0.04 **18.** 0.2426 **19.** 593.44 **20.** 5967.5 m

Calculator Corner, p. 203

1. 14.3 **2.** 2.56 **3.** 200 **4.** 0.75 **5.** 20 **6.** 0.064 **7.** 15.7 **8.** 75.8

Exercise Set 3.4, p. 207

1. 2.99 **3.** 23.78 **5.** 7.48 **7.** 7.2 **9.** 1.143 **11.** 4.041 **13.** 0.07 **15.** 70 **17.** 20 **19.** 0.4 **21.** 0.41 **23.** 8.5 **25.** 9.3 **27.** 0.625 **29.** 0.26 **31.** 15.625 **33.** 2.34 **35.** 0.47 **37.** 0.2134567 **39.** 21.34567 **41.** 1023.7 **43.** 9.3 **45.** 0.0090678 **47.** 45.6 **49.** 2107 **51.** 303.003 **53.** 446.208 **55.** 24.14 **57.** 13.0072 **59.** 19.3204 **61.** 473.188278 **63.** 10.49 **65.** 911.13 **67.** 205 **69.** $1288.36 **71.** $206.34 billion **73.** D$_W$ **75.** $\frac{6}{7}$ **76.** $\frac{7}{8}$ **77.** $\frac{19}{73}$ **78.** $\frac{19}{73}$ **79.** $2 \cdot 2 \cdot 3 \cdot 3 \cdot 19$, or $2^2 \cdot 3^2 \cdot 19$ **80.** $2 \cdot 3 \cdot 3 \cdot 3 \cdot 3$, or $2 \cdot 3^4$ **81.** $3 \cdot 3 \cdot 223$, or $3^2 \cdot 223$ **82.** $5 \cdot 401$ **83.** $15\frac{1}{8}$ **84.** $5\frac{7}{8}$ **85.** 6.254194585 **87.** 1000 **89.** 100

Margin Exercises, Section 3.5, pp. 211–215

1. 0.8 **2.** 0.45 **3.** 0.275 **4.** 1.32 **5.** 0.4 **6.** 0.375 **7.** $0.1\overline{6}$ **8.** $0.\overline{6}$ **9.** $0.\overline{45}$ **10.** $1.\overline{09}$ **11.** $0.\overline{428571}$ **12.** 0.7; 0.67; 0.667 **13.** 0.8; 0.81; 0.808 **14.** 6.2; 6.25; 6.245 **15.** 0.510 **16.** 24.2 mpg **17.** 12.1 million digital cameras **18.** 0.72 **19.** 0.552 **20.** 9.6575

Exercise Set 3.5, p. 216

1. 0.23 **3.** 0.6 **5.** 0.325 **7.** 0.2 **9.** 0.85 **11.** 0.375 **13.** 0.975 **15.** 0.52 **17.** 20.016 **19.** 0.25 **21.** 1.16 **23.** 1.1875 **25.** $0.2\overline{6}$ **27.** $0.\overline{3}$ **29.** $1.\overline{3}$ **31.** $1.1\overline{6}$ **33.** $0.\overline{571428}$ **35.** $0.91\overline{6}$ **37.** 0.3; 0.27; 0.267 **39.** 0.3; 0.33; 0.333 **41.** 1.3; 1.33; 1.333 **43.** 1.2; 1.17;

1.167 **45.** 0.6; 0.57; 0.571 **47.** 0.9; 0.92; 0.917 **49.** 0.2; 0.18; 0.182 **51.** 0.3; 0.28; 0.278 **53.** **(a)** 0.429; **(b)** 0.75; **(c)** 0.571; **(d)** 1.333 **55.** 15.8 mpg **57.** 17.8 mpg **59.** 15.2 mph **61.** $24.5625; $24.56 **63.** $3.734375; $3.73 **65.** $59.875; $59.88 **67.** 11.06 **69.** 8.4 **71.** $417.51\overline{6}$ **73.** 0 **75.** 2.8125 **77.** 0.20425 **79.** 317.14 **81.** 0.1825 **83.** 18 **85.** 2.736 **87.** DW **89.** 21 **90.** $238\frac{7}{8}$ **91.** 10 **92.** $\frac{43}{52}$ **93.** $50\frac{5}{24}$ **94.** $30\frac{7}{10}$ **95.** $1\frac{1}{2}$ **96.** $14\frac{13}{24}$ **97.** $1\frac{1}{24}$ cups **98.** $1\frac{33}{100}$ in. **99.** 270 **100.** 792 **101.** $0.\overline{142857}$ **103.** $0.\overline{428571}$ **105.** $0.\overline{714285}$ **107.** $0.\overline{1}$ **109.** $0.\overline{001}$

Margin Exercises, Section 3.6, pp. 220–222

1. (b) **2.** (a) **3.** (d) **4.** (b) **5.** (a) **6.** (d) **7.** (b) **8.** (c) **9.** (b) **10.** (b) **11.** (c) **12.** (a) **13.** (c) **14.** (c)

Exercise Set 3.6, p. 223

1. (d) **3.** (c) **5.** (a) **7.** (c) **9.** 1.6 **11.** 6 **13.** 60 **15.** 2.3 **17.** 180 **19.** (a) **21.** (c) **23.** (b) **25.** (b) **27.** 1800 ÷ 9 = 200 posts; answers may vary **29.** DW **31.** Repeating **32.** Multiple **33.** Factors **34.** Solution **35.** Multiplicative **36.** Reciprocals **37.** Denominator; multiple **38.** Divisible; divisible **39.** Yes **41.** No **43.** **(a)** +, ×; **(b)** +, ×, −

Margin Exercises, Section 3.7, pp. 226–232

1. $2.5 billion **2.** 199.1 lb **3.** $51.26 **4.** $368.75 **5.** 96.52 cm^2 **6.** $1.33 **7.** 28.6 mpg **8.** $716,667 **9.** $158,760

Translating for Success, p. 233

1. I **2.** C **3.** N **4.** A **5.** G **6.** B **7.** D **8.** O **9.** F **10.** M

Exercise Set 3.7, p. 234

1. $17.8 billion **3.** $19.15 **5.** 102.8°F **7.** $64,333,333.33 **9.** Area: 8.125 cm^2; perimeter: 11.5 cm **11.** 22,691.5 **13.** 20.2 mpg **15.** $30 **17.** 11.9752 ft^3 **19.** 78.1 cm **21.** 28.5 cm **23.** 2.31 cm **25.** 876 calories **27.** $1171.74 **29.** 227.75 ft^2 **31.** 0.335 **33.** $53.04 **35.** 2152.56 yd^2 **37.** 10.8¢ **39.** $906.50 **41.** 5.665 billion **43.** 1.4°F **45.** $266,791 **47.** $165,565 **49.** $1,131,429 **51.** DW **53.** 6335 **54.** $\frac{31}{24}$ **55.** $6\frac{5}{6}$ **56.** $\frac{23}{15}$ **57.** $\frac{1}{24}$ **58.** 2803 **59.** $\frac{2}{15}$ **60.** $1\frac{5}{6}$ **61.** $\frac{129}{251}$ **62.** $\frac{5}{16}$ **63.** $\frac{13}{25}$ **64.** $\frac{25}{19}$ **65.** 28 min **66.** $7\frac{1}{5}$ min **67.** 186 calories **68.** 30 calories **69.** $17.28

Concept Reinforcement, p. 241

1. False **2.** True **3.** True **4.** False **5.** True

Summary and Review: Chapter 3, p. 241

1. 6,590,000 **2.** 6,900,000 **3.** Three and forty-seven hundredths **4.** Thirty-one thousandths **5.** Twenty-seven and eleven hundred-thousandths **6.** Seven millionths **7.** $\frac{9}{100}$ **8.** $\frac{4561}{1000}$ **9.** $\frac{89}{1000}$ **10.** $\frac{30,227}{10,000}$ **11.** 0.034 **12.** 4.2603 **13.** 27.91 **14.** 867.006 **15.** 0.034 **16.** 0.91 **17.** 0.741 **18.** 1.041 **19.** 17.4 **20.** 17.43 **21.** 17.429 **22.** 17 **23.** 574.519 **24.** 0.6838 **25.** 229.1 **26.** 45.551 **27.** 29.2092 **28.** 790.29 **29.** 29.148 **30.** 70.7891 **31.** 12.96 **32.** 0.14442 **33.** 4.3 **34.** 0.02468 **35.** 7.5 **36.** 0.45 **37.** 45.2 **38.** 1.022 **39.** 0.2763 **40.** 1389.2 **41.** 496.2795 **42.** 6.95 **43.** 42.54 **44.** 4.9911 **45.** $15.52 **46.** 1.9 lb **47.** $912.68 **48.** $307.49 **49.** 14.5 mpg **50.** **(a)** 102.6 lb; **(b)** 14.7 lb **51.** 272 **52.** 216 **53.** 4 **54.** $125 **55.** 2.6 **56.** 1.28 **57.** 2.75 **58.** 3.25 **59.** $1.1\overline{6}$ **60.** $1.\overline{54}$ **61.** 1.5 **62.** 1.55 **63.** 1.545 **64.** $82.73 **65.** $4.87 **66.** 2493¢ **67.** 986¢ **68.** 1.8045 **69.** 57.1449 **70.** 15.6375 **71.** $41.537\overline{3}$
72. DW Multiply by 1 to get a denominator that is a power of 10:

$$\frac{44}{125} = \frac{44}{125} \cdot \frac{8}{8} = \frac{352}{1000} = 0.352.$$

We can also divide to find that $\frac{44}{125} = 0.352$.
73. DW Each decimal place in the decimal notation corresponds to one zero in the power of ten in the fraction notation. When the fractions are multiplied, the number of zeros in the denominator of the product is the sum of the number of zeros in the denominators of the factors. So the number of decimal places in the product is the sum of the number of decimal places in the factors.
74. **(a)** $2.56 \times 6.4 \div 51.2 - 17.4 + 89.7 = 72.62$; **(b)** $(11.12 - 0.29) \times 3^4 = 877.23$
75. $\frac{1}{3} + \frac{2}{3} = 0.33333333\ldots + 0.66666666\ldots$
$= 0.99999999\ldots.$
Therefore, $1 = 0.99999999\ldots$ because $\frac{1}{3} + \frac{2}{3} = 1$.
76. $2 = 1.\overline{9}$ **77.** $a = 5$, $b = 9$

Test: Chapter 3, p. 244

1. [3.3b] 8,900,000,000 **2.** [3.3b] 3,756,000 **3.** [3.1a] Two and thirty-four hundredths **4.** [3.1a] One hundred five and five ten-thousandths **5.** [3.1b] $\frac{91}{100}$ **6.** [3.1b] $\frac{2769}{1000}$ **7.** [3.1b] 0.074 **8.** [3.1b] 3.7047 **9.** [3.1b] 756.09 **10.** [3.1b] 91.703 **11.** [3.1c] 0.162 **12.** [3.1c] 0.078 **13.** [3.1c] 0.9 **14.** [3.1d] 6 **15.** [3.1d] 5.68 **16.** [3.1d] 5.678 **17.** [3.1d] 5.7 **18.** [3.2a] 0.7902 **19.** [3.2a] 186.5 **20.** [3.2a] 1033.23 **21.** [3.2b] 48.357 **22.** [3.2b] 19.0901 **23.** [3.2b] 152.8934 **24.** [3.3a] 0.03 **25.** [3.3a] 0.21345 **26.** [3.3a] 73,962 **27.** [3.4a] 4.75 **28.** [3.4a] 30.4 **29.** [3.4a] 0.19 **30.** [3.4a] 0.34689 **31.** [3.4a] 34,689 **32.** [3.4b] 84.26 **33.** [3.2c] 8.982 **34.** [3.7a] $314.99 **35.** [3.7a] 28.3 mpg **36.** [3.7a] $6572.45 **37.** [3.7a] $181.93 **38.** [3.7a] 58.24 million passengers **39.** [3.6a] 198 **40.** [3.6a] 4 **41.** [3.5a] 1.6 **42.** [3.5a] 0.88 **43.** [3.5a] 5.25 **44.** [3.5a] 0.75

45. [3.5a] $1.\overline{2}$ **46.** [3.5a] $2.\overline{142857}$ **47.** [3.5b] 2.1 **48.** [3.5b] 2.14 **49.** [3.5b] 2.143 **50.** [3.3b] $9.49 **51.** [3.4c] 40.0065 **52.** [3.4c] 384.8464 **53.** [3.5c] 302.4 **54.** [3.5c] $52.339\overline{4}$ **55.** [3.7a] $35 **56.** [3.1b, c] $\frac{2}{3}, \frac{5}{7}, \frac{15}{19}, \frac{11}{13}, \frac{17}{20}, \frac{13}{15}$

CHAPTER 4

Margin Exercises, Section 4.1, pp. 248–258

1. $\frac{5}{11}$, or 5:11 **2.** $\frac{57.3}{86.1}$, or 57.3:86.1 **3.** $\frac{6\frac{3}{4}}{7\frac{2}{5}}$, or $6\frac{3}{4}:7\frac{2}{5}$ **4.** $\frac{739}{12}$ **5.** $\frac{12}{14}$ **6.** $\frac{44}{211.1}; \frac{211.1}{44}$ **7.** $\frac{38.2}{56.1}$ **8.** $\frac{205}{278}; \frac{278}{205}; \frac{278}{483}$ **9.** 3.6 is to 12 as 3 is to 10 **10.** 1.2 is to 1.5 as 4 is to 5 **11.** $\frac{9}{16}$ **12.** 5 mi/hr, or 5 mph **13.** 12 mi/hr, or 12 mph **14.** $\frac{89}{13}$ km/h, or 6.85 km/h **15.** 1100 ft/sec **16.** 4 ft/sec **17.** $\frac{121}{8}$ ft/sec, or 15.125 ft/sec **18.** 250 ft/sec **19.** $\frac{714 \text{ home runs}}{1330 \text{ strikeouts}}$; approximately 0.537 home run per strikeout **20.** Yes **21.** No **22.** No **23.** 14 **24.** $11\frac{1}{4}$ **25.** 10.5 **26.** 2.64 **27.** 10.8 **28.** 445 calories **29.** 15 gal **30.** 9.5 in. **31.** 2074 deer

Calculator Corner, p. 254

1. 27.5625 **2.** 25.6 **3.** 15.140625 **4.** 40.03952941

Exercise Set 4.1, p. 259

1. $\frac{4}{5}$ **3.** $\frac{56.78}{98.35}$ **5.** $\frac{356}{100{,}000}; \frac{173}{100{,}000}$ **7.** $\frac{3}{4}$ **9.** $\frac{7}{9}$ **11.** $\frac{478}{213}; \frac{213}{478}$ **13.** 40 km/h **15.** 7.48 mi/sec **17.** 23 mpg **19.** About 18.3 points/game **21.** 0.623 gal/ft^2 **23.** 25 beats/min **25.** No **27.** Yes **29.** Yes **31.** No **33.** 45 **35.** 10 **37.** 20 **39.** 18 **41.** 0.06 **43.** 5 **45.** 1 **47.** 12.5725 **49.** 880 calories **51.** 177 million, or 177,000,000 **53.** 13,500 mi **55.** 120 lb **57.** 3360 mi **59.** **(a)** 1254.66 Mexican Pesos; **(b)** 344.32 **61.** 175 bulbs **63.** 2975 ft^2 **65.** 100 oz **67.** 954 deer **69.** ^{D}W **71.** 50 **72.** 9.5 **73.** 14.5 **74.** 152 **75.** $6\frac{7}{20}$ cm **76.** $17\frac{11}{20}$ cm **77.** 2150 earned runs

Margin Exercises, Section 4.2, pp. 264–267

1. $\frac{70}{100}; 70 \times \frac{1}{100}; 70 \times 0.01$ **2.** $\frac{23.4}{100}; 23.4 \times \frac{1}{100}; 23.4 \times 0.01$ **3.** $\frac{100}{100}; 100 \times \frac{1}{100}; 100 \times 0.01$ **4.** 0.34 **5.** 0.789 **6.** 0.06625 **7.** 0.18 **8.** 0.0008 **9.** 24% **10.** 347% **11.** 100% **12.** 32.1% **13.** 25.3%

Calculator Corner, p. 265

1. 0.14 **2.** 0.00069 **3.** 0.438 **4.** 1.25

Exercise Set 4.2, p. 268

1. $\frac{90}{100}; 90 \times \frac{1}{100}; 90 \times 0.01$ **3.** $\frac{12.5}{100}; 12.5 \times \frac{1}{100}; 12.5 \times 0.01$ **5.** 0.67 **7.** 0.456 **9.** 0.5901 **11.** 0.1 **13.** 0.01 **15.** 2 **17.** 0.001 **19.** 0.0009 **21.** 0.0018 **23.** 0.2319 **25.** 0.14875 **27.** 0.565 **29.** 0.09; 0.58 **31.** 0.44 **33.** 0.36 **35.** 47% **37.** 3% **39.** 870% **41.** 33.4% **43.** 75% **45.** 40% **47.** 0.6% **49.** 1.7% **51.** 27.18% **53.** 2.39% **55.** 26%; 38% **57.** 17.7% **59.** 21.5% **61.** ^{D}W **63.** $33\frac{1}{3}$ **64.** $37\frac{1}{2}$ **65.** $9\frac{3}{8}$ **66.** $18\frac{9}{16}$ **67.** $5\frac{11}{14}$ **68.** $111\frac{2}{3}$ **69.** $0.\overline{6}$ **70.** $0.\overline{3}$ **71.** $0.8\overline{3}$ **72.** $1.41\overline{6}$ **73.** $2.\overline{6}$ **74.** 0.9375

Margin Exercises, Section 4.3, pp. 272–274

1. 25% **2.** 62.5%, or $62\frac{1}{2}\%$ **3.** $66.\overline{6}\%$, or $66\frac{2}{3}\%$ **4.** $83.\overline{3}\%$, or $83\frac{1}{3}\%$ **5.** 57% **6.** 76% **7.** $\frac{3}{5}$ **8.** $\frac{13}{400}$ **9.** $\frac{2}{3}$

10.

FRACTION NOTATION	$\frac{1}{5}$	$\frac{5}{6}$	$\frac{3}{8}$
DECIMAL NOTATION	0.2	$0.83\overline{3}$	0.375
PERCENT NOTATION	20%	$83.\overline{3}\%$, or $83\frac{1}{3}\%$	$37\frac{1}{2}\%$

Calculator Corner, p. 272

1. 52% **2.** 38.46% **3.** 110.26% **4.** 171.43% **5.** 59.62% **6.** 28.31%

Calculator Corner, p. 275

1. 30.54; 1.31% **2.** 32.05; 1.20% **3.** 34.47; 1.19% **4.** 26.47; 1.00% **5.** 11.98; 4.32% **6.** 17.52; 0.89%

Exercise Set 4.3, p. 276

1. 41% **3.** 5% **5.** 20% **7.** 30% **9.** 50% **11.** 87.5%, or $87\frac{1}{2}\%$ **13.** 80% **15.** $66.\overline{6}\%$, or $66\frac{2}{3}\%$ **17.** $16.\overline{6}\%$, or $16\frac{2}{3}\%$ **19.** 18.75%, or $18\frac{3}{4}\%$ **21.** 81.25%, or $81\frac{1}{4}\%$ **23.** 16% **25.** 5% **27.** 34% **29.** 40%; 18% **31.** 22% **33.** 5% **35.** 9% **37.** $\frac{17}{20}$ **39.** $\frac{5}{8}$ **41.** $\frac{1}{3}$ **43.** $\frac{1}{6}$ **45.** $\frac{29}{400}$ **47.** $\frac{1}{125}$ **49.** $\frac{203}{800}$ **51.** $\frac{176}{225}$ **53.** $\frac{711}{1100}$ **55.** $\frac{3}{2}$ **57.** $\frac{13}{40{,}000}$ **59.** $\frac{1}{3}$ **61.** $\frac{2}{25}$ **63.** $\frac{3}{5}$ **65.** $\frac{1}{50}$ **67.** $\frac{7}{20}$ **69.** $\frac{47}{100}$

71.

FRACTION NOTATION	DECIMAL NOTATION	PERCENT NOTATION
$\frac{1}{8}$	0.125	12.5%, or $12\frac{1}{2}\%$
$\frac{1}{6}$	$0.1\overline{6}$	$16.\overline{6}\%$, or $16\frac{2}{3}\%$
$\frac{1}{5}$	0.2	20%
$\frac{1}{4}$	0.25	25%
$\frac{1}{3}$	$0.\overline{3}$	$33.\overline{3}\%$, or $33\frac{1}{3}\%$
$\frac{3}{8}$	0.375	37.5%, or $37\frac{1}{2}\%$
$\frac{2}{5}$	0.4	40%
$\frac{1}{2}$	0.5	50%

73.

FRACTION NOTATION	DECIMAL NOTATION	PERCENT NOTATION
$\frac{1}{2}$	0.5	50%
$\frac{1}{3}$	$0.\overline{3}$	$33.\overline{3}\%$, or $33\frac{1}{3}\%$
$\frac{1}{4}$	0.25	25%
$\frac{1}{6}$	$0.1\overline{6}$	$16.\overline{6}\%$, or $16\frac{2}{3}\%$
$\frac{1}{8}$	0.125	12.5%, or $12\frac{1}{2}\%$
$\frac{3}{4}$	0.75	75%
$\frac{5}{6}$	$0.8\overline{3}$	$83.\overline{3}\%$, or $83\frac{1}{3}\%$
$\frac{3}{8}$	0.375	37.5%, or $37\frac{1}{2}\%$

75. $\mathrm{D_W}$ **77.** 70 **78.** 5 **79.** 400 **80.** 18.75
81. 23.125 **82.** 25.5 **83.** 4.5 **84.** $8\frac{3}{4}$ **85.** $33\frac{1}{3}$
86. $37\frac{1}{2}$ **87.** $83\frac{1}{3}$ **88.** $20\frac{1}{2}$ **89.** $43\frac{1}{8}$ **90.** $62\frac{1}{6}$
91. $18\frac{3}{4}$ **92.** $7\frac{4}{9}$ **93.** $\frac{18}{17}$ **94.** $\frac{209}{10}$ **95.** $\frac{203}{2}$ **96.** $\frac{259}{8}$
97. $11.\overline{1}\%$ **99.** $257.\overline{46317}\%$ **101.** $0.01\overline{5}$
103. $1.04\overline{142857}$

Margin Exercises, Section 4.4, pp. 280–283

1. $12\% \times 50 = a$ **2.** $a = 40\% \times 60$ **3.** $45 = 20\% \times t$
4. $120\% \times y = 60$ **5.** $16 = n \times 40$ **6.** $b \times 84 = 10.5$
7. 6 **8.** $35.20 **9.** 225 **10.** $50 **11.** 40%
12. 12.5%

Calculator Corner, p. 284

1. 1.2 **2.** $5.04 **3.** 48.64 **4.** $22.40 **5.** 0.0112
6. $29.70 **7.** 450 **8.** $1000 **9.** 2.5% **10.** 12%

Exercise Set 4.4, p. 285

1. $y = 32\% \times 78$ **3.** $89 = a \times 99$ **5.** $13 = 25\% \times y$
7. 234.6 **9.** 45 **11.** $18 **13.** 1.9 **15.** 78%
17. 200% **19.** 50% **21.** 125% **23.** 40 **25.** $40
27. 88 **29.** 20 **31.** 6.25 **33.** $846.60 **35.** $\mathrm{D_W}$
37. $\frac{9}{100}$ **38.** $\frac{179}{100}$ **39.** $\frac{875}{1000}$, or $\frac{7}{8}$ **40.** $\frac{125}{1000}$, or $\frac{1}{8}$
41. $\frac{9375}{10{,}000}$, or $\frac{15}{16}$ **42.** $\frac{6875}{10{,}000}$, or $\frac{11}{16}$ **43.** 0.89 **44.** 0.07
45. 0.3 **46.** 0.017 **47.** $800 (can vary); $843.20
49. $10,000 (can vary); $10,400 **51.** $1875

Margin Exercises, Section 4.5, pp. 288–290

1. $\frac{12}{100} = \frac{a}{50}$ **2.** $\frac{40}{100} = \frac{a}{60}$ **3.** $\frac{130}{100} = \frac{a}{72}$ **4.** $\frac{20}{100} = \frac{45}{b}$
5. $\frac{120}{100} = \frac{60}{b}$ **6.** $\frac{N}{100} = \frac{16}{40}$ **7.** $\frac{N}{100} = \frac{10.5}{84}$ **8.** $225
9. 35.2 **10.** 6 **11.** 50 **12.** 30% **13.** 12.5%

Exercise Set 4.5, p. 291

1. $\frac{37}{100} = \frac{a}{74}$ **3.** $\frac{N}{100} = \frac{4.3}{5.9}$ **5.** $\frac{25}{100} = \frac{14}{b}$ **7.** 68.4
9. 462 **11.** 40 **13.** 2.88 **15.** 25% **17.** 102%
19. 25% **21.** 93.75% **23.** $72 **25.** 90 **27.** 88
29. 20 **31.** 25 **33.** $780.20 **35.** $\mathrm{D_W}$ **37.** 8
38. 4000 **39.** 8 **40.** 2074 **41.** 100 **42.** 15
43. $8.0\overline{4}$ **44.** $\frac{3}{16}$, or 0.1875 **45.** $\frac{43}{48}$ qt **46.** $\frac{1}{8}$ T
47. $1134 (can vary); $1118.64

Margin Exercises, Section 4.6, pp. 294–299

1. About 9.5% **2.** 14,560,000 workers **3.** **(a)** \$1475; **(b)** \$38,350 **4.** **(a)** \$9218.75; **(b)** \$27,656.25 **5.** About 2.9% **6.** About 76.6%

Translating for Success, p. 300

1. J **2.** M **3.** N **4.** E **5.** G **6.** H **7.** O **8.** C **9.** D **10.** B

Exercise Set 4.6, p. 301

1. About 13,247 wild horses **3.** 3 yr: \$21,080; 5 yr: \$17,680 **5.** Overweight: 176.4 million people; obese: 73.5 million people **7.** Acid: 20.4 mL; water: 659.6 mL **9.** About 1808 mi **11.** About 39,867,000 people **13.** 36.4 items correct; 3.6 items incorrect **15.** 95 items **17.** 25% **19.** 166; 156; 146; 140; 122 **21.** 8% **23.** 20% **25.** About 86.5% **27.** \$30,030 **29.** \$16,174.50; \$12,130.88 **31.** About 27% **33.** 34.375%, or $34\frac{3}{8}\%$ **35.** **(a)** 7.5%; **(b)** 879,675 **37.** 71% **39.** \$1560 **41.** 80% **43.** 98,775; 18.0% **45.** 799,065; 14.8% **47.** 4,550,688; 38.1% **49.** \$36,400 **51.** 40% **53.** ^{D}W **55.** $2.\overline{27}$ **56.** 0.44 **57.** 3.375 **58.** $4.\overline{7}$ **59.** 0.92 **60.** $0.8\overline{3}$ **61.** 0.4375 **62.** 2.317 **63.** 3.4809 **64.** 0.675 **65.** About 5 ft 6 in. **67.** $83\frac{1}{3}\%$

Margin Exercises, Section 4.7, pp. 307–315

1. \$48.50; \$717.45 **2.** \$5.39; \$140.14 **3.** 6% **4.** \$999 **5.** \$5628 **6.** 12.5%, or $12\frac{1}{2}\%$ **7.** \$1675 **8.** \$180; \$360 **9.** 20% **10.** \$301 **11.** \$225.75 **12.** **(a)** \$33.53; **(b)** \$4833.53 **13.** \$2376.20 **14.** \$7690.94

Calculator Corner, p. 315

1. \$16,357.18 **2.** \$12,764.72

Exercise Set 4.7, p. 316

1. \$2.65 **3.** \$19.53 **5.** \$16.39; \$361.39 **7.** 5% **9.** \$2000 **11.** \$800 **13.** \$719.86 **15.** 5.6% **17.** \$2700 **19.** 5% **21.** \$980 **23.** \$5880 **25.** \$420 **27.** \$30; \$270 **29.** \$2.55; \$14.45 **31.** \$125; \$112.50 **33.** \$40%; \$360 **35.** \$30; 16.7% **37.** \$8 **39.** \$84 **41.** \$113.52 **43.** \$925 **45.** \$671.88 **47.** **(a)** \$80.14; **(b)** \$6580.14 **49.** **(a)** \$147.95; **(b)** \$10,147.95 **51.** **(a)** \$46.03; **(b)** \$5646.03 **53.** \$441 **55.** \$2802.50 **57.** \$7853.38 **59.** \$99,427.40 **61.** \$4243.60 **63.** \$28,225.00 **65.** \$9270.87 **67.** \$129,871.09 **69.** \$4101.01 **71.** ^{D}W **73.** Reciprocals **74.** Divisible by 6 **75.** Additive **76.** Quotient **77.** Perimeter **78.** Divisible by 3 **79.** Prime **80.** Proportional **81.** He bought the plaques for $\$166\frac{2}{3} + \$250$, or $\$416\frac{2}{3}$, and sold them for \$400, so he lost money. **83.** 9.38%

Margin Exercises, Section 4.8, pp. 322–326

1. **(a)** \$97; **(b)** interest: \$86.40; amount applied to principal: \$10.60; **(c)** interest: \$55.17; amount applied to principal: \$41.83; **(d)** At 13.6%, the principal was decreased by \$31.23 more than at the 21.3% rate. The interest at 13.6% is \$31.23 less than at 21.3%. **2.** **(a)** Interest: \$91.78; amount applied to principal: \$229.22; **(b)** \$58; **(c)** \$4080 **3.** Interest: \$843.94; amount applied to principal: \$135.74 **4.** **(a)** Interest: \$844.69; amount applied to principal: \$498.64; **(b)** \$88,799.40; **(c)** The Sawyers will pay \$110,885.40 less in interest with the 15-yr loan than with the 30-yr loan. **5.** **(a)** Interest: \$701.25; amount applied to principal: \$548.89; **(b)** \$72,025.20; **(c)** The Sawyers will pay \$16,774.20 less in interest with the 15-yr loan at $5\frac{1}{2}\%$ than with the 15-yr loan at $6\frac{5}{8}\%$.

Exercise Set 4.8, p. 327

1. **(a)** \$98; **(b)** interest: \$86.56; amount applied to principal: \$11.44; **(c)** interest: \$51.20; amount applied to principal: \$46.80; **(d)** At 12.6%, the principal is decreased by \$35.36 more than at the 21.3% rate. The interest at 12.6% is \$35.36 less than at 21.3%. **3.** **(a)** Interest: \$125.14; amount applied to principal: \$312.79; **(b)** \$51.24; **(c)** \$7991.60, \$11,504, \$3512.40 **5.** **(a)** Interest: \$854.17; amount applied to principal: \$155.61; **(b)** \$199,520.80; **(c)** new principal: \$163,844.39; interest: \$853.36; amount applied to principal: \$156.42 **7.** **(a)** Interest: \$854.17; amount applied to principal: \$552; **(b)** \$89,110.60; **(c)** The Martinez family will pay \$110,410.20 less in interest with the 15-yr loan than with the 30-yr loan. **9.** \$99,917.71; \$99,834.94 **11.** \$99,712.04; \$99,422.15 **13.** \$149,882.75; \$149,764.79 **15.** \$199,382.07; \$198,760.41 **17.** **(a)** \$2395, \$21,555; **(b)** \$52.09, \$401.97; **(c)** \$239.88 **19.** **(a)** \$595; \$11,305; **(b)** interest: \$87.61; amount applied to principal: \$273.47; **(c)** \$1693.88 **21.** ^{D}W **23.** ^{D}W **25.** 18 **26.** $\frac{22}{7}$ **27.** 265.625 **28.** 1.1 **29.** $0.\overline{5}$ **30.** $2.\overline{09}$ **31.** $0.91\overline{6}$ **32.** $1.\overline{857142}$ **33.** $2.\overline{142857}$ **34.** $1.58\overline{3}$ **35.** 4,030,000,000,000 **36.** 5,800,000 **37.** 42,700,000 **38.** 6,090,000,000,000

Concept Reinforcement, p. 330

1. True **2.** False **3.** True **4.** False

Summary and Review: Chapter 4, p. 330

1. $\frac{47}{84}$ **2.** $\frac{46}{1.27}$ **3.** $\frac{83}{100}$ **4.** $\frac{0.72}{197}$ **5.** $\frac{3}{4}$ **6.** $\frac{9}{16}$ **7.** 26 mpg **8.** 6300 rpm **9.** 0.638 gal/ft^2 **10.** 0.72 serving/lb **11.** No **12.** Yes **13.** 32 **14.** 7 **15.** $\frac{1}{40}$ **16.** 24 **17.** \$4.45 **18.** 27 circuits **19.** 832 mi **20.** Approximately 3,293,558 kg **21.** **(a)** 206.1 Euros; **(b)** \$60.65 **22.** 27 acres **23.** 6 in. **24.** Approximately 2096 lawyers **25.** 56% **26.** 1.7% **27.** 37.5% **28.** $33.\overline{3}\%$, or $33\frac{1}{3}\%$ **29.** 0.735 **30.** 0.065 **31.** $\frac{6}{25}$ **32.** $\frac{63}{1000}$ **33.** $30.6 = p \times 90$; 34% **34.** $63 = 84\% \times n$; 75

35. $y = 38\frac{1}{2}\% \times 168$; 64.68   36. $\frac{24}{100} = \frac{16.8}{b}$; 70

37. $\frac{42}{30} = \frac{N}{100}$; 140%   38. $\frac{10.5}{100} = \frac{a}{84}$; 8.82

39. 223 students; 105 students   40. 42%   41. 2500 mL
42. 12%   43. 92   44. $14.40   45. 5%   46. 11%
47. $42; $308   48. 14%   49. $2940
50. Approximately 18.4%   51. $36   52. (a) $394.52;
(b) $24,394.52   53. $121   54. $7727.26   55. $9504.80
56. (a) $129; (b) interest: $100.18; amount applied to principal: $28.82; (c) interest: $70.72; amount applied to principal: $58.28; (d) At 13.2%, the principal is decreased by $29.46 more than at the 18.7% rate. The interest at 13.2% is $29.46 less than at 18.7%.
57. No; the 10% discount was based on the original price rather than on the sale price.
58. A 40% discount is better. When successive discounts are taken, each is based on the previous discounted price rather than on the original price. A 20% discount followed by a 22% discount is the same as a 37.6% discount off the original price.
59. Finishing paint: 11 gal; primer: 16.5 gal   60. $66\frac{2}{3}\%$

Test: Chapter 4, p. 334

1. [4.1a] $\frac{85}{97}$ 2. [4.1a] $\frac{0.34}{124}$ 3. [4.1a] $\frac{9}{10}$ 4. [4.1a] $\frac{25}{32}$
5. [4.1b] 22 mpg   6. [4.1b] $1\frac{1}{3}$ servings/lb   7. [4.1c] Yes
8. [4.1c] No   9. [4.1d] 12   10. [4.1d] 360
11. [4.1e] 4.8 min   12. [4.1e] 525 mi
13. [4.1e] (a) 3490.425 Hong Kong dollars; (b) $102.49
14. [4.1e] About 86,151 arrests   15. [4.2b] 0.064
16. [4.2b] 38%   17. [4.3a] 137.5%
18. [4.3b] $\frac{13}{20}$ 19. [4.4a, b] $a = 40\% \cdot 55$; 22
20. [4.5a, b] $\frac{N}{100} = \frac{65}{80}$; 81.25%   21. [4.6a] 400 passengers; 575 passengers   22. [4.6a] 654   23. [4.6b] $50.\overline{90}\%$
24. [4.6a] 5.5%   25. [4.7a] $16.20; $340.20
26. [4.7b] $630   27. [4.7c] $40; $160   28. [4.7d] $8.52
29. [4.7d] $5356   30. [4.7e] $1110.39
31. [4.7e] $11,580.07   32. [4.6b] 2.9, 26.1%; 0.6, 37.5%; 2.0, 20%; 2.1, 0.4   33. [4.7c] $275; about 14.1%
34. [4.8a] $119,909.14; $119,817.72   35. [4.7b] $194,600
36. [4.7b, e] $2546.16

CHAPTER 5

Margin Exercises, Section 5.1, pp. 340–345

1. 75   2. 54.9   3. 81   4. 19.4   5. 37   6. 34 mpg
7. 2.5   8. 94   9. 17   10. 17   11. 91   12. $3700
13. 67.5   14. 45   15. 34, 67   16. No mode exists.
17. (a) 17 g; (b) 18 g; (c) 19 g
18. Wheat A: average stalk height ≈ 25.21 in.; wheat B: average stalk height ≈ 22.54 in.; wheat B is better.

Calculator Corner, p. 342

1. 285.5   2. 75; 54.9; 81; 19.4   3. $202.\overline{3}$

Exercise Set 5.1, p. 346

1. Average: 21; median: 18.5; mode: 29
3. Average: 21; median: 20; modes: 5, 20
5. Average: 5.2; median: 5.7; mode: 7.4
7. Average: 239.5; median: 234; mode: 234
9. Average: $23.\overline{8}$; median: 15; mode: 1   11. 31 mpg
13. 2.7   15. Average: $8.19; median: $8.49; mode: $6.99
17. 90   19. 263 days
21. Bulb A: average time = 1171.25 hr; bulb B: average time ≈ 1251.58 hr; bulb B is better   23. DW   25. 196
26. $\frac{4}{9}$ 27. 1.96   28. 1.999396   29. 225.05
30. 126.0516   31. $\frac{3}{35}$ 32. $\frac{14}{15}$ 33. 118.75%
34. 68.75%   35. 51.2%   36. 97.81%   37. 182
39. 10 home runs   41. 58

Margin Exercises, Section 5.2, pp. 349–352

1. Zone (Men's Menu)   2. Both Ornish (Eat More, Weigh Less) and Zone (Men's Menu) have 17 servings of fruits and vegetables   3. Ornish (Eat More, Weigh Less)
4. About 43 g   5. About 24.5 g   6. 22 g
7. Average: about 9.3 g; median: 7 g; mode: 7 g
8. 20 calories   9. Yes   10. 560 mg   11. 24%
12. 14 g   13. 2100 rhinos   14. There are twelve times as many black rhinos as Sumatran rhinos, or 3300 more.
15. 3537 rhinos; answers may vary
16. 795 cups; answers may vary
17. 750 cups; answers may vary
18. 1830 cups; answers may vary

Exercise Set 5.2, p. 353

1. 483,612,200 mi   3. Neptune   5. All
7. 11 Earth diameters   9. Average: $27{,}884.\overline{1}$ mi; median: 7926 mi; no mode exists   11. 92°   13. 108°
15. 3   17. 90° and higher   19. 30% and higher
21. 50%   23. 1940: 1976; 1980: 3849; 94.8%
25. 1920; 3186; 1266   27. 1.0 billion   29. 2070
31. 1650 and 1850   33. 3 billion; 75%   35. Africa
37. 470,000 gal   39. 340,000 gal   41. DW
43. Cabinets: $13,444; countertops: $4033.20; appliances: $2151.04; fixtures: $806.64
44. Cabinets: $3597.55; countertops: $1276.55; labor: $2901.25; flooring: $696.30   45. $\frac{6}{25}$ 46. $\frac{9}{20}$
47. $\frac{6}{125}$ 48. $\frac{8}{125}$ 49. $\frac{531}{1000}$ 50. $\frac{873}{1000}$ 51. 1   52. $\frac{1}{50}$
53.

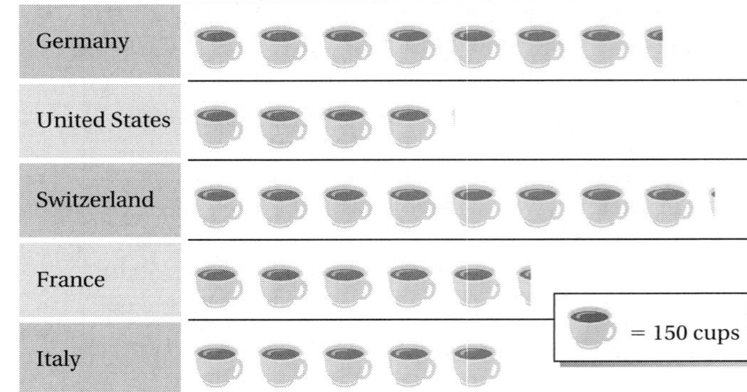

Margin Exercises, Section 5.3, pp. 358–363

1. About 77 million **2.** NASCAR **3.** NFL, NBA, and MLB **4.** 60 women **5.** 85+ **6.** 60–64 **7.** Yes **8.**

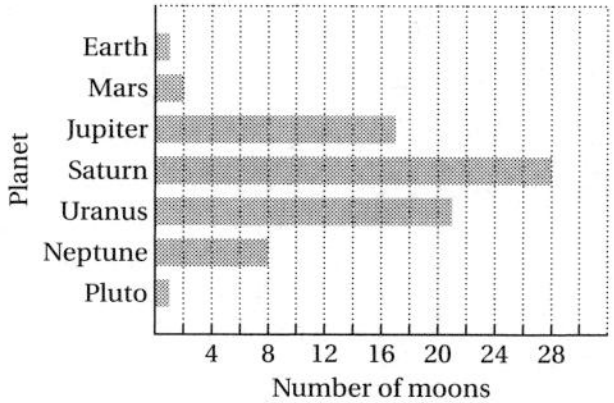

9. Month 7, June **10.** Months 1 and 2, December and January; months 4 and 5, March and April; months 6 and 7, May and June; and months 11 and 12, October and November **11.** Month 2, January, and month 8, July **12.** $900 **13.** 40 yr **14.** $1300 **15.**

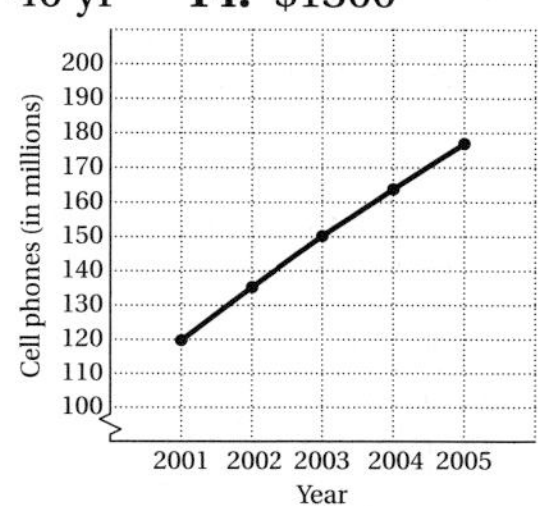

Exercise Set 5.3, p. 364

1. 190 calories **3.** 1 slice of chocolate cake with fudge frosting **5.** 1 cup of premium chocolate ice cream **7.** About 120 calories **9.** About 920 calories **11.** About 28 lb **13.** 1970: $11,000; 2002: $58,000; $47,000; 427% **15.** 1970: $6000; 2002: $25,000; $19,000; 317% **17.** $4000 **19.** $11,000 **21.**

23. Indianapolis **25.** $28.7\overline{3}$ min **27.** 1995 and 1996, 1996 and 1997, 1998 and 1999 **29.** 51% **31.** $3.9 billion **33.** 30.4 yd **35.** 1988 and 1995 **37.**

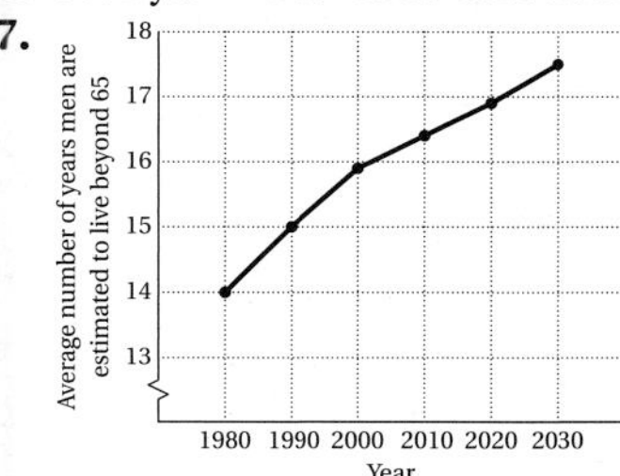

39. 25%0 **41.** 10.1% **43.** 1995 and 1996 **45.** About 308 murders **47.** 14.1% **49.** DW **51.** Natural **52.** Add; divide **53.** Simple **54.** Marked price; rate of discount; discount; sale price **55.** Compound **56.** Repeating **57.** Subtrahend **58.** Terminating **59.**

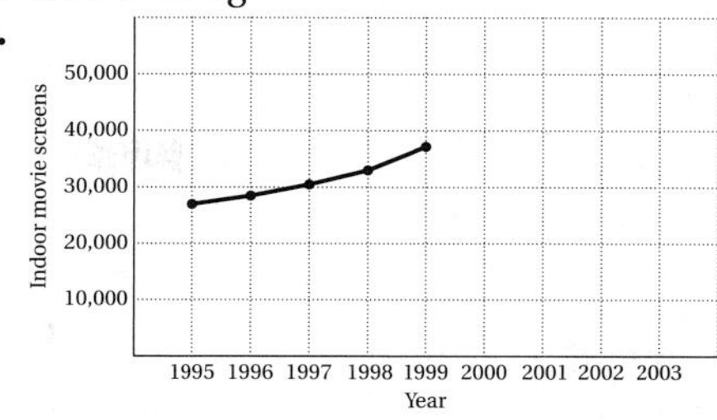

Using the line through the points for 1996 and 1998 (instead of a line through the point for 1999) probably gives more realistic estimates of the number of movie screens for the years 2000, 2001, and 2003. 2000: 37,500; 2001: 40,000; 2003: 44,500. Answers may vary.

Margin Exercises, Section 5.4, pp. 369–370

1. Spaying or neutering **2.** 87% **3.** $1122 **4.** 8% + 3%,or 11% **5.**

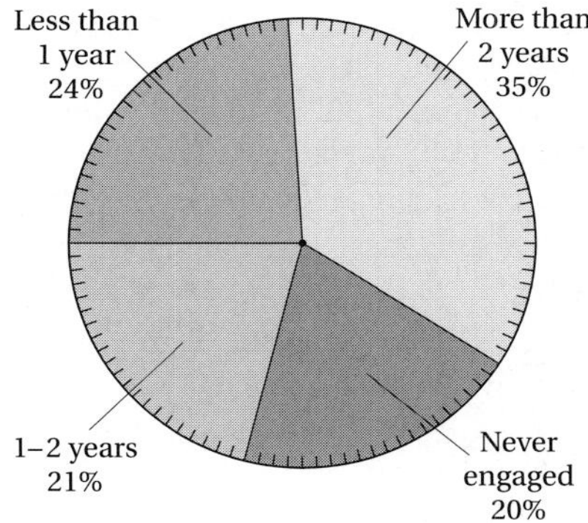

Translating for Success, p. 371

1. D **2.** B **3.** J **4.** K **5.** I **6.** F **7.** N **8.** E **9.** L **10.** M

Exercise Set 5.4, p. 372

1. 17.1% **3.** About 613 students **5.** 86.1% **7.** Food **9.** 14% **11.**

13.

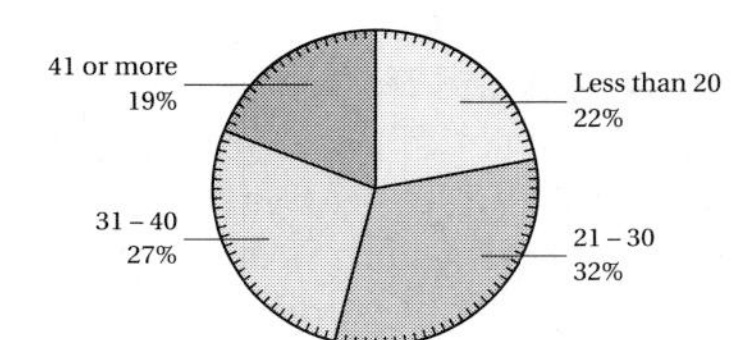

15. 
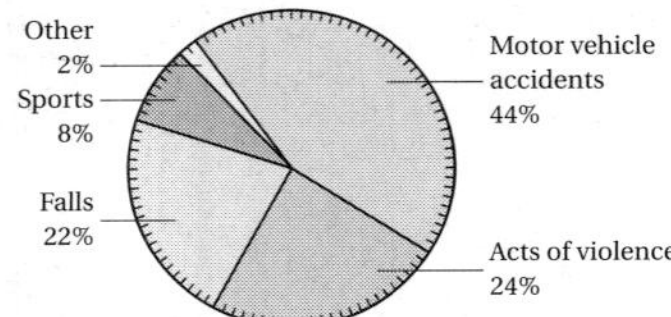

17. DW

Concept Reinforcement, p. 375

1. True **2.** False **3.** True **4.** True

Summary and Review: Chapter 5, p. 375

1. \$32.25 **2.** \$47.00 **3.** \$20.95 **4.** \$25.40 **5.** No **6.** \$19.75 **7.** 14,000 officers **8.** Los Angeles **9.** Houston **10.** 12,500 officers **11.** 26 **12.** 11 and 17 **13.** 0.2 **14.** 700 and 800 **15.** \$17 **16.** 20 **17.** \$110.50; \$107 **18.** 33 mpg **19.** 96 **20.** About 420 calories **21.** About 440 calories **22.** Big Bacon Classic **23.** Plain Single **24.** Plain Single **25.** Chicken Club **26.** About 50 calories **27.** About 220 calories **28.** Under 20 **29.** 12 accidents **30.** 13 accidents per 100 drivers **31.** 45–74 **32.** 11 accidents per 100 drivers **33.** Under 20 **34.** 22% **35.** 11% **36.** 1600 travelers **37.** 25%

38. 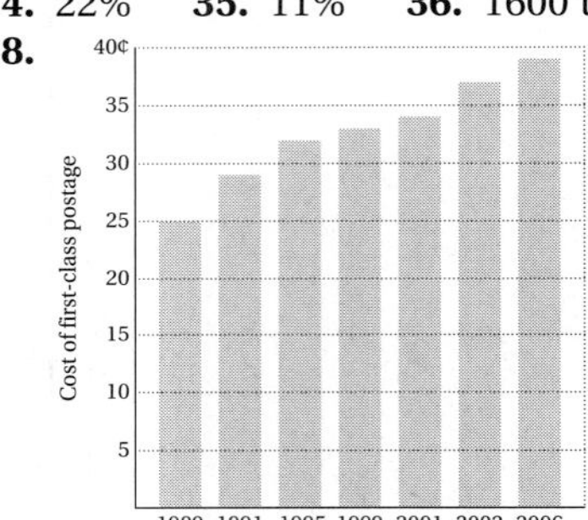

39. 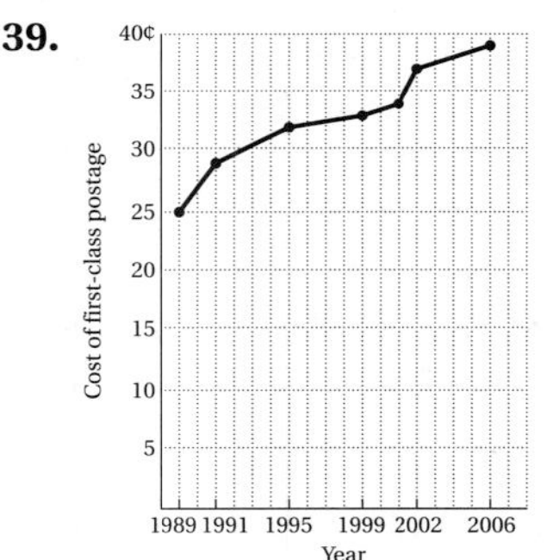

40. Battery A: average $\approx$ 43.04 hr; battery B: average $=$ 41.55 hr; battery A is better. **41.** 38.5 **42.** 13.4 **43.** 1.55 **44.** 1840 **45.** \$16.$\overline{6}$ **46.** 321.$\overline{6}$ **47.** 38.5 **48.** 14 **49.** 1.8 **50.** 1900 **51.** \$17 **52.** 375 **53.** 3.1

54. DW The average, the median, and the mode are "center points" that characterize a set of data. You might use the average to find a center point that is midway between the extreme values of the data. The median is a center point that is in the middle of all the data. That is, there are as many values less than the median than there are values greater than the median. The mode is a center point that represents the value or values that occur most frequently. **55.** DW The equation could represent a person's average income during a 4-yr period. Answers may vary. **56.** $a = 316$, $b = 349$

Test: Chapter 5, p. 379

1. [5.2a] 179 lb **2.** [5.2a] 111 lb **3.** [5.2a] 5 ft 3 in.; medium frame **4.** [5.2a] 5 ft 11 in.; medium frame **5.** [5.2b] Japan **6.** [5.2b] United States **7.** [5.2b] 800 lb **8.** [5.2b] 400 lb **9.** [5.1a] 49.5 **10.** [5.1a] 2.6 **11.** [5.1a] 15.5 **12.** [5.1b, c] 50.5; 52 **13.** [5.1b, c] 3; 1 and 3 **14.** [5.1b, c] 17.5; 17 and 18 **15.** [5.1a] 27 mpg **16.** [5.1a] 76 **17.** [5.3c] 53% **18.** [5.3c] 41% **19.** [5.3c] 1967 **20.** [5.3c] 2006 **21.** [5.3b]

22. [5.2a], [5.3a] 197 mph **23.** [5.2a], [5.3a] No; the greyhound can run 14 mph faster than a human. **24.** [5.2a], [5.3a] 89 mph **25.** [5.2a], [5.3a] 61 mph **26.** [5.4b]

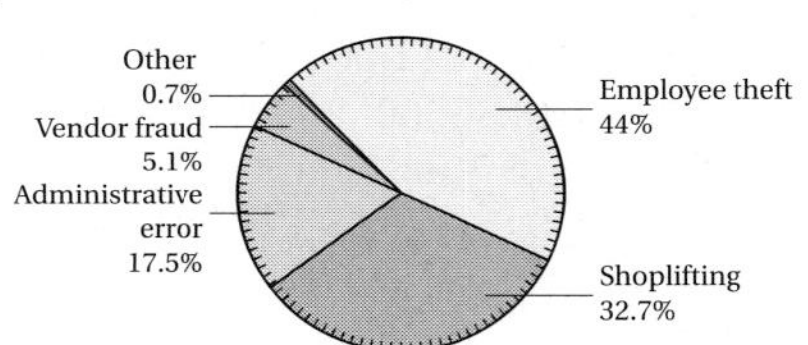

27. [5.2a], [5.4a] Employee theft: \$10.12 billion; shoplifting: \$7.521 billion; administrative error: \$4.025 billion; vendor fraud: \$1.173 billion; other: \$0.161 billion

28. [5.3b] 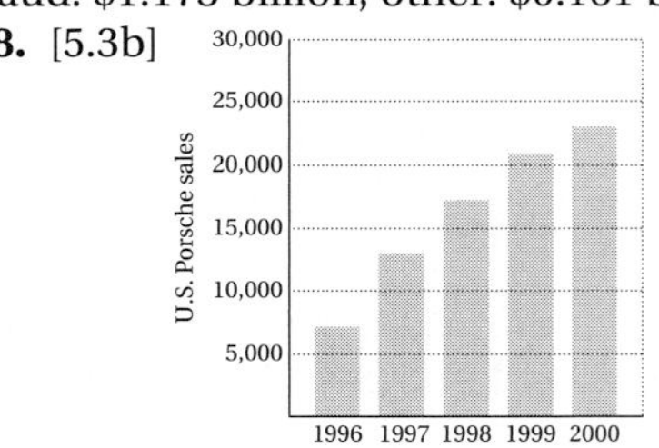

29. [5.3d] 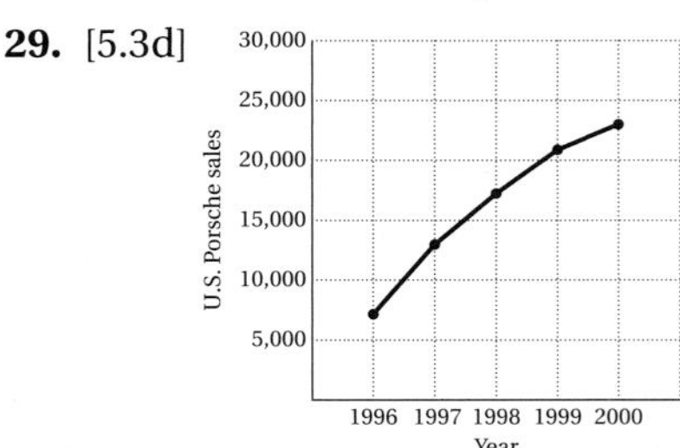

30. [5.1d] Bar A: average $\approx$ 8.417; bar B: average $\approx$ 8.417; equal quality **31.** [5.1a] 2.9 **32.** [5.1a, b] $a = 74$, $b = 111$

CHAPTER 6

Margin Exercises, Section 6.1, pp. 384–390

1. (a) and (b) E F (c) $\overline{EF}, \overline{FE}$ **2.** P Q
3. P Q **4.** P Q, P
5. P Q, Q **6.** R S
7. R S, R and S **8.** R S, R
9. R S, S **10.** R S, no endpoints **11.** $\overleftrightarrow{RS}, \overleftrightarrow{SR}, \overleftrightarrow{RT}, \overleftrightarrow{TR}, \overleftrightarrow{ST}, \overleftrightarrow{TS}, n$
12. Angle *DEF*, angle *FED*, $\angle DEF$, $\angle FED$, or $\angle E$
13. Angle *PQR*, angle *RQP*, $\angle PQR$, $\angle RQP$, or $\angle Q$ **14.** 127°
15. 33° **16.** Right **17.** Acute **18.** Obtuse
19. Straight **20.** Not perpendicular **21.** Perpendicular
22. (a) $\triangle ABC$; (b) $\triangle ABC, \triangle MPN$; (c) $\triangle DEF, \triangle GHI, \triangle JKL, \triangle QRS$ **23.** Yes **24.** No **25.** (a) $\triangle DEF$; (b) $\triangle GHI, \triangle QRS$; (c) $\triangle ABC, \triangle MPN, \triangle JKL$ **26.** Quadrilateral
27. Hexagon **28.** Triangle **29.** Quadrilateral
30. Dodecagon **31.** Octagon **32.** 180° **33.** 64°
34. (a) 3; (b) 180°; (c) 3, 540° **35.** 1080° **36.** 4140°

Exercise Set 6.1, p. 391

1. G H, $\overline{GH}, \overline{HG}$ **3.** Q D, $\overrightarrow{QD}$ **5.** $\overleftrightarrow{DE}, \overleftrightarrow{ED}, \overleftrightarrow{DF}, \overleftrightarrow{FD}, \overleftrightarrow{EF}, \overleftrightarrow{FE}, l$ **7.** Angle *GHI*, angle *IHG*, $\angle GHI$, $\angle IHG$, or $\angle H$ **9.** 10° **11.** 180° **13.** 130°
15. Obtuse **17.** Acute **19.** Straight **21.** Obtuse
23. Acute **25.** Obtuse **27.** Not perpendicular
29. Perpendicular **31.** Scalene; obtuse **33.** Scalene; right **35.** Equilateral; acute **37.** Scalene; obtuse
39. Quadrilateral **41.** Pentagon **43.** Triangle
45. Pentagon **47.** Hexagon **49.** 1440° **51.** 900°
53. 2160° **55.** 3240° **57.** 46° **59.** 120° **61.** 43°
63. DW **65.** \$160 **66.** \$22.50 **67.** \$148
68. \$1116.67 **69.** \$33,597.91 **70.** \$413,458.31
71. \$641,566.26 **72.** \$684,337.34 **73.** $m\angle 2 = 67.13°$; $m\angle 3 = 33.07°$; $m\angle 4 = 79.8°$; $m\angle 5 = 67.13°$
75. $m\angle ACB = 50°$; $m\angle CAB = 40°$; $m\angle EBC = 50°$; $m\angle EBA = 40°$; $m\angle AEB = 100°$; $m\angle ADB = 50°$

Margin Exercises, Section 6.2, pp. 396–398

1. 26 cm **2.** 46 in. **3.** 12 cm **4.** 17.5 yd **5.** $27\frac{5}{6}$ in.
6. 40 km **7.** 21 yd **8.** 31.2 km **9.** 70 ft; \$346.50

Exercise Set 6.2, p. 399

1. 17 mm **3.** 15.25 in. **5.** 18 km **7.** 30 ft
9. 79.14 cm **11.** 88 ft **13.** 182 mm
15. 826 m; \$6021.54 **17.** 122 cm **19.** (a) 228 ft; (b) \$1046.52 **21.** DW **23.** \$19.20 **24.** \$96
25. 1000 **26.** 1331 **27.** 225 **28.** 484 **29.** 49
30. 64 **31.** 5% **32.** 11% **33.** 9 ft

Margin Exercises, Section 6.3, pp. 401–406

1. 8 cm² **2.** 56 km² **3.** $18\frac{3}{8}$ yd² **4.** 144 km²
5. 118.81 m² **6.** $12\frac{1}{4}$ yd² **7.** 43.8 cm² **8.** 12.375 km²
9. 96 m² **10.** 18.7 cm² **11.** 100 m² **12.** 88 cm²
13. 228 in²

Exercise Set 6.3, p. 407

1. 15 km² **3.** 1.4 in² **5.** $6\frac{1}{4}$ yd² **7.** 8100 ft² **9.** 50 ft²
11. 169.883 cm² **13.** $41\frac{2}{9}$ in² **15.** 484 ft²
17. 3237.61 km² **19.** $28\frac{57}{64}$ yd² **21.** 32 cm² **23.** 60 in²
25. 104 ft² **27.** 45.5 in² **29.** 8.05 cm² **31.** 297 cm²
33. 7 m² **35.** 1197 m² **37.** (a) About 8473 ft²; (b) about \$102 **39.** 630.36 ft² **41.** (a) 819.75 ft²; (b) 10 gal; (c) \$179.50 **43.** 80 cm² **45.** 675 cm²
47. 21 cm² **49.** 852.04 ft² **51.** DW **53.** Three
54. Parallel **55.** Perpendicular **56.** Prime
57. Angle **58.** Cents; dollars **59.** Perimeter
60. Multiplicative; additive **61.** 16,914 in²

Margin Exercises, Section 6.4, pp. 412–416

1. 9 in. **2.** 5 ft **3.** 62.8 m **4.** 88 m **5.** 34.296 yd
6. $78\frac{4}{7}$ km² **7.** 339.62 cm² **8.** 12-ft–diameter flower bed, by about 13.04 ft²

Exercise Set 6.4, p. 417

1. 14 cm; 44 cm; 154 cm² **3.** $1\frac{1}{2}$ in.; $4\frac{5}{7}$ in.; $1\frac{43}{56}$ in²
5. 16 ft; 100.48 ft; 803.84 ft² **7.** 0.7 cm; 4.396 cm; 1.5386 cm² **9.** 3 cm; 18.84 cm; 28.26 cm² **11.** 153.86 ft²
13. 2.5 cm; 1.25 cm; 4.90625 cm² **15.** 3.454 ft
17. 65.94 yd² **19.** 45.68 ft **21.** 26.84 yd **23.** 45.7 yd
25. 100.48 m² **27.** 6.9972 cm² **29.** 64.4214 in²
31. DW **33.** 16 **34.** 289 **35.** 125 **36.** 64
37. 37.5% **38.** 62.5% **39.** $66.\overline{6}$% **40.** 20%
41. 41.01% **42.** \$730 **43.** 10 cans **44.** 5 lb
45. 3.1416 **47.** $3d$; πd; circumference of one ball, since $\pi > 3$

Margin Exercises, Section 6.5, pp. 421–426

1. 12 cm³ **2.** 3087 in³ **3.** 128 ft³ **4.** 38.4 m³; 75.2 m²
5. $1\frac{7}{8}$ ft³; $10\frac{1}{4}$ ft² **6.** 785 ft³ **7.** 67,914 m³
8. $91{,}989\frac{1}{3}$ ft³ **9.** 38.77272 cm³ **10.** 1695.6 m³
11. 528 in³ **12.** $83.7\overline{3}$ mm³

Calculator Corner, p. 425

1. Left to the student **2.** Left to the student

Exercise Set 6.5, p. 427

1. 768 cm³; 512 cm² **3.** 45 in³; 87 in²
5. 75 m³; 145 m² **7.** $357\frac{1}{2}$ yd³; $311\frac{1}{2}$ yd² **9.** 803.84 in³
11. 353.25 cm³ **13.** 41,580,000 yd³
15. $4{,}186{,}666\frac{2}{3}$ in³ **17.** 124.72 m³ **19.** $1950\frac{101}{168}$ ft³
21. 113,982 ft³ **23.** 24.64 cm³ **25.** 4747.68 cm³
27. 367.38 m³ **29.** 143.72 cm³ **31.** 32,993,441,150 mi³
33. 646.74 cm³ **35.** 61,600 cm³ **37.** 5832 yd³

39. DW **41.** $\frac{7}{20}$ **42.** $\frac{171}{200}$ **43.** $\frac{3}{8}$ **44.** $\frac{2}{3}$ **45.** $\frac{5}{6}$ **46.** $\frac{1}{6}$ **47.** 7500 **48.** 46; 2 cc left over **49.** About 57,480 in^3 **51.** 9.425 L **53.** The diameter of the earth at the equator is about 7930 mi. The diameter of the earth between the north and south poles is about 7917 mi. If we use the average of these two diameters (7923.5 mi), the volume of the earth is $\frac{4}{3} \cdot \pi \cdot \left(\frac{7923.5}{2}\right)^3$, or about 260,000,000,000 mi^3.
55. 0.331 m^3

Margin Exercises, Section 6.6, pp. 432–439

1. $\angle 1$ and $\angle 2$; $\angle 1$ and $\angle 4$; $\angle 2$ and $\angle 3$; $\angle 3$ and $\angle 4$
2. 45° **3.** 72° **4.** 5° **5.** $\angle 1$ and $\angle 2$; $\angle 1$ and $\angle 4$; $\angle 2$ and $\angle 3$; $\angle 3$ and $\angle 4$ **6.** 142° **7.** 23° **8.** 90° **9.** Not congruent **10.** Congruent **11.** Not congruent **12.** Congruent **13.** $m\angle 1 = 10°, m\angle 3 = 129°, m\angle 5 = 41°, m\angle 6 = 129°$ **14.** $\angle 1$ and $\angle 3$, $\angle 2$ and $\angle 4$, $\angle 5$ and $\angle 7$, $\angle 6$ and $\angle 8$ **15.** $\angle 2, \angle 3, \angle 6$, and $\angle 7$ **16.** $\angle 2$ and $\angle 7$, $\angle 6$ and $\angle 3$ **17.** $m\angle 7 = m\angle 1 = m\angle 5 = 51°, m\angle 8 = m\angle 2 = m\angle 6 = m\angle 4 = 129°$
18. $\angle CED \cong \angle BEA, \angle ECD \cong \angle EBA, \angle EDC \cong \angle EAB, \angle CEA \cong \angle BED$ **19.** $\angle TPQ \cong \angle TRS, \angle TQP \cong \angle TSR$

Exercise Set 6.6, p. 440

1. 79° **3.** 23° **5.** 32° **7.** 61° **9.** 177° **11.** 41° **13.** 95° **15.** 78° **17.** Not congruent **19.** Congruent **21.** $m\angle 2 = 67°, m\angle 3 = 33°, m\angle 4 = 80°, m\angle 6 = 33°$
23. **(a)** $\angle 1$ and $\angle 3$, $\angle 2$ and $\angle 4$, $\angle 8$ and $\angle 6$, $\angle 7$ and $\angle 5$; **(b)** $\angle 2, \angle 3, \angle 6$, and $\angle 7$; **(c)** $\angle 2$ and $\angle 6$, $\angle 3$ and $\angle 7$
25. $m\angle 6 = m\angle 2 = m\angle 8 = 125°, m\angle 5 = m\angle 3 = m\angle 7 = m\angle 1 = 55°$ **27.** $\angle ABE \cong \angle DCE$, 95°; $\angle BAE \cong \angle CDE$; $\angle AEB \cong \angle DEC$; $\angle BED \cong \angle AEC$ **29.** $\angle AEC \cong \angle DCE$, 50°; $\angle BED \cong \angle EDC$, 41° **31.** $\frac{45}{4}$, or $11\frac{1}{4}$ **32.** $\frac{129}{8}$, or $16\frac{1}{8}$
33. 118 **34.** $\frac{44}{3}$, or $14\frac{2}{3}$

Margin Exercises, Section 6.7, pp. 443–448

1. $\angle A \cong \angle D, \angle B \cong \angle E, \angle C \cong \angle F; \overline{AB} \cong \overline{DE}, \overline{AC} \cong \overline{DF}, \overline{BC} \cong \overline{EF}$ **2.** $\angle N \cong \angle P, \angle M \cong \angle R, \angle O \cong \angle Q; \overline{NM} \cong \overline{PR}, \overline{NO} \cong \overline{PQ}, \overline{MO} \cong \overline{RQ}$ **3.** (a), (c) **4.** (a)
5. (b) **6.** None **7.** SAS **8.** ASA **9.** SSS **10.** SAS
11. $\triangle SRW \cong \triangle STV$ by ASA; $\angle RSW \cong \angle TSV, \overline{RS} \cong \overline{TS}, \overline{SW} \cong \overline{SV}$ **12.** $\triangle GKP \cong \triangle PTR$ by ASA. Thus corresponding parts $\overline{GP}$ and $\overline{PR}$ are congruent, and P is the midpoint of $\overline{GR}$. **13.** $m\angle C = 27°, m\angle B = m\angle D = 153°$
14. $m\angle S = 114°, m\angle P = m\angle R = 66°$ **15.** $QR = 10, SR = 8$ **16.** $EF = 13.6, DE = GF = 20.4$

Exercise Set 6.7, p. 449

1. $\angle A \cong \angle R, \angle B \cong \angle S, \angle C \cong \angle T; \overline{AB} \cong \overline{RS}, \overline{AC} \cong \overline{RT}, \overline{BC} \cong \overline{ST}$ **3.** $\angle D \cong \angle G, \angle E \cong \angle H, \angle F \cong \angle K; \overline{DE} \cong \overline{GH}, \overline{DF} \cong \overline{GK}, \overline{EF} \cong \overline{HK}$ **5.** $\angle X \cong \angle U, \angle Y \cong \angle V, \angle Z \cong \angle W; \overline{XY} \cong \overline{UV}, \overline{XZ} \cong \overline{UW}, \overline{YZ} \cong \overline{VW}$ **7.** $\angle A \cong \angle F, \angle C \cong \angle D, \angle B \cong \angle E; \overline{AC} \cong \overline{FD}, \overline{AB} \cong \overline{FE}, \overline{CB} \cong \overline{DE}$ **9.** $\angle M \cong \angle Q, \angle N \cong \angle P, \angle O \cong \angle S; \overline{MN} \cong \overline{QP}, \overline{MO} \cong \overline{QS}, \overline{NO} \cong \overline{PS}$
11. No **13.** Yes **15.** Yes **17.** No **19.** Yes
21. Yes **23.** Yes **25.** Yes **27.** Yes **29.** ASA

31. SAS **33.** SSS or SAS **35.** $\overline{PR} \cong \overline{TR}, \overline{SR} \cong \overline{QR}$, $\angle PRQ \cong \angle TRS$ (vertical angles); $\triangle PRQ \cong \triangle TRS$ by SAS
37. $m\angle GLK = m\angle GLM = 90°, \angle GLK \cong \angle GLM, \overline{GL} \cong \overline{GL}, \overline{KL} \cong \overline{ML}$; $\triangle KLG \cong \triangle MLG$ by SAS **39.** $\overline{AE} \cong \overline{CD}, \overline{AB} \cong \overline{CB}, \overline{EB} \cong \overline{DB}$; $\triangle AEB \cong \triangle CDB$ by SSS **41.** $\triangle LKH \cong \triangle GKJ$ by SAS; $\angle HLK \cong \angle JGK, \angle LHK \cong \angle GJK, \overline{LH} \cong \overline{GJ}$
43. $\triangle PED \cong \triangle PFG$ by ASA. As corresponding parts, $\overline{EP} \cong \overline{FP}$; thus P is the midpoint of $\overline{EF}$. **45.** $m\angle A = 70°, m\angle D = m\angle B = 110°$ **47.** $m\angle M = 71°, m\angle J = m\angle L = 109°$ **49.** $TU = 9, NU = 15$ **51.** $KL = 3\frac{1}{2}, ML = JK = 7\frac{1}{2}$ **53.** $AC = 28, ED = 38$ **55.** 45.2%
56. $33\frac{1}{3}\%$ **57.** 55% **58.** 88% **59.** $\frac{2.7}{13.1}; \frac{13.1}{2.7}$ **60.** $\frac{1}{4}; \frac{3}{4}$
61. 1.75 **62.** 2.34 **63.** 0.234 **64.** 0.0234 **65.** 13.85

Margin Exercises, Section 6.8, pp. 454–458

1. (a), (b), (d) **2.** $\overline{PQ} \leftrightarrow \overline{GH}, \overline{QR} \leftrightarrow \overline{HK}, \overline{PR} \leftrightarrow \overline{GK}$, $\angle P \leftrightarrow \angle G, \angle Q \leftrightarrow \angle H, \angle R \leftrightarrow \angle K$ **3.** $\angle J \cong \angle A, \angle K \cong \angle B, \angle L \cong \angle C; \frac{JK}{AB} = \frac{JL}{AC} = \frac{KL}{BC}$ **4.** $\frac{PN}{TS} = \frac{PM}{TR} = \frac{MN}{RS}$
5. $BT = 6\frac{3}{4}, CT = 9$ **6.** $QR = 10$ **7.** 24.75 ft
8. 7.5 ft

Translating for Success, p. 459

1. K **2.** G **3.** B **4.** H **5.** O **6.** M **7.** E
8. A **9.** D **10.** I

Exercise Set 6.8, p. 460

1. $\angle R \leftrightarrow \angle A, \angle S \leftrightarrow \angle B, \angle T \leftrightarrow \angle C, \overline{RS} \leftrightarrow \overline{AB}, \overline{RT} \leftrightarrow \overline{AC}, \overline{ST} \leftrightarrow \overline{BC}$ **3.** $\angle C \leftrightarrow \angle W, \angle B \leftrightarrow \angle J, \angle S \leftrightarrow \angle Z, \overline{CB} \leftrightarrow \overline{WJ}, \overline{CS} \leftrightarrow \overline{WZ}, \overline{BS} \leftrightarrow \overline{JZ}$ **5.** $\angle A \cong \angle R, \angle B \cong \angle S, \angle C \cong \angle T; \frac{AB}{RS} = \frac{AC}{RT} = \frac{BC}{ST}$ **7.** $\angle M \cong \angle C, \angle E \cong \angle L, \angle S \cong \angle F; \frac{ME}{CL} = \frac{MS}{CF} = \frac{ES}{LF}$ **9.** $\frac{PS}{ND} = \frac{SQ}{DM} = \frac{PQ}{NM}$
11. $\frac{TA}{GF} = \frac{TW}{GC} = \frac{AW}{FC}$ **13.** $QR = 10, PR = 8$
15. $EC = 18$ **17.** 36 ft **19.** 100 ft **21.** DW
23. $\frac{147}{5}$, or $29\frac{2}{5}$ **24.** 0.244 **25.** 78 **26.** 61.1611

Concept Reinforcement, p. 462

1. False **2.** True **3.** False **4.** True **5.** False

Summary and Review: Chapter 6, p. 463

1. 54° **2.** 180° **3.** 140° **4.** 90° **5.** Acute
6. Straight **7.** Obtuse **8.** Right **9.** 60°
10. Scalene **11.** Right **12.** 720° **13.** 23 m
14. 4.4 m **15.** 228 ft; 2808 ft^2 **16.** 36 ft; 81 ft^2
17. 17.6 cm; 12.6 cm^2 **18.** 60 cm^2 **19.** 35 mm^2
20. 22.5 m^2 **21.** 29.64 cm^2 **22.** 88 m^2 **23.** $145\frac{5}{9}$ in^2
24. 840 ft^2 **25.** 8 m **26.** $\frac{14}{11}$ in., or $1\frac{3}{11}$ in. **27.** 14 ft
28. 20 cm **29.** 50.24 m **30.** 8 in. **31.** 200.96 m^2
32. $5\frac{1}{11}$ in^2 **33.** 1038.555 ft^2 **34.** 93.6 m^3; 150 m^2
35. 193.2 cm^3; 240.4 cm^2 **36.** 31,400 ft^3 **37.** 4.71 in^3
38. $33.49\overline{3}$ cm^3 **39.** 942 cm^3 **40.** 8° **41.** 85°

42. 147° **43.** 47° **44.** $m\angle 2 = 105°$, $m\angle 3 = 37°$, $m\angle 4 = 38°$, $m\angle 6 = 37°$ **45.** **(a)** $\angle 1$ and $\angle 5$, $\angle 4$ and $\angle 8$, $\angle 3$ and $\angle 7$, $\angle 2$ and $\angle 6$; **(b)** $\angle 4$, $\angle 5$, $\angle 2$, and $\angle 7$; **(c)** $\angle 4$ and $\angle 7$, $\angle 2$ and $\angle 5$ **46.** $m\angle 1 = m\angle 3 = m\angle 7 = m\angle 5 = 45°$, $m\angle 6 = m\angle 2 = m\angle 8 = 135°$
47. $\angle D \cong \angle R$, $\angle H \cong \angle Z$, $\angle J \cong \angle K$; $\overline{DH} \cong \overline{RZ}$, $\overline{DJ} \cong \overline{RK}$, $\overline{HJ} \cong \overline{ZK}$ **48.** $\angle A \cong \angle G$, $\angle B \cong \angle D$, $\angle C \cong \angle F$, $\overline{AB} \cong \overline{GD}$, $\overline{AC} \cong \overline{GF}$, $\overline{BC} \cong \overline{DF}$ **49.** ASA **50.** SSS
51. None **52.** $\overline{IJ} \cong \overline{KJ}$, $\angle HJI \cong \angle LJK$, $\angle HIJ \cong \angle LKJ$; $\triangle JIH \cong \triangle JKL$ by ASA **53.** $m\angle C = 63°$, $m\angle B = m\angle D = 117°$; $BC = 23$, $CD = 13$ **54.** $\angle C \cong \angle F$, $\angle Q \cong \angle A$, $\angle W \cong \angle S$; $\frac{CQ}{FA} = \frac{CW}{FS} = \frac{QW}{AS}$ **55.** $MO = 14$
56. $\mathbf{D_W}$ Linear measure is one-dimensional, area is two-dimensional, and volume is three-dimensional.
57. $\mathbf{D_W}$ Volume of two spheres, each with radius r: $2(\frac{4}{3}\pi r^3) = \frac{8}{3}\pi r^3$; volume of one sphere with radius $2r$: $\frac{4}{3}\pi(2r)^3 = \frac{32}{3}\pi r^3$. The volume of the sphere with radius $2r$ is four times the volume of the two spheres, each with radius r: $\frac{32}{3}\pi r^3 = 4 \cdot \frac{8}{3}\pi r^3$. **58.** 100 ft^2
59. 7.83998704 m^2 **60.** 42.05915 cm^2

Test: Chapter 6, p. 467

1. [6.1b] 90° **2.** [6.1b] 35° **3.** [6.1b] 180°
4. [6.1b] 113° **5.** [6.1c] Right **6.** [6.1c] Acute
7. [6.1c] Straight **8.** [6.1c] Obtuse **9.** [6.1f] 35°
10. [6.1e] Isosceles **11.** [6.1e] Obtuse **12.** [6.1f] 540°
13. [6.2a], [6.3a] 32.82 cm; 65.894 cm^2
14. [6.2a], [6.3a] $19\frac{1}{2}$ in.; $23\frac{49}{64}$ in^2 **15.** [6.3b] 25 cm^2
16. [6.3b] 12 m^2 **17.** [6.3b] 18 ft^2 **18.** [6.4a] $\frac{1}{4}$ in.
19. [6.4a] 9 cm **20.** [6.4b] $\frac{11}{14}$ in. **21.** [6.4c] 254.34 cm^2
22. [6.4d] 65.46 km; 103.815 km^2 **23.** [6.5a] 84 cm^3; 142 cm^2 **24.** [6.5e] 420 in^3 **25.** [6.5b] 1177.5 ft^3
26. [6.5c] $4186.\overline{6}$ yd^3 **27.** [6.5d] 113.04 cm^3
28. [6.6a] 149° **29.** [6.6a] 11° **30.** [6.6c] $m\angle 2 = 110°$, $m\angle 3 = 8°$, $m\angle 4 = 62°$, $m\angle 6 = 8°$ **31.** [6.6d] $m\angle 6 = m\angle 2 = m\angle 8 = 120°$, $m\angle 5 = m\angle 3 = m\angle 7 = m\angle 1 = 60°$
32. [6.7a] $\angle C \cong \angle A$, $\angle W \cong \angle T$, $\angle S \cong \angle Z$, $\overline{CW} \cong \overline{AT}$, $\overline{WS} \cong \overline{TZ}$, $\overline{SC} \cong \overline{ZA}$ **33.** [6.7a] SAS **34.** [6.7a] None
35. [6.7a] ASA **36.** [6.7a] None
37. [6.7b] $m\angle G = 105°$, $m\angle D = m\angle F = 75°$; $EF = 11$, $DE = GF = 20$ **38.** [6.7b] $LJ = 6.4$, $KM = 6$
39. [6.8a] $\angle E \cong \angle T$, $\angle R \cong \angle G$, $\angle S \cong \angle F$; $\frac{ER}{TG} = \frac{RS}{GF} = \frac{SE}{FT}$
40. [6.8b] $EK = 18$, $ZK = 27$ **41.** [6.3a] 2 ft^2
42. [6.3b] 1.875 ft^2 **43.** [6.5a] 0.65 ft^3
44. [6.5d] 0.033 ft^3 **45.** [6.5b] 0.055 ft^3

CHAPTER 7

Margin Exercises, Section 7.1, pp. 472–476

1. $14{,}410 + x = 15{,}300$; 890 ft **2.** 64 **3.** 28 **4.** 60
5. 192 ft^2 **6.** 25 **7.** 16 **8.** 12 hr **9.** $x - 8$
10. $y + 8$, or $8 + y$ **11.** $m - 4$ **12.** $\frac{1}{2}p$
13. $6 + 8x$, or $8x + 6$ **14.** $a - b$
15. $59\%x$, or $0.59x$ **16.** $xy - 200$ **17.** $p + q$

Calculator Corner, p. 474

1. 56 **2.** 11.9 **3.** 1.8 **4.** 34,427.16 **5.** 20.1
6. 29.9

Exercise Set 7.1, p. 477

1. 32 min; 69 min; 81 min **3.** 1935 m^2 **5.** 260 mi
7. 24 ft^2 **9.** 56 **11.** 8 **13.** 1 **15.** 6 **17.** 2
19. $b + 7$, or $7 + b$ **21.** $c - 12$ **23.** $q + 4$, or $4 + q$
25. $a + b$, or $b + a$ **27.** $x \div y$, or $\frac{x}{y}$, or x/y, or $x \cdot \frac{1}{y}$
29. $x + w$, or $w + x$ **31.** $n - m$ **33.** $x + y$, or $y + x$
35. $2z$ **37.** $3m$ **39.** $4a + 6$, or $6 + 4a$ **41.** $xy - 8$
43. $2t - 5$ **45.** $3n + 11$, or $11 + 3n$
47. $4x + 3y$, or $3y + 4x$
49. $89\%s$, or $0.89s$, where s is the salary **51.** $s + 0.05s$
53. $65t$ miles **55.** $\$50 - x$ **57.** $\mathbf{D_W}$ **59.** $2 \cdot 3 \cdot 3 \cdot 3$
60. $2 \cdot 2 \cdot 2 \cdot 2 \cdot 2$ **61.** $2 \cdot 2 \cdot 3 \cdot 3 \cdot 3$
62. $2 \cdot 2 \cdot 2 \cdot 2 \cdot 2 \cdot 2 \cdot 3$ **63.** $3 \cdot 11 \cdot 31$ **64.** 18
65. 96 **66.** 60 **67.** 96 **68.** 396 **69.** $\frac{1}{4}$ **71.** 0

Margin Exercises, Section 7.2, pp. 482–488

1. 8; −5 **2.** 125; −50 **3.** −3 **4.** −10; 156
5. −120; 50; −80 **6.**

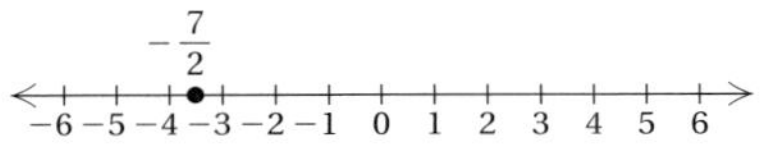

7.

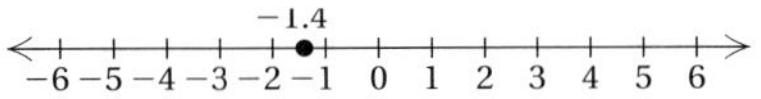

8.

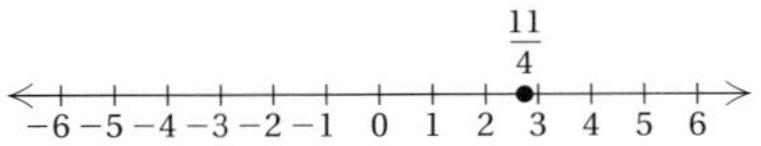

9. −0.375

10. $-0.\overline{54}$ **11.** $1.\overline{3}$ **12.** < **13.** < **14.** > **15.** >
16. > **17.** < **18.** < **19.** > **20.** $7 > -5$
21. $4 < x$ **22.** False **23.** True **24.** True **25.** 8
26. 9 **27.** $\frac{2}{3}$ **28.** 5.6

Calculator Corner, p. 483

1. −0.75 **2.** −0.45 **3.** −0.125 **4.** −1.8 **5.** −0.675
6. −0.6875 **7.** −3.5 **8.** −0.76

Calculator Corner, p. 484

1. 8.717797887 **2.** 17.80449381 **3.** 67.08203932
4. 35.4807407 **5.** 3.141592654 **6.** 91.10618695
7. 530.9291585 **8.** 138.8663978

Calculator Corner, p. 487

1. 5 **2.** 17 **3.** 0 **4.** 6.48 **5.** 12.7 **6.** 0.9
7. $\frac{5}{7}$ **8.** $\frac{4}{3}$

Exercise Set 7.2, p. 489

1. −34,000,000 **3.** 24; −2 **5.** 950,000,000; −460
7. Alley Cats: −34; Strikers: 34
9.

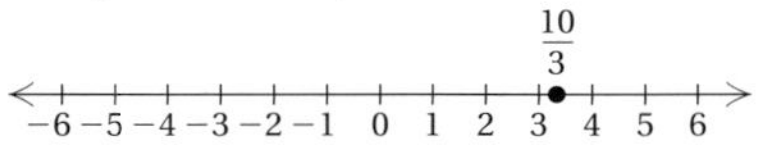

11.

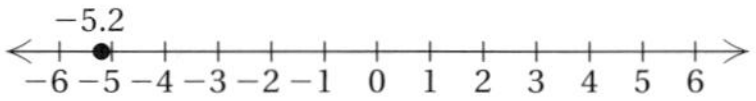

13.

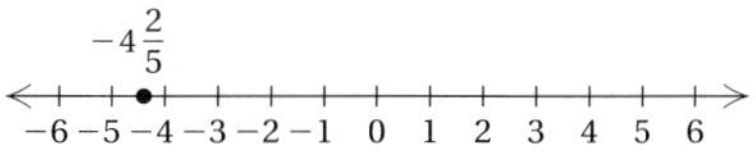

15. −0.875
17. $0.8\overline{3}$ **19.** $-1.1\overline{6}$ **21.** $0.\overline{6}$ **23.** 0.1 **25.** −0.5
27. 0.16 **29.** > **31.** < **33.** < **35.** < **37.** >
39. < **41.** > **43.** < **45.** < **47.** > **49.** <
51. < **53.** True **55.** False **57.** $x < -6$
59. $y \geq -10$ **61.** 3 **63.** 10 **65.** 0 **67.** 30.4
69. $\frac{2}{3}$ **71.** 0 **73.** $3\frac{5}{8}$ **75.** DW **77.** 0.63
78. 0.238 **79.** 1.1 **80.** 0.2276 **81.** 52% **82.** 125%
83. $83.\overline{3}\%$, or $83\frac{1}{3}\%$ **84.** 59.375%, or $59\frac{3}{8}\%$
85. $-\frac{5}{6}, -\frac{3}{4}, -\frac{2}{3}, \frac{1}{6}, \frac{3}{8}, \frac{1}{2}$
87. −8.76, −−5.16, −4.24, −2.13, 1.85, 5.23
89. $\frac{1}{9}$ **91.** $5\frac{5}{9}$, or $\frac{50}{9}$

Margin Exercises, Section 7.3, pp. 492–496

1. −3 **2.** −3 **3.** −5 **4.** 4 **5.** 0 **6.** −2 **7.** −11
8. −12 **9.** 2 **10.** −4 **11.** −2 **12.** 0 **13.** −22
14. 3 **15.** 0.53 **16.** 2.3 **17.** −7.7 **18.** −6.2
19. $-\frac{2}{9}$ **20.** $-\frac{19}{20}$ **21.** −58 **22.** −56 **23.** −14
24. −12 **25.** 4 **26.** −8.7 **27.** 7.74 **28.** $\frac{8}{9}$ **29.** 0
30. −12 **31.** −14; **32.** −1; 1 **33.** 19; −19
34. 1.6; −1.6 **35.** $-\frac{2}{3}; \frac{2}{3}$ **36.** $\frac{9}{8}; -\frac{9}{8}$ **37.** 4
38. 13.4 **39.** 0 **40.** $-\frac{1}{4}$ **41.** −2 students

Exercise Set 7.3, p. 497

1. −7 **3.** −6 **5.** 0 **7.** −8 **9.** −7 **11.** −27
13. 0 **15.** −42 **17.** 0 **19.** 0 **21.** 3 **23.** −9
25. 7 **27.** 0 **29.** 35 **31.** −3.8 **33.** −8.1
35. $-\frac{1}{5}$ **37.** $-\frac{7}{9}$ **39.** $-\frac{3}{8}$ **41.** $-\frac{19}{24}$ **43.** $\frac{1}{24}$
45. $\frac{8}{15}$ **47.** $\frac{16}{45}$ **49.** 37 **51.** 50 **53.** −1409
55. −24 **57.** 26.9 **59.** −8 **61.** $\frac{13}{8}$ **63.** −43
65. $\frac{4}{3}$ **67.** 24 **69.** $\frac{3}{8}$ **71.** 13,796 ft **73.** −3°F
75. −$20,300 **77.** He owes $85. **79.** DW **81.** 0.57
82. 0.713 **83.** 0.238 **84.** 0.92875 **85.** 125%
86. 12.5% **87.** 52% **88.** 40.625%
89. All positive **91.** (b)

Margin Exercises, Section 7.4, pp. 500–502

1. −10 **2.** 3 **3.** −5 **4.** −1 **5.** 2 **6.** −4 **7.** −2
8. −11 **9.** 4 **10.** −2 **11.** −6 **12.** −16 **13.** 7.1
14. 3 **15.** 0 **16.** $\frac{3}{2}$ **17.** −8 **18.** 7 **19.** −3
20. −23.3 **21.** 0 **22.** −9 **23.** $-\frac{6}{5}$ **24.** 12.7
25. 214°F higher

Exercise Set 7.4, p. 503

1. −7 **3.** −6 **5.** 0 **7.** −4 **9.** −7 **11.** −6
13. 0 **15.** 14 **17.** 11 **19.** −14 **21.** 5 **23.** −1
25. 18 **27.** −3 **29.** −21 **31.** 5 **33.** −8 **35.** 12
37. −23 **39.** −68 **41.** −73 **43.** 116 **45.** 0
47. −1 **49.** $\frac{1}{12}$ **51.** $-\frac{17}{12}$ **53.** $\frac{1}{8}$ **55.** 19.9
57. −8.6 **59.** −0.01 **61.** −193 **63.** 500
65. −2.8 **67.** −3.53 **69.** $-\frac{1}{2}$ **71.** $\frac{6}{7}$ **73.** $-\frac{41}{30}$
75. $-\frac{2}{15}$ **77.** $-\frac{1}{48}$ **79.** $-\frac{43}{60}$ **81.** 37 **83.** −62
85. −139 **87.** 6 **89.** 108.5 **91.** $\frac{1}{4}$ **93.** 2319 m
95. $347.94 **97.** **(a)** 77; **(b)** −41 **99.** 381 ft **101.** DW
103. 100.5 **104.** 226 **105.** 13 **106.** 50 **107.** $\frac{11}{12}$
108. $\frac{41}{64}$ **109.** False; $3 - 0 \neq 0 - 3$ **111.** True
113. True

Margin Exercises, Section 7.5, pp. 507–510

1. 20; 10; 0; −10; −20; −30 **2.** −18 **3.** −100 **4.** −80
5. $-\frac{5}{9}$ **6.** −30.033 **7.** $-\frac{7}{10}$ **8.** −10; 0; 10; 20; 30
9. 27 **10.** 32 **11.** 35 **12.** $\frac{20}{63}$ **13.** $\frac{2}{3}$ **14.** 13.455
15. −30 **16.** 30 **17.** 0 **18.** $-\frac{8}{3}$ **19.** 0
20. 0 **21.** −30 **22.** −30.75 **23.** $-\frac{5}{3}$ **24.** 120
25. −120 **26.** 6 **27.** 4; −4 **28.** 9; −9 **29.** 48; 48
30. 55°C

Exercise Set 7.5, p. 511

1. −8 **3.** −48 **5.** −24 **7.** −72 **9.** 16 **11.** 42
13. −120 **15.** −238 **17.** 1200 **19.** 98 **21.** −72
23. −12.4 **25.** 30 **27.** 21.7 **29.** $-\frac{2}{5}$ **31.** $\frac{1}{12}$
33. −17.01 **35.** $-\frac{5}{12}$ **37.** 420 **39.** $\frac{2}{7}$ **41.** −60

43. 150 **45.** $-\frac{2}{45}$ **47.** 1911 **49.** 50.4 **51.** $\frac{10}{189}$
53. -960 **55.** 17.64 **57.** $-\frac{5}{784}$ **59.** 0 **61.** -720
63. $-30{,}240$ **65.** 1 **67.** 16, -16; 16, -16
69. 441; -147 **71.** 20; 20 **73.** -2; 2 **75.** -20 lb
77. $-54°$C **79.** \$12.71 **81.** -32 m **83.** D_W
85. 180 **86.** $2 \cdot 2 \cdot 2 \cdot 2 \cdot 2 \cdot 2 \cdot 2 \cdot 2 \cdot 2 \cdot 3 \cdot 3$ **87.** $\frac{2}{3}$
88. $\frac{8}{9}$ **89.** $\frac{6}{11}$ **90.** $\frac{41}{265}$ **91.** $\frac{11}{32}$ **92.** $\frac{37}{67}$ **93.** $\frac{1}{24}$
94. 6 **95.** (a)

97.
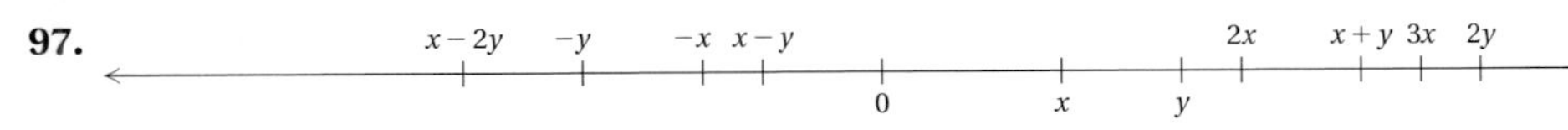

Exercise Set 7.6, p. 520

1. -8 **3.** -14 **5.** -3 **7.** 3 **9.** -8 **11.** 2
13. -12 **15.** -8 **17.** Not defined **19.** 0 **21.** $\frac{7}{15}$
23. $-\frac{13}{47}$ **25.** $\frac{1}{13}$ **27.** $\frac{1}{4.3}$ **29.** -7.1 **31.** $\frac{q}{p}$
33. $4y$ **35.** $\frac{3b}{2a}$ **37.** $4 \cdot \left(\frac{1}{17}\right)$ **39.** $8 \cdot \left(-\frac{1}{13}\right)$
41. $13.9 \cdot \left(-\frac{1}{1.5}\right)$ **43.** $x \cdot y$ **45.** $(3x + 4)\left(\frac{1}{5}\right)$
47. $(5a - b)\left(\frac{1}{5a + b}\right)$ **49.** $-\frac{9}{8}$ **51.** $\frac{5}{3}$ **53.** $\frac{9}{14}$
55. $\frac{9}{64}$ **57.** -2 **59.** $\frac{11}{13}$ **61.** -16.2 **63.** Not defined
65. 44.3% **67.** -5.1% **69.** D_W **71.** 33 **72.** 129
73. 1 **74.** 1296 **75.** $\frac{22}{39}$ **76.** 0.477 **77.** 87.5%
78. $\frac{2}{3}$ **79.** $\frac{9}{8}$ **80.** $\frac{128}{625}$ **81.** $\frac{1}{-10.5}$; -10.5, the reciprocal of the reciprocal is the original number
83. Negative **85.** Positive **87.** Negative

Margin Exercises, Section 7.6, pp. 514–519

1. -2 **2.** 5 **3.** -3 **4.** 8 **5.** -6 **6.** $-\frac{30}{7}$
7. Not defined **8.** 0 **9.** $\frac{3}{2}$ **10.** $-\frac{4}{5}$ **11.** $-\frac{1}{3}$
12. -5 **13.** $\frac{1}{1.6}$ **14.** $\frac{2}{3}$
15.

NUMBER	OPPOSITE	RECIPROCAL
$\frac{2}{3}$	$-\frac{2}{3}$	$\frac{3}{2}$
$-\frac{5}{4}$	$\frac{5}{4}$	$-\frac{4}{5}$
0	0	Not defined
1	-1	1
-8	8	$-\frac{1}{8}$
-4.5	4.5	$-\frac{1}{4.5}$

16. $\frac{4}{7} \cdot \left(-\frac{5}{3}\right)$ **17.** $5 \cdot \left(-\frac{1}{8}\right)$ **18.** $(a - b) \cdot \left(\frac{1}{7}\right)$
19. $-23 \cdot a$ **20.** $-5 \cdot \left(\frac{1}{7}\right)$ **21.** $-\frac{20}{21}$ **22.** $-\frac{12}{5}$
23. $\frac{16}{7}$ **24.** -7 **25.** $\frac{5}{-6}, -\frac{5}{6}$ **26.** $\frac{-8}{7}, \frac{8}{-7}$
27. $\frac{-10}{3}, -\frac{10}{3}$ **28.** $-3.4°$F per minute

Calculator Corner, p. 519

1. -4 **2.** -0.3 **3.** -12 **4.** -9.5 **5.** -12 **6.** 2.7
7. -2 **8.** -5.7 **9.** -32 **10.** -1.8 **11.** 35
12. 14.44 **13.** -2 **14.** -0.8 **15.** 1.4 **16.** 4

Margin Exercises, Section 7.7, pp. 523–531

1.

Value	$x + x$	$2x$
$x = 3$	6	6
$x = -6$	-12	-12
$x = 4.8$	9.6	9.6

2.

Value	$x + 3x$	$5x$
$x = 2$	8	10
$x = -6$	-24	-30
$x = 4.8$	19.2	24

3. $\frac{6}{8}$ **4.** $\frac{3t}{4t}$ **5.** $\frac{3}{4}$ **6.** $-\frac{4}{3}$ **7.** $\frac{x}{8}$ **8.** $\frac{3}{4q}$ **9.** 1; 1
10. -10; -10 **11.** $9 + x$ **12.** qp
13. $t + xy$, or $yx + t$, or $t + yx$ **14.** 19; 19 **15.** 150; 150
16. $(r + s) + 7$ **17.** $(9a)b$
18. $(4t)u$, $(tu)4$, $t(4u)$; answers may vary
19. $(2 + r) + s$, $(r + s) + 2$, $s + (r + 2)$; answers may vary
20. **(a)** 63; **(b)** 63 **21.** **(a)** 80; **(b)** 80 **22.** **(a)** 28; **(b)** 28
23. **(a)** 8; **(b)** 8 **24.** **(a)** -4; **(b)** -4 **25.** **(a)** -25; **(b)** -25

26. $5x, -8y, 3$ **27.** $-4y, -2x, 3z$ **28.** $3x - 15$
29. $5x + 5$ **30.** $\frac{3}{5}p + \frac{3}{5}q - \frac{3}{5}t$ **31.** $-2x + 6$
32. $5x - 10y + 20z$ **33.** $-5x + 10y - 20z$
34. Associative law of multiplication
35. Identity property of 1
36. Commutative law of addition
37. Distributive law of multiplication over addition
38. Identity property of 0
39. Commutative law of multiplication
40. Associative law of addition
41. $6(x - 2)$ **42.** $3(x - 2y + 3)$ **43.** $b(x + y - z)$
44. $2(8a - 18b + 21)$ **45.** $\frac{1}{8}(3x - 5y + 7)$
46. $-4(3x - 8y + 4z)$ **47.** $3x$ **48.** $6x$ **49.** $-8x$
50. $0.59x$ **51.** $3x + 3y$ **52.** $-4x - 5y - 7$
53. $-\frac{2}{3} + \frac{1}{10}x + \frac{7}{9}y$

Exercise Set 7.7, p. 532

1. $\frac{3y}{5y}$ **3.** $\frac{10x}{15x}$ **5.** $\frac{2x}{x^2}$ **7.** $-\frac{3}{2}$ **9.** $-\frac{7}{6}$ **11.** $\frac{4s}{3}$
13. $8 + y$ **15.** nm **17.** $xy + 9$, or $9 + yx$
19. $c + ab$, or $ba + c$ **21.** $(a + b) + 2$ **23.** $8(xy)$
25. $a + (b + 3)$ **27.** $(3a)b$
29. $2 + (b + a), (2 + a) + b, (b + 2) + a$; answers may vary **31.** $(5 + w) + v; (v + 5) + w; (w + v) + 5$; answers may vary
33. $(3x)y, y(x \cdot 3), 3(yx)$; answers may vary
35. $a(7b), b(7a), (7b)a$; answers may vary **37.** $2b + 10$
39. $7 + 7t$ **41.** $30x + 12$ **43.** $7x + 28 + 42y$
45. $7x - 21$ **47.** $-3x + 21$ **49.** $\frac{2}{3}b - 4$
51. $7.3x - 14.6$ **53.** $-\frac{3}{5}x + \frac{3}{5}y - 6$
55. $45x + 54y - 72$ **57.** $-4x + 12y + 8z$
59. $-3.72x + 9.92y - 3.41$ **61.** $4x, 3z$ **63.** $7x, 8y, -9z$
65. $2(x + 2)$ **67.** $5(6 + y)$ **69.** $7(2x + 3y)$
71. $5(x + 2 + 3y)$ **73.** $8(x - 3)$
75. $4(-y + 8)$, or $-4(y - 8)$ **77.** $2(4x + 5y - 11)$
79. $a(x - 1)$ **81.** $a(x - y - z)$
83. $6(-3x + 2y + 1)$, or $-6(3x - 2y - 1)$
85. $\frac{1}{3}(2x - 5y + 1)$ **87.** $19a$ **89.** $9a$ **91.** $8x + 9z$
93. $7x + 15y^2$ **95.** $-19a + 88$ **97.** $4t + 6y - 4$
99. b **101.** $\frac{13}{4}y$ **103.** $8x$ **105.** $5n$ **107.** $-16y$
109. $17a - 12b - 1$ **111.** $4x + 2y$ **113.** $7x + y$
115. $0.8x + 0.5y$ **117.** $\frac{35}{6}a + \frac{3}{2}b - 42$ **119.** D_W
121. 144 **122.** 72 **123.** 144 **124.** 60 **125.** 32
126. 72 **127.** 90 **128.** 108 **129.** $\frac{89}{48}$ **130.** $\frac{5}{24}$
131. $-\frac{5}{24}$ **132.** 30% **133.** Not equivalent; $3 \cdot 2 + 5 \neq 3 \cdot 5 + 2$ **135.** Equivalent; commutative law of addition **137.** $q(1 + r + rs + rst)$

Margin Exercises, Section 7.8, pp. 536–540

1. $-x - 2$ **2.** $-5x - 2y - 8$ **3.** $-6 + t$ **4.** $-x + y$
5. $4a - 3t + 10$ **6.** $-18 + m + 2n - 4z$ **7.** $2x - 9$
8. $3y + 2$ **9.** $2x - 7$ **10.** $3y + 3$ **11.** $-2a + 8b - 3c$
12. $-9x - 8y$ **13.** $-16a + 18$ **14.** $-26a + 41b - 48c$
15. $3x - 7$ **16.** $-18.6x - 19y$ **17.** 2 **18.** 18 **19.** 6
20. 17 **21.** $5x - y - 8$ **22.** -1237 **23.** 8 **24.** 4
25. 317 **26.** -12

Calculator Corner, p. 540

1. -11 **2.** 9 **3.** 114 **4.** 117,649 **5.** $-1{,}419{,}857$
6. $-1{,}124{,}864$ **7.** $-117{,}649$ **8.** $-1{,}419{,}857$
9. $-1{,}124{,}864$ **10.** -4 **11.** -2 **12.** 787

Translating for Success, p. 541

1. N **2.** I **3.** A **4.** K **5.** J **6.** F **7.** M
8. B **9.** G **10.** E

Exercise Set 7.8, p. 542

1. $-2x - 7$ **3.** $-8 + x$ **5.** $-4a + 3b - 7c$
7. $-6x + 8y - 5$ **9.** $-3x + 5y + 6$ **11.** $8x + 6y + 43$
13. $5x - 3$ **15.** $-3a + 9$ **17.** $5x - 6$ **19.** $-19x + 2y$
21. $9y - 25z$ **23.** $-7x + 10y$ **25.** $37a - 23b + 35c$
27. 7 **29.** -40 **31.** 19 **33.** $12x + 30$ **35.** $3x + 30$
37. $9x - 18$ **39.** $-4x - 64$ **41.** -7 **43.** -7
45. -16 **47.** -334 **49.** 14 **51.** 1880 **53.** 12
55. 8 **57.** -86 **59.** 37 **61.** -1 **63.** -10
65. -67 **67.** -7988 **69.** -3000 **71.** 60 **73.** 1
75. 10 **77.** $-\frac{13}{45}$ **79.** $-\frac{23}{18}$ **81.** -122 **83.** D_W
85. Integers **86.** Additive inverses
87. Commutative law **88.** Identity property of 1
89. Associative law **90.** Associative law
91. Multiplicative inverses **92.** Identity property of 0
93. $6y - (-2x + 3a - c)$ **95.** $6m - (-3n + 5m - 4b)$
97. $-2x - f$ **99.** **(a)** 52; 52; 28.130169; **(b)** -24; -24; -108.307025 **101.** -6

Concept Reinforcement, p. 546

1. True **2.** True **3.** False **4.** True **5.** False
6. True **7.** False

Summary and Review: Chapter 7, p. 546

1. 4 **2.** $19\%x$, or $0.19x$ **3.** $-45, 72$ **4.** 38
5. -2.5
$-6\ -5\ -4\ -3\ -2\ -1\ 0\ 1\ 2\ 3\ 4\ 5\ 6$
6. $\frac{8}{9}$
$-6\ -5\ -4\ -3\ -2\ -1\ 0\ 1\ 2\ 3\ 4\ 5\ 6$
7. $<$ **8.** $>$
9. $>$ **10.** $<$ **11.** -3.8 **12.** $\frac{3}{4}$ **13.** $\frac{8}{3}$ **14.** $-\frac{1}{7}$
15. 34 **16.** 5 **17.** -3 **18.** -4 **19.** -5 **20.** 1
21. $-\frac{7}{5}$ **22.** -7.9 **23.** 54 **24.** -9.18 **25.** $-\frac{2}{7}$

26. -210 27. -7 28. -3 29. $\frac{3}{4}$ 30. 40.4
31. -2 32. 2 33. -9 34. 8-yd gain 35. $-\$130$
36. \$4.64 37. \$18.95 38. $15x - 35$ 39. $-8x + 10$
40. $4x + 15$ 41. $-24 + 48x$ 42. $2(x - 7)$
43. $6(-x + 1)$, or $-6(x - 1)$ 44. $5(x + 2)$
45. $3(-x + 4y - 4)$, or $-3(x - 4y + 4)$ 46. $7a - 3b$
47. $-2x + 5y$ 48. $5x - y$ 49. $-a + 8b$ 50. $-3a + 9$
51. $-2b + 21$ 52. 6 53. $12y - 34$ 54. $5x + 24$
55. $-15x + 25$ 56. True 57. False 58. $x > -3$
59. D_W If the sum of two numbers is 0, they are opposites, or additive inverses of each other. For every real number a, the opposite of a can be named $-a$, and $a + (-a) = (-a) + a = 0$. 60. D_W No; $|0| = 0$, and 0 is not positive. 61. $-\frac{5}{8}$ 62. -2.1 63. 1000
64. $4a + 2b$

Test: Chapter 7, p. 549

1. [7.1a] 6 2. [7.1b] $x - 9$ 3. [7.1a] 240 ft^2
4. [7.2d] $<$ 5. [7.2d] $>$ 6. [7.2d] $>$ 7. [7.2d] $<$
8. [7.2e] 7 9. [7.2e] $\frac{9}{4}$ 10. [7.2e] 2.7 11. [7.3b] $-\frac{2}{3}$
12. [7.3b] 1.4 13. [7.3b] 8 14. [7.6b] $-\frac{1}{2}$
15. [7.6b] $\frac{7}{4}$ 16. [7.4a] 7.8 17. [7.3a] -8
18. [7.3a] $\frac{7}{40}$ 19. [7.4a] 10 20. [7.4a] -2.5
21. [7.4a] $\frac{7}{8}$ 22. [7.5a] -48 23. [7.5a] $\frac{3}{16}$
24. [7.6a] -9 25. [7.6c] $\frac{3}{4}$ 26. [7.6c] -9.728
27. [7.8d] -173 28. [7.8d] -5 29. [7.4b] 14°F
30. [7.3c], [7.4b] Up 15 points 31. [7.5b] 16,080
32. [7.6d] $\frac{33}{35}$°C per minute 33. [7.7c] $18 - 3x$
34. [7.7c] $-5y + 5$ 35. [7.7d] $2(6 - 11x)$
36. [7.7d] $7(x + 3 + 2y)$ 37. [7.4a] 12
38. [7.8b] $2x + 7$ 39. [7.8b] $9a - 12b - 7$
40. [7.8c] $68y - 8$ 41. [7.8d] -4 42. [7.8d] 448
43. [7.2d] $-2 \geq x$ 44. [7.2e], [7.8d] 15
45. [7.8c] $4a$ 46. [7.7e] $4x + 4y$

CHAPTER 8

Margin Exercises, Section 8.1, pp. 552–555

1. False 2. True 3. Neither 4. Yes 5. No
6. No 7. Yes 8. 9 9. -13 10. 22 11. 13.2
12. -6.5 13. -2 14. $\frac{31}{8}$

Exercise Set 8.1, p. 556

1. Yes 3. No 5. No 7. Yes 9. No 11. No
13. 4 15. -20 17. -14 19. -18 21. 15
23. -14 25. 2 27. 20 29. -6 31. $6\frac{1}{2}$ 33. 19.9
35. $\frac{7}{3}$ 37. $-\frac{7}{4}$ 39. $\frac{41}{24}$ 41. $-\frac{1}{20}$ 43. 5.1 45. 12.4
47. -5 49. $1\frac{5}{6}$ 51. $-\frac{10}{21}$ 53. D_W 55. -11 56. 5
57. $-\frac{5}{12}$ 58. $\frac{1}{3}$ 59. $-\frac{3}{2}$ 60. -5.2 61. $-\frac{1}{24}$
62. 172.72 63. $\$83 - x$ 64. $65t$ miles 65. 342.246
67. $-\frac{26}{15}$ 69. -10 71. All real numbers 73. $-\frac{5}{17}$
75. 13, -13

Margin Exercises, Section 8.2, pp. 558–561

1. 15 2. $-\frac{7}{4}$ 3. -18 4. 10 5. 10 6. $-\frac{4}{5}$
7. 7800 8. -3 9. 28

Exercise Set 8.2, p. 562

1. 6 3. 9 5. 12 7. -40 9. 1 11. -7 13. -6
15. 6 17. -63 19. 36 21. -21 23. $-\frac{3}{5}$ 25. $-\frac{3}{2}$
27. $\frac{9}{2}$ 29. 7 31. -7 33. 8 35. 15.9 37. -50
39. -14 41. D_W 43. $7x$ 44. $-x + 5$ 45. $8x + 11$
46. $-32y$ 47. $x - 4$ 48. $-5x - 23$ 49. $-10y - 42$
50. $-22a + 4$ 51. $8r$ miles 52. $\frac{1}{2}b \cdot 10\ m^2$, or $5b\ m^2$
53. -8655 55. No solution 57. No solution
59. $\frac{b}{3a}$ 61. $\frac{4b}{a}$

Margin Exercises, Section 8.3, pp. 564–570

1. 5 2. 4 3. 4 4. 39 5. $-\frac{3}{2}$ 6. -4.3 7. -3
8. 800 9. 1 10. 2 11. 2 12. $\frac{17}{2}$ 13. $\frac{8}{3}$
14. $-\frac{43}{10}$, or -4.3 15. 2 16. 3 17. -2 18. $-\frac{1}{2}$
19. Yes 20. Yes 21. Yes 22. Yes 23. No
24. No 25. No 26. No 27. All real numbers
28. No solution

Calculator Corner, p. 571

1. Left to the student 2. Left to the student

Exercise Set 8.3, p. 572

1. 5 3. 8 5. 10 7. 14 9. -8 11. -8 13. -7
15. $\frac{2}{3}$ 17. 6 19. 4 21. 6 23. -3 25. 1
27. 6 29. -20 31. 7 33. 2 35. 5 37. 2
39. 10 41. 4 43. 0 45. -1 47. $-\frac{4}{3}$ 49. $\frac{2}{5}$
51. -2 53. -4 55. $\frac{4}{5}$ 57. $-\frac{28}{27}$ 59. 6 61. 2
63. No solution 65. All real numbers 67. 6 69. 8
71. 1 73. All real numbers 75. No solution
77. 17 79. $-\frac{5}{3}$ 81. -3 83. 2 85. $\frac{4}{7}$
87. No solution 89. All real numbers 91. $-\frac{51}{31}$
93. D_W 95. -6.5 96. -75.14 97. $7(x - 3 - 2y)$
98. $8(y - 11x + 1)$ 99. -160 100. $-17x + 18$
101. $91x - 242$ 102. 0.25 103. $-\frac{5}{32}$ 105. $\frac{52}{45}$

Margin Exercises, Section 8.4, pp. 576–579

1. 2.8 mi 2. 280,865 socks 3. 341 mi 4. $q = 3B$
5. $r = \frac{d}{t}$ 6. $I = \frac{E}{R}$ 7. $x = y - 5$ 8. $x = y + 7$

9. $x = y + b$ **10.** $y = \frac{5x}{9}$, or $\frac{5}{9}x$ **11.** $p = \frac{bq}{a}$
12. $x = \frac{y - b}{m}$ **13.** $Q = \frac{a + p}{t}$ **14.** $D = \frac{C}{\pi}$
15. $c = 4A - a - b - d$

Exercise Set 8.4, p. 580

1. **(a)** 57,000 Btu's; **(b)** $a = \frac{B}{30}$ **3.** **(a)** $1\frac{3}{5}$ mi; **(b)** $t = 5M$
5. **(a)** 1423 students; **(b)** $n = 15f$
7. 10.5 calories per ounce **9.** 42 games **11.** $x = \frac{y}{5}$
13. $c = \frac{a}{b}$ **15.** $x = y - 13$ **17.** $x = y - b$
19. $x = 5 - y$ **21.** $x = a - y$ **23.** $y = \frac{5x}{8}$, or $\frac{5}{8}x$
25. $x = \frac{By}{A}$ **27.** $t = \frac{W - b}{m}$ **29.** $x = \frac{y - c}{b}$
31. $b = 3A - a - c$ **33.** $t = \frac{A - b}{a}$ **35.** $h = \frac{A}{b}$
37. $w = \frac{P - 2l}{2}$, or $\frac{1}{2}P - l$ **39.** $a = 2A - b$
41. $a = \frac{F}{m}$ **43.** $c^2 = \frac{E}{m}$ **45.** $x = \frac{c - By}{A}$ **47.** $t = \frac{3k}{v}$
49. DW **51.** 0.92 **52.** -90 **53.** -9.325 **54.** 44
55. -13.2 **56.** $-21a + 12b$ **57.** 0.031 **58.** 0.671
59. $\frac{1}{6}$ **60.** $-\frac{3}{2}$
61. **(a)** 1901 calories;
(b) $a = \frac{917 + 6w + 6h - K}{6}$;
$h = \frac{K - 917 - 6w + 6a}{6}$;
$w = \frac{K - 917 - 6h + 6a}{6}$
63. $b = \frac{Ha - 2}{H}$, or $a - \frac{2}{H}$; $a = \frac{2 + Hb}{H}$, or $\frac{2}{H} + b$
65. A quadruples. **67.** A increases by $2h$ units.

Margin Exercises, Section 8.5, pp. 584–587

1. $13\% \cdot 80 = a$ **2.** $a = 60\% \cdot 70$ **3.** $43 = 20\% \cdot b$
4. $110\% \cdot b = 30$ **5.** $16 = n \cdot 80$ **6.** $n \cdot 94 = 10.5$
7. 1.92 **8.** 115 **9.** 36% **10.** 111,416 mi^2
11. About 1.2 million **12.** About 58%

Exercise Set 8.5, p. 588

1. 20% **3.** 150 **5.** 546 **7.** 24% **9.** 2.5 **11.** 5%
13. 25% **15.** 84 **17.** 24% **19.** 16% **21.** $46\frac{2}{3}$
23. 0.8 **25.** 5 **27.** 40 **29.** \$198 **31.** \$1584
33. \$528 **35.** Japan: 44.3%; Germany: 24.9%
37. About 603 at-bats **39.** \$195 **41.** **(a)** 16%; **(b)** \$29
43. **(a)** \$3.75; **(b)** \$28.75 **45.** **(a)** \$28.80; **(b)** \$33.12
47. 200 women **49.** About 31.5 lb **51.** \$655; 196%
53. \$2190; \$1455 **55.** \$3935; 261% **57.** DW
59. 181.52 **60.** 0.4538 **61.** 12.0879 **62.** 844.1407

63. $a + c$ **64.** $7x - 9y$ **65.** -3.9 **66.** $-6\frac{1}{8}$
67. Division; subtraction **68.** Exponential; division; subtraction **69.** 6 ft 7 in.

Margin Exercises, Section 8.6, pp. 593–602

1. $62\frac{2}{3}$ mi **2.** Jenny: 8 in.; Emma: 4 in.; Sarah: 6 in.
3. 313 and 314 **4.** 60,417 copies
5. Length: 84 ft; width: 50 ft **6.** First: 30°; second: 90°; third: 60° **7.** Second: 80; third: 89 **8.** \$8400 **9.** \$658

Translating for Success, p. 603

1. B **2.** H **3.** G **4.** N **5.** J **6.** C **7.** L **8.** E
9. F **10.** D

Exercise Set 8.6, p. 604

1. 180 in.; 60 in. **3.** \$4.29 **5.** \$6.3 billion **7.** $699\frac{1}{3}$ mi
9. 1204 and 1205 **11.** 41, 42, 43 **13.** 61, 63, 65
15. Length: 48 ft; width: 14 ft **17.** \$75 **19.** \$85
21. 11 visits **23.** 28°, 84°, 68° **25.** 33°, 38°, 109°
27. \$350 **29.** \$852.94 **31.** 12 mi **33.** \$36
35. \$25 and \$50 **37.** DW **39.** $-\frac{47}{40}$ **40.** $-\frac{17}{40}$
41. $-\frac{3}{10}$ **42.** $-\frac{32}{15}$ **43.** -10 **44.** 1.6 **45.** 409.6
46. -9.6 **47.** -41.6 **48.** 0.1 **49.** 120 apples
51. About 0.65 in. **53.** \$9.17, not \$9.10

Margin Exercises, Section 8.7, pp. 609–616

1. **(a)** No; **(b)** no; **(c)** no; **(d)** yes; **(e)** no; **(f)** no
2. **(a)** Yes; **(b)** yes; **(c)** yes; **(d)** no; **(e)** yes; **(f)** yes
3. $x \leq 4$
4. $x > -2$
5. $-2 < x \leq 4$
6. $\{x \mid x > 2\}$;
7. $\{x \mid x \leq 3\}$;
8. $\{x \mid x < -3\}$;
9. $\{x \mid x \geq \frac{2}{15}\}$ **10.** $\{y \mid y \leq -3\}$
11. $\{x \mid x < 8\}$;
12. $\{y \mid y \geq 32\}$;
13. $\{x \mid x \geq -6\}$ **14.** $\{y \mid y < -\frac{13}{5}\}$
15. $\{x \mid x > -\frac{1}{4}\}$ **16.** $\{y \mid y \geq \frac{19}{9}\}$ **17.** $\{y \mid y \geq \frac{19}{9}\}$
18. $\{x \mid x \geq -2\}$ **19.** $\{x \mid x \geq -4\}$ **20.** $\{x \mid x > \frac{8}{3}\}$

Exercise Set 8.7, p. 617

1. **(a)** Yes; **(b)** yes; **(c)** no; **(d)** yes; **(e)** yes
3. **(a)** No; **(b)** no; **(c)** no; **(d)** yes; **(e)** no
5. $x > 4$
7.
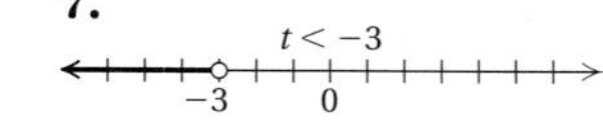

9.
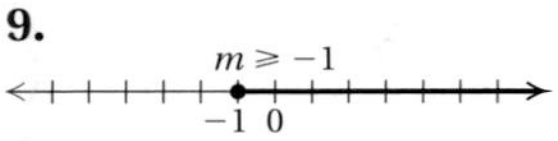

11.
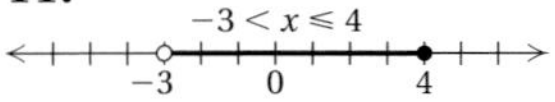

13.
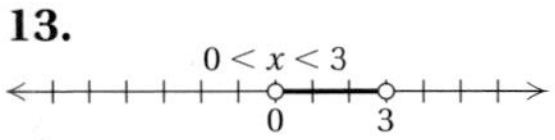

15. $\{x \mid x > -5\}$;
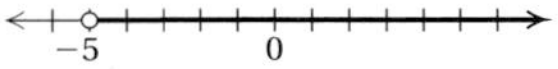

17. $\{x \mid x \leq -18\}$;
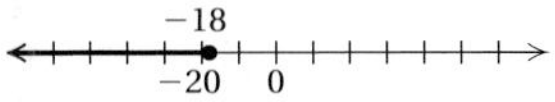

19. $\{y \mid y > -5\}$

21. $\{x \mid x > 2\}$ **23.** $\{x \mid x \leq -3\}$ **25.** $\{x \mid x < 4\}$ **27.** $\{t \mid t > 14\}$ **29.** $\{y \mid y \leq \frac{1}{4}\}$ **31.** $\{x \mid x > \frac{7}{12}\}$ **33.** $\{x \mid x < 7\}$;
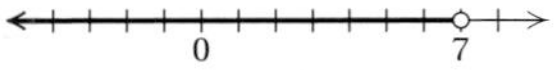

35. $\{x \mid x < 3\}$;
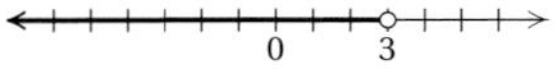

37. $\{y \mid y \geq -\frac{2}{5}\}$ **39.** $\{x \mid x \geq -6\}$ **41.** $\{y \mid y \leq 4\}$ **43.** $\{x \mid x > \frac{17}{3}\}$ **45.** $\{y \mid y < -\frac{1}{14}\}$ **47.** $\{x \mid x \leq \frac{3}{10}\}$ **49.** $\{x \mid x < 8\}$ **51.** $\{x \mid x \leq 6\}$ **53.** $\{x \mid x < -3\}$ **55.** $\{x \mid x > -3\}$ **57.** $\{x \mid x \leq 7\}$ **59.** $\{x \mid x > -10\}$ **61.** $\{y \mid y < 2\}$ **63.** $\{y \mid y \geq 3\}$ **65.** $\{y \mid y > -2\}$ **67.** $\{x \mid x > -4\}$ **69.** $\{x \mid x \leq 9\}$ **71.** $\{y \mid y \leq -3\}$ **73.** $\{y \mid y < 6\}$ **75.** $\{m \mid m \geq 6\}$ **77.** $\{t \mid t < -\frac{5}{3}\}$ **79.** $\{r \mid r > -3\}$ **81.** $\{x \mid x \geq -\frac{57}{34}\}$ **83.** $\{x \mid x > -2\}$ **85.** **DW** **87.** -74 **88.** 4.8 **89.** $-\frac{5}{8}$ **90.** -1.11 **91.** -38 **92.** $-\frac{7}{8}$ **93.** -9.4 **94.** 1.11 **95.** 140 **96.** 41 **97.** $-2x - 23$ **98.** $37x - 1$ **99. (a)** Yes; **(b)** yes; **(c)** no; **(d)** no; **(e)** no; **(f)** yes; **(g)** yes **101.** No solution

Margin Exercises, Section 8.8, pp. 621–623

1. $m \geq 92$ **2.** $c \geq 4000$ **3.** $p \leq 21{,}900$ **4.** $45 < t < 55$ **5.** $d > 15$ **6.** $w < 110$ **7.** $n > -2$ **8.** $c \leq 12{,}500$ **9.** $d \leq 11.4\%$ **10.** $s \geq 23$ **11.** $\frac{9}{5}C + 32 < 88$; $\{C \mid C < 31\frac{1}{9}°\}$

12. $\dfrac{91 + 86 + 89 + s}{4} \geq 90$; $\{s \mid s \geq 94\}$

Exercise Set 8.8, p. 624

1. $n \geq 7$ **3.** $w > 2$ kg **5.** 90 mph $< s <$ 110 mph **7.** $a \leq 1{,}200{,}000$ **9.** $c \geq \$1.50$ **11.** $x > 8$ **13.** $y \leq -4$ **15.** $n \geq 1300$ **17.** $A \leq 500$ L **19.** $3x + 2 < 13$ **21.** $\{x \mid x \geq 84\}$ **23.** $\{C \mid C < 1063°\}$ **25.** $\{Y \mid Y \geq 1935\}$ **27.** $\{L \mid L \geq 5$ in.$\}$ **29.** 15 or fewer copies **31.** 5 min or more **33.** 2 courses **35.** 4 servings or more **37.** Lengths greater than or equal to 92 ft; lengths less than or equal to 92 ft **39.** Lengths less than 21.5 cm **41.** The blue-book value is greater than or equal to \$10,625. **43.** It has at least 16 g of fat. **45.** Dates at least 6 weeks after July 1 **47.** Heights greater than or equal to 4 ft **49.** 21 calls or more **51.** **DW** **53.** Even **54.** Odd **55.** Additive **56.** Multiplicative **57.** Equivalent **58.** Addition principle **59.** Multiplication principle; is reversed **60.** Solution **61.** Temperatures between $-15°$C and $-9\frac{4}{9}°$C **63.** They contain at least 7.5 g of fat per serving.

Concept Reinforcement, p. 630

1. True **2.** True **3.** True **4.** False **5.** True **6.** False

Summary and Review: Chapter 8, p. 630

1. -22 **2.** 1 **3.** 25 **4.** 9.99 **5.** $\frac{1}{4}$ **6.** 7 **7.** -192 **8.** $-\frac{7}{3}$ **9.** $-\frac{15}{64}$ **10.** -8 **11.** 4 **12.** -5 **13.** $-\frac{1}{3}$ **14.** 3 **15.** 4 **16.** 16 **17.** All real numbers **18.** 6 **19.** -3 **20.** 28 **21.** 4 **22.** No solution **23.** Yes **24.** No **25.** Yes **26.** $\{y \mid y \geq -\frac{1}{2}\}$ **27.** $\{x \mid x \geq 7\}$ **28.** $\{y \mid y > 2\}$ **29.** $\{y \mid y \leq -4\}$ **30.** $\{x \mid x < -11\}$ **31.** $\{y \mid y > -7\}$ **32.** $\{x \mid x > -\frac{9}{11}\}$ **33.** $\{x \mid x \geq -\frac{1}{12}\}$

34.
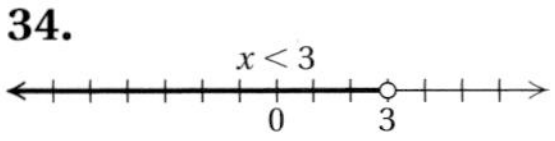

35.
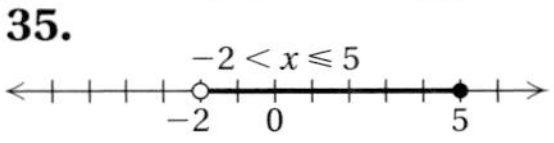

36.
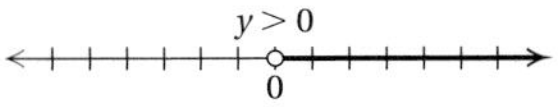

37. $d = \dfrac{C}{\pi}$ **38.** $B = \dfrac{3V}{h}$

39. $a = 2A - b$ **40.** $x = \dfrac{y - b}{m}$ **41.** Length: 365 mi; width: 275 mi **42.** 345, 346 **43.** \$2117 **44.** 27 appliances **45.** 35°, 85°, 60° **46.** 15 **47.** 18.75% **48.** 600 **49.** About 26% **50.** \$220 **51.** \$53,400 **52.** \$138.95 **53.** 86 **54.** $\{w \mid w > 17$ cm$\}$ **55.** **DW** The end result is the same either way. If s is the original salary, the new salary after a 5% raise followed by an 8% raise is $1.08(1.05s)$. If the raises occur the other way around, the new salary is $1.05(1.08s)$. By the commutative and associative laws of multiplication, we see that these are equal. However, it would be better to receive the 8% raise first, because this increase yields a higher salary initially than a 5% raise. **56.** **DW** The inequalities are equivalent by the multiplication principle for inequalities. If we multiply both sides of one inequality by -1, the other inequality results. **57.** 23, -23 **58.** 20, -20

59. $a = \dfrac{y - 3}{2 - b}$

Test: Chapter 8, p. 633

1. [8.1b] 8 **2.** [8.1b] 26 **3.** [8.2a] -6 **4.** [8.2a] 49 **5.** [8.3b] -12 **6.** [8.3a] 2 **7.** [8.3a] -8 **8.** [8.1b] $-\frac{7}{20}$ **9.** [8.3c] 7 **10.** [8.3c] $\frac{5}{3}$ **11.** [8.3b] $\frac{5}{2}$ **12.** [8.3c] No solution **13.** [8.3c] All real numbers **14.** [8.7c] $\{x \mid x \leq -4\}$ **15.** [8.7c] $\{x \mid x > -13\}$ **16.** [8.7d] $\{x \mid x \leq 5\}$ **17.** [8.7d] $\{y \mid y \leq -13\}$ **18.** [8.7d] $\{y \mid y \geq 8\}$ **19.** [8.7d] $\{x \mid x \leq -\frac{1}{20}\}$ **20.** [8.7e] $\{x \mid x < -6\}$ **21.** [8.7e] $\{x \mid x \leq -1\}$

22. [8.7b]
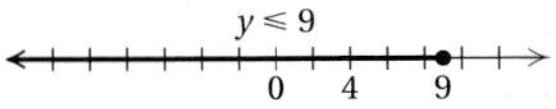

23. [8.7b, e]
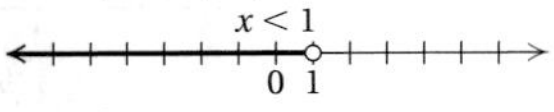

24. [8.7b]
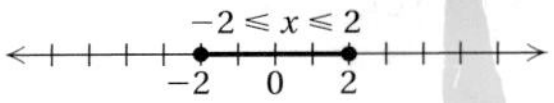

25. [8.5a] 18 **26.** [8.5a] 16.5% **27.** [8.5a] 40,000

28. [8.5a] About 53.4% **29.** [8.6a] Width: 7 cm; length: 11 cm **30.** [8.5a] About \$240.7 billion
31. [8.6a] 2509, 2510, 2511 **32.** [8.6a] \$880
33. [8.6a] 3 m, 5 m **34.** [8.8b] $\{l \mid l \geq 174 \text{ yd}\}$
35. [8.8b] $\{b \mid b \leq \$105\}$ **36.** [8.8a] $\{c \mid c \leq 143{,}750\}$
37. [8.4b] $r = \dfrac{A}{2\pi h}$ **38.** [8.4b] $x = \dfrac{y - b}{8}$
39. [8.4b] $d = \dfrac{1 - ca}{-c}$, or $\dfrac{ca - 1}{c}$
40. [7.2e], [8.3a] 15, −15 **41.** [8.6a] 60 tickets

CHAPTER 9

Margin Exercises, Section 9.1, pp. 636–646

1.–8

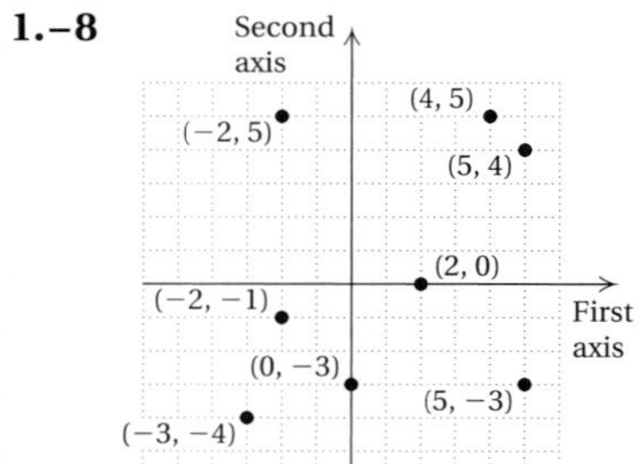

9. Both are negative numbers. **10.** First, positive; second, negative **11.** I **12.** III **13.** IV **14.** II **15.** On an axis **16.** A: $(-5, 1)$; B: $(-3, 2)$; C: $(0, 4)$; D: $(3, 3)$; E: $(1, 0)$; F: $(0, -3)$; G: $(-5, -4)$ **17.** No **18.** Yes **19.** $(-2, -3)$, $(1, 3)$; answers may vary

20.

x	y	(x, y)
−3	6	(−3, 6)
−1	2	(−1, 2)
0	0	(0, 0)
1	−2	(1, −2)
3	−6	(3, −6)

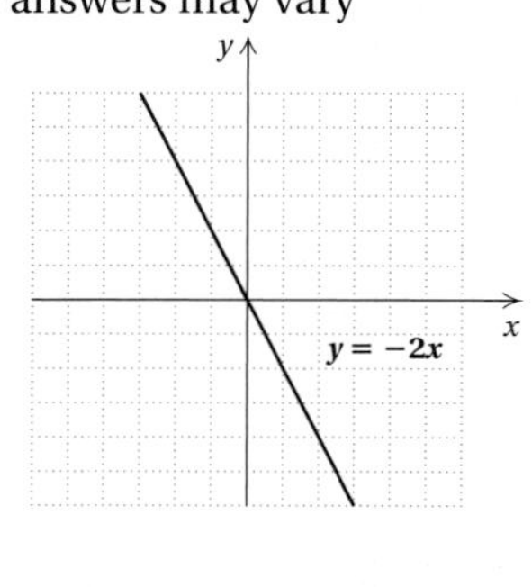

21.

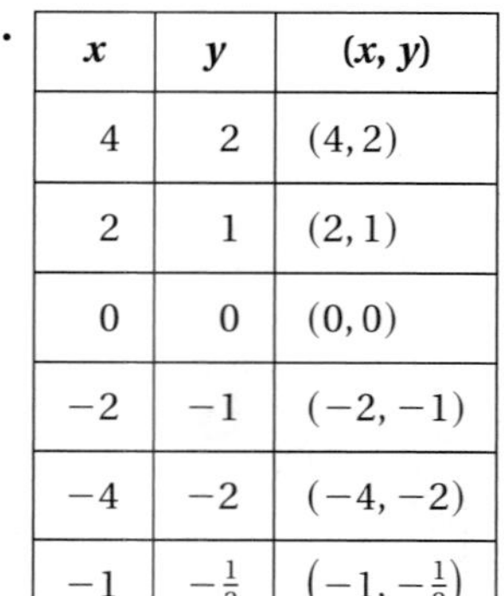

x	y	(x, y)
4	2	(4, 2)
2	1	(2, 1)
0	0	(0, 0)
−2	−1	(−2, −1)
−4	−2	(−4, −2)
−1	$-\frac{1}{2}$	$\left(-1, -\frac{1}{2}\right)$

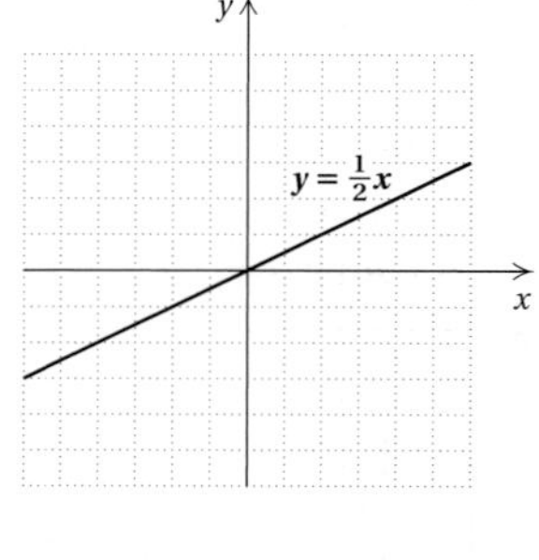

22.

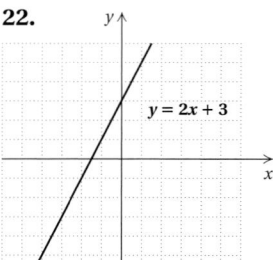

23.

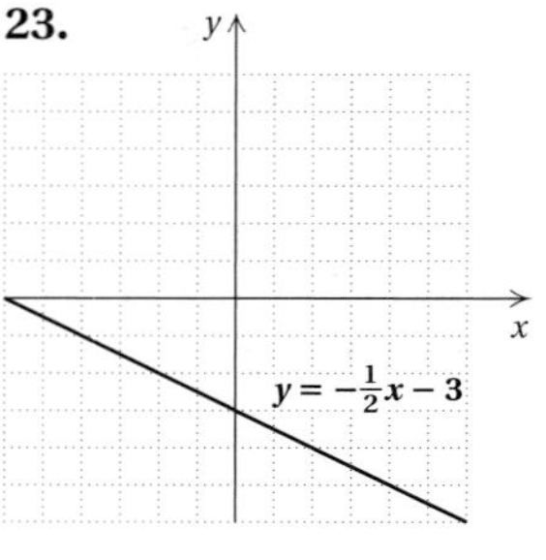

24.

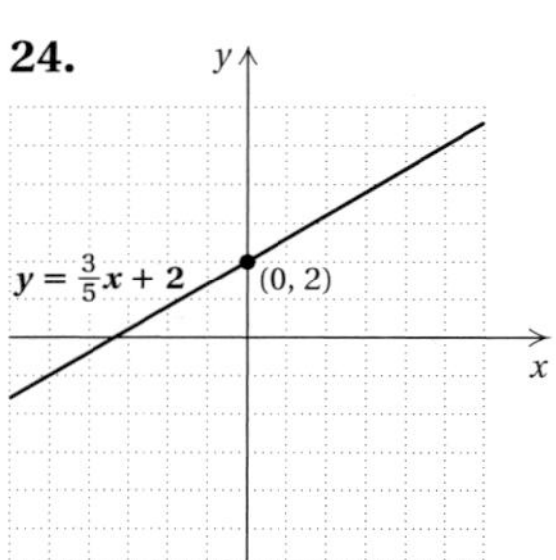

25.

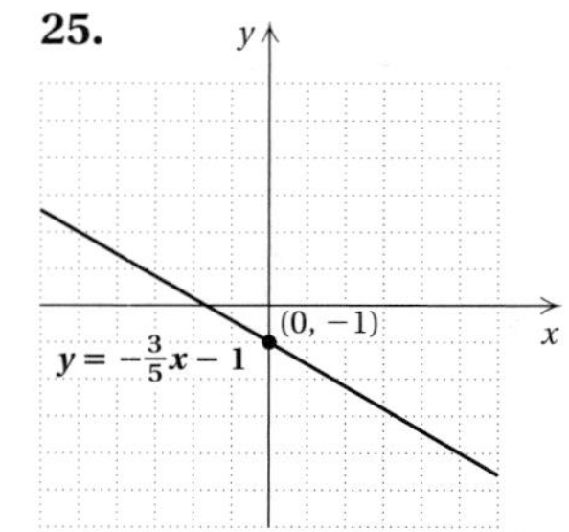

26.

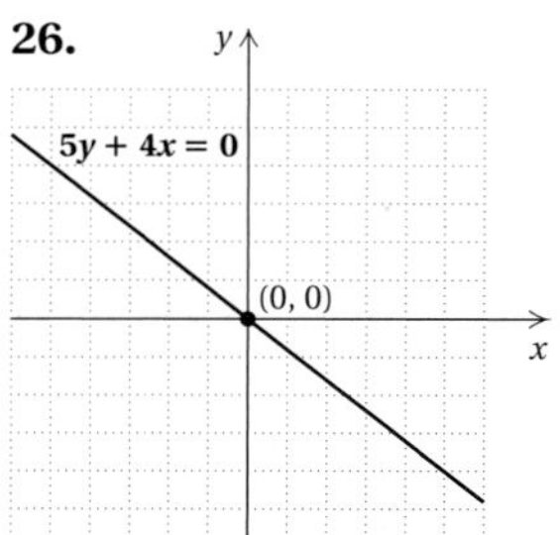

27.

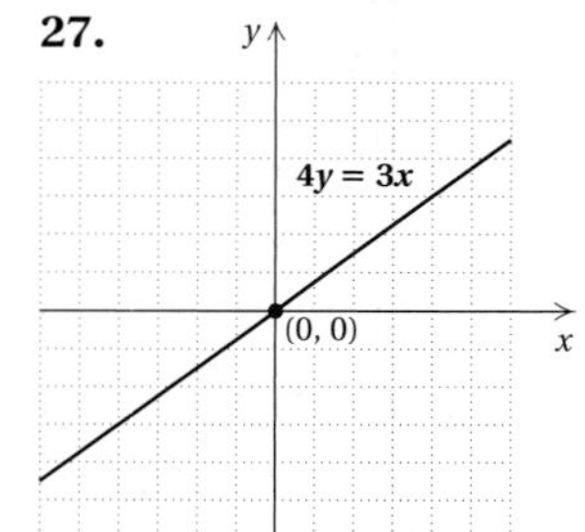

28.

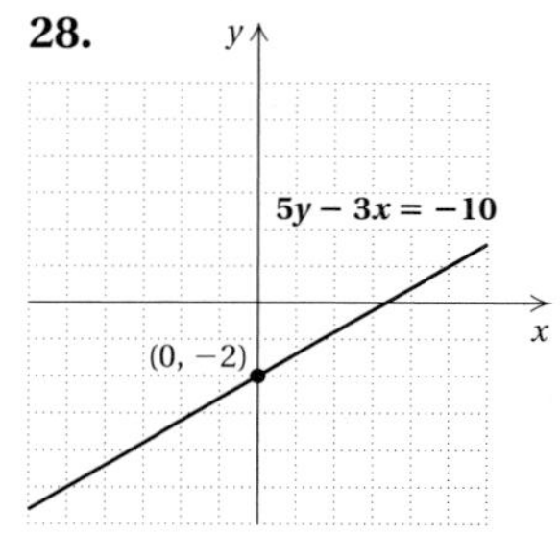

29.

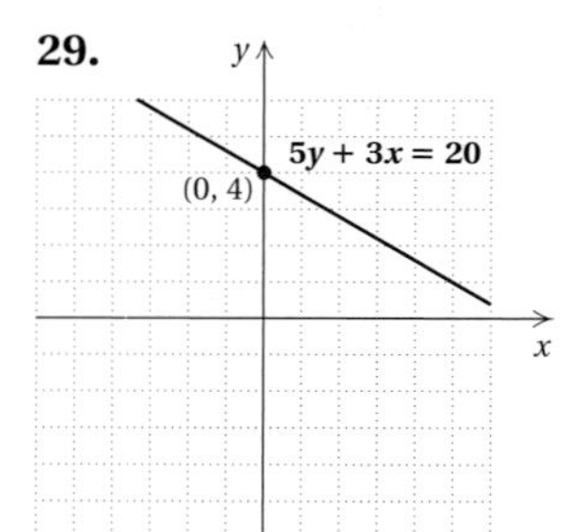

30. (a) \$2720; \$2040; \$680; \$0; **(b)** about \$1700; **(c)** about 2.8 yr

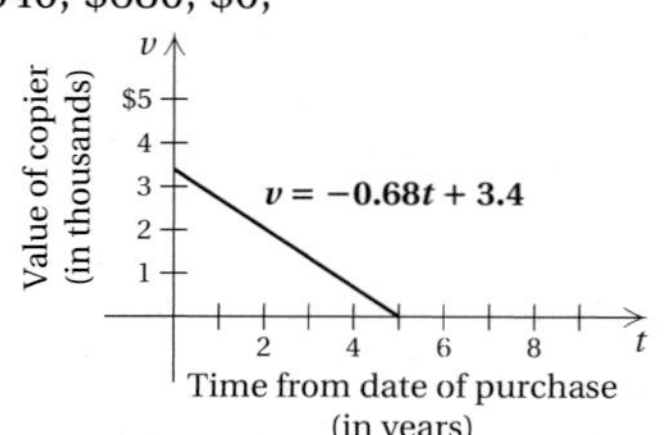

Calculator Corner, p. 642

1. Left to the student

Calculator Corner, p. 647

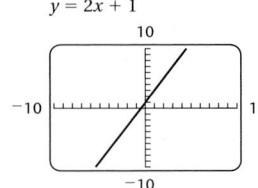

1. $y = 2x + 1$

2. $y = -3x + 1$

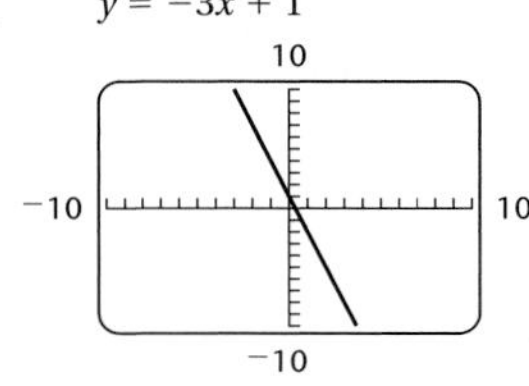

3. $y = -5x + 3$

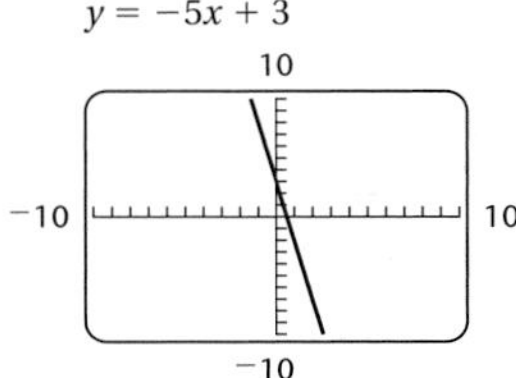

4. $y = 4x - 5$

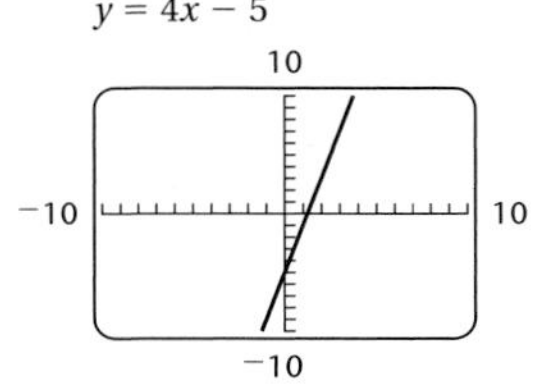

5. $y = \frac{4}{5}x + 2$

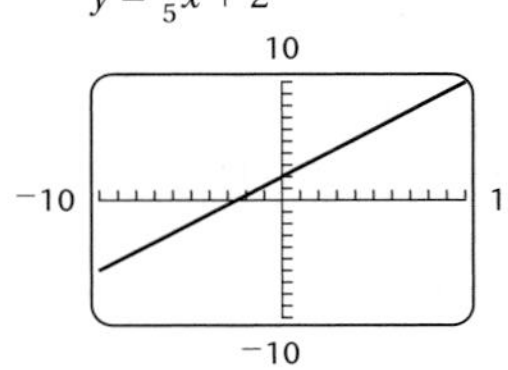

6. $y = -\frac{3}{5}x - 1$

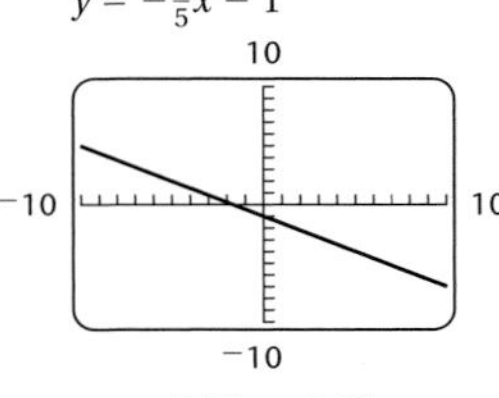

7. $y = 2.085x + 5.08$

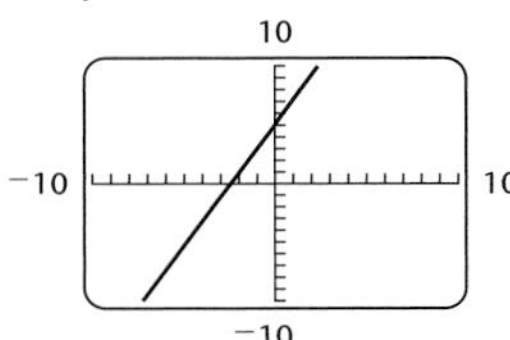

8. $y = -3.45x - 1.68$

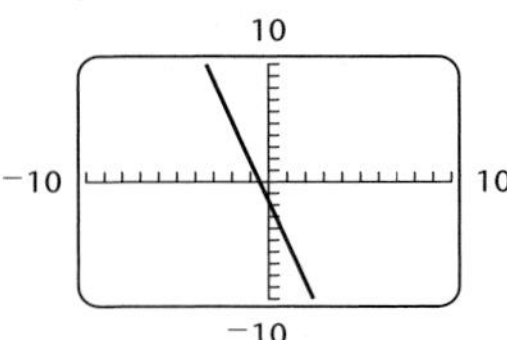

Exercise Set 9.1, p. 648

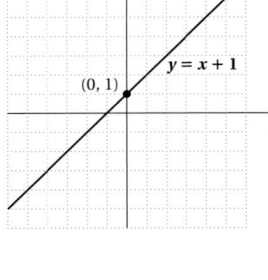

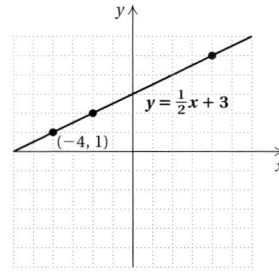

1.

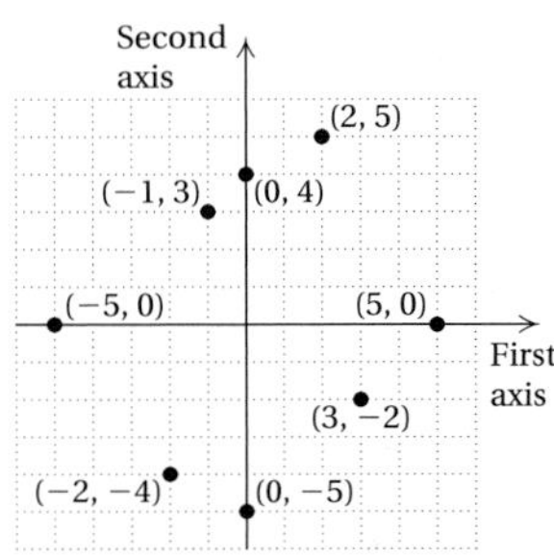

3. II **5.** IV **7.** III **9.** On an axis **11.** II **13.** IV **15.** II **17.** I, IV **19.** I, III

21. A: (3, 3); B: (0, −4); C: (−5, 0); D: (−1, −1); E: (2, 0)

23. No **25.** No **27.** Yes

29.

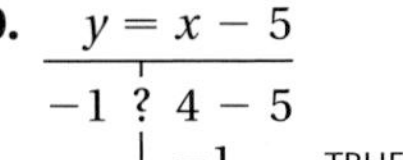

$y = x - 5$	
-1 ?	$4 - 5$
	-1 TRUE

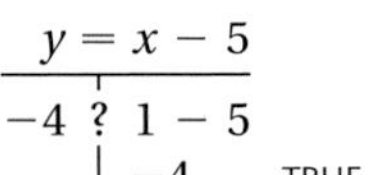

$y = x - 5$	
-4 ?	$1 - 5$
	-4 TRUE

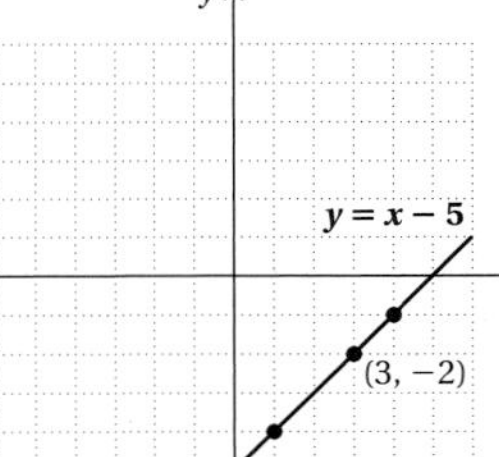

31.

$y = \frac{1}{2}x + 3$	
5 ?	$\frac{1}{2} \cdot 4 + 3$
	$2 + 3$
	5 TRUE

$y = \frac{1}{2}x + 3$	
2 ?	$\frac{1}{2}(-2) + 3$
	$-1 + 3$
	2 TRUE

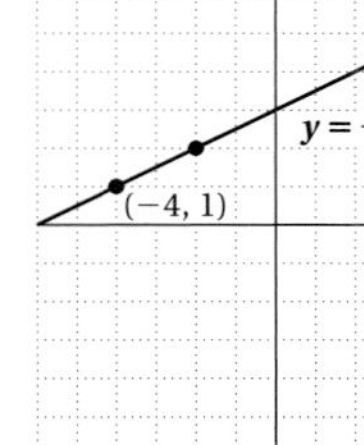

33.

$4x - 2y$	$= 10$
$4 \cdot 0 - 2(-5)$?	10
$0 + 10$	
10	TRUE

$4x - 2y$	$= 10$
$4 \cdot 4 - 2 \cdot 3$?	10
$16 - 6$	
10	TRUE

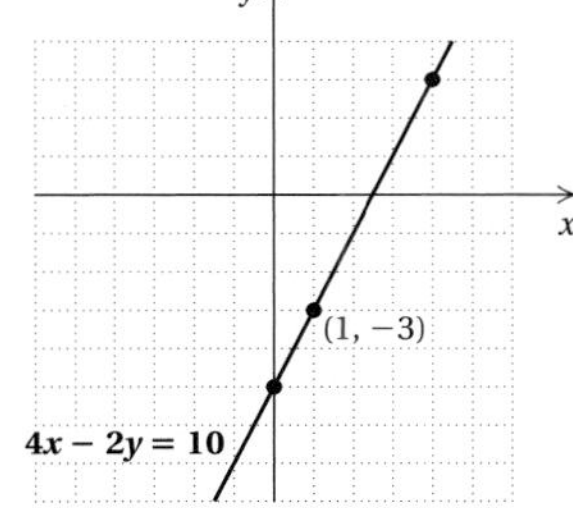

35.

x	y
−2	−1
−1	0
0	1
1	2
2	3
3	4

37.

x	y
−2	−2
−1	−1
0	0
1	1
2	2
3	3

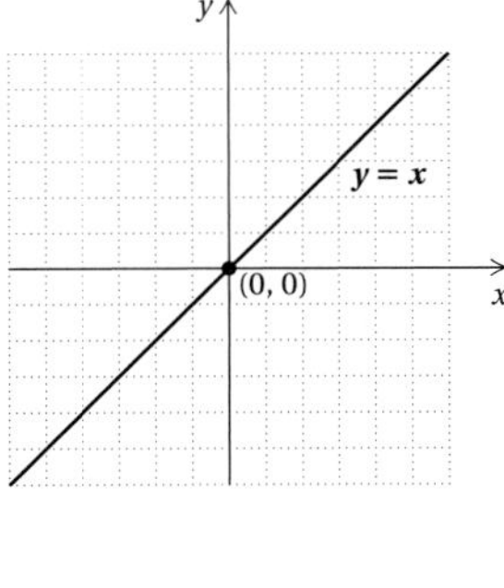

39.

x	y
−2	−1
0	0
4	2

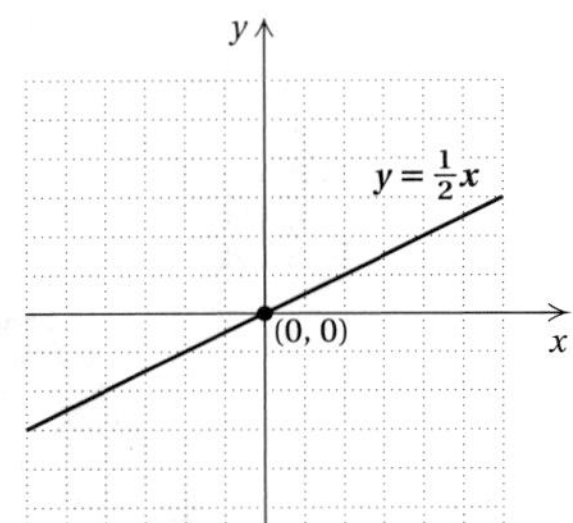

41.

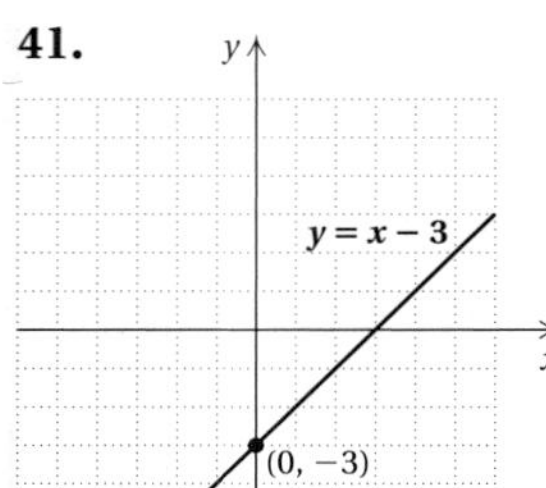

43.

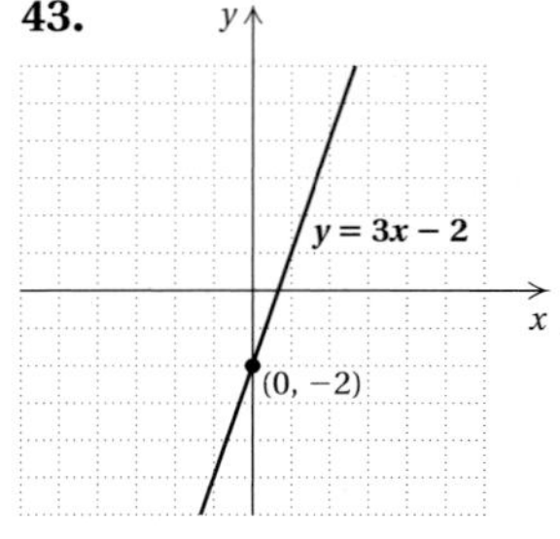

45.

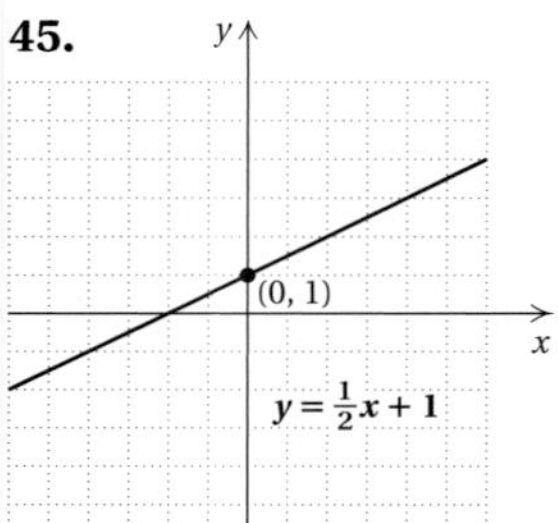

47.

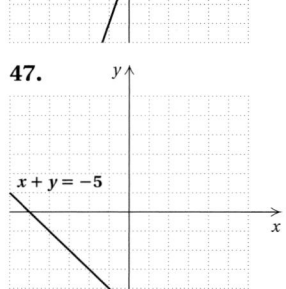

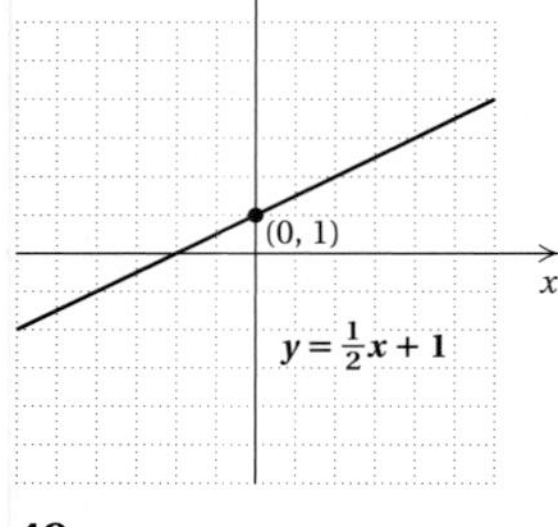

49.

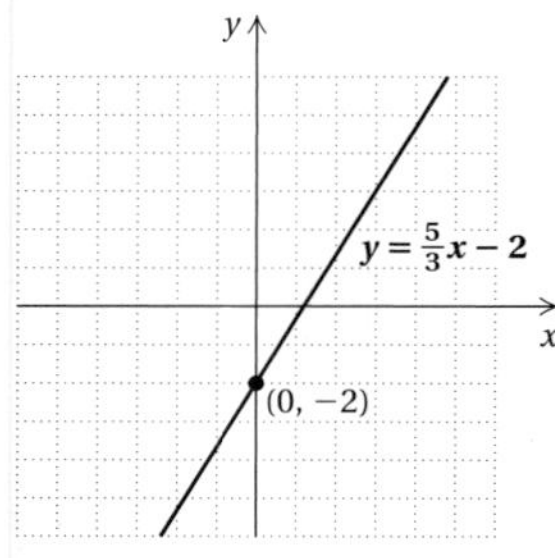

51.

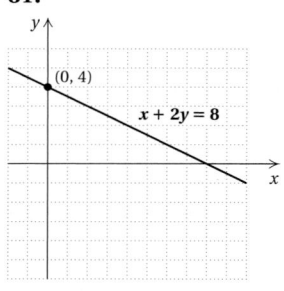

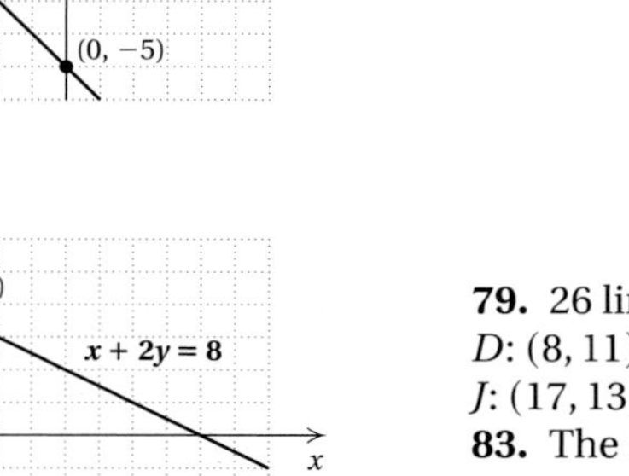

53.

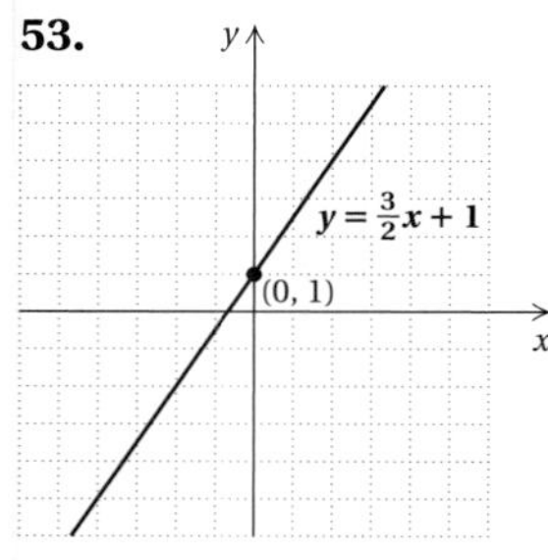

55.

57.

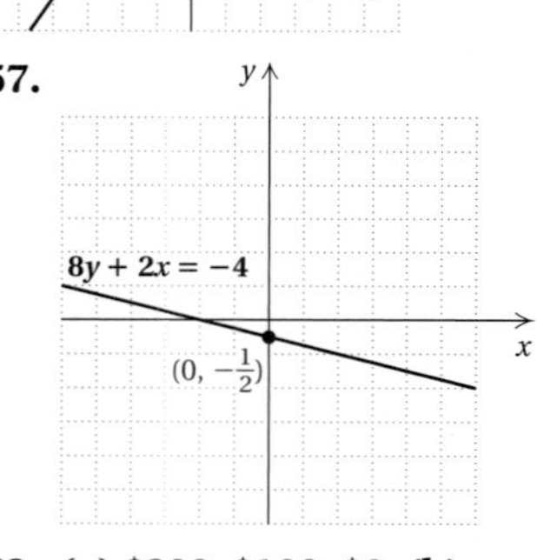

59. **(a)** \$300, \$100, \$0; **(b)** \$50; **(c)** 3 yr

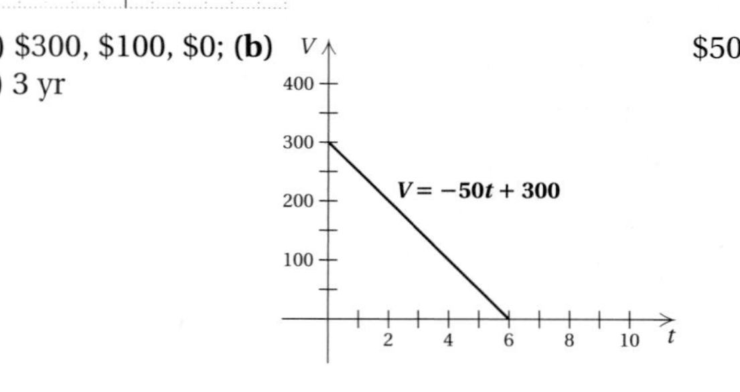

61. **(a)** 22 gal, 25.2 gal, 30 gal, 33.2 gal; **(b)** 27 gal; **(c)** in 13 yr, or in 2008

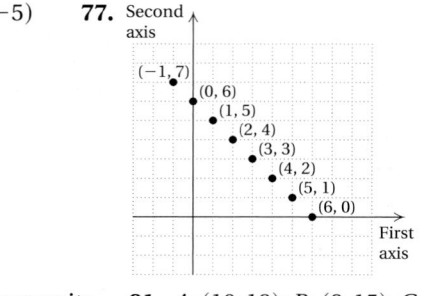

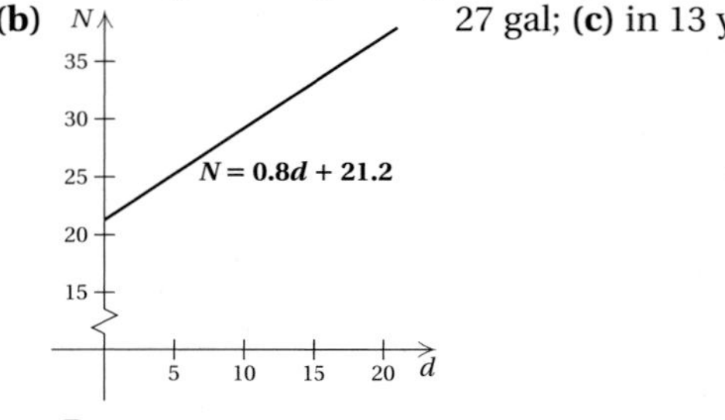

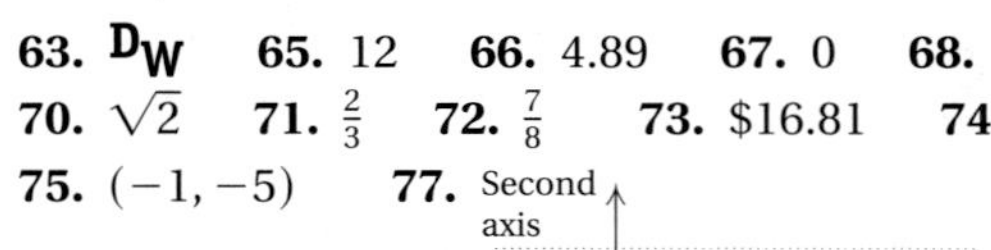

63. $\mathbf{D_W}$ **65.** 12 **66.** 4.89 **67.** 0 **68.** $\frac{4}{5}$ **69.** 3.4 **70.** $\sqrt{2}$ **71.** $\frac{2}{3}$ **72.** $\frac{7}{8}$ **73.** \$16.81 **74.** \$18.40 **75.** $(-1, -5)$ **77.**

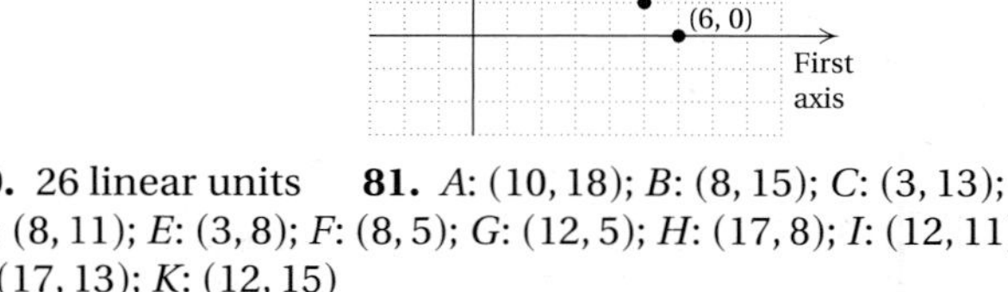

79. 26 linear units **81.** A: (10, 18); B: (8, 15); C: (3, 13); D: (8, 11); E: (3, 8); F: (8, 5); G: (12, 5); H: (17, 8); I: (12, 11); J: (17, 13); K: (12, 15)
83. The figure is translated 3 units down.

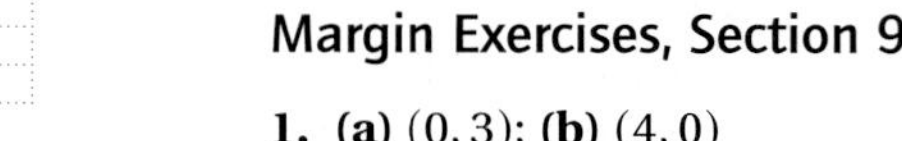

Margin Exercises, Section 9.2, pp. 655–659

1. **(a)** (0, 3); **(b)** (4, 0)

2.

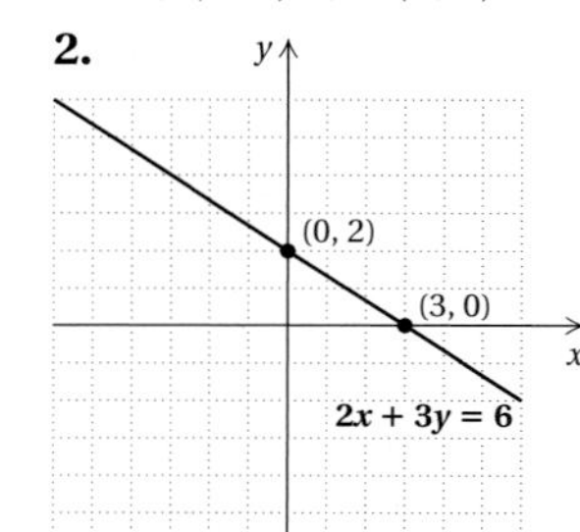

3.

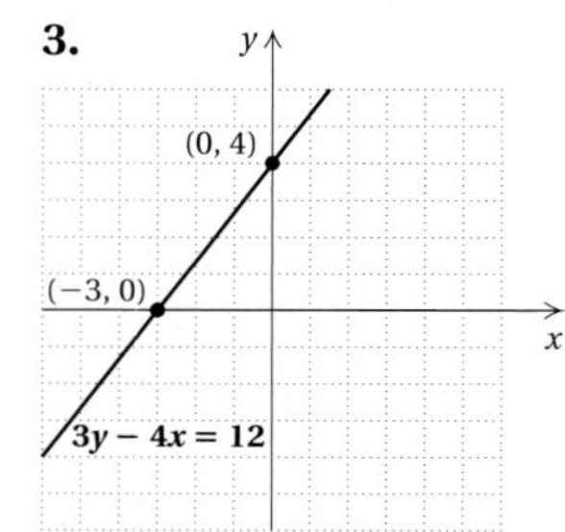

4.

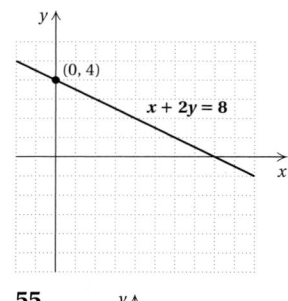

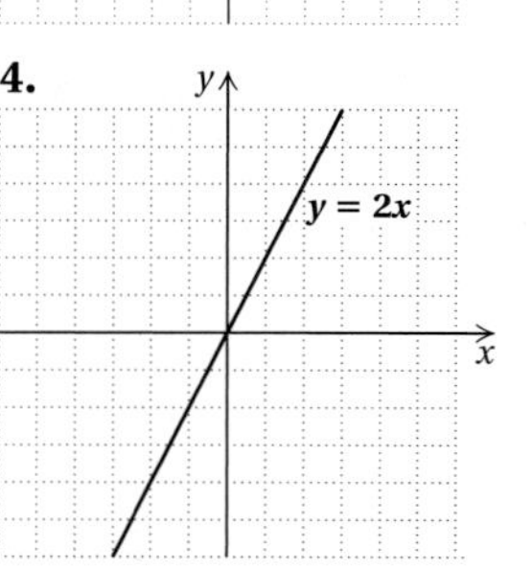

5.

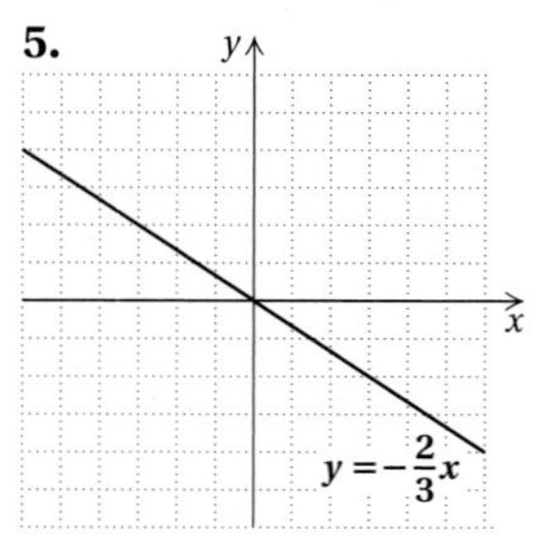

6.

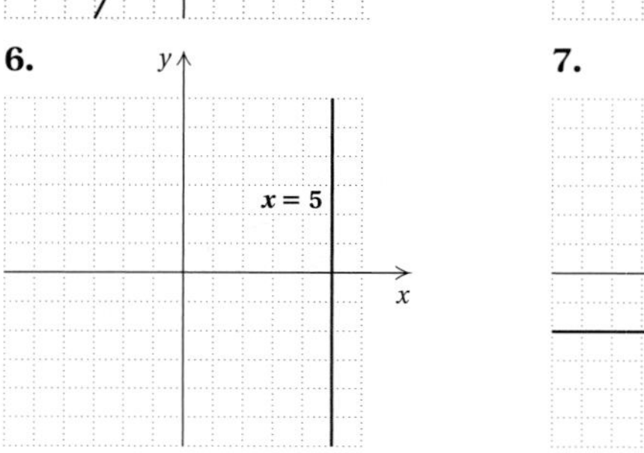

7.

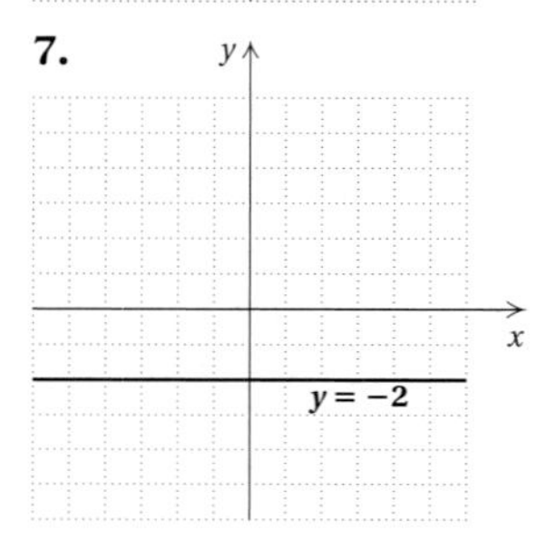

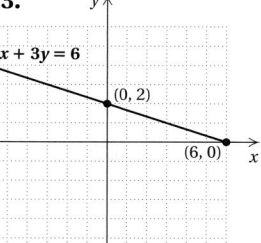

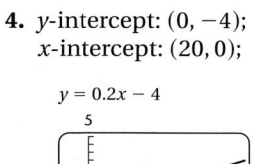

8.

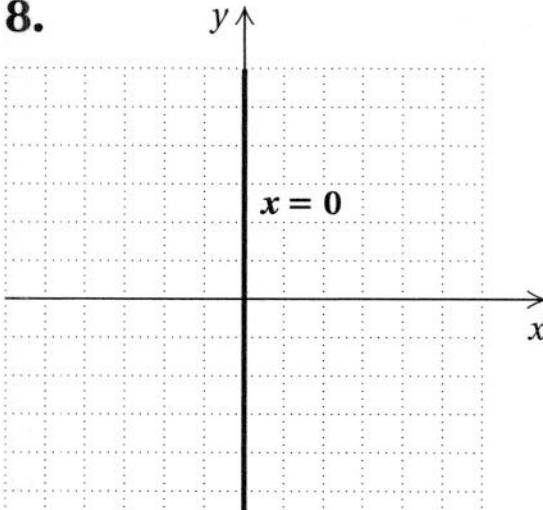

9.

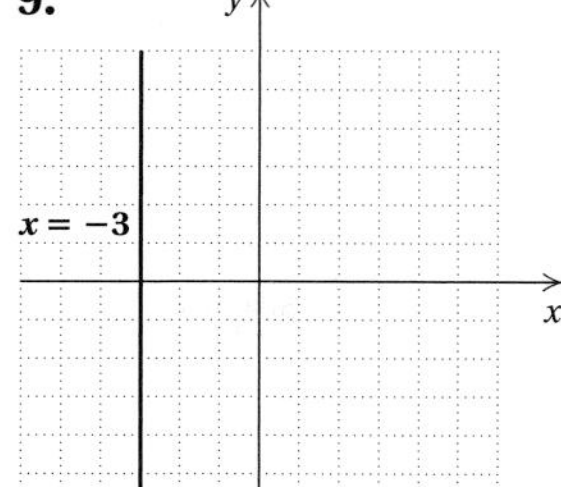

Calculator Corner, p. 657

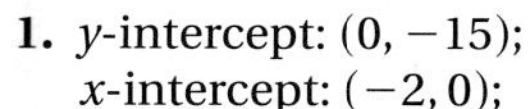

1. y-intercept: $(0, -15)$; x-intercept: $(-2, 0)$;

$y = -7.5x - 15$

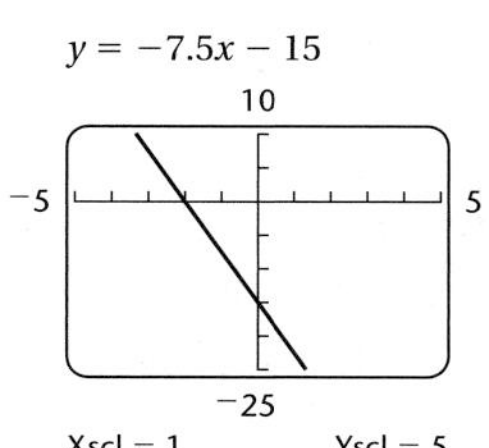

2. y-intercept: $(0, 43)$; x-intercept: $(-20, 0)$;

$y = 2.15x + 43$

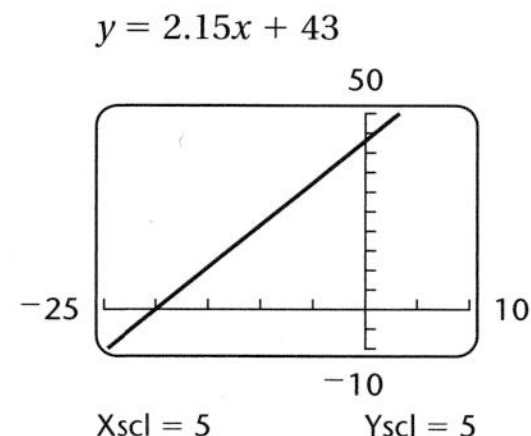

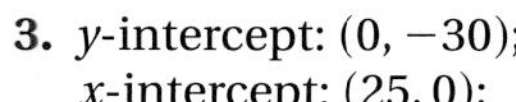

3. y-intercept: $(0, -30)$; x-intercept: $(25, 0)$;

$y = (6x - 150)/5$

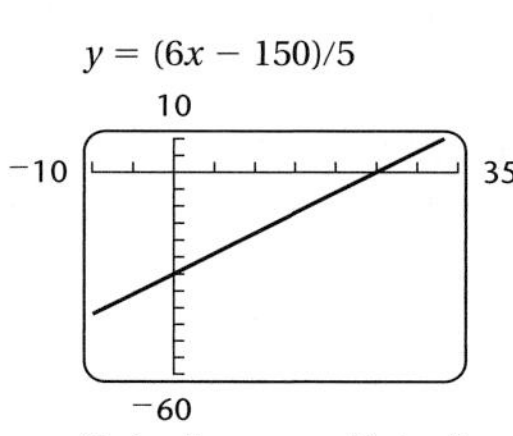

4. y-intercept: $(0, -4)$; x-intercept: $(20, 0)$;

$y = 0.2x - 4$

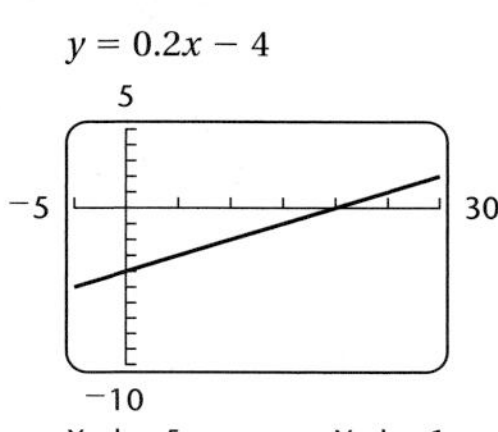

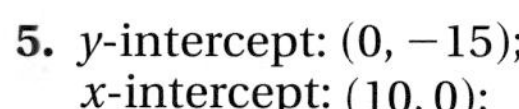

5. y-intercept: $(0, -15)$; x-intercept: $(10, 0)$;

$y = 1.5x - 15$

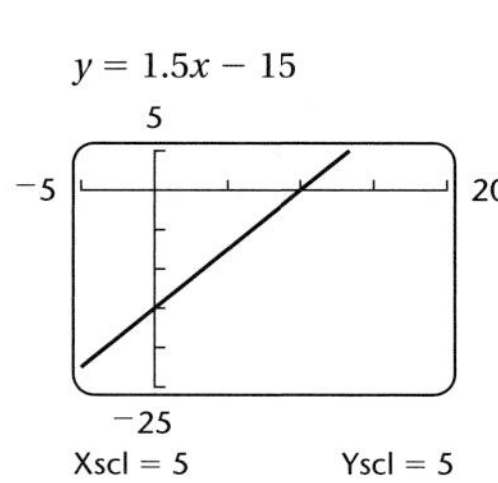

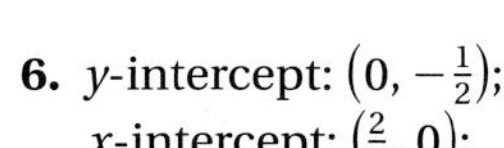

6. y-intercept: $\left(0, -\frac{1}{2}\right)$; x-intercept: $\left(\frac{2}{5}, 0\right)$;

$y = (5x - 2)/4$

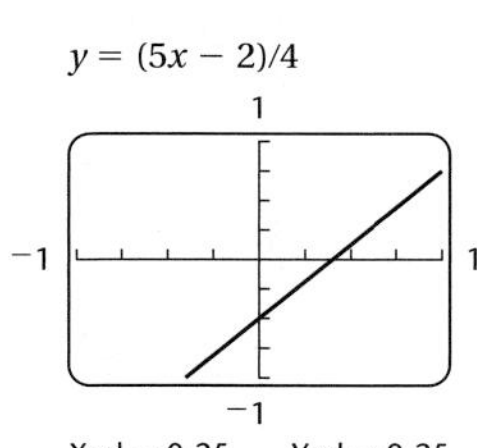

Visualizing for Success, p. 660

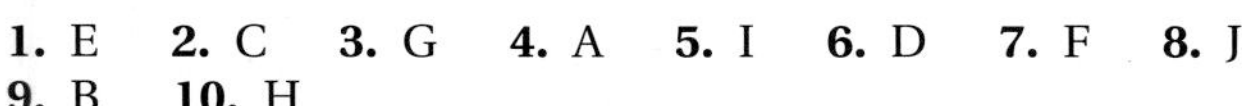

1. E **2.** C **3.** G **4.** A **5.** I **6.** D **7.** F **8.** J **9.** B **10.** H

Exercise Set 9.2, p. 661

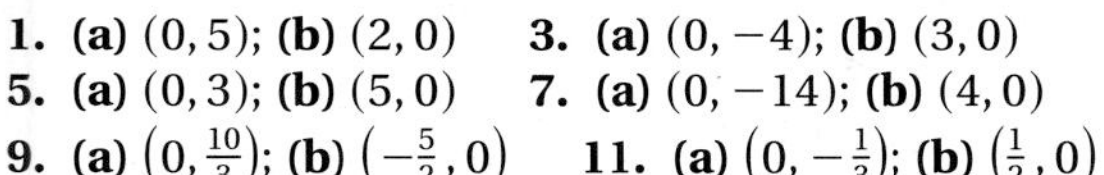

1. (a) $(0, 5)$; **(b)** $(2, 0)$ **3. (a)** $(0, -4)$; **(b)** $(3, 0)$ **5. (a)** $(0, 3)$; **(b)** $(5, 0)$ **7. (a)** $(0, -14)$; **(b)** $(4, 0)$ **9. (a)** $\left(0, \frac{10}{3}\right)$; **(b)** $\left(-\frac{5}{2}, 0\right)$ **11. (a)** $\left(0, -\frac{1}{3}\right)$; **(b)** $\left(\frac{1}{2}, 0\right)$

13.

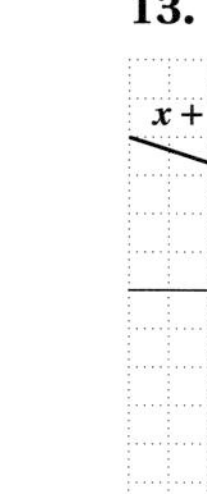

15.

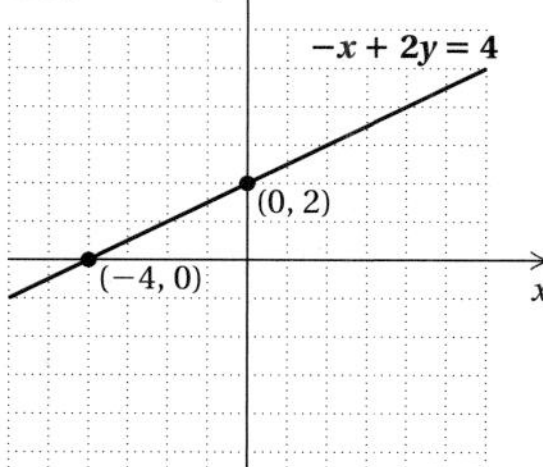

17.

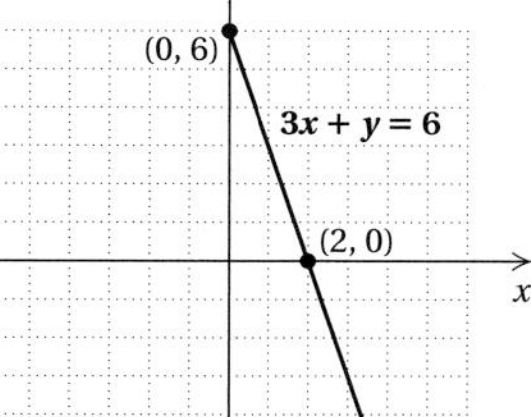

19.

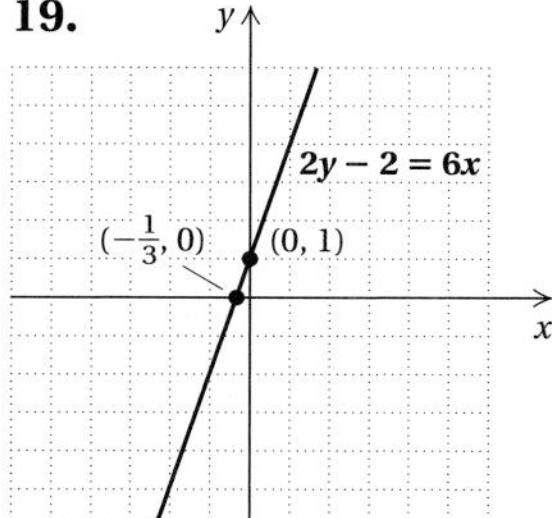

21.

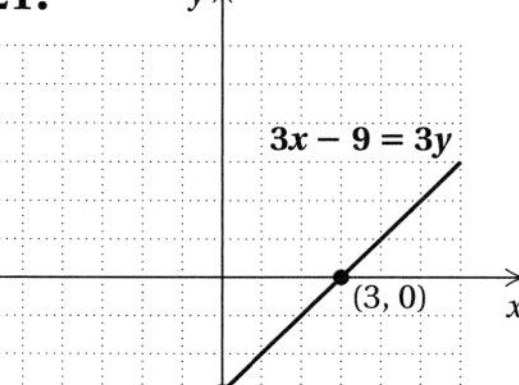

23.

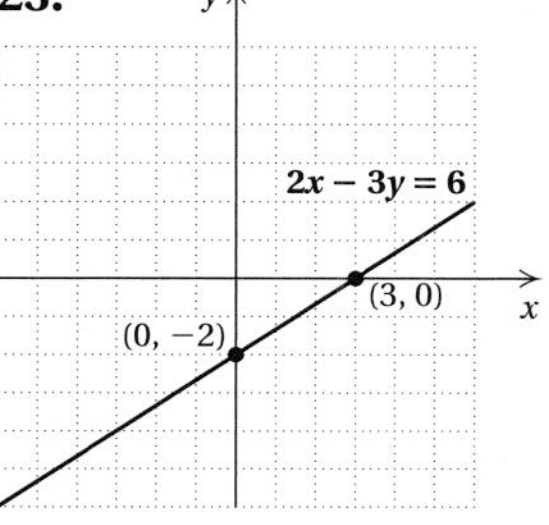

25.

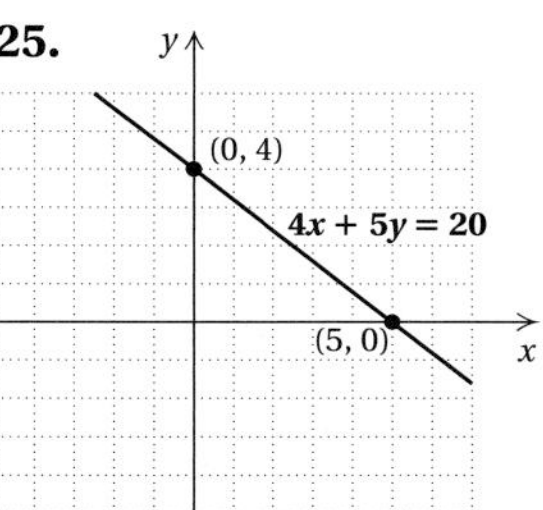

27.

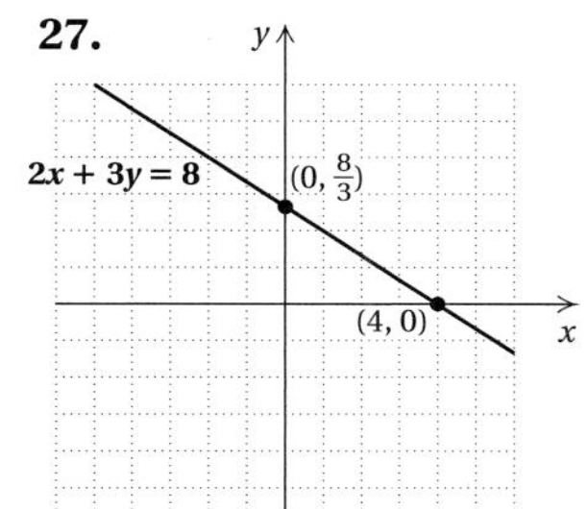

29.

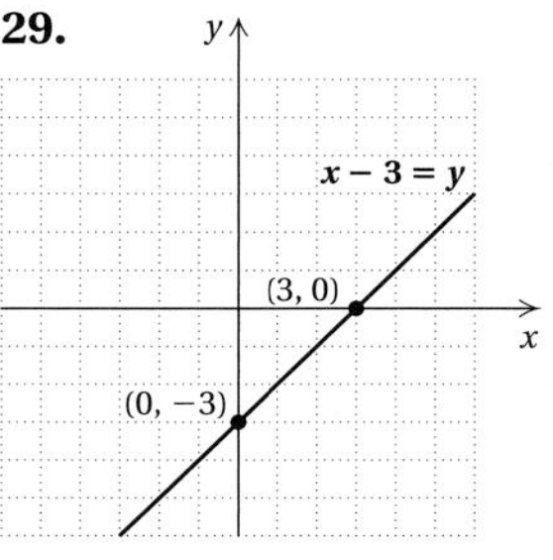

31.

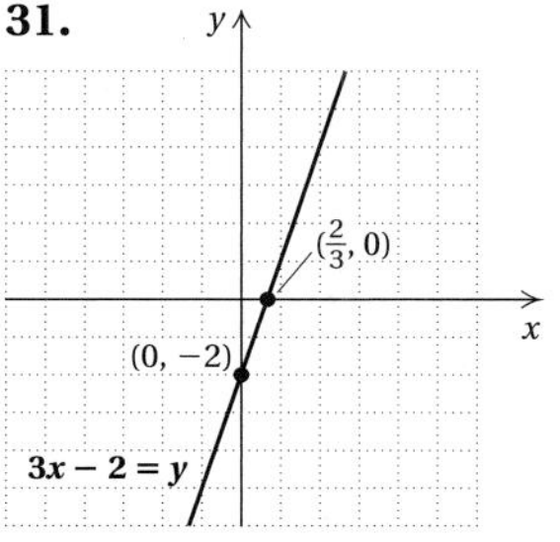

33.

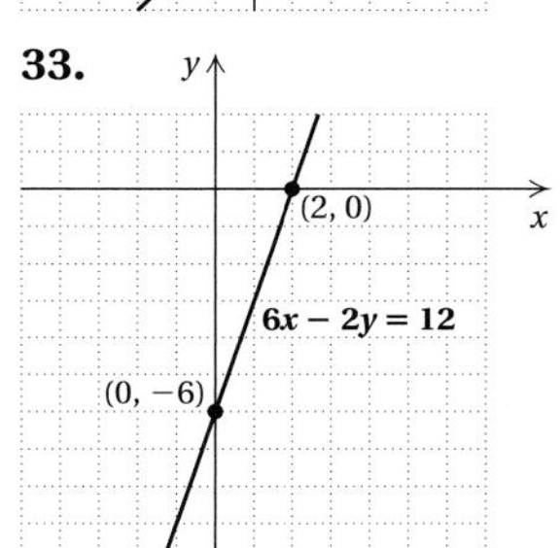

35.

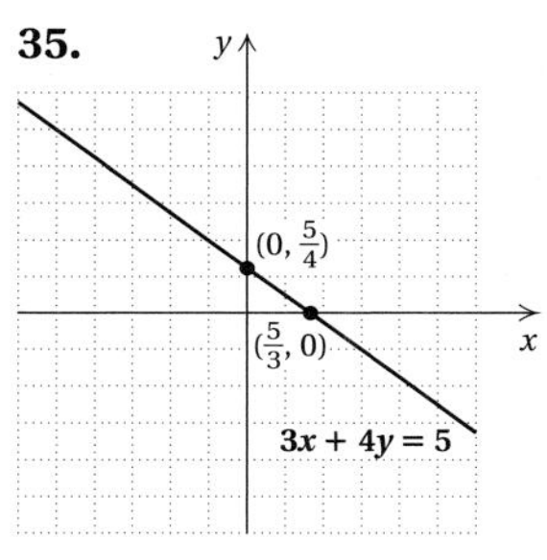

37.

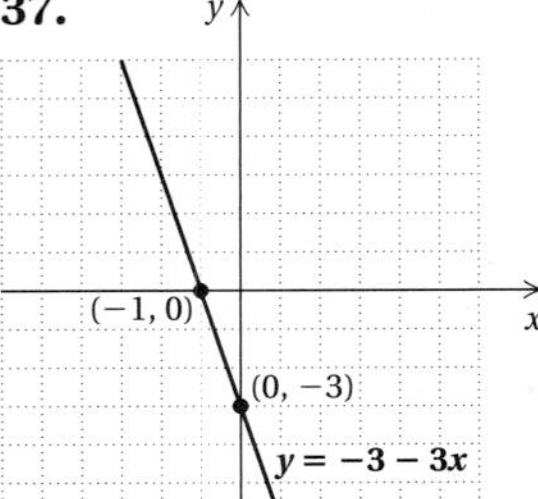

39.

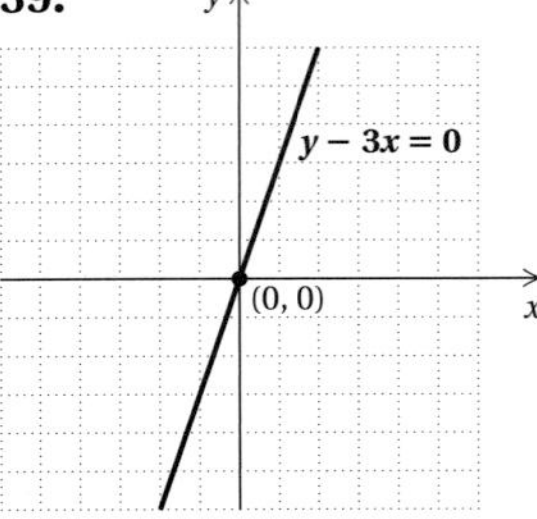

41.

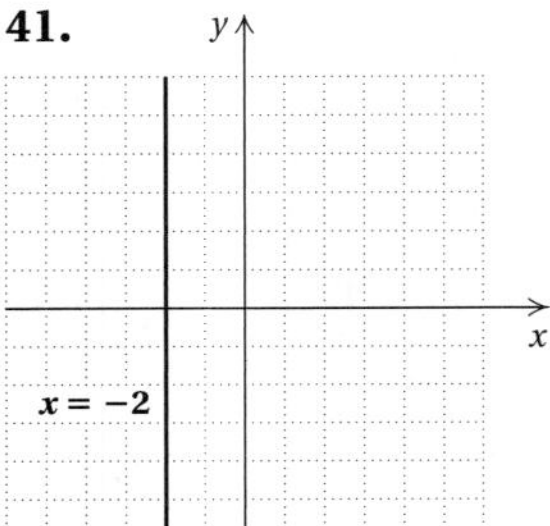

43.

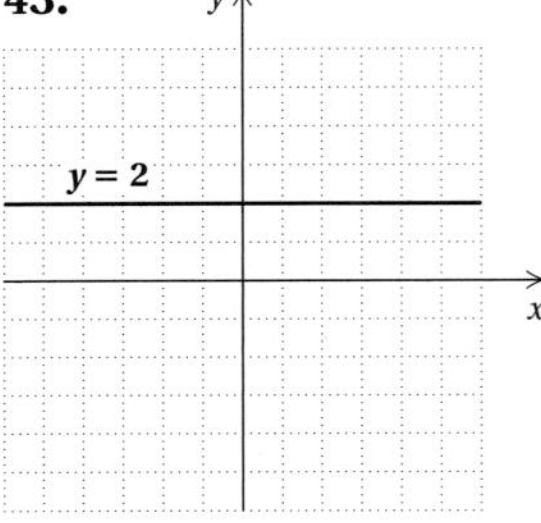

45.

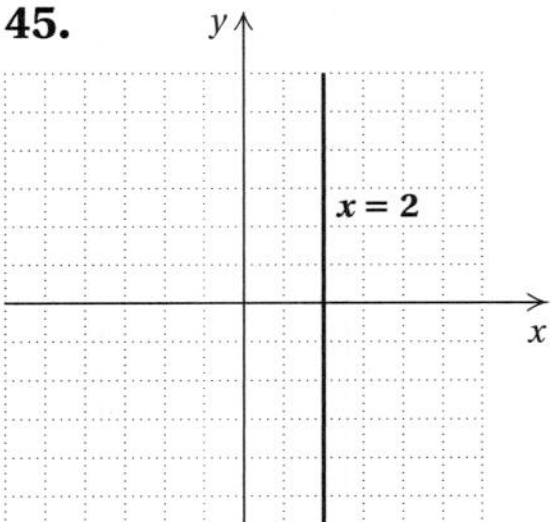

47.

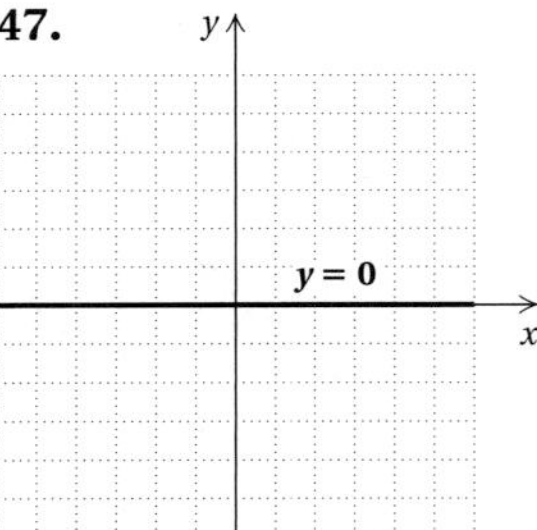

49.

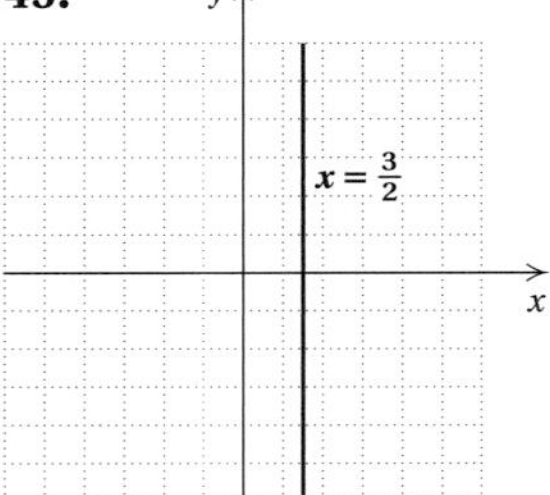

51.

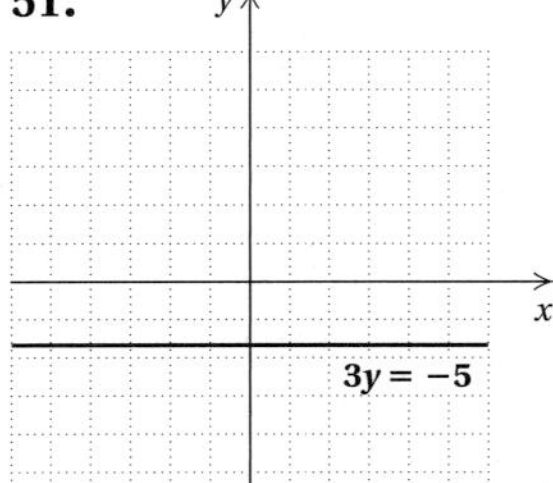

53.

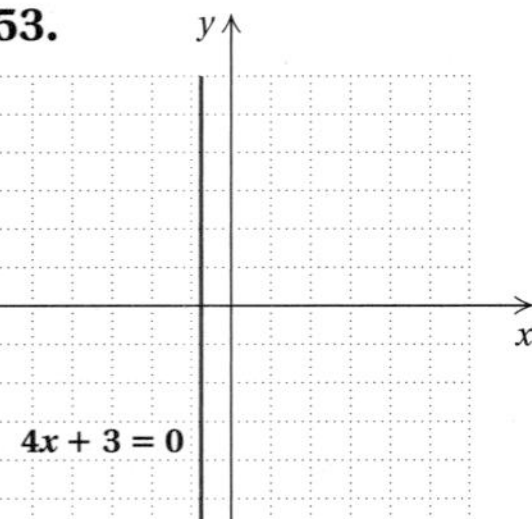

55.

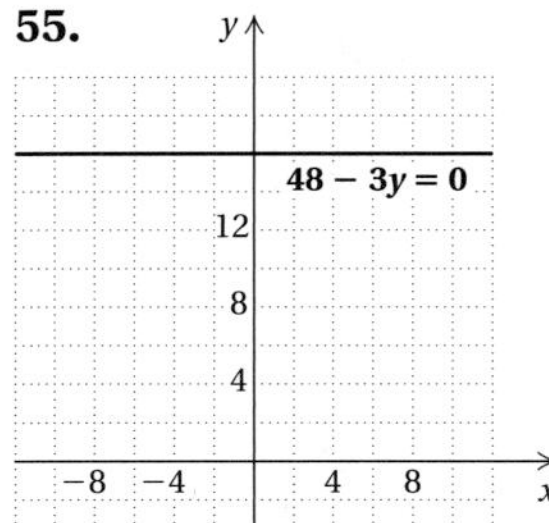

57. $y = -1$ **59.** $x = 4$ **61.** $^{\mathrm{D}}_{\mathrm{W}}$ **63.** 16% **64.** \$32.50 **65.** $\{x|x > -40\}$ **66.** $\{x|x \leq -7\}$ **67.** $\{x|x < 1\}$ **68.** $\{x|x \geq 2\}$ **69.** $y = -4$ **71.** $k = 12$

Margin Exercises, Section 9.3, pp. 668–673

1. $\frac{2}{5}$

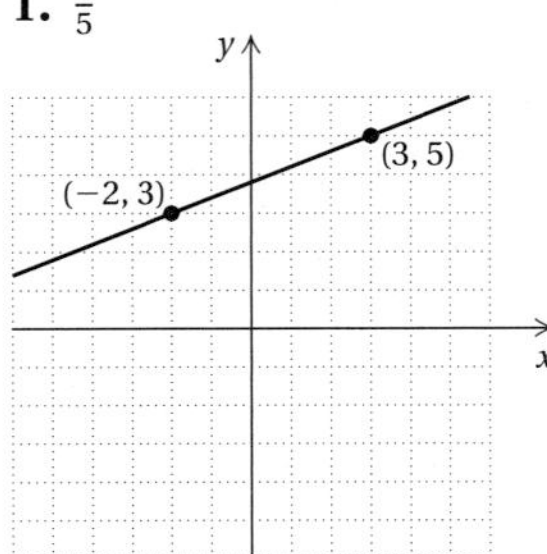

2. $-\frac{5}{3}$

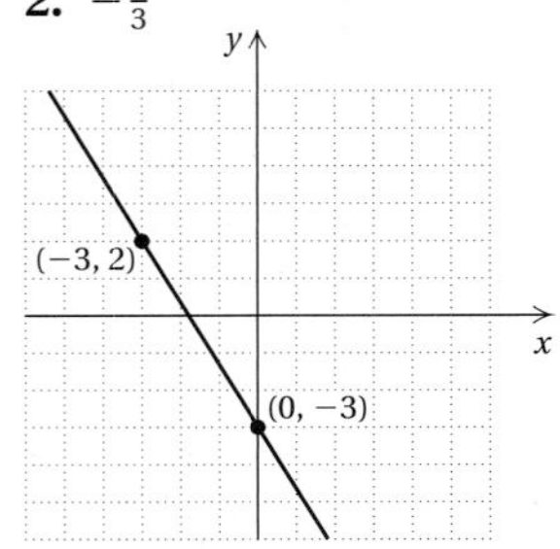

3. $63\frac{7}{11}\%$, or $63.\overline{63}\%$ **4.** 4.5 cents per minute **5.** −1700 deaths by firearms per year **6.** 4 **7.** −17 **8.** −1 **9.** $\frac{2}{3}$ **10.** −1 **11.** $\frac{5}{4}$ **12.** Not defined **13.** 0

Calculator Corner, p. 672

1. This line will pass through the origin and slant up from left to right. This line will be steeper than $y = 10x$.
2. This line will pass through the origin and slant up from left to right. This line will be less steep than $y = \frac{5}{32}x$.

Calculator Corner, p. 673

1. This line will pass through the origin and slant down from left to right. This line will be steeper than $y = -10x$.
2. This line will pass through the origin and slant down from left to right. This line will be less steep than $y = -\frac{5}{32}x$.

Exercise Set 9.3, p. 674

1. $-\frac{3}{7}$ **3.** $\frac{2}{3}$ **5.** $\frac{3}{4}$ **7.** 0

9. $-\frac{4}{5}$;

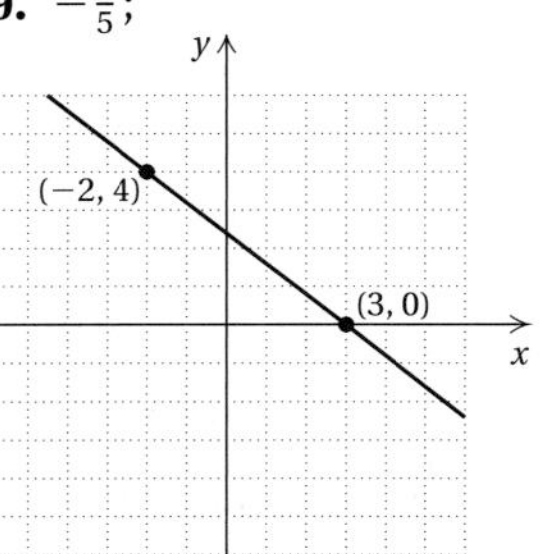

11. 3;

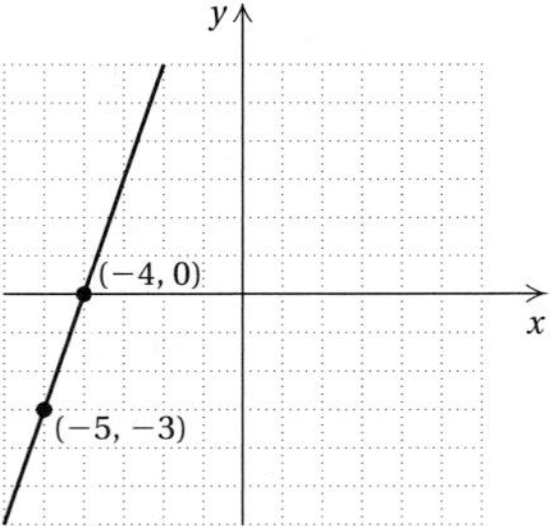

13. $-\frac{5}{6}$;

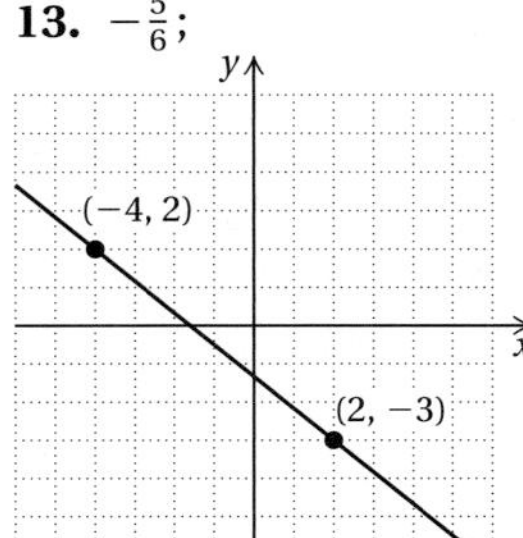

15. $\frac{7}{8}$;

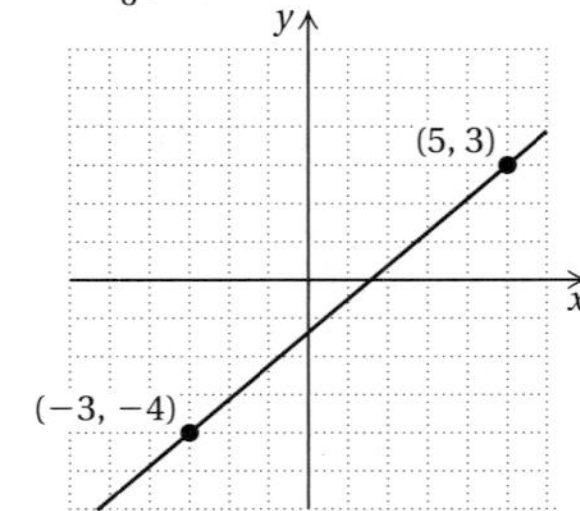

17. $\frac{2}{3}$ **19.** Not defined **21.** $-\frac{5}{13}$ **23.** 0 **25.** $\frac{12}{41}$
27. $\frac{28}{129}$ **29.** About 29.4% **31.** 25 miles per gallon
33. $-\$500$ per year **35.** About 7600 people per year
37. -10 **39.** 3.78 **41.** 3 **43.** $-\frac{1}{5}$ **45.** $-\frac{3}{2}$
47. Not defined **49.** -2.74 **51.** 3 **53.** $\frac{5}{4}$ **55.** 0
57. DW **59.** $\frac{4}{25}$ **60.** $\frac{1}{3}$ **61.** $\frac{3}{8}$ **62.** $\frac{3}{4}$ **63.** \$3.57
64. \$48.60 **65.** 20% **66.** \$18 **67.** \$45.15 **68.** \$55
69. $y = -x + 5$ **71.** $y = x + 2$
73.

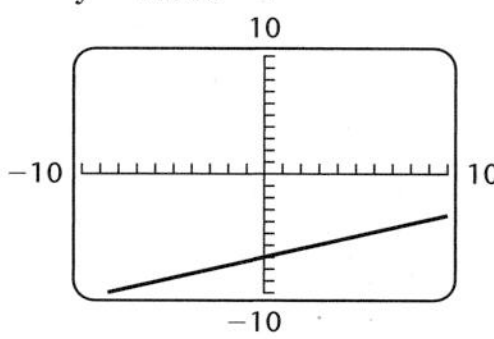

X	Y1
−10	−10.5
−9.9	−10.47
−9.8	−10.43
−9.7	−10.4
−9.6	−10.36
−9.5	−10.33
−9.4	−10.29

X = −10

75.

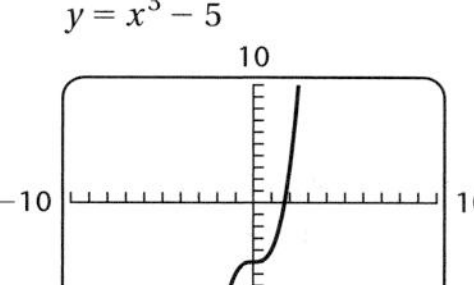

X	Y1
−10	−1005
−9.9	−975.3
−9.8	−946.2
−9.7	−917.7
−9.6	−889.7
−9.5	−862.4
−9.4	−835.6

X = −10

Margin Exercises, Section 9.4, pp. 679–682

1. Slope: 5; y-intercept: (0, 0)
2. Slope: $-\frac{3}{2}$; y-intercept: (0, −6)
3. Slope: $-\frac{3}{4}$; y-intercept: $\left(0, \frac{15}{4}\right)$
4. Slope: 2; y-intercept: $\left(0, -\frac{17}{2}\right)$
5. Slope: $-\frac{7}{5}$; y-intercept: $\left(0, -\frac{22}{5}\right)$
6. $y = 3.5x - 23$ **7.** $y = 13$ **8.** $y = -7.29x$
9. $y = 5x - 18$ **10.** $y = -3x - 5$ **11.** $y = 6x - 13$
12. $y = -\frac{2}{3}x + \frac{14}{3}$ **13.** $y = x + 2$ **14.** $y = 2x + 4$

Exercise Set 9.4, p. 683

1. Slope: −4; y-intercept: (0, −9)
3. Slope: 1.8; y-intercept: (0, 0)
5. Slope: $-\frac{8}{7}$; y-intercept: (0, −3)
7. Slope: $\frac{4}{9}$; y-intercept: $\left(0, -\frac{7}{9}\right)$
9. Slope: $-\frac{3}{2}$; y-intercept: $\left(0, -\frac{1}{2}\right)$
11. Slope: 0; y-intercept: (0, −17) **13.** $y = -7x - 13$
15. $y = 1.01x - 2.6$ **17.** $y = -2x - 6$
19. $y = \frac{3}{4}x + \frac{5}{2}$ **21.** $y = x - 8$ **23.** $y = -3x + 3$
25. $y = x + 4$ **27.** $y = -\frac{1}{2}x + 4$ **29.** $y = -\frac{3}{2}x + \frac{13}{2}$
31. $y = -4x - 11$ **33.** **(a)** $T = -0.75a + 165$;
(b) −0.75 beat per minute per year; **(c)** 127.5 beats per minute **35.** DW **37.** $\frac{53}{7}$ **38.** $\frac{3}{8}$ **39.** $\frac{24}{19}$
40. $\frac{125}{7}$ **41.** $\frac{1}{3}$ **42.** $-\frac{1}{12}$ **43.** $\frac{42}{25}$ **44.** $\frac{5}{7}$
45. $y = 3x - 9$ **47.** $y = \frac{3}{2}x - 2$

Margin Exercises, Section 9.5, pp. 685–687

1.

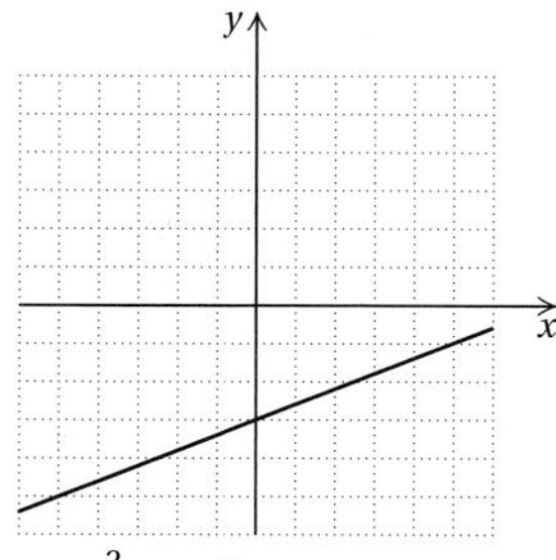

$y = \frac{2}{5}x - 3$

2.

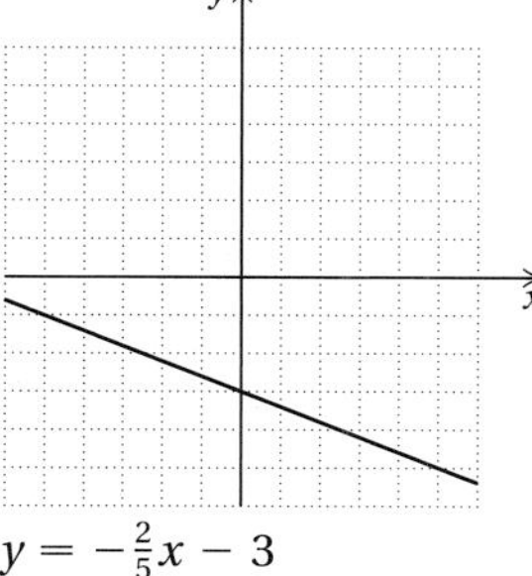

$y = -\frac{2}{5}x - 3$

3.

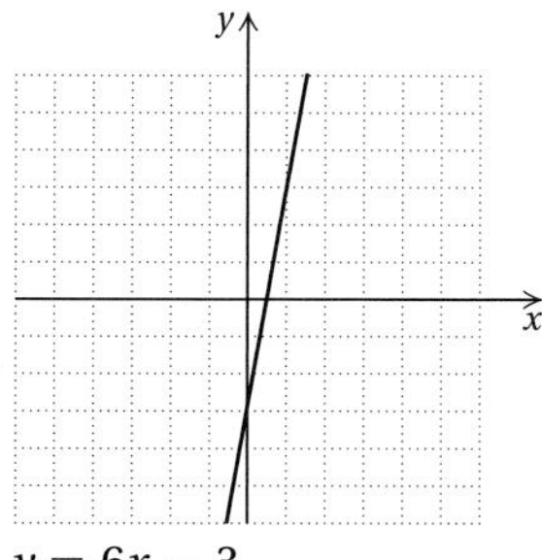

$y = 6x - 3$

4.

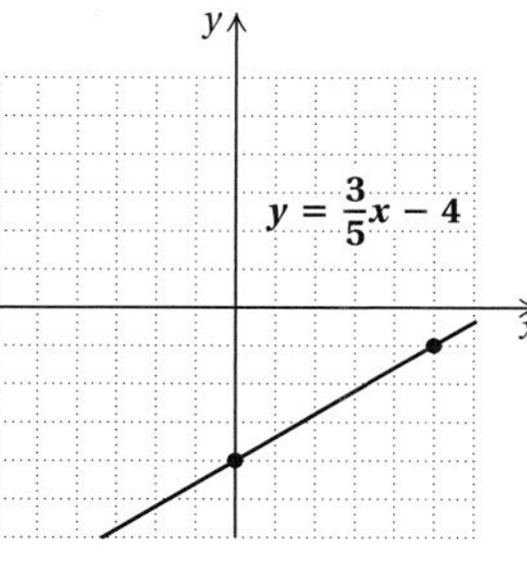

5.

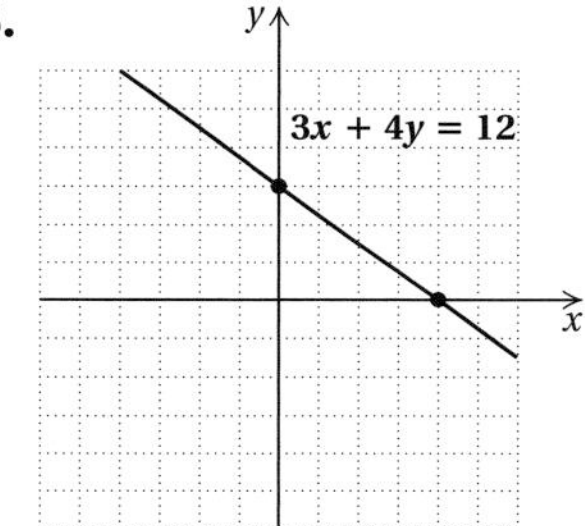

Exercise Set 9.5, p. 688

1.

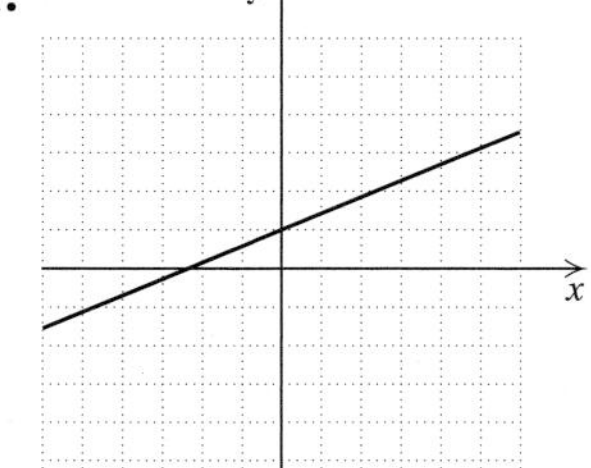

3.

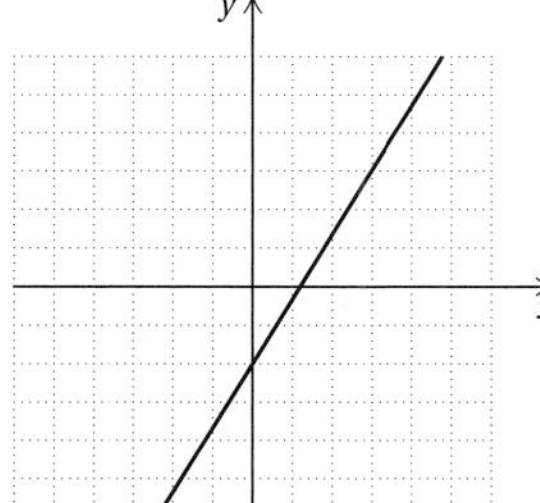

5.

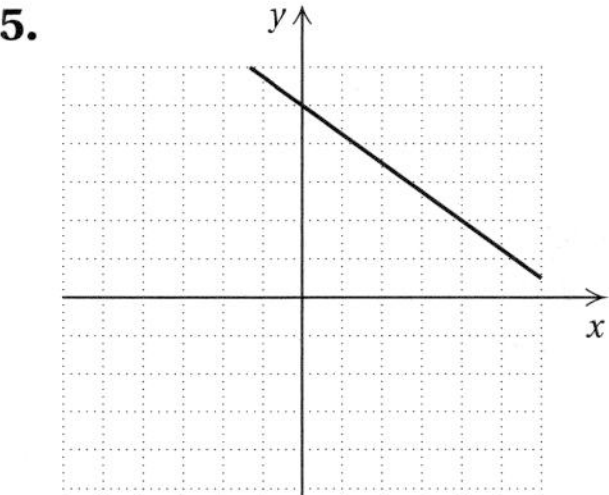

7.

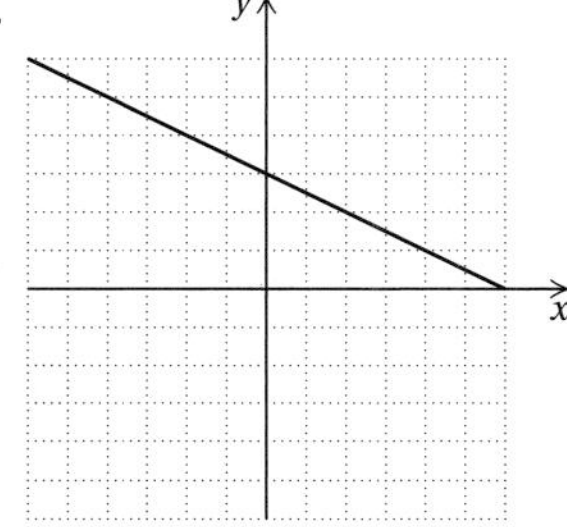

9.

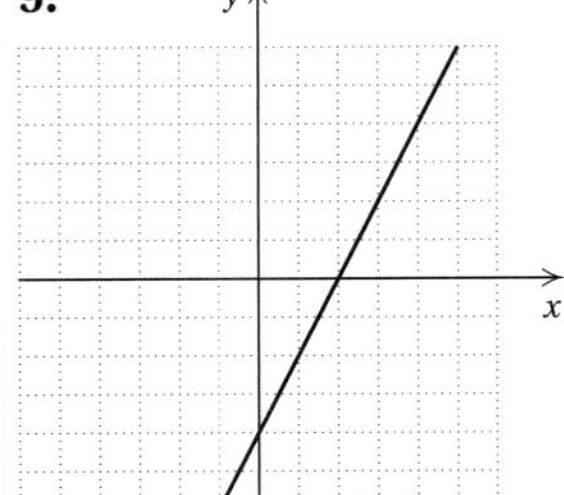

11.

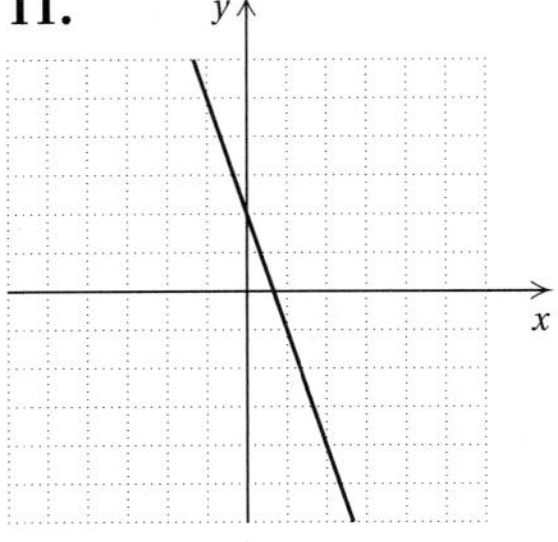

13.

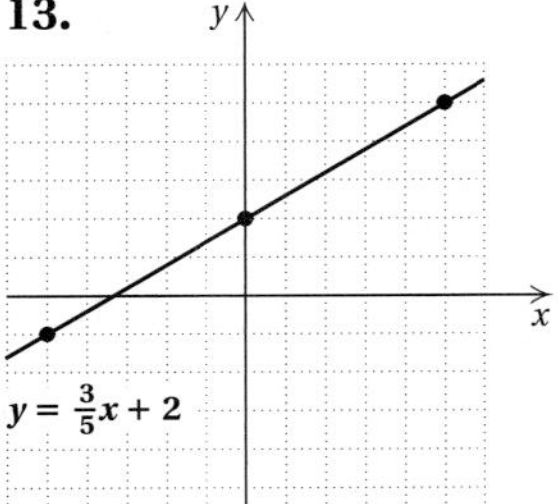

15.

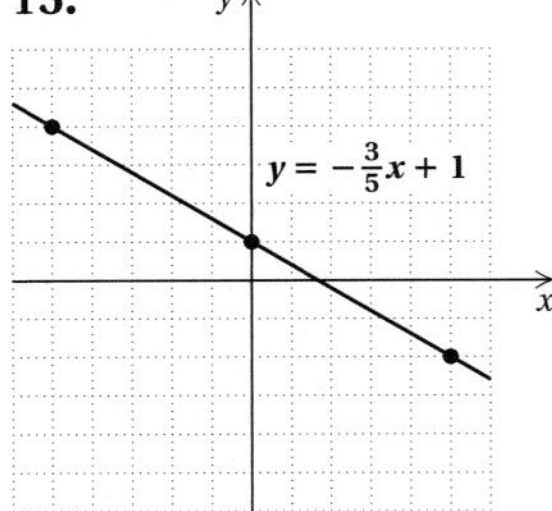

17.

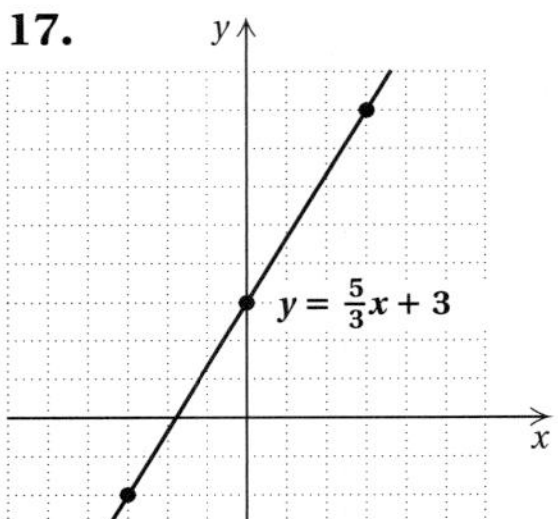

19.

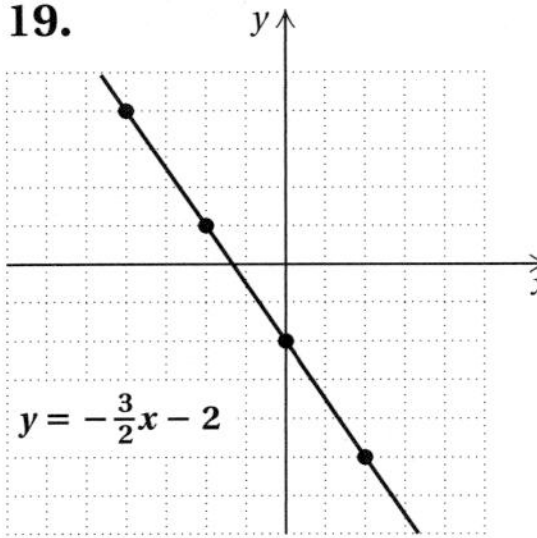

21.

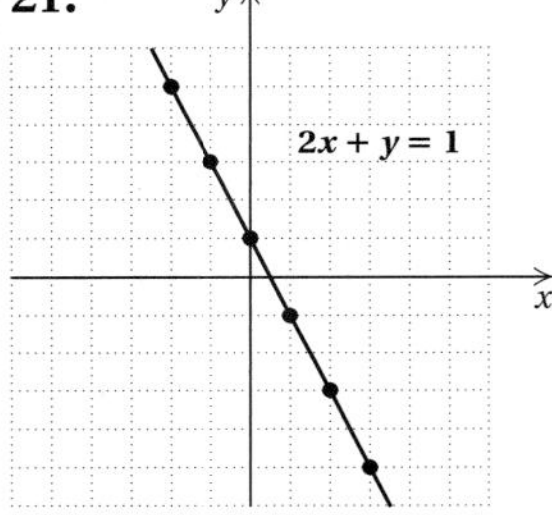

23.

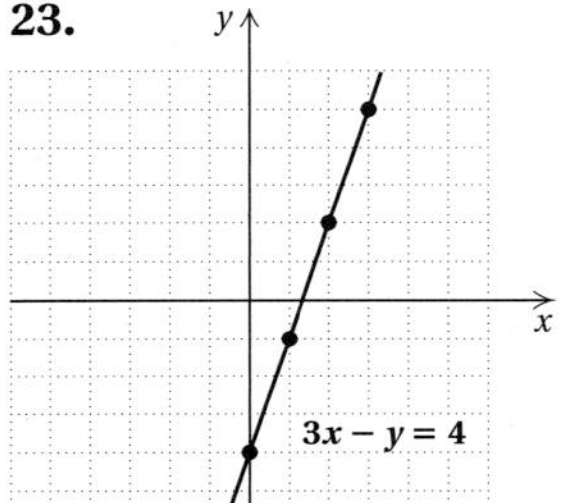

25.

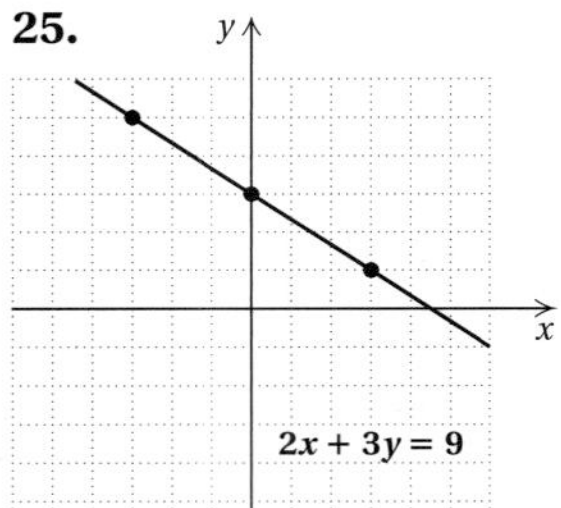

27.

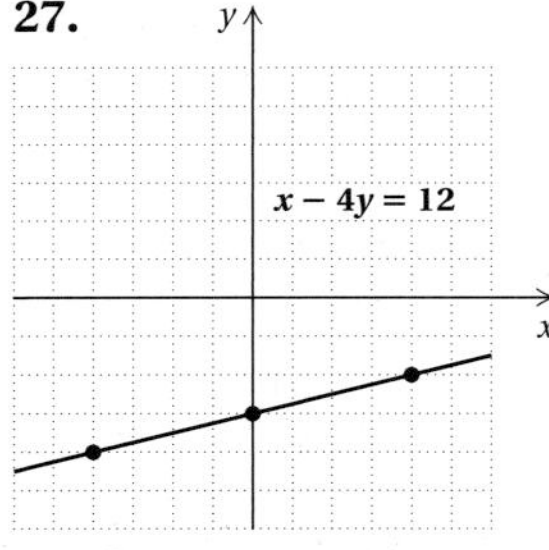

29.

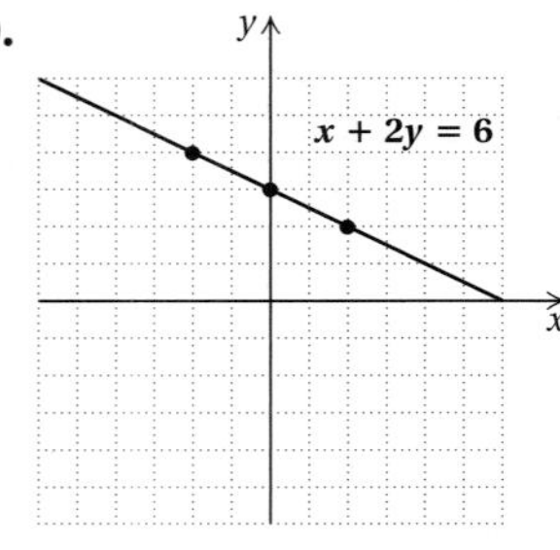

31. DW **33.** $\frac{13}{10}$

34. $-\frac{1}{6}$ **35.** $\frac{69}{100}$, or 0.69 **36.** $-\frac{3}{5}$, or -0.6 **37.** 0
38. Not defined **39.** Not defined **40.** 0
41. Increase of about 416 kidney transplants per year; 416
42. Increase of about 264 liver transplants per year; 264
43. $y = 1.5x + 16$ **45.**

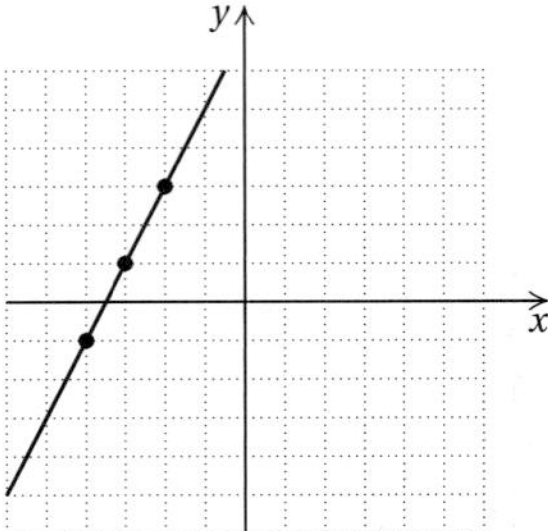

Margin Exercises, Section 9.6, p. 693

1. No **2.** Yes **3.** Yes **4.** No

Exercise Set 9.6, p. 695

1. Yes **3.** No **5.** No **7.** No **9.** Yes **11.** Yes
13. No **15.** Yes **17.** Yes **19.** Yes **21.** No
23. Yes **25.** Parallel **27.** Neither **29.** DW
31. Equivalent equations **32.** Addition principle
33. Multiplication principle **34.** Horizontal
35. Vertical **36.** Slope **37.** x-intercept
38. y-intercept **39.** $y = 3x + 6$
41. $y = -3x + 2$ **43.** $y = \frac{1}{2}x + 1$ **45.** 16
47. A: $y = \frac{4}{3}x - \frac{7}{3}$; B: $y = -\frac{3}{4}x - \frac{1}{4}$

Margin Exercises, Section 9.7, pp. 697–700

1. No **2.** No **3.**

$y < x$

4.

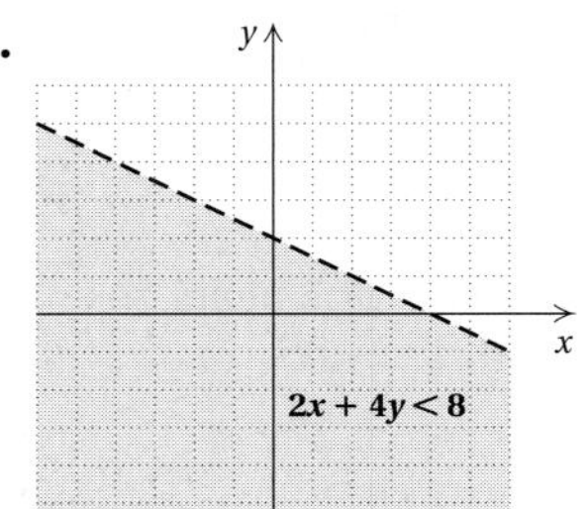

5.

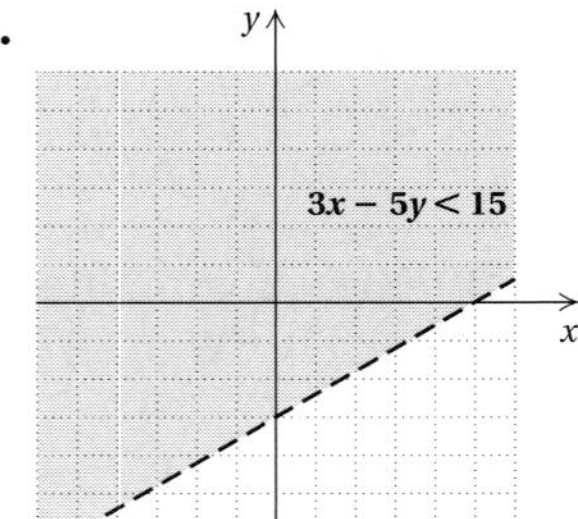

6. 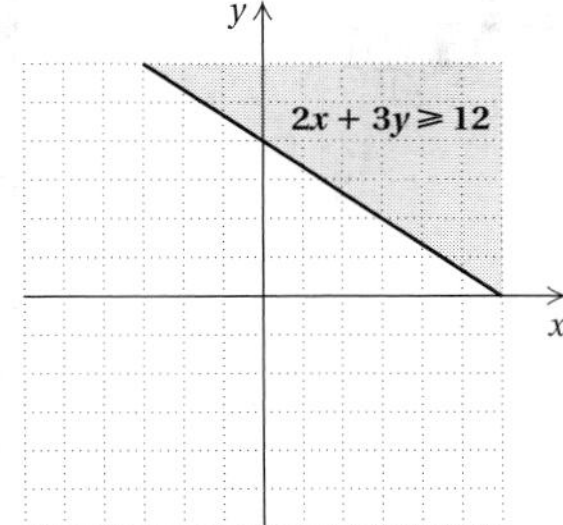

7. 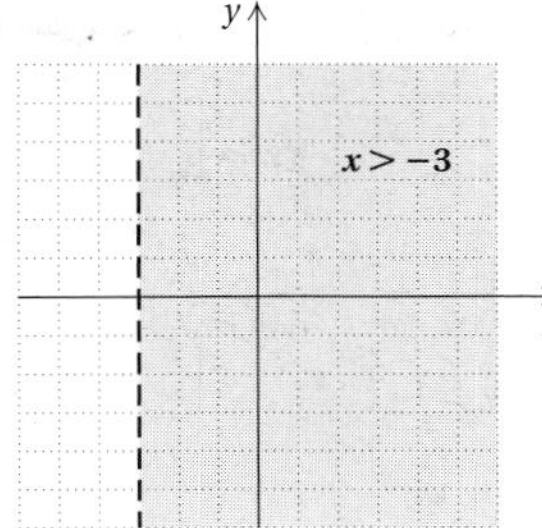

8. 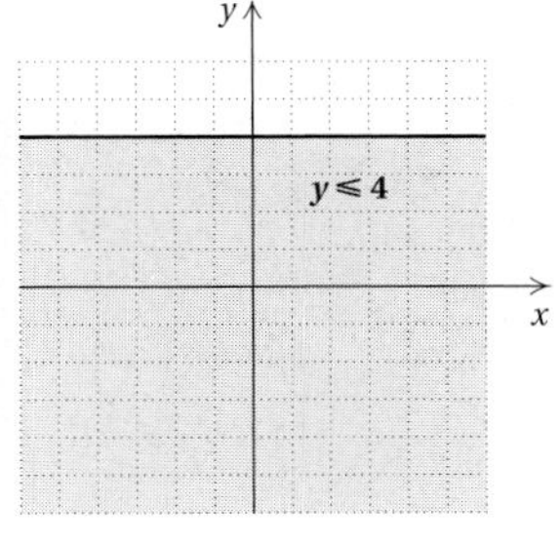

15. 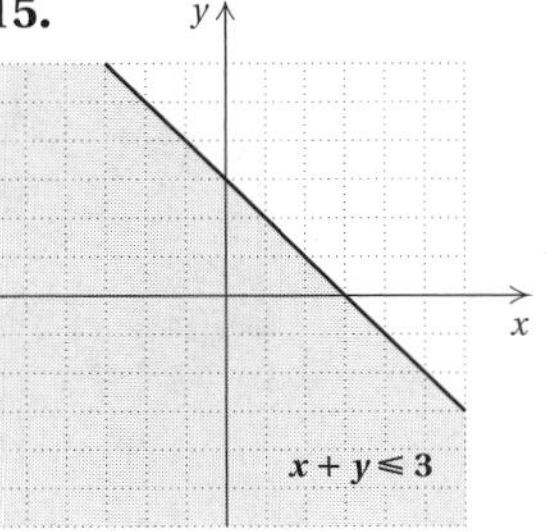

17.

19. 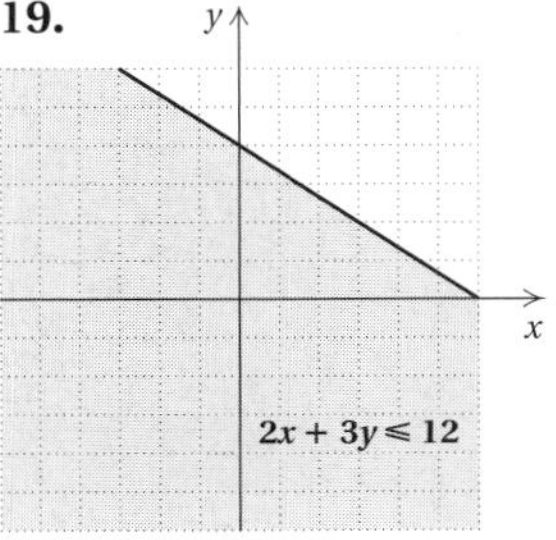

21. 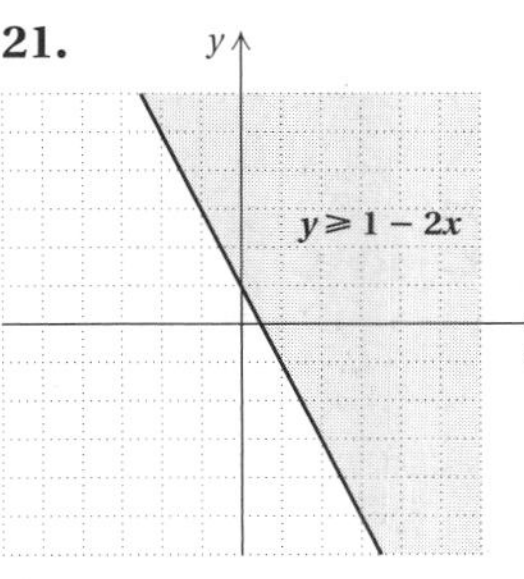

23. 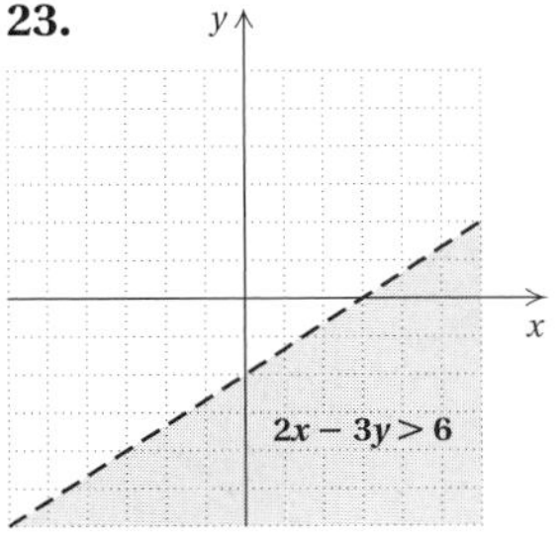

25. 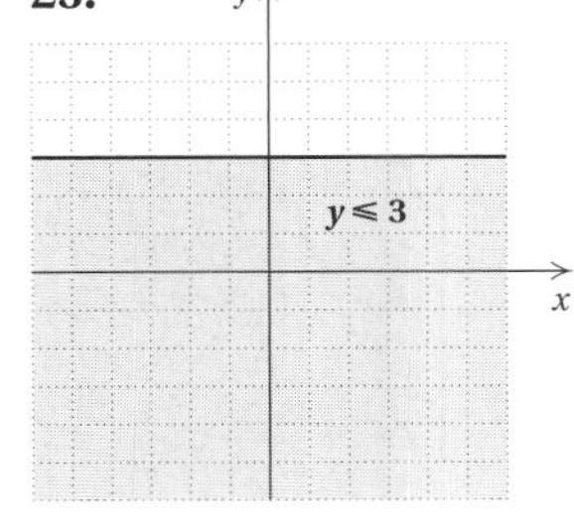

27. 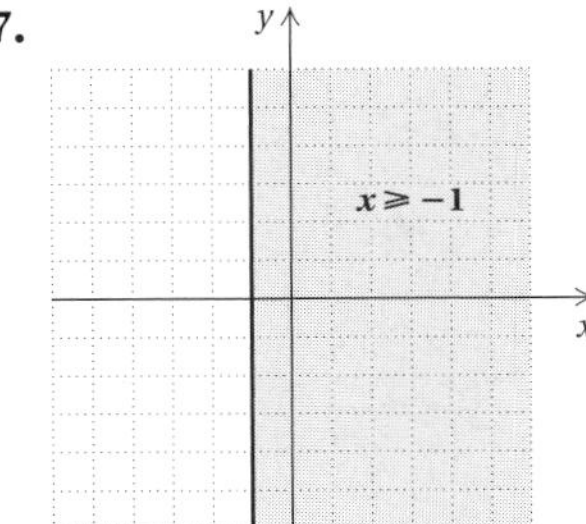

Calculator Corner, p. 701

1. Left to the student **2.** Left to the student

Visualizing for Success, p. 702

1. D **2.** H **3.** E **4.** A **5.** J **6.** F **7.** C **8.** B **9.** I **10.** G

Exercise Set 9.7, p. 703

1. No **3.** Yes **5.** 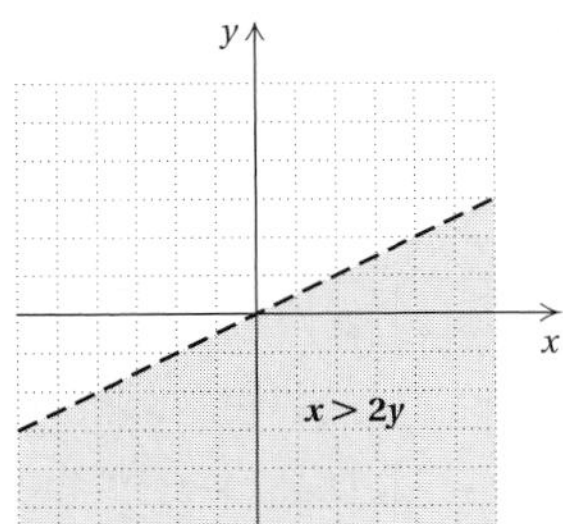

7. 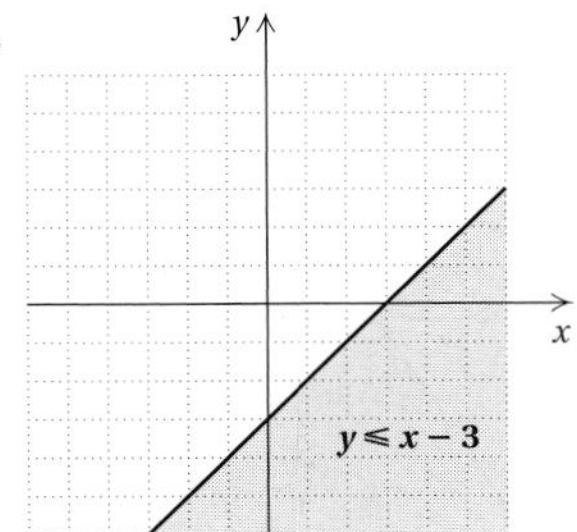

9. 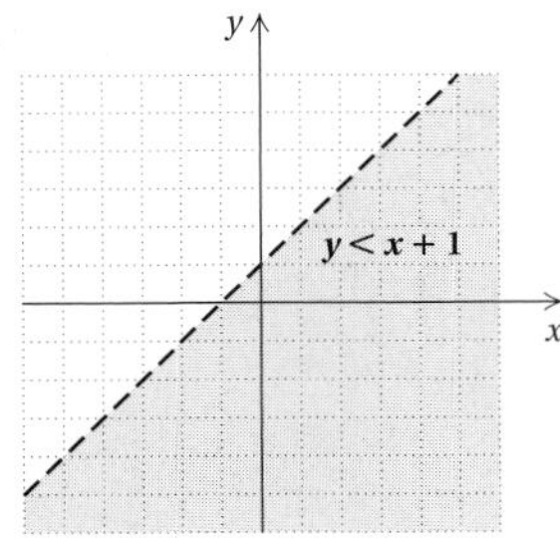

11. 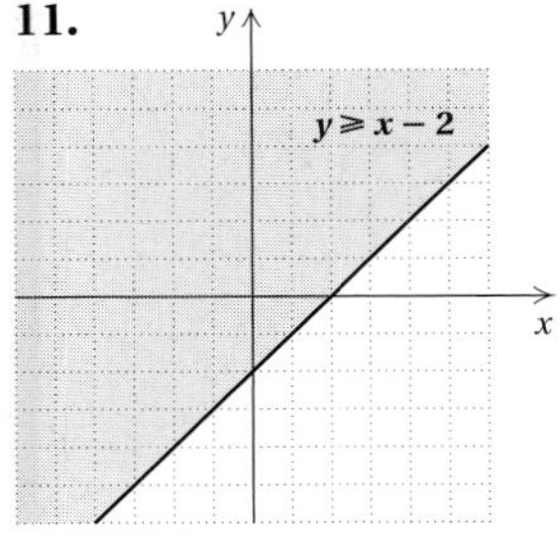

13. 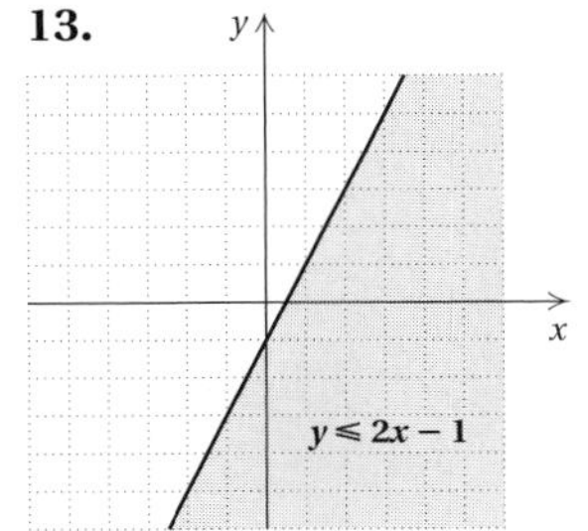

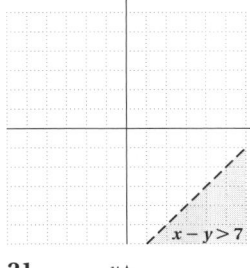

29. ^{D}W **31.** Parallel **32.** Perpendicular **33.** Neither **34.** Neither **35.** $35c + 75a > 1000$

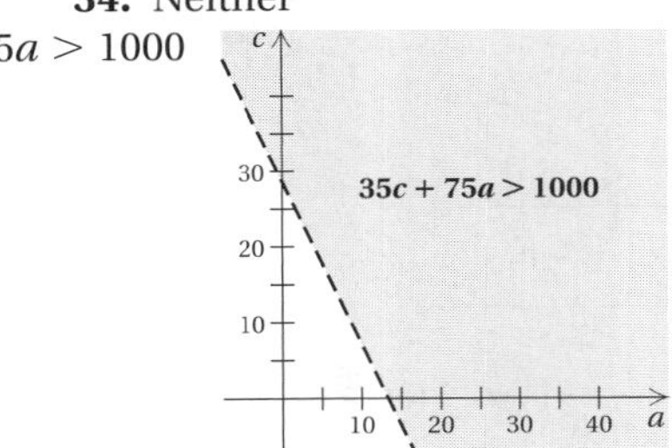

Concept Reinforcement, p. 705

1. False **2.** True **3.** True **4.** True **5.** False **6.** True

Summary and Review: Chapter 9, p. 705

1. $(-5, -1)$ **2.** $(-2, 5)$ **3.** $(3, 0)$

4.–6.

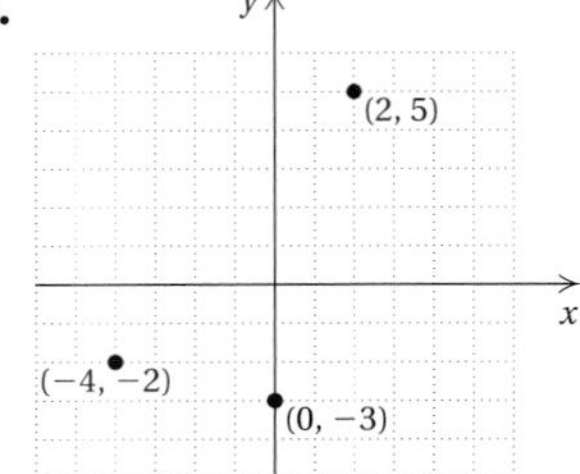

7. IV **8.** III **9.** I **10.** No **11.** Yes

12.

$$\begin{array}{c|c} 2x - y & 3 \\ \hline 2 \cdot 0 - (-3) & ?\ 3 \\ 0 + 3 & \\ 3 & \end{array} \quad \text{TRUE}$$

$$\begin{array}{c|c} 2x - y & 3 \\ \hline 2 \cdot 2 - 1 & ?\ 3 \\ 4 - 1 & \\ 3 & \end{array} \quad \text{TRUE}$$

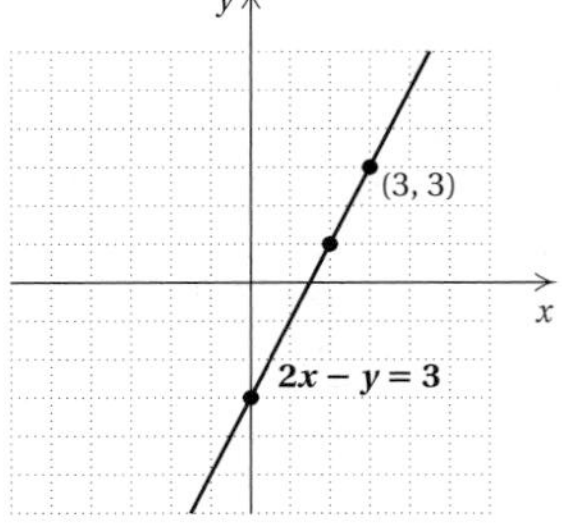

13.

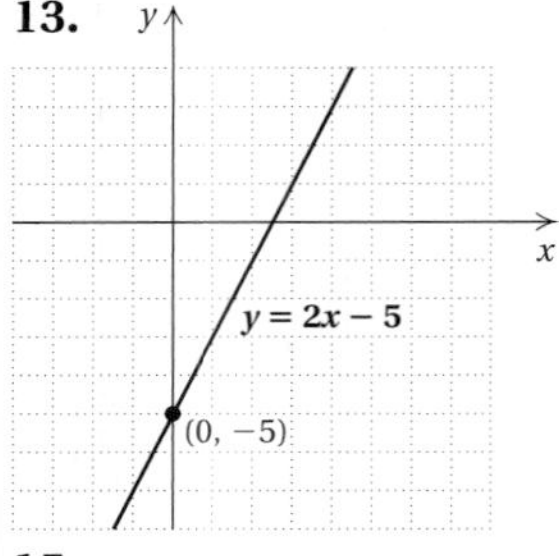

14.

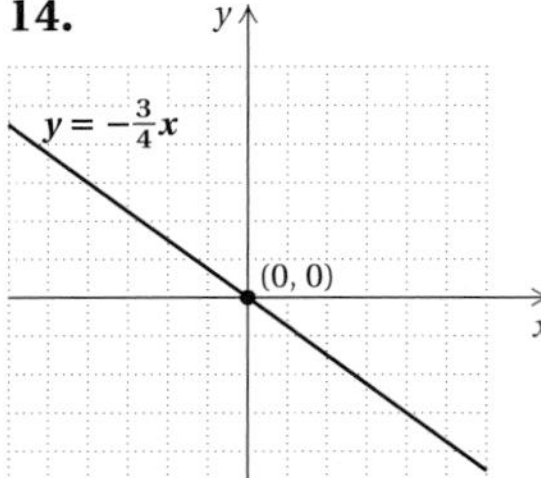

15.

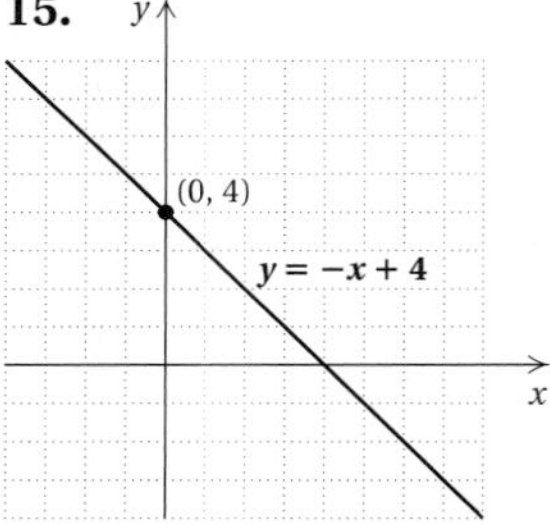

16.

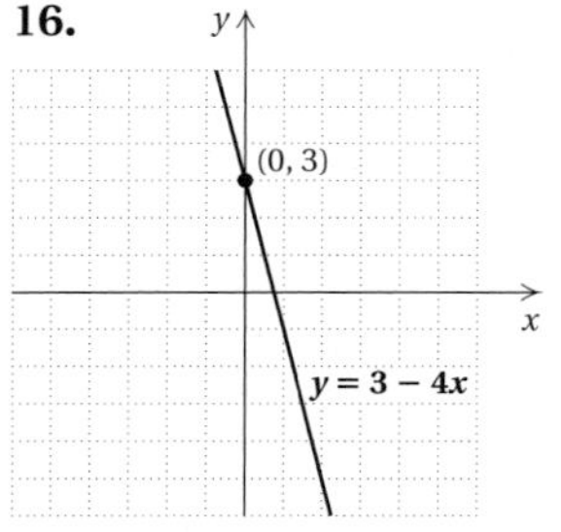

17.

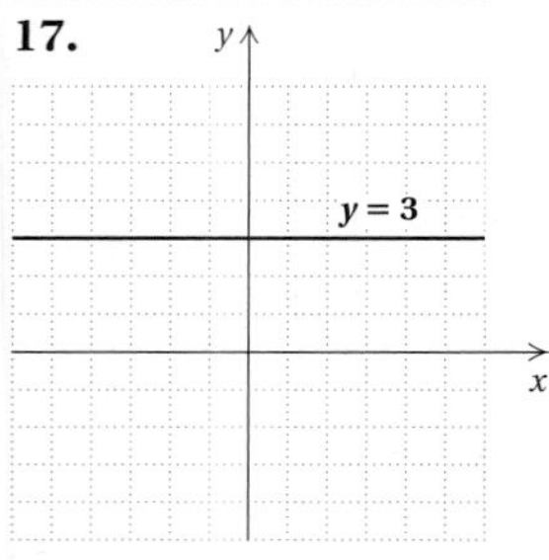

18.

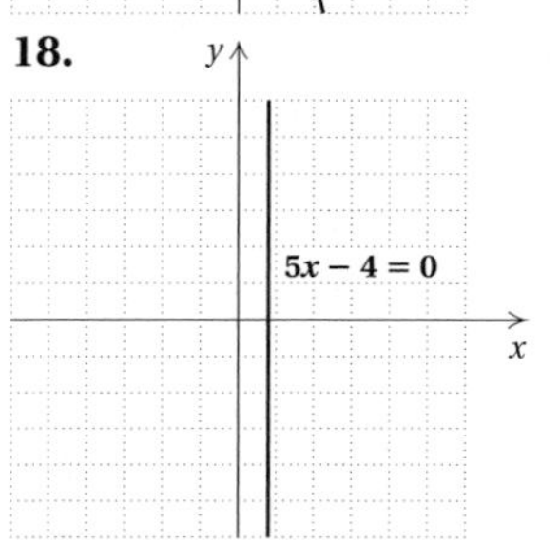

19.

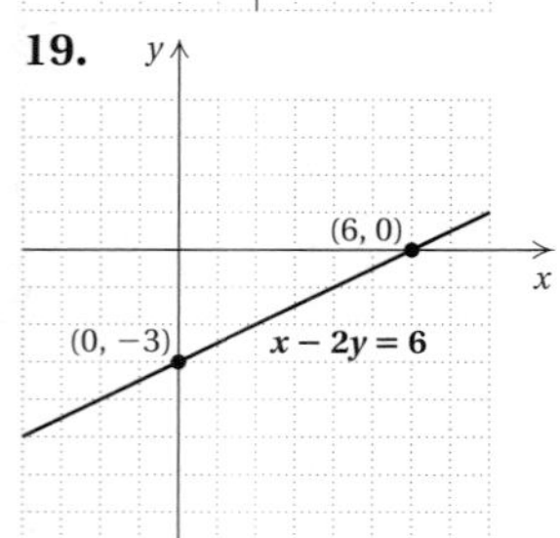

20.

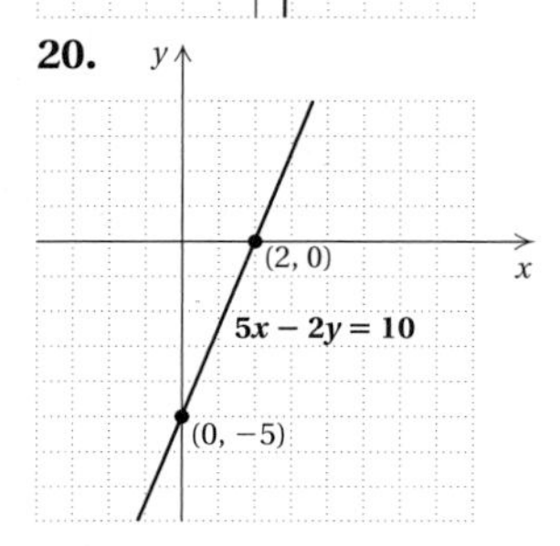

21. **(a)** $14\frac{1}{2}$ ft^3, 16 ft^3, $20\frac{1}{2}$ ft^3, 28 ft^3;

(b)

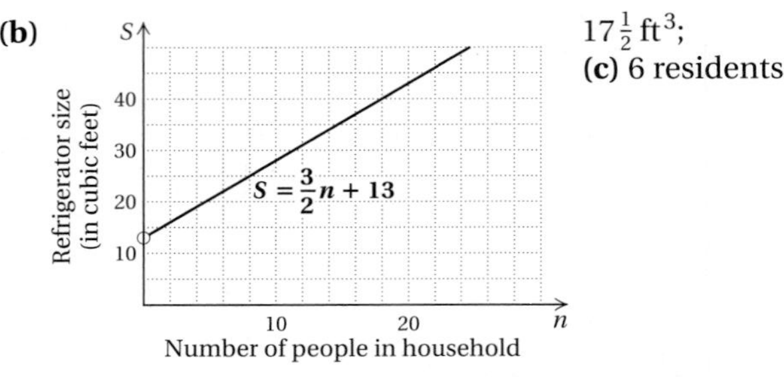

$17\frac{1}{2}$ ft^3; **(c)** 6 residents

22. **(a)** 2.4 driveways per hour; **(b)** 25 minutes per driveway

23. 4 manicures per hour **24.** $\frac{1}{3}$ **25.** $-\frac{1}{3}$

26. $\frac{3}{5}$;

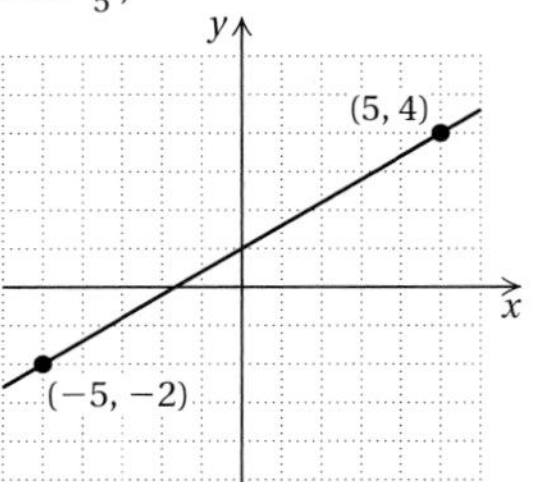

27. -1;

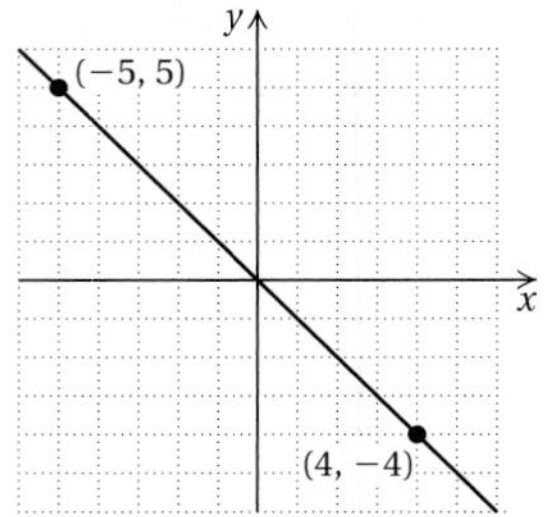

28. 7% **29.** $-\frac{5}{8}$ **30.** $\frac{1}{2}$ **31.** Not defined **32.** 0

33. Slope: -9; y-intercept: $(0, 46)$ **34.** Slope: -1; y-intercept: $(0, 9)$ **35.** Slope: $\frac{3}{5}$; y-intercept: $\left(0, -\frac{4}{5}\right)$

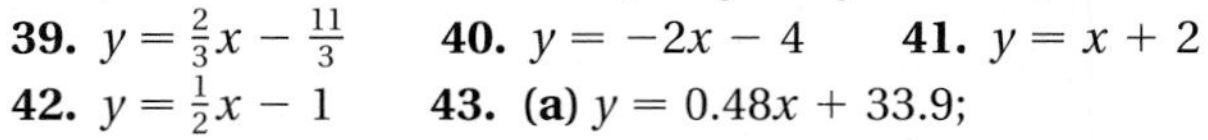

36. $y = -2.8x + 19$ **37.** $y = \frac{5}{8}x - \frac{7}{8}$ **38.** $y = 3x - 1$

39. $y = \frac{2}{3}x - \frac{11}{3}$ **40.** $y = -2x - 4$ **41.** $y = x + 2$

42. $y = \frac{1}{2}x - 1$ **43.** **(a)** $y = 0.48x + 33.9$; **(b)** 0.48 percent per year; **(c)** 60.3%

44.

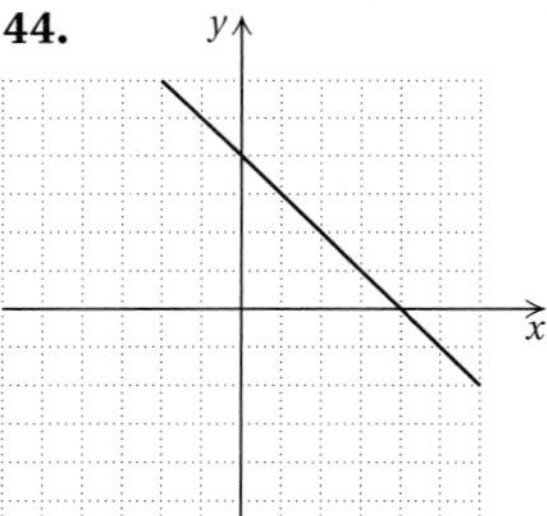

45.

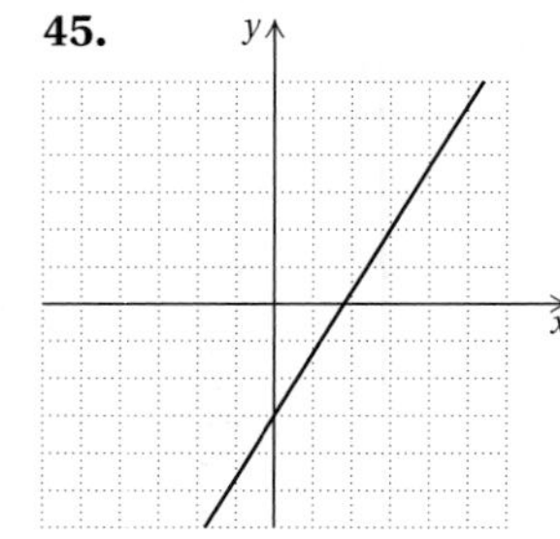

46.

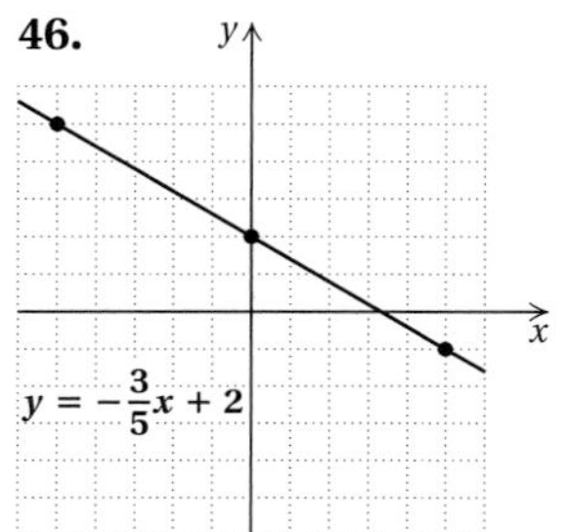

47.

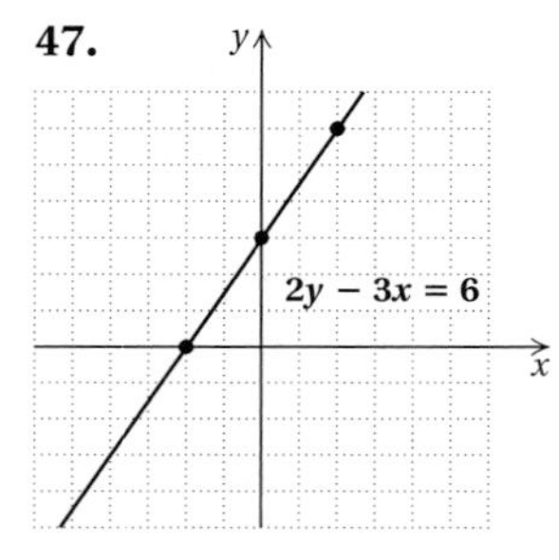

48. Parallel **49.** Perpendicular **50.** Parallel

51. Neither **52.** No **53.** No **54.** Yes

55.

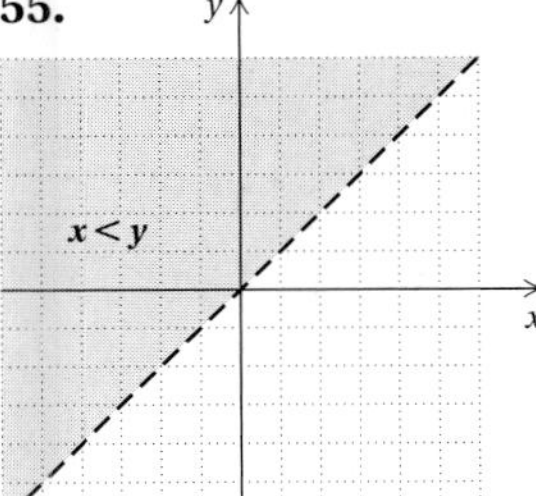

56.

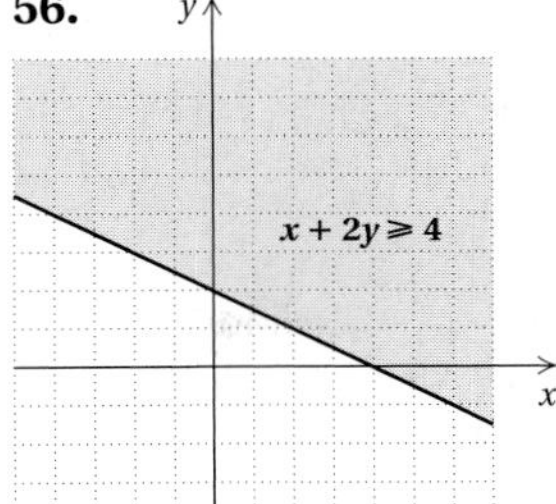

57.

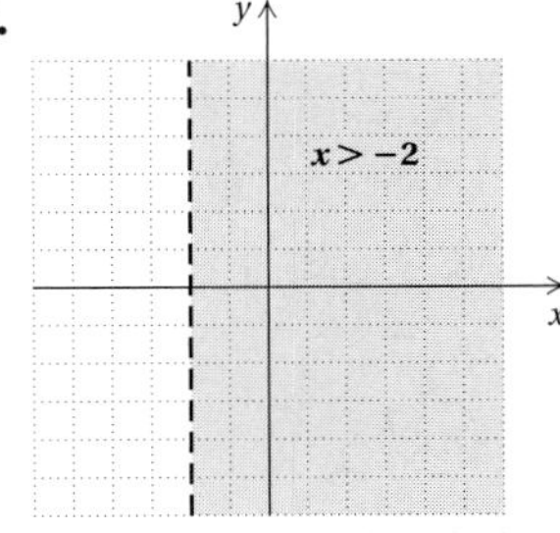

58. DW The y-intercept is the point at which the graph crosses the y-axis. Since a point on the y-axis is neither left nor right of the origin, the first or x-coordinate of the point is 0.

59. DW

The graph of $x < 1$ on a number line consists of the points in the set $\{x|x < 1\}$. The graph of $x < 1$ on a plane consists of the points, or ordered pairs, in the set $\{(x, y)|x + 0 \cdot y < 1\}$. This is the set of ordered pairs with first coordinate less than 1.

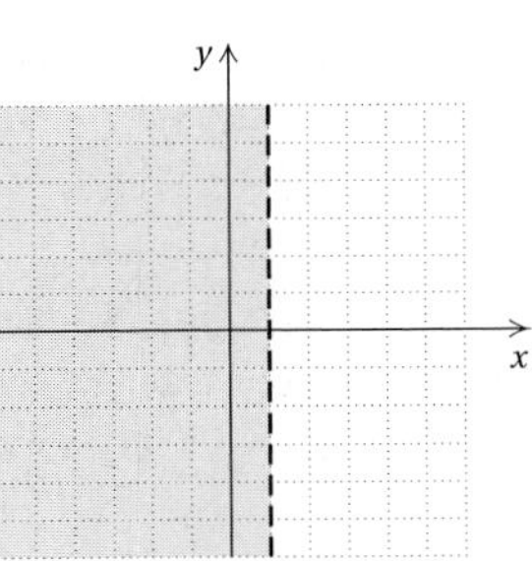

60. DW First plot the y-intercept, (0, 2458). Then, thinking of the slope as $\frac{37}{100}$, plot a second point on the line by moving up 37 units and to the right 100 units from the y-intercept. Next, thinking of the slope as $\frac{-37}{-100}$, start at the y-intercept and plot a third point by moving down 37 units and to the left 100 units. Finally, draw a line through the three points.
61. $m = -1$ **62.** 45 square units; 28 linear units
63. **(a)** 3.709 feet per minute; **(b)** about 0.2696 minute per foot **64.** $-\frac{1}{2}, \frac{1}{2}, -2, 2$

Test: Chapter 9, p. 711

1. [9.1a] II **2.** [9.1a] III **3.** [9.1b] (3, 4)
4. [9.1b] (0, −4)
5. [9.1c]

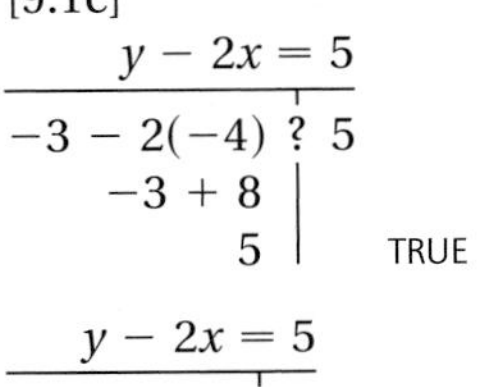

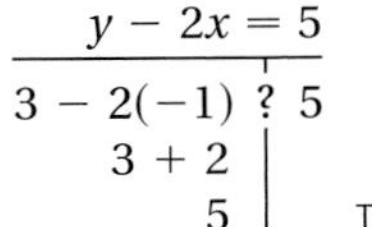

$y - 2x = 5$	
$-3 - 2(-4) \; ? \; 5$	
$-3 + 8$	
5	TRUE

$y - 2x = 5$	
$3 - 2(-1) \; ? \; 5$	
$3 + 2$	
5	TRUE

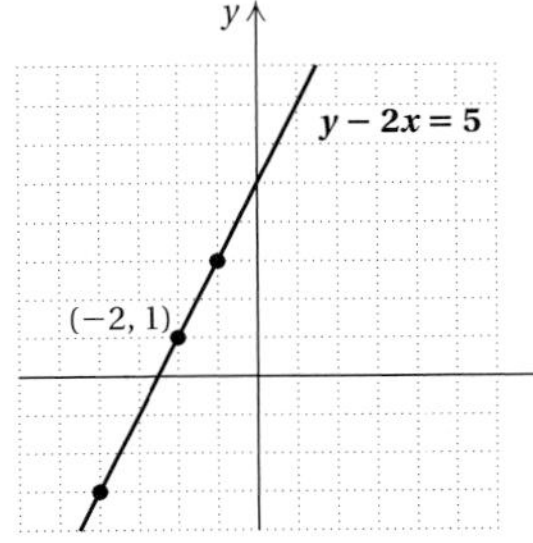

6. [9.1d]

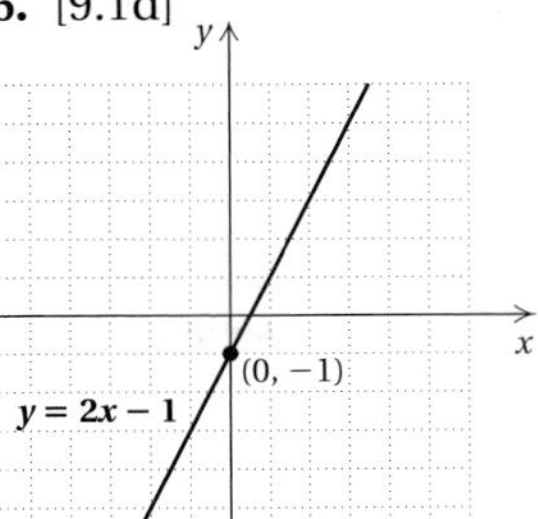

7. [9.1d]

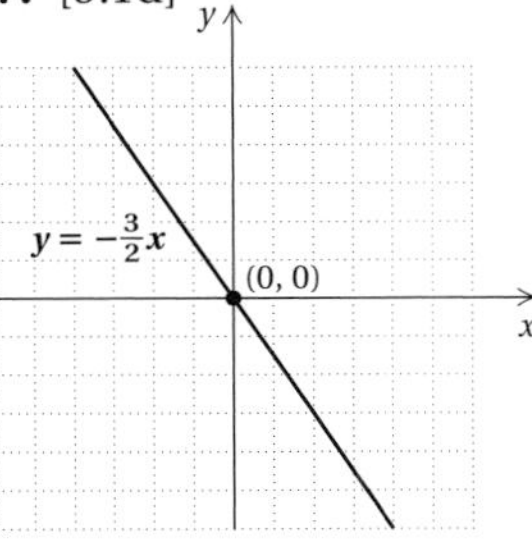

8. [9.2b]

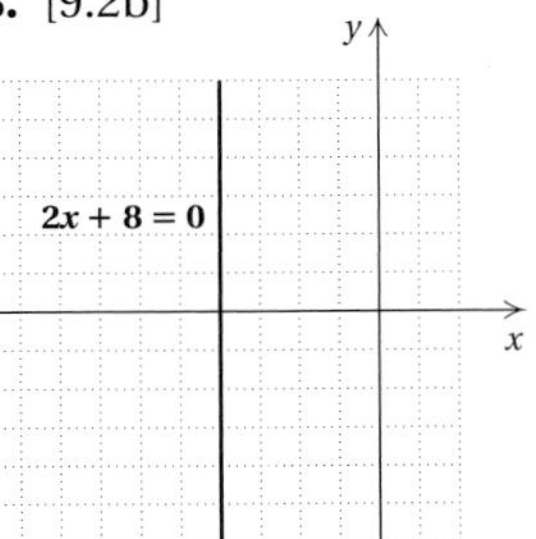

9. [9.2b]

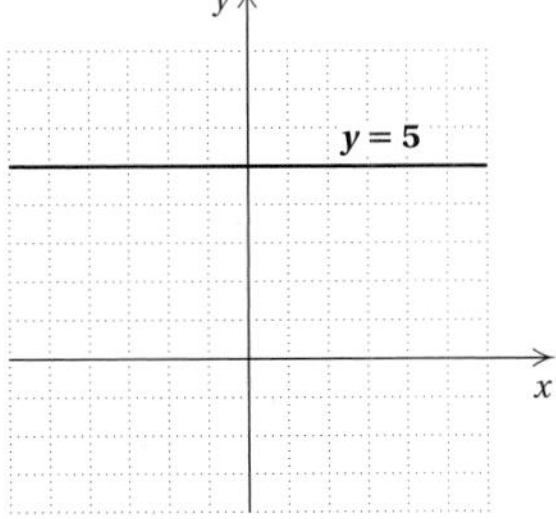

10. [9.2a]

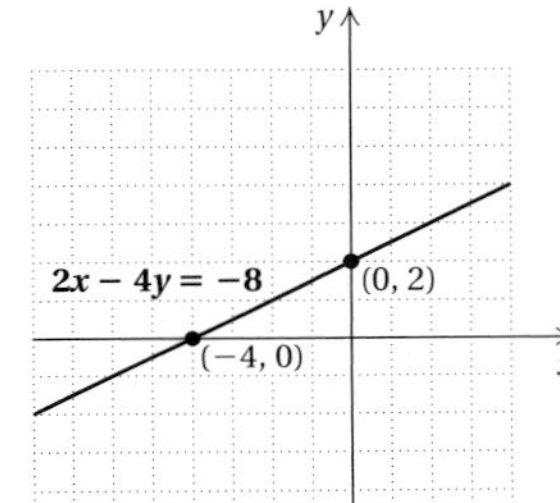

11. [9.2a]

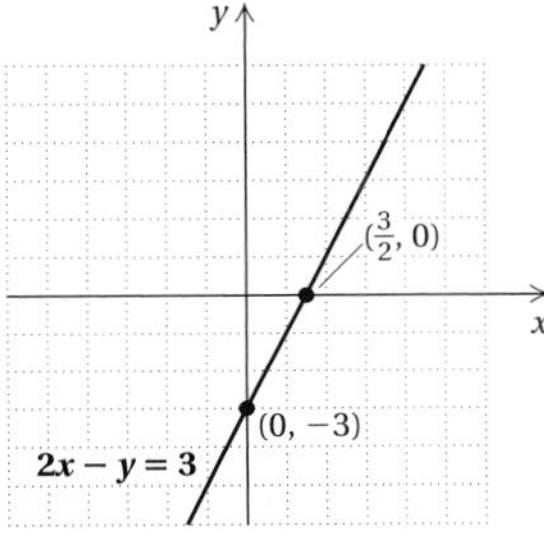

12. [9.1e] **(a)** \$5000; \$11,600; \$14,000; \$16,400; \$17,000;
(b)

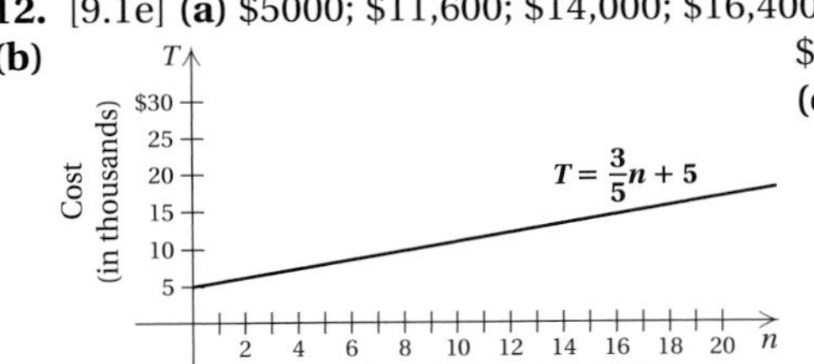

(c) 2015

13. [9.3b] **(a)** 14.5 floors per minute; **(b)** $4\frac{4}{29}$ seconds per floor **14.** [9.3b] 87.5 miles per hour **15.** [9.3a] −2
16. [9.3a] $\frac{3}{8}$;

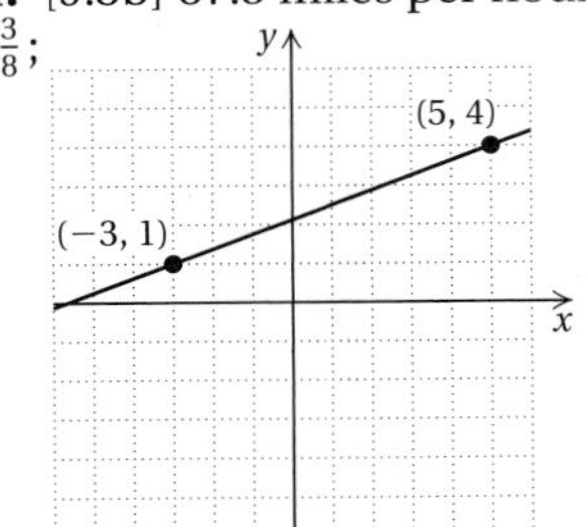

17. [9.3b] $-\frac{1}{20}$ **18.** [9.3c] **(a)** $\frac{2}{5}$; **(b)** not defined

19. [9.5a]

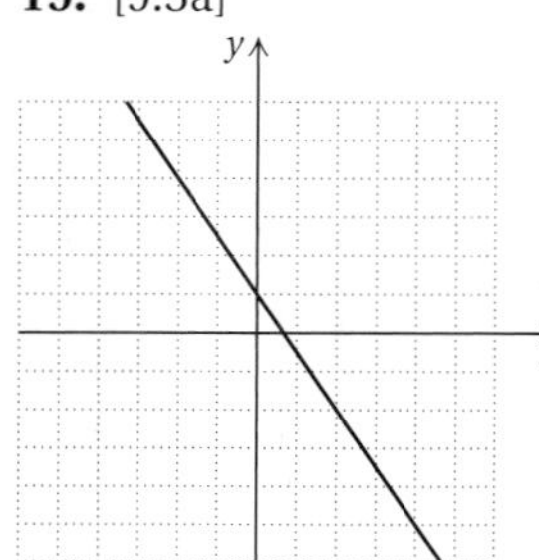

20. [9.5a]

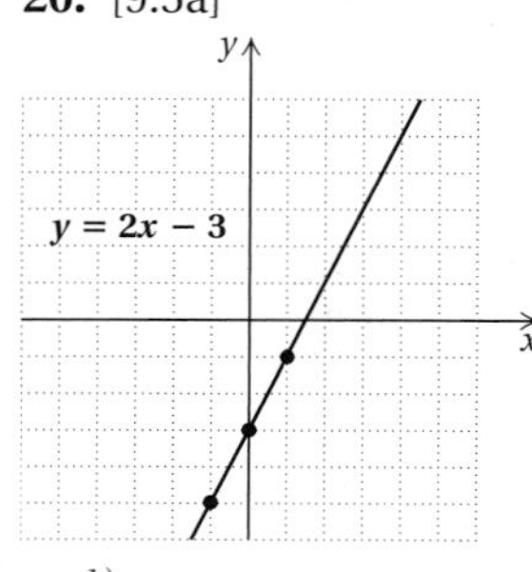

21. [9.4a] Slope: 2; y-intercept: $\left(0, -\frac{1}{4}\right)$
22. [9.4a] Slope: $\frac{4}{3}$; y-intercept: $(0, -2)$
23. [9.4a] $y = 1.8x - 7$ **24.** [9.4a] $y = -\frac{3}{8}x - \frac{1}{8}$
25. [9.4b] $y = x + 2$ **26.** [9.4b] $y = -3x - 6$
27. [9.4c] $y = -3x + 4$ **28.** [9.4c] $y = \frac{1}{4}x - 2$
29. [9.4c] **(a)** $y = 75.3x + 52.4$; **(b)** about 75 lung transplants per year; **(c)** about 1333 lung transplants
30. [9.6a, b] Parallel **31.** [9.6a, b] Neither
32. [9.6a, b] Perpendicular **33.** [9.7a] No
34. [9.7a] Yes **35.** [9.7b]

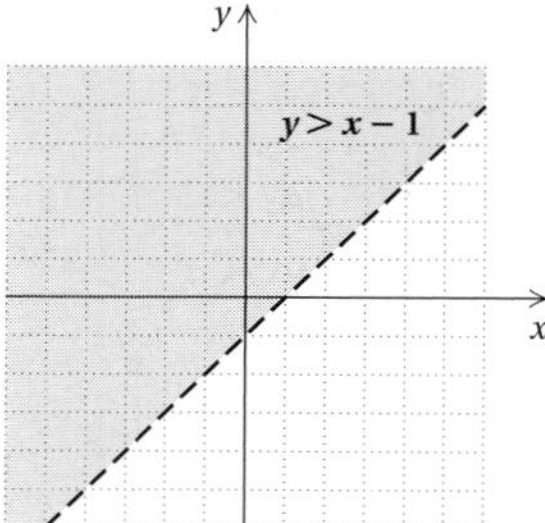

36. [9.7b]

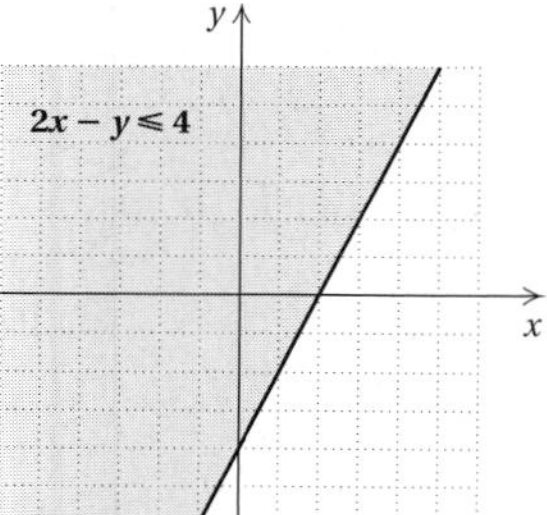

37. [9.1a] 25 square units; 20 linear units **38.** [9.6b] 3

CHAPTER 10

Margin Exercises, Section 10.1, pp. 716–721

1. $5 \cdot 5 \cdot 5 \cdot 5$ **2.** $x \cdot x \cdot x \cdot x \cdot x$ **3.** $3t \cdot 3t$ **4.** $3 \cdot t \cdot t$ **5.** $(-x) \cdot (-x) \cdot (-x) \cdot (-x)$ **6.** $-1 \cdot y \cdot y \cdot y$ **7.** 6 **8.** 1 **9.** 8.4 **10.** 1 **11.** -1.4 **12.** 0 **13.** 125 **14.** 160 **15.** 3215.36 cm^2 **16.** 119 **17.** 3; -3 **18.** **(a)** 144; **(b)** 36; **(c)** no **19.** 3^{10} **20.** x^{10} **21.** p^{24} **22.** x^5 **23.** a^9b^8 **24.** 4^3 **25.** y^4 **26.** p^9 **27.** a^4b^2 **28.** $\frac{1}{4^3} = \frac{1}{64}$ **29.** $\frac{1}{5^2} = \frac{1}{25}$ **30.** $\frac{1}{2^4} = \frac{1}{16}$ **31.** $\frac{1}{(-2)^3} = -\frac{1}{8}$ **32.** $\frac{4}{p^3}$ **33.** x^2 **34.** 5^2 **35.** $\frac{1}{x^7}$ **36.** $\frac{1}{7^5}$ **37.** b **38.** t^6

Exercise Set 10.1, p. 722

1. $3 \cdot 3 \cdot 3 \cdot 3$ **3.** $(-1.1)(-1.1)(-1.1)(-1.1)(-1.1)$ **5.** $\left(\frac{2}{3}\right)\left(\frac{2}{3}\right)\left(\frac{2}{3}\right)\left(\frac{2}{3}\right)$ **7.** $(7p)(7p)$ **9.** $8 \cdot k \cdot k \cdot k$ **11.** $-6 \cdot y \cdot y \cdot y \cdot y$ **13.** 1 **15.** b **17.** 1 **19.** -7.03 **21.** 1 **23.** ab **25.** a **27.** 27 **29.** 19 **31.** -81 **33.** 256 **35.** 93 **37.** 136 **39.** 10; 4 **41.** 3629.84 ft^2 **43.** $\frac{1}{3^2} = \frac{1}{9}$ **45.** $\frac{1}{10^3} = \frac{1}{1000}$ **47.** $\frac{1}{7^3} = \frac{1}{343}$ **49.** $\frac{1}{a^3}$ **51.** $8^2 = 64$ **53.** y^4 **55.** z^n **57.** 4^{-3} **59.** x^{-3} **61.** a^{-5} **63.** 2^7 **65.** 8^{14} **67.** x^7 **69.** 9^{38} **71.** $(3y)^{12}$ **73.** $(7y)^{17}$ **75.** 3^3 **77.** $\frac{1}{x}$ **79.** x^{17} **81.** $\frac{1}{x^{13}}$ **83.** $\frac{1}{a^{10}}$ **85.** 1 **87.** 7^3 **89.** 8^6 **91.** y^4 **93.** $\frac{1}{16^6}$ **95.** $\frac{1}{m^6}$ **97.** $\frac{1}{(8x)^4}$ **99.** 1 **101.** x^2 **103.** x^9 **105.** $\frac{1}{z^4}$ **107.** x^3 **109.** 1 **111.** $5^2 = 25$; $5^{-2} = \frac{1}{25}$; $\left(\frac{1}{5}\right)^2 = \frac{1}{25}$; $\left(\frac{1}{5}\right)^{-2} = 25$; $-5^2 = -25$; $(-5)^2 = 25$; $-\left(-\frac{1}{5}\right)^2 = -\frac{1}{25}$; $\left(-\frac{1}{5}\right)^{-2} = 25$ **113.** D_W **115.** 8 in., 4 in. **116.** 228, 229 **117.** 25,543.75 ft^2 **118.** 51°, 27°, 102° **119.** $\frac{23}{14}$ **120.** $\frac{11}{10}$ **121.** $4(x - 3 + 6y)$ **122.** $2(128 - a - 2b)$ **123.** No **125.** No **127.** y^{5x} **129.** a^{4t} **131.** 1 **133.** $>$ **135.** $<$ **137.** $-\frac{1}{10{,}000}$

Margin Exercises, Section 10.2, pp. 726–733

1. 3^{20} **2.** $\frac{1}{x^{12}}$ **3.** y^{15} **4.** $\frac{1}{x^{32}}$ **5.** $\frac{16x^{20}}{y^{12}}$ **6.** $\frac{25x^{10}}{y^{12}z^6}$ **7.** x^{74} **8.** $\frac{27z^{24}}{y^6x^{15}}$ **9.** $-\frac{1}{y^{24}}$ **10.** $-\frac{1}{8x^{12}}$ **11.** $-\frac{y^{15}}{27x^6}$ **12.** $\frac{x^{12}}{25}$ **13.** $\frac{8t^{15}}{w^{12}}$ **14.** $\frac{9}{x^8}$ **15.** 5.17×10^{-4} **16.** 5.23×10^8 **17.** 689,300,000,000 **18.** 0.0000567 **19.** 5.6×10^{-15} **20.** 7.462×10^{-13} **21.** 2.0×10^3 **22.** 5.5×10^2 **23.** 1.884672×10^{11} L **24.** The mass of Saturn is 9.5×10 times the mass of Earth.

Calculator Corner, p. 731

1. 1.3545×10^{-4} **2.** 9.044×10^5 **3.** 3.2×10^5 **4.** 3.6×10^{12} **5.** 3×10^{-6} **6.** 4×10^5 **7.** 8×10^{-26} **8.** 3×10^{13}

Exercise Set 10.2, p. 734

1. 2^6 **3.** $\frac{1}{5^6}$ **5.** x^{12} **7.** $\frac{1}{a^{18}}$ **9.** t^{18} **11.** $\frac{1}{t^{12}}$ **13.** x^8 **15.** a^3b^3 **17.** $\frac{1}{a^3b^3}$ **19.** $\frac{1}{m^3n^6}$ **21.** $16x^6$ **23.** $\frac{9}{x^8}$ **25.** $\frac{1}{x^{12}y^{15}}$ **27.** $x^{24}y^8$ **29.** $\frac{a^{10}}{b^{35}}$ **31.** $\frac{25t^6}{r^8}$ **33.** $\frac{b^{21}}{a^{15}c^6}$ **35.** $\frac{9x^6}{y^{16}z^6}$ **37.** $\frac{16x^6}{y^4}$ **39.** $a^{12}b^8$ **41.** $\frac{y^6}{4}$

43. $\frac{a^8}{b^{12}}$ **45.** $\frac{8}{y^6}$ **47.** $49x^6$ **49.** $\frac{x^6y^3}{z^3}$ **51.** $\frac{c^2d^6}{a^4b^2}$
53. 2.8×10^{10} **55.** 9.07×10^{17} **57.** 3.04×10^{-6}
59. 1.8×10^{-8} **61.** 10^{11} **63.** 2.96×10^{8} **65.** 10^{-7}
67. 87,400,000 **69.** 0.00000005704 **71.** 10,000,000
73. 0.00001 **75.** 6×10^{9} **77.** 3.38×10^{4}
79. 8.1477×10^{-13} **81.** 2.5×10^{13} **83.** 5.0×10^{-4}
85. 3.0×10^{-21} **87.** Approximately 1.325×10^{14} ft^3
89. The mass of Jupiter is 3.18×10^{2} times the mass of Earth.
91. 1×10^{22} **93.** The mass of the sun is 3.33×10^{5} times the mass of Earth. **95.** 4.375×10^{2} days **97.** DW
99. $9(x - 4)$ **100.** $2(2x - y + 8)$ **101.** $3(s + t + 8)$
102. $-7(x + 2)$ **103.** $\frac{7}{4}$ **104.** 2 **105.** $-\frac{12}{7}$ **106.** $-\frac{11}{2}$
107.

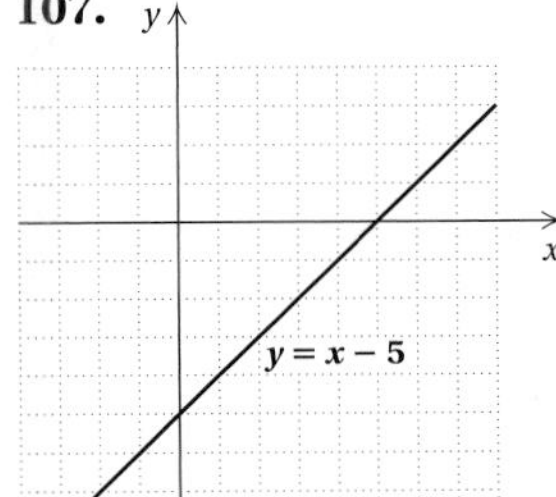

108.

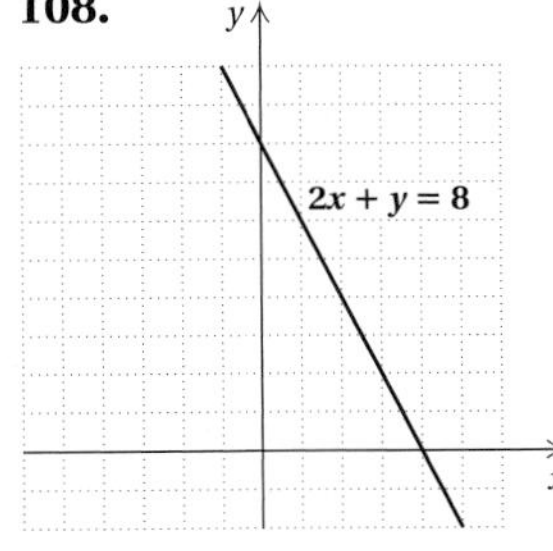

109. 2.478125×10^{-1} **111.** $\frac{1}{5}$ **113.** 3^{11} **115.** 7
117. $\frac{1}{0.4}$, or 2.5 **119.** False **121.** False **123.** True

Margin Exercises, Section 10.3, pp. 739–746

1. $4x^2 - 3x + \frac{5}{4}$; $15y^3$; $-7x^3 + 1.1$; answers may vary
2. -19 **3.** -104 **4.** -18 **5.** 21 **6.** 6; -4
7. 132 games **8.** 360 ft **9.** **(a)** 7.55 parts per million; **(b)** When $t = 3$, $C \approx 7.5$; so the value found in part (a) appears to be correct. **10.** 20 parts per million
11. $-9x^3 + (-4x^5)$ **12.** $-2y^3 + 3y^7 + (-7y) + (-9)$
13. $3x^2, 6x, \frac{1}{2}$ **14.** $-4y^5, 7y^2, -3y, -2$ **15.** $4x^3$ and $-x^3$
16. $4t^4$ and $-7t^4$; $-9t^3$ and $10t^3$ **17.** $5x^2$ and $7x^2$; $3x$ and $-8x$; -10 and 11 **18.** 2, -7, -8.5, 10, -4 **19.** $8x^2$
20. $2x^3 + 7$ **21.** $-\frac{1}{4}x^5 + 2x^2$ **22.** $-4x^3$ **23.** $5x^3$
24. $25 - 3x^5$ **25.** $6x$ **26.** $4x^3 + 4$
27. $-\frac{1}{4}x^3 + 4x^2 + 7$ **28.** $x^5 - \frac{9}{20}x^4 + \frac{49}{5}$
29. $6x^7 + 3x^5 - 2x^4 + 4x^3 + 5x^2 + x$
30. $7x^5 - 5x^4 + 2x^3 + 4x^2 - 3$
31. $14t^7 - 10t^5 + 7t^2 - 14$ **32.** $-2x^2 - 3x + 2$
33. $10x^4 - 8x - \frac{1}{2}$ **34.** 4, 2, 1, 0; 4 **35.** 0, 3, 6, 5; 6
36. x **37.** x^3, x^2, x, x^0 **38.** x^2, x **39.** x^3
40. $2x^3 + 4x^2 + 0x - 2$; $2x^3 + 4x^2 \quad - 2$
41. $a^4 + 0a^3 + 0a^2 + 0a + 10$; $a^4 \quad + 10$
42. Monomial **43.** None of these **44.** Binomial
45. Trinomial

Calculator Corner, p. 742

1. 3; 2.25; -27 **2.** 44; 0; 9.28 **3.** 13; -3.32; 7
4. -1; -7; -40.6

Exercise Set 10.3, p. 747

1. -18; 7 **3.** 19; 14 **5.** -12; -7 **7.** $\frac{13}{3}$; 5 **9.** 9; 1
11. 56; -2 **13.** 1112 ft **15.** **(a)** 1138.34 billion kilowatt-hours, 1499.46 billion kilowatt-hours, 1950.86 billion kilowatt-hours, 2402.26 billion kilowatt-hours, 3305.06 billion kilowatt-hours; **(b)** left to the student
17. \$18,750; \$24,000 **19.** -4, 4, 5, 2.75, 1
21. 1,820,000; 3,660,000 **23.** 9 words **25.** 6 **27.** 15
29. $2, -3x, x^2$ **31.** $-2x^4, \frac{1}{3}x^3, -x, 3$ **33.** $6x^2$ and $-3x^2$
35. $2x^4$ and $-3x^4$; $5x$ and $-7x$ **37.** $3x^5$ and $14x^5$; $-7x$ and $-2x$; 8 and -9 **39.** -3, 6 **41.** 5, $\frac{3}{4}$, 3
43. -5, 6, -2.7, 8, -2 **45.** $-3x$ **47.** $-8x$
49. $11x^3 + 4$ **51.** $x^3 - x$ **53.** $4b^5$ **55.** $\frac{3}{4}x^5 - 2x - 42$
57. x^4 **59.** $\frac{15}{16}x^3 - \frac{7}{6}x^2$ **61.** $x^5 + 6x^3 + 2x^2 + x + 1$
63. $15y^9 + 7y^8 + 5y^3 - y^2 + y$ **65.** $x^6 + x^4$
67. $13x^3 - 9x + 8$ **69.** $-5x^2 + 9x$ **71.** $12x^4 - 2x + \frac{1}{4}$
73. 1, 0; 1 **75.** 2, 1, 0; 2 **77.** 3, 2, 1, 0; 3 **79.** 2, 1, 6, 4; 6
81.

TERM	COEFFICIENT	DEGREE OF THE TERM	DEGREE OF THE POLYNOMIAL
$-7x^4$	-7	4	
$6x^3$	6	3	
$-3x^2$	-3	2	4
$8x$	8	1	
-2	-2	0	

83. x^2, x **85.** x^3, x^2, x^0 **87.** None missing
89. $x^3 + 0x^2 + 0x - 27$; $x^3 \quad - 27$
91. $x^4 + 0x^3 + 0x^2 - x + 0x^0$; $x^4 \quad - x$
93. None missing **95.** Trinomial **97.** None of these
99. Binomial **101.** Monomial **103.** DW
105. 27 apples **106.** -19 **107.** $-\frac{17}{24}$ **108.** $\frac{5}{8}$
109. -2.6 **110.** $\frac{15}{2}$ **111.** $b = \frac{C + r}{a}$ **112.** 45%; 37.5%; 17.5% **113.** $3(x - 5y + 21)$ **115.** $3x^6$ **117.** 10
119. -4, 4, 5, 2.75, 1 **121.** 1,820,000; 3,660,000

Margin Exercises, Section 10.4, pp. 753–756

1. $x^2 + 7x + 3$ **2.** $-4x^5 + 7x^4 + x^3 + 2x^2 + 4$
3. $24x^4 + 5x^3 + x^2 + 1$ **4.** $2x^3 + \frac{10}{3}$ **5.** $2x^2 - 3x - 1$
6. $8x^3 - 2x^2 - 8x + \frac{5}{2}$ **7.** $-8x^4 + 4x^3 + 12x^2 + 5x - 8$
8. $-x^3 + x^2 + 3x + 3$ **9.** $-4x^3 + 6x - 3$
10. $-5x^4 - 3x^2 - 7x + 5$
11. $-14x^{10} + \frac{1}{2}x^5 - 5x^3 + x^2 - 3x$ **12.** $2x^3 + 2x + 8$
13. $x^2 - 6x - 2$ **14.** $-8x^4 - 5x^3 + 8x^2 - 1$
15. $x^3 - x^2 - \frac{4}{3}x - 0.9$ **16.** $2x^3 + 5x^2 - 2x - 5$
17. $-x^5 - 2x^3 + 3x^2 - 2x + 2$ **18.** Sum of perimeters: $13x$; sum of areas: $\frac{7}{2}x^2$ **19.** $(x^2 - 64)$ ft^2

Calculator Corner, p. 756

1. Yes **2.** Yes **3.** No **4.** Yes **5.** No **6.** Yes

Exercise Set 10.4, p. 757

1. $-x + 5$ **3.** $x^2 - \frac{11}{2}x - 1$ **5.** $2x^2$ **7.** $5x^2 + 3x - 30$
9. $-2.2x^3 - 0.2x^2 - 3.8x + 23$ **11.** $6 + 12x^2$
13. $-\frac{1}{2}x^4 + \frac{2}{3}x^3 + x^2$
15. $0.01x^5 + x^4 - 0.2x^3 + 0.2x + 0.06$
17. $9x^8 + 8x^7 - 6x^4 + 8x^2 + 4$
19. $1.05x^4 + 0.36x^3 + 14.22x^2 + x + 0.97$ **21.** $5x$
23. $x^2 - \frac{3}{2}x + 2$ **25.** $-12x^4 + 3x^3 - 3$ **27.** $-3x + 7$
29. $-4x^2 + 3x - 2$ **31.** $4x^4 - 6x^2 - \frac{3}{4}x + 8$
33. $7x - 1$ **35.** $-x^2 - 7x + 5$ **37.** -18
39. $6x^4 + 3x^3 - 4x^2 + 3x - 4$
41. $4.6x^3 + 9.2x^2 - 3.8x - 23$ **43.** $\frac{3}{4}x^3 - \frac{1}{2}x$
45. $0.06x^3 - 0.05x^2 + 0.01x + 1$ **47.** $3x + 6$
49. $11x^4 + 12x^3 - 9x^2 - 8x - 9$ **51.** $x^4 - x^3 + x^2 - x$
53. $\frac{23}{2}a + 12$ **55.** $5x^2 + 4x$ **57.** $(r + 11)(r + 9)$; $9r + 99 + 11r + r^2$, or $r^2 + 20r + 99$
59. $(x + 3)(x + 3)$, or $(x + 3)^2$; $x^2 + 3x + 9 + 3x$, or $x^2 + 6x + 9$ **61.** $\pi r^2 - 25\pi$ **63.** $18z - 64$ **65.** $^{\mathbf{D}}\mathbf{W}$
67. 6 **68.** -19 **69.** $-\frac{7}{22}$ **70.** 5 **71.** 5 **72.** 1
73. $\frac{39}{2}$ **74.** $\frac{37}{2}$ **75.** $\{x|x \geq -10\}$ **76.** $\{x|x < 0\}$
77. $20w + 42$ **79.** $2x^2 + 20x$ **81.** $y^2 - 4y + 4$
83. $12y^2 - 23y + 21$ **85.** $-3y^4 - y^3 + 5y - 2$

Margin Exercises, Section 10.5, pp. 761–764

1. $-15x$ **2.** $-x^2$ **3.** x^2 **4.** $-x^5$ **5.** $12x^7$ **6.** $-8y^{11}$
7. $7y^5$ **8.** 0 **9.** $8x^2 + 16x$ **10.** $-15t^3 + 6t^2$
11. $-5x^6 - 25x^5 + 30x^4 - 40x^3$
12. **(a)** $(y + 2)(y + 7) = y \cdot (y + 7) + 2(y + 7)$
$= y \cdot y + y \cdot 7 + 2 \cdot y + 2 \cdot 7$
$= y^2 + 7y + 2y + 14$
$= y^2 + 9y + 14$;
(b) $(y + 2)(y + 7)$, or $y^2 + 2y + 7y + 14$, or $y^2 + 9y + 14$
13. $x^2 + 13x + 40$ **14.** $x^2 + x - 20$
15. $5x^2 - 17x - 12$ **16.** $6x^2 - 19x + 15$
17. $x^4 + 3x^3 + x^2 + 15x - 20$
18. $6y^5 - 20y^3 + 15y^2 + 14y - 35$
19. $3x^3 + 13x^2 - 6x + 20$
20. $20x^4 - 16x^3 + 32x^2 - 32x - 16$
21. $6x^4 - x^3 - 18x^2 - x + 10$

Calculator Corner, p. 764

1. Correct **2.** Correct **3.** Not correct **4.** Not correct

Exercise Set 10.5, p. 765

1. $40x^2$ **3.** x^3 **5.** $32x^8$ **7.** $0.03x^{11}$ **9.** $\frac{1}{15}x^4$ **11.** 0
13. $-24x^{11}$ **15.** $-2x^2 + 10x$ **17.** $-5x^2 + 5x$
19. $x^5 + x^2$ **21.** $6x^3 - 18x^2 + 3x$ **23.** $-6x^4 - 6x^3$
25. $18y^6 + 24y^5$ **27.** $x^2 + 9x + 18$ **29.** $x^2 + 3x - 10$
31. $x^2 - 7x + 12$ **33.** $x^2 - 9$ **35.** $25 - 15x + 2x^2$
37. $4x^2 + 20x + 25$ **39.** $x^2 - \frac{21}{10}x - 1$
41. $x^2 + 2.4x - 10.81$ **43.** $(x + 2)(x + 6)$, $x^2 + 8x + 12$
45. $(x + 1)(x + 6)$, $x^2 + 7x + 6$ **47.**

49. **51.**

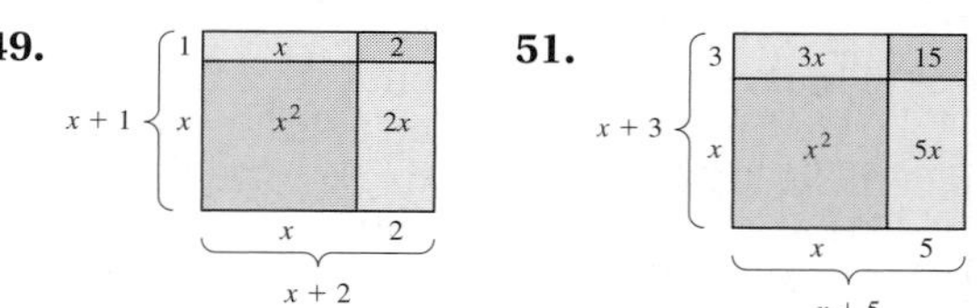

53. $x^3 - 1$ **55.** $4x^3 + 14x^2 + 8x + 1$
57. $3y^4 - 6y^3 - 7y^2 + 18y - 6$ **59.** $x^6 + 2x^5 - x^3$
61. $-10x^5 - 9x^4 + 7x^3 + 2x^2 - x$
63. $-1 - 2x - x^2 + x^4$ **65.** $6t^4 + t^3 - 16t^2 - 7t + 4$
67. $x^9 - x^5 + 2x^3 - x$ **69.** $x^4 - 1$
71. $x^4 + 8x^3 + 12x^2 + 9x + 4$
73. $2x^4 - 5x^3 + 5x^2 - \frac{19}{10}x + \frac{1}{5}$ **75.** $^{\mathbf{D}}\mathbf{W}$ **77.** $-\frac{3}{4}$
78. 6.4 **79.** 96 **80.** 32 **81.** $3(5x - 6y + 4)$
82. $4(4x - 6y + 9)$ **83.** $-3(3x + 15y - 5)$
84. $100(x - y + 10a)$ **85.**

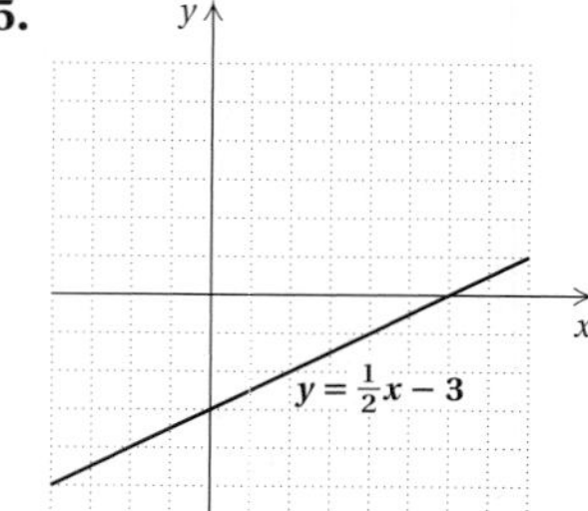

86. $\frac{23}{19}$ **87.** $75y^2 - 45y$ **89.** $V = (4x^3 - 48x^2 + 144x)\text{ in}^3$; $S = (-4x^2 + 144)\text{ in}^2$ **91.** 5 **93.** $(x^3 + 2x^2 - 210)\text{ m}^3$
95. 0 **97.** 0

Margin Exercises, Section 10.6, pp. 769–773

1. $x^2 + 7x + 12$ **2.** $x^2 - 2x - 15$ **3.** $2x^2 - 9x + 4$
4. $2x^3 - 4x^2 - 3x + 6$ **5.** $12x^5 + 10x^3 + 6x^2 + 5$
6. $y^6 - 49$ **7.** $t^2 + 8t + 15$ **8.** $-2x^7 + x^5 + x^3$
9. $x^2 - \frac{16}{25}$ **10.** $x^5 + 0.5x^3 - 0.5x^2 - 0.25$
11. $8 + 2x^2 - 15x^4$ **12.** $30x^5 - 27x^4 + 6x^3$
13. $x^2 - 25$ **14.** $4x^2 - 9$ **15.** $x^2 - 4$ **16.** $x^2 - 49$
17. $36 - 16y^2$ **18.** $4x^6 - 1$ **19.** $x^2 - \frac{4}{25}$
20. $x^2 + 16x + 64$ **21.** $x^2 - 10x + 25$ **22.** $x^2 + 4x + 4$
23. $a^2 - 8a + 16$ **24.** $4x^2 + 20x + 25$
25. $16x^4 - 24x^3 + 9x^2$ **26.** $60.84 + 18.72y + 1.44y^2$
27. $9x^4 - 30x^2 + 25$ **28.** $(A + B)^2$ represents the area of the large square. This includes all four sections. $A^2 + B^2$ represents only two of the sections. **29.** $x^2 + 11x + 30$
30. $t^2 - 16$ **31.** $-8x^5 + 20x^4 + 40x^2$
32. $81x^4 + 18x^2 + 1$ **33.** $4a^2 + 6a - 40$
34. $25x^2 + 5x + \frac{1}{4}$ **35.** $4x^2 - 2x + \frac{1}{4}$
36. $x^3 - 3x^2 + 6x - 8$

Visualizing for Success, p. 774

1. E, F **2.** B, O **3.** S, K **4.** R, G **5.** D, M **6.** J, P **7.** C, L **8.** N, Q **9.** A, H **10.** I, T

Exercise Set 10.6, p. 775

1. $x^3 + x^2 + 3x + 3$ **3.** $x^4 + x^3 + 2x + 2$ **5.** $y^2 - y - 6$ **7.** $9x^2 + 12x + 4$ **9.** $5x^2 + 4x - 12$ **11.** $9t^2 - 1$ **13.** $4x^2 - 6x + 2$ **15.** $p^2 - \frac{1}{16}$ **17.** $x^2 - 0.01$ **19.** $2x^3 + 2x^2 + 6x + 6$ **21.** $-2x^2 - 11x + 6$ **23.** $a^2 + 14a + 49$ **25.** $1 - x - 6x^2$ **27.** $\frac{9}{64}y^2 - \frac{5}{8}y + \frac{25}{36}$ **29.** $x^5 + 3x^3 - x^2 - 3$ **31.** $3x^6 - 2x^4 - 6x^2 + 4$ **33.** $13.16x^2 + 18.99x - 13.95$ **35.** $6x^7 + 18x^5 + 4x^2 + 12$ **37.** $8x^6 + 65x^3 + 8$ **39.** $4x^3 - 12x^2 + 3x - 9$ **41.** $4y^6 + 4y^5 + y^4 + y^3$ **43.** $x^2 - 16$ **45.** $4x^2 - 1$ **47.** $25m^2 - 4$ **49.** $4x^4 - 9$ **51.** $9x^8 - 16$ **53.** $x^{12} - x^4$ **55.** $x^8 - 9x^2$ **57.** $x^{24} - 9$ **59.** $4y^{16} - 9$ **61.** $\frac{25}{64}x^2 - 18.49$ **63.** $x^2 + 4x + 4$ **65.** $9x^4 + 6x^2 + 1$ **67.** $a^2 - a + \frac{1}{4}$ **69.** $9 + 6x + x^2$ **71.** $x^4 + 2x^2 + 1$ **73.** $4 - 12x^4 + 9x^8$ **75.** $25 + 60t^2 + 36t^4$ **77.** $x^2 - \frac{5}{4}x + \frac{25}{64}$ **79.** $9 - 12x^3 + 4x^6$ **81.** $4x^3 + 24x^2 - 12x$ **83.** $4x^4 - 2x^2 + \frac{1}{4}$ **85.** $9p^2 - 1$ **87.** $15t^5 - 3t^4 + 3t^3$ **89.** $36x^8 + 48x^4 + 16$ **91.** $12x^3 + 8x^2 + 15x + 10$ **93.** $64 - 96x^4 + 36x^8$ **95.** $t^3 - 1$ **97.** 25; 49 **99.** 56; 16 **101.** $a^2 + 2a + 1$ **103.** $t^2 + 10t + 24$ **105.** $\mathbf{D_W}$ **107.** Lamps: 500 watts; air conditioner: 2000 watts; television: 50 watts **108.** $\frac{28}{27}$ **109.** $-\frac{41}{7}$ **110.** $\frac{27}{4}$ **111.** $y = \dfrac{3x - 12}{2}$, or $y = \dfrac{3}{2}x - 6$ **112.** $a = \dfrac{5d + 4}{3}$, or $a = \dfrac{5}{3}d + \dfrac{4}{3}$ **113.** $30x^3 + 35x^2 - 15x$ **115.** $a^4 - 50a^2 + 625$ **117.** $81t^{16} - 72t^8 + 16$ **119.** -7 **121.** First row: 90, -432, -63; second row: 7, -18, -36, -14, 12, -6, -21, -11; third row: 9, -2, -2, 10, -8, -8, -8, -10, 21; fourth row: -19, -6 **123.** Yes **125.** No

Margin Exercises, Section 10.7, pp. 779–782

1. -7940 **2.** -176 **3.** 1889 calories **4.** $-3, 3, -2, 1, 2$ **5.** 3, 7, 1, 1, 0; 7 **6.** $2x^2y + 3xy$ **7.** $5pq - 8$ **8.** $-4x^3 + 2x^2 - 4y + 2$ **9.** $14x^3y + 7x^2y - 3xy - 2y$ **10.** $-5p^2q^4 + 2p^2q^2 + 3p^2q + 6pq^2 + 3q + 5$ **11.** $-8s^4t + 6s^3t^2 + 2s^2t^3 - s^2t^2$ **12.** $-9p^4q + 9p^3q^2 - 4p^2q^3 - 9q^4 + 5$ **13.** $x^5y^5 + 2x^4y^2 + 3x^3y^3 + 6x^2$ **14.** $p^5q - 4p^3q^3 + 3pq^3 + 6q^4$ **15.** $3x^3y + 6x^2y^3 + 2x^3 + 4x^2y^2$ **16.** $2x^2 - 11xy + 15y^2$ **17.** $16x^2 + 40xy + 25y^2$ **18.** $9x^4 - 12x^3y^2 + 4x^2y^4$ **19.** $4x^2y^4 - 9x^2$ **20.** $16y^2 - 9x^2y^4$ **21.** $9y^2 + 24y + 16 - 9x^2$ **22.** $4a^2 - 25b^2 - 10bc - c^2$

Exercise Set 10.7, p. 783

1. -1 **3.** -15 **5.** 240 **7.** -145 **9.** 3.715 liters **11.** 92.4 m **13.** 44.46 in^2 **15.** 63.78125 in^2 **17.** Coefficients: 1, -2, 3, -5; degrees: 4, 2, 2, 0; 4 **19.** Coefficients: 17, -3, -7; degrees: 5, 5, 0; 5 **21.** $-a - 2b$ **23.** $3x^2y - 2xy^2 + x^2$ **25.** $20au + 10av$ **27.** $8u^2v - 5uv^2$ **29.** $x^2 - 4xy + 3y^2$ **31.** $3r + 7$ **33.** $-b^2a^3 - 3b^3a^2 + 5ba + 3$ **35.** $ab^2 - a^2b$ **37.** $2ab - 2$ **39.** $-2a + 10b - 5c + 8d$ **41.** $6z^2 + 7zu - 3u^2$ **43.** $a^4b^2 - 7a^2b + 10$ **45.** $a^6 - b^2c^2$ **47.** $y^6x + y^4x + y^4 + 2y^2 + 1$ **49.** $12x^2y^2 + 2xy - 2$ **51.** $12 - c^2d^2 - c^4d^4$ **53.** $m^3 + m^2n - mn^2 - n^3$ **55.** $x^9y^9 - x^6y^6 + x^5y^5 - x^2y^2$ **57.** $x^2 + 2xh + h^2$ **59.** $r^6t^4 - 8r^3t^2 + 16$ **61.** $p^8 + 2m^2n^2p^4 + m^4n^4$ **63.** $4a^6 - 2a^3b^3 + \frac{1}{4}b^6$ **65.** $3a^3 - 12a^2b + 12ab^2$ **67.** $4a^2 - b^2$ **69.** $c^4 - d^2$ **71.** $a^2b^2 - c^2d^4$ **73.** $x^2 + 2xy + y^2 - 9$ **75.** $x^2 - y^2 - 2yz - z^2$ **77.** $a^2 - b^2 - 2bc - c^2$ **79.** $3x^4 - 7x^2y + 3x^2 - 20y^2 + 22y - 6$ **81.** $\mathbf{D_W}$ **83.** IV **84.** III **85.** I **86.** II

87.

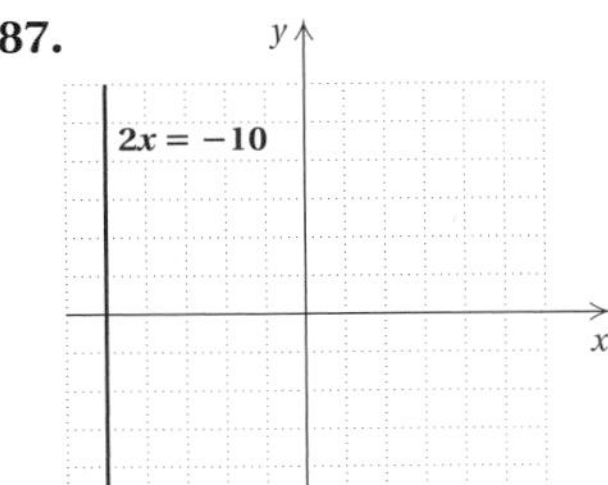

88.

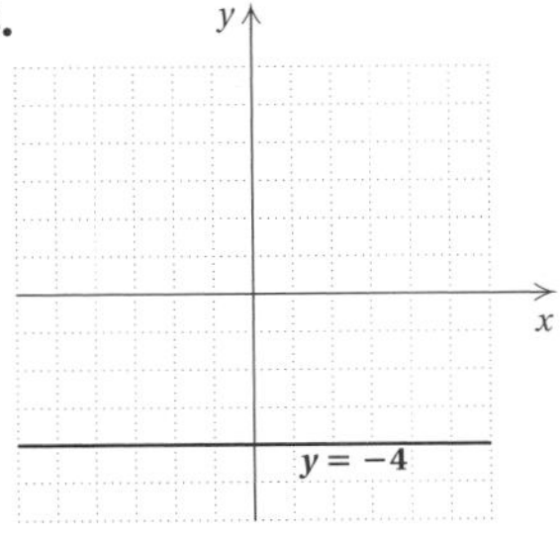

89.

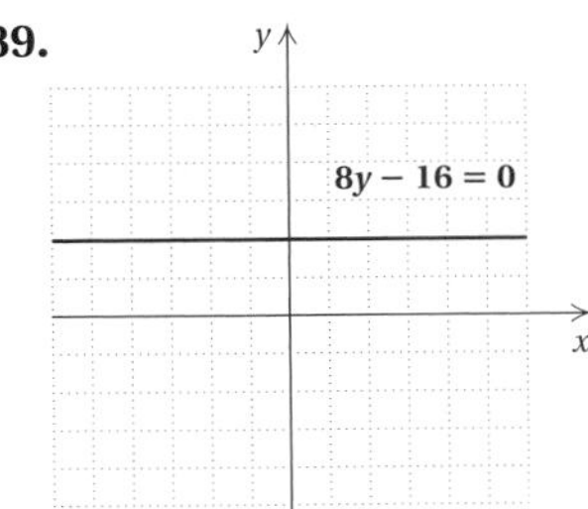

90.

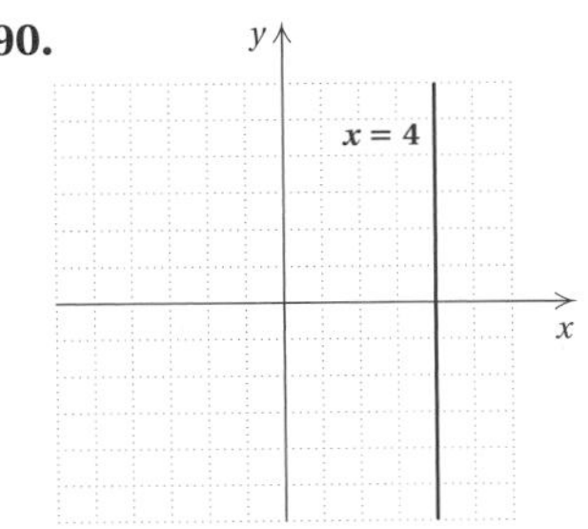

91. $4xy - 4y^2$ **93.** $2xy + \pi x^2$ **95.** $2\pi nh + 2\pi mh + 2\pi n^2 - 2\pi m^2$ **97.** 16 gal **99.** \$15,638.03

Margin Exercises, Section 10.8, pp. 788–791

1. $4x^2$ **2.** $-7x^{11}$ **3.** $-28p^3q$ **4.** $\frac{1}{4}x^4$ **5.** $7x^4 + 8x^2$ **6.** $x^2 + 3x + 2$ **7.** $2x^2 + x - \frac{2}{3}$ **8.** $4x^2 - \frac{3}{2}x + \frac{1}{2}$ **9.** $2x^2y^4 - 3xy^2 + 5y$ **10.** $x - 2$ **11.** $x + 4$ **12.** $x + 4$, R -2, or $x + 4 + \dfrac{-2}{x + 3}$ **13.** $x^2 + x + 1$ **14.** $2x^3 - x^2 + 3x - 1$ with R = 11; or $2x^3 - x^2 + 3x - 1 + \dfrac{11}{4x + 2}$

Exercise Set 10.8, p. 792

1. $3x^4$ **3.** $5x$ **5.** $18x^3$ **7.** $4a^3b$
9. $3x^4 - \frac{1}{2}x^3 + \frac{1}{8}x^2 - 2$ **11.** $1 - 2u - u^4$
13. $5t^2 + 8t - 2$ **15.** $-4x^4 + 4x^2 + 1$
17. $6x^2 - 10x + \frac{3}{2}$ **19.** $9x^2 - \frac{5}{2}x + 1$
21. $6x^2 + 13x + 4$ **23.** $3rs + r - 2s$ **25.** $x + 2$
27. $x - 5 + \frac{-50}{x - 5}$ **29.** $x - 2 + \frac{-2}{x + 6}$ **31.** $x - 3$
33. $x^4 - x^3 + x^2 - x + 1$ **35.** $2x^2 - 7x + 4$
37. $x^3 - 6$ **39.** $3x^2 + x + 2 + \frac{10}{5x + 1}$ **41.** $t^2 + 1$
43. $\mathbf{D_W}$ **45.** Product **46.** Monomial
47. Multiplication; equivalent **48.** $x = a$
49. Trinomial **50.** Quotient **51.** Absolute value
52. Slope **53.** $x^2 + 5$ **55.** $a + 3 + \frac{5}{5a^2 - 7a - 2}$
57. $2x^2 + x - 3$ **59.** $a^5 + a^4b + a^3b^2 + a^2b^3 + ab^4 + b^5$
61. -5 **63.** 1

Concept Reinforcement, p. 795

1. False **2.** True **3.** False **4.** True **5.** False
6. True

Summary and Review: Chapter 10, p. 795

1. $\frac{1}{7^2}$ **2.** y^{11} **3.** $(3x)^{14}$ **4.** t^8 **5.** 4^3 **6.** $\frac{1}{a^3}$ **7.** 1
8. $9t^8$ **9.** $36x^8$ **10.** $\frac{y^3}{8x^3}$ **11.** t^{-5} **12.** $\frac{1}{y^4}$
13. 3.28×10^{-5} **14.** 8,300,000 **15.** 2.09×10^4
16. 5.12×10^{-5} **17.** 4.4676×10^9 gal **18.** 10
19. $-4y^5, 7y^2, -3y, -2$ **20.** x^2, x^0 **21.** 3, 2, 1, 0; 3
22. Binomial **23.** None of these **24.** Monomial
25. $-2x^2 - 3x + 2$ **26.** $10x^4 - 7x^2 - x - \frac{1}{2}$
27. $x^5 - 2x^4 + 6x^3 + 3x^2 - 9$
28. $-2x^5 - 6x^4 - 2x^3 - 2x^2 + 2$ **29.** $2x^2 - 4x$
30. $x^5 - 3x^3 - x^2 + 8$ **31.** Perimeter: $4w + 6$; area: $w^2 + 3w$ **32.** $(t + 3)(t + 4)$, $t^2 + 7t + 12$
33. $x^2 + \frac{7}{6}x + \frac{1}{3}$ **34.** $49x^2 + 14x + 1$
35. $12x^3 - 23x^2 + 13x - 2$ **36.** $9x^4 - 16$
37. $15x^7 - 40x^6 + 50x^5 + 10x^4$ **38.** $x^2 - 3x - 28$
39. $9y^4 - 12y^3 + 4y^2$ **40.** $2t^4 - 11t^2 - 21$ **41.** 49
42. Coefficients: 1, −7, 9, −8; degrees: 6, 2, 2, 0; 6
43. $-y + 9w - 5$
44. $m^6 - 2m^2n + 2m^2n^2 + 8n^2m - 6m^3$
45. $-9xy - 2y^2$ **46.** $11x^3y^2 - 8x^2y - 6x^2 - 6x + 6$
47. $p^3 - q^3$ **48.** $9a^8 - 2a^4b^3 + \frac{1}{9}b^6$ **49.** $5x^2 - \frac{1}{2}x + 3$
50. $3x^2 - 7x + 4 + \frac{1}{2x + 3}$ **51.** 0, 3.75, −3.75, 0
52. $\mathbf{D_W}$ 578.6×10^{-7} is not in scientific notation because 578.6 is larger than 10. **53.** $\mathbf{D_W}$ A monomial is an expression of the type ax^n, where n is a whole number and a is a real number. A binomial is a sum of two monomials and has two terms. A trinomial is a sum of three monomials and has three terms. A general polynomial is a monomial or a sum of monomials and has one or more terms.

54. $\frac{1}{2}x^2 - \frac{1}{2}y^2$ **55.** $400 - 4a^2$ **56.** $-28x^8$ **57.** $\frac{94}{13}$
58. $x^4 + x^3 + x^2 + x + 1$ **59.** 16 ft by 8 ft

Test: Chapter 10, p. 798

1. [10.1d, f] $\frac{1}{6^5}$ **2.** [10.1d] x^9 **3.** [10.1d] $(4a)^{11}$
4. [10.1e] 3^3 **5.** [10.1e, f] $\frac{1}{x^5}$ **6.** [10.1b, e] 1 **7.** [10.2a] x^6
8. [10.2a, b] $-27y^6$ **9.** [10.2a, b] $16a^{12}b^4$ **10.** [10.2b] $\frac{a^3b^3}{c^3}$
11. [10.1d], [10.2a, b] $-216x^{21}$
12. [10.1d], [10.2a, b] $-24x^{21}$
13. [10.1d], [10.2a, b] $162x^{10}$ **14.** [10.1d], [10.2a, b] $324x^{10}$
15. [10.1f] $\frac{1}{5^3}$ **16.** [10.1f] y^{-8} **17.** [10.2c] 3.9×10^9
18. [10.2c] 0.00000005 **19.** [10.2d] 1.75×10^{17}
20. [10.2d] 1.296×10^{22} **21.** [10.2e] 1.5×10^4 files
22. [10.3a] −43 **23.** [10.3d] $\frac{1}{3}, -1, 7$ **24.** [10.3g] 3, 0, 1, 6; 6
25. [10.3i] Binomial **26.** [10.3e] $5a^2 - 6$
27. [10.3e] $\frac{7}{4}y^2 - 4y$ **28.** [10.3f] $x^5 + 2x^3 + 4x^2 - 8x + 3$
29. [10.4a] $4x^5 + x^4 + 2x^3 - 8x^2 + 2x - 7$
30. [10.4a] $5x^4 + 5x^2 + x + 5$
31. [10.4c] $-4x^4 + x^3 - 8x - 3$
32. [10.4c] $-x^5 + 0.7x^3 - 0.8x^2 - 21$
33. [10.5b] $-12x^4 + 9x^3 + 15x^2$ **34.** [10.6c] $x^2 - \frac{2}{3}x + \frac{1}{9}$
35. [10.6b] $9x^2 - 100$ **36.** [10.6a] $3b^2 - 4b - 15$
37. [10.6a] $x^{14} - 4x^8 + 4x^6 - 16$
38. [10.6a] $48 + 34y - 5y^2$
39. [10.5d] $6x^3 - 7x^2 - 11x - 3$
40. [10.6c] $25t^2 + 20t + 4$
41. [10.7c] $-5x^3y - y^3 + xy^3 - x^2y^2 + 19$
42. [10.7e] $8a^2b^2 + 6ab - 4b^3 + 6ab^2 + ab^3$
43. [10.7f] $9x^{10} - 16y^{10}$ **44.** [10.8a] $4x^2 + 3x - 5$
45. [10.8b] $2x^2 - 4x - 2 + \frac{17}{3x + 2}$
46. [10.3a] 3, 1.5, −3.5, −5, −5.25 **47.** [10.4d] $28a + 90$
48. [10.4d] $(t + 2)(t + 2)$, $t^2 + 4t + 4$
49. [10.5b], [10.6a] $V = l^3 - 3l^2 + 2l$
50. [8.3b], [10.6b, c] $-\frac{61}{12}$

CHAPTER 11

Margin Exercises, Section 11.1, pp. 803–808

1. 20 **2.** 7 **3.** 72 **4.** 1 **5.** $4x^2$ **6.** y^2 **7.** $4mn^2$
8. $7x^3$ **9. (a)** $3x + 6$; **(b)** $3(x + 2)$
10. (a) $2x^3 + 10x^2 + 8x$; **(b)** $2x(x^2 + 5x + 4)$
11. $x(x + 3)$ **12.** $y^2(3y^4 - 5y + 2)$
13. $3x^2y(3x^2y - 5x + 1)$ **14.** $\frac{1}{4}(3t^3 + 5t^2 + 7t + 1)$
15. $7x^3(5x^4 - 7x^3 + 2x^2 - 9)$ **16.** $28(3x^2 - 2x + 1)$
17. $(x^2 + 3)(x + 7)$ **18.** $(x^2 + 2)(a + b)$
19. $(x^2 + 3)(x + 7)$ **20.** $(2t^2 + 3)(4t + 1)$
21. $(3m^3 + 2)(m^2 - 5)$ **22.** $(3x^2 - 1)(x - 2)$
23. $(2x^2 - 3)(2x - 3)$ **24.** Not factorable using factoring by grouping

Exercise Set 11.1, p. 809

1. x **3.** x^2 **5.** 2 **7.** $17xy$ **9.** x **11.** x^2y^2
13. $x(x-6)$ **15.** $2x(x+3)$ **17.** $x^2(x+6)$
19. $8x^2(x^2-3)$ **21.** $2(x^2+x-4)$
23. $17xy(x^4y^2+2x^2y+3)$ **25.** $x^2(6x^2-10x+3)$
27. $x^2y^2(x^3y^3+x^2y+xy-1)$
29. $2x^3(x^4-x^3-32x^2+2)$
31. $0.8x(2x^3-3x^2+4x+8)$
33. $\frac{1}{3}x^3(5x^3+4x^2+x+1)$ **35.** $(x^2+2)(x+3)$
37. $(5a^3-1)(2a-7)$ **39.** $(x^2+2)(x+3)$
41. $(2x^2+1)(x+3)$ **43.** $(4x^2+3)(2x-3)$
45. $(4p^2+1)(3p-4)$ **47.** $(5x^2-1)(x-1)$
49. $(x^2-3)(x+8)$ **51.** $(2x^2-9)(x-4)$ **53.** DW
55. $\{x|x > -24\}$ **56.** $\{x|x \le \frac{14}{5}\}$ **57.** 27
58. $p = 2A - q$ **59.** $y^2+12y+35$ **60.** $y^2+14y+49$
61. y^2-49 **62.** $y^2-14y+49$

63.

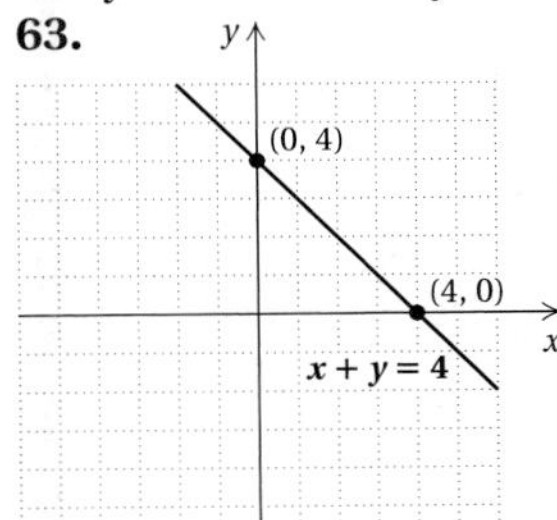

64.

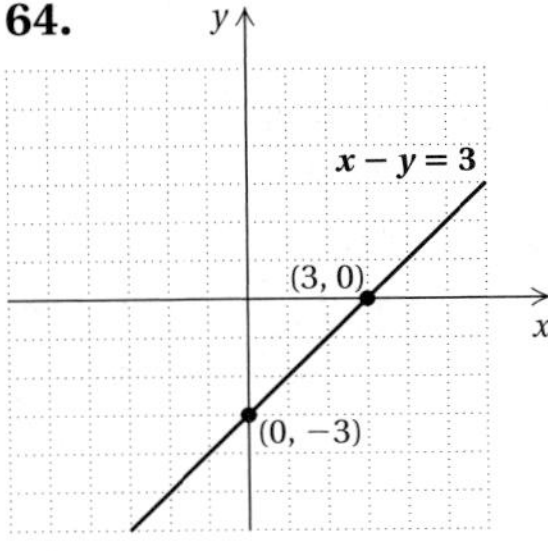

65.

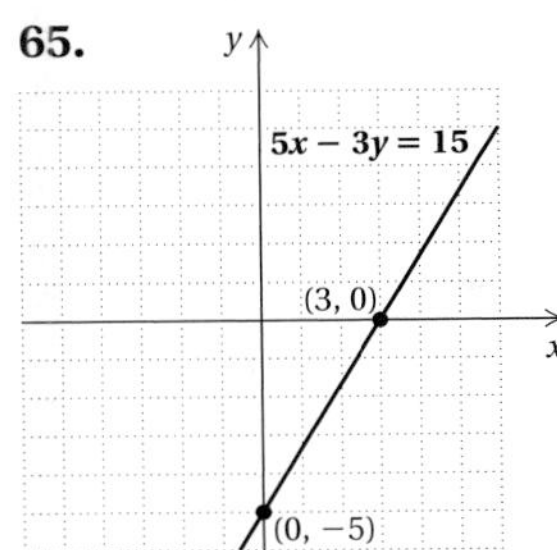

66.

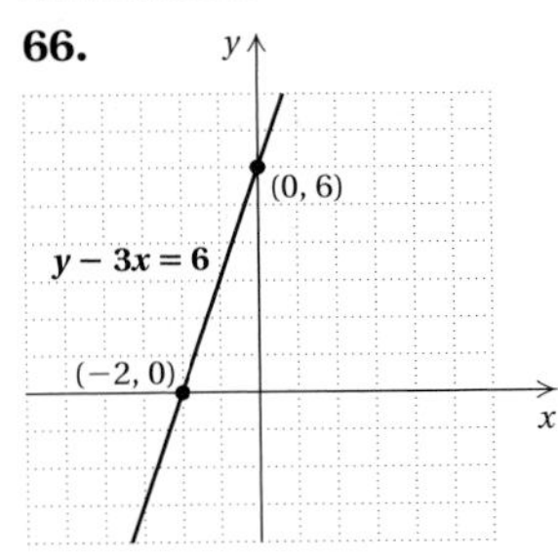

67. $(2x^3+3)(2x^2+3)$ **69.** $(x^7+1)(x^5+1)$
71. Not factorable by grouping

Margin Exercises, Section 11.2, pp. 811–816

1. (a) $-13, 8, -8, 7, -7$; **(b)** 13, 8, 7; both 7 and 12 are positive; **(c)** $(x+3)(x+4)$ **2.** $(x+9)(x+4)$
3. The coefficient of the middle term, -8, is negative.
4. $(x-5)(x-3)$ **5.** $(t-5)(t-4)$ **6. (a)** 23, 10, 5, 2; the positive factor has the larger absolute value; **(b)** $-23, -10, -5, -2$; the negative factor has the larger absolute value; **(c)** $(x+3)(x-8)$ **7. (a)** $-23, -10, -5, -2$; the negative factor has the larger absolute value; **(b)** 23, 10, 5, 2; the positive factor has the larger absolute value;
(c) $(x-2)(x+12)$ **8.** $(a-2)(a+12)$
9. $(t+2)(t-12)$ **10.** $(y-6)(y+2)$
11. $(t^2+7)(t^2-2)$ **12.** Prime **13.** $x(x+6)(x-2)$
14. $p(p-q-3q^2)$ **15.** $3x(x+4)^2$
16. $-1(x+2)(x-7)$, or $(-x-2)(x-7)$, or $(x+2)(-x+7)$ **17.** $-1(x+3)(x-6)$, or $(-x-3)(x-6)$, or $(x+3)(-x+6)$

Exercise Set 11.2, p. 817

1.

PAIRS OF FACTORS	SUMS OF FACTORS
1, 15	16
−1, −15	−16
3, 5	8
−3, −5	−8

$(x+3)(x+5)$

3.

PAIRS OF FACTORS	SUMS OF FACTORS
1, 12	13
−1, −12	−13
2, 6	8
−2, −6	−8
3, 4	7
−3, −4	−7

$(x+3)(x+4)$

5.

PAIRS OF FACTORS	SUMS OF FACTORS
1, 9	10
−1, −9	−10
3, 3	6
−3, −3	−6

$(x-3)^2$

7.

PAIRS OF FACTORS	SUMS OF FACTORS
−1, 14	13
1, −14	−13
−2, 7	5
2, −7	−5

$(x+2)(x-7)$

9.

PAIRS OF FACTORS	SUMS OF FACTORS
1, 4	5
−1, −4	−5
2, 2	4
−2, −2	−4

$(b+1)(b+4)$

11.

PAIRS OF FACTORS	SUMS OF FACTORS
$\frac{1}{3}, \frac{1}{3}$	$\frac{2}{3}$
$-\frac{1}{3}, -\frac{1}{3}$	$-\frac{2}{3}$
$1, \frac{1}{9}$	$\frac{10}{9}$
$-1, -\frac{1}{9}$	$-\frac{10}{9}$

$(x+\frac{1}{3})^2$

13. $(d-2)(d-5)$ **15.** $(y-1)(y-10)$ **17.** Prime **19.** $(x-9)(x+2)$ **21.** $x(x-8)(x+2)$ **23.** $y(y-9)(y+5)$ **25.** $(x-11)(x+9)$ **27.** $(c^2+8)(c^2-7)$ **29.** $(a^2+7)(a^2-5)$ **31.** $(x-6)(x+7)$ **33.** Prime **35.** $(x+10)^2$ **37.** $-1(x-10)(x+3)$, or $(-x+10)(x+3)$, or $(x-10)(-x-3)$ **39.** $-1(a-2)(a+12)$, or $(-a+2)(a+12)$, or $(a-2)(-a-12)$ **41.** $x^2(x-25)(x+4)$ **43.** $(x-24)(x+3)$ **45.** $(x-9)(x-16)$ **47.** $(a+12)(a-11)$ **49.** $(x-15)(x-8)$ **51.** $-1(x+12)(x-9)$, or $(-x-12)(x-9)$, or $(x+12)(-x+9)$ **53.** $(y-0.4)(y+0.2)$ **55.** $(p+5q)(p-2q)$ **57.** $-1(t+14)(t-6)$, or $(-t-14)(t-6)$, or $(t+14)(-t+6)$ **59.** $(m+4n)(m+n)$ **61.** $(s+3t)(s-5t)$ **63.** $6a^8(a+2)(a-7)$ **65.** DW **67.** DW **69.** $16x^3-48x^2+8x$ **70.** $28w^2-53w-66$ **71.** $49w^2+84w+36$ **72.** $16w^2-88w+121$ **73.** $16w^2-121$ **74.** y^3-3y^2+5y **75.** $6x^2+11xy-35y^2$ **76.** $27x^{12}$ **77.** $\frac{8}{3}$ **78.** $-\frac{7}{2}$ **79.** 29,555 **80.** 100°, 25°, 55° **81.** 15, −15, 27, −27, 51, −51 **83.** $\left(x+\frac{1}{4}\right)\left(x-\frac{3}{4}\right)$ **85.** $(x+5)\left(x-\frac{5}{7}\right)$ **87.** $(b^n+5)(b^n+2)$ **89.** $2x^2(4-\pi)$

Margin Exercises, Section 11.3, pp. 822–825

1. $(2x+5)(x-3)$ **2.** $(4x+1)(3x-5)$ **3.** $(3x-4)(x-5)$ **4.** $2(5x-4)(2x-3)$ **5.** $(2x+1)(3x+2)$ **6.** $-1(2x-1)(3x+2)$, or $(2x-1)(-3x-2)$, or $(-2x+1)(3x+2)$ **7.** $-2(3x-4)(x+1)$, or $2(-3x+4)(x+1)$, or $2(3x-4)(-x-1)$ **8.** $(2a-b)(3a-b)$ **9.** $3(2x+3y)(x+y)$

Calculator Corner, p. 826

1. Correct **2.** Correct **3.** Not correct **4.** Not correct **5.** Not correct **6.** Correct **7.** Not correct **8.** Correct

Exercise Set 11.3, p. 827

1. $(2x+1)(x-4)$ **3.** $(5x+9)(x-2)$ **5.** $(3x+1)(2x+7)$ **7.** $(3x+1)(x+1)$ **9.** $(2x-3)(2x+5)$ **11.** $(2x+1)(x-1)$ **13.** $(3x-2)(3x+8)$ **15.** $(3x+1)(x-2)$ **17.** $(3x+4)(4x+5)$ **19.** $(7x-1)(2x+3)$ **21.** $(3x+2)(3x+4)$ **23.** $(3x-7)^2$ **25.** $(24x-1)(x+2)$ **27.** $(5x-11)(7x+4)$ **29.** $-2(x-5)(x+2)$, or $2(-x+5)(x+2)$, or $2(x-5)(-x-2)$ **31.** $4(3x-2)(x+3)$ **33.** $6(5x-9)(x+1)$ **35.** $2(3y+5)(y-1)$ **37.** $(3x-1)(x-1)$ **39.** $4(3x+2)(x-3)$ **41.** $(2x+1)(x-1)$ **43.** $(3x+2)(3x-8)$ **45.** $5(3x+1)(x-2)$ **47.** $p(3p+4)(4p+5)$ **49.** $-1(3x+2)(3x-8)$, or $(-3x-2)(3x-8)$, or $(3x+2)(-3x+8)$ **51.** $-1(5x-3)(3x-2)$, or $(-5x+3)(3x-2)$, or $(5x-3)(-3x+2)$ **53.** $x^2(7x-1)(2x+3)$ **55.** $3x(8x-1)(7x-1)$ **57.** $(5x^2-3)(3x^2-2)$ **59.** $(5t+8)^2$ **61.** $2x(3x+5)(x-1)$ **63.** Prime **65.** Prime **67.** $(4m+5n)(3m-4n)$ **69.** $(2a+3b)(3a-5b)$ **71.** $(3a+2b)(3a+4b)$ **73.** $(5p+2q)(7p+4q)$ **75.** $6(3x-4y)(x+y)$ **77.** DW **79.** $q=\frac{A+7}{p}$ **80.** $x=\frac{y-b}{m}$ **81.** $y=\frac{6-3x}{2}$ **82.** $q=p+r-2$ **83.** $\{x|x>4\}$ **84.** $\{x|x\le\frac{8}{11}\}$ **85.**

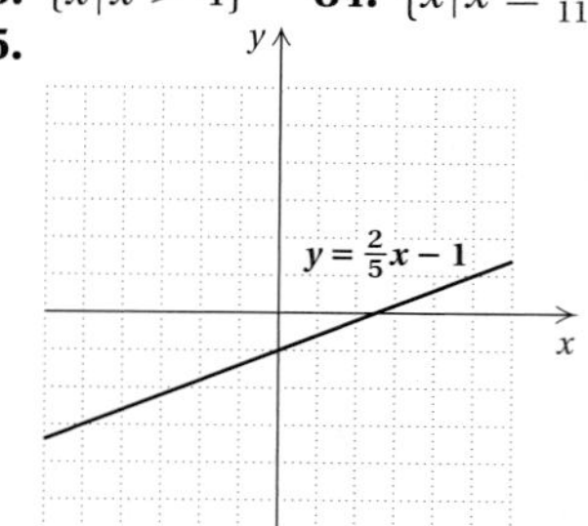

86. y^8 **87.** y-intercept: $(0,-4)$; x-intercept: $(16,0)$ **88.** y-intercept: $(0,4)$; x-intercept: $(16,0)$ **89.** y-intercept: $(0,-5)$; x-intercept: $(6.5,0)$ **90.** y-intercept: $\left(0,\frac{2}{3}\right)$; x-intercept: $\left(\frac{5}{8},0\right)$ **91.** y-intercept: $(0,4)$; x-intercept: $\left(\frac{4}{5},0\right)$ **92.** y-intercept: $(0,-5)$; x-intercept: $\left(\frac{5}{2},0\right)$ **93.** $(2x^n+1)(10x^n+3)$ **95.** $(x^{3a}-1)(3x^{3a}+1)$ **97.–105.** Left to the student

Margin Exercises, Section 11.4, p. 831

1. $(2x+1)(3x+2)$ **2.** $(4x+1)(3x-5)$ **3.** $3(2x+3)(x+1)$ **4.** $2(5x-4)(2x-3)$

Exercise Set 11.4, p. 832

1. $(x+7)(x+2)$ **3.** $(x-1)(x-4)$ **5.** $(2x+3)(3x+2)$ **7.** $(x-4)(3x-4)$ **9.** $(5x+3)(7x-8)$ **11.** $(2x-3)(2x+3)$ **13.** $(2x^2+5)(x^2+3)$ **15.** $(2x-1)(x+4)$ **17.** $(3x+5)(x-3)$ **19.** $(2x+7)(3x+1)$ **21.** $(3x-1)(x-1)$ **23.** $(2x+3)(2x-5)$ **25.** $(2x-1)(x+1)$ **27.** $(3x+2)(3x-8)$ **29.** $(3x-1)(x+2)$ **31.** $(3x-4)(4x-5)$ **33.** $(7x+1)(2x-3)$ **35.** $(3x+2)(3x+4)$ **37.** $(3x-7)^2$ **39.** $(24x+1)(x-2)$ **41.** $-1(3a-1)(3a+5)$, or $(-3a+1)(3a+5)$, or $(3a-1)(-3a-5)$ **43.** $-2(x-5)(x+2)$, or $2(-x+5)(x+2)$, or $2(x-5)(-x-2)$ **45.** $4(3x-2)(x+3)$ **47.** $6(5x-9)(x+1)$ **49.** $2(3y+5)(y-1)$ **51.** $(3x-1)(x-1)$ **53.** $4(3x+2)(x-3)$ **55.** $(2x+1)(x-1)$ **57.** $(3x-2)(3x+8)$ **59.** $5(3x+1)(x-2)$ **61.** $p(3p+4)(4p+5)$ **63.** $-1(5x-4)(x+1)$, or $(-5x+4)(x+1)$, or $(5x-4)(-x-1)$ **65.** $-3(2t-1)(t-5)$, or $3(-2t+1)(t-5)$, or $3(2t-1)(-t+5)$ **67.** $x^2(7x-1)(2x+3)$ **69.** $3x(8x-1)(7x-1)$ **71.** $(5x^2-3)(3x^2-2)$ **73.** $(5t+8)^2$ **75.** $2x(3x+5)(x-1)$ **77.** Prime **79.** Prime **81.** $(4m+5n)(3m-4n)$ **83.** $(2a+3b)(3a-5b)$ **85.** $(3a-2b)(3a-4b)$

87. $(5p + 2q)(7p + 4q)$ 89. $6(3x - 4y)(x + y)$
91. $-6x(x - 5)(x + 2)$, or $6x(-x + 5)(x + 2)$, or $6x(x - 5)(-x - 2)$ 93. $x^3(5x - 11)(7x + 4)$ 95. ^{D}W
97. $\{x|x < -100\}$ 98. $\{x|x \geq 217\}$ 99. $\{x|x \leq 8\}$
100. $\{x|x < 2\}$ 101. $\{x|x \geq \frac{20}{3}\}$ 102. $\{x|x > 17\}$
103. $\{x|x > \frac{26}{7}\}$ 104. $\{x|x \geq \frac{77}{17}\}$
105. About 6369 km, or 3949 mi 106. 40°
107. $(3x^5 - 2)^2$ 109. $(4x^5 + 1)^2$
111.–119. Left to the student

Margin Exercises, Section 11.5, pp. 837–841

1. Yes 2. No 3. No 4. Yes 5. No 6. Yes
7. No 8. Yes 9. $(x + 1)^2$ 10. $(x - 1)^2$
11. $(t + 2)^2$ 12. $(5x - 7)^2$ 13. $(7 - 4y)^2$
14. $3(4m + 5)^2$ 15. $(p^2 + 9)^2$ 16. $z^3(2z - 5)^2$
17. $(3a + 5b)^2$ 18. Yes 19. No 20. No 21. No
22. Yes 23. Yes 24. Yes 25. $(x + 3)(x - 3)$
26. $4(t + 4)(t - 4)$ 27. $(a + 5b)(a - 5b)$
28. $x^4(8 + 5x)(8 - 5x)$ 29. $5(1 + 2t^3)(1 - 2t^3)$
30. $(9x^2 + 1)(3x + 1)(3x - 1)$
31. $\left(4 + \frac{1}{9}y^4\right)\left(2 - \frac{1}{3}y^2\right)\left(2 + \frac{1}{3}y^2\right)$
32. $(7p^2 + 5q^3)(7p^2 - 5q^3)$

Exercise Set 11.5, p. 842

1. Yes 3. No 5. No 7. No 9. $(x - 7)^2$
11. $(x + 8)^2$ 13. $(x - 1)^2$ 15. $(x + 2)^2$ 17. $(q^2 - 3)^2$
19. $(4y + 7)^2$ 21. $2(x - 1)^2$ 23. $x(x - 9)^2$
25. $3(2q - 3)^2$ 27. $(7 - 3x)^2$ 29. $5(y^2 + 1)^2$
31. $(1 + 2x^2)^2$ 33. $(2p + 3q)^2$ 35. $(a - 3b)^2$
37. $(9a - b)^2$ 39. $4(3a + 4b)^2$ 41. Yes 43. No
45. No 47. Yes 49. $(y + 2)(y - 2)$
51. $(p + 3)(p - 3)$ 53. $(t + 7)(t - 7)$
55. $(a + b)(a - b)$ 57. $(5t + m)(5t - m)$
59. $(10 + k)(10 - k)$ 61. $(4a + 3)(4a - 3)$
63. $(2x + 5y)(2x - 5y)$ 65. $2(2x + 7)(2x - 7)$
67. $x(6 + 7x)(6 - 7x)$ 69. $\left(\frac{1}{4} + 7x^4\right)\left(\frac{1}{4} - 7x^4\right)$
71. $(0.3y + 0.02)(0.3y - 0.02)$ 73. $(7a^2 + 9)(7a^2 - 9)$
75. $(a^2 + 4)(a + 2)(a - 2)$ 77. $5(x^2 + 9)(x + 3)(x - 3)$
79. $(1 + y^4)(1 + y^2)(1 + y)(1 - y)$
81. $(x^6 + 4)(x^3 + 2)(x^3 - 2)$ 83. $\left(y + \frac{1}{4}\right)\left(y - \frac{1}{4}\right)$
85. $\left(5 + \frac{1}{7}x\right)\left(5 - \frac{1}{7}x\right)$ 87. $(4m^2 + t^2)(2m + t)(2m - t)$
89. ^{D}W 91. -11 92. 400 93. $-\frac{5}{6}$ 94. -0.9
95. 2 96. -160 97. $x^2 - 4xy + 4y^2$ 98. $\frac{1}{2}\pi x^2 + 2xy$
99. y^{12} 100. $25a^4b^6$ 101.

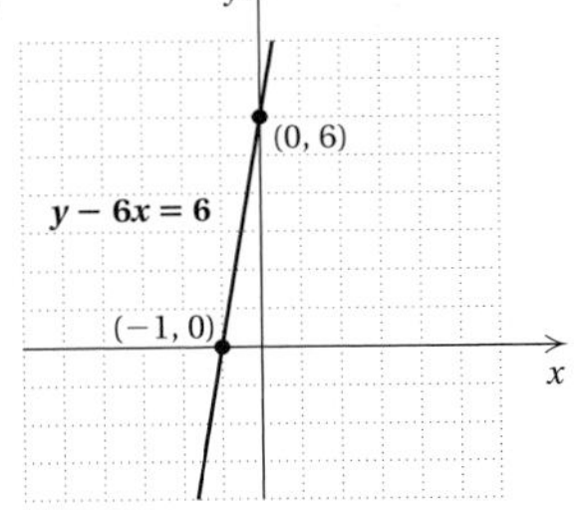

102.

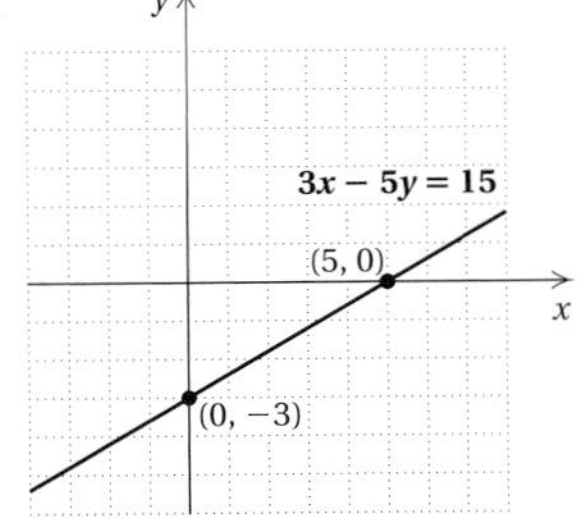

103. Prime

105. $(x + 11)^2$ 107. $2x(3x + 1)^2$
109. $(x^4 + 2^4)(x^2 + 2^2)(x + 2)(x - 2)$
111. $3x^3(x + 2)(x - 2)$ 113. $2x\left(3x + \frac{2}{5}\right)\left(3x - \frac{2}{5}\right)$
115. $p(0.7 + p)(0.7 - p)$ 117. $(0.8x + 1.1)(0.8x - 1.1)$
119. $x(x + 6)$ 121. $\left(x + \dfrac{1}{x}\right)\left(x - \dfrac{1}{x}\right)$
123. $(9 + b^{2k})(3 - b^k)(3 + b^k)$ 125. $(3b^n + 2)^2$
127. $(y + 4)^2$ 129. 9 131. Not correct
133. Not correct

Margin Exercises, Section 11.6, pp. 847–849

1. $3(m^2 + 1)(m + 1)(m - 1)$ 2. $(x^3 + 4)^2$
3. $2x^2(x + 1)(x + 3)$ 4. $(3x^2 - 2)(x + 4)$
5. $8x(x - 5)(x + 5)$ 6. $5(3x^2 - 2y)(x^2 + y)$
7. $2p^4q^2(5p^2 + 2pq + q^2)$ 8. $(a - b)(2x + 5 + y^2)$
9. $(a + b)(x^2 + y)$ 10. $(x^2 + y^2)^2$ 11. $(xy + 1)(xy + 4)$
12. $(p^2 + 9q^2)(p + 3q)(p - 3q)$

Exercise Set 11.6, p. 850

1. $3(x + 8)(x - 8)$ 3. $(a - 5)^2$ 5. $(2x - 3)(x - 4)$
7. $x(x + 12)^2$ 9. $(x + 2)(x - 2)(x + 3)$
11. $3(4x + 1)(4x - 1)$ 13. $3x(3x - 5)(x + 3)$
15. Prime 17. $x(x - 3)(x^2 + 7)$ 19. $x^3(x - 7)^2$
21. $-2(x - 2)(x + 5)$, or $2(-x + 2)(x + 5)$, or $2(x - 2)(-x - 5)$ 23. Prime
25. $4(x^2 + 4)(x + 2)(x - 2)$
27. $(1 + y^4)(1 + y^2)(1 + y)(1 - y)$ 29. $x^3(x - 3)(x - 1)$
31. $\frac{1}{9}\left(\frac{1}{3}x^3 - 4\right)^2$ 33. $m(x^2 + y^2)$ 35. $9xy(xy - 4)$
37. $2\pi r(h + r)$ 39. $(a + b)(2x + 1)$
41. $(x + 1)(x - 1 - y)$ 43. $(n + p)(n + 2)$
45. $(3q + p)(2q - 1)$ 47. $(2b - a)^2$, or $(a - 2b)^2$
49. $(4x + 3y)^2$ 51. $(7m^2 - 8n)^2$ 53. $(y^2 + 5z^2)^2$
55. $\left(\frac{1}{2}a + \frac{1}{3}b\right)^2$ 57. $(a + b)(a - 2b)$
59. $(m + 20n)(m - 18n)$ 61. $(mn - 8)(mn + 4)$
63. $r^3(rs - 2)(rs - 8)$ 65. $a^3(a - b)(a + 5b)$
67. $\left(a + \frac{1}{5}b\right)\left(a - \frac{1}{5}b\right)$ 69. $(x + y)(x - y)$
71. $(4 + p^2q^2)(2 + pq)(2 - pq)$
73. $(1 + 4x^6y^6)(1 + 2x^3y^3)(1 - 2x^3y^3)$
75. $(q + 1)(q - 1)(q + 8)$ 77. $(7x + 8y)^2$ 79. ^{D}W
81. $y = -3x - 12$ 82. $y = 8x - 4$ 83. $y = -\frac{2}{3}x + \frac{7}{3}$
84. $y = -0.28x - 1.16$ 85. $-\frac{14}{11}$ 86. $25x^2 - 10xt + t^2$
87. $X = \dfrac{A + 7}{a + b}$ 88. $\{x|x < 32\}$ 89. $(a + 1)^2(a - 1)^2$
91. $(3.5x - 1)^2$ 93. $(5x + 4)(x + 1.8)$
95. $(y + 3)(y - 3)(y - 2)$ 97. $(a^2 + 1)(a + 4)$
99. $(x + 2)(x - 2)(x - 1)$ 101. $(y - 1)^3$
103. $(y + 4 + x)^2$

Margin Exercises, Section 11.7, pp. 855–858

1. $3, -4$ **2.** $7, 3$ **3.** $-\frac{1}{4}, \frac{2}{3}$ **4.** $0, \frac{17}{3}$ **5.** $-2, 3$ **6.** $-4, 7$ **7.** 3 **8.** $0, 4$ **9.** $-\frac{4}{3}, \frac{4}{3}$ **10.** $3, \frac{7}{2}$ **11.** $-5, 2$ **12.** $-3, 3$ **13.** $(-5, 0), (1, 0)$ **14.** $0, 3$

Calculator Corner, p. 859

1. Left to the student

Exercise Set 11.7, p. 860

1. $-4, -9$ **3.** $-3, 8$ **5.** $-12, 11$ **7.** $0, -3$ **9.** $0, -18$ **11.** $-\frac{5}{2}, -4$ **13.** $-\frac{1}{5}, 3$ **15.** $4, \frac{1}{4}$ **17.** $0, \frac{2}{3}$ **19.** $-\frac{1}{10}, \frac{1}{27}$ **21.** $\frac{1}{3}, -20$ **23.** $0, \frac{2}{3}, \frac{1}{2}$ **25.** $-5, -1$ **27.** $-9, 2$ **29.** $3, 5$ **31.** $0, 8$ **33.** $0, -18$ **35.** $-4, 4$ **37.** $-\frac{2}{3}, \frac{2}{3}$ **39.** -3 **41.** 4 **43.** $0, \frac{6}{5}$ **45.** $-1, \frac{5}{3}$ **47.** $-\frac{1}{4}, \frac{2}{3}$ **49.** $-1, \frac{2}{3}$ **51.** $-\frac{7}{10}, \frac{7}{10}$ **53.** $-2, 9$ **55.** $\frac{4}{5}, \frac{3}{2}$ **57.** $(-4, 0), (1, 0)$ **59.** $\left(-\frac{5}{2}, 0\right), (2, 0)$ **61.** $(-3, 0), (5, 0)$ **63.** $-1, 4$ **65.** $-1, 3$ **67.** $\mathbf{D}_\mathbf{W}$ **69.** $(a + b)^2$ **70.** $a^2 + b^2$ **71.** -16 **72.** -4.5 **73.** $-\frac{10}{3}$ **74.** $\frac{3}{10}$ **75.** $-5, 4$ **77.** $-3, 9$ **79.** $-\frac{1}{8}, \frac{1}{8}$ **81.** $-4, 4$ **83.** Answers may vary. **(a)** $x^2 - x - 12 = 0$; **(b)** $x^2 + 7x + 12 = 0$; **(c)** $4x^2 - 4x + 1 = 0$; **(d)** $x^2 - 25 = 0$; **(e)** $40x^3 - 14x^2 + x = 0$ **85.** 2.33, 6.77 **87.** 0, 2.74

Margin Exercises, Section 11.8, pp. 863–868

1. Length: 24 in.; width: 12 in. **2.** Height: 25 ft; width: 10 ft **3. (a)** 342 games; **(b)** 9 teams **4.** 22 and 23 **5.** 24 ft **6.** 3 m, 4 m

Translating for Success, p. 869

1. O **2.** M **3.** K **4.** I **5.** G **6.** E **7.** C **8.** A **9.** H **10.** B

Exercise Set 11.8, p. 870

1. Length: 6 cm; width: 4 cm **3.** Length: 12 ft; width: 2 ft **5.** Height: 4 cm; base: 14 cm **7.** Base: 8 m; height: 16 m **9.** 182 games **11.** 12 teams **13.** 4950 handshakes **15.** 25 people **17.** 20 people **19.** 14 and 15 **21.** 12 and 14; -12 and -14 **23.** 15 and 17; -15 and -17 **25.** Hypotenuse: 17 ft; leg: 15 ft **27.** 32 ft **29.** 9 ft **31.** Dining room: 12 ft by 12 ft; kitchen: 12 ft by 10 ft **33.** 4 sec **35.** 5 and 7 **37.** $\mathbf{D}_\mathbf{W}$ **39.** Factor **40.** Factor **41.** Product **42.** Common factor **43.** Trinomial **44.** Quotient rule **45.** y-intercept **46.** Slope **47.** 35 ft **49.** 5 ft **51.** 30 cm by 15 cm **53.** 7 ft

Concept Reinforcement, p. 876

1. True **2.** False **3.** True **4.** False **5.** True

Summary and Review: Chapter 11, p. 876

1. $5y^2$ **2.** $12x$ **3.** $5(1 + 2x^3)(1 - 2x^3)$ **4.** $x(x - 3)$ **5.** $(3x + 2)(3x - 2)$ **6.** $(x + 6)(x - 2)$ **7.** $(x + 7)^2$ **8.** $3x(2x^2 + 4x + 1)$ **9.** $(x^2 + 3)(x + 1)$ **10.** $(3x - 1)(2x - 1)$ **11.** $(x^2 + 9)(x + 3)(x - 3)$ **12.** $3x(3x - 5)(x + 3)$ **13.** $2(x + 5)(x - 5)$ **14.** $(x^3 - 2)(x + 4)$ **15.** $(4x^2 + 1)(2x + 1)(2x - 1)$ **16.** $4x^4(2x^2 - 8x + 1)$ **17.** $3(2x + 5)^2$ **18.** Prime **19.** $x(x - 6)(x + 5)$ **20.** $(2x + 5)(2x - 5)$ **21.** $(3x - 5)^2$ **22.** $2(3x + 4)(x - 6)$ **23.** $(x - 3)^2$ **24.** $(2x + 1)(x - 4)$ **25.** $2(3x - 1)^2$ **26.** $3(x + 3)(x - 3)$ **27.** $(x - 5)(x - 3)$ **28.** $(5x - 2)^2$ **29.** $(7b^5 - 2a^4)^2$ **30.** $(xy + 4)(xy - 3)$ **31.** $3(2a + 7b)^2$ **32.** $(m + t)(m + 5)$ **33.** $32(x^2 - 2y^2z^2)(x^2 + 2y^2z^2)$ **34.** $1, -3$ **35.** $-7, 5$ **36.** $-4, 3$ **37.** $\frac{2}{3}, 1$ **38.** $-4, \frac{3}{2}$ **39.** $-2, 8$ **40.** Height: 6 cm; base: 5 cm **41.** -18 and -16; 16 and 18 **42.** -19 and -17; 17 and 19 **43.** 3 ft **44.** 6 km **45.** $(-5, 0), (-4, 0)$ **46.** $\left(-\frac{3}{2}, 0\right), (5, 0)$ **47.** $\mathbf{D}_\mathbf{W}$ Answers may vary. The area of a rectangle is 90 m^2. The length is 1 m greater than the width. Find the length and the width. **48.** $\mathbf{D}_\mathbf{W}$ Because Sheri did not first factor out the largest common factor, 4, her factorization will not be "complete" until she removes a common factor of 2 from each binomial. **49.** 2.5 cm **50.** 0, 2 **51.** Length: 12; width: 6 **52.** No solution **53.** $2, -3, \frac{5}{2}$ **54.** $-2, \frac{5}{4}, 3$ **55.** $(\pi - 2)x^2$

Test: Chapter 11, p. 879

1. [11.1a] $4x^3$ **2.** [11.2a] $(x - 5)(x - 2)$ **3.** [11.5b] $(x - 5)^2$ **4.** [11.1b] $2y^2(2y^2 - 4y + 3)$ **5.** [11.1c] $(x^2 + 2)(x + 1)$ **6.** [11.1b] $x(x - 5)$ **7.** [11.2a] $x(x + 3)(x - 1)$ **8.** [11.3a], [11.4a] $2(5x - 6)(x + 4)$ **9.** [11.5d] $(2x + 3)(2x - 3)$ **10.** [11.2a] $(x - 4)(x + 3)$ **11.** [11.3a], [11.4a] $3m(2m + 1)(m + 1)$ **12.** [11.5d] $3(w + 5)(w - 5)$ **13.** [11.5b] $5(3x + 2)^2$ **14.** [11.5d] $3(x^2 + 4)(x + 2)(x - 2)$ **15.** [11.5b] $(7x - 6)^2$ **16.** [11.3a], [11.4a] $(5x - 1)(x - 5)$ **17.** [11.1c] $(x^3 - 3)(x + 2)$ **18.** [11.5d] $5(4 + x^2)(2 + x)(2 - x)$ **19.** [11.3a], [11.4a] $(2x + 3)(2x - 5)$ **20.** [11.3a], [11.4a] $3t(2t + 5)(t - 1)$ **21.** [11.2a] $3(m + 2n)(m - 5n)$ **22.** [11.7b] $-4, 5$ **23.** [11.7b] $-5, \frac{3}{2}$ **24.** [11.7b] $-4, 7$ **25.** [11.8a] Length: 8 m; width: 6 m **26.** [11.8a] Height: 4 cm; base: 14 cm **27.** [11.8a] 5 ft **28.** [11.7b] $(-5, 0), (7, 0)$ **29.** [11.7b] $\left(\frac{2}{3}, 0\right), (1, 0)$ **30.** [11.8a] Length: 15; width: 3 **31.** [11.2a] $(a - 4)(a + 8)$ **32.** [11.7b] $-\frac{8}{3}, 0, \frac{2}{5}$ **33.** [10.6b], [11.5d] (d)

CHAPTER 12

Margin Exercises, Section 12.1, pp. 882–887

1. 3 **2.** $-8, 3$ **3.** None **4.** $\frac{(2x + 1)x}{(3x - 2)x}$

5. $\frac{(x+1)(x+2)}{(x-2)(x+2)}$ **6.** $\frac{(x-8)(-1)}{(x-y)(-1)}$ **7.** 5 **8.** $\frac{x}{4}$
9. $\frac{2x+1}{3x+2}$ **10.** $\frac{x+1}{2x+1}$ **11.** $x+2$ **12.** $\frac{y+2}{4}$
13. -1 **14.** -1 **15.** -1 **16.** $\frac{a-2}{a-3}$ **17.** $\frac{x-5}{2}$

Calculator Corner, p. 888

1. Correct **2.** Correct **3.** Not correct **4.** Not correct
5. Not correct **6.** Not correct **7.** Correct **8.** Correct

Exercise Set 12.1, p. 889

1. 0 **3.** 8 **5.** $-\frac{5}{2}$ **7.** $-4, 7$ **9.** $-5, 5$ **11.** None
13. $\frac{(4x)(3x^2)}{(4x)(5y)}$ **15.** $\frac{2x(x-1)}{2x(x+4)}$ **17.** $\frac{-1(3-x)}{-1(4-x)}$
19. $\frac{(y+6)(y-7)}{(y+6)(y+2)}$ **21.** $\frac{x^2}{4}$ **23.** $\frac{8p^2q}{3}$ **25.** $\frac{x-3}{x}$
27. $\frac{m+1}{2m+3}$ **29.** $\frac{a-3}{a+2}$ **31.** $\frac{a-3}{a-4}$ **33.** $\frac{x+5}{x-5}$
35. $a+1$ **37.** $\frac{x^2+1}{x+1}$ **39.** $\frac{3}{2}$ **41.** $\frac{6}{t-3}$
43. $\frac{t+2}{2(t-4)}$ **45.** $\frac{t-2}{t+2}$ **47.** -1 **49.** -1 **51.** -6
53. $-x-1$ **55.** $\frac{56x}{3}$ **57.** $\frac{2}{dc^2}$ **59.** $\frac{x+2}{x-2}$
61. $\frac{(a+3)(a-3)}{a(a+4)}$ **63.** $\frac{2a}{a-2}$ **65.** $\frac{(t+2)(t-2)}{(t+1)(t-1)}$
67. $\frac{x+4}{x+2}$ **69.** $\frac{5(a+6)}{a-1}$ **71.** $\mathrm{D_W}$ **73.** 18 and 20; -18 and -20 **74.** 3.125 L **75.** $(x-8)(x+7)$
76. $(a-8)^2$ **77.** $x^3(x-7)(x+5)$
78. $(2y^2+1)(y-5)$ **79.** $(2+t)(2-t)(4+t^2)$
80. $10(x+7)(x+1)$ **81.** $(x-7)(x-2)$ **82.** Prime
83. $(4x-5y)^2$ **84.** $(a-7b)(a-2b)$ **85.** $x+2y$
87. $\frac{(t-9)^2(t-1)}{(t^2+9)(t+1)}$ **89.** $\frac{x-y}{x-5y}$
91.

$$\frac{5(2x+5)-25}{10} = \frac{10x+25-25}{10} = \frac{10x}{10} = x$$

You get the same number you selected. To do a number trick, ask someone to select a number and then perform these operations. The person will probably be surprised that the result is the original number.

Margin Exercises, Section 12.2, pp. 893–895

1. $\frac{2}{7}$ **2.** $\frac{2x^3-1}{x^2+5}$ **3.** $\frac{1}{x-5}$ **4.** x^2-3 **5.** $\frac{6}{7}$
6. $\frac{5}{8}$ **7.** $\frac{(x-3)(x-2)}{(x+5)(x+5)}$ **8.** $\frac{3a^2}{a-5}$ **9.** $\frac{x-3}{x+2}$
10. $\frac{(x-3)(x-2)}{x+2}$ **11.** $\frac{y+1}{y-1}$

Exercise Set 12.2, p. 896

1. $\frac{x}{4}$ **3.** $\frac{1}{x^2-y^2}$ **5.** $a+b$ **7.** $\frac{x^2-4x+7}{x^2+2x-5}$ **9.** $\frac{3}{10}$
11. $\frac{1}{4}$ **13.** $\frac{b}{a}$ **15.** $\frac{(a+2)(a+3)}{(a-3)(a-1)}$ **17.** $\frac{(x-1)^2}{x}$
19. $\frac{1}{2}$ **21.** $\frac{15}{8}$ **23.** $\frac{15}{4}$ **25.** $\frac{a-5}{3(a-1)}$ **27.** $\frac{(x+2)^2}{x}$
29. $\frac{3}{2}$ **31.** $\frac{c+1}{c-1}$ **33.** $\frac{y-3}{2y-1}$ **35.** $\frac{x+1}{x-1}$ **37.** $\mathrm{D_W}$
39. $\{x \mid x \geq 77\}$ **40.** Height: 7 in.; base: 10 in.
41. $8x^3-11x^2-3x+12$ **42.** $-2p^2+4pq-4q^2$
43. $\frac{4y^8}{x^6}$ **44.** $\frac{125x^{18}}{y^{12}}$ **45.** $\frac{4x^6}{y^{10}}$ **46.** $\frac{1}{a^{15}b^{20}}$ **47.** $-\frac{1}{b^2}$
49. $\frac{a+1}{5ab^2(a^2+4)}$

Margin Exercises, Section 12.3, pp. 898–899

1. 144 **2.** 12 **3.** 10 **4.** 120 **5.** $\frac{35}{144}$ **6.** $\frac{1}{4}$ **7.** $\frac{11}{10}$
8. $\frac{9}{40}$ **9.** $60x^3y^2$ **10.** $(y+1)^2(y+4)$
11. $7(t^2+16)(t-2)$ **12.** $3x(x+1)^2(x-1)$

Exercise Set 12.3, p. 900

1. 108 **3.** 72 **5.** 126 **7.** 360 **9.** 500 **11.** $\frac{65}{72}$
13. $\frac{29}{120}$ **15.** $\frac{23}{180}$ **17.** $12x^3$ **19.** $18x^2y^2$
21. $6(y-3)$ **23.** $t(t+2)(t-2)$
25. $(x+2)(x-2)(x+3)$ **27.** $t(t+2)^2(t-4)$
29. $(a+1)(a-1)^2$ **31.** $(m-3)(m-2)^2$
33. $(2+3x)(2-3x)$ **35.** $10v(v+4)(v+3)$
37. $18x^3(x-2)^2(x+1)$ **39.** $6x^3(x+2)^2(x-2)$
41. $\mathrm{D_W}$ **43.** $(x-3)^2$ **44.** $2x(3x+2)$
45. $(x+3)(x-3)$ **46.** $(x+7)(x-3)$ **47.** $(x+3)^2$
48. $(x-7)(x+3)$ **49.** $120x^4$; $8x^3$; $960x^7$
50. $48x^6$; $16x^5$; $768x^{11}$ **51.** $20x^2$; $10x$; $200x^3$
52. $48ab^3$; $4ab$; $192a^2b^4$ **53.** $120x^3$; $2x^2$; $240x^5$
54. a^{15}; a^5; a^{20} **55.** 24 min

Margin Exercises, Section 12.4, pp. 902–905

1. $\frac{7}{9}$ **2.** $\frac{3+x}{x-2}$ **3.** $\frac{6x+4}{x-1}$ **4.** $\frac{10x^2+9x}{48}$ **5.** $\frac{9x+10}{48x^2}$
6. $\frac{4x^2-x+3}{x(x-1)(x+1)^2}$ **7.** $\frac{2x^2+16x+5}{(x+3)(x+8)}$
8. $\frac{8x+88}{(x+16)(x+1)(x+8)}$ **9.** $\frac{x-5}{4}$ **10.** $\frac{x-1}{x-3}$
11. $\frac{-2x-11}{3(x+4)(x-4)}$

Exercise Set 12.4, p. 906

1. 1 **3.** $\frac{6}{3+x}$ **5.** $\frac{2x+3}{x-5}$ **7.** $\frac{2x+5}{x^2}$ **9.** $\frac{41}{24r}$
11. $\frac{4x+6y}{x^2y^2}$ **13.** $\frac{4+3t}{18t^3}$ **15.** $\frac{x^2+4xy+y^2}{x^2y^2}$
17. $\frac{6x}{(x-2)(x+2)}$ **19.** $\frac{11x+2}{3x(x+1)}$ **21.** $\frac{x^2+6x}{(x+4)(x-4)}$

23. $\frac{6}{z+4}$ **25.** $\frac{3x-1}{(x-1)^2}$ **27.** $\frac{11a}{10(a-2)}$

29. $\frac{2x^2+8x+16}{x(x+4)}$ **31.** $\frac{7a+6}{(a-2)(a+1)(a+3)}$

33. $\frac{2x^2-4x+34}{(x-5)(x+3)}$ **35.** $\frac{3a+2}{(a+1)(a-1)}$ **37.** $\frac{1}{4}$

39. $-\frac{1}{t}$ **41.** $\frac{-x+7}{x-6}$ **43.** $y+3$ **45.** $\frac{2b-14}{b^2-16}$

47. $a+b$ **49.** $\frac{5x+2}{x-5}$ **51.** -1 **53.** $\frac{-x^2+9x-14}{(x-3)(x+3)}$

55. $\frac{2x+6y}{(x+y)(x-y)}$ **57.** $\frac{a^2+7a+1}{(a+5)(a-5)}$

59. $\frac{5t-12}{(t+3)(t-3)(t-2)}$ **61.** $\mathbf{D_W}$ **63.** x^2-1

64. $13y^3-14y^2+12y-73$ **65.** $\frac{1}{8x^{12}y^9}$ **66.** $\frac{x^6}{25y^2}$

67. $\frac{1}{x^{12}y^{21}}$ **68.** $\frac{25}{x^4y^6}$ **69.** 

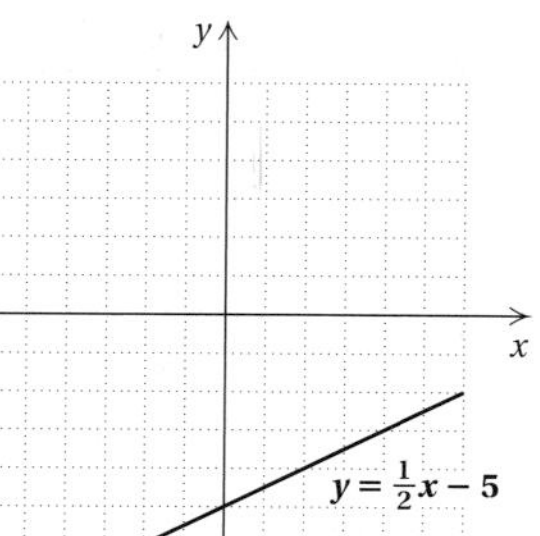

70.

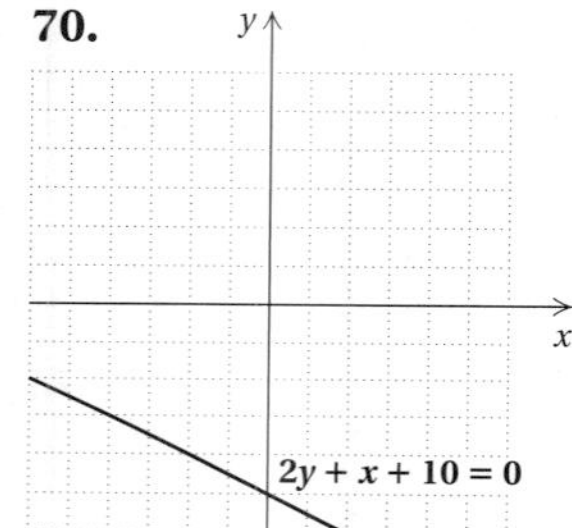

71.

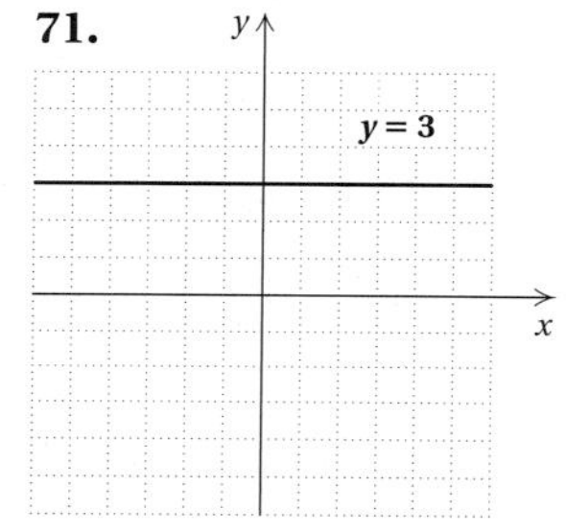

72.

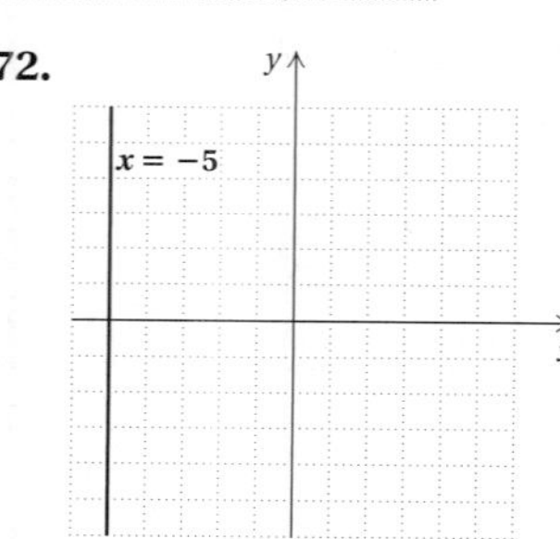

73. -8 **74.** $\frac{5}{6}$

75. 3, 5 **76.** $-2, 9$

77. Perimeter: $\frac{16y+28}{15}$; area: $\frac{y^2+2y-8}{15}$

79. $\frac{(z+6)(2z-3)}{(z+2)(z-2)}$

81. $\frac{11z^4-22z^2+6}{(z^2+2)(z^2-2)(2z^2-3)}$

83.–85. Left to the student

Margin Exercises, Section 12.5, pp. 910–913

1. $\frac{4}{11}$ **2.** $\frac{5}{y}$ **3.** $\frac{x^2+2x+1}{2x+1}$ **4.** $\frac{-x-7}{15x}$

5. $\frac{x^2-48}{(x+7)(x+8)(x+6)}$ **6.** $\frac{3x-1}{3}$ **7.** $\frac{4x-3}{x-2}$

8. $\frac{-8y-28}{(y+4)(y-4)}$ **9.** $\frac{x-13}{(x+3)(x-3)}$ **10.** $\frac{6x^2-2x-2}{3x(x+1)}$

Exercise Set 12.5, p. 914

1. $\frac{4}{x}$ **3.** 1 **5.** $\frac{1}{x-1}$ **7.** $\frac{-a-4}{10}$ **9.** $\frac{7z-12}{12z}$

11. $\frac{4x^2-13xt+9t^2}{3x^2t^2}$ **13.** $\frac{2x-40}{(x+5)(x-5)}$ **15.** $\frac{3-5t}{2t(t-1)}$

17. $\frac{2s-st-s^2}{(t+s)(t-s)}$ **19.** $\frac{y-19}{4y}$ **21.** $\frac{-2a^2}{(x+a)(x-a)}$

23. $\frac{8}{3}$ **25.** $\frac{13}{a}$ **27.** $\frac{8}{y-1}$ **29.** $\frac{x-2}{x-7}$ **31.** $\frac{4}{a^2-25}$

33. $\frac{2x-4}{x-9}$ **35.** $\frac{9x+12}{(x+3)(x-3)}$ **37.** $\frac{1}{2}$

39. $\frac{x-3}{(x+3)(x+1)}$ **41.** $\frac{18x+5}{x-1}$ **43.** 0 **45.** $\frac{-9}{2x-3}$

47. $\frac{20}{2y-1}$ **49.** $\frac{2a-3}{2-a}$ **51.** $\frac{z-3}{2z-1}$ **53.** $\frac{2}{x+y}$

55. $\mathbf{D_W}$ **57.** x^5 **58.** $30x^{12}$ **59.** $\frac{b^{20}}{a^8}$ **60.** $18x^3$

61. $\frac{6}{x^3}$ **62.** $\frac{10}{x^3}$ **63.** $x^2-9x+18$ **64.** $(4-\pi)r^2$

65. $\frac{30}{(x-3)(x+4)}$ **67.** $\frac{x^2+xy-x^3+x^2y-xy^2+y^3}{(x^2+y^2)(x+y)^2(x-y)}$

69. Missing side: $\frac{-2a-15}{a-6}$; area: $\frac{-2a^3-15a^2+12a+90}{2(a-6)^2}$

71.–73. Left to the student

Margin Exercises, Section 12.6, pp. 918–921

1. $\frac{33}{2}$ **2.** $\frac{3}{2}$ **3.** 3 **4.** $-\frac{1}{8}$ **5.** 1 **6.** 2 **7.** 4

Calculator Corner, p. 922

1.–12. Left to the student

Study Tip, p. 923

1. Rational expression **2.** Solutions **3.** Rational expression **4.** Rational expression **5.** Rational expression **6.** Solutions **7.** Rational expression **8.** Solutions **9.** Solutions **10.** Solutions **11.** Rational expression **12.** Solutions **13.** Rational expression

Exercise Set 12.6, p. 924

1. $\frac{6}{5}$ **3.** $\frac{40}{29}$ **5.** $\frac{47}{2}$ **7.** -6 **9.** $\frac{24}{7}$ **11.** $-4, -1$
13. $-4, 4$ **15.** 3 **17.** $\frac{14}{3}$ **19.** 5 **21.** 5 **23.** $\frac{5}{2}$
25. -2 **27.** $-\frac{13}{2}$ **29.** $\frac{17}{2}$ **31.** No solution **33.** -5
35. $\frac{5}{3}$ **37.** $\frac{1}{2}$ **39.** No solution **41.** No solution
43. 4 **45.** No solution **47.** $-2, 2$ **49.** 7
51. $\mathbf{D_W}$ **53.** Quotient **54.** Product **55.** Reciprocals
56. Factoring **57.** Greatest **58.** Not **59.** Subtract
60. Additive inverses **61.** $-\frac{1}{6}$ **63.** Left to the student

Margin Exercises, Section 12.7, pp. 930–935

1. $3\frac{3}{7}$ hr **2.** Greg: 40 mph; Nancy: 60 mph **3.** 58 km/L
4. 0.28 **5.** 124 km/h **6.** 2.4 fish/yd^2

7. About 34.6 gal **8.** 90 whales **9.** Yes; approximately 224 walks **10.** 24.75 ft **11.** About 34.9 ft

Translating for Success, p. 936

1. K **2.** E **3.** C **4.** N **5.** D **6.** O **7.** F **8.** H **9.** B **10.** A

Exercise Set 12.7, p. 937

1. $2\frac{2}{9}$ hr **3.** $25\frac{5}{7}$ min **5.** $3\frac{15}{16}$ hr **7.** $22\frac{2}{9}$ min **9.** $3\frac{3}{4}$ min **11.** Sarah: 30 km/h; Rick: 70 km/h **13.** Passenger: 80 mph; freight: 66 mph **15.** 20 mph **17.** Hank: 14 km/h; Kelly: 19 km/h **19.** Ralph: 5 km/h; Bonnie: 8 km/h **21.** 3 hr **23.** $\frac{5}{9}$ divorce/marriage **25.** 2.3 km/h **27.** 66 g **29.** 1.92 g **31.** 1.75 lb **33.** $1\frac{11}{39}$ kg **35. (a)** 0.332; **(b)** 243 hits; **(c)** 232 hits **37.** 22 in.; 55.8 cm **39.** $7\frac{1}{4}$; 57.9 cm **41.** $7\frac{1}{2}$; $23\frac{3}{5}$ in. **43.** 287 trout **45.** 200 duds **47. (a)** 4.8 tons; **(b)** 48 lb **49.** $\frac{21}{2}$ **51.** $\frac{8}{3}$ **53.** $\frac{35}{3}$ **55.** 15 ft **57.** DW **59.** x^{11} **60.** x **61.** $\frac{1}{x^{11}}$ **62.** $\frac{1}{x}$

63.

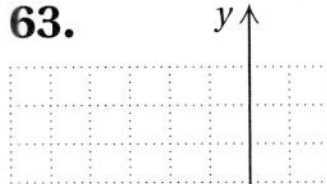
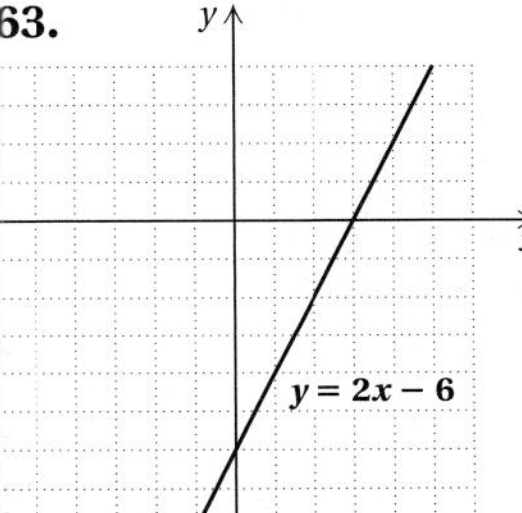

64.

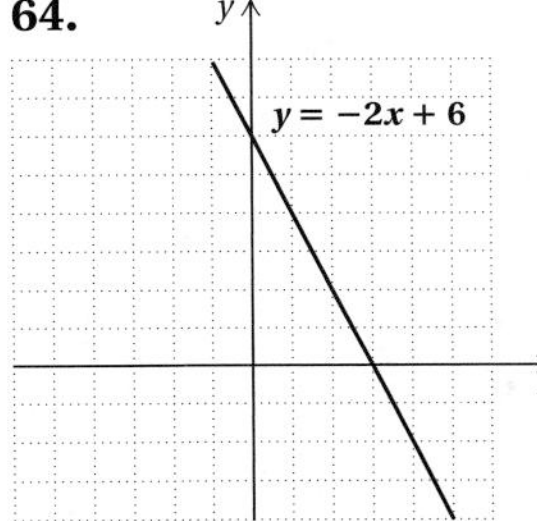

65.

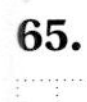
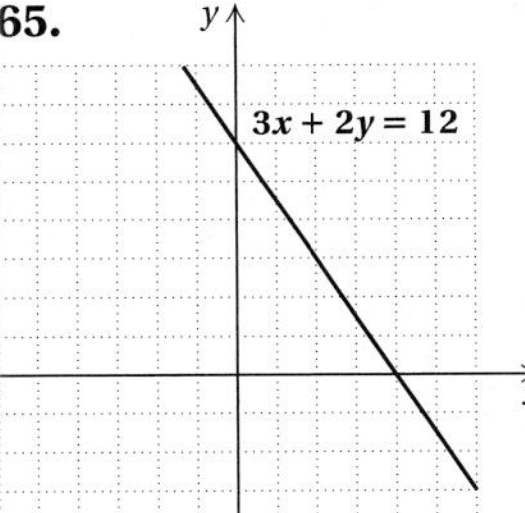

66.

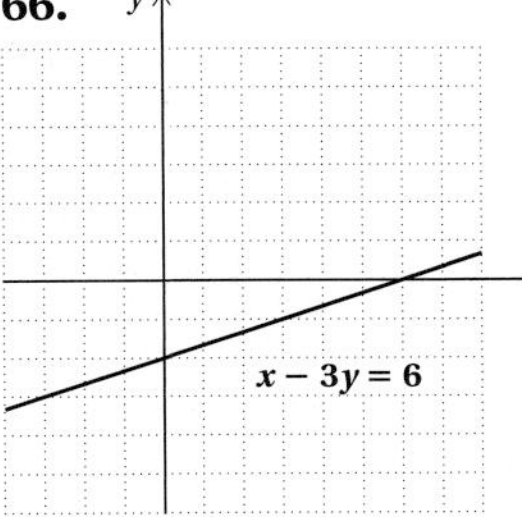

67.

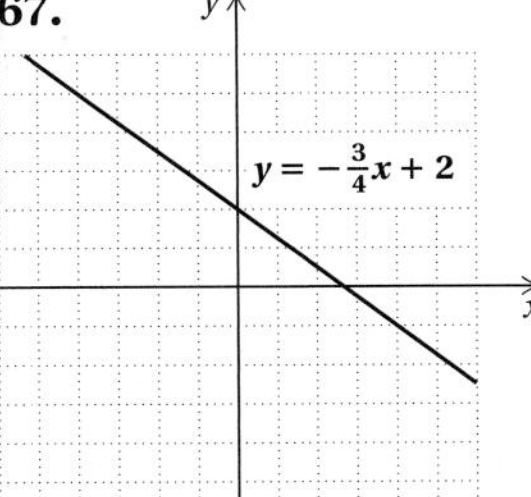

68.

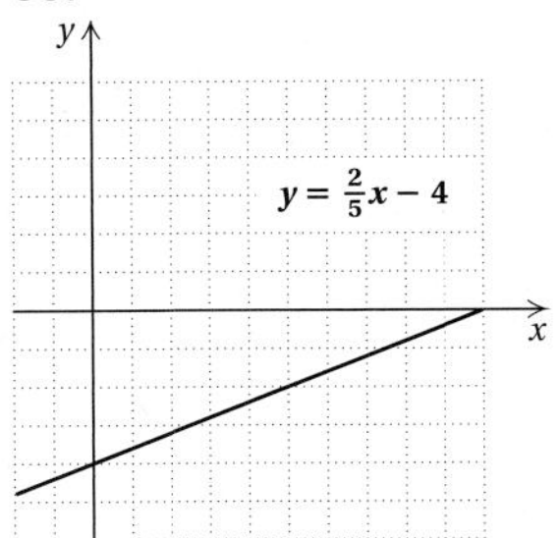

69. Ann: 6 hr; Betty: 12 hr **71.** $27\frac{3}{11}$ min **73.** $t = \frac{ab}{b + a}$

Margin Exercises, Section 12.8, pp. 945–947

1. $\frac{136}{5}$ **2.** $\frac{7x^2}{3(2 - x^2)}$ **3.** $\frac{x}{x - 1}$ **4.** $\frac{136}{5}$ **5.** $\frac{7x^2}{3(2 - x^2)}$ **6.** $\frac{x}{x - 1}$

Exercise Set 12.8, p. 948

1. $\frac{25}{4}$ **3.** $\frac{1}{3}$ **5.** -6 **7.** $\frac{1 + 3x}{1 - 5x}$ **9.** $\frac{2x + 1}{x}$ **11.** 8 **13.** $x - 8$ **15.** $\frac{y}{y - 1}$ **17.** $-\frac{1}{a}$ **19.** $\frac{ab}{b - a}$ **21.** $\frac{p^2 + q^2}{q + p}$ **23.** $\frac{2a^2 + 4a}{5 - 3a^2}$ **25.** $\frac{60 - 15a^3}{126a^2 + 28a^3}$ **27.** $\frac{ac}{bd}$ **29.** 1 **31.** $\frac{4x + 1}{5x + 3}$ **33.** DW **35.** $4x^4 + 3x^3 + 2x - 7$ **36.** 0 **37.** $(p - 5)^2$ **38.** $(p + 5)^2$ **39.** $50(p^2 - 2)$ **40.** $5(p + 2)(p - 10)$ **41.** 14 yd **42.** 12 ft, 5 ft **43.** $\frac{(x - 1)(3x - 2)}{5x - 3}$ **45.** $\frac{5x + 3}{3x + 2}$

Margin Exercises, Section 12.9, pp. 951–955

1. $y = 7x$; 287 **2.** $y = \frac{5}{8}x$; $\frac{25}{2}$ **3. (a)** $C = \frac{7}{360}n$; **(b)** \$0.4667; \$0.0194 **4. (a)** $V = 0.88E$; **(b)** 174.24 lb **5.** $y = \frac{63}{x}$; 3.15 **6.** $y = \frac{900}{x}$; 562.5 **7.** 8 hr **8. (a)** $t = \frac{300}{r}$; **(b)** 7.5 hr

Exercise Set 12.9, p. 956

1. $y = 4x$; 80 **3.** $y = \frac{8}{5}x$; 32 **5.** $y = 3.6x$; 72 **7.** $y = \frac{25}{3}x$; $\frac{500}{3}$ **9. (a)** $P = 5.6H$; **(b)** \$196 **11. (a)** $C = 11.25S$; **(b)** \$101.25 **13. (a)** $M = \frac{1}{6}E$; **(b)** $18.\overline{3}$ lb; **(c)** 30 lb **15. (a)** $N = 80{,}000S$; **(b)** 16,000,000 instructions/sec **17.** $93\frac{1}{3}$ servings **19.** $y = \frac{75}{x}$; $\frac{15}{2}$, or 7.5 **21.** $y = \frac{80}{x}$; 8 **23.** $y = \frac{1}{x}$; $\frac{1}{10}$ **25.** $y = \frac{2100}{x}$; 210 **27.** $y = \frac{0.06}{x}$; 0.006 **29. (a)** Direct; **(b)** $69\frac{3}{8}$ players **31. (a)** Inverse; **(b)** $4\frac{1}{2}$ hr **33. (a)** $N = \frac{280}{P}$; **(b)** 10 gal **35. (a)** $I = \frac{1920}{R}$; **(b)** 32 amperes **37. (a)** $m = \frac{40}{n}$; **(b)** 10 questions **39.** 8.25 ft **41.** DW **43.** DW **45.** $\frac{8}{5}$ **46.** 11 **47.** 9, 16 **48.** $-12, -9$ **49.** $\frac{2}{5}, \frac{4}{7}$ **50.** $-\frac{1}{7}, \frac{3}{2}$ **51.** $\frac{1}{3}$ **52.** $\frac{47}{20}$ **53.** -1 **54.** 49 **55.** The y-values become larger. **57.** $P^2 = kt$ **59.** $P = kV^3$

Concept Reinforcement, p. 961

1. False **2.** True **3.** True **4.** True **5.** False

Summary and Review: Chapter 12, p. 961

1. 0 **2.** 6 **3.** −6, 6 **4.** −6, 5 **5.** −2 **6.** None **7.** $\frac{x-2}{x+1}$ **8.** $\frac{7x+3}{x-3}$ **9.** $\frac{y-5}{y+5}$ **10.** $\frac{a-6}{5}$ **11.** $\frac{6}{2t-1}$ **12.** $-20t$ **13.** $\frac{2x^2-2x}{x+1}$ **14.** $30x^2y^2$ **15.** $4(a-2)$ **16.** $(y-2)(y+2)(y+1)$ **17.** $\frac{-3x+18}{x+7}$ **18.** −1 **19.** $\frac{2a}{a-1}$ **20.** $d+c$ **21.** $\frac{4}{x-4}$ **22.** $\frac{x+5}{2x}$ **23.** $\frac{2x+3}{x-2}$ **24.** $\frac{-x^2+x+26}{(x-5)(x+5)(x+1)}$ **25.** $\frac{2(x-2)}{x+2}$ **26.** $\frac{z}{1-z}$ **27.** $c-d$ **28.** 8 **29.** −5, 3 **30.** $5\frac{1}{7}$ hr **31.** 240 km/h, 280 km/h **32.** 95 mph, 175 mph **33.** 160 defective calculators **34.** **(a)** $\frac{12}{13}$ c; **(b)** $4\frac{1}{5}$ c; **(c)** $9\frac{1}{3}$ c **35.** 10,000 blue whales **36.** 6 **37.** $y=3x$; 60 **38.** $y=\frac{1}{2}x$; 10 **39.** $y=\frac{4}{5}x$; 16 **40.** $y=\frac{30}{x}$; 6 **41.** $y=\frac{1}{x}$; $\frac{1}{5}$ **42.** $y=\frac{0.65}{x}$; 0.13 **43.** \$288.75 **44.** 1 hr **45.** DW $\frac{5x+6}{(x+2)(x-2)}$; used to find an equivalent expression for each rational expression with the LCM as the least common denominator **46.** DW $\frac{3x+10}{(x-2)(x+2)}$; used to find an equivalent expression for each rational expression with the LCM as the least common denominator **47.** DW 4; used to clear fractions **48.** DW $\frac{4(x-2)}{x(x+4)}$; method 1: used to multiply by 1 using LCM/LCM; method 2: LCM of the denominators in the numerator used to subtract in the numerator and LCM of the denominators in the denominator used to add in the denominator **49.** $\frac{5(a+3)^2}{a}$ **50.** $\frac{10a}{(a-b)(b-c)}$ **51.** They are equivalent proportions.

Test: Chapter 12, p. 964

1. [12.1a] 0 **2.** [12.1a] −8 **3.** [12.1a] −7, 7 **4.** [12.1a] 1, 2 **5.** [12.1a] 1 **6.** [12.1a] None **7.** [12.1c] $\frac{3x+7}{x+3}$ **8.** [12.1d] $\frac{a+5}{2}$ **9.** [12.2b] $\frac{(5x+1)(x+1)}{3x(x+2)}$ **10.** [12.3c] $(y-3)(y+3)(y+7)$ **11.** [12.4a] $\frac{23-3x}{x^3}$ **12.** [12.5a] $\frac{8-2t}{t^2+1}$ **13.** [12.4a] $\frac{-3}{x-3}$ **14.** [12.5a] $\frac{2x-5}{x-3}$ **15.** [12.4a] $\frac{8t-3}{t(t-1)}$ **16.** [12.5a] $\frac{-x^2-7x-15}{(x+4)(x-4)(x+1)}$ **17.** [12.5b] $\frac{x^2+2x-7}{(x-1)^2(x+1)}$ **18.** [12.8a] $\frac{3y+1}{y}$ **19.** [12.6a] 12 **20.** [12.6a] −3, 5 **21.** [12.9a] $y=2x$; 50 **22.** [12.9a] $y=0.5x$; 12.5 **23.** [12.9c] $y=\frac{18}{x}$; $\frac{9}{50}$ **24.** [12.9c] $y=\frac{22}{x}$; $\frac{11}{50}$ **25.** [12.9b] 240 km **26.** [12.9d] $1\frac{1}{5}$ hr **27.** [12.7b] 16 defective spark plugs **28.** [12.7b] 50 zebras **29.** [12.7a] 12 min **30.** [12.7a] Craig: 65 km/h; Marilyn: 45 km/h **31.** [12.7b] 15 **32.** [12.7a] Rema: 4 hr; Reggie: 10 hr **33.** [12.8a] $\frac{3a+2}{2a+1}$

CHAPTER 13

Margin Exercises, Section 13.1, pp. 969–972

1. Yes **2.** No **3.** (2, −3) **4.** (−4, 3) **5.** No solution **6.** Infinite number of solutions **7.** **(a)** 3; **(b)** 3; **(c)** same **8.** **(a)** 3; **(b)** same

Calculator Corner, p. 972

1. (−1, 3) **2.** (0, 5) **3.** (2, −3) **4.** (−4, 1) **5.** (3, 2) **6.** $\left(\frac{7}{5}, -\frac{9}{5}\right)$

Exercise Set 13.1, p. 973

1. Yes **3.** No **5.** Yes **7.** Yes **9.** Yes **11.** (4, 2) **13.** (4, 3) **15.** (−3, −3) **17.** No solution **19.** (2, 2) **21.** $\left(\frac{1}{2}, 1\right)$ **23.** Infinite number of solutions **25.** (5, −3) **27.** DW **29.** $\frac{2x^2-1}{x^2(x+1)}$ **30.** $\frac{-4}{x-2}$ **31.** $\frac{9x+12}{(x-4)(x+4)}$ **32.** $\frac{2x+5}{x+3}$ **33.** Trinomial **34.** Binomial **35.** Monomial **36.** None of these **37.** $A=2$, $B=2$ **39.** $x+2y=2$, $x-y=8$ **41.–47.** Left to the student

Margin Exercises, Section 13.2, pp. 976–978

1. (3, 2) **2.** (3, −1) **3.** $\left(\frac{24}{5}, -\frac{8}{5}\right)$ **4.** Length: 25 m; width: 21 m

Exercise Set 13.2, p. 979

1. (1, 9) **3.** (2, −4) **5.** (4, 3) **7.** (−2, 1) **9.** (2, −4) **11.** $\left(\frac{17}{3}, \frac{16}{3}\right)$ **13.** $\left(\frac{25}{8}, -\frac{11}{4}\right)$ **15.** (−3, 0) **17.** (6, 3) **19.** Length: 94 ft; width: 50 ft **21.** Length: $3\frac{1}{2}$ in.; width: $1\frac{1}{2}$ in. **23.** Length: 365 mi; width: 275 mi **25.** Length: 40 ft; width: 20 ft **27.** Length: 110 yd; width: 60 yd **29.** 16 and 21 **31.** 12 and 40 **33.** 20 and 8 **35.** DW

37. 
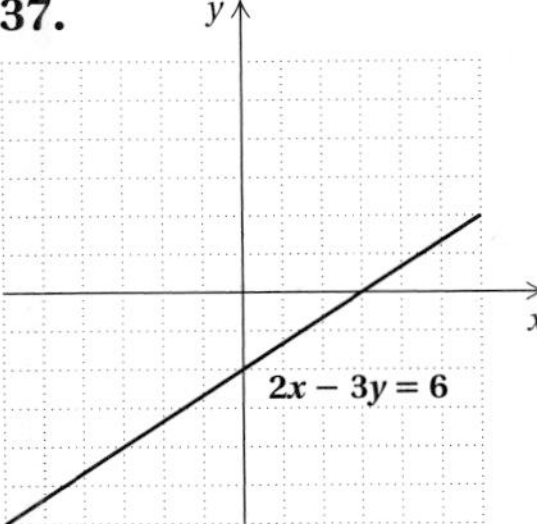

38. 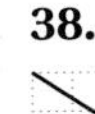
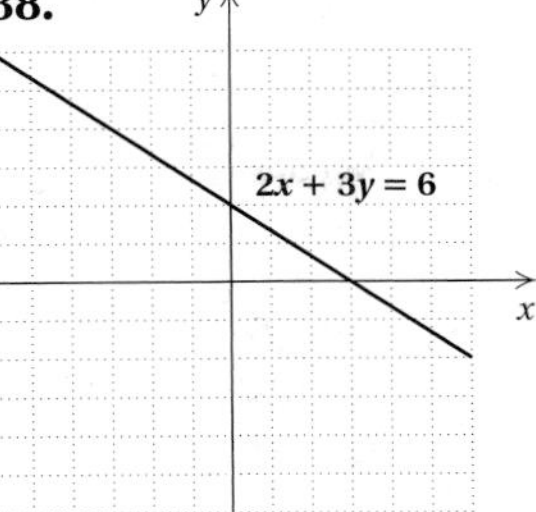

39. 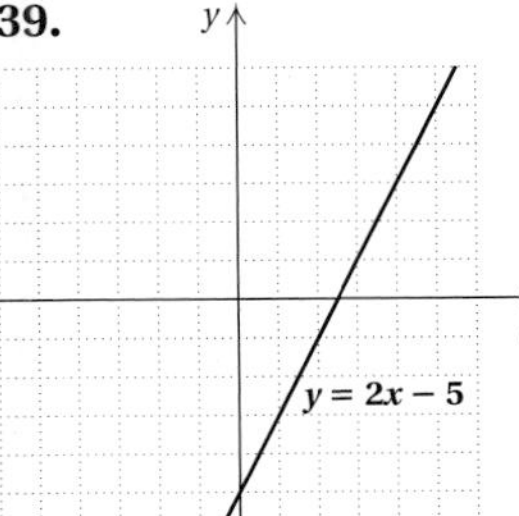

40. 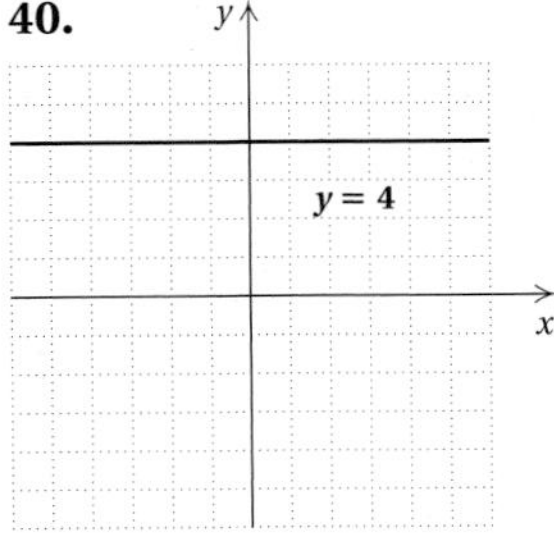

41. $(3x - 2)(2x - 3)$ **42.** $(4p + 3)(p - 1)$
43. Not factorable **44.** $(3a - 5)(3a + 5)$ **45.** x^3
46. x^7 **47.** $\dfrac{1}{x^7}$ **48.** $\dfrac{b^3}{a^3}$ **49.** $(5.\overline{6}, 0.\overline{6})$ **51.** $(4.38, 4.33)$
53. Baseball: 30 yd; softball: 20 yd

Margin Exercises, Section 13.3, pp. 983–988

1. $(3,2)$ **2.** $(1,-2)$ **3.** $(1,4)$ **4.** $(-8,3)$ **5.** $(1,1)$
6. $(-2,-1)$ **7.** $\left(\frac{17}{13}, -\frac{7}{13}\right)$ **8.** No solution
9. Infinite number of solutions **10.** $(1,-1)$
11. $(2,-2)$

Calculator Corner, p. 987

1. We get equations of two lines with the same slope but different y-intercepts. This indicates that the lines are parallel, so the system of equations has no solution.
2. We get equivalent equations. This indicates that the lines are the same, so the system of equations has an infinite number of solutions.

Exercise Set 13.3, p. 989

1. $(6,-1)$ **3.** $(3,5)$ **5.** $(2,5)$ **7.** $\left(-\frac{1}{2}, 3\right)$
9. $\left(-1, \frac{1}{5}\right)$ **11.** No solution **13.** $(-1,-6)$ **15.** $(3,1)$
17. $(8,3)$ **19.** $(4,3)$ **21.** $(1,-1)$ **23.** $(-3,-1)$
25. $(3,2)$ **27.** $(50,18)$ **29.** Infinite number of solutions
31. $(2,-1)$ **33.** $\left(\frac{231}{202}, \frac{117}{202}\right)$ **35.** $(-38,-22)$ **37.** D_W
39. Slope; y-intercepts **40.** Perpendicular **41.** Solution
42. Direct **43.** Horizontal **44.** Inverse
45. Slope–intercept **46.** Graph
47.–65. Left to the student **67.** $(5,2)$ **69.** $(0,-1)$
71. $(0,3)$ **73.** $x = \dfrac{c-b}{a-1}, y = \dfrac{ac-b}{a-1}$

Margin Exercises, Section 13.4, pp. 992–997

1. Hamburger: \$1.99; chicken: \$1.70 **2.** Adults: 125; children: 41 **3.** 50% alcohol: 22.5 L; 70% alcohol: 7.5 L
4. Seed A: 30 lb; seed B: 20 lb **5.** Quarters: 7; dimes: 13

Exercise Set 13.4, p. 999

1. Two-pointers: 32; three-pointers: 7
3. 24-exposure: 12; 36-exposure: 5 **5.** Hay: 10 lb; grain: 5 lb **7.** \$50 bonds: 13; \$100 bonds: 6
9. Cardholders: 128; non-cardholders: 75
11. Adults: 1300; children: 330
13. Solution A: 40 L; solution B: 60 L **15.** Dimes: 70; quarters: 33 **17.** Brazilian: 200 lb; Turkish: 100 lb
19. 28% fungicide: 100 L; 40% fungicide: 200 L
21. Large type: 4 pages; small type: 8 pages
23. 70% cashews: 36 lb; 45% cashews: 24 lb
25. Type A: 12; type B: 4; 180 **27.** Kuyatts': 32 yr; Marconis': 16 yr **29.** Randy: 24; Mandy: 6 **31.** 50°, 130°
33. 28°, 62° **35.** 87-octane: 12 gal; 93-octane: 6 gal
37. Dr. Zeke's $53\frac{1}{3}$ oz; Vitabrite: $26\frac{2}{3}$ oz **39.** D_W
41. $(5x + 9)(5x - 9)$ **42.** $(6 + a)(6 - a)$
43. $4(x^2 + 25)$ **44.** $4(x + 5)(x - 5)$

45. 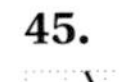
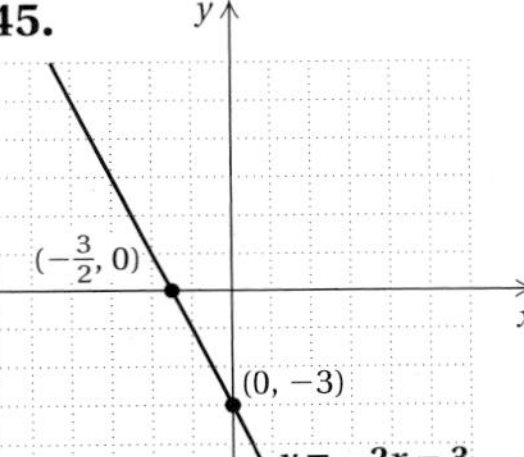

46. 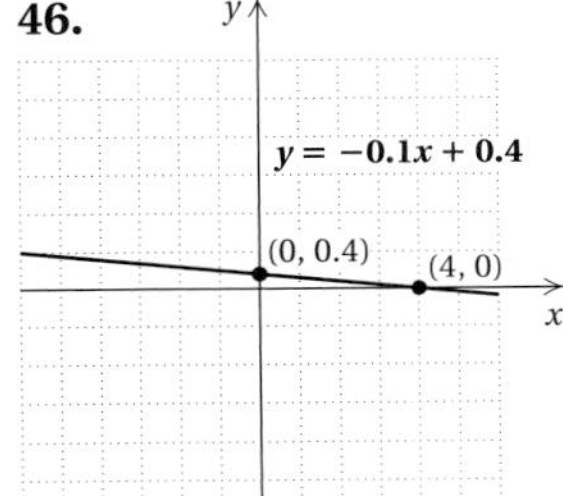

47. 
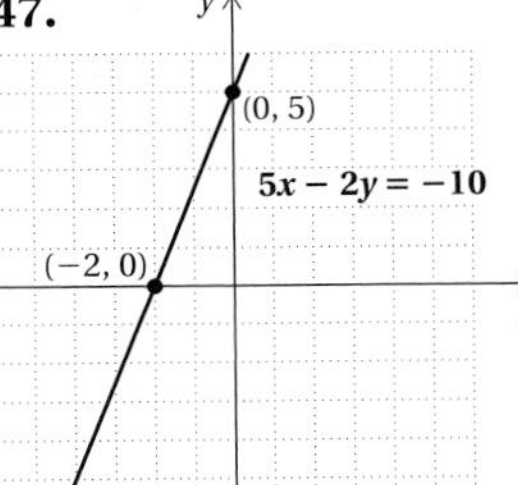

48. 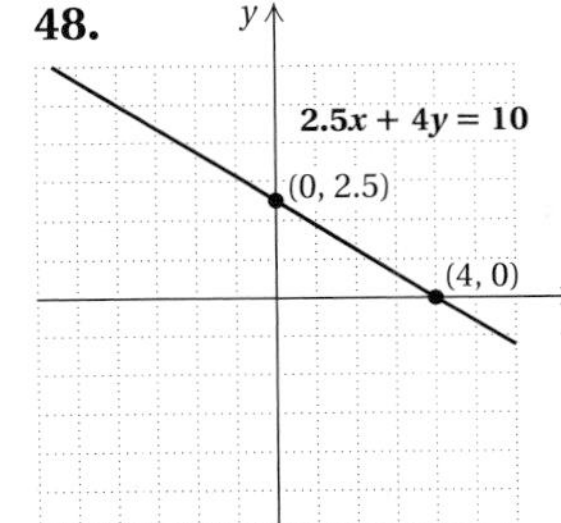

49. $\dfrac{x-3}{x+2}$ **50.** $\dfrac{x+5}{x-5}$ **51.** $\dfrac{-x^2 - 7x + 23}{(x+3)(x-4)}$
52. $\dfrac{-2x + 10}{(x+1)(x-1)}$ **53.** 43.75 L **55.** $4\frac{4}{7}$ L **57.** 54

Margin Exercises, Section 13.5, pp. 1006–1008

1. 324 mi **2.** 3 hr **3.** 168 km **4.** 275 km/h

Translating for Success, p. 1009

1. C **2.** A **3.** G **4.** E **5.** J **6.** I **7.** B **8.** L
9. O **10.** F

Exercise Set 13.5, p. 1010

1.

SPEED	TIME
30	t
46	t

4.5 hr

3.

SPEED	TIME	
72	$t + 3$	$\rightarrow d = 72(t + 3)$
120	t	$\rightarrow d = 120t$

$4\frac{1}{2}$ hr

5.

SPEED	TIME	
$r + 6$	4	$\rightarrow d = (r + 6)4$
$r - 6$	10	$\rightarrow d = (r - 6)10$

14 km/h

7. 384 km **9.** **(a)** 24 mph; **(b)** 90 mi
11. $1\frac{23}{43}$ min after the toddler starts running, or $\frac{23}{43}$ min after the mother starts running **13.** 15 mi **15.** DW
17. $\frac{x}{3}$ **18.** $\frac{x^5y^3}{2}$ **19.** $\frac{a + 3}{2}$ **20.** $\frac{x - 2}{4}$ **21.** 2
22. $\frac{1}{x^2 + 1}$ **23.** $\frac{x + 2}{x + 3}$ **24.** $\frac{3(x + 4)}{x - 1}$ **25.** $\frac{x + 3}{x + 2}$
26. $\frac{x^2 + 25}{x^2 - 25}$ **27.** $\frac{x + 2}{x - 1}$ **28.** $\frac{x^2 + 2}{2x^2 + 1}$
29. Approximately 3603 mi **31.** $5\frac{1}{3}$ mi

Concept Reinforcement, p. 1012

1. False **2.** True **3.** True **4.** False **5.** False

Summary and Review: Chapter 13, p. 1012

1. No **2.** Yes **3.** Yes **4.** No **5.** $(6, -2)$ **6.** $(6, 2)$
7. $(0, 5)$ **8.** No solution **9.** $(0, 5)$ **10.** $(-3, 9)$
11. $(3, -1)$ **12.** $(1, 4)$ **13.** $(-2, 4)$ **14.** $(1, -2)$
15. $(3, 1)$ **16.** $(1, 4)$ **17.** $(5, -3)$ **18.** $(-2, 4)$
19. $(-2, -6)$ **20.** $(3, 2)$ **21.** $(2, -4)$
22. Infinite number of solutions **23.** $(-4, 1)$
24. Length: 37.5 cm; width: 10.5 cm
25. Orchestra: 297; balcony: 211 **26.** 40 L of each
27. Asian: 4800 kg; African: 7200 kg
28. Peanuts: 8 lb; fancy nuts: 5 lb **29.** About 371 min
30. 87-octane: 2.5 gal; 95-octane: 7.5 gal **31.** Jeff: 39; his son: 13 **32.** 32°, 58° **33.** 77°, 103° **34.** 135 km/h
35. 412.5 mi **36.** DW The equations have the same slope but different y-intercepts, so they represent parallel lines. Thus the system of equations has no solution.
37. DW Translating. It is difficult to define the variables and write equations that represent the mathematical relationships in the exercises. Answers may vary.
38. \$960 **39.** $C = 1, D = 3$ **40.** $(2, 0)$
41. $y = -x + 5, y = \frac{2}{3}x$ **42.** $x + y = 4, x + y = -3$
43. Rabbits: 12; pheasants: 23

Test: Chapter 13, p. 1015

1. [13.1a] No
2. [13.1b]

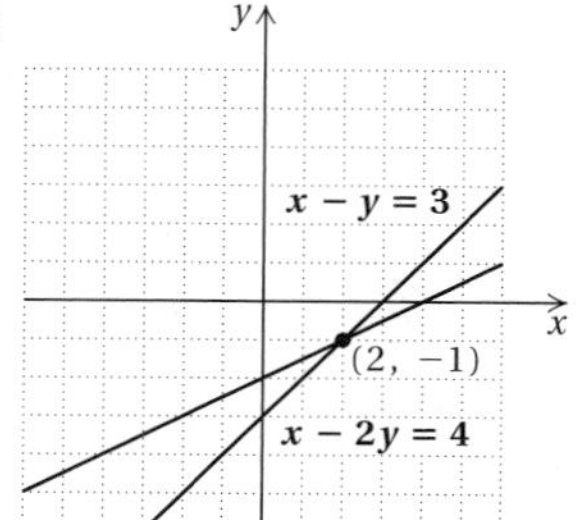

3. [13.2a] $(8, -2)$ **4.** [13.2b] $(-1, 3)$
5. [13.2a] No solution **6.** [13.3a] $(1, -5)$
7. [13.3b] $(12, -6)$ **8.** [13.3b] $\left(\frac{1}{2}, -\frac{2}{3}\right)$
9. [13.3b] $(5, 1)$ **10.** [13.2c] Length: 2108.5 yd; width: 2024.5 yd **11.** [13.4a] Solution A: 40 L; solution B: 20 L **12.** [13.5a] 40 km/h
13. [13.2c] Concessions: \$2850; rides: \$1425
14. [13.2c] Hay: 415 acres; oats: 235 acres
15. [13.2c] 45°, 135° **16.** [13.4a] 87-octane: 4 gal; 93-octane: 8 gal **17.** [13.4a] 20 min
18. [13.5a] 11 hr **19.** [13.1a] $C = -\frac{19}{2}; D = \frac{14}{3}$
20. [13.4a] 5 **21.** [9.4c], [13.1b] $y = \frac{1}{5}x + \frac{17}{5}, y = -\frac{3}{5}x + \frac{9}{5}$
22. [9.4c], [13.1b] $x = 3, y = -2$

CHAPTER 14

Margin Exercises, Section 14.1, pp. 1018–1022

1. 6, −6 **2.** 8, −8 **3.** 11, −11 **4.** 12, −12 **5.** 4
6. 7 **7.** 10 **8.** 21 **9.** −7 **10.** −13 **11.** 3.873
12. 5.477 **13.** 31.305 **14.** −25.842 **15.** 0.816
16. −1.732 **17.** **(a)** About 28.3 mph; **(b)** about 49.6 mph
18. 227 **19.** $45 + x$ **20.** $\frac{x}{x + 2}$ **21.** $x^2 + 4$ **22.** Yes
23. No **24.** No **25.** Yes **26.** 13 **27.** $|7w|$ **28.** $|xy|$
29. $|xy|$ **30.** $|x - 11|$ **31.** $|x + 4|$ **32.** xy **33.** xy
34. $x - 11$ **35.** $x + 4$ **36.** $5y$ **37.** $\frac{1}{2}t$

Calculator Corner, p. 1019

1. 6.557 **2.** 10.050 **3.** 102.308 **4.** 0.632
5. −96.985 **6.** −0.804

Exercise Set 14.1, p. 1023

1. 2, −2 **3.** 3, −3 **5.** 10, −10 **7.** 13, −13
9. 16, −16 **11.** 2 **13.** −3 **15.** −6 **17.** −15
19. 19 **21.** 2.236 **23.** 20.785 **25.** −18.647
27. 2.779 **29.** −120 **31.** **(a)** 13 spaces; **(b)** 24 spaces

33. 0.833 sec **35.** 0.911 sec **37.** 200 **39.** $a - 4$
41. $t^2 + 1$ **43.** $\frac{3}{x + 2}$ **45.** No **47.** Yes **49.** No
51. c **53.** $3x$ **55.** $8p$ **57.** ab **59.** $34d$ **61.** $x + 3$
63. $a - 5$ **65.** $2a - 5$ **67.** D_W **69.** 61°, 119°
70. 38°, 52° **71.** \$10,660 **72.** $\frac{1}{x + 3}$ **73.** 1
74. $\frac{(x + 2)(x - 2)}{(x + 1)(x - 1)}$ **75.** 1.7, 2.2, 2.6 **77.** 16, −16
79. 7, −7

Margin Exercises, Section 14.2, pp. 1026–1029

1. **(a)** 20; **(b)** 20 **2.** $\sqrt{33}$ **3.** 5 **4.** $\sqrt{\frac{30}{77}}$
5. $\sqrt{x^2 + x}$ **6.** $\sqrt{x^2 - 4}$ **7.** $4\sqrt{2}$ **8.** $2\sqrt{23}$
9. $8\sqrt{2t}$ **10.** $11\sqrt{3q}$ **11.** $3x\sqrt{7}$ **12.** $x + 7$ **13.** $9m$
14. $\sqrt{3}(x - 10)$ **15.** t^2 **16.** t^{10} **17.** h^{23} **18.** $x^3\sqrt{x}$
19. $2x^5\sqrt{6x}$ **20.** $3\sqrt{2}$ **21.** 10 **22.** $4x^3y^2$
23. $5xy^2\sqrt{2xy}$ **24.** $14q^2r^4\sqrt{3q}$

Calculator Corner, p. 1029

1. False **2.** False **3.** False **4.** True

Exercise Set 14.2, p. 1030

1. $2\sqrt{3}$ **3.** $5\sqrt{3}$ **5.** $2\sqrt{5}$ **7.** $10\sqrt{6}$ **9.** $9\sqrt{6}$
11. $3\sqrt{x}$ **13.** $4\sqrt{3x}$ **15.** $4\sqrt{a}$ **17.** $8y$ **19.** $x\sqrt{13}$
21. $2t\sqrt{2}$ **23.** $6\sqrt{5}$ **25.** $12\sqrt{2y}$ **27.** $2x\sqrt{7}$
29. $x - 3$ **31.** $\sqrt{2}(2x + 1)$ **33.** $\sqrt{y}(6 + y)$ **35.** t^3
37. x^6 **39.** $x^2\sqrt{x}$ **41.** $t^9\sqrt{t}$ **43.** $(y - 2)^4$
45. $2(x + 5)^5$ **47.** $6m\sqrt{m}$ **49.** $2a^2\sqrt{2a}$
51. $2p^8\sqrt{26p}$ **53.** $8x^3y\sqrt{7y}$ **55.** $3\sqrt{6}$ **57.** $3\sqrt{10}$
59. $6\sqrt{7x}$ **61.** $6\sqrt{xy}$ **63.** 13 **65.** $5b\sqrt{3}$ **67.** $2t$
69. $a\sqrt{bc}$ **71.** $2xy\sqrt{2xy}$ **73.** 18 **75.** $\sqrt{10x - 5}$
77. $x + 2$ **79.** $6xy^3\sqrt{3xy}$ **81.** $10x^2y^3\sqrt{5xy}$
83. $33p^4q^2\sqrt{2pq}$ **85.** $16a^3b^3c^5\sqrt{3abc}$ **87.** D_W
89. (−2, 4) **90.** $\left(\frac{1}{8}, \frac{9}{8}\right)$ **91.** (2, 1) **92.** (10, 3)
93. 360 ft^2 **94.** Adults: 350; children: 61 **95.** 80 L of 30%; 120 L of 50% **96.** 10 mph **97.** $\sqrt{5}\sqrt{x - 1}$
99. $\sqrt{x + 6}\sqrt{x - 6}$ **101.** $x\sqrt{x - 2}$ **103.** 0.5
105. $4y\sqrt{3}$ **107.** $18(x + 1)\sqrt{y(x + 1)}$ **109.** $2x^3\sqrt{5x}$

Margin Exercises, Section 14.3, pp. 1034–1037

1. 4 **2.** 5 **3.** $x\sqrt{6x}$ **4.** $\frac{4}{3}$ **5.** $\frac{1}{5}$ **6.** $\frac{6}{x}$ **7.** $\frac{3}{4}$
8. $\frac{15}{16}$ **9.** $\frac{7}{y^5}$ **10.** **(a)** $\frac{\sqrt{15}}{5}$; **(b)** $\frac{\sqrt{15}}{5}$ **11.** $\frac{\sqrt{10}}{4}$
12. $\frac{10\sqrt{3}}{3}$ **13.** $\frac{\sqrt{21}}{7}$ **14.** $\frac{\sqrt{5r}}{r}$ **15.** $\frac{8y\sqrt{7}}{7}$

Exercise Set 14.3, p. 1038

1. 3 **3.** 6 **5.** $\sqrt{5}$ **7.** $\frac{1}{5}$ **9.** $\frac{2}{5}$ **11.** 2 **13.** $3y$
15. $\frac{4}{7}$ **17.** $\frac{1}{6}$ **19.** $-\frac{4}{9}$ **21.** $\frac{8}{17}$ **23.** $\frac{13}{14}$ **25.** $\frac{5}{x}$
27. $\frac{3a}{25}$ **29.** $\frac{5}{y^5}$ **31.** $\frac{x^9}{7}$ **33.** $\frac{\sqrt{10}}{5}$ **35.** $\frac{\sqrt{14}}{4}$
37. $\frac{\sqrt{3}}{6}$ **39.** $\frac{\sqrt{10}}{6}$ **41.** $\frac{3\sqrt{5}}{5}$ **43.** $\frac{2\sqrt{6}}{3}$ **45.** $\frac{\sqrt{3x}}{x}$
47. $\frac{\sqrt{xy}}{y}$ **49.** $\frac{x\sqrt{5}}{10}$ **51.** $\frac{\sqrt{14}}{2}$ **53.** $\frac{3\sqrt{2}}{4}$ **55.** $\frac{\sqrt{6}}{2}$
57. $\sqrt{2}$ **59.** $\frac{\sqrt{55}}{11}$ **61.** $\frac{\sqrt{21}}{6}$ **63.** $\frac{\sqrt{6}}{2}$ **65.** 5
67. $\frac{\sqrt{3x}}{x}$ **69.** $\frac{4y\sqrt{5}}{5}$ **71.** $\frac{a\sqrt{2a}}{4}$ **73.** $\frac{\sqrt{42x}}{3x}$
75. $\frac{3\sqrt{6}}{8c}$ **77.** $\frac{y\sqrt{xy}}{x}$ **79.** $\frac{3n\sqrt{10}}{8}$ **81.** D_W
83. (4, 2) **84.** (10, 30) **85.** No solution
86. Infinite number of solutions **87.** $\left(-\frac{5}{2}, -\frac{9}{2}\right)$
88. $\left(\frac{26}{23}, \frac{44}{23}\right)$ **89.** $\frac{(x + 7)^2}{x - 7}$ **90.** $\frac{(x - 2)(x - 5)}{(x - 3)(x - 4)}$
91. $\frac{a - 5}{2}$ **92.** $\frac{x - 3}{x - 2}$ **93.** $9x^2 - 49$ **94.** $16a^2 - 25b^2$
95. $21x - 9y$ **96.** $14a - 6b$ **97.** 1.57 sec; 3.14 sec; 8.88 sec; 11.10 sec **99.** 1 sec
101. $\frac{\sqrt{5}}{40}$ **103.** $\frac{\sqrt{5x}}{5x^2}$ **105.** $\frac{\sqrt{3ab}}{b}$ **107.** $\frac{3\sqrt{10}}{100}$
109. $\frac{y - x}{xy}$

Margin Exercises, Section 14.4, pp. 1042–1045

1. $12\sqrt{2}$ **2.** $5\sqrt{5}$ **3.** $-12\sqrt{10}$ **4.** $5\sqrt{6}$ **5.** $\sqrt{x + 1}$
6. $\frac{3}{2}\sqrt{2}$ **7.** $\frac{8\sqrt{15}}{15}$ **8.** $\sqrt{15} + \sqrt{6}$
9. $4 + 3\sqrt{5} - 4\sqrt{2} - 3\sqrt{10}$ **10.** $2 - a$
11. $25 + 10\sqrt{x} + x$ **12.** 2 **13.** $7 - \sqrt{5}$
14. $\sqrt{5} + \sqrt{2}$ **15.** $1 + \sqrt{x}$ **16.** $\frac{21 - 3\sqrt{5}}{44}$
17. $-6 - \sqrt{35}$ **18.** $\frac{7 + 7\sqrt{x}}{1 - x}$

Exercise Set 14.4, p. 1046

1. $16\sqrt{3}$ **3.** $4\sqrt{5}$ **5.** $13\sqrt{x}$ **7.** $-9\sqrt{d}$ **9.** $25\sqrt{2}$
11. $\sqrt{3}$ **13.** $\sqrt{5}$ **15.** $13\sqrt{2}$ **17.** $3\sqrt{3}$ **19.** $2\sqrt{2}$
21. 0 **23.** $(2 + 9x)\sqrt{x}$ **25.** $(3 - 2x)\sqrt{3}$
27. $3\sqrt{2x + 2}$ **29.** $(x + 3)\sqrt{x^3 - 1}$
31. $(4a^2 + a^2b - 5b)\sqrt{b}$ **33.** $\frac{2\sqrt{3}}{3}$ **35.** $\frac{13\sqrt{2}}{2}$
37. $\frac{\sqrt{6}}{6}$ **39.** $\sqrt{15} - \sqrt{3}$ **41.** $10 + 5\sqrt{3} - 2\sqrt{7} - \sqrt{21}$
43. $9 - 4\sqrt{5}$ **45.** −62 **47.** 1 **49.** $13 + \sqrt{5}$
51. $x - 2\sqrt{xy} + y$ **53.** $-\sqrt{3} - \sqrt{5}$ **55.** $5 - 2\sqrt{6}$
57. $\frac{4\sqrt{10} - 4}{9}$ **59.** $5 - 2\sqrt{7}$ **61.** $\frac{12 - 3\sqrt{x}}{16 - x}$
63. $\frac{24 + 3\sqrt{x} + 8\sqrt{2} + \sqrt{2x}}{64 - x}$ **65.** D_W **67.** $\frac{5}{11}$
68. $-\frac{38}{13}$ **69.** −1, 6 **70.** 2, 5 **71.** $\frac{x^6}{3}$ **72.** $\frac{x - 3}{4(x + 3)}$
73. Jolly Juice: 1.6 L; Real Squeeze: 6.4 L **74.** $\frac{1}{3}$ hr
75. −9, −2, −5, −17, −0.678375 **77.** Not equivalent
79. Not correct **81.** $11\sqrt{3} - 10\sqrt{2}$
83. True; $(3\sqrt{x + 2})^2 = (3\sqrt{x + 2})(3\sqrt{x + 2}) = (3 \cdot 3)(\sqrt{x + 2} \cdot \sqrt{x + 2}) = 9(x + 2)$

Margin Exercises, Section 14.5, pp. 1050–1054

1. $\frac{64}{3}$ **2.** 2 **3.** $\frac{3}{8}$ **4.** 4 **5.** 1 **6.** 4 **7.** About 276 mi **8.** About 9 mi **9.** About 52 ft

Calculator Corner, p. 1053

1. Left to the student **2.** Left to the student

Exercise Set 14.5, p. 1055

1. 36 **3.** 18.49 **5.** 165 **7.** $\frac{621}{2}$ **9.** 5 **11.** 3 **13.** $\frac{17}{4}$ **15.** No solution **17.** No solution **19.** 9 **21.** 12 **23.** 1, 5 **25.** 3 **27.** 5 **29.** No solution **31.** $-\frac{10}{3}$ **33.** 3 **35.** No solution **37.** 9 **39.** 1 **41.** 8 **43.** 256 **45.** About 232 mi **47.** 16,200 ft **49.** 211.25 ft; 281.25 ft **51.** $^{\mathrm{D}}\mathrm{W}$ **53.** Slope; *y*-intercepts **54.** Square root **55.** Principal square root **56.** Positive; direct **57.** Quotient **58.** Positive; inverse **59.** Quotient **60.** Product **61.** −2, 2 **63.** $-\frac{57}{16}$ **65.** 13 **67.** Left to the student **69.** Left to the student

Margin Exercises, Section 14.6, pp. 1060–1061

1. $\sqrt{65} \approx 8.062$ **2.** $\sqrt{75} \approx 8.660$ **3.** $\sqrt{10} \approx 3.162$ **4.** $\sqrt{175} \approx 13.229$ **5.** $\sqrt{325}$ ft ≈ 18.028 ft

Translating for Success, p. 1062

1. J **2.** K **3.** N **4.** H **5.** G **6.** E **7.** O **8.** D **9.** B **10.** C

Exercise Set 14.6, p. 1063

1. 17 **3.** $\sqrt{32} \approx 5.657$ **5.** 12 **7.** 4 **9.** 26 **11.** 12 **13.** 2 **15.** $\sqrt{2} \approx 1.414$ **17.** 5 **19.** 3 **21.** $\sqrt{211{,}200{,}000}$ ft ≈ 14,533 ft **23.** 240 ft **25.** $\sqrt{18}$ cm ≈ 4.243 cm **27.** $\sqrt{208}$ ft ≈ 14.422 ft **29.** $^{\mathrm{D}}\mathrm{W}$ **31.** $\left(-\frac{3}{2}, -\frac{1}{16}\right)$ **32.** $\left(\frac{8}{5}, 9\right)$ **33.** $\left(-\frac{9}{19}, \frac{91}{38}\right)$ **34.** $(-10, 1)$ **35.** $-\frac{1}{3}$ **36.** $\frac{5}{8}$ **37.** $12 - 2\sqrt{6} \approx 7.101$

Concept Reinforcement, p. 1066

1. True **2.** True **3.** True **4.** False **5.** False **6.** False **7.** True **8.** True

Summary and Review: Chapter 14, p. 1066

1. 8, −8 **2.** 20, −20 **3.** 6 **4.** −13 **5.** 1.732 **6.** 9.950 **7.** −17.892 **8.** 0.742 **9.** −2.055 **10.** 394.648 **11.** $x^2 + 4$ **12.** $5ab^3$ **13.** No **14.** Yes **15.** No **16.** No **17.** No **18.** No **19.** m **20.** $x - 4$ **21.** $\sqrt{21}$ **22.** $\sqrt{x^2 - 9}$ **23.** $-4\sqrt{3}$ **24.** $4t\sqrt{2}$ **25.** $\sqrt{t-7}\sqrt{t+7}$ **26.** $x + 8$ **27.** x^4 **28.** $m^7\sqrt{m}$ **29.** $2\sqrt{15}$ **30.** $2x\sqrt{10}$ **31.** $5xy\sqrt{2}$ **32.** $10a^2b\sqrt{ab}$ **33.** $\frac{5}{8}$ **34.** $\frac{2}{3}$ **35.** $\frac{7}{t}$ **36.** $\frac{\sqrt{2}}{2}$ **37.** $\frac{\sqrt{2}}{4}$ **38.** $\frac{\sqrt{5y}}{y}$ **39.** $\frac{2\sqrt{3}}{3}$ **40.** $\frac{\sqrt{15}}{5}$ **41.** $\frac{x\sqrt{30}}{6}$ **42.** $8 - 4\sqrt{3}$ **43.** $13\sqrt{5}$ **44.** $\sqrt{5}$ **45.** $\frac{\sqrt{2}}{2}$ **46.** $7 + 4\sqrt{3}$ **47.** 1 **48.** 52 **49.** No solution **50.** 0, 3 **51.** 9 **52.** 20 **53.** $\sqrt{3} \approx 1.732$ **54.** $\sqrt{2{,}600{,}000{,}000}$ ft ≈ 50,990 ft **55.** 9 ft **56. (a)** About 63 mph; **(b)** 405 ft **57.** $^{\mathrm{D}}\mathrm{W}$ It is incorrect to take the square roots of the terms in the numerator individually—that is, $\sqrt{a+b}$ and $\sqrt{a} + \sqrt{b}$ are not equivalent. The following is correct:

$$\sqrt{\frac{9+100}{25}} = \frac{\sqrt{9+100}}{\sqrt{25}} = \frac{\sqrt{109}}{5}.$$

58. $^{\mathrm{D}}\mathrm{W}$ **(a)** $\sqrt{5x^2} = \sqrt{5}\sqrt{x^2} = \sqrt{5} \cdot |x| = |x|\sqrt{5}$. The given statement is correct.
(b) Let $b = 3$. Then $\sqrt{b^2 - 4} = \sqrt{3^2 - 4} = \sqrt{9-4} = \sqrt{5}$, but $b - 2 = 3 - 2 = 1$. The given statement is false.
(c) Let $x = 3$. Then $\sqrt{x^2 + 16} = \sqrt{3^2 + 16} = \sqrt{9 + 16} = \sqrt{25} = 5$, but $x + 4 = 3 + 4 = 7$. The given statement is false.
59. $\sqrt{1525}$ mi ≈ 39.051 mi **60.** 2 **61.** $b = \pm\sqrt{A^2 - a^2}$ **62.** 6

Test: Chapter 14, p. 1069

1. [14.1a] 9, −9 **2.** [14.1a] 8 **3.** [14.1a] −5 **4.** [14.1b] 10.770 **5.** [14.1b] −9.349 **6.** [14.1b] 4.127 **7.** [14.1d] $4 - y^3$ **8.** [14.1e] Yes **9.** [14.1e] No **10.** [14.1f] a **11.** [14.1f] $6y$ **12.** [14.2c] $\sqrt{30}$ **13.** [14.2c] $\sqrt{x^2 - 64}$ **14.** [14.2a] $3\sqrt{3}$ **15.** [14.2a] $5\sqrt{x-1}$ **16.** [14.2b] $t^2\sqrt{t}$ **17.** [14.2c] $5\sqrt{2}$ **18.** [14.2c] $3ab^2\sqrt{2}$ **19.** [14.3b] $\frac{3}{2}$ **20.** [14.3b] $\frac{12}{a}$ **21.** [14.3c] $\frac{\sqrt{10}}{5}$ **22.** [14.3c] $\frac{\sqrt{2xy}}{y}$ **23.** [14.3a, c] $\frac{3\sqrt{6}}{8}$ **24.** [14.3a] $\frac{\sqrt{7}}{4y}$ **25.** [14.4a] $-6\sqrt{2}$ **26.** [14.4a] $\frac{6\sqrt{5}}{5}$ **27.** [14.4b] $21 - 8\sqrt{5}$ **28.** [14.4b] 11 **29.** [14.4c] $\frac{40 + 10\sqrt{5}}{11}$ **30.** [14.6a] $\sqrt{80} \approx 8.944$ **31.** [14.5a] 48 **32.** [14.5a] −2, 2 **33.** [14.5b] −3 **34.** [14.5c] **(a)** About 237 mi; **(b)** 34,060.5 ft **35.** [14.6b] $\sqrt{15{,}700}$ yd ≈ 125.300 yd **36.** [14.1a] $\sqrt{5}$ **37.** [14.2b] y^{8n}

CHAPTER 15

Margin Exercises, Section 15.1, pp. 1072–1078

1. (a) *y*-intercept: (0, −3); *x*-intercept: $\left(-\frac{9}{2}, 0\right)$;

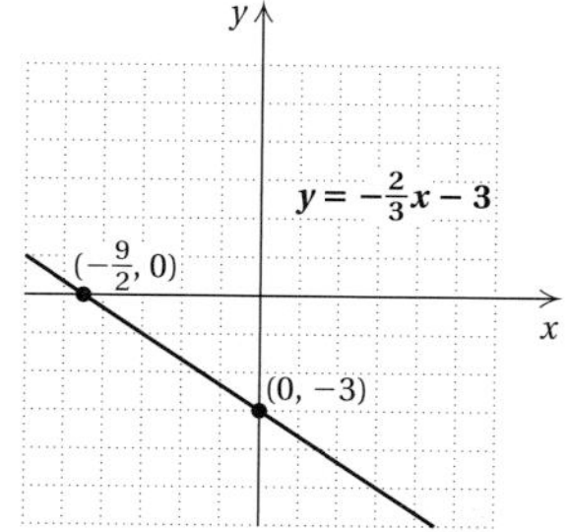

(b) $-\frac{9}{2}$; **(c)** $-\frac{9}{2}$; first coordinate; $\left(-\frac{9}{2}, 0\right)$

2. (a) $-1, 3$; (b) $-1, 3$; first coordinates, $(-1, 0), (3, 0)$
3. $6x^2 + 7x - 3 = 0; a = 6, b = 7, c = -3$
4. $y^2 - 8y = 0; a = 1, b = -8, c = 0$
5. $x^2 + 9x - 3 = 0; a = 1, b = 9, c = -3$
6. $4x^2 + x + 4 = 0; a = 4, b = 1, c = 4$
7. $5x^2 - 21 = 0; a = 5, b = 0, c = -21$ 8. $0, -4$
9. $0, \frac{3}{5}$ 10. $-2, \frac{3}{4}$ 11. $1, 4$ 12. $5, 13$
13. (a) 14 diagonals;

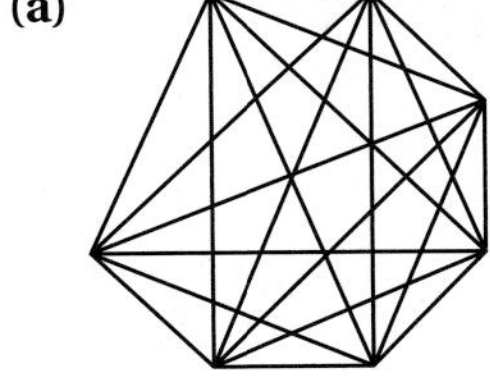

(b) 14 diagonals; (c) 11 sides

Calculator Corner, p. 1078

1. $0.6, 1$ 2. $-1.5, 5$ 3. $3, 8$ 4. $2, 4$

Exercise Set 15.1, p. 1079

1. $x^2 - 3x + 2 = 0; a = 1, b = -3, c = 2$
3. $7x^2 - 4x + 3 = 0; a = 7, b = -4, c = 3$
5. $2x^2 - 3x + 5 = 0; a = 2, b = -3, c = 5$
7. $0, -5$ 9. $0, -2$ 11. $0, \frac{2}{5}$ 13. $0, -1$
15. $0, 3$ 17. $0, \frac{1}{5}$ 19. $0, \frac{3}{14}$ 21. $0, \frac{81}{2}$ 23. $-12, 4$
25. $-5, -1$ 27. $-9, 2$ 29. $3, 5$ 31. -5 33. 4
35. $-\frac{2}{3}, \frac{1}{2}$ 37. $-\frac{2}{3}, 4$ 39. $-1, \frac{5}{3}$ 41. $-5, -1$
43. $-2, 7$ 45. $-5, 4$ 47. 4 49. $-2, 1$
51. $-\frac{2}{5}, 10$ 53. $-4, 6$ 55. 1 57. $2, 5$
59. No solution 61. $-\frac{5}{2}, 1$ 63. 35 diagonals
65. 7 sides 67. D_W 69. 8 70. -13 71. $2\sqrt{2}$
72. $2\sqrt{3}$ 73. $2\sqrt{5}$ 74. $2\sqrt{22}$ 75. $9\sqrt{5}$
76. $2\sqrt{255}$ 77. 2.646 78. 4.796 79. 1.528
80. 22.908 81. $-\frac{1}{3}, 1$ 83. $0, \frac{\sqrt{5}}{5}$ 85. $-1.7, 4$
87. $-1.7, 3$ 89. $-2, 3$ 91. 4

Margin Exercises, Section 15.2, pp. 1081–1086

1. $\sqrt{10}, -\sqrt{10}$ 2. 0 3. $\frac{\sqrt{6}}{2}, -\frac{\sqrt{6}}{2}$ 4. $7, -1$
5. $-4 \pm \sqrt{11}$ 6. $-5, 11$ 7. $1 \pm \sqrt{5}$ 8. $2, 4$
9. $-10, 2$ 10. $6 \pm \sqrt{13}$ 11. $-2, 5$ 12. $\frac{-3 \pm \sqrt{33}}{4}$
13. About 7.5 sec

Exercise Set 15.2, p. 1087

1. $11, -11$ 3. $\sqrt{7}, -\sqrt{7}$ 5. $\frac{\sqrt{15}}{5}, -\frac{\sqrt{15}}{5}$ 7. $\frac{5}{2}, -\frac{5}{2}$
9. $\frac{7\sqrt{3}}{3}, -\frac{7\sqrt{3}}{3}$ 11. $\sqrt{3}, -\sqrt{3}$ 13. $\frac{8}{7}, -\frac{8}{7}$
15. $-7, 1$ 17. $-3 \pm \sqrt{21}$ 19. $-13 \pm 2\sqrt{2}$
21. $7 \pm 2\sqrt{3}$ 23. $-9 \pm \sqrt{34}$ 25. $\frac{-3 \pm \sqrt{14}}{2}$
27. $-5, 11$ 29. $-15, 1$ 31. $-2, 8$ 33. $-21, -1$
35. $1 \pm \sqrt{6}$ 37. $11 \pm \sqrt{19}$ 39. $-5 \pm \sqrt{29}$
41. $\frac{7 \pm \sqrt{57}}{2}$ 43. $-7, 4$ 45. $\frac{-3 \pm \sqrt{17}}{4}$
47. $\frac{-3 \pm \sqrt{145}}{4}$ 49. $\frac{-2 \pm \sqrt{7}}{3}$ 51. $-\frac{1}{2}, 5$
53. $-\frac{5}{2}, \frac{2}{3}$ 55. About 9.6 sec 57. About 4.4 sec
59. D_W 61. Product 62. Quadratic; equivalent
63. Principal square root 64. Square root
65. Quotient 66. Product 67. Quotient
68. Power; multiply 69. $-12, 12$ 71. $-16\sqrt{2}, 16\sqrt{2}$
73. $-2\sqrt{c}, 2\sqrt{c}$ 75. $49.896, -49.896$ 77. $-9, 9$

Margin Exercises, Section 15.3, pp. 1091–1094

1. $-4, \frac{1}{2}$ 2. $-2, 5$ 3. $-2 \pm \sqrt{11}$
4. No real-number solutions 5. $\frac{4 \pm \sqrt{31}}{5}$
6. $-0.3, 1.9$

Calculator Corner, p. 1093

1. The equations $x^2 + x = -1$ and $x^2 + x + 1 = 0$ are equivalent. The graph of $y = x^2 + x + 1$ has no x-intercepts, so the equation $x^2 + x = -1$ has no real-number solutions.
2. Yes 3. No

Calculator Corner, p. 1094

1. $-1.3, 5.3$ 2. $-5.3, 1.3$ 3. $-0.3, 1.9$

Exercise Set 15.3, p. 1095

1. $-3, 7$ 3. 3 5. $-\frac{4}{3}, 2$ 7. $-\frac{5}{2}, \frac{3}{2}$ 9. $-3, 3$
11. $1 \pm \sqrt{3}$ 13. $5 \pm \sqrt{3}$ 15. $-2 \pm \sqrt{7}$
17. $\frac{-4 \pm \sqrt{10}}{3}$ 19. $\frac{5 \pm \sqrt{33}}{4}$ 21. $\frac{1 \pm \sqrt{3}}{2}$
23. No real-number solutions 25. $\frac{5 \pm \sqrt{73}}{6}$
27. $\frac{3 \pm \sqrt{29}}{2}$ 29. $-\sqrt{5}, \sqrt{5}$ 31. $-2 \pm \sqrt{3}$
33. $\frac{5 \pm \sqrt{37}}{2}$ 35. $-1.3, 5.3$ 37. $-0.2, 6.2$
39. $-1.2, 0.2$ 41. $0.3, 2.4$ 43. D_W 45. $3\sqrt{10}$
46. $\sqrt{6}$ 47. $2\sqrt{2}$ 48. $(9x - 2)\sqrt{x}$ 49. $4\sqrt{5}$
50. $3x^2\sqrt{3x}$ 51. $30x^5\sqrt{10}$ 52. $\frac{\sqrt{21}}{3}$ 53. $y = \frac{141}{x}$
54. $3\frac{1}{3}$ hr 55. $0, 2$ 57. $\frac{3 \pm \sqrt{5}}{2}$ 59. $\frac{-7 \pm \sqrt{61}}{2}$
61. $\frac{-2 \pm \sqrt{10}}{2}$ 63.–69. Left to the student

Margin Exercises, Section 15.4, pp. 1097–1099

1. (a) $I = \frac{9R}{E}$; (b) $R = \frac{EI}{9}$ 2. $x = \frac{y - 5}{a - b}$ 3. $f = \frac{pq}{q + p}$
4. $L = \frac{r^2}{20}$ 5. $L = \frac{T^2 g}{4\pi^2}$ 6. $m = \frac{E}{c^2}$ 7. $r = \sqrt{\frac{A}{\pi}}$
8. $n = \frac{1 + \sqrt{1 + 4N}}{2}$

Exercise Set 15.4, p. 1100

1. $I = \frac{VQ}{q}$ **3.** $m = \frac{Sd^2}{kM}$ **5.** $d^2 = \frac{kmM}{S}$
7. $W = \sqrt{\frac{10t}{T}}$ **9.** $t = \frac{A}{a+b}$ **11.** $x = \frac{y-c}{a+b}$
13. $a = \frac{bt}{b-t}$ **15.** $p = \frac{qf}{q-f}$ **17.** $b = \frac{2A}{h}$
19. $h = \frac{S - 2\pi r^2}{2\pi r}$, or $h = \frac{S}{2\pi r} - r$ **21.** $R = \frac{r_1 r_2}{r_2 + r_1}$
23. $Q = \frac{P^2}{289}$ **25.** $E = \frac{mv^2}{2g}$ **27.** $r = \frac{1}{2}\sqrt{\frac{S}{\pi}}$
29. $A = \frac{-m + \sqrt{m^2 + 4kP}}{2k}$ **31.** $a = \sqrt{c^2 - b^2}$
33. $t = \frac{\sqrt{s}}{4}$ **35.** $r = \frac{-\pi h + \sqrt{\pi^2 h^2 + \pi A}}{\pi}$
37. $v = 20\sqrt{\frac{F}{A}}$ **39.** $a = \sqrt{c^2 - b^2}$ **41.** $a = \frac{2h\sqrt{3}}{3}$
43. $T = \frac{2 + \sqrt{4 - a(m-n)}}{a}$ **45.** $T = \frac{v^2 \pi m}{8k}$
47. $x = \frac{d\sqrt{3}}{3}$ **49.** $n = \frac{1 + \sqrt{1 + 8N}}{2}$ **51.** $b = \frac{a}{3S - 1}$
53. $B = \frac{A}{QA + 1}$ **55.** $n = \frac{S + 360}{180}$, or $n = \frac{S}{180} + 2$
57. $t = \frac{A - P}{Pr}$ **59.** $D = \frac{BC}{A}$ **61.** $a = \frac{-b}{C - K}$, or
$a = \frac{b}{K - C}$ **63.** DW **65.** $\sqrt{65} \approx 8.062$
66. $\sqrt{75} \approx 8.660$ **67.** $\sqrt{41} \approx 6.403$ **68.** $\sqrt{44} \approx 6.633$
69. $\sqrt{1084} \approx 32.924$ **70.** $\sqrt{5} \approx 2.236$
71. $\sqrt{424}$ ft ≈ 20.591 ft **72.** $\sqrt{12{,}500}$ yd ≈ 111.803 yd
73. $3x\sqrt{2}$ **74.** $8x^2\sqrt{3x}$ **75.** $3t$ **76.** $x^3\sqrt{x}$
77. (a) $r = \frac{C}{2\pi}$; (b) $A = \frac{C^2}{4\pi}$ **79.** $\frac{1}{3a}, 1$

Margin Exercises, Section 15.5, pp. 1103–1105

1. Length: 14.8 yd; width: 4.6 yd **2.** 3.8 yd, 5.8 yd
3. 3 km/h

Translating for Success, p. 1106

1. M **2.** G **3.** F **4.** L **5.** D **6.** N **7.** J
8. E **9.** B **10.** C

Exercise Set 15.5, p. 1107

1. Width: about 61 in.; height: about 34 in.
3. Length: 10 ft; width: 7 ft **5.** 16 in.; 24 in.
7. Length: 20 cm; width: 16 cm
9. 4.6 m; 6.6 m **11.** Length: 5.6 in.; width: 3.6 in.
13. Length: 6.4 cm; width: 3.2 cm **15.** 3 cm
17. 7 km/h **19.** 0 km/h (no wind) or 40 km/h
21. 0 km/h (stream is still) or 4 km/h **23.** 8 mph
25. 1 km/h **27.** DW **29.** $8\sqrt{2}$ **30.** $12\sqrt{10}$
31. $(2x - 7)\sqrt{x}$ **32.** $-\sqrt{6}$ **33.** $\frac{3\sqrt{2}}{2}$ **34.** $\frac{2\sqrt{3}}{3}$
35. $5\sqrt{6} - 4\sqrt{3}$ **36.** $(9x + 2)\sqrt{x}$
37. $12\sqrt{2}$ in. ≈ 16.97 in.; two 12-in. pizzas

Margin Exercises, Section 15.6, pp. 1113–1114

1. $(0, -3)$

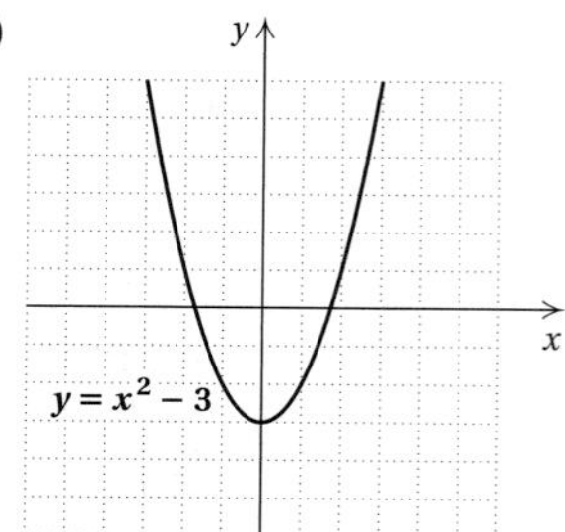

2. $(1, 3)$

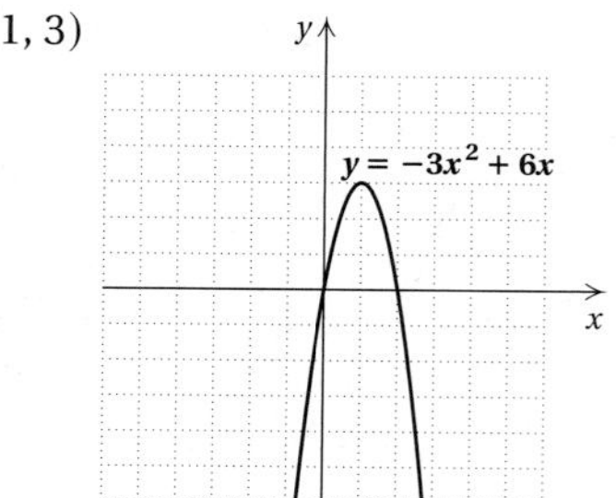

3. $(2, 0)$

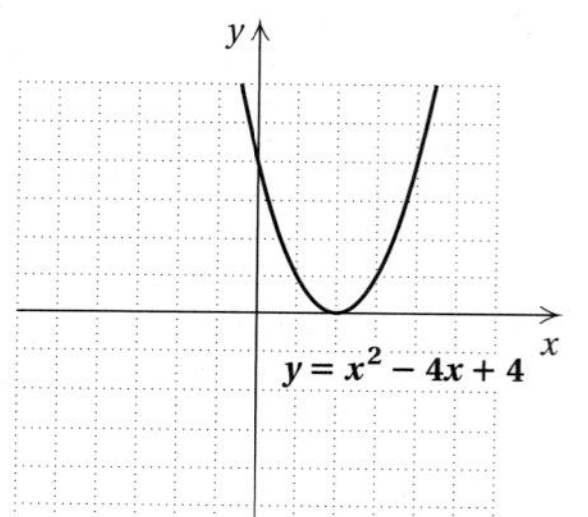

4. $(-\sqrt{3}, 0)$; $(\sqrt{3}, 0)$

5. $(-4, 0)$; $(-2, 0)$ **6.** $\left(\frac{-2 - \sqrt{6}}{2}, 0\right)$; $\left(\frac{-2 + \sqrt{6}}{2}, 0\right)$
7. None

Calculator Corner, p. 1112

Left to the student

Calculator Corner, p. 1114

Left to the student

Visualizing for Success, p. 1115

1. J **2.** F **3.** H **4.** G **5.** B **6.** E **7.** D
8. I **9.** C **10.** A

Exercise Set 15.6, p. 1116

1.

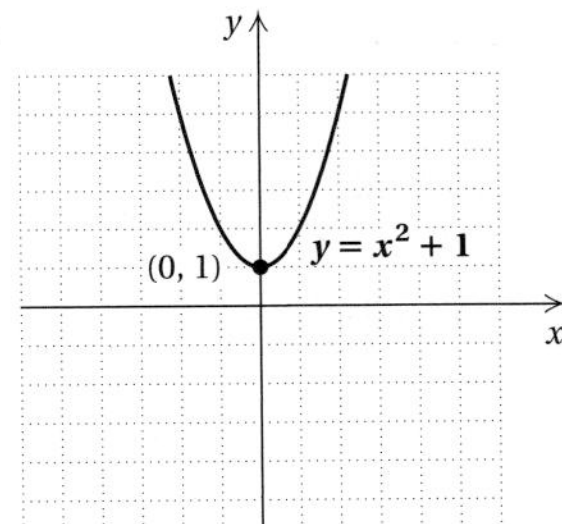

3.

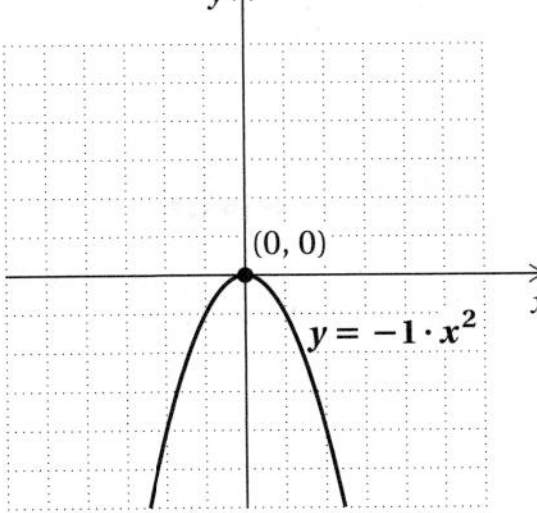

5.

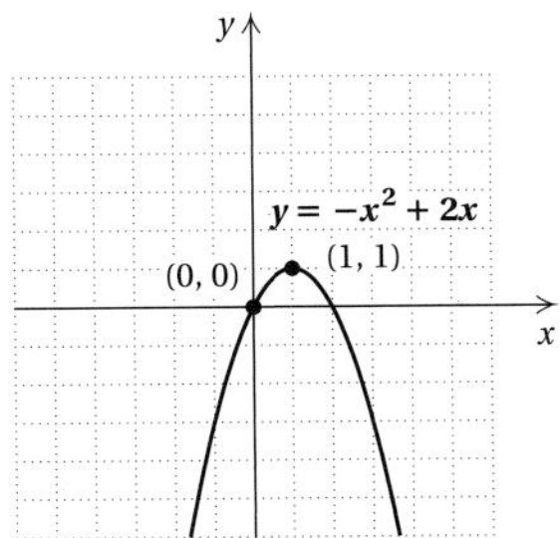

7.

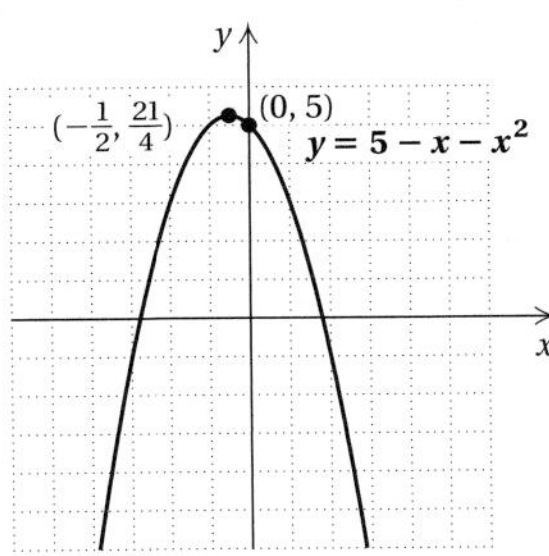

9.

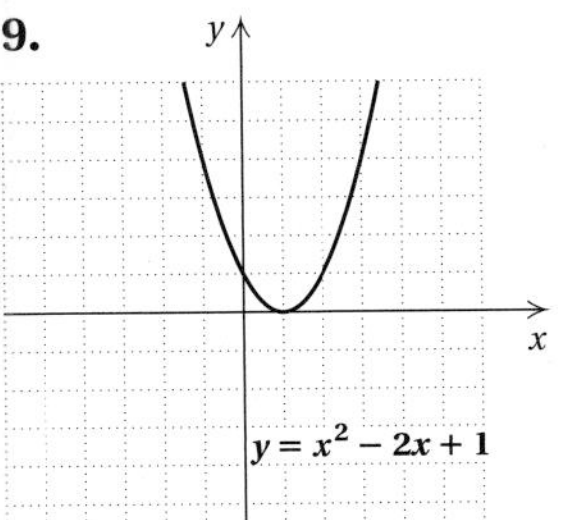

11.

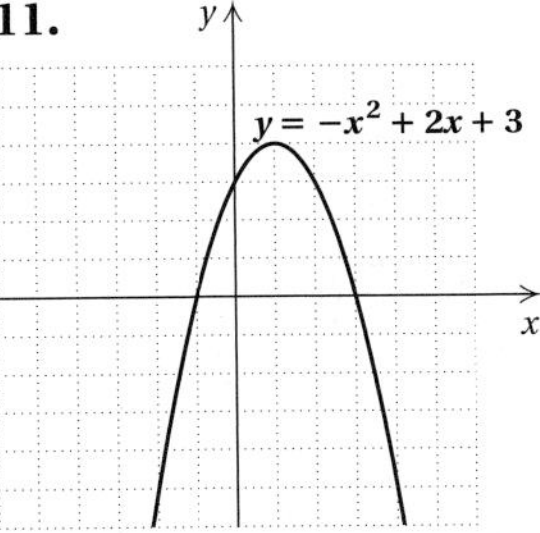

13.

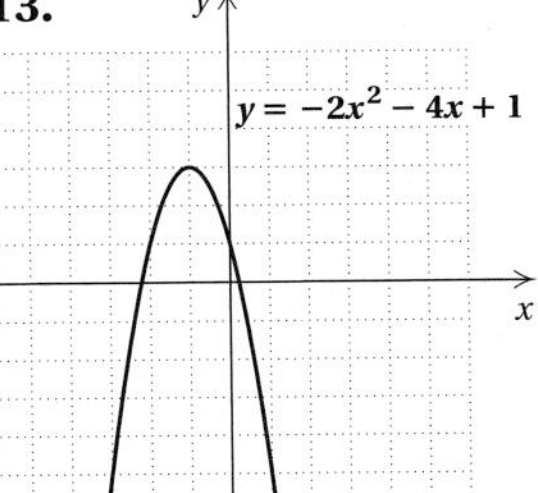

15.

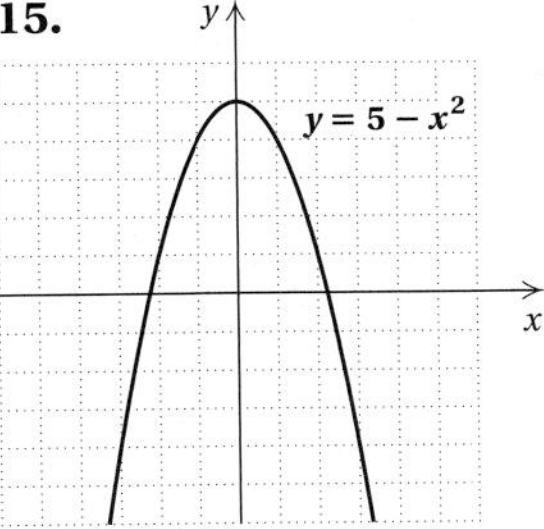

17.

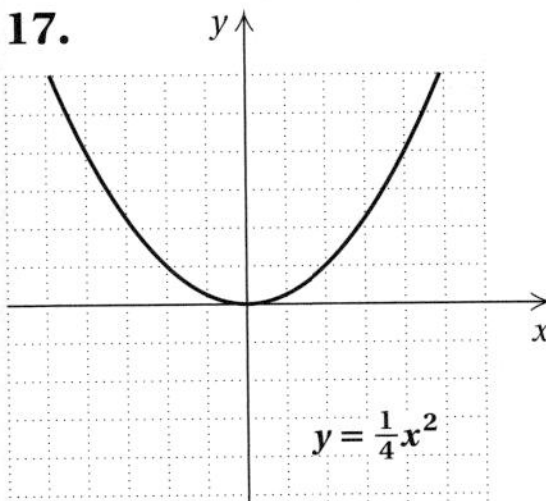

19.

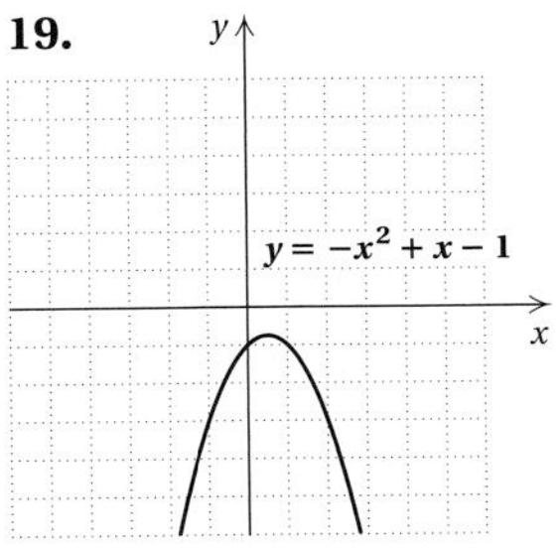

21.

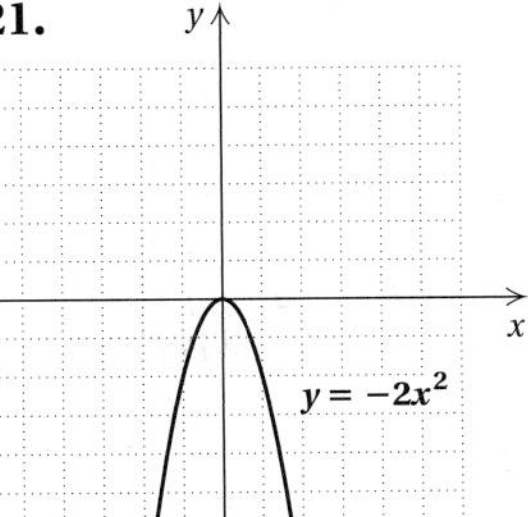

23.

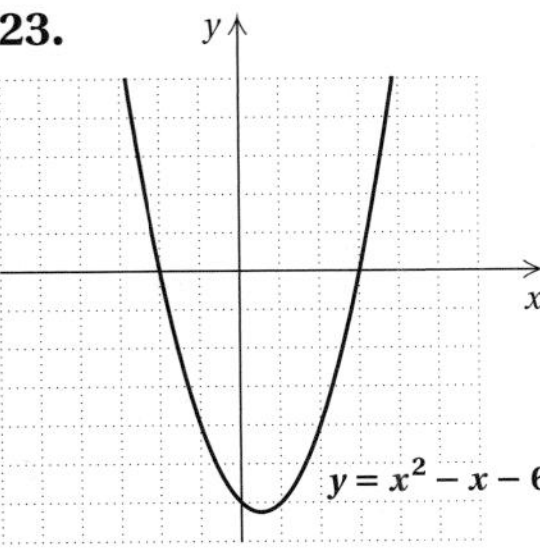

25. $(-\sqrt{2}, 0)$; $(\sqrt{2}, 0)$ **27.** $(-5, 0)$; $(0, 0)$
29. $\left(\frac{-1 - \sqrt{33}}{2}, 0\right)$; $\left(\frac{-1 + \sqrt{33}}{2}, 0\right)$ **31.** $(3, 0)$
33. $(-2 - \sqrt{5}, 0)$; $(-2 + \sqrt{5}, 0)$ **35.** None
37. $\mathbf{D_W}$ **39.** $22\sqrt{2}$ **40.** $25y^2\sqrt{y}$ **41.** $y = \frac{29.76}{x}$
42. 35 **43.** **(a)** After 2 sec; after 4 sec; **(b)** after 3 sec; **(c)** after 6 sec **45.** 16; two real solutions
47. -161.91; no real solutions

Margin Exercises, Section 15.7, pp. 1120–1125

1. Yes **2.** No **3.** Yes **4.** No **5.** Yes **6.** Yes
7. **(a)** -33; **(b)** -3; **(c)** 2; **(d)** 97; **(e)** -9
8. **(a)** 21; **(b)** 9; **(c)** 14; **(d)** 69

9.

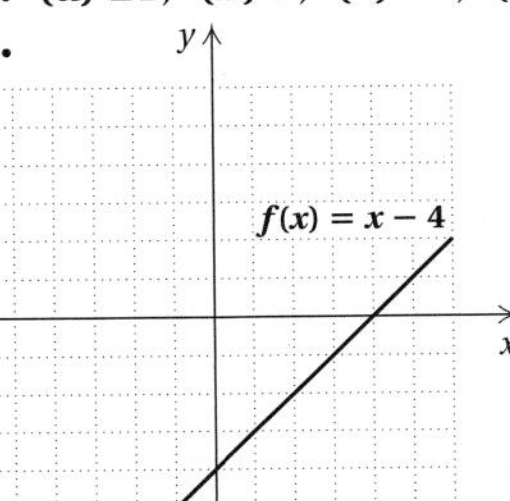

10.

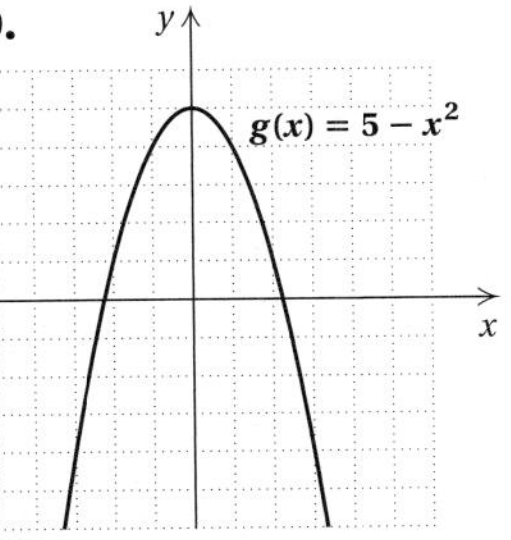

11.

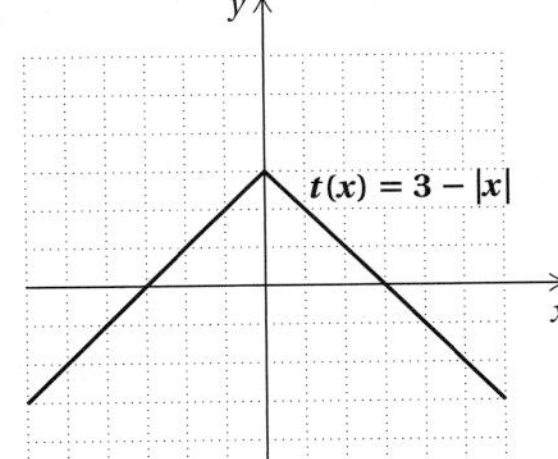

12. Yes **13.** No

14. No **15.** Yes **16.** About \$43 million
17. About \$6 million

Calculator Corner, p. 1122

1. -12.8 **2.** -9.2 **3.** -2 **4.** 20

Exercise Set 15.7, p. 1126

1. Yes **3.** Yes **5.** No **7.** Yes **9.** Yes **11.** Yes **13.** A relation but not a function **15.** (a) 9; (b) 12; (c) 2; (d) 5; (e) 7.4; (f) $5\frac{2}{3}$ **17.** (a) -21; (b) 15; (c) 42; (d) 0; (e) 2; (f) -162.6 **19.** (a) 7; (b) -17; (c) 24.1; (d) 4; (e) -26; (f) 6 **21.** (a) 0; (b) 5; (c) 2; (d) 170; (e) 65; (f) 230 **23.** (a) 1; (b) 3; (c) 3; (d) 4; (e) 11; (f) 23 **25.** (a) 0; (b) -1; (c) 8; (d) 1000; (e) -125; (f) -1000 **27.** (a) 159.48 cm; (b) 153.98 cm **29.** $1\frac{20}{33}$ atm; $1\frac{10}{11}$ atm; $4\frac{1}{33}$ atm **31.** 1.792 cm; 2.8 cm; 11.2 cm

33.

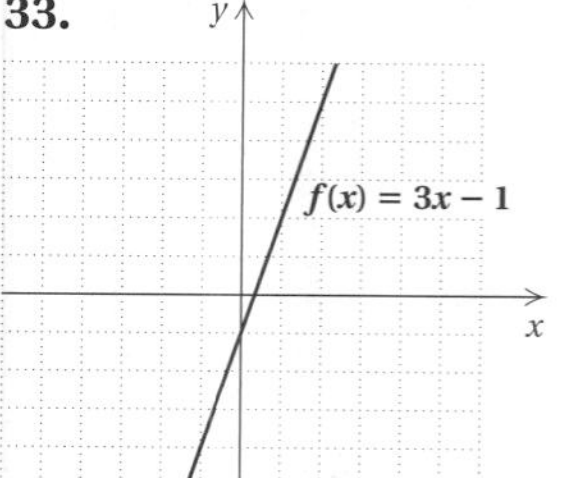

35.

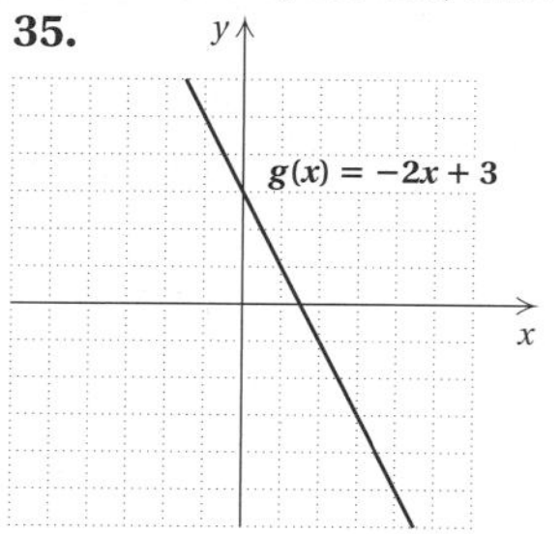

37.

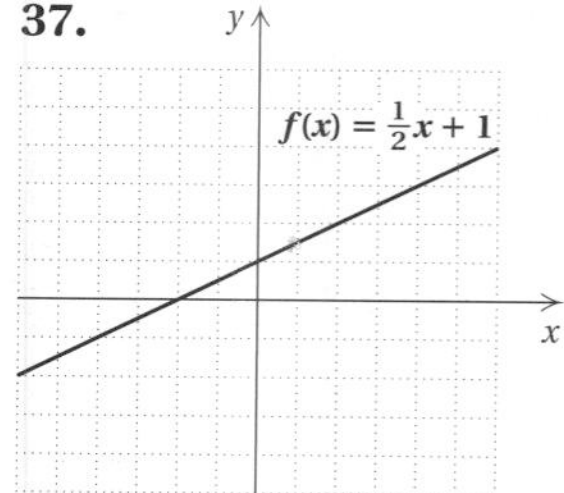

39.

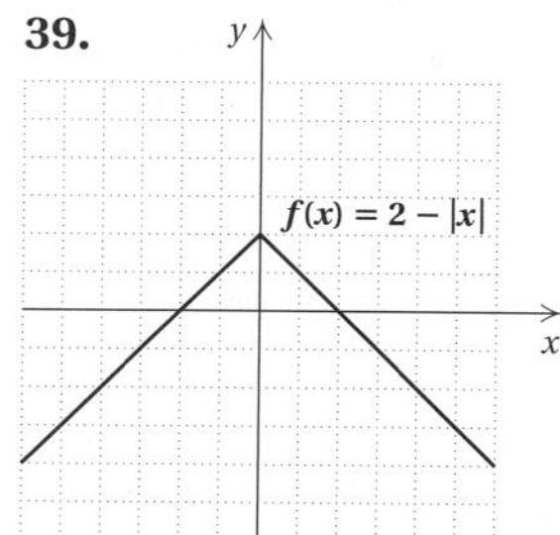

41.

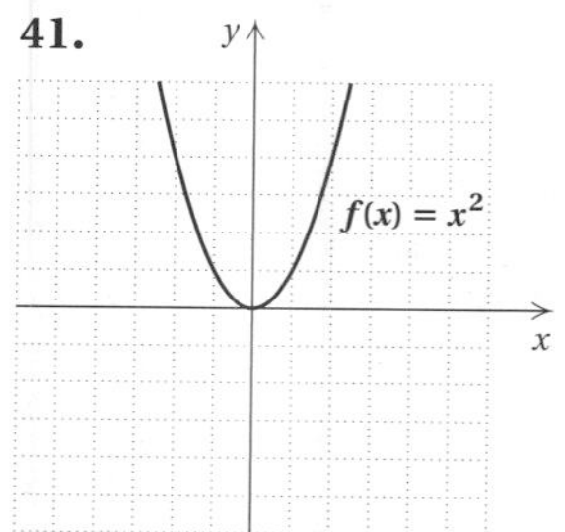

43.

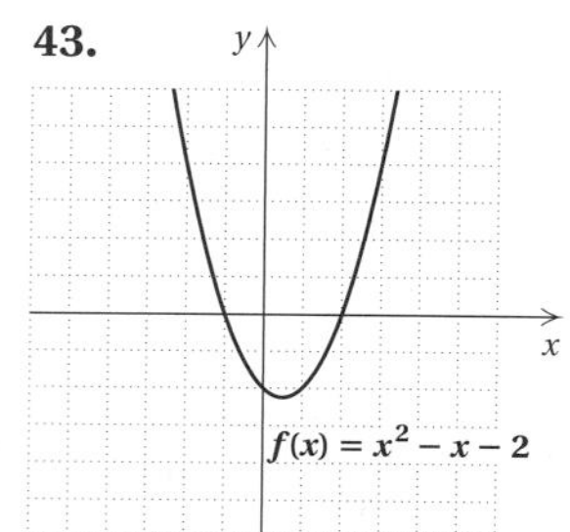

45. Yes **47.** Yes **49.** No **51.** No **53.** About 75 per 10,000 men **55.** $\mathbf{D_W}$ **57.** No **58.** Yes **59.** No solution **60.** Infinite number of solutions

61.

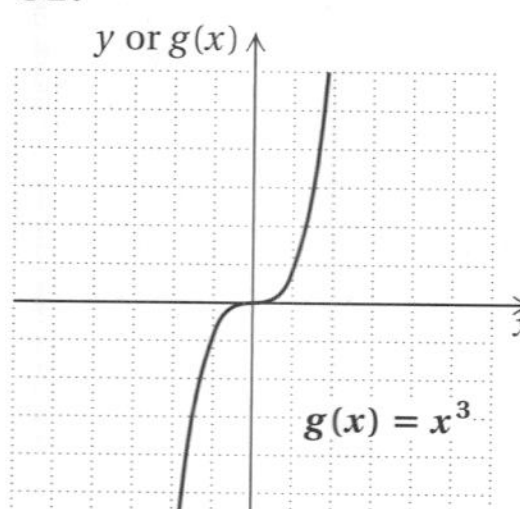

63.

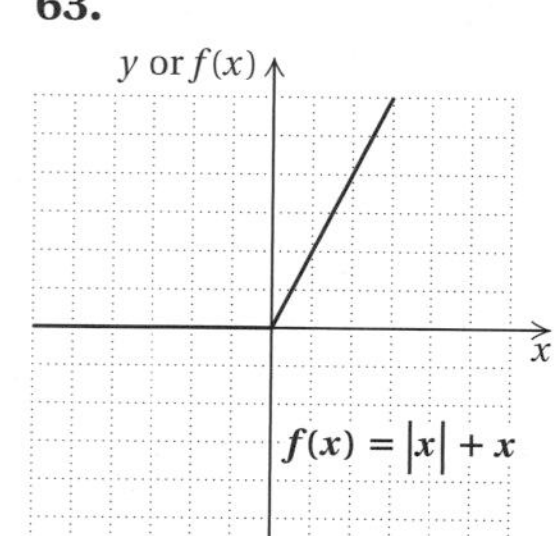

Concept Reinforcement, p. 1130

1. True **2.** False **3.** True **4.** True **5.** True **6.** False

Summary and Review: Chapter 15, p. 1130

1. $-\sqrt{3}, \sqrt{3}$ **2.** $-2\sqrt{2}, 2\sqrt{2}$ **3.** $\frac{3}{5}, 1$ **4.** $-2, \frac{1}{3}$ **5.** $-8 \pm \sqrt{13}$ **6.** 0 **7.** $0, \frac{7}{5}$ **8.** $1 \pm \sqrt{11}$ **9.** $\frac{1 \pm \sqrt{10}}{3}$ **10.** $-3 \pm 3\sqrt{2}$ **11.** $\frac{2 \pm \sqrt{3}}{2}$ **12.** $\frac{3 \pm \sqrt{33}}{2}$ **13.** No real-number solutions **14.** $0, \frac{4}{3}$ **15.** $-5, 3$ **16.** 1 **17.** $\frac{5 \pm \sqrt{17}}{2}$ **18.** $-1, \frac{5}{3}$ **19.** 0.4, 4.6 **20.** $-1.9, -0.1$ **21.** $T = L(4V^2 - 1)$

22.

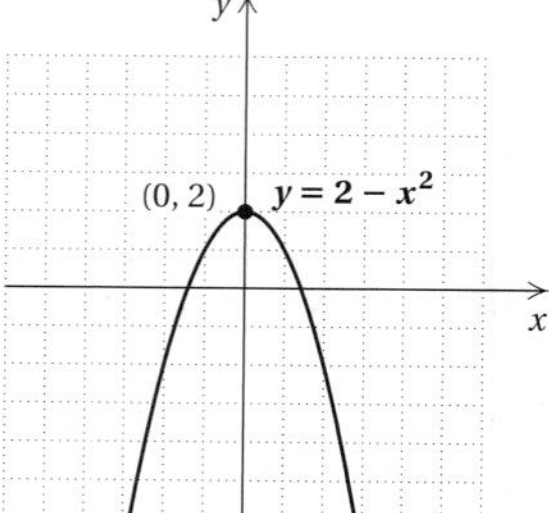

23.

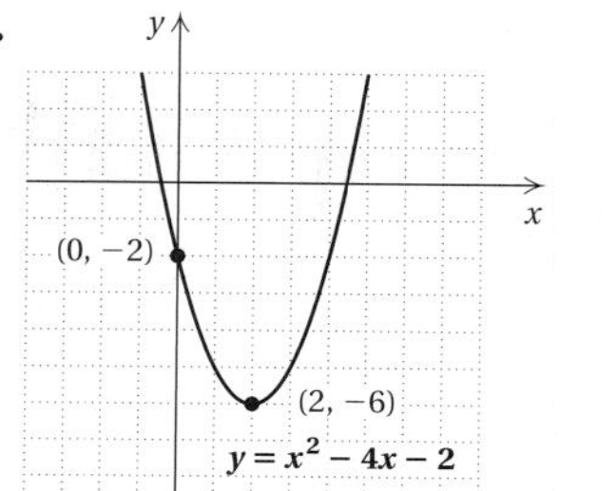

24. $(-\sqrt{2}, 0); (\sqrt{2}, 0)$ **25.** $(2 - \sqrt{6}, 0); (2 + \sqrt{6}, 0)$ **26.** 4.7 cm, 1.7 cm **27.** 15 ft **28.** About 6.3 sec **29.** $-1, -7, 2$ **30.** 0, 0, 19 **31.** 2700 calories

32.

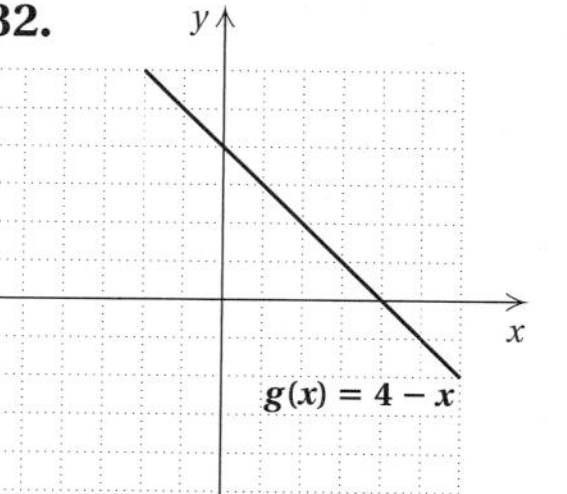

33.

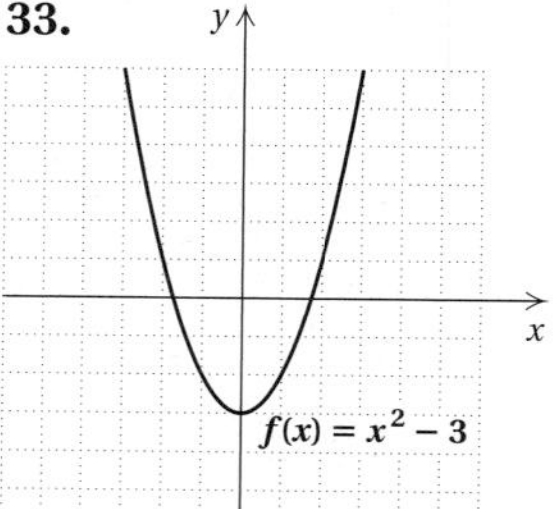

34.

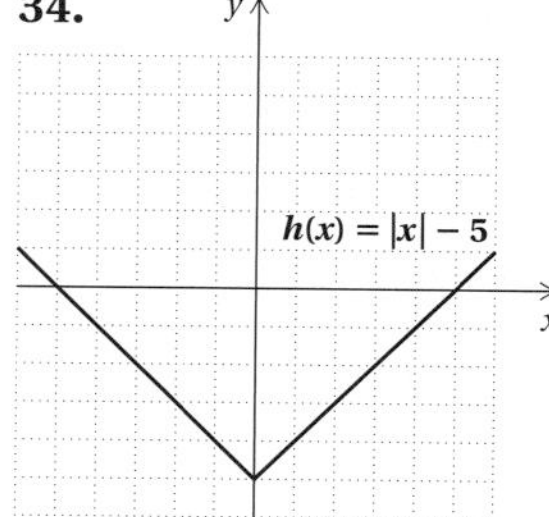

35.

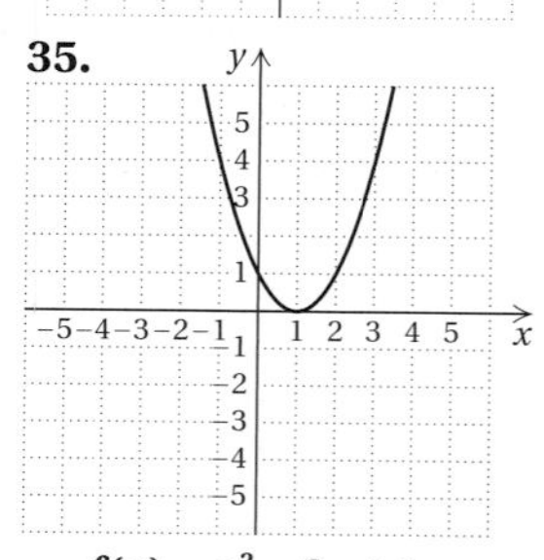

36. No **37.** Yes

38. $^{\mathrm{D}}\mathrm{W}$

Equation	Form	Example
Linear	Reducible to $x = a$	$3x - 5 = 8$
Quadratic	$ax^2 + bx + c = 0$	$2x^2 - 3x + 1 = 0$
Rational	Contains one or more rational expressions	$\frac{x}{3} + \frac{4}{x-1} = 1$
Radical	Contains one or more radical expressions	$\sqrt{3x-1} = x - 7$
Systems of equations	$Ax + By = C$, $Dx + Ey = F$	$4x - 5y = 3$, $3x + 2y = 1$

39. $^{\mathrm{D}}\mathrm{W}$ **(a)** The third line should be $x = 0$ *or* $x + 20 = 0$; the solution 0 is lost in the given procedure. Also, the last line should be $x = -20$. **(b)** The addition principle should be used at the outset to get 0 on one side of the equation. Since this was not done in the given procedure, the principle of zero products was not applied correctly.
40. 31 and 32; -32 and -31 **41.** $5\sqrt{\pi}$ in., or about 8.9 in.
42. 25 **43.** $-4, -2$ **44.** $-5, -1$ **45.** $-6, 0$
46. -3

Test: Chapter 15, p. 1133

1. [15.2a] $-\sqrt{5}, \sqrt{5}$ **2.** [15.1b] $-\frac{8}{7}, 0$
3. [15.1c] $-8, 6$ **4.** [15.1c] $-\frac{1}{3}, 2$ **5.** [15.2b] $8 \pm \sqrt{13}$
6. [15.3a] $\frac{1 \pm \sqrt{13}}{2}$ **7.** [15.3a] $\frac{3 \pm \sqrt{37}}{2}$
8. [15.3a] $-2 \pm \sqrt{14}$ **9.** [15.3a] $\frac{7 \pm \sqrt{37}}{6}$
10. [15.1c] $-1, 2$ **11.** [15.1c] $-4, 2$
12. [15.2c] $2 \pm \sqrt{14}$ **13.** [15.3b] $-1.7, 5.7$
14. [15.4a] $n = \frac{-b + \sqrt{b^2 + 4ad}}{2a}$
15. [15.6b] $\left(\frac{1 - \sqrt{21}}{2}, 0\right), \left(\frac{1 + \sqrt{21}}{2}, 0\right)$
16. [15.6a]

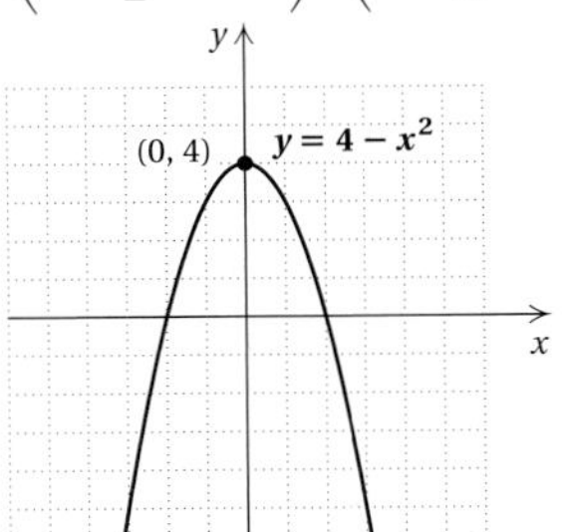

17. [15.6a]

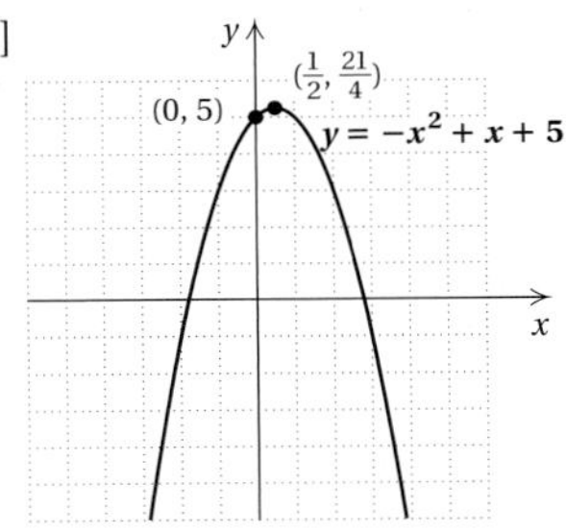

18. [15.7b] 1; $1\frac{1}{2}$; 2 **19.** [15.7b] 1; 3; -3
20. [15.5a] Length: 6.5 m; width: 2.5 m
21. [15.5a] 24 km/h **22.** [15.7e] 25.98 min
23. [15.7c]

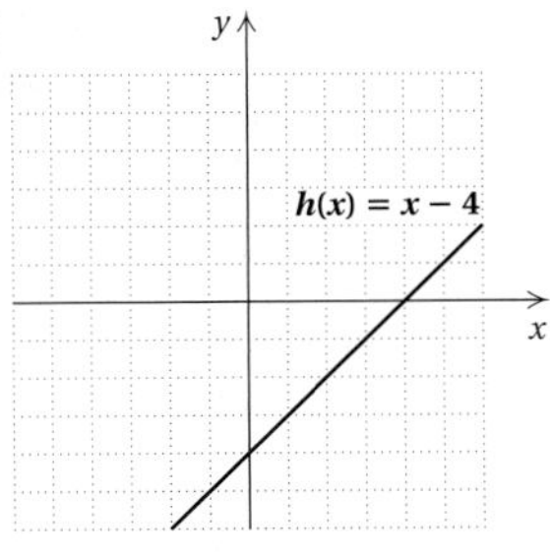

24. [15.7c]

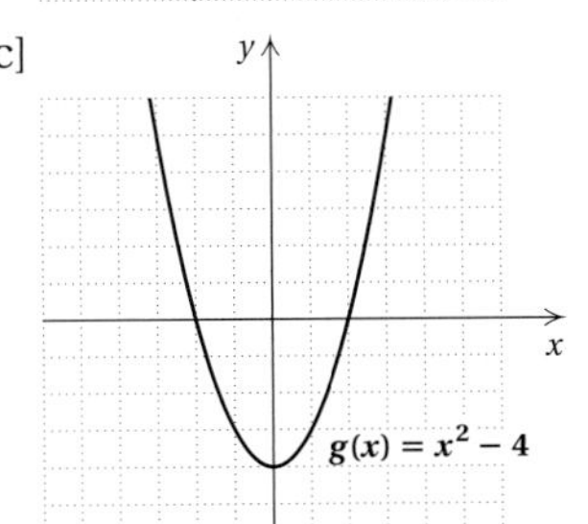

25. [15.7d] Yes

26. [15.7d] No **27.** [15.5a] $5 + 5\sqrt{2}$
28. [14.2b], [15.3a] $1 \pm \sqrt{5}$

APPENDIXES

Margin Exercises, Appendix A, pp. 1136–1146

1. 2 **2.** 3 **3.** $1\frac{1}{2}$ **4.** $2\frac{1}{2}$ **5.** 288 **6.** $8\frac{1}{2}$
7. 240,768 **8.** 6 **9.** $1\frac{5}{12}$ **10.** 8 **11.** $11\frac{2}{3}$, or $11.\overline{6}$
12. 5 **13.** 0.5 **14.** 2 cm, or 20 mm **15.** 2.3 cm, or 23 mm **16.** 4.4 cm, or 44 mm **17.** cm **18.** km
19. mm **20.** m **21.** cm **22.** m **23.** 23,000
24. 400 **25.** 178 **26.** 9040 **27.** 7.814 **28.** 781.4
29. 0.967 **30.** 8,900,000 **31.** 6.78 **32.** 97.4 **33.** 0.1
34. 8.451 **35.** 91.4 **36.** 804.5 **37.** 1479.843
38. 913.912 **39.** 343.735 m **40.** 0.125

Exercise Set A, p. 1147

1. 12 **3.** $\frac{1}{12}$ **5.** 5280 **7.** 108 **9.** 7 **11.** $1\frac{1}{2}$
13. 26,400 **15.** $5\frac{1}{4}$, or 5.25 **17.** $3\frac{1}{3}$ **19.** 37,488
21. $1\frac{1}{2}$ **23.** $1\frac{1}{4}$, or 1.25 **25.** 110 **27.** 2 **29.** 300
31. 30 **33.** $\frac{1}{36}$ **35.** 126,720 **37.** **(a)** 1000; **(b)** 0.001
39. **(a)** 10; **(b)** 0.1 **41.** **(a)** 0.01; **(b)** 100 **43.** **(a)** 6700; **(b)** Substitute **45.** **(a)** 0.98; **(b)** Multiply by 1 **47.** 8.921
49. 0.05666 **51.** 566,600 **53.** 4.77 **55.** 688 **57.** 0.1
59. 100,000 **61.** 142 **63.** 0.82 **65.** 450
67. 0.000024 **69.** 0.688 **71.** 230 **73.** 3.92 **75.** 180; 0.18 **77.** 278; 27.8 **79.** 48,440; 48.44 **81.** 4000; 400
83. 0.027; 0.00027 **85.** 442,000; 44,200 **87.** 100.65
89. 727.4394 **91.** 104.585 **93.** 289.62
95. 112.63 **97.** 9.14 **99.** 87.8896 **101.** 1250.061
103. 19.05 **105.** 1376.136 **107.** 0.51181
109.–119. Answers may vary, depending on the conversion factor used.

	yd	cm	in.	m	mm
109.	0.2604	23.8	9.37006	0.238	238
111.	0.2361	21.59	$8\frac{1}{2}$	0.2159	215.9
113.	52.95934	4844	1907.0828	48.44	48,440
115.	4	365.6	144	3.656	3656
117.	0.000295	0.027	0.0106299	0.00027	0.27
119.	483.548	44,200	17,401.54	442	442,000

121. 1 in. = 25.4 mm

Margin Exercises, Appendix B, pp. 1151–1155

1. 80 **2.** 4.32 **3.** 32,000 **4.** kg **5.** kg **6.** mg **7.** g **8.** t **9.** 6200 **10.** 3.048 **11.** 77 **12.** 234.4 **13.** 6700 **14.** 0.001 **15.** 0.5 mg

Exercise Set B, p. 1156

1. 2000 **3.** 3 **5.** 64 **7.** 12,640 **9.** 0.1 **11.** 5 **13.** 26,000,000,000 lb **15.** 1000 **17.** 10 **19.** $\frac{1}{100}$, or 0.01 **21.** 1000 **23.** 10 **25.** 234,000 **27.** 5.2 **29.** 6.7 **31.** 0.0502 **33.** 8.492 **35.** 58.5 **37.** 800,000 **39.** 1000 **41.** 0.0034 **43.** 1000 **45.** 0.325 **47.** 125 mcg **49.** 0.875 mg; 875 mcg **51.** 4 tablets **53.** 8 cc **55.** 144 packages

Margin Exercises, Appendix C, pp. 1159–1161

1. 40 **2.** 20 **3.** mL **4.** mL **5.** L **6.** L **7.** 970 **8.** 8.99 **9.** 2.4 L **10.** **(a)** About 59.14 mL; **(b)** about 0.059 L

Exercise Set C, p. 1162

1. 1000; 1000 **3.** 87,000 **5.** 0.049 **7.** 0.000401 **9.** 78,100 **11.** 320 **13.** 10 **15.** 32 **17.** 20 **19.** 14 **21.** 88

	gal	qt	pt	cups	oz
23.	1.125	4.5	9	18	144
25.	16	64	128	256	2048
27.	0.25	1	2	4	32
29.	0.1171875	0.46875	0.9375	1.875	15

	L	mL	cc	cm^3
31.	2	2000	2000	2000
33.	64	64,000	64,000	64,000
35.	0.443	443	443	443

37. 2000 mL **39.** 0.32 L **41.** 59.14 mL **43.** 500 mL **45.** 125 mL/hr **47.** 9 **49.** $\frac{1}{5}$ **51.** 6 **53.** 15 **55.** $0.21\overline{3}$ oz

Margin Exercises, Appendix D, pp. 1164–1166

1. 120 **2.** 1461 **3.** 1440 **4.** 1 **5.** 80°C **6.** 0°C **7.** −20°C **8.** 80°F **9.** 100°F **10.** 50°F **11.** 176°F **12.** 95°F **13.** 35°C **14.** 45°C

Exercise Set D, p. 1167

1. 24 **3.** 60 **5.** $365\frac{1}{4}$ **7.** 0.05 **9.** 8.2 **11.** 6.5 **13.** 10.75 **15.** 336 **17.** 4.5 **19.** 56 **21.** 86,164.2 sec **23.** 77°F **25.** 104°F **27.** 186.8°F **29.** 136.4°F **31.** 35.6°F **33.** 41°F **35.** 5432°F **37.** 30°C **39.** 55°C **41.** $81.\overline{1}$°C **43.** 60°C **45.** 20°C **47.** $6.\overline{6}$°C **49.** 37°C **51.** **(a)** 136°F = $57.\overline{7}$°C, $56\frac{2}{3}$°C = 134°F; **(b)** 2°F **53.** About 0.03 yr

Margin Exercises, Appendix E, pp. 1169–1170

1. $\{0, 1, 2, 3, 4, 5, 6, 7\}$ **2.** $\{-5, 5\}$ **3.** True **4.** True **5.** True **6.** False **7.** True **8.** $\{-2, -3, 4, -4\}$ **9.** $\{a, i\}$ **10.** $\{\ \}$, or $\varnothing$ **11.** $\{-2, -3, 4, -4, 8, -5, 7, 5\}$ **12.** $\{a, e, i, o, u, m, r, v, n\}$ **13.** $\{a, b, c, d, e, f, g, h, i, j, k, l, m, n, o, p, q, r, s, t, u, v, w, x, y, z\}$

Exercise Set E, p. 1171

1. $\{3, 4, 5, 6, 7, 8\}$ **3.** $\{41, 43, 45, 47, 49\}$ **5.** $\{-3, 3\}$ **7.** False **9.** True **11.** True **13.** True **15.** True **17.** False **19.** $\{c, d, e\}$ **21.** $\{1, 10\}$ **23.** $\{\ \}$, or $\varnothing$ **25.** $\{a, e, i, o, u, q, c, k\}$ **27.** $\{0, 1, 7, 10, 2, 5\}$ **29.** $\{a, e, i, o, u, m, n, f, g, h\}$ **31.** $\{x | x \text{ is an integer}\}$ **33.** $\{x | x \text{ is a real number}\}$ **35.** $\{\ \}$, or $\varnothing$ **37.** **(a)** A; **(b)** A; **(c)** A; **(d)** $\{\ \}$, or $\varnothing$ **39.** True

Margin Exercises, Appendix F, pp. 1173–1174

1. $(x - 3)(x^2 + 3x + 9)$ **2.** $(4 - y)(16 + 4y + y^2)$ **3.** $(y + 2)(y^2 - 2y + 4)$ **4.** $(5 + t)(25 - 5t + t^2)$ **5.** $(3x - y)(9x^2 + 3xy + y^2)$ **6.** $(2y + z)(4y^2 - 2yz + z^2)$ **7.** $(m + n)(m^2 - mn + n^2)(m - n)(m^2 + mn + n^2)$ **8.** $2xy(2x^2 + 3y^2)(4x^4 - 6x^2y^2 + 9y^4)$ **9.** $(3x + 2y)(9x^2 - 6xy + 4y^2)(3x - 2y)(9x^2 + 6xy + 4y^2)$ **10.** $(x - 0.3)(x^2 + 0.3x + 0.09)$

Exercise Set F, p. 1175

1. $(z + 3)(z^2 - 3z + 9)$ **3.** $(x - 1)(x^2 + x + 1)$ **5.** $(y + 5)(y^2 - 5y + 25)$ **7.** $(2a + 1)(4a^2 - 2a + 1)$ **9.** $(y - 2)(y^2 + 2y + 4)$ **11.** $(2 - 3b)(4 + 6b + 9b^2)$ **13.** $(4y + 1)(16y^2 - 4y + 1)$ **15.** $(2x + 3)(4x^2 - 6x + 9)$ **17.** $(a - b)(a^2 + ab + b^2)$ **19.** $\left(a + \frac{1}{2}\right)\left(a^2 - \frac{1}{2}a + \frac{1}{4}\right)$ **21.** $2(y - 4)(y^2 + 4y + 16)$ **23.** $3(2a + 1)(4a^2 - 2a + 1)$ **25.** $r(s + 4)(s^2 - 4s + 16)$ **27.** $5(x - 2z)(x^2 + 2xz + 4z^2)$ **29.** $(x + 0.1)(x^2 - 0.1x + 0.01)$ **31.** $8(2x^2 - t^2)(4x^4 + 2x^2t^2 + t^4)$ **33.** $2y(y - 4)(y^2 + 4y + 16)$ **35.** $(z - 1)(z^2 + z + 1)(z + 1)(z^2 - z + 1)$ **37.** $(t^2 + 4y^2)(t^4 - 4t^2y^2 + 16y^4)$ **39.** 1; 19; 19; 7; 1 **41.** $(x^{2a} + y^b)(x^{4a} - x^{2a}y^b + y^{2b})$ **43.** $3(x^a + 2y^b)(x^{2a} - 2x^ay^b + 4y^{2b})$ **45.** $\frac{1}{3}\left(\frac{1}{2}xy + z\right)\left(\frac{1}{4}x^2y^2 - \frac{1}{2}xyz + z^2\right)$ **47.** $y(3x^2 + 3xy + y^2)$ **49.** $4(3a^2 + 4)$

Margin Exercises, Appendix G, pp. 1178–1179

1. $y = -3x - 11$ **2.** $y = 5x + 11$ **3.** $y = 6x - 23$ **4.** $y = \frac{2}{3}x + \frac{8}{3}$ **5.** $y = -\frac{9}{4}x + \frac{7}{4}$ **6.** $y = -9x - 16$

Exercise Set G, p. 1180

1. $y = 4x - 18$ **3.** $y = -2x + 12$ **5.** $y = 3x + 4$ **7.** $y = -3x - 6$ **9.** $y = 4$ **11.** $y = -\frac{4}{5}x + \frac{23}{5}$ **13.** $y = x + 3$ **15.** $y = x$ **17.** $y = \frac{5}{3}x - 5$ **19.** $y = 3x + 5$ **21.** $y = -\frac{7}{4}x$ **23.** $y = \frac{2}{11}x + \frac{91}{66}$ **25.** $y = 5x + 3$

Margin Exercises, Appendix H, pp. 1181–1182

1. $\{6, -6\}$;

−6 0 6

2. $\varnothing$ **3.** $\{0\}$ **4.** $\{2, -2\}$ **5.** $\left\{\frac{17}{4}, -\frac{17}{4}\right\}$ **6.** $\{4, -4\}$ **7.** $\{3, 5\}$ **8.** $\left\{-\frac{13}{3}, 7\right\}$ **9.** $\varnothing$

Exercise Set H, p. 1183

1. $\{3, -3\}$ **3.** $\varnothing$ **5.** $\{0\}$ **7.** $\{15, -9\}$ **9.** $\left\{\frac{7}{2}, -\frac{1}{2}\right\}$ **11.** $\left\{\frac{23}{4}, -\frac{5}{4}\right\}$ **13.** $\{11, -11\}$ **15.** $\{291, -291\}$ **17.** $\{8, -8\}$ **19.** $\{7, -7\}$ **21.** $\{2, -2\}$ **23.** $\{8, -7\}$ **25.** $\{2, -12\}$ **27.** $\left\{\frac{7}{2}, -\frac{5}{2}\right\}$ **29.** $\varnothing$ **31.** $\left\{-\frac{13}{54}, -\frac{7}{54}\right\}$ **33.** $\{x \mid x \geq -5\}$ **35.** $\left\{1, -\frac{1}{4}\right\}$

Margin Exercises, Appendix I, pp. 1184–1185

1. 14 **2.** 11 **3.** 10 **4.** $\sqrt{37} \approx 6.083$ **5.** $\left(\frac{3}{2}, -3\right)$ **6.** $(9, -5)$

Exercise Set I, p. 1186

1. 5 **3.** $\frac{9}{4}$ **5.** $\sqrt{29} \approx 5.385$ **7.** $\sqrt{648} \approx 25.456$ **9.** 7.1 **11.** $\frac{\sqrt{41}}{7} \approx 0.915$ **13.** $\sqrt{6970} \approx 83.487$ **15.** $\sqrt{a^2 + b^2}$ **17.** $\sqrt{17 + 2\sqrt{14} + 2\sqrt{15}} \approx 5.677$ **19.** $\sqrt{9{,}672{,}400} \approx 3110.048$ **21.** $\frac{2}{5}$ **23.** $\left(\frac{3}{2}, \frac{7}{2}\right)$ **25.** $\left(0, \frac{11}{2}\right)$ **27.** $\left(-1, -\frac{17}{2}\right)$ **29.** $(-0.25, -0.3)$ **31.** $\left(-\frac{1}{12}, \frac{1}{24}\right)$ **33.** $\left(\frac{\sqrt{2} + \sqrt{3}}{2}, \frac{3}{2}\right)$ **35.** $\sqrt{49 + k^2}$ **37.** $8\sqrt{m^2 + n^2}$ **39.** Yes **41.** $(2, 4\sqrt{2})$

Margin Exercises, Appendix J, pp. 1188–1194

1. 3, −3 **2.** 6, −6 **3.** 11, −11 **4.** 1 **5.** 6 **6.** $\frac{9}{10}$ **7.** 0.08 **8.** **(a)** 4; **(b)** −4; **(c)** does not exist as a real number **9.** **(a)** 7; **(b)** −7; **(c)** does not exist as a real number **10.** **(a)** 12; **(b)** −12; **(c)** does not exist as a real number **11.** $\frac{5}{8}$ **12.** −0.9 **13.** 1.2 **14.** $28 + x$ **15.** $\frac{y}{y + 3}$ **16.** $|y|$ **17.** 24 **18.** $5|y|$ **19.** $4|y|$ **20.** $|x + 7|$ **21.** $2|x - 2|$ **22.** $7|y + 5|$ **23.** $|x - 3|$ **24.** −4 **25.** $3y$ **26.** $2(x + 2)$ **27.** $-\frac{7}{4}$ **28.** 3 **29.** −3 **30.** x **31.** y **32.** 0 **33.** $-2x$ **34.** $3x + 2$ **35.** 3 **36.** −3 **37.** Does not exist as a real number **38.** 0 **39.** $2|x - 2|$ **40.** $|x|$ **41.** $|x + 3|$ **42.** $x + 3$ **43.** $3x$ **44.** $\sqrt[4]{y}$ **45.** $\sqrt{3a}$ **46.** 2 **47.** 5 **48.** $\sqrt[5]{a^3b^2c}$ **49.** $(19ab)^{1/3}$ **50.** $19(ab)^{1/3}$ **51.** $\left(\frac{x^2y}{16}\right)^{1/5}$ **52.** $7(2ab)^{1/4}$ **53.** $\sqrt[5]{x^3}$ **54.** 4 **55.** 32 **56.** $(7abc)^{4/3}$ **57.** $6^{7/5}$ **58.** $\frac{1}{2}$ **59.** $\frac{1}{(3xy)^{7/8}}$ **60.** $\frac{1}{27}$ **61.** $\frac{7p^{3/4}}{q^{6/5}}$ **62.** $\left(\frac{7n}{11m}\right)^{2/3}$ **63.** $7^{14/15}$ **64.** $5^{1/3}$ **65.** $9^{2/5}$ **66.** $\frac{q^{1/8}}{p^{1/3}}$

Exercise Set J, p. 1195

1. 4, −4 **3.** 12, −12 **5.** 20, −20 **7.** $-\frac{7}{6}$ **9.** 14 **11.** 0.06 **13.** 18.628 **15.** 1.962 **17.** $y^2 + 16$ **19.** $\frac{x}{y - 1}$ **21.** $4|x|$ **23.** $12|c|$ **25.** $|p + 3|$ **27.** $|x - 2|$ **29.** 3 **31.** $-4x$ **33.** −6 **35.** $0.7(x + 1)$ **37.** −5 **39.** −1 **41.** $-\frac{2}{3}$ **43.** $|x|$ **45.** 6 **47.** y **49.** $x - 2$ **51.** $\sqrt[7]{y}$ **53.** 2 **55.** $\sqrt[5]{a^3b^3}$ **57.** 8 **59.** 343 **61.** $17^{1/2}$ **63.** $18^{1/3}$ **65.** $(xy^2z)^{1/5}$ **67.** $(3mn)^{3/2}$ **69.** $(8x^2y)^{5/7}$ **71.** $\frac{1}{x^{1/4}}$ **73.** $\left(\frac{8yz}{7x}\right)^{3/5}$ **75.** $5^{7/8}$ **77.** $7^{1/4}$ **79.** $4.9^{1/2}$ **81.** $a^{8/3}b^{5/2}$ **83.** $a^{23/12}$

Margin Exercises, Appendix K, pp. 1197–1200

1. $\{x \mid x < -3 \text{ or } x > 1\}$ **2.** $\{x \mid -3 < x < 1\}$ **3.** $\{x \mid -3 \leq x \leq 1\}$ **4.** $\{x \mid x < -4 \text{ or } x > 1\}$ **5.** $\{x \mid -4 \leq x \leq 1\}$ **6.** $\{x \mid x < -1 \text{ or } 0 < x < 1\}$

7.

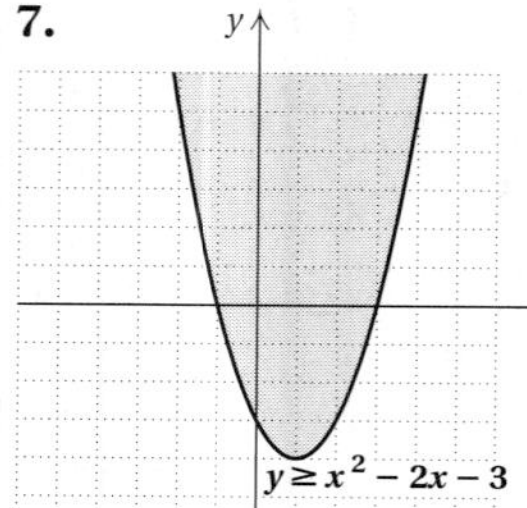

8.

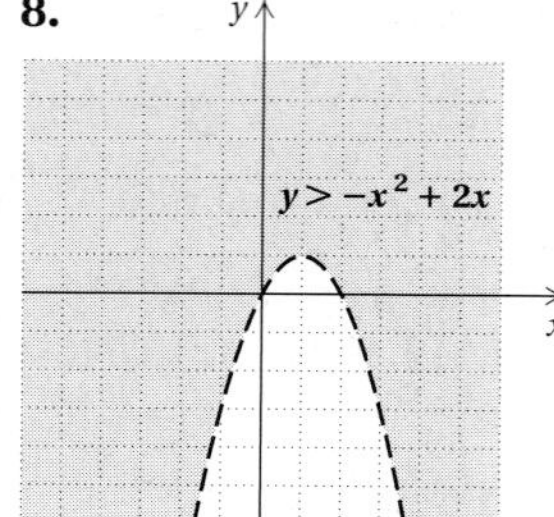

Exercise Set K, p. 1201

1. $\{x | x < -2 \text{ or } x > 6\}$ 3. $\{x | -2 \leq x \leq 2\}$
5. $\{x | -1 \leq x \leq 4\}$ 7. $\{x | -1 < x < 2\}$
9. All real numbers 11. $\{x | 2 < x < 4\}$
13. $\{x | x < -2 \text{ or } 0 < x < 2\}$
15. $\{x | -9 < x < -1 \text{ or } x > 4\}$
17. $\{x | x < -3 \text{ or } -2 < x < 1\}$

19.

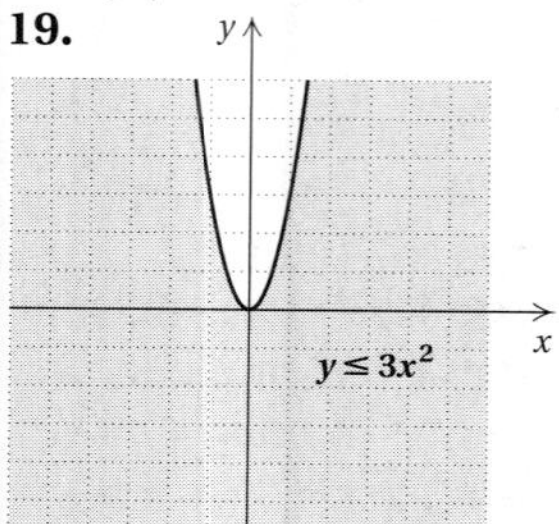

21.

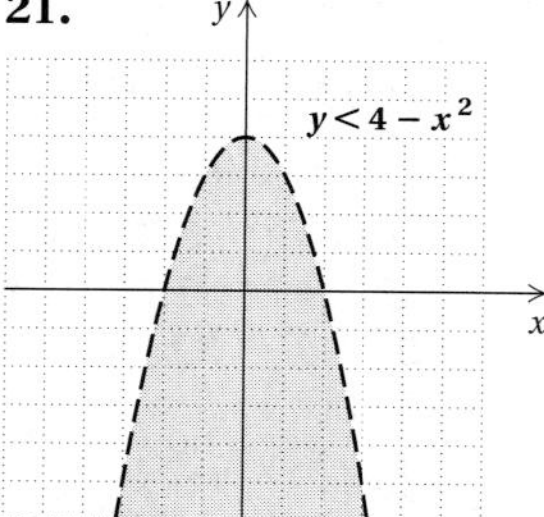

23.

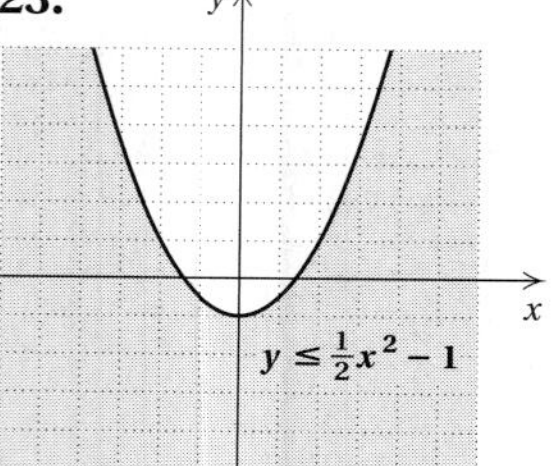

25.

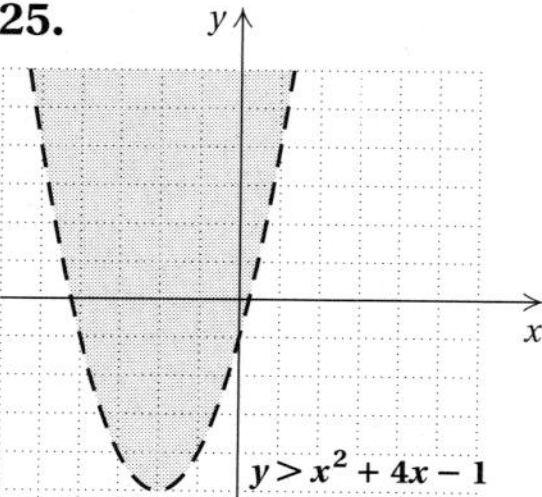

27.

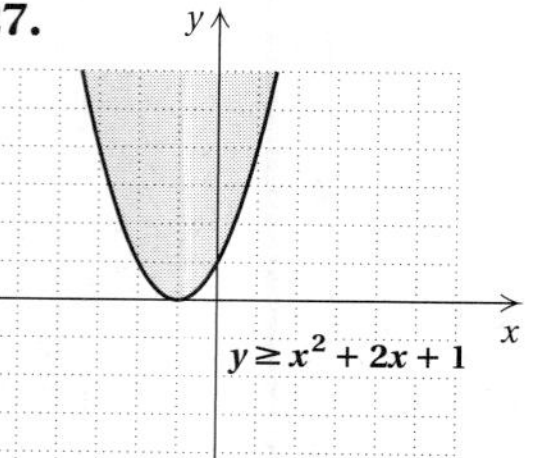

29.

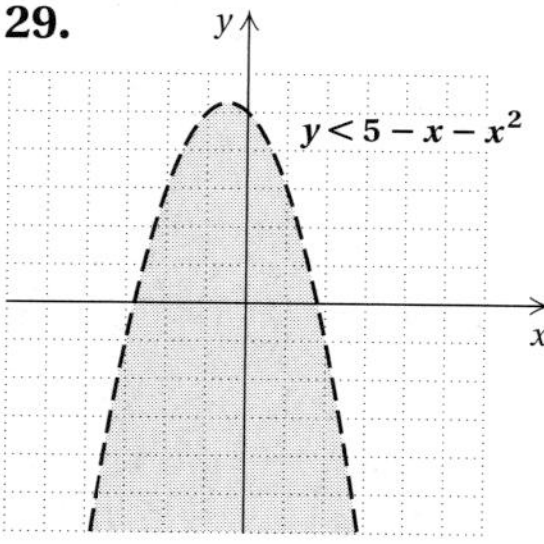

31.

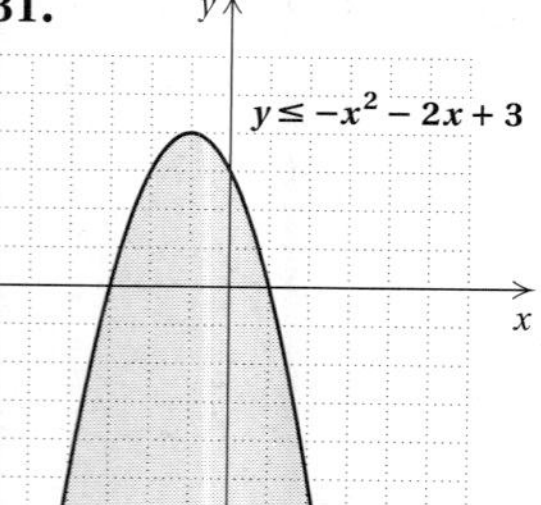

33.

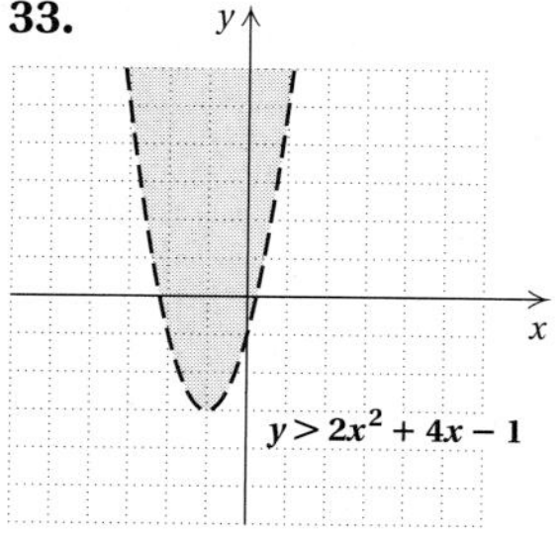

Margin Exercises, Appendix L, pp. 1204–1206

1.

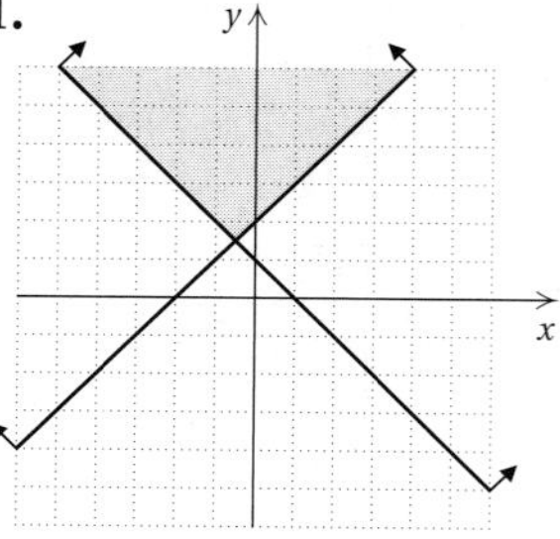

2.

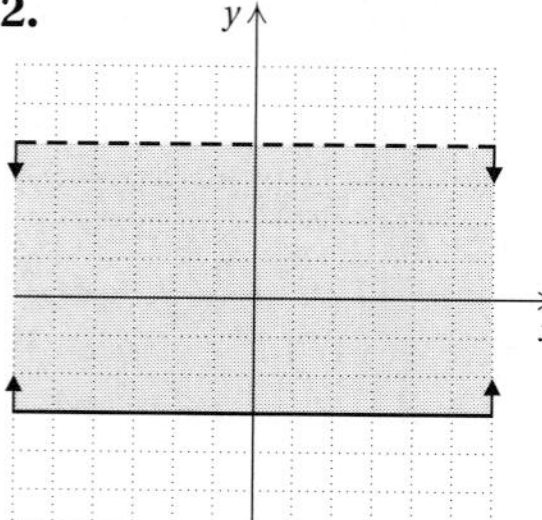

3.

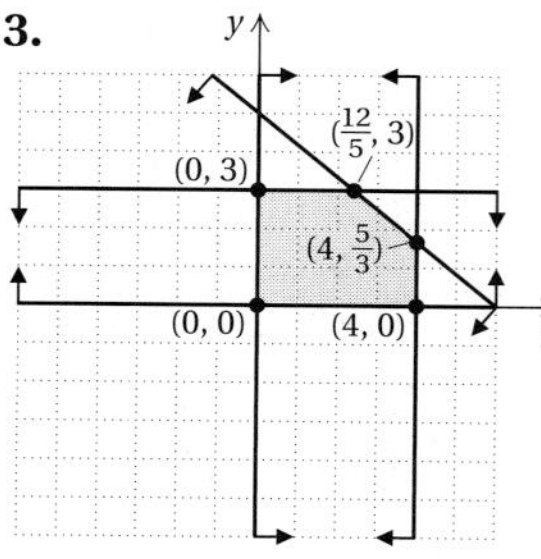

4.

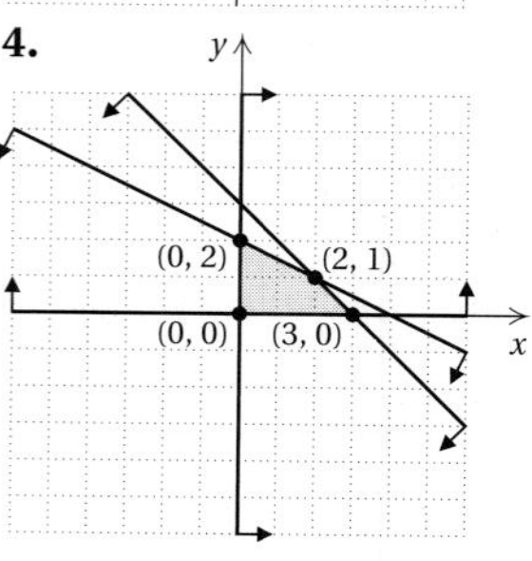

Exercise Set L, p. 1207

1.

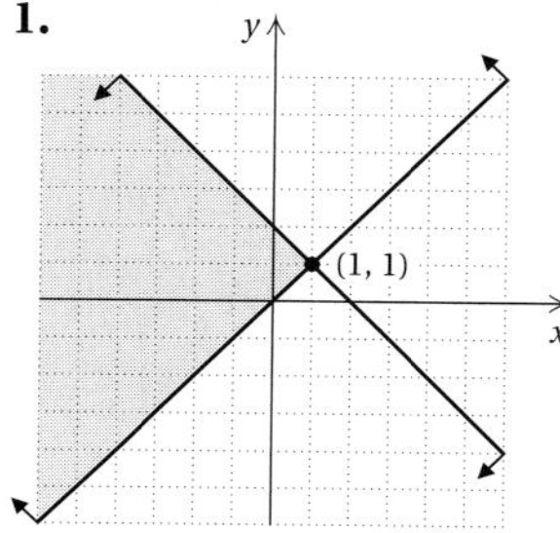

3.

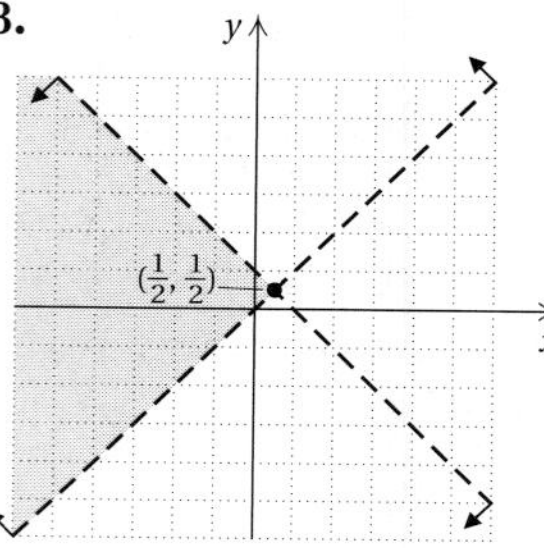

5.

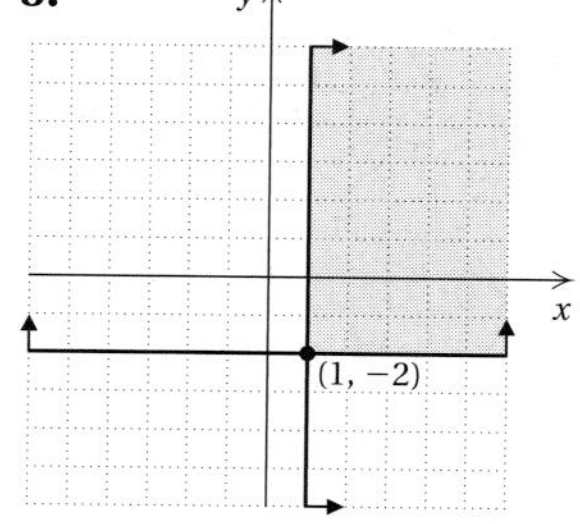

7.

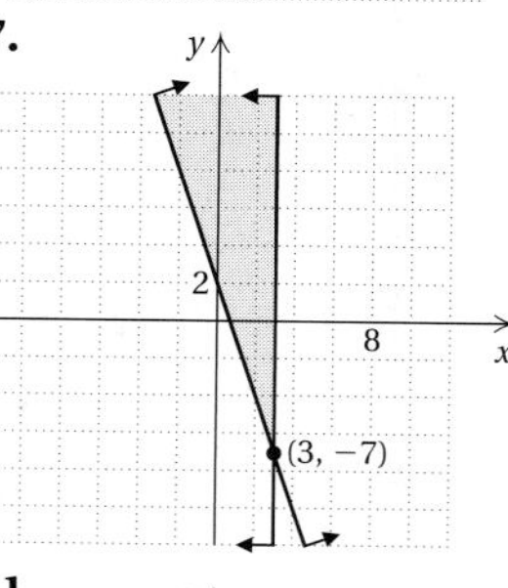

9.

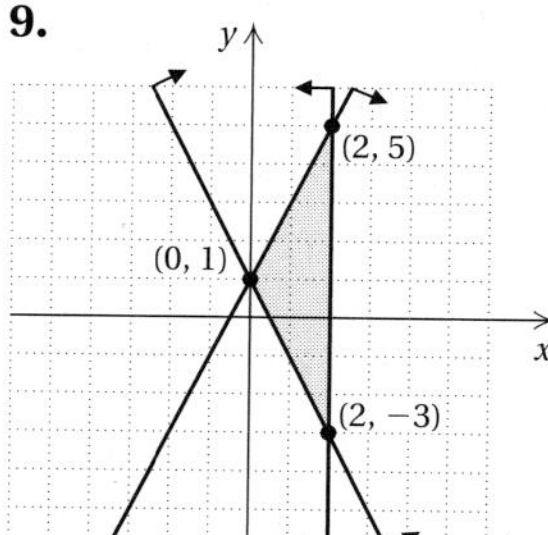

11.

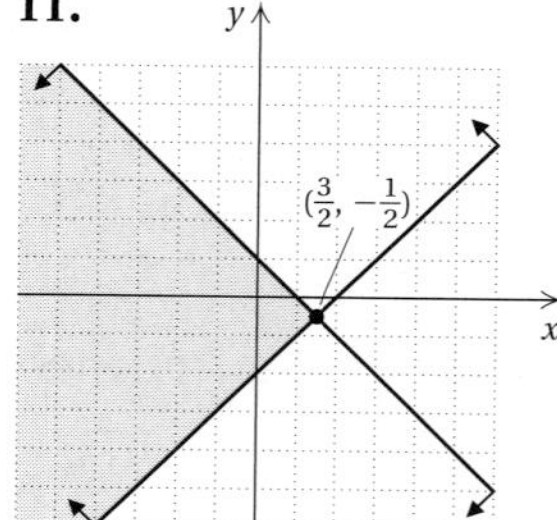

13.

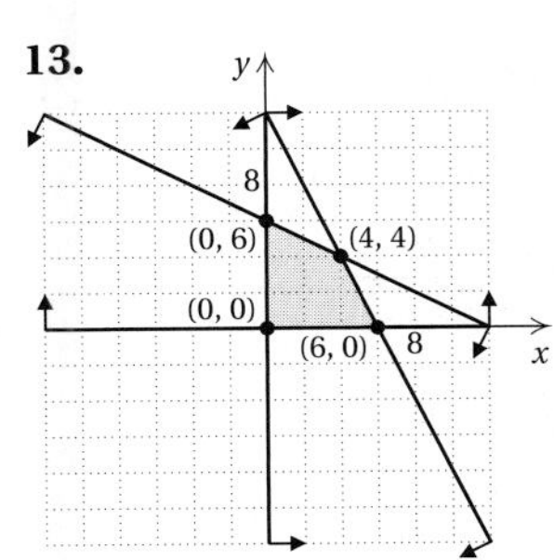

15.

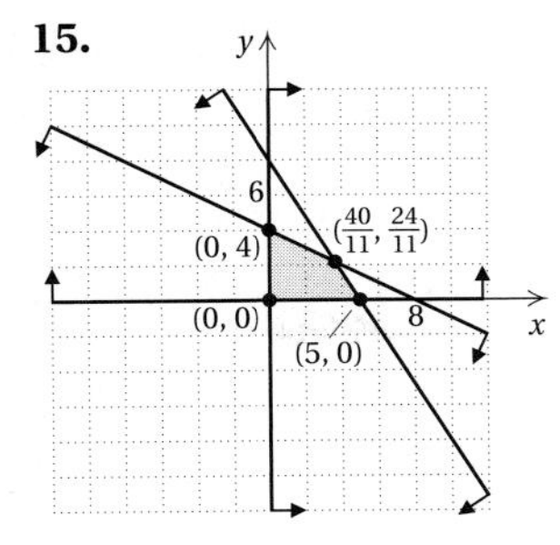

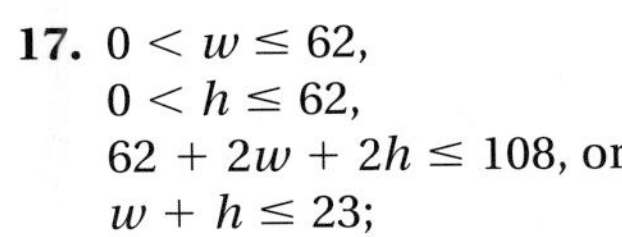

17. $0 < w \leq 62$,
$0 < h \leq 62$,
$62 + 2w + 2h \leq 108$, or
$w + h \leq 23$;

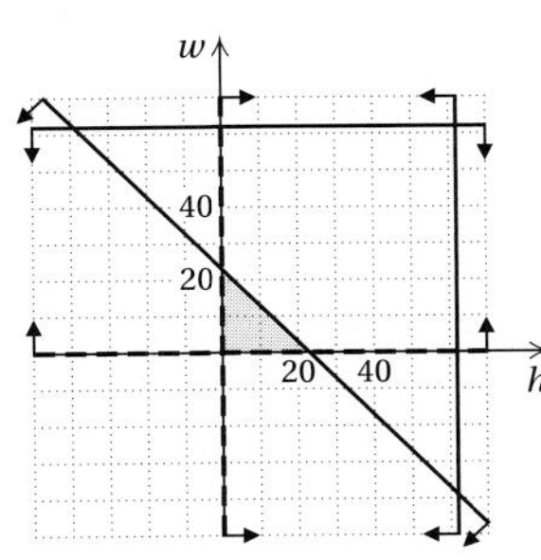

Margin Exercises, Appendix M, pp. 1210–1212

1. $\frac{82}{100}$ **2.** 14.7%, 50.4% **3.** $\frac{1}{2}$ **4. (a)** $\frac{1}{13}$; **(b)** $\frac{1}{4}$; **(c)** $\frac{1}{2}$; **(d)** $\frac{2}{13}$ **5.** $\frac{6}{11}$ **6.** 0 **7.** 1

Exercise Set M, p. 1213

1. 0.57, 0.43 **3.** 0.075, 0.134, 0.057, 0.071, 0.030 **5.** 0.633 **7.** 52 **9.** $\frac{1}{4}$ **11.** $\frac{1}{2}$ **13.** $\frac{2}{13}$ **15.** $\frac{2}{7}$ **17.** 0 **19.** $\frac{5}{36}$ **21.** $\frac{5}{36}$ **23.** $\frac{1}{36}$

Margin Exercises, Appendix N, pp. 1215–1218

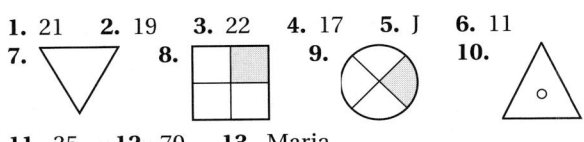

1. 21 **2.** 19 **3.** 22 **4.** 17 **5.** J **6.** 11 **7.** **8.** **9.** **10.** **11.** 35 **12.** 70 **13.** Maria

Exercise Set N, p. 1219

1. 19 **3.** V **5.** 43 **7.** **9.** **11.** 17

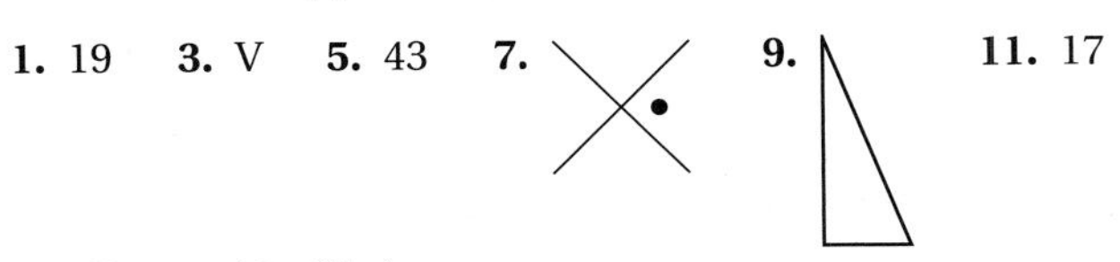

13. Bev **15.** Chris

Margin Exercises, Appendix O, p. 1222

1. \$110.30 **2.** \$809.92

Exercise Set O, p. 1223

1. \$12 **3.** \$10.91 **5.** \$611.50 **7.** \$727.50 **9.** 28.4 mpg **11.** 137,500 lb **13.** \$31.80 **15.** \$83.93 **17.** \$591.75 **19.** 4 ft

Exercise Set P, p. 1225

1. B **2.** B **3.** D **4.** A **5.** C **6.** B **7.** C **8.** D **9.** C **10.** C **11.** A **12.** A **13.** D **14.** D **15.** B **16.** A **17.** B **18.** C **19.** A **20.** B **21.** C **22.** D **23.** B **24.** A **25.** C **26.** A **27.** D **28.** A **29.** B **30.** D **31.** C **32.** B **33.** A **34.** D **35.** A **36.** C **37.** D **38.** C **39.** C **40.** C **41.** B **42.** C **43.** A **44.** B **45.** A **46.** B

Glossary

A

Abscissa The first coordinate in an ordered pair of numbers
Absolute value The distance that a number is from 0 on the number line
***ac*-method** A method for factoring trinomials of the type $ax^2 + bx + c, a \neq 1$, involving the product, ac, of the leading coefficient a and the last term c
Acute angle An angle whose measure is greater than 0° and less than 90°
Acute triangle A triangle in which all angles are acute
Addends In addition, the numbers being added
Additive identity The number 0
Additive inverse A number's opposite; two numbers are additive inverses of each other if their sum is zero.
Additive inverse of a polynomial Two polynomials are additive inverses, or opposites, of each other if their sum is zero.
Algebraic expression A number or variable or a collection of numbers and variables on which operations are performed
Angle A set of points consisting of two rays (half-lines) with a common endpoint (vertex)
Area The number of square units that fill a plane region
Ascending order When a polynomial is written with the exponents of the variable increasing as read from left to right, it is said to be in ascending order.
Associative law of addition The statement that when three numbers are added, regrouping the addends gives the same sum
Associative law of multiplication The statement that when three numbers are multiplied, regrouping the factors gives the same product
Average A center point of a set of numbers found by adding the numbers and dividing by the number of items of data; also called the *mean*
Axes Two perpendicular number lines used to identify points in a plane

B

Bar graph A graphic display of data using bars proportional in length to the numbers represented
Base In exponential notation, the number being raised to a power
Binomial A polynomial containing two terms

C

Celsius A temperature scale in which water freezes at 0° and water boils at 100°
Circumference The distance around a circle
Coefficient The numeric multiplier of a variable
Commission A percent of total sales paid to a salesperson
Commutative law of addition The statement that when two numbers are added, changing the order in which the numbers are added does not affect the sum
Commutative law of multiplication The statement that when two numbers are multiplied, changing the order in which the numbers are multiplied does not affect the product
Complementary angles Two angles for which the sum of their measures is 90°
Completing the square Adding a particular constant to an expression so that the resulting sum is a perfect square
Complex fraction expression A rational expression that has one or more rational expressions within its numerator and/or denominator
Complex-number system A number system that contains the real-number system and is designed so that negative numbers have square roots
Complex rational expression A rational expression that has one or more rational expressions within its numerator and/or denominator
Composite number A natural number, other than 1, that is not prime
Compound interest Interest computed on the sum of an original principal and the interest previously accrued by that principal
Congruent angles Two angles that have the same measure
Congruent segments Two line segments that have the same length
Congruent triangles Triangles in which corresponding angles and sides are congruent
Conjugates Pairs of radical terms, like $\sqrt{a} + \sqrt{b}$ and $\sqrt{a} - \sqrt{b}$ or $c + \sqrt{d}$ and $c - \sqrt{d}$, for which the product does not have a radical term

Consecutive even integers Even integers that are two units apart
Consecutive integers Integers that are one unit apart
Consecutive odd integers Odd integers that are two units apart
Constant A number or letter that stands for just one number
Constant of proportionality The constant in an equation of direct or inverse variation
Coordinates The numbers in an ordered pair
Coplanar lines Lines in the same plane
Cube root The number c is called a cube root of a if $c^3 = a$.

D

Decimal notation A representation of a number containing a decimal point
Deductive reasoning Reasoning used when conclusions are made on the basis of a set of facts
Degree of a polynomial The degree of the term of highest degree in a polynomial
Degree of a term The sum of the exponents of the variables
Denominator The number below the fraction bar in a fraction
Descending order When a polynomial is written with the powers of the variable decreasing as read from left to right, it is said to be in descending order.
Diagonal of a quadrilateral A line segment that joins two opposite vertices
Diameter A line segment that passes through the center of a circle and has its endpoints on the circle
Difference The result of subtracting one number from another
Difference of cubes An expression that can be written in the form $A^3 - B^3$
Difference of squares An expression that can be written in the form $A^2 - B^2$
Digit A number 0, 1, 2, 3, 4, 5, 6, 7, 8, or 9 that fills a place-value location
Direct variation A situation that translates to an equation of the form $y = kx$, with k a positive constant
Discount The amount subtracted from the original price of an item to find the sale price
Discriminant The radicand, $b^2 - 4ac$, from the quadratic formula
Distance formula The equation $d = \sqrt{(x_2 - x_1)^2 + (y_2 - y_1)^2}$ that represents the distance between two points, (x_1, y_1) and (x_2, y_2), on the coordinate plane
Distributive law of multiplication over addition The statement that multiplying a factor by the sum of two numbers gives the same result as multiplying the factor by each of the two numbers and then adding
Distributive law of multiplication over subtraction The statement that multiplying a factor by the difference of two numbers gives the same result as multiplying the factor by each of the two numbers and then subtracting
Dividend In division, the number being divided
Divisible The number b is said to be divisible by another number a if b is a multiple of a.
Divisor In division, the number dividing another number
Domain The set of all first coordinates of the ordered pairs in a function

E

Elimination method An algebraic method that uses the addition principle to solve a system of equations
Empty set The set without members
Equation A number sentence that says that the expressions on either side of the equals sign, =, represent the same number
Equation of direct variation An equation, described by $y = kx$ with k a positive constant, used to represent direct variation
Equation of inverse variation An equation, described by $y = \dfrac{k}{x}$ with k a positive constant, used to represent inverse variation
Equiangular triangle A triangle in which all angles are congruent
Equilateral triangle A triangle in which all sides are the same length
Equivalent equations Equations with the same solutions
Equivalent expressions Expressions that have the same value for all allowable replacements
Equivalent inequalities Inequalities that have the same solution set
Evaluate To substitute a value for each occurrence of a variable in an expression
Even root A root with an even index
Event A set of outcomes
Exponent In expressions of the form a^n, the number n is an exponent.
Exponential notation A representation of a number using a base raised to a power

F

Factor *Verb*: to write an equivalent expression that is a product; *Noun*: a multiplier
Factoring Writing an expression as a product
Factorization A number expressed as a product of two or more numbers
Factorization of a polynomial An expression that names the polynomial as a product
Fahrenheit A temperature scale in which water freezes at 32° and water boils at 212°
FOIL To multiply two binomials by multiplying the First terms, the Outside terms, the Inside terms, and then the Last terms
Formula An equation that uses numbers or letters to represent a relationship between two or more quantities
Fraction equation An equation containing one or more rational expressions; also called a *rational equation*
Fraction expression A quotient, or ratio, of two polynomials; also called a *rational expression*

Fraction notation A number written using a numerator and a denominator

Function A correspondence that assigns to each member of a set called the domain exactly one member of a set called the range

G

Grade The measure of a road's steepness

Graph A picture or diagram of the data in a table; a line, curve, or collection of points that represents all the solutions of an equation

Greatest common factor (GCF) The common factor of a polynomial with the largest possible coefficient and the largest possible exponent(s)

H

Hypotenuse In a right triangle, the side opposite the right angle

I

Identity property of 1 The statement that the product of a number and 1 is always the original number

Identity property of 0 The statement that the sum of a number and 0 is always the original number

Index In the radical $\sqrt[n]{a}$, the number n is called the index.

Inductive reasoning Reasoning used when conclusions are made on the basis of observations

Inequality A mathematical sentence using $<$, $>$, $\leq$, $\geq$, or $\neq$

Input A member of the domain of a function

Integers The whole numbers and their opposites

Intercept The point at which a graph intersects the x- or y-axis

Interest A percentage of an amount invested or borrowed

Interest rate The percent at which interest is calculated on a principal

Intersecting lines Lines that cross each other at a common point

Intersection of sets *A* and *B* The set of all elements that are common to *both* A and B

Inverse variation A situation that translates to an equation of the form $y = \dfrac{k}{x}$, with k a positive constant

Irrational number A real number that cannot be named as a ratio of two integers

Isosceles triangle A triangle in which two or more sides are the same length

L

Leading coefficient The coefficient of the term of highest degree in a polynomial

Leading term The term of highest degree in a polynomial

Least common denominator (LCD) The least common multiple of the denominators of two or more fractions

Least common multiple (LCM) The smallest number that is a multiple of two or more numbers

Legs In a right triangle, the two sides that form the right angle

Like radicals Radicals that have the same index and the same radicand

Like terms Terms that have exactly the same variable factors

Line Two rays that have common points and continue in opposite directions

Line of symmetry A line that can be drawn through a graph such that the part of the graph on one side of the line is an exact reflection of the part on the opposite side

Line graph A graph in which quantities are represented as points connected by straight-line segments

Linear equation Any equation that can be written in the form $Ax + By + C$, where x and y are variables

Linear function A function that can be described by an equation of the form $y = mx + b$, where x and y are variables

Linear inequality An inequality whose related equation is a linear equation

M

Marked price The original price of an item

Mean A center point of a set of numbers found by dividing the sum of the numbers by the number of items of data; also called the *average*

Median In a set of data listed in order from smallest to largest, the middle number if there is an odd number of data items, or the average of the two middle numbers if there is an even number of data items

Metric system A measurement system used in most countries of the world, but very little in the United States

Minuend The number from which another number is being subtracted

Mixed numeral A number represented by a whole number and a fraction less than 1

Mode The number or numbers that occur most often in a set of data

Monomial A constant, a variable, or a product of a constant and one or more variables

Motion problem A problem that deals with distance, speed (or rate), and time

Multiple of a number A product of the number and some natural number

Multiplication property of 0 The statement that the product of 0 and any real number is 0

Multiplicative identity The number 1

Multiplicative inverses Reciprocals; two numbers whose product is 1

N

Natural numbers The counting numbers: 1, 2, 3, 4, 5, . . .

Negative integers Integers to the left of zero on the number line

Numerator The number above the fraction bar in a fraction

O

Obtuse angle An angle whose measure is greater than 90° and less than 180°
Obtuse triangle A triangle in which one angle is an obtuse angle
Odd root A root with an odd index
Opposite The opposite, or additive inverse, of a number x is written $-x$. Opposites are the same distance from 0 on the number line but on different sides of 0.
Opposite of a polynomial Two polynomials are opposites, or additive inverses, of each other if their sum is zero.
Ordered pair A pair of numbers of the form (a, b) for which the order in which the numbers are listed is important
Ordinate The second coordinate in an ordered pair of numbers
Origin The point $(0, 0)$ on a graph where the two axes intersect
Original price The price of an item before a discount is deducted
Outcome The result of an experiment
Output A member of the range of a function

P

Palindrone prime A prime number that becomes a prime number when its digits are reversed
Parabola The graph of a quadratic equation
Parallel Coplanar lines that do not intersect
Parallel lines Lines in the same plane that never intersect; two lines are parallel if they have the same slope.
Parallelogram A four-sided polygon with two pairs of parallel sides
Percent notation A representation of a number as n parts per 100; $n\%$
Perfect square A rational number p for which there exists a number a for which $a^2 = p$
Perfect-square trinomial A trinomial that is the square of a binomial
Perimeter The distance around an object or the sum of the lengths of its sides
Periods Groups of three digits, separated by commas
Perpendicular Two lines are perpendicular if they intersect to form a right angle.
Perpendicular lines Two lines that intersect to form a right angle; two lines are perpendicular if the product of their slopes is -1.
Pi (π) The number that results when the circumference of a circle is divided by its diameter; $\pi \approx 3.14$, or $\frac{22}{7}$
Pictograph A graphic means of displaying information using symbols to represent the amounts
Point–slope equation The equation $y - y_1 = m(x - x_1)$, where x_1, y_1, and m are real numbers and m is the slope and (x_1, y_1) is a point that lies on the graph of the equation
Polygon A closed geometric figure with three or more line segments as sides
Polynomial A monomial or a sum of monomials
Polynomial equation An equation in which two polynomials are set equal to each other
Positive integers Integers to the right of zero on the number line
Prime factorization A factorization of a composite number as a product of prime numbers
Prime number A natural number that has exactly two different factors: itself and 1
Prime polynomial A polynomial that cannot be factored using only integer coefficients
Principal An amount of money that is invested or borrowed
Principal square root The nonnegative square root of a number
Principle of zero products The statement that an equation $ab = 0$ is true if and only if $a = 0$ is true or $b = 0$ is true, or both are true
Product The result when one number is multiplied by another
Proportion An equation stating that two ratios are equal
Proportional numbers Two pairs of numbers having the same ratio
Protractor A device used to measure angles
Purchase price The price of an item before sales tax is added
Pythagorean equation The equation $a^2 + b^2 = c^2$, where a and b are lengths of the legs of a right triangle and c is the length of the hypotenuse
Pythagorean theorem In any right triangle, if a and b are the lengths of the legs and c is the length of the hypotenuse, then $a^2 + b^2 = c^2$.

Q

Quadrants The four regions into which the axes divide a plane
Quadratic equation An equation of the form $ax^2 + bx + c = 0$, where $a \neq 0$
Quadratic formula The solutions of $ax^2 + bx + c = 0$, $a \neq 0$, are given by the equation $x = \dfrac{-b \pm \sqrt{b^2 - 4ac}}{2a}$.
Quadratic function A second-degree polynomial function in one variable
Quadratic inequality An inequality whose related equation is a quadratic equation
Quotient The result when one number is divided by another

R

Radical equation An equation in which a variable appears in a radicand
Radical expression An algebraic expression in which a radical symbol appears
Radical sign The symbol $\sqrt{}$
Radicand The expression under the radical sign
Radius A line segment with one endpoint on the center of a circle and the other endpoint on the circle
Range The set of all second coordinates of the ordered pairs in a function
Rate A ratio used to compare two different kinds of measure

Ratio The quotient of two quantities; for example, the ratio of a to b is $\frac{a}{b}$, also written $a:b$
Rational equation An equation containing one or more rational expressions; also called a *fraction equation*
Rational expression A quotient, or ratio, of two polynomials; also called a *fraction expression*
Rational numbers Any number that can be written as the ratio of two integers $\frac{a}{b}$, where $b \neq 0$
Rationalizing the denominator A procedure for finding an equivalent expression without a radical in the denominator
Real numbers All rational and irrational numbers
Reciprocal A multiplicative inverse; two numbers are reciprocals if their product is 1
Rectangle A four-sided polygon with four 90° angles
Relation A correspondence between a first set, the domain, and a second set, the range, such that each member of the domain corresponds to at least one member of the range
Repeating decimal A decimal in which a number pattern repeats indefinitely
Right angle An angle whose measure is 90°
Right triangle A triangle in which one angle is a right angle
Rise The change in the second coordinate between two points on a line
Roster notation A way of naming sets by listing all the elements in the set
Rounding Approximating the value of a number; used when estimating
Run The change in the first coordinate between two points on a line

S

Sale price The price of an item after a discount has been deducted
Sales tax A tax added to the purchase price of an item
Sample space The set of all possible outcomes
Scalene triangle A triangle in which all sides are of different lengths
Scientific notation A number written in the form $M \times 10^n$, where n is an integer, $1 \leq M < 10$, and M is expressed in decimal notation
Segment A geometric figure consisting of two points, called endpoints, and all points between them
Set A collection of objects
Set-builder notation The naming of a set by describing basic characteristics of the elements in the set
Similar figures Figures with the same shape, but not necessarily the same size
Similar triangles Triangles in which corresponding sides are proportional and corresponding angles are congruent
Simple interest A percentage of an amount P invested or borrowed for t years, computed by calculating principal × interest rate × time in years
Simplify To rewrite an expression in an equivalent, abbreviated form
Slope The ratio of the rise to the run for any two points on a line
Slope–intercept equation An equation of the form $y = mx + b$, where x and y are variables, the slope is m, and the y-intercept is $(0, b)$
Solution A replacement for the variable that makes an equation true
Solution set The set of all solutions of an equation, an inequality, or a system of equations or inequalities
Solve To find all solutions of an equation, an inequality, or a system of equations or inequalities; to find the solution(s) of a problem
Sphere The set of all points in space that are a given distance (radius) from a given point (center)
Square A four-sided polygon with four right angles and all sides of equal length
Square of a number A number multiplied by itself
Square root of a number The number c is a square root of a if $c^2 = a$
Standard form of a linear equation An equation written in the form $Ax + By = C$
Standard form of a quadratic equation An equation written in the form $ax^2 + bx + c = 0$ with $a > 0$
Statistic A number that describes a set of data
Straight angle An angle whose measure is 180°
Subsets Sets that are parts of other sets
Substitute To replace a variable with a number
Substitution method An algebraic method for solving systems of equations
Subtrahend In subtraction, the number being subtracted
Sum The result in addition
Sum of cubes An expression that can be written in the form $A^3 + B^3$
Sum of squares An expression that can be written in the form $A^2 + B^2$
Supplementary angles Two angles for which the sum of their measures is 180°
Surface area The sum of the areas of all of the faces of a three-dimensional figure
System of equations A set of two or more equations that are to be solved simultaneously

T

Term A number, a variable, or a product or a quotient of numbers and/or variables
Terminating decimal A decimal that can be written using a finite number of decimal places
Total price The sum of the purchase price of an item and the sales tax on the item
Transversal A line that intersects two or more coplanar lines in different points
Trapezoid A polygon with four sides, two of which, the bases, are parallel to each other
Triangle A three-sided polygon
Trinomial A polynomial containing three terms
Trinomial square The square of a binomial expressed as three terms

U

Union of sets *A* and *B* The set of all elements belonging to *either A* or *B*
Unit price The ratio of price to the number of units

V

Value The numerical result after a number has been substituted into an expression
Variable A letter that represents an unknown number
Variable expression An expression containing a variable
Variation constant The constant in an equation of direct or inverse variation
Vertex The common endpoint of two rays that form an angle; the point at which the graph of a quadratic equation crosses its axis of symmetry
Vertical angles Two non-straight angles formed by two pairs of opposite rays
Vertical-line test The statement that a graph represents a function if it is impossible to draw a vertical line that intersects the graph more than once
Volume The number of cubic units needed to fill a three-dimensional figure

W

Whole numbers The natural numbers and 0: 0, 1, 2, 3, 4, 5, . . .

X

***x*-intercept** The point at which a graph crosses the x-axis

Y

***y*-intercept** The point at which a graph crosses the y-axis

Index

C

H

I

O

P

T